Catalogue of Palaearctic Coleoptera

Volume 4

DEDICATION

We dedicate the present volume to the late Karel Majer (12.V.1949-8.VIII.2000), Brno, Czech Republic. A scientific illustrator by profession, he was deeply involved in the study of Coleoptera. His enthusiasm and untiring interest in the taxonomy and phylogeny allowed him to become one of the significant contributors to the modern study of Cleroidea. He was involved in the work on the Catalogue from the very beginning of the project, but his premature death prevented him to accomplish the task.

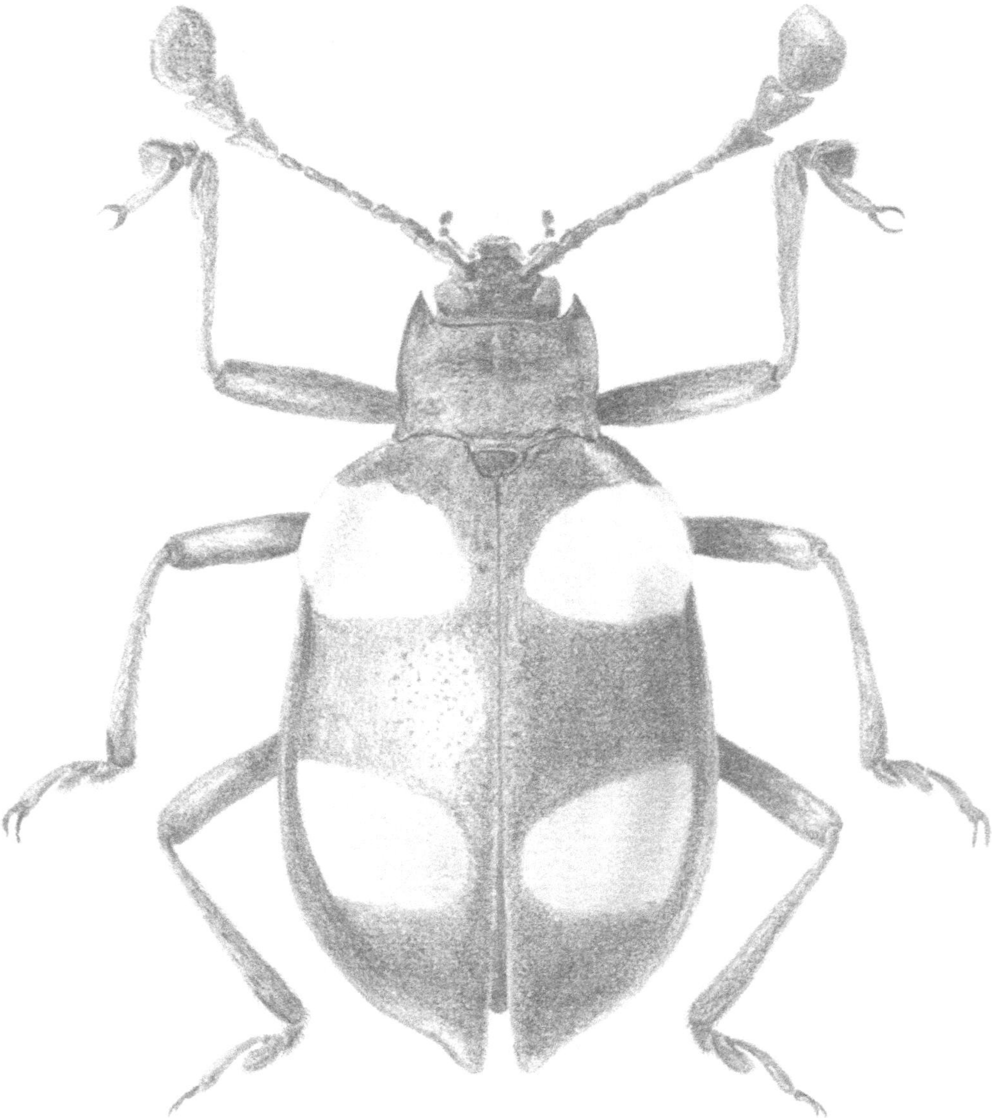

Eutectanus coelestinus (Gerstaecker, 1857)

Catalogue of Palaearctic Coleoptera

Volume 4

Elateroidea - Derodontoidea - Bostrichoidea
Lymexyloidea - Cleroidea - Cucujoidea

Edited by
I. LÖBL & A. SMETANA

Apollo Books
Stenstrup, 2007

Edited by: I. Löbl & A. Smetana.

Printed by: Vinderup Bogtrykkeri A/S.

Published by:
Apollo Books
Kirkeby Sand 19
DK-5771 Stenstrup
Denmark
Telephone + 45 62 26 37 37
Telefax + 45 62 26 37 80
E-mail: apollobooks@vip.cybercity.dk
Web site: www.apollobooks.com

Chapters from this publication should be cited as follows:
Muona, J.: Eucnemidae, pp. 81-87. – In I. Löbl & A. Smetana (ed.):
Catalogue of Palaearctic Coleoptera, Vol. 4. Stenstrup: Apollo Books, 935 pp.

ISBN 87-88757-67-6 (Vol. 4)
ISBN 87-88757-71-4 (Vols 1-8)

Date of publication: 30.06.2007.

CONTENTS

INTRODUCTION

Structured knowledge requires categorisation of observations and phenomena, and one of these phenomena is life. Structured knowledge of life, as it evolved and is present on our planet, requires cataloguing.

Taxonomy has a profound impact in many fields, including genetics, physiology, ecology, plant protection, nature conservation, and legislation. At the core of taxonomic information are taxonomic catalogues that serve as guides to the diversity of life. They summarize the work in systematics and taxonomy that enable assessment of faunal diversity and classification, and provide information on historical taxonomic actions. Catalogues are also essential in achieving a universal language in taxonomy that assures unambiguous transfer of biological information.

The ever increasing number of taxa described since the tenth edition of Linné's *Systema Naturae* was probably the main motivation for compiling catalogues in the past. Since the middle of the 19th century numerous catalogues were published, usually covering restricted parts of the world and served as checklists rather than works that we would qualify today as full catalogues. While knowledge is dynamic, catalogues are static and reflect a time restricted information. Therefore, there is always a need of newer, up-to-date, and complete catalogues. The first world catalogue of Coleoptera was Gemminger & Harold's *Catalogus Coleopterorum* issued from 1868 to 1874. It was followed by Junk & Schenkling's (editors) *Coleopterorum Catalogus*, published in separate parts from 1910 to 1940. At present, even though it is desperately obsolete, the *Coleopterorum Catalogus* remains the only source of comprehensive taxonomic information on a world basis for many families of beetles. The only catalogue covering all groups of Coleoptera occurring in the Palaearctic region is Winkler's 1924 – 1932 *Catalogus Coleopterorum regionis palaearcticae*. However, the Junk & Schenkling and Winkler catalogues differ. The former one is approaching the definition of a "full" catalogue, while the latter is rather a checklist. They both share two serious omissions. The primary sources of taxonomic acts were not consistently checked, and the information on distribution was not detailed enough to satisfy the needs of biologists. In addition, many rules subsequently adopted by the *International Code of Zoological Nomenclature* were not respected. Due to the lack of modern catalogues, the language in taxonomic literature remains inadequate in many respects. In particular, unavailable names are regularly used as available and sometimes even regarded as valid, various spellings are used for the same taxon, and references to taxonomic actions are inexact to a high degree for papers published during the 20th century.

Undoubtedly, the Coleoptera of the Palaearctic Region were over the years studied by more professional and amateur biologists than those in other parts of the World, and the results may exceed those of the study of many other arthropods. It is therefore remarkable that an overview of the present knowledge of the Palaearctic Coleoptera is badly lagging behind other groups, such as the Nearctic Coleoptera, or Palaearctic Diptera. Systematic studies of many groups of Palaearctic Coleoptera, even in view of their popularity for biologists, have to start with searches for data scattered throughout the literature of the past 250 years.

A full catalogue provides information concerning all published names within a group, their classification, taxonomic history, and so on. Ideally, the present Catalogue should include such data. However, considering the number of taxa and the low number of present day taxonomists, along with difficulties in accessing primary literature, a full catalogue of all Palaearctic beetles would not be completed within a reasonable period of time.

Having this in mind, the aim of this Catalogue is to provide a tool that meets the most urgent needs: 1) a list of available names, both valid and invalid, of taxa occurring in the Palaearctic Region, in their verified orthography and with correct publication dates, 2) a complete list of verified references to primary descriptions, and 3) informative distributions of the species and subspecies. Thus, the Catalogue is expected to respond to questions related to biodiversity and to the stability of the taxonomic nomenclature.

This Catalogue is a collective work involving many authors. For its realisation, it was necessary to achieve a reasonable consensus in very diverse fields. The Editors appreciate the understanding of the need of consistency by the contributing authors. The most controversial points concern the spelling of names, and the delimitation of the Palaearctic Region and its main subdivisions.

The first point is purely formal. The alternatives adopted by the Editors and presented below result from the consensus achieved during the discussions with colleagues. The Editors are aware of the difficulty in satisfying all opinions and appeal to the understanding of the user.

The second point pertains to the presentation of distributional data. The Palaearctic Region, as it is defined for the purpose of this Catalogue, includes some areas that are usually considered to belong to the Afrotropical, Oriental, and Pacific Regions respectively, i.e. the south of the Arabian Peninsula and Suqutra, all of Pakistan, the Himalayan part of India including Arunachal Pradesh, Bhutan, and Nepal, the south of China, and the Pacific Islands of Japan. The main reasons for inclusion of these areas are as follows:

Old records of "Arabia" and "China" may pertain to any states on the Arabian Peninsula and to Jordan, or to any part of the People's Republic of China, respectively.

1. Recent field work in the Himalaya and in the mountains of mainland China and Taiwan provides evidence of altitudinal faunal transition in Coleoptera, as well as in other insect groups. While subtropical climate with dominant Oriental taxa prevails at low elevations, significant faunal changes are found already at elevations of 1500 m, and almost "pure Palaearctic" taxa are present at elevations of 2500 m and above. Thus, it appears illusory to draw simple biogeographical frontiers in any of these parts of the world.
2. The Catalogue includes more information. This is an obvious advantage in the absence of modern catalogues covering the Afrotropical and Oriental regions.

The Catalogue of Palaearctic Coleoptera is published in a number of independent volumes, each having its own Reference section and Index. The taxonomic, geographical and bibliographical information in this volume is presented as in the two previous Volumes.

The present volume treats the polyphagous superfamilies Elateroidea, Derodontoidea, Bostrichoidea, Lymexyloidea, Cleroidea, and Cucujoidea. As in the first three volumes, the adopted arrangement of families and subfamilies within the superfamilies is consistent with J. F. Lawrence & A. F. Newton's 1995 classification (in J. Pakaluk & S. A. Ślipiński (eds): *Biology, phylogeny, and classification of Coleoptera. Papers celebrating the 80th birthday of Roy A. Crowson*. Volume 2. Warszawa: Muzeum i Instytut Zoologii PAN). The more important changes concern the classification of the cleroids, and the treatement of languriines as a subfamily of Erotylidae.

The volume includes about 28.900 names of taxa and 6.905 primary references to the genus- and species-groups names.

The Editors adopted the policy that the gap between the publication date of each volume of the Catalogue and the date of the corresponding entry deadline should not exceed three years. While the first Volume contains the available genus- and species-groups names in Archostemata, Myxophaga and Adephaga published before January 1, 2000, the second Volume contains the available names in Staphyliniformia published before January 1, 2002, the third Volume contains the available names in Scarabaeoidea, Dascilloidea, Buprestoidea and Byrrhoidea published before January 1, 2004, the present Volume is extended to contain the available Elateroidea, Derodontoidea, Bostrichoidea, Lymexyloidea, Cleroidea, and Cucujoidea names published before January 1, 2005.

An index of the species-group names was not provided for practical reasons associated with the printing and bookbinding costs. Therefore, as for the previous volumes, an electronic version of species-group names indexes is available on the web sites of the Apollo Books (www.apollobooks.com) and of the Muséum d'histoire naturelle in Geneva (www.ville-ge.ch/musinfo/mhng).

The Editors were repeatedly asked to make the Catalogue available electronically, either as an online interactive database, or as CD-Rom. After careful consideration they decided to publish the work only in book form for the following reasons: Information in taxonomy, unlike that in most other biological fields, remains useful for extremely long periods of time, extending over centuries. Taxonomic work deals commonly with sources published throughout the 19th and 20th centuries and often even earlier. Consequently, catalogues that summarize taxonomic work are consulted over a very long period of time. The simple, although often ignored, reason for the durability of taxonomic data is in the fact that taxonomy provides the language indispensable for unambiguous conveyance of biological information. The bulk of correctly recorded data in Catalogues remains informative even under a continuous inflow of additional new taxa, and even if the assignments and ranks of taxa and the validity or invalidity of names are changing, and the known distributions of species become gradually more precise. It is therefore necessary to insure access to taxonomic work, including Catalogues, not only in spatial but also in temporal dimensions. Experience shows that printed texts may remain available for centuries, while life expectancy of electronic information is unknown. Ideally, works like this Catalogue, should be available in both print- and electronic formats; however, the real world is not ideal and the production of printed Catalogues is associated with considerable costs. The consequence of providing low-cost online or CD-Rom editions would have a serious negative economic impact on the production of printed version. We believe that under the globally inadequate institutional interest for the needs of taxonomy, the livelihood of those who are willing to accept the commercially hazardous production of printed taxonomic publications should be protected. We advocate the growth and continuation of taxonomy, not its demise.

TAXONOMIC INFORMATION

The present Catalogue includes all available names, both valid and invalid, of extant beetle taxa described before January 1, 2005 and known to occur in the Palaearctic Region, as it is defined below. The higher classification, from suborder down to subfamily, is based on the work of Lawrence & Newton (1995): *Families and subfamilies of Coleoptera (with selected genera, notes, references and data on family-group names)*. However, the classification proposed in this work is not taken as dogma, and changes are accepted when considered well founded. All taxa below subfamily rank are arranged alphabetically within the higher taxon and the synonyms follow the respective valid name alphabetically.

Extinct taxa, names rejected by the ICZN (*International Code of Zoological Nomenclature, Fourth Edition.* London: International Trust for Zoological Nomenclature, 1999), misspellings, misidentifications and other nomina nuda are not included in the Catalogue. However, concepts that are important for nomenclatorial purposes, may be included. Similarly, all infrasubspecific names, such as those established as "morpha", "natio", or "race", "subvariety" and "aberration", or proposed as variety and form of a subspecies or another variety, and names published, e.g., as *A-us b-us c-us* , but specified in the text that they are actually proposed for a "natio" or "race", etc., are not considered subspecific and are therefore excluded from the Catalogue. Names proposed as varieties and forms before 1961 are included, if deemed subspecific under the provisions of the ICZN, Article 45.6. Unjustified emendations may be included.

The currently valid names of the family-group taxa include the name of the author and the year of the publication. Their synonyms are not listed.

The names of the genus-group taxa are given with the name of the author, and the year and page of publication. The page given is the page where the name and the actual description of the taxon is printed. The type species of all genus-group names are given in their original combination. If the type species is currently regarded as a junior synonym, the valid senior synonym is given in brackets in its original combination.

The names of the species-group taxa are given with the name of the author, and the year and page of publication. The page given is the page where the name and the actual description of the taxon is printed. In bi-languaged Chinese/English and Japanese/English publications both respective pages of the actual description may be given. For species-group taxa subsequently transferred to another genus, the name of the original genus is given in parentheses, following the page of publication.

Some authors (e.g., V. Apfelbeck, H. John) published the same description twice, or even more times, in separate papers. Such publications produce, de facto, primary homonyms and objective synonyms. The first publication in such cases is referred to as indicated above, followed by the mark = , the year and first page of the subsequent description/s in square brackets. This is particularly important for taxa that are erroneously associated with their junior description.

The following symbols, all given in square brackets following the page of publication, or the original combination when applicable, are used for taxonomic information : HN for homonyms, RN for replacement names, NO for nomina oblita, NP for nomina protecta, DA for doubtful assignment, and EA for erroneous assignment.

Taxa considered incertae sedis and nomina dubia are listed separately at the end of the nearest applicable taxon.

Taxonomic and nomenclarorial acts published after December 31, 2004 are considered only when they concern taxa described on or before that date.

DISTRIBUTIONAL INFORMATION

The limits of the Palaearctic region, as those of other biogeographical regions, are arbitrarily defined (Map 1). For practical reasons, the boundaries of the Palaearctic Region, as they were established for the Catalogue (see above), usually follow national boundaries. The region includes Europe, Africa north of the Sahara, and Asia except for the part that is arbitrarily defined as belonging to the Oriental Region.

For each species and subspecies an outline of its present distribution is given. Fossil records are not considered. The information is given by means of symbols, presented at three levels.

The first level is the subdivision of the Palaearctic Region into three main parts, Europe (letter **E**, bold), North Africa (letter **N**, bold) and Asia (letter **A**, bold).

Europe includes the Azores, Iceland and Turkey west of the Bosporus. The eastern boundaries are a matter of controversy. In the Catalogue, Europe includes Russia west of the main ridge of the Ural mountains, the Permsk Oblast, Bashkortostan Republic and Orenburskaya Oblast, and the small part of Kazakhstan west of the Ural River. It includes the Caucasian republics of Georgia, Armenia and Azerbaijan. The south-eastern boundaries are the political boundaries of the Asian part of Turkey, Iran, Kazakhstan, and the Caspian and Black seas.

Yugoslavia has been recently split into two independent countries: Serbia and Montenegro. At the time of the split almost all distributional records for the respective species were finished, using the symbol YU. It would have been an unreasonably time consuming effort to retroactively differentiate these records, therefore, as a compromise, the term Yugoslavia is no longer used in the list of geographical symbols on page 14, but records for Serbia and Montenegro are still listed under the symbol YU in the body of the Catalogue.

North Africa includes Morocco (incl. Western Sahara), Algeria, Tunisia, Libya and Egypt west of the Suez Canal, and the Canary and Madeira islands.

Asia includes Sinai and the Arabian Peninsula (including Suqutra), Turkey east of the Bosporus, the Middle East and Central Asian countries, Russia east of the main ridge of the Ural mountains, Korea, Japan (including Ryukyu [= Nansei] Islands and the Japanese Pacific Islands), the entire People's Republic of China, Taiwan, Bhutan, Nepal, North India along the base of the Himalaya (Arunachal Pradesh, Uttaranchal, Uttar Pradesh, Himachal Pradesh), Jammu & Kashmir and all of Pakistan. Thus, India is the only state for which the strict political boundaries are not respected. Large parts of Uttar Pradesh south of Nepal are overpopulated plains. Information on Coleoptera from this North Indian state is based almost exclusively on its Himalayan districts lying west of Nepal, which are in the present state Uttaranchal.

The second level of the geographic information is provided by two-letter symbols for countries, major areas of Russia and North Indian states, and by three-letter symbols for provinces of mainland China and for Taiwan (see Table 1, Map 3).

The symbols are arranged in alphabetical order within the first-level subdivision. Some, usually older, distributional records cannot be accommodated within the structure of the Catalogue (e.g., Arabia, Caucasus, North India, Siberia). Such information is given in quotation marks (e.g., "Caucasus") behind the last symbol of the respective first-level symbol. Russia is subdivided into six major sub-regions (Map 2), each of which has its own two-letter symbol. These, as well as the symbol RU for Russia, are used only when more detailed information is not available. Similarly, the symbol CH for the People's Republic of China, as well as the seven two-letter symbols for China's major regions, are used only in the absence of more detailed geographical information.

In general, the published distributional information is based on both identified material in collections and on published records, scattered in an enormous number of taxonomic and faunistic papers that are virtually impossible to review in their entirety. Revisions of collections reveal a high proportion of misidentifications, attaining 40% of specimens in some large museums. The degree of identification reliability and of the records derived from identifications, is a function of faunal diversity and quality of systematic revisions. Thus, the reliability in general increases from south to north and from poorly studied groups to "popular" groups. At present, a number of modern catalogues or check-lists, covering the beetle faunas of many European and some extra-European countries or archipelagos, are available. The use of data contained in these and other faunistic works is left to the discretion of the authors, who may also add unpublished information available to them. The second-level geographic information is not necessarily exhaustive, it should rather be considered as a base for future faunal research.

The third-level geographic information concerns species and subspecies with restricted distribution. Taxa of this category may be strict endemites, or taxa comparatively widely distributed in one area but restricted in another area. For example, the distributional record of a species widely distributed in North Africa with isolated occurrence on Pantelleria would appear as follows: **E**: IT (Pantelleria) **N**: AG MO TU. The third-level information is facultative. It is given in parentheses after the respective second-level symbol. The official language of the respective state is used for records in languages using the Latin alphabet, or it is transliterated from the Cyrillic alphabet. Records in languages using non-Latin or Cyrillic characters (e.g. Chinese or Japanese pictographs) are translated into English, and the translated geographical terms are spelled as closely as possible to those used in the Times Atlas®, or in other well-known sources. Detailed geographical information may refer to natural geographical features such as islands, mountains, lakes, valleys, caves, or to administrative entities, such as districts.

The extralimital distribution of some Palaearctic species is indicated by three letter symbols in bold, cosmopolitan species by the symbol **COS**, all located at the end of the respective geographical information (see Table 1). Introductions are indicated by the letter "i" (e.g., **E**i: GB). The extralimital regions for the needs of the Catalogue are defined as follows (see Map 1): Nearctic (**NAR**): north of Mexico; Neotropical (**NTR**): south of the United States; Afrotropical (**AFR**): south of the North African states included in the Palaearctic Region; Oriental (**ORR**): areas south of the People's Republic of China and Taiwan, areas south of the Himalaya in India, the Philippines, Malaysia and Indonesia south to the Lydekker line; Australian (**AUS**): south of the Lydekker line, Pacific.

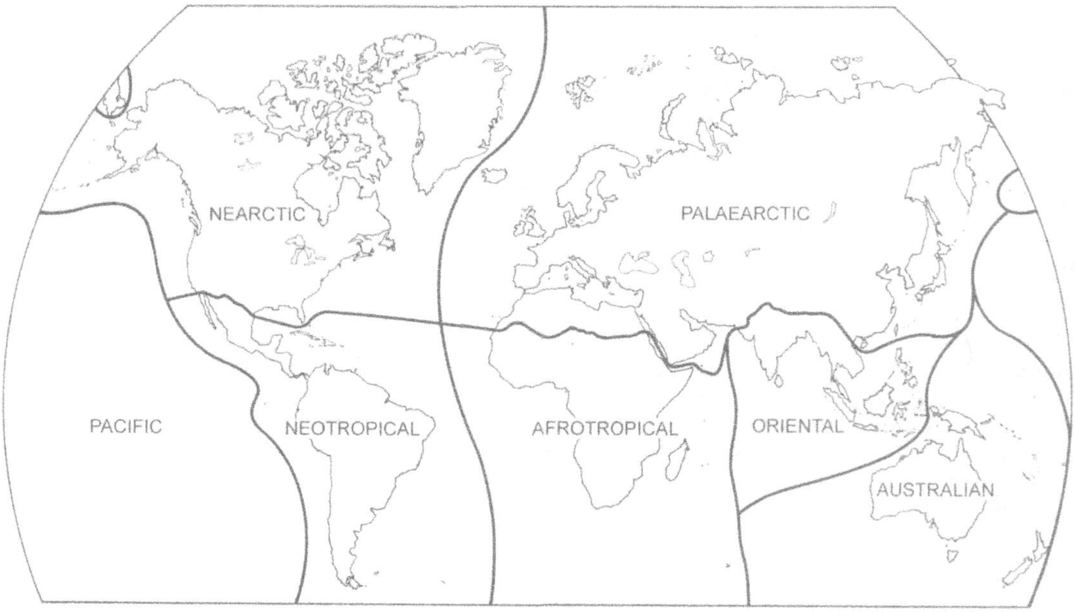

Map 1. The limits of the geographic Regions as defined for the purpose of this Catalogue

Map 2. Subdivisions of Russia

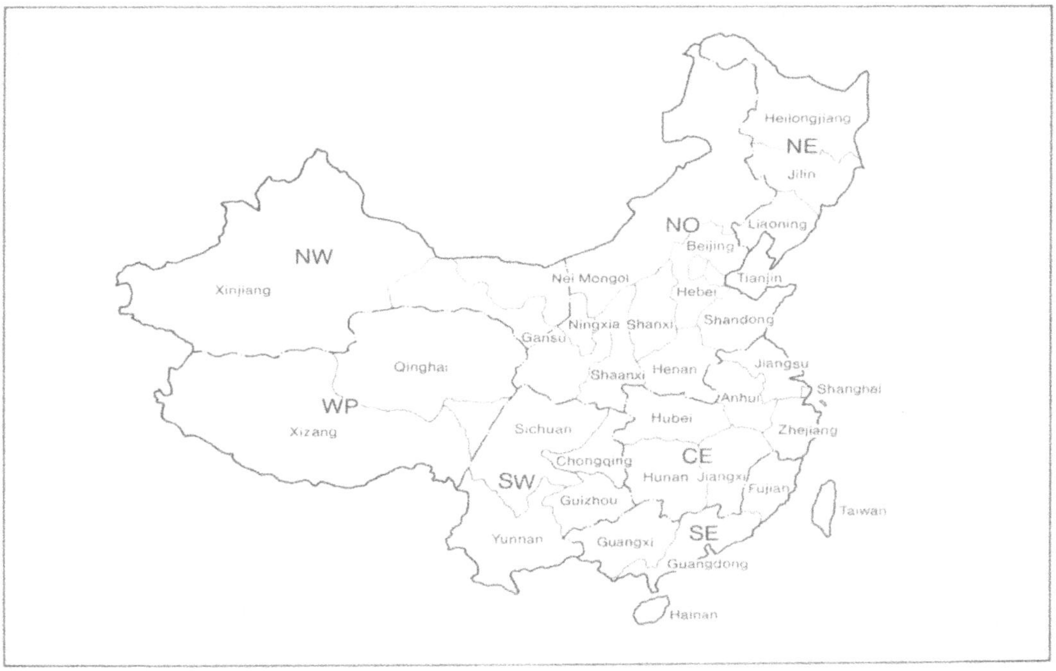

Map 3. Subdivisions and provinces of the People's Republic of China

Table 1: GEOGRAPHICAL SYMBOLS

E	**Europe**
AB	Azerbaijan
AL	Albania
AN	Andorra
AR	Armenia
AU	Austria
AZ	Azores
BE	Belgium
BH	Bosnia Herzegovina
BU	Bulgaria
BY	Belarus
CR	Croatia
CT	Russia: Central European Territory
CZ	Czech Republic
DE	Denmark
EN	Estonia
FA	Faeroe Islands
FI	Finland
FR	France (incl. Corsica, Monaco)
GB	Great Britain (incl. Channel Islands)
GE	Germany
GG	Georgia
GR	Greece (incl. Crete)
HU	Hungary
IC	Iceland
IR	Ireland
IT	Italy (incl. Sardinia, Sicily, San Marino)
KZ	Kazakhstan
LA	Latvia
LS	Liechtenstein
LT	Lithuania
LU	Luxembourg
MA	Malta
MC	Macedonia
MD	Moldavia
NL	The Netherlands
NR	Norway
NT	Russia: North European Territory
PL	Poland
PT	Portugal
RO	Romania
RU	Russia
SK	Slovakia
SL	Slovenia
SP	Spain (incl. Gibraltar)
SR	Svalbard (Spitzbergen)
ST	Russia: South European Territory
SV	Sweden
SZ	Switzerland
TR	Turkey
UK	Ukraine
YU	Serbia and Montenegro

N	**North Africa**
AG	Algeria
CI	Canary Islands
EG	Egypt
LB	Libya
MO	Morocco (incl. Western Sahara)
MR	Madeira Archipelago
TU	Tunisia

A	**Asia**
AE	Arab Emirates
AF	Afghanistan
AP	India: Arunachal Pradesh
BA	Bahrain
BT	Bhutan
CE	China: Central Territory
CH	China
CY	Cyprus
ES	Russia: East Siberia
FE	Russia: Far East
HP	India: Himachal Pradesh
IN	Iran
IQ	Iraq
IS	Israel
JA	Japan
JO	Jordan
KA	India: Kashmir
KI	Kyrgyzstan
KU	Kuwait
KZ	Kazakhstan
LE	Lebanon
MG	Mongolia
NE	China: Northeast Territory
NC	North Korea
NO	China: Northern Territory
NP	Nepal
NW	China: Northwest Territory
OM	Oman
PA	Pakistan

QA Qatar (incl. United Arab Emirates)
RU Russia
SA Saudi Arabia
SC South Korea
SD India: Sikkim, Darjeeling District
SE China: Southeastern Territory (incl. Macao, Hongkong)
SI Egypt: Sinai
SW China: Southwestern Territory
SY Syria
TD Tajikistan
TM Turkmenistan
TR Turkey
UP India: Uttaranchal, Uttar Pradesh
UZ Uzbekistan
WP China: Western Plateau
WS Russia: west Siberia
YE Yemen (incl. Suqutra)

CHINA: PROVINCES, AUTONOMOUS REGIONS OR MUNICIPALITIES, AND TAIWAN

ANH Anhui (Anhwei)
BEI Beijing (Peking or Peiping)
CHQ Chongqing
FUJ Fujian (Fukien)
GAN Gansu (Kansu)
GUA Guandong (Kwantung)
GUI Guizhou (Kweichow)
GUX Guangxi (Kwangsi)
HAI Hainan
HEB Hebei (Hopeh)
HEI Heilongjiang (Heilungkiang)
HEN Henan (Honana)
HKG Hongkong
HUB Hubei (Hupeh)
HUN Hunan
JIA Jiangsu (Kiangsu)
JIL Jilin (Kirin)
JIX Jiangxi (Kiangsi)
LIA Liaoning
MAC Macao
NIN Ningxia (Ningsia)
NMO Nei Mongol (Inner Mongolia)
QIN Qinghai (Tsinghai)
SCH Sichuan (Szechwan)
SHA Shaanxi (Shensi)
SHG Shanghai
SHN Shandong (Shantung)
SHX Shanxi (Shansi)
TAI Taiwan (Formosa)
TIA Tianjin (Tsiensin)
XIN Xinjiang (Sinkiang)
XIZ Xizang (Tibet)
YUN Yunnan
ZHE Zhejiang (Chekiang)

WORLD ZOOGEOGRAPHICAL REGIONS

AFR Afrotropical Region
AUR Australian Region
NAR Nearctic Region
NTR Neotropical Region
ORR Oriental Region

BIBLIOGRAPHIC INFORMATION

References are given to, and only to, primary sources of the genus-group and species-group names included in the Catalogue. No bibliographical reference is given to family-group names, to designation of type-species, to secondary taxonomic acts, or to sources of any other information. Names of all authors, if applicable, are given for each reference. An effort was put forth to unify the spellings of the names. For names that appeared in publications in two or more different orthographies, the one used in the Catalogue is that which was considered to be correct after thorough verification (e.g., the spelling Josef Müller for the author who also published as Giuseppe Müller). In such cases the alternative spelling follows the original one in brackets. The original name is used for authors that subsequently adopted additional names, or honorary epithets (e.g., A. Semenov, and not A. Semenov-Tian-Shanskyi). Initials are given for all authors, even if they are missing in the original publication. Names published in two or more alternative spellings, that seem to be equally correct, are spelled following Schmitt, Hübner & Gaedike (1998): *Nomina Auctorum. Auflösung von Abkürzungen taxonomischer Autoren-Namen. Nova Supplementa Entomologica* **11**: 3-189. An exception is made for F. L. Castelnau de Laporte, who is referred to as Laporte in recent major works. Therefore, the name Laporte is arbitrarily used in the Catalogue. Names in Cyrillic are transliterated as for any other Cyrillic text, unless an alternative spelling was preferred by the author, or the alternative name is generally used in the literature (e.g., Jakobson, Tschitschérine). It should be noted that lowercase prepositions such as French "de" and German "von" are not given with the author's name in the list of the taxa.

Titles of references are given in full, i.e. not abbreviated, with subtitles, as originally published. Text written in Cyrillic is transliterated. Titles in languages using non-Latin or non-Cyrillic characters are translated into English and are presented in square brackets, followed by the indication of the original language in parentheses.

Titles of periodicals are given in full. Series numbers are given in parentheses, and always precede the boldface volume number. The numbers of separate issues are only provided when they are paginated separately, in which case they are in parentheses following the volume number. Unnumbered pages are indicated in square brackets. The titles of some periodicals vary over the years. For these periodicals the titles are given as they appeared for the respective items referred to (e.g., *Coleopterologische Rundschau* / *Koleopterologische Rundschau*). Contrarily, the journal *Entomologische Blätter* had several subtitles during its existence, the last of them being *Entomologische Blätter für Biologie und Systematik der Käfer*. Since there is no danger of confusion with any other periodical, the journal is cited simply as *Entomologische Blätter*. If the year of the volume differs from the actual year of publication, the former follows the volume number and is given is square brackets. If periodicals have no volume number, the year of the volume is given.

The *Bulletin de la Société Entomologique de France* is one of the more important periodicals in the field of systematics. The title changed almost continuously for more than 60 years, appearing usually as *Bulletin des séances*, *Bulletin trimestriel* or *Bulletin entomologique*. In addition, the first numbered volume, 37, was published 100 years after the creation of the *Annales* and the *Bulletin* of the French Entomological Society in 1832. To avoid confusion, this periodical is referred to as *Bulletin de la Société Entomologique de France,* and, from 1832 to 1931, is identified by the year in which it was published. From the year 1932, the volume number is given.

The *Bulletin de la Société Entomologique de France* was published from 1873 to 1894, in addition to the well known quarterly version, in a bi-mensuel version "*Bulletin Bimensuel*". The pagination of the two versions are not identical, the publication of the bi-mensuel version preceded the quarterly one. The *Bulletin Bimensuel* is ignored by almost all coleopterists, and is not respected in the Catalogue.

For ease of access, references of the same author and year were placed in order based on the first page number of the paper, with Roman numerals being placed last, with respect to the number of the respective issue. Books were placed first for a given year, unless a page range was cited, in which case the first-page rule was followed. Papers of the same year and with the same first page numbers were placed in order of their last page number. Thus the references are not arranged chronologically within each year.

ACKNOWLEDGEMENTS

As for the previous volumes of the Catalogue, the assistance of libraries, and of colleagues who had access to such libraries, was one of the prerequisites for realisation of the present work. The most important libraries to be mentioned are: Muséum d'histoire naturelle, Genève, Switzerland (Mrs. Christele Mougin); Agriculture and Agri-Food Canada (Entomological Library) Ottawa, Canada (Mrs Patricia Madaire and Mr. Steve Gamman); Deutsches Entomologisches Institut, Müncheberg, Germany (Mrs Jutta Moebert, Monika Schiele, and Helga Wehrens, Mr. Reinhard Gaedike and Lothar Zerche), and Zoologisches Museum, Berlin, Germany (Mr. Johannes Frisch).

Special thanks are due to Lee H. Herman, New York, USA, and Mikael Sörensson, Lund, Sweden whose assistance with rare publications was essential. Mr. H. W. Rücker, Neuwied, Germany, made available his own specialist knowledge on Latridiidae and completed the distributional data for many taxa. Mr. Y. Okushima, Kurashiki, Japan, provided data on Japanese Cantharidae and Antonio Sánchez-Ruiz provided data on Elateridae of North Africa and the Canary Islands. Otto Merkl, Budapest, Hungary, verified the correct distribution in Hungary and Romania in a large number of cases, in particular for Cleroidea, and Miguel Angel Alonso Zarazaga, Madrid, Spain, corrected Spanish references.

In addition, the Editors and authors of the present volume are indebted to many individuals who commented on different parts of the Catalogue, added distributional information, completed references, and helped in various ways. These include also coauthors, and are as follows: R. Allemand, Lyon, France; F. Angelini, Brindisi, Italy; A. Assen, Leiden, Netherlands; M. Barclay, London, Great Britain; W. Barries, Wien, Austria; N. Berti, Paris, France; C. Besuchet, Genève, Switzerland; O. Biström, Helsinki, Finland; R. G. Booth, London, Great Britain; L. Borowiec, Wrocław, Poland; B. Brugge, Amsterdam, Netherlands; Y. Bousquet, Ottawa, Canada; M. Brancucci, Basel, Switzerland; M. Budenz, Staten Island, USA; J. Clary, Lyon, France; R. Constantin, Saint-Lô, France; J. Cooter, Hereford, Great Britain; M. Danilevsky, Moscow, Russia; J. Esser, Berlin, Germany; G. Gillerfors, Varberg. Sweden; N. Evenhuis, Honolulu, USA; N. F. Heal, Faversham, Great Britain; J. Hájek, Prague, Czech Republic; H. E. J. Hammond, Edmonton, Canada; B. Jaeger, Berlin, Germany; J. Jelínek, Prague, Czech Republic; B. Jaeger, Berlin, Germany; P. Johnson, La Manga, Spain; C. Johnson, Glossop, Great Britain; S. Jonsson, Uppsala, Sweden; late Z. Kaszab, Budapest, Hungary; I. Kershner, St. Petersburg, Russia; S. Kirejtschuk, St. Petersburg, Russia; J. Kolibáč, Brno, Czech Republic; F. T. Krell, London, Great Britain; H. Labrique, Lyon, France; B. Landry, Genève, Swirtezland; J. F. Lawrence, Gympie, Australia; L. Leseigneur, Meylan, France; R. A. B. Leschen, Auckland, New Zealand; L. Leseigneur, Meylan, France; G. Liberti, Uboldo, Italy; D. V. Logunov, Manchester, Great Britain; I. J. López-Colón, Rivas-Vaciamadrid, Spain; A. Machado Carillo, La Laguna-Tenerife, Spain; D. J. Mann, Oxford, Great Britain; J. Mawdsley, Washington, D. C., USA; O. Merkl, Budapest, Hungary; J. Mertlik, Hradec Králové, Czech Republic; H. Meybohm, Stelle, Germany; G. Nardi, Marmiloro, Italy; H. Mühle, München, Germany; Y. Muona, Helsinki, Finland; K. Needham, Vancouver, Canada; N. Nikitsky, Moscow, Russia; M. V. Nabozhenko, Rostov on Don, Russia; P. Oromi, La Laguna, Canary Isles, Spain; C. G. Paz, Santiago de Compostela, Spain; M. Perreau, Paris, France; G. Platia, Forli, Italy; R. Poggi, Genova, Italy; late A. de Porochin, Tockport, Great Britain; V. Puthz, Schlitz, Germany; S. Reid, York, Great Britain; A. Rey, Genova, Italy; P. Richoux, Lyon, France; G. de Rougemont, Dieppe, France; S. Saluk, Minsk, Belorus; W. Schawaller, Stuttgart, Germany; H. Schoenmann, wien, Austria; R. Schuh, Wien, Austria; S. Shute, London, Great Britain; H. Silfverberg, Helsinki, Finland; P. Skidmore, Elkesley, Great Britain; S. A. Ślipiński, Canberra, Australia; E. Sprecher-Uebersax, Basel, Switzerland; V. Švihla, Prague, Czech Republic; A. Taghavian, Paris, France; M. C. Thomas, Gainesville, USA; T. A. Trikhleb, Donetsk, Ukraine; S. E. Tshernyshev, Novosibirsk, Russia; M. Uhlig, Berlin, Germany; L. Vilhelmsen, Copenhagen, Denmark; M. G. Volkovitsh, St. Petersburg, Russia; O. Vorst, Utrecht, Netherlands; S. A.Williams, Faversham, Great Britain; L. Zerche, Müncheberg, Germany.

The following colleagues identified errors that appeared in the previous volumes of the Catalogue: Miguel A. Alongo Zarazaga, Madrid, Spain; Aleš Bezděk, České Budějovice, Czech Republic; Tristao V. Branco, Porto, Portugal; Robert Constantin, Saint-Lô, France; Anthony E. Davies, Ottawa, Canada; Eduard Jendek, Bratislava, Slovakia; Colin Johnson, Glossop, Great Britain; Mark J. Kalashian, Erevan, Armenia; Vítězslav Kubáň, Brno, Czech Republic; Hans Mühle, München, Germany; Mark G. Volkovitsh, St. Petersburg, Russia; Chi-Feng Lee,

Taipei, Taiwan; Andreas Pütz, Eisenhüttenstadt, Germany; Antonio Rey, Genova, Italy, and Stefano Ziani, Meldola, Italy.

The frontispiece illustration was made by Piotr Wegrzynowicz and provided by Wioletta K. Tomaszewska, both from Warszawa, Poland, and the photography of K. Majer was provided by Jiří Kolibáč, Brno, Czech Republic.

Mr. Daniel Löbl, Genève, Switzerland, developed the software tools needed to generate the Indexes, and was helpful in solving a of number computer problems.

Ivan Löbl is grateful to his wife Daniela for her continuous support.

THE EDITORS

Ivan Löbl
Muséum d'histoire naturelle

Genève, Switzerland

Aleš Smetana
Agriculture and
Agri-Food Canada
Ottawa, Canada

COORDINATOR OF THIS VOLUME

Ivan Löbl, Muséum d'histoire naturelle, Genève, Switzerland

AUTHORS

superfamily ELATEROIDEA Leach, 1815

family ARTEMATOPODIDAE Lacordaire, 1857	L. BOCÁK
family CEROPHYTIDAE Latreille, 1834	L. BOCÁK
family EUCNEMIDAE Eschscholtz, 1829	J. MUONA
family THROSCIDAE Laporte, 1840	L. LESEIGNEUR
family ELATERIDAE Leach, 1815 (- Cebrioninae, Lissominae, Subprotelaterinae)	P. C. CATE
subfamily Cebrioninae Latreille, 1812	A. SÁNCHEZ-RUIZ, I. LÖBL,
subfamily Subprotelaterinae Fleutiaux, 1916	I. LÖBL
subfamily Lissominae Laporte, 1835	A. SMETANA
family PLASTOCERIDAE Crowson, 1972	L. BOCÁK
family DRILIDAE Blanchard, 1845	L. BOCÁK
family OMALISIDAE Lacordaire, 1857	L. BOCÁK
family LYCIDAE Laporte, 1838	M. BOCÁKOVÁ, L. BOCÁK
family PHENGODIDAE LeConte, 1861	L. BOCÁK
family LAMPYRIDAE Latreille, 1817	M. GEISTHARDT, M. SATÔ
family OMETHIDAE LeConte, 1861	L. BOCÁK
family CANTHARIDAE Imhoff, 1856	S. KAZANTSEV, M. BRANCUCCI

superfamily DERODONTOIDEA LeConte, 1861

family DERODONTIDAE LeConte, 1861	J. HÁVA

superfamily BOSTRICHOIDEA Latreille, 1802

family NOSODENDRIDAE Erichson, 1846	F. STEINHAMMER
family DERMESTIDAE Latreille, 1804 (- Thorictinae)	J. HÁVA
subfamily Thorictinae Agassiz, 1846	I. LÖBL
family ENDECATOMIDAE LeConte, 1861	I. LÖBL
family BOSTRICHIDAE Latreille, 1802	J. BOROWSKI
family PTINIDAE Latreille, 1802	
subfamily Gibbiinae Mulsant & Rey, 1868	J. BOROWSKI
subfamily Ptininae Latreille, 1802	J. BOROWSKI
remaining subfamilies	P. ZAHRADNÍK

superfamily Bostrichoidea, incertae sedis

family JACOBSONIIDAE Heller, 1926	I. LÖBL

superfamily LYMEXYLOIDEA Fleming, 1821

family LYMEXYLIDAE Fleming, 1821	G. CUCCODORO

superfamily CLEROIDEA Latreille, 1802

family PHLOIOPHILIDAE Kiesenwetter, 1863	A. MAYOR

family TROGOSSITIDAE Latreille, 1802	J. KOLIBÁČ
family THANEROCLERIDAE Chapin, 1924	J. KOLIBÁČ
family CLERIDAE Latreille, 1802	I. LÖBL, J. ROLČÍK, J. KOLIBÁČ, R. GERSTMEIER
family ACANTHOCNEMIDAE Crowson, 1964	A. MAYOR
family PRIONOCERIDAE Lacordaire, 1857	A. MAYOR
family MELYRIDAE Leach, 1815	A. MAYOR
family DASYTINAE Laporte, 1840	A. MAYOR
family MALACHIIDAE Fleming, 1821	A. MAYOR

superfamily CUCUJOIDEA Latreille, 1802

family SPHINDIDAE Jacquelin du Val, 1860	J. JELÍNEK
family KATERETIDAE Erichson, 1840	J. JELÍNEK
family NITIDULIDAE Latreille, 1802	J. JELÍNEK, P. AUDISIO
family MONOTOMIDAE Laporte, 1840	J. JELÍNEK
family HELOTIDAE Reitter, 1876	P. WEGRZYNOWICZ
family PHLOEOSTICHIDAE Reitter, 1911	P. WEGRZYNOWICZ
family SILVANIDAE Kirby, 1837	D. G. H. HALSTEAD, I. LÖBL, J. JELÍNEK
family PASSANDRIDAE Erichson, 1845	I. LÖBL
family CUCUJIDAE Latreille, 1802	P. WEGRZYNOWICZ
family LAEOPHLOEIDAE Ganglbaeur, 1899	P. WEGRZYNOWICZ
family PHALACRIDAE Leach, 1815	Z. ŠVEC
family CRYPTOPHAGIDAE Kirby, 1837	C. JOHNSON, J. C. OTERO, R. A. B. LESCHEN
family EROTYLIDAE Latreille, 1802	P. WEGRZYNOWICZ
family BYTURIDAE Jacquelin du Val, 1858	I. LÖBL
family BIPHYLLIDAE LeConte, 1861	J. JELÍNEK
family BOTHRIDERIDAE Erichson, 1845	A. ŚLIPIŃSKI
family CERYLONIDAE Billberg, 1820	A. ŚLIPIŃSKI
family ALEXIIDAE Imhoff, 1856	W. K. TOMASZEWSKA
family DISCOLOMATIDAE Horn, 1878	W. SCHAWALLER
family ENDOMYCHIDAE Leach, 1815	
subfamily Merophysiinae Seidlitz, 1872	H. W. RÜCKER, I. LÖBL
remaining subfamilies	W. K. TOMASZEWSKA
family COCCINELLIDAE Latreille, 1807	I. KOVÁŘ
family CORYLOPHIDAE LeConte, 1852	S. BOWESTEAD
family LATRIDIIDAE Erichson, 1842	C. JOHNSON

MAILING AND E-MAIL ADDRESSES OF THE AUTHORS

AUDISIO Paulo, Dipartemento di Biologia Animale e dell'Uomo, Università di Roima "La Sapienza", viale dell'Università 32, I-00185 Roma, Italy. E-mail: paulo.audisio@uniroma1.it

BOCÁK Ladislav, Dept. Zoology, Faculty of Sciences UP, tř. Svobody 26, CZ-771 46 Olomouc, Czech Republic. E-mail: lbocak@lycos.com; ladislav.bocak@upol.cz

BOCÁKOVÁ Milada, Dept. of Biology, Pedagogical Faculty, Palacký University, Purkrabská 2, CZ-771 00 Olomouc, Czech Republic. E-mail: milada.bocakova@upol.cz

BOROWSKI Jerzy, Warszaw Agricultural University, Department of forerst Protection, Rakowiecka 26/30, 02-528 Warszawa, Poland. E-mail: ptinusik@poczta.onet.pl

BOWESTEAD Stanley, 48 Grimshaw Lane, Ormskirk, Lancashire L39 1PD, Great Britain.

BRANCUCCI Michel, Naturhistorisches Museum, Augustinergasse 2, CH-4001 Basel, Switzerland. E-mail: michel.brancucci@bs.ch

CATE C. Peter, Hebragasse 4/18, A-1090 Wien, Austria. E-mail: peter.cate@utanet.at

CUCCODORO Giulio, Muséum d'histoire naturelle, Case postale 6343, 1211 Genève 6, Switzerland. E-mail: giulio.cuccodoro@ ville-ge.ch

GEISTHARDT Michael, Auringer Straße 22, D-65207 Wiesbaden, Germany. E-mail: michael.geisthardt@t-online.de; geisthardt@entomology-iev.de

GERSTMEIER Roland, Technische Universität Muenchen, Lehrstuhl für Tierökologie, Am Hochanger 13, D-85350 Freising, Germany. E-mail: r.gerstmeier@wzw.tum.de

HALSTEAD David G. H., 57 Meadow Way, Old Windsor, Berks SL4 2NY, England

HÁVA Jiří, Private Entomological Laboratory and Collection, Únitice u Prahy, CZ-252 62 Praha-západ, Czech Republic. E-mail: jh.dermestidae@volny.cz

JELÍNEK Josef, The Natural History Museum, Kunratice 1, 148 00 Praha 4, Czech Republic. E-mail: jj.nitidula@seznam.cz

JOHNSON Colin, 17 Peaknaze Close, Simmondley, Glossop, Derbyshire SK 13 9UN, Great Britain.

KAZANTSEV Sergey, Donetskaya 13-326, 109651 Moscow, Russia. E-mail: kazantss@yahoo.com

KOLIBÁČ Jiří, Moravské zemské muzeum, Entomologické odd., Hviezdoslavova 29a, CZ-62700 Brno, Czech Republic. E-mail: ento.kol@volny.cz

KOVÁŘ Ivo, The Natural History Museum, Kunratice 1, 148 00 Praha 4, Czech Republic. E-mail: entomologie@volny.cz

LESCHEN Richard A. B., New Zealand Arthropod Collection, Private Bag 92170, Auckland, New Zealand. E-mail: LeschenR@landcareresearch.co.nz

LESEIGNEUR Lucien, rue des Aiguinards, F-38240 Meylan, France. E-mail: lucien.leseigneur@free.fr

LÖBL Ivan, Muséum d'histoire naturelle, Case postale 6343, 1211 Genève 6, Switzerland. E-mail: ivan.lobl@bluewin.ch

MAYOR Adrien J., Great Smoky Mountains National Park, 107 Park Headquarters Road, Gatlinburg, TN 37738, USA. E-mail: adrien_mayor@nps.gov

MUONA Jyrki, Zoological museum, Division of entomology, P. O. Box 17, FIN 00014 University of Helsinki, Finland. E-mail: jyrki.muona@helsinki.fi

OTERO J. Carlos, Departemente de Biologia Animal, Facultad de Biologia, 15782 Santiago de Compostela, Spain. E-mail: baotero@usc.es

ROLČÍK Jakub, Slavíkova 11, CZ-120 00 Praha 2, Czech Republic. E-mail: jrolcik@mail.ti.cz

RÜCKER H. Wolfgang, Von Ebner-Eschenbach-Strasse 12, D-56567 Neuwied, Germany. E-mail: info@latridiidae.de; wolfgang@wruecker.de

ŚLIPIŃSKI Adam, CSIRO, Canberra, Australia. E-mail: adam.slipinski@csiro.au

SÁNCHEZ-RUIZ Antonio, Depto. de Biodiversidad y Biología Evolutiva, Museo Nacional de Ciencias Naturales, Jose Gutierrez Abascal, 2, E-28006 Madrid, Spain. E-mail: a.s.r@wanadoo.es

† SATÔ Masataka, Dia Cuore Tokushige 306, Kamegahora 3-1404, Midoriku, Nagoya, 458-0804 Japan

SCHAWALLER Wolfgang, Staatliches Museum für Naturkunde, Rosenstein 1, D-70191 Stuttgart, Germany. E-mail: schawaller.smns@naturkundemuseum-bw.de

ŚLIPIŃSKI Adam S., Division of entomology, CSIRO, GOP Box 1700, Canberra, ACT "601, Australia. E-mail: Adam.Slipinski@csiro.au

SMETANA Aleš, Biodiversity, Systematic entomology, Agriculture and Agri-Food Canada, Ottawa, Canada. E-mail: smetanaa@agr.gc.ca

ŠVEC Zdĕnek, Žerotinská 4, CZ 170 00 Praha 7, Czech Republic. E-mail: zdenek.svec@volny.cz; zdenek.svec@szpi.gov.cz

STEINHAMER Frank, Sylter Strasse 18, D-90425 Nürnberg, Germany

TOMASZEWSKA Wioletta K., Museum & Institute of Zoology, Polish Academy of Sciences, ul. Wilcza 64, 00-679 Warszawa, Poland. E-mail: wiola@wa.home.pl; wiolkat@miiz.waw.pl

WEGRZYNOWICZ Piotr, Museum & Institute of Zoology, Polish Academy of Sciences, ul. Wilcza 64, 00-679 Warszawa, Poland. E-mail: piotr@ miiz.waw.pl

ZAHRADNÍK Peter, Forestry and Game Management Institute, Jíloviště-Strnady, CZ 156 04 Praha 5 – Zbraslav, Czech Republic. E-mail: zahradnik@vulhm.cz

ERRATA for VOLUME 1

p. 35, add: SZ to *Hygrobia hermanni* Fabricius
p. 78, add: SP to *Rhysodes sulcatus* Fabricius
p. 524, correct spelling: *Amphyginus* Haliday, from *Amphiginus* Haliday

ERRATA for VOLUME 2

p. 19, "p. 455: *Agonum thoracicum*", replace "Dejean's name" with "Fischer von Waldheim's name"
p. 24 and index, *Eupachylopus ripae* (Lewis, 1885), change [2x] to: *Eopachylopus*
p. 26, *Euconnus henrii* Davies, nom. nov. for *Euconnus coiffati* Franz, 1985, change [2x] to *coiffaiti*
p. 26, *Euconnus rufescens* Ganglbauer, 1899 syn. nov. of *Euconnus motschulskii* Motschulsky, 1837, change to: (Motschulsky, 1837)
p. 27, *Stenichnus protervus*, synonym *Stenichnus trunctatus* (Coquerel, 1860), change to: *truncatus*
p. 28, *Bythinus grandipalpis* Stephens, 1832 syn. nov., change to *grandipalpus* Stephens, 1835
p. 29, *Tychus bysantinicus* Karaman, 1955, change to: (Karaman, 1955)
p. 51, *Enochrus nigriceps* L. Redtenbacher, 1867: 26 (*Philhydrus*) [HN], correct year: 1868
p. 64, *Cercyon sternalis* Sharp, 1918: 276 and *pumilio* Sharp, 1918: 277, remove (*Cercyon*) from both
p. 87, *Eblisia*, *cavifrnos* Leoni, 1907, change to: *cavifrons*; *marginata* C.G. Thomson, 1867a: 397 (*Platysomai*), change to: (*Platysoma*)
p. 90, *Glymma candezii* Marseul, delete synonym: *candezei* Jacquelin du Val, 1858: 31 [RN] [should be "candezii" pl. 31, fig. 153, and not RN]
p. 92, genus *Gnathoncus* Jacquelin du Val, 1858: 112, correct year: 1857
p. 126, *Ptilium haemorrhoidale* Boheman, 1844: 156, remove (*Ptilium*)
p. 127, *Cissidium matthewsi* Johnson was erroneously given as a replacement name for *Camptodium adustipenne* A. Matthews, 1884. In fact, the latter is a misidentification of *C. adustipenne* Motschulsky, 1855.
Cissidium matthewsi Johnson **sp. nov**. was characterized in Matthews, 1884: 83. Holotype male: [Japan] Manazuru, Kanagawa Pref., 3-Nov. 1967, M. Kubota leg./ / Manchester Museum Holotype/ (in Manchester University Museum, MMUE).
Paratypes: Japan: 1 male, 1 female, same data; 2 females, same data, Odawara, 3-7.iv.1946; 3, same data, 28.iii. 1971; 1 male, same data, 21.ii.1970 (in Manchester University Museum, MMUE).

p. 130, *Acrotrichis* (*A.*) *williamsi* Johnson was erroneously given as a replacement name for *Acratrichis fascicularis* Wollaston, 1854. In fact, the latter is a misidentification of *A. fascicularis* (Herbst, 1793).
Acrotrichis (*A.*) *williamsi* Johnson, **sp. nov**. resembles *A. danica* Sundt, 1958 in size, colour, eye facets, mentum, structure and pale colour of spermatheca, but differs in fine (not coarse) pronotal granulation, pronotal side-edge from 45 degrees more rounded (not feebly curved) and approaching *fascicularis* (Herbst, 1793), elytra straight sided in both sexes (not apically narrowed in male) and in slight apical (not strong) narrowing of male aedeagus.
Holotype male: Madeira, iv.72, Levada de Toros, S.A. Williams 800 m/ /Manchester Museum Holotype / (in Manchester University Museum, MMUE).
Paratypes: Madeira: 3 males, 5 females, same data; 1 male, 1 female, same data, Encumeada, 1000 m; 1 female, same data, Funchal; 3 females, Ribeiro Frio, Balcões, 18.iv.1981, laurisilva litter, leg. J.G. Blower; 1 male, same locality, iv.1981, leg. R.R. Askew; 1 female, Faja de Noguiero, 17.iv.1981, Ocatea litter, leg. R. R. Askew (in Manchester University Museum, MMUE).

p. 165, *Drimeotus viehmanni* Ienistea, change to: Ieniştea
p. 169, *Pholeuon comani* Ienistea and *knirschi serbani* Ienistea, change to: Ienіştea
p. 172, *Soprochaeta dacica* and *orghidani* Ienistea, change to: Ienіştea
p. 181, tribe Agathidiini Westwood, 1840, correct date: 1838
p. 206, genus *Etelea*, correct spelling: *tenuis* Petri
p. 211, *Euconnus coiffati* Franz, 1962: 228, change to: *coiffaiti*

p. 213, add *Euconnus chinensis* Li J. & Chen P., 1993: 163 **A**: "China"
p. 217, *Neuraphes diocletianus* Reitter, 1882e: 566, remove (*Neuraphes*)
p. 224, *Scydmaenus reversus* Sharp, 1874c: 128, remove (*Scydmaenus*)
p. 225, *Scydmaenus antidotus* Schaum, 1841: , remove (*Scydmaenus*); add: page 25
p. 230, *Eusilpha jakowlewi similator* Shibata, 1969: 51, remove (*Eusilpha*)
p. 234, *Eonecrophorus tenuicornis* Kurosawa, 1985: 47, remove (*Eonecrophorus*)
p. 252, *Trichodromeus penicillatus* Reitter, synonym *sogdianus* A. Bordoni, remove (*Trichodromeus*)
p. 265, *Phyllodrepa atricapilla* Bernhauer, 1903c: 191, remove (*Phyllodrepa*)
p. 300, *Prosthecarthron sauteri* Raffray, 1914a: 3, remove (*Prosthecarthron*)
p. 304 and 310, remove *bulgaricus* Reitter, 1880f: 535 (*Bythinus*) from page 304 and place it in synonymy with *simoni* Reitter, p. 310
p. 327, *Pselaphus pilicollis* Reitter, 1883d: 293, remove (*Pselaphus*)
p. 340, *Lamprinodes hammarstroemi* Luze, 1901a: 182, remove (*Lamprinodes*)
p. 341, *Sepedophilus bipunctatus*, synonym *corybantes* Gistel, change (*Tchyporus*) to: *Tachyporus*
p. 367, *Alevonota sericata* Pace, 1998e: 947, remove (*Alevonota*)
p. 375, *Atheta hypnorum* Kiesenwetter, synonym *rufipes* J. Sahlberg, 1876, remove (*Atheta*)
p. 385, *Atheta glabriculoides* A. Strand, 1958: 355, remove (*Atheta*)
p. 391, *Atheta ingenua* Pace, 1993b: 104, remove (*Atheta*)
p. 393, *Atheta sororcula* Cameron, synonym *pseudocaesula* Cameron, 1944a, remove (*Atheta*)
p. 394, *Thinobaena* Thomson, 1859: 39 (= *Aleochara vestita* Gravenhorst, 1806), change to: = *Paederus vestitus*
p. 394, *vestita* Gravenhorst, 1806: 140 (*Aleochara*), change to: (*Paederus*)
p. 409, *Hydrosmecta tenuis* Cameron, 1939a: 283, remove (*Hydrosmecta*)
p. 446, *Leptarthrophaena* Scheerpeltz, 1948b: 164, change to: 1948a: 170
p. 483, *Oxypoda orousetti* Pace, 1992e: 321 **A**: NP, change to: *orousseti*
p. 496, *Scaphidium cinnamoneum* Champion, 1927c: 270, change to: *cinnamomeum*
p. 534, *Thinobius* [incertae sedis] *angusticeps* Fauvel, 1889a: 83, remove (*Thinobius*)
p. 545, *Stenus prometheus* Puthz, 1967d: 232, delete ": TR"
p. 602, *Tetartopeus sibiricum* Eppelsheim, 1876, change to: *sibiricus*
p. 603, *Tetartopeus gracile* C. Hampe, 1867: 371, change to: *gracilis*
gracile Poppius, 1909a: 22, change to: *gracilis*
punctatum Zetterstedt, 1828: 84, change to: *punctatus*
p. 613, *Paederus gottschei* H. Kolbe, 1886: 179, remove (*Paederus*)
p. 630, *Bisnius protenus* Schubert, 1906: 382, add: (*Philonthus*)
p. 640, *Philonthus aberrans* Cameron, 1932: 111, remove (*Philonthus*)
p. 679, *Ocypus queinneci* Coiffait, 1984d: 379, remove (*Ocypus*)
p. 681, *Platydracus perreaui* Coiffait, 1984d: 378, remove (*Platydracus*)
plagiicollis Fairmaire, 1891: cxci, add: (*Ocypus*)
p. 690, *Leptacinus gracilis* Fauvel, 1895b: 240, remove (*Leptacinus*)
p. 693, *hongkongensis* L. Redtenbacher, 1867: 28 (*Xantholinus*), correct year: 1868
p. 751, add reference: Falcoz L. 1931. Notes pour la connaissance de la faune française. Coléoptères. *Annales de la Société Linnéenne de Lyon* **75** [1930]: 133-138.
p. 772, Heller K. M. 1910, [correct to:] Fünfter Beitrag zur Papuanischen Käferfauna ... *Abhandlungen und Berichte des Königlichen Zoologischen und Anthropologisch-Ethnographischen Museums zu Dresden*
p. 781, Jacquelin du Val P. N. C. 1856 [correct to:] Pp. 1-40. In: Famille des staphylinides. In: *Manuel Entomologique. Genera des coléoptères d'Europe comprenant leur classification en familles naturelles, la description de tous les genres, des tableaux synoptiques destinés à faciliter l'étude, le catalogue de toutes les espèces de nombreux dessins au trait de caractères. Tome Deuxième.* Paris: A. Deyrolle, 285 + [2] pp.
p. 782, Jacquelin du Val P. N. C. 1857b [correct to:] Pp. 41-128, pls. 1-67. In: *Manuel Entomologique. Genera des coléoptères d'Europe.... Tome Deuxième.* Paris: A. Deyrolle, 285 + [2] pp.
delete reference: Jacquelin du Val P. N. C. 1858: *Genera des coléoptères d'Europe...*
p. 782, Jacquelin du Val P. N. C. 1868. *Genera des coléoptères d'Europe...*, add "*Tome deuxième*"
p. 802, LePeletier A. & Serville J. 1825: [new taxa]., add: "In: Pp. 1-344." and change "344 pp." to "832 pp."
p. 838, Pace R. 1992e: Aleocharinae nepalasi del Museo di Genevra, change to: nepalesi

p. 856, Redtenbacher L. 1867: Reise der Österreichischen Fregatte Novara..., correct year: 1868 [see: Higgins, 1964: *J. Soc. Bibl. Nat. Hist.* 4: 158]
p. 878, Scheerpeltz O. 1948a: , change "[new taxa]" to: "Familie: Staphylinidae. (Kurzflügler, Raubkäfer, "Stapheln") (45. Beitrag zur Kenntnis der paläarktischen Staphyliniden.). Pp. 149-220."
p. 910, Westwood J. O. 1840, change to: Synopsis of the genera of British insects. Pp. 81-158. In: An introduction... [only pp. 81-158 are 1840]
p. 915 [index], Agathidiini Westwood, 1840, correct date: 1838
p. 927 [index], *Gnathoncus* Jacquelin du Val, 1858, correct year: 1857

ERRATA for VOLUME 3

Due to a problem in formatting, a number of special characters have not been taken over in the definitive pdf file. The affected author names are:
Bezdek/Bezděk, Ciampor/Čiampor, Endrodi/Endrődi, Graells/Graëlls, Jirí Hájek/Jiří Hájek, Holynski/Hołyński, Vítezslav Kubán/ Vítězslav Kubáň, Mikšic/Mikšić, Rakovic/Rakovič, Ružicka/Růžička, Tesar/Tesař, Uhagon/Uhagón, Všetecka/Všetečka.

Correct years of taxa published by W. W. Saunders: from 1872 to 1852, from 1874 to 1854, and from 1879 to 1839.

p. 19, correct spelling: Euchirinae Burmeister, 1842
p. 20, correct authorship: the first author of Julodinae is V. KUBÁŇ
p. 20, correct authorship of Agrilinae for V. KUBÁŇ to: (without genus *Agrilus* and subtribe Aphanisticina)
p. 20, add as authors of the subfamily Agrilinae subtribe Aphanisticina: V. KUBÁŇ, M. J. KALASHIAN
p. 21, correct e-mail for M. J. Kalashian: mkalashian@yahoo.com
p. 51, correct under *Buprestis haemorrhoidalis sibirica* Fleischer: Art. 23.9.5
p. 63, add: WS to *Ceruchus chrysomelinus* Hochenwart
p. 63, add: FE (Kuril) to *Ceruchus lignarius lignarius* Lewis
p. 63, add: **A**: KZ MG to *Sinodendron cylindricum* Linnaeus
p. 65, add: TM, delete RU for *Lucanus ibericus* Motschulsky
p. 65, add: FE (Kuril) to *Lucanus maculifemoratus maculifemoratus* Motschulsky
p. 66, *Pseudolucanus*, correct original combination and date of type species: *Scarabaeus capreolus* Linnaeus, 1764
p. 67, correct year for *coerulescens* Uddman (in synonymy with *Platycerus caraboides*) to 1790
p. 67, add: FE (Skovorodino District) to *Platycerus caprea* De Geer
p. 72, add: **E**: KZ to *Dorcus parallelipipedus* Linnaeus
p. 82, correct year for *Bolbaffroides validus* (Klug) to 1845
p. 82, *Bolbelasmus*, correct spelling of type species: *Bolboceras gallicum* Mulsant, 1842 (the gender of *Bolboceras* is neuter)
p. 86, *Chelotrupes*, correct type species: *Scarabaeus momus* Fabricius, 1792 (= *Scarabaeus momus* A. G. Olivier, 1789)
p. 86, subgenus *Typhaeus*, delete from the list of synonyms:
- "*Armidens* Boucomont, 1902: 12 [unjustified emendation]". *Armidens* Boucomont, 1912 (not 1902): 12 is an incorrect subsequent spelling and, as such, an unavailable name.
- "*Minotaurus* Jekel, 1866: 546 [HN] type species: *Scarabaeus typhoeus* Linnaeus, 1758". Jekel (1866) credited the name *Minotaurus* to Mulsant, obviously an error for Mulsant & Godart, and designated *Scarabaeus typhoeus* Linnaeus, 1758 as its type species

p. 89, genus *Sericotrupes*, correct original combination of type species: *Scarabaeus niger* Marsham, 1802
p. 89, *Thorectes*, correct: type species *Scarabaeus laevigatus* sensu Mulsant, 1842 (= *Scarabaeus intermedius* O. G. Costa, 1839)
p. 89, add: *Zuninoeus* López-Colón, 1989 (p. 92) in synonymy of *Thorectes*
p. 90, add: *hoppei* Sturm & Hagenbach, 1825: 486 (*Geotrupes*) **E**: CR IT SK SL, and its synonyms (from p. 92)
p. 92, delete: *Zuninoeus* López-Colón, 1989

p. 95, correct page for *Codocera ferruginea* Eschscholtz, 1818 to 451
p. 95, genus *Ochodaeus*, correct type species: *Melolontha chrysomelina* Fabricius, 1775 (= *Scarabaeus chrysomeloides* Schrank, 1771)
p. 95, add: **A**: "North China" to *Codocera ferruginea chinensis* Balthasar
p. 100, add: **A**: TR to *Glaphyrus cinnanerinus* Schweiger
p. 101, add: **A**: TR to *Glaphyrus ornatus* Schweiger
p. 101, add: **A**: IQ to *Glaphyrus pubescens* Schweiger
p. 103, *Aegialia*, correct type species: *Scarabaeus globosus* Kugelann, 1794 (= *Scarabaeus arenarius* Fabricius, 1787)
p. 110, *Amidorus*, correct type species: *Aphodius sericatus* W. Schmidt, 1840 (= *Scarabaeus obscurus* Fabricius, 1792)
p. 123, *Eudolus*, correct date of type species: *Scarabaeus quadriguttatus* Herbst, 1783
p. 128, *Mendidaphodius*, correct original combination of valid name of type species: (= *Mendidius brancsiki* Reitter, 1899)
p. 136, *Planolinus*, correct type species: *Scarabaeus foetidus* Fabricius, 1792 (= *Scarabaeus fasciatus* A. G. Olivier, 1789)
p. 138, *Subrinus*, correct authorship of type species: *Aphodius illigeri* Mulsant & Rey, 1870
p. 141, *Oxyomus*, correct type species: *Scarabaeus porcatus* Fabricius, 1775 (= *Scarabaeus sylvestris* Scopoli, 1763)
p. 143, *Theotimius* Huchet, 2000 is erroneously listed within Didactyliina Pittino, 1985. It should be placed in Chironinae Blanchard, 1845 (p. 150)
p. 144, *Parataenius*, add to type species: (= *Aphodius derbesis* Solier, 1851)
p. 144, *Brindalus*, correct type species: *Phycochus azoricus* Landin, 1960 (= *Aphodius porcicollis* Illiger, 1803)
p. 145, *Psammodius*, correct type species: *Aphodius sulicollis* Illiger, 1802 (= *Scarabaeus asper* Fabricius, 1775)
p. 147, *Platytomus*, correct type species: *Platytomus sabulosus* Mulsant, 1842 (= *Scarabaeus tibialis* Fabricius, 1798)
p. 147, add: *jailensis* Apostolov & Maltzev, 1986: 88 (*Pleurophorus*) in synonymy of *Platynotus variolosus* Kolenati, 1846
p. 147, *Pleurophorus*, correct authorship of type species: *Scarabaeus caesus* Panzer, 1796
p. 148, *Rhyssemus*, correct type species: *Scarabaeus asper* Fabricius, 1775 sensu Mulsant, 1842 (= *Ptinus germanus* Linnaeus, 1767)
p. 150, add: *Rhyparus nepalensis* Balthasar, 1971a: 19 **A**: NP
p. 150, add: *Rhyparus schachti* Balthasar, 1971a: 21 **A**: NP
p. 154, *Gymnopleurus*, correct type species: *Scarabaeus geoffroyi* Fuessly, 1775
p. 158, *Bubas*, correct authorship and date: *Bubas* Dejean, 1833: 143
p. 159, *Onitis*, correct type species: *Scarabaeus sphinx* A. G. Olivier, 1789 [HN] (= *Onitis belial* Fabricius, 1798)
p. 161, correct spelling: Cambefort
pp. 176 and 177, *Scarabaeus*, the validly designated type species is *Scarabaeus hercules* Linnaeus, 1758. However, it is recommended that current usage, i.e., type species *Scarabaeus sacer* Linnaeus, 1758, should be maintained until a case is made to the International Commission on Zoological Nomenclature and a ruling is published
p. 176, *Ateuchetus*, correct type species: *Scarabaeus laticollis* Linnaeus, 1758
p. 176, correct year for *reflexicornis* L. Redtenbacher, 1867 (in synonymy with *Onthophagus spinifex*) to 1868
p. 177, *Scarabaeus*, correct name of third synonym: *Heliocantharus* MacLeay, 1821
p. 181, correct spelling: Euchirinae Burmeister, 1842
p. 184, correct year for *Ectinohoplia* L. Redtenbacher and *E. sulphuriventris* L. Redtenbacher to: 1868
pp. 184 and 186, *Hoplia*, correct type species: *Scarabaeus farinosus* Linnaeus, 1761 (= *Scarabaeus argenteus* Poda von Neuhaus, 1761)

p. 184, *Decamera*, correct type species: *Melolontha pulverulenta* Fabricius, 1775 sensu Mulsant, 1842 (= *Scarabaeus philanthus* Fuessly, 1775)
p. 186, add: **E**: GR to *Hoplia caucasica danieli* Tesař
p. 186, delete: *Echyra* Erichson, 1847 from synonymy with *Hoplia*; *Echyra* is a valid Madagascar taxon
p. 191, *Anoxia*, correct date and page of synonym: *Catalasis* Dejean, 1833: 159
p. 191, add: **A**: HAI to *Malaisius hainanensis* Y. Zhang
p. 192, *Mesanoxia*, correct authorship of type species: *Melolontha australis* Schönherr, 1817
p. 201, add: **A**: AF to *Hemictenius petrovitzi* Gusakov
p. 214, add: KZ to *Chioneosoma komarowi* Brenske
p. 218, correct year for *Ancylonycha nigra* L. Redtenbacher to 1868
p. 218, correct year *Holochelus* (*Miltotrogus*) *vernus* Germar to 1824
p. 222, *Monotropus*, delete synonym; *Paramonotropus* is not an available name
p. 226, correct year for *jeannei* Baraud (synonym with *Rhizotrogus monticola*) to: 1970b
p. 229, *Apotriodonta*, correct original combination of type species: *Triodonta hispanica* Baraud, 1962
p. 239, add: **N**: MO to *Neomaladera neglecta* Baraud
p. 248, add: **A**: TD to *Trochaloschema valentini* Novikov
p. 251, add: KZ to *Pseudadoretus phthisicus* Dohrn
p. 251, *Anisoplia*, correct authorship and date: *Anisoplia* Schönherr, 1817: 186. Futhermore, de validly designated type species is *Scarabaeus horticola* Linnaeus, 1758; however, it is recommended that current usage, i.e., type species *Scarabaeus agricola* Poda von Neuhaus, 1761, should be maintained until a case is made to the International Commission on Zoological Nomenclature and a ruling is published
p. 256, correct year for *Phyllopertha chinense* L. Redtenbacher to 1868
p. 258, add: **A**: HAI to *Anomala cariniventris* Lin
p. 260, add: **A**: HAI to *Anomala flavilimbata* Lin
p. 261, add: **A**: HAI to *Anomala myriopila* Lin
p. 261, add: **A**: HAI to *Anomala olivea* Lin
p. 262, add: **A**: HAI to *Anomala pilicauda* Lin
p. 262, add: **A**: HAI to *Anomala profundisulca* Lin
p. 262, add: **A**: HAI to *Anomala projecta* Lin
p. 271, delete "KZ" from *Trigonocnemis lanuginosa* Semenov
p. 273, correct year for *Popilia cerchnopyga* Lin to 1981
p. 277, correct year for *Myrina pfeifferi* L. Redtenbacher to 1868 (2x)
p. 281, add: **A**: HAI to *Microryctes hainanensis* Y. Zhang
p. 280, correct year for *Heteronychus lioderes* L. Redtenbacher to 1868
p. 282, *Phyllognathus*, correct type species: *Scarabaeus silenus* Fabricius, 1775 (= *Scarabaeus excavatus* Forster, 1771)
p. 282, *Oryx* (synonym of *Phyllognathus*), delete type species; no type species has ever been validly designated
p. 284, add: KZ UZ to *Aethiessa szekessyi* Brasavola de Massa
p. 284, correct year and page for *Cetonia aurata aurata* Linnaeus to: 1758: 352
p. 284, *Cetonia*, correct date of type species: *Scarabaeus auratus* Linnaeus, 1758
p. 289, *Protaetia*, correct type species: *Cetonia spectabilis* Schaum, 1841
p. 289, *Cetonischema*, correct date of type species: *Scarabaeus aeruginosus* Drury, 1773
p. 289, *Cetonischema*, add as synonym: *Potosia* Mulsant & Rey, 1871: 413 type species *Cetonia speciosissima* Scopoli, 1876 (= *Scarabaeus aeruginosus* Drury, 1773)
p. 294, delete subgenus *Potosia*
pp. 294-296, move species listed under subgenus *Potosia* to the subgenus *Netocia* (p. 291)
p. 296, add: [HN] to *metallica* Paykull, 1799: 203 (*Cetonia*)
pp. 297 and 298, *Tropinota*, correct type species: *Tropinota reyi* Mulsant, 1842 (= *Scarabaeus squallidus* Scopoli, 1763)
p. 298, *Oxythyrea*, correct date of type species: *Scarabaeus sticticus* Linnaeus, 1767
p. 309, add: FE (Kuril) to *Gnorimus viridiopacus* Lewis
p. 311, correct year for *mandarinus* L. Redtenbacher (in synonymy with *Trichius bowringii* J. Thomson) to 1868
p. 312, *Dasyvalgus formosanus* Moser, 1915 is a correctly in synonym with *D. luzonicus* Kraatz, 1883; delete it

from list of valid names
p. 314, correct year for *Nycteus meridionalis* Laporte to: 1838
p. 324, add to Dascillidae, Karumiinae:

tribe Emmitini Escalera, 1914

Emmita Escalera, 1914: 349 [RN] type species *Emmita serricornis* Escalera, 1913
Emma Escalera, 1913: 319 [HN]
serricornis Escalera, 1913: 320 (*Emma*) **N**: MO

p. 325, add: superfamily Dascilloidea, incertae sedis
Spermodenops mollipennis Abeille de Perrin, 1881: 87 **A**: "Palestina"
p. 326, correct reference for *Julodella cymbiformis* Bílý to: 1990b
p. 329, delete [DA] from *Julodis zablodskii*
p. 330, correct authorship for *Acmaeodera chalcithorax* to: Volkovitsh, 1986a: 19
p. 330, correct spelling for *Acmaeodera moralesi* Espanol & Mateu to: *M. moralesi* Español & Mateu
p. 331, correct authorship for *Acmaeodera crinita crinita* to: Gory, 1840: 41
pp. 331, 337, 366, change for *Acmaeodera pilosellae persica*, *Acmaeoderella vetusta*, *Sphenoptera melitta* and *S. rangnovi* AM to AR
p. 331, 339, delete MN from *Acmaeodera pilosellae pilosellae* and *Acmeoderella adspersula adspersula*
p. 332, correct reference for *Acmaeodera ottomana ottomana* Frivaldszky von Frievald to: 1837: 175
p. 332, add *Acmaeodera deprivata* Volkovitsh, 1983: 564 as synonym with *A. simulans simulans* Abeille de Perrin, 1891
p. 332, correct spelling of Abeile to: Abeille
p. 333, delete [RN] from *vaillant* Spinola, 1838
p. 333, delete *kureimatica* Cobos, 1958 from synonymy of *Acmaeodera nigellata* Abeille de Perrin, 1904
p. 334, correct spelling for Thery to: Théry
p. 334, delete *punctatissima* J. Thomson, 1878 from synonymy of *Acmaeodera polita polita* Klug, 1829
p. 334, delete *turanica* Obenberger from synonymy of *Acmaeoderella caspica suturifera* Reitter, 1904
p. 334, correct reference for *Acmaeoderella caspica turkestanica* Obenberger to: 1934o
p. 334, correct reference for *faldermanni* Obenberger and *florilega* Obenberger (synonyms with *Acmaeoderella circassica*) to: 1934o
p. 335, correct spelling for *Acmaeoderella albosetosa* Rosenhouer to *A. albosetosa* Rosenhauer
p. 335, correct authorship and data for *Acmaeoderella jonica* to: Volkovitsh, 1989a: 58
p. 335, correct page for *Acmaeoderella dermestoides* Frivaldszky von Frivald to: 180
p. 336, change for *Acmaeoderella flavofasciata flavofasciata* SR to YU, and delete MN
p. 336, correct reference for *Acmaeoderella dilatisquamis* Obenberger to: 1928b
p. 337, correct spelling for *Acmaeoderella cyrta* Iablokov-Khnzorian to: Iablokoff-Khnzorian
p. 337, correct year for *Acmaeoderella squamosa* Théry to: 1914
p. 337, correct reference for *Acmaeoderella henoni* to: [= 1894a: 48]
p. 337, correct spelling for *Acmaeoderella vetusta* Menetries to: *A. vetusta* Ménétriés
p. 338, correct spelling from *Acmaeoderella aleshi* Obenberger to: *A. alesi* Obenberger
p. 338, delete [= 1894a: 48] from *ordinata* Abeille de Perrin and add it to *regularis* Abeille de Perrin (in synonymy with *Acmaeoderella rufomarginata*)
p. 341, correct author for *Thurntaxisia cottyi* from Fairmaire & Coquerel, 1866 to: Fairmaire, 1866a
p. 343, correct spelling and data for *Chrysochroa fulgidissima alternans* Watterhouse to: *C. fulgidissima alternans* Waterhouse, 1888b
p. 344, change for *Steraspis squamosa squamosa* AE to EG
p. 344, correct spelling for *Chalcophora japonica chinensis* Schauffuss to *C. japonica chinensis* Schaufuss
p. 344, add original combination (*Buprestis*) to *Chalcophora japonica japonica* Gory, 1840
p. 344, correct reference to [*Steraspis*] *intermedia* Kerremans, 1891a: clvii to 1891: clvii
pp. 344, 346, 372, change for *Chalcophora detrita marani*, *Capnodis porosa porosa* and *Anthaxia lucens lucens* BG to BU
p. 345, correct reference for *Sapaia* Bílý and *S. brodskyi* Bílý from 1994c to: 1994a
p. 345, add original combination (*Chalcophora*) to *Chalcophorella quadrioculata* L. Redtenbacher
p. 346, change for *Capnodis semisuturalis* LI to LE

p. 347, correct spelling *Cyphosoma lausoniae* to: *C. lawsoniae*
p. 348, add original combination (*Buprestis*) to *Lampetis mimosae interrupta* Laporte & Gory
p. 348, correct year for *Lampetis nigritorum* Laporte & Gory to 1837
p. 351, correct spelling *Lamprodila savoi* to: *L. savioi*
p. 351, correct reference for *ottomana* Obenberger (synonym with *Lamprodila balcanica*) to: 1952e
p. 351, delete *bella* Gory, 1840 from *Palmar* and add it as nomen dubium in *Palmar* to p. 352
p. 351, add: **A**: JA TAI to *Lamprodila davidis intermedia* Kurosawa, 1956
p. 351, delete JA TAI from *Lamprodila davidis davidis* Fairmaire
p. 352, correct reference for *beata* Obenberger (in synonymy with *Lamprodila virgata*) to: 1952e
p. 352, add: **A**: "India" to *Buprestis plebeja*
p. 352, correct reference for *anchorifera* Bílý (in synonymy with *Sphenoptera anchorata* Bílý) to 1994c
p. 355, correct authors and refence for *henonii* Fairmaire, 1866a (in synonymy with *Sphenoptera cylindricollis* Marseul) to: *henonii* Fairmaire & Coquerel, 1866
p. 355, change for *Sphenoptera puberula* TA to TD
pp. 355, 373, change for *Sphenoptera puberula* and *Anthaxia richteri* KG to KI
p. 356, add to subgenus *Chrysoblemma*: *mutilata* Théry, 1929b: 130 **N**: EG
p. 356, correct spelling of *Chrysoblemma scovtzi alfierii* to *scovitzii alfierii*
p. 358, add to subgenus *Deudora*: *antoinei* Théry, 1930a: 182 **N**: MO
p. 366, add *darius* Obenberger, 1927a: 81as synonym with *Sphenoptera repetekensis* Obenberger, 1927
p. 368, correct reference for *paradoxa* Kerremans (in synonymy with *Sphenoptera konbirensis*) to Kerremans, 1913c
p. 369, add: genus *Sphenoptera*, nomen dubium
parvula Gory & Laporte, 1839: 20 [HN] **E**: IT
p. 369, correct spelling and reference *hilaris* Mulsant & Godard, 1855 (in synonymy with *Anthaxia anatolica* Chevrolat) to Mulsant & Godart, 1855a
p. 370, delete *hirticollis* Abeille de Perrin, 1893 and *perrini* Obenberger, 1918 from synonymy of *Anthaxia cypraea* Abeille de Perrin
p. 371, correct spelling *Anthaxia maschelli* to: *A. maschelii*
p. 372, correct page for *Anthaxia cyanipennis* Gory, 1841 from 268 to: 286
p. 372, delete *lyciae* Magnani, 1996 from subgenus *Anthaxia* Eschscholtz, 1829
p. 373, correct spelling for *Anthaxia plavilschikovi* Obenberger to: *plaviscikovi* Obenberger
p. 373, correct reference for *Anthaxia plicata plicata* Kiesenwetter to: 1859b
p. 373, correct spelling for *Anthaxia sedilloti rabinowitschi* Théry to: *sedilloti rabinovitchi* Théry
p. 373, correct spelling for *Anthaxia shirazensis* Obenberger to: *shirasensis* Obenberger
p. 374, change spelling for *Anthaxia mirabilis* Zikharev, 1918 to: Zhicharev, 1918
p. 378, correct reference for *Anthaxia ursulae* Niehuis to: 1989b
p. 378, correct reference for *Haplanthaxia sinica* Bílý to 1994b
p. 379, add *frankenbergeri* Obenberger, 1914b: 14 as synonym with *Anthaxia godeti* Gory & Laporte, 1839
p. 379, delete AU from distribution of *Anthaxia kubani*
pp. 379, 380, correct spelling Bercedo-Párano to: Bercedo-Páramo
p. 380, correct reference for *Anthaxia pinquis* Kirsch to: 1880a
p. 380, correct year for *Anthaxia potanini* Ganglbauer to: 1890
p. 380, correct data for *Anthaxia pinguis* Kirsch to 1880a
p. 382, *Buprestis novemmaculata novemmaculata* Linnaeus, correct **AFR** to: **AFR**i
p. 383, add **E**: IT to *Eurythyrea fastidiosa* (P. Rossi)
p. 384, correct reference for *Chrysobothris andamana* Kerremans, 1891a: clxii to: 1891: clxii
p. 385, correct reference for *beesoni* Obenberger (in synonymy with *Chrysobothris parvipunctata* Obenberger) to 1928a
p. 386, correct reference for *quadridens* Rey (in synonymy with *Chrysobothris solieri* Laporte & Gory) to 1890b
p. 386, add to *Chrysobothris*: *znojkoi* Semenov & Richter, 1934: 93 **A**: KA
p. 388, change for *Agrilus cuprescens chrysoderes* **E**: TR to **A**: TR
p. 389, add **A**: TR to *Agrilus integerrimus*
p. 389, add **A**: TR to *Agrilus lineola lineola*
p. 389, add **A**: TR to *Agrilus salicis*

p. 389, change for *Agrilus tschitscherini* **E**: KZ to **A**: KZ
p. 390, add **A**: KZ to *Agrilus validiusculus*
p. 390, add **A**: TR to *Agrilus convexicollis*
p. 390, add **A**: KZ TR to *Agrilus cyanescens cyanescens*
p. 391, change for *Agrilus ganglbaueri* **E**: KZ to **A**: KZ
p. 391, add **A**: TR to *Agrilus buresi*
p. 392, add: **A**: TR to *Agrilus grandiceps hemiphanes*
p. 392, add GE to *Agrilus hastulifer hastulifer*
p. 392, add: **E**: KZ **A**: KZ to *Agrilus sulcicollis*
p. 393, change for *Agrilus auricollis korbi* **E**: TR to **A**: TR
p. 393, add **A**: KZ to *Agrilus betuleti*
p. 394, add **A**: KZ to *Agrilus subauratus subauratus*
p. 394, add: **A**: TR to *Agrilus marozzinii*
p. 394, add: **A**: TR to *Agrilus roscidus*
p. 394, add: **A**: TR to *Agrilus viridicaerulans rubi*
p. 394, change for *Agrilus viridicaerulans viridicaerulans* **E**: TR to **A**: TR
p. 394, add: **A**: KZ TR to *Agrilus sinuatus sinuatus*
p. 395, change for *Agrilus constantini* **E**: KZ to **A**: KZ
p. 395, add: **A**: TR to *Agrilus hyperici*
p. 395, add **A**: KZ to *Agrilus lopatini*
p. 395, add: **A**: TR to *Agrilus macroderus*
p. 395, change for *Agrilus sulcifer* **E**: TR to **A**: TR
p. 395, change for *Agrilus unguiculosus* **E**: KZ to **A**: KZ
p. 395, change for *Agrilus ustjurti* **E**: KZ to **A**: KZ
p. 395, add **A**: KZ to *Agrilus vaginalis philippovi*
p. 395, change for *Agrilus vaginalis vaginalis* **E**: KZ to **A**: KZ
p. 395, add: **A**: KZ TR to *Agrilus ater*
p. 395, add: **A**: KZ TR to *Agrilus albogularis albogularis*
p. 396, add: **A**: KZ to *Agrilus sericans kuldjensis*
p. 396, add: **A**: KZ TR to *Agrilus sericans sericans*
p. 396, change for *Agrilus adlbaueri* **E**: TR to **A**: TR
p. 397, delete: *Agrilus binotatus* Gory, 1841: 236 [ND] **E**: IT
p. 397, change for *Agrilus candidiventris* **E**: KZ to **A**: KZ
p. 397, change for *Agrilus chlorophyllus* **E**: TR to **A**: TR
p. 397, delete *Agrilus confinis* Faldermann, 1835a: 24 **A**: "China boreali"
p. 398, change for *Agrilus fleischeri fleischeri* **E**: KZ to **A**: KZ
p. 398, change for *Agrilus foliicornis* **E**: TR to **A**: TR
p. 398, change for *Agrilus gerschuni* **E**: KZ to **A**: KZ
p. 398, change for *Agrilus hittita* **E**: TR to **A**: TR
p. 399, change for *Agrilus kutahyanus* **E**: TR to **A**: TR
p. 399, add **A**: TR to *Agrilus litura*
p. 399, change for *Agrilus lugovoi* **E**: KZ to **A**: KZ
p. 400, change for *Agrilus morawitzi* **E**: KZ to **A**: KZ
p. 400, change for *Agrilus nivosus* **E**: KZ to **A**: KZ
p. 400, change for *Agrilus oliveri* **E**: TR to **A**: TR
p. 400, delete *Agrilus pisanus* Curletti, 1980 as valid species, it is a synonym of *A. paganettii* Obenberger, 1913
p. 401, delete nominal subspecies of *Agrilus quadrisignatus* Marseul, 1866
p. 401, remouve *Agrilus quadristictulus* Obenberger, 1940 from synonymy with *A. quadrisignatus* Marseul, 1866 and place it as synonym of *A. sospes* Lewis, 1893
p. 401, change for *Agrilus rhoos* **E**: TR to **A**: TR
p. 401, add: *quadristictulus* Obenberger, 1940: 179 in synonymy with *Agrilus sospes* Lewis
p. 402, delete: *sperkii* Solsky, 1873: 159 [ND] **E**: BY PL UK
p. 402, change for *Agrilus suvorovi* **E**: TR to **A**: KZ TR
p. 402, change for *Agrilus turcicus* **E**: TR to **A**: TR

p. 402, change for *Agrilus uzbekistanus* **E**: KZ to **A**: KZ
p. 403, change for *Agrilus voriseki* **E**: KZ to **A**: KZ
p. 403, correct spelling of authors of *Agrilus zanthoxylumi*: Zhang R. & Wang
p. 403, add as nomina dubia:
Agrilus binotatus Gory, 1841: 236 **E**: IT
Agrilus confinis Faldermann, 1835a: 24 **A**: "China boreali"
Agrilus sperkii Solsky, 1873: 159 **E**: BY PL UK
p. 405, add PL and SK to *Aphanisticus emarginatus*
p. 405, correct for *Aphanisticus peninsulae* Obenberger, 1924j to Obenberger, 1924h
p. 407, correct for *Paracylindromorphus pinguis rabinovitchi* Théry, 1835a: to 1935b:
p. 407, correct for *Paracylindromorphus richteri* Théry, 1837f: to Théry, 1937b:
p. 407/408, delete *Amorphosoma* from Clematina (p. 408) and add them under Amorphosomatina (p. 407)
pp. 407, 408, 415, correct authorship for Majer, 2000 to Majer in Kubáň et al. 2000
p. 408, correct reference for *Coraegrilus* Fairmaire, 1889b to 1889a
p. 410, correct spelling Bonelii to Bonelli
p. 411, correct reference for *Coraebus sakagutii sakagutii* Kurosawa from 1962a to 1963a
p. 412, *Coraebus yuanyunae*: delete HUN
p. 412, replace *Coraebus yanshanensis* Peng, 1990: 11, 13 **A**: HUB SCH SHX by *C. zhangi* Peng, 1990: 11, 13 **A**: HUN JIX
p. 414, add CZ to *Meliboeus fulgidicollis*
p. 414, correct reference for *Meliboeus impressithorax* Pic to 1924c: 54
p. 415, delete from subgenus *Meliboeus*: *tribulis* Faldermann, 1835a: 157 (*Agrilus*) [DA] **E**: "Transcaucasus", and add it as a nomen dubium in the genus *Meliboeus*
p. 418, correct *Trachys goberti* Gozis, 1889b to Gozis, 1889
p. 419, add to *Trachys*:
menthae Bedel, 1921: 225 **E**: AU CR FR IT PT SL SP SZ
subacuminata Pic, 1922c: 29
p. 420, change for *Trachys problematica* Obenberger DE to GE
p. 420, delete *alpina* Obenberger, *menthae* Bedel and *subacuminata* Pic from synonymy of *Trachys problematica* Obenberger
p. 421, add under genus *Trachys*, nomina dubia
alpina Obenberger, 1918a: 60
p. 421, delete line: *viridis* Samouelle, 1819b: 42 [DA] **E**: GB; it is listed below as a nomen dubium
p. 421, add as nomina dubia, probably family Buprestidae
Buprestis dubia Piller & Mitterpacher, 1783: 38, 113 **E**: SL
Buprestis duodevigintipunctata Piller & Mitterpacher, 1783: 38 **E**: SL
p. 439, for *Stenelmis zhangi*, correct spelling of autors to Jäch & Kodada
p. 441, correct authorship of *Curimopsis eos* to: Fabbri, 2003
p. 442, correct authorship of *Curimopsis saeticirrata* to: Fabbri, 2003
p. 453, correct year for Ptilodactylidae Laporte to: 1838
p. 454, correct year for Ptilodactylinae Laporte to: 1838
p. 462, add: Apostolov L. G. & Malcev I. V. 1986: Plastinchatoucye zhuki (Coleoptera, Scarabaeidae) Kryma. Pp. 88-97. In: *Prirodookhrannye issledovaniya ekosistem gornogo Kryma*. Simferopol.
p. 477, add: Bílý S. 1994a: Sapaia brodskyi gen. n., sp. n. from Vietnam (Coleoptera, Buprestidae, Chalcophorini). *Folia Heyrovskyana* **2**: 8-11.
p. 477, correct reference Bílý S. 1994 to Bílý S. 1994b
p. 477, add: Bílý S. 1994c: Sphenoptera (Chrysoblemma) anchorifera sp. n. from Sahara (Coleoptera: Buprestidae). *European Journal of Entomology* **91**: 327-329.
p. 503, add: Escalera M. M. de la 1913: Dos nuevos tribus de Drílidos. *Boletín de la Real Societad Española de Historia Natural* **13**: 318-320.
p. 503, add: Escalera M. M. de la 1914: Rectificación sinonímica. *Boletín de la Real Societad Española de Historia Natural* **14**: 349.
p. 520, add: Gory H. L. 1840: *Histoire naturelle et iconographie des insectes Coléoptères. Monographie des*

buprestides, Supplément. Tome IV, livraisons 36-42. Paris: P. Duménil, pp. 1-72, 79-106, 111-126, 141-142.

p. 520, complete pagination in reference Gory H. L. 1841 by: 73-78, and 143-356 + 1-7 + 1-2 + 1.

p. 521, add: Gozis M. des. 1889: Description d'un Trachys de France. *Revue d'Entomologie* **8**: 203-205.

p. 523, correct year for Guérin-Méneville from 1850 to 1859.

p. 524, add: Gusakov A. A. 2003: A new species of the genus Hemictenius from Afghanistan (Coleoptera: Scarabaeidae: Pachydeminae). *Russian Entomological Journal* **12**: 69-70.

p. 527, add: Herman G. 1966a: Première contribution à la connaissance des Anthaxia. *Bulletin & Annales de la Société Royale d'Entomologie de Belgique* **102**: 241-247.

p. 527, add: Herman G. 1966b: Deuxième contribution à la connaissance des Anthaxia. *Bollettino dell'Associazione Romana di Entomologia* **21**: 40-42.

p. 541, add: Kerremans C. 1891: Buprestides nouveaux et remarques synonymiques. *Bulletin ou Comptes-Rendus des Séances de la Société Entomologique de Belgique* **35**: clvi-clxiv.

p. 542, add: Kerremans C. 1913c: *Monographie des buprestides. Tome VII. Sphenoptera* (*fin*). Bruxelles: A. Breuer, pp. 1-96 (livraison 1-3).

p. 559, add: Laporte [de Castelnau] F. L. N. Caumont de. 1835: Études entomologiqes, ou descriptions d'insectes nouveaux et observations sur la synonymie. *Revue Entomologique* (Silbermann) **3**: 157-181.

p. 565, add: Lin P. 2002: Coleoptera: Rutelidae. Pp. 345-355. In: Huang F. (ed.): *Forest Insects of Hainan.* National Natural Science Foundation of China, xv + 1064 pp., 31 pls (in Chinese and English).

p. 559, correct year for Laporte [de Castelnau] F. L. N. Caumont de. 1836: Etudes entomologiques, … to 1838

p. 569, correct spelling MacLeay W. J. 1864, and not MacLeay W. S. 1864

p. 594, add: Novikov O. A. 1999: Novyy vid roda Trochaloschema Reitter (Coleoptera, Scarabaeidae) s Karateginskogo khrebta Tadzhikistana. *Izvestiya Kharkovskogo Entomologicheskogo Obshchestva* **7** (1): 30-33.

p. 608, add: Panzer G. W. F. 1796: *Faunae Insectorum Germanicae initia; oder Deutschlands Insecten. Fuenfter Iahrgang.* [Heft 35]. Nürnberg: Felsecker, 24 pp. + 24 pls.

p. 856, correct year for Redtenbacher L. 1867: Reise der Österreichischen Fregatte Novara..., to: 1868 [see: Higgins, 1964: *J. Soc. Bibl. Nat. Hist.* 4: 158]

p. 639, add: Schaefer L. 1935: A propos de "Julodis onopordi" F. *Bulletin Mensuel de la Société Linnéenne de Lyon* **4**: 46-47.

p. 643, complete reference Schröter J. S. 1776 by: 488 pp., 2 pls.

p. 643, add: Schweiger H. 1949: Neue westasiatische Glaphyrus-Arten (Col. Scarabaeidae). *Entomon* **1**: 114-115.

p. 649, add: Spinola M. 1838: Essai sur les espèces des genres Steraspis et Acmaeodera. *Annales de la Société Entomologique de France* **7**: 303-394.

p. 656, change Théry A. 1930 to Théry A. 1930b

p. 658, correct title for Thomson J. 1879b to: Revue du groupe des Psiloptérides de la famille des Buprestides.

p. 658, add: Tôyama M. 1986b: The buprestid genus Chalcophorella Kerremans and its related genera (Coleoptera, Buprestidae). Pp. 188-193. In: Ueno S. I. (ed.): *Entomological papers presented to Yoshihiko Kurosawa on the occasion of his retirement.* Tokyo: Coleopterists' Association of Japan, 342 pp.

p. 660, add: Villa A. & Villa G. B. 1835: *Coleopterorum species novae in supplemento satutatae diagnosibus atque observationibus illustratae.* Mediolani, 47-50 pp.

p. 668, add: Zhang R. & Wang T. 1992: Agrilus zanthoxylumi Hou. P. 402-403. In: Xiao Gangrou (ed.): *Forest Insects of China. The second edition* (*Revised & enlarged*). Beijing: China forestry publishing house, 1362 pp.

p. 668, add: Zhang Y. 2002: Coleoptera: Scarabaeoidea (Parts). Pp. 326-336. In: Huang F. (ed.): *Forest Insects of Hainan.* National Natural Science Foundation of China, xv + 1064 pp., 31 pls (in Chinese and English).

p. 682, add: *Kubaniellus* Kalashian, 1997 … 406

p. 688, add: *Steraspis* Dejean, 1833 ……..... 344

NEW NOMENCLATORIAL and TAXONOMIC ACTS, and COMMENTS

Eucnemidae

Jyrki Muona & Diane Alaruikka

Comments

The oldest available name for the family is Melasidae, but as it would not promote stability to change the name again, an application has been sent to the ICZN to suppress the use of Melasidae for this purpose. The proper spelling for the family name would be Eucnemididae, the plural stem being Eucnemides, but as the family group names are not changed usually for this reason only, we have continued to use the familiar Eucnemidae.

Elater buprestoides Fabricius, 1792 is placed as a nomen dubium, it may eventually be senior synonym of *Nematodes filum* Fabricius, 1801

Missidentification

Epiphanis cornutus Eschscholtz is often considered an introduction in Europe, but we see no reason to adopt this view. The species has been known from Japan for a long time, the records from the Mediterranean region are old and there is an unconfirmed record from China as well.

New combinations

Dirhagus marginatus Fleutiaux, 1923 and *Dirrhagus morimotoi* Hisamatsu, 1960 were earlier placed in the genus *Microrhagus* Dejean, but are here included in the genus *Rhagomicrus* Fleutiaux, this being the first time this world-wide genus is reported from the Palearctic.

New synonymy

Microrhagus palmi (Olexa, 1963) is regarded a subjective junior synonym of *Microrhagus pygmaeus* (Fabricius, 1792). No diagnostic genitalic features have been reported and the described differences in the antennae and the last ventrite are subject to considerable variation in all *Microrhagus* species. This view was shared by our late collague Willy Lucht as well.

Infrasubspecific name

Dirrhagus yasumatsui f. *unicolor* Hisamatsu, 1960 was clearly proposed as an infrasubspecific name and has been omitted from the list.

Elateridae: Cebrioninae

I. Löbl

Unavailable genus-group name and designation of type species

Paulusiella: Mandl (1974: 640) stated he is establishing his new genus *Paulusiella* for *Escalerina serraticornis* Paulus, 1972, and a new species, *P. fossulatipennis* Mandl. In the same paper he also described *P. richteri*, with a subspecies, *P. richteri acuticollis*, and *P. pallida*. A type species was not designated and *Paulusiella* Mandl is according to the ICZN Art. 13.3 unavailable. The description given by Mandl, 1974 characterizes the group. *Paulusiella fossulatipennis* Mandl, 1974 is here designated as the type species.

Elateridae

P. C. Cate

Comments to nomenclature

Gistel (sometimes spelled in the literature "Gistl") published unnecessary replacement names and based new family-group names on some of them. These names were ignored for about a century and half. Nevertheless, some recent authors used them as valid. The resurrections of Gistel's names may be accepted according to the ICZN Art. 40.2, 40.2.1. They affect the stability of the nomenclature and are in contradiction to the fundamental aim of the zoological nomenclature (ICZN, Introduction). In elaterids, the following, well established family-group names were recently replaced in some works:

Adrastini Candéze, 1863 by Synaptini Gistel, 1856
Ctenicerini Fleutiaux, 1936 by Prosternini Gistel, 1856
Conoderini Fleutiaux, 1919 by Oophorini Gistel, 1856
Denticollinae Stein & J. Weise, 1877 by Dendrometrinae Gistel, 1856

The suppression of Gistel's senior synonyms is pending on a ruling of the ICZN. These names are not accepted in the present work.

Ampedus aethiops (Lacordaire, 1835), published as *Elater aethiops*, is a primary junior homonom of *Elater aethiops* Herbst, 1806 (placed in *Pristilophus* by Germar, 1844 and since not included in the same genus). It is used as available, according to the ICZN Art. 23.9.

Ampedus elongatulus (Fabricius): Fabricius, 1787: 174 described his new species as *Elater elongantulus*, an incorrect original spelling corrected by himself (Fabricius, 1792: 229). *Elater elongatus* Gmelin, 1790: 1907 is an unjustified emendation of *elongatulus* Fabricius. *Elater elongatulus* has been used for two centuries for a common, widely distributed species, at present placed in *Ampedus*. *Elater glycereus* Herbst, 1784 was considered as an invalid synonym of *E. elongatulus* (at least since Candèze, 1859). As the name *Ampedus glycereus* (Herbst, 1784) was recently used as valid, it is in the interest of the nomenclatural stability of this common taxon to have it suppressed. *Ampedus elongatulus* (Fabricius) is used here furtheron as valid, pending on a ruling of the ICZN.

Ampedus ochropterus Germar, 1844 is a junior secondary homonym of *Elater ochropterus* Eschscholtz, 1830, now placed in synonymy of *Ampedus pomorum* Herbst, 1784. *Ampedus ochropterus* Germar has been in use for over a century and a half. It is in the interest of nomenclatorial stability to retain this name. *Ampedus ochropterus* Germar is used here furtheron as valid, pending on a ruling of the ICZN.

Cardiophorus rufipes (Goeze, 1777), a well known species, was originally published in *Elater* and was there a primary junior homonym of *E. rufipes* DeGeer, 1774 (presently and at least since Gemminger & Harold, 1869 placed in synonymy with *Prosternon tessellatum* (Linnaeus, 1758)). *Elater rufipes* Goeze was considered unavailable and unnecessarily replaced by a new name, *Cardiophorus goezei* Sánchez-Ruiz, 1996. It is used as available, according to the ICZN, Art. 23.9.

Cryptalaus Ôhira, 1967 was published in May and has priority over *Paracalais* Neboiss, 1967 published in July.

Dicronychus nigromaculatus (Motschulsky, 1860): The description of *Melanoxanthus nigromaculatus* Motschulsky, 1860 was published on 21st April and has priopríty over *Cardiophorus nigropunctatus* Candèze, 1860, published in July

Selatosomus latus (Fabricius): *Elater latus* Fabricius, 1801b: 232 is a junior primary homonym of *Elater latus* Fuessely, 1775: 16 and *Elater latus* Sulzer, 1776: 53, both nomina dubia. The name of Fabricius has been in use for two centuries. It is therefore in the interest of nomenclatorial stability to retain this name. *Selatosomus latus* (Fabricius) is used here furtheron as valid, pending on a ruling of the ICZN.

Zorochros dermestoides: Herbst, 1806 published his new species in *Elater*, spelled in three different ways: *demustoides* (p. 85), *dermestoides* (pl. 165), and *dermesorides* (Verzeichnis, p. ii). The spelling *dermestoides* is obviously the correct one and was in general use for almost two ceturies. Thre remaining alternative original spellings are not available (ICZN, Art. 19.3, 32.4). The name *dermestoides* was in *Elater* a primary junior homonym of *Elater dermestoides* Linnaeus, 1758 (at present in Throscidae). Both species were placed in different genera already by Latreille (Latreille, 1807a: 37) and were not included in the same genus since. Nevertheless, some authors replaced *dermestoides* by the incorrect alternative original spelling *demustoides*. This is unnecessary and not accepted in the present work.

New synonyms

Actenicerus yaku var. *koyamai* Nakane & Kishii, 1958 **syn. nov.** of *Actenicerus* (*Acnitecerus*) *yaku* Nakane & Kishii, 1958
Adrastus lacertosus var. *guillebeaui* Buysson, 1896 **syn. nov.** of *Adrastus lacertosus* Erichson, 1841
Adrastus limbatus var. *porrectifrons* Gozis, 1882 **syn. nov.** of *Adrastus axillaris* Erichson, 1841
Adrastus axillaris var. *nigrinus* Schilsky, 1890 **syn. nov.** of *Adrastus axillaris* Erichson, 1841
Adrastus axillaris var. *pseudolimbatus* Binaghi, 1936 **syn. nov.** of *Adrastus axillaris* Erichson, 1841
Adrastus montanus var. *exutus* Buysson, 1926 **syn. nov.** of *Adrastus montanus* (Scopoli, 1763)
Adrastus pallens var. *paykulli* Szombathy, 1910 **syn. nov.** of *Adrastus Adrastus rachifer* (Geoffroy, 1785)
Adrastus pallens var. *siculus* Ragusa, 1896 **syn. nov.** of *Adrastus Adrastus rachifer* (Geoffroy, 1785)
Adrastus rachifer var. *humerotinctus* Buysson, 1926 **syn. nov.** of *Adrastus rachifer* (Geoffroy, 1785)
Adrastus rachifer var. *nigratorius* Buysson, 1926 **syn. nov.** of *Adrastus rachifer* (Geoffroy, 1785)
Adrastus rachifer var. *reitteri* Buysson, 1896 **syn. nov.** of *Adrastus rachifer* (Geoffroy, 1785)
Aeoloides waltlii Candèze, 1859 var. *isabellinus* Buysson, 1912 **syn. nov.** of *Aeoloides waltlii* Candèze, 1859
Aeolus atricapillus var. *mosquensis* Ivanov, 1902 **syn. nov.** of *Drasterius atricapillus* (Germar, 1824)
Aeolus candezei var. *laeticolor* Reitter, 1891 **syn. nov.** of *Aeoloides figuratus* (Germar, 1844)
Aeolus candezei var. *obscuricolor* Reitter, 1891 **syn. nov.** of *Aeoloides figuratus* (Germar, 1844)
Aeolus cruciatus var. *humeralis* Reitter, 1891 **syn. nov.** of *Drasterius atricapillus* (Germar, 1824)
Aeolus cruciatus var. *nigripennis* Reitter, 1891 **syn. nov.** of *Drasterius atricapillus* (Germar, 1824)
Aeoloderma crucifer var. *flavofactus* Buysson, 1912 **syn. nov.** of *Aeoloderma crucifer* (Rossi, 1790)
Aeoloderma crucifer var. *lateralipictus* Buysson, 1912 **syn. nov.** of *Aeoloderma crucifer* (Rossi, 1790)
Aeoloderma crucifer var. *paludum* Buysson, 1912 **syn. nov.** of *Aeoloderma crucifer* (Rossi, 1790)
Aeoloderma crucifer var. *m-disjunctus* Buysson, 1912 **syn. nov.** of *Aeoloderma crucifer* (Rossi, 1790)
Aeoloides waltli var. *isabellinus* Buysson, 1912 **syn. nov.** of *Aeoloides waltli* Candèze, 1859
Agonischius altus var. *striatus* Fleutiaux, 1918 **syn. nov.** of *Vuilletus altus* (Candèze, 1863)
Agonischius cyaneus var. *nitidus* Fleutiaux, 1895 **syn. nov.** of *Ludioschema cyaneum* (Candèze, 1863)
Agonischius cyaneus var. *thoracicus* Fleutiaux, 1895 **syn. nov.** of *Ludioschema cyaneum* (Candèze, 1863)
Agonischius florentini **syn. nov.** of *Ludioschema delauneyi* (Fleutiaux, 1887)
Agonischius obscuripes var. *brunneus* Miwa, 1928 **syn. nov.** of *Ludioschema obscuripes* (Gyllenhal, 1817)
Agonischius obscuripes var. *candezellus* Miwa, 1928 **syn. nov.** of *Ludioschema obscuripes* (Gyllenhal, 1817)
Agonischius obscuripes var. *ferrugineus* Fleutiaux, 1918 **syn. nov.** of *Ludioschema obscuripes* (Gyllenhal, 1817)
Agonischius obscuripes var. *ruficephalus* Miwa, 1928 **syn. nov.** of *Ludioschema okinawense* (Miwa, 1927a)
Agonischius sulcicollis var. *obscurus* Fleutiaux, 1895 **syn. nov.** of *Ludioschema sulcicolis* (Candèze, 1878b)
Agriotes acuminatus var. *integricollis* Reitter, 1911 **syn. nov.** of *Agriotes acuminatus* (Stephens, 1830)
Agriotes acuminatus var. *pallidicolor* Pic, 1910b **syn. nov.** of *Agriotes acuminatus* (Stephens, 1830)
Agriotes algirinus var. *obscurior* Pic, 1916 **syn. nov.** of *Megathous algirinus* (Candèze, 1860)
Agriotes bevis var. *nigricollis* Schwarz, 1891 **syn. nov.** of *Agriotes bevis* Candèze, 1863
Agriotes bevis var. *nigripennis* Schwarz, 1891 **syn. nov.** of *Agriotes bevis* Candèze, 1863
Agriotes corsicus var. *fuscipennis* Fiori, 1899 **syn. nov.** of *Agriotes infuscatus* Desbrochers des Loges, 1870
Agriotes corsicus var. *nigripennis* Fiori, 1899 **syn. nov.** of *Agriotes infuscatus* Desbrochers des Loges, 1870
Agriotes corsicus var. *nigrithorax* Fiori, 1899 **syn. nov.** of *Agriotes infuscatus* Desbrochers des Loges, 1870
Agriotes gallicus var. *atacinus* Buysson, 1906 **syn. nov.** of *Agriotes gallicus* Lacordaire, 1835
Agriotes gallicus var. *robigus* Buysson, 1906 **syn. nov.** of *Agriotes gallicus* Lacordaire, 1835
Agriotes gallicus var. *semibrunneus* Pic, 1910 **syn. nov.** of *Agriotes gallicus* Lacordaire, 1835
Agriotes infuscatus var. *synaptoides* Reitter, 1906 **syn. nov.** of *Agriotes infuscatus* Desbrochers des Loges, 1870
Agriotes infuscatus var. *elegans* Schwarz, 1891a **syn. nov.** of *Agriotes infuscatus* Desbrochers des Loges, 1870
Agriotes litigiosus var. *fulvescens* Candèze, 1863 **syn. nov.** of *Agriotes litigiosus* (Rossi, 1792)
Agriotes litigiosus var. *rufipennis* Buysson, 1893 **syn. nov.** of *Agriotes litigiosus* (Rossi, 1792)
Agriotes litigiosus var. *vesubiensis* Pic, 1914 **syn. nov.** of *Agriotes litigiosus* (Rossi, 1792)

Agriotes obscurus var. *cinnanomeus* Buysson, 1893 **syn. nov.** of *Agriotes obscurus* (Linnaeus, 1758)
Agriotes obscurus var. *radhosticola* Fleischer, 1910 **syn. nov.** of *Agriotes obscurus* (Linnaeus, 1758)
Agriotes elongatus var. *humerosignatus* Buysson, 1926 **syn. nov.** of *Agriotes pilosellus* (Schönherr, 1817)
Agriotes elongatus var. *simulator* Buysson, 1893 **syn. nov.** of *Agriotes pilosellus* (Schönherr, 1817)
Agriotes pallidulus var. *gabilloti* Pic, 1910 **syn. nov.** of *Agriotes pallidulus* (Illiger, 1807)
Agriotes pilosellus var. *buyssoni* Méquignon, 1927 **syn. nov.** of *Agriotes pilosellus* (Schönherr, 1817)
Agriotes rufipalpis var. *suturalis* Schwarz, 1891 **syn. nov.** of *Agriotes rufipalpis* Brullé, 1832
Agriotes sonanis var. *brunneipennis* Miwa, 1931 **syn. nov.** of *Ectinus sonanis* (Miwa, 1928)
Agriotes sonanis Miwa, 1928b var. *nigrinus* Miwa, 1931 **syn. nov.** of *Ectinus sonanis* (Miwa, 1928)
Agriotes sordidus var. *buyssoni* Pic, 1895 **syn. nov.** of *Agriotes sordidus* (Illiger, 1807)
Agriotes sordidus (var. *circumcinctus* Fuente, 1903 **syn. nov.** of *Agriotes sordidus* (Illiger, 1807)
Agriotes sordidus var. *clarior* Buysson, 1924 **syn. nov.** of *Agriotes sordidus* (Illiger, 1807)
Agriotes sordidus var. *neglectus* Schwarz, 1891 **syn. nov.** of *Agriotes sordidus* (Illiger, 1807)
Agriotes sordidus var. *nigriceps* Daniel, 1903 **syn. nov.** of *Agriotes sordidus* (Illiger, 1807)
Agriotes sordidus var. *nigroscutellatus* Buysson, 1924 **syn. nov.** of *Agriotes sordidus* (Illiger, 1807)
Agriotes sordidus var. *parallelus* Schwarz, 1891 **syn. nov.** of *Agriotes sordidus* (Illiger, 1807)
Agriotes sordidus var. *raduloides* Buysson, 1924 **syn. nov.** of *Agriotes sordidus* (Illiger, 1807)
Agriotes sordidus var. *rusticus* Schwarz, 1891 **syn. nov.** of *Agriotes sordidus* (Illiger, 1807)
Agriotes sordidus var. *schwarzi* Buysson, 1893 **syn. nov.** of *Agriotes sordidus* (Illiger, 1807)
Agriotes sordidus var. *seminiger* Buysson, 1893 **syn. nov.** of *Agriotes sordidus* (Illiger, 1807)
Agriotes sordidus var. *sericans* Rey, 1891 **syn. nov.** of *Agriotes sordidus* (Illiger, 1807)
Agriotes sputator var. *depactus* Buysson, 1926 **syn. nov.** of *Agriotes sputator* (Linnaeus, 1758)
Agriotes sputator var. *negatus* Buysson, 1895 **syn. nov.** of *Agriotes sputator* (Linnaeus, 1758)
Agriotes sputator var. *productus* Rey, 1891 **syn. nov.** of *Agriotes sputator* (Linnaeus, 1758)
Agriotes starcki var. *productus* Rey, 1891b **syn. nov.** of *Agriotes sputator* (Linnaeus, 1758)
Agriotes tauricus var. *rufangulus* Buysson, 1896 **syn. nov.** of *Agriotes tauricus* Heyden, 1882b
Agriotes tauricus var. *rufus* Schwarz, 1891 **syn. nov.** of *Agriotes tauricus* Heyden, 1882b
Agriotes ustulatus var. *bucovinensis* Hormuzachi, 1901 **syn. nov.** of *Agriotes ustulatus* (Schaller, 1783)
Agriotes ustulatus var. *foveolatus* Rey, 1891 **syn. nov.** of *Agriotes ustulatus* (Schaller, 1783)
Agriotes ustulatus var. *inustus* Buysson, 1906 **syn. nov.** of *Agriotes ustulatus* (Schaller, 1783)
Agriotes ustulatus var. *martialis* Pic, 1914 **syn. nov.** of *Agriotes ustulatus* (Schaller, 1783)
Agriotes ustulatus var. *piceus* Buysson, 1890 **syn. nov.** of *Agriotes ustulatus* (Schaller, 1783)
Agriotes ustulatus var. *subrectus* Rey, 1891 **syn. nov.** of *Agriotes ustulatus* (Schaller, 1783)
Agriotes ustulatus var. *tristis* Schilsky, 1888 **syn. nov.** of *Agriotes ustulatus* (Schaller, 1783)
Athous austriacus var. *flecki* Reitter, 1906 **syn. nov.** of *Athous* (*Haplathous*) *austriacus* Desbrochers des Loges, 1873
Athous bicolor var. *gracilis* Buysson, 1905 **syn. nov.** of *Athous* (*Orthathous*) *bicolor* (Goeze, 1777)
Athous castanescens var. *luctuosus* Buysson, 1906 **syn. nov.** of *Athous* (*Euplathous*) *frigidus* Mulsant & Guillbeau, 1855
Athous castanescens var. *rubiginosus* Buysson, 1906 **syn. nov.** of *Athous* (*Euplathous*) *frigidus* Mulsant & Guillbeau, 1855
Athous circassicus var. *atripes* Reitter, 1905 **syn. nov.** of *Athous* (*Orthathous*) *circassicus* Reitter, 1888a
Athous circumscriptus var. *flaveolus* Buysson, 1925 **syn. nov.** of *Athous* (*Haplathous*) *austriacus* Desbrochers des Loges, 1873b
Athous circumscriptus var. *flavidus* Buysson, 1925 **syn. nov.** of *Athous* (*Haplathous*) *austriacus* Desbrochers des Loges, 1873
Athous circumscriptus var. *humeropictus* Buysson, 1925 **syn. nov.** of *Athous* (*Haplathous*) *austriacus* Desbrochers des Loges, 1873
Athous circumscriptus var. *lemovicium* Buysson, 1925 **syn. nov.** of *Athous* (*Haplathous*) *austriacus* Desbrochers des Loges, 1873
Athous circumscriptus var. *neomelanoderes* Buysson, 1925 **syn. nov.** of *Athous* (*Haplathous*) *austriacus* Desbrochers des Loges, 1873
Athous circumscriptus var. *subrubricus* Reitter, 1905 **syn. nov.** of *Athous* (*Haplathous*) *austriacus* Desbrochers des Loges, 1873

Athous emaciatus var. *pallidior* Buysson, 1925 **syn. nov.** of *Athous* (*Haplathous*) *emaciatus* Candèze, 1860
Athous dejeani var. *mariei* Méquignon, 1927 **syn. nov.** of *Athous* (*Euplathous*) *dejeani* Laporte, 1840
Athous dejeani var. *pallidior* Buysson, 1925 **syn. nov.** of *Athous* (*Haplathous*) *emaciatus* Candèze, 1860
Athous difficilis var. *divergens* Reitter, 1905 **syn. nov.** of *Athous* (*Orthathous*) *difficilis* Dufour, 1843
Athous difformis var. *fallaciosus* Buysson, 1924a **syn. nov.** of *Athous* (*Orthathous*) *campyloides* Newman, 1833
Athous difformis var. *punctipennis* Reitter, 1905 **syn. nov.** of *Athous* (*Orthathous*) *campyloides* Newman, 1833
Athous frigidus var. *capucinus* Buysson, 1906 **syn. nov.** of *Athous* (*Euplathous*) *frigidus* Mulsant & Guillbeau, 1855
Athous frigidus var. *leptotropus* Buysson, 1925 **syn. nov.** of *Athous* (*Euplathous*) *frigidus* Mulsant & Guillbeau, 1855
Athous godarti var. *hispanicus* Pic, 1910 **syn. nov.** of *Athous* (*Pleurathous*) *godarti* Mulsant & Guillebeau, 1856
Athous godarti var. *varensis* Reitter, 1905 **syn. nov.** of *Athous* (*Pleurathous*) *godarti* Mulsant & Guillebeau, 1856
Athous godarti var. *xambeui* Pic, 1910 **syn. nov.** of *Athous* (*Pleurathous*) *godarti* Mulsant & Guillebeau, 1856
Athous haemorrhoidalis var. *asiaticus* Buysson, 1905 **syn. nov.** of *Athous* (*Athous*) *haemorrhoidalis* (Fabricius, 1801)
Athous haemorrhoidalis var. *croissendeaui* Buysson, 1891 **syn. nov.** of *Athous* (*Athous*) *haemorrhoidalis* (Fabricius, 1801)
Athous haemorrhoidalis var. *faeculentus* Buysson, 1888 **syn. nov.** of *Athous* (*Athous*) *haemorrhoidalis* (Fabricius, 1801)
Athous haemorrhoidalis var. *faeculentus* Buysson, 1888 **syn. nov.** of *Athous* (*Athous*) *haemorrhoidalis* (Fabricius, 1801)
Athous haemorrhoidalis var. *interpositus* Rey, 1891 **syn. nov.** of *Athous* (*Athous*) *haemorrhoidalis* (Fabricius, 1801)
Athous haemorrhoidalis var. *lacordairei* Méquignon, 1929 **syn. nov.** of *Athous* (*Athous*) *haemorrhoidalis* (Fabricius, 1801)
Athous haemorrhoidalis var. *lacordairei* Méquignon, 1929 **syn. nov.** of *Athous* (*Athous*) *haemorrhoidalis* (Fabricius, 1801)
Athous haemorrhoidalis jureceki Jagemann, 1939 **syn. nov.** of *Athous* (*Athous*) *haemorrhoidalis* (Fabricius, 1801)
Athous haemorrhoidalis var. *multiimpressus* Pic, 1910 **syn. nov.** of *Athous* (*Athous*) *haemorrhoidalis* (Fabricius, 1801)
Athous herbigradus var. *gabilloti* Pic, 1910 **syn. nov.** of *Athous* (*Orthathous*) *herbigradus* Mulsant & Guillebeau, 1855a
Athous hirtus var. *balthasari* Roubal, 1935 **syn. nov.** of *Hemicrepidius hirtus* (Herbst, 1784)
Athous hirtus var. *daliberti* Buysson, 1926 **syn. nov.** of *Hemicrepidius hirtus* (Herbst, 1784)
Athous hirtus var. *lubini* Buysson, 1926 **syn. nov.** of *Hemicrepidius hirtus* (Herbst, 1784)
Athous hirtus var. *meles* Buysson, 1906 **syn. nov.** of *Hemicrepidius hirtus* (Herbst, 1784)
Athous laevigatus var. *fortimargo* Buysson, 1905 **syn. nov.** of *Athous* (*Haplathous*) *laevistriatus* Dufour, 1851
Athous niger var. *leporinus* Buysson, 1916 **syn. nov.** of *Hemicrepidius* (*Haplathous*) *niger* (Linnaeus, 1758)
Athous niger var. *schaumi* Buysson, 1901 **syn. nov.** of *Hemicrepidius* (*Haplathous*) *niger* (Linnaeus, 1758)
Athous niger var. *rubrocallatus* Roubal, 1936 **syn. nov.** of *Hemicrepidius* (*Haplathous*) *niger* (Linnaeus, 1758)
Athous obscurus var. *deubeli* Buysson, 1916 **syn. nov.** of *Athous* (*Athous*) *haemorrhoidalis* (Fabricius, 1801)
Athous obscurus var. *jouvenoti* Buysson, 1926 **syn. nov.** of *Athous* (*Athous*) *haemorrhoidalis* (Fabricius, 1801)
Athous puncticollis var. *punctithorax* Reitter, 1905 **syn. nov.** of *Athous* (*Athous*) *vittatus* (Fabricius, 1792)
Athous puncticollis var. *ravouxi* Pic, 1909 **syn. nov.** of *Athous* (*Athous*) *vittatus* (Fabricius, 1792)
Athous puncticollis var. *testaceicolor* Pic, 1906 **syn. nov.** of *Athous* (*Athous*) *vittatus* (Fabricius, 1792)
Athous rhombeus var. *obscuratus* Pic, 1904 **syn. nov.** of *Stenagostus rhombeus* (A. G. Olivier, 1790)
Athous rufangulus var. *ruficollis* Miwa, 1930 **syn. nov.** of *Hemicrepidius rufangulus rufangulus* (Miwa, 1928)
Athous subfuscus var. *kafkanus* Reitter, 1905 **syn. nov.** of *Athous* (*Haplathous*) *subfuscus* O. F. Müller, 1764
Athous subfuscus var. *polonicus* Reitter, 1905 **syn. nov.** of *Athous* (*Haplathous*) *subfuscus* O. F. Müller, 1764
Athous sylvaticus var. *laevigatus* Desbrochers des Loges, 1871 **syn. nov.** of *Athous* (*Haplathous*) *sylvaticus* Mulsant & Guillebeau, 1855a
Athous undulatus var. *simplicitus* Heyden, 1884 **syn. nov.** of *Diacanthus undulatus* (DeGeer, 1774)

Athous (*Pleurathous*) *uncicollis* var. *gredosanus* Reitter, 1904 **syn. nov.** of *Athous* (*Pleurathous*) *uncicollis* Perris, 1864
Athous vittatus var. *brevitinctus* Buysson, 1924 **syn. nov.** of *Athous* (*Athous*) *vittatus* (Fabricius, 1792)
Athous vittatus var. *decipiens* Rye, 1891 **syn. nov.** of *Athous* (*Athous*) *vittatus* (Fabricius, 1792)
Athous vittatus var. *exaequatus* Buysson, 1924 **syn. nov.** of *Athous* (*Athous*) *vittatus* (Fabricius, 1792)
Athous vittatus var. *filicti* Buysson, 1891c **syn. nov.** of *Athous* (*Athous*) *vittatus* (Fabricius, 1792)
Athous vittatus var. *inopinatus* Buysson, 1891c **syn. nov.** of *Athous* (*Athous*) *vittatus* (Fabricius, 1792)
Athous vittatus var. *lanatus* Rey, 1891 **syn. nov.** of *Athous* (*Athous*) *vittatus* (Fabricius, 1792)
Athous vittatus var. *ocskayi* Kiesenwetter, 1858 **syn. nov.** of *Athous* (*Athous*) *vittatus* (Fabricius, 1792)
Athous vittatus var. *stephensi* Buysson, 1891 **syn. nov.** of *Athous* (*Athous*) *vittatus* (Fabricius, 1792)
Athous vittatus var. *tenuevittatus* Reitter, 1905 **syn. nov.** of *Athous* (*Athous*) *vittatus* (Fabricius, 1792)
Athous undulatus var. *mediofasciatus* Pic, 1903 **syn. nov.** of *Diacanthous undulatus* (DeGeer, 1774)
Athous undulatus var. *simplicitus* Heyden, 1884 **syn. nov.** of *Diacanthous undulatus* (DeGeer, 1774)
Athous zebei Bach, 1852 var. *balmei* Pic, 1910 **syn. nov.** of *Athous* (*Haplathous*) *zebei* Bach, 1852
Athous (*Crepidophorus*) *mutilatus* var. *barbarinus* Pic, 1916 **syn. nov.** of *Crepidophorus mutilatus* (Rosenhauer, 1847)
Betarmon vittatus Candèze, 1880 **syn. nov.** of *Idolus picipennis* (Bach, 1852)
Calambus bipustulatus var. *diapurus* Buysson, 1928 **syn. nov.** of *Calambus bipustulatus* (Linnaeus, 1767)
Calambus bipustulatus var. *epipleuralis* Buysson, 1905 **syn. nov.** of *Calambus bipustulatus* (Linnaeus, 1767)
Calambus bipustulatus var. *perferrugineus* Buysson, 1928 **syn. nov.** of *Calambus bipustulatus* (Linnaeus, 1767)
Campsosternus auratus var. *niger* Fleutiaux, 1918 **syn. nov.** of *Campsosternus auratus* (Drury, 1773)
Campylus flabellatus var. *obscuripennis* Reitter, 1906 **syn. nov.** of *Denticollis flabellatus* Reitter, 1906
Campylus rubens var. *abchasiaticus* Buysson, 1906 **syn. nov.** of *Denticollis rubens* Piller & Mitterpacher, 1783
Campylus rubens var. *dichromus* Candèze, 1863 **syn. nov.** of *Denticollis rubens* Piller & Mitterpacher, 1783
Campylus rubens var. *moestus* Buysson, 1906 **syn. nov.** of *Denticollis rubens* Piller & Mitterpacher, 1783
Campylus rubens var. *mangonis* Buysson, 1929 **syn. nov.** of *Denticollis rubens* Piller & Mitterpacher, 1783
Cardiotarsus fulvipes var. *difficilis* Fleutiaux, 1947 **syn. nov.** of *Cardiotarsus fulvipes* Fleutiaux, 1918
Cardiophorus anticus var. *sidonius* Pic, 1912e **syn. nov.** of *Cardiophorus anticus* Erichson, 1840
Cardiophorus argiolus var. *andreini* Buysson, 1922 **syn. nov.** of *Cardiophorus argiolus* Gené, 1836
Cardiophorus argiolus var. *angularoides* Buysson, 1913 **syn. nov.** of *Cardiophorus argiolus* Gené, 1836
Cardiophorus argiolus var. *anticecoloratus* Buysson, 1922 **syn. nov.** of *Cardiophorus argiolus* Gené, 1836
Cardiophorus argiolus var. *involutus* Buysson, 1922 **syn. nov.** of *Cardiophorus argiolus* Gené, 1836
Cardiophorus argiolus var. *conjonctellus* Buysson, 1922 **syn. nov.** of *Cardiophorus argiolus* Gené, 1836
Cardiophorus argiolus var. *posticeminiatus* Buysson, 1913 **syn. nov.** of *Cardiophorus argiolus* Gené, 1836
Cardiophorus argiolus var. *rubricollis* Schwarz, 1893 **syn. nov.** of *Cardiophorus argiolus* Gené, 1836
Cardiophorus argiolus var. *rufimembris* Buysson, 1913 **syn. nov.** of *Cardiophorus argiolus* Gené, 1836
Cardiophorus argiolus var. *sardeus* Candèze, 1860 **syn. nov.** of *Cardiophorus argiolus* Gené, 1836
Cardiophorus argiolus var. *tibialoides* Buysson, 1913 **syn. nov.** of *Cardiophorus argiolus* Gené, 1836
Cardiophorus beduinus var. *mekaliensis* Pic, 1912 **syn. nov.** of *Cardiophorus beduinus* Buysson, 1902
Cardiophorus beduinus var. *retrominiatus* Buysson, 1911 **syn. nov.** of *Cardiophorus beduinus* Buysson, 1902
Cardiophorus beduinus var. *tebessensis* Pic, 1912 **syn. nov.** of *Cardiophorus beduinus* Buysson, 1902
Cardiophorus biguttatus (var. *familiaris* Buysson, 1902 **syn. nov.** of *Cardiophorus biguttatus* (A. G. Olivier, 1790)
Cardiophorus biguttatus var. *suturatus* Buysson, 1902 **syn. nov.** of *Cardiophorus biguttatus* (A. G. Olivier, 1790)
Cardiophorus bimaculatus (Fabricius) [error for *Cardiophorus bipunctatus* (Fabricius, 1798)] var. *perrisi* Pic, 1893 **syn. nov.** of *Cardiophorus bipunctatus* (Fabricius, 1798)
Cardiophorus bipunctatus var. *alcalanus* Pic, 1923 **syn. nov.** of *Cardiophorus bipunctatus* (Fabricius, 1798)
Cardiophorus bipunctatus var. *almeiensis* Cobos, 1950 **syn. nov.** of *Cardiophorus bipunctatus* (Fabricius, 1798)
Cardiophorus bipunctatus var. *boeticus* Buysson, 1902 **syn. nov.** of *Cardiophorus bipunctatus* (Fabricius, 1798)
Cardiophorus bipunctatus var. *conjunctus* Pic, 1912 **syn. nov.** of *Cardiophorus bipunctatus* (Fabricius, 1798)
Cardiophorus bipunctatus var. *faroensis* Pic, 1912 **syn. nov.** of *Cardiophorus bipunctatus* (Fabricius, 1798)
Cardiophorus bipunctatus var. *lateobliteratus* Pic, 1902 **syn. nov.** of *Cardiophorus bipunctatus* (Fabricius, 1798)

Cardiophorus bipunctatus var. *nigrovelatus* Buysson, 1912 **syn. nov.** of *Cardiophorus bipunctatus* (Fabricius, 1798)
Cardiophorus bipunctatus var. *quadristigmosus* Buysson, 1912 **syn. nov.** of *Cardiophorus bipunctatus* (Fabricius, 1798)
Cardiophorus bipunctatus var. *semilimbatus* Pic, 1912 **syn. nov.** of *Cardiophorus bipunctatus* (Fabricius, 1798)
Cardiophorus bipunctatus var. *sexstigmosus* Buysson, 1912 **syn. nov.** of *Cardiophorus bipunctatus* (Fabricius, 1798)
Cardiophorus bipunctatus var. *semiobliteratus* Pic, 1902 **syn. nov.** of *Cardiophorus bipunctatus* (Fabricius, 1798)
Cardiophorus bipunctatus var. *smithi* Fuente, 1920 **syn. nov.** of *Cardiophorus bipunctatus* (Fabricius, 1798)
Cardiophorus bipunctatus var. *stigmosulus* Buysson, 1912e **syn. nov.** of *Cardiophorus bipunctatus* (Fabricius, 1798)
Cardiophorus bonnairei var. *laghouatensis* Pic, 1902 **syn. nov.** of *Cardiophorus bonnairei* Buysson, 1893a
Cardiophorus brunneus var. *extinctoides* Buysson, 1912 **syn. nov.** of *Paracardiophorus brunneus* (Erichson, 1840)
Cardiophorus brunneus var. *stilpnithorax* Buysson, 1912 **syn. nov.** of *Paracardiophorus brunneus* (Erichson, 1840)
Cardiophorus cinereus var. *gabrieli* Gerhardt, 1898 **syn. nov.** of *Dicronychus cinereus* (Herbst, 1784)
Cardiophorus collaris var. *paganettii* Pic, 1907 **syn. nov.** of *Cardiophorus collaris* Erichson, 1840
Cardiophorus collaris var. *retropictus* Buysson, 1911 **syn. nov.** of *Cardiophorus collaris* Erichson, 1840
Cardiophorus conformis var. *caroli* Pic, 1893 **syn. nov.** of *Cardiophorus letourneuxi* Buysson, 1891
Cardiophorus crux var. *alfieri* Pic, 1931 **syn. nov.** of *Cardiophorus crux* Erichson, 1840
Cardiophorus crux var. *shimonius* Fleutiaux, 1919 **syn. nov.** of *Cardiophorus crux* Erichson, 1840
Cardiophorus cyanipennis var. *delangrangei* Pic, 1899 **syn. nov.** of *Cardiophorus cyanipennis* Mulsant & Wachanru, 1852
Cardiophorus decorus var. *melanothorax* Candèze, 1860 **syn. nov.** of *Cardiophorus decorus* Faldermann, 1835
Cardiophorus eleonorae var. *buckhardti* Pic, 1910 **syn. nov.** of *Cardiophorus eleonorae* (Gené, 1836)
Cardiophorus eleonorae var. *humeralis* Bickhardt, 1908 **syn. nov.** of *Cardiophorus eleonorae* (Gené, 1836)
Cardiophorus exaratus var. *strigipennis* Buysson, 1902 **syn. nov.** of *Cardiophorus exaratus* Erichson, 1840
Cardiophorus hauseri var. *brevescapulatus* Pic, 1912 **syn. nov.** of *Cardiophorus hauseri* Schwarz, 1900
Cardiophorus hauseri var. *subunicolor* Pic, 1902 **syn. nov.** of *Cardiophorus hauseri* Schwarz, 1900
Cardiophorus hiekei **syn. nov.** of *Cardiophorus ebeninus* (Germar, 1843)
Cardiophorus hipponensis var. *rubidiventris* Buysson, 1919 **syn. nov.** of *Cardiophorus vestigialis* Erichson, 1840
Cardiophorus hoploderus var. *neosuturalis* Pic, 1902 **syn. nov.** of *Cardiophorus hoploderus* Candèze, 1860
Cardiophorus hoploderus var. *vittatithorax* Pic, 1902 **syn. nov.** of *Cardiophorus hoploderus* Candèze, 1860
Cardiophorus hoploderus var. *vittatithorax* Pic, 1902 **syn. nov.** of *Cardiophorus hoploderus* Candèze, 1860
Cardiophorus insignis var. *manueli* Pic, 1910 **syn. nov.** of *Cardiophorus insignis* Desbrochers des Loges, 1875
Cardiophorus insignis var. *sanguinicollis* Sahlberg, 1913 **syn. nov.** of *Cardiophorus insignis* Desbrochers des Loges, 1875
Cardiophorus nigerrimus var. *villosior* Buysson, 1911 **syn. nov.** of *Cardiophorus nigerrimus* Erichson, 1840
Cardiophorus nigropunctatus var. *cruciatus* Faust, 1877 **syn. nov.** of *Platynychus* (*Platynychus*) *nigropunctatus* (Candèze, 1860)

Cardiophorus nigropunctatus var. *duchoni* Pic, 1914 **syn. nov.** of *Platynychus* (*Platynychus*) *nigropunctatus* (Candèze, 1860)
Cardiophorus nigropunctatus var. *tedshenensis* Pic, 1914 **syn. nov.** of *Platynychus* (*Platynychus*) *nigropunctatus* (Candèze, 1860)
Cardiophorus ovipennis var. *mimicus* Buysson, 1911 **syn. nov.** of *Cardiophorus ovipennis* Desbrichers des Logis, 1875
Cardiophorus peyerimhoffi var. *ietenensis* Buysson, 1902 **syn. nov.** of *Cardiophorus peyerimhoffi* Buysson, 1902
Cardiophorus pharaonum var. *intrusus* Buysson, 1912 **syn. nov.** of *Cardiophorus pharaonum* Buysson, 1911

Cardiophorus (*Platynychus*) *rubripes* var. *flavipubens* Reitter, 1911 **syn. nov.** of *Dicronychus rubripes* (Germar, 1824)
Cardiophorus ruficruris var. *clarior* Buysson, 1912 **syn. nov.** of *Cardiophorus ruficruris* Brullé, 1832
Cardiophorus ruficruris var. *depictus* Buysson, 1913 **syn. nov.** of *Cardiophorus ruficruris* Brullé, 1832
Cardiophorus rufipes var. *persicus* Schwarz, 1892 **syn. nov.** of *Cardiophorus vestigialis* Erichson, 1840
Cardiophorus rufipes var. *pueli* Buysson, 1906 **syn. nov.** of *Cardiophorus goezei* Sánchez-Ruiz, 1996
Cardiophorus rufipes var. *ragusae* Ragusa, 1911 **syn. nov.** of *Cardiophorus vestigialis* Erichson, 1840
Cardiophorus sacratus var. *augustalisi* Pic, 1913 **syn. nov.** of *Cardiophorus sacratus* Erichson, 1840
Cardiophorus sacratus var. *sinuatevittatus* Pic, 1928 **syn. nov.** of *Cardiophorus discicollis* (Herbst, 1806)
Cardiophorus sacratus var. *tokatensis* Pic, 1913 **syn. nov.** of *Cardiophorus sacratus* Erichson, 1840
Cardiophorus scapulatus var. *exoplatus* Buysson, 1912 **syn. nov.** of *Cardiophorus scapulatus* Candèze, 1878
Cardiophorus scapulatus var. *gafsanus* Pic, 1909 **syn. nov.** of *Cardiophorus scapulatus* Candèze, 1878
Cardiophorus scapulatus var. *gafsensis* Buysson, 1912 **syn. nov.** of *Cardiophorus scapulatus* Candèze, 1878
Cardiophorus signatus var. *astorganus* Pic, 1912 **syn. nov.** of *Cardiophorus signatus* (A. G. Olivier, 1790)
Cardiophorus signatus var. *granjasensis* Pic, 1899 **syn. nov.** of *Cardiophorus signatus* (A. G. Olivier, 1790)
Cardiophorus signatus var. *manzanalensis* Pic, 1912 **syn. nov.** of *Cardiophorus signatus* (A. G. Olivier, 1790)
Cardiophorus signatus var. *moreli* Pic, 1923 **syn. nov.** of *Cardiophorus signatus* (A. G. Olivier, 1790)
Cardiophorus signatus var. *nigrithorax* Fuente, 1909 **syn. nov.** of *Cardiophorus signatus* (A. G. Olivier, 1790)
Cardiophorus signatus var. *semilutescens* Pic, 1912 **syn. nov.** of *Cardiophorus signatus* (A. G. Olivier, 1790)
Cardiophorus stoliger var. *henoni* Pic, 1912 **syn. nov.** of *Cardiophorus stoliger* Buysson, 1902b
Cardiophorus syriacus var. *fulvofasciatus* Reitter, 1894 **syn. nov.** of *Cardiophorus syriacus* (Linnaeus, 1758)
Cardiophorus syriacus var. *obscurofasciatus* Reitter, 1894 **syn. nov.** of *Cardiophorus syriacus* (Linnaeus, 1758)
Cardiophorus ulcerosus var. *acupictus* Buysson, 1922 **syn. nov.** of *Cardiophorus ulcerosus* Gené, 1836
Cardiophorus ulcerosus var. *anguloseminiatus* Buysson, 1922 **syn. nov.** of *Cardiophorus ulcerosus* Gené, 1836
Cardiophorus ulcerosus var. *anguloseminiatus* Buysson, 1922 **syn. nov.** of *Cardiophorus ulcerosus* Gené, 1836
Cardiophorus ulcerosus var. *antecenotatus* Ragusa, 1911 **syn. nov.** of *Cardiophorus ulcerosus* Gené, 1836
Cardiophorus ulcerosus var. *conjontucus* Buysson, 1922 **syn. nov.** of *Cardiophorus ulcerosus* Gené, 1836
Cardiophorus ulcerosus var. *extensenigratus* Buysson, 1922 **syn. nov.** of *Cardiophorus ulcerosus* Gené, 1836
Cardiophorus ulcerosus var. *infimus* Buysson, 1906 **syn. nov.** of *Cardiophorus vestigialis* Erichson, 1840
Cardiophorus ulcerosus var. *kabylianus* Pic, 1899 **syn. nov.** of *Cardiophorus ulcerosus* Gené, 1836
Cardiophorus ulcerosus var. *neoposticeseminiatus* Buysson, 1922 **syn. nov.** of *Cardiophorus ulcerosus* Gené, 1836
Cardiophorus ulcerosus var. *peronatus* Buysson, 1913 **syn. nov.** of *Cardiophorus ulcerosus* Gené, 1836
Cardiophorus ulcerosus var. *posticepictus* Buysson, 1913 **syn. nov.** of *Cardiophorus ulcerosus* Gené, 1836
Cardiophorus variipennis var. *humeronotatus* Pic, 1902 **syn. nov.** of *Cardiophorus hauseri* Schwarz, 1900
Cardiophorus (*Platynychus*) *cinereus* var. *cephalonicus* Buysson, 1913 **syn. nov.** of *Dicronychus cinereus* (Herbst, 1784)
Cardiophorus (*Platynychus*) *cinereus* var. *lusitanicus* Buysson, 1913 **syn. nov.** of *Dicronychus cinereus* (Herbst, 1784)
Cardiophorus (*Platynychus*) *versicolor* (Herbst, 1784) var. *lusitanicus* Buysson, 1913 **syn. nov.** of *Dicronychus cinereus* (Herbst, 1784)
Chiagosnius delauneyi var. *deminutus* Fleutiaux, 1939 **syn. nov.** of *Ludioschema delauneyi* (Fleutiaux, 1887)
Chiagosnius suturalis var. *gagates* Fleutiaux, 1939 **syn. nov.** of *Ludioschema suturale suturale* (Candèze, 1863)

Corymbites (*Selatosomus*) *aeneus* (var. *luctuosus* Depoli, 1928 **syn. nov.** of *Selatosomus* (*Selatosomus*) *aeneus* (Linnaeus, 1758)
Corymbites aeneus var. *mutator* Rey, 1891 **syn. nov.** of *Selatosomus* (*Selatosomus*) *aeneus* (Linnaeus, 1758)
Corymbites aeneus var. *subrugosicollis* Rey, 1891 **syn. nov.** of *Selatosomus* (*Selatosomus*) *aeneus* (Linnaeus, 1758)
Corymbites bipustulatus var. *flavescens* Schilsky, 1888 **syn. nov.** of *Calambus bipustulatus* (Linnaus, 1767)
Corymbites bipustulatus var. *rubidus* Wimmel, 1894 **syn. nov.** of *Calambus bipustulatus* (Linnaus, 1767)
Corymbites bipustulatus var. *tenebricans* Buysson, 1890 **syn. nov.** of *Calambus bipustulatus* (Linnaus, 1767)
Corymbites cuprascens LeConte, 1853 **syn. nov.** of *Actenicerus* (*Actenicerus*) *sjaelandicus* (O. F. Müller, 1764)
Corymbites cupreus var. *brevelineatus* Pic, 1926 **syn. nov.** of *Ctenicera cuprea* (Fabricius, 1775)

Corymbites cupreus var. *discoidalis* Pic, 1913 **syn. nov.** of *Ctenicera cuprea* (Fabricius, 1775)
Corymbites cupreus var *hilfi* Reitter, 1902 **syn. nov.** of *Ctenicera cuprea* (Fabricius, 1775)
Corymbites cupreus var. *reductus* Pic, 1923 **syn. nov.** of *Ctenicera cuprea* (Fabricius, 1775)
Corymbites cupreus var. *transsylvanicus* Szombathy, 1910 **syn. nov.** of *Ctenicera cuprea* (Fabricius, 1775)
Corymbites (*Anostirus*) *haemapterus* var. *araxis* Reitter, 1914 **syn. nov.** of *Anostirus eschscholtzi* (Faldermann, 1835b)
Corymbites haematodes var. *flavus* Gredler, 1863 **syn. nov.** of *Anostirus purpureus* (Poda, 1761)
Corymbites impressus var. *rufipes* Schilsky, 1888 **syn. nov.** of *Paraphotistus* (*Mosotalesus*) *impressus impressus* (Fabricius, 1792)
Corymbites quercus var. *testaceus* **syn. nov.** of *Pheletes quercus* (A. G. Olivier, 1790)
Corymbites (*Metactenicerus*) *speciosus* var. *viridus* Miwa, 1934 **syn. nov.** of *Corymbitodes speciosus* (Miwa, 1928)
Corymbites (*Actenicerus*) *tessellatus* var. *jurae* Pic, 1913 **syn. nov.** of *Actenicerus* (*Actenicerus*) *sjaelandicus* (O. F. Müller, 1764)
Corymbites (*Actenicerus*) *tessellatus* var. *lineatus* Pic, 1927 **syn. nov.** of *Actenicerus* (*Actenicerus*) *sjaelandicus* (O. F. Müller, 1764)
Corymbites tessellatus var. *nigropilosus* Gerhardt, 1912 **syn. nov.** of *Actenicerus* (*Actenicerus*) *sjaelandicus* (O. F. Müller, 1764)
Corymbites tessellatus var. *strigatus* Gerhardt, 1897 **syn. nov.** of *Actenicerus* (*Actenicerus*) *sjaelandicus* (O. F. Müller, 1764)
Corymbites virens var. *stramineus* Candèze, 1892 **syn. nov.** of *Ctenicera virens* (Schrank, 1781)
Corymbitodes gratus var. *koikei* Kishii & Ôhira, 1956 **syn. nov.** of *Corymbitodes gratus* (Lewis, 1894)
Corymbitodes gratus var. *tenuithorax* Kishii & Ôhira, 1956 **syn. nov.** of *Corymbitodes gratus* (Lewis, 1894)
Crepidophorus mutilatus var. *barbarinus* Pic, 1916 **syn. nov.** of *Crepidophorus mutilatus* (Rosenhauer, 1847)
Cryptohypnus crux var. *reductus* Buysson, 1912 **syn. nov.** of *Zorochros crux* (Küster, 1849)
Cryptohypnus pulchellus var. *bipunctatus* Schilsky, 1888 **syn. nov.** of *Negastrius pulchellus* (Linnaeus, 1761)
Cryptohypnus pulchellus var. *ripicola* Friederichs, 1901 **syn. nov.** of *Negastrius pulchellus* (Linnaeus, 1761)
Cryptohypnus riparius var. *tournieri* Pic, 1910 **syn. nov.** of *Zorochros riparius* (Fabricius, 1792)
Cryptohypnus sabulicola var. *contentus* Friederichs, 1901 **syn. nov.** of *Negastriua riparius sabulicola* (Boheman, 1852)
Cryptohypnus sabulicola var. *laetus* Friederichs, 1901 **syn. nov.** of *Negastriua riparius sabulicola* (Boheman, 1852)
Cryptohypnus sabulicola var. *maestus* Friederichs, 1901 **syn. nov.** of *Negastriua riparius sabulicola* (Boheman, 1852)
Cryptohypnus sabulicola var. *modestus* Friederichs, 1901 **syn. nov.** of *Negastriua riparius sabulicola* (Boheman, 1852)
Cryptohypnus tetragraphus var. *bipustulatus* Schilsky, 1888 **syn. nov.** of *Zorochros dermestoides* (Herbst, 1806)
Ctenicera cuprea var. *brevipectinata* Binaghi, 1940 **syn. nov.** of *Ctenicera cuprea* (Fabricius, 1775)
Ctenicera (*Anostirus*) *purpurea* var. *lecomtei* Méquignon, 1927 **syn. nov.** of *Anostirus purpureus* (Poda, 1761)
Dolopius marginatus var. *gyllenhali* Buysson, 1926 **syn. nov.** of *Dalopius marginatus* (Linnaeus, 1758)
Denticollis acuticollis var. *amurensis* Pic, 1916 **syn. nov.** of *Denticollis acuticollis* (Motschulsky, 1860b)
Denticollis inaequalis var. *limbatipennis* Pic, 1916 **syn. nov.** of *Denticollis inaequalis* (Candèze, 1879a)
Denticollis linearis var. *alsaticus* Pic, 1916 **syn. nov.** of *Denticollis linearis* (Linnaeus, 1758)

Denticollis linearis var. *jacqueti* Pic, 1916 **syn. nov.** of *Denticollis linearis* (Linnaeus, 1758)
Denticollis linearis var. *maculithorax* Pic, 1916 **syn. nov.** of *Denticollis linearis* (Linnaeus, 1758)
Denticollis linearis var. *simulatrix* Roubal, 1941 **syn. nov.** of *Denticollis linearis* (Linnaeus, 1758)
Denticollis (*Campylus*) *rubens* var. *notatithorax* Pic, 1916 **syn. nov.** of *Denticollis rubens* Piller & Mitterpacher, 1783
Diacanthus bipustulatus var. *semiflavus* Fleischer, 1886 **syn. nov.** of *Calambus bipustulatus* (Linnaeus, 1767)
Drasterius brahminus var. *delauneyi* Fleutiaux, 1918 **syn. nov.** of *Drasterius prosternalis* Candèze, 1878b
Drasterius bimaculatus (Rossi, 1790) var. *notatulus* Buysson, 1912 **syn. nov.** of *Drasterius aegyptiacus* Buysson, 1905
Drasterius brahminus var. *delauneyi* Fleutiaux, 1918 **syn. nov.** of *Drasterius brahminus* Candèze, 1859

Drasterius collaris var. *nigricollis* Fleutiaux, 1918 **syn. nov.** of *Drasterius collaris* Candèze, 1859
Drasterius bimaculatus var. *aegyptiacus* Buysson, 1912 **syn. nov.** of *Drasterius bimaculatus* (Rossi, 1790)
Drasterius bimaculatus var. *angulosepictus* Buysson, 1900 **syn. nov.** of *Drasterius bimaculatus* (Rossi, 1790)
Drasterius bimaculatus var. *anticenotatus* Buysson, 1912 **syn. nov.** of *Drasterius bimaculatus* (Rossi, 1790)
Drasterius bimaculatus var. *variegatulus* Buysson, 1912 **syn. nov.** of *Drasterius aegyptiacus* Buysson, 1905
Drasterius bimaculatus var. *anticus* Reitter, 1890 **syn. nov.** of *Drasterius bimaculatus* (Rossi, 1790)
Drasterius bimaculatus var. *apicalis* Reitter, 1890 **syn. nov.** of *Drasterius bimaculatus* (Rossi, 1790)
Drasterius bimaculatus var. *basalis* Reitter, 1890 **syn. nov.** of *Drasterius bimaculatus* (Rossi, 1790)
Drasterius bimaculatus var. *bengasinus* Binaghi, 1941 **syn. nov.** of *Drasterius bimaculatus* (Rossi, 1790)
Drasterius bimaculatus var. *bisbisignatus* Buysson, 1913 **syn. nov.** of *Drasterius bimaculatus* (Rossi, 1790)
Drasterius bimaculatus var. *cantheriatus* Buysson, 1900 **syn. nov.** of *Drasterius bimaculatus* (Rossi, 1790)
Drasterius bimaculatus var. *diluticollis* Reitter, 1890 **syn. nov.** of *Drasterius bimaculatus* (Rossi, 1790)
Drasterius bimaculatus var. *dorsalis* Reitter, 1890 **syn. nov.** of *Drasterius bimaculatus* (Rossi, 1790)
Drasterius bimaculatus var. *grossus* Sahlberg, 1913 **syn. nov.** of *Drasterius bimaculatus* (Rossi, 1790)
Drasterius bimaculatus var. *immaculatus* Oliveira, 1893 **syn. nov.** of *Drasterius bimaculatus* (Rossi, 1790)
Drasterius bimaculatus var. *latepictus* Buysson, 1900 **syn. nov.** of *Drasterius bimaculatus* (Rossi, 1790)
Drasterius bimaculatus var. *lateralis* Reitter, 1890 **syn. nov.** of *Drasterius bimaculatus* (Rossi, 1790)
Drasterius bimaculatus var. *lineatobasalis* Fuente, 1903 **syn. nov.** of *Drasterius bimaculatus* (Rossi, 1790)
Drasterius bimaculatus var. *montandoni* Buysson, 1896 **syn. nov.** of *Drasterius bimaculatus* (Rossi, 1790)
Drasterius bimaculatus var. *pseudobiskrensis* Buysson, 1912 **syn. nov.** of *Drasterius aegyptiacus* Buysson, 1905
Drasterius bimaculatus var. *sexmaculata* Ragusa, 1893 **syn. nov.** of *Drasterius bimaculatus* (Rossi, 1790)
Drasterius bimaculatus var. *sexsignatus* Buysson, 1896 **syn. nov.** of *Drasterius bimaculatus* (Rossi, 1790)
Drasterius bimaculatus var. *trisignatus* Chobaut, 1897 **syn. nov.** of *Drasterius bimaculatus* (Rossi, 1790)
Drasterius bimaculatus wachsmanni Reitter, 1905 **syn. nov.** of *Drasterius bimaculatus* (Rossi, 1790)
Ectinus aterrimus var. *cavillator* Buysson, 1926 **syn. nov.** of *Ectinus aterrimus* (Linnaeus, 1761)
Ectinus subaeneus var. *xanthodon* Maerkel, 1847 **syn. nov.** of *Sericus* (*Sericoderma*) *subaeneus* (W. Redtenbacher, 1842)
Elater aethiops var. *dilutimembris* Buysson, 1911 **syn. nov.** of *Ampedus aethiops* (Lacordaire, 1835)
Elater aethiops var. *rugosissimus* Tanzer, 1929 **syn. nov.** of *Ampedus aethiops* (Lacordaire, 1835)
Elater (*Cardiophorus*) *argiolus* var. *fascicollis* Gené, 1836 **syn. nov.** of *Cardiophorus argiolus* Gené, 1836
Elater assimilis **syn. nov.** of *Actenicerus* (*Actenicerus*) *sjaelandicus* (O. F. Müller, 1764)
Elater balteatus var. *sorsakoskii* Hellén, 1934 **syn. nov.** of *Ampedus lepidus* (Mäklin, 1876)
Elater cinnabarinus [sic] var. *dimorphus* Reitter, 1918 **syn. nov.** of *Ampedus cinnabarinus* (Eschscholtz, 1829)
Elater crocatus var. *nigrolineatus* Pic, 1920 **syn. nov.** of *Ampedus nigroflavus* (Goeze, 1777)
Elater crocatus var. *podolicus* Reitter, 1889 **syn. nov.** of *Ampedus nigroflavus* (Goeze, 1777)
Elater elegantulus var. *seidlitzi* Buysson, 1896 **syn. nov.** of *Ampedus elegantulus* (Schönherr, 1817)
Elater elongatulus var. *discanicus* Buysson, 1896 **syn. nov.** of *Ampedus elongatulus* (Herbst, 1784)
Elater elongatulus var. *balteatulus* Reitter, 1889 **syn. nov.** of *Ampedus elongatulus* (Herbst, 1784)
Elater ferrugineus var. *occitanicus* Villers, 1789 **syn. nov.** of *Elater ferrugineus* Linnaeus, 1758
Elater ferrugatus var. *fulvilegulus* Reitter, 1889 **syn. nov.** of *Ampedus pomorum* (Herbst, 1784)
Elater ferrugatus var. *nigriventris* Reitter, 1889 **syn. nov.** of *Ampedus pomorum* (Herbst, 1784)
Elater fulgens A. G. Olivier, 1790, is synonym with *Campsosternus auratus* (Drury, 1773), **syn. reafirmed**
Elater humeralis Dufour, 1843 **syn. nov.** of *Ctenicera cuprea* (Fabricius, 1775)

Elater lateralis Fabricius, 1794 **syn. nov.** of *Ampedus tristis* (Linnaeus, 1758)
Elater lenkoranus var. *caspicus* Reitter, 1918 **syn. nov.** of *Ampedus lenkoranus* (Reitter, 1889)
Elater lythropterus var. *atropilosus* B. G. Rye, 1903 **syn. nov.** of *Ampedus cinnabarinus* Eschscholtz, 1829
Elater nigrinus var. *fulvangulus* Csiki, 1940 **syn. nov.** of *Ampedus nigrinus* (Herbst, 1874)
Elater pomonae var. *gerardi* Candèze, 1890 **syn. nov.** of *Ampedus pomonae* (Stephens, 1830)
Elater pomonae var. *paraminiatus* Buysson, 1906 **syn. nov.** of *Ampedus sanguinolentus sanguinolentus* (Schrank, 1776)
Elater pomonae var. *paraminiatus* Buysson, 1906 **syn. nov.** of *Ampedus sanguinolentus sanguinolentus* (Schrank, 1776)
Elater pomorum var. *apicalis* Schilsky, 1892 **syn. nov.** of *Ampedus pomorum* (Herbst, 1784)

Elater praeustus var. *laetulus* Reitter, 1918 **syn. nov.** of *Ampedus circassicus* (Reitter, 1887)
Elater praeustus var. *talyschensis* Reitter, 1889 **syn. nov.** of *Ampedus circassicus* (Reitter, 1887)
Elater praeustus var. *tauricus* Reitter, 1889 **syn. nov.** of *Ampedus praeustus* (Fabricius, 1792)
Elater pulcher var. *pulcherrimus* Reitter, 1889 **syn. nov.** of *Ampedus pulcher* (Baudi di Selve, 1871)
Elater pulcher var. *araxidis* Reitter, 1889 **syn. nov.** of *Ampedus circassicus* (Reitter, 1887)
Elater pulcher var. *pulcherrimus* Reitter, 1889 **syn. nov.** of *Ampedus pulcher* (Baudi di Selve, 1871)
Elater rivularius var. *alpestris* Zetterstedt, 1828 **syn. nov.** of *Hypnoidus rivularius rivularius* (Gyllenhal, 1808)
Elater ruficeps var. *semiobscurus* Pic, 1902 **syn. nov.** of *Brachygonus ruficeps* (Mulsant & Guillebeau, 1855a)
Elater sanguineus var. *semisanguineus* Reitter, 1889 **syn. nov.** of *Ampedus sanguineus* (Linnaues, 1758)
Elater sanguinolentus var. *paleatus* Candèze, 1893 **syn. nov.** of *Ampedus sanguinolentus sanguinolentus* (Schrank, 1776)
Elater satrapa var. *intermedius* Buysson, 1896 **syn. nov.** of *Ampedus rufipennis* (Stephens, 1830)
Elater satrapa var. *pandellei* Buysson, 1896 **syn. nov.** of *Ampedus rufipennis* (Stephens, 1830)
Elater semiruber Stephens, 1830 **syn. nov.** of *Ampedus cinnabarinus* (Eschscholtz, 1829)
Elater tristis var. *amurensis* Pic, 1909 **syn. nov.** of *Ampedus sobrinus* (Motschulsky, 1860)
Elater (*Cardiophorus*) *ulcerosus* var. *angularis* Gené, 1836 **syn. nov.** of *Cardiophorus ulcerosus* (Gené, 1836)
Harminius castaneus var. *ferrugineus* Binaghi, 1972, **syn nov**. of *Harminius spiniger* Candèze, 1860
Harminius undulatus triornatus Tanzer, 1934 **syn. nov.** of *Diacanthous undulatus* (DeGeer, 1774)
Hemicrepidius hirtus var. *maculicollis* Pic, 1897 **syn. nov.** of *Hemicrepidius niger* (Herbst, 1784)
Heteroderes algirinus var. *infuscatus* Sahlberg, 1903 **syn. nov.** of *Heteroderes algirinus* (Lucas, 1846)
Heteroderes crucifer var. *multidisjunctus* Pic, 1912 **syn. nov.** of *Aeoloderma crucifer* (Rossi, 1790)
Hypnoidus dermestoides var. *diluviatus* Buysson, 1913 **syn. nov.** of *Zorochros dermestoides* (Herbst, 1806)
Hypnoidus dermestoides var. *humeropictus* Buysson, 1900 **syn. nov.** of *Zorochros dermestoides* (Herbst, 1806)
Hypnoidus dermestoides var. *laevithorax* Buysson, 1906 **syn. nov.** of *Zorochros dermestoides* (Herbst, 1806)
Hypnoidus dermestoides (var. *pseudotetragraphus* Binaghi, 1933 **syn. nov.** of *Zorochros dermestoides* (Herbst, 1806)
Hypnoidus dermestoides var. *submaculatus* Reitter, 1895 **syn. nov.** of *Zorochros dermestoides* (Herbst, 1806)
Hypnoidus dermestoides var. *subnotatus* Reitter, 1895 **syn. nov.** of *Zorochros dermestoides* (Herbst, 1806)
Hypnoidus flavipes var. *mangonis* Buysson, 1913 **syn. nov.** of *Zorochros flavipes* (Aubè, 1850)
Hypnoidus frigidus var. *granulosus* Buysson, 1896 **syn. nov.** of *Hypnoidus rivularius rivularius* Gyllenhal, 1808
Hypnoidus meridionalis var. *levantinus* K. & J. Daniel, 1903 **syn. nov.** of *Zorochros meridionalis* (Laporte, 1840)
Hypnoidus pulchellus var. *olivieri* Buysson, 1896 **syn. nov.** of *Negastrius pulchellus* (Linnaeus, 1758)
Hypnoidus pulchellus var. *panzeri* Buysson, 1896 **syn. nov.** of *Negastrius pulchellus* (Linnaeus, 1758)
Idolus picipennis var. *angulatus* Rey, 1891 **syn. nov.** of *Idolus picipennis* (Bach, 1852)
Idolus picipennis var. *moreli* Pic, 1926 **syn. nov.** of *Idolus picipennis* (Bach, 1852)
Idolus picipennis var. *transpictus* Buysson, 1896 **syn. nov.** of *Idolus picipennis* (Bach, 1852)
Ischnodes languidus var. *syriacus* Pic, 1904 **syn. nov.** of *Haterumelater languidus* (Buysson, 1891b)
Limonius lythrodes var. *nigricollis* Schilsky, 1888 **syn. nov.** of *Limonius quercus* (A. G. Olivier, 1790)
Limonius minutus var. *cyanichrous* Buysson, 1926 **syn. nov.** of *Limonius minutus* (Linnaeus, 1758)
Limonius parvulus var. *serosus* Buysson, 1902 **syn. nov.** of *Nothodes parvulus* (Panzer, 1799)
Limonius pilosus var. *cyanescens* Buysson, 1902 **syn. nov.** of Cidnopus *pilosus* (Leske, 1785)
Limonius pilosus var. *ferrugineus* Buysson, 1902 **syn. nov.** of Cidnopus *pilosus* (Leske, 1785)
Limonius parvulus (var. *dilutus* Buysson, 1926 **syn. nov.** of *Nothodes parvulus* (Panzer, 1799)
Ludius nigropilosus Schwarz, 1902 **syn. nov.** of *Anostirus eschscholtzi* (Faldermann, 1835)
Ludius(*Corymbites*) *cupreus* var. *diffusus* Buysson, 1929 **syn. nov.** of *Ctenicera cuprea* (Fabricius, 1775)
Ludius(*Hypoganus*) *cinctus* var. *akbesianus* Pic, 1903 **syn. nov.** of *Hypoganus inunctus* (Lacordaire, 1835)
Ludius(*Hypoganus*) *cinctus* var. *semitestaceus* Pic, 1905 **syn. nov.** of *Hypoganus inunctus* (Lacordaire, 1835)
Ludius (*Liotrichus*) *affinis* var. *inhumanus* Pic, 1916 **syn. nov.** of *Liotrichus affinis* (Payküll, 1800)
Ludius ferrugineus var. *morio* Schilsky, 1888 **syn. nov.** of *Elater ferrugineus ferrugineus* Linnaeus, 1758
Megapenthes volxemi var. *retowskii* Heyden, 1882 **syn. nov.** of *Procraerus carinifrons* (Desbrochers des Loges 1875)
Melanotus badachshanicus Dolin, 1988 **syn. nov.** of *Melanotus indigenus* Gurjeva, 1988

Melanotus castanipes var. *sublaevicollis* Rey, 1891 **syn. nov.** of *Melanotus* (*Melanotus*) *castanipes* (Paykull, 1800)
Melanotus darvasicus Dolin, 1988 **syn. nov.** of *Melanotus nitidicornis* Gurjeva, 1988
Melanotus dichrous var. *incertus* Schwarz, 1892 **syn. nov.** of *Melanotus* (*Melanotus*) *dichrous* (Erichson, 1841)
Melanotus rufipes var. *dichroicolor* Buysson, 1914 **syn. nov.** of *Melanotus* (*Melanotus*) *villosus* (Geoffroy, 1785)
Melanotus rufipes var. *subrugatus* Rey, 1891 **syn. nov.** of *Melanotus* (*Melanotus*) *crassicollis* Erichson, 1841
Melanotus rufipes var. *subfus* Schwarz, 1892 **syn. nov.** of *Melanotus* (*Melanotus*) *villosus* (Geoffroy, 1785)
Melanotus sogdianus Dolin & Latifi, 1988 **syn. nov.** of *Melanotus kryzhanovskyi* Gurjeva, 1988
Metaricus viridis subopacus Nakane, 1959 **syn. nov.** of *Vuilletus viridis* (Lewis, 1894)
Negastrius pulchellus var. *exiguus* (Randall, 1838) **syn. nov.** of *Negastrius pulchellus* (Linnaeus, 1761)
Neodiploconus elegans var. *takanoi* Miwa, 1930 **syn. nov.** of *Priopus superbus* (Fleutiaux, 1895)
Orithales serraticornis var. *separatus* Pic, 1929 **syn. nov.** of *Orithales serraticornis* (Paykull, 1800)
Paranomus guttatus var. *biguttatus* Pic, 1929 **syn. nov.** of *Eanus* (*Eanus*) *guttatus* (Germar, 1839)
Paranomus guttatus var. *innotatus* Pic, 1929 **syn. nov.** of *Eanus* (*Eanus*) *guttatus* (Germar, 1839)
Paranomus guttatus var. *uniguttatus* Pic, 1929 **syn. nov.** of *Eanus* (*Eanus*) *guttatus* (Germar, 1839)
Pheletes quercus var. *solarii* Buysson, 1902 **syn. nov.** of *Pheletes quercus* (A. G. Olivier, 1790)
Porthmidius austriacus var. *agnusi* Méquignon, 1927 **syn. nov.** of *Porthmidius austriacus* (Schrank, 1781)
Prodrasterius sexpunctatus var. *immaculatus* Miwa, 1929 **syn. nov.** of *Babadrasterius sexpunctatus* (Miwa, 1927)
Prodrasterius sexpunctatus var. *quadrimaculatus* Miwa, 1929b **syn. nov.** of *Babadrasterius sexpunctatus* (Miwa, 1927)
Prosternon holosericeus var. *semirufus* Pic, 1909 **syn. nov.** of *Prosternon tessellatum* (Linnaeus, 1758)
Selatosomus aeneus var. *marginatus* Papp, 1943 **syn. nov.** of *Selatosomus* (*Selatosomus*) *aeneus* (Linnaeus, 1758)
Selatosomus impressus var. *alpicola* Szombathy, 1910 **syn. nov.** of *Paraphotistus* (*Mosotalesus*) *impressus impressus* (Fabricius, 1792)
Selatosomus cruciatus var. *bifenestratus* Pic, 1912c **syn. nov.** of *Selatosomus* (*Pristilophus*) *cruciatus* (Linnaeus, 1758)
Selatosomus guttatus var. *seminotatus* Pic, 1919 **syn. nov.** of *Eanus* (*Eanus*) *guttatus* (Germar, 1839a)
Selatosomus juldanus var. *diversipennis* Reitter, 1910 **syn. nov.** of *Pseudanostirus juldanus* Reitter, 1910
Selatosomus rugosus var. *carreti* Buysson, 1894 **syn. nov.** of *Selatosomus* (*Selatosomus*) *confluens rugosus* Germar, 1825
Senodonia quadricollis Laporte, 1838 var. *sinensis* Jagemann, 1943 **syn. nov.** of *Senodonia quadricollis* Laporte, 1838
Sericus brunneus var. *obscurior* Pic, 1913 **syn. nov.** of *Sericus* (*Sericus*) *brunneus brunneus* (Linnaeus, 1758)
Sericus brunneus var. *sulcipennis* Buysson, 1893 **syn. nov.** of *Sericus* (*Sericus*) *brunneus brunneus* (Linnaeus, 1758)
Sericus subaeneus var. *planeti* Buysson, 1926 **syn. nov.** of *Sericus* (*Sericoderma*) *subaeneus* (W. Redtenbacher, 1842)
Silesis dimidiatipennis var. *basalis* Pic, 1912 **syn. nov.** of *Peripontius dimidiatipennis* (Reiche & Saulcy, 1857)
Silesis dimidiatipennis var. *mimicus* Pic, 1912 **syn. nov.** of *Peripontius dimidiatipennis* (Reiche & Saulcy, 1857)
Silesis musculus var. *flavicollis* Fleutiaux, 1902 **syn. nov.** of *Silesis musculus musculus* Candèze, 1873
Silesis rutilipennis var. *hipponensis* Pic, 1912 **syn. nov.** of *Peripontius rutilipennis* (Illiger, 1807)

Silesis rutilipennis var. *postobscurus* Pic, 1912 **syn. nov.** of *Peripontius rutilipennis* (Illiger, 1807)
Silesis terminatus var. *indistinctus* Buysson, 1912 **syn. nov.** of *Peripontius terminatus* (Erichson, 1841)
Sinaptus [sic for *Synaptus*] *filiformis* var. *atripes* Pic, 1941 **syn. nov.** of *Synaptus filiformis* (Fabricius, 1781)
Sinaptus [sic for *Synaptus*] *filiformis* var. *leprieuri* Pic, 1941 **syn. nov.** of *Synaptus filiformis* (Fabricius, 1781)
Synaptus filiformis var. *levantinus* Sahlberg, 1913 **syn. nov.** of *Synaptus filiformis* (Fabricius, 1781)

New combinations and assignments
Aeoloides imitator (Reitter, 1891) **comb. nov**. from *Aelous*

Adelocera ichihashii Ôhira, 1978 from *Adelocera* (*Brachylacon*) *ichihashi* Ôhira, 1978, **new assignment**
Adelocera omotoensis Ôhira, 1978 from *Adelocera* (*Brachylacon*) *omotoensis* Ôhira, 1978, **new assignment**
Adelocera shibatai Kishii, 1996 from *Adelocera* (*Brachylacon*) *shibatai* Kishii, 1996, **new assignment**
Adelocera takasago Kishii, 1996 from *Adelocera* (*Brachylacon*) *takasago* Kishii, 1996, **new assignment**
Agrypnus amamianus (Kishii, 1974) from *Agrypnus* (*Colaulon*) *amamianus* (Kishii, 1974)), **new assignment**
Agrypnus babanus Kishii, 1990 from *Agrypnus* (*Colaulon*) *babanus* Kishii, 1990c), **new assignment**
Agrypnus heianus Kishii, 1985 from *Agrypnus* (*Colaulon*) *heianus* Kishii, 1985, *Compsolacon* **new assignment**
Agrypnus korotjaevi (Gurjeva, 1987) **comb. nov**. from *Compsolacon*
Agrypnus (s. str.) *kusuii* Ôhira, 1993 from *Agrypnus* (*Colaulon*) *kusuii* Ôhira, 1993, **new assignment**
Agrypnus (s. str.) *ryukyuensis* Kishii, 1985 from *Agrypnus* (*Sabikikorius*) *ryukyuensis* Kishii, 1985), **new assignment**
Agrypnus takasago Kishii, 1990c from *Agrypnus* (*Colaulon*) *takasago* Kishii, 1990c), **new assignment**
Agrypnus tsukamotoi (Kishii, 1956) from *Agrypnus* (*Colaulon*) *tsukamotoi* (Kishii, 1956), **new assignment**
Agrypnus tsushimensis kusuii Ôhira, 2004 from *Agrypnus* (*Colaulon*) *tsushimensis kusuii* Ôhira, 2004, **new assignment**
Agrypnus tsushimensis tsushimensis Ôhira, 1986 from *Agrypnus* (*Colaulon*) *tsushimensis* Ôhira, 1986, **new assignment**
Dicronychus consentaneus (Kollar, 1848) **comb. nov.** from *Cardiophorus* (based on study of the type material)
Dicronychus devius (Candèze, 1882) **comb. nov.** from *Cardiophorus*
Dicronychus hippanicus (Orlov, 1997) **comb. nov.** from *Cardiophorus*
Dicronychus pseudobesus (Orlov, 1993) **comb. nov.** from *Cardiophorus*
Dicronychus tzkheniensis (Orlov, 1997) **comb. nov.** from *Cardiophorus*
Drasterius agnatus (Candèze, 1873) **comb. nov.** from *Prodrasterius* [originally *Aeolus*]
Lacon (s. str.) *kikuchii* (Miwa, 1929) from *Lacon* (*Alaotypus*) *kikuchii* (Miwa, 1929), **new assignment**
Lacon (s. str.) *kintaroui* Kishii, 1990 from *Lacon* (*Alaotypus*) *kintaroui* Kishii, 1990), **new assignment**
Lacon (s. str.) *kushihige* Kishii, 1990 from *Lacon* (*Alaotypus*) *kushihige* Kishii, 1990), **new assignment**
Lacon (s. str.) *puriensis* Kishii, 1991 from *Lacon* (*Alaotypus*) *puriensis* Kishii, 1991), **new assignment**
Lacon (s. str.) *ramatasenseni* (Miwa, 1934) from *Lacon* (*Alaotypus*) *ramatasenseni* Miwa, 1934), **new assignment**
Lacon (s. str.) *rotundicollis* Kishii & Jiang, 1994 from *Lacon* (*Alaotypus*) *rotundicollis* Kishii & Jiang, 1994), **new assignment**
Ludioschema akashii (Ôhira, 1969) **comb. nov.** from *Chiagosnius*
Ludioschema atkinsoni (Candèze, 1878) **comb. nov.** from *Agonischius*
Ludioschema cyaneum (Candèze, 1863) **comb. nov.** from *Chiagosnius*[originally *Agonischius*]
Ludioschema delauneyi (Fleutiaux, 1887) **comb. nov.** from *Chiagosnius* [originally *Agonischius*]
Ludioschema dorsale (Candèze, 1878) **comb. nov.** from *Chiagosnius* [originally *Agonischius*]
Ludioschema massiei (Fleutiaux, 1939) **comb. nov.** from *Chiagosnius*
Ludioschema minor (Fleutiaux, 1903) **comb. nov.** from *Chiagosnius* [originally *Agonischius*]
Ludioschema nigripenne (Fleutiaux, 1895) **comb. nov.** from *Chiagosnius*) [originally *Agonischius*]
Ludioschema paludum (Fleutiaux, 1939) **comb. nov.** from *Chiagosnius*
Ludioschema simulator (Candèze, 1897) **comb. nov**. from *Chiagosnius* [originally *Agonischius*]
Phorocardius vicinus (Kollar, 1848) **comb. nov.** from *Cardiophorus* (based on study of the type material)
Podeonius aquilus formosensis (Kishii, 1989) **comb. nov**. from *Akitsu*
Podeonius castelnaui (Candèze, 1878) **comb. nov**. from *Akitsu* [originally *Anchastus*]

Podeonius taiwanus (Ôhira, 1972) **comb. nov**. from *Akitsu*) [originally *Anchastus*]

Replacement names

Adelocera manillensis **nom. nov.** for *Adelocera oblonga* (Fleuriaux, 1934) [originally *Brachylacon oblongus* Fleutiaux, 1934] (nec *Lacon punctatus oblongus* (Della Beffa, 1931) [originally *Adelocera punctata oblonga* Della Beffa, 1931]
Meristhus sikkimensis **nom. nov.** for *Meristhus crenulatus* Mukhopadhyay & Chakraborty, 2003 [nec *Meristhus crenulatus* Vats & Kashyap, 1996]

Silesis jagemanni **nom. nov.** for *Silesis sinensis* Jagemann, 1934 (December) [nec *Silesis sinensis* Fleutiaux, 1934 (August), at present in *Lanecarus*,)
Hemiops germari **nom. nov.** for *Hemiops nigripes* Germar, 1843 [nec *Hemiops nigripes* Laporte, 1838]

Changes in rank

Ampedus straminipennis (Heyden, 1880) is tentatively elevated to species rank from *Elater ochropterus* Küster [sic] var. *straminipennis* Heyden, 1880, pending examination of type specimens
Campsosternus dohrnii fairmairei Fleutiaux, 1918 is tentatively elevated to subspecies from *Campsosternus mouhoti* var. *fairmairei* Fleutiaux, 1918 pending examination of type specimens
Lacon punctatus oblongus (Della Beffa, 1931) is tentatively elevated to subspecies from *Adelocera punctatus* (Herbst) var. *oblonga* Della Beffa, 1931 pending examination of type specimens

Resurrection

Athous fausti Reitter, 1890: This species was described from the Caucasus, but hat been sunk into synonymy of *Megathous dauricus* (Mannerheim, 1852), an eastern Sibirian species not occurring in the Caucasus. For this reason *Athous fausti* Reitter, 1890 is resurrected from synonymy until examination of the type material can clarify its status.

Spelling

Ampedus cinnaberinus (Eschscholtz, 1829) is the original but unused original spelling. The name of this species is commonly spelled *A. cinnabarinus*. According to the ICZN, Art. 33.3.1 the subsequent spelling is to be preserved.
Ampedus turkmenicus Dolin: Dolin, 1977b: 357 described his new species as *Ampedus turcmenicus*, an incorrect original spelling corrected by himself (Dolin, 1977c: 360).
Mulsanteus guillebelli Mulsant & Godart, 1853 is an incorrect original spelling, as the authors remark they are dedicating the species to Mr. Guillebeau. The correct spelling *guillebeaui* is used in this catalogue.
Phorotarsus Motschulsky, 1861 is a spelling used by Motschulsky and subsequent authors for *Pharotarsus* Motschulsky, 1859.
Szombatya Platia is obviously an incorrect original spelling and emended to *Szombathya*.
Zorochrus C. G. Thomson, 1864 is a misspelling of *Zorochros* C. G. Thomson, 1859. Although the spelling *Zorochrus* was used more frequently in the 20 century, recent authors use *Zorochros*. This original spelling is retained in the present work.

Comments

Ampedus sanguinolentus sanguinolentus (Schrank, 1776) var. *immaculatus* (Schaufuss, 1882) is an infrasubspecific name used currently to denote sexually modified males. The original description of *immaculatus* Schaufuss was not seen.
Drasterius aegyptiacus Buysson, 1905: This species and its synonyms *aegyptiacoides* Buysson, 1912, *assuanensis* Pic, 1913, *biskrensis* Desbrochers des Loges, 1870, and *kebiliensis* Buysson, 1912 were originally described as varieties of *Aeoloides figuratus* (Germar, 1844) or later considered such. Platia & Schimmel, 1997 examined the type series of *Cryptohypnus figuratus* from Mesopotamia and Egypt and consider the single specimen from Egypt to belong to *Drasterius aegyptiacus*. The above-mentioned synonyms are described from Egypt and are therefore tentatively assigned to *Drasterius aegyptiacus*. *Aeoloides figuratus* (Germar, 1844) is not known to occur in Egypt.
Lanelater tomentosus Fabricius, 1798: Vats & Kashyap, 1993: 218 report this species from India (Uttar Pradesh). This is almost certainly a misidentification. The species is known only from the Philippine Islands.
Metasternum: Stibick 1979 and 1980 distinguishes in *Hypnoidus* two groups, *Hypnoidus* s. str. and *Metasternus*. He does not assign a type species to the *Metasternus* group. *Metasternus* is a nomen nudum.

Agajev, 1988 proposed an infrasubspecific name *Athous jugicola* Pérez Arcas, 1872 ab. *lenkoranus*. The name may denote an undescribed species in *Hemicrepidius* but is unavailable and therefore omitted from the list.

Elateridae: Lissominae

A. Smetana

Designation of type species

Lissodes Berthold, 1827: 335 type species *Elater equestris* Fabricius, 1798 (= *Drapetes mordelloides* Host, 1789), by present designation. Berthold (l.c.) did not include any species. Latreille (1829: 452) was the first one to list several species under *Lissodes*, including *Elater equestris*.

Lycidae

M. Bocáková & L. Bocák

Reassignment

Benibotarus (s. str.) *longicornis* (Reiche, 1878) from the subgenus *Sibetarus* Bocák & Bocáková, 1991, placed there by Kazantsev in 1995. *B. longicornis* has the characters typical for *Benitotarsus* s. str. (apical portion of parameres emarginate dorsally, phallus straight or slightly bent in lateral view) and lacks the characters of *Sibetarus* (apical portion of parameres emarginate ventrally, phallus S-shaped in lateral view).

Designation of type species

Paralopheros Kazantsev, 1993 was established for two species *Conderis diversipennis* Pic, 1921 and *Pyropterus fukiensis* Kleine, 1940. The designated type species was misspelled as *Conderis diversicornis* Pic, 1921. The correct spelling is *Conderis diversipennis* Pic, 1921.

Dumbrellia Lea, 1909 was established for two species, and was not yet fixed by a type species. We designate here one of the originaly included species, *Calochromus brevicornis* Lea, 1898, as the type species.

Lampyridae

M. Geisthardt

Comments

Baolacus Pic, 1915 (with type species *Baolacus lajoyei* Pic, 1915) is currently placed in Drilidae. *Baolacus nepalensis* Wittmer, 1979 was placed in Lampyridae without any justification.

Lampyris platyptera Fairmaire, 1887 was overlooked by E. Olivier (1907b) and McDermott (1966).

Pyractonema pugnax from Japan was transferred to *Lucidina* by E. Olivier in 1907. E. Olivier (1907b: 24) and McDermott (1966: 71) give as reference to this name E. Olivier, 1902 (in Bull. Mus. Hist. Nat. 8: 189). Their statement is incorrect and it is assumed that *P. pugnax* was described elsewhere, in an unknown paper.

Rhagophthalmus ohbai Wittmer, 1994 is placed in Lampyridae, based on molecular phylogenetic analysis (H. Suzuki, 1997).

Stenocladius was transferred to Phengodidiae, and *Drilaster* placed as incertae sedid within elateroids by Branham & Wenzel, 2001. This opinion is not followed in the present Catalogue.

New synonyms

Diaphanes citrinus E. Olivier, 1911 var. *notatithorax* Pic, 1955 **syn. nov.** of *Diaphanes citrinus* E. Olivier, 1911

Pyrocoelia analis (Fabricius, 1801) var. *insidiosa* E. Olivier, 1886 **syn. nov.** of *Pyrocoelia analis* (Fabricius, 1801)

Pyrocoelia motschulskyi (Motschulsky, 1854) var. *sinuatocollis* E. Olivier, 1886 **syn. nov.** of *Pyrocoelia motschulskyi* (Motschulsky, 1854)

Pyrocoelia signaticollis E. Olivier, 1886 var. *diversicollis* Pic, 1937 **syn. nov.** of *Pyrocoelia signaticollis* E. Olivier, 1886
Lamprohiza mulsanti Kiesenwetter, 1850 var. *peltata* Rey, 1891 **syn. nov.** of *Lamprohiza mulsanti* Kiesenwetter, 1850
Pristolycus sagulatus Gorham, 1883 var. *diversecostatus* Pic, 1915 **syn. nov.** of *Pristolycus sagulatus* Gorham, 1883
Lampyroidea graeca Laporte, 1833 var. *euboica* Pic, 1914 **syn. nov.** of *Lampyroidea graeca* Laporte, 1833
Lampyroidea quadrinotata Wittmer, 1935 var. *binotata* Wittmer, 1935 **syn. nov.** of *Lampyroidea quadrinotata* Wittmer, 1935
Lampyroidea syriaca A. Costa, 1875 var. *innotaticollis* Pic, 1914 **syn. nov.** of *Lampyroidea syriaca* A. Costa, 1875

Cantharidae

S. Kazantsev

Designation of type species

Hapaloderus Motschulsky, 1853. *Hapaloderus ustulatus* Motschulsky, 1853, the first species assigned to the genus, is here designated as type species.
Dalthomes Fiori, 1905. *Malthodes penninus* Baudi di Selve, 1871, the first species assigned to the genus, is here designated as type species.
Mesomalthodes Fiori, 1905. *Malthodes dispar* Germar, 1824, the first species assigned to the genus, is here designated as type species.
Promalthodes Fiori, 1905. *Malthodes simplex* Kiesenwetter, 1872, the first species assigned to the genus, is here designated as type species.
Malthodinus Fiori, 1905. *Malthodes facetus* Kiesenwetter, 1863, the first species assigned to the genus, is here designated as type species.
Prothalmodes Fiori, 1905. *Malthodes cruciatus* Baudi di Selve, 1859, the first species assigned to the genus, is here designated as type species.
Podistrina Seidlitz, 1889. *Malthodes apterus* Mulsant, 1862 is here designated as type species.

New synonyms

Wittmercantharis Satô, 1986 **syn. nov.** of *Cyrtomoptila* Motschulsky, 1859. There definitely seems to be no difference between the two taxa.

Note. The following available names were described or placed as varieties before 1961. They are synonymized to comply with provisions of the ICZN. This is done without the study of relevant types, and it is possible that some of them might be revalidated in future.
Podabrus procerulus var. *bosnica* Pic, 1903 **syn. nov.** of *Dichelotarsus procerulus* (Kiesenwetter, 1860)
Podabrus heydeni var. *immaculicollis* Pic, 1905 **syn. nov.** of *Hatchiana heydeni* (Kiesenwetter, 1879)
Podabrus alpinus var. *anatolica* Pic, 1920 **syn. nov.** of *Podabrus alpinus Paykull, 1798*
Podabrus alpinus var. *apicalis* Pic, 1906 **syn. nov.** of *Podabrus alpinus* Paykull, 1798
Podabrus alpinus var. *apicenigra* Roubal, 1936 **syn. nov.** of *Podabrus alpinus* Paykull, 1798
Podabrus alpinus var. *falzonii* Burlini, 1942 **syn. nov.** of *Podabrus alpinus* Paykull, 1798
Podabrus alpinus var. *lateralis* Erichson, 1839 **syn. nov.** of *Podabrus alpinus* Paykull, 1798
Podabrus alpinus var. *mocquerysi* Reiche, 1878 **syn. nov.** of *Podabrus alpinus* Paykull, 1798
Podabrus alpinus var. *rubens* Fabricius, 1801 **syn. nov.** of *Podabrus alpinus* Paykull, 1798
Podabrus alpinus var. *ruficeps* Gabriel, 1905 **syn. nov.** of *Podabrus alpinus* Paykull, 1798
Podabrus alpinus var. *semibrunnea* Pic, 1914 **syn. nov.** of *Podabrus alpinus* Paykull, 1798
Podabrus flavus var. *latithorax* Pic, 1926 **syn. nov.** of *Podabrus flavus* Pic, 1926
Ancistronycha abdominalis var. *cyannipennis* Bach, 1854 **syn. nov.** of *Ancistronycha abdominalis* (Fabricius, 1798)
Ancistronycha abdominalis var. *passeriana* Gredler, 1854 **syn. nov.** of *Ancistronycha abdominalis* (Fabricius,

1798)
Ancistronycha erichsonii var. *inapicalis* Pic, 1902 **syn. nov.** of *Ancistronycha erichsonii erichsonii* Bach, 1852
Ancistronycha erichsonii var. *neapolitana* Pic, 1918 **syn. nov.** of *Ancistronycha erichsonii erichsoni* Bach, 1852
Ancistronycha rufidens var. *fiorii* Pic, 1914 **syn. nov.** of *Ancistronycha rufidens* Marseul, 1864
Ancistronycha violacea var. *innotaticeps* Pic, 1907 **syn. nov.** of *Ancistronycha violacea* (Paykull, 1798)
Ancistronycha violacea var. *moraviensis* Pic, 1918 **syn. nov.** of *Ancistronycha violacea* (Paykull, 1798)
Ancistronycha violacea var. *notaticeps* Pic, 1907 **syn. nov.** of *Ancistronycha violacea* (Paykull, 1798)
Ancistronycha violacea var. *piceobinotata* Pic, 1922 **syn. nov.** of *Ancistronycha violacea* (Paykull, 1798)
Ancistronycha violacea var. *tigurina* Dietrich, 1857 **syn. nov.** of *Ancistronycha violacea* (Paykull, 1798)
Athemus multistriatus var. *atricollis* Pic, 1937 **syn.nov.** of *Lycocerus multistriatus* (Pic, 1906)
Athemus multistriatus var. *brevenotata* Pic, 1924 **syn. nov.** of *Lycocerus multistriatus* (Pic, 1906)
Athemus multistriatus var. *lineaticeps* Pic, 1924 **syn. nov.** of *Lycocerus multistriatus* (Pic, 1906)
Athemus flavonotaticeps var. *obliterata* Pic, 1937 **syn. nov.** of *Lycocerus flavonotaticeps* Pic, 1937
Athemus atripennis var. *griseopubens* Pic, 1928 **syn. nov.** of *Lycocerus atripennis* (Pic, 1922)
Athemus japonicus var. *edmei* Pic, 1932 **syn. nov.** of *Lycocerus japonicus japonicus* (Kiesenwetter, 1874)
Armidia signata var. *apfelbecki* Pic, 1902, **syn. nov.** of *Armidia signata* (Ahrens, 1814)
Armidia signata var. *insignata* Pic, 1902, **syn. nov.** of *Armidia signata* (Ahrens, 1814)
Armidia signata var. *paulosignata* Pic, 1902 **syn. nov.** of *Armidia signata* (Ahrens, 1814)
Cantharis dimidiatipes var. *detectiventris* Pic, 1904 **syn. nov.** of *Boveycantharis dimidiatipes* (Reiche & Saulcy, 1857)
Cantharis alluaudi var. *macrocephala* Kocher, 1954 **syn. nov.** of *Cantharis alluaudi* Pic, 1922
Cantharis annularis var. *bisbigemina* Pic, 1917 **syn. nov.** of *Cantharis annularis* Ménétriés, 1836
Cantharis annularis var. *gredosensis* Pic, 1918 **syn. nov.** of *Cantharis annularis* Ménétriés, 1836
Cantharis annularis var. *immaculatithorax* Pic, 1900, **syn.nov.** of *Cantharis annularis* Ménétriés, 1836
Cantharis annularis var. *insignaticollis* Pic, 1900h **syn. nov.** of *Cantharis annularis* Ménétriés, 1836
Cantharis annularis var. *longitarsis* Pandelle, 1867 **syn. nov.** of *Cantharis annularis* Ménétriés, 1836
Cantharis annularis var. *pueli* Pic, 1913 **syn. nov.** of *Cantharis annularis* Ménétriés, 1836
Cantharis annularis var. *quadrinotatithorax* Pic, 1914 **syn. nov.** of *Cantharis annularis* Ménétriés, 1836
Cantharis annularis var. *simoni* Pic, 1922 **syn. nov.** of *Cantharis annularis* Ménétriés, 1836
Cantharis annularis var. *trinotatithorax* Pic, 1914 **syn. nov.** *Cantharis annularis* Ménétriés, 1836
Cantharis atlasiensis var. *amismizana* Pic, 1932 **syn. nov.** *Cantharis atlasiensis* Pic, 1932
Cantharis convexicollis var. *kabyliana* Pic, 1900 **syn. nov.** *Cantharis convexicollis* (Fairmaire, 1875)
Cantharis deportata var. *rotroui* Pic, 1924 **syn. nov.** *Cantharis deportata* Fairmaire, 1876
Cantharis figurata var. *cruachana* Chitty, 1893 **syn. nov.** *Cantharis figurata* Mannerheim, 1843
Cantharis figurata var. *fumata* Everts, 1921 **syn. nov.** *Cantharis figurata* Mannerheim, 1843
Cantharis figurata var. *infigurata* Pic, 1914 **syn. nov.** *Cantharis figurata* Mannerheim, 1843
Cantharis figurata var. *luteata* Schilsky, 1888 **syn. nov.** *Cantharis figurata* Mannerheim, 1843
Cantharis figurata var. *scoticus* Sharp, 1867 **syn. nov.** *Cantharis figurata* Mannerheim, 1843
Cantharis flavilabris var. *affinis* (Stephens, 1830) **syn. nov.** of *Cantharis flavilabris* Fallén, 1807
Cantharis flavilabris var. *maculata* Schilsky, 1888 **syn. nov.** of *Cantharis flavilabris* Fallén, 1807
Cantharis flavilabris var. *nigricollis* Haenel, 1940 **syn. nov.** of *Cantharis flavilabris* Fallén, 1807
Cantharis fusca var. *basithorax* Pic, 1902 **syn. nov.** of *Cantharis fusca* Linnaeus, 1758
Cantharis fusca var. *conjuncta* Schilsky, 1888 **syn. nov.** of *Cantharis fusca* Linnaeus, 1758
Cantharis fusca var. *digoniensis* Pic, 1914 **syn. nov.** of *Cantharis fusca* Linnaeus, 1758
Cantharis fusca var. *fracassii* Leoni, 1905 **syn. nov.** of *Cantharis fusca* Linnaeus, 1758
Cantharis fusca var. *immaculicollis* (Laporte, 1840) **syn. nov.** of *Cantharis fusca* Linnaeus, 1758
Cantharis fusca var. *marciana* Leoni, 1905 **syn. nov.** of *Cantharis fusca* Linnaeus, 1758
Cantharis fusca var. *teutonica* Pic, 1913 **syn. nov.** of *Cantharis fusca* Linnaeus, 1758
Cantharis fuscipennis var. *ambigua* Fiori, 1914 **syn. nov.** of *Cantharis fuscipennis* (Mulsant, 1862)
Cantharis fuscipennis var. *rufifrons* (Marseul, 1864) **syn. nov.** of *Cantharis fuscipennis* (Mulsant, 1862)
Cantharis inculta var. *genei* Gené, 1839 **syn. nov.** of *Cantharis inculta* Gené, 1839
Cantharis inculta var. *laconiensis* Pic, 1918 **syn. nov.** of *Cantharis inculta* Gené, 1839
Cantharis inculta var. *leveillei* Pic, 1927 **syn. nov.** of *Cantharis inculta* Gené, 1839
Cantharis inculta var. *sardiniensis* Pic, 1918 **syn. nov.** of *Cantharis inculta* Gené, 1839

Cantharis italica var. *decolor* Fiori, 1899 **syn. nov.** of *Cantharis italica* Fiori, 1914
Cantharis italica var. *imitans* Fiori, 1914 **syn. nov.** of *Cantharis italica* Fiori, 1914
Cantharis livida var. *bicolorata* Ragusa, 1893 **syn. nov.** of *Cantharis livida* Linnaeus, 1758
Cantharis livida var. *distinguendus* (Gautier des Cottes, 1872) **syn. nov.** of *Cantharis livida* Linnaeus, 1758
Cantharis livida var. *eremita* Rosenhauer, 1856 **syn. nov.** of *Cantharis livida* Linnaeus, 1758
Cantharis livida var. *ferdinandi* Pic, 1909 **syn. nov.** of *Cantharis livida* Linnaeus, 1758
Cantharis livida var. *inscapularis* Pic, 1909 **syn. nov.** of *Cantharis livida* Linnaeus, 1758
Cantharis livida var. *luteiceps* Schilsky, 1888 **syn. nov.** of *Cantharis livida* Linnaeus, 1758
Cantharis livida var. *mendax* (Rey, 1891) **syn. nov.** of *Cantharis livida* Linnaeus, 1758
Cantharis livida var. *menetriesi* Faldermann, 1838 **syn. nov.** of *Cantharis livida* Linnaeus, 1758
Cantharis livida var. *minutehumeralis* Pic, 1953 **syn. nov.** of *Cantharis livida* Linnaeus, 1758
Cantharis livida var. *mostarensis* Pic, 1926 **syn. nov.** of *Cantharis livida* Linnaeus, 1758
Cantharis livida var. *nigrifrons* (Stephens, 1833) **syn. nov.** of *Cantharis livida* Linnaeus, 1758
Cantharis livida var. *nudipennis* Reitter, 1911 **syn. nov.** of *Cantharis livida* Linnaeus, 1758
Cantharis livida var. *paganosa* Pic, 1942 **syn. nov.** of *Cantharis livida* Linnaeus, 1758
Cantharis livida var. *predispar* Pic, 1942 **syn. nov.** of *Cantharis livida* Linnaeus, 1758
Cantharis livida var. *rufescens* Trella, 1925 **syn. nov.** of *Cantharis livida* Linnaeus, 1758
Cantharis livida var. *ruficeps* Reitter, 1911 **syn. nov.** of *Cantharis livida* Linnaeus, 1758
Cantharis livida var. *rufipes* Herbst, 1783 **syn. nov.** of *Cantharis livida* Linnaeus, 1758
Cantharis livida var. *scapularis* L. Redtenbacher, 1858 **syn. nov.** of *Cantharis livida* Linnaeus, 1758
Cantharis livida var. *sicula* Bourgeois, 1893 **syn. nov.** of *Cantharis livida* Linnaeus, 1758
Cantharis livida var. *sulcicornis* Reitter, 1911 **syn. nov.** of *Cantharis livida* Linnaeus, 1758
Cantharis livida var. *unicoloriceps* Kaszab, 1955 **syn. nov.** of *Cantharis livida* Linnaeus, 1758
Cantharis livida var. *varendorffi* Reitter, 1904 **syn. nov.** of *Cantharis livida* Linnaeus, 1758
Cantharis maroccana var. *semiobliterata* Pic, 1935 **syn. nov.** of *Cantharis maroccana* Bedel, 1921
Cantharis maura var. *azrouana* Pic, 1923 **syn. nov.** of *Cantharis maura* Pic, 1921
Cantharis maura var. *curtehumeralis* Pic, 1935 **syn. nov.** of *Cantharis maura* Pic, 1921
Cantharis mauritanica var. *atroscutellaris* Pic, 1900 **syn. nov.** of *Cantharis mauritanica* P. H. Lucas, 1846
Cantharis mauritanica var. *vaucheri* Pic, 1911 **syn. nov.** of *Cantharis mauritanica* P. H. Lucas, 1846
Cantharis monacha var. *apennina* Fiori, 1909 **syn. nov.** of *Cantharis monacha* Moscardini, 1962
Cantharis monacha var. *bimaculata* Fiori, 1914 **syn. nov.** of *Cantharis monacha* Moscardini, 1962
Cantharis monacha var. *morio* Baudi di Selve, 1873 **syn. nov.** of *Cantharis monacha* Moscardini, 1962
Cantharis nigra var. *discotestacea* Pic, 1905 **syn. nov.** of *Cantharis nigra* (DeGeer, 1774)
Cantharis nigra var. *suturalis* Schilsky, 1890 **syn. nov.** of *Cantharis nigra* (DeGeer, 1774)
Cantharis nigra var. *theresae* Pic, 1900 **syn. nov.** of *Cantharis nigra* (DeGeer, 1774)
Cantharis nigricans var. *arvernica* Buysson, 1908 **syn. nov.** of *Cantharis nigricans* O. F. Müller, 1776
Cantharis nigricans var. *bipallida* Kaszab, 1955 **syn. nov.** of *Cantharis nigricans* O. F. Müller, 1776
Cantharis nigricans var. *discoidea* (Stephens, 1839) **syn. nov.** of *Cantharis nigricans* O. F. Müller, 1776
Cantharis nigricans var. *hospes* Rosenhauer, 1847 **syn. nov.** of *Cantharis nigricans* O. F. Müller, 1776
Cantharis nigricans var. *luteipes* Schilsky, 1888 **syn. nov.** of *Cantharis nigricans* O. F. Müller, 1776
Cantharis nigricans var. *martialis* Pic, 1908 **syn. nov.** of *Cantharis nigricans* O. F. Müller, 1776
Cantharis nigricans var. *monachi* Pic, 1918 **syn. nov.** of *Cantharis nigricans* O. F. Müller, 1776
Cantharis nigricans var. *nigrescens* Trella, 1925 **syn. nov.** of *Cantharis nigricans* O. F. Müller, 1776
Cantharis nigricans var. *unicoloripes* Kaszab, 1955 **syn. nov.** of *Cantharis nigricans* O. F. Müller, 1776
Cantharis nigricans var. *vittigera* Baudi di Selve, 1871 **syn. nov.** of *Cantharis nigricans* O. F. Müller, 1776
Cantharis obscura var. *baudii* Pic, 1909 **syn. nov.** of *Cantharis obscura* Linnaeus, 1758
Cantharis obscura var. *kaszabi* Vosyka, 1957 **syn. nov.** of *Cantharis obscura* Linnaeus, 1758
Cantharis pallida var. *maritima* Fiori, 1914 **syn. nov.** of *Cantharis pallida* Goeze, 1777
Cantharis pallida var. *pseudoassimilis* Pic, 1917 **syn. nov.** of *Cantharis pallida* Goeze, 1777
Cantharis pallida var. *ustulata* Kiesenwetter, 1851 **syn. nov.** of *Cantharis pallida* Goeze, 1777
Cantharis pellucida var. *croissandeaui* Pic, 1927 **syn. nov.** of *Cantharis pellucida* Fabricius, 1792
Cantharis pellucida var. *dietzi* Roelofs, 1946 **syn. nov.** of *Cantharis pellucida* Fabricius, 1792
Cantharis pellucida var. *meridionalis* Fiori, 1914 **syn. nov.** of *Cantharis pellucida* Fabricius, 1792
Cantharis pellucida var. *mirabilis* Dalla Torre, 1880 **syn. nov.** of *Cantharis pellucida* Fabricius, 1792

Cantharis pellucida var. *obscuripes* Everts, 1921 **syn. nov.** of *Cantharis pellucida* Fabricius, 1792
Cantharis pellucida var. *rauterbergi* Reitter, 1893 **syn. nov.** of *Cantharis pellucida* Fabricius, 1792
Cantharis peninsularis var. *bisignata* Fiori, 1914 **syn. nov.** of *Cantharis peninsularis* Fiori, 1914
Cantharis peninsularis var. *parens* Fiori, 1914 **syn. nov.** of *Cantharis peninsularis* Fiori, 1914
Cantharis praecox var. *vittatocollis* (Reiche, 1861) **syn. nov.** of *Cantharis praecox* Gené, 1836
Cantharis pulicaria var. *curtipennis* Rey, 1891 **syn. nov.** of *Cantharis pulicaria* Fabricius, 1781
Cantharis puncticollis var. *notaticeps* Pic, 1903 **syn. nov.** of *Cantharis puncticollis* Levrat, 1857
Cantharis puncticollis var. *obscuripennis* Pic, 1907 **syn. nov.** of *Cantharis puncticollis* Levrat, 1857
Cantharis puncticollis var. *paulonotata* Pic, 1903 **syn. nov.** of *Cantharis puncticollis* Levrat, 1857
Cantharis pyrenaea var. *correai* Pic, 1927 **syn. nov.** of *Cantharis pyrenaea* Pic, 1906
Cantharis pyrenaea var. *getschmanni* Heyden, 1880, **syn.n.***Cantharis pyrenaea* Pic, 1906
Cantharis pyrenaea var. *plurivittata* Pic, 1906 **syn. nov.** of *Cantharis pyrenaea* Pic, 1906
Cantharis quadripunctata var. *bakonyensis* Kaszab, 1955 **syn. nov.** of *Cantharis quadripunctata* (O. F. Müller, 1776)
Cantharis quadripunctata var. *bipuncticollis* Kaszab, 1955 **syn. nov.** of *Cantharis quadripunctata* (O. F. Müller, 1776)
Cantharis quadripunctata var. *bosnica* Pic, 1906 **syn. nov.** of *Cantharis quadripunctata* (O. F. Müller, 1776)
Cantharis quadripunctata var. *cincticollis* Pic, 1911 **syn. nov.** of *Cantharis quadripunctata* (O. F. Müller, 1776)
Cantharis quadripunctata var. *fulvipennis* Germar, 1824 **syn. nov.** of *Cantharis quadripunctata* (O. F. Müller, 1776)
Cantharis quadripunctata var. *fulvithorax* Kaszab, 1955 **syn. nov.** of *Cantharis quadripunctata* (O. F. Müller, 1776)
Cantharis quadripunctata var. *mihalovicsi* Kaszab, 1955 **syn. nov.** of *Cantharis quadripunctata* (O. F. Müller, 1776)
Cantharis quadripunctata var. *moreli* Pic, 1923 **syn. nov.** of *Cantharis quadripunctata* (O. F. Müller, 1776)
Cantharis quadripunctata var. *rosenhaueri* Kaszab, 1955 **syn. nov.** of *Cantharis quadripunctata* (O. F. Müller, 1776)
Telephorus reichei var. *hispanica* (Reiche, 1878) **syn. nov.** of *Cantharis reichei* Mulsant, 1862
Cantharis reichei var. *reynosana* Pic, 1922 **syn. nov.** of *Cantharis reichei* Mulsant, 1862
Cantharis rufa var. *alexandris* Pic, 1914 **syn. nov.** of *Cantharis rufa* Linnaeus, 1758
Cantharis rufa var. *liturata* Fallén, 1807 **syn. nov.** of *Cantharis rufa* Linnaeus, 1758
Cantharis rufa var. *padana* Fiori, 1914 **syn. nov.** of *Cantharis rufa* Linnaeus, 1758
Cantharis rufa var. *roelofsi* Pic, 1913 **syn. nov.** of *Cantharis rufa* Linnaeus, 1758
Cantharis rustica var. *agnani* Pic, 1913 **syn. nov.** of *Cantharis rustica* Fallén, 1807
Cantharis rustica var. *leonardi* Pic, 1913 **syn. nov.** of *Cantharis rustica* Fallén, 1807
Cantharis rustica var. *lokvensis* Stiller, 1926 **syn. nov.** of *Cantharis rustica* Fallén, 1807
Cantharis rustica var. *nigripalpis* Rey, 1891 **syn. nov.** of *Cantharis rustica* Fallén, 1807
Cantharis rustica var. *rufolimbaticollis* Pic, 1914 **syn. nov.** of *Cantharis rustica* Fallén, 1807
Cantharis rustica var. *subcincticollis* Pic, 1913 **syn. nov.** of *Cantharis rustica* Fallén, 1807
Cantharis rustica var. *tenuipes* Rey, 1891 **syn. nov.** of *Cantharis rustica* Fallén, 1807
Cantharis rustica var. *trimaculatithorax* Pic, 1914, **syn.n.***Cantharis rustica* Fallén, 1807
Cantharis rustica var. *vesubiensis* Pic, 1922 **syn. nov.** of *Cantharis rustica* Fallén, 1807
Cantharis sublivida var. *mardinensis* Pic, 1906 **syn. nov.** of *Cantharis sublivida* Pic, 1904
Cantharis tunisea var. *pauli* Pic, 1915 **syn. nov.** of *Cantharis tunisea* Pic, 1900
Cantharis versicolor var. *rufithorax* Delkeskamp, 1939 **syn. nov.** of *Cantharis versicolor* (Baudi di Selve, 1871)
Cantharis xanthoporpa var. *signaticollis* Mulsant, 1862 **syn. nov.** of *Cantharis xanthoporpa* Kiesenwetter, 1860
Cantharis lateralis var. *andrei* Pic, 1935 **syn. nov.** of *Cantharis lateralis* Linnaeus, 1758
Cantharis lateralis var. *nigricollis* Trella, 1932 **syn. nov.** of *Cantharis lateralis* Linnaeus, 1758
Cantharis lateralis var. *nigrinotata* Pic, 1905 **syn. nov.** of *Cantharis lateralis* Linnaeus, 1758
Cantharis lateralis var. *notaticollis* Schilsky, 1890 **syn. nov.** of *Cantharis lateralis* Linnaeus, 1758
Cantharis lateralis var. *sexnotata* Roubal, 1934 **syn. nov.** of *Cantharis lateralis* Linnaeus, 1758

Cratosilis denticollis var. *punctipennis* (Kiesenwetter, 1851) **syn. nov.** of *Cratosilis denticollis* (Schummel, 1844)
Cratosilis distinguenda var. *saasensis* Pic, 1923 **syn. nov.** of *Cratosilis distinguenda* (Baudi di Selve, 1859)
Cratosilis laeta var. *discolor* (Baudi di Selve, 1873) **syn. nov.** of *Cratosilis laeta* (Fabricius, 1792)
Cratosilis laeta var. *semiobscura* Pic, 1901 **syn. nov.** of *Cratosilis laeta* (Fabricius, 1792)
Metacantharis clypeata var. *rubens* Marseul, 1864 **syn. nov.** of *Metacantharis clypeata* (Illiger, 1798)
Metacantharis discoidea var. *carreti* Bourgeois, 1900 **syn. nov.** of *Metacantharis discoidea* (Ahrens, 1812)
Metacantharis discoidea var. *flavicollis* Gerhardt, 1909 **syn. nov.** of *Metacantharis discoidea* (Ahrens, 1812)
Metacantharis discoidea var. *fontisbellaquei* Pic, 1914 **syn. nov.** of *Metacantharis discoidea* (Ahrens, 1812)
Metacantharis discoidea var. *gaberti* Roubal, 1919 **syn. nov.** of *Metacantharis discoidea* (Ahrens, 1812)
Metacantharis discoidea var. *indiscoidea* Pic, 1914 **syn. nov.** of *Metacantharis discoidea* (Ahrens, 1812)
Metacantharis discoidea var. *liturata* (L. Redtenbacher, 1849) **syn. nov.** of *Metacantharis discoidea* (Ahrens, 1812)
Metacantharis discoidea var. *telonensis* Pic, 1914 **syn. nov.** of *Metacantharis discoidea* (Ahrens, 1812)
Metacantharis picciolii var. *fraudulenta* Fiori, 1914 **syn. nov.** of *Metacantharis picciolii* (Ragusa, 1870)
Metacantharis picciolii var. *picticollis* Ragusa, 1893 **syn. nov.** of *Metacantharis picciolii* (Ragusa, 1870)
Metacantharis taurigrada var. *obscuripes* J. R. Sahlberg, 1913 **syn. nov.** of *Metacantharis taurigrada* Bourgeois, 1900
Podabrus novemexcavatus var. *testaceicollis* Wittmer, 1951 **syn. nov.** of *Micropodabrus novemexcavatus* (Wittmer, 1951)
Occathemus tarsalis var. *oculatifer* (Pic, 1900) **syn. nov.** of *Occathemus tarsalis* (Mulsant, 1862)
Rhagonycha aetolica var. *subaptera* Pic, 1909 **syn. nov.** of *Rhagonycha aetolica* (Kiesenwetter, 1859)
Rhagonycha atra var. *fauconetti* Pic, 1914 **syn. nov.** of *Rhagonycha atra* (Linnaeus, 1767)
Rhagonycha atra var. *glacialis* Bourgeois, 1909 **syn. nov.** of *Rhagonycha atra* (Linnaeus, 1767)
Rhagonycha atra var. *rhaetica* Stierlin, 1862 **syn. nov.** of *Rhagonycha atra* (Linnaeus, 1767)
Rhagonycha atra var. *sabauda* Pic, 1914 **syn. nov.** of *Rhagonycha atra* (Linnaeus, 1767)
Rhagonycha atra var. *subrobusta* Pic, 1923 **syn. nov.** of *Rhagonycha atra* (Linnaeus, 1767)
Rhagonycha barbara var. *bettingeri* (Pic, 1927) **syn. nov.** of *Rhagonycha barbara* (Fabricius, 1801)
Rhagonycha barbara var. *sulcata* Mulsant, 1862 **syn. nov.** of *Rhagonycha barbara* (Fabricius, 1801)
Rhagonycha chlorotica var. *corsica* Reiche, 1861 **syn. nov.** of *Rhagonycha chlorotica* (Gené, 1839)

Rhagonycha fossulata var. *scutellaris* P. H. Lucas, 1846 **syn. nov.** of *Rhagonycha fossulata* (P. H. Lucas, 1846)
Rhagonycha fugax var. *infuscata* J. R. Sahlberg, 1871 **syn. nov.** of *Rhagonycha fugax fugax* Mannerheim, 1843
Rhagonycha fulva var. *cailloli* Chobaut, 1914 **syn. nov.** of *Rhagonycha fulva* (Scopoli, 1763)
Rhagonycha fulva var. *curtithorax* Pic, 1920 **syn. nov.** of *Rhagonycha fulva* (Scopoli, 1763)
Rhagonycha fulva var. *delahoni* Schilski, 1908 **syn. nov.** of *Rhagonycha fulva* (Scopoli, 1763)
Rhagonycha fulva var. *inapicalis* Fiori, 1914 **syn. nov.** of *Rhagonycha fulva* (Scopoli, 1763)
Rhagonycha fulva var. *usta* (Gemminger, 1870) **syn. nov.** of *Rhagonycha fulva* (Scopoli, 1763)
Rhagonycha herbea var. *pallidithorax* Pic, 1910 **syn. nov.** of *Rhagonycha herbea* Marseul, 1864
Rhagonycha hesperica var. *georgi* Pic, 1909 **syn. nov.** of *Rhagonycha hesperica* Baudi di Selve, 1859
Rhagonycha hesperica var. *inapicalis* Pic, 1903 **syn. nov.** of *Rhagonycha hesperica* Baudi di Selve, 1859
Rhagonycha hesperica var. *kochi* Pic, 1935 **syn. nov.** of *Rhagonycha hesperica* Baudi di Selve, 1859
Rhagonycha hesperica var. *lanjaroni* Pic, 1952 **syn. nov.** of *Rhagonycha hesperica* Baudi di Selve, 1859
Rhagonycha hesperica var. *lineatipennis* Pic, 1908 **syn. nov.** of *Rhagonycha hesperica* Baudi di Selve, 1859
Rhagonycha hesperica var. *obscurimembris* Pic, 1952 **syn. nov.** of *Rhagonycha hesperica* Baudi di Selve, 1859
Rhagonycha hesperica var. *oliveti* Kiesenwetter, 1866 **syn. nov.** of *Rhagonycha hesperica* Baudi di Selve, 1859
Rhagonycha hesperica var. *spinifera* Pic, 1903 **syn. nov.** of *Rhagonycha hesperica* Baudi di Selve, 1859
Rhagonycha kuleghana var. *bodoi* Pic, 1911 **syn. nov.** of *Rhagonycha kuleghana* (Marseul, 1868)
Rhagonycha kuleghana var. *deyrollei* Pic, 1902 **syn. nov.** of *Rhagonycha kuleghana* (Marseul, 1868)
Rhagonycha licenti var. *latiorithorax* Pic, 1938 **syn. nov.** of *Rhagonycha licenti* Pic, 1938 =.
Rhagonycha licenti var. *unicolorithorax* Pic, 1938 **syn. nov.** of.*Rhagonycha licenti* Pic, 1938

Rhagonycha lignosa var. *berolinensis* (Gmelin, 1790) **syn. nov.** of *Rhagonycha lignosa* (O. F. Müller, 1764)
Rhagonycha lignosa var. *pallida* (Fabricius, 1787) **syn. nov.** of *Rhagonycha lignosa* (O. F. Müller, 1764)
Rhagonycha lignosa var. *subabbreviata* Rey, 1891 **syn. nov.** of *Rhagonycha lignosa* (O. F. Müller, 1764)
Rhagonycha lignosa var. *subnigrofemoralis* Pic, 1914 **syn. nov.** of *Rhagonycha lignosa* (O. F. Müller, 1764)
Podabrus lutea var. *banatica* Rosenhauer, 1847 **syn. nov.** of *Rhagonycha lutea* (O. F. Müller, 1764)
Rhagonycha lutea var. *maerkeli* Kiesenwetter, 1851 **syn. nov.** of *Rhagonycha lutea* (O. F. Müller, 1764)
Rhagonycha lutea var. *nubila* Baudi di Selve, 1871 **syn. nov.** of *Rhagonycha lutea* (O. F. Müller, 1764)
Rhagonycha milleri var. *netolitzkyi* Penecke, 1912 **syn. nov.** of *Rhagonycha milleri* Kiesenwetter, 1860
Rhagonycha milleri var. *bravardi* Pic, 1917 **syn. nov.** of *Rhagonycha milleri* Kiesenwetter, 1860
Rhagonycha morio var. *doctoris* Pic, 1909 **syn. nov.** of *Rhagonycha morio* Kisenwetter, 1851
Rhagonycha nigriceps var. *atricapilla* Kiesenwetter, 1850 **syn. nov.** of *Rhagonycha nigriceps* (Waltl, 1838)
Rhagonycha nigriceps var. *cantabrica* Heyden, 1880 **syn. nov.** of *Rhagonycha nigriceps* (Waltl, 1838)
Rhagonycha nigriceps var. *nigricornis* Petri, 1912 **syn. nov.** of *Rhagonycha nigriceps* (Waltl, 1838)
Rhagonycha nigripes var. *herzegivina* Pic, 1911 **syn. nov.** of *Rhagonycha nigripes* W. Redtenbacher, 1842
Rhagonycha nigripes var. *saturipennis* Pic, 1909 **syn. nov.** of *Rhagonycha nigripes* W. Redtenbacher, 1842
Rhagonycha nigriventris var. *femorata* Rey, 1891 **syn. nov.** of *Rhagonycha nigriventris* Motschulsky, 1860
Rhagonycha nigriventris var. *innotatithorax* Pic, 1902 **syn. nov.** of *Rhagonycha nigriventris* Motschulsky, 1860
Rhagonycha nigriventris var. *signicollis* Rey, 1891 **syn. nov.** of *Rhagonycha nigriventris* Motschulsky, 1860
Rhagonycha ornaticollis var. *nigropygidialis* Pic, 1954 **syn. nov.** of *Rhagonycha ornaticollis* Marseul, 1864
Rhagonycha peyroni var. *discobirufa* Pic, 1946 **syn. nov.** of *Rhagonycha peyroni* (Marseul, 1864)
Rhagonycha quadricollis var. *brevinotata* Pic, 1908 **syn. nov.** of *Rhagonycha quadricollis* Kiesenwetter, 1851
Rhagonycha quadricollis var. *marginella* Baudi di Selve, 1859 **syn. nov.** of *Rhagonycha quadricollis* Kiesenwetter, 1851
Rhagonycha querceti var. *bugnioni* Bourgeois, 1903 **syn. nov.** of *Rhagonycha querceti* Kiesenwetter, 1865
Rhagonycha querceti var. *pardalensis* Pic, 1908 **syn. nov.** of *Rhagonycha querceti* Kiesenwetter, 1865
Rhagonycha tangeriana var. *marocana* Pic, 1949 **syn. nov.** of *Rhagonycha tangeriana* Pic, 1903
Rhagonycha testacea var. *obscurithorax* Pic, 1902 **syn. nov.** of *Rhagonycha testacea* (Linnaeus, 1758)
Rhagonycha testacea var. *taurinensis* Pic, 1905 **syn. nov.** of *Rhagonycha testacea* (Linnaeus, 1758)
Rhagonycha tripunctata var. *unistigmata* Pic, 1900 **syn. nov.** of *Rhagonycha tripunctata* (Reiche & Saulcy, 1857)
Rhagonycha varians var. *pici* Jacobson, 1911 **syn. nov.** of *Rhagonycha varians* (Rosenhauer, 1856)
Rhagonycha varians var. *subnotaticeps* Pic, 1935 **syn. nov.** of *Rhagonycha varians* (Rosenhauer, 1856)
Rhagonycha viduata var. *soror* Kiesenwetter, 1860 **syn. nov.** of *Rhagonycha viduata* (Kuester, 1854)
Themus leechianus var. *aeneipennis* (Gorham, 1889) **syn. nov.** of *Themus leechianus* (Gorham, 1889)
Themus praelongus var. *babaulti* Pic, 1945 **syn. nov.** of *Themus praelongus* Champion, 1926
Themus subviridipennis var. *nonfriedi* Pic, 1934 **syn. nov.** of *Themus subviridipennis* Pic, 1916
Silis ruficollis var. *gallica* Pic, 1927 **syn. nov.** of *Crudosilis ruficollis* (Fabricius, 1775)
Silis ruficollis var. *hungarica* Pic, 1910 **syn. nov.** of *Crudosilis ruficollis* (Fabricius, 1775)
Silis ruficollis var. *signicollis* Everts, 1920 **syn. nov.** of *Crudosilis ruficollis* (Fabricius, 1775)
Laemoglyptus bomfordii var. *innotaticollis* Pic, 1937 **syn. nov.** of *Laemoglyptus bomfordii* Fairmaire, 1897
Malthinus aegyptiacus var. *piceonotatus* Pic, 1916 **syn. nov.** of *Malthinus aegyptiacus* (Pic, 1899)
Malthinus axillaris var. *moricei* Pic, 1930 **syn. nov.** of *Malthinus axillaris* Kiesenwetter, 1852
Malthinus axillaris var. *nigrithorax* Wittmer, 1936 **syn. nov.** of *Malthinus axillaris* Kiesenwetter, 1852
Malthinus balteatus var. *kruperi* Pic, 1901 **syn. nov.** of *Malthinus balteatus* Suffrian, 1851
Malthinus balteatus var. *obscuriceps* Pic, 1900 **syn. nov.** of *Malthinus balteatus* Suffrian, 1851
Malthinus balteatus var. *sinuatipes* Pic, 1908 **syn. nov.** of *Malthinus balteatus* Suffrian, 1851
Malthinus deceptor var. *tiburtina* Baudi di Selve, 1893 **syn. nov.** of *Malthinus deceptor* Baudi di Selve, 1893
Malthinus facialis var. *luteola* Trella, 1925 **syn. nov.** of *Malthinus facialis* C. G. Thomson, 1864
Malthinus facialis var. *sabaudus* Pic, 1901 **syn. nov.** of *Malthinus facialis* C. G. Thomson, 1864
Malthinus flaveolus var. *griseipennis* Pic, 1913 **syn. nov.** of *Malthinus flaveolus* (Herbst, 1784)
Malthinus flaveolus var. *immunis* (Marsham, 1802) **syn. nov.** of *Malthinus flaveolus* (Herbst, 1784)
Malthinus flaveolus var. *subfusca* Pic, 1906 **syn. nov.** of *Malthinus flaveolus* (Herbst, 1784)
Malthinus incrassicornis var. *diversicollis* Pic, 1906, **syn.n.***Malthinus incrassicornis* (Pic, 1900)

Malthinus insignipes var. *indisjuncta* Pic, 1910 **syn. nov.** of *Malthinus insignipes* Pic, 1907
Malthinus laticollis var. *vitalei* Pic, 1916 **syn. nov.** of *Malthinus laticollis* (Pic, 1900)
Malthinus longipennis var. *biskrensis* (Pic, 1900) **syn. nov.** of *Malthinus longipennis* P. H. Lucas, 1846
Malthinus longipennis var. *flavicans* (Bourgeois, 1903) **syn. nov.** of *Malthinus longipennis* P. H. Lucas, 1846
Malthinus longipennis var. *flavipennis* (Pic, 1903) **syn. nov.** of *Malthinus longipennis* P. H. Lucas, 1846
Malthinus longipennis var. *nigripes* Bourgeois, 1897 **syn. nov.** of *Malthinus longipennis* P. H. Lucas, 1846
Malthinus maculithorax var. *bordei* Pic, 1911 **syn. nov.** of *Malthinus maculithorax* Pic, 1898
Malthinus maculithorax var. *lineata* Pic, 1899 **syn. nov.** of *Malthinus maculithorax* Pic, 1898
Malthinus olcesei var. *griseonotata* Pic, 1951 **syn. nov.** of *Malthinus olcesei* (Pic, 1898)
Malthinus olcesei var. *limbatus* (Pic, 1900) **syn. nov.** of *Malthinus olcesei* (Pic, 1898)
Malthinus olcesei var. *maculaticeps* Pic, 1920 **syn. nov.** of *Malthinus olcesei* (Pic, 1898)
Malthinus pici var. *nigrina* Wittmer, 1936 **syn. nov.** of *Malthinus pici* Wittmer, 1936
Malthinus scriptus var. *diversithorax* Pic, 1901 **syn. nov.** of *Malthinus scriptus* Kiesenwetter, 1851
Malthinus scriptus var. *filicornis* Kiesenwetter, 1851 **syn. nov.** of *Malthinus scriptus* Kiesenwetter, 1851
Malthinus scriptus var. *marginipennis* Bourgeois, 1888 **syn. nov.** of *Malthinus scriptus* Kiesenwetter, 1851
Malthinus seriepunctatus var. *binotaticollis* Pic, 1913 **syn. nov.** of *Malthinus seriepunctatus* Kiesenwetter, 1851
Malthinus seriepunctatus var. *distinctipes* Pic, 1901 **syn. nov.** of *Malthinus seriepunctatus* Kiesenwetter, 1851
Malthinus seriepunctatus var. *mediocrenotatus* Pic, 1914 **syn. nov.** of *Malthinus seriepunctatus* Kiesenwetter, 1851
Malthinus seriepunctatus var. *mimica* Pic, 1906 **syn. nov.** of *Malthinus seriepunctatus* Kiesenwetter, 1851
Malthinus seriepunctatus var. *multonotata* Pic, 1910 **syn. nov.** of *Malthinus seriepunctatus* Kiesenwetter, 1851
Malthinus seriepunctatus var. *pallidithorax* Pic, 1899 **syn. nov.** of *Malthinus seriepunctatus* Kiesenwetter, 1851
Malthinus seriepunctatus var. *raveli* Pic, 1906 **syn. nov.** of *Malthinus seriepunctatus* Kiesenwetter, 1851
Malthinus seriepunctatus var. *subopacus* Pic, 1900 **syn. nov.** of *Malthinus seriepunctatus* Kiesenwetter, 1851
Malthinus sordidus var. *brunneosulcata* Pic, 1940 **syn. nov.** of *Malthinus sordidus sordidus* Kiesenwetter, 1871
Malthinus sulcicollis var. *mutata* Pic, 1910 **syn. nov.** of *Malthinus sulcicollis* Marseul, 1878
Malthinus theresae var. *smyrnensis* Pic, 1915 **syn. nov.** of *Malthinus theresae* Pic, 1899
Malthinus theryi var. *longevittatus* Pic, 1953 **syn. nov.** of *Malthinus theryi* Pic, 1920
Malthinus tuniseus var. *diversicolor* Pic, 1905 **syn. nov.** of *Malthinus tuniseus terrestris* Pic, 1905
Malthodes argivus var. *hymettensis* Pic, 1903 **syn. nov.** of *Malthodes argivus* Kiesenwetter, 1859
Malthodes atroaptera var. *algirica* (Bourgeois, 1900) **syn. nov.** of *Malthodes atroapterus* Fairmaire, 1884
Malthodes atroapterus var. *alata* Pic, 1918 **syn. nov.** of *Malthodes atroapterus* Fairmaire, 1884
Malthodes atroapterus var. *diversistylus* Pic, 1908 **syn. nov.** of *Malthodes atroapterus* Fairmaire, 1884
Malthodes barbarus var. *obscurior* Pic, 1935 **syn. nov.** of *Malthodes barbarus* Pic, 1903
Malthodes batillifer var. *semilimbatus* Pic, 1903 **syn. nov.** of *Malthodes batillifer* Bourgeois, 1901
Malthodes berytensis var. *thoracicus* J. R. Sahlberg, 1902 **syn. nov.** of *Malthodes berytensis* Reiche & Saulcy, 1857
Malthodes bifurcatus var. *malcolmi* Pic, 1904 **syn. nov.** of *Malthodes bifurcatus* Kiesenwetter, 1852
Malthodes cognatus var. *andreae* Ganglbauer, 1906 **syn. nov.** of *Malthodes cognatus* A. Costa, 1858
Malthodes cognatus var. *nigra* Fiori, 1905 **syn. nov.** of *Malthodes cognatus* A. Costa, 1858
Malthodes cognatus var. *nigripes* Mulsant, 1862 **syn. nov.** of *Malthodes cognatus* A. Costa, 1858
Malthodes dispar var. *arvernica* Pic, 1910 **syn. nov.** of *Malthodes dispar* (Germar, 1824)
Malthodes dispar var. *decolorata* (Fiori, 1905) **syn. nov.** of *Malthodes dispar* (Germar, 1824)
Malthodes dispar var. *jurassica* Pic, 1914 **syn. nov.** of *Malthodes dispar* (Germar, 1824)
Malthodes dispar var. *laterufescens* Pic, 1914 **syn. nov.** of *Malthodes dispar* (Germar, 1824)
Malthodes dispar var. *neglecta* Mulsant, 1862 **syn. nov.** of *Malthodes dispar* (Germar, 1824)
Malthodes dispar var. *nigrina* (Fiori, 1905) **syn. nov.** of *Malthodes dispar* (Germar, 1824)
Malthodes dispar var. *noualhieri* Bourgeois, 1894 **syn. nov.** of *Malthodes dispar* (Germar, 1824)
Malthodes dispar var. *putoni* Bourgeois, 1902 **syn. nov.** of *Malthodes dispar* (Germar, 1824)
Malthodes dispar var. *subobscuripes* Pic, 1914 **syn. nov.** of *Malthodes dispar* (Germar, 1824)
Malthodes distinctithorax var. *chromodera* Pic, 1912 **syn. nov.** of *Malthodes distinctithorax* Pic, 1901
Malthodes flavoguttatus var. *inapicalis* Pic, 1910 **syn. nov.** of *Malthodes flavoguttatus* Kiesenwetter, 1852
Malthodes flavoguttatus var. *obscurimembris* Pic, 1933 **syn. nov.** of *Malthodes flavoguttatus* Kiesenwetter, 1852

Malthodes forcipipher var. *ponferradanus* Pic, 1910 **syn. nov.** of *Malthodes forcipipher* (Kiesenwetter, 1851)
Malthodes ganglbaueri var. *rufithorax* Falzoni, 1924 **syn. nov.** of *Malthodes ganglbaueri* Fiori, 1905
Malthodes gratiosus var. *simbruinica* Luigioni, 1919|1920 **syn. nov.** of *Malthodes gratiosus* Ganglbauer, 1906
Malthodes guttifer var. *guttegenus* Ganglbauer, 1911 **syn. nov.** of *Malthodes guttifer* Kiesenwetter, 1852
Malthodes guttifer var. *robustior* Pic, 1903 **syn. nov.** of *Malthodes guttifer* Kiesenwetter, 1852
Malthodes hexacanthus var. *inepta* (Fiori, 1908) **syn. nov.** of *Malthodes hexacanthus* Kiesenwetter, 1852
Malthodes hexacanthus var. *tetracanthus* Kiesenwetter, 1871 **syn. nov.** of *Malthodes hexacanthus* Kiesenwetter, 1852
Malthodes laciniatus var. *ragusai* Fiori, 1904 **syn. nov.** of *Malthodes laciniatus laciniatus* Kiesenwetter, 1852
Malthodes licenti var. *subguttata* Pic, 1938 **syn. nov.** of *Malthodes licenti* Pic, 1938
Malchinus liegeli var. *brancsiki* Reitter, 1903 **syn. nov.** of *Malthodes liegeli* J. Weise, 1890
Malthodes ludwigi var. *abdominalis* Pic, 1915 **syn. nov.** of *Malthodes ludwigi ludwigi* (Pic, 1901)
Malthodes ludwigi var. *olympica* Pic, 1901 **syn. nov.** of *Malthodes ludwigi ludwigi* (Pic, 1901)
Malthodes marginatus var. *constricta* Fiori, 1905 **syn. nov.** of *Malthodes marginatus* (Latreille, 1806)
Malthodes maurus var. *atrostylus* Pic, 1914 **syn. nov.** of *Malthodes maurus* (Laporte, 1840)
Malthodes minimus var. *marginicollis* Schilsky, 1892 **syn. nov.** of *Malthodes minimus* (Linnaeus, 1758)
Malthodes montandoni var. *comanensis* Pic, 1922 **syn. nov.** of *Malthodes montandoni* (Pic, 1922)
Malthodes montandoni var. *vlascana* Pic, 1922 **syn. nov.** of *Malthodes montandoni* (Pic, 1922)
Malthodes mysticus var. *atrithorax* Pic, 1926 **syn. nov.** of *Malthodes mysticus mysticus* Kiesenwetter, 1852
Malthodes mysticus var. *megalophtalmus* Ganglbauer, 1911 **syn. nov.** of *Malthodes mysticus mysticus* Kiesenwetter, 1852
Malthodes mysticus var. *obscuriuscula* Dietrich, 1857 **syn. nov.** of *Malthodes mysticus mysticus* Kiesenwetter, 1852
Malthodes numidicus var. *variicollis* Pic, 1912 **syn. nov.** of *Malthodes numidicus* Pic, 1909
Malthodes pictipennis var. *pallidicolor* Pic, 1903 **syn. nov.** of *Malthodes pictipennis* Pic, 1903
Malthodes prodigiosus var. *servilis* Pic, 1915 **syn. nov.** of *Malthodes prodigiosus* Kiesenwetter, 1852
Malthodes simplex var. *romanus* Pic, 1903 **syn. nov.** of *Malthodes simplex* Kiesenwetter, 1872
Malthodes spathifer var. *mionopsis* Ganglbauer, 1911 **syn. nov.** of *Malthodes spathifer* Kiesenwetter, 1852
Malthodes spathifer var. *tetra* (Fiori, 1908) **syn. nov.** of *Malthodes spathifer* Kiesenwetter, 1852
Malthodes spretus var. *guttulifer* Ganglbauer, 1911 **syn. nov.** of *Malthodes spretus* Kiesenwetter, 1852
Malthodes subincisicollis var. *lineolatus* Pic, 1915 **syn. nov.** of *Malthodes subincisicollis* Pic, 1915
Malthodes theryi var. *meyeri* Pic, 1953 **syn. nov.** of *Malthodes theryi* Pic, 1918
Malthodes tingutanus var. *basicornis* Pic, 1918 **syn. nov.** of *Malthodes tingutanus* Pic, 1912
Malthodes tingutanus var. *escorialensis* Pic, 1904 **syn. nov.** of *Malthodes tingutanus* Suffrian, 1851
Malchinus carinicollis var. *scutellaris* Pic, 1899 **syn. nov.** of *Macrocerus carinicollis* (J. Weise, 1895)
Malchinus tunicatus var. *abeillei* Pic, 1911 **syn. nov.** of *Macrocerus tunicatus* (Kiesenwetter, 1863)
Trypherus m-signatus var. *bimaculata* Pic, 1919 **syn. nov.** of *Trypherus m-signatus* Pic, 1919

New replacement names

Lycocerus harmandianus Kazantsev, **nom. nov.** for *Lycocerus harmandi* (Pic, 1926), nec Pic, 1925
Lycocerus limatus Kazantsev, **nom. nov.** for *Lycocerus limbatipennis* (Pic, 1926), nec Pic, 1906
Lycocerus purpureus Kazantsev, **nom. nov.** for *Lycocerus purpurascens* (Wittmer, 1978), nec Pic, 1911
Lycocerus rubrispinis Kazantsev, **nom. nov.** for *Lycocerus rubripennis* (Pic, 1906), nec Hope, 1831
Lycocerus wittmerorum Kazantsev, **nom. nov.** for *Lycocerus wittmeri* (Švihla, 2004), nec Kazantsev, 1999
Cantharis occidentalis Kazantsev, **nom. nov.** for *Cantharis occipitalis* (Fairmaire, 1876), nec Rosenhauer, 1847
Macrohabronychus obscuricolorissimus Kazantsev, **nom. nov.** for *Macrohabronychus obscuricolor* (Pic, 1935), nec Pic, 1905
Themus (s. str.) *hmong* Kazantsev, **nom. nov.** for *Themus* (s. str.) *rufoscutus* (Pic, 1926), nec Pic, 1921
Micropodabrus wittmeri Kazantsev, **nom. nov.** for *Micropodabrus formosanus* Wittmer, 1954, nec Pic, 1910

Spelling

Cantharis xantholoma Gebler, 1833, though originally described as *C. cantholoma*, should preserve its commonly used name, in accordance with the ICZN, Art. 33.2.3.1.

Cantharidae
M. Brancucci

Comments and new combinations

Habronychus Wittmer, 1981c: 399 is currently used as valid although it is unavailable. The name was made available in Wittmer, 1982c: 124. *Macrohabronychus* Wittmer, 1981c: 400 was published as a subgenus of the unavailable genus name *Habronychus*. Hence, *Macrohabronychus* Wittmer, 1981 must be used as the valid genus name while *Habronychus* Wittmer, 1982 is one of its valid subgeneric names, resulting in following new combinations:

Macrohabronychus (s. str.) *aritai* (Satô, 1986) **comb. nov**. from *Habronychus*
Macrohabronychus (s. str.) *biungulatus* (Champion, 1925) **comb. nov**. from *Habronychus*
Macrohabronychus (s. str.) *kurosawai* (Wittmer, 1982) **comb. nov**. from *Habronychus*
Macrohabronychus (s. str.) *miyatakei* (Ishida, 1986) **comb. nov**. from *Habronychus*
Macrohabronychus (s. str.) *nantouanus* (Wittmer, 1982) **comb. nov**. from *Habronychus*
Macrohabronychus (s. str.) *parallelicollis* (Pic, 1921) **comb. nov**. from *Habronychus*
Macrohabronychus (s. str.) *providus* (Kiesenwetter, 1874) **comb. nov**. from *Habronychus*
Macrohabronychus (s. str.) *sauteri* (Pic, 1934) **comb. nov**. from *Habronychus*
Macrohabronychus (s. str.) *semimetallicus* (Wittmer, 1997) **comb. nov**. from *Habronychus*
Macrohabronychus (s. str.) *szechwanus* (Wittmer, 1988) **comb. nov**. from *Habronychus*
Macrohabronychus (*Monohabronychus*) *furcatus* (Wittmer, 1986) **comb. nov**. from *Habronychus*
Macrohabronychus (*Monohabronychus*) *intermixtus* (Wittmer, 1954) **comb. nov**. from *Habronychus*
Macrohabronychus (*Monohabronychus*) *lalashanensis* (Okushima & Satô, 1999) **comb. nov**. from *Habronychus*
Macrohabronychus (*Monohabronychus*) *longicornis* (Okushima & Satô, 1999) **comb. nov**. from *Habronychus*
Macrohabronychus (*Monohabronychus*) *multilimbatus* (Pic, 1910) **comb. nov**. from *Habronychus*

Bostrichidae
J. Borowski

Designation of type species

Bostrychoplites Lesne, 1899 was established for 11 species including *Bostrichus cornutus* A. G. Olivier, 1790, but not fixed by a type species. *Bostrichus cornutus* A. G. Olivier, 1790 is here designated as the type species.

Xyloperthodes Lesne, 1906 was established for 12 species including *Sinoxylon nitidipenne* Murray, 1867, but not fixed by a type species. *Sinoxylon nitidipenne* Murray, 1867 is here designated as the type species.

Ptinidae
J. Borowski

New synonyms

Dignomus brevipilis (Desbrochers des Loges, 1875) **syn. nov.** of *Dignomus carinatus* (P. H. Lucas, 1846)

Note. The description of *Ptinus brevipilis* Desbrochers des Loges, 1875 was based on specimens coming from Corsica and Sicily. A study of the type material (Muséum National d'Histoire Naturelle, Paris; Hungarian Natural History Museum, Budapest; Naturhistorisches Museum Bern) showed that the specimens from Sicily are males of *Dignomus carinatus* (P. H. Lucas, 1846), while the specimens from Corsica are *Dignomus lusitanus* (Illiger, 1807).

Sundaptinus Bellés, 1991 **syn. nov.** of *Myrmecoptinus* Wasmann, 1916, based on study of type material of *Myrmecoptinus* from the Wasmann's collection (Natuurhistorisch Museum Maastrich).

New combinations

Myrmecoptinus deplanatus (Pic, 1954) **comb. nov**. from *Hanumanus*

Myrmecoptinus kuronis (Ohta, 1930) **comb. nov**. from *Sundaptinus*
Myrmecoptinus sauteri (Pic, 1914) **comb. nov**. from *Sundaptinus*

Ptinidae

P. Zahradník

New replacement name

Xyletinus (s. str.) *nardii* Zahradník, **nom. nov**. for *Xyletinus* (s. str.) *bicolor* Zahradník, 1996

Designation of type species

Dorcadolomum Gistel, 1856 was established for three three species including *Dorcatoma bistriata* Paykull, 1798; *D. bistriata* Paykull is here designated as the type species of *Dorcadolomum*.

New synonyms

Anobium alexandri Logvinovskiy, 1981 **syn. nov.** of *Hemicoelus ussuriensis* Karapetyan, 1980
Dorcadolomum Gistel, 1856 **syn. nov.** of *Dorcatoma* Herbst, 1792
Dorcatoma chrysomelina var. *diversesculpturata* Roubal, 1934 **syn. nov.** of *Dorcatoma chrysomelina* Sturm, 1837
Dorcatoma furcatum Logvinovskiy, 1980 **syn. nov.** of *Dorcatoma japonica* Pic, 1937
Dryophilus excelsus Gistel, 1857 **syn. nov.** of *Dryophilus pusillus* (Gyllenhal, 1808)
Dryophilus pusilloides Obenberger, 1917 **syn. nov.** of *Homophthalmus rugicollis* (Mulsant & Rey, 1853)
Ernobius kiesenwetteri var. *lateobscurus* Pic, 1921 **syn. nov.** of *Ernobius kiesenwetteri* Schilsky, 1898
Falsogastrallus elongatus var. *bicolor* Pic, 1931 **syn. nov.** of *Falsogastrallus elongatus* Pic, 1931
Hemicoelus khnozoriani Karapetyan, 1980 **syn. nov.** of *Hemicoelus fulvicornis* Sturm, 1837
Lasioderma semirufulum var. *leveillei* Pic, 1927 **syn. nov.** of *Lasioderma semirufulum* Reitter, 1897
Lasioderma semirufulum var. *vacaresiense* Chobaut, 1924 **syn. nov.** of *Lasioderma semirufulum* Reitter, 1897
Mizodorcatoma sibirica var. *subrufescens* Pic, 1914 **syn. nov.** of *Mizodorcatoma sibirica* Reitter, 1879
Ochina bicolor Csiki, 1903 **syn. nov.** of *Lasioderma thoracicum* Morawitz, 1861
Priobium dendrobiformis Reitter, 1901 **syn. nov.** of *Priobium carpini* Herbst, 1793
Stagetodes breviusculus var. *luteopubens* Pic, 1909 **syn. nov.** of *Stagetodes breviusculus* Fairmaire, 1886
Xyletinus ruficollis var. *mariei* Méquignon, 1940 **syn. nov.** of *Xyletinus ruficollis* Gebler, 1833

New combinations

Cacotemnus kompanzevi (Logvinovskiy, 1979) **comb. nov**. from *Hemicoelus*
Pseudomesothes pulverulentus latior (Pic, 1954) **comb. nov**. from *Mesothes*
Stagetus sydowi (Reitter, 1913) **comb. nov**. from *Lasioderma*

Ptinidae

J. Jelínek, P. Zahradník

Anobium nitidum Fabricius, 1792 belongs to the family Ciidae. *Hemicoelus nitidus* auct. in Anobiidae is a misinterpretation dating back to Herbst (1793); it is necessary to use for this species the oldest available name, *Hemicoelus canaliculatus* (C. G. Thomson, 1863).

Ptinidae/Chrysomelidae
M. Döberl

New synonymy

Epitrix canariensis Franz, 1996 **syn. nov.** of *Neodryophilus cryptophagoides* (Wollaston, 1864), based on study of type material.

Dermestidae
J. Háva

New synonyms

Eunorops Gistel, 1856 **syn. nov.** of *Attagenus* Latreille, 1802
Megalotomum Gistel, 1856 **syn. nov.** of *Megatoma* Herbst, 1792
Trogodermum Gistel, 1856 **syn. nov.** of *Trogoderma* Dejean, 1821
Anthrenus beali modestus Zhantiev, 2004 **syn. nov.** of *A. beali* Zhantiev, 2004
Anthrenus bleusei Pic, 1899 (type material studied) **syn. nov.** of *A. aegyptiacus* Pic, 1899
Attagenus arrowi Kalík, 1954 (type material studied) **syn. nov.** of *A. sinensis* Pic, 1927
Attagenus duplex octopunctatus Mroczkowski, 2002 (type material studied) **syn. nov.** of *A. duplex* Reitter, 1890
Attagenus euryovatus H. Li, 1988 (type material studied) **syn. nov.** of *A. sparsutus* Reitter, 1881
Attagenus hirtulus var. *confusus* (Pic, 1951) (type material studied) **syn. nov.** of *A. obtusus* Gyllenhal, 1808
Attagenus picipennis Pic, 1894 (type material studied) **syn. nov.** of *A. brunneus* Faldermann, 1835
Attagenus quisquiliorum Gistel, 1857 **syn. nov.** of *A. pellio* (Linnaeus, 1758)
Attagenus schaefferi (C. Sahlberg, 1819) **syn. nov.** of *A. unicolor* (Brahm)
Attagenus subdiurnus Gistel, 1857 **syn. nov.** of *A. punctatus* (Scopoli, 1772)
Attagunus urbicola Gistel, 1857 **syn. nov.** of *A. unicolor unicolor* Brahm, (1790)
Dermestes caninus Gravenhorst, 1807 **syn. nov.** of *D. undulatus* Brahm, 1790
Dermestes diadema Motschulsky, 1849 (type material studied) **syn. nov.** of *D. mustelinus* Erichson, 1848
Dermestes geltzermani Kozminykh, 1997 (type material studied) **syn. nov.** of *D. palmi* Sjöberg, 1950
Phradonoma aegyptiacum (Pic, 1901) (type material studied) **syn. nov.** of *P. cercyonoides* Reitter, 1886
Telopes senegalensis Pic, 1915 (type material studied) **syn. nov.** of *A. cyphonoides* Reitter, 1881

New combinations and assignments (type material studied)

Anthrenus (*Anthrenodes*) *aegyptiacus* Pic, 1899 from *Anthrenus* s. str.
Anthrenus (*Anthrenodes*) *jordanicus* Pic, 1934 from *Anthrenus* s. str.
Evorinea flava (Motschulsky, 1863) **comb. nov**. from *Trinodes* Dejean, 1821
Ctesias maculifasciata (Reitter, 1899) **comb. nov**. from *Trogoderma* Dejean, 1821

Changes in rank

Anthrenus (*Anthrenus*) *delicatus armstrongi* Mroczkowski, 1952 from species rank
Attagenus barbieri Pic, 1946 as valid species, from variety of *Attagenus trifasciatus* Fabricius, 1787 (type material studied)
Attagenus addendus J. R. Sahlberg,1903 as valid species, from variety of *Attagenus fallax* Gené, 1839 (type material studied)
Globicornis semilimbata Pic, 1906 as valid species, from variety of *Globicornis corticalis* Eichhoff, 1863

Lymexylidae
G. Cuccodoro

Designation of type species

Calolymus was published by Thunberg, 1791 as a genus group name that remained without any included species. *Lymexylon navale* (Linnaeus, 1758) is here designated at the type species of *Calolymus*.

Cleridae
J. Rolčík and J. Kolibáč

New synonyms

Clerus funebralis Gistel, 1857 **syn. nov.** of *Tarsostenus univittatus* (P. Rossi, 1792); based on the original description, the type material is lost.
Clerus vermivorus Gistel, 1857 **syn. nov.** of *Thanasimus formicarius* (Linnaeus, 1758); based on the original description, the type material is lost.

Melyridae
A. J. Mayor

Designations of type species

Melyrosoma oceanicum Wollaston, 1854 is designated as type species of *Melyrosoma* Wollaston, 1854
Opatrum granulatum Fabricius, 1792 is designated as type species of *Melyris* (*Melyridella*) Champion, 1919

New synonyms

Anthodromius apicalis var. *basalis* (Pic, 1908) **syn. nov.** of *Anthodromius apicalis* (Schilsky, 1897)
Anthodromius rufolimbatus var. *ludovici* (Pic, 1911) **syn. nov.** of *Anthodromius rufolimbatus* (Pic, 1911)
Anthodromius turkestanicus var. *flavipes* Schilsky, 1899 **syn. nov.** *Anthodromius turkestanicus* Heyden, 1886
Anthodromius turkestanicus var. *haplocnemoides* Reitter, 1902 **syn. nov.** of *Anthodromius turkestanicus* Heyden, 1886
Anthodromius turkestanicus var. *nigripes* Schilsky, 1899 **syn. nov.** *Anthodromius turkestanicus* Heyden, 1886
Ceralus varians var. *maculicollis* Schilsky, 1897 **syn. nov.** of *Ceralus varians* (Morawitz, 1861)
Ceralus varians var. *rufifrons* Schilsky, 1897 **syn. nov.** of *Ceralus varians* (Morawitz, 1861)
Procerallus aplocnemiformis var *diversepunctatus* (Pic, 1920) **syn. nov.** of *Procerallus aplocnemiformis* (Pic, 1915)
Falsomelyris rotundicollis var. *pallidipes* (Pic, 1898) **syn. nov.** of *Falsomelyris rotundicollis* (Pic, 1894)
Falsomelyris ruficrus var. *aristidis* Pic, 1894 **syn. nov.** of *Falsomelyris ruficrus* (Fairmaire, 1882)
Melyris klugii var. *dubia* Schilsky, 1897 **syn. nov.** of *Melyris klugii* (Baudi, 1873)
Melyris oblonga var. *limbata* Pic, 1895 **syn nov.** of *Melyris oblonga* (Fabricius, 1775)
Melyris oblonga var. *lineatocollis* (Pic, 1899) **syn nov.** of *Melyris oblonga* (Fabricius, 1775)
Melyris oblonga var. *maculicollis* (Fuente, 1913) **syn nov.** of *Melyris oblonga* (Fabricius, 1775)
Melyris oblonga var. *mulsanti* (Schilsky, 1897) **syn nov.** of *Melyris oblonga* (Fabricius, 1775)
Melyris reitteri var. *marginata* (Heyden, 1890) **syn. nov.** of *Melyris reitteri* (Heyden, 1890)

Dasytidae
A. J. Mayor

Designations of type species

Dasytes imperialis Gené, 1836 designated as type species of *Cosmiocomus* Küster, 1850
Dolichosoma (*Dolichophron*) *cylindromorphum* Kiesenwetter, 1867 designated as type species of *Dolichophron* Kiesenwetter, 1867
Dermestes niger Linnaeus, 1761 designated as type species of *Enodius* Laporte, 1838

Dasytes subaeneus Schönherr, 1817 designated as type species of *Pseudodasytes* Mulsant & Rey, 1868
Dermestes niger Linnaeus, 1761 designated as type species of *Psilocorse* C. G. Thomson, 1864

New replacement name

Aplocnemus sahlbergi **nom. nov.** for *Aplocnemus puncticollis* (Sahlberg, 1913), preoccupied by *Aplocnemus subcostatus* var. *puncticollis* (Schilsky, 1894) = *Aplocnemus subcostatus* Schilsky, 1894.

New synonyms and reconfirmed synonymy

Allotarsus baudii var. *canescens* Pardo Alcaide, 1954 **syn. nov.** of *Allotarsus baudii* (Bourgeois, 1888)
Allotarsus praticola var. *nigripes* Schilsky, 1896 **syn. nov.** of *Graellsinus brachialis* (Jacquelin du Val, 1860)
Allotarsus senescens var. *fulvimanus* Escalera, 1927 **syn. nov.** of *Allotarsus senescens* (Jacquelin du Val, 1860)
Amauronia (*Aphyctus*) *megacephala* var. *robusta* Pic, 1923 **syn. nov.** of *Amauronia subaenea* Westwood, 1839
Amauronia elegans var. *infasciata* Pic, 1922 **syn. nov.** of *Mauroania elegans* (Kiesenwetter, 1867)
Amauronia elegans var. *viridescens* Pic, 1894 **syn. nov.** of *Mauroania elegans* (Kiesenwetter, 1867)
Danacaea (sic) *hispana* var. *heydeni* Schilsky, 1897 **syn. nov.** of *Danacea hispanica* Gougelet & H. Barneville, 1859
Danacaea acutangula var. *nevadensis* Pic, 1904 **syn. nov.** of *Danacea acutangula* Schilsky, 1897
Danacaea angulata var. *ochripes* Schilsky, 1897 **syn. nov.** of *Danacea angulata* Küster, 1850
Danacaea atripes var. *ornata* Procházka, 1894 **syn. nov.** of *Danacea atripes* Graëlls, 1858
Danacaea atripes var. *tibialis* Pic, 1894 **syn. nov.** of *Danacea atripes* Graëlls, 1858
Danacaea cervina var. *diversipes* Pic, 1936 **syn. nov.** of *Danacea cervina* Küster, 1850
Danacaea championi var. *fallax* Schilsky, 1897 **syn. nov.** *Danacea championi* Marseul, 1878
Danacaea championi var. *fallax* Schilsky, 1897 **syn. nov.** of *Danacea championi* Marseul, 1878
Danacaea corsica var. *luteipes* Schilsky, 1897 **syn. nov.** of *Danacea corsica* Kiesenwetter, 1871
Danacaea cypria var. *lineatithorax* Pic, 1910 **syn. nov.** of *Danacea cypria* Schilsky, 1897
Danacaea cypria var. *livadiensis* Pic, 1910 **syn. nov.** of *Danacea cypria* Schilsky, 1897
Danacaea cypria var. *maculaticeps* Pic, 1910 **syn. nov.** of *Danacea cypria* Schilsky, 1897
Danacaea dentatithorax var. *simplicithorax* Pic, 1909 **syn. nov.** of *Danacea dentatithorax* Pic, 1909
Danacaea genistae var. *berytensis* Pic, 1902 **syn. nov.** of *Danacea genistae* Marseul, 1868
Danacaea imperialis var. *testaceoapicalis* Pic 1947 **syn. nov.** of *Danacea imperialis* Gené, 1836
Danacaea imperialis var. *uniformis* Schilsky, 1897 **syn. nov.** of *Danacea imperialis* Gené, 1836
Danacaea imperialis var. *versicolor* Schilsky, 1897 **syn. nov.** of *Danacea imperialis* Gené, 1836
Danacaea induta var. *atricornis* Pic, 1922 **syn. nov.** of *Danacea induta* Schilsky, 1897
Danacaea induta var. *simulatrix* Schilsky, 1897 **syn. nov.** of *Danacea induta* Schilsky, 1897
Danacaea krüperi var. *varipes* Schilsky, 1897 **syn. nov.** of *Danacea kreuperi* Schilsky, 1897
Danacaea leprieuri var. *picipes* Schilsky, 1897 **syn. nov.** of *Danacea leprieuri* Pic, 1894
Danacaea leprieuri var. *ruficornis* Schilsky, 1897 **syn. nov.** of *Danacea leprieuri* Pic, 1894
Danacaea lusitana var. *testaceipes* Pic, 1897 **syn. nov.** of *Danacea hispanica* Gougelet & H. Barneville, 1859
Danacaea lysholmi var. *subrobusta* Pic, 1903 **syn. nov.** of *Danacea lysholmi* Pic, 1900
Danacaea marginata var. *graeca* Procházka, 1894 **syn. nov.** of *Danacea marginata* Küster, 1851
Danacaea marginata var. *maculipennis* Pic, 1899 **syn. nov.** of *Danacea marginata* Küster, 1851
Danacaea marginata var. *rufula* Schilsky, 1897 **syn. nov.** of *Danacea marginata* Küster, 1851

Comments

Camptolegnum Schilsky, 1894 is currently listed as a junior synonym of *Divales* Laporte, 1838. According to synonymy listing in Majer (1984), it is implied a misidentification. The case has not yet been adequately treated.

Malachiidae

A. J. Mayor

Designations of type species

Antidipnis palpator Marseul, 1868 (= *Colotes cinctus* Motschulsky, 1854) designated as the type species of *Colotes* (*Heterodipnis*) Peyron, 1877
Apalochrus (*Spinapalochrus*) *rufofasciatus* Pic, 1919 designated as the type species of *Opisthapalochrus* (*Spinapalochrus*) Pic, 1919
Chionotopus damascaenus Abeille de Perrin, 1881 designated as the type species of *Chionotopus* Abeille de Perrin, 1881
Colotes flavocinctus Marseul, 1868 designated as the type species of *Colotes* (*Gynedipnis*) Abeille de Perrin, 1890
Cyrtosus (*Homogine*) *uhagoni* Escalera, 1913 designated as the type species of *Apalochroides* Escalera, 1913
Ebaeus (*Mixis*) *hystrix* Abeille de Perrin, 1885 designated as the type species of *Mixis* Abeille de Perrin, 1885
Hypebaeus vitticollis Abeille de Perrin, 1881 designated as the type species of *Hypebaeus* (*Allogynes*) Abeille de Perrin, 1891
Malachius apicalis Say, 1825 designated as the type species of *Pseudebaeus* Horn, 1872
Malachius lobatus A. G. Olivier, 1790 designated as the type species of *Sphinginus* Mulsant & Rey, 1867
Malachius bipunctatus Say, 1823 designated as the type species of *Collops* Erichson, 1840
Scalopterus infuscatus Motschulsky, 1860 designated as the type species of *Scalopterus* Moschulsky, 1860
Pseudocerapheles superbus Pic, 1914 designated as the type species of *Pseudocerapheles* Pic, 1914

New synonyms and reconfirmed synonymy

Anthocomus cardinalis var. *mesopotamicus* Pic, 1901 **syn. nov.** of *Anthocomus cardinalis* Abeille de Perrin, 1881
Anthocomus coccineus var. *bedeli* Méquignon, 1948 **syn. nov.** of *Anthocomus coccineus* (Schaller, 1783)
Anthocomus coccineus var. *subscutellaris* Pic, 1924 **syn. nov.** of *Anthocomus coccineus* (Schaller, 1783)
Anthocomus fasciatus var. *latior* Abeille de Perrin, 1891 **syn. nov.** of *Anthocomus fasciatus* (Linnaeus, 1758)
Attalus aenescens var. *puncticollis* Wollaston, 1862 **syn. nov.** of *Attalus aenescens* Wollaston, 1862
Attalus aenescens var. *umbrinus* Wollaston, 1862 **syn. nov.** of *Attalus aenescens* Wollaston, 1862
Attalus chamaeleon var. *niger* Escalera, 1921 **syn. nov.** of *Fortunatius chamaeleon* (Escalera, 1921)
Attalus chrysanthemi var. *dasytoides* Wollaston, 1862 **syn. nov.** of *Attalus chrysanthemi* Wollaston, 1862
Attalus gracilis var. *lugubris* Abeille de Perrin, 1891 **syn. nov.** of *Attalus gracilis* Kiesenwetter, 1865
Attalus gracilis var. *miser* Abeille de Perrin, 1891 **syn. nov.** of *Attalus gracilis* Kiesenwetter, 1865
Attalus heydeni var. *notaticeps* Pic, 1949 **syn. nov.** of *Attalus heydeni* Abeille de Perrin, 1887
Attalus lateralis var. *corsicus* Pic, 1954 **syn. nov.** of *Attalus minimus* (P. Rossi, 1790)
Attalus lutatus var. *tournieri* Pic, 1914 **syn. nov.** of *Attalus lutatus* Abeille de Perrin, 1890
Attalus mackiei var. *cockerelli* Pic, 1931 **syn. nov.** of *Attalus mackiei* Pic, 1931
Attalus mackiei var. *kocheri* Pic, 1949 **syn. nov.** of *Attalus mackiei* Pic, 1931
Attalus marraquensis var. *ornatus* Escalera, 1914 **syn. nov.** of *Attalus marraquensis* Escalera, 1914
Attalus metallicus var. *luteotibialis* Evers, 1959 **syn. nov.** of *Fortunatius metallicus* (Wollaston, 1862)
Attalus metallicus var. *similis* (Wollaston, 1862) **syn. nov.** of *Fortunatius metallicus* (Wollaston, 1862)
Attalus omophloides var. *convexifrons* Abeille de Perrin, 1891 **syn. nov.** of *Attalus omophloides* Abeille de Perrin, 1882
Attalus oudjanus var. *santiagoi* Pardo Alcaide, 1955 **syn. nov.** of *Attalus oudjanus* Pic, 1920
Attalus ovatipennis var. *wolastoni* Pic, 1954 **syn. nov.** of *Attalus ovatipennis* Wollaston, 1862
Attalus parietariae var. *limbatus* Kiesenwetter, 1863 **syn. nov.** of *Attalus parietariae* (Erichson, 1840)
Attalus pellucidus var. *felschei* Evers, 1960 **syn. nov.** of *Attalus pellucidus* (Wollaston, 1854)
Attalus pellucidus var. *rufipes* Evers, 1960 **syn. nov.** of *Attalus pellucidus* (Wollaston, 1854)
Attalus reductus var. *reductissimus* Pic, 1936 **syn. nov.** of *Attalus reductus* Pic, 1936
Attalus rostratus var. *floricola* (Wollaston, 1854) **syn. nov.** of *Attalus rostratus* (Wollaston, 1854)
Attalus ruficollis var. *pauperculus* Wollaston, 1862 **syn. nov.** of *Attalus ruficollis* Wollaston, 1862
Attalus rufoapicalis var. *inapicalis* Pic, 1917 **syn. nov.** of *Attalus rufoapicalis* Pic, 1917

Attalus semitogatus var. *henoni* Pic, 1904 **syn. nov.** of *Attalus semitogatus* (Fairmaire, 1863)
Attalus sicanus var. *kabylianus* Pic, 1898 **syn. nov.** of *Attalus sicanus* Erichson, 1840
Attalus tuberculatus var. *pallidipes* Escalera, 1921 **syn. nov.** of *Attalus tuberculatus* Wollaston, 1862
Attalus varitarsis var. *gavoyi* Pic, 1927 **syn. nov.** of *Attalus varitarsis* (Kraatz, 1862)
Attalus varitarsis var. *subdisjunctus* Pic, 1927 **syn. nov.** of *Attalus varitarsis* (Kraatz, 1862)
Axinotarsus algericus var. *chobauti* Escalera, 1926 **syn. nov.** of *Axinotarsus algiricus* Abeille de Perrin, 1896
Axinotarsus brevicornis var. *curtelineatus* Pic, 1927 **syn. nov.** of *Axinotarsus brevicornis* (Kraatz, 1862)
Axinotarsus brevicornis var. *normandi* Escalera, 1926 **syn. nov.** of *Axinotarsus brevicornis* (Kraatz, 1862)
Axinotarsus brevicornis var. *orientalis* Escalera, 1926 **syn. nov.** of *Axinotarsus brevicornis* (Kraatz, 1862)
Axinotarsus marginalis var. *infuscatus* Abeille de Perrin, 1891 **syn. nov.** of *Axinotarsus marginalis* (Laporte, 1840)
Axinotarsus pallitarsus var. *bleusei* Pic, 1927 **syn. nov.** of *Axinotarsus pallitarsus* (Fairmaire, 1880)
Axinotarsus pallitarsus var. *vaulogeri* Pic, 1927 **syn. nov.** of *Axinotarsus pallitarsus* (Fairmaire, 1880)
Axinotarsus pulicarius var. *martialis* Pic, 1918 **syn. nov.** of *Axinotarsus pulicarius* (Fabricius, 1777)
Axinotarsus pulicarius var. *obscuratus* Escalera, 1926 **syn. nov.** of *Axinotarsus pulicarius* (Fabricius, 1777)
Axinotarsus pulicarius var. *truncatellus* (Fallén, 1807) **syn. nov.** of *Axinotarsus pulicarius* (Fabricius, 1777)
Axinotarsus rissor var. *escalerai* Pic, 1927 **syn. nov.** of *Axinotarsus algiricus* Abeille de Perrin, 1896
Axinotarsus robustus var. *parvisignatus* Escalera, 1914 **syn. nov.** of *Axinotarsus robustus* Pic, 1905
Axinotarsus robustus var. *simulans* Escalera, 1926 **syn. nov.** of *Axinotarsus robustus* Pic, 1905
Axinotarsus tristis var. *risor* Abeille de Perrin, 1891 **syn. nov.** of *Axinotarsus algiricus* Abeille de Perrin, 1896
Axinotarsus varius var. *uhagoni* Escalera, 1926 **syn. nov.** of *Axinotarsus varius* Uhagón, 1901
Brachemys brevipennis var. *nigripes* (Wittmer, 1935) **syn. nov.** of *Brachemys brevipennis* (Laporte, 1836)
Cephaloncus albozonatus var. *atticus* (Abeille de Perrin, 1891) **syn. nov.** of *Cephaloncus albozonatus* (Abeille de Perrin, 1883)
Cephaloncus albozonatus var. *cyaneicollis* (Abeille de Perrin, 1891) **syn. nov.** of *Cephaloncus albozonatus* (Abeille de Perrin, 1883)
Cephaloncus rhinoceros var. *bicoloriceps* (Pic, 1917) ***syn. nov.*** of *Troglops rhinoceros* (Marseul, 1868)
Cerapheles terminatus var. *obscuricornis* Pic, 1906 **syn. nov.** of *Cerapheles terminatus* (Ménétriés, 1832)
Charopus nitidus var. *macroderus* Abeille de Perrin, 1890 **syn. nov.** of *Charopus rotundatus* Erichson, 1840
Chionotopus bimaculatus var. *inhumeralis* Pic, 1911 **syn. nov.** of *Chionotopus bimaculatus* Pic, 1911
Clanoptilus (*Hypoptilus*) *curticornis* var. *subelongatus* (Pic, 1911) **syn. nov.** of *Clanoptilus* (*Hypoptilus*) *curticornis* (Kiesenwetter, 1865)
Clanoptilus abdominalis var. *cyrenaicus* (Pic, 1925) **syn. nov.** of *Clanoptilus abdominalis* (Fabricius, 1798)
Clanoptilus abdominalis var. *mozabitus* (Pic, 1898) **syn. nov.** of *Clanoptilus abdominalis* (Fabricius, 1798)
Clanoptilus abdominalis var. *pseudobimaculatus* (Pic, 1925) **syn. nov.** of *Clanoptilus abdominalis* (Fabricius, 1798)
Clanoptilus abdominalis var. *subcrucifer* (Pic, 1904) **syn. nov.** of *Clanoptilus abdominalis* (Fabricius, 1798)
Clanoptilus affinis var. *concolor* (Krauss, 1902) **syn. nov.** of *Clanoptilus affinis* (Ménétriés, 1832)
Clanoptilus elegans var. *haemorrhoidalis* (Abeille de Perrin, 1891) **syn.** of *Clanoptilus elegans* (A. G. Olivier, 1790)
Clanoptilus elegans var. *purpureonotatus* (Pic, 1952) **syn. nov.** of *Clanoptilus elegans* (A. G. Olivier, 1790)
Clanoptilus geniculatus var. *obconicus* (Abeille de Perrin, 1891) **syn. nov.** of *Clanoptilus geniculatus* (Germar, 1824)
Clanoptilus heliophilus var. *breviruber* (Pic, 1914) **syn. nov.** of *Clanoptilus heliophilus* (Peyron, 1877)
Clanoptilus laticollis var. *peninsularis* (Evers, 1948) **syn.** of *Clanoptilus laticollis* (Rosenhauer, 1856)
Clanoptilus marginellus var. *atriceps* (Abeille de Perrin, 1891) **syn. nov.** of *Clanoptilus marginellus* (A. G. Olivier, 1790)
Clanoptilus marginellus var. *lambessanus* (Pic, 1951) **syn. nov.** of *Clanoptilus marginellus* (A. G. Olivier, 1790)
Clanoptilus marginellus var. *pseudosardous* (Reclaire and van der Wiel, 1932) **syn. nov.** of *Clanoptilus marginellus* (A. G. Olivier, 1790)
Clanoptilus marginellus var. *subdilatatus* (Pic, 1923) **syn. nov.** of *Clanoptilus marginellus* (A. G. Olivier, 1790)
Clanoptilus parilis var. *commissus* (Abeille de Perrin, 1891) **syn. nov.** of *Clanoptilus parilis* (Erichson, 1840)
Clanoptilus spinosus var. *clorizans* (Erichson, 1840) **syn. nov.** of *Clanoptilus spinosus* (Erichson, 1840)
Clanoptilus versicolor var. *erivanicus* (Pic, 1912) **syn. nov.** of *Clanoptilus versicolor* (Faldermann, 1835)

Clanoptilus versicolor var. *punctonotatus* (Pic, 1912) **syn. nov.** of *Clanoptilus versicolor* (Faldermann, 1835)
Clanoptilus versicolor var. *viridibasis* (Reitter, 1890) **syn. nov.** of *Clanoptilus versicolor* (Faldermann, 1835)
Colotes cinctus var. *inlateralis* Pic, 1938 **syn. nov.** of *Colotes cinctus* Motschulsky, 1854
Colotes ebaeiformis var. *obscuriceps* Pic, 1908 **syn. nov.** of *Colotes ebaeiformis* Abeille de Perrin, 1900
Colotes flavocinctus var. *atricollis* Abeille de Perrin, 1890 **syn. nov.** of *Colotes flavocinctus* Marseul, 1868
Colotes flavocinctus var. *signatipennis* Pic, 1938 **syn. nov.** of *Colotes flavocinctus* Marseul, 1868
Colotes javeti var. *rufithorax* Jacquelin du Val, 1852 **syn. nov.** of *Colotes javeti* Jacquelin du Val, 1852
Colotes maculatus var. *semireductus* Pic, 1915 **syn. nov.** of *Colotes maculatus* (Laporte, 1836)
Colotes obsoletus var. *innotaticeps* Pic, 1938 **syn. nov.** of *Colotes obsoletus* Erichson, 1840
Colotes ogieri var. *unicolor* Pic, 1895 **syn. nov.** of *Colotes ogieri* (Fairmaire, 1863)
Colotes punctatus var. *colon* Abeille de Perrin, 1890 **syn. nov.** of *Colotes punctatus* (Erichson, 1840)
Colotes punctatus var. *inmaculatus* Escalera, 1914 **syn. nov.** of *Colotes punctatus* (Erichson, 1840)
Colotes punctatus var. *pygmaeus* Abeille de Perrin, 1890 **syn. nov.** of *Colotes punctatus* (Erichson, 1840)
Cordylepherus viridis var. *concolor* (Schilsky, 1888) **syn. nov.** of *Cordylepherus viridis* (Fabricius, 1787)
Cordylepherus viridis var. *pseudobipustulatus* (Evers, 1944) **syn. nov.** of *Cordylepherus viridis* (Fabricius, 1787)
Cordylepherus viridis var. *semiviridis* (Pic, 1940) **syn. nov.** of *Cordylepherus viridis* (Fabricius, 1787)
Cyrtosus anceyi var. *cruralis* Abeille de Perrin, 1891 **syn. nov.** *Cyrtosus anceyi* (Abeille de Perrin, 1869)
Cyrtosus cyaneipennis var. *bifrons* Abeille de Perrin, 1891 **syn. nov.** of *Cyrtosus cyanipennis* (Erichson, 1840)
Cyrtosus cyaneipennis var. *nigricornis* Evers 1948 **syn. nov.** of *Cyrtosus cyanipennis* (Erichson, 1840)
Cyrtosus cyanipennis var. *angustimargo* (Uhagón, 1894) **syn. nov.** of *Cyrtosus cyanipennis* (Erichson, 1840)
Cyrtosus dispar var. *variicollis* Abeille de Perrin, 1891 **syn. nov.** of *Cyrtosus dispar* (Fairmaire, 1859)
Cyrtosus meridionalis var. *parvulus* Pic, 1904 **syn. nov.** of *Cyrtosus meridionalis* Abeille de Perrin, 1885
Cyrtosus nitidicollis var. *romaniensis* Pic, 1901 ***syn. nov.*** of *Cyrtosus nitidicollis* (Chevrolat, 1873)
Cyrtosus ovalis var. *cyaneicollis* Abeille de Perrin, 1894 **syn. nov.** of *Cyrtosus ovalis* (Laporte, 1836)
Cyrtosus ovalis var. *maculicollis* Abeille de Perrin, 1894 **syn. nov.** of *Cyrtosus ovalis* (Laporte, 1836)
Cyrtosus ovalis var. *variicollis* Abeille de Perrin, 1894 **syn. nov.** of *Cyrtosus ovalis* (Laporte, 1836)
Cyrtosus rufifrons var. *cyanipes* Abeille de Perrin, 1885 **syn. nov.** of *Cyrtosus rufifrons* Abeille de Perrin, 1885
Cyrtosus rufifrons var. *diversipes* Pic, 1912 **syn. nov.** of *Cyrtosus ferreri* (Pardo Alcaide, 1966)
Cyrtosus rufifrons var. *tibiellus* Abeille de Perrin, 1885 **syn. nov.** of *Cyrtosus rufifrons* Abeille de Perrin, 1885
Cyrtosus sinuaticollis var. *moutoni* (Pic, 1937) **syn. nov.** of *Cyrtosus sinuaticollis* (Pic, 1913)
Ebaeus abietinus var. *elzeari* Pic, 1911 **syn. nov.** of *Ebaeus abietinus* Abeille de Perrin, 1869
Ebaeus limbellus ssp. *tianshanensis* Wittmer, 1970 **syn.** of. *Ebaeus limbellus* Peyron, 1877
Ebaeus luctuosus var. *koenigi* Pic, 1912 **syn. nov.** of *Ebaeus luctuosus* Abeille de Perrin, 1890
Ebaeus pedicularius var. *montenegrinus* Pic, 1915 **syn. nov.** of *Ebaeus pedicularius* (Fabricius, 1775)
Ebaeus rubetorum var. *viridipennis* Pic, 1908 **syn. nov.** of *Ebaeus rubetorum* Peyron, 1877
Ebaeus rufoapicalis var. *paulonotatus* Pic, 1948 **syn. nov.** of *Ebaeus rufoapicalis* Pic, 1926
Haplomalachius diversicornis var. *maculaticollis* (Pic in Pic and Lindberg, 1932) **syn. nov.** of *Haplomalachius diversicornis* (Pic, 1932)
Hypebaeus albifrons var. *hypocrita* Abeille de Perrin, 1891 **syn. nov.** of *Hypebaeus albifrons* (Fabricius, 1775)
Hypebaeus flavicollis var. *obscurithorax* Pic 1903 **syn. nov.** of *Hypebaeus flavicollis* (Erichson, 1840)
Hypebaeus tripartitus var. *tauricola* Pic, 1911 **syn. nov.** of *Hypebaeus tripartitus* (Marseul, 1868)
Intybia blaisei var. *albonotatus* (Pic, 1926) **syn. nov.** of *Intybia duplex* (Champion, 1921)
Intybia viridithorax var. *atricollis* (Pic, 1921) **syn. nov.** of *Intybia viridithorax* (Pic, 1921)
Laius sikkimensis var. *testaceilabris* Pic, 1914 **syn. nov.** of *Laius sikkimensis* Pic, 1914
Laius venustus var. *bedforti* Pic, 1928 **syn. nov.** of *Laius venustus* Erichson, 1840
Laius venustus var. *letourneuxi* Pic, 1928 **syn. nov.** of *Laius venustus* Erichson, 1840
Malachiomimus mirandus var. *transversus* (Abeille de Perrin, 1891) **syn. nov.** of *Malachiomimus mirandus* (Abeille de Perrin, 1885)
Malachius aeneus var. *aeneoides* Abeille de Perrin, 1891 **syn. nov.** of *Malachius aeneus* (Linnaeus, 1758)
Malachius basalis var. *cruciger* Abeille de Perrin, 1890 **syn. nov.** of *Malachius basalis* Abeille de Perrin, 1890
Malachius bipustulatus var. *aetolicus* Kiesenwetter, 1866 **syn. nov.** of *Malachius bipustulatus* (Linnaeus, 1758)
Malachius bipustulatus var. *immaculatus* Abeille de Perrin 1891 **syn. nov.** of *Malachius bipustulatus* (Linnaeus, 1758)

Malachius bipustulatus var. *immaculicollis* Mulsant & Rey, 1867 **syn. nov.** of *Malachius bipustulatus* (Linnaeus, 1758)
Malachius caramanicus var. *anticenotatus* Pic, 1904 **syn. nov.** of *Malachius caramanicus* Pic, 1912
Malachius carnifex var. *latenotatus* Pic, 1914 **syn. nov.** of *Malachius carnifex* Erichson, 1840
Malachius carnifex var. *naxosensis* Pic, 1923 **syn. nov.** of *Malachius carnifex* Erichson, 1840
Malachius coccineus var. *baudueri* Pic, 1930 **syn. nov.** of *Malachius coccineus* Waltl, 1838
Malachius ensiculus var. *delagrangei* Pic 1898 **syn. nov.** of *Malachius ensiculus* Abeille de Perrin, 1891
Malachius ensiculus var. *metallescens* Abeille de Perrin, 1891 **syn. nov.** of *Malachius ensiculus* Abeille de Perrin, 1891
Malachius fausti var. *crucifer* Reitter, 1888 **syn. nov.** of *Malachius fausti* Abeille de Perrin, 1885
Malachius fausti var. *miniatus* Reitter, 1888 **syn. nov.** of *Malachius fausti* Abeille de Perrin, 1885
Malachius foveifrons var. *drouardi* Pic, 1923 **syn. nov.** of *Malachius prolongatus* Motschulsky, 1866
Malachius gethsemaniensis var. *rubidiformis* Pic, 1899 **syn. nov.** of *Malachius gethsemaniensis* Abeille de Perrin, 1881
Malachius heydeni var. *chehirensis* Pic, 1923 **syn. nov.** of *Malachius heydeni* Abeille de Perrin, 1882
Malachius lusitanicus var. *amasiae* Abeille de Perrin, 1891 **syn. nov.** of *Malachius lusitanicus* Erichson, 1840
Malachius montanus var. *viridistriatus* Bourgeois, 1900 **syn. nov.** of *Malachius caramanicus* Pic, 1912
Malachius rubidus var. *mazarakii* Baeckman, 1903 **syn. nov.** of *Malachius rubidus* Erichson, 1840
Malachius rubidus var. *sublateralis* Pic, 1912 **syn. nov.** of *Malachius rubidus* Erichson, 1840
Malachius rubidus var. *valesiacus* Pic, 1929 **syn. nov.** of *Malachius rubidus* Erichson, 1840
Malachius vittatus var. *subhumeralis* Pic, 1911 **syn. nov.** of *Malachius vittatus* Ménétriés, 1832
Microlipus rosti var. *quercicola* (Champion, 1927) **syn. nov.** of *Microlipus Rosti* (Pic, 1914)
Microlipus rosti var. *rectifasciatus* (Champion, 1927) **syn. nov.** of *Microlipus Rosti* (Pic, 1914)
Nepachys cardiacae var. *punctipennis* (Dodero, 1929) **syn. nov.** of *Nepachys cardiacae* (Linnaeus, 1761)
Nudopectinus flabellicornis var. *bijunctus* (Pic, 1899) **syn. nov.** of *Nudopectinus flabellicornis* (Erichson, 1840)
Nudopectinus flabellicornis var. *sudanicus* (Pic, 1931) **syn. nov.** of *Nudopectinus flabellicornis* (Erichson, 1840)
Pelochrus pallidulus var. *gallicus* Uhagón, 1900 **syn. nov.** of *Pelochrus pallidulus* (Erichson, 1840)
Pelochrus pallidulus var. *maculicollis* Pic, 1903 **syn. nov.** of *Pelochrus pallidulus* (Erichson, 1840)
Protapalochrus festivus var. *paulojunctus* (Pic, 1953) **syn. nov.** of *Protapalochrus festivus* (Zoubkoff, 1833)
Troglops angustatus var. *normandi* Pic, 1919 **syn. nov.** of *Troglops angustatus* Reitter, 1872
Troglops basicollis var. *ruficollis* Uhagón, 1900 **syn. nov.** of *Troglops basicollis* Fairmaire, 1877
Troglops ferrugineus var. *nigritarsis* Escalera, 1914 **syn. nov.** of *Troglops ferrugineus* Escalera, 1914
Troglops orientalis var. *laevissimus* Abeille de Perrin, 1890 **syn. nov.** of *Troglops orientalis* Abeille de Perrin, 1881
Troglops palaestinus var. *robusticeps* Pic, 1924 **syn. nov.** of *Troglops palaestinus* Pic, 1902
Troglops pyriventris var. *theryi* Pic, 1917 **syn. nov.** of *Troglops pyriventris* Abeille de Perrin, 1885

Nomina dubia

The name *Malachius annulatus* Gebler, 1829 is currently placed as an "in part" synonym of *Clanoptilus elegans* (A. G. Olivier, 1790) and of *Clanoptilus affinis* (Ménétriés, 1832). It is considered to be a nomen dubium.
The name *Malachius bellieri* Peyron, 1877 is currently placed as an "in part" synonym of *Clanoptilus geniculatus* (Germar, 1824) and of *Clanoptilus dissimilis* (Baudi di Selve, 1873). It is considered to be a nomen dubium.
The name *Malachius triangularis* J. R. Sahlberg, 1908 is currently placed in synonymy *Malachius securiclatus* Baudi di Selve, 1873 (♂) and of *Malachius aeneus* (Linnaeus, 1758) (♀). It is placed as a nomen dubium.

Monotomidae/Latridiidae

C. Johnson

New transfer and new synonym

Dermestes fenestralis Linnaeus, 1758. The type specimen in the Linnaean collection, London, examined by C. Johnson and R.G. Booth, belongs not to the genus *Corticaria* Marsham, 1802 (Latridiidae) but to the family Monotomidae and is identical with *Rhizophagus parvulus* (Paykull, 1800). The type specimen (examined on 3rd March 1999) is pinned through the middle of the elytra which are left gaping, is staged on a white new continental mount, bearing a large brown label /fenestralis/ in Linnaeus's hand, as well as a modern white label and the number /15/.

Monotomidae
J. Jelínek

Comments

Holocephala Fairmaire, 1886, which was erroneously considered homonym of *Holocephalus* Hope, 1838, is in fact the oldest available name for *Shoguna* Lewis, 1889, but it has not been used for about 100 years. Even though it was not possible to find a sufficient number of papers necessary to conserve *Shoguna* as nomen protectum (ICZN, Art. 23.9.1.2), *Shoguna* is here listed as valid. The case was submitted to the International Commission of Zoological Nomenclature.

Sphindidae
J. Jelínek

Unavailable name

Sphindus gyllenhalli Chevrolat, 1833 is currently listed as junior synonym of *Sphindus dubius* (Gyllenhal, 1808); the name was published in synonymy with "*Nitidula dubia*" Gyllenhall and is therefore unavailable.

Silvanidae
D. G. H. Halstead

New synonyms

Silvanoprus prolixicornis Sengupta & Pal, 1996 **syn. nov.** of *Silvanoprus grouvellei* (Reitter, 1876)

Designation of type species

Monanops Grouvelle, 1913, a subgenus of *Monanus*, originally included the following species: *Silvanus albertisi* Grouvelle, 1883, *Monanus* (*Monanops*) *telephanoides* Grouvelle, 1913, *Monanus* (*Monanops*) *bouchardi* Grouvelle, 1913, *Monanus* (*Monanops*) *longipennis* Grouvelle, 1913 and *Monanus* (*Monanops*) *insolitus* Grouvelle, 1913. *Monanus* (*Monanops*) *bouchardi* Grouvelle, 1913 is hereby designated as the type species.

Microsilvanus Grouvelle, 1913, a subgenus currently not in use, originally included *Silvanus* (*Microsilvanus*) *pumilus* Grouvelle, 1913, *Silvanus* (*Microsilvanus*) *minimus* Grouvelle, 1913 and *Silvanus vitulus* Grouvelle, 1883. *Silvanus* (*Microsilvanus*) *pumilus* Grouvelle, 1913 is hereby designated as the type species.

Silvanidae
J. Jelínek

New synonymy

Silvanus westehauseri Gistel, 1857 **syn. nov.** of *Silvanus bidentatus* (Fabricius, 1792)

Phalacridae
Z. Švec

Designation of type species

Parischius Guillebeau, 1896 was established for two new species, *P. auluaudi* Guillebeau, and *P. perparvulus* Guillebeau. *Parischius auluaudi* Guillebeau, 1896 is here designated as the type species.

New synonyms

Note. The following available names were described as variaties or forms before 1961. They are synonymized to comply with the provisions of the ICZN. This is done without the study of relevant types, and it is possible that some of them might be revalidated in future.

Merobrachys koltzei var. *fenestratus* (Reitter, 1887) **syn. nov.** of *Merobrachys koltzei* (Reitter, 1887)
Olibrosoma sinuaticolle var. *dorsale* (Guillebeau, 1892) **syn. nov.** of *Olibrosoma sinuaticolle* (Guillebeau, 1892)
Olibrus aenescens var. *lucidus* Guillebeau, 1892 **syn. nov.** of *Olibrus aenescens* Küster, 1852
Olibrus aenescens var. *subniger* Guillebeau, 1892 **syn. nov.** of *Olibrus aenescens* Küster, 1852
Olibrus bicolor var. *dohrni* Flach, 1888 **syn. nov.** of *Olibrus bicolor* (Fabricius, 1792)
Olibrus bisignatus var. *fauveli* Guillebeau, 1892 **syn. nov.** of *Olibrus bisignatus* (Ménétriés, 1849)
Olibrus flavicornis var. *perfidus* Flach, 1888 **syn. nov.** of *Olibrus flavicornis* (Sturm, 1807)
Olibrus particeps var. *castanopterus* Guillebeau, 1892 **syn. nov.** of *Olibrus particeps* Mulsant & Rey, 1861
Phalacrus corruscus var. *punctipennis* J. R. Sahlberg,1913 **syn. nov.** of *Phalacrus corruscus* (Panzer, 1797)
Phalacrus maximus var. *ambiguous* Guillebeau, 1892 **syn. nov.** of *Phalacrus maximus* Fairmaire, 1852
Phalacrus maximus var. *reitteri* Flach, 1888 **syn. nov.** of *Phalacrus maximus* Fairmaire, 1852
Phalacrus substriatus var. *fortestriatus* J. R. Sahlberg,1889 **syn. nov.** of *Phalacrus substriatus* Gyllenhal, 1813
Tolyphus punctulatus var. *flavipes* A. Fleischer, 1928 **syn. nov.** of *Tolyphus punctulatus* Rosenhauer, 1856

Cryptophagidae/Ptiliidae

C. Johnson

New assignment and synonymy

Silpha evanescens Marsham, 1802, currently placed in Cryptophagidae, is a senior synonym of *Ptenidium* (*Ptenidium*) *pusillum* (Gyllenhal, 1808).

Cryptophagidae

C. Johnson

Note. New nomenclatural and taxonomic acts are primarily based upon personal studies of type material, genitalia included. If type material is unknown or was exceptionally not available, then the acts are based on descriptions alone.

Designation of type species

Atomaria Stephens, 1829. The earliest cited type species is *Dermestes nigripennis* Paykull, 1798 (Westwood, 1839) [= *Latridius nigripennis* Kugelann, 1794], a short and curved member of the genus, not a parallel species. If this implementation was carried out, it would change every name in this species-rich genus. In order to maintain the accepted sub-generic names used for over a century in Europe, an application will be made to the ICZN to conserve current usage. *Atomaria linearis* Stephens, 1830, is suggested as the type species.

Elachistus A. Villa & J. B. Villa, 1833. The name was established for *Cryptophagus brunnipes* Gyllenhal, 1808 and *Dermestes globulum* [sic] Paykull, 1798. *Dermestes globulus* Paykull, 1798 is here designated as the type species.

Hydrophytophagus Shuckard, 1839 was established for two species, *Ips caricis* A. G. Olivier, 1790 and *Cryptophagus typhae* Fallén, 1802. *Cryptophagus typhae* Fallén, 1802 is here designated as the type species (*Telmatophilus typhae* (Fallén, 1802)).

Limnesius Sturm, 1843. Six species are listed in this genus which was assigned (incorrectly) to Erichson. These are *brunneus* (Gyllenhal, 1808) (= *Zavaljus brunneus* (Gyllenhal), Languriidae), *sparganii* (Ahrens, 1812), *caricis* (Gyllenhal) (= A. G. Olivier, 1790), *affinis* Sturm (? identity uncertain), *typhae* Gyllenhal (= Fallén, 1802) and *schonherrii* (Gyllenhal, 1808). As type species is here designated *Cryptophagus sparganii* Ahrens, 1812 (= *Telmatophilus sparganii* (Ahrens, 1812)).

Psychidium Heer, 1841 was established for "*exigius* Ullr.", "*globulus* Pk." and "*atomus* Müll.". *Dermestes globulus* Paykull, 1798 is here designated as the type species.

New replacement names

Atomaria (*Anchicera*) *nikitskyi* Johnson, **nom. nov.** for *A.* (*A.*) *lateralis* Reitter, 1888 (nec Gistel, 1857).

New taxa

Cryptophagus cylindrellus Johnson, **sp. nov.** for *C. cylindrus* sensu auctorum, nec *C. cylindrus* Kiesenwetter, 1858. Holotype male: [Switzerland] Corbeyrier 5/69 Vd 1000 Toumayeff/ /HOLOTYPE/ (in Muséum d'Histoire Naturelle, Geneva). Paratypes: 2, Slovakia or. Lanartov u Bard. O. Kavan leg./ (in Manchester University Museum, MMUE). Note: this species may be readily distinguished (e. g., as characterized in Reška, 1994).

Cryptophagus lyubarskyi Johnson, **sp. nov.** for *C. robustus* sensu Lyubarsky, 1999 (misidentification; nec *C. robustus* Bruce, 1959). Holotype female: /Dolpo, Ringmo, Phoksumdo-See 10.-15.vi.1973, 3600-3900 m/, /NEPAL-Expeditionen, Jochen Martens/, /Cryptophagus robustus Bruce det. Lyubarsky/ (in Staatliches Museum für Naturkunde, Stuttgart).

Micrambe (*Micrambe*) *woodroffei* Johnson, **sp. nov.** for *M.* (*M.*) *villosus* sensu auctorum. Holotype male: Harz German/, /Ex.coll. J. Klapperich/, /villosa Heer N. Bruce det./ Manchester Museum Holotype/ (in Manchester University Museum, MMUE).Paratypes: 1 male, same data; 1 female, Italy, Mte Gargano, Ex.coll. J. Klapperich; 1 female, Portugal, Coimbra, 27.5.1959, leg. Lindberg; 1 female, Turkey, Korikos, Silifka, 25.4.1967, J. Klapperich; 1 female, same data, Ephes, 2.5.1967 (in Manchester University Museum, MMUE). Note: This species is characterized in many works and may be readily distinguished (see Reška 1994).

Resurrections

Antherophagus similis Curtis, 1835 resurrected from synonymy with *A. pallens* (Linné, 1758).
Cryptophagus ariadne Otero & González, 1980 resurrected from synonymy with *C. pallidus* Sturm, 1845
Cryptophagus gracilis Reitter, 1875 resurrected from synonymy with *C. baldensis* Erichson, 1846
Cryptophagus insulicola Roubal, 1919 resurrected from synonymy with *C. dentatus* (Herbst 1793)
Cryptophagus jakowlewi Reitter, 1888 resurrected from synonymy with *C. pallidus* Sturm, 1845
Cryptophagus puncticollis P. H. Lucas, 1846 resurrected from synonymy with *C. pilosus* (Gyllenhal, 1827)
Cryptophagus punctipennis C. N. F. Brisout de Barneville, 1863 (= *C. pilosus* sensu auctorum) resurrected from synonymy with *C. pilosus* (Gyllenhal, 1827)
Cryptophagus reflexus Rey, 1889 (= *C. pallidus* 'sensu auctorum) resurrected from synonymy with *C. scanicus* (Linnaeus, 1758)
Cryptophagus ulicis Stephens, 1830 resurrected from synonymy with *C. vini* (Panzer, 1797)
Cryptophagus uncinatus Stephens, 1830 resurrected from synonymy with *C. acutangulus* Gyllenhal, 1827
Himascelis nigratus Sengupta, 1978 resurrected from synonymy with *H. brunneus* Gyllenhal, 1827
Micrambe pilosula (Erichson, 1846) resurrected from synonymy with *M. villosus* Sengupta, 1978
Micrambinus Reitter, 1906 resurrected as valid subgenus from synonymy with *Micrambe* C. G. Thomson, 1863

Changes in rank

Micrambe occidentalis (Wollaston, 1863) upgraded to specific rank, from subspecies of *M. hesperia* (Wollaston, 1863)

New combination

Spaniophaenus fraudulentus (Lyubarsky, 1997) **comb. nov.** from *Micrambe*

New synonyms

Antherophagus nigricornis (Fabricius, 1787) **syn. nov.** of *Antherophagus pallens* (Linnaeus, 1758). Based on descriptions.
Antherophagus nigricornis var. *fasciatus* Roubal, 1935 **syn. nov.** of *Antherophagus pallens* (Linnaeus, 1758). Based on descriptions.
Antherophagus nigricornis var. *fasciferus* Roubal, 1935 **syn. nov.** of *Antherophagus pallens* (Linnaeus, 1758). Based on descriptions.
Antherophagus nigricornis var. *maculipennis* A. Villa & G. B. Villa, 1833 **syn. nov.** of *Antherophagus pallens* (Linnaeus, 1758). Based on descriptions.
Atomaria angellata Lyubarsky, 1995 **syn. nov.** of *Atomaria* (*Anchicera*) *obliqua* Johnson, 1971
Atomaria attila var. *flavicolor* Roubal, 1931 **syn. nov.** of *Atomaria* (*Anchicera*) *attila* Reitter, 1878
Atomaria barani C. N. F. Brisout de Barneville, 1863 **syn. nov.** of *Atomaria* (*Atomaria*) *pulchra* Erichson, 1846
Atomaria frugi Lyubarsky, 1997 **syn. nov.** of *Atomaria* (*Anchicera*) *plecta* Lyubarsky, 1995
Atomaria linearis var. *coriacea* Petri, 1926 **syn. nov.** of *Atomaria* (*Atomaria*) *linearis* Stephens, 1830. Based on descriptions.
Atomaria linearis var. *distincticornis* Roubal, 1913 **syn. nov.** of *Atomaria* (*Atomaria*) *linearis* Stephens, 1830. Based on descriptions.
Atomaria lineola (Notman, 1920) **syn. nov.** of *Atomaria* (*Atomaria*) *wollastoni* Sharp, 1867
Atomaria melanica Hatch, 1962 **syn. nov.** of *Atomaria* (*Atomaria*) *atrata* Reitter, 1875
Atomaria mentita Lyubarsky, 1999 **syn. nov.** of *Atomaria* (*Anchicera*) *khumbuensis* Johnson, 1971
Atomaria ochracea Zimmermann, 1869 **syn. nov.** of *Atomaria* (*Anchicera*) *fuscata* (Schönherr, 1808)
Atomaria pilosula Wollaston, 1864 **syn. nov.** of *Atomaria* (*Anchicera*) *fasciata* Kolenati, 1846
Atomaria semitestacea Reitter, 1888 **syn. nov.** of *Atomaria* (*Anchicera*) *fulvipennis* Mannerheim, 1846
Cryptophagus acuminatus Coombs & Woodroffe, 1955 **syn. nov.** of *Cryptophagus insulicola* Roubal, 1919
Cryptophagus albipilus Rey, 1889 **syn. nov.** of *Cryptophagus pilosus* Gyllenhal, 1827
Cryptophagus altaicus Bruce, 1938 **syn. nov.** of *Cryptophagus conicollis* Reitter, 1896
Cryptophagus amfidoxus Bruce, 1963 **syn. nov.** of *Cryptophagus hauseri* Reitter, 1890
Cryptophagus amplus Grouvelle, 1912 **syn. nov.** of *Cryptophagus saginatus* Sturm, 1845. Based on descriptions.
Cryptophagus bedeli Bruce, 1936 **syn. nov.** of *Cryptophagus gracilis* Reitter, 1875
Cryptophagus bengalensis Sengupta, 1980 **syn. nov.** of *C. johnsoni* Sengupta, 1980. Based on descriptions.
Cryptophagus bettae Macquart, 1831 **syn. nov.** of *Atomaria* (*Atomaria*) *linearis* Stephens, 1830. Based on descriptions.
Cryptophagus bipedes Bruce, 1963 **syn. nov.** of *Cryptophagus dilatipennis* Reitter, 1888
Cryptophagus callosipennis Grouvelle, 1919 **syn. nov.** of *Cryptophagus decoratus* Reitter, 1874
Cryptophagus carinatus Bruce, 1936 **syn. nov.** of *Cryptophagus circassicus* Reitter, 1888. Based on descriptions.
Cryptophagus confertus Casey, 1900 **syn. nov.** of *Cryptophagus jakowlewi* Reitter, 1888d.
Cryptophagus croaticus var. *tridentatus* Roubal, 1936 **syn. nov.** of *Cryptophagus montanus* C. N. F. Brisout de Barneville, 1863. Based on descriptions.
Cryptophagus deceptor Grouvelle, 1919 **syn. nov.** of *Cryptophagus dilutus* Reitter, 1874
Cryptophagus demarzoi Otero & Angelini, 1981 **syn. nov.** of *Cryptophagus badius* Sturm, 1845
Cryptophagus domesticus Hubenthal, 1920 **syn. nov.** of *Cryptophagus pubescens* Sturm, 1845. Based on descriptions.
Cryptophagus elegans Grouvelle, 1919 **syn. nov.** of *Cryptophagus decoratus* Reitter, 1874
Cryptophagus erichsoni Reitter, 1888, doubtful **syn. nov.** of *Cryptophagus intermedius* Bruce, 1934b. Based on descriptions.
Cryptophagus flavipennis Faldermann, 1838 doubtful **syn. nov.** of *Cryptophagus dilutus* Reitter, 1874. Based on descriptions.
Cryptophagus freyi Bruce, 1941 **syn. nov.** of *Cryptophagus reflexus* Rey, 1889

Cryptophagus galilei Otero, 1997 **syn. nov.** of *Cryptophagus laticollis* P. H. Lucas, 1846
Cryptophagus heteroclitus Lyubarsky, 1997 **syn. nov.** of *Cryptophagus castanescens* Grouvelle, 1916
Cryptophagus hirtulus Kraatz, 1858 **syn. nov.** of *Cryptophagus uncinatus* Stephens, 1830
Cryptophagus impressicollis Tournier, 1872 **syn. nov.** of *Cryptophagus acutangulus* Gyllenhal, 1827
Cryptophagus kamtschaticus Lyubarsky, 1992 **syn. nov.** of *Cryptophagus bidentatus* Mäklin, 1853. Based on descriptions.
Cryptophagus kochi Bruce, 1943 **syn. nov.** of *Cryptophagus quadrimaculatus* Reitter, 1877
Cryptophagus lateralis Rey, 1889 **syn. nov.** of *Cryptophagus reflexus* Rey, 1889
Cryptophagus laticollis L. Miller, 1858 **syn. nov.** of *Catopochrotides termitophilus* Kieseritzky & Reichardt, 1936. Based on descriptions.
Cryptophagus lindbergorum Bruce, 1934 **syn. nov.** of *Paramecosoma pilosula* Erichson, 1846
Cryptophagus lomus Sengupta, 1980 **syn. nov.** of *Dermestes cellaris* Scopoli, 1763. Based on descriptions.
Cryptophagus longipennis Grouvelle, 1919 **syn. nov.** of *Cryptophagus decoratus* Reitter, 1874
Cryptophagus longiventris Grouvelle, 1919 **syn. nov.** of *Cryptophagus latangulus* Reitter 1889
Cryptophagus ludovici Breit, 1908 **syn. nov.** of *Cryptophagus durus* Reitter, 1878
Cryptophagus madadorum Roubal, 1945 **syn. nov.** of *Cryptophagus montanus* C. N. F. Brisout de Barneville, 1863
Cryptophagus marshami Stephens, 1830 **syn. nov.** of *Dermestes lycoperdi* Scopoli, 1763
Cryptophagus martensi Sengupta, 1980 **syn. nov.** of *Cryptophagus johnsoni* Sengupta, 1980
Cryptophagus mazetieri Falcoz, 1929 **syn. nov.** of *Cryptophagus setulosus* Sturm, 1845
Cryptophagus niger C. N. F. Brisout de Barneville, 1863 **syn. nov.** of *Dermestes scanicus* Linnaeus, 1758
Cryptophagus ophthalmicus Bruce, 1938 **syn. nov.** of *Cryptophagus quadrimaculatus* Reitter, 1877
Cryptophagus outereloi Otero & González, 1983 **syn. nov.** of *Cryptophagus skalitzkyi* Reitter, 1875
Cryptophagus podolkai Roubal, 1916 **syn. nov.** of *Cryptophagus circassicus* Reitter, 1888. Based on descriptions.
Cryptophagus posticus Reitter, 1888 **syn. nov.** of *Cryptophagus axillaris* Reitter, 1875. Based on descriptions.
Cryptophagus reichardti Bruce, 1936 **syn. nov.** of *Cryptophagus dorsalis* C. Sahlberg, 1819. Based on descriptions.
Cryptophagus robustus Bruce, 1959 **syn. nov.** of *Cryptophagus himalaicus* Bruce, 1953
Cryptophagus rotundatus Coombs & Woodroffe, 1955 **syn. nov.** of *Cryptophagus puncticollis* P. H. Lucas, 1846.
Cryptophagus rufulus J. R. Sahlberg,1926 **syn. nov.** of *Cryptophagus lapponicus* Gyllenhal, 1827. Based on descriptions.
Cryptophagus rutae Bruce, 1940 **syn. nov.** of *Cryptophagus nitiduloides* Wollaston, 1854
Cryptophagus schuberti Bruce, 1963 **syn. nov.** of *Cryptophagus dilatipennis* Reitter, 1888
Cryptophagus sculpturatus Bruce, 1941 **syn. nov.** of *Cryptophagus japonicus* Reitter, 1874
Cryptophagus subelongatus Rey, 1889 **syn. nov.** of *Cryptophagus puncticollis* P. H. Lucas, 1846
Cryptophagus subinermis Grouvelle, 1919 **syn. nov.** of *Cryptophagus dilutus* Reitter, 1874.
Cryptophagus subtilis C. G. Thomson, 1873 **syn. nov.** of *Cryptophagus cellaris* (Scopoli, 1793). Based on descriptions.
Cryptophagus szekessyi Bruce, 1938 **syn. nov.** of *Cryptophagus hauseri* Reitter, 1890
Cryptophagus thoracicus J. R. Sahlberg,1903 **syn. nov.** of *Cryptophagus punctipennis* C. N. F. Brisout de Barneville, 1863
Cryptophagus uralensis Bruce, 1963 **syn. nov.** of *Cryptophagus jakowlewi* Reitter, 1888
Cryptophagus vagus Bruce, 1938 **syn. nov.** of *Cryptophagus insulicola* Roubal, 1919
Cryptophagus varians Grouvelle, 1919 **syn. nov.** of *Cryptophagus decoratus* Reitter, 1874
Cryptophagus vaulogeri Grouvelle, 1912 **syn. nov.** of *Cryptophagus punctipennis* C. N. F. Brisout de Barneville, 1863. Based on descriptions.
Cryptophagus wollastoni Bruce, 1941 **syn. nov.** of *Micrambe hesperia* (Wollaston, 1863)
Dermestes vini Panzer, 1797 [HN] **syn. nov.** of *Micrambe ulicis* (Stephens, 1830)
Ephistemus punctatus Blatchley, 1925 **syn. nov.** of *Curelius japonicus* (Reitter, 1878). Based on descriptions.
Henoticus japonicus Nakane & Hisamatsu, 1963 **syn. nov.** of *Henoticus sinensis* Bruce, 1943
Henoticus serratus nepalensis Johnson, 1975 **syn. nov.** of *Henoticus sinensis* Bruce, 1943
Himascelis gracilis Lyubarsky, 1999 **syn. nov.** of *Himascelis brunneus* Sengupta, 1978

Himascelis nepalensis Lyubarsky, 1999 **syn. nov.** of *Himascelis nigratus* Sengupta, 1978
Hypocoprus formicetorum Motschulsky, 1839 **syn. nov.** of *Hypocoprus latridioides* Motschulsky, 1839
Hypocoprus latridioides var. *pici* Lavagne, 1920 **syn. nov.** of *Hypocoprus latridioides* Motschulsky, 1839. Based on descriptions.
Micrambe duclouxi Grouvelle, 1910 **syn. nov.** of *Micrambinus bimaculatus* (Panzer, 1798). Based on descriptions.
Paramecosoma infuscatum Halbert, 1910 **syn. nov.** of *Paramecosoma melanocephalum* (Herbst, 1793). Based on descriptions.
Sternodea dichroa Reitter, 1896 **syn. nov.** of *Sternodea lederi* Reitter, 1876
Sternodea haroldi Reitter, 1876 **syn. nov.** of *Sternodea lederi* Reitter, 1876. Based on descriptions.
Telmatophilus bicolor Reitter, 1913 **syn. nov.** of *Telmatophilus schonherrii* (Gyllenhal, 1808)
Thelmatophilus balcanicus Karaman, 1960 **syn. nov.** of *Telmatophilus brevicollis* Aubé, 1862. Based on descriptions.

Designations of lectotypes

Cryptophagus acutangulus Gyllenhal, 1827. Lectotype female [long-pinned through right elytron]: [blank, dirty pale square]/ /"a" [red label]/ /Uppsala Univ. Zool. Mus. Gyllenhals saml. TYP nr 1326 [black print on red label]/ (in University Museum of Evolution, Zoology, Uppsala). Paralectotype: 1 female, /"b"/ same labels (in University Museum of Evolution, Zoology, Uppsala).
Cryptophagus marshami Stephens, 1830. Lectotype female [short-pinned through right elytron and staged]: Marshami/ /BRITISH ISLES, J. Stephens Coll., BM 1853 – 46/ /Standing as Cryptophagus marshami in Stephens Coll./ (in Natural History Museum, London).
Cryptophagus pilosus Gyllenhal, 1827. Lectotype female [long-pinned through right elytron]: [blank dirty pale square]/ /"c" [red label]/ /Uppsala Univ. Zool. Mus. Gyllenhals saml. TYP nr 1327 [black print on red label]/ (in University Museum of Evolution, Zoology, Uppsala). Paralectotypes: 4 females: "a"/ /18./ /[blank dirty pale square]/; /"d"/ /[blank dirty pale square]/, /"e"/, /"g"/ /[blank dirty pale square]/ all with same label as lectotype (in University Museum of Evolution, Zoology, Uppsala).
Cryptophagus subdepressus Gyllenhal, 1827. Lectotype female [long-pinned through right elytron]: [blank dirty whitish square]/ /"c" [red label]/ /Uppsala Univ. Zool. Mus. Gyllenhals saml. TYP nr 1328 [black print on red label]/ (in University Museum of Evolution, Zoology, Uppsala).
Cryptophagus uncinatus Stephens, 1830. Lectotype female [glued on small narrow point and staged]: BRITISH ISLES, J. Stephens Coll., BM 1853 – 46/ /Standing as Cryptophagus uncinatus in Stephens Coll./ (in Natural History Museum, London). Paralectotype: 1 female, same labels as lectotype (in Natural History Museum, London).
Kateretes dentatus Herbst, 1793. Lectotype male [carded and aedeagus dissected out]: /Doublette von Erichson/ /Lectotypus Kateretes dentatus HERBST, 1793 design. Esser 2004/ (in Museum für Naturkunde, Berlin). Paralectotypes: see Esser (in press).
Tenebrio pallens Linné, 1758. Lectotype female [lacking most antennae, pinned through right elytron and staged]: /9 pallens/ [Linnaeus's hand], /10/ [recent printed white label] (In Linnean Society of London).

Cryptophagidae

C. Johnson & J. Esser

Note. New taxonomic and nomenclatural acts are based upon type specimens, dissected where appropriate.

Resurrection

Antherophagus silaceus (Herbst, 1792), as valid species, from synonymy of *Antherophagus nigricornis* (Fabricius, 1787)

New synonyms

Antherophagus canescens Grouvelle, 1916 **syn. nov.** of *Antherophagus silacea* (Herbst 1792)
Atomaria cognata Erichson, 1846 **syn. nov.** of *Atomaria* (*Anchicera*) *analis* Erichson, 1846
Cryptophagus rufus Waltl, 1838 **syn. nov.** of *Atomaria* (*Atomaria*) *zetterstedti* (Zetterstedt, 1838)

Cryptophagus thomsoni Reitter, 1875 **syn. nov.** of *Cryptophagus pallidus* Sturm, 1845

Cryptophagidae

C. Johnson & R. A. B. Leschen

Note. All Rey types (in Lyon, France) were identified by Bruce during the 1950s and his notes form the basis for new synonyms. Grouvelle and Casey types were examined by Leschen in Paris and Washington.

New synonyms

Cryptophagus alpestris Rey, 1889 **syn. nov.** of *Cryptophagus lapponicus* Gyllenhal, 1827
Cryptophagus convexus Rey, 1889 **syn. nov.** of *Cryptophagus distinguendus* Sturm, 1845
Cryptophagus corticum Rey, 1889 **syn. nov.** of *Cryptophagus pilosus* Gyllenhal, 1827
Cryptophagus curtulus Rey, 1889 **syn. nov.** of *Cryptophagus acutangulus* Gyllenhal, 1827
Cryptophagus germaini Grouvelle, 1905 **syn. nov.** of *Cryptophagus saginatus* Sturm, 1845
Cryptophagus obtusidens Rey, 1889 **syn. nov.** of *Cryptophagus pilosus* Gyllenhal, 1827
Cryptophagus rugicollis Rey, 1889 **syn. nov.** of *Cryptophagus distinguendus* Sturm, 1845
Cryptophagus subhirtus Rey, 1889 **syn. nov.** of *Cryptophagus distinguendus* Sturm, 1845
Cryptophagus subinermis Rey, 1889 **syn. nov.** of *Cryptophagus pallidus* Sturm, 1845
Cryptophagus substriatus Rey, 1889 **syn. nov.** of *Cryptophagus pilosus* Gyllenhal, 1827
Cryptophagus suturalis Rey, 1889 **syn. nov.** of *Cryptophagus scanicus* (Linnaeus, 1758)
Cryptophagus tamaricis Rey, 1889 **syn. nov.** of *Cryptophagus pilosus* Gyllenhal, 1827
Cryptophagus verrucifer Rey, 1889 **syn. nov.** of *Cryptophagus distinguendus* Sturm, 1845
Cryptophagus vestitus Rey, 1889 **syn. nov.** of *Cryptophagus distinguendus* Sturm, 1845
Hypocoprus tenuis Casey, 1916 **syn. nov.** of *Hypocoprus latridioides* Motschulsky, 1839
Pteryngium duclouxi Grouvelle, 1916 **syn. nov.** of *Micrambe bimaculatus* (Panzer, 1798)

Cryptophagidae

C. Johnson & J. C. Otero

Note. New taxonomic and nomenclatural acts are based upon type specimens, dissected where appropriate.

New synonyms

Cryptophagus bimacularis Grouvelle, 1916 **syn. nov.** of *Cryptophagus castanescens* Grouvelle, 1916
Cryptophagus binotatus Grouvelle, 1916 **syn. nov.** of *Cryptophagus castanescens* Grouvelle, 1916
Cryptophagus malaisei Bruce, 1944 **syn. nov.** of *Cryptophagus curtus* Grouvelle, 1916
Cryptophagus subnudus Grouvelle, 1916 **syn. nov.** of *C. auropubens* Grouvelle, 1916
Cryptophagus vicinus Grouvelle, 1916 is tentatively placed in synonymy of *Cryptophagus castanescens* Grouvelle, 1916, it was erroneously placed in synonymy of *Cryptophagus pumilus* Reitter, 1875

Corylophidae/Cryptophagidae

C. Johnson

New synonymy

Ephistemus nigriclavis Stephens, 1829 **syn. nov.** of *Orthoperus brunnipes* (Gyllenhal, 1808).
Lectotype [glued on small oblong card and staged]: BRITISH ISLES, J. Stephens Coll., BM 1853 – 46/ /Standing as *Ephistemus nigriclavis* in Stephens Coll./ (in Natural History Museum, London).

Mycetophagidae/Cryptophagidae

C. Johnson

New synonym

Cryptophagus humeralis Stephens, 1830 **syn. nov.** of *Pseudotriphyllus suturalis* (Fabricius, 1801) (Mycetophagidae). Syntypes in Natural History Museum, London examined.

Endomychidae: Merophysiinae
W. Rücker

Holoparamecus azoricus Méquignon, 1942 **syn. nov.** of *H. singularis* (Beck, 1817)

Endomychidae: Lycoperdininae and Stenotarsinae
W. K. Tomaszewska

Lycoperdina bovistae var. *leprieuri* Pic, 1943 **syn. nov.** of *Lycoperdina bovistae* Fabricius, 1792
Lycoperdina bovistae var. *rufescens* Pic, 1943 **syn. nov.** of *Lycoperdina bovistae* Fabricius, 1792
Perrisina brevis var. *theresae* Pic, 1944 **syn. nov.** of *Perrisina brevis* (Perris, 1864)

Coccinellidae
I. Kovář

New replacement names
Calvia sichuanica **nom. nov.** for *C. sicardi* Mader, 1930, preoccupied by *C. sicardi* (Nunenmacher, 1912) (originally assigned to *Agrabia*)
Nephus (s. str.) *ganglbaueri* **nom. nov.** for *N. pictus* Ganglbauer, 1899, preoccupied by *N. pictus* (Gorham, 1897) (originally assigned to *Scymnus*)
Pseudaspidimerus hoangi **nom. nov.** for *P. limbatus* Hoang, 1982, preoccupied by *P limbatus* Kapur, 1948
Scymnus kapuri **nom. nov.** for *S. rubicundus* (Kapur, 1951), (originally assigned to *Pullus*), preoccupied by *S. rubicundus* Erichson, 1847
Scymnus marinellus **nom. nov.** for *S. marinus* H. Kamiya, 1961, preoccupied by *S. marinus* (Mulsant, 1850) (originally assigned to *Rhizobius*). The type material is in the coll. Helfer, National Museum in Prague and was studied.

Priorities
Scymnus (s. str.) *frontalis* (Fabricius, 1787) is a junior synonym of *S. alticus* (Schrank, 1781). The synonymy of the two names was established by Schönherr, 1808: 204. However, the senior name was not used as valid while the junior name was consistently considered as valid. As the condition to declare *Scymnus frontalis* (Fabricius, 1787) a nomen protectum are not met, the case is to be submitted to the International Commission of Zool. Nomenclature.
Scymnus (s. str.) *nigrinus* Kugelann, 1794 is a junior synonym of *S. minimus* (O. F. Müller, 1776). The synonymy of the two names was given already by Schönherr, 1808: 207 and Stephens, 1829: 238. The senior name was not used as valid while the junior name was consistently considered as valid. As the condition to declare *Scymnus nigrinus* Kugelann, 1794 a nomen protectum are not met, the case is to be submitted to the International Commission of Zool. Nomenclature.
Scymnus (*Pullus*) *suturalis* Thunberg, 1781 is a junior synonym of *S. pusillus* (O. F. Müller, 1776). The synonymy of the two names was established already by Mulsant, 1846: 256. The senior name was not used as valid while the junior name was consistently considered as valid. As the condition to declare *Scymnus suturalis* Thunberg, 1781 a nomen protectum are not met, the case is to be submitted to the International Commission of Zool. Nomenclature.

Changes in rank
Scymnus (s. str.) *magnomaculatus* Fürsch, 1958 from *S. frontalis* f. *magnomaculatus* Fürsch, 1958

New combinations
Aspidimerus blandus (Mader, 1933) **comb. nov.** from *Cryptogonus*
Aspidimerus nigritus (X. Pang & Mao, 1979) **comb. nov.** from *Cryptogonus*
Exochomus discors (Barovskiy, 1927) **comb. nov.** from *Brumus*
Exochomus gebleri J. Weise, 1885 **comb. nov.** from *Brumus*

Exochomus jacobsoni (Barovskiy, 1927) **comb. nov.** from *Brumus*
Exochomus mongolicus (A. Fleischer, 1900) **comb. nov.** from *Brumus*
Exochomus octosignatus (Gebler, 1830) **comb. nov.** from *Brumus*
Illeis nevili (Dohrn, 1882: 373) **comb. nov.** from *Bulaea*
Paracoelopterus berytensis (J. Weise, 1884) **comb. nov.** from *Pharus*
Nephus (s. str.) *jacobsoni* (Barovskiy, 1906) **comb. nov.** from *Scymnus*
Oenopia excelens (Crotch, 1874) **comb. nov.** from *Dysis*
Phymatosternus huangea (Cao & Xiao, 1984), from *Platynaspis*
Platynaspidius angulimaculatus (Mader, 1938) **comb. nov.** from *Platynaspis*
Platynaspidius sexmaculatus (Cao & Xiao, 1984) **comb. nov.** from *Platynaspis*
Platynaspidius variegatus (Crotch, 1874) **comb. nov.** from *Platynaspis*, **comb. n.**
Scymnus (*Mimopullus*) *marinus* (Mulsant, 1850) **comb. nov.** from *Rhizobius* (= *Rhyzobius*)
Scymniscus anomus (Mulsant & Rey, 1852) **comb. nov.** from *Nephus* (*Sidis*)
Scymniscus armeniacus (Canepari, 1979) **comb. nov.** from *Nephus* (*Sidis*)
Scymniscus biflammulatus (Motschulsky, 1837) **comb. nov.** from *Nephus* (*Sidis*)
Scymniscus biguttatus (Mulsant, 1850) **comb. nov.** from *Nephus* (*Sidis*)
Scymniscus castanopterus (Mulsant, 1850) **comb. nov.** from *Scymnus*
Scymniscus fuerschi (Plaza, 1981) **comb. nov.** from *Nephus* (*Sidis*))
Scymniscus helgae (Fürsch, 1965) **comb. nov.** from *Nephus* (*Sidis*)
Scymniscus horioni (Fürsch, 1965) **comb. nov.** from *Nephus* (*Sidis*)
Scymniscus kahleni (Fürsch, 1997) **comb. nov.** from *Nephus* (*Sidis*)
Scymniscus macilentus (Fürsch, 1965) **comb. nov.** from *Nephus* (*Sidis*)
Scymniscus mainanderi (Fürsch, 1965) **comb. nov.** from *Nephus* (*Sidis*)
Scymnuscus marikovski (Savoiskaya, 1971) **comb. nov.** from *Scymnus*
Scymniscus tristiculus (J. Weise, 1929) **comb. nov.** from *Nephus* (*Sidis*)
Scymniscus spilotus (J. Weise, 1901) **comb. nov.** from *Nephus* (*Sidis*)
Scymniscus splendidulus (Steinius, 1952) **comb. nov.** from *Nephus* (*Sidis*)
Stethorus pusillus (Herbst, 1897) **comb. nov.** from *Scymnus*
Tetrabrachys arnoldi (Iablokoff-Khnzorian, 1969) **comb. nov.** from *Lithophilus*
Tetrabrachys klapperichi (Iablokoff-Khnzorian, 1974) **comb. nov.** from *Lithophilus*
Tetrabrachys naviauxi (Duverger, 1983) **comb. nov.** from *Lithophilus*
Tetrabrachys rifensis (Iablokoff-Khnzorian, 1974) **comb. nov.** from *Lithophilus*
Tetrabrachys smarti (Iablokoff-Khnzorian, 1974) **comb. nov.** from *Lithophilus*

Reconfirmed combinations
Exochomus cedri J. R. Sahlberg,1913b, from *Brumus*
Exochomus kiritshenkoi Barovskiy, 1922 from *Brumus*
Exochomus mongol Barovskiy, 1922 from *Brumus*
Exochomus quadriguttatus A. Fleischer, 1900 from *Brumus*
Exochomus quadripustulatus (Linnaeus, 1758), from *Brumus*
Exochomus septemmaculatus J. Weise, 1885, from *Brumus*
Exochomus undulatus J. Weise, 1878, from *Brumu*

New synonyms:
Diloponis Pope, 1962 **syn. nov.** of *Paracoelopterus* Normand, 1936
Diloponis fuerschi Yazdani & Ahmadi, 1992 **syn. nov.** of *Paracoelopterus berytensis* (J. Weise, 1884)
Adonia Mulsant, 1846 **syn. nov.** of *Hippodamia* (s. str.) Dejean, 1836 [nec Chevrolat, 1836, through type species fixation]
Anchilleis Iablokoff-Khnzorian, 1979 **syn. nov.** of *Macroilleis* Miyatake, 1965
Mononeda (*Paramicraspis*) Hoang, 1983 **syn. nov.** of *Micraspis* Dejean, 1836 [nec Chevrolat, 1836]
Nedina Hoang, 1983 **syn. n.** of *Protothea* J. Weise, 1898
Paraplatynaspis Hoang, 1983 **syn. nov.** of *Platynaspidius* Miyatake, 1961
Barovskya Iablokoff-Khnzorian, 1979 **syn. nov.** of *Tytthaspis* Crotch, 1874
Anisolemnia laichauana Hoang, 1983 **syn. nov.** of *Coelophora duvaucelii* (Mulsant, 1850a)

Aspidimerus dongpaoensis Hoang, 1982 **syn. nov.** of *Aspidimerus nigritus* X. Pang & Mao, 1979
Aspidimerus sexmaculatus X. Pang & Mao, 1979 **syn. nov.** of *Aspidimerus mouhotii* Crotch, 1874
Coccinella caucasica Motschulsky, 1837 **syn. nov.** of *Oenopia impustulata* (Linnaeus, 1758) [exhibits gradual variability]
Diloponis fuerschi Yazdani & Ahmadi, 1992 **syn. nov.** of *Paracoelopterus berytensis* (J. Weise, 1884)
Exochomus saharae Sicard, 1929c **syn. nov.** of *Parexochomus pubescens* (Küster, 1848) [type material studied]
Nephus patagiatus montanus Hoang, 1982 **syn. nov.** of *Nephus levaillanti* (Mulsant, 1850)
Nephus rutaneni Fürsch, 1986 **syn. nov.** of *Nephus jacobsoni* (Barovskiy, 1906)
Oeniopia guttata (T. Blackburn, 1892) **syn. nov.** of *Oenopia excelens* Crotch, 1874
Paracoelopterus rufinus Normand, 1936 **syn. nov.** of *Paracoelopterus berytensis* (J. Weise, 1884)
Platynaspis bimaculatus X. Pang & Mao, 1979 **syn. nov.** of *Platynaspidius variegatus* (Crotch, 1874)
Paraplatynaspis bimaculatus Hoang, 1983 **syn. nov.** of *Platynaspidius variegatus* (Crotch, 1874)
Scymnus bellophalus Capra, 1925 **syn. nov.** of *Scymnus latemaculatus* Motschulsky, 1858
Scymnus changajensis Bielawski, 1965 **syn. nov.** of *Scymniscus zaisanensis* Dobrzhanskiy, 1926
Scymnus obsoletus J. Weise, 1890 **syn. nov.** of *Scymnus castanopterus* Mulsant, 1850
Scymnus silesiacus lipicensis G. Müller, 1924 **syn. nov.** of *Scymnus* (*Pullus*) *silesiacus* J. Weise, 1902

Valid names for particular species

Chilocorus bivulnerus Mulsant, 1850 is a **valid name** for *stigma* Say, 1835, preoccupied by A. G. Olivier, 1808 (in *Coccinella*) **NAR**
Diomus messorius Brèthes, 1925 is a **valid name** for *pusillus* Berg, 1874, preoccupied by Herbst, 1797 (in *Scymnus*) **NTR**
Hippodamia (*Hemispherica*) *lengi* Johnson, 1910 is a **valid name** for *apicalis* Casey, 1899, preoccupied by Wimmel, 1897 **NAR**
Epilachna immaculicollis Chevrolat, 1834 is a **valid name** for *tredecimnotata* Latreille, 1833 (*Epilachna*), preoccupied by Thunberg, 1784 (*Coccinella*, now *Epilachna*) **NAR, NTR**
Hyperaspis maculifera Melsheimer, 1847 is a **valid name** for *undulata* Say, 1824, preoccupied by Thunberg, 1781, Schaler, 1783, and Herbst, 1786 (in *Coccinella*) **NAR**
Psyllobora octodecimmaculata Thunberg, 1795 is a **valid name** for *variegata* Fabricius, 1781, preoccupied by Goeze, 1777 (in *Coccinella*) **AFR**.
Scymnus (s. str.) *magnomaculatus* Fürsch, 1958 is a **valid name** for *quadriguttatus* Capra, 1924, preoccupied by A. Fleischer, 1900 (in *Scymnus*) **PAR**

Comments

Coccinella nepalensis Iablokoff-Khnzorian, 1982 is extralimital. Its description was based on material of *Lioadalia flavomaculata* (DeGeer, 1778), stated to be from Nepal but probably from Natal.

Introduced species

Following species are now established in the Palaearctic region, as defined in the present Catalogue:
Cryptolaemus montrouzieri Mulsant, 1853, originally from **AUR**
Delphastus catalinae (Horn, 1895), originally from **NAR, NTR**
Lindorus forestieri (Mulsant, 1853), originally from **AUR**
Lindorus lophantae (Blaisdell, 1892), originally from **AUR**
Serangium montazeri Fürsch, 1995, originally from **NAR**
Serangium parcesetosum Sicard, 192, originally from **ORR**
Sticholotis madagassa J. Weise, 1909, originally from **AFR**
Nephus (*Geminosipho*) *reunioni* (Fürsch, 1974), originally from **AFR**
Rodolia cardinalis (Mulsant, 1850), originally from **AUR**

Following species were introduced to Europe but may not be established:

Chilocorus bivulnerus Mulsant, 1850 (as *C. stigma* Say, 1835), **E**: FR
Chilocorus kuwanae Silvestri, 1909, **E**: FR IT
Delphastus catalinae (Horn, 1895), **E**: GB NL
Harmonia axyridis (Pallas, 1775), **E**: AZ CZ GE
Hippodamia glacialis (Fabricius, 1775), **E**: SV
Lindorus lophantae (Blaisdell, 1892), **E**: BE
Nephus (Bipunctatus) conjunctus (Wollaston, 1870), **E**: NL

Species reported from, but not established in the Palaearctic region, as defined in the present Catalogue:
Chilocorus cacti (Linnaeus, 1767), **N**: MO from **NAR NTR**
Harmonia octomaculata (Fabricius, 1775), **E**: FI originally from **AUR** and **ORR**
Hippodamia convergens (Guérin-Méneville, 1844), **E**: FR IT from **NAR** and **NTR**
Hippodamia glacialis (Fabricius, 1775), **E**: SV from **NAR**
Eriopis connexa (Germar, 1824), **E**: GB SV from **NTR**; the material from GB is likely *Eriopis connexa chilensis* Hofmann, 1972
Hyperaspis arrowi Brèthes, 1925, **N**: CI from **NTR**
Hyperaspis silvestrii J. Weise, 1909, **E**: IT from **NTR**
Platynaspis atrata (Thunberg, 1820) [as *Platynaspis pilosa* Sicard, 1930], **N**: MO from **AFR**
Procula douei Mulsant, 1850, **E**: GB from **NTR**
Psyllobora (Psyllobora) vigintimaculata (Say, 1824), **E**: GE from **NAR**

Corylophidae

S. Bowestead

Comments

The type material of east Palaearctic species described by Reitter, Matthews and Paulian was examined (see Bowestead, 1999; 2003; Leschen and Bowestead, 2001). Species described by Nakane have been interpreted from the coloured plates in Nakane (1963), since the types were not available for study.

New combinations

Clypastraea polita (Reitter, 1878) **comb. nov.** from *Arthrolips*, based on study of type material.
Arthrolips brittoni (Paulian, 1948) **comb. nov.** from *Sacium*, based on study of type material.
Arthrolips galloisis (Paulian, 1962) **comb. nov.** from *Parmulus*, based on study of type material.
Arthrolips japonica (Paulian, 1962) **comb. nov.** from *Sacinops*, based on study of type material.
Arthrolips major (Paulian, 1962) **comb. nov.** from *Sacinops*, based on study of type material.

New synonyms

Sacium politum Matthews, 1899 **syn. nov.** of *Clypastraea polita* (Reitter, 1878), based on study of type material.
Corylophus japonicus Matthews, 1899 **syn. nov.** of *Corylophus cassidoides* (Marsham 1802), based on study of type material.

Latridiidae

C. Johnson

New nomenclatural and taxonomic acts are primarily based upon personal studies of type material, genitalia included. If type material is unknown or was exceptionally not available, then it is based on descriptions.

Designation of type species

Abothria Belon, 1897. This subgenus of *Corticaria* Marsham, 1802 is based upon Belon's group 9 and includes three species: *Latridius elongatus* Gyllenhal, 1827, *Corticaria japonica* Reitter 1877a and *C. ferruginea* Marsham, 1802. *Latridius elongatus* Gyllenhal, 1827 is here designated as the type species (*Corticaria elongata* (Gyllenhal, 1827)).

Adasia Belon, 1897. This subgenus of *Corticaria* Marsham, 1802 is based upon Belon's groups 5, 6 and 7 and includes *Ips impressa* A. G. Olivier, 1790 from group 5; *Corticaria saginata* Mannerheim, 1844, *Dermestes serratus* Paykull, 1798, *C. weisei* Reitter, 1875c and *C. obscura* C. N. F. Brisout de Barneville, 1863 from group 6; and 8 other species in group 7. *Dermestes serratus* Paykull, 1798 from group 6 is here designated as the type species (*Corticaria serrata* (Paykull, 1798)).

Cartoderema Reitter, 1911. This subgenus of *Cartodere* "sensu auctorum" is based on two species, *Latridius elongatus* Curtis, 1830 and *Corticaria ruficollis* Marsham, 1802. *Latridius elongatus* Curtis, 1830 [HN] is here designated as the type species (*Dienerella vincenti* Johnson, sp. nov.).

Corticarina Reitter, 1881 is based upon *Latridius similatus* Gyllenhal, 1827, *Melanophthalma ovalipennis* Reitter, 1875c, *L. fusculus* Hummel (= *fusculus* Gyllenhal, 1827), *L. fulvipes* Comolli, 1837 and *Corticaria truncatella* Mannerheim, 1844. *Corticarina truncatella* Mannerheim, 1844 is here designated as the type species (*Corticarina truncatella* (Mannerheim, 1844)).

Dienerella Reitter, 1911. This subgenus of *Cartodere* "sensu auctorum" is based on two species, *Lathridius elegans* Aubé, 1850 and *Cartodere pilifera* Reitter, 1875. *Lathridius elegans* Aubé, 1850 is here designated as the type species.

Isidius Motschulsky, 1866 is based upon *Lathridius gemellatus* Mannerheim, 1844, *L. quadricollis* Mannerheim, 1844, *L. sobrinus* Mannerheim, 1844, *L. protensicollis* Mannerheim, 1843 and two new species, *Isidius recticollis* Motschulsky and *I. kamtschaticus* Motschulsky. *Latridius protensicollis* Mannerheim, 1843 is here designated as the type species.

Melanopsis Gozis, 1875 is based upon *Latridius gibbosus* Herbst, 1793, *Corticaria hortensis* Motschulsky, 1867 (= Mannerheim, 1844), *brevicollis* (Villa ?) [?, uncertain identity], *Melanophthalma cylindricollis* Motschulsky, 1866 and *Latridius distinguendus* Villa (= Comolli, 1837). *Corticaria hortensis* Mannerheim, 1844 is here designated as the type species (*Melanophthalma transversalis* (Gyllenhal, 1827)).

Oropsime Gozis, 1881 is based upon *Corticaria psammeticha* Motschulsky, 1867, *C. subrugosa* Motschulsky, 1867, *C. pilitecta* Motschulsky, 1867 and *Melanophthalma carinulata* Motschulsky, 1867. *Melanophthalma carinulata* Motschulsky, 1867 is here designated as the type species (*Corticarina curta* (Wollaston, 1854)).

Permidius Motschulsky, 1866 is based upon 33 species, a mixture of old and new ones, including *Latridius rugosus* Herbst, 1793 on page 248. *Latridius rugosus* Herbst, 1793 is here designated as the type species (*Enicmus rugosus* (Herbst, 1793)).

Resurrections

Corticaria angulata Wollaston, 1871 from synonymy with *Melanophthalma taurica* (Mannerheim, 1844).

Corticaria curta Wollaston, 1854 from synonymy with *Corticarina fulvipes* (Comolli, 1837).

Corticaria parvicollis Mannerheim, 1844 from synonymy with *M. taurica* (Mannerheim, 1844).

Corticaria parvula Mannerheim, 1844 from synonymy with *Corticarina similata* (Gyllenhal, 1827).

Corticaria pharaonis Motschulsky, 1867 from synonymy with *C. fulva* (Comolli, 1837).

Corticaria psammeticha Motschulsky, 1867 from synonymy with *Migneauxia crassiuscula* (Aubé, 1850).

Corticaria quadrimaculata Mannerheim, 1844 from synonymy with *C. illaesa* Mannerheim, 1844.

Dermestes minutus Fabricius, 1792 from synonymy with *Cortinicara gibbosa* (Herbst, 1793) (*Corticarina minuta* (Fabricius, 1792)

Lathridius assimilis Mannerheim, 1844 from synonymy with *Latridius minutus* (Linné, 1767).

Lathridius caucasicus Mannerheim, 1844 from synonymy with *Stephostethus sinuatocollis* (Faldermann, 1837).

Lathridius gemellatus Mannerheim, 1844 from synonymy with *Latridius minutus* (Linné, 1767).

[*Lathridius reflexus* LeConte, 1855 from synonymy with *Latridius minutus* (Linné, 1767) (*Latridius minutus* 'sensu auctorum Americana'). Based on descriptions. **NAR**].

Latridius porcatus Herbst, 1793 upgraded to species from *Latridius minutus* (Linné, 1767).

Latridius transversalis Gyllenhal, 1827, from synonymy with *Cortinicara gibbosa* (Herbst, 1793) (*Melanophthalma transversalis* (Gyllenhal, 1827)).

Melanophthalma algirina Motschulsky, 1866 from synonymy with *M. taurica* (Mannerheim, 1844).

Melanophthalma extensa Rey, 1889 from synonymy with *M. taurica* (Mannerheim, 1844).

Neotype designation

Latridius transversalis Gyllenhal, 1827 (= *Melanophthalma transversalis* (Gyllenhal, 1827)):

Type material of this species, the type species of *Melanophthalma* Motschulsky, 1866, is lost, being missing from its old card mount and the inside of the drawer. For the stability of the nomenclature it is necessary to fix the name by a neotype. A dissected ♂ from Sweden is here designated as a neotype. It is labeled: S.Sk. Södra Sandby, 21.6.1966, G.Gillerfors/, /NEOTYPE/, /NEOTYPE ♂, Latridius transversalis Gyllenhal, 1827. det. C. Johnson 2005/ (to Gyllenhal's *Insecta Svecica* collection, University of Uppsala, Sweden)). Gyllenhal's name now takes precedence over *M. curticollis* (Mannerheim, 1844).

Lectotype designation

Dermestes longicornis Herbst, 1784.

Lectotype female [long-pinned through right elyton]: /53214 [printed]/, /impressa Oliv n.sp. longicornis m. Sahlby [handwritten]/, /longicornis Ht. Dermest. fenestra-lis Pk. Vix Fab [very large handwritten label with black border]/, /Hist.- Coll. (Coleoptera) Nr. 53214 Corticaria longicornis Herbst. German.? Zool. Mus. Berlin [small printed label with black-edged border]/, /Corticaria abietorum Motschulsky, 1867 det. W. Rücker/ (in Museum für Naturkunde, Berlin).

Paralectotypes [long-pinned through right elyton]: 2 females, /Hist.- Coll. 53214 German.?/, /Zool. Mus. Belin/, /Hist.- Coll. (Coleoptera) Nr. 53214 Corticaria longicornis Herbst. German.? Zool. Mus. Berlin [small printed label with black-edged border]/ (in Museum für Naturkunde, Berlin).

New species

Corticaria peezi Johnson, **sp. nov.** This species is *C. weisei* sensu auctorum, a misidentification; nec *weisei* Reitter, 1875).

Holotype male: [Greece] umg. Olympias, Chalkidike/, /N-Griechenland, Mazedonien, lg. H. Franz, Mai 1979/, /Manchester Museum HOLOTYPE/ (in Manchester University Museum, MMUE).

Paratypes: 3, same data; 2 males, 1 female, /GREECE: 14.v.84, Corinth, S.A.Williams/; 1 male, 3, /Graecia Parnass./, /CORTICARIA weisei Reitter det. W. Rücker, 1982/; 1 male, [Austria] /OST-TIROL, leg. A. Kofler/, /U-Prin.la..[?], 6.3.77/, /2 hi-Fö/, /Corticaria weisei Reitt., det. C. Johnson, 1980/ (all with Manchester Museum PARATYPE labels, in Manchester University Museum, MMUE). This species is characterized in many works, e.g. Peez (1967).

Corticaria porochini Johnson, **sp. nov.** This species is *C. longicornis* sensu auctorum, a misidentification; nec *longicornis* Herbst, 1793).

Holotype male: [Finland] Samorovo/, /Sundman/, /Corticaria longicornis Hbst. C.Johnson det./, /HOLOTYPE/ (in Zoological Museum of Helsinki University).

Paratypes: 1 female, /H:fors/, /J.Sahlb./, /C.longicornis/, /Corticaria longicornis Hbst. C.Johnson det./; 2, /Sakkola/, /U. Sahlb./; 1, /862/, /Suomi PK Hammas-lahti 12.7.1950, R. Linnavuori leg./, /longicornis det. A.Strand/; 1, ČSSR – Silesia or., Šenov u Ostravy, P. Nohel 4.XII.65/ (all with Manchester Museum PARATYPE labels, in Manchester University Museum, MMUE). This species is characterized in many works, e.g. Peez (1967).

Corticarina saluki Johnson, **sp. nov.** This species is *C. strandi* sensu Saluk, 1992 (a misidentification; nec *strandi* Johnson, 1972).

Holotype male: [Russia] S. Primorie, Anisimovka, 60 km W of Partizansk, 21.VII.1988 in litter, lgt. S. Kurbatov [in Russian]/, /Corticarina strandi Johnson ♂ det. S. Saluk, 1988/, /Corticarina ? pr. strandi Johnson S. Saluk det. 2005/, /HOLOTYPE/ (in Zoological Museum of Moscow University).

Notes. Features of small head and pronotum, amply curved elytra, brown colour, antennal length and structure, long and narrow aedeagus all show this species to be closely allied to the section which includes *similata* (Gyllenhal), not *strandi* Johnson. It differs from *similata* by being broadest at the basal third of the elytra and by its bifurcate aedeagal apex. The aedeagal figure (Saluk, 1992) is taken from my Indian *strandi* figure.

Dienerella (*Cartoderema*) *vincenti* Johnson, **sp. nov.** This species is *Latridius elongatus* Curtis, 1830 (nec *L. elongatus* Gyllenhal, 1827 = *Corticaria elongata* (Gyllenhal, 1827))

Holotype (unsexed): [Austria] Nordtirol R.Wohlmann/, Kranebitter Klam 15.10.36/, /Cartodere elongatus/, /Manchester Museum HOLOTYPE/ (in Manchester University Museum, MMUE).

Paratypes: 1, Finland 682:20 St Luvia, 22.7.1977, T. Clayhills/, /D.elongata (Curtis) det. C. Johnson, 1979/; 2, FRANCE: Hte. Savoie, 24.7.87, J. Steffen/, /sous une souche/; 1, FRANCE: 15-29.8.82, Ariège: Lanoux, S. Bowestead/, woodland litter/; 1, [Germany] DDR: 11.iv.1977, Waltersdorf b. Zittau, M. Sieber/, /Dienerella elongata Curt. det. C. Johnson/; 1, Feldkirchen /Neuwied Dat. 6.4.1985, leg. W. Rücker/, /DIENERELLA elongata (Curtis, 1830) det. W. Rücker/; 1, Germania, Neuwied 12 Rheintal in Küche, Datum: 22.5.1963 leg. Edeltrud Rücker/, /DIENERELLA elongata (Curtis, 1830) det. W. Rücker/; Switzerland: 1, Berne, Heimiswil, Dreienbärgli, 740 m, 4.1994, leg. E. Kobel; 6, Genève, Nant d'Aisy, 18.6.1988, vieille paille, leg. C. Besuchet; 6, same data, iii-vii.1989, piège bière-sel; 2, Grisons, Martina, 26.8.1968, feuilles mortes, leg. C. Besuchet; 2, Grisons, Bregaglia Soglio, 900 m, 8.6.1992, écorces châtaignier, leg. C. Besuchet; 4, Tessin, Curtina, Val Colla, 10.8.1976, leg. S. Vit; 5, Valais, Grimisuat, 900 m, 13.5.1992, leg. C.Johnson (all with Manchester Museum PARATYPE labels, in Manchester University Museum, MMUE). This common species is characterized in many works, e.g. Peez (1967), although type material is absent from the Curtis collection (Vincent, 1990a).

Enicmus rueckeri Johnson, **sp. nov.** This species is *E. dubius* sensu Rücker, 1992 (a misidentification; nec *dubius* Mannerheim, 1844).

Holotype male: Mongolia bor., Reitter/, /Enicmus dubius Mannh., Coll. Reitter/, /dubius Mannh. det. Rücker, 1982/, /HOLOTYPE/ (in Hungarian Natural History Museum, Budapest, HNHMB).

Paratypes: 1 female, /MONGOLIA: Suchebaator aimak, 44 km SSW von Baruun urt, 1050 m, Exp. Dr. Z. KASZAB, 1965/, /nr. 350, 2.VIII.1965/, /dubius Mannh. det. Rücker, 1982/; 1, Central aimak, SO von Somon Bajanzogt, 1600 m, Exp. Dr. Z. KASZAB, 1966/, /nr. 529, 11.VI.1966/; 1, 1 male, Bulgan aimak, 11 km W von Somon Bajanuur am See, Bajan nuur, 1000 m, Exp. Dr. Z. KASZAB, 1968/, /nr. 958,14.VI-24.VII.1968/; 1, 1 male, Central aimak, 11 km OSO von Somon Bajanzogt, 1600-1700 m, Exp. Dr. Z. KASZAB, 1968/, /nr. 1150, 26.VII.1968/; 1 female, [Russia] Quellgebiet des Irkut, Leder/, /dubius Mannh., Coll. Reitter/ (all with PARATYPE labels, in Hungarian Natural History Museum, Budapest, HNHMB).

Migneauxia ottoi Johnson, **sp. nov.** for *M. lederi* 'sensu auctorum' (misidentification; nec Reitter, 1875c).

Holotype male: [Italy] /SICILY leg. Melly/, /Coll. Melly/, /HOLOTYPE/ (in Muséum d'Histoire Naturelle, Geneva).

Paratypes: 2, /ITALY SICILIA, Castello (TP) 9.XII.1993 Baia Castellammare del Golfo prato, leg. Sabella/, /LATRIDIIDAE Migneauxia lederi Reitter, 1875, det. W. Rücker, 1996/; 2, /ITALIA Calabria, Aspromonte, 13.IV.1997 Melitto (RC), Chorio flum Melitto, sotto pietre in formic, leg. F. ANGELINI/, /LATRIDIIDAE Migneauxia lederi Reitter, 1875, det. W. Rücker, 2001/ (all with Manchester Museum PARATYPE labels, in Manchester University Museum, MMUE). This species is characterized in many works, e.g. Otto (1979).

Migneauxia phili Johnson, **sp. nov.** for *M. inflata* 'sensu auctorum' (misidentification; nec Rosenhauer, 1856).

Holotype male: [Spain] Hof Alamaja, Sierra Negra, 21.2.1951/, /Umg. Carmona, Andalusien, leg. H. Franz/, /Migneauxia inflata Rosh. det. H. Franz/ , /Manchester Museum HOLOTYPE/ (in Manchester University Museum, MMUE).

Paratypes: 1 male, [Spain] Cortijo de Maribañetz, 22.2.1951/, /Umg. Sevilla, Andalusien, leg. H. Franz/; 1, /Umg. Chechaouen, Nordmarokko, lg. H. Franz/; 1 female, /Maroc, 26.1.62, Fouarat, R. Mussard/; 1 male, /Maroc, 17.v.68, Rharb Azbaoua – Fakra, leg. Lapin/ (all with Manchester Museum PARATYPE labels, in Manchester University Museum, MMUE). This species is characterized in many works, e.g. Otto (1979).

New synonyms

Cartodere costipennis Reitter, 1877 **syn. nov.** of *Dienerella elegans* (Aubé, 1850). Based on descriptions
Cartodere depilis Belon, 1895 **syn. nov.** of *Dienerella elegans* (Aubé, 1850). Based on descriptions
Cartodere elegantula Reitter, 1890 **syn. nov.** of *Dienerella elegans* (Aubé, 1850)
Cartodere forticostis Reitter, 1911 **syn. nov.** of *Dienerella elegans* (Aubé, 1850)

Cartodere strupii Hölzel, 1944 **syn. nov.** of *Dienerella schueppeli* (Reitter, 1881). Based on descriptions.
Cartodere subcostella Reitter, 1915 **syn. nov.** of *Dienerella filiformis* (Gyllenhal, 1827)
Coricaria fuscotestacea Motschulsky, 1861 **syn. nov.** of *Cortinicara gibbosus* (Herbst, 1793)
Corticaria abietorum Motschulsky, 1867 **syn. nov.** of *Corticaria longicornis* (Herbst, 1784)
Corticaria adelaidae Blackburn, 1891 **syn. nov.** of *Corticaria japonica* Reitter, 1877
Corticaria attenuata Motschulsky, 1867 **syn. nov.** of *Corticaria fulva* (Comolli, 1837)
Corticaria boreophila Motschulsky, 1867 **syn. nov.** of *Corticaria ferruginea* Marsham, 1802
Corticaria carinulata Motschulsky, 1867 **syn. nov.** of *Corticarina curta* (Wollaston, 1854)
Corticaria ciliata Motschulsky, 1867 **syn. nov.** of *Corticaria fulva* (Comolli, 1837)
Corticaria concinnula Mannerheim, 1844 **syn. nov.** of *Corticaria pubescens* (Gyllenhal, 1827)
Corticaria curticollis Mannerheim, 1844 **syn. nov.** of *Melanophthalma transversalis* (Gyllenhal, 1827)
Corticaria diluta Mannerheim, 1844 **syn. nov.** of *Corticaria pubescens* (Gyllenhal, 1827)
Corticaria fagi H. Brisout de Barneville, 1881 [HN] **syn. nov.** of *Corticaria inconspicua* Wollaston, 1860. Based on descriptions.
Corticaria flavifrons Motschulsky, 1867 **syn. nov.** of *Corticaria serratus* (Paykull, 1798)
Corticaria illaesiformis Reitter, 1898 **syn. nov.** of *Corticaria subpilosula* Reitter, 1898
Corticaria intricata Mannerheim, 1844 **syn. nov.** of *Corticaria pubescens* (Gyllenhal, 1827)
Corticaria kaufmanni Reitter, 1881 **syn. nov.** of *Corticaria diecki* Reitter, 1875. Based on descriptions.
Corticaria limbicollis Motschulsky, 1867 **syn. nov.** of *Corticaria rubripes* Mannerheim, 1844
Corticaria melanocara Gistel, 1857 **syn. nov.** of *Corticarina minutus* (Fabricius, 1792). Based on descriptions
Corticaria melanophthalma Mannerheim, 1844 **syn. nov.** of *Corticaria serrata* (Paykull, 1798)
Corticaria metallica Reitter, 1874 **syn. nov.** of *Corticaria fulva* (Comolli, 1837)
Corticaria monticola H. Brisout de Barneville, 1881 **syn. nov.** of *Corticaria illaesa* Mannerheim, 1844
Corticaria obscuripes Reitter, 1898 **syn. nov.** of *Corticaria rubripes* Mannerheim, 1844. Based on descriptions.
Corticaria pallida Marsham, 1802 **syn. nov.** of *Cortinicara gibbosa* (Herbst, 1793). Based on descriptions.
Corticaria perpulchra Reitter, 1892 **syn. nov.** of *Corticaria quadrimaculata* Mannerheim, 1844
Corticaria piligera Mannerheim, 1844 **syn. nov.** of *Corticaria punctulata* Marsham, 1802
Corticaria pilitecta Motschulsky, 1867 **syn. nov.** of *Corticarina curta* (Wollaston, 1854)
Corticaria planula Fall, 1899 **syn. nov.** of *Corticaria cucujiformis* Reitter, 1881
Corticaria psammeticha Motschulsky, 1867 **syn. nov.** of *Migneauxia subrugosa* (Motschulsky, 1867)
Corticaria punctatissima Motschulsky, 1867 **syn. nov.** of *Corticaria illaesa* Mannerheim, 1844
Corticaria resecta Walker, 1859 **syn. nov.** of *Cortinicara gibbosa* (Herbst, 1793)
Corticaria rudebecki Dajoz, 1970 **syn. nov.** of *Corticaria japonica* Reitter, 1877
Corticaria setosa Motschulsky, 1867 **syn. nov.** of *Corticaria quadrimaculata* Mannerheim, 1844
Corticaria solarii Reitter, 1904 **syn. nov.** of *Corticaria beloni* Reitter, 1889
Corticaria strandi Roubal, 1934 **syn. nov.** of *Corticaria aequalis* Reitter, 1898
Corticaria subcordata Rey, 1889 **syn. nov.** of *Corticaria serrata* (Paykull, 1798)
Corticaria subfusca Sharp, 1902 **syn. nov.** of *Corticarina cavicollis* (Mannerheim, 1844)
Corticaria subpicea Motschulsky, 1867 **syn. nov.** of *Corticaria obscura* C. N. F. Brisout de Barneville, 1863
Corticaria subrugosa Motschulsky, 1867 **syn. nov.** of *Migneauxia psammeticha* (Motschulsky, 1867)
Corticaria tenuicornis Motschulsky, 1867 **syn. nov.** of *Corticaria serrata* (Paykull, 1798)
Corticaria tincta Mannerheim, 1844 **syn. nov.** of *Corticaria pubescens* (Gyllenhal, 1827)
Corticaria transversicollis Motschulsky, 1867 **syn. nov.** of *Corticaria pharaonis* Motschulsky, 1867
Corticaria unicarinulata Motschulsky, 1867 **syn. nov.** of *Corticaria fulva* (Comolli, 1837)
Corticaria weisei Reitter, 1875 **syn. nov.** of *Corticaria serrata* (Paykull, 1798)
Corticarina obfuscata A. Strand, 1937 **syn. nov.** of *Corticaria parvula* Mannerheim, 1844
Dermestes marginatus Paykull, 1798 doubtful **syn. nov.** of *Latridius porcatus* Herbst, 1793. Based on descriptions.
Dienerella tanakai Hisamatsu, 1985 **syn. nov.** of *Dienerella lurida* Rücker,1983. Based on descriptions.
Enicmus alius J. Weise, 1972 **syn. nov.** of *Latridius porcatus* Herbst, 1793
Enicmus anatolicus Reitter, 1903 **syn. nov.** of *Enicmus varendorffi* Reitter, 1903
Enicmus distincticollis Roubal, 1933 **syn. nov.** of *Latridius hirtus* Gyllenhal, 1827. Based on descriptions.
Enicmus lederi Reitter, 1875b **syn. nov.** of *Enicmus testaceus* (Stephens, 1830)
Enicmus nidicola Palm, 1944 **syn. nov.** of *Latridius gemellatus* (Mannerheim, 1844). Based on descriptions.

Enicmus pseudominutus A. Strand, 1958 **syn. nov.** of *Latridius assimilis* (Mannerheim, 1844). Based on descriptions.
Enicmus pseudotestaceus Franz, 1974 **syn. nov.** of *Latridius testaceus* Stephens, 1830
Enicmus ventriosus Peyerimhoff, 1929 **syn. nov.** of *Enicmus histrio* Joy & Tomlin, 1910
Lathridius aegyptiacus Motschulsky, 1866 **syn. nov.** of *Cartidere constricta* (Gyllenhal 1827). Based on descriptions.
Lathridius anthracinus Mannerheim, 1844 **syn. nov.** of *Latridius porcatus* Herbst, 1793
Lathridius campicola Gerhardt, 1911 **syn. nov.** of *Stephostethus rugicollis* (A. G. Olivier, 1790). Based on descriptions.
Lathridius lenensis Poppius, 1903 **syn. nov.** of *Stephostethus attenuatus* (Mannerheim, 1844)
Lathridius obtusatus Roubal, 1951 **syn. nov.** of *Stephostethus angusticollis* (Gyllenhal, 1827). Based on descriptions.
Lathridius parallelipennis Solsky, 1876 **syn. nov.** of *Dienerella filiformis* (Gyllenhal, 1827). Based on descriptions.
Lathridius perpusillus Walker, 1855 **syn. nov.** of *Cartodere constricta* (Gyllenhal, 1827). Based on descriptions.
Lathridius subbrevis Motschulsky, 1866 **syn. nov.** of *Stephostethus lardarius* (DeGeer, 1775)
Lathridius subnudus Reitter, 1889 **syn. nov.** of *Stephostethus pandellei* C. N. F. Brisout de Barneville, 1863
Latridius creticus Rücker, 1986 **syn. nov.** of *Latridius amplus* Johnson, 1977. Based on descriptions.
Latridius depressus Grimmer, 1841 **syn. nov.** of *Enicmus rugosus* (Herbst, 1793). Based on descriptions.
Latridius fulvipes Comolli, 1837 doubtful **syn. nov.** of *Corticarina similata* (Gyllenhal, 1827). Based on descriptions.
Latridius fusculus Gyllenhal, 1827 **syn. nov.** of *Corticarina minuta* (Fabricius, 1792)
Latridius herbaceus Gistel, 1857 **syn. nov.** of *Cortinicara gibbosa* (Herbst, 1793). Based on descriptions.
Latridius klapperichi C. Johnson, 1977 **syn. nov.** of *Lathridius peacockae* Sengupta, 1976
Latridius latissimus Rücker, 1986 **syn. nov.** of *Latridius amplus* C. Johnson, 1977. Based on descriptions.
Latridius lignarius Gistel, 1857 **syn. nov.** of *Stephostethus rugicollis* (A. G. Olivier, 1790). Based on descriptions.
Latridius ovatus Grimmer, 1841 **syn. nov.** of *Stephostethus rugicollis* (A. G. Olivier, 1790). Based on descriptions.
Latridius ruficornis Kugelann, 1794 **syn. nov.** of *Enicmus rugosus* (Herbst, 1793). Based on descriptions.
Latridius subterraneus Gistel, 1857 **syn. nov.** of *Enicmus brevicornis* (Mannerheim, 1844). Based on descriptions.
Melanophthalma ceylanica Dajoz, 1975 **syn. nov.** of *Melanophthalma angulicollis* Motschulsky, 1866
Melanophthalma gibbula Motschulsky, 1866 **syn. nov.** of *Cortinicara gibbosa* (Herbst, 1793)
Melanophthalma hexagona Motschulsky, 1866 **syn. nov.** of *Melanophthalma angulicollis* Motschulsky, 1866
Melanophthalma ovalipennis Reitter, 1875 **syn. nov.** of *Corticarina minuta* (Fabricius, 1792)
Melanophthalma prominens Johnson, 1972a **syn. nov.** of *Melanophthalma angulicollis* Motschulsky, 1866
Melanophthalma terricula Fall, 1899 **syn. nov.** of *Corticarina latipennis* (J. R. Sahlberg,1871)
Metophthalmus brenskei Reitter, 1884 **syn. nov.** of *Metophthalmus lacteolus* Motschulsky, 1866
Metophthalmus convexiusculus J. R. Sahlberg,1913 **syn. nov.** of *Metophthalmus ragusae* Reitter, 1875
Metophthalmus creticus Reitter, 1908 **syn. nov.** of *Metophthalmus lacteolus* Motschulsky, 1866
Metophthalmus heiligi Reitter, 1908 **syn. nov.** of *Metophthalmus ragusae* Reitter, 1875
Metophthalmus hungaricus Reitter, 1884 **syn. nov.** of *Metophthalmus lacteolus* Motschulsky, 1866
Metophthalmus intermedius Binaghi, 1946 **syn. nov.** of *Metophthalmus niveicollis* (Jacquelin du Val, 1859)
Metophthalmus obesus Reitter, 1881 **syn. nov.** of *Metophthalmus niveicollis* (Jacquelin du Val, 1859)
Metophthalmus jordanensis J. Sahlberg 1913 **syn. nov.** of *Metophthalmus judaicus* J. R. Sahlberg,1913
Metophthalmus siculus Binaghi, 1946 **syn. nov.** of *Metophthalmus ragusae* Reitter, 1875
Metophthalmus solarii Binaghi, 1946 **syn. nov.** of *Metophthalmus lacteolus* Motschulsky, 1866
Metophthalmus syriacus Reitter, 1884 **syn. nov.** of *Metophthalmus ragusae* Reitter, 1875. Based on descriptions.
Migneauxia orientalis Reitter 1877 **syn. nov.** of *Migneauxia lederi* Reitter, 1875
Migneauxia parvicollis Peyerimhoff, 1929 **syn. nov.** of *Migneauxia lederi* Reitter, 1875
Migneauxia sudanensis C. Johnson, 1976 **syn. nov.** of *Migneauxia psammeticha* (Motschulsky, 1867)
Permidius basalis Motschulsky, 1866 **syn. nov.** of *Latridius consimilis* (Mannerheim, 1844)
Permidius flavicornis Motschulsky, 1866 **syn. nov.** of *Latridius assimilis* (Mannerheim, 1844)

Permidius inflaticeps Motschulsky, 1866 **syn. nov.** of *Dienerella elegans* (Aubé, 1850)
Revelieria globosa J. R. Sahlberg,1913 **syn. nov.** of *Revelieria genei* (Aubé, 1850)

Comments
Based on descriptions, following names are tentatively placed in synonymy:
Corticaria clairi H. Brisout de Barneville, 1881 = *Dermestes serratus* Paykull, 1798
Enicmus kolbei Wanka, 1929 = *Lathridius gemellatus* Mannerheim, 1844
Lathridius sulcifrons Broun, 1895 = *Lathridius assimilis* Mannerheim, 1844
Lathridius scitus Mannerheim, 1844 = *Lathridius assimilis* Mannerheim, 1844
Latridius herborisans Gistel, 1857 = *Lathridius assimilis* Mannerheim, 1844

CATALOGUE

series **ELATERIFORMIA** Crowson, 1960
superfamily **ELATEROIDEA** Leach, 1815
family ARTEMATOPIDAE Lacordaire, 1857
subfamily **Artematopodinae** Lacordaire, 1857

genus *Eurypogon* Motschulsky, 1860d: 363 type species *Ochina nigra* Melsheimer, 1846
Pseudodactylus Hampe, 1866: 374 type species *Pseudodactylus cribratus* Hampe, 1866
brevipennis Sakai, 1982b: 59 **A**: JA
cribratus Hampe, 1866: 374 (*Pseudodactylus*) **E**: IT
granulatus Sakai, 1982b: 57 **A**: JA
hisamatsui Sakai, 1982b: 60 **A**: JA
japonicus Sakai, 1982b: 52 **A**: JA
ocularis Sakai, 1982b: 56 **A**: JA

genus *Macropogon* Motschulsky, 1845a: 38 type species *Macropogon sibiricum* Motschulsky, 1845
pubescens Motschulsky, 1860d: 362 **A**: FE (Kamchatka)
sibiricus Motschulsky, 1845a: 38 **A**: ES

family CEROPHYTIDAE Latreille, 1834

genus *Cerophytum* Latreille, 1809: 379 type species *Cerophytum elateroides* Latreille, 1809
elateroides Latreille, 1809: 375 **E**: AU BU CR CZ FR GE HU LA LU NL PL RO SK SL SZ
japonicum Sasaji, 1999: 97 **A**: JA

family EUCNEMIDAE Eschscholtz, 1829
subfamily **Phyllocerinae** Reitter, 1905
tribe Anelastini Reitter, 1921

genus *Anelastes* Kirby, 1819: 384 type species *Anelastes drurii* Kirby, 1819
Silenus Latreille, 1834: 128 type species *Silenus brunneus* Latreille, 1834
barbarus P. H. Lucas, 1846: 167 **E**: CR FR GR IT SP **N**: AG LB MO TU **A**: IN TM TR
crenulatus Bonvouloir, 1872: 695 (*Hypocaelus*) **A**: AF **ORR**

tribe Phyllocerini Reitter, 1905

genus *Phyllocerus* Le Peletier de Saint-Fargeau & Audinet-Serville, 1825: 116 type species *Phyllocerus flavipennis* Le Peletier de Saint-Fargeau & Audinet-Serville, 1825
Cephalodendron Latreille, 1834: 166 type species *Eucnemis ramicornis* Latreille, 1834
Isosoma Ménétriés, 1832: 160 [HN] type species *Isosoma elateroides* Ménétriés, 1832
apicalis Pic, 1929d: 5 **N**: LB
diversicornis Pic, 1905l: 181 **E**: TR
elateroides Ménétriés, 1832: 160 (*Isosoma*) **E**: GR "Caucasus"
herculeanus Schaufuss, 1863: 185
flavipennis Le Peletier de Saint-Fargeau & Audinet-Serville, 1825: 116 **E**: CR IT
fulvipennis Germar, 1839b: 196
grahmanni Spinola, 1838: xlii
spinolae Guérin-Méneville, 1838b: 13
ghilarovi Gurjeva & Valiakhmedow, 1982: 95 **A**: TD
longipennis Ferrari, 1869: 193 **E**: GG
oreadis Wurst, 1996: 52 **A**: TR
ullmanni Kirchsberg, 1898: 677 **E**: IT SL

subfamily **Melasinae** Fleming, 1821

tribe Calyptocerini Muona, 1993

genus *Otho* Lacordaire, 1857: 113 type species *Melasis sphondyloides* Germar, 1818
amamiensis Nakane, 1987c: 171 **A**: JA (Amami-Ôshima)
himalayensis Lucht, 1989: 175 **A**: NP SD
nipponicus Fleutiaux, 1923b: 318 **A**: JA
sphondyloides Germar, 1818: 235 (*Melasis*) **E**: AU BU BY CR CT FE GE HU LA RO UK **A**: JA
sibiricus Motschulsky, 1845a: 34 (*Hypocoelus*)

genus *Calyptocerus* Guérin-Ménèville, 1843: 177 type species *Calyptocerus leboucherii* Guérin-Ménèville, 1843
favipunctatus Lucht, 1986: 103 **A**: TAI

tribe Dirhagini Reitter, 1911

genus *Balistica* Motschulsky, 1861c: 116 type species *Balistica picipes* Motschulsky, 1861
elongata Fleutiaux, 1923b: 314 **A**: JA

genus *Cafolus* Bonvouloir, 1871: 75 type species *Cafolus moestus* Bonvouloir, 1875
Arhagus Fleutiaux, 1921b: 72 type species *Arhagus bakeri* Fleutiaux, 1921
carinicollis Fleutiaux, 1923b: 307 (*Arhagus*) **A**: JA

genus *Clypeorhagus* Olexa, 1975: 157 type species *Microrhagus clypeatus* Hampe, 1850
clypeatus Hampe, 1850: 350 (*Microrhagus*) **E**: CT GE HU LA PL RO SK UK

genus *Farsus* Jacquelin du Val, 1860a: 116 type species *Hylochares unicolor* Latreille, 1834 (= *Elater dubius* Piller & Mitterpacher, 1783)
Rhagomerus Iablokoff-Khnzorian, 1964: 162 type species *Rhagomerus obesus* Iablokoff-Khnzorian, 1964
ainu Fleutiaux, 1923b: 306 **A**: JA
dubius Piller & Mitterpacher, 1783: 83 (*Elater*) **E**: AU CZ FR HU IT RO SP ST "Yugoslavia" **A**: SY
alticollis Lacordaire, 1857: 115 (*Hylochares*)
siculus Ragusa, 1893: 302
unicolor Latreille, 1834: 128 (*Hylochares*)
obenbergeri Roubal, 1941a: 12 **A**: SCH
obesus Iablokoff-Khnzorian, 1964: 164 (*Rhagomerus*) **E**: GG

genus *Fryanus* Fleutiaux, 1902d: 663 type species *Compsocnemis dohertyi* Fleutiaux, 1899
japonicus Hisamatsu, 1957: 46 **A**: JA

genus *Microrhagus* Dejean, 1833: 85 type species *Elater pygmaeus* Fabricius, 1792
Arhagus Méquignon, 1925b: 187 [HN] type species *Microrhagus pyrenaeus* Bonvouloir, 1872
Aulacostenus Motschulsky, 1870: 117 type species *Aulacostenus pavidus* Motschulsky, 1870
Dichodirhagus Méquignon, 1925: 240 [RN] type species *Microrhagus pyrenaeus* Bonvouloir, 1872
Dirhagus Latreille, 1834: 130 type species *Elater pygmaeus* Fabricius, 1792
Emyirhagus Olexa, 1975: 162 type species *Microrhagus emyi* Rouget, 1856
emyi Rouget, 1856: 193 (*Elater*) **E**: AU CR CT CZ FR IT GE HU PL RO SK SP SV SZ UK "Yugoslavia"
foveolatus Fleutiaux, 1923b: 308 (*Dirhagus*) **A**: FE JA
elateroides Fursov, 1936a: 348 (*Dirrhagus*)
hummleri Reitter, 1911a: 204 (*Microrrhagus*) **E**: IT
klapperichi Lucht, 1984: 289 (*Dirhagus*) **A**: FUJ
lepidus Rosenhauer, 1847: 14 (*Microrhagus*) **E**: AU BE BH BY CT CZ FI FR GE HU IT LA LT NL NT PL RO SK SV
manueli Fairmaire, 1856a: 530
nitidus A. Costa, 1847a: 101
lindbergi Palm, 1958: 35 (*Dirrhagus*)
miyatakei Hisamatsu, 1955: 101 (*Balistica*) **A**: JA

mystagogus Fleutiaux, 1923b: 309 (*Dirhagus*) **A**: FE JA
nigrinus Hisamatsu, 1960: 92 (*Dirrhagus*) **A**: JA
pectinicornis Hisamatsu, 1960: 94 (*Dirrhagus*) **A**: JA
pygmaeus Fabricius, 1792b: 234 (*Elater*) **E**: BE BH BU BY CT CR CZ DE EN FI FR GB GE HU IT LA LT NL NT NR PL RO SK SZ SV
brevis Hampe, 1863: 288
chevrolati Stierlin, 1864: 148
nominandus Roubal, 1915: 170 (*Dirrhagus*)
palmi Olexa, 1963: 88 (*Dirhagus*)
pyrenaeus Bonvouloir, 1872: 528 **E**: BU FR GE
ramosus Fleutiaux, 1902a: 24 **A**: JA
savioi Fleutiaux, 1925b: 177 (*Arhagus*) **A**: FUJ
towadensis Hisamatsu, 1960: 94 (*Dirrhagus*) **A**: JA

genus *Dirrhagofarsus* Fleutiaux, 1935a: 15 type species *Microrhagus lewisi* Fleutiaux, 1900
Attenuorhagus Olexa, 1975: 161 type species *Hypocoelus attenuatus* Mäklin, 1845
attenuatus Mäklin, 1845: 547 (*Hypocoelus*) **E**: AU CT CR FI HU GE NT PL RO SK
longicornis Hampe, 1852: 160 (*Microrhagus*)
hungaricus Pic, 1918a: 1 (*Dirrhagus*)
lewisi Fleutiaux, 1900: 358 (*Microrhagus*) **A**: JA **NAR**
modestus Fleutiaux, 1923b: 309 (*Dirhagus*) **A**: FE JA

genus *Rhagomicrus* Fleutiaux, 1902d: 658 type species *Microrhagus meticulosus* Bonvouloir, 1872 (= *Eucnemis humeralis* Say, 1836)
marginatus Fleutiaux, 1923b: 308 (*Dirhagus*) **A**: FE JA
morimotoi Hisamatsu, 1960: 96 (*Dirrhagus*) **A**: JA

genus *Rhacopus* Hampe, 1855b: 256 type species *Rhacopus cinnamomeus* Hampe, 1855
Eurhagus Méquignon, 1922: 372 type species *Eucnemis sahlbergi* Mannerheim, 1823
Hyporhagus Reitter, 1921a: 81 type species *Eucnemis sahlbergi* Mannerheim, 1823
Pseudorhacopus Olexa, 1975: 158 type species *Dirrhagus ferrugineus* Reitter, 1889
bodemeyeri Olexa, 1962: 43 (*Dirrhagus*) **A**: IN
ferrugineus Reitter, 1889m: 282 (*Dirrhagus*) **A**: IN
yasumatsui Hisamatsu, 1960: 99 (*Dirrhagus*) **A**: JA
unicolor Hisamatsu, 1960: 100 (*Dirrhagus*)
olexai Hisamatsu, 1963: 29 (*Dirrhagus*) **A**: JA
sahlbergi Mannerheim, 1823: 27 (*Eucnemis*) **E**: AU BH BY CR CT EN FI FR GE GR HU IT LA NT PL RO SK SV **A**: MG WS
cinnamomeus Hampe, 1855b: 256

genus *Sarpedon* Bonvouloir, 1871: 71 type species *Sarpedon scabrosus* Bonvouloir, 1875
atratus Fleutiaux, 1923b: 317 **A**: FE JA

tribe Epiphanini Muona, 1993

genus *Hylis* Gozis, 1886: 21 type species *Eucnemis procerulus* Mannerheim, 1823
Elatocoelus Hyslop, 1921: 644 [RN] type species *Hypocaelus matthiesseni* Reitter, 1906
Hypohylis Reitter, 1911a: 203 type species *Hypocaelus matthiesseni* Reitter, 1906
cariniceps Reitter, 1902g: 208 (*Hypocoelus*) **E**: AU BH CZ CR FI FR GB GR GE HU IT LS LU NL NR PL RO SK SV SZ
foveicollis C. G. Thomson, 1874: 550 (*Nematodes*) **E**: AU BH BE CR CT CZ DE GE FI FR HU IT LA LS LU NL NT PL RO SK SV SZ
fleischeri Olexa, 1954b: 67 (*Hypocoelus*)
harmandi Fleutiaux, 1923b: 326 (*Hypocoelus*) **A**: FE JA "NE China"
japonicus Fleutiaux, 1902a: 24 (*Hypocoelus*) **A**: FE JA
matthiesseni Reitter, 1906e: 239 (*Hypocoelus*) **E**: "Caucasus"

olexai Palm, 1955: 212 (*Hypocoelus*) **E**: AU BH BE CR CT CZ DE GE FR GB IT LA LS LU NL PL RO SK SV
elaterinus A. Villa & J. B. Villa, 1838: 62 (*Eucnemis*) [DA]
musculus Rosenhauer, 1847: 98 (*Nematodes*) [DA]
procerulus Mannerheim, 1823: 32 (*Eucnemis*) **E**: AU BE CT CZ SZ EN FI FR LA LT NR NT SV UK
simplicifrons Reitter, 1902g: 209 (*Hypocoelus*)
slipinskii Burakowski & Buchholz, 1991: 122 **E**: BH
simonae Olexa, 1970: 91 (*Hypocoelus*) **E**: BU FR HU SP
yanoi Hisamatsu, 1955: 98 (*Hypocoelus*) **A**: JA
yoshidai Nakane, 1987c: 175 (*Hypocoelus*) [DA] **A**: JA

genus *Epiphanis* Eschscholtz, 1829: 35 type species *Epiphanis cornutus* Eschscholtz, 1829
Ceratotaxia Sharp, 1908: 399 type species *Ceratotaxia tristis* Sharp, 1908
Prosopotropis Abeille de Perrin, 1898a: 35 type species *Prospotropis devillei* Abeille de Perrin, 1898 (= *Epiphanis cornutus* Eschscholtz, 1829)
cornutus Eschscholtz, 1829: 11 **E**: BU FR IT **N**: AG **A**: CH FE JA **NAR**
cristatus LeConte, 1852a: 46
devillei Abeille de Perrin, 1898a: 36 (*Prosopotropis*)

tribe Hylocharini Jacquelin du Val, 1859

genus *Hylochares* Latreille, 1834: 127 type species *Elater cruentatus* Gyllenhal, 1808
cruentatus Gyllenhal, 1808a: 435 (*Elater*) **E**: BY CT EN FI LA NT PL **A**: WS
harmandi Fleutiaux, 1900: 360 **A**: FE JA

tribe Melasini Fleming, 1821

genus *Isorhipis* Lacordaire, 1835: 622 type species *Isorhipis lepaigei* Lacordaire, 1835 (= *Tharops melasoides* Laporte, 1835)
Tharops Laporte, 1835b: 168 [HN] type species *Tharops melasoides* Laporte, 1835
banghaasi Reitter, 1899c: 217 (*Tharops*) **A**: FE JA
krivolutzhki L. N. Medvedev, 1970: 167
foveata Hisamatsu, 1955: 96 **A**: JA
marmottani Bonvouloir, 1871: 99 (*Tharops*) **E**: AU BH BU CT CZ EN FR GE NR PL RO SK SV
melasoides Laporte, 1835b: 169 (*Tharops*) **E**: AU BH BU CR CT CZ FR GE HU IT PL RO SK SV TR
lepaigei Lacordaire, 1835: 623 (*Tharops*)
strepens W. Redtenbacher, 1842: 9 (*Tharops*)
minuta L. N. Medvedev, 1970: 167 **A**: FE
nigriceps Mannerheim, 1823: 35 (*Eucnemis*) **E**: AB AR BH BU CR FR HU RO SL ST **A**: TR
potanini Semenov, 1891: 345 (*Tharops*) **A**: GAN

genus *Melasis* A. G. Olivier, 1790a: no 30: 1 type species *Hispa flabellicornis* Fabricius, 1775 (= *Elater buprestoides* Linnaeus, 1761)
buprestoides Linnaeus, 1761: 209 (*Elater*) **E**: AB AU BE BH BU BY CR CT CZ DE FI FR GB GE GR HU IR IT LU MC NR PL RO SK SL SP ST SV SZ TR UK YU **N**: AG **A**: "Siberia"
cylindricus Harrer, 1784: 144 (*Ptilinus*)
dentatus Geoffroy, 1785: 34 (*Cucujus*)
elateroides Illiger, 1806: 227
flabellicornis Fabricius, 1775: 70 (*Hispa*)
rufipennis Pic, 1907e: 122 [HN]
fermini Sánchez-Ruiz & Rosa, 2003: 1 **E**: SP
japonica Hisamatsu, 1963: 30 **A**: JA
sachalinensis Suzuki & Lucht, 1983: 41 **A**: FE
sinensis Lucht, 1982: 15 **A**: TAI
tibialis Lucht, 1982: 17 **A**: TAI

tribe Xylobiini Reitter, 1911

genus *Bioxylus* Fleutiaux, 1923b: 320 type species *Xylobius japonensis* Fleutiaux, 1900
bidentatus Hisamatsu, 1956: 157 (*Xylobius*) **A**: JA

galloisi Fleutiaux, 1923b: 320 **A**: JA
japonensis Fleutiaux, 1900: 359 (*Xylobius*) **A**: FE JA
laticollis Hisamatsu, 1963: 33 **A**: JA
personatus Mamaev, 1976: 149 **A**: FE
pilosellus Hisamatsu, 1959b: 23 **A**: JA
shimoyamai Hisamatsu, 1959b: 26 **A**: JA

genus *Proxylobius* Fleutiaux, 1900: 360 type species *Proxylobius helleri* Fleutiaux, 1900
Phizoschilus Fleutiaux, 1930a: 273 type species *Phizoschilus gardneri* Fleutiaux, 1930
galloisi Fleutiaux, 1923b: 304 **A**: JA
helleri Fleutiaux, 1900: 361 **A**: FE JA
longicornis Hisamatsu, 1958: 118 **A**: JA

genus *Xylophilus* Mannerheim, 1823: 14 type species *Elater alni* Fabricius, 1801 (= *Elater corticalis* Paykull, 1800)
Xylobius Latreille, 1834: 124 type species *Elater alni* Fabricius, 1801
Xyloecus Dejean, 1833: 85 type species *Elater alni* Fabricius, 1801
ainu Fleutiaux, 1923b: 321 (*Xylobius*) **A**: FE JA
corticalis Paykull, 1800: 43-44 (*Elater*) **E**: AU BY CR CZ CT DE EN FI FR GE IT LA PL RO SK SV "Yugoslavia"
alni Fabricius, 1801b: 246 (*Elater*)
humeralis Dufour, 1851: 24 (*Xylobius*)
inhumeralis Pic, 1907e: 122 (*Xylobius*)
limbatus Pic, 1907e: 122 (*Xylobius*)
leseigneuri Olexa, 1960: 55 **E**: GG
minimus Hisamatsu, 1959b: 30 (*Xylobius*) **A**: JA
rufomarginatus Fleutiaux, 1923b: 321 (*Xylobius*) **A**: JA
talyschensis Olexa, 1969: 61 **E**: AB
testaceus Herbst, 1806: 101 (*Elater*) **E**: CR CZ FR HU IT PL RO SK SZ "Yugoslavia"
seiditzi Csiki, 1906: 406 [RN]
ussuriensis Mamaev, 1976: 149 **A**: FE JA

subfamily **Eucneminae** Eschscholtz, 1829

tribe Dyscharachthini Muona, 1993

genus *Dyscharachthis* Blackburn, 1900b: 217 type species *Dyscharachthis brevipennis* Blackburn, 1900
Galloisius Fleutiaux, 1923b: 294 type species *Galloisius amplicollis* Fleutiaux, 1923
amplicollis Fleutiaux, 1923b: 295 (*Galloisius*) **A**: JA SE TAI

tribe Eucnemini Eschscholtz, 1829

genus *Eucnemis* Ahrens, 1812: 38 type species *Eucnemis capucina* Ahrens, 1812
capucina Ahrens, 1812: 38 **E**: AU BE BG BH BY CR CT CZ DE GE FI FR HU IT LA LS LT LU NL NR NT PL RO SP SK SZ SV WS **N**: AG
ater Geoffroy, 1785: 128 (*Cucujus*)
carinatus Billberg, 1820: 21 (*Eucnemis*)
deflexicollis Mannerheim, 1823: 23 (*Elater*)
macrotis Beck, 1817: 18 (*Elater*)
zaitzevi Mamaev, 1976: 137 **A**: ES FE

genus *Poecilochrus* Bonvouloir, 1871: 148 type species *Poecilochrus piceus* Bonvouloir, 1874
japonicus Fleutiaux, 1923b: 293 **A**: JA

tribe Galbitini Muona, 1991

genus *Galbites* Fleutiaux, 1918b: 59 type species *Galba marmorata* Guérin-Ménèville, 1830
Galba Latreille, 1829a: 451 [HN] type species *Galba marmorata* Guérin-Ménèville, 1830
australiae Lea, 1919: 741 (*Galba*) **A**: BH TAI **AUR ORR**

bakeri Fleutiaux, 1926a: 36 (*Pterotarsus*)
mouhoti Fleutiaux, 1923c: 303 (*Pterotarsus*)
borealis Hisamatsu, 1980: 27 (*Pterotarsus*) **A**: JA
chevrolati Fleutiaux, 1923c: 304 (*Pterotarsus*) **A**: TAI **ORR**
chrysocoma Hope, 1845a: 14 (*Galba*) **A**: CH **ORR**
sauteri Muona, 1991: 27 **A**: TAI **ORR**
tomentosa Montrouzier, 1855: 13 (*Galba*) **A**: TAI **ORR**
tuberculata L. Redtenbacher, 1868: 90 (*Galba*) **A**: TAI **AUR ORR**
wallacei Perroud & Montrousier, 1864: 98 (*Galba*) **A**: CH **AUR ORR**

subfamily **Macrulacinae** Fleutiaux, 1922
tribe Anelastidini Muona, 1993

genus *Anelastidius* Jacquelin du Val, 1860a: 122 type species *Anelastidius ineditus* Jacquelin du Val, 1863 (= *Eucnemis feisthameli* Graells, 1847)
feisthameli Graells, 1847: 307 (*Eucnemis*) **E**: FR SP **N**: AG MO TU
algericus Pic, 1894n: 105 (*Fornax*)
ineditus Jacquelin du Val, 1863: pl. 28
zanzii Lacordaire, 1857: 117 (*Anelastes*)

tribe Euryptychini Mamaev, 1976

genus *Dromaeolus* Kiesenwetter, 1858a: 197 type species *Eucnemis barnabita* A. Villa & J. B. Villa, 1838
Megathambus Reitter, 1911a: 201 type species *Dromaeolus maronita* Bonvouloir, 1871
Melanus Broun, 1881: 676 type species *Melanus sculptus* Broun, 1881 (= *Elater nigellus* White, 1846)
barnabita A. Villa & J. B. Villa, 1838: 62 (*Euchnemis* [sic]) **E**: AU BH CR CT CZ FR GE HU IT PL RO SK SV **N**: AG
brevipes Fleutiaux, 1923b: 299 **A**: JA
cariniceps Fleutiaux, 1923b: 299 **A**: JA
gracilicornis Roubal, 1912: 124 **E**: "Caucasus"
japonensis Fleutiaux, 1923b: 299 **A**: JA
lewisi Fleutiaux, 1923b: 299 **A**: JA
maronita Bonvouloir, 1871: 201 **E**: ST "Caucasus"
nipponensis Fleutiaux, 1923b: 299 **A**: JA
rufulus Fleutiaux, 1923b: 299 **A**: JA

genus *Euryptychus* LeConte, 1852a: 46 type species *Eucnemis heterocerus* Say, 1836
Dyscolocerus Bonvouloir, 1871: 73 type species *Dyscolocerus subnitidus* Bonvouloir, 1872
ishiharai Hisamatsu, 1994: 151 **A**: JA (Ogasawara)
konigi Reitter, 1885i: 379 (*Phlegon*) **E**: UK
lewisi Fleutiaux, 1923b: 324 **A**: FE JA
vicinus Fleutiaux, 1923b: 324 **A**: FE JA
kishkini Fursov, 1936a: 348 (*Phlegon*)
ussuriensis L. N. Medvedev, 1970: 168

tribe Macraulacini Fleutiaux, 1922

genus *Asiocnemis* Mamaev, 1976: 139 type species *Dromaeolus morawitzi* Semenov, 1891
morawitzi Semenov, 1891: 346 (*Dromaeolus*) **A**: ES TM

genus *Fornax* Laporte, 1835b: 172 type species *Fornax ruficollis* Laporte, 1835
Filifornax Fleutiaux, 1945: 188 type species *Fornax leseleuci* Fleutiaux, 1899
Monilifornax Fleutiaux, 1945: 189 type species *Fornax lambertoni* Fleutiaux, 1926
consobrinus Hisamatsu, 1963: 26 **A**: JA
hachijonis Hisamatsu, 1963: 28 **A**: JA
hisamatsui Mamaev, 1976: 142 **A**: FE
lewisi Fleutiaux, 1923b: 295 **A**: JA
nipponicus Fleutiaux, 1923b: 295 **A**: JA

victor Fleutiaux, 1923b: 295 **A**: JA

genus *Heterotaxis* Bonvouloir, 1871: 75 type species *Heterotaxis myrmidon* Bonvouloir, 1875
nipparensis Hisamatsu, 1957: 45/46 **A**: JA

genus *Procladidus* Fleutiaux, 1921a: 80 type species *Dicladus favrei* Fleutiaux, 1912
coomani Fleutiaux, 1927d: 134 **A**: TAI **ORR**

genus *Pseudochapianus* Lucht, 1987: 38 type species *Pseudochapianus lopatini* Lucht, 1987
lopatini Lucht, 1987: 38 **A**: TD
rugosus Cobos, 1964: 363 (*Chapianus*) **A**: NO

genus *Raapia* Fleutiaux, 1899: 569 type species *Raapia galboides* Fleutiaux, 1899
sauteri Fleutiaux, 1929: 206 **A**: TAI

genus *Scython* Laporte, 1835b: 169 type species *Scython bicolor* Laporte, 1835
coloratus Bonvouloir, 1872: 493 **A**: TAI **ORR**
maculicollis Bonvouloir, 1872: 494 **A**: HKG **ORR**

genus *Serrifornax* Fleutiaux, 1926a: 45 type species *Fornax subflabellatus* Fairmaire, 1880
tumidicollis Redtenbacher, 1867: 92 (*Fornax*) **A**: TAI **ORR**

genus *Spinifornax* Fleutiaux, 1926: 46 type species *Fornax senex* Fleutiaux, 1926
miurai Nakane, 1987b: 10 **A**: JA (Yakushima)
oshimanus Nakane, 1987c: 174 [DA] **A**: JA (Amami-Ôshima)

genus *Thambus* Bonvouloir, 1871: 160 type species *Thambus agilis* Bonvouloir, 1871
friwaldszkyi Bonvouloir, 1871: 161 **E**: AU BU CR HU RO SK

genus *Torigaia* Hisamatsu & Satô, 1959: 16 type species *Torigaia bicolor* Hisamatsu & Satô, 1959
bicolor Hisamatsu & Satô, 1959: 16 **A**: JA

tribe Nematodini Leiler, 1976

genus *Nematodes* Berthold, 1827: 335 type species *Elater filum* Fabricius, 1801
Emathion Laporte, 1835b: 171 type species *Emathion cylindricum* Laporte, 1834
Hypocaelus Dejean, 1833: 85 type species *Elater filum* Fabricius, 1801
Sphaerocephalus Eschoscholtz, 1836: pl. type species *Sphaerocephalus brasiliensis* Germar, 1839
filum Fabricius, 1801b: 240 (*Elater*) **E**: AU FR GE HU IT PT RO SE SK UK
watanabei Hisamatsu, 1963: 31 **A**: JA

family Eucnemidae, nomen dubium
Elater buprestoides Fabricius, 1792b: 234

family THROSCIDAE Laporte, 1840
subfamily **Throscinae** Laporte, 1840
tribe Throscini Laporte, 1840

genus *Aulonothroscus* Horn, 1890: 193, 208 type species *Throscus validus* LeConte, 1868
brevicollis Bonvouloir, 1859: 8 (*Throscus*) **E**: AU BE BH BU BY CR CT CZ FI FR GB GE GR HU IR IT LS NL PL PT RO SK SP ST SV SZ TR UK "Caucasus" **N**: AG **A**: IN SY
dilatatus Reitter, 1901c: 60 (*Throscus*) **E**: AB
elongatulus Wollaston, 1865: 189, Append.: 30 (*Throscus*) **N**: CI (Gomera)
grancanariae Franz, 1982: 55 (*Throscus*) **N**: CI (Gran Canaria)
incrassatus Peyerhimoff, 1949: 272 **N**: MO

intactus Peyerhimoff, 1949: 271 **N**: MO
integer Wollaston, 1857a: 82 (*Throscus*) **N**: MR
kocheri Cobos, 1959: 155 **N**: MO
laticollis Rybinsky, 1897: 56 (*Throscus*) **E**: BY CR CT FI FR NR NT PL **A**: FE
major Barovskij, 1933: 192
latiusculus Wollaston, 1865: 189, Append.: 30 (*Throscus*) **N**: CI (Hierro)
longulus J. Weise, 1879a: 148 (*Throscus*) **A**: FE JA
maximus Pic, 1938a: 1 (*Trixagus*) **A**: TR
nikitskyi Egorov, 1992 **A**: FE
nipponicus Cobos, 1969: 50 **A**: JA
tenerifae Franz, 1982: 53 (*Throscus*) **N**: CI (Tenerife)
wollastoni Franz, 1982: 55 (*Throscus*) **N**: CI (Palma)

genus *Trixagus* Kugelann, 1794: 534 type species *Elater dermestoides* Linnaeus, 1767
Throscus Latreille, 1797: 42 type species *Elater dermestoides* Linnaeus, 1767
algiricus Bonvouloir, 1862: 350 (*Throscus*) **E**: FR IT PT **N**: AG MO MR
corsicus Reitter, 1889b: 35
angulatus Reitter, 1921a: 73 **N**: AG MO
asiaticus Bonvouloir, 1859: 9 (*Throscus*) **E**: AB AR GG IT **A**: SY
astruci Pic, 1920a: 6 **A**: SD
atticus Reitter, 1921a: 69 **E**: DE FI FR GE GR IT RO ST SV SZ **A**: AF CY TM
caucasicus Reitter, 1921a: 70
hauseri Cobos, 1963: 5 (*Throscus*)
bachofeni Reitter, 1905b: 91 (*Throscus*) **E**: ST **A**: TM
carinifrons Bonvouloir, 1859: 9 (*Throscus*) **E**: AB BE BH BU BY CT CZ DE FI FR GB GE GR HU IT LA LS LU NL NR NT PL RO SK SP ST SV SZ TR UK "Caucasus" **A**: CY FE
elateroides L. Redtenbacher, 1858: 402
seriatus Blair, 1942: 287
similis Baudi di Selve, 1870: 89 (*Throscus*)
chinensis Cobos, 1966a: 345 (*Throscus*) **A**: ZHE
dermestoides Linnaeus, 1767: 656 (*Elater*) **E**: AU AZ BU BY CR CT CZ DE EN FI FR GB GE HU IR IT LA LS LT LU NL NR NT PL PT RO SK SP ST SV SZ TR UK **A**: CY ES FE KZ TR
adstrictor Herbst, 1792: 140 (*Dermestes*)
clavicornis Geoffroy, 1785: 39 (*Elater*)
planifrons Rey, 1891a: 27
sulcifrons Rey, 1891a: 27
vastator Kugelann, 1792: 479 (*Dermestes*)
difficilis Pic, 1906j: 65 (*Throscus*) **N**: EG
duvalii Bonvouloir, 1859: 10 (*Throscus*) **E**: AU CZ DE FI FR GB GE GR HU IT PL RO SK SL SP ST SV **A**: CY
dohrni Bethe, 1868: 36
elateroides elateroides Heer, 1841: 443 (*Throscus*) **E**: AB AZ CR CT CZ FR GE GB GR HU IT LA NL PL RO SK SP ST UK YU **N**: AG MO MR **A**: SY TR
modestus J. Weise, 1877: 23
obtusus Kiesenwetter, 1858a: 183
elateroides simplicifrons Reitter, 1921a: 70 **E**: CR "Caucasus"
exul Bonvouloir, 1859: 10 (*Throscus*) **E**: CT FR GE GR HU PT RO SZ TR
aurociliatus Reitter, 1885h: 314
gauthieri Cobos, 1969: 45 (*Throscus*) **N**: AG
gracilis Wollaston, 1854: 237 **E**: AR AU AZ BH BU CR FR GB GE GR HU IT MD PT SP ST SZ **N**: AG MO MR **A**: CY TR
latior Pic, 1908d: 42 **A**: UZ (Andizhan)
lesegneuri Muona, 2002: 187 **E**: HU
maai Cobos, 1966a: 343 (*Throscus*) **A**: FUJ
mesopotamicus Pic, 1906j: 65 (*Throscus*) **A**: IQ
micado Reitter, 1921a: 68 **A**: FE JA

minutus Rey, 1891a: 27 **E**: IT FR SP **A**: MO AG
moderatus Schenkling, 1928: 20 (*Throscus*) [RN] **E**: AB **A**: SY
bonvouloirii Desbrochers des Loges, 1875a: 37 (*Throscus*) [HN]
rufulus Reitter, 1921a: 71
niger Gebler, 1830: 86 (*Throscus*) **A**: WS
obtusus Curtis, 1827: 163 (*Throscus*) **E**: AB BE CZ DE FR GE GB GR HU IT LS NL NR PL PT RO SL SK SP ST SZ UK **N**: AG EG MO MR **A**: CY TR
ciliaris Rey, 1891a: 27 (*Throscus*)
pusillus Heer, 1841: 443 (*Throscus*)
orientalis Bonvouloir, 1859: 28 (*Throscus*) **E**: TR "Caucasus" **A**: CY ES FE IN KZ ST SY TD TR "Siberia"
ovalis Reitter, 1921a: 73 **E**: AB ST
penicillus Reitter, 1921a: 70 **N**: AG TU
perversus Cobos, 1958: 3 (*Throscus*) **A**: AF TM
rougeti Fauvel, 1885: 333 (*Throscus*) **E**: FR IT SP SV "Kaspimeergebiet"
schenklingi Reitter, 1921a: 71 **A**: JA
syriacus Reitter, 1921a: 71 **A**: SY
turkestanus Reitter, 1921a: 71 **E**: AB **A**: "Turkestan" "Transcaspie"

family ELATERIDAE Leach, 1815
subfamily **Cebrioninae** Latreille, 1802

genus *Analestesa* Leach, 1824: 35 type species *Analestesa testacea* Leach, 1824
Cebriognathus Chobaut, 1899a: 22 type species *Cebriognathus desertorum* Chobaut, 1899
arabica Paulus, 1981: 261 (*Cebriognathus*) **A**: SA
persica Paulus, 1981: 261 (*Cebriognathus*) **A**: IN
testacea Leach, 1824: 36 **N**: AG
desertorum Chobaut, 1899a: 23 (*Cebriognathus*)

genus *Cebrio* A. G. Olivier, 1790a: no 30bis: 1 type species *Cistela gigas* Fabricius, 1787
Hammonia Latreille, 1817: 235 type species *Cebrio brevicornis* A. G. Olivier, 1790
Tibesia Leach, 1824: 41 type species *Cebrio ruficollis* Fabricius, 1798
abdominalis Laporte, 1840a: 253 **N**: AG
algerica Dalla Torre, 1911a: 4 [RN] **N**: AG
deformis Chevrolat, 1874c: 383 [HN]
amori Graells, 1851: 123 **E**: SP
amphimalloides Escalera, 1914: 218 **N**: MO
amplicollis Fairmaire, 1866a: 39 **N**: AG
andalusicus Jacquelin du Val, 1860d: 120 **E**: SP
angustatus Kocher, 1961a: 47 **N**: MO
angusticornis Fairmaire, 1873: 328 **N**: "North Africa"
annulicornis Chevrolat, 1874c: 410 **N**: AG
antennatus Chevrolat, 1874c: 392 **E**: GR
anthracinus Chevrolat, 1874c: 403 **E**: SP
apicalis Chevrolat, 1882: 4 **E**: SP
ariasi ariasi Escalera, 1914: 222 **N**: MO
nigrescens Escalera, 1914: 223
varipennis Escalera, 1914: 223
ariasi outkensis Kocher, 1967: 281 **N**: MO
aterrimus Chevrolat, 1874c: 396 **N**: AG
atlasicus atlasicus Kocher, 1952: 318 **N**: MO
atlasicus panousei Kocher, 1952: 319 **N**: MO
atricapillus Chevrolat, 1874c: 412 **N**: AG
atriceps Chevrolat, 1874c: 407 **N**: AG
attenuatus P. H. Lucas, 1846: 172 **N**: AG
barbarus P. H. Lucas, 1846: 170 **N**: AG

basicornis Chevrolat, 1874a: 31 **N**: AG
bayonnei Chobaut, 1897c: 204 **N**: AG
benedicti Fairmaire, 1849c: 420 **E**: IT
bipartitus Chevrolat, 1874c: 376 **N**: AG
biskrensis Fairmaire, 1876c: 93 **N**: AG
boucadensis Chevrolat, 1874a: 37 **N**: AG
bruleriei Heyden, 1870: 122 **E**: PT
cantabricus Bercedo-Páramo & López-Colón, 2003: 17 **E**: SP
capitatus Fairmaire, 1866a: 40 **N**: AG
carbonarius carbonarius Chevrolat, 1874a: 25 **E**: SP
carbonarius oropensis Caminero, 1983b: 135 **E**: SP
carinicollis Chevrolat, 1874c: 372 **N**: AG
carrenii Graells, 1846: 218 **E**: PT SP
catoxanthus Chevrolat, 1874c: 400 **N**: AG
  melanocephala Gandolphe, 1865: 12
cincticollis Chevrolat, 1874c: 365 **N**: AG
cinctiventris Chevrolat, 1874c: 379 **N**: AG
collaris Fairmaire, 1873: 337 **N**: AG
compactilis Chevrolat, 1874c: 398 **N**: AG
comptus Chevrolat, 1874c: 413 **N**: AG
conformis Chevrolat, 1874c: 368 **N**: AG
consimilis Chevrolat, 1874c: 368 **N**: AG
constantinensis Chevrolat, 1874c: 381 **N**: AG
convexiusculus Fairmaire, 1876c: 93 **N**: MO
cordubensis Perez Arcas, 1865: 32 **E**: SP
corsicus Jacquelin du Val, 1860d: 111 **E**: FR IT
coxalis Chevrolat, 1874c: 393 **N**: AG
crassus Fairmaire, 1871: 382 **N**: "Barbarie"
curvipes Pic, 1921a: 3 **N**: MO
cyrenaicus Pic, 1935h: 148 **N**: LB (Cyrenaica)
debduensis Kocher, 1967: 285 **N**: MO
decipiens Fairmaire, 1866a: 37 **N**: AG
decolor Chobaut, 1896b: 400 **N**: AG
deformis Chevrolat, 1874a: 38 **N**: AG
denominandus Chevrolat, 1874a: 37 **N**: AG
  confusus Fairmaire, 1866a: 36
dimidiatus P. H. Lucas, 1846: 171 **N**: AG MO
  costicollis Fairmaire, 1867: 402
  hirundinis Chevrolat, 1876b: clxxxvi
distinguendus Fairmaire, 1866a: 36 **N**: AG
divisus Chevrolat, 1874c: 375 **N**: AG
dubitabilis Fairmaire, 1871: 381 **N**: AG
dubius P. Rossi, 1790: 234 **E**: IT
  brevicornis A. G. Olivier, 1790a: no 30bis: 5
  nigricornis Leoni, 1906b: 204
dufouri Graells, 1851: 122 **E**: SP
elenacompteae Compte, 1988: 57 **E**: SP
elenae Fairmaire, 1882b: 446 **N**: TU
ernesti Chevrolat, 1874c: 408 **N**: AG
erythrogonus Chevrolat, 1874c: 369 **N**: AG
erythropterus Chevrolat, 1874c: 380 **N**: AG
fabricii Leach, 1824: 40 **E**: FR IT PT SP
  xanthomerus Germar, 1844: pl. 8
fairmairii Raffray, 1873: 373 **N**: AG
falsicolor Fairmaire, 1871: 381 **N**: AG

favieri favieri Fairmaire, 1884e: cxviii **N**: MO
favieri zaerensis Kocher, 1952: 313 **N**: MO
filicornis Fairmaire, 1873: 338 **N**: AG
fiorii Leoni, 1906b: 212 **E**: IT
fossulatus Perris, 1865: 508 **E**: FR
frater Jacquelin du Val, 1860d: 115 **E**: PT
fuscatus A. Costa, 1847b: 141 **E**: IT
fusciventris Chevrolat, 1874a: 33 **N**: AG
gandolphei Guérin-Méneville, 1859: clxxxvi **N**: AG
geminus Chevrolat, 1874c: 415 **N**: AG
 numidicus Fairmaire, 1866a: 40 [HN]
geniculatus Chevrolat, 1874c: 417 **N**: AG
germari Jacquelin du Val, 1860d: 127 **E**: PT SP
 melanocephalus Germar, 1839a: 17 [HN]
 nigricans Ragusa, 1894: 10
getschmanni Chevrolat, 1872: 409 **E**: SP
gigas Fabricius, 1787a: 84 (*Cistela*) **E**: FR IT SP
 latreillei Leach, 1824: 43
 longicollis A. G. Olivier, 1790a: no 30bis: 5
 promelus Leach, 1824: 39
grandipennis Fairmaire, 1871: 383 **N**: AG
guyonii Guérin-Méneville, 1844b: 403 **N**: AG
 obscurior Pic, 1905h: 154
gypsicola Graells, 1858: 48 **E**: SP
holofulvus Kocher, 1967: 286 **N**: MO
humerosus Chevrolat, 1874c: 413 **N**: AG
impresicollis Chevrolat, 1874c: 392 **E**: SP
impressifrons Fairmaire, 1867: 402 **N**: AG
infuscatus Chevrolat, 1874c: 384 **N**: AG
insignitus Jacquelin du Val, 1860d: 131 **E**: SP
insularis Chevrolat, 1874c: 42 **E**: AL CR GR
 alleonis Fairmaire, 1880d: 240
intermedius Chevrolat, 1874c: 383 **N**: AG
kruegeri Pic, 1935h: 149 **N**: LB (Cyrenaica)
 gridellii Pic, 1935h: 149
juvencus Chevrolat, 1874c: 377 **N**: AG
labouli Pic, 1935c: 6 **N**: AG
lanuginosus Chevrolat, 1874c: 371 **N**: AG
latericollis Chevrolat, 1874c: 382 distribution unknown, possibly **N**: AG
laticornis Chevrolat, 1874c: 409 **N**: MO
levaillanti Chevrolat, 1874c: 401 **N**: AG
longipennis Fairmaire, 1866a: 31 **N**: AG
lucasi Fairmaire, 1866a: 39 **N**: AG MO
 melanocephalus P. H. Lucas, 1846: 172 [HN]
luctuosus Fairmaire, 1867: 400 **N**: AG
luteolus Fairmaire, 1873: 339 **N**: AG
macilentus Chevrolat, 1874c: 382 **N**: AG
maculicollis maculicollis Fairmaire, 1856b: 531 **N**: MO
 mogadoricus Escalera, 1914: 217
maculicollis susicus Kocher, 1961a: 40 **N**: MO
malaccensis Dieck, 1870: 124 **E**: SP
marginipennis Fairmaire, 1866a: 43 **N**: "North Africa"
maroccanus Chevrolat, 1874c: 418 **N**: MO
marraquense Escalera, 1914: 219 **N**: MO
mediatlantis mediatlantis Kocher, 1952: 320 **N**: MO

mesatlanticus Kocher, 1956a: 83 [unjustified emendation]
mediatlantis rufisternis Kocher, 1952: 321 **N**: MO
melanocephalus Leach, 1824 (*Hammonia*) **E**: PT SP
melas Fairmaire, 1866a: 29 **N**: AG
melillense Escalera, 1914: 223 **N**: MO
obscuratus Escalera, 1914: 224
monastirensis Pic, 1935i: 158 **E**: MC
morio Leach, 1824: 40 **E**: PT SP
moyses Fairmaire, 1852: 82 **E**: PT SP
mussardi Kocher, 1969: 107 **N**: MO
muticus Chevrolat, 1874c: 386 **N**: AG
neapolitanus A. Costa, 1847b: 137 **E**: IT
nigricans P. H. Lucas, 1846: 175 **N**: AG
nigriceps Fairmaire, 1873: 338 **N**: AG
nigricollis Laporte, 1838: 16 **A**: TR
nigricornis Leoni, 1906b: 204 **E**: IT
numida Fairmaire, 1866a: 31 **N**: AG
numidicus P. H. Lucas, 1846: 114 **N**: AG
obscuripes Chevrolat, 1874c: 394 **N**: AG
barbarus Gandolphe, 1865: 14 [HN]
obtusangulus Fairmaire, 1866a: 28 **N**: AG
oranensis Chevrolat, 1874c: 402 **N**: AG MO
pachycephalus Chevrolat, 1874: 416 **N**: AG
pallidipennis Chevrolat, 1874a: 35 **N**: AG
parvicollis Dieck, 1870: 122 **E**: SP
patruelis Fairmaire, 1866a: 33 **N**: MO
pectoralis Chevrolat, 1875: 538 **N**: AG
pelissieri Pic, 1905h: 154 **N**: AG
pellucidus Fairmaire, 1894: 310 **N**: AG
personatus Chevrolat, 1874c: 419 **E**: SP
peyerimhoffi Kocher, 1952: 316 **N**: MO
piceiventris Fairmaire, 1894: 310 **N**: AG
picipennis Chevrolat, 1874c: 378 **N**: AG
pilifrons Fairmaire, 1867: 401 **N**: MO
nigrinus Escalera, 1914: 218
nigrosuturalis Escalera, 1914: 218
poupillieri Chevrolat, 1874c: 370 **N**: AG
puberulus Chevrolat, 1874c: 422 **E**: SP
pubicollis Pic, 1935c: 5 **N**: AG
pubicornis Fairmaire, 1869a: 233 **E**: PT
quadraticollis Chevrolat, 1874c: 385 **N**: AG
raynali Kocher, 1954c: 283 **N**: MO
reichei Fairmaire, 1867: 401 **N**: AG
reymondi Kocher, 1952: 309 **N**: MO
rifensis Kocher, 1963: 193 **N**: MO
roberti Pic, 1921a: 3 **N**: MO
robustus Pic, 1935c: 6 **N**: AG
rotroui Kocher, 1961a: 43 **N**: MO
rozasi Cobos, 1985: 331 **E**: SP
rubicundus Jacquelin du Val, 1860d: 126 **E**: SP
rubrocinctus Chevrolat, 1874c: 371 **N**: AG
rufangulus Chevrolat, 1874c: 366 **N**: AG
ruficollis Fabricius, 1798: 102 **N**: MO
reymondi Kocher, 1952: 309
rufifrons Graells, 1849: 620 **E**: SP

rufipes Chevrolat, 1861: 267 **N**: AG
rugecostatus Chevrolat, 1874a: 34 **N**: AG
rungsi Kocher, 1952: 322 **N**: MO
saintpierrei Chevrolat, 1874c: 395 **N**: AG
sardous Perris, 1869: 16 **E**: IT
satanas Kocher, 1952: 315 **N**: MO
scutellaris Fairmaire, 1867: 403 **N**: AG
sefrensis Chobaut, 1896b: 399 **N**: AG
segmentatus Chevrolat, 1874c: 390 **N**: MO
cinnabaripennis Chevrolat, 1874c: 391
duodecimarticulatus Escalera, 1914: 220
rufinus Escalera, 1914: 221
seguranus Caminero, 1983a: 27 **E**: SP
semimarginatus Chevrolat, 1874a: 33 **N**: AG
semistriatus Chevrolat, 1874a: 36 **N**: AG
seoanei Perez Arcas, 1865: 30 **E**: SP
spurcaticollis Fairmaire, 1871: 380 **N**: AG
striatifrons Fairmaire, 1876c: 93 **N**: MO
strictus Gené, 1836: 177 **E**: IT
doderoi Leoni, 1906b: 214
varicolor Perris, 1869: 17
subattenuatus Pic, 1905h: 154 **N**: AG
subgracilis Kocher, 1961a: 44 **N**: MO
sulcicollis Chevrolat, 1874a: 30 **N**: AG
superbus Jacquelin du Val, 1860d: 119 **E**: SP
suturalis Boisduval, 1835: 102 **E**: SP
tarifensis Dieck, 1870: 121 **E**: SP
testaceus Laporte, 1840a: 253 **E**: SP
theryi Pic, 1935: 159 **N**: AG
tibialis Chevrolat, 1874a: 32 **N**: AG
tiziifrensis Cobos, 1961: 73 **N**: MO
transversalis Chevrolat, 1874c: 411 **N**: AG
tricolor Graells, 1858: 47 **E**: SP
ventralis Chevrolat, 1874c: 397 **N**: AG
wohlberedti Pic, 1935: 150 **N**: LB (Cyrenaica)
xanthoderus Fairmaire, 1866a: 44 **N**: AG
vaulogeri Pic, 1905h: 153
xanthognathus Chevrolat, 1874a: 31 **N**: AG
xanthopus Fairmaire, 1866a: 32 **N**: MO
gracilissimus Fairmaire, 1876c: 93
ysernii Graells, 1858: 48 **E**: SP

genus *Cebriorhipis* Chevrolat, 1875: 525 type species *Cebriorhipis elongatus* Chevrolat, 1875
striatulus Bourgeois, 1896: 117 **A**: TAI **ORR**
uniformis Fairmaire, 1887b: 145 **A**: FUJ (Xiamen)

genus *Paulusiella* Löbl, 2007 [see New acts] type species *Paulusiella fossulatipennis* Mandl, 1974
fossulatipennis Mandl, 1974: 636 **A**: PA
richteri acuticollis Mandl, 1974: 642 **A**: IN (Belutschistan)
richteri richteri Mandl, 1974: 641 **A**: IN
pallida Mandl, 1974: 642 **A**: IN (Makran)
serraticornis Paulus, 1972: 38 (*Escalerina*) **A**: IN

subfamily **Oxynopterinae** Candèze, 1857
tribe Campsosternini Fleutiaux, 1927

genus *Campsosternus* Latreille, 1834: 141 type species *Elater fulgens* A. G. Olivier, 1790 (= *Elater auratus* Drury, 1773)
auratus Drury, 1773: 65 (*Elater*) **A**: FUJ GUA GUI GUX HAI HEN HKG HUB HUN JIX SCH SHG TAI YUN ZHE **ORR**
fulgens A. G. Olivier, 1790: no 31: 12 (*Elater*)
niger Fleutiaux, 1918b: 200
bimaculatus Jiang, 1991: 276 **A**: SCH
brunneicornis Candèze, 1889b: 82 **A**: SD
cantori Hope, 1842b: 453 [= 1843b: 288] **A**: BT **ORR**
csikii Szombathy, 1909: 120 **A**: GUI JIA SCH SHG
davidis Fairmaire, 1887a: 120 **A**: FUJ **ORR**
dohrnii dohrnii Westwood, 1848: 71 **A**: GUX YUN **ORR**
dohrnii fairmairei Fleutiaux, 1918b: 199 **A**: GUX **ORR**
elongatus Fleutiaux, 1926b: 58 **A**: JIX SCH XIZ YUN **ORR**
fruhstorferi Schwarz, 1902b: 318 **A**: GUX HUB HUN JIA JIX YUN **ORR**
gemma Candèze, 1857: 344 **A**: ANH FUJ GUA GUI HUB HUN JIA JIX SCH SHG ZHE
guishuni Ôhira, 1977b: 34 **A**: TAI
guizhouensis Jiang, 1991: 277 **A**: GUI
malaisianus Candèze, 1865: 19 **A**: "China" **ORR**
matsumurae Miwa, 1929c: 344 [RN] **A**: JA (Ishigaki-jima, Iriomote-jima)
quadrimaculatus Miwa, 1927a: 14 [HN]
mirabilis Fleutiaux, 1930b: 409 **A**: HUN TAI
nobuoi Ôhira, 1966a: 8 **A**: JA (Yonaguni-jima)
quadrimaculatus Fleutiaux, 1926b: 59 **A**: YUN
rasilus Schimmel, 1994: 48 **A**: NP SD
saundersi Candèze, 1874: 203 **A**: YUN
sobrinus Candèze, 1874: 201 **A**: YUN **ORR**
stephensii Hope, 1831: 25 (*Elater*) **A**: NP XIZ **ORR**
watanabei Miwa, 1929b: 241 **A**: TAI
yasuakii W. Suzuki, 2002: 326 **A**: TAI

tribe Oxynopterini Candèze, 1857

genus *Ceropectus* Fleutiaux, 1927a: 117 type species *Pectocera messi* Candèze, 1874
messi Candèze, 1874: 207 (*Pectocera*) **A**: FUJ GUA GUX HKG HUB **ORR**

genus *Oxynopterus* Hope, 1842a: 77 type species *Elater mucronatus* Olivier, 1792
subgenus Oxynopterus Hope, 1842a: 77 type species *Elater mucronatus* Olivier, 1792
Leptophyllus Hope, 1842a: 79 type species *Leptophyllus Strachani* Hope, 1842
annamensis Fleutiaux, 1918c: 212 **A**: YUN **ORR**

genus *Sternocampsus* Fleutiaux, 1927c: 104 type species *Sternocampsus villosus* Fleutiaux, 1927
castaneus Jiang, 1999: 34 [DA] **A**: YUN

tribe Pectocerini Gurjeva, 1974

genus *Ceroleptus* Fleutiaux, 1927a: 121 type species *Pectocera sulcata* Fleutiaux, 1903
Neopectocera Ôhira, 1971c: 207 type species *Pectocera nivea* Fleutiaux, 1918
brevicollis Candèze, 1878b: lvi (*Pectocera*) **A**: GUA

genus *Pectocera* Hope, 1842a: 79 type species *Pectocera Cantori* Hope, 1842
amamiinsulana amamiinsulana Nakane, 1957: A238 **A**: JA (Amami-Ôshima)
amamiinsulana ogatai Kishii, 1996b: 13 **A**: JA (Toku-no-shima)
babai Kishii, 1990c: 19, 31 **A**: TAI

cantori Hope, 1842a: 79 **A**: FUJ HUN NP SCH SD XIZ YUN **ORR**
mellii Hope, 1842a: 79
chaurengiensis Schimmel, 1994: 56 **A**: UP
deboisae Schimmel, 1994: 58 **A**: SD
delessertii Guérin-Méneville, 1840: 38 (*Elater*) **A**: NP **ORR**
fleutiauxi Schimmel, 1997: 47 **A**: YUN **ORR**
formosana Kishii, 1990c: 18, 29 **A**: TAI
fortunei Candèze, 1873: 6 **A**: FUJ HAI HUB JIA JIX NC SC SCH ZHE **ORR**
hige hige Kishii, 1993b: 2 **A**: JA
hige yakuana Kishii, 1993b: 4 **A**: JA (Yaku-shima)
holzschuhi Schimmel, 1994: 59 **A**: NP
indica Schimmel, 1994: 62 **A**: UP
ingridae Schimmel, 1994: 63 **A**: NP SD
jiangxiana Kishii & Jiang, 1994: 96 **A**: JIX
kashmirensis Schimmel, 1997: 48 **A**: KA
kobayashii Kishii, 1996e: 103 **A**: TAI
maruyamai Kishii, 2001b: 41 **A**: JA (Kume-jima)
sechuana Schimmel, 2003b: 286 **A**: SCH
sikkimensis Schimmel, 1994: 65 **A**: SD
tonkinensis Fleutiaux, 1918b: 202 **A**: SCH YUN ZHE **ORR**
yaeyamana W. Suzuki, 1976: 263 **A**: JA (Iriomote-jima, Ishigaki-jima) TAI
yonaha Kishii, 1996b: 10 **A**: JA (Okinawa-hontô)

subfamily **Tetralobinae** Laporte, 1840
tribe Tetralobini Laporte, 1840
subtribe Tetralobina Laporte, 1840

genus *Sinelater* Laurent, 1967: 94 type species *Tetralobus perroti* Fleutiaux, 1941
perroti Fleutiaux, 1941: 107 (*Tetralobus*) **A**: FUJ GUA GUI GUX HAI HUB HUN JIX SCH ZHE **ORR**

subfamily **Pityobiinae** Hyslop, 1917
tribe Adzusini Kishii, 1989

genus *Adzusa* Kishii, 1957a: 10 type species *Adzusa inexpecta* Kishii, 1957
inexpecta Kishii, 1957a: 11 **A**: JA

tribe Rostricephalini Fleutiaux, 1947

genus *Rostricephalus* Fleutiaux, 1918b: 252 type species *Rostricephalus vitalisi* Fleutiaux, 1918
vitalisi Fleutiaux, 1918b: 253 **A**: GUA TAI **ORR**

subfamily **Agrypninae** Candèze, 1857
tribe Agrypnini Candèze, 1857

genus *Adelocera* Latreille, 1829a: 451 type species *Elater ovalis* Germar, 1824
Aganolacon Ôhira, 1967a: 55 type species *Brachylacon* (*Aganolacon*) *shirozui* Ôhira, 1967
Brachylacon Motschulsky, 1858c: 60 type species *Brachylacon microcephalus* Motschulsky, 1858
Pericus Candèze, 1857: 167 type species *Pericus nitidus* Candèze, 1857
Prolacon Fleutiaux, 1934a: 179 type species *Prolacon alluaudi* Fleutiaux, 1934
Trachylacon Motschulsky, 1858c: 61 type species *Trochylacon* [sic!] *fulvicollis* Motschulsky, 1858
adspersus Candèze, 1857: 115 (*Lacon*) **A**: "North India"
aethiopicus Candèze, 1882: 7 (*Lacon*): **A**: YE **AFR**
afghana Platia & Gudenzi, 2002b: 75 **A**: AF
ami Kishii, 1996a: 98 **A**: TAI
brunneus Lewis, 1894a: 29 (*Lacon*) **A**: JA
iriei Ôhira, 1978a: 32
difficilis Lewis, 1894a: 29 (*Lacon*) **A**: JA TAI
fulvicollis Motschulsky, 1858c: 61 (*Trochylacon* [sic!]) **A**: HKG **ORR**

distinctus Fleutiaux, 1935b: 14 (*Trachylacon*)
gressitti Ôhira, 1972a: 3 (*Brachylacon*) **A**: TAI
ichihashii Ôhira, 1978a: 31 **A**: JA
imfoveatus Vats & Kashyap, 1992a: 8 **A**: UP
nitidus Candèze, 1857: 167 (*Pericus*) **A**: "North India"
sanguinolentus Candèze, 1893b: 172 (*Pericus*)
omotoensis Ôhira, 1978a: 33 **A**: JA (Ishigaki-jima)
ovalis Germar, 1824: 49 (*Elater*) **A**: IN
pentagonalus Vats & Kashyap, 1992a: 5 **A**: UP **ORR**
pretzmanni Platia & Gudenzi, 2002b: 75 **A**: IN
pygmaeus Baudi di Selve, 1871: 49 (*Lacon*) **A**: CY IS LB TR
adanensis Jagemann, 1944: 333
schuberti Platia & Gudenzi, 2002b: 76 **A**: TR
shibatai Kishii, 1996a: 95 **A**: TAI
shirozui ishigakiensis Ôhira, 1967b: 105 (*Aganolacon*) **A**: JA (Ishigaki-jima)
shirozui shirozui Ôhira, 1967a: 55 (*Aganolacon*) **A**: TAI
takasago Kishii, 1996a: 99 **A**: TAI
tumens Candèze, 1873: 4 (*Lacon*) **A**: FUJ GUA JA TAI
yemenita Platia & Schimmel, 1997: 293 **A**: YE

genus *Agraeus* Candèze, 1857: 165 type species *Agraeus mannerheimii* Candèze, 1857
Cavicoxum Pic, 1928b: 21 type species *Cavicoxum monstrosum* Pic, 1928
bhasini Fleutiaux, 1932b: 228 **A**: NP UP
excavatus Fleutiaux, 1918c: 207 **A**: NP
coomani Fleutiaux, 1927a: 97
falsus Fleutiaux, 1927a: 97 **A**: HAI **ORR**
feai Fleutiaux, 1935b: 23 **A**: HP SD UP **ORR**

genus *Agrypnus* Eschscholtz, 1829: 32 type species *Elater murinus* Linnaeus, 1758
Archontas Gozis, 1886: 23 type species *Elater murinus* Linnaeus, 1758
Archontoides Cobos, 1966b: 651 type species *Archontoides pretoriensis* Cobos, 1966
Centrostethus Schwarz, 1898c: 414 [RN] type species *Elater* (*Conoderus*) *cuspidatus* Klug, 1833
Colaulon Arnett, 1952: 116 type species *Elater rectangularis* Say, 1825
Compsolacon Reitter, 1905a: 6 type species *Elater crenicollis* Ménétriés, 1832
Cryptolacon Nakane & Kishii, 1955b: 1 type species *Cryptolacon miyamotoi* Nakane & Kishii, 1955
Enoploderes Schwarz, 1898a: 131 [HN] type species *Elater* (*Conoderus*) *cuspidatus* Klug, 1833
Homeolacon Blackburn, 1890: 90 type species *Homeolacon gracilis* Blackburn, 1890
Lobotarsus Schwarz, 1898a: 131 type species *Lobotarsus decoratus* Schwarz, 1898
Mecynocanthus Hope, 1837: 53 type species *Mecynocanthus unicolor* Hope, 1837
Myrmodes Candèze, 1857: 168 type species *Myrmodes akidiformis* Candèze, 1857
Neolacon Miwa, 1929b: 234 type species *Neolacon formosanus* Miwa, 1929
Paralacon Reitter, 1905a: 6 type species *Elater cinnamomeus* Candèze, 1874
Pseudolacon Blackburn, 1890: 89 type species *Pseudolacon rufus* Blackburn, 1890
Pyrganus Golbach, 1968: 198 type species *Lacon tuspanensis* Candèze, 1857
Sabikikorius Nakane & Kishii, 1955b: 3 type species *Lacon fuliginosus* Candèze, 1865
Sagojyo Kishii, 1964: 30 type species *Colaulon yuppe* Kishii, 1964
Tylotarsus Germar, 1840: 247 type species *Tylotarsus cinctipes* Germar, 1840
acristatus Vats & Kashyap, 1992c: 107 **A**: UP
acuminipennis Fairmaire, 1878b: 109 (*Lacon*) **A**: GUI GUX HUB JIX SCH YUN **ORR**
acutangulus Fleutiaux, 1936a: 15 (*Compsolacon*) **A**: GAN SCH
aequalis Candèze, 1900: 78 (*Lacon*) **A**: KA PA
agrestis Vats & Kashyap, 1992c: 123 **A**: UP
amamianus Kishii, 1974: 3 (*Colaulon*) **A**: JA
amamiensis amamiensis Miwa, 1934: 248 (*Paralacon*) **A**: JA (Amami-Ôshima, Toku-no-shima, Kika-jima)
amamiensis arenicola Nakane & Kishii, 1955b: 3 **A**: JA (Takara-jima)
amamiensis okinawanus Ôhira, 1967b: 101 (*Adelocera*) **A**: JA (Okinawa-hontô)
ami Kishii, 1995: 4 **A**: TAI

anathesinus Candèze, 1897: 10 (*Lacon*) **A**: "China"
angustus Fleutiaux, 1942: 1 **A**: YUN **ORR**
argentosquamus Vats & Kashyap, 1992c: 102 **A**: UP
argillaceus argillaceus Solsky, 1871: 360 (*Lacon*) **A**: BEI GAN GUX HAI HEN HUB FE JIL LIA NC NMO SC SCH TAI XIZ YUN **ORR**
cinnamomeus Candèze, 1874: 76 (*Lacon*)
argillaceus shirozui Ôhira, 1966c: 216 (*Adelocera*) **A**: TAI
arorai Vats & Kashyap, 1992c: 105 **A**: HP
babanus Kishii, 1990c: 12, 21 **A**: TAI
baibaranus Hayek, 1973: 129 [RN] **A**: FUJ GUI JIX SCH TAI XIZ YUN
formosanus Miwa, 1929b: 235 (*Neolacon*) [HN]
bengalensis Garg, Saini & Vasu, 1997b: 284 **A**: SD
bilyi Cate, Platia & Schimmel, 2002: 26 **A**: IN
binodulus binodulus Motschulsky, 1861a: 8 (*Lacon*) **A**: FE JA JIX NC SC
albomaculatus Miwa, 1934: 68 (*Lacon*)
binodulus coreanus Kishii, 1961: 24 **A**: GAN HEN HUB JA (Tsushima) JIX LIA NC SC
binus Candèze, 1889b: 73 (*Lacon*) **A**: HP
bipapulatus Candèze, 1865: 11 (*Lacon*) **A**: FUJ GUI GUX HEI HEN HUB JA JIA JIL JIX LIA NMO SCH TAI YUN
blackburni Hayek, 1973: 132 [RN] **A**: SD
duplex Candèze, 1897: 9 (*Lacon*) [HN]
blairei Fleutiaux, 1927a: 77 (*Adelocera*) **A**: GUA **ORR**
boomensis Vats & Kashyap, 1992c: 101 **A**: UP **ORR**
brachychaetus Kollar, 1844: 506 (*Lacon*) **A**: KA PA
brunnipennis Candèze, 1857: 133 (*Lacon*) **A**: NP "North India"
buyssoni Jagemann, 1944: 332 (*Lacon*) **A**: XIZ
campestris Vats & Kashyap, 1992c: 116 **A**: UP
cashmiriensis Della Beffa, 1931: 181 (*Compsolacon*) **A**: KA PA
coarctatus Candèze, 1874: 70 (*Lacon*) **A**: NP "Himalaya"
coenosus Hope, 1831: 25 (*Elater*) **A**: NP "North India" **ORR**
nepalensis Candèze, 1871: 72 (*Lacon*)
colonicus Candèze, 1882: 8 (*Lacon*) **A**: YUN **ORR**
confertus Vats & Kashyap, 1992c: 140 **A**: HP UP
consobrinus Candèze, 1857: 125 (*Lacon*) **A**: HP UP
cordicollis Candèze, 1865: 9 (*Lacon*) **A**: JA JIX "Korea"
costicollis Candèze, 1857: 116 (*Lacon*) **A**: NP UP YUN **ORR**
crenicollis Ménétriés, 1832: 156 (*Elater*) **E**: AB AR BH BU FR GG GR IT ST UK YU **A**: IN TR
davidis Fairmaire, 1878b: 109 (*Lacon*) **A**: FUJ HUB HUN GAN GUI JIL LIA NIN SCH SHX XIZ YUN **ORR**
depressus Candèze, 1874: 77 (*Lacon*) **A**: NC SC
dilatipennis Kishii, 1995: 6 **A**: TAI
ellipticus Candèze, 1857: 135 (*Lacon*) **A**: AF HP PA UP "Himalaya"
propinquus Candèze, 1857: 135 (*Lacon*)
formosanus formosanus Bates, 1866: 348 (*Lacon*) **A**: TAI
formosanus lanyuanus Kishii, 1995: 4 **A**: TAI (Lanyu Is.)
fuliginosus Candèze, 1865: 10 (*Lacon*) **A**: JA SC HEB
fulvastra Fleutiaux, 1940c: 91 (*Adelocera*) **A**: XIZ
fuscoluridus Vats & Kashyap, 1992c: 96 **A**: UP **ORR**
gypsatus Candèze, 1891b: 773 (*Lacon*) **A**: GUI YUN **ORR**
haedulus Candèze, 1857: 122 (*Lacon*) **A**: "North India"
heianus Kishii, 1985: 5 **A**: JA (Toku-no-shima)
herzi Koenig, 1887: 353 (*Lacon*) **A**: NC SC "North China"
himalayanus Jagemann, 1944: 334 (*Compsolacon*) **A**: NP XIZ
hispidulus Candèze, 1857: 125 (*Lacon*) **A**: HAI **ORR**
holontelius Vats & Kashyap, 1992c: 121 **A**: HP UP
hypnicola Kishii, 1964: 33 (*Colaulon*) **A**: JA SC

scrofilia Ôhira, 1969b: 43 (*Colaulon*)
impectinatus Vats & Kashyap, 1992c: 119 **A**: HP
indianus Garg, Saini & Vasu, 1997b: 285 **A**: SD **ORR**
indosinensis Fleutiaux, 1927a: 86 (*Adelocera*) **A**: GUA HAI **ORR**
inlustris Vats & Kashyap, 1992c: 126 **A**: UP **ORR**
jacksoni Vats & Kashyap, 1992c: 132 **A**: UP
jagemanni Jiang, 1993: 138 [RN] **A**: XIZ
schwarzi Jagemann, 1944: 333 (*Lacon*] [HN]
judex Candèze, 1874: 62 (*Lacon*) **A**: HUB JIA JIX NMO SHG
jurulosus Candèze, 1889b: 73 (*Lacon*) **A**: SD HP
kawamurae Miwa, 1929b: 230 (*Lacon*) **A**: JIX SCH TAI
korotjaevi Gurjeva, 1987a: 41 (*Compsolacon*) **A**: TD
kuluensis Vats & Kashyap, 1992c: 110 **A**: HP
kumayuni Garg, Saini & Vasu, 1997b: 287 **A**: UP
kusuii Ôhira, 1993: 249 **A**: JA
lapideus Candèze, 1857: 141 (*Lacon*) **A**: YUN **ORR**
lameyi Fleutiaux, 1927a: 85 (*Adelocera*) **A**: GUA **ORR**
liukueiensis Kishii, 1990c: 12, 22 **A**: TAI
lucidus Vats & Kashyap, 1992c: 115 **A**: UP
lupinosus Candèze, 1857: 130 (*Lacon*) **A**: "North India"
lutosus Candèze, 1857: 119 (*Lacon*) **A**: NP
maisus Vats & Kashyap, 1992c: 130 **A**: HP UP
mikawaensis Ôhira, 1986b: 213 **A**: JA
miyakei miyakei Ôhira, 1967b: 100 **A**: JA (Amami-Ôshima, Toku-no-shima)
miyakei seinoi Kishii, 1985: 2 **A**: JA (Okinawa-hontô)
miyamotoi erabu Kishii, 1996b: 4 **A**: JA (Kuchino-erabu)
miyamotoi haterumanus Ôhira, 1967b: 104 (*Colaulon*) **A**: JA (Ryukyus)
miyakoensis Ôhira, 1969a: 93 (*Colaulon*)
miyamotoi ihai Ôhira, 1967b: 104 (*Colaulon*) **A**: JA (Ryukyus)
miyamotoi ishiharai Ôhira, 1994a: 285 **A**: JA (Miyake-jima)
miyamotoi kikai Kishii, 1985: 5 **A**: JA (Kikai-ga-shima)
miyamotoi kume Kishii, 1996b: 6 **A**: JA (Kume-shima)
miyamotoi miyamotoi Nakane & Kishii, 1955b: 2 (*Cryptolacon*) **A**: JA (Tokara, Yaku-shima) SC (Cheju)
yaku Kishii, 1996b: 4
miyamotoi ogatai Kishii, 1996b: 4 **A**: JA (Toku-no-shima)
miyamotoi senkakuanus Ôhira, 2004: 70 **A**: JA (Senkaku)
miyamotoi tosa Kishii, 1996b: 4 **A**: JA (Okino-shima)
modestus Candèze, 1857: 118 (*Lacon*) **A**: JIX TAI **ORR**
major Fleutiaux, 1889: 139 (*Lacon*)
mohanensis Vats & Kashyap, 1992c: 99 **A**: UP
montanus Miwa, 1929b: 229 (*Lacon*) **A**: HUB SCH TAI
murinus Linnaeus, 1758: 406 (*Elater*) **E**: AB AL AN AR AU BE BH BU BY CR CT CZ DE EN FI FR GE GB GG GR HU IR IT LA LS LT LU MC MD NL NR NT PL RO SK SL SP ST SV SZ TR UK YU **A**: ES FE IN KZ TR WS XIN
kokeilli Küster, 1845: 6 (*Lacon*)
mucorea, LeConte, 1853c: 491 (*Adelocera*)
nebulosus Razoumowsky, 1789: 161 (*Elater*)
sordidus Candèze, 1857: 114 (*Lacon*)
muscroides Kishii, 1995: 10 **A**: TAI
musculus Candèze, 1857: 141 (*Lacon*) **A**: FUJ GUA HAI HKG HUB JIA JIX SCH ZHE
muticus Herbst, 1806: 93 (*Elater*) **A**: UP **ORR**
nagaoi Ôhira, 1966a: 7 (*Colaulon*) **A**: JA (Ryukyus) TAI
niger Schwarz, 1905: 258 (*Lacon*) **A**: SD
oinodon Wurst, Schimmel & Platia, 2001: 542 **A**: YE
omanensis Platia & Schimmel, 1997: 293 **A**: OM

orientalis Hope, 1843a: 63 [= 1845a: 9] **A**: ZHE
oshimanus Ôhira, 1969a: 91 **A**: JA (Amami-Ôshima)
pardalinus Candèze, 1882: 9 (*Lacon*) **A**: SD "Himalaya"
pecirkai Jagemann, 1944: 333 (*Lacon*) **A**: XIZ
planocorpus Vats & Kashyap, 1992c: 138 **A**: UP
plantaferusus Vats & Kashyap, 1992c: 123 **A**: UP
planus Vats & Kashyap, 1992c: 111 **A**: UP **ORR**
polishaensis Ôhira, 1977b: 33 **A**: TAI
rameshi Vats & Kashyap, 1992c: 129 **A**: HP UP
repercussus Vats & Kashyap, 1992c: 131 **A**: UP
ryukyuensis Kishii, 1985: 3 **A**: JA (Ryukyus)
sakaguchii Miwa, 1927a: 14 (*Lacon*) **A**: JA (Ryukyus) TAI
sakishimanus sakishimanus Ôhira, 1967b: 100 **A**: JA (Ryukyus)
sakishimanus tamurai Kishii, 1985: 3 **A**: JA (Okinawa-hontôs)
saltatus Vats & Kashyap, 1992c: 138 **A**: HP UP
sauteri Ôhira, 1970f: 208 (*Adelocera*) **A**: GUI TAI
scrofa Candèze, 1873: 4 (*Lacon*) **A**: JA JIX SC
 katoi Kishii, 1992b: 2 **A**: JA
scutellaris hamai Ôhira, 1967b: 99 **A**: JA (Ryukyus)
scutellaris scutellaris Candèze, 1895: 9 (*Lacon*) **A**: JA (Tokara, Amami, Okinawa-hontô)
scutellatus Candèze, 1857: 111 (*Lacon*) **A**: UP "China" **ORR**
serrula Candèze, 1857: 122 (*Lacon*) **A**: GUA NP "North India" **ORR**
setiger Bates, 1866: 348 (*Lacon*) **A** GUA JA (Iriomote-jima) SC TAI YUN **ORR**
shirakii Matsumura, 1910: 39 (*Lacon*) [= 1911b: 146] **A**: TAI
sinensis Candèze, 1857: 139 (*Lacon*) **A**: GUA XIZ YUN **ORR**
 massiei Fleutiaux, 1895: 685 (*Lacon*)
 tonkinensis Jagemann, 1944: 334 (*Compsolacon*)
solanensis Vats & Kashyap, 1992c: 127 **A**: HP
souslapisus Vats & Kashyap, 1992c: 135 **A**: HP
squamafraxineus Vats & Kashyap, 1992c: 103 **A**: UP
sternoviridis Vats & Kashyap, 1992c: 97 **A**: UP
subargillus Vats & Kashyap, 1992c: 128 **A**: HP
subfaenum Vats & Kashyap, 1992c: 133 **A**: HP UP **ORR**
sublapideus Vats & Kashyap, 1992c: 134 **A**: UP
substramentum Vats & Kashyap, 1992c: 92 **A**: UP
taciturnus koshunensis Miwa, 1929b: 233 (*Paralacon*) **A**: TAI
taciturnus taciturnus Candèze, 1874: 60 (*Lacon*) **A**: FUJ HAI HUB JIX TAI **ORR**
taiwanus Miwa, 1927a: 13 (*Lacon*) **A**: GUI TAI
takasago Kishii, 1990c: 12, 19 **A**: TAI
thibetanus Reitter, 1913e: 658 (*Compsolacon*) **A**: AF IN PA XIZ
thuligadensis Vats & Kashyap, 1992c: 94 **A**: UP
tostus Candèze, 1857: 129 (*Lacon*) **A**: FUJ HKG JIX TAI UP XIZ **ORR**
transversus Candèze, 1857: 123 (*Lacon*) **A**: UP "North India" **ORR**
triticumunis Vats & Kashyap, 1992c: 124 **A**: HP UP
truncatus Herbst, 1806: 95 (*Elater*) **A**: NP SD XIZ "North India" **ORR**
 cylindricus Candèze, 1857: 121 (*Lacon*)
tsukamotoi Kishii, 1956: 18 (*Cryptolacon*) **A**: JA
 satsuma Kishii, 1996b: 3 **A**: JA
tsushimensis tsushimensis Ôhira, 1986c: 13 **A**: JA
 sado Kishii, 1996b: 3 (*Colaulon*) **A**: JA
tsushimensis kusuii Ôhira, 2004: 68 **A**: JA
tuberculipennis Miwa, 1929b: 230 (*Lacon*) **A**: TAI
tuberculosus Vats & Kashyap, 1992c: 100 **A**: UP
tuberosus Vats & Kashyap, 1992c: 136 **A**: HP UP **ORR**

turkestanicus Schwarz, 1902a: 198 (*Lacon*) **A**: TD UZ
upadhyai Vats & Kashyap, 1992c: 106 **A**: UP
uraiensis Miwa, 1929b: 231 (*Lacon*) **A**: GUI TAI
vatsai Vats & Kashyap, 1992c: 119 **A**: HP
yuppe Kishii, 1964: 31 (*Colaulon*) **A**: JA (Ryukyus)

genus *Anathesis* Candèze, 1865: 21 type species *Anathesis laconoides* Candèze, 1865
laconoides Candèze, 1865: 21 **A**: HAI TAI YUN **ORR**
laticornis Kirsch, 1875: 33 (*Penia*)

genus *Christinea* Gurjeva, 1987a: 39 type species *Christinea mirabilis* Gurjeva, 1987
mirabilis Gurjeva, 1987a: 39 **A**: TM UZ

genus *Danosoma* C. G. Thomson, 1859: 103 type species *Elater conspersus* Gyllenhal, 1808
Delox Quelle, 1932: 208 type species *Elater conspersus* Gyllenhal, 1808
conspersa Gyllenhal, 1808a: 377 (*Elater*) **E**: BY CT EN FI LA NR NT PL SV **A**: ES FE MG NC WS "Korea"
fasciata Linnaeus, 1758: 406 (*Elater*) **E**: AU BU BY CT CZ EN FI FR GE IT LA LT NR NT PL RO SK SL SP SV SZ UK YU **A**: ES FE GAN JA KZ MG TR WS XIN "North China" "Korea"
angustata J. R. Sahlberg, 1903a: 25 (*Adelocera*) [HN]
inaequalis DeGeer, 1774: 148 (*Elater*)
sachalinensis Miwa, 1927a: 13 (*Adelocera*)
sahlbergi Schwarz, 1906: 315 (*Adelocera*) [RN]

genus *Hemicleus* Candèze, 1857: 180 type species *Hemicleus caffer* Candèze, 1857
ferrantei Buysson, 1912a: 19 **N**: EG **A**: SA
normandi Buysson, 1920a: 12 **N**: TU

genus *Lacon* Laporte, 1838: 11 type species *Elater atomarius* Fabricius, 1798: 139 (= *Elater punctatus* Herbst, 1784)
Alaotypus Schwarz, 1902b: 307 type species *Alaotypus subpectinatus* Schwarz, 1902
Arnettia Golbach, 1969: 155 type species *Adelocera aberrans* Candèze, 1874
Aulacon Arnett, 1952: 112 type species *Adelocera nobilis* Fall, 1932
Cornilacon Golbach, 1969: 158 type species *Adelocera longicornis* Champion, 1894
Diphyaulon Arnett, 1952: 111 type species *Adelocera pyrsolepis* LeConte, 1866
Kobulacon Chûjô & Ôhira, 1965: 2 type species *Lacon quadrinodatus* Lewis, 1894
Latilacon Golbach, 1969: 158 type species *Adelocera laticollis* Candèze, 1857
Lepidelater Smith, 1969: 11 type species *Lepidelater misticius* Mignot, 1969
Lepidotus Stephens, 1830: 374 [HN] type species *Elater varius* A. G. Olivier, 1790 (= *Elater querceus* Herbst, 1784)
Monocyrton Golbach, 1969: 156 type species *Adelocera chabannei* Guérin-Méneville, 1829
Ocneus Candèze, 1857: 84 type species *Ocneus limbatus* Candèze, 1857
Pangaura Gistel, 1856: 366 type species *Elater varius* A. G. Olivier, 1790 (= *Elater querceus* Herbst, 1784)
Scelisus Candèze, 1863: 327 type species *Scelisus sanguineus* Candèze, 1863
Sulcilacon Fleutiaux, 1927a: 65 type species *Adelocera geographica* Candèze, 1865
Zalepia Arnett, 1953: 7 [RN] type species *Elater varius* A. G. Olivier, 1790 (= *Elater querceus* Herbst, 1784)
altaicus Candèze, 1882: 2 (*Adelocera*) **A**: ES FE KZ MG WS XIN
atayal Kishii, 1991c: 11, 19 **A**: TAI
aurosquamosus Jagemann, 1944: 332 (*Adelocera*) **A**: IN TR
bicolor Fleutiaux, 1940c: 90 **A**: YUN
boninensis Ôhira, 1970e: 11 (*Alaotypus*) **A**: JA (Ogasawara)
caeruleus Schimmel, 1998d: 234 **A**: NP
candezei Desbrochers des Loges, 1875a: 37 (*Adelocera*) **A**: IS JO TR
kraatzi Schwarz, 1893d: 192 (*Adelocera*)
carinatus Vats & Kashyap, 1992b: 15 **A**: UP **ORR**
churakagi Ôhira, 1971a: 39 (*Alaotypus*) **A**: JA (Ryukyus)
collisus Candèze, 1889b: 70 (*Adelocera*) **A**: SD
delagrangei Buysson, 1891d: cxxxviii **A**: TR

drusa Marseul, 1870: 380 (*Adelocera*) **A**: LB SY
bruleriei Desbrochers des Loges, 1875a: 37 (*Adelocera*)
flavopilosus Vats & Kashyap, 1992b: 18 **A**: UP
funebris Solsky, 1881: 231 (*Adelocera*) **A**: IN TD TM UZ
tristis Kraatz, 1882c: 319 (*Adelocera*)
ganglbaueri Schwarz, 1894: 145 (*Adelocera*)
gillerforsi Platia & Schimmel, 1994: 78 **E**: GR **A**: CY TR
graecus Candèze, 1857: 61 (*Adelocera*) **E**: GR **A**: TR
griseus Schwarz, 1900: 98 (*Adelocera*) **A**: UZ
incomptus Kraatz, 1882c: 319 (*Adelocera*) **A**: UZ
kapleri Platia & Schimmel, 1994: 79 **E**: GR **A**: TR
kikuchii Miwa, 1929b: 226 (*Adelocera*) **A**: TAI
kintaroui Kishii, 1990c: 14, 24 **A**: TAI
kushihige Kishii, 1990c: 14, 26 **A**: TAI
kryzhanovskyi Dolin & Atamuradov, 1989: 33 **A**: TM
lepidopterus Panzer, 1800a: 4 (*Elater*) **E**: AU BU CR CT CZ DE FI FR GE GG GR IT LA NT PL RO SK SL ST SV SZ UK **A**: IN WS
chrysoprasus Herbst, 1806: 18 (*Elater*)
grafii F. J. Schmidt, 1832: 83 (*Elater*)
lithophilus Candèze, 1857: 60 (*Adelocera*) **N**: EG "North Africa"
maeklinii maeklinii Candèze, 1865: 6 (*Adelocera*) **A**: GUX JA "Korea"
antennata Lewis, 1894a: 27 (*Adelocera*)
maeklinii maenamii Ôhira, 1969b: 41 (*Alaotypus*) **A**: JA
mausoni Hayek, 1973: 71 [RN] **A**: TAI **ORR**
sanguinea Fleutiaux, 1908: 164 (*Adelocera*) [HN]
mekrani Candèze, 1889b: 71 (*Adelocera*) **A**: IN PA
modestus Boisduval, 1835: 108 (*Elater*) [NP] **A**: IN JA (Iwo-jima) OM QA SA TAI **AFR AUR NTR ORR**
castanidens Gyllenhal, 1817: 134 (*Elater*) [NO]
coomani Fleutiaux, 1927a: 66
nigroplagiatus Blanchard, 1853: 85 (*Agrypnus*)
pruinosus Fairmaire, 1849a: 35 (*Agrypnus*) [= 1849b: 359 (*Adelocera*)]
squalidus Fairmaire, 1849a: 35 (*Agrypnus*) [= 1849b: 359 (*Adelocera*)]
quadulpensis Fleutiaux & Sallé, 1889: 407 (*Adelocera*)
tessellata Candèze, 1895: 6 (*Adelocera*)
vicina Candèze, 1891b: 772 (*Adelocera*)
monticola Candèze, 1897: 6 (*Adelocera*) **A**: "Himalaya"
nigrofucus Vats & Kashyap, 1992b: 16 **A**: UP
obscurus Fleutiaux, 1940c: 89 **A**: YUN
olivieri Candèze, 1874: 20 (*Adelocera*) **A**: SD
parallelus amamiensis Ôhira, 1967b: 98 **A**: JA
parallelus parallelus Lewis, 1894a: 28 (*Adelocera*) **A**: JA JIL "Korea"
punctatus oblongus Della Beffa, 1931: 181 (*Adelocera*) **A**: KA
punctatus punctatus Herbst, 1779: 316 (*Elater*) **E**: AB AR BH BU CR CT CZ FR GE GG GR HU IT MA PT RO SK SL SP ST SZ TR UK YU **A**: CY JO SY TR **N**: MO
atomarius Fabricius, 1798: 139 (*Elater*)
carbonarius Schrank, 1781a: 184 (*Elater*)
pulverulentus Panzer, 1795: 235 (*Elater*)
puriensis Kishii, 1991d: 2 [RN] **A**: TAI
formosanus Kishii, 1966a: 7 (*Compsolacon*) [HN]
quadrinodatus Lewis, 1894a: 28 **A**: JA
querceus Herbst, 1784b: 113 (*Elater*) **E**: AU BU CT CZ FR GE GB HU IT LT PL RO SK SP SV SZ UK
varius A. G. Olivier, 1790: no 31: 32 (*Elater*)
ramatasenseni Miwa, 1934: 180 (*Adelocera*) **A**: TAI
robustus Fleutiaux, 1902c: 213 (*Adelocera*) **A**: "Himalaya" **ORR**
rotundicollis Kishii & Jiang, 1994: 89 **A**: GAN
rubripennis Fleutiaux, 1940c: 89 **A**: YUN

sanguineus Candèze, 1863: 328 (*Scelisus*) **A**: AP NP PA "North India"
solai Platia & Gudenzi, 2000f: 66 **A**: TR
unicolor Candèze, 1874: 21 (*Adelocera*) **A**: IN
yayeyamanus Miwa, 1934: 246 (*Adelocera*) **A**: JA (Ryukyus)
yunnanus Fleutiaux, 1940c: 90 **A**: YUN

genus *Lanelater* Arnett, 1952: 105 type species *Agrypnus schotti* LeConte, 1853
Amaurus Laporte, 1840a: 237 [HN] type species *Amaurus senegalensis* Laporte, 1840 (= *Elater notodonta* Latreille, 1827)
aequalis Candèze, 1857: 25 (*Agrypnus*) **A**: GUA GUX HAI JA (Ishigaki-jima) JIX SHA SHX TAI YUN **AFR ORR**
insularis Fairmaire, 1891c: lxx (*Agrypnus*)
afoveatus Vats & Kashyap, 1993: 220 **A**: UP **ORR**
arabicus Candèze, 1874: 6 (*Agrypnus*) **A**: IN SA YE
arenarum Platia & Schimmel, 1997: 294 **A**: OM
bartoni Fleutiaux, 1902b: 163 (*Agrypnus*) **A**: IN **ORR**
bipunctatus Candèze, 1857: 29 (*Agrypnus*) **A**: IN PA UP **ORR**
buettikeri Chassain, 1983: 130 **A**: OM SA YE
cinereus Candèze, 1857: 40 (*Agrypnus*) **A**: SD UP **ORR**
densus Vats & Kashyap, 1993: 219 **A**: UP
fuscipes Fabricius, 1775: 211 (*Elater*) **A**: NP SD UP **AFR ORR**
convexicollis Fleutiaux, 1918b: 182 (*Agrypnus*)
gigas Fleutiaux, 1918b: 182 (*Agrypnus*)
fusiformis Candèze, 1857: 39 (*Agrypnus*) **A**: GUA HAI **ORR**
glabrosus Vats & Kashyap, 1993: 220 **A**: UP **ORR**
judaicus Reiche & Saulcy, 1857a: 418 (*Agrypnus*) **A**: IQ IS JO LE SY TR
lacertosus Candèze, 1857: 38 (*Agrypnus*) **A**: NP PA **ORR**
lucerus Vats & Kashyap, 1993: 216 **A**: UP **ORR**
luridus Fabricius, 1781: 265 (*Elater*) **A**: "West Hindustan" **AFR ORR**
gilvus Candèze, 1865: 5 (*Agrypnus*)
moestus Candèze, 1857: 34 (*Agrypnus*)
punctatus Candèze, 1857: 26 (*Agrypnus*)
sondaicus Candèze, 1857: 33 (*Agrypnus*)
naviauxi Chassain, 1983: 130 **A**: OM SA YE
notodonta Latreille, 1827: 275 (*Elater*) **E**: IT (Sicily) **N**: AG EG LB MO TU **A**: OM SA YE **AFR**
himerensis Ragusa, 1881: 8 (*Agrypnus*)
senegalensis Laporte, 1840a: 237 (*Amaurus*)
olcesii Fairmaire, 1884b: 446 (*Agrypnus*) **N**: MO
parvus Platia & Schimmel, 1997: 296 **A**: AE OM
persicus Candèze, 1874: 6 (*Agrypnus*) **A**: IN TR
pescadorensis Miwa, 1934: 178 (*Agrypnus*) **A**: TAI (Pescadores)
politus Candèze, 1857: 43 (*Agrypnus*) **A**: GUA HAI JIX TAI **ORR**
pubescens Candèze, 1857: 31 (*Agrypnus*) **N**: EG **AFR**
bocandei Candèze, 1857: 32 (*Agrypnus*)
saudarabicus Chassain, 1983: 131 **A**: AE IN OM SA
scorteccii Binaghi, 1941a: 73 (*Agrypnus*) **N**: LB
scutopentagonus Vats & Kashyap, 1993: 216 **A**: UP
vishvai Vats & Kashyap, 1993: 221 **A**: UP **ORR**
wittmeri Chassain, 1983: 135 **A**: SA

genus *Meristhus* Candèze, 1857: 162 type species *Elater lepidotus* Palisot de Beauvois, 1805
subgenus *Meristhus* Candèze, 1857: 162 type species *Elater lepidotus* Palisot de Beauvois, 1805
crenulatus Vats & Kashyap, 1996: 71 [DA] **A**: UP
lepidotus Palisot de Beauvois, 1805: 11 (*Elater*) **A**: SA **AFR**
sabensis Wurst, Schimmel & Platia, 2001: 543 **A**: YE
sikkimensis Cate, **nom. nov.** [see New acts] **A**: SD

crenulatus Mukhopadhyay & Chakraborty, 2003: 225 [HN]
subgenus *Sulcimerus* Fleutiaux, 1947: 255 type species *Meristhus quadripunctatus* Candèze, 1857
afghanus Platia & Gudenzi, 2002b: 78 **A**: AF
babai Kishii, 1991c: 11, 17 **A**: TAI
niponensis Lewis, 1894a: 30 **A**: JA SC
quadripunctatus Candèze, 1857: 163 **A**: FUJ GUX HUB JIX NP SCH TAI UP **ORR**

genus *Octocryptus* Candèze, 1892b: 486 type species *Octocryptes cardoni* Candèze, 1892
babaulti Fleutiaux, 1944b: 148 **A**: HP
cardoni Candèze, 1892b: 487 **A**: NP "North India" **ORR**
kalesarensis Vats & Kashyap, 1995b: 215 **A**: UP **ORR**
maindroni Fleutiaux, 1944b: 148 **A**: OM
wittmeri Chassain, 1979b: 197 **A**: IN SA

genus *Rismethus* Fleutiaux, 1947: 257 type species *Meristhus scobinula* Candèze, 1857
longicollis Candèze, 1897: 11 (*Meristhus*) **A**: NP **ORR**
lotharensis Ôhira & Becker, 1973b: 464 (*Meristhus*) **A**: NP
nepalensis Ôhira & Becker, 1973b: 462 (*Meristhus*) **A**: NP
pistrinarius Candèze, 1857: 164 (*Meristhus*) **A**: NP UP "North India" **ORR**
ryukyuensis Ôhira, 1999: 409 **A**: JA TAI (Lukao Is.)
scobinula Candèze, 1857: 164 (*Meristhus*) **A**: CH JA TAI **NAR ORR**
setarius Chevrolat, 1867: 596 (*Meristhus*)
texanus Horn, 1871: 300 (*Meristhus*)

genus *Saudilacon* Chassain, 1983: 138 type species *Saudilacon ochraceus* Chassain, 1983
balcis Wurst, Schimmel & Platia, 2001: 543 **A**: YE
ochraceus Chassain, 1983: 138 **A**: SA

tribe Agrypnini, species incertae sedis
Lacon variegatus Motschulsky, 1854c: 45 **A**: BEI

tribe Conoderini Fleutiaux, 1919

genus *Aeoloderma* Fleutiaux, 1928a: 135 type species *Elater crucifer* P. Rossi, 1790
babylonicum Candèze, 1859: 227 (*Monocrepidius*) **A**: "Mesopotamia"
brachmana Candèze, 1859: 345 (*Aeolus*) **A**: FUJ GUA GUX HUB JA JIX SCH TAI "North India" **ORR**
ancorale Schwarz, 1901b: 24 (*Heteroderes*)
multilineatum Candèze, 1878a: 118 (*Heteroderes*)
parduum Candèze, 1859: 345 (*Aeolus*)
tessellatum Motschulsky, 1860e: 518 (*Aeolus*)
vittatum Matsumura, 1910: 37 (*Aeolus*) [= 1911a: 144] [HN]
crucifer P. Rossi, 1790: 183 (*Elater*) **E**: AB AL AR AZ BU CT FR GG GR IT MC PT RO SP ST TR UK **N**: AG EG LB MO TU **A**: CY IN KI KZ PA SY TM TR
flavofactum Buysson, 1912a: 34
lateralipictum Buysson, 1912a: 32
m-disjunctum Buysson, 1912a: 33
multidisjunctum Pic, 1912g: 25 (*Heteroderes*)
paludum Buysson, 1912a: 32
obliteratum Candèze, 1859: 356 (*Heteroderes*) **A**: IQ TR "Palestine"
savioi Fleutiaux, 1936a: 16 **A**: HEI JIA MG SHG
sillemi Fleutiaux, 1935c: 296 **A**: XIZ
sinense Candèze, 1859: 346 (*Aeolus*) **A**: HAI HKG GUA GUX SCH **ORR**
stigmaticum Candèze, 1895: 33 (*Drasterius*) **A**: YE

genus *Aeoloides* Schwarz, 1906: 109 type species *Heteroderes sequester* Candèze, 1859
agriotides Wurst, Schimmel & Platia, 2001: 544 **A**: YE

figuratus Germar, 1844: 148 (*Cryptohypnus*) **E**: AB AR GG ST **A**: AF IQ IN KI KU KZ PA QA SA SY TD TM TR UZ "Palestine"

candezei Reitter, 1891c: 146 (*Aeolus*)
laeticolor Reitter, 1891c: 147 (*Aeolus*)
obscuricolor Reitter, 1891c: 147 (*Aeolus*)
persicus Motschulsky, 1860e: 518 (*Aeolus*)
turkomanus Candèze, 1889b: 95 (*Heteroderes*)

filipponii Platia & Schimmel, 1997: 298 **A**: SA YE
grisescens Germar, 1844: 151 (*Cryptohypnus*) **E**: AB AR CT GG GR ST UK **N**: EG LB MO **A**: AF CY IN IQ KI KZ MG OM NMO PA QA SA SY TD TM TR UZ WS **AFR**

bicarinatus Reitter, 1891c: 148 (*Aeolus*)
grayii Wollaston, 1857c: 505 (*Monocrepidius*)

hauseri Reitter, 1896b: 35 (*Aeolus*) **A**: AF KZ MG TM UZ
holzschuhi Platia & Schimmel, 1997: 299 **A**: IN OM PA
imitator Reitter, 1891c: 148 (*Aeolus*) **N**: TU
iranicus Platia & Gudenzi, 2002b: 80 **A**: IN
spissus Candèze, 1889b: 95 (*Heteroderes*) **A**: YE **AFR**
waltlii Candèze, 1859: 380 (*Heteroderes*) **N**: EG

isabellinus Buysson, 1912a: 36

genus *Aeolosomus* Dolin, 1982a: 106 type species *Cryptohypnus rossii* Germar, 1844
rossii Germar, 1844: 148 (*Cryptohypnus*) **E**: AB AR GG ST UK **A**: IN KI KZ TD TM TR UZ

genus *Aeolus* Eschscholtz, 1829: 33 type species *Elater scriptus* Fabricius, 1801

Oophorus Dejean, 1833: 93 type species *Elater elegans* Fabricius, 1792

melliculus moreleti Tarnier, 1860: 94 **E**: AZ **NTR**

tristis Fleutiaux, 1911: 257

genus *Babadrasterius* Ôhira, 1994b: 224, 230 type species *Babadrasterius urabensis* Ôhira, 1994
sexpunctatus Miwa, 1927a: 15 (*Heteroderes*) **A**: JIA TAI

immaculatus Miwa, 1929b: 246 (*Prodrasterius*)
quadrimaculatus Miwa, 1929b: 245 (*Prodrasterius*)
sacchari Miwa, 1927a: 15 (*Heteroderes*)

triangularis Eschscholtz, 1822: 73 [= 1823: 129] (*Elater*) **A**: GUA HAI JA (Ryukyus) JIA **ORR**

proximus Fleutiaux, 1895: 686 (*Heteroderes*)

urabensis Ôhira, 1994b: 224, 230 **A**: JA (Ryukyus)

genus *Conoderus* Eschscholtz, 1829: 31 type species *Conoderus fuscofasciatus* Eschscholtz, 1829

Monocrepidius Eschscholtz, 1829: 31 type species *Monocrepidius pallipes* Eschscholtz, 1829

bicarinatus Vats & Chauhan, 1992c: 29 **A**: UP
chitrakutensis Vats & Chauhan, 1992c: 34 **A**: HP UP
crocopus Hope, 1843a: 63 [= 1845a: 9] (*Ludius*) **A**: ZHE
dohrni Candèze, 1878b: lxxvii **A**: YUN "Himalaya" **ORR**
elegans Candèze, 1878b: lxxvii **A**: GUA HAI YUN "North India" **ORR**
garhwalensis Vats & Chauhan, 1992c: 32 **A**: UP
longipennis Vats & Chauhan, 1992c: 33 **A**: UP
nigripes Vats & Chauhan, 1992c: 30 **A**: HP UP
nigrolineatus Vats & Chauhan, 1992c: 35 **A**: UP
nigromaculosus Vats & Chauhan, 1992c: 35 **A**: UP
posticus Eschscholtz, 1822: 76 [= 1823: 132] **E**: AZ **NTR**
pallipes Eschscholtz, 1829: 31 (*Monocrepidius*) **A**: JA (Iwo-jima) **AUR**

rufangulus Fairmaire, 1849a: 35 [= 1849b: 355] (*Monocrepidius*)
taitensis Boheman, 1858: 68 (*Elater*)

productus arabicus Chassain, 1979b: 200 (*Prodrasterius*) **A**: IN OM QA SA YE
productus productus Peyerimhoff, 1931b: 66 **N**: AG EG
tapovanensis Vats & Chauhan, 1992c: 32 **A**: UP

tonkinensis Fleutiaux, 1895: 686 (*Monocrepidius*) **A**: ANH GUI **ORR**

genus *Drasterius* Eschscholtz, 1829: 33 type species *Elater bimaculatus* P. Rossi, 1790
Epistrophus Gistel, 1834: 12 type species *Elater bimaculatus* P. Rossi, 1790
Prodrasterius Fleutiaux, 1927b: 91 type species *Drasterius brahminus* Candèze, 1859
aegyptiacus Buysson, 1905a: 18 **N**: AG EG LB MO TU **A**: SA YE
aegyptiacoides Buysson, 1912c: 86 (*Drasterius*)
assuanensis Pic, 1913h: 121 (*Drasterius*)
biskrensis Desbrochers des Loges, 1870a: 68 (*Heteroderes*)
kebiliensis Buysson, 1912c: 86 (*Drasterius*)
riyadhensis Chassain, 1979b: 201
notatulus Buysson, 1912a: 38
pseudobiskrensis Buysson, 1912a: 41
variegatulus Buysson, 1912a: 39
agnatus Candèze, 1873: 8 (*Aeolus*) **A**: GAN HUB FE JA JIX LIA NC SC "Northeast China"
atricapillus Germar, 1824: 40 (*Elater*) **E**: AB AR CT GG ST TR **A**: IN KI KZ TD TM UZ XIN
ballioni Heyden, 1885: 287 (*Aeolus*)
cruciatus Candèze, 1859: 330 (*Aeolus*)
humeralis Reitter, 1891c: 147 (*Aeolus*)
mosquensis Ivanov, 1902: 32 (*Aeolus*)
nigripennis Reitter, 1891c: 147 (*Aeolus*)
sarmaticus Motschulsky, 1860e: 517 (*Aeolus*)
bimaculatus P. Rossi, 1790: 182 (*Elater*) **E**: AB AL AR AU BE BH BU CR CT CZ FR GE GG GR HU IT MA MC MD PL PT RO SK SL SP ST SZ UK YU **N**: AG CI (Lanzarote) EG LB MO TU **A**: AF CY IN IQ IS JO KI KZ SY TM TR UZ
angulosepictus Buysson, 1900a: 261
anticenotatus Buysson, 1912a: 38
anticus Reitter, 1890b: 171
apicalis Reitter, 1890b: 172
basalis Reitter, 1890b: 171
bengasinus Binaghi, 1941c: 177
binotatus P. Rossi, 1792: 59 (*Elater*)
bisbisignatus Buysson, 1913b: 74
breviusculus Desbrochers des Loges, 1869: 46 [= 1870a: 121] (*Agriotes*)
cantheriatus Buysson, 1900a: 262
diluticollis Reitter, 1890b: 171
dorsalis Reitter, 1890b: 171
fenestratus Küster, 1851a: 45 (*Cryptohypnus*)
grossus J. R. Sahlberg, 1913b: 16
immaculatus Oliveira, 1893: 189
latepictus Buysson, 1900a: 261
lateralis Reitter, 1890b: 171
lineatobasalis Fuente, 1903: 342
montandoni Buysson, 1895a: 39
pallipes Küster, 1848c: 19 (*Cryptohypnus*)
propinquus Desbrochers des Loges, 1869: 30 [= 1870a: 106] (*Cryptohypnus*)
quadrisignatus Küster, 1849b: 16 (*Cryptohypnus*)
sexmaculata Ragusa, 1893: 304
sexsignatus Buysson, 1895a: 39
trisignatus Chobaut, 1897a: 38
variegatus Küster, 1851a: 44 (*Cryptohypnus*)
wachsmanni Reitter, 1905b: 91
brahminus Candèze, 1859: 426 **A**: FUJ GUA GUX HUB JA (Ryukyus) JIX PA SCH UP "Himalaya" **ORR**
fougueti Fleutiaux, 1918b: 212
candezei Fleutiaux, 1918b: 213 (*Prodrasterius*) **A**: GUA **ORR**
collaris asaokai Ôhira, 1994b: 219, 229 (*Prodrasterius*) **A**: JA (Ryukyus)
collaris collaris Candèze, 1859: 427 **A**: NP PA UP **ORR**
nigricollis Fleutiaux, 1918b: 212
collaris taiwanus Kishii, 1996d: 89 (*Prodrasterius*) **A**: TAI

csorbai Platia & Gudenzi, 1997b: 408 **A**: PA
makrisi Preiss & Platia, 2003: 101 **A**: CY
okinawensis Ôhira, 2000: 365 **A**: JA
prosternalis Candèze, 1878b: cxxxv **A**: JA (Okinawa) "Himalaya" **ORR**
annamensis Schwarz, 1902b: 322 (*Heteroderes*)
delauneyi Fleutiaux, 1918b: 213
drasteroides Fleutiaux, 1895: 686 (*Heteroderes*)
schatzmayri Binaghi, 1941c: 180 **A**: SI
sulcatulus Candèze, 1859: 427 **A**: NP PA UP "Himalaya" **ORR**
indus Candèze, 1859: 428

genus *Heteroderes* Latreille, 1834: 155 type species *Heteroderes fuscus* Latreille, 1834
Exaeolus Gozis, 1886: 22 type species *Oophorus algirinus* Lucas, 1846
abyssinus Candèze, 1859: 366 **N**: EG **AFR**
albicans Candèze, 1878b: lxxxiii **A**: GUA HAI HUB SCH TAI YUN XIZ **ORR**
algirinus Lucas, 1846: 166 (*Oophorus*) **E**: FR SP **N**: AG CI (La Gomera) EG LB MO TU
infuscatus J. R. Sahlberg, 1903d: 41
modestus Candèze, 1859: 370
amaculatus Vats & Chauhan, 1993c: 170 **A**: UP **ORR**
approximatus Desbrochers des Loges, 1875a: 37 **A**: IS SY
azoricus Tarnier, 1860: 93 (*Oophorus*) **E**: AZ
atlanticus Candèze, 1865: 26
biellipticus Buysson, 1890c: ccxiii **A**: SY "Palestine"
changi Ôhira, 1967a: 57 **A**: TAI
cinerascens Candèze, 1859: 362 **A**: NP
farus Vats & Chauhan, 1993c: 172 **A**: UP **ORR**
gallagheri Platia & Schimmel, 1997: 301 **A**: AE IN OM PA
grancanariensis Palm, 1976b: 72 **A**: CI (Grand Canaria)
heydeni Reitter, 1891c: 148 (*Aeolus*) [RN] **A**: AF IN TD TM PA UZ
sericeus Reitter, 1887h: 512 (*Aeolus*) [HN]
inexpectatus Kishii, 1969: 4 **A**: JA
intermedius Candèze, 1895: 32 (*Heteroderes*) **A**: TAI **ORR**
kordofanus Candèze, 1859: 364 **N**: EG **AFR**
mitigatus Candèze, 1859: 365
multimaculatus Schwarz, 1898b: 189
kuluensis Jagemann, 1945: 40 **A**: HP
kusuii Ôhira, 1994b: 225, 231 **A**: JA (Ryukyus)
lenis Candèze, 1859: 357 **A**: HP KA UP **ORR**
luristanicus Pic, 1916i: 6 **A**: IN
macroderes Candèze, 1859: 358 **A**: NP SD YUN **ORR**
musculus Germar, 1844: 151 (*Cryptohypnus*) **N**: AG EG **A**: SA YE
aegyptiacus Motschulsky, 1860e: 518 (*Aeolus*)
oblitus Candèze, 1859: 361 **A**: GUI JIX TAI SCH **ORR**
sacchari Matsumura, 1910: 37 (*Agriotes*) [= 1911a: 144]
omanensis Platia & Schimmel, 1997: 303 **A**: OM
ornatus Candèze, 1888: 675 **A**: HAI **ORR**
pecirkai Jagemann, 1945: 39 **A**: KA
ruteri Chassain, 1979b: 204 **A**: OM SA
sericeus Candèze, 1859: 357 **A**: UP "Himalaya"
tabularis Vats & Chauhan, 1993c: 172 **A**: UP **ORR**
vagus Candèze, 1895: 32 **Ei**: FR SP **Ai**: SA **NTR**

genus *Melanthoides* Candèze, 1865: 23 type species *Melanthoides latimanus* Candèze, 1865
partitus Candèze, 1897: 31 **A**: GUA **ORR**

genus *Nanseia* Kishii, 1985: 5 type species *Prodrasterius erabuensis* Kishii, 1966
erabuensis Kishii, 1966b: 8 (*Prodrasterius*) **A**: JA (Ryukyus, Satsunans, Sakishima) **ORR**
iriomotecola Kishii, 1972: 7 (*Ampedus*)

genus *Neodrasterius* Kishii, 1996d: 90 type species *Neodrasterius kiyoyamai* Kishii, 1996
hisamatsui hisamatsui Ôhira & Satô, 1964: 28 (*Prodrasterius*) **A**: JA (Ryukyus)
hisamatsui yaeyamensis Kishii, 1972: 5 (*Prodrasterius*) **A**: JA (Ryukyus)
kiyoyamai Kishii, 1996d: 91 **A**: TAI

genus *Nipponodrasterius* Kishii, 1966b: 9 type species *Nipponodrasterius alpicola* Kishii, 1966
alpicola Kishii, 1966b: 9 **A**: JA

genus *Pachyderes* Guérin-Ménéville, 1830: pl. 13 type species *Pachyderes ruficollis* Guérin-Méneville, 1830
coccineus Candèze, 1859: 174 **A**: NP **ORR**
niger Candèze, 1878b: 14 **A**: NP YUN **ORR**

tribe Hemirhipini Candèze, 1857

subtribe Chalcolepidiina Lacordaire, 1857

genus *Calais* Laporte, 1838: 9 type species *Calais senegalensis* Laporte, 1838 (= *Elater excavatus* Fabricius, 1801)
afghanicus Chassain, 1991: 214 **A**: AF PA
brandti Platia & Gudenzi, 1999b: 118 **A**: IN
parreysii Steven, 1829: 285 (*Elater*) [= 1830: 134; 1832: 81] **E**: CR GR ST UK **A**: AF CY IN TR
goryi Gory, 1832: pl. IX
persicus Chassain, 1991: 211 **A**: IN
sulcicollis Gahan, 1900a: 11 (*Alaus*) **A**: JE (Suqutra)
wittmeri Chassain, 1983: 136 **A**: SA

genus *Cryptalaus* Ôhira, 1967b: 97 type species *Alaus putridus* Candèze, 1857 sensu Ôhira, 1967b (= *Alaus larvatus* Candèze, 1874)
Paracalais Neboiss, 1967: 261 type species *Alaus suboculatus* Candèze, 1857
berus Candèze, 1865: 15 (*Alaus*) **A**: FUJ GUA GUX HAI HUN JA JIX SC TAI YUN ZHE **ORR**
cenchris Candèze, 1857: 231 (*Alaus*) **A**: SD **ORR**
chinensis Ôhira, 1970f: 216 (*Paracalais*) **A**: GUI YUN
elongatus Miwa, 1929b: 237 (*Alaus*) **A**: TAI
eryx Candèze, 1874: 140 (*Alaus*) **A**: NP SD **ORR**
larvatus larvatus Candèze, 1874: 141 (*Alaus*) **A**: FUJ GUA GUX HAI HUN JA (Ryukyu, Ogasawara) JIA JIX SCH SHG TAI YUN ZHE **ORR**
larvatus pini Lewis, 1894a: 30 (*Alaus*) **A**: JA JIX SCH
satoi Ôhira, 1964c: 235 (*Alaus*)
lynceus Candèze, 1874: 139 (*Alaus*) **A**: "North India"
mus Miwa, 1929b: 238 (*Alaus*) **A**: TAI
abrotanus Ôhira, 1966c: 215 (*Alaus*)
nubilus Candèze, 1857: 230 (*Alaus*) **A**: YUN **ORR**
podagrus Candèze, 1874: 125 (*Alaus*) **A**: TAI **ORR**
sculptus Westwood, 1848: 72 (*Alaus*) **A**: GUA HUN SD XIZ YUN "North India" **ORR**
coomani Fleutiuax, 1927: 107 (*Alaus*)
shibatai Kishii, 1996a: 104 **A**: TAI
sordidus Westwood, 1848: 72 (*Alaus*) **A**: HAI NP SD SCH YUN XIZ "North India" **ORR**
yamato Nakane, 1957: A237 (*Alaus*) **A**: JA

tribe Pseudomelanactini Arnett, 1967

genus *Anthracalaus* Fairmaire, 1889a: 349 type species *Alaus westermanni* Candèze, 1857
Pseudomelanactes Mathieu, 1961: 474 type species *Melanactes agrypnoides* Van Dyke, 1932

chujoi Miwa, 1939: 65 (*Pristilophus*) **A**: TAI
moricii Fairmaire, 1889a: 349 **A**: FUJ GUA GUX JIA JIX SCH YUN **ORR**
sakaguchii Miwa, 1929c: 348 (*Pristilophus*) **A**: JA (Ryukyus)
magnus Miwa, 1929c: 349 (*Pristilophus*)

subfamily Agrypninae, incertae sedis
genus *Tetrigus* Candèze, 1857: 254 type species *Tetrigus parallelus* Candèze, 1857
Elatrigus Reitter, 1905a: 10 type species *Tetrigus cyprius* Baudi di Selve, 1871
Ludioctenus Fairmaire, 1893b: lxviii type species *Ludioctenus akbesianus* Fairmaire, 1893
Nycterilampus Montrouzier, 1860: 258 type species *Nycterilampus lifuanus* Montrouzier, 1860
babai Kishii, 1990c: 16, 28 **A**: TAI
cyprius Baudi di Selve, 1871: 50 **E**: GR **A**: CY LE SY TR
akbesianus Fairmaire, 1893b: lxix (*Ludioctenus*)
flabellatus Germar, 1844: 186 (*Aphanobius*) **A**: HEB SHG SC **ORR**
pectinicornis Motschulsky, 1854f: 63 (*Agriotes*)
fujitai Ôhira, 1978b: 553 **A**: JA (Ogasawara)
kusuii Ôhira, 1974b: 24 **A**: JA (Ogasawara)
lewisi Candèze, 1873: 6 **A**: FUJ GAN HEB HEN HUB HUN JA JIX SC SHG SHN SHX TAI ZHE **ORR**
grandis Lewis, 1879b: 155
okinawensis Ôhira, 1967b: 96 **A**: JA (Ryukyus)
parallelus Candèze, 1857: 255 **A**: JIA **ORR**
submontanus Vats & Kashyap, 1992d: 197 **A**: HP **ORR**
taiwanus Kishii, 1996a: 105 **A**: TAI

subfamily **Elaterinae** Leach, 1815
tribe Adrastini Candèze, 1863
genus *Adrastus* Eschscholtz, 1829: 35 type species *Elater limbatus* Fabricius, 1777
anatolicus Platia & Schimmel, 1991b: 138 **A**: TR
axillaris Erichson, 1841a: 122 **E**: AL AU BE BU CZ DE FR GE GR HU IT LS RO SK SL SZ UK
alpicola Rey, 1891c: 85
nigrinus Schilsky, 1890: 188
porrectifrons Gozis, 1882: 197
pseudolimbatus Binaghi, 1936: 222
turcicus Stierlin, 1861: 219
circassicus Reitter, 1896d: 158 **E**: AB AR GG ST **A**: TR
dolini Wellschmied, 1978: 121 **E**: AB GG ST **A**: IN TR
fraterculus Gurjeva, 1976c: 96 **E**: AB AR GG ST
gurjevae Penev, 1983: 181 **E**: BU
juditae Laibner, 1991: 327 **E**: CZ
kryshtali Dolin, 1988a: 175 **E**: CZ GE SK UK
lacertosus Erichson, 1841a: 126 **E**: AU BU CZ FR GE HU IT PL RO SK SP SZ UK
guillebeaui Buysson, 1896b: 156
limbatus Fabricius, 1777: 235 (*Elater*) **E**: AL AU BE BH BU BY CR CZ FR GE GR HU IT LT LU MC MD NL PL RO SK SL SP ST SZ UK YU **A**: TR
longicornis Gurjeva, 1976c: 96 **E**: AB GG ST **A**: TR
melonii Platia & Gudenzi, 2000f: 75 **E**: IT (Sardegna)
miegi Graells, 1858: 46 **E**: SP
montanus Scopoli, 1763: 92 (*Elater*) **E**: AL AU BE BH BU CR CZ FR GE GR HU IT LU MD NT PL PT RO SK SL SP SZ UK **A**: TR
exutus Buysson, 1926b: 179
humilis Erichson, 1841a: 127
rudis Küster, 1848b: 34 (*Dolopius*)
montenegroensis Laibner, 1987: 393 **E**: YU (Crna Gora)
pallens Fabricius, 1792b: 232 (*Elater*) **E**: AU BE BU BY CT CZ DE EN FI FR GE GB GR HU IR IT LA LS LT LU MC NL NR NT PL PT RO SK SL SP ST SV SZ UK YU **A**: TR
nitidulus Marsham, 1802: 380 (*Elater*) [HN]

protractus Roubal, 1924b: 81 **E**: GG ST **A**: TR
 lateralis Herbst, 1806: 101 (*Elater*)
 maurus Gurjeva, 1976c: 97
 paykulli Szombathy, 1910d: 582
rachifer Geoffroy, 1785: 39 (*Elater*) **E**: AL AR AU BE BU CR CZ EN FR GB GE GR HU IT LS LT LU MC MD NL PL RO SK SL SP SZ TR YU **A**: TR
 humerotinctus Buysson, 1926b: 178
 luteipennis Erichson, 1841a: 125
 nanus Herbst, 1806: 96 (*Elater*)
 nigratorius Buysson, 1926b: 178
 parvus Schönherr, 1817: 317 (*Elater*) [RN]
 pusillus Fabricius, 1801b: 246 (*Elater*)
 reitteri Buysson, 1896b: 159
 siculus Ragusa, 1896a: 141
samedovi Dolin & Agajev, 1974: 54 **E**: AB **A**: IN
sbordonii Guglielmi & Platia, 1985: 200 **E**: GR
sekerae Reitter, 1910b: 173 **E**: IT
 binaghii Leseigneur, 1969: 93
temperei Leseigneur, 1974: 112 **E**: FR (Corse)

genus *Ctenoplus* Candèze, 1863: 463 type species *Ctenoplus javanensis* Candèze, 1863
coomani Fleutiaux, 1940b: 20 **A**: FUJ GUA **ORR**
 coomani Fleutiaux, 1940b: 22 (*Silesis*)
 topali Ôhira, 1973f: 333

genus *Glyphonyx* Candèze, 1863: 451 type species *Glyphonyx gundlachii* Candèze, 1863
subgenus *Glyphonyx* Candèze, 1863: 451 type species *Glyphonyx gundlachii* Candèze, 1863
acristatus Vats & Chauchan, 1992g: 205 **A**: UP
ami Kishii, 1991b: 21, 33 **A**: TAI
apayao Kishii, 1991b: 24, 46 **A**: TAI
arisanus Miwa, 1931a: 206 **A**: TAI
atayal Kishii, 1991b: 22, 36 **A**: TAI
bicolor bicolor Candèze, 1895: 66 **A**: JA
bicolor yaku Kishii, 1976a: 13 **A**: JA
brunneipennis Miwa, 1931a: 206 **A**: TAI
castaneus Miwa, 1931a: 207 **A**: TAI
chipenensis Kishii, 1994d: 13 **A**: TAI
dalopioides Nakane, 1959c: 92 **A**: JA
 babai Kishii & Ôhira, 1956: 77, 84
depremerus Vats & Chauchan, 1992g: 203 **A**: UP
esakii Miwa, 1934: 263 **A**: JA
exilis Kishii, 1991b: 24, 53 **A**: TAI
flavicollis Kishii, 1991b: 23, 43 **A**: TAI
formosanus Ôhira, 1972a: 12 **A**: TAI
fuscicollis Kishii, 1991b: 23, 38 **A**: TAI
godivariensis Ôhira & Becker, 1972a: 135 **A**: NP
grossus Kishii, 1994d: 11 **A**: TAI
haterumarum haterumarum Ôhira, 1968a: 130 **A**: JA (Iriomote-jima)
haterumarum takahashii Kishii, 1974: 16 **A**: JA (Ishigake-jima)
housaii Kishii, 1991b: 24, 48 **A**: TAI
ihai Ôhira, 1968a: 127 **A**: JA (Okinawa-hontô I.)
illepidus Candèze, 1873: 31 **A**: JA
iriomotensis Miwa, 1934: 263 **A**: JA (Iriomote-jima, Ishigaki-jima, Yonaguni-jima)
kankaui Miwa, 1931a: 206 **A**: TAI
kintaroui Kishii, 1991b: 24, 51 **A**: TAI
kishiii Ôhira, 1968a: 128 **A**: JA (Amami-Ôshima)

koshikiensis Ôhira, 2003b: 361 **A**: JA (Kamikoshiki-jima)
kurosawai Nakane & Kishii, 1958: A299 **A**: JA (Yakushima)
liukuiensis Kishii, 1989b: 39, 50 **A**: TAI
longicornis Kishii, 1989b: 40, 54 **A**: TAI
longipennis Ôhira, 1966e: 335 **A**: TAI
longulus Miwa, 1930c: 94 **A**: TAI
makiharai Ôhira, 1973c: 31 **A**: JA (Okinawa-hontô)
marginalis Kishii, 1994c: 13 **A**: TAI
matobai Kishii, 1974: 18 **A**: JA
muneaka Kishii, 1991b: 22, 35 **A**: TAI
nepalensis Ôhira & Becker, 1972a: 137 **A**: NP
nitidicollis Kishii, 1991b: 23, 41 **A**: TAI
nomurai Ôhira, 1968a: 128 **A**: JA (Okinawa-hontô)
oiwakensis Kishii, 1991d: 12 **A**: TAI
okinawanus Chûjô, 1959: 5 **A**: JA (Amami-Ôshima, Okinawa-hontô)
paiwan Kishii, 1991b: 23, 42 **A**: TAI
pallidipes Miwa, 1934: 264 **A**: JA (Ishigaki-jima, Iriomote-jima)
parallelaris Kishii, 1991d: 17 **A**: TAI
rasariti Ôhira & Becker, 1972a: 136 **A**: NP
rubiginosus Kishii, 1991b: 24, 49 **A**: TAI
rubricollis Miwa, 1928b: 49 **A**: TAI
rufithorax Kishii, 1991b: 23, 41 **A**: TAI
sauteri Miwa, 1931c: 207 **A**: TAI
scutellecrista Vats & Chauchan, 1992g: 204 **A**: UP
shibatai Ôhira, 1968a: 129 **A**: JA (Amami-Ôshima)
shirouzuanus Kishii, 1991d: 14 **A**: TAI
taiwan Kishii, 1991b: 24, 53 **A**: TAI
taiwanus Kishii, 1991d: 13 **A**: TAI
takasago Kishii, 1989b: 40, 52 **A**: TAI
tamurai tamurai Kishii, 1985: 24 **A**: JA (Iriomote-jima)
tamurai yamayai Kishii, 1986: 51 **A**: JA (Okinawa-hontô)
uedai erabuanus Kishii, 1988d: 180 **A**: JA (Kuchierabu-jima)
uedai uedai Kishii, 1977a: 50 **A**: JA (Yaku-shima)
vunun Kishii, 1991b: 23, 43 **A**: TAI
yona Kishii, 1996b: 37 **A**: JA (Okinawa-hontô)
yonaguni Kishii, 1974: 14 **A**: JA (Ishigaki-jima,Yonaguni-jima)
yoshimotoi Ôhira, 1971b: 542 **A**: JA (Amami-Ôshima)
yuwancola yambarus Kishii, 1986: 51 **A**: JA (Okinawa-hontô)
yuwancola yuwancola Ôhira, 1971b: 542 **A**: JA (Amami-Ôshima)

genus *Kinokunia* Kishii, 1998c: 81 type species *Kinokunia yoshidai* Kishii, 1998
yoshidai Kishii, 1998c: 82 **A**: JA

genus *Lanecarus* Ôhira, 1962a: 199 type species *Agriotes palustris* Lewis, 1894
amianus Miwa, 1934: 204 (*Melanoxanthus*) **A**: TAI
babai Kishii, 1991b: 20, 28 **A**: TAI
ihai Ôhira, 1962a: 199 **A**: JA (Amami-Ôshima, Okinawa-hontô, Tokuno-shima)
katsuyai Ôhira, 2003b: 363 **A**: JA
lineatus Kishii, 1991b: 20, 30 **A**: TAI
modestus Candèze, 1878b: cxcvii (*Silesis*) **A**: GUA HKG SCH
palustris Lewis, 1894a: 314 (*Agriotes*) **A**: JA
    japonicus Nakane, 1957: A238 (*Synaptus*)
sinensis Fleutiaux, 1934a: 185 (*Silesis*)
    unicus Fleutiaux, 1936a: 21 (*Silesis*) **A**: GAN SCH

genus *Nepalinus* Platia & Gudenzi, 2000b: 265 type species *Nepalinus unicus* Platia & Gudenzi, 2000
unicus Platia & Gudenzi, 2000b: 265 **A**: NP

genus *Peripontius* Gurjeva, 1979a: 314 type species *Cratonychus dimidiatipennis* Reiche & Saulcy, 1857
concolor Desbrochers des Loges, 1875a: 42 (*Silesis*) **N**: TU **A**: SY TR
cordubensis Heyden, 1882a: 45 (*Silesis*) **E**: SP
crassus Buysson, 1906a: 20 [= 1912b: 42] (*Silesis*) **A**: IN TR
dentatus Platia & Schimmel, 1991b: 137 **A**: TR
dimidiatipennis Reiche & Saulcy, 1857a: 416 (*Cratonychus*) **E**: GR **N**: AG MO TU
basalis Pic, 1912n: 65 (*Silesis*)
fiduciarius Buysson, 1906b: 468 [= Buysson, 1912f: 133]
mimicus Pic, 1912n: 65 (*Silesis*)
vaulogeri Buysson, 1905b: 110 (*Silesis*)
elegantulus Platia, 2004b: 85 **A**: IQ TR
ingridae Schimmel, 1996d: 121 **A**: CY
jagemanni Platia & Gudenzi, 2000f: 76 **A**: SY TR
juvencus Buysson, 1887a: 141 (*Silesis*) **A**: LE
oertzeni Schwarz, 1897a: 136 (*Silesis*) **E**: GR (Kríti) **A**: TR
omissus Buysson, 1889a: 212 (*Silesis*) **A**: SY TR
rutilipennis Illiger, 1807a: 6 (*Elater*) **E**: FR IT PT SP **N**: AG MO
bicolor P. H. Lucas, 1846: 169 (*Adrastus*)
hipponensis Pic, 1912n: 65 (*Silesis*)
postobscurus Pic, 1912n: 65 (*Silesis*)
saroveci Platia, 2004b: 89 **A**: TR
schimmeli Zeising & Brunne, 2003: 193 **A**: TR
terminatus Erichson, 1841a: 118 (*Adrastus*) **E**: AL BU CR FR GR IT MC RO SP TR **N**: MO **A**: CY IS JO LE SY TR
indistinctus Buysson, 1912f: 133
turcicus Platia & Tarnawski, 1998: 271 **A**: IN IQ TR

genus *Silesis* Candèze, 1863: 458 type species *Silesis hilaris* Candèze, 1863
Okinawana Kishii, 1976c: 54 type species *Silesis hatayamai* Kishii, 1975
Parasilesis Ôhira, 1990: 75 type species *Silesis musculus* Candèze, 1873
absimilis Candèze, 1863: 463 **A**: FUJ GUA GUI HKG JIX TAI **ORR**
atramentarius Platia & Schimmel, 1991a: 108 **A**: NP SD
beckeri Platia & Schimmel, 1993b: 117 **A**: SD
besucheti Platia & Schimmel, 1993b: 118 **A**: BT NP
bhaktai Platia & Schimmel, 1991a: 110 **A**: SD
bhutanensis Platia & Schimmel, 1996a: 245 **A**: BT
brancuccii Platia & Schimmel, 1991a: 113 **A**: NP SD
brevicarinatus Garg, Saini & Vasu, 1998: 462 **A**: SD
brevis Platia & Schimmel, 1991a: 114 **A**: NP
brignolii Platia & Schimmel, 1991a: 115 **A**: NP SD **ORR**
canaliculatus Platia & Schimmel, 1991a: 115 **A**: NP SD
candezei Platia & Schimmel, 1991a: 116 **A**: NP SD
carinoscutellis Garg & Vasu, 1997: 55 **A**: AP SD
catei Platia & Schimmel, 1991a: 118 **A**: NP SD UP
almorensis Vats & Chauhan, 1993b: 39
coactus Garg & Vasu, 1997: 56 **A**: SD
convexicollis Platia & Schimmel, 1991a: 120 **A**: SD
convexus Platia & Schimmel, 1991a: 121 **A**: SD
cylindriformis Ôhira & Becker, 1971: 580 **A**: NP SD
davidi Platia & Schimmel, 1991a: 122 **A**: BEI
dilatatus Platia & Schimmel, 1993b: 120 **A**: SD
ecarinatus Garg, Saini & Vasu, 1998: 463 **A**: AP

excavatus Platia & Schimmel, 1991a: 123 **A**: SD
exiguus Platia & Schimmel, 1991a: 124 **A**: NP
florentini Fleutiaux, 1895: 690 **A**: SHN TAI **ORR**
fluens Platia & Schimmel, 1991a: 125 **A**: NP SD
girardi Platia & Schimmel, 1991a: 129 **A**: SCH
granarius Candèze, 1895: 68 **A**: NP SCH SD YUN **ORR**
grisescens Candèze, 1895: 68 **A**: NP SD
hatayamai amami Kishii, 1976c: 54 **A**: JA (Amami-Ôshima)
hatayamai hatayamai Kishii, 1975: 6 **A**: JA (Okinawa-hontô, Tokuno-shima)
holzschuhi Platia & Schimmel, 1991a: 132 **A**: NP
humilis Platia & Schimmel, 1991a: 133 **A**: NP SD
improvisus Gurjeva, 1976a: 36 **A**: FE NC "Korea"
incognitus Platia & Schimmel, 1991a: 134 **A**: SD
iriomotensis Ôhira, 1968a: 127 **A**: JA (Iriomote-jima)
jagemanni Cate, **nom. nov.** [see New acts] **A**: SCH YUN
  sinensis Jagemann, 1934: 124 [HN]
jendeki Platia & Schimmel, 1996a: 248 **A**: YUN
krali Platia & Schimmel, 1996a: 249 **A**: YUN **ORR**
kubani Platia & Schimmel, 1993b: 121 **A**: XIZ YUN
lebischi Platia & Schimmel, 1991a: 139 **A**: NP SD **ORR**
loebli Platia & Schimmel, 1993b: 121 **A**: NP
longipennis Schwarz, 1902b: 347 **A**: BT NP SD UP YUN
  cylindricus Schwarz, 1902b: 348 [DA]
martensi Platia & Schimmel, 1991a: 141 **A**: NP
melanocephalus Fleutiaux, 1918b: 269 **A**: YUN **ORR**
migliaccioi Platia & Schimmel, 1991a: 143 **A**: NP SD
musculus musculus Candèze, 1873: 31 **A**: FE (Kurile) JA "Korea"
  candezei Lewis, 1879a: 16 (*Agriotes*) [RN]
  crocatus Candèze, 1895: 68
  ferrugineipennis Motschulsky, 1866a: 166 (*Agriotes*) [HN]
  flavicollis Fleutiaux, 1902a: 23
  flavipennis Lewis, 1894a: 315
  harmandi Fleutiaux, 1900: 358
musculus nomurai Kishii, 1993b: 12 **A**: JA (Hachijô-jima)
musculus quelpartianus Kishii, 2002a: 16 **A**: SC (Cheju)
musculus tsushimensis Kishii, 1976c: 51 **A**: JA (Tsushima)
mutabilis Bates, 1866: 349 **A**: TAI
  formosanus Matsumura, 1910: 36 [= 1911a: 143] (*Agriotes*)
  taichuensis Matsumura, 1910: 36 [= 1911a: 143] (*Agriotes*)
nepalensis Ôhira & Becker, 1971: 579 **A**: NP SD
nigriceps Candèze, 1892b: 495 **A**: BT FUJ GUA HAI SCH SHA **ORR**
  duporti Fleutiaux, 1918b: 268
numerosus Platia & Schimmel, 1991a: 145 **A**: NP SD
okinawensis erabuanus Kishii, 1979a: 8 **A**: JA (Okinoerabu-jima)
okinawensis okinawensis Miwa, 1928b: 50 **A**: JA (Ryukys)
okinawensis seinoi Kishii, 1976a: 13 **A**: JA (Nakano-shima)
parvulus Platia & Schimmel, 1991a: 146 **A**: SD
pasangi Platia & Schimmel, 1991a: 148 **A**: NP SD UP
  portulacis Vats & Chauhan, 1993b: 42
pekinensis Platia & Schimmel, 1991a: 149 **A**: BEI FUJ SCH
probsti Platia & Schimmel, 1991a: 150 **A**: NP
punctatus Platia & Schimmel, 1991a: 152 **A**: NP
pusillus Platia & Schimmel, 1993b: 122 **A**: NP
raii Platia & Schimmel, 1991a: 153 **A**: NP SD
rufipes Candèze, 1897: 79 **A**: HUB HUN SCH ZHE **ORR**

rugosus Platia & Schimmel, 1991a: 156 **A**: NP SD
sabatinelli Platia & Schimmel, 1991a: 157 **A**: NP SD UP
sanguinicollis Candèze, 1863: 461 **A**: HP NP PA SCH UP
sauteri babai Kishii, 1985: 24 **A**: JA (Iriomote-jima, Ishigaki-jima)
sauteri sauteri Miwa, 1930c: 95 **A**: FUJ GAN SCH TAI
scabripennis Lewis, 1894a: 315 **A**: JA
schawalleri Platia & Schimmel, 1991a: 160 **A**: SD
schuhi Platia & Schimmel, 1993b: 123 **A**: UP
schwarzi Platia & Schimmel, 1991a: 160 **A**: SCH
sericeus Candèze, 1878b: cxcvii **A**: ANH NP SD UP
  protuberancis Vats & Chauhan, 1993b: 39
sherpana Ôhira & Becker, 1971: 577 **A**: NP SD
shirozui Kishii, 1959b: 62 **A**: JA (Ryukyus)
simulatus Candèze, 1863: 462 **A**: HP NP SD UP XIZ **ORR**
singularis Platia & Schimmel, 1991a: 166 **A**: NP
smetanai Platia & Schimmel, 1993b: 124 **A**: NP
spiniformis Platia & Schimmel, 1991a: 166 **A**: NP SD
subcarinatus Platia & Schimmel, 1996a: 250 **A**: NP
subrugosus Platia & Schimmel, 1993b: 124 **A**: SD NP **ORR**
suturalis Candèze, 1892b: 494 (*Glyphonyx*) **A**: NP UP **ORR**
  inficetus Candèze, 1892b: 494
  nigrostriatus Vats & Chauhan, 1993b: 41
umbilicatus Platia & Schimmel, 1991a: 167 **A**: NP
vatsi Garg, Saini & Vasu, 1998: 460 **A**: SD
vignai Platia & Schimmel, 1991a: 168 **A**: NP SD
wewalkai Platia & Schimmel, 1991a: 169 **A**: NP SD
wittmeri Platia & Schimmel, 1991a: 170 **A**: SD **ORR**
yaku Kishii, 1976c: 51 **A**: JA (Kuro-shima, Yaku-shima)
yunnanensis Platia & Schimmel, 1996a: 251 **A**: YUN

genus *Synaptus* Eschscholtz, 1829: 32 type species *Elater filiformis* Fabricius, 1781
  Ctenonychus Stephens, 1830: 272 type species *Ctenonychus hirsutus* Stephens, 1830 (= *Elater filiformis* Fabricius, 1781)
filiformis Fabricius, 1781: 270 (*Elater*) **E**: AB AL AR AU BE BH BU BY CR CT CZ EN FR GB GE GG GR HU IT LA LS LT LU MC MD NL PL PT RO SK SL SP ST SZ UK YU **A**: IN KZ SY TR
  atripes Pic, 1941b: 1 (*Sinaptus* [sic])
  cinereus Illiger, 1807a: 4 (*Elater*) [HN]
  erivanus Faldermann, 1835a: 160
  hirsutus Stephens, 1830: 273 (*Ctenonychus*)
  incola Herbst, 1806: 80 (*Elater*)
  leprieuri Pic, 1941b: 1 (*Sinaptus* [sic])
  levantinus J. R. Sahlberg, 1913d: 134
  unguliserris Gyllenhal, 1817: 136 (*Elater*)

genus *Tolphorea* Gurjeva, 1983: 81 type species *Tolphorea volans* Gurjeva, 1983
bicolor Cate, Platia & Schimmel, 2002: 41 **A**: IN
kurdistanicus Platia & Gudenzi, 1996a: 142 **A**: TR
luristanicus Pic, 1920h: 22 (*Agriotes*) **A**: IN
pallida Platia, 2004c: 105 **A**: TR
turcica Platia & Gudenzi, 2000f: 76 **A**: TR
volans Gurjeva, 1983: 82 **A**: IN

tribe Adrastini, species incertae sedis
Adrastus aequalis Vats & Chauhan, 1992e: 181 **A**: UP
Adrastus rougeatrus Vats & Chuahan, 1992e: 182 **A**: UP

tribe Agriotini Champion, 1894

genus *Agriotes* Eschscholtz, 1829: 34 type species *Elater sputator* Linnaeus, 1758

subgenus *Agriotes* Eschscholtz, 1829: 34 type species *Elater sputator* Linnaeus, 1758

Agriodrastus Reitter, 1911a: 222 type species *Elater pallidulus* Illiger, 1807
Cataphagus Stephens, 1830: 247 type species *Elater lineatus* Linnaeus, 1767
Fructuarius Gistel, 1848: xi [unjustified RN]
Hemirhipus Dillwyn, 1829: 31 [HN] type species *Elater lineatus* Linnaeus, 1767
Lepidotus Gistel, 1834: 12 [HN] type species *Elater sputator* Linnaeus, 1758
Pedetes Kirby, 1837: 145 [HN] type species *Elater obscurus* Linnaeus, 1758

acuminatus Stephens, 1830: 248 (*Cataphagus*) **E**: AU BE BU CZ DE FR GB GE GR HU IT LS LT MC MD NL PL RO SK SL SP SZ UK **A**: TR
pallidicolor Pic, 1910m: 49
sobrinus Kiesenwetter, 1858a: 258
adanensis Pic, 1910q: 65 **A**: TR
adlbaueri Platia & Schimmel, 1992: 91 **A**: TR
aequalis Schwarz, 1891a: 107 **E**: IT
aquilus Platia, 2003a: 15 **A**: TR
anachoretus Ménétriés, 1832: 159 (*Elater*) **E**: AB
anatolicus Platia, 2003a: 14 **A**: TR
andalusiacus Franz, 1967b: 77 **E**: SP
asaokai Ôhira, 1994d: 317 **A**: JA (Ryukyus)
audisioi Platia & Schimmel, 1992: 89 **A**: TR
babanus Kishii, 1989b: 37, 43 **A**: TAI
bagherii Platia, Furlan & Gudenzi, 2002: 68 **A**: IN
barkulensis Jagemann, 1942a: 30 **A**: XIN
barriesi Cate & Platia, 1997: 109 **A**: TR
bogatschevi bogatschevi Dolin, 1969: 124 **E**: GG ST
bogatschevi rugosus Gurjeva, 1979a: 405 **E**: AB ST
bonnairei Buysson, 1890c: ccxiv **N**: AG
brevis Candèze, 1863: 390 **E**: AB AU AR AU BH BU CR CZ FR GE GG GR HU IT MD PL SK SL SP ST UK YU **A**: IN TM TR
nigricollis Schwarz, 1891a: 101
nigripennis Schwarz, 1891a: 101
caspicus Heyden, 1883c: 358 **E**: AB AR ST **A**: AF IN KI KZ TM UZ
cilicensis Platia & Gudenzi, 1998c: 634 **A**: TR
colchicus Gurjeva, 1975c: 250 **E**: GG
colonnellii Guglielmi & Platia, 1985: 202 **A**: TR
connexivus Schwarz, 1891a: 107 **N**: MO
conspicuus Schwarz, 1891a: 86 **A**: TR
constrictus Reitter, 1900b: 86 **A**: IN SY TR
corsicus Candèze, 1863: 379 **E**: AL FR (Corse) IT GR PT SP
crassiusculus Platia & Gudenzi, 1998c: 634 [RN] **A**: TR
collaris Schwarz, 1891a: 96 (*Agriotes*) [HN]
cribrithorax Pic, 1910q: 65 **A**: TR
curtus curtus Candèze, 1878b: cxciii **E**: PT SP **N**: MO
curtus kocheri Franz, 1967b: 76 **N**: MO
cylindricollis Schwarz, 1902a: 299 **A**: FUJ
defreinai Platia & Gudenzi, 1998a: 57 **A**: TR
depressus Schwarz, 1891a: 96 **E**: "Caucasus" **A**: TR
desbrochersi Buysson, 1890d: clvii **N**: AG
dilataticoxis Buysson, 1904b: 42 **N**: MO
dusaneki Platia & Gudenzi, 1998c: 633 **A**: TR
elegantulus Lewis, 1894a: 313 **A**: JA
yasumatsui Miwa, 1933b: 71
ellenicus Cate & Platia, 1997: 113 **E**: GR

emaciatus Platia & Gudenzi, 1997a: 205 **A**: SY
flavobasalis Heyden, 1889a: 328 **E**: PT
paulinoi Candèze, 1893a: 195
fulgens Ôhira, 1966e: 332 **A**: TAI
furlani Platia, 2003a: 16 **A**: TR
fusiformis Candèze, 1876b: cxciv **A**: HEN NO SE
gallicus Lacordaire, 1835: 670 **E**: AU BE BH BU CR CZ FR GE GR IT LA LU LS NL PL SK SP SZ
atacinus Buysson, 1906b: 460
robigus Buysson, 1906b: 460
semibrunneus Pic, 1910m: 49
ganglbaueri Schwarz, 1891a: 86 **A**: LE SY
praetermissus Schwarz, 1891a: 87
goksunensis Platia, 2004c: 101 **A**: TR
gracilis Miwa, 1931c: 205 (*Agonischius*) **A**: TAI
graecus Franz, 1967b: 83 **E**: GR MC TR
grandinii Candèze, 1863: 381 **N**: AG MO TU
granulosus Platia & Schimmel, 1992: 90 **A**: TR
gurgistanus Faldermann, 1835a: 161 (*Synaptus*) **E**: AB AR BU GG GR MC RO ST UK **A**: KZ TR
karabachensis Kolenati, 1846b: 39
monachus Mulsant & Godart, 1855b: 265
hedini Fleutiaux, 1936a: 20 **A**: GAN
heydeni Schwarz, 1891a: 85 **A**: TR
hirayamai Miwa, 1934: 261 **A**: JA (Ryukyus)
incognitus Schwarz, 1891a: 110 **E**: ST
informis Schwarz, 1891a: 94 **A**: TR
edmondi Pic, 1906a: 1
infuscatus Desbrochers des Loges, 1870b: 117 **E**: AB AR CR FR GG IT ST **A**: IN TR
elegans Schwarz, 1891a: 98
fuscipennis Fiori, 1899b: 163
nigripennis Fiori, 1899b: 163
nigrithorax Fiori, 1899b: 163
synaptoides Reitter, 1906e: 240
integricollis Reitter, 1911a: 222 **E**: AR **A**: TR
iranicus Platia, Furlan & Gudenzi, 2002: 69 **A**: IN
iraqensis Platia & Gudenzi, 1997a: 208 **A**: IQ
izmirensis Cate & Platia, 1997: 111 **A**: TR
karsantianus Pic, 1910q: 65 **A**: TR
kinzelbachi Platia & Schimmel, 1994: 81 **A**: SY
kirghisicus Iablikoff-Khnzorian, 1970c: 179 **A**: KI
kraatzi Schwarz, 1891a: 85 **A**: SY TR
kubani Platia & Gudenzi, 1997a: 202 **A**: SY
laevicarinatus Platia & Gudenzi, 1999a: 27 **E**: GR
lapicida Faldermann, 1835a: 177 **E**: AB AR GG ST **A**: IN
leinfesti Platia & Gudenzi, 1998a: 57 **A**: TR
leucophaeatus Candèze, 1873: 30 **A**: FE JA
leuthneri Platia & Gudenzi, 1997a: 205 **A**: SY
lineatus Linnaeus, 1767: 653 (*Elater*) **E**: AB AL AR AU BE BH BU BY CR CT CZ DE EN ES FI FR GB GE GG GR HU IR IT LA LS LT LU MC MD NL NR NT PL PT RO SK SL SP ST SV SZ TR UK YU **A**: ES GAN IN JIL KI KZ LIA SY TM TR XIN **AURi** **NARi**
segetis Bjerkander, 1779: 285 (*Elater*)
suecicus Gmelin, 1790: 1907 (*Elater*)
striatus Fabricius, 1775: 215 (*Elater*) [HN]
striolatus Fabricius, 1803: 50 (*Elater*)
suecicus Gmelin, 1790: 1907 (*Elater*)
litigiosus P. Rossi, 1792: 58 (*Elater*) **E**: AU BU FR (Corse) GR IT MC SL SP SZ
attenuatus Desbrochers des Loges, 1869: 30 [= 1870b: 119]

fulvescens Candèze, 1863: 394 [HN]
laichartingi Gredler, 1863: 219
meridionalis Desbrochers des Loges, 1869: 30 [= 1870b: 119]
murinus L. Miller, 1862: 343
rufipennis Buysson, 1893a: 43
vesubiensis Pic, 1916j: 13
liukuiensis Kishii, 1989b: 36, 40 **A**: TAI
lizleri Platia, 2003a: 17 **A**: TR
lundbergi Platia, 1989: 83 **A**: CY TR
luteonotatus Pic, 1913i: 137 **A**: LE
maceki Platia & Gudenzi, 1997a: 201 **A**: SY
magnanii Platia & Gudenzi, 1996a: 145 **A**: CY
medvedevi Dolin, 1960: 196 **E**: AB AU HU MD RO SK ST UK
mertliki Platia, 2003a: 17 **A**: TR
meticulosus Candèze, 1863: 396 **E**: AB AR GG ST **A**: AF IN IQ KZ MG PA SA SY TD TM UZ
minor Schwarz, 1891a: 95
modestus Kiesenwetter, 1858a: 259 **E**: AU BU CZ FR GR IT HU MD RO SK SP ST UK **A**: IN KZ SY
incognitus Schwarz, 1891a: 110
ponticus Stepanov, 1935: 199
nadezhdae Cherepanov, 1965: 35 **A**: MG WS XIN
nigricans Platia & Gudenzi, 1997a: 204 **A**: SY
nigropubens Reitter, 1904e: 158 **A**: TR
nigror Platia, 2003a: 20 **A**: TR
nuceus Fairmaire, 1866b: 260 **A**: TR
obscurus Linnaeus, 1758: 406 (*Elater*) **E**: AB AR AU BE BU BY CT CZ DE EN FI GB GE GG HU IR IT LA LS LT NR NT PL RO SK SL SP ST SV SZ YU **A**: GAN ES FE KZ MG SHA SHX WS XIN **NAR**i
badius O. F. Müller, 1764: 17 (*Elater*)
cinnanomeus Buysson, 1893a: 41
fuscusminor DeGeer, 1774: 147 (*Elater*)
hirtellus Herbst, 1806: 94 (*Elater*)
nigrinus Kolbe, 1924: 53
obtusus DeGeer, 1774: 147 (*Elater*)
radhosticola A. Fleischer, 1910b: 329
variabilis Fabricius, 1792b: 230 (*Elater*)
olivieri Desbrochers des Loges, 1875b: cxxxviii **N**: AG MO TU
hipponensis Schwarz, 1891a: 111
oxianus Iablokoff-Khnzorian, 1970a: 59 **A**: AF TD TM
pallidulus Illiger, 1807a: 6 (*Elater*) **E**: AU BE CZ FR GB GE HU IR IT LU NL PT RO SK SP SZ UK
gabilloti Pic, 1910m: 49
picipennis Bach, 1896: 287
rufulus Lacordaire, 1835: 673
umbrinus Germar, 1824: 46 (*Elater*)
paludum Kiesenwetter, 1859a: 18 **E**: AL BG GR MC TR YU **A**: IS JO SY TR
filiformis Candèze, 1863: 397
pamirensis Gurjeva, 1978b: 31 **A**: TD
passosi Platia & Serrano, 2002: 318 **E**: PT
pavesii Platia & Gudenzi, 2001: 310 **E**: GR
peregrinus Gurjeva, 1972c: 866 **A**: SCH
petterssoni Platia & Gudenzi, 2000f: 74 **A**: TR
pilosellus Schönherr, 1817: 307 (*Elater*) [RN] **E**: AL AU BE BH BU BY CR CT CZ DE FR GE GR HU IT LS LU MC MD NL PL RO SK SL SP ST SV SZ TR UK YU **N**: AG MO **A**: IN TR
buyssoni Méquignon, 1927: 23
elongatus Marsham, 1802: 385 (*Elater*)
flavescens Laporte, 1840a: 246 (*Elater*)
humerosignatus Buysson, 1926b: 198
simulator Buysson, 1893a: 34
panzeri Dufour, 1851: 326 (*Elater*)

vilis Illiger, 1807a: 5 (*Elater*) [HN]
propleuralis Platia & Gudenzi, 1998: 635 **A**: TR
proximoides Platia, Furlan & Gudenzi, 2002: 70 **A**: IN
proximus Schwarz, 1891a: 113 **E**: AB AU BU GR HU IT MD PT SK SP ST UK **N**: MO **A**: JO TR
lineatoides Reitter, 1911a: 222
radula Desbrochers des Loges, 1875a: 41 **A**: IS LE SY TR
brevicollis Schwarz, 1891a: 104
reitteri Schwarz, 1891a: 101 **E**: GG ST
riesei Platia, 2004a: 42 **A**: TR
rufipalpis Brullé, 1832 141 **E**: BH BU CR GR HU SK YU
piceolus Küster, 1853: 77
rubiginosus Candèze, 1863: 389
suturalis Schwarz, 1891: 106
rugipennis Schwarz, 1891a: 113 **A**: ES FE HEB MG
samai Platia, Furlan & Gudenzi, 2002: 72 **A**: IN
sameki Platia, 2003a: 21 **A**: TR
schuberti Platia, 2004c: 104 **A**: TR
schurmanni Platia & Gudenzi, 1998a: 58 **A**: TR
sericatus Schwarz, 1891a: 113 **A**: ANH FE FUJ HEB HEN HUN JIA MG SCH SHA SHN ZHE "Korea"
sexualis Platia, 2004c: 104 **A**: TR
siciliensis Pic, 1912s: 83 [RN] **E**: IT (Sicilia)
buyssoni Ragusa, 1911: 191 [HN]
shirozui Kishii, 1991d: 10 **A**: TAI
shykshanus Kishii, 1989b: 37, 46 **A**: TAI
soederbomi Fleutiaux, 1936a: 20 **A**: MG NMO
solai Platia, 2004a: 42 **A**: TR
sordidus Illiger, 1807a: 7 (*Elater*) **E**: AU BE FR GB GE GR IT PT RO SP SZ **N**: AG CI MO
buyssoni Pic, 1895a: 137
circumcinctus Fuente, 1903: 343
clarior Buysson, 1924b: 206
hispanicus Desbrochers des Loges, 1871: 369
italicus Baudi di Selve, 1871: 75
marginipennis P. H. Lucas, 1846: 168 (*Dolopius*)
neglectus Schwarz, 1891a: 110
nigriceps Daniel, 1903a: 251
nigroscutellatus Buysson, 1924b: 206
parallelus Schwarz, 1891a: 108
raduloides Buysson, 1924b: 206
rusticus Schwarz, 1891a: 108
schwarzi Buysson, 1893a: 49
scutellatus Schaufuss, 1882b: 622
seminiger Buysson, 1893a: 50
sericans Rey, 1891c: 85
sputator Linnaeus, 1758: 405 (*Elater*) **E**: AB AL AR AU BE BH BU BY CR CT CZ DE EN FR GB GE GG GR HU IT LA LS LT LU MC MD NL NR NT PL RO SL SP ST SV SZ UK YU **N**: AG **A**: ES FE IN KZ MG NO SY TR WS XIN **NAR**i
brunnicornis Gebler, 1830: 80
corallifer Eschscholtz, 1830a: 18
cribrosus Eschscholtz, 1830a: 18
depactus Buysson, 1926b: 196
fusculus Illiger, 1805: 101 (*Elater*)
graminicola L. Redtenbacher, 1849: 311 [RN]
melanocephalus Desbrochers des Loges, 1870b: 119
negatus Buysson, 1895: cclxxxiii
productus Rey, 1891c: 85
squalidus squalidus Schwarz, 1891a: 104 **A**: IN KI KZ MG
nadari Buysson, 1899a: 21
squalidus vesperalis Gurjeva, 1972c: 870 **A**: KI UZ XIN

starcki Schwarz, 1891a: 94 **E**: AB GG ST **A**: TR
niger Schwarz, 1891a: 95
pectoralis Schwarz, 1891a: 95
schneideri Schwarz, 1891a: 95
stepanovorum Orlov, 1997b: 275 **E**: ST
strigosus Kiesenwetter, 1858a: 261 **E**: GR **A**: CY IS SY TR
subsulcatus Pic, 1913i: 137 **A**: TR
subvittatus fuscicollis Miwa, 1928b: 44 **A**: JA "NE China"
subvittatus hegurensis Kishii, 1964: 28 **A**: JA (Hegura-jima)
subvittatus ogurae Lewis, 1894a: 313 **A**: JA "Korea"
nakayamai Ôhira, 1976: 50
yamadai Ôhira, 1976: 49
subvittatus subvittatus Motschulsky, 1860e: 490 **A**: ANH FE FUJ GAN GUX HEB HEI HEN HUB JIA JIL LIA NC NIN NMO QIN SC SCH SHA SHN SHX XIN ZHE
rubidicinctus Buysson, 1905a: 16
sylviae Cate & Platia, 1997: 111 **A**: TR
syriacus Platia & Gudenzi, 1997a: 204 **A**: SY
tadzhikistanicus Gurjeva, 1967b: 201 **A**: KI TD TM UZ
takasago Kishii, 1989b: 37, 47 **A**: TAI
tauricus Heyden, 1882b: 155 **E**: AB AR GG ST UK **A**: TR
rufangulus Buysson, 1896a: 42
rufus Schwarz, 1891a: 93
tonkinensis Fleutiaux, 1895: 690 (*Silesis*) **A**: GUI JIX TAI YUN ZHE **ORR**
colonus Fleutiaux, 1895: 690 (*Silesis*)
formosanus Miwa, 1928b: 46 (*Agriotides*)
nodieri Fleutiaux, 1918b: 265 (*Agonischius*)
pallidiangulus Miwa, 1928b: 46 (*Agriotes*)
transcaucasicus Dolin, 2003: 36 **E**: ST
tristiculus Reitter, 1906a: 390 [RN] **A**: TR
tristis Schwarz, 1891a: 105 [HN]
turcicus Candèze, 1863: 379 **E**: AB AR GG ST TR **A**: TR
turkmenicus Platia & Gudenzi, 1999a: 26 **A**: TM
unicolor Koenig, 1889b: 538 **A**: MG NMO XIN
ustulatus Schaller, 1783: 303 (*Elater*) **E**: AL AU BE BH BU BY CR CZ DE EN FI FR GE GG GR HU IT LS LT LU MC MD NL PL RO SK SL SP ST SZ UK **N**: MO TU **A**: IN TR **NTR**i
blandus Germar, 1824: 48 (*Elater*)
bucovinensis Hormuzachi, 1901: 103
confusus Bach, 1852: 47
flavicornis Panzer, 1799a: 6 (*Elater*)
foveolatus Rey, 1891c: 85
fusculus Laporte, 1840a: 248
gilvellus Lacordaire, 1835: 670
inustus Buysson, 1906b: 460
martialis Pic, 1914o: 58
nitidicollis L. Miller, 1881: 220
piceus Buysson, 1890a: 18
subrectus Rey, 1891c: 85
tristis Schilsky, 1888: 190
ustulus Schönherr, 1817: 303 (*Elater*)
vastus Gurjeva, 1972c: 871 **A**: KZ
werneri Platia, 2003a: 22 **A**: TR
witzgalli Platia & Schimmel, 1993a: 52 **A**: TR
yushihanus Kishii, 1989b: 37, 42 **A**: TAI
zinovjevi Gurjeva, 1967a: 137 **A**: ES FE MG NC NO
subgenus *Tinecoides* Gurjeva, 1979a: 376 type species *Agriotes koltzei* Reitter, 1890
koltzei Reitter, 1890a: 147 **A**: ES FE JIL NMO "Korea"

genus *Agriotes*, species incertae sedis
dehiscans Gistel, 1857: 3 **E:** "Styria"
oreas Gistel, 1857: 3 **E:** "Styria"
sareptanus Desbrochers des Loges, 1869: 46

genus *Chatanayus* Fleutiaux, 1940a: 124 type species *Agonischius ruficollis* Fleutiaux, 1903 (= *Agonischius fleutiauxi* Schwarz, 1906)
Neoagriotes Ôhira, 1962b: 263 type species *Agriotes insularis* Miwa, 1934
fuscus Miwa, 1930c: 93 (*Agriotes*) **A**: TAI
insularis insularis Miwa, 1934: 260 (*Agriotes*) **A**: JA
insularis isaoi Kishii, 1973: 16 (*Neoagriotes*) **A**: JA
ishiharai ishiharai Nakane & Kishii, 1954a: 14 (*Agriotes*) **A**: JA
ishiharai seinoi Kishii, 1984: 20 **A**: JA
taiyarui Miwa, 1934: 232 (*Agriotes*) **A**: TAI

genus *Dalopius* Eschscholtz, 1829: 34 type species *Elater marginatus* Linnaeus, 1758
Dolopius Eschscholtz, 1838: pl. post p. 60 [unjustified emendation]
ainu Kishii, 1962: 26 **A**: FE JA JIL
bizen Kishii, 1984: 28 **A**: JA
exilis Kishii, 1956: 20 **A**: FE JA
japonicus Kishii, 1984: 25 **A**: JA
kushiroensis Ôhira, 1997b: 25 **A**: JA
marginatus Linnaeus, 1758: 405 (*Elater*) **E**: AL AU BY CT CZ DE EN FI FR GB GE GR HU IR IT LA LS LT NR NT PT RO SL SP ST SV SZ **A**: ES HEB GAN KZ MG NMO TR WS
depressus Eschscholtz, 1818: 479 (*Elater*)
fulvus Marsham, 1802: 379 (*Elater*)
gyllenhali Buysson, 1926b: 188
lateralis A. G. Olivier, 1790: no 31: 50 (*Elater*) [HN]
lineatus DeGeer, 1774: 158 (*Elater*) [HN]
lineolus Panzer, 1795: 238 (*Elater*)
sticticus Panzer, 1800b: 9 (*Elater*)
suturalis Marsham, 1802: 379 (*Elater*)
miwai Ôhira, 1972c: 45 **A**: JA
mutsuensis Ôhira, 1989c: 61 **A**: JA
naomii Kishii, 1981: 19 **A**: JA
niponensis Ôhira, 1970a: 11 **A**: JA
obscuricollis Jiang, 1999: 126, 186 **A**: HUB
patagiatus Lewis, 1894a: 315 (*Adrastus*) **A**: JA
lewisi Fleutiaux, 1900: 358 (*Dolopius*)
puerilis Candèze, 1879a: 282 (*Agriotes*) **A**: ES FE HEI MG
radiculosus Gurjeva, 1964: 150 **A**: KZ TR
singeri Brancsik, 1914: 58 (*Dolopius*) **A**: ES WS
tamui Kishii, 1957c: 86 **A**: JA
hirasanus Kishii, 1966a: 38
yakuensis Kishii, 1975: 6 **A**: JA (Yaku-shima)

genus *Ectinoides* Kishii, 1966a: 40 type species *Agriotes leucophaeatus* sensu Kishii, 1966 (= *Megapenthes insignitus* Lewis, 1894)
insignitus insignitus Lewis, 1894a: 46 (*Megapenthes*) **A**: FE JA
insignitus ogatai Kishii, 1986 49 **A**: JA (Yaku-shima)

genus *Ectinus* Eschscholtz, 1829: 34 type species *Elater aterrimus* Linnaeus, 1761
Eumenus Gistel, 1834: 12 type species *Elater aterrimus* Linnaeus, 1761
aterrimus Linnaeus, 1761: 206 (*Elater*) **E**: AU BE BU BY CT CZ DE EN FI FR GE IT LA LS LT LU NL NR NT PL RO SP SV SZ **A**: WS
atratus Illiger, 1805: 100 (*Elater*)

cavillator Buysson, 1926b: 205
nudus Küster, 1853: 74 [erroneous p. 84]
volhyniensis Fischer von Waldheim, 1824: 202 (*Elater*)
dahuricus dahuricus Candèze, 1863: 394 (*Agriotes*) **A**: FE HEI JA LIA NC
dahuricus persimilis Lewis, 1894a: 312 (*Agriotes*) **A**: FE JA JIL NE
exulatus Candèze, 1873: 29 (*Agriotes*) **A**: JA
hidaensis Ôhira, 1998a: 161 **A**: JA
higonius Lewis, 1894a: 43 (*Megapenthes*) **A**: JA
insidiosus Lewis, 1894a: 44 (*Megapenthes*) **A**: JA
hattorii Ôhira, 1954: 45 (*Agriotes*)
koshiki Kishii, 1984: 46 **A**: JA
longicollis Lewis, 1894a: 311 (*Agriotes*) **A**: JA JIL "Korea"
miyakei Ôhira, 1964c: 233 **A**: JA
nipponicus Kishii, 1979b: 36 **A**: JA
nokozanus Miwa, 1928b: 45 (*Agriotes*) **A**: TAI
obakoae Kishii, 1979b: 34 **A**: HEI JA
obscurolineatus Kishii, 1979b: 38 **A**: JA
peregrinus Gurjeva, 1979a: 361 **A**: BEI SCH
piloselloides Schwarz, 1891a: 88 (*Agriotes*) **A**: FE
puberulus Miwa, 1928b: 44 (*Agriotes*) **A**: JA
saitoanus Kishii, 1987: 11 **A**: JA
sepes Lewis, 1879b: 157 (*Agriotes*) **A**: JA
sericeus babai Kishii, 1961: 47 **A**: JA (Tsushima)
sericeus sericeus Candèze, 1878b: cxciii (*Agriotes*) **A**: FE FUJ GUI HEB HEN HUB HUN JA JIL SCH SHN "Korea"
sericans Lewis, 1879b: 157 (*Agriotes*)
sonanis Miwa, 1928b: 45 (*Agriotes*) **A**: HUB TAI
brunneipennis Miwa, 1931b: 78
nigrinus Miwa, 1931b: 78
tamnaensis Kishii, 2002a: 14 **A**: SC (Cheju)
yushiroi W. Suzuki, 1999: 189 [RN] **A**: GUI SCH TAI
angustatus Miwa, 1928b: 46 (*Agriotes*) [HN]

genus *Insuliectinus* Kishii, 1984: 12 type species *Insuliectinus amami* Kishii, 1984
amami Kishii, 1984: 13 **A**: JA (Amami-Ôshima)

genus *Pseudagriotes* Schwarz, 1896b: 103 type species *Pseudagriotes holtzi* Schwarz, 1896
holtzi Schwarz, 1896b: 104 **A**: TR

genus *Sadoganus* Ôhira, 1956: 11 type species *Sadoganus babai* Ôhira, 1956
babai Ôhira, 1956: 12 **A**: JA

tribe Ampedini Gistel, 1856

genus *Alestrus* Méquignon, 1942b: 10 type species *Elastrus dolosus* Crotch, 1867
dolosus Crotch, 1867: 386 (*Elastrus*) **E**: AZ

genus *Ampedus* Dejean, 1833: 92 type species *Elater sanguineus* Linnaeus, 1758
subgenus *Ampedus* Dejean, 1833: 92 type species *Elater sanguineus* Linnaeus, 1758
Lewisiellus Kishii, 1955b: 81 type species *Elater gracilipes* Lewis, 1894
abdominalis Schimmel, 1993a: 114 **A**: HP
adlbaueri Schimmel, 1990: 76 **A**: TR
adrastoides Reitter, 1896g: 289 (*Elater*) **A**: FE
adventicus Schimmel, 2003b: 267 **A**: HEB
aequicollis Schimmel, 1993a: 115 **A**: SCH

aethiops Lacordaire, 1835: 657 (*Elater*) **E**: AL AU BH BU CR CZ FR GE IT LS PL RO SK SL SP SZ UK YU **A**: TR
  dilutimembris Buysson, 1911a: 17 (*Elater*)
  foveicollis Germar, 1844: 174
  rugosissimus Tanzer, 1929: 76 (*Elater*)
  scrofa Germar, 1844: 173
agajewi Dolin, 1970: 15 **E**: AB **A**: TR
agilis Platia, 2001a: 66 **A**: TR
agnatus Schimmel, 1993a: 116 **A**: TAI
ainu ainu Lewis, 1894a: 35 (*Elater*) **A**: FE JA
ainu hondoensis Kishii, 1985: 17 **A**: JA
aizu Kishii, 1996b: 35 **A**: JA
akihikoi Kishii, 1985: 21 **A**: JA
alpinus Kishii, 1992b: 18 **A**: JA
  lewisi Kishii, 1955b: 82
alticola alticola Silfverberg, 1977: 92 [RN] **A**: FE JA JIL
  montanus Lewis, 1894a: 36 (*Elater*) [HN]
alticola tokoro Kishii, 1998a: 18 **A**: JA
altiusculus Schimmel, 1998d: 233 **A**: XIZ
amakazaricola Kishii & Ôhira, 1956: 75, 83 **A**: JA
amamiensis Ôhira, 1968c: 140 **A**: JA (Ryukyus)
amicus Gurjeva & Dolin, 1970: 19 **A**: TD UZ
anatolicus Platia & Gudenzi, 2002b: 86 **A**: TR
antoniae Reitter, 1889f: 113 (*Elater*) **E**: AR ST
apicalis Reitter, 1889f: 113 (*Elater*) **E**: GG UK
ariharai Kishii, 1985: 20 **A**: JA
armeniacus Dolin & Agajev, 1993: 67 **E**: AR
ashiaka Kishii, 1998a: 16 **A**: JA
ashiunus Kishii, 1976a: 11 **A**: JA
assingi Schimmel, 1996b: 63 **A**: CY
atripes Reitter, 1896g: 288 (*Elater*) **A**: HEI KI KZ XIN
  deplanantus Reitter, 1918b: 92 (*Elater*)
  subbasalis Pic, 1916i: 9 (*Elater*)
auranticulus Reitter, 1887f: 213 (*Elater*) **E**: ST
aureopilosus Kishii, 1988a: 17 **A**: JA
aureovestitus aureovestitus Kishii, 1966a: 36 **A**: JA
  oiwakensis Ôhira, 1974a: 57 **A**: JA
aureovestitus katoi Kishii, 1998a: 19 **A**: JA
aurilegulus Schaufuss, 1863: 200 (*Elater*) **E**: FR PT SP **N**: AG MO TU
auripes Reitter, 1895c: 200 (*Elater*) **E**: AU CZ FR GE IT SK SZ
aurosericeus Gurjeva, 1957: 468 (*Elater*) [= 1959: 210] **A**: KI KZ UZ XIN
azurescens azurescens Candèze, 1865: 29 (*Elater*) **A**: FE (Kunashir) JA (Hokkaido)
  pirika Kishii, 1985: 17
azurescens scutellaris Lewis, 1894a: 35 (*Elater*) **A**: JA
babai Kishii, 1966b: 6 **A**: JA
babaulti Schimmel, 1993a: 117 **A**: HP PA (Punjab)
balcanicus Dolin, 1983: 30 **E**: BH
balteatus Linnaeus, 1758: 405 (*Elater*) **E**: AU BE BH BU BY CT CZ DE EN FI FR GB GE HU IR IT LA LS LT LU NR NT PL RO SK SL SP ST SV SZ UK YU **A**: ES KZ MG TR WS
  flavescens Geoffroy, 1785: 35 (*Elater*)
  histrio Gmelin, 1790: 1914 (*Elater*)
basalis Mannerheim, 1852b: 291 **A**: ES FE JIL MG
becvari Schimmel, 2003b: 267 **A**: YUN
beppin Kishii, 1992b: 11 **A**: JA
besucheti Schimmel, 1993a: 118 **A**: PA

beybienkoi Dolin, 1976: 1728 **A**: FE (Sachalin)
biformis Dolin, 1970: 17 **E**: AB **A**: IN
bilyi Schimmel, 1993a: 119 **A**: NP
bocakorum Schimmel, 2003b: 268 **A**: MG
bonianus Schimmel, 2003b: 269 **A**: XIZ
boquilobensis Platia & Serrano, 2002: 320 **E**: PT
bouweri Schimmel, 1984: 9 **E**: AU CZ GE LA
brancuccii Schimmel, 1993a: 121 **A**: NP SD
brunnicornis Germar, 1844: 173 **E**: AU CZ FR GE IT SK
fontisbellaquei Iablokoff, 1937: 64
callegarii Platia & Gudenzi, 2000c: 11 **E**: IT
canaliculatus Reitter, 1918b: 86 (*Elater*) **E**: GR **A**: TR
cardinalis Schiödte, 1865: 534 (*Elater*) **E**: AL AU BE BH CR CT CZ DE FR GE GB GR HU IT LA NR PL SK SP ST SV UK
coccinatus E. C. Rye, 1867: 249 (*Elater*)
titanus Reitter, 1918a: 43 (*Elater*)
carinthiacus Bouwer, 1984a: 3 **E**: AU SL
catei Schimmel, 1998a: 217 **A**: GAN SCH
chegdeus Schimmel, 2003b: 269 **A**: HEB
chigo Kishii, 1992b: 14 **A**: JA
chlamydatus Lewis, 1894a: 36 (*Elater*) **A**: JA
chokai Kishii, 1985: 16 **A**: JA
chopalensis Schimmel, 1993a: 123 **A**: HP
cinnabarinus Eschscholtz, 1829: 33 (*Elater*) **E**: AB AR AU BE BH BU BY CR CT CZ DE EN FI FR GB GE GG GR HU IT LA LT LU MD NL NR NT PL PT RO SK SL SP ST SV SZ UK **A**: ES FE IN KZ TR WS
angusticollis Heyden, 1886d: 278 (*Elater*)
atropilosus B. G. Rye, 1903: 80 (*Elater*)
dimorphus Reitter, 1918b: 84 (*Elater*)
lythropterus Germar, 1844: 156
pictaviensis Buysson, 1887b: xc (*Elater*)
semiruber Stephens, 1830: 257 (*Elater*)
circassicus Reitter, 1887f: 212 (*Elater*) **E**: AB AR GG ST **A**: TR
araxidis Reitter, 1889f: 113 (*Elater*)
laetulus Reitter, 1918b: 97 (*Elater*)
talyschensis Reitter, 1889f: 111 (*Elater*)
coenobita A. Costa, 1881: 34 (*Elater*) **E**: AR GG GR IT ST
ochripennis Reitter, 1887f: 212 (*Elater*)
cognatus Gurjeva, 1977a: 45 **A**: FE
commutabilis Schimmel, 1993a: 124 **A**: YUN
convexicollis Lewis, 1894a: 34 (*Elater*) **A**: JA
corsicus Reitter, 1918b: 91 (*Elater*) **E**: FR (Corse)
coruscus Gurjeva, 1977b: 800 **E**: AB GG
cretensis Platia, 2004a: 40 **E**: GR (Kríti)
cyanicollis Dolin, 1970: 16 **E**: AB **A**: IN
decumanus Gurjeva, 1977b: 803 **E**: AB **A**: TR
demaisoni Buysson, 1899c: 367 (*Elater*) **A**: TR
dewanus Kishii, 1979b: 32 **A**: JA
dilutipes Motschulsky, 1860b: 111 (*Elater*) **A**: FE NE
doii Miwa, 1927a: 17 (*Elater*) **A**: FE (Kunashir, Iturup)
dosanko Kishii, 1987: 5 **A**: JA
dundai Schimmel, 1993a: 126 **A**: SCH
echigo Kishii, 1991d: 4 **A**: JA
echizen Kishii, 2002b: 113 **A**: JA (Honshu)
elegantulus Schönherr, 1817: 303 (*Elater*) [RN] **E**: AB AU BE BH BU BY CR CZ DE FR GE GR HU IT LA MD PL RO SK SL SP ST SV SZ UK **A**: IN TR
adustus Eschscholtz, 1829: 33 (*Elater*)

austriacus Laporte, 1840a: 244 (*Elater*) [HN]
lederi Reitter, 1888g: 426 (*Elater*)
seidlitzi Buysson, 1896b: 222 (*Elater*)
elongatulus Fabricus, 1787a: 174 (*Elater*) **E**: AB AR AU BE BU BY CR CT CZ EN FR GE GB GG GR HU IR IT LA LT MD NL PL RO SK SL SP SP ST SZ UK YU **A**: TR
balteatulus Reitter, 1889: 113 (*Elater*)
bodemeyeri Reitter, 1906c: 449 (*Elater*)
discanicus Buysson, 1896b: 200 (*Elater*)
elongantulus Fabricus, 1787a: 174 (*Elater*) [incorrect orig. spelling]
gliscereus Panzer, 1795: 241 (*Elater*)
glycereus Herbst, 1784b: 112 (*Elater*)
emishi Kishii, 1982: 45 **A**: JA
erythrogonus P. W. Müller, 1821: 186 (*Elater*) **E**: AU BH BU BY CR CT CZ DE EN FI FR GE GG HU IT LA LT MD NT PL RO SK SL ST SV SZ UK YU **A**: TR
concolor Stierlin, 1862: 36 (*Elater*)
etorupensis Miwa, 1929d: 454 (*Elater*) **A**: FE (Iturup)
fagi fagi Lewis, 1894a: 34 (*Elater*) **A**: JA
fagi kitami Kishii, 1998a: 15 **A**: JA
ferganensis Stepanov, 1935: 196 (*Elater*) **A**: TD TM UZ
flavovestitus Kishii, 1982: 43 **A**: JA
formosensis Miwa, 1929e: 487 (*Elater*) **A**: TAI
forticornis Schwarz, 1900: 99 (*Elater*) **E**: AL AU BH BU CZ GE GR HU IT RO SL SK YU
foveolatus Schimmel, 1993a: 128 **A**: UP
francolinus Bouwer, 1984c: 84 **E**: GE
fuentei Sánchez-Ruiz, 1996: 143 [RN] **E**: SP
rufiventris Candèze, 1859: 450 (*Elater*) [HN]
fulvipennis Ôhira, 1967a: 59 **A**: TAI
fulvipes Motschulsky, 1860b: 111 (*Elater*) **A**: FE
gagatinus Candèze, 1895: 33 (*Elater*) **A**: ES FE HEI KZ WS
candezei Kraatz, 1879a: 123 (*Elater*) [HN]
compactus Candèze, 1891a: 91 (*Elater*) [RN, HN]
galii Gurjeva, 1979a: 253 **A**: FE
gallicus Bouwer, 1990: 60 **E**: FR
ganeshanus Schimmel, 2003b: 270 **A**: NP
ganglbaueri Reitter, 1889f: 113 (*Elater*) **E**: AB AR GG ST **A**: TR
gaoligongshanus Schimmel, 1996a: 161 **A**: YUN
geiko Kishii, 1996b: 33 **A**: JA
girardi Schimmel, 1993a: 129 **A**: FUJ
gozaishi Kishii, 1990b: 7 **A**: JA
gracilipes Lewis, 1894a: 42 (*Elater*) **A**: JA
galloisi Miwa, 1934: 77 (*Elater*)
gurjevae Dolin & Ôhira, 1976: 389 **A**: FE
gyojya Kishii, 2004b: 71 **A**: JA
hayekae Schimmel, 1993a: 130 **A**: UP
hebeiensis Schimmel, 2003b: 271 **A**: BEI HEB
hirticollis Semenov, 1891: 348 (*Elater*) **E**: AB GG ST **A**: TR
hispanicus Platia & Gudenzi, 1999a: 29 **E**: SP
hjorti B. G. Rye, 1905: 356 (*Elater*) **E**: AU CZ DE GE GR HU LA LT NR PL SK SV **A**: IN
honguanus Ôhira, 1962b: 262 **A**: JA
hosodai Kishii, 1990b: 13 **A**: JA
houwau Kishii, 1990b: 6 **A**: JA
hypogastricus hypogastricus Candèze, 1873: 9 (*Elater*) **A**: JA SC
miwai Kishii, 1958: 29
hypogastricus kosugi Kishii, 1982: 46 **A**: JA
impressicollis Bouwer, 1984c: 86 **E**: GE
iranicus Gurjeva, 1972a: 305 **A**: IN

ishikawai Kishii, 1998a: 10 **A**: JA
ivanovi Jakobson, 1913: 750 (*Elater*) [RN] **A**: JA
nigroventris Lewis, 1894a: 38 (*Elater*) [HN]
japonicus japonicus Silfverberg, 1977: 92 [RN] **A**: JA
rufipes Lewis, 1894a: 40 (*Elater*) [HN]
japonicus kosugiensis Kishii, 1983a: 33 **A**: JA
japonicus okhotskanus Kishii, 1997a: 10 **A**: JA
jendeki Schimmel, 1993a: 132 **A**: YUN
jiangxiensis Schimmel, 1996a: 162 **A**: JIX
jindrai Schimmel, 2003b: 271 **A**: HEB
johannae Zeising & Sieg, 1992: 110 **E**: AB
juldusanus Reitter, 1918b: 102 (*Elater*) **A**: KZ XIN
kai Kishii, 1986: 47 **A**: JA
kamuro Kishii, 1996b: 30 **A**: JA
karneri Schimmel, 1996a: 162 **E**: GR
karpathicus Buysson, 1886a: ccxvi (*Elater*) **E**: BY MD RO UK
kasugensis kasugensis Kishii, 1966b: 4 **A**: JA
yoshidai Ôhira, 1974a: 58
kasugensis yakushimensis Kishii, 1983a: 35 **A**: JA
kibane Kishii, 1994d: 6 **A**: JA
kibunensis Kishii, 1966a: 32 **A**: JA
kikuchii W. Suzuki, 1985a: 82 **A**: JA
kintaroi Ôhira, 1989a: 14 **A**: JA
kintaroui Kishii, 1991a: 9, 25 **A**: TAI
kirgisicus Platia, 2001a: 72 **A**: KI
kiso Kishii, 1990b: 11 **A**: JA
kisuji Kishii, 2004b: 74 **A**: JA
kobiki Kishii, 2002b: 109 **A**: JA
koenigi Semenov, 1891: 348 (*Elater*) **E**: AB GG ST
koltzei Reitter, 1901h: 179 (*Elater*) **A**: KI KZ
bicoloratus Buysson, 1912e: 130 (*Elater*)
koschwitzi Schimmel, 1990: 81 **E**: SP
kozoanus Kishii, 1998a: 13 **A**: JA
kucerai Schimmel, 2003b: 272 **A**: SCH SHA
kurama Kishii, 1983b: 51 **A**: JA
kurilensis Dolin & Ôhira, 1976: 390 **A**: FE JA
lateritius Dolin & Protzenko, 1965: 19 (*Elater*) **A**: KI TD UZ
islamicus Bouwer, 1984b: 7
latiusculus Reitter, 1889l: 260 (*Elater*) **A**: FE
lenkoranus Reitter, 1889f: 111 (*Elater*) **E**: AB AR ST **A**: IN
caspicus Reitter, 1918b: 91 (*Elater*)
lepidus Mäklin, 1878: 30 (*Elater*) **E**: CT FI SV **A**: ES FE WS
eurinus Cherepanov, 1957: 205 (*Elater*)
sorsakoskii Hellén, 1934: 85 (*Elater*)
leptogracilis Ôhira, 1967a: 58 **A**: TAI
lijiangensis Schimmel, 1996a: 163 **A**: YUN
lini Kishii, 1990b: 10 **A**: TAI
lubricus Buysson, 1891d: cxli **A**: TR
luguanus Schimmel, 2003b: 273 **A**: YUN
macedonicus Schimmel, 1996b: 65 **E**: GR
magistrettii Platia & Schimmel, 1988: 453 **E**: IT
mamajevi Dolin & Ôhira, 1976: 388 **A**: FE
maroccanus Schimmel, Platia & Martin, 1992: 173 **N**: MO
martini Zeising & Sieg, 1992: 126 **N**: TU
maruyamanus Ôhira, 1992d: 83 **A**: JA

masamichii Kishii, 1990a: 17 **A**: TAI
masatoi Kishii, 2002b: 115 **A**: JA
masculatus Ôhira, 1966d: 269 **A**: TAI
matobai matobai Kishii, 1982: 42 **A**: JA
matobai yamayai Kishii, 1987: 8 **A**: JA
medvedevi Gurjeva, 1977b: 799 **A**: TM
melanurus Mulsant & Guillebeau, 1855b: 319 **E**: CR CZ BU FR GE GR IT SK
menoko Kishii, 1994d: 4 **A**: JA
meybohmi Zeising & Brunne, 2003: 190 **N**: MO
mica Gurjeva, 1977a: 45 **A**: FE HEI
miyatai Ôhira, 1992d: 81 **A**: JA
minos Wurst, 1997a: 2 **E**: GR (Kríti)
mizunoanus Kishii, 1988a: 15 **A**: JA
mizunoi Kishii, 1976a: 11 **A**: JA
motoshigei Kishii, 1999a: 1 **A**: JA
motschulskyi Gurjeva, 1979a: 214 **A**: FE
mutsu Kishii, 1991d: 3 **A**: JA
nanus Silfverberg, 1977: 92 [RN] **A**: JA
parvulus Lewis, 1894a: 41 (*Elater*) [HN]
nemoralis Bouwer, 1980: 328 **E**: AU CZ GE IT LA SK SL SZ
nepalensis Schimmel, 1993a: 135 **A**: NP
nigerrimus Lacordaire, 1835: 657 (*Elater*) **E**: AU BE BH BU CR CT CZ DE FR GB GE GR HU IT MD NL PL RO SK SL SP ST SV SZ UK "Caucasus" **A**: TR
obsidianus Germar, 1844: 174
rufitarsis Desvignes, 1842: 326 (*Elater*)
nigrinus Herbst, 1784b: 114 (*Elater*) **E**: AU BE BU BY CR CT CZ DE EN FI FR GB GE GG IT LA LT NL NR NT PL RO SK SL SP ST SV SZ UK **A**: ES FE JIL KZ LIA MG TR WS **NAR**
anthracinus LeConte, 1869: 378 (*Elater*)
fulvangulus Csiki, 1951: 147 (*Elater*)
lugubris O. F. Müller, 1776: 60 (*Elater*) [DA]
nigrinoides Reitter, 1918b: 101 (*Elater*)
pilosulus Herbst, 1806: 69 (*Elater*)
nigroflavus Goeze, 1777: 568 (*Elater*) **E**: AL AU BE BU BY CT CZ DE EN FI FR GE GR HU IT LA LU MD NL NR NT PL RO SK SL SP ST SV SZ UK **A**: KZ TR WS
crocatus Lacordaire, 1835: 655 (*Elater*)
erubescens Eschscholtz, 1829: 33 (*Elater*)
nigrolineatus Pic, 1920h: 21 (*Elater*)
podolicus Reitter, 1889f: 114 (*Elater*)
nigromaculatus Schimmel, 1993a: 136 **A**: XIZ
nigror Reitter, 1896g: 288 (*Elater*) **A**: ES FE WS
nubatama Kishii, 1988a: 12 **A**: JA
ocellatus Buysson, 1891d: cxl (*Elater*) **A**: IN TR
fragarinus Wurst, 1995c: 149
ochrinulus Reitter, 1887f: 213 (*Elater*) **E**: ST UK
ochropterus Germar, 1844: 160 **E**: AB AR BH BU GG GR ST UK **A**: TR
ogatai Kishii, 1983a: 33 **A**: JA
ogloblini Denisova, 1948: 47 (*Elater*) **A**: KI TD UZ
ohsawai Kishii, 2004b: 74 **A**: JA
ohtsukai Ôhira, 1988: 156 **A**: JA
okudai Kishii, 1998a: 21 **A**: JA
optabilis awa Kishii, 1994d: 3 **A**: JA
optabilis kusuii Ôhira, 1989a: 5 **A**: JA
optabilis optabilis Lewis, 1894a: 33 (*Elater*) **A**: FE JA
ordinarius Gurjeva, 1970: 20 **A**: FE
orientalis Lewis, 1894a: 33 (*Elater*) **A**: FE JA

otome otome Kishii, 1992b: 8 **A**: JA
otome shikokuanus Kishii, 2002b: 112 **A**: JA
pachycollis Ôhira, 1973b: 119 **A**: JA
pakistanicus Platia, 1988a: 9 **A**: KA PA
pallipes Kraatz, 1879a: 124 (*Elater*) **A**: ES FE JIL MG NC NW WS
patricius Gurjeva, 1957: 466 [= 1959: 205] (*Elater*) **A**: FE
gracilis Gurjeva, 1957: 466 [= 1959: 205] (*Elater*)
pauxillus Lewis, 1894a: 42 (*Elater*) **A**: JA JIL LIA
brunneisuturalis Kishii, 1966b: 5
persicus Gurjeva, 1972a: 304 **E**: AB ST **A**: IN
persimilis Dolin & Katyukha, 1987: 15 **A**: FE
phoenicium Platia, 2001a: 68 **A**: LE
picitarsis Motschulsky, 1860b: 110 (*Elater*) **A**: FE
simillimus Kraatz, 1879a: 123 (*Elater*)
platiai Schimmel, 1990: 78 **A**: TR
pomonae Stephens, 1830: 257 (*Elater*) **E**: AB AR AU BE BH BU BY CR CT CZ DE EN FI FR GE GG HU IR IT LA LT LU MD NL NR NT PL PT RO SK SL SP ST SV SZ UK YU **A**: ES FE IN KZ MG TR WS
gerardi Candèze, 1890: cxxi (*Elater*)
pomonaeformis Buysson, 1887b: xc (*Elater*)
pomorum Herbst, 1784b: 112 (*Elater*) **E**: AB AL AR AU BE BH BU BY CR CT CZ DE EN FI FR GB GE GG GR HU IR IT LA LS LT LU MC MD NL NR NT PL RO SK SL SP ST SV SZ UK YU **A**: ES FE JIL HEI KZ MG WS
adumbratus Buysson, 1896b: 205
apicalis Schilsky, 1892: 198
brigittae Bouwer, 1980: 326
crocatus Stephens, 1839: 179 (*Elater*) [HN]
ferrugatulus Reitter, 1889f: 112 (*Elater*)
ferrugatus Lacordaire, 1835: 655 (*Elater*)
fulvilegulus Reitter, 1889f: 112 (*Elater*)
jakowlewi Semenov, 1891: 350 (*Elater*)
ochropterus Eschscholtz, 1830a: 17 (*Elater*) [HN]
nigriventris Reitter, 1889f: 112 (*Elater*)
robustus Bouwer, 1980: 330
pooti Wurst, 1995c: 150 **E**: PT
praeustus Fabricius, 1792b: 229 (*Elater*) **E**: AB AR AU BE BH BU BY CT CR CZ DE EN FI FR GE GR HU IR IT LA LT LU MD NL NR NT PL PT RO SK SL SP ST SV SZ UK YU **N**: MO **A**: ES KZ TR WS
aster B. G. Rye, 1905: 355 (*Elater*)
aurileguloides Buysson, 1886b: xxviii (*Elater*)
exsanguis Eschscholtz, 1830a: 17 (*Elater*)
subdepressus Rey, 1887: 78 (*Elater*)
tauricus Reitter, 1889f: 111 (*Elater*)
probsti Zeising & Sieg, 1992: 112 **A**: NP
pseudoflavobasalis Schimmel, 1993a: 139 **A**: TAI
pulcher Baudi di Selve, 1871: 51 (*Elater*) **A**: CY TR
pulcherrimus Reitter, 1889f: 113 (*Elater*)
punctatus Schwarz, 1893d: 192 (*Elater*) **A**: SY TR
purus Gurjeva, 1977b: 802 **A**: ES FE
pyrenaeus Zeising, 1981: 17 **E**: SP
quadrisignatus Gyllenhal, 1817: 139 (*Elater*) **E**: AU BH CR CZ FR GE HU FR IT RO SK UK YU
chalusi Guérin-Méneville, 1847: 7
quercicola Buysson, 1887b: xc (*Elater*) **E**: AB AU BE CZ DE FR GB GE HU IT RO SK SP ST SZ **N**: TU
miniatus Gorham, 1892b: 304 (*Elater*)
quercicollis Schimmel, 1993a: 140 **A**: KA PA
rabusculus Dolin & Ôhira, 1976: 389 **A**: FE
radula Candèze, 1900: 87 (*Elater*) **A**: KA PA UP
rasilis Schimmel, 1993a: 142 **A**: YUN
reitteri Semenov, 1891: 348 (*Elater*) **A**: TD

ressli Platia & Gudenzi, 1999a: 29 **A**: TR
rimmae Dolin & Chantladze, 1989: 169 **A**: GG
rubellus Gurjeva, 1977b: 800 **E**: AB GG ST
rubridorsus Lewis, 1879b: 155 (*Elater*) **A**: JA
rufipennis Stephens, 1830: 256 (*Elater*) **E**: AB AR AU BH BU CR CZ DE FR GB GE GG GR HU IT MD PL PT RO SK SL SP ST SV UK YU **N**: AG MO **A**: IN
dibaphoides Buysson, 1887b: xc (*Elater*)
dibaphus Schiödte, 1865: 533 (*Elater*)
intermedius Buysson, 1896b: 218 (*Elater*)
pandellei Buysson, 1896b: 218 (*Elater*)
satrapa Kiesenwetter, 1858a: 338 (*Elater*)
rugosus Schimmel, 1982: 183 **E**: CZ GE SK
saghaliensis Dolin & Ôhira, 1976: 387 **A**: FE
samai Schimmel & Platia, 1988: 38 **N**: AG MO
samedovi Dolin & Agajev, 1983: 32 **E**: AB AR ST **A**: TR
sanguineus Linnaeus, 1758: 405 (*Elater*) **E**: AB AL AU BE BH BU BY CR CT CZ DE EN FI FR GE GG GR HU IT LA LS LT LU MD NL NR NT PT RO SK SL SP ST SV SZ UK **N**: AG MO **A**: ES FE GAN IN KA KZ MG TR WS **ORR**
burdigalensis Buysson, 1886: xxvii (*Elater*)
coccineus Schiødte, 1870: 533 (*Elater*) [HN]
haemelyter Schrank, 1798: 590 (*Elater*)
rubidus Candèze, 1859: 447 (*Elater*)
semisanguineus Reitter, 1889f: 110 (*Elater*)
sanguinolentoides Platia, Furlan & Gudenzi, 2002: 73 **A**: IN
sanguinolentus nippon Kishii, 1982: 42 **A**: JA
sanguinolentus sanguinolentus Schrank, 1776: 69 (*Elater*) **E**: AB AL AR AU BE BH BU BY CR CT CZ DE EN FI GB GE GG GR HU IT LA LS LT LU MC MD NL NR NT PL PT RO SK SL SP ST SV SZ UK YU **N**: MO **A**: ES FE HEI IN JIL KZ MG NMO TR WS
ephippium A. G. Olivier, 1790: no 31: 41 (*Elater*)
haemopterus Gmelin, 1790: 1917 (*Elater*)
irreductus Buysson, 1911a: 27 (*Elater*)
immaculatus Schaufuss, 1882a: 1 (*Elater*)
paleatus Candèze, 1893c: 286
paraminiatus Buysson, 1906b: 469 (*Elater*)
ryei Lewis, 1879b: 155 (*Elater*)
sanorum Kishii, 2002a: 12 **A**: SC (Cheju)
sapporoensis W. Suzuki, 1985a: 84 [RN] **A**: JA
rotundicollis Miwa, 1927a: 16 (*Elater*) [HN]
satrapoides Zeising & Sieg, 1992: 123 **E**: "Caucasus"
sausai Schimmel, 1996a: 164 **A**: YUN
sawadai Kishii, 1985: 18 **A**: JA
schawalleri Schimmel, 1993a: 144 **A**: KA PA
schoettlei Zeising & Brunne, 2003: 192 **A**: TR
serenus Gurjeva, 1977b: 804 **E**: AB GG ST
shakotanensis Miwa, 1927a: 16 (*Elater*) **A**: FE
shinoharai Kishii, 1977b: 32 **A**: JA
shotakahashii Kishii, 1991d: 8 **A**: JA
sichuanensis Schimmel, 1993a: 146 **A**: SCH
sikkimensis Schimmel, 1993a: 147 **A**: NP SD
silvaticus Gurjeva, 1977b: 802 **A**: FE
sinanoensis Ôhira, 1974a: 56 **A**: JA
sinuatus Germar, 1844: 170 **E**: AB AR AU BH BU CR CZ FR GE GG GR HU IT MD PL RO SK SL ST SZ UK YU **A**: IN TR
cuneiformis Hampe, 1866: 373
sobrinus Motschulsky, 1860b: 111 (*Elater*) **A**: ES FE MG WS
amurensis Pic, 1909d: 105 (*Elater*)

christophi Jakobson, 1913: 750 [RN] (*Elater*)
compactus Reitter, 1889f: 116 (*Elater*) [HN]
straminipennis Heyden, 1880b: 289 (*Elater*) **E**: SP
subglabratus Schimmel, 1993a: 147 **A**: SCH
suecicus Palm, 1976c: 140 [RN] **E**: AU BY CT CZ FI NR NT SK SV UK
borealis Palm, 1947a: 164 (*Elater*) [HN]
taichangi Ôhira, 1967a: 57 **A**: TAI
takahachi Kishii, 1985: 19 **A**: JA
takeuchii Kishii, 1955b: 81 **A**: JA
insignitus Kishii, 1955b: 81
takaoensis Ôhira, 1973b: 118 **A**: JA
talamellii Platia & Gudenzi, 2000a: 148 **E**: SP
tamba Kishii, 1976a: 12 **A**: JA
tatsumii Ôhira, 1992d: 84 **A**: JA
tattakensis Ôhira, 1966d: 270 **A**: TAI
tenuistriatus Lewis, 1894a: 39 (*Elater*) **A**: JA
hakusanus Ôhira, 1954: 42
serionus Kishii, 1966a: 35
tibetanus Schimmel, 1993a: 148 **A**: XIZ YUN
tingitanus Wurst, 1995a: 30 **N**: MO
tokachi Kishii, 1991d: 7 **A**: JA
tokugoensis W. Suzuki, 1985a: 84 [RN] **A**: JA
longipennis Miwa, 1934: 76 (*Elater*) [HN]
towadensis Kishii, 1987: 4 **A**: JA
triangulum Dorn, 1924: 129 (*Elater*) **E**: AU CZ DE GE IT SK SL SV UK
triangulus Bouwer, 1980: 333
tristis Linnaeus, 1758: 406 (*Elater*) **E**: AU BU BY CT CZ EN FI GB GE IT LA NR NT PL RO SK SV UK **A**: WS
lateralis Fabricius, 1794: 452 (*Elater*)
trossulus Gurjeva, 1979a: 261 **A**: FE (Kunashir)
tsugaru Kishii, 1992b: 6 **A**: JA
turanus Semenov, 1891: 349 (*Elater*) **A**: TD UZ
juglandis Iablokoff-Khnzorian, 1964: 159 (*Elater*)
turkmenicus Dolin, 1977b: 357 **A**: TM
turcmenicus Dolin, 1977c: 360 [incorrect original spelling]
uralensis Gurjeva, 1954: 198 (*Elater*) **A**: KZ
vandalitae Lohse, 1976: 91 **E**: AU BU CZ GE LA PL SK
verae Gurjeva, 1977b: 801 **E**: AR
vestitus vestitus Lewis, 1894a: 39 (*Elater*) **A**: JA SC
vestitus watanabei Kishii, 1979a: 7 **A**: JA
vestitus yakuinsulanus Kishii, 1983b: 53 [RN] **A**: JA
yakuensis Kishii, 1979a: 7 [HN]
vignai Guglielmi & Platia, 1985: 197 **E**: GR
wachtangi Dolin, 1970: 13 **E**: AB GG ST **A**: TR
wittmeri Schimmel, 1993a: 150 **A**: HP
yagii Kishii, 1990a: 19 **A**: TAI
yagishiriensis Ôhira, 1986a: 88 **A**: JA
yaku ohdai Kishii, 1983b: 53 **A**: JA
yaku yaku Kishii, 1969: 7 **A**: JA
yakuensis Kishii, 1976a: 10 **A**: JA (Yaku-shima)
yamato Kishii, 1988a: 14 **A**: JA
yezo Kishii, 1968: 13 **A**: JA
yousukei Arimoto, 2004: 1 **A**: JA (Ishigaki-jima)
yukara Kishii, 1997a: 7 **A**: JA
yukoanus Kishii, 1962: 25 **A**: JA

yulongshanus Schimmel, 1993a: 151 **A**: YUN
yunnanus Schimmel, 1993a: 152 **A**: YUN
ziegleri Zeising & Sieg, 1983: 26 **E**: GE
subgenus *Miwaelater* W. Suzuki, 1999: 142 [RN] type species *Elater habunensis* Miwa, 1934
Pseudelater Miwa, 1934: 20 [HN] type species *Elater habunensis* Miwa, 1934
aritai aritai Ôhira & Satô, 1964: 27 **A**: JA (Amami-Ôshima)
aritai kuniyoshii Ôhira, 1968c: 141 **A**: JA (Okinawa-hontô)
aritai mikyoensis Ôhira, 1998a: 164 **A**: JA (Tokuno-shima)
carbunculus Lewis, 1879b: 156 (*Elater*) **A**: JA SC
fleutiauxi Schimmel, 1993a: 127 **A**: SCH
habunensis Miwa, 1934: 197 (*Elater*) **A**: TAI
loebli Schimmel, 1993a: 133 **A**: PA
mirificus Schimmel, 1993a: 134 **A**: PA
nikkoensis Ôhira, 1973b: 117 (*Pseudelater*) **A**: JA
niponicus Lewis, 1894a: 32 (*Elater*) **A**: JA JIL
otobei Kishii, 1986: 46 **A**: JA
soboensis kiianus Kishii, 1986: 45 **A**: JA
soboensis soboensis Ôhira, 1963b: 176 **A**: JA
tsuneoi Ôhira, 1998a: 163 **A**: JA
viridipennis Kishii, 1994c: 9 **A**: TAI
subgenus *Parelater* Candèze, 1882: 70 type species *Elater coccineus* Candèze, 1865
Pademus Kishii, 1957b: 70 type species *Elater canalicollis* Lewis, 1894
analogicus Kolbe, 1886: 198 (*Elater*) **A**: SC
canalicollis canalicollis Lewis, 1894a: 38 (*Elater*) **A**: JA
canalicollis shidoui Ôhira, 1993: 252 **A**: JA
coccineus Candèze, 1865: 29 (*Elater*) **A**: BT SD UP **ORR**
miles Lewis, 1894a: 37 (*Elater*) **A**: JA
nomurai Ôhira, 1968c: 141 **A**: JA (Amami-Ôshima)
puniceus Lewis, 1879b: 156 (*Elater*) **A**: JA "Korea"
shirozui Ôhira, 1966d: 267 **A**: TAI
subcostatus Kolbe, 1886: 197 (*Elater*) **A**: SC
yushiroi W. Suzuki, 1985a: 85 [RN] **A**: TAI
bimaculatus Miwa, 1929e: 487 (*Elater*) [HN]

genus *Ampedus*, species incertae sedis
nigrotinctus Gistel, 1857: 4, 36 **E**: GE (Bayern)

genus *Brachygonus* Buysson, 1912g: 139 type species *Elater megerlei* Lacordaire, 1835
bouyoni Chassain, 1992: 327 (*Ampedus*) **E**: AU CZ FR IT HU PT SK SP
campadellii Platia & Gudenzi, 2000c: 13 **E**: IT
frater Wurst, 1995c: 147 **A**: TR
gratiosus Platia & Schimmel, 1991b: 139 **E**: GR **A**: TR
hadullanus Wurst, 1995a: 31 **A**: TR
megerlei Lacordaire, 1835: 656 (*Elater*) **E**: AU BE BU CR CZ FR GE GR HU IT PL RO SK SP ST SZ UK **A**: TR
melanotites Gredler, 1863: 230 (*Elater*)
meraculus Reitter, 1889f: 116 (*Elater*) **E**: GR
ruficeps Mulsant & Guillebeau, 1855a: 29 (*Ampedus*) **E**: AU BU CZ FR GB GE GR HR HU IT PT RO SK SP ST SZ UK
pallidus L. Redtenbacher, 1858: 500 (*Ampedus*)
semiobscurus Pic, 1902l: 72
ruficepsoides Platia & Gudenzi, 2002b: 87 **A**: TR

genus *Haterumelater* Ôhira, 1968c: 143 type species *Elater bicarinatus* Candèze, 1873
bicarinatus bicarinatus Candèze, 1873: 9 (*Elater*) **A**: JA

hachijoensis Kishii, 1985: 14 **A**: JA
insulanus Kishii, 1976c: 48 **A**: JA
bicarinatus heianus Kishii, 1976c: 49 **A**: JA (Kuchi-erabu-jima)
bicarinatus shibatai Ôhira, 1968c: 143 **A**: JA (Tokara, Ryukyus)
bicarinatus yaku Kishii, 1976c: 49 **A**: JA (Tokara)
bifoveolatus Miwa, 1927a: 17 (*Elater*) **A**: JA (Ogasawara)
bimaculatus Schimmel, 1998a: 218 **A**: IN
fulvago Marseul, 1868b: 176 (*Elater*) **E**: BU GR IT **A**: CY IS LB TR
julianum Penev, 1987: 158
picinus Buysson, 1891d: cxlii (*Ischnodes*)
preissi Schimmel, 1995: 49
languidus Buysson, 1891d: cxlii (*Ischnodes*) **E**: GR **A**: TR
syriacus Pic, 1904l: 81 (*Ischnodes*)
schembrii Platia, 1985b: 94 (*Ischnodes*) **E**: MA
tauricola Gurjeva, 1957: 460 [= 1959: 201] (*Elater*) **E**: UK

genus *Ischnodes* Germar, 1844: 180 type species *Elater sanguinicollis* Panzer, 1793
kadleci Mertlik, 2005: 171 **A**: TR
sanguinicollis maiko W. Suzuki, 1985a: 85 **A**: JA
otome Kishii, 1985: 14
sanguinicollis sanguinicollis Panzer, 1793a: 13 (*Elater*) **E**: AB AU BE BH CR CZ DE FR GB GE GR HU IT MD PL RO SK SL SP SV SZ UK **A**: IN JA SY
fulvicollis Stephens, 1830: 252 (*Elater*)
nigernitens Voet, 1791: 109 (*Erythrotus*)
sibiricus Cherepanov, 1966b: 27 **A**: FE WS

genus *Kometsukia* Kishii, 1957a: 11 type species *Kometsukia vesticornis* Kishii, 1957
vesticornis Kishii, 1957a: 11 **A**: JA

genus *Reitterelater* Platia & Cate, 1990: 111 [RN] type species *Elater fulvus* Reitter, 1891
amamiensis Ôhira, 1968c: 142 (*Ectamenogonus*) **A**: JA
dubius Platia & Cate, 1990: 113 **E**: AU CZ FR GE GR HU IT SK SL SV UK **A**: TR
fulvus Reitter, 1891g: 234 (*Elater*) **A**: KI KZ UZ
kuriharai Ôhira, 2000: 366 **A**: JA
miyako Kishii, 1969: 9 (*Ectamenogonus*) **A**: JA
rugipennis Lewis, 1894a: 41 (*Elater*) **A**: JA "Korea"

tribe Dicrepidiini Candèze, 1859

genus *Adiaphorus* Candèze, 1859: 47 type species *Adiaphorus gracilicornis* Candèze, 1859
elevatus Vats & Chauhan, 1992d: 177 **A**: UP
levisus Vats & Chauhan, 1992d: 178 **A**: UP
modestus Candèze, 1892b: 488 **A**: NP **ORR**

genus *Ischiodontus* Candèze, 1859: 90 type species *Ischiodontus pinguis* Candèze, 1859
daitoensis Ôhira, 1969a: 96 **A**: JA (Ryukyus)
kawaii Ôhira, 1967d: 27 **A**: JA (Ryukyus)
langfordi Van Zwaluwenburg, 1957: 14 (*Propsephus*) **A**: JA (Ogasawara)

genus *Neopsephus* Kishii, 1990a: 12 type species *Neopsephus takasago* Kishii, 1990
takasago Kishii, 1990a: 14 **A**: TAI

genus *Propsephus* Hyslop, 1921: 666 [RN] type species *Psephus beniniensis* Candèze, 1859
Psephus Candèze, 1859: 19 [HN] type species *Psephus beniniensis* Candèze, 1859
claudei Wurst, Schimmel & Platia, 2001: 545 **A**: YE
cyaneus Candèze, 1878b: lvii **A**: SD **ORR**

occupator Wurst, Schimmel & Platia, 2001: 545 **A**: YE

genus *Sephilus* Candèze, 1878a: 108 type species *Sephilus frontalis* Candèze, 1878
formosanus Schwarz, 1902b: 319 **A**: TAI
okinawensis Kishii, 1992b: 2 **A**: JA (Ryukyus)
shibatai Kishii, 1999b: 6 **A**: TAI
tanakai Kishii, 1992b: 5 **A**: JA (Ishigaki-jima I.)

tribe Elaterini Leach, 1815

subtribe Elaterina Leach, 1815

genus *Agonischius* Candèze, 1863: 407 type species *Agonischius pectoralis* Candèze, 1863
Agriotides Schwarz, 1907: 273 type species *Agonischius mutabilis* Schwarz, 1898
breviusculus Candèze, 1863: 426 **A**: HKG XIZ
chamoliensis Vats & Chauchan, 1994: 145 [DA] **A**: UP
chauhani Vats & Chauchan, 1994: 148 **A**: UP
cyanipennis Candèze, 1863: 418 **A**: XIZ **ORR**
cyanoreflexus Vats & Chauchan, 1994: 147 **A**: UP
francki Fleutiaux, 1934a: 185 **A**: SCH
indicus Fleutiaux, 1933a: 15 **A**: UP **ORR**
monachus Candèze, 1878b: cxcvi **A**: GUA
monticola Candèze, 1895: 62 **A**: SD
pantolamprus Candèze, 1863: 419 **A**: "Himalaya"
quadrilineatus Hope, 1843a: 63 [= 1845a: 10] (*Ludius*) **A**: ZHE
scapularis Candèze, 1863: 425 **A**: HKG
speculifer Fairmaire, 1889b: 34 **A**: CE YUN
subsericeus Kolbe, 1886: 197 **A**: SC

genus *Aphanobius* Eschscholtz, 1829: 33 type species *Aphanobius longicollis* Eschscholtz, 1829
alaomorphus Candèze, 1863: 319 **A**: FUJ HEN HUN JIA JIX SD YUN ZHE **ORR**
melanurus Schwarz, 1896b: 102 **A**: "Himalaya"
touffus Vats & Chauhan, 1992f: 189 [DA] **A**: HP UP

genus *Campylomorphus* Jacquelin du Val, 1860a: 130 type species *Elater homalisinus* Illiger, 1807
homalisinus Illiger, 1807a: 14 (*Elater*) **E**: FR IT PT SP
suturanigra Chevrolat, 1840: 15 (*Athous*)

genus *Dolerosomus* Motschulsky, 1860d: 382 type species *Dolerosomus flavipennis* Motschulsky, 1860 (= *Elater silaceus* Say, 1825)
gracilis Candèze, 1873: 10 (*Megapenthes*) **A**: FE (Kurile) JA

genus *Elater* Linnaeus, 1758: 404 type species *Elater ferrugineus* Linnaeus, 1758
Ludius Berthold, 1827: 336 type species *Elater ferrugineus* Linnaeus, 1758
Morous Gistel, 1834: 12 type species *Elater ferrugineus* Linnaeus, 1758
Steatoderus Dejean, 1833: 94 type species *Elater ferrugineus* Linnaeus, 1758
asmodaius Wurst, 1994b: 118 **E**: GR
brancuccii Schimmel, 1996c: 122 **A**: SD
businskyi Schimmel, 2003b: 287 **A**: HUB
ferrugineus ferrugineus Linnaeus, 1758: 405 **E**: AL AR AU BE BH BU CT CZ DE FR GB GE GG GR HU IT MD NL RO SK SP ST SV SZ UK
morio Schilsky, 1888: 187 (*Ludius*)
occitanicus Villers, 1789: 319 (*Elater*)
occitanus Buysson, 1893a: 61 (*Ludius*) [unjustified emendation]
ruber Geoffroy, 1785: 35 (*Elater*)
ferrugineus lenkoranus Gurjeva, 1974a: 70 **E**: AB **A**: IN TM
luctuosus Solsky, 1871: 364 (*Ludius*) **A**: FE NC

niponensis Lewis, 1894a: 264 (*Ludius*) **A**: HEI JA "Korea"
rubiginosus Candèze, 1889a: 96 (*Ludius*) **A**: GUI HUB HUN **ORR**
splendens Gurjeva, 1974a: 71 **E**: AB
tauricus Schwarz, 1897c: 400 (*Ludius*) **A**: TR

genus *Girardelater* Schimmel, 1999: 27 type species *Megapenthes porrectus* Fleuteaux, 1918
werneri Schimmel, 1999: 32 **A**: NP **ORR**

genus *Ludigenoides* Platia, 2004d: 702 type species *Ludigenus minor* Candèze, 1888
melantoides Fleutiaux, 1940d: 194 (*Neotrichophorus*) **A**: YUN **ORR**
schawalleri Platia, 2004d: 707 **A**: NP **ORR**

genus *Ludigenus* Candèze, 1863: 325 type species *Ludigenus politus* Candèze, 1863
politus Candèze, 1863: 326 **A**: HAI NP YUN **ORR**

genus *Ludioschema* Reitter, 1891g: 238 type species *Ludioschema emerichi* Reitter, 1891
Chiagosnius Fleutiaux, 1940a: 136 type species *Elater obscuripes* Gyllenhal, 1817
akashii Ôhira, 1969a: 99 (*Chiagosnius*) **A**: JA (Ryukyus)
atkinsoni Candèze, 1895: 63 (*Agonischius*) **A**: SD
apayao Kishii, 1993a: 24 (*Chiagosnius*) **A**: TAI
cyaneum Candèze, 1863: 418 (*Agonischius*) **A**: ZHE **ORR**
fruhstorferi Schwarz, 1902b: 340 (*Agonischius*)
nitidum Fleutiaux, 1895: 689 (*Agonischius*)
thoracicum Fleutiaux, 1895: 689 (*Agonischius*)
delauneyi Fleutiaux, 1887: 64 (*Agonischius*) **A**: GUA **ORR**
decoratum Candèze, 1889b: 121 (*Agonischius*)
deminutum Fleutiaux, 1940a: 143 (*Chiagosnius*)
florentini Fleutiaux, 1895: 688 (*Agonischius*)
fusciventre Fleutiaux, 1895: 688 (*Agonischius*)
dorsale Candèze, 1878b: cxcv (*Agonischius*) **A**: FUJ GUA GUX HAI HKG **ORR**
brunneipes Fleutiaux, 1918b: 259 (*Agonischius*)
emerichi Reitter, 1891g: 239 **E**: "Caucasus"
longum Candèze, 1895: 63 (*Agonischius*) **A**: SD **ORR**
*marginicolle*Miwa, 1928b: 48 (*Agonischius*) **A**: TAI
massiei Fleutiaux, 1940a: 140 (*Chiagosnius*) **A**: HAI **ORR**
metallicum Candèze, 1895: 62 (*Agonischius*) **A**: FUJ NP SD ZHE **ORR**
minor Fleutiaux, 1903a: 576 (*Agonischius*) **A**: GUA HAI **ORR**
nigripenne Fleutiaux, 1895: 689 (*Agonischius*) **A**: GUA GUI GUX **ORR**
obscuripes Gyllenhal, 1817: 131 (*Elater*) **A**: FUJ GUA GUX HEB HKG HUB HUN JA JIA JIX KA SC SCH TAI XIZ YUN ZHE **ORR**
brunneum Miwa, 1928b: 48 (*Chiagosnius*)
candezellus Miwa, 1928b: 48 (*Chiagosnius*)
cashmirense Kollar, 1844: 507 (*Ludius*)
ferrugineum Fleutiaux, 1918b: 262 (*Chiagosnius*)
obscuricolle Fleutiaux, 1895: 688 (*Agonischius*)
okinawense Miwa, 1927a: 18 (*Megapenthes*) **A**: JA (Ryukyus) TAI
iwasakii Miwa, 1927a: 20 (*Megapenthes*)
ruficephalum Miwa, 1928b: 48 (*Agonischius*)
opacipenne Kishii, 1993a: 25 (*Chiagosnius*) **A**: TAI
pallidum Fleutiaux, 1940a: 146 (*Chiagosnius*) **A**: GUA **ORR**
simulator Candèze, 1897: 73 (*Agonischius*) **A**: NP SD
sulcicolle Candèze, 1878b: cxcv (*Agonischius*) **A**: JIA ZHE JIX GUX HAI **ORR**
niger Fleutiaux, 1895: 689 (*Agonischius*)
obscurum Fleutiuax, 1895: 689 (*Agonischius*)
suturale maculicolle Fleutiaux, 1940a: 141 (*Chiagosnius*) **A**: GUX YUN **ORR**
suturale suturale Candèze, 1863: 416 (*Agonischius*) **A**: HAI **ORR**

candezei Fleutiaux, 1903a: 576 (*Agonischius*)
dussauti Fleutiaux, 1918b: 260 (*Agonischius*)
gagates Fleutiuax, 1939: 141 (*Chiagosnius*)
taikozanum Miwa, 1928b: 49 (*Agonischius*) **A**: TAI
vittiger formosanum Miwa, 1927a: 19 (*Megapenthes*) **A**: TAI
vittiger fuscomarginatum Lewis, 1896c: 337 (*Aphanobius*) **A**: HUB JA (Ryukyus) SCH
vitticolle Miwa, 1927a: 18 (*Megapenthes*)
vittiger vittiger Heyden, 1887: 267 (*Agonischius*) **A**: FUJ GUX SC
yushiroi W. Suzuki, 1999: 179 [RN] **A**: JA TAI
sanguinicolle Miwa, 1928b: 48 [HN]

genus *Mulsanteus* Gozis, 1875: 50 [RN] type species *Trichophorus guillebeaui* Mulsant & Godart, 1853
Nairus Iablokoff-Khnzorian, 1974a: 52 type species *Nairus dux* Iablokoff-Khnzorian, 1974
Neotrichophorus Jakobson, 1913: 742 [RN] type species *Trichophorus guillebeaui* Mulsant & Godart, 1853
Trichophorus Mulsant & Godart, 1853: 181 [HN] type species *Trichophorus guillebeaui* Mulsant & Godart, 1853
anchastinus Candèze, 1882: 104 (*Ludius*) **A**: SCH
dux Iablokoff-Khnzorian, 1974a: 54 (*Nairus*) **E**: AR
guillebeaui Mulsant & Godart, 1853: 183 (*Trichophorus*) **E**: BU FR GR IT SP **A**: CY IN IQ SY TR "Palestine"
guillebelli Mulsant & Godart, 1853: 183 (*Trichophorus*) [incorrect original spelling]
grandis Faldermann, 1836: 368 (*Ludius*) **A**: TM
hirtellus Candèze, 1863: 303 **A**: NP **ORR**
junior junior Candèze, 1873: 27 (*Ludius*) **A**: JA SC (Cheju)
junior yakuensis Kishii: 1983a: 37 (*Neotrichophorus*) **A**: JA (Yaku-shima)
linteatus aureopilosus Miwa, 1934: 259 (*Neotrichophorus*) **A**: JA (Ryukyus)
linteatus linteatus Candèze, 1873: 28 (*Ludius*) **A**: JA "Korea"
ligatus Schwarz, 1906: 259 (*Trichophorus*)
yamamotoi Nakane & Kishii, 1955a: 44 (*Neotrichophorus*)
manuelae Platia & Gudenzi, 1998a: 59 **E**: GR **A**: TR
rubuginosus Ôhira, 1966d: 272 (*Neotrichophorus*) **A**: JA (Ryukyus) TAI
schaumi Candèze, 1882: 103 (*Ludius*) **E**: GR **A**: SY TR "Palestine"
taygetanus Reitter, 1887h: 514 (*Ludius*)
shirozui Ôhira, 1966d: 272 (*Neotrichophorus*) **A**: TAI
turanicus Reitter, 1887h: 513 (*Ludius*) **A**: AF KI KZ TD TM UZ XIN

genus *Orthostethus* Lacordaire, 1857: 207 type species *Aphanobius infuscatus* Germar, 1844
Nipponoelater Kishii, 1985: 23 type species *Ludius sieboldi* Candèze, 1873
amami Kishii, 1987: 12 (*Elater*) **A**: JA (Amami-Ôshima)
babai babai Kishii, 1987: 14 (*Elater*) **A**: JA (Iriomote-jima, Ishigake-jima, Okinawa-honto)
babai seinoi Kishii, 1987: 16 (*Elater*) **A**: JA (Amami-Ôshima, Tokuno-shima)
babai taiwanus Kishii, 1989b: 37, 49 **A**: TAI "mainland China"
kometsuki Kishii, 1985: 23 (*Elater*) **A**: JA
sieboldi sieboldi Candèze, 1873: 27 (*Ludius*) **A**: FUJ GAN GUX HEB HEN JA JIX SCH ZHE "Korea"
unicolor Fleutiaux, 1900: 357 (*Aphanobius*)
sieboldi tsushimensis Ôhira, 1997c: 37,40 **A**: JA (Tsushima)
sinensis Candèze, 1882: 103 (*Ludius*) **A**: FUJ GUA GUX JIX SCH

genus *Parallelostethus* Schwarz, 1906: 258 type species *Elater attenuatus* Say, 1825
acutus Candèze, 1863: 299 (*Ludius*) **A**: SD **ORR**
georgelewisi W. Suzuki, 1985a: 88 [RN] **A**: JA
candezei Lewis, 1894a: 265 (*Ludius*) [HN]
sakishimensis Ôhira, 1967d: 30 **A**: JA (Iriomote-jima, Ishigake-jima)
thoracicus Fleutiaux, 1918b: 256 (*Ludius*) **A**: TAI **ORR**

genus *Pittonotus* Jacquelin du Val, 1860a: 128 type species *Elater theseus* Germar, 1817
simoni Stierlin, 1879b: 511 (*Ectinus*) **A**: SY
theseus Germar, 1817: 218 (*Elater*) **E**: BU CR FR GR **N**: EG **A**: CY IN IS JO LE TR

genus *Taiwanostethus* Kishii, 1994d: 7 type species *Taiwanostethus tanidai* Kishii, 1994
tanidai Kishii, 1994d: 9 **A**: TAI

genus *Vuilletus* Fleutiaux, 1940a: 123 type species *Agonischius altus* Candèze, 1889
Metaricus Nakane & Kishii, 1958: A295 type species *Sericosomus viridis* Lewis, 1894
altus Candèze, 1889b: 120 (*Agonischius*) **A**: NP SD **ORR**
striatus Fleutiaux, 1918b: 259
amamiensis okinawanus Ôhira, 1971b: 535 **A**: JA (Okinawa-hontô)
babai Kishii, 1991b: 9, 25 **A**: TAI
crebrepunctatus Nakane, 1959c: 94 (*Metricarus*) **A**: JA
elongatus Nakane & Kishii, 1958: A296 (*Metaricus*) **A**: JA (Yakushima)
gemmula Candèze, 1878b: cxciv (*Agonischius*) **A**: HKG "China"
mushanus Miwa, 1928b: 47 (*Agonischius*) **A**: TAI
peropacus Nakane, 1959c: 94 (*Metaricus*) **A**: JA
potanini Gurjeva, 1972a: 306 **A**: SCH
viridis Lewis, 1894a: 314 (*Sericosomus*) **A**: GUI JA SCH
nitidicollis Kishii & Ôhira, 1956: 74 (*Metaricus*)
subopacus Nakane, 1959c: 94 (*Metaricus*)
yagii Kishii, 1992a: 5 **A**: TAI
yamazakii Ôhira, 1987b: 285 **A**: JA

subtribe Sericosomina Hyslop, 1917

genus *Sericus* Eschscholtz, 1829: 34 type species *Elater brunneus* Linnaeus, 1758
subgenus *Sericoderma* Dolin, 1973: 84 type species *Ectinus subaeneus* W. Redtenbacher, 1842
subaeneus W. Redtenbacher, 1842: 12 (*Ectinus*) **E**: AU BU CZ FR GE IT MD PL RO SK SL SZ UK
jucundus Märkel, 1847: 80 (*Ectinus*)
micans Mulsant & Guillebeau, 1855a: 31 (*Diacanthus*)
planeti Buysson, 1926b: 187
xanthodon Märkel, 1847: 79 (*Ectinus*)
subgenus *Sericus* Eschscholtz, 1829: 34 type species *Elater brunneus* Linnaeus, 1758
Amphilabris Gistel, 1834: 12 type species *Elater brunneus* Linnaeus, 1758
Atractopterus LeConte, 1853c: 454 type species *Atractopterus fusiformis* LeConte, 1853
Sericosomus Dejean, 1833: 96 type species *Elater brunneus* Linnaeus, 1758
bifoveolatus Lewis, 1894a: 45 (*Megapenthes*) **A**: JA
nobuchii Kishii, 1955c: 110 (*Vuilletus*)
brunneus brunneus Linnaeus, 1758: 404 (*Elater*) **E**: AU BE BU BY CT CZ DE EN FI FR GB GE HU IR IT LA LT LU NL NR NT PL RO SK SL SV SZ UK **A**: ES FE JIL KZ MG WS
affinis Motschulsky, 1859b: 228 [= 1860f: 544] (*Sericosomus*)
brunnipennis Stephens, 1839: 177 (*Sericosomus*)
fugax Fabricius, 1801b: 237 (*Elater*)
maculosus Gmelin, 1790: 1915 (*Elater*)
nigromaculatus DeGeer, 1774: 158 (*Elater*)
obscurior Pic, 1913k: 153
rubidus Candèze, 1863: 428 (*Sericosomus*)
sulcipennis Buysson, 1893a: 65
tibialis Laporte, 1840a: 248 (*Agriotes*)
vulgaris Herbst, 1784b: 114 (*Elater*)
brunneus sachalinensis Miwa, 1928b: 40 (*Pittonotus*) **A**: FE JA
flavipennis Miwa, 1928b: 41 (*Pittonotus*)
montanus Miwa, 1928b: 41 (*Pittonotus*)
clarus Gurjeva, 1972a: 305 **E**: AU CZ FR SK SZ **A**: ES
fujisanus Lewis, 1894a: 44 (*Megapenthes*) **A**: FE JA
aenescens Miwa, 1928b: 42 (*Pittonotus*)
grandis Nakane & Kishii, 1954b: 739
hiramatsui Ôhira, 1995d: 190 (*Shirozulus*) **A**: JA
ishidai Kishii, 1955b: 82 (*Vuilletus*) **A**: JA

jezoensis Ôhira, 1992e: 92 **A**: JA
lahaulensis Vats & Chauhan, 1992f: 190 [DA] **A**: HP
subgenus *Shirozulus* Ôhira, 1966d: 270 type species *Shirozulus formosanus* Ôhira, 1966
formosanus Ôhira, 1966d: 271 **A**: TAI

tribe Megapenthini Gurjeva, 1973

genus *Abelater* Fleutiaux, 1947: 384 type species *Melanoxanthus rubiginosus* Candèze, 1878
babanus Kishii, 1989a: 25, 29 **A**: TAI
bousaianus Kishii, 1991a: 9, 21 **A**: TAI
brancuccii Schimmel, 2004b: 51 **A**: NP
himalayanus Schimmel, 2004: 109 **A**: NP SD
iriomotensis Kishii, 1985: 13 **A**: JA (Iriomote-jima)
makiharai Ôhira, 1970d: 20 **A**: TAI
nigrolineatus Kishii, 1991a: 9, 23 **A**: TAI
pulcherus Miwa, 1933a: 8 (*Melanoxanthus*) **A**: JA (Ryukyus)
rubiginosus Candèze, 1878b: cxxxviii (*Melanoxanthus*) **A**: SD
satoi Ôhira, 1968c: 137 (*Melanoxanthus*) **A**: JA (Ryukyus)
shirozui Kishii, 1959b: 60 (*Melanoxanthus*) **A**: JA (Amami-Ôshima, Yaku-shima)
sinensis Schimmel, 2004b: 202 **A**: YUN
taiwanus Kishii, 1989a: 30 **A**: TAI
trivittatus Schwarz, 1901b: 27 (*Melanoxanthus*) **A**: NP **ORR**
  andrewesi Fleutiaux, 1933a: 10 (*Melanoxanthus*)

genus *Agaripenthes* Ôhira, 1970c: 84 type species *Agriotes helvolus* Candèze, 1873
babai Kishii, 1989a: 25, 27 (*Procraerus*) **A**: TAI
helvolus Candèze, 1873: 30 (*Agriotes*) **A**: JA LIA
  flavus Fleutiaux, 1902a: 19 (*Megapenthes*)
hoshidatensis Ôhira, 1968c: 136 (*Procraerus*) **A**: JA (Iriomote-jima)
nagaoi Ôhira, 1968c: 136 (*Procraerus*) **A**: JA (Ryukyus)
satoi Ôhira, 2003a: (*Procraerus*) 243 **A**: JA (Okinawa)
shibatai Kishii, 1969: 5 (*Procraerus*) **A**: JA (Yaku-shima)
tsutsuii Nakane & Kishii, 1955b: 5 (*Procraerus*) **A**: JA (Nakano-shima)
watanabei Ôhira, 1960: 24 (*Procraerus*) **A**: JA (Miyake, Mikura)

genus *Amamipenthes* Kishii, 1973: 6 type species *Amamipenthes matobai* Kishii, 1973
formosensis Kishii, 1991a: 8, 17 **A**: TAI
gardneri Schimmel, 2004b: 241 **A**: SD
harmandi Schimmel, 2004b: 242 **A**: SD
matobai Kishii, 1973: 7 **A**: JA (Amami-Ôshima)

genus *Cateanus* Schimmel, 2004b: 243 type species *Melanoxanthus bakeri* Fleutiaux, 1914
brancuccii Schimmel, 2004b: 248 **A**: NP
horni Schwarz, 1901b: 26 (*Melanoxanthus*) **A**: NP **ORR**
lijiangensis Schimmel, 2004b: 258 **A**: YUN
probsti Schimmel, 2004b: 265 **A**: NP

genus *Dolinolus* Schimmel, 1999: 160 type species *Megapenthes carinatus* Candèze, 1878
malus Fleutiaux, 1928a: 164 (*Megapenthes*) **A**: GUA **ORR**

genus *Ectamenogonus* Buysson, 1894b: cccxiv type species *Ludius montandoni* Buysson, 1889
  Penthelater Ôhira, 1970a: 9 type species *Ludius plebejus* Candèze, 1873
  Rhodopenthes Gurjeva, 1973: 450 type species *Ludius plebejus* Candèze, 1873
bispinicollis Miwa, 1929e: 488 (*Elater*) **A**: TAI
lehmanni Schimmel, 1999: 179 **A**: FUJ GUA SHG

matobai Kishii, 1973: 12 (*Penthelater*) **A**: JA (Ryukyus)
melanotoides Reitter, 1891g: 234 (*Elater*) **E**: AB AR GG ST
montandoni Buysson, 1889b: ccii (*Ludius*) **E**: BU CZ FR GR IT RO SK SP **A**: CY IN SY TR
  vassilevi Gurjeva & Penev, 1983: 736 (*Penthelater*)
plebejus Candèze, 1873: 28 (*Ludius*) **A**: JA SC SCH TAI
  fusconiger Kolbe, 1886: 199 (*Megapenthes*)
robustus Kishii, 1966a: 28 (*Homotechnes*) **A**: JA
russicus Gurjeva, 1972a: 303 **A**: TD
schoedei Schimmel, 1999: 180 **A**: BEI SHG YUN
umber Bates, 1866: 349 (*Melanotus*) **A**: TAI
yakuensis Ôhira, 1970a: 9 (*Penthelater*) **A**: JA

genus *Gamepenthes* Fleutiaux, 1928a: 148 type species *Megapenthes octomaculatus* Schwarz, 1898
antennatus Miwa, 1934: 224 (*Athous*) **A**: TAI
  ruficollis Miwa, 1934: 224 (*Athous*)
bengalensis Schimmel, 2004b: 443 **A**: SD
brancucci Schimmel, 2003a: 32 **A**: NP
holzschuhi Schimmel, 2003a: 36 **A**: SCH YUN
hubeiensis Schimmel, 2003a: 38 **A**: ANH HUB
jiangxiensis Schimmel, 2004b: 445 **A**: JIX
montivagus Miwa, 1929e: 491 (*Melanoxanthus*) **A**: TAI
octoguttatus Candèze, 1882: 66 (*Megapenthes*) **A**: YUN **ORR**
octomaculatus Schwarz, 1898a: 144 (*Megapenthes*) **A**: BT NP SD
ornatus Lewis, 1894a: 47 (*Megapenthes*) **A**: JA
pictipennis Lewis, 1894a: 48 (*Melanoxanthus*) **A**: JA
sausai Schimmel, 2003a: 43 **A**: YUN **ORR**
sichuanensis Schimmel, 2003a: 44 **A**: SCH
similis Lewis, 1894a: 182 (*Melanoxanthus*) **A**: JA
versipellis Lewis, 1894a: 47 (*Megapenthes*) **A**: FUJ HUB JA ZHE **ORR**
  shirozui Kishii, 1958: 29
yoshidai Ôhira, 1995c: 27 **A**: JA

genus *Ganoxanthus* Fleutiaux, 1928a: 148 type species *Melanoxanthus virgatus* Candèze, 1891
virgatus Candèze, 1891b: 777 (*Melanoxanthus*) **A**: XIZ YUN **ORR**

genus *Hayekpenthes* Ôhira, 1970c: 87 type species *Megapenthes pallidus* Lewis, 1894
babakinsanus Kishii, 1991a: 8, 19 **A**: TAI
flavus Schimmel, 2004b: 291 **A**: YUN
girardi Schimmel, 2004b: 292 **A**: SD
indicus Schimmel, 2004b: 293 **A**: SD
pallidus masakii Kishii, 1979a: 8 **A**: JA (Ishigake-jima, Iriomote-jima)
pallidus okinawensis Kishii, 1979a: 8 **A**: JA (Okinawa)
pallidus pallidus Lewis, 1894a: 46 (*Megapenthes*) **A**: JA SCH
pallidus tokunoshimensis Kishii, 1979a: 8 **A**: JA (Toku-no-shima)
parallelaris Miwa, 1927a: 21 (*Megapenthes*) **A**: TAI

genus *Houwau* Kishii, 1993c: 87 type species *Houwau alpicola* Kishii, 1993
alpicola Kishii, 1993c: 88 **A**: JA (Mt. Houwau-zan, Yamanashi Pref.)

genus *Megapenthes* Kiesenwetter, 1858a: 353 type species *Elater lugens* W. Redtenbacher, 1842
  Dolopiosomus Motschulsky, 1860d: 380 type species *Dolopiosomus aterrimus* Motschulsky, 1860 (= *Trichophorus tartareus* LeConte, 1859)
  Horizoteichos Van Zwaluwenburg, 1931: 432 type species *Horizoteichos papuensis* Zwaluwenburg, 1931
azumai Arimoto, 1990: 89 **A**: JA
bengalensis Schimel, 1998b: 296 **A**: SD **ORR**

dajuensis Schimmel, 2004b: 297 **A**: YUN
dongchuanensis Schimmel, 2004b: 298 **A**: YUN
funebrioides Schimmel, 2004b: 299 **A**: YUN
funebris Candèze, 1882: 65 **A**: SCH YUN
holzschuhi Schimmel, 2004b: 301 **A**: NP SD
hummeli Fleutiaux, 1936a: 16 **A**: GAN JIA JIX SCH
japonicus Fleutiaux, 1902a: 18 **A**: JA
jizuensis Schimmel, 2004b: 303 **A**: SCH YUN
kubani Schimmel, 2004b: 305 **A**: YUN
kurosawai W. Suzuki, 1986: 207 **A**: JA
kusuii Ôhira, 1973d: 327 (*Procraerus*) **A**: JA (Ogasawara)
linearis Candèze, 1865: 30 **A**: SD
lugens W. Redtenbacher, 1842: 11 (*Sericosomus*) **E**: AB AL AR AU BH BU BY CZ FR GB GE GG GR HU IT MC MD RO SK SP ST SZ UK YU **N**: AG **A**: TR
  gagates Curtis, 1854: 12 (*Ectinus*)
luteipes Hope, 1843a: 63 [= 1845a: 9] (*Ludius*) **A**: ZHE
oblongicollis Miwa, 1929e: 492 (*Pengamenthes*) **A**: TAI
opacus Candèze, 1873: 10 **A**: JA
pallidulus Cate, Platia & Schimmel, 2002: 39 **A**: IN
pierrei Leseigneur, 1955: 88 **N**: MO
rugosus Schimmel, 2004b: 311 **A**: YUN
rutilipennis Candèze, 1859: 502 **E**: AB ST **A**: IN TR
sechuanensis Schimmel, 2004b: 313 **A**: SCH
semelai Schimmel, 2004b: 314 **A**: YUN
shirozui shirozui Kishii, 1959a: 14 **A**: JA
shrozui taichii Kishii, 1975: 3 **A**: JA (Ryukyus)
tattakensis Ôhira, 1966d: 266 (*Melanoxanthus*) **A**: TAI
tetricus Candèze, 1859: 504 **A**: HKG
yunnanus Schimmel, 2004b: 316 **A**: YUN

genus *Megapenthoides* Gurjeva & Dolin, 1972: 615 type species *Megapenthes gussakovskii* Gurjeva, 1961
gussakovskii Gurjeva, 1961: 885 (*Megapenthes*) **A**: TD

genus *Melanoxanthus* Eschscholtz, 1838: table after page 60 type species *Elater melanocephalus* Fabricius, 1781
  Pacificola Van Zwaluwenburg, 1932: 131 type species *Pacificola obscura* Van Zwaluwenburg, 1932
almaca Wurst, Schimmel & Platia, 2001: 550 **A**: YE
anticus Candèze, 1892b: 490 **A**: NP **ORR**
bimaculatus Fleutiaux, 1918b: 222 **A**: NP **ORR**
doriae Candèze, 1878a: 125 **A**: AP **ORR**
flavithorax Schimmel, 2004b: 333 **A**: NP SD
illustris Fleutiaux, 1921c: 73 **A**: HAI **ORR**
ingridae Schimmel, 2004b: 339 **A**: NP
lansbergi Candèze, 1882: 67 **A**: AP NP SD **ORR**
melanocephalus Fabricius, 1781: 272 (*Elater*) **A**: CH HAI JA NP TAI **AFR NTR ORR**
  nigroterminatus W. S. MacLeay, 1826: 441 (*Elater*)
melanurus Candèze, 1878b: cxxxvi **A**: HAI
mirabilis Fleutiaux, 1928a: 156 (*Ganoxanthus*) **A**: NP SD **ORR**
nigripennis Fleutiaux, 1928a: 154 **A**: TAI ORR
schawalleri Schimmel, 2004b: 355 **A**: NP **ORR**
sonani Miwa, 1934: 252 **A**: JA
wenzelorum Schimmel, 2004b: 359 **A**: BT
yushiroi W. Suzuki, 1999: 158 [RN] **A**: TAI
  quadriguttatus Miwa, 1927a: 21

genus *Neopenthes* Kishii, 1973: 9 type species *Neopenthes pallidihumeralis* Kishii, 1973
horishanus Miwa, 1929e: 489 (*Elater*) **A**: TAI
pallidihumeralis Kishii, 1973: 10 **A**: JA (Ryukyus)

genus *Parapenthes* Jiang, 1999: 111, 185 type species *Parapenthes varimaculatus* Jiang, 1999
varimaculatus Jiang, 1999: 112, 186 **A**: FUJ

genus *Platianeus* Schimmel, 2004b: 366 type species *Platianeus kubani* Schimmel, 2004
holzschuhi Schimmel, 2004b: 368 **A**: NP

genus *Procraerus* Reitter, 1905a: 11 type species *Elater tibialis* Lacordaire, 1835
basilaris Kishii, 1991a: 7, 15 **A**: TAI
bhutanensis Schimmel, 1999: 99 **A**: BT
bicolor Platia & Gudenzi, 2002b: 89 **A**: TR
brancuccii Schimmel, 1999: 53 **A**: NP
cariniceps Lewis, 1894a: 45 (*Megapenthes*) **A**: FE JA
carinifrons Desbrochers des Loges, 1875a: 41 (*Agriotes*) **E**: AB AR BU GG ST UK
fulvescens Reitter, 1887h: 512 (*Megapenthes*)
retowskii Heyden, 1882b: 155 (*Megapenthes*)
transcaucasicus Stierlin, 1875: 474 (*Athous*)
volxemi Candèze, 1878b: cxxxv (*Megapenthes*)
cretensis Platia & Gudenzi, 1996a: 147 **E**: GR (Kríti)
kadesanus Ôhira, 1969c: 30
kathmandulis Ôhira & Becker, 1974a: 389 (*Megapenthes*) **A**: NP SD
koshunensis Miwa, 1929e: 494 (*Megapenthes*) **A**: TAI
ligatus Candèze, 1878a: 124 (*Melanoxanthus*) **A**: FUJ GUA HKG HUB JIX ZHE **ORR**
nepalensis Ôhira & Becker, 1974a: 387 **A**: NP
opacofulvus Reitter, 1898g: 181 (*Megapenthes*) **E**: AB AR GG ST **A**: KZ UZ
rougemonti Schimmel, 2003a: 238 **A**: HKG
sauteri Schimmel, 1999: 88 **A**: TAI
sinensis Schimmel, 1999: 135 **A**: GUA GUX HEB ZHE
sonami Miwa, 1929e: 494 (*Megapenthes*) **A**: TAI
stepanovi Iablokoff-Khnzorian, 1959a: 59 **E**: AB **A**: IN
suturalis Matsumura, 1910: 39 (*Ludius*) [= 1911a: 145] **A**: TAI
tibialis Lacordaire, 1835: 658 (*Elater*) **E**: AU BE BH BU CR CT CZ DE FI FR GB GE GR HU IT LT LV MD NR PL RO SK SL SP SV ST SZ UK YU **N**: AG **A**: IN
cothurnatus Curtis, 1854: 16 (*Aplotarsus*)
divaricatus Desbrochers des Loges, 1865: 208 (*Megapenthes*)
subcarinatus Germar, 1844: 177 (*Ampedus*)
variegatus Candèze, 1878b: cxc (*Ludius*) **A**: GUA JA (Ryukyus) TAI **AUR ORR**
brunniventris Candèze, 1892c: 804 (*Megapenthes*)
wittmeri Schimmel, 1999: 101 **A**: BT NP
yagii Kishii, 1994c: 7 **A**: TAI

genus *Sawadapenthes* Ôhira, 1970c: 85 type species *Gamepenthes amami* Kishii, 1959
amami amami Kishii, 1959b: 59 (*Gamepenthes*) **A**: JA (Amami-Ôshima)
amami masakii Kishii, 1975: 2 **A**: JA (Okinawa)
himalayansis Schimmel, 2004b: 381 **A**: SD
nepalensis Schimmel, 2004b: 391 **A**: NP
rugosus Schimmel, 2004b: 395 **A**: SD **ORR**

genus *Simodactylus* Candèze, 1859: 169 type species *Elater cinnamomeus* Boisduval, 1835
rougemonti Schimmel, 2003a: 163 **A**: HKG
yamianus Ôhira, 1970d: 19 **A**: TAI

genus *Xanthopenthes* Fleutiaux, 1928a: 160 type species *Megapenthes birmanicus* Candèze, 1888
Xanthelater Miwa, 1931d: 259 type species *Elater granulipennis* Miwa, 1929
bicarinatus Lewis, 1894a: 43 (*Megapenthes*) **A**: JA TAI
birmanicus Candèze, 1888: 677 (*Megapenthes*) **A**: GUA **ORR**
fuscus Fleutiaux, 1928a: 168 (*Megapenthes*)
melanoxanthoides Fleutiaux, 1918b: 219 (*Megapenthes*)
moderatus Candèze, 1888: 678 (*Megapenthes*)
obscurus Fleutiuax, 1918a: 219 (*Megapenthes*)
tonkinensis Fleutiaux, 1928a: 167 (*Megapenthes*)
vicinus Fleutiaux, 1889: 142 (*Megapenthes*)
brancuccii Schimmel, 1999: 207 **A**: NP
brendelli Schimmel, 1998d: 247 **A**: HKG "China"
dignoscus Schimmel, 2003a: 247 **A**: HKG
friedrichi Schimmel, 1999: 287 **A**: HKG
granulipennis Miwa, 1929e: 489 (*Elater*) **A**: FUJ GUA GUI HUB JA (Ryukyus) JIA JIX SCH TAI ZHE
hiramatsui Ôhira, 1995c: 28 **A**: JA (Okinawa)
hummeli Fleutiaux, 1936a: 16 (*Megapenthes*) **A**: GAN JIA JIX SCH
konoi Nakane & Kishii, 1955b: 6 **A**: JA
kucerai Schimmel, 2004b: 484 **A**: UP
lugubris Candèze, 1888: 678 (*Megapenthes*) **A**: GUA **ORR**
minimus Schimmel, 2004b: 485 **A**: YUN
modestus Candèze, 1859: 507 (*Megapenthes*) **A**: NP **ORR**
parvulus Fleutiaux, 1928a: 172 **A**: YUN **ORR**
probsti Schimmel, 1999: 217 **A**: NP
robustus Miwa, 1929e: 490 (*Elater*) **A**: GUI HUB HUN TAI ZHE
rufus Candèze, 1859: 503 (*Megapenthes*) **A**: NP **ORR**
sinensis Schimmel, 1999: 254 **A**: HKG
vagus Schimmel, 1999: 195 **A**: GUA SCH SHX **ORR**
werneri Schimmel, 1999: 220 **A**: UP **ORR**
wittmeri Schimmel, 1999: 257 **A**: SD **ORR**

tribe Megapenthini, species incertae sedis
Melanoxanthus agranulosus Vats & Chauhan, 1992a: 22 **A**: UP
Megapenthes variabilis Vats & Chauhan, 1992a: 20 **A**: HP UP

tribe Physorhinini Candèze, 1859

genus *Anchastelater* Fleutiaux, 1928a: 144 type species *Anchastelater ornatus* Fleutiuax, 1928
ornatus Fleutiaux, 1928a: 144 **A**: YUN **ORR**

genus *Anchastus* LeConte, 1853c: 459 type species *Anchastus digitatus* LeConte, 1853
birganjensis Ôhira & Becker, 1974a: 385 **A**: NP
davidi Candèze, 1878b: lxxxiv **A**: JIX
falsus Candèze, 1897: 47 **A**: NP **ORR**
humeralis Candèze, 1859: 411 **A**: UP **ORR**
parvicollis Candèze, 1897: 46 (*Porthmidius*) **A**: SD
sinensis Candèze, 1859: 410 **A**: GUA HKG
vanharteni Wurst, Schimmel & Platia, 2001: 551 **A**: YE

genus *Astanchus* Gurjeva, 1979a: 124 type species *Anchastus ussuriensis* Gurjeva, 1975
sinensis Schimmel, 2004a: 208 **A**: HEB
ussuriensis Gurjeva, 1975d: 311 (*Anchastus*) **A**: FE MG

genus *Chastanus* Dolin & Gurjeva, 1976a: 645 type species *Anchastus rosti* Schwarz, 1897
afghanus Platia & Gudenzi, 2002b: 90 **A**: AF

castaneus Miwa, 1934: 253 (*Anchastus*) **A**: JA (Ryukyus)
rosti Schwarz, 1897a: 128 (*Anchastus*) **E**: AB **A**: IN

genus *Podeonius* Kiesenwetter, 1858: 229 type species *Elater acuticornis* Germar, 1824
Akitsu Kishii, 1985: 22 type species *Anchastus mus* Lewis, 1894
acuticornis Germar, 1824: 57 (*Elater*) **E**: AL AU CR CZ FR GE GR HU IT PL RO SK SP SZ UK YU **A**: TR
amamiensis Ôhira, 1996b: 7, 13 (*Anchastus*) **A**: JA (Amami-Ôshima)
aquilus aquilus Candèze, 1873: 8 (*Anchastus*) **A**: JA SC
aquilus formosensis Kishii, 1989a: 31 (*Akitsu*) **A**: TAI
aquilis ryukyuensis Ôhira, 1968c: 134 (*Anchastus*) **A**: JA (Ryukyus)
castelnaui Candèze, 1878b: lxxxiv (*Anchastus*) **A**: GUA TAI ZHE **ORR**
fujianus Schimmel, 2003b: 279 **A**: FUJ
himalayanus Schimmel, 2003b: 279 **A**: NP SD
mus Lewis, 1894a: 31 (*Anchastus*) **A**: JA
nagaoi Ôhira, 1968c: 135 (*Anchastus*) **A**: JA (Ryukyus)
rufipes Lewis, 1894a: 32 (*Anchastus*) **A**: JA
shaanxianus Schimmel, 2003b: 281 **A**: SHA
subcyaneus Kiesenwetter, 1858a: 351 (*Brachycrepis*) **E**: GR
taiwanus Ôhira, 1972a: 9 (*Anchastus*) **A**: TAI
yamamotoi Ôhira, 1996b: 10, 14 (*Anchastus*) **A**: JA (Ogasawara)

genus *Porthmidius* Germar, 1847: 7 type species *Ampedus fulvus* W. Redtenbacher, 1842 (= *Elater austriacus* Schrank, 1781)
austriacus Schrank, 1781a: 186 (*Elater*) **E**: AU BE BH BU CR CZ FR GE GR HU IT MD MK RO SK SL SZ UK YU **A**: TR
agnusi Méquignon, 1927: 23
fulvus W. Redtenbacher, 1842: 11 (*Ampedus*)
gelineki Reitter, 1904d: 148 (*Elater*) **E**: BH

genus *Suzukielater* Kishii, 1987: 8 type species *Sphenomerus babai* Kishii & Ôhira, 1956
babai Kishii & Ôhira, 1956: 74, 82 (*Sphenomerus*) **A**: JA

tribe Pomachiliini Candèze, 1859

genus *Betarmon* Kiesenwetter, 1858a: 265 type species *Elater bisbimaculatus* Fabricius, 1803
bisbimaculatus Fabricius, 1803: 48 (*Elater*) **E**: AB AR AU BU CR CZ FR GE GG GR HU IT PL RO SK SL SP SZ UK YU **A**: IN TR
ferruginatus Schönherr, 1817: 317 (*Elater*) [RN]
ferrugineus Scopoli, 1763: 95 (*Elater*) [HN]
quadrimaculatus Fabricius, 1792b: 233 (*Elater*) [HN]

genus *Idiotarmon* Binaghi, 1940a: 99 type species *Betarmon quadrivittatus* Ragusa, 1893
quadrivittatus Ragusa, 1893: 305 (*Betarmon*) **E**: IT
gestroi Buysson, 1893b: 417 (*Idolus*)

genus *Idolus* Desbrochers des Loges, 1875a: 42 [RN] type species *Athous picipennis* Bach, 1852
Agriotella Brown, 1933: 179 type species *Elater bigeminatus* Randall, 1838
Logesius Gozis, 1875: 51 [RN, unjustified]
Metopius Desbrochers des Loges, 1870b: 122 [HN]
adrastoides Reitter, 1888d: 180 (*Logesius*) **E**: AB AR GG ST **A**: TR
picipennis Bach, 1852: 32 (*Athous*) **E**: AU BE BH BU CR CZ FR GE HU IT LS MC NL PL PT RO SK SL SP SZ UK YU
angulatus Rey, 1891c: 85
mocsaryi Szombathy, 1910a: 190
moreli Pic, 1926f: 9
scapulatus Candèze, 1863: 400 (*Agriotes*)
styriacus L. Redtenbacher, 1858: 512 (*Dolopius*)

transpictus Buysson, 1896b: 165
vittatus Candèze, 1880: 297 (*Betarmon*)

genus *Idotarmonides* Agajev, 1985a: 66 type species *Betarmon anatolicus* Candèze, 1882
anatolicus Candèze, 1882: 74 (*Betarmon*) **E**: AB GG ST **A**: TR
quadrisignatus Reitter, 1894i: 302 (*Betarmon*)
bicolor Platia & Gudenzi, 1999b: 118 **A**: IQ
bagdadensis Buysson, 1904a: 6 (*Athous*) **A**: IQ OM
rydhi Platia & Schimmel, 1992: 87 **E**: GR **A**: TR

genus *Isidus* Mulsant & Rey, 1875b: 405 type species *Isidus moreli* Mulsant & Rey, 1875
appendiculatus Pic, 1912g: 25 **N**: AG
letourneuxi Pic, 1902k: 64 **N**: EG
moreli Mulsant & Rey, 1875b: 406 **E**: FR IT MA SP ST **N**: AG EG MO TU **A**: CY TR

genus *Parabetarmon* Ôhira, 1970b: 106 type species *Neotrichophorus carinicephalus* Miwa, 1931
carinicephalus Miwa, 1931b: 205 (*Neotrichophorus*) **A**: JA (Iriomote-jima) TAI

subfamily Elaterinae, species incertae sedis
Dolerosomus sericarius Motschulsky, 1866a: 166 **A**: JA
Dolopius contiguus Faldermann, 1835b: 178

subfamily **Melanotinae** Candèze, 1859 (1856)

genus *Melanotus* Eschscholtz, 1829: 32 type species *Elater fulvipes* Herbst, 1806 (= *Elater villosus* Geoffroy, 1785)
subgenus *Melanotus* Eschscholtz, 1829: 32 type species *Elater fulvipes* Herbst, 1806 (= *Elater villosus* Geoffroy, 1785)
Cratonychus Dejean, 1833: 87 type species *Elater obscurus* sensu A. G. Olivier, 1790 [HN]
Cremnostethus Schwarz, 1901c: 197 type species *Cremnostethus nigricollis* Schwarz, 1901
Kensakulus Chûjô & Ôhira, 1965: 24 type species *Melanotus invectitius* Candèze, 1865
Natomelus Dolin, 1979: 71 type species *Natomelus arcanus* Dolin, 1979
Perimecus Dillwyn, 1829: 32 type species *Elater fulvipes* Herbst, 1806 (= *Elater villosus* Geoffroy, 1785)
Xanthus Gistel, 1834: 11 type species *Elater niger* Fabricius, 1792 (= *Melanotus punctolineatus* Pelerin, 1829)
abdominalis Platia & Schimmel, 2001: 168 **A**: FUJ
acuminatus Reitter, 1891g: 237 **E**: ST **A**: KI KZ TD TM UZ
adlbaueri Platia & Schimmel, 2001: 68 **A**: NP SD
admirabilis Dolin, 1994: 107 **A**: TM
aemulus Erichson, 1841a: 97 (*Cratonychus*) **E**: GR TR **A**: IN SY TR "Palestine"
afghanicus Gurjeva, 1990: 41 **A**: AF
agilis Platia & Schimmel, 2001: 287 **A**: GUX
amianus Kishii, 1992b: 21 **A**: TAI
anachoretus Ménétriés, 1832: 159 (*Elater*) **E**: AB
angustus Platia & Schimmel, 2002: 318 **A**: YUN
ankangensis Platia & Schimmel, 2004: 108 **A**: HUB SHA
annosus Candèze, 1865: 48 **A**: JA JIL SC ZHE
arcanus Dolin, 1979: 72 **A**: TD TM
arctoides Kishii, 2002a: 21 **A**: SC
arctus Candèze, 1882: 89 **A**: GUI JIA JIX SCH YUN
argodi Platia & Schimmel, 2001: 170 **A**: HKG
armeniacus Schwarz, 1892b: 365 **E**: AR
arnoldii Dolin, 1990: 117 **A**: UZ
asper Platia & Schimmel, 2001: 170 **A**: GAN
atkinsoni Platia & Schimmel, 2001: 71 **A**: UP
atratus Fleutiaux, 1933b: 222 **A**: YUN **ORR**
vuilleti Fleutiaux, 1933b: 224

atricapillus Reitter, 1891g: 238 **E**: AR **A**: TR
auberti Platia & Schimmel, 2001: 171 **A**: HKG HUN SCH YUN
augustianus Lee, 1998: 106 **A**: SC
augustus Platia & Schimmel, 2002: 318 **A**: YUN
auriculus Platia & Schimmel, 2001: 71 **A**: NP
ausobskyi Platia & Schimmel, 2001: 72 **A**: KA NP SD
avitus Candèze, 1886: 186 **A**: KI KZ TD TM
babaulti Platia & Schimmel, 2001: 74 **A**: HP UP **ORR**
bacoti Platia & Schimmel, 2001: 172 **A**: XIZ
badchysicus Dolin & Atamuradov, 1980: 83 **A**: TM
badgleyi Platia & Schimmel, 2001: 75 **A**: NP SD **ORR**
bajulus Erichson, 1841a: 98 (*Cratonychus*) **A**: SY TR
bannaensis Ôhira, 1978b: 554 **A**: JA
barbatus Gurjeva, 1990: 47 **A**: AF
beckeri Platia & Schimmel, 2001: 76 **A**: NP
becvari Platia & Schimmel, 2002: 319 **A**: YUN
beesoni Platia & Schimmel, 2001: 77 **A**: KA UP
bequesti Platia & Schimmel, 2001: 78 **A**: HP KA
benesi Platia & Schimmel, 2002: 319 **A**: SCH
beshkenticus Dolin & Latifi, 1989: 21 **A**: TD
besucheti Platia & Schimmel, 2001: 79 **A**: PA
bhaktai Platia & Schimmel, 2001: 80 **A**: BT NP SD **ORR**
bhutanensis Platia & Schimmel, 2001: 81 **A**: BT SD
bieti Platia & Schimmel, 2001: 173 **A**: GAN SCH XIZ
bifoveatus Vats & Chauhan, 1991a: 191 **A**: UP
bilyi Platia & Schimmel, 2001: 82 **A**: NP
binaghii Platia & Schimmel, 2001: 174 **A**: SCH XIZ YUN **ORR**
blumenthali Platia & Gudenzi, 2002b: 92 **A**: AF
bocaki Platia & Schimmel, 2001: 176 **A**: SCH YUN
bogatschevi Dolin, 1978: 96 **A**: TD
boninensis Ôhira, 1974b: 25 **A**: JA (Ogasawara)
brancuccii Platia & Schimmel, 2001: 84 **A**: NP
breuili Platia & Schimmel, 2001: 85 **A**: KA
brevithorax Kishii & Platia, 1993a: 67 **A**: JA (Ishigaki)
brignolii Guglielmi & Platia, 1985: 206 **E**: GR
bronsteinae Dolin, 1990: 118 **A**: UZ
brunneolus Dolin & Latifi, 1989: 23 **A**: TD
brunnicornis Schwarz, 1892c: 164 **A**: BEI HEB HEN
brunniopacus Kishii, 1989d: 33, 47 **A**: TAI
brunnipes Germar, 1824: 41 (*Elater*) **E**: AB AR AU BE BH BU BY CR CT CZ FR GE GG GR HU IT MC MD PL RO SK SL ST SZ TR UK **A**: IN KA TR XIN
 fascicularis Küster, 1851b: 37 (*Cratonychus*)
 subvestitus Brullé, 1832: 137
burakowskii Buchholz & Dolin, 1990: 713 **A**: TD
businskyi Platia & Schimmel, 2001: 178 **A**: QIN XIZ
carbonarius Candèze, 1882: 88 **A**: JIA LIA SC SHG
carinicollis Schwarz, 1894: 148 **A**: TD
 maurus Gurjeva, 1979c: 70
carinulatus Platia & Schimmel, 2001: 180 **A**: YUN
carnosus Platia & Schimmel, 2001: 180 **A**: YUN
castanipes Paykull, 1800: 23 (*Elater*) **E**: AU BE BH BU BY CR CT CZ DE EN FI FR GB GE GG GR HU IT LA LS LT LU NR NT PL RO SK SL SP ST SV SZ UK YU **A**: ES FE IN JIL PA SCH SHX TR UP WS YUN **NAR**
 bernhardinus Stierlin, 1879a: 439
 fulvipes Gyllenhal, 1808a: 407 (*Elater*) [HN]

fuscusmajor DeGeer, 1774: 146 (*Elater*)
inaequalis LeConte, 1853c: 476 (*Cratonychus*)
insulsus Buysson, 1914b: 83
longipennis Küster, 1848b: 25 (*Cratonychus*)
neopunctaticollis Buysson, 1920b: 37
paradoxus F. E. Melsheimer, 1844: 152
picipes Schrank, 1789: 78 (*Elater*)
rufescens Fallén, 1802: 17 (*Elater*)
scrobicollis LeConte, 1853c: 476 (*Cratonychus*)
sublaevicollis Rey, 1891b: 68
texanus Candèze, 1860: 351
ussuricus Brancsik, 1914: 58
catei Platia & Schimmel, 2001: 181 **A**: SCH
caudex Lewis, 1879b: 156 **A**: JA HEI JIL LIA HEB SHX SHA HEN HUB HUN GUX QIN
cavaleriei Platia & Schimmel, 2001: 182 **A**: SCH YUN
cechovskyi Platia & Schimmel, 2001: 183 **A**: SCH
cerberus Gurjeva, 1989b: 26 **A**: KZ
cervenkai Platia & Schimmel, 2002: 320 **A**: GUI
championi Platia & Schimmel, 2001: 88 **A**: KA
chatlonicus Dolin & Cate, 1990: 113 **A**: TD
chejuensis Lee, 1999: 42 **A**: SC
chengi Kishii, 1994d: 17 **A**: TAI
christophi Dolin & Gurjeva, 1994: 112 **A**: TM
cinerascens Küster, 1851b: 34 (*Cratonychus*) **E**: AL BU CR GR IT RO SL YU
combyi Platia & Schimmel, 2001: 184 **A**: YUN
compactus Candèze, 1860: 317 **N**: AG MO
corticeus Fairmaire, 1871: 380 (*Athous*)
comptus Platia & Schimmel, 2001: 185 **A**: AHN JIX YUN
conicicollis Reitter, 1891g: 235 **A**: KI KZ XIN
convexiusculus Platia & Schimmel, 2001: 89 **A**: PA
coolsi Platia & Schimmel, 2001: 90 **A**: BT NP SD
coomani Platia & Schimmel, 2001: 186 **A**: YUN **ORR**
indosinensis Fleutiaux, 1933b: 228
coreanus Lee, 1999: 40 **A**: LIA SC
correctus correctus Candèze, 1865: 48 **A**: JA
correctus issikii Miwa, 1929c: 346 **A**: JA (Yaku-shima, Kuchierabu-jima)
crassicollis Erichson, 1841a: 98 (*Cratonychus*) **E**: AU BA BU BY CR CT CZ EN FR GE GR HU IT LA MC PL RO SK SL SP ST SZ TR YU **A**: TR
brunnipes Lacordaire, 1835: 633 (*Cratonychus*) [HN]
declivis Rey, 1891b: 68
parumpunctatus Schwarz, 1892c: 155
subrugatus Rey, 1891b: 68
tristis Küster, 1851b: 32 (*Cratonychus*)
crassus Platia & Schimmel, 2001: 91 **A**: NP SD **ORR**
daccordii Platia & Schimmel, 2002: 321 **A**: XIZ
davidi Platia & Schimmel, 2001: 187 **A**: SCH
danilevskyi Dolin & Cate, 1994: 122 [= 1995: 169] **A**: TM
daruma Kishii, 2004d: 170 **A**: JA (Kume-jima Is.)
dejeani Platia & Schimmel, 2001: 188 **A**: YUN
depressicollis Fleutiaux, 1933b: 229 **A**: YUN **ORR**
desbrochersi Candèze, 1882: 87 **E**: AZ
dichroides Platia & Gudenzi, 1999b: 119 **A**: IN
dichrous Erichson, 1841a: 93 (*Cratonychus*) **E**: AZ BU CR FR GR IT MC PT SP TR **A**: TR "Palestina" **N**: AG CI (Fuerteventura, La Palma) EG MO TU **NARi**
amplithorax Mulsant & Guillebeau, 1855a: 17 (*Cratonychus*)
incertus Schwarz, 1892c: 162

marocanus Buysson, 1918b: 111
mauritanicus P. H. Lucas, 1846: 162 (*Cratonychus*)
sublucens Abeille de Perrin, 1872a: 32
dignus Platia & Schimmel, 2001: 310 **A**: YUN **ORR**
dilaticollis Reitter, 1891g: 236 **A**: KI
dilutus Platia & Schimmel, 2001: 190 **A**: YUN
discrepans Dolin & Latifi, 1989: 24 **A**: TD
dolini Atamuradov, 1990: 82 **A**: UZ
dubernardi Platia & Schimmel, 2001: 191 **A**: YUN
dubiosus Platia & Schimmel, 2001: 192 **A**: YUN **ORR**
duchainei Platia & Schimmel, 2001: 193 **A**: GUA HAI HKG JIX YUN
dundai Platia & Schimmel, 2001: 194 **A**: SCH
dureli Platia & Schimmel, 2001: 94 **A**: SD **ORR**
dusaneki Platia & Schimmel, 2002: 322 **A**: SCH
dussauti Fleutiaux, 1918b: 240 **A**: YUN **ORR**
elongatus Hope, 1831: 25 (*Elater*) **A**: NP SD
elsasseri Platia & Schimmel, 2001: 98 **A**: KA
erythropygus erythropygus Candèze, 1873: 20 **A**: JA NC SC "NE China"
erythropygus sado Kishii, 1994d: 21 **A**: JA (Sado)
erythropygus taishu Kishii, 1994d: 21 **A**: JA
escalerai Platia & Schimmel, 2001: 196 **A**: YUN
excavatus Dolin & Latifi, 1997: 85 **A**: TD
excelsus Platia & Schimmel, 2001: 197 **A**: YUN
excoffieri Platia & Schimmel, 2001: 198 **A**: YUN
exularis Dolin, 1990: 119 **A**: KZ UZ
fairmairei Fleutiaux, 1918b: 239 **A**: ZHE **ORR**
farkaci Platia & Schimmel, 2001: 200 **A**: YUN
fergussoni Platia & Schimmel, 2001: 201 **A**: SCH
ferreroi Platia & Schimmel, 2001: 315 **A**: YUN **ORR**
ferrugineus Schwarz, 1892b: 366 **E**: CR GR
fiumii Platia & Schimmel, 2001: 201 **A**: SHA SHX
flavus Platia & Schimmel, 2001: 202 **A**: XIN
fleutiauxi Platia & Schimmel, 2001: 203 **A**: YUN ZHE
fortnumi fortnumi Candèze, 1878b: clxvii **A**: JA
fortnumi yezoensis Ôhira, 1992c: 35 **A**: JA
fortunati Platia & Schimmel, 2001: 204 **A**: GUI SCH YUN ZHE
fougueorum Platia & Schimmel, 2004: 109 **A**: YUN
fragilis Schwarz, 1892c: 160 **A**: UZ
fragilloides Dolin, 1988c: 80 **A**: TD
fraseri Platia & Schimmel, 1993a: 55 **A**: SY TR
frontalis Platia & Schimmel, 2001: 205 **A**: YUN ZHE **ORR**
fruhstorferi Platia & Schimmel, 2001: 206 **A**: GUA YUN
fulvus Reitter, 1891g: 237 **A**: IN IQ TM
fusciceps Gyllenhal, 1817: 135 (*Elater*) **E**: AB AR AZ BU CR CT GG GR MC MD RO ST UK **A**: CY KZ IN IQ IS LE SY TR
conformis Faldermann, 1835a: 163 (*Cratonychus*)
fuscatocollis Brullé, 1832: 137
persicus Ménétriés, 1832: 157 (*Elater*)
robustus Erichson, 1841a: 99 (*Cratonychus*)
torosus Erichson, 1841a: 100 (*Cratonychus*)
villosus Brullé, 1832: 137 [HN]
gardneri Platia & Schimmel, 2001: 100 **A**: HP
ghilarovi Dolin & Atamuradov, 1989: 39 **A**: TM
girardi Platia & Schimmel, 2001: 207 **A**: SCH XIZ YUN
glanei Platia & Schimmel, 2001: 208 **A**: YUN

globosus Platia & Schimmel, 2002: 323 **A**: HKG
gononi Platia & Schimmel, 2002: 324 **A**: XIZ
gracilifasciatus Platia & Schimmel, 2001: 101 **A**: NP SD
gracilipennis Kishii & Platia, 1993b: 119 **A**: TAI
graecus Platia & Schimmel, 1993a: 56 **E**: GR
gratus Platia & Schimmel, 2001: 103 **A**: NP
gudenzii Platia & Schimmel, 2001: 209 **A**: GAN GUI HUN SCH SHA
gurjevae Dolin, 1997b: 352 **A**: UZ
hachijoensis Ôhira, 1969b: 46 **A**: JA (Hachijô-jima)
hainanensis Platia & Schimmel, 2001: 210 **A**: HAI
hallasanae Lee, 1998: 106 **A**: SC
hartmanni Platia & Schimmel, 2004: 101 **A**: NP
hauseri Dolin, 1971: 650 **A**: TD UZ
hausschweisi Platia & Schimmel, 2001: 211 **A**: SCH
hebeiensis Platia & Schimmel, 2004: 110 **A**: HEB
heydeni Schwarz, 1892b: 365 **A**: KI TD UZ
hiekei Platia & Schimmel, 2001: 212 **A**: XIZ YUN
himalayanus Platia & Schimmel, 2001: 104 **A**: HP KA
hirticornis Herbst, 1806: 47 (*Elater*) **A**: GUI HAI NP SCH YUN **ORR**
    fuscus Erichson, 1841a: 106 (*Cratonychus*)
hissaricus Dolin & Latifi, 1988a: 58 **A**: TD
holzschuhi Platia & Schimmel, 2001: 106 **A**: NP SD **ORR**
hopei Platia & Schimmel, 2001: 108 **A**: BT NP SD
horai Platia & Schimmel, 2001: 109 **A**: HP
hourai Kishii, 1989d: 34, 49 **A**: TAI
humilis Schwarz, 1892c: 156 **A**: KI KZ XIN
hunanensis Platia & Schimmel, 2001: 213 **A**: HUN SHA
imitator Dolin, 1997b: 353 **A**: KZ TD UZ
indicus Platia & Schimmel, 2001: 110 **A**: UP
indigenus Gurjeva, 1988b: 338 **A**: TD
    badachshanicus Dolin, 1988b: 59
insularis Lee & Woo, 1999: 15 **A**: SC
intermedius Fleutiaux, 1940d: 193 **A**: "China" **ORR**
iranicus Platia & Gudenzi, 1999b: 119 **A**: IN
ishigakianus Kishii, 1974: 12 **A**: JA (Ishigaki-jima, Iriomote-jima)
jansoni Platia & Schimmel, 2001: 111 **A**: KA
jendeki Platia & Schimmel, 2001: 215 **A**: GAN SCH
jucundus Platia & Schimmel, 2001: 216 **A**: FUJ
kabakovi Gurjeva, 1990: 46 **A**: AF
kabateki Platia & Schimmel, 2002: 324 **A**: SCH YUN
kakusatoi Kishii, 1998a: 25 **A**: SC
kasyi Platia & Gudenzi, 2002b: 92 **A**: AF
kaulbacki Platia & Schimmel, 2001: 217 **A**: SCH
kiungdoni Platia & Schimmel, 2001: 218 **A**: XIZ YUN
kintaroui Kishii, 1989d: 31, 37 **A**: TAI
kirghizicus Dolin, 1969: 132 **A**: KI
kliri Cate, Platia & Schimmel, 2002: 43 **A**: IN
kolibaci Platia & Schimmel, 2002: 325 **A**: GAN HEN SHA
kolthoffi Platia & Schimmel, 2001: 219 **A**: HUN
kopetdaghensis Dolin & Atamuradov, 1994: 118 **A**: TM
kopetzi Platia & Schimmel, 2004: 102 **A**: NP
kourai Kishii, 1994d: 18 **A**: NC
kraatzi Schwarz, 1892c: 155 **A**: GR

kryzhanovskyi Gurjeva, 1988b: 343 **A**: TD
sogdianus Dolin & Latifi, 1988b: 825
kubani Platia & Schimmel, 2001: 220 **A**: YUN
kucerai Platia & Schimmel, 2002: 326 **A**: YUN
kumaunensis Vats & Chauhan, 1991a: 188 **A**: UP
kumensis Ôhira, 1999: 414 **A**: JA
lameyi Fleutiaux, 1918b: 238 **A**: FUJ GUA GUI GUX HEN SCH SHA TAI ZHE **ORR**
languidus Platia & Schimmel, 2001: 329 **A**: YUN **ORR**
latissimus Platia & Schimmel, 2001: 113 **A**: NP SD "Himalaya"
latus Platia & Schimmel, 2001: 222 **A**: FUJ
lebischi Platia & Schimmel, 2001: 114 **A**: BT NP SD **ORR**
legatoides Kishii, 1975: 5 **A**: JA
legatus erabu Kishii, 1988c: 126 **A**: JA (Okierabu-jima)
legatus leeanus Kishii, 2002a: 20 **A**: SC (Cheju)
legatus legatus Candèze, 1860: 323 **A**: FUJ GAN GUA GUX FE JA NC NE SC SHN YUN
laticollis Motschulsky, 1861a: 9
legatus masakianus Kishii, 1977a: 50 **A**: JA
legatus ogatai Kishii, 1988c: 124 **A**: JA (Tokara, Ôsumi)
akusekianus Ôhira, 1997d: 346 **A**: JA
lehmanni Platia & Schimmel, 2001: 224 **A**: FUG GUA GUX **ORR**
leningeri Platia & Schimmel, 2001: 115 **A**: NP
lewisi lewisi Schenkling, 1927: 277 [RN] **A**: JA
longipennis Lewis, 1894a: 192 [HN]
lewisi tsushimensis Kishii, 1986: 53 **A**: JA (Tsushima)
liui Platia & Schimmel, 2001: 225 **A**: YUN
liukueiensis Kishii, 1989d: 33, 45 **A**: TAI
loebli Platia & Schimmel, 2001: 116 **A**: HP KA PA UP
longicornis Candèze, 1860: 332 **A**: SD
longipes Gurjeva, 1988b: 342 **A**: KI
longipilosus Vats & Chauhan, 1991a: 191 **A**: HP
longus Gurjeva, 1990: 44 **A**: AF
lutaoanus Kishii, 1994d: 14 **A**: TAI
macrofasciatus Platia & Schimmel, 2001: 118 **A**: NP **ORR**
maculicornis Buysson, 1912e: 130 **A**: KI UZ
maestus Platia & Schimmel, 2002: 327 **A**: SCH
magnus Platia & Schimmel, 2001: 226 **A**: ZHE
malaisei Fleutiaux, 1942: 12 **A**: YUN **ORR**
mamillanus Gurjeva, 1989b: 30 **A**: KZ
manifestus Gurjeva, 1989b: 31 **A**: KI
marchandi Platia & Schimmel, 2001: 227 **A**: FUJ HKG HUN ZHE
martensi Platia & Schimmel, 2001: 119 **A**: NP
masamichii Kishii, 2004d: 172 **A**: JA (Kume-jima)
matsumurai Schenkling, 1927: 277 [RN] **A**: FE JA NC SC "Manshukuo"
carinatus Matsumura, 1911a: 121 [HN]
medianus Platia & Schimmel, 2002: 328 **A**: YUN
medvedevi Dolin, 1997b: 350 **A**: UZ
melli Platia & Schimmel, 2001: 229 **A**: ANH GUI GUX ZHE
mellonii Platia & Schimmel, 2001: 120 **A**: NP
mertliki Platia & Schimmel, 2001: 230 **A**: YUN
michailovi Dolin, 1988c: 78 **A**: TD
microaugustus Platia & Schimmel, 2002: 329 **A**: XIZ
minimus Dolin, 1997b: 354 **A**: TD
mirus Platia & Schimmel, 2001: 121 **A**: NP XIZ
mniszechi Platia & Schimmel, 2001: 231 **A**: XIZ YUN
moffartsi Platia & Schimmel, 2001: 122 **A**: KA "India"

mongolicus Gurjeva, 1968: 295 **A**: KZ MG NMO
monticola Ménétriés, 1832: 156 (*Elater*) **E**: AB AR BU GG ST **A**: IN TR
morbosus Candèze, 1886: 186 **A**: KI TD TM UZ
mursini Dolin, 1990: 121 **A**: KZ
murzini Platia & Schimmel, 2002: 330 **A**: YUN
mutatus Harold, 1869: 125 [RN] **A**: IN SY
persicus Faldermann, 1835a: 161 [HN] (*Cratonychus*)
mutilatus Platia & Schimmel, 2001: 231 **A**: ZHE
narukawai Kishii, 1996c: 71 **A**: JA
nepalensis Ôhira & Becker, 1974b: 560 **A**: NP
niger Lee, 1998: 105 **A**: SC
nigricollis Schwarz, 1901c: 197 (*Cremnostethus*) **A**: SD **ORR**
nikodymi Platia & Schimmel, 2001: 232 **A**: GAN QIN
niponicus Ôhira, 1964b: 35 **A**: JA
nitidicornis Gurjeva, 1988b: 343 **A**: TD
darvasicus Dolin, 1988d: 826
notabilis Dolin, 1994: 117 **A**: AF TM
nuceus Candèze, 1882: 89 **A**: GUA HUN JIX SC SCH **ORR**
oblongissimus Platia & Schimmel, 2002: 317 [RN] **A**: SD
oblongulus Platia & Schimmel, 2001: 124 [HN]
oblongulus Kishii, 1989d: 32, 39 **A**: TAI
ocellatopunctatus Lewis, 1894a: 193 **A**: JA
ocellatus Platia & Schimmel, 2001: 125 **A**: PA
opacus Platia & Schimmel, 2001: 234 **A**: GUI HUN JIX SCH YUN ZHE
opicus Candèze, 1900: 93 **A**: KA PA
osellai Guglielmi & Platia, 1985: 210 **A**: TR
oshimanus Ôhira, 1967d: 34 **A**: JA (Amami-Ôshima)
otobeanus Kishii, 1988b: 103 **A**: JA
ovtschinnikovi Dolin, 1997b: 352 **A**: UZ
pakistanicus Platia & Schimmel, 2001: 127 **A**: PA
pansanus Kishii, 1998a: 22 **A**: NC
paradoxus Dolin, 1990: 120 **A**: UZ
parallelicollis Platia & Schimmel, 2001: 235 **A**: SCH
parphianus Dolin & Atamuradov, 1989: 39 **A**: TM
passerculus Gurjeva, 1988b: 345 **A**: TM
paulusi Platia & Schimmel, 2002: 331 **A**: XIZ
pearsoni Platia & Schimmel, 2001: 128 **A**: BT NP
pekarovici Platia & Schimmel, 2002: 331 **A**: GAN
pendulus Platia & Schimmel, 2001: 236 **A**: SCH
perlongus Candèze, 1900: 94 **A**: KA PA **ORR**
perplexus Platia & Schimmel, 2001: 237 **A**: XIZ
persimilis Dolin & Latifi, 1988a: 56 **A**: TD
perturbatus Platia & Schimmel, 2004: 112 **A**: HUB
piceipennis Platia & Schimmel, 2004: 111 **A**: SHX
piceus Ménétriés, 1849: 51 (*Adrastus*) **A**: KI TM UZ
subtilis Schwarz, 1892c: 159
pichoni Platia & Schimmel, 2001: 238 **A**: FUJ GUX HUB JIX ZHE
pieli Platia & Schimmel, 2001: 239 **A**: ANH FUJ SHA
pishanensis Kishii, 1989d: 34, 50 **A**: TAI
planipennis Candèze, 1900: 93 **A**: HP KA
planus Platia & Schimmel, 2001: 240 **A**: GUA
platypus Vats & Chuahan, 1991 193 **A**: HP UP
porioni Platia & Schimmel, 2001: 131 **A**: KA PA
problematicus Platia & Schimmel, 2001: 241 **A**: GUI JIA ZHE **ORR**

probsti Platia & Schimmel, 2001: 132 **A**: NP
propexus Candèze, 1860: 326 **A**: JIA NC SC ZHE
prosternalis Platia & Schimmel, 2001: 243 **A**: XIZ
pseudoalburnus Platia & Schimmel, 2001: 244 **A**: YUN
pseudoarctus Platia & Schimmel, 2001: 245 **A**: BEI SCH SHA
pseudolegatus Platia & Schimmel, 2001: 246 **A**: FUJ SHA
pseudoregalis Platia & Schimmel, 2001: 247 **A**: ANH FUJ GUA GUX HUB JIX SCH
pullatus Platia & Schimmel, 2001: 248 **A**: YUN
pulverosus Hope, 1831: 26 (*Elater*) **A**: BT NP SD
pulvinatus Gurjeva, 1990: 43 **A**: AF
pucticollis Platia & Schimmel, 2001: 249 **A**: XIZ
punctatostriatus Schwarz, 1892b: 365 **A**: SY
punctolineatus Pelerin, 1829: 524 (*Elater*) **E**: AL AU BE BH BU CR CT CZ DE FR GB GE GR HU IT LA MC MD NL PL PT RO SK SL SP SV SZ UK YU **A**: SY TR
ater Eschscholtz, 1829: 32
niger Fabricius, 1792b: 221 (*Elater*)
punctosinus Cate, Platia & Schimmel, 2002: 44 **A**: IN
punctosus Walker, 1858: 280 (*Athous*) **A**: KA NP PA UP **ORR**
foliatus Vats & Chauhan, 1991a: 195
vermiculatus Candèze, 1860: 336
pusillus Platia & Schimmel, 2001: 249 **A**: SCH YUN
pygmaeus Stepanov, 1930a: 91 **A**: FE
raii Platia & Schimmel, 2001: 136 **A**: BT NP SD
ramitensis Dolin, 1997b: 354 **A**: TD
rectus Gurjeva, 1990: 48 **A**: AF
regalis Candèze, 1860: 325 **A**: FUJ GUA GUI GUX HUB HUN JIA JIX SCH SHG **ORR**
reitteri Platia & Schimmel, 2002: 332 **A**: XIN
remissus Gurjeva, 1989b: 34 **A**: TD
riesei Platia & Schimmel, 2001: 137 **A**: PA
roannei Platia & Schimmel, 2001: 251 **A**: YUN
ronkayi Platia & Schimmel, 2002: 311 **A**: NP
rollei Platia & Schimmel, 2001: 138 **A**: KA (Jammu)
rougemonti Platia & Schimmel, 2002: 333 **A**: HKG
rufinoides Platia & Schimmel, 2001: 139 **A**: NP **ORR**
rufiventris Miwa, 1930b: 64 **A**: TAI
rugosipennis Fleutiaux, 1933b: 231 **A**: YUN **ORR**
rustamovi Dolin & Atamuradov, 1987: 74 **A**: TM
sabatinellii Platia & Schimmel, 2001: 141 **A**: HP UP
satoi okinawensis Ôhira, 1982: 21 **A**: JA (Ryukyus)
satoi sakishimensis Ôhira, 1982: 24 **A**: JA (Ryukyus)
satoi satoi Ôhira, 1967d: 35 **A**: JA
savioi Platia & Schimmel, 2001: 253 **A**: FUJ JIA JIX SCH SHG YUN ZHE
schawalleri Platia & Schimmel, 2001: 142 **A**: HP KA NP
schmidti Platia & Schimmel, 2004: 103 **A**: NP
schoenmanni Platia & Schimmel, 2001: 254 **A**: SCH YUN
schwarzi Platia & Schimmel, 2001: 255 **A**: ZHE
scissus Platia & Schimmel, 2001: 256 **A**: BEI
sciurus Candèze, 1882: 89 **A**: BEI GAN SCH
scutellatus Platia & Schimmel, 2004: 104 **A**: NP
seniculus Candèze, 1873: 21 **A**: JA
candezellus Miwa, 1934: 100
senilis senilis Candèze, 1865: 47 **A**: JA
invectitius Candèze, 1865: 47 **A**: JA
senilis yakuinsulanus Kishii, 1986: 52 **A**: JA (Yaku-shima)
seriatus Platia & Schimmel, 2001: 143 **A**: KA

sichuanensis Platia & Schimmel, 2001: 258 **A**: SCH
sikkimensis Platia & Schimmel, 2001: 144 **A**: SD
similaris Dolin & Latifi, 1989: 20 **A**: TD
sinensis Platia & Schimmel, 2001: 259 **A**: GUA JIA
siniaevi Platia & Schimmel, 2002: 334 **A**: SCH
singularis Platia & Schimmel, 2001: 260 **A**: SCH
siskai Platia & Schimmel, 2002: 335 **A**: GAN
skopini Dolin, 1971: 652 **A**: KI
sladkovi Dolin & Atamuradov, 1986: 71 **A**: TM
slipinskyi Platia & Schimmel, 2001: 145 **A**: NP
smetanai Platia & Schimmel, 2001: 146 **A**: KA (Jammu)
sobrinus Ménétriés, 1832: 157 (*Elater*) **E**: AB AR GG ST **A**: IN TM
    menetriesi Faldermann, 1835a: 162 (*Cratonychus*)
solus Platia & Schimmel, 2001: 261 **A** JIX SCH YUN
sordidatus Gurjeva, 1989b: 35 **A**: KI
sordidus Platia & Schimmel, 2001: 262 **A**: JIX SCH XIZ YUN
souliei Platia & Schimmel, 2001: 263 **A**: YUN
spernendus kosugi Kishii, 1975: 4 **A**: JA (Yaku-shima)
spernendus spernendus Candèze, 1873: 21 **A**: JA
spevari Platia & Schimmel, 2001: 264 **A**: SCH YUN
splendidus Platia & Schimmel, 2001: 265 **A**: FUJ GUA JIX **ORR**
splichali Platia & Schimmel, 2002: 335 **A**: XIZ
stegnus Gurjeva, 1990: 45 **A**: AF
stepari Platia & Schimmel, 2002: 336 **A**: GAN
stotzneri Platia & Schimmel, 2001: 266 **A**: SCH
striatulus Platia & Schimmel, 2001: 267 **A**: GUX ZHE
strnadi Platia & Schimmel, 2001: 362 **A**: HUN **ORR**
subcarinatus Platia & Schimmel, 2001: 146 **A**: NP UP
subglabratus Platia & Schimmel, 2001: 147 **A**: NP
subspinosus Platia & Schimmel, 2001: 268 **A**: HUB NP SCH SHA YUN
sukenagai Kishii, 1986: 55 **A**: JA (Ishigaki-jima)
sugonjaevi Gurjeva, 1990: 40 **A**: AF
suwonensis Lee, 1998: 103 **A**: SC
svihlai Platia & Schimmel, 2002: 337 **A**: GAN
taiwanus Kishii, 1989d: 32, 43 **A**: TAI
takahashii Kishii, 1974: 12 **A**: JA (Ryukyus)
takaoanus Kishii, 1989d: 32, 41 **A**: TAI
takasago Kishii, 1989d: 33, 44 **A**: TAI
tamsuyensis Bates, 1866: 349 **A**: FUJ SCH TAI
tauricola Dolin, 1980: 79 **E**: UK (Krym)
tenebrosus Erichson, 1841a: 93 (*Cratonychus*) **E**: AR AU BH CR CZ FR GG GR HU IT MK PT RO SK SL SP ST SZ
thomasi Platia & Schimmel, 2001: 148 **A**: UP
tibetanus Platia & Schimmel, 2001: 269 **A**: SCH XIZ
tristis Platia & Schimmel, 2001: 270 **A**: SCH YUN
tryznai Platia & Schimmel, 2002: 338 **A**: SCH
turcicus Platia & Gudenzi, 2002b: 94 **A**: TR
turkestanicus Schwarz, 1892c: 152 **A**: KI
turkmenicus Dolin & Atamuradov, 1986: 70 **A**: TM
    lanatus Gurjeva, 1988b: 339
turnai Platia & Schimmel, 2001: 271 **A**: GAN SCH
umbrosus Platia & Schimmel, 2001: 272 **A**: YUN
unicolor Fleutiaux, 1930a: 274 **A**: NP SD **ORR**
vaillanti Platia & Schimmel, 2001: 273 **A**: YUN

valentini Gurjeva, 1989b: 37 [RN] **A**: TD
 michailovi Gurjeva, 1988b: 341 [HN]
validus Schwarz, 1892b: 366 **A**: KI
variabilis Platia & Schimmel, 2001: 274 **A**: GUX SCH XIZ YUN
vartiani Platia & Gudenzi, 2002b: 93 **A**: AF
venalis Candèze, 1860: 323 **A**: HUN JIX YUN ZHE
ventralis Candèze, 1860: 324 **A**: ANH FUJ HEN JIA JIX SCH SHG ZHE
vesperatus Gurjeva, 1989b: 38 **A**: TD
vetustus Schwarz, 1894: 147 **A**: TD
vicinus Gurjeva, 1989b: 39 **A**: TD
vidualis Gurjeva, 1988b: 344 **A**: TD
vignai Platia & Schimmel, 2001: 277 **A**: GUI SCH ZHE
vilis Gurjeva, 1989b: 40 **A**: TM
villosus Geoffroy, 1785: 40 (*Elater*) **E**: AB AL AR AU BE BH BU BY CR CT CZ DE EN FI FR GB GE GG GR HU IR IT LA LS LT LU MD MK NL NR NT PL PT RO SK SL SP SV ST SZ TR UK YU **N**: AG MO **A**: IN TR YUN "Manchuria"
 aspericollis Mulsant & Guillebeau, 1855a: 18
 atripennis Eschscholtz, 1818: 480 (*Elater*)
 bicolor Fabricius, 1801b: 234 (*Elater*)
 candezei Stierlin, 1864: 149
 discolor Gmelin, 1790: 1914 (*Elater*)
 dichroicolor Buysson, 1914b: 82
 erythropus Gmelin, 1790: 1912 (*Elater*)
 fulvipes Herbst, 1806: 46 (*Elater*)
 hidalgoi Perez Arcas, 1874: 131
 picticornis Heyden, 1870: 119
 punctatocollis C. N. F. Brisout de Barneville, 1862: 600 (*Cratonychus*)
 rufipes Herbst, 1784b: 113 (*Elater*) [HN]
 rugosus Marsham, 1802: 381 (*Elater*)
 subrufus Schwarz, 1892c: 152
waltersi Platia & Schimmel, 2001: 151 **A**: UP
watanabei Ôhira, 2002a: 337 **A**: JA (Kuro-shima)
weigeli Platia & Schimmel, 2002: 312 **A**: NP
weiperti Platia & Schimmel, 2004: 105 **A**: NP
wittmeri Platia & Schimmel, 2001: 152 **A**: PA
woonhahi Lee & Woo, 1999: 17 **A**: SC
wrasei Platia & Schimmel, 2002: 339 **A**: SCH
wui Platia & Schimmel, 2001: 279 **A**: YUN
yagianus Kishii, 1990a: 23 **A**: TAI
yamayai Kishii, 1985: 26 **A**: JA
yangi Platia & Schimmel, 2001: 280 **A**: FUJ
yayeyamacola Kishii, 1974: 9 **A**: JA (Ryukyus)
yunnanensis Platia & Schimmel, 2001: 281 **A**: YUN
zerchei Platia & Schimmel, 2001: 153 **A**: SD **ORR**
zethneri Platia & Schimmel, 2001: 154 **A**: PA (Murree)
zhantievi Dolin, 1969: 133 **A**: TD UZ
zhilongensis Platia & Schimmel, 2001: 282 **A**: SCH
subgenus *Nishimelanotus* Ôhira, 2001: 20 type species *Melanotus masakii* Kishii, 1975
masakii Kishii, 1975: 3 **A**: JA (Okinawa-hontô)
subgenus *Spheniscosomus* Schwarz, 1892a: 132 type species *Melanotus cuneiformis* Baudi di Selve, 1871
 Melanotopsis Lewis, 1894a: 191 type species *Melanotus cete* Candèze, 1860
amamiensis amamiensis Ôhira, 1967d: 31 **A**: JA (Amami-Ôshima, Okinawa-hontô)
amamiensis yambarus Kishii, 1985: 25 **A**: JA (Okinawa-hontô)
apicalis Fleutiaux, 1889: 144 **A**: NP **ORR**
atayal Kishii & Platia, 1991 **A**: TAI
babai Kishii, 1989d: 31, 35 **A**: TAI

bertrandi Platia & Schimmel, 2001: 156 **A**: NP "Himalaya" **ORR**
cete cete Candèze, 1860: 332 (*Melanotus*) **A**: JA SC "China"
amussitatus Candèze, 1873: 19
cete tokarensis Ôhira, 1997d: 345 **A**: JA (Nakano-shima)
cribricollis Faldermann, 1835b: 361 (*Ludius*) **A**: BEI FUJ GAN GUA GUI HEB HUB JA JIA JIX LIA NC NMO SC SCH SHA SHG SHN SHX YUN ZHE
restrictus Candèze, 1865: 47
cuneiformis Baudi di Selve, 1871: 53 **E**: FR GR IT SP **N**: MO **A**: CY TR
deyrollei Platia & Schimmel, 2002: 313 **A**: SD **ORR**
frequens Miwa, 1930b: 61 (*Speniscosomus*) **A**: TAI
haemorrhous Candèze, 1860: 333 **A**: UP **ORR**
punctorotundus Vats & Chauhan, 1991a: 194
horishanus Miwa, 1927b: 111 (*Speniscosomus*) **A**: TAI
housaii Kishii, 1991c: 6, 12 **A**: TAI
ignotus Platia & Schimmel, 2001: 159 **A**: NP XIZ
ingens Platia & Schimmel, 2001: 286 **A**: AHN JIX ZHE
jagemanni Platia & Schimmel, 2002: 340 **A**: SCH YUN
japonicus Ôhira, 1974b: 26 **A**: JA
kawakatsui Kishii, 1990b: 14 **A**: TAI
koikei Kishii & Ôhira, 1956: 76, 83 **A**: JA
loochooensis loochooensis Miwa, 1929c: 347 **A**: JA (Ryukyus)
loochooensis tanchamelis Ôhira, 1967d: 33 **A**: JA (Ryukyus)
melanotoides melanotoides Miwa, 1930b: 61 (*Spheniscosomus*) **A**: TAI
melanotoides saitorum W. Suzuki, 1984: 38 **A**: JA (Yonaguni-jima)
oberthueri Platia & Schimmel, 2001: 162 **A**: BT NP
omotoensis Ôhira, 1966a: 9 **A**: JA (Ishigaki-jima, Iriomote-jima)
pilosulus Miwa, 1927b: 110 (*Spheniscosomus*) **A**: TAI
procerus Platia & Schimmel, 2001: 163 **A**: NP
shinoharai Kishii & Platia, 1993b: 117 **A**: TAI
sulcatus Candèze, 1860: 338 **A**: BT SD
sulcicollis Mulsant & Guillebeau, 1855a: 19 (*Cratonychus*) **E**: CR FR GR IT SP **N**: MO
telum Candèze, 1895: 49 **A**: BT SD
vunun Kishii, 1991c: 6, 12 **A**: TAI
werneri Platia & Schimmel, 2001: 167 **A**: NP SD UP **ORR**

genus *Melanotus*, species incerae sedis
abietinus Gistel, 1857: 43 (Cratonychus) **E**: GE
piger Motschulsky, 1854c: 45 (*Cratonychus*) **A**: SC

genus *Metriaulacus* Schwarz, 1901c: 198 type species *Metriaulacus nigrolaterus* Schwarz, 1901
badiipennis Candèze, 1860: 334 (*Melanotus*) **A**: BT NP **ORR**
formosanus Miwa, 1927b: 113 **A**: TAI **ORR**
gobius Candèze, 1860: 329 (*Melanotus*) **A**: HP NP SD **ORR**
angustiformis Vats & Chauhan, 1991a: 191 (*Melanotus*)
permacer Candèze, 1900: 94 (*Spheniscosomus*)

genus *Priopus* Laporte, 1840a: 251 type species *Priopus frontalis* Laporte, 1840
Diploconus Candèze, 1860: 290 [HN] type species *Diploconus peregrinus* Candèze, 1860
Neodiploconus Hyslop, 1921: 658 [RN] type species *Diploconus peregrinus* Candèze, 1860
Ploconides Fleutiaux, 1933b: 208 type species *Diploconus spiloderes* Candèze, 1865
Pulchronotus Fleutiaux, 1933b: 206 type species *Diploconus ornatus* Candèze, 1891
Thaumastiellus Schwarz, 1902b: 335 type species *Thaumastiellus bioculatus* Schwarz, 1902
angulatus Candèze, 1860: 297 (*Diploconus*) **A**: FUJ GAN GUA GUI GUX HAI HEN HKG HUB HUN JIA JIX SCH TAI YUN **ORR**
basilaris Schenkling, 1927: 266 [RN] **A**: YUN **ORR**

basalis Fleutiaux, 1918b: 233 [HN] (*Diploconus*)
bibari Kishii, 2002a: 17 **A**: SC (Cheju)
boninsis Zwaluwenburg, 1957: 64 (*Neodiploconus*) **A**: JA (Ogasawara)
brevis Candèze, 1897: 61 (*Diploconus*) **A**: YUN **ORR**
castaneus Miwa, 1930c: 92 (*Neodiploconus*) **A**: TAI
desgodinsi Platia & Schimmel, 1995: 55 **A**: BT
diversus Fleutiaux, 1933b: 210 (*Neodiploconus*) **A**: YUN **ORR**
sublaevis Fleutiaux, 1918b: 233 (*Diploconus*)
dureli Platia & Schimmel, 1995: 58 **A**: AP BT SD
dirangensis Garg & Vasu, 1996: 260 (*Neodiploconus*)
ferrugineipennis ferrugineipennis Miwa, 1927b: 110 (*Neodiploconus*) **A**: JA SC (Cheju)
ferrugineipennis kuniyoshii Ôhira, 1967d: 37 (*Neodiploconus*) **A**: JA (Okinawa-hontô)
ishigakiensis Ôhira, 1967d: 37 (*Neodiploconus*) **A**: JA (Ishigaki-jima)
kubotai W. Suzuki, 1978: 400 (*Thaumastiellus*) **A**: FUJ HUN TAI
kume Kishii, 2004d: 169 **A**: JA (Kume-jima)
miniaticollis Hope, 1831: 25 (*Elater*) **A**: BT NP SD **ORR**
nitidus Candèze, 1860: 334 (*Melanotus*)
miniatus Hope, 1831: 25 (*Elater*) **A**: BT NP SD
mirabilis Fleutiaux, 1923a: 12 **A**: YUN **ORR**
minidiversus Platia & Schimmel, 1996b: 204 **A**: SD **ORR**
nigerrimus Fleutiaux, 1903a: 575 (*Diploconus*) **A**: FUJ GUA GUI HAI **ORR**
obayashii Miwa, 1934: 215 (*Neodiploconus*) **A**: TAI
ornatus Candèze, 1891b: 782 (*Diploconus*) **A**: YUN **ORR**
pulchellus Fleutiaux, 1923a: 13 (*Diploconus*) **A**: SCH YUN **ORR**
rufulus Candèze, 1891b: 781 (*Diploconus*) **A**: GUA HAI TAI **ORR**
russatus Fleutiaux, 1933b: 211 (*Neodiploconus*)
rugosus Fleutiaux, 1918b: 232 (*Diploconus*) **A**: SCH **ORR**
sanguinicollis Miwa, 1930b: 59 (*Neodiploconus*) **A**: TAI
satoi matobai Kishii, 1973: 16 (*Neodiploconus*) **A**: JA (Iriomote-jima)
satoi satoi Nakane & Kishii, 1956: A157 (*Neodiploconus*) **A**: JA (Miyako-jima)
superbus Fleutiaux, 1895: 687 (*Diploconus*) **A**: TAI **ORR**
elegans Szombathy, 1910b: 358 (*Diploconus*)
takanoi Miwa, 1930b: 59 (*Neodiploconus*)
tonkinensis Fleutiaux, 1933b: 213 (*Neodiploconus*) **A**: FUJ **ORR**
trimaculatus Hope, 1831: 25 (*Elater*) **A**: BT NP SD UP
neeraae Garg & Vasu, 1996: 261 (*Neodiploconus*)
vafer Erichson, 1841a: 108 (*Cratonychus*) **A**: SD FUJ HAI **ORR**
coracinus Candèze, 1860: 494 (*Diploconus*)
yagianus Kishii, 1993a: 29 **A**: TAI
yonaguni Kishii, 1982: 47 (*Neodiploconus*) **A**: JA

genus *Szombathya* Platia, 1986: 2 type species *Psellis formosana* Szombathy, 1910
formosana Szombathy, 1910c: 446 (*Psellis*) **A**: TAI

subfamily **Hypnoidinae** Schwarz, 1906
tribe Hypnoidini Schwarz, 1906

genus *Ascoliocerus* Méquignon, 1930a: 94 [RN] type species *Scoliocerus basalis* Motschulsky, 1859
Colioascerus Ôhira, 1968d: 11 type species *Hypolithus fluviatilis* Lewis, 1894
Scoliocerus Motschulsky, 1859b: 227 [= 1860f: 543] [HN] type species *Scoliocerus basalis* Motschulsky, 1859
barbatus J. R. Sahlberg, 1887a: 30 (*Cryptohypnus*) **A**: FE **NAR**
basalis Motschulsky, 1859b: 227 [= 1860f: 543] (*Scoliocerus*) **A**: CH ES FE WS
fluviatilis Lewis, 1894a: 184 (*Hypolithus*) **A**: JA
saxatilis expansicornis Lewis, 1894a: 184 (*Hypolithus*) **A**: JA
saxatilis saxatilis Lewis, 1894a: 183 (*Hypolithus*) **A**: FE JA

genus *Berninelsonius* Leseigneur, 1970: 20 type species *Elater hyperboreus* Gyllenhal, 1827
hyperboreus Gyllenhal, 1827: 350 (*Elater*) **E**: FI FR IT NR NT SV SZ **A**: ES FE JIL MG NC NMO SCH WS **NAR**
planatus Eschscholtz, 1829: 34 (*Hypolithus*)

genus *Homotechnes* Candèze, 1882: 71 type species *Homotechnes corymbitoides* Candèze, 1882
brunneofuscus Nakane, 1954: 14 (*Hypnoidus*) **A**: JA
corymbitoides Candèze, 1882: 71 **A**: HEN HUB JIX SCH
motschulskyi aizu Kishii, 1994d: 2 **A**: JA
motschulskyi akitensis Kishii, 1979b: 31 (*Hypolithus*) **A**: JA
motschulskyi alpicola Kishii, 1968: 5 (*Hypolithus*) **A**: JA (Asahi-dake Mts. in Niigata Pref.)
motschulskyi babai Kishii, 1957b: 73 (*Hypolithus*) **A**: JA (Niigata & Yamagata Pref.)
motschulskyi echigoanus Kishii & Ôhira, 1956: 74, 81 (*Hypolithus*) **A**: JA (Niigara & Nagano Pref.)
motschulskyi etchu Kishii, 1996b: 16 **A**: JA (Toyama & Ishikawa Pref.)
motschulskyi fleutiauxi Ôhira, 1968e: 30 (*Hypolithus*) **A**: JA (Nagano Pref.)
motschulskyi fukuinosaitoui Kishii, 1987: 1 (*Hypolithus*) **A**: JA (Mt. Kammuri)
motschulskyi hideoanus Kishii, 1990b: 2 (*Hypolithus*) **A**: JA
motschulskyi hirayuanus Kishii, 2001d: 55 **A**: JA (Gifu Prefecture)
motschulskyi hiranoi Ôhira, 1993: 250 (*Hypolithus*) **A**: JA (Hakone Mts. in Kanagawa Pref.)
motschulskyi hiromasai Ôhira, 1968e: 29 (*Hypolithus*) **A**: JA
motschulskyi ibukianus Ôhira, 1997b: 26 **A**: JA
motschulskyi ishizuchi Kishii, 1985: 6 (*Hypolithus*) **A**: JA
motschulskyi ikutanii Nakane, 1963c: 222 (*Hypolithus*) **A**: JA (Tate-yama Mt.)
motschulskyi iwatensis Ôhira, 1968d: 11 (*Hypolithus*) **A**: JA
motschulskyi izucola Kishii, 2001d: 52 **A**: JA (Mt. Manjiro-dake in Shizuoka Pref.)
motschulskyi izumii Kishii, 1985: 6(*Hypolithus*) **A**: JA
motschulskyi kai Kishii, 1996b: 14 **A**: JA
motschulskyi kaniei Ôhira, 1996c: 314 (*Hypolithus*) **A**: JA
motschulskyi karasawa Kishii, 1999c: 21 **A**: JA
motschulskyi kawasei Ôhira, 1995a: 96 (*Hypolithus*) **A**: JA (Mie Pref.)
motschulskyi kongoensis Kishii, 1969: 1 (*Hypolithus*) **A**: JA
motschulskyi kuratai Kishii, 1968: 5 (*Hypolithus*) **A**: JA
motschulskyi kurotai Ôhira, 1994d: 318 (*Hypolithus*) **A**: JA
motschulskyi monnaiensis Ôhira, 1968e: 29 (*Hypolithus*) **A**: JA
motschulskyi morii Kishii, 2004c: 164 **A**: JA
motschulskyi motschulskyi Fleutiaux, 1902a: 22 (*Corymbites*) **A**: JA
motschulskyi mutsuurai Nakane, 1963c: 222 (*Hypolithus*) **A**: JA (Haku-san Mts.)
motschulskyi narukawai Kishii, 1998b: 77 **A**: JA
motschulskyi oginoi Kishii, 2001d: 57 **A**: JA (Toyama Prefecture)
motschulskyi ohbayashii Ôhira, 1968b: 74 (*Hypolithus*) **A**: JA (Ohyama Mt. in Kanagawa Pref.)
motschulskyi ohkawai Kishii, 1986: 38 (*Hypolithus*) **A**: JA
motschulskyi ohmi Kishii, 1996b: 15 **A**: JA (Niigata Prefecture)
motschulskyi oikensis Kishii, 2002d: 139 **A**: JA (Mt. Oike-date in Shiga & Mie Pref.)
motschulskyi okudai Kishii, 1998b: 79 **A**: JA
motschulskyi ondai Kishii, 1990b: 3 (*Hypolithus*) **A**: JA
motschulskyi ontakecola Kishii, 1968: 8 (*Hypolithus*) **A**: JA
motschulskyi sawadai Ôhira, 1963a: 16 (*Hypolithus*) **A**: JA
motschulskyi shimotsuke Kishii, 1986: 39 (*Hypolithus*) **A**: JA
motschulskyi shimoyamai Ôhira, 1984: 54 (*Hypolithus*) **A**: JA
motschulskyi tachikawai Ôhira, 1968e: 30 (*Hypolithus*) **A**: JA
motschulskyi taichii Nakane, 1963c: 222 (*Hypolithus*) **A**: JA (Ohdaigahara Mts. in Mie Pref.)
motschulskyi tsurugi Ôhira, 1963a: 17 (*Hypolithus*) **A**: JA
motschulskyi yamizo Kishii, 1989f: 100 (*Hypolithus*) **A**: JA
motschulskyi yokozekii Kishii, 2001d: 53 **A**: JA (Mt. Shirouma-dake in Nagano Prefecture)

motschulskyi yoshizawai Ôhira, 1996c: 312 (*Hypolithus*) **A**: JA
nodai Ôhira, 1996a: 25 (*Hypolithus*) **A**: JA (Tsushima)
ogatai Kishii, 1983a: 30 (*Hypolithus*) **A**: JA (Yaku-shima)
yahikoanus Kishii, 1986: 41 (*Hypolithus*) **A**: JA

genus *Hypnoidus* Dillwyn, 1829: 32 type species *Elater riparius* Fabricius, 1792
Amphius Gistel, 1834: 12 type species *Elater riparius* Fabricius, 1792
Cryptohypnus Eschscholtz, 1830a: 17 type species *Elater riparius* Fabricius, 1792
affinis Dolin & Cate, 2002: 66 **A**: SCH
alpinus Dolin, 1999b: 83 **A**: XIZ
alticola Gurjeva, 1963a: 30 **A**: KI
anatolicus Platia & Gudenzi, 2000f: 67 **A**: TR
audryae Stibick, 1980: 253 **A**: UP
auricomus Reitter, 1910b: 175 (*Cryptohypnus*) **A**: KI XIZ
badachschanicus Dolin, 1998a: 8 **A**: TD
balassogloi Candèze, 1889b: 104 (*Cryptohypnus*) **A**: KI KZ
beckeri Stibick, 1980: 254 **A**: NP SD XIZ
beneschi Dolin, 1999b: 84 **A**: XIZ
bicolor Eschscholtz, 1829: 33 (*Hypolithus*) **A**: FE **NAR**
brevicollis Dolin & Cate, 2002: 65 **A**: SHA YUN
brevis Dolin & Cate, 2001: 122 **A**: NP
brezinai Dolin & Cate, 1998: 46 **A**: SCH
businskyi Dolin & Cate, 1998: 48 **A**: HUB
butti Stibick, 1980: 269 **A**: PA XIZ
cachemirensis Candèze, 1897: 52 (*Cryptohypnus*) **A**: HP KA PA
himalayensis Stibick, 1980: 261
canaliculatus Gebler, 1841b: 583 (*Cryptohypnus*) **A**: ES WS XIN XIZ
candezei Fleutiaux, 1905: 326 [RN] **A**: HP
indicus Candèze, 1900: 92 (*Cryptohypnus*) [HN]
carinatissimus Cherepanov, 1957: 224 **A**: KI KZ
conformis Dolin & Cate, 2003: 42 **A**: YUN
consobrinus Mulsant & Guillebeau, 1855a: 30 (*Cryptohypnus*) **E**: AU CZ FR IT SK SZ
valesiacus Stierlin, 1879a: 440 (*Cryptohypnus*)
costae Stibick, 1980: 268 **A**: NP
deuvei Dolin & Cate, 2001: 124 **A**: KA
dzhungaricus Dolin, 1998a: 6 **A**: KZ
fuchsi Stibick, 1979: 292 **A**: XIN
ghilarovi Dolin & Bessolitzina, 1992: 846 **A**: KZ
gibbus Gebler, 1847b: 421 (*Cryptohypnus*) **A**: ES WS
gruenwaldti Dolin, 1998a: 5 **A**: UZ
gurjevae Stibick, 1979: 295 **A**: KI
guttatus Dolin & Cate, 2002: 68 **A**: GAN
haplonotus Reitter, 1910b: 176 (*Cryptohypnus*) **A**: KI KZ
heinrichi Stibick, 1980: 270 **A**: NP SD XIZ
jeffreyi Dolin & Cate, 2002: 62 **A**: NP SCH YUN XIZ
koltzei Reitter, 1910b: 176 (*Cryptohypnus*) **A**: FE MG
kopetdaghensis Dolin, 1998a: 4 **A**: TM
kuldzhensis Dolin & Cate, 1998: 46 **A**: XIN
leseigneuri Stibick, 1980: 266 **A**: SD
mendeli Dolin & Cate, 2001: 125 **A**: HP
murzini Dolin & Cate, 2002: 64 **A**: YUN
narynensis Dolin & Cate, 1998: 46 **A**: KI
nepalensis Ôhira & Becker, 1973a: 74 (*Hypolithus*) **A**: NP XIZ
spillmani Stibick, 1980: 265
nitidicollis Koenig, 1889b: 533 (*Cryptohypnus*) **A**: XIN XIZ

nitidus Fleutiaux, 1902a: 19 (*Cryptohypnus*) **A**: JA
obovatus Dolin & Cate, 2003: 40 **A**: SHA
olivaceus Eschscholtz, 1829: 33 (*Hypolithus*) [DA] **A**: FE (Kamtschatka)
opacicollis Reitter, 1910b: 174 (*Cryptohypnus*) **A**: UZ
parallelocollis Dolin & Cate, 2002: 66 **A**: XIZ
patoni Dolin, 1999b: 81 **A**: XIZ
persimilis Dolin & Cate, 2001: 127 **A**: KA
pictus Bessolitzina & Dolin, 1992: 847 **A**: KZ WS
problematicus Dolin & Cate, 2003: 40 **A**: UP
riparius Fabricius, 1792b: 232 (*Elater*) **E**: AN AU BE BU BY CT CZ DE EN FA FI FR GB GE IC IR IT LA LT NL NR NT PL RO SK SL SP ST SV SZ UK **A**: ES FE WS XIN
littoreus Herbst, 1806: 86 (*Elater*)
meyeri Stierlin, 1862: 36 (*Cryptorhypnus*)
politus Fabricius, 1792b: 234 (*Elater*)
tournieri Pic, 1910m: 50 (*Cryptohypnus*)
rivularius ainu Ôhira & Yasuda, 1994: 6 **A**: JA
rivularius rebunensis Ôhira & Yasuda, 1994: 7 **A**: JA (Rebun-to Is.)
rivularius rivalis Lewis, 1894a: 184 (*Cryptohypnus*) **A**: FE JA
rivularius rivularius Gyllenhal, 1808a: 403 (*Elater*) **E**: AB AU BU BY CZ EN FI FR GE GG IT LA NR NT PL RO SK SL ST SV SZ UK **A**: CH ES FE MG WS
alpestris Zetterstedt, 1828: 250 (*Elater*)
frigidus Kiesenwetter, 1858a: 361 (*Cryptohypnus*)
granulosus Buysson, 1900a: 255
riparius Panzer, 1796b: 12 (*Elater*) [HN]
rufescens Gebler, 1847b: 423 (*Cryptohypnus*) **A**: ES WS XIN
sachalinensis Dolin, 1987b: 348 **A**: FE
sajanus Bessolitzina & Dolin, 1991: 850 **A**: ES
semiaeneus Pic, 1916a: 1 **A**:"Siberia"
shanskyi Stibick, 1979: 302 **A**: FE
stibicki Dolin & Bessolitzina, 1992: 848 **A**: KZ WS
tibetanus Dolin & Cate, 1998: 49 **A**: XIZ
tilloae Stibick, 1980: 264 **A**: NP or UP "Rapti River"
topali Dolin & Cate, 2001: 128 **A**: KA
tscherepanovi Stibick, 1979: 306 **A**: WS (Tuva)
tuvensis Dolin & Bessolitzina, 1992: 850 **A**: ES
vonhayeka Stibick, 1980: 263 **A**: XIZ

genus *Hypolithus* Eschscholtz, 1829: 33 type species *Hypolithus littoralis* Eschscholtz, 1829
Yezodima Miwa, 1928b: 40 type species *Yezodima convexum* Miwa, 1928
littoralis convexus Miwa, 1928b: 39 (*Yezodima*) **A**: FE JA
littoralis littoralis Eschscholtz, 1829: 33 **A**: FE **NAR**
littoralis michinoku Kishii, 2004c: 161 **A**: JA

genus *Ligmargus* Stibick, 1976: 210 type species *Cryptohypnus funebris* Candèze, 1860
aeneoniger aeneoniger Miwa, 1928b: 39 (*Yezodima*) **A**: FE (Kunashir) JA
kurilensis Gurjeva, 1972a: 301 (*Hypolithus*)
aeneoniger emishi Kishii, 2004c: 162 (*Ligmargus*) **A**: JA
depressus Gebler, 1847b: 420 (*Cryptohypnus*) **A**: CH ES FE HEI MG WS
kalabi Mertlik, 2001: 157 **A**: PA
margae Stibick, 1980: 248 **A**: PA UP

subfamily **Pleonominae** Semenov & Pjatakova, 1936
tribe Pleonomini Semenov & Pjatakova, 1936
genus *Clon* Semenov, 1900a: 594 type species *Clon cerambycinus* Semenov, 1900
cerambycinus Semenov, 1900a: 596 **A**: UZ

umbratus Semenov, 1900a: 597

genus *Gurjevelater* Platia & Gudenzi, 2000f: 80 type species *Gurjevelater iranicus* Platia & Gudenzi, 2000
iranicus Platia & Gudenzi, 2000f: 81 **A**: IN
vartianorum Platia & Gudenzi, 2000f: 81 **A**: AF PA

genus *Lomopneus* Gurjeva, 1976b: 78 type species *Lomopneus gracilis* Gurjeva, 1976
gracilis Gurjeva, 1976b: 79 **A**: AF
kasyi Platia & Gudenzi, 2000f: 78 **A**: AF
mundus Gurjeva, 1976b: 79 **A**: IN

genus *Pleonomus* Ménétriés, 1849: 48 type species *Pleonomus tereticollis* Ménétriés, 1849
Ictus Candèze, 1863: 240 type species *Ictus sinensis* Candèze, 1863
angusticollis Reitter, 1894g: 242 **A**: AF KI TM
tschitscherini Semenov, 1895: 137
canaliculatus Faldermann, 1835b: 362 (*Cratonychus*) **A**: ANH FUJ GAN GUI GUX HEB HEI HEN HUB HUN JIA JIL LIA MG NMO SHA SHN SHX ZHE
acutidens Fairmaire, 1878b: 110 (*Athous*)
sinensis Candèze, 1863: 240 (*Ictis*)
spinicollis Faldermann, 1835b: 414 (*Serropalpus*)
laticornis Reitter, 1900c: 88 **A**: AF KI TD TM UZ
makiharai Ôhira, 1970b: 104 **A**: JA (Iriomote-jima)
tereticollis Ménétriés, 1849: 49 **A**: KI KZ
rotundicollis Kraatz, 1881b: 325

subfamily **Denticollinae** Stein & J. Weise, 1877 (1856)
tribe Denticollini Stein & J. Weise, 1877 (1856)
subtribe Athouina Candèze, 1859

genus *Arctapila* Candèze, 1891a: 119 type species *Cryptohypnus bruckii* Candèze, 1865
bruckii Candèze, 1865: 38 (*Cryptohypnus*) **E**: FR

genus *Athous* Eschscholtz, 1829: 33 type species *Elater vittatus* Fabricius, 1792
subgenus *Alcimathous* Reitter, 1905a: 32 type species *Athous sacheri* Kiesenwetter, 1858
mokrzeckii Łomnicki, 1923: 53 **E**: BU
sacheri Kiesenwetter, 1858a: 320 **E**: RO
subgenus *Athous* Eschscholtz, 1829: 33 type species *Elater vittatus* Fabricius, 1792
Anathrotus Dillwyn, 1829: 33 type species *Elater vittatus* Fabricius, 1792
Gripus Gistel, 1834: 11 type species *Elater vittatus* Fabricius, 1792
Grypocrarus C. G. Thomson, 1859: 104 type species *Elater haemorrhoidalis* Fabricius, 1801
cingulatus L. Miller, 1881: 219 **E**: CR (Dalmatia) YU (Crna Gora)
densatus Reitter, 1905a: 43 **E**: AL BH CR YU
haemorrhoidalis Fabricius, 1801b: 235 (*Elater*) **E**: AL AR AU AZi BE BU BY CR CT CZ DE EN FI FR GB GE GR HU IR IT LA LS LT LU MC MD NL NR PL PT RO SK SL SP ST SV SZ TR UK YU **A**: CY KZ TR
asiaticus Buysson, 1905d: 388
cariniscapus Buysson, 1916: 108
croissendeaui Buysson, 1891a: 244
deubeli Buysson, 1916: 108
faeculentus Buysson, 1888b: 1
interpositus Rey, 1891b: 68
jouvenoti Buysson, 1926a: 48
jureceki Jagemann, 1939: 70
lacordairei Méquignon, 1929: 276
leucophaeus Lacordaire, 1835: 641
multiimpressus Pic, 1910m: 50
nigropiceus Ivanov, 1902: 35
ruficaudis Gyllenhal, 1808a: 409 (*Elater*)

sicardi Buysson, 1891a: 244
ineptus Candèze, 1860: 142 **E**: FR IT
crenatostriatus Reiche, 1869: 383
depressifrons Desbrochers des Loges, 1869: 45 [= 1870b: 113]
limoniiformis Candèze, 1865: 50 **E**: IT
nigricornis C. N. F. Brisout de Barneville, 1866: 377 **E**: PT SP
rubrotestaceus Desbrochers des Loges, 1871: 370 **E**: PT SP
vittatoides Reitter, 1905a: 46 **E**: AL GR
vittatus Fabricius, 1792b: 224 (*Elater*) **E**: AL AR AU BE BH BU BY CR CT CZ DE EN FR GB GE GR HU IT LA LS LT LU MC MD NL NR PL PT RO SK SL SP ST SV SZ UK YU **A**: TR
analis Mulsant & Guillebeau, 1856: 89 [HN]
angularis Stephens, 1830: 277 (*Anathrotus*)
balteatus Geoffroy, 1785: 38 (*Elater*) [HN]
brevitinctus Buysson, 1924a: 114
conicicollis Desbrochers des Loges, 1870b: 109
decipiens Rey, 1891b: 68
denigratus Buysson, 1895: cclxxxiii
dimidiatus Drapiez, 1820: 223 (*Elater*)
exaequatus Buysson, 1924a: 113
filicti Buysson, 1891a: 243
impallens Buysson, 1887c: xcvii
inopinatus Buysson, 1891a: 242
lanatus Rey, 1891b: 68
lineatus Villers, 1789: 304 (*Elater*) [HN]
murinus Reiche, 1869: 380
niger Fiori, 1899b: 162
ocskayi Kiesenwetter, 1858a: 317
puncticollis Kiesenwetter, 1858a: 316 [RN]
punctatithorax Buysson, 1916: 109
punctithorax Reitter, 1905a: 44
ravouxi Pic, 1909d: 105
semipallens Mulsant & Guillebeau, 1855a: 28
stephensi Buysson, 1891a: 243
tenuevittatus Reitter, 1905a: 47
testaceicolor Pic, 1906g: 33
subgenus *Euplathous* Reitter, 1905a: 34 type species *Elater canus* Dufour, 1843
canus Dufour, 1843a: 40 (*Elater*) **E**: FR
dejeanii Laporte, 1840a: 244 (*Elater*) **E**: FR
cervinus Laporte, 1840: 244 (*Elater*)
dejeani Mulsant & Guillebeau, 1856: 64 [HN]
fuscicornis Mulsant & Guillebeau, 1856: 67
mariei Méquignon, 1928: 133
obtusifrons Desbrochers des Loges, 1871: 371
subrubiginosus Buysson, 1906b: 439
lugubris Buysson, 1906b: 439
vestitus Mulsant & Guillebeau, 1856: 79
frigidus Mulsant & Guillebeau, 1855a: 21 **E**: FR IT
castanescens Mulsant & Guillebeau, 1855a: 27
capucinus Buysson, 1906b: 435
cognatus Buysson, 1906b: 435
leptotropus Buysson, 1925: 160
luctuosus Buysson, 1906b: 432
oblongicollis Desbrochers des Loges, 1871: 370
peragalloi Reiche, 1864: 247
rubiginosus Buysson, 1906b: 432
saturiator Buysson, 1906b: 435
iablokoffi Leseigneur, 1972: 236 **E**: FR
mandibularis Dufour, 1843a: 41 (*Elater*) **E**: FR SP
dejeani Dufour, 1843a: 41 (*Elater*) [HN]

titanus Mulsant & Guillebeau, 1856: 60
proximus Hampe, 1864: 190 **E**: BU GR RO
haemus Buysson, 1890b: ccvi
spectabilis Hampe, 1864: 191
villardi Carret, 1904: 170 **E**: IT
subgenus *Haplathous* Reitter, 1905a: 33 type species *Athous circumductus* Ménétriés, 1832
Exanathrotus Méquignon, 1930b: 320 type species *Elater subfuscus* O. F. Müller, 1764
abdurachmanovi Dolin & Penev, 2004: 90 **E**: ST
alpestris Orlov, 1994: 684 **E**: ST
angulifrons Reitter, 1905a: 66 **E**: AU CR IT SL
stellatus Reitter, 1905a: 67
austriacus Desbrochers des Loges, 1873a: 360 **E**: AU BH BU CR CZ HU IT MC RO SL SZ UK
circumscriptus Candèze, 1860: 466
flaveolus Buysson, 1925: 147
flavidus Buysson, 1925: 147
flecki Reitter, 1906e: 240
humeropictus Buysson, 1925: 147
lemovicium Buysson, 1925: 148
neomelanoderes Buysson, 1925: 147
subrubricus Reitter, 1905a: 72
barthei Leseigneur, 1958: 88 **E**: FR
belloi Guglielmi & Platia, 1985: 173 **E**: GR
brevicornis Desbrochers des Loges, 1871: 369 **E**: FR (Corse) IT (Sardegna)
cachecticus Candèze, 1860: 455 **E**: IT
carpathicus Reitter, 1905a: 71 **E**: RO
cavifrons L. Redtenbacher, 1858: 493 **E**: AU CR IT SL
milleri Reitter, 1905a: 65
circassiensis baksanicus Dolin & Penev, 2004: 94 **E**: ST
circassiensis circassiensis Reitter, 1905a: 73 **E**: AR GG ST
circassiensis diplogrammus Orlov, 1994: 686 **E**: GG
circassiensis frontalis Dolin & Penev, 2004: 2004: 94 **E**: ST
circumductus Ménétriés, 1832: 158 (*Elater*) **E**: AB AR GG ST
corcyreus Reitter, 1905a: 62 **E**: GR
corsicus Reiche, 1861b: 207 **E**: FR (Corse) IT (Sardegna)
crassicornis Candèze, 1860: 456 **E**: FR IT
brunneicollis Boubers & Leseigneur, 1956: 38
laticornis Boubers & Leseigneur, 1956: 36
csikii Platia, 2001b: 522 **E**: MC
curtus Dolin & Penev, 2004: 91 **E**: ST
daghestanicus Reitter, 1890d: 246 **E**: AB ST
delmastroi Platia & Gudenzi, 1998b: 268 **E**: IT
dilaticornis Reitter, 1905a: 68 **E**: ST
pliginskyi Reitter, 1910b: 177
emaciatus Candèze, 1860: 454 **E**: FR IT SZ
flavescens Mulsant & Guillebeau, 1855a: 25
pallidipennis Desbrochers des Loges, 1870b: 110 [HN]
pallidior Buysson, 1925: 145
euxinus Buysson, 1912g: 142 **E**: RO
filicornis Dufour, 1851: 328 (*Elater*) **E**: FR
flavipennis Candèze, 1860: 453 **E**: FR IT
delphinas Reiche, 1869: 385
lavergnei Reiche, 1869: 384
hilfi Reitter, 1912d: 317 **E**: BU
incognitus Platia, 1988b: 159 **E**: IT
iristonicus subalpinus Dolin & Penev, 2004: 95 **E**: ST
iristonicus iristonicus Dolin, 1971: 649 **E**: ST

kerkyranus Reitter, 1905a: 63 **E**: GR
kobachidzei Dolin & Chantladze, 1982: 161 **E**: AB GG ST
korabicus Csiki, 1940a: 250 **E**: AL MC
laevistriatus Dufour, 1851: 327 (*Elater*) **E**: FR SP
fortimargo Buysson, 1905d: 400
laevigatus Desbrochers des Loges, 1871: 371
sylvaticus Mulsant & Guillebeau, 1855a: 23
latior Orlov, 1994: 683 **E**: ST
lepontinus Schwarz, 1900: 106 **E**: IT
villigeroides Reitter, 1905a: 55
leprieuri Desbrochers des Loges, 1870a: 72 **N**: AG
lgockii Dolin, 1983: 29 **E**: ST
minusculus Orlov, 1994: 682
longicornis Candèze, 1865: 50 **E**: FR
magnanii Guglielmi & Platia, 1985: 176 **E**: GR
marginicollis Reitter, 1890d: 245 **E**: GG GR IT **A**: TR
melanoderes Mulsant & Guillebeau, 1855a: 22 **E**: FR IT
agnatus Reiche, 1869: 387
fallax Desbrochers des Loges, 1869: 33 [= 1870b: 115]
pallens Mulsant & Guillebeau, 1855a: 24
quadricollis Desbrochers des Loges, 1870b: 134
mollis Reitter, 1889s: 372 **E**: RO
monilicornis Schwarz, 1897a: 134 **E**: BU GR MC
osellai Guglielmi & Platia, 1985: 177 **E**: GR
pacei Guglielmi & Platia, 1985: 178 **E**: GR
panellai Platia & Schimmel, 1991b: 134 **E**: GR
patoni Dolin & Penev, 2004: 92 **E**: ST
pfefferi Roubal, 1932d: 282 **E**: BU
protoracicus Platia & Schimmel, 1991b: 135 **E**: GR
putschkovi Dolin & Penev, 2004: 93 **E**: ST
pyrenaeus Candèze, 1865: 50 **E**: FR SP
reflexicollis Dolin & Penev, 2004: 93 **E**: ST
reynosae C. N. F. Brisout de Barneville, 1866: 376 **E**: PT SP
ruteri Chassain, 1985a: 66 **E**: FR
settei Guglielmi & Platia, 1985: 175 **E**: GR
sosybius Reitter, 1905a: 73 **E**: ST
spalatrensis Reitter, 1894f: 102 **E**: BH CR
subfuscus O. F. Müller, 1764: 17 (*Elater*) **E**: AB AL AR AU BE BH BU BY CR CT CZ DE EN FA FI FR GB GE GG GR HU IR IT LA LS LT LU MC MD NL NR NT PL PT RO SK SL SP ST SV SZ TR UK YU **A**: IN TR
analis Fabricius, 1792b: 230 (*Elater*)
derisorius Buysson, 1925: 153
dilutus Illiger, 1807a: 4 (*Elater*)
jugoslavicus Dolin, 1983: 30
kafkanus Reitter, 1905a: 52
montanus Candèze, 1860: 463
polonicus Reitter, 1905a: 52
tricolor Gmelin, 1790: 1917 (*Elater*)
talamellii Platia & Gudenzi, 1998c: 630 **E**: GR
tauricola Reitter, 1905a: 68 **E**: UK
tauricus Candèze, 1860: 470 **E**: UK
transsylvanicus Platia, 2001b: 523 **E**: RO
tschukini Reitter, 1910b: 177 **E**: ST
turcicus Reitter, 1905a: 75 **E**: AL BU TR **A**: TR
utschderensis Reitter, 1890d: 246 **E**: AB GG ST
kovali Orlov, 1994: 685

villiger Mulsant & Guillebeau, 1855a: 20 **E**: FR
villosulus Desbrochers des Loges, 1875a: 41 **E**: FR (Corse) IT (Sardegna)
zebei Bach, 1852: 32 **E**: AU CZ FR GE GR IT LS PL RO SK SZ UK
balmei Pic, 1910u: 89
biformis L. Redtenbacher, 1858: 493
leucophaeus Kiesenwetter, 1858a: 321 [HN]
subgenus *Neonomopleus* Schenkling, 1927: 308 [RN] type species *Pleonomus strictus* Candèze, 1863 (= *Athous elongatus* C. N. F. Brisout de Barneville, 1866)
Nomopleus Reitter, 1891a: 210 [HN] type species *Pleonomus strictus* Candèze, 1863 (= *Athous elongatus* C. N. F. Brisout de Barneville, 1866)
cobosi Platia, 2003b: 34 **E**: SP
discors Reitter, 1904f: 237 **E**: SP
elongatus C. N. F. Brisout de Barneville, 1866: 380 **E**: PT SP
strictus Candèze, 1863: 498 (*Pleonomus*) [HN]
koschwitzi Platia, 2003b: 34 **E**: SP
longissimus Reitter, 1904f: 238 **E**: PT SP
martinae Platia, 2003b: 34 **E**: SP
martinezi Reitter, 1904f: 238 **E**: SP
masoae Platia, 2003b: 36 **E**: SP
procerus Illiger, 1807a: 1 (*Elater*) **E**: PT
chloroticus Candèze, 1865: 51
recticollis Graells, 1858: 83 **E**: SP
escorialensis Mulsant & Guillebeau, 1856: 71
kiesenwetteri Schaufuss, 1863: 199 (*Campylus*)
schimmeli Platia, 2003b: 36 **E**: SP
schurmanni Platia & Serrano, 2002: 313 **E**: PT
tenuis C. N. F. Brisout de Barneville, 1866: 379 **E**: SP
subgenus *Orthathous* Reitter, 1905a: 34 type species *Athous jejunus* Kiesenwetter, 1858
Grypathous Reitter, 1905a: 34 type species *Elater longicollis* A. G. Olivier, 1790 (= *Elater bicolor* Goeze, 1777)
acutangulus Fairmaire, 1866b: 261 **E**: GR **A**: TR
thessalonicus Reiche, 1869: 386
albanicus Csiki, 1940a: 249 **E**: AL
anatolicus Platia, 1989: 82 **A**: TR
angustulus Candèze, 1860: 443 **E**: PT SP
apfelbecki Reitter, 1905a: 86 **E**: BU HU RO YU
astrabadensis Faust, 1877b: 309 **A**: IN
audisioi Guglielmi & Platia, 1985: 180 **A**: TR
azoricus Platia & Gudenzi, 2002b: 81 **N**: AZ
balcanicus Reitter, 1905a: 86 **E**: BU
barriesi Platia & Gudenzi, 1996b: 236 **E**: BU TR
benedikti Platia, 2003c: 263 **A**: TR
bicolor Goeze, 1777: 568 (*Elater*) **E**: AU BE BU CR CZ FR GB GE HU IT LU NL PL RO SK SL SP ST SZ UK YU
crassicollis Lacordaire, 1835: 642
gracilis Buysson, 1905d: 374
lateralis Marsham, 1802: 380 (*Elater*) [HN]
longicollis A. G. Olivier, 1790: no 31: 38 (*Elater*)
marginellus Herbst, 1806: 75 (*Elater*)
melanophtalmus Geoffroy, 1785: 39 (*Elater*)
suturalis Panzer, 1795: 239 (*Elater*)
unicolor Marsham, 1802: 379 (*Elater*)
binaghii Platia, 1984a: 1 **E**: IT
bolivari Reitter, 1904f: 238 **E**: SP
bolognai Guglielmi & Platia, 1985: 181 **E**: GR
brevipennis Schwarz, 1897a: 135 **E**: GG ST
mesmini Buysson, 1900b: 231

brisouti Sánchez-Ruiz, 1996: 92 [RN] **E**: FR SP
lateralis C. N. F. Brisout de Barneville, 1866: 378 [HN]
bulgaricus Platia, 2001b: 523 **E**: BU
campyloides Newman, 1833: 509 **E**: BE FR GB GEi SP **NAR**i
affinis Cooper, 1865: 61
bonvouloiri Reiche, 1869: 386
cylindricollis Mulsant & Guillebeau, 1856: 82
difformis Lacordaire, 1835: 640
fallaciosus Buysson, 1924a: 143
interstitialis Desbrochers des Loges, 1873a: 366
punctipennis Reitter, 1905a: 83
cantabricus Schaufuss, 1862b: 48 **E**: SP
carpathophilus Reitter, 1905a: 106 **E**: MD RO UK
cavazzutii Platia, 2003c: 264 **A**: TR
caviformis Reitter, 1905a: 90 **E**: BH CR HU
cavulus Reitter, 1905a: 91 **E**: UK
cavus Germar, 1817: 218 (*Elater*) **E**: BH BU CR HU MC RO
cervicolor Heyden, 1880b: 289 **E**: PT SP
chamboveti Mulsant & Godart, 1868: 409 **E**: FR
croaticus Platia & Gudenzi, 2002b: 81 **E**: CR
curtulus Desbrochers des Loges, 1873a: 360 **E**: SP
daccordii Guglielmi & Platia, 1985: 182 **A**: TR
dasycerus Buysson, 1890d: clvii **E**: BU TR **A**: TR
debilis Reiche, 1869: 382 **E**: SP
difficilis Dufour, 1843a: 41 (*Elater*) **E**: FR SP
divergens Reitter, 1905a: 108
distinctithorax Desbrochers des Loges, 1873a: 361 **E**: SP
dorgaliensis Buysson, 1912f: 131 **E**: FR (Corse) IT (Sardegna)
durazzoi Platia, 1985a: 1 **E**: AL
dusaneki Platia, 2003c: 265 **A**: TR
eckerleini Platia & Gudenzi, 2000a: 150 **E**: FR
edirnensis Platia & Gudenzi, 2000f: 71 **E**: TR
epirus Stierlin, 1875: 475 **E**: AL GR **A**: TR
calosirus Reitter, 1895c: 201
freudei Platia, 1989: 79 **E**: GR MC **A**: TR
frontalis Platia & Schimmel, 1991b: 136 **A**: TR
gagliardii Platia, 1988c: 305 **E**: CR
galiberti Buysson, 1918a: 27 **E**: FR
gallicus Platia & Gudenzi, 2000f: 70 **E**: FR
ganglbaueri Schwarz, 1897a: 133 **E**: CR IT SL
gerezianus Reitter, 1905a: 99 **E**: SP
gianassoi Platia & Gudenzi, 1996b: 236 **A**: TR
gobanzi Reitter, 1905a: 105 **E**: BH CR
goriciensis Reitter, 1905a: 110 **E**: IT SL
gottwaldi Lohse, 1978: 26 **E**: SK
graecus Platia, 1989: 80 **E**: GR **A**: TR
gudenzii Gulglielmi & Platia, 1985: 183 **A**: TR
harmodius Reitter, 1905a: 85 **E**: GR
herbigradus Mulsant & Guillebeau, 1855a: 26 **E**: FR
basalis Candèze, 1860: 472
gabilloti Pic, 1910m: 50
hetzeli Platia, 2004c: 99 **A**: TR
holtzi Reitter, 1905a: 85 **E**: GR
insularis Desbrochers des Loges, 1869: 33 **E**: FR (Corse)
iranicus Platia, 2004c: 100 **A**: IN

jejunus Kiesenwetter, 1858a: 326 **E**: CR MD PL UK
judicariensis Schwarz, 1900: 107 **E**: IT
kaszabi Dolin, 1986: 145 **E**: BH CR HU RO YU
kovancii Platia, 2003c: 265 **A**: TR
kruegeri Reitter, 1905a: 88 **E**: SP
lambeleti Leseigneur, 2004b: 237 **E**: FR SP
lassallei Platia & Gudenzi, 1996b: 237 **A**: TR
leonhardi Reittter, 1905: 81 **E**: BH
lomnickii Reitter, 1905a: 92 **E**: MD PL RO UK
luigionii Platia, 1988c: 303 **E**: IT
malmusii Platia & Gudenzi, 2000f: 72 **A**: TR
margheritae Guglielmi & Platia, 1985: 184 **A**: TR
melonii Platia, 1984b: 120 **E**: IT
mendesi Platia & Serrano, 2002: 316 **E**: PT
mertliki Platia & Gudenzi, 2002b: 83 **A**: TR
meuseli Reitter, 1905a: 91 **E**: BH YU (Serbia)
mingrelicus Reitter, 1890d: 247 **E**: AB AR GG ST
nadari Buysson, 1904c: 60 **E**: FR
naseri J. Müller, 1916: 89 **E**: SL
novaki Penecke, 1907: 13 **E**: CR
obsoletus Illiger, 1807a: 2 (*Elater*) **E**: PT
olbiensis Mulsant & Guillebeau, 1856: 93 **E**: FR IT
    hispidus Candèze, 1860: 474
olcesei Buysson, 1905c: 211 **N**: MO
paflagonensis Platia & Gudenzi, 1998c: 632 **A**: TR
pallidus Platia & Gudenzi, 2002b: 84 **E**: BH
pedemontanus Platia, 1988b: 161 **E**: IT
penevi Platia, 2001b: 526 **E**: BU
picipennis Reitter, 1905a: 106 **E**: RO
plagipennis Reitter, 1905a: 112 **E**: CR IT RO SL
pomboi Platia & Borges, 2002: 92 **E**: AZ
propinquus Buysson, 1890b: ccv **E**: BU TR
ruffoi Guglielmi & Platia, 1985: 186 **A**: TR
sabatinellii Guglielmi & Platia, 1985: 187 **A**: TR
samai Guglielmi & Platia, 1985: 188 **E**: BU GR
sameki Platia, 2003c: 266 **A**: TR
senaci Buysson, 1890b: cciv **E**: TR **A**: TR
    ottomanus Reitter, 1905a: 88
serbicus Reitter, 1905a: 105 **E**: BH BU YU
silicensis Laibner, 1975: 412 **E**: HU SK SL
singularis Reitter, 1905a: 89 **E**: CR
sinuatocollis Desbrochers des Loges, 1869: 33 [= 1870b: 112] **E**: FR
    ecoffeti Reiche, 1863: 80
snizeki Platia, 2004c: 100 **A**: TR
stoimenovae Platia, 2001b: 526 **E**: BU
subtruncatus Mulsant & Guillebeau, 1856: 85 **E**: FR
    grandini Desbrochers des Loges, 1870b: 111
    strictus Reiche, 1863: 81
    virgatus Reiche, 1863: 82
subvirgatus Daniel, 1904: 79 **E**: IT
svihlai Platia & Gudenzi, 1998c: 632 **A**: TR
szombathyi Schenkling, 1927: 340 [RN] **E**: CR HU
    discrepans Szombáthy, 1910d: 588 [HN]
tekkirazicus Platia, 2003c: 266 **A**: TR
tomentosus Mulsant & Guillebeau, 1855a: 24 **E**: FR

tribertii Guglielmi & Platia, 1985: 189 **A**: TR
vicinus Desbrochers des Loges, 1873a: 361 **E**: FR PT SP
vomeroi Platia, 1988c: 306 **E**: CR
warchalowskii Platia & Tarnawski, 1998: 270 **A**: TR
wewalkai Platia, 1989: 81 **A**: TR
zanettii Guglielmi & Platia, 1985: 190 **A**: TR
zappiorum Platia, 1985b: 93 **E**: IT
subgenus *Pleurathous* Reitter, 1905a: 33 type species *Athous godarti* Mulsant & Guillebeau, 1856
circassicus Reitter, 1888d: 180 **E**: AR GG ST
atripes Reitter, 1905a: 50
godarti Mulsant & Guillebeau, 1856: 75 **E**: FR SP
hispanicus Pic, 1910u: 89
varensis Reitter, 1905a: 50
xambeui Pic, 1910u: 89
rosinae Reitter, 1899b: 203 **E**: AB ST
uncicollis Perris, 1864: 285 **E**: PT SP
gredosanus Reitter, 1904f: 237
uhagoni Buysson, 1904a: 7

genus *Athous*, species incertae sedis
alnicola Gistel, 1857: 12 **E**: GE (Bayern)
corticeus Fairmaire, 1870: 380 **N**: MO
dimidiatus Gistel, 1857: 51 **E**: GE (Bayern)
fausti Reitter, 1890d: 244 **E**: "Caucasus"
francoisi Fleutiaux, 1928a: 116 **A**: HAI **ORR**
humeralis Fischer von Waldheim, 1824: 203 (*Elater*) **E**: ST
kubani Schimmel, 1998c: 149 **A**: YUN
opacus Fleutiaux, 1934a: 183 **A**: SCH
parallelopipedus Brullé, 1832: 139 **E**: GR
roralis Gistel, 1857: 40
rufithorax Miwa, 1930c: 92 **A**: TAI
takasago Kishii, 1991a: 6, 13 (*Sphenomerus*)
shirozui Ôhira, 1966c: 218 **A**: TAI
strictus Fischer von Waldheim, 1824: 204 (*Elater*) **E**: ST
temperatus Miwa, 1930b: 66 **A**: TAI
weigeli Schimmel, 1998c: 151 **A**: NP

genus *Athousius* Reitter, 1905a: 112 type species *Athous holdereri* Reitter, 1900
holdereri Reitter, 1900e: 159 (*Athous*) **A**: GAN XIN XIZ
humeralis Miwa, 1927a: 21 (*Megapenthes*) **A**: JA "Korea"
rufipennis Kishii, 1958: 29 (*Athous*)
isamutanakai Kishii, 2004a: 61 **A**: JA
kitanoi Ôhira, 1972b: 19 **A**: JA
obakoensis Kishii, 1976a: 9 **A**: JA
yukikoa Ôhira, 1976: 47 (*Grypathous*) **A**: JA
wudanganus Kishii & Jiang, 1996: 135 **A**: HUB

genus *Chinathous* Kishii & Jiang, 1996: 136 type species *Chinathous robustus* Kishii & Jiang, 1996
robustus Kishii & Jiang, 1996: 137 **A**: GUX JIX

genus *Cidnopus* C. G. Thomson, 1859: 106 type species *Elater nigripes* Gyllenhal, 1808 (= *Elater pilosus* Leske, 1785)
aeruginosus A. G. Olivier, 1790: no 31: 33 (*Elater*) **E**: AU BE BU BY CT CZ DE EN FI FR GB GE HU IR IT LA LS LT LU MC NL NR NT PL RO SK SL SP SV SZ UK YU **A**: "Central Asia" "Siberia"
cylindricus P. Rossi, 1792: 58 (*Elater*)

obsoletus Marsham, 1802: 387 (*Elater*)
crassipes Schwarz, 1900: 105 (*Limonius*) **A**: TR
hubeiensis Kishii & Jiang, 1996: 131 **A**: HUB
koltzei Reitter, 1895d: 256 (*Limonius*) **A**: MG ES FE NMO WS
macedonicus Cate & Platia, 1989: 185 **E**: GR MC
marginellus brusteli Leseigneur, 2004a: 231 **E**: FR
marginellus marginellus Perris, 1864: 284 (*Limonius*) **E**: FR PT SP
marginipennis Lewis, 1894a: 195 (*Limonius*) **A**: JA
obienesis Cherepanov, 1966a: 23 (*Limonius*) **A**: WS
parallelus Motschulsky, 1860b: 111 (*Limonius*) **A**: ES FE MG WS
pilosus Leske, 1785: 11 (*Elater*) **E**: AL AU BE BU BY CR CT CZ DE EN FR GE GR HU IT LA LS LT LU NL NR PT RO SK SL SP ST SV SZ UK YU **A**: CH TR
cyanescens Buysson, 1902b: 274 (*Limonius*)
ferrugineus Buysson, 1902b: 274 (*Limonius*)
nigripes Gyllenhal, 1808a: 395 (*Elater*) [HN]
platiai Mertlik, 1996: 203 **E**: HU SK
pseudopilosus Platia & Gudenzi, 1985: 21 **E**: IT MC SP
ruzenae Laibner, 1977: 388 (*Limonius*) **E**: HU PL SK
schurmanni Platia & Gudenzi, 1998a: 53 **E**: AL GR **A**: TR
scutellaris Dolin, 2003: 36 **A**: FE
turcicus Platia, 2004a: 40 **A**: TR

genus *Elathous* Reitter, 1890e: 247 type genus *Elathous buyssoni* Reitter, 1890
Athoina Reitter, 1906a: 399 type species *Elathous perrisi* Desbrochers, 1873
Elathouina Reitter, 1905a: 19 type species *Athous revelierei* Mulsant & Rey, 1875 (= *Elathous perrisi* Desbrochers, 1873)
Melanathous Reitter, 1905a: 24 type species *Melanathous sequensi* Reitter, 1905
brucki Candèze, 1878b: clxviii (*Athous*) **E**: GR
sequensi Reitter, 1905a: 25 (*Melanathous*)
brunneus Lewis, 1894a: 195 (*Limonius*) **A**: JA
buyssoni Reitter, 1890e: 248 **E**: AB AR
candezei Reitter, 1890e: 248 **E**: UK ST
chiarae Guglielmi & Platia, 1985: 192 **E**: GR
crovatoi Platia & Schimmel, 1993a: 52 **E**: GR
cyprius Baudi di Selve, 1871: 54 (*Athous*) **A**: CY
depressus Platia & Schimmel, 1991b: 131 **A**: TR
ekaterinae Preiss, 2003: 55 **A**: SY
emgei Schwarz, 1893c: 191 (*Leptoschema*) **E**: GR
impressifrons Hampe, 1866: 373 (*Athous*) **E**: CR
mertliki Platia & Schimmel, 1992: 93 **A**: TR
niger Schwarz, 1897a: 130 **E**: AB GR
nigricans Platia & Schimmel, 1991b: 132 **E**: GR (Kríti) **A**: TR
pavesii Platia & Schimmel, 1992: 94 **A**: TR
perrisii Desbroches des Loges, 1873a: 366 (*Athous*) **E**: FR (Corse) IT (Sardegna)
amicus Perris, 1875: 4 (*Athous*)
revelierei Mulsant & Rey, 1875a: 81 (*Athous*)
rufobasalis Wurst, 1994a: 62 **A**: TR
rufus Candèze, 1860: 384 (*Limonius*) **E**: PT SP
schwarzi Reitter, 1898h: 346 **A**: TR
smyrnensis Schwarz, 1903: 207 **A**: TR
syriacus Schwarz, 1903: 208 **A**: LE SY
vasquezi Cobos, 1969: 11 (*Leptoschema*) **N**: MO
yamamotoi Ôhira, 1995d: 189 **A**: JA

genus *Jonthadocerus* Buysson, 1918b: 109 type species *Jonthadocerus theryi* Buysson, 1918 (= *Athous rufescens* Escalera, 1914)

rufescens Escalera, 1914: 214 (*Athous*) **N**: MO
theryi Buysson, 1918b: 110

genus *Limoniscus* Reitter, 1905a: 14 type species *Elater violaceus* P. W. J. Müller, 1821
ainu Ôhira, 1968d: 9 (*Gambrinus*) **A**: JA
amamiensis Ôhira, 1966a: 10 (*Gambrinus*) **A**: JA (Amami-Ôshima)
atricolor Lewis, 1879b: 157 (*Limonius*) **A**: JA
elegans Buysson, 1891d: cxxxvii (*Limonius*) **A**: SY TR
hinakurai Kishii, 1998a: 2 **A**: JA
hiramatsui Ôhira, 1976: 48 (*Gambrinus*) **A**: JA
hosodai Kishii, 1989e: 65 (*Gambrinus*) **A**: JA
imitans Lewis, 1894a: 196 (*Limonius*) **A**: JA
katoi Kishii, 2002c: 129 **A**: JA
kawaharai Kishii, 2002c: 132 **A**: JA
kraatzi kraatzi Candèze, 1879a: 281 (*Limonius*) **A**: FE JIL SC
kraatzi nihonicus Kishii, 1966a: 46 (*Gambrinus*) **A**: JA
limbatipennis Nakane & Kishii, 1955a: 43 (*Gambrinus*) **A**: JA
montivagus Lewis, 1894a: 194 (*Limonius*) **A**: JA
naomii Kishii, 1997a: 4 **A**: JA
niponensis Lewis, 1894a: 194 (*Limonius*) **A**: JA
ogatai Kishii, 1985: 8 (*Gambrinus*) **A**: JA
rufipennis Lewis, 1894a: 197 (*Limonius*) **A**: JA SC
rufovittatus Ôhira, 1963a: 15 (*Gambrinus*) **A**: JA
suturalis Gebler, 1844: 99 (*Ampedus*) **E**: UK ST **A**: KI KZ
takabai Kishii, 1997a: 1 **A**: JA
violaceus P. W. J. Müller, 1821: 184 (*Elater*) **E**: AU CZ DE EN FR GB GE GR HU PL RO SK SP **A**: TR
vittatus Candèze, 1873: 22 (*Limonius*) **A**: HKG JA "Korea"
flavolineatus Miwa, 1927a: 19 (*Megapenthes*)
wittmeri Chassain, 1998: 120 **A**: IN
yamato Kishii, 1998a: 5 **A**: JA

genus *Limonius* Eschscholtz, 1829: 33 type species *Elater minutus* Linnaeus, 1758
Dendrometrus Gistel, 1848: xi [unjustified RN]
Kibunea Kishii, 1966: 46 type species *Limonius eximius* Lewis, 1894
approximans Lewis, 1894a: 198 **A**: JA
baconi Candèze, 1860: 412 **A**: "Himalaya"
eximia Lewis, 1894a: 318 **A**: JA
purpureipennis Miwa, 1927a: 20 (*Megapenthes*)
ignicollis Lewis, 1894a: 197 **A**: JA
kouichiana Kishii, 1989e: 63 (*Kibunea*) **A**: JA
minutus Linnaeus, 1758: 406 (*Elater*) **E**: AB AR AL AU BE BH BU BY CR CT CZ DE EN FI FR GB GE GG GR HU IT LA LS LT LU MC MD NL PL RO SK SL SP ST SV SZ UK YU **A**: IN KZ SY TR
aereus Brullé, 1832: 139
angustus Herbst, 1806: 98 (*Elater*)
crenulatus Faldermann, 1935b: 169 (*Limonius*)
cyanichrous Buysson, 1926b: 166
forticornis Bach, 1852: 34
nigroaeneus Marsham, 1802: 382 (*Elater*)
nitidicollis Laporte, 1840a: 242 (*Elater*)
narukawai Kishii, 1993d: 151 **A**: JA
nigrinus Candèze, 1900: 94 **A**: KA
substriatus Gebler, 1833: 279 (*Elater*) **A**: "Siberia"

genus *Nothodes* LeConte, 1861: 171 type species *Limonius dubitans* LeConte, 1853
marginicollis Lewis, 1894a: 196 (*Limonius*) **A**: JA

parvulus Panzer, 1799a: 7 (*Elater*) **E**: AL AU BE BH BU CR CT CZ EN FR GE GR HU IT LA LS LU MD MK NL PL PT RO SK SL SP ST SZ TR YU **A**: SY TR

dilutus Buysson, 1926a: 48 (*Limonius*)
mus Illiger, 1807a: 12 (*Elater*)
serosus Buysson, 1902b: 280 (*Limonius*)

genus *Neoathousius* Schimmel & Platia, 1991b: 881 type species *Neoathousius loebli* Schimmel & Platia, 1991
brancuccii Schimmel & Platia, 1991b: 884 **A**: UP
decorus Gurjeva, 1975b: 169 (*Athousius*) **A**: AF
exiguus Schimmel & Platia, 1991b: 885 **A**: UP
ferrugineus Schimmel & Platia, 1991b: 888 **A**: PA
hartmanni Schimmel, 1998c: 151 **A**: NP
hiermeieri Schimmel & Platia, 1991b: 888 **A**: UP
kubani Schimmel, 1996a: 172 **A**: YUN
loebli Schimmel & Platia, 1991b: 890 **A**: PA
longicornis Schimmel & Platia, 1991b: 890 **A**: PA
manaliensis Schimmel & Platia, 1991b: 892 **A**: HP
miandamensis Schimmel & Platia, 1991b: 892 **A**: PA
minusculus Schimmel, 1998b: 101 **A**: KA
recticornis Schimmel & Platia, 1991b: 893 **A**: UP
riesei Platia, 1988a: 9 (*Athousius*) **A**: PA
rusticus Schimmel & Platia, 1991b: 894 **A**: PA
sinensis Schimmel & Platia, 1994: 233 **A**: SCH YUN
singularis Schimmel & Platia, 1991b: 894 **A**: PA
wittmeri Schimmel & Platia, 1991b: 895 **A**: PA

genus *Paradima* Miwa, 1928b: 38 type species *Paradima tattakensis* Miwa, 1928
tattakensis Miwa, 1928b: 38 (*Paradima*) **A**: TAI

genus *Parathous* Fleutiaux, 1918b: 242 type species *Parathous sanguineus* Fleutiaux, 1918
sanguineus Fleutiaux, 1918b: 242 **A**: HUB SCH **ORR**
sulcicollis Miwa, 1928a: 136 (*Athous*) **A**: TAI

genus *Pheletes* Kiesenwetter, 1858a: 328 type species *Elater bructeri* Panzer, 1795 (= *Eater aeneoniger* DeGeer, 1774)

Hammionus Gistel, 1834: 11 [HN] type species *Elater bructeri* Panzer, 1795 (= *Eater aeneoniger* DeGeer, 1774)

aeneoniger DeGeer, 1774: 159 (*Elater*) **E**: AU BE BU CR CT CZ DE EN FI FR GE HU IT LU MD NL NR NT PL RO SK SL SV SZ UK

bructeri Panzer, 1795: 243 (*Elater*)
minutus Paykull, 1800: 40 (*Elater*) [HN]
nitidus Gmelin, 1790: 1915 (*Elater*)
pusillus J. R. Sahlberg, 1900: 17 (*Limonius*)

quercus A. G. Olivier, 1790: no 31: 51 (*Elater*) **E**: AU BH BU BY CR CZ DE FR GE GR HU IT LU MD PT RO SK SL SP SZ UK **A**: SY

candezei Buysson, 1888a: ccxii
lythrodes Germar, 1813: 129 (*Elater*)
nigricollis Schilsky, 1888: 187 (*Limonius*)
solarii Buysson, 1902b: 284
testaceus Schilsky, 1888: 187 (*Corymbites*)

reitteri Gurjeva, 1976d: 195 (*Pheletes*) **A**: MG NMO

genus *Solskyana* Dolin, 1978: 45 type species *Limonius villiger* Solsky, 1881
hirta Dolin, 1978: 48 **A**: KZ UZ
villiger Solsky, 1881: 237 (*Limonius*) **E**: AR **A**: KI KZ UZ

genus *Subathous* Fleutiaux, 1918b: 203 type species *Subathous tonkinensis* Fleutiaux, 1918
formosensis Kishii, 1993a: 16 **A**: TAI

genus *Taiwanathous* Miwa, 1930b: 68 type species *Taiwanathous arisanus* Miwa, 1930
arisanus Miwa, 1930b: 68 **A**: TAI

genus *Tropihypnus* Reitter, 1905a: 9 type species *Paracardiophorus bimargo* Reitter, 1896
Crypnoidus Fleutiaux, 1928b: 252 type species *Quasimus setosus* Buysson, 1914 (= *Paracardiophorus bimargo* Reitter, 1896)
bimargo Reitter, 1896e: 235 (*Paracardiophorus*) **A**: KI KZ XIN
ferruginosus Buysson, 1914a: 43 (*Quasimus*)
setosus Buysson, 1914a: 42 (*Quasimus*)
chatterjeei Fleutiaux, 1928b: 253 (*Crypnoidus*) **A**: UP
gardneri Fleutiaux, 1928b: 253 (*Crypnoidus*) **A**: AF HP UP
namsooa Stibick, 1968: 178 **A**: SD
punjabae Stibick, 1968: 180 **A**: SD **ORR**
rungbongi Stibick, 1968: 172 **A**: NP SD SHA
unicolor Gurjeva, 1987b: 50 **A**: TD

subtribe Denticollina Stein & J. Weise, 1877 (1859)

genus *Campylomimus* G. Müller, 1960: 72 type species *Campylomimus brelihi* G. Müller, 1960
brelihi G. Müller, 1960: 72 **E**: BH

genus *Denticollis* Piller & Mitterpacher, 1783: 86 type species *Denticollis rubens* Piller & Mitterpacher, 1783
Campylus Fischer von Waldheim, 1824: 153 type species *Elater linearis* Linaeus, 1758
Exophthalmus Berthold, 1827: 336 type species *Elater linearis* Linaeus, 1758
Hammionus Charpentier, 1825: 191 type species *Elater denticollis* Fabricius, 1794 (= *Denticollis rubens* Piller & Mitterpacher, 1783)
Lepturoides Herbst, 1784b: 103 type species *Elater linearis* Linnaeus, 1758
Notopeda Gistel, 1834: 11 type species *Elater linearis* Linnaeus, 1758
acutangulus J. R. Sahlberg, 1903a: 26 (*Campylus*) **A**: ES
acuticollis Motschulsky, 1860b: 113 (*Campylus*) **A**: FE
amurensis Pic, 1916l: 21
borealis Paykull, 1800: 6 (*Elater*) **E**: BY CZ EN FI FR GE LA NR NT PL SV UK **A**: ES FE MG WS
caucasicola Roubal, 1914: 153 (*Campylus*) **E**: ST
cinctus Candèze, 1863: 479 (*Campylus*) **A**: ES FE JIL MG WS
davidis Fairmaire, 1889b: 34 (*Campylus*) **A**: HUB SCH
dilutiangulus Motschulsky, 1860b: 112 (*Campylus*) **A**: FE
erebus Gurjeva, 1987a: 42 **A**: FE
flabellatus Reitter, 1906f: 274 (*Campylus*) **E**: ST
obscuripennis Reitter, 1906f: 274 (*Campylus*)
flavipes Germar, 1846: 155 (*Campylus*) **A**: FE
formosus Gurjeva, 1987a: 43 **A**: FE
inaequalis Candèze, 1879a: 282 (*Lepturoides*) **A**: FE GUX
limbatipennis Pic, 1916l: 21
interpositus Roubal, 1941b: 30 **E**: CZ PL SK
jacobsoni Reitter, 1903e: 280 **A**: ES
korbi Pic, 1904d: 25 (*Campylus*) **A**: "Siberia"
linearis Linnaeus, 1758: 404 (*Elater*) **E**: AU BE BH BU BY CR CT CZ DE EN FI FR GB GE GR HU IR IT LA LS LT LU MD NL NR NT PL RO SK SL SP SV SZ UK YU **A**: ES FE HEI IN KZ MG TR WS
alsaticus Pic, 1916l: 21
bicolor Piller & Mitterpacher, 1783: 86 (*Denticollis*)
cantharoides A. G. Olivier, 1790: no 31: 36 (*Elater*)
chrysomelinus O. F. Müller, 1776: 60 (*Elater*)
dispar Paykull, 1800: 37 (*Elater*)
jacqueti Pic, 1916l: 21

livens Linnaeus, 1761: 206 (*Elater*)
lividus P. L. S. Müller, 1802: 320 (*Elater*)
lucens Fabricius, 1775: 213 (*Elater*)
maculithorax Pic, 1916l: 21
marginellus Fabricius, 1787a: 174 (*Elater*)
mesomelus Linnaeus, 1758: 405 (*Elater*)
simulatrix Roubal, 1941b: 32
subcantharoides Buysson, 1906b: 452
suturalis Dufour, 1851: 328 (*Elater*)
variabilis DeGeer, 1774: 154
miniatus Candèze, 1885: 121 (*Lepturoides*) **A**: GUX JA SC
oculatus Lewis, 1894a: 317 (*Lepturoides*)
scutellaris Lewis, 1894a: 317 (*Lepturoides*)
mongolicus Motschulsky, 1860b: 113 (*Campylus*) **A**: JIL MG NMO
mounaldaui Miwa, 1931b: 83 **A**: TAI
nigricollis Gebler, 1830: 79 (*Campylus*) **A**: ES MG
nipponensis ainu Kishii, 1987: 2 **A**: JA
nipponensis nipponensis Ôhira, 1973e: 35 **A**: JA
parallelicollis Aubé, 1850: 336 (*Campylus*) **E**: AB AR GG ST **A**: TR
pectinatus Reitter, 1906f: 273 (*Campylus*) **E**: GG
puerilis Miwa, 1931b: 83 **A**: TAI
rubens Piller & Mitterpacher, 1783: 86 **E**: AU BE BH BU BY CR CT CZ DE EN FR GE GG HU IT LA LS LU MD NR PL RO SK SL ST SV SZ UK YU
abchasiaticus Buysson, 1906b: 449 (*Campylus*)
denticollis Fabricius, 1794: 451 (*Elater*)
dichromus Candèze, 1863: 476 (*Campylus*)
mangonis Buysson, 1929: 254 (*Campylus*)
moestus Buysson, 1906b: 449 (*Campylus*)
notatithorax Pic, 1916l: 21
pyropterus A. G. Olivier, 1790: no 31: 35 (*Elater*)
varians shirozui Ôhira, 1963b: 178 **A**: JA
varians varians Germar, 1846: 152 (*Campylus*) **A**: ES FE JIL MG **NAR**
fulvus Motschulsky, 1860c: 386 (*Campylus*)
sahlbergi Germar, 1846: 151 (*Campylus*)
variabilis Eschscholtz, 1829: 33 [HN]
versicolor Lewis, 1894a: 316 (*Lepturoides*) **A**: JA
yasumii Ôhira, 1993: 251 **A**: JA

genus *Denticolloides* Gurjeva, 1963b: 1410 type species *Denticolloides paradoxus* Gurjeva, 1963
bajtenovi Gurjeva & Tugusheva, 1967: 64 **A**: KI KZ
fulvus Dolin, 1969: 129 **A**: KI XIN
hosokawai Arimoto, 2000: 369 **A**: JA
oblongicollis Dolin & Cate, 1999: 26 **A**: KI
paradoxus Gurjeva, 1963b: 1410 **A**: KZ XIN
sinensis Dolin & Cate, 1999: 28 **A**: XIN

genus *Odontoderus* Schwarz, 1894: 149 type species *Odontoderus spinicollis* Schwarz, 1894
Diadochus Buysson, 1896c: cdvi type species *Diadochus antigai* Buysson, 1895
antigai Buysson, 1896c: cdvii (*Diadochus*) **E**: CR IT SP
gurjevae Orlov, 1995: 619 **E**: GG ST
spinicollis Schwarz, 1894: 149 **A**: SY

subtribe Hemicrepidiina Campion, 1894

genus *Crepidophorus* Mulsant & Guillebeau, 1853: 189 type species *Crepidophorus anthracinus* Mulsant & Guillebeau, 1854 (= *Athous mutilatus* Rosenhauer, 1847)
expolitus Gurjeva, 1987a: 44 **A**: FE

laetus Candèze, 1879a: 281 (*Athous*) **A**: FE
jocosus Candèze, 1882: 90 (*Athous*)
pictus Kraatz, 1891: 125 (*Athous*)
mutilatus Rosenhauer, 1847: 16 (*Athous*) **E**: AU BU BY CT CZ DE FI FR GE HU LA LT PL SK SL SV SZ UK
anthracinus Mulsant & Guillebeau, 1853: 191
barbarinus Pic, 1916a: 1
foveolatus Hampe, 1850: 351 (*Athous*)
rufiventris Eschscholtz, 1822: 71 [= 1823: 127] (*Elater*) **A**: ES FE WS **NAR**

genus *Diacanthous* Reitter, 1905a: 25 type species *Elater undulatus* DeGeer, 1774
ainu Miwa, 1928a: 135 (*Harminius*) **A**: JA
amurensis Platia & Gudenzi, 1999a: 24 **A**: FE
antennatus Kishii, 1957a: 10 (*Harminius*) **A**: JA
dshesynensis Cherepanov, 1957: 257 **A**: WS
ontakeanus Kishii, 1969: 2 (*Harminius*) **A**: JA
undosus Lewis, 1894a: 256 (*Athous*) **A**: FE JIL JA NMO
undulatus DeGeer, 1774: 155 (*Elater*) **E**: AU BY CR CT CZ EN FI FR GB GE IT LA LS LT NR NT PL RO SK SL SP SV SZ UK **A**: ES FE MG WS XIZ **NAR**
bifasciatus Gyllenhal, 1808a: 383 (*Elater*)
cinereofasciatus Eschscholtz, 1829: 33 (*Athous*)
cinereofasciatus Motschulsky, 1860b: 108 (*Athous*) [HN]
limbaticollis Motschulsky, 1859b: 227 [= 1860f: 543, 1860b: 108] (*Athous*)
major Reitter, 1905a: 28 (*Harminius*)
mediofasciatus Pic, 1903n: 153 (*Athous*)
simplicitus Heyden, 1884: 290 (*Athous*)
trifasciatus Panzer, 1792b: 14 (*Elater*)
triornatus Tanzer, 1934: 127 (*Harminius*)
undatus Gmelin, 1790: 1915 (*Elater*)
unifasciatus Motschulsky, 1859b: 227 [= 1860f: 543] (*Athous*)

genus *Harminius* Fairmaire, 1851: 527 [= 1852: 80] type species *Athous spiniger* Candèze, 1860
Pseudocorymbites Fiori, 1898: 162 type species *Athous spiniger* Candèze, 1860
florentinus Desbrochers des Loges, 1870b: 114 (*Athous*) **E**: IT
galloisi Miwa, 1928a: 134 **A**: JA
gigas Reitter, 1890f: 249 (*Athous*) **E**: AL BH CR YU
kurotai Ôhira, 1997d: 343 **A**: JA
nihonicus Kishii, 1979a: 4 **A**: JA
nikkoensis Miwa, 1928a: 134 **A**: JA
singularis chishimanus Miwa, 1928a: 134 **A**: FE
singularis hondoensis Kishii, 1985: 9 **A**: JA
singularis singularis Lewis, 1894a: 201 (*Athous*) **A**: FE JA SC
spiniger Candèze, 1860: 460 (*Athous*) **E**: IT MA **N**: AG
castaneus Fairmaire, 1851: 528 [= 1852: 81] (*Athous*)
ferrugineus Binaghi, 1972: 144

genus *Hemicrepidius* Germar, 1839b: 212 type species *Hemicrepidius thomasi* Germar, 1839 (= *Hemicrepidius memnonius* Herbst, 1806)
subgenus *Hemicrepidius* Germar, 1839b: 212 type species *Hemicrepidius thomasi* Germar, 1839 (= *Hemicrepidius memnonius* Herbst, 1806)
Heterathous Reitter, 1905a: 32 type species *Athous sanguinicollis* J. Frivaldszky, 1892 (= *Psephus desertor* Candèze, 1873)
Asaphes Kirby, 1837: 146 [HN] type species *Pedetes* (*Asaphes*) *ruficornis* Kirby, 1837 (= *Hemicrepidius memnonius* Herbst, 1806)
Pseudathous Méquignon, 1930a: 95 type species *Elater hirtus* Herbst, 1784
carbonarius Stepanov, 1935: 194 (*Harminius*) **E**: ST
chinensis Kishii & Jiang, 1996: 145 **A**: FUJ HUB
colchicus Iablokoff-Khnzorian, 1962a: 114 (*Athous*) **E**: GG ST
consors Heyden, 1884: 293 **A**: FE

coreanus Kishii, 2002a: 6 **A**: SC (Cheju)
corvinus Reitter, 1905a: 38 (*Athous*) **A**: SY TR
cotesi Candèze, 1895: 49 (*Athous*) **A**: "Hindoustan"
cylindricus Kishii & Jiang, 1996: 140 **A**: SCH
decorus Fleutiaux, 1918b: 245 (*Athous*) **A**: JIA JIX **ORR**
desertor amamicola Kishii, 1969: 3 **A**: JA (Amami-Ôshima)
desertor chejuanus Kishii, 2002a: 9 **A**: SC (Cheju)
desertor desertor Candèze, 1873: 7 (*Psephus*) **A**: JA
brunneipennis Lewis, 1894a: 200 (*Athous*)
flavipennis Miwa, 1934: 113 (*Athous*)
isshikii Miwa, 1928a: 142 (*Corymbites*)
sanguinicollis J. Frivaldszky, 1892: 124 (*Athous*)
shirozui Kishii, 1958: 28
virens Candèze, 1873: 24 (*Athous*)
desertor taishu Kishii, 2001c: 49 **A**: JA (Tsushima)
desertor yakuensis Kishii, 1969: 3 **A**: JA (Yaku-shima)
rufoabdominalis Nakane & Kishii, 1958: A297
flavipennis Cherepanov, 1957: 258 (*Harminius*) **A**: KZ WS
guizhouensis Kishii & Jiang, 1996: 141 **A**: GUI
hirtus Herbst, 1784b: 114 (*Elater*) **E**: AB AL AU BE BH BU BY CR CT CZ DE EN FR GB GE GG GR HU IR IT LA LS LT LU MC MD NL NR PL PT RO SK SL SP ST SV SZ UK YU **A**: CY IN TR
ater Geoffroy, 1785: 38 (*Elater*)
balthasari Roubal, 1935b: 61 (*Athous*)
daliberti Buysson, 1926a: 47 (*Athous*)
lublini Buysson, 1926a: 47 (*Athous*)
maculicollis Pic, 1897b: 5
meles Buysson, 1906b: 369 (*Athous*)
nigrinus Marsham, 1802: 389 (*Elater*) [HN]
porrectus C. G. Thomson, 1868: 356 (*Athous*)
rambouseki Roubal, 1925b: 61 (*Athous*)
jugicola Perez Arcas, 1872: 101 (*Athous*) **E**: PT SP
kibane Kishii, 1989c: 7 **A**: JA
koenigi Schwarz, 1897a: 132 (*Athous*) **E**: AR
kumaso Kishii, 2001c: 46 **A**: JA (Kyushu)
niger Linnaeus, 1758: 406 (*Elater*) **E**: AB AL AU BE BU BY CR CT CZ DE EN FI FR GE GG HU IT LA LS LT LU NL NR NT PL RO SL
aeneithorax Desbrochers des Loges, 1870b: 108 (*Athous*)
alpinus L. Redtenbacher, 1858: 491 (*Athous*)
deflexus C. G. Thomson, 1868: 355 (*Athous*)
laesus Germar, 1824: 56 (*Elater*)
leporinus Buysson, 1916: 97 (*Athous*)
pubescens Mannerheim, 1827: 6 (*Elater*)
rubrocallatus Laco, 1936: 54
schaumi Buysson, 1901b: 101 (*Athous*)
scrutator Herbst, 1806: 73 (*Elater*)
semirufus Desbrochers des Loges, 1873a: 356 (*Athous*)
nigripennis Miwa, 1928a: 136 (*Athous*) **A**: TAI
nigritulus Reitter, 1890d: 244 **E**: AB AR GG **A**: IN TR
oblongus Solsky, 1871: 362 (*Athous*) **A**: FE JIL LIA NC
insulsus Buysson, 1904c: 59 (*Athous*)
plicatulus Reitter, 1905a: 40 (*Athous*)
pullus Reitter, 1905a: 40 **A**: TR (Kars)
raddei Faust, 1877b: 307 (*Athous*) **E**: GG
rufangulus rufangulus Miwa, 1928a: 137 (*Athous*) **A**: TAI
nigricollis Miwa, 1928a: 137 (*Athous*)
ruficollis Miwa, 1930b: 68 (*Athous*)
rufangulus tricolor Kishii & Jiang, 1996: 144 **A**: SCH ZHE

schneideri Kiesenwetter, 1878: 203 (*Athous*) **E**: AR GG **A**: TR
perniger Reitter, 1896b: 35 (*Athous*)
secessus hallaensis Kishii, 2002a: 8 **A**: SC (Cheju)
secessus izuinsulanus Kishii, 1977a: 49 (*Pseudathous*) **A**: JA (Izu-shoto Is.)
secessus kuniyoshii Ôhira, 1967c: 46 **A**: JA (Okinawa-hontô)
secessus okadomei Ôhira, 1966a: 11 **A**: JA (Kuchinoérabu-jima)
secessus secessus Candèze, 1873: 23 (*Athous*) **A**: JA SC
secessus yakuanus Kishii, 2001c: 45 **A**: JA (Yakushima)
sinuatus insularis Kishii, 2002a: 9 **A**: SC (Cheju)
sinuatus satouchii Kishii, 1961: 50 **A**: JA (Yakushima)
sinuatus sinuatus Lewis, 1894a: 199 (*Athous*) **A**: JA
harmandi Fleutiaux, 1902a: 21 (*Athous*)
subopacus Kishii & Jiang, 1996: 148 **A**: JIX
tartarus Candèze, 1860: 426 (*Athous*) **E**: UK
terukoanus Kishii, 1961: 36 **A**: JA
variabilis Fleutiaux, 1918b: 246 (*Athous*) **A**: GUI HUB HUN SCH **ORR**
vulpeculus Reitter, 1890d: 245 **E**: AR GG **A**: TR

subgenus *Miwacrepidius* Ôhira, 1962b: 263 type species *Athous subcyaneus* Motschulsky, 1866
subcyaneus Motschulsky, 1866a: 166 (*Athous*) **A**: FE JA
praenobilis Lewis, 1895b: 407 (*Athous*)

genus *Medakathous* Kishii, 1964: 17 type species *Athous jactatus* Lewis, 1894
amami Kishii, 1985: 10 (*Hemicrepidius*) **A**: JA
yamawakii Ôhira, 1987b: 287 (*Hemicrepidius*) **A**: JA
jactatus babai Ôhira, 1956: 17 **A**: JA (Awa-shima)
jactatus jactatus Lewis, 1894a: 255 (*Athous*) **A**: JA "Korea"

genus *Megathous* Reitter, 1905a: 25 type species *Athous menetriesi* Reitter, 1890
Harminathous Kishii, 1955b: 78 type species *Harminathous nakanei* Kishii, 1955 (= *Athous suturalis* Candèze, 1873)
algirinus Candèze, 1860: 451 (*Athous*) **N**: AG MO TU
obscurior Pic, 1916a: 1
raffrayi Desbrochers des Loges, 1873a: 364 (*Athous*)
altaicus Schwarz, 1900: 108 (*Athous*) **A**: KZ WS XIN
satanulus Reitter, 1902d: 187 (*Athous*)
barrosi Méquignon, 1932: 123 (*Harminius*) **E**: PT
dauricus Mannerheim, 1852b: 292 (*Athous*) **A**: FE HEI JIL MG NMO
eximius Buysson, 1904c: 58 (*Athous*)
urgensis Schwarz, 1900: 109 (*Athous*)
escalerai Cobos, 1958: 55 (*Athous*) **N**: MO
ficcuzzensis Buysson, 1912f: 132 (*Athous*) **E**: IT (Sicilia)
fiorii Platia & Marini, 1990: 35 **E**: FR
jacobsoni Reitter, 1905a: 29 (*Harminius*) **A**: GAN NMO
menetriesi Reitter, 1890d: 243 (*Athous*) **E**: AB **A**: IN
martini Buysson, 1901c: 125 (*Athous*)
nigerrimus Desbrochers des Loges, 1869: 33 [= 1870b: 106] (*Athous*) **E**: FR IT SZ
langsdorfi Stierlin, 1880: 590 (*Athous*)
peyerimhoffi Buysson, 1917: 14 **N**: AG
pici Buysson, 1898b: 238 (*Athous*) **N**: MO
rifensis Cobos, 1969: 15 (*Harminius*) **N**: MO
sedakovii Mannerheim, 1852b: 293 (*Athous*) **A**: ES MG WS
suturalis Candèze, 1873: 23 (*Athous*) **A**: JA "Korea"
nakanei Kishii, 1955b: 79 (*Harminathous*)

genus *Mucromorphus* Ôhira, 1962b: 263 type species *Athous montanus* Miwa, 1934 (=*Mucromorphus miwai miwai* Kishii, 1962)
miwai miwai Kishii, 1962: 24 **A**: FE JA
montanus Miwa, 1934: 113 (*Athous*) [HN]
miwai yushiroi W. Suzuki, 1985a: 89 **A**: FE JA

genus *Pseudocrepidophorus* Dolin & Agajev, 1988: 5 type species *Elater flavescens* Eschscholtz, 1818
concavifrons Wurst, 1994a: 59 **A**: TR
filicollis Reitter, 1890d: 147 (*Athous*) **E**: UK (Odessa)
filicollis sensu auct. **A**: FE "NE China"
flavescens Eschscholtz, 1818: 480 (*Elater*) **E**: AB BU GG ST **A**: IN TR
cavatus Candèze, 1860: 431 (*Limonius*)

genus *Scutellathous* Kishii, 1955b: 79 type species *Athous comes* Lewis, 1894
comes Lewis, 1894a: 200 (*Athous*) **A**: JA
fujisanus Ôhira, 1963a: 16 **A**: JA
horioi Kishii, 1955b: 80 **A**: JA
ozakii Ôhira, 1992a: 225 **A**: JA
porrecticollis Lewis, 1894a: 201 (*Athous*) **A**: JA
daisetsusanus Miwa, 1928a: 136 (*Athous*)
sasajii Kishii, 2001a: 209 **A**: JA
shikokuanus Kishii, 1985: 9 **A**: JA
seinoi Kishii, 2001a: 212 **A**: JA
yakuensis Nakane & Kishii, 1958: A296 **A**: JA (Yaku-shima)
yamashitai Arimoto, 1992: 73 **A**: TAI

genus *Stenagostus* C. G. Thomson, 1859: 104 type species *Elater rufus* DeGeer, 1774
Eschscholtzia Laporte, 1840a: 232 [HN] type species *Elater rhombeus* A. G. Olivier, 1790
laufferi Reitter, 1904f: 236 (*Athous*) **E**: PT SP
probosus Buysson, 1901c: 126 (*Athous*) **E**: AB AR GG
rhombeus A. G. Olivier, 1790: no 31: 22 (*Elater*) **E**: AU BE BH BU BY CR CZ DE FR GB GE HU IT LS MD NL PL PT RO SK SL SP ST SV SZ UK **A**: TR
obscuratus Pic, 1904i: 61 (*Athous*)
pubescens Stephens, 1830: 274 (*Anathrotus*)
robustus Stierlin, 1863b: 91 [= 1864: 149] (*Athous*)
simplex Méquignon, 1930b: 311 (*Athous*)
villosus Geoffroy, 1785: 40 (*Elater*)
rosti Schwarz, 1897a: 131 (*Athous*) **E**: AB GG ST
caucasicus Buysson, 1900b: 232
rufus DeGeer, 1774: 144 (*Elater*) **E**: AU BE BH BU BY CR CT CZ EN FR GE GR HU IT LA LT NR PL RO SK SL SP ST SV SZ UK
melanophthalmos Gmelin, 1790: 1914 (*Elater*)
testaceus Frölich, 1792: 161 (*Elater*)
sardiniensis Reitter, 1914b: 266 (*Athous*) **E**: IT (Sardegna)
umbratilis Lewis, 1894a: 198 (*Athous*) **A**: JA SC
obscuratus Nakane, 1958b: 86
zuercheri Reitter, 1909b: 57 (*Athous*) **E**: GR **A**: SY TR

genus *Yukara* Kishii, 1962: 21 type species *Athous inornatus* Lewis, 1894
inornata Lewis, 1894a: 255 (*Athous*) **A**: FE JA
fluvipennis Matsumura, 1911a: 122 (*Athous*)
testaceipennis Miwa, 1928a: 138 (*Athous*)

tribe Ctenicerini Fleutiaux, 1936 (1863)

genus *Actenicerus* Kiesenwetter, 1858a: 285 type species *Elater tessellatus* sensu Germar, 1843 (= *Elater sjaelandicus* O. F. Müller, 1764) [nec *Elater tesselatus* Linnaeus, 1758]

subgenus *Acnitecerus* Gurjeva, 1989a: 78 type species *Corymbites infirmus* Reitter, 1892

aerosus aerosus Lewis, 1879b: 157 (*Athous*) **A**: FE JP SC

minoensis Miwa, 1934: 116 (*Corymbites*)

modestus Lewis, 1894a: 258 (*Corymbites*)

otsukae Miwa, 1928a: 142 (*Corymbites*)

aerosus miyanouranus Kishii, 1968: 9 (*Malloea*) **A**: JA (Yaku-shima)

alternatus Heyden, 1886c: 272 (*Ctenicerus*) **A**: ES FE HUN GUI JIL NC NO SC

koraiensis Kishii, 1978: 19

ruptivittis Denisova, 1948: 40 (*Corymbites*)

ashiaka Kishii, 1985: 12 **A**: JA

athoides Kishii, 1955b: 78 (*Malloea*) **A**: JA

chokai Kishii, 1966b: 2 (*Malloea*) **A**: JA

defloratus Schwarz, 1902a: 291 (*Ludius*) **A**: GUA SCH **ORR**

formosensis Miwa, 1928a: 139 (*Corymbites*) **A**: TAI

fruhstorferi Schwarz, 1902a: 291 (*Ludius*) **A**: FUJ GUX HUB HUN JIX SCH **ORR**

giganteus Kishii, 1975: 1 **A**: JA

infirmus Reitter, 1892b: 152 (*Corymbites*) **A**: FE HUB JIL NC SC

jeanvoinei Fleutiaux, 1936b: 280 **A**: GUA YUN **ORR**

kiashianus Miwa, 1928a: 139 (*Corymbites*) **A**: JA

yamashitai Ôhira, 1968d: 10

kunimi Kishii, 1966b: 3 (*Malloea*) **A**: JA

kurilensis Dolin, 1987b: 347 **A**: FE

maculipennis Schwarz, 1902a: 289 (*Ludius*) **A**: ANH FUJ GUA GUX HUB HUN JIX SCH YUN ZHE **ORR**

montanus Kishii, 1998a: 9 **A**: JA

mushanus Miwa, 1928a: 139 (*Corymbites*) **A**: TAI

nagaoi Ôhira, 1967c: 47 **A**: JA (Amami-Ôshima)

naomii Kishii, 1996b: 19 **A**: JA

nempta Kishii, 1996b: 18 **A**: JA

octomaculatus Kishii, 1978: 25 **A**: JA

odaisanus Miwa, 1928a: 140 (*Corymbites*) **A**: FUJ JA SC

ohbayashii Ôhira, 1964a: 13 **A**: JA

orientalis Candèze, 1889b: 114 (*Corymbites*) **A**: JA JIL LIA "Korea"

akitu Kishii, 1955a: 20

praenobilis Lewis, 1894a: 259 (*Corymbites*)

pruinosus Motschulsky, 1861a: 9 **A**: FE HEB HEI HUB JIL JA NC SC ZHE

suzukii hegurensis Kishii, 1964: 26 (*Malloea*) **A**: JA (Hegura-jima)

suzukii suzukii Miwa, 1928a: 140 (*Corymbites*) **A**: JA

taishu Kishii, 1996b: 21 **A**: JA (Tsushima)

takeshii Arimoto, 1992: 75 **A**: TAI

toyoshimai Arimoto & Watanabe, 1993: 245 **A**: JA

tsugaru Kishii, 1978: 17 **A**: JA

yaku Nakane & Kishii, 1958: A 294 **A**: JA (Yaku-shima)

koyamai Nakane & Kishii, 1958: A 295

yamashiro Kishii, 1998a: 7 **A**: JA

subgenus *Actenicerus* Kiesenwetter, 1858a: 285 type species *Elater tessellatus* Linnaeus, 1758 sensu Germar, 1843 (= *Elater siaelandicus* O. F. Müller, 1764) [nec *Elater tesselatus* Linnaeus, 1758]

Malloea Arnett, 1955: 600 type species *Elater siaelandicus* O. F. Müller, 1764

paulinoi Desbrochers des Loges, 1873b: cxviii (*Corymbites*) **E**: PT SP

sjaelandicus O. F. Müller, 1764: 17 (*Elater*) **E**: AU BE BH BU BY CT CZ DE EN FI FR GB GE GR HU IR IT LA LS LT LU MD NL NR NT PL PT RO SK SL SP ST SV SZ UK **A**: ES FE KZ NO WS "Korea" **NAR**

assimilis Gyllenhal, 1808a: 394 (*Elater*)

cuprascens LeConte, 1853c: 444 (*Corymbites*)

jurae Pic, 1913k: 153 (*Corymbites*)
lineatus Pic, 1927g: 9 (*Corymbites*)
micans Germar, 1843a: 62 (*Corymbites*)
nigropilosus Gerhardt, 1912: 463 (*Corymbites*)
nubilus Schrank, 1786: 46 (*Elater*)
siaelandicus O. F. Müller, 1764: 17 (*Elater*) [incorrect original spelling]
strigatus Gerhardt, 1897: 205 (*Corymbites*)
viridis Say, 1825b: 255 (*Elater*)

genus *Aganohypoganus* Ôhira, 1966b: 42 type species Hypoganus mirabilis Miwa, 1934
mirabilis Miwa, 1934: 125 (*Hypoganus*) **A**: JA

genus *Anostirus* C. G. Thomson, 1859: 103 type species *Elater castaneus* Linnaeus, 1758
Calostirus C. G. Thomson, 1864: 67 type species *Elater castaneus* Linnaeus, 1758
Ipostirus Binaghi, 1940c: 198 type species *Ludius parumcostatus* Buysson, 1894
Parastirus Binaghi, 1940c: 197 type species *Elater purpureus* Poda, 1761
Pseudostirus Binaghi, 1940c: 198 type species *Corymbites sulphuripennis* Germar, 1843
alexandri Wurst, 1995c: 152 **A**: TR
ataturki Platia & Gudenzi, 2000f: 68 **A**: TR
atropilosus Buysson, 1898a: 296 **E**: RO SK
boeberi Germar, 1824: 51 (*Elater*) **A**: ES FE HEI KZ MG NC NMO WS XIN
humeralis Motschulsky, 1860b: 110 (*Ctenocera* [sic!])
immaculatus Miwa, 1928a: 138 (*Corymbites*)
kamtschaticus Jakobson, 1913: 738 (*Corymbites*) [RN]
pacificus Motschulsky, 1860b: 110 (*Ctenocera* [sic!])
quadripustulatus Fischer von Waldheim, 1824: 202 (*Elater*)
testaceipennis Motschulsky, 1860b: 109 (*Ctenocera* [sic!]) [HN]
castaneus castaneus Linnaeus, 1758: 405 (*Elater*) **E**: AU BE BU BY CT CZ DE EN FI FR GB GE HU IT LA LT LU NL NR NT PL RO SK SL SV UK YU **A**: ES FE KZ MG WS
flavopectinicornis DeGeer, 1774: 153 (*Elater*)
castaneus japonicus Kishii & Ôhira, 1956: 73, 81 **A**: FE (Kunashir) JA
cerrutii Binaghi, 1940c: 210 **E**: IT
daimio Lewis, 1894a: 256 (*Corymbites*) **A**: FE (Kunashir) JA "Korea"
dalmatinus G. Müller, 1925: 26 **E**: BH CR
decorus Schwarz, 1906: 223 (*Ludius*) [RN] **A**: KI
decoratus Schwarz, 1897a: 136 [HN] (*Diacanthus*)
machnovskyi Dolin, 1969: 127
eschscholtzii Faldermann, 1835a: 173 (*Ludius*) **E**: AB AR GG ST **A**: IN TR
araxis Reitter, 1914a: 195 (*Corymbites*)
nigropilosus Schwarz, 1902a: 288 (*Ludius*)
gabilloti Pic, 1907k: 153 (*Corymbites*) **E**: FR IT
ghilarovi Gurjeva, 1988a: 30 **E**: GG ST
gracilicollis Stierlin, 1896: 329 (*Corymbites*) **E**: AU CZ FR GE HU IT LS PL RO SK SL SZ UK **A**: TR
gudenzii Platia, 1983: 143 **E**: IT
haemapterus Illiger, 1807a: 13 (*Elater*) **E**: PT SP
hirculus Gurjeva, 1988a: 32 **A**: KI
holtzi Schwarz, 1902a: 289 (*Corymbites*) **A**: TR
incostatus Pic, 1905j: 169 (*Ludius*) **A**: SY TR
jarmilae Cechovský & Platia, 1991: 13 **E**: GR
lauianus Wurst, 1995d: 264 **E**: AU
lederi Heyden, 1878: 204 (*Corymbites*) **E**: AB AR GG ST **A**: IN TR
marginatus Pic, 1931j: 159 (*Ctenicera*) **E**: IT
melas Koenig, 1887: 353 (*Corymbites*) **E**: ST
nubilosus Wurst & Schimmel, 1995: 25 **A**: TR
parumcostatus Buysson, 1894a: 80 (*Ludius*) **E**: FR SP
plagifer Reitter, 1914a: 195 (*Corymbites*) **A**: KI XIN
altaicus Gurjeva, 1978a: 138

scherdlini Reitter, 1914a: 196 (*Corymbites*)
pseudosulphuripennis Binaghi, 1940c: 216 **E**: FR
pullatus Gurjeva, 1989a: 151 **A**: KZ
pulchellus Denisova, 1948: 41 (*Corymbites*) **A**: KZ
purpureus Poda, 1761: 41 (*Elater*) **E**: AU BE BU CT CZ DE EN FR GE GG HU IT LA LS LT LU NL PL RO SK SL SP ST SZ UK TR YU **A**: TR
flavus Gredler, 1863: 226 (*Corymbites*)
haematodes Fabricius, 1777: 235 (*Elater*)
istriacus Scopoli, 1763: 94 (*Elater*)
scopoli Fuessly, 1775: 16 (*Elater*) [RN]
sulcatus Razoumowsky, 1789: 162 (*Elater*)
lecomtei Méquignon, 1927: 22 (*Ctenicera*)
reissi Reitter, 1914a: 196 (*Corymbites*) **E**: IT
richterae Gurjeva, 1988a: 31 **E**: AB
saltinii Cate, Platia & Schimmel, 2002: 34 **A**: IN TR
semiaurantiacus Fairmaire, 1891f: cxxxvii (*Diacanthus*) **A**: KA
stramineipennis Binaghi, 1940c: 214 **E**: RO
sulphuripennis Germar, 1843a: 55 (*Corymbites*) **E**: AU FR IT GE RO SL SZ
affinis Laporte, 1840a: 239 (*Ctenicera*)
penninus Binaghi, 1940c: 226
suvorovi Reitter, 1910b: 173 (*Corymbites*) **A**: KI KZ
teheranus Binaghi, 1940c: 207 **A**: IN
trivialis Gurjeva, 1988a: 32 **E**: GG ST **A**: TR
turcicus Platia & Mertlik, 1996: 225 **A**: TR
turkestanicus Stepanov, 1935: 192 (*Corymbites*) **A**: KI KZ UZ
venustus Gurjeva, 1988a: 34 **A**: TD
zenii binaghianus Wurst, 1995b: 88 **E**: SL
zenii zenii Rosenhauer, 1856: 138 (*Corymbites*) **E**: AU IT SL

genus *Aplotarsus* Stephens, 1830: 271 type species *Elater rufipes* sensu Stephens, 1830 (= *Elater incanus* Gyllenhal, 1827)
angustulus Kiesenwetter, 1858a: 291 (*Corymbites*) **E**: AB AU BE BU CZ FR GB GE PL RO SK ST SZ UK
ganglbaueri Rybiński, 1902: 11 (*Ludius*)
moreli Pic, 1933e: 10 (*Corymbites*)
putonis Desbrochers des Loges, 1873b: cxviii (*Corymbites*)
subobscuripes Pic, 1933e: 10 (*Corymbites*)
turdus Candèze, 1860: 393 (*Limonius*)
imperceptus Gurjeva, 1987a: 46 **A**: FE
incanus Gyllenhal, 1827: 358 (*Elater*) **E**: AU BE BY CZ DE EN FI FR GB GE IR IT LA LS LU NL NR NT PL RO SK SP SV SZ UK **N**: AG **A**: KZ
gyllenhalii Reiche, 1860: CIX (*Corymbites*)
henoni Candèze, 1895: 65 (*Sericus*)
ochropterus Stephens, 1830: 272
sachalinensis Dolin & Katyukha, 1987: 11 (*Haplotarsus*) **A**: FE
semipalatinus Jakobson, 1913: 740 (*Selatosomus*) [RN] **A**: KZ XIZ
altaicus Schwarz, 1900: 110 (*Corymbites*) [HN]
sladkovi Dolin & Katyukha, 1987: 13 (*Haplotarsus*) **A**: FE
subdepressus Denisova, 1948: 44 **A**: FE (*Selatosomus*)
tibialis Schwarz, 1900: 111 (*Corymbites*) **A**: FE ES XIN
tibiellus Chevrolat, 1865: 351 (*Athous*) **E**: SP
nivicola Kiesenwetter, 1866a: 391 (*Corymbites*)

genus *Calamboganus* Gurjeva, 1989a: 96 type species *Calambus abscondidus* Gurjeva, 1987
abscondidus Gurjeva, 1987a: 47 (*Calambus*) **A**: FE

genus *Calambus* C. G. Thomson, 1859: 103 type species *Elater bipustulatus* Linnaeus, 1767
bipustulatus Linnaeus, 1767: 652 (*Elater*) **E**: AU BE BH BU CZ DE FR GB GE HU IR IT LA LT LU NL NR PL RO SK SL SP SV SZ UK YU **A**: WS
bimaculatus Geoffroy, 1785: 38 (*Elater*) [HN]
diapurus Buysson, 1928: 235
epipleuralis Buysson, 1905a: 17
ferrugineus Buysson, 1906b: 465
flavescens Schilsky, 1888: 190 (*Corymbites*)
perferrugineus Buysson, 1928: 235 (*Ludius*)
rubidus Wimmel, 1894: 6 (*Corymbites*)
semiflavus A. Fleischer, 1886: 235 (*Diacanthus*)
tenebricans Buysson, 1890a: 18
japonicus Fleutiaux, 1902a: 22 (*Corymbites*) **A**: JA
jureceki Stepanov, 1930b: 135 **A**: FE
kusuii Ôhira, 1994a: 285 **A**: JA
mundulus Lewis, 1879b: 157 (*Corymbites*) **A**: FE JA
ussuriensis Denisova, 1948: 45 (*Selatosomus*) **A**: ES FE WS

genus *Corymbitodes* Buysson, 1904c: 58 type species *Corymbitodes longicollis* Buysson, 1904 (= *Corymbites christophi* Kiesenwetter, 1879)
Metactenicerus Miwa, 1934: 37 type species *Corymbites gratus* Lewis, 1894
captiosus Gurjeva, 1985b: 782 **A**: FE
christophi Kiesenwetter, 1879a: 145 (*Corymbites*) **A**: FE
atricornis Heyden, 1884: 278 (*Corymbites*)
longicollis Buysson, 1904c: 58
concolor Lewis, 1894a: 263 (*Corymbites*) **A**: JA
lewisi Fleutiaux, 1902a: 23 (*Corymbites*)
corybas Gurjeva, 1982a: 20 **A**: TD
epipedus Gurjeva, 1985b: 784 **A**: FE
excellens Gurjeva, 1962: 252 **A**: KI
fuscipes Miwa, 1931b: 72 **A**: TAI
gratus Lewis, 1894a: 262 (*Corymbites*) **A**: FE JA JIL JIX SCH
koikei Kishii & Ôhira, 1956: 73, 81
tenuithorax Kishii & Ôhira, 1956: 73, 81
kiiensis Ôhira, 1999: 410 **A**: JA
nikkoensis Jakobson, 1913: 740 (*Selatosomus*) [RN] **A**: JA SC
ferrugineipennis Lewis, 1894a: 262 (*Corymbites*) [HN]
obscuripes Lewis, 1894a: 264 (*Corymbites*) **A**: JA
rubripennis Lewis, 1894a: 263 (*Corymbites*) **A**: JA
rufus Jiang, 1993: 147 **A**: HUB
speciosus Miwa, 1928a: 144 (*Corymbites*) **A**: TAI
viridus Miwa, 1934: 227 (*Corymbites*)
takabai Kishii, 2004a: 64 **A**: JA

genus *Ctenicera* Latreille, 1829a: 454 type species *Elater pectinicornis* Linnaeus, 1758
Clenioceerus Stephens, 1829: 10 type species *Elater pectinicornis* Linnaeus, 1758
Corymbites Latreille, 1834: 150 type species *Elater pectinicornis* Linnaeus, 1758
Ludius Eschscholtz, 1829: 34 [HN] type species *Elater pectinicornis* Linnaeus, 1758
bonomii Binaghi, 1940b: 106 **E**: IT
bosnica Apfelbeck, 1894: 217 [= 1896: 556] (*Ludius*) **E**: BH BU
cuprea Fabricius, 1775: 212 (*Elater*) **E**: AL AU BE BH BU BY CR CT CZ EN FI FR GB GE HU IR IT LS LT LU MC MD NL NR NT PL PT RO SK SL SP ST SV SZ UK YU **A**: ES KZ MG WS XIN
aeruginosus Fabricius, 1798: 139 (*Elater*)
brevelineatus Pic, 1926f: 9 (*Corymbites*)
brevipectinata Binaghi, 1940b: 111
chalybeus Sowerby, 1806: pl. 72 (*Elater*)
diffusus Buysson, 1929: 251 (*Ludius*)

discoidalis Pic, 1913k: 153 (*Corymbites*)
hilfi Reitter, 1902a: 6 (*Ludius*)
humeralis Dufour, 1843a: 42 (*Elater*)
pyraeneus Charpentier, 1825: 189 (*Elater*)
reductus Pic, 1923i: 10 (*Corymbites*)
transsylvanicus Szombathy, 1910d: 577 (*Corymbites*)
doderoi Binaghi, 1940b: 117 **E**: IT
heyeri Saxesen, 1838: 805 (*Elater*) **E**: AU BE BH CZ FR GE IT LU PL RO SK SL ST SZ UK
croaticus Germar, 1843a: 59 (*Corymbites*)
regalis Szombathy, 1911: 182 (*Ctenicera*)
kiesenwetteri C. N. F. Brisout de Barneville, 1866: 375 (*Corymbites*) **E**: SP
pectinicornis Linnaeus, 1758: 406 (*Elater*) **E**: AR AU BE BU BY CT CZ DE EN FI FR GB GE GG HU IT LA LS LT LU MC NL NR NT PL RO SK SL SP SV SZ UK **A**: ES JIL KZ TR SHX WS
aeneopectinicornis DeGeer, 1774: 145 (*Elater*)
schneebergi Roubal, 1932d: 284 (*Corymbites*) **E**: BU RO
virens Schrank, 1781b: 317 (*Elater*) **E**: AU BE BH BU CZ EN FR GE HU IT LS PL RO SK SL SZ UK **NAR**
aeneicollis Olivier, 1792: 264 (*Elater*)
aulicus Panzer, 1800b: 6 (*Elater*)
inaequalis A. G. Olivier, 1790: no 31: 22 (*Elater*)
signatus Panzer, 1800b: 5 (*Elater*)
stramineus Candèze, 1892a: 228 (*Corymbites*)
virescens Gmelin, 1790: 1906 (*Elater*)

genus *Eanus* LeConte, 1861: 171 type species *Limonius estriatus* LeConte, 1853
subgenus *Eanus* LeConte, 1861: 171 type species *Limonius estriatus* LeConte, 1853
Gyrus Jakobson, 1913: 740 [RN] type species *Ludius guttatus* Germar, 1839
Paranomus Kiesenwetter, 1858a: 303 [HN] type species *Ludius guttatus* Germar, 1839
guttatus Germar, 1839a: pl. 5 (*Ludius*) **E**: AU CR CZ FR PL RO SL UK
biguttatus Pic, 1929a: 1
innotatus Pic, 1929a: 1
seminotatus Pic, 1919f: 13
uniguttatus Pic, 1929a: 1
singularis Mannerheim, 1852b: 290 (*Diacanthus*) **A**: FE
subgenus *Pareanus* Gurjeva, 1989a: 90 type species *Elater costalis* Paykull, 1800
costalis costalis Paykull, 1800: 37 (*Elater*) **E**: FI NR NT SV **A**: ES FE WS **NAR**
parvicollis Mannerheim, 1853: 230 (*Diacanthus*)
costalis konishii Ôhira, 1962a: 198 **A**: JA
costalis shibatai Kishii, 1968: 11 **A**: JA

genus *Hypoganus* Kiesenwetter, 1858a: 299 type species *Elater cinctus* Paykull, 1800 (= *Elater inunctus* Lacordaire, 1835)
inunctus Lacordaire, 1835: 642 (*Elater*) **E**: AB AU BE BH BU CR CZ DE FR GE GG GR HU IT LS LU MD NL NR PL RO SK SL ST SV SZ UK **A**: TR
akbesianus Pic, 1903k: 129
cinctus Paykull, 1800: 10 (*Elater*) [HN]
semitestaceus Pic, 1905m: 181
miyatakei Ôhira, 1966b: 41 **A**: JA JIL
stepanovi Denisova, 1948: 46 **E**: AR GG ST **A**: IN TR
tibetis Čechovský & Kubáň, 1997: 115 **A**: YUN

genus *Liotrichus* Kiesenwetter, 1858a: 288 type species *Elater affinis* Paykull, 1800
subgenus *Acteniceromorphus* Kishii, 1955a: 19 type species *Corymbites selectus* Candèze, 1865
chlamydates Lewis, 1894a: 263 (*Corymbites*) **A**: JA "Korea"
ferrugineipennis Candèze, 1879a: 281 (*Corymbites*) **A**: FE
patruus Heyden, 1884: 294 (*Corymbites*)
fulvipennis Lewis, 1894a: 258 (*Corymbites*) **A**: JA "Korea"
himalayanus Ôhira & Becker, 1973c: 565 **A**: NP

hiramatsui Ôhira, 1996c: 311 **A**: JA
kabakovi Gurjeva, 1986: 557 **A**: AF
kurofunei Miwa, 1934: 121 (*Corymbites*) **A**: JA
kochianus Ôhira, 1963b: 177
minoensis Ôhira, 1954: 44 (*Ctenicera*) **A**: JA
nepalensis Ôhira & Becker, 1973c: 563 **A**: NP
nipponensis Ôhira, 1973d: 326 **A**: JA
nishidai Kishii, 1986: 43 **A**: JA
ozeanus Nakane & Kishii, 1954b: 737 (*Actenicerus*) **A**: JA
sasakii Ôhira, 1970a: 10 **A**: JA
selectus Candèze, 1865: 53 (*Corymbites*) **A**: FE (Kunashir) JA SC
singularis Gurjeva, 1986: 557 **A**: KI
subopacus Kishii, 1982: 35 **A**: JA
tengu Miwa, 1934: 122 (*Corymbites*) **A**: JA
tsukamotoi Kishii, 1966a: 52 **A**: JA
subgenus *Liotrichus* Kiesenwetter, 1858a: 288 type species *Elater affinis* Paykull, 1800
affinis Paykull, 1800: 12 (*Elater*) **E**: AU BH BU BY CT CZ FI FR GE IT LA LT NR NT PL RO SK SL SV SZ UK YU **A**: ES FE HEI JIL KZ MG WS
dauricus Motschulsky, 1860b: 110 (*Pristilophus*)
erythropus Fischer von Waldheim, 1824: 203 (*Elater*) [HN]
inhumanus Pic, 1916l: 21 (*Ludius*)
sibiricus Germar, 1824: 58 (*Elater*)
bessolitzinae Gurjeva, 1986: 556 **A**: ES
hypocrita hypocrita Lewis, 1894a: 259 (*Corymbites*) **A**: JA
alpensis Miwa, 1928a: 143 (*Corymbites*)
hypocrita kurilensis Miwa, 1928a: 144 (*Corymbites*) **A**: FE
ligneus Candèze, 1879a: 281 (*Corymbites*) **A**: FE

genus *Hypoganomorphus* Dolin, 1975a: 115 type species *Diacanthus laevicollis* Mannerheim, 1852
laevicollis Mannerheim, 1852b: 288 (*Diacanthus*) **A**: ES FE JIL MG NMO WS
gracilis Motschulsky, 1860b: 109 (*Selatosomus*)
suzuyamanus Matsumura, 1911a: 121 (*Agriotes*)

genus *Metanomus* Buysson, 1887d: cxxxii type species *Corymbites montivagus* Rosenhauer, 1847 (= *Athous infuscatus* Eschscholtz, 1829)
badrinathensis Vats & Chauhan, 1992b: 28 (*Ctenicera*) [DA] **A**: HP UP
grouvellei Buysson, 1900a: 250 (*Hypnoidus*) **E**: FR IT
indiana Garg & Saini, 1996b: 321 (*Ctenicera*) [DA] **A**: HP
infuscatus Eschscholtz, 1829: 33 (*Athous*) **E**: AU BH CZ FR GE HU IT RO SK SL UK **A**: ES FE MG WS **NAR**
acutus Mulsant & Guillebeau, 1855a: 20 (*Athous*)
gebleri Gebler, 1847b: 412 (*Athous*)
lutescens Fall, 1910: 134 (*Corymbites*)
montivagus Rosenhauer, 1847: 15 (*Pristilophus*)

genus *Neopristilophus* Buysson, 1894a: 78 type species *Elater depressus* Germar, 1823 (= *Elater insitivus* Germar, 1824)
gougeleti Fairmaire, 1859d: cli (*Pristilophus*) **E**: SP
hispanicus Sáez-Bolaño, 1989: 23 (*Harminius*)
insitivus Germar, 1824: 44 (*Elater*) **E**: AL AU BH BU CR CZ GE GR HU PL RO SK UK YU
depressus Germar, 1823b: pl. 9 (*Elater*)
parvus Vats & Chauhan, 1992b: 25 (*Ctenicera*) [DA] **A**: UP
serrifer Candèze, 1873: 25 (*Corymbites*) **A**: JA SC
yakuanus Kishii, 1982: 40 **A**: JA (Yakushima)

genus *Orithales* Kiesenwetter, 1858a: 287 type species *Elater serraticornis* Paykull, 1800
serraticornis serraticornis Paykull, 1800: 21 (*Elater*) **E**: AU CT CZ EN FI FR GE LA NR NT PL RO SK SV SZ UK **A**: ES FE JA KZ WS
longulus Gyllenhal, 1808a: 399 (*Elater*)
parvulus Herbst, 1806: 98 (*Elater*) [HN]
separatus Pic, 1929a: 1
serricornis Herbst, 1806: 100 (*Elater*)
serraticornis yatsuensis Kishii, 1966b: 1 **A**: JA
serraticornis yezoensis Miwa, 1934: 126 **A**: JA

genus *Paraphotistus* Kishii, 1966a: 53 type species *Corymbites notabilis* Candèze, 1873
subgenus *Mosotalesus* Kishii, 1977b: 31 type species *Elater impressus* Fabricius, 1792
Haplotrichus Dolin, 1982a: 238 type species *Elater impressus* Fabricius, 1792
adiposus Gurjeva, 1978a: 139 (*Selatosomus*) **A**: TD
capillatus Gurjeva, 1979b: 47 (*Selatosomus*)
auronebulosus Reitter, 1896g: 290 (*Selatosomus*) **A**: GAN KI KZ XIN
baerii Kuschakewitsch, 1861: 55 (*Corymbites*) **A**: ES FE NC WS
elevatus Matsumura, 1911a: 122 (*Corymbetes* [sic])
cordialis Gurjeva, 1989a: 197 **A**: JIL QIN XIZ
imitatus Gurjeva, 1989a: 193 **A**: KI
impressoides Reitter, 1901d: 76 (*Ludius*) **A**: KI TD UZ
impressus impressus Fabricius, 1792b: 223 (*Elater*) **E**: AU BE BU BY CH CT CZ DE EN FI FR GB GE HU IT LA LS LT NL NR NT PL RO SK SL SV SZ UK **A**: ES FE JA KZ MG WS "Korea"
alpicola Szombáthy, 1910d: 578 (*Selatosomus*)
rufipes Schilsky, 1888: 187 (*Corymbites*)
impressus sachalinensis Miwa, 1928a: 141 (*Corymbites*) **A**: FE JA "Korea"
insularis Platia & Gudenzi, 1998c: 629 **E**: GR
kabakovi Gurjeva, 1989a: 190 **A**: AF
obscuroaeneus Koenig, 1889b: 536 (*Corymbites*) **A**: GAN NMO QIN XIZ
picticollis Fairmaire, 1891f: cxxxviii (*Diacanthus*) **A**: KA
przewalskyi Koenig, 1889b: 535 (*Corymbites*) **A**: GAN NMO QIN XIZ
roubali Jagemann, 1942a: 29 (*Corymbites*) **A**: SCH
rufolimbatus Denisova, 1948: 41 (*Selatosomus*)
semenovi Koenig, 1889b: 536 (*Corymbites*) **A**: GAN QIN XIZ
subalpinus Gurjeva, 1979b: 45 (*Selatosomus*) **A**: KI
subgenus *Paraphotistus* Kishii, 1966a: 53 type species *Corymbites notabilis* Candèze, 1873
iyoensis Kishii, 1982: 37 **A**: JP
niger Miwa, 1928a: 142 (*Corymbites*) **A**: JA
notabilis notabilis Candèze, 1873: 25 (*Corymbites*) **A**: FE JA SC
notabilis yagii Kishii, 1982: 39 **A**: JA (Yakushima)
subgenus *Setasomus* Gurjeva, 1985a: 576 type species *Elater nigricornis* Panzer, 1799
nigricornis Panzer, 1799a: 5 (*Elater*) **E**: AU BE BY CT CZ DE EN FI FR GB GE HU IT LA LT LU MD NL NR NT PL RO SK SL SP SV SZ UK **A**: ES KZ WS
metallicus Paykull, 1800: 19 (*Elater*)

genus *Poemnites* Buysson, 1894a: 78 type species *Diacanthus aeratus* Mulsant & Guillebeau, 1856
subgenus *Gnathodicrus* Fleutiaux, 1934a: 183 type species *Gnathodicrus francki* Fleutiaux, 1934
babai Kishii, 1991a: 4, 10 **A**: TAI
francki Fleutiaux, 1934a: 183 **A**: SCH
perpendicularis Fleutiaux, 1918b: 248 (*Corymbites*) **A**: YUN ZHE **ORR**
tonkinensis Fleutiaux, 1918b: 249 (*Corymbites*) **A**: SCH YUN **ORR**
subgenus *Poemnites* Buysson, 1894a: 78 type species *Diacanthus aeratus* Mulsant & Guillebeau, 1856
aeratus Mulsant & Guillebeau, 1856: 97 (*Diacanthus*) **E**: FR
exotans Candèze, 1900: 96 (*Corymbites*) [DA] **A**: PA **ORR**
famelicus Candèze, 1889b: 113 (*Corymbites*) [DA] **A**: SD

fraudator Candèze, 1897: 66 (*Corymbites*) [DA] **A**: SD
hamirensis Cherepanov, 1957: 310 (*Selatosomus*) **A**: KZ MG WS
speculifer Candèze, 1889b: 114 (*Corymbites*) [DA] **A**: YUN
thibeticus Candèze, 1900: 95 (*Corymbites*) [DA] **A**: XIZ
himalayana Garg & Saini, 1996b: 318 (*Ctenicera*) [DA] **A**: HP **ORR**
longiantennus Vats & Chauhan, 1992b: 27 (*Ctenicera*) [DA] **A**: HP UP
subgenus *Thacana* Fleutiaux, 1936b: 282 type species *Corymbites cambodiensis* Fleutiaux, 1918
cambodiensis Fleutiaux, 1918b: 250 (*Corymbites*) **A**: GUI JIX **ORR**

genus *Prosternon* Latreille, 1834: 151 type species *Elater holosericeus* A. G. Olivier, 1790 (= *Elater tessellatum* Linnaeus, 1758)
Lepidotus Curtis, 1838: 694 [HN] type species *Elater holosericeus* A. G. Olivier, 1790 (= *Elater tessellatum* Linnaeus, 1758)
Propheletes Fiori, 1905b: 131 type species *Limonius grandii* Fiori, 1905
Tactocomus Kiesenwetter, 1858a: 301 type species *Elater holosericeus* A. G. Olivier, 1790 (= *Elater tessellatum* Linnaeus, 1758)
admirabile Gurjeva, 1984b: 552 **E**: AB **A**: IN
dolini Agajev, 1985c: 63
aurichalceum Stepanov, 1930a: 90 **A**: FE JIL
chrysocomum Germar, 1843a: 69 (*Diacanthus*) **E**: BU CZ HU GR RO SK
egregium Denisova, 1948: 45 **A**: NMO
montanum Gurjeva, 1980: 93 **A**: MG MNO WS
sericeum Gebler, 1824: 203 (*Elater*) **A**: ES FE KZ MG WS XIN **NAR**
uncinatus Eschscholtz, 1829: 34 (*Ludius*)
syriacum Buysson, 1891d: cxxxvii **A**: SY TR
anatolicus Reitter, 1901f: 100 (*Ludius*)
tessellatum Linnaeus, 1758: 406 (*Elater*) **E**: AB AL AR AU BE BH BU BY CR CT CZ DE EN FI FR GB GE GG GR HU IR IT LA LS LT LU MD NL NR NT PL PT RO SK SL SP ST SV SZ UK YU **A**: ES JIL KZ NC SC TR WS
bifasciatus Küster, 1847b: 38 (*Diacanthus*)
grandii Fiori, 1905: 131 (*Limonius*)
holosericeus A. G. Olivier, 1790: no 31: 27 (*Elater*)
nubilipennis Germar, 1843a: 69 (*Diacanthus*)
rufipes DeGeer, 1774: 150 (*Elater*)
rufounguiculatus DeGeer, 1774: 148 (*Elater*)
semirufus Pic, 1909d: 105
undulatus J. C. Schaeffer, 1766a: pl. 4, fig. 7 (*Elater*)

genus *Pseudanostirus* Dolin, 1964: 120 type species *Diacanthous globicollis* Germar, 1843
adustus Gurjeva, 1982b: 27 **A**: KI
altaicus altaicus Eschscholtz, 1829: 34 (*Ludius*) **A**: KZ MG XIN
depressus Candèze, 1879b: 283 (*Corymbites*)
tarbagataicus Jakobson, 1913: 739 (*Selatosomus*) [RN]
altaicus kasachstanicus Gurjeva, 1964: 151 (*Selatosomus*) **A**: KZ
amurensis Jagemann, 1942b: 596 (*Corymbites*) **A**: FE
alpinus Gurjeva, 1978a: 144 (*Calambus*) **A**: TD
bicolor Dolin & Gurjeva, 1988: 63 **A**: KZ
densatus Reitter, 1910b: 172 (*Selatosomus*) **A**: KZ XIN
dilatatus Miwa, 1928a: 143 (*Corymbites*) **A**: FE JA
ecarinatus Stepanov, 1930a: 88 (*Selatosomus*) **A**: ES FE HEI MG WS
furfurosus Gurjeva, 1982b: 24 **A**: KI
globicollis Germar, 1843a: 79 (*Diacanthus*) **E**: BY CT CZ FI GE HU MD NT PL RO SK UA
sphaerothorax Reitter, 1890h: 265 (*Diacanthus*)
jasoni Dolin & Chantladze, 1987: 621 **E**: GG ST **A**: TR
juldanus Reitter, 1910b: 172 (*Selatosomus*) **A**: XIN
diversipennis Reitter, 1910b: 172 (*Selatosomus*)

montivagus Gurjeva, 1982b: 25 **A**: KI
orodromus Gurjeva, 1982b: 25 **A**: KI
pippingskoeldii Mannerheim, 1852b: 285 (*Corymbites*) **A**: FE
risillus Gurjeva, 1978a: 140 (*Selatosomus*) **A**: KZ
vicinus Gurjeva, 1984a: 295 **A**: MG WS

genus *Selatosomus* Stephens, 1830: 268 type species *Elater aeneus* Linnaeus, 1758
subgenus *Pristilophus* Latreille, 1834: 151 type species *Elater melancholicus* Fabricius, 1798
cruciatus Linnaeus, 1758: 404 (*Elater*) **E**: AU BE BY CT CZ DE EN FI FR GE HU IT LA LT LU MD NL NR NT PL RO SK SL ST SV SZ UK **A**: KZ
    bifenestratus Pic, 1912i: 33
melancholicus alpestris Ménétriés, 1832: 155 (*Elater*) **E**: AB AR GG ST **A**: TR
    profugus Faldermann, 1835a: 176 (*Ludius*)
melancholicus melancholicus Fabricius, 1798: 139 (*Elater*) **E**: AU BY CT FI FR GE IR IT NR NT RO SV SZ UK **A**: ES FE KI KZ MG NMO
    anxius Gebler, 1843: 38 (*Ludius*)
    cenisius Pic, 1909d: 105 (*Ludius*)
    coerulescens Motschulsky, 1860b: 109 (*Selatosomus*)
    depressicornis Motschulsky, 1860b: 109 (*Selatosomus*)
    robustus Stierlin, 1898: 49 (*Diacanthus*)
    scabricollis Eschscholtz, 1822: 68 [= 1823: 124] (*Elater*)
    simplonicus Stierlin, 1884: 44 (*Corymbites*)
melancholicus tianshanicus Denisova, 1948: 42 **A**: KI KZ XIN
onerosus Lewis, 1894a: 260 (*Corymbites*) **A**: JA SCH SHX
pacatus Lewis, 1894a: 261 (*Corymbites*) **A**: JA NIN
punctatissimus Ménétriés, 1851: 53 (*Diacanthus*) **A**: ES FE MG XIN "NE China"
    atriceps J. R. Sahlberg, 1903a: 24 (*Ludius*)
    nigrita Mannerheim, 1852b: 289 (*Diacanthus*)
    punctatissimus Mannerheim, 1852b: 287 (*Diacanthus*)
    spretus Mannerheim, 1852b: 285 (*Diacanthus*)
vagepictus Lewis, 1894a: 261 (*Corymbites*) **A**: JP
subgenus *Selatapteria* Tarnawski, 1995: 83 type species *Selatosomus messorobius* Dolin, 1971
jakobsoni Stepanov, 1930a: 87 **A**: UZ
messorobius Dolin, 1971: 647 **A**: KZ
subgenus *Selatosomus* Stephens, 1830: 268 type species *Elater aeneus* Linnaeus, 1758
    Aphotistus Kirby, 1837: 149 type species *Elater aeneus* Linnaeus, 1758
    Diacanthus Latreille, 1834: 151 type species *Elater aeneus* Linnaeus, 1758
    Eanoides Kishii, 1966a: 50 type species *Corymbites puerilis* Candèze, 1873
aeneomicans Fairmaire, 1889b: 33 (*Corymbites*) **A**: GAN GUI HUB JIA SCH
aeneus Linnaeus, 1758: 406 (*Elater*) **E**: AL AR AU BE BH BU BY CR CT CZ DE EN FI FR GB GE GG GR HU IT LA LS LT LU MC MD NL NR NT PL RO SK SL SP ST SV SZ UK YU **A**: ES HEI JIL KZ MG TR WS XIN
    aeneusrufipes DeGeer, 1774: 149 (*Elater*)
    bicolor Depoli, 1913: 22
    coeruleus Herbst, 1784b: 111 (*Elater*)
    cyaneus Marsham, 1802: 388 (*Elater*)
    germanus Linnaeus, 1761: 207 (*Elater*)
    hispanicus Reitter, 1910b: 167
    luctuosus Depoli, 1928: 230 (*Corymbites*)
    marginatus Papp, 1943: 209
    mutator Rey, 1891b: 85 (*Corymbites*)
    nitens Scopoli, 1763: 91 (*Elater*)
    subpuberulus Reitter, 1910b: 166
    subrugosicollis Rey, 1891b: 85 (*Corymbites*)
    superbus Daniel, 1903a: 252 (*Ludius*)
    viridescens Depoli, 1928: 229 (*Corymbites*)
    viturati Pic, 1910r: 67

albipubens Reitter, 1910b: 166 **A**: QIN SCH XIZ
alekseevi Dolin & Penev, 1988: 25 **E**: ST
alpinus Vats & Chauhan, 1992b: 30 (*Ctenicera*) [DA] **A**: HP
ampliatus Fairmaire, 1891e: cxxvii (*Diacanthus*) **A**: KA
amplicollis Germar, 1843a: 80 (*Diacanthus*) **E**: AL BU FR GR IT MC RO SL SP **A**: SY TR
buyssoni Szombathy, 1910d: 579
lugubricus Reitter, 1910b: 169
lugubris Dufour, 1851: 326 (*Ludius*)
montenegrinus Reitter, 1910b: 168
paganetii Reitter, 1910b: 168
pyrenaeus Laporte, 1840a: 241 (*Ludius*) [HN]
rufoabdominalis Xambeu, 1909: 4 (*Corymbites*)
armeniacus Dolin, 1982b: 20 **E**: AB AR **A**: TR
beysehiricus Mertlik, 2000: 4 **A**: TR
caucasicus Ménétriés, 1832: 155 (*Elater*) **E**: AB AR ST **A**: TR
circassicus Reitter, 1910b: 169
plorator Candèze, 1889b: 114 (*Corymbites*)
violaceus Candèze, 1889b: 115 (*Corymbites*)
centralis Candèze, 1882: 96 (*Corymbites*) **A**: GUX JIL MG NE NMO SCH
confluens confluens Gebler, 1830: 80 (*Elater*) **A**: ES KZ MG WS
confluens rugosus Germar, 1836: pl. 7 (*Ludius*) **E**: AU FR GE IT PL SK SL SZ
carreti Buysson, 1894a: 98
coreanus Miwa, 1928a: 141 (*Corymbites*) **A**: ES FE HEI JIL LIA MGNMO NC SC
reichardti Denisova, 1948: 42
gloriosus Kishii, 1955b: 78 (*Diacanthus*) **E**: NT **A**: ES FE JIL JA NC WS **NAR**
graecus Tarnawski, 1995: 37 **E**: GR
jailensis Dolin, 1971: 645 **E**: UK (Krym)
latissimus Reitter, 1910b: 171 **A**: TR
latus Fabricius, 1801b: 232 (*Elater*) **E**: AB AL AR AU BE BH BU BY CR CT CZ EN FR GE GG GR HU IT LA LT LU MC MD NL PL PT RO SK SL SP ST SZ UK YU **A**: ES FE HEI IN JIL KZ MG NIN NMO SY TR WS
corpulentus Candèze, 1879b: 283 (*Corymbites*)
gravidus Germar, 1843a: 78 (*Diacanthus*)
milo Germar, 1843a: 78 (*Diacanthus*)
pectinicornis Geoffroy, 1785: 37 (*Elater*) [HN]
subrubripes Pic, 1909d: 105 (*Ludius*)
sulcatus Candèze, 1882: 96 (*Corymbites*)
taygetanus Pic, 1905j: 170 (*Ludius*)
luanus Kishii & Jiang, 1999: 15 **A**: JIX
miyajimanus Ôhira, 1971a: 40 (*Aphotistus*) **A**: JA
pecirkanus Reitter, 1910: 167 **E**: AB **A**: TR
puberulus Candèze, 1879a: 282 (*Corymbites*) **A**: BEI FE HUB JIL LIA NC SC
puerilis Candèze, 1873: 26 (*Corymbites*) **A**: JA
puncticollis Motschulsky, 1866a: 167 **A**: FE GAN HUB JA JIL LIA NC SC
punctipennis Reitter, 1910b: 169 **A**: TR
roborowskyi Koenig, 1889b: 538 (*Corymbites*) **A**: XIZ
saginatus Ménétriés, 1832: 154 (*Elater*) **E**: AB AR ST **A**: IN TR
pasticus Ménétriés, 1832: 155 (*Elater*)
salebrosus Gurjeva, 1989a: 257 **A**: NE (Da Hinggan Ling)
sameki Mertlik, 2000: 6 **A**: TR
songoricus Kraatz, 1879b: 283 (*Corymbites*) **A**: KZ MG
tauricus Dolin, 1975b: 1619 **E**: UK (Krym)
whitii Candèze, 1863: 171 (*Corymbites*) **A**: KA "Himalaya"
subgenus *Warchalowskia* Tarnawski, 1995: 64 type species *Corymbites atratus* Ballion, 1878
acceptus Gurjeva, 1989a: 262 **E**: AB **A**: TM
ampliformis Reitter, 1910b: 171 **E**: AB AR

atratus Ballion, 1878: 293 (*Corymbites*) **A**: KI KZ
macropalpus Reitter, 1910b: 170
barlaensis Mertlik, 2000: 2 **A**: TR
denisovae Gurjeva, 1978a: 142 **A**: UZ
informis Kraatz, 1879c: 287 (*Corymbites*) **A**: KI KZ
fraterculus Gurjeva, 1978a: 143
paradoxus Koenig, 1887: 354 (*Corymbites*)
karabachensis Dolin, 1982b: 24 **E**: AB
lemniscatus Denisova, 1948: 43 **A**: KI
logvinenkoae Dolin, 1982b: 21 **E**: AB AR
mirificus Gurjeva, 1972a: 301 **A**: TD
mirus Gurjeva, 1972a: 299 **A**: TD
persimilis Dolin, 1982b: 22 **A**: TD
theresae Tarnawski, 1995: 79 **A**: KZ
turkestanicus Tarnawski, 1995: 81 **A**: "Turkestan"
victor Gurjeva, 1982c: 545 **A**: KI

genus *Selatosomus*, species incertae sedis
nanus Gurjeva, 1975b: 248 **A** or **E**: "Armenia"

genus *Taiyalu* Kishii, 1993b: 11 type species *Thacana shinoharai* Kishii, 1990
shinoharai Kishii, 1990b: 4 (*Thacana*) **A**: TAI

tribe Ctenicerini, species incertae sedis
Corymbites kiesenwetteri Gistel, 1857: 12 [HN] **E**: AU (Schärding) GE (Passau)

subfamily **Diminae** Candèze, 1863

tribe Dimini Candèze, 1863

genus *Brancuccia* Schimmel & Platia, 1991a: 267 type species *Brancuccia atramentaria* Schimmel & Platia, 1991
atramentaria Schimmel & Platia, 1991a: 268 **A**: NP
cava Schimmel & Platia, 1991a: 270 **A**: NP

genus *Csikia* Szombathy, 1910b: 359 type species *dimatoides* Szombathy, 1910
dimatoides Szombathy, 1910b: 360 **A**: TAI
rougemonti Schimmel, 1998d: 251 **A**: HKG
sausai Schimmel, 1996a: 166 **A**: JIX

genus *Dima* Charpentier, 1825: 191 type species *Dima elateroides* Charpentier, 1825
Celox Schaufuss, 1863: 201 type species *Celox dima* Schaufuss, 1863
aspera Schimmel & Platia, 1991a: 276 **A**: NP
beckeri Schimmel & Platia, 1991a: 277 **A**: NP
bhutanensis Schimmel, 1996e: 135 **A**: BT
brancuccia Schimmel & Platia, 1991a: 279 **A**: NP
brunnea Schimmel & Platia, 1991a: 280 **A**: NP
caetrata Schimmel & Platia, 1991a: 282 **A**: NP
canalicollis Schimmel & Platia, 1991a: 283 **A**: NP
convexiculata Schimmel & Platia, 1991a: 284 **A**: SD
dalmatina Küster, 1844: 13 **E**: CR GR
dima Schaufuss, 1863: 202 (*Celox*) **E**: SP
assoi Perez Arcas, 1872: 131
perezii Seidlitz, 1867: 178
dolini Schimmel, 1996a: 167 **A**: YUN
elateroides Charpentier, 1825: 191 **E**: AU BH CR IT SL RO YU

ferruginea Schimmel & Platia, 1991a: 285 **A**: NP
friedrichi Schimmel, 1998d: 251 **A**: SCH
ganeshana Schimmel, 2002: 207 **A**: NP
hayekae Schimmel & Platia, 1991a: 287 **A**: NP
hladilorum Schimmel, 1987: 42 **E**: GR
holzschuhi Schimmel, 1996e: 141 **A**: NP
isabellae Dajoz, 1973a: 103 **E**: GR
jaegeri Schimmel, 2002: 208 **A**: NP
katomandulia Ôhira & Becker, 1972b: 531 **A**: NP
klausnitzeri Schimmel, 1996a: 167 **A**: YUN
kubani Schimmel, 1996a: 168 **A**: YUN
leistoides Candèze, 1863: 234 (*Penia*) **A**: "Hindustan"
lijiangensis Schimmel & Cate, 1991: 137 **A**: YUN
loebli Schimmel & Platia, 1991a: 289 **A**: NP
longicornis Schimmel & Platia, 1991a: 290 **A**: NP
macedonica Schimmel, 1993b: 247 **E**: AL MC
martensi Schimmel & Platia, 1991a: 291 **A**: NP
nebriomorpha W. Suzuki, 1979: 207 **A**: TAI
niehuisi Schimmel & Platia, 1992a: 297 **A**: SCH
oberthueri Schimmel, 1993b: 247 **A**: YUN
olympica Meschnigg, 1934: 67 **E**: GR
pakistanicus W. Suzuki, 1985b: 183 **A**: PA
pecoudi Fleutiaux, 1944a: 40 **E**: GR
probsti Schimmel & Platia, 1991a: 293 **A**: NP
pseudoleistoides Schimmel & Platia, 1991a: 294 **A**: SD
raineri Wurst, 1997a: 4 **E**: GR MC
rugosicollis Schimmel, 2002: 209 **A**: NP
schawalleri Schimmel & Platia, 1991a: 295 **A**: NP
smetanai Schimmel & Platia, 1991a: 297 **A**: NP
spicata Schimmel, 1998d: 252 **A**: SCH
yunnana Fleutiaux, 1916a: 256 **A**: YUN

genus *Neocsikia* Ôhira & Becker, 1972b: 535 type species *Neocsikia nepalensis* Ôhira & Becker, 1972
krishna W. Suzuki, 1982a: 79 **A**: NP SD
nepalensis Ôhira & Becker, 1972b: 535 **A**: NP

genus *Neodima* Schimmel & Platia, 1992a: 299 type species *Neodima sichuanensis* Schimmel & Platia, 1992
cechovskyi Schimmel, 1998a: 222 **A**: SCH
sichuanensis Schimmel & Platia, 1992a: 300 **A**: SCH

genus *Paracsikia* Schimmel & Platia, 1991a: 300 type species *Paracsikia nigerrima* Schimmel & Platia, 1991
brunnea Schimmel & Platia, 1991a: 302 **A**: NP
indica Fleutiaux, 1916a: 256 (*Dima*) **A**: SD
minima Schimmel & Platia, 1991a: 336 (*Penia*) **A**: SD
nigerrima Schimmel & Platia, 1991a: 303 **A**: NP
parvula Schimmel & Platia, 1991a: 305 **A**: SD
schmidti Schimmel, 1995: 52 **A**: NP
sulphurea Schimmel, 2002: 211 **A**: NP

genus *Parapenia* W. Suzuki, 1982b: 84 type species *Parapenia nigroapicalis* W. Suzuki, 1982
jagemanni Schimmel, 2002: 211 **A**: YUN
significata Schimmel, 1998b: 103 **A**: YUN
taiwana Miwa, 1930c: 93 (*Csikia*) **A**: TAI
tonkinensis Fleutiaux, 1918c: 226 (*Penia*) **A**: YUN **ORR**

yunnana Schimmel, 1993b: 249 **A**: YUN

genus *Penia* Laporte, 1838: 11 type species *Elater eschscholtzi* Hope, 1831
annapurnica Schimmel, 2002: 217 **A**: NP
babai Kishii, 1994b: 211 **A**: TAI
bicolorata Schimmel & Platia, 1991a: 313 **A**: BT
bimaculatus Schwarz, 1905: 262 **A**: NP **ORR**
brancuccii Schimmel, 1996e: 164 **A**: BT
brevis Candèze, 1863: 231 **A**: "Himalaya" "India or."
canaliculata Candèze, 1863: 235 **A**: NP SD
cincta Schimmel & Platia, 1991a: 316 **A**: NP
comosa Schimmel, 1993b: 246 (*Dima*) **A**: ZHE
concolor Schimmel & Platia, 1991a: 318 **A**: SD
dolini Schimmel, 1998d: 255 **A**: GUI
eschscholtzi Hope, 1831: 26 (*Elater*) **A**: NP SD
falcata Schimmel, 2003b: 288 **A**: SCH
girardi Schimmel & Platia, 1991a: 321 **A**: SD
gracilis Candèze, 1863: 235 **A**: NP "India"
hayekae Schimmel & Platia, 1991a: 324 **A**: NP SD
hirtella Candèze, 1863: 231 **A**: NP SD
  chotarea Ôhira & Becker, 1972b: 533
holzschuhi Schimmel & Platia, 1991a: 326 **A**: NP SD
hubeiana Schimmel, 2003b: 289 **A**: HUB
imprimosa Schimmel, 2002: 221 **A**: NP
jansoni Schimmel & Platia, 1991a: 328 **A**: SD
jendeki Schimmel, 1998a: 223 **A**: GUI
kasantsevi Schimmel, 2002: 221 **A**: SCH
kaszabi Dolin & W. Suzuki, 1987: 33 **A**: SD
lateralis Schwarz, 1905: 261 **A**: NP SD
longicollis Dolin, 1987a: 37 **A**: NP
longipes Candèze, 1863: 232 **A**: NP SD
longiuscula Schimmel & Platia, 1991a: 333 **A**: NP SD
marmorata Schwarz, 1905: 262 **A**: BT SD **ORR**
martensi Schimmel & Platia, 1991a: 335 **A**: NP SD
nebrioides Candèze, 1863: 233 **A**: SD
ohirai W. Suzuki & Dolin, 1987: 34 **A**: NP SD
omeishanensis Schimmel, 1996a: 169 **A**: YUN
opatroides Candèze, 1895: 54 **A**: BT SD
plagiata Candèze, 1891b: 786 **A**: NP SD **ORR**
proxima Schimmel & Platia, 1991a: 342 **A**: NP
pseudolateralis Schimmel & Platia, 1991a: 344 **A**: NP SD
pseudoplagiata Schimmel & Platia, 1992a: 304 **A**: UP
raii Schimmel & Platia, 1991a: 345 **A**: SD
rugosa Schimmel & Platia, 1991a: 346 **A**: SD
rustica Schimmel & Platia, 1991a: 348 **A**: SD
sachtlebeni Dolin & W. Suzuki, 1987: 32 **A**: NP SD
sausai Schimmel, 1996a: 169 **A**: YUN
semiruber Schimmel & Platia, 1991a: 349 **A**: NP
sichuana Schimmel, 2002: 225 **A**: SCH
similis Schimmel & Platia, 1991a: 351 **A**: NP SD
singularis Schimmel & Platia, 1991a: 353 **A**: NP
striata Schimmel, 1993b: 250 **A**: ZHE
suzukii Schimmel & Platia, 1991a: 355 **A**: SD
takasago Kishii, 1997b: 12 **A**: TAI

tomentosa Candèze, 1863: 232 **A**: NP SD
turnai Schimmel, 2003b: 290 **A**: HUB
wangdiensis Schimmel, 1996e: 190 **A**: BT
wittmeri W. Suzuki & Dolin, 1984: 168 **A**: NP SD
wudangana Kishii & Jiang, 1996: 149 **A**: HUB
yunnana Schimmel, 1993b: 251 **A**: YUN
zeisingi Schimmel, 2002: 227 **A**: SD

genus *Pseudocsikia* Schimmel & Platia, 1991a: 357 type species *Pseudocsikia rustica* Schimmel & Platia, 1991
formosana Ôhira, 1972a: 8 (*Csikia*) **A**: TAI
laticollis Schimmel & Platia, 1991a: 358 **A**: NP
gaoligongshana Schimmel, 1996a: 172 **A**: YUN
rustica Schimmel & Platia, 1991a: 360 **A**: NP

tribe Senodoniini Dolin, 2000

genus *Senodonia* Laporte, 1838: 12 type species *Senodonia quadricollis* Laporte, 1838
Allotrius Laporte, 1840a: 231 [HN] type species *Allotrius quadricollis* Laporte, 1838
Hemiolimerus Candèze, 1863: 227 type species *Hemiolimerus emodi* Candèze, 1863
Orientis Vats & Kashyap, 1992f: 252 type species *Orientis montanus* Vats & Kashyap, 1992
brancuccii Schimmel & Platia, 1992b: 240 **A**: BT NP SD
emodi Candèze, 1863: 183 (*Hemiolimerus*) **A**: BT NP SD
flagellaris Garg & Saini, 1996a: 91 (*Orientis*) **A**: SD **ORR**
hiekei Schimmel & Platia, 1992b: 246 **A**: SE
montanus Vats & Kashyap, 1992f: 252 (*Orientis*) **A**: HP
quadricollis Laporte, 1838: 12 (*Semiotus*) **A**: GUA ZHE **ORR**
sinensis Jagemann, 1943: 100
sculpticollis Fairmaire, 1889a: 350 (*Allotrius*) **A**: GUX **ORR**

subfamily **Hemiopinae** Fleutiuax, 1941

tribe Hemiopini Fleutiuax, 1941

genus *Hemiops* Laporte, 1838 15 type species *Hemiops flava* Laporte, 1838
Oxysternus Latreille, 1834: 164 [HN] type species *Elater crassus* Gyllenhal, 1817
alternata Fairmaire, 1878b: 110 **A**: "Chine centrale"
chinensis Germar, 1843a: 53 **A**: CH
flava Laporte, 1838: 15 **A**: GUA HAI HUN NP SD TAI UP YUN ZHE **ORR**
lutea Germar, 1843a: 52
plana Laporte, 1840a: 254
germari Cate, **nom. nov.** [see New acts] **A**: FUJ GUA GUX HAI HUB HUN JIA JIX SCH XIZ ZHE **ORR**
nigripes Germar, 1843a: 52 [HN]
nigripennis Candèze, 1893b: 178 **A**: CH **ORR**
sinensis Candèze, 1882: 106 **A**: GUA SHG XIN
substriata Fleutiaux, 1903a: 580 **A**: FUJ HUN YUN
assamensis Schwarz: 1905: 264

genus *Parhemiops* Candèze, 1878b: cxcviii type species *Parhemiops palliatus* Candèze, 1878
nepalensis Ôhira & Becker, 1972a: 138 **A**: NP

genus *Plectrosternus* Lacordaire, 1857: 227 type species *Plectrosternus rufus* Lacordaire, 1857
Legna Walker, 1858: 281 type species *Legna idonea* Walker, 1858
convexus Vats & Kashyap, 1992e: 247 **A**: UP
rufus Lacordaire, 1857: 228 **A**: TAI **ORR**

subfamily **Negastriinae** Nakane & Kishii, 1956

genus *Arhaphes* Candèze, 1860: 98 type species *Arhaphes diptychus* Candèze, 1860

Pharotarsus Motschulsky, 1861c: 119 type species *Pharotarsus picturatus* Motschulsky, 1861 (= *Arhaphes diptychus* Candèze, 1860)
Phorotarsus Motschulsky, 1860d: 358 [unused original spelling]
opacus Candèze, 1878b: cxlii (*Arrhaphes*) **A**: NP "Himalaya" **ORR**
takasago Kishii, 1994a: 179, 188 **A**: TAI

genus *Buffeventius* Fleutiaux, 1925a: 102 type species *Buffeventius lividus* Fleutiaux, 1925
somnambulus Wurst, Schimmel & Platia, 2001: 553 **A**: YE

genus *Cardiohypnus* Fleutiaux, 1928c: 282 type species *Cardiophorus mirabilis* Candèze, 1860
beesoni Fleutiaux, 1928c: 286 **A**: UP
bhutanicus Dolin, 1992b: 225 **A**: BT
brancuccii Dolin, 1992b: 218 **A**: NP UP
coomani Fleutiaux, 1925c: 185 (*Cardiophorus*) **A**: TAI **ORR**
sonani Miwa, 1930a: 11
kabakovi Dolin, 1992b: 221 **A**: AF PA
mirabilis Candèze, 1860: 141 (*Cardiophorus*) **A**: NP **ORR**
nepalensis Dolin, 1992b: 223 **A**: BT NP
stephensi Fleutiaux, 1928c: 286 **A**: SD NP
stevensi Fleutiaux, 1928c: 286 [incorrect original spelling]
wittmeri Dolin, 1992b: 215 **A**: KA PA

genus *Fleutiauxellus* Méquignon, 1930a: 95 type species *Aplotarsus maritimus* Curtis, 1840
subgenus *Fleutiauxellus* Méquignon, 1930a: **95** type species *Aplotarsus maritimus* Curtis, 1840
Curtisius Miwa, 1934: 24 type species *Aplotarsus maritimus* Curtis, 1840
algidus J. R. Sahlberg, 1883: 28 (*Negastrius*) **E**: FI NR NT SV
grandiniger Han, 2000: 22 **A**: SC
maritimus Curtis, 1840: 278 (*Aplotarsus*) **E**: AU CZ FI FR GB GE IT NR SK SL SV SZ UK
gracilis Mulsant & Guillebeau, 1855a: 30 (*Cryptohypnus*)
morio Kiesenwetter, 1858a: 362 (*Cryptohypnus*)
scotus Candèze, 1860: 62 (*Cryptohypnus*)
parvus Han, 2000: 24 **A**: SC
sabulo Han, 2000: 27 **A**: SC
yezoensis yezoensis Ôhira, 1973a: 97 **A**: JA
yezoensis tsugaru Ôhira, 1994c: 153 **A**: JA
subgenus *Migiwa* Kishii, 1966a: 14 type species *Negastrius niponicus* Kishii, 1957b
awaensis Ôhira, 1994d: 315 **A**: JA
cruciatus Candèze, 1873: 11 (*Cryptohypnus*) **A**: JA
curatus coreaensis Han, 2000: 20 **A**: SC
curatus curatus Candèze, 1873: 12 (*Cryptohypnus*) **A**: JA JIL
curatus okinawanus Ôhira, 1967c: 45 **A**: JA
curatus septentrionalis Kishii, 1976b: 31 **A**: JA
echigoanus Kishii, 1976b: 32 **A**: JA
insulsus Candèze, 1873: 13 (*Cryptohypnus*) **A**: JA
ishidai Kishii, 1976b: 36 **A**: JA
katamon Kishii, 1985: 28 **A**: JA
kishiii Ôhira, 1967c: 44 **A**: JA
modestus Lewis, 1894a: 186 (*Cryptohypnus*) **A**: JA
niponicus Kishii, 1957b: 69 (*Negastrius*) **A**: JA
ohmi Kishii, 1976b: 35 **A**: JA
ohkawai Kishii, 1998a: 25 **A**: JA
shirahatai Kishii, 1976b: 34 **A**: JA
obscurimaculatus Kishii, 1957b: 69 (*Negastrius*)
tsukadai Kishii, 1987: 17 **A**: JA
tsushimensis Kishii, 1976b: 31 **A**: JA

tutus Lewis, 1894a: 186 (*Cryptohypnus*) **A**: JA
  kurotai Ôhira, 1992a: 223
yezo Kishii, 1976b: 37 **A**: JA
yotsuboshi Kishii, 1976b: 33 **A**: FE JA SC
subgenus *Neomigiwa* Kishii, 1976b: 38 type species *Migiwa* (*Neomigiwa*) *nikkoensis* Kishii, 1976
amamiensis Ôhira, 1967c: 44 (*Migiwa*) **A**: JA
babanus Kishii, 1994a: 180, 190 **A**: TAI
houwau Kishii, 1992b: 28 **A**: JA
interstinctus Lewis, 1894a: 185 (*Cryptohypnus*) **A**: JA
nikkoensis Kishii, 1976b: 38 (*Migiwa*) **A**: JA
optatus Lewis, 1894a: 185 (*Cryptohypnus*) **A**: JA
quadrillum Candèze, 1873: 11 (*Cryptohypnus*) **A**: HEI JA JIL "Korea"
sugiurai Kishii, 1993b: 17 **A**: JA

genus *Futomigiwa* Kishii, 1976b: 39 type species *Negastrius rivalioides* Kishii, 1962
rivalioides Kishii, 1962: 19 (*Negastrius*) **A**: JA

genus *Hisago* Kishii, 1985: 7 type species *Hypolithus korobokkurus* Kishii, 1968
korobokkurus Kishii, 1968: 1 (*Hypolithus*) **A**: JA

genus *Madadicus* Stibick, 1971: 380 type species *Monadicus alluaudi* Fleutiaux, 1932
sulcator Wurst, Schimmel & Platia, 2001: 552 **A**: YE

genus *Menoko* Kishii, 1976b: 42 type species *Menoko pallidula* Kishii, 1976
alpense Kishii, 1987: 19 **A**: JA
difficile Lewis, 1894a: 187 (*Cryptohypnus*) **A**: JA
kawaharai Kishii, 1993b: 13 **A**: JA
nigrum Kishii, 1985: 26 **A**: JA
nitida Fleutiaux, 1902a: 19 (*Cryptohypnus*) **A**: JA
ozakii Ôhira, 1995c: 25 (*Oedostethus*) **A**: JA
pallidula pallidula Kishii, 1976b: 42 **A**: JA
pallidula yasudai Ôhira, 1995b: 6 (*Oedostethus*) **A**: JA
pirika Kishii, 1994d: 22 **A**: JA
tanagami Kishii, 1985: 27 **A**: JA

genus *Microhypnus* Kishii, 1976b: 43 type species *Cryptohypnus agilis* Lewis, 1894
agilis Lewis, 1894a: 189 (*Cryptohypnus*) **A**: JA
  gifuensis Ôhira, 1973a: 97 (*Migiwa*)

genus *Negastrius* C. G. Thomson, 1859: 106 type species *Elater pulchellus* Linnaeus, 1761
  Yezostrius Kishii, 1976b: 30 type species *Negastrius aino* Kishii, 1957
aino Kishii: 1957a: 12 **A**: JA
arenicola Boheman, 1854: 76 (*Cryptohypnoidus*) **E**: CT DE GE NR SV PL
hieroglyphicus Pic, 1899h: 139 (*Cryptohypnus*) **N**: EG
nadezhdae Dolin, 1971: 642 **A**: KI
pulchellus Linnaeus, 1761: 209 (*Elater*) **E**: AU BE BH BG BY CT CZ DE EN FI FR GB GE HU IT LA LT LU NL NR NT PL PT RO SK SL SP ST SV SZ UK **A**: ES FE KZ MG WS **NAR**
  bipunctatus Schilsky, 1888: 187 (*Cryptohypnus*)
  exiguus Randall, 1838b: 35 (*Elater*)
  guttatulus Melsheimer, 1845: 214 (*Cryptohypnus*)
  olivieri Buysson, 1900a: 237 (*Hypnoidus*)
  panzeri Buysson, 1900a: 237 (*Hypnoidus*)
  quadrilunulatus Buysson, 1900a: 237 (*Hypnoidus*)
  ripicola Friederichs, 1901: 82 (*Cryptohypnus*)
  trimaculatus Fabricius, 1792b: 233 (*Elater*)

pagmanus Nakane, 1966c: 237 **A**: AF
sabulicola Boheman, 1854: 74 (*Cryptohypnoidus*) **E**: AU BE BG BU BY CT CZ DE FR GB GE GR HU IT LA LT NL NT PL RO SK SV UK
contentus Friederichs, 1901: 82 (*Cryptohypnus*)
laetus Friederichs, 1901: 82 (*Cryptohypnus*)
maestus Friederichs, 1901: 82 (*Cryptohypnus*)
modestus Friederichs, 1901: 82 (*Cryptohypnus*)
yosiii Nakane, 1966c: 236 **A**: AF

genus *Neohypdonus* Stibick, 1971: 374 type species *Cryptohypnus gentilis* LeConte, 1866
arcticus altaicus Cherepanov, 1957: 230 (*Hypnoidus*) **A**: ES KZ MG
arcticus arcticus Candèze, 1860: 66 (*Cryptohypnus*) **E**: FI NR NT SV **A**: WS
boreaphilus C. G. Thomson, 1864: 112 (*Negastrius*)
nepalensis Ôhira & Becker, 1973a: 72 **A**: NP
tumescens LeConte, 1853c: 486 (*Cryptohypnus*) **A**: ES **NAR**

genus *Oedostethus* LeConte, 1853c: 489 type species *Oedostethus femoralis* LeConte, 1853
aeneocollis Dolin, 1987b: 346 **A**: FE
aerarius Reitter, 1895b: 89 (*Hypnoidus*) **A**: ES MG WS
gurjevae Dolin & Bessolitzina, 1982: 633
ainu Ôhira, 1995b: 7 **A**: JA
ambiguus Buysson, 1914a: 41 (*Hypnoidus*) **A**: KI KZ XIN
apterus Dolin & Bessolitzina, 1990: 23 **A**: ES FE
babai Ôhira, 1991: 12, 16 **A**: JA
bodoanus Reitter, 1913e: 658 (*Hypnoidus*) **A**: KI XIN
tianshanicus Dolin, 1978: 20 (*Negastrius*)
canicapillus Kim, Han & Lee, 2000: 73 **A**: SC
convexus Cherepanov, 1957: 232 (*Hypnoidus*) **A**: WS
cryptohypnoidus Miwa, 1930a: 6 (*Paracardiophorus*) **A**: TAI
curticornis Dolin & Bessolitzina, 1982: 632 **A**: ES
ghilarovi Dolin, 1978: 20 (*Negastrius*) **A**: FE
graniger Cherepanov, 1956: 45 (*Hypnoidus*) **A**: ES MG WS
hosodai Kishii, 1992b: 24 **A**: JA
insolentus Dolin & Bessolitzina, 1990: 26 **A**: FE
kadowakii Kishii, 1976b: 41 (*Neohypdonus*) **A**: JA
kanmiyai Ôhira, 1991: 9, 15 **A**: JA
kaszabi Gurjeva, 1968: 293 (*Cryptohypnus*) **A**: MG NMO
kohnoi Ôhira, 1991: 12, 16 **A**: JA
kolymensis Dolin & Bessolitzina, 1990: 25 **A**: FE
koreanus Kim, Han & Lee, 2000: 71 **A**: SC
latissimus Cherepanov, 1957: 229 (*Hypnoidus*) **A**: ES
matisi Dolin, 1992a: 13 **A**: ES FE
mediocris Gurjeva, 1972b: 457 (*Cryptohypnus*) **A**: MG
minutulus Cherepanov, 1957: 232 (*Hypnoidus*) **A**: WS
mystax Gurjeva, 1971: 83 (*Cryptohypnus*) **A**: ES MG
nazarenkoi Dolin, 1992a: 18 **A**: FE
nitobei Miwa, 1930a: 2 (*Hypnoidus*) **A**: TAI
novaki Binaghi, 1935: 22 (*Hypnoidus*) **E**: IT
nubilus Bessolitzina, 1974: 81 (*Negastrius*) **A**: ES
ovalatus Kim, Han & Lee, 2000: 72 **A**: SC
petrenkoi Dolin, 1992a: 15 **A**: ES FE HEI KZ WS
problematicus Dolin, 1992a: 16 **A**: FE
punctulatus Dolin & Katyukha, 1977: 178, 181 (*Negastrius*) **A**: FE
quadripustulatus Fabricius, 1792b: 235 (*Elater*) **E**: AU BE BY CT CZ DE EN FR GB GE HU IT LA LT NL NT PL RO SK SP SV SZ UK **A**: ES FE HEI WS **NAR**

agricola Zetterstedt, 1824: 155 (*Elater*)
quadrum Gyllenhal, 1827: 357 (*Elater*)
hoepfneri Germar, 1844b: 143 (*Cryptohypnus*)
sachalinensis Dolin & Katyukha, 1977: 177, 180 (*Negastrius*) **A**: FE
similarius Dolin & Medvedev, 2002: 95 **E**: NT
simplicipunctatus Cherepanov, 1956: 45 (*Hypnoidus*) **A**: ES WS
sincerus Gurjeva, 1975a: 47 (*Negastrius*) **A**: ES
submontanus Dolin & Bessolitzina, 1982: 633 **A**: ES
telluris Lewis, 1879b: 156 (*Cryptohypnus*) **A**: JA
tenuicornis Germar, 1824: 60 (*Elater*) **E**: AU BU BY CR CZ EN GE IT HU LA LT NT PL RO SK SL SZ **A**: WS
elongatus L. Redtenbacher, 1849: 305 (*Cryptohypnus*)
minusculus Bach, 1852: 35 (*Limonius*)
tenuis Gurjeva, 1963a: 25 (*Cryptohypnus*) **A**: KI
ugo Kishii, 1992b: 26 **A**: JA
varians Gurjeva, 1968: 293 (*Cryptohypnus*) **A**: MG NMO
yamamotoi Kishii, 1976b: 40 (*Neohypdonus*) **A**: JA
zherichini Dolin, 1985b: 144 **A**: ES WS

genus *Quasimus* Gozis, 1886: 22 type species *Elater minutissimus* Germar, 1823
subgenus *Miquasus* Kishii, 1959a: 9 type species *Cryptohypnus luteipes* Candèze, 1873
arimotoi Ôhira, 2002b: 483 **A**: JA (Amami-Ôshima)
atayal Kishii, 1994a: 182, 193 **A**: TAI
besucheti Dolin, 2001: 137 **A**: SD
cariosus Dolin, 1997a: 152 **A**: NP
convexipennis Kishii, 1982: 48 **A**: JA
formosanus Ôhira, 1968f: 364 **A**: TAI
kai Kishii, 1994d: 26 **A**: JA
luteipes Candèze, 1873: 15 (*Cryptohypnus*) **A**: JA
sobosanus Kishii, 1970: 10
ohkurai Ôhira, 1992b: 123 **A**: JA (Yakushima)
tenuis Kishii, 1994d: 24 **A**: JA (Ryukyus)
subgenus *Quasimus* Gozis, 1886: 22 type species *Elater minutissimus* Germar, 1823
abruptus Dolin, 1997a: 144 **A**: HP SD **ORR**
ami Kishii, 1994a: 184, 195 **A**: TAI
anhuiensis Li, 1993: 164 **A**: ANH
atomarius Candèze, 1882: 76 (*Hypnoidus*) **A**: SD "Himalaya"
babai Kishii, 1970: 26 **A**: JA
bengalicus Dolin, 2001: 132 **A**: SD **ORR**
brancuccii Dolin, 1993: 191 **A**: NP SD
brevis Kishii, 1980: 88 **A**: NP **ORR**
chibi Kishi, 1976b: 21 **A**: JA
cordatus Miwa, 1934: 254 **A**: JA (Ryukyus)
echigoanus Kishii, 1976a: 7 **A**: JA
elongatissimus Dolin, 1993: 186 **A**: KA PA
exilis Motschulsky, 1858c: 58 (*Cryptorhypnus* [sic!]) **A**: BT NP YUN
hiroyoshii Ôhira, 1998b: 373 **A**: JA (Ryukyus)
holzschuhi Dolin, 2001: 133 **A**: KA NP PA SD
horishanus Miwa, 1930a: 3 **A**: TAI
imasakai Kishii, 1976b: 22 **A**: JA (Okinawa-honto)
inopinatus Dolin, 2001: 135 **A**: SD **ORR**
ishigakianus Kishii, 1976b: 19 **A**: JA (Ishigaki-jima)
issunboushi Kishii, 1966a: 10 **A**: JA
japonicus Kishii, 1959a: 9 **A**: JA
kiiensis Kishii, 1976a: 8 **A**: JA

kintaroui Kishii, 1982: 50 **A**: JA (Ishigaki-jima)
korellipticus Han, 2000: 38 **A**: SC
kyotoensis Kishii, 1966a: 12 **A**: JA
liliputanus Germar, 1844b: 145 (*Cryptohypnus*) **E**: IT (Sicily) **N**: AG TU **A**: CY
longulus Kishii, 1970: 26 **A**: JA
minutissimus Germar, 1823a: pl. 8 (*Elater*) **E**: AB AL AR AU BE BH BU CR CT CZ FR GE GG GR HU IT MC NL PL RO SK SL SP ST SZ UK YU **A**: IN TR "Siberia"
minutissimus Laporte, 1840a: 246 (*Elater*) [HN]
miwai Ôhira, 1966c: 217 **A**: TAI
miyakonis Kishii, 1976b: 23 **A**: JA (Miyako-jima)
obovatus Dolin, 1993: 183 **A**: NP SD UP
okinawensis Ôhira, 1998b: 375 **A**: JA (Okinawa)
ovalioides Kishii, 1970: 15 **A**: JA (Tsushima)
ovalis Candèze, 1873: 15 (*Cryptohypnus*) **A**: FE JA "China"
pakistanicus Dolin, 1993: 185 **A**: PA "India"
parallelus Schwarz, 1902b: 332 (*Hypnoidus*) **A**: NP SD **ORR**
parvulus Kishii, 1970: 12 **A**: JA
persimilis Dolin, 1993: 188 **A**: NP SD
petrimagni Dolin & Latifi, 1997: 83 **A**: TD
pseudovalis Kishii, 1994d: 29 **A**: NC
ranzanus Kishii, 1970: 17 **A**: JA
reclinatus Ôhira, 1966c: 217 **A**: TAI
satoi kimurai Kishii, 1970: 12 **A**: JA (Yaku-shima)
satoi ogatai Kishii, 1985: 30 **A**: JA (Amami-Ôshima)
satoi satoi Ôhira, 1967c: 43 **A**: JA (Tokara, Amami-Ôshima)
sausai Dolin, 2001: 122 **A**: SD **ORR**
shaxianensis Jiang, 1999: 156, 186 **A**: FUJ
shibatai matobai Kishii, 1974: 4 **A**: JA (Amami-Ôshima)
isaoi Kishii, 1979a: 3
shibatai shibatai Kishii, 1970: 19 **A**: JA
shimabarensis Kishii, 1979a: 3 **A**: JA
shirakii Miwa, 1930a: 6 **A**: TAI
shouichii Kishii, 1979a: 2 **A**: JA (Miyako-jima)
subovalis Fleutiaux, 1930c: 643 **A**: HKG YUN **ORR**
takahashii Miwa, 1934: 254 **A**: JA (Ishigaki)
takakurai Kishii, 1976a: 8 **A**: JA
tsurugi Kishii, 1976b: 20 **A**: JA
tsushimensis Kishii, 1970: 22 **A**: JA (Tsushima)
uguriensis heianus Kishii, 1970: 25 **A**: JA (Kuchi-no-erabu)
uguriensis okicola Kishii, 1970: 24 **A**: JA (Oki-no-shima)
uguriensis uguriensis Kishii, 1970: 24 **A**: JA (Uguri-jima)
vunun Kishii, 1994a: 184, 197 **A**: TAI
wittmeri Dolin, 1993: 189 **A**: BT NP
yakuensis Kishii, 1959a: 11 **A**: JA
yamayai Kishii, 1985: 29 **A**: JA (Iriomote-jima)
yasuii Kishii, 1970: 16 **A**: JA
subgenus *Wittmeroquasimus* Dolin, 1993: 195 type speces *Quasimus* (*Wittmeroquasimus*) *ocellatus* Dolin, 1993
interpositus Dolin, 1993: 193 **A**: BT "India"
ocellatus Dolin, 1993: 195 **A**: BT "India"
paradoxus Dolin, 1993: 197 **A**: NP

genus *Yukoana* Kishii, 1959a: 7 type species *Cryptohypnus ellipticus* Candèze, 1873
amamiensis Ôhira, 1967c: 43 **A**: JA (Amami-Ôshima)
angustatus Miwa, 1927b: 107 **A**: JA

bhutanicus Dolin, 1993: 198 **A**: BT (*Quasimus*)
carinicollis Lewis, 1894a: 188 (*Cryptohypnus*) **A**: JA
elongata amimicola Kishii, 1970: 6 **A**: JA (Amami-Ôshima)
elongata elongata Kishii, 1970: 6 **A**: JA (Yaku-shima)
elongata okinawana Ôhira, 1971b: 534 **A**: JA (Okinawa-honto)
elliptica Candèze, 1873: 14 (*Cryptohypnus*) **A**: JA
punctatus Miwa, 1927b: 107 (*Quasimus*)
formosana Ôhira, 1966c: 217 **A**: TAI
housaiana Kishii, 1994: 188, 199 **A**: TAI
iramatsui Ôhira, 1978b: 552 **A**: JA
kashmirensis Dolin, 1993: 199 (*Quasimus*) **A**: KA
monticola Kishii, 1961: 33 **A**: JA (Yaku-shima)
shirozuana Kishii, 1961: 29 **A**: JA SC
taiwana Ôhira, 1968f: 365 **A**: TAI
takasago Kishii, 1994a: 188, 201 **A**: TAI
tamui Kishii, 1959a: 7 **A**: JA
terukoe Kishii, 1961: 32 **A**: JA

genus *Zorochros* C. G. Thomson, 1859: 106 type species *Elater dermestoides* Herbst, 1806
subgenus *Pronegastrius* Ôhira, 1963c: 275 type species *Cryptohypnus humeralis* Candèze, 1873
humeralis humeralis Candèze, 1873: 13 (*Cryptohypnus*) **A**: JA JIL
humeralis tsushimensis Kishii, 1976b: 30 **A**: JA (Tsushima)
humeralis yakuensis Kishii, 1976b: 29 **A**: JA (Yakushima)
subgenus *Thurana* Stibick, 1971: 384 type species *Cryptohypnus scitus* Candèze, 1897
iriei Ôhira, 1977a: 11 **A**: JA (Ishigaki)
perpusillus Candèze, 1878b: cxlii (*Cryptohypnus*) **A**: NP SD „West Himalaya“
sausai Dolin, 2002a: 39 **A**: YUN
subgenus *Yamatostrius* Kishii, 1976b: 26 type species *Cryptohypnus albipilis* Candèze, 1873
albipilis Candèze, 1873: 12 (*Cryptohypnus*) **A**: JA
brevis Miwa, 1927b: 107 (*Hypnoidus*)
osawai Ôhira, 1972b: 20 (*Negastrius*) **A**: JA
subgenus *Zorochros* C. G. Thomson, 1859: 106 type species *Elater dermestoides* Herbst, 1806
Chibistrius Kishii, 1976b: 27 type species *Hypnoidus lewisi* Schwarz, 1906
aenigmatosus Dolin, 1995: 20 **A**: TD UZ
aequicollis Reitter, 1895b: 89 (*Hypnoidus*) **E**: AB AR ST **A**: KI TD TM UZ
alysidotus Kiesenwetter, 1858a: 368 (*Cryptohypnus*) **E**: AB AR BU CR FR GR IT MC **A**: CY SY TR
amalec Peyerimhoff, 1907: 22 (*Hypnoidus*) **A**: IS SI
amamiensis Arimoto, 1987: 133 (*Yamatostrius*) **A**: JA (Amami-Ôshima)
anatolicus Dolin & Mertlik, 2002: 91 (*Zorochrus*) **E**: AR GG **A**: PA TR
angularis Candèze, 1869: 122 (*Cryptohypnus*) [RN] **E**: PT SP **N**: AG MO
exilis Candèze, 1860: 84 (*Cryptohypnus*) [HN]
araxicola Reitter, 1895b: 91 (*Hypnoidus*) **E**: AR GG **A**: IN TR
birganjianus Ôhira & Becker, 1973a: 70 (*Zorochrus*) **A**: NP
boubersi Leseigneur, 1970: 30 **E**: AU CZ IT SL
brancuccii Dolin, 1999a: 32 (*Zorochrus*) **A**: NP
capillatus Dolin, 1999a: 38 (*Zorochrus*) **A**: FUJ
caucasicus Jagemann, 1939: 68 (*Hypnoidus*) **E**: GG ST **A**: TR
cinefactus Lewis, 1894a: 188 (*Cryptohypnus*) **A**: JA
cretensis Platia, 2004c: 106 **E**: GR (Kríti)
crux Küster, 1849b: 15 (*Cryptohypnus*) **E**: FR (Corse) IT (Sardegna)
reductus Buysson, 1912f: 134 (*Cryptohypnus*)
curtus Germar, 1844b: 141 (*Cryptohypnus*) **E**: FR IT PT SP **N**: AG MO TU
davidianus Candèze, 1882: 75 (*Cryptohypnus*) **A**: SCH
dermestoides Herbst, 1806: 85 (*Elater*) **E**: AN AR AU BE BH BU BY CR CT CZ DE EN FI FR GE GB GG GR IR IT LT MC MD NL NR NT PL PT RO SK SL SP ST SV SZ UK YU **N**: MO

bipustulatus Schilsky, 1888: 188 (*Cryptohypnus*)
diluviatus Buysson, 1913b: 61 (*Hypnoidus*)
humeropictus Buysson, 1900a: 241 (*Hypnoidus*)
laevithorax Buysson, 1906b: 470 (*Hypnoidus*)
minimus Lacordaire, 1835: 660 (*Elater*)
pseudotetragraphus Binaghi, 1933: 210 (*Hypnoidus*)
submaculatus Reitter, 1895b: 88 (*Hypnoidus*)
subnotatus Reitter, 1895b: 88 (*Hypnoidus*)
dufouri Buysson, 1900a: 239 (*Hypnoidus*) **E**: AU BE BH CZ FR GE MC SK SL SZ
flavipes Aubé, 1850: 338 (*Cryptohypnus*) **E**: AU BE CZ FR GE IT HU NR PL RO SK SL SP ST SZ
fauconneti Pic, 1910v: 234 (*Hypnoidus*)
mangonis Buysson, 1913b: 63 (*Hypnoidus*)
quadrum Dufour, 1851: 327 (*Elater*)
viturati Pic, 1910v: 234 (*Hypnoidus*)
freyi Dolin, 1996: 267 (*Zorochrus*) **A**: IQ
georgi Dolin, 1996: 268 (*Zorochrus*) **A**: IQ
georgicus Dolin & Chantladze, 1980: 189 (*Zorochrus*) **E**: GG **A**: TR
godavariensis Ôhira & Becker, 1974b: 557 (*Zorochrus*) **A**: NP **ORR**
gurjevae Dolin, 1995: 24 (*Zorochrus*) **A**: KZ
heyrovskyi Roubal, 1940: 52 (*Hypnoidus*) **A**: IQ SY TR
hispidus Dolin & Atamuradov, 1987: 75 (*Zorochrus*) **A**: TM
housaianus Kishii, 1994a: 181, 192 (*Zorochrus*) **A**: TAI
hummeli Fleutiaux, 1936a: 17 (*Zorochrus*) **A**: CH MG
ibericus Franz, 1967a: 34 (*Hypnoidus*) **E**: AU BU CZ FR GG IT MC SK SP ST UK **N**: AG TU
ponticus Dolin, 1977a: 21 (*Zorochrus*)
indicus Motschulsky, 1858c: 57 (*Cryptorhypnus* [sic!]) **A**: NP **ORR**
fouqueti Fleutiaux, 1918b: 224 (*Hypnoidus*)
interpositus Dolin & Agajev, 1993: 65 (*Zorochrus*) **A**: AB GG TM
iranicus Dolin, 2002b: 92 (*Zorochrus*) **A**: IN PA
kopeckyi Platia & Gudenzi, 1998c: 636 **A**: TR
kopetdagensis Dolin, 1994: 44 (*Zorochrus*) **A**: IN TM
kourili Roubal, 1936b: 65 (*Hypnoidus*) **E**: BU GR MC
kurdicus Dolin & Mertlik, 2002: 90 (*Zorochrus*) **A**: TR
lazorkoi Dolin, 2003: 30 (*Zorochrus*) **A**: UZ
lewisi Schwarz, 1906: 316 (*Hypnoidus*) [RN] **A**: JA
atomarius Lewis, 1894a: 187 (*Cryptohypnus*) [HN]
ivanovi Jakobson, 1913: 753 (*Zorochrus*) [RN]
lotharensis Ôhira & Becker, 1973a: 71 (*Zorochrus*) **A**: NP
mansusanensis Han, 2000: 35 (*Zorochrus*) **A**: SC
meridionalis Laporte, 1840a: 246 (*Elater*) **E**: AB AU BH BU CR CZ FR GE GG GR HU IT MC NL PL RO SK SL SP ST SZ **A**: TR
lapidicola Germar, 1844b: 144 (*Cryptohypnus*)
levantinus K. Daniel & J. Daniel, 1903b: 333 (*Hypnoidus*)
troglodytes Dufour, 1851: 327 (*Elater*)
merkli Mertlik, 1998: 75 (*Zorochros*) **A**: GR
mesasiaticus Dolin, 1995: 21 (*Zorochrus*) **A**: KI KZ TD UZ
murinoides Gurjeva, 1963a: 26 (*Cryptohypnus*) **A**: KI KZ TD TM UZ
murinus Reitter, 1895b: 91 (*Hypnoidus*) **E**: AB AR GG ST **A**: KI TM TR
nanus Gurjeva, 1963a: 23 (*Cryptohypnus*) **A**: KI
nikkoensis Kishii, 1976b: 27 (*Pronegastrius*) **A**: JA
oblongus Fleutiaux, 1934c: 476 (*Zorochrus*) **A**: HP **ORR**
pakistanicus Dolin, 1999a: 34 (*Zorochrus*) **A**: PA
palii Dolin & Atamuradov, 1987: 75 (*Zorochrus*) **A**: TM
pallicrus Desbrochers des Loges, 1875a: 37 (*Cryptohypnus*) **A**: IN IQ SY TR
cryptohypnoides Candèze, 1882: 57 (*Heteroderes*)
delicatulus Schwarz, 1897a: 129 (*Hypnoidus*)

penevi Dolin, 2003: 31 (*Zorochrus*) **E**: ST
persimilis Dolin, 1996: 270 (*Zorochrus*) **A**: LE
pictus Candèze, 1892b: 491 (*Cryptohypnus*) **A**: UP **ORR**
pilosellus Reitter, 1895b: 90 (*Hypnoidus*) **E**: AB AR GR **A**: IN TM TR
pumilio Kiesenwetter, 1858a: 368 (*Cryptohypnus*) **E**: GR FR IT
quadriguttatus Laporte, 1840a: 245 (*Elater*) **E**: AR AU BU CZ FR GR HU IT MC PL PT RO SK SP ST
tetragraphus Germar, 1844b: 143 (*Cryptohypnus*)
quadrinaevus Reitter, 1895b: 87 (*Hypnoidus*) **E**: AB AR ST **N**: AG **A**: IN TR
recellentus Dolin, 1995: 19 (*Zorochrus*) **E**: AB AR **A**: IN TR
sericeus Candèze, 1892b: 490 (*Cryptohypnus*) **A**: PA **ORR**
sinensis Fleutiaux, 1936a: 17 (*Zorochrus*) **A**: GAN
stibicki Leseigneur, 1970: 30 (*Zorochrus*) **E**: AU GE IT HU SL YU **A**: TR
suzukii Ôhira, 1995d: 187 (*Zorochrus*) **A**: JA
tarbinskyi Dolin, 1998b: 93 (*Zorochrus*) **A**: KI
trigonochirus Binaghi, 1933: 205 (*Hypnoidus*) **E**: FR IT
truncatus Han, 2000: 33 (*Zorochrus*) **A**: NC SC
tschatkalensis Dolin, 1971: 641 (*Zorochrus*) **A**: KI
turkmenicus Dolin & Atamuradov, 1994: 48 (*Zorochrus*) **A**: TM
wrasei Dolin, 1999a: 36 (*Zorochrus*) **A**: SHA
yunnanus Fleutiaux, 1940c: 100 (*Zorochrus*) **A**: BT GUI NP YUN **ORR**
yunanus Fleutiaux, 1940c: 100 (*Zorochrus*) [incorrect original spelling]

genus *Zorochros* species incertae sedis
Cryptohypnus tetratoma Rosenhauer, 1856: 136 **E**: SP
Elater quadripustulatus Preyssler, 1792: 8 **E**: CZ (Prag)

subfamily Negastriinae, species incertae sedis
Cryptohypnus advena Gistel, 1857: 40 **E**: GE (Bayern)
Cryptohypnus niphe Gistel, 1857: 40 **E**: GE
Hypnoides quadrimaculatus Vats & Chauhan, 1991b: 294 [DA] **A**: UP
Monadicus amriki Garg, Vasu & Vats, 1995: 364 [EA] **A**: SD
Monadicus colocassius Vats & Chauhan, 1991b: 296 [EA] **A**: UP
Monadicus pyrusus Vats & Chauhan, 1991b: 297 [EA] **A**: UP

subfamily **Cardiophorinae** Candèze, 1860
tribe Cardiophorini Candèze, 1860

genus *Allocardiophorus* Ôhira, 1989b: 79 type species *Paracardiophorus nigroapicalis* Miwa, 1927
nigroapicalis Miwa, 1927b: 109 (*Paracardiophorus*) **A**: JA (Ryukyus)
shirozui Ôhira, 1968f: 368 (*Paracardiophorus*) **A**: HUB SCH TAI

genus *Blaiseus* Fleutiaux, 1931: 307 type species *Blaiseus bedeli* Fleutiaux, 1931
bedeli Fleutiaux, 1931: 308 **A**: GUA **ORR**

genus *Cardiophorus* Eschscholtz, 1829: 34 type species *Elater gramineus* Scopoli, 1763
subgenus *Cardiophorus* Eschscholtz, 1829: 34 type species *Elater gramineus* Scopoli, 1763
Caloderus Stephens, 1830: 269 type species *Elater thoracicus* Fabricius, 1775 (= *Elater gramineus* Scopoli, 1763)
Lasiocerus Buysson, 1912e: 129 type species *Cardiophorus schusteri* Buysson, 1912
Melanotus Gistel, 1834: 11 [HN] type species *Elater thoracicus* Fabricius, 1775 (= *Elater gramineus* Scopoli, 1763)
Nyctor Semenov & Pjatakova, 1936: 101 type species *Nyctor expallidus* Semenov & Pjatakova, 1936
Zigocardiophorus Mardjanian, 1981: 247 type species *Cardiophorus nigratissimus* Buysson, 1891
abdominalis Aubé, 1850: 337 **N**: AG TU
acuminatus Platia & Gudenzi, 2002a: 28 **A**: TR
aemulus Platia & Gudenzi, 2002a: 28 **A**: TR
aeneomicans Gurjeva, 1966: 82 **A**: TM UZ

aeneoniger Gurjeva, 1966: 81 **A**: TD TM UZ
aenigmatosus Dolin & Šauša, 1997: 44 **A**: FE
agilis Platia & Gudenzi, 2002a: 30 **A**: TR
albofasciatus Schwarz, 1893a: 1 **E**: IT (Sicilia)
alepensis Pic, 1914d: 10 **A**: SY
analis Schwarz, 1892d: 383 **A**: TR
angustiformis Buysson, 1914a: 41 **N**: MO
annulicornis Desbrochers des Loges, 1875a: 40 **E**: SP
anticus Erichson, 1840: 285 **E**: BU FR GR HU IT SK **A**: TR "Palestine"
  sidonius Pic, 1912n: 66
antoinei Buysson, 1924b: 207 **N**: MO
apterus Platia & Gudenzi, 2002a: 31 **A**: TR
arabs Wurst, Schimmel & Platia, 2001: 548 **A**: YE
araxicola Iablokoff-Khnzorian, 1970c: 177 **E**: AB AR **A**: TR
arcanus Dolin, 1985a: 135 **E**: AB AR
argiolus Gené, 1836: 174 (*Elater*) **E**: FR (Corse) IT (Sardegna) MA
  maculicrus Desbrochers des Loges, 1869: 34 [= 1870b: 100]
  andreini Buysson, 1922: 255
  angularoides Buysson, 1913b: 49
  anticecoloratus Buysson, 1922: 255
  conjunctellus Buysson, 1922: 255
  fascicollis Gené, 1836: 174 (*Elater*)
  involutus Buysson, 1922: 255
  posticeminiatus Buysson, 1913b: 49
  rufimembris Buysson, 1913b: 49
  sardeus Candèze, 1860: 131
  tibialoides Buysson, 1913b: 50
asellus Erichson, 1840: 300 **E**: AU BE BU BY CT CZ DE EN FI FR GB GE GR HU LA LU MC NL NT PL RO SK SP ST SV UK
asper Gurjeva, 1966: 76 **A**: KZ UZ
aussanicus Wurst, Schimmel & Platia, 2001: 547 **A**: YE
azerbaidzhanicus Agajev, 1985b: 56 **E**: AB
badchysicus Dolin, 1977b: 354 [= 1977c: 357] **A**: TM
baibaranus Miwa, 1930a: 10 **A**: TAI
balchanensis Dolin & Atamuradov, 1989: 35 **A**: TM
balearicus Platia & Gudenzi, 1999a: 18 **E**: SP (Baleares)
balkenohli Platia & Schimmel, 1997: 304 **A**: SA
barrosi Guérin, 1893: 63 [RN] **E**: PT SP
  reitteri Guérin, 1893: 31 [HN]
beduinus Buysson, 1902c: 431 **N**: AG MO TU
  mekaliensis Pic, 1912n: 65
  retrominiatus Buysson, 1911b: 145
  tebessensis Pic, 1912n: 65
bellus Platia & Gudenzi, 2000e: 603 **A**: TR
belonis Desbrochers des Loges, 1870b: 101 **E**: MA
bilyi Platia & Gudenzi, 2000d: 136 **E**: GR
biguttatus A. G. Olivier, 1790: no 31: 47 (*Elater*) **E**: BE FR IT SP **N**: AG
  familiaris Buysson, 1902b: 313
  farinesi Villa & Villa, 1838: 62
  ornatus Candèze, 1860: 136
  pictus Laporte, 1840a: 249 (*Caloderus*)
  submaculatus Laporte, 1840a: 249 (*Caloderus*)
  suturatus Buysson, 1902b: 313
bipunctatus Fabricius, 1798: 140 (*Elater*) **E**: PT SP **N**: AG MO
  alcalanus Pic, 1923i: 10
  almeriensis Cobos, 1950b: 155
  boeticus Buysson, 1902c: 439

conjunctus Pic, 1912g: 25
faroensis Pic, 1912l: 57
lateobliterus Pic, 1902n: 306
nigrovelatus Buysson, 1912d: 100
perrisi Pic, 1893e: 122
quadristigmosus Buysson, 1912d: 101
semilimbatus Pic, 1912g: 25
semiobliteratus Pic, 1902n: 306
sexstigmosus Buysson, 1912d: 100
smithi Fuente, 1920: 151
stigmosulus Buysson, 1912d: 101
bogatschevi Dolin, 1985a: 134 **E**: AB **A**: IN KZ TM
bucculatus Candèze, 1860: 211 **A**: UP "Hindustan" **ORR**
bucharensis Schwarz, 1900: 102 **A**: TD UZ
candezei Koenig, 1889b: 534 **A**: XIN
carnosus Platia & Gudenzi, 2002a: 32 **A**: IN IQ TR
castillanus Buysson, 1902c: 435 **E**: FR IT PT SP **N**: AG MO
arribasi Cobos, 1961: 75
caucasicus Desbrochers des Loges, 1875a: 40 **E** "Caucasus"
chivensis Stepanov, 1935: 198 **A**: TD TM UZ
chobauti Buysson, 1899d: 213 **N**: AG
collaris Erichson, 1840: 283 **E**: FR (Corse) IT
paganettii Pic, 1907g: 129
retropictus Buysson, 1911a: 35
commiphorae Wurst, Schimmel & Platia, 2001: 549 **A**: YE
convexithorax Desbrochers des Loges, 1869: 34 [= 1870b: 99] **N**: EG **A**: IS SI SY
costicollis Vats & Chauhan, 1991c: 21 **A**: UP **ORR**
crux Erichson, 1840: 291 **N**: EG **AFR**
alfierii Pic, 1931i: 136
gaertneri Schwarz, 1901a: 16
raffrayi Schwarz, 1896a: 92
shimonius Fleutiaux, 1919: 93
cyanipennis Mulsant & Wachanru, 1852a: 2 **A**: TR
delagrangei Pic, 1899h: 139
decoratus Fleutiaux, 1936a: 18 **A**: GAN
demaisoni Buysson, 1898c: 323 **N**: EG
discicollis Herbst, 1806: 92 (*Elater*) **E**: AU BH BU CR CT CZ GE GG GR HU IT MC MD PL RO SK SL ST SZ UK YU **A**: KZ SY TD TR WS XIN "Palestine"
brevinotaticollis Pic, 1913m: 169
ganglbaueri Buysson, 1897: 259
pleuralis Buysson, 1899d: 212
sinuatevittatus Pic, 1928e: 9
volgensis Fischer von Waldheim, 1824: 202 (*Elater*)
doggeri Vats & Chauhan, 1991c: 13 **A**: HP
dolini Mardjanian, 1985: 137 **E**: AR BU GR HU MC RO ST TR UK **A**: TR
ebeninus Germar, 1824: 58 (*Elater*) **E**: AU BU BY CR CT CZ EN FR GE GG GR HU IT LA LT MC PL RO SK ST SZ UK **A**: ES KZ MG TR
breviatus Desbrochers des Loges, 1875a: 39
hiekei Mardjanian, 1983: 143
ecorculatus Gurjeva, 1972a: 307 **A**: TD TM
eleonorae Gené, 1836: 177 (*Elater*) **E**: FR IT
bickhardti Pic, 1910m: 49 [RN]
humeralis Bickhardt, 1908: 202 [HN]
eliasi Pic, 1904p: 298 **A**: TR
elongatissimus Escalera, 1914: 212 **N**: MO
exaratus exaratus Erichson, 1840: 304 **E**: FR GR IT PT SP **N**: AG TU
pusillus Desbrochers des Loges, 1869: 46 [= 1870b: 103]

strigipennis Buysson, 1902b: 317
exaratus tenueseriatus Cobos, 1958: 53 **N**: MO
eximius Candèze, 1860: 140 **A**: AF NP PA SD
expallidus Semenov & Pjatakova, 1935: 102 (*Nyctor*) **A**: TD TM UZ
ferganicus Dolin, 2003: 31 **A**: KI
ferrugatipes Candèze, 1895: 45 [EA] **A**: SD
flavocapillus Vats & Chauhan, 1991c: 22 **A**: UP
flavus Erichson, 1840: 304 **A**: YE "Arabia"
flexus Vats & Chauhan, 1991c: 15 **A**: UP **ORR**
formosus Curtis, 1840: 278 **A**: XIZ **ORR**
anaticus Candèze, 1860: 140
fossulatus Platia & Gudenzi, 2002a: 32 **A**: TR
foveiventris Schwarz, 1900: 103 **E**: GR
foveolatus Schwarz, 1900: 103 **E**: GR
frequens Platia & Gudenzi, 2002a: 34 **A**: TR
freudei Platia & Gudenzi, 2000d: 138 **E**: GR
frigidus Cobos, 1961: 79 **E**: SP
gebleri gebleri Candèze, 1860: 137 **A**: KZ MG XIN
gebleri meridionalis Gurjeva, 1966: 80 **A**: TD TM UZ
gebleri rutilipes Gurjeva, 1966: 80 **A**: KI
georgioui Preiss & Platia, 2003: 118 **A**: CY
getschmanni Candèze, 1880: 297 **E**: PT SP
glasunovi Gurjeva, 1966: 82 **A**: TD
graciliformis Dolin & Atamuradov, 1989: 38 **A**: TM
gramineus Scopoli, 1763: 95 (*Elater*) **E**: AB AR AU BE BH BU CR CT CZ FR GE GG GR HU IT LA LU MD NL PL PT RO SK SL SP ST SV SZ UK YU **A**: TR "Palestine"
ruficollis Geoffroy, 1785: 36 (*Elater*) [HN]
thoracicus Fabricius, 1775: 214 (*Elater*)
granulicollis Dolin, 1998b: 97 **A**: KZ
haridwarensis Vats & Chauhan, 1991c: 25 **A**: UP
hauseri Schwarz, 1900: 100 **A**: AF IN KI KZ TM UZ XIN
humeronotatus Pic, 1902l: 72
brevescapulatus Pic, 1912g: 25
subunicolor Pic, 1902l: 72
variipennis Schwarz, 1900: 100
hayekae Platia & Gudenzi, 2002a: 35 **A**: TR
heydeni Buysson, 1915: 9 **A**: FE
hiemalis Dolin & Atamuradov, 1980: 81 **A**: TM
hinkei I. Frivaldszky, 1837: 176 **E**: AB AR BU GR **A**: CY TR
fulvofasciatus Reitter, 1894h: 254
obscurofasciatus Reitter, 1894h: 254
hispanicus Cobos, 1961: 78 **E**: SP
hissaricus Dolin & Latifi, 1989: 19 **A**: TD
humilis Erichson, 1840: 308 **N**: EG **AFR**
hummeli Fleutiaux, 1936a: 18 **A**: GAN
imitator Platia & Gudenzi, 2002a: 36 **A**: TR
impressiventris Schwarz, 1900: 102 **A**: TR
boyadjiani Pic, 1908g: 57
indentilis Vats & Chauhan, 1991c: 20 **A**: UP
indicus Garg, Saini & Vasu, 1997a: 65 **A**: SD
inermis Schwarz, 1900: 101 **A**: KI UZ
insignis Desbrochers des Loges, 1875a: 39 **A**: IS SY "Palestine"
manueli Pic, 1910m: 50
sanguinicollis J. R. Sahlberg, 1913d: 135
italicus Platia & Bartolozzi, 1988: 88 **E**: IT

jelineki Cate, Platia & Schimmel, 2002: 47 **A**: IN
jermolenkoi Dolin, 1969: 123 **E**: AB ST UK
juniperinus Orlov, 1993: 96 **E**: ST
kantneri Platia & Gudenzi, 2002b: 95 **A**: SY
kaszabi Gurjeva, 1969: 193 **A**: MG
keyserlingi Koenig, 1889b: 534 **A**: JIA MG XIN XIZ
kindermannii Candèze, 1860: 128 **E**: AR **A**: LE SY TR
  akbesianus Pic, 1899h: 139
kinzelbachi Chassain, 1979a: 70 **A**: SY TR
klapperichi Gurjeva, 1975b: 170 **A**: AF PA
kopetdaghensis Dolin & Atamuradov, 1994: 91 **A**: TM
koschwitzi Platia & Gudenzi, 1999a: 20 **E**: SP
kronbladi Platia & Gudenzi, 2000d: 138 **E**: GR (Kríti)
kruegeri Pic, 1928i: 104 **N**: LB
kryzhanovskyi Dolin & Chantladze, 1980: 190 **E**: AB AR GG ST
kugitangi Dolin & Atamuradov, 1986: 69 **A**: TM
kurashvili Dolin & Chantladze, 1987: 622 **A**: GG
kuschkensis Dolin & Atamuradov, 1989: 37 **A**: TM
kuwaitianus Platia & Schimmel, 1997: 305 **A**: KU
latelimbatus Pic, 1905c: 105 **N**: MO
  circumpictus Buysson, 1924b: 207
levis Platia & Gudenzi, 2002a: 36 **A**: TR
lineatus Gurjeva, 1964: 149 **A**: KZ
linnavuorii Platia & Schimmel, 1997: 306 **A**: IQ OM YE
ljudmilae Gurjeva, 1987a: 48 **A**: TM
lompei Zeising & Brunne, 2003: 195 **E**: PT
longulus Erichson, 1840: 306 **A**: KI UZ
lundbergi Platia & Gudenzi, 2000e: 604 **A**: TR
maculatus Cate, Platia & Schimmel, 2002: 48 **A**: IN
maculicollis Reiche & Saulcy, 1857a: 420 **E**: GR **A**: TR
magnanii Platia, Furlan & Gudenzi, 2002: 76 **A**: IN
maritimus Dolin, 1971: 644 **E**: BU CZ RO ST UK **A**: KZ TR
marmottani Buysson, 1912f: 129 **N**: MO
medvedevi Gurjeva, 1966: 77 **A**: TM
megathorax Faldermann, 1835a: 171 **E**: AB AR GG GR (Samos) **A**: IN IS JO LE SY TR
  antennalis Desbrochers des Loges, 1875a: 39 **E**: GG **A**: IS TR
melampus Illiger, 1807a: 8 (*Elater*) **E**: FR PT SP **N**: MO
  deflexus Schaufuss, 1862d: 310
  flavicornis Rey, 1891b: 68
  graellsii Candèze, 1860: 160
  longicornis Schaufuss, 1862b: 48 [= 1862d: 310]
  vaucheri Buysson, 1919: 115
mesasiaticus Dolin & Latifi, 1997: 84 **A**: KI TD UZ
michai Platia & Gudenzi, 2000d: 140 **E**: GR
microcephalus Candèze, 1882: 77 **N**: MO
miniaticollis Candèze, 1860: 122 **E**: AR GR **A**: IN LE SY TR "Palestine"
moralesi Cobos, 1958: 52 **N**: MO
motschulskyi Heyden, 1880a: 121 [RN] **A**: FE JIL
  curtulus Motschulsky, 1860b: 112 [HN]
mus Candèze, 1889b: 106 **A**: ES
mutabilis Gurjeva, 1966: 75 **A**: KI TD TM UZ
nanus Gurjeva, 1966: 82 **A**: TM
nathaliae Platia & Schimmel, 1988: 455 **E**: IT
neoalienus Platia & Gudenzi, 1999b: 121 [RN] **E**: AB AR **A**: IQ
  alienus Dolin, 1969: 126 [HN]

neoarnoldii Platia & Gudenzi, 1999b: 125 [RN] **E**: ST UK
arnoldii Dolin, 1960: 190 [HN]
neuberti Platia & Gudenzi, 2002a: 38 **A**: TR
nigerrimus Erichson, 1840: 296 **E**: AU BE BU CT CZ FR GE GR HU IT MD NL PL RO SK SP ST SZ UK **A**: TR
villosior Buysson, 1911a: 47
nigratissimus Buysson, 1891d: cxxxix **E**: AB AR GG **A**: IN IQ SY TM TR
nigricans Vats & Chauhan, 1991c: 25 **A**: UP **ORR**
nigricollis Erichson, 1840: 306 **A**: KI TD UZ
nigroaeneus Dolin & Protzenko, 1965: 22 **A**: KI UZ
nigrocyaneus Buysson, 1912e: 129 **A**: TD
nigrosuturus Vats & Chauhan, 1991c: 28 **A**: UP **ORR**
niponicus Lewis, 1894a: 190 **A**: JA
nitidus Candèze, 1878b: clxv **A**: "North India"
nubilosus Schwarz, 1895a: 37 **E**: ST
numida Candèze, 1860: 132 **N**: AG TU
cloueti Buysson, 1899d: 213
henoni Candèze, 1889b: 105
lethierryi Desbrochers des Loges, 1875a: 39
obscurior Buysson, 1899d: 212
obliteratus Dolin & Šauša, 1997: 41 **A**: FE NC
oertzeni Schwarz, 1900: 104 **E**: GR (Kríti)
olgae Solsky, 1881: 233 **A**: KI KZ TM UZ
margelanicus Heyden, 1881: 325
opacus Gurjeva, 1966: 87 **A**: TD
ordubadensis Agajev, 1985b: 57 **E**: AB
ovipennis Desbrichers des Logis, 1875b: 38 **A**: CY
cyprius Pic, 1910i: 33
mimicus Buysson, 1911c: 167
parvulus Platia & Gudenzi, 2000d: 142 **E**: GR **A**: TR
pavesii Platia & Gudenzi, 2000e: 604 **A**: TR
pellitoides Gurjeva, 1966: 86 **A**: KI
pellitus Schwarz, 1892d: 383 **A**: KI TM UZ
persianus Cate, Platia & Schimmel, 2002: 49 **A**: IN
peyerimhoffi Buysson, 1902c: 426 **A**: "Arabie Petrée"
ietenensis Buysson, 1902c: 427
picinus Platia & Gudenzi, 2002a: 39 **A**: IN TR
picticollis Kraatz, 1882c: 320 **A**: TD UZ
pinguis Lewis, 1894a: 189 **A**: JA "Korea" "NE China"
platiai Chassain, 1985b: 142 [RN] **E**: GR (Santorin)
spatulatus Chassain, 1984: 95 [HN]
plurimus Platia & Gudenzi, 2002a: 39 **A**: TR
poncyi Buysson, 1903: 129 **E**: SP **N**: AG MO
tingitanus Buysson, 1903: 129
procerulus Kiesenwetter, 1859a: 21 **E**: GR
kiesenwetteri Candèze, 1860: 156
procerus Platia & Gudenzi, 2002a: 40 **A**: TR
proprius Platia & Gudenzi, 2002a: 42 **A**: TR
pruinosus Buysson, 1902c: 428 **A**: IN
przewalskii Gurjeva, 1966: 78 **A**: KI KZ
pseudogramineus Mardjanian, 1983: 141 **E**: AR
pubicollis Buysson, 1918b: 111 **N**: MO
pubicornis Buysson, 1912f: 131 (*Lasiocerus*) **N**: TU
quadriflavonotatus Vats & Chauhan, 1991c: 27 **A**: UP
qatarensis Platia, 2004a: 43 **A**: QA

ressli Platia & Gudenzi, 2000e: 605 **A**: TR
riedeli Platia & Gudenzi, 2000d: 144 **E**: GR
roberi Platia & Gudenzi, 2002b: 96 **E**: GR (Samos)
rorulentus Gurjeva: 1966: 84 **A**: TD
rotundicollis I. Frivaldszky, 1845: 181 **E**: GR **A**: TR
turgescens Buysson, 1891d: cxxxix
ruficollis Linnaeus, 1758: 405 (*Elater*) **E**: AU BE BU BY CT CZ DE EN FI FR GE GR HU IT LA LT LU NL NR NT PL RO SK SP SV SZ UK **A**: WS
ruficruris Brullé, 1832: 140 **E**: AR IT GR SP **N**: MO **A**: CY IN TR
clarior Buysson, 1912f: 130
depictus Buysson, 1913a: 14
pythiae Buysson, 1902c: 436
tibialis Erichson, 1840: 295
rufipes Goeze, 1777: 569 (*Elater*) **E**: AB AR AU BE BH BU CR CT CZ FR GE GR HU IT LA MC MD PL PT RO SK SP ST SV SZ **N**: AG EG MO TU **A**: IN KZ TM TR WS
goezei Sánchez-Ruiz, 1996: 61 [RN]
pallipes Brullé, 1832: 141 [HN]
pueli Buysson, 1906b: 471
ruizi Platia & Gudenzi, 1999a: 21 **E**: SP
sacratus Erichson, 1840: 284: **E**: AB AR GG ST **N**: EG LB **A**: CY IQ IN IS JO LE SY TR TM
augustalisi Pic, 1913m: 169
nigricornis Baudi di Selve, 1859b: 342
tokatensis Pic, 1913m: 169
samai Platia & Gudenzi, 2000e: 605 **A**: TR
saudarabicus Platia & Schimmel, 1997: 306 **A**: SA
sbeitlensis Buysson, 1905c: 210 **N**: TU
scapulatus Candèze, 1878b: clxi **N**: MO TU
exoplatus Buysson, 1912f: 129
gafsanus Pic, 1909m: 153
gafsensis Buysson, 1912f: 129
scheuerni Chassain, 1979a: 73 **A**: IS JO
schuberti Platia & Gudenzi, 2002a: 42 **A**: TR
schurmanni Platia & Gudenzi, 2002a: 44 **A**: TR
schusteri Buysson, 1912e: 128 **A**: TM
sericapilosus Vats & Chauhan, 1991c: 23 **A**: UP
serratus Mukhopadhyay & Chakraborty, 2003: 233 **A**: SD
signatus A. G. Olivier, 1790: no 31: 48 (*Elater*) **E**: PT SP **N**: AG
astorganus Pic, 1912g: 25
granjasensis Pic, 1899h: 140
manzanalensis Pic, 1912g: 25
moreli Pic, 1923i: 10
nigrithorax Fuente, 1909: 307
semilutescens Pic, 1912g: 25
sexmaculatus P. H. Lucas, 1846: 164
sexpunctatus Latreille, 1804a: 31 (*Elater*) [RN]
sikkimensis Garg, Saini & Vasu, 1997a: 67 **A**: SD
silesioides Escalera, 1914: 213 **N**: MO
silferbergi Platia & Schimmel, 1997: 307 **A**: YE
sinoconstrictus Vats & Chauhan, 1991c: 17 **A**: UP **ORR**
skulei Platia & Schimmel, 1997: 307 **A**: KU OM SA YE
somcheticus Schwarz, 1896c: 148 **E**: AB AR GG **A**: TR
sonani Miwa, 1930a: 9 **A**: FUJ TAI
stussineri Buysson, 1913a: 14 **E**: AB GR MC **A**: CY TR
syriacoides Platia & Cate, 2001: 118 **E**: AB AR **A**: IN SY TD TM
syriacus Linnaeus, 1758: 404 (*Elater*) **E**: AB AR GG **A**: IN SY TR TM
pictus Faldermann, 1835a: 170
tadzhikistanicus Gurjeva, 1966: 73 **A**: TD

tarbinskyi Dolin, 1998b: 95 **A**: KI
tenelloides Dolin, 1960: 189 **E**: UK
tenellus Reiche & Saulcy, 1857a: 421 **N**: EG **A**: IS JO LE SY
terskeyensis Dolin, 1998b: 96 **A**: KI
thebaicus Candèze, 1860: 177 **N**: EG
tokgajevi Dolin, 1977b: 356 [= 1977c: 359] **A**: TM
triangularis Vats & Chauhan, 1991c: 26 **A**: HP UP
tricolor Reitter, 1896g: 289 **A**: KI KZ
trimaculatus Schwarz, 1894: 146 **A**: IN
turgidus Erichson, 1840: 295 **A**: SY TR
tzkheniensis Orlov, 1997a: 174 (*Cardiophorus*) **E**: GG
ulcerosus Gené, 1836: 175 (*Elater*) **E**: IT MA **N**: AG MO TU
acupictus Buysson, 1922: 257
angularis Gené, 1836: 176 (*Elater*)
anguloseminiatus Buysson, 1922: 258
anticenotatus Ragusa, 1911: 193
apicalis Rey, 1891b: 68
circinatopictus Buysson, 1922: 257
conjonctus Buysson, 1922: 257
extensenigratus Buysson, 1922: 257
kabylianus Pic, 1899h: 139
neoposticeseminiatus Buysson, 1922: 256
peronatus Buysson, 1913b: 50
posticepictus Buysson, 1913b: 50
rubricollis Schwarz, 1893b: 3
uncinatus Platia & Gudenzi, 2002a: 46 **A**: TR
unicus Vats & Chauhan, 1991c: 24 **A**: UP
uzbek Dolin, 2003: 35 **A**: KI UZ
valichanovi Dolin, 2003: 34 **A**: KZ
varius Cate, Platia & Schimmel, 2002: 50 **A**: IN
vespertinus Dolin & Atamuradov, 1989: 36 **A**: TM
vestigialis Erichson, 1840: 293 **E**: AB AL AR AU BE BH BU CR CT CZ FI FR GB GE GG GR HU IT MD NR NT PL PT RO SK SL SP SV ST SZ UK YU **N**: AG EG LB MO TU **A**: CY ES FE IN KZ MG TM TR WS
atramentarius Erichson, 1840: 297
atripes Buysson, 1902c: 425
captiosus Normand, 1941: 333
curtulus Mulsant & Guillebeau, 1855b: 317
erichsoni Buysson, 1901c: 125
hipponensis Desbrochers des Loges, 1870b: 102
infimus Buysson, 1906b: 471
laetipes Buysson, 1902c: 425
maurus Desbrochers des Loges, 1870a: 71 [RN]
mauritanicus Desbrochers des Loges, 1869: 34 [= 1870b: 103] [HN]
nijazi Dolin & Agajev, 1983: 31
neotericus Buysson, 1902a: 228
persicus Schwarz, 1892d: 384
ragusae Ragusa, 1911: 194
rubidiventris Buysson, 1919: 116
vexillarius Candèze, 1889b: 110 **A**: KI KZ TM UZ
russowi Heyden, 1889a: 328
viberti Buysson, 1912f: 130 **N**: TU
vulgaris Motschulsky, 1860b: 111 **A**: FE MG
widenfalki Leiler, 1967: 106 **E**: SV
winkleri Platia & Gudenzi, 2002a: 47 **A**: SY
witzgalli Platia & Gudenzi, 2002a: 46 **A**: TR
zeisingi Platia, 2004c: 106 **A**: TR
zianii Platia & Gudenzi, 2002b: 96 **E**: GR

subgenus *Coptostethus* Wollaston, 1854: 238 type species *Coptostethus femoratus* Wollaston, 1854
abora Wurst & Cate, 1994: 419 **N**: CI (Grand Canaria)
arozarenai Cobos, 1970: 76 **N**: CI (Tenerife)
auarita Liberto & Wurst, 1999: 333 **N**: CI (LaPalma)
brunneipennis Wollaston, 1863: 218 (*Coptostethus*) **N**: CI (Tenerife)
cabrerai Cobos, 1970: 50 **N**: CI (Fuenteventura)
canariensis Wollaston, 1858a: 196 (*Coptostethus*) **N**: CI (Tenerife)
cobossanchezi Díaz de Castro & Sánchez-Ruiz, 2002: 190 [RN] **N**: CI (La Gomera)
differens Cobos, 1983: 239 [HN]
crassiusculus Wollaston, 1864: 213 (*Coptostethus*) **N**: CI (Gran Canaria)
divergens Cobos, 1983: 238 **N**: CI (El Hierro)
femoratus Wollaston, 1854: 240 (*Coptostethus*) **N**: MR
fernandezi Cobos, 1970: 77 **N**: CI (Tenerife)
globulicollis Wollaston, 1862d: 440 (*Coptostethus*) **N**: CI (Tenerife)
gomerensis Cobos, 1970: 80 **N**: CI (La Gomera)
gracilis Wollaston, 1864: 211 (*Coptostethus*) **N**: CI (Tenerife)
guanche Cobos, 1970: 88 **N**: CI (Gran Canaria)
guayote Wurst & Cate, 1994: 422 **N**: CI (Gran Canaria)
hayeki Cobos, 1970: 90 **N**: CI (Tenerife)
hierrensis Franz, 1980: 114 **N**: CI (El Hierro)
inflatus Cobos, 1970: 94 **N**: CI (Tenerife)
lindbergi Cobos, 1970: 62 **N**: CI
longicornis Lindberg, 1953: 7 (*Coptostethus*) **N**: CI (Tenerife)
machadoi Cobos, 1983: 241 **N**: CI (La Gomera)
mateui Cobos, 1970: 82 **N**: CI (Tenerife)
mendizabali Cobos, 1970: 70 **N**: CI (Tenerife)
minutus Har. Lindberg, 1951: 5 (*Coptostethus*) **N**: CI (Gran Canaria)
obscurus Wollaston, 1864: 210 (*Coptostethus*) **N**: CI (El Hierro)
obtusus Wollaston, 1864: 213 (*Coptostethus*) **N**: CI (Tenerife)
oromii Cobos, 1978: 147 **N**: MR
palmensis Cobos, 1970: 84 **N**: CI (La Palma)
platai Cobos, 1979: 199 **N**: CI (Lanzarote)
taylori Cobos, 1970: 60 **N**: CI (Tenerife)
vestitus Har. Lindberg, 1951: 5 (*Coptostethus*) **N**: CI (Gan Canaria)
vittatus Har. Lindberg, 1953: 7 (*Coptostethus*) **N**: CI (Gan Canaria)
wollastoni Cobos, 1970: 74 **N**: CI (Tenerife)
subgenus *Metacardiophorus* Gurjeva, 1966: 91 type species *Cardiophorus sogdianus* Gurjeva, 1966
ineptus Gurjeva, 1966: 92 **A**: UZ
sogdianus Gurjeva, 1966: 91 **A**: UZ
valentini Gurjeva, 1978a: 145 **A**: TD
subgenus *Perrinellus* Buysson, 1899f: 282 type species *Athous argentatus* Abeille de Perrin, 1894 (= Cardiphorus *reitteri* Schwarz, 1891)
aegyptiacus Candèze, 1860: 125 **N**: EG TU
bonnairei Buysson, 1893c: vii **N**: AG
laghouatensis Pic, 1902n: 306
bousaadensis Buysson, 1899e: 213 **N**: AG
pallidipennis Pic, 1897a: 2 [HN]
disgamus Buysson, 1902c: 431 **N**: AG TU
impexus Buysson, 1905c: 209
gracilis Gurjeva, 1966: 90 **A**: TD
hoploderus Candèze, 1860: 175 **N**: AG MO TU **AFR**
neosuturalis Pic, 1902l: 72
vittatithorax Pic, 1902l: 72
letourneuxi Buysson, 1891c: lxviii **N**: EG
caroli Pic, 1893e: 122

lopatini Gurjeva, 1966: 88 **A**: TD
pharaonum Buysson, 1911b: 140 **N**: AG EG **A**: CY
intrusus Buysson, 1912a: 22
reitteri Schwarz, 1891b: 151 **A**: IS SY
argentatus Abeille de Perrin, 1894b: 92 (*Athous*)
sinaitus Pic, 1919f: 14 (*Perinellus*) **A**: SI
stoliger Buysson, 1902c: 429 **N**: AG MO TU
henonianus Schenkling, 1927: 540 [RN]
henoni Pic, 1912n: 65 [HN]

genus *Cardiophorus*, nomen dubium
undinarum Gistel, 1857: 7 **E**: "Styria"

genus *Cardiotarsus* Eschscholtz, 1838: [pl.] type species *Cardiophorus capensis* Candèze, 1860
babai Kishii, 1994b: 213 **A**: TAI
fulvipes Fleutiaux, 1918c: 223 **A**: GUA **ORR**
difficilis Fleutiaux, 1947: 372
housaianus Kishii, 1994d: 32 **A**: TAI
humeralis Miwa, 1930a: 6 **A**: FUJ TAI
longipennis Schwarz, 1902a: 261 **A**: "Himalaya"
pallidipes Miwa, 1930a: 5 **A**: GUI TAI
rarus Miwa, 1927b: 108 **A**: FUJ TAI
rotundicollis Fleutiaux, 1931: 329 **A**: GUA **ORR**
sinensis Candèze, 1860: 230 **A**: HKG
taiwanus Kishii, 1994d: 30 **A**: TAI
tonkinensis Fleutiaux, 1931: 330 **A**: GUA **ORR**
yagii Kishii, 1992a: 8 **A**: TAI
yamazakii Ôhira, 1968a: 134 **A**: JA (Okinawa-hontô, Amami-Ôshima)

genus *Craspedostethus* Schwarz, 1898c: 414 type species *Craspedonotus rufiventris* Schwarz, 1898
Craspedonotus Schwarz, 1898a: 148 type species *Craspedonotus rufiventris* Schwarz, 1898
Tropidiplus Fleutiaux, 1903b: 251 type species *Tropidiplus tellinii* Fleutiaux, 1903
antennalis Cate, Platia & Schimmel, 2002: 52 **A**: AF IN
buettikeri Platia & Schimmel, 1997: 315 **A**: SA
dilutus Erichson, 1840: 307 (*Cardiophorus*) **N**: AG EG **A**: IN OM SA SI YE **AFR**
jocularis Candèze, 1889b: 108 (*Cardiophorus*)
wittmeri Chassain, 1979b: 207
flavescens Chassain, 1979b: 209 **A**: OM SA YE
iranicus Platia & Gudenzi, 1999b: 121 **A**: IN
linnavuorii Platia & Gudenzi, 1999b: 122 **A**: IQ
longicornis Cate, Platia & Schimmel, 2002: 53 **A**: IN
myrrhilegius Wurst, Schimmel & Platia, 2001: 550 **A**: YE
permodicus Faldermann, 1835a: 172 (*Drasterius*) **E**: AR ST **A**: IN
fulvescens Reitter, 1891c: 145 (*Aeolus*)
saudianus Chassain, 1979b: 209 **A**: SA
schusteri Schwarz, 1897a: 129 (*Paracardiophorus*) **A**: IQ IN

genus *Dicronychus* Brullé, 1832: 138 [RN] type species *Dicronychus obesus* Brullé, 1832 (= *Dicronychus brullei* Platia & Gudenzi, 2003)
Gauroderus C. G. Thomson, 1859: 104 type species *Elater cinereus* Herbst, 1784
Paramecus Dillwyn, 1829: 32 [HN] type species *Paramecus cordiger* Dillwyn, 1829 (= *Elater equiseti* Herbst, 1806)
abyssinus Candèze, 1889b: 109 (*Cardiophorus*) **N**: EG **A**: SI **AFR**
adanensis Pic, 1908d: 43 (*Cardiophorus*) **A**: TR
aenescens Platia & Gudenzi, 2003: 28 **E**: GR
apteriformis Platia & Gudenzi, 2004: 10 **A**: TR

asperulus Candèze, 1860: 194 (*Cardiophorus*) **E**: PT
bactrianus Gurjeva, 1966: 92 (*Cardiophorus*) **A**: TD TM
beauchenei Fleutiaux, 1918c: 226 (*Cardiophorus*) **A**: HAI **ORR**
berrai Platia & Gudenzi, 2003: 29 **E**: GR
bicolor Platia & Gudenzi, 2004: 10 **A**: TR
blandus Solsky, 1881: 235 **A**: KI TD TM UZ
bozdagensis Platia & Gudenzi, 2004: 10 **A**: TR
brancuccii Platia & Schimmel, 1997: 309 **A**: IQ KU OM QA SA
brullei Platia & Gudenzi, 2003: 34 [RN] **E**: GR
obesus Brullé, 1832: 138 (*Elater*) [HN]
buettikeri Platia & Schimmel, 1997: 310 **A**: SA YE
bureschi Roubal, 1936b: 66 (*Cardiophorus*) **E**: BU
campadelli Platia & Gudenzi, 2003: 30 **E**: GR
chassaini Platia & Schimmel, 1997: 311 **A**: SA
cinereus Herbst, 1784b: 114 (*Elater*) **E**: AB AR AU BE BU BY CR CT CZ DE FR GE GG GR HU IT LA LU LT MC MD NL PL PT RO SK SL SP ST SV ST SZ TR **A**: CH LE KZ TR
agnatus Candèze, 1860: 189 (*Cardiophorus*)
cephalonicus Buysson, 1913b: 52 (*Cardiophorus*)
equiseti Herbst, 1806: 67 (*Elater*) [HN]
gabrieli Gerhardt, 1898a: 14
lusitanicus Buysson, 1913b: 52 (*Cardiophorus*)
suturalis Eschscholtz, 1829: 34 (*Cardiophorus*)
testaceus Fabricius, 1792b: 229 (*Elater*)
weberi Waltl, 1838b: 269 (*Elater*)
conductus Erichson, 1840: 316 (*Cardiophorus*) **A**: JIX **ORR**
consentaneus Kollar, 1848: 508 (*Cardiophorus*) **A**: KA
dahuricus Candèze, 1889b: 110 (*Cardiophorus*) **A**: ES
davidianus Candèze, 1882: 80 (*Cardiophous*) **A**: SCH
decorus Faldermann, 1835a: 179 (*Dolopius*) **E**: AB AR GG ST **A**: KZ TR
melanothorax Candèze, 1860: 192 (*Cardiophorus*)
deminuatus Platia & Gudenzi, 2004: 12 **A**: TR
desbrochersi Platia & Gudenzi, 2004: 12 **A**: IN TR
devius Candèze, 1882: 81 (*Cardiophorus*) **A**: SC SCH
drurei Pic, 1908d: 43 (*Cardiophorus*) **A**: IQ
equiseti Herbst, 1784b: 114 (*Elater*) **E**: AB AR AU BE BU BY CT CZ EN FR GB GE GG GR HU IT LA LT NL NR NT PL PT RO SK SP ST SZ UK YU **A**: IN KZ TR
cordiger Dillwyn, 1829: 32 (*Paramecus*)
luridipes Lacordaire, 1835: 653 (*Elater*)
equisetioides Lohse, 1976: 101 **E**: AU BE CZ DE FR GE HU IT LA PL SK SV
exstinctus Erichson, 1840: 317 (*Cardiophorus*) **N**: EG **A**: "Mesopotamia"
ferruginosus Platia & Schimmel, 1997: 311 **A**: AE OM
fulvivellus Candèze, 1860: 214 (*Cardiophorus*) **A**: NP **ORR**
fusivittatus Platia & Gudenzi, 1999b: 122 **A**: IN IQ
gillerforsi Platia & Gudenzi, 2004: 12 **A**: TR
graingeri Platia & Schimmel, 1997: 312 **A**: SA
haematomus Candèze, 1860: 198 **N**: AG MO
hedenborgii Candèze, 1860: 197 (*Cardiophorus*) **N**: EG **A**: SI **AFR**
hippanicus Orlov, 1997a: 172 (*Cardiophorus*) **E**: ST
hoberlandti Cate, Platia & Schimmel, 2002: 55 **A**: IN
iconiensis Pic, 1908d: 43 (*Cardiophorus*) **A**: TR
incanus Erichson, 1840: 311 (*Cardiophorus*) **E**: IT (Sardegna)
intermedius Platia & Gudenzi, 2003: 30 **E**: GR
involucer Platia & Gudenzi, 2004: 16 **A**: SY
larseni Platia & Schimmel, 1997: 312 **A**: OM
latescapulatus Buysson, 1906a: 19 [= 1912b: 40] (*Cardiophorus*) **E**: AR **A**: IN

eustictus Buysson, 1906a: 20 [= 1912b: 41] (*Cardiophorus*)
luciae Laibner, 1985: 131 **E**: BU
macedonicus Platia & Gudenzi, 2002a: 33 **E**: MC
marani Roubal, 1936b: 67 (*Cardiophorus*) **E**: BU
martini Platia & Gudenzi, 2003: 33 **E**: GR
mauritanicus Motschulsky, 1858c: 59 (*Platynchus*) **N**: AG
merkli Pic, 1910m: 50 (*Cardiophorus*) **E**: TR **A**: CY TR
mesopotamicus Platia & Gudenzi, 1999b: 125 **A**: IQ IN
mixtus Fleutiaux, 1931: 326 (*Platynychus*) **A**: HAI **ORR**
mossulensis Pic, 1912n: 65 **A**: IQ (*Cardiophorus*)
mutilus Platia & Gudenzi, 2004: 14 **A**: TR
nebulosus Motschulsky, 1858: 59 (*Platynychus*) **A**: HAI YUN **ORR**
nigropunctatus Motschulsky, 1860a: 58 (*Melanoxanthus*) **E**: AR ST **A**: KI KZ TM
cruciatus Faust, 1877b: 307
duchoni Pic, 1914d: 10
nigropunctatus Candèze, 1860: 137 (*Cardiophorus*)
tedschensis Pic, 1914d: 10
obesus Krynicki, 1832: 95 (*Elater*) **E**: UK
crassicollis Erichson, 1840: 313 (*Cardiophorus*)
geminatus Erichson, 1840: 314 (*Cardiophorus*)
obscuripennis Pic, 1899h: 140 (*Cardiophorus*) **E**: TR **A**: TR
omanensis Platia & Schimmel, 1997: 314 **A**: OM
perstriatus Buysson, 1912c: 87 (*Cardiophorus*) **N**: AG
pici Platia & Gudenzi, 2004: 14 **A**: TR
pseudobesus Orlov, 1993: 98 (*Cardiophorus*) **E**: ST
pseudofebriens Buysson, 1912a: 25 **N**: EG **A**: SI
saalensis Buysson, 1902c: 426
quadrinaevus Reitter, 1891b: 25 **E**: AB **A**: TM
rubripes Germar, 1824: 59 (*Elater*) **E**: AB AR AU BU CR CZ GE GR HU IT MC MD PL RO SP SK SL ST SZ TR YU **A**: IN KZ TR
flavipubens Reitter, 1911a: 232 (*Cardiophorus*)
messenicus Brullé, 1832: 138 (*Elater*)
permixtus Buysson, 1912f: 131 (*Platynychus*)
pollux Germar, 1824: 59 (*Elater*)
sahlbergi Schwarz, 1900: 104 **A**: KI TM
schmalfussi Chassain, 1984: 96 **E**: GR (Santorin)
senaci Desbrochers des Loges, 1869: 34 [= 1870b: 104] (*Cardiophorus*) **E**: GR RO TR **A**: TR
septentrionalis Fleutiaux, 1931: 318 (*Platynychus*) **A**: GUA **ORR**
submontanus Vats, 1984: 8 **A**: UP
subulipennis Faldermann, 1835b: 363 **A**: MG "North China"
jucundus Gurjeva, 1974b: 146
talhouki Platia & Schimmel, 1997: 314 **A**: SA YE
turcicus Platia & Gudenzi, 2004: 16 **A**: TR
variatus Desbroches des Loges, 1875a: 38 (*Cardiophorus*) **A**: SY TR
versicolor Mulsant & Guillebeau, 1856: 95 (*Cardiophorus*) **E**: AU BU FR GR IT SP **A**: TR
gallicus Pic, 1923i: 10 (*Cardiophorus*)
wagneri Pečírka, 1926: 67 (*Cardiophorus*) **E**: SP
wittmeri Platia & Schimmel, 1997: 315 **A**: OM SA YE

genus *Neocardiophorus* Gurjeva, 1966: 95 type species *Neocardiophorus mamajevi* Gurjeva, 1966
fausti Gurjeva, 1966: 96 **A**: TM
mamajevi Gurjeva, 1966: 95 **A**: UZ

genus *Paracardiophorus* Schwarz, 1895b: 40 type species *Cardiophorus musculus* Erichson, 1840
amami Kishii, 1977a: 54 **A**: JA (Amami-Ôshima)
brunneus Erichson, 1840: 307 **N**: AG EG **A**: SI "Palestine" **AFR**

febriens Candèze, 1860: 197
conformis Desbrochers des Loges, 1875a: 38 (*Cardiophorus*)
extinctoides Buysson, 1912a: 27
stilpnithorax Buysson, 1912a: 26
buettikeri Chassain, 1979b: 210 **A**: OM SA
carduelis Candèze, 1865: 39 (*Cardiophorus*) **A**: SE TAI **ORR**
rufithorax Miwa, 1930a: 10
confalonierii Pic, 1928i: 104 **N**: LB
coreanus coreanus Kishii, 1977a: 55 **A**: HEI SC NE
coreanus tsushimensis Kishii, 1977a: 56 **A**: JA
devastans Matsumura, 1910: 38 (*Cardiophorus*) [= 1911a: 145] **A**: JIA TAI ZHE
erythrurus Candèze, 1882: 78 (*Cardiophorus*) **A**: FE MG NE SCH
flavobasalis Schwarz, 1902a: 260 **A**: FUJ HUB
granarius Candèze, 1889b: 106 (*Cardiophorus*) **A**: CH
imasakai Ôhira, 1995a: 95 **A**: JA
jenisseensis J. R. Sahlberg, 1903a: 27 **A**: ES
kaszabi Gurjeva, 1968: 296 **A**: MG NMO
leprieuri Pic, 1902n: 305 (*Cardiophorus*) **N**: AG
lewisi Fleutiaux, 1902a: 20 (*Cardiophorus*) **A**: JA
musculus musculus Erichson, 1840: 299 (*Cardiophorus*) **E**: AU BU CT CZ FR GE GG GR HU IT LT PL RO SK SP ST SZ UK **N**: AG TU **A**: ES WS
aenescens Desbrochers des Loges, 1875a: 40 (*Cardiophorus*)
concolor Desbrochers des Loges, 1875a: 38 (*Cardiophorus*)
gaullei Buysson, 1902c: 434 (*Cardiophorus*)
nakanei hondoensis Ôhira, 1997a: 2,9 **A**: JA
nakanei nakanei Ôhira, 1986b: 215 **A**: JA (Yakushima)
opacus Lewis, 1894a: 190 (*Cardiophorus*) **A**: JA
pullatus hachijo Kishii, 1994d: 34 **A**: JA (Hachijo-jima)
pullatus pullatus Candèze, 1873: 16 (*Cardiophorus*) **A**: FE JA MG NC SC
pullatus yasudai Ôhira, 1986b: 215 (*Paracardiophorus*) **A**: JA (Hokkaido)
saharae Leseigneur, 1955: 90 **N**: MO
sequens purpuratus Kishii, 1977a: 55 **A**: JA (Yaku-shima)
sequens sequens Candèze, 1873: 16 (*Cardiophorus*) **A**: JA FUJ ZHE SHA SHX "Korea"
subaeneus Fleutiaux, 1902a: 20 (*Cardiophorus*)
tokara erabu Kishii, 1977a: 53 **A**: JA (Oki-no-erabu-jima, Yoron)
tokara heianus Kishii, 1977a: 54 **A**: JA (Tokuno-shima)
tokara kikai Kishii, 1977a: 54 **A**: JA (Kikai-ga-shima, Amami-Ôshima)
tokara tokara Nakane & Kishii, 1955b: 7 **A**: JA (Ryukyus)
tokara yakuensis Ôhira, 1970a: 12 **A**: JA (Yaku-shima)

genus *Phorocardius* Fleutiaux, 1931: 308 type species *Cardiophorus florentini* Fleutiaux, 1895
bombycinus Candèze, 1895: 46 (*Cardiophorus*) **A**: SD
comptus Candèze, 1860: 202 (*Cardiophorus*) **A**: NP "India" **ORR**
magnus Fleutiaux, 1931: 310 **A**: HAI **ORR**
moorii Candèze, 1860: 206 (*Cardiophorus*) **A**: NP SD **ORR**
unguicularis Fleutiaux, 1918c: 222 (*Cardiophorus*) **A**: YUN **ORR**
sobrinus Laporte, 1840a: 250 (*Caloderus*) **A**: NP **ORR**
systens Candèze, 1860: (*Cardiophorus*) **A**: NP **ORR**
vicinus Kollar, 1848: 507 (*Cardiophorus*) **A**: KA
yanagiharae Miwa, 1927b: 109 (*Cardiophorus*) **A**: SCH TAI

genus *Platynychus* Motschulsky, 1858c: 58 type species *Platynychus indicus* Motschulsky, 1858
subgenus *Displatynychus* Ôhira, 1987a: 92 type species *Cardiophorus adjutor* Candèze, 1873
adjutor adjutor Candèze, 1873: 17 (*Cardiophorus*) **A**: CH JA SC
adjutor tamurai Kishii, 1974: 17 (*Dicronychus*) **A**: JA (Ryukyus)

taiwanus Kishii, 1991c: 9, 15 **A**: TAI
subgenus *Platynychus* Motschulsky, 1858c: 58 type species *Platynychus indicus* Motschulsky, 1858
Paradicronychus Dolin, 1975a: 116 type species *Cardiophorus inflatus* Candèze, 1882
amamianus Kishii, 1979a: 10 (*Dicronychus*) **A**: JA (Amami-Ôshima)
anpingensis Miwa, 1930c: 91 (*Cardiophorus*) **A**: TAI
formosanus Matsumura, 1910: 38 (*Cardiophorus*) [= 1911a: 145] **A**: JGAN HEN HUB A (Ryukyus) JIX TAI
imasakai Kishii, 1979a: 9 (*Dicronychus*) **A**: TAI
inflatus Candèze, 1882: 78 (*Cardiophorus*) **A**: FE NE
javanus Candèze, 1860: 205 (*Cardiophorus*) **A**: HAI **ORR**
loochooensis Kishii, 1979a: 10 (*Dicronychus*) **A**: JA (Okinawa-hontô, Ishigaki-jima & Iriomote-jima)
masafumii Ôhira, 1973d: 328 (*Dicronychus*) **A**: JA (Iriomote-jima)
nebulosus Motschulsky, 1858c: 59 **A**: YUN **ORR**
nothus Candèze, 1865: 43 (*Cardiophorus*) **A**: FUJ GAN GUI HEB HUN HUB JA JIA JIX SCH
ferrugineus Lewis, 1894a: 191 (*Cardiophorus*) [RN]
pauper Candèze, 1873: 17 (*Cardiophorus*)
rameus Lewis, 1879b: 156 (*Cardiophorus*)
systenus Candèze, 1860: 210 (*Cardiophorus*) **A**: "India" **ORR**

genus *Ryukyucardiophorus* Ôhira, 1973c: 32 type species *Paracardiophorus loochooensis* Miwa, 1934
babai Kishii, 1991d: 19 **A**: TAI
loochooensis Miwa, 1934: 255 (*Paracardiophorus*) **A**: JA (Ryukyus) TAI
mertliki Platia & Gudenzi, 1999a: 23 **A**: TR
tokarensis Kishii, 1976a: 13 **A**: JA (Nakano-shima)

subfamily **Physodactylinae** Fleutiaux, 1892

genus *Idiotropa* Schwarz, 1897b: 63 type species *Athous henoni* Abeille de Perrin, 1894
henoni Abeille de Perrin, 1894b: 91 (*Athous*) **N**: AG

genus *Toxognathus* Fairmaire, 1878c: 271 type species *Toxognathus costulatus* Fairmaire, 1878
beauchenei Fleutiaux, 1918b: 276 **A**: GUA HKG **ORR**
fairmairei Fleutiaux, 1940c: 104 **A**: FUJ

subfamily **Subprotelaterinae** Fleutiaux, 1920

genus *Subprotelater* Fleutiaux, 1916a: 387 type species *Subprotelater bakeri* Fleutiaux, 1916
hisamatsui Nakane, 1987c: 172 **A**: JA (Ogasawara)
japonicus Nakane & Hisamatsu, 1991: 115 **A**: JA (Nara)

subfamily **Lissominae** Laporte, 1835

genus *Drapetes* Dejean, 1821: 34 type species *Elater equestris* Fabricius, 1798 (= *Elater mordelloides* Host, 1789)
Lissodes Berthold, 1827: 335 type species *Elater equestris* Fabricius, 1798 (= *Elater mordelloides* Host, 1789)
abei Nakane, 1985b: 31 **A**: JA
flavipes flavipes Baudi di Selve, 1870: 87 **A**: CY
flavipes subparallelus Pic, 1908b: 9 **A**: TR (Toros Dağlari)
jansoni Bonvouloir, 1895: 355 **A**: TAI **ORR**
mordelloides Host, 1789: 298 (*Elater*) **E**: AU AZ BY CR CT CZ EN FI FR GE GG HU IT LA LS LT NT PL RO SK ST SV SZ UK **A**: FE KZ SY TR WS
angustatus Pic, 1897j: 220
biguttatus Piller & Mitterpacher, 1783: 38 (*Buprestis*) [HN]
bipustulatus Gerstaecker, 1860: 164
caucasicus Ménétriés, 1832: 160
cinctus Panzer, 1796a: 21 (*Elater*)
equestris Fabricius, 1798: 140 (*Elater*)
immaculatus A. Fleischer, 1889: 90
theresae Reitter, 1908h: 296

sulcatus Rost, 1892: 402 **E**: ST (Kavkaz)
talyshensis Dolin & Agajev, 1985: 1746 **E**: AB
torigaii Nakane, 1985b: 30 **A**: JA

genus *Drapetes*, nomen dubium
heydeni Heyden, 1867: 378 patria unknow

family Elateridae, species incertae sedis
Chalcolepis kashyapi Vats & Kashyap, 1992d: 196 [EA] **A**: UP
Chalcolepis pannus Vats & Kashyap, 1992d: 194 [EA] **A**: HP UP
Chalcolepis rotundoextremus Vats & Kashyap, 1992d: 195 [EA] **A**: UP
Tetralobus quadrifoveatus Vats & Kashyap, 1995a: 171 [EA] **A**: UP
Cardiorhinus truncatus Vats & Chauhan, 1993a: 40 [EA] **A**: UP
Ctenicera montana Vats & Chauhan, 1992a: 28 **A**: HP
Ctenicera subnivosus Vats & Chauhan, 1992a: 31 **A**: HP
Glyphochilus spissus Vats & Chauhan, 1993a: 35 [EA] **A**: UP **ORR** [probably *Pachyderes*]
Horistonotus spatulatus Vats & Chauhan, 1993a: 37 [EA] **A**: HP KA UP [perhaps Negatriinae]
Orientis bicoloris Vats & Chauhan, 1993a: 39 **A**: UP [possibly *Hemirhaphes*]
Panspaeus guttatus Sharp, 1877: 409 **Ei**: GB [introduced from New Zealand]

family Elateridae, nomina dubia
Elater advena Fabricius, 1787: 174 [HN] **N**: TU
Elater advena Scopoli, 1763: 94 [HN] **E**: "Carniola"
Elater aterrimus Paykull, 1800: 6 [HN] **E**: SV
Elater brevicollis Schrank, 1789: 80 [HN] **E**: "Österreich"
Elater corticalis Paykull, 1800: 41 **E**: "Westrogothia"
Elater crocopus Hope, 1831: 26 **A**: NP
Elater cruentus A. G. Olivier, 1790: no 31: 45 **N**: "Côte de Barbarie"
Elater degener Scopoli, 1763: 95 **E**: "Carniola"
Elater depressus Fischer von Waldheim, 1824: 202 [HN] **E**: "Volhynia"
Elater dorsalis Paykull, 1800: 4 [HN] **E**: SV
Elater dubius Mayer, 1779: 183 **E**: CZ
Elater equiseti Illiger, 1807: 8 [HN] **E**: PT
Elater equestris Fabricius, 1798: 140 **E**: "Germania"
Elater europaeus Herbst, 1806: 102 **E**: "Europa"
Elater flavatus J. C. Schaeffer, 1766a: pl. 31
Elater flavipes O. F. Müller, 1776: 61 [HN] **E**: DE or NR
Elater flexuosus A. G. Olivier, 1790: 31: 53 **E**: IT (Pisa)
Elater fuscus Schrank, 1781a: 189 [HN] **E**: AU ("Lincii")
Elater griseus Herbst, 1784: 114 [HN] **E**: "Berlin"
Elater haematoceras Schrank, 1781a: 185 **E**: AU ("Vienna")
Elater inunctus Panzer, 1795: 241 **E**: "Brunsvigiae"
Elater laticollis Fischer von Waldheim, 1824: 203 **E**: "prope Volgam"
Elater latus Fuessly, 1775: 16 [HN] **E**: SZ
Elater latus Sulzer, 1776: 53 [HN] **E**: SZ
Elater lineolus Panzer, 1795: 239 **E**: "Germania"
Elater maculatus P. Rossi, 1790: 182 [HN] **E**: "Etrusca"
Elater maurus Herbst, 1806: 69 **E**: "Deutschland"
Elater modestus Herbst, 1806: 103 [HN] **E**: "Deutschland"
Elater nigricornis Herbst, 1806: 66 [HN] **E**: "Deutschland"
Elater nigrinus Paykull, 1800: 39 [HN] **E**: "Uplandia"
Elater picipes Panzer, 1795: 242 [HN] **E**: "Germania"
Elater planatus Charpentier, 1825: 190 **E**: "Pyrenean Mts"
Elater prolifericornis P. Rossi, 1790: 180 **E**: IT ("Etrusca")

Elater pullus O. F. Müller, 1776: 59 [HN] **E**: DE or NR
Elater purpureus Schrank, 1781a: 187 [HN] **E**: AU ("Lincii")
Elater pygmaeus Schrank, 1798: 595 **E**: GE ("Baiern")
Elater rufipalpis Schrank, 1789: 79 **E**: "Österreich"
Elater rufiventris Gebler, 1833: 278 [HN] **A**: WS
Elater serraticornis Scopoli, 1763: 91 [HN] **E**: "Carniola"
Elater sexmaculatus O. F. Müller, 1764: 17 **E**: "Friedrichsdal"
Elater signatulus Herbst, 1806: 78 **E**: "Europa"
Elater socialis Linnaeus, 1767: 1067 **E**: PT
Elater taeniata Panzer, 1795: 236 **E**: "Germania"
Elater tetraspilotus Hope, 1831: 25 **A**: NP
Elater thoracicus Scopoli, 1763: 92 **E**: "Carniolia"
Elater ustulatus Paykull, 1800: 32 [HN] **E**: SV
Elater vilis Herbst, 1806: 103 **E**: "Europa"
Limonius araratensis Faldermann, 1835b: 168

family PLASTOCERIDAE Crowson, 1972

genus *Plastocerus* Schaum, 1852: 49 type species *Callirhipis angulosus* Germar, 1844
Pseudophyllocerus Reitter, 1896e: 234 type species *Pseudophyllocerus atricolor* Reitter, 1896
angulosus Germar, 1844a: pl. 5 (*Callirhipis*) **A**: TR
atricolor Reitter, 1896e: 234 (*Pseudophylocerus*)

family DRILIDAE Blanchard, 1845

genus *Drilus* G. A. Olivier, 1790a: no 23: 1 type species *Drilus flavescens* G. A. Olivier, 1790
Cochleoctonus Mielzinsky, 1824: 74 type species *Cochleoctonus vorax* Mielzinsky, 1824
amabilis Schaufuss, 1867: 85 **E**: SP (Baleares)
attenuatus Pic, 1914l: 50 **E**: GR (Rodos)
subparalleleus Pic, 1934b: 25
bicolor Schaufuss, 1867: 83 **A**: SY
concolor Ahrens, 1812: 13 **E**: AU BH BU CR CZ DE EN FI GE HU IT LS NR PL RO RU SK SV UK YU
ater Audouin, 1824: 459
pectinatus Gyllenhal, 1817: 12 (*Dasytes*)
creticus Pic, 1905j: 170 **E**: GR (Kríti)
distincticollis Pic, 1907g: 129 **E**: GR (Zakynthos)
flavescens G. A. Olivier, 1790a: no 23: 4 **E**: AU BE GE FR IT MA SP SV SZ
flabellatus Kiesenwetter, 1859c: 158
parisinus Thunberg, 1792: 93
serraticornis Marsham, 1802: 374 (*Cantharis*)
vorax Mielzinsky, 1824: 75 (*Cochleoctonus*)
frontalis Schaufuss, 1867: 84 **A**: TR
fulvicollis Audouin, 1824: 460 **E**: CR FR
ruficollis Guérin de Méneville, 1829: 48
venustus Guérin de Méneville, 1829: 48
fulvicornis Kiesenwetter, 1859c: 159 **E**: GR
fulvitarsis Baudi di Selve, 1872: 120 **E**: "Caucasus"
funebris Reitter, 1884c: 79 **E**: GR (Kríti)
humeralis Pic, 1931d: 9 **A**: CY
iljini Barovskij, 1922a: 25 **A**: IN
latithorax Pic, 1902c: 24 **A**: TR
longulus Kiesenwetter, 1859c: 159 **E**: GR (Kríti)
mauritanicus P. H. Lucas, 1842: 1187 **E**: SP **N**: AG MO
tangerianus Escalera, 1914: 538
novoathonius Sumakow, 1903: 48 **E**: "Caucasus"

obscuricornis Pic, 1899j: 205 **A**: TR
posticus Schaufuss, 1867: 83 **A**: SY
rectus Schaufuss, 1867: 83 **A**: SY
rufipes Baudi di Selve, 1871: 62 (*Malacogaster*) **A**: CY SY
reitteri Bourgeois, 1908b: 241
schwarzi Reitter, 1891h: 246 **E**: CR
testaceipes Pic, 1933e: 9 **A**: CY
truquii Baudi di Selve, 1871: 62 (*Malacogaster*) **A**: CY

genus *Euanoma* Reitter, 1889e: 98 type species *Euanoma starcki* Reitter, 1889
elongata Pic, 1932e: 21 **E**: "Caucasus, Ossetia"
starcki Reitter, 1889e: 99 **E**: GG ST

genus *Malacogaster* Bassi, 1834: pl. 99 type species *Malacogaster passerinii* Bassi, 1834
adustus Chevrolat, 1854: 433 **A**: SY
akbesianus Fairmaire, 1895b: cx **A**: SY
bassii P. H. Lucas, 1870: 58 **N**: AG LB
ruficollis Dodero, 1925: 7
holomelas Peyerimhoff, 1949: 266 **N**: AG
maculiventris Reitter, 1894a: 4 **E**: SP (Andalucía)
nigripes Schaufuss, 1867: 85 **E**: SP **N**: AG
heydeni Reitter, 1894a: 4
passerinii Bassi, 1834: pl. 99 **E**: IT (Sicilia)
thoracicus L. Redtenbacher, 1858: 525
rubripes Peyerimhoff, 1949: 265 **N**: AG
rutllanti Alcaide, 1946: 457 **N**: MO
tilloides Wollastone, 1864: 215 **N**: CI

genus *Paradrilus* Kiesenwetter, 1866a: 369 type species *Paradrilus opacus* Kiesenwetter, 1866
opacus Kiesenwetter, 1866a: 369 **E**: SP

genus *Pseudeuanoma* Pic, 1901e: 33 type species *Pseudeuanoma obscura* Pic, 1901
jonica jonica Pic, 1901f: 49 **E**: GR (Zakynthos)
ionica semiobscura Pic, 1901b: 49 **E**: GR (Kefalonia)
obscura Pic, 1901e: 33 **E**: TR
reitteri Pic, 1901f: 49 **E**: GR (Kerkyra, Zakynthos)
semitestacea Pic, 1907k: 153 **E**: TR
subimpressa Pic, 1901i: 74 **E**: GR

genus *Selasia* Laporte, 1838: 19 type species *Selasia rhipiceroides* Laporte, 1838
arabica Geisthardt, 2003: 100 **A**: YE
bleusei E. Olivier, 1913a: 92 **E**: GR (Rodos)
homhilia Geisthardt, 2003: 104 **A**: YE (Suqutra)

family OMALISIDAE Lacordaire, 1857

genus *Omalisus* Geoffroy, 1762: 179 type species *Omalisus fontisbellaquei* Geoffroy, 1785
subgenus *Omalisus* Geoffroy, 1762: 179 type species *Omalisus fontisbellaquei* Geoffroy, 1785
Homalisus Illiger, 1801b: 139 [unjustified emendation]
Omalysus Hope, 1831: 26 [incorrect subsequent spelling]
fontisbellaquei Geoffroy, 1785: 64 **E**: AU BE BH BY CR CZ BU FR GE HU IT LS NL PL RO SL SK SP SZ UK **A**: TR
nigricans Schilsky, 1888: 188 (*Homalisus*)
rougeti J. E. Olivier, 1884: 42 (*Phosphaenus*)
suturalis G. A. Olivier, 1790a: no 24: 4

nicaeensis Lesne, 1921a: 181 **E**: FR
sanguinipennis Laporte, 1840a: 263 **E**: CR IT
taurinensis Baudi di Selve, 1872: 93 **E**: FR
victoris Mulsant, 1852: 79 **E**: FR
subgenus *Phaeopterus* A. Costa, 1857: 220 type species *Phaeopterus unicolor* A. Costa, 1857
flavangulus Spaeth, 1898: 657 (*Homalisus*) **E**: CR
graecus Pic, 1901f: 49 **E**: GR
minutus Pic, 1938g: 13 (*Homalisus*) **E**: IT
nigricornis Reitter, 1881d: 220 **E**: BH CR YU
unicolor A. Costa, 1857: 220 **E**: FR (Corse) IT

genus *Thilmanus* Gemminger, 1869: 1683 type species *Thilmanus obscurus* Baudi di Selve, 1872
longipennis Pic, 1912i: 41 **E**: FR IT (Sardegna)
  latipes Pic, 1913o: 185
obscurus Baudi di Selve, 1872: 96 **E**: FR (Corse) IT (Sardegna)
  fuscus Baudi di Selve, 1872: 97

family LYCIDAE Laporte, 1838
subfamily **Lycinae** Laporte, 1838
tribe Lycini Laporte, 1838

genus *Lipernes* C. O. Waterhouse, 1879: 9 type species *Lipernes perspectus* C. O. Waterhouse, 1879
perspectus C. O. Waterhouse, 1879: 9 **A**: CE SE **ORR**

genus *Lycostomus* Motschulsky, 1861c: 136 type species *Lycus coccineus* Motschulsky, 1861 (= *Lycus similis* Hope, 1831)
subgenus *Haplolycus* Bourgeois, 1883c: lxii type species *Lycus congener* Gerstaecker, 1871
platypterus Bourgeois, 1884a: 65 (*Lycus*) **A**: SA YE
subgenus *Lycostomus* Motschulsky, 1861c: 136 type species *Lycus coccineus* Motschulsky, 1861 (= *Lycus similis* Hope, 1831)
aequalis C. O. Waterhouse, 1879: 10 **A**: CH
afghan Kleine, 1936b: 180 **A**: AF "Himalayas"
ambiguus C. O. Waterhouse, 1879: 11 **A**: KA PA **ORR**
  acutecostatus Fairmaire, 1891d: lxxxix
  orientalis Kleine, 1930a: 94
atricollis Kraatz, 1879a: 125 (*Dictyoptera*) **A**: FE SC "Siberia"
atrimembris Pic, 1926h: 31 **A**: TAI **ORR**
  jeanvoinei Pic, 1927d: 4
auriculatus Fairmaire, 1891d: xc **A**: KA
bourgeoisi bourgeoisi Reitter, 1910c: 205 **A**: JA
bourgeoisi rufofemoratus Reitter, 1910c: 205 **A**: JA
crassus Kleine, 1926c: 97 **A**: TAI
darjeelensis Kazantsev, 1991: 162 **A**: SD **ORR**
davidi Fairmaire, 1878b: 118 (*Porrostoma*) **A**: CH
  debilis C. O. Waterhouse, 1879: 15
  flavipes Pic, 1937j: 147
  sublineatus Pic, 1937k: 169
decorus Kleine, 1935: 77 **A**: UP
delectus Kazantsev, 1991: 163 **A**: SD
delicuus Kazantsev, 1991: 163 **A**: SD **ORR**
formosanus Pic, 1937j: 147 **A**: JA SE TAI
hedini Kleine, 1928: 225 **A**: CH "Tibet"
hirtus Kleine, 1940b: 109 **A**: ZHE (Mokan Shan)
honestus Bourgeois, 1885b: 73 (*Lycus*) **A**: SE TAI NP **ORR**
  modestus C. O. Waterhouse, 1879: 11

klapperichi Kleine, 1940a: 90 **A**: FUJ
koreanus Kleine, 1936a: 13 **A**: NC SC
koshimizui Nakane, 1993: 156 **A**: JA
kraatzi Bourgeois, 1882a: 117 **A**: TR
  rufiventris Kraatz, 1879a: 125 (*Dictyoptera*)
luridus Kleine, 1940b: 109 **A**: JIA
modestus Kiesenwetter, 1874b: 250 (*Lycus*) **A**: CH (East Coast) FE (Sakhalin) JA SC
  rugipennis Matsumura, 1911a: 123 (*Lycus*)
moupinensis Fairmaire, 1889b: 36 **A**: CH
nigricollis Bourgeois, 1905b: 46 **A**: NP **ORR**
nigripes Fabricius, 1787a: 163 (*Pyrochroa*) **A**: BT NP SD **ORR**
pallidenitens Kleine, 1940b: 108 **A**: ZHE
patruelis Bourgeois, 1883a: 625 **A**: SD
placidus C. O. Waterhouse, 1879: 14 **A**: HGK TAI
porphyrophorus Solsky, 1871: 366 (*Dictyoptera*) **A**: CH NC SC "Siberia"
potanini Kazantsev, 2002b: 257 **A**: SCH
praeustus Fabricius, 1792b: 107 (*Lycus*) **A**: KA NP SD UP **ORR**
  cardinalis Kleine, 1926a: 2
  laosensis Pic, 1937j: 148
  lateritius Gorham, 1895a: 299
  legitimus Kleine, 1939b: 19
purpureivestis Gorham, 1895a: 299 **A**: KA
restitutus Bourgeois, 1885b: 75 **A**: NP SD UP
rubrocinctus rubrocinctus Fairmaire, 1886c: 338 **A**: SD XIZ YUN **ORR**
rufocostatus Kleine, 1940b: 107 **A**: FUJ ZHE
semiellipticus kumamotonis Nakane, 1969a: 12 **A**: JA
semiellipticus semiellipticus Reitter, 1910c: 204 **A**: FE JA
similis Hope, 1831: 26 (*Lycus*) **A**: CH KA NP SD TAI UP **ORR**
  cinnabarinus Candèze, 1861: 29 (*Lycus*)
  coccineus Motschulsky, 1861: 136 (*Lycus*)
  geminus Walker, 1858: 281 (*Lycus*)
  suturalis L. Redtenbacher, 1848: 508 (*Lycus*)
  triangularis Hope, 1831: 26 (*Lycus*)
stevensi Kleine, 1928: 225 **A**: "Upper Assam"
stoetzneri Kleine, 1936a: 13 **A**: SCH
thoracicus C. O. Waterhouse, 1879: 16 **A**: BT SD **ORR**
  tonkineus Pic, 1923g: 8
tsilingensis Kazantsev, 2002b: 256 **A**: GAN SHA
villosus Kleine, 1940b: 110 **A**: HAI
vaucheri Bourgeois, 1905e: 291 **N**: AG MO

tribe Macrolycini Kleine, 1928

genus *Dilophotes* C. O. Waterhouse, 1879: 75 type species *Lycus exilis* C. O. Waterhouse, 1878
  Stenolycus Ohbayashi, 1956: 58 type species *Stenolycus ohirai* Ohbayashi, 1956
atricollis Pic, 1926i: 68 **A**: TAI
atrorufus atrorufus Kiesenwetter, 1879b: 305 (*Eros*) **A**: JA
diversipennis Pic, 1922d: 13 (*Macrolycus*) **A**: YUN
libnetoides Nakane, 1971: 138 **A**: SE TAI
lizipingensis Bic, 2002: 19 **A**: GUI SCH
ohirai Ohbayashi, 1956: 58 (*Stenolycus*) **A**: JA (Honshu) TAI
particularis Pic, 1928c: 32 **A**: NP **ORR**
vandykei Nakane, 1970b: 358 (*Stenolycus*) **A**: JA (Hokkaido)

genus *Macrolycus* C. O. Waterhouse, 1878: 96 type species *Macrolycus coccineus* C. O. Waterhouse, 1878 (=*Macrolycus bowringi* C. O. Waterhouse, 1878)
Cerceros Kraatz, 1879a: 126 type species *Cerceros pectinicornis* Kraatz, 1879
aemulus Barovskij, 1930c: 580 **A**: ES FE JA (Honshu) NC NE SC
kleinei Nakane, 1967c: 71
laticollis Pic, 1935g: 110
alishanus Nakane, 1967c: 72 **A**: TAI
atronotatus Pic, 1939h: 165 **A**: SCH SHX
aurantiacus Kazantsev, 2001b: 102 **A**: YUN
bocakorum Kazantsev, 2001b: 104 **A**: SCH SHA
bowringi C. O. Waterhouse, 1878: 105 **A**: SD **ORR**
praedicabilis Kleine, 1939a: 17 (*Ponyalis*)
confusus Nakane, 1994a: 6 **A**: JA (Honshu)
crassicornis Nakane, 1967c: 72 **A**: SE TAI
dominator dominator Kleine, 1925c: 327 **A**: TAI
dominator ishigakianus Satô & Ohbayashi, 1968: 65 **A**: JA (Yayeyama)
erythropterus Nakane, 1967c: 72 **A**: SE TAI
excellens Nakane, 1969a: 31 **A**: JA
flabellatus Motschulsky, 1860b: 114 (*Lygistopterus*) **A**: ES FE JA MG NC NE NO SC SE TAI
kurilensis Medvedev, 1966: 35
pectinicornis Kraatz, 1879a: 127 (*Cerceros*)
pectinifer Kiesenwetter, 1874b: 251 (*Celetes*)
galinae Kazantsev, 2001b: 105 **A**: SCH
gansuensis Kazantsev, 2002b: 253 **A**: GAN
hyugaensis Nakane, 1994a: 6 **A**: JA (Kyushu)
kazantsevi Chernyshev, 1996: 1093 **A**: FE (Kuril, Kunashir)
kotuensis Kazantsev, 1993d: 105 **A**: JA
laetus Nakane, 1967c: 72 **A**: SE TAI
medvedevi Kazantsev, 1993d: 103 **A**: FE
montanus Nakane, 1967c: 71 **A**: JA (Honshu)
multicostatus Kazantsev, 2002b: 255 **A**: SCH
murzini Kazantsev, 2001b: 107 **A**: SHA
nagaii Nakane, 1994a: 5 **A**: JA (Kyushu)
okinawanus Satô & Ohbayashi, 1968: 66 **A**: JA (Okinawa)
oreophilus Kazantsev, 2002b: 255 **A**: SCH YUN
ozeanus Nakane, 1994a: 6 **A**: JA (Honshu)
pubescens Barovskij, 1929a: 231 **A**: ES FE
pusillus Kazantsev, 2001b: 103 **A**: FUJ
shaanxiensis Kazantsev, 2001b: 102 **A**: SHA
shirakii Nakane, 1961b: 11 **A**: JA (Amami-Ôshima)
similaris chugokuensis Nakane, 1994a: 4 **A**: JA (Honshu)
similaris similaris Nakane, 1969a: 26 **A**: JA
spinicollis Fairmaire, 1891b: xiv **A**: HUB
inaequalis Pic, 1922d: 13
submontanus Nakane, 1994a: 5 **A**: JA (Honshu)
taiwanus Nakane, 1967b: 72 **A**: TAI
ussuricus Kazantsev, 1993d: 102 **A**: FE (Ussuri)
wrasei Kazantsev, 2002b: 255 **A**: SCH
yunnanus Kazantsev, 2001b: 100 **A**: SCH

genus *Mesolycus* Gorham, 1883d: 398 type species *Mesolycus puniceus* Gorham, 1883
Biphilodes Kazantsev, 2000b: 329 type species *Dilophotes ilyai* Kazantsev, 2000
Flabellodilophotes Pic, 1912e: 9 type species *Flabellodilophotes obscurus* Pic, 1912
berezowskii Kazantsev, 2000b: 331 (*Dilophotes*) **A**: HUB SCH SHA YUN
bhutanensis Bic, 2002: 10 (*Dilophotes*) **A**: BT

holzschuhi Bic, 2002: 11 **A**: BT **ORR**
hubeicus Kazantsev, 2004: 19 **A**: HUB
ilyai Kazantsev, 2000b: 329 (*Dilophotes*) **A**: SCH
moxiensis Bic, 2002: 17 (*Dilophotes*)
mediozonatus Nakane, 1955a: 35 **A**: JA
puniceus Gorham, 1883d: 399
qinlinganus Kazantsev, 2000b: 330 **A**: GAN
tibetanus Kazantsev, 2000b: 330 **A**: SCH YUN

subfamily **Erotinae** LeConte, 1881
tribe Erotini LeConte, 1881

genus *Benibotarus* Kôno, 1932: 56 type species *Eros spinicoxis* Kiesenwetter, 1874
subgenus *Benibotarus* Kôno, 1932: 56 type species *Eros spinicoxis* Kiesenwetter, 1874
alternatus Fairmaire, 1856a: 531 (*Dictyopterus*) **E**: FR PT SP
decipiens Marseul, 1875: xlii (*Dictyopterus*)
spinicoxis Kiesenwetter, 1874b: 254 (*Eros*) **A**: FE (Kunashir) JA SC TAI
krivolutzkii Medvedev, 1966: 36
longicornis Reiche, 1878b: xxvii (*Eros*) **E**: GG
schelkovnikovi Barovskij, 1930d: 615 (*Pyropterus*)
subgenus *Sibetarus* Bocák & Bocáková, 1991: 319 type species *Dictyopterus taygetanus* Pic, 1905
arnoldi Barovskij, 1932: 139 (*Pyropterus*) **E**: AB
taygetanus Pic, 1905n: 185 (*Dictyoptera*) **E**: AU BH CR CZ FR GE GR HU PL RO SK
fiedleri Reitter, 1907a: 25 (*Dictyopterus*)

genus *Chinotaphes* Bocák & Bocáková, 1999: 48 type species *Chinotaphes weibaoshanensis* Bocák & Bocáková, 1999
weibaoshanensis Bocák & Bocáková, 1999: 48 **A**: YUN

genus *Dictyoptera* Latreille, 1829a: 464 type species *Cantharis sanguineus* sensu Linnaeus, 1761 (= *Pyrochroa aurora* Herbst, 1784)
Dictyopterus Mulsant, 1838: 80 type species *Pyrochroa aurora* Herbst, 1784
aurora Herbst, 1784b: 105 (*Pyrochroa*) **E**: AL AU BE BH BU BY CR CZ DE EN FI FR GE GB GR HU IT LA LS LT MC NR PL RO RU SK SL SP SV SZ UK YU **A**: ES FE JA NC SC WS **N**: AG **NAR**
caprai Puppin, 1974: 40
coccinea Linnaeus, 1767: 648 (*Lampyris*)
coccinata Say, 1835: 155 (*Omalisus*)
hybrida Mannerheim, 1843c: 88 (*Dictyopterus*)
limbaticollis Pic, 1914l: 50
sanguinea sensu Linnaeus, 1761: 202 (*Cantharis*)
superba Motschulsky, 1860b: 115 (*Dictyopterus*)
bicoloricornis Pic, 1933i: 110 (*Pyropterus*) **A**: SCH
elegans Nakane & J. R. Winkler, 1952: 133 **A**: JA (Honshu)
formosana Nakane, 1968c: 217 **A**: TAI
forticornis Pic, 1942a: 1 (*Dictyopterus*) **A**: CH
gansuensis Kazantsev, 2004: 25 **A**: GAN
gorhami Kôno, 1932: 57 (*Dictyopterus*) **A**: JA (Honshu, Kyushu, Shikoku) TAI
erythropterus Gorham, 1883d: 400 (*Eros*)
marginicollis Boheman, 1858: 73 **A**: HKG
sapporensis Kôno, 1932: 57 (*Dictyopterus*) **A**: JA
ohbayashii Nakane, 1969a: 129
speciosa Ohbayashi, 1954: 19 **A**: JA (Honshu, Kyushu)
velata Gorham, 1883d: 402 (*Eros*) **A**: FE (Kunashiri) JA (Hokkaido, Honshu)
pseudovelata Medvedev, 1966: 36

genus *Eropterus* Green, 1951: 14 type species *Dictyopterus trilineatus* Melsheimer, 1846
Asioplateros Nakane, 1969b: 9 type species *Asioplateros flavipennis* Nakane, 1969

aritai Satô & Ohbayashi, 1968: 66 (*Aplatopterus*) **A**: JA (Yayeyama) TAI
flavipennis Nakane, 1969b: 9 (*Asioplateros*) **A**: TAI
nothus Kiesenwetter, 1874b: 258 (*Eros*) **A**: JA
ohkurai Matsuda, 1992: 94 **A**: SE TAI
taiwanus Matsuda, 1992: 92 **A**: SE TAI
yakushimaensis Ohbayashi, 1954: 22 (*Plateros*) **A**: JA (Yakushima)

genus *Erotides* C. O. Waterhouse, 1879: 37 type species *Eros oblitus* Newman, 1838
subgenus *Glabroplatycis* Pic, 1914l: 50 type species *Dictyopterus cosnardi* Chevrolat, 1831
cosnardi Chevrolat, 1831: livr. 14: pl. 45 (*Dictyopterus*) **E**: AL AU BE BH BU CR CZ GE DE FI FR GB GR HU LS MC NL PL RO SK SV SZ UK YU **A**: WS
 flavescens L. Redtenbacher, 1849: 319 (*Dictyopterus*)
 mercki Mulsant, 1838: 80 (*Dictyopterus*)
kanoi Nakane, 1967e: 285 (*Platycis*) **A**: TAI
matsudai Bocák, 1996: 1 (*Platycis*) **A**: YUN
nasutus Kiesenwetter, 1874b: 255 (*Eros*) **E**: CT **A**: ES FE (Kunashir) JA SC TAI WS ZHE
 raddensis Pic, 1904d: 25 (*Platycis*)
schneideri Kiesenwetter, 1878: 206 (*Eros*) **E**: AB (Talysh)
 abdominalis Reiche, 1878b: xxviii (*Eros*)
taiwanus Kôno, 1932: 59 (*Platycis*) **A**: TAI

genus *Greenarus* Kazantsev, 1995a: 40 type species *Eros thoracicus* Randall, 1838
nigripennis Nakane & Ohbayashi, 1958: 79 **A**: JA

genus *Helcophorus* Fairmaire, 1891e: cxxix type species *Helcophorus miniatus* Fairmaire, 1891
brevilobus Bocáková, 1994: 133 (*Pyropterus*) **A**: KA
delicatus Kazantsev, 1996a: 84 (*Pyropterus*) **A**: HUB
gobindanus Kazantsev, 2004: 28 **A**: UP
miniatus Fairmaire, 1891e: cxxix **A**: KA PA
 gilvus Kleine, 1928: 226 (*Pyropterus*)
 miniatus Pic, 1928j: 267 (*Dictyopterus*) [HN]
 unicolor Gorham, 1903: 324 (*Xylobanus*)
murzini Kazantsev, 2004: 26 **A**: YUN
ostindicus Kazantsev, 1996a: 83 (*Pyropterus*) **A**: SD
tricolor Kazantsev, 2000a: 80 (*Pyropterus*) **A**: YUN

genus *Laterialis* Kazantsev, 1990: 14 type species *Eros oculatus* Gorham, 1883
subgenus *Laterialis* Kazantsev, 1990: 14 type species *Eros oculatus* Gorham, 1883
oculatus Gorham, 1883d: 401 (*Eros*) **A**: FE JA TAI
 diversicollis Pic, 1942a: 2 (*Pyropterus*)
 sanguinipennis Nakane & Ohbayashi, 1958: 80 (*Benibotarus*)
subgenus *Tricostaeptera* Kazantsev, 1997a: 160 type species *Kolibaceum* [sic!] *sanguinea* Kazantsev, 1997
sanguinea Kazantsev, 1997a: 160 (*Kolibacium*) **A**: CE
shirozui Nakane, 1968c: 218 (*Benibotarus*) **A**: SE TAI

genus *Lopheros* LeConte, 1881: 23 type species *Lycus fraternus* Randall, 1838
subgenus *Eulopheros* Kazantsev, 1995a: 41 type species *Dictyopterus harmandi* Bourgeois, 1902
harmandi harmandi Bourgeois, 1902b: 89 (*Dictyopterus*) **A**: FE (Sakhalin, Kuril) JA (Hokkaido, Honshu)
 nakanei J. R. Winkler, 1953: 407 (*Aplatopterus*)
harmandi hiroshimensis Nakane, 1969a: 109 **A**: JA (Honshu)
harmandi shikokuanus Nakane, 1969a: 110 **A**: JA (Shikoku)
subgenus *Lopheros* LeConte, 1881: 23 type species *Lycus fraternus* Randall, 1838
 Aplatopterus Reitter, 1911a: 250 type species *Lycus rubens* Gyllenhal, 1817
 Eudictyoptera Barovskij, 1930a: 287 type species *Eudictyoptera brevicornis* Barovskij, 1930
 Falsoconderis Pic, 1926h: 22 type species *Falsoconderis pubescens* Pic, 1926

brevicornis Barovskij, 1930a: 288 (*Eudictyoptera*) **A**: FE
imanensis Medvedev, 1970: 163 (*Aplatopterus*)
brunneorubens Medvedev, 1966: 37 (*Aplatopterus*) **A**: FE (Kuril) JA
konoi Nakane, 1969a: 115
crassipalpis Nakane, 1969a: 117 **A**: JA (Hokkaido, Honshu)
lineatus Gorham, 1883d: 406 (*Plateros*) **E**: PL **A**: FE (Ussuri) JA (Honshu) NE
mamaevi Medvedev, 1979: 240 (*Aplatopterus*)
motschoulskii Barovskij, 1930b: 359 (*Aplatopterus*)
mandshuricus Kleine, 1940b: 112 (*Eudictyoptera*) **A**: NE
minimus Nakane, 1969a: 120 **A**: JA (Honshu)
rubens Gyllenhal, 1817: 31 (*Lycus*) **E**: AU BY CZ FI FR GE HU IT LA LS PL RO RU SK SL SP SV SZ UK YU
sanguinea Scopoli, 1763: 38 (*Cassida*)
vavrai Roubal, 1929a: 31 (*Dictyopterus*)
septentrionalis Kôno, 1932: 58 (*Aplatopterus*) **A**: FE (Sakhalin, Kuril) JA (Hokkaido, Honshu)
japonicus Pic, 1942a: 2 (*Aplopterus*)

genus *Platycis* C. G. Thomson, 1864: 162 type species *Pyrochroa minuta* Fabricius, 1787
subgenus *Konoplatycis* Nakane, 1969a: 155 type species *Platycis otome* Kôno, 1932
otome Kôno, 1932: 59 (*Platycis*) **A**: JA
subgenus *Platycis* C. G. Thomson, 1864: 162 type species *Pyrochroa minuta* Fabricius, 1787
consobrinus Bourgeois, 1902b: 90 **A**: JA (Honshu, Kyushu)
minutus Fabricius, 1787a: 163 (*Pyrochroa*) **E**: AU BE BH CZ DE GE FI FR GB HU IR IT LA LS NR PL RO RU SK SL SP SV SZ UK YU **A**: ES FE JA WS
nigrorubra DeGeer, 1774: 46 (*Lampyris*)
pusilla Gmellin, 1790: 1886 (*Lampyris*)
siculus Pic, 1914l: 50

genus *Propyropterus* Nakane, 1968c: 219 type species *Propyropterus pygidialis* Nakane, 1968
subgenus *Paralopheros* Kazantsev, 1993b: 64 type species *Conderis diversipennis* Pic, 1921
diversipennis Pic, 1921b: 7 (*Conderis*) **A**: YUN
fukiensis Kleine, 1940a: 91 (*Pyropterus*) **A**: FUJ
subgenus *Propyropterus* Nakane, 1968c: 219 type species *Propyropterus pygidialis* Nakane, 1968
plateroides Kazantsev, 2004: 32 **A**: TAI
pygidialis Nakane, 1968c: 219 **A**: SE TAI
kanoi Nakane, 1968c: 222 **A**: SE TAI

genus *Punicealis* Kazantsev, 1990: 11 type species *Dictyoptera* (*Punicealis*) *medvedevi* Kazantsev, 1990
barovskii Medvedev, 1966: 36 (*Pyropterus*) **A**: FE (Kunashir)
martensi Kazantsev, 1993b: 61 (*Dictyoptera*) **A**: NP
medvedevi Kazantsev, 1990: 12 (*Dictyoptera*) **A**: FE (Kunashir, Primorsky Kray)
mirandus Barovskij, 1930b: 360 (*Dictyoptera*) **A**: FE (Kunashir)

genus *Pyropterus* Mulsant, 1838: 81 type species *Lycus affinis* Paykull, 1799
Hiekeolycus Winkler, 1987: 66 type species *Hiekeolycus berendti* Winkler, 1987
Xylobanoides Kleine, 1928: 237 type species *Xylobanus unicolor* Gorham, 1903
atricollis Motschulsky, 1860b: 115 (*Dictyopterus*) **A**: FE (Amur)
insularis Kazantsev, 1995a: 35 **A**: FE (Kuril Is.)
maritimus Kazantsev, 1995a: 37 **A**: FE (Ussuri)
nigroruber DeGeer, 1774: 46 (*Lampyris*) **E**: AU BH BU BY CR CZ DE EN FI FR GB GE GR HU IR IT LA LT MC NR PL RO RU SK SL SP SV SZ UK YU **A**: ES FE JA WS
affinis Paykull, 1799: 176 (*Lycus*)
aurora Fabricius, 1801b: 116 (*Lycus*)
nigrorubra Goeze, 1777: 525 (*Lampyris*)
rubripes Pic, 1897j: 220 (*Eros*) **N**: AG

tribe Taphini Bocák & Bocáková, 1990

genus *Coloberos* Bourgeois, 1885b: 82 type species *Coloberos tenellus* Bourgeois, 1885
tenellus Bourgeois, 1885b: 83 **A**: SD

genus *Taphes* C. O. Waterhouse, 1878: 102 type species *Taphes brevicollis* C. O. Waterhouse, 1878
brevicollis C. O. Waterhouse, 1878: 111 **A**: FUJ UP **ORR**

genus *Parataphes* Kleine, 1925b: 103 type species *Parataphes dohertyi* Kleine, 1925
nanus Kleine, 1939a: 9 **A**: CH **ORR**

genus *Lycoprogentes* Pic, 1915f: 6 type species *Lycoprogentes pouilloni* Pic, 1915
Parapyropterus Kleine, 1926c: 104 type species *Parapyropterus nigrostriatus* Kleine, 1926
Protaphes Kleine, 1926e: 363 type species *Protaphes confertus* Kleine, 1926
Pseudosynchonnus Pic, 1922d: 13 type species *Pseudosynchonnus testaceus* Pic, 1922
himalejicus Bourgeois, 1882b: cxliv (*Pyropterus*) **A**: SD
nigrostriatus Kleine, 1926c: 104 (*Parapyropterus*) **A**: TAI
bistriatus Kleine, 1943: 148 (*Protaphes*)
pouilloni Pic, 1915f: 6 **A**: SD

subfamily **Platerodinae** Kleine, 1928
tribe Libnetini Bocák & Bocáková, 1990

genus *Dihammatus* C. O. Waterhouse, 1879: 29 type species *Dihammatus cribripennis* C. O. Waterhouse, 1879
atricolor Pic, 1921b: 1 **A**: TAI
brancuccii Kazantsev, 1992b: 262 **A**: NP
chinensis Bocáková, 2003: 174 **A**: YUN
holzschuhi Bocáková, 2003: 174 **A**: NP
parvus Nakane, 1971: 151 **A**: SE TAI
wittmeri Kazantsev, 1992b: 261 **A**: NP SD **ORR**
xunyanbanensis Bocáková, 2003: 174 **A**: SHA

genus *Libnetis* C. O. Waterhouse, 1879: 77 type species *Libnetus pumilio* C. O. Waterhouse, 1878
Libnetomimus Kleine, 1927a: 1 type species *Libnetomimus setosus* Kleine, 1927
Libnetus C. O. Waterhouse, 1878: 104 type species *Libnetus pumilio* C. O. Waterhouse, 1878
bicoloratus Kleine, 1939a: 16 **A**: SD
birmanensis Kleine, 1928: 255 **A**: GUI **ORR**
chinensis Bocáková, 2000: 225 **A**: GUI
confucius Kazantsev, 2000a: 90 **A**: JIA
edentatus Bocáková, 2000: 226 **A**: GUI
fodingshanensis Bocáková, 2000: 226 **A**: GUI
granicollis Kiesenwetter, 1879b: 305 (*Eros*) **A**: JA
kazuoi Satô & Ohbayashi, 1968: 71 **A**: JA (Ishigaki)
leei Kazantsev & P. S. Yang, 1999: 243 **A**: TAI
monachus Kazantsev, 2000a: 88 **A**: JIA
opacus Pic, 1921d: 14 **A**: TAI **ORR**
piceovittatus Kazantsev, 1992b: 264 **A**: NP
sauteri Kazantsev, 1993c: 72 **A**: TAI
similis Kleine, 1928: 255 **A**: UP
sinica Kazantsev & Yang, 1999: 243 **A**: FUJ
tenebrans Kleine, 1950: 22 (*Plateros*) [HN]
taiwanus Kazantsev, 2002b: 262 **A**: TAI
xilingensis Kazantsev, 2000a: 88 **A**: SCH
xunyanbanensis Bocáková, 2003: 177 **A**: SHA
yunnanensis Bocáková, 2000: 228 **A**: YUN

genus *Libnetisia* Pic, 1921b: 1 type species *Libnetisia atricornis* Pic, 1921
latrunculus Kazantsev, 2000a: 90 (*Libnetis*) **A**: SCH
xunyanbanensis Bocáková, 2004: 62 **A**: SHA
yunnanensis Bocáková, 2004: 61 **A**: YUN

genus *Prolibnetis* Bocáková, 2004: type species *Prolibnetis nepalensis* Bocáková, 2004
nepalensis Bocáková, 2004: 58 **A**: NP

tribe Lyponiini Bocák & Bocáková, 1990

***genus Lyponia* C. O. Waterhouse, 1878: 99** type species *Lyponia debilis* C. O. Waterhouse, 1878
subgenus *Lyponia* C. O. Waterhouse, 1878: 99 type species *Lyponia debilis* C. O. Waterhouse, 1878
debilis C. O. Waterhouse, 1878: 107 **A**: FUJ GUI
  pieli Pic, 1937k: 169
delicatula Kiesenwetter, 1874b: 254 (*Eros*) **A**: JA (Honshu, Shikoku, Kyushu, Tsushima)
kuatunensis Bocák, 1999: 79 **A**: FUJ
nepalensis Nakane, 1983a: 116 **A**: NP
osawai Nakane, 1969a: 89 **A**: FE (Kunashir) JA (Honshu)
shaanxiensis Kazantsev, 2002a: 202 **A**: SHA
subgenus *Poniella* Kazantsev, 2002a: 199 type species *Lyponia gongashanica* Bocák, 1999
gongashanica Bocák, 1999: 84 **A**: SCH
kleinei Nakane, 1967e: 286 **A**: TAI
minuta Bocák, 1999: 82 **A**: YUN
nigroscutellaris Ohbayashi, 1956: 59 (*Plateros*) **A**: JA (Honshu, Shikoku, Kyushu)
  subpectinata Nakane, 1969a: 86
palpalis Nakane, 1971: 140 **A**: TAI
shennongjinensis Bocák, 1999: 82 **A**: HUB
subgenus *Weiyangia* Bocák, 1999: 74 type species *Lyponia brevicollis* Fairmaire, 1889
brevicollis Fairmaire, 1889b: 37 **A**: SCH
pertica Bocák, 1999: 75 **A**: SCH YUN
taliensis Bocák, 1999: 75 **A**: YUN

genus *Ponyalis* Fairmaire, 1900: 623 type species *Ponyalis laticornis* Fairmaire, 1900
cincinatus Kazantsev, 2002a: 203 **A**: TAI
chifenglee Kazantsev, 2002a: 203 **A**: TAI
daucinus Kazantsev, 2002a: 203 **A**: TAI
dolosa Kleine, 1924: 173 (*Lyponia*) **A**: TAI
fukiensis Bocák, 1999: 92 (*Lyponia*) **A**: FUJ
gestroi Pic, 1912d: 2 (*Lyponia*) **A**: JA (Yayeyama) TAI
  formosana Kleine, 1924: 174 (*Lyponia*)
gracilis Bocák, 1999: 89 (*Lyponia*) **A**: FUJ
himalejica Bourgeois, 1885b: 79 (*Lyponia*) **A**: SD **ORR**
  aurantiaca Pic, 1927d: 5 (*Lyponia*)
  ochraceicollis Pic, 1923g: 9 (*Lyponia*)
  robusticollis Pic, 1939h: 165 (*Lyponia*)
  waterhousei Gorham, 1890c: 543 (*Lyponia*)
ishigakiana Nakane, 1961b: 14 (*Lyponia*) **A**: FUJ JA (Ishigaki)
klapperichi Bocák, 1999: 100 (*Lyponia*) **A**: FUJ
laticornis Fairmaire, 1900: 623 **A**: GUI YUN **ORR**
  diversicornis Pic, 1926j: 70 (*Lyponia*)
  guerryi Pic, 1939h: 165 (*Lyponia*)
  limbaticollis Pic, 1926j: 70 (*Lyponia*)
  patruelis Kleine, 1939a: 18 (*Lyponia*)
  robusta Pic, 1922d: 13 (*Lyponia*)
nigrohumeralis Pic, 1939i: 220 (*Lyponia*) **A**: HUB SCH SHA
oshimana Nakane, 1961b: 14 (*Lyponia*) **A**: JA (Amami-Ôshima)

quadricollis Kiesenwetter, 1874b: 252 (*Celetes*) **A**: JA SC
militans Kiesenwetter, 1874b: 253 (*Eros*)
sichuanensis Bocák, 1999: 88 (*Lyponia*) **A**: SCH
tryznai Bocák, 1999: 95 (*Lyponia*) **A**: SCH

tribe Platerodini Kleine, 1928

genus *Plateros* Bourgeois, 1879a: xix type species *Eros brasiliensis* P. H. Lucas, 1857
Calleros Gorham, 1881: 25 type species *Calleros puniceus* Gorham, 1881
Calloplateros Pic, 1923k: 30 type species *Calloplateros particularis* Pic, 1923
Costatoplateros Pic, 1949a: 5 type species *Plateros* (*Costatoplateros*) *fortecostatus* Pic, 1949
Ditoneces C. O. Waterhouse, 1879: 31 type species *Lycus punctipennis* C. O. Waterhouse, 1878
Cautirodes Pic, 1921b: 3 type species *Ditoneces malaccanus* Pic, 1921
Falsotrichalus Pic, 1921b: 3 type species *Falsotrichalus gracilicornis* Pic, 1921
Graciloplateros Pic, 1921b: 1 type species *Graciloplateros sapientis* Pic: 1921
Libnetomorphus Pic, 1921b: 2 type species *Libnetomorphus cephalotes* Pic, 1921
Melampyrus C. O. Waterhouse, 1879: 30 type species *Lycus alternans* C. O. Waterhouse, 1878
Microditoneces Pic, 1921b: 5 type species *Microditoneces brevelineatus* Pic, 1921
Microplateros Pic, 1921b: 2 type species *Microplateros diversithorax* Pic, 1921
Neoceletes Leng & Shoemaker, 1915: 55 type species *Neoceletes crateracollis* Leng & Shoemaker, 1915
Planeteros Gorham, 1883e: 591 type species *Planeteros ochropterus* Gorham, 1883
Tolianus Pic, 1921b: 3 type species *Tolianus diversithorax* Pic, 1921
alishanus Nakane, 1969c: 473 **A**: SE TAI
annapurnensis Bocák & Bocáková, 1999: 55 **A**: NP
apicicornis Pic, 1925h: 19 (*Ditoneces*) **A**: TAI
atsushii Nakane, 1995b: 45 (*Melaneros*) **A**: JA
businskyi Bocáková, 1997b: 184 (*Melaneros*) **A**: XIZ
berezovskii Kazantsev, 2001a: 15 (*Melaneros*) **A**: SCH
bhaktai Kazantsev, 1991: 176 **A**: NP
birganjensis Kazantsev, 2001a: 13 (*Melaneros*) **A**: NP
brancuccii Kazantsev, 1991: 173 **A**: NP SD
brevehumeralis Pic, 1928b: 16 (*Ditoneces*) **A**: SD **ORR**
brevelineatus Pic, 1922k: 73 **A**: SE TAI
brevenotatus Pic, 1921b: 6 **A**: SE TAI
chinensis C. O. Waterhouse, 1879: 29 **A**: HKG TAI SW **ORR**
flavomarginatus Kleine, 1936c: 264
formosanus Pic, 1921b: 7
sycophanta Fairmaire, 1889a: 352
cimicoides Kleine, 1928: 242 **A**: NP SD UP **ORR**
cinereus Kazantsev, 1991: 166 **A**: NP
coccinipennis Nakane, 1971: 142 **A**: SE TAI
compressus Bocáková, 1997a: 5 (*Melaneros*) **A**: NP
confrater Kleine, 1928: 241 **A**: UP
coracinus Kiesenwetter, 1874b: 257 (*Eros*) **A**: JA
curtelineatus Pic, 1926h: 31 **A**: YUN
dentifer Kazantsev, 1991: 177 **A**: NP
dubius Bocáková, 1997b: 187 (*Melaneros*) **A**: YUN
gomadanus Nakane, 1969a: 72 **A**: JA (Honshu)
flavonotatus Kleine, 1928: 242 **A**: UP **ORR**
fukienensis Bocáková, 1997b: 178 (*Melaneros*) **A**: FUJ
fuscodiscus Kazantsev, 1993a: 29 **A**: "Himalaya"
gurkha Kazantsev, 2001a: 14 (*Melaneros*) **A**: NP UP
harmandi Bourgeois, 1902c: 92 **A**: NP SD YUN
sikkimanus Kleine, 1928: 240
hasegawai Nakane & Baba, 1961: 5 **A**: JA
holzschuhi Bocáková, 1997a: 4 (*Melaneros*) **A**: BT
hyuganus Nakane, 1995b: 45 (*Melaneros*) **A**: JA

ignius M. Chûjô & Satô, 1970: 27 **A**: JA (Honshu)
ikarianus Nakane, 1961b: 14 **A**: JA (Amami-Ôshima)
imasakai Matsuda, 1985: 29 **A**: JA (Yayeyama)
incurvus Bocáková, 1997b: 178 (*Melaneros*) **A**: FUJ
infuscatus Bocáková, 1997b: 185 (*Melaneros*) **A**: YUN
jageri Bocák & Bocáková, 1999: 53 **A**: NP
japonicus Nakane, 1969a: 77 **A**: JA (Honshu, Kyushu)
jenisi Bocáková, 1997a: 5 (*Melaneros*) **A**: NP
jizushanensis Bocáková, 1997b: 188 (*Melaneros*) **A**: YUN
kanoi Nakane, 1969c: 471 (*Ditoneces*) **A**: SE TAI
kirishima Nakane, 1975: 30 **A**: JA (Kyushu)
kleineanus Nakane, 1971: 150 **A**: TAI
kosakai Nakane, 1977a: 96 **A**: JA
krivolutzkii Medvedev, 1966: 38 **A**: FE (Kuril)
kubani Bocáková, 1997b: 188 (*Melaneros*) **A**: YUN
kumatai Nakane, 1983a: 118 **A**: NP
kurentzovi Medvedev, 1970: 165 **A**: FE (Ussuri)
laeticeps Pic, 1921b: 4 (*Ditoneces*) **A**: TAI
latenigrescens Nakane, 1983a: 121 **A**: NP
laterculus Kazantsev, 1991: 174 **A**: NP
latevittatus Nakane, 1983a: 120 **A**: NP
leechi Nakane, 1971: 145 **A**: SE TAI
macrosuturalis Kazantsev, 1991: 170 **A**: NP
maculatithorax Pic, 1921b: 4 (*Ditoneces*) **A**: TAI
  griseostriatus Kleine, 1926c: 101(*Ditoneces*)
mahabharaticus Kazantsev, 2001a: 14 (*Melaneros*) **A**: NP UP
manaliensis Bocáková, 1997a: 2 (*Melaneros*) **A**: HP
marginicollis Nakane, 1969a: 70 **A**: JA (Honshu)
miwai M. Satô & N. Ohbayashi, 1968: 68 **A**: JA (Amami-Ôshima)
moxiensis Bocáková, 1997b: 182 (*Melaneros*) **A**: SCH
nakachii Nakane, 1961b: 12 (*Ditoneces*) **A**: JA (Okinawa)
nepalensis Kazantsev, 1991: 172 **A**: NP
nigrosulcatus Kleine, 1928: 250 **A**: SD **ORR**
nozirianus Nakane, 1969a: 211 **A**: JA (Honshu)
pallidus Pic, 1921b: 5 (*Ditoneces*) **A**: CH
phoeniceus Kazantsev, 2001a: 14 (*Melaneros*) **A**: SCH
picianus Nakane, 1971: 146 **A**: TAI
  rubripennis Pic, 1921b: 5 (*Microditoneces*) [HN]
planatus obconiceps Pic, 1925h: 19 (*Ditoneces*) **A**: SE TAI
planatus planatus C. O. Waterhouse, 1879: 27 **A**: NP SCH YUN **ORR**
  fulgens Kleine, 1933: 20
  incisicollis Pic, 1921b: 5 (*Ditoneces*)
  sulcatithorax Pic, 1925h: 18 (*Ditoneces*)
  tuberculatus Pic, 1921b: 6
ponderosus Kleine, 1940: 114 (*Ditoneces*) **A**: ZHE
propinquus C. O. Waterhouse, 1879: 32 (*Ditoneces*) **A**: CH
proximus Bourgeois, 1902c: 93 **A**: SD
purpurivestis Gorham, 1883d: 406 **A**: JA (Honshu)
purus Kleine, 1926c: 99 **A**: FUJ JA (Honshu) NC SC TAI
  koreanus Kleine, 1936c: 263
rubens Bocáková, 1997b: 186 (*Melaneros*) **A**: YUN
rufomarginatus M. Chûjô & Satô, 1970: 28 **A**: JA (Iriomote)
sakishimana Nakane, 1985c: 155 **A**: JA (Ryukyu)
sauteri Pic, 1925h: 19 (*Ditoneces*) **A**: TAI
  gressitti Nakane, 1971: 147

schawalleri Kazantsev, 1993a: 29 **A**: NP
sherpa Kazantsev & P. S. Yang, 1999: 243 (*Melaneros*) **A**: SD
demissus Kazantsev, 1991: 176 [HN]
shibatai Nakane, 1961b: 13 **A**: JA (Amami-Ôshima) TAI
shirakii Nakane, 1961b: 13 **A**: JA (Amami-Ôshima)
sichuanensis Bocáková, 1997b: 182 (*Melaneros*) **A**: SCH
sincerus Kleine, 1928: 247 **A**: NP UP
smetanai Kazantsev, 2001a: 13 (*Melaneros*) **A**: NP
sordidus Fairmaire, 1889b: 36 **A**: YUN
suturalis Pic, 1919a: 21 **A**: NP UP
tryznai Bocáková, 1997b: 186 (*Melaneros*) **A**: SCH
tsinensis Pic, 1926h: 23 **A**: YUN
udoensis Nakane, 1985c: 154 **A**: JA (Kyushu)
ussuriensis Barovskij, 1931c: 160 **A**: FE (Ussuri)
viduus Nakane, 1971: 149 **A**: SE TAI
weibaoshanensis Bocáková, 1997b: 186 (*Melaneros*) **A**: YUN
wittmerianus Kazantsev, 1991: 173 **A**: NP
yaku Nakane, 1969a: 73 **A**: JA (Kyushu, Yakushima)
yayeyamanus M. Chûjô & Satô, 1970: 30 **A**: JA (Yayeyama)
yunnanensis Bocáková, 1997b: 184 (*Melaneros*) **A**: YUN

subfamily **Lyropaeinae** Bocák & Bocáková, 1989

genus *Lyropaeus* C. O. Waterhouse, 1878: 104 type species *Lycus fallax* Walker, 1858
Paralycus Medvedev & Kazantsev, 1992: 55 type species *Paralycus helvius* Medvedev & Kazantsev, 1992
laticollis Pic, 1926d: 4 **A**: NP

subfamily **Miniduliticolinae** Kazantsev, 2003

genus *Platerodrilus* Pic, 1921d: 13 type species *Platerodrilus sinuatus* Pic, 1921
Duliticola Mjoeberg, 1925: 133 type species *Duliticola paradoxa* Mjoeberg, 1925
Macrolibnetis Pic, 1938m: 280 type species *Macrolibnetis depressus* Pic, 1938
indicus Wittmer, 1966b: 229 **A**: NP SD

genus *Sinodulia* Kazantsev, 2003: 14 type species *Sinodulia gressiti* Kazantsev, 2003
gressitti Kazantsev, 2003: 16 **A**: HUB

subfamily **Ateliinae** Kleine, 1928

genus *Atelius* C. O. Waterhouse, 1878: 104 type species *Lycus expansicornis* Walker, 1858
wittmeri Bocák, 1997 **A**: HAI

subfamily **Metriorrhynchinae** Kleine, 1926

tribe Conderini Bocák & Bocáková, 1990

genus *Conderis* C. O. Waterhouse, 1879: 59 type species *Calopteron signicolle* Kirsch, 1875
Pseudoconderis Pic, 1921b: 8 type species *Pseudoconderis gorhami* Pic, 1921
chujoi Ohbayashi, 1954: 20 **A**: JA (Shikoku)
longipennis Pic, 1921b: 8 (*Pseudoconderis*) **A**: CH
major C. O. Waterhouse, 1879: 60 **A**: SD
orientis Gorham, 1883d: 403 **A**: JA (Honshu)
pictus Gorham, 1883d: 404 **A**: JA (Honshu, Shikoku, Kyushu)
pseudoconderoides Kazantsev, 2002b: 257 **A**: GAN SHA
rufohumeralis Nakane, 1958b: 88 **A**: JA (Honshu)
signicollis Kirsch, 1875: 36 (*Calopteron*) **A**: CH NP SD **ORR**
festivus C. O. Waterhouse, 1878: 110
velutinus C. O. Waterhouse, 1878: 110
shirozui Nakane, 1968c: 222 **A**: SE TAI

waterhousei Bourgeois, 1885b: 81 **A**: NP SD UP
wolongensis Kazantsev, 2002b: 257 **A**: SCH

genus *Xylobanellus* Kleine, 1930b: 171 type species *Xylobanellus atricolor* Kleine, 1930
Chuzenjianus Kôno, 1932: 61 type species *Chuzenjianus tenuis* Kôno, 1932
Rossioptera Kazantsev, 1988: 169 type species *Dictyoptera erythroptera* Baudi di Selve, 1871
erythropterus Baudi di Selve, 1871: 93 (*Eros*) **E**: BY CT FI LA PL **A**: FE (Amur, Ussuri)
amurensis Pic, 1932e: 21 (*Pyropterus*)
amurensis Kleine, 1933: 10 [HN]
wankowiczi Bourgeois, 1879b: xxxix (*Eros*)
gansuensis Kazantsev, 2000a: 81 **A**: GAN
tenuis Kôno, 1932: 61 (*Chuzenjianus*) **A**: FE (Kunashir) JA (Hokkaido, Honshu)

tribe Metriorrhynchini Kleine, 1926

genus *Bulenides* C. O. Waterhouse, 1879: 34 type species *Lycus* [gen. 21] *obsoletus* C. O. Waterhouse, 1878
purpureus Pic, 1922k: 74 **A**: SE TAI **ORR**

genus *Cautires* C. O. Waterhouse, 1879: 36 type species *Lycus excellens* C. O. Waterhouse, 1878
amamiensis Nakane, 1961b: 12 **A**: JA (Amami-Ôshima)
bannanus Nakane, 1969a: 197 **A**: JA (Ishigaki)
bulenoides Kazantsev, 2000a: 81 **A**: TAI
callimarginatus Kazantsev, 2002b: 259 **A**: HUB
certus Kleine, 1926c: 103 (*Cladophorus*) **A**: TAI
taiwanensis Jeng & P.-S. Yang, 2002: 183 [RN]
chui Kazantsev & P.-S. Yang, 1999: 240 **A**: SE TAI
diversicollis Pic, 1925h: 8 (*Cladophorus*) **A**: TAI
inlimbatus Pic, 1925h: 18
dubatolovi Kazantsev, 1995a: 42 **A**: FE (South Ussuri)
fainanensis Pic, 1925h: 8 (*Cladophorus*) **A**: TAI
singularicollis Pic, 1925h: 18 (*Cladophorus*)
galae Kazantsev, 2000a: 86 **A**: GUI
geometricus Kiesenwetter, 1874b: 256 (*Eros*) **A**: FE (Kunashir) JA (Honshu, Shikoku, Kyushu)
bourgeoisi Harold, 1879b: 333 (*Caenia*)
incompositus Ohbayashi, 1954: 21 (*Cladophorus*)
hsui Kazantsev & P.-S. Yang, 1999: 240 **A**: SE TAI
hyonosen Nakane, 1969a: 201 **A**: JA (Honshu)
ilanensis Kazantsev & P.-S. Yang, 1999: 242 **A**: SE TAI
incorruptus Kleine, 1928: 229 **A**: NP SD **ORR**
jengi Kazantsev, 2000a: 82 **A**: TAI
yengi Kazantsev, 2000a: 82 [alternative original spelling]
katoi Nakane, 1969b: 11 **A**: SE TAI
kazuoi Satô & Ohbayashi, 1968: 67 **A**: JA (Ishigaki, Iriomote)
klapperichi Bocák & Bocáková, 1987: 284 **A**: TAI
kurilensis Medvedev, 1966: 37 (*Cladophorus*) **A**: JA (Hokkaido) FE (Kunashir)
leei Kazantsev, 2002b: 259 **A**: TAI
lineatus Hope, 1831: 26 (*Omalysus*) **A**: NP **ORR**
mao Kazantsev, 2000a: 84 **A**: JIA
miaoensis Kazantsev, 2002b: 259 **A**: GUI
nakanei nakanei J. R. Winkler, 1953: 405 (*Cladophorus*) **A**: JA (Honshu, Shikoku)
nakanei yakushimanensis Nakane, 1969a: 208 **A**: JA (Yakushima)
nigrostriatus Kleine, 1928: 228 **A**: SD
postmaculatus Kleine, 1926d: 141 **A**: NP UP **ORR**
procautiroides Kazantsev, 2002b: 259 **A**: GAN HUB SCH SHA
rubromarginatus Kleine, 1926d: 136 **A**: HAI
sanguinolentus Kleine, 1940b: 113 **A**: ZHE

sichuanensis Kazantsev, 2000a: 86 **A**: SCH
takakurai Nakane, 1981c: 127 **A**: JA (Kyushu)
tengjiensis Kazantsev, 2002b: 261 **A**: TAI
vigens Kazantsev, 2000a: 84 **A**: FUJ JIA
yuasai Nakane, 1969a: 208 **A**: JA (Honshu)
zahradniki ohdaisanus Nakane, 1969a: 199 **A**: JA (Honshu: Kii Peninsula)
zahradniki zahradniki J. R. Winkler, 1953: 406 (*Cladophorus*) **A**: JA (Honshu)

genus *Cladophorus* Guérin-Méneville, 1831a: pl. II type species *Cladophorus formosus* Guérin-Méneville, 1831
Odontocerus Guérin-Méneville, 1838a: 72 [HN]
Spacekia E. Strand, 1936: 168 type species *Cladophorus formosus* Guérin-Méneville, 1830
incompletus Pic, 1939h: 172 **A**: YUN
kansuensis Kleine, 1940b: 113 **A**: GAN
rudeplicatus Pic, 1925h: 8 **A**: SD
sinensis Pic, 1925h: 9 **A**: CH

genus *Metanoeus* C. O. Waterhouse, 1879: 73 type species *Lycus* [gen. 37] *conformis* C. O. Waterhouse, 1878
formosanus Pic, 1926i: 67 **A**: SE TAI

genus *Metriorrhynchus* Gemminger & Harold, 1869: 1629 type species *Lycus parallelus* Guérin-Méneville, 1838
dessumi Pic, 1949a: 13 **A**: CH

genus *Microtrichalus* Pic, 1921b: 9 type species *Microtrichalus singularis* Pic, 1921
Falsoenylus Pic, 1926h: 29 type species *Falsoenylus basipennis* Pic, 1926
fuliginosus Bourgeois, 1883a: 646 (*Trichalus*) **A**: YUN **ORR**
rianganus Pic, 1927i: 41 (*Trichalus*)

genus *Procautires* Kleine, 1925a: 32 type species *Procautires toxopei* Kleine, 1925
socius Kleine, 1926c: 104 **A**: SE TAI

genus *Xylobanus* C. O. Waterhouse, 1879: 38 type species *Lycus costifer* Walker, 1858
alveolus Bourgeois, 1885b: 80 **A**: SD **ORR**
angusticollis Motschulsky, 1860b: 115 (*Dictyoptera*) **A**: FE (Sakhalin, Kuriles) JA (Hokkaido)
ogumae Matsumura, 1911a: 124 (*Eros*)
kunasirensis Medvedev, 1966: 38
sutschaensis Kleine, 1933: 11
approximans Bourgeois, 1885b: 80 **A**: NP **ORR**
austrogansuensis Kazantsev, 2002b: 261 **A**: GAN
basivittatus Nakane, 1969a: 187 **A**: JA
chinensis Kleine, 1927b: 56 **A**: HAI
fastidiosus C. O. Waterhouse, 1879: 39 **A**: "Himalaya" **ORR**
japonicus Bourgeois, 1902b: 91 **A**: FE (Kunashir) JA (Hokkaido, Honshu)
rubrocostatus Pic, 1921: 16
leechi Nakane, 1971: 153 **A**: SE TAI
macrolycoides Kazantsev, 2002b: 261 **A**: TAI
minshanicus Kazantsev, 2002b: 261 **A**: SCH
montanus Kleine, 1928: 234 **A**: NP **ORR**
montiphionus Kazantsev, 2000a: 87 **A**: SCH
niger Ohbayashi, 1954: 21 **A**: JA (Honshu)
nigrimembris Pic, 1922d: 16 **A**: SE TAI
noacki Kleine, 1950: 22 **A**: CH
putris Kleine, 1926b: 7 **A**: UP

subfamily **Calochrominae** Lacordaire, 1857

genus *Adoceta* Bourgeois, 1882a: 96 type species *Adoceta caroli* Bourgeois, 1882
alluaudi Peyerimhoff, 1925b: 55 **N**: MO
caroli Bourgeois, 1882a: 97 **N**: AG

genus *Calochromus* Guérin-Méneville, 1833: 158 type species *Calochromus glaucopterus* Guérin-Méneville, 1833
Micronychus Motschulsky, 1861c: 138 type species *Micronychus bimaculatus* Motschulsky, 1861
apicalis Hope, 1831: 26 (*Omalysus*) **A**: NP
costulatus Kleine, 1928: 260 **A**: KA **ORR**
darjeelinensis Bourgeois, 1883a: 622 **A**: SD
formosanus Nakane, 1969b: 13 **A**: SE TAI
impressipes Pic, 1916e: 14 **A**: SD **ORR**
kaschmirensis Kleine, 1928: 259 **A**: KA UP **ORR**
monticola Kleine, 1930c: 287 **A**: XIZ
nagaii Nakane, 1961a: 5 **A**: JA (Honshu)
nigrovittatus Pic, 1925h: 12 **A**: SD **ORR**
orbatus C. O. Waterhouse, 1877b: 197 **A**: SD **ORR**
quadraticollis Bourgeois, 1889: 672 **A**: GAN
robusticollis Pic, 1942b: 4 **A**: XIZ
rubrovestitus Nakane & Ohbayashi, 1955: 29 **A**: JA (Honshu, Shikoku, Kyushu)
sanguinosus Kleine, 1950: 24 **A**: CH
serraticornis Kleine, 1928: 263 **A**: UP
shirozui Nakane, 1969b: 15 **A**: SE TAI
singaporensis Pic, 1925h: 12 **A**: UP **ORR**
subparallelus Pic, 1925h: 17 **A**: TAI
sungkangensis Nakane, 1969b: 12 **A**: SE TAI
testaceocinctus Pic, 1938k: 160 **A**: YUN **ORR**
thibetanus Pic, 1916d: 20 **A**: NP XIZ
triguttatus Kleine, 1928: 260 **A**: NP UP

genus *Dumbrellia* Lea, 1909: 99 type species *Calochromus brevicornis* Lea, 1898
ramicornis Pic, 1928c: 31 (*Calochromus*) **A**: CH **ORR**

genus *Lygistopterus* Mulsant, 1838: 79 type species *Cantharis sanguineus* Linnaeus, 1758
anorachilus Ragusa, 1883: 251 **E**: IT (Sicilia)
cobosi Pardo Alcaide, 1961: 70 **N**: MO
sanguineus escalerai Pic, 1942c: 1 **A**: TR
sanguineus sanguineus Linnaeus, 1758: 401 (*Cantharis*) **E**: AL AU BE BH BU BY CR CZ GE DE EN FI FR GR HU IT LT MD MC NL NR PL PT RO RU SL SK SV TR UK YU SP SZ **A**: ES FE WS
coccineus Goeze, 1777: 523 (*Lampyris*)
collaris Motschulsky, 1860b: 115 (*Lygistopterus*)
crassicornis Matsumura, 1911a: 124 (*Dictyopterus*)
rufipennnis Latreille, 1806: 256 (*Lycus*)
villosus DeGeer, 1774: 47 (*Lampyris*)

family PHENGODIDAE LeConte, 1861

subfamily **Rhagophthalminae** E. Olivier, 1907

genus *Cydistus* Bourgeois, 1885d: xxxvii type species *Cydistus reitteri* Bourgeois, 1885
chindaaricus Bolívar y Pieltain, 1913: 317 **A**: IN
escalerai Bolívar y Pieltain, 1913: 316 **A**: SY
minor Bolívar y Pieltain, 1913: 317 **A**: IN
nigripennis Wittmer, 1979e: 338 **A**: IN

persicus Bolívar y Pieltain, 1913: 316 **A**: IN
reitteri Bourgeois, 1885d: cccviii **A**: SY
zurcheri Bourgeois, 1908a: 211 **A**: TR
reitteri Reitter, 1908e: 133

genus *Dodecatoma* Westwood, 1843: 6: no 16 type species *Dodecatoma bicolor* Westwood, 1843
gracilis Wittmer, 1995a: 110 **A**: NP
riedeli Wittmer, 1995a: 112 **A**: UP
parvicornis Wittmer, 1979b: 89 **A**: AF

genus *Menghouius* Kawashima, 2000: 132 type species *Rhagophthalmus ingens* Fairmaire, 1896
giganteus Fairmaire, 1888a: 25 (*Rhagophthalmus*) **A**: YUN
ingens Fairmaire, 1897b: 227 (*Rhagophthalmus*) **A**: CH **ORR**

***genus Ochotyra* Pascoe, 1862: 323** type species *Ochotyra semiusta* Pascoe, 1862
longipennis Pic, 1923g: 29 [= 1925h: 17] (*Rhagophthalmus*) **A**: CH

genus *Rhagophthalmus* Motschulsky, 1853a: 45 type species *Rhagophthalmus scutellatus* Motschulsky, 1853
angulatus Wittmer, 1997f: 258 **A**: HUB
elongatus Wittmer, 1994b: 348 **A**: GUX
formosanus Kawashima & Sugaya, 2003: 354 **A**: TAI
gibbosulus Fairmaire, 1900: 624 **A**: CH
jeniferae Kawashima & Satô, 2001: 423 **A**: TAI
kiangsuensis Wittmer, 1994b: 346 **A**: JIA
motschulskyi E. Olivier, 1911c: 472 **A**: HKG
ohbai Wittmer, 1994b: 344 **A**: JA (Okinawa)
sausai Wittmer, 1997f: 257 **A**: GUI
scutellatus Motschulsky, 1854c: 45 **A**: CH
semisulcatus Wittmer, 1997f: 259 **A**: YUN
sulcatus Pic, 1925i: 72 **A**: SD
sulcicollis bhutanensis Wittmer, 1997f: 261 **A**: BT
sulcicollis sulcicollis E. Olivier, 1911c: 471 **A**: XIZ
tienmushanensis Wittmer, 1994b: 346 **A**: ZHE
xanthogenus E. Olivier, 1911c: 471 **A**: CH

family LAMPYRIDAE Latreille, 1817
subfamily **Amydetinae** E. Olivier, 1907
tribe Amydetini E. Olivier, 1907

genus *Vesta* Laporte, 1833: 133 type species *Vesta chevrolati* Laporte, 1833
Cratolampis Motschulsky, 1853a: 43 type species *Cratolampis flavicollis* Motschulsky, 1854
chevrolati Laporte, 1833: 133 **A**: TAI **ORR**
davidis Fairmaire, 1878b: 114
formosana Pic, 1944c: 2 **A**: TAI
impressicollis Fairmaire, 1891b: xv **A**: GUI TAI
rufiventris Motschulsky, 1854a: 43 (*Cratolampis*) **A**: GUI TAI

subfamily **Cyphonocerinae** Crowson, 1972

genus *Cyphonocerus* Kiesenwetter, 1879b: 311 type species *Cyphonocerus ruficollis* Kiesenwetter, 1879
harmandi E. Olivier, 1903: 19 (*Psilocladus*) **A**: NP SD **ORR**
hwadongensis Jeng, P.-S. Yang & Satô, 1998: 393 **A**: TAI
inelegans Nakane, 1967a: 7 **A**: JA
jenniferi Jeng & Satô, 1999: 406 **A**: TAI
marginatus Lewis, 1895a: 115 **A**: JA

okinawanus amamianus Jeng, P.-S. Yang & Satô, 1998: 386 **A**: JA
okinawanus okinawanus Nakane, 1983b: 142 **A**: JA
ruficollis Kiesenwetter, 1879b: 312 **A**: JA
variolosus E. Olivier, 1902b: 188 (*Psilocladus*)
sanguineus klapperichi Pic, 1955c: 25 **A**: FUJ
sanguineus sanguineus Pic, 1911p: 143 **A**: CH TAI
formosanus Satô, 1970: 22
taiwanus Nakane, 1967a: 8 **A**: TAI
watarii Satô, 1991: 191 **A**: JA
yayeyamensis Satô, 1976: 59 **A**: JA

subfamily **Lampyrinae** Latreille, 1817
tribe Lampyrini Latreille, 1817

genus *Diaphanes* Motschulsky, 1853a: 54 type species *Diaphanes luniger* Motschulsky, 1853
Lychnebius Motschulsky, 1853a: 44 type species *Lychnebius adustus* Motschulsky, 1853
Phaenopyrus E. Olivier, 1885: 347 type species *Phaenopyrus birmaensis* E. Olivier, 1885
cheni Jeng, 2001: 224 **A**: TAI
citrinus E. Olivier, 1911b: 147 **A**: CH TAI
notatithorax Pic, 1955c: 26
watanabei Matsumura, 1918: 89 (*Lampyris*)
elongatus Pic, 1927k: 106 **A**: XIZ
exsanguis E. Olivier, 1909a: 248 **A**: GUI
flavilateralis Jeng & P.-S. Yang, 2001: 218 **A**: TAI
formosus E. Olivier, 1910b: 286 **A**: TAI
harmandi Pic, 1925i: 73 **A**: SD
indicus Motschulsky, 1854e: 15 **A**: SD
lampyroides E. Olivier, 1891: 598 (*Pyrocoelia*) **A**: TAI **ORR**
longithorax Pic, 1933b: 6 **A**: SCH
marginellus Hope, 1831: 26 (*Lampyris*) **A**: NP
niveus Jeng & Satô, 2001: 227 **A**: TAI
nubilus Jeng & Lai, 2001: 230 **A**: TAI
pallidus Pic, 1938b: 15 **A**: JIA
serotinus E. Olivier, 1907c: 178 **A**: CH
sinuatus E. Olivier, 1908b: 263 **A**: XIZ
splichali E. Olivier, 1912: 172 **A**: XIZ

genus *Lampyris* Geoffroy, 1762: 165 type species *Cantharis noctiluca* Linnaeus, 1767
Lamprotomus Motschulsky, 1853a: 46 type species *Lampyris orientalis* Faldermann, 1835
algerica algerica Ancey, 1870: 87 **N**: AG
mucronata E. Olivier, 1884: 25
algerica attenuata Fairmaire, 1875a: 512 **N**: TU LB (Cyrenaica)
algerica levigata Geisthardt, 1983a: 30 **N**: AG TU
algerica occidentalis Geisthardt, 1983a: 29 **N**: MO
ambigena Jacquelin du Val, 1860c: 99 **E**: IT (Sicilia)
mutabilis E. Olivier, 1884: 31
angustata Motschulsky, 1854e: 17 (*Lampronetes*) **E**: GG "Caucasus"
angustula angustula Fairmaire, 1895b: cx **A**: SY TR
angustula iraqi Geisthardt, 1999: 99 **A**: IQ
berytensis Fairmaire, 1866b: 263 **A**: LE SY
brevicollis Motschulsky, 1854e: 21 **E**: GG
brutia A. Costa, 1882: 34 **E**: IT
caspica Motschulsky, 1854e: 17 (*Lampronetes*) **E**: AB GG KZ ST **A**: IN TM TR
cincta Motschulsky, 1854e: 20 **A**: ES
costalis Motschulsky, 1854e: 20 **E**: AR

depressiuscula Motschulsky, 1854e: 19 **E**: GG
exilis E. Olivier, 1894: 135 **N**: AG MO TU
numidica Normand, 1935: 235
fuscata Geisthardt, 1987: 98 **E**: IT
fuscata apuliae Geisthardt, 1987: 99 **E**: IT (Apulia)
germariensis Jacquelin du Val, 1860c: 97 **E**: BH CR
zenkeri Schaum, 1862: 61
gridelli Pic, 1935h: 147 **A**: LB (Cyrenaica)
hellenica Geisthardt, 1983a: 40 **E**: GR (Peloponnisos)
hummeli Pic, 1933b: 6 **A**: GAN SCH
lareynii Jacquelin du Val, 1859b: 12 **E**: FR (Corse)
lareynii maculata Geisthardt, 1987: 93 **E**: IT (Giglio)
limbata Motschulsky, 1854e: 21 **E**: GG
lobata Motschulsky, 1854e: 18 (*Lampronetes*) **A**: MG
membranacea Motschulsky, 1854e: 16 (*Lampronetes*) **E**: GG
monticola Geisthardt, 2000: 175 **E**: GR
nervosa E. Olivier, 1883b: lxx **N**: AG **A**: IS JO SY
barbara E. Olivier, 1884: 30
noctiluca Linnaeus, 1767: 643 (*Cantharis*) **E**: AB AL AN AR AU BE BH BU CR CT CZ DE EN FI FR GB GE GG GR HU IT LA LS LT LU MC MD NL NR NT PL PT RO RU SK SL SP ST SV SZ UK YU **A**: ES FE MG
bellieri Reiche, 1858: 155
carreti E. Olivier, 1895a: 212
longipennis Motschulsky, 1854e: 21
minor E. Olivier, 1901: 129
obscurella Motschulsky, 1854e: 22
parvicollis E. Olivier, 1901: 129
sibirica Gebler, 1847b: 429
submucronata Rey, 1891d: 114
thoracica Motschulsky, 1854e: 22
olivieriana Heyden, 1890b: 359 **A**: KZ XIN
olivieriana afghana Nakane, 1966c: 235 **A**: AF
orientalis Faldermann, 1835a: 184 **E**: AB GG "Caucasus"
pallida Geisthardt, 1987: 97 **E**: MA
pseudozenkeri Geisthardt, 1999: 95 **E**: TR
raymondi Mulsant & Rey, 1859a: 158 **E**: FR (Alpes) IT SP
lusitanica Motschulsky, 1854e: 19 [HN]
sardiniae brunnea Geisthardt, 1987: 95 **E**: IT (Asinara)
sardiniae sardiniae Geisthardt, 1987: 94 **E**: IT (Sardegna)
satoi Geisthardt, 2003: 67 **A**: NP
setosa Geisthardt, 1983a: 36 **A**: JO SY
spinifer Pic, 1923i: 9 **N**: EG
turkestanica Heyden, 1881: 326 **A**: KZ TD TM UZ
zenkeri Germar, 1817: 211 **E**: BU BH CR GR MC RO YU
soror Schaum, 1862: 61
zenkeri liebegotti Geisthardt, 1985: 288 **E**: GR (Kyra Panagiá)
zenkeri sulpuri Geisthardt, 1999: 98 **E**: GR (west Peloponnisos)

genus *Nyctophila* E. Olivier, 1884: 1 type species *Lampyris reichii* Jacquelin du Val, 1859
anatolica Geisthardt, 1982b: 127 **A**: TR
bonvouloirii Jacquelin du Val, 1860c: 101 (*Lampyris*) **E**: IT (Sicilia)
calabriae Geisthardt, 1983b: 41 **E**: IT (Calabria)
caucasica caucasica Motschulsky, 1854e: 19 (*Lamprotomus*) **A**: IN
caucasica incisa E. Olivier, 1884: 14 (*Lampyris*) **A**: IN TM UZ
caucasica insignis Geisthardt, 1988: 474 **A**: IN
caucasica lenkorani Geisthardt, 1982a: 69 **E**: AB

caucasica morgani E. Olivier, 1907a: 28 **A**: IN
caucasica nigrita Geisthardt, 1982a: 70 **A**: IN
caucasica scutata Geisthardt, 1983b: 55 **A**: IN
colorata Geisthardt, 1983b: 44 **E**: GR (Thíra [= Santorin])
confusa confusa Geisthardt, 1982b: 122 [RN] **N**: MO
mauritanica G. A. Olivier, 1790: 13 (*Lampyris*) [HN]
confusa pygidialis Geisthardt, 1982b: 124 **N**: MO
confusa variata Geisthardt, 1982b: 124 **N**: MO
graeca Geisthardt, 1990: 2 **E**: GR
heydeni E. Olivier, 1884: 9 (*Lampyris*) **E**: SP (Baleares) **A**: MO
incognita Geisthardt, 1982b: 127 **A**: SY
invenusta Geisthardt, 1983b: 41 **A**: SY
libani Laporte, 1833: 139 (*Lampyris*) **A**: CY LE SY TR
insignis Ancey, 1870: 86 (*Lampyris*)
maculicollis Fairmaire, 1866b: 262 (*Lampyris*) **E**: MC TR **A**: IN SY
molesta Jacquelin du Val, 1860c: 102 (*Lampyris*) **E**: IT (Liguria)
pseudocaucasica Geisthardt, 1982a: 72 **E**: "Caucasus" **A**: IN
reichii brullei Reiche, 1863: 129 (*Lampyris*) **E**: GR
zencheri Brullé, 1832: 143 [RN]
reichii reichii Jacquelin du Val, 1859b: 13 (*Lampyris*) **E**: FR LS PT SP TR YU
bicarinata Mulsant & Revelière, 1860: 130 (*Lampyris*)
bidens Rey, 1891d: 114
hispanica E. Olivier, 1884: 17
mauritanica G. A. Olivier, 1790a: no 28: 13 (*Lampyris*)
riegeri Geisthardt, 1990: 5 **E**: GR (Parnon)
scabripennis E. Olivier, 1907b: 45 **E**: TR
syriaca montana Geisthardt, 1983b: 50 **A**: TR
syriaca septentrionalis Geisthardt, 1983b: 51 **A**: TR
syriaca syriaca Baudi di Selve, 1871: 58 (*Lampyris*) **A**: IS SY

genus *Pelania* Mulsant, 1860: 120 type species *Lampyris mauritanica* Linnaeus, 1767
Lampronetes Motschulsky, 1853a: 45 type species *Lampyris mauritanica* Linnaeus, 1767
mauritanica Linnaeus, 1767: 645 (*Cantharis*) **E**: FR PT SP **N**: AG MO TU
angustipennis E. Olivier, 1883b: lxix
imperfecta E. Olivier, 1899: 92
obtusa Fairmaire, 1866a: 44
scutellata Fairmaire, 1884c: xxxv

genus *Pyrocoelia* Gorham, 1880: 91 type species *Lampyris bicolor* Fabricius, 1801
Eulampyris Fairmaire, 1897d: 250 type species *Eurylampyris bivitrea* Fairmaire, 1897
Eurylampyris Fairmaire, 1897a: 69 type species *Eurylampyris bivitrea* Fairmaire, 1897
Lychnocrepis Motschulsky, 1853a: 36 type species *Lychnocrepis motschulskyi* Motschulsky, 1853
Lychnuris Dejean, 1833: 101 type species *Lampyris bicolor* Fabricius, 1801
Stroggulus Motschulsky, 1845a: 36 type species *Stroggulus auritus* Motschulsky, 1845
Strongylomorphus Motschulsky, 1853a: 26 type species *Stroggulus auritus* Motschulsky, 1845
amplissima E. Olivier, 1886: 202 **A**: GAN GUI SCH XIZ
analis Fabricius, 1801b: 100 (*Lampyris*) **A**: FUJ TAI **ORR**
foochowensis Gorham, 1880: 93
insidiosa E. Olivier, 1886: 204
atripennis Lewis, 1896c: 337 **A**: JA TAI
bivitrea Fairmaire, 1897a: 69 (*Eurylampyris*)
iwasakii Matsumura, 1918: 88 (*Lucernula*)
atripes Pic, 1937j: 141 (*Pyrocaelia*) **A**: XIZ
aurita Motschulsky, 1845a: 36 (*Stroggulus*) **A**: MG
discicollis Kiesenwetter, 1874b: 258 (*Lucernula*) **A**: JA
kawamurae Matsumura, 1918: 88 (*Lucernula*)
enervis E. Olivier, 1909a: 248 (*Vesta*) **A**: GUI

flaviventris Fairmaire, 1878b: 114 (*Lucernuta*) **A**: CE
formosana E. Olivier, 1911b: 146 **A**: CH TAI
fumata Fairmaire, 1868: 336 (*Lucernuta*) **A**: CH
fumosa Gorham, 1883d: 409 (*Lucidota*) **A**: JA
　umbrosa E. Olivier, 1902b: 189
grandicollis Fairmaire, 1891b: xvi **A**: CH
iriomotensis Nakane, 1985c: 157 **A**: JA
kanamarui Kishida, 1936: 18 **A**: GUA NMO
matsumurai abdominalis Nakane, 1977a: 88 [RN] **A**: JA
　rufiventris Nakane, 1963b: 45
matsumurai kumejimensis M. Chûjô & Satô, 1972: 23 (*Lychnuris*) **A**: JA
matsumurai matsumurai Nakane, 1963b: 45 **A**: JA
miyako Nakane, 1981b: 127 **A**: JA
motschulskyi Motschulsky, 1854a: 34 (*Lychnocrepis*) **A**: BEI SCH
　nigroflava Fairmaire, 1878b: 113 (*Lucernuta*)
　sinuatocollis E. Olivier, 1886: 202
moupinensis Fairmaire, 1889b: 37 (*Pyrocaelia*) **A**: SCH
opaca E. Olivier, 1885: 349 **A**: SCH **ORR**
　basicrus Fairmaire, 1897d: 250 (*Eulampyris*)
oshimana Nakane, 1985c: 157 **A**: JA
pectoralis E. Olivier, 1883a: 328 **A**: BEI WP
pekinensis Gorham, 1880: 94 **A**: BEI
　coreana Kishida, 1934: 116
praetexta E. Olivier, 1911: 66 **A**: TAI
　rostriformis E. Olivier, 1911: 66
　tappana Matsumura, 1918: 88 (*Lucernula*)
prolongata Jeng & Lai, 1999: 358 **A**: TAI
pygidialis Pic, 1926h: 36 **A**: YUN
rufa E. Olivier, 1886: 200 **A**: JA NC SC ZHE
　coreana Matsumura, 1928: 67
　luteicollis Fairmaire, 1897a: 69 (*Eulampyris*)
　saishutonis Matsumura, 1928: 67
　tsushimana Matsumura, 1928: 65
sanguiniventer E. Olivier, 1911b: 145 **A**: TAI
scutellaris Pic, 1926h: 36 **A**: CH
signaticollis E. Olivier, 1886: 199 **A**: CH
　diversicollis Pic, 1937j: 141 (*Pyrocaelia*)
thibetana E. Olivier, 1886: 203 **A**: XIZ

tribe Photinini E. Olivier, 1907

genus *Lamprigera* Motschulsky, 1853a: 47 type species *Lamprigera boyei* Motschulsky, 1854
　Lamprophorus Gemminger & Harold, 1869: 1646 type species *Lamprigera boyei* Motschulsky, 1854
angustior Fairmaire, 1886c: 337 (*Lamprophorus*) **A**: YUN
marussii Pic, 1955d: 47 (*Lamprophorus*) **A**: NMO XIN
morator E. Olivier, 1891: 596 (*Lamprophorus*) **A**: GUI **ORR**
nepalensis Hope, 1831: 26 (*Lampyris*) **A**: NP
nitidicollis Fairmaire, 1891d: xc (*Lamprophorus*) **A**: KA
scutatus Fairmaire, 1897d: 250 (*Lamprophorus*) **A**: SCH
yunnana Fairmaire, 1897c: 216 (*Lamprophorus*) **A**: GUI TAI YUN

genus *Lamprohiza* Motschulsky, 1853a: 47 type species *Lampyris splendidula* Linnaeus, 1767
boieldieui Jacquelin du Val, 1859b: 17 **E**: FR IT
delarouzei Jacquelin du Val, 1859b: 18 **E**: FR (Alpes-Maritimes)
foliacea Baudi di Selve, 1872: 99 **E**: IT (Sardegna)
germari Küster, 1844: 17 (*Lampyris*) **E**: CR

morio Baudi di Selve, 1875: 37 (*Lamprorhiza*) **E**: IT
mulsanti Kiesenwetter, 1850: 224 **E**: FR SP
farinesi A. Villa & J. B. Villa, 1838: 62 (*Lampyris*)
peltata Rey, 1891d: 114
paulinoi E. Olivier, 1884: 34 **E**: PT
splendidula Linnaeus, 1767: 644 (*Lampyris*) **E**: AU BE BH CR FR GE GR HU IT LS MC NL SZ UK YU
antiqua Brullé, 1832: 143 (*Lampyris*)
sencki Villaret, 1833: 352 (*Lampyris*)

genus *Lucidina* Gorham, 1883d: 408 type species *Lucidina accensa* Gorham, 1883
accensa Gorham, 1883d: 408 **A**: JA
harmandi E. Olivier, 1902b: 188 (*Pyractonema*)
biplagiata Motschulsky, 1866a: 167 (*Lichnuris*) **A**: FE JA NC SC
angusticollis Kiesenwetter, 1874b: 261 (*Lucidota*)
puerile E. Olivier, 1902b: 189 (*Pyractonema*)
tabida Kiesenwetter, 1874b: 259 (*Lucidota*)
vulnerata Kiesenwetter, 1874b: 260 (*Lucidota*)
glaber Kleine, 1950: 22 (*Plateros*) **A**: XIN
klapperichi Pic, 1955c: 25 **A**: FUJ TAI
natsumiae M. Chûjô & Satô, 1972: 24 **A**: JA
okadai Nakane & Ohboyashi, 1949: 34 **A**: JA
roseonotata Pic, 1917b: 5 **A**: TAI
tonkinea Pic, 1917b: 5 **A**: CH **ORR**
vitalisi Pic, 1917b: 4 **A**: TAI **ORR**
vitalisi trinotaticollis Pic, 1937k: 170 **A**: ZHE
vitalisi truncatithorax Pic, 1955c: 25 **A**: FUJ XIN TAI

genus *Lucidotopsis* McDermott, 1960: 18 type species *Lucidota cruenticollis* Fairmaire, 1889
carinicollis Fairmaire, 1889b: 38 (*Lucidota*) **A**: JIX
cruenticollis Fairmaire, 1889b: 38 (*Lucidota*) **A**: SCH

genus *Phosphaenopterus* Schaufuss, 1870: 60 type species *Phosphaenopterus metzneri* Schaufuss, 1870
metzneri Schaufuss, 1870: 61 **E**: FR PT
montandoni Bourgeois, 1900c: 337 **E**: RO

genus *Phosphaenus* Laporte, 1833: 138 type species *Phosphaenus hemipterus* Geoffroy, 1762
Geopyris Dejean, 1833: 103 type species *Lampyris hemiptera* Geoffroy, 1762
hemipterus Geoffroy, 1762: 168 (*Lampyris*) **E**: AL AU BE BH CR CZ DE FR GE IT NL LA LS LT LU MC PL RO RU SK SL SP SV SZ
brachypterus Motschulsky, 1854e: 23
coarctatus E. Olivier, 1884: 39
flavocinctus E. Olivier, 1884: 39
macropterus E. Olivier, 1884: 39

genus *Pristolycus* Gorham, 1883d: 407 type species *Pristolycus sagulatus* Gorham, 1883
diversecostatus Pic, 1915f: 6 **A**: SCH
kanoi Nakane, 1967e: 289 **A**: TAI ZHE
nigronotatus Pic, 1916e: 17 **A**: YUN
sagulatus adachii Satô, 1986c: 261 **A**: FE (Kunashir) JA
sagulatus amami Nakane, 1961b: 15 **A**: JA (Amami-Ôshima)
sagulatus sagulatus Gorham, 1883d: 407 **A**: JA ZHE
shikokensis Ohbayashi & Satô, 1963: 127 **A**: JA
tricolor Pic, 1928d: 4 **A**: YUN **ORR**

subfamily **Luciolinae** Lacordaire, 1857
tribe Curtosini McDermott, 1964

genus *Curtos* Motschulsky, 1845a: 36 type species *Curtos mongolicus* Motschulsky, 1845
bilineatus Pic, 1927i: 37 **A**: CH
costipennis Gorham, 1880: 102 (*Luciola*) **A**: CH JA TAI
iwasakii Matsumura, 1918: 85
elongatus Jeng & P.-S. Yang, 1998: 335 **A**: TAI
fulvocapitalis Jeng & Satô, 1998: 337 **A**: TAI
impolitus E. Olivier, 1913b: 271 (*Luciola*) **A**: TAI
mongolicus Motschulsky, 1845a: 36 **A**: MG NO
mundulus E. Olivier, 1913b: 271 (*Luciola*) **A**: TAI
koshunensis Matsumura, 1918: 86
obscuricolor Jeng & Lai, 1998: 339 **A**: TAI
okinawanus Matsumura, 1918: 85 **A**: JA
okinawae Matsumura, 1928: 64 [RN]
pseudosauteri Geisthardt, 2004: 2 **A**: SCH
ruficollis Jeng & Chang, 1998: 342 **A**: TAI
sauteri E. Olivier, 1913b: 272 **A**: TAI

tribe Luciolini Lacordaire, 1857

genus *Lampyroidea* A. Costa, 1875: 169 type species *Lampyroidea syriaca* A. Costa, 1875
achaiaca Geisthardt, 1999: 107 **E**: GR (Achaia)
antennalis Geisthardt, 1988: 31 **A**: IN
dispar Fairmaire, 1857b: 739 (*Luciola*) **E**: BU TR
greaca Laporte, 1833: 147 (*Luciola*) **E**: GR (Naxos) TR
euboica Pic, 1914p: 74
lucifer Reiche & Saulcy, 1857b: 169 (*Luciola*) **A**: CY SY
lucifera E. Olivier, 1884: 46
maculata Geisthardt, 2004: 2 **A**: IN
maculicollis Mulsant & Wachanru, 1852a: 3 (*Colophotia*) **A**: CY SY TR
maculithorax Gemminger, 1870: 120 (*Luciola*)
nigrescens E. Olivier, 1884: 46 **A**: SY
persica E. Olivier, 1907a: 28 **A**: IN
quadrinotata binotata Wittmer, 1935d: 248 **E**: GR (Aegean Islands)
quadrinotata quadrinotata Wittmer, 1935d: 247 **E**: GR (Aegean Islands)
schneideri Geisthardt, 1999: 104 **A**: IQ
syriaca A. Costa, 1875: clxix **A**: IS SY TR
innotaticollis Pic, 1914p: 74
tenenbaumi Pic, 1928a: 1 **A**: IS

genus *Luciola* Laporte, 1833: 146 type species *Luciola pedemontana* Motschulsky, 1853 (= *Lampyris italica* Linnaeus, 1767)
subgenus *Hotaria* Yuasa, 1937: 107 type species *Luciola parvula* Kiesenwetter, 1874
papariensis Doi, 1932: 63 (*Luciola*) **A**: CH JA
tsushimana Nakane, 1970a: 286 (*Luciola*)
parvula Kiesenwetter, 1874b: 263 (*Luciola*) **A**: JA
ibukiyamana Nawa, 1921: 256 (*Luciola*)
subgenus *Luciola* Laporte, 1833: 146 type species *Luciola pedemontana* Motschulsky, 1853
Atyphella Olliff, 1890: 645 type species *Atyphella lychnus* Olliff, 1890
Delopleurus Motschulsky, 1853a: 50 type species *Delopleurus fuscus* Motschulsky, 1854
Delopyrus Motschulsky, 1853a: 50 type species *Delopyrus dregei* Motschulsky, 1853
Pygolampis Kirby & Spence, 1828: 407 type species *Lampyris italica* Linnaeus, 1767
anceyi E. Olivier, 1883a: 330 **A**: GUI TAI **ORR**
bourgeoisi E. Olivier, 1895b: cxlviii **A**: CH

cerata E. Olivier, 1911b: 147 **A**: TAI
formosana Pic, 1916d: 9
formosana Matsumura, 1918: 84 [HN]
chinensis Linnaeus, 1767: 645 (*Lampyris*) [DA] **A**: FUJ GUI TAI
vespertina Fabricius, 1801b: 103 (*Lampyris*)
cruciata cruciata Motschulsky, 1854e: 53 **A**: ES JA
towadensis Nakane, 1987d: 173
curtithorax Pic, 1927i: 49 **A**: TAI **ORR**
davidis E. Olivier, 1895b: cxlix **A**: JIX
ficta E. Olivier, 1909a: 249 **A**: GUI HKG SCH TAI
filiformis filiformis E. Olivier, 1913b: 271 **A**: JA TAI
filiformis yayeyamana Matsumura, 1918: 84 **A**: JA
fissicollis Fairmaire, 1891b: xvi **A**: CH
flava Pic, 1929i: 73 **A**: CH
flavida Hope, 1845a: 10 (*Colophotia*) **A**: SHG (Zhoushan)
fletcheri Pic, 1935a: 10 **A**: SD
fukienensis Pic, 1955c: 29 **A**: FUJ
hydrophila Jeng, P.-S. Yang & Lai, 2003: 544 **A**: TAI
impedita E. Olivier, 1902a: 80 **A**: JIX
antica Fairmaire, 1889b: 39 [HN]
italica Linnaeus, 1767: 645 (*Lampyris*) **E**: CR GR IT MC RO SL SZ TR YU
illyrica Motschulsky, 1854e: 54
minuta Motschulsky, 1854e: 51
obtusangula Motschulsky, 1854e: 52
pedemontana Motschulsky, 1853a: 55
suturalis Motschulsky, 1854e: 55
japonica Thunberg, 1784b: 79 (*Lampyris*) **A**: TAI **ORR**
judaica E. Olivier, 1884: 48 **A**: SY
kagiana Matsumura, 1928: 63 **A**: TAI
kervillei E. Olivier, 1909b: 398 **A**: LE SY
klapperichi Pic, 1955b: 26 **A**: FUJ
kuroiwae Matsumura, 1918: 84 **A**: JA
lateralis Motschulsky, 1860b: 114 **A**: ES JA
vitticollis Kiesenwetter, 1874b: 261
laticollis Gorham, 1883a: 4 **A**: CH
limbalis Fairmaire, 1889b: 38 **A**: JIX
lusitanica Charpentier, 1825: 194 (*Lampyris*) **E**: AU BH BU CR FR GR HU IT MC PT RO SL SP ST TR UK YU
caucasica Motschulsky, 1854e: 52
collaris Motschulsky, 1854e: 51
erythrocephalus E. Olivier, 1885: 362
italica G. A. Olivier, 1790a: no 28: 17 (*Lampyris*)
papii Mikšić, 1969: 45
mehadiensis Motschulsky, 1854e: 50
mentonensis Peragallo, 1863: 665
mingrelica Motschulsky, 1845e: 36
minor Baudi di Selve, 1873: 229
pontica Motschulsky, 1854e: 55
mongolica Motschulsky, 1860b: 114 **A**: ES MG
nitescens E. Olivier, 1903: 19 **A**: SD
novaki G. Müller, 1946: 112 **E**: YU
ovalis Hope, 1831: 26 **A**: GUI NP TAI
owadai Satô & Kimura, 1994: 160 **A**: JA
picticollis Kiesenwetter, 1874b: 262 **A**: JA
cruciata Harold, 1877: 357
pieli Pic, 1939f: 135 **A**: HAI

praeusta Kiesenwetter, 1874b: 263 **A**: TAI **ORR**
affinis Gorham, 1880: 101
gorhami Ritsema, 1883: 4
roseicollis Pic, 1933b: 6 **A**: GAN JIA
satoi Jeng & P.-S. Yang, 2003: 251 **A**: TAI
stigmaticollis Fairmaire, 1887a: 123 **A**: YUN
substriata Gorham, 1880: 100 **A**: JIA TAI
terminalis E. Olivier, 1883a: 330 **A**: TAI **ORR**
trilucida Jeng & Lai, 2003: 248 **A**: TAI
unmunsana Doi, 1931: 54 **A**: NC

subfamily **Ototretinae** McDermott, 1964

genus *Drilaster* Kiesenwetter, 1879b: 310 type species *Drilaster axillaris* Kiesenwetter, 1879
Ototreta E. Olivier, 1900: 237 type species *Ototreta weyersi* E. Olivier, 1900
akusekianus Nakane, 1983b: 143 **A**: JA (Tokora)
atricollis Nakane, 1977a: 90 **A**: TAI
axillaris Kiesenwetter, 1879b: 311 **A**: JA
bipustulata E. Olivier, 1908a: 114 (*Ototreta*)
bicolor Satô, 1968: 25 **A**: JA
flavicollis Nakane, 1977a: 96 **A**: JA TAI
flavipennis Nakane, 1977a: 90 **A**: JA TAI
fuscicollis Nakane, 1977a: 88 **A**: JA
iokii Satô, 1968: 26 **A**: JA
kimotoi Nakane, 1977a: 93 **A**: TAI
ohbayashii Satô, 1968: 24 **A**: JA
okinawensis Nakane, 1977a: 91 **A**: JA
olivieri Pic, 1911b: 36 (*Ototreta*) **A**: TAI
parvus Nakane, 1977a: 94 **A**: TAI
purpureicollis Pic, 1911b: 36 (*Ototreta*) **A**: TAI
robusta Pic, 1922b: 29 (*Ototreta*) **A**: CH
rollei Pic, 1911b: 36 (*Ototreta*) **A**: TAI
shibatai Satô, 1968: 23 **A**: JA
anomalus Nakane, 1977a: 92
takashii Nakane, 1977a: 89 **A**: TAI
takahashii Lai, Satô & P.-S. Yang, 1998: 208
unicolor Lewis, 1895a: 115 **A**: JA

genus *Stenocladius* Fairmaire, 1878b: 112 type species *Stenocladius davidis* Fairmaire, 1878
Hydaspes Bourgeois, 1890a: 171 type species *Hydaspes fairmairei* Bourgeois, 1890
azumai azumai Nakane, 1981b: 126 **A**: JA
azumai iheyanus Kawashima & Satou, 2004: 391 **A**: JA (Iheya-jima)
azumai keramenis Kawashima & Satou, 2004: 394 **A**: JA (Tokashiki-jima)
bicoloripes Pic, 1918c: 3 **A**: TAI
chinensis Geisthardt, 2004: 5 **A**: SCH
davidis Fairmaire, 1878b: 113 **A**: CE
flavipennis Kawashima, 1999: 148 **A**: JA
rufithorax Wittmer, 1995a: 113 **A**: NP
yoshikawai Nakane, 1981b: 125 **A**: JA
bicoloripes Satô, 1985a: 122 [HN]
shirakii Nakane, 1981b: 125
yoshimasai Kawashima, 1999: 144 **A**: JA

family Lampyridae, species incertae sedis
Baolacus nepalensis Wittmer, 1979b: 90 **A**: NP
Lampyris letourneuxi E. Olivier, 1884: 24 **A**: IS

Lampyris platyptera Fairmaire, 1887a: 122 **A**: YUN
Rhagophthalmus ohbai Wittmer, 1994b: 344 **A**: JA

family OMETHIDAE LeConte, 1861
subfamily **Driloniinae** Crowson, 1972

genus *Drilonius* Kiesenwetter, 1874b: 282 type species *Drilonius striatulus* Kiesenwetter, 1874
Curtodrilus Pic, 1921d: 13 type species *Curtodrilus costatus* Pic, 1921
chinensis Wittmer, 1995a: 116 **A**: YUN
osawai Nakane, 1950b: 30 **A**: JA
rufithorax Wittmer, 1995a: 115 **A**: SCH
striatulus Kiesenwetter, 1874b: 282 **A**: JA

subfamily **Omethinae** LeConte, 1861

genus *Omethes* LeConte, 1861: 187 type species *Omethes marginatus* LeConte, 1861
Elianus Lewis, 1895a: 113 type species *Elianus rugiceps* Lewis, 1895
rugiceps Lewis, 1895a: 113 (*Elianus*) **A**: JA

family CANTHARIDAE Imhoff, 1856 (1815)
subfamily **Cantharinae** Imhoff, 1856 (1815)
tribe Podabrini LeConte, 1881

genus *Asiopodabrus* Wittmer, 1982d: 121 type species *Podabrus taiwanus* Wittmer, 1982
ainu ainu Nakane & Makino, 1990: 184 (*Podabrus*) **A**: FE (Kurils, Sakhalin) JA (Hokkaido, Honshu)
ainu shirakawanus Takahashi & Kiriyama, 2000: 130 **A**: JA (Honshu)
aomoriensis Takahashi & Okushima, 1999: 24 **A**: JA (Honshu)
asperipunctatus Kang & Okushima, 2003: 346 **A**: SC
azusa Nakane & Makino, 1990: 185 (*Podabrus*) **A**: JA (Honshu)
bilineatus bilineatus Pic, 1904f: 33 (*Podabrus*) **A**: JA (Sado)
bilineatus nozirianus Nakane & Makino, 1990: 186 (*Podabrus*) **A**: JA (Honshu)
bilineatus oisoanus Takahashi, 1992: 96 (*Podabrus*) **A**: JA (Honshu)
circumangulatus Kang & Kim, 2000b: 209 **A**: SC
falciformis Takahashi & Okushima, 1999: 30 **A**: JA (Honshu)
fragiliformis Kang & Kim, 2000b: 208 **A**: SC
fragilis Nakane & Makino, 1989: 12 (*Podabrus*) **A**: JA (Shikoku, Kyushu)
fukudai Nakane & Makino, 1990: 191 (*Podabrus*) **A**: JA (Honshu)
gifuensis Takahashi & Kiriyama, 2000: 123 **A**: JA (Honshu)
guangxiensis Wittmer, 1997a: 33 **A**: GUX
gyotokui Nakane & Makino, 1989: 8 (*Podabrus*) **A**: JA (Honshu, Kyushu)
hana Nakane & Makino, 1990: 187 (*Podabrus*) **A**: JA (Honshu)
hidaensis Takahashi & Kiriyama, 2000: 129 **A**: JA (Honshu)
hinakurai Takahashi, 1996: 21 **A**: JA (Honshu)
hinokiboranus Takahashi, 1992: 102 (*Podabrus*) **A**: JA (Honshu)
hiranoi Takahashi, 1992: 97 (*Podabrus*) **A**: JA (Honshu, Izu-oshima)
hiwadanus Takahashi & Kiriyama, 2000: 142 **A**: JA (Honshu)
horikawai Takahashi & Okushima, 1999: 27 **A**: JA (Honshu)
hoshiensis Nakane, 1990: 190 (*Podabrus*) **A**: JA (Honshu)
hosodai Takahashi & Kiriyama, 2000: 130 **A**: JA (Honshu)
hozumii Nakane & Makino, 1990: 190 (*Podabrus*) **A**: JA (Honshu)
hyogoensis Nakane & Makino, 1989: 8 (*Podabrus*) **A**: JA (Honshu)
ichihashii Takahashi & Kiriyama, 2003: 293 **A**: JA (Honshu)
ichitai Takahashi & Okushima, 1999: 22 **A**: JA (Honshu)
ihai Wittmer, 1960b: 107 (*Podabrus*) **A**: JA (Ryukyus)
inexpectus Takahashi, 1992: 102 (*Podabrus*) **A**: JA (Honshu)

inugoejianus Takahashi, 1992: 96 **A**: JA (Honshu)
ishidai Nakane & Makino, 1990: 188 (*Podabrus*) **A**: JA (Honshu)
iwanadomensis Takahashi & Kiriyama, 2000: 132 **A**: JA (Honshu)
kadowakii Nakane & Makino, 1989: 4 (*Podabrus*) **A**: JA (Honshu, Shikoku, Oki)
kamikochianus kamikochianus Nakane & Makino, 1990: 185 (*Podabrus*) **A**: JA (Honshu)
kamikochianus kanagawaensis Takahashi, 1992: 94 (*Podabrus*) **A**: JA (Honshu)
kamitakaranus Takahashi & Kiriyama, 2000: 126 **A**: JA (Honshu)
kansaiensis Nakane & Makino, 1990: 186 (*Podabrus*) **A**: JA (Honshu)
kasugensis Nakane & Makino, 1989: 5 (*Podabrus*) **A**: JA (Honshu, Shikoku)
kashiensis Takahashi, 2002a: 50 **A**: JA (Honshu)
kawaianus Takahashi & Kiriyama, 2000: 138 **A**: JA (Honshu)
kiiensis hakonensis Takahashi, 1992: 100 (*Podabrus*) **A**: JA (Honshu)
kiiensis kiiensis Nakane & Nakino, 1989: 10 (*Podabrus*) **A**: JA (Honshu)
kintokisanus Takahashi, 1992: 95 (*Podabrus*) **A**: JA (Honshu)
kiriyamai Takahashi, 2002a: 53 **A**: JA (Honshu)
kishii Nakane & Makino, 1989: 5 (*Podabrus*) **A**: JA (Honshu)
kiso imasakai Takahashi & Okushima, 1999: 28 **A**: JA (Honshu)
kiso kiso Nakane, 1989b: 3 (*Podabrus*) **A**: JA (Honshu)
 haroldi Nakane, 1990: 186 (*Podabrus*)
komatsui Nakane, 1990: 189 (*Podabrus*) **A**: JA (Honshu)
kouichii Takahashi & Kiriyama, 2000: 133 **A**: JA (Honshu)
kurbatovi Kazantsev, 1998b: 573 **A**: FE
kuriyamanus Takahashi, 2002a: 54 **A**: JA (Honshu)
kuroisawanus Takahashi & Kiriyama, 2000: 135 **A**: JA (Honshu)
kusamai Takahashi, 1999a: 71 **A**: JA (Honshu)
lictorius Lewis, 1895a: 109 (*Podabrus*) **A**: JA (Honshu, Shikoku)
 yanoi Nakane, 1989b: 3 (*Podabrus*)
longilaterophyses longilaterophyses Takahashi & Kiriyama, 2000: 140 **A**: JA (Honshu)
longilaterophyses mitsukanus Takahashi & Kiriyama, 2000: 140 **A**: JA (Honshu)
macilentus Kiesenwetter, 1874b: 264 (*Podabrus*) **A**: JA (Honshu, Shikoku)
maedai Nakane, 1989b: 12 (*Podabrus*) **A**: JA (Kyushu)
malthinoides hayato Nakane, 1989b: 4 (*Podabrus*) **A**: JA (Kyushu)
malthinoides malthinoides Kiesenwetter, 1874b: 265 (*Podabrus*) **A**: JA (Honshu, Shikoku)
 sulcithorax Pic, 1904d: 26 (*Podabrus*)
masaakii Nakane, 1990: 183 (*Podabrus*) **A**: JA (Honshu)
masatakai Takahashi & Kiriyama, 2003: 290 **A**: JA (Honshu)
masatoi Takahashi, 1999a: 72 **A**: JA (Honshu)
micronigerinus Takahashi, 1992: 99 (*Podabrus*) **A**: JA (Honshu)
mikunisanus Takahashi, 1992: 92 (*Podabrus*) **A**: JA (Honshu)
nakanei Takahashi & Okushima, 1999: 32 **A**: JA (Honshu)
nakaoi Nakane, 1990: 183 (*Podabrus*) **A**: JA (Honshu, Kyushu)
nanasawaensis Takahashi, 1992: 101 (*Podabrus*) **A**: JA (Honshu)
naratougeanus Takahashi & Kiriyama, 2000: 135 **A**: JA (Honshu)
neglectus minus Takahashi, 1992: 98 (*Podabrus*) **A**: JA (Honshu)
neglectus neglectus Nakane, 1989b: 13 (*Podabrus*) **A**: JA (Shikoku, Kyushu)
nenoharanus Takahashi & Kiriyama, 2000: 138 **A**: JA (Honshu)
nigripalustris Takahashi, 2002a: 57 **A**: JA (Honshu)
ninoseanus Nakane & Makino, 1990: 184 (*Podabrus*) **A**: JA (Honshu)
ochraceus Kiesenwetter, 1874b: 265 (*Podabrus*) **A**: JA (Kyushu)
ohshirakawanus Takahashi & Kiriyama, 2000: 127 **A**: JA (Honshu)
okadai Nakane & Makino, 1990: 188 (*Podabrus*) **A**: JA (Honshu)
okianus Nakane & Makino, 1990: 184 (*Podabrus*) **A**: JA (Oki)
okuhidanus Takahashi & Kiriyama, 2000: 139 **A**: JA (Honshu)
oreumsensis Kang & Okushima, 2003: 342 **A**: SC

owarianus Nakane & Makino, 1990: 184 (*Podabrus*) **A**: JA (Honshu)
ozeanus Nakane & Makino, 1990: 187 (*Podabrus*) **A**: JA (Honshu)
pseudolictrius Takahashi, 1992: 103 (*Podabrus*) **A**: JA (Honshu)
sakaii Wittmer, 1982d: 124 **A**: TAI
sannouanus sannouanus Nakane & Makino, 1989: 11 (*Podabrus*) **A**: JA (Honshu)
sannouanus tomokoae Satô, 1989: 182 **A**: JA (Honshu)
shimashimanus Nakane & Makino, 1989: 11 (*Podabrus*) **A**: JA (Honshu)
shiobaranus Takahashi, 2002a: 50 **A**: JA (Honshu)
shiraisawanus Takahashi & Kiriyama, 2000: 144 **A**: JA (Honshu)
shirakomanus Takahashi & Okushima, 1999: 31 **A**: JA (Honshu)
simillimus Nakane & Nakino, 1989: 10 (*Podabrus*) **A**: JA (Honshu)
syozoi palustris Nakane & Makino, 1989: 7 (*Podabrus*) **A**: JA (Honshu)
syozoi syozoi Nakane & Makino, 1989: 7 (*Podabrus*) **A**: JA (Honshu)
taiwanus Wittmer, 1982d: 123 **A**: TAI
tajimensis Nakane & Nakino, 1990: 187 (*Podabrus*) **A**: JA (Honshu)
takaii Nakane & Makino, 1990: 189 (*Podabrus*) **A**: JA (Honshu)
takaosanus Nakane & Nakino, 1990: 189 (*Podabrus*) **A**: JA (Honshu)
tanasawanoatamanus Takahashi, 1992: 93 (*Podabrus*) **A**: JA (Honshu)
taoi Takahashi, 1992: 100 (*Podabrus*) **A**: JA (Honshu)
temporalis Harold, 1878: 73 (*Podabrus*) **A**: JA (Honshu)
tochigiensis Takahashi, 2002b: 199 **A**: JA (Honshu)
tokugoanus Nakane & Makino, 1989: 9 (*Podabrus*) **A**: JA (Honshu)
tokui Satô, 1989: 182 **A**: JA (Honshu)
towadanus Takahashi & Okushima, 1999: 25 **A**: JA (Honshu)
tryznai Švihla, 2004b: 157 **A**: SHA
tsubonae Takahashi & Kiriyama, 2000: 143 **A**: JA (Honshu)
tsuchikawai Takahashi, 1999a: 74 **A**: JA (Honshu)
tsuyukii Takahashi, 1992: 98 (*Podabrus*) **A**: JA (Honshu)
uchigadaninus Takahashi & Kiriyama, 2000: 126 **A**: JA (Honshu)
uedai hisashii Nakane, 1990: 190 (*Podabrus*) **A**: JA (Honshu, Oki)
uedai uedai Nakane & Makino, 1989: 13 (*Podabrus*) **A**: JA (Kyushu)
umasakanus Takahashi, 2002a: 51 **A**: JA (Honshu)
yama Nakane & Makino, 1990: 187 (*Podabrus*) **A**: JA (Honshu)
yamatonus Takahashi & Kiriyama, 2000: 124 **A**: JA (Honshu)
yokonensis Takahashi, 2002a: 52 **A**: JA (Honshu)
yoshidai fujisanus Nakane & Makino, 1989: 6 (*Podabrus*) **A**: JA (Honshu)
yoshidai yoshidai Nakane, 1989b: 2 (*Podabrus*) **A**: JA (Honshu)
yunishigawanus Takahashi, 2002a: 56 **A**: JA (Honshu)
yushinensis Takahashi, 1992: 94 (*Podabrus*) **A**: JA (Honshu)

genus *Dichelotarsus* Motschulsky, 1860d: 400 type species *Rhagonycha piniphila* Eschscholtz, 1830
  Anolisus Mulsant, 1862b: 195 type species *Cantharis lapponica* Gyllenhal, 1810
angusticollis Motschulsky, 1860b: 117 **A**: ES FE JA (Hokkaido) MG
barkalovi Kazantsev, 1998b: 572 **A**: ES
cheni Pic, 1932i: 207 (*Anolisus*) **A**: JIX
davidi Pic, 1926r: 357 (*Anolisus*) **A**: SCH
flavimanus Motschulsky, 1860b: 116 **E**: AU FI NR NT SV **A**: ES FE MG WS
  callosus J. R. Sahlberg, 1887a: 31 (*Podabrus*)
  obscuripes J. R. Sahlberg, 1871: 435 (*Podabrus*)
  ratteri Reitter, 1906b: 35 (*Podabrus*)
kolymensis Kazantsev, 1992c: 276 **A**: FE
lapponicus Gyllenhal, 1810: xv (*Cantharis*) **E**: FI NR NT SV **A**: ES FE JA (Hokkaido) MG WS
  amurensis Heyden, 1880a: 127 (*Podabrus*) [RN]
  bytonii Mannerheim, 1849: 231 (*Cantharis*)
  flavipes Motschulsky, 1860b: 116 [HN]

karelini Marseul, 1864: 12 (*Podabrus*)
labiatus Motschulsky, 1860b: 117
nervosus Motschulsky, 1860b: 117
luteonotatus Pic, 1902i: 55 (*Podabrus*) **E**: GR
procerulus Kiesenwetter, 1860: 470 (*Cantharis*) **E**: BH BU FR IT **A**: TR
bosnica Pic, 1903g: 115
recticollis Motschulsky, 1860b: 117 **A**: FE
septemnotatus Pic, 1907p: 173 (*Podabrus*) **A**: NP "India"

genus *Dichelotarsus*, species incertae sedis
lunulatus Fischer de Waldheim, 1844: 34 (*Podabrus*) **E**: "Ross. merid. "
nigriventris Fischer de Waldheim, 1844: 33 (*Podabrus*) **E**: UK
vittatus Fischer de Waldheim, 1844: 34 (*Podabrus*) **A**: "Tartaria magna"

genus *Hatchiana* Fender, 1966: 321 type species *Hatchiana arizonensis* Fender, 1966
abei Nakane & Makino, 1981: 59 (*Podabrus*) **A**: JA (Honshu)
glochidiata Kazantsev, 1996b: 206 **A**: "Korea"
heydeni Kiesenwetter, 1879b: 306 (*Podabrus*) **A**: JA (Honshu, Shikoku, Kyushu)
immaculicollis Pic, 1905d: 113 (*Podabrus*)
reini Heyden, 1879: 351 (*Podabrus*)
hikosana Nakane & Makino, 1981: 60 (*Podabrus*) **A**: JA (Honshu, Kyushu)
hirugatakensis Takahashi, 1992: 92 (*Podabrus*) **A**: JA (Honshu)
ishiharai Satô, 1986c: 255 (*Podabrus*) **A**: JA (Shikoku)
jirisanensis Kang & Kim, 2000b: 207 **A**: SC
kyushuensis Takakura, 1988b: 152 **A**: JA (Kyushu)
osawai Nakane & Makino, 1981: 61 (*Podabrus*) **A**: JA (Honshu)
rosinae Pic, 1904d: 25 (*Podabrus*) **A**: FE HEI NC SC

genus *Podabrus* Westwood, 1838: 27 type species *Cantharis alpinus* Paykull, 1798
Brachynotus Kirby, 1837: 249 type species *Brachynotus benneti* Kirby, 1837
Malthacus Kirby, 1837: 249 type species *Malthacus puncticollis* Kirby, 1837
alpinus Paykull, 1798: 259 (*Cantharis*) **E**: AU BE BY CT CZ DE EN FI FR GB GE HU IR IT LA LT LU NR NT PL RO SK SV SZ UK **A**: ES FE MG WS
anatolica Pic, 1920h: 21
apicalis Pic, 1906k: 81
apicenigra Roubal, 1936c: 197
falzonii Burlini, 1942: 13
iljini Barovskij, 1908a: 174
lateralis Erichson, 1839: 369
melancholica Dalla Torre, 1879: 5
mocquerysi Reiche, 1878a: 383
necrophora Dalla Torre, 1879: 5
rubens Fabricius, 1801a: 297 (*Cantharis*)
ruficeps Gabriel, 1905a: 142
semibrunnea Pic, 1914r: 166
annulatus Mannerheim, 1825: 28 (*Cantharis*) **E**: NT **A**: ES FE MG WS HEI "Korea"
altaicus Pic, 1902g: 47
ochoticus Motschulsky, 1860b: 115
pallidior Pic, 1902g: 47
dilaticollis Motschulsky, 1860b: 115 **A**: ES FE
distincticollis Pic, 1905d: 113
exophtalmus Fairmaire, 1888b: 122 **A**: BEI
flavus Pic, 1926r: 356 **A**: SCH
latithorax Pic, 1926r: 356
longissimus Pic, 1905d: 113 **A**: FE (Sakhalin, Kurils) JA (Hokkaido, Honshu) "Korea"
sapporensis Matsumura, 1911a: 125 (*Telephorus*)
magnus S. Wang & C. Yang, 1990: 239 **A**: NMO

marginellus Motschulsky, 1860b: 116 **A**: ES FE MG
maritimus Kazantsev, 1996b: 205 **A**: FE
rotundimargo Kazantsev, 1996b: 203 **A**: FE

tribe Podabrini, species incertae sedis
Podabrus chinensis J. Li, 1992: 101 **A**: HEI
Podabrus paraheydeni J. Li & Chen, 1993: 165 **A**: ANH

tribe Cantharini Imhoff, 1856 (1815)

genus *Ancistronycha* Märkel, 1852: 589 type species *Cantharis abdominalis* Fabricius, 1798
abdominalis Fabricius, 1798: 67 (*Cantharis*) **E**: AN AU BU CZ FR GB GE IT LS PL RO SK SP SZ UK
consobrina Märkel, 1852: 589
cyanea Curtis, 1828: 215 (*Telephorus*)
cyanea Dietrich, 1857: 119 (*Telephorus*) [HN]
cyannipennis Bach, 1854: 64
maculithorax Pic, 1904a: 2
passeriana Gredler, 1854: 6 (*Cantharis*)
astur Heyden, 1880b: 298 **E**: PT SP
erichsonii erichsonii Bach, 1852: 68 (*Telephorus*) **E**: AU CZ FR GE HU IT PL RO SK SZ UK **A**: TR
inapicalis Pic, 1902k: 63
neapolitana Pic, 1918h: 22
rotundicollis Dietrich, 1857: 131
erichsonii kurbatovi Kazantsev, 1989: 242 **E**: GG ST
lucens Moscardini, 1967: 32 **E**: CR IT
occipitalis Rosenhauer, 1847: 18 (*Cantharis*) **E**: BU PL SK UK
taygetana Pic, 1902k: 63 **E**: GR (Taygetos)
violacea Paykull, 1798: 260 (*Cantharis*) **E**: AU BE CR CT CZ DE EN FI FR GE HU IT LA LT LU NR NT PL RO SK ST SV SZ UK **A**: WS
innotaticeps Pic, 1907d: 113
moraviensis Pic, 1918h: 21
notaticeps Pic, 1907d: 113
piceobinotata Pic, 1922g: 25 (*Cantharis*)
tigurina Dietrich, 1857: 119
violacea Dietrich, 1857: 118

genus *Armidia* Mulsant, 1862a: 283 type species *Cantharis signata* Ahrens, 1814
gracilis Pandellé, 1867: 179 (*Rhagonycha*) **E**: SP
notaticornis Pic, 1927c: 10 (*Rhagonycha*)
lohsei Wittmer, 1975b: 262 **E**: SP
nobilissima Reitter, 1884c: 78 **E**: GR
jonica Pic, 1901f: 50
moricei Pic, 1901f: 50
signata Ahrens, 1814: pl. 10 (*Cantharis*) **E**: AL AU BH CR GE GR HU IT
apfelbecki Pic, 1902i: 55
breviventris Pacher, 1853: 45 (*Telephorus*)
insignata Pic, 1902i: 56
paulosignata Pic, 1902i: 56
unicolor Dufour, 1851: 329 (*Rhagonycha*) **E**: FR SP
ericeti Kiesenwetter, 1860: 508 (*Cantharis*)

genus *Bactrocantharis* Barovskij, 1926c: 235 type species *Bactrocantharis kaznakovi* Barovskij, 1926
bogatchevi Kazantsev, 1989: 240 **A**: UZ
ciliatocollis Pic, 1906k: 83 **A**: TD
kaznakovi Barovskij, 1926c: 236
kuchleri Pic, 1913o: 187
ruzickai Švihla, 1992: 330 **A**: UZ
sohnsteini Kazantsev, 1998d: 865 **A**: TD

genus *Bactronycha* Kazantsev, 2001d: 1077 type species *Bactronycha tamerlan* Kazantsev, 2001
tamerlan Kazantsev, 2001d: 1078 **A**: UZ

genus *Boveycantharis* Wittmer, 1969b: 78 type species *Metacantharis rufimanoides* Wittmer, 1969
akschehirensis Wittmer, 1969b: 85 **A**: TR
deserticola Wittmer, 1969b: 83 **A**: IS JO SY TR
dimidiatipes Reiche & Saulcy, 1857b: 172 (*Telephorus*) **A**: CY JO LE SY TR
detectiventris Pic, 1904j: 71 (*Cantharis*)
varicolor Wittmer, 1969b: 84
funesta Faldermann, 1835a: 190 (*Cantharis*) **E**: AB AR GG
hetitica Wittmer, 1969b: 81 **A**: TR
holzschuhi Švihla, 1999: 158 **A**: TR
malatiensis Heyden, 1886d: 279 (*Cantharis*) **A**: TR
mersinensis Švihla, 1999: 158 **A**: TR
olympica Marseul, 1864: 57 (*Telephorus*) **E**: GR
phoeniciensis Wittmer, 1969b: 82 **A**: SY
reitteri Pic, 1906k: 82 (*Cantharis*) **E**: GR
rufimana Ménétriés, 1832: 161 (*Cantharis*) **E**: AB AR GG ST **A**: TR
rufimanoides Wittmer, 1969b: 80 **A**: TR
tauricola Švihla, 1999: 156 **A**: TR
tokatensis Pic, 1898h: 138 (*Telephorus*) **A**: TR
unguiculata Baudi di Selve, 1872: 104 (*Telephorus*) **E**: SP
species incertae sedis
tatvanensis Wittmer, 1997e: 299 **A**: TR

genus *Cantharis* Linnaeus, 1758: 400 type species *Cantharis fusca* Linnaeus, 1758
subgenus *Cantharis* Linnaeus, 1758: 400 type species *Cantharis fusca* Linnaeus, 1758
Oripa Motschulsky, 1860d: 398 type species *Oripa transmarina* Motschulsky, 1860
Silotrachelus Solsky, 1881: 31 type species *Silotrachelus semirufus* Solsky, 1881
Telephorus J. C. Schaeffer, 1766b: pl. CXXIII type species *Cantharis fusca* Linnaeus, 1758
allosensis Pic, 1924d: 25 **E**: FR
alluaudi Pic, 1922h: 30 **N**: MO
macrocephala Kocher, 1954a: 151
annularis Ménétriés, 1836: 150 **E**: AL BU CR CT CZ DE FR GE GG GR HU MC RO SK SP ST SZ UK YU **A**: ES IN KZ MG TR WS
bisbigemina Pic, 1917g: 13
gredosensis Pic, 1918h: 22
illyrica Mulsant, 1861: 188
immaculatithorax Pic, 1900j: 85
insignaticollis Pic, 1900j: 85
longitarsis Pandellé, 1867: 177
nigrolabra Pic, 1915h: 21
oculata Gebler, 1817: 316 [HN]
pueli Pic, 1913l: 161
quadrinotatithorax Pic, 1914m: 52
simoni Pic, 1922e: 19
subgemina Rey, 1891e: 115
trinotatithorax Pic, 1914m: 52
antennalis Marseul, 1864: 40 (*Telephorus*) **E**: SP
araticollis Fairmaire, 1897c: 216 (*Telephorus*) **A**: JIX
ariasi Mulsant, 1862a: 277 (*Telephorus*) **E**: SP
assimilis Paykull, 1798: 261 **E**: CZ HU SV YU
atlasiensis Pic, 1932g: 26 [= 1932c: 5] **N**: MO
amismizana Pic, 1932g: 26 [=1932c: 5]
atrata Marseul, 1864: 27 (*Telephorus*) **E**: SP

atrifrons Fairmaire, 1889b: 40 (*Telephorus*) **A**: SCH
atripalpis Kocher, 1954a: 153 **N**: MO
basithorax Pic, 1902g: 47 **E**: BU **A**: TR
  bulgarica Švihla, 1983: 2
beckeri Pic, 1902i: 56 (*Rhagonycha*) **E**: ST
bilunata Marseul, 1864: 22 (*Telephorus*) **A**: IS SY TR
  dilutipes Pic, 1906k: 82
  distincticollis Pic, 1906k: 82
biimpressiceps Pic, 1930i: 78 **A**: CH
biplagiata Ballion, 1871: 351 (*Telephorus*) **A**: ES KI KZ MG UZ XIN
  auliensis Pic, 1914a: 4
  bimaculifera Heyden, 1888a: 41
  biplagiaticollis Pic, 1914a: 5
  chianschanensis Pic, 1913o: 187
  dia Reitter, 1898e: 123
  funestula Pic, 1906k: 82
  kiritschenkoi Barovskij, 1926c: 240
  musarti Pic, 1913o: 187
  uniplagiata Pic, 1914a: 5
boroveci Švihla, 1999: 146 **E**: AR
bouvieri Pic, 1924j: 479 **A**: SD
brancuccii Švihla, 1992: 303 **A**: XIN
brevicornis Kiesenwetter, 1852: 595 (*Telephorus*) **E**: FR SP
brevipennis Faldermann, 1835a: 189 **E**: AR **A**: IQ SY TR
  akbesiana Pic, 1906m: 96
  brachyptera Holdhaus, 1920: 50
brullei Marseul, 1864: 49 (*Telephorus*) **E**: GR
brunneipennis Heyden, 1889b: 673 **A**: FE HEI GAN MG SCH SHA SHX
  luteolimbata Pic, 1902g: 47
bubsequa Gozis, 1882: 198 **E**: AR
carinulata Kocher, 1954a: 156 **N**: MO
cedricola Wittmer, 1971d: 230 **A**: CY IN SY TR
  cyprogenia Švihla, 1983: 1
ciliciensis Bourgeois, 1900a: 152 **E**: AB **A**: IN TR
colona Erichson, 1841b: 169 **N**: AG
confossicollis Fairmaire, 1891f: ccviii (*Telephorus*) **A**: HUB (Chang-yang)
convexicollis Fairmaire, 1875a: 513 (*Rhagonycha*) **N**: AG TU
  kabyliana Pic, 1900j: 85
coriacea Fairmaire, 1891d: xci (*Telephorus*) **A**: KA
cornix Abeille de Perrin, 1869b: 42 (*Telephorus*) **E**: FR
coronata Gyllenhal, 1808b: 62 **E**: PT SP **N**: MO
corvina Moscardini, 1962: 112 **E**: IT
cretica Wittmer, 1971d: 230 **E**: GR (Kríti)
cryptica Ashe, 1947: 59 **E**: BE DE GB GE FR IR LU PL PT SP SV SZ
curta Marseul, 1864: 43 (*Telephorus*) **N**: AG TU
cypria Marseul, 1864: 36 (*Telephorus*) **A**: CY
darwiniana Sharp, 1867: 436 **E**: BE DE FR GB GE LU NL
daurica Gebler, 1832: 45 **A**: ES FE MG
  mannerheimi Pic, 1902k: 63 [HN]
decipiens Baudi di Selve, 1872: 112 **E**: AU BE CR CZ DE FR GE HU IT LU NR PL SK SL SP SV SZ YU **N**: AG **A**: TR
decolorans Brullé, 1832: 147 **E**: GR
  kalavrytana Pic, 1909q: 169
  latemaculata Pic, 1905i: 161
  parnassica Pic, 1903g: 115
delagrangei Delkeskamp, 1939: 75 **E**: BU GR **A**: TR

ottoi Delkeskamp, 1939: 75
trisignaticollis Delkeskamp, 1939: 75
deportata Fairmaire, 1876a: 38 **N**: AG MO
rotroui Pic, 1924c: 21
dissipata Gemminger, 1870: 120 [RN] **E**: FR (Corse) IT (Sardegna)
binominata Pic, 1906k: 81
dichroa Marseul, 1864: 53 (*Telephorus*) [HN]
dichroma Reiche, 1861b: 207 (*Telephorus*)
dobrzhanskii Barovskij, 1926c: 238 **A**: KI
monticola Barovskij, 1926c: 239
edentula Baudi di Selve, 1872: 106 **E**: ST
emilieae Švihla, 1992: 319 **A**: UZ
ephippigera Brullé, 1832: 147 (*Telephorus*) **E**: GR
europea Pic, 1921d: 28 [RN] **E**: IT (Sicilia)
insularis Fiori, 1914b: 71 [HN]
fallaciosa Barovskij, 1926c: 237 **A**: KI
angustiplaga Barovskij, 1926c: 237
semispoliata Barovskij, 1926c: 237
spoliata Barovskij, 1926c: 237
falzonii Fiori, 1914a: 16 **E**: IT
figurata Mannerheim, 1843c: 92 **E**: AR AU BE BU BY CT CZ DE EN FI FR GB GE GG HU IR IT LA LS LT LU NR NT PL RO SK ST SV SZ UK **A**: ES FE JA KZ UZ
bicolor Fabricius, 1798: 69 [HN]
cruachana Chitty, 1893: 143
fumata Everts, 1921: 307
haemorrhoidalis C. G. Thomson, 1864: 183
infigurata Pic, 1914r: 178
liturata Dietrich, 1857: 126 (*Telephorus*)
luteata Schilsky, 1888: 189
scotica Sharp, 1867: 437 (*Telephorus*)
flavilabris Fallén, 1807: 12 **E**: AR AU BE BY CT CZ EN FR GB GE GR HU IT LT NT PL RO SK ST UK **A**: TR
affinis Stephens, 1830: 294 (*Telephorus*)
atra Stephens, 1830: 294 (*Telephorus*)
discotestacea Pic, 1905n: 186
fiscicollis Kiesenwetter, 1852: 598 (*Telephorus*)
fulvicollis Fabricius, 1792a: 218 [HN]
maculata Schilsky, 1888: 189
nigricollis Haenel, 1940: 189
nivalis Ahrens, 1814: pl. 10 (*Telephorus*)
thoracica L. Redtenbacher, 1858: 527
foetida Ménétriés, 1832: 162 **E**: ST
forticornis Heyden, 1885: 287 **A**: KI UZ
andischanensis Pic, 1914a: 4
arisi Pic, 1906k: 82
nigropubescens Barovskij, 1926c: 239
franciana Kiesenwetter, 1866b: 248 **E**: SP
fusca Linnaeus, 1758: 401 **E**: AU BE BY CT CZ EN FR GB GE GR HU IT LS LT LU NT PL RO SK ST SZ UK **A**: WS
antica Laporte, 1840a: 271 (*Telephorus*)
aurita Rey, 1891e: 114 (*Telephorus*)
basithorax Pic, 1902g: 47
conjuncta Schilsky, 1888: 188
curticornis Rey, 1891e: 115 (*Telephorus*)
digoniensis Pic, 1914m: 52
fracassii Leoni, 1905: 130
immaculicollis Laporte, 1840a: 272 (*Telephorus*)
marciana Leoni, 1905: 130
rugifrons Laporte, 1840a: 241 (*Telephorus*)

teutonica Pic, 1913l: 161
fuscipennis Mulsant, 1862a: 187 (*Telephorus*) **E**: CR IT (Sicilia)
ambigua Fiori, 1914a: 4
neptunia Baudi di Selve, 1893: 38
rufifrons Marseul, 1864: 25 (*Telephorus*)
heleocharis Satô, Okushima & Ishida, 2002: 210 **A**: JA
ictaria Fiori, 1914b: 63 **E**: IT
ifranensis Pic, 1952a: 65 **N**: MO
inculta Gené, 1839: 15 **E**: IT (Sardegna) FR (Corse)
genei Gené, 1839: 15
laconiensis Pic, 1918h: 22
leveillei Pic, 1927g: 9
sardiniensis Pic, 1918h: 22
inpectoralis Pic, 1908j: 73 **A**: TR (Taurus)
instabilis Kiesenwetter, 1866b: 246 **E**: PT SP
iranica Wittmer, 1975b: 261 **A**: IN
italica Fiori, 1914a: 12 **E**: IT
decolor Fiori, 1899a: 139
imitans Fiori, 1914a: 13
jacobsoni Barovskij, 1926c: 237 **A**: KZ
heptapotamica Barovskij, 1928a: 97
jindrai Švihla, 2004b: 173 **A**: SHA
kafkai Švihla, 1999: 149 **A**: IN
kervillei Pic, 1932d: 17 **E**: AU CZ GE HU PL RO SK YU **A**: TR
muelleri Hicker, 1955: 55
knizeki Švihla, 2004b: 174 **A**: HEB
kugartensis Barovskij, 1926c: 239 **A**: KI
liburnica Depoli, 1912: 101 **E**: AU AL BH CZ FR GE GR HU IT PL RO SK UK YU
csikii Stiller, 1926: 136
pulicaroides Depoli, 1921: 144
szegediensis Stiller, 1942: 223
livida Linnaeus, 1758: 401 **E**: AB AL AR AU BE BH BU BY CR CT CZ DE EN FI FR GB GE GG GR HU IT LS LT LU MC NT PL PT RO SK SP ST SV SZ UK YU **N**: MO **A**: IS TR WS **NAR**
bicolorata Ragusa, 1894: 39
canescens Schaller, 1783: 302
dispar Fabricius, 1792a: 213
dispar Paykull, 1798: 259 [HN]
distinguenda Gautier des Cottes, 1872: 224 (*Telephorus*)
eremita Rosenhauer, 1856: 142
femoralis Baudi di Selve, 1872: 108
ferdinandi Pic, 1909q: 169
flava DeGeer, 1774: 70 (*Cicindela*)
fusca Scopoli, 1763: 38
hungarica Csiki, 1903: 443
inscapularis Pic, 1909q: 169
latescapularis Pic, 1953a: 10
luteiceps Schilsky, 1888: 189
media Gmelin, 1790: 1891
mendax Rey, 1891e: 115 (*Telephorus*)
menetriezi Faldermann, 1839: 84
minutehumeralis Pic, 1953a: 10
mostarensis Pic, 1926f: 9
nigrifrons Stephens, 1833: 45 (*Telephorus*)
nigripennis Schilsky, 1888: 189
nigripes Schilsky, 1889: 342
nudipennis Reitter, 1911a: 258
opaca O.F. Müller, 1776: 61 (*Telephorus*)
paganosa Pic, 1942c: 7
predispar Pic, 1942c: 7

rufescens Trella, 1925: 125
ruficeps Reitter, 1911a: 259
rufipes Herbst, 1784b: 107
scapularis L. Redtenbacher, 1858: 527
sicula Bourgeois, 1893: 17
sulcicornis Reitter, 1911a: 258
unicoloriceps Kaszab, 1955: 294
varendorffi Reitter, 1904e: 159
macrosilesiformis Pic, 1923f: 7 **A**: SCH
aethalina Barovskij, 1926c: 242
marginiventris funebris Marseul, 1864: 29 (*Telephorus*) **A**: IN TR
marginiventris marginiventris Reiche & Saulcy, 1857b: 171 (*Telephorus*) **A**: IN IS JO LE SY TR
bloudana Pic, 1915j: 29
insignatithorax Delkeskamp, 1939: 9
postarcuata Delkeskamp, 1939: 9
mariae Pic, 1952a: 66 **N**: MO
maroccana Bedel, 1921: 59 **N**: MO
semiobliterata Pic, 1935f: 13
massurae Levrat, 1859: 33 **N**: TU
maura Pic, 1921m: 146 **N**: MO
azouana Pic, 1923h: 9
curtehumeralis Pic, 1935f: 13
lindbergi Pic, 1932g: 26 [= 1932c: 5]
mauritanica P. H. Lucas, 1846: 186 **N**: MO TU
atroscutellaris Pic, 1900j: 86
vaucheri Pic, 1911j: 113
melaspis Chevrolat, 1854: 435 (*Telephorus*) **E**: AR GG **A**: IN IQ LE SY TR
atrofoveolata Pic, 1918h: 22
melaspoides Wittmer, 1971d: 233 **A**: JO LE SY
merula Moscardini, 1962: 107 **E**: IT
metallipennis Wittmer, 1997e: 289 **A**: YUN
minutemaculata Wittmer, 1997e: 286 **A**: SHA SHX
monacha Moscardini, 1962: 116 **E**: IT
apennina Fiori, 1909a: 24
bimaculata Fiori, 1914a: 9
morio Baudi di Selve, 1873: 230
montana Stierlin, 1889: 76 **E**: AR AU CZ FR GE IT LS SZ
morio Fabricius, 1792a: 220 **E**: GE
nevadensis Pic, 1908l: 89 **E**: SP
nigra DeGeer, 1774: 72 (*Telephorus*) **E**: AU BE BY CR CT CZ EN FI FR GB GE GR HU IR IT NT PL RO SK SV SZ SP UK "Balkans" **A**: WS
bicolor Herbst, 1784b: 108
coerulocephala Thunberg, 1784a: 18
discotestacea Pic, 1905n: 186
fulva Gmelin, 1790: 1896
fulvicollis L. Redtenbacher, 1858: 257
melanoptera Eschscholtz, 1818: 463
ruficollis P. Rossi, 1790: 169
suturalis Schilsky, 1890: 178
theresae Pic, 1900j: 86
thoracica A. G. Olivier, 1790a: no 26: 12 (*Telephorus*)
nigricans O.F. Müller, 1776: 61 **E**: AR AU BE BU BY CR CT CZ DE EN FR FI GB GE GG HU IR IT LS LT LU NT PL RO SK SL ST SV SZ UK **A**: TR
alpestris Fiori, 1914a: 11
arvernica Buysson, 1908: 72
bipallida Kaszab, 1955: 295
discoidea Stephens, 1839: 190 (*Telephorus*)
hospes Rosenhauer, 1847: 19

immaculata Schilsky, 1888: 189
luteipes Schilsky, 1888: 189
martialis Pic, 1908f: 49
monachi Pic, 1918h: 22
nigrescens Trella, 1925: 125
pallidosignata Pic, 1904a: 2
picipes Schilsky, 1908: 602
plumbea Eschscholtz, 1818: 461
unicoloripes Kaszab, 1955: 295
vittigera Baudi di Selve, 1872: 107
nigricolor Pic, 1906k: 83 **A**: FE GAN "Korea"
nigricornis Laporte, 1840a: 274 (*Telephorus*) **E**: GE
obscura Linnaeus, 1758: 402 **E**: AL AU BE BH BU BY CR CT CZ EN FR GB GE GR HU IT LT LU MC NR PL NT RO SK SL ST SV SZ UK YU **A**: WS
baudii Pic, 1909o: 164
bicolor Fiori, 1899a: 139
kaszabi Vosyka, 1957: 123
occidentalis Kazantsev, **nom. nov.** [see New acts] **N**: MO TU
occipitalis Fairmaire, 1876a: 38 (*Telephorus*) [HN]
ochreata Reiche, 1878a: 384 (*Telephorus*) **E**: FR (Corse)
ochracea Fiori, 1914b: 80
oculimarginalis Hicker, 1935: 292 **A**: CH ("Turkestan") UZ
ferganica Švihla, 1992: 302
orgivasa Pic, 1954f: 96 **E**: SP
paganettii Flach, 1907: 16 (*Telephorus*) **E**: IT
pakistana Wittmer, 1997e: 291 **A**: PA
palliata Gyllenhal, 1808b: 63 **E**: SP
pallida Goeze, 1777: 541 **E**: AU BE BY CR CT CZ DE GB GE FR HU IR IT LA LS LT NT PL PT RO SK SP ST SV SZ UK YU **A**: ES FE JA MG NC WS "Korea"
baikalica Pic, 1906k: 83
bicolor Creutzer, 1797: 13
lutescens Geoffroy, 1785: 61
maritima Fiori, 1914b: 70
melanogastrica Motschulsky, 1860b: 117
pectoralis Motschulsky, 1860b: 118
pseudoassimilis Pic, 1917g: 13
rufa Gyllenhal, 1808a: 350 [HN]
ustulata Kiesenwetter, 1852: 596
paludosa Fallén, 1807: 12 **E**: AU BE CT CZ DE EN FI FR GB GE HU IR IT NR NT PL SK SZ UK **A**: ES WS
aethiops Curtis, 1840: 279 (*Telephorus*)
atra Thunberg, 1792: 106 [HN]
boreella Zetterstedt, 1838: 83
nigritula Dietrich, 1857: 129
pamphylica Wittmer, 1971d: 236 **A**: IN TR
paradoxa Hicker, 1960: 78 **E**: AL AU BE BH CR CZ GE GR HU IT NL PL SK YU
cinctithorax Moscardini, 1963: 109
paulinoi Kiesenwetter, 1870: 125 **E**: FR PT SP
schrammi Pic, 1907b: 104
pellucida Fabricius, 1792a: 214 **E**: AL AU BE BH BU BY CZ CT EN FR GB GE HU IR IT LS LT LU NT PL RO SK SL ST SZ UK YU **A**: WS
baudii Fiori, 1914b: 47
cantiata Stephens, 1830: 301 (*Telephorus*)
croissandeaui Pic, 1927g: 9
dietzi Roelofs, 1946: 240
meridionalis Fiori, 1914b: 49
mirabilis Dalla Torre, 1880: 4 (*Telephorus*)
obscuripes Everts, 1921: 306
rauterbergi Reitter, 1893b: 176

peninsularis Fiori, 1914b: 47 **E**: IT
bisignata Fiori, 1914b: 73
parens Fiori, 1914b: 75
piceogeniculata Pic, 1926p: 236 **A**: YUN
plebeja Kiesenwetter, 1874b: 278 **A**: JA
podabriformis Pic, 1917j: 160 **A**: "North China"
praecox Gené, 1836: 178 **E**: FR (Corse) IT (Sardegna, Sicilia)
vittatocollis Reiche, 1861b: 208 (*Telephorus*)
prusiensis Marseul, 1864: 33 (*Telephorus*) **E**: BU **A**: TR
anatolica Bourgeois, 1900a: 151
atrocapitata Pic, 1903m: 145
eurynota Bourgeois, 1900a: 153
pseudoquadripunctata Vosyka, 1957: 123 **E**: SK
pulicaria Fabricius, 1781: 260 **E**: AL AU BU CR CZ FR GE GR HU IT LA LU MC PL PT SK SP SZ UK **A**: TR
curticollis Pic, 1906k: 82
curtipennis Rey, 1891e: 115
detectiventris Pic, 1904j: 71
discicollis Brullé, 1832: 144
opaca Germar, 1824: 68
pumilio pumilio Heyden, 1885: 288 **A**: UZ
pumilio reductipennis Wittmer, 1966c: 491 **A**: KI
puncticollis Levrat, 1857: 418 **N**: AG TU
notaticeps Pic, 1903b: 146
obscuripennis Pic, 1907d: 114
paulonotata Pic, 1903b: 146
pyrenaea Pic, 1906l: 89 **E**: FR PT SP
bivittata Marseul, 1864: 57 (*Telephorus*)
correai Pic, 1927g: 9
getschmanni Heyden, 1880b: 291
lineata Kiesenwetter, 1852: 594 (*Telephorus*)
plurivittata Pic, 1906l: 89
quadripunctata O.F. Müller, 1776: 63 (*Telephorus*) **E**: AR AU BH BU BY CT CZ DE EN FI FR GE GR HU IT LT NT PL RO SK ST SV SZ UK **A**: FE IN TR WS
adusta Bourgeois, 1900a: 152
bakonyensis Kaszab, 1955: 294
bipuncticollis Kaszab, 1955: 294
bosnica Pic, 1906k: 82
cincticollis Pic, 1911u: 169
dilatata L. Redtenbacher, 1849: 324 (*Telephorus*)
fulvipennis Germar, 1824: 68
fulvithorax Kaszab, 1955: 294
mihalovicsi Kaszab, 1955: 294
moreli Pic, 1923h: 9
nigricornis Letzner, 1847: 74
rosenhaueri Kaszab, 1955: 294
sumovica Roubal, 1913c: 109
reichei Mulsant, 1862a: 186 **E**: FR PT SP
andalusiaca Pic, 1900j: 85
hispanica Reiche, 1878a: 383 (*Telephorus*)
reynosana Pic, 1922e: 19
rifensis Kocher, 1961b: 147 **N**: MO
rufa Linnaeus, 1758: 401 **E**: AU BE BU BY CT CZ DE EN FI FR GB GE HU IR IT LS LT LU NT PL RO SK SP ST SV SZ UK YU **A**: AF ES FE HEI KI KZ MG NC QIN SC TD UZ WS XIN **NAR**
alexandris Pic, 1914m: 52
analis Stephens, 1829: 11
conditiva Pic, 1912r: 82
confinis Stephens, 1830: 302 (*Telephorus*)

griseipennis Stephens, 1835: 416 (*Telephorus*)
korbi Pic, 1912r: 82
litterata Eschscholtz, 1818: 461
liturata Fallén, 1807: 18
maculicollis Stephens, 1835: 415 (*Telephorus*)
padana Fiori, 1914b: 66
roelofsi Pic, 1913k: 153
rufescens Dietrich, 1857: 124 (*Telephorus*) [HN]
tenuilimbatus Ballion, 1871: 352 (*Telephorus*)
testaceipes Stephens, 1830: 303 (*Telephorus*)
turkestanica Pic, 1913o: 187

rufonotaticeps Pic, 1921d: 29 **A**: CH
rufoscuta Pic, 1922b: 31 **A**: YUN
rustica Fallén, 1807: 9 **E**: AL AU BE BH BU BY CR CT CZ DE EN FR GB GE GR HU IT LS LT LU NT PL RO SL SP SV SK ST SZ UK YU **N**: EG **A**: SY TR WS

agnani Pic, 1913j: 145
albanica Pic, 1947: 5
asignata Trella, 1925: 125
leonardi Pic, 1913i: 145
lokvensis Stiller, 1926: 137
mimithorax Pic, 1910m: 49
nigripalpis Rey, 1891e: 115
rubrocollis Rüschkamp, 1928: 150
rufolimbaticollis Pic, 1914r: 174
subcincticollis Pic, 1913k: 153
tenuipes Rey, 1891e: 115
trimaculatithorax Pic, 1914m: 52
vesubiensis Pic, 1922e: 19

satoi Wittmer, 1997a: 37 **A**: GUX
schoeni Švihla, 1992: 319 **A**: KI TD
seidlitzi Kiesenwetter, 1866a: 393 **E**: SP
semimetallica Pic, 1932j: 250 **A**: SCH
semirufa Solsky, 1881: 33 (*Silotrachelus*) **A**: KI
sicula Pic, 1906n: 197 **E**: IT (Sicilia)
smyrnensis Marseul, 1864: 109 (*Telephorus*) **E**: AL BU GR **A**: IS LE SY TR

eduardi Pic, 1914m: 52
graecoturca Wittmer, 1986a: 101
hellenica Heyden, 1883b: 310
morfini Pic, 1901c: 25
ruficeps Kiesenwetter, 1859a: 26 (*Telephorus*) [HN]
rufocapitatus Gemminger, 1870: 120 (*Telephorus*)

soeulensis Pic, 1922c: 1 **A**: SC
sublivida Pic, 1904c: 9 **A**: IQ SY TR

mardinensis Pic, 1906k: 82

swampingana Pic, 1916f: 4 **A**: CH ("Swamping")
teinturieri Mulsant & Godart, 1866: 447 (*Telephorus*) **N**: AG
terminata Faldermann, 1835a: 192 **E**: AB AR AU BE CR CT CZ FR GE GG GR HU IT LS MD PL RO SK SL ST SZ UK **A**: IN LE TR

dahlgreni Wittmer, 1984a: 1
fissicollis Fairmaire, 1884a: 179 (*Telephorus*)
melanoscelis Kolenati, 1846b: 41
sudetica Letzner, 1847: 75
vesubiella Bourgeois, 1893: 19

thibetana Gorham, 1889: 102 (*Telephorus*) **A**: SCH
thibetanomima Wittmer, 1997e: 294 **A**: SCH YUN
torretassoi Wittmer, 1935a: 72 **E**: IT
trabeata Faimaire, 1896: 93 (*Telephorus*) **A**: SD
tristis Fabricius, 1798: 68 **E**: AU CZ FR GE HU IT PL RO SK SZ UK

tunisea Pic, 1900j: 86 **N**: AG TU
pauli Pic, 1915j: 29
tunisiensis Pic, 1953a: 10 **N**: TU
versicolor Baudi di Selve, 1872: 102 (*Telephorus*) **E**: IT
hummleri Pic, 1908f: 49
rufithorax Delkeskamp, 1939: 106
voriseki Švihla, 1992: 307 **A**: KZ
xantholoma Gebler, 1833: 280 **A**: WS
xanthoporpa Kiesenwetter, 1860: 484 **E**: FR SP
signaticollis Mulsant, 1862a: 234
xantholoma Kiesenwetter, 1852: 590 (*Telephorus*) [HN]
ziganadagensis Wittmer, 1971d: 238 **A**: TR
zolotikhini Kazantsev, 1994a: 15 **A**: FE
subgenus *Cyrtomoptila* Motschulsky, 1860d: 398 type species *Cantharis lateralis* Linnaeus, 1758
Absidiella Wittmer, 1972b: 130 type species *Pseudoabsidia kaszabi* Wittmer, 1971
Gymnocantharis Wittmer, 1979e: 328 type species *Gymnocantharis himalaica* Wittmer, 1979
Wittmercantharis Satô, 1986c: 259 type species *Telephorus vulcanus* Lewis, 1895
biocellata Fairmaire, 1891e: cxxx (*Telephorus*) **A**: KA PA "Tibet"
bipuncticollis Gorham, 1895a: 312
curtata Kiesenwetter, 1874b: 273 **A**: JA (Hokkaido, Honshu, Shikoku)
fibulata Märkel, 1852: 593 **E**: AU CR CZ FR PL SK SL
gemina Dahlgren, 1974: 145 **E**: AU IT SZ
lateralis Linnaeus, 1758: 402 **E**: AU BE BH BY CR CT CZ DE EN FI FR GB GE GG GR HU IR IT LA LT LU MC NT PL RO SK SP ST SZ UK YU **N**: MO **A**: AF KI KZ MG TR
afghana Švihla, 1992: 316
andrei Pic, 1935f: 13
marginata Stephens, 1830: 296 (*Telephorus*)
nigricollis Trella, 1932: 225
nigrinotata Pic, 1905n: 186
notaticollis Schilsky, 1890: 178
oralis Ahrens, 1814: pl. 11
sexnotata Roubal, 1934c: 206
lucida Pic, 1906l: 89 **A**: AF TD
klapperichi Wittmer, 1956a: 156
pagana Rosenhauer, 1847: 20 **E**: AU BE CZ FR GE IT LU PL SK SZ UK
albomarginata Märkel, 1852: 592
plagiata Heyden, 1889b: 675 **A**: FE GAN JA NC SC
argentosa Barovskij, 1929b: 266
guilleti Pic, 1905d: 113
inlateralis Pic, 1902i: 55
raddensis Pic, 1904d: 26
vulcana Lewis, 1895a: 112 (*Telephorus*)
podistroides Švihla, 1992: 312 **A**: UZ
pygmaea mongolica Pic, 1906k: 83 **A**: AF KA KZ MG
himalaica Wittmer, 1979e: 329 (*Gymnocantharis*)
kaszabi Wittmer, 1971c: 192 (*Pseudoabsidia*)
pygmaea pygmaea Ménétriés, 1832: 162 **E**: AB **A**: IN TM UZ
inforticornis Pic, 1913o: 187
subgenus *Taiwanocantharis* Wittmer, 1984c: 147 type species *Cantharis tripunctata* Wittmer, 1984
pallidithorax Wittmer, 1984c: 148 **A**: TAI
tripunctata Wittmer, 1984c: 147 **A**: TAI

genus *Cantharomorphus* Fiori, 1914b: 82 type species *Cantharomorphus longipes* Fiori, 1914
longipes Fiori, 1914b: 83 **E**: IT (Sicilia)
rufidens Marseul, 1864: 20 (*Ancistronycha*) **E**: FR (Corse) IT (Sardegna)
fiorii Pic, 1914m: 52 (*Ancistronycha*)

genus *Cordicantharis* Švihla, 1999: 160 type species *Cantharis cordicollis* Küster, 1854
acutangula Fairmaire, 1884a: 169 (*Telephorus*) **A**: TR
amanicola Pic, 1903r: 169 (*Cantharis*)
bodemeyeri Bourgeois, 1900a: 152 (*Cantharis*) **A**: TR
cordicollis Küster, 1854: 61 (*Cantharis*) **E**: AB AR GG ST **A**: TR
cyanipennis Faldermann, 1835a: 191 (*Cantharis*) **A**: IN
caspica Reitter, 1898e: 122 (*Cantharis*)
diabolica Reitter, 1895a: 82 (*Cantharis*) **A**: TR
drahuska Švihla, 2004b: 176 **A**: SHA
iliaca Marseul, 1864: 41 (*Telephorus*) **A**: TR
chehirensis Pic, 1918h: 22 (*Cantharis*)
pontica Pic, 1904c: 9 (*Cantharis*)
longicollis Kiesenwetter, 1859a: 24 (*Telephorus*) **E**: AL AU BH BU CR CZ GR HU IT MA PL RO SK SL UK YU **A**: TR
similis Švihla, 1999: 163 **A**: TR
talyschensis Pic, 1900n: 182 (*Cantharis*) **E**: AB
wittmeri Švihla, 1999: 162 **A**: LE

genus *Cratosilis* Motschulsky, 1860d: 399 type species *Rhagonycha redtenbacheri* Märkel, 1852 (= *Cantharis denticollis* Schummel, 1844)
Pygidia Mulsant & Rey, 1861e: 191 type species *Pygidia hypocrita* Mulsant & Rey, 1861 (= *Cantharis denticollis* Schummel, 1844)
denticollis Schummel, 1844: 194 (*Cantharis*) **E**: AU BE CZ FR GE HU IT LU PL RO SK SZ UK
hypocrita Mulsant & Rey, 1861e: 192 [pl.] (*Pygidia*)
nivalis L. Redtenbacher, 1849: 324 (*Rhagonycha*)
planicollis Kiesenwetter, 1852: 601 (*Rhagonycha*)
punctipennis Kiesenwetter, 1852: 600 (*Rhagonycha*)
redtenbacheri Märkel, 1852: 601 (*Rhagonycha*)
distinguenda Baudi di Selve, 1859a: 295 (*Rhagonycha*) **E**: FR IT SZ
laricicola Kiesenwetter, 1861: 381 (*Cantharis*)
saasensis Pic, 1923h: 9
laeta Fabricius, 1792a: 218 (*Cantharis*) **E**: FR (Alpes Maritimes) IT SZ
discolor Baudi di Selve, 1873: 235 (*Rhagonycha*)
semiobscura Pic, 1901h: 66 (*Pygidia*)
sicula Marseul, 1864: 100 **E**: IT
theresae Pic, 1901f: 50 **E**: GR

genus *Cyrebion* Fairmaire, 1891f: ccvii type species *Cyrebion laticorne* Fairmaire, 1891
laticorne Fairmaire, 1891f: ccvii **A**: HUB (Chang-yang)

genus *Falsopodabrus* Pic, 1927i: 40 type species *Podabrus refossicollis* Pic, 1907
himalaicus Wittmer, 1974e: 631 **A**: BH SD
martensi Wittmer, 1979e: 331 (*Stenothemus*) **A**: NP SD
particularis Pic, 1931g: 77 (*Stenothemus*) **A**: YUN
refossicollis Pic, 1907q: 175 (*Podabrus*) **A**: NP SD XIZ

genus *Islamocantharis* Wittmer & Magis, 1978: 134 type species *Islamocantharis orientalis* Wittmer & Magis, 1978
subgenus *Islamocantharis* Wittmer & Magis, 1978: 134 type species *Islamocantharis orientalis* Wittmer & Magis, 1978
auroraensis Kazantsev, 1989: 244 **E**: AB AR **A**: IN
businskae Wittmer, 1997e: 300 **A**: HUB SHA
diffusa Wittmer & Magis, 1978: 136 **A**: IN
orientalis Wittmer & Magis, 1978: 134 **E**: AR **A**: IN TR
subgenus *Unicantharis* Švihla, 1999: 160 type species *Cantharis cilicia* Pic, 1904
cilicia Pic, 1904c: 9 (*Cantharis*) **A**: TR

johanidesi Švihla, 2004a: 77 **A**: TR

genus *Leiothorax* Wittmer, 1978e: 350 type species *Leiothorax kaschmirensis* Wittmer, 1978
kashmirensis Wittmer, 1978e: 352 **A**: KA
kopetzi Švihla, 2004b: 208 **A**: NP
minutus Wittmer, 1974c: 57 (*Stenothemus*) **A**: NP

genus *Lycocerus* Gorham, 1889: 108 type species *Lycocerus serricornis* Gorham, 1889 (= *Omalisus maculicollis* Hope, 1831)
Andrathemus Wittmer, 1978c: 155 type species *Andrathemus purpurascens* Wittmer, 1978 (= *Lycocerus purpureus* Kazantsev [RN])
Athemellus Wittmer, 1972b: 124 type species *Athemellus maculithorax* Wittmer, 1972
Athemus Lewis, 1895a: 110 type species *Telephorus suturellus* Motschulsky, 1860
Isathemus Wittmer, 1995c: 185 type species *Athemus pallidulus* Wittmer, 1995
Mikadocantharis Wittmer & Magis, 1978: 137 type species *Cantharis japonica* Kiesenwetter, 1874
adusticollis Kiesenwetter, 1874b: 274 (*Cantharis*) **A**: FE (Kurils) JA
aegrotus Kiesenwetter, 1874b: 275 (*Cantharis*) **A**: JA
aenescens Fairmaire, 1889b: 40 (*Podabrus*) **A**: CH (Koui-Tcheou)
akemiae Satô & Okushima, 1992: 861 (*Andrathemus*) **A**: JA (Ryukyus)
alpicolus Nakane, 1992: 77 (*Mikadocantharis*) **A**: JA (Honshu)
amplus Wittmer, 1995c: 278 (*Athemus*) **A**: SHG
angulatus Wittmer, 1995c: 198 (*Andrathemus*) **A**: YUN
arisanensis Wittmer, 1954d: 278 (*Athemus*) **A**: TAI
asperipennis Fairmaire, 1891f: ccvii (*Telephorus*) **A**: SCH YUN
limbatipennis Pic, 1906k: 83 (*Cantharis*)
stotzneri Pic, 1926m: 154 (*Cantharis*)
atricolor Pic, 1922b: 31 (*Cantharis*) **A**: "China"
atripennis Pic, 1922b: 30 (*Malthinus*) **A**: CH TAI
griseopubens Pic, 1928g: 55 (*Malthinus*)
nigripennis Pic, 1938j: 121 (*Malthinus*) [RN]
taihokuensis Wittmer, 1954d: 276 (*Podabrinus*)
attristatus Kiesenwetter, 1874b: 272 (*Cantharis*) **A**: JA (Honshu)
niger Wittmer, 1983b: 225 (*Themus*)
pseudosuturellus Yajima & Nakane, 1969b: 186 (*Athemus*)
roninus Lewis, 1895a: 110 (*Athemus*)
atronotatus Pic, 1932b: 4 (*Rhagonycha*) **A**: SCH YUN
atroopacus Pic, 1937h: 62 (*Athemus*) **A**: TAI
babai Ishida, 1986: 199 (*Andrathemus*) **A**: FE (Kurils) JA (Honshu, Kyushu)
balangensis Wittmer, 1997e: 275 (*Athemus*) **A**: SCH
basicrus Fairmaire, 1886c: 339 (*Athemus*) **A**: JIX
benesi Švihla, 2004b: 186 (*Andrathemus*) **A**: SCH
bhutanensis Wittmer, 1978c: 152 (*Athemus*) **A**: BT
bicoloripennis Pic, 1924j: 478 (*Cantharis*) **A**: NP SD
bilamellatus Wittmer, 1995c: 245 (*Athemus*) **A**: SD
bilineatus Wittmer, 1995c: 275 (*Isathemus*) **A**: HUB (Chingkiang)
bilyi Švihla, 2004b: 188 (*Andrathemus*) **A**: NP
bipartitus Wittmer, 1972b: 128 (*Athemellus*) **A**: TAI
bisbicostatus Pic, 1924j: 478 (*Cantharis*) **A**: BT SD **ORR**
brahminicus Gorham, 1889: 98 (*Telephorus*) **A**: BT **ORR**
brancuccii Švihla, 2004b: 194 **A**: NP
brevelineatus Pic, 1916k: 15 **A**: YUN
atriceps Pic, 1923g: 33 [HN]
nigriceps Pic, 1938k: 158 [RN]
brunneonotaticeps Pic, 1922b: 32 (*Cantharis*) **A**: SHN (Jiao Xian)
brunneoreductus Pic, 1926l: 143 (*Cantharis*) **A**: SCH
brunneus Wittmer, 1955: 40 (*Podabrinus*) **A**: TAI

canthariformis Ishida, 1986: 210 (*Athemellus*) **A**: JA (Honshu)
carolusi Wittmer, 1995c: 242 (*Athemus*) **A**: YUN
centrochinensis Švihla, 2004b: 182 (*Athemus*) **A**: HUB SHA
chinensis Wittmer, 1982e: 344 (*Athemellus*) **A**: ZHE (Tienmu Shan)
chosokeiensis Pic, 1937h: 63 (*Andrathemus*) **A**: TAI
chujoi Wittmer, 1972b: 128 (*Athemellus*) **A**: TAI
cinctiventris Pic, 1937k: 170 (*Athemus*) **A**: BEI
cordieri Pic, 1915e: 15 **A**: NP SD UP
maculaticeps Pic, 1921e: 5
trilineatus Pic, 1921e: 5
variipes Champion, 1926c: 256
costulatus Wittmer, 1972b: 134 (*Lycocerus*) **A**: TAI
curtipennis Švihla, 2004b: 183 (*Andrathemus*) **A**: XIZ
curvatus Wittmer, 1995c: 260 (*Isathemus*) **A**: SCH
cyanurus Hope, 1831: 26 (*Telephorus*) **A**: NP SD **ORR**
duchoni Pic, 1911p: 143 (*Cantharis*)
mangini Pic, 1926r: 354 (*Themus*)
oberthuri Pic, 1955a: 11 (*Cantharis*)
rubricollis Hope, 1831: 26 (*Telephorus*)
solanensis Pic, 1955a: 11 (*Cantharis*)
viator Gorham, 1889: 98 (*Telephorus*)
daitoensis Kinoda & Nakane, 1993a: 6 (*Andrathemus*) **A**: JA (Honshu)
delicatulus Wittmer, 1995c: 197 (*Andrathemus*) **A**: SCH
dongchuanus Wittmer, 1995c: 231 (*Athemus*) **A**: YUN
elongaticeps Wittmer, 1972b: 127 (*Athemellus*) **A**: TAI
elongaticollis Pic, 1932b: 3 (*Podabrinus*) **A**: CH
elongatissimus Wittmer, 1997e: 281 (*Athemus*) **A**: YUN
elongatus Pic, 1921e: 5 (*Lycocerus*) **A**: CH
flavonotaticeps Pic, 1937h: 63 (*Athemus*) **A**: TAI
obliterata Pic, 1937h: 63 (*Athemus*)
genaemaculatus Wittmer, 1995c: 259 (*Isathemus*) **A**: FUJ (Shaowu)
gibbicollis Fairmaire, 1887a: 124 (*Telephorus*) **A**: FUJ
sinensis Gorham, 1889: 106 (*Telephorus*)
gracilitarsis Pic, 1921d: 30 (*Cantharis*) **A**: TAI
gressitti Wittmer, 1954d: 279 (*Athemus*) **A**: TAI
guerryi Pic, 1906k: 83 (*Cantharis*) **A**: CH **ORR**
jeanvoinei Pic, 1927d: 2 (*Cantharis*)
fainanus Pic, 1910t: 77 (*Cantharis*) **A**: TAI
flavonotaticeps Pic, 1937h: 63 (*Athemus*)
hickeri Pic, 1934d: 54 (*Themus*)
hanatanii Okushima, 1996: 118 (*Athemellus*) **A**: JA (Ryukyus)
hansi Švihla, 2004b: 179 (*Athemellus*) [RN] **A**: TAI
sauteri Pic, 1926i: 68 (*Cantharis*)
harmandi Pic, 1925i: 72 **A**: NP SD
harmandianus Kazantsev, **nom. nov.** [see New acts] **A**: SD
harmandi Pic, 1926r: 355 (*Rhagonycha*) [HN]
hasegawai Kinoda & Nakane, 1993b: 159 (*Andrathemus*) **A**: JA (Honshu)
hedini Pic, 1933b: 3 (*Themus*) **A**: GAN SCH YUN
hiroshii Satô & Okushima, 2001: 117 **A**: JA (Ryukyu)
hubeiensis Švihla, 2004b: 191 (*Athemus*) **A**: HUB
humerolineatus Wittmer, 1997a: 36 (*Athemus*) **A**: GUX
incisus Wittmer, 1995c: 250 (*Athemus*) **A**: NP
indentatus Wittmer, 1995c: 233 (*Athemus*) **A**: NP
infuscatus Yajima & Nakane, 1969b: 188 (*Athemus*) **A**: JA (Honshu, Shikoku, Kyushu)
inopaciceps Pic, 1926m: 153 (*Cantharis*) **A**: SCH
insulsus insulsus Harold, 1878: 76 (*Telephorus*) **A**: JA (Honshu)

insulsus lewisi Pic, 1906k: 81 (*Cantharis*) [RN] **A**: JA (Honshu, Shikoku, Kyushu)
dichroa Lewis, 1895a: 111 (*Telephorus*) [HN]
irrawadicus Švihla, 2004b: 184 (*Andrathemus*) **A**: XIZ
ishidai ishidai Nakane & Kinoda, 1992: 75 (*Andrathemus*) **A**: JA (Hokkaido, Honshu)
ishidai kotokuanus Nakane & Kinoda, 1992: 76 (*Andrathemus*) **A**: JA (Honshu)
ishidai ozeanus Nakane & Kinoda, 1992: 74 (*Andrathemus*) **A**: JA (Honshu)
ishiharai Ishida, 1986: 203 (*Andrathemus*) **A**: JA (Honshu, Shikoku, Kyushu)
hisamatsui Ishida, 1986: 201 (*Andrathemus*)
japonicus japonicus Kiesenwetter, 1874b: 266 (*Cantharis*) **A**: JA (Honshu, Shikoku, Kyushu)
edmei Pic, 1932a: 1 (*Cantharis*)
japonicus takaosanus Nakane, 1992: 77 (*Mikadocantharis*) **A**: JA (Honshu)
jelineki Švihla, 2004b: 189 (*Andrathemus*) **A**: HUB SHA
jejuensis Kang & Okushima, 2003: 347 (*Andrathemus*) **A**: SC
jindrai Švihla, 2004b: 185 (*Andrathemus*) **A**: SCH
kamjeensis Wittmer, 1978c: 152 (*Athemus*) **A**: BT
kejvali Švihla, 2004b: 193 **A**: SHA
kiontochananus Pic, 1921e: 4 (*Cantharis*) **A**: JIX SHN
atropygidialis Pic, 1937j: 143 (*Cantharis*)
perroudi Pic, 1937j: 143 (*Cantharis*)
kuatunensis Wittmer, 1995c: 224 (*Athemus*) **A**: FUJ
kubani Švihla, 2004b: 190 (*Athemus*) **A**: SHA
kunigamiensis Ishida, 1989: 83 (*Andrathemus*) **A**: JA (Ryukyus)
kurokamiensis Kinoda & Nakane, 1993b: 160 (*Andrathemus*) **A**: JA (Honshu)
kyushuensis Takakura, 1989: 157 (*Andrathemus*) **A**: JA (Kyushu)
lalaensis Wittmer, 1983a: 159 (*Athemus*) **A**: TAI
lateritius Gorham, 1889: 109 **A**: BT NP "India"
ruficeps Pic, 1915b: 3
indicus Pic, 1921e: 5
limatus Kazantsev, **nom. nov**. [see New acts] **A**: YUN
limbatipennis Pic, 1926p: 235 (*Podabrus*) [HN]
limbatus Pic, 1915l: 78 **A**: SCH YUN
hirsutus Pic, 1921e: 4
notaticollis Pic, 1921e: 4
lineatipennis Wittmer, 1953a: 35 (*Andrathemus*) **A**: JA (Honshu)
lineatithorax Wittmer, 1995c: 244 (*Athemus*) **A**: NP
lineatocollis Hope, 1831: 26 (*Omalisus*) **A**: NP SD
lohitensis Wittmer, 1960a: 101 (*Andrathemus*) **A**: "Tibet"
longicornis Wittmer, 1995c: 247 (*Athemus*) **A**: NP
longipilis longipilis Wittmer, 1995c: 218 (*Andrathemus*) **A**: SCH YUN
longipilis subincisus Wittmer, 1995c: 219 (*Andrathemus*) **A**: YUN
maculiceps Wittmer, 1983a: 161 (*Athemus*) **A**: TAI
maculicollis Hope, 1831: 26 (*Omalisus*) **A**: NP SD
serricornis Gorham, 1889: 109
maculielytris Ishida, 1986: 197 (*Athemus*) **A**: JA (Honshu, Shikoku, Kyushu)
maculithorax Wittmer, 1972b: 126 (*Athemellus*) **A**: TAI
magniceps Wittmer, 1995c: 237 (*Athemus*) **A**: SD
magnius Ishida, 1986: 204 (*Andrathemus*) **A**: JA (Honshu)
martae Švihla, 2004b: 178 (*Athemellus*) **A**: SCH
matsunagai Imasaka & Okushima, 2000: 374 (*Andrathemus*) **A**: JA (Honshu, Shikoku, Kyushu)
metallescens fukienensis Wittmer, 1954c: 109 (*Themus*) **A**: FUJ
metallescens metallescens Gorham, 1889: 107 (*Telephorus*) **A**: FUJ (Kiukiang) TAI
metallicipennis Fairmaire, 1887a: 124 (*Telephorus*) **A**: YUN
miekoae Takakura, 1987: 179 (*Athemellus*) **A**: JA (Kyushu)
milosi Švihla, 2004b: 180 (*Athemus*) **A**: SCH
minutonitidus Wittmer, 1995c: 188 (*Athemellus*) **A**: SCH

minusculus Barovskij, 1926c: 242 (*Cantharis*) **A**: SCH
moupinensis Pic, 1926r: 355 (*Cantharis*) **A**: SCH
moxiensis Wittmer, 1995c: 202 (*Andrathemus*) **A**: SCH
mugangensis Švihla, 2004b: 184 (*Andrathemus*) **A**: SCH
multiimpressus Wittmer, 1997e: 285 (*Athemus*) **A**: SHA
multistriatus Pic, 1906k: 82 (*Absidia*) **A**: SD
atricollis Pic, 1937h: 64 (*Athemus*)
brevenotata Pic, 1924j: 480 (*Athemus*)
lineaticeps Pic, 1924j: 480 (*Athemus*)
nakanei akiensis Nakane & Kinoda, 1992: 75 (*Andrathemus*) **A**: JA (Honshu)
nakanei nakanei Wittmer, 1953a: 35 (*Cantharis*) **A**: JA (Honshu, Kyushu)
nanpingensis Wittmer, 1995c: 204 (*Andrathemus*) **A**: SCH
nanshanensis Wittmer, 1983a: 162 (*Athemus*) **A**: TAI
naokii Okushima & Imasaka, 2000: 383 (*Andrathemus*) **A**: JA (Ryukyus)
nepalensis Wittmer, 1958: 115 (*Athemus*) **A**: NP
makiharai Wittmer, 1985b: 381 (*Athemus*)
newtoni Pic, 1921d: 30 (*Cantharis*) **A**: NP SD
granigera Champion, 1926b: 201 (*Cantharis*)
nigerrimus Yajima & Nakane, 1969b: 187 (*Athemus*) **A**: JA (Honshu)
nigriceps Wittmer, 1955: 39 (*Podabrinus*) **A**: GUA
nigricollis Wittmer, 1984c: 141 (*Lycocerus*) **A**: TAI
nigricolor Wittmer, 1951: 99 (*Podabrinus*) **A**: FUJ
nigrimembris Kazantsev, 1994a: 13 (*Athemellus*) **A**: FE "Korea"
nigripes Wittmer, 1995c: 209 (*Athemus*) **A**: SCH
nigrithorax Wittmer, 1995c: 194 (*Athemus*) **A**: YUN
nigroannulatus Pic, 1922b: 30 (*Cantharis*) **A**: JIX ZHE
nigrobilineatus Pic, 1916k: 16 (*Lycocerus*) **A**: YUN
nigroverticalis Fairmaire, 1887a: 123 (*Telephorus*) **A**: YUN **ORR**
choganhensis Pic, 1923g: 40 (*Athemus*)
humerotestacea Pic, 1938i: 84 (*Athemus*) [RN]
matensis Pic, 1923g: 61 (*Cantharis*)
monostigma Champion, 1926b: 197 (*Athemus*)
nigroabdominalis Pic, 1914b: 6 (*Cantharis*)
obconicicollis Pic, 1914b: 6 (*Cantharis*)
rufithorax Pic, 1923g: 40 (*Cantharis*)
subarcuatipes Pic, 1916e: 12 (*Cantharis*)
subelongtipennis Pic, 1921d: 29 (*Cantharis*)
subnitida Pic, 1906k: 83 (*Cantharis*)
testaceohumeralis Pic, 1936a: 1 (*Cantharis*) [HN]
trinotatithorax Pic, 1921d: 29 (*Cantharis*)
oberthueri Gorham, 1889: 102 (*Telephorus*) **A**: SCH
cordimacula Barovskij, 1926c: 243 (*Ancistronycha*)
distinctipes Pic, 1955b: 17 (*Themus*)
unimaculata Pic, 1920g: 19 (*Cantharis*)
oedemeroides Kiesenwetter, 1874b: 268 (*Cantharis*) **A**: JA (Honshu, Shikoku, Kyushu)
okabei Takahashi, 1992: 105 (*Andrathemus*) **A**: JA (Honshu, Izu Is.)
okinawanus Ishida, 1986: 196 (*Athemus*) **A**: JA (Ryukyus)
okuyugawaranus imasakai Kinoda & Nakane, 1993b: 158 (*Andrathemus*) **A**: JA (Honshu, Shikoku)
okuyugawaranus okuyugawaranus Takahashi, 1992: 106 (*Andrathemus*) **A**: JA (Honshu)
ondreji Švihla, 2004b: 181 (*Athemus*) **A**: SCH
orientalis Gorham, 1889: 105 (*Telephorus*) **A**: FUJ JIX
bigibbulus Fairmaire, 1900: 627 (*Telephorus*)
dimidiaticrus Fairmaire, 1889b: 41 (*Telephorus*)
subrufofemoralis Pic, 1921e: 4 (*Cantharis*)
oudai Švihla, 2004b: 179 (*Athemus*) **A**: SCH
pallicolor Wittmer, 1951: 98 (*Podabrinus*) **A**: FUJ
pallidipes Wittmer, 1997e: 278 (*Andrathemus*) **A**: YUN

paviei Bourgeois, 1890b: 186 (*Cantharis*) **A**: YUN **ORR**
  miniatus Champion, 1926c: 251
phulchokiensis Wittmer, 1995c: 248 (*Athemus*) **A**: NP
pictus Wittmer, 1983a: 170 (*Athemellus*) **A**: TAI
pieli Pic, 1937k: 171 (*Athemus*) **A**: SCH
pilipes Pic, 1911y: 271 (*Podabrus*) **A**: TAI
pluricostatus Fairmaire, 1887a: 125 (*Telephorus*) **A**: FUJ TAI
  militaris Gorham, 1889: 110 (*Lycocerus*)
podabroides Champion, 1926b: 199 (*Andrathemus*) **A**: SD
posticelimbatus Pic, 1926p: 236 (*Cantharis*) **A**: YUN
postobscuripes Wittmer, 1995c: 196 (*Andrathemus*) **A**: SCH
prattianus Gorham, 1889: 105 (*Athemus*) **A**: FUJ (Kiukiang)
pubescens Wittmer, 1983a: 167 (*Athemellus*) **A**: TAI
pubicollis Heyden, 1889b: 674 (*Cantharis*) **A**: GAN HEB SCH
  benardi Pic, 1924j: 479 (*Cantharis*)
  norensis Pic, 1913j: 145 (*Cantharis*)
  subaeneipennis Pic, 1917j: 161 (*Cantharis*)
pulchellus Wittmer, 1995c: 215 (*Athemus*) **A**: SCH
purpureus Kazantsev, **nom. nov**. [see New acts] **A**: BT
  purpurascens Wittmer, 1978c: 155 (*Andrathemus*) [HN]
reductus Wittmer, 1995c: 227 (*Athemus*) **A**: NP
rhagonychiformis Wittmer, 1983a: 169 (*Athemellus*) **A**: TAI
rufocapitatus Kazantsev, 1999: 128 **A**: YUN
rubripennis Hope, 1831: 26 (*Telephorus*) **A**: BT KA NP SD **ORR**
  purpurascens Pic, 1911p: 143 (*Cantharis*)
rubrispinis Kazantsev, **nom. nov**. [see New acts] **A**: SD
  rubripennis Pic, 1906k: 83 (*Cantharis*) [HN]
rubropilosus Wittmer, 1995c: 266 (*Andrathemus*) **A**: YUN
rugulicollis Fairmaire, 1886c: 340 (*Telephorus*) **A**: JIX
  bartoni Gorham, 1889: 106 (*Telephorus*)
ryukyuanus Wittmer, 1970b: 23 (*Pseudoabsidia*) **A**: JA (Ryukyus)
sabdeensis Wittmer, 1995c: 199 (*Andrathemus*) **A**: SCH
safraneki Švihla, 2004b: 187 (*Andrathemus*) **A**: XIZ
sanguineus brevemarginatus Wittmer, 1983a: 169 (*Athemellus*) **A**: TAI
sanguineus sanguineus Wittmer, 1951: 98 (*Podabrinus*) **A**: FUJ
sannenensis Pic, 1926h: 29 (*Absidia*) **A**: YUN
sauteri Pic, 1926i: 68 (*Cantharis*) **A**: TAI
schneideri Švihla, 2004b: 177 (*Athemellus*) **A**: SCH
semiarcuatipes Pic, 1927e: 36 (*Podabrinus*) **A**: ZHE
semicyaneus Wittmer, 1983a: 164 (*Athemellus*) **A**: TAI
semiextensus Wittmer, 1995c: 264 (*Andrathemus*) **A**: BT NP SD
semimetallicus Pic, 1932j: 250 (*Cantharis*) **A**: SCH
shigeii Kinoda & Nakane, 1993b: 159 (*Andrathemus*) **A**: JA (Honshu)
shikokensis Ishida, 1986: 209 (*Athemellus*) **A**: JA (Shikoku)
shimomurai Wittmer, 1983a: 160 (*Athemus*) **A**: TAI
sichuanus Wittmer, 1995c: 186 (*Athemus*) **A**: SCH
simplicicornis Pic, 1938k: 152 (*Podabrus*) **A**: CH
simulator extremus Wittmer, 1997e: 280 (*Athemus*) **A**: YUN
simulator simulator Wittmer, 1997e: 278 (*Athemus*) **A**: YUN
singularicollis Pic, 1933b: 5 (*Podabrinus*) **A**: GAN
striatipennis Wittmer, 1995c: 228 (*Athemus*) **A**: NP
striatus Kazantsev, 1994a: 11 (*Athemellus*) **A**: FE
suturalis Pic, 1908m: 95 (*Rhagonycha*) **A**: YUN
suturellus izuensis Okushima & Satô, 2000b: 321 (*Athemus*) **A**: JA (Honshu)
suturellus luteipennis Kiesenwetter, 1874b: 273 (*Cantharis*) **A**: JA (Honshu, Shikoku, Kyushu)

melanopus Harold, 1878: 74 (*Telephorus*)
yakushimanus Yajima & Nakane, 1969b: 186 (*Athemus*)
suturellus suturellus Motschulsky, 1861a: 10 (*Telephorus*) **A**: FE JA (Hokkaido)
honshuensis Yajima & Nakane, 1969b: 185 (*Athemus*)
insuturellus Yajima & Nakane, 1969b: 185 (*Athemus*)
nikkoensis Yajima & Nakane, 1969b: 185 (*Athemus*)
suzukii Wittmer, 1995c: 200 (*Athemus*) **A**: NP SD
taoyuanus Wittmer, 1983a: 158 (*Athemus*) **A**: TAI
angusticollis Pic, 1955b: 16 (*Themus*) [HN]
angustithorax Wittmer, 1975b: 260 (*Athemus*) [RN]
terricola Champion, 1926b: 194 (*Podistra*) **A**: XIZ
teruhisai Okushima, 1991a: 35 (*Andrathemus*) **A**: JA (Ryukyus)
testaceilabris Pic, 1926r: 357 (*Podabrinus*) **A**: SCH
testaceipes Pic, 1932b: 4 (*Podabrinus*) **A**: CH
tcheonanus Pic, 1922b: 31 (*Cantharis*) **A**: SCH YUN
tienmushanus Wittmer, 1995c: 239 (*Isathemus*) **A**: FUJ
tobiranus Okushima & Imasaka, 2000: 381 (*Andrathemus*) **A**: JA (Honshu)
trimaculatus Hope, 1831: 26 (*Telephorus*) **A**: NP UP
rufohumeralis Pic, 1926h: 29 (*Cantharis*)
tryznai Švihla, 2004b: 177 (*Athemellus*) **A**: SCH
tsuifengensis Wittmer, 1983a: 166 (*Athemellus*) **A**: TAI
tsuyukii Takahashi, 1992: 104 (*Athemus*) **A**: JA (Honshu)
ueharaensis Okushima, 1996: 121 (*Athemellus*) **A**: JA (Ryukyus)
varipubens Wittmer, 1978c: 157 (*Andrathemus*) **A**: NP
varithorax Wittmer, 1997a: 36 (*Athemus*) **A**: GUX
viridinitidus Pic, 1906o: 218 (*Cantharis*) **A**: YUN
atripes Pic, 1906o: 218 (*Cantharis*)
vitellinus Kiesenwetter, 1874b: 277 (*Cantharis*) **A**: JA SC
wulaianus Wittmer, 1983a: 163 (*Athemus*) **A**: TAI
walteri Švihla, 2004b: 183 (*Athemus*) [RN] **A**: ZHE
testaceipes Pic, 1937k: 171 (*Athemus*) [HN]
wangi Švihla, 2004b: 183 (*Athemus*) [RN] **A**: HUB (Wuling Mts)
maculithorax S. Wang & C. Yang, 1993: 264 (*Athemus*) [HN]
watanabei Ishida & Satô, 1993: 47 (*Athemellus*) **A**: JA (Honshu)
wittmeri Kazantsev, 1999: 126 **A**: "East Tibet"
wittmerorum Kazantsev, **nom. nov.** [see New acts] **A**: NP
nepalensis Wittmer, 1981e: 409 (*Podistra*) [HN]
wittmeri Švihla, 2004b: 190 (*Andrathemus*) [HN]
xanthogaster Champion, 1926b: 199 (*Athemus*) **A**: UP
yaeyamanus Okushima, 1996: 115 (*Athemellus*) **A**: JA (Ryukyus)
yato Takahashi, 1992: 107 (*Cantharis*) **A**: JA (Honshu)
yonaensis Satô & Okushima, 1992: 858 (*Athemus*) **A**: JA (Ryukyus)
yunnanus Fairmaire, 1886c: 338 (*Lipernes*) **A**: SCH YUN **ORR**
curticollis Pic, 1921e: 5
fumatus Champion, 1926c: 253
pubicollis Pic, 1915l: 78
pyrochroides Champion, 1926c: 255
tibetanus Pic, 1921e: 5
zdeneki Švihla, 2004b: 192 (*Athemus*) **A**: SCH

genus *Macrohabronychus* Wittmer, 1981d: 400 type species *Anolisus parallelus* Champion, 1926
subgenus *Habronychus* Wittmer, 1982d: 124 type species *Malthinus sauteri* Pic, 1934
aritai Satô, 1986c: 256 **A**: JA (Ryukyus)
biungulatus Champion, 1925d: 264 (*Anolisus*) **A**: UP
kurosawai Wittmer, 1982d: 126 (*Habronychus*) **A**: TAI
miyatakei Ishida, 1986: 218 (*Habronychus*) **A**: JA (Yaku-shima)

nantouanus Wittmer, 1982d: 125 (*Habronychus*) **A**: TAI
obscuricolorissimus Kazantsev, **nom. nov**. [see New acts] **A**: JA
obscuricolor Pic, 1935a: 12 (*Podabrinus*) [HN]
parallelicollis Pic, 1921e: 5 (*Lycocerus*) **A**: YUN
notatithorax Pic, 1947: 8 (*Lycocerus*)
providus Kiesenwetter, 1874b: 269 (*Cantharis*) **A**: FE JA
modesta Kiesenwetter, 1874b: 276 (*Cantharis*)
obscuricolor Pic, 1905e: 121 (*Absidia*)
sauteri Pic, 1934d: 46 (*Malthinus*) **A**: TAI
semimetallicus Wittmer, 1997e: 361 (*Habronychus*) **A**: SD
szechwanus Wittmer, 1988c: 366 (*Habronychus*) **A**: SCH
subgenus *Macrohabronychus* Wittmer, 1981d: 400 type species *Anolisus parallelus* Champion, 1926
biungulatus Champion, 1925d: 264 (*Anolisus*) **A**: BT NP
himalaicus Wittmer, 1965c: 85 (*Podabrus*)
chaoi Wittmer, 1997e: 363 **A**: YUN
harmandi Pic, 1924j: 481 (*Podabrus*) **A**: SD
hieroglyphicus Champion, 1926a: 124
inaequalis inaequalis Pic, 1924j: 480 (*Podabrus*) **A**: NP SD
inaequalis limbatus Wittmer, 1981d: 402 **A**: NP
multilaculatus lopchuensis Wittmer, 1981d: 403 **A**: NP SD
multimaculatus multimaculatus Pic, 1916d: 9 (*Cantharis*) **A**: UP
parallelus Champion, 1926a: 123 (*Anolisus*) **A**: SD
septemnotatus Pic, 1907p: 173 (*Anolisus*) **A**: BT NP SD
subgenus *Monohabronychus* Okushima & Satô, 1999b: 394 type species *Podabrinus intermixtus* Wittmer, 1954
furcatus Wittmer, 1986b: 131 (*Stenothemus*) **A**: TAI
intermixtus Wittmer, 1954d: 275 (*Podabrinus*) **A**: TAI
lalashanensis Okushima & Satô, 1999b: 400 (*Habronychus*) **A**: TAI
longicornis Okushima & Satô, 1999b: 398 (*Habronychus*) **A**: TAI
multilimbatus Pic, 1910x: 271 (*Podabrus*) **A**: FUJ TAI
humeralis Wittmer, 1951: 100 (*Podabrinus*)

genus *Malchinomorphus* Pic, 1922i: 31 type species *Malchinomorphus longiceps* Pic, 1922
longiceps Pic, 1922i: 31 **A**: CH

genus *Metacantharis* Bourgeois, 1886: 103 type species *Cantharis discoidea* Ahrens, 1812
araxicola Reitter, 1891b: 25 (*Cantharis*) **E**: AR **A**: IN TR
disparipennis Bourgeois, 1900a: 153
attalia Pic, 1902i: 55 (*Cantharis*) **A**: TD UZ
bucharica Pic, 1914a: 5 (*Cantharis*)
diversenotata Pic, 1913o: 187 (*Cantharis*)
korzhinskii Barovskij, 1926c: 243
senckenbergi Pic, 1913o: 187 (*Cantharis*)
balcanograeca Wittmer, 1969b: 76 **E**: AL GR YU
brachyelytrata brachyelytrata Švihla, 1992: 321 **A**: AF
brachyelytrata walteri Švihla, 1999: 152 **A**: TR
clypeata Illiger, 1798: 299 **E**: AL AR AU BU CR CT CZ FR GB GE GG GR HU IT LT MC PL RO SK SP ST SZ TR UK YU **N**: AG **A**: IN
haemorrhoidalis Fabricius, 1792a: 216 (*Cantharis*) [HN]
inlimbata Pic, 1902k: 63
meziendi Marseul, 1864: 61 (*Telephorus*)
nivea Panzer, 1798a: 5 (*Cantharis*)
rubens Marseul, 1864: 63
signata Faldermann, 1835a: 190 (*Cantharis*)
testacea Scopoli, 1763: 39 (*Cantharis*)
discoidea Ahrens, 1812: 12 (*Cantharis*) **E**: AL AU BH CR CZ FR GE GR HU IT PL RO SK SL SZ
carreti Bourgeois, 1900b: 181

deserta Dietrich, 1857: 131 (*Telephorus*)
flavicollis Gerhardt, 1909: 417
fontisbellaquei Pic, 1914m: 52
humeralis L. Redtenbacher, 1849: 323 (*Telephorus*)
indiscoidea Pic, 1914m: 52
lineata Bach, 1854: 68 (*Telephorus*)
liturata L. Redtenbacher, 1849: 323 (*Telephorus*)
notata Waltl, 1838b: 269 (*Cantharis*)
semidiscoidalis Pic, 1909q: 169
telonensis Pic, 1914r: 179
fulvoides Pic, 1904c: 9 (*Cantharis*) **A**: TR
kostali Wittmer, 1997e: 297
katarensis Dahlgren, 1985b: 163 (*Cantharis*) **E**: GR
keiseri Wittmer, 1969b: 75 **E**: GR
peloponnesica Wittmer, 1974a: 1 **E**: GR
picciolii Ragusa, 1870: 316 (*Telephorus*) **E**: IT
fraudulenta Fiori, 1914b: 81
picticollis Ragusa, 1894: 40
raptor Ballion, 1871: 351 (*Telephorus*) **A**: AF KZ TD TM UZ
samarkandensis Pic, 1913o: 185
submarginalis Ballion, 1871: 351 (*Telephorus*)
rosinae Pic, 1902c: 24 (*Cantharis*) **A**: TR
taurigrada Bourgeois, 1900a: 154 **A**: SY TR
angorensis Pic, 1903m: 146
obscuripes J. R. Sahlberg, 1913d: 140
torosensis Švihla, 1999: 152 **A**: TR
turcica Marseul, 1864 60 (*Telephorus*) **E**: GR
apicalis Reiche & Saulcy, 1857b: 176 (*Telephorus*)
wittmeri brachyelytrata Švihla, 1992: 325 **A**: AF
wittmeri wittmeri Švihla, 1992: 325 **A**: AF PA
zarudnyi Barovskij, 1928a: 98 **A**: IN

genus *Micropodabrus* Pic, 1920a: 6 type species *Micropodabrus obscurus* Pic, 1920
Cephalomalthinus Pic, 1921f: 5 type species *Cephalomalthinus ocularis* Pic, 1921
Fissocantharis Pic, 1921d: 27 type species *Fissocantharis opaca* Pic, 1921
Fissopodabrus Pic, 1927d: 2 type species *Fissopodabrus gracilipes* Pic, 1927
Harmonycha Wittmer, 1938b: 302 type species *Rhagonycha sulcicornis* Pic, 1939
Javaesilis Pic, 1955a: 7 type species *Javaesilis specialicornis* Pic, 1955
Kandyosilis Pic, 1929i: 70 type species *Kandyosilis bryanti* Pic, 1929
Stenopodabrus Nakane, 1992: 78 type species *Rhagonycha longipes* Wittmer, 1953
abdominalis Pic, 1906k: 81 (*Podabrus*) **A**: BT SD TR
ocularis Pic, 1921f: 5 (*Cephalomalthinus*)
ocularis Champion, 1926a: 124 (*Anolisus*)
ruficollis Pic, 1921d: 28 (*Podabrus*)
sikkimensis Pic, 1924j: 480 (*Podabrus*)
albonotatus Pic, 1924j: 479 (*Rhagonycha*) **A**: SD
albovarius Wittmer, 1983c: 240 **A**: SD
angustus Fairmaire, 1900: 624 (*Podabrus*) **A**: FUJ
arunensis Wittmer, 1983c: 253 **A**: NP
bicoloriceps Wittmer, 1995b: 117 **A**: TAI
bidifformis Wittmer, 1988c: 350 **A**: GUX
biimpressus Wittmer, 1982d: 138 (*Kandyosilis*) **A**: TAI
bothriderus Fairmaire, 1887a: 125 (*Rhagonycha*) **A**: FUJ
nigricoloriceps Pic, 1928g: 55
carinatus Wittmer, 1989: 225 (*Kandyosilis*) **A**: NP
chaoi Wittmer, 1988c: 362 **A**: YUN
chinensis Wittmer, 1993a: 219 (*Fissocantharis*) **A**: YUN

chujoi Wittmer, 1972b: 136 (*Kandyosilis*) **A**: TAI
cicatricosus Wittmer, 1988c: 360 **A**: FUJ GUX
crassicornis Pic, 1922c: 4 (*Podabrus*) **A**: CH
curvicornis Wittmer, 1997e: 311 [RN] **A**: TAI
    uenoi Wittmer, 1995f: 407 (*Kandyosilis*) [HN]
denominatus Wittmer, 1997e: 310 [RN] **A**: TAI
    formosanus Wittmer, 1954d: 274 (*Podabrus*) [HN]
    taiwanus Wittmer, 1982d: 130 [HN]
diffusus Wittmer, 1983c: 252 **A**: SD **ORR**
dromedarius Champion, 1926a: 120 (*Podabrus*) **A**: NP "Tibet"
eburneus Bourgeois, 1902: 426 (*Podabrus*) **A**: SD
fenchihuensis Wittmer, 1982d: 131 **A**: TAI
fissus Wittmer, 1997e: 313 **A**: SCH SHA
flavicornis Gorham, 1889: 108 (*Telephorus*) **A**: FUJ GUX
flavimembris Wittmer, 1951: 96 (*Podabrus*) **A**: FUJ
flavofacialis Pic, 1926h: 29 (*Podabrus*) **A**: FUJ
    denticornis Wittmer, 1951: 96 (*Podabrus*)
formosanus Pic, 1910x: 271 (*Rhagonycha*) **A**: TAI
    curvatipes Pic, 1922c: 4 (*Podabrus*)
    obscurioripes Pic, 1926p: 235 (*Rhagonycha*)
fumidiformis Wittmer, 1983c: 243 **A**: NP SD
fumidus Champion, 1926a: 125 (*Anolisus*) **A**: NP UP "Tibet"
fuscolineatus Champion, 1926a: 121 (*Podabrus*) **A**: NP UP
gaoligongensis Wittmer, 1997e: 315 **A**: YUN
grahami Wittmer, 1997e: 316 **A**: SCH YUN
gressitti Wittmer, 1954d: 272 (*Podabrus*) **A**: TAI
gyalpo Wittmer, 1983c: 241 **A**: BT
hualiensis Wittmer, 1995a: 112 **A**: TAI
hupeanus Wittmer, 1956b: 304 (*Kandyosilis*) **A**: HUB
incrassatus Wittmer, 1988c: 354 **A**: YUN
kurosawai Wittmer, 1982d: 132 **A**: TAI
lesnei Pic, 1924j: 479 (*Cantharis*) **A**: SD
lineolatus Pic, 1924j: 481 (*Podabrus*) **A**: NP SD "Tibet"
    annulicornis Champion, 1926a: 121 (*Podabrus*)
lishanensis Wittmer, 1982d: 131 **A**: TAI
liuchowensis Wittmer, 1989: 212 **A**: GUX
longiceps Pic, 1908m: 95 (*Rhagonycha*) **A**: YUN
longipes Wittmer, 1953a: 36 (*Rhagonycha*) **A**: JA (Honshu, Shikoku)
maculivertex Ishida, 1989: 82 **A**: JA (Ryukyus)
marmoratus Bourgeois, 1902: 426 (*Podabrus*) **A**: SD
minor Wittmer, 1997e: 318 **A**: TAI
minutissimus Pic, 1933i: 117 (*Kandyosilis*) **A**: GAN
minutulus Wittmer, 1995b: 142 (*Kandyosilis*) **A**: YUN
mucronatus Wittmer, 1979e: 333 (*Kandyosilis*) **A**: JA (Ryukyus) TAI
multicostatus Wittmer, 1982d: 136 (*Kandyosilis*) **A**: TAI
multiexcavatus Wittmer, 1988c: 361 **A**: GUX **ORR**
nantouensis Wittmer, 1995b: 113 **A**: TAI
nodicornis Wittmer, 1982d: 137 (*Kandyosilis*) **A**: TAI
notatithorax Pic, 1922c: 4 (*Podabrus*) **A**: YUN
    diversipennis Pic, 1922c: 4 (*Podabrus*)
    particularicornis Pic, 1926h: 29 (*Podabrus*)
novemexcavatus Wittmer, 1951: 97 (*Podabrus*) **A**: FUJ
    testaceicollis Wittmer, 1951: 97
pallidiceps Pic, 1911y: 271 (*Rhagonycha*) **A**: ZHE
paralleloides Wittmer, 1997e: 312 [RN] **A**: SCH

parallelus Wittmer, 1993a: 211 [HN]
parallelus Wittmer, 1983c: 243 **A**: BT
pauloincrassatus Wittmer, 1951: 97 (*Podabrus*) **A**: FUJ
pieli Pic, 1937k: 172 (*Lycocerus*) **A**: FUJ TAI
piluchiensis Wittmer, 1995b: 114 **A**: TAI
pingtungensis Wittmer, 1982d: 129 **A**: TAI
pseudolongiceps Wittmer, 1988c: 363 **A**: YUN
pseudonotatithorax Wittmer, 1988c: 363 **A**: YUN
purpurascens Hope, 1831: 26 (*Telephorus*) **A**: HP NP SD UP
rollei Pic, 1907p: 172 (*Rhagonycha*)
rufolineatus Wittmer, 1983c: 254 **A**: SD
safraneki Švihla, 2004b: 162 **A**: SHA
satoi Wittmer, 1982d: 133 **A**: TAI
semifumatus Fairmaire, 1889b: 39 (*Podabrus*) **A**: GAN SCH
bicoloricornis Pic, 1926r: 356 (*Podabrus*)
hummeli Pic, 1933b: 4 (*Rhagonycha*)
nigrosubapicalis Pic, 1926c: 5 (*Rhagonycha*)
shaanxiensis Wittmer, 1995b: 121 **A**: SHA
shimomurai Wittmer, 1995f: 410 (*Kandyosilis*) **A**: NP
shinborii Takahashi, 1992: 104 **A**: JA (Honshu)
sikkimensis Wittmer, 1989: 225 (*Kandyosilis*) **A**: SD
similis Wittmer, 1951: 97 (*Podabrus*) **A**: FUJ
simplicicornis Wittmer, 1993a: 209 **A**: TAI
sinensis Wittmer, 1988c: 353 **A**: GUX
specialithorax Pic, 1916e: 12 (*Cantharis*) **A**: SD
ssulingensis Wittmer, 1995b: 116 **A**: TAI
sublineolatus Wittmer, 1983c: 244 **A**: BT
subniger Wittmer, 1972b: 132 (*Bisadia*) **A**: TAI
tachulanensis Wittmer, 1988c: 358 **A**: FUJ
taipeianus Wittmer, 1982d: 134 **A**: TAI
taiwanus Wittmer, 1979e: 332 (*Kandyosilis*) **A**: TAI
taoyuanus ilanensis Wittmer, 1995b: 117 **A**: TAI
taoyuanus taoyuanus Wittmer, 1995b: 116 **A**: TAI
tenuicollis Champion, 1926a: 122 (*Podabrus*) **A**: BT NP SD
tigerhillensis Wittmer, 1995f: 410 (*Kandyosilis*) **A**: SD
tridifformis Wittmer, 1988c: 349 **A**: SCH
uenoi Wittmer, 1982d: 130 **A**: TAI
uninodulatus Wittmer, 1997a: 34 **A**: GUX
viaticus Lewis, 1895a: 111 (*Telephorus*) **A**: JA (Honshu)
particularicornis Pic, 1938k: 152 (*Podabrus*)
yayeyamanus Satô, 1986: 256 **A**: JA (Ryukyus)
yunnanus Wittmer, 1988c: 357 **A**: YUN

genus *Mimopodabrus* Wittmer, 1997e: 320 type species *Micropodabrus obscurior* Wittmer, 1954
jendeki Wittmer, 1997e: 323 **A**: GUI
obscurior Wittmer, 1954d: 272 (*Podabrus*) **A**: TAI
oudai Švihla, 2004b: 166 **A**: SCH
satoi Wittmer, 1997a: 34 **A**: GUX
singularis Wittmer, 1997e: 323 **A**: SCH
yunnanus Wittmer, 1993a: 214 (*Kandyosilis*) **A**: YUN

genus *Occathemus* Švihla, 1999: 168 type species *Cantharis tarsalis* Mulsant, 1862
tarsalis Mulsant, 1862c: 79 (*Cantharis*) [RN] **A**: CY IN SY TR
disjuncta Barovskij, 1926c: 241 (*Cantharis*)
nigritarsis Mulsant & Wachanru, 1852b: 163 (*Telephorus*) [HN]

oculatifer Pic, 1900j: 86 (*Cantharis*)
pulverulenta Barovskij, 1926c: 241 (*Cantharis*)

genus *Pacificanthia* Kazantsev, 2001c: 43 type species *Telephorus consors* LeConte, 1881
knirschi Pic, 1929b: 5 (*Cantharis*) **A**: FE

genus *Pakabsidia* Wittmer, 1972b: 124 type species *Pakabsidia swatensis* Wittmer, 1972
brevipennis Švihla, 2004b: 171 **A**: NP
carinthiaca Wittmer, 1973a: 118 (*Pseudoabsidia*) **E**: AU
heinzi Wittmer, 1978e: 349 **A**: TR
himachalica Wittmer, 1997e: 307 **A**: HP
kuluana Wittmer, 1979c: 134 **A**: HP
ladakhensis Wittmer, 1997e: 307 **A**: KA (Ladakh)
lateriemarginata Wittmer, 1997e: 309 **A**: KA (Ladakh)
mixtus Wittmer, 1993a: 253 (*Habronychus*) **A**: NP
olexai Švihla, 1983: 3 **E**: GG
quadricollis Wittmer, 1993a: 251 (*Habronychus*) **A**: NP
semiopaca Pic, 1909b: 37 (*Cantharis*) **A**: AF HP PA **ORR**
pooanus Pic, 1921e: 4 (*Cantharis*)
testaceitincta Wittmer, 1979c: 134
seximpressus Pic, 1921d: 29 (*Cantharis*) **A**: NP SD
sikkimensis Wittmer, 1997e: 360 (*Habronychus*) **A**: SD
sintanana Pic, 1938i: 84 (*Cantharis*) [RN] **A**: KA PA
kaganensis Wittmer, 1981e: 407
kaschmirensis Pic, 1916e: 12 (*Cantharis*) [HN]
swatensis Wittmer, 1972b: 135 **A**: PA
thomasi Wittmer, 1981e: 408 **A**: PA

genus *Podabrinus* Fairmaire, 1896a: 92 type species *Podabrinus oedemeroides* Fairmaire, 1896
oedemeroides Fairmaire, 1896a: 92 **A**: SD

genus *Podistra* Motschulsky, 1839: 78 type species *Podistra alpina* Motschulsky, 1839
subgenus *Absidia* Mulsant, 1862a: 277 type species *Cantharis pilosa* Paykull, 1798 (= *Podistra schoenherri* Dejean, 1836)
pentheri Ganglbauer, 1922: 70 **E**: BH CR YU
rufotestacea Letzner, 1845: 72 (*Cantharis*) **E**: AU BE CR CT CZ FI FR GE HU IT NT PL RO SK SL SV SZ UK **A**: ES WS
angularis J. R. Sahlberg, 1871: 363 (*Cantharis*)
discreta Bourgeois, 1887: 134
schoenherri Dejean, 1836: 118 (*Cantharis*) [RN] **E**: AU BE BY CT CZ EN FI FR HU GE IT LS LT NR NT RO SK ST SV SZ UK **A**: ES FE WS
pilosa Paykull, 1798: 264 (*Cantharis*) [HN]
subgenus *Hemipodistra* Ganglbauer, 1922: 70 type species *Hemipodistra rupicola* Kiesenwetter, 1863
birnbacheri Krauss, 1894: 180 **E**: AU IT
saxicola Krauss, 1894: 181
motschulskyi Kazantsev, 2001d: 1079 **E**: ST
rupicola Kiesenwetter, 1863a: 729 **E**: AU IT RO SL
savitskiorum Kazantsev, 2001d: 1080 **E**: ST
subgenus *Podistra* Motschulsky, 1839: 78 type species *Podistra alpina* Motschulsky, 1839
alpina Motschulsky, 1839: 79 **E**: ST
circassicola Reitter, 1889e: 100 **E**: ST
piceolineata Pic, 1922e: 18 **E**: ST
rufomaculata Kazantsev, 1998c: 692 **E**: ST
starcki Ganglbauer, 1922: 72 **E**: ST
subgenus *Pseudoabsidia* Wittmer, 1969c: 128 type species *Telephorus prolixus* Märkel, 1852

arnoldii Kazantsev, 1998c: 692 **E**: ST
caucasica Ganglbauer, 1922: 70 **E**: ST
chekanovskii Kazantsev, 1994: 16 **A**: ES
cherkessica Kazantsev, 1998c: 693 **E**: ST
cicatriculata Wittmer, 1973a: 119 **E**: AB or AR "Araxestal"
ossetica Kazantsev, 1998c: 693 **E**: ST
prolixa Märkel, 1852: 598 (*Telephorus*) **E**: AU CZ FR IT SK SL SZ
sulcifrons Märkel, 1852: 599 (*Cantharis*)
unicolor Faldermann, 1835aa: 195 (*Cantharis*) **E**: ST
ussurica Wittmer, 1979c: 138 **A**: FE WS "Korea"
wittmeri Kazantsev, 1992a: 44 **E**: GG

genus *Podistra*, species incertae sedis
beauprei Pic, 1906k: 81 [RN] **N**: AG
latiuscula Bourgeois, 1906: 97 [HN]
japonica Kiesenwetter, 1874b: 279 **A**: FE JA
longicornis Witmer, 1978b: 348 **A**: KA
nonveilleri G. Müller, 1946: 113 **E**: YU
vaulogeri Bourgeois, 1905c: 106 **N**: AG

genus *Prothemellus* Švihla, 1992: 287 type species *Absidia afghana* Wittmer, 1956
afghanus Wittmer, 1956a: 157 (*Absidia*) **A**: AF PA
kunarensis Wittmer, 1997e: 301 **A**: PA

genus *Prothemus* Champion, 1926b: 195 type species *Prothemus neglectus* Champion, 1926
benesi Švihla, 2004b: 208 **A**: SCH
bimaculaticollis Pic, 1937k: 171 (*Cantharis*) **A**: CH
chaoi Wittmer, 1987a: 84 **A**: SCH
chinensis Wittmer, 1987a: 75 **A**: SHA TAI
ciusianus Kiesenwetter, 1874b: 267 (*Cantharis*) **A**: JA (Honshu, Shikoku, Kyushu)
emeiensis Wittmer, 1993a: 215 **A**: SCH
grouvellei Pic, 1906k: 83 (*Cantharis*) **A**: YUN
hisamatsui Wittmer, 1984c: 144 **A**: TAI
impressiventris Fairmaire, 1891f: ccix (*Telephorus*) **A**: HUB (Chang-yang)
kanoi Wittmer, 1984c: 146 **A**: JA (Ryukyus) TAI
kiukianganus Gorham, 1889: 107 (*Telephorus*) **A**: FUJ
limbolarius Fairmaire, 1900: 628 (*Telephorus*) **A**: CH TAI
fainanensis Pic, 1916f: 4 (*Cantharis*)
maculithorax Wittmer, 1997e: 302 **A**: HUB
minor Wittmer, 1984c: 145 **A**: FUJ TAI
monochrous Fairmaire, 1900: 627 (*Telephorus*) **A**: TAI
mupinensis Wittmer, 1987a: 78 **A**: SCH
neglectus Champion, 1926b: 195 **A**: NP SD
oberthuri Pic, 1955b: 18 (*Cantharis*)
neimongolanus S. Wang & C. Yang, 1990: 240 **A**: NMO
nigrihumeralis Okushima & Satô, 1997b: 335 **A**: JA (Ryukyus)
notsui Wittmer, 1984c: 144 **A**: TAI
opacipennis Pic, 1906k: 84 (*Cantharis*) **A**: YUN
piluensis Wittmer, 1987a: 80 **A**: TAI
purpureipennis Gorham, 1889: 107 (*Telephorus*) **A**: FUJ SHA
reinii Kiesenwetter, 1879b: 307 (*Cantharis*) **A**: JA (Honshu)
enokidoi Takahashi, 1992: 108
ryukyuanus Wittmer, 1975b: 260 **A**: JA (Ryukyus)
sanguinosus Fairmaire, 1900: 629 (*Telephorus*) **A**: CH
angustioripennis Pic, 1927l: 132 (*Cantharis*)

semimetallicus Wittmer, 1993a: 215 **A**: SCH
similithorax Pic, 1922b: 31 (*Cantharis*) **A**: YUN
subobscurus Pic, 1906k: 84 (*Cantharis*) **A**: YUN
fouana Pic, 1930i: 78 (*Cantharis*)
szechwanus Wittmer, 1987a: 74 **A**: SCH
varicolor Wittmer, 1954d: 280 **A**: TAI
venustus Wittmer, 1997a: 38 **A**: GUX
vrianatangus Pic, 1921d: 28 (*Cantharis*) [RN] **A**: "Tibet"
thibetanus Pic, 1916e: 12 (*Cantharis*) [HN]
vuilleti Pic, 1914b: 8 (*Cantharis*) **A**: YUN **ORR**
minutenotata Pic, 1914b: 8 (*Cantharis*)
rubronotaticeps Pic, 1914b: 8 (*Cantharis*)
watanabei Okjushima & Satô, 2002: 348 **A**: TAI
yunnanus Wittmer, 1987a: 77 **A**: YUN

***genus Pseudopodabrus* Pic, 1906k: 81** type species *Pseudopodabrus impressiceps* Pic, 1906
atriceps Pic, 1922c: 4 (*Podabrus*) **A**: CH
kubani Wittmer, 1995b: 124 **A**: YUN
latoimpressus Wittmer, 1995b: 127 **A**: YUN
semicircularis Wittmer, 1995b: 125 **A**: YUN

genus *Rhagonycha* Eschscholtz, 1830b: 64 type species *Cantharis fulva* Scopoli, 1763
subgenus *Rhagonycha* Eschscholtz, 1830b: 64 type species *Cantharis fulva* Scopoli, 1763
Nastonycha Motschulsky, 1853c: 77 type species *Nastonycha brachyptera* Motschulsky, 1853
Pseudocratosilis Moscardini & Sassi, 1970: 192 type species *Pygidia graeca* Pic, 1901 (= *Rhagonycha corcyrea* Pic, 1901)
abchasica Pic, 1900n: 183 **E**: AR GG ST
adanensis Dahlgren, 1968: 116 **A**: TR
addenda Dahlgren, 1972: 136 **E**: SP
aetolica Kiesenwetter, 1859a: 25 (*Telephorus*) **E**: AL GR
subaptera Pic, 1909u: 185
aibgaena Kazantsev, 1992a: 47 **E**: ST
alagoesa Reitter, 1893a: 111 (*Nastonycha*) **E**: AR ST
albolimbata Pic, 1926c: 5 **A**: SCH
aliena Dahlgren, 1972: 131 **E**: GR (Rodos, Samos) **N**: AG **A**: CY IQ LE SY TR
chevrolati Dahlgren, 1968: 99 [HN]
alpicola Barovskij, 1928a: 98 **A**: ES KI KZ MG UZ
kiritshenkoi Barovskij, 1929b: 267
shavrovi Barovskij, 1928a: 99
amaguensis Švihla, 1995: 74 **A**: ES
anapensis Dahlgren, 1985a: 88 **E**: ST
anatolica Dahlgren, 1968: 110 **A**: TR
andalusica Dahlgren, 1975: 104 **E**: SP
angulatocollis A. Costa, 1858: 25 **E**: IT
angulosa Kazantsev, 1994b: 77 **A**: ES
angusta Marseul, 1864: 74 (*Armidia*) **A**: TR
longula Dahlgren, 1968: 122 [HN]
wittmeri Dahlgren, 1968: 167 [RN]
antennata Švihla, 2004a: 78 **A**: TR
approximana Fairmaire, 1884a: 169 (*Telephorus*) **E**: TR **A**: LE SY
brumanensis Pic, 1901g: 57
cryptica Dahlgren, 1968: 96
arakawadakensis Satô, 1976: 55 **A**: JA (Honshu)
arrounoana Pic, 1932g: 26 [= 1932c: 7] **N**: MO
asiatica Wittmer, 1971c: 195 **A**: FE "Korea"
atra Linnaeus, 1767: 649 (*Cantharis*) **E**: AU BE BY CT CZ EN FI FR GE HU IT LT NT PL RO SK ST SZ UK **A**: NC WS

fauconetti Pic, 1914m: 52
glacialis Bourgeois, 1909: 391
plumbea P. Rossi, 1792: 56
rhaetica Stierlin, 1862: 37
sabauda Pic, 1914m: 52
subrobusta Pic, 1923a: 1
atriceps Pic, 1922e: 18 **E**: ST
atrovaria atrovaria Wittmer, 1971c: 196 **A**: ES FE KZ MG
atrovaria iremelica Kazantsev, 1994b: 95 **A**: WS
balcanica Pic, 1901g: 57 **E**: BU **A**: TR
helleni Dahlgren, 1968: 119
osmana Wittmer, 1972a: 76 (*Cratosilis*)
vicina Dahlgren, 1968: 118 [HN]
barbara Fabricius, 1801a: 299 (*Cantharis*) **N**: AG MO TU
bettingeri Pic, 1927a: 1
sulcata Mulsant, 1862b: 343
basarukini Kazantsev, 1994b: 94 **A**: FE JA (Hokkaido)
benedikti Švihla, 2002b: 308 **A**: SY
berbera Švihla, 1995: 85 [RN] **N**: MO
marocana Švihla, 1990a: 196 [HN]
bernhaueri Wittmer, 1972a: 75 **A**: TR
bicallosithorax Pic, 1938k: 161 **A**: CH ("Tchen Ly")
bicolor Takahashi & Imasaka, 1997: 80 **A**: JA (Shimokoshiki-jima)
bilimbata Wittmer, 1954c: 110 **A**: FUJ
bohaci Švihla, 1990a: 197 **E**: ST
brachyptera Motschulsky, 1853c: 77 (*Nastonycha*) **E**: ST
brancuccii Švihla, 1993: 274 **A**: TR
bythinica Marseul, 1864: 95 **A**: TR
carolusi Švihla, 1993: 270 **A**: TR
carpathica Ganglbauer, 1896: 183 **E**: SK UK
caspica Wittmer, 1972a: 65 **E**: GG ST **A**: IN
tbilisiana Švihla, 1977: 178
catei Švihla, 1993: 271 **A**: TR
caucasica Wittmer, 1972a: 65 **E**: AR GG ST
cembricola Eschscholtz, 1822: 62 (*Cantharis*) **A**: ES FE MG SC
sjoestedti Pic, 1926b: 2
chevrolati Marseul, 1864: 80 **E**: GG ST **A**: CY IS JO LE SY TR **N**: AG
apicalis Chevrolat, 1854: 434 (*Telephorus*)
orientalis Dahlgren, 1968: 100
chirorodakensis Kazantsev & Takahashi, 2001: 275 **A**: JA (Hokkaido)
chlorotica Gené, 1839: 16 (*Cantharis*) **E**: FR (Corse) IT (Sardegna, Sicilia)
corsica Reiche, 1861b: 208
circassicola Reitter, 1888f: 207 **E**: ST
compacta Wittmer, 1972a: 75 **E**: GG
complicans Dahlgren, 1979: 62 **E**: YU
confusa Dahlgren, 1975: 103 **E**: FR SP
consociata Heyden, 1883b: 311 **A**: TR
corcyrea Pic, 1901f: 50 **E**: AL BU GR MC YU
graeca Pic, 1901h: 66 (*Pygidia*)
coreana Pic, 1921d: 28 **A**: FE JA (Hokkaido) SC
croceipennis Fairmaire, 1884e: cxix **N**: MO
atlasica Pic, 1925f: 6
cruentata Reiche, 1862: 79 (*Telephorus*) **E**: GR **A**: TR
hoemorroidalis Reiche & Saulcy, 1857b: 173 [HN]
dahlgreni Wittmer, 1972a: 63 **A**: TR
decorata Pic, 1912l: 57 **E**: GR YU

delagrangei Pic, 1898h: 139 (*Armidia*) **A**: IQ SY TR
drurei Pic, 1905i: 161
nigrocapitata Bourgeois, 1900a: 155
nigroterminata Reitter, 1898h: 346 (*Armidia*)
diversipennis Pic, 1921d: 27 **A**: YUN
diversipes Pic, 1905n: 186 **E**: GR RO
deubeli Petri, 1912: 139
divisa Dahlgren, 1972: 144 **E**: SP
dolini Švihla, 1995: 78 **A**: ES FE
drienensis Dahlgren, 1978: 13 **E**: BH
duplicata Dahlgren, 1968: 115 **A**: TR
dvoraki Švihla, 1995: 73 **A**: TR
elbursiaca Wittmer, 1972c: 195 **A**: IN
elongata Fallén, 1807: 11 (*Cantharis*) **E**: AL AU BE BH BU BY CT CZ EN FR GB GE HU LT LU NR NT PL RO SK SV SZ UK YU **A**: ES WS
atra Paykull, 1798: 266 [HN]
paludosa L. Redtenbacher, 1849: 325 [HN]
elongatipes Wittmer, 1972a: 71 **A**: TR
erevanensis Dahlgren, 1985a: 89 **E**: AR
erimoensis Kazantsev & Takahashi, 2001: 272 **A**: JA (Hokkaido)
esfandiarii Wittmer, 1972c: 197 **A**: IN
falcifera Dahlgren, 1972: 140 **E**: SP
femoralis Brullé, 1832: 148 (*Telephorus*) **E**: AL BH BU GR HU RO YU
albanica Pic, 1927q: 372
flava Pic, 1926r: 356 (*Podabrus*) **A**: SCH
flavicornis Kazantsev, 1992a: 46 **E**: GG
flavotibialis Medvedev & Ryvkin, 1989: 138 **A**: FE MG
fonticola Kazantsev, 1994b: 83 **A**: FE
fossulata P. H. Lucas, 1846: 187 (*Telephorus*) **N**: AG
scutellaris P. H. Lucas, 1846: 186
fugax fugax Mannerheim, 1843c: 89 **E**: AU BY CZ FI IT LA LT NT PL RO SK SZ
infuscata J. R. Sahlberg, 1871: 364
scopolii Gredler, 1866: 245
fugax strejceki Švihla, 1993: 268 **E**: ST
fulva Scopoli, 1763: 39 (*Cantharis*) **E**: AB AL AU BE BH BU BY CR CT CZ DE EN FR GB GE GG GR HU IR IT LA LS LT LU MC NL PL PT RO SK SP ST SV SZ TR UK **N**: MO **A**: IN SY TM
bimaculata De Geer, 1774: 71 (*Telephorus*)
cailloli Chobaut, 1914: 31
curtithorax Pic, 1920f: 17
delahoni Schilski, 1908: 602
inapicalis Fiori, 1914b: 87
maculata Fourcrois, 1785: 60 (*Cicindela*)
melanura A. G. Olivier, 1790a: no 26: 8 (*Telephorus*)
terminalis L. Redtenbacher, 1849: 324
usta Gemminger, 1870: 121 (*Telephorus*)
fulvaliena Švihla, 1995: 85 **E**: GR **A**: TR
fulvonigra Dahlgren, 1972: 148 **N**: MO
furcata Wittmer, 1997e: 327 **A**: YUN
furcatiformis Wittmer, 1997e: 328 **A**: SCH
fuscitibia Rey, 1891e: 115 **E**: AL AU BH CR FR GR HU IT PT RO SK SZ UK
improvisa Dahlgren, 1976: 88
maculata Schilsky, 1890: 180 [HN]
maritima Pic, 1902h: 53
gallica Pic, 1923a: 1 **E**: AU BE BHCR CZ FR GE HU IT LU NL PL RO SK SZ UK YU **A**: TR
redtenbacheri Kaszab, 1955: 295
gansuensis Švihla, 2002b: 310 **A**: GAN SCH SHA
geniculata Gebler, 1832: 46 **A**: ES FE JA (Hokkaido) MG

amurensis Pic, 1902d: 25
genistae Kiesenwetter, 1866a: 392 **E**: PT SP
georgiana Kazantsev, 1992a: 49 **E**: GG
ghilanensis Dahlgren, 1985a: 89 **E**: AR
gilleforsi Švihla, 1993: 264 **A**: TR
gilvipennis Rosenhauer, 1856: 141 (*Podabrus*) **E**: SP
gressitti Wittmer, 1956b: 304 **A**: FUJ
gruziana Wittmer, 1972a: 69 **E**: GG **A**: TR
hagai Kazantsev & Takahashi, 2001: 270 **A**: JA (Hokkaido)
herbea Marseul, 1864: 82 **E**: FR **N**: AG TU **A**: SY
pallidithorax Pic, 1910j: 37
hesperica Baudi di Selve, 1859a: 296 **E**: PT SP
georgi Pic, 1909u: 185
inapicalis Pic, 1903i: 122
kochi Pic, 1935d: 11
lanjaroni Pic, 1952b: 9
lineatipennis Pic, 1908l: 89
obscurimembris Pic, 1952b: 9
oliveti Kiesenwetter, 1866b: 251
semilimbipennis Pic, 1917h: 17
spinifera Pic, 1903i: 122
hetitica Wittmer, 1972a: 68 **A**: TR
hispanica Pic, 1932g: 26 [= 1932c: 8] **E**: SP
hokkaidoensis Kazantsev & Takahashi, 2001: 272 **A**: JA (Hokkaido)
holtzi Pic, 1905n: 186 **E**: GR
moreana Pic, 1909g: 185
holzschuhi Wittmer, 1972a: 66 **A**: TR
hubeiana Wittmer, 1997e: 326 **A**: HUB
hyperborea Kazantsev, 1994b: 84 **A**: FE
hyrcana Švihla, 2002b: 310 **A**: IN
iberica Dahlgren, 1975: 103 **E**: PT SP
icelica Švihla, 1995: 71 **A**: TR
indistincta Medvedev & Ryvkin, 1989: 140 **A**: ES FE MG "Korea"
intermedia Wittmer, 1972a: 64 **A**: TR
interposita Dahlgren, 1978: 12 **E**: AU CR PL SL YU
iranica Wittmer, 1972c: 198 **A**: IN
itoi Kazantsev & Takahashi, 2001: 273 **A**: JA (Hokkaido)
jindrai Švihla, 2002b: 311 **A**: SHA
kabakovi Kazantsev, 1994b: 81 **A**: FE
kadleci Švihla, 2002b: 313 **A**: TR
kanwonensis Švihla, 1995: 77 **A**: "Korea"
karsensis Švihla, 1995: 81 **A**: IN TR
kefallinica Dahlgren, 1975: 109 **E**: GR (Kefallonia)
kiesenwetteri kiesenwetteri Marseul, 1864: 90 (*Armidia*) **E**: GR (Samos) **A**: CY TR
bodoi Pic, 1911t: 167
deyrollei Pic, 1902g: 48
kuleghana Marseul, 1868a: 117 (*Armidia*)
kiesenwetteri piciana Dahlgren, 1975: 105 **A**: IS JO LE SY TR
kobiensis alexeevi Kazantsev, 1992a: 50 **E**: ST
kobiensis kobiensis Dahlgren, 1975: 111 **E**: ST
koreaensis Kang & Kim, 2000a: 160 **A**: SC
kronbladi Švihla, 1993: 267 **A**: TR
kubanensis Pic, 1900n: 184 **E**: ST
kurbatovi Kazantsev, 1994b: 73 **A**: FE
kurdistana Švihla, 1983: 4 **A**: IN IQ
kurilica Wittmer, 1971c: 197 **A**: FE GAN JA (Hokkaido)

latipennis Pic, 1898h: 138 **A**: SY
latiuscula J. R. Sahlberg, 1887a: 32 **A**: ES FE JA MG WS
caroli Pic, 1905e: 121
distinctipes Pic, 1905e: 121
lederi Pic, 1909s: 177 **A**: ES FE "Korea"
lencoranica Kazantsev, 1992a: 50 **E**: AB
licenti Pic, 1938k: 161 **A**: GAN MNO
latiorithorax Pic, 1938k: 161
unicolorithorax Pic, 1938k: 161
lignosa O.F. Müller, 1764: 16 (*Cantharis*) **E**: AL AU BE BY BH BU CR CT CZ EN FR GB GE GR HU IR IT LT LU NT PL RO SK SL SZ UK YU **A**: TR
berolinensis Gmelin, 1790: 1896 (*Cantharis*)
melanocephala Herbst, 1784b: 108 (*Cantharis*)
pallida Fabricius, 1787a: 167 (*Cantharis*)
pallipes Fabricius, 1781: 259 (*Cantharis*)
subabbreviata Rey, 1891e: 115
subnigrofemoralis Pic, 1914m: 52
limbatella Wittmer, 1997e: 334 **A**: SCH
lundbergi Švihla, 1993: 258 **A**: TR
luristana Švihla, 1995: 81 **A**: IN TR
lutea O.F. Müller, 1764: 16 (*Cantharis*) **E**: AL AU BE BH BU CR CZ DE GB GE GR FR HU IR IT LU NR PL RO SK SL SP SV SZ TR UK YU
apicalis Curtis, 1840: 279 (*Telephorus*)
banatica Rosenhauer, 1847: 17 (*Podabrus*)
fuscicornis A. G. Olivier, 1790a: no 26: 11 (*Telephorus*)
maerkeli Kiesenwetter, 1852: 606
melanocephala Creutzer, 1797: 12 (*Cantharis*)
nubila Baudi di Selve, 1872: 111
lyncea Bourgeois, 1900a: 155 (*Armidia*) **A**: TR
bodemeyeri Dahlgren, 1968: 115
macedonica Dahlgren, 1985a: 85 **E**: YU
maceki Švihla, 1983: 5 **A**: IQ
machulkai Švihla, 1993: 260 **E**: BU
maculicollis Märkel, 1852: 607 **E**: AU BU CZ FR GR IT MA PL RO SK SZ UK YU
meisteri Gredler, 1857: 24 (*Cantharis*)
malthinoides Pic, 1921e: 4 **A**: "India"
mandchurica Pic, 1911o: 137 **A**: NE
mandibularis siberiana Kazantsev, 1994b: 93 **A**: ES FE
marginithorax Wittmer, 1972a: 74 **A**: TR
martini Pic, 1908l: 89 **E**: PT SP
matisi Kazantsev, 1994b: 85 **A**: FE
melanodera Eschscholtz, 1818: 462 (*Cantharis*) **E**: ST
meridionalis Dahlgren, 1975: 103 **E**: FR
milleri Kiesenwetter, 1860: 516 **E**: AU BU IT MC YU
bravardi Pic, 1917h: 17
netolitzkyi Penecke, 1912: 240
mimica Medvedev & Ryvkin, 1989: 140 **A**: FE
minima Pic, 1921e: 4 **A**: "India"
mlikovskyi Švihla, 1995: 74 **A**: ES "Korea"
mongolica Wittmer, 1971c: 197 **A**: MG
montana Biggs, 1947: 9 **N**: MO
monticola Dahlgren, 1985b: 163 **E**: GR
morio Kiesenwetter, 1852: 609 **E**: CZ FR RO SK SP UK YU
doctoris Pic, 1909s: 177
morvani Wittmer, 1974a: 2 **A**: IN
neglecta Dahlgren, 1975: 105 **E**: IT

nevadensis Švihla, 1995: 82 **E**: SP
nigriceps Waltl, 1838b: 269 (*Cantharis*) **E**: AU CR CZ FR GE HU IT PL RO SK SL SP SZ UK **A**: TR
atricapilla Kiesenwetter, 1850: 224
boops Kiesenwetter, 1852: 602
cantabrica Heyden, 1880b: 298
forticornis Pic, 1905k: 177
nigricornis Petri, 1912: 139
styriaca Baudi di Selve, 1872: 111
nigricollis Motschulsky, 1849: 83 **E**: IT SP
guadarramensis Graells, 1858: 49 (*Telephorus*)
semiflava Küster, 1854: 69
nigripes W. Redtenbacher, 1842: 13 (*Cantharis*) **E**: AU BU CT CZ FR GE HU IT LT NT PL RO SK SP SZ UK YU
barbara Bach, 1854: 71
herzegovina Pic, 1911n: 129
melanocera Schummel, 1844: 193
saturipennis Pic, 1909u: 185
nigritarsis Brullé, 1832: 145 **E**: GR IT (Sicilia) MA **A**: IN IS SY
fuscipes P. H. Lucas, 1853: 569 (*Telephorus*)
sulcicollis Brullé, 1832: 146 (*Telephorus*)
tibialis Brullé, 1832: 145 (*Telephorus*)
nigriventris Motschulsky, 1860b: 117 **E**: AU BE BH BY CR CT CZ EN FE FI FR GB GE HU KZ LA LT LU NR NT RO SK SL ST SV SZ UK YU **A**: ES FE KZ MG WS
femorata Rey, 1891e: 115
innotatithorax Pic, 1902h: 54
limbata C. G. Thomson, 1864: 191
nigrofemorata Schilsky, 1890: 178
signicollis Rey, 1891e: 115
testacea Gyllenhal, 1808a: 355
nigroimpressa Pic, 1922b: 31 (*Cantharis*) **A**: ZHE (Hangzou) TAI
limbatipennis Wittmer, 1956b: 303
nigrosuta A. Fiori, 1900: 108 **E**: IT
italica Pic, 1908a: 2
nitida Baudi di Selve, 1859a: 296 **E**: IT (Sardegna) SP
nopporensis Wittmer, 1971c: 199 **A**: JA (Hokkaido)
notaticollis Rosenhauer, 1856: 143 **E**: SP
nurdagensis Wittmer, 1972a: 66 **A**: TR
oboensis Kazantsev, 1994b: 72 **A**: MG
opaca Mulsant, 1862a: 325 **E**: FR PT SP
ornaticollis Marseul, 1864: 93 **E**: PT SP **N**: AG MO
nigropygidialis Pic, 1954b: 4
osellai Švihla, 1993: 261 **A**: TR
pacifica Kazantsev, 1994b: 80 **A**: FE
pamphylica Wittmer, 1972a: 74 **A**: TR
paflagonica Švihla, 2004a: 79 **A**: TR
parviocellata Kang & Kim, 2000a: 161 **A**: SC
patricia Kiesenwetter, 1866a: 375 (*Cantharis*) **E**: PT SP
dalmatina Pic, 1915g: 17
granatensis Pic, 1908l: 89
pedemontana Baudi di Selve, 1872: 113 **E**: FR IT SZ
pekinensis Švihla, 2002b: 311 **A**: BEI
persica Pic, 1933e: 10 **E**: AB GG ST **A**: IN TM
pfefferi Švihla, 1977: 179
peteri Švihla, 1995: 80 **A**: TM
peyroni Marseul, 1864: 74 (*Armidia*) **A**: SY TR
discobirufa Pic, 1946: 5
sanguinolenta Reiche, 1878a: 384

plagiella Marseul, 1864: 85 **E**: SP **N**: MO
planicollis Kazantsev, 1994b: 76 **A**: FE
prijutensis Kazantsev, 1992a: 45 **E**: GG
prudeki Švihla, 2002b: 309 **N**: MO
przewalskii Barovskij, 1929b: 267 **A**: NMO
pseudoconsociata Dahlgren, 1968: 121 **A**: TR
pseudogeniculata Kazantsev & Takahashi, 2001: 277 **A**: JA (Hokkaido)
pseudoreitteri Kazantsev, 2001d: 1080 **E**: GG
pseudorossica Kazantsev, 1992a: 48 **A**: TR
quadricollis Kiesenwetter, 1852: 607 **E**: PT SP **N**: AG TU
　brevinotata Pic, 1908l: 90
　fedjensis Pic, 1901c: 25
　limbipennis Marseul, 1864: 85
　marginella Baudi di Selve, 1859a: 297
querceti Kiesenwetter, 1866a: 383 (*Cantharis*) **E**: PT SP **N**: AG MO
　bugnioni Bourgeois, 1903a: 74
　pardalensis Pic, 1908l: 90
rambouseki Švihla, 1993: 265 **E**: MC
rassouli Wittmer, 1981e: 406 **A**: IQ TR
reflexa Wittmer, 1972a: 71 **A**: TM
reitteri Dahlgren, 1975: 109 **E**: ST
richteri Wittmer, 1972c: 196 **A**: IN
robusticornis Wittmer, 1972a: 69 **A**: TR
robustula Wittmer, 1951: 101 **A**: FUJ
rorida Kiesenwetter, 1867a: 46 (*Cantharis*) **E**: AU BH CR CZ HU RO SK UK YU
rossica Wittmer, 1972a: 73 **E**: ST
rosti Pic, 1900n: 183 **E**: ST
roubali Švihla, 1977: 180 **E**: ST
rufopallida Wittmer, 1972a: 73 **A**: IS
ruzickai Švihla, 2002b: 307 **A**: SY TR
rydhi Švihla, 1993: 268 **A**: TR
safraneki Švihla, 2002b: 312 **A**: SCH SHA
sareptana Marseul, 1868b: 178 **E**: ST UK
saurica Kazantsev, 1994b: 82 **A**: KZ
sibirica Wittmer, 1971c: 198 **A**: FE JA (Hokkaido)
sihotana Kazantsev, 1994b: 75 **A**: FE
similata Dahlgren, 1976: 88 **E**: GR
spinosa Wittmer, 1997e: 329 **A**: SCH
stanislavi Švihla, 2002b: 308 **A**: TR
straminea Kiesenwetter, 1859a: 27 **E**: GR
striatofrons Dahlgren, 1972: 137 **E**: PT SP
stusaki Švihla, 1992: 286 **A**: ES MG
syriaca Dahlgren, 1968: 110 **A**: IS JO LE SY
taiwanonigra Wittmer, 1982d: 135 **A**: TAI
takizawai Kazantsev & Takahashi, 2001: 273 **A**: JA (Hokkaido)
talyschensis Khnzorjan, 1959: 58 **E**: AB
tangeriana Pic, 1903i: 122 **N**: MO
　lindbergi Pic, 1932g: 26 [= 1932c: 8]
　marocana Pic, 1949a: 16
testacea Linnaeus, 1758: 403 (*Cantharis*) **E**: AU BE BY CR CT CZ DE EN FI FR GB GE HU IT LA LS LT LU NR NT PL RO SK ST SZ UK **A**: ES WS
　florida Schioedte, 1872: 8
　obscurithorax Pic, 1902h: 54
　ochropa Stephens, 1830: 296 (*Telephorus*)
　pellucida Brahm, 1790: 110 (*Cantharis*)

taurinensis Pic, 1905c: 105
testaceopallida Wittmer, 1997e: 325 **A**: HUB
theryi Pic, 1935f: 13 **N**: MO
tibetana Švihla, 2002b: 305 **A**: XIZ
tongluensis Champion, 1926c: 244 **A**: SD
transbaikalica Pic, 1906l: 89 **A**: ES KZ MG
jakovlevi Barovskij, 1908b: 293
smirnovi Barovskij, 1908b: 294
transita Wittmer, 1971c: 195 **A**: ES FE JA (Tsushima) "Korea"
translucida Krynicki, 1832: 99 (*Cantharis*) **E**: AU BE BH CR CZ FR GB GE HU IR IT LS PL RO SK SL SZ UK YU
concolor Märkel, 1852: 605
rufescens Letzner, 1847: 76 (*Cantharis*)
tridentata Wittmer, 1972a: 66 **A**: TR
tripunctata Reiche & Saulcy, 1857b: 175 (*Telephorus*) **A**: CY IS JO LE SY TR
chtaurana Švihla, 1977: 182
unistigmata Pic, 1900j: 86
tryznai Švihla, 2002b: 306 **A**: SCH
tseiana Kazantsev, 1992a: 46 **E**: ST
turcica Wittmer, 1972a: 68 **A**: TR
ulaensis Kazantsev, 1994b: 78 **A**: ES JA (Hokkaido) MG
upepesankensis Kazantsev & Takahashi, 2001: 275 **A**: JA (Hokkaido)
uralensis Dahlgren, 1972: 131 **A**: ES
varians Rosenhauer, 1856: 140 (*Podabrus*) **E**: FR PT SP
diversipes Pic, 1908l: 90 [HN]
fairmairei Marseul, 1864: 91
galiciana Gougelet, 1859: 238
galloisi Pic, 1908l: 89
heteronota Pandellé, 1867: 178
manzanalensis Pic, 1917h: 17
manzanalensis Pic, 1927f: 5 [HN]
pici Jakobson, 1911b: 680 [RN]
subnotaticeps Pic, 1935d: 11
vartiani Švihla, 1993: 273 **A**: SY
vicina Baudi di Selve, 1872: 108 **E**: UK
viduata Küster, 1854: 71 (*Cantharis*) **E**: AL BU GR RO YU
soror Kiesenwetter, 1860: 518
xanthopa Kiesenwetter, 1859a: 27 (*Telephorus*)
vitticollis Ménétriés, 1832: 162 **E**: AR AB GG ST **A**: IN
obscurior Pic, 1900n: 184
voriseki Švihla, 1977: 179 **E**: ST
walteri Švihla, 1993: 262 **A**: TR
weichowensis Wittmer, 1997e: 331 **A**: SCH
xanthochroina Fairmaire, 1884a: 180 **A**: LE SY TR
libanicola Pic, 1901b: 17
yunnana Wittmer, 1997e: 332 **A**: YUN
zwicki Wittmer, 1972a: 70 **A**: TR
subgenus *Spartiolepta* Bedel & Bourgeois, 1901: 177 type species *Rhagonycha marseuli* Bedel & Bourgeois, 1901
lucasi Jakobson, 1911b: 679 [RN] **N**: AG
geniculata P. H. Lucas, 1846: 188 [HN]
marseuli Bedel & Bourgeois, 1901: 177 **N**: AG
geniculata Marseul, 1864: 100 [HN]
subgenus *Ussurycha* Kazantsev, 1995c: 91 type species *Rhagonycha lederi* sensu Kazantsev, 1994 (= *Rhagonycha kazantsevi* Švihla, 1995)
kazantsevi Švihla, 1995: 76 **A**: FE

genus *Sidabia* Švihla, 1994: 116 type species *Sidabia bilyi* Švihla, 1994
bilyi Švihla, 1994: 116 **N**: AG

genus *Silicantharis* Švihla, 1992: 326 type species *Silicantharis walteri* Švihla, 1992
walteri Švihla, 1992: 326 **A**: CH ("Turkestan")

genus *Sinometa* Wittmer, 1969b: 86 type species *Sinometa besucheti* Wittmer, 1969
besucheti Wittmer, 1969b: 86 **E**: GR **A**: TR

genus *Sogdocantharis* Kazantsev, 1993e: 154 type species *Sogdocantharis chimtargensis* Kazantsev, 1993
chimtargensis Kazantsev, 1993e: 155 **A**: TD

genus *Stenothemus* Bourgeois, 1907: 292 type species *Themus harmandi* Bourgeois, 1902
angulatus Wittmer, 1974c: 50 **A**: SD
badius Kiesenwetter, 1874b: 278 (*Cantharis*) **A**: FE (Kuril) JA
hilgendorfi Harold, 1878: 75 (*Telephorus*)
benesi benesi Švihla, 2004b: 202 **A**: SCH
benesi shaanxiensis Švihla, 2004b: 202 **A**: SHA
decolor Bourgeois, 1907: 292 **A**: SD
dentatus alexandrae Švihla, 2004b: 196 **A**: SD
diffusus Wittmer, 1974c: 58 **A**: SCH **ORR**
dundai Švihla, 2004b: 198 **A**: GAN SCH XIN
expansus Wittmer, 1974c: 55 **A**: SD **ORR**
fukiensis Wittmer, 1974c: 60 **A**: FUJ
ganeshai Švihla, 2004b: 197 **A**: NP
grahami Wittmer, 1974c: 60 **A**: SCH
harmandi Bourgeois, 1902: 427 (*Themus*) **A**: SD **ORR**
innotaticollis Pic, 1920a: 6
jindrai Švihla, 2004b: 200 **A**: SCH
kansuensis Pic, 1933b: 4 **A**: GAN
kuatunensis Wittmer, 1979e: 330 **A**: FUJ
mamorui Okushima & Satô, 1999a: 137 **A**: TAI
nepalensis Wittmer, 1974c: 57 **A**: NP
owadai Okushima & Satô, 1997a: 85 **A**: TAI
piligerus Champion, 1925d: 266 **A**: SD **ORR**
schneideri Švihla, 2004b: 201 **A**: YUN
separatus Wittmer, 1974c: 58 **A**: NP
taiwanus Okushima & Satô, 1997a: 88 **A**: TAI
tryznai Švihla, 2004b: 197 **A**: SCH
volaticomimus Wittmer, 1974c: 54 **A**: UP
volaticus Champion, 1925d: 265 **A**: NP "India"
wittmeri Okushima & Satô, 1999a: 133 **A**: TAI
yunnanus Švihla, 2004b: 198 **A**: YUN

genus *Themus* Motschulsky, 1858b: 27 type species *Themus cyanipennis* Motschulsky, 1858
subgenus *Gallerucocantharis* Pic, 1913a: 3 type species *Gallerucocantharis subtumida* Pic, 1913
tincticornis Wittmer, 1997e: 272 **A**: SD
subgenus *Haplothemus* Wittmer, 1973b: 186 type species *Themus gracilis* Wittmer, 1973
bagmatiensis Wittmer, 1995b: 137 **A**: NP
bicoloriceps Kopetz, 2004: 130 **A**: SD
bidentatus Wittmer, 1997e: 262 **A**: NP
bilyi Wittmer, 1993a: 206 **A**: NP
brahmaputrensis Švihla, 2004b: 170 **A**: XIZ
brunescens Wittmer, 1965c: 87 **A**: NP

chumbiensis Champion, 1926: 201 (*Cantharis*) **A**: BT "Tibet"
cultellatus Wittmer, 1973b: 200 **A**: NP SD **ORR**
dhaulagiriensis Kopetz, 2004: 132 **A**: NP
excavatus Kopetz, 2004: 137 **A**: AP BT NP SD **ORR**
explanaticollis Pic, 1917e: 5 **A**: TAI
ferganensis Kazantsev, 2001d: 1076 **A**: KI
fuliginosus fuliginosus Bourgeois, 1891: cxxxix **A**: NP SD **ORR**
fuliginosus puncticeps Champion, 1926b: 134 **A**: NP "Tibet"
gnatongensis Wittmer, 1981c: 392 **A**: SD
gracilimimus Wittmer, 1983b: 237 **A**: NP
gracilis Wittmer, 1973b: 196 **A**: SCH
granulipennis Pic, 1931g: 77 **A**: YUN
hackeli Švihla, 2004b: 171 **A**: GAN SHA
hartmanni Kopetz, 2004: 141 **A**: NP
inequalithorax Pic, 1916d: 9 **A**: SD "Tibet"
ishigakiensis Okushima, 1991b: 196 **A**: JA (Ryukyus)
jaegeri Kopetz, 2004: 134 **A**: NP
kolibaci Wittmer, 1997e: 263 **A**: NP
lamellatus Wittmer, 1973b: 198 **A**: SD **ORR**
lycoceriformis Pic, 1916d: 10 (*Polemius*) **A**: YUN
makiharai Wittmer, 1983b: 237 **A**: NP
mediofasciatus Pic, 1934f: 129 **A**: SCH
niisatoi Satô & Takahashi, 1989: 215 **A**: JA (Honshu)
obscurior Wittmer, 1981c: 387 **A**: NP
ohkawai Satô, 1976: 53 **A**: JA (Honshu, Kyushu)
pakistanus Wittmer, 1960a: 102 (*Podistra*) **A**: NP PA
  lulusarensis Wittmer, 1981c: 391 [RN]
pallidipes Wittmer, 1954d: 277 **A**: TAI
pallidobrunneus Wittmer, 1973b: 201 **A**: SCH
pokaliensis Wittmer, 1983b: 234 **A**: NP
probsti Wittmer, 1993a: 208 **A**: NP
recurvus Wittmer, 1973b: 200 **A**: SD
reflexus Wittmer, 1995b: 139 **A**: SD **ORR**
rugosocyaneus Fairmaire, 1889b: 43 (*Telephorops*) **A**: SCH YUN "Tibet"
  diversipes Pic, 1947: 7 (*Cantharis*)
  magnifica Pic, 1906k: 83 (*Cantharis*)
satoi Wittmer, 1983a: 149 **A**: TAI
schawalleri Kopetz, 2004: 138 **A**: NP
schmidti Kopetz, 2004: 140 **A**: NP
schneideri Švihla, 2004b: 169 **A**: SCH SHA
szeschwanensis Wittmer, 1973b: 218 **A**: SCH
simlanus Pic, 1924j: 477 (*Cantharis*) **A**: BT NP UP
talianus Pic, 1917c: 5 (*Cantharis*) **A**: YUN
tigerensis Wittmer, 1987b: 102 **A**: SD
violatipennis S. Wang & J. Yang, 1993: 265 **A**: HUB (Wuling Mts)
subgenus *Telephorops* Fairmaire, 1886c: 339 type species *Telephorops impressipennis* Fairmaire, 1886
  Tryblius Fairmaire, 1897b: 228 type species *Tryblius cavipennis* Fairmaire, 1897
bicoloricornis Wittmer, 1983a: 153 **A**: TAI
cavipennis Fairmaire, 1897b: 228 **A**: BT NP SD
  ancoralis Champion, 1926a: 128
chalybeipennis Gorham, 1889: 96 **A**: PA "North India" **ORR**
coelestis Gorham, 1889: 104 (*Telephorus*) **A**: FUJ JIX
  rugosus Pic, 1929e: 8
crassimargo Champion, 1926a: 127 **A**: BT SD
crassipes Pic, 1929e: 8 **A**: YUN **ORR**

impressipennis Fairmaire, 1886c: 339 (*Telephorops*) **A**: GUI TAI
 violaceipennis Gorham, 1889: 105 (*Telephorus*)
laboissierei Pic, 1929l: 195 (*Triblius*) **A**: YUN **ORR**
minor Wittmer, 1997e: 272 **A**: YUN
nepalensis Hope, 1931: 26 (*Telephorus*) **A**: NP
sauteri Pic, 1912j: 46 (*Cantharis*) **A**: TAI
separandus Wittmer, 1975a: 252 **A**: BT SD
subcaeruleiformis Wittmer, 1983b: 199 **A**: SCH
subcaeruleus Pic, 1911n: 132 (*Triblius*) **A**: YUN
uncinatus Wittmer, 1983b: 200 **A**: SCH
subgenus *Themus* Motschulsky, 1858b: 27 type species *Themus cyanipennis* Motschulsky, 1858
atripes Pic, 1937k: 170 **A**: FUJ
benesi Švihla, 2004b: 167 **A**: SCH
bieti Gorham, 1889: 101 (*Telephorus*) **A**: SCH YUN
bilituratus Barovskij, 1909c: 326 (*Cantharis*) **A**: AF PA TD
 nuristanus Hicker, 1937: 180
 subannulicornis Pic, 1915b: 3 (*Cantharis*)
bimaculiceps Wittmer, 1958: 114 **A**: SCH
bitinctus uniformis Wittmer, 1983b: 218 **A**: FUJ (Yen-Ping)
cavaleriei Pic, 1926n: 211 (*Cantharis*) **A**: HUB
chaoi Wittmer, 1983b: 219 **A**: YUN
chatengensis Wittmer, 1973b: 206 **A**: SD
comans Champion, 1926: 137 **A**: UP
coomani Pic, 1923g: 60 (*Cantharis*) **A**: YUN **ORR**
 brevemaculiceps Pic, 1927l: 132 (*Cantharis*)
corayi Wittmer, 1983b: 216 **A**: FUJ SHA
coriaceipennis Fairmaire, 1889b: 41 (*Telephorus*) **A**: SCH
 licenti Pic, 1938k: 161
cribripennis Wittmer, 1983a: 151 **A**: TAI
curticornis Wittmer, 1995b: 131 **A**: SCH
cyanipennis Motschulsky, 1858b: 28 **A**: FE JA
 venatrix Kiesenwetter, 1874b: 271 (*Cantharis*)
 viridipennis Kisenwetter, 1874: 271 (*Cantharis*)
davidis Fairmaire, 1889b: 42 (*Telephorops*) **A**: SCH
deuvei Wittmer, 1993a: 204 **A**: GAN
eberti Wittmer, 1965c: 86 **A**: NP
elongatior elongatior Pic, 1955b: 16 **A**: NP SD
episcopalis episcopalis Kiesenwetter, 1874b: 269 (*Cantharis*) **A**: JA (Kyushu)
episcopalis purpureoaeneus Nakane, 1988: 78 **A**: JA (Honshu, Shikoku)
episcopalis satsumanus Yajima & Nakane, 1988: 78 **A**: JA (Kyushu)
formosanus Wittmer, 1954d: 277 **A**: TAI
foveicollis Fairmaire, 1900: 628 (*Telephorus*) **A**: FUJ
 testaceithorax Pic, 1938k: 155
galloisi Pic, 1932a: 1 **A**: JA
glasunovi alaicus Kazantsev, 2001d: 1076 **A**: KI
glasunovi glasunovi Barovskij, 1909: 325 (*Cantharis*) **A**: AF KA KI PA TD
guizhouensis Wittmer, 1997e: 264 **A**: GUI
dhudkundensis Kopetz, 2004: 114 **A**: NP
dolphukangensis Kopetz, 2004: 115 **A**: NP
hedini Pic, 1933b: 3 **A**: SCH
hmong Kazantsev, **nom. nov.** [see New acts] **A**: YUN **ORR**
 rufoscutus Pic, 1926h: 35 [HN]
hobsoni himalaicus Wittmer, 1973b: 215 **A**: UP "Tibet"
hobsoni hobsoni Champion, 1926: 136 **A**: "Tibet"
hobsoni tenuilimbatus Wittmer, 1973b: 213 **A**: NP "Tibet"

hypopelius Fairmaire, 1900: 625 (*Telephorus*) **A**: FUJ
imperialis Gorham, 1889: 102 **A**: FUJ JIA SCH YUN **ORR**
imperator Pic, 1906k: 81 [RN]
regalis Gorham, 1889: 103 (*Telephorus*) [HN]
kabakovi Wittmer, 1979c: 133 **A**: AF
kambaiticomimus Wittmer, 1997e: 261 **A**: YUN
kaschmirensis Pic, 1909e: 108 (*Cantharis*) **A**: PA
kazuoi azumai Kinoda & Nakane, 1988: 78 **A**: JA (Ryukyus)
kazuoi kazuoi Ohbayashi & Satô, 1968: 33 **A**: JA (Ryukyus)
kerstinae Kopetz, 2004: 120 **A**: NP
khasianus Gorham, 1889: 97 (*Telephorus*) **A**: UP NP SD BT **ORR**
kuatunensis Wittmer, 1983b: 209 **A**: ANH FUJ JIX
kumaonensis Wittmer, 1973b: 207 **A**: NP PA UP
kurosawai Satô, 1982: 35 **A**: JA (Ryukyus)
lahoulensis Wittmer, 1973b: 211 **A**: "Himalaya: Lahoul"
larrygrayi Wittmer, 1982a: 29 **A**: FUJ
leechianus Gorham, 1889: 104 (*Telephorus*) **A**: FUJ TAI
aeneipennis Gorham, 1889: 104 (*Telephorus*)
hemixanthus Fairmaire, 1900: 626 (*Telephorus*)
limbatus Wittmer, 1983b: 202 **A**: YUN
lineatofemoralis Pic, 1917c: 4 (*Cantharis*) **A**: BT NP SD UP **ORR**
longipennis Fairmaire, 1891d: xci (*Telephorus*) **A**: KA
luteipes Pic, 1938k: 161 **A**: GAN HEB SHA
maximus afghanus Wittmer, 1973b: 215 **A**: AF
maximus karnaliensis Kopetz, 2004: 122 **A**: NP
maximus maximus Pic, 1907l: 158 (*Cantharis*) **A**: UP
menieri dequinensis Wittmer, 1997e: 264 **A**: YUN
menieri menieri Wittmer, 1981c: 388 **A**: SD
micronotatus Pic, 1920g: 19 (*Cantharis*) **A**: YUN
midas Kiesenwetter, 1874b: 270 (*Cantharis*) **A**: JA
montanus Wittmer, 1983b: 204 **A**: GAN SHA SCH
muganglingensis Švihla, 2004b: 168 **A**: SCH
namnasae Wittmer, 1973b: 216 **A**: SD
nigropolitus Wittmer, 1995b: 128 **A**: YUN
nobilis nobilis Gorham, 1889: 103 (*Telephorus*) **A**: FUJ
confusus Fairmaire, 1900: 625 (*Telephorus*)
generosus Pic, 1906l: 89 (*Cantharis*)
fraternus Fairmaire, 1900: 626 (*Telephorus*)
reductenotatus Pic, 1922b: 31
nobilis reitteri Hicker, 1960: 78 **A**: FUJ
nobuoi Okushima & Satô, 2000a: 147 **A**: TAI
omeiensis Wittmer, 1983b: 214 **A**: SCH
parallelus Wittmer, 1973b: 218 **A**: GAN "East Tibet"
particularis Pic, 1929i: 75 **A**: YUN
paulometallicus Wittmer, 1995b: 134 **A**: NP
pindarae dentatus Kopetz, 2004: 124 **A**: NP
pindarae pindarae Champion, 1926: 136 **A**: UP
pindaraemimus difficilis Kopetz, 2004: 125 **A**: NP
pindaraemimus khumjungensis Wittmer, 1973b: 222 **A**: NP
pindaraemimus pindaraemimus Wittmer, 1973b: 222 **A**: SD
praelongus Champion, 1926: 135 **A**: NP UP
babaulti Pic, 1945: 12
purpuratus Wittmer, 1983a: 151 **A**: TAI
reductus Wittmer, 1983b: 212 **A**: GUI
reymondi Bourgeois, 1907: 291 **A**: NP

robustus Wittmer, 1973b: 219 **A**: KA
ruficollis Wittmer, 1995b: 130 **A**: SCH
rufocapitatus Wittmer, 1995b: 136 **A**: FUJ
safedkoensis Wittmer, 1973b: 221 **A**: AF
saipalensis Kopetz, 2004: 127 **A**: NP
semistrangulatus Pic, 1917c: 4 (*Cantharis*) **A**: SD
senensis Pic, 1922b: 32 (*Cantharis*) **A**: YUN **ORR**
infraterna Pic, 1922c: 1 (*Cantharis*)
shensanus Wittmer, 1983b: 211 **A**: SHA
sichuanus Wittmer, 1997e: 266 **A**: SCH
sikkimensis Pic, 1911h: 109 (*Cantharis*) **A**: BT NP SD "Tibet" **ORR**
semicyaneus Pic, 1913d: 99 (*Tylocerus*)
sobosanus kitakyushuensis Takakura, 1989: 157 **A**: JA (Kyushu)
sobosanus sobosanus Satô & Ishida, 1982: 34 **A**: JA (Kyushu)
stevensi Champion, 1926: 132 **A**: SD
stigmaticus Fairmaire, 1888b: 123 (*Telephorus*) **A**: BEI GAN HEB JIA NMO QIN SHA
subopacipennis Pic, 1910t: 77 (*Cantharis*) **A**: TAI
subrufolineatus Wittmer, 1995b: 134 **A**: YUN
subviridipennis flavoscutellaris Pic, 1936a: 1 **A**: UP
subviridipennis subviridipennis Pic, 1916e: 13 **A**: NP SD
nonfriedi Pic, 1934d: 54
subviridipennis vicinus Wittmer, 1973b: 227 **A**: "NW India"
tabulatus Wittmer, 1973b: 225 **A**: HP KA UP
taiwanus Wittmer, 1983a: 148 **A**: TAI
testaceicollis Wittmer, 1983b: 225 **A**: GAN SCH SHA
thudamensis Wittmer, 1981c: 390 **A**: NP
tryznai Švihla, 2004b: 167 **A**: SCH
tsushimensis Satô & Ishida, 1982: 32 **A**: JA (Tsushima)
tumlonganus Pic, 1916k: 16 (*Cantharis*) **A**: BT NP SD
chrysocephalus Champion, 1926: 133
vastiorum Wittmer, 1983b: 206 **A**: FUJ
venningi Champion, 1926: 130 **A**: "NE India" **ORR**
viridipennis Kiesenwetter, 1874b: 271 (*Telephorus*) **A**: JA
viridissimus Pic, 1906k: 84 (*Cantharis*) **A**: YUN
walteri Švihla, 2004b: 169 [RN] **A**: PA
pakistanus Wittmer, 1973b: 203 [HN]
yunnanus Wittmer, 1983b: 221 **A**: YUN

genus *Themus*, species incertae sedis
kingsiensis Wittmer, 1983b: 225 **A**: GUX
pallidocincticollis Pic, 1917b: 7 (*Cantharis*) **A**: "Tibet"
quadratus Wittmer, 1983b: 224 **A**: SCH

genus *Walteriella* Kazantsev, 2001d: 1081 [RN] type species *Lycocerus particularis* Pic, 1923
Walteria Kazantsev, 1999: 117 [HN] type species *Lycocerus particularis* Pic, 1923
decipiens Gorham, 1889: 109 (*Lycocerus*) **A**: NP
taiwana Kazantsev, 1999: 120 **A**: TAI

genus *Yukikoa* Satô, 1976: 51 type species *Themus wittmeri* Nakane, 1963
akitai Takahashi, 2003: 88 **A**: JA (Honshu)
kamezawai Takahashi, 2003: 92 **A**: JA (Honshu)
kanekoi Takahashi, 2003: 94 **A**: JA (Honshu)
masatakai Takahashi, 2003: 90 **A**: JA (Honshu)
mizunoi Takahashi, 2003: 89 **A**: JA (Honshu)
onzuiensis Takahashi, 2003: 86 **A**: JA (Honshu)

wittmeri Nakane, 1963b: 46 (*Themus*) **A**: JA (Honshu)

subfamily **Silinae** Mulsant, 1862
tribe Silini Mulsant, 1862

genus *Asiosilis* Wittmer, 1977a: 274 type species *Asiosilis cochleata* Wittmer, 1974
birmanica Pic, 1911v: 173 (*Silis*) **A**: BT SD **ORR**
dehraduna Wittmer, 1957a: 30 (*Silis*) **A**: UP
himalaica Wittmer, 1977a: 282 **A**: HP NP UP
infrapunctata infrapunctata Wittmer, 1977a: 280 **A**: BT
newtoni Pic, 1927m: 140 (*Silis*) **A**: NP SD **ORR**
obscuritarsis Pic, 1907u: 195 (*Silis*) **A**: NP "Himalaya"
pectinicornis Champion, 1924c: 253 (*Silis*) **A**: UP **ORR**
singularicornis Pic, 1924j: 477 (*Silis*) **A**: BT SD **ORR**
staehlini Wittmer, 1977a: 282 **A**: BT
unilamellata Champion, 1926b: 208 (*Silis*) **A**: BT **ORR**
wuermlii Wittmer, 1977a: 282 **A**: BT

genus *Cordylocera* Guérin-Méneville, 1838a: 77 type species *Tylocerus antennatus* Guérin-Méneville, 1838
livida Hope, 1831: 26 (*Anisotelus*) **A**: NP

genus *Crudosilis* Kazantsev, 1994c: 102 type species *Crudosilis altaica* Kazantsev, 1994
alaica Kazantsev, 1998a: 289 **A**: KI
altaica Kazantsev, 1994c: 110 **A**: WS
barovskii Kazantsev, 1994c: 113 **A**: KZ
beringiana Kazantsev, 1994c: 110 **A**: FE
dowgailoi Kazantsev, 1998a: 290 **A**: KI
euphrosyne Barovskij, 1928b: 219 (*Silis*) **A**: TD
ferganica Kazantsev, 1998a: 292 **A**: KI
lomakini Chernyshev, 1998: 240 **A**: ES
magadanica Kazantsev, 1998a: 286 **A**: FE
mingrelica Kazantsev, 1994c: 109 **E**: GG
mongolica Kazantsev, 1998a: 286 **A**: ES MG
motschulskii Barovskij, 1928b: 218 (*Silis*) **A**: ES
  bidentata Motschulsky, 1859d: 568 (*Silis*) [HN]
ruficollis Fabricius, 1775: 206 (*Cantharis*) **E**: AU BE BY CT CZ DE FI FR GB GE HU IR IT LA LS LT NL PL RO SK SV SZ UK **A**: KZ WS
  bidentata O.F. Müller, 1776: 62 (*Telephorus*)
  gallica Pic, 1927h: 13 (*Silis*)
  hungarica Pic, 1910i: 33 (*Silis*)
  rubricollis Charpentier, 1825: 195 (*Silis*)
  rugicollis Gebler, 1833: 279 (*Silis*)
  signicollis Everts, 1920: 230 (*Silis*)
  torquata Gyllenhal, 1827: 340 (*Cantharis*)
rugimana Kazantsev, 1994c: 107 **A**: FE
sandalica Kazantsev, 1998a: 292 **A**: KI
sexdentata Mannerheim, 1853: 295 (*Silis*) **A**: CH ES FE MG
sinica Kazantsev, 1998a: 288 **A**: QIN
tchernyshevi Kazantsev, 1998a: 287 **A**: ES
tolli Barovskij, 1928b: 220 (*Silis*) **A**: FE
turkestanica Kazantsev, 1998a: 291 **A**: KI
wittmeri Kazantsev, 1998a: 288 **A**: GAN
xanthoderus Solsky, 1881: 35 (*Silotrachelus*) **A**: KI

genus *Eusilis* Reitter, 1887e: 233 type species *Eusilis przewalskii* Barovskij, 1928
przewalskii Barovskij, 1928b: 216 **A**: ZIN

genus *Laemoglyptus* Fairmaire, 1886c: 334 type species *Laemoglyptus fissiventris* Fairmaire, 1886
atripes Pic, 1921e: 7 **A**: TAI
bimaculatus Wittmer, 1956b: 305 **A**: HUB
bomfordii Fairmaire, 1897b: 225 **A**: NP UP
innotaticollis Pic, 1937l: 189
chujoi Wittmer, 1972b: 138 **A**: TAI
fissiventris Fairmaire, 1886c: 335 **A**: SCH
grandis Pic, 1921e: 7 **A**: TAI
iriomotoensis Wittmer, 1972b: 137 **A**: JA (Ryukyus)
nakanei Wittmer, 1974a: 4 **A**: JA (Tokara)
pectinatus Lewis, 1895a: 112 (*Silis*) **A**: JA
pingtungensis Wittmer, 1984c: 150 **A**: TAI
ramifer Champion, 1926b: 204 (*Silis*) **A**: SD
rubrithorax Pic, 1912c: 13 (*Drilosilis*) **A**: TAI
sauteri Pic, 1911x: 188 (*Drilosilis*) **A**: TAI
subspinosus Pic, 1910w: 259 **A**: JA (Ryukyus) TAI
taihorinensis Wittmer, 1984c: 152 **A**: TAI
vicinus Pic, 1921e: 6 **A**: SD

genus *Macrosilis* Pic, 1911v: 173 type species *Macrosilis fortunei* Pic, 1911
brevior Pic, 1917e: 5 **A**: TAI
fortunei Pic, 1911v: 173 **A**: "South China" **ORR**
laticollis Boheman, 1858: 79 (*Cantharis*) **A**: HKG

genus *Paradiscodon* Wittmer, 1954a: 24 type species *Cantharis vanicorensis* Boisduval, 1835
donckieri Pic, 1906k: 83 (*Cantharis*) **A**: "China"

genus *Podosilis* Wittmer, 1978c: 159 type species *Silis fruhstorferi* Pic, 1906
afghana Wittmer, 1978c: 161 **A**: AF
amplilobata Wittmer, 1997e: 243 **A**: HUB
apicecarinata Wittmer, 1997e: 233 **A**: YUN
bicoloriceps Wittmer, 1997: 255 **A**: SD **ORR**
circumcincta Wittmer, 1997e: 249 **A**: SCH YUN
distenda Wittmer, 1997e: 246 **A**: YUN
donckieri Pic, 1906k: 84 (*Silis*) **A**: YUN
convexa Wittmer, 1992a: 2
diversihamata Pic, 1922c: 2 (*Silis*)
elongaticornis Wittmer, 1997e: 228 **A**: YUN
fruhstorferi Pic, 1906b: 91 (*Silis*) **A**: BT SD
fukiena Wittmer, 1997e: 357 **A**: FUJ
jendeki Wittmer, 1997e: 359 **A**: GUI
jizushanensis Wittmer, 1997e: 226 **A**: YUN
langana Pic, 1923g: 43 (*Silis*) **A**: YUN **ORR**
laokaiensis Pic, 1914b: 11 (*Silis*) **A**: YUN **ORR**
nitidissima Pic, 1922b: 30 (*Silis*) **A**: YUN
obscurissima Pic, 1906k: 84 (*Silis*) **A**: YUN
omissa Wittmer, 1954c: 111 (*Silis*) **A**: JA (Honshu)
shirahatai Nakane, 1963a: 172 (*Silis*)
pallidiventris Fairmaire, 1889b: 35 (*Silis*) **A**: SCH
sichuana Wittmer, 1997e: 359 **A**: SCH
sinensis Pic, 1906k: 84 (*Silis*) **A**: YUN
thibetana Pic, 1916h: 8 (*Silis*) **A**: "Tibet"
yunnana Wittmer, 1997e: 245 **A**: YUN

genus *Silis* Charpentier, 1825: 194 type species *Cantharis spinicollis* Charpentier, 1825 (= *Silis nitidula* Fabricius, 1792)
bianchii Barovskij, 1926a: 13 **A**: ES
bifossicollis Wittmer, 1963: 114 **A**: "Central China"
dzungarica Kazantsev, 1994c: 102 **A**: KZ
jacutica Barovskij, 1926a: 13 **A**: ES FE
nitidula Fabricius, 1792a: 220 (*Cantharis*) **E**: AU CT CZ EN FR GE HU LA LT PL RO SK UK
excisa Germar, 1824: 71 (*Telephorus*)
spinicollis Charpentier, 1825: 194
olschwangi Kazantsev, 1994c: 99 **A**: WS
telephoroides Pic, 1913d: 98 **A**: PA
triimpressa Pic, 1926a: 1 **A**: FE "Korea"
urjanhaica Kazantsev, 1997b: 1216 **A**: ES

genus *Sphaerarthrum* Waterhouse, 1884b: 281 type species *Telephorus praeustus* Guérin-Méneville, 1838
stygianum Gorham, 1889: 99 (*Telephorus*) **A**: SD **ORR**
nuguei Pic, 1910s: 70 (*Silis*)

subfamily **Malthininae** Kiesenwetter, 1852
tribe Malthinini Kiesenwetter, 1852

genus *Caccodes* Sharp, 1885b: 157 type species *Malthodes oceaniae* Bourgeois, 1884
kanekoi Takahashi, 1999b: 203 **A**: JA (Ogasawara)
niponicus formosanus Pic, 1937h: 64 **A**: TAI
niponicus niponicus Kiesenwetter, 1874b: 281 (*Malthodes*) **A**: JA (Honshu, Kyushu)
shirahatai Ohbayashi, 1955: 51 (*Maltypus*)
oceaniae Bourgeois, 1884b: 288 (*Malthodes*) **A**: JA **AUR ORR**
debilis Sharp, 1885b: 157

genus *Malthinellus* Kiesenwetter, 1874b: 280 type species *Malthinellus bicolor* Kiesenwetter, 1874
bicolor Kiesenwetter, 1874b: 281 **A**: JA (Honshu, Kyushu) "Korea"
chinensis Wittmer, 1956b: 307 **A**: HUB
chujoi Wittmer, 1961: 1 (*Malthinus*) **A**: JA (Ryukyus)

genus *Malthinus* Latreille, 1806: 261 type species *Cantharis flaveola* Herbst, 1784
subgenus *Indomalthinus* Brancucci, 1978: 305 type species *Malthinus bhaktai* Brancucci, 1978
albidipennis Pic, 1907p: 172 **A**: HP NP PA UP **ORR**
duplolineatus Champion, 1926b: 209
bengalensis Brancucci, 1978: 314 **A**: SD
bhaktai Brancucci, 1978: 312 **A**: SD
ganesh Brancucci, 1981: 54 [RN] **A**: HP KA UP
multinotatus Pic, 1929i: 75 (*Anolisus*) [HN]
spissicornis Champion, 1926b: 208 **A**: UP
tantricus Brancucci, 1978: 316 **A**: SD
subgenus *Malthinus* Latreille, 1806: 261 type species *Cantharis flaveola* Herbst, 1784
Apteromalthinus Escalera, 1913b: 322 type species *Apteromalthinus pithanoides* Escalera, 1913
Malachidius Motschulsky, 1860a: 62 type species *Malachidius rufifrons* Motschulsky, 1860
Progeutes Abeille de Perrin, 1894b: 92 type species *Malthinus longipennis* P. H. Lucas, 1846
Ymnis Des Gozis, 1886: 23 type species *Malthinus flaveolus* Herbst, 1784
aegyptiacus Pic, 1899n: 372 (*Progeutes*) **N**: EG
aegyptiacus Ferrante, 1908: 26
piceonotatus Pic, 1916g: 7
aetolicus Wittmer, 1974d: 363 **E**: GR
alemdagensis Wittmer, 1974d: 409 **A**: TR
aliceae Švihla, 1990a: 199 **E**: AB GG
angulatus Wittmer, 1997a: 39 **A**: GUX

angusticollis Pic, 1898f: 120 **N**: AG MO
anselmoi Wittmer, 1971: 180 **N**: MO
armipes Kiesenwetter, 1871: 76 **E**: IT (Sardegna, Sicilia)
artvinensis Wittmer, 1978e: 356 **A**: TR
aspoecki Wittmer, 1974a: 4 **E**: SP
atlasicus Wittmer, 1940b: 546 **N**: MO
axillaris Kiesenwetter, 1852: 256 **E**: GR (Rodos) **A**: CY IS SY TR
    moricei Pic, 1930d: 5
    nigrithorax Wittmer, 1936c: 677
axillaroides Wittmer, 1936c: 678 **A**: JO
balteatus Suffrian, 1851: 50 **E**: AU CR CZ FR GB GE GR HU IR PT SK SP SV SZ TR UK **A**: TR
    fasciatus C. G. Thomson, 1864: 194
    kruperi Pic, 1901f: 51
    lateralis Motschulsky, 1853a: 4
    merkli J. Weise, 1895e: 214
    obscuriceps Pic, 1900h: 61
    praetextus Bourgeois, 1899b: 370
    sinuatipes Pic, 1908e: 46
    taygetanus Pic, 1902k: 63
bandamensis Palm, 1975b: 128 **N**: CI
baroniurbanii Wittmer, 1974e: 630 **A**: BT
basantapurensis Wittmer, 1984b: 96 **A**: NP
basimarginatus Wittmer, 1966b: 234 **A**: AP
bezdeki Švihla, 2002a: 127 **E**: TR
bhaumiki Wittmer, 1978e: 355 **A**: SD
bicoloriceps Wittmer, 1997e: 342 **A**: YUN
biglianii Fiori, 1915a: 45 **E**: FR IT
biguttatus Linnaeus, 1758: 402 (*Cantharis*) **E**: AU BE CT CZ EN FR GE HU IT LS NT PL RO SK SZ UK
    biguttulus Paykull, 1800: 445 (*Cantharis*)
    striatulus Mulsant, 1862a: 367
bilineatus Kiesenwetter, 1852: 257 **E**: BE FR GE IT LU SZ
bitincticollis Wittmer, 1997e: 343 **A**: KI
brancuccii Wittmer, 1978e: 359 **A**: IN TR
brevereductus Pic, 1953a: 10 (*Malachidius*) **N**: MO
brunneomarginatus Wittmer, 1978e: 354 **A**: SD
bulgaricus Švihla, 1990a: 200 **E**: BU
byzantinus Wittmer, 1974d: 410 **A**: TR
cartennaeensis Constantin, 1979: 631 **N**: AG
caucasicus Wittmer, 1974d: 420 **E**: AR ST **A**: TR
centrocarinatus Wittmer, 1969d: 174 **N**: AG
centrolineatus Wittmer, 1995b: 157 **A**: FUJ
chobauti Bourgeois, 1897: 41 (*Progeutes*) **N**: AG
cincticollis Kiesenwetter, 1866a: 393 **E**: PT SP
complexus Wittmer, 1993b: 287 **A**: TR
conspicuus conspicuus Kiesenwetter, 1852: 250 **A**: IN IQ IS JO LE SY TR
    rufithorax Pic, 1903i: 122 (*Progeutes*)
conspicuus triggiber Marseul, 1878a: 16 **A**: IN IS SY
    abdelkader Marseul, 1868b: 180
    brevenotaticeps Pic, 1923i: 9
    lateflavicans Pic, 1910o: 57 (*Progeutes*)
    notaticollis Pic, 1900j: 86 (*Progeutes*)
    sahlbergi Pic, 1909j: 137 (*Progeutes*)
    simoni Pic, 1913c: 97
    unionis Pic, 1909j: 137 (*Progeutes*)
cordobensis Wittmer, 1971b: 175 **E**: SP
cornutus Wittmer, 1974d: 395 **A**: TR

corsicanus Constantin, 1975: 85 **E**: FR (Corse)
corylicus Wittmer, 1974d: 415 **A**: TR
crassifemur Wittmer, 1974a: 5 **A**: NP
crassipes Pic, 1899n: 373 **A**: IS LE SY TR
schatzmayri Pic, 1951a: 9
crenaticeps Wittmer, 1974d: 413 **A**: TR
crenulatomimus Wittmer, 1984c: 154 **A**: TAI
crenulatus Wittmer, 1979e: 335 **A**: TAI
creticus Wittmer, 1971e: 368 **E**: GR (Kríti)
croceicollis Wollaston, 1862c: 426 **N**: CI
flammeicollis Wollaston, 1865: 194
dalmatinus Pic, 1902k: 63 **E**: BH CR YU
angustipennis Pic, 1908i: 65
damascenus Wittmer, 1936c: 678 **A**: SY TR
danielssoni Wittmer, 1995b: 156 **E**: GR
deceptor Baudi di Selve, 1893: 43 **E**: IT
tiburtina Baudi di Selve, 1893: 43
dentatofemoralis Wittmer, 1987b: 105 **A**: SD
dentipes Wittmer, 1979e: 337 **A**: NP
depauperatus Wollaston, 1862c: 425 **N**: CI (Lanzarote, Fuertaventura)
devillei Abeille de Perrin, 1898b: 92 **E**: FR IT
diffusus Kiesenwetter, 1866a: 369 **E**: SP **N**: MO
dimorphus cilicius Švihla, 2002a: 123 **E**: TR
dimorphus dimorphus Wittmer, 1935d: 249 **E**: GR (Rodos)
dimorphus phrygius Švihla, 2002a: 124 **E**: TR
dogueti Constantin, 1979: 622 **N**: AG
dromioides Bourgeois, 1885e: cxxvi **E**: GR
drurei Pic, 1904g: 49 (*Progeutes*) **A**: IQ
dryocoetes Rottenberg, 1871: 242 **E**: IT
tibiellus Pic, 1918e: 10
dubius Motschulsky, 1853a: 4 **A**: TR
elbursensis Wittmer, 1978e: 360 **A**: IN
elongaticollis Pic, 1907c: 105 **N**: AG
emeiensis Wittmer, 1993b: 292 **A**: SCH
espadanensis Wittmer, 1981e: 411 **E**: SP
euboeicus Wittmer, 1974d: 364 **E**: GR
evidentissimus Wittmer, 1974d: 415 **E**: AR
excisipes Pic, 1899n: 376 **A**: SY
maculipennis Pic, 1900b: 15
externus Pic, 1900d: 37 **N**: AG TU
facialis C. G. Thomson, 1864: 195 **E**: AB AR AU BE CR CT CZ FR GE GR HU IT PL NT RO SK SL ST SZ **A**: TR
delicatulus Bourgeois, 1899b: 370
luteola Trella, 1925: 126
maritimus Pic, 1899a: 19
sabaudus Pic, 1901j: 80
fasciatus A. G. Olivier, 1790a: no 26: 18 (*Telephorus*) **E**: AU BE CR CT CZ FR GB GE HU IT PL RO SK SV SZ UK
humeralis Marsham, 1802: 374 (*Cantharis*)
luteolus Stephens, 1830: 306
subopaciceps Pic, 1915h: 21
fenchihuensis Wittmer, 1979e: 335 **A**: TAI
filiformis Wittmer, 1997e: 340 **A**: YUN
flaveoloides Švihla, 1997b: 123 **E**: GG ST **A**: TR

flaveolus Herbst, 1786: 171 (*Cantharis*) **E**: AU BE BY CR CT CZ EN FR GB GE GR IR HU IT LT PL NT RO SK SL SP ST SZ UK
  collaris Latreille, 1806: 262 (*Cantharis*)
  flavus Latreille, 1806: 262 (*Cantharis*)
  griseipennis Pic, 1913c: 97
  immunis Marsham, 1802: 374 (*Cantharis*)
  minimus A. G. Olivier, 1790a: no 26: 17 (*Telephorus*)
  punctatus Geoffroy, 1785: 174 (*Necydalis*) [HN]
  subfusca Pic, 1906d: 24
foliiformis Wittmer, 1975b: 263 **E**: CR YU
forcepiformis Wittmer, 1978e: 356 **E**: GR
frantiseki Švihla, 2002a: 129 **A**: SY
freyi Wittmer, 1974d: 404 **A**: TR
frontalis Marsham, 1802: 373 (*Cantharis*) **E**: AU BE CZ FR GB GE IT SP SZ
  concolor Stephens, 1830: 310 (*Malthinus*)
  immaculatus Stephens, 1830: 307 (*Malthinus*)
  longipennis Motschulsky, 1853a: 5
  pinicola Stephens, 1830: 308 (*Malthinus*)
fuerteventurensis Palm, 1975b: 124 **N**: CI
furcatus Wittmer, 1974d: 425 **A**: TR
gafsaensis Wittmer, 1969d: 172 **N**: TU
ganglbaueri Wittmer, 1974d: 419 **E**: GR MC
gardneri Pic, 1929i: 75 **A**: "India"
garganicus Fiori, 1915a: 50 **E**: IT
geigyi Wittmer, 1971: 175 **E**: SP
geniculatus Kiesenwetter, 1859a: 27 **E**: GR IT
glabellus Kiesenwetter, 1852: 254 **E**: BE CZ FR GE IT SK SP SZ
graecus Wittmer, 1986a: 102 **E**: GR
gratiosus Pic, 1901f: 51 **E**: GR (Samos)
guruguensis Wittmer, 1971: 187 **N**: MO
gyalpo Wittmer, 1974e: 628 **A**: BT NP
hetiticus Wittmer, 1974d: 412 **A**: TR
hofferi Švihla, 1980a: 242 **N**: AG
humeralis Kiesenwetter, 1874b: 280 **A**: JA (Honshu)
iblanensis Wittmer, 1971: 177 **N**: MO
ikauensis Wittmer, 1978e: 362 **N**: MO
ilisicus Wittmer, 1974d: 363 **E**: GR (Purgos)
impressicollis Fairmaire, 1884d: lxiv **N**: AG TU
  batnensis Pic, 1900b: 15
  tamarisci Pic, 1900f: 55
impressithorax Švihla, 1997b: 125 **E**: SP **N**: AG
inaequalithorax Pic, 1933i: 117 **A**: JIA
incanus Kiesenwetter, 1852: 258 **A**: TR
incrassicornis Pic, 1900j: 86 (*Progeutes*) **N**: AG
  diversicollis Pic, 1906c: 17
incrassipes Wittmer, 1956a: 160 **A**: AF
inflavus Marseul, 1878a: 36 **N**: AG TU
  algericus Pic, 1900b: 19
insignipes Pic, 1907a: 97 **E**: IT
  calaber Pic, 1909q: 169
  indisjuncta Pic, 1910a: 1
intermedius Wittmer, 1971: 177 **N**: MO
interruptus Wittmer, 1974d: 423 **A**: TR
ionicus Pic, 1901f: 51 **E**: GR
  corcyreus Pic, 1901f: 51
  insularis Pic, 1915g: 18

rufescens Pic, 1901f: 51
israelsoni Palm, 1975b: 127 **N**: CI
japonicus Ohbayashi, 1949: 21 **A**: JA (Honshu)
jordanicus Wittmer, 1974d: 396 **A**: JO
kadleci Švihla, 2002a: 133 **A**: SY
kamengensis Wittmer, 1966b: 235 **A**: AP
kafkai Švihla, 2002a: 125 **A**: CY
kaschmirensis Wittmer, 1974d: 418 **A**: KA
kaszabi Wittmer, 1974d: 418 **E**: "Caucasus" **A**: TR
khuzistanicus Wittmer, 1972c: 199 **A**: IN
kiesenwetteri C. N. F. Brisout de Barneville, 1863: 84 **E**: FR
klapperichi Wittmer, 1951: 101 **A**: FUJ
kobensis Lewis, 1895a: 115 (*Malthodes*) **A**: JA (Honshu)
kraatzi Pic, 1900h: 61 **E**: SP
kumejimensis Satô, 1986c: 258 **A**: JA (Ryukyus)
lacteifrons Marseul, 1878a: 43 **E**: IT (Sicilia) **N**: AG TU
laesus Kiesenwetter, 1871b: 76 **N**: AG TU
flavicollis Pic, 1904b: 4
grandiceps Pic, 1901p: 112
normandi Pic, 1900j: 87
obscurithorax Pic, 1900b: 15
laevicollis Kiesenwetter, 1859a: 28 **E**: GR
langtangensis Wittmer, 1993a: 222 **A**: NP
lasistanicus Wittmer, 1974d: 409 **A**: TR
lasithiensis Wittmer, 1975b: 263 **E**: GR (Kríti)
laticollis Pic, 1900b: 14 (*Progeutes*) **E**: IT (Sicilia) **N**: AG
vitalei Pic, 1916g: 6
latirufus Pic, 1901b: 17 (*Malthodes*) **N**: TU
henrici Pic, 1901b: 17 (*Malthodes*)
libenae Švihla, 2002a: 130 **A**: SY
lindbergi Palm, 1975b: 121 **N**: CI (Lanzarote)
lineatocollis Champion, 1920b: 247 **A**: NP UP "India"
lituratus Motschulsky, 1853a: 4 **E**: SP
longicornis Kiesenwetter, 1866a: 374 **E**: SP
longipennis P. H. Lucas, 1846: 189 **N**: AG MO TU
biskrensis Pic, 1900j: 86 (*Progeutes*)
flavicans Bourgeois, 1903a: 75 (*Progeutes*)
flavipennis Pic, 1903o: 156 (*Progeutes*)
nigripes Bourgeois, 1897: 40
longithorax Pic, 1899n: 373 **A**: IS SY
caifensis Pic, 1900d: 37
curtenotatus Pic, 1931c: 5
ruficeps Pic, 1899n: 373
lopchuensis Wittmer, 1978e: 353 **A**: SD
lundbergi Švihla, 1994: 117 **A**: TR
luteus Pic, 1935f: 13 (*Apteromalthinus*) **N**: MO
maculithorax Pic, 1898b: 28 **N**: MO
bordei Pic, 1911l: 121
lineata Pic, 1899n: 373
madoniensis Švihla, 2002a: 122 **E**: IT (Sicilia)
magisi Wittmer, 1974d: 400 **A**: TR
magniceps Wittmer, 1997e: 337 **A**: SCH
malkini Wittmer, 1986a: 102 **E**: GR
manaliensis Wittmer, 1995b: 151 **A**: HP
mandli Wittmer, 1966c: 493 **A**: IN
marashensis Wittmer, 1974d: 395 **A**: TR

marginicollis Ganglbauer, 1906a: 49 **E**: IT (Sicilia)
marocanus Wittmer, 1971: 183 **N**: MO
maspalomensis Palm, 1975b: 130 **N**: CI
medianus Wittmer, 1974d: 389 **A**: TR
mediterraneus Wittmer, 1974d: 390 **A**: TR
mesopotamicus Wittmer, 1974d: 414 **A**: TR
mexensis Wittmer, 1969d: 178 **N**: MO
mimosinensis Wittmer, 1988c: 368 **A**: GUX
minimus Palm, 1975b: 122 **N**: CI
moravicus Švihla, 1997a: 101 **E**: CZ
mraceki Švihla, 2002a: 121 **N**: MO
muchei Wittmer, 1981e: 412 **A**: TD
mucoreus Kiesenwetter, 1879b: 309 **A**: JA
muehlei Švihla, 1990a: 197 **A**: SY
multimaculatus Wittmer, 1995b: 153 **A**: YUN
multinotatus Pic, 1929j: 94 **N**: LB
multispinosus Wittmer, 1995b: 155 **A**: GUA
mutabilis Wollaston, 1862c: 424 **N**: CI
mutilatus Kiesenwetter, 1879b: 308 **A**: JA
nabataeicus Wittmer, 1974d: 391 **A**: JO
nakanei Wittmer, 1955: 41 **A**: JA (Honshu, Kyushu)
nankingensis Wittmer, 1997e: 339 **A**: JIA
neapolitanus Pic, 1905n: 186 **E**: IT
neglectus Palm, 1975b: 129 **N**: CI
nepalensis Wittmer, 1966b: 235 **A**: NP
nigerrimus Constantin, 1975: 80 **E**: SP
nigrescens Palm, 1975b: 124 **N**: CI
nitidithorax Pic, 1905g: 145 **N**: AG TU
notaticeps Wittmer, 1956a: 159 **A**: AF
notsui Wittmer, 1984c: 158 **A**: TAI
obscuripennis Pic, 1898j: 153 **N**: AG
obscuripes Kiesenwetter, 1866b: 256 **E**: SP
ohbai Wittmer, 1984c: 158 **A**: TAI
okinawanus Wittmer, 1970b: 23 **A**: JA (Ryukyus)
olcesei Pic, 1898f: 119 **N**: MO
  griseonotata Pic, 1951a: 8
  limbatus Pic, 1900j: 86 (*Progeutes*)
  maculaticeps Pic, 1920b: 7
olympiacus Wittmer, 1974d: 364 **E**: GR
ornatus Rosenhauer, 1856: 146 **E**: PT SP
otini Wittmer, 1971b: 172 **N**: MO
oujdaensis Wittmer, 1969d: 175 **N**: MO
palingensis Wittmer, 1993b: 291 **A**: TAI
pallicolor Fairmaire, 1883d: xcii **N**: AG TU
pallidipennis Pic, 1900b: 15 **N**: AG MO TU
  brevior Pic, 1900f: 54
  femoralis Pic, 1900j: 87
  laterufus Pic, 1900f: 54
  semiobscurus Pic, 1900f: 54
pallidipes Fairmaire, 1884d: lxiv **E**: IT **N**: AG TU
  atriceps Pic, 1900f: 54
  biinterruptus Pic, 1903r: 169
pamphylicus Wittmer, 1974d: 417 **A**: TR
panachaicus Wittmer, 1974d: 364 **E**: GR
parallelus Wittmer, 1974d: 365 **E**: GR

pardoi Wittmer, 1969d: 175 **N**: MO
parnassicus Wittmer, 1974d: 385 **E**: GR
paulocrassus Wittmer, 1974d: 388 **A**: KA
paulomaculatus Wittmer, 1997e: 337 **A**: YUN
persicus Pic, 1901d: 34 **A**: IN
petraensis Wittmer, 1971b: 185 **A**: JO
peyerimhoffi Constantin, 1979: 630 **N**: MO
pici Wittmer, 1936c: 678 **A**: IS
  nigrina Wittmer, 1936c: 678
pithanoides Escalera, 1913b: 323 (*Apteromalthinus*) **N**: MO
planulus Wittmer, 1969d: 173 **N**: AG MO
  horaki Švihla, 1980a: 243
planus Wittmer, 1988c: 368 **A**: GUX
prudeki Švihla, 2002a: 126 **E**: TR
pseudobiguttatus Constantin, 1975: 81 **E**: FR IT
pseudoflaveolus Wittmer, 1974d: 422 **E**: GG ST **A**: TR
pseudointerruptus Wittmer, 1974: 101 **A**: TR
pseudopersicus Wittmer, 1972c: 201 **A**: IN
pseudoreflexus Švihla, 1994: 118 **E**: IT (Sicilia)
pseudoscriptus Wittmer, 1971b: 170 **E**: IT
quadrinodosus Champion, 1926b: 209 **A**: "Tibet"
quadratipennis Kim & Kang, 2000: 115 **A**: SC
quedenfeldti Pic, 1910o: 57 (*Progeutes*) **N**: MO
ramamensis Wittmer, 1978e: 354 **A**: SD
rauschi Wittmer, 1978e: 362 **A**: IN
reductelineatus Pic, 1922b: 30 **A**: TAI
reductipennis Wittmer, 1971b: 185 **A**: JO
reflexus Wittmer, 1974d: 406 **E**: IT
ressli Wittmer, 1978e: 358 **A**: TR
rhaphidiceps Kiesenwetter, 1852: 261 **E**: BU **A**: IN TR
  obscuricollis Fairmaire, 1883c: xxxiv
robusticornis Wittmer, 1993b: 293 **A**: SCH
robustus Motschulsky, 1853a: 4 **E**: ST
rosinae Pic, 1900d: 37 **A**: SY TR
  delagrangei Pic, 1900h: 61
rubricollis Baudi di Selve, 1859a: 297 **E**: BU CR FR GR IT **N**: TU **A**: SY
  pyrrhoderus Fairmaire, 1883c: xxxiv
rufifrons Motschulsky, 1860a: 63 (*Malachidius*) **A**: CY LE SY TR
  sericellus Marseul, 1868b: 179
rydhi Wittmer, 1993b: 285 **A**: TR
sagamiensis Takahashi, 1992: 109 **A**: JA (Honshu)
samsunicus Wittmer, 1974d: 414 **A**: TR
sanpedroensis Wittmer, 1971b: 183 **E**: SP
scapularis Marseul, 1878a: 23 **E**: IT (Sicilia) MA
schaubi Wittmer, 1969d: 178 **N**: MO
schoeni Švihla, 2002a: 128 **E**: GR (Rodos)
scriptipennis Pic, 1900j: 87 **E**: SP **N**: AG MO
  tenietensis Pic, 1900j: 87
scriptoides Wittmer, 1971b: 171 **E**: SP
scriptus Kiesenwetter, 1852: 611 **E**: FR GR IT (Sicilia) PT SP **N**: AG MO MR
  diversithorax Pic, 1901f: 51
  filicornis Kiesenwetter, 1852: 611
  maculiventris Fairmaire, 1884d: lxv
  marginipennis Bourgeois, 1888b: 164
scutellaris Rosenhauer, 1856: 145 **E**: SP **N**: MO

seriepunctatus Kiesenwetter, 1852: 610 **E**: AU BE BH CR CZ FR GE HU IR IT PT RO SK SL SP SZ YU **N**: AG
binotaticollis Pic, 1913o: 185
distinctipes Pic, 1901c: 25
mediocrenotatus Pic, 1914r: 187
mimica Pic, 1906f: 31
multonotata Pic, 1910a: 1
pallidithorax Pic, 1899j: 207
raffrayi Pic, 1910a: 1
raveli Pic, 1906f: 31
subopacus Pic, 1900j: 87
setulosus Wittmer, 1988c: 370 **A**: HKG
sexpustulatus Wittmer, 1995b: 147 **A**: SCH
shimomurai Wittmer, 1984c: 157 **A**: TAI
sibilleae Wittmer, 1969d: 179 **A**: JO
sicanus Kiesenwetter, 1871: 77 **E**: IT (Sicilia)
sikkimensis Pic, 1924j: 477 **A**: SD
simplicipes Pic, 1899n: 374 **E**: GR **A**: TR
sinensis Pic, 1910p: 61 **A**: HKG SCH TAI
sordidus novaki Reitter, 1889s: 373 **E**: SL
sordidus sordidus Kiesenwetter, 1871: 76 **E**: FR (Alpes-Maritimes) IT SZ
brunneosulcata Pic, 1940b: 10
ssulingensis Wittmer, 1993b: 289 **A**: TAI
stanislavi Švihla, 2004a: 80 **A**: TR
stemmleri Wittmer, 1974e: 629 **A**: BT
stigmatias Kiesenweter, 1866: 255 **E**: SP
strangulatus Wittmer, 1974d: 395 **A**: TR
subgeniculatus Pic, 1930f: 14 [RN] **N**: MO
geniculatus Pic, 1921m: 146 [HN]
sulcicollis Marseul, 1878a: 24 **N**: AG TU
collaris J. R. Sahlberg, 1903d: 43 [HN]
mutata Pic, 1910a: 1 [RN]
swaneticus Wittmer, 1974d: 420 **E**: GG ST
swatensis Wittmer, 1987b: 104 **A**: PA
syriacus Marseul, 1868b: 178 **A**: JO SY
maximiceps Pic, 1917g: 13
tadzhikistanicus Švihla, 1980a: 244 **A**: TD
taforaltensis Wittmer, 1969d: 175 **N**: MO
taiwanoniger Wittmer, 1979e: 336 **A**: TAI
taiwanus Wittmer, 1984c: 157 **A**: TAI
tauri anatolicus Wittmer, 1974d: 389 **A**: TR
tauri tauri Pic, 1907m: 161 **A**: TR
nigronotatus Pic, 1908i: 65
temperei Constantin, 1975: 85 **E**: FR (Corse)
theresae Pic, 1899n: 373 **A**: TR
smyrnensis Pic, 1915h: 21
theryi Pic, 1920b: 8 **N**: MO
longevittatus Pic, 1953e: 117
torretassoi Pic, 1934h: 449 **N**: EG
trinotaticeps Pic, 1951a: 9 **E**: SP
tuniseus terrestris Pic, 1905g: 146 **N**: AG
diversicolor Pic, 1905g: 146
tuniseus tuniseus Fairmaire, 1875a: 513 (*Malthodes*) **N**: AG MO TU
fuscipes Bourgeois, 1897: 42
mathieui Pic, 1900b: 14
nigribuccis Marseul, 1878a: 34

obscurior Bourgeois, 1897: 42
obscuricolor Pic, 1920f: 17
pubescens Pic, 1911l: 121
tangerianus Pic, 1902b: 18
tortiscelis Pic, 1898j: 154
varus Bourgeois, 1900a: 94
turcicus Pic, 1899j: 206 **E**: AU CZ GG GR HU RO SK **A**: TR
montandoni Pic, 1907j: 137
turcmenicus Švihla, 1980a: 244 **A**: TM
uniformis Wittmer, 1997e: 342 **A**: YUN
validiceps Pic, 1928a: 2 **A**: IN
vandykei Wittmer, 1956b: 305 **A**: ZHE
vartiani Wittmer, 1966c: 492 **A**: IN
vaucheri Pic, 1901p: 111 **N**: MO
kocheri Pic, 1953e: 117
vaulogeri Pic, 1915h: 21 **N**: AG
versatilis Delkeskamp, 1939: 239 [RN] **E**: IT (Sicilia)
laticeps Fiori, 1915a: 48 [HN]
verticalis Pic, 1930d: 5 **E**: IT (Sicilia)
vitellinus Kiesenwetter, 1866a: 374 **E**: SP
vixlimbatus Wittmer, 1984c: 155 **A**: TAI
vonschulthessi Wittmer, 1974e: 627 **A**: BT
walteri Švihla, 2002a: 132 **E**: TR
wewalkai Wittmer, 1980b: 410 **E**: GR
wittmeri Švihla, 1998: 227 **A**: TR
yangmingensis Wittmer, 1984c: 156 **A**: TAI
yunnanus Wittmer, 1995b: 150 **A**: YUN
zahradniki Švihla, 2002a: 131 **E**: TR

tribe Malthodini Böving & Craighead, 1930

genus *Frostia* Fender, 1951: 524 type species *Malthodes transversus* LeConte, 1866 (= *Frostia laticollis* LeConte, 1861)
Stenomaltypus Wittmer, 1978d: 298 type species *Maltypus bhutanensis* Wittmer, 1975
abdominalis Wittmer, 1995b: 167 **A**: YUN
bhutanensis Wittmer, 1975a: 256 (*Maltypus*) **A**: BT
cornuta Wittmer, 1997e: 353 **A**: YUN
pectinata Wittmer, 1992d: 101 **A**: NP
peculiaria Wittmer, 1993b: 281 **A**: SCH
pseudobhutanensis Brancucci, 1984: 188 **A**: NP
semiinflata Wittmer, 1993b: 279 **A**: YUN

genus *Malthodes* Kiesenwetter, 1852: 242 type species *Malthinus marginatus* Latreille, 1806
subgenus *Malthodes* Kiesenwetter, 1852: 242 type species *Malthinus marginatus* Latreille, 1806
Apodistrus Reitter, 1882: 30 type species *Malthodes brachypterus* Kiesenwetter, 1863
Dalthomes Fiori, 1905c: 239 type species *Malthodes penninus* Baudi di Selve, 1872
Hapaloderus Motschulsky, 1853a: 6 type species *Hapaloderus ustulatus* Motschulsky, 1853
Maltharchus J. Weise, 1892e: 408 type species *Cantharis brevicollis* Paykull, 1798
Malthinellus Seidlitz, 1889b: 510 [HN]
Malthodellus Bourgeois, 1894a: 174 [RN]
Malthodinus Fiori, 1905c: 236 type species *Malthodes facetus* Kiesenwetter, 1863
Mesomalthodes Fiori, 1905c: 236 type species *Malthodes dispar* Germar, 1824
Paramalthodes J. Weise, 1893b: 364 type species *Malthodes nigricollis* Baudi di Selve, 1859
Podistrina Seidlitz, 1889b: 510 [HN] type species *Malthodes apterus* Mulsant, 1862 (= *Malthodes lobatus* Kiesenwetter, 1852)
Promalthodes Fiori, 1905c: 235 type species *Malthodes simplex* Kiesenwetter, 1872
Prothalmodes Fiori, 1905c: 239 type species *Malthodes cruciatus* Baudi di Selve, 1859

abandensis Wittmer, 1970a: 48 **A**: TR
abeillei Bourgeois, 1894d: cclxxxix (*Podistrina*) **N**: AG
abkhasicus Wittmer, 1979c: 141 **E**: GG ST
acutopygus Wittmer, 1966c: 499 **A**: JO
adjaricus Wittmer, 1992c: 29 **E**: GG
aegyptiacus Wittmer, 1937: 167 **N**: EG
aemulus Kiesenwetter, 1861: 382 **E**: FR IT SZ
akbesianus Pic, 1901e: 36 **A**: SY TR
akshehirensis Wittmer, 1970a: 34 **A**: TR
alanyaensis Wittmer, 1970a: 28 **A**: TR
alatauicus Kazantsev, 1995d: 104 **A**: KI
albanicus Wittmer, 1970a: 56 **E**: AL
alemdagensis Wittmer, 1970a: 44 **A**: TR
alfierii Pic, 1926o: 228 **N**: EG
allenspachi Wittmer, 1980b: 406 **E**: GR
alpicola Kiesenwetter, 1852: 289 **E**: AU CZ GE IT SZ
alpinus Motschulsky, 1853a: 11 (*Hapaloderus*) **E**: AU SZ
amasiae Wittmer, 1971e: 363 **A**: TR
amplithorax Wittmer, 1992c: 26 **E**: ST
analis Gebler, 1842: 38 (*Malthinus*) **A**: "Siberia"
anatolicus Wittmer, 1966c: 496 **A**: TR
andalusiacus Wittmer, 1970a: 18 **E**: SP
angorensis Pic, 1908d: 42 **A**: TR
angulatus Wittmer, 1970a: 50 **A**: TR
angulipygus Wittmer, 1980b: 395 **E**: GR
angulistilus Wittmer, 1978e: 365 **A**: IN
angusticollis Motschulsky, 1853a: 9 (*Hapaloderus*) **E**: SZ
ansariensis Švihla, 1994: 121 **A**: SY
antiochiacus Wittmer, 1970a: 35 **A**: TR
appendicifer Bourgeois, 1903d: 219 **N**: AG MO TU
arbaensis Pic, 1904b: 3 (*Podistrina*) **N**: TU
arbustorum Kiesenwetter, 1866b: 262 **E**: SP
argivus Kiesenwetter, 1859a: 29 **E**: GR
  hymettensis Pic, 1903q: 161
armitagei Pic, 1914q: 77 **N**: MO
atlasicus Pic, 1932g: 26 [= 1932c: 10] **N**: MO
atratus atratus Baudi di Selve, 1859a: 299 **E**: FR SP
  meloiformis Linder, 1863: 120
atratus guadarramensis Wittmer, 1970a: 71 **E**: SP
atratus samniticus Fiori, 1908: 2 **E**: IT
atroapterus Fairmaire, 1884d: lxvi **N**: AG TU
  alata Pic, 1918g: 13
  algirica Bourgeois, 1900d: 284 (*Podistrina*)
  diversistylus Pic, 1909a: 328
austriacus Hicker, 1960: 80 **E**: AU
baicalicus Wittmer, 1970a: 15 **A**: ES FE
balfourbrownei Wittmer, 1966c: 494 **E**: GG **A**: TR
barbarus Pic, 1903h: 157 **N**: AG MO TU
  obscurior Pic, 1935j: 238
batillifer Bourgeois, 1901: 143 **N**: AG TU
  semilimbatus Pic, 1903c: 90
beauprei Pic, 1909c: 97 **N**: AG
benedikti Švihla, 2002a: 135 **A**: SY
berberidis Kiesenwetter, 1866b: 261 **E**: SP
bertolinii Fiori, 1905c: 248 (*Dalthomes*) **E**: IT SZ

berytensis Reiche & Saulcy, 1857b: 177 **A**: CY IS LE SY
  grypopygus Pic, 1912m: 58
  thoracicus J. R. Sahlberg, 1903c: 23
besucheti besucheti Wittmer, 1970a: 99 **A**: TR
besucheti bucakensis Wittmer, 1970a: 39 **E**: GR
bidens Wittmer, 1970a: 58 **E**: BU
bidentatopygus Pic, 1935j: 239 **N**: AG TU
bifurcatus Kiesenwetter, 1852: 297 **E**: IT (Sicilia) MA
  cameroni Pic, 1903h: 155
  malcolmi Pic, 1904g: 49
binotaticollis Pic, 1935f: 14 **N**: MO
biroi Kaszab, 1955: 300 **E**: RO (Carpathians)
bohaci Švihla, 2002a: 137 **E**: GG
boicus Kiesenwetter, 1863a: 723 **E**: AU GE IT (Alps)
borsaensis Švihla, 1980a: 247 **E**: RO
bosnicus Wittmer, 1980b: 411 **E**: BH
bourgeoisi Reitter, 1888e: 210 (*Malchinus*) **E**: GG ST **A**: IN TR
  rhaphidostylus Pic, 1912m: 59
brachythorax Motschulsky, 1853a: 7 (*Hapaloderus*) **E**: AU
brancuccii Wittmer, 1984c: 160 **A**: TAI
brevepigus Pic, 1935f: 14 **N**: MO
brevicollis Paykull, 1798: 269 (*Cantharis*) **E**: AU BE BU BH CT CZ FR GB GE IT NL NT PL SK SV SZ UK **A**: ES WS
  lunifer Czwalina, 1884: 272
  nigellus Kiesenwetter, 1852: 307
brevicornis Fiori, 1906a: 6 (*Maltharcus*) **E**: IT
brevifurcatus Pic, 1912m: 59 **E**: AU IT
brevistylus Fiori, 1906a: 15 (*Maltharcus*) **E**: IT
brignolii Wittmer, 1974b: 21 **A**: LE
bugnioni Pic, 1902l: 71 **N**: TU
camptopygus Pic, 1912m: 61 **N**: AG
canariensis Escalera, 1921: 297 **N**: CI
cantabricus Wittmer, 1958: 118 **E**: SP
carinatopygus Wittmer, 1971f: 518 **E**: AL BH
carmelicus Wittmer, 1966a: 3 **A**: IS
caspicus Wittmer, 1972c: 202 **A**: IN
castaneicollis Reitter, 1888e: 208 **E**: ST
caucasicus Wittmer, 1958: 120 **E**: ST
caudatomimicus Wittmer, 1970a: 80 **E**: AU IT
caudatus J. Weise, 1892e: 410 **E**: AU CZ GE HU SZ
cavicornis Wittmer, 1982e: 345 **A**: HNG
cerambyciformis Normand, 1950a: 67 **N**: TU
championi Holdhaus, 1923: 131 **E**: FR (Corse)
chelifer Kiesenwetter, 1852: 614 (*Malthinus*) **E**: FR SP
chorezmicus Švihla, 1980a: 247 **A**: UZ
chvojkai Švihla, 1997b: 125 **E**: AL
cinctellus Motschulsky, 1853a: 7 (*Hapaloderus*) **E**: FR
circassicus Švihla, 1980a: 249 **E**: ST
clermonti Pic, 1912m: 58 **A**: TM
cochlearius Pic, 1909q: 170 **E**: FR (Hautes-Pyrénées) SP (Asturia)
cognatus A. Costa, 1858: 25 **E**: IT
  andreae Ganglbauer, 1906b: 54
  nigra Fiori, 1906a: 7
  nigriceps Mulsant, 1862a: 411
comptus Bourgeois, 1894a: 188 **E**: FR (Alpes-Maritimes)

concoloripennis Pic, 1923a: 2 **A**: TR
conicus Wittmer, 1970a: 66 **E**: IT
constantini Wittmer, 1971e: 365 **E**: IT (Elba)
continuus Bourgeois, 1899a: 340 **E**: FR
corcyreus Bourgeois, 1903b: 152 **E**: GR
cordiger Kiesenwetter, 1863b: 429 **E**: FR (Corse) IT (Sardegna)
cordigerioides Pic, 1915k: 46 **E**: IT
cornutopygus Wittmer, 1966a: 1 **A**: IS
corsicus Kiesenwetter, 1871: 79 **E**: FR (Corse) IT (Sardegna)
cortadesi Pic, 1905g: 146 **N**: AG
crassicornis Maeklin, 1846: 179 (*Malthinus*) **E**: AU CT CZ FI FR GB GE GG HU NT ST SV SZ UK **A**: IN TR
  helveticus Kiesenwetter, 1852: 279
  moesiacus J. Weise, 1895e: 215 (*Maltharcus*)
  subunicolor Pic, 1911o: 138
  vosykai Wittmer, 1964: 100
creticus Kiesenwetter, 1859a: 29 **E**: GR (Kríti)
croceicollis Motschulsky, 1853a: 7 (*Hapaloderus*) **E**: GE SZ
cruciatus Baudi di Selve, 1859a: 301 **E**: FR (Corse) IT (Sardegna)
crucicollis Fairmaire, 1884d: lxv **N**: AG
cruciferarum Kiesenwetter, 1866b: 261 **E**: SP
curticornis Wittmer, 1997a: 40 **A**: GUX
curtipennis Motschulsky, 1853a: 8 (*Hapaloderus*) **E**: GG
curvifurca Hicker, 1955: 58 **E**: GR (Kerkyra)
cyphonurus Kiesenwetter, 1861: 383 **E**: AU FR IT SZ
cypricus Pic, 1911i: 137 **A**: CY
debilis Kiesenwetter, 1852: 296 **E**: AU BE BH CR CT CZ FR GE HU IT NT PL RO SK SL ST SZ UK
  volgensis Kiesenwetter, 1872c: 377
delagrangei Pic, 1901e: 35 **A**: SY
delkeskampi Wittmer, 1968b: 278 [RN] **A**: FUJ
  longipennis Wittmer, 1951: 102 [HN]
demaisoni Pic, 1909s: 177 **A**: TR
denizlianus bergamensis Švihla, 2002a: 141 **E**: TR
denizlianus denizlianus Wittmer, 1970a: 24 **A**: TR
desbrochersi Pic, 1903p: 157 **N**: AG
deubeli Wittmer, 1970a: 16 **E**: RO
dieneri Kaszab, 1955: 300 **E**: HU RO SK **A**: TR
dilizhanus Švihla, 1983: 9 **E**: AR
dilutus Motschulsky, 1853a: 12 (*Hapaloderus*) **E**: GE
dimidiaticollis dimidiaticollis Rosenhauer, 1847: 20 (*Malthinus*) **E**: AU CT CZ GE HU SK ST UK **A**: IS JO KZ LE TR
  auritus Motschulsky, 1853a: 9 (*Hapaloderus*)
  pulicarius L. Redtenbacher, 1849: 327 (*Malthinus*)
  signaticollis Barovskij, 1929b: 268
dimidiaticollis seleukos Wittmer, 1972c: 203 **A**: IN
dimidiaticollis zagrosensis Wittmer, 1978e: 370 **A**: IN
discicollis Baudi di Selve, 1859a: 302 **E**: FR IT
  meridianus Mulsant, 1862a: 402
dispar Germar, 1824: 73 (*Malthinus*) **E**: AU BE BU BH CZ FR GB GE GR HU IR IT NL PL SK SZ UK
  arvernica Pic, 1910r: 67
  decolorata Fiori, 1905c: 250 (*Mesomalthodes*)
  decora Bourgeois, 1903f: 336
  jurassica Pic, 1914r: 193
  laetus Waltl, 1838b: 270 (*Malthinus*)
  laterufescens Pic, 1914m: 53
  neglecta Mulsant, 1862b: 396
  noualhieri Bourgeois, 1894a: 191

putoni Bourgeois, 1902d: 182
subobscuripes Pic, 1914m: 52
distinctithorax Pic, 1901e: 36 **A**: TR
binotaticollis Pic, 1911q: 145
chromodera Pic, 1912m: 58
hickeri Wittmer, 1958: 119
distinctus Pic, 1901b: 17 **N**: AG TU
diversiventris Pic, 1903p: 157 **N**: AG
doctoris Pic, 1935j: 239 **N**: AG
doderoi Bourgeois, 1903c: 210 (*Podistrina*) **E**: IT (Sardegna)
elaboratus Pic, 1903p: 156 **N**: TU
elongaticornis Pic, 1918e: 10 **N**: AG
emiliae Pliginsky, 1923: 141 **E**: UK (Krym)
ensifer Kiesenwetter, 1871: 80 **E**: FR (Corse) IT (Sardegna)
escorialinus Pic, 1910a: 1 [RN] **E**: SP
escorialensis Pic, 1909q: 170 [HN]
espanoli Wittmer, 1958: 115 **E**: SP
euboicus euboicus Pic, 1903l: 138 **E**: GR (Evvoia)
euboicus gerhardti Pic, 1912m: 59 **E**: GR (Kerkyra)
europaeus Wittmer, 1970a: 81 **E**: AU CZ HU GE PL SK SZ
facetus Kiesenwetter, 1863b: 429 **E**: FR IT
favarcqi Pic, 1903r: 169 **N**: AG
fenchihuensis Wittmer, 1984c: 160 **A**: TAI
fibulatus Kiesenwetter, 1852: 300 **E**: AU BE CT CZ GB GE HU IT NL NT PL SK SV SZ UK YU
fiorii Ganglbauer, 1906: 50 **E**: FR (Alpes-Maritimes) IT YU
flavoguttatus Kiesenwetter, 1852: 294 **E**: AU BE BU CT CZ FR GB GE HU IR IT NT SK SP SV SZ UK
alpinus Mulsant, 1862a: 399
inapicalis Pic, 1910f: 67
limbiventris C. G. Thomson, 1864: 206
minimus Gyllenhal, 1808a: 344 (*Cantharis*) [HN]
minimus Stephens, 1830: 309 (*Malthinus*) [HN]
nigricollis Dufour, 1851: 329
obliquus J. Weise, 1887a: 53
obscurimembris Pic, 1933f: 13
forcipifer Kiesenwetter, 1852: 614 (*Malthinus*) **E**: FR PT SP
ponferradanus Pic, 1910a: 1
fortesulcatus Pic, 1938k: 161 **A**: CH ("Tchen Ly")
frantiseki Švihla, 2002a: 134 **A**: SY
frater Wittmer, 1970a: 26 **A**: TR
freyi Wittmer, 1939a: 3 **E**: IT
furcatistylus Pic, 1919d: 9 **N**: TU
furcatopygus Wittmer, 1954b: 51 **A**: JA (Honshu)
furthi Švihla, 1994: 122 **A**: IS
fuscus Waltl, 1838b: 270 (*Malthinus*) **E**: AU BE CT CZ GB GE FI FR IR IT NL NT SK ST SZ UK
pellucidus Kiesenwetter, 1852: 280
ganglbaueri Fiori, 1906a: 4 **E**: IT
rufithorax Falzoni, 1924: 109
gavoyi Bourgeois, 1905a: 21 **E**: FR (Pyrénées)
genistae Kiesenwetter, 1866b: 260 **E**: SP
gibbiceps Reitter, 1889c: 68 (*Malchinus*) **E**: AB
gillerforsi Wittmer, 1993b: 293 **A**: TR
goliath Pic, 1903p: 158 **N**: AG
gracilipygus Pic, 1919d: 9 **N**: AG
graecus Kiesenwetter, 1872c: 374 **E**: GR
spinosostylus Pic, 1915k: 45
winkleri Pic, 1912m: 59

grandicornis Pic, 1905c: 106 **N**: AG
gratiosus Ganglbauer, 1906b: 51 **E**: IT
 simbruinica Luigioni, 1920: 206
grigorjevi Barovskij, 1929d: 487 **A**: TD
 esgaudensis Wittmer, 1993b: 299
guttatoapicalis Pic, 1938k: 161 **A**: CH (Tcheou)
guttifer Kiesenwetter, 1852: 287 **E**: AU BE BH CT CZ FI FR GB GE IR IT NT PL SK SL SP SZ UK **A**: MG SCH YUN
 guttegenus Ganglbauer, 1911: 269
 robustior Pic, 1903f: 154
gyropygus Pic, 1912m: 58 **A**: LE
hagai Takahashi, 2002c: 279 **A**: JA (Honshu)
halabensis Švihla, 2002a: 136 **A**: SY
hamatus Kiesenwetter, 1852: 615 **E**: FR (Corse) IT (Sardegna)
hastatostylus Wittmer, 1966c: 498 **A**: LE
hastulifer Kiesenwetter, 1871: 79 **E**: IT (Sicilia)
heinertzi Wittmer, 1979c: 142 **A**: TR
hellenicus Wittmer, 1980b: 393 **E**: GR
herzegovinus Wittmer, 1970a: 56 **E**: BH
hetiticus hetiticus Wittmer, 1970a: 29 **A**: TR
hetiticus karatepeensis Wittmer, 1970a: 30 **A**: TR
hexacanthus Kiesenwetter, 1852: 303 **E**: AU BE BH CZ FR GE HU IT RO SK SZ YU
 funditor J. Weise, 1887a: 56
 inepta Fiori, 1908: 11 (*Maltharcus*)
 ornatula Bourgeois, 1894a: 203
 tetracanthus Kiesenwetter, 1871: 79
hikosanus Takakura, 1988c: 153 **A**: JA (Honshu, Kyushu)
hipponensis Pic, 1903p: 155 **N**: AG
hiranoi Nakane, 1977a: 97 **A**: JA (Hachijô-jima)
hladili Švihla, 1980a: 245 **E**: YU
hoberlandti Švihla, 1983: 6 **A**: IN
holdhausi Kaszab, 1955: 298 **E**: AU CZ HU RO SK SZ
holtzi Pic, 1904c: 9 **E**: GR (Kríti)
holzschuhi Wittmer, 1978e: 364 **A**: IN
ibericus Rosenhauer, 1856: 147 **E**: SP
ilicicola Pic, 1912m: 60 **N**: AG
ilvensis Holdhaus, 1923: 131 **E**: IT (Elba, Giglio)
impennis Pic, 1901d: 35 (*Podistrina*) **N**: TU
implicatus Wittmer, 1963: 117 **A**: JO
impressithorax Wittmer, 1997e: 349 **A**: YUN
inapicalis Pic, 1905g: 146 **N**: AG
inapicipennis Pic, 1923j: 19 **N**: MO
incilostilus Wittmer, 1970a: 45 **A**: TR
incostatus Wittmer, 1972c: 204 **A**: IN
insularis Kiesenwetter, 1871: 80 **E**: FR (Corse) IT (Sardegna)
insulcatus Pic, 1903q: 161 **E**: AL GR (Kerkyra)
jaromiri Švihla, 2002a: 138 **E**: GG
jelineki Švihla, 2002a: 144 **E**: SP
jordanicus Wittmer, 1963: 115 **A**: JO
kahleni Wittmer, 1982e: 345 **E**: AU CZ SK
kantnerorum Švihla, 2002a: 145 **A**: SY
kazantsevi Wittmer, 1992c: 34 **E**: ST
kazakhstanicus Wittmer, 1979c: 141 **A**: KZ
kerzhneri kerzhneri Wittmer, 1979e: 338 **A**: FE (Kurils) JA (Hokkaido)
 nikitskyi Kazantsev, 1995d: 100

kerzhneri morishamai Takahashi, 2002c: 277 **A**: JA (Honshu)
kiesenwetteri Wollaston, 1854: 243 **N**: MR
kinugasai Takahashi, 2002c: 276 **A**: JA (Honshu)
kirisensis Švihla, 2004a: 82 **A**: TR
kiritshenkoi Barovskij, 1931a: 101 **A**: FE MG
kaszabi Wittmer, 1965b: 84
kisilensis Wittmer, 1970a: 60 **A**: TM (Kopet-Dag)
klapperichi assyrius Švihla, 2002a: 142 **E**: TR
klapperichi klapperichi Wittmer, 1963: 116 **A**: JO
klapperichi mesopotamicus Švihla, 1983: 8 **A**: IQ
kobiensis Wittmer, 1970a: 62 **E**: GG ST
kochi Wittmer, 1935e: 460 **E**: SP
koestlini Wittmer, 1980b: 411 **E**: GR (Kríti)
kopeckyi Švihla, 2002a: 140 **E**: TR
kopetzi Švihla, 2002a: 141 **E**: TR
kopetdagensis Wittmer, 1970a: 59 **A**: TM
kouichii Takahashi, 2002c: 275 **A**: JA (Honshu)
kuatunensis Wittmer, 1951: 102 **A**: FUJ
kurbatovi Kazantsev, 1995d: 102 **A**: FE
kurosawai Wittmer, 1954b: 48 **A**: JA (Hokkaido, Honshu)
kyushuensis Takakura, 1988c: 154 **A**: JA (Honshu, Kyushu)
laciniatus agrigentinus Fiori, 1915a: 55 **E**: IT (Sicilia)
laciniatus laciniatus Kiesenwetter, 1852: 304 **E**: IT (Sicilia, Sardegna)
messenius Fiori, 1904: 74
ragusai Fiori, 1904: 75
lamelliger Wittmer, 1993b: 302 **A**: SCH
lameyi Pic, 1909c: 98 **N**: AG
latepallidus Pic, 1902g: 48 (*Podistrina*) **N**: AG
latialis Fiori, 1906a: 13 **E**: IT
latithorax Pic, 1903g: 114 **N**: AG
lautus lautus Kiesenwetter, 1852: 285 **E**: BH CR IT SL
nigricollis Baudi di Selve, 1859a: 298
schreiberi Reitter, 1891i: 259
lautus narentinus Wittmer, 1970a: 15 **E**: BH
lederi Pic, 1912m: 59 **E**: ST
leveincisus Wittmer, 1966a: 2 **A**: IS
libenae Švihla, 2002a: 144 **A**: SY
licenti Pic, 1938k: 161 **A**: NMO
subguttata Pic, 1938k: 161
liegeli J. Weise, 1890b: 348 **E**: AU BU CR GE GR IT
brancsiki Reitter, 1903b: 232 (*Malchinus*)
sulcata Bourgeois, 1903b: 154 (*Malthodellus*)
ligulatistylus Wittmer, 1960b: 108 **E**: GG
lividianus Pic, 1911w: 177 **A**: CY TR
lobatus Kiesenwetter, 1852: 317 **E**: AU CZ FR GE IT SK
apterus Mulsant, 1862a: 411
brachypterus Kiesenwetter, 1863a: 725
massiliensis Abeille de Perrin, 1896a: 155
lobicollis Reitter, 1888e: 208 (*Apodistrus*) **E**: "Caucasus"
lokveanus Wittmer, 1970a: 55 **E**: HU IT
longipygus Wittmer, 1954b: 49 **A**: JA (Hokkaido, Honshu)
longistylus Pic, 1909c: 97 **N**: AG
lucernensis Wittmer, 1981a: 95 **E**: CZ SZ
ludwigi ludwigi Pic, 1901f: 50 (*Podistrina*) **E**: GR
abdominalis Pic, 1915k: 45

cephalenicus Pic, 1915k: 45
olympica Pic, 1901f: 51
ludwigi patrasicus Wittmer, 1980b: 400 **E**: GR
luteostylus Pic, 1919d: 9 **N**: TU
lyriformis Wittmer, 1992c: 29 **E**: ST
maceki Švihla, 1994: 119 **A**: SY
maderi Hicker, 1955: 59 **E**: AL BH YU
makiharai Wittmer, 1984c: 162 **A**: TAI
malickyi Wittmer, 1980b: 413 **A**: CY
manubriatus Kiesenwetter, 1863b: 430 **E**: IT
marashensis Wittmer, 1970a: 51 **A**: TR
marginatus Latreille, 1806: 261 (*Malthinus*) **E**: AU BE BH CT CZ FR GB GE IR IT NT PL SK SL SP SV SZ UK YU
biguttatus A. G. Olivier, 1790a: no 26: 16 (*Telephorus*) [HN]
constrictus Fiori, 1905c: 241
nigricollis Stephens, 1830: 308 (*Malthinus*)
marocanus Pic, 1930f: 13 **N**: MO
maronitus Pic, 1903q: 162 **A**: SY
mathieui Pic, 1903p: 156 **N**: AG
mauritanicus Pic, 1912m: 60 **N**: MO TU
maurus Laporte, 1840a: 277 (*Malthinus*) **E**: AL AU BE CZ FR GB IT GE SK SV SZ UK YU
atrostylus Pic, 1914r: 194
carbonarius Waltl, 1838b: 269 (*Malthinus*)
distans C. G. Thomson, 1864: 204
profanus Kiesenwetter, 1852: 302
unicolor Dufour, 1851: 329 (*Malthinus*)
mediocricornis Pic, 1918g: 13 **N**: AG
medvedevi Wittmer, 1992c: 22 **E**: GG ST
megesticus Wittmer, 1935d: 249 **E**: GR (Castelrosso)
mendax Kiesenwetter, 1871: 77 **E**: IT (Sardegna) FR (Corse)
mersinensis Pic, 1901e: 35 **A**: TR
micropygus Wittmer, 1970a: 28 **A**: TR
minimus Linnaeus, 1758: 402 (*Cantharis*) **E**: AU BE BH BU CT CZ FR GB GE HU IR IT LU NL NT RO SK ST SV SZ UK YU **A**: ES WS
cognatus Stephens, 1830: 309 (*Malthinus*)
fulvicollis Stephens, 1830: 309 (*Malthinus*)
marginicollis Schilsky, 1892: 198
melanocephalus Marsham, 1802: 374 (*Cantharis*)
ruficollis Latreille, 1806: 261 (*Malthinus*)
sanguinicollis Fallén, 1807: 15 (*Cantharis*)
sanguinolentus Gyllenhal, 1808a: 343
minoicus Wittmer, 1971e: 366 **E**: GR (Kríti)
minutopygus Wittmer, 1954b: 48 **A**: JA (Honshu)
misellus Kiesenwetter, 1852: 299 **E**: AU BE FR GE IT LU SZ
moczarskii Ganglbauer, 1912: 181 **E**: UK (Krym)
tauricus Istomina, 1969: 1104
modestus Kiesenwetter, 1852: 616 **E**: FR SP
monstrosus Pic, 1909c: 98 **N**: AG
montandoni Pic, 1922g: 26 **E**: MC RO
comanensis Pic, 1922g: 26
vlascana Pic, 1922g: 26
montanus Kiesenwetter, 1863a: 723 **E**: AU BE GE HU SZ
morio Motschulsky, 1853a: 11 (*Hapaloderus*) **E**: AU
muraniensis Švihla, 2004a: 82 **E**: SK SV
murgianus Pic, 1914q: 77 **E**: IT
mushensis Wittmer, 1997e: 345 **A**: TR

mutatus Wittmer, 1970a: 40 **E**: ST
mysticus mysticus Kiesenwetter, 1852: 282 **E**: AL AU BE CZ FR GB GE IR IT LS RO SK SL SV SZ UK **A**: FE NC
atrithorax Pic, 1926b: 1
megalophtalmus Ganglbauer, 1911: 268
obscuriuscula Dietrich, 1857: 132
mysticus reductostylus Wittmer, 1980b: 400 **E**: GR YU
nantouensis nantouensis Wittmer, 1984c: 162 **A**: TAI
nantouensis taoyuanus Wittmer, 1984c: 163 **A**: TAI
neoganglbaueri Wittmer, 1970a: 57 **A**: LE SY
nevadensis Constantin, 1971: 961 **E**: SP
nigerrimus Pic, 1903l: 139 **E**: GR
nigrithorax Pic, 1909c: 99 **N**: AG
niitakaensis Wittmer, 1955: 42 **A**: TAI
notaticollis Bourgeois, 1902a: 60 (*Podistrina*) **N**: AG
novaki Reitter, 1894e: 48 (*Podistrina*) **E**: CR
numidicus Pic, 1909c: 97 **N**: AG MO TU
algiricus Pic, 1907e: 121
variicollis Pic, 1912m: 60
nurdagensis Wittmer, 1970a: 42 **E**: TR
nyholmi Wittmer, 1970a: 52 **E**: GG ST **A**: TR
obscuricollis Motschulsky, 1853a: 10 (*Hapaloderus*) **E**: UK (Krym)
obscuricolor Pic, 1903p: 156 **N**: AG TU
occultistylus Wittmer, 1958: 116 **E**: SP
ohbayashii Wittmer, 1954b: 48 **A**: JA (Honshu)
olympicus Švihla, 2004a: 84 **A**: TR
opaciceps Pic, 1901f: 50 (*Podistrina*) **E**: GR
oranensis Pic, 1901b: 18 **N**: AG
ordubadensis iranicus Wittmer, 1972c: 202 **A**: IN
ordubadensis ordubadensis Reitter, 1889a: 24 **E**: AR **A**: TR
orientalicus Wittmer, 1970a: 52 **E**: ST **A**: TR
osellai Wittmer, 1974b: 23 **A**: LE
paganettii Pic, 1912m: 60 **E**: GR YU
palaestinus Pic, 1901e: 36 **A**: IS
pallidipennis Pic, 1903p: 157 **N**: AG
pamphylicus Wittmer, 1970a: 32 **E**: TR
paradoxus Hicker, 1915a: 252 **E**: IT
parthenias Kiesenwetter, 1871: 78 **E**: IT
partipygus Wittmer, 1939a: 1 **E**: IT
parvus Kazantsev, 1995d: 102 **A**: FE
peneri Wittmer, 1966a: 4 **A**: IS
peniculatus Wittmer, 1970a: 49 **A**: TR
penninus icaricus Wittmer, 1940b: 545 **E**: AU IT SZ
penninus penninus Baudi di Selve, 1872: 119 **E**: IT SZ
penninus raeticus Wittmer, 1970a: 72 **E**: SZ
pergamonensis Wittmer, 1970a: 33 **A**: TR
pergensis Wittmer, 1993b: 296 **A**: TR
phoenicus Wittmer, 1970a: 41 **A**: TR
picticollis Kiesenwetter, 1871: 80 **E**: IT **N**: AG
pictipennis Pic, 1903p: 156 **N**: AG TU
pallidicolor Pic, 1903m: 156
pinnatus Kiesenwetter, 1871: 79 **E**: IT
marchii Pic, 1922g: 25 **E**: IT
pirazzolii Pic, 1922g: 26 **E**: IT
planatus Wittmer, 1970a: 30 **A**: TR

prescutellaris Pic, 1938k: 162 **A**: CH ("Tcheu")
procerulus Kiesenwetter, 1852: 306 **E**: FR IT
prodigiosus Kiesenwetter, 1852: 308 **A**: TR
servilis Pic, 1915a: 1
proximus Wittmer, 1968b: 276 **A**: IS JO
prudeki Švihla, 2002a: 139 **E**: TR
pseudobesucheti Wittmer, 1970a: 61 **E**: ST **A**: TR
pseudolobatus Wittmer, 1956a: 161 **A**: AF
pseudonipponicus Takahashi, 1996: 32 **A**: JA (Honshu)
pseudoparthenias Wittmer, 1980b: 404 **E**: GR
pulchellus P. H. Lucas, 1846: 189 (*Malthinus*) **N**: AG
pumilus Brébisson, 1835: 153 (*Malthinus*) **E**: BE CZ FR GB GE IR SK SP SV SZ UK **A**: IN TR
atomus C. G. Thomson, 1864: 202
brevicollis Kiesenwetter, 1852: 312 [HN]
pusillopygus Wittmer, 1965a: 82 **A**: CY
pygomelas Bourgeois, 1902a: 62 **N**: AG
pyrenaeus Constantin, 1971: 963 **E**: FR
quadrispinus Kiesenwetter, 1852: 613 **E**: SP
quadristilus Wittmer, 1992c: 32 **E**: ST
rabatensis Pic, 1920f: 17 (*Podistrina*) **N**: MO
rangonii Fiori, 1912a: 37 **E**: IT
rauschorum Wittmer, 1997e: 345 **A**: TR
raymondi Kiesenwetter, 1871: 77 **E**: IT (Sardegna)
recurvus Baudi di Selve, 1872: 118 (*Malthinus*) **E**: IT FR
mancinii Pic, 1935c: 5
reductopygus Pic, 1935d: 240 **N**: AG
reitteri Bourgeois, 1895: cclxvi (*Podistrina*) **A**: SY
rhadamantys Švihla, 1983: 7 **E**: GR (Kríti)
robustipygus Wittmer, 1997e: 347 **A**: YUN
robustithorax Pic, 1919f: 13 **N**: AG
rolciki Švihla, 1998: 228 **A**: TR
romadinae Kazantsev, 1995d: 106 **A**: TM
rosmarini Kiesenwetter, 1866b: 261 **E**: SP
rugosipennis Pic, 1925b: 1 (*Podistrina*) **N**: LB (Cyrenaica)
ruralis Kiesenwetter, 1871: 80 **E**: IT (Sicilia)
baudii Fiori, 1905: pl. 3
ruzickai Švihla, 2002a: 135 **E**: TR
rydhi Švihla, 1994: 122 **A**: TR
sahlbergi Bourgeois, 1903b: 153 **E**: GR (Kríti)
sardous Motschulsky, 1853a: 9 (*Hapaloderus*) **E**: IT (Sardegna)
sassarensis Strassen, 1954: 269 **E**: IT (Sardegna)
schatzmayri Wittmer, 1935d: 250 **E**: GR (Rodos)
schuberti Wittmer, 1966c: 495 **A**: TR
sedileformis Wittmer, 1993b: 297 **A**: GR
seleucianus euphraticus Švihla, 1994: 121 **A**: TR
seleucianus seleucianus Wittmer, 1970a: 25 **A**: TR
serbotae macedonicus Švihla, 1980a: 245 **E**: MC
serbotae occidentalis Wittmer, 1970a: 71 **E**: FR
serbotae serbotae Petri, 1906: 224 (*Podistrella*) **E**: RO (Carpatii merid.)
setifer Baudi di Selve, 1872: 117 **E**: FR IT SP SZ
sibilleae Wittmer, 1970a: 42 **A**: TR
sichuanus Wittmer, 1993b: 300 **A**: SCH
siculus Kiesenwetter, 1852: 310 **E**: FR IT SZ
silvicola Bourgeois, 1903e: 247
sidonius Pic, 1911q: 145 **A**: SY

signatus Chevrolat, 1861: 268 (*Malthinus*) **N**: AG
simplex Kiesenwetter, 1872c: 372 **E**: IT
romanus Pic, 1903p: 157
simpliciventris Pic, 1909c: 97 **N**: TU
simplipygus Wittmer, 1954b: 50 **A**: JA (Kyushu)
sinensis Pic, 1932i: 206 **A**: CH
solarii Pic, 1912m: 60 **E**: IT
sotschienesis Wittmer, 1970a: 63 **E**: ST
spalatrensis J. Weise, 1893b: 365 **E**: CR
spathifer Kiesenwetter, 1852: 313 **E**: AU BE CT CZ FR GE HU IT LU NT SK SP SV SZ UK
mionopsis Ganglbauer, 1911: 273
tetra Fiori, 1908: 9 (*Maltharcus*)
spatulistilus Wittmer, 1970a: 43 **A**: TR
spectabilis Kiesenwetter, 1871: 78 **E**: IT
sporadicus Wittmer, 1980b: 392 **E**: GR
spretus Kiesenwetter, 1852: 290 **E**: AU CZ FR GE HU IT NL RO SK SL SZ
affinis Mulsant, 1862a: 415
guttulifer Ganglbauer, 1911: 269
stanislavi Švihla, 2004a: 84 **A**: TR
stoltzi Hicker, 1915b: 380 **E**: IT
strejceki Švihla, 1990a: 200 **E**: GG ST
stylifer Kiesenwetter, 1866b: 262 **E**: SP
stylopygus Pic, 1912m: 59 **E**: BH CR
subsericeus Baudi di Selve, 1872: 116 [HN]
subdiversus Pic, 1903p: 157 **N**: AG
subductus Kiesenwetter, 1863a: 722 **E**: AU GE IT
subincisicollis Pic, 1915k: 46 **E**: CR
lineolatus Pic, 1915k: 46
subsericeus Kiesenwetter, 1852: 319 **E**: AU CR IT YU
cephalotes Baudi di Selve, 1859a: 298
tristis Kiesenwetter, 1872c: 373
sulcicollis Kiesenwetter, 1879b: 309 **A**: JA (Honshu)
sulphuribasis Reitter, 1888e: 209 **E**: AR
sulphuriguttatus Motschulsky, 1853a: 8 (*Hapaloderus*) **E**: GG
sveci Švihla, 1997b: 125 **E**: HU
svihlai Wittmer, 1995b: 169 [RN] **A**: IN
iranicus Wittmer, 1978e: 365 [HN]
szekessyi Wittmer, 1970a: 58 **E**: YU
taeniatus Bourgeois, 1903d: 220 **N**: TU
taipehanus Wittmer, 1956b: 307 **A**: TAI
taiwanus Wittmer, 1984c: 161 **A**: TAI
talyshensis Wittmer, 1972c: 203 **A**: IN
tanurensis Švihla, 1994: 121 **A**: IS
tartusensis Wittmer, 1981e: 412 **A**: SY
tauricus Pic, 1917g: 13 [RN] **E**: UK (Krym)
pumilus Motschulsky, 1853a: 13 (*Hapaloderus*) [HN]
taygetanus Švihla, 1983: 5 **E**: GR
tectosagum Pic, 1909c: 97 **E**: FR (Pyrénées)
tenax Kiesenwetter, 1871: 97 **E**: IT (Sardegna) FR (Corse)
tenietensis Pic, 1903p: 157 **N**: AG
tetraglyphis Hicker, 1953: 130 **E**: IT (Adamello)
theryi Pic, 1918e: 10 **N**: MO
meyeri Pic, 1953a: 3
tianshanicus Kazantsev, 1995d: 106 **A**: UZ
tingitanus Pic, 1912m: 60 **N**: MO

basicornis Pic, 1918g: 13
tokatensis Pic, 1911h: 108 **A**: TR
tordi Wittmer, 1970a: 52 **E**: GG **A**: TR
torretassoi Wittmer, 1935e: 459 **A**: LE
transeuropaeus Wittmer, 1970a: 83 **E**: AL AU CZ GE IT SK SZ
trapezoidalis Bourgeois, 1903d: 219 **N**: AG TU
tricuspidatus Wittmer, 1965a: 83 **A**: TR
trifurcatus atramentarius Kiesenwetter, 1852: 286 **E**: AU GE FR HU IT SK SZ YU
chobauti Bourgeois, 1899a: 338
manevali Pic, 1944b: 1
minor Stierlin, 1863a: 60
obscurior Pic, 1901o: 105 (*Podistrina*)
villardi Bourgeois, 1893: 32 (*Podistrina*)
trifurcatus peyerimhoffi Bourgeois, 1900d: 378 (*Podistrina*) **E**: FR (Alpes-Maritime)
trifurcatus trifurcatus Kiesenwetter, 1852: 284 **E**: AU CZ GE FR IT SZ YU
quadrifidus J. Weise, 1887a: 58
trifurcatus uralicus Kazantsev, 1995d: 99 **A**: WS (Ural)
tropidipygus Pic, 1912m: 58 **A**: SY
truncatopygus Pic, 1935f: 14 **N**: MO
tungluanus Wittmer, 1956b: 308 **A**: ZHE
turcicus Kiesenwetter, 1872c: 375 **A**: TR
umbrosus Kiesenwetter, 1871: 78 **E**: IT
ungulopygus Pic, 1911o: 137 **A**: CY
unicolor Motschulsky, 1853a: 9 (*Hapaloderus*) **E**: GG
ustulatus Motschulsky, 1853a: 6 (*Hapaloderus*) **E**: AU
validicornis Suffrian, 1851: 55 **E**: PT SP
escorialensis Pic, 1904c: 9
hispanicus Baudi di Selve, 1859a: 300
vaucheri Abeille de Perrin, 1896a: 155 (*Podistrina*) **N**: MO
vavrai Švihla, 1998: 230 **A**: TR
ventralis Motschulsky, 1853a: 11 (*Hapaloderus*) **E**: AU SZ
vincens Gredler, 1870: 10 **E**: AU IT
viridiventris Motschulsky, 1853a: 12 (*Hapaloderus*) **E**: NT (Ladoga)
viturati Pic, 1902g: 48 (*Podistrina*) **N**: AG
walteri Švihla, 2002a: 143 **E**: TR
weigeli Švihla, 2004a: 85 **A**: TR
wewalkai Wittmer, 1970a: 46 **A**: TR
wittmeri Constantin, 1971: 959 **E**: SP
yukihikoi Takahashi, 1996: 35 **A**: JA (Honshu)
yunnanus Wittmer, 1993b: 303 **A**: YUN
zacynthius Wittmer, 1980b: 390 **E**: GR
zahradniki Švihla, 1998: 230 **A**: TR
zangherii Hicker, 1958: 26 **E**: IT
zurcheri Pic, 1911w: 177 **A**: CY
x-notatus Pic, 1914q: 77
subgenus *Podistrina* Fairmaire, 1875a: 514 type species *Podistrina doriae* Fairmaire, 1875
doriae Fairmaire, 1875a: 514 **N**: TU
forficula Normand, 1935: 241 **N**: TU
lindbergi Pic, 1932g: 27 [= 1932c: 10] **N**: MO
normandi Pic, 1900i: 77 **N**: TU
pici Wittmer, 1940b: 544 **N**: MO
reducta Pic, 1935f: 14 **N**: MO
supersedens Peyerimhoff, 1925b: 55 **N**: AG MO

genus *Maltypus* Motschulsky, 1860a: 61 type species *Maltypus niger* Motschulsky, 1860
bengalicus Wittmer, 1978d: 303 **A**: SD
chanakhae Wittmer, 1975a: 256 **A**: BT
chinensis Wittmer, 1995b: 161 **A**: YUN
dilatatus compressus Wittmer, 1984b: 105 **A**: NP
dilatatus dilatatus Wittmer, 1984b: 105 **A**: NP
holzschuhi Wittmer, 1984b: 100 **A**: NP
kamjeensis Wittmer, 1975a: 254 **A**: BT
kugleri Wittmer, 1975a: 253 **A**: BT
ovalis Wittmer, 1984b: 102 **A**: NP
papunensis Wittmer, 1984b: 98 **A**: NP
pulchoki Wittmer, 1984b: 102 **A**: NP
ryukyuanus Wittmer, 1970b: 24 **A**: JA (Ryukyus) TAI
tongsae Wittmer, 1975a: 254 **A**: BT
torsae Wittmer, 1975a: 254 **A**: BT

genus *Protomaltypus* Wittmer, 1978d: 297 type species *Protomaltypus ghumensis* Wittmer, 1978
basicrassa Wittmer, 1997e: 355 **A**: YUN
expansicornis Wittmer, 1997e: 355 **A**: YUN
ghumensis Wittmer, 1978d: 302 **A**: SD
godavariensis Wittmer, 1978d: 300 **A**: NP
indicus Wittmer, 1957b: 575 (*Mimomalthinus*) **A**: SD
kurbatovi Wittmer, 1997e: 356 **A**: SCH
nepalensis Wittmer, 1978d: 298 **A**: NP
semiinflatus Wittmer, 1995b: 169 **A**: YUN

tribe Malchinini Brancucci, 1980

genus *Macrocerus* Motschulsky, 1845a: 38 type species *Macrocerus oculatus* Motschulsky, 1845
Malchinus Kiesenwetter, 1863a: 726 type species *Malthodes sinuatocollis* Kiesenwetter, 1852
carinatus Pic, 1901d: 34 (*Malchinus*) **A**: TR
carinicollis J. Weise, 1895e: 213 (*Malchinus*) **E**: GR (Chios) **A**: TR
bodemeyeri Pic, 1911w: 177 (*Malchinus*)
malchinoides Pic, 1899m: 357 (*Podistrella*)
scutellaris Pic, 1899m: 358 (*Malchinus*)
holomelas Fairmaire, 1887c: clxii (*Malchinus*) **E**: AL BU
kadleci Švihla, 2002a: 120 **E**: BU MC
nigrinus Schaufuss, 1866: 180 (*Malchinus*) **E**: CT HU RO UK
obscuripes obscuripes Pic, 1900j: 87 (*Malchinus*) **A**: TR
obscuripes soguksuensis Brancucci, 1980: 335 (*Malchinus*) **A**: TR
obscuripes tokatensis Pic, 1903d: 97 (*Malthodes*) **A**: TR
oculatus Motschulsky, 1845a: 38 **E**: AB AR GG ST **A**: TR
circassicus Reitter, 1889c: 67 (*Malchinus*)
rosti Pic, 1900n: 184 (*Malchinus*)
similis Barovskij, 1927a: 17 (*Malchinus*)
sinuatocollis chvojkai Švihla, 2002a: 121 **E**: AL
sinuatocollis sinuatocollis Kiesenwetter, 1852: 274 (*Malchinus*) **E**: AL BH BU CR GE GR HU IT MC RO SL YU
demissus Kiesenwetter, 1863a: 727 (*Malchinus*)
pseudorhagonycha Reitter, 1889c: 67 (*Malchinus*)
subelongatus Pic, 1901f: 50 (*Malchinus*)
tunicatus Kiesenwetter, 1863a: 728 (*Malchinus*) **E**: FR IT
abeillei Pic, 1911w: 177 (*Malchinus*)
majori Piccioli, 1877: 230 (*Podabrus*)
telephoroides Abeille de Perrin, 1869a: 39 (*Malchinus*)

subfamily **Chauliognathinae** LeConte, 1861

tribe Ichthyurini Champion, 1915

genus *Ichthyurus* Westwood, 1848: 83 type species *Ichtyurus lateralis* Westwood, 1848
Biurus Motschulsky, 1853a: 13 type species *Biurus apicalis* Motschulsky, 1853
Diurus Gemminger, 1869: 1684 [RN] type species *Biurus apicalis* Motschulsky, 1853
Dissacurus Gemminger, 1872: 207 [RN] type species *Biurus apicalis* Motschulsky, 1853
abnormis Gestro, 1892: 1046 **A**: SD
annulatus Champion, 1924c: 260 **A**: SD
bidens Champion, 1925c: 177 **A**: UP
bilineatus Pic, 1915f: 6 **A**: TAI
nigrosignatus Pic, 1954c: 15
bourgeoisi Gestro, 1892: 1023 **A**: CH
davidi Gestro, 1892: 1038 **A**: JIX
distorticornis Champion, 1925c: 178 **A**: UP
gianellii Gestro, 1888: 127 **A**: SD
rankhensis Champion, 1924c: 257
heidiae Brancucci, 1983: 265 **A**: TAI
himalaicis Champion, 1920b: 247 **A**: "Himalaya"
klapperichi Brancucci, 1983: 261 **A**: TAI
maculifrons Champion, 1920b: 248 **A**: "India"
malthiniformis Champion, 1924c: 260 **A**: UP
nigripennis Pic, 1913b: 17 **A**: TAI
pici Delkeskamp, 1939: 210 [RN]
sauteri Pic, 1937h: 65 [HN]
oblongonotatus Champion, 1927: 80 **A**: UP
pieli Pic, 1924i: 89 **A**: SHG
ranikhetensis Champion, 1924c: 257 **A**: UP
saccularis Champion, 1927: 80 **A**: SD
savioi Pic, 1928b: 14 **A**: CH
tridens Champion, 1924c: 258 **A**: UP
vandepolli Gestro, 1892: 1029 **A**: "South China"

genus *Microichthyurus* Pic, 1919e: 14 type species *Microichthyurus impressicollis* Pic, 1919
Satoichthyurus Wittmer, 1978b: 85 type species *Ichthyurus yukikoae* Satô, 1976
apicipennis Pic, 1913b: 18 (*Ichthyurus*) **A**: TAI
haennii Brancucci, 1983: 268 **A**: TAI
ilanensis Brancucci, 1985a: 240 **A**: TAI
minutulus Gestro, 1906: 276 (*Ichthyurus*) **A**: JA (Ryukyus)
yukihoae Satô, 1976: 57 (*Ichthyurus*)
pennatus Lewis, 1895a: 114 (*Biurus*) **A**: JA (Honshu, Shikoku)
satoi Brancucci, 1983: 273 **A**: TAI
shimomurai Brancucci, 1983: 271 **A**: TAI
yukikoae Satô, 1976: 57 (*Ichthyurus*) **A**: JA (Ryukyus)

genus *Trypheridium* Brancucci, 1985b: 245 type species *Trypherus nuristanicus* Wittmer, 1956
nuristanicus Wittmer, 1956a: 158 (*Trypherus*) **A**: AF PA TD
lopatini Medvedev, 1964a: 44 (*Trypherus*)

genus *Trypherus* LeConte, 1851: 346 type species *Malthinus latipennis* Germar, 1824
subgenus *Trypherus* LeConte, 1851: 346 type species *Malthinus latipennis* Germar, 1824
Lygerus Kiesenwetter, 1852: 242 type species *Malthinus latipennis* Germar, 1824
Molorchus Say, 1824c: 292 [HN] type species *Molorchus marginalis* Say, 1824 (= *Trypherus latipennis* Germar, 1824)
Trypheromimus Pic, 1919e: 12 type species *Trypherus nankineus* Pic, 1913
atratulus Brancucci, 1985c: 266 **A**: JA (Honshu)
babai Brancucci, 1985c: 276 **A**: JA (Honshu)

charbinensis Brancucci, 1985c: 283 **A**: HEI
chujoi Brancucci, 1983: 295 **A**: TAI
fenchihuensis Brancucci, 1983: 282 **A**: TAI
kanoi Brancucci, 1983: 276 **A**: TAI
makiharai Satô, 1976: 56 **A**: JA (Ryukyus)
m-signatus Pic, 1919i: 24 **A**: TAI
bimaculata Pic, 1919i: 24
mutilatus Kiesenwetter, 1879b: 308 (*Malthinus*) **A**: JA (Honshu)
nakanei Brancucci, 1983: 278 **A**: TAI
nankineus Pic, 1913n: 182 **A**: JIA
obscurimembris Pic, 1933b: 5 (*Trypherus*)
nantouensis Brancucci, 1983: 289 **A**: TAI
nigrinus Brancucci, 1985c: 273 **A**: JA (Honshu)
niponicus Lewis, 1879d: 463 (*Ichthyurus*) **A**: FE JA
atriceps Lewis, 1895a: 114 (*Ichthyurus*)
limbatus Pic, 1919e: 12
ohbayashii Brancucci, 1985c: 264 **A**: TAI
parilis Brancucci, 1983: 283 **A**: TAI
perplexus Brancucci, 1983: 292 **A**: TAI
plagiocephalus Brancucci, 1983: 286 **A**: TAI
pseudoparilis Brancucci, 1983: 279 **A**: TAI
rossicus Barovskij, 1922b: 276 (*Ichthyurus*) **A**: FE
sauteri Gestro, 1911: 50 (*Ichthyurus*) **A**: TAI
similis Brancucci, 1983: 296 **A**: TAI
simulator Brancucci, 1983: 288 **A**: TAI
taihorinensis Brancucci, 1983: 291 **A**: TAI
taiwanensis Brancucci, 1983: 293 **A**: TAI

superfamily Elateroidea, taxon incertae sedis
genus *Porostenus* Motschulsky, 1853a: 5 type species *Porostenus pediloides* Motschulsky, 1853
pediloides Motschulsky, 1853a: 5 **E**: FR

series **BOSTRICHIFORMIA** Forbes, 1926
superfamily **DERODONTOIDEA** LeConte, 1861
family DERODONTIDAE LeConte, 1861
subfamily **Derodontinae** LeConte, 1861

genus *Derodontus* LeConte, 1861: 100 type species *Cryptophagus maculatus* Melsheimer, 1844
Mycetomychus J. Frivaldszky, 1865: 192 type species *Corticaria macularis* Fuss, 1850
japonicus Hisamatsu, 1964: 52 **A**: JA (Shikoku)
longiclavis Nikitsky, 1987: 34 **A**: FE
macularis Fuss, 1850: 127 (*Corticaria*) **E**: AU BH CR CZ FR GE HU LS PL RO SK ST SZ
osseticus Nikitsky, 1993: 152 **E**: ST (Osetia)
raffrayi Grouvelle, 1916d: 296 **E**: IT (Sicilia)
tuberosus Hisamatsu & Sakai, 1985: 133 **A**: JA (Shikoku)

subfamily **Laricobiinae** Mulsant & Rey, 1863

genus *Laricobius* Rosenhauer, 1846: 5 type species *Laricobius erichsoni* Rosenhauer, 1846
caucasicus Rost, 1893: 341 **E**: "W. Caucasus"
erichsoni Rosenhauer, 1846: 7 **E**: AU BE BU CR CZ DE FR GB GE HU LS IT NL PL RO SK SV SZ YU **NAR**
lituratus Gorham, 1878: 156
niger Pic, 1895f: 88
kovalevi Nikitsky, 1992: 44 **A**: FS

loebli Jelínek & Háva, 2001: 150 **A**: NP
minutus Nikitsky, 1992: 43 **A**: FE
mirabilis Háva & Jelínek, 1999: 115 **A**: SCH SHA
sahlbergi Reitter, 1883c: 42 **A**: ES MG
schawalleri Háva & Jelínek, 2000: 184 **A**: NP

subfamily **Peltasticinae** LeConte, 1861

genus *Peltastica* Mannerheim, 1852a: 334 type species *Peltastica tuberculata* Mannerheim, 1852
amurensis Reitter, 1879e: 220 **A**: FE (Kuril) JA
reitteri Lewis, 1883a: 79

superfamily **BOSTRICHOIDEA** Latreille, 1802

family NOSODENDRIDAE Erichson, 1846

genus *Nosodendron* Latreille, 1804d: 146 type species *Byrrhus fascicularis* Olivier, 1790
Dendrodipnis Wollaston, 1873b: 33 type species *Dendrodipnis coenosus* Wollaston, 1873
subgenus *Nosodendron* Latreille, 1804d: 146 type species *Byrrhus fascicularis* Olivier, 1790
asiaticum Lewis, 1889a: 229 **A**: JA FE
coenosum Wollaston, 1873b: 34 (*Dendrodipnis*) **A**: JA TAI
disjectum Champion, 1923d: 580 **A**: AP
fasciculare A. G. Olivier, 1790a: no 13: 8 (*Byrrhus*) **E**: AU BE BY CR CT CZ DE FR GB GE HU IT LS LT NL PL RO SK SL SV SZ UK
hispidum Champion, 1923d: 590 **A**: AP NP PA SD **ORR**
irregulare Pic, 1928c: 2
marginatum marginatum Reitter, 1886: 218 (*Dendrodipnis*) **A**: SD
marginatum rugiferum Champion, 1923d: 588 **A**: SD **ORR**
nepalense Háva & Farkač, 2003: 245 **A**: NP
punctulatum Reitter, 1886: 217 (*Dendrodipnis*) **A**: SD **ORR**
sikkimense Champion, 1923d: 585 **A**: SD
subgenus *Nosoglobulus* Háva, 2003e: 247 type species *Nosodendron loebli* Háva, 2003
loebli Háva, 2003e: 247 **A**: HUB YUN
smetanai Háva, 2003e: 247 **A**: NP

family DERMESTIDAE Latreille, 1804

subfamily **Dermestinae** Latreille, 1804

tribe Dermestini Latreille, 1804

genus *Dermestes* Linnaeus, 1758: 342 type species *Dermestes lardarius* Linnaeus, 1758
subgenus *Dermestes* Linnaeus, 1758: 342 type species *Dermestes lardarius* Linnaeus, 1758
ater DeGeer, 1774: 223 **E**: AB AL AN AR AU AZ BE BH BU BY CR CT CZ DE FI FR GB GE GG GR HU IC IR IT KZ LA LT LU MA MC NL NR NT PL PT RO SK SL SP ST SV SZ TR UK YU **N**: AG CI EG LB MO MR TU **A**: AE AF AP CE CH CY ES FE HP IN IQ IS JA JO KA KI KZ LE MG NC NE NO NP NW OM PA RU SA SC SE SI SW SY TD TM TR UP UZ WP WS YE **COS**
cadaverinus Fabricius, 1775: 55
chinensis Motschulsky, 1866a: 168
domesticus Germar, 1824: 83
favarcqui Godard, 1883: 383
felinus Fabricius, 1787a: 34
hispidulus Montrouzier, 1860: 264
noxius Mulsant & Rey, 1868a: 55
piceus Thunberg, 1781: 8
rufoapicalis Pic, 1951a: 7
subcostatus Murray, 1867b: 22
subsulcatus Ballion, 1878: 277
bicolor bicolor Fabricius, 1781: 64 **E**: AU BU BY CR CT CZ FR GB GE GR HU IT KZ MC NL PL PT RO RU

SK SL SP ST SV SZ TR UK YU **A**: AF IN IQ JA KZ MG NC PA SY TM UZ "Siberia"
bicolor impressus Pic, 1898a: 2 **N**: AG EG MO TU
bicolor sinensis Háva, 2004f: 230 **A**: XIN
freudei Kalík & Ohbayashi, 1982: 65 **A**: HEB HEI HEN JA NMO RU SHA
haemorrhoidalis Küster, 1852: 64 **E**: AU CZ DE FR GB GE GR HU IR IT KZ NL PL RO RU SK SP SV **N**: EG MO TU **A**: JA MG RU **COS**
gulo Mulsant & Godart, 1855b: 275
hispanicus Kalík, 1952b: 75 **E**: IT (Sardegna) SP **N**: AG MO TU
lardarius Linnaeus, 1758: 354 **E**: AB AL AN AR AU AZ BE BH BU BY CR CT CZ DE FI FR GB GE GG GR HU IC IR IT KZ LA LT LU MA MC NL NR NT PL PT RO RU SK SL SP ST SV SZ TR UK YU **N**: AG CI EG LB MO MR TU **A**: AE AF AP CE CY ES FE HP IN IQ IS JA JO KA KI KZ LE MG NP NE NC NO NW OM PA RU SA SC SE SI SW SY TD TM TR UP UZ WP WS YE **COS**
atrobasalis Pic, 1951a: 7
conjunctus Reitter, 1887d: 173
luganensis Stierlin, 1902: 425
proximus Reitter, 1887d: 173
leechi Kalík, 1952a: 45 **E**: CT GBi NT SP ST **N**: EG MO TU **A**: AF ES IN NP PA TD TM UZ WS "India"
nidum Arrow, 1915b: 426 **A**: FE NC MG TAI **NAR**
elongatus LeConte, 1854b: 109 [HN]
palmi Sjöberg, 1950: 162 **E**: ES FE FI SV WS **A**: WS
geltzermani Kozminykh, 1997: 77
peruvianus Laporte, 1840b: 33 **E**: CZ GE GB FI IT NL PL RU SP SV **N**: CI EG MO TU **A**: TAI **NAR NTR**
angustatus C. F. A. Schaeffer, 1931: 174
angustus T. L. Casey, 1900: 143
peruanus Erichson, 1847: 95
schneideri Háva, 2002a: 35 **A**: MG
vorax Motschulsky, 1860b: 123 **A**: ES FE GAN GUX HEB HEI JA JIL LIA SHN XIN ZHE
albofasciatus Matsumura & Yokoyama, 1928b: 51
subgenus *Dermestinus* Zhantiev, 1967: 1352 type species *Dermestes murinus* Linnaeus, 1758
aurichalceus Küster, 1846b: 50 **E**: FR IT SP SZ **N**: MO TU
holosericeus Tournier, 1860: 318 [HN]
mustelinus Perris, 1853: 638 [HN]
carnivorus Fabricius, 1775: 55 **E**: CT DE FR GB GE IR IT NL NT PL RU SP ST SV SZ **N**: EG MO TU **A**: FUJ GUA HKG SCH TAI "N. India" **COS**
doemmlingi Meier, 1899b: 99
impressicollis Reitter, 1881a: 29
mucoreus LeConte, 1854b: 108
muscoreus Reitter, 1881a: 28
sobrinus LeConte, 1854b: 108
unicolor Lepesme, 1950: 44
versicolor Laporte, 1840b: 33
coarctatus Harold, 1877: 347 **A**: GAN HEI JA JIL LIA NMO SHA "East Siberia"
coronatus Steven, 1808: 85 **E**: ST TR **A**: AF IN MG UP XIN
dimidiatus, 1808: 89 **E**: MD ST **A**: KZ MG NO XIZ "Siberia"
orientalis Kuznecova, 1928: 20
rosea Kuznecova, 1928: 22
elegans Gebler, 1830: 99 **A**: KI KZ MG NO TD TM UZ
elegans Solsky, 1876: 269 [HN]
nadezhdae Plavilshchikov, 1924: 231
erichsoni Ganglbauer, 1904: 16 **E**: AB AL AR AU BE BH BU BY CR CT CZ DE FR GE GR HU IT KZ LA MC NT PL RO RU SK SL SP ST SZ TR UK YU **N**: AG MO TU **A**: WS
tessellatus Erichson, 1846: 432 [HN]
fasciventris Reitter, 1881a: 28 **A**: KZ RU
frischii Kugelann, 1792: 478 **E**: AB AL AN AR AU AZ BE BH BU BY CR CT CZ DE FI FR GB GE GG GR HU IC IR IT KZ LA LT LU MA MC NL NR NT PL PT RO RU SK SL SP ST SV SZ TR UK YU **N**: AG CI EG LB MO MR TU **A**: AE AF AP CE CH CY ES FE HP IN IQ IS JA JO KA KI KZ LE MG NP NE NC NO NW OM PA SA SC SE SI SW SY TD TM TR UP UZ WP WS YE **COS**

heyrovskyi Obenberger, 1917a: 32
pollinctus Hope, 1834: 54
rufimembris Pic, 1951a: 7
sternimaculatus Marcu, 1957: 531
vulpinus Herbst, 1792: 124 [HN]

fulvicollis Reitter, 1881a: 29 **A**: UP

gyllenhalii gyllenhalii Laporte, 1840b: 34 **E**: AB AU BU BY CR CT CZ DE EN FI FR GE GR HU IT KZ LA LT MC NR NT PL RO RU SK SL SP ST SV SZ TR UK YU **A**: IN KZ WS

atomarius Erichson, 1846: 431

gyllenhali helmi Reitter, 1889m: 279 **E**: CR DE GE PL SV

hankae Háva, 1999b: 143 **E**: FR IT SP **N**: AG

intermedius intermedius Kalík, 1951b: 41 **E**: AU BU CZ GR GE HU IT MC RO SK TR **A**: CY

intermedius iranicus Háva & Kalík, 1999: 444 **A**: IN IQ SY

kafkai Háva & Kalík, 1999: 445 **A**: IN

kaszabi Kalík, 1950: 64 **E**: AB AR BH BU CR CZ GE GR HU RO SK TR YU **A**: IN

laniarius Illiger, 1801a: 85 **E**: AB AL AR AU BE BH BU BY CR CT CZ DE EN FI FR GE GR HU IT KZ LA LT MC NL NR NT PL PT RO RU SK SL SP ST SV SZ TR UK YU **N**: MO TU **A**: CY IN KI KZ NW PA RU SCH SHA TD TM TR UZ

affinis Gyllenhal, 1808a: 148
catta Duftschmid, 1825: 34 [HN]
macellarius Herbst, 1792: 126 [HN]
murinus Herbst, 1792: 126 [HN]

leopardinus Mulsant & Godart, 1855b: 273 **E**: BU CR TR UK YU **A**: RU

loebli Háva, 2002f: 539 **A**: IS SY

maculatus DeGeer, 1774: 223 **E**: AB AL AN AR AU AZ BE BH BU BY CR CT CZ DE EN FI FR GB GE GG GR HU IC IR IT KZ LA LT LU MA MC NL NR NT PL PT RO RU SK SL SP ST SV SZ TR UK YU **N**: AG CI EG LB MO MR TU **A**: AE AF AP CE CH CY ES FE HP IN IQ IS JA JO KA KI KZ LE MG NP NE NC NO NW OM PA RU SA SC SE SI SW SY TD TM TR UP UZ WP WS YE **COS**

cinereus L. Redtenbacher, 1868: 43
cyprius Pic, 1951a: 7
elongatus Hope, 1834: 55
kurseongensis Lepesme, 1939: 192
lupinus Mannerheim, 1843b: 257
marginatus Thunberg, 1781: 7
pakistanicus Havelka, 1951: 143
rattulus Mulsant & Rey, 1868a: 30
roei Hope, 1834: 55
semistriatus Boheman, 1851: 583
senex Germar, 1824: 84
sudanicus Gredler, 1878: 506
truncatus T. L. Casey, 1916: 180
vulpinus Fabricius, 1781: 64

marocanus Háva, 1999a: 134 **N**: MO

murinus auriceps Reitter, 1905c: 244 **A**: MG NC RU

murinus murinus Linnaeus, 1758: 356 **E**: AB AL AR AU BE BH BU BY CR CT CZ DE EN FI FR GE GR HU IT KZ LA LT MC NL NR NT PL PT RO RU SK SL SP ST SV SZ TR YU **N**: CI MO TU **A**: CY IN KI KZ NW PA RU TD TM TR UZ

catta Panzer, 1789: 123
marinus Fabricius, 1787a: 35
nebulosus DeGeer, 1774: 197
roseiventris Laporte, 1840b: 34

mustelinus Erichson, 1846: 433 **E**: FR GE IT PL PT SP

diadema Motschulsky, 1849: 89
tessellatus A. G. Olivier, 1790a: no 9: 9 [HN]

normandi Kalík, 1951c: 154 **N**: AG TU

pardalis Billberg, 1808: 90 **E**: FR GE PT IT SP **N**: AG MO TU

thoracicus Gené, 1836: 22

planus Mroczkowski, 1960: 210 **A**: TM UZ
sardous asiaticus Háva, 2002f: 540 **A**: IN JO SY TR
sardous fulvofasciatus Ganglbauer, 1904: 18 **N**: AG EG LB MO TU
striatellus Reitter, 1908b: 45
sardous sardous Küster, 1846b: 49 **E**: FR IT PT SP **N**: AG EG LB MO TU
shaanxiensis Z. Cao, 1987: 271 **A**: SHA TAI
sibiricus Erichson, 1846: 427 **E**: AR TR **A**: ES HEB HEI IN KZ MG NMO TM ZU XIN
solskyi Dalla Torre, 1911b: 49 [RN] **A**: IN
ratus Solsky, 1876: 269 [HN]
szekessyi Kalík, 1950: 61 **E**: AB AU BH BY CR CT CZ DE FI FR GE GR HU IT KZ MC NR NT PL RO RU SK SL SP ST SV SZ TR UK YU **A**: KZ MG WS
tessellatocollis pici Mroczkowski, 1963a: 377 **A**: HP NO NP SD UP
tessellatocollis tessellatocollis Motschulsky, 1860b: 124 **A**: ES FUJ GAN GUI GUX HEB HEI HEN HUN JA JIA JIL NIN NMO QIN WS SHA SHN SHX XIZ YUN
kaliki Havelka, 1957: 106
undulatus Brahm, 1790: 114 **E**: AB AL AR AU BE BH BU BY CR CT CZ DE FI FR GE GR HU IT KZ LA LT MC NL NR NT PL PT RO RU SK SL SP ST SV SZ TR YU **N**: AG CI EG LB MO TU **A**: CY GUX HEB IN JIL KI KZ NMO PA QIN RU TD TM TR UZ XIN **NAR**
caninus Gravenbhorst, 1807: 114
cylindricus T. L. Casey, 1916: 181
dissector Kirby, 1837: 115
tessellatus Illiger, 1801a: 84 [HN]
variegatus Brullé, 1832: 163
vulpecula Herbst, 1792: 123
subgenus *Montandonia* Jacquet, 1886b: 320 type species *Montandonia catopoides* Jacquet, 1886 (= *Dermestes latissimus* Bielz, 1850)
depressus Gebler, 1830: 99 (*Montandonia*) **A**: ES KZ
fuliginosus P. Rossi, 1792: 17 **E**: AU BH BU BY CR CT CZ FR GE GR HU IT MC NT PL RO RU SK SL SP ST TR **A**: IN
ater Duftschmid, 1825: 32 [HN]
hirticollis Fabricius, 1792a: 229 **E**: FR PT IT (Sardegna) SP **N**: AG EG MO TU
latissimus Bielz, 1850: 180 **E**: HU RO
catopoides Jacquet, 1886b: 319 (*Montandonia*)
olivieri Lepesme, 1939: 192 **E**: AR AU BH BU BY CR CT CZ FR GE GG GR HU IT MC NT PL RO RU SK SL SP ST TR **A**: IN SY TM TR
ater A. G. Olivier, 1790a: no 9: 9 [HN]
roubali Kalík, 1951c: 152 (*Montandonia*) **A**: ES KZ TM
sichuanicus Háva, 1999b: 147 **A**: GAN QIN SCH

tribe Marioutini Jakobson, 1913

genus *Mariouta* Pic, 1898d: 73 type species *Mariouta letourneuxi* Pic, 1898
letourneuxi Pic, 1898d: 73 **N**: AG EG MO
stangei Reitter, 1910d: 239 **A**: OM SA TM

genus *Rhopalosilpha* Arrow, 1929a: 97 type species *Rhopalosilpha wasmanni* Arrow, 1929
wasmanni Arrow, 1929a: 97 **A**: IN SA

subfamily **Thorictinae** Agassiz, 1846

tribe Thaumaphrastini Anderson, 1949

genus *Thorictodes* Reitter, 1875d: 45 type species *Thorictodes heydeni* Reitter, 1875
Thaumaphrastus Blaisdell, 1927: 123 type species *Thaumaphrastus karanisensis* Blaisdell, 1927
brevipennis Zhang & Liu, 1986: 67 [= 1988: 27] **A**: YUN
erraticus Champion, 1922: 73 **A**: UP
heydeni Reitter, 1875d: 46 **E**: FR IT RU SP **N**: AG EG MO **A**: IS JA JIA PA TR **AFR NAR NTR ORR**
karanisensis Blaisdell, 1927: 123 (*Thaumaphrastus*)

tribe Thorictini Agassiz, 1846

genus *Thorictus* Germar, 1834: 15 type species *Thorictus castaneus* Germar, 1834
Cyphomanes Gistel, 1848: xi [RN]
Sphaerophorus Waltl, 1835: 64 type species *Sphaerophorus castaneus* Waltl, 1835
Xylonotrogus Motschulsky, 1839: 70 type species *Xylonotrogus laticollis* Motschulsky, 1839
aequabilis John, 1965e: 325 **N**: MO
andreinii Escalera, 1942b: 383 **N**: LB
angustus John, 1964d: 334 **A**: AF
bacunus John, 1971: 145 **E**: AB
badadjanidis John, 1971: 147 **E**: AB
basalis John, 1968: 30 **N**: MO
baudii baudii Reitter, 1881f: 92 **A**: CY LE "Syria" **N**: EG
baudii ganglbaueri Wasmann, 1897: 242 **A**: LE
baudoni John, 1967: 150 [= 1969: 127] **N**: MO
becharensis Chobaut, 1924b: 105 **N**: MO
bifoveolatus Reitter, 1887g: 286 **A**: KZ
bonnairei bonnairei Wasmann, 1894: 219 **N**: AG MO TU
nivellei Escalera, 1925c: 311
silvestrii Escalera, 1923: 379
bonnairei nigroides Wasmann, 1925: 98 **E**: PT
brevipennis John, 1963b: 206 **A**: "Syria"
buigasi Escalera, 1923: 379 **N**: MO
canariensis Wollaston, 1862d: 439 **N**: CI
obsitus Wollaston, 1864: 185
castaneus castaneus Germar, 1834: no 15 **N**: AG EG LB MO **A**: "Syria"
peyroni Pic, 1924f: 29
castaneus leonhardi Reitter, 1913e: 654 **E**: GR (Zakynthos)
ciliatus Reitter, 1881f: 91 **A**: IS
cirenaicus John, 1965e: 338 **N**: LB
cobosi John & Andreae, 1967: 149 **N**: CI
cobosi John, 1969: 126
consimilis John, 1965e: 339 **E**: SP
crassus John, 1965e: 339 **A**: TM
deviedmai John, 1965f: 509 [= 1965e: 340] **N**: CI
dilatipennis Reitter, 1881f: 94 **N** EG **A**: "Syria"
dimidiatus Peyron, 1857a: 707 **E**: GR **N**: EG **A**: CY IS LE TR "Syria"
dispar Baudi di Selve, 1870: 58
doderoi Escalera, 1942b: 381 **N**: LB
doriae Escalera, 1942b: 386 **A**: IN
ehlersii ehlersii Perez Arcas, 1872: 95 **E**: SP
ehlersii therondi Chobaut, 1924b: 106 **N**: AG MO
escalerai John, 1965f: 508 [= 1965e: 342] **E**: SP
escorialanus John, 1965e: 329 **E**: SP
fairmairii fairmairii Raffray, 1873: 369 **N**: AG MO
angulatus Escalera, 1914: 134
fairmairii tunisius Chobaut, 1924c: 216 **N**: TU
tunisius John, 1965e: 358 [HN]
foreli Wasmann, 1894: 219 [= 1895a: 44] **N**: AG MO TU
foveicollis Reitter, 1880g: 545 **E**: AB ST **A**: IS UZ
franzi John, 1967: 463 **N**: CI
fuhesanus John, 1964d: 336 **A**: JO
gestroi Escalera, 1942b: 385 **N**: LB
gigas Wollaston, 1862d: 439 **N**: CI
grandiceps Pic, 1931i: 134 **N**: EG

grandicollis grandicollis Germar, 1842: 4 **E**: AR CR FR GR IT MA PT SP ST UK **N**: AG EG MO TU **A**: SY TR
atticus Rey, 1889i: 58
bicolor Kraatz, 1858: 141
castaneus Waltl, 1835: 64 (*Sphaerophorus*) [HN]
corsicus Rey, 1889i: 58
creticus Obenberger, 1917b: 77
gallicus Peyron, 1857a: 713
germari P. H. Lucas, 1846: 234
hauseri John, 1971: 145 [HN]
laticollis Motschulsky, 1839: 71
loricatoides Wasmann, 1897: 243
loricatus Peyron, 1857a: 710
rufithorax Rey, 1889i: 58
grandicollis hungaricus J. Weise, 1875: 101 **E**: HU UK
grandicollis westwoodi Wollaston, 1854: 220 **N**: MR
hauseri John, 1965e: 343 **A**: TD
helleri John, 1963b: 216 **N**: CI
hendeli Reitter, 1910a: 128 **A**: "Transkaspien"
hilfi John, 1965e: 344 **E**: GR (Zakynthos)
hoppi John, 1971: 152 **E**: PT
immutatus John, 1963b: 195 **N**: AG
impressithorax Pic, 1930h: 62 **N**: EG
incisicollis Pic, 1924f: 29 **N**: EG
incultus John, 1971: 150 **E**: FR (Corse)
insulcatus Pic, 1924f: 29 **N**: EG
irakensis John, 1963b: 201 **A**: IQ
kabulanus John, 1964d: 333 **A**: AF
kandaharicus John, 1962: 284 [= 1963: 212] **A**: AF
kaznakovi R. Schmidt, 1904: 389 **A**: UZ
khinjanus John, 1964d: 332 **A**: AF
klapperichi John, 1964d: 336 **A**: JO
kocheri John, 1965e: 346 [= 1965a: 29] **N**: MO
kochi John, 1964b: 163 **N**: LB
koenigi Reitter, 1887g: 286 **A**: TM UZ
impressibasis Reitter, 1910a: 128
kraatzi Wasmann, 1895b: 292 **N**: AG
kurdistanus John, 1965b: 51 [= 1965e: 347] **A**: "Kurdistan"
lederi Reitter, 1881f: 90 **E**: AB AR **A**: IN UZ
lethierryi Fairmaire, 1875b: 495 **N**: AG
lameerei Wasmann, 1902: 159
lindbergi John, 1963b: 216 **N**: CI
longipennis Coye, 1870: 374 **A**: LE
lucasi John, 1965e: 331 **N**: MO
lucusensis Escalera, 1923: 377 **N**: MO
manni Reichenspenger, 1926: 105 **A**: JO
marginicollis Schaum, 1859b: 73 **N**: EG
martinsi Wasmann, 1925: 95 **E**: PT
mauritanicus P. H. Lucas, 1846: 235 **E**: IT SP **N**: AG MO TU **A**: SY
nivalis Wasmann, 1895b: 291
mogadoricus Escalera, 1914: 132 **N**: MO
munganasti Reitter, 1908b: 43 **N**: EG LB
myrmecophilus Reitter, 1881f: 93 **E**: AB AR
nabeulanus John, 1964c: 191 **N**: TU
nanus John, 1964d: 338 **A**: JO
normandi Chobaut, 1924c: 215 **N**: TU

obenbergeri John, 1965e: 349 **E**: AB AR
orientalis Peyron, 1857a: 706 **E**: GR "Caucasus" **A**: IQ IS SY TR "Transkaukasien"
circumflexus Coye, 1870: 371
duplimargo Reitter, 1909f: 317
lebedewi Reitter, 1909f: 317
spectabilis Kraatz, 1858: 142
paganettii Obenberger, 1917b: 77 **E**: GR (Kríti)
cretensis John, 1971: 148
palmi John, 1964c: 190 **A**: CY
parnassus John, 1971: 148 **E**: GR
pauciseta Wasmann, 1894: 219 [= 1895a: 43] **N**: TU
persicus Reitter, 1881f: 91 **A**: IN
petranus John, 1965f: 507 **A**: JO SY
patranus John, 1965e: 350
peyerimhoffi Chobaut, 1904: 243 **N**: AG **A**: "Arabie: Kasr-er-Rabbat"
pilosus Peyron, 1857a: 702 **E**: GR **N**: EG LB TU **A**: CY IQ IS LE SY TR UZ
mesopotamicus Escalera, 1923: 377
piliger Schaum, 1858: 141
pubescens Coye, 1870: 373
rufocinctus Pic, 1924f: 29
politus Wasmann, 1895b: 292 **E**: GR
postangulis Reitter, 1895e: 296 **A**: UZ
procerus John, 1962: 283 [= 1963: 208] **A**: "Syria"
puncticollis P. H. Lucas, 1846: 236 **N**: AG
leboulei Escalera, 1923: 378
quiquesulcatus John, 1964d: 335 **A**: AF
rectangulatus John, 1965f: 509 [= 1965e: 353; 1965a: 31] **N**: MO
reitteri John, 1963b: 206 **A**: "Turkestan"
rollei John, 1971: 152 **N**: EG
rotroui John, 1965e: 333 **N**: MO
rotundithorax Escalera, 1942b: 388 **E**: PT
ruzskii Semenov, 1903: 353 **A**: UZ
schatzmayri John, 1965d: 67 [= 1965e: 353] **N**: LB
seriesetosus seriesetosus Fairmaire, 1870: 373 **N**: AG MO
arganicus Escalera, 1923: 380
rugulosus Reitter, 1881f: 91
seriesetosus tritomentosus John, 1965e: 336 **N**: MO
sicilianus John, 1965e: 354 [= 1966: 19] **E**: IT (Sicilia) **N**: MO
simillimus Escalera, 1914: 133 **N**: MO
simoni John, 1965e: 334 **N**: AG
striatus Reitter, 1889m: 278 **E**: AR
stricticollis minutalis Normand, 1950a: 75 **N**: TU (Le Kef)
stricticollis stricticollis Kraatz, 1859: 73 **E**: CR GR IT **A**: "Syria" **N**: AG MO
nigripennis Reitter, 1881f: 95
studti John, 1971: 149 **E**: IT (Sicilia)
subcastaneus Chobaut, 1898: 74 **N**: AG
subpusillus John, 1965f: 511 [= 1965e: 355] **N**: CI
sulcicollis Perez Arcas, 1868: 492 **E**: SP **N**: AG TU
tamadabanus John, 1964a: 135 **N**: CI
tejedanus John & Andreae, 1967: 149 **N**: CI
tejedanus John, 1969: 125
tenerifanus John, 1965f: 512 [= 1965e: 356] **N**: CI
testaceus Pic, 1930h: 62 **N**: EG
theryi Chobaut, 1924b: 107 **N**: MO
tripolitanus Escalera, 1942b: 383 **N**: LB
trisulcatus mecheriensis John, 1965e: 332 **N**: MO

trisulcatus trisulcatus Reitter, 1881f: 88 **N**: AG TU
tuberosus Reitter, 1881f: 93 **A**: CY "Palestina"
vaulogeri Escalera, 1923: 378 **N**: MO
vaucheri Chobaut, 1924b: 105 **N**: MO
vestitus Wollaston, 1864: 186 **N**: CI
villosissimus Escalera, 1923: 377 **N**: MO
vonoertzeni John, 1971: 150 **E**: GR (Kríti)
walanganus John, 1964d: 331 **A**: AF
wasmanni Reitter, 1895e: 296 **A**: UZ
wollastoni John, 1963b: 215 **N**: CI

subfamily **Orphilinae** LeConte, 1861

genus *Orphilus* Erichson, 1846: 424 type species *Byrrhus niger* P. Rossi, 1790
beali Zhantiev, 2001: 612 **E**: FR GR SP TR **A**: CY IS LE SY TR
niger P. Rossi, 1790: 39 (*Byrrhus*) **E**: AB AL AR AU BE BH BU BY CR CT CZ FR GE GR HU IT KZ MC NT PL PT RO SK SL SP SZ TR UK YU **N**: AG EG MO TU **A**: TD UZ
glaber Creutzer, 1796a: 11 (*Anthrenus*)
glabratus Fabricius, 1801a: 109 (*Anthrenus*)

subfamily **Trinodinae** T. L. Casey, 1900

tribe Thylodriadini Mroczkowski, 1954

genus *Thylodrias* Motschulsky, 1839: 75 type species *Thylodrias contractus* Motschulsky, 1839
Ignotus Slosson, 1903: 187 type species *Ignotus aenigmaticus* Slosson, 1903
Hospitopterus Pic, 1921l: 54 type species *Hospitopterus efflatouni* Pic, 1921
Thelodrias Agassiz, 1846a: 161 type species *Thylodrias contractus* Motschulsky, 1839
Thelydrias Agassiz, 1846b: 368 type species *Thylodrias contractus* Motschulsky, 1839
contractus Motschulsky, 1839: 76 **E**: CZ DE FI FR GE IT NR PL SP SV **N**: AG EG TU **A**: JA RU **AUR NAR NTR**
aenigmaticus Slosson, 1903: 187 (*Ignotus*)
efflatouni Pic, 1921l: 54 (*Hospitopterus*)

tribe Trinodini T. L. Casey, 1900

genus *Evorinea* Beal, 1961: 118 type species *Apsectus iotus* Arrow, 1915
bicolorata Pic, 1939g: 144 (*Trinodes*) **N**: EG
curta Pic, 1894s: 119 (*Trinodes*) **N**: AG
flava Motschulsky, 1863: 433 (*Trinodes*) **N**: EG **A**: IS SY **AFR**
hirtella Walker, 1858: 207 (*Trinodes*) **A**: NP **ORR**
indica Arrow, 1915b: 450 (*Apsectus*) **A**: JA TAI **ORR**
hisamatsui Ohbayashi, 1977: 357
iota Arrow, 1915b: 449 (*Apsectus*) **A**: JA TAI **AFR ORR**
minor Pic, 1922b: 6 (*Trinodes*)

genus *Trinodes* Dejean, 1821: 47 type species *Nitidula hirta* Fabricius, 1781
amamiensis Ohbayashi, 1977: 355 **A**: JA
bicoloratus Pic, 1939g: 144 **N**: EG
curtus Pic, 1894s: 119 **N**: AG
hirtus Fabricius, 1781: 93 (*Nitidula*) **E**: AR AU BH BU BY CR CZ DE FR GE GR HU IT KZ LA LT MC NT PL RO SK SP SV SZ TR UK YU **N**: AG TU **A**: IS TM RU
ater Dalla Torre, 1879: 100
castanea Dalla Torre, 1879: 100
testacea Dalla Torre, 1879: 100
insulanus Zhantiev, 1988: 1422 **A**: FE (Sakhalin) JA (Honshu)
niger Matsumura & Yokoyama, 1928b: 53 **A**: TAI
rufescens Reitter, 1877i: 376 **A**: JA NC
sinensis Fairmaire, 1886c: 319 **A**: JIA SCH SHX

subfamily **Attageninae** Laporte, 1840

tribe Attagenini Laporte, 1840

genus *Attagenus* Latreille, 1802: 121 type species *Dermestes pellio* Linnaeus, 1758

subgenus *Aethriostoma* Motschulsky, 1858c: 47 type species *Aethriostoma undulata* Motschulsky, 1858

Brachysphyrus Blackburn, 1903: 159 type species *Brachysphyrus irroratus* Blackburn, 1903

Pseudotelopes Pic, 1916c: 2 type species *Attagenus simoni* Pic, 1916

sparsutus Reitter, 1881a: 36 (*Aethriostoma*) **A**: NO NP XIZ

euryovatus H. Li, 1988: 388

rufescens Pic, 1933g: 71 (*Aethriostoma*)

undulatus Motschulsky, 1858c: 47 (*Aethriostoma*) **A**: GUA GUX YUN "India" **AFR ORR**

rufipes Walker, 1859a: 53 (*Attagenus*)

simoni Pic, 1916c: 2

testaceipes Pic, 1916c: 2

subgenus *Attagenus* Latreille, 1802: 121 type species *Dermestes pellio* Linnaeus, 1758

Eunorops Gistel, 1856: 361 type species *Dermestes pellio* Linnaeus, 1758

Lanorus Mulsant & Rey, 1868a: 68 type species *Dermestes vigintiguttatus* Fabricius, 1775

Megatoma Kugelann, 1792: 479 type species *Megatoma schaefferi* Herbst, 1792

Telopes L. Redtenbacher, 1843b: 976 type species *Telopes dispar* L. Redtenbacher, 1843

abbreviatus Heer, 1856: 140 **N**: CI

addendus J. R. Sahlberg, 1903d: 34 **N**: AG MO TU **A**: IS

afghanus Háva, 2000b: 163 **A**: AF

antennatus Laporte, 1840b: 36 **E**: SP **N**: EG TU

anthrenoides Wollaston, 1864: 159 (*Telopes*) **N**: CI

apicalis Pic, 1942b: 7 **A**: SA **AFR**

aristidis Pic, 1894s: 119 (*Telopes*) **N**: AG EG **A**: IN

astacurus Peyerimhoff, 1931b: 64 **N**: AG "Central Sahara"

assuanensis Pic, 1899l: 254 (*Telopes*) **N**: AG EG

atripennis Pic, 1938e: 9 **A**: SA YE **AFR**

attenuatus Pic, 1894x: 210 (*Telopes*) **N**: AG EG **A**: IS SY

augustatus Ballion, 1871: 330 **A**: MG SCH TD TM XIZ YUN

angustatus Solsky, 1876: 271

aurantiacus aurantiacus Reitter, 1900h: 274 **E**: TR **A**: IQ IS SY

aurantiacus erevanicus Zhantiev, 1963a: 417 **E**: AR

barbieri Pic, 1946: 6 **N**: EG LB **A**: IS

bifasciatus bifasciatus A. G. Olivier, 1790a: no 9: 13 (*Dermestes*) **E**: AL BU CR CT FR GR IT NT PT SL SP ST TR YU **N**: AG EG LB MO TU **A**: AF IN IQ IS LE SY TM

brunneonotatus Pic, 1893b: 60 (*Telopes*)

chakouri Pic, 1907k: 153

pulcher Faldermann, 1835a: 226

unicolor Alfieri, 1976: 117

wachanrui Mulsant & Rey, 1868a: 91

bifasciatus decoloratus Mulsant & Rey, 1868a: 96 **E**: SP **N**: AG EG MO TU

bifasciatus incertus Mulsant & Rey, 1868a: 96 **E**: SP **N**: AG EG MO TU

bifasciatus tigrinus Fabricius, 1792a: 229 (*Dermestes*) **E**: AL BU CR CT FR GR IT PT SL SP ST TR YU **N**: AG EG LB MO TU **A**: AF IN IQ IS LE SY TM

poecilus Germar, 1839a: 9

bifasciatus wegeneri Pic, 1899l: 252 **N**: TU **A**: IS LE

birmanicus Arrow, 1915b: 428 **A**: YUN "NE India" **ORR**

biskrensis Pic, 1904a: 2 **N**: AG MO **AFR**

brittoni Kalík, 1954: 368 **A**: IN

brunneopunctatus Pic, 1893e: 122 (*Telopes*) **N**: EG **A**: IS SY

impunctatus Pic, 1894r: 116 (*Telopes*)

brunnescens Pic, 1904k: 73 (*Telopes*) **E**: TR

brunneus Faldermann, 1835a: 227 **E**: AU BH BU CT CZ FR GE GR HU IT LA NT PL RO SK SP ST TR YU **N**: AG EG MO TU **A**: AF ES FE HEB HEN IN IS NMO PA TM TR UZ WS XIN **NAR**

bicolor T. L. Casey, 1900: 147

dalmatinus Küster, 1847b: 44
elongatulus T. L. Casey, 1900: 147
elongatus T. L. Casey, 1916: 183
extricatus T. L. Casey, 1900: 146
longicornis Pic, 1894f: 66
marginicollis Handschuch, 1847b: 76
obliquus Rey, 1889i: 59
picipennis Pic, 1894f: 66
sordidus Heer, 1841: 438
calabricus Reitter, 1881b: 77 **E**: IT (Sicilia)
caracal Zhantiev, 1963b: 456 **A**: NO NW UZ
civetta Mulsant & Rey, 1868a: 104 (*Telopus*) **E**: MA SP **N**: AG LB MO TU **A**: IS
coquereli Mulsant & Rey, 1868a: 102 (*Telopus*) **N**: AG TU
curvicornis J. R. Sahlberg, 1913a: 26 **N**: EG
cyphonoides Reitter, 1881a: 34 **E**: FR IT SP RO RU TR **N**: EG MO TU **A**: AF HEB HEN HUN IN NMO RU XIN XIZ YUN"India" **COS**
alfierii Pic, 1910e: 17
senegalensis Pic, 1915d: 2
decoloratus Mulsant & Rey, 1868a: 96 **E**: SP **N**: AG EG MO TU
dichrous Roth, 1851: 122 **A**: SA **AFR**
dispar L. Redtenbacher, 1843b: 976 (*Telopes*) **E**: GR TR **N**: EG **A**: IN SY
redtenbacheri Peyron, 1857b: 720
duplex Reitter, 1890i: 359 (*Telopes*) **A**: IQ MG TD TM UZ
octopunctatus Mroczkowski, 2002: 29
eremivagus Peyerimhoff, 1943: 11 **N**: TU "Sahara"
escalerai Mader, 1954: 53 **N**: MO
antennatus Escalera, 1914: 137 [HN]
escalerai Kocher, 1956a: 25 [RN, HN]
fallax Gené, 1839: 19 **E**: FR GR IT PT SP **N**: AG EG LB MO TU **A**: IS
distinctus Mulsant & Rey, 1868a: 85
octomaculatus Pic, 1894s: 119
sexguttatus Sturm, 1843: 94
fasciatus Thunberg, 1795: 105 (*Anthrenus*) **E**: CT DE FI FR GB GE IT NT SP ST SV **N**: AG EG MO TU **A**: AF ES FE FUJ IN IS JA KI KZ MG NP OM SA SY TD TM TR UZ WS YE YUN **COS**
annulifer Laporte, 1840b: 36
cinnamomeus Roth, 1851: 122
gloriosae Fabricius, 1798: 76 (*Anthrenus*)
plebeius Sharp, 1885b: 147
subfasciatum Chevrolat, 1864: 617 (*Trogoderma*)
unifasciatus Fairmaire, 1860: 168
fasciolatus Solsky, 1876: 275 (*Telopes*) **A**: KZ MG SA TD TM UZ
gobicola J. Frivaldszky, 1892: 122 **A**: AF GAN KA KI KZ MG NMO QIN SD TD TM UP XIN
heydeni Reitter, 1881a: 35 (*Telopes*) **N**: AG EG MO **A**: IS
opacus Pic, 1946: 6 (*Telopes*)
rotroui Pic, 1927f: 5 (*Telopes*)
hirtulus Rosenhauer, 1856: 109 **E**: PT SP **N**: MO TU
incognitus Háva, 2003a: 28 **E**: SP
indicus Kalík, 1954: 368 **A**: NP UP
jacobsoni Zhantiev, 1963a: 417 **A**: IN SY TD
jelineki Háva, 2004b: 9 **A**: IN TR
karnali Háva, 2001d: 211 **A**: NP
lepidus Háva & Kovařík, 2004: 83 **N**: MO
leprieuri Reitter, 1887a: 53 (*Telopes*) **N**: EG **A**: IS
aegyptiacus Pic, 1893e: 122 (*Telopes*)
lineatus Pic, 1894g: 71 (*Telopes*) **N**: AG EG MO TU
lobatus Rosenhauer, 1856: 108 **E**: FR IT RO RU SP TR **N**: AG EG MO TU **A**: AF BEI IN IQ MG SA TD TM UP UZ XIN

byturoides Solsky, 1876: 272
longipennis Pic, 1904f: 33 **A**: KZ MG NMO TD XIN
lynx Mulsant & Rey, 1868a: 105 (*Telopus*) **E**: ST **N**: EG **A**: IN IS KZ MG SA SY TD TM TR UZ
sareptanus Reitter, 1878a: 178
maltensis Pic, 1894f: 66 **E**: MA
maritimus Gené, 1839: 20 (*Telopes*) **E**: IT SP **N**: AG TU
molitor Reitter, 1889t: 557 **A**: IN KZ MG XIN
mongolicus Zhantiev, 1973a: 188 **A**: MG
multifasciatus Wollaston, 1863: 218 (*Telopes*) **N**: CI
nepalensis Háva, 2001d: 209 **A**: NP XIZ
obtusus Gyllenhal, 1808a: 88 (*Dermestes*) **E**: AL AR BU CR FR GR IT MA PT SP ST SVi TR **N**: AG CI EG MO TU **A**: IN IQ IS SA
confusus Pic, 1951c: 5 (*Telopes*)
holosericeus C. R. Sahlberg, 1819: 52 (*Dermestes*)
orientalis Reitter, 1878a: 178 **E**: "Caucasus" **N**: TU **A**: IN IQ SY TR
pantherinus Ahrens, 1814: 11 (*Dermestes*) **E**: AU CT CZ FR GE HU IT NT PL SK ST SZ YU TR
pellio Linnaeus, 1758: 355 (*Dermestes*) **E**: AB AL AN AR AU AZ BE BH BU BY CR CT CZ DE EN FI FR GB GE GG GR HU IC IR IT KZ LA LT LU MA MC NL NR NT PL PT RO SK SL SP ST SV SZ TR UK YU **N**: AG CI EG LB MO MR TU **A**: AE AF AP CE CH CY ES FE HP IN IQ IS JA JO KA KI KZ LE MG NP NE NC NO NW OM PA SA SC SE SI SW SY TD TM TR UP UZ WP WS YE **COS**
atra Herbst, 1792: 95 (*Megatoma*)
bipunctatus DeGeer, 1774: 197 (*Dermestes*)
cylindricornis Schrank, 1785: 315 (*Dermestes*)
fruteti Grimmer, 1841: 39
macellarius Fabricius, 1781: 63 (*Dermestes*)
pilosissimus Roubal, 1932b: 66
quisquiliorum Gistel, 1857: 71
schranki Kugelann, 1792: 480 (*Megatoma*)
persicus Reitter, 1881b: 78 **E**: GR **A**: IN
pulcher Faldermann, 1835a: 226
pictus Ballion, 1871: 330 **A**: AF IN IS KZ TD TM UZ
placitus Normand, 1950a: 72 **N**: TU
posticalis Fairmaire, 1879a: 169 **E**: SP **N**: AG EG MO TU **A**: AE IS OM SA SY YE **AFR**
subelongatus Pic, 1929f: 9 (*Telopes*)
pseudomolitor Zhantiev, 1963b: 454 **A**: MG TM UZ
pubescens Pic, 1894f: 66 (*Telopes*) **N**: EG
punctatus Scopoli, 1772: 87 (*Dermestes*) **E**: AU CT CZ DE GE HU NT PL SK ST SV
multiguttatus Schrank, 1798: 427 (*Dermestes*)
quadripunctatus Fuessly, 1775: 4 (*Dermestes*)
subdiurnus Gistel, 1857: 5
vigintiguttatus Fabricius, 1775: 56 (*Dermetes*)
vigintipunctatus Haberfelner, 1889: 390
quadricolor Sumakow, 1907: 15 (*Trogoderma*) **A**: KI KZ MG TD TM TR UZ
soederbomi Pic, 1933b: 2
quadrimaculatus Kraatz, 1858: 145 **E**: AL BH BU CR FR GR SL TR YU **A**: IS JO
bivittatus Mulsant & Rey, 1868a: 97
quadrinotatus Pic, 1938g: 10 **A**: YUN
quadritinctus Reitter, 1889m: 280 **E**: AR **A**: IN
reitteri Mroczkowski, 1968: 90 (*Telopes*) [RN] **E**: PT SP **N**: AG MO TU
antennatus Reitter, 1887a: 54 [HN]
robustus Pic, 1899l: 253 **N**: EG **A**: JO
rossii Ganglbauer, 1904: 26 [RN] **E**: BU GR IT SP MA **N**: AG TU
bifasciatus P. Rossi, 1794: 79 (*Dermestes*) [HN]
ruficolor Pic, 1918b: 1 **A**: YUN
rufomaculatus Pic, 1907v: 274 (*Telopes*) **N**: AG
scalaris Pic, 1893b: 60 (*Telopes*) **N**: EG LB

schaefferi schaefferi Herbst, 1792: 93 (*Megatoma*) **E**: AU BU CR CZ FI FR GB GE GR HU IT LA LT RO RU SK SP SZ TR **N**: CI **A**: NMO WS **NAR**

bicolor Harold, 1868: 104

dichrous LeConte, 1854b: 110

sparsus T. L. Casey, 1916: 184

seminiger Fairmaire, 1864: 639 (*Telopes*) **N**: AG EG TU

seniculus Solsky, 1876: 276 (*Telopes*) **A**: AF IN MG TD TM UZ

sericeus Guérin-Méneville, 1844a: 65 (*Megatoma*) **N**: AG EG

sieversi Reitter, 1896g: 286 **A**: IN TM

sikkimensis Pic, 1942b: 7 **A**: SD

silvaticus Zhantiev, 1976: 126 **E**: CZ HU SK "Caucasus" **A**: IN "Siberia"

similaris Mulsant & Rey, 1868a: 85 **N**: AG LB TU

simonis Reitter, 1881a: 34 **N**: EG **A**: IS JO SY

sexmaculatus J. R. Sahlberg, 1913d: 107

simplex Reitter, 1881b: 79 **E**: IT **N**: LB MO TU **A**: IS

rufipennis Mulsant & Rey, 1868a: 100 [HN]

sinensis Pic, 1927c: 3 **A**: YUN

arrowi Kalík, 1954: 367

smirnovi Zhantiev, 1973b: 283 **E**: AU CZ DE FI EN GE GB IR LA PL RO RU SK SZ **N**: MO **A**: OM RU SA **AFR ORR**

steinbergi Zhantiev, 1963a: 418 **A**: IN KZ TD TM

suspiciosus Solsky, 1876: 274 **A**: AF IN KI KZ MG TD TM UZ XIN

tessellatus Reitter, 1887a: 54 (*Telopes*) **N**: EG **A**: IS SY

trifasciatus Fabricius, 1787a: 34 (*Dermestes*) **E**: BH FR GBi GEi IT SP SZi PT **N**: AG EG MO TU **A**: IS

turcomanus Zhantiev, 1963b: 455 **A**: TM

unicolor japonicus Reitter, 1877i: 375 **Ei**: RO SZ **A**: JA MG NC NO "Siberia"

amurensis Pic, 1942b: 7

canadensis T. L. Casey, 1916: 183

nankineus Pic, 1916d: 15

unicolor simulans Solsky, 1876: 272 **A**: AF KZ MG NMO TD TM UZ XIN

unicolor unicolor Brahm, 1790: 144 (*Dermestes*) **E**: AB AL AN AR AU AZ BE BH BU BY CR CT CZ DE EN FI FR GB GE GG GR HU IC IR IT KZ LA LT LU MA MC NL NR NT PL PT RO SK SL SP ST SV SZ TR UK YU **N**: AG CI EG LB MO MR TU **A**: AE AF AP CE CH CY ES FE HP IN IQ IS JA JO KA KI KZ LE MG NC NE NO NP NW OM PA SA SC SE SI SW SY TD TM TR UP UZ WP WS YE **COS**

brevicornis Herbst, 1792: 95 (*Megatoma*)

cylindricornis Schrank, 1798: 446 (*Nitidula*)

cylindricornis Say, 1825a: 185 [HN]

deficiens T. L. Casey, 1900: 146

floricola Say, 1825a: 185

fulvipes Mulsant & Rey, 1868a: 71

macellarius Duftschmid, 1825: 39 (*Dermestes*)

megatoma Fabricius, 1798: 71 (*Megatoma*)

megadoma Sturm, 1826: 98

obscurus Say, 1825a: 185

piceus A. G. Olivier, 1790a: no 9: 10 (*Dermestes*) [HN]

schaefferi C. Sahlberg, 1819: 50 (*Dermestes*)

stygialis Mulsant &Rey, 1868a: 73

urbicola Gistel, 1857: 67

uniformis Fairmaire, 1860: 169 (*Telopes*) **E**: IT (Sicilia) **N**: AG EG LB MO TU **A**: IQ SY

vagepictus Fairmaire, 1889b: 13 **A**: GUX SCH YUN XIZ

wollastoni Mroczkowski, 1964: 180 [RN] **N**: CI

fasciatus Wollaston, 1863: 218 (*Telopes*) [HN]

woodroffei Halstead & Green, 1979: 97 **E**: DE FI GE SV

genus *Sefrania* Pic, 1899c: 28 type species *Sefrania bleusei* Pic, 1899

bleusei Pic, 1899c: 29 **Ei**: FR PL **N**: AG MO TU

tribe Egidyellini Semenov, 1916

genus *Egidyella* Reitter, 1899e: 285 type species *Egidyella prophetea* Reitter, 1899

Trichopleurias J. R. Sahlberg, 1903a: 35 type species *Trichopleurias deserti* J. R. Sahlberg, 1903 (= *Egidyella prophetea* Reitter, 1899)

prophetea Reitter, 1899e: 285 **A**: TM UZ

deserti J. R. Sahlberg, 1903a: 36 (*Trichopleurias*)

subfamily **Megatominae** Leach, 1815

tribe Anthrenini T. L. Casey, 1900

genus *Anthrenus* Geoffroy, 1762: 113 type species *Dermestes scrophulariae* Linnaeus, 1758

subgenus *Anthrenodes* Chobaut, 1898: 76 type species *Anthrenus umbellatarum* Chobaut, 1898

aegyptiacus Pic, 1899l: 255 **N**: EG

bleusei Pic, 1899f: 85

amoenulus Reitter, 1896g: 286 **E**: TR **A**: IN TM TR "Caspian Lowlands"

buettikeri Mroczkowski, 1980: 126 **A**: SA

ineptus Háva & Tezcan 2004: 53 **A**: IN IQ TR

israelicus Háva, 2004d: 91 **A**: IS

jordanicus Pic, 1934c: 29 **A**: IS JO

maculifer Reitter, 1881a: 58 **A**: GUA HKG NP TAI XIN **ORR**

albonotatus Pic, 1922b: 6

globiger Arrow, 1915b: 447

multimaculatus Pic, 1918b: 2

malkini Mroczkowski, 1980: 125 **A**: AE OM SA YE

minor Wollaston, 1865: 163 [RN] **E**: GR **N**: AG CI EG MO TU **A**: SA

albidoflavus Reitter, 1881b: 91

claviger Wollaston, 1861b: 161 [HN]

pulchellus Gestro, 1889: 37 **A**: SA YE

sarnicus Mroczkowski, 1963b: 697 **E**: GB NLi

semenovi semenovi Zhantiev, 1976: 161 **A**: "Pamir"

semenovi rupestris Zhantiev, 1976: 162 **A**: TD

umbellatarum Chobaut, 1898: 76 **N**: AG EG MO **A**: IS

albidipennis Chobaut, 1898: 76

wittmeri Mroczkowski, 1980: 125 **A**: SA **AFR**

subgenus *Anthrenops* Reitter, 1881b: 86 type species *Anthrenus coloratus* Reitter, 1881

bellulus bellulus Chobaut, 1897e: 277 **N**: AG

bellulus ornatus Kocher, 1955b: 40 **E**: SP **N**: MO

coloratus Reitter, 1881b: 91 **E**: BU FR GE GR IT RU SP TR YU **N**: AG CI EG MO TU **A**: AE AF IS KZ SY TD TM TR "India" **AFR NAR**

gobicus Zhantiev, 2004b: 2 **A**: MG

hissaricus Mroczkowski, 1961b: 591 **A**: TD

longus Arrow, 1915b: 448 **A**: YE (Suqutra) **AFR**

longipennis Pic, 1937d: 6

nahiricus Zhantiev, 1976: 88 **E**: AR

parthicus Zhantiev, 1976: 88 **A**: IN TM (Kopet)

rufescens Pic, 1923d: 3 **A**: NP "North India"

subclaviger Reitter, 1881a: 59 **A**: SA UP YE

zagrosensis Háva, 2004d: 90 **A**: IN

zebra Reitter, 1889m: 280 **E**: AB AR **A**: AF IN TM

araraticus Zhantiev, 1976: 161

tigrinus Reitter, 1889m: 281

subgenus *Anthrenus* Geoffroy, 1762: 113 type species *Dermestes scrophulariae* Linnaeus, 1758

Anthrenus J. C. Schaeffer, 1766: pl. xxvi type species *Dermestes scrophulariae* Linnaeus, 1758

Byrrhus Sulzer, 1776: 25 type species *Dermestes scrophulariae* Linnaeus, 1758

Hypoceuthes Gerstaecker, 1871: 46 type species *Hypoceuthes aterrimus* Gerstaecker, 1871

angustefasciatus Ganglbauer, 1904: 42 **E**: IT PT SP **N**: MO

bilyi Háva, 2000b: 168 **A**: KI
crustaceus Reitter, 1881a: 58 **N**: EG **A**: IS SA SY TR YE **AFR**
delicatus armstrongi Mroczkowski, 1952: 27 **E**: BU GR TR YU
delicatus delicatus Kiesenwetter, 1852: 579 **E**: AL BH BU CR FR IT MA MC PT SP TR YU **N**: AG EG MO TU **A**: IN IS SY TR
 isabellinae Mulsant & Rey, 1868a: 154
dorsatus Mulsant & Rey, 1868a: 154 **E**: MA **N**: AG TU
festivus Erichson, 1846: 458 **E**: FR PT IT (Sardegna, Sicilia) SP (Baleares) **N**: MO TU **A**: CY IS
 funestus Mulsant & Rey, 1868a: 145
flavidulus Reitter, 1889a: 23 **E**: AR **A**: IN JO
flavipes albopunctatus Pic, 1895o: 142 **A**: IN SY TM
 albopunctatus Pic, 1895o: 142
 isabellae Reitter, 1899c: 216
flavipes flavipes LeConte, 1854b: 112 **E**: CZ DE FI FR GB GE IT NL SP **N**: EG MO TU **A**: AF GUA HKG IN IS JA RU SA TD **COS**
 alboimpletus J. R. Sahlberg, 1903d: 34
 cinnamomeus Gredler, 1878: 507
 importatus Pic, 1952: 1
 seminiveus T. L. Casey, 1916: 199
 vorax Waterhouse, 1883: 61
goliath Saulcy, 1868: 140 **E**: AL BU CR FR GR HU IT MC PT RO SP TR YU **N**: AG EG MO TU **A**: SY
kaliki Pic, 1952: 1 **N**: AG
latefasciatus Reitter, 1892a: 134 **A**: AF IN KI KZ MG NC SY TD TM UZ XIN
lindbergi Mroczkowski, 1959: 100 **A**: AF PA
lopatini Zhantiev, 1976: 158 **A**: TD TM
mesopotamicus Háva, 2001c: 65 **A**: IN IQ SY
miniatulus Reitter, 1899c: 216 **A**: TR SY
miniopictus Bedel, 1884: 21 **E**: SP **N**: AG MO TU
munroi Hinton, 1943: 14 **E**: BU FR IT SP TR **N**: AG LB MO **A**: CY IS JO SY
nipponensis Kalík & Ohbayashi, 1985: 77 **A**: ES FE HEB HEI HEN JA LIA NMO SCH SHA SHN
oceanicus Fauvel, 1903: 338 **Ei**: CZ GB **Ni**: EG **A**: GUA **AUR NAR ORR**
 basifasciatus Pic, 1935e: 13
 fasciatus Herbst, 1797: 337 [HN]
 latebasalis Pic, 1927j: 66
pfefferi Kalík, 1954: 370 **E**: GR
picturatus hintoni Mroczkowski, 1952: 29 **A**: CH
picturatus makolskii Mroczkowski, 1950: 187 **E**: PL SK
picturatus melanoleucus Solsky, 1876: 279 **A**: AF KI KZ TD UZ
picturatus picturatus Solsky, 1876: 280 **A**: AF GG IN KI KZ TD TM TR UZ **AFR**i
 incertus Mulsant & Rey, 1868a: 154 **E**: TR
pimpinellae isabellinus Küster, 1848a: 38 **E**: FR GR IT SP **N**: AG MO TU **NAR**i
 niveipennis J. R. Sahlberg, 1903d: 34
 niveus Reitter, 1881b: 88
pimpinellae mroczkowskii Kalík, 1954: 369 **E**: AL BH BU FR (Corse) CR GR IT SL **N**: AG
pimpinellae pimpinellae Fabricius, 1775: 61 (*Byrrhus*) **E**: AB AR AU BE BH BU BY CR CT CZ DE EN FI FR GB GE GG GR HU IT KZ LA LS LT LU NL NR PL RO SK SP SV SZ TR UK YU **N**: EG MO **A**: KI MG RU TD **COS**
 dimidiatus Hauser, 1894: 23
 scrophulariae Geoffroy, 1785: 27 [HN]
rauterbergi Reitter, 1908b: 46 **N**: EG **A**: SA
rotundulus Reitter, 1889a: 23 **A**: AF IN TM TR
 toleranda Pic, 1908d: 41
scrophulariae albidus Brullé, 1832: 164 **E**: AL BH BU CR GR MC TR YU **A**: SY UZ
 gravidus Küster, 1848a: 37
 hilairei Pic, 1952: 1
 proteus Kraatz, 1858: 147

senex Kraatz, 1858: 147
signatus Reitter, 1881b: 90
scrophulariae scrophulariae Linnaeus, 1758: 356 (*Dermestes*) **E**: AB AR AU BE BH BU BY CR CT CZ DE EN FI FR GB GE GG GR HU IT KZ LA LS LT LU NL NR PL RO SK SP SV SZ TR UK YU **N**: EG MO **A**: IN LE KI MG RU TD **COS**
histrio Fabricius, 1792a: 264
insulicola Obenberger, 1917a: 32
rubricollis J. R. Sahlberg, 1913d: 109
suecicus Palm, 1940: 1
variegatus Scopoli, 1763: 16
verbasci Herbst, 1797: 328
simonis Reitter, 1881a: 59 **N**: EG TU **A**: IN IS SY TR
x-signum Reitter, 1881b: 88 **E**: FR (Corse) **N**: AG EG MO TU
subgenus *Florilinus* Mulsant & Rey, 1868a: 141 type species *Dermestes museorum* Linnaeus, 1761
araxensis Zhantiev, 1976: 89 **E**: AR
bactrianus Zhantiev, 2004a: 4 **A**: TD
bajtenovi Sokolov, 1974: 92 **A**: KZ
caucasicus Reitter, 1881b: 92 **E**: AR AU PL **A**: GG IN TM
armeniacus Zaitzev, 1919: 177
coreanus Mroczkowski, 1966: 440 **A**: NC
flavidus Solsky, 1876: 281 **E**: PL **A**: AF KZ KI NO SHX TD TR TM UZ XIN
rufulus Solsky, 1876: 281
gracilis Zhantiev, 2004b: 1 **A**: KZ
hartmanni Háva, 2000b: 170 **A**: NP
japonicus Ohbayashi, 1985: 59 **A**: JA
kaszabi Zhantiev, 1973a: 193 **A**: MG
kompantzevi Zhantiev, 2004a: 3 **A**: KI
mongolicus Zhantiev, 1973a: 193 **A**: MG
mugodsharicus Sokolov, 1974: 91 **A**: KZ
museorum Linnaeus, 1761: 145 (*Dermestes*) **E**: AB AL AN AR AU AZ BE BH BU BY CR CT CZ DE EN FI FR GB GE GG GR HU IC IR IT KZ LA LT LU MA MC NL NR NT PL PT RO SK SL SP ST SV SZ TR UK YU **N**: AG CI EG LB MO MR TU **A**: AE AF AP CE CH CY ES FE HP IN IQ IS JA JO KA KI KZ LE MG NC NE NO NP NW OM PA SA SC SE SI SW SY TD TM TR UP UZ WP WS YE **COS**
muscorum Fabricius, 1787a: 39 [HN]
pellio Thunberg, 1815: 151
vagus Gmelin, 1790: 1615
varius Stephens, 1830: 130
verbasci Fabricius, 1775: 61
oberthueri Reitter, 1881b: 91 **E**: FR IT **N**: AG TU
olgae Kalík, 1946: 134 **E**: AU BU CZ FI GB GE HU LA PL SK SV UK
pallidus Sokolov, 1974: 95 **A**: TD
qinlingensis Háva, 2004f: 227 **A**: SHX
shikokensis Ohbayashi, 1985: 64 **A**: JA (Shikoku)
solskianus Sokolov, 1974: 95 **A**: KZ
sordidulus Reitter, 1889j: 256 **E**: GR SP TR **N**: EG **A**: CY SY
sveci Háva, 2004a: 1 **E**: GR
tanakai Ohbayashi, 1985: 61 **A**: JA SC
tuvensis Zhantiev, 1976: 88 **A**: ES (Tuva)
ussuricus Zhantiev, 1988: 1422 **A**: FE
subgenus *Helocerus* Mulsant & Rey, 1868a: 141 type species *Anthrenus fuscus* A. G. Olivier, 1790
fuscus A. G. Olivier, 1790b: 152 **E**: AL AR AU BH BU BY CR CZ DE EN FR GB GE GR HU IT LA LT NL PL RO RU SK SP SV SZ TR UK YU **N**: MO **NAR ORR**
claviger Erichson, 1846: 458
minutus Erichson, 1846: 459 **E**: FR IT PT SP
flavosquamosus Pic, 1903i: 121
polonicus Mroczkowski, 1951: 253 **E**: BU BY EN GE HU LA PL RO SK UK YU

subgenus *Nathrenus* T. L. Casey, 1900: 161 type species *Byrrhus verbasci* Linnaeus, 1767
biskrensis Reitter, 1887a: 69 **E**: MA IT (Sicilia) **N**: AG LB MO TU
bomiensis Háva, 2004b: 4 **A**: XIZ
bulirschi Háva, 2000b: 170 **A**: IS SY TR
exilis Mulsant & Rey, 1868a: 158 **N**: AG EG
jakli Háva, 2001a: 1 **A**: OM
knizeki Háva, 2004f: 226 **A**: HEB
liliputianus Mulsant & Rey, 1868a: 159 **N**: EG
molitor Aubé, 1850: 335 **E**: AL BU FR GR IT MA SP TR YU **N**: AG EG MO TU **A**: IS
albidus Laporte, 1840b: 37 [HN]
ochraceus Reiche, 1868a: 161
nocivus Mulsant & Godart, 1872c: 240 **N**: AG
pilosus Pic, 1923d: 3 **A**: XIZ
pubifer Reitter, 1899c: 217 **A**: IN SY TR
akbesianus Reitter, 1899c: 217
sichuanicus Háva, 2004f: 225 **A**: SCH
signatus Erichson, 1846: 456 **E**: AU BH BU CR CZ GR HU IT SL TR YU
apicalis Küster, 1847b: 48
incanus J. Frivaldszky, 1882: 28
transcaspicus Mroczkowski, 1960: 217 **A**: IN TM
tryznai Háva, 2001b: 61 **A**: XIZ
turnai Háva, 2004f: 226 **A**: SCH
verbasci Linnaeus, 1767: 568 (*Byrrhus*) **E**: AB AL AN AR AU AZ BE BH BU BY CR CT CZ DE EN FI FR GB GE GG GR HU IC IR IT KZ LA LT LU MA MC NL NR NT PL PT RO SK SL SP ST SV SZ TR UK YU **N**: AG CI EG LB MO MR TU **A**: AE AF AP CE CH CY ES FE HP IN IQ IS JA JO KA KI KZ LE MG NE NC NO NP NW OM PA SA SC SE SI SW SY TD TM TR UP UZ WP WS YE **COS**
adspersus Herbst, 1797: 332
caseyi Hinton, 1945: 328
confusus Reitter, 1886a: 69
destructor Melsheimer, 1844: 116
florilegus Geoffroy, 1785: 27
funebris Reitter, 1889j: 256
nebulosus Reitter, 1886a: 69
nitidulus Küster, 1847b: 47
pictus Germar, 1813: 121
pistor T. L. Casey, 1900: 161
substriatus T. L. Casey, 1900: 161
tomentosus Thunberg, 1815: 151
tricolor Herbst, 1797: 333
varius Fabricius, 1775: 60 (*Bostrichus*)
versicolor Reitter, 1887a: 70
vorax T. L. Casey, 1900: 161
zahradniki Háva, 2003d: 225 **E**: TR **A**: TR
subgenus *Peacockia* Menier & Villemant, 1993: 61 type species *Anthrenus vladimiri* Menier & Villemant, 1993
vladimiri Menier & Villemant, 1993: 62 **N**: MO
subgenus *Ranthenus* Mroczkowski, 1962: 255 type species *Anthrenus alatauensis* Mroczkowski, 1962
alatauensis Mroczkowski, 1962: 254 **A**: KZ
kryzhanovskii Sokolov, 1979: 924 **A**: KI
prudeki Háva, 2002b: 45 **A**: IN TR
zeravshanicus Sokolov, 1979: 925 **A**: TD
subgenus *Solskinus* Mroczkowski, 1962: 252 type species *Anthrenus leucogrammus* Solsky, 1876
assimilis Zhantiev, 1976: 168 **A**: KI
auratus Zhantiev, 1979: 68 **A**: UZ
beali Zhantiev, 2004a: 1 **A**: KI
modestus Zhantiev, 2004a: 2
becvari Háva, 2004f: 227 **A**: YUN

bucharicus Zhantiev, 1976: 166 **A**: TM UZ
dsungaricus Mroczkowski, 1962: 251 **A**: KZ
heptamerus Peyerimhoff, 1924: 62 **N**: MO
jacobsoni Zhantiev, 1976: 165 **A**: TD
leucogrammus Solsky, 1876: 282 **A**: TD UZ
milkoi Zhantiev, 2004a: 2 **A**: KZ
similis Zhantiev, 1976: 167 **A**: TD "Tian shan"
sinensis Arrow, 1915b: 445 **A**: AP FE NO UP
sogdianus Zhantiev, 1976: 166 **A**: TD
tadzhicus Mroczkowski, 1961b: 592 **A**: TD
talassicus Sokolov, 1980: 62 **A**: KZ

genus *Anthrenus*, species incertae sedis
palaeoaegyptiacus Grüss, 1930: 181 **N**: EG

tribe Megatomini Leach, 1815

genus *Anthrenocerus* Arrow, 1915: 443 type species *Anthrenus australis* Hope, 1843
australis Hope, 1843c: 319 (*Anthrenus*) **E**i: AU BE CZ FR GB GE IT NL SZ **AUR**

genus *Ctesias* Stephens, 1830: 124 type species *Dermestes serra* Fabricius, 1792
subgenus *Ctesias* Stephens, 1830: 124 type species *Dermestes serra* Fabricius, 1792
Ctenias Reitter, 1911a: 151 type species *Dermestes serra* Fabricius, 1792
Tiresias Stephens, 1835: 410 type species *Dermestes serra* Fabricius, 1792
hebei Háva, 2004f: 230 **A**: HEB
orientalis Zhantiev, 1988: 1421 **A**: FE
serra Fabricius, 1792a: 234 (*Dermestes*) **E**: AU BE BH BU BY CR CT CZ DE EN FI FR GB GE GR HU IT LA LS LT LU MC NL NR NT PL RO SK SL SP ST SV SZ UK YU "Caucasus"
viennensis Herbst, 1797: 336 (*Anthrenus*)
subgenus *Decemctesias* Háva, 2004e: 154 type species *Ctesias syriaca* Ganglbauer, 1904
intermedia Mroczkowski, 1961a: 195 **A**: TM
kaliki Mroczkowski, 1961a: 193 **A**: AF
maculifasciata Reitter, 1899e: 283 (*Trogoderma*) **A**: SY TR
syriacum Dalla Torre, 1911b: 70 [RN]
morocco Háva, 2000b: 165 **N**: MO
syriaca Ganglbauer, 1904: 37 **A**: SY TR IS
sogdiana Zhantiev, 1975: 310 **A**: UZ
subgenus *Novemctesias* Háva, 2004e: 154 type species *Ctesias tschuiliensis* Sokolov, 1972
fasciata Zhantiev, 1975: 310 **A**: KZ (Karatau: Balamur)
nuratavica Sokolov, 1983: 803 **A**: UZ (Nuratau)
tschuiliensis Sokolov, 1972: 127 **A**: KZ
subgenus *Tiresiomorpha* Pic, 1954c: 11 type species *Tiresiomorpha klapperichi* Pic, 1954
gemma Zhantiev, 1976: 152 **A**: TD
klapperichi Pic, 1954c: 11 (*Tiresiomorpha*) **A**: FUJ HUB
schawalleri Háva, 2002c: 157 **A**: NP

genus *Globicornis* Latreille, 1829a: 511 type species *Dermestes rufitarsis* Creutzer, 1796 (= *Dermestes nigripes* Fabricius, 1792)
subgenus *Globicornis* Latreille, 1829a: 511 type species *Dermestes rufitarsis* Creutzer, 1796 (= *Dermestes nigripes* Fabricius, 1792)
alpina Pic, 1912q: 75 **E**: FR
bifasciata Perris, 1866: 186 (*Hadrotoma*) **E**: FR IT (Sicilia) SP **N**: AG TU
breviclavis Reitter, 1878a: 179 (*Hadrotoma*) **E**: AR BU TR
fasciata Fairmaire & Brisout de Barneville, 1859: 45 (*Hadrotoma*) **E**: BE DE FR GE IT SP SZ
kafkai Háva, 2000c: 369 **A**: IN TR

mongolicus Zhantiev, 1973a: 191 **A**: MG
nigripes Fabricius, 1792a: 233 (*Dermestes*) **E**: AB AR AU BU CR CZ FR GB GE GR HU IR IT KZ PL RO RU SK SP SV SZ YU
plantaris Curtis, 1838: 682 (*Dermestes*)
rufitarsis Creutzer, 1796a: 6 (*Dermestes*)
picta Küster, 1851a: 39 (*Trogoderma*) **E**: AL BH BU CR FR GR HU IT RO SK SP TR YU
quadrinaeva Reitter, 1908f: 213 **A**: KZ TD
rufoguttata Pic, 1899l: 254 **A**: IS SY
unifasciata Pic, 1908d: 41
signatipennis Pic, 1899l: 255 **A**: CY IS JO LE SY
bicruciata J. R. Sahlberg, 1913a: 28
tristis Reitter, 1881b: 84 (*Hadrotoma*) **E**: GR IT
variegata Küster, 1851a: 40 (*Trogoderma*) **E**: FR GR IT SP
sicula Reitter, 1881b: 76 (*Attagenus*)
subgenus *Elania* Mulsant & Rey, 1868a: 111 type species *Hadrotoma depressa* Mulsant & Rey, 1868
depressa Mulsant & Rey, 1868a: 118 (*Hadrotoma*) **E**: FR
vaulogeri Pic, 1900f: 53 **N**: MO TU
brunneithorax Pic, 1900f: 53
subgenus *Hadrotoma* Erichson, 1846: 424 type species *Dermestes marginatus* Paykull, 1798
Mesalia Mulsant & Rey, 1868a: 106 type species *Mesalia guillebelli* Mulsant & Rey, 1868 (= *Dermestes emarginata* Gyllenhal, 1808)
antoniae Reitter, 1884a: 7 (*Hadrotoma*) **E**: AB **A**: IN
corticalis Eichhoff, 1863: 437 (*Hadrotoma*) **E**: AU BU CR CZ DE FI FR GB GE GR HU IT PL RO RU SK SP SV SZ TR YU **A**: KZ
hispanica Pic, 1908k: 81 **E**: SP
emarginata Gyllenhal, 1808a: 153 (*Dermestes*) **E**: AU BU CR CZ DE EN FI FR GB GE GR HU IT KZ LA LT NR PL RO SK SP SV SZ TR YU "Caucasus" **A**: WS
guillebelli Mulsant & Rey, 1868a: 109 (*Mesalia*)
marginata Paykull, 1798: 280 (*Dermestes*) [HN]
semilimbata Pic, 1906b: 9 **E**: GR
sulcata C. N. F. Brisout de Barneville, 1866: 371 (*Hadrotoma*) **E**: FR IT SP
subgenus *Pseudomesalia* Ganglbauer, 1900: 149 type species *Hadrotoma bodemeyeri* Ganglbauer, 1900
bodemeyeri Ganglbauer, 1900: 148 (*Hadrotoma*) **A**: IN SY TR
maculatus Háva, 2004d: 92 **A**: AF IN
quadriguttatus Reitter, 1878f: 95 (*Hadrotoma*) **E**: AR **A**: AF IQ KZ TD TM UZ
kulti Kalík, 1951a: 2
solskyi Reitter, 1881a: 38 (*Trogoderma*)
subfasciata Dalla Torre, 1911b: 64 [RN]
sunica Iablokoff-Khnzoryan, 1961: 49
unifasciatus Pic, 1908d: 41
quadripunctata Zhantiev, 1975: 309 **A**: KZ

genus *Hirtomegatoma* Pic, 1931c: 5 type species *Hirtomegatoma infasciata* Pic, 1931
infasciata Pic, 1931c: 5 **A**: JO
nigra Háva, 2003b: 174 **A**: IN

genus *Megatoma* Herbst, 1792: 92 type species *Dermestes undatus* Linnaeus, 1758
subgenus *Caucasotoma* Mroczkowski, 1967: 5 type species *Megatoma riedeli* Mroczkowski, 1967
riedeli Mroczkowski, 1967: 11 **E**: "Caucasus"
subgenus *Megatoma* Herbst, 1792: 92 type species *Dermestes undatus* Linnaeus, 1758
Aprogramme Reitter, 1886a: 40 type species *Dermestes undatus* Linnaeus, 1758
Asprogramme Gozis, 1886: 10 type species *Dermestes undatus* Linnaeus, 1758
Megalotomum Gistel, 1856: 361 type species *Dermestes undatus* Linnaeus, 1758
pubescens Zetterstedt, 1828: 136 (*Dermestes*) **E**: FI NR SV LT **A**: MG WS **NAR**
fuscus Gebler, 1841b: 585 (*Anthrenus*)
innotata Pic, 1902g: 48

prolixa Beal, 1967: 299
ruficornis Aubé, 1866: 161 **E**: FR GR HU IT SK TR YU
pici Kalík, 1952b: 76
undata undata Linnaeus, 1758: 355 (*Dermestes*) **E**: AB AR AU CZ DE EN FI FR GB GE GR HU IT KZ LA LT PL RO RU SK SP SV SZ UK YU "Caucasus"
glaber C. R. Sahlberg, 1819: 50 (*Dermestes*)
undulatus Linnaeus, 1761: 141 (*Dermestes*)
unifasciata Ganglbauer, 1904: 28
undata ussuriensis Mroczkowski, 1967: 7 **A**: ES FE
subgenus *Pseudohadrotoma* Kalík, 1951a: 4 type species *Globicornis obenbergeri* Kalík, 1951 (= *Globicornis friebi* Pic, 1938)
conspersa Solsky, 1876: 275 **A**: AF KZ KI NO UZ
maculosa Reitter, 1887a: 55 (*Approgramme*)
mnotata Pic, 1938a: 1
friebi Pic, 1938a: 1 (*Globicornis*) **A**: ES MG WS
obenbergeri Kalík, 1951a: 3 (*Globicornis*)
graeseri Reitter, 1887a: 56 (*Approgramme*) **E**: BY CT **A**: KI MG "Siberia"
latenotata Pic, 1902g: 48
indica Háva, 2000b: 165 **A**: KA (Ladakh)
tianschanica Sokolov, 1972: 129 **A**: KI KZ MG XIN

genus *Orphinus* Motschulsky, 1858c: 48 type species *Orphinus haemorrhoidalis* Motschulsky, 1858
subgenus *Curtophinus* Pic, 1954c: 11 type species *Curtophinus bicolor* Pic, 1954
bicolor Pic, 1954c: 11 **A**: FUJ
subgenus *Falsoorphinus* Pic, 1931a: 4 type species *Falsoorphinus notatipennis* Pic, 1931
ovalis Arrow, 1915: 436 (*Thaumaglossa*) **A**: CH
quadrimaculatus Matsumura & Yokoyama, 1928b: 52 (*Trinodes*) **A**: JA
notatipennis Pic, 1931a: 4
subgenus *Orphinus* Motschulsky, 1858c: 48 type species *Orphinus haemorrhoidalis* Motschulsky, 1858
Aethriosia Reitter, 1908b: 45 type species *Aethriosia globulicornis* Reitter, 1908
bipunctutus Matsumura & Yokoyama, 1928b: 53 (*Orphilus*) **A**: JA
chinensis Háva, 2004f: 228 **A**: HEB
fasciatus Matsumura & Yokoyama, 1928b: 54 (*Orphilus*) **A**: TAI
fulvipes Guérin-Méneville, 1838d: 138 (*Globicornis*) **Ei**: GB GE **AFR AUR NAR NTR ORR**
brasiliensis Reitter, 1881a: 37 (*Trogoderma*)
brevicorne Sharp, 1885b: 150 (*Cryptorhopalum*)
unicolor Kolbe, 1910: 24 (*Trogoderma*)
globulicornis Reitter, 1908b: 46 (*Aethriosia*) **N**: EG
boehmi Reitter, 1908b: 46
ferrantei Reitter, 1908b: 46
rubromaculatum J. R. Sahlberg, 1913a: 29 (*Trogoderma*)
hartmanni Háva, 2001d: 213 **A**: BT NP SD
japonicus Arrow, 1915: 438 **A**: JA NC SC
pseudoovalis Háva, 2004f: 228 **A**: HUB
sikkimensis Háva & Herrmann, 2004: 201 **A**: SD
kresli Háva, 2003f: 251 **A**: NP
tadzhicus Zhantiev, 1975: 310 **A**: TD
yunnanus Háva, 2004f: 229 **A**: YUN

genus *Phradonoma* Jacquelin du Val, 1859a: 257 type species *Anthrenus villosulus* Duftschmid, 1825
Orbeola Mulsant & Rey, 1868a: 134 type species *Trogoderma hirsutula* Mulsant & Rey, 1868
amoenulum Reitter, 1889t: 557 (*Trogoderma*) **A**: MG
calliesi Peyerimhoff, 1928: 304 (*Trogoderma*) **N**: AG MO TU
cercyonoides Reitter, 1887a: 62 (*Trogoderma*) **N**: EG LB MO **A**: IS SY
aegyptiacum Pic, 1901d: 33 (*Trogoderma*)
demaisoni Reitter, 1889g: 127 (*Trogoderma*)

dichroum Reitter, 1900f: 219 (*Trogoderma*) **A**: KI KZ
hirsutulum Reiche, 1868a: 134 (*Trogoderma*) **A**: AE IS SA SY TM YE **AFR**
maculifasciatum Reitter, 1887e: 205 (*Trogoderma*) **A**:XIN
nobile Reitter, 1881b: 85 (*Trogoderma*) **E**: GR PT SP **N**: AG EG LB MO TU **A**: AE AF CY IN IQ IS JO SA SY TD TM UP UZ **AFR**
distincta Kalík, 1954: 369
paulonotata Pic, 1908d: 41 (*Trogoderma*)
oculata Háva, 2004c: 78 **A**: YE
parthicum Zhantiev, 1976: 83 **A**: TM
tricolor Arrow, 1915b: 431 (*Trogoderma*) **A**: OM SA UP YE
trizonatum Fairmaire, 1883d: clvi (*Trogoderma*) **N**: AG MO
major Pic, 1895e: 79 (*Trogoderma*)
tamaricis Pic, 1894v: 180 (*Trogoderma*)
turcomanicum Mroczkowski, 1960: 214 **A**: IN TM
villosulum Duftschmid, 1825: 28 (*Anthrenus*) **E**: AR AU BU BY CR CZ DE FI FR GB GE GR HU IT PL RO SK SP SZ TR YU **A**: IN KI SY TR

genus *Reesa* Beal, 1967: 310 type species *Perimegatoma vespulae* Milliron, 1939
vespulae Milliron, 1939: 570 (*Perimegatoma*) **E**: AU CZ DE FI FR GB GE HU NL NR PL RU SK SV SZ **N**: AG EG MO TU **A**: AF JA RU **NAR AUR**

genus *Thaumaglossa* L. Redtenbacher, 1868: 43 type species *Thaumaglossa ruficapillata* L. Redtenbacher, 1868
Axinocerus Jayne, 1882: 347 type species *Axinocerus americanus* Jayne, 1882
Orphiloides Matsumura & Yokoyama, 1928a: 131 type species *Orphiloides ovivorus* Matsumura & Yokoyama, 1928
Phradonomorphus Pic, 1916c: 3 type species *Aethriostoma rufofasciata* Pic, 1915
Thaumatoglossa T. L. Casey, 1900: 145 type species *Axinocerus americanus* Jayne, 1882
bimaculata Arrow, 1915a: 113 **A**: UP **AFR**
chujoi Ohbayashi, 1982: 41 **A**: JA **ORR**
herrmanni Háva, 2003c: 206 **A**: HKG
hilleri Reitter, 1881a: 42 **A**: JA NO NP TAI UP **ORR**
indiana indiana Veer, Chauhan & Singh, 2004: 389 **A**: UP
laeta Arrow, 1915b: 435 **A**: CH JA TAI **ORR**
bimaculatus Matsumura & Yokoyama, 1928b: 53 (*Orphilus*)
rufomaculata Pic, 1938f: 12 (*Aethriostoma*)
rufocapillata L. Redtenbacher, 1868: 44 **E**: GEi NLi **A**: CH JA NP **AFR ORR**
africana Kalík, 1955: 321
atricolor Pic, 1915d: 2 (*Aethriostoma*)
ovalis Fleutiaux, 1887: 62 (*Anthrenus*)
ovivorus Matsumura & Yokoyama, 1928a: 131 (*Orphiloides*)
sinensis Pic, 1951c: 5
sumatrana Pic, 1951c: 5
uninotata Pic, 1954d: 56 (*Orphinus*) **A**: FUJ
yeti Háva, 2003f: 251 **A**: AF NP

genus *Trogoderma* Dejean, 1821: 46 type species *Anthrenus elongatulus* Fabricius, 1801 (= *Anthenus glabrus* Herbst, 1783)
Acolpus Jayne, 1882: 347 type species *Acolpus primus* Jayne, 1882
Asidora Mulsant & Rey, 1868a: 122 type species *Trogoderma fuscicornis* Mulsant & Rey, 1868
Ditorines Normand, 1936a: 298 type species *Ditorines denticornis* Normand, 1936
Entomotrogus Ganglbauer, 1904: 27 type species *Entomotrogus megatomoides* Reitter, 1881
Eucnocerus Sharp, 1902: 648 type species *Eucnocerus dispar* Sharp, 1902
Eurhopalus Solier, 1849: 372 type species *Eurhopalus variegatus* Solier, 1849
Macroprion Hope, 1840: 68 type species *Anthrenus serraticornis* Fabricius, 1792
Ocelliger R. A. Philippi & F. Philippi, 1864: 283 type species *Ocelliger ater* R. A. Philippi & F. Philippi, 1864
Psacus Pascoe, 1866c: 446 type species *Psacus attagenoides* Pascoe, 1866
Pseudomegatoma Pic, 1915d: 3 type species *Pseudomegatoma boliviensis* Pic, 1915

Trogoderma Berthold, 1827: 352 [HN] type species *Anthrenus elongatulus* Fabricius, 1801
Trogoderma Latreille, 1829a: 511 [HN] type species *Anthrenus elongatulus* Fabricius, 1801
Trogodermum Gistel, 1856: 361 type species *Anthrenus elongatulus* Fabricius, 1801
albonotatum Reiche, 1868a: 128 **E**: "Pyrénées"
angustum Solier, 1849: 374 (*Eurhopalus*) **E**: AU CZ DE FI FR GB GE HU IT LA NL NR PL RU SV **NAR NTR ORR**
bifasciatum L. Redtenbacher, 1868: 44
boliviensis Pic, 1915d: 4 (*Pseudomegatoma*)
arcanum Zhantiev, 2002: 1 **E**: GG
bactrianum Zhantiev, 1976: 3 [= 1973: 148] **A**: TD TM
denticornis Normand, 1936a: 299 (*Ditorines*) **N**: TU
glabrum Herbst, 1783: 26 (*Anthrenus*) **E**: AU CZ DE FI FR GB GE GR HU IR IT LA NL PL RO RU SK SP SZ YU "Caucasus" **N**: AG MO TU **A**: IN KZ MG OM "Siberia" **NAR**
boron Beal, 1954: 53
elongatulus Fabricius, 1801a: 106 (*Anthrenus*)
fuscicornis Mulsant & Rey, 1868a: 122
limbatum Pic, 1934b: 26
niger Herbst, 1797: 338 (*Anthrenus*)
ruficornis Latreille, 1807a: 39 (*Anthrenus*)
subfasciatus Gyllenhal, 1808a: 155 (*Dermestes*)
versicolor Illiger, 1801a: 86 (*Dermestes*)
granarium Everts, 1898: 610 **E**: AU CZ DE FR GB GE NL NR PL RO SK SP SZ SV **N**: EG MO TU **A**: AF CH CY IN IS KI KZ RU SA TD UZ YE **COS**
afrum Priesner, 1951: 133
khapra Arrow, 1917: 481
inclusum LeConte, 1854b: 110 **E**: CR FR GB GR HU IT NL RU SP TR YU **N**: AG EG MO TU **A**: IN IS JA RU TR UP **AFR NAR**
advena T. L. Casey, 1900: 154
brunnescens T. L. Casey, 1916: 189
frosti T. L. Casey, 1916: 188
meridionalis Kraatz, 1858: 146
nigrescans T. L. Casey, 1916: 188
obsolescens T. L. Casey, 1900: 154
versicolor Mutchler & Weiss, 1927: 12
irroratum Reitter, 1881a: 41 **N**: EG
quinquefasciata Leesberg, 1906: 79 [HN]
koenigi Pic, 1954d: 55 **A**: FUJ
longisetosum Chao & Lee, 1966: 245 **E**: CZi **A**: JA SHX
megatomoides Reitter, 1881b: 85 **E**: AU CZ FR GB GE HU NL PL SK SV **N**: AG **A**: IN **NAR**
filipes Sharp, 1902: 647
gravidum Sharp, 1902: 648
nigripubens Reitter, 1911d: 291 (*Entomotrogus*)
mongolicum Zhantiev, 1973a: 190 **A**: MG
octaedron Peyerimhoff, 1943: 12 **N**: "Sahara"
ornatum Say, 1825a: 185 (*Megatoma*) **A**: JA **NAR**
dispar Sharp, 1902: 649 (*Eucnocerus*)
pallipes Ziegler, 1845: 269
parvula T. L. Casey, 1916: 186
procera T. L. Casey, 1916: 185
pusillum LeConte, 1854b: 111
serriger T. L. Casey, 1900: 153 [HN]
serriferum Dalla Torre, 1911b: 70 [RN]
tarsale Melsheimer, 1844: 116
quinquefasciatum Jacquelin du Val, 1859a: 257 **E**: FR
sinense Pic, 1927c: 6 **A**: HUB
teukton Beal, 1956: 560 **A**: CH IN JA KI KZ SC TD **NAR**
oothecophilum Chao & Lee, 1966: 245

variabile Ballion, 1878: 277 **E**: AR CZi FI GB IT LA NLi SV **N**: MO **A**: AF IN KZ MG NE SA TD TM UZ **COS**
parabile Beal, 1954: 53
persica Pic, 1914e: 10
scabripennis T. L. Casey, 1916: 189
turkestanicum Pic, 1914e: 10
varium Matsumura & Yokoyama, 1928b: 51 (*Megatoma*) **A**: HUN JA SC SCH
laticorne Chao & Lee, 1966: 247
versicolor Creutzer, 1799: 117 (*Anthrenus*) **E**: AU BU CR CT CZ DE FI FR GB GE GR HU IR IT KZ LA NL PL PT RO RU SK SP SV SZ TR UK YU **N**: AG GE MO TU **A**: AF CH IN IS JA KI KZ MG RU SA SY TD TM UZ WS **COS**
costae Mulsant & Rey, 1868a: 127
elongatulus Duftschmid, 1825: 26 (*Anthrenus*)
flexuosa C. G. Thomson, 1862: 207
grandis Thunberg, 1815: 151 (*Anthrenus*)
hieroglyphica Abeille de Perrin, 1872b: xlii
reductum Pic, 1919h: 17
testaceicorne Perris, 1862: 196
yunnaneunsis Liu & Zhang, 1986: 178 **A**: YUN
yunnaneunsis Zhang, Liu & Jing, 1999: 1765 [HN]

genus *Turcicornis* Háva, 2000a: 115 type species *Turcicornis kopeckyi* Háva, 2000
kopeckyi Háva, 2000a: 116 **A**: TR

genus *Zhantievus* Beal, 1992: 403 type species *Zhantievus lymantriae* Beal, 1992
lymantriae Beal, 1992: 407 **N**: MO

family ENDECATOMIDAE LeConte, 1861

genus *Endecatomus* Mellié, 1847: 108 type species *Anobium reticulatum* Herbst, 1793
Dictyalotus L. Redtenbacher, 1849: 348 type species *Anobium reticulatum* Herbst, 1793
lanatus Lesne, 1934a: 174 **A**: FE JA JIL
reticulatus Herbst, 1793: 70 (*Anobium*) **E**: AU BE CZ FR GE HU IT NL PL RO RU SK SL SZ **A**: ES FE "Ural"

family BOSTRICHIDAE Latreille, 1802
subfamily **Bostrichinae** Latreille, 1802
tribe Apatini Jacquelin du Val, 1861

genus *Apate* Fabricius, 1775: 54 type species *Ligniperda terebrans* Pallas, 1772
monachus Fabricius, 1775: 54 **E**: ABi FR GBi GEi GGi IT NLi PT SP SVi SZi **N**: AG EG LB MO TU **A**: CY IS LE SY YE **AFR NTR**i
carmelita Fabricius, 1801b: 379
francisca Fabricius, 1801b: 379
gibba Fabricius, 1798: 156 (*Synodendron*)
mendica A. G. Olivier, 1790c: 108 (*Bostrichus*)
rufiventris P. H. Lucas, 1843c: xxv
semicostata J. Thomson, 1858: 83
sennii De Stefani-Perez, 1911: 62
terebrans Pallas, 1772: 7 (*Ligniperda*) **E**i: AB AT GB GG SP **A**i: IS YE **AFR NTR**i
dispar Fåhraeus, 1871: 663
muricatus Fabricius, 1775: 54 (*Apate*)

genus *Phonapate* Lesne, 1895: 178 type species *Apate uncinata* Karsch, 1881
chan Semenov, 1891: 351 (*Apate*) **A**: KZ TM UZ
deserti Semenov, 1891: 351 (*Apate*) **A**: KZ TM UZ

fimbriata Lesne, 1909: 541 **A**: SE HKG **ORR**
uncinata africana Vrydagh, 1961: 9 **A**: SA YE **AFR**
uncinata moghrebica Lesne, 1934b: 218 **E**: SP **N**: AG EG LB MO TU **AFR**
uncinata Lesne, 1902a: 123 [HN]
uncinata uncinata Karsch, 1881: 46 (*Apate*) **N**: EG **A**: AE AF BA CY IN IQ OM QA SA TR YE **AFR ORR**
arabs Lesne, 1909: 552
ganglbaueri Zoufal, 1894: 37 (*Ligniperda*)
nitidipennis Waterhose, 1881: 472 (*Apate*)

genus *Xylomedes* Lesne, 1902a: 118 type species *Apate rufocoronata* Fairmaire, 1892
carbonnieri Lesne, 1897b: 235 (*Apate*) **N**: AG LB TU
cornifrons Baudi di Selve, 1874b: 334 (*Apate*) **A**: CY IS LE SY TR
coronata Marseul, 1883: 183 (*Apate*) **N**: AG EG MO TU
sericans Marseul, 1883: 183 (*Apate*)
laticornis Lesne, 1895: 178 (*Apate*) **A**: YE **N**: EG **AFR**
rufocoronata Fairmaire, 1892a: 104 (*Apate*) **N**: AG EG LB MO TU **A**: SA YE **AFR**
turcica Lesne, 1941a: 134 **E**: SY TR

tribe Bostrichini Latreille, 1802

genus *Amphicerus* LeConte, 1861: 208 type species *Apate bicaudatus* Say, 1824
subgenus *Amphicerus* LeConte, 1861: 208 type species *Apate bicaudatus* Say, 1824
Schistoceros Lesne, 1899b: 442 type species *Bostrichus bimaculatus* A. G. Olivier, 1790
cornutus cornutus Pallas, 1772: 8 (*Ligniperda*) **E**i: PL **NAR NTR**
bicornutus Latreille, 1813: 65 (*Bostrichus*)
consanguineus Lesne, 1899b: 513 (*Schistoceros*)
maritimus Casey, 1898: 68
migrator Sharp, 1885b: 160 (*Bostrichus*)
peregrinus Erichson, 1847: 87 (*Bostrichus*)
punctipennis LeConte, 1858: 73 (*Apate*)
hamatus Fabricius, 1787a: 33 (*Apate*) **E**i: GE **NAR**
aspericollis Germar, 1824: 465 (*Apate*)
bicaudatus Say, 1824: 320 (*Apate*)
gracilis Casey, 1898: 64
serricollis Germar, 1824: 464 (*Apate*)
subgenus *Caenophrada* Waterhouse, 1888: 350 type species *Caenophrada anobioides* Waterhouse, 1888
anobioides Waterhouse, 1888: 350 (*Caenophrada*) **A**: AF PA SA UP YE **AFR ORR**
bimaculata A. G. Olivier, 1790c: 109 (*Bostrichus*) **E**: AF AL BH BU CR FR GR HU IT KZ MC PT RO SP ST TR UK YU **N**: AG MO TU **A**: CY IS KI SY TD TM TR **NAR**i
auritus Frivaldszky von Frivald, 1835: 267 (*Apate*)

genus *Bostrichus* Geoffroy, 1762: 301 type species *Dermestes capucinus* Linnaeus, 1758
capucinus Linnaeus, 1758: 355 (*Dermestes*) **E**: AB AL AR AU BE BH BU BE BY CR CT CZ DE FI FR GB GE GG GR HU IR IT KZ LA LT MC MD NL NT PL PT RO SK SL SP ST SV SZ TR UK YU **N**: AG EG LB MO TU **A**: CH CY ES IN IS KZ LE NW SI SY TM TR WS
luctuosus A. G. Olivier, 1790c: 109
nigriventris P. H. Lucas, 1843c: xxv (*Apate*)
ruber Geoffroy, 1785: 133
rubriventris Zoufal, 1894: 40 (*Bostrychus*)
rugosus Fabricius, 1801b: 380 (*Apate*)

genus *Bostrychoplites* Lesne, 1899b: 443 type species *Bostrichus cornutus* A. G. Olivier, 1790
arabicus Lesne, 1935a: 271 **A**: YE
cornutus A. G. Olivier, 1790c: 108 (*Bostrichus*) **E**i: AU FI GB GE IT SP SV **N**i: EG MO **A**: YE **AFR NAR**i
abyssinicus Murray, 1867b: 90 (*Bostrichus*)
destructor Burchell, 1822: 325 (*Bostrichus*)
normandi normandi Lesne, 1897b: 236 (*Bostrychus*) **N**: AG EG TU **AFR**
normandi yemenensis Lesne, 1935a: 272 **A**: YE

zickeli Marseul, 1867: xxxiv (*Apate*) **Ei**: GE **N**: AG EG LB MO TU **A**: YE **AFR**
hamaticollis Fairmaire, 1874: 407 (*Apate*)

genus *Bostrychopsis* Lesne, 1899b: 444 type species *Bostrichus cephalotes* A. G. Olivier, 1795
bengalensis Lesne, 1895: 174 (*Bostrychus*) **A**: UP **ORR**
cristaticeps Lesne, 1906a: 403 **A**: YE (Suqutra)
jesuita Fabricius, 1775: 54 (*Apate*) **Ei**: GE **AUR**
canarii Nördlinger, 1880: 66 (*Apate*)
parallela Lesne, 1895: 174 (*Bostrychus*) **Ei**: AU GE **A**: GUA HAI HKG JA SE TAI UP **AUR NARi ORR**
affinis Lesne, 1899b: 536
reichei Marseul, 1867: xxxv (*Apate*) **N**: AG EG LB TU **AFR**
roonwali Rai, 1966: 576 **A**: UP
valida valida Lesne, 1899b: 541 **Ei**: GE **NTR**

genus *Dolichobostrychus* Lesne, 1899b: 583 type species *Bostrychus angustus* Steinheil, 1872
yunnanus Lesne, 1913a: 191 **A**: SCH YUN

genus *Heterobostrychus* Lesne, 1899b: 443 type species *Bostrichus aequalis* Waterhouse, 1884
aequalis Waterhouse, 1884a: 215 (*Bostrichus*) **Ei**: GB GE IT NL SV **A**: BT HP ISi TAI UP YUN **AFRi AUR NARi ORR**
papuensis W. J. MacLeay, 1886: 154 (*Rhizopertha*)
uncipennis Lesne, 1895: 173 (*Bostrychus*)
ambigenus Lesne, 1920: 295 **A**: YUN
brunneus Murray, 1867b: 92 (*Bostrichus*) **Ei**: BE FR GB GE IT PL SP **Ni**: EG **AFR AURi NARi**
grayanus Wollaston, 1867: 109 (*Bostrychus*)
picipennis Fåhraeus, 1871: 669 (*Bostrichus*)
hamatipennis Lesne, 1895: 173 (*Bostrychus*) **Ei**: GE **A**: BT HP JA SD SHG TAI **AFR ORR**
carinipennis Matsumura, 1915a: 128 (*Apate*) [HN]
nipponensis Lewis, 1896c: 339 (*Apate*)

genus *Lichenophanes* Lesne, 1899b: 443 type species *Bostrichus tristis* Fåhraeus, 1871
carinipennis Lewis, 1896c: 338 (*Apate*) **A**: JA TAI **ORR**
carinatus Lewis, 1896c: 339 (*Apate*)
guttatus Matsumura, 1915a: 129 (*Bostrychus*)
khmerensis Lesne, 1896b: 511 (*Bostrychus*)
numida Lesne, 1899b: 460 **E**: IT PT SP **N**: AG MO TU **AFR**
varius Illiger, 1801c: 172 (*Apate*) **E**: AB AL AR AU BH BU CR CT CZ EN FR GE GG GR HU IT MC MD PL PT RO SK SL SP ST SZ TR UK YU **N**: AG EG LB MO TU **A**: CY IN SY TM TR
dufourii Latreille, 1807b: 7 (*Bostrichus*)
gallicus Panzer, 1806: 17 (*Apate*)

genus *Micrapate* Casey, 1898: 66 type species *Sinoxylon dinoderoides* Horn, 1878
Bostrychulus Lesne, 1899b: 444 type species *Sinoxylon dinoderoides* Horn, 1878
brunnipes Fabricius, 1801b: 383 (*Apate*) **Ei**: FR (Corse) GE **NTR**
puncticollis Kiesenwetter, 1877: 39 (*Apate*)
xyloperthoides Jacquelin du Val, 1859c: 40 (*Apate*) **E**: FR IT PT SP **N**: AG MO TU **NARi**

genus *Parabostrychus* Lesne, 1899b: 442 type species *Bostrychus elongatus* Lesne, 1895
acuticollis Lesne, 1913a: 192 **A**: GUA NP TAI **ORR**

tribe Sinoxylonini Lesne, 1899

genus *Calopertha* Lesne, 1906b: 446 type species *Sinoxylon subretusum* Ancey, 1881
costatipennis Lesne, 1906b: 457 **A**: YE **AFR**
subretusa Ancey, 1881: 509 (*Sinoxylon*) **N**: EG MO **A**: SA YE **AFR**
truncatula Ancey, 1881: 509 (*Sinoxylon*) **N**: EG MO **A**: IN PA SA YE **AFR**

genus *Sinoxylon* Duftschmid, 1825: 85 type species *Dermestes muricatus* Linnaeus, 1767
Apatodes Blackburn, 1889b: 1429 type species *Apatodes macleayi* Blackburn, 1889 (= *Sinoxylon anale* Lesne, 1897)
Trypocladus Guérin-Méneville, 1845: 17 type species *Bostrichus sexdentatum* A. G. Olivier, 1790
anale Lesne, 1897a: 21 **E**i: AU BE FI GB GE PL **A**: CE HP IQ NP PA SE UP YUN **AFR**i **AUR**i **NAR**i **NTR**i **ORR**
geminatum Schilsky, 1899: 80
macleayi Blackburn, 1889b: 1429 (*Apatodes*)
atratum kohlarianum Lesne, 1906b: 540 **A**: HP
birmanum Lesne, 1906b: 474 **E**i: GE **ORR**
capillatum Lesne, 1895: 175 **A**: KA
ceratoniae Linnaeus, 1758: 353 (*Scarabaeus*) **E**i: GE SV **N**: AG EG LB MO TU **A**: AE OM SA YE **AFR**
bicuspidatum Ancey, 1879: 139
crassum crassum Lesne, 1897a: 21 **E**i: GE PL **A**: HP IN NP OMi UP **ORR**
cucumella Lesne, 1906b: 474 **A**: BT **ORR**
eucerum Lesne, 1932: 657 **A**: SHA
flabrarius Lesne, 1906b: 474 **E**i: GB **A**: BT HKG SW **ORR**
japonicum Lesne, 1895: 175 **A**: JA **NAR**i
mangiferae M. Chûjô, 1936b: 407 **A**: TAI
muricatum Linnaeus, 1767: 562 (*Dermestes*) **E**: AB AL AU BH BU CR FR GE GR HU IT PT SL SP ST TR UK **N**: AG CI EG LB MO TU **A**: CY IS SY TR **NAR**i
bidens Fabricius, 1798: 157 (*Bostrichus*)
chalcographum Panzer, 1793e: 4 (*Bostrichus*)
muricatum Fabricius, 1792b: 359 (*Synodendron*)
sexdentatum A. G. Olivier, 1790c: 110 (*Bostrichus*)
oleare Lesne, 1932: 655 **A**: HP
perforans Schrank, 1789: 64 (*Bostrichus*) **E**: AB AL AR AU BH BU CR CZ FR GE GG GR HU IT MC PL PT RO SK SL SP ST SZ TR UK YU **A**: CY IS SY TM TR
bispinosum A. G. Olivier, 1790c: 110 (*Bostrichus*)
pugnax Lesne, 1904: 159 **A**: AF IN KA PA **ORR**
rejectum Hope, 1845a: 16 (*Apate*) **A**: GUA
senegalense Karsch, 1881: 42 (*Synoxylon*) **E**i: GE UK **N**: AG EG LB MO **AFR**
coronatum Zoufal, 1894: 38
vermiculatum Lesne, 1906b: 501
succisum Lesne, 1895: 176 **N**: MO TU **AFR**
sudanicum Lesne, 1895: 176 **E**i: GE **N**: EG **A**: YE **AFR** **ORR**
tignarium Lesne, 1902a: 111 **A**: SCH YUN **ORR**
unidentatum Fabricius, 1801b: 377 (*Sinodendron*) **E**i: GB GE IT RU **A**: CH YE **AFR** **AUR**i **NAR**i **NTR**i **ORR**
conigerum Gerstaecker, 1855: 268

genus *Xyloperthodes* Lesne, 1906b: 447 type species *Sinoxylon nitidipennis* Murray, 1867
nitidipennis nitidipennis Murray, 1867b: 94 (*Sinoxylon*) **E**i: GB GE **AFR**
plagiatus Fåhraeus, 1871: 666 (*Xylopertha*)
politus Quedenfeldt, 1886: 327 (*Xylopertha*)

tribe Xyloperthini Lesne, 1921

genus *Calophagus* Lesne, 1902a: 108 type species *Calophagus pekinensis* Lesne, 1902
pekinensis Lesne, 1902a: 109 **A**: BEI

genus *Enneadesmus* Mulsant, 1851c: 208 type species *Bostrichus trispinosus* A. G. Olivier, 1795
auricomus Reitter, 1898a: 13 (*Xylopertha*) **A**: IN SY TM UZ
bidentatus A. G. Olivier, 1795b: no 77: 16 (*Bostrichus*) [HN]
forficula Fairmaire, 1883b: 95 (*Apate*) **E**: GR (Kríti) **N**: AG EG LB MO TU **A**: IN IS JO OM PA SI SA YE **AFR** **ORR**
auberti Chobaut, 1897b: 170 (*Xylopertha*)
capensis Lesne, 1901b: 607

obtusidentatus Lesne, 1899a: 228 (*Xylopertha*) **Ei**: GE **N**: EG **A**: IQ SA SY YE **AFR**
scopini Fursov, 1936a: 349 **A**: UZ (Tashkent)
minor Fursov, 1936a: 349
trispinosus A. G. Olivier, 1795b: no 77: 16 (*Bostrichus*) **E**: CR FR IT PT SP **N**: AG CI EG LB MO TU **A**: IN IQ
lesnei Chobaut, 1898: 81 (*Xylopertha*)

genus *Octodesmus* Lesne, 1901b: 479 type species *Octodesmus episternalis* Lesne, 1901
parvulus Lesne, 1897a: 19 (*Xylopertha*) **Ei**: GB **A**: UP **ORR**

genus *Paraxylion* Lesne, 1941b: 150 type species *Xylion bifer* Lesne, 1932
bifer Lesne, 1932: 659 (*Xylion*) **A**: HP UP **ORR**

genus *Paraxylogenes* Damoiseau, 1968: 2 type species *Paraxylogenes pistaciae* Damoiseau, 1968
pistaciae Damoiseau, 1968: 4 **A**: IN IQ PA

genus *Psicula* Lesne, 1941b: 146 type species *Psicula heterogama* Lesne, 1941
heterogama Lesne, 1941b: 147 **A**: SD **ORR**

genus *Scobicia* Lesne, 1901b: 479 type species *Apate chevrieri* A. Villa & J. B. Villa, 1835
Xylopertha Casey, 1898: 66 [HN]
barbata Wollaston, 1860b: 359 (*Enneadesmus*) **E**: AZ **N**: MR **NAR**i
barbifrons Wollaston, 1864: 252 (*Xylopertha*) **N**: CI
chevrieri A. Villa & J. B. Villa, 1835: 49 (*Apate*) **E**: AB AL AU BH BU CR FR GG GR HU IT PT SL SP ST SZ TR UK YU **N**: AG EG LB MO TU **A**: CY IN IS LE SY TR **NAR**i
foveicollis Allard, 1869: 468 (*Xylopertha*)
ficicola Wollaston, 1865: 231 (*Xylopertha*) **N**: CI
pustulata Fabricius, 1801b: 381 (*Apate*) **E**: AB AU CR FR GEi GG GR HU IT PT RO SP ST SZ UK **N**: AG EG LB MO TU **A**: CY SY TR
humeralis P. H. Lucas, 1843c: xxv (*Apate*)

genus *Tetrapriocera* Horn, 1878: 541 type species *Bostrichus longicornis* A. G. Olivier, 1795
longicornis A. G. Olivier, 1795b: no 77: 15 (*Bostrichus*) **Ei**: GE **NTR**
schwarzi Horn, 1878: 545

genus *Xylion* Lesne, 1901b: 478 type species *Xylion securifer* Lesne, 1901
adustus Fåhreus, 1871: 667 (*Xylopertha*) **Ei**: GB **AFR**
pierronii Fairmaire, 1880e: 246 (*Xylopertha*)

genus *Xylobiops* Casey, 1898: 66 type species *Apate basilaris* Say, 1824
basilaris Say, 1824a: 321 (*Apate*) **Ei**: GB GE **NAR**

genus *Xylodectes* Lesne, 1901b: 478 type species *Xylopertha ornata* Lesne, 1897
ornatus Lesne, 1897a: 19 (*Xylopertha*) **A**: SD TAI **ORR**

genus *Xylogenes* Lesne, 1901b: 479 type species *Rhizoperta dilatata* Reitter, 1889
dilatata Reitter, 1889d: 112 (*Rhizoperta*) **E**: AB **A**: IN IQ JO SY TM
mesopotamicus Lesne, 1937: 195 **E**: AR **A**: IQ SY
semenovi Lesne, 1904: 157 **A**: TM XIN
sindicola Lesne, 1936: 136 **A**: PA

genus *Xylopertha* Guérin-Méneville, 1845: 17 type species *Apate sinuata* Fabricius, 1792 (= *Botrichus retusa* A. G. Olivier, 1790)
Xylonites Lesne, 1901b: 478 type species *Bostrichus retusus* A. G. Olivier, 1790
dunensis Rai & Chatterjee, 1964: 122 **A**: "North India"

praeusta Germar, 1817: 226 (*Apate*) **E**: CR FR IT PT SP **N**: AG MO TU
appendiculata P. H. Lucas, 1847: 446
reflexicauda Lesne, 1937: 199 **A**: IN PA (*Xylonites*)
retusa A. G. Olivier, 1790c: 110 (*Bostrichus*) **E**: AB AL AU BE BH BU CR CZ FR GE GG GR HU IT LU MC NL PL PT RO SK SL SP ST SZ TR UK YU **N**: AG LB MO TU **A**: CY ES TR WS
aterrima Faldermann, 1837: 250 (*Apate*)
sinuata Fabricius, 1792b: 362 (*Apate*)

genus *Xyloperthella* Fisher, 1950: 116 type species *Bostrichus piceus* A. G. Olivier, 1790
crinitarsis Imhoff, 1843: 177 (*Apate*) Ei: FR GB GE NL **AFR NAR**i
pubescens Murray, 1867b: 93 (*Sinoxylon*)
picea picea A. G. Olivier, 1790c: 110 (*Bostrichus*) **E**: GBi GEi IT NLi SP **N**: AG EG LB MO TU **A**: IS SA SY YE **AFR NTRi**
cultrata J. Thomson, 1858: 83 (*Apate*)
femorata Klug, 1833: 203 (*Apate*)
fumata Murray, 1867b: 94 (*Sinoxylon*)
heydeni Schilsky, 1899: 92 (*Xylopertha*)

genus *Xylopsocus* Lesne, 1901b: 479 type species *Apate capucinus* Fabricius, 1781
bicuspis Lesne, 1901b: 629 **A**: JA TAI
capucinus Fabricius, 1781: 62 (*Apate*) **A**: CH JA SD TAI **AFR AUR NAR**i **NTR ORR**
eremitus A. G. Olivier, 1790c: 110 (*Bostrichus*)
marginata Fabricius, 1801b: 382 (*Apate*)
nicobaricus L. Redtenbacher, 1868: 114 (*Enneadesmus*)
distinctus Rai, 1967: 140 **A**: UP
galloisi Lesne, 1937: 198 **A**: JA

genus *Xylothrips* Lesne, 1901b: 479 type species *Apate flavipes* Illiger, 1801
cathaicus Reichardt, 1966: 82 **A**: HEN HUB
flavipes Illiger, 1801c: 171 (*Apate*) **E**i: GB **N**: AG **A**: BT JA SD TAI UP **AFR AUR NAR**i **ORR**
dominicana Fabricius, 1801b: 380 (*Apate*)
iracundus Snellen van Vollenhoven, 1869: 10 (*Bostrichus*)
mutilatus Walker, 1858: 286 (*Bostrichus*)
religiosae Fairmaire, 1850: 50 (*Apate*) [HN]
sinuata Stephens, 1830: 351 (*Apate*) [HN]

subfamily **Dinoderinae** C. G. Thomson, 1863

genus *Dinoderopsis* Lesne, 1906a: 400 type species *Dinoderopsis escharipora* Lesne, 1906
escharipora Lesne, 1906a: 401 **A**: YE (Suqutra)
serriger Lesne, 1923: 56 **A**: YE **AFR**

genus *Dinoderus* Stephens, 1830: 352 type species *Dinoderus ocellaris* Stephens, 1830
subgenus *Dinoderastes* Lesne, 1914: 244 type species *Dinoderastes scabricauda* Lesne, 1914
exilis Lesne, 1932: 653 **A**: BT SD
japonicus Lesne, 1895: 170 **E**i: AU GE NL SV **A**: CH JA **AUR**i **NAR**i
pubicollis Van Dyke, 1923: 45
tsugae Matsumura, 1915b: 183
speculifer Lesne, 1895: 169 **A**: JA JIL
subgenus *Dinoderus* Stephens, 1830: 352 type species *Dinoderus ocellaris* Stephens, 1830
Patea Casey, 1898: 66 type species *Dinoderus brevis* Horn, 1878
bifoveolatus Wollaston, 1858c: 409 (*Rhyzopertha*) **E**i: AU BE CR CZ DE FI FR GB GE IT NL PL SK SP SZ **N**i: EG MR **A**: CH JA SA YE **COS**
perpunctatus Lesne, 1895: 170
brevis Horn, 1878: 549 **E**i: AZ FI GE GR NT PL SV **A**: CH **AFR**i **NAR**i **NTR**i **ORR**
distinctus Lesne, 1898b: 325 **E**i: GE **NTR**i **ORR**

minutus Fabricius, 1775: 54 (*Apate*) **E**i: AU BE CZ DE FI FR GB GE GR IT NL NR PL SK SV **N**i: EG MO **A**: CY JA SY TAI ZHE **COS**

japonicus Matsumura, 1915b: 184 [HN]
sicula Baudi di Selve, 1874b: 336 (*Rhizopertha*)
substriatus Stephens, 1830: 352 [HN]
vertens Walker, 1859a: 260 (*Bostrichus*)

ocellaris ocellaris Stephens, 1830: 352 **E**i: BE BH FR GB GE NL PL SV **NAR**i **ORR**

pilifrons Lesne, 1895: 170

piceolus Lesne, 1933: 259 **A**: HKG

genus *Prostephanus* Lesne, 1898b: 342 type species *Apate punctatus* Say, 1826
truncatus Horn, 1878: 549 (*Dinoderus*) **E**i: FR GE **A**i: CH IQ IS **AFR**i **NAR** **NTR**

genus *Rhyzopertha* Stephens, 1830: 354 type species *Synodendron pusillum* Fabricius, 1798 (= *Synodendron dominicum* Fabricius, 1792)

Rhizopertha Guérin-Méneville, 1845: 17 type species *Synodendron dominicum* Fabricius, 1792

dominica Fabricius, 1792b: 359 (*Synodendron*) **E**i: AR AU AZ BE BY CR CZ DE FI FR GB GE GR HU IR IT LA NL NR NT PL SK SP SV **N**i: AG CI EG MR TU **A**i: CY HKG IQ JA SY TAI YE **COS**

fissicornis Marsham, 1802: 82 (*Ptinus*)
frumentaria Nördlinger, 1855: 238 (*Apate*)
moderata Walker, 1859a: 260 (*Bostrichus*)
piceus Marsham, 1802: 88 (*Ptinus*)
pusillum Fabricius, 1798: 156 (*Synodendron*)
rufa Hope, 1845a: 17 (*Apate*)

genus *Stephanopachys* Waterhouse, 1888: 349 type species *Apate substriatus* Paykull, 1800
brunneus Wollaston, 1862d: 440 (*Dinoderus*) **N**: CI
himalayanus Lesne, 1932: 651 **A**: HP
linearis Kugelann, 1792: 495 (*Apate*) **E**: AB AR AU BY CT CZ DE EN FI FR (Corse) GE GG IT LA LT NR NT PL SL SV SZ UK **A**: CH ES FE WS

elongatus Paykull, 1800: 143 (*Apate*)

quadricollis Fairmaire, 1878a: 83 (*Dinoderus*) **E**: CR FR GR IT PT SP SVi UK **N**: AG MO TU **A**: SY TR
sachalinensis Matsumura, 1911a: 126 (*Rhyzorertha*) **A**: FE
substriatus Paykull, 1800: 142 (*Apate*) **E**: AL AU BH BU BY CR CT CZ EN ES FI FR GB GE GR HU IT LA LT MC MD NR NT PL RO SK SL SV SZ UK YU **A**: ES FE SY WS **NAR**

pacificus Casey, 1898: 73 (*Dinoderus*)
rhapontica Motschulsky, 1853b: 22 (*Apate*)

subfamily **Dysidinae** Lesne, 1921

genus *Apoleon* Gorham, 1885a: 51 type species *Apoleon edax* Gorham, 1885
edax Gorham, 1885a: 52 **A**i: GUA **ORR**

spineus Chen & Yin, 2003: 113 (*Dysides*)

subfamily **Lyctinae** Billberg, 1820

tribe Lyctini Billberg, 1820

genus *Acantholyctus* Lesne, 1924: 79 type species *Lyctus cornifrons* Lesne, 1898
cornifrons cornifrons Lesne, 1898a: 139 (*Lyctus*) **N**: AG EG MO TU **AFR**

genus *Lyctoxylon* Reitter, 1879c: 196 type species *Minthea dentata* Pascoe, 1866
beesonianum Lesne, 1936: 133 **A**: HP
convictor Lesne, 1936: 132 **A**: HP
dentatum Pascoe, 1866b: 141 (*Minthea*) **E**i: GB GE NL **A**: CH JA **AFR**i **AUR**i **NAR**i **NTR**i **ORR**

japonum Reitter, 1879c: 199
seriehispidum Kiesenwetter, 1879b: 319 (*Lyctus*)

genus *Lyctus* Fabricius, 1792b: 502 type species *Lyctus canaliculatus* Fabricius, 1792 (= *Dermestes linearis* Goeze, 1777)

subgenus *Lyctus* Fabricius, 1792b: 502 type species *Lyctus canaliculatus* Fabricius, 1792 (= *Dermestes linearis* Goeze, 1777)

Eulyctus Jakobson, 1915: 896 type species *Lyctus suturalis* Faldermann, 1837

asiaticus Iablokoff-Khnzorian, 1976a: 96 **A**: KI TD

carbonarius Waltl, 1834: 167 **E**i: AU BE FI FR GB GE IT NL SV SZ **A**i: IS JA **N**i: EG **AUR**i **NAR NTR ORR**i

leacocianus Wollaston, 1860a: 256

modestus Lesne, 1911b: 534

planicollis LeConte, 1858: 74

cavicollis LeConte, 1866a: 103 **E**i: AU GB GE SZ **AUR**i **NAR**

linearis Goeze, 1777: 148 (*Dermestes*) **E**: AB AL AU AZ BE BH BU BY CR CT CZ DE FI FR GB GE GG GR HU ICi IR IT KZ LA LS LT LU MC MD NL NR PL PT RO SK SL SP ST SV SZ TR UK YU **N**: AG CI EG MO TU **A**: CY JA KZ SY TR TM UZ **AUR**i **NAR NTR**i

axillaris Melsheimer, 1844: 113

canaliculatus Fabricius, 1792b: 504

crassicornis Lesne, 1916: 96

duftschmidi Gozis, 1881b: cxxxv

fuscus Paykull, 1800: 332

oblongus Geoffroy, 1785: 19 (*Dermestes*)

striatus Melsheimer, 1844: 112

unipunctatus Herbst, 1783: 40 (*Dermestoides*)

parallelocollis Blackburn, 1888: 266 **A**i: IS **AUR**

pubescens Panzer, 1792c: 17 **E**: AB AL AR AU BE BH BU CR CT CZ FR GB GE GR HU IT KZ LS LU MC MD NL PL PT RO SK SL SP ST SZ TR UK YU **A**: CY TR

bicolor Comolli, 1837: 41

caucasicus Tournier, 1874: 412

shestakovi Fursov, 1939: 89 **E**: ST

simplex Reitter, 1879c: 198 **E**i: GE **NTR**

sinensis Lesne, 1911a: 48 **E**i: GB **A**: JA NE NO "Korea" **AUR**i

suturalis Faldermann, 1837: 255 **E**: AB AR BY GG RU UK **A**: KA KZ TM UZ

deyrollei Tournier, 1874: 411

turkestanicus Lesne, 1935b: 300 **A**: TM UZ XIN

turkestanicus Fursov, 1939: 89 [HN]

subgenus *Xylotrogus* Stephens, 1830: 116 type species *Xylotrogus brunneus* Stephens, 1830

africanus africanus Lesne, 1907: 302 **E**i: BE FR GB GE IT PL SP SZ TR **N**: AG EG MO **A**: IS JA SY **AFR NAR**i **ORR**i

politus Kraus, 1911: 118

brunneus Stephens, 1830: 117 (*Xylotrogus*) **E**i: BE CZ DE FI FR GB GE IC IR IT NL NR PT SK SP SV SZ **N**i: LB MO MR TU **A**: JA TAI UZ "Korea" **COS**

carolinae Casey, 1891: 13

costatus Blackburn, 1888: 265

disputans Walker, 1858: 206

glycyrrhizae Chevrolat, 1844: 191

jatrophae Wollaston, 1867: 112

retractus Walker, 1858: 206

rugulosus Montrouzier, 1861: 266

discedens Blackburn, 1888: 267 **E**i: GE **AUR ORR**

malayanus Lesne, 1910: 254

hipposideros Lesne, 1908: 356 **E**i: FI IT SZ **N**: AG MO **AFR**

genus *Minthea* Pascoe, 1866a: 97 type species *Minthea squamigera* Pascoe, 1866

Eulachus Blackburn, 1885: 141 type species *Ditoma rugicollis* Walker, 1858

Lyctopholis Reitter, 1879c: 196 type species *Minthea squamigera* Pascoe, 1866

obsita Wollaston, 1867: 112 (*Lyctus*) **E**i: AU **AFR NAR**i **NTR**i

reticulata Lesne, 1931: 98 **E**i: GB **AUR NAR**i **ORR**

rugicollis Walker, 1858: 206 (*Ditoma*) **E**i: AU BE DE FI FR GB GE IT (Sicilia) NL SV **A**: CH JA **COS**
foveicollis Reitter, 1879c: 199 (*Lyctopholis*)
hispida Blackburn, 1885: 141 (*Eulachus*)
similata Pascoe, 1866b: 141
squamigera Pascoe, 1866a: 97 **E**i: AU GB GR **NTR**
stichothrix Reitter, 1879c: 199 (*Lyctopholis*)

tribe Trogoxylini Lesne, 1921

genus *Trogoxylon* LeConte, 1862: 209 type species *Xylotrogus parallelopipedus* Melsheimer, 1844
aequale Wollaston, 1867: 111 (*Lyctus*) **E**i: BE NL SZ **AFR**i **NAR NTR ORR**i
californicus Casey, 1891: 13 (*Lyctus*)
curtulum Casey, 1891: 13 (*Lyctus*)
auriculatum Lesne, 1932: 654 **A**: HP
impressum Comolli, 1837: 40 (*Lyctus*) **E**: AL AR AU BH BU CR CZ DEi FIi FR GE GG GR HU IT MC NRi PT RO SK SL SP ST SVi SZ TR UK YU **N**: AG CI EG LB MO TU **A**: CY IN SY TM TR **NAR**i **NTR**i
capitalis Schaufuss, 1882a: 534 (*Lyctus*)
laevipennis Faldermann, 1837: 256 (*Lyctus*)
laevis Galeazzi, 1854: 64 (*Lyctus*)
parallelopipedum Melsheimer, 1844: 112 (*Xylothrogus*) **E**i: GB IT **AUR**i **NAR**
praeustum Erichson, 1847: 88 (*Lyctus*) **E**i: FI GE SZ **NTR ORR**i
prostomoides Gorham, 1883c: 212 (*Lyctus*)

subfamily **Polycaoninae** Lesne, 1896

genus *Polycaon* Laporte, 1838: 30 type species *Polycaon chiliensis* Laporte, 1838
Allaeocnemis LeConte, 1853a: 233 type species *Allaeocnemis stoutii* LeConte, 1853
Exops Curtis, 1839: 203 type species *Exops bevani* Curtis, 1839
stoutii LeConte, 1853a: 233 (*Allaeocnemis*) **E**i: IT (Sicilia) PL **NAR NTR**
ovicollis LeConte, 1859a: 49 (*Exops*)

subfamily **Psoinae** Blanchard, 1851
tribe Psoini Blanchard, 1851

genus *Coccographis* Lesne, 1901a: 349 type species *Coccographis nigrorubra* Lesne, 1901
nigrorubra Lesne, 1901a: 349 **A**: CH **ORR**

genus *Psoa* Herbst, 1797: 214 type species *Psoa viennensis* Herbst, 1797
subgenus *Psoa* Herbst, 1797: 214 type species *Psoa viennensis* Herbst, 1797
dubia P. Rossi, 1792: 17 (*Dermestes*) **E**: AL AU BH BU CR FR GE GR HU IT PT SL SP TR UK YU **N**: AG EG LB MO TU **A**: CY TR
herbsti Küster, 1847a: 45
italica Le Peletier de Saint-Fargeau & Audinet-Serville, 1825: 224
sanguinea Giorna, 1792: 48 (*Attelabus*)
viennensis Herbst, 1797: 215 **E**: AB AL AR AU BH BU CR CZ GE GG GR HU IT KZ MC RO SK SL ST TR UK YU **A**: CY TR
grandis Motschulsky, 1845a: 92

genus *Stenomera* P. H. Lucas, 1850: 38 type species *Stenomera blanchardii* P. H. Lucas, 1850
assyria Lesne, 1895: 169 **A**: IN IQ SY TR
blanchardii P. H. Lucas, 1850: 41 **N**: AG MO TU
nigrina Pic, 1893e: 122

family PTINIDAE Latreille, 1802
subfamily **Gibbiinae** Mulsant & Rey, 1868
tribe Gibbiini Mulsant & Rey, 1868

genus *Gibbium* Scopoli, 1777: 505 type species *Ptinus scotias* Fabricius, 1781 (= *Scotias psylloides* Czenpinski, 1778)

Scotias Czenpinski, 1778: 51 type species *Scotias psylloides* Czenpinski, 1778
aequinoctiale Boieldieu, 1854: lxxxiv **E**i: AB AR AU BE FR GB GE GR IT PL SP TR **N**i: AG CI EG LB TU **A**i: AP BT GUA HEN HKG HP IN JA LE NP PA SC SD SY TM TR UP YE YUN **AFR AUR NAR NTR ORR COS**
aegyptiacum Pic, 1894w: 203
chevrolatii Boieldieu, 1854: lxxxv
einsteini Bellés, 1980: 845
boieldieui Levrat, 1857: 419 **E**: AB AR GG GR RU UK **A**: IN TR
laevigena Reitter, 1884h: 296
psylloides Czenpinski, 1778: 51 (*Scotias*) **E**: AB AL AR AU BE BH BU CR CZ DEi FIi FR GB GE GG GR HU IRi IT MA MC MD NL NTi PL PT SK SL SP ST SVi SZ TR UK YU **N**: AG EG LB MO MR TU **A**: CY FEi IN IQ IS JAi LE LIAi SA SCi SY TAIi TR
apterus Geoffroy, 1785: 57 (*Bruchus*)
longicorne Reitter, 1884h: 296
scotias Fabricius, 1781: 74 (*Ptinus*)
seminulum Schrank, 1781a: 36 (*Ptinus*)

tribe Meziini Bellés, 1985

genus *Mezium* Curtis, 1828: 232 type species *Ptinus sulcatus* Fabricius, 1781
affine Boieldieu, 1856c: 674 **E**: AL AUi BE BH CR DEi FIi FR GBi GEi GG GR IRi IT MA MC NLi PT SL SP SVi TR YU **N**: AG EG LB MO MR TU **A**: CY SY TR **AUR**i **NAR**i
hirtipenne Reiche, 1864: 241
americanum Laporte, 1840a: 297 (*Gibbium*) **E**i: AU AZ FR GR IT MA SP **N**: AGi CI MOi MR **AUR**i **NAR**i **NTR**i
arachnoides Desbrochers des Loges, 1875a: 50
nitidipenne Germain, 1856: 395 (*Gibbium*)
giganteum Escalera, 1914: 258 **E**: SP **N**: MO
horridum Har. Lindberg, 1951: 8 **N**: CI
sulcatum Fabricius, 1781: 73 (*Ptinus*) **E**: AUi AZ CRi DEi FRi ITi PT SP SVi **N**: CI MR **NAR**i
cristatum Boheman, 1858: 86
hirticolle Laporte, 1840a: 297 (*Gibbium*)

genus *Stethomezium* Hinton, 1943: 50 type species *Stethomezium squamosum* Hinton, 1943
Falsomezium Pic, 1950b: 1 type species *Mezium notiale* Péringuey, 1899
notiale Péringuey, 1899: 243 (*Mezium*) **E**i: GB **N**i: EG **AFR**
nooitgedag Irish, 1996: 319
squamosum Hinton, 1943: 51

subfamily **Ptininae** Latreille, 1802

tribe Ptinini Latreille, 1802

genus *Casapus* Wollaston, 1862b: 194 type species *Casapus bonvouloirii* Wollaston, 1862
subgenus *Casapus* Wollaston, 1862b: 194 type species *Casapus bonvouloirii* Wollaston, 1862
Casopus Wollaston, 1864: 237 type species *Casopus bonvouloirii* Wollaston, 1862
alticola Wollaston, 1862b: 198 **N**: CI
barbicollis Israelson, 1993: 113 (*Piotes*) **N**: CI
bonvouloirii Wollaston, 1862b: 196 **N**: CI
dilaticollis Wollaston, 1862b: 197 **N**: CI
longicornis Palm, 1976a: 31 (*Casopus*)
pedatus Wollaston, 1865: 211 (*Casopus*)
hirsutus Bellés, 1996a: 81 **N**: CI
radiosus Wollaston, 1862b: 199 **N**: CI
wollastoni Uyttenboogaart, 1935: 9
subcalvus Wollaston, 1862b: 200 **N**: CI
subgenus *Piotes* Wollaston, 1862b: 211 type species *Piotes inconstans* Wollaston, 1862
inconstans Wollaston, 1862b: 212 (*Piotes*) **N**: CI
lanata Wollaston, 1862b: 212 (*Piotes*)

vestita Wollaston, 1862b: 213 (*Piotes*) **N**: CI

genus *Cyphoniptus* Bellés, 1992: 257 type species *Ptinus sulcithorax* Pic, 1899
sulcithorax Pic, 1899b: 28 (*Ptinus*) **A**: AP BT SCH SD XIZ YUN
thibetanus Pic, 1916f: 5 (*Ptinus*)

genus *Dignomus* Wollaston, 1862b: 201 type species *Dignomus gracilipes* Wollaston, 1862
Eutaphrus Mulsant & Rey, 1868b: 53 type species *Ptinus irroratus* Kiesenwetter, 1852
Pseudentaphrus Pic, 1896i: 109 type species *Ptinus fairmairei* Pic, 1895
Pseudeutaphrus Pic, 1930a: 1 type species *Ptinus tewfiki* Pic, 1930
albipilis Reitter, 1884h: 318 (*Bruchus*) **E**: FR SP
atticus Pic, 1902e: 32 (*Ptinus*) **E**: GR **A**: TR
aureopilis Desbrochers des Loges, 1875a: 49 (*Ptinus*) **N**: EG **A**: IS LE SY
morfini Pic, 1929g: 14 (*Ptinus*)
obscuripennis Pic, 1930e: 10 (*Ptinus*)
beauprei Pic, 1907c: 105 (*Ptinus*) **N**: TU
bedeli Pic, 1895c: 51 (*Ptinus*) **N**: AG TU
brevicrinitus Desbrochers des Loges, 1871: 359 (*Ptinus*) **E**: FR (Corse) IT (Sardegna)
canaliculatus Reitter, 1884c: 85 (*Bruchus*) **E**: GR
graecus Pic, 1896c: 41 (*Eutaphrus*)
carinatus P. H. Lucas, 1846: 210 (*Ptinus*) **E**: IT MA **N**: AG MO TU
brevipilis Desbrochers des Loges, 1875a: 49 (*Ptinus*)
foveolatus Boieldieu, 1854: lxxix (*Ptinus*)
siculus Kiesenwetter, 1877: 52 (*Ptinus*)
damascenus Baudi di Selve, 1873: 256 (*Ptinus*) **A**: IS LE SY TR
ganglbaueri Reitter, 1884h: 320 (*Bruchus*)
desertorum Reitter, 1887b: 29 (*Bruchus*) **E**: AB ST **N**: EG **A**: SI SY
damascenus Reitter, 1884h: 320 (*Bruchus*) [HN]
dilophus Illiger, 1807a: 23 (*Ptinus*) **E**: PT SP **N**: AG MO TU
diversicollis Pic, 1940a: 2 (*Ptinus*) **N**: MO
panousei Pic, 1948a: 1 (*Ptinus*)
elongatior Pic, 1919f: 14 (*Ptinus*) **N**: EG
fairmairei Pic, 1895i: 102 (*Ptinus*) **A**: YE
fossulatus P. H. Lucas, 1846: 208 (*Ptinus*) **E**: PT SP **N**: AG MO
abbreviatus Boieldieu, 1854: lxxix (*Ptinus*)
agricultor Rosenhauer, 1856: 173 (*Ptinus*)
algericus Pic, 1894b: 27 (*Ptinus*)
frivaldszkyi Reitter, 1884h: 319 (*Bruchus*) **E**: GR TR **A**: CY IS LE SY TR
gibbicollis P. H. Lucas, 1846: 210 (*Ptinus*) **E**: PT SP **N**: AG MO
farinosus Boieldieu, 1854: lxxviii (*Ptinus*)
globipennis Reitter, 1884h: 321 (*Bruchus*) **N**: AG
gracilipes Wollaston, 1862b: 202 **N**: CI MO
nivosus Escalera, 1914: 263 (*Eutaphrus*)
impressithorax Pic, 1896a: 6 (*Ptinus*) **A**: KA UP
irroratus Kiesenwetter, 1852: 622 (*Ptinus*) **E**: CR FR GE GR IT PL PT SP
alpinus Boieldieu, 1854: lxxviii (*Ptinus*)
porosensis Pic, 1929g: 14 (*Ptinus*)
quadridens Chevrolat, 1863: 86 (*Ptinus*)
jacqueti Pic, 1895i: 102 (*Eutaphrus*) **E**: FR IT SP
klapperichi Pic, 1956: 5 (*Ptinus*) **A**: AF
kochi Pic, 1934c: 30 (*Ptinus*) **A**: YE **N**: EG
kroliki Borowski, 2002: 155 **A**: TR
lesinae Reitter, 1884c: 85 (*Bruchus*) **E**: CR
longipennis Pic, 1897k: 295 (*Ptinus*) **N**: AG TU
lusitanus Illiger, 1807a: 21 (*Ptinus*) **E**: FR (Corse) IT (Sardegna) PT SP **N**: AG MO TU
loboderus Mulsant & Rey, 1868b: 60 (*Ptinus*)

major Pic, 1901j: 79 (*Ptinus*) **N**: TU
marraquensis Escalera, 1914: 264 (*Eutaphrus*) **N**: MO
mesopotamicus Pic, 1894f: 66 (*Ptinus*) **A**: IN IQ SA YE
nitidus Duftschmid, 1825: 65 (*Ptinus*) **E**: AU BU CZ HU MD PL RO SK SL ST UK KZ **A**: KZ
omanensis Borowski, 2000: 125 **A**: OM
pardoi Pic, 1953a: 9 (*Ptinus*) **N**: AG MO
pici Grandjean, 1895: 70 (*Ptinus*) **N**: AG MO
tondui Pic, 1930e: 10 (*Ptinus*)
pubens Abeille de Perrin, 1897: 131 (*Ptinus*) **E**: FR IT SP
hispanicus Pic, 1907a: 99 (*Ptinus*)
pulverulentus Boieldieu, 1854: lxxviii (*Ptinus*) **N**: EG
quadricornis Pic, 1895i: 103 (*Eutaphrus*) **E**: GR **A**: CY IS LE SY TR
quedenfeldti Reitter, 1891b: 29 (*Eutaphrus*) **N**: AG EG LB TU
soubironi Pic, 1895i: 102 (*Eutaphrus*)
reichei Boieldieu, 1854: lxxix (*Ptinus*) **E**: AL CR GR
monstrositarsis Pic, 1896c: 42 (*Eutaphrus*)
reitteri Bellés, 1996b: 216 [RN] **N**: MO
gracilipes Reitter, 1891b: 28 (*Eutaphrus*) [HN]
rotundipennis Pic, 1897g: 120 (*Ptinus*) **A**: SY
tewfiki Pic, 1930a: 1 (*Ptinus*) **N**: EG
theryi Pic, 1894y: cccxlviii [= 1894d: 49] (*Ptinus*) **N**: AG LB TU
tomentifer Pic, 1896i: 109 (*Ptinus*)
tuniseus Pic, 1897d: 61 (*Ptinus*) **N**: TU
urbanus Borowski, 1999: 1 **A**: SA
xylopertha Boieldieu, 1856a: 310 (*Ptinus*) **E**: GR **N**: AG EG LB TU **A**: CY IS LE SY TR
fulvohirtus Reitter, 1891b: 29 (*Eutaphrus*)

genus *Epauloecus* Mulsant & Rey, 1868b: 49 type species *Ptinus crenatus* Fabricius, 1792 (=*Ptinus unicolor* Piller & Mitterpacher, 1783)
Tipnus C. G. Thomson, 1859: 88 [HN] type species *Ptinus crenatus* Fabricius, 1792 (= *Ptinus unicolor* Piller & Mitterpacher, 1783)
unicolor Piller & Mitterpacher, 1783: 67 (*Ptinus*) **E**: AB AU AZ BE BH BU CR CT CZ DE EN FA FI FR GB GE GG GR HU ICi IR IT LA LS LT LU MC MD NL NR NT PL PT RO SK SL SP ST SV SZ UK YU **NAR**i
cerevisiae Marsham, 1802: 90 (*Ptinus*)
crenatus Fabricius, 1792a: 240 (*Ptinus*)
fuscus Gradl, 1881: 306 (*Niptus*)
lutosus Mulsant & Rey, 1868b: 203
minutus Panzer, 1795c: 114 (*Ptinus*)
ovatus Marsham, 1802: 90 (*Ptinus*)
salinus Schilling, 1843: 175 (*Ptinus*)

genus *Eurostodes* Reitter, 1884h: 299 type species *Ptinus* (*Eurostus*) *quisquiliarum* Baudi di Selve, 1874
quisquiliarum Baudi di Selve, 1874b: 327 (*Ptinus*) **E**: IT

genus *Eurostoptinus* Pic, 1895e: 80 type species *Eurostus algericus* Pic, 1895
algericus Pic, 1895e: 80 (*Eurostus*) **N**: AG
mazuri Borowski, 1995: 372 **N**: AG

genus *Hanumanus* Bellés, 1991: 83 type species *Hanumanus sumateraensis* Bellés, 1991
senilis Kiesenwetter, 1879b: 317 (*Ptinus*) **A**: JA **ORR**

genus *Kedirinus* Bellés, 1991: 72 type species *Kedirinus papuanus* Bellés, 1991
albidiceps Pic, 1914d: 9 (*Ptinus*) **A**: TAI **ORR**

genus *Lapidoniptus* Bellés, 1981: 70 type species *Piarus hispanicus* Pic, 1953
hispanicus Pic, 1953f: 144 (*Piarus*) **E**: SP

genus *Mezioniptus* Pic, 1944c: 5 type species *Mezioniptus impressicollis* Pic, 1944
fausti Heyden, 1886b: 187 (*Niptus*) **A**: TM
impressicollis Pic, 1944c: 5 **A**: NIN

genus *Myrmecoptinus* Wasmann, 1916: 206 type species *Myrmecoptinus butteli* Wasmann, 1916
Sundaptinus Bellés, 1991: 76 type species *Sundaptinus planophthalmus* Bellés, 1991
deplanatus Pic, 1954e: 63 (*Ptinus*) **A**: FUJ
kuronis Ohta, 1930b: 108 (*Ptinus*) **A**: TAI
sauteri Pic, 1914d: 8 (*Ptinus*) **A**: JA TAI

genus *Niptodes* Reitter, 1884h: 299 type species *Niptus elongatus* Boieldieu, 1854
subgenus *Niptodes* Reitter, 1884h: 299 type species *Niptus elongatus* Boieldieu, 1854
Microptinus Kiesenwetter, 1877: 79 [HN] type species *Niptus elongatus* Boieldieu, 1854
atricornis Pic, 1912u: 219 **N**: MO
brunneus Pic, 1898d: 73 (*Microptinus*) **N**: TU
carbonarius Rosenhauer, 1856: 176 (*Ptinus*) **E**: PT SP
constrictus Kiesenwetter, 1867b: 134 (*Niptus*)
elongatus Boieldieu, 1854: lxxxiii (*Niptus*) **E**: IT **N**: AG MO
ferrugulus Reitter, 1884h: 300 (*Niptus*) **E**: SP
lusitanus Reitter, 1888g: 427 (*Niptus*) **E**: PT
melitensis Pic, 1903q: 162 (*Microptinus*) **E**: MA
minimus Heyden, 1870: 128 (*Eurostus*) **E**: PT SP
nitidus Pic, 1895i: 100 (*Microptinus*) **E**: SP **N**: MO
nobilis Reitter, 1884h: 300 (*Niptus*) **E**: IT
reitteri Pic, 1895i: 99 (*Microptinus*) **E**: IT
rotundicollis P. H. Lucas, 1846: 209 (*Ptinus*) **N**: AG TU
hispidus Pic, 1895i: 100 (*Microptinus*)
rotundipennis Reitter, 1884h: 301(*Niptus*) **N**: AG MO
tournieri Pic, 1895i: 100 (*Microptinus*) **N**: AG
subgenus *Paraniptus* Kiesenwetter, 1877: 79 type species *Ptinus globulus* Illiger, 1807
globulus Illiger, 1807a: 26 (*Ptinus*) **E**: GEi PT SP **N**: MO
subgenus *Pseudoniptus* Reitter, 1884h: 298 type species *Niptus globipennis* Reitter, 1884
globipennis Reitter, 1884h: 299 (*Niptus*) **N**: MO
ovipennis Reitter, 1884h: 300 (*Niptus*) **N**: MO

genus *Niptus* Boieldieu, 1856a: 290 type species *Ptinus hololeucus* Faldermann, 1835
hololeucus Faldermann, 1835a: 214 (*Ptinus*) **E**: AL AR AU AZ BE BH BU BY CR CT CZ DE EN FI FR GB GE GG GR HU ICi IR IT KZ LA LS LT LU MA MC MD NL NR NT PL PT RO SK SL SP ST SV SZ TR UK YU **N**: AG CI EG LB MO TU **A**: AF AP CE CY ES FE HKG HP IN IQ IS JA JO KI KU KZ LE MG NP NE NC NO NW PA RU SC SHG SI SY TD TM TR UZ WS **COS**
brevesetosus Pic, 1956: 5

genus *Piarus* Wollaston, 1862b: 209 type species *Piarus basalis* Wollaston, 1862
basalis Wollaston, 1862b: 210 **N**: CI

genus *Pseudeurostus* Heyden, 1906: 424 [RN] type species *Ptinus frigidus* Boieldieu, 1854
Eurostus Mulsant & Rey, 1868b: 49 [HN] type species *Ptinus frigidus* Boieldieu, 1854
Niptinus C. Yang, 1980: 26 [HN] type species *Niptus hilleri* Reitter, 1877
Niptomimus Pic, 1950c: 10 type species *Niptus subelongatus* Pic, 1950
anemophilus Chobaut, 1901: 298 (*Eurostus*) **E**: FR IT SZ
apenninus Baudi di Selve, 1874b: 328 (*Ptinus*) **E**: IT
bordei Sainte-Claire Deville, 1921: 394 (*Niptus*) **E**: FR (Corse)
frigidus Boieldieu, 1854: lxxxii (*Ptinus*) **E**: AU FR IT SL SZ
fagniezi Méquignon, 1948: 77
helveticus Pic, 1902b: 18 (*Eurostus*)

otti Bremi-Wolf, 1855b: 329 (*Ptinus*)
testaceus Pic, 1937g: 9 (*Eurostus*)
hilleri Reitter, 1877i: 378 (*Niptus*) **E**i: DE FI GB GE IR SV **A**: ES FE JA NO NE NW SC SCH **NAR**i
alienus Brown, 1940: 116 (*Eurostus*)
sahoroensis Ohta, 1930b: 106 (*Eurostus*)
subelongatus Pic, 1950c: 10 (*Niptus*)
yezoensis Ohta, 1930b: 107 (*Eurostus*)
submetallicus Fairmaire, 1862a: 583 (*Ptinus*) **E**: FR SP

genus *Ptinus* Linnaeus, 1767: 537 type species *Cerambyx fur* Linnaeus, 1758
subgenus *Bruchoptinus* Reitter, 1884h: 303 type species *Ptinus rufipes* A. G. Olivier, 1790
antennatus Pic, 1896d: 43 **E**: BU TR
biformis Reitter, 1880e: 476 **E**: AR GG ST UK (Krym)
pallidior Pic, 1895j: 107
brevivittis Reitter, 1881d: 221 **E**: AL BH CR GR
caucasicus Pic, 1897f: 102 (*Pseudoptinus*) **E**: AB AR GG
eduardi Pic, 1910o: 57 **A**: TR
elbrusicola A. Fleischer, 1915: 130 **A**: IN
femoralis Reitter, 1884h: 306 (*Bruchus*) **E**: GR IT SP **N**: AG TU **A**: CY
madoni Pic, 1934c: 30
raveli Pic, 1901j: 79
henoni Pic, 1897g: 119 **E**: AB GR **A**: TR
italicus Aragona, 1830: 17 **E**: AU FR IT
neapolitanus Pic, 1925a: 1
ivanensis Reitter, 1902f: 193 (*Bruchoptinus*) **E**: BH YU
libanicus Pic, 1899e: 82 **A**: LE SY
brevior Pic, 1907c: 105
palliatus Perris, 1847: 465 **E**: AU FR GB GE IT PT SP YU
germanus Fabricius, 1781: 72 [HN]
pellitus Desbrochers des Loges, 1875a: 50 **E**: GR (Kríti) **A**: CY LE TR SY
rufipes A. G. Olivier, 1790a: no 17: 8 **E**: AB AL AU BE BH BU BY CR CT CZ DE EN FI FR GB GE GR HU IT LA LT LU MC MD NL NR NT PL RO SK SL SP ST SV SZ TR UK YU **A**: CY TR
brunnescens Pic, 1896d: 43
elegans Illiger, 1798: 346
montandoni Pic, 1911w: 177
humeralis Pic, 1895j: 107
obscurithorax Pic, 1895j: 107
subelongatus Pic, 1901g: 57
vicinus Pic, 1897f: 102 (*Bruchoptinus*)
schatzmayeri Pic, 1934c: 29 **E**: MC
syriacus Pic, 1896g: 87 **A**: SY
torretassoi Pic, 1934c: 29 **E**: GR **A**: TR
subgenus *Cyphoderes* Mulsant & Rey, 1868b: 54 type species *Ptinus bidens* A. G. Olivier, 1790
bidens A. G. Olivier, 1790a: no 17: 8 **E**: AU CR FR GR IT PT SL SP SVi SZ **N**: CIi
minutus Laporte, 1840a: 296
quercus Kiesenwetter, 1867b: 129 [HN]
sycophanta Illiger, 1807a: 25
catalonicus Bellés, 2002: 99 **E**: IT SP
hirticornis Kiesenwetter, 1867b: 130 **E**: IT PT SP **N**: MO
logesi Pic, 1894h: 71
japonicus Reitter, 1877i: 377 **E**i: FR **A**: CE ES FE GAN JA NC NE SC SD
raptor Sturm, 1837c: 53 **E**: AU BE BH BU BY CR CT CZ DE EN FI FR GB GE GR HU IT KZ LA LT MD NL NR NT PL RO RU SK SL ST SV SZ UK YU **A**: ES FE KZ MG WS **NAR**i
obscuripennis Pic, 1929g: 13
transbaicalicus Pic, 1929g: 13
schlerethi Reitter, 1884h: 317 (*Bruchus*) **E**: AU BH BU CR CZ GR HU MD PL RO SK SL ST UK YU
subgenus *Gynopterus* Mulsant & Rey, 1868b: 53 type species *Ptinus sexpunctatus* Panzer, 1789

Ptinobruchus Brenske & Reitter, 1884: 83 type species *Ptinus crassicornis* Kiesenwetter, 1877
aubei Boieldieu, 1854: lxxxi **E**: CR FR IT SL SP **N**: AG TU
formosus Rosenhauer, 1856: 175
semiunicolor Pic, 1911a: 145
subunicolor Pic, 1911q: 145
barrosi Pic, 1905j: 170 **E**: PT
crassicornis Kiesenwetter, 1877: 50 **E**: GR IT
forticornis Reitter, 1878e: 94
desertor Reitter, 1896c: 73 (*Bruchus*) **A**: TM TR UZ
diversipennis Pic, 1907k: 154 **A**: SY
dubius Sturm, 1837c: 75 **E**: AB AL AU BE BH BU BY CR CT CZ DE EN FI FR GB GE GG GR HU IT LA LT LU MC NL NR NT PL PT RO SK SL SP ST SV SZ TR UK YU **N**: AG CI MO **A**: CY TR
intonsus Pic, 1907r: 177 **N**: AG
peringueyi Pic, 1896b: 30 [RN] **N**: EG **AFR**
affinis Péringuey, 1888: 113 [HN]
dilataticornis Pic, 1929d: 5
maculifer Pic, 1896a: 6
latipennis Pic, 1914n: 56
zavattarii Pic, 1952: 25
pyrenaeus Pic, 1897d: 61 **E**: FR PT SP
martinezi Escalera, 1917: 234 (*Ptinobruchus*)
superbus Abeille de Perrin, 1897: 132
quadrisignatus Ménétriés, 1832: 165 **E**: AB AR GG **A**: FE KA KZ TR UZ
sexsignatus Faldermann, 1835a: 215
salvatori Pic, 1901f: 52 **E**: GR
sexpunctatus Panzer, 1789: 11 **E**: AB AL AU BE BH BU BY CR CT CZ DE FI FR GB GE GR HU IT LA LU MC NL NR NT PL PT RO SK SL SP ST SV SZ UK YU
basicornis Reitter, 1881e: 222
dispar Pic, 1895i: 102 (*Gynopterus*)
gavoyi Pic, 1929g: 13
massiliensis Pic, 1911q: 145
subroseus Reitter, 1888g: 430 (*Bruchus*) **E**: GR
tokatensis Pic, 1901g: 58 **A**: TR
variegatus P. Rossi, 1792: 20 **E**: AB AL AU BE BH BU CR CT CZ FR GE GG GR HU IT MA MC MD PL PT RO SK SL SP ST SZ TR UK YU **N**: AG EG LB MO MR TU **A**: CY IS LE SY TR **NARi**
duvali Lareynie, 1852: xc
insularis Desbrochers des Loges, 1871: 342 [HN]
mauritanicus P. H. Lucas, 1846: 208
vaulogeri Pic, 1894e: 61 **N**: AG
subgenus *Pseudoptinus* Reitter, 1884h: 303 type species *Ptinus coarcticollis* Sturm, 1837
anatolicus Pic, 1900e: 48 **E**: BU **A**: TR
cylindricornis Reitter, 1904c: 81 (*Eurostus*)
arragonicus Brenske & Reitter, 1884: 82 (*Bruchus*) **E**: FR SP
auberti Abeille de Perrin, 1869b: 42 **E**: CR FR IT PT SL SP
dalmatinus Pic, 1895j: 107
rufipennis Pic, 1929g: 14
biscrensis Pic, 1905c: 106 **N**: AG
capellae Reitter, 1880b: 52 **E**: AU BH BU CR CZ HU IT SL SZ YU
austriacus Reitter, 1890j: 391 (*Pseudoptinus*)
rufipennis Pic, 1932e: 22
coarcticollis Sturm, 1837c: 77 **E**: AR AU BH BU BY CT CZ EN FI FR GE HU IT LA LT MD NL PL RO RU SK SP ST SV TR UK
debilicornis Boieldieu, 1856b: 488
gracilicornis Pic, 1896c: 41 (*Pseudoptinus*)
indutus Abeille de Perrin, 1894c: xxv
cumaniensis Pic, 1896d: 42 **E**: GR
grandjeani Pic, 1895p: cxxix **N**: AG TU

dayensis Pic, 1895c: 50
kutzschenbachi Reitter, 1878a: 217 **E**: AB AR GG ST (Kavkaz) UK (Krym)
lederi Reitter, 1884h: 305 (*Bruchus*) **E**: AB AR
*levantinus J. R. Sahlberg, 19*13a: 48 **A**: LE SY TR (*Pseudoptinus*)
lichenum Marsham, 1802: 89 **E**: AU BE CZ DE FR GB GE IT LS LU NL PL RO SP SV SZ YU
fuscus Sturm, 1837c: 62
lepidus A. Villa & J. B. Villa, 1838: 62
ornatus P. W. J. Müller, 1821: 218
similis Marsham, 1802: 90
maculosus Abeille de Perrin, 1895a: ccxlvii **E**: FR SP **N**: AG
subcarinatus Pic, 1900f: 55
madoni Pic, 1932e: 22 **E**: SP
martini Pic, 1893f: lxxxvii **N**: AG TU
kabylianus Pic, 1895e: 80
martini Pic, 1893d: 112 [HN]
obscurior Pic, 1895e: 81
rubi Pic, 1900f: 55
micans Brenske & Reitter, 1884: 81 (*Bruchus*) **E**: AR ST (Kavkaz)
nikitanus Brenske & Reitter, 1884: 82 (*Bruchus*) **E**: CR YU
obscuricollis Pic, 1895q: cxci **N**: AG
semirufulus Pic, 1896i: 108
oertzeni Reitter, 1888g: 428 (*Bruchus*) **E**: GR
leucaspis Reitter, 1888g: 428 (*Bruchus*)
peyroni Pic, 1899e: 82 **A**: LE SY TR
rufolimbatus Pic, 1908d: 42 **E**: BU
rugosicollis Desbrochers des Loges, 1875a: 50 (*Niptus*) **E**: GR TR **N**: EG LB TU **A**: CY IS LE SY TR
berytensis Pic, 1899e: 82
*curtipilis*Pic, 1911s: 154
*longipilis*Pic, 1896h: 93
obscuripes J. R. Sahlberg, 1913d: 155
sanninensis Pic, 1911s: 154 **A**: LE TR
spissicornis Abeille de Perrin, 1894c: xxv **E**: FR IT
subaeneus Brenske & Reitter, 1884: 81 (*Bruchus*) **E**: GR TR **A**: CY IS LE SY TR
pistaciae Pic, 1901f: 51
tauricus Brenske & Reitter, 1884: 82 (*Bruchus*) **E**: UK (Krym)
subgenus *Ptinus* Linnaeus, 1767: 537 type species *Cerambyx fur* Linnaeus, 1758
Bruchus O. F. Müller, 1764: 15 type species *Cerambyx fur* Linnaeus, 1758
Heteroplus Mulsant & Rey, 1868b: 54 type species *Ptinus pusillus* Sturm, 1837
Heteroptinus Reitter, 1884h: 303 type species *Ptinus obesus* P. H. Lucas, 1846
Pseudobruchus Pic, 1894w: 203 type species *Ptinus pustuliferus* Pic, 1894
affinis Desbrochers des Loges, 1871: 341 **E**: IT MA **N**: TU
argolisanus Reitter, 1884h: 310 (*Bruchus*) **E**: GR
atricapillus Kiesenwetter, 1877: 56 **E**: CR IT **N**: AG TU
bicinctus Sturm, 1837c: 57 **E**: AU BH BU CR CZ HU MC RO SL YU
bescidicus Reitter, 1906g: 282
calcaratus Kiesenwetter, 1877: 57 **E**: AL AR AU BH BU CR CZ GG GR HU IT MC MD PL RO SK SL ST SZ UK YU
edmundi Abeille de Perrin, 1894d: cxxxi
calcarifer Reitter, 1888g: 429 (*Bruchus*) **E**: GR
circassicus Pic, 1900e: 48 **E**: AR GG ST (Kavkaz)
comptus Chevrolat, 1873: 204 **A**: SY
corsicus Kiesenwetter, 1877: 55 **E**: FR (Corse) IT
desbrochersi Reitter, 1884h: 313 (*Bruchus*)
validicornis Reitter, 1884h: 309 (*Bruchus*)
dauricus Reitter, 1906g: 283 **A**: ES FE WS
explanatus Fauvel, 1891: 59 **E**: FR
ferrantei Pic, 1912o: 72 **N**: EG

fimicola Desbrochers des Loges, 1875a: 50 **N**: AG EG LB TU **A**: JO SI **AFR**
olivieri Pic, 1893a: 33
fringillae Pic, 1927b: 1 **N**: EG
fulvocapillus Pic, 1901j: 79 **N**: AG MO TU
fur Linnaeus, 1758: 393 (*Cerambyx*) **E**: AB AL AR AU AZ BE BH BU BY CR CT CZ DE EN FA FI FR GB GE GG GR HU ICi IR IT KZ LA LS LT LU MA MC MD NL NR NT PL PT RO SK SL SP ST SV SZ TR UK YU **N**: AG CI EG LB MO MR TU **A**: AF AP BT CE CY ES FE HP IN IQ IS JA JO KI KZ LE MG NP NC PA SC SI SY TAI TD TM TR WS **COS**
furunculus O. F. Müller, 1776: 57 (*Bruchus*)
germanus Goeze, 1777: 159 [HN]
humeralis Say, 1835: 165
longipes P. Rossi, 1792: 20
pulex Goeze, 1776: 90
rapax DeGeer, 1774: 231
striatus Fabricius, 1787a: 40
gylippus Reitter, 1906g: 283 **E**: GR
hirsutus Pic, 1895i: 101 **A**: IS LE SY
kiesenwetteri Reitter, 1884h: 83 (*Bruchus*) **E**: GR TR **A**: SY TR
creticus Iablokoff-Khnzorian & Karapetyan, 1991: 33
cytherias Pic, 1903r: 170
leuthneri Pic, 1900e: 49
kocheri Pic, 1953c: 19 **N**: MO
kruperi Pic, 1929g: 14 **E**: GR
latro Fabricius, 1775: 63 **E**: AB AL AR AU AZ BE BH BU BY CR CT CZ DE EN FI FR GB GE GG GR HU IR IT KZ LA LS LT LU MA MC MD NL NR NT PL PT RO SK SL SP ST SV SZ TR UK YU **N**: AG CI EG LB MO MR TU **A**: AF AP BT CE CY ES FE HP IN IQ IS JA JO KI KZ LE MG NP NC PA SC SI SY TD TM TR UZ WS **COS**
advena Wollaston, 1854: 261
brunneus Duftschmid, 1825: 65
brevipennis Pic, 1896i: 108
clavipes Panzer, 1805: 4
corticinus Rottenberg, 1871: 245
densepubens Pic, 1950c: 11
furoides Escalera, 1914: 261
hirtellus Sturm, 1837c: 80 [HN]
hirticollis P. H. Lucas, 1846: 212
letourneuxi Pic, 1894e: 60
lucasii Boieldieu, 1856c: 636 [RN]
mobilis Moore, 1957: 201 [RN]
moorei Iablokoff-Khnzorian & Karapetyan, 1991: 41
obsoletus Baudi di Selve, 1874b: 326
pueli Pic, 1936d: 25
rufus P. H. Lucas, 1846: 207 [HN]
testaceus A. G. Olivier, 1790a: no 17: 9 [HN]
leprieuri Pic, 1896i: 108 **E**: IT **N**: AG TU
calabricus Pic, 1908d: 42
meisteri Reitter, 1884h: 309 (*Bruchus*) **E**: AB AR GG ST **A**: TR
merkli Pic, 1895i: 101 **E**: TR **A**: TR
montanus Escalera, 1914: 262 **N**: MO
nigripennis Comolli, 1837: 18 **E**: IT
obesus P. H. Lucas, 1846: 211 **E**: AU BH CR FR GR IT PT SL SP **N**: AG EG LB MO TU **A**: CY
pulchellus Boieldieu, 1854: lxxxi
reflexus Rey, 1891f: 131
solitarius Rosenhauer, 1856: 172
perrini Reitter, 1884h: 312 (*Bruchus*) **E**: FR IT SP
ellipticus Reitter, 1894i: 304 (*Bruchus*)
perplexus Mulsant & Rey, 1868b: 160 **E**: FR IT
phlomidis Boieldieu, 1854: lxxx **E**: BH BU CR GR HU IT TR YU **A**: TR

kaufmanni Reitter, 1881e: 223
turcicus Pic, 1895b: 29
pilosus P. W. J. Müller, 1821: 220 **E**: AB AL AR AU BH BU CR CZ FR GE GG GR HU IT LA MC MD PL RO SK SL SP ST SV SZ UK YU **A**: CY TR
hungaricus Iablokoff-Khnzorian & Karapetyan, 1991: 33
intermedius Boieldieu, 1856c: 646
pallidus Duftschmid, 1825: 66
pallipes Duftschmid, 1825: 66
podolicus Iablokoff-Khnzorian & Karapetyan, 1991: 32 **E**: AB AU BE BH BU BY CR CT CZ DE EN FI FR GB GE HU IT KZ LA LT LU MC MD NL NR NT PL PT RO SK SL SP ST SV SZ UK YU **N**: AG **NARi**
fontisbellaquei Iablokoff-Khnzorian & Karapetyan, 1991: 32
pusillus Sturm, 1837c: 65 **E**: AU BE CZ DE FR GB GE HU IR IT LS LU NL SK SL SP SV SZ UK
quadripunctatus Gebler, 1833: 282 **E**: AB **A**: ES FE KZ WS
oblongulus Motschulsky, 1860b: 154
sexornatus Reitter, 1896c: 72 (*Bruchus*)
reitteri Pic, 1894g: 71 **E**: IT MA **N**: AG TU
leonhardi Breit, 1908: 160
obscurus Pic, 1896i: 109
rufus Brullé, 1832: 157 **E**: GR
separatus Pic, 1896k: 201 **N**: AG
spitzyi A. Villa & G. B. Villa, 1838: 62 **E**: AU BH CR FR IT MA PT SP SZ **N**: AG MO TU
cisti Chevrolat, 1865: 352
pustuliferus Pic, 1894w: 203
robustior Pic, 1929g: 13
ruber Rosenhauer, 1856: 175
subpillosus Sturm, 1837c: 82 **E**: AB AL AR AU BH BU BY CR CT CZ DE EN FI FR GB GE GR HU IR IT LA LT MC MD NR PL RO SK SL SP ST SV SZ TR UK YU
swaneticus Reitter, 1906g: 284 **E**: AR GG ST (Kavkaz)
tarsalis Reitter, 1884c: 84 (*Bruchus*) **E**: GR IT **A**: CY
timidus C. N. F. Brisout de Barneville, 1866: 382 **E**: PT SP
moroderi Reitter, 1906g: 281
villiger Reitter, 1884h: 311 (*Bruchus*) **E**: AB AU BH BU BY CR CT CZ DE EN FI GBi GE GG GR HU KZ LA LT MD NR NT PL RO SK SL ST SV UK YU **A**: ES FE KZ TM WS **NARi**
balticus Iablokoff-Khnzorian & Karapetyan, 1991: 48
subgenus *Tectoptinus* Iablokoff-Khnzorian & Karapetyan, 1986: 513 type species *Ptinus tectus* Boieldieu, 1856
exulans Erichson, 1842: 147 **E**i: AU AZ BH CR CZ DE FI FR GB GE GR IT NL SP ST SV UK **N**i: AG EG MO **A**i: CY ES FE HKG NP SY **AFR**i **AUR** **NAR**i **NTR**i **ORR**i
tectus Boieldieu, 1856c: 652 **E**i: AU BE BH BU BY CR CT CZ DE EN FA FI FR GB GE GR HU IC IR IT KZ LA LT LU MC MD NL NR NT PL PT SK SL SP ST SV SZ TR UK YU **N**i: AG EG MO TU **A**i: AF CH CY ES FE IN IS JA KZ NO NP NC SC SD SI SY TR **AFR**i **AUR** **NAR**i **NTR**i **ORR**i **COS**
ocellus Brown, 1929: 109
pilosus White & Butler, 1846: 8 [HN]

genus *Ptinus*, nomina dubia
abbreviatus Schrank, 1798: 542
agarici Schrank, 1798: 542
chrysomeloides Kugelann, 1792: 502

tribe Sphaericini Bellés, 1982

genus *Sphaericus* Wollaston, 1854: 263 type species *Ptinus dawsoni* Wollaston, 1854
subgenus *Doramasus* Bellés, 1994: 64 type species *Sphaericus crotchianus* Wollaston, 1864
crotchianus Wollaston, 1864: 242 **N**: CI (Gomera)
franzi Leiler, 1984: 163 **N**: CI (Gran Canaria: Isleta near Las Palmas)
langer Erber & Hinterseher, 2002: 166 **N**: CI (Tenerife)
marmoratus Wollaston, 1865: 219 **N**: CI (Gomera)
orzolensis Leiler, 1984: 166 **N**: CI

pseudofranzi Erber & Hinterseher, 2002: 176 **N**: CI (Gran Canaria)
rotundatus Wollaston, 1865: 34 **N**: CI (Hierro)
rotundicollis Israelson, 1980: 190 **N**: CI (Lanzarote)
thurepalmi Leiler, 1984: 165 **N**: CI (Tenerife)
subgenus *Nitpus* Jacquelin du Val, 1860e: 138 type species *Nitpus gonospermi* Jacquelin du Val, 1860
Microptinus Wollaston, 1865: 215 type species *Nitpus gonospermi* Jacquelin du Val, 1860
ptinoides Boieldieu, 1854: lxxxiii (*Trigonogenius*) **N**: AG CI MO **A**: CY SY
gonospermi Jacquelin du Val, 1860e: 138 (*Nitpus*)
subgenus *Sphaericus* Wollaston, 1854: 263 type species *Ptinus dawsoni* Wollaston, 1854
Outipnus Gozis, 1886: 24 type species *Trigonogenius exiguus* Boieldieu, 1854
Ptinodes Wollaston, 1865: 219 [HN] type species *Ptinus fragilis* Wollaston, 1854
Tipnus Jacquelin du Val, 1860e: 137 [HN] type species *Trigonogenius gibboides* Boieldieu, 1854
Wollastonella R. Lucas, 1920: 559 type species *Ptinus fragilis* Wollaston, 1854
albopictus albopictus Wollaston, 1854: 267 (*Ptinus*) **N**: MR
albopictus albosquamosus Erber & Hinterseher, 2002: 261 **N**: MR (Illheu Chão)
albopictus brevinasus Erber & Hinterseher, 2002: 254 **N**: MR
albopictus carinasus Erber & Hinterseher, 2002: 253 **N**: MR (Deserta Grande)
albopictus flavotarsus Erber & Hinterseher, 2002: 255 **N**: MR
albopictus minutus Erber & Hinterseher, 2002: 259 **N**: MR (Illheu Chão)
albopictus plantaginis Erber & Hinterseher, 2002: 257 **N**: MR (Porto Santo)
ambiguus Wollaston, 1865: 253 **N**: MR
ater Leiler, 1984: 168 **N**: MR
bicolor Bellés, 1982: 104 **N**: MR (Selvagens)
dawsoni Wollaston, 1854: 263 (*Ptinus*) **N**: MR (Desertas)
erinaceus Erber & Hinterseher, 2002: 225 **N**: MR (Illheu Chão)
exiguus Boieldieu, 1854: lxxxiv (*Trigonogenius*) **E**: FR IT PT SP **N**: AG MO
feloi Erber & Hinterseher, 2002: 228 **N**: CI (La Palma)
flavosquamosus Erber & Hinterseher, 2002: 234 **N**: MR
fragilis Wollaston, 1854: 271 (*Ptinus*) **N**: MR
gibbicollis Wollaston, 1862b: 208 **N**: CI MO
leileri Erber & Hinterseher, 2002: 232 **N**: MR (Porto Santo)
longicornis Wollaston, 1854: 270 (*Ptinus*) **N**: MR
machadoi Bellés, 1994: 72 **N**: CI (Lanzarote)
naviculiformis Erber & Hinterseher, 2002: 223 **N**: MR
nigrescens Wollaston, 1857a: 91 (*Trigonogenius*) **N**: MR
niveus Boieldieu, 1854: lxxxiv (*Trigonogenius*) **N**: AG CI MO **A**: IS SY
nodulus Wollaston, 1854: 265 (*Ptinus*) **N**: MR
obscurus Erber & Hinterseher, 2002: 230 **N**: MR
orbatus Wollaston, 1854: 264 (*Ptinus*) **N**: MR
pilula Wollaston, 1854: 266 (*Ptinus*) **N**: MR
pinguis Wollaston, 1854: 264 (*Ptinus*) **E**: AZ BE CZi DEi FIi FR GBi GEi GRi HUi IRi IT NLi PLi PT RUi SP SVi UKi **N**: AG CI MO MR **Ai**: CH CY **COS**
africanus Pic, 1896a: 6
brasiliensis Pic, 1928h: 74
densepunctatus Pic, 1936f: 131
gibboides Boieldieu, 1854: lxxxiv (*Trigonogenius*)
grandjeani Pic, 1896a: 6
impunctipennis Wollaston, 1864: 241
madecassus Pic, 1909e: 108
saetiger Erber & Hinterseher, 2002: 218 **N**: MR
selvagensis Bellés, 2001: 10 **N**: MR (Selvagens)
simplex Wollaston, 1862b: 207 **N**: CI
orbicularis Lindberg, 1953: 8
truncatus basibulbosus Erber & Hinterseher, 2002: 216 **N**: MR (main island)
truncatus interpositus Erber & Hinterseher, 2002: 218 **N**: MR (main island)
truncatus truncatus Erber & Hinterseher, 2002: 214 **N**: MR (Illheu Chão)

velhocabrali Israelson, 1984a: 1 **E**: AZ
ventriculus Erber & Hinterseher, 2002: 201 **N**: MR

genus *Stereocaulophilus* Bellés, 1995: 43 type species *Stereocaulophilus volcanius* Bellés, 1995
volcanius Bellés, 1995: 44 **N**: CI

genus *Trigonogenius* Solier, 1849: 464 type species *Ptinus globulus* Solier, 1849
globosus Solier, 1849: 464 (*Ptinus*) **E**i: AU BE CZ DE FI FR GB GE GR IT PL RU SP SV UK **N**i: AG EG MO **A**i: CH FE IS JA **COS**
curtus Pic, 1947: 6
farctus LeConte, 1866a: 100
nigronotatus Pic, 1901q: 278
tournoueri Pic, 1917l: 258 (*Niptus*)
tropicus Kirsch, 1889: 12

subfamily **Anobiinae** Fleming, 1821
tribe Anobiini Fleming, 1821

genus *Anobium* Fabricius, 1775: 62 type species *Anobium punctatum* DeGeer, 1774
Byrrhus O. F. Müller, 1776: xxii [HN]
cymoreki Español, 1963a: 81 **E**: SP **N**: MO
hederae Ihssen, 1949: 107 **E**: AU BE BU CR CZ FR GE GG IT LS PT SL SP SZ YU
inexspectatum Lohse, 1954: 213 **E**: AB AU AZ BE CZ FR GB GE IT NL NT SL SP SZ
punctatum DeGeer, 1774: 230 (*Ptinus*) **E**: AB AU BE BU BY CR CT CZ DE EN FA FI FR GB GE GG HU ICi IR IT LA LT NL NR NT PL PT RO SK SL SP SV SZ TR **N**: AG CI EG MO MR TU **A**: CY ES FE HKG IN IS TR **COS**
amplicolle Broun, 1880: 339
caelatum Mulsant & Rey, 1864: 81
cylindricum Marsham, 1802: 83 (*Ptinus*)
domesticus Geoffroy, 1785: 26 (*Birrhus*)
latreillei Dufour, 1843a: 48
pubescens Herbst, 1793: 67
pumilum LeConte, 1865: 232 (*Hadrobregmus*)
ruficolle Herbst, 1793: 65
ruficorne Broun, 1880: 340
striatum A. G. Olivier, 1790a: no 16: 9

genus *Cacotemnus* LeConte, 1861: 204 type species *Anobium errans* Melsheimer, 1846
demokidovi Toskina, 2004a: 57 **E**: ST
kompanzevi Logvinovskiy, 1979a: 42 (*Hemicoelus*) **A**: KI
panfilovi Toskina, 2004a: 59 **E**: ST
parfentievi Toskina, 2004a: 61 **A**: UZ
rufipes Fabricius, 1792a: 236 (*Anobium*) **E**: AU BE BY CR CT CZ EN FI FR GB GE GG HU IT LA LT NL NR NT PL RO SK ST SV SZ UK YU **A**: ES FE KZ WS TR
albidus Toskina, 1976: 1730
captiosus Toskina, 1976: 1730
cinnamomeum Sturm, 1837b: 115 (*Anobium*)
elongatus Paykull, 1798: 303 (*Anobium*)
juglandis Herbst, 1793: 61 (*Anobium*)
uralensis Pic, 1920h: 22 (*Hadrobregmus*)
thomsoni Kraatz, 1881a: 302 (*Hadrobregmus*) **E**: AU BY CT CZ EN FI GE GG IT LA NR NT PL SK ST SV **A**: ES WS
zavadskii Toskina, 2004a: 61 **E**: AB

genus *Hemicoelus* LeConte, 1861: 204 type species *Anobium gibbicollis* LeConte, 1859
canaliculatus C. G. Thomson, 1863: 160 (*Hadrobregmus*) **E**: AB AU BE BY CR CT CZ DE EN FI FR GB GE GR IT LA MA NL NR NT PL RO SK SP ST SV SZ UK YU **A**: FE IN

costatus Aragona, 1830: 16 (*Anobium*) **E**: AL AR AU BE BU CR CZ DE FR GE GG GR HU IT LS MC NL PL PT RO SK SL SP SV SZ UK **A**: TR
fagi Mulsant & Rey, 1863d: 72 (*Anobium*) [HN]
fagicola Mulsant & Rey, 1864: 89 (*Anobium*) [RN]
fulvicornis Sturm, 1837b: 114 (*Anobium*) **E**: AB AR AU BE BU CR CT CZ DE FI FR GB GE GG GR HU IT NL PL RO SK SL SP ST SV SZ UK **A**: IN TR
khnzoriani Karapetyan, 1980: 301 (*Anobium*)
morio A. Villa & G. B. Villa, 1835: 48 (*Anobium*)
ruber Reitter, 1897e: 216 (*Byrrhus*)
sericeus C. G. Thomson, 1863: 160 (*Hadrobregmus*)
rufipennis Duftschmid, 1825: 56 (*Anobium*) **E**: AU BU CR CZ FR GE GG GR HU IT PL RO SK ST **A**: TR
ussuriensis Karapetyan, 1980: 303 (*Anobium*) **A**: FE
alexandri Logvinovskiy, 1981: 33 (*Anobium*)

genus *Microbregma* Seidlitz, 1889a: 501 type species *Anobium emarginatum* Duftschmid, 1825
emarginatum Duftschmid, 1825: 54 (*Anobium*) **E**: AU BE BY CT CZ FI FR GE HU IT LA LT NL NR NT PL RO SK SP SV SZ **A**: ES FE HEI JA JIL TR **NAR**
foveatum Kirby, 1837: 190 (*Anobium*)
granicollis Fall, 1905: 185

tribe Gastrallini White, 1982

genus *Falsogastrallus* Pic, 1914d: 10 type species *Falsogastrallus sauteri* Pic, 1914
Gastrallomimum Pic, 1939a: 7 type species *Gastralomimum unistriatum* Pic, 1939
Neogastrallus Fisher, 1938: 43 type species *Neogastrallus librinocens* Fisher, 1938
curtus Toskina, 2003: 195 **A**: JO
elongatus Pic, 1931b: 1 **A**: YUN
bicolor Pic, 1931b: 1
sauteri Pic, 1914d: 10 **A**: GUA GUX HKG JA JIA JIX SCH TAI **NAR**i
skopini Español & Bellés, 1984: 26 **A**: KZ TD UZ
stemmleri Español, 1977c: 306 **A**: BT
unistriatus Zoufal, 1897: 206 (*Gastrallus*) **E**: AU CR GR HU IT YU **A**: LE SY TD TR

genus *Gastrallus* Jacquelin du Val, 1860f: 142 type species *Anobium immarginatum* P. W. J. Müller, 1821
affinis Sakai, 1984b: 19 **A**: JA
clematorum Fursov, 1958: 144 **A**: UZ
corsicus Schilsky, 1898: 51 **E**: CR FR GR IT PT SP **N**: AG MO **A**: IS SY
coarcticollis Schilsky, 1898: 51
cymoreki Español, 1990a: 47 **A**: NP
dimidiatus Sakai, 1984b: 22 **A**: JA
erdosi Toskina, 2003: 187 **A**: IS
immarginatus P. W. J. Müller, 1821: 196 (*Anobium*) **E**: AR AU BE BU CR CZ DE FR GB GE GR HU IT PL RO SK SP SV SZ UK **N**: TU **A**: HEI JA JIL LE LIA **ORR**
cornicularium Beck, 1817: 14 (*Anobium*)
exilis Gyllenhal, 1827: 325 (*Anobium*)
sericatum Laporte, 1840a: 294 (*Anobium*)
indicus Reitter, 1913a: 16 **A**: UP **ORR**
insuetus Logvinovskiy, 1978: 28 **A**: KZ
insulcatus Pic, 1937i: 124 **A**: UP **ORR**
multistriatus Pic, 1943a: 9
knizeki Zahradník, 1996b: 267 **E**: AU CZ FR GE SK
kocheri Español, 1963b: 15 **E**: GR (Kríti) IT SP **N**: MO TU
laevigatus A. G. Olivier, 1790a: no 16: 12 (*Anobium*) **E**: AU BE BH BU CR CZ FR GE GG GR HU IT NL PL PT RO SK SP ST SZ UK YU **N**: AG MO TU **A**: CY IS JA SY TR
parallelum Küster, 1849c: 39 (*Anobium*)
lyctoides Wollaston, 1865: 35 (*Anobium*) **N**: CI
mauritanicus Español, 1963b: 18 **E**: IT **N**: MO TU **A**: TR

ornatulus Toskina, 2003: 189 **E**: AB ST
phloeophagus Iablokoff-Khnzorian, 1960: 149 **E**: AR
pubens Fairmaire, 1875a: 515 **E**: AB SP **N**: EG TU **A**: IS JO LE SY TM **AFR**
striatus Zoufal, 1897: 206
subtuberculatus Pic, 1948b: 16 [= 1949: 87]
rollei Reitter, 1912a: 104 **E**: SP
subtilis Toskina, 1998: 52 **A**: SA
testaceicornis Pic, 1922b: 7 **A**: TAI
tuberculatus Pic, 1914d: 10 **A**: TAI **ORR**
wittmeri Español, 1977c: 306 **A**: BT NP

genus *Mimogastrallus* Sakai, 2003: 333 type species *Mimogastrallus hideoi* Sakai, 2003
hideoi Sakai, 2003: 335 **A**: JA

tribe Hadrobregmini White, 1982

genus *Hadrobregmus* C. G. Thomson, 1859: 89 type species *Anobium denticolle* Creutzer, 1796
Coelostethus LeConte, 1861: 204 type species *Anobium notatum* Say, 1825
Dendrobium Mulsant & Rey, 1863d: 58 type species *Anobium denticolle* Creutzer, 1796
bicolor Español, 1990b: 172 **E**: SP **N**: MO
carpetanus Heyden, 1870: 129 (*Anobium*) **E**: SP **N**: MO
confusus Kraatz, 1881a: 301 (*Anobium*) **E**: CT FI LA NT PL SV **A**: ES WS
denticollis Creutzer, 1796a: 8 (*Anobium*) **E**: AU BE BU CT CZ DE FR GB GE HU IT LA LT NL PL RO SK SP ST SV SZ
ishiharai Sakai, 1994: 289 **A**: JA
nagaii Sakai, 1978: 36 **A**: JA
pertinax Linnaeus, 1758: 355 (*Dermestes*) **E**: AU BEi BH BU BY CT CZ DE EN FI FR GB GE GG GR HU IT LA LT NLi NR NT PL RO SK SP SV SZ UK **A**: ES FE HEI JA JIL LIA TR WS "Korea"
bipunctatus Schrank, 1776: 61 (*Dermestes*)
fagi Herbst, 1783: 26 (*Anobium*)
rufus Gmelin, 1790: 1606 (*Ptinus*)
rugosus Gmelin, 1790: 1604 (*Ptinus*)
striatus Fabricius, 1787a: 39 (*Anobium*)
pineti cedretorum Peyerimhoff, 1920: 267 (*Anobium*) **N**: AG
pineti pineti Peyerimhoff, 1920: 266 (*Anobium*) **N**: AG
vulsus Reichardt & Toskina, 1973: 1335 (*Priobium*) **A**: FE

genus *Priobium* Motschulsky, 1845a: 35 type species *Anobium carpini* Herbst, 1793
Trypopitys L. Redtenbacher, 1849: 346 type species *Anobium serricorne* Duftschmid, 1825
carpini Herbst, 1793: 58 (*Anobium*) **E**: AB AR AU BE BH BU BY CT CZ DE FI FR GE GG HU IT LA LT NL NR NT PL PT RO SK SP ST SV SZ UK YU **A**: CY ES FE KZ LE SY TR WS **NAR**i
dendrobiformis Reitter, 1901a: 25 (*Trypopithys*)
excisum Mannerheim, 1843c: 93 (*Anobium*)
serricorne Duftschmid, 1825: 50 (*Anobium*)
cylindricum Nakane, 1963b: 47 (*Trypopitys*) **A**: JA
inermicollis Pic, 1907p: 173 (*Trypopithys*) **A**: PA
luteopilosum Pic, 1901n: 94 (*Trypopitys*) **A**: UP **ORR**
persimile Nakane, 1989a: 4 **A**: JA
sulcicolle Chevrolat, 1860: 76 (*Xyletinus*) **N**: AG
yamashitai Nakane, 1989a: 3 **A**: JA

tribe Nicobiini White, 1982

genus *Nicobium* LeConte, 1861: 204 type species *Anobium hirtum* Illiger, 1807 (= *Anobium castaneum* A. G. Olivier, 1790)
Neobium Mulsant & Rey, 1863d: 58 type species *Anobium hirtum* Illiger, 1807 (= *Anobium castaneum* A. G. Olivier, 1790)
auroguttatum Reitter, 1908a: 21 **A**: SY

castaneum A. G. Olivier, 1790a: no 16: 7 (*Anobium*) **E**: AB AU AZ BH CR CZ FR GEi GR IT MA PT RO SL SP ST SZi UK **N**: AG CI EG LB MO MR **A**: CY GUI IS JA JIA LIA SY TAI TR **NARi NTRi**
fasciatum Dufour, 1843a: 47 (*Anobium*)
hirtum Illiger, 1807a: 19 (*Anobium*)
tomentosum Mulsant & Rey, 1863d: 81 (*Anobium*)
schneideri Reitter, 1878a: 218 **E**: AB AZ GG ST UK **N**: EG **A**: IN TR
velatum Wollaston, 1854: 276 (*Anobium*) **N**: MR
villosum Brullé, 1838: 60 (*Anobium*) **E**: PT SP **N**: CI MO **AFRi NTRi**
lineaticolle Pic, 1937d: 4
subparallelum Saraiva, 1957: 48

genus *Trichodesma* LeConte, 1861: 204 type species *Anobium gibbosum* Say, 1825
subgenus *Trichodesma* LeConte, 1861: 204 type species *Anobium gibbosum* Say, 1825
Ptinodes LeConte, 1861: 204 type species *Anobium setifer* LeConte, 1858
fasciculare Reitter, 1877i: 378 (*Nicobium*) **A**: JA
japonica Pic, 1906a: 2 **A**: JA
kirishimana Nakane, 1979: 133 **A**: JA
kurosawai Sakai, 1986: 211 **A**: TAI
lewisii Kiesenwetter, 1879b: 315 **A**: JA
pulchra Pic, 1903h: 119 **A**: SD
regalis Pic, 1900g: 57 **A**: NP
uruma Sakai, 1985: 27 **A**: JA

tribe Stegobiini White, 1982

genus *Neoligomerus* Español, 1981a: 50 type species *Neoligomerus elongatus* Español, 1981
elongatus Español, 1981a: 52 **A**: KA

genus *Oligomerus* L. Redtenbacher, 1849: 347 type species *Anobium brunneum* A. G. Olivier, 1790
Octophorus Iablokoff-Khnzorian, 1960: 151 type species *Oligomerus fraxini* Iablokoff-Khnzorian, 1960
Oligomerinus Portevin, 1931: 484 type species *Anobium ptilinoides* Wollaston, 1854
brunneus A. G. Olivier, 1790a: no 16: 8 (*Anobium*) **E**: AB AU BE BU CR CT CZ FR GE HU IT LA LT NTi PL RO SK SL SP ST SV SZ UK **A**: JA TD WS
oblongum Laporte, 1840a: 293 (*Anobium*)
chujoi Sakai, 1982a: 48 **A**: JA
disruptus Baudi di Selve, 1874b: 329 (*Anobium*) **E**: FR (Corse) GR IT (Sardegna) **N**: AG **A**: CY
explanatus Sakai, 1982a: 45 **A**: JA
fraxini Iablokoff-Khnzorian, 1960: 151 **E**: AR
japonicus Sakai, 1982a: 46 **A**: JA LIA "Korea"
ptilinoides Wollaston, 1854: 278 (*Anobium*) **E**: AB AUi CR FR GEi GR HU IT MA PLi PT RO SK SL SP ST SZ TR UK YU **N**: AG CI EG MO MR TU **A**: CY HEI IN IS JA JIL KZ LIA SY TR
brunneus Duftschmid, 1825: 54 (*Anobium*)
oculatum Wollaston, 1865: 36 (*Anobium*)
reyi C. N. F. Brisout de Barneville, 1867b: 180 (*Anobium*)
retowskii Schilsky, 1898: 32 **E**: AR BU CT CZ GG GR SK ST UK **A**: TR

genus *Paroligomerus* Logvinovskiy, 1979a: 42 type species *Paroligomerus costatus* Logvinovskiy, 1979
costatus Logvinovskiy, 1979a: 44 **A**: TD

genus *Priartobium* Reitter, 1901a: 23 type species *Priartobium serrifunis* Reitter, 1901
leonhardi Roubal, 1917b: 52 **E**: IT (Sardegna)
serrifunis Reitter, 1901a: 24 **E**: BU CR FR (Corse) IT UK **N**: AG **A**: IS

genus *Pseudoligomerus* Pic, 1933b: 3 type species *Pseudoligomerus hummeli* Pic, 1933
hummeli Pic, 1933b: 3 **A**: GAN

genus *Stegobium* Motschulsky, 1860b: 154 type species *Dermestes paniceus* Linnaeus, 1758
Artobium Mulsant & Rey, 1863d: 58 type species *Dermestes panniceus* Linnaeus, 1758
Sitodrepa C. G. Thomson, 1863: 166 type species *Dermestes paniceus* Linnaeus, 1758
paniceum Linnaeus, 1758: 357 (*Dermestes*) **E**: AB AU AZ BE CR CT CZ DE EN FI FR GB GE GG GR HU IR IT LA LT MA MC NL NR NT PL PT RO SK SP ST SV SZ UK **N**: AG CI EG LB MO MR TU **A**: AHN BEI CY FE FUJ GAN GUA GUI GUX HAI HEB HEI HEN HKG HUB HUN IS JA JIA JIL JIX JO KI KZ LIA MAC NIN NMO NP QIN SCH SHA SHG SHN SHX TAI TIA TM TR UP XIN XIZ YUN ZHE **AFR**i **AUR**i **NAR**i **NTR**i **ORR**i
bonariensis Steinheil, 1873: 575 (*Cis*)
ferrugineum Herbst, 1783: 27 (*Anobium*)
ireos A. Villa & G. B. Villa, 1833: 33 (*Anobium*)
minutum Fabricius, 1792a: 238 (*Anobium*)
nanum Küster, 1849c: 45 (*Anobium*)
obesum Melsheimer, 1846: 309 (*Anobium*)
rubellum Marsham, 1802: 85 (*Ptinus*)
striatopunctata Steinheil, 1873: 574 (*Cis*)
tenuicorne Marsham, 1802: 84 (*Ptinus*)
tenuestriatum Say, 1825a: 173 (*Anobium*)
testaceus Thunberg, 1784a: 6 (*Ptinus*)
upsaliensis Gmelin, 1790: 1608 (*Ptinus*)
villosum Melsheimer, 1846: 309 (*Anobium*)

tribe Stegobiini, genus incertae sedis
genus *Holcobius* Sharp, 1881: 520 type species *Holcobius major* Sharp, 1881
japonicus Pic, 1903u: 183 (*Xyletinus*) **A**: HUB JA

subfamily **Dorcatominae** C. G. Thomson, 1859

tribe Calymmaderini White, 1982

genus *Calymmaderus* Solier, 1849: 472 type species *Calymmaderus capucinus* Solier, 1849
subgenus *Calymmaderus* Solier, 1849: 472 type species *Calymmaderus capucinus* Solier, 1849
Byrrhocerus Brèthes, 1919: 26 type species *Byrrhocerus newmani* Brèthes, 1919
Eutheca Kiesenwetter, 1877: 155 type species *Eutheca solida* Kiesenwetter, 1877
Phoberus Kirsch, 1874: 400 [HN] type species *Phoberus byrrhoides* Kirsch, 1874
Thaptor Gorham, 1883c: 205 type species *Thaptor pupatus* Gorham, 1883
oblongus Gorham, 1883c: 206 (*Thaptor*) **E**i: AZ **NTR**
solidus Kiesenwetter, 1877: 155 (Eutheca) **E**: PT SP **N**: MR

tribe Cryptoramorphini White, 1982

genus *Cryptoramorphus* White, 1966: 959 type species *Cryptoramorphus floridanus* White, 1966
boninensis Ford, 1958: 80 (*Mirosternus*) **A**: JA
longiusculus Reitter, 1877i: 380 (*Mesocoelopus*) **A**: JA

tribe Dorcatomini C. G. Thomson, 1859

genus *Anitys* C. G. Thomson, 1863: 175 type species *Dorcatoma rubens* J. J. Hoffmann, 1803
Amblytoma Mulsant & Rey, 1864: 381 type species *Dorcatoma rubens* J. J. Hoffamnn, 1803
rubens J. J. Hoffmann, 1803: 103 (*Dorcatoma*) **E**: AU BE CR CZ DE FI FR GB GE GR IT NR PL RU SK SL SV UK **N**: EG
cognata Mulsant & Rey, 1864: 388 (*Amblytoma*)

genus *Byrrhodes* LeConte, 1878a: 412 type species *Byrrhodes setosus* LeConte, 1878
Eutylistus Fall, 1905: 264 type species *Dorcatoma incomptum* LeConte, 1865
Fachus R. Lucas, 1920: 294 [RN] type species *Byrrhodes setosus* LeConte, 1878
Notiotheca Brèthes, 1923c: 458 type species *Dorcatoma bimaculata* Phillippi, 1864
Priotoma Gorham, 1886: 350 type species *Dorcatoma quadrimaculata* Gorham, 1883
baroniurbanii Español, 1977c: 315 **A**: BT
irregularis Sakai, 1984a: 11 **A**: JA

nipponicus Sakai, 1984a: 13 **A**: JA
tomokunii Sakai, 1996: 277 **A**: TAI

genus *Caenocara* C. G. Thomson, 1859: 90 type species *Dorcatoma bovistae* J. J. Hoffmann, 1803
Enneatoma Mulsant & Rey, 1864: 367 [HN] type species *Dorcatoma affine* Sturm, 1837
Tylistus LeConte, 1861: 203 type species *Dorcatoma simile* Say, 1835
affine Sturm, 1837c: 15 (*Dorcatoma*) **E**: AU BE BH BU BY CR CT CZ FR GB GE GG HU IT LA LT NL PL RO SK SP SZ UK YU **A**: FE
anatolica Zahradník, 1998: 282 **A**: TR
bovistae J. J. Hoffmann, 1803: 100 (*Dorcatoma*) **E**: AU BE CT CZ DE FI FR GB GE HU IT LA LT NL NR NT PL RO SK SL SP ST SV SZ UK "Caucasus" **A**: ES FE MG WS
bicolor Halbherr, 1894: 14
castaneae Gredler, 1863: 263 (*Dorcatoma*)
meridionale Laporte, 1840a: 294 (*Dorcatoma*)
subalpina Bonelli, 1812: 162 (*Dorcatoma*)
ganglbaueri Schilsky, 1899: 65 **A**: TRi
indica Español, 1977b: 183 **A**: SD
rufitarsis Reitter, 1878d: 90 **A**: JA NC
granulum Kiesenwetter, 1879b: 318 (*Dorcatoma*)
grouvellei Pic, 1937d: 4
subglobosum Mulsant & Rey, 1864: 377 (*Enneatoma*) **E**: AU BE BU CR CT CZ FR GE GG GR IT NL PL RO SK SP ST SZ UK **N**: TU **A**: ES FE MG SC
atrata Schilsky, 1899: 67
tsuchiguri Sakai, 1984a: 15 **A**: JA

genus *Dorcatoma* Herbst, 1792: 103 type species *Dorkatoma dresdensis* Herbst, 1792
subgenus *Dorcatoma* Herbst, 1792: 103 type species *Dorkatoma dresdensis* Herbst, 1792
Dorcadolomum Gistel, 1856: 368 type species *Dorcatoma bistriata* Herbst, 1798
Serrocerus Kugelann, 1792: 486 type species *Serrocerus glaber* Kugelann, 1792 (= *Dorkatoma dresdensis* Herbst, 1792)
dresdensis Herbst, 1792: 104 **E**: AU BE BH BU BY CT CZ DE EN FI FR GB GE GG HU IT LA LT NL NR NT PL RO SK SL SP SV SZ UK YU **A**: FE WS
bistriata Paykull, 1798: 318
dorcatoma Fabricius, 1798: 73 (*Ptilinus*)
glaber Kugelann, 1792: 486 (*Serrocerus*)
sinuata C. G. Thomson, 1863: 173
striatus Kugelann, 1792: 487 (*Serrocerus*)
farbiaki Zahradník, 1998: 280 **A**: TR
lomnickii Reitter, 1903d: 239 **E**: BY CT NT PL SK
obtrita Logvinovskiy, 1980: 151 **A**: FE
palmi Zahradník, 1996a: 98 **E**: GR **A**: CY
punctulata Mulsant & Rey, 1864: 346 **E**: AU BU CT CZ FI FR GE IT NR NT PL SK SV **A**: ES FE WS
elongata Baudi di Selve, 1873: 263
robusta A. Strand, 1938: 80 **E**: AB AU BY CT CZ FI GE HU LA NR NT PL SK SV
subgenus *Pilosodorcatoma* Borowski, 1999: 128 type species *Dorcatoma substriata* Hummel, 1829
ambjoerni Baranowski, 1985: 205 **E**: CZ GB SV
androgyna Büche, 2001: 229 **E**: AU FR GE
chrysomelina Sturm, 1837c: 7 **E**: AU BE BY CT CZ DE EN FI FR GB GE HU IT LA LT NL NR NT PL RO SK SL SP ST SV SZ UK **A**: WS
dorcatoma Illiger, 1798: 334 (*Anobium*)
diversesculpturata Roubal, 1934a: 37
polita Rey, 1892: 3
simplex Rey, 1892: 3
janssoni Büche & Lundberg, 2002: 79 **E**: LA SV
minor Zahradník, 1993: 80 **E**: AU CZ FR GE HU PL RO SK SV
setosella lucens Peyerimhoff, 1926c: 348 **N**: AG
setosella setosella Mulsant & Rey, 1864: 356 **E**: AU BE CR CZ FR GE GR HU IT NL PL SZ UK **A**: TR

substriata Hummel, 1829: 24 **E**: AU BE CT CZ DE FI FR GB GE HU IT NT PL PT RO SK SP SV SZ UK
serra Panzer, 1795a: 10 (*Dermestes*) [HN]
suspecta Rey, 1892: 3
subgenus *Sternitodorcatoma* Borowski, 1999: 127 type species *Bruchus flavicornis* Fabricius, 1792
flavicornis Fabricius, 1792b: 374 (*Bruchus*) **E**: AR AU BE BY CT CZ DE FI FR GB GE GG HU IT LA LT NL NR PL RO SK SV SZ UK **A**: CY
murina Baudi di Selve, 1874b: 334

genus *Dorcatoma*, species incertae sedis
externa Mulsant & Rey, 1864: 367 **Ei**: FR
gardneri Pic, 1937i: 126 **A**: UP
hattorii Kôno, 1938: 175 **A**: FE JA
japonica Pic, 1937d: 4 **A**: FE JA
furcatum Logvinovskiy, 1980: 152
rimosa Hayashi, 1951: 7
lanuginosa Baudi di Selve, 1873: 263 **E**: IT **A**: AG
osensis Hayashi, 1951: 6 **A**: JA
pilosella Reitter, 1901a: 44 **A**: FE
polypori Hayashi, 1954: 90 **A**: JA
shigaensis Hayashi, 1951: 8 **A**: JA
shigensis Hayashi, 1954: 93
vaulogeri agenjoi Español, 1978b: 82 **E**: SP
vaulogeri latior Pic, 1948a: 14 **N**: MO
vaulogeri vaulogeri Pic, 1927g: 9 **N**: AG TU
bisulcata Pic, 1927g: 9

genus *Mizodorcatoma* Hayashi, 1955: 21 type species *Mizodorcatoma pinicolae* Hayashi, 1955
densepunctata Español, 1977b: 189 **A**: SD **ORR**
dommeri Rosenhauer, 1856: 171 (*Dorcatoma*) **E**: FR IT SP **N**: AG MO TU
dichroa Boieldieu, 1859: 471 (*Dorcatoma*)
obscuricolor Pic, 1927g: 9 (*Dorcatoma*)
magojyakushi Sakai, 1996: 279 **A**: JA
pinicola Hayashi, 1951: 7 (*Dorcatoma*) **A**: JA
fometaria Hayashi, 1951: 7 (*Dorcatoma*)
pinicolae Hayashi, 1955: 22
pulcherima Sakai, 1979: 134 **A**: JA
sibirica Reitter, 1879e: 226 (*Dorcatoma*) **A**: FE JA TAI
subrufescens Pic, 1914d: 11 (*Dorcatoma*)
trisulcata Champion, 1924c: 246 (*Caenocara*) **A**: UP
uenoi Sakai, 1983: 25 **A**: NP SD

tribe Petaliini White, 1982

genus *Synanobium* Schilsky, 1898: 22 type species *Synanobium ganglbaueri* Schilsky, 1898 (= *Rhadine parmatum* Baudi di Selve, 1874)
subgenus *Synanobium* Schilsky, 1898: 22 type species *Synanobium ganglbaueri* Schilsky, 1898 (= *Rhadine parmatum* Baudi di Selve, 1874)
Rhadine Baudi di Selve, 1874b: 331 [HN] type species *Rhadine parmata* Baudi di Selve, 1874
angustissimum Pic, 1903s: 172 (*Rhadine*) **A**: SA **AFR**
parmatum Baudi di Selve, 1874b: 332 (*Rhadine*) **N**: EG **A**: CY IS LE SY **AFR**
ganglbaueri Schilsky, 1898: 22

tribe Prothecini White, 1982

genus *Methemus* Broun, 1882: 409 [RN] type species *Capnodes griseipilus* Broun, 1881
Capnodes Broun, 1881: 690 [HN] type species *Capnodes griseipilus* Broun, 1881
Capnodula E. Strand, 1928: 2 [RN] type species *Capnodes griseipilus* Broun, 1881
Dicoelocephalus Lea, 1924: 45 type species *Dicoelocephalus granipennis* Lea, 1924

pendleburyi Pic, 1936e: 54 (*Falsogastrallus*) **A**: BT **AFR ORR**

genus *Peritheca* Logvinovskiy, 1978: 29 type species *Peritheca gyrosa* Logvinovskiy, 1978
gyrosa Logvinovskiy, 1978: 29 **A**: TD TM
phototropus Logvinovskiy, 1978: 28 (*Stagetus*) **A**: KZ TM UZ

genus *Protheca* LeConte, 1865: 241 type species *Protheca hispida* LeConte, 1865
Picatoma Lepesme, 1947l: 224 type species *Leptobia guadalupensis* Pic, 1909
plicatipennis Pic, 1912j: 46 (*Theca*) **A**: TAI

genus *Sculptotheca* Schilsky, 1900: 16 type species *Theca hilleri* Schilsky, 1900
Paratheca Scott, 1924: 423 type species *Paratheca borneensis* Scott, 1924
hilleri Schilsky, 1900: 16 (*Theca*) **A**: JA

genus *Stagetodes* Español, 1970b: 159 type species *Theca breviuscula* Fairmaire, 1886
breviusculus Fairmaire, 1886a: 447 (*Theca*) **A**: SA **AFR**
africanus Pic, 1929e: 7 (*Theca*)
luteopubens Pic, 1909l: 147 (*Stagetus*)
denticornis Champion, 1919c: 242 (*Stagetus*) **A**: UP
tetsuyai Sakai, 1995a: 243 **A**: JA
zozu Sakai, 1995a: 246 **A**: JA

genus *Stagetus* Wollaston, 1861a: 11 type species *Stagetus hirtulus* Wollaston, 1861
Anomotheca Schilsky, 1899: 40 type species *Theca reitteri* Schilsky, 1899
Stagetomimus Pic, 1923d: 13 type species *Stagetomimus minor* Pic, 1923
Theca Mulsant & Rey, 1861f: 293 [= 1861a: 40] [HN] type species *Theca byrrhoides* Mulsant & Rey, 1861
andalusiacus andalusiacus Aubé, 1861a: 96 (*Theca*) **E**: PT SP **N**: MO
obscuriceps Pic, 1907c: 106 (*Theca*)
andalusiacus convexus Pic, 1921h: 9 (*Theca*) **N**: AG MO
andalusiacus cribricollis Aubé, 1861c: xlv (*Theca*) **E**: GR IT **N**: AG MO TU
ausonicus Obenberger, 1917a: 35
batnensis Pic, 1902m: 80 (*Theca*)
punctatus Pic, 1921h: 9 (*Theca*)
remotus Brenske & Reitter, 1884: 87 (*Theca*)
arabicus Español, 1981b: 269 **A**: SA
beesoni Pic, 1937i: 125 (*Theca*) **A**: KA
borealis Israelson, 1971a: 70 **E**: CT FI GEi NR NT SV
byrrhoides Mulsant & Rey, 1861f: 295 [= 1861a: 42] (*Theca*) **E**: AB AR FR GE IT PL ST **N**: AG MO TU **A**: KZ TM TR UZ
byrrhoides Aubé, 1861a: 95 (*Theca*) [HN]
robustus Pic, 1921h: 9 (*Theca*)
championi Schilsky, 1899: 41 (*Theca*) **E**: SP **N**: AG MO
numidica Pic, 1903a: 81 (*Theca*)
laticollis Pic, 1921h: 9 (*Theca*)
marginatus Pic, 1921h: 9 (*Theca*)
conicicollis Schilsky, 1899: 48 (*Theca*) **E**: GR
curimoides curimoides Reitter, 1884c: 86 (*Theca*) **E**: GR **A**: IS LE SY
curimoides exiguus Roubal, 1917b: 53 **E**: GR
dorcatomoides Brenske & Reitter, 1884: 87 (*Theca*) **A**: CY IS LE TR
elongatus Mulsant & Rey, 1861f: 297 [= 1861a: 44] (*Theca*) **E**: AU FR GE GR IT SL SP **N**: AG MO TU **A**: CY IS TR
inplicata Pic, 1910c: 10 (*Theca*)
raphaelensis Aubé, 1861a: 96 (*Theca*)
rufescens Pic, 1902m: 80 (*Theca*)
euphorbiae Israelson, 1971b: 292 **N**: CI
ferreri Español, 1994: 151 **E**: SP

ferrugineus Español, 1969a: 53 **A**: IS JO SA
franzi Español, 1969b: 117 **E**: FR GR HU SP **N**: AG MO TU **A**: IS SY TR
gardneri Pic, 1937i: 125 (*Theca*) **A**: UP **ORR**
hirtulus brachypilosus Israelson, 1971b: 291 **N**: CI
hirtulus crenatus Wollaston, 1861a: 13 **N**: CI
hirtulus hirtulus Wollaston, 1861a: 12 **N**: CI
　intermedius Wollaston, 1865: 222
hirtulus orientalis Israelson, 1974: 87 **N**: CI
indicus Pic, 1937i: 124 (*Theca*) **A**: UP
italicus Reitter, 1885i: 380 (*Theca*) **E**: GR IT
khnzoriani Español, 1969a: 56 **A**: TD UZ
latior Pic, 1899j: 207 (*Theca*) **A**: IS LE SY TR
　dorcatomoides Reitter, 1885i: 381 (*Theca*) [HN]
　reitteri Schilsky, 1899: 40 (*Theca*) [RN]
madoni Pic, 1932b: 13 (*Theca*) **A**: SA **AFR**
makarovi Zahradník, 1997: 111 **A**: TM
montanus Toskina, 1998: 87 **A**: OM SA
pellitus Chevrolat, 1859: 384 (*Xyletinus*) **E**: FR PT **N**: AG MO TU **A**: SA **AFR**
　nodieri Pic, 1932b: 13 (*Theca*)
pilula Aubé, 1861a: 95 (*Theca*) **E**: AU CR CZ FR GE GR HU IT MC PL SK SL SP UK **N**: AG EG **A**: IS KZ SY TD TR
　obscurior Pic, 1908n: 118
　rufonotatus Pic, 1910c: 9 (*Theca*)
propinquus Obenberger, 1917a: 36 **E**: GR
proximus Pic, 1937i: 125 (*Theca*) **A**: UP
puncticollis Brenske & Reitter, 1884: 86 (*Theca*) **E**: GR PT **N**: AG MO TU **A**: CY IS TR
　*laterufus*Pic, 1921h: 9 (*Theca*)
　moreanus Pic, 1921h: 9 (*Theca*)
rufonotatus Pic, 1956: 6 (*Theca*) **A**: AF
sardous Reitter, 1915d: 116 (*Theca*) **E**: IT (Sardegna) SP
striatulus Schilsky, 1899: 51 (*Theca*) **A**: TR
sydowi Reitter, 1913d: 291 (*Mesothes*) **A**: JIA SHG
thurepalmi Israelson, 1971b: 291 **N**: CI
uenoi Sakai, 1995b: 419 **A**: JA
vicinus Pic, 1921h: 9 (*Theca*) **A**: IS TR
wollastoni Español, 1969b: 110 **N**: MO
xyletinus Reitter, 1889m: 283 (*Theca*) **E**: AR
yunnanus Pic, 1911m: 122 **A**: YUN

genus *Striatheca* White, 1973: 48 type species *Striatheca lineata* White, 1973
cariniceps Sakai, 1983: 23 **A**: NP

tribe Prothecini, genera incertae sedis
genus *Dimorphotheca* Español, 1977b: 205 type species *Dimorphotheca oculata* Español, 1977
albomarginata Sakai, 1996: 275 **A**: JA

genus *Macrodorcatoma* Pic, 1937i: 125 type species *Macrodorcatoma multistriata* Pic, 1937
multistriata Pic, 1937i: 126 **A**: UP

subfamily **Dryophilinae** LeConte, 1861
tribe Ptilineurini White, 1982
genus *Ptilinastes* Lesne, 1913b: 293 type species *Ptilinastes gerardi* Lesne, 1913
gerardi Lesne, 1913b: 294 **A**: JA

genus *Ptilineurus* Reitter, 1901a: 24 type species *Ptilinus marmoratus* Reitter, 1877

marmoratus Reitter, 1877i: 379 (*Ptilinus*) **E**: SVi **A**: AHN GUI GUX HEB HEN HUB JA JIA JIL JIX LIA NMO SHA SCH SHN TAI YUN **NAR**i **ORR**
ramicornis Kiesenwetter, 1879b: 317 (*Ptilinus*)
pictipennis Fairmaire, 1895a: 175 (*Trachelobrachys*) **A**: JIA SCH SHA

tribe Dryophilini LeConte, 1861

genus *Dryophilus* Chevrolat, 1832: 3 type species *Dryophilus anobioides* Chevrolat, 1832
anobioides Chevrolat, 1832: 3 **E**: AU BE CR CZ FR GB GE HU IT MD NL PL SK SL SP SZ UK **N**: AG TU
compressicornis Mulsant & Rey, 1853c: 17 (*Anobium*)
curticornis Pic, 1921a: 2 **N**: AG
densipilis Abeille de Perrin, 1872a: 34 **E**: FR GR IT PT SP **N**: AG
forticornis Abeille de Perrin, 1875: 212 **E**: IT **A**: CY IS LE SY TR
longicollis Mulsant & Rey, 1853c: 14 (*Anobium*) **E**: AU CR CZ FR GE GR IT PL PT SK SP SZ UK YU **N**: AG MO TU
elongatus Pic, 1921a: 2
luigionii Pic, 1921a: 2 **E**: IT
pusillus Gyllenhal, 1808a: 294 (*Anobium*) **E**: AU BE CR CZ DE EN FI FR GB GE GR HU IR IT LA LT NL NR PL PT RO SK SL SP SV SZ UK "Caucasus" **N**: EG **A**: CY
excelsus Gistel, 1857: 12
semipallidus Pic, 1902m: 79
striatellum Beck, 1817: 15 (*Anobium*)
rufescens Pic, 1921a: 2 **E**: GR (Kríti)
siculus Ragusa, 1896b: 90 **E**: IT
subimpressum Pic, 1921a: 2 **N**: TU

genus *Grynobius* C. G. Thomson, 1859: 89 type species *Anobium excavatus* Kugelann, 1792 (= *Anobium planus* Fabricius, 1787)
planus Fabricius, 1787a: 39 (*Anobium*) **E**: AN AU BE CT DE FI FR GB GE GG HU IR IT NL NR NT PL SP ST SV SZ UK **N**: AG **A**: TR
crenulatus Laporte, 1840a: 293 (*Anobium*)
eichboffii Seidlitz, 1889a: 498 (*Priobium*)
excavatus Kugelann, 1792: 488 (*Anobium*)
ferrugineus Gmelin, 1790: 1606 (*Ptinus*)
kiesenwetteri Edwards, 1921: 102 (*Priobium*)
tricolor A. G. Olivier, 1790a: no 16: 10 (*Anobium*)

genus *Homophthalmus* Abeille de Perrin, 1875: 209 type species *Anobium rugicollis* Mulsant & Rey, 1853
rugicollis Mulsant & Rey, 1853c: 19 (*Anobium*) **E**: AU BH BU CR FR GE GR IT MC SL SP ST SZ UK **A**: TR
pusilloides Obenberger, 1917a: 35 (*Dryophillus*)
rydhi Zahradník, 1996a: 93 **A**: TR

genus *Neodryophilus* Español & Bellés, 1981: 69 type species *Anobium cryptophagoides* Wollaston, 1864
apicalis Pic, 1932g: 26 [= 1932c: 2] (*Dryophilus*) **N**: MO
cryptophagoides Wollaston, 1864: 250 (*Anobium*) **N**: CI
canariensis Har. Lindberg, 1951: 8 (*Dryophilus*)
canariensis Franz, 1996: 133 (*Epitrix*) [HN]

genus *Pseudodryophilus* Heyden, 1891b: 230 [RN] type species *Dryophilus raphaelensis* Mulsant & Rey, 1861 (= *Anobium paradoxus* Rosenhauer, 1856)
Ptinodes Mulsant & Rey, 1864: 32 [HN] type species *Dryophilus raphaelensis* Mulsant & Rey, 1861
paradoxus Rosenhauer, 1856: 177 (*Anobium*) **E**: FR IT SP "Caucasus" **N**: AG MO
raphaelensis Mulsant & Rey, 1861g: 333 [= 1861b: 80] (*Dryophilus*)
succinctus Chevrolat, 1861: 153 (*Hedobia*)

subfamily **Ernobiinae** Pic, 1912
tribe Ernobiini Pic, 1912

genus *Episernomorphus* Roubal, 1917b: 50 type species *Episernomorphus leonhardi* Roubal, 1917
leonhardi Roubal, 1917b: 50 **E**: BH

genus *Episernus* C. G. Thomson, 1863: 151 type species *Episernus angulicollis* C. G. Thomson, 1863
Amphibolus Mulsant & Rey, 1863d: 139 [HN] type species *Anobium gentile* Rosenhauer, 1847
Claudius Gozis, 1882: 200 [HN] type species *Anobium gentile* Rosenhauer, 1847
analis Kiesenwetter, 1879b: 319 **A**: JA
angulicollis C. G. Thomson, 1863: 151 **E**: AU FI IT NR NT SV SZ **A**: ES
acutangulus Leinberg, 1904: 20
striatulus Leinberg, 1904: 20
tenuicollis Leinberg, 1904: 21
ganglbaueri Schilsky, 1898: 18 **E**: AU
gentilis Rosenhauer, 1847: 21 (*Anobium*) **E**: AU FR GE IT SZ
obscurior Pic, 1899b: 25
pallidior Pic, 1912k: 49
semirufus Pic, 1930f: 13 (*Gastrallus*)
granulatus J. Weise, 1887a: 59 **E**: AU BH CZ FR GE HU IT NT PL SK SZ **A**: ES
sulcatus Leinberg, 1904: 18
henschi Reitter, 1901a: 12 **E**: BH
hispanus Kiesenwetter, 1877: 100 **E**: FR SP
diversus Pic, 1930f: 13 (*Gastrallus*)
pyrenaeus Mařan, 1941: 123 **E**: FR
striatellus C. N. F. Brisout de Barneville, 1863: 87 (*Gastrallus*) **E**: AU CR CZ FR GE IT PL SK SP SZ
achillis Gozis, 1882: 201 (*Claudius*)
testaceus Pic, 1912k: 49
sulcicollis Schilsky, 1898: 16 **E**: AU
tatarinovae Toskina & Nikitsky, 2003: 1126 **E**: NT
taygetanus Mařan, 1941: 123 **E**: GR
turcicus Zahradník, 1998: 269 **A**: TR

genus *Ernobius* C. G. Thomson, 1859: 88 type species *Dermestes mollis* Linnaeus, 1758
Conophoribium Chevrolat, 1862: 391 type species *Dermestes mollis* Linnaeus, 1758
Liozoum Mulsant & Rey, 1863d: 92 type species *Dermestes mollis* Linnaeus, 1758
Philoxylon LeConte, 1861: 205 type species *Anobium convexifrons* Melsheimer, 1846
abietinus Gyllenhal, 1808a: 298 (*Anobium*) **E**: AU BE BY CR CT CZ DE FI FR GE GG GR IT LA NL NR NT PL RO SK SV SZ UK
abietis Fabricius, 1792a: 238 (*Anobium*) **E**: AU BE BU BY CR CT CZ DE EN FI FR GB GE GG GR IT LA LS LT NL NR NT PL RO SK SL SV SZ UK **N**: AG **A**: JA
abieticola C. G. Thomson, 1863: 149
brevicorne Bach, 1852: 106 (*Anobium*)
laevis Marsham, 1802: 84 (*Ptinus*)
anatolicus C. Johnson, 1975c: 86 **A**: TR
angelinii Lohse, 1991: 49 **E**: IT
angusticollis Ratzeburg, 1837: 45 (*Anobium*) **E**: AU BE CT CZ DE FI FR GB GE GR HU IT LA LT NL NR NT PL RO SK SL SV SZ TR
parvicollis Mulsant & Rey, 1863d: 121 (*Liozoum*)
tabidus Kiesenwetter, 1877: 120, 126
besucheti Zahradník, 2000: 133 **E**: SZ
cupressi Chobaut, 1899b: 104 **E**: FR GR SP **A**: CY
curticollis Pic, 1921g: 6 **A**: JA
explanatus explanatus Mannerheim, 1843c: 94 (*Anobium*) **E**: BY CT EN FI NR NT SV
frigidus C. G. Thomson, 1863: 148
explanatus phobos Gottwald, 1971: 169 **E**: AU GE PL SK SZ **A**: ES
deimos Gottwald, 1971: 170

freudei Lohse, 1970: 99 **E**: AU IT SZ
fulvus C. Johnson, 1975c: 89 **E**: FR (Corse) IT (Sicilia)
galasjevae Toskina, 2002b: 387 **A**: ES
gallicus C. Johnson, 1975c: 89 **E**: FR SP
gigas Mulsant & Rey, 1863d: 113 (*Liozoum*) **E**: FR GB GE IT LA LT NT PT SP ST SZ UK **N**: AG **A**: TR
mulsantianus Sharp, 1916b: 221
subelongatus Pic, 1914h: 34
impressithorax Pic, 1902m: 79 **N**: AG
indicus Pic, 1943a: 10 **A**: SD
jaroslavli Logvinovskiy, 1977a: 126 **E**: CT
juniperi Chobaut, 1899c: 117 **E**: FR IT SP **N**: AG MO
bedeli Reitter, 1901a: 16
kailidisi C. Johnson, 1975c: 86 **E**: GR **A**: TR
kiesenwetteri Schilsky, 1898: 50 **E**: AU CR CT CZ FR GG GR HU IT LA LT MC PL RO ST SZ **A**: TR
lateobscurus Pic, 1921a: 2
schilskyi Reitter, 1901a: 15
laticollis Pic, 1927g: 9 **E**: AU FR GR IT SP SZ **N**: TU
granulosus Pic, 1939b: 17
lenkae Zahradník, 1998: 267 **A**: TR
longicornis Sturm, 1837b: 124 (*Anobium*) **E**: AU BE BH BY CT CZ DE EN FI FR GE IT LA LT NL NR NT PL RO SK SV SZ **A**: ES
baudueri Pic, 1915g: 17
biimpressus Pic, 1937b: 1
densicornis Mulsant & Rey, 1863d: 128 (*Liozoum*)
dilutipennis Gerhardt, 1905: 222
longicornis Ratzeburg, 1837: 44 (*Anobium*) [HN]
microtomus J. R. Sahlberg, 1871: 431
sericeum Duftschmid, 1825: 53 (*Anobium*)
quittardi Pic, 1916a: 3
lucidus Mulsant & Rey, 1863d: 109 (*Liozoum*) **E**: FR GE PT SP
lucidipennis Pic, 1914h: 34
madoni Pic, 1930f: 13 **A**: CY
minutissimus Sakai, 2002: 128 **A**: JA
mollis espanoli C. Johnson, 1975c: 82 **N**: CI
mollis mollis Linnaeus, 1758: 355 (*Dermestes*) **E**: AR AU AZ BE BY CR CT CZ DE EN FA FI FR GB GE GG GR HU IC IR IT LA LT MC NL NR NT PL PT RO SK SL SP ST SV SZ UK **N**: EG MO MR TU **A**: ES FE HEI IS JA JIL LIA NC TAI TM TR WS **AFRi** **AURi** **NARi** **NTRi** **ORRi**
consimilis Mulsant & Rey, 1863d: 117 (*Liozoum*)
convexifrons Melsheimer, 1846: 309 (*Anobium*)
laetus Mulsant & Rey, 1864: 171 (*Liozoum*)
reversus Sharp, 1916b: 222
sulcatulus Mulsant & Rey, 1863d: 111 (*Liozoum*)
sybaris Kugelann, 1792: 490 (*Anobium*)
tarsatus Kraatz, 1881a: 302
testaceus Kugelann, 1792: 489 (*Anobium*)
mulsanti fructuum Peyerimhoff, 1920: 266 **N**: AG
mulsanti mulsanti Kiesenwetter, 1877: 119 **E**: AU FR GE IT PL **N**: TU
anabaptista Gozis, 1882: 199
nigrinus Sturm, 1837b: 126 (*Anobium*) **E**: AB AR AU BE BH BU BY CR CT CZ DE EN FI FR GB GE GG GR HU IT LA LT NL NR NT PL RO SK SL SP SV SZ UK **N**: CI **A**: TR
berardi Lavagne, 1914: 137
canaliculatus C. G. Thomson, 1871: 380
fuscus Mulsant & Rey, 1863d: 131 (*Liozoum*)
mayeti Pic, 1914h: 27
nigriclava Roubal, 1917b: 51
nigrinus Ratzeburg, 1837: 45 (*Anobium*) [HN]
politus L. Redtenbacher, 1849: 346 (*Anobium*)
pueli Lavagne, 1914: 137

rufescens Pic, 1916a: 3
oculeus Toskina, 2002b: 388 **E**: FR
oertzeni Schilsky, 1900: 14 **E**: CY GR IT **A**: IS
schilskyanus Sharp, 1916b: 221
oxycedri Reitter, 1901a: 18 **N**: AG MO
tiaretensis Pic, 1921a: 2
pallidipennis Pic, 1902m: 79 **E**: GR SP **N**: AG MO
diversepunctatus Pic, 1914h: 33
pardoi Español, 1962: 66 **N**: MO
parens Mulsant & Rey, 1863d: 118 (*Liozoum*) **E**: FR GE HU IT PT SP UK **N**: AG MO TU
bipartitus Pic, 1935c: 5
crassicorne Mulsant & Rey, 1864: 175 (*Liozoum*)
obscuriventris Pic, 1902m: 79
pini pini Sturm, 1837b: 121 (*Anobium*) **E**: AU BE CR CT CZ DE EN FI FR GB GE GG GR HU IT LA NL NT PL RO SK SP SV SZ **N**: AG TU **A**: CY TR
crassiusculus Mulsant & Rey, 1864: 175 (*Liozoum*)
galiberti Pic, 1918e: 9
oblitus Sharp, 1916a: 179
pini Ratzeburg, 1837: 43 (*Anobium*) [HN]
tauri Pic, 1907s: 185
pini tauricus Toskina, 2002b: 391 **E**: UK
pruinosus Mulsant & Rey, 1863d: 101 (*Liozoum*) **E**: FR GE GR IT SK SP **N**: AG MO TU **A**: LE
beauprei Pic, 1908g: 57
cedri Reitter, 1901a: 18
incisus Pic, 1904a: 2
minutus Pic, 1921a: 2
normandi Pic, 1899b: 24
theryi Pic, 1902m: 79
tuniseus Pic, 1921a: 2
reflexus Mulsant & Rey, 1863d: 96 (*Liozoum*) **E**: FR SP **N**: AG MO TU
algiricus Pic, 1915g: 17
reitteri Pic, 1902m: 79 **A**: TR
robusticornis Mařan, 1941: 124 **E**: GR **A**: TR
rufus Illiger, 1807a: 20 (*Anobium*) **E**: IT PT SP **N**: MO MR **AFRi**
nitidulus Wollaston, 1871: 262 (*Anobium*)
simillimus Sakai, 2002: 126 **A**: JA
subopacus Pic, 1904a: 2 **E**: FR
syriacus Pic, 1899b: 25 **A**: SY TR
pubescens Pic, 1921a: 2
vinolasi Novoa & Baselga, 2000: 403 **E**: SP
watanabei Sakai, 2002: 123 **A**: JA
wittmeri Español, 1977a: 6 **A**: TR

tribe Ozognathini White, 1982

genus *Ozognathus* LeConte, 1861: 205 type species *Anobium cornutum* LeConte, 1859
Durangoum Pic, 1903u: 182 type species *Durangoum mexicanum* Pic, 1903
Micranobium Gorham, 1883c: 202 type species *Micranobium exiguum* Gorham, 1883
cornutus LeConte, 1859b: 87 (*Anobium*) **E**i: MA **N**i: MR TU **NAR**
misellus LeConte, 1865: 226

tribe Xestobiini White, 1982

genus *Hyperisus* Mulsant & Rey, 1863d: 90 type species *Anobium plumbeum* Illiger, 1801
caucasicum Logvinovskiy, 1977a: 128 (*Xestobium*) **E**: AR GG ST UK
declive Dufour, 1843a: 47 (*Anobium*) **E**: FR IT SP SZ
velutinum Mulsant & Rey, 1863d: 88
filicorne Israelson, 1974: 74 (*Xestobium*) **N**: CI
impressum Wollaston, 1865: 35 (*Anobium*) **N**: CI

pacificus Sakai, 2001: 328 **A**: JA
plumbeum Illiger, 1801a: 87 (*Anobium*) **E**: AR AU BE BH BU BY CR CZ DE FR GB GE GG GR HU IT MC NL PL RO RU SK SL SP SZ UK YU **A**: CY SY TR
aeneicolle Bach, 1852: 105 (*Anobium*)
bicoloripenne Pic, 1933f: 13
circassicum Pic, 1922e: 18
coerulescens Geoffroy, 1785: 27 (*Birrhus*)
ernobiforme Reitter, 1901a: 13
erythropum Stephens, 1830: 342 (*Anobium*)
graecum Mařan, 1941: 122
politum Duftschmid, 1825: 53 (*Anobium*)
rufonotatum Pic, 1923a: 1
subaeneum Reitter, 1897e: 216
syriacum Pic, 1897h: 194
thoracicum P. Rossi, 1790: 41 (*Anobium*)

genus *Xestobium* Motschulsky, 1845a: 35 type species *Dermestes tessellatus* Villers, 1789 (= *Ptinus rufovillosum* DeGeer, 1774)
Cnecus C. G. Thomson, 1859: 88 type species *Dermestes tessellatus* Villers, 1789 (=*Ptinus rufovillosum* DeGeer, 1774)
africanum Español, 1964: 130 **E**: SP **N**: AG MO TU
austriacum Reitter, 1890j: 392 **E**: AU CZ GE PL SK UK
elegans Logvinovskiy, 1981: 32 **A**: FE
luriei Toskina, 2002b: 393 **A**: ES
pleshanovi Toskina, 2002b: 395 **A**: FE
punctatum Logvinovskiy, 1977a: 129 **A**: FE
rafesi Toskina, 2002b: 395 **A**: FE
rufovillosum DeGeer, 1774: 230 (*Ptinus*) **E**: AR AU BE BH BU BY CT CZ DE FI FR GB GE GR HU IR IT LA LT MD NL NR NT PL PT RO SK SL ST SV SZ TR UK YU **A**: ES FE HKG TM TR WS **AUR**i **NAR**i
faber Thunberg, 1784a: 6 (*Ptinus*)
fatidicus Shaw, 1792: 104
fuscus Gmelin, 1790: 1606 (*Ptinus*)
pulsator Schaller, 1783: 249 (*Ptinus*)
pulsatorium Scriba, 1790b: 156 (*Anobium*)
rubiginosus O. F. Müller, 1776: 57 (*Byrrhus*)
squalidum LeConte, 1874: 64
tessellatum A. G. Olivier, 1790a: no 16: 6 (*Anobium*)
tesselatum Villers, 1789: 54 (*Dermestes*)
shibatai Sakai, 1979: 131 **A**: JA
subincanum Reitter, 1878a: 219 **E**: AB AR GG IT ST UK **A**: TR
circassicum Reitter, 1890j: 393

tribe Xestobiini, taxa incertae sedis
genus *Ochina* Dejean, 1821: 40 type species *Ptilinus hederae* P. W. J. Müller, 1821 (= *Crioceris ptinoides* Marsham, 1802)
subgenus *Dulgieris* Gozis, 1886: 25 type species *Ptilinus latreillii* Bonelli, 1812
latreillii Bonelli, 1812: 167 (*Ptilinus*) **E**: AR AU BE CR CZ FR GE GG GR HU IT RO SK SP ST SZ UK YU **A**: TR
sanguinicollis Duftschmid, 1825: 56 (*Anobium*)
subgenus *Ochina* Dejean, 1821: 40 type species *Ptilinus hederae* P. W. J. Müller, 1821 (= *Crioceris ptinoides* Marsham, 1802)
Cittobium Mulsant & Rey, 1864: 237 type species *Crioceris ptinoides* Marsham, 1802
ferruginea Schilsky, 1899: 11 **E**: AB GR IT
italica Pic, 1908d: 42
pallidicolor Pic, 1908d: 42
hirsuta Seidlitz, 1889b: 540 **E**: FR IT SP **N**: AG MO TU
africana Pic, 1897i: 203
leveillei Sainte-Claire Deville, 1910: 315 **E**: FR (Corse) **A**: MO

numidica Bedel, 1890: cxxxv **N**: AG MO
ptinoides Marsham, 1802: 228 (*Crioceris*) **E**: AU BE CR CZ DE FR GB GE GR IR IT LA LS LT NL PT SP SZ TR **N**: AG **A**: TR
fasciata Geoffroy, 1785: 26 (*Birrhus*) [HN]
hederae P. W. J. Müller, 1821: 193 (*Ptilinus*)

subfamily **Eucradinae** LeConte, 1861
tribe Hedobiini White, 1982

genus *Anhedobia* Nakane, 1963b: 48 type species *Hedobia capucina* Reitter, 1877
capucina Reitter, 1877i: 376 (*Hedobia*) **A**: JA LIA
cristata Kiesenwetter, 1879b: 316 (*Hedobia*)

genus *Clada* Pascoe, 1887: 8 type species *Clada waterhousei* Pascoe, 1887
subgenus *Clada* Pascoe, 1887: 8 type species *Clada waterhousei* Pascoe, 1887
Paranobium Gahan, 1900b: 104 type species *Paranobium posticum* Gahan, 1900
Pseudoclada M. Pic & T. Pic, 1901: 170 type species *Hedobia quinquecostata* Schilsky, 1898
Xystrophorus Lesne, 1902b: 476 type species *Xystrophorus horni* Lesne, 1902
alternata Fairmaire, 1886a: 448 (*Anobium*) **N**: AG **AFR**
babai Sakai, 1987: 70 **A**: TAI
denticornis barbara Pic, 1903d: 97 (*Xystrophorus*) **N**: AG EG MO TU
vaulogeri Pic, 1921a: 2 (*Xystrophorus*)
villiersi Pic, 1950f: 181
denticornis insularis Israelson, 1974: 72 **N**: CI
denticornis maroccana Español, 1968: 81 **N**: MO
espagnoli Israelson, 1974: 71 **N**: CI
fernandezi Español, 1968: 75 **N**: CI
forbesii Gahan, 1903: 290 (*Paranobium*) **A**: YE (Abd el Kuri)
formosana Sakai, 1987: 69 **A**: TAI
halperini Español, 1978a: 1 **A**: IS
insulcata Pic, 1933b: 2 **A**: ANH GAN YUN
latipennis Pic, 1943a: 9 (*Hedobia*) **N**: MO
maxima Pic, 1903c: 106 (*Hedobia*) **A**: YUN
mouhoti Pic, 1903b: 99 (*Xystrophorus*) **A**: "Asia c."
okinawana Sakai, 1991: 167 **A**: JA
oromii Español, 1978c: 152 **N**: MR (Selvagens)
pohli Toskina, 2004b: 460 **A**: YE
sarracena Español, 1979: 251 **A**: SA
tricostata Baudi di Selve, 1874b: 325 (*Hedobia*) **A**: CY IS LE SY
obscuricolor Pic, 1909s: 178 (*Hedobia*)
quinquecostata Schilsky, 1898: 55 (*Hedobia*)
vittula Sakai, 1987: 72 **A**: TAI
subgenus *Taiwanoclada* Sakai, 1987: 66 type species *Clada shibatai* Sakai, 1987
shibatai Sakai, 1987: 66 **A**: TAI

genus *Hedobia* Dejean, 1821: 41 type species *Ptinus pubescens* A. G. Olivier, 1790
atricolor Pic, 1926c: 4 **A**: YUN
minor Pic, 1926c: 4 **A**: HUN
multipunctata Pic, 1944c: 5 **A**: GAN
pubescens A. G. Olivier, 1790a: no 17: 5 (*Ptinus*) **E**: AB AU CR CZ FR GE GG GR HU IT PL RO SK SP ST UK YU **A**: TR
nigrina Reitter, 1897e: 215
unicolor Reitter, 1898f: 137 [HN]
unicolor Pic, 1897h: 194 **E**: AB AR GG ST **A**: IN SA TR

genus *Ptinomorphus* Mulsant & Rey, 1868b: 31 type species *Ptinus imperialis* Linnaeus, 1767

angustatus C. N. F. Brisout de Barneville, 1862: 602 (*Hedobia*) **E**: FR IT SP **N**: AG MO
corsicus Pic, 1903l: 138 (*Hedobia*)
aquilus Toskina, 2001: 143 **E**: SV
exilis Kiesenwetter, 1879b: 316 (*Hedobia*) **A**: JA
harmandi Pic, 1901r: 343 (*Hedobia*)
fursovi Toskina, 2001: 146 **A**: WS
imperialis Linnaeus, 1767: 565 (*Ptinus*) **E**: AB AU BE BH BU CR CZ DE EN FI FR GB GE GG GR HU IR IT NL NR PL PT RO SK SL SP ST SV SZ UK **A**: TR
alpinus Pic, 1898k: 156 (*Hedobia*)
angustior Pic, 1896f: 71 (*Hedobia*)
cruciatus Geoffroy, 1785: 58 (*Bruchus*)
dayremi Pic, 1942c: 2
imperialis O. F. Müller, 1776: 57 (*Bruchus*)
interruptus Pic, 1893c: 111 (*Hedobia*)
luteosuturalis Pic, 1925g: 13
maculatus Pic, 1922e: 18 (*Hedobia*)
nigronotatus Pic, 1942c: 2
senex Kraatz, 1876: 151
theresae Pic, 1933f: 13
x-notatus Pic, 1942c: 2
magnificus angustior Pic, 1896f: 71 (*Hedobia*) **E**: GG ST
magnificus magnificus Reitter, 1880e: 476 (*Hedobia*) **E**: AL AR GG GR ST UK **A**: TR
caucasicus Pic, 1901g: 58
tauricola Pic, 1906p: 228 (*Hedobia*)
perpulchrus Obenberger, 1917a: 34 (*Hedobia*) **E**: GR
regalis Duftschmid, 1825: 61 (*Ptinus*) **E**: AB AU BH CR CZ FR GE GG GR HU IT NL PL RO SK SP ST SZ UK YU **A**: IN TR
aureopilosus Pic, 1901g: 58
belius Reitter, 1898f: 139 (*Hedobia*)
dispersus Pic, 1930e: 10 (*Hedobia*)
circassicus Pic, 1896f: 72 (*Hedobia*)
satanulus Reitter, 1901a: 8 (*Hedobia*)
rosti Pic, 1896f: 71 (*Hedobia*) **E**: AR GG ST UK **A**: TR
sericeus Toskina, 2001: 146 **E**: ST
tatjanae Logvinovskiy, 1978: 26 **E**: ST **A**: KI TR

subfamily **Mesocoelopodinae** Mulsant & Rey, 1864
tribe Mesocoelopodini Mulsant & Rey, 1864

genus *Hisamatsua* Sakai, 1977: 119 type species *Hisamatsua japonica* Sakai, 1977
japonica Sakai, 1977: 120 **A**: JA

genus *Mesocoelopus* Jacquelin du Val, 1860f: 143 type species *Ptilinus niger* P. W. J. Müller, 1821
brevistriatus Leiler, 1979: 93 **E**: GR **A**: TR
buettikeri Español, 1981b: 268 **A**: OM SA
collaris Mulsant & Rey, 1864: 324 **E**: AU CR FR GB GE IT MA PT SP **NARi**
creticus Fairmaire, 1880d: 241 **E**: GR
espagnoli Toskina, 2002a: 208 [RN] **A**: SA
grandis Toskina, 1998: 72 [HN]
gallagheri Toskina, 1998: 71 **A**: OM
ingibbosus Pic, 1924f: 29 **N**: EG **A**: IS OM SA TR YE **AFR**
leileri Israelson, 1976: 13 **N**: CI
mimeuri Pic, 1953c: 19 **N**: MO
niger P. W. J. Müller, 1821: 191 (*Ptilinus*) **E**: AU BE CR CZ FR GE GR HU IT LS PL PT SK SP ST SZ TR UK YU **N**: AG MO **A**: TR
hederae Dufour, 1843b: 323 (*Xyletinus*) [HN]
hederae Laporte, 1840a: 294 (*Dorcatoma*)
murinus Sturm, 1837b: 88 (*Xyletinus*)

pubescens Duftschmid, 1825: 46 (*Ptilinus*)
substriatus Schilsky, 1900: 73 **A**: SA

genus *Mesothes* Mulsant & Rey, 1864: 311 type species *Xyletinus ferrugineus* Mulsant & Rey, 1861
canariensis Español & Oromí, 1984: 90 **N**: CI
ferrugineus Mulsant & Rey, 1861g: 336 [= 1861b: 83] (*Xyletinus*) **E**: BU CR FR GR IT SK SL SP ST UK YU **A**: IS LE SY TR
granulatus Pic, 1907s: 185 **A**: TR
substriatus Pic, 1938b **A**: JIX

genus *Microcoelopus* Toskina, 1998: 75 type species *Microcoelopus raripilosus* Toskina, 1998
arabicus Toskina, 1998: 79 **A**: SA
crassicornis Toskina, 1998: 81 **A**: SA
curtus Toskina, 1998: 81 **A**: SA
productus Toskina, 1998: 83 **A**: SA
raripilosus Toskina, 1998: 77 **A**: SA
talhouki Toskina, 1998: 84 **A**: SA
tenebrosus Toskina, 1998: 85 **A**: SA

genus *Nesocoelopus* Español, 1977c: 312 type species *Mesocoelopus indicus* Pic, 1914
indicus Pic, 1914c: 8 (*Mesocoelopus*) **A**: BT HP KA
miyatakei Sakai, 1979: 132 **A**: JA

genus *Pseudomesothes* Español, 1967: 45 type species *Lasioderma pulverulenta* Reitter, 1877
pulverulentus latior Pic, 1954d: 55 (*Mesothes*) **A**: FUJ
pulverulentus pulverulentus Reitter, 1877i: 379 (*Lasioderma*) **A**: JA
ussuriensis Zahradník, 1997: 107 **A**: FE

tribe Tricorynini White, 1982

genus *Rhamna* Peyerimhoff, 1913: 518 type species *Rhamna semen* Peyerimhoff, 1913
kinoshitai Sakai, 1993: 571 **A**: JA
saudita Español, 1981b: 268 **A**: SA
semen Peyerimhoff, 1913: 519 **E**: SP **N**: MO **AFR**
yaeyamana Sakai, 1993: 572 **A**: JA

genus *Stagetomorphus* Pic, 1914c: 7 type species *Stagetomorphus indicus* Pic, 1914
subgenus *Stagetomorphus* Pic, 1914c: 7 type species *Stagetomorphus indicus* Pic, 1914
Aulacanobium Lea, 1924: 44 type species *Dorcatoma lanigerum* Olliff, 1888
amamiensis Sakai, 1979: 137 **A**: JA
indicus Pic, 1914c: 8 **A**: HP UP **ORR**

genus *Tricorynus* G. R. Waterhouse, 1849: lxviii type species *Tricorynus zeae* G. R. Waterhouse, 1849
Catorama Guérin-Méneville, 1850: 431 type species *Catorama tabaci* Guérin-Méneville, 1850
Cathorama Gemminger & Harold, 1869: 1784 type species *Catorama tabaci* Guérin-Ménevile, 1850
Hemiptychus LeConte, 1865: 239 type species *Hemiptychus punctatus* LeConte, 1865
Xylotheca Reitter, 1897d: 204 type species *Xylotheca meieri* Reitter, 1897
herbarius Gorham, 1883c: 207 (*Cathorama*) **Ei**: GE NL **AUR NAR NTR**
claveri Pic, 1923c: 5 (*Catorama*)
geminatum Fall, 1905: 251 (*Catorama*)
goudoti Pic, 1927c: 8 (*Catorama*)
latipennis Pic, 1927c: 9 (*Catorama*)
maroniensis Pic, 1927c: 8 (*Catorama*)
meieri Reitter, 1897d: 205 (*Xylotheca*)
mexicanum Sharp, 1885b: 160 (*Catorama*)
minasensis Pic, 1927c: 8 (*Catorama*)
minor Meier, 1899a: 72 (*Xylotheca*)

pusillus Sharp, 1885b: 160 (*Catorama*)
sharpi Pic, 1912v: 265 (*Catorama*)
venezuelensis Pic, 1927c: 9 (*Catorama*)
tabaci Guérin-Méneville, 1850: 437 (*Catorama*) **Ei**: CR DE FR GE HU **NAR NTR**
impressifrons Fall, 1905: 234 (*Catorama*)
puncticeps Gorham, 1886: 348 (*Cathorama*)

subfamily **Ptilininae** Shuckard, 1840
tribe Ptilinini Shuckard, 1840

genus *Indanobium* Español, 1970a: 98 type species *Ptilinus tonkineus* Pic, 1917
binodulus Motschulsky, 1858c: 65 (*Ptilinus*) **A**: HP UP **ORR**
formosanus Kôno & Kim, 1937: 30 (*Ptilinus*) **A**: TAI
kyushuensis Nakane, 1989a: 1 (*Paroligomerus*) **A**: JA

genus *Nepalanobium* Sakai, 1983: 20 type species *Nepalanobium quadricostatum* Sakai, 1983
quadricostatum Sakai, 1983: 21 **A**: NP

genus *Plumilus* White, 1974: 445 [RN] type species *Ptilinus grandicollis* Ménétriés, 1832
Plumaria Reitter, 1883a: 123 [HN] type species *Ptilinus grandicollis* Ménétriés, 1832
grandicollis Ménétriés, 1832: 165 (*Ptilinus*) **E**: AB AR GEi GG RO ST **A**: IN IS TM TR

genus *Ptilinus* Geoffroy, 1762: 64 type species *Ptilinus fuscus* Geoffroy, 1785
asiaticus Toskina, 1995: 15 **A**: IN TM
cercidiphylli Kôno & Kim, 1937: 29 **A**: JA
cylindripennis Wollaston, 1854: 285 **E**: AZ **N**: MR
fuscus Geoffroy, 1785: 133 (*Bostrichus*) **E**: AU BE BU BY CT CZ EN FI FR GB GE GR HU IT MC NL NR NT PL RO SK SP SV SZ UK "Caucasus" **N**: AG **A**: AF ES FE GAN JA KZ LIA NMO QIN TD TM TR UZ WS
aspericollis Mulsant & Rey, 1853a: 7 [HN]
asperulus Gemminger, 1870: 121 [RN]
costatus Gyllenhal, 1827: 329
flabelicornis Le Peletier de Saint Fargeau & Audinet-Serville, 1825: 238
flavescens Laporte, 1840a: 295
flavescens Geoffroy, 1785: 19 (*Dermestes*)
mulsanti Marseul, 1886: 296 [RN]
galasjevi Toskina, 1995: 17 **A**: FE
galloisi Kôno & Kim, 1937: 30 **A**: JA
iranicus Toskina, 1995: 21 **A**: IN
lepidus Wollaston, 1864: 251 **N**: CI
longicornis Toskina, 1995: 21 **E**: AR GG ST
nikitskii Toskina, 1995: 23 **A**: FE
panfilovi Toskina, 1995: 23 **A**: KI
pectinicornis Linnaeus, 1758: 355 (*Dermestes*) **E**: AB AR AU AZ BH BU BY CR CT CZ DE FR GB GE GG HU IR IT LA LT NL NR NT PL RO SK SP ST SV SZ TR UK YU **N**: MO MR **A**: ES FE HUB JIX SY TM TR UP WS **NAR**i
aspericollis Ménétriés, 1832: 166
cylindricus O. F. Müller, 1776: 81
discolor Faldermann, 1839: 5 (*Xyletinus*)
impressifrons Küster, 1847a: 42
pectinatus Laicharting, 1781: 68 (*Bostrichus*)
serraticornis Marsham, 1802: 82 (*Ptinus*)
phellodendri Logvinovskiy, 1979b: 69 **A**: FE
pleshanovi Toskina, 1995: 25 **A**: ES
reichardti Persov, 1964: 169 **A**: FE
sinaitus Pic, 1939g: 144 **A**: SI
sulcatus Pic, 1937i: 124 **A**: UP

genus *Yunnanobium* Español, 1965: 56 type species *Oligomerus longicorne* Pic, 1907
longicorne Pic, 1907p: 173 (*Oligomerus*) **A**: SCH **ORR**

subfamily **Xyletininae** Gistel, 1856
tribe Lasiodermini White, 1982

genus *Lasioderma* Stephens, 1835: 417 type species *Ptilinus testaceum* Duftschmid, 1825 (=*Ptinus serricorne* Fabricius, 1792)
Hypora Mulsant & Rey, 1864: 294 type species *Ptinus serricorne* Fabricius, 1792
Pseudochina Jacquelin du Val, 1860f: 143 type species *Ptilinus haemorrhoidalis* Illiger, 1807
aethiops Iablokoff-Khnzorian, 1976c: 124 **A**: TD
amabile Sakai, 1975: 101 **A**: JA
anatolicum Zahradník, 1996a: 96 **A**: TR
antoinei Peyerimhoff, 1949: 272 **N**: MO
aterrimum Roubal, 1916: 48 **E**: CR CZ GE GR SK ST
atrorubrum Toskina, 1999: 24 **E**: ST
atrum arnoldi Toskina, 1999: 26 **E**: ST
atrum atrum Toskina, 1999: 25 **E**: ST
babadjanidesi Reitter, 1913d: 291 **E**: AB TM
baudii Schilsky, 1899: 25 **E**: CR FR GR IT PT SP **N**: AG CI EG LB TU **A**: CY IS LE SY
vestitum Schilsky, 1899: 25
bubalus bubalus Fairmaire, 1860: 631 **E**: AB FR IT MA PT SP **N**: AG LB **A**: CY IS SA
apicatum Mulsant & Rey, 1864: 295 (*Pseudochina*)
bubalus phelipaearum Peyerimhoff, 1926c: 345 **N**: AG MO TU
buettikeri Toskina, 1998: 60 **A**: SA
corsicum Schilsky, 1899: 24 **E**: FR (Corse)
costulatum Schilsky, 1899: 28 **E**: AB AR **A**: KZ MG TD TM UZ
desectum Wollaston, 1861a: 13 (*Xyletinus*) **N**: CI
excavatum Wollaston, 1861a: 15 (*Xyletinus*) **N**: CI
festai Pic, 1925b: 3 **N**: AG LB TU
flavicollis Wollaston, 1865: 34 (*Xyletinus*) **N**: CI
formosanum Pic, 1943a: 10 **A**: TAI
fuscum Rey, 1892: 3 (*Pseudochina*) **E**: FR GR
gallagheri Toskina, 1998: 62 **A**: OM
haemorrhoidale Illiger, 1807a: 18 (*Ptilinus*) **E**: AZ BU CR FR GR HU IT MA PT SP ST SZ UK **N**: AG CI EG LB MO TU **A**: CY IS MG SY TR "Korea" **NAR**i
bicolor Schaufuss, 1882b: 622
florieni Pic, 1911a: 11
punctulatissimus Wollaston, 1861a: 17 (*Notiomimus*)
unicolor Schilsky, 1899: 33
villosum Laporte, 1840a: 295 (*Anobium*)
impexum Toskina, 1998: 63 **A**: SA
impunctatum Schilsky, 1899: 26 **N**: AG
japonica Pic, 1943a: 10 **A**: JA
kaszabi Español, 1971: 52 **A**: MG
kiesenwetteri Schilsky, 1899: 31 **E**: AB AR CR CZ FR GG GR IT SP ST UK **A**: IS KZ SY TM TR UZ
kirgizicum Toskina, 1999: 30 **A**: KI
kryzhanovskii Logvinovskiy, 1977b: 286 **E**: AB
kurbatovi Toskina, 1999: 31 **A**: UZ
laeve Illiger, 1807a: 17 (*Ptilinus*) **E**: AU FR HU IT PT SP SZ
latifrons Toskina, 1998: 64 **A**: KU
latitans Wollaston, 1861a: 14 (*Xyletinus*) **N**: CI MO
logvinovskii Toskina, 1998: 60 [RN] **A**: KZ
minutum Logvinovskiy, 1977b: 285 [HN]
luminosum Logvinovskiy, 1977b: 287 **E**: AR **A**: ES

mazokhini Toskina, 1999: 31 **E**: AR
melanocephalum Schilsky, 1899: 32 **E**: IT SP **A**: AG MO
micans Mannerheim, 1829: 24 (*Dorcatoma*) **E**: NTi
mikheechevi Toskina, 1999: 32 **E**: AB
minutum Har. Lindberg, 1951: 8 **N**: CI
mongolicum Español, 1971: 54 **A**: MG
mulsanti Schilsky, 1899: 29 **E**: FR SP **N**: AG TU
chobauti Reitter, 1901a: 35
multipunctatum Toskina, 1999: 35 **E**: ST
nikitskii Toskina, 1999: 35 **A**: FE
obscurum Solsky, 1868: 36 (*Pseudochina*) **E**: AB AU BH BU CR FR GG GR HU IT RO SK ST UK YU **A**: ES IN MG KZ TR
oculeum Toskina, 1999: 36 **E**: AB
punctulatum Reitter, 1884e: 254 **E**: UK **N**: LB **A**: IS LE MG SY
pusillum Español, 1979: 254 **A**: SA
redtenbacheri redtenbacheri Bach, 1852: 115 (*Xyletinus*) **E**: AB AR AU BE BU CR CT CZ FR GE GG GR HU IT MC NL NT PL PT RO SK SP ST SZ UK **N**: EG LB TU **A**: CY IQ IS KZ MG SI SY TM TR UZ WS
caucasicum Pic, 1904f: 34
cyphonoides F. Moravitz, 1861: 290 (*Xyletinus*)
fulvescens Mulsant & Rey, 1864: 301 (*Pseudochina*)
redtenbacheri stepposum Toskina, 1999: 37 **E**: ST
rufomarginatum Pic, 1923b: 6 **A**: "Korea"
semirufulum Reitter, 1897f: 217 **E**: FR IT **N**: AG TU
leveillei Pic, 1927g: 9
vacaresiense Chobaut, 1924a: 103
serricorne Fabricius, 1792a: 241 (*Ptinus*) **E**: AB AR AU AZ BE BU CR CT CZ DE EN FI FR GB GE GG GR HU IR IT LA NL NR NT PL PT SK SP ST SV SZ TR UK **N**: AG CI EG LB MR TU **A**: AHN BT CY ES FE FUJ GUA GUI GUX HEI HEN HKG HUB IS JA JIL JIX KZ LIA MG NP SA SY TAI TR UP WS ZHE "Korea" **AFR**i **AUR**i **NAR**i **NTR**i **ORR**i
brevis Wollaston, 1861a: 15 (*Xyletinus*)
castaneum Melsheimer, 1846: 308
testaceum Duftschmid, 1825: 46 (*Ptilinus*)
testaceum Stephens, 1835: 417 [HN]
splendidum Toskina, 1998: 66 **A**: OM
striola Rey, 1892: 3 (*Pseudochina*) **E**: IT (Sicilia)
tadzhika Zahradník, 1997: 108 **A**: TD
tenebricum Toskina, 1998: 66 **A**: OM
tenuicorne Iablokoff-Khnzorian, 1964: 165 **A**: TD
thoracicum Morawitz, 1861: 289 (*Xyletinus*) **E**: CR HU SK SL ST UK **A**: KZ TR
bicolor Csiki, 1903: 445 (*Ochina*)
torquatum Chevrolat, 1859: 384 (*Xyletinus*) **E**: FR GR IT **N**: AG TU
triste Roubal, 1919: 38 **E**: GR
turkestanicum Reitter, 1901a: 36 **E**: ST **A**: KZ TM UZ
turkmenicum Toskina, 1999: 41 **A**: TM

genus *Megorama* Fall, 1905: 206 type species *Catorama simplex* LeConte, 1865
Catoramina Pic, 1951b: 6 type species *Catoramina signata* Pic, 1951
Falsoptilinus Pic, 1950c: 10 type species *Falsoptilinus modestus* Pic, 1950
densepunctatum Pic, 1923b: 6 (*Lasioderma*) **A**: BT **ORR**
japonicola Sakai, 1975: 103 **A**: JA
subserratum Israelson, 1974: 80 **N**: CI

genus *Pseudolasioderma* Logvinovskiy, 1978: 26 type species *Lasioderma brevis* Reitter, 1891
brevis Reitter, 1891b: 30 (*Lasioderma*) **A**: KZ TM UZ

tribe Xyletinini Gistel, 1856

genus *Neoxyletinus* Español, 1983: 83 type species *Xyletinus angustatus* Pic, 1907
angustatus Pic, 1907p: 173 (*Xyletinus*) **A**: YUN **ORR**
assamensis Español, 1983: 85 **A**: SD **ORR**
nepalensis Español, 1983: 84 **A**: NP
sikkimensis Español, 1983: 84 **A**: NP SD **ORR**
tibetanus Gottwald, 1977: 176 (*Xyletinus*) [RN] **A**: XIZ
striatus Pic, 1922a: 4 (*Xyletinus*) [HN]

genus *Trachelobrachys* Gemminger, 1870: 124 [RN] type species *Brachytrachelus kiesenwetteri* Morawitz, 1861
Brachytrachelus Morawitz, 1861: 290 [HN] type species *Brachytrachelus kiesenwetteri* Morawitz, 1861
kiesenwetteri Morawitz, 1861: 290 (*Brachytrachelus*) **E**: ST UK "Caucasus" **A**: KZ
obscurior Pic, 1902l: 72 (*Xyletinus*)

genus *Xyletinus* Latreille, 1809: 376 type species *Ptilinus ater* Creutzer, 1796
subgenus *Calypterus* Mulsant & Godart, 1859: 181 type species *Xyletinus sericans* Mulsant & Godart, 1859
Notiomimus Wollaston, 1861a: 15 type species *Notiomimus fimicola* Wollaston, 1861
bucephalus bucephalus Illiger, 1807a: 16 (*Ptilinus*) **E**: FR GR IT PT SP ST **N**: EG LB MO TU **A**: CY IQ IS LE SA SI SY **AFR**
bicoloricornis Pic, 1924f: 29
peregrinus Chevrolat, 1861: 154
sericans Mulsant & Godart, 1859: 181 (*Calypterus*)
striatipennis Fairmaire, 1857a: 638
bucephalus omanus Toskina, 1998: 55 **A**: OM
bucephalus theanus Reitter, 1901a: 29 **E**: BU GR **N**: EG **A**: IS SI SY TR
brunnescens Obenberger, 1917a: 36
megatomoides Roubal, 1917a: 11
elongatus Logvinovskiy, 1977: 415 **A**: MG
fimicola Wollaston, 1861a: 17 (*Notiomimus*) **N**: CI MO
materliki Toskina, 2002a: 207 **A**: YE
tadzhika Iablokoff-Khnzorian, 1978c: 61 **A**: TD
wollastoni Gottwald, 1977: 162 [RN] **N**: CI MO
holosericeus Wollaston, 1861a: 17 (*Notiomimus*) [HN]
subgenus *Evaniocerius* Gottwald, 1983: 136 type species *Xyletinus semilimbatus* Pic, 1899
semilimbatus Pic, 1899j: 208 **N**: EG LB MO TU **A**: IS JO SI
laterufus Pic, 1920i: 48
poliphylloides Escalera, 1914: 256
testaceipennis Pic, 1905b: 97
subgenus *Pseudocalypterus* Gottwald, 1977: 159 type species *Xyletinus pectiniferus* Fairmaire, 1879
pectiniferus Fairmaire, 1879b: 258 **E**: IT **N**: AG MO
alienus Lohse, 1957: 31
pruinosus Desbrochers des Loges, 1889: clxxxvi
subgenus *Xeronthobius* Morawitz, 1863: 164 type species *Ptilinus pallens* Germar, 1824
bucephaloides Reitter, 1901a: 28 **N**: EG LB TU **A**: CY IN OM TR
desertus Barovskij, 1929c: 276 (*Xeronthobius*) **A**: KZ
kozlovi Emetz, 1977: 414 **A**: NMO
lecerfi Kocher, 1956b: 43 **E**: SP **N**: MO
medvedevi Logvinovskiy, 1977: 414 **A**: KZ
ocularis Reitter, 1901a: 28 **E**: ST **A**: KZ MG NMO
pallens Germar, 1824: 79 (*Ptilinus*) **E**: AB CT ST UK **A**: IN KZ PA TM TR UZ WS
pallidus Laporte, 1840a: 295
subgenus *Xyletinus* Latreille, 1809: 376 type species *Ptilinus ater* Creutzer, 1796
Sternoplus Mulsant & Rey, 1864: 263 type species *Xyletinus ater* Creutzer, 1796
asiaticus Reitter, 1901a: 33 **A**: MG NMO

ater Creutzer, 1796a: 9 (*Ptilinus*) **E**: AR AU BE BU CR CT CZ DE EN FI FR GE GG HU IT LA LT NL NR NT PL RO SK SP ST SV SZ UK **A**: ES FE KZ MG TR WS
brevitarsis Schilsky, 1898: 68
serratus Fabricius, 1798: 73 (*Ptilinus*)
striatus Kugelann, 1792: 487 (*Anobium*)
tenebricosus Solsky, 1868: 35
thomsoni Schilsky, 1906b: 23
baicalicus Zahradník, 1997: 104 **A**: ES
balcanicus Gottwald, 1977: 168 **E**: BH BU GR IT MC YU **A**: TR
cylindricus Kofler, 1970b: 144 **A**: CY TR
distinguendus Kofler, 1970a: 65 **E**: AU CZ HU SK
dolini Zahradník, 1997: 102 **A**: KZ
emetzi Gottwald, 1977: 171 **A**: MG
excellens Kofler, 1970c: 365 **E**: CR CT
fibyensis Lundblad, 1949: 4 **E**: AU BE BU BY CT CZ FI FR GE HU IT LA PL RO SK ST SV SZ UK **A**: WS
groenblomi Kangas, 1955: 3
formosus Mannerheim, 1849: 232 **E**: CT SP ST UK **A**: ES KZ MG TM WS
haemorrhoidalis Español, 1971: 51 **A**: MG
hanseni Jansson, 1947: 118 **E**: AU BE DE FI GE HU NR NT SV
chinensis Frivaldszky, 1892: 124 **A**: CH
interpositus Gottwald, 1977: 166 **E**: ST
kofleri Gottwald, 1977: 175 [RN] **E**: GR
robustus Kofler, 1970a: 69 [HN]
laticollis Duftschmid, 1825: 46 (*Ptilinus*) **E**: AB AR AU BH BU BY CR CT CZ DE EN FR GE GR HU IT LA LT NL NR PL PT RO SK SP ST SV SZ UK YU **N**: EG LB MO **A**: ES FE IN IS JO KZ LE SY TD TR WS
flavicornis Rey, 1892: 3
flavipes Laporte, 1840a: 295
fulvicollis Reitter, 1890c: 196
holosericeus Dufour, 1854: 304
kocheri Pic, 1952: 65
oblongulus Mulsant & Rey, 1864: 279
robustus Pic, 1924f: 29
rufescens Schilsky, 1899: 13
simplex Rey, 1892: 3
latiusculus Kofler, 1970a: 72 **E**: BU GG
longitarsis longitarsis Jansson, 1942: 23, 26 **E**: AU BE BU BY CR CT CZ DE FI GB GE GG HU IT LA NR NT SK ST SV SZ UK **A**: KZ MG TR
fulvipes Arnoldi, 1965: 252
longitarsis teberdaensis Gottwald, 1977: 173 **E**: GG ST
lundbergi Zahradník, 1996a: 96 **A**: TR
maculatus Kiesenwetter, 1877: 146 **E**: BU CT GG HU RO ST UK **A**: KZ TR WS
moraviensis Gottwald, 1977: 165 **E**: AU CZ HU
muehlei Gottwald, 1983: 133 **E**: SP
nardii Zahradník, **nom. nov**. [see New acts] **E**: AB GR
bicolor Zahradník, 1997: 105 [HN]
ornatus Germar, 1842: 24 **E**: CT FR GR HU ST UK **A**: ES KZ MG TR WS
amphitelus Arnoldi, 1965: 250
castor Arnoldi, 1965: 249
mongolicus Español, 1971: 46
purpuricollis Reitter, 1901a: 30
renovatus Gozis, 1882: 201
rufus Schilsky, 1899: 19
sericeus Moravitz, 1861: 289
superbus Roubal, 1942: 24
pectinatus pectinatus Fabricius, 1792a: 244 (*Ptilinus*) **E**: AB AU BE BY CT CZ EN FI FR GE GR HU IT LA LT NL NR NT PL RO SK ST SV SZ UK **A**: ES FE SI WS
denticornis Fabricius, 1792a: 240 (*Ptinus*)

flabellicornis Herbst, 1793: 66 (*Anobium*)
gagliardii Pic, 1933f: 13
schaelkownikowi Pic, 1904f: 34
scutellaris Schilsky, 1899: 17
serraticornis P. Rossi, 1790: 42 (*Anobium*)
pectinatus tapanus Pic, 1933f: 13 **E**: "Caucasus"
planicollis Lohse, 1957: 31, 36 **E**: AU BE CT CZ FI GE LA PL SK ST SV UK **A**: ES MG WS
commodus Arnoldi, 1965: 252
kaszabi Español, 1971: 48
suecicus Lundberg, 1972: 57
pseudoblongulus Gottwald, 1977: 174 **E**: AU CZ SK UK
reichardti Iablokoff-Khnzorian, 1970a: 60 **A**: WS
ruficollis Gebler, 1833: 282 **E**: AU BU CT CZ FR GR HU IT RO SP ST UK **A**: CY ES FE KZ TR WS
discicollis Moravitz, 1861: 288
mariei Méquignon, 1940: 27
rufithorax Lareynie, 1852: xc
thoracicus Frivaldszky von Frivald, 1835: 256
sareptanus Kiesenwetter, 1877: 146 **E**: AB AR BU CR CT FR GG GR HU MC ST UK **A**: ES FE KI KZ TM TR UZ WS
subrotundatus Lareynie, 1852: xci **E**: AU BH BU CR CT CZ FI FR GE GG GR HU IT MC RO SK SP ST SV UK **A**: IS KZ SY TR
tibialis Rey, 1892: 3
tibialis Schilski, 1899: 15 [HN]
tomentosus Sakai, 1975: 97 **A**: JA
tremulicola Kangas, 1958: 28 **E**: FI SV
vaederoeensis Lundberg, 1969: 211 **E**: BU CZ GE SK SV **A**: CY
wewalkai Gottwald, 1977: 169 **A**: IS TR
wuermlii Español, 1977c: 308 **A**: BT
subgenus *Xyletomimus* Reitter, 1901a: 28 type species *Xyletinus sanguineocinctus* Fairmaire, 1859
leprieuri Chobaut, 1894: 107 **E**: SP **N**: AG MO TU
flabellatus Normand, 1950a: 76
sanguineocinctus Fairmaire, 1859c: cv **E**: AB AR CT FR GG PT SP ST UK **N**: AG MO **A**: TR
disconiger Pic, 1906a: 3
miniaticollis Pic, 1902l: 72

genus *Xyletinus*, species incertae sedis
longipennis Gebler, 1833: 281 **A**: WS

tribe Xyletinini, genera incertae sedis
genus *Deroptilinus* Lea, 1924: 40 type species *Deroptilinus granicollis* Lea, 1924
Flabellomesothes Pic, 1936g: 419 type species *Flabellomesothes duboisi* Pic, 1936
alcobei Español, 1974: 58 **A**: BT **ORR**
obscurus Sakai, 1975: 107 **A**: JA

genus *Metholcus* Jacquelin du Val, 1860f: 142 type species *Ptilinus cylindricus* Germar, 1817 (= *Trypopitys phoenicis* Fairmaire, 1859)
arabicus Español, 1979: 254 **A**: SA
buettikeri Toskina, 1998: 55 **A**: SA
gracilicornis Español, 1972b: 68 **N**: MO
gracilipes gracilipes Israelson, 1974: 81 **N**: CI (Gran Canaria)
gracilipes substriatus Israelson, 1974: 84 **N**: CI (Fuerteventura)
phoenicis Fairmaire, 1859a: 53 (*Trypopitys*) **E**: AU BU CR FR GR IT MA SP ST SZ YU **N**: AG EG TU **A**: CY IS LE SY TR **AFR**
abyssinicus Pic, 1903a: 81
cylindricus Germar, 1817: 202 (*Ptilinus*) [HN]
longipennis Chevrolat, 1860: 75 (*Xyletinus*)

madoni Pic, 1930f: 13
raymondi Mulsant & Godart, 1859: 177 (*Trypopitys*)
rotundicollis Schilsky, 1898: 72 **E**: GR **N**: EG **A**: IS LE SY TR
subseriatus J. R. Sahlberg, 1913d: 156

genus *Paraxyletinus* Español, 1972a: 60 type species *Paraxyletinus israelsoni* Español, 1972
israelsoni Español, 1972a: 63 **N**: CI
oculatissimus Israelson, 1974: 86 **N**: CI
ornatus fernandezi Español & Oromí, 1984: 89 **N**: CI
ornatus ornatus Israelson, 1974: 84 **N**: CI

genus *Pseudoptilinus* Leiler, 1969: 57 type species *Ptilinus fissicollis* Reitter, 1877
fissicollis Reitter, 1877d: 24 (*Ptilinus*) **E**: AU BU CT CZ FR GE HU PL RO SK ST SV UK **A**: KI KZ MG TR WS
banatensis Pic, 1912k: 49 (*Ptilinus*)
bernhaueri Reitter, 1916: 85 (*Ptilinus*)
punctatostriatum Faust, 1877a: 35 (*Ptilinus*)

family JACOBSONIIDAE Heller, 1926

genus *Sarothrias* Grouvelle, 1918: 7 type species *Sarothrias eximius* Grouvelle, 1918
Jacobsonium Heller, 1926: 126 type species *Jacobsonium dimerum* Heller, 1926
hygrophilus Pal, 1998: 3 **A**: AP

genus *Derolathrus* Sharp, 1908: 430 type species *Derolathrus atomus* Sharp, 1908
Gomya Dajoz, 1973c: 1051 type species *Gomya insularis* Dajoz, 1973
Lathridiomorphus Franz, 1969a: 136 type species *Lathridiomorphus anophthalmus* Franz, 1969
anophthalmus Franz, 1969a: 137 (*Lathridiomorphus*) **N**: CI
parvulus Rücker, 1983c: 685 (*Gomya*) **N**: MR

series **CUCUJIFORMIA** Lameere, 1938
superfamily **LYMEXYLOIDEA** Fleming, 1821
family LYMEXYLIDAE Fleming, 1821
subfamily **Hylecoetinae** Gistel, 1856

genus *Elateroides* J. C. Schaeffer, 1777: pl. cxxxix type species *Elateroides primus* Schaller, 1783 (= *Cantharis dermestoides* Linnaeus, 1761)
Hylecoetus Latreille, 1806: 266 type species *Cantharis dermestoides* Linnaeus, 1761
Hylecerus Jacquelin du Val, 1860a: 206 type species *Lymexylon flabellicorne* D. H. Schneider, 1791
dermestoides Linnaeus, 1761: 201 (*Cantharis*) **E**: AU BE BY CT CZ DE EN FI FR GB GE HU IR IT LS LT NL NR NT PL RO SK SV SZ **A**: SCH ES FE JA WS
aestivalis Gistel, 1857: 14 (*Hylecoetus*)
aprilinus Gistel, 1857: 14 (*Hylecoetus*)
barbatus Schaller, 1783: 322 (*Mordella*)
cossis Lewis, 1896b: 65 (*Hylecoetus*)
cribricollis Fairmaire, 1889b: 45 (*Hylecoetus*)
francofurthana Herbst, 1784b: 145 (*Lytta*)
fulvipennis Reitter, 1911a: 299 (*Hylecoetus*)
marci Linnaeus, 1767: 681 (*Meloe*)
matsushitai Kôno, 1938: 199 (*Hylecoetus*)
morio Fabricius, 1787a: 165 (*Lymexylon*)
nycturgus Gistel, 1857: 25 (*Hylecoetus*)
primus Schaller, 1783: 322 (*Elateroides*)
proboscideus Fabricius, 1781: 256 (*Lymexylon*)
flabellicornis D. H. Schneider, 1791: 109 (*Lymexylon*) **E**: BY CT EN FI GE LA LT NR NT PL RO **A**: ES FE JA TAI "Korea"
formosanus Miwa, 1935: 456 (*Hylecoetus*)

orientalis Fursov, 1935b: 92 (*Hylecoetus*)

subfamily **Atractocerinae** Laporte, 1840

genus *Arractocetus* Kurosawa, 1985: 112 type species *Atractocerus nipponicus* Nakane, 1984
bruijnii Gestro, 1874: 545 (*Atractocerus*) **A**: SE **ORR**
monticola Kurosawa, 1985: 114 (*Atractocerus*) **A**: TAI
nipponicus Nakane, 1984: 596 (*Atractocerus*) **A**: JA TAI

genus *Atractocerus* Palisot de Beauvois, 1801: 516 type species *Necydalis brevicornis* Linnaeus, 1767
Macrogaster Thunberg, 1805: 281 type species *Necydalis brevicornis* Linnaeus, 1767
reversus Walker, 1858: 285 **A**: HKG **ORR**

genus *Hymaloxylon* Kurosawa, 1985: 112 type species *Atractocerus quercus* Gardner, 1935
aspoecki Paulus, 2004: 282 **A**: YUN
quercus Gardner, 1935: 69 (*Atractocerus*) **A**: NP UP

genus *Raractocetus* Kurosawa, 1985: 111 type species *Atractocerus emarginatus* Laporte, 1838
emarginatus Laporte, 1838: 59 (*Atractocerus*) **A**: TAI **ORR**
celebensis Gestro, 1874: 545 (*Atractocerus*)
debilis Walker, 1858: 285 (*Atractocerus*)
fissicollis Fairmaire, 1885: cviii (*Atractocerus*)
horni Bourgeois, 1905d: 133 (*Atractocerus*)
luteolus Fairmaire, 1882a: 217 (*Atractocerus*)
mirabilis Miwa, 1935: 457 (*Atractocerus*)

genus *Urtea* Paulus, 2004: 280 type species *Urtea graeca* Paulus, 2004
graeca Paulus, 2004: 281 **E**: GR

subfamily Atractocerinae, species incertae sedis
Atractocerus atricollis Pic, 1955c: 21 **A**: FUJ

subfamily **Lymexylinae** Fleming, 1821

genus *Lymexylon* Fabricius, 1775: 204 type species *Cantharis navalis* Linnaeus, 1758
Calolymus Thunberg, 1791: 26 type species *Cantharis navalis* Linnaeus, 1758
Pterophorus Herbst, 1784b: 105 [HN]
amamianum Kurosawa, 1985: 116 **A**: JA TAI
miyakei Nakane, 1985a: 28 **A**: JA (Amami-Ôshima) TAI
navale Linnaeus, 1758: 403 (*Cantharis*) **E**: AU BE BY CZ DE EN FR GE GB GG GR HU IT LA LT NL PL RO SK SV ST SZ TR UK **N**: MR **A**: TR "Korea"
flavipes Fabricius, 1787a: 165
saxonica Gmelin, 1790: 1900 (*Cantharis*)
persicum Fursov, 1935b: 92 **E**: ST
ruficolle Kurosawa, 1949: 27 **A**: JA

subfamily **Melittommatinae** Wheeler, 1986

genus *Melittomma* Murray, 1867b: 314 type species *Hylecaetus brasiliensis* Laporte, 1832
Neohylecoetus Kurosawa, 1956: 80 type species *Hylecaetus javanicus* Chevrolat, 1844
oculare Nakane, 1963c: 221 (*Lymexylon*) **A**: JA

superfamily **CLEROIDEA** Latreille, 1802

family PHLOIOPHILIDAE Kiesenwetter, 1863

genus *Phloiophilus* Stephens, 1830: 81 type species *Phloiophilus edwardsii* Stephens, 1830
edwardsii Stephens, 1830: 81 **E**: AU BE CZ DE FR GB GE HU IR NL PL SK SV SZ
bimaculatus Stephens, 1830: 82

cooperi Stephens, 1830: 82

family TROGOSSITIDAE Latreille, 1802
subfamily **Trogossitinae** Latreille, 1802
tribe Calityni Winkler, 1924

genus *Calitys* C. G. Thomson, 1859: 71 type species *Hispa scabra* Thunberg, 1784
scabra Thunberg, 1784a: 15 (*Hispa*) **E**: AU BE BU CR CT CZ EN FI FR GE GR IT LA LT LU NL NR NT PL RO RU SK SL SV SZ UK YU **N**: "North Africa" **A**: ES FE WS

tribe Gymnochilini Lacordaire, 1854

genus *Anacypta* Illiger, 1807b: 301 type species *Nitidula punctata* Fabricius, 1801
Acrops Dalman, 1824: 15 type species *Acrops metallica* Dalman, 1824 (= *Nitidula punctata* Fabricius, 1801)
cicatricosa Reitter, 1880a: 29 (*Acrops*) **A**: "Himalaya" **ORR**
rugosa Léveillé, 1888: 606 (*Acrops*)
higonia Lewis, 1888: 220 (*Acrops*) **A**: JA

genus *Leperina* Erichson, 1844: 453 type species *Trogosita decorata* Erichson, 1842
Onyschomorpha Arrow, 1900: 94 type species *Onyschomorpha marmorata* Arrow, 1900
squamulata Gebler, 1830: 97 (*Peltis*) **A**: ES FE JA MG NMO NC NE
tibialis Reitter, 1889i: 217 **A**: JA

genus *Xenoglena* Reitter, 1876a: 40 type species *Xenoglena deyrollei* Reitter, 1876
quadrisignata Mannerheim, 1852b: 303 (*Gymnochila*) **A**: FE JA MG NC NE
yunnanensis Léveillé, 1907: 406 **A**: YUN

tribe Trogossitini Latreille, 1802

genus *Alindria* Erichson, 1844: 451 type species *Trogossita grandis* Le Peletier de Saint-Fargeau & Audinet-Serville, 1825
auberti Léveillé, 1905: 92 **A**: SCH
orientalis L. Redtenbacher, 1844: 549 (*Trogosita*) **A**: KA
parallela Léveillé, 1889: 411
virescens Léveillé, 1907: 401 **A**: "India"

genus *Leipaspis* Wollaston, 1862a: 140 type species *Leipaspis lauricola* Wollaston, 1862
caulicola caulicola Wollaston, 1862a: 142 **N**: CI
caulicola oceanica Wollaston, 1865: 116 **N**: MR (Selvagens)
lauricola gomerensis Plata & Prendes, 1981: 229 **N**: CI (El Hierro, La Gomera)
lauricola lauricola Wollaston, 1862a: 142 **N**: CI (La Palma, Tenerife)
pinicola Wollaston, 1862a: 143 **N**: CI

genus *Melambia* Erichson, 1844: 450 type species *Trogosita gigas* Fabricius, 1798
gautardi Tournier, 1872b: 441 **N**: EG
tekkensis Koenig, 1889a: 301 **A**: "Transcaspia"

genus *Nemozoma* Latreille, 1804b: 239 type species *Dermestes elongatus* Linnaeus, 1761
Aponemosoma Lepesme & Paulian, 1944: 141 type species *Nemosoma caucasicum* Ménétriés, 1832
Cylidrella Sharp, 1891: 389 type species *Cylidrella mollis* Sharp, 1891
Monesoma Léveillé, 1895: cxliv type species *Nemosoma cornutum* Sturm, 1826
Nemosomia Reitter, 1876a: 11 type species *Nemosomia vorax* Reitter, 1876
Paranemosoma Lepesme & Paulian, 1944: 140 type species *Nemosoma punctatum* Léveillé, 1886
Pseudalindria Fall, 1910: 126 type species *Pseudalindria fissiceps* Fall, 1910
Sturmia Ragusa, 1892: 194 type species *Nemosoma cornutum* Sturm, 1826
caucasicum Ménétriés, 1832: 224 **E**: AU PL ST SK UK
fasciicole Hampe, 1864: 193

cornutum Sturm, 1826: 77 **E**: ST UK
elongatum Linnaeus, 1761: 41 (*Dermestes*) **E**: AU BE BH BU BY CR CT CZ DE EN FI FR GE GR HU IT LA LT NR PL PT RO RU SK SL SP SV SZ UK YU **N**: TU **A**: SY TR
corsicum Reitter, 1875c: 13
fasciatum Panzer, 1796a: 22 (*Colydium*)
siculum Ragusa, 1891: 193
syriacum Pic, 1901e: 33
tuniseum Pic, 1900h: 61 (*Nemosoma*)

genus *Seidlitzella* Jakobson, 1915: 893 [RN] type species *Peltis procera* Kratz, 1858
Cymba Seidlitz, 1875: 34 [HN] type species *Peltis procera* Kratz, 1858
Peltocymba Reitter, 1920: 220 type species *Peltis procera* Kratz, 1858
procera Kraatz, 1858: 136 (*Peltis*) **E**: GR **A**: CY TR

genus *Temnoscheila* Westwood, 1830: 231 type species *Trogossita caerulea* A. G. Olivier, 1790
Temnochila Erichson, 1844: 449 [unjustified emendation]
caerulea A. G. Olivier, 1790a: no 19: 6 (*Trogossita*) **E**: AU BH BU CR FR GR PT SK SP ST SV SZ **N**: AG CI TU **A**: CH
gemella Bedel, 1900: 335 (*Trogosita*)
pini Brullé, 1838: 70 (*Trogosita*)
rogenhoferi Reitter, 1875c: 19 (*Trogosita*)
japonica Reitter, 1875c: 20 (*Trogosita*) **A**: FE JA NC NE
tristis Mulsant & Rey, 1853b: 8 (*Trogosita*) **E**: IT **NTR**
cribricollis Reitter, 1875c: 14 (*Trogosita*)

genus *Tenebroides* Piller & Mitterpacher, 1783: 87 type species *Tenebrio mauritanicus* Linnaeus, 1758
latens Wollaston, 1862a: 143 **N**: CI
maroccanus Reitter, 1884e: 254 (*Tenebrioides*) **E**: AZ FR SP **N**: MO
mauritanicus Linnaeus, 1758: 417 (*Tenebrio*) **E**: AU AZ BE BU BY CR CZ DE EN FI FR GB GE GR HU IC IT LA LT NL NR PL PT RO RU SK SL SP SV SZ UK YU **N**: AG CI EG LB MO TU **A**: ANH ES FE FUJ GUA GUI HEN HKG HUB JA JIA JIX MNO SCH TAI WS **AFR AUR NTR ORR**
bucephalus Herbst, 1784b: 141 (*Carabus*)
caraboides Linnaeus, 1758: 418 (*Tenebrio*)
complanatus Piller & Mitterpacher, 1783: 87
dubius Scriba, 1790a: 42 (*Lucanus*)
fuscus Preyssler, 1790: 6 (*Lucanus*)
nitidus Horn, 1862: 83 (*Trogosita*)
piceus Dalla Torre, 1879: 32
striatus Geoffroy, 1785: 3 (*Platycerus*)
rectus Wollaston, 1862a: 144 **N**: CI
turkestanicus Ballion, 1871: 330 (*Trogosita*) **A**: "Turkestan"

subfamily **Peltinae** Kirby, 1837

tribe Ancyronini Kolibáč, 2006

genus *Ancyrona* Reitter, 1876a: 51 type species *Ancyrona lewisi* Reitter, 1876
diversa Pic, 1921a: 1 (*Ostoma*) **A**: FE
haroldi Reitter, 1877i: 375 **A**: JA
japonica Reitter, 1889i: 217 (*Ostoma*) **E**: BU HU SK **A**: JA
lewisi Reitter, 1876a: 52 **A**: JA
shibatai Nakane, 1963b: 46 **A**: JA

tribe Lophocaterini Crowson, 1964

genus *Grynocharis* C. G. Thomson, 1859: 71 type species *Silpha oblonga* Linnaeus, 1758
oblonga Linnaeus, 1758: 362 (*Silpha*) **E**: AU BU BY CR CT CZ EN FI FR GE GR IT LA LT NR NT PL RO RU SK SL SV SZ UK YU
pubescens Erichson, 1844: 457 (*Peltis*) **E**: GG ST (Kavkaz) UK (Krym)

genus *Grynocharis*, nomen dubium
caucasica Motschulsky, 1863: 508 **E**: "Caucasus"

genus *Lophocateres* Olliff, 1883b: 180 type species *Lophocateres nanus* Olliff, 1883 (= *Peltis pusillus* Klug, 1833)
pusillus Klug, 1833: 71 (*Peltis*) **E**: BE FR GE NL PL SP **N**: AG CI EG **A**: GUA GUI GUX HUJ HEB HUB IS JIA JIX SC SCH SD TAI YUN ZHE
nanus Olliff, 1883b: 180 (*Lophocateres*)

tribe Peltini Kirby, 1837

genus *Peltis* O. F. Müller, 1764: 13 type species *Silpha grossa* Linnaeus, 1758
Gaurambe C. G. Thomson, 1862: 189 type species *Silpha ferruginea* Linnaeus, 1758
Ostoma Laicharting, 1781: 102 type species *Silpha ferruginea* Linnaeus, 1758
Zimioma Gozis, 1886: 11 type species *Silpha grossa* Linnaeus, 1758
ferruginea Linnaeus, 1758: 361 (*Silpha*) **E**: AU BY CT CZ DE EN FI FR GE HU LA LS LT NR NT PL RO RU SK SL SP ST SV SZ UK **A**: ES FE JA NC WS **NAR**
cimicoides DeGeer, 1774: 183 (*Silpha*)
fraterna Randall, 1838a: 17 (*Peltis*)
nigricans Dalla Torre, 1879: 32
nigrina Casey, 1916: 284
rubicunda Laicharting, 1781: 102
septentrionalis Randall, 1838a: 17 (*Peltis*)
gigantea Reitter, 1882b: 147 (*Ostoma*) **A**: ES FE JA NE
grossa Linnaeus, 1758: 361 (*Silpha*) **E**: AU BY CT CZ EN FI FR GE LA LT NR NT PL RO SK SL SP ST SV SZ UK
jakowlewi Semenov, 1898c: 283 (*Ostoma*) **E**: ST
valida Lewis, 1894b: 34 **A**: JA

tribe Thymalini Léveillé, 1888

genus *Thymalus* Latreille, 1802: 133 type species *Peltis brunnea* Thunberg, 1794 (= *Cassida limbata* Fabricius, 1787)
Thymalops Iablokoff-Khnzorian, 1962b: 424 type species *Cassida limbata* Fabricius, 1787
aubei Léveillé, 1877: 155 **E**: GG "Caucasus"
subtilis Reitter, 1889m: 278
chinensis Fairmaire, 1900: 616 **A**: FUJ
laticeps Lewis, 1894b: 33 **A**: JA
limbatus Fabricius, 1787a: 63 (*Cassida*) **E**: AU BE BU BY CR CT CZ DE EN FI FR GE IT LA LS LT NL NR NT PL PT RO SK SL SP SV SZ UK **N**: TU
brunneus Thunberg, 1794: 104 (*Peltis*)
rubiginosus Gmelin, 1790: 1643 (*Cassida*)
oblongus Reitter, 1889m: 278 **E**: CT NT SV **A**: ES
parviceps Lewis, 1894b: 33 **A**: JA
punctidorsum Lewis, 1894b: 33 **A**: JA

family THANEROCLERIDAE Chapin, 1924

tribe Thaneroclerini Chapin, 1924

subtribe Isoclerina Kolibáč, 1992

genus *Isoclerus* Lewis, 1892: 191 type species *Isoclerus pictus* Lewis, 1892
Lyctosoma Lewis, 1892: 192 type species *Lyctosoma parallelum* Lewis, 1892
disinlei Kolibáč, 1992: 325 **A**: TAI
elongatus Schenkling, 1906: 278 (*Thaneroclerus*) **A**: YUN
parallelus Lewis, 1892: 192 (*Lyctosoma*) **A**: JA SCH **ORR**
parvum Schenkling, 1916: 222 (*Thaneroclerus*)
pictus Lewis, 1892: 191 **A**: JA

subtribe Thaneroclerina Chapin, 1924

genus *Neoclerus* Lewis, 1892: 190 type species *Neoclerus ornatulus* Lewis, 1892
ornatulus Lewis, 1892: 190 **A**: JA TAI

genus *Thaneroclerus* Lefebvre, 1838: 13 type species *Clerus buquet* Lefebvre, 1835
aino Lewis, 1892: 190 **A**: JA
buquet Lefebvre, 1835: 577 (*Clerus*) **E**: BE GE HU PL NL **N**: EG **A**: AP BEI FUJ GUA GUX HEN HUN JA KA LIA MNO RU SA SCH SD TAI YUN ZHE
  dermestoides Klug, 1842: 310 (*Clerus*)
  girodi Chevrolat, 1880: 31
  tabacci Matsumura, 1935: 234 (*Metademius*)
quasitardatus Corporaal, 1939c: 349 **A**: KA

family CLERIDAE Latreille, 1802
subfamily **Tillinae** Fleming, 1821

genus *Cladiscus* Chevrolat, 1843: 33 type species *Cladiscus strangulatus* Chevrolat, 1843
ater Pic, 1926h: 21 **A**: CH
attenuatus Gorham, 1893: 569 **A**: YUN **ORR**
obeliscus Lewis, 1892: 185 **A**: JA NP SCH YUN **ORR**
  strangulata Kiesenwetter, 1879b: 314 (*Cymatodera*) [HN]
pallidicornis Corporaal & van der Wiel, 1949: 192 **A**: TAI
pectinicornis Klug, 1842: 270 (*Tillus*) **A**: UP
sanguinicollis Spinola, 1844a: 125 (*Tillus*) **A**: SD TAI **ORR**
sauteri Schenkling, 1912b: 322 **A**: TAI
terminalis Schenkling, 1912b: 322 **A**: TAI
yunnanus Corporaal & van der Wiel, 1949: 192 **A**: YUN

genus *Cylidroctenus* Kraatz, 1899b: 214 type species *Tillus chalybaeus* Westwood, 1849
chalybaeum Westwood, 1849: 51 (*Tillus*) **A**: NP SD **ORR**

genus *Cylidrus* Latreille, 1829a: 476 type species *Clerus cyaneus* Fabricius, 1787
beesoni Corporaal, 1926: 209 **A**: UP **ORR**
cyaneus Fabricius, 1787a: 126 (*Clerus*) **A**: JA TAI **AFR AUR ORR**
  alcyoneus Pascoe, 1860: 44
  coeruleus Spinola, 1844a: 85
  pallipes Chevrolat, 1874b: 280
  vescoi Fairmaire, 1849b: 361
megacephalus Spinola, 1844a: 88/89 **N**: EG **A**: IS **AFR**

genus *Denops* Fischer von Waldheim, 1829b: 65 type species *Denops longicollis* Fischer von Waldheim, 1829 (= *Tillus albofasciatus* Charpentier, 1825)
  Dinops Agassiz, 1846: 119 [unjustified emendation]
albofasciatus Charpentier, 1825: 198 (*Tillus*) **E**: AR AU BU CR CZ FR GE GG GR HU IT PT RO SK SL SP ST UK (Krym) YU "Caucasus" **N**: AG EG LB MO TU **A**: IS LE
  agilis P. H. Lucas, 1843c: xxv (*Cylidrus*)
  longicollis Fischer von Waldheim, 1829b: 65
  personatus Aragona, 1830: 14 (*Tillus*)
  rubrofasciatus Pic, 1894b: 27
  spinolai Ragusa, 1896b: 82
  tenuicollis Spinola, 1844a: 90 (*Tillus*)
canariensis Palm, 1978: 36 **N**: CI (Gran Canaria)

genus *Dedana* Fairmaire, 1888a: 26 type species *Dedana rufodorsata* Fairmaire, 1888
rufodorsata Fairmaire, 1888a: 26 **A**: FUJ

genus *Diplopherusa* Heller, 1922: 530 type species *Diplopherusa tumidipes* Heller, 1922
kitamurai Nakane, 1989c: 149 **A**: JA
rosti Schenkling, 1908a: 362 (*Cladiscus*) **A**: HP
shibatai Nakane, 1963b: 46 (*Ancyrona*) **A**: JA

genus *Elasmocylidrus* Corporaal, 1939a: 18 type species *Cylidrus tricolor* Corporaal, 1926
tricolor Corporaal, 1926: 210 (*Cylidrus*) **A**: UP

genus *Eucymatodera* Schenkling, 1899a: 333 type species *Eucymatodera variabilis* Schenkling, 1899
senegalensis Laporte, 1832: 399 (*Tilloidea*) **N**: AG EG **A**: IN SA **AFR**
hottentotta Kuwert, 1893a: 469 (*Cymatodera*)
pubescens Laporte, 1838: 37 (*Tilloidea*)
rubricollis Guérin-Méneville, 1835a: pl. 15 (*Tillus*)
rufoscutus Pic, 1932f: 25 (*Tillus*)

genus *Falsotillus* Gerstmeier & Kuff, 1992: 70 type species *Tillus igarashii* Kôno, 1930
igarashii Kôno, 1930: 134 (*Tillus*) **A**: FE JA "Korea"

genus *Flabellotilloidea* Gerstmeier & Kuff, 1992: 67 type species *Tillus palaestinus* Pic, 1900
bayonnei Chobaut, 1897d: 241 (*Phloeocopus*) **N**: AG MO
palaestina Pic, 1900a: 2 (*Tillus*) **A**: IS JO TR
vaulogeri Pic, 1906e: 25 (*Tillus*) **N**: TU

genus *Gastrocentrum* Gorham, 1876: 63 type species *Gastrocentrum pauper* Gorham, 1876
nitidum Schenkling, 1916: 117 **A**: TAI
unicolor White, 1849: 56 (*Notoxus*) **A**: TAI **ORR**
pauper Gorham, 1876: 63

genus *Gracilotillus* Pic, 1933c: 16 type species *Tillus semicyaneus* Pic, 1933
fasciatus Schenkling, 1908a: 362 (*Cladiscus*) **A**: HP UP
shibatai Miyatake, 1965a: 21 **A**: JA (Ryukyus)

genus *Nodepus* Gorham, 1892c: 719 type species *Nodepus conicicollis* Gorham, 1892
conicicollis Gorham, 1892c: 720 **A**: UP "India" **ORR**

genus *Prioceromorphus* Pic, 1950d: 23 type species *Priocleromorphus flavofasciatus* Pic, 1950
flavofasciatus Pic, 1950d: 24 **N**: EG

genus *Orthocladiscus* Corporaal & van der Wiel, 1948: 181 type species *Cladiscus longipennis* Westwood, 1849
longipennis Westwood, 1849: 52 (*Cladiscus*) **A**: SD

genus *Paracladiscus* Miyatake, 1965a: 19 type species *Paracladiscus atricolor* Miyatake, 1965
atricolor Miyatake, 1965a: 20 **A**: JA (Ryukyus)

genus *Spinoza* Lewis, 1892: 184 type species *Spinoza caerulea* Lewis, 1892
Melanoclerus Chapin, 1919: 226 type species *Melanoclerus parva* Chapin, 1919
caerulea Lewis, 1892: 185 **A**: JA

genus *Teloclerus* Schenkling, 1903b: 12 type species *Tillus uniformis* Westwood, 1852
Hendecatus Schenkling, 1906: 246 type species *Tillus subnotatus* Westwood, 1852
compressicornis Klug, 1842: 273 (*Tillus*) **E**: IT (Sardegna) PT SP **N**: AG MO TU **A**: IS JA SA **AFR**
belamyi Fairmaire, 1891a: 247 (*Macrotelus*)
pallidicolor Reitter, 1894d: 46 (*Phloeocopus*)
rufopiceus Schenkling, 1899a: 332 (*Tillus*)

subnotatus Westwood, 1852: 37 (*Tillus*)

genus *Tillodenops* Hintz, 1905: 305 type species *Tillodenops brunneus* Hintz, 1905
bimaculatus Schenkling, 1899b: 545 (*Denops*) **N**: EG **A**: SA **AFR**
angustatus Pic, 1922f: 21 (*Cylidrus*)
plagiatus Fairmaire, 1892a: 102 (*Denops*) **A**: IN OM SA **AFR**

genus *Tilloidea* Laporte, 1832: 398 type species *Clerus unifasciatus* Fabricius, 1787
birmanica Gorham, 1892c: 729 (*Tillus*) **A**: CH TAI **ORR**
iranica Gerstmeier & Kuff, 1992: 63 **A**: IN
laevigata Peyerimhoff, 1927: 237 (*Tillus*) **N**: AG MO
munin Nakane, 1977b: 157 **A**: JA (Bonin Is.)
notata Klug, 1842: 276 (*Tillus*) **A**: FUJ GUA GUX HEB HEN HUB JA MNO NP SCH SD TAI UP ZHE "Korea" **ORR**
basithorax Pic, 1927c: 1 (*Tillus*)
bipartita Blanchard, 1853: 59 (*Tillus*)
distincta White, 1849: 49 (*Tillus*)
lewisii Gorham, 1878: 160 (*Tillus*)
lewisii Kiesenwetter, 1879b: 313 (*Tillus*) [HN]
multicolor Fairmaire, 1886b: 43 (*Tillus*)
multipartita Fairmaire, 1889a: 353 (*Tillicera*)
quadricolor Heller, 1921: 534 (*Tillus*)
semperana Gorham, 1876: 62
tristis Schenkling, 1899a: 332 (*Tillus*)
transversalis Charpentier, 1825: 199 (*Clerus*) **E**: BU HU CZ FR GR IT MA MC PT SK SP **N**: AG EG MO TU **A**: TR
myrmecodes Spinola, 1844: 102 (*Tillus*)
unifasciata Fabricius, 1787a: 125 (*Clerus*) **E**: AB AR AU BE BH BU CR CT CZ FR GE GG GR HU IT MC PT RO SK SL SP ST SZ UK YU **N**: EG **A**: IN TR UP
cingulata Corporaal, 1939a: 22 (*Tillus*)
elbursiaca Winkler, 1982a: 7
fasciata Geoffroy, 1785: 135 (*Clerus*) [HN]
laterufa Corporaal, 1949b: 420 (*Tillus*) [RN]
minor Sulzer, 1776: 42 (*Attelabus*)
serraticornis Villers, 1789: 222 (*Attelabus*)
tricolor Spinola, 1844a: 97 (*Tillus*) [HN]

genus *Tillus* A. G. Olivier, 1790a: no 22: 3 type species *Chrysomela elongata* Linnaeus, 1758
Cylinder Voet, 1796: 78 type species *Cylinder coeruleus* Voet, 1796
Thanasimomorpha Blackburn, 1891b: 304 type species *Tillus bipartitus* Blanchard, 1853
concolor Nakane, 1996: 134 **A**: TAI
discoidalis Fairmaire, 1891f: ccx **A**: HUB
elongatus Linnaeus, 1758: 377 (*Chrysomela*) **E**: AB AR AU BE BH BU BY CR CT CZ DE EN FI FR GB GE GG HU IR IT LA LS LT NL NR NT PL RO SK SL SP ST SV SZ UK YU (Crna Gora) "Caucasus" **A**: FE KZ
ambulans Fabricius, 1787a: 93 (*Lagria*)
ater Fabricius, 1792b: 80 (*Lagria*)
aterrimus Eschscholtz, 1818: 459
bimaculatus Donovan, 1807: 50
coeruleus Voet, 1796: 78 (*Cylinder*)
comma Petri, 1912: 146
fleischeri Roubal, 1936a: 29
hyalinus Sturm, 1826: 202 [RN]
marchiae Gmelin, 1790: 1731 (*Lagria*)
niger Ragusa, 1896b: 82
ruficollis Herbst, 1783: 68 (*Lagria*)
rugulosus Henschel, 1861: 45
ibericus Bahillo de la Puebla, López-Colón & García-París, 2003: 58 **E**: SP
flabellicornis Fairmaire, 1866a: 22 **N**: AG MO

mozabitus Chobaut, 1897d: 240 **E**: AG
nakamurai Nakane, 1996: 133 **A**: TAI
pallidipennis espinosai Winkler, 1985: 119 **E**: IT (Sicilia)
pallidipennis pallidipennis Bielz, 1850: 179 **E**: AU BU GR HU IT RO SK "Caucasus" **A**: TR
pectinicornis Abeille de Perrin, 1893a: ccxxxvi **E**: GR (Kríti) **A**: IN TR
holtzi Hintz, 1902: 228
succinta Spinola, 1844a: 100 **A**: UP "Indes Orientales"
vicarius Corporaal, 1935: 34 **A**: HP

genus *Wittmeridecus* Winkler, 1981: 252 type species *Phloeocopus mediozonatus* Fairmaire, 1892
mediozonatus Fairmaire, 1892a: 103 (*Phloeocopus*) **E**: IT SP **N**: AG MO **A**: IS SA SY **AFR**
apicipennis Pic, 1932b: 8 (*Tillus*)

subfamily **Hydnocerinae** Spinola, 1844

tribe Hydnocerini Spinola, 1844

genus *Emmepus* Motschulsky, 1845a: 41 type species *Emmepus arundinis* Motschulsky, 1845
Afghanobaenus Winkler, 1974: 397 type species *Afghanobaenus jarmilae* Winkler, 1974
Brachyclerus Fairmaire, 1883d: clvii type species *Brachyclerus bonnairii* Fairmaire, 1883
Kaszabaenus Winkler, 1974: 392 *Kaszabaenus sensillifer* Winkler, 1974
Sinobaenus Winkler, 1978: 247 type species *Sinobaenus sedlaceki* Winkler, 1978
arundinis Motschulsky, 1845a: 42 **E**: AR **A**: KZ MG TM
bonnairii Fairmaire, 1883d: clvii (*Brachyclerus*) **N**: AG EG
juncicola Abeille de Perrin, 1898c: 255
nigroapicalis Pic, 1934h: 454
elegans Fursov, 1936b: 352 **A**: TD
emiliae Winkler, 1974: 404 (*Afghanobaenus*) **A**: AF
jarmilae Winkler, 1974: 398 (*Afghanobaenus*) **A**: AF
maowenensis Kolibáč, 1998: 139 **A**: SCH
ottonis Winkler, 1974: 401 (*Afghanobaenus*) **A**: AF
pallidipennis Pic, 1898a: 2 **N**: AG
pilgrim Kolibáč, 1998: 141 **A**: TAI
sedlaceki Winkler, 1978: 247 (*Sinobaenus*) **A**: ZHE
sensillifer Winkler, 1974: 394 (*Kaszabaenus*) **A**: AF

genus *Neohydnus* Gorham, 1892c: 742 type species *Neohydnus despectus* Gorham, 1892
Laiomorphus Pic, 1927e: 20 type species *Laiomorphus bicolor* Pic, 1927
aritai Miyatake, 1985b: 154 **A**: JA
coomani Pic, 1934f: 133 **A**: GUX **ORR**
formosanus Miyatake, 1965c: 158 **A**: TAI
hozumii Nakane, 1981b: 128 **A**: JA
jiuzhaigouensis Kolibáč, 1998: 139 **A**: SCH
latior Pic, 1940b: 7 **A**: CH
shirozui Miyatake, 1965c: 157 **A**: TAI
sinensis Pic, 1954e: 62 **A**: FUJ

tribe Callimerini Kolibáč, 1998

genus *Callimerus* Gorham, 1876: 64 type species *Xylobius dulcis* Westwood, 1852
Brachycallimerus Chapin, 1924: 180 type species *Callimerus latifrons* Chapin, 1924
Caloclerus Kuwert, 1893b: 480 type species *Caloclerus mirabilis* Kuwert, 1893
Crassocallimerus Pic, 1929c: 16 type species *Callimerus latesignatus* Gorham, 1892
Cucujocallimerus Pic, 1929c: 16 type species *Callimerus coomani* Pic, 1925
Stenocallimerus Corporaal & Pic, 1940: 189 type species *Callimerus prasinatus* Lewis, 1896
albofasciatus J. Thomson, 1860: 61 (*Lemidia*) **A**: "North India"
albovarius Westwood, 1849: 50 (*Xylobius*) **A**: "Himalaya"

amabilis Gorham, 1876: 66 **A**: "Himalaya" **ORR**
scutellaris Schenkling, 1901: 105 **A**: SD
apicalis Pic, 1954: 61 **A**: FUJ
benedictus Gorham, 1893: 573 **A**: "Himalaya"
pernix Schenkling, 1915: 112
bicoloripes Pic, 1925e: 32 **A**: TAI
bicoloritarsis Pic, 1925e: 32 **A**: YUN
chinensis Schenkling, 1915: 111 **A**: GUA GUI JIX SHN
cyaneus J. Thomson, 1860: 62 (*Lemidia*) **A**: CH
decoratus Gorham, 1893: 570 (*Lemidia*) **A**: BT SD **ORR**
tredecimpunctatus Kuwert, 1893b: 484
distinctus Corporaal, 1939a: 29 **A**: UP
dorsalis J. Thomson, 1860: 61 (*Lemidia*) **A**: "North India"
duodecimpustulatus Schenkling, 1912b: 323 **A**: TAI "North India"
grandjeani Schenkling, 1901: 104 **A**: "Himalaya" **ORR**
hydnoceroides Gorham, 1893: 574 **A**: SD **ORR**
albosuturatus Kuwert, 1893b: 482
insignitus Lesne, 1928: 35 **A**: YUN
ishigakensis Miyatake, 1985b: 155 **A**: JA
koenigi Pic, 1954d: 58 **A**: FUJ
subattenuatus Pic, 1954d: 59
maderi Corporaal, 1939b: 189 **A**: ZHE
micans Corporaal & Pic, 1940: 190 (*Stenocallimerus*) **A**: GUX **ORR**
nigroapicalis Pic, 1955c: 23 **A**: FUJ
okinawanus Yajima & Nakane, 1969a: 5 (*Stenocallimerus*) **A**: JA (Okinawa)
perroti Pic, 1939c: 19 **A**: GUX **ORR**
prasinatus Lewis, 1896c: 338 (*Callimerus*) **A**: CH JA
nigripes Corporaal, 1926: 177 (*Callimerus*)
ryukyuensis Miyatake, 1985b: 154 **A**: JA
sauteri Corporaal & Pic, 1940: 192 **A**: TAI
taiwanus Miyatake, 1965c: 159 (*Stenocallimerus*) **A**: TAI
tenuissimus Corporaal, 1926: 176 (*Callimerus*) **A**: SD

subfamily **Clerinae** Latreille, 1802

genus *Allonyx* Jacquelin du Val, 1860a: 196 type species *Attelabus qudrimaculatus* Schaller, 1783
quadrimaculatus Schaller, 1783: 288 (*Attelabus*) **E**: AU BE CR CZ FR GE GR HU IT MC NL PL PT SK SP SZ **A**: CY TR
cilicensis Pic, 1917h: 17
frontalis Méquignon, 1949: 114
maculatus Geoffroy, 1785: 136 (*Clerus*)

genus *Anthicoclerus* Schenkling, 1906: 264 type species *Clerus anthicoides* Westwood, 1849
anthicoides Westwood, 1849: 59 (*Clerus*) **A**: SD TAI **ORR**
ruficollis Corporaal, 1926: 214 **A**: UP
ruficeps Corporaal, 1926: 214

genus *Astigmus* Kuwert, 1894c: 399 type species *Astigmus ventralis* Kuwert, 1894
delatouchei Fairmaire, 1887a: 126 (*Stigmatium*) [DA] **A**: FUJ

genus *Bousquetoclerus* Menier, 1997: 158 type species *Bouquetoclerus arachnoides* Menier, 1997
arachnoides Menier, 1997: 158 **A**: YUN

genus *Canariclerus* Winkler, 1982b: 475 type species *Clerus paivae* Wollaston, 1862
paivae Wollaston, 1862a: 163 (*Clerus*) **N**: CI

genus *Clerus* Geoffroy, 1762: 34 type species *Clerus mutillarius* Fabricius, 1775
Pseudoclerops Jacquelin du Val, 1860a: 196 type species *Clerus mutillarius* Fabricius, 1775
dealbatus Kraatz, 1879a: 128 (*Pseudoclerops*) **A**: FE FUJ JIL SC SHG **ORR**
moutoni Pic, 1936b: 3 (*Thanasimus*)
formosanus Nakane, 1996: 136 **A**: TAI
guishanensis Nakane, 1996: 138 **A**: TAI
intermedius Nakane, 1996: 138 **A**: TAI
klapperichi Pic, 1954e: 61 (*Pseudoclerops*) **A**: FUJ
mutillaeformis Reitter, 1895a: 83 **A**: TR
mutillarius africanus Kocher, 1955a: 39 **N**: AG MO TU
mutillarius mutillarius Fabricius, 1775: 157 **E**: AU BE BH BU CR CT CZ FR GE GG GR HU IT PL PT RO SK SL SP ST SZ UK YU
alboapicalis Pic, 1950b: 7 (*Pseudoclerops*)
fasciatus Geoffroy, 1785: 135
formicaroides Schrank, 1776: 65 (*Attelabus*)
mutillaroides Reitter, 1894d: 47 **E**: AB AR **A**: TR
mendax Kuwert, 1894c: 404 (*Pseudoclerops*)
sinae Chapin, 1927: 20 (*Pseudoclerops*) **A**: FUJ ZHE
luteofasciatus Pic, 1940b: 7 (*Pseudoclerops*)
sinensis Pic, 1931a: 7 (*Pseudoclerops*)
thanasimoides Chevrolat, 1874b: 313 **A**: AF HKG YUN **ORR**

genus *Coptoclerus* Chapin, 1924: 209 type species *Coptoclerus sericeus* Chapin, 1924
gressitti Miyatake, 1985b: 157 **A**: JA

genus *Ekisius* Winkler, 1987: 170 type species *Ekisius vitreus* Winkler, 1987
vitreus Winkler, 1987: 170 **A**: YUN

genus *Jenjouristia* Fursov, 1936b: 351 type species *Jenjouristia philippovi* Fursov, 1936
philippovi Fursov, 1936b: 351 **A**: FE

genus *Omadius* Laporte, 1838: 48 type species *Omadius indicus* Laporte, 1838
Falsomadius Gerstmeier, 2002: 110 type species *Ommadius aurulentus* Heller, 1921
Ommadius Gorham, 1895a: 296 [unjustified emendaton]
mediofasciatus Westwood, 1852: 44 (*Clerus*) **A**: HAI SD **ORR**
coomani Pic, 1933c: 21 (*Ommadius*)
latifascia Kuwert, 1894b: 79 (*Ommadius*)
nigropunctatus Chevrolat, 1874b: 318
posticalis Gorham, 1876: 105
nigromaculatus Lewis, 1892: 187 **A**: JA TAI
tricinctus Gorham, 1892c: 742 (*Ommadius*) **A**: SD TAI **ORR**
pectoralis Schenkling, 1912b: 326 (*Ommadius*)
zebratus Westwood, 1852: 43 (*Clerus*) **A**: "India Orientali" **ORR**
genus *Omadius*, species incertae sedis
alishanus Nakane, 1996: 138 **A**: TAI

genus *Opilo* Latreille, 1802: 111 type species *Attelabus mollis* Linnaeus, 1758
Eupocus Illiger, 1807c: 341 [RN]
Notoxus Fabricius, 1775: 158 type species *Attellabus mollis* Linnaeus, 1758
Opilus Latreille, 1806: 272 type species *Attelabus mollis* Linnaeus, 1758
abeillei Korge, 1960: 186 [RN] **E**: PT SP **N**: AG
barbarus Abeille de Perrin, 1893d: clxxvii [RN] **N**: AG EG MO TU
foveicollis Abeille de Perrin, 1893a: ccxxxvi [HN]
theryi Pic, 1912l: 57
cilicicus Hintz, 1902: 229 **A**: TR
communimacula Fairmaire, 1888b: 124 (*Clerus*) **A**: BEI MG SHX

licenti Pic, 1938h: 15
difficilis Schenkling, 1912b: 324 **A**: TAI
domesticus Sturm, 1837b: 16 (*Notoxus*) **E**: AU AZ BE CR CZ DE FR GB GE GG GR HU IT MA NL NR PL PT RO SK SL SP SV SZ UK "Caucasus" **N**: MO MR TU **A**: BEI "Korea" **NAR NTR**
hladilorum Winkler, 1984b: 194
subdelectus Pic, 1903n: 154
subfasciatus Spinola, 1844a: 222/224 (*Notoxus*)
vittatus Leoni, 1906a: 64
fenestratus Pic, 1953g: 321 **A**: "China: Shan-Hai-Kouan"
formosanus Schenkling, 1912b: 324 **A**: JA (Ryukyus) TAI
germanus Chevrolat, 1843: 37 (*Opilus*) **E**: FR GEi PL PT **N**: AG TU
impressipennis Gorham, 1903: 334 **A**: HP **ORR**
klapperichi Winkler, 1974: 376 **A**: AF
longipilis Fairmaire, 1892a: 105 **N**: AG **A**: SA **AFR**
luteonotatus Pic, 1926c: 7 **A**: HUN JIX ZHE
miyatakei Yoshimichi, 2004: 113 **A**: JA
mollis Linnaeus, 1758: 388 (*Attelabus*) **E**: AU AZ BE BH BU CR CZ DE EN FR GB GE GG HU IT LA LS LT MC NL NR NT PT RO SK SL SP ST SZ **A**: CH JA TAI "Korea" **COS**
aequalis Reitter, 1894d: 44
carinatus Lewis, 1892: 186
cruciger Geoffroy, 1785: 135 (*Clerus*)
fuscofasciatus DeGeer, 1775: 159 (*Clerus*)
guerini Pic, 1911o: 138
sexmaculatus Pic, 1912l: 57
viturati Pic, 1900h: 62
niponicus Lewis, 1892: 186 **A**: FE JA "Korea" **ORR**
pallidus A. G. Olivier, 1795b: 76: 11 (*Clerus*) **E**: AU BE BH BU CR CZ FR GE GR HU IT MD PL RO SK SL SP SZ UK YU **N**: AG **A**: CY TR
centromaculatus Klug, 1842: 320 (*Opilus*)
schaedia P. Rossi, 1790: 140 (*Notoxus*)
rugulicollis Schenkling, 1906: 260 **A**: SD
sinensis Pic, 1926h: 21 **A**: SCH
grahami Chapin, 1927: 19
shirozui Miyatake, 1965c: 161 **A**: TAI
sordidus Westwood, 1852: 42 (*Opilus*) **A**: NP SD UP
taeniatus Klug, 1842: 320 (*Opilus*) **E**: AB AL AU BH BU CZ GG GR HU IT MC RO SK SL ST TR **N**: LB **A**: CY TR
cruentatus Spinola, 1844: 225 (*Notoxus*)
fallax Chevrolat, 1876a: 11
frontalis Klug, 1842: 321 (*Opilus*)
mimonti Boieldieu, 1859: 471 (*Opilus*)
rubrofasciatus Kolenati, 1846b: 46 (*Tillus*)
rufofasciatus Schaufuss, 1872: 374 (*Opilus*)
scuttelatum Schaufuss, 1862d: 311 (*Enoplium*)
syriacus Spinola, 1844a: 225 (*Notoxus*)
thoracicus Klug, 1842: 321 (*Opilus*)
torretassoi Wittmer, 1935a: 65
testaceipennis Pic, 1927c: 1 **A**: CH
theresae Pic, 1951b: 6 **A**: SD (Padong)
dureli Pic, 1951b: 6
tilloides Chevrolat, 1876a: 12 **A**: "Syria"
triangulus Schenkling, 1902b: 322 **A**: JIX ZHE **ORR**
variipennis Nakane, 1996: 135 **A**: TAI

genus *Orthrius* Gorham, 1876: 74 type species *Orthrius cylindricus* Gorham, 1876
Tillopilo Winkler, 1958: 245 type species *Tillopilo corporaali* Winkler, 1958
abdominalis Germar, 1824: 80 (*Clerus*) **A**: SCH "Indes Orientalis" **ORR**

latior Pic, 1932b: 9
abruptepunctatus Schenkling, 1932: 24 **A**: HUN SCH YUN
binotatus Fischer von Waldheim, 1829a: 44 (*Clerus*) **A**: JA TAI "Bengal" **ORR**
andamanensi Schenkling, 1906: 268
sellatus Westwood, 1852: 42 (*Thanasimus*)
brachialis Gorham, 1893: 577 **A**: UP **ORR**
carinifrons Schenkling, 1900: 17 **A**: JA TAI
corporaali Winkler, 1958: 246 (*Orthrius*) **A**: ZHE
disjunctus Pic, 1926c: 6 **A**: YUN
dorsalis Schenkling, 1902b: 324 **A**: SD
particularis Pic, 1926c: 6 **A**: "Korea"
pieli Pic, 1936c: 15 **A**: FUJ HUN JIX ZHE
prolongatus Schenkling, 1912b: 325 **A**: TAI
rufotestaceus Schenkling, 1902b: 323 **A**: SD UP
sibiricus Pic, 1933d: 5 **A**: FE
sinensis Gorham, 1876: 70 (*Opilo*) **A**: GUI
stevensi Corporaal, 1926: 180 **A**: SD
striatopunctatus Schenkling, 1901: 106 **A**: HP SD UP **ORR**
bimaculatus Schenkling, 1901: 106
brunneus Corporaal, 1926: 180
striatulus Motschulsky, 1860b: 113 (*Opilus*) **A**: FE "Korea"
subscalaris Pic, 1954e: 62 **A**: FUJ
sulcatus Pic, 1926c: 6 **A**: CH **ORR**
tuberculicollis Schenkling, 1902b: 323 **A**: SD

genus *Pieleus* Pic, 1940b: 4 type species *Orthrius irregularis* Pic, 1940
irregularis Pic, 1940b: 4 (*Orthrius*) **A**: CH

genus *Phloiocopus* Guérin-Méneville, 1835b: pl. 40 type species *Phloiocopus tricolor* Guérin-Méneville, 1835
Phloeocopus Lacordaire, 1857: 437 [unjustified emendation]
Phloiocopus Spinola, 1841: 73 [HN] type species *Phloiocopus tricolor* Guérin-Méneville, 1835
andresi Schenkling, 1912a: 38 (*Phloeocopus*) **N**: EG **A**: LE
rufus Pic, 1935b: 3 (*Phloeocopus*)
arabicus Corporaal, 1941: 359 (*Phloeocopus*) **A**: SA YE
basalis Klug, 1842: 323 (*Opilus*) **N**: EG **A**: CY SI TR
tricolor Guérin-Méneville, 1835b: pl. 40 **A**: SA SY YE **AFR**
discolor Desmarest, 1860: pl. 20
inaequalis Reitter, 1894d: 46 (*Phloeocopus*)
interruptus Klug, 1842: 323 (*Opilus*)
lelieurii Spinola, 1844: 338
lesueurii Spinola, 1844: 338

genus *Sedlacekius* Winkler, 1972: 38 type species *Sedlacekius rex* Winkler, 1972
pountianus Menier, 1981: 108 **A**: YE (Aden)

genus *Stigmatium* Gray, 1832: 375 type species *Stigmatium cicindeloides* Gray, 1832
Cyclotomocerus Kuwert, 1894c: 400 type species *Cyclotomocerus gemmatus*, 1894
Dasyceroclerus Kuwert, 1894c: 399 type species *Dasyceroclerus viridis* Kuwert, 1894
Microastigmus Pic, 1938c: 2 type species *Microastigmus maindroni* Pic, 1938
Phaeocyclotomus Kuwert, 1894c: 400 type species *Stigmatium dorsiger* Westwood, 1852
Rhytidoclerus Kuwert, 1894c: 400 type species *Stigmatium rufiventris* Westwood, 1849
Xestoclerus Chapin, 1924: 399 type species *Xestonotus eximius* Kuwert, 1894
birmanicum Gorham, 1892c: 740 **A**: UP **ORR**
ceramboides Motschulsky, 1861c: 125 (*Stigmatium*) **A**: TAI **ORR**
cicindeloides Gray, 1832: 376 **A**: SD **ORR**
diversipes Schenkling, 1916: 118 **A**: TAI

formosanum Pic, 1938c: 2 **A**: TAI
igai Nakane, 1963b: 46 **A**: JA
indicum Pic, 1938c: 6 **A**: SD
mutillaecolor White, 1849: 51 (*Tillicera*) **A**: HUN JIA NP SD YUN ZHE **ORR**
dimidiatum Chevrolat, 1874b: 314
nakanei Iga, 1949: 37 **A**: FE (Kunashir) JA
pilosellum Gorham, 1878: 163 (*Thanasimus*) **A**: JA TAI
pilosellum Kiesenwetter, 1879b: 312 (*Clerus*) [HN]
postmaculatus Nakane, 1963b: 46 (*Clerus*) **A**: JA (Ryukyu)
rufiventris Westwood, 1849: 54 (*Rhytidoclerus*) **A**: FUJ YUN
ryukyuense Miyatake, 1985b: 157 **A**: JA
sauteri Pic, 1938c: 2 **A**: TAI
torulentus Westwood, 1855: 21 (*Stigmatium*) **A**: UP

genus *Thanasimodes* Murray, 1867b: 319 type species *Thanasimodes metallicus* Murray, 1867
desertorum Lesne, 1905: 388 (*Opilo*) **N**: AG **AFR**
dorsalis P. H. Lucas, 1843b: 159 (*Opilo*) [= 1843c: xxiv (*Opilus*)] **N**: AG MO TU **AFR**
luteofasciatus Pic, 1932h: 30
gigas Laporte, 1838: 42 (*Notoxus*) **N**: TU **AFR**
abdominalis Fairmaire, 1891f: ccxci (*Opilo*)
tropicus Klug, 1842: 322 (*Opilus*)
insignis Escalera, 1914: 249 (*Opilo*) **N**: MO

genus *Thanasimus* Latreille, 1806: 270 type species *Attelabus formicarius* Linnaeus, 1758
Cleroides J. C. Schaeffer, 1777: pl. 137
Pseudoclerus Jacquelin du Val, 1860a: 196 type species *Clerus brevicollis* Spinola, 1844
albomaculatus Lewis, 1892: 187 **A**: JA
carbonarius Gorham, 1893: 577 **A**: SD **ORR**
femoralis Zetterstedt, 1828: 122 (*Clerus*) **E**: AB AR AU BH BY CR CT CZ DE EN FI FR GB GE GG HU IT LA LS LT NL NR NT PL RO SK SP (Pirineos) ST SV SZ UK "Caucasus" **N**: AG **A**: ES KZ MG WS
austriacus Reitter, 1894d: 48
nigricollis Seidlitz, 1889a: 493
pectoralis Fuss, 1863: 67 (*Clerus*)
rufipes Brahm, 1797: 136 (*Clerus*) [HN]
formicarius formicarius Linnaeus, 1758: 387 (*Attelabus*) **E**: AL AR AU BE BU BY CR CT CZ DE EN FI FR GB GE HU IR IT LA LS LT NL NR NT PL PT RO SK SL SP ST SV SZ UK **N**: MO TU **A**: ES FE GAN NMO "Korea" **NAR**i
assimilis Méquignon, 1948: 77
fasciatus Geoffroy, 1785: 135 (*Clerus*)
brioni Pic, 1946: 5
hilaris Buysson, 1901a: 95
laetipes Reitter, 1894d: 48
roubali Roubal, 1936a: 30
schaeferi Méquignon, 1949: 114
semifasciatus A. Fleischer, 1887: 237 (*Cleroides*)
vermivorus Gistel, 1857: 32 (*Clerus*)
formicarius mascarensis Corporaal, 1932: 137 **N**: AG MO
fukiensis Corporaal, 1948: 242 **A**: FUJ
himalayensis Stebbing, 1914: 18 [fig. 11] **A**: HP UP
karafutonis Kôno, 1930: 135 **A**: FE "Korea"
lewisi Jakobson, 1911b: 717 (*Cleroides*) [RN] **A**: HEN JA JIL QIN SHN "Korea"
nigricollis Lewis, 1892: 189 [HN]
pieli Pic, 1940b: 7
nigriventralis Krivolutskaya, 1981: 34 **A**: FE
subscutellaris Westwood, 1852: 44 **A**: HP UP
substriatus mimolewisi Satô, 1982: 395 **A**: JA

substriatus substriatus Gebler, 1832: 47 (*Clerus*) **A**: ES FE HEN JA MG WS "Korea"

genus *Tillicera* Spinola, 1841: 73 type species *Tillicera javana* Spinola, 1844
soror Schenkling, 1902a: 127 **A**: BT **ORR**

genus *Trichodes* Herbst, 1792: 154 type species *Attelabus apiarius* Linnaeus, 1758
Pachyscelis Hope, 1840: 137 type species *Clerus ammios* Fabricius, 1787
aberrans Winkler, 1974: 389 **A**: AF
affinis Chevrolat, 1843: 38 **E**: GR **N**: EG **A**: IS JO SY TR
antiquus Klug, 1842: 334
apicalis Champenois, 1900: 14 (*Clerus*) [HN]
illepidus Walker, 1871: 14 (*Clerus*)
metasternalis Reitter, 1894d: 65
sagittifer Champenois, 1900: 15 (*Clerus*)
subapicalis Chevrolat, 1874b: 306
albanicus Winkler & Žirovnický, 1980: 469 **E**: AL
alberi Escherich, 1894: 198 **A**: IQ TR
cappadocicus Ganglbauer, 1905: 275
korbianus Ganglbauer, 1905: 274
alvearius Fabricius, 1792a: 209 (*Clerus*) **E**: AB AU BE CR CZ FR GB GE HU IT LS NL PL SK SL SP SZ UK **N**: AG MO
affinis Spinola, 1844a: 330
alveolarius Latreille, 1804a: 154 (*Clerus*)
apiarius O.F. Müller, 1776: 95 (*Attelabus*) [HN]
dahlii Spinola, 1844: 300
lugdunensis Pic, 1919c: 7
postinterruptus Pic, 1939e: 25
ravouxi Pic, 1903m: 146
theresae Pic, 1902f: 46
ammios Fabricius, 1787a: 126 (*Clerus*) **E**: GR IT SP **N**: AG MO TU
arthriticus Spinola, 1844a: 322/324
flavicornis Germar, 1838: 4
gracilior Corporaal & Vári, 1946: 84
omoplatus Spinola, 1844a: 322/225
sipylus Fabricius, 1775: 158 (*Clerus*) [HN]
subfasciatus Spinola, 1844a: 322/324
apiarius Linnaeus, 1758: 388 (*Attelabus*) **E**: AB AL AU BE BH BU BY CR CT CZ EN FR GB GE GG GR HU IT KZ LA LS LT NL NT PL PT RO SK SL SP ST SZ TR UK YU **A**: KZ TR
angusticolor Reitter, 1894d: 61
apiaroides Reitter, 1894d: 57
apicida Klug, 1842: 331
apivore Walckenaer, 1802: 78
arcuatus Spinola, 1844: 307
binotatus Pic, 1912l: 57
bisbiinteruptus Pic, 1917g: 13
corallinus Ménétriés, 1832: 166
crassipedarius Chevrolat, 1876a: 6
elegans Spinola, 1844a: 307
georgianus Chevrolat, 1874b: 306
grandjeani Pic, 1932f: 25
interruptus Klug, 1842: 331
jacqueti Pic, 1912l: 57
pannonicus Spinola, 1844a: 305/307
quinquenotatus Pic, 1917g: 13
reductus Pic, 1917g: 13
subtrifasciatus Klug, 1842: 331
taygetanus Hintze, 1902: 229
trellai Corporaal, 1931: 138
tricolor Roubal, 1931b: 183

unifasciatus Klug, 1842: 331
axillaris Fischer von Waldheim, 1842: 9 **A**: AF KI KZ TD TM UZ XIN
apicalis Brancsik, 1899: 99
bucharicus Fursov, 1936b: 352
confusus Mader, 1931a: 193
deliquus Reitter, 1894d: 72
hintzi Pic, 1945: 14 [RN]
humeralis Hintz, 1904: 422 [HN]
interruptus Hintz, 1904: 422 [HN]
immarginatus Hintz, 1904: 422
jucundus Reitter, 1894d: 72
kraatzi Reitter, 1894h: 252 [RN]
nobiliformis Reitter, 1894d: 72
opacus Fursov, 1936b: 154
sanguineosignatus Escherich, 1894: 191 [HN]
violaceus Heyden, 1892: 105
audouini dubreuili Reymond, 1956: 42 **N**: MO
bitaeniatus Kraatz, 1894: 128 **A**: IN
crabroniformis Fabricius, 1787a: 126 (*Clerus*) **E**: AR BU CR FR GR IT (Sicilia) MC SL **A**: TR
gulo Chevrolat, 1874b: 304
hidalgo Winkler, 1959: 71
lepidus Brullé, 1832: 154 (*Clerus*)
creticus Brodský, 1982: 189 **E**: GR (Kríti)
cruciatus Fursov, 1939: 40 **E**: AR
cyprius Reitter, 1894d: 77 **A**: CY IS
cyprius Champenois, 1900: 23 (*Clerus*) [HN]
davidis Deyrolle, 1878: 119 **A**: FUJ JIX **ORR**
davidi Schenkling, 1903a: 68
inapicipennis Pic, 1955c: 23
dilatipennis Reitter, 1894d: 75 **A**: IN IS SY
doriae Baudi di Selve, 1874b: 323 **A**: IN IQ
bipunctatus Reitter, 1894d: 60
olivieri Chevrolat, 1843: 38 [HN]
ephippiger Chevrolat, 1874b: 307 **E**: AR **A**: IN IQ IS SY TR
jactans Reitter, 1899d: 276
leucippus Reitter, 1899d: 276
theophilei Chevrolat, 1876a: 18
escherichi Reitter, 1893d: 303 **A**: TD TM UZ
kulaebensis Reitter, 1894d: 73
favarius Illiger, 1801a: 80 (*Clerus*) **E**: AL AR AU BH BU CR GR HU MC PL RO SK SL ST TR UK YU "Caucasus" **A**: CY TR
axillaris Spinola, 1844: 313
elengatulus Spinola, 1844a: 313/315
hispanicus Spinola, 1844a: 313/315
interruptefasciatus Ronchetti, 1910: 201
nigratus Heyden, 1894: 88 DEZ
obliquatus Brullé, 1832: 155 (*Clerus*)
schreberi Gerhadt, 1908: 210
flavocinctus Spinola, 1844a: 322/324 **E**: FR IT PT SP **N**: AG LB MO TU
dauci Spinola, 1844: 322/324
hispanus Chevrolat, 1876a: 19
podagricus Chevrolat, 1876a: 18
x-littera Chevrolat, 1874b: 309
flavotarsis Gerstmeier, 1985: 130 **A**: IQ
galbula Reitter, 1894h: 252 **A**: TM
ganglbaueri Escherich, 1894: 166 **A**: CY IN IS TR
israelicus Winkler, 1963: 890
turcicus Winkler & Žirovnický, 1980: 463

graecus Winkler & Žirovnický, 1980: 466 **E**: GR
hauseri Escherich, 1894: 192 **A**: KZ TD TM UZ
- *maculifer* Champenois, 1900: 20 (*Clerus*)
- *pilosus* Fursov, 1936b: 353
- *proximus* Champenois, 1900: 20 (*Clerus*)
- *semilimbatus* Champenois, 1900: 20 (*Clerus*)

heydeni Escherich, 1892: 230 **E**: AB AR **A**: IN IQ TR
- *anatolicus* Reitter, 1894d: 78

hofferi Žirovnický, 1974: 316 **N**: AG
holtzi Hintz, 1902: 230 **A**: TR
- *tauricus* S. Zimmermann, 1971: 604

inapicalis Pic, 1945: 14 **A**: IN
inapicipennis Pic, 1955c: 23 **A**: FUJ
inermis Reitter, 1894d: 76 **E**: AB/AR TR **A**: TR
- *akbesianus* Reitter, 1915e: 124 [RN]
- *femininus* Champenois, 1900: 25 (*Clerus*) [RN]
- *imitator* Reitter, 1899d: 276 [HN]

insignis Fischer von Waldheim, 1829a: 68 **E**: AB AR **A**: IN TR
- *araxicola* Reitter, 1894d: 69
- *armeniacus* Baudi di Selve, 1873: 254
- *caucasicus* Chevrolat, 1874b: 305
- *caucasicus* Kraatz, 1882c: 320 [HN]
- *phedinus* Spinola, 1844: 313

ircutensis Laxmann, 1770: 595 (*Attelabus*) **E**: PL RO SK SZ UK **A**: CY ES FE JIL MG NE NMO TR WS "Korea"
- *bangi* Reitter, 1894d: 62
- *bifasciatus* Fabricius, 1781: 202 (*Clerus*)
- *humeralis* Kraatz, 1894: 135

jelineki Brodský & Winkler, 1988: 297 **A**: IN
jubilarii Corporaal, 1932: 138 **A**: AF
kabakovi Melnik, 2004: 147 **A**: AF
kirghizicus Fursov, 1939: 89 **A**: KZ
laminatus Chevrolat, 1843: 40 **E**: AB GR UK (Krym) **A**: CY IN IQ IS JO SY TR
- *abeillei* Reitter, 1894d: 76
- *angustus* Chevrolat, 1843: 41
- *fallax* Escherich, 1894: 189
- *flavocinctus* Chevrolat, 1876a: 19
- *quadrigutullus* Escherich, 1892: 244
- *religatus* Champenois, 1900: 23 (*Clerus*)
- *smyrnensis* Spinola, 1884: 322
- *viridis* Rey, 1891f: 131 (*Clerus*)
- *visnagae* Spinola, 1844: 322/325

leucopsideus A. G. Olivier, 1795b: 76: 8 (*Clerus*) **E**: FR IT PT SP **N**: AG LB MO TU
- *annae* Coreia de Barros, 1929: 12
- *fuscicornis* Spinola, 1844a: 319/320
- *hypocritus* Chevrolat, 1861: 151
- *sanguinosus* Chevrolat, 1861: 150
- *xanthoceros* Champenois, 1900: 5 (*Clerus*)

longicollis Gerstmeier, 1985: 129 **A**: IN
longissimus Abeille de Perrin, 1881: 99 (*Clerus*) **E**: AB AR BU GR **N**: EG **A**: CY SY TR
- *angustifrons* Abeille de Perrin, 1881: 100 (*Clerus*)
- *discendens* Champenois, 1900: 8 (*Clerus*) [RN]

martini Fairmaire, 1896b: 223 **A**: YE **AFR**
moghrebicus Lesne, 1925: 15 **N**: MO (Moyen Atlas)
namanganensis Fursov, 1939: 89 **A**: UZ
nobilis Klug, 1842: 335 **E**: AB AR GR TR **A**: IN KZ SY TD TM TR UZ
- *carceli* Chevrolat, 1843: 39
- *consanguineus* Kraatz, 1894: 130

guttifer Reitter, 1893c: 258
illustris Spinola, 1844a: 311
normalis Reitter, 1894d: 70
quadrilineolatus Kraatz, 1894: 128
reitteri Champenois, 1900: 19 (*Clerus*) [RN]
ruber Escherich, 1897: 30
sanguineosignatus Reitter, 1894: 83
suturangulus Reitter, 1893c: 258
variabilis Spinola, 1844: 311
oberthueri Champenois, 1900: 16 (*Clerus*) **A**: TR
sbordonii Audisio, 1975: 236
octopunctatus Fabricius, 1787a: 126 (*Clerus*) **E**: FR IT PT SP **N**: AG MO
andalusiacus Pic, 1895l: 114
fasciatus Champenois, 1900: 10
octomaculatus Villers, 1789: 222 (*Attelabus*)
quadrimaculatus Kraatz, 1894: 117
sexpunctatus Rey, 1891f: 130 (*Clerus*)
sexpustulatus Barros, 1929: 13
suturifer Champenois, 1900: 11
persicus Kraatz, 1894: 126 **E**: AB AR **A**: IN TM
pulcherrimus Escherich, 1892: 227 **A**: IN TR
punctatus punctatus Fischer von Waldheim, 1829a: 68 **E**: BU GR MC TR UK (Krym) "Caucasus" **A**: CY IN IS JO SY TR
bosphoranus Champenois, 1900: 14 (*Clerus*)
illustris Spinola, 1844: 315
lafertei Chevrolat, 1843: 39
senilis Klug, 1842: 333
vicinus Spinola, 1844a: 313
viridanus Kraatz, 1894: 126
viridifasciatus Chevrolat, 1843: 40
quadriguttatus Adams, 1817: 313 (*Clerus*) **E**: AB AR BU GG GR KZ ST TR UK (Krym) "Caucasus" **A**: IN JO KZ SY TR
quadripunctatus Spinola, 1844a: 325
quadripustulatus Brullé, 1832: 156 (*Clerus*)
rectilinea Reitter, 1894d: 71 **A**: TM
escherichi Kraatz, 1894: 132
signum Reitter, 1894d: 71
reichei Mulsant & Rey, 1863a: 95 (*Clerus*) **E**: AR **A**: IN SY TR "Palestina"
biniterruptus Pic, 1945: 14
conjunctus Escherich, 1892: 229
imitator Reitter, 1894d: 56
klugii Kraatz, 1873b: 242 [RN]
maximus Escherich, 1892: 229
minimus Kraatz, 1894: 121
olivierii Klug, 1842: 336 [HN]
rubrolimbatus Chevrolat, 1876a: 20 **A**: IS LE SY TR
difficilis Escherich, 1894: 199
sedlaceki intercalaris Winkler, 1974: 386 **A**: AF (Panchir)
sedlaceki sedlaceki Winkler, 1974: 383 **A**: AF (Nuristan)
sexpustulatus Chevrolat, 1874b: 306 **E**: AR **A**: SY TR
erratus Champenois, 1898: 115 [RN]
quadrimaculatus Escherich, 1894: 188
triangulus Reitter, 1894d: 68
varifasciatus Kraatz, 1894: 127
similis Kraatz, 1894: 124 **E**: GR
sinae Chevrolat, 1874b: 303 **A**: BEI FE HEB HEN HUB HUN GAN JIL JIX LIA MG NMO QIN SC SCH SHA SHX XIN XIZ ZHE
frivaldszkyi Reitter, 1894d: 61
interruptus Kraatz, 1894: 122 [HN]

pekinensis Pic, 1895f: 88 [RN]
spinolai Kolbe, 1886: 199
thibetanus Kraatz, 1894: 122
sipylus atticus Chevrolat, 1874b: 308 **E**: GR
sipylus praetentus S. Zimmermann, 1973: 417 **A**: IN TR
sipylus sipylus Linnaeus, 1758: 387 (*Attelabus*) **E**: AB GR ST "Caucasus" **A**: IN IS LE TR
ammios Spinola, 1844: 325
frater Kraatz, 1893: 80
fraterculus Zaitzev, 1915: 45
subfasciatus Kraatz, 1873b: 239
spectabilis Kraatz, 1883: 346 **A**: KZ TD TM TR UZ
brancsiki Reitter, 1899d: 276
imperfectus Breit, 1912: 7
kuwerti Reitter, 1893d: 303
subvittatus Champion, 1900: 4
suspectus Escherich, 1892: 226 **A**: TR
suturalis Seidlitz, 1889b: 527 **E**: SP
syriacus Spinola, 1844: 316 **A**: IS JO SY TR
kindermanni Chevrolat, 1874b: 307
olivieri Spinola, 1844: 147 [HN]
talyshensis Zaitzev, 1915: 45 **E**: AB **A**: IN
turkestanicus Kraatz, 1882b: 113 **A**: KZ TM UZ
bimaculatus Kraatz, 1894: 131
immarginatus Kraatz, 1894: 127
latefasciatus Kraatz, 1894: 131
quadrimaculatus Kraatz, 1894: 135
trinominatus Pic, 1895l: 114
umbellatarum A. G. Olivier, 1795b: no 76: 5 (*Clerus*) **E**: FR PT SP **N**: AG LB MO TU
aindrahamensis Bodemeyer, 1929: 42
balearicus Pic, 1920d: 10
batnensis Pic, 1920f: 17
bisbidivisus Pic, 1945: 14
gabilotti, Pic, 1920: 17
gabilloti Pic, 1945: 14 [HN]
galfanus Bodemeyer, 1929: 18
leprieuri Pic, 1920d: 10
maroccanus Kraatz, 1894: 123
multifasciatus Pic, 1920d: 10
subobliteratus Pic, 1920e: 13
umbellatarum Spinola, 1844: 29
viridiaureus Abeille de Perrin, 1881: 101 (*Clerus*) **A**: IS
winkleri Žirovnický, 1976: 15 **E**: GR
zaharae Chevrolat, 1861: 152 **N**: AG MO TU **AFR**
punctonotus Pic, 1895f: 88
zebra Faldermann, 1835a: 207 **E**: AB AR **A**: IN
rufitarsis Chevrolat, 1874b: 302
zhelochovtsevi Fursov & Rikhter, 1958: 708 **A**: KZ TD TM UZ

genus *Winklerius* Menier, 1986: 224 type species *Opilo grandis* Stierlin, 1868
grandis Stierlin, 1868: 345 (*Opilus*) **A**: IN IQ SA
albonotatus Pic, 1901a: 10 (*Opilo*)

genus *Xenorthrius* Gorham, 1892c: 733 type species *Xenorthrius mouhoti* Gorham, 1892
elongatus Miyatake, 1985b: 156 **A**: JA
impressicollis Pic, 1954e: 62 **A**: FUJ
incarinipes Miyatake, 1965c: 163 **A**: TAI
mouhoti Gorham, 1892c: 734 **A**: "India" **ORR**
robustus Corporaal, 1939a: 25 **A**: SD UP

simplex Schenkling, 1903c: 16 **A**: HKG TAI
umbratus Schenkling, 1912b: 325 **A**: HKG TAI
simplex Schenkling, 1912b: 326 [HN]

subfamily **Korynetinae** Laporte, 1838

genus *Allochotes* Westwood, 1875: 241 type species *Allochotes bicolor* Westwood, 1875
Sisyrnophorus C. O. Waterhouse, 1876b: 125 type species *Sisyrnophorus maculatus* C. O. Waterhouse, 1876
amamioshimanus Miyatake, 1965a: 23 **A**: JA (Ryukyu)
dichroa Lewis, 1891: 210 (*Sisynophorus*[sic]) **A**: JA TAI
bicolor Lewis, 1895a: 122 (*Sisyrnophorus*) [HN]
punctipennis Champion, 1923c: 79 **A**: UP
sauteri Schenkling, 1912b: 328 **A**: TAI
violaceipennis Schenkling, 1912b: 328 **A**: TAI
yunnensis Schenkling, 1908c: 702 **A**: YUN
yuwanensis Yajima & Nakane, 1969a: 6 **A**: JA (Ryukyus)

genus *Corynetinus* Reitter, 1894d: 82 type species *Corynetes fimetarius* Wollaston, 1862
fimetarius Wollaston, 1862d: 440 (*Corynetes*) **N**: AG CI EG LB SI TU
desbrochersi Reitter, 1897b: 79 (*Corynetes*)
fulgidicollis Desbrochers des Loges, 1897: 167 (*Corynetes*)

genus *Dermestoides* J. C. Schaeffer, 1771: pl. ccxx type species *Dermestes sanguinicollis* Fabricius, 1787
Aporthopleura Barr, 1977: 32 type species *Dermestes sanguinicollis* Fabricius, 1787
Eurenoplium Žirovnický, 1977: 45 type species *Dermestes sanguinicollis* Fabricius, 1787
Orthopleura Spinola, 1844: 80 type species *Dermestes sanguinicollis* Fabricius, 1787
sanguinicollis Fabricius, 1787a: 35 (*Dermestes*) **E**: BH BY CZ DE FR GE GG GR IT PL RO SK SP ST SV SZ UK "Caucasus"
cornigera Harrer, 1784: 145 (*Ptillinus*)
dulcis Ledoux, 1833: 474 (*Enoplium*)
weberi Fabricius, 1798: 118 (*Tillus*)

genus *Enoplium* Latreille, 1802: 111 type species *Tillus serraticornis* A. G. Olivier, 1790
Pelonides Kuwert, 1894a: 8 type species *Enoplium quadripunctatum* Say, 1823
balteatum Fairmaire, 1889a: 354 **A**: "S. China" **ORR**
doderoi Luigioni, 1926: 51 **E**: GR (Kríti) IT (Sardegna, Sicilia)
serraticorne A. G. Olivier, 1790a: no 22: 4 (*Tillus*) **E**: AL FR GEi GR MC IT PLi SL SP UK (Krym) **N**: "North Africa"
dentanum P. Rossi, 1790: 34 (*Dermestes*)
praeustum Bigliani, 1907: 156

genus *Iwawakia* Miyatake, 1985b: 158 type species *Iwawakia femorata* Miyatake, 1985
femorata Miyatake, 1985b: 158 **A**: JA
trimaculata Satô, 2002: 112 **A**: JA

genus *Korynetes* Herbst, 1792: 148 type species *Clerus caeruleus* DeGeer, 1775
[*Corynetes* Paykull, 1798: suppressed]
Corynetops Jacquelin du Val, 1860a: 201 type species *Clerus caeruleus* DeGeer, 1775
caeruleus DeGeer, 1775: 163 (*Clerus*) **E**: AB AL AU BE BU CR CT CZ DE FI FR GB GE GR HU IT LA MC NL NR PL RO SK SV SZ "Caucasus" **A**: JA TAI
geoffroyanus Laicharting, 1783: 247 (*Attelabus*)
meridionalis Obenberger, 1916: 23 (*Corynetes*)
nigrocoeruleus Geoffroy, 1799: 304 (*Clerus*)
obenbergeri Jansson, 1936: 42 [RN]
unicolor Chevrolat, 1876a: 43 (*Corynetes*)
violaceus Linnaeus, 1758: 356 (*Dermestes*)
coxalis Reitter, 1894d: 84 **E**: GR (Kríti) **A**: JO SY

geniculatus Klug, 1842: 347 (*Corynetes*) **E**: FR PT SP **N**: AG MO TU
pexicollis Fairmaire, 1868: 484 (*Corynetes*)
pusillus Klug, 1842: 347 (*Corynetes*) **E**: AL FR IT
ruficornis Sturm, 1837b: 42 (*Corynetes*) **E**: AU CR CZ FR GE HU PL RO SK SV SZ "Caucasus"
quintus Klug, 1842: 347
violaceus Scopoli, 1763: 18 (*Dermestes*)
rugipennis Ballion, 1871: 353 (*Corynetes*) **A**: TD

genus *Necrobia* A. G. Olivier, 1795b: no 76bis: 4 type species *Dermestes violaceus* Linnaeus, 1758
kelescenyi Gebien, 1899: 97 **E**i: GE (Hamburg)
konowi E. Hoffmann, 1887: 96 (*Agonolia*) **E**: IT
ruficollis Fabricius, 1775: 57 (*Dermestes*) **E**: AU AZ BE BU BY CZ DE EN FI FR GE GB HU IR IT LA LS LT NL NR NT PT SK SL SP SV SZ **N**: CI EG MO TU **A**: ANH FUJ GUA GUI GAN HAI HEI HUB HUN JA LIA SCH SHA SHG SHN TAI YUN ZHE "Korea" **COS**
rufipes DeGeer, 1775: 165 (*Clerus*) **E**: AB AR AU AZ BE BY CR CZ DE EN FI FR GB GE GR HU IR IT LA LS LT NL NT PT RO SK SL SP ST SV SZ **N**: AG CI EG LB MO TU **A**: ANH FE FUJ GAN GUA GUI HAI HUB HUN IN JA MG NMO SA SCH SHA SHN SHX TD TR UP XIN YUN ZHE "Korea" **COS**
aeneipennis Csiki, 1900a: 124
amethystina Stephens, 1835: 417
aspera Walker, 1858: 283
cupreonitens Lauffer, 1905: 406
dermestoides Piller & Mitterpacher, 1783: 68 (*Tenebrio*)
flavipes Klug, 1842: 353 (*Corynetes*)
foveicollis Schenkling, 1900: 20
glabra Jurine, 1814: 44 (*Corynetes*)
mumiarum Hope, 1834: 54
pilifera Reitter, 1894d: 85
violacea Linnaeus, 1758: 356 (*Dermestes*) **E**: AB AU BE BH BU BY CR CZ DE EN FI FR GB GE IR IT LA LT NL NR NT PT RO SK SL SP ST SV SZ YU **N**: MO **A**: FE GUA GUX IN MG NMO QIN WS TAI XIN **COS**
angustata Faldermann, 1835a: 210 (*Corynetes*)
chalybea Sturm, 1837b: 43 (*Corynetes*)
dalmatina Obenberger, 1916: 22 (*Corynetes*)
errans Melsheimer, 1846: 307
jablanicensis Obenberger, 1916: 22 (*Corynetes*)
pygmaea L. Redtenbacher, 1858: 552 (*Corynetes*)
quadra Marsham, 1802: 323 (*Clerus*)

genus *Necrobinus* Reitter, 1894: 86 type species *Corynetes defunctorum* Waltl, 1835
defunctorum Waltl, 1835: 63 (*Corynetes*) **E**: FR PT SP **N**: AG EG MO TU **A**: IS
anthracinus Bedel, 1899: 115 [RN]
carbonarius Spinola, 1844: 108 (*Necrobia*)
defunctorum Reitter, 1894d: 86 [HN]
frater Reitter, 1894d: 86
laufferi Reitter, 1902h: 213
lugubris Gistel, 1857: 34 (*Necrobia*)
obscuripes Pic, 1898e: 98 (*Opetiopalpus*)
subterraneus Chevrolat, 1876a: 43 (*Necrobia*)

genus *Opetiopalpus* Spinola, 1844: 110 type species *Clerus scutellaris* Panzer, 1797
Opetiopselaphus Gemminger & Harold, 1869: 1759 [unjustified emendation]
bicolor Laporte, 1838: 50 (*Corynetes*) **E**: PT SP **N**: MO
hybridus Baudi di Selve, 1874b: 324 (*Corynetes*) **E**: SP
obesus Westwood, 1849: 63 **A**: GUA HEN JA SHG YUN **ORR**
morulus Kiesewetter, 1879: 314 (*Corynetes*)
sabulosus Motschulsky, 1840: 178 (*Corynetes*) **E**: AR MD UK "Balkan" "Caucasus" **A**: ES MG TM "NE China"

martynovi Plavilshchikov, 1924: 231
obscuricollis Schenkling, 1898: 368
scutellaris Panzer, 1797b: 19 (*Clerus*) **E**: AB AU BH CZ EN GE HU IT LT PL RO SK ST **N**: "North Africa" **A**: KZ MG **AFR AUR**
semiruber Bedel, 1899: 115 **N**: AG
tuniseus Pic, 1906e: 26 **N**: TU
wagneri Schenkling, 1926: 102 **E**: SP

genus *Platytenerus* Miyatake, 1985b: 158 type species *Tenerus castaneus* Kôno, 1930
castaneus Kôno, 1930: 136 (*Tenerus*) **A**: JA

genus *Syriopelta* Winkler, 1984a: 178 type species *Orthopleura funebris* Fairmaire, 1892
funebris Fairmaire, 1892b: 148 (*Orthopleura*) **A**: TR

genus *Tarsostenus* Spinola, 1844: 287 type species *Clerus univittatus* P. Rossi, 1792
Paratillus Gorham, 1876: 62 type species *Clerus carus* Newman, 1840
carus Newman, 1840: 15 (*Clerus*) **E**i: FR GB **AFR**i **AUR NAR**i
univittatus P. Rossi, 1792: 44 (*Clerus*) **E**: AB AR BE CZ FR GR HU IT NT PL PT RO SK SL SP ST SZ **N**: AG EG TU **A**: GUA GUI GUX HAI HEB HUB JA SCH SHA TAI UP YUN "Korea" **COS**
albofasciatus Melsheimer, 1846: 306 (*Opilus*)
biguttatus Montrouzier, 1860: 260 (*Clerus*)
fasciatellus Spinola, 1844: 172 (*Dupontiella*)
fasciatus Curtis, 1829: 270 (*Opilus*)
funebralis Gistel, 1857: 46 (*Clerus*)
incertus W. J. MacLeay, 1872: 269 (*Opilus*)
maerens Westwood, 1849: 57 (*Notoxus*)
moerens Westwood, 1852: 50 (*Opilus*)
picipennis Westwood, 1849: 48 (*Tillus*)
succintus Chevrolat, 1842: 277 (*Tillus*)

genus *Teneroides* Gahan, 1910: 69 type species *Teneroides tavoyanus* Gahan, 1910
maculicollis Lewis, 1892: 189 **A**: JA TAI
higonius Lewis, 1892: 189 (*Tenerus*)

genus *Teneropsis* Chapin, 1924: 275 type species *Teneropsis sibuyanus* Chapin, 1924
atricornis Pic, 1926g: 16 (*Pelonium*) **A**: SD
discrepans Gorham, 1892c: 745 (*Pelonium*) **A**: SD UP **ORR**
tristis Corporaal, 1926: 189 (*Pelonium*)
extraneus Gorham, 1877: 424 (*Pelonium*) **A**: SD **ORR**
formosanus Schenkling, 1912b: 329 (*Pelonium*) **A**: TAI
lividipennis Schenkling, 1912b: 330 (*Pelonium*) **A**: TAI
mundus Schenkling, 1912b: 329 (*Pelonium*) **A**: TAI

genus *Tenerus* Laporte, 1838: 43 type species *Tenerus praeustus* Laporte, 1838
Cylistus Klug, 1842: 354 type species *Cylistus variabilis* Klug, 1842
Prionophorus Blanchard, 1853: 64 type species *Prionophorus bicolor* Blanchard, 1853
curticollis Schenkling, 1906: 305 **A**: NP SD
cyaneus Gorham, 1877: 404 **A**: CH "India" **ORR**
femoralis Corporaal, 1939a: 33 **A**: UP
flavicollis Gorham, 1877: 404 **A**: UP **ORR**
formosanus Schenkling, 1912b: 329 **A**: JA TAI
hilleri Harold, 1877: 357 **A**: JA TAI "Korea"
javanus Gorham, 1877: 404 **A**: "India" **ORR**
lewisi Lohde, 1899: 305 [RN] **A**: FE JA
cyaneus Lewis, 1892: 189 [HN]
melanurus Gorham, 1877: 409 **A**: UP **ORR**

parryanus Gorham, 1877: 402 **A**: "Indian Archipelago"
praeustus Laporte, 1838: 43 **A**: UP **ORR**
proximus Schenkling, 1908c: 703 **A**: UP **ORR**
quadrimaculatus Corporaal, 1926: 187 **A**: UP **ORR**
savioi Pic, 1925e: 32 **A**: CH
signaticollis Laporte, 1838: 44 **A**: TAI "India" **ORR**
atripennis Pic, 1926q: 301
binotatus Chevrolat, 1874b: 323
discoidalis Pic, 1928k: 377
gorhami Pic, 1945: 14 [HN]
gorhamianus Corporaal, 1949b: 357 [RN]
marginipennis Gestro, 1882: 316
notatipennis Pic, 1926h: 22
reductesignatus Pic, 1945: 14
robustus Schenkling, 1902b: 330
subfasciatus Pic, 1945: 14
viridipennis J. Thomson, 1860: 64 **A**: "India"

family ACANTHOCNEMIDAE Crowson, 1964

genus *Acanthocnemus* Perris, 1866: 187 type species *Acanthocnemus ciliatus* Perris, 1866
Eurema Abeille de Perrin, 1894b: 93 type species *Eurema dilutum* Abeille de Perrin, 1894
Hovacnemus Fairmaire, 1898: 232 type species *Hovacnemus pallitarsis* Fairmaire, 1898
nigricans Hope, 1845b: 105 (*Dasytes*) **E**i: FR IT **N**: AG **A**: CY **AFR**i **ORR**i **AUR**
ciliatus Perris, 1866: 188
dilutus Abeille de Perrin, 1894b: 93 (*Eurema*)
fauveli Bourgeois, 1884b: 289
fuscipennis Hope, 1845b: 105 (*Dasytes*)
immaturus Schilsky, 1896b: 361
kraatzi Schilsky, 1896b: 361
pallitarsis Fairmaire, 1898: 233 (*Hovacnemus*)
truquii Baudi di Selve, 1874b: 321

family PRIONOCERIDAE Lacordaire, 1857

tribe Lobonychini Majer, 1987

genus *Lobonyx* Jacquelin du Val, 1859d: 42 type species type species *Lagria aenea* Fabricius, 1787
Eulobonyx Kraatz, 1882a: 97 type species *Eulobonyx turkestanicus* Kraatz, 1882
aeneus Fabricius, 1787a: 94 (*Lagria*) **E**: IT (Sardegna, Sicilia) PT SP **N**: "North Africa"
aurarius Illiger, 1807b: 302 (*Dasytes*)
barbarus Gmelin, 1790: 1731 (*Cryptocephalus*)
ciliatus Gräells, 1842: 221 (*Dasytes*)
arisi Schilsky, 1906b: 11 (*Eulobonyx*) **A**: "Turkestan"
bicoloripes Pic, 1924j: 476 **A**: SD
bioculatus Abeille de Perrin, 1894b: 93 (*Psilothrix*) **A**: SY
curticollis Pic, 1924e: 27 (*Eulobonyx*) **A**: "Indes"
exasperatus Champion, 1920a: 72 **A**: UP
gorhami Pic, 1924e: 27 (*Eulobonyx*) **A**: "Indes"
gracilis Reitter, 1872b: 180 **E**: SP **N**: AG MO
ruficollis Raffray, 1873: 376
granulatus Pic, 1920a: 8 (*Eulobonyx*) **A**: "Indes"
intuberculatus Pic, 1924e: 27 (*Eulobonyx*)
guerryi Pic, 1920a: 8 (*Eulobonyx*) **A**: CH
inaequalis Pic, 1920a: 8 (*Eulobonyx*) **A**: PA
kashmirensis Fairmaire, 1891e: cxxxi **A**: KA
haemorrhoidalis Schilsky, 1906b: 12 (*Eulobonyx*)
scheibei Blair, 1937: 181 (*Eulobonyx*) **A**: AF (Nuristan)

sericeus Champion, 1920a: 71 (*Eulobonyx*) **A**: "NW Himalaya"
thoracicus Majer, 1990: 159 **A**: IN
turkestanicus Kraatz, 1882a: 98 (*Eulobonyx*) **A**: KZ "Turkestan"
varipes Schilsky, 1894a: 53

tribe Prionocerini Lacordaire, 1857

genus *Idgia* Laporte, 1838: 27 type species *Idgia terminata* Laporte, 1838
Deromma Kollar & L. Redtenbacher, 1848: 511 type species *Deromma melanura* Kollar & L. Redtenbacher, 1848
Diprosopus Mulsant, 1851d: 209 type species *Diprosopus melanurus* Mulsant, 1851
Thaccona Walker, 1859a: 260 type species *Thaccona dimelaena* Walker, 1859
abori Pic, 1913p: 199 **A**: AP
amplipennis Pic, 1908h: 59 **A**: GUI **ORR**
obscurimembris Pic, 1923l: 73
arabica Champion, 1919a: 336 **A**: YE
ardesiaca Pic, 1908h: 59 **A**: CH
asirensis Wittmer, 1980a: 117 **A**: SA
assimilis Hope, 1831: 26 (*Telephorus*) **A**: NP UP
bimaculata Pic, 1910t: 75 **A**: TAI
circumdata Pic, 1909n: 160 **A**: "India"
costulata Pic, 1941a: 16 **A**: HAI
curticeps Pic, 1920c: 8 **A**: "India"
cyaneipennis Pic, 1920c: 7 **A**: "India"
deusta Fairmaire, 1878b: 118 **A**: FUJ GUA JIA JIX SHG TAI YUN **ORR**
femorata Champion, 1919a: 346 **A**: HAI **ORR**
flavicollis L. Redtenbacher, 1868: 111 **A**: HKG TAI
flavirostris Pascoe, 1860: 43 **A**: SE
foveifrons Fairmaire, 1900: 629 **A**: TAI
grandis Pic, 1933b: 5 **A**: GAN
granulipennis Fairmaire, 1891f: ccix **A**: GUI (Chang-yang)
abdominalis Pic, 1920c: 7
haemorrhoidalis Pic, 1906o: 217 **A**: YUN
hoffmanni Gressitt, 1939: 191 **A**: GUA
huegeli L. Redtenbacher, 1868: 110 (*Prionocerus*) **A**: CH
incerta Pic, 1920c: 8 **A**: "India"
iriomoteana Nakane, 1981c: 129 **A**: JA (Irimote, Yaeyama)
laticornis Champion, 1919a: 337 **A**: YE
lineata Pic, 1907f: 125 **A**: "Indes"
confusa Pic, 1920c: 7
major Pic, 1908o: 128 **A**: YUN
melanura Kollar & L. Redtenbacher, 1848: 512 (*Deromma*) **A**: HP KA SD
moupinensis Fairmaire, 1889b: 44 **A**: SCH
nitida Champion, 1919a: 364 **A**: PA
oculata L. Redtenbacher, 1868: 110 **A**: HAI
particularicornis Pic, 1939g: 145 **A**: SI
subparallela Pic, 1920c: 8 **A**: "India"
testaceipes Pic, 1908m: 95 **A**: YUN
thibetanus Obenberger, 1918: 60 (*Prionocerus*) **A**: "West Himalaya"
ungulata Champion, 1919a: 339 **A**: FUJ JIX GUA HAI HKG SE
virescens Champion, 1919a: 340 **A**: SCH
viridescens Gorham, 1895a: 319 **A**: HP SD UP **ORR**
metallescens Fairmaire, 1896a: 94 (*Prionocerus*)

genus *Prionocerus* Perty, 1831: 33 type species *Prionocerus coeruleipennis* Perty, 1831
bicolor L. Redtenbacher, 1868: 109 **A**: SD TAI **ORR**
caeruleipennis Perty, 1831: 33 **A**: HAI JA SE **AFR AUR ORR**

brevicornis Schaufuss, 1887: 126
fuscipennis Lewis, 1879d: 464
tereticollis Sturm, 1826: 170

family MELYRIDAE Leach, 1815

tribe Cerallini Pic, 1929

genus *Anthodromius* L. Redtenbacher, 1850: 48 type species *Anthodromius variabilis* L. Redtenbacher, 1850
apicalis Schilsky, 1897b: 82 (*Cerallus*) **A**: "Turkestan"
basalis Pic, 1908d: 42 (*Cerallus*)
bicolor Kiesenwetter, 1872b: 317 (*Cerallus*) **A**: TD "Russia" "Turkestan"
longulus Reitter, 1896b: 38 (*Cerallus*)
carinulatus Reitter, 1902j: 229 (*Cerallus*) **A**: UZ
fulvipes Reitter, 1902j: 229 (*Cerallus*) **A**: UZ
hispanicus Kiesenwetter, 1872b: 317 (*Cerallus*) **E**: SP
inexpectatus Majer, 1990: 180 **A**: IN
kiesenwetteri Kraatz, 1882a: 98 (*Cerallus*) **A**: ES
laticollis Majer, 1990: 179 **A**: IN
marginatus Baudi di Selve, 1874a: 298 **A**: SY
nigrinus Reitter, 1902j: 228 (*Cerallus*) **A**: UZ
nigropilosus Reitter, 1902j: 228 (*Cerallus*) **A**: TM
pekinensis Fairmaire, 1889c: cci (*Cerallus*) **A**: BEI
piceus Majer, 1990: 176 **A**: IN
reticulatus Majer, 1990: 176 **A**: IN
robustus Majer, 1990: 180 **A**: IN
rufipennis Majer, 1990: 174 **A**: IN
rufolimbatus Pic, 1911u: 169 (*Cerallus*) **A**: SY
ludovici Pic, 1911u: 170 (*Cerallus*)
turkestanicus Heyden, 1886b: 186 **A**: AF TD UZ "Turkestan"
flavipes Schilsky, 1899: 35
haplocnemoides Reitter, 1902j: 229
nigripes Schilsky, 1899: 35
smaragdinus Reitter, 1896b: 37 (*Cerallus*)
variabilis L. Redtenbacher, 1850: 48 **A**: IN

genus *Cerallus* Jacquelin du Val, 1859d: 43 type species *Dasytes rubidus* Gyllenhal, 1817
Pristochira Morawitz, 1862: 316 type species *Pristochira varians* Morawitz, 1862
analis Gebler, 1830: 90 (*Dasytes*) **A**: "Turkestan"
brevicollis Kiesenwetter, 1872b: 315 **E**: ST **A**: "Turkestan"
coriaceus Majer, 1990: 171 **A**: IN
flavipennis Reitter, 1889m: 283 **E**: AB
interpositus Majer, 1990: 169 **A**: IN
luteus Kiesenwetter, 1872b: 315 **E**: TR
concolor Kiesenwetter, 1872b: 316
pallidipennis Majer, 1990: 165 **A**: IN
pilosus Reitter, 1889m: 283 **E**: AB GG HU ST "Caucasus" **A**: TM
rubidus Gyllenhal, 1817: 12 (*Dasytes*) **E**: HU IT SK ST YU
schilskyi Reitter, 1896b: 38 (*Cerallus*) **E**: AB
similis Majer, 1990: 167 **A**: IN
tataricus Baudi di Selve, 1874a: 299 **A**: "Russia" "Tatarstan"
varians Morawitz, 1862: 317 (*Pristochira*) **E**: HU SK ST **A**: TR
maculicollis Schilsky, 1897b: 77
rufifrons Schilsky, 1897b: 77

genus *Procerallus* Champion, 1922b: 129 type species *Dasytes aplocnemiformis* Pic, 1915
altivolans Champion, 1925c: 178 **A**: UP XIZ

aplocnemiformis Pic, 1915m: 97 (*Dasytes*) **A**: KA UP **ORR**
diversepunctatus Pic, 1920a: 8 (*Dasytes*)
picipes Champion, 1927: 85 **A**: HP

tribe Melyrini Leach, 1815

genus Falsomelyris Pic, 1913b: 15 type species *Melyris niger* Fabricius, 1781
Melyridella Champion, 1919b: 160 type species *Opatrum granulatum* Fabricius, 1792
crenicollis Pic, 1904m: 90 **N**: AG
granulatus Fabricius, 1792a: 90 (*Opatrum*) **E**: FR (Corse) IT (Sicilia) SP **N**: MO
andalusiaca Waltl, 1835: 62 (*Dasytes*)
opaca Schilsky, 1897b: 100 **E**: FR (Corse) IT (Sardegna, Sicilia) **N**: MO TU
rotundicollis Pic, 1894a: 13 **N**: AG
pallidipes Pic, 1898l: 273
rubripes P. H. Lucas, 1846: 201 **N**: AG MO TU
ruficrus Fairmaire, 1883e: clxxxix (*Melyris*) **N**: AG EG LB
amaliae Heyden, 1886a: 56 (*Melyris*)
aristidis Pic, 1894a: 13

genus *Flabellomelyris* Pic, 1942d: 77 type species *Flabellomelyris bicoloripennis* Pic, 1942
bicoloripennis Pic, 1942d: 77 **N**: "Central Sahara"

genus *Melyresthes* Reitter, 1891b: 27 type species *Melyresthes cardinalis* Reitter, 1891
cardinalis Reitter, 1891b: 27 **E**: AB

genus *Melyris* Fabricius, 1775: 58 type species *Melyris viridis* Fabricius, 1775
Zygia Fabricius, 1775: 126 type species *Zygia oblonga* Fabricius, 1775
bicalcarata Champion, 1919b: 209 **A**: SA
bicolor Fabricius, 1801a: 311 **E**: IT (Sicilia) **N**: EG **A**: SA SY
cucullata Champion, 1919b: 210 **A**: YE
cyanopectus Champion, 1919b: 211 **A**: YE
itigiensis Champion, 1919b: 190 **A**: YE (Suqutra)
klugii Baudi di Selve, 1874a: 315 (*Zygia*) **N**: EG **A**: SA **AFR**
dubia Schilsky, 1897b: 91 **N**: EG
longicollis Schilsky, 1897b: 88 (*Zygia*) **A**: CY SY
oblonga Fabricius, 1775: 126 (*Zygia*) **E**: IT (Sicilia) SP **N**: AG TU **A**: IQ SY "Persian Gulf" **NARi**
limbata Pic, 1895n: 126
lineatocollis Pic, 1899g: 113
maculicollis Fuente, 1913: 473
mulsanti Schilsky, 1897b: 94
reitteri Heyden, 1890a: 73 (*Zygia*) **N**: LB TU
marginata Heyden, 1890a: 73
rostrata Reiche & Saulcy, 1857b: 184 (*Zygia*) **A**: IS SY
melanocephala L. Redtenbacher, 1858: 548 (*Zygia*)
ruficolor Chobaut, 1897d: 239 **N**: AG
rufoscutellaris Pic, 1935b: 3 **E**: TR
scutellaris Mulsant, 1851b: 190 (*Zygia*) **N**: AG EG TU
shaumariensis Peacock, 1980: 236 **A**: JO
sinaita Pic, 1912r: 82 (*Zygia*) **A**: SI
versicolor Chevrolat, 1844: 50 **E**: FR (Corse) IT (Sardegna) SP TR **N**: AG EG **A**: SY "Arabia"

genus *Melyrodasytes* Pic, 1950e: 92 type species *Melyrodasytes costatus* Pic, 1950
costatus Pic, 1950e: 92 **N**: MO

genus *Melyrosoma* Wollaston, 1854: 253 type species *Melyrosoma oceanicum* Wollaston, 1854
abdominale Wollaston, 1857a: 87 **N**: MR

artemisiae Wollaston, 1854: 254 **N**: MR
blackmorei Wollaston, 1875: 218 **N**: MO
costipenne Wollaston, 1862c: 448 **N**: CI
flavescens Wollaston, 1862c: 449 **N**: CI
hirtum Wollaston, 1862c: 449 **N**: CI
oceanicum Wollaston, 1854: 253 **N**: MR

family DASYTIDAE Laporte, 1840

subfamily **Chaetomalachiinae** Majer, 1987

genus *Achaetomalachius* Majer, 1989e: 751 type species *Chaetomalachius tibialis* Iablokoff-Khnzorian, 1966
aeneonitens Champion, 1922b: 146 (*Dasytes*) **A**: HP NP UP
bengalensis Pic, 1907p: 172 (*Dasytiscus*) **A**: PA
cinereus Majer, 1996a: 9 **A**: KA
clarus Majer, 1996a: 16 **A**: AF
coloratus Majer, 1996a: 14 **A**: HP
confusus Majer, 1996a: 10 **A**: AF
cupreatus Majer, 1996a: 13 **A**: AF
danacaeoides Champion, 1922b: 148 (*Dasytes*) **A**: HP KA
discretus Gorham, 1895a: 323 (*Dasytes*) **A**: HP PA
kangraensis Champion, 1925c: 180 (*Dasytes*) **A**: HP
kashmirensis Champion, 1922b: 148 (*Dasytes*) **A**: KA
  crenulatus Pic, 1922l: 156 (*Dasytes*)
  dusuensis Pic, 1922l: 156 (*Dasytes*)
kumaonensis Champion, 1922b: 149 (*Dasytes*) **A**: UP
mutabilis Champion, 1922b: 146 (*Dasytes*) **A**: UP
nepalensis Majer, 1996a: 14 **A**: NP
praetinctus Majer, 1996a: 10 **A**: AF
rosti Majer, 1996c: 177 **A**: HP
scabricollis Champion, 1922b: 147 (*Dasytes*) **A**: HP KA
tenuicornis Majer, 1996a: 13 **A**: NP UP
tibetensis Champion, 1925c: 179 (*Dasytes*) **A**: XIZ
tibialis Iablokoff-Khnzorian, 1966a: 79 (*Chaetomalachius*) **A**: AF
vagus Majer, 1996a: 6 **A**: PA

genus *Asiothrix* Majer, 1996d: 472 type species *Dasytiscus minutus* Wittmer, 1954
minuta Wittmer, 1954e: 322 (*Dasytiscus*) **A**: YE

genus *Brodskyana* Majer, 1996d: 484 type species *Dasytiscus semipallidus* Reitter, 1899
furcilla Majer, 1996d: 487 **A**: JO
juncta Majer, 1996d: 485 **A**: IN
lubrica Majer, 1996d: 485 **A**: IN
semipallida Reitter, 1899d: 275 (*Dasytiscus*) **E**: AR
  semipallens Schilsky, 1900: 7 (*Dasytiscus*)

genus *Chaetomalachius* Kraatz, 1882a: 96 type species *Chaetomalachius dasytoides* Kraatz, 1882
  Chaetomalachius Majer, 1989d: 349 type species *Chaetomalachius staudingeri* Schilsky, 1900
  Trithrix Schilsky, 1896a: 32 HN type species *Dasytiscus longipilis* Reitter, 1889
acutipennis Majer, 1989d: 347 **A**: TD
aeneolus Majer, 1989d: 325 **A**: AF IN TD TM
aeneus Schilsky, 1900: 5 **A**: UZ KI TD TM
australis Majer, 1988b: 426 **A**: AF
bactrianus Iablokoff-Khnzorian, 1966a: 83 **A**: KI TD UZ
bilyi Majer, 1989d: 345 **A**: TD

cyanellus Kiesenwetter, 1863a: 625 (*Dasytiscus*) **A**: LE SY TR
dasytoides Kraatz, 1882a: 97 **A**: IN KI KZ TD TM UZ
    transcaspicus Bourgeois, 1885c: 256 (*Dasytiscus*)
dispar bodemeyeri Majer, 1989d: 317 **A**: IN SY
dispar dispar Schilsky, 1896a: 54 **A**: IN SY
forticornis Schilsky, 1896a: 50 **A**: KI KZ UZ
hauseri Schilsky, 1894a: 31 (*Dasytes*) **A**: TD KZ UZ
insidiosus Iablokoff-Khnzorian, 1966a: 81 **A**: AF
jelineki Majer, 1989d: 319 **A**: IN
klapperichi Iablokoff-Khnzorian, 1966a: 80 **A**: AF
kotschyi Schilsky, 1896a: 51 **A**: IN
largus Majer, 1988b: 422 **A**: AF
longipilis Reitter, 1889d: 111 (*Dasytiscus*) **A**: TM
    pilipes Reitter, 1889d: 111 (*Dasytiscus*)
    ruficornis Schilsky, 1896a: 52
marginicollis Reitter, 1891b: 26 (*Dasytes*) **E**: AB AR **A**: TR IN TM
minutipenis Tshernyshev, 2002: 212 **A**: KZ
pulcher Majer, 1988b: 424 **A**: AF
punctatus Schilsky, 1896a: 55 **N**: EG
reitteri Majer, 1988b: 421 **E**:: AB AR **A**: IN TR
staudingeri Schilsky, 1900: 6 **A**: TD UZ
    crinipes Iablokoff-Khnzorian, 1966a: 83
wittmeri Majer, 1989d: 329 **A**: IN KI TM

genus *Danaceothrix* Majer, 1989e: 755 type species *Danaceothrix murina* Majer, 1989
glaberrima Majer, 1996d: 470 **A**: SCH
monilicornis Champion, 1922b: 150 (*Dasytes*) **A**: UP
    brevis Champion, 1925c: 169 (*Dasytes*)
    breviusculus Champion, 1922b: 150 (*Dasytes*) [HN]
murina Majer, 1989e: 756 **A**: AF

genus *Dasythrix* Majer, 1996d: 462 type species *Dasytidius flavoniger* Majer, 1991
flavoniger Majer, 1991b: 1 (*Dasytidius*) **A**: IN

genus *Dasytidius* Schilsky, 1896a: 32 H type species *Dasytiscus indutus* Kiesenwetter, 1859
addillaensis Wittmer, 1979d: 191 (*Dasytiscus*) **A**: SA YE
alfierii Wittmer, 1935c: 188 (*Dasytiscus*) **A**: IQ IS JO SI SY
    atratus Majer, 1991a: 2
aurescens Majer, 1991a: 4 **A**: SY
bourgeoisi Schilsky, 1896a: 72 (*Dasytiscus*) **N**: MO
    otini Majer, 1990a: 47
brevicornis Majer, 1996c: 164 **A**: TR
candidus Majer, 1991b: 7 **A**: AF
congruens Majer, 1989c: 143 **E**: GR (Rodos, Sporades) **A**: TR
constantini Majer, 1990a: 46 **N**: MO
crassicornis Majer, 1996c: 162 **A**: SY
crenulatus Pic, 1924a: 3 (*Danacea*) **N**: LB
    convexus Pic, 1928i: 103 (*Dasytiscus*)
deportatus Peyerimhoff, 1929c: 194 (*Dasytiscus*) **N**: AG
diversimembris Pic, 1937a: 52 (*Dasytiscus*) [RN] **N**: AG MO
    diversipes Pic, 1922h: 30 (*Dasytiscus*) [HN]
emgei Reitter, 1884c: 79 (*Dasytiscus*) **E**: GR
ethologus Majer, 1989c: 145 **E**: GR (Kríti)
fulvipes Reitter, 1885e: 244 (*Dasytiscus*) **A**: IS LE SY TR
funebris Majer, 1989c: 153 **A**: TR

gracilis Escalera, 1914: 246 (*Dasytiscus*) **N**: AG MO
hauseri Reitter, 1890i: 360 (*Dasytiscus*) **A**: UZ
impar Majer, 1996c: 171 **A**: TR
inchoatus Majer, 1991a: 5 **A**: CY
indutus aegaeicus Liberti, 1986: 188 (*Dasytiscus*) **E**: GR (Aegean Islands, Kríti, Rodos)
indutus dalmatinus Majer, 1989c: 152 **E**: CR
indutus indutus Kiesenwetter, 1859c: 179 **E**: GR
indutus similis Schilsky, 1896a: 77 (*Dasytiscus*) **A**: CY TR
  nigricornis Schilsky, 1896a: 66
infinitus Majer, 1991a: 11 **A**: IS JO SY TR
insularis Majer, 1996c: 162 **E**: GR (Lesvos)
kalalovae Majer, 1991a: 3 **A**: IN IQ TR
kubani Majer, 1991b: 10 **E**: AR
laticollis Bourgeois, 1885c: 267 (*Dasytiscus*) **A**: IQ
latissimus Majer, 1989c: 147 **A**: TR
  avius Majer, 1989c: 149
licenti Pic, 1938k: 162 (*Dasytiscus*) **A**: CH
longiventris Majer, 1991b: 10 **A**: IN
maceki Majer, 1991a: 9 **A**: IQ
malkini Majer, 1996c: 171 **A**: TR
mantici Majer, 1996b: 88 **N**: MO
margelanus Majer, 1991b: 5 **A**: TD
marsaleki Majer, 1996c: 164 **A**: KI
medius Rottenberg, 1871: 244 (*Dasytiscus*) **E**: IT (Sicilia) **N**: AG MO TU
  beckeri Kiesenwetter, 1871: 86 (*Dasytiscus*)
  obesus Kiesenwetter, 1871: 85 (*Dasytiscus*)
  pexus Kiesenwetter, 1871: 85 (*Dasytiscus*)
  sedilloti Bourgeois, 1885c: 268 (*Dasytiscus*)
  squamatus Kiesenwetter, 1871: 86 (*Dasytiscus*)
  theresae Pic, 1896e: 48 (*Dasytiscus*)
melitensis Bourgeois, 1885c: 270 (*Dasytiscus*) **E**: MA
muehlei Majer, 1996c: 168 **A**: YE
nigripes Pic, 1894p: 112 (*Dasytiscus*) **E**: AB GR (Rodos) **A**: CY IN IQ SY TR
  attenuatus Majer, 1991a: 7
  obscuripes Pic, 1896e: 48 (*Dasytiscus*)
  svihlai Majer, 1991b: 6
normandi Majer, 1990a: 43 **N**: TU
opertus Majer, 1991b: 15 **A**: IN
optivus Majer, 1989c: 143 **A** TR
pardoi Majer, 1990a: 49 **N**: MO
persicus Pic, 1926a: 1 (*Dasytiscus*) **A**: IN
petrowi Pic, 1923i: 9 (*Dasytiscus*) **N**: EG LB
princeps Majer, 1991a: 14 **A**: IN IQ
prosperus Majer, 1989c: 141 **A**: TR
quadricollis Schilsky, 1896a: 75 (*Dasytiscus*) **A**: IN
quaesitus Majer, 1991b: 15 **A**: IN
ragusae Procházka, 1895: 139 (*Dasytiscus*) **E**: IT (Sicilia) **N**: LB TU
  gestroi Schilsky, 1897b: 75 (*Dasytiscus*)
recticollis Majer, 1996c: 163 **A**: SY
robustus Majer, 1991b: 12 **A**: IN
rufimanus Bourgeois, 1885c: 269 (*Dasytiscus*) **A**: TD UZ
  atripes Schilsky, 1896a: 79 (*Dasytiscus*)
sequensi Reitter, 1902g: 210 (*Dasytiscus*) **E**: AR **A**: KZ
sparsepubens Pic, 1924a: 2 (*Dasytiscus*) **N**: LB
subnudus Majer, 1991a: 9 **A**: IQ IS JO SY

subsyriacus Majer, 1991a: 7 **A**: JO SY
syriacus Reitter, 1885e: 245 (*Dasytiscus*) **A**: CY IS JO LE SY TR
*obscuripes J. R. Sahlberg, 1913*d: 151 (*Dasytiscus*)
syrticus Bourgeois, 1885c: 266 (*Dasytiscus*) **N**: AG TU
neglectus Schilsky, 1897a: 84 (*Dasytiscus*)
tajikistanus Majer, 1991b: 9 **A**: TD
transversus Majer, 1991b: 14 **A**: IN
turnai Majer, 1996c: 169 **A**: YUN
ugamicus Majer, 1991b: 3 **A**: UZ
vestitus Kiesenwetter, 1863a: 625 (*Dasytiscus*) **N**: AG MO TU
henoni Pic, 1900j: 88 (*Dasytiscus*)
virescens Baudi di Selve, 1873b: 319 (*Dasytiscus*) **A**: CY
wartmanni Reitter, 1897f: 219 (*Dasytiscus*) **N**: AG
nigrofemoralis Pic, 1894p: 112 (*Dasytiscus*)
wittmeri Majer, 1991a: 13 **A**: SA

genus *Dasytiscus* Kiesenwetter, 1859c: 179 type species *Dasyes rufitarsis* P. H. Lucas, 1853
abeillei Bourgeois, 1885c: 261 **A**: IS JO
concolor Schilsky, 1896a: 83 (*Dasytiscus*)
affinis Morawitz, 1862: 318 **E**: AB AL AR BU CR GG GR MC MD RO ST TR UK YU **A**: TR
rotundicollis Reitter, 1885e: 246
amasyanus Majer, 1996d: 482 **A**: TR
fallax Majer, 1988a: 140 **A**: IS JO
flaveolus Reitter, 1889j: 257 **E**: GR (Rodos Sporades) **A**: TR
graminis Majer, 1996d: 480 [RN] **E**: CR GR
rufitarsis P. H. Lucas, 1853: 571 (*Dasytes*) [HN]
hebraicus Bourgeois, 1883b: liii **A**: IS LE
hybridus Reitter, 1885e: 244
heydeni Reitter, 1891f: 226 [RN] **A**: KZ TD UZ
analis Reitter, 1890i: 361 [HN]
hissarus Majer, 1988a: 133 **A**: TD
hladili Majer, 1988a: 137 **A**: TR
impressicollis Reitter, 1885e: 245 **E**: AL CR GR YU
jordanicus Majer, 1988a: 142 **A**: JO
klapperichi Majer, 1988a: 129 **A**: AF
minimus J. R. Sahlberg, 1903a: 32 (*Dasytes*) **A**: IN
aenescens J. R. Sahlberg, 1903a: 32 (*Dasytes*)
bicoloriceps Pic, 1926a: 1 (*Dasytiscus*)
collaris J. R. Sahlberg, 1903a: 32 (*Dasytes*)
infuscatus J. R. Sahlberg, 1903a: 32 (*Dasytes*)
verticalis J. R. Sahlberg, 1903a: 32 (*Dasytes*)
minotaurus Majer, 1988a: 132 **E**: GR (Kríti)
pallens Majer, 1988a: 136 **A**: IN
praecox Küster, 1851a: 57 (*Dasytes*) **E**: GR **A**: TR
rufipennis Majer, 1996d: 480 **A**: TR
schereri Majer, 1988a: 138 **A**: IS JO
simulator Majer, 1988a: 140 **A**: IS
strejcekorum Majer, 1996c: 180 **E**: AR

genus *Dasytiscus*, nomen dubium
graminicola Kiesenwetter, 1859c: 180

genus *Euthrix* Majer, 1996d: 476 type species *Euthrix sulcicollis* Majer, 1996
lata Majer, 1996d: 477 **A**: AF
sulcicollis Majer, 1996d: 477 **A**: PA

genus *Haplothrix* Schilsky, 1896a: 32 H type species *Dasytiscus aequalis* Reitter, 1885
aequalis Reitter, 1885e: 246 (*Dasytiscus*) **E**: GR
armeniaca Kiesenwetter, 1878: 215 (*Dasytiscus*) **E**: AR
captiosa Majer, 1989b: 38 **A**: TR
chalif Majer, 1989b: 41 **A**: IN IQ
effusa Majer, 1989b: 35 **A**: IN
leucosa Majer, 1989b: 37 **A**: AF
mera Majer, 1989b: 39 **A**: TR
pospisili Majer, 1989b: 36 **E**: GR (Rodos, Sporades)
prudeki Majer, 1996b: 90 **A**: TR
pusilla Schilsky, 1896a: 85 (*Dasytiscus*) **A**: TR
ruficollis Reitter, 1889a: 25 (*Dasytiscus*) **E**: AR GG **A**: TR
femoralis Schilsky, 1896a: 59
nigricollis Schilsky, 1896a: 59
sanguinicollis Majer, 1989b: 33 **A**: AF
soluta Majer, 1989b: 34 **A**: IN
subtilis Reitter, 1885e: 245 (*Dasytiscus*) **E**: GR (Kríti, Rodos) **A**: CY TR
puberulus Bourgeois, 1885c: 262 (*Dasytiscus*)
vana Majer, 1989b: 33 **A**: HP

genus *Hladilium* Majer, 1996d: 443 type species *Dasytes cambiensis* Gorham, 1895
amplexum Champion, 1925c: 179 (*Dasytes*) **A**: SD
cambiense Gorham, 1895a: 323 (*Dasytes*) **A**: HP UP **ORR**
assamensis Pic, 1922l: 155 (*Dasytes*)
kuluensis Pic, 1922l: 155 (*Dasytes*)
himalayanum Pic, 1911n: 133 (*Dasytes*) **A**: KA
insulcatum Pic, 1922l: 156 (*Dasytes*) **A**: HP
molestum Majer, 1996d: 455 **A**: NP
obstinatum Majer, 1996d: 455 **A**: NP
perplexum Majer, 1996d: 458 **A**: NP
perpolitum Majer, 1996d: 449 **A**: KA
speculiferum Champion, 1922b: 149 (*Dasytes*) **A**: UP

genus *Mimothrix* Majer, 1989e: 752 type species *Mimothrix olivacea* Majer, 1989
agnoscenda Pic, 1907p: 172 (*Dasytiscus*) **A**: YUN
atrotibialis Pic, 1917e: 5 (*Dasytiscus*) **A**: YUN
diversipes Pic, 1917e: 5 (*Dasytiscus*)
baishuiensis Majer, 1995: 188 **A**: YUN
bella Majer, 1995: 195 **A**: AF
cinis Majer, 1995: 183 **A**: AF TD
crassifemur Majer, 1995: 197 **A**: BT
cyanea Majer, 1995: 185 **A**: NP
cybaea Majer, 1995: 179 **A**: AF
donjuan Majer, 1996b: 88 **A**: YUN
extranea Majer, 1995: 198 **A**: AF
fraterna Majer, 1995: 186 **A**: AF
histrio Majer, 1995: 197 **A**: SD
jendeki Majer, 1995: 188 **A**: YUN
kubani Majer, 1995: 191 **A**: YUN
maowensis Majer, 1995: 192 **A**: SCH
matsumurai Majer, 1995: 186 **A**: NP
menieri Majer, 1995: 194 **A**: YUN
montivaga Majer, 1995: 202 **A**: YUN
olivacea Majer, 1989e: 753 **A**: BT NP SD

pamirensis Majer, 1996c: 177 **A**: TD
rhododendri Majer, 1996b: 88 **A**: YUN
roshtkalensis Majer, 1995: 181 **A** TD
scortea Majer, 1995: 185 **A**: KA
sparsehirsuta Pic, 1907p: 172 (*Dasytiscus*) **A**: YUN
trapezicollis Majer, 1995: 183 **A**: KA (Ladakh)

genus *Neothrix* Majer, 1989e: 754 type species *Dasytes aspericollis* Champion, 1925
aspericollis Champion, 1925c: 180 (*Dasytes*) **A**: UP

genus *Parathrix* Majer, 1989e: 748 type species *Dasytiscus plumbeus* Kiesenwetter, 1878
aeneola Reitter, 1889a: 25 (*Dasytiscus*) **E**: AB AR **A**: IN
anatolica Reitter, 1902g: 209 (*Dasytiscus*) **A**: TR
attonita Majer, 1990a: 17 **A**: IN
bimetallica Abeille de Perrin, 1894b: 93 (*Dasytiscus*) **A**: IQ
facunda Majer, 1990a: 19 **A**: SY
gravida Majer, 1990a: 16 **E**: AB
plumbea Kiesenwetter, 1878: 214 (*Dasytiscus*) **E**: AB
tristis Majer, 1996d: 462 **A**: IN

subfamily **Danaceinae** C. G. Thomson, 1859

tribe Amauroniodini Majer, 1987

genus *Amauronia* Westwood, 1839: 174 type species *Amauronia subaenea* Westwood, 1839
    Aphyctus Jacquelin du Val, 1860f: 142 type species *Amauronia megacephala* Kiesenwetter, 1859
brenskei Reitter, 1884c: 80 (*Aphyctus*) **E**: GR
    bicoloriceps Pic, 1923a: 2
brevicornis Schilsky, 1897b: 71 (*Aphyctus*) **E**: BU MC RO TR UK **A**: TR
insignis Majer, 1997: 377 **A**: JO
maritima Majer, 1997: 373 **E**: GR (Kríti, Lesbos, Naxos)
pendula Majer, 1997: 375 **E**: RO
procera Majer, 1997: 371 **E**: CR
saginata Kiesenwetter, 1873: 31 (*Aphyctus*) **E**: GR
subaenea Westwood, 1839: 175 **E**: BU GR MC RO TR YU
    megacephala Kiesenwetter, 1859c: 182
    muelleri Kiesenwetter, 1873: 32 (*Aphyctus*)
    robusta Pic, 1923a: 2

genus *Indamauronia* Majer, 1997: 397 type species *Amauronia westwoodi* Champion, 1922
westwoodi Champion, 1922b: 130 (*Amauronia*) **A**: HP NP UP
    limbatithorax Pic, 1907p: 172 (*Dasytiscus*)

genus *Mauroania* Majer, 1997: 385 type species *Amauronia picta* Kiesenwetter, 1859
bettingeri Pic, 1912f: 17 (*Amauronia*) **N**: AG MO
bourgeoisi Pic, 1894q: 114 (*Amauronia*) **E**: PT SP
elegans Kiesenwetter, 1867b: 127 (*Amauronia*) **E**: SP
    infasciata Pic, 1922g: 26 (*Amauronia*)
    viridescens Pic, 1894q: 114 (*Amauronia*)
hispana Kiesenwetter, 1859c: 181 (*Amauronia*) **E**: SP
juniperi Bourgeois, 1894c: cciv (*Amauronia*) **N**: AG MO
longula Desbrochers des Loges, 1889: clxxxv (*Amauronia*) **E**: SP **N**: AG TU
picta Kiesenwetter, 1859c: 181 (*Amauronia*) **E**: AL GR **A**: TR

genus *Perisphinctus* Schilsky, 1897b: 68 type species *Aphyctus picticollis* Heyden, 1889
picticollis Heyden, 1889a: 328 (*Aphyctus*) **A**: KZ UZ
    charopoides Reitter, 1891b: 27 (*Aphyctus*)

turanicus Majer, 1997: 369 **A**: KI TD UZ

genus *Pseudamauronia* Pic, 1915k: 45 type species *Amauronia violacea* Abeille de Perrin, 1894
Amauronioides Champion, 1923b: 13 type species *Amauronioides schilskyi* Champion, 1923
danacaeoides Reitter, 1885e: 245 (*Dasytiscus*) **E**: GR
schilskyi Champion, 1923b: 13
hirtula Reitter, 1885e: 247 (*Dasytiscus*) **E**: AZ **A**: IN
brevicollis Pic, 1926a: 1 (*Amauronia*)
caspicus Schilsky, 1896a: 45 (*Dasytes*)
violacea Abeille de Perrin, 1894b: 93 (*Amauronia*) **A**: LE SY TR

tribe Danacaeomimini Majer, 1987

genus *Cyrdobaldus* Majer, 1989a: 4 type species *Cyrdobaldus nigrofemoralis* Majer, 1989
hybridus Majer, 1989a: 6 **A**: IN
nigrofemoralis Majer, 1989a: 6 **A**: IN

genus *Cyrdothrix* Majer, 1989a: 25 type species *Cyrdothrix hirta* Majer, 1989
atra Majer, 1989a: 26 **A**: KA
hirta Majer, 1989a: 28 **A**: UZ
simplicitarsis Champion, 1925c: 180 (*Dasytes*) **A**: XIZ

genus *Danacaeomimus* Champion, 1922b: 130 type species *Danacaeomimus nigropectus* Champion, 1922
albellus Majer, 1989a: 12 **A**: IN
danilevskii Majer, 1989a: 14 **A**: TD
fausti Reitter, 1889a: 26 (*Dasytiscus*) **A**: KZ UZ
glaber Majer, 1989a: 17 **A**: AF
gracilis Majer, 1989a: 17 **A**: AF
haemorrhoidalis Ballion, 1878: 296 (*Danacea*) **A**: "West China"
kolibaci Majer, 1996b: 86 **A**: SCH
licenti Majer, 1996b: 85 **A**: SHX
longicollis Majer, 1989a: 14 **A**: AF
maleci Majer, 1989a: 20 **A**: KI KZ UZ
nadari Pic, 1896j: 148 (*Danacaea*) **A** UZ
diluticornis Reitter, 1902g: 211 (*Dasyticus*)
nigropectus Champion, 1922b: 131 **A**: UP
obenbergeri Reitter, 1915b: 43 (*Danacaeina*) **A**: TD UZ
peacockae Majer, 1989a: 11 **A**: PA
pubescens Pic, 1895m: 122 (*Danacaea*) **A**: KI TD UZ WS
sahlbergi Pic, 1898f: 118 (*Danacaea*)
turkestanicus Schilsky, 1897b: 74 (*Dasytiscus*)
puellaris Majer, 1989a: 17 **A**: IN
sericeus Majer, 1989a: 12 **A**: IN
shivae Majer, 1989a: 13 **A**: AF
songpanensis Majer, 1996b: 86 **A**: SCH

genus *Piechockia* Majer, 1986: 307 type species *Piechockia ingens* Majer, 1986
ingens Majer, 1986: 307 **A**: MG

genus *Schoenia* Majer, 1989a: 23 type species *Schoenia grandis* Majer, 1989
grandis Majer, 1989a: 25 **A**: UZ
korbunovi Majer, 1989a: 24 **A**: KI

tribe Danaceini Kiesenwetter, 1863

genus *Danacea* Laporte, 1838: 31 type species *Lagria pallipes* Panzer, 1793
subgenus *Allodanacaea* Liberti, 1985: 336 type species *Danacaea romana* Pic, 1902

caneparii Liberti, 1985: 346 **E**: IT (Pantelleria) **N**: TU (Archipel La Galite)
caprariae Liberti, 1985: 338 **E**: IT (Capraia, Gorgona)
constantini Liberti, 1985: 351 **E**: FR (Corse)
elongatipennis Pic, 1917i: 22 **E**: IT (Sicilia)
ilicis Liberti, 1985: 341 **E**: IT (Giannutri Giglio Mte Argentario)
jonica Liberti, 1985: 343 **E**: IT (Calabria)
milleri Schilsky, 1897a: 49 **E**: FR (Corse) IT (Sardegna)
nympha Liberti, 1985: 348 **E**: IT (Sardegna)
oreas Liberti, 1985: 350 **E**: IT (Sardegna)
picicornis picicornis Küster, 1850: 18 (*Cosmiocomus*) **E**: IT (Sardegna)
picicornis supramontana Liberti, 1985: 354 **E**: IT (Sardegna)
misella Baudi di Selve, 1873a: 314
poggii Liberti, 1985: 340 **E**: IT (Montecristo)
romana Pic, 1902e: 31 **E**: IT
sicana Liberti, 1985: 345 **E**: IT (Sicilia)
sulcitana Liberti, 1985: 356 **E**: IT (Sardegna)
thymi Liberti & Schembri, 2002: 180 **E**: MA
wittmeri Liberti, 1985: 342 **E**: IT (Campania, Capri, Ischia)
subgenus *Danacea* Laporte, 1838: 31 type species *Lagria pallipes* Panzer, 1793
Cosmiocomus Küster, 1850: 10 type species *Dasytes imperialis* Gené, 1836
Dermatoma Motschulsky, 1845a: 37 type species *Lagria pallipes* Panzer, 1793
acutangula Schilsky, 1897a: 15 **E**: SP
nevadensis Pic, 1904l: 81
aenea Morawitz, 1862: 319 **E**: UK
albanica Apfelbeck, 1911: 217 **E**: AL
albella Reitter, 1900b: 86 **A**: TR
amabilis J. R. Sahlberg, 1903a: 29 **E**: GR (Kríti)
ambigua Mulsant & Rey, 1868d: 282 **E**: AU CR FR IT SL SP SZ
reyi Chobaut, 1895: clvi
strangulaticollis Pic, 1937b: 1
subacutipennis Pic, 1914p: 73
tyrolensis Procházka, 1894: 28
anatolica Schilsky, 1897a: 58 **A**: TR
angulata Küster, 1850: 10 (*Cosmiocomus*) **E**: BH CR YU
incana Reitter, 1884f: 256
ochripes Schilsky, 1897a: 23
apicalis Pic, 1894u: 135 **A**: SY
abeillei Chobaut, 1895: clvi
argentata J. R. Sahlberg, 1903a: 30 **A**: UZ
atripes Graells, 1858: 57 **E**: PT SP
ornata Procházka, 1894: 11
tibialis Pic, 1894m: 100
aurichalcea Küster, 1850: 17 (*Cosmiocomus*) **E**: CR IT
barrosi Pic, 1907o: 169 **E**: PT
baudii Schilsky, 1897a: 36 **E**: UK
biskrensis Pic, 1894m: 106 **N**: AG
bleusei Pic, 1895m: 122 **E**: GR (Rodos)
bosnica Pic, 1913f: 105 **E**: BH
bourgeoisi Zurcher, 1911a: 242 **A**: CY
brevipennis Pic, 1932h: 29 **A**: CY
bucciarellii Liberti, 1979: 36 **E**: IT (Sicilia)
bulgarica Pic, 1902m: 79 **E**: BU
cartennaeensis Constantin, 1988: 350 **N**: AG
caucasica Schilsky, 1897a: 32 **E**: GG ST
cavifrons Pic, 1895m: 122 **A**: SY

cervina cervina Küster, 1850: 12 (*Cosmiocomus*) **E**: MC RO YU
cervina montenegrina Apfelbeck, 1911: 218 **E**: YU
cervina vulpina Apfelbeck, 1911: 218 **E**: CR
diversipes Pic, 1936a: 3
championi Marseul, 1878b: 55 **E**: GR
fallax Schilsky, 1897a: 24
citrina Procházka, 1894: 31 **A**: LE
clavipes Schilsky, 1897a: 61 **E**: SP
coiffaiti Constantin, 1990: 63 **E**: FR
concii Liberti, 1984a: 170 **E**: IT
confinis Schilsky, 1897a: 54 **E**: SP
conicicollis Schilsky, 1897a: 26 **A**: SY
consimilis Schilsky, 1897a: 69 **E**: GR (Rodos)
corsica Kiesenwetter, 1871: 86 **E**: FR (Corse) IT (Sardegna)
luteipes Schilsky, 1897a: 33
crenulata Pic, 1924a: 3 **N**: LB
cretica Kiesenwetter, 1859c: 185 **E**: GR (Kríti)
cusanensis A. Costa, 1847a: 105 (*Dasytes*) **E**: IT
cylindricollis Schilsky, 1897a: 70 **E**: GR
cypria Schilsky, 1897a: 3 **A**: CY
lineatithorax Pic, 1910k: 41
livadiensis Pic, 1910k: 41
maculaticeps Pic, 1910k: 41
dauci Liberti, 1984a: 176 **E**: IT (Giglio and Giannutri, Mte Argentario)
delagrangei Pic, 1895m: 121 **A**: TR
delphini Liberti, 1984a: 172 **E**: IT (Liguria)
dentatithorax Pic, 1907k: 153 **A**: TR
simplicithorax Pic, 1909f: 113
denticollis Baudi di Selve, 1861: 182 **E**: FR IT SZ
denticulata Pic, 1903t: 177 **E**: SP
distincta P. H. Lucas, 1846: 201 (*Dasytes*) **E**: IT (Pantelleria) **N**: AG TU
latior Pic, 1894u: 135
regalis Schilsky, 1897a: 9
viridescens Pic, 1894m: 105 (*Danacaea*)
eludens Liberti & Schembri, 2002: 179 **E**: MA
femoralis Schilsky, 1907: 100 **E**: SP
flava Procházka, 1894: 31 **E**: AR GG **A**: TR
fuscoaenea Fairmaire, 1880d: 241 **E**: SP
ganglbaueri Procházka, 1894: 31 **E**: GR (Kríti)
genistae Marseul, 1868b: 190 **A**: LE SY TR
berytensis Pic, 1902e: 31
grandiceps, Pic, 1927h: 13 **A**: TR
gridellii Pic, 1955e: 131 **N**: LB
hispanica Gougelet & H. Brisout de Barneville, 1859: ccxxxvii (*Cosmiocomus*) **E**: SP
heydeni Schilsky, 1897a: 4
testaceipes Pic, 1897e: 95
holtzi Pic, 1904b: 3 **A**: TR
ciliciensis Zurcher, 1911b: 255
hypoleuca Kiesenwetter, 1859c: 184 **E**: GR **A**: CY
imperialis Gené, 1836: 180 (*Dasytes*) **E**: IT (Sardegna)
solarii Pic, 1895m: 122
testaceoapicalis Pic, 1947: 6
uniformis Schilsky, 1897a: 13
versicolor Schilsky, 1897a: 13
induta Schilsky, 1897a: 29 **E**: GG ST **A**: TR
atricornis Pic, 1922e: 17

simulatrix Schilsky, 1897a: 29
iners Kiesenwetter, 1859c: 184 **E**: GR
insularis Schilsky, 1897a: 28 **E**: GR (Dodekanisos)
intermedia Apfelbeck, 1911: 218 **E**: BH
kiesenwetteri Heyden, 1870: 127 **E**: SP **A**: AG
klapperichi Constantin, 1983: 331 **A**: TR
korbi Schilsky, 1897a: 73 **E**: SP **N**: AG
kraatzi Pic, 1895m: 121 **N**: TU
krueperi Schilsky, 1897a: 22 **E**: GR
varipes Schilsky, 1897a: 22
lata Kiesenwetter, 1867b: 125 **E**: SP
latipennis Pic, 1903m: 146 **E**: GR (Kríti)
leonardii Liberti, 1979: 42 **E**: FR (Corse) IT (Arcipelago Toscana)
leprieuri Pic, 1894m: 105 **N**: AG TU
obscuripes Schilsky, 1897a: 33 O
picipes Schilsky, 1897a: 48
ruficornis Schilsky, 1897a: 48
ligurica Liberti, 1984a: 174 **E**: IT (Liguria)
limbata Schilsky, 1897a: 62 **E**: GR
longepubens Pic, 1927h: 13 **N**: AG
longiceps Mulsant & Rey, 1868d: 305 **E**: FR SP
amica Pic, 1895d: 68
cusanensis Mulsant & Rey, 1868d: 301 [HN]
paradoxa Mulsant & Rey, 1868d: 309
semiflava Pic, 1895d: 68
longipilis Pic, 1894m: 101, 105 **N**: AG
lucana Liberti, 1989: 290 **E**: IT (south)
luctuosa Desbrochers des Loges, 1889: clxxxv **N**: AG
luigionii Pic, 1920d: 9 **E**: IT
lusitana Heyden, 1870: 127 **E**: PT
lusitanica Procházka, 1894: 14
luteipalpis Schilsky, 1907: 99 **E**: CR
luteopubens Pic, 1922g: 26 **E**: SP
lysholmi Pic, 1900c: 27 **A**: CY SY TR
subrobusta Pic, 1903q: 162
maculicornis Pic, 1914p: 73 **E**: GR (Rodos)
major Pic, 1902i: 56 **E**: GR
marginata Küster, 1851a: 58 (*Cosmiocomus*) **E**: AL BH BU CR GR HU IT MC SK SL TR YU **A**: CY SY TR
fulvescens Brancsik, 1910: 188
graeca Procházka, 1894: 16
maculipennis Pic, 1899j: 207
reitteri Procházka, 1894: 33
rufula Schilsky, 1897a: 47
thoracica Schilsky, 1897a: 47
martini Pic, 1904b: 2 **E**: SP
pardalensis Pic, 1904b: 2
mecheriensis Pic, 1898f: 117 **N**: AG
nigrofemorata Pic, 1898f: 118
micans Procházka, 1894: 32 **E**: AR GG **A**: TR
astrabadensis Pic, 1922e: 18
minuta Pic, 1894m: 102, 105 **N**: AG MO
aphyctoides Procházka, 1894: 12
batnensis Schilsky, 1897a: 76
bicoloripes Pic, 1946: 5
dubia Schilsky, 1897a: 72
obscuripes Pic, 1895d: 68
mitis Küster, 1850: 11 (*Cosmiocomus*) **E**: IT (Sardegna)

laevicollis Baudi di Selve, 1873a: 313
monastirensis Pic, 1917i: 22 **E**: MC (Bitola)
jupillei Pic, 1927q: 375
montivaga Mulsant & Rey, 1868d: 284 **E**: FR IT SZ
moreana Pic, 1905i: 161 **E**: GR
morosa Kiesenwetter, 1863a: 664 **E**: AU BH HU IT PL RO SK SL
infuscata Schilsky, 1897a: 56
murina Küster, 1850: 15 (*Cosmiocomus*) **E**: CR YU
macrocephala Schaufuss, 1862a: 19
mutata Pic, 1895c: 69 **E**: GR
cephalonica Pic, 1901m: 89
tibialis Procházka, 1894: 20
nana Kiesenwetter, 1863a: 661 **E**: SP
neglecta Schilsky, 1897a: 35 **E**: PT
nigripalpis Fiori, 1912b: 133 **E**: IT (Sicilia) MA
nigritarsis alpina Pic, 1894m: 102 **E**: FR IT SL SZ
nigritarsis brunneipes Pic, 1927h: 13 **E**: IT
nigritarsis ingauna Liberti, 1984a: 165 **E**: FR IT
nigritarsis nigritarsis Küster, 1850: 14 (*Cosmiocomus*) **E**: AU BE CZ FR GE HU NL PL RO SK SP SV SZ UK
antennata Schilsky, 1897a: 74
nova Jakobson, 1911b: 710 [RN]
obscuritarsis Pic, 1895d: 68
opacula Abeille de Perrin, 1894b: 94
tomentosa Panzer, 1797d: 12 (*Byturus*) [HN]
nigritarsis zolotarewi Pic, 1910i: 34 **E**: GG ST
nitidissima Pic, 1904k: 73 **A**: TR
normandi Constantin, 1988: 351 **N**: TU
obscura Schilsky, 1897a: 27 **E**: GR
ochroleuca Schilsky, 1897a: 20 **A**: TR
oertzeni Schilsky, 1897a: 68 **E**: GR (Dodekanisos)
olivacea Baudi di Selve, 1873a: 311 **A**: IN TR
olympiaca Schilsky, 1897a: 53 **E**: GR
opulenta Schilsky, 1897a: 14 **E**: GR MC
macedonica Pic, 1909f: 113
pallidipalpis Abeille de Perrin, 1894b: 93 **E**: AL CR GR HU IT RO SL YU
merditana Apfelbeck, 1911: 218
pallidipennis Schilsky, 1896b: 364
reyi Procházka, 1894: 33
violaceicollis Chobaut, 1895: clv
pallipes Panzer, 1793: 11 (*Lagria*) **E**: AN AU BY CZ DE EN FR GE HU IT PL RO SK SL SP SV SZ UK
codinai Pic, 1937b: 1
flavipes Panzer, 1793a: 11 (*Lagria*)
lethierryi Pic, 1918f: 9
reyi Pic, 1894h: 72
parnassia Schilsky, 1897a: 78 **E**: GR
particeps Mulsant & Rey, 1868c: 282 **E**: SP
heydeni Procházka, 1894: 32
particularipennis Pic, 1935b: 3 **A**: TR
persica Pic, 1922e: 18 **A**: IN
pesruchesi Schilsky, 1897a: 57Q **N**: AG
peyroni Pic, 1902e: 31 **A**: SY
pici Bleuse, 1896: 250 **E**: SP
pickai Nigrin, 1984: 318 **N**: TU
plumbea kabyliana Constantin, 1988: 346 **N**: AG
plumbea plumbea Desbrochers des Loges, 1889: clxxxv **N**: AG
femorata Schilsky, 1897a: 9

posterecta Pic, 1902e: 31 **E**: GR
poupillieri Gougelet & H. Brisout de Barneville, 1859: ccxxxviii (*Cosmiocomus*) **N**: AG TU
prochazkae Pic, 1895e: 79 **N**: AG TU
purkynei Obenberger, 1917a: 33 **E**: BU
pygmaea Schaufuss, 1869: 20 **E**: SP (Baleares)
quadricollis Schilsky, 1897a: 71 **N**: AG TU
tunisea Pic, 1922g: 26
quedenfeldti Schilsky, 1897a: 64 **N**: MO
pedestris Schilsky, 1897a: 64
rambouseki Roubal, 1909: 341 **E**: BU
reissi Roubal, 1911: 101 **E**: AR GG
retowskii Reitter, 1890c: 195 **E**: UK
aequilata Procházka, 1894: 31
nigripes Schilsky, 1897a: 40
rostrata Procházka, 1894: 34 **E**: GR (Tinos)
rusicadensis Constantin, 1988: 350 **N**: AG
sardoa declivis Liberti, 1989: 286 **E**: IT (Sardegna)
sardoa mancinii Pic, 1927h: 13 **E**: FR (Corse) IT (Capraia, Elba, Gorgona)
sardoa sardoa Kisenwetter, 1871: 86 **E**: IT (Sardegna)
satanas Procházka, 1894: 34 **E**: AR GG ST
scyllea Liberti, 1981: 420 **E**: IT (Calabria)
sequensi Reitter, 1901f: 100 **A**: TR
serbica Kiesenwetter, 1863a: 661 **E**: BH BU CR CZ GR HU LA MC SK UK
taygetana Pic, 1902i: 56
sericea Schilsky, 1897a: 65 **N**: TU
shardaghensis Apfelbeck, 1918: 101 **E**: AL
sicula Pic, 1895m: 122 **E**: IT (Sicilia)
spinicollis Schilsky, 1900: 10 **A**: SY
splendida Pic, 1903q: 162 **A**: TR
strangulata Pic, 1922g: 26 **N**: TU
subelongata Pic, 1901a: 10 **N**: TU
syrensis Pic, 1910i: 33 **E**: GR (Siros)
syriaca Schilsky, 1897a: 11 **A**: SY
taurica Baudi di Selve, 1873a: 311 **E**: UK
tauricola Pic, 1904k: 73 **A**: TR
temporalis Schilsky, 1897a: 16 **E**: IT (Sicilia)
thessalonicensis Apfelbeck, 1911: 217 **E**: GR
tokatensis Pic, 1901m: 89 **A**: TR
trinacriae Liberti, 1979: 43 **E**: IT (Sicilia)
tripolitana Schilsky, 1897a: 63 **N**: LB TU
tristis Schilsky, 1900: 9 **A**: KI
valida Heyden, 1878: 215 **E**: AR GG ST
fuscata Schilsky, 1897a: 6
robusta Procházka, 1894: 33
valonensis Apfelbeck, 1911: 219 **E**: AL
variepubens Pic, 1903t: 177 **N**: AG TU
vitticollis Schilsky, 1897a: 45 **E**: GR (Naxos) **A**: CY
dubiosa Jakobson, 1911b: 709 [RN]
dubia Schilsky, 1907: [Nachträge und Berichtigungen]
winneguthi Apfelbeck, 1911: 219 **E**: AL GR
zahradniki Nigrin, 1986: 359 **A**: IS LE
ziczac Schaufuss, 1869: 19 **E**: SP (Baleares)
zingara Liberti & Baviera, 2004: 150: **E**: IT (Sicilia)

subfamily Danaceinae, genus incertae sedis

genus *Bilyella* Majer, 1988d: 436 type species *Dasytiscus gracilitarsis* Reitter, 1902
apis Majer, 1988d: 451 **A**: IN
basiunguis Majer, 1988d: 442 **A**: AF
bipartita Majer, 1988d: 441 **A**: AF
brevispina Majer, 1988d: 460 **A**: KA
breviunguis Majer, 1988d: 444 **A**: AF
felis Majer, 1988d: 446 **A**: AF
femina Majer, 1988d: 459 **A**: AF
furcata Majer, 1988d: 463 **A**: PA
gracilitarsis Reitter, 1902g: 210 (*Dasytiscus*) **A**: KZ TD UZ
javrozae Majer, 1988d: 451 **A**: TD
maleci Majer, 1988d: 449 **A**: TD
pecten Majer, 1988d: 457 **A**: AF
saccata Majer, 1988d: 452 **A**: AF
simplex Majer, 1988d: 463 **A**: AF PA
spinosa Majer, 1988d: 461 **A**: PA
stilet Majer, 1988d: 454 **A**: PA
unispina Majer, 1988d: 441 **A**: AF
ursus Majer, 1988d: 444 **A**: AF IN
vaginata Majer, 1988d: 453 **A**: KA PA

subfamily **Dasytinae** Laporte, 1840
tribe Dasytini Laporte, 1840

genus *Allotarsodasytes* Pic, 1911e: 97 type species *Allotarsus monstrosipes* Pic, 1909
atrimembris Pic, 1931c: 5 **E**: PT
monstrosipes Pic, 1909s: 179 (*Allotarsus*) **E**: SP

genus *Allotarsus* Graells, 1858: 55 type species *Allotarsus melaleucotrichos* Graells, 1858
baudii Bourgeois, 1888a: 28 (*Henicopus*) **E**: SP
canescens Pardo Alcaide, 1954b: 97
cobosi Pardo Alcaide, 1956: 21 **E**: SP
dentipes Raffray, 1873: 375 (*Henicopus*) **N**: AG MO
physomerus Fairmaire, 1883d: cxii (*Henicopus*)
kiesenwetteri Reitter, 1872b: 181 (*Henicopus*) **N**: AG
luteonotatus Pic, 1953a: 5 **N**: MO
melaleucotrichos Graells, 1858: 55 **E**: SP
senex Rosenhauer, 1856: 167 (*Enicopus*)
peyerimhoffi Pardo Alcaide, 1950: 21 **N**: MO
senescens Jacquelin du Val, 1860b: 87 (*Henicopus*) **E**: SP
fulvimanus Escalera, 1927: 15
vidualis Escalera, 1927: 16 **E**: SP

genus *Dasytes* Paykull, 1799: 156 type species *Dermestes niger* Linnaeus, 1761
subgenus *Anthoxenus* Motschulsky, 1845a: 37 type species *Dasytes subaeneus* Schönherr, 1817
Hapalogluta C. G. Thomson, 1864: 148 type species *Dasytes subaeneus* Schönherr, 1817
Pseudodasytes Mulsant & Rey, 1868d: 76 type species *Dasytes subaeneus* Schönherr, 1817
albosetosus Schilsky, 1896a: 46 **E**: SP
nigropunctatus Kiesenwetter, 1863a: 641
crassitarsis Peyerimhoff, 1925a: 6 **N**: MO
dilutipes Reitter, 1896b: 36 **A**: "Turkestan"
gilvipes Schilsky, 1903: 96 **A**: UZ
microps G. Müller, 1935: 154 **E**: AL
mixtus Schilsky, 1896a: 47 **A**: SY
nigropunctatus Küster, 1849c: 27 **E**: PT SP
asperulus Graells, 1858: 56

provincialis Abeille de Perrin, 1907b: xx **E**: FR IT
pyrrhostoma Kolenati, 1846b: 45 **E**: AB
sefrensis Pic, 1894p: 112 **N**: AG
  semiauratus Pic, 1922g: 27
seriatopunctatus Schilsky, 1897a: 82 **A**: SY
subaeneus Schönherr, 1817: 15 **E**: AN AU BH BU CR CZ DE FR GE GR HU IT LU MC PL RO SK SL SP SV SZ TR YU
  aeneus A. G. Olivier, 1790a: no 21: 11 (*Melyris*) [HN]
  scaber Suffrian, 1843b: 336
terminalis Jacquelin du Val, 1863: pl. 45 **E**: FR PT SP **N**: AG MO
  innotaticollis Pic, 1937a: 86
  nigrithorax Fuente, 1897: 130
vulpinus L. Redtenbacher, 1843b: 984 **A**: SY
subgenus *Dasytes* Paykull, 1799: 156 type species *Dermestes niger* Linnaeus, 1761
  Enodius Laporte, 1838: 31 type species *Dermestes niger* Linnaeus, 1761
  Psilocorse C. G. Thomson, 1864: 151 type species *Dermestes niger* Linnaeus, 1761
abeillei Schilsky, 1897b: 5 **A**: IQ SY
acutangulus Schilsky, 1895: 10 **N**: AG
acutipennis Baudi di Selve, 1873a: 307 **N**: AG
  acutus Schilsky, 1895: 9
aequalis Schilsky, 1897b: 12 **E**: GR
algiricus P. H. Lucas, 1846: 199 **N**: AG
  confinis Schilsky, 1897a: 79
altaicus Schilsky, 1900: 2 **A**: MG WS
atrimembris Pic, 1924b: 17 **A**: "Siberia"
aurasiacus Peyerimhoff, 1920: 244 **N**: AG
aurescens Champion, 1922b: 151 **A**: UP
blascoi Constantin, 1992: 401 **E**: SP
brevesulcatus Pic, 1931e: 13 **A**: TR
brevicollis Schilsky, 1897b: 14 **A**: IN IQ TR
brevioripennis Pic, 1939d: 21 **N**: MO
brutus Pic, 1937a: 65 [RN] **N**: LB TU
  breviusculus Schilsky, 1897b: 13 [HN]
buphtalmus Baudi di Selve, 1873a: 307 **E**: AL BH BU CR CZ GR HU IT MC SK TR YU
  gelineki Reitter, 1913d: 289
  oertzeni Schilsky, 1895: 13
  viertli Schilsky, 1897b: 19
carbonarius Schilsky, 1896a: 41 **E**: AB AR **A**: IN TR
chalybeus Ménétriés, 1832: 164 **E**: AB
  obscureaenea Ménétriés, 1832: 164
circassicola Reitter, 1896b: 37 **E**: ST
corallipes Reitter, 1902g: 211 (*Dasytiscus*) **A**: KZ
crassicornis Schilsky, 1896a: 40 **A**: SY
creticus Constantin, 1983: 329 **E**: GR (Kríti)
delagrangei Pic, 1893g: clxi **A**: SY
  transversefasciatus Reitter, 1896a: 18
depexus Peyerimhoff, 1925b: 58 **N**: MO
diabolicus Chobaut, 1898: 79 **N**: AG
dilectus Normand, 1948: 88 **N**: TU
distinguendus Pic, 1930f: 14 **N**: AG
doderoi Pic, 1924h: 80 **E**: IT (Sardegna)
dolens Rosenhauer, 1856: 163 **E**: SP
edoughensis Pic, 1897j: 219 **N**: AG
emarginatus Schilsky, 1895: 8 **N**: AG MO
eximius Majer, 1986: 314 **A**: MG
fausti Schilsky, 1896a: 39 **E**: ST

festucarum Peyerimhoff, 1925b: 57 **A**: MO
flavescens Gené, 1839: 57 **E**: FR (Corse) IT MA **N**: AG TU
apicalis Ragusa, 1896b: 72 (*Divales*)
biskrensis Pic, 1895c: 80
faillae Ragusa, 1896b: 72
inapicalis Pic, 1894p: 112
nigriceps Schilsky, 1896a: 35
parvulus Schilsky, 1894a: 32
pectoralis Baudi di Selve, 1873a: 297
posticus Solsky, 1868: 34 (*Dasytiscus*)
rufotestaceus Reitter, 1889s: 373 (*Dasytiscus*)
scutellaris Solsky, 1868: 34-35 (*Dasytiscus*)
unicolor Schilsky, 1894a: 32
forticulus Normand, 1948: 91 **N**: TU
fuscipes Brullé, 1832: 153 **E**: GR
gonocerus Mulsant & Rey, 1868d: 95 **E**: FR (Pyrénées) SP (Pirineos)
sericatus Mulsant & Rey, 1868d: 100
graeculus Kiesenwetter, 1863a: 634 **E**: GR
gridellii Pic, 1955e: 131 **N**: LB
griseus Küster, 1849c: 26 **E**: CR
hickeri Kaszab, 1955: 305 **E**: HU
hirsutulus Faldermann, 1835a: 205 **E**: GG
incanus Rosenhauer, 1856: 155 **E**: SP
nigropilosus Schilsky, 1896a: 57 (*Dasytiscus*)
nitidicollis Schilsky, 1896a: 43
incrassatus Schilsky, 1900: 1 **N**: AG
interjectus Peyerimhoff, 1925a: 3 **N**: AG
korbi Schilsky, 1897b: 10 **E**: SP
laevicollis Schilsky, 1897a: 81 **E**: BU **A**: SY
laeviusculus Schilsky, 1896a: 42 (*Dasytes*) **E**: AB AR **A**: IN
latrunculus Normand, 1948: 92 **N**: TU
lepineyi Peyerimhoff, 1935: 193 **N**: MO
loboensis Mayor, 2004: 85 [RN] **E**: SP
nevadensis Pic, 1954f: 98 [HN]
longipennis Gebler, 1833: 281 **A**: "Siberia"
marginatus Fischer von Waldheim, 1842: 9 **A**: "Siberia"
analis Fischer von Waldheim, 1844: 38
mathieui Pic, 1925f: 5 **N**: AG
meridionalis Schilsky, 1894a: 19 **E**: "Caucasus"
minimus J. R. Sahlberg, 1903a: 32 **E**: GG
aenescens J. R. Sahlberg, 1903a: 32
collaris J. R. Sahlberg, 1903a: 32
infuscatus J. R. Sahlberg, 1903a: 32
verticalis J. R. Sahlberg, 1903a: 32
minutissimus Pic, 1894i: 76 **A**: SY
modestus Pic, 1934b: 25 **E**: GR
moniliatus Kiesenwetter, 1863a: 633 **E**: CR GR UK YU **A**: TR
luteicornis Schilsky, 1894a: 30
montanus Mulsant & Rey, 1868d: 86 **E**: FR (Massif Central)
caesicornis Mulsant & Rey, 1868d: 90
incertus Schilsky, 1895: 3
opacithorax Pic, 1908e: 46
moreli Schilsky, 1899: 1 **E**: FR (Pyrénées-Orientales) SP (Pirineos)
murinus Schilsky, 1894a: 28 **E**: GR
niger Linnaeus, 1761: 146 (*Dermestes*) **E**: AN AU BE BH BY CT CZ DE EN FI FR GB GE HU IT LA LT NL NR PL RO SK SL SP SV SZ UK YU **A**: MG "Siberia"
palustris Fiori, 1906b: 262

perplexus Mulsant & Rey, 1868d: 83
villosus A. G. Olivier, 1790a: no 21: 9 (*Melyris*)
nigritus Kiesenwetter, 1859c: 169 **E**: GR
nigropilosus Reitter, 1885e: 244 (*Dasytiscus*) **E**: AN PT SP **N**: AG MO
nimbosus Majer, 1986: 311 **A**: MG
occiduus Mulsant & Rey, 1870: 120 **E**: FR
orientalis Schilsky, 1894a: 23 **A**: TR
otini Peyerimhoff, 1948d: 270 **N**: MO
pauperculus Laporte, 1840a: 282 **E**: FR IT SL SP
mesmini Pic, 1908e: 46
pilicornis Kiesenwetter, 1864: 388
ponferradanus Pic, 1913f: 105 (*Haplocnemus*)
persimilis Pic, 1930f: 14 **N**: MO
pilosus Germar, 1824: 75 **A**: "Siberia"
praecelsus Peyerimhoff, 1925b: 56 **N**: MO
recticollis Schilsky, 1895: 6 **A**: SY
reticulatus Schilsky, 1894a: 20 **N**: LB
robusticornis Pic, 1924b: 17 **A**: "Turkestan"
setosus Waltl, 1835: 61 **E**: SP
similis Schilsky, 1895: 5 **N**: AG
staudingeri Schilsky, 1897a: 80 **E**: SP
striatulus Brullé, 1832: 153 **E**: AL BU CR GR IT SK
dalmatinus Baudi di Selve, 1873a: 305
nudus Schilsky, 1894a: 26
subacuminatus Pic, 1938k: 154 **E**: GR (Rodos)
subdepressus Schilsky, 1895: 7 **N**: AG TU
subemarginatus Pic, 1932g: 27 [= 1932c: 21] **N**: MO
subimpressus Pic, 1932g: 27 [= 1932c: 21] **N**: MO
submissus Normand, 1948: 90 **N**: TU
taguchii Nakane, 1981b: 128 **A**: JA
tardus Schaufuss, 1876: 384 [RN] **E**: AL BH BU CR GR IT SL YU
brevicornis Kiesenwetter, 1863: 730 [HN]
tarsalis Zetterstedt, 1838: 86 **E**: FI NT SV
testaceipes Pic, 1924b: 17 **E**: GG
thoracicus lucanus Wittmer, 1935a: 75 **E**: BU GR IT (Basilicata)
thoracicus thoracicus Mulsant & Rey, 1868: 83 **E**: FR IT SZ
timidus Baudi di Selve, 1873b: 318 **A**: CY
tomokunii Satô, 1986b: 253 **A**: FE JA
tondui Pic, 1925f: 5 **N**: AG
torretassoi Pic, 1934b: 25 **E**: GR (Dodecanisos: Meyisti) **A**: TR
trifasciatus L. Redtenbacher, 1850: 47 **A**: IN
tripubescens Pic, 1925f: 5 **N**: AG
troglavensis Apfelbeck, 1911: 216 **E**: BH CR
tuniseus Pic, 1925f: 10 **N**: TU
vernalis Normand, 1935: 246 **N**: TU
villiger Pic, 1895r: cccl **A**: SY
delagrangei Schilsky, 1894a: 17 [HN]
gracilis Schilsky, 1897e: 16
vittiger Reitter, 1896a: 18
xanthoceros Gebler, 1830: 89 **A**: "Siberia"
xanthocnemus Kolenati, 1846b: 46 **E**: AB GG ST **A**: TR
cyprius Baudi di Selve, 1873b: 317
lederi Schilsky, 1897b: 34
subgenus *Hypodasytes* Mulsant & Rey, 1868d: 76 type species *Dasytes obscurus* Gyllenhal, 1813
alpigradus Kiesenwetter, 1863a: 633 **E**: AU CZ EN GE IT PL RO SK SL SZ
cyaneus Baudi di Selve, 1889: 131 [HN]

fiorii Jakobson, 1911b: 705 [RN]
montanus Gredler, 1866: 254
coerulescens Küster, 1852: 81 **E**: FR (Corse) IT (Sardegna)
sardous Rosenhauer, 1856: 164
erratus Schilsky, 1895: 2 **E**: AN FR IT SP SZ
apenninus Schilsky, 1896a: 37
intermedius Fiori, 1909b: 32
valesiacus Pic, 1910u: 89
grenieri Kiesenwetter, 1871: 83 **E**: FR (Corse)
laufferi Schilsky, 1906a: 98 **E**: SP
lombardus Fiori, 1909b: 33 **E**: IT SZ
metallicus Fabricius, 1792b: 81 (*Lagria*) **E**: IT (Sicilia) **N**: AG TU
montivagus Schilsky, 1907: 97 **E**: SP
obscurus Gyllenhal, 1813: 685 **E**: AU BY CT CZ EN FI FR GE IT LA LT NL NR NT PL RU SK SV SZ **A**: "Siberia"
borealis C. G. Thomson, 1864: 151
letzneri J. Weise, 1887a: 55
rugipennis C. G. Thomson, 1870a: 139
productus Schilsky, 1894a: 22 **E**: IT (Calabria, Sicilia) **N**: TU
subalpinus austriacus Lohse, 1977: 183 **E**: AU
subalpinus subalpinus Baudi di Selve, 1873a: 304 **E**: AU GE IT FR SZ
subalpinus vesubiensis Pic, 1946: 9 **E**: FR
subfasciatus Kiesenwetter, 1867b: 118 **E**: SP
tristiculus Mulsant & Rey, 1868d: 100 **E**: FR (Corse) IT SP
baudii Jakobson, 1911b: 705 [RN]
mulsanti Schilsky, 1894a: 25
seriatus Mulsant & Rey, 1868d: 106
sublaevis Mulsant &Rey, 1868d: 104
subgenus *Mesodasytes* Mulsant & Rey, 1868d: 76 type species *Telephorus plumbeus* O.F. Müller, 1776
aeneiventris Küster, 1850: 6 **E**: FR GR IT SP **N**: AG MO TU
roberti Abeille de Perrin, 1907b: xx
aeratus Stephens, 1830: 319 **E**: AU BE BH BU CR CZ DE EN FR GB GE GR HU IR IT NL NR PL PT RO SK SL SP SV SZ UK **A**: TR
aerosus Kiesenwetter, 1867b: 116
serricornis Stephens, 1830: 320
subaeneus C. G. Thomson, 1864: 149 [HN]
variicornis Schilsky, 1894a: 38
akbesianus Pic, 1894i: 76 **A**: SY
albipilis Pic, 1894p: 112 **N**: AG
alticola Palm, 1974: 41 **N**: CI (Tenerife)
atricornis Pic, 1931e: 13 **A**: TR
basicornis Nakane, 1963b: 47 **A**: JA
bourgeoisi Schilsky, 1903: 95 **E**: RO
brenskei Schilsky, 1894a: 42 **E**: GR
brevitarsis Normand, 1935: 247 **N**: AG TU
canariensis Har. Lindberg, 1951: 4 **N**: CI
croceipes Kiesenwetter, 1866a: 366 **E**: FR IT SP **N**: AG LB MO TU
corsicus Schilsky, 1894c: 228 [RN]
cruralis Mulsant & Rey, 1868d: 132 [HN]
dispar Wollaston, 1862c: 445 **N**: CI (Gran Canaria)
diversimembris Pic, 1925f: 5 **N**: TU
gemmatus Liberti, 1984b: 325 **A**: TAI
inexspectatus Palm, 1974: 43 **N**: CI (Tenerife)
israelsoni Palm, 1974: 41 **N**: CI
iteratus Peyerimhoff, 1925a: 5 **E**: IT (Sardegna) **N**: AG TU
japonicus Kiesenwetter, 1874b: 288 **A**: FE JA

kocheri Pic, 1950e: 92 **N**: MO
lanzarotensis Palm, 1974: 44 **N**: CI
macrophtalmus Liberti, 1982: 377 **A**: TAI
nigroaeneus Küster, 1850: 7 **E**: GR IT FR PT SP ST **N**: AG MO **A**: SY
tibiellus Mulsant & Rey, 1868d: 126
oculatus Kiesenwetter, 1867b: 115 **E**: SP
opacicornis Liberti, 1982: 378 **A**: TAI
pallidohirtus Pic, 1899l: 258 **A**: LE SY
plumbeus O. F. Müller, 1776: 576 (*Telephorus*) **E**: AN AL AU BE BH BU BY CR CT CZ DE EN FI FR GB GE GR HU IT LA LS LT LU MC NL NR NT PL RO SK SL SP SV SZ UK YU **A**: "Siberia"
atripes Sainte-Claire Deville, 1908: 216
coerulescens Schilsky, 1898: 2
coxalis Mulsant & Rey, 1868d: 126
flavipes Fabricius, 1787a: 94 (*Lagria*)
nigrinus Schilsky, 1898: 2
nigrofemoralis Schilsky, 1898: 2
pueli Fagniez, 1946: 27
ragusae Schilsky, 1894b: 15
punctulatus Schilsky, 1897a: 83 **N**: AG
shimomurai Liberti, 1982: 374 **A**: TAI
subaenescens Wollaston, 1862c: 444 **N**: CI
taiwanensis Liberti, 1982: 375 **A**: TAI
variolosus Baudi di Selve, 1873a: 307 **E**: SP
viberti Pic, 1922g: 27 **N**: AG
virens Marsham, 1802: 230 (*Tillus*) **E**: AN AU BE BH CZ EN FR GB GE GR HU IT NL PL RO SK SL SP SZ **N**: AG **A**: IN
dubius Schilsky, 1894c: 232
flavipes A. G. Olivier, 1790a: no 21: 12 (*Melyris*) [HN]
nigripes Schilsky, 1894a: 37
ochsi Fagniez, 1946: 20, 27
plumbeus Illiger, 1798: 310
pullus Abeille de Perrin, 1907b: xxi
puncticollis Reitter, 1888e: 211
tibialis Zetterstedt, 1828: 119
vulgaris Nakane, 1963b: 47 **A**: JA
subgenus *Metadasytes* Mulsant & Rey, 1868d: 76 type species *Telephorus caeruleus* DeGeer, 1774
aeneithorax Pic, 1925f: 5 **N**: TU
baborensis Pic, 1903j: 128 **N**: AG
caeruleus DeGeer, 1774: 76 (*Telephorus*) **E**: AU BE BH BU CR CT CZ EN DE FI FR GE GR HU IT LA LT LU MC NL NR PL RO SK SL SP SV SZ YU
bulgaricus Pic, 1903k: 129 **E**: BU
cyaneus Fabricius, 1775: 63 (*Anobium*)
virescens Westhoff, 1881: 174
fusculus Illiger, 1801a: 82 (*Melyris*) **E**: AU BU BY CT CZ EN DE EN FI GE HU IT LA LS LT NR NT PL RO SK ST SV SZ UK "Caucasus" **A**: MG WS
femoralis Krynicki, 1832: 101
immaturus Schilsky, 1897b: 23
pallipes Faldermann, 1832: 202
jurjurensis Pic, 1898f: 119 **N**: AG
nigrocyaneus Mulsant & Rey, 1868d: 143 **E**: FR GE HU SK SZ
varipes Schilsky, 1894a: 36
genus *Dasytes*, species incertae sedis
calabrus A. Costa, 1847b: 143 **E**: IT (Calabria) [probable synonym with *Aplocnemus jejunus* Kiesenwetter, 1863]

genus *Divales* Laporte, 1838: 31 type species *Hispa bipustulata* Fabricius, 1781
argyrostictus Peyerimhoff, 1948d: 268 **N**: MO

bipustulatus Fabricius, 1781: 82 (*Hispa*) **E**: BU FR GE GR IT SZ **N**: AG MO TU
angustijunctus Pic, 1937f: 5
baudii Pic, 1937a: 57
bimaculatus P. Rossi, 1790: 35 (*Melyris*)
cruentus Pic, 1893b: 46
erythromelas Küster, 1852: 80 (*Dasytes*)
fallax Schilsky, 1894a: 10
nigricollis Ragusa, 1896b: 71
nigromaculatus P. H. Lucas, 1846: 197 (*Dasytes*)
portai Pic, 1937a: 57
qudrimaculatus Baudi di Selve, 1873a: 294 (*Dasytes*)
quadrisignatus Escalera, 1914: 243
rufulus Schilsky, 1894a: 10
subobliteratus Pic, 1937f: 5
sulcicollis Schilsky, 1897: 34C
variegatus P. H. Lucas, 1846: 196 (*Dasytes*)
cinctus Gené, 1839: 37 (*Dasytes*) **E**: FR (Corse) IT (Sardegna, Sicilia)
affinis Schilsky, 1897b: 3
apicatus Schilsky, 1897e: 3
ater Schilsky, 1888: 189 (*Dasytes*)
atratulus Schilsky, 1897b: 3
communimacula A. Costa, 1847b: 142
conjunctus Schilsky, 1894a: 15
discendens Schilsky, 1897b: 3
ephippiatus Schilsky, 1897b: 3
flavipennis Baudi di Selve, 1873a: 296 (*Dasytes*)
gemellatus Schilsky, 1897: 34
notaticollis Schilsky, 1894a: 15
quadrinotatus Schilsky, 1897b: 3
reyanus Gozis, 1881b: cxxxv (*Dasytes*) [RN]
tibialis Mulsant & Revelière, 1861: 10 (*Dasytes*) [HN]
densepunctatus Pic, 1903t: 177 **N**: AG
haemorrhoidalis Fabricius, 1798: 118 (*Lagria*) **E**: SP **N**: AG MO
carbonarius Schilsky, 1894a: 11
kraatzi Schilsky, 1894a: 12
lineatocollis Schilsky, 1894a: 12
maculipennis Schilsky, 1894a: 11
melyroides Pic, 1896e: 47
quadrisignatus Schilsky, 1894a: 11
rosenhaueri Schilsky, 1894a: 11
rotroui Pic, 1933f: 14
signatus Schilsky, 1897b: 34
uninotatus Pic, 1932g: 27 [= 1932c: 21]
mauritanicus P. H. Lucas, 1846: 196 (*Dasytes*) **E**: SP **N**: AG
apex Schilsky, 1897b: 34
nigripennis Schilsky, 1894a: 9
rufiventris Schilsky, 1894a: 9
quadrimaculatus A. G. Olivier, 1790a: no 21: 10 (*Melyris*) **E**: FR IT SZ
decoloratus Schilsky, 1894a: 14
mequignoni Fagniez, 1946: 22
quadripustulatus Fabricius, 1798: 116 (*Hispa*)
semiornatus Schilsky, 1894a: 14
similis Schilsky, 1894a: 14
rufitarsis Baudi di Selve, 1873a: 295 (*Dasytes*) **N**: AG
amplipennis Baudi di Selve, 1873a: 295
uhligi Majer, 1984: 291 **E**: IT
weisei Schilsky, 1894a: 13 **E**: SP **N** AG MO

genus *Dolichophron* Kiesenwetter, 1867c: 138 type species *Dolichosoma cylindromorphum* Kiesenwetter, 1867

cylindromorphum Kiesenwetter, 1867c: 138 (*Dolichosoma*) **A**: CY IS LE SY
hartungii Wollaston, 1862c: 446 (*Dolichosoma*) **N**: CI
kiesenwetteri Reitter, 1891b: 26 (*Dolichophron*) **N**: LB
semisericeum Abeille de Perrin, 1907b: xxi **N**: LB

genus *Dolichosoma* Stephens, 1830: 318 type species *Lagria linearis* P. Rossi, 1794
Linotoma Motschulsky, 1845a: 37 type species *Lagria linearis* P. Rossi, 1794
filum Fairmaire, 1860: 630 **E**: FR (Corse)
lineare P. Rossi, 1794: 92 (*Lagria*) **E**: AN AU BE BH BY CR CT CZ DE EN GB FI FR GE HU IT LA LT NL NR NT PL RO SK SP ST SV SZ UK **A**: ES KZ
filiforme Panzer, 1799b: 17 (*Lagria*)
subdensatum Mulant & Rey, 1868: 273
submicaceum Mulant & Rey, 1868: 274
subnodosum Mulant & Rey, 1868: 274
maxima Schilsky, 1894a: 52 **E**: UK
simile Brullé, 1832: 154 (*Dasytes*) **E**: CR FR (Bouches-du-Rhône) GR HU IT **A**: TR
cinereum Ménétriés, 1832: 165

genus *Enicopus* Stephens, 1830: 318 type species *Dermestes hirtus* Linnaeus, 1767
subgenus *Dasysoma* Schilsky, 1896a: 32 type species *Henicopus privignus* Kiesenwetter, 1867
privignus Kiesenwetter, 1867b: 113 (*Henicopus*) **E**: SP
corvinus Reitter, 1903a: 43 (*Dasytiscus*)
subgenus *Enicopus* Stephens, 1830: 318 type species *Dermestes hirtus* Linnaeus, 1767
Henicopus Agassiz, 1846: 138 [unjustidied emendation]
andradei Pardo Alcaide, 1966c: xxiii (*Henicopus*) **E**: PT
armatus P. H. Lucas, 1846: 198 (*Dasytes*) [EA] **N**: AG
armipes Jacquelin du Val, 1860b: 74 (*Henicopus*)
calcaratus Kiesenwetter, 1859c: 165 (*Henicopus*) **E**: SP
bonvouloiri Boieldieu, 1859: 468
confusus Jacquelin du Val, 1860b: 66 (*Henicopus*) **E**: SP
distinguendus Jacquelin du Val, 1860b: 66 (*Henicopus*) **E**: SP
gracilis Schilsky, 1896: 20a (*Henicopus*) **E**: SP
heydeni Kiesenwetter, 1870: 126 (*Henicopus*) **E**: SP
latior Bourgeois, 1888a: 26 (*Henicopus*)
hirtus Linnaeus, 1767: 563 (*Dermestes*) **E**: AL AU BU CZ FR GE GR CR HU IT MC RO SK SL SV SZ **A**: MG "Turkestan" "Korea"
truncatus Fairmaire, 1859a: 51
turkestanicus Schilsky, 1896a: 21a (*Henicopus*)
hoplotarsus Jacquelin du Val, 1860b: 84 (*Henicopus*) **E**: PT SP
ibericus Jacquelin du Val, 1860b: 75 (*Henicopus*) **E**: SP
scutellaris Küster, 1849c: 14 (*Dasytes*)
korbi Schilsky, 1896a: 19 (*Henicopus*) **E**: SP
longimanus Kiesenwetter, 1859c: 168 **E**: FR (Pyrénées) SP
parnassi Kiesenwetter, 1859c: 166 (*Henicopus*) **E**: GR
orientalis Fairmaire, 1859a: 51
pilosus Scopoli, 1763: 41 (*Cantharis*) **E**: FR IT SL SP SZ
acutatus Boieldieu, 1859: 469
ater Fabricius, 1787a: 94 (*Lagria*)
fairmairei Schilsky, 1896a: 21 (*Henicopus*)
falculifer Fairmaire, 1859a: 53 (*Henicopus*)
niger Petagna, 1787: 12 (*Lagria*)
steveni Motschulsky, 1849: 82
subvittatus Fairmaire, 1859a: 52
plumbeus Schilsky, 1896a: 22 (*Henicopus*) **A**: TR
pyrenaeus Fairmaire, 1859a: 50 **E**: FR (Pyrénées) SP (Pirineos)
rugosicollis Jacquelin du Val, 1860b: 71 (*Henicopus*) **E**: PT SP
dubius Schilsky, 1896a: 16 (*Henicopus*)

hirsutus Schilsky, 1896a: 16 (*Henicopus*)
sarae Bahillo de la Puebla & López-Colón, 2004: 55 **E**: SP
scutellaris Fabricius, 1792b: 80 (*Lagria*) **E**: SP
amphicoma Graells, 1858: 59 (*Enodius*)
spiniger Jacquelin du Val, 1860b: 83 (*Henicopus*) **E**: PT SP
spinosus Schilsky, 1896a: 10 (*Henicopus*) **E**: BU
transcaucasicus Schilsky, 1896a: 25 (*Henicopus*) **E**: GG
subgenus *Parahenicopus* Portevin, 1931: 453 type species *Henicopus vittatus* Kiesenwetter, 1859
espanoli Pardo Alcaide, 1966a: 179 (*Henicopus*) **E**: SP
lagari Pardo Alcaide, 1966a: 177 (*Henicopus*) **E**: SP
paulinoi Bourgeois, 1884c: lxiv (*Henicopus*) **E**: PT
perezi Kiesenwetter, 1867b: 111 (*Henicopus*) **E**: PT SP
simplicipes Jacquelin du Val, 1860b: 92 (*Henicopus*) **E**: SP
vittatus Kiesenwetter, 1859c: 165 (*Henicopus*) **E**: SP
vestitus Schilsky, 1896a: 2 (*Henicopus*)

genus *Graellsinus* Escalera, 1927: 22 type species *Dasytes praticola* Waltl, 1835
barrosi Escalera, 1927: 25 **E**: PT
brachialis Jacquelin du Val, 1860b: 95 (*Henicopus*) **E**: PT SP
nigripes Schilsky, 1896a: 33a (*Allotarsus*)
praticola Waltl, 1835: 61 (*Dasytes*) **E**: PT SP
sartorii Schilsky, 1896a: 34 (*Allotarsus*)

genus *Hauseria* Reitter, 1896c: 70 type species *Hauseria acmaeopsiformis* Reitter, 1896
acmaeopsiformis Reitter, 1896c: 71 **A**: AF

genus *Psilothrix* Küster, 1850: 9 type species *Dasytes protensus* Gené, 1836
subgenus *Dolichomorphus* Fiori, 1905a: 82 type species *Psilothrix rufimanus* Fiori, 1905
Dolichosomina Reitter, 1911a: 284 type species *Psilothrix femorale* Morawitz, 1862
femoralis Morawitz, 1862: 317 (*Dolichosoma*) **E**: BU HU IT RO ST UK YU **A**: KZ MG "Siberia"
rufimanus Fiori, 1905a: 82
subgenus *Psilothrix* Küster, 1850: 9 type species *Dasytes protensus* Gené, 1836
Lasius Motshulsky, 1845: 37 [HN] type species *Melyris nobilis* Illiger, 1798
albilanea Peyerimhoff, 1925a: 7 (*Psilotrix* [sic]) **N**: AG MO
aureola Kiesenwetter, 1859c: 178 (*Dolichosoma*) **E**: FR (Corse) GR IT MA SP (Baleares) **N**: AG TU
splendidum Schaufuss, 1867: 81 (*Dolichosoma*)
cyrtosina Peyerimhoff, 1935: 193 **N**: MO
illustris Wollaston, 1854: 252 (*Dasytes*) **E**: PT **N**: MO MR
kurdistanica Schilsky, 1897b: 25 **E**: AB
latipennis Pic, 1900i: 77 **E**: GR (Rodos)
melanostoma Brullé, 1832: 150 (*Dasytes*) **E**: GR IT (Puglia)
lucidula Brullé, 1832: 150 (*Dasytes*)
pharaonum Kiesenwetter, 1867c: 139 (*Dolichosoma*) **N**: EG
protensa Gené, 1836: 179 (19, Separate) (*Dasytes*) **E**: FR (Corse) IT (Sardegna, Sicilia) **N**: AG TU
severa Kiesenwetter, 1859c: 178 (*Dolichsoma*) **E**: BH BU CR FR (Alpes-de-Haute-Provence)
smaragdina P. H. Lucas, 1846: 195 (*Dasytes*) **E**: IT **N**: TU
ultramarina Schaufuss, 1867: 82 (*Dolichosoma*) **E**: GR (Rodos)
viridicoerulea Geoffroy, 1785: 63 (*Cicindela*) **E**: BE CR DE FR GB GE GR IR IT LA MA MC NL PL PT RO SL SP SV SZ TR YU **N**: AG CI LB MO TU **A**: TR
cyaneus A. G. Olivier, 1790a: no 21: 8 (*Melyris*)
fulminans Schilsky, 1894a: 46
nobilis Illiger, 1798: 308 (*Melyris*)
violaceipennis Pic, 1908e: 47
viridis P. Rossi, 1792: 35 (*Lagria*)

genus *Trochantodon* Escalera, 1927: 9 type species *Enicopus tibiellus* Rosenhauer, 1856

tibiellus Rosenhauer, 1856: 165, 167 (*Enicopus*) **E**: SP **N**: MO
trichomelas Pardo Alcaide, 1966a: 189

tribe Dasytini, genus incertae sedis
genus *Apovates* Majer, 1986: 309 type species *Apovates impeditus* Majer, 1986
impeditus Majer, 1986: 309 **A**: MG

subfamily **Gietellinae** Constantin & Menier, 1987
genus *Gietella* Constantin & Menier, 1987: 53 type species *Gietella fortunata* Constantin & Meier, 1987
faialensis Menier & Constantin, 1988: 30 **N**: AZ (Faial)
fortunata Constantin & Menier, 1987: 55 **N**: CI (Hierro, LaPalma, Lanzarote)

subfamily **Listrinae** Majer, 1990
genus *Danacaeina* Reitter, 1887e: 234 type species *Danacaeina bicolor* Reitter, 1887
bicolor Reitter, 1887e: 234 **A**: KI MG XIN
villosus Schilsky, 1907: 98 (*Dasytiscus*)

genus *Picolistrus* Majer, 1990b: 377 type species *Dasytiscus inhirsutus* Pic, 1922
inhirsutus Pic, 1922l: 156 (*Dasytiscus*) **A**: YUN

genus *Sinolistrus* Majer, 1990b: 376 type species *Dasytiscus sinensis* Pic, 1907
sinensis Pic, 1907q: 175 (*Dasytiscus*) **A**: YUN

subfamily **Rhadalinae** LeConte, 1862
tribe Aplocnemini Majer, 1987
genus *Aplocnemus* Stephens, 1830: 316 type species *Crioceris impressa* Marsham, 1802
subgenus *Aplocnemus* Stephens, 1830: 316 type species *Crioceris impressa* Marsham, 1802
Elicopis Stephens, 1829: 136 type species *Crioceris impressa* Marsham, 1802
Haplocnemus Agassiz, 1846: 29 [unjustified emendation]
Pseudaphyctus Pic, 1896e: 47 type species *Haplocnemus tournieri* Pic, 1896 (= *Aplocnemus tumidus* Kiesenwetter, 1863)
acutangulus Schilsky, 1897a: 92 (*Haplocnemus*) **E**: IT (Sicilia) **N**: AG TU
lacoi Majer, 1985: 35
adanensis Pic, 1908f: 50 **A**: TR
albipilis Kiesenwetter, 1863a: 651 (*Haplocnemus*) **E**: PT SP
algiricus Schilsky, 1894a: 73 (*Haplocnemus*) **N**: AG
alpestris Kiesenwetter, 1861: 385 (*Haplocnemus*) **E**: AU CZ FR IT SK SP SZ UK
theresae Pic, 1914o: 57
alternatus Peyerimhoff, 1931a: 1 (*Haplocnemus*) [RN] **N**: AG MO
cribripennis Pic, 1922g: 27 (*Haplocnemus*) [HN]
nigrescens Pic, 1949b: 74 (*Haplocnemus*)
andalusicus Rosenhauer, 1856: 158 (*Dasytes*) **E**: PT SP
diversus Schilsky, 1897b: 49 (*Haplocnemus*)
korbi Schilsky, 1897b: 49 (*Haplocnemus*)
laetus Schilsky, 1897b: 34 (*Haplocnemus*)
nevadensis Pic, 1908a: 90 (*Haplocnemus*)
angelinii Liberti, 1995: 162 **E**: AL GR IT
aubei Kiesenwetter, 1867b: 122 (*Haplocnemus*) **E**: SP
escalerai Pic, 1908l: 90 (*Haplocnemus*)
colasi Pic, 1954f: 97 (*Haplocnemus*)
uhagoni Schilsky, 1897a: 86 (*Haplocnemus*)
baborensis Pic, 1908p: 200 (*Haplocnemus*) **N**: AG
basalis Küster, 1849c: 19 (*Dasytes*) **E**: CR GR RO
distinctipes Pic, 1908f: 50 (*Haplocnemus*)
beauprei Pic, 1909d: 105 (*Haplocnemus*) **N**: TU
berytensis J. R. Sahlberg, 1913: 45 (*Haplocnemus*) **A**: SY

bicoloratus Pic, 1935ff: 15 (*Haplocnemus*) **N**: MO
theryi Pic, 1935f: 15 (*Haplocnemus*)
biskraensis Schilsky, 1897a: 96 (*Haplocnemus*) **N**: AG
gridellii Pic, 1928i: 103 (*Haplocnemus*)
brevis Rosenhauer, 1856: 161 (*Dasytes*) **E**: PT SP **N**: MO
pellucens Kiesenwetter, 1866a: 383 (*Haplocnemus*)
brevissimus Pic, 1908a: 50 (*Haplocnemus*) **E**: GR
holtzi Pic, 1908f: 50 (*Haplocnemus*)
caelatus Brullé, 1832: 150 (*Dasytes*) **E**: GR
calidus Mulsant & Rey, 1868d: 239 (*Haplocnemus*) **E**: FR SP
caramanicus J. R. Sahlberg, 1913a: 46 (*Haplocnemus*) **A**: TR
caroli Pic, 1941b: 1 (*Haplocnemus*) **N**: MO
caucasicus Schilsky, 1897a: 89 (*Haplocnemus*) **E**: AR GG ST
chalconatus Germar, 1817: 209 (*Dasytes*) **E**: AU CR FR IT SL SZ
aestivus Kiesenwetter, 1863a: 657 (*Haplocnemus*)
fauconneti Pic, 1914o: 57 (*Haplocnemus*)
pinicola Kiesenwetter, 1863a: 658 (*Haplocnemus*)
raffrayi Pic, 1908f: 51 (*Haplocnemus*)
subviolaceus Pic, 1922g: 27 (*Haplocnemus*)
chlorosomus P. H. Lucas, 1846: 199 (*Dasytes*) **N**: AG
cupreus P. H. Lucas, 1846: 199 (*Dasytes*)
coeruleatus Rosenhauer, 1856: 160 (*Dasytes*) **E**: FR (Var) SP **N**: AG MO
barnevillei Kiesenwetter, 1867b: 121 (*Haplocnemus*)
mirabilis Schilsky, 1897a: 99 (*Haplocnemus*)
consobrinus Rosenhauer, 1856: 157 (*Dasytes*) **E**: PT SP
cobosi Pic, 1953f: 144 (*Haplocnemus*)
corcyricus L. Miller, 1866: 818 (*Haplocnemus*) **E**: IT GR (Kerkyra)
ragusae Schilsky, 1894a: 68 (*Haplocnemus*)
cribrarius Brullé, 1832: 151 (*Dasytes*) **E**: GR
cribricollis Mulsant & Rey, 1868d: 234 (*Haplocnemus*) **E**: FR (Corse) IT
elongatus Pic, 1921a: 3 (*Haplocnemus*)
erosus Mulsant & Rey, 1868d: 233 (*Haplocnemus*)
cribripennis Pic, 1921a: 3 (*Haplocnemus*) **E**: GR
cribrosus Schilsky, 1897b: 44 (*Haplocnemus*) **N**: AG MO
festai Pic, 1925b: 2 (*Haplocnemus*)
robustior Pic, 1928i: 103 (*Haplocnemus*)
croceicornis Kiesenwetter, 1863a: 649 (*Haplocnemus*) **E**: UK
cupreatus Schilsky, 1897a: 88 (*Haplocnemus*) **E**: GG
curtus Pic, 1921a: 3 (*Haplocnemus*) **N**: AG
cylindricus Kiesenwetter, 1863a: 651 (*Haplocnemus*) **E**: FR
cyrenaicus Pic, 1955e: 132 (*Haplocnemus*) **N**: LB
dentatus Schilsky, 1897a: 100 (*Haplocnemus*) **N**: AG
depressicollis Schilsky, 1897b: 52 (*Haplocnemus*) **E**: GG
diaphanus Schilsky, 1897a: 94 (*Haplocnemus*) **N**: AG
difficilis Holdhaus, 1923: 137 (*Haplocnemus*) **E**: FR (Corse) IT (Arcipelago Toscana)
mancinii Pic, 1927h: 14 (*Haplocnemus*)
flavicornis Schilsky, 1897a: 87 (*Haplocnemus*) **E**: AB AR **A**: TR
areschanus Pic, 1908k: 81 (*Haplocnemus*)
fortepunctatus Pic, 1935f: 15 (*Haplocnemus*) **N**: MO
gestroi Schilsky, 1897b: 58 (*Haplocnemus*) **N**: AG TU
grancanariensis Lindberg, 1953: 7 (*Haplocnemus*) **N**: CI (Gran Canaria)
heydeni Schilsky, 1894a: 81 (*Haplocnemus*) **N**: AG
hickeri Pic, 1935k: 256 (*Haplocnemus*) **A**: TR
impressus Marsham, 1802: 226 (*Crioceris*) **E**: AU BE BH CR CZ CT DE EN FI FR GB GE HU IT LA LS LT NL NR PL RU SK SL SP ST SV SZ UK
castiliensis Schilsky, 1897b: 41 (*Haplocnemus*)

pini L. Redtenbacher, 1849: 335 (*Dasytes*)
serratus L. Redtenbacher, 1849: 335 (*Dasytes*)
serrulatus Schilsky, 1906a: 100 (*Haplocnemus*)
incognitus Faldermann, 1835a: 204 (*Dasytes*) **E**: AB GG ST
integer Baudi di Selve, 1873a: 302 (*Haplocnemus*) **E**: BU HU IT RO SK SZ
jejunus Kiesenwetter, 1863a: 652 **E**: FR IT
capillicornis Abeille de Perrin, 1907b: xxi (*Haplocnemus*)
kaszabi Majer, 1982: 431 **E**: HU
kiesenwetteri Schilsky, 1897b: 50 (*Haplocnemus*) **E**: GR
koziorowiczi Desbrochers des Loges, 1870b: 122 (*Haplocnemus*) **E**: FR (Corse) IT (Capraia, Elba)
capraianus Pic, 1931j: 159 (*Haplocnemus*)
xanthopus Kiesenwetter, 1871: 85 (*Haplocnemus*)
kruegeri Pic, 1929j: 94 (*Haplocnemus*) **N**: LB
kubanensis Pic, 1909c: 99 (*Haplocnemus*) **E**: ST
lateralis Schilsky, 1894a: 71 (*Haplocnemus*) **E**: AB
latior Pic, 1908f: 50 (*Haplocnemus*) **E**: GR
libanicus Pic, 1901a: 9 (*Haplocnemus*) **A**: SY
laetipes Abeille de Perrin, 1907b: xxii (*Haplocnemus*)
limbipennis Kiesenwetter, 1866a: 383 (*Haplocnemus*) **E**: PT SP
longicornis Kocher, 1962: 196 (*Haplocnemus*) **N**: MO
longulus Schilsky, 1897a: 91 (*Haplocnemus*) **N**: AG TU
brunnescens Pic, 1922g: 27 (*Haplocnemus*)
macedonicus Pic, 1922e: 17 (*Haplocnemus*) **E**: GR
marginatus Rottenberg, 1871: 243 (*Haplocnemus*) **E**: GR IT (Sicilia)
impressipennis Pic, 1921f: 5 (*Haplocnemus*)
marginicollis Schilsky, 1897b: 39 (*Haplocnemus*) **N**: AG
maurus Schilsky, 1895: 18 (*Haplocnemus*) **N**: AG
mohammed Chobaut, 1898: 78 (*Haplocnemus*) **N**: AG
nigricornis garganicus Liberti, 1995: 171 **E**: IT (Puglia)
nigricornis nigricornis Fabricius, 1792b: 81 (*Lagria*) **E**: AU BE BH BY CR CT CZ DE EN FI FR GB GE GR HU IT LA LT NL NR PL RO SK SL SP ST SV SZ UK
chalybeus Germar, 1824: 78 (*Dasytes*)
femoralis Illiger, 1807b: 302 (*Dasytes*)
femoratus Schilsky, 1897b: 46 (*Haplocnemus*)
fuscitibia Mulsant & Rey, 1868d: 247 (*Haplocnemus*)
punctatus Germar, 1824: 77 (*Dasytes*)
rufolateralis Pic, 1914o: 57 (*Haplocnemus*)
testaceofemoralis Pic, 1917i: 22
variolatus A. Costa, 1882: 35 (*Haplocnemus*)
viertli Schilsky, 1897b: 46 (*Haplocnemus*)
nigripes Mulsant & Rey, 1868d: 223 (*Haplocnemus*) **N**: AG TU
palaestinus Baudi di Selve, 1873b: 319 (*Haplocnemus*) **A**: IS LE SY
panalpinus Liberti, 1995: 189 **E**: IT
parumpunctatus Schilsky, 1897b: 53 (*Haplocnemus*) **E**: GG
pectinatus Küster, 1849c: 18 (*Dasytes*) **E**: FR (Corse) IT (Sardegna, Sicilia) MA **N**: AG TU
eumerus Mulsant & Rey, 1868d: 195 (*Haplocnemus*)
flavipes Schilsky, 1897b: 59 (*Haplocnemus*)
fulvipes Schilsky, 1894c: 230 (*Haplocnemus*)
marchali Pic, 1914o: 57 [DA]
melitensis Schilsky, 1897b: 60 (*Haplocnemus*)
obscuripes Schilsky, 1897b: 59 (*Haplocnemus*)
siculus Kiesenwetter, 1863a: 654 (*Haplocnemus*)
pectinicornis P. H. Lucas, 1846: 200 (*Dasytes*) **N**: AG
pertusus Kiesenwetter, 1859c: 172 (*Haplocnemus*) **E**: GR TR **A**: CY
adaliensis Pic, 1908f: 50 (*Haplocnemus*)
elongatior Pic, 1902e: 32 (*Haplocnemus*)
pesruchesi Schilsky, 1897a: 90 (*Haplocnemus*) **N**: AG

pulverulentus Küster, 1849c: 17 (*Dasytes*) **E**: BH BU CR HU SL YU
obscurus Germar, 1817: 209 (*Dasytes*) [HN]
punctiger Schilsky, 1897a: 98 (*Haplocnemus*) **E**: "Caucasus"
quercicola Mulsant & Rey, 1868d: 243 (*Haplocnemus*) **E**: FR
ramicornis Kiesenwetter, 1863a: 651 **E**: UK
raymondi Sante-Claire Deville, 1908a: 218, 219 (*Haplocnemus*) **E**: FR (Corse)
rufipes L. Miller, 1862: 345 (*Haplocnemus*) **E**: GR MC
semicaeruleus Pic, 1922e: 17 (*Haplocnemus*)
taygetana Pic, 1908f: 50 (*Haplocnemus*)
rufomarginatus Perris, 1869: 18 (*Haplocnemus*) **E**: IT (Sardegna) **N**: AG
edoughensis Pic, 1897c: 41 (*Haplocnemus*)
rugicollis Schilsky, 1897a: 95 (*Haplocnemus*) **N**: AG
rugulosus Rosenhauer, 1856: 159 (*Dasytes*) **E**: SP
mateui Pic, 1954f: 97 (*Haplocnemus*)
subcoeruleus Pic, 1908l: 90 (*Haplocnemus*)
sculpturatus Wollaston, 1862c: 447 (*Haplocnemus*) **N**: CI MR
tenerifensis Pic, 1922e: 17 (*Haplocnemus*)
serbicus Kiesenwetter, 1863a: 653 (*Haplocnemus*) **E**: CR RO UK YU
tibiellus Schilsky, 1897b: 34 (*Haplocnemus*)
serratus Brullé, 1832: 152 (*Dasytes*) **E**: GR
similis Schilsky, 1906a: 99 (*Haplocnemus*) **A**: TR
smyrnensis Pic, 1901b: 18 (*Haplocnemus*) **A**: TR
strandi Marcu, 1936: 518 (*Haplocnemus*) **E**: RO
subinteger Pic, 1902j: 60 (*Haplocnemus*) **A**: IQ SY
tarsalis C. R. Sahlberg, 1822: 113 (*Dasytes*) **E**: AU CT CZ DE EN FI GE IT LA LT NR NT PL SK SV SZ
kuesteri Schilsky, 1894d: 331 (*Haplocnemus*)
pandellei Pic, 1918d: 7
rufitarsis C. R. Sahlberg, 1822: 113 (*Dasytes*)
tarsicola J. R. Sahlberg, 1913a: 43 (*Haplocnemus*) **A**: SY
thalensis Pic, 1930b: 1 (*Haplocnemus*) **N**: TU
trinacriensis Ragusa, 1872: 83 (*Haplocnemus*) **E**: IT (Sicilia) MA **N**: TU
tuberculifer Motschulsky, 1850: 364 (*Haplocnemus*) **E**: GE
tumidus Kiesenwetter, 1863a: 650 **E**: SP **N**: MO
tournieri Pic, 1896e: 47 (*Pseudaphyctus*)
tuniseus Pic, 1908f: 50 (*Haplocnemus*) **N**: TU
venustulus Schilsky, 1897a: 93 (*Haplocnemus*) **E**: ST
vestitus Wollaston, 1862c: 447 **N**: CI
nigrofemoratus Pic, 1894p: 112 (*Haplocnemus*)
virens aerosus Schilsky, 1897a: 97 (*Haplocnemus*) **E**: SP
virens virens Suffrian, 1843b: 337 (*Dasytes*) **E**: AU FR GE IT RO SP SZ YU
ahenus Kiesenwetter, 1863a: 656 (*Haplocnemus*)
hispanicus Pic, 1953f: 143 (*Haplocnemus*)
orientalis Lohse, 1977: 181 (*Haplocnemus*)
tibialis Schilsky, 1894d: 331 (*Haplocnemus*)
viridescens Pic, 1935f: 15 **N**: MO
subgenus *Diplambe* Schilsky, 1894c: 234 type species *Dasytes montivagus* Rosenhauer, 1856
abietum Kiesenwetter, 1859c: 171 (*Haplocnemus*) **E**: GR
afer Schilsky, 1897b: 33 (*Haplocnemus*) **N**: AG
akbesianus Pic, 1896e: 48 (*Haplocnemus*) **A**: CY SY
anatolicus Schilsky, 1903: 97 (*Haplocnemus*) **A**: TR
subelongatus Pic, 1910c: 9 (*Haplocnemus*)
atricornis Pic, 1921a: 3 (*Haplocnemus*) **E**: GR
breviusculus Schilsky, 1897b: 31 **N**: AG
crenicollis Kiesenwetter, 1863a: 654 (*Haplocnemus*) **E**: IT (Sicilia) **N**: TU
curtipennis Pic, 1908p: 200 (*Haplocnemus*) **N**: AG
delagrangei Pic, 1902e: 32 (*Haplocnemus*) **A**: SY

desertorum Pic, 1896e: 48 (*Haplocnemus*) **N**: AG
rufofemoratus Pic, 1908f: 49 (*Haplocnemus*)
duplicatus Kiesenwetter, 1871: 85 (*Haplocnemus*) **E**: IT (Sardegna)
geniculatus Schilsky, 1903: 98 (*Haplocnemus*) **A**: TR
hebraicus Schilsky, 1906b: 13 (*Haplocnemus*) **A**: IS
imperforatus Pic, 1908f: 49 (*Haplocnemus*) **N**: TU
latipennis Pic, 1911q: 146 (*Haplocnemus*) **A**: LE SY
rechmayanus Pic, 1911u: 169 (*Haplocnemus*)
montbabor Mayor, 2004: 85 [RN] **N**: AG
baborensis Pic, 1922g: 26 (*Haplocnemus*) [HN]
montivagus Rosenhauer, 1856: 156 (*Dasytes*) **E**: SP
perforatus Schilsky, 1897b: 35 (*Haplocnemus*) **N**: AG
peyroni Pic, 1899l: 259 (*Haplocnemus*) **A**: SY
testaceipes Pic, 1932g: 27 [= 1832c: 24] (*Haplocnemus*) **N**: MO
thessalicus Pic, 1908f: 49 (*Haplocnemus*) **E**: GR
subgenus *Holcopleura* Schilsky, 1894c: 234 type species *Haplocnemus reitteri* Schilsky, 1894
pristocerus Kiesenwetter, 1859c: 170 (*Haplocnemus*) **E**: GR
reitteri Schilsky, 1894a: 59 (*Haplocnemus*) **A**: TR
sahlbergi Mayor, 2007 **nom. nov**. [see New acts] **A**: TR
puncticollis J. R. Sahlberg, 1913a: 40 (*Haplocnemus*) [HN]
turcicus Schilsky, 1897b: 30 (*Haplocnemus*) **A**: TR
subgenus *Ischnopalpus* Schilsky, 1894c: 235 type species *Haplocnemus subcostatus* Schilsky, 1894
curticornis Pic, 1910c: 9 (*Haplocnemus*) **N**: MO
gracilicornis Schilsky, 1897b: 36 (*Haplocnemus*) **E**: SP **N**: TU
alluaudi Pic, 1900m: 165 (*Haplocnemus*)
griseopubescens Pic, 1899l: 259 (*Haplocnemus*) **A**: IS
hierichunticus J. R. Sahlberg, 1913a: 42 (*Haplocnemus*) **A**: IS
morio Schönherr, 1817: 12 (*Dasytes*) **N**: MO
nitidus Majer, 1990d: 197 **A**: IN
persicus Schilsky, 1897b: 37 (*Haplocnemus*) **A**: IN
rufipedes Escalera, 1914: 244 (*Haplocnemus*) **N**: MO
russicus Reitter, 1890a: 148 (*Haplocnemus*) **E**: AB AR **A**: IN
sanctus Pic, 1902e: 32 (*Haplocnemus*) **A**: IS
subcostatus Schilsky, 1894a: 62 (*Haplocnemus*) **N**: AG MO
bonnairei Pic, 1894p: 112 (*Xaplochnemus* [sic])
puncticollis Schilsky, 1894a: 62 (*Haplocnemus*)
viberti Pic, 1922g: 27 (*Haplocnemus*)
suggara alfierii Wittmer, 1937: 169 (*Haplocnemus*) **N**: EG
suggara suggara Peyerimhoff, 1929c: 193 (*Haplocnemus*) **N**: AG
syriacus Schilsky, 1894a: 63 (*Haplocnemus*) **A**: CY IS SY
caerulescens Schilsky, 1897b: 34 (*Haplocnemus*)
tingitanus Baudi di Selve, 1873a: 300 (*Haplocnemus*) **N**: MO
lixensis Bourgeois, 1911: 158 (*Haplocnemus*)
subacuminatus Pic, 1934a: 21 (*Haplocnemus*)

genus *Kubanius* Majer, 1983: 385 type species *Kubanius elegans* Majer, 1983
aurescens Champion, 1922b: 151 (*Dasytes*) **A**: UP
elegans Majer, 1983: 387 **A**: KZ
wittmeri Majer, 1988c: 431 **A**: PA

genus *Trichoceble* C. G. Thomson, 1859: 109 type species *Melyris floralis* Oliver, 1790
Julistus Kiesenwetter, 1859c: 174 type species *Julistus funera* Kiesenwetter, 1859
albohirta Majer, 1990d: 194 **A**: IN
araratica Iablokoff-Khnzorian, 1978f: 243 **E**: AR
arbustorum Kiesenwetter, 1859c: 176 (*Julistus*) **E**: GR

cincta Pic, 1929a: 1 **A**: "Kurdistan"
convexa Pic, 1909f: 113 **A**: SY
curta Baudi di Selve, 1873a: 297 (*Julistus*) **E**: CR
curvipennis Majer, 1990d: 193 **A**: IN
elongata Schilsky, 1897a: 85 **A**: TR
floralis A. G. Olivier, 1790a: no 21: 11 (*Melyris*) **E**: AU BE CZ DE EN FR GE HU IT LT NL NR PL RU SK SV SZ "Caucasus" **A**: "Siberia"
    floricola Kiesenwetter, 1861: 386 (*Julistus*)
fulvopilis Reitter, 1889a: 25 (*Julistus*) **E**: AB AR **A**: IN
funera Kiesenwetter, 1859c: 175 (*Julistus*) **E**: GR **A**: TR
    brunneipennis Reitter, 1902k: 258
    caucasica Reitter, 1902k: 259
grandis Schilsky, 1896a: 90 **A**: SY
griseohirta Reitter, 1885i: 380 (*Julistus*) **E**: GR (Kríti)
heydeni Schilsky, 1897c: 155 **A**: BEI
immarginata Reitter, 1902k: 257 **A**: TR (Eski-Chéhir)
laeta Majer, 1986: 303 **A**: MG
lederi Schilsky, 1896a: 92 **E**: AR GG
longicornis Kiesenwetter, 1863a: 647 (*Julistus*) **E**: GR (Kyklades: Siros)
major Pic, 1922e: 17 **E**: GR
mediocris Majer, 1986: 306 **A**: MG
memnonia Kiesenwetter, 1861: 385 (*Julistus*) **E**: AU CZ DE FR GE HU IT LT NR PL RO SK SV SZ "Caucasus"
    fulvohirta C. N. F. Brisout de Barneville, 1862: 601 (*Julistus*)
    semirufescens Pic, 1915h: 22 (*Julistus*)
nigra Pic, 1921a: 4 **E**: GR
oculata Schilsky, 1896a: 98 **E**: GR
oertzeni Schilsky, 1896a: 91 **E**: GR (Kríti)
pallidipes Pic, 1944c: 12 (*Julistus*) **A**: CH
ramicornis Schilsky, 1900: 4 **A**: KI
schatzmayeri Pic, 1909f: 113 **E**: GR
schilskyi Reitter, 1902k: 259 **E**: GG
sparsepunctata Pic, 1921a: 4 **A**: TR
subcoerulea Pic, 1921a: 4 **E**: GR
subcoriacea Reitter, 1902k: 260 **E**: GR
testaceipes Pic, 1921a: 4 **A**: CY
torretassoi Wittmer, 1935d: 254 **E**: GR (Rodos)
unguicularis Reitter, 1902k: 258 **A**: TR
unicolor Pic, 1932j: 251 **A**: SCH

tribe Microjulistini Majer, 1987

genus *Flavojulistus* Majer, 1990d: 198 type species *Flavojulistus melanophthalmus* Majer, 1990
melanophthalmus Majer, 1990d: 199 **A**: IN

genus *Microjulistus* Reitter, 1889d: 111 type species *Microjulistus fulvus* Reitter, 1889
    Ceralliscus Bourgeois, 1894b: 121 type species *Ceralliscus raffrayi* Bourgeois, 1894
chobauti Pic, 1895e: 79 (*Ceralliscus*) **N**: AG
duboisi Peyerimhoff, 1931b: 51 **N**: AG (Hoggar)
fulvus Reitter, 1889d: 112 **E**: AB AR **A**: IN MG TM TR
    obscurithorax Pic, 1905b: 97
    moreli Pic, 1927g: 10 (*Ceralliscus*)
    pectoralis Pic, 1927g: 10 (*Ceralliscus*)
    rufithorax Pic, 1927g: 10 (*Cerailiscus*)
gibbipennis Chobaut, 1898: 80 **N**: AG
laticollis Pic, 1894l: 95 (*Cerailiscus*) **N**: AG

lysholmi Pic, 1899i: 170 **A**: SY
abdominalis Pic, 1899i: 170, 171
nigripennis Pic, 1899i: 170, 171
meynieri Peyerimhoff, 1931b: 51 **N**: AG (Hoggar)
minutus Pic, 1927f: 5 [EA] **E**: FR (Corse)
nigricollis Pic, 1894l: 95 (*Ceralliscus*) **N**: AG
raffrayi Bourgeois, 1894b: 122 (*Ceralliscus*) **N**: AG
nigrifrons Pic, 1895e: 80 (*Ceralliscus*)
rubricollis Abeille de Perrin, 1907b: xxii (*Ceralliscus*) **A**: YE
wegeneri Pic, 1899i: 170 **N**: EG
obscurus Pic, 1899i: 170, 171

genus *Semijulistus* Schilsky, 1894c: 227 type species *Dasytes callosus* Solsky, 1868
Celsus Lewis, 1895a: 118 type species *Celsus spectabilis* Lewis, 1895
Eurelymis Casey, 1895: 602 type species *Eurelymis speculifer* Casey, 1895
callosus Solsky, 1868: 32 (*Dasytes*) **A**: KZ WS
elongatus Nakane, 1981b: 129 (*Celsus*) **A**: JA
spectabilis Lewis, 1895a: 119 (*Celsus*) **A**: JA

tribe Rhadalini LeConte, 1862

genus *Rhadalus* LeConte, 1852b: 212 type species *Rhadalus testaceus* LeConte, 1852
Cymbolus Gorham, 1886: 324 type species *Cymbolus rufopiceus* Gorham, 1886
asiaticus Pic, 1951c: 13 (*Cymbolus*) **A**: "Tuhang"

tribe Rhadalini, genera incertae sedis
genus *Antinea* Peyerimhoff, 1929c: 191 type species *Antinea saxicola* Peyerimhoff, 1929
saxicola Peyerimhoff, 1929c: 191 **N**: AG (Hoggar)

genus *Jelinekius* Majer, 1990d: 183 type species *Jelinekius persicus* Majer, 1990
brunneus Majer, 1990d: 190 **A**: IN
minor Majer, 1990d: 192 **A**: IN
persicus Majer, 1990d: 185 **A**: IN

family MALACHIIDAE Fleming, 1821

subfamily **Carphurinae** Champion, 1923

genus *Carphuroides* Champion, 1923a: 45 type species *Helcogaster pectinatus* Sharp, 1885
Asiocarphurus Wittmer, 1981b: 229 type species *Asiocarphurus bhaktai* Wittmer, 1981
almorensis Champion, 1925d: 268 **A**: AF UP
astruci Pic, 1918c: 4 (*Carphurus*) **A**: NP SD
bhaktai Wittmer, 1981b: 231 (*Asiocarphurus*) **A**: SD
brendelli Wittmer, 1995e: 340 **A**: UP
chinensis Wittmer, 1999: 214 **A**: YUN
emeiensis Wittmer, 1994a: 15 **A**: SCH
formosanus Wittmer, 1982f: 351 **A**: TAI
grossepunctatus Wittmer, 1981b: 225 **A**: SD
latemarginatus Wittmer, 1981b: 227 **A**: BT
lopchuensis Wittmer, 1981b: 228 **A**: SD
nuristanus Wittmer, 1981b: 220 **A**: AF
pectinatus Sharp, 1885b: 157 (*Helcogaster*) **A**: CH **ORR**
ellingensis Wittmer, 1981b: 227 **A**: SD
ilosicornis Wittmer, 1982f: 351 **A**: TAI
lagiatus Kiesenwetter, 1874b: 287 (*Carphurus*) **A**: FE JA
obustithorax Wittmer, 1981b: 231 (*Asiocarphurus*) **A**: BT
rosti Pic, 1902d: 25 (*Attalus*) **A**: ES (Transbaikal) FE (Amur)

hemipterus Evers, 1964: 16 (*Attalus*)
transbaicalicus Wittmer, 1954c: 111 (*Carphurus*)
sikkimensis Wittmer, 1981b: 223 **A**: BT SD
smetaceki Wittmer, 1981b: 223 **A**: UP
talhouki Wittmer, 1984d: 245 **A**: SA

genus *Carphurus* Erichson, 1840: 132 type species *Carphurus dispar* Erichson, 1840
nigritulus Champion, 1923a: 30 **A**: UP **ORR**
rubroannulatus Motschulsky, 1860a: 64 **A**: CH **ORR**
rufifrons Pic, 1921d: 20 **A**: TAI
sivalikensis Wittmer, 1981b: 218 **A**: UP
topali Wittmer, 1978a: 63 **A**: TAI **ORR**
varicolor Wittmer, 1983a: 311 **A**: NP **ORR**

genus *Telocarphurus* Wittmer, 1939b: 30 type species *Telocarphurus drescheri* Wittmer, 1939
eglini Wittmer, 1981b: 229 **A**: SD
taiwanus Wittmer, 1982f: 350 **A**: TAI

subfamily **Malachiinae** Fleming, 1821

tribe Apalochrini Mulsant & Rey, 1867

genus *Apalochrus* Erichson, 1840: 50 type species *Apalochrus femoralis* Erichson, 1840
Hapalochrus Agassiz, 1846: 27 [unjustified emendation]
Paratinus Abeille de Perrin, 1891b: 220 type species *Apalochrus femoralis* Erichson, 1840
femoralis Erichson, 1840: 53 **E**: AU CZ DE EN FI FR HU LA NT LA LT PL RU SK ST SV UK YU **A**: "Turkestan"
flavicollis Schaufuss, 1871: 157 **E**: ST **A**: "Turkestan"
flavicornis Abeille de Perrin, 1890a: 50 **E**: ST
fulvicollis Gebler, 1844: 99 (*Malachius*) **E**: ST
maculicollis Peyron, 1877: 274
maculicollis Motschulsky, 1860b: 118
notatithorax Pic, 1936a: 2 **E**: ST
notatus mongolicus Evers, 1968b: 124 **A**: MG
notatus notatus Zubkov, 1833: 349 (*Malachius*) **A**: MG UZ
pallipes Motschulsky, 1860b: 118 **E**: ST
turkestanicus Pic, 1907a: 99 **A**: "Turkestan"

genus *Collops* Erichson, 1840: 54 type species *Malachius bipunctatus* Say, 1823
alatauensis Wittmer, 1992b: 15 [DA] **A**: KZ
bicoloripennis Pic, 1920h: 21 (*Laius*) [DA] **A**: "Siberia"
nigerrimus Wittmer, 1992b: 14 [DA] **A**: XIZ
susamirensis Wittmer, 1992b: 12 [DA] **A**: KI

genus *Dromanthomorphus* Pic, 1921g: 7 type species *Dromanthomorphus cyaneus* Pic, 1921
Flabellapalochrus Pic, 1923g: 49, 51 type species *Hapalochrous mirabilis* Pic, 1907
armenus Iablokoff-Khnzorian, 1978b: 175 (*Apalochrus*) **E**: AR
excavatus Wittmer, 1995e: 348 **A**: YUN
omega Evers, 1965: 149 (*Apalochrus*) **A**: MG
boops Evers, 1968a: 33
pectinicornis Erichson, 1840: 53 (*Apalochrus*) **E**: ST

genus *Hadrocnemus* Kraatz, 1895a: 59 type species *Hadrocnemus conradti* Kraatz, 1895 (= *Hapalochrus malachioides* Fairmaire, 1887)
Amanicollops Pic, 1908c: 25 type species *Amanicollops obscurus* Pic, 1908
cavaleriei Pic, 1923o: 376 (*Apalochrus*) **A**: CH
eversi Iablokoff-Khnzorian, 1988: 46 (*Troglocollops*) **A**: ES (Altai)

fernandoi Pic, 1925d: 4 (*Apalochrus*) **A**: YE
maindroni Abeille de Perrin, 1900a: 21 (*Hapalochrous*) **A**: PA

genus *Intybia* Pascoe, 1866c: 448 type species *Intybia guttata* Pascoe, 1866
birmanica Champion, 1921b: 338 (*Laius*) **A**: YUN **ORR**
duplex Champion, 1921b: 339 (*Laius*) **A**: TAI **ORR**
albonotatus Pic, 1926k: 79 (*Laius*)
blaisei Pic, 1926k: 79 (*Laius*)
eversi Hicker, 1948: 61 (*Laius*) **A**: FUJ **ORR**
fouqueti Pic, 1944a: 6 (*Laius*) **A**: FUJ **ORR**
gressitti Wittmer, 1956b: 313 (*Laius*) **A**: HUB
hamata Wittmer, 1996: 321 (*Laius*) **A**: NP
histrio Kiesenwetter, 1874b: 283 (*Lajus* [sic]) **A**: FE (Kunashir) JA SHG
flavicornis Kiesenwetter, 1874b: 283 (*Lajus* [sic]) [HN]
kiesenwetteri Lewis, 1893b: 151 [RN]
kawasakii Nakane, 1956b: 162 (*Laius*) **A**: JA
kishiii Nakane, 1955a: 33 (*Laius*) **A**: JA
latefasciata Pic, 1919a: 12 (*Laius*) **A**: TAI
niponicus Lewis, 1895a: 116 (*Laius*) **A**: FE JA
pellegrini longior Wittmer, 1997d: 193 **A**: SC
pellegrini pellegrini Pic, 1910p: 62 (*Laius*) **A**: FE (Kunashir) JA (Tsushima, Yedo)
planicornis Wittmer, 1982f: 370 (*Laius*) **A**: FUJ TAI
polititthorax Wittmer, 1997d: 198 **A**: CH
producta Wittmer, 1995e: 356 **A**: YUN
rotundula Wittmer, 1997d: 200 **A**: FUJ
sinensis Pic, 1910t: 87 (*Laius*) **A**: HKG **ORR**
swatowensis Wittmer, 1956b: 312 (*Laius*) **A**: GUA
taiwana Wittmer, 1997d: 203 **A**: TAI
takaraensis Nakane, 1955b: 376 (*Laius*) **A**: JA (Tokara)
tsushimensis Satô & Ohbayashi, 1968: 5 (*Laius*) **A**: JA (Tsushima)
viridithorax shaanxiensis Wittmer, 1997d: 195 **A**: SHA
viridithorax viridithorax Pic, 1921d: 19 (*Laius*) **A**: SHG
atricollis Pic, 1921d: 19

genus *Laius* Guérin-Méneville, 1838a: 78 type species *Laius cyaneus* Guérin-Méneville, 1838
Megadeuterus Westwood, 1833: 678 type species *Paussus flavicornis* Fabricius, 1801
Rotomalachius Kôno, 1937: 121 type species *Rotomalachius esakii* Kôno, 1937
afghanistanicus Wittmer, 1956a: 155 **A**: AF
asahinai Nakane, 1955b: 375 **A**: JA (Tokara)
aurantiacus arabicus Wittmer, 1954e: 322 **A**: SA YE
bengalensis Wittmer, 1986c: 230 **A**: NP
bivittatus Wittmer, 1955: 45 **A**: YUN
brancuccii Wittmer, 1989: 234 **A**: UP
bulbifer Champion, 1921b: 326 **A**: UP
caeruleipennis Pic, 1914a: 16 **A**: AF HP KA (Jammu)
canaliculatus Wittmer, 1986c: 222 **A**: NP
dimidiatus Plavilshchikov, 1924: 230 (*Collops*) **A**: NE
erectodentatus Wittmer, 1982f: 368 **A**: TAI
fenchihuensis Wittmer, 1982f: 370 **A**: TAI
flavicornis Fabricius, 1801b: 75 (*Paussus*) **A**: CH **ORR**
foveicornis Pic, 1917d: 5 **A**: UP
klapperichi Hicker, 1948: 60 **A**: FUJ
lunatus Champion, 1921b: 328 **A**: UP
miyamotoi Nakane, 1955b: 373 **A**: JA (Tokara, Ryukyus)
nitidicollis Wittmer, 1994a: 13 **A**: SCH

nodifrons Champion, 1921b: 327 **A**: UP
nuristanicus Wittmer, 1989: 232 **A**: AF
pakistanicus Wittmer, 1986c: 225 **A**: PA
partepolitus Wittmer, 1986c: 223 **A**: UP
pashtun Wittmer, 1989: 231 **A**: PA
pelegrini Pic, 1910p: 62 **A**: JA
pici Miwa, 1931a: 108 [RN] **A**: TAI
  sexmaculatus Pic, 1926k: 78 [HN]
rubithorax Pic, 1907t: 190 **A**: YUN
savioi Pic, 1927n: 152 **A**: SHG
schuhi Wittmer, 1994a: 12 **A**: UP
sexmaculatus Pic, 1919a: 12 **A**: TAI
sikkimensis Pic, 1914c: 16 **A**: SD
  testaceilabris Pic, 1914c: 16
theresae Pic, 1944a: 6 **A**: YUN
venustus Erichson, 1840: 63 **N**: EG **A**: SA YE **AFR**
  abyssinicus Pic, 1917d: 5
  bedforti Pic, 1928e: 5
  bifasciatus Laporte, 1840a: 280
  letourneuxi Pic, 1928e: 5
  rugosiceps Pic, 1922m: 159

genus *Myrmecophasma* Bourgeois, 1885f: clxxvii [RN] type species *Myrmecospectra nietneri* Motschulsky, 1858
  Myrmecodes Gemminger & Harold, 1869: 1777 [HN, RN] type species *Myrmecospectra nietneri* Motschulsky, 1858
  Myrmecospectra Motschulsky, 1858c: 65 [HN] type species *Myrmecospectra nietneri* Motschulsky, 1858
fukienensis Evers, 1948b: 155 **A**: FUJ

genus *Nudopectinus* Evers, 1987: 14 type species *Apalochrus flabellicornis* Erichson, 1840
flabellicornis Erichson, 1840: 52 (*Apalochrus*) **N**: EG **AFR**
  bijunctus Pic, 1899k: 231
  sudanicus Pic, 1931h: 101

genus *Opisthapalochrus* Evers, 1987: 14 type species *Cantharis fasciata* Fabricius, 1792
subgenus *Opisthapalochrus* Evers, 1987: 14 type species *Cantharis fasciata* Fabricius, 1792
flavosellatus Fairmaire, 1887a: 125 (*Hapalochrus*) **A**: ZHE
subgenus *Spinapalochrus* Pic, 1919a: 12 type species *Apalochrus rufofasciatus* Pic, 1919
rufofasciatus Pic, 1919a: 12 (*Apalochrus*) **A**: TAI **ORR**

genus *Paratinoides* Medvedev, 1964b: 157 type species *Paratinoides apterus* Medvedev, 1964
apterus Medvedev, 1964b: 158 **A**: KZ
arcticus Medvedev, 1958: 469 (*Collops*) **A**: FE (Chukotskiy Okrug)

genus *Protapalochrus* Evers, 1987: 13 type species *Apalochrus flavolimbatus* Mulsant & Rey, 1853
chamaeleon Evers, 1971d: 233 (*Apalochrus*) **A**: MG
fedtschenkoi Solsky, 1882: 39 (*Apalochrus*) **A**: AF MG UZ "Turkestan"
festivus Zubkov, 1833: 348 (*Malachius*) **E**: ST "Caucasus" **A**: TM **AFR**
  paulojunctus Pic, 1953h: 1530
  variegatus Erichson, 1840: 52 (*Apalochrus*)
flavolimbatus Mulsant & Rey, 1853b: 8 **E**: FR GR HU IT SP "Caucasus" **N**: AG MO TU **A**: CY
  cretica Pic, 1903t: 177
  tricolor Kiesenwetter, 1859a: 30 (*Apalochrus*)
persicus Champion, 1920c: 324 (*Hapalochrus*) **A**: IN TR "Kurdistan"
pictus Motschulsky, 1860b: 118 (*Apalochrus*) **E**: ST "Ural"
  vittatus Morawitz, 1861: 286
puncticollis Wittmer, 1970c: 96 (*Apalochrus*) **A**: MG

vaulogeri Abeille de Perrin, 1898c: 257 (*Hapalochrous*) **N**: AG **AFR**
grandicornis Pic, 1942c: 8 (*Apalochrus*)
viridimetallicus Pic, 1927i: 39 (*Troglops*)

genus *Protocollops* Evers, 1991b: 108 type species *Apalochrus apicalis* Ballion, 1878
apicalis Ballion, 1878: 294 (*Apalochrus*) **A**: XIN
abrinoides Abeille de Perrin, 1885c: cxvi (*Collops*)
bulganensis Evers, 1968a: 31 (*Collops*) **A**: MG

genus *Simoderus* Abeille de Perrin, 1891b: 210 type species *Malachius reflexicollis* Gebler, 1843
reflexicollis Gebler, 1843: 38 (*Malachius*) **A**: "Siberia"

genus *Troglocollops* Wittmer, 1965d: 143 type species *Collops obscuricornis* Motschulsky, 1860
subgenus *Alexeus* Tshernyshev, 1994: 71 type species *Troglocollops tshemerisi* Tshernyshev, 1994
dlabolai Wittmer, 1968a: 21 **A**: MG
logunovi Tshernyshev, 1994: 76 **A**: ES (Tuva)
mongolicus Wittmer, 1965d: 143 **A**: MG
tshemerisi Tshernyshev, 1994: 71 **A**: ES (Altai)
subgenus *Troglocollops* Wittmer, 1965d: 143 type species *Collops obscuricornis* Motschulsky, 1860
betpakdalensis Medvedev, 1964b: 157 (*Collops*) **A**: KZ
bituberculatus Wittmer, 1979c: 153 **A**: MG
coronatus Wittmer, 1992b: 15 **A**: KZ (Saur)
gobiensis Evers, 1987: 18 **A**: MG
mordkovitshi Tshernyshev, 1994: 60 **A**: ES WS
niger Wittmer, 1979c: 154 **A**: MG
obscuricornis Motschulsky, 1860b: 118 (*Collops*) **A**: ES FE KZ MG WS
sulcithorax Pic, 1914p: 73 (*Collops*)
unicolor Peyron, 1877: 272 (*Apalochrus*)
turanicus Evers, 1966c: 105 **A**: "Turkestan"
wittmeri Tshernyshev, 1994: 60 **A**: KZ
subgenus *Troglocollopsoides* Wittmer, 1979c: 151 type species *Troglocollops medvedevi* Wittmer, 1979
bythinocerus Reitter, 1902d: 188 (*Cerapheles*) **A**: MG
cinctus Gebler, 1847b: 434 (*Ebaeus*) **A**: ES FE MG
limbatus Motschulsky, 1860b: 118 (*Collops*)
kaszabi Evers, 1965: 148 **A**: ES MG WS
latecavus Wittmer, 1969e: 515 **A**: MG
latecavus medvedevi Wittmer, 1979c: 151 **A**: MG
latecavus urianchaicus Tshernyshev, 1994: 66 **A**: ES (Tuva)

tribe Attalini Abeille de Perrin, 1890

genus *Attalus* Erichson, 1840: 89 type species *Attalus lusitanicus* Erichson, 1840
subgenus *Abrinus* Mulsant & Rey, 1867: 153 type species *Malachius limbatus* Fabricius, 1798
amictus Erichson, 1840: 105 (*Anthocomus*) **E**: FR IT PT SP SZ
obliteratithorax Pic, 1911g: 106
analis Panzer, 1798a: 6 (*Malachius*) **E**: AU CZ FR GE GR HU IT PL PT SK SP SZ TR UK
fagi Motschulsky, 1854d: 56 (*Anthocomus*)
innotatus Pic, 1911g: 105
limbatus Fabricius, 1798: 71 (*Malachius*) **E**: FR PT SP (Baleares) **N**: AG MO
ulicis Erichson, 1840: 105 (*Anthocomus*)
parietariae Erichson, 1840: 103 (*Anthocomus*) **E**: IT SP (Baleares)
limbatus Kiesenwetter, 1863a: 270
pictus Kiesenwetter, 1850: 224 (*Anthocomus*) **E**: FR SP
ragusae Schaufuss, 1871: 157 (*Axinotarsus*) **E**: IT (Sicilia)
panormitanus Ragusa, 1871: 282 (*Attalus*)
subgenus *Antholinus* Mulsant & Rey, 1867: 150 type species *Anthocomus varitarsis* Kraatz, 1862

bayonnei Chobaut, 1898: 77 **N**: AG
elzeari Uhagón, 1901b: 360 **E**: FR IT SP **N**: AG MO
abeillei Uhagón, 1900: 352 [HN]
fageli Evers, 1955: 76 **N**: AG
hauseri Abeille de Perrin, 1887a: 324 **A**: KZ
iguermalensis Escalera, 1936: 434 **N**: MO
jocosus Erichson, 1840: 101 (*Anthocomus*) **E**: FR IT PT SP **N**: AG
kokandicus Solsky, 1882: 41 (*Ebaeus*) **A**: UZ
labilis Erichson, 1840: 106 (*Anthocomus*) **E**: FR (Corse) IT (Sardegna)
marraquensis Escalera, 1914: 237 **N**: MO
ornatus Escalera, 1914: 238
melanopus Evers, 1966b: 102 **N**: MO
minimus P. Rossi, 1790: 170 (*Cantharis*) **E**: AZ FR IT PT SP SZ **N**: AG MO TU
aemulus Erichson, 1840: 102 (*Anthocomus*)
collaris Laporte, 1840a: 279 (*Malachius*)
corsicus Pic, 1954a: 1 [RN]
inlateralis Pic, 1911g: 105 [HN]
lateralis Erichson, 1840: 101 (*Anthocomus*)
militaris Wollaston, 1854: 245 (*Malachius*)
mirleftensis Evers, 1962: 158 **N**: MO
reitteri Abeille de Perrin, 1891a: 582 **N**: MO
sericans Erichson, 1840: 102 (*Anthocomus*) **E**: FR (Corse) IT SP
moreli Pic, 1927f: 5
tenietensis Abeille de Perrin, 1882a: 139 (*Antholinus*) **N**: AG **A**: IS
varitarsis Kraatz, 1862c: 270 (*Anthocomus*) **E**: FR IT PT SP **N**: AG MO TU
distinctus Mulsant & Rey, 1867: 154 (*Antholinus*)
gavoyi Pic, 1927f: 5
subdisjunctus Pic, 1927f: 5
tarsalis Perris, 1864: 287 (*Malachius*)
subgenus *Attalus* Erichson, 1840: 89 type species *Attalus lusitanicus* Erichson, 1840
Acletus LeConte, 1852d: 167 type species *Acletus nigrellus* LeConte, 1852
Flabellattalus Pic, 1923n: 80 type species *Attalus eichelbaumi* Pic, 1908
Pecteropus Wollaston, 1854: 245 type species *Pecteropus maderensis* Wollaston, 1854
Scalopterus Motschulsky, 1860d: 406 type species *Scalopterus infuscatus* Motschulsky, 1860
abeillei Pic, 1900f: 53 **N**: TU
lineatithorax Pic, 1900f: 53
obliterata Pic, 1900f: 54
acaciae Peyerimhoff, 1929b: 110 **N**: AG
acusae Evers, 1988a: 77 **N**: CI (Gran Canaria)
aegyptiacus Pic, 1911c: 73 **N**: EG
aenescens Wollaston, 1862c: 438 **N**: CI
puncticollis Wollaston, 1862c: 438
umbrinus Wollaston, 1862c: 438
aetnensis Abeille de Perrin, 1891b: 228 **E**: IT (Sicilia)
afourerus Evers, 1995b: 92 **N**: MO
aldeae Evers, 1961a: 9 **N**: CI (Gran Canaria)
alluaudi Pic, 1903g: 113 **N**: TU
alpinus Giraud, 1851: 132 (*Ebaeus*) **E**: AU FR IT SZ
alyssi Peyerimhoff, 1925c: 267 **N**: MO
amami Satô, 1986a: 235 **A**: JA (Ryukyus)
anagaensis Evers, 1961a: 11 **N**: CI (Tenerife)
angustifrons Wollaston, 1862c: 427 (*Pecteropus*) **N**: CI (Gomera)
anthicoides Wollaston, 1862c: 435 **N**: CI (Fuerteventura, Lanzerote)
anticus Kiesenwetter, 1866a: 393 **E**: PT SP
ariasi Escalera, 1933c: 359 **N**: MO
atrilabris Pic, 1925h: 14 **A**: CH

atrocyaneus Uhagón, 1898: 41 **E**: SP
baguenai Escalera, 1942a: 133 **E**: SP
bandamae Evers, 1961a: 8 **N**: CI (Gran Canaria)
betancuriensis Evers, 1959: 225 **N**: CI (Fuerteventura)
bisculpturatus Wollaston, 1862c: 430 **N**: CI (Fuerteventura)
bobadillae Evers, 1960: 168 **N**: CI (Gomera)
bolivari Escalera, 1933a: 220 **N**: MO
bonnairei Abeille de Perrin, 1885b: 144 **N**: AG MO
calcarifer Champion, 1922c: 219 **A**: SD
caroli Pic, 1934b: 26 [RN] **N**: MO
  alluaudi Pic, 1933h: 102 [HN]
  atlasicus Escalera, 1933d: 395
cerastes Abeille de Perrin, 1893c: clxxvi **N**: TU
chinensis A. Fleischer, 1919: 211 **A**: SHN TAI
  incertus Pic, 1925h: 13
  luperoides Pic, 1933i: 112
chloroticus Fairmaire, 1864: 642 (*Ebaeus*) **N**: AG TU
chrysanthemi Wollaston, 1862c: 432 **N**: CI (Fuerteventura, Lanzerote)
  dasytoides Wollaston, 1862c: 432
chujoanus Wittmer, 1982b: 53 **A**: JA (Ryukyus)
coeruleipennis Har. Lindberg, 1951: 3 **N**: CI (Gran Canaria)
coloratus Abeille de Perrin, 1882a: 140 **E**: SP (Baleares)
columbi Evers, 1960: 167 **N**: CI (Gomera)
commixtus Wollaston, 1862c: 433 **N**: CI (Fuerteventura, Lanzerote)
concoloricollis Pic, 1932h: 29 **A**: "Palestine"
convolvuli Abeille de Perrin, 1882a: 145 **N**: MO
crispus Lindberg, 1953: 6 **N**: CI (Hierro)
cupreomicans Abeille de Perrin, 1882b: 181 **N**: AG MO
  emiliae Pic, 1953e: 117
  rotroui Peyerimhoff, 1935: 191
cupreus Escalera, 1914: 236 **N**: MO
curticornis Pic, 1936a: 1 **N**: MO
curtipennis Pic, 1938d: 5 **A**: "Palestine"
cyaneus Fabricius, 1787a: 169 (*Malachius*) **E**: FR IT MA SP SZ **N**: AG TU
  erythroderus Erichson, 1840: 90
  genei Küster, 1848a: 13
  luxurians Erichson, 1840: 90
dalmatinus Erichson, 1840: 91 **E**: CR GR IT (Sicilia) TR UK YU
dasytiformis Abeille de Perrin, 1885b: 153 [RN] **N**: AG MO
  dasytoides Abeille de Perrin, 1882b: 180 [HN]
demoflysi Normand, 1951: 28 **N**: TU
discoluteus Pic, 1925h: 13 **A**: SHX **ORR**
diversicollis Chobaut, 1898: 77 **N**: AG
donceeli Pic, 1908q: 228 **A**: HP
dumonti Peyerimhoff, 1929a: 37 **N**: AG
elongatulus Lewis, 1895a: 117 **A**: JA
euphorbiae Escalera, 1921: 306 **N**: CI (Tenerife)
eversi Plata Negrache, 1987a: 157 **N**: CI (La Palma)
felix Evers, 1971c: 68 **N**: CI (Gran Canaria)
fernandezi Evers, 1961b: 174 **N**: CI (Gran Canaria)
festai Pic, 1925b: 1 **N**: LB
flavoapicalis Pic, 1933i: 112 **A**: JA
flavocinctus Pic, 1930e: 9 **A**: "Turkestan"
fortunatus Evers, 1971c: 72 **N**: CI (El Hierro)
fuentei Pardo Alcaide, 1975a: 96 **N**: MO

ghardaiensis Wittmer, 1994a: 8 **N**: AG
gineri Escalera, 1942a: 134 **E**: SP (Baleares)
gomerensis Evers, 1961a: 13 **N**: CI (Gomera)
gracilentus Mulant & Rey, 1867: 204 **E**: FR **N**: AG MO TU
  barbarus Motschulsky, 1854d: 55
gracilis Kiesenwetter, 1866a: 372 **E**: SP **N**: AG MO
  lugubris Abeille de Perrin, 1891b: 227
  miser Abeille de Perrin, 1891a: 575
guanche Evers, 1988a: 77 **N**: CI (Gran Canaria)
guilinensis Wittmer, 1999: 242 **A**: GUX
hariensis Evers, 1959: 226 **N**: CI (Lanzarote)
hasimoufensis Evers, 1995b: 93 **A**: MG
huashanensis Wittmer, 1999: 242 **A**: SHX
incisopygus Wittmer, 1956b: 309 **A**: TAI
insusetus Evers, 1959: 226 **N**: CI (Lanzarote)
interjunctus Peyerimhoff, 1925c: 267 **N**: MO
jordanicus Wittmer, 1979c: 149 **A**: JO
kaimon Nakane, 1987a: 5 **A**: JA
klapperichi emeiensis Wittmer, 1999: 240 **A**: SCH
klapperichi klapperichi Evers, 1948a: 52 **A**: FUJ
kuluensis Pic, 1910j: 37 **A**: HP
kurosawai ejimai Nakane, 1989c: 151 **A**: JA
kurosawai kurosawai Satô, 1986a: 236 **A**: JA (Ryukyus)
laevicollis Wollaston, 1862c: 434 **N**: CI MO
  pallidicolor Escalera, 1940: 259
latopygus Wittmer, 1972b: 139 **A**: TAI
laufferi Uhagón, 1898: 41 **E**: SP
leptocephalus Abeille de Perrin, 1885b: 144 **N**: MO
longicornis Pic, 1948b: 15 **A**: JIA
luctuosus Har. Lindberg, 1953: 6 [RN] **N**: CI (Fuerteventura)
  lindbergius Pic, 1954a: 1 [RN]
  tristis Har. Lindberg, 1951: 3 [HN]
lusitanicus didymae Evers, 1962: 162 **N**: MO **AFR**
lusitanicus lusitanicus Erichson, 1840: 89 **E**: AZ FR IT PT SP **N**: AG MO MR TU **AFR**
  miniaticollis Tarnier, 1860: 95
lutatus Abeille de Perrin, 1890a: 46 **N**: EG
  tournieri Pic, 1914l: 50
mackiei Pic, 1931f: 60 **N**: MO
  cockerelli Pic, 1931f: 60
  kocheri Pic, 1949a: 5 **N**: MO
maderensis Wollaston, 1854: 247 (*Pecteropus*) **N**: MR
mairei Peyerimhoff, 1926b: 193 **N**: MO
makarios Evers, 1971c: 70 **N**: CI (La Gomera)
mascae Evers, 1961a: 10 **N**: CI (Tenerife)
mateui Pardo Alcaide, 1954c: 105 **E**: SP
melittensis Peyron, 1877: 159 **E**: IT MA
mitlaensis Wittmer, 1937: 168 **A**: SI
mokattamensis Pic, 1912i: 33 **N**: EG
montanus Pic, 1953c: 18 **N**: MO
monticola Evers, 1961a: 7 **N**: CI (Tenerife)
mozabita Chobaut, 1897e: 278 **N**: AG EG MO **A**: IS JO
nanpingensis Wittmer, 1999: 244 **A**: SCH
nigropustulatus Champion, 1925d: 266 **A**: UP
nourrichelii nourrichelii Laporte, 1838: 29 (*Malachius*) **E**: IT
  palliatus Baudi di Selve, 1872: 127

transfuga Kiesenwetter, 1859b: 58 (*Anthocomus*)
obscuripennis Pic, 1905h: 154 **N**: AG
obscurithorax Pic, 1898j: 153 **N**: AG
obscurus Wollaston, 1862c: 437 **N**: CI (Gran Canaria)
oceanicus Evers, 1971a: 175 **N**: CI (Graciosa, Lanzarote)
ochraceus Har. Lindberg, 1951: 3 **N**: CI (Graciosa, Lanzarote)
okinawanus Wittmer, 1960b: 110 **A**: JA (Ryukyus)
olivensis Evers, 1959: 225 **N**: CI (Fuerteventura)
omophloides Abeille de Perrin, 1882b: 181 **N**: AG MO
convexifrons Abeille de Perrin, 1891a: 577
ornatissimus Wollaston, 1862c: 431 **N**: CI (Gomera, La Palma)
oudjanus Pic, 1920f: 17 **N**: MO
santiagoi Pardo Alcaide, 1955: 46
ouinensis Evers, 1995b: 94 **N**: MO
ovatipennis Wollaston, 1862c: 429 **N**: CI
excelsus Wollaston, 1862c: 430
gracilipes Wollaston, 1862c: 430
subopacus Wollaston, 1862c: 430
wolastoni [sic] Pic, 1954a: 1 [RN]
paelex Evers, 1991a: 102 **N**: CI (La Palma)
pallidior Escalera, 1921: 305 **N**: CI (Tenerife)
pallipes Wollaston, 1864: 220 **N**: CI (Gomera, Tenerife)
palmensis Har. Lindberg, 1951: 3 **N**: CI (La Palma)
panurgus Evers, 1991a: 105 **N**: CI (La Palma)
papilioformis Wittmer, 1997b: 46 **A**: JA
paradoxus Abeille de Perrin, 1882b: 182 **E**: SP **N**: AG MO
pardoalcaidei Plata Negrache, 1990a: 51 **N**: CI (La Palma)
pardoi Escalera, 1933c: 360 **N**: MO
pellucidus Wollaston, 1854: 247 (*Pecteropus*) **N**: AG CI (Tenerife)
felschei Evers, 1960: 168
perforatus Abeille de Perrin, 1882a: 140 **N**: AG TU
petiti Pic, 1906c: 17 **N**: AG
pini Evers, 1971c: 69 **N**: CI (Tenerife)
platai Santiago, 1987: 289 **N**: MO
plocamae Evers, 1961a: 12 **N**: CI (Gran Canaria)
postremus Abeille de Perrin, 1882c: cxxv [RN] **E**: IT (Sicilia)
ragusae Abeille de Perrin, 1882a: 146 [HN]
propinquus Evers, 1988a: 77 **N**: CI (Gran Canaria)
punctifer Abeille de Perrin, 1885b: 145 **N**: AG
pygidialis bicornutus Wittmer, 1982f: 360 **A**: TAI
pygidialis pygidialis Wittmer, 1982f: 359 **A**: TAI
pygorufotinctus Wittmer, 1941: 1125 [RN] **A**: "Palestine"
rufopygus Pic, 1932e: 21 [HN]
reductus Pic, 1936a: 1 **N**: MO
reductissimus Pic, 1936a: 1
remanei Evers, 1986a: 91 **N**: MR (Porto Novo)
rosenhaueri rosenhaueri Evers, 1962: 185 **E**: PT SP **N**: MO
cyaneus Rosenhauer, 1856: 152 (*Ebaeus*)
rosenhaueri zarcoi Pardo Alcaide, 1958: 170 **E**: SP
rostratus Wollaston, 1854: 250 (*Pecteropus*) **N**: MR (Bugio Dezerta Grande, Porto Santos)
floricola Wollaston, 1854: 250
rufescens Escalera, 1914: 234 **N**: MO
ruficollis Wollaston, 1862c: 428 **N**: CI (La Palma, Tenerife)
pauperculus Wollaston, 1862c: 428
rufoapicalis Pic, 1917e: 6 **A**: TAI
inapicalis Pic, 1917e: 6

rugifrons Wollaston, 1862c: 431 **N**: CI (Gomera)
rugosus Wollaston, 1854: 249 (*Pecteropus*) **N**: MR
santiagoi Evers, 1988c: 172 **E**: SP
schatzmayri Wittmer, 1934: 453 **N**: EG
scitulus Wollaston, 1864: 218 (*Pecteropus*) **N**: CI (Gomera)
segurensis Escalera, 1926b: 429 **E**: SP
semitogatus Fairmaire, 1864: 641 (*Ebaeus*) **E**: IT (Sicilia) **N**: AG TU
 henoni Pic, 1904m: 90
shimomurai Wittmer, 1982f: 356 **A**: TAI
sicanus Erichson, 1840: 91 **E**: AU CR GR IT (Sicilia, Sardegna) MA SP (Baleares) TR YU **N**: AG TU
 kabylianus Pic, 1898h: 137
 obscuripes Pic, 1928e: 9
 testaceipes Pic, 1903r: 169
 ventralis Abeille de Perrin, 1883: 55
 viridifrons Schaufuss, 1869: 18 (*Ebaeus*)
simplicissimus Escalera, 1933d: 395 **N**: MO
smaragdinus Abeille de Perrin, 1894b: 92 (*Malachius*) **N**: AG
stotzneri Pic, 1926h: 35 **A**: XIZ
subaenescens Escalera, 1921: 304 **N**: CI (Tenerife)
sublaevis Lindberg, 1953: 7 **N**: CI (La Palma)
subopacus Wollaston, 1862c: 437 **N**: CI (Fuertaventura, Lanzerote)
susicola Evers, 1962: 160 **N**: MO
taiwanohirsutus Wittmer, 1982f: 357 **A**: TAI
tamriensis Evers, 1962: 161 **N**: MO
tarfayensis Evers, 1966b: 99 **N**: MO
testaceimembris Wittmer, 1954e: 320 **A**: YE
theryi Pic, 1920d: 9 **N**: MO
tizritensis Evers, 1995b: 91 **N**: MO
tristis P. H. Lucas, 1846: 193 (*Malachius*) **E**: IT (Sicilia) PT SP **N**: AG MO TU
trochantinus Wittmer, 1982b: 53 **A**: JA (Ryukyus)
 tokara Satô, 1986a: 239
tuberculatus Wollaston, 1862c: 436 **N**: CI (Gomera, Tenerife)
 pallidipes Escalera, 1921: 308
victoriensis Wittmer, 1982e: 346 **A**: HKG
vidualis Pardo Alcaide, 1968: 271 **E**: IT (Sicilia)
viridescens Pic, 1932g: 27 [= 1932c: 15] **N**: MO
 masculiantennatus Escalera, 1933d: 394
walteri Satô, 1986a: 240 **A**: JA (Ryukyus)
walterwittmeri Santiago, 1987: 216 **N**: MO
xanthopus Kiesenwetter, 1879a: 146 **A**: ES FE (Amur)
 azureus Abeille de Perrin, 1885b: 143 (*Anthocomus*)
xauensis Escalera, 1933a: 219 **N**: MO
yemenitus Wittmer, 1982c: 109 **A**: YE
yonaguni Satô, 1986a: 240 **A**: JA (Ryukyus)
zillae Peyerimhoff, 1929b: 109 **N**: AG
subgenus *Ebaeimorphus* Abeille de Perrin, 1891a: 401 type species *Malachius maculicollis* P. H. Lucas, 1846
escaleraianus Pardo Alcaide, 1954a: 81 **E**: SP
 escalerai Pardo Alcaide, 1948: 25 [HN]
maculicollis P. H. Lucas, 1846: 194 (*Malachius*) **E**: FR MA **N**: AG MO TU
 ecaudatus Peyron, 1877: 135 (*Axinotarsus*)
 immaculicollis Pic, 1903g: 113
subgenus *Maurattalus* Pardo Alcaide, 1975a: 99 type species *Attalus ferreri* Escalera, 1940
bremondi Peyerimhoff, 1935: 192 **N**: MO
escalerai escalerai Wittmer, 1948: 173 [RN] **N**: MO
 nigricornis Escalera, 1940: 257 [HN]
escalerai susicus Escalera, 1940: 258 **N**: MO (Igli en el Sus)

ferreri Escalera, 1940: 255 **N**: MO
hamadae Evers, 1962: 159 **N**: MO
heydeni Abeille de Perrin, 1887b: 447 **N**: MO
  notaticeps Pic, 1949a: 5
  uniformis Abeille de Perrin, 1891a: 567
neoterminalis Wittmer, 1948: 173 [RN] **N**: MO
  terminalis Escalera, 1940: 256 [HN]
panousei Pic, 1950e: 93 **N**: MO
poweli Pic, 1933d: 5 **N**: MO
regalis Evers, 1966b: 101 **N**: MO

genus *Attalus*, species incertae sedis
fruhstorferi Pic, 1905a: 92 **A**: SD
genualifer Abeille de Perrin, 1900a: 19 **A**: UZ (Buchara)
niponensis Pic, 1907i: 134 **A**: JA
nuptiabilis Abeille de Perrin, 1900a: 22 **A**: SY
pici Abeille de Perrin, 1894b: 92 **N**: AG

genus *Fortunatius* Evers, 1971b: 27 type species *Micromimetes jucundus* Wollaston, 1862
canariensis Lindberg, 1953: 6 (*Sphinginus*) **N**: CI (Hierro, Tenerife)
chamaeleon adejensis Evers, 1961a: 6 (*Sphinginus*) **N**: CI (Tenerife)
chamaeleon chamaeleon Escalera, 1921: 302 (*Attalus*) **N**: CI (Tenerife)
  niger Escalera, 1921: 303
jubilatus Evers, 1971b: 28 **N**: CI (Gran Canaria)
jucundus Wollaston, 1862c: 441 (*Micromimetes*) **N**: CI (Gran Canaria)
laurisilvae Evers, 1988a: 73 **N**: CI (La Palma)
lozanocaboi Plata Negrache, 1990b: 60 **N**: CI (La Gomera)
mencey mencey Evers, 1971b: 29 **N**: CI (La Gomera)
mencey nigerrimus Evers, 1995b: 95 **N**: CI (La Palma)
metallicus Wollaston, 1862c: 438 (*Attalus*) **N**: CI (Lanzerote, Tenerife)
  luteotibialis Evers, 1959: 226 (*Attalus*)
  similis Wollaston, 1862c: 438
posticus Wollaston, 1862c: 434 (*Attalus*) **N**: CI (Fuerteventura)
sparsepubens Wittmer, 1940b: 546 (*Attalus*) **N**: CI (Tenerife)
subextinctus Evers, 1990: 161 **N**: CI (La Palma)

genus *Ifnidius* Escalera, 1940: 260 type species *Ifnidius microphthalmus* Escalera, 1935
atlanticus Evers, 1981: 155 **N**: MR (Selvagens)
microphthalmus Escalera, 1940: 261 **N**: MO
milleri Wollaston, 1867: 97 (*Pecteropus*) **N**: CI (Saint Vincent)
petricola Plata Negrache, 1987b: 162 **N**: CI (El Golfo, Lanzarote)

genus *Macrotrichopherus* Evers, 1962: 156 type species *Macrotrichopherus spectabilis* Evers, 1962
spectabilis Evers, 1962: 156 **N**: MO (Sahara Occidental)

genus *Malachiomimus* Champion, 1921a: 108 type species *Malachiomimus latifrons* Champion, 1921
banghaasi Pic, 1907p: 171 (*Attalus*) **A**: PA
bucharicus Pic, 1907a: 97 (*Attalus*) **A**: UZ (Buchara)
buettikeri Wittmer, 1980a: 116 (*Attalus*) **A**: SA
erinaceus Abeille de Perrin, 1890a: 45 (*Attalus*) **A**: IN
espanoli Pardo Alcaide, 1945: 460 (*Attalus*) **N**: MO
eximius Peyron, 1877: 175 (*Ebaeus*) **A**: IS
ferganicus Wittmer, 1970c: 92 (*Attalus*) **A**: UZ
impressifrons Pic, 1919a: 16 (*Attalus*) **A**: AF "Indes" **ORR**
issenensis Evers, 1962: 163 (*Attalus*) **N**: MO

kabulensis Wittmer, 1979c: 150 (*Attalus*) **A**: AF
latifrons Champion, 1921a: 108 **A**: UP
luteipes Champion, 1921a: 109 **A**: UP
mirandus Abeille de Perrin, 1885b: 148 (*Ebaeus*) **A**: MG "Siberia" "Turkestan"
transversus Abeille de Perrin, 1891a: 590
multicolor Reitter, 1889d: 109 (*Nepachys*) **A**: TM
pectinatus Kiesenwetter, 1866b: 268 (*Attalus*) **E**: SP
sericeus Wittmer, 1957b: 578 **A**: XIZ
sinaitus Pic, 1921j: 13 (*Attalus*) **A**: SI
subunicolor Pic, 1948a: 16 **A**: YUN

genus *Nepachys* C. G. Thomson, 1859: 112 type species *Cantharis cardiacae* Linnaeus, 1761
Ebaeomorphus Pic, 1904e: 28 type species *Ebaeomorphus ramicornis* Pic, 1904
afghanistanicus Wittmer, 1956a: 150 **A**: AF PA
amaenus Peyron, 1877: 170 (*Attalus*) **A**: TR SY
cardiacae Linnaeus, 1761: 205 (*Cantharis*) **E**: BY DE FI FR GE IT NR PL SK SV SZ RU "Caucasus" **A**: "Siberia" "Turkestan"
punctipennis Dodero, 1929: 230
coreanus Wittmer, 1989: 228 **A**: NC
dentipennis Wittmer, 1984f: 255 **A**: PA
drouardi Pic, 1923o: 375 (*Attalus*) **A**: JA
formosanus formosanus Wittmer, 1956b: 309 (*Attalus*) **A**: TAI (Mt. Lishan)
formosanus latus Wittmer, 1982f: 361 (*Attalus*) **A**: TAI (Bi-kei)
formosanus recurvus Wittmer, 1982f: 362 (*Attalus*) **A**: TAI (Mt. Nanfengshan)
formosanus subtilis Wittmer, 1982f: 361 (*Attalus*) **A**: TAI (Urai, Taipei Pref.)
impressus Wittmer, 1972b: 139 (*Attalus*) **A**: TAI
inermis Wittmer, 1972b: 141 (*Attalus*) **A**: TAI
iranicus Švihla, 1987: 279 **A**: IN
japonicus Kiesenwetter, 1874b: 286 (*Attalus*) **A**: FE JA
kashmirensis Wittmer, 1984f: 259 **A**: KA
kurokawensis Nakane, 1987a: 3 (*Attalus*) **A**: JA
paghmanensis Wittmer, 1974a: 11 (*Attalus*) **A**: AF
peucedani Abeille de Perrin, 1885a: 10 (*Attalus*) **E**: FR SP **N**: MO
pulchellus Mulsant & Rey, 1861b: 78 (*Anthocomus*) [HN]
rubroapicalis Pic, 1936a: 1 **N**: MO
ryukyuanus danjoanus Nakane, 1989c: 151 **A**: JA (Meshima, Danjo-gunto)
ryukyuanus ryukyuanus Wittmer, 1982b: 51 **A**: JA (Ryukyus)
ryukyuanus yamanei Nakane, 1984: 590, 596 **A**: JA (Yaku-shima)
spinosus Wittmer, 1984f: 258 **A**: AF
subimpressus Wittmer, 1992b: 25 **A**: SHG
swatensis Wittmer, 1984f: 263 **A**: PA
tsushimanus Nakane, 1981b: 128 **A**: JA

genus *Sphinginus* Mulsant & Rey, 1867: 154 type species *Malachius lobatus* A. G. Olivier, 1790
coarctatus Erichson, 1840: 104 (*Anthocomus*) **E**: AU FR GR HU IT SK SZ YU **N**: AG MO TU
riffensis Escalera, 1933b: 255
constrictus Erichson, 1840: 104-105 (*Anthocomus*) **E**: FR IT (Sardegna) **N**: AG TU
corcyreus Pic, 1901h: 66
sanguineicollis Abeille de Perrin, 1891b: 407
tangerianus Escalera, 1933b: 255
lobatus apicalis Perris, 1864: 289 (*Attalus*) **E**: FR (Corse) IT (Sardegna) MA
limbatus Mink, 1853: 59 (*Troglops*)
posticus Mulsant & Rey, 1867: 184 (*Antholinus*)
lobatus lobatus A. G. Olivier, 1790a: no 27: 12 (*Malachius*) **E**: GB PT SP **N**: AG TU

genus *Sternodeattalus* Wittmer, 1970b: 25 type species *Sternodeattalus chujoi* Wittmer, 1970
chujoi Wittmer, 1970b: 25 **A**: JA (Ryukyus)

tribe Colotini Abeille de Perrin, 1890

genus *Asiapelochrus* Wittmer, 1996: 305 type species *Abeillea delerei* Wittmer, 1956
delerei Wittmer, 1956a: 121 (*Abeillea*) **A**: AF
khaganensis Wittmer, 1996: 307 **A**: PA
kutiauensis Wittmer, 1996: 305 **A**: AF

genus *Centrocolotes* Wittmer, 1987c: 649 type species *Pseudocolotes testaceipes* Pic, 1928
testaceipes Pic, 1928f: 38 (*Pseudocolotes*) **A**: "SW Arabia" **AFR**
    arambourgi Pic, 1935l: 295 (*Pseudocolotes*)

genus *Colotes* Erichson, 1840: 129 type species *Malachius maculatus* Laporte, 1838
    Pseudocolotes Abeille de Perrin, 1900b: 166 type species *Pseudocolotes cribripennis* Abeille de Perrin, 1900
subgenus *Allodipnis* Abeille de Perrin, 1890c: 258 type species *Ebaeus ogieri* Fairmaire, 1864
hispanus Uhagón, 1900: 39 **E**: SP
ogieri Fairmaire, 1864: 641 (*Ebaeus*) **N**: AG EG MO TU
    unicolor Pic, 1895p: cxxvii
uhagoni Abeille de Perrin, 1890a: 49 **E**: SP
subgenus *Antidipnis* Wollaston, 1858b: 337 type species *Colotes rubripes* Jacquelin du Val, 1859
punctatus Erichson, 1840: 122 (*Charopus*) **E**: FR GR IT MA PT SP YU **N**: AG CI MO TU
    colon Abeille de Perrin, 1890c: 259
    inmaculatus Escalera, 1914: 233
    obscurithorax Pic, 1914g: 24
    pygmaeus Abeille de Perrin, 1890c: 259
    rubripes Jacquelin du Val, 1852: 707 (*Colotes*)
scutellaris Pic, 1920i: 46 **N**: EG
subgenus *Colotes* Erichson, 1840: 129 type species *Malachius maculatus* Laporte, 1838
almoraensis Wittmer, 1977c: 324 **A**: UP
afghanistanicus Wittmer, 1981b: 187 **A**: AF
atrofemoratus Champion, 1922c: 217 **A**: UP
banihalensis Wittmer, 1981b: 184 **A**: KA (Jammu)
benardi Wittmer, 1970c: 75 **A**: IN
burakowskii Wittmer, 1969a: 54 **A**: MG
cabrerensis Tenenbaum, 1915: 1 **E**: SP (Baleares)
dollfusi Pic, 1938l: 251 **A**: SI
dushambicus Wittmer, 1979c: 144 **A**: TD
ebaeiformis Abeille de Perrin, 1900a: 20 **A**: AF TM
    levecrassus Wittmer, 1970c: 76
emeljanovi Wittmer, 1976: 171, 180 **A**: MG
flavoscutellatus Wittmer, 1981b: 184 **A**: AF
flexuosus Wittmer, 1981b: 182 **A**: AF
gobiensis Evers, 1964: 15 **A**: MG
gorhami Champion, 1921a: 71 [RN] **A**: HP UP
    deplanaticeps Pic, 1929b: 3 (*Laius*)
    punctatus Gorham, 1895a: 320 (*Periebaeus*) [HN]
hampei L. Redtenbacher, 1874: 28 **E**: AU CZ HU SK
iranicus Wittmer, 1981b: 185 **A**: IN
impressicornis Wittmer, 1994a: 3 **A**: KA
kovari Švihla, 1987: 277 **A**: IN
longior Pic, 1938e: 10 **N**: EG
maculatus Laporte, 1838: 29 (*Malachius*) **E**: AL AU BU CR FR GR IT MA MC PT SK SP TR YU "Caucasus" **N**: AG LB MO TU **A**: SY
    semireductus Pic, 1915i: 25

suturalis Motschulsky, 1854d: 56
trinotatus Erichson, 1840: 130
mongolicus Wittmer, 1976: 173 **A**: MG
nepalensis Wittmer, 1977c: 323 **A**: NP
nigricollis Wittmer, 1976: 174 **A**: MG
nigrosignatus Pic, 1933c: 25 **N**: EG
protrudioculatus Wittmer, 1965c: 87 **A**: NP
simlaensis Wittmer, 1977c: 321 **A**: HP
tedjenensis Wittmer, 1970c: 77 **A**: TM
thuloensis Wittmer, 1977c: 323 **A**: NP
triangulatus Wittmer, 1977c: 322 **A**: BT
uncifer Abeille de Perrin, 1903: 335 **N**: AG MO TU
subgenus *Gynedipnis* Abeille de Perrin, 1890c: 257 type species *Colotes flavocinctus* Marseul, 1868
anax Iablokoff-Khnzorian, 1978b: 172 **E**: AR
anthicinus Baudi di Selve, 1871: 70 **A**: CY JO SA SY TM "Palestine"
flavocinctus Marseul, 1868b: 189 **E**: ST TR **A**: CY IN SY "Turkestan"
atricollis Abeille de Perrin, 1890c: 260
maculatus Kiesenwetter, 1864: 387 (*Antidipnis*)
signatipennis Pic, 1938e: 10
kasachstanicus Medvedev, 1964b: 155 **A**: KZ
subgenus *Heterodipnis* Peyron, 1877: 261 type species *Antidipnis palpator* Marseul, 1868 (= *Colotes cinctus* Motshulsky, 1854)
cinctus cinctus Motschulsky, 1854d: 56 **N**: EG **A**: CY SY
inlateralis Pic, 1938e: 10
palpator Marseul, 1868b: 188 (*Antidipnis*)
cinctus suturellus Wittmer, 1935c: 186 **A**: SI
subgenus *Homoeodipnis* Jacquelin du Val, 1859e: 47 type species *Colotes javeti* Jacquelin du Val, 1859
javeti Jacquelin du Val, 1852: 705 **E**: FR IT PT SP **N**: AG EG MO TU
rufithorax Jacquelin du Val, 1852: 706
subgenus *Micromimetes* Wollaston, 1862c: 439 type species *Micromimetes alutaceus* Wollaston, 1862
alutaceus Wollaston, 1862c: 441 **N**: CI (Gran Canaria)
ebaeiformis Abeille de Perrin, 1900a: 20 **A**: "Transcaspian"
obscuriceps Pic, 1908d: 41
mauritii Abeille de Perrin, 1900a: 23 **A**: "Cisjordania"
subgenus *Psauter* Marseul, 1890: 257 type species *Colotes obsoletus* Erichson, 1840
florieni Pic, 1911l: 121 **A**: SY
obsoletus Erichson, 1840: 130 **E**: GR TR YU **A**: CY IS LE SY
innotaticeps Pic, 1938e: 10
heteropalpus Marseul, 1868b: 186 (*Antidipnis*)
nigripennis Küster, 1848a: 18
punctipennis Kiesenwetter, 1859a: 32 (*Ebaeus*)
subfasciatus Pic, 1920d: 10 **N**: MO
subgenus *Pseudodipnis* Abeille de Perrin, 1890c: 257 type species *Antidipnis galbula* Kiesenwetter, 1864
galbula Kiesenwetter, 1864: 387 (*Antidipnis*) **E**: ST
flavomaculata Becker, 1864: 477 (*Antidipnis*)

genus *Colotrema* Wittmer, 1979d: 190 type species *Colotes asirensis* Wittmer, 1979
asirensis Wittmer, 1979d: 190 (*Colotes*) **A**: SA
impressus Wittmer, 1954e: 316 (*Colotes*) **A**: YE
testaceus testaceus Wittmer, 1979d: 191 (*Colotes*) **A**: SA
testaceus wajjensis Wittmer, 1980a: 115 (*Colotes*) **A**: SA (Wadi Wajj)

genus *Juengerius* Evers, 1995a: 2 [RN] type species *Abeillea minutula* Wittmer, 1956
Wittmerius Evers, 1988b: 168 [HN] type species *Abeillea minutula* Wittmer, 1956
minutula Wittmer, 1956a: 122 (*Abeillea*) **A**: AF IN

genus *Pachyebaeus* Wittmer, 1970d: 152 type species *Pachyebaeus nodulosus* Wittmer, 1970
cameroni Pic, 1903t: 177 (*Hypebaeus*) **A**: "Arabia" **AFR**
genus *Pardonius* Evers, 1988a: 72 type species *Attalus rufomaculatus* Wittmer, 1941
rufomaculatus Wittmer, 1941: 1125 (*Attalus*) [RN] **N**: MO
marocanus Pic, 1944a: 3 [RN]
rufoapicalis Pic, 1933h: 103 [HN]
rufoposticalis Pic, 1950a: 15 [RN]

genus *Pelochrus* Mulsant & Rey, 1867: 188 type species *Anthocomus pallidulus* Erichson, 1840
atriceps Pic, 1910e: 17 **E**: "Caucasus"
brevicollis Abeille de Perrin, 1885b: 145 (*Attalus*) **E**: "Caucasus"
fusculus Peyron, 1877: 157 (*Attalus*) **A**: LE SY
fusculus Pic, 1910o: 58 (*Attalus*)
moroderi Pic, 1915i: 25 **E**: SP
pallidulus Erichson, 1840: 107 (*Anthocomus*) **E**: FR IT PT SP **N**: AG MO
gallicus Uhagón, 1900: 46
maculicollis Pic, 1903q: 162
pallidus Abeille de Perrin, 1882a: 146
scutellaris Abeille de Perrin, 1890a: 46 **N**: AG MO TU

genus *Transvestitus* Evers, 1988a: 72 type species *Attalus infernalis* Evers, 1961
infernalis Evers, 1961a: 9 (*Attalus*) **N**: CI (Tenerife)

tribe Ebaeini Portevin, 1931

genus *Ebaeus* Erichson, 1840: 113 type species *Cantharis pedicularius* Linnaeus, 1758
subgenus *Ebaeus* Erichson, 1840: 113 type species *Cantharis pedicularius* Linnaeus, 1758
abietinus Abeille de Perrin, 1869a: 44 **E**: FR
elzeari Pic, 1911g: 106
affinis P. H. Lucas, 1846: 193 (*Malachius*) **N**: AG MO TU
albopictus Wittmer, 1970c: 90 **A**: TM
alboterminatus Champion, 1921a: 78 (*Hypebaeus*) **A**: HP UP
aleppensis Wittmer, 1970c: 79 **A**: SY
altaiensis Evers, 1968a: 26 **A**: MG
alutaceus Wittmer, 1986b: 133 **A**: TAI
ampliterminatus Wittmer, 1968b: 279 **A**: KA
amurensis Pic, 1904g: 50 **A**: FE (Amur)
anatolicus Wittmer, 1986a: 110 **E**: TR
appendiculatus Erichson, 1840: 116 **E**: AU BH BU CZ GE HU IT PL RO RU SK SP (Baleares) SZ UK **A**: TR
mediterraneus Abeille de Perrin, 1882a: 148 (*Malachius*)
argenteonotatus Pic, 1948a: 13 **A**: SD
ater Kiesenwetter, 1863a: 608 **E**: AU BU CZ FR GR HU SK UK
pannonicus Švihla, 1990b: 358
bagdadensis Wittmer, 1966c: 501 **A**: IQ
basipes Abeille de Perrin, 1891a: 383 **A**: TM
battonii Pardo Alcaide, 1962: 274 **E**: AU IT SZ
baudueri Peyron, 1877: 184 **E**: TR
bifurcatus Champion, 1927: 83 **A**: HP
bilineatus Champion, 1927: 83 **A**: HP
bipartitus Wittmer, 1970c: 89 **A**: TD UZ
bleusei Abeille de Perrin, 1900a: 19 **E**: GR (Rodos) TR
bulbifer Kolenati, 1846b: 43 (*Malachius*) **E**: GG
caspius caspius Peyron, 1877: 177 **E**: ST **A**: TR
caspius flavolimbatus Wittmer, 1968c: 207 **A**: IN TM
cedrorum Wittmer, 1981b: 213 **A**: NP
chibaensis Nakane, 1989c: 150 **A**: JA

chopalensis Wittmer, 1981b: 207 **A**: HP
coerulescens Erichson, 1840: 115 **E**: AU BH BU CR CZ GE GR IT MC SK YU **A**: LE SY TR
nigricollis Küster, 1846a: 49 (*Attalus*)
pugio Marseul, 1868b: 185
collaris collaris Erichson, 1840: 117 **E**: AU CR FR GR IT (Sardegna) PT RU SP SZ
congressarius Fairmaire, 1857a: 637
flavifrons Baudi di Selve, 1872: 126
collaris princeps Abeille de Perrin, 1881: 115 **N**: AG TU
contemptus Pardo Alcaide, 1954a: 81 **N**: AG MO TU
cordiger Evers, 1968a: 28 **A**: KZ MG
creticus Wittmer, 1968c: 209 **E**: GR (Kríti)
cyaneoviridis Wittmer, 1970c: 82 **A**: TD
decorus Abeille de Perrin, 1885a: 12 **A**: AF UZ "Central Asia"
tricolor Ballion, 1871: 353
delicatulus Wittmer, 1956a: 144 **A**: AF
dolini Wittmer, 1995e: 374 **A**: TM
elburzensis Wittmer, 1986a: 109 **A**: IN
erythropus Peyron, 1877: 183 **A**: ES KI KZ MG
fischeri A. Fleischer, 1909a: 244 **A**: KZ "Turkestan"
ferganicus Wittmer, 1970c: 86 **A**: UZ
flavicornis Erichson, 1840: 114 **E**: AU CR CZ FR GE HU IT MC PL RO RU SK SP (Baleares) SV SZ UK **A**: MG TR "Siberia"
apicalis Reichenbach, 1826: 167
flavobullatus flavobullatus Marseul, 1868b: 184 **A**: CY IS LE SY
flavobullatus orientalis Wittmer, 1979c: 145 **A**: TR
flavomarginatus Pic, 1929g: 13 (*Hypebaeus*) **A**: IS
fukienensis Evers, 1948a: 50 **A** FUJ
gibbosus Wittmer, 1956a: 141 **A**: AF
gibbus Drapiez, 1819: 133 (*Malachius*) **E**: FR IT MA SZ
cyaneus Laporte, 1838: 29 (*Malachius*)
taeniatus Mulsant & Rey, 1867: 223
glabricollis Mulsant & Rey, 1867: 229 **E**: FR PT SP
guilanensis Wittmer, 1986a: 111 **A**: IN
hartmanni Evers, 1996: 91 (*Kuatunia*) **A**: NP
hissarensis Wittmer, 1992b: 21 **A**: TD
horaianus Nakane, 1991b: 114 **A**: JA
humilis Erichson, 1840: 116 **E**: IT (Sardegna) SP **N**: AG MO TU
adolescens Abeille de Perrin, 1883: 34
iranicus Wittmer, 1970c: 83 **A**: IN
judaicus Evers, 1948a: 49 **A**: IS JO
jumlaensis Wittmer, 1981b: 206 **A**: NP
kaganensis Wittmer, 1981b: 211 **A**: PA
karadjensis Wittmer, 1970c: 88 **A**: IN
karakumensis Wittmer, 1970c: 91 **A**: TM
kasachstanicus Švihla, 1980b: 396 **A**: KZ UZ
kaszabi Evers, 1968a: 30 **A**: MG
kocheri Pardo Alcaide, 1962: 278 **N**: MO
kukkulensis Švihla, 1987: 280 **A**: UZ
lahinvanensis Wittmer, 1981b: 205 **A**: KA
lapplandicus Evers, 1993: 157 **E**: LA NR SV
laterinodosus Wittmer, 1936a: 153 (*Hypebaeus*) **A**: SI
lepidus Abeille de Perrin, 1895b: cccliv **E**: GG
levesulcatus Wittmer, 1992b: 19 **A**: AF
limbellus limbellus Peyron, 1877: 189 **A**: ES KI MG
altaicus Iablokoff-Khnzorian, 1978b: 174 (*Hypebaeus*)

epipleuralis Abeille de Perrin, 1885b: 146
flavoniger Evers, 1971d: 228
sailugemensis Wittmer, 1992b: 23
tianshanensis Wittmer, 1970c: 80
limbellus nigrotibialis Tshernyshev, 2003: 288 **A**: ES (Altai)
lopatini Dowgailo, 1997: 376 **A**: KZ
luctuosus Abeille de Perrin, 1890a: 48 **E**: AR "Caucasus"
koenigi Pic, 1912b: 9
lunulatus Wittmer, 1958: 124 **A**: TAI
maculiceps Dowgailo, 1997: 373 **A**: UZ
maculithorax Wittmer, 1970c: 89 **A**: "Turkestan"
majeri Švihla, 1987: 280 **A**: UZ
malickyi Wittmer, 1992b: 24 **E**: GR
mandchuricus Pic, 1911h: 107 **A**: NE
medvedevi Wittmer, 1970c: 80 **A**: IN TR
mendax Kiesenwetter, 1866b: 270 **E**: FR SP
nitidulus Fabricius, 1792a: 225 (*Malachius*)
miyamotoi Wittmer, 1955: 42 **A**: TAI
modestus Abeille de Perrin, 1885a: 11 **A**: ES MG
formaneki Reitter, 1894g: 240
mongolicus Evers, 1968a: 27 **A**: MG
montanus Wittmer, 1981b: 205 **A**: PA
nankinensis Pic, 1908m: 96 **A**: JIA
nanschanensis Reitter, 1900e: 160 **A**: XIZ (Nanschan)
nantouensis Wittmer, 1986b: 134 **A**: TAI
naranensis Wittmer, 1981b: 204 **A**: PA
nigerrimus Wittmer, 1981b: 211 **A**: UP
nigrocaudatus Abeille de Perrin, 1885a: 11 **E**: IT **A**: MG
nodifer Wittmer, 1981b: 209 **A**: KA (Jammu)
oculifer Abeille de Perrin, 1885b: 149 **A**: "Turkestan"
ornaticeps Wittmer, 1956a: 145 **A**: AF
ovosimilis Wittmer, 1964: 103 **E**: AR
pallidiceps Wittmer, 1956a: 143 **A**: AF
pallidipes Wittmer, 1992b: 23 **A**: UZ
pathanus Wittmer, 1981b: 210 **A**: PA
patricius Abeille de Perrin, 1900a: 19 **A**: MG
pedicularius atrotibialis Tshernyshev, 2003: 294 **A**: ES (Gornyy Altai)
pedicularius pedicularius Linnaeus, 1758: 402 (*Cantharis*) **E**: AU BE BU BY CZ DE FI FR GE HU IT NL PL RO RU SK SP SV SZ UK YU (Montenegro) **A**: TM "Turkestan"
montenegrinus Pic, 1915i: 22
praeoccupatus Gemminger & Harold, 1869: 1699 (*Malachius*) [RN]
praeustus Gyllenhal, 1808a: 364 (*Malachius*)
productus Stephens, 1830: 315 (*Malachius*)
personatus Abeille de Perrin, 1885b: 149 **A**: "Turkestan"
pinorum Wittmer, 1981b: 207 **A**: HP
poecilus Iablokoff-Khnzorian, 1978b: 174 **A**: TM
pseudothoracicus Wittmer, 1978e: 376 **A**: KA (Ladakh)
reductemaculatus Wittmer, 1956a: 148 **A**: AF
robusticornis Wittmer, 1964: 104 **A**: KI
rosali Pardo Alcaide, 1958: 170 [= 1960: 126] **E**: PT SP
rosinae Pic, 1902a: 9 **E**: TR
rubetorum Peyron, 1877: 187 **E**: TR "Caucasus" **A**: TR
viridipennis Pic, 1908b: 9
ruffoi Pardo Alcaide, 1962: 288 **E**: IT MA
rufipes Morawitz, 1861: 288 **E**: RU ST UK **A**: KZ TM

rufoapicalis Pic, 1926s: 454 **A**: CH
  paulonotatus Pic, 1948c: 16
rufoterminatus Wollaston, 1865: 195 (*Malachius*) **N**: CI
schawalleri Wittmer, 1988a: 25 **A**: IS
scheibei Wittmer, 1956a: 142 **A**: AF
scutellaris Pic, 1908d: 41 **A**: AF TM
  parvantai Wittmer, 1956a: 147
semilateralis Pic, 1938k: 162 **A**: SHX
seawardii Tshernyshev, 2003: 297 **A**: KZ
shansiensis Wittmer, 1992b: 17 **A**: SHX
shirahatai Wittmer, 1953b: 71 **A**: SHX
sillemi Hicker, 1935: 294 **A**: MG
speciosus Iablokoff-Khnzorian, 1959b: 69 **E**: AR
spinifer Wittmer, 1970c: 81 **A**: KI
spinimargo Wittmer, 1970c: 88 **A**: TM
srinagarensis Wittmer, 1981b: 212 **A**: KA
strejceki Švihla, 1980b: 394 **A**: TD
suarezi Pardo Alcaide, 1958: 169 [= 1960: 124] **E**: SP
szechuanensis Wittmer, 1954c: 112 **A**: SCH
tadjikus Wittmer, 1999: 172 **A**: TD
tener Dohrn, 1873: 246 **A**: IN
tenuicornis Champion, 1922c: 264 **A**: UP
tenuis Wittmer, 1986a: 113 **A**: IN
tergalis Wittmer, 1992b: 17 **A**: TM
testaceipes Wittmer, 1981b: 213 **A**: PA
thoracicus Geoffroy, 1785: 63 (*Cicindela*) **E**: BE CZ FR GE IT NL PL PT SP SZ
  rubroniger Goeze, 1777: 544 (*Malachius*)
tibiaemaculatus Wittmer, 1956a: 146 **A**: AF
transbaikalicus korsuni Tshernyshev, 2003: 290 **A**: ES
transbaikalicus transbaikalicus Pic, 1912i: 33 **A**: ES MG
trimaculatus Gebler, 1830: 88 **A**: MG "Siberia"
tripictus Abeille de Perrin, 1890a: 47 **A**: AF UZ (Buchara)
tritinctus Wittmer, 1956c: 1077 [RN] **A**: AF
  tricoloratus Wittmer, 1956a: 140 [HN]
tshernyshevi Wittmer, 1995d: 271 **A**: TM
turcmenicus Wittmer, 1995e: 376 **A**: TM
turcicus Wittmer, 1968c: 208 **E**: TR
turkestanicus Abeille de Perrin, 1885b: 146 **A**: AF KI TD UZ
uzbekus Wittmer, 1999: 174 **A**: UZ
velatus Peyron, 1877: 192 **A**: SY
velazquezi Pardo Alcaide, 1952: 79 **N**: MO
vinculatus Wittmer, 1981b: 210 **A**: HP
wahrmani Wittmer, 1966c: 502 **A**: IS
wittmeri Pardo Alcaide, 1962: 282 **N**: MO
wittmerorum Švihla, 1980b: 397 **A**: TD UZ
yamawakii Nakane, 1985c: 158 **A**: JA
zoltani Wittmer, 1969e: 513 **A**: MG
subgenus *Majerus* Wittmer, 1995d: 268 type species *Ebaeus impressus* Wittmer, 1995
alexandri Tshernyshev, 2003: 284 **A**: UZ
impressus Wittmer, 1995d: 268 **A**: TM

genus *Epiebaeus* Wittmer, 1995d: 283 type species *Epiebaeus glabricollis* Wittmer, 1995
glabricollis Wittmer, 1995d: 283 **A**: NP
yunnanus Wittmer, 1995d: 285 **A**: YUN

genus *Holzschuhus* Wittmer, 1996: 297 type species *Ebaeus pauloimpressus* Pic, 1948
chunkingensis Wittmer, 1996: 299 **A**: SCH
jizushanensis Wittmer, 1999: 178 **A**: YUN
longicornis Wittmer, 1981b: 188 (*Abeillea*) **A**: NP
nigriceps Wittmer, 1999: 176 **A**: YUN
pauloimpressus Pic, 1948a: 15 (*Ebaeus*) **A**: YUN
pseudochunkingensis Wittmer, 1999: 180 **A**: YUN
spatulifer Wittmer, 1996: 301 **A**: SCH

genus *Hypebaeina* Wittmer, 1995d: 280 type species *Hypebaeus discifer* Abeille de Perrin, 1881
alaiensis Pic, 1906b: 10 (*Attalus*) **A**: KI (Alai Range)
andreevae Wittmer, 1999: 187 **A**: UZ
badakshana Wittmer, 1999: 186 **A**: AF
crassitibialis Wittmer, 1966a: 5 (*Hypebaeus*) **A**: IS
discifer Abeille de Perrin, 1881: 118 (*Hypebaeus*) **A**: IS
viduus Abeille de Perrin, 1881: 115 (*Attalus*)
hauseri Wittmer, 1999: 184 **A**: UZ (East Buchara, Tschitschantan)
kutteri Wittmer, 1971e: 369 (*Hypebaeus*) **A**: JO
milkoi Tshernyshev, 1999b: 367 **A**: KI
nabataeica Wittmer, 1971e: 369 (*Hypebaeus*) **A**: JO
nainiensis Champion, 1922c: 267 **A**: UP
taplejungana Wittmer, 1997c: 161 **A**: NP
torretassoi Wittmer, 1935c: 187 (*Hypebaeus*) **A**: SI
triemarginata Wittmer, 1999: 186 **A**: KI
viridimetallica Wittmer, 1994a: 4 (*Hypebaeus*) **A**: BT
waygalensis Wittmer, 1999: 188 **A**: AF

genus *Hypebaeus* Kiesenwetter, 1863a: 610 type species *Malachius flavipes* Fabricius, 1787
subgenus *Alloceps* Abeille de Perrin, 1891a: 345 type species *Malachius albifrons* Fabricius, 1775
albifrons albifrons Fabricius, 1775: 225 (*Malachius*) **E**: FR IT PT SP SZ **N**: AG MO
anticus Laporte, 1838: 29 (*Malachius*)
bejarensis Pic, 1911g: 106
hypocrita Abeille de Perrin, 1891a: 352
nigritulus Dufour, 1843a: 45
albifrons atlasicus Evers, 1966b: 99 **N**: MO (Atlas Amizmiz, Atlas Reraia)
alicianus Jacquelin du Val, 1859c: 38 (*Ebaeus*) **E**: FR SP
cedrorum Abeille de Perrin, 1885b: 150 (*Hypebaeus*) **N**: AG MO TU
subgenus *Allogynes* Abeille de Perrin, 1891a: 345 type species *Hypebaeus vitticollis* Abeille de Perrin, 1881
albofacialis J. R. Sahlberg, 1908: 71 **A**: JO SY "Palestine"
jordanicus Wittmer, 1960b: 109
bucharicus Pic, 1907g: 130 **A**: UZ (Buchara)
cyanipennis Baudi di Selve, 1871: 68 **A**: SY
grandiceps Pic, 1907a: 97 **A**: UZ (Buchara)
nodipennis Krynicki, 1832: 100 (*Malachius*) **E**: GR TR **A**: CY IQ IS SY
cordicollis Kiesenwetter, 1859a: 33 (*Ebaeus*)
scitulus Erichson, 1840: 122 (*Charopus*)
vitticollis Abeille de Perrin, 1881: 117 **A**: IS
obscuriceps Pic, 1899l: 257
subgenus *Hypebaeus* Kiesenwetter, 1863a: 610 type species *Malachius flavipes* Fabricius, 1787
Lobatomixis Pic, 1915c: 11 type species *Lobatomixis cavifrons* Pic, 1915
Pseudebaeus Horn, 1872: 109, 118 type species *Malachius apicalis* Say, 1825
adumbratus Champion, 1921a: 76 **A**: UP
albobinotatus Wittmer, 1956a: 130 **A**: AF
albocaudatus Champion, 1921a: 78 **A**: UP
albomaculatus Pic, 1924j: 476 **A**: SD

annulifer Pic, 1907p: 171 (*Attalus*) **N**: EG
apiceimpressus Wittmer, 1981b: 193 **A**: KA
apicetestaceus Wittmer, 1955: 43 (*Lobatomixis*) **A**: ZHE
appendicifer Wittmer, 1981b: 197 **A**: BT
atronotatus Champion, 1925d: 267 **A**: UP
bactrianus Švihla, 1980b: 397 **A**: TD
basimaculatus Wittmer, 1958: 123 **A**: TD UZ
bicornutus Wittmer, 1981b: 199 **A**: HP
bilyi Švihla, 1980b: 399 **A**: TD
binotatus Wittmer, 1956a: 136 **A**: AF
bodenheimeri Pic, 1932h: 29 **A**: "Palestine"
borgorum Wittmer, 1987b: 110 **A**: NP
brisouti Mulsant & Rey, 1867: 235 **E**: FR PT SP
cerastes Champion, 1921a: 106 **A**: UP
chlorizans Kiesenwetter, 1874b: 286 (*Ebaeus*) **A**: JA SHX
clavatus Champion, 1922c: 266 **A**: UP
confusus Pic, 1925h: 14 [RN] **A**: SD
    diversipennis Champion, 1922: 270 [HN]
coreanus Wittmer, 1975b: 265 **A**: "Korea"
coronatus coronatus Wittmer, 1956a: 134 **A**: AF
coronatus himachalicus Wittmer, 1981b: 192 **A**: HP PA
cyathiger Wittmer, 1956a: 131 **A**: AF
dalailamai Wittmer, 1981b: 198 **A**: SD
darjeelingensis Wittmer, 1981b: 200 **A**: SD
drassicus drassicus Wittmer, 1978e: 372 **A**: KA (Ladakh)
drassicus fatulaensis Wittmer, 1978e: 374 **A**: KA
elongaticornis Wittmer, 1960b: 110 **N**: EG
fissomimus Wittmer, 1981b: 191 **A**: PA
fissus Champion, 1927: 84 **A**: PA (Punjab)
flavicollis Erichson, 1840: 117 (*Ebaeus*) **E**: IT MA SP (Baleares) TR **N**: AG
    italicus Reitter, 1904e: 159 (*Ebaeus*)
    obscurithorax Pic, 1903k: 129
flavipes Fabricius, 1787a: 169 (*Malachius*) **E**: AU BE BH CR CZ FR GE HU IT LA LU PL RU SK SP SV SZ **A**: ES FE WS
    ochropus Gmelin, 1790: 1899 (*Malachius*)
    perspicillatus Bremi-Wolf, 1855a: 199 (*Ebaeus*)
    posticus Kiesenwetter, 1866a: 393 (*Hypebaeus*)
    praeustus Fabricius, 1792a: 224 (*Malachius*)
    productus A. G. Olivier, 1790a: no 27: 13 (*Malachius*)
flavocaudatus Nakane, 1956b: 163 (*Ebaeus*) **A**: JA
flexuosus Champion, 1922c: 269 **A**: UP
foveatus Champion, 1925d: 267 **A**: UP
gestroi Abeille de Perrin, 1890b: 90 **A**: IN
hikosanus Takakura, 1988a: 151 **A**: JA
impressipennis Wittmer, 1981b: 196 **A**: NP
inapicalis Pic, 1899j: 205 **E**: GR (Rodos)
incisus Wittmer, 1984a: 2 **A**: "Siberia" ("Altai: Koscha-Agatsch")
ingeae Wittmer, 1978e: 374 **A**: KA
inglanensis Wittmer, 1981b: 189 **A**: PA
intermixtus Wittmer, 1956a: 124 **A**: AF
kafiristanicus Wittmer, 1956a: 125 **A**: AF
karatshaicus Wittmer, 1979c: 148 **A**: KI
kiotoensis Pic, 1954b: 4 **A**: JA
kirgisicus Wittmer, 1970c: 78 **A**: KI
klapperichi Wittmer, 1956a: 137 **A**: AF

krokeri Evers, 1986b: 101 **A**: KA
lenis Champion, 1922c: 269 **A**: UP
levecyaneus Wittmer, 1982f: 354 **A**: TAI
libanus Peyron, 1877: 199 **E**: EG **A**: LE SY
lindemannae Wittmer, 1960a: 104 **A**: PA
lopchuensis Wittmer, 1981b: 201 **A**: SD
minutus Wittmer, 1956a: 133 **A**: AF
miyakoanus Nakane, 1985c: 160 **A**: JA (Miyako)
mongolicus Evers, 1968a: 25 **A**: MG
mylabrinus Baudi di Selve, 1871: 69 **A**: CY TR SY
nepalensis Wittmer, 1981b: 196 **A**: NP
niger Wittmer, 1955: 44 (*Lobatomixis*) **A**: ZHE
nigrithorax Wittmer, 1956a: 123 **A**: AF
nuristanicus Wittmer, 1956a: 126 **A**: AF
ohbayashii Wittmer, 1954c: 113 (*Ebaeus*) **A**: FE JA
okinawensis Wittmer, 1958: 123 **A**: JA
pakistanicus Wittmer, 1960a: 105 **A**: PA
pallidipennis Wittmer, 1988a: 24 **A**: IS
pallidulus Wittmer, 1956a: 132 **A**: AF
peyerimhoffi Abeille de Perrin, 1907a: 18 **A**: SI
 alfierii Pic, 1920i: 45
picticollis Kiesenwetter, 1874b: 287 (*Ebaeus*) **A**: JA
pius Kiesenwetter, 1866a: 375 **E**: PT SP
pseudolamellatus Wittmer, 1981b: 195 **A**: UP
pseudonepalensis Wittmer, 1987b: 109 **A**: NP UP
pygialis Abeille de Perrin, 1883: 34 (*Ebaeus*) **N**: AG
quadrisignatus Champion, 1921a: 105 **A**: UP
quinqueplagiatus Champion, 1921a: 75 **A**: HP
ramirezi Pardo Alcaide, 1958: 169 [= 1960: 121] **E**: SP
satoi Wittmer, 1971a: 16 **A**: JA (Ryukyus)
schereri Wittmer, 1966b: 236 **A**: SD
schmidi Wittmer, 1960a: 103 **A**: PA
scutellomaculatus Wittmer, 1956a: 135 **A**: AF
senaci Abeille de Perrin, 1890a: 48 (*Ebaeus*) **A**: TR
simillimus Wittmer, 1956a: 127 **A**: AF
spinicornis Champion, 1921a: 105 **A**: UP
stevensi Champion, 1922c: 268 **A**: SD
subfractus J. R. Sahlberg, 1908: 71 **A**: IS SY
suffusus Champion, 1922c: 268 **A**: UP
sulcicauda Champion, 1921a: 104 **A**: UP
tibiellus Wittmer, 1979c: 145 **A**: TD
tincticollis Champion, 1921a: 74 **A**: UP
tokunoshimanus Nakane, 1985c: 159 **A**: JA
towadensis Nakane, 1985c: 159 **A**: JA
triangularis Champion, 1922c: 267 **A**: UP
triguttatus Champion, 1921a: 75 **A**: HP
tripartitus Marseul, 1868b: 183 (*Anthocomus*) **A**: IS SY TR
 tauricola Pic, 1911g: 106
umbilicatus J. R. Sahlberg, 1908: 72 **A**: IS "Palestine"
uncatus Champion, 1921a: 77 **A**: UP
vaillanti Pic, 1923o: 374 **A**: CH (Nan Shan
vesiculiger Marseul, 1868b: 183 (*Anthocomus*) **A**: CY IS LE SY TR
vicinus Peyron, 1877: 205 **A**: IS TR
virgineus Abeille de Perrin, 1885a: 13 **N**: AG MO

wittmeri Kaszab, 1955a: 302 **E**: SL
yoronensis Nakane, 1985c: 160 **A**: JA

genus *Hypomixis* Wittmer, 1995d: 286 type species *Hypebaeus lucifer* Wittmer, 1956
acutispina Wittmer, 1995d: 295 **A**: SCH
afghanistanica Wittmer, 1956a: 129 (*Hypebaeus*) **A**: AF PA
azureipennis Wittmer, 1995d: 290 **A**: YUN
bituberculata Wittmer, 1995d: 289 **A**: YUN
bivittata Wittmer, 1995d: 291 **A**: YUN
brancuccii Wittmer, 1999: 181 **A**: UP
cuspidata Champion, 1922c: 220 (*Attalus*) **A**: NP UP
erectus Wittmer, 1999: 183 **A**: YUN
hickeriana Wittmer, 1955: 44 (*Lobatomixis*) **A**: FUJ
intusdirigeta Wittmer, 1995d: 293 **A**: YUN
kalimpongensis Wittmer, 1995d: 293 **A**: SD
kryzhanovskiji Wittmer, 1979c: 147 **A**: TD
ladakhensis Wittmer, 1978e: 370 (*Hypebaeus*) **A**: KA (Ladakh)
lamellata Champion, 1921a: 76 (*Hypebaeus*) **A**: UP
lucifer Wittmer, 1956a: 130 (*Hypebaeus*) **A**: AF
prescutellaris Pic, 1938k: 162 (*Hypebaeus*) **A**: SHX
serpentinus Wittmer, 1999: 182 **A**: YUN
tinctoapicalis Wittmer, 1994a: 5 (*Hypebaeus*) **A**: NW
transversus Wittmer, 1994a: 6 (*Hypebaeus*) **A**: SD
triangulimaculata Wittmer, 1995d: 289 **A**: SCH

genus *Hypotroglops* Wittmer, 1983d: 311 type species *Hypotroglops nepalensis* Wittmer, 1983
gaheshensis Wittmer, 1999: 190 **A**: NP
gorapaniensis Wittmer, 1983d: 317 **A**: NP
impressus Wittmer, 1983d: 315 **A**: SD
nepalensis Wittmer, 1983d: 313 **A**: NP
planiceps Wittmer, 1983d: 312 **A**: NP
uenoi Wittmer, 1983d: 316 **A**: NP

genus *Incisomalachius* Pic, 1914f: 15 type species *Incisomalachius notaticeps* Pic, 1914
notaticeps Pic, 1914f: 15 **A**: UP

genus *Indiattalus* Pic, 1915c: 11type species *Attalus diversipennis* Pic, 1915
diversipennis Pic, 1915c: 12 (*Attalus*) **A**: "Himalaya"

genus *Indiebaeus* Wittmer, 1957a: 44 type species *Indiebaeus crassitibia* Wittmer, 1957
cyaneonotatus Pic, 1903i: 122 (*Attalus*) **A**: SD UP
  crassitibia Wittmer, 1957a: 44
dentistylus Wittmer, 1957a: 44 **A**: SD
metallipennis Wittmer, 1994a: 8 **A**: SD
spinosus Wittmer, 1995d: 273 **A**: CH

genus *Kuatunia* Evers, 1948a: 51 type species *Kuatunia klapperichi* Evers, 1948
bullosa Wittmer, 1995d: 279 **A**: NP
emeiensis Wittmer, 1995d: 278 **A**: SCH
guilinensis Wittmer, 1995d: 275 **A**: GUX
klapperichi Evers, 1948a: 51 **A**: FUJ
nigroopaca Wittmer, 1995d: 276 **A**: SCH
oblongula chujoi Wittmer, 1954c: 113 (*Ebaeus*) **A**: JA (Kamikochi, Shinano)
oblongula oblongula Kiesenwetter, 1874b: 287 (*Hypebaeus*) **A**: FE JA

oloyensis Wittmer, 1999: 176 **A**: ES
sichuana Wittmer, 1995d: 278 **A**: SCH

genus *Mixis* Abeille de Perrin, 1885b: 148 type species *Ebaeus hystrix* Abeille de Perrin, 1885
brittoni Wittmer, 1954e: 319 (*Lobatomixis*) **A**: YE
curvispinus Wittmer, 1954e: 318 (*Ebeaus*) **A**: YE
hystrix Abeille de Perrin, 1885b: 147 (*Ebaeus*) **A**: "Turkestan"
perforatus Wittmer, 1954e: 320 (*Lobatomixis*) **A**: YE
pilosoplicatus Wittmer, 1954e: 318 **A**: YE
scotti Wittmer, 1954e: 317 (*Ebaeus*) **A**: SA YE

genus *Platyebaeus* Wittmer, 1995d: 282 type species *Platyebaeus bellulus* Wittmer, 1995
bellulus Wittmer, 1995d: 283 **A**: YUN
kubani Wittmer, 1999: 174 **A**: YUN

genus *Sceloattalus* Wittmer, 1966b: 236 type species *Sceloattalus nepalensis* Wittmer, 1966
nepalensis Wittmer, 1966b: 237 **A**: NP

genus *Tropiebaeus* Wittmer, 1957b: 576 type species *Attalus carinatifer* Pic, 1919
carinatifer Pic, 1919a: 16 (*Attalus*) **A**: AF UP
  auritus Champion, 1921a: 73
carinatipennis Pic, 1905a: 92 (*Ebaeus*) **A**: BT SD **ORR**
cavernosus Champion, 1922c: 264 (*Hypebaeus*) **A**: UP
externecostatus Pic, 1925h: 14 (*Hypebaeus*) **A**: PA UP
garhwalensis Wittmer, 1957b: 577 **A**: PA UP
kuluensis Pic, 1910j: 37 (*Attalus*) **A**: HP PA
  brevecarinatus Pic, 1921g: 7 (*Hypebaeus*)
nepalensis Wittmer, 1980c: 468 **A**: NP
puncticeps Wittmer, 1980c: 471 **A**: SD
rubroapicalis Pic, 1907f: 126 (*Ebaeus*) **A**: YUN
satoi Wittmer, 1981b: 214 **A**: NP
sawadai Wittmer, 1995d: 273 **A**: XIZ

genus *Urodactylus* J. Thomson, 1858: 80 type species *Urodactylus bicaudatus* J. Thomson, 1858
vastiorum Wittmer, 1995e: 311 **A**: OM

genus *Yunnanebaeus* Wittmer, 1996: 326 type species *Yunnanebaeus costipennis* Wittmer, 1996
costipennis Wittmer, 1996: 327 **A**: YUN

tribe Illopini Jakobson, 1911

genus *Condylops* L. Redtenbacher, 1850: 47 type species *Condylops erichsonii* L. Redtenbacher, 1850
  Diabolattalus Pic, 1920a: 7 type species *Diabolattalus diversipes* Pic, 1920
  Flabellohedybius Pic, 1917d: 3 type species *Flabellohedybius maculatipennis* Pic, 1917
  Hedybioattalus Pic, 1917e: 6 type species *Hedybioattalus testaceicornis* Pic, 1917
  Hedybiinus Champion, 1921c: 454 type species *Hedybius clypeolus* Erichson, 1840
arabicus Wittmer, 1979d: 186 **A**: SA
basiexcavatus Wittmer, 1996: 316 **A**: YUN
bocaki Wittmer, 1994a: 11 **A**: YUN
costatus Wittmer, 1983d: 324 **A**: "W. China"
cristatus Champion, 1921a: 110 (*Hedybius*) **A**: UP
diversipedoides Wittmer, 1996: 312 **A**: YUN
dongchuanensis Wittmer, 1995e: 361 **A**: YUN
erichsonii L. Redtenbacher, 1850: 47 **A**: IN
flabellatus Champion, 1925d: 268 (*Hedibiinus*) **A**: UP
fossulifer Wittmer, 1996: 314 **A**: YUN

guerryi Pic, 1903b: 81 (*Attalus*) **A**: YUN
habashanensis Wittmer, 1996: 317 **A**: YUN
impressus Wittmer, 1995e: 367 **A**: YUN
inimpressus Wittmer, 1995e: 369 **A**: YUN
iranicus Wittmer, 1979e: 343 **A**: IN
jizushanensis Wittmer, 1995e: 361 **A**: YUN
leveexcavatus Wittmer, 1996: 311 **A**: YUN
libanicus Wittmer, 1965a: 85 (*Hedybiinus*) **A**: LE
lijiangensis Wittmer, 1995e: 370 **A**: YUN
lopatini Wittmer, 1992b: 32 **A**: UZ
maculatipennis Pic, 1917d: 3 (*Flabellohedybius*) **A**: AF
oculilobatus Wittmer, 1996: 316 **A**: YUN
pasynkouensis Wittmer, 1983d: 324 **A**: "W China"
paulocostatus Wittmer, 1983d: 321 **A**: XIZ
pauloimpressus Wittmer, 1995e: 365 **A**: YUN
potanini Wittmer, 1983d: 325 **A**: XIZ
sellatus Wittmer, 1995e: 362 **A**: YUN
semilimbatus Pic, 1906c: 17 (*Attalus*) **A**: YUN
shimomurai Wittmer, 1982f: 363 (*Hedybioattalus*) **A**: TAI
sinensis Pic, 1903f: 105 (*Attalus*) **A**: YUN
subcyaneus Pic, 1948c: 16 (*Attalus*) **A**: YUN
tcheouensis Pic, 1925h: 14 (*Attalus*) **A**: HEN
testaceicornis Pic, 1917e: 6 (*Hedybius*) **A**: CH
triangulatus Wittmer, 1996: 314 **A**: SCH
turkmenicus Wittmer, 1970c: 95 **A**: TM

genus *Hedybius* Erichson, 1840: 92 type species *Hedybius smaragdulus* Erichson, 1840
nigroplagiatus Champion, 1922c: 262 (*Hedybius*) [DA] **A**: UP

genus *Penhedybius* Wittmer, 1985a: 343 type species *Penhedybius damaricus* Wittmer, 1985
marginatus Wittmer, 1979e: 342 (*Flabellohedybius*) **A**: IN
tuberculatus Wittmer, 1965a: 86 (*Flabellohedybius*) **A**: AF

tribe Malachiini Fleming, 1821

genus *Acromalachius* Evers, 1985: 9 type species *Malachius flexicornis* Baudi di Selve, 1873
bicinctus Chevrolat, 1854: 436 (*Anthocomus*) **A**: IS JO LE
  crassicornis Baudi di Selve, 1871: 67 (*Malachius*)
clavicornis Peyron, 1877: 60 (*Malachius*) **A**: IS SY TR
flexicornis Baudi di Selve, 1873: 241 (*Malachius*) **A**: SY TR

genus *Anhomodactylus* Mayor & Wittmer, 1981: 416 type species *Malachius eximius* Lewis, 1895
albilabrus Pic, 1914p: 73 (*Anthocomus*) **A**: FE (Amur)
biimpressifrons Pic, 1932j: 252 (*Malachius*) **A**: SCH
eximius Lewis, 1895a: 117 (*Malachius*) **A**: JA
gotoi Nakane, 1956b: 161 (*Malachius*) **A**: JA
ohbayashii Wittmer, 1953a: 37 (*Malachius*) **A**: JA

genus *Anthocomus* Erichson, 1840: 97 type species *Cantharis fasciatus* Linnaeus, 1758
subgenus *Anthocomus* Erichson, 1840: 97 type species *Cantharis fasciatus* Linnaeus, 1758
  Paremballus Abeille de Perrin, 1891b: 195 type species *Anthocomus fenestratus* Linder, 1864
abdominalis Pic, 1903f: 106 **A**: SHG
apalochroides Abeille de Perrin, 1890b: 89 **A**: IN
  hapalochroides Abeille de Perrin, 1891b: 195 [unjustified emendation]
bilineatus Pic, 1907a: 98 **A**: TM UZ

tedschenensis Evers, 1985: 14 (*Clanoptilus*)
coreanus Pic, 1911h: 107 **A**: "Korea"
dux Abeille de Perrin, 1890a: 44 **A**: AF
fasciatus Linnaeus, 1758: 402 (*Cantharis*) **E**: AU BE BH BU BY CR CT CZ DE EN FI FR GB GE GR HU IR IT LA LS LT MC NL NR PL RO RU SK SP SV SZ YU (Serbia) "Caucasus" **A**: SY TR
bituberculatus Stephens, 1830: 313 (*Malachius*)
gouani Linnaeus, 1767: 1068 (*Meloë*)
latemaculatus Pic, 1920h: 21
latior Abeille de Perrin, 1891b: 199
regalis Charpentier, 1818: 232 (*Malachius*)
fenestratus Linder, 1864: 251 **E**: FR PT SP
flavicrus Abeille de Perrin, 1883: 30 (*Malachius*) **E**: AB
gratissimus Abeille de Perrin, 1882a: 137 **A**: SY
miniaceus Gemminger & Harold, 1869: 1691 (*Malachius*) [RN] **A**: SY "Palestine"
miniatus Reiche & Saulcy, 1857b: 178 (*Malachius*) [HN]
pulcher Pic, 1921m: 147 **N**: MO
rufus rufus Herbst, 1784b: 108 (*Malachius*) **E**: AU BE BY CZ DE EN FI FR GB GE HU IT LA LS LT PL SK SP SV SZ UK "Caucasus" **A**: IN SY TM TR
bedeli Méquignon, 1948: 77
coccineus Schaller, 1783: 303 (*Cantharis*) [HN]
erythromelas Gmelin, 1790: 1899 (*Malachius*)
sanguinolentus Fabricius, 1787a: 169 (*Malachius*)
subscutellaris Pic, 1924g: 80
rufus grossicornis Abeille de Perrin, 1891b: 190 **A**: "Persia"
semipolitus Abeille de Perrin, 1882a: 138 **A**: SY
tricoloripennis Evers, 1948a: 58 **E**: TR
subgenus *Celidus* Mulsant & Rey, 1867: 131 type speces *Malachius equestris* Fabricius, 1781
Neotrotus Abeille de Perrin, 1891b: 187 type species *Malachius bipunctatus* Harrer, 1784
cardinalis Abeille de Perrin, 1881: 111 **E**: ST (Astrakhan) **A**: IQ SY
mesopotamicus Pic, 1901a: 10
damascenus Pic, 1915i: 25 **A**: SY
doriae Baudi di Selve, 1873: 245 **A**: IN SY
equestris Fabricius, 1781: 500 (*Malachius*) **E**: AU BE BU CR CZ EN FI FR GE GR HU IT LA LS LT NL NR PL RO RU SK SP SV SZ "Caucasus" **NAR**i
bipunctatus Harrer, 1784: 83 (*Malachius*)
herbsti Gmelin, 1790: 1899 (*Malachius*)
quadripustulatus Illiger, 1798: 306 (*Malachius*)
humeralis Morawitz, 1862: 314 **E**: AU BU CZ GR RO SK ST SZ **A**: TR
miniatus Kolenati, 1846b: 43 (*Malachius*) **E**: GR ST "Caucasus" **A**: SY TR "Palestine"
sellatus Solsky, 1866a: 91
pupillatus Abeille de Perrin, 1890a: 44 **E**: "Caucasus" **A**: SY TR
subgenus *Omphalius* Abeille de Perrin, 1891b: 187 type species *Malachius haeres* Abeille de Perrin, 1883
haeres Abeille de Perrin, 1883: 31 (*Malachius*) **E**: GR
thalassinus Abeille de Perrin, 1883: 31 (*Attalus*) **E**: AU BU HU **A**: SY TR
amanicola Abeille de Perrin, 1900a: 17 (*Malachius*)
infasciatus Pic, 1901b: 18 (*Anthocomus*)

genus *Anthocomus*, species incertae sedis
bipartitus Wittmer, 1982f: 365 **A**: TAI
bowringi Pic, 1948a: 16 **A**: HKG
creticus Evers, 1994: 77 **E**: GR (Kríti)
cyaneipennis Wittmer, 1940a: 100, 103 (*Anthocomus*) **A**: JIA SHA
denticornis Wittmer, 1986b: 134 **A**: TAI
flaveolus Abeille de Perrin, 1893b: clxv **E**: RO (Dobrudscha)
kinzelbachi Wittmer, 1981e: 415 **A**: SY
limbellus Wittmer, 1953b: 73 (*Malachius*) **A**: SHX

lineatipennis Wittmer, 1995e: 372 **A**: YUN
miandamensis Wittmer, 1986a: 120 **A**: PA
mongolicus Wittmer, 1969a: 57 **A**: MG
pseudococcineus Wittmer, 1974a: 19 **A**: PA
quadrimaculatus Pic, 1933e: 9 **A**: IS
similicornis Wittmer, 1999: 195 **A**: YUN
smyrnensis Wittmer, 1971e: 373 **A**: TR
swatensis Wittmer, 1986a: 119 **A**: PA
testaceoterminalis Wittmer, 1995e: 374 **A**: YUN

genus *Axinotarsus* Motschulsky, 1854d: 55 type species *Malachius pulicarius* Fabricius, 1777
subgenus *Axinotarsus* Motschulsky, 1854d: 55 type species *Malachius pulicarius* Fabricius, 1777
Adenophorus C. G. Thomson, 1859: 111 type species *Cantharis rubricollis* Marsham, 1802
africanus Escalera, 1926a: 271 **N**: AG MO TU
alticola Abeille de Perrin, 1882a: 138 **A**: SY
amilcar Escalera, 1926a: 273 **N**: TU
anatolicus Evers, 1961b: 176 **E**: TR
bicoloriceps Pic, 1932g: 27 [= 1932c: 18] **N**: MO
championi Escalera, 1926a: 275 **N**: TU
cobosi Pardo Alcaide, 1966b: 527 **N**: MO
doderoi Escalera, 1926a: 236 **E**: FR (Corse) IT (Sardegna)
figuerasi Pardo Alcaide, 1950: 17 **N**: MO
freydenbergi Escalera, 1926a: 265 **N**: MO
insularis Abeille de Perrin, 1881: 113 **E**: FR (Corse) IT **N**: AG
intermedius Escalera, 1914: 239 **N**: AG MO
nigricollis Escalera, 1926a: 268
italicus Escalera, 1926a: 258 **E**: IT
justei Pardo Alcaide, 1971: 247 **N**: MO
longicornis Kiesenwetter, 1859a: 34 (*Malachius*) **E**: IT **N**: AG
rufithorax Kiesenwetter, 1859b: 59 (*Anthocomus*)
marginalis Laporte, 1840a: 279 (*Malachius*) **E**: AU BE CZ DE FR GE HU IT LA NL PL PT RO RU SK SP SZ "Caucasus"
infuscatus Abeille de Perrin, 1891a: 616
massinisa Escalera, 1926a: 277 **N**: LB
nigritarsis Abeille de Perrin, 1885a: 10 **E**: PT SP **N**: AG
pallitarsus Fairmaire, 1880a: 6 (*Malachius*) **E**: FR (Corse) **N**: AG MO TU
bleusei Pic, 1927f: 5
vaulogeri Pic, 1927o: 167
peninsularis Abeille de Perrin, 1881: 112 **E**: FR (Corse) SP **N**: AG TU
pulicarius Fabricius, 1777: 234 (*Malachius*) **E**: AU BE CZ DE FR GB GE HU IT LA MA NL NR PL PT SK SP ST SV SZ UK "Caucasus" **N**: AG MO
leucogastra Preyssler, 1790: 58 (*Cantharis*)
martialis Pic, 1918g: 17 (*Axynotarsus* [sic])
nemoralis Gmelin, 1790: 1898 (*Cantharis*)
obscuratus Escalera, 1926a: 244
truncatellus Fallén, 1807: 25 (*Malachius*)
risor Abeille de Perrin, 1891a: 614 **N**: AG
algiricus Abeille de Perrin, 1896b: 392
chobauti Escalera, 1926a: 248
escalerai Pic, 1927f: 5
ruficollis A. G. Olivier, 1790a: no 27: 9 (*Malachius*) **E**: AU BE CZ DE FR GB GE HU IT LS NL PL PT RO SK SP SV SZ **N**: AG TU
rubricollis Marsham, 1802: 371 (*Cantharis*)
siciliensis Escalera, 1926a: 246 **E**: IT (Sicilia)
transcaspicus Pic, 1907a: 98 **A**: TM
tristiculus Kiesenwetter, 1866b: 267 **E**: IT (Sardegna) SP

tristis Perris, 1864: 287 (*Malachius*) **E**: SP **N**: AG MO
tunezinus Escalera, 1926a: 257 **N**: TU
tunetanus Dodero, 1926: 136 [injustified emendation]
varius nevadensis Escalera, 1926a: 240 **E**: SP (Sierra Nevada)
varius varius Uhagón, 1901a: 28 **E**: PT SP
uhagoni Escalera, 1926a: 239
subgenus *Maurocerus* Pardo Alcaide, 1971: 232, 241 type species *Axinotarsus wifaqi* Pardo Alcaide, 1971
violaceus Abeille de Perrin, 1890a: 44 **N**: MO
compressicornis Escalera, 1914: 238
wifaqi Pardo Alcaide, 1971: 241 **N**: MO
subgenus *Tarxinosus* Pardo Alcaide, 1971: 231, 239 type species *Malachius brevicornis* Kraatz, 1862
brevicornis Kraatz, 1862c: 270 (*Malachius*) **E**: IT (Sardegna) SP **N**: AG MO TU
curtelineatus Pic, 1927o: 167
normandi Escalera, 1926a: 224
orientalis Escalera, 1926a: 223
implicatus Escalera, 1926a: 225 **N**: AG MO TU
robustus Pic, 1905c: 106 **N**: MO
parvisignatus Escalera, 1914: 238

genus *Cerapheles* Mulsant & Rey, 1867: 141 type species *Malachius lateplagiatus* Fairmaire, 1862
Diaphonus Mulsant & Rey, 1867: 144 type species *Malachius terminatus* Ménétriés, 1832
hauseri Pic, 1907a: 99 **A**: UZ (Buchara)
lateplagiatus Fairmaire, 1862c: 550 (*Malachius*) **E**: FR PT SP
notaticollis Pic, 1907j: 138 **E**: "Caucasus"
terminatus Ménétriés, 1832: 164 (*Malachius*) **E**: AB AU BE CZ FR GB GE GR HU IT NL PL SK SZ TR "Caucasus" **A**: IQ SY TM TR **NAR**i
festivus W. Redtenbacher, 1842: 14 (*Anthocomus*)
obscuricornis Pic, 1906h: 41
venustus Guérin-Méneville, 1844a: 48 (*Melyris*)
utebensis Hodgson Torres & Plata Negrache, 1987: 173 **E**: SP
zoelleri Flach, 1895: 45 **E**: RO TR

genus *Ceratistes* Fischer von Waldheim, 1844: 35 type species *Malachius cornutus* Gebler, 1825
Hammamalachius Evers, 1985: 9 type species *Malachius dilaticornis* Germar, 1824
cervulus Reitter, 1894f: 103 (*Malachius*) **E**: BU RO
cornutus Gebler, 1825: 47 [= 1859: 45] (*Malachius*) **A**: ES KZ LE WS
dentifrons Erichson, 1840: 73 (*Malachius*) **E**: FR IT SP
dilaticornis Germar, 1824: 74 (*Malachius*) **E**: BH BU CR FR GR IT MC RO SZ TR YU

genus *Charopus* Erichson, 1840: 119 type species *Malachius pallipes* A. G. Olivier, 1790
apicalis Kiesenwetter, 1859a: 33 **E**: FR (Corse) GR IT MA **N**: AG MO TU
bicolor Peyron, 1877: 221 **A**: LE
bonadonai Pardo Alcaide, 1962: 294 **E**: FR
concolor Fabricius, 1801a: 310 (*Malachius*) **E**: AL AU BU CR CZ FR (Corse) GE GR HU IT PT RO SK SP (Baleares) ST SZ TR YU (Montenegro) "Caucasus" **N**: AG **A**: TR
furcatipennis A. Villa & G. B. Villa, 1837: 17 (*Malachius*)
nigripes Kolenati, 1846b: 44 (*Ebaeus*)
diversipennis Pic, 1906h: 41 **E**: TR
docilis Kiesenwetter, 1852: 619 **E**: FR IT SP SZ **N**: AG
hamifer Kiesenwetter, 1866a: 387
flavipes Paykull, 1798: 274 (*Malachius*) **E**: AU BE BH BU CZ DE FI FR GE HU IT LA NL NR PL RO RU SK SV SZ UK
plumbeocupreus Goeze, 1777: 545 (*Cantharis*)
plumbeonigra Goeze, 1777: 544 (*Cantharis*)
laptelensis Champion, 1925b: 108 **A**: UP
madidus Kiesenwetter, 1863a: 614 **E**: AU CR IT SZ

nigricans Peyron, 1877: 220 **A**: IS SY
nubilus Hodgson Torres & Plata Negrache, 1987: 177 **E**: SP
pallipes A. G. Olivier, 1790a: no 27: 11 (*Malachius*) **E**: BE FR GE IT NL PT SP SZ **N**: AG
bicoloripes Pic, 1914q: 77
grandicollis Kiesenwetter, 1852: 620
rotundatus Geoffroy, 1785: 64 (*Cicindela*) [HN]
varipes Baudi di Selve, 1872: 128
philoctetes Abeille de Perrin, 1885a: 13 **E**: GG HU IT MC TR "Caucasus" **A**: TM
rotundatus Erichson, 1840: 121 **E**: AL FR (Corse) GE IT (Sardegna, Sicilia) PT SP **N**: AG MO TU
macroderus Abeille de Perrin, 1890c: 253
multicaudis Kiesenwetter, 1866a: 383
nitidus Küster, 1849b: 18
saginatus Kiesenwetter, 1852: 621
rubronotatus Pic, 1911g: 105 **E**: "Caucasus"
thoracicus Morawitz, 1861: 287 **E**: HU ST
tibetanus Champion, 1925b: 109 **A**: XIZ

genus *Chionotopus* Abeille de Perrin, 1881: 126 type species *Chionotopus damascaenus* Abeille de Perrin, 1881
bimaculatus Pic, 1911g: 107 **A**: IN
inhumeralis Pic, 1911d: 91
damascaenus Abeille de Perrin, 1881: 125 **A**: SY
humeralis Reiche & Saulcy, 1857b: 180 (*Malachius*) **A**: IS SY TR
mossulensis Abeille de Perrin, 1881: 126 **A**: IQ

genus *Clanomalachius* Tshernyshev, 2000: 209 type species *Malachius resecatus* Wittmer, 1979
debrosi Wittmer, 1979a: 5 (*Malachius*) **A**: KI TD UZ
dolini Wittmer, 1992b: 29 (*Clanoptilus*) **A**: KI KZ
insignicornis Pic, 1909h: 129 (*Malachius*) **A**: AF KI TD TM
kiesenwetteri Solsky, 1881: 242 (*Malachius*) **A**: KI TD UZ "Turkestan"
kuehnelti Hicker, 1959: 71 (*Malachius*) **A**: KI UZ
resecatus Wittmer, 1979a: 3 (*Malachius*) **A**: UZ

genus *Clanoptilus* Motschulsky, 1854b: 32 type species *Clanoptilus antennatus* Motshulsky, 1854
subgenus *Clanoptilus* Motschulsky, 1854b: 32 type species *Clanoptilus antennatus* Motshulsky, 1854
Falsomicrinus Pic, 1949b: 73 type species *Malachius concoloricollis* Pic, 1949
abbreviatus Tshernyshev, 1999a: 144 **A**: KI
abdominalis abdominalis Fabricius, 1798: 71 (*Malachius*) **N**: AG EG LB MO TU **A**: AF SY
bimaculatus Pic, 1951b: 4 (*Malachius*)
coeruleus Erichson, 1840: 82 (*Malachius*)
cyrenaicus Pic, 1925b: 2
hilaris Rosenhauer, 1856: 150 (*Malachius*)
illusus Abeille de Perrin, 1890a: 41 (*Malachius*)
lippus Chevrolat, 1866a: 102 (*Malachius*)
marginicollis P. H. Lucas, 1846: 191 (*Malachius*)
mozabitus Pic, 1898e: 97
pseudobimaculatus Pic, 1954a: 1
serricornis Ballion, 1871: 352 (*Malachius*)
semilimbatus Fairmaire, 1862c: 550 (*Malachius*)
subcrucifer Pic, 1904m: 90
abdominalis limbifer Kiesenwetter, 1850: 224 (*Malachius*) **E**: FR PT SP
abdominalis rufimanus Kocher, 1954b: 207 **N**: MO
aegyptiacus Evers, 1985: 13 **N**: EG
affinis Ménétriés, 1832: 164 (*Malachius*) **E**: AU BH BU CR FR GR HU IT MC PL RU SK SP SZ TR UK "Caucasus" **A**: ES IN XIZ "Turkestan"
concolor Krauss, 1902a: 21
gracilis L. Miller, 1857: 138 (*Malachius*)

obtusicornis Solsky, 1881: 248 (*Malachius*)
aktaschensis Švihla, 1987: 284 **A**: UZ
ambiguus Peyron, 1877: 102 (*Malachius*) **E**: AL AR AU BH BU CR GR HU IT RO RU TR YU "Caucasus" **A**: TM "Turkestan"
aeneocupreus Peyron, 1877: 102
armeniacus Ménétrés, 1832: 199 (*Malachius*)
anatolicus Wittmer, 1986a: 117 **E**: TR
angulicornis Pic, 1949b: 73 (*Malachius*) **N**: MO
arnaizi Pardo Alcaide, 1966: 34 (*Malachius*) **E**: IT FR SP
ayatollai Wittmer, 1994a: 26 **A**: IN
bedeli Abeille de Perrin, 1881: 110 (*Malachius*) **N**: AG
brodskyi Švihla, 1987: 284 **E**: GR (Kríti)
calabrus Baudi di Selve, 1873: 244 (*Malachius*) **E**: IT
capricornis Peyron, 1877: 94 (*Malachius*) **A**: SY
cephalotes Wittmer, 1956a: 151 (*Malachius*) **A**: AF
colasi Wittmer, 1967: 1090 (*Malachius*) **A**: IN TR
serricornis Wittmer, 1988d: 403
crux Abeille de Perrin, 1883: 28 (*Malachius*) **A**: IS
danieli Abeille de Perrin, 1893b: clxiv (*Malachius*) **E**: TR
tokatensis Abeille de Perrin, 1900a: 17 (*Malachius*)
dilatipes Wittmer, 1974a: 13 (*Malachius*) **A**: KA
druzicus Wittmer, 1986a: 115 **A**: IS
dupplicatus Kolenati, 1846b: 44 (*Malachius*) **E**: "Caucasus"
debilis Kiesenwetter, 1878: 211 (*Malachius*)
opacus Kiesenwetter, 1878: 210 (*Malachius*)
durandi Pardo Alcaide, 1970: 352 (*Malachius*) **E**: BH BU GG GR RO "Caucasus" **A**: TR
elegans A. G. Olivier, 1790a: no 27: 6 (*Malachius*) **E**: AL AU BE BH BU CZ FR GB GE GR IT MC NL PL PT RU SK SP SZ TR "Caucasus" **A**: IN TR "Siberia"
haemorrhoidalis Abeille de Perrin, 1891b: 140
purpureonotatus Pic, 1952b: 9 (*Malachius*)
emarginatus Krauss, 1902a: 24 (*Malachius*) **E**: AU FR IT GE SZ
falcifer Abeille de Perrin, 1882a: 114 (*Malachius*) **E**: BU HU IT ST TR "Caucasus"
fissipennis Wittmer, 1956b: 311 (*Malachius*) **A**: ZHE
flabellicornis Solsky, 1881: 240 (*Malachius*) **E**: ST **A**: "Turkestan"
banghaasi Abeille de Perrin, 1900a: 16 (*Malachius*)
vestalis Abeille de Perrin, 1900a: 17 (*Malachius*)
fractus Wittmer, 1988d: 401 **A**: SY
geniculatus Germar, 1824: 73 (*Malachius*) **E**: AL AR AU BH BU CR CZ FR GE GR HU IT MC PL PT RU SK SP SZ TR UK YU "Caucasus" **A**: IN MG SY TM "Siberia" "Turkestan"
annulatus: Mulsant & Rey, 1867: 60 (*Malachius*)
guttifer Abeille de Perrin, 1893b: clxiii (*Malachius*)
obconicus Abeille de Perrin, 1891b: 141
pulcherrimus Pic, 1912b: 1 (*Chionotopus*)
uncicornis Solsky, 1881: 245 (*Malachius*)
heliophilus heliophilus Peyron, 1877: 84 (*Malachius*) **A**: IN IQ SY TR
breviruber Pic, 1914h: 33
heliophilus kervillei Pic, 1932d: 17 (*Malachius*) **A**: TR
ibex Abeille de Perrin, 1890a: 42 (*Malachius*) **E**: AB AR GG ST
imperialis Morawitz, 1862: 315 (*Anthocomus*) **E**: ST **A**: AF LE TD TM
coccineovarius Solsky, 1881: 249 (*Anthocomus*)
rubromarginatus Reitter, 1889d: 110 (*Malachius*)
indicus Pic, 1913e: 101 (*Malachius*) **A**: UP
insignis Buquet, 1840: 242 (*Malachius*) **N**: AG TU
iridicollis Marseul, 1868b: 182 (*Malachius*) **A**: SY
italicus Pardo Alcaide, 1967: 99 (*Malachius*) **E**: IT
jordanicus Wittmer, 1965a: 84 (*Malachius*) **A**: JO

karpathosensis Wittmer, 1988a: 28 **E**: GR
kaszabi Wittmer, 1958: 127 (*Malachius*) **E**: AR
kopetensis Wittmer, 1994a: 19 **A**: TM
kruegeri Pic, 1934g: 147 (*Malachius*) **N**: LB
laterimaculatus Wittmer, 1966c: 502 (*Malachius*)
kruperi Pic, 1909j: 137 (*Malachius*) **E**: TR
laticollis Rosenhauer, 1856: 149 (*Malachius*) **E**: SP (Andalusía)
peninsularis Evers, 1948a: 57
lineaticollis Pic, 1917e: 6 (*Malachius*) **A**: TAI
maculiventris Chevrolat, 1854: 435 (*Malachius*) **A**: CY IS SY TR
malatyensis Wittmer, 1988d: 401 **E**: TR
marginellus A. G. Olivier, 1790a: no 27: 6 (*Malachius*) **E**: AU BE BH BU BY CR CZ DE FR GB GE HU IT LA LT MC NL PL PT RO RU SK SP SV SZ UK **N**: AG MO TU **AFR**
angustimarginalis Donisthorpe, 1933: 25 (*Malachius*)
atriceps Abeille de Perrin, 1891a: 670
bispinosus Curtis, 1827: 167 (*Malachius*)
dentipennis Laporte, 1838: 28 (*Malachius*)
lambessanus Pic, 1951b: 4 (*Malachius*)
oxys Abeille de Perrin, 1890a: 40 (*Malachius*)
pseudosardous Reclaire & van der Wiel, 1932: 426 (*Malachius*)
subdilatatus Pic, 1923i: 9 (*Malachius*)
milkoi Tshernyshev, 1999a: 142 **A**: KI KZ UZ
minutemarginatus Wittmer, 1992b: 25 **A**: TD
miricollis Peyerimhoff, 1948d: 268 (*Malachius*) **N**: MO
concoloricollis Pic, 1949b: 73 (*Malachius*)
mosulensis Wittmer, 1988d: 400 **A**: IQ
nigroplumbeus Pic, 1926a: 1 (*Malachius*) **A**: "Turkestan"
oranicus Evers, 1966a: 45 (*Malachius*) **N**: AG
otini Peyerimhoff, 1948d: 267 (*Malachius*) **N**: MO
oxyacanthae Peyron, 1877: 286 (*Malachius*) **A**: LE SY
paludosus Peyron, 1877: 110 (*Malachius*) **A**: SY TR
parilis Erichson, 1840: 80 (*Malachius*) **E**: IT (Sardegne, Sicilia)
commissus Abeille de Perrin, 1891b: 141 (*Malachius*)
peyroni Abeille de Perrin, 1881: 108 (*Malachius*) **A**: SY
phereclus Reitter, 1898e: 123 (*Malachius*) **E**: AB/AR "Araxesthal bei Ordubad"
piceoterminatus Pic, 1911q: 145 (*Malachius*) **E**: TR
politus Wittmer, 1988d: 402 **E**: TR
pseudoserrifer Wittmer, 1994a: 20 **A**: IN
pseudoturanicus Wittmer, 1979a: 7 (*Malachius*) **A**: AF
rectispinus Wittmer, 1992b: 28 **A**: TM
reichenspergeri Wittmer, 1956a: 154 (*Malachius*) **A**: AF
robusticornis Wittmer, 1994a: 17 **A**: IN
rufithorax Wittmer, 1956a: 153 (*Malachius*) **A**: AF
rufus A. G. Olivier, 1790a: no 27: 4 (*Malachius*) **E**: FR IT SP SZ **N**: AG MO TU
savioi Pic, 1937k: 172 (*Malachius*) **A**: JIX "Korea"
senylia Tshernyshev, 2000: 207 **A**: KI UZ
serrifer Pic, 1911a: 107 (*Malachius*) **A**: IN
sexplagiatus Abeille de Perrin, 1882a: 115 (*Malachius*) **A**: IN SI SY
fourtaui Pic, 1921k: 42 (*Malachius*)
sikkimensis Pic, 1907p: 172 (*Malachius*) **A**: SD UP
sinensis Abeille de Perrin, 1900a: 17 (*Malachius*) **A**: CH
spinipennis Germar, 1824: 75 (*Malachius*) **E**: AU BH BU CR GR IT MC PL RO RU SZ TR UK "Caucasus" **A**: SY "Western Asia"
antennatus Motschulsky, 1854b: 32 (*Clanoptilus*)
brevispina Kiesenwetter, 1871: 81 (*Malachius*)
cruentatus Motschulsky, 1854b: 31 (*Malachius*)

muticus Baudi di Selve, 1872: 123 (*Malachius*)
spinosus Brullé, 1832: 149 (*Malachius*)
spinosus Erichson, 1840: 81-82 (*Malachius*) **E**: AR AU AZ BH FR GE GR IT MC PL PT RO RU SK SP SZ UK **N**: AG MO TU
angustatus Ménétriés, 1832: 163 (*Malachius*)
chlorizans Erichson, 1840: 81 (*Malachius*)
condrillieri Pic, 1952b: 9 (*Malachius*)
strangulatus Abeille de Perrin, 1885b: 143 (*Cyrtosus*) **E**: AU CZ FR GB GE HU IT NL RU SK TR UK "Caucasus" **A**: IN KI KZ UZ
pseudospinosus Medvedev, 1964b: 156 (*Malachius*)
schelkownikowi Pic, 1911j: 113 (*Malachius*)
vulneratus Abeille de Perrin, 1891b: 135 (*Malachius*)
turanicus Reitter, 1889d: 109 (*Malachius*) **A**: TM "Turkestan"
turcmenicus Abeille de Perrin, 1890a: 39 (*Malachius*)
valentinae Tshernyshev, 1999a: 142 **A**: UZ
versicolor Faldermann, 1835a: 195 (*Malachius*) **E**: AB AR GG **A**: IN "Turkestan"
erivanicus Pic, 1912t: 89
punctonotatus Pic, 1912t: 89 (*Malachius*)
tricolor Baudi di Selve, 1873: 242 (*Malachius*)
viridibasis Reitter, 1890b: 172
wittmeri Hicker, 1959: 69 (*Malachius*) **A**: SCH
subgenus *Hypoptilus* Mulsant & Rey, 1867: 38 type species *Malachius barnevillei* Puton, 1865
assimilis Erichson, 1871: 65 (*Malachius*) **E**: GR TR ST **A**: IS SY
barnevillei Puton, 1865: 131 (*Malachius*) **E**: DE FR GB IT SP SV SZ
bielawskii Wittmer, 1969a: 55 (*Malachius*) **A**: MG
curticornis Kiesenwetter, 1866a: 388 (*Malachius*) **E**: SP
subelongatus Pic, 1911k: 119
dissimilis Baudi di Selve, 1873: 241 (*Malachius*) **E**: IT (Sicilia) MA "Caucasus" **A**: SY TM
dlabolai Wittmer, 1979e: 344 (*Malachius*) **A**: IN
dubatolovi Tshernyshev, 1999a: 139 **A**: TM
dudkoi Tshernyshev, 1999a: 137 **A**: ES (Chita area)
flavipalpis Abeille de Perrin, 1891b: 137 (*Malachius*) **A**: TM
foveatus Medvedev, 1964b: 156 (*Malachius*) **A**: KZ MG RU
hajazensis Pic, 1929b: 1 (*Malachius*) **A**: "Arabia"
jarantajensis Wittmer, 1977b: 282 (*Malachius*) **A**: MG
judex Abeille de Perrin, 1885a: 7 (*Malachius*) **E**: AR (Baku) "Caucasus"
lutescens Krauss, 1902a: 22 (*Malachius*) **A**: TD TM UZ
pseudolutescens Tshernyshev, 2000: 203 **E**: AB GG
sajnensis Wittmer, 1969a: 57 (*Malachius*) **A**: MG
sardous Erichson, 1840: 76 (*Malachius*) **E**: FR (Corse) IT (Sardegna)
schaeferi Pic, 1953b: 7 (*Malachius*)
transadriaticus Evers, 1969b: 241 (*Malachius*) **E**: AL IT (Calabria, Sicilia)
viridanus Mulsant & Wachanru, 1852a: 4 [= 1852b: 164] (*Malachius*) **A**: LE SY TR
distinctipes Pic, 1908b: 9

nomina dubia, genus *Clanoptilus*
annulatus Gebler, 1830: 86 (*Malachius*)
bellieri Peyron, 1877: 53 (*Malachius*)

genus *Cordylepherus* Evers, 1985: 8 type species *Malachius viridis* Fabricius, 1787
davidi Pic, 1922j: 51 (*Malachius*) **A**: LIA
facialis Gebler, 1832: 47 (*Malachius*) **A**: ES MG NE
persimplex Reitter, 1908c: 129 (*Malachius*)
faustus Erichson, 1840: 74 (*Malachius*) **A**: CH ES MG
aurichalceus Gebler, 1832: 46 (*Malachius*)
profugus Reitter, 1894g: 241 (*Malachius*)

konigi Abeille de Perrin, 1900a: 18 (*Malachius*) **E**: AR **A**: TR
linearis Morawitz, 1861: 288 (*Malachius*) **E**: ST **A**: KZ LE
moupinensis Pic, 1922j: 50 (*Malachius*) **A**: SCH
oberthuerii Uhagón, 1879: 209 (*Malachius*) **E**: PT SP
sibiricus Kiesenwetter, 1879a: 146 (*Malachius*) **A**: ES MG NE "Korea"
 quadricollis Abeille de Perrin, 1885a: 5 (*Malachius*)
sierranus Evers, 1948a: 55 (*Malachius*) **E**: SP
 suarezi Pardo Alcaide, 1956: 19
sponsus Abeille de Perrin, 1883: 25 (*Malachius*) **A**: CY
vaillanti Pic, 1923o: 376 (*Malachius*) **A**: "Central Asia"
viridis Fabricius, 1787a: 169 (*Malachius*) **E**: AL AU BE BH BY CR CZ EN FI FR GB GE HU IT LA LS NL LT NR NT PL RO SK SP SV SZ UK "Caucasus" **N**: AG **A**: AF MG "Siberia" "Turkestan"
 apicalis A. Villa & G. B. Villa, 1833: 33
 concolor Schilsky, 1888b: 189
 cyanea Thunberg, 1787: 55 (*Cantharis*)
 pseudobipustulatus Evers, 1944: 138
 semiviridis Pic, 1940b: 10
walteri Evers, 1961b: 177 (*Malachius*) **A**: AF
xantholoma Kiesenwetter, 1874b: 284 (*Malachius*) **A**: FE (Kunashir) JA (Hokkaido)

genus *Cyrtosus* Motschulsky, 1854d: 55 type species *Cyrtosus nodicornis* Motshulsky, 1854
subgenus *Cyrtosus* Motschulsky, 1854d: 55 type species *Cyrtosus nodicornis* Motshulsky, 1854
 Anthodytes Kiesenwetter, 1863a: 591 type species *Malachius cyanipennis* Erichson, 1840
abeillei Dodero, 1922: 51 **E**: IT
afer Abeille de Perrin, 1885b: 141 **N**: AG TU
annulipes Pic, 1936a: 2 **N**: MO
baborensis Pic, 1905e: 121 **N**: AG
christophi Kiesenwetter, 1879a: 146 (*Malachius*) **A**: FE CH
codinai Pardo Alcaide, 1962: 297 **N**: MO
cyanipennis Erichson, 1840: 86 (*Malachius*) **E**: FR (Corse) IT PT SP **N**: AG MO TU
 angustimargo Uhagón, 1901a: 91 (*Malachius*)
 bifrons Abeille de Perrin, 1891b: 172
 nigricornis Evers, 1948a: 57
diversicornis Pic, 1904f: 33 **N**: AG
dolorosus Abeille de Perrin, 1885a: 4 **N**: AG
eversi Pardo Alcaide, 1966b: 532 **N**: MO
ferreri Pardo Alcaide, 1966b: 533 **E**: SP **N**: MO
 diversipes Pic, 1912u: 221
flavilabris Waltl, 1835: 60 (*Malachius*) **E**: PT SP **N**: AG MO
kaszabi Pardo Alcaide, 1966b: 532 **N**: MO
lohsei Evers, 1995b: 86 **N**: MO
luteotibialis Evers, 1962: 165 **N**: MO
mauritanicus P. H. Lucas, 1846: 192 (*Malachius*) **E**: SP **N**: AG MO TU
 angusticollis P. H. Lucas, 1846: 191, 192 (*Malachius*)
 nodicornis Motschulsky, 1854d: 55
meridionalis Abeille de Perrin, 1885b: 140 **E**: SP **N**: AG
 parvulus Pic, 1904j: 90
rufifrons Abeille de Perrin, 1885b: 140 **N**: AG MO **A**: SY
 curtelineatus Pic, 1936a: 2 **N**: MO
 cyanipes Abeille de Perrin, 1885b: 140
 tibiellus Abeille de Perrin, 1885b: 140
semimarginatus Fairmaire, 1880a: 5 (*Malachius*) **N**: AG MO
seraphini Pardo Alcaide, 1966b: 533 **N**: AG
sinuaticollis Pic, 1913d: 99 (*Malachius*) **A**: SC ZHE
 moutoni Pic, 1937k: 172
tibialis Abeille de Perrin, 1900a: 18 **A**: SY

vaulogeri Abeille de Perrin, 1900a: 19 **N**: AG
viridescens Pic, 1936a: 2 **N**: MO
vitticollis Kiesenwetter, 1874b: 285 (*Malachius*) **A**: CH JA NC
subgenus *Oogynes* Mulsant & Rey, 1867: 33 type species *Oogynes signicollis* Mulsant & Rey, 1867
Apalochroides Escalera, 1913a: 34 type species *Cyrtosus reitteri* Abeille de Perrin, 1890
aestivus Peyron, 1877: 18 **A**: SY
anceyi Abeille de Perrin, 1869c: 404 (*Oogynes*) **E**: "Caucasus" **A**: SY TR
cruralis Abeille de Perrin, 1891b: 163
cerealis Peyron, 1877: 19 **A**: LE SY
corniculatus Kraatz & Kiesenwetter, 1864: 308 (*Anthodytes*) **E**: GR TR
cornifrons Kiesenwetter, 1878: 213
dispar dispar Fairmaire, 1859b: 277 (*Charopus*) **E**: FR (Corse) IT
variicollis Abeille de Perrin, 1891b: 230
dispar unicolor Dodero, 1922a: 50-51 IT (Sardegna)
frigidus frigidus Peyron, 1877: 24 **A**: LE
frigidus syriacus Pic, 1904b: 3 **A**: SY
longicollis Erichson, 1840: 86 (*Malachius*) **E**: FR (Corse) IT (Sardegna) SP
bicolor Perris, 1864: 288 (*Malachius*)
signicollis Mulant & Rey, 1867: 33 (*Oogynes*)
nitidicollis Chevrolat, 1873: 204 (*Malachius*) **E**: TR **N**: LB **A**: SY
caelatus Peyron, 1877: 23
romaniensis Pic, 1901b: 18
ovalis Laporte, 1838: 28 (*Malachius*) **E**: GR IT SP SZ TR
armifrons Kraatz, 1862c: 269 (*Malachius*)
cyaneicollis Abeille de Perrin, 1894a: 91
maculicollis Abeille de Perrin, 1894a: 91
parallelus Küster, 1848a: 14 (*Anthocomus*)
variicollis Abeille de Perrin, 1894a: 90
reitteri Abeille de Perrin, 1890a: 35 **E**: "Caucasus"
schneideri Abeille de Perrin, 1890a: 35 **E**: "Caucasus"
subcylindricus Pic, 1904b: 3 **A**: SY
uhagoni Abeille de Perrin, 1894a: 92 **E**: SP

genus *Haplomalachius* Evers, 1985: 8 type species *Malachius conformis* Erichson, 1840
subgenus *Flabellomalachius* Evers, 1985: 8 type species *Malachius flabellatus* Frivaldszky von Frivald, 1835
akbesianus Pic, 1898h: 138 (*Malachius*) **A**: SY
deferreri Wittmer, 1994a: 10 **E**: SP
diversicornis Pic, 1932g: 27 [= 1932c: 19] (*Malachius*) **N**: MO
maculaticollis Pic, 1932c: 19
flabellatus Frivaldszky von Frivald, 1835: 255 (*Malachius*) **E**: GR TR **A**: CY SY
strigicollis Motschulsky, 1854b: 32 (*Clanoptilus*)
hispanus Perris, 1864: 286 (*Malachius*) **E**: SP **N**: AG MO
ishiharai ishiharai Satô & Wittmer, 1989: 57 **A**: JA
ishiharai kasantsevi Wittmer, 1996: 307 **A**: FE
transbaikalensis Tshernyshev, 1999a: 130 **A**: ES
subgenus *Haplomalachius* Evers, 1985: 8 type species *Malachius conformis* Erichson, 1840
bocharensis Abeille de Perrin, 1900a: 17 (*Malachius*) **A**: UZ (Buchara)
ceraphelоides Wittmer, 1971e: 372 (*Malachius*) **A**: JO
conformis Erichson, 1840: 76 **E**: TR **A**: IQ SY
karsensis Wittmer, 1988a: 29 **E**: TR
melanorhynchus Abeille de Perrin, 1890a: 43 (*Malachius*) **E**: AB or AR "Araxesthal"
potanini Tshernyshev, 1999a: 133 **A**: GAN
pseudomelanorhynchus Tshernyshev, 1999a: 133 **A**: IN
shirazensis Wittmer, 1988a: 30 **A**: IN
simplex Hicker, 1959: 72 (*Malachius*) **N**: MO
strigicrus Fairmaire, 1889b: 44 (*Malachius*) **A**: XIZ

syriacus Wittmer, 1999: 194 **A**: SY
tibetanus Evers, 1985: 12 **A**: XIZ

genus *Malachius* Fabricius, 1775: 207 type species *Cantharis aeneus* Linnaeus, 1758
subgenus *Malachius* Fabricius, 1775: 207 type species *Cantharis aeneus* Linnaeus, 1758
Hapalorhinus LeConte, 1859b: 74 type species *Hapalorhinus mirandus* LeConte, 1859
aeneus Linnaeus, 1758: 402 (*Cantharis*) **E**: AU BE BH BU BY CT CZ DE EN FI FR GB GE GR HU IT LA LT MC NL NR NT PL PT RO SK SP SV SZ TR UK YU (Serbia) "Caucasus" **A**: IN TR "Siberia" **NAR**i
aeneoides Abeille de Perrin, 1891a: 649
agenjoi Pardo Alcaide, 1975b: 236 **E**: GR
alepensis Pic, 1914h: 33 **A**: SY
artvinensis Wittmer, 1974a: 15 **E**: TR
australis Mulsant & Rey, 1867: 93 **E**: FR IT MA SZ
basalis Abeille de Perrin, 1890a: 37 **E**: "Caucasus"
cruciger Abeille de Perrin, 1890a: 37
bicornutus Fairmaire, 1889b: 43 **A**: XIZ
bilyi Švihla, 1987: 282 **E**: GR (Kríti)
bingoelensis Wittmer, 1999: 192 **E**: TR
bipustulatus Linnaeus, 1758: 402 (*Cantharis*) **E**: AU BE BH BU BY CR CT CZ DE EN FI FR GB GE GR HU IR IT LA LS LT NL NR NT PL PT RO RU SK SP ST SV SZ TR UK **A**: ES JA TR WS
aetolicus Kiesenwetter, 1866b: 265
immaculatus Abeille de Perrin, 1891a: 680
immaculicollis Mulsant & Rey, 1867: 93
laticeps Stephens, 1835: 416
capricornis Peyron, 1877: 94 **A**: SY
caramanicus Pic, 1912b: 9 [RN] **A**: SY TR
anticenotatus Pic, 1904h: 57
montanus Peyron, 1877: 66 [HN]
viridistriatus Bourgeois, 1900a: 156
cardinalis Pic, 1906i: 49 **E**: ST **A**: AF
badachshanensis Iablokoff-Khnzorian, 1986c: 223
carnifex Erichson, 1840: 69 **E**: GR IT ST "Caucasus" **A**: CY TR
fallaciosus Baudi di Selve, 1871: 64
latenotatus Pic, 1914h: 33
naxosensis Pic, 1923a: 2
stolatus Mulsant & Godart, 1855a: 168
suturalis Motschulsky, 1854b: 31
suturellus Kiesenwetter, 1859b: 31
cavifrons anatolicus Wittmer, 1974a: 14 **A**: TR
cavifrons cavifrons Kiesenwetter, 1878: 209 **E**: "Caucasus"
cedricola Peyron, 1877: 285 **A**: LE SY
coccineus Waltl, 1838c: 445 **E**: GR **A**: CY SY TR
baudueri Pic, 1930d: 5
coeruleoscutatus Fairmaire, 1891e: cxxx **A**: AF KA
cressius Pic, 1904c: 10 **E**: GR (Kríti)
cyprius Baudi di Selve, 1871: 64 **A**: CY
dama Abeille de Perrin, 1888: 70 **A**: SY TR
demaisoni Abeille de Perrin, 1900a: 16 **E**: TR
diversipes Abeille de Perrin, 1893b: clxii **A**: IN
ensiculus Abeille de Perrin, 1891b: 405 **A**: SY
delagrangei Pic, 1898b: 28
metallescens Abeille de Perrin, 1891b: 405
ephippiger L. Redtenbacher, 1843b: 983 **E**: AR TR **A**: IN SY
cervicornis Abeille de Perrin, 1890a: 36
procerus J. R. Sahlberg, 1908: 67
erythropterus Erichson, 1840: 68 **A**: IQ SY

extensicornis Tshernyshev, 1999a: 135 **A**: TM UZ
faldermanni Faldermann, 1835a: 196 **E**: AB AR GG TR **A**: IN IQ (Kurdistan)
allochromus Abeille de Perrin, 1885a : 9
ornatus Faldermann, 1835a: 198
fausti Abeille de Perrin, 1885°: 7 **E**: AB AR **A**: AF IN TM
crucifer Reitter, 1888g: 427
miniatus Reitter, 1888g: 427
flammeus flammeus Abeille de Perrin, 1882a: 110 **A**: IS JO
flammeus reducticornis Wittmer, 1966a: 8 **A**: IS
fucatus Peyron, 1877: 58 **E**: "Caucasus" **A**: IN SY TR
angustatus Motschulsky, 1854b: 32 (*Clanoptilus*) [HN]
carinifrons Baudi di Selve, 1873: 240
macer Kiesenwetter, 1878: 208
gethsemaniensis Abeille de Perrin, 1881: 106 **A**: IS
rubidiformis Pic, 1899l: 258
gigas Evers, 1948a: 53 **E**: AR
graecus Kraatz, 1862a: 122 **E**: GR (Corfu, Kríti) IT TR "Caucasus" **A**: CY
elaphus Abeille de Perrin, 1890a: 42
submarginatus Motschulsky, 1854b: 31
hastulifer Solsky, 1881: 238 **A**: IN KZ
heydeni Abeille de Perrin, 1882: 113 **E**: TR **A**: IN
chehirensis Pic, 1923a: 2
junceus Peyron, 1877: 62 **A**: "Syria"
kasosensis Wittmer, 1988a: 26 **E**: GR
kraussi Reitter, 1902f: 192 **E**: TR
stylifer Sahlberg, 1908: 68
sultani Pic, 1912t: 89
kumaonensis Evers, 1985: 12 **A**: UP
labiatus Brullé, 1832: 149 **E**: GR **A**: LE "Turkestan"
nigrosetosus Motschulsky, 1854b: 32
tenellus Kraatz, 1862a: 123
lusitanicus atlasicus Escalera, 1914: 240 **N**: MO
lusitanicus lusitanicus Erichson, 1840: 72 **E**: FR IT MA PT SP **N**: MO **A**: TR
amasiae Abeille de Perrin, 1891: 679
escalerai Wittmer, 1941: 1125
maculithorax Pic, 1898b: 28 **N**: MO
manschukuoensis Hicker, 1959: 70 **A**: HEI
mariae Abeille de Perrin, 1885°: 6 **E**: TR
monticola Kiesenwetter, 1878: 210 **E**: AR
kulzeri Evers, 1948°: 55
okinawaensis Nakane, 1985c: 157 **A**: JA
palaestinus Peyron, 1877: 93 **A**: IS
pickai Švihla, 1987: 281 **E**: GR
pilosulus Wittmer, 1999: 192 **A**: IQ
prolongatus Motschulsky, 1866: 167 **A**: FE JA
drouardi Pic, 1923o: 376
foveifrons Kiesenwetter, 1874b: 284
rubidus Erichson, 1840: 7 **E**: AL AU BH BU CZ FR GE GR HU IT LT MC PL SK SZ TR **A**: SY
fallax Strübing, 1854: 198
mazarakii Baeckman, 1903: 92
sublateralis Pic, 1912t: 89
valesiacus Pic, 1929a: 1
sculptifrons Abeille de Perrin, 1883: 27 **A**: SY "Caspian sea" "Middle East"
scutellaris Erichson, 1840: 67 **E**: AU BE BH BU CZ FR GE HU LT PL RU SK SZ TR **A**: SY
securiclatus Baudi di Selve, 1873: 237 **A**: SY TR
semiaeneus Abeille de Perrin, 1891°: 639 **E**: TR "Caucasus" **A**: IN
ornatus Baudi di Selve, 1872: 121

faldermanni Kiesenwetter, 1878: 212
truncaticornis Abeille de Perrin, 1890°: 36 **A**: SY
vittatus Ménétriés, 1832: 163 **E**: "Caucasus"
subhumeralis Pic, 1911j: 113
subgenus *Malachius*, nomen dubium *triangularis* J. R. Sahlberg, 1908: 69
subgenus *Protomalachius* Evers, 1985: 9 type species *Malachius abeillei* Bauduer, 1874
abeillei Bauduer, 1874: clxiv **A**: IS
holzschuhi Wittmer, 1974a: 16 **A**: AF

genus *Malachius*, species incertae sedis
alfierii Pic, 1919f: 14 **N**: EG
arctelimbatus Pic, 1904o: 287 **A**: JA
armitagei Pic, 1923a: 2 **A**: IN
atticus Pic, 1902c: 24 **E**: GR
becquarti Wittmer, 1940a: 99, 102 **A**: JIL
ilicis Laporte, 1840a: 280 **E**: PT
immaculithorax Pic, 1908c: 27 **A**: SY
japonicus Pic, 1904o: 288 **A**: JA
latenotatus Pic, 1928a: 1 **N**: MO
opacipennis Abeille de Perrin, 1882a: 111 **A**: LE SY
perrini Jakobson, 1911b: 696 [RN] **E**: ST **A**: UZ "Turkestan"
marginicollis Kraatz, 1882b: 320 (*Attalus*) [HN]
subangustatus Pic, 1910e: 18 **E**: "Caucasus" **A**: TM
testaceipes Wittmer, 1960a: 106 **A**: UP
tokmakensis Evers, 1969b: 240 **A**: "Turkestan"
yunnanus Pic, 1926s: 454 **A**: YUN

genus *Malachius*, nomen dubium
Cantharis chrysomeloides Thunberg, 1784a: 17 **E**: FR

genus *Micrinus* Mulsant & Rey, 1867: 39 type species *Malachius inornatus* Küster, 1846
dimorphus Abeille de Perrin, 1881: 105 (*Malachius*) **E**: FR SP
heteromorphus Abeille de Perrin, 1869a: 42 (*Malachius*) **E**: FR IT SZ
laticollis Baudi di Selve, 1873: 245 (*Malachius*)
inornatus Küster, 1846b: 38 (*Malachius*) **E**: AU FR IT SP SZ
cyanescens Mulsant & Rey, 1852a: 219

genus *Microlipus* LeConte, 1852d: 168 type species *Microlipus laticeps* LeConte, 1852
Abeillea J. R. Sahlberg, 1908: 73 type species *Hypebaeus tenuicollis* Abeille de Perrin, 1882
Homogynes Abeille de Perrin, 1891b: 157 type species *Cyrtosus lethieryi* Peyron, 1877
Pseudocerapheles Pic, 1914f: 15 type species *Pseudocerapheles superbus* Pic, 1914
anulatus Evers, 1988b: 166 **A**: TD
asiaticus Wittmer, 1982f: 367 **A**: TAI
baudii Abeille de Perrin, 1891b: 229 (*Cyrtosus*) **E**: SP
citrinoguttatus Reiche & Saulcy, 1857b: 181 (*Anthocomus*) **A**: IS
forticornis Wittmer, 1984a: 3 **A**: MG
inimpressus Wittmer, 1982f: 364 (*Anthocomus*) **A**: TD
lethierryi Peyron, 1877: 17 (*Cyrtosus*) **E**: PT SP
limbicollis Abeille de Perrin, 1882a: 112 (*Malachius*)
longipennis Champion, 1922c: 261 (*Pseudocerapheles*) **A**: UP
manueli Evers, 1948c: 190 [RN] **N**: MO
uhagoni Escalera, 1913a: 34 (*Cyrtosus*) [HN]
marmottani Peyron, 1877: 145 (*Attalus*) **N**: AG MO
oculatus Wittmer, 1999: 196 **A**: YUN
rosti Pic, 1914f: 16 (*Pseudocerapheles*) **A**: HP

quercicola Champion, 1927: 84
rectifasciatus Champion, 1927: 84
rubrifrons J. R. Sahlberg, 1908: 73 (*Pelochrous*) **A**: IS
senguptai Wittmer, 1974a: 17 (*Malachius*) **A**: KA
superbus Pic, 1914f: 15 (*Pseudocerapheles*) **A**: HP
tenuicollis Abeille de Perrin, 1882a: 149 (*Hypebaeus*) **A**: IS
usbekistanicus Evers, 1988b: 167 **A**: UZ
vandykei Wittmer, 1956b: 310 (*Malachius*) **A**: JIA "Korea"

tribe Troglopini Mulsant & Rey, 1867

genus *Attalusinus* Leng, 1918: 206 type species *Ebaeus submarginatus* LeConte, 1852
alfierii Pic, 1909u: 185 (*Callotroglops*) **N**: EG
holzschuhi Wittmer, 1984d: 243 **A**: SA
petraensis Wittmer, 1986a: 107 **A**: JO

genus *Brachemys* Abeille de Perrin, 1890c: 206 [RN] type species *Atelestus hemipterus* Erichson, 1840
subgenus *Atelestodes* Abeille de Perrin, 1890c: 207 type species *Brachemys peragalloi* Abeille de Perrin, 1890
heydeni Flach, 1895: 45 **E**: GR (Lesbos) TR
peragalloi Perris, 1866: 186 **E**: FR IT
subgenus *Brachemys* Abeille de Perrin, 1890c: 206 [RN] type species *Atelestus hemipterus* Erichson, 1840
Atelestus Erichson, 1840: 122 [HN] type species *Atelestus hemipterus* Erichson, 1840
Neatelestus Bedel, 1906a: 297 [RN] type species *Atelestus hemipterus* Erichson, 1840
brevipennis brevipennis Laporte, 1838: 29 (*Malachius*) **E**: FR (Corse) GR IT (Sardegna) SP
hemipterus Erichson, 1840: 123 (*Atelestus*)
nigriceps Wittmer, 1935d: 252 (*Brachemis* [sic])
brevipennis dalmatinus J. Müller, 1908: 315 (*Neatelestus*) **E**: CR
demelti Evers, 1961b: 174 **E**: TR
erichsonii Küster, 1844: 20 (*Atelestus*) **E**: CR

genus *Brachyattalus* Wittmer, 1988e: 591 type species *Attalusinus leveimpressus* Wittmer, 1985
anastasei Pic, 1931i: 135 (*Callotroglops*) **N**: EG
arabicus Wittmer, 1980a: 115 (*Carphuroides*) **A**: SA
tenuipubens Wittmer, 1986a: 114 (*Attalus*) **A**: IN

genus *Callotroglops* Abeille de Perrin, 1890c: 220 type species *Troglops bisignatus* Abeille de Perrin, 1885
atrithorax Pic, 1921l: 54 **N**: EG
basipictus Peyerimhoff, 1929b: 107 **N**: AG
bisignatus Abeille de Perrin, 1885b: 151 (*Troglops*) **N**: AG MO TU
convexicollis Wittmer, 1935c: 185 **A**: SI
descarpentriesi Peyerimhoff, 1948b: 73 **N**: AG
eburifer Peyron, 1877: 244 (*Troglops*) **A**: SY
mesatlanticus Peyerimhoff, 1948b: 72 **N**: MO
niveus Abeille de Perrin, 1898c: 256 **N**: AG
planicollis Abeille de Perrin, 1885b: 152 (*Troglops*) **A**: KZ
remanei Evers, 1995b: 87 **N**: MO

genus *Cephalogonia* Wollaston, 1862c: 442 type species *Cephalogonia cerasina* Wollaston, 1862
autumnalis Evers, 1986c: 143 **N**: CI (Gran Canaria)
cerasina Wollaston, 1862c: 444 **N**: CI (La Palma, Tenerife)
mephistopheles Escalera, 1921: 298 (*Cephalogonia*)
fortunata Evers, 1986c: 137 **N**: CI (La Palma)
gautardi Abeille de Perrin, 1881: 123 **N**: CI (Gran Canaria)
euphorbiae Uyttenboogaart, 1929: 158
satanas Escalera, 1921: 300 (*Cephalogonia*) **N**: CI (Tenerife)
scurra Evers, 1986c: 141 **N**: CI (Gran Canaria)

genus *Cephaloncus* Westwood, 1863: 178 type species *Cephaloncus capito* Westwood, 1863
subgenus *Cephaloncus* Westwood, 1863: 178 type species *Cephaloncus capito* Westwood, 1863
Chalicoroides Champion, 1922d: 323 type species *Chalicorus triguttulus* Abeille de Perrin, 1900
Trogliscus Peyron, 1877: 232 type species *Troglops rhinoceros* Marseul, 1868
aegyptiacus Abeille de Perrin, 1900a: 20 (*Troglops*) **N**: EG
albozonatus albozonatus Abeille de Perrin, 1883: 36 (*Troglops*) **E**: GR **N**: AG **A**: CY
atticus Abeille de Perrin, 1891b: 226
cyaneicollis Abeille de Perrin, 1891b: 226
albozonatus tauricus Wittmer, 1988a: 23 **E**: TR (Taurus Mts)
bispinus Wittmer, 1935c: 181 **A**: IS JO SI
capito Westwood, 1863: 178 **N**: CI (Gran Canaria)
capitulo Evers, 1986c: 129 **N**: CI (La Palma)
gracilispinus Wittmer, 1956a: 119 **A**: AF TM
jelineki Wittmer, 1979e: 340 **A**: IN
karsensis Wittmer, 1995e: 377 **E**: TR
khnzoriani Wittmer, 1964: 101 **A**: TD TM
klapperichianus Wittmer, 1966c: 500 **A**: JA
kryzhanovskyi Tshernyshev, 1997: 676 **A**: TM (Uzboi)
kubani Švihla, 1984: 251 **A**: TM
kumaonensis Wittmer, 1979e: 342 **A**: UP
kunarensis Wittmer, 1958: 121 **A**: AF
laureatus Evers & Plata, 1990: 159 **N**: CI (Gran Canaria)
lindbergi Evers, 1961b: 172 **N**: CI (Tenerife)
microcephalus Pic, 1952d: 65 (*Troglops*) **N**: MO
mordkovitshi Tshernyshev, 1997: 678 **A** KI TD UZ
nuristanicus Wittmer, 1956a: 118 **A**: AF
palmensis García & Evers, 1997: 87 **N**: CI (La Palma)
polevodi Tshernyshev, 2001: 157 **A**: TM
reducticeps Pic, 1932e: 21 (*Troglops*) **A**: TM
reitteri afghanistanus Wittmer, 1956a: 121 **A**: AF
reitteri reitteri A. Fleischer, 1910a: 146 (*Troglops*) **A**: KZ
rhinoceros Marseul, 1868b: 189 (*Troglops*) **A**: LE SY TR
bicoloriceps Pic, 1917a: 2 (*Traugliscus*)
torretassoi Wittmer, 1936b: 375 **N**: LB
yemenitus Wittmer, 1988b: 47 **A**: YE
subgenus *Traugliscolus* Pic, 1933a: 1 type species *Traugliscus diversus* Pic, 1933
diversus Pic, 1933a: 1 (*Traugliscus*) **A**: IS

genus *Embrocerus* Peyron, 1877: 225 type species *Embrocerus variegatus* Peyron, 1877
biguttatus Abeille de Perrin, 1881: 119 (*Psiloderes*) **A**: SY
dowgailoi Wittmer, 1995e: 379 **A**: UZ
jerashensis Wittmer, 1995e: 382 **A**: JO
klapperichi Wittmer, 1956a: 117 **A**: AF
turcicus Wittmer, 1964: 102 (*Cephaloncus*) **A**: TR
variegatus Peyron, 1877: 227 **A**: LE

genus *Psiloderes* Peyron, 1877: 223 type species *Charopus formicarius* Reiche & Saulcy, 1857
Mendizabalina Pardo Alcaide, 1956: 17 type species *Mendizabalina zariquieyi* Pardo Alcaide, 1956
cyrtosoides Abeille de Perrin, 1883: 35 (*Troglops*) **N**: AG
diabolicus Abeille de Perrin, 1885b: 150 (*Troglops*) **N**: AG TU
flexicornis Wittmer, 1938a: 86 **N**: LB
formicarius Reiche & Saulcy, 1857b: 182 (*Charopus*) **A**: IS JO SY
judaeicus Wittmer, 1985c: 417 **A**: IS
marraquensis Escalera, 1914: 232 (*Troglops*) **N**: MO

nasiraensis Wittmer, 1985c: 417 **A**: IS
pluriarmatus Belon, 1878: xxviii **A**: IQ
sinaiticus Wittmer, 1935c: 182 **A**: SI
tuniseus Wittmer, 1985c: 414 **N**: TU
viridiceps Pic, 1929k: 108 (*Troglops*) **N**: EG
zariquieyi Pardo Alcaide, 1956: 18 (*Mendizabalina*) **E**: SP

genus *Staphylotroglops* Evers, 1966b: 96 type species *Staphylotroglops amabilis* Evers, 1966
amabilis Evers, 1966b: 97 **N**: MO

genus *Troglomorphus* Wittmer, 1996: 303 type species *Troglomorphus omanus* Wittmer, 1996
omanus Wittmer, 1996: 303 **A**: OM

genus *Troglops* Erichson, 1840: 125 type species *Cantharis albicans* Linnaeus, 1767
albicans Linnaeus, 1767: 649 (*Cantharis*) **E**: AL AU BE BH BU CR CZ FR GE GR HU IT PL RU SK SZ "Caucasus" **A**: SY
  angulatus Fabricius, 1792a: 223 (*Malachius*)
alfierianus Wittmer, 1934: 450 **N**: EG
alonsoi López-Colón, 1997: 445 **E**: SP
angustatus Reitter, 1872b: 181 **N**: AG TU
  normandi Pic, 1919d: 9
apterus Wittmer, 1935c: 184 **N**: EG
arabicus Wittmer, 1979d: 188 **A**: SA
ardaensis Wittmer, 1985c: 406 **A**: JO
assaensis Evers, 1962: 154 **N**: MO
atlasicus Evers, 1995b: 88 **N**: MO
atriceps Pic, 1917a: 2 **N**: AG
auriculatus Peyerimhoff, 1929b: 109 **N**: AG
barbieri Wittmer, 1984e: 250 **N**: AG
basicollis Fairmaire, 1877a: 141 **E**: SP **N**: AG
  corallifer Fairmaire, 1877a: 141
  ruficollis Uhagón, 1900: 25
berber Wittmer, 1985c: 402 **N**: MO
bolivari Escalera, 1937: 47 **N**: MO
bourgeoisi Abeille de Perrin, 1885a: 15 **N**: AG MO
brevis Erichson, 1840: 128 **E**: IT (Sardegna, Sicilia) SP **N**: AG MO
  glaber Kiesenwetter, 1866a: 371 (*Charopus*)
  nigripes Waltl, 1835: 60 (*Malachius*)
buettikeri Wittmer, 1979d: 187 **A**: SA
canaliculatus Abeille de Perrin, 1890a: 49 **N**: AG MO
capitatus Erichson, 1840: 127 **E**: FR (Corse) PT SP **N**: AG MO
  algiricus Perris, 1873: 6
  corsicus Perris, 1864: 290
  verticalis Erichson, 1840: 128 (*Troglops*)
carbonarius Pic, 1917a: 2 **N**: AG MO
caucasicus Pic, 1917a: 2 **E**: "Caucasus"
cephalotes A. G. Olivier, 1790a: no 27: 12 (*Malachius*) **E**: AU BH CR FR GB IT PL SZ **N**: AG MO **NAR**i
  bicornis A. Costa, 1847a: 103 (*Malachius*)
  corniger Kiesenwetter, 1863a: 729
  cruentus Kiesenwetter, 1863a: 618 (*Troglops*)
  dufouri Perris, 1857: 128
colasi Wittmer, 1968b: 278 **E**: GG **A**: TR
constantini Wittmer, 1985c: 399 **N**: MO
diminutus Abeille de Perrin, 1885a: 14 **E**: FR GR YU **N**: AG
exophthalmus Fairmaire, 1880a: 6 **N**: AG MO
  laminatus Pic, 1917a: 1

fascipennis Reitter, 1896b: 36 **E**: AB AR
ferrugineus Escalera, 1914: 230 **N**: MO
  alluaudi Pic, 1953d: 61
  nigritarsis Escalera, 1914: 231
furcatus Abeille de Perrin, 1885b: 151 **E**: PT SP **N**: AG MO
  barrosi Pic, 1917a: 1 (*Troglops*)
  laticeps Pic, 1922h: 30
ganglbaueri Abeille de Perrin, 1895b: cccliii **A**: SY
gobiensis Pic, 1923o: 372 **A**: MG (Gobi)
hatayensis Wittmer, 1974a: 10 **E**: TR
horridus Escalera, 1914: 229 **N**: MO
  panousei Pic, 1953d: 61
ignoratus Evers, 1989: 97 **E**: SP
incisus Pic, 1917a: 1 **N**: AG
inexpectus Abeille de Perrin, 1900a: 20 **N**: AG
infurcatus Pic, 1912u: 219 **N**: MO
  xauensis Escalera, 1937: 48
italicus Wittmer, 1984e: 245 **E**: IT
khotanensis Hicker, 1935: 292 **A**: UZ
latifrons Peyron, 1877: 235, 238-239 **E**: IT (Sicilia) **N**: AG MO TU
  innotaticollis Pic, 1917a: 1
maculatus Wittmer, 1954e: 315 **A**: YE
marginatus Waltl, 1835: 60 (*Malachius*) **E**: IT PT SP **N**: AG MO TU **A**: CY SY
  aterrimus Allard, 1869: 467 (*Charopus*)
  martini Pic, 1917a: 2
muehlei Wittmer, 1988b: 49 **A**: YE
nasutus Wittmer, 1988b: 50 **A**: YE
olympicus Wittmer, 1988a: 21 **E**: GR
orientalis Abeille de Perrin, 1881: 121 **E**: TR **N**: AG EG **A**: IS SY
  laevissimus Abeille de Perrin, 1890c: 240
palaestinus Pic, 1902e: 31 **N**: EG **A**: IS
  robusticeps Pic, 1924c: 21
pistaciae Evers, 1995b: 90 **N**: MO
punctatulus Abeille de Perrin, 1885a: 16 **E**: PT SP **N**: MO
pyriventris notaticeps Pic, 1917a: 1 **N**: TU
pyriventris pyriventris Abeille de Perrin, 1885a: 15 **N**: AG TU
  theryi Pic, 1917a: 5
rabinovitchi Wittmer, 1934: 450 **N**: EG
rubrifrons J. R. Sahlberg, 1908: 73 (*Pelochrus*) **A**: IS
ruficolor Pic, 1930c: 4 **N**: EG **AFR**
semiinermis Wittmer, 1970c: 74 **E**: TR
silo Erichson, 1840: 127 **E**: FR IT SZ
  impressifrons Ponza, 1805: 76 (*Cantharis*)
sonyae Wittmer, 1982c: 108 **A**: SA
spinifrons Abeille de Perrin, 1894b: 92 **N**: AG
  cultricornis Abeille de Perrin, 1909b: 179
susicus Evers, 1962: 152 **N**: MO
tafraoutensis Wittmer, 1985c: 398 **N**: MO
taurulus Abeille de Perrin, 1900a: 20 **N**: AG
tizintestus Evers, 1995b: 89 **N**: MO
trinotaticollis Pic, 1933h: 102 **N**: MO
vestitus Abeille de Perrin, 1898c: 256 **N**: AG
yusi Pardo Alcaide, 1975a: 89 **E**: SP

superfamily **CUCUJOIDEA** Latreille, 1802
family SPHINDIDAE Jacquelin du Val, 1860
subfamily **Sphindinae** Jacquelin du Val, 1860

genus *Sphindus* Dejean, 1821: 102 type species *Nitidula dubia* Gyllenhal, 1808
Coniophagus Mink, 1853: 58 type species *Coniophagus humeralis* Mink, 1853 (= *Nitidula dubia* Gyllenhal, 1808)
brevis Reitter, 1879d: 202 **A**: FE JA
obesus Kiesenwetter, 1879b: 318
castaneipennis Reitter, 1879d: 201 **A**: JA
dubius Gyllenhal, 1808a: 243 (*Nitidula*) **E**: AB AU AZ BH BY CR CT CZ DE EN FI FR GB GE GR HU IT LT NR NT PL PT RO SK SL SP ST SV SZ UK **N**: AG CI
gyllenhalii Chevrolat, 1833: pl. 6
humeralis Mink, 1853: 59 (*Coniophagus*)
rufipennis Roubal, 1931b: 184
himalayaensis Lesne, 1925: 511 **A**: SD
semirufus Champion, 1922a: 74 **A**: UP **ORR**

genus *Odontosphindus* LeConte, 1878b: 601 type species *Odontosphindus denticollis* LeConte, 1878
grandis Hampe, 1861: 67 (*Sphindus*) **E**: BH CR GR RO SK

subfamily **Aspidiphorinae** Kiesenwetter, 1877

genus *Aspidiphorus* Dejean, 1821: 47 type species *Nitidula orbiculata* Gyllenhal, 1808
Coniporus C. G. Thomson, 1863: 167 type species *Nitidula orbiculata* Gyllenhal, 1808
asiaticus Champion, 1924c: 263 **A**: UP
bhuswargabasi Sengupta & Pal, 1982: 390 **A**: KA
bhutia Sengupta & Pal, 1982: 390 **A**: BT
confusus Reitter, 1902b: 139 (*Aspidophorus*) **E**: AB AR
dravida Sengupta & Pal, 1982: 390 **A**: HP **ORR**
japonicus Reitter, 1879d: 202 (*Aspidophorus*) **A**: FE JA
lareyniei Jacquelin du Val, 1859c: 39 **E**: AB AU BH BU CR CZ FR IT RO SK SL ST
maheswar Pal & Sengupta, 2003: 98 **A**: HP
orbiculatus Gyllenhal, 1808a: 242 (*Nitidula*) **E**: AB AU BE BH BU BY CR CT CZ DE EN FI FR GB GE GG HU IR IT LT NL NR NT PL RO SK SL SP SV ST SZ UK **N**: AG **A**: FE
striatopunctatus Laporte, 1840a: 294 (*Dorcatoma*)
sakaii Sasaji, 1993: 18 **A**: JA
uma Pal & Sengupta, 2003: 100 **A**: HP

family KATERETIDAE Erichson, 1846

genus *Anamartus* Jelínek, 1976: 19 type species *Anamartus sinuatus* Jelínek, 1976
appli Ganglbauer, 1900: 147 (*Amartus*) **E**: AB AR **A**: LE TR
aurosericeus Reitter, 1873d: 169 (*Brachypterus*) **A**: IS JO LE SY TR
dilutitarsis Solsky, 1876: 247 (*Brachypterus*) **A**: AF TD TM UZ
incognitus Jelínek, 1976: 25 **E**: GG **A**: IS JO LE SY TR
jelineki Audisio, 1980: 21 **A**: IS JO
olexai Jelínek, 1976: 23 **E**: AR **A**: TR
opacus Abeille de Perrin, 1892: 66 (*Brachypterus*) **A**: IS JO LE SY
sinuatus Jelínek, 1976: 29 **A**: TR
strobli Reitter, 1885i: 376 (*Amartus*) **A**: IQ IS JO SY TR
testaceimembris Pic, 1919b: 5 (*Amartus*)

genus *Brachyleptus* Motschulsky, 1845a: 54 type species *Brachyleptus canescens* Motschulsky, 1845 (= *Brachypterus quadratus* Sturm, 1844)
algiricus Grouvelle, 1912c: 97 **N**: AG MO TU **A**: IS JO SI
argenteolus Reitter, 1896h: 295 **A**: AF KI KZ TD UZ

auripubens Reitter, 1896h: 294 **A**: IS JO SY TR
notativentris Reitter, 1902e: 188
aurosus Reitter, 1885i: 377 **E**: AB AR BU GR ST **N**: EG **A**: AF IN IQ IS JO LE SY TM TR
reitteri Ganglbauer, 1899a: 457
tomentiventris Reitter, 1896h: 294
bicoloratus Reitter, 1896h: 295 **A**: AF KZ TD TM UZ
discolor Reitter, 1896h: 295 **E**: AB AR GG ST **A**: CY IQ IS JO LE SY TR
rufipennis Pic, 1919b: 5
papaveris Grouvelle, 1912c: 96 **N**: AG MO TU
quadratus Sturm, 1844: 19 (*Brachypterus*) **E**: AB AR AU BH BU CR GG GR HU MC RO SK ST TR UK YU **A**: CY IQ IS JO LE SY TR
canescens Motschulsky, 1845a: 54

genus *Brachypterolus* Grouvelle, 1913d: 387 [RN] type species *Catheretes gravidus* Illiger, 1798 (= *Dermestes pulicarius* Linnaeus, 1758)
Heterostomus Jacquelin du Val, 1858: 138 [HN] type species *Catheretes gravidus* Illiger, 1798 (= *Dermestes pulicarius* Linnaeus, 1758)
antirrhini Murray, 1864: 248 (*Brachypterus*) **E**: AL AU AZ BE BU CZ FR GB GE GR HU IT LS NL PL PT RO SK SP SZ TR UK **N**: AG LB MO MR TU **A**: CY IS JO KZ LE SI SY TR
plumbeus Abeille de Perrin, 1892: 67 (*Brachypterus*)
villiger Reitter, 1885h: 316 (*Brachypterus*)
cinereus Heer, 1841: 413 (*Cateretes*) **E**: FR PT SP SZ **N**: MO
rufilabris Reitter, 1919: 14
cylindricus Normand, 1936b: 82 **N**: TU
drusus Abeille de Perrin, 1892: 67 (*Brachypterus*) **A**: LE
linariae Stephens, 1830: 53 (*Cateretes*) **E**: AU BH BU CT DE EN FI FR GB GE GG GR HU IT LA MC NL NR NT PL RO SP ST SV YU **A**: ES FE MG TR
autumnalis Gistel, 1857: 574 (*Brachypterus*)
cornelii Spornraft, 1966: 16 [RN]
dilutipes Reitter, 1909e: 303 (*Heterostomus*)
immundus Reitter, 1899b: 199 (*Brachypterus*)
linariae Cornelius, 1863: 113 (*Brachypterus*) [HN]
longulus Reitter, 1885i: 377(*Heterostomus*) **E**: FR PT SP **N**: AG MO
pumilio Reitter, 1919: 14
pumilus Rey, 1889c: 3 (*Brachypterus*)
nanulus Reitter, 1919: 14 **A**: IS
pulicarius Linnaeus, 1758: 357 (*Dermestes*) **E**: AU AZ BH BU BY CR CT DE EN FI FR GB GE HU IT LA LT NL NR NT PL ST SV SZ UZ YU **A**: ES FE JA KZ MG NW SC TR WS "Korea" **NAR**
gravidus Illiger, 1798: 395 (*Catheretes*)
flaviclavis Rey, 1889: 51 (*Brachypterus*)
fuscopubens Reitter, 1919: 15
laticollis Küster, 1848c: 35 (*Brachypterus*)
lepturoides Scopoli, 1793: 17 (*Dermestes*)
mordelloides Notman, 1920: 29 (*Heterostomus*)
scutellatus Panzer, 1792c: 11 (*Scaphidium*)
shimoyamai Hisamatsu, 1985: 177 **A**: JA
vestitus Kiesenwetter, 1850: 223 (*Brachypterus*) **E**: AU CZ FR GB GE IT SK SP SZ **N**: MO
obtusus Guillebeau, 1886: 2 (*Brachypterus*)

genus *Brachypterus* Kugelann, 1794: 560 type species *Dermestes urticae* Fabricius, 1792
Heterobrachium Wollaston, 1864: 108 type species *Heterobrachium longimanum* Wollaston, 1864
Parabrachypterus Hisamatsu, 1976: 19 type species *Parabrachypterus yukikoae* Hisamatsu, 1976
Virbius Gozis, 1886: 11 [RN] type species *Dermestes urticae* Fabricius, 1792
aeneomicans Wollaston, 1865, Appendix: 16 **N**: CI (Gomera, Gran Canaria, Tenerife)
pallens Har. Lindberg, 1953: 8
curtulus Wollaston, 1864: 110 **E**: IT PT SP **N**: AG CI (Fuerteventura, Lanzarote) MO TU
nigriclavis Guillebeau, 1896b: 50

fulvipes Erichson, 1843b: 231 **E**: AU BE BY CR CZ EN FR GE HU LT NL PL RO SK SL ST SV SZ UK **A**: ES FE MG TR WS
sibiricus Csiki, 1901b: 106
glaber Newman, 1834: 200 (*Catheretes*) **E**: AB AZ BH BU BY CR DE EN FI FR GB GE GR HU IR IT LA LT MA MC NL NR PL SP SV TR **N**: AG MO TU **A**: CY TR
ater Motschulsky, 1849: 93
collinus Gistel, 1857: 590
lucasii Murray, 1864: 244 [RN]
meridionalis Küster, 1848c: 38
pubescens Erichson, 1843b: 231
pubescens P. H. Lucas, 1846: 219 (*Cercus*) [HN]
unicolor Küster, 1849b: 29
horii Hisamatsu, 1976: 22 (*Parabrachypterus*) **A**: TAI
kaszabi Jelínek, 1965a: 136 **A**: MG
kraatzi Grouvelle, 1903a: 108 (*Heterhelus*) **A**: SD
labiatus Erichson, 1843b: 232 **E**: FR IT (Sardegna) PT SP **N**: AG LB MO MR TU
metallescens Schaufuss, 1869: 13 (*Cercus*)
pallipes Murray, 1864: 243
sydowi Schilsky, 1910: 100
longicollis Jelínek, 1978a: 176 **A**: BT
longimanus Wollaston, 1864: 109 (*Heterobrachium*) **N**: CI (Gomera, La Palma, Tenerife)
proximus Kirejtshuk, 1989b: 148 **A**: FE
rotundicollis Murray, 1864: 244 **E**: AL BH BU GR MC YU **A**: CY IS JO LE SY TR
urticae Fabricius, 1792a: 235 (*Dermestes*) **E**: AB AL AR AU BH BU BY CR CT CZ DE EN FI FR GB GE GG GR HU IR IT LA LT MC NL NR NT PL RO SK SL SP ST SV SZ TR UK YU **A**: ES FE JA KI KZ NO SC TR WS "Korea" **NAR**
abbreviatus Herbst, 1792: 190 (*Strongylus*)
affinis Heer, 1841: 411 (*Cateretes*)
erythropus Marsham, 1802: 132 (*Nitidula*)
flavicornis Küster, 1848c: 40
pelusius Gistel, 1857: 582
pusillus Melsheimer, 1844: 105 (*Cercus*)
solaris Gistel, 1857: 577
velatus Wollaston, 1863: 217 **N**: CI (Gomera, Hierro, Tenerife)
viridinitens Har. Lindberg, 1951: 6 **N**: CI
yukikoae Hisamatsu, 1976: 20 (*Parabrachypterus*) **A**: TAI

genus *Heterhelus* Jacquelin du Val, 1858: 137 type species *Cercus sambuci* Erichson, 1843 (= *Cateretes scutellaris* Heer, 1841)
Boreades Parsons, 1943: 139 type species *Cercus abdominalis* Erichson, 1843
angusticollis Reitter, 1878j: 167 (*Amartus*) **A**: JA
scutellaris Heer, 1841: 412 (*Cateretes*) **E**: AU BH BY CR CZ DE EN FI FR GE HU IT LA LS MC NL NR NT PL SK SV SZ UK **A**: ES FE JA MG "Korea"
japonicus Reitter, 1878j: 166 (*Amartus*)
longipennis Murray, 1864: 234 (*Cercus*)
morio Reitter, 1878j: 166 (*Amartus*)
sambuci Erichson, 1843b: 229 (*Cercus*)
solani Heer, 1841: 412 (*Cateretes*) **E**: AU BH BU CZ FR GE HU NL PL RO SK SL SP SZ UK YU **A**: ES FE JA MG SC
rhenanus Bach, 1866: 480 (*Cercus*)
rubiginosus Erichson, 1843b: 232 (*Brachypterus*)
spireae Märkel, 1857: 177 (*Cercus*)

genus *Kateretes* Herbst, 1793: 11 type species *Dermestes pedicularius* Linnaeus, 1758
Ammalaptus Gistel, 1856: 362 type species *Cercus rufilabris* Latreille, 1807
Anisocera Stephens, 1835: 438 type species *Anisocera spireae* Stephens, 1835 (= *Dermestes pedicularius* Linnaeus, 1758)

Anomaeocera Shuckard, 1840: 25 type species *Anisocera spireae* Stephens, 1835 (= *Dermestes pedicularius* Linnaeus, 1758)
Cateretes Illiger, 1798: 306 type species *Dermestes pedicularius* Linnaeus, 1758
Catheretes Gyllenhal, 1808a: 245 type species *Dermestes pedicularius* Linnaeus, 1758 (=*Dermestes pedicularius* Linnaeus, 1758)
Cercus Latreille, 1797: 68 type species *Cercus pedicularius* Latreille, 1804 (=*Dermestes pedicularius* Linnaeus, 1758)
Pulion Gozis, 1886: 12 type species *Cercus rufilabris* Latreille, 1807
dalmatinus Sturm, 1844: 12 (*Cercus*) **E**: AB AU CR GR HU IT **A**: IN IS TR
testaceus Murray, 1864: 236 (*Cercus*)
flavicans Fairmaire, 1860: 166 (*Cercus*) **N**: AG MO TU
japonicus Hisamatsu, 1985: 177 **A**: JA (Honshu)
mixtus Kirejtshuk, 1989b: 145 **E**: HU RO SK UK **A**: KZ
ornatus Jelínek, 1978a: 175 (*Cateretes*) **A**: HP UP
pedicularius Linnaeus, 1758: 357 (*Dermestes*) **E**: AU BY CR CT CZ DE EN FI FR GB GE HU IR IT LA LS LT MC NL NR NT PL SK SL SP ST SV SZ UK YU **A**: WS
alandicus Leinberg, 1900: 186 (*Cateretes*)
arundinis Laporte, 1840b: 12 (*Cercus*)
glaberrimus Paykull, 1798: 287 (*Dermestes*)
nectarineus Gistel, 1857: 578 (*Cercus*)
nigriventris Leinberg, 1900: 186 (*Cateretes*)
pallens Rey, 1889c: 3 (*Cercus*)
scutellaris Leinberg, 1900: 186 (*Cateretes*)
similis Marsham, 1802: 127 (*Silpha*)
spireae Stephens, 1835: 438 (*Anisocera*)
truncatus Fabricius, 1792a: 261 (*Nitidula*)
unicolor Marsham, 1802: 127 (*Silpha*)
pusillus Thunberg, 1794: 69 (*Nitidula*) **E**: AU CR CT EN FI FR GB GE HU IR IT LA LT NR NT PL ST SV UK YU **A**: ES FE WS **NAR**
bipustulatus Paykull, 1798: 286 (*Dermestes*) [HN]
inglorius Gozis, 1882: 196 (*Cercus*)
lapponicus Leinberg, 1900: 185, 187 (*Cateretes*)
niger Seidlitz, 1888a: 208 (*Cercus*)
nigricollis Leinberg, 1900: 185, 187 (*Cateretes*)
ochraceus Murray, 1864: 233 (*Cercus*)
punctulatus Marsham, 1802: 126 (*Silpha*)
scissus Parsons, 1943: 138 (*Cateretes*)
suturalis Murray, 1864: 233 (*Cercus*)
tripunctatus Illiger, 1798: 387 (*Nitidula*)
uliginosus Gistel, 1857: 591 (*Cercus*)
rufilabris Latreille, 1807a: 16 (*Cercus*) **E**: AU BH CR CZ DE FR GB GE GR HU IR IT LA LS MA NL PL RO SK SL SP SV SZ TR UK YU **N**: AG MO TU **A**: TR
caricis Stephens, 1830: 54 (*Cateretes*)
confusus Méquignon, 1948: 77 (*Cateretes*)
fulvus Erichson, 1843b: 232 (*Cercus*)
junci Stephens, 1830: 54 (*Cateretes*)
pallidus Heer, 1841: 411 (*Cateretes*)
rubicundus Heer, 1841: 411 (*Cateretes*)

genus *Platamartus* Reitter, 1892b: 151 type species *Platamartus jakowlewi* Reitter, 1892
jakowlewi Reitter, 1892b: 151 **A**: ES
solskyi Kirejtshuk, 1984b: 517 **A**: FE

genus *Sibirhelus* Kirejtshuk, 1989b: 147 type species *Heterhelus corpulentus* Reitter, 1900
corpulentus Reitter, 1900g: 231 (*Heterhelus*) **A**: ES FE JA

family Kateretidae, nomen dubium
Silpha histeroides Thunberg, 1784a: 8 **E**: SV

family NITIDULIDAE Latreille, 1802

subfamily **Amphicrossinae** Kirejtshuk, 1986

genus *Amphicrossus* Erichson, 1843b: 346 type species *Nitidula ciliata* A. G. Olivier, 1811
Cametis Motschulsky, 1863: 440 type species *Cametis picea* Motschulsky, 1863
Lobostoma Fairmaire, 1892a: 90 [HN] type species *Lobostoma picea* Fairmaire, 1892 (= *Amphicrossus accidentus* Kirejtshuk, 1997)
Nitidopecten Reichensperger, 1913: 199 type species *Nitidopecten comes* Reichensperger, 1913 (= *Amphicrossus parallelus* Grouvelle, 1912)
Rhacostoma Berg, 1898: 18 [RN] type species *Lobostoma picea* Fairmaire, 1892 (= *Amphicrossus accidentus* Kirejtshuk, 1997)
discolor Erichson, 1843b: 348 **A**: SD SE TAI **ORR**
hisamatsui Jelínek, 1993: 36 **A**: FE JA
japonicus Reitter, 1873d: 100 **A**: FE HP JA TAI ZHE "Korea" **ORR**
korschefskyi Bollow, 1941: 236
nigrinus Bollow, 1941: 235
lewisi Reitter, 1873d: 100 **A**: JA SCH TAI "Korea" **ORR**
subvittatus Heller, 1923: 68

subfamily **Epuraeinae** Kirejtshuk, 1986

genus *Amystrops* Grouvelle, 1906c: 312 type species *Amystrops modiglianii* Grouvelle, 1906
Haptognathus Gillogly, 1962: 160 type species *Haptognathus pacificus* Gillogly, 1962
Platychorinus Grouvelle, 1906b: 201 type species *Platychorinus dilutus* Grouvelle, 1906
Platychoropsis Grouvelle, 1913d: 398 [RN] type species *Platychorinus dilutus* Grouvelle, 1906
Propetes Reitter, 1875e: 62 type species *Epuraea nigripennis* L. Redtenbacher, 1868
nigripennis L. Redtenbacher, 1868: 34 (*Epuraea*) **A**: JA (Ryukyus) TAI **ORR**
epuraeoides Grouvelle, 1914: 38 (*Amystrops*)
formosianus Grouvelle, 1914: 39 (*Amystrops*)
montanus Grouvelle, 1913a: 103 (*Amystrops*)
remotus Reitter, 1873d: 178 (*Haptoncus*)

genus *Carpocryraea* Kirejtshuk, 1998: 358 type species *Carpophilus familiaris* Grouvelle, 1897
familiaris Grouvelle, 1897: 353 (*Carpophilus*) **A**: SD **ORR**

genus *Epuraea* Erichson, 1843b: 267 type species *Nitidula silacea* Herbst, 1784
subgenus *Aphenolia* Reitter, 1884g: 259 [= 1885c: 16] type species *Aphenolia pseudosoronia* Reitter, 1884
pseudosoronia Reitter, 1884g: 262 [= 1885c: 16] (*Aphenolia*) **A**: FE JA "Korea"
subgenus *Epuraea* Erichson, 1843b: 267 type species *Nitidula silacea* Herbst, 1784
Dadopora C. G. Thomson, 1859: 68 type species *Nitidula decemguttata* Fabricius, 1792 (= *Nitidula guttata* A. G. Olivier, 1811)
acea Kirejtshuk, 1998: 122 **A**: UP
acelsa Kirejsthuk, 1998: 125 **A**: UP
aestiva Linnaeus, 1758: 574 (*Silpha*) **E**: AB AL AN AR AU AZ BE BH BU BY CR CT CZ DE EN FI FR GB GE GG GR HU IR IT LA LS LT LU MC MD NL NR NT PL PT RO SK SL SP ST SV SZ TR UK YU **A**: IN JA TR WS **NAR**
bisignata Sturm, 1844: 80
convexiuscula Mannerheim, 1843b: 255 (*Nitidula*)
depressa Illiger, 1798: 386 (*Nitidula*)
grandiclava Roubal, 1939: 6
obsoleta Herbst, 1793: 240 (*Nitidula*) [HN]
ochracea Sturm, 1844: 54
testacea A. G. Olivier, 1790b: 160 (*Anthribus*)
vaccinii Gistel, 1857: 542 (*Nitidula*)
villosa Thunberg, 1794: 70 (*Nitidula*)
alpicola Nakane, 1966b: 65 **A**: JA
angustula Sturm, 1844: 74 **E**: AU BE BH BU BY CR CT CZ DE EN FI FR GB GE GR HU IR IT LA LT MC MD NL NR NT PL RO SK ST SL SP SV SZ UK YU **A**: ES FE TR WS
parallela Reitter, 1875a: 6

apposita Reitter, 1884g: 302 **A**: FE JA
argus Reitter, 1894c: 25 **E**: AU CR CT MD PL RO SL SZ UK **A**: ES FE JA NE WS
argoides Sjöberg, 1939b: 118
nobilis Reitter, 1894c: 26
biguttata Thunberg, 1784a: 9 (*Silpha*) **E**: AL AN AU AZ BE BH BU BY CR CT CZ DE EN FI FR GB GE GR HU IT LA LS LT LU MC MD NL NR NT PL SK SL SP ST SV SZ TR UK YU
bipunctata Heer, 1841: 398 (*Nitidula*) [HN]
binotata Reitter, 1873a: 15 **E**: AU BE BH BU BY CR CT CZ DE EN FI FR GB GE GR HU IT LA LT MC MD NR NT PL RO SK SL ST SV SZ UK YU **A**: ES MG WS
nana Reitter, 1873d: 19
birmanica Grouvelle, 1892c: 840 **A**: BT FUJ SD TAI **ORR**
andrewesi Grouvelle, 1908c: 350
arcuata Grouvelle, 1908c: 350
zurstrasseni Kirejtshuk, 1987a: 67
boreella Zetterstedt, 1828: 152 (*Nitidula*) **E**: AL AU BE BH BU BY CR CT CZ EN FI FR GE GR HU IT LA LS LT MC MD NR NT PL RO SK SL ST SV SZ UK YU **A**: ES FE TR WS **NAR**
brunnescens Motschulsky, 1860b: 128
nigra Mäklin, 1853: 204
picea Mäklin, 1853: 205
cameroni Kirejsthuk, 1998: 137 **A**: SD
carpathica Reitter, 1878a: 51 **E**: RO ST UK **A**: FE JA
quadrimaculata Reitter, 1884g: 260
championi Kirejtshuk, 1998: 141 **A**: UP
compacta Kirejtshuk, 1998: 143 **A**: NP ORR
contractula J. R. Sahlberg, 1889: 102 **E**: CT FI NR NT **A**: ES FE (Kamchatka, Magadan)
helleni Sjöberg, 1939b: 117
wuorentausi Sjöberg, 1939b: 119
contraria Kirejtshuk, 1998: 146 **A**: SD
cribrata Grouvelle, 1903a: 111 **A**: SD
curtula Reitter, 1894c: 25 **A**: ES FE MG
cyclops Jelínek, 1978a: 185 **A**: BT NP SD **ORR**
decolor Reitter, 1884g: 261 **A**: FE JA
densepunctata Nakane & Hisamatsu, 1967: 67 **A**: JA
deplanata Motschulsky, 1860b: 127 **A**: ES
deubeli Reitter, 1898h: 340 **E**: AU CT CZ DE EN FI GE NR NT PL RO SK SV UK **A**: WS
lysholmi Munster, 1933: 275
distincta Grimmer, 1841: 38 (*Nitidula*) **E**: AU BH BY CR CT CZ FI FR GB GE HU IT LA MD NL PL RO SK SL ST SZ UK **A**: ES FE JA KZ
polyporina Gistel, 1857: 542 (*Nitidula*)
domina Reitter, 1873d: 31 **A**: JA
obnoxia Reitter, 1873d: 32
drapeta Reitter, 1909c: 101 **A**: IN TD TM UZ
durula Reitter, 1894c: 25 **A**: ES FE
durula Reitter, 1919: 66 [HN]
sibirica J. R. Sahlberg, 1903a: 17 (*Omosiphora*)
excisicollis Reitter, 1873a: 18 **E**: DE GE NT
danica Sjöberg, 1939b: 115
fageticola Audisio, 1991: 520 [RN] **E**: AB AL AU BE BH BU BY CR CT CZ DE EN FR GE GR HU IT LA LT MC MD NL NT PL RO SK SL SP ST SZ UK YU
castanea Duftschmid, 1825: 135 (*Nitidula*) [HN]
foveicollis Reitter, 1873d: 33 **A**: JA
funeraria Reitter, 1884g: 260 **A**: FE JA NC SE
fungicola Hisamatsu, 1968b: 29 **A**: FE JA
fuscicollis Stephens, 1835: 406 (*Nitidula*) **E**: AL AN AU BE BH BU CR CT CZ DE EN FR GB GE GR HU IT LU MC MD NL PL PT RO SK SL SP ST SV SZ TR UK YU **N**: AG MO TU **A**: CY IN
diffusa C. N. F. Brisout de Barneville, 1863: 46

guttifera Reitter, 1887c: 83
ragusae Reitter, 1906c: 219
fussi Reitter, 1875a: 7 **E**: AU BY CT EN FI NR NT SV **A**: ES MG
interjecta Sjöberg, 1939b: 113
guttata A. G. Olivier, 1811: 214 (*Nitidula*) **E**: AB AL AN AR AU BE BH BU BY CR CT CZ DE EN FI FR GB GE GG GR HU IT LA LS LT LU MC MD NL NR NT PL PT RO SK SL SP ST SV SZ TR UK YU **A**: IN TR
decemguttata Fabricius, 1792a: 258 (*Nitidula*) [HN]
hilleri Reitter, 1877f: 109 **E**: CT FI NT **A**: FE JA MG WS
concurrens Sjöberg, 1939a: 98
hisamatsui Nakane, 1966b: 66 **A**: JA "Korea"
intercalatis Kirejtshuk, 1995a: 6 **A**: FE JA
kaszabi Kirejtshuk, 1992: 157 **E**: CT **A**: FE JA
korotyaevi Kirejtshuk, 1992: 154 **A**: FE (Kurily Is.)
laeta Kirejtshuk, 1998: 155 **A**: SD
laeviuscula Gyllenhal, 1827: 299 (*Nitidula*) **E**: AU BY CT CZ DE EN FI FR GE IT LA LT NR NT PL RO SK SL SV SZ UK **A**: FE WS
laferi Kirejtshuk, 1992: 143 **A**: FE
latipes Grouvelle, 1896c: 278 **N**: AG MO **A**: TR (Toros Dağlari)
latissima Reitter, 1880c: 42 **A**: UP **ORR**
linearis Mäklin, 1853: 205 **A**: FE (Magadan) **NAR**
longiclavis Sjöberg, 1939b: 116 **E**: AU BY CT CZ DE EN FI FR GE IT LA LS LT NR NT RO SK SL SV SZ UK
longipennis Sjöberg, 1939b: 113 **E**: NR NT SV **A**: ES
dolosa Kirejtshuk, 1995b: 279
longiungulata Kirejtshuk, 1998: 160 **A**: NP **ORR**
longula Erichson, 1845: 154 **E**: AL AN AU AZ BE BH BU BY CR CT CZ DE EN FI FR GB GE GG GR HU IR IT LA LS LT LU MC MD NL PL RO SK SL SP ST SV SZ TR UK YU **A**: JO TR
apicipennis Reitter, 1909c: 101 [RN]
erichsonis Reitter, 1894c: 30 [HN]
ornata Reitter, 1873a: 16
subdepressa Rey, 1889c: 4
marseuli Reitter, 1873a: 20 **E**: AB AL AN AU BE BH BU BY CR CT CZ DE EN FI FR GB GE GG GR HU IR IT LA LS LT LU MC MD NL NR NT PL PT RO SK SL SP ST SV SZ TR UK YU **A**: ES IN MG TR
acuta Biström, 1977: 17 [RN
bickhardti Sainte-Claire Deville, 1906: 294
lenkorana Méquignon, 1945: 20
pusilla Illiger, 1798: 386 (*Nitidula*) [HN]
melina Erichson, 1843b: 269 **E**: AL AN AU BE BH BU BY CR CT CZ DE EN FI FR GB GE GG GR HU IR IT LA LS LT LU MC MD NL NR NT PL RO SK SL SP ST SV SZ TR UK YU **A**: TR WS
mestsheryakovi Kirejtshuk, 1992: 137 **A**: FE
muehli Reitter, 1908g: 245 **E**: AU BE BH BY CR CT CZ DE EN FI GE HU IT LA NR NT PL RO SK SL SV SZ UK **A**: FE TR
nepalica Kirejtshuk, 1998: 164 **A**: NP
oblonga Herbst, 1793: 245 (*Nitidula*) **E**: AU BY CT EN FI FR GE IT LA LT NR NT RO SK SL SV SZ UK **A**: ES FE JA "Korea" **NAR**
parilis Reitter, 1873d: 37
truncatella Mannerheim, 1846: 514
opalizans J. R. Sahlberg, 1889: 104 **E**: BY DE EN FI NR NT SV
rugulosa J. R. Sahlberg, 1889: 105
pallescens Stephens, 1835: 406 (*Nitidula*) **E**: AL AN AU BE BH BU BY CR CT CZ DE EN FI FR GB GE GG GR HU IR IT LA LS LT LU MC MD NL NR NT PL PT RO SK SL SP ST SV SZ TR UK YU **A**: AF ES FE JA KZ MG TR WS "Korea"
abietina J. R. Sahlberg, 1889: 108
florea Erichson, 1845: 155
pellax Reitter, 1873d: 33 **A**: JA

placida Mäklin, 1853: 203 **E**: AU BH BY CT DE EN FI IT LA LT NR NT RO SV SZ UK **A**: ES FE GAN MG **NAR**
incompleta Motschulsky, 1860b: 126
lapponica Reitter, 1894c: 24 [RN]
opacula Poppius, 1904: 3
rogalandica Helliesen, 1913: 5
polina Kirejtshuk, 1987a: 66 **A**: FUJ SD
propria Kirejtshuk, 1998: 173 **A**: SD
pseudorapax Kirejtshuk, 1995a: 8 **A**: FE JA
pumila Grouvelle, 1897: 359 **A**: BT HP NP PA SD UP
apicalis Grouvelle, 1903a: 110
pygmaea Gyllenhal, 1808a: 225 (*Nitidula*) **E**: AL AU BE BH BU BY CR CT CZ DE EN FI FR GE GR IT LA LS LT MC MD NL NR NT PL RO SK SL ST SV SZ TR UK YU **A**: ES FE TR
planidorsis Reitter, 1894c: 32
rubromarginata Reitter, 1873a: 20
quadrangula Motschulsky, 1860b: 127 **A**: ES FE
rapax Reitter, 1884g: 261 [= 1885c: 15] **A**: FE JA
reichardti Sjöberg, 1939b: 118 **A**: FE
riedeli Kirejtshuk, 1998: 178 **A**: HP NP UP
rifensis Audisio, 1983: 119 **E**: SP **N**: AG MO
rubronotata Reitter, 1873d: 37 **A**: FE JA
rufobrunnea Sjöberg, 1939b: 111 **E**: CT FI NT SK **A**: FE WS
rufomarginata Stephens, 1830: 41 (*Nitidula*) **E**: AU BE BH BU BY CR CT CZ DE EN FI FR GB GE GR HU IR IT LA LT MC MD NL NR NT PL RO SK SL ST SV SZ UK YU **A**: FE JA MG WS **NAR**
boreades Parsons, 1967: 736
inaequalis Rey, 1889c: 4
limbata C. R. Sahlberg, 1820b: 74 (*Nitidula*) [HN]
parvula Sturm, 1844: 67
testacea Rey, 1889c: 4 [HN]
sengleti Audisio, 1991: 518 **A**: IN
silacea Herbst, 1783: 36 (*Nitidula*) **E**: AL AU BE BH BU BY CR CT CZ DE EN FI FR GB GE GG GR HU IR IT LA LT MC MD NL NR NT PL RO SK SL ST SV SZ UK YU **A**: IN MG WS
deleta Sturm, 1844: 56
fagi C. N. F. Brisout de Barneville, 1874: lxxi
laricina Motschulsky, 1860b: 127
silesiaca Reitter, 1873a: 9 **E**: BY CT CZ EN FI LT NT PL SV UK **A**: ES WS
similis Reitter, 1873d: 32 **A**: JP
simplissima Kirejtshuk, 1998: 181 **A**: NP
sinicola Jelínek, 1980: 93 **A**: SCH
subnitida Kirejsthuk, 1998: 184 **A**: NP
subparallela Grouvelle, 1896c: 278 **A**: TR (Nur Dağlari)
tenuis Jelínek, 1978a: 179 **A**: NP PA
terminalis Mannerheim, 1843c: 95 (*Nitidula*) **E**: AL AU BE BH BU BY CR CT CZ DE EN FI FR GB GE GR HU IR IT LA LT MC MD NL NR NT PL PT RO SK SL SP ST SV SZ UK YU **A**: CH ES JA MG TR WS **NAR**
adumbrata Mannerheim, 1852a: 336
immunda Sturm, 1844: 59
infuscata Mäklin, 1853: 206
nigricans Schilsky, 1892: 196
seidlitzi Schilsky, 1893: 354
tenenbaumi Sjöberg, 1939b: 115
thoracica Tournier, 1872a: 250 **E**: AU BH BU BY CR CT CZ DE EN FI FR GB GE GR HU IR IT LA LT MC MD NL NR NT PL RO SK SL ST SV SZ UK YU **A**: ES FE TR
sericata Reitter, 1873d: 21
suturalis Reitter, 1873d: 22
titana Kirejtshuk, 1998: 188 **A**: SCH

unicolor A. G. Olivier, 1790a: no 12: 17 (*Nitidula*) **E**: AB AL AN AU AZ BE BH BU BY CR CT CZ DE EN FI FR GB GE GG GR HU IR IT LA LS LT LU MC MD NL NR NT PL PT RO SK SL SP ST SV SZ TR UK YU **N**: AG CI MO MR **A**: CY ES FE JA KZ MG TR WS

heeri Tournier, 1872b: 439
maculata Dalla Torre, 1879: 87
marginata Dalla Torre, 1879: 87
obsoleta Fabricius, 1792a: 256 (*Nitidula*)
subangulata Motschulsky, 1860b: 127
trapezicollis Motschulsky, 1860b: 127
x-rubrum J. R. Sahlberg, 1911: 49

variegata Herbst, 1793: 245 (*Nitidula*) **E**: AB AL AU BE BH BU BY CR CT CZ DE EN FI FR GB GE GR HU IT LA LS LT MC MD NL NR NT PL RO SK SL SP ST SV SZ TR UK YU **A**: ES

carinthiaca Hölzel, 1960: 34
fennica J. R. Sahlberg, 1889: 103
monochroa Reitter, 1894c: 24
nigropunctata Reitter, 1919: 65
variabilis Reitter, 1894c: 24

vicaria Kirejtshuk & Kvamme, 2001: 411 **A**: FE (Sakhalin)
waterhousei Grouvelle, 1908c: 348 **A**: "NW Himalaya"

subgenus *Epuraeanella* Crotch, 1874b: 76 type species *Omosita castanea* Melsheimer, 1844 (= *Epuraea nearctica* Kirejtshuk & Kvamme, 2001)

Omosiphora Reitter, 1875e: 56 type species *Nitidula rufa* Say, 1825

amurensis Kirejtshuk, 1992: 132 **A**: FE MG
fossicollis Grouvelle, 1908c: 347 **A**: NP PA SD
georgica Reitter, 1877h: 296 (*Omosiphora*) **E**: AR GG
hammondi Kirejtshuk, 1992: 130 **A**: SHA YUN
limbata Fabricius, 1787a: 52 (*Nitidula*) **E**: AL AN AU BE BH BU BY CR CT CZ DE EN FI FR GB GE GR HU IT LA LT LU MD NL NR NT PL RO SK SL SP ST SV SZ TR UK YU **A**: WS

skalitzkyi Reitter, 1875e: 57 (*Omosiphora*)
testaceidorsis Roubal, 1936a: 138

martensi Kirejtshuk, 1998: 291 [RN] **A**: NP SD UP

ornata Grouvelle, 1903a: 112 (*Omosita*) [HN]
ornatula Kirejtshuk, 1984a: 178 [HN]

neglecta Heer, 1841: 396 (*Nitidula*) **E**: AL AU BE BH BU BY CR CT CZ DE EN FI FR GB GE HU IT LA LS LT LU MC MD NL NR NT PL RO SK SL SP ST SV SZ UK YU **A**: ES FE TR TR

convergens Reitter, 1894c: 23

nigerrima Kirejtshuk, 1998: 294 **A**: YUN **ORR**
nikitskyi Kirejtshuk, 1992: 129 **A**: FE

subgenus *Haptoncurina* Jelínek, 1977: 376 type species *Epuraea angustula* Motschulsky, 1863 (= *Epuraea motschulskii* Reitter, 1873)

motschulskii Reitter, 1873d: 29 [RN] **A**: BT CH FUJ JA NC SCH SD TAI UP ZHE **AFR ORR NAR**i

angustula Motschulsky, 1863: 439 [HN]
insularis Grouvelle, 1906c: 319 (*Haptoncus*)
rhombotela Gillogly, 1982: 286 (*Haptoncus*)
weisei Grouvelle, 1909b: 132

paulula Reitter, 1873d: 36 **A**: FUJ HEI HEN JA NC TAI **ORR**
reflexicollis Motschulsky, 1863: 438 **A**: UP **ORR**

subgenus *Haptoncus* Murray, 1864: 401 type species *Haptoncus tetragonus* Murray, 1864 (= *Epuraea ocularis* Fairmaire, 1849)

Haptoncura Reitter, 1875e: 61 type species *Epuraea luteola* Erichson, 1843

concolor Murray, 1864: 402 (*Haptoncus*) **A**: CH JA (Tokara) NP PA TAI UP "Korea" **AFR AUR NAR NTR ORR**

mellitula Reitter, 1873d: 40
minuta Reitter, 1873d: 40 [HN]
munda Sharp, 1878: 139 (*Haptoncus*)
nitescens Grouvelle, 1897: 362 (*Haptoncura*)
perminuta Reitter, 1919: 67 [RN]

uenoi Nakane, 1959a: 56 (*Haptoncus*)
fallax Grouvelle, 1897: 361 (*Micrurula*) **A**: FE HP JA NE SD SE UP TAI "Korea" **AUR ORR**
opaca Grouvelle, 1897: 362 (*Haptoncura*) [HN]
opacula Grouvelle, 1906c: 317 (*Haptoncus*)
murrayi Grouvelle, 1906c: 318 (*Haptoncus*)
epuraeoides Grouvelle, 1908c: 344 (*Haptoncus*)
luteola Erichson, 1843b: 272 **E**i: AL AN AZ BH BU CR FR GE GR IT MA MC PT RO SL SP TR UK YU **N**i: CI MR **A**: AE AFi ANH CYi FUJ GUA HEN HUB HUN ISi IQi JA JIX NP PA SA SCH SD TAI TRi UP ZHE "Korea" **COS**
albertisi Reitter, 1880f: 455 (*Haptoncus*)
floreola Sharp, 1890: 305 (*Haptoncus*)
intendens Walker, 1858: 206 (*Nitidula*)
paupercula Reitter, 1873d: 179 (*Haptoncus*)
pubescens Murray, 1864: 403 (*Haptoncus*)
submaculata Walker, 1859a: 52 (*Nitidula*)
subquadrata Reitter, 1877e: 22 (*Haptoncura*)
texana Crotch, 1874b: 76
testacea Murray, 1864: 403 (*Haptoncus*)
vulpecula L. Redtenbacher, 1868: 34
ocularis Fairmaire, 1849b: 363 **E**i: AU FR GE IT PT SL SP SZ **N**i: CI **A**: AP JA TAI TRi "Korea" **COS**
barbula Gillogly, 1962: 172 (*Haptoncus*)
bifasciata Kraatz, 1895b: 148
bisignata Boheman, 1851: 565 [HN]
bohemani Plavilshchikov, 1924: 232 [RN]
decorata Reitter, 1873d: 41
significans Walker, 1858: 206 (*Nitidula*)
tetragona Murray, 1864: 401 (*Haptoncus*)
thiemei Reitter, 1873d: 41
subgenus *Micruria* Reitter, 1875e: 58 type species *Epuraea mandibularis* Reitter, 1873
Micrurula Reitter, 1884g: 259 [RN] type species *Epuraea mandibularis* Reitter, 1873
adolescens Kirejtshuk, 1998: 215 **A**: XIZ
aldridgei Kirejtshuk, 1997: 255 **A**: SHA
atra Jelínek, 1978a: 195 **A**: BT NP **ORR**
auripubens Reitter, 1901d: 70 **A**: CE MG SE WS
bergeri Sjöberg, 1939b: 110 **A**: CH FE JA "Korea"
biplagiata Kirejtshuk, 1998: 225 **A**: SD
calcarifera Kirejtshuk, 1998: 229 **A**: NP **ORR**
cerina Grouvelle, 1894b: 579 **A**: BT SD **ORR**
subochracea Grouvelle, 1903a: 111
transversecollis Jelínek, 1978a: 198
commutata Grouvelle, 1913d: 395 [RN] **A**: FE (Kuril) JA
japonica Reitter, 1873d: 39 [HN]
consanguinea Grouvelle, 1914: 41 **A**: TAI
convexa Grouvelle, 1908c: 354 **A**: BT HP NP PA SD UP
affinis Grouvelle, 1908c: 353 [HN]
hisamatsui Jelínek, 1978a: 193 [HN]
persimilis Grouvelle, 1913b: 253 [RN]
curvipes Hisamatsu, 1961: 26 **A**: JA
dentipes Hisamatsu, 1961: 27 **A**: JA TAI
dura Reitter, 1884g: 261 [= 1885c: 16] (*Micrurula*) **A**: JA
grouvellei Jelínek, 1978a: 197 [RN] **A**: BT NP SD TAI **ORR**
gangamoria Gillogly, 1982: 284 (*Haptoncus*)
pusilla Grouvelle, 1894b: 585 (*Macroura*) [HN]
haptoncoidea Kirejtshuk, 1992: 134 **A**: FE
harmandi Grouvelle, 1902: 16 **A**: CH FE JA
himalayaensis Kirejtshuk, 1998: 247 **A**: NP
insolita Grouvelle, 1908c: 353 **A**: PA UP
japonica Motschulsky, 1861a: 12 (*Pria*) **A**: FE JA

fusciceps Reitter, 1884g: 261 [= 1885c: 15] (*Micrurula*)
klapperichi Kirejtshuk, 1998: 252 **A**: FUJ
lisa Kirejsthuk, 1987b: 70 **A**: SCH UP
mandibularis Reitter, 1873d: 38 **A**: HEI JA SHA SC TAI "Korea"
melanocephala Marsham, 1802: 136 (*Nitidula*) **E**: AB AL AN AR AU BE BH BU BY CR CT CZ DE EN FI FR GB GE GG GR HU IR IT LA LS LT LU MC MD NL NR NT PL PT RO SK SL SP ST SV SZ TR UK YU **N**: AG MO TU **A**: IN TR
affinis Stephens, 1830: 40 (*Nitidula*)
brunnea Heer, 1841: 400 (*Nitidula*) [HN]
discolor Waltl, 1838b: 270 (*Nitidula*)
ferruginea Heer, 1841: 400 (*Nitidula*) [HN]
heterochroa Gistel, 1857: 542 (*Nitidula*)
truncata Stephens, 1830: 39 (*Nitidula*) [HN]
potaninorum Kirejtshuk, 1987a: 71 **A**: SCH
punctata Kirejtshuk, 1992: 128 **A**: FUJ JA SCH
reticulata Grouvelle, 1892c: 841 **A**: BT NP PA SD UP **ORR**
braeti Grouvelle, 1894b: 578
confinis Jelínek, 1978a: 183
exculta Grouvelle, 1913d: 394 [RN]
omissa Grouvelle, 1908c: 349
ornata Grouvelle, 1903a: 109 [HN]
parabraeti Jelínek, 1978a: 180
rotundula Kirejtshuk, 1998: 268 **A**: NP
specialis Kirejsthuk, 1998: 272 **A**: NP
submicrurula Reitter, 1884g: 261 [= 1885c: 15] **A**: FE (Kuril) JA NC
subita Kirejtshuk, 1992: 125 **A**: FE
subtilis Grouvelle, 1894b: 579 **A**: AP BT NP PA SD UP **ORR**
descarpentriesi Jelínek, 1978a: 188
diluticollis Grouvelle, 1913a: 102 (*Pria*)
tshistyakovae Kirejtshuk, 1987a: 72 **A**: GAN SCH
vulpina Kirejtshuk, 1998: 278 **A**: SCH
wittmeri Jelínek, 1978a: 190 **A**: BT FUJ JA SD
accurata Kirejtshuk, 1987a: 69
subgenus *Polinexa* Kirejtshuk, 1989a: 70 type species *Epuraea decellei* Jelínek, 1977
rougemonti Jelínek, 1988: 44 **A**: YE

genus *Grouvellia* Kirejtshuk, 1984a: 172 type species *Haptoncus piceus* Reitter, 1873
picea Reitter, 1873d: 178 (*Haptoncus*) **A**: SD **ORR**
mirmidon Grouvelle, 1913a: 101 (*Pria*)

genus *Tetrisus* Murray, 1864: 404 type species *Tetrisus cholevoides* Murray, 1864
subgenus *Trimenus* Murray, 1864: 405 type species *Trimenus adpressus* Murray, 1864 (= *Epuraea parallelopipeda* Motschulsky, 1863)
accomodus Kirejtshuk, 1984a: 173 (*Trimenus*) **A**: NP **ORR**
curvipes Grouvelle, 1908c: 338 **A**: NP SD **ORR**
confusus Kirejtshuk, 1984a: 174 (*Trimenus*)
epuraeoides Grouvelle, 1892c: 839 **A**: SD **ORR**
hydroporoides Murray, 1864: 405 **A**: UP **ORR**
parallelopipedus Motschulsky, 1863: 438 (*Epuraea*) **A**: JA SD SE UP **AUR ORR**
adpressus Murray, 1864: 406 (*Trimenus*)
angustatus Murray, 1864: 407 (*Trimenus*)
opacipennis Nakane, 1959a: 56 (*Trimenus*)

genus *Taenioncus* Kirejtshuk, 1984a: 192 type species *Carpophilus cylindricus* Murray, 1864
cylindricus Murray, 1864: 382 (*Carpophilus*) **A**: UP **AUR ORR**
tenuis Murray, 1864: 382 (*Carpophilus*) **A**: JA SE UP "Korea" **AUR ORR**

hana Nakane, 1959a: 54 (*Carpophilus*)

subfamily **Carpophilinae** Erichson, 1842

genus *Carpophilus* Stephens, 1830: 50 type species *Dermestes hemipterus* Linnaeus, 1758

subgenus *Carpophilus* Stephens, 1830: 50 type species *Dermestes hemipterus* Linnaeus, 1758

Tribrachys LeConte, 1861: 83 type species *Carpophilus caudalis* LeConte, 1859 (= *Carpophilus discoideus* LeConte, 1858)

amplus Kirejtshuk, 1992: 164 **A**: FE

bifenestratus Murray, 1864: 364 **E**: CR FR GR (Kríti) IT **N**: AG CI EG MO MR TU **A**: CY IS KZ SA SD SY TR **AFR ORR**

biguttatus Klug, 1862: 208 [HN]

sanguineofasciatus Gemminger & Harold, 1868: 813 [RN]

tenenbaumi Roubal, 1926a: 171

tersus Wollaston, 1865: Appendix: 16

bipustulatus Heer, 1841: 416 (*Ips*) **E**: AU BH BU CR FR GE GG GR HU IT NL RO SK SP ST SZ UK **A**: AF TR SY

chalybeus Murray, 1864: 369 **A**: BEI FE JA NC NE TAI "Korea"

canus Murray, 1864: 369

heterostomoides Reitter, 1877i: 371 (*Heterhelus*)

punctatissimus Reitter, 1877i: 372

delkeskampi Hisamatsu, 1963: 60 **A**: FE FUJ GAN GUA GUI GUX HAI HEB HEI HEI JA (Amami-Ôshima, Iriomote-Jima, Okinawa-Jima) JIX SD TAI XIN YUN "Korea" **AFR ORR**

funereus Murray, 1864: 358 **A**: JA SD TAI **ORR**

hemipterus Linnaeus, 1758: 358 (*Dermestes*) **E**: AB AL AU AZ BE BH BY CR CZ DE FI FR GB GE GR HU IC IR IT LA MA MD NL PL PT SK SP SV SZ **N**: AG CI EG LB MR TU **A**: FUJ GUA GUX HUB HUN IN IQ IS JA JIA JIX JO LE SA SCH TAI TM TR YUN **COS**

aterrimus W. J. MacLeay, 1871: 161

bimaculatus Linnaeus, 1767: 569 (*Silpha*)

brevipennis Germain, 1856: 397

cadaverinus Fabricius, 1801a: 354 (*Nitidula*)

circumdatus Ragusa, 1892: 187

dimidiatus Heer, 1841: 413 (*Cateretes*) [HN]

ficus Fabricius, 1801b: 603 (*Stenus*)

flexuosus Herbst, 1790: 246 (*Nitidula*) [HN]

pictus Heer, 1841: 413 (*Cateretes*)

quadriguttatus Thunberg, 1794: 70 (*Nitidula*)

quadratus Fabricius, 1798: 74 (*Nitidula*)

humerosus Reitter, 1873d: 176 **A**: FE JA

indicus Hisamatsu, 1963: 61 **A**: IN IQ JO KU PA SA TAI **ORR**

jelineki Audisio & Kirejtshuk, 1989: 791 [RN] **A**: IN KU SA **ORR**

pictus Jelínek, 1986: 456 [HN]

marginellus Motschulsky, 1858c: 40 **Ei**: AU AZ BE BU BY CT CZ DE FI GB GE GR IT NL NR SK SP SV "Caucasus" **Ni**: CI **A**: FUJ GUX HKG HUB ISi JA JIX NMO NP SCH SD SHN TAI TRi YUN ZHE "Korea" **COS**

nitens Fall, 1910: 125

oblongopunctatus Grouvelle, 1903a: 108 **A**: SD

obsoletus Erichson, 1843b: 259 **Ei**: CR GR IT MA PT SP **Ni**: AG EG MO TU **A**: AE ANH AP CH CYi FUJ GUA GUX HEB HEN HKG HUB HUN INi IQi ISi JA JIA JIX LEi LIA SAi SCH SD SHA SYi TAI TRi YE YUN ZHE **AFR ORR**

cribellatus Motschulsky, 1858c: 41

immaculatus P. H. Lucas, 1846: 217

sericeus Motschulsky, 1858c: 41

strigipennis Motschulsky, 1858c: 41

quadrisignatus Erichson, 1843b: 257 **E**: AU AZ FR GR IT SP UK "Caucasus" **N**: AG CI (Tenerife) EG MR **A**: IN IS SA SY TR YE

subcalvus Kirejtshuk, 1984b: 520 **A**: FE JA

succisus Erichson, 1843b: 259 **Ei**: AZ (Saõ Miguel) **NTR**

subgenus *Ecnomorphus* Motschulsky, 1858c: 42 type species *Nitidula sexpustulata* Fabricius, 1791
acutangulus Reitter, 1884g: 258 **A**: FE JA NC SC TAI
cingulatus Reitter, 1884g: 258
biguttatus Motschulsky, 1858c: 43 (*Ecnomorphus*) **A**: SD **ORR**
cingulatoides Nakane, 1959a: 53 **A**: FE (Kuril) JA (Tokara-Rettō)
ligneus Murray, 1864: 351 **Ei**: EN GR **Ni**: CI MO **Ai**: TR "Korea" **NAR**
decipiens Horn, 1879: 279
sexpustulatus Fabricius, 1792a: 260 (*Nitidula*) **E**: AB AU BE BH BU CR CZ FR GB GR GG GR HU ITNL PL PT RO SK SL SP ST SZ UK **N**: AG MO TU **A**: LE SY TR
abbreviatus Panzer, 1794b: 21 (*Lyctus*)
quadrimaculatus Ragusa, 1892: 187
sibiricus Reitter, 1879e: 215 **A**: FE JA
signatus Grouvelle, 1908c: 333 **A**: SD
ornatus Grouvelle, 1908c: 334
subgenus *Eidocolastus* Murray, 1864: 399 type species *Colastus plagiatipennis* Motschulsky, 1858
plagiatipennis Motschulsky, 1858c: 39 (*Colastus*) **A**: SD **ORR**
dilutus Motschulsky, 1858c: 39 (*Colastus*)
subgenus *Megacarpolus* Reitter, 1919: 56 type species *Carpophilus grandis* Motschulsky, 1860
grandis Motschulsky, 1861b: 40 **A**: FE
triton Murray, 1864: 357 **A**: FE JA TAI "Korea"
titanus Reitter, 1884g: 259
subgenus *Myothorax* Murray, 1864: 372 type species *Carpophilus mutilatus* Erichson, 1843
dimidiatus Fabricius, 1792a: 261 (*Nitidula*) **Ei**: AZ GR IT NL **Ni**: CI EG **Ai**: AE ANH CY FUJ GAN GUA GUX HEB HEN HKG HUB HUN IQ JA JIA JIX JO SCH SHN SHX SY TR YUN **COS**
auropilosus Wollaston, 1854: 117
pusillus Stephens, 1830: 51
contengens Walker, 1858: 206 (*Nitidula*) **A**: JA NP TAI **AFR AUS ORR**
fumatus Boheman, 1851: 564 (*Brachypeplus*) **Ei**: AZ IT (Sicilia) PT **Ni**: MO **Ai**: IQ **COS**
ochropterus Klug, 1862: 209
lewisi Reitter, 1884g: 258 **A**: JA
maculatus Murray, 1864: 372 **A**: GUX HEN HKG HUB JA **AUR ORR**
vittiger Murray, 1864: 373
mutilatus Erichson, 1843b: 258 **E**: AL AZ FR (Corse) GR IT MA NL PT SP **N**: CI (Tenerife) EG MO **A**: AF CY IN IS JA JO LE SA SY TAI TR **COS**
luridus Murray, 1864: 377
pilosellus Motschulsky, 1858c: 41
nepos Murray, 1864: 381, 397 **Ei**: AZ CR GR IT PT SP **Ni**: CI EG MO TU **Ai**: IQ JA JO SA SY TR **COS**
freemani Dobson, 1956: 41
schioedtei Murray, 1864: 381 **A**: TAI **ORR**
truncatus Murray, 1864: 381, 397 **Ei**: AU CR CZ FR GE GR IT PL SK SL **Ni**: EG MO **A**: FUJ GAN GUA GUI GUX HAI HEB HEI HEN HKG JA JIL LIA NMO SHA SHN SHX TAI Tri YUN **COS**
floridanus Fall, 1910: 122
halli Dobson, 1954: 299
zeaphilus Dobson, 1969: 99 **Ei**: IT (Sicilia) PT SP **A**: JO TR YE **AFR**
subgenus *Stauroglossicus* Murray, 1864: 398 type species *Stauroglossicus terminalis* Murray, 1864
dubitabilis Grouvelle, 1897: 355 (*Stauroglossicus*) **A**: BT **ORR**

genus *Nitops* Murray, 1864: 382 type species *Carpophilus ophthalmicus* Murray, 1864
Endomerus Murray, 1864: 384 type species *Carpophilus piger* Murray, 1864
Heterodontus Murray, 1864: 342 type species *Carpophilus piger* Murray, 1864
pubescens Murray, 1864: 383 (*Carpophilus*) **A**: YUN **ORR**

genus *Urophorus* Murray, 1864: 342 type species *Ips rubripennis* Heer, 1841
subgenus *Anophorus* Kirejtshuk, 1990b: 87 type species *Nitidula humeralis* Fabricius, 1798
adumbratus Murray, 1864: 344 (*Carpophilus*) **A**: FUJ GUI ZHE **ORR**
conspicuus Murray, 1864: 345 (*Carpophilus*)

humeralis Fabricius, 1798: 74 (*Nitidula*) **E**i: AL AN AU BH BU CR CY SP FR GB GR IT MC MA MR PT RO SP TR UK YU **N**i: AG CI EG LB MO TU **A**i: AE CY FUJ GUA GUI GUX IN IS JA SA SCH TAI TR YE YUN ZHE **COS**

punctatus Fleutiaux, 1887: 61 (*Carpophilus*)

rickseckeri Fall, 1910: 124 (*Carpophilus*)

foveicollis Murray, 1864: 344 (*Carpophilus*) **A**: JA (Ryukyus) NP SD TAI YUN **ORR**

prodicus Hinton, 1944: 172 (*Carpophilus*) **A**: HEB PA **ORR**

subgenus *Urophorus* Murray, 1864: 342 type species *Ips rubripennis* Heer, 1841

aria Audisio, Jelínek & Kirejtshuk, 1990: 84 **A**: IN

colonnellii Audisio & Kirejtshuk, 1989: 791 **A**: TR

kryzhanovskiji Kirejtshuk, 1984b: 522 (*Carpophilus*) **A**: TD UZ

maghrebinus Audisio, 1993a: 173 **N**: AG

rubripennis Heer, 1841: 417 (*Ips*) **E**: AL AR AU BH BU CR FR GR HU IT MT RO SK SP SZ UK

castanopterus Erichson, 1843b: 256 (*Carpophilus*)

yakushenkoi Audisio & Kirejtshuk, 1989: 792 **E**: AR **N**: AG **A**: AF CY JO KI KZ LE PA TD TM TR UZ

turanicus Audisio & Kirejtshuk, 1989: 793

subfamily **Cryptarchinae** C. G. Thomson, 1859

genus *Cryptarcha* Shuckard, 1840: 165 type species *Nitidula strigata* Fabricius, 1787

Africanips Lechanteur, 1959: 107 type species *Africanips niger* Lechanteur, 1959 (= *Cryptarcha lechanteuri* Kirejtshuk, 1997)

Cryptarchina Iablokoff-Khnzorian, 1966b: 313 type species *Cryptarcha incognita* Iablokoff-Khnzorian, 1966

Cryptarchula Ganglbauer, 1899a: 550 type species *Nitidula undata* A. G. Olivier, 1790

Cryptarchus Heer, 1841: 409 type species *Nitidula strigata* Fabricius, 1787

Lepiarcha Sharp, 1891: 385 type species *Cryptarcha omositoides* Reitter, 1873

bifasciata Baudi di Selve, 1870: 52 **N**: EG **A**: CY LE TR

dubia Grouvelle, 1890: 126 **A**: SD **ORR**

incognita Iablokoff-Khnzorian, 1966b: 311 **E**: AR **A**: IN

inhalita Reitter, 1884g: 269 [= 1885c: 79] **A**: FE JA SHA **ORR**

kapfereri Reitter, 1913b: 123 **A**: FE

lewisii Reitter, 1873d: 152 **A**: FE FUJ HEB JA YUN "Korea"

maculata Reitter, 1873d: 151 **A**: JA SD NP SE TAI "Korea" **ORR**

bonifacyi Grouvelle, 1912h: 502

fraterna Grouvelle, 1908c: 395

lesnei Grouvelle, 1903a: 118

longipennis Hisamatsu, 1961: 31

ritsemae Olliff, 1884: 246

okinawensis Hisamatsu, 1985: 195 **A**: JA (Okinawa-Jima)

strigata Fabricius, 1787a: 51 (*Nitidula*) **E**: AU BH BY CT CZ DE EN FI FR GB GE GR HU IT LA LT MC NL NR NT PL PT SK SL SP ST SV SZ TR UK **A**: FE HUB JA SC SHA SHX WS "Korea"

circassica Reitter, 1887g: 287

graphica Schrank, 1796: 160 (*Dermestes*)

lateralis C. R. Sahlberg, 1820b: 77 (*Nitidula*)

punctatissima Boieldieu, 1859: 467

quadrisignata Küster, 1845: 15

verbasci Thunberg, 1794: 71 (*Nitidula*)

tuberculifera Jelínek, 1964: 2 **A**: AF PA

undata A. G. Olivier, 1790a: no 12: 14 (*Nitidula*) **E**: AB AU BH BY CT CZ DE FR GB GE GR HU IT LA LT NL PL RO SK SL SP ST SV SZ TR UK YU

imperialis Fabricius, 1792a: 257 (*Nitidula*)

nebulosa Marsham, 1802: 134 (*Nitidula*)

genus *Eucalosphaera* Jelínek, 1978a: 217 [RN] type species *Cryptarcha ocularis* Reitter, 1873

Calosphaera Jelínek, 1974: 192 [HN] type species *Cryptarcha ocularis* Reitter, 1873

feae Grouvelle, 1892c: 855 (*Cryptarcha*) **A**: BT NP **ORR**

genus *Glischrochilus* Reitter, 1873d: 162 type species *Silpha quadripustulata* Linnaeus, 1761 (= *Silpha quadripunctata* Linnaeus, 1758)
subgenus *Cephalips* Arrow, 1937: 101 type species *Librodor egregius* Grouvelle, 1892
egregius Grouvelle, 1892c: 856 (*Librodor*) **A**: HUN NP SD XIN ZHE **ORR**
cyclops Jelínek, 1975b: 137
monticola Jelínek, 1975b: 137
subgenus *Glischrochilus* Reitter, 1873d: 162 type species *Silpha quadripustulata* Linnaeus, 1761 (= *Silpha quadripunctata* Linnaeus, 1758)
Ips Fabricius, 1777: 23 [HN] type species *Ips quadriguttatus* Fabricius, 1777
biguttulus Motschulsky, 1860b: 129 (*Ips*) **A**: FE (Kamchatka)
angusticollis Motschulsky, 1860b: 129 (*Ips*)
brevenotatus Pic, 1926b: 1
cruciatus Motschulsky, 1860b: 129 (*Ips*) **A**: ES FE JA MG
quadripunctatus Linnaeus, 1758: 359 (*Silpha*) **E**: AU BE BH BU BY CT CZ DE EN FI FR GB GE GR IR IT LA LT MC NL NR NT PL SK SP ST SV SZ UK YU **A**: ES MG WS
niger J. R. Sahlberg, 1889: 85 (*Ips*)
quadripustulatus Linnaeus, 1761: 148 (*Silpha*)
subgenus *Librodor* Reitter, 1884g: 269 type species *Ips quadriguttatus* Fabricius, 1777
Cryptarchips Reitter, 1911a: 37 type species *Cryptarcha ipsoides* Reitter, 1879
affinis Kirejtshuk, 1984b: 529 **A**: ES FE
audisioi Jelínek, 1999a: 201 **A**: NP
becvari Jelínek, 1999a: 204 **A**: TAI YUN
binaevus Reitter, 1879e: 218 (*Cryptarcha*) **A**: FE JA
christophi Reitter, 1879e: 219 (*Ips*) **A**: FE JA SC
clavatus Reitter, 1884g: 270 [= 1885c: 80] (*Librodor*) **A**: JA
fasciatus A. G. Olivier, 1790a: No.12: 7 (*Nitidula*) **Ei**: GE SZ **NAR**
geminatus Melsheimer, 1844: 108 (*Ips*)
quadrimaculosus Melsheimer, 1844: 107 (*Ips*)
flavipennis Reitter, 1875i: 120 (*Cryptarcha*) **A**: BT NP SD UP YUN **ORR**
flavoguttatus Reitter, 1875i: 120 (*Cryptarcha*) **A**: BT UP
forcipatus Fairmaire, 1889b: 12 (*Librodor*) **A**: SHN
formosus Jelínek, 1999a: 207 **A**: YUN
grandis Tournier, 1872b: 440 (*Ips*) **E**: AR BY CT GG LA ST UK
latefasciatus Reitter, 1883c: 41 (*Ips*)
hortensis Geoffroy, 1785: 21 (*Dermestes*) **E**: AU BE BH BU BY CR CT CZ DE EN FI FR GB GE GR HU IR IT LA LS LT NL NR NT PL RO SK SP ST SV SZ UK **A**: ES FE TR WS
olivieri Bedel, 1891: 153 [RN]
puncticollis Trella, 1923: 112
quadripunctatus A. G. Olivier, 1790a: no 12: 9 (*Nitidula*) [HN]
ipsoides Reitter, 1879e: 218 (*Cryptarcha*) **A**: FE JA SC "Korea"
japonius Motschulsky, 1858b: 28 (*Ips*) **A**: ANH FE GUI HUB JA JIA NC NP SC SCH SHA SHN SHX TAI YUN **ORR**
chinensis Reitter, 1873d: 160 (*Ips*)
davidis Fairmaire, 1878b: 93 (*Ips*)
nankineus Fairmaire, 1878b: 93 (*Ips*)
superbus Jelínek, 1975b: 143
klapperichi Jelínek, 1975b: 139 **A**: FUJ
latior Jelínek, 1999a: 210 **A**: YUN
luteoniger Jelínek, 1982a: 195 **A**: NP
pallidescriptus Jelínek, 1999a: 212 **A**: HEN SCH SHA YUN
pantherinus Reitter, 1879e: 217 (*Cryptarcha*) **A**: FE JA SC
parvipustulatus Kolbe, 1886: 180 (*Ips*) **A**: FE GAN GUI HUB JA SC SCH YUN "Korea"
pubescens Jelínek, 1982a: 192
popei Jelínek, 1975b: 138 **A**: XIZ **ORR**
pulcher Jelínek, 1975b: 132 **A**: NP YUN

quadriguttatus Fabricius, 1777: 214 (*Ips*) **E**: AL AU BH BY CR CT CZ DE EN FI FR GB GE GR HU IR IT LA LT MC NL NR PL RO SK SL SP ST SV SZ UK YU TR
bidisjunctus Pic, 1917: 21
decemguttatus A. G. Olivier, 1790a: no 12: 10 (*Nitidula*) [HN]
diversenotatus Pic, 1917: 21
quadrinotatus Scriba, 1790b: 165 (*Ips*)
subinterruptus Pic, 1917: 21
quadrisignatus Say, 1835: 169 (*Ips*) **Ei**: AU BU BY CR CZ GE HU IT MD NL PL RO SK SL ST SZ TR UK YU **Ai**: KZ **NAR**
bipunctatus Melsheimer, 1844: 108 (*Ips*)
bipustulatus Melsheimer, 1844: 108 (*Ips*)
canadensis Brown, 1932: 259
sexpustulatus Reitter, 1873d: 161 (*Ips*) [HN]
similis Melsheimer, 1844: 108 (*Ips*)
rufiventris Reitter, 1879e: 219 (*Ips*) **A**: FE JA SC
subcylindricus Reitter, 1884g: 271 [= 1885c: 101] (*Librodor*) **A**: FE (Kuril) JA

genus *Pityophagus* Shuckard, 1840: 171 type species *Dermestes ferrugineus* Linnaeus, 1761
basalis Reitter, 1874c: 514 **A**: JA
ferrugineus Linnaeus, 1761: 145 (*Dermestes*) **E**: AL AU BE BH BU BY CR CT CZ DE EN FI FR GB GE GR HU IR IT LA LT NL NR NT PL SK SL SP ST SV SZ UK YU **N**: AG MO **A**: FE TR
dermestoides Panzer, 1793c: 15 (*Lyctus*)
linearis Latreille, 1807a: 11 (*Nitidula*)
laevior Abeille de Perrin, 1872b: xlii **E**: AU BH BU CR CZ FR GE HU IT PL PT RO SK SL SP **N**: MR
deubeli Ganglbauer, 1899b: 532
quercus Reitter, 1877d: 17 **E**: BH BU CZ GR HU MC PL RO SK SP UK

subfamily **Nitidulinae** Latreille, 1802

genus *Aethina* Erichson, 1843b: 306 type species *Aethina pubescens* Erichson, 1843
subgenus *Aethina* Erichson, 1843b: 306 type species *Aethina pubescens* Erichson, 1843
Aethinopa Reitter, 1875i: 109 type species *Aethinopa fulvovestita* Reitter, 1875
Meligethopsis Rebmann, 1944: 25 type species *Meligethopsis singularis* Rebmann, 1944 (= *Aethina aeneipennis* Reitter, 1873)
Pseudomystrops Grouvelle, 1913d: 389 type species *Idaethina humeralis* Grouvelle, 1890
aeneipennis Reitter, 1873d: 85 **A**: FUJ JA SCH **ORR**
singularis Rebmann, 1944: 26 (*Meligethopsis*)
argus Grouvelle, 1890: 125 **A**: TAI **ORR**
berezovskyi Kirejtshuk, 1986b: 56 **A**: SCH
concinna Jelínek & Kirejtshuk, 1986b: 67 **A**: TAI
cyaneipennis Grouvelle, 1903a: 116 **A**: SD
dundai Jelínek, 1995: 116 **A**: SCH
elongata Reitter, 1875i: 109 **A**: UP
flavicollis Reitter, 1884g: 266 [= 1885c: 41] **A**: JA SCH TAI "Korea" **ORR**
maculicollis Reitter, 1884g: 266 [= 1885c: 41]
humeralis Grouvelle, 1890: 124 (*Idaethina*) **A**: AP **ORR**
basalis Grouvelle, 1897: 367
concolor Grouvelle, 1913d: 390 (*Pseudomystrops*)
convexa Grouvelle, 1913d: 391 (*Pseudomystrops*)
orientalis Olliff, 1884: 245 (*Mystrops*)
inconspicua Nakane, 1967b: 68 **A**: FE FUJ JA NP PA SCH TAI UP YUN "Korea" **ORR**
mantici Jelínek, 1995: 109 **A**: NP
ocularis Jelínek, 1978a: 211 **A**: BT SD
planta Kirejtshuk, 1986b: 55 **A**: GAN
rufobasalis Jelínek, 1995: 110 **A**: SCH SHA **ORR**
scutellaris Jelínek & Kirejtshuk, 1986: 61 **A**: FUJ
testacea Jelínek & Kirejtshuk, 1986: 63 **A**: FUJ GUA

tumida Murray, 1867a: 177 **N**: EGi **AFR AUS**i **NAR**i
vicina Grouvelle, 1894b: 585 **A**: SD **ORR**
nigrocastanea Grouvelle, 1903a: 115
subgenus *Circopes* Reitter, 1873d: 79 type species *Pocadius subquadratus* Motschulsky, 1858
abdita Kirejtshuk, 1986b: 72 **A**: FUJ **ORR**
atratula Jelínek & Kirejtshuk, 1986: 74 [RN] **A**: BT NP SD
atra Jelínek, 1978a: 209 [HN]
jelineki Kirejtshuk, 1986b: 76 [RN] **A**: NP TAI
basalis Hisamatsu, 1978: 8 (*Circopes*) [HN]
marginata Grouvelle, 1894b: 584 (*Circopes*) **A**: SD
suturalis Reitter, 1884g: 266 [= 1885c: 41] **A**: FE FUJ JA "Korea"
subgenus *Cleidorura* Kirejtshuk & Lawrence, 1999: 239 type species *Aethina transfusa* Kirejtshuk & Lawrence, 1999
obscura Reitter, 1873d: 86 **A**: "Himalaya" **ORR**
subgenus *Idaethina* Gemminger & Harold, 1868: 824 type species *Carpophilus longipennis* Motschulsky, 1858 (= *Meligethes orientalis* Nietner, 1856)
Macroura Reitter, 1873d: 80 [HN] type species *Macroura nigra* Reitter, 1873 (= *Nitidula abbreviata* Fabricius, 1781)
Olliffura Jelínek & Kirejtshuk, 1986: 47 [RN] type species *Macroura meligethoides* Reitter, 1873
meligethoides Reitter, 1873d: 82 (*Macroura*) **A**: "Himalaya"
orientalis Nietner, 1856: 534 (*Meligethes*) **A**: UP **ORR**
longipennis Motschulsky, 1858c: 40 (*Carpophilus*)
respondens Walker, 1859a: 53 (*Meligethes*)
subrugosa Grouvelle, 1894a: 461 (*Macroura*) **A**: NP SHN UP **ORR**
dubia Grouvelle, 1894a: 461 (*Macroura*)
subgenus *Ithyra* Reitter, 1873d: 78 type species *Ithyra hirsutula* Reitter, 1873
hirsutula Reitter, 1873d: 78 **E**: IT (Sicilia) **A**: SA YE **AFR**

genus *Aethinopsis* Grouvelle, 1908c: 379 type species *Aethinopsis antennata* Grouvelle, 1908
antennata Grouvelle, 1908c: 379 **A**: NP SD

genus *Amphotis* Erichson, 1843b: 290 type species *Nitidula marginata* Fabricius, 1781
Oostomum Gistel, 1856: 362 type species *Nitidula marginata* Fabricius, 1781
marginata Fabricius, 1781: 91 (*Nitidula*) **E**: AU BE BH BU BY CR CT CZ DE FR GB GE GG GR HU IT LA LT MD NL PL PT RO SK SL SP ST SV SZ UK YU **A**: TR
biloba Herbst, 1793: 238 (*Nitidula*)
martini C. N. F. Brisout de Barneville, 1878: lxiii **E**: PT SP **N**: AG CI MO TU
tenorioi Plata-Negrache & Prendes-Ayala, 1980: 191
orientalis Reiche, 1861a: 1 **E**: AL BH BU CR GR IT SP TR UK YU **N**: MO **A**: CY IS LE SY TR

genus *Anister* Grouvelle, 1901: 102 type species *Anister raffrayi* Grouvelle, 1901
Oturovana Reitter, 1915f: 135 type species *Oturovana carpophiloides* Reitter, 1915 (= *Anister raffrayi* Grouvelle, 1901)
raffrayi Grouvelle, 1901: 102 **N**: EG LB TU **A**: IS SA **AFR**
ares Hinton, 1942a: 126
carpophiloides Reitter, 1915f: 136 (*Oturovana*)

genus *Atarphia* Reitter, 1884g: 260 type species *Atarphia fasciculata* Reitter, 1884
fasciculata Reitter, 1884g: 263 [= 1885c: 39] **A**: CH FE JA
quadripunctata Reitter, 1884g: 263 [= 1885c: 39] **A**: CH FE JA

genus *Cychramus* Kugelann, 1794: 543 type species *Strongylus quadripunctatus* Herbst, 1792
Campta Stephens, 1830: 44 type species *Sphaeridium luteum* Fabricius, 1787
chloroticus Fairmaire, 1860: 168 **N**: AG TU
algericus Pic, 1894t: 134
fairmairei Pic, 1894c: 44
henoni Fairmaire, 1867: 395

luteus Fabricius, 1787b: 378 (*Sphaeridium*) **E**: AU BH BU BY CR CT CZ DE EN FI FR GB GE GG GR HU IR IT LA LT NL NR NT PL PT RO SK SL SP ST SV SZ UK YU **A**: ES FE GAN JA MG SCH SHA TR WS YUN "Korea" **ORR**
alutaceus Reitter, 1875l: 360
dorsalis Reitter, 1884g: 266 [= 1885c: 42]
floricola Reitter, 1884g: 266 [= 1885c: 43]
fungicola Heer, 1841: 408
latus Scriba, 1790b: 175 (*Nitidula*)
montandoni Pic, 1893h: cclxxxviii
pubescens Pic, 1894t: 133
subopacus Reitter, 1884g: 266 [= 1885c: 42]
unicolor A. G. Olivier, 1790b: 160 (*Anthribus*)
verris P. Rossi, 1790: 60 (*Nitidula*)
plagiatus Reitter, 1884g: 266 [= 1885c: 42] **A**: JA
lewisi Reitter, 1884g: 266 [= 1885c: 42]
variegatus Herbst, 1792: 184 (*Strongylus*) **E**: AU BE BH BY CT CZ DE EN FI FR GE HU IT LA LS LT NL NR NT PL RO SK ST SV SZ UK YU **A**: ES FE JA SCH SHA YUN "Korea" **ORR**
colon Fabricius, 1792a: 78 (*Sphaeridium*)
quadripunctatus Herbst, 1792: 185 (*Strongylus*)

genus *Cyllodes* Erichson, 1843b: 342 type species *Strongylus ater* Herbst, 1792
Pseudocamptodes Grouvelle, 1896b: 76 type species *Pseudocamptodes africanus* Grouvelle, 1896
Strongylus Herbst, 1792: 179 [HN] type species *Strongylus ater* Herbst, 1792
angustipes Grouvelle, 1908c: 385 **A**: SD
ater Herbst, 1792: 188 (*Strongylus*) **E**: AR BH BU BY CR CT CZ DE FI FR GE GG HU IT LA MC NT PL RO SK SL SP ST SV SZ UK YU **A**: ES FE JA TR WS "Korea"
aterrimus Reitter, 1879e: 210 (*Strongylus*)
glabratus Panzer, 1795: 26 (*Sphaeridium*)
morio Kugelann, 1794: 537 (*Volvoxis*)
tenebrio Gistel, 1857: 536 (*Cychramus*)
tenebrio Gistel, 1857: 552 (*Pocadius*) [HN]
wankowiczi Hochhut, 1872: 212 (*Agathidium*)
bifascies Walker, 1859a: 259 (*Tritoma*) **A**: FE JA TAI "Korea" **ORR**
binotatus Reitter, 1879e: 217 (*Strongylus*)
dorsalis Reitter, 1884g: 268 [= 1885c: 77] (*Strongylus*)
opponens Walker, 1859b: 219 (*Chilocorus*)
ornatus Motschulsky, 1863: 442 (*Camptodes*)
notatus Reitter, 1873d: 129 (*Strongylus*)
breviusculus Reitter, 1884g: 268 [= 1885c: 77] (*Strongylus*) **A**: JA
dubius Reitter, 1877i: 374 (*Strongylus*) **A**: FE JA "Korea" **ORR**
excellens Reitter, 1884g: 268 [= 1885c: 78] (*Strongylus*) **A**: JA **ORR**
indicus Grouvelle, 1894a: 462 **A**: KA **ORR**
montanus Grouvelle, 1903b: 126 (*Pseudocamptodes*)
literatus Reitter, 1878c: 89 (*Strongylus*) **A**: JA TAI **ORR**
punctidorsum Nakane & Hisamatsu, 1955: 55 **A**: JA SCH SHA
nakanei Hisamatsu, 1961: 28 **A**: JA TAI
multimaculatus Grouvelle, 1914: 45 **A**: TAI
semiglobosus Reitter, 1884g: 268 [= 1885c: 77] (*Strongylus*) **A**: JA
tigrinus Grouvelle, 1914: 44 **A**: TAI

genus *Ecnomaeus* Erichson, 1843b: 264 type species *Ecnomaeus planus* Erichson, 1843
haroldi Reitter, 1873d: 182 **A**: "Himalaya"
waterhousei Grouvelle, 1908c: 342 **A**: "NW India"

genus *Hebasculinus* Kirejtshuk, 1992: 195 type species *Hebascus hilleri* Reitter, 1877 (=*Hebascus japonus* Reitter, 1877)
japonus Reitter, 1877i: 372 (*Hebascus*) **A**: FE JA

hilleri Reitter, 1877i: 373 (*Hebascus*)
cybaeus Kirejsthuk, 1992: 195 **A**: FE

genus *Ipidia* Erichson, 1843b: 289 type species *Ips quadrinotata* Fabricius, 1798 (= *Ipidia binotata* Reitter, 1875)
subgenus *Hemipidia* Kirejtshuk, 1992: 188 type species *Nitidula sexguttata* R. F. Sahlberg, 1834
chujoi Hisamatsu, 1982: 55 **A**: TAI
octusis Reitter, 1919: 76 (*Stelidota*) **A**: FE JA
sexguttata R. F. Sahlberg, 1834a: 6 (*Nitidula*) **E**: AU BY CT FI GE GG LT PL RO SK SL ST
integra Wankowicz, 1867: 252
lata Aubé, 1850: 328
sibirica Reitter, 1879e: 216 (*Stelidota*) **A**: FE JA SCH **ORR**
sjoebergi Jelínek, 1978a: 199 **A**: SD
subgenus *Ipidia* Erichson, 1843b: 289 type species *Ips quadrinotata* Fabricius, 1798 (= *Ipidia binotata* Reitter, 1875)
binotata Reitter, 1875h: 88 **E**: AL AU BH BU BY CR CT CZ EN FI FR GE GR HU IT LA LT NR NT PL RO SK SL SP ST SV SZ UK YU **A**: WS
quadrimaculata Quensel, 1790: 12 (*Silpha*) [HN]
quadrinotata Fabricius, 1798: 179 (*Ips*) [HN]
quadriplagiata Biström, 1978: 88 [RN]
variolosa latior Grouvelle, 1914: 41 **A**: AP JA (Yaku-Shima) SHA TAI **ORR**
variolosa variolosa Reitter, 1879e: 215 **A**: ES FE JA NE "Korea"
hattorii Kôno, 1940: 57 (*Stelidota*)

genus *Megauchenia* W. S. MacLeay, 1825: 42 type species *Megauchenia setipennis* W. S. MacLeay, 1825
Ischaena Erichson, 1843b: 287 type species *Ischaena angustata* Erichson, 1843
angustata Erichson, 1843b: 288 (*Ischaena*) **A**: AP JA (Ryukyus) NP TAI **ORR**
peninsularis Grouvelle, 1890: 123, 124 (*Ischaena*)
quadricollis rotundata Kirejtshuk, 1990a: 82 **A**: TAI **ORR**
setipennis W. S. MacLeay, 1825: 46 **A**: TAI **ORR**
elongata Erichson, 1843b: 288 (*Ischaena*)

genus *Neopallodes* Reitter, 1884g: 264 type species *Pallodes hilleri* Reitter, 1877
clavatus Reitter, 1884g: 269 [= 1885c: 78] **A**: FE JA
falsus Grouvelle, 1913d: 398 (*Pallodes*) [RN] **A**: NP SD **ORR**
harmandi Grouvelle, 1903a: 117 (*Pallodes*) [HN]
lindskogi Kirejtshuk, 1987c: 158
globosus Kirejtshuk, 1994b: 239 **A**: NP
hilleri Reitter, 1877i: 374 (*Pallodes*) **A**: FE JA TAI YUN
bouvieri Grouvelle, 1902: 16 (*Pallodes*)
circumflexus Reitter, 1879e: 210 (*Pallodes*)
inermis Reitter, 1884g: 269 [= 1885c: 79] **A**: FE JA YUN
harmandi Grouvelle, 1902: 17
omogonis Hisamatsu, 1953: 141 **A**: JA
turulosus Kirejtshuk, 1994b: 246 **A**: NP
vietnamicus Kirejtshuk, 1987c: 152 **A**: SD **ORR**
vicinus Grouvelle, 1892c: 850 **A**: FE JA TAI **ORR**

genus *Nitidula* Fabricius, 1775: 77 type species *Silpha bipustulata* Linnaeus, 1761 (= *Silpha bipunctata* Linnaeus, 1758)
Theridiosmum Gistel, 1856: 361 type species *Silpha bipustulata* Linnaeus, 1761 (= *Silpha bipunctata* Linnaeus, 1758)
bipunctata Linnaeus, 1758: 359 (*Silpha*) **E**: AU BH BY CT CZ DE EN FI FR GB GE HU IR IT LA LT NL NR NT PL PT RO SK SL SP ST SV **A**: ES FE HEI MG TR WS **NAR**
bipustulata Linnaeus, 1761: 148 (*Silpha*)
impustulata Ganglbauer, 1899a: 491
nigra Ragusa, 1892: 189

scarabaeoides Scopoli, 1763: 15 (*Dermestes*)
carnaria Schaller, 1783: 257 (*Silpha*) **E**: AB AL AR AZ BH BU BY CR CZ DE EN FR GB GE GR HU IT LA LT MC MD NL PL RO SK SL SP ST SV UK YU **N**: CI EG MR **A**: CY ES GAN HEI HEN IN JIA KI KZ LE LIA MG NIN NMO QIN SHA TM TR UZ XIN **NAR**
flavipennis Heer, 1841: 400
guttalis Herbst, 1793: 247
quadripustulata Fabricius, 1792a: 255
variata Stephens, 1830: 36
eremita Audisio, 1990: 25 [RN] **N**: AG EG LB TU **A**: AE IN IQ IS SA
ciliata Erichson, 1843b: 275 [HN]
flavomaculata P. Rossi, 1790: 58 **E**: BE BU CR CT FR GE GR HU IT MA PT SP ST UK YU **N**: AG CI EG MO MR TU **A**: AF CY IN IS JO LE SY TD TM TR UZ
flexuosa A. G. Olivier, 1790a: no 12: 7
latiplaga Solsky, 1876: 248
fusula Gebler, 1833: 283 **E**: ST **A**: ES KZ MG TM WS
regalis Zubkov, 1833: 320
maculosa Fairmaire, 1866a: 19 **N**: AG EG LB **A**: IQ IS JO
mollicella Reitter, 1873d: 44
rufipes Linnaeus, 1767: 573 (*Silpha*) **E**: BH BU BY CR CZ DE EN FI FR GB GE HU IT LA LT NL NR NT PL PT SK SL SP ST SV UK YU **A**: FE KZ MG NE TR UZ WS **NAR**
bicolor Dalla Torre, 1879: 87
castanea C. R. Sahlberg, 1820b: 74
fulvipes Geoffroy, 1785: 22 (*Dermestes*)
marginata Dalla Torre, 1879: 87
obscura Fabricius, 1777: 215
ossium Kirby, 1837: 106

genus *Omosita* Erichson, 1843b: 298 type species *Silpha depressa* Linnaeus, 1758
Saprobia Ganglbauer, 1899a: 489 type species *Silpha colon* Linnaeus, 1758
Scatocharis Gistel, 1856: 362 type species *Silpha depressa* Linnaeus, 1758
colon Linnaeus, 1758: 362 (*Silpha*) **E**: AB AU AZ BH BU BY CR CT CZ DE EN FI FR GB GE HU IR IT LA LT MD MR NL NR NT PL RO SK SL SP SV TR UK YU **A**: BEI ES FE GAN HEB HEI HEN JA JIA JIL KZ MG NC NIN NMO QIN SHA SHN SHX TR WS XIN YUN ZHE "Korea" **AUR NAR**
bipartita Trost, 1801: 18 (*Nitidula*)
haemorrhoidalis Fabricius, 1777: 216 (*Nitidula*)
viana Gistel, 1857: 587
depressa Linnaeus, 1758: 362 (*Silpha*) **E**: AU BH BU BY CR CT CZ DE EN FI FR GB GE HU IR IT LA LT NL NR NT PL SK SL SP ST SV SZ UK YU **A**: ES FE KZ MG WS
colon Herbst, 1793: 234 (*Nitidula*) [HN]
ferruginea Herbst, 1786: 159 (*Nitidula*) [HN]
immaculata A. G. Olivier, 1790a: no 12: 11 (*Nitidula*)
sordida Fabricius, 1792a: 258 (*Nitidula*)
varia A. G. Olivier, 1790a: no 12: 12 (*Nitidula*)
discoidea Fabricius, 1775: 78 (*Nitidula*) **E**: AR AU AZ BH BU BY CR CT CZ DE EN FR GB GE GG HU IR IT LA LT MC MD MR NL NR NT PL RO SK SL SP ST SV SZ UK YU **A**: FE GUX HUN IN JA SCH **NAR NTR**
cincta Heer, 1841: 396 (*Nitidula*)
inversa LeConte, 1857: 36
funesta Reitter, 1873d: 45 **E**: SPi **NTR**
japonica Reitter, 1874c: 510 **A**: FE FUJ JA SCH "Korea"
smetanai Jelínek, 1999b: 252 **A**: GAN SCH

genus *Oxystrongylus* Reitter, 1911a: 26 type species *Oxystrongylus seriepunctatus* Reitter, 1911 (= *Xenostrongylus ovulum* Fairmaire, 1875)
ovulum Fairmaire, 1875a: 504 (*Xenostrongylus*) **E**: SP **N**: AG EG LB MO TU
seriepunctatus Reitter, 1911a: 26
sanctissimus Roubal, 1927c: 134 (*Xenostrongylus*) **A**: IS JO SI

palaestinensis Amsel & Hering, 1931: 133 (*Xenostrongylus*)

genus *Pallodes* Erichson, 1843b: 348 type species *Strongylus annulifer* Laporte, 1840
subgenus *Coxollodes* Kirejtshuk, 1987c: 168 type species *Pallodes cyrtusoides* Reitter, 1884
cyrtusoides Reitter, 1884g: 269 [= 1885c: 79] **A**: JA TAI **ORR**
subgenus *Pallodes* Erichson, 1843b: 348 type species *Strongylus annulifer* Laporte, 1840
alutaceus Kirejtshuk, 1987c: 160 **A**: SD
amamiensis Hisamatsu, 1956: 166 **A**: JA (Amami-Shotō)
kaszabi Kirejtshuk, 1987c: 162 **A**: NP
oleosus Kirejtshuk, 1987c: 159 **A**: SD
umbratilis Reitter, 1873d: 134 **A**: FE (Kuril) JA NC "Korea"

genus *Parametopia* Reitter, 1884g: 259 type species *Parametopia x-rubrum* Reitter, 1884
indica Grouvelle, 1894b: 581 **A**: SD
rotundata Reitter, 1881c: 75 (*Prometopia*) **A**: TAI **ORR**
x-rubrum Reitter, 1884g: 262 [= 1885c: 15] **A**: JA TAI

genus *Phenolia* Erichson, 1843b: 299 type species *Nitidula grossa* Fabricius, 1801
subgenus *Lasiodites* Jelínek, 1999b: 276 type species *Nitidula picta* MacLeay, 1825
amplificator Hisamatsu, 1956: 165 (*Lasiodactylus*) **A**: JA (Amami-Ôshima) NP TAI **ORR**
borealis Hisamatsu, 1968a: 11 (*Lasiodactylus*) **A**: JA
inaequalis Grouvelle, 1914: 42 (*Lasiodactylus*) **A**: NP TAI **ORR**
limbata tibialis Boheman, 1851: 568 (*Soronia*) **Ei**: AZ MD **AFR**
acutipennis Grouvelle, 1915b: 521 (*Lasiodactylus*)
biplicata Fairmaire, 1880c: 181 (*Lordites*)
caffra Boheman, 1851: 567 (*Soronia*)
curvitibia Kraatz, 1895b: 151 (*Lordites*)
dorsalis Grouvelle, 1915a: 112 (*Lasiodactylus*)
grammica Klug, 1855: 649 (*Lordites*)
maculipennis Kraatz, 1895b: 152 (*Lordites*)
sinuatipennis Fairmaire, 1880c: 181 (*Lordites*)
monticola Grouvelle, 1910b: 250 (*Lasiodactylus*) **A**: NP TAI YUN **ORR**
nepalensis Hisamatsu, 1978: 8 (*Lasiodactylus*) **A**: NP
picta W. S. MacLeay, 1825: 40 (*Nitidula*) **A**: JA PA SHA TAI "Korea" **AFR AUS ORR**
costulata Fairmaire, 1869b: 775 (*Lordites*)
glabricola Candèze, 1861: 340 (*Lordites*)
testudinaria Reitter, 1873d: 87 (*Lasiodactylus*)
pubescens Grouvelle, 1908c: 362 (*Lasiodactylus*) **A**: PA
tuberculifera Reitter, 1873d: 87 (*Lasiodactylus*) **A**: JA

genus *Physoronia* Reitter, 1884g: 260 [= 1885c: 18] type species *Physoronia explanata* Reitter, 1884
subgenus *Lordyrodes* Reitter, 1884g: 264 [= 1885c: 43] type species *Lordyrodes latipes* Reitter, 1884
dentipes Jelínek, 1978a: 204 (*Lordyrodes*) **A**: BT
kaszabi Kirejtshuk, 1984a: 182 (*Lordyrodes*) **A**: SD
latipes Reitter, 1884g: 267 [= 1885c: 44] (*Lordyrodes*) **A**: JA
subgenus *Physoronia* Reitter, 1884g: 260 [= 1885c: 18] type species *Physoronia explanata* Reitter, 1884
Osotima Rebmann, 1944: 22 type species *Osotima klapperichi* Rebmann, 1944 (= *Physoronia explanata* Reitter, 1884)
brunnea Kirejtshuk, 1984a: 181 (*Atarphia*) **A**: SCH YUN **ORR**
caudata Jelínek, 1999b: 266 **A**: YUN
explanata Reitter, 1884g: 263 [= 1885c: 18] **A**: FE FUJ JA
klapperichi Rebmann, 1944: 24 (*Osotima*)
olexai Jelínek, 1999b: 265 **A**: TAI **ORR**
pseudoexplanata Jelínek, 1999b: 270 **A**: SCH
schneideri Jelínek, 1999b: 272 **A**: SCH TAI
subgenus *Pocadiodes* Ganglbauer, 1899a: 543 type species *Pocadius wajdelota* Wankowicz, 1869
gracilis Jelínek, 1978a: 207 (*Pocadiodes*) **A**: UP

harmandi Grouvelle, 1903a: 114 (*Pocadius*) **A**: SD
japonica Reitter, 1873d: 94 (*Pocadius*) **A**: FE JA "Korea"
  rufimargo Reitter, 1884g: 267 [= 1885c: 43] (*Pocadius*)
  unicolor Reitter, 1884g: 267 [= 1885c: 43 (*Pocadius*)
wajdelota Wankowicz, 1869: 415 (*Pocadius*) **E**: AU BY CR GE LT PL SK ST SZ

genus *Platychora* Erichson, 1843b: 284 type species *Platychora polita* Erichson, 1843
  Pherocopis J. Thomson, 1858: 42 type species *Pherocopis ebena* J. Thomson, 1858
hololeptoides Nakane, 1967b: 67 **A**: JA
insularis Hisamatsu, 1964: 54 **A**: JA

genus *Pocadites* Reitter, 1884g: 264 type species *Stelidota dilatimana* Reitter, 1877
chujoi Hisamatsu, 1959c: 30 **A**: JA (Ryukyus) TAI
corpulentus Reitter, 1884g: 267 **A**: FE JA
dilatimanus Reitter, 1877f: 110 (*Stelidota*) **A**: FE JA
  dorsiger Reitter, 1884g: 267
oviformis Reitter, 1884g: 267 **A**: JA
rufobasalis Reitter, 1884g: 267 **A**: FE JA
sauteri Grouvelle, 1914: 43 **A**: TAI

genus *Pocadius* Erichson, 1843b: 318 type species *Nitidula ferruginea* Fabricius, 1775
  Holopterus Gistel, 1856: 361 type species *Nitidula ferruginea* Fabricius, 1775
adustus Reitter, 1888d: 175 **E**: AU BY CT CZ DE FR GB GE GR NL PL SK SP ST SZ **A**: TR
  lanuginosus Franz, 1969b: 146
ferrugineus Fabricius, 1775: 77 (*Nitidula*) **E**: AU BY CT CZ DE EN FI FR GB GE GR HU IR IT LA LT MD NL NR NT PL SK SP ST SV SZ **A**: ES FE (Kamchatka) IN TR WS
  fulvus Marsham, 1802: 136 (*Nitidula*)
  pilosus P. Rossi, 1794: 84 (*Sphaeridium*)
  striatus A. G. Olivier, 1790a: no 12: 14 (*Nitidula*)
  thoracicus Reitter, 1888d: 176
  rottrautae Kniephof, 1913: 187
nobilis Reitter, 1873d: 95 **A**: FE JA YUN "Korea" **ORR**
  yunnanensis Grouvelle, 1910b: 251

genus *Prometopia* Erichson, 1843b: 279 type species *Nitidula sexmaculata* Say, 1825
brevis Jelínek, 1978a: 201 **A**: BT NP **ORR**
gracilis Grouvelle, 1908c: 358 **A**: NP **ORR**
quadrimaculata Motschulsky, 1863: 440 **A**: BT JA NP SD TAI **ORR**
unidentata Hisamatsu, 1959a: 21 **A**: FE JA

genus *Psilopyga* LeConte, 1853b: 286 type species *Psilopyga histrina* LeConte, 1853
  Eugoniopus Reitter, 1884g: 264 type species *Eugoniopus lewisi* Reitter, 1884
lewisi Reitter, 1884g: 267 [= 1885c: 76] (*Eugoniopus*) **A**: JA

genus *Soronia* Erichson, 1843b: 277 type species *Silpha grisea* Linnaeus, 1758
  Norosia Portevin, 1931: 149 type species *Nitidula punctatissima* Illiger, 1794
braeti Grouvelle, 1894b: 580 **A**: NP SD
elongata Cameron, 1903: 97 **E**: AR **N**: MO **A**: CY JO LE SY TR
fracta Reitter, 1884g: 263 [= 1885c: 18] **A**: CH FE JA "Korea"
  maxima Heller, 1923: 67
grisea Linnaeus, 1758: 362 (*Silpha*) **E**: AL AU BH BU BY CR CT CZ DE EN FI FR GB GE GR HU IR IT LA LS LT MD NL NR NT PL PT RO SK SP ST SV UK YU **A**: ES FE JA MG "Korea" **NAR**
  colon A. G. Olivier, 1811: 210 (*Nitidula*) [HN]
  conicicollis Roubal, 1923: 181
  japonica Reitter, 1873d: 47
  maculata DeGeer, 1774: 184 (*Silpha*)

varia Fabricius, 1781: 92 (*Nitidula*)
variegata Geoffroy, 1785: 20 (*Dermestes*)
imperialis Grouvelle, 1903a: 113 **A**: SD
laevigata Kirejtshuk, 1984b: 519 **A**: SCH
lewisi Reitter, 1884g: 263 [= 1885c: 17] **A**: FE JA
minima Grouvelle, 1903a: 113 **A**: SD
oblonga C. N. F. Brisout de Barneville, 1863: 46 **E**: BH BU CR FR GR IT SL SP YU **N**: AG MO TU
punctatissima Illiger, 1794: 598 (*Nitidula*) **E**: AN BH BY CR CT CZ DE EN FI FR GB HU IR IT LA LT MD NL NR NT PL SK SL SP ST SV SZ UK YU

genus *Stelidota* Erichson, 1843b: 300 type species *Nitidula geminata* Say, 1825
franzi Jelínek, 1984b: 147 **A**: NP SD UP **ORR**
geminata Say, 1825a: 181 (*Nitidula*) **Ei**: AU AZ BE FR IT SL SZ TR **NAR NTR**
multiguttata Reitter, 1877f: 110 **A**: JA NP PA SCH TAI UP **AFR ORR**
nepalensis Jelínek, 1984b: 145 **A**: NP SD UP

genus *Thalycra* Erichson, 1843b: 305 type species *Strongylus sericeus* Sturm, 1839 (= *Nitidula fervida* A. G. Olivier, 1790)
Perthalycra Horn, 1879: 309 type species *Perthalycra murrayi* Horn, 1879
emmanueli Auroux, 1967: 91 **E**: FR SP
fervida A. G. Olivier, 1790a: no 12: 15 (*Nitidula*) **E**: AL AR BE BH CT CZ DE FI FR GB GE GG GR HU IT LA LT NL NR NT PT PL SK SP ST SV SZ UK **A**: ES TR
aridosa Gistel, 1857: 590
sericea Sturm, 1839: 114 (*Strongylus*)
wittmeri Jelínek, 1982a: 179 **A**: PA

genus *Triacanus* Erichson, 1843b: 352 type species *Nitidula apicalis* Erichson, 1834
japonicus Hisamatsu, 1985: 193 (*Tricanus*) **A**: JA (Honshu, Shikoku)
sauteri Grouvelle, 1914: 46 (*Tricanus*) **A**: TAI

genus *Ussuriphia* Kirejtshuk, 1992: 185 type species *Soronia hilleri* Reitter, 1877
hilleri Reitter, 1877f: 109 (*Soronia*) **A**: FE JA NC SCH YUN **ORR**
excellentis Kirejtshuk, 1984a: 178 (*Omosita*)

genus *Xenostrongylus* Wollaston, 1854: 127 type species *Xenostrongylus histrio* Wollaston, 1854
Strongyllodes Kirejtshuk, 1992: 196 type species *Xenostrongylus variegatus* Fairmaire, 1891
Strongylolasius Reitter, 1911a: 26 type species *Xenostrongylus arcuatus* Liesenwetter, 1859
arcuatus Kiesenwetter, 1859b: 57 **E**: AU BH CR FR HU IT (Sardegna) SL SZ
canariensis Wollaston, 1854: 127 **N**: CI
deyrollei Jacquelin du Val, 1860f: 140 **E**: IT PT SP **N**: AG MO TU
hirsutus Fairmaire, 1860: 167
obsoletus Chevrolat, 1861: 264
histrio Wollaston, 1854: 128 **N**: MR
lateralis Chevrolat, 1861: 264 **E**: IT **N**: AG TU
siculus Ragusa, 1892: 192
levantinus Audisio & Jelínek, 2001: 492 **A**: TR (Nur Dağlari)
plazae Audisio & Kirejtshuk, 1989: 796 **E**: SP **N**: MO
titanus Audisio, 1986a: 11 **N**: CI (Tenerife, La Gomera)
truncatus Kiesenwetter, 1870: 104 **E**: SP **N**: AG CI MO
variegatus Fairmaire, 1891f: cxcii **A**: BEI FE HEN SHA

subfamily **Cillaeinae** Kirejtshuk & Audisio, 1986

genus *Brachypeplus* Erichson, 1842: 148 type species *Brachypeplus planus* Erichson, 1843
Nitidulopsis Walker, 1858: 206 type species *Nitidulopsis aequalis* Walker, 1858
Onicotis Murray, 1864: 288 type species *Brachypeplus auritus* Murray, 1864
Selis Murray, 1864: 302 type species *Brachypeplus cuneatus* Murray, 1864

Tasmus Murray, 1864: 287 type species *Brachypeplus planus* Erichson, 1842
amplus Grouvelle, 1914: 34 **A**: TAI
deyrollei Murray, 1864: 294 **Ei**: FR **AFR**
dorsalis Grouvelle, 1897: 343 **A**: JA **ORR**
mauli Gardner & Classey, 1962: 153 **E**: AZ **N**: MR
pudicus Kirejtshuk & Gillerfors, 1987: 2

genus *Cillaeus* Laporte, 1835a: 133 type species *Cillaeus castaneus* Laporte, 1835
ryukyuensis Hisamatsu, 1985: 178 **A**: JA (Ryukyus)

genus *Ithyphenes* Murray, 1864: 316 type species *Ithyphenes gnatho* Murray, 1864
sauteri Grouvelle, 1914: 35 **A**: TAI

genus *Platynema* Ritsema, 1885: 29 [RN] type species *Orthogramma fuscipennis* Murray, 1864
Orthogramma Murray, 1864: 318 [HN] type species *Orthogramma fuscipennis* Murray, 1864
japonica Hisamatsu, 1985: 178 **A**: JA (Ryukyus)

subfamily **Meligethinae** C. G. Thomson, 1859

genus *Cyclogethes* Kirejtshuk, 1979d: 359 type species *Cyclogethes orientalis* Kirejtshuk, 1979
abnormis Kirejtshuk, 1979d: 365 **A**: SD **ORR**
aldridgei Kirejtshuk, 1980b: 849 **A**: NP UP YUN **ORR**

genus *Kabakovia* Kirejtshuk, 1979d: 356 type species *Pria latipes* Grouvelle, 1908
latipes Grouvelle, 1908c: 366 (*Pria*) **A**: NP SD UP **ORR**

genus *Meligethes* Stephens, 1830: 30 type species *Nitidula atrata* A. G. Olivier, 1790
Acanthogethes Reitter, 1871: 155 type species *Sphaeridium fuscum* A. G. Olivier, 1790
Astylogethes Kirejtshuk, 1992: 168 type species *Nitidula subrugosa* Gyllenhal, 1808
Chromogethes Kirejtshuk, 1989a: 85 type species *Meligethes splendidulus* Reitter, 1872
Clypeogethes Scholz, 1932: 97 type species *Meligethes leonhardi* Scholz, 1932 (= *Meligethes elongatus* Rosenhauer, 1856)
Idiogethes Kirejtshuk, 1977b: 626 type species *Idiogethes angustitarsis* Kirejtshuk, 1977
Lariopsis Kirejtshuk, 1989a: 86 type species *Meligethes variabilis* Reitter, 1872
Odontogethes Reitter, 1871: 154 type species *Meligethes hebes* Erichson, 1845 (= *Nitidula denticulata* Heer, 1841)
abiens Kirejtshuk, 1979c: 64 **E**: CT
accentus Kirejtshuk, 1978: 578 **A**: TD
acicularis C. N. F. Brisout de Barneville, 1863: 58 **E**: AB AL AR AU BE BH BU BY CR CZ FR GE GG GR HU IT MC RO SK SL SP ST SZ TR UK YU **A**: IN TR
angulosus Rebmann, 1940b: 579
adenesis Easton, 1954a: 346 **A**: YE
aeneoviridinitens Audisio, 1993b: 188 **E**: AR **A**: TR (Karadeniz Dağlari)
aeneus Fabricius, 1775: 78 (*Nitidula*) **E**: AB AL AN AU AZ BE BH BU BY CR CT CZ DE EN FI FR GB GE GG GR HU IR IT LA LS LT LU MA MC MD NL NR NT PL PT RO SK SL SP ST SV SZ TR UK YU **N**: AG CI EG LM MO TU **A**: AF CY ES FE IN IQ IS JO KA KI KZ LE MG NE NC NO NW PA SI SY TD TM TR UZ WS **NAR**
alpestris Heer, 1841: 405 (*Nitidula*)
australis Küster, 1848c: 42
boops Easton, 1957b: 386
brassicae Reitter, 1875a: 16
bonvouloiri C. N. F. Brisout de Barneville, 1872: 10
californicus Reitter, 1871: 68
coeruleus Marsham, 1802: 132 (*Nitidula*)
dauricus Motschulsky, 1849: 93
florilegus Geoffroy, 1785: 12 (*Scarabaeus*)
kelchii Gistel, 1857: 530
latipes Marsham, 1802: 13 (*Nitidula*)
minor Rey, 1889d: 27
moerens LeConte, 1857: 37

mutatus Gemminger & Harold, 1868: 827 [RN]
nigricornis Stephens, 1830: 48
nigrinus Marsham, 1802: 120 (*Nitidula*)
peristericus Roubal, 1943: 65
psyllius Herbst, 1784a: 21 (*Dermestes*)
pubens Rey, 1889d: 27
rotundangulus Ganglbauer, 1899a: 506
rubripennis Reitter, 1871: 68
ruficornis LeConte, 1859: 6 [HN]
rufimanus LeConte, 1857: 37
semiaeneus Ganglbauer, 1899a: 506
subtilis Waltl, 1838b: 270 (*Nitidula*)
urticae Stephens, 1830: 47
viridipennis Motschulsky, 1860b: 129

affinis Jelínek, 1982c: 295 **A**: FUJ SCH
amei Audisio & Kirejtshuk, 1988: 560 **E**: GG ST **A**: TR (Karadeniz Dağlari)
ancestor Kirejtshuk, 1980b: 843 **A**: BEI SHX
angustatus Küster, 1848c: 44 **E**: AU (Kärnten) CR FR (Var) IT RO SL SZ (Ticino)
liguricus Reitter, 1872a: 128
angustitarsis Kirejtshuk, 1977b: 626 (*Idiogethes*) **A**: TM UZ
anthracinus C. N. F. Brisout de Barneville, 1863: 48 **E**: AB AL AR AU BE BH BU BY CR CZ FR GE GG GR HU IT MC MD NL PL PT SK SL SP ST SZ TR UK YU **A**: IN TR
arabicus Jelínek, 1988: 47 **A**: SA YE
armeniacus Audisio, Jelínek & Stevanović, 1999: 68 **E**: AR GG ST (Kavkaz) **A**: TR (eastern provinces)
assamensis Kirejtshuk, 1980b: 845 **A**: AP **ORR**
assimilis Sturm, 1845: 33 **E**: AL AN AU BE BH BU BY CR CT CZ FR GE GR HU IT LS LU MC MD NL PL RO SK SL SP (Pirineos) ST SZ TR UK YU **A**: KZ TR
fibularis Erichson, 1845: 197
astacus Easton, 1957c: 413 **A**: JA
ater C. N. F. Brisout de Barneville, 1863: 50 **E**: AL AU BH BU CR FR (Var) GE (Bayern) GR HU IT (Trieste) PL RO SL ST TR UK YU **A**: IN IQ SY TR
atramentarius Förster, 1849: 19 **E**: AU BE BH BU BY CR CT CZ DE FR GB GE GR (Makedonia) HU IT LU MC MD NL PL RO SK SL SV (Skåne) ST SZ UK YU
diecki Reitter, 1871: 116
kirschi Reitter, 1871: 115
atratus A. G. Olivier, 1790a: no 12: 18 (*Nitidula*) **E**: AL AN AU BE BH BU BY CR CT CZ EN FI FR GB GE HU IR IT LA LS LU MC MD NL NR NT PL PT RO SK SL SP ST SV SZ TR UK YU **N**i: TU **A**: ES FE IN KI KZ MG NE NO NW UZ WS **AFR**i **NAR**i
rufipes Marsham, 1802: 130 (*Nitidula*) [HN]
atrovirens Jelínek, 1982b: 205 **E**: AB AR GG ST **A**: IN TR
audisioi Jelínek, 1997: 132 **A**: XIZ
auricomus Rebmann, 1956b: 131 **A**: FUJ SCH
auripilis Reitter, 1889t: 558 **A**: GAN JIA SHA SCH SHX XIZ
bactrianus Jelínek, 1982c: 292 **A**: TD TM
bidens C. N. F. Brisout de Barneville, 1863: 49 **E**: AL AN AR AU BE BH BU BY CR CT CZ FR GB GE GG GR HU IR IT LS LU MC MD NL PL PT RO SK SL SP ST SV SZ TR UK YU **A**: TR
bidentatus C. N. F. Brisout de Barneville, 1863: 61 **E**: AL AN AU BE BH BU BY CR CT CZ DE FR GB GE GR HU IR IT LS LU MC MD NL PL PT RO SK SL SP ST SZ UK YU **N**: AG MO TU **A**: TR (eastern provinces)
corsicus Sainte-Claire Deville, 1908: 232
gresseri Bach, 1872: 35
obtusus Rey, 1889d: 28
binotatus Grouvelle, 1894a: 460 **A**: NE YUN **ORR**
biondii Audisio, 1988: 213 **A**: TR (eastern provinces)
bithynicus Audisio, 1988: 192 **A**: TR (Karadeniz Dağlari)
bolognai Audisio, 1977b: 23 **E**: AB GG **A**: TR (eastern provinces)

bourdilloni Easton, 1968: 46 **A**: NE YUN
brachialis Erichson, 1845: 195 **E**: AB AL AN AR AU BE BH BU CR CZ FR GE GG GR HU IT MC MD NL PL RO SK SL SP (Pirineos) ST SZ TR UK YU **A**: IN KZ TR WS
dives Reitter, 1872a: 127
braeti Grouvelle, 1894b: 583 **A**: SD
brevipilus Kirejtshuk, 1980b: 837 **A**: SCH YUN
brevis Sturm, 1845: 57 **E**: AL AN AR AU BE BH BU BY CR CZ FR GB GE GG GR HU IT LS LU MC MD NL PL PT RO SK SL SP ST SZ TR UK YU **A**: TR
khevenhuelleri L. Miller, 1852: 112
lobanovi Kirejtshuk, 1984b: 529
mutabilis Rosenhauer, 1856: 102
pictus E. C. Rye, 1871: 75
brisouti Reitter, 1871: 138 **E**: FR (Hérault) PT SP **N**: MO
therondi Easton, 1957a: 88
brittoni Easton, 1954a: 349 **A**: YE
brunnicornis Sturm, 1845: 27 **E**: AB AL AN AR AU BE BH BU BY CR CT CZ EN FI FR GB GE GG GR HU IR IT LA LS LT LU MC MD NL NR NT PL PT RO SK SL SP ST SV SZ TR UK YU **N**: LB (Cyrenaica) **A**: ES (Altai) KI KZ NW TR UZ WS
letzneri Reitter, 1873d: 74
bucciarellii Audisio, 1976a: 9 **E**: AL AR BU GG GR IT MC TR YU **A**: IS LE JO SY TR
buduensis Ganglbauer, 1899a: 524 **E**: AL AR AU (Neusiedler See) BH BU CR GG HU IT (southern Apennines) MC RO SK ST TR UK YU **A**: CY IN IQ IS KZ LE JO SY TR UZ
jejunus Reitter, 1913b: 122
buyssoni C. N. F. Brisout de Barneville, 1882: xxix **E**: AU BU CR CZ FR GE GG HU PL RO SK SL ST SZ TR UK YU **A**: TR
wankae Reitter, 1919: 33
canariensis Kirejtshuk, 1997: 257 **N**: CI
capucinus Robert, 1909: 42 [RN] **N**: AG
bicolor Reitter, 1896f: 268 [HN]
carinulatus Förster, 1849: 16 **E**: AL AN AR AU BE BH BU BY CR CT CZ EN FI FR GB GE GG GR HU IR IT LA LT LU MC MD NL NR NT PL PT RO SK SL SP ST SV SZ TR UK YU **A**: IN JO LE SY TR
binodulus Rey, 1889d: 28
carbonarius Förster, 1849: 5
coriaceus Rey, 1889d: 28
crenulatus Rey, 1889d: 28
erythropus Gyllenhal, 1808a: 237 (*Nitidula*) [HN]
misellus Rey, 1889d: 28
pusillus Rey, 1889d: 28
teter Rey, 1889d: 28
carpathicus Audisio, Jelínek & Stevanović, 1999: 68 **E**: RO (Transsylvanian Alps)
castanescens Grouvelle, 1903a: 114 **A**: GUI SD TAI YUN
caudatus Guillebeau, 1897c: 226 **E**: AU BY CZ DE FR IT PL NR RO SZ UK (Karpaty) YU **A**: TR
chinensis Kirejtshuk, 1979c: 65 **A**: SCH YUN
chlorocyaneus Jelínek & Audisio, 1977: 3 **E**: AU BH BU CZ IT (Alpi Graie) SL
ciliatus Easton, 1956a: 58 **N**: AG TU
cinereus Jelínek, 1978a: 213 **A**: BT
circularis J. R. Sahlberg, 1903a: 18 **A**: ES FE MG NC NE NO
coeruleovirens Förster, 1849: 2 **E**: AU BE BH BU BY CR CT CZ DE EN FI FR GE GR (Makedonia) HU IT (Alpi Giulie) LA LT NL NR NT PL RO SK SL SV ST SZ UK YU
rhenanus Reitter, 1872a: 126
seidlitzi Plavilshchikov, 1924: 232
subrubicundus Reitter, 1871: 63
coerulescens Kraatz, 1858: 134 **E**: GR
conjungens Grouvelle, 1910b: 246 **A**: YUN
cooteri Audisio, 1989a: 25 **E**: YU (Crna Gora)

coracinus Sturm, 1845: 17 **E**: AB AL AN AR AU BE BH BU BY CR CT CZ EN FI FR GB GE GG GR (north) HU IT (north) LA LT LU MC MD NL NR NT PL PT (north) RO SK SL SP (north) ST SV SZ TR UK YU **A**: ES FE IN (north) KI KZ MG NW TD TM TR UZ WS

basalis Reitter, 1919: 22

decoloratus Förster, 1849: 18

epuraeoides Reitter, 1875a: 10

pumilus Erichson, 1845: 176

coronillae Easton, 1962: 74 **E**: SP (Andalucía) **N**: AG MO TU

solskyi Kirejtshuk, 1980a: 353

corvinus Erichson, 1845: 177 **E**: AL AU BE BH BU BY CR CT CZ EN FI FR GB GE HU IT LA LT LU MC MD NL NR NT PL RO SK SL ST (Kavkaz) SZ UK YU **A**: ES FE JA KI KZ MG NC NO NW WS

placidus Easton, 1957c: 401

ventralis Baudi di Selve, 1870: 50

cyaneus Easton, 1957c: 408 **A**: JA

cyrenaicus Rebmann, 1940b: 576 **N**: LB (Cyrenaica)

czwalinai Reitter, 1871: 71 **E**: AU BH BY CR CZ FR GE HU IT PL RO SK SL SP SV (Skåne) SZ UK

denticulatus Heer, 1841: 402 (*Nitidula*) **A**: AL AN AU BE BH BU BY CR CT CZ EN FI FR GE GG GR HU IT LA LS LT LU MC MD NL NR NT PL PT RO SK SL SP ST SV SZ UK YU **A**: ES FE JA KI KZ MG NE NC NO NW TR UZ WS

hebes Erichson, 1845: 172

honshuensis Easton, 1956b: 559

marginatus Gredler, 1870: 7

marginalis Motschulsky, 1860b: 120

olivaceus Heer, 1841: 402 (*Nitidula*) [HN]

devillei Grouvelle, 1912d: 168 **E**: AR AU BU (Rodopi) FR (Alpes) GE (Alpen) GG IT ST SZ UK YU **A**: ES (Altai) KZ MG TR (Karadeniz Dağlari) WS

dieckmanni Audisio & Jelínek, 1984: 568 **E**: AB AR GG ST (Kavkaz) **A**: IN (northern region) TR (Karadeniz Dağlari)

difficilis Heer, 1841: 403 (*Nitidula*) **E**: AB AL AN AU BE BH BU BY CR CT CZ DE EN FI FR GB GE GG GR HU IR IT LA LS LT LU MA MC MD NL NR NT PL PT RO SK SL SP ST SV SZ TR UK YU **A**: ES FE IN JA KI KZ NE NC NO NW TD TR UZ WS

luctifer Reitter, 1872a: 129

milleri Reitter, 1872a: 130

parcus Rey, 1889d: 28

schenklingi Reitter, 1919: 31

dilutipes Easton, 1957b: 395 **A**: AF TD

discoideus Erichson, 1845: 200 **E**: AB AR AU BU CR CZ GG GR HU PL RO SK ST TR UK YU **A**: IN IQ KZ TR

anatolicus Schilsky, 1903: 94

glaucii Kolenati, 1846b: 53 (*Nitidula*)

maculatus Schilsky, 1894: 7

distinctus Sturm, 1845: 59 **E**: AB AL AR AU BE BH BU BY CR CT CZ DE EN FI FR GE GG GR HU IT LA LS LT MC MD NL PL PT RO SK SL SP SV ST SZ TR UK YU **A**: IN IQ TR

confusus C. N. F. Brisout de Barneville, 1872: 24

minimus Rey, 1889d: 27

diversus Schilsky, 1893: 354 **E**: AB AR GG ST **A**: ES (Altai) IN KZ TM TR (eastern provinces)

tschistyakovae Kirejtshuk, 1990b: 101

yakushenkoi Kirejtshuk, 1992: 175

egenus Erichson, 1845: 202 **E**: AB AL AR AU BE BH BU BY CR CT CZ DE EN FI FR GR GG GR HU IT LA LS LT MC MD NL PL RO SK SL SP (north eastern) ST SV (Skåne) ST SZ TR UK YU **A**: CY IN IQ IS JO KZ LE SA SY TM TR

alutaceus Rey, 1889d: 27

menthae C. N. F. Brisout de Barneville, 1863: 57

mucronatus Rey, 1889d: 28

elongatus Rosenhauer, 1856: 98 **E**: FR (Pyrénées-Orientales) IT (Pantelleria) PT SP **N**: AG LB MO TU **A**: IS SY

aenescens Fairmaire, 1875a: 504

crotchi C. N. F. Brisout de Barneville, 1872: 3
freyi Rebmann, 1940a: 548
leonhardi Scholz, 1932: 97
transgrediens Rebmann, 1940a: 550
verrucicollis Jelínek, 1978b: 331
erichsonii C. N. F. Brisout de Barneville, 1863: 59 **E**: AL AU BH BU CR FR GB GE GR HU IT RO SK SL SP SZ TR YU **N**: AG MO TU
stierlini Reitter, 1872c: 268
erysimicola Audisio & De Biase, 2001: 377 [RN] **E**: AB AL AR AU BH BU CR CZ FR (southern provinces) GE GG GR HU IT RO SK SL SP ST (Kavkaz) SZ TR UK YU **A**: IN LE SY TR
aenescens Ganglbauer, 1899a: 504 [HN]
exilis Sturm, 1845: 53 **E**: AB AL AN AR AU BE BH BU BY CR CT CZ DE EN FI FR GB GE GG GR HU IR IT LA LS LT LU MC MD NL PL PT RO SK SL SP ST SV SZ TR UK YU **N**: MO **A**: TR
explanatus Reitter, 1900a: 52 **A**: KZ SY TD UZ
ferrugineus Reitter, 1873c: 51 **A**: SD
flavicollis Reitter, 1873d: 76 **A**: CE FE JA NC NE NO SC SE ZHE
perversecoloratus Roubal, 1943: 65
semirufus Reitter, 1879e: 216
flavimanus Stephens, 1830: 45 **E**: AB AL AN AR AU BE BH BU BY CR CT CZ EN FI FR GB GE GG GR HU IR IT LA LS LT LU MC MD NL NR NT PL PT RO SK SL SP ST SV SZ TR UK **A**: ES FE KI KZ MG NC NE NO NW TD TM TR UZ WS
asperrimus Guillebeau, 1897c: 225
borealis Motschulsky, 1845b: 364
foersteri Reitter, 1871: 54
foveifrons Reitter, 1871: 55
lumbaris Sturm, 1845: 7
ogumae Matsum ura, 1911: 116
sibiricus Rebmann, 1956b: 127
fulvipes C. N. F. Brisout de Barneville, 1863: 49 **E**: AN BE FR GB GE (Rheinland) IT (Piemonte, Liguria) NL PT SP SZ (Jura) **N**: AG MO TU
fulvicornis Rey, 1889d: 27
rubripes Mulsant & Rey, 1863c: 4
fumatus Erichson, 1845: 205 **E**: AL AU BH CR FR GR HU IT MC SL YU
funereus Jelínek, 1967: 26 **E**: AB AL AR BH BU CR GG GR IT MC SL ST TR UK YU **A**: IN TR
fuscus A. G. Olivier, 1790a: no 12: 10 (*Sphaeridium*) **E**: AN FR GR IT MA PT SP SZ **N**: AG MO TU
barbarus P. H. Lucas, 1846: 216 (*Cercus*)
bicolor P. H. Lucas, 1846: 215 (*Cercus*)
pinguis Horn, 1879: 314
pyrenaicus Laporte, 1840b: 9 (*Strongylus*)
gagathinus Erichson, 1845: 201 **E**: AL AN AU BE BH BU BY CR CT CZ DE EN FI FR GB GE GG GR HU IT LA LS LT LU MC MD NL PL RO SK SL SP ST SV SZ UK YU **A**: ES FE HEI JA KZ NC NE NO SC TR (Artvin) WS
cristatus Förster, 1849: 9
mikado Reitter, 1884g: 265 [= 1885c: 40]
gracilis C. N. F. Brisout de Barneville, 1863: 48 **E**: BE FR GE (Rheinland) IT PT SP SZ
grenieri C. N. F. Brisout de Barneville, 1872: 7 **E**: FR IT MA PT SP **N**: AG LB MO TU
unidentatus Schilsky, 1894: 5
griseus Jelínek, 1978a: 216 **A**: BT
gurjevae Kirejtshuk, 1984b: 526 **A**: ES (Yakutia) MG
haemorrhoidalis Förster, 1849: 4 **E**: AB AL AN AR AU BE BH BU BY CR CT CZ DE FR GB GE GG HU IT LA LU MC MD NL PL RO SK SL SP ST SV SZ TR UK YU **A**: IN TR
floribundus Reitter, 1877h: 293
praestans Kirejtshuk, 1977b: 637
hammondi Kirejtshuk, 1980b: 840 **A**: SHA
haroldi Reitter, 1877i: 372 **A**: FE JA
longipennis Easton, 1957c: 405
hercules Audisio, 1986b: 551 **N**: CI (Tenerife)

himalayaensis Kirejtshuk, 1980b: 841 **A**: UP
hladili Jelínek, 1982d: 306 **E**: AR GG **A**: IN IQ SY TR
hoffmanni Reitter, 1871: 92 **E**: AB AL AR AU BH BU BY CR CZ FR GR HU IT MA MC MD PT RO SK SL SP ST SV (Öland) TR UK YU **N**: AG MO **A**: CY ES FE IN JA KZ LE SY TR UZ WS
japonicus Reitter, 1873d: 73
niger C. N. F. Brisout de Barneville, 1863: 50
parvulus C. N. F. Brisout de Barneville, 1872: 28
peneckei Marcu, 1939: 374
holzschuhi Jelínek & Spornraft, 1979: 9 **A**: TR (Toros Dağlari)
humerosus Reitter, 1871: 60 **E**: AU CR CZ IT (Alpi Graie) RO SK UK (Karpaty)
subalpinus Reitter, 1878b: 52
immundus Kraatz, 1858: 135 **E**: AL BH BU (Pontic areas) CR FR (Provence) GR IT MA PT RO (Pontic areas) YU (Crna Gora) **N**: AG LB MO TU **A**: CY IS JO LE TR
castaneus C. N. F. Brisout de Barneville, 1863: 55
picipennis Mulsant & Rey, 1863c: 6
incanus Sturm, 1845: 42 **E**: AL AR AU AZ BE BH BU BY CR CZ FR GB GE GG GR HU IT MC MD NL PL PT RO SK SL SP SZ TR UK YA **N**: AG MO **A**: IN IQ IS JO LE SY TR
gredleri Reitter, 1873d: 72
initialis Kirejtshuk, 1979c: 59 **A**: LIA SCH SHX
interjectus Jelínek & Spornraft, 1979: 7 **E**: AR **A**: TR (Karadeniz Dağlari)
intermedius Kirejtshuk, 1979c: 53 **A**: KZ
intermixtus Kirejtshuk, 1979c: 61 **A**: KI
isoplexidis Wollaston, 1854: 123 **N**: MR
echii Wollaston, 1857a: 40 [RN]
jelineki Audisio, 1976b: 299 **E**: AU BE BH BU CR CZ FR GE HU IT PL RO SL SP ST SZ UK YU
jordanis Jelínek & Spornraft, 1979: 10 **E**: AR ST (Karadeniz Dağlari) UK (Krym) **A**: IQ IS JO LE SY TR
kasparyani Kirejtshuk, 1984b: 525 **A**: BEI FE (Primorie) HEB
kaszabi Audisio, 1979: 134 **E**: GG
khnzoriani Kirejtshuk, 1979e: 679 **E**: AR **A**: TR
kirejtshuki Audisio, 1979: 130 **E**: GG ST (Kavkaz) **A**: TR (Karadeniz Dağlari)
klapperichi Easton, 1957b: 389 **A**: AF IN KA PA TD TM UZ
ahriman Jelínek, 1981a: 111
kraatzii Reitter, 1871: 135 **E**: AB AL AR AU BU CR CZ GG GR HU PL RO SK ST TR UK YU **A**: CY IN IQ TR
assyricus Obenberger, 1914: 104
basimargo Reitter, 1919: 49
frivaldszkyi Reitter, 1875h: 88
kunzei Erichson, 1845: 182 **E**: AL AN AU BE BH BU BY CR CT CZ FR GB GE GG HU IT LU MC MD NL PL RO SK SL SP (Pirineos) ST SZ TR UK YU **A**: TR (Karadeniz Dağlari)
blandulus Reitter, 1872a: 130
kvaki Kirejtshuk, 1977a: 42 **A**: KI KZ TD TM
lamii Rosenhauer, 1856: 101 **E**: PT SP **N**: AG MO TU
ampulla Easton, 1955: 236
laguncula Easton, 1955: 232
tridens Easton, 1955: 235
leati Easton, 1956a: 49 **E**: AB AR GG IT (Sicilia) ST **N**: AG LB TU **A**: IN
lederi Reitter, 1871: 85 **E**: PT SP **N**: AG MO TU
marmottani C. N. F. Brisout de Barneville, 1872: 19
lepidii L. Miller, 1851: 111 **E**: AB AL AR AU BH BU BZ CR CT CZ FR (Pyrénées) GE GR HU IT MC MD PL SK SL SP ST SZ TR UK YU **A**: AF CY ES FE IN IQ IS JO KI LE SY TD TM TR UZ WS
chalybaeus Reitter, 1871: 131
hypocrita C. N. F. Brisout de Barneville, 1867a: 117 [RN]
solitarius Reitter, 1872a: 131
subtilis C. N. F. Brisout de Barneville, 1866: 368 [HN]
lindbergi Rebmann, 1940b: 578 **E**: AL CR FR GR IT MA SL YU **N**: AG TU
lloydi Easton, 1968: 44 **A**: NE TAI XIZ YUN

longulus Schilsky, 1894: 8 **E**: AR **A**: TR (eastern provinces)
lugubris Sturm, 1845: 49 **E**: AB AL AN AR AU BE BH BU BY CR CT CZ DE EN FI FR GB GE GG GR HU IR IT LA LS LT LU MC MD NL PL RO SK SL SP ST SV SZ TR UK YU **A**: ES FE KZ MG TR WS
ebeninus Förster, 1849: 15
mellitulus Reitter, 1872a: 132
lunariae Audisio & De Biase, 1999: 101 **E**: GR IT
mandibularis J. R. Sahlberg, 1913a: 24 **E**: GG ST (Kavkaz) **A**: TR
nitidissimus Roubal, 1943: 66
robustus Jelínek & Marek, 1966: 455
maroccanus Easton, 1956a: 56 **N**: MO
matronalis Audisio & Spornraft, 1990: 72 **E**: AB AU BE CR CZ DE FR GE HU IT PL SP SL SV SZ **A**: TR (Karadeniz Dağlari)
maurus Sturm, 1845: 36 **E**: AB AL AN AR AU BE BH BU BY CR CT CZ FR GE GG GR HU IT LA LS LT LU MC MD NL NR PL PT RO SK SL SP ST SV SZ TR UK YU **A**: CY ES IN IQ IS KI KZ NW SY TM TR UZ WS
calvus Reitter, 1871: 99
durus Reitter, 1871: 99
meridionalis Reitter, 1871: 99
rufimanus J. R. Sahlberg, 1913d: 102 [HN]
medvedevi Kirejtshuk, 1978: 583 **E**: GG ST (Kavkaz) **A**: TR (Erzurum Prov.)
minutus C. N. F. Brisout de Barneville, 1863: 60 **E**: FR IT (Sicilia) PT SP SZ GR (Kríti) **N**: AG MO TU
parallelus Reitter, 1871: 92
mirae Audisio, Jelínek & Stevanović, 1999: 62 **E**: YU (Crna Gora)
mirator Kirejtshuk, 1979a: 71 **A**: FE (Primorie)
mithra Jelínek, 1978b: 333 **A**: IS JO SY TR
morosus Erichson, 1845: 184 **E**: AB AL AN AU BE BH BU BY CR CT CZ DE EN FI FR GB GE GG GR HU IR IT LA LS LT LU MY MC MD NL NR NT PL PT RO SK SL SP ST SV SZ TR UK YU **N**: MO **A**: ES FE IN JA KI KZ MG NE NC NO NW TD TM TR UZ WS
lewisii Reitter, 1873d: 74
memnonius Erichson, 1845: 183
mus Reitter, 1884g: 265 [= 1885c: 40] **A**: JA
nakanei Easton, 1957c: 418 **A**: BEI CE ES FE HEB JA MG NC NE NO SC SW
eastoni Jelínek, 1965a: 138
nanus Erichson, 1845: 193 **E**: AB AL AN AR AU BE BH BU BY CR CT CZ FR GB GE GG GR HU IT LU MC MD NL PL PT RO SK SL SP ST SV (Skåne) SZ TR UK YE **A**: CY IN IQ IS JO KZ LE SY TR
dalmatinus Reitter, 1871: 89
marrubii C. N. F. Brisout de Barneville, 1863: 54
rugicollis Rey, 1889d: 27
schneideri Reitter, 1878a: 157
spinipes Reitter, 1871: 87
nepalensis Easton, 1968: 42 **A**: NE
nigerrimus Rosenhauer, 1856: 99 **E**: AN FR (south) IT (Sardegna) PT SP **N**: AG MO TU
infidus Grouvelle, 1913d: 393
tropicus Reitter, 1871: 129
tropicus Rey, 1889d: 28 [HN]
nigrescens Stephens, 1830: 47 **E**: AB AL AN AU BE BH BU BY CR CT CZ DE EN FI FR GB GE GG GR HU IR IT LA LS LT LU MA MC MD NL NR NT PL PT RO SK SL SP ST SV SZ TR UK YU **N**: AG EG LB MO MR TU **A**: AE AF CY ES FE IN IQ IS JO KA KI KZ LE MG NE NC NO NW PA SA SI SY TD TM TR UZ WS **NARi**
funebris Förster, 1849: 23
medius Rey, 1889d: 28
pallipes Rey, 1889d: 28
picipes Sturm, 1845: 47
saulcyi Reitter, 1872a: 133
seminulum LeConte, 1857: 37
subrugosus Stephens, 1830: 48
subsimilis Rey, 1889d: 28

xanthocerus Stephens, 1830: 48
nigritus P. H. Lucas, 1846: 216 (*Epuraea*) **E**: FR IT PT SP **N**: AG MO TU
nitidicollis Reitter, 1873c: 49 **A**: ES FE JA NC
normandi Easton, 1954b: 382 **E**: FR (Landes) PT SP **N**: AG MO TU
norvegicus Easton, 1959a: 50 **E**: CT NR
nuragicus Audisio & Jelínek, 1990: 80 **E**: FR (Corse) IT (Sardegna)
obscurus Erichson, 1845: 203 **E**: BE DE FR GB GE IR IT LU NL NR PT SP SV SZ **N**: AG MO
palmatus Erichson, 1845: 204
ochropus Sturm, 1845: 23 **E**: AU BE BH BU BY CR CT CZ DE EN FI FR GB GE HU IT LA LT LU MD NL NR PL PT RO SK SL SP ST SV SZ UK YU **A**: ES FE KZ MG NC NE NO NW WS
abbreviatus Reitter, 1919: 31
quadristriatus Förster, 1849: 6
opacus Rosenhauer, 1856: 100 **E**: FR (Provence) IT (Sardegna) PT SP UK (Krym) **N**: AG MO TU
natricis C. N. F. Brisout de Barneville, 1863: 175
subopacus Reitter, 1874c: 511
oreophilus Audisio, 1984: 368 **E**: IT SZ (Graubünden, Ticino)
osellai Audisio & Jelínek, 2000: 312 **A**: TR (Yozgat, Nevşehir)
otini Easton, 1954b: 388 **E**: PT SP **N**: AG MO TU
ovatus Sturm, 1845: 44 **E**: AL AU BE BH BU BY CR CT CZ EN FI FR GB GE HU IR IT LA LT LU MC MD NL NR NT PL RO SK SL ST (Kavkaz) SV SZ UK YU
ovalis C. G. Thomson, 1862: 159
pamirensis Kirejtshuk, 1979c: 55 **A**: PA TD
paschalis Spornraft, 1975: 13 **E**: IT
pectinatus Schilsky, 1894: 6 **E**: AR GG **A**: TR
pectoralis Rebmann, 1956a: 47 **A**: FUJ GUI TAI ZHE
pedicularius Gyllenhal, 1808a: 236 (*Nitidula*] **E**: AB AL AN AR AU BE BH BU BY CR CT CZ DE EN FI FR GB GE GG GR HU IR IT LA LS LT LU MC MD NL NR NT PL PT RO SK SL SP ST SV SZ TR UK YU **N**: AG **A**: ES FE IN KI KZ MG NW TR UZ WS
aestimabilis Reitter, 1872a: 133
austriacus Reitter, 1871: 124
bituberculatus Förster, 1849: 8
fuliginosus Erichson, 1845: 196
luctuosus Förster, 1849: 15
melanarius Förster, 1849: 24
viduatus Heer, 1841: 403 (*Nitidula*)
perceptus Jelínek & Spornraft, 1979: 6 **A**: IN (Zagros)
persicus Faldermann, 1835a: 225 **E**: AB AL AN AR AU BE BH BU BY CR CT CZ FR GB GE GG GR HU IR IT LS LU MC MD NL PL RO SK SL SP ST SV SZ TR UK YU **A**: ES KI KZ NW TD TR UZ WS XIN
sinuans Rey, 1889d: 27
tenebrosus Förster, 1849: 13
pharetra Easton, 1957b: 394 **A**: AF IN KA PA TM UZ
planiusculus Heer, 1841: 404 (*Nitidula*) **E**: AB AL AN AR AU BE BH BU BY CR CT CZ FR GB GE GG GR HU IR IT LS LU MA MC MD NL PL PT RO SK SL SP ST SZ TR UK YU **N**: AG EG LB MO MR TU **A**: CY IN IQ IS JO KZ LE SI SY TR
murinus Erichson, 1845: 191
ruficornis Heer, 1841: 404 (*Nitidula*) [HN]
seniculus Erichson, 1845: 192
potanini Kirejtshuk, 1979c: 57 **A**: SCH
praetermissus Easton, 1957c: 403 **A**: FE JA HEI NC
primoriensis Kirejtshuk, 1987a: 88 **A**: FE (Primorie)
privus Kirejtshuk, 1977a: 44 **A**: TD
prometheus Jelínek, 1982b: 202 **E**: GG ST (Kavkaz) **A**: IN (north) TR (Karadeniz Dağlari)
ponticus Jelínek, 1982b: 204
punctatissimus Reitter, 1897h: 312 **E**: AR **A**: TR
punctatus C. N. F. Brisout de Barneville, 1863: 56 **E**: AL BH BU CR FR GR IT MA MC PT RO (Karadeniz Dağlari) SL SP SZ TR YU (Crna Gora) **A**: IS JO LE SY TR

ciliaris Rey, 1889d: 28
bruckii Reitter, 1871: 149
subregularis Rey, 1889d: 28
rebmanni Easton, 1957b: 392 **A**: AF PA TD
reitteri Schilsky, 1894: 4 **E**: AB AN AR AU BE BH FR GE GG HU IT (Monti Albani) RO SP (Pirineos, Cordillera Cantábrica) ST (Kavkaz) SZ UK (Karpaty) **A**: ES KI KZ (Altai) MG (Altai) TR WS
reyi Guillebeau, 1885b: 2 **E**: AL AN AR BH BU CR FR GG GR HU IT MC PT RO SL SP ST SZ TR UK YU **A**: TR
rosenhaueri Reitter, 1871: 84 **E**: AB AL AR AU BH BU BY CR CZ GE GG GR HU IT PL RO SK SL ST TR UK YU **A**: KZ SY TR
kirschenblatti Kirejtshuk, 1979b: 39
krueperi Reitter, 1873d: 75
rotroui Easton, 1952: 189 **N**: AG MO TU
rotundicollis C. N. F. Brisout de Barneville, 1863: 56 **E**: AL AN BE BH BU CR FR GB GE GG GR HU IT MA MC NL PT RO SL SP ST (Pontic areas) SZ TR UK YU **N**: AG EG LB TU MO **A**: CY IQ (northwest areas) IS JO LE SI SY TR
obsoletus Rey, 1889d: 28
ruficornis Marsham, 1802: 131 (*Nitidula*) **E**: AB AL AN AU BE BH BU BY CR CT CZ DE EN FI FR GB GE GG GR HU IR IT LA LS LT LU MA MC MD NT PL PT RO SK SL SP ST SV SZ TR UK YU **N**: AG TU **A**: CY IN IQ IS JO KZ LE SI SY TD TM TR UZ WS
flavicornis L. Miller, 1852: 111
flavipes Sturm, 1845: 46
moestus Erichson, 1845: 194
pruinosus Rey, 1889d: 27
salvan Audisio, De Biase & Antonini, 2003: 123 **E**: IT (Alpi Maritime)
saxatilis Audisio, 1988: 202 **A**: IN TR
schilskyi Reitter, 1897h: 312 **A**: AF TD TM UZ
scholzi Easton, 1960: 337 **E**: AL BH CR GR IT (southern Apennines, Sicilia) YU (Crna Gora)
scotti Easton, 1954a: 343 **A**: SA YE **AFR**
semenovi Kirejtshuk, 1979a: 66 **A**: FE (Ussuri) SCH SHA
serripes Gyllenhal, 1827: 301 (*Nitidula*) **E**: AN AU BE BH BU BY CR CT CZ DE EN FI FR GB GE GG HU IT LA LS LT LU MD NL PL RO SK SL SP ST SV SZ UK YU **A**: ES (Altai) KI KZ WS
alpigradus Reitter, 1871: 82
exaratus Förster, 1849: 12
quadridens Förster, 1849: 21
shimoyamai Hisamatsu, 1964: 56 **A**: JA
shirakii Hisamatsu, 1956: 168 **A**: JA SCH TAI
shirozui Hisamatsu, 1965: 137 **A**: TAI
simillimus Kirejtshuk, 1984b: 522 **A**: FE LIA
simplex Kraatz, 1858: 135 **E**: GR
solidus Kugelann, 1794: 530 (*Sphaeridium*) **E**: AB AL AN AR AU BE BH BU BY CR CT CZ DE EN FR GB GE GG GR HU IT LA LS LT LU MC MD NL PL RO SK SL SP ST SV SZ TR UK YU **A**: TR
spadix Easton, 1954a: 341 **A**: YE
spornrafti Audisio, 1977a: 15 **E**: FR (Alpes-Maritimes) IT YU
squamosus Jelínek & Marek, 1966: 453 **E**: BU (Rila, Rodopi)
strejceki Jelínek, 1982c: 297 **A**: UZ
subaeneus Sturm, 1845: 11 **E**: AL AN AU BE BH BU BY CR CT CZ DE EN FI FR GE GR (Pindos) HU IT LA LT LU MC MD NL NR NT PL RO SK SL SP ST SV SZ UK YU
subater Kirejtshuk, 1980b: 844 **A**: HEI SHX
subfumatus Ganglbauer, 1899a: 532 **E**: FR IT
submetallicus Sainte-Claire Deville, 1908: 231 **E**: AL BH BU CR FR GR HU IT MC SL ST TR UK YU **A**: ES KZ TR WS
subrugosus Gyllenhal, 1808a: 236 (*Nitidula*) **E**: AB AL AN AU AZ BE BH BU BY CR CT CZ DE EN FI FR GB GE GG GR HU IR IT LA LS LT LU MC MD NL NR NT PL PT RO SK SL SP ST SV SZ TR UK YU **N**:

AG CI EG LB MO TU **A**: AF CY ES FE IN IQ IS JA JO KA KI KZ LE MG NE NC NO NW PA SC SHA SI SY TD TM TR UZ WS

substrigosus Erichson, 1845: 178

turbidescens Easton, 1957c: 400

subtristis Easton, 1957b: 398 **A**: AF TD UZ

tristoides Easton, 1957b: 399

sulcatus C. N. F. Brisout de Barneville, 1863: 51 **E**: AL AN AR AU BE BH BU BY CR CT CZ DE EN FI FR GE GG GR HU IT LA LS LT LU MY MD NL NT PL RO SK SL SP ST SZ TR UK YU **A**: ES (Altai Mts) KI KZ NW TR WS

melancholicus Reitter, 1871: 103

moraviacus Reitter, 1871: 76

ranunculi Reitter, 1872a: 127

schweigeri Jelínek, 1968: 1

symphyti Heer, 1841: 405 (*Nitidula*) **E**: AB AL AN AR AU BE BH BU BY CR CT CZ EN FR GE GG GR HU IT LA LS LT LU MC MD NL NT PL PT RO SK SL SP ST SZ TR UK YU **N**: AG (Djurdjura) **A**: IN TR

glabripennis Roubal, 1943: 67

reclairei Scholz, 1933: 77

syriacus C. N. F. Brisout de Barneville, 1872: 20 **A**: IS JO SY

tatjanae Kirejtshuk, 1982: 35 **A**: UZ

tauricus Jelínek & Spornraft, 1979: 8 **E**: AB AR UK (Krym) **A**: IN IQ TR

tener Reitter, 1873c: 49 **E**: CR (Dalmacija) GR MC IT (Monte Gargano, Gran Sasso) ST (Kavkaz) YU (Montenegro) **A**: TR

tilmani Easton, 1968: 40 **A**: NE PA UP XIZ

topali Kirejtshuk, 1988: 66 **A**: "Northeast India" **ORR**

torquatus Jelínek, 1997: 128 **A**: TAI

transmissus Kirejtshuk, 1988: 62 **A**: SCH YUN

tristis Sturm, 1845: 40 **E**: AL AN AU BE BH BU BY CR CT CZ DE EN FI FR GE GR HU IT LA LS LT LU MC MD NL NR PL PT RO SK SL SP ST SV SZ TR UK YU

turcicus Jelínek, 1982b: 207 **A**: TR

umbrosus Sturm, 1845: 38 **E**: AB AL AN AR AU BE BH BU BY CR CT CZ DE EN FI FR GB GE GG GR HU IR IT LA LS LT LU MC MD NL NR PL PT RO SK SL SP ST SV SZ TR UK YU **N**: TU **A**: KI KZ TR WS

transsylvanicus Petri, 1912: 154

varicollis Wollaston, 1854: 126 **E**: IT (Sardegna) PT SP **N**: AG CI MO MR TU

erythropus Hartung, 1857: 140

metallicus Rosenhauer, 1856: 96

ryei Wollaston, 1871: 238

theryi Guillebeau, 1894c: cccxxvii

theryi Pic, 1894j: 77 [HN]

variolosus Easton, 1964: 3 **E**: AU BU CR FR (Alpes-Maritimes) GG GR HU RO SK ST (Kavkaz) TR UK YU **A**: TR

villosus C. N. F. Brisout de Barneville, 1863: 53 **E**: AL AN AU BH BU CR FR GE GR HU IT NL PL PT RO SL SP SZ TR YU **N**: MO

hispanicus Reitter, 1871: 86

violaceus Reitter, 1873d: 71 **A**: CE FE JA NE NO SE ZHE

viridescens Fabricius, 1787a: 52 (*Nitidula*) **E**: AB AL AN AR AU BE BH BU BY CR CT CZ DE EN FI FR GB GE GG GR HU IR IT LA LS LT LU MC MD NL NR PL PT RO SK SL SP ST SV SZ TR UK YU **N**: AG MO TU **A**: ES (Altai Mts) IN IS JO KZ LE SY TR WS XIN (Tian Shan) **NAR**i

azureus Heer, 1841: 406 (*Nitidula*)

bohemicus Obenberger, 1914: 104

coloreus Rey, 1889d: 27

discolor Reitter, 1872c: 265

germanicus Reitter, 1871: 69

perviridis Reitter, 1896f: 268

pratincola Gistel, 1857: 600

rufipes Fabricius, 1775: 78 (*Nitidula*) [HN]

virescens C. G. Thomson, 1862: 154

vomer Kirejtshuk, 1978: 590 **E**: BU RO ST (Pontic areas) TR UK (Krym) **A**: TR

vulpes Solsky, 1876: 251 **A**: AF KI PA TD TM UZ
lutra Solsky, 1876: 252
melanocephalus Rebmann, 1956b: 132
subopacus Reitter, 1891b: 24 (*Micrurula*) [HN]
transmutatus Grouvelle, 1913d: 393 [RN]
wagneri Rebmann, 1956a: 45 **A**: FUJ
wittmeri Jelínek &Audisio, 1977: 6 **E**: AR **A**: TR (Karadeniz Dağlari)
wollastoni Easton, 1950: 309 [RN] **N**: CI
virescens Wollaston, 1864: 113 [HN]
xanthopus Solsky, 1876: 290 **A**: UZ
yemenensis Easton, 1954a: 338 **N**: EG **A**: IS JO SA SI YE **AFR**
zapparolii Audisio, 1989b: 179 **A**: IN SY TR
zarudnyi Kirejtshuk, 1984b: 527 **A**: IN (northwestern region) TR (southeastern provinces)

genus *Meligethinus* Grouvelle, 1906b: 202 type species *Meligethinus humeralis* Grouvelle, 1906
Prianella Reitter, 1919: 16 type species *Pria pallidula* Erichson, 1843
apicalis Grouvelle, 1894b: 582 (*Pria*) **A**: CH SD **ORR**
gedrosiacus Jelínek, 1981a: 106 **A**: AE IN SA
pallidulus Erichson, 1843b: 308 (*Pria*) **E**: CRi FR (Mediterranean coast, Corse) GRi IT (western coast, Sardegna, Sicilia) PT SLi SP TRi **N**: AG LB MO TU
prioides Reitter, 1875n: 393 (*Meligethes*)
plagiatus Grouvelle, 1894b: 582 (*Pria*) **A**: SD TAI
elegans Grouvelle, 1914: 37 (*Pria*)
quadricollis Kirejtshuk, 1987a: 87 **A**: UP
tschungseni Kirejtshuk, 1987a: 85 **A**: FUJ SCH YUN

genus *Pria* Stephens, 1830: 49 type species *Silpha truncatella* Marsham, 1802 (= *Laria dulcamarae* Scopoli, 1763)
Cormyphora Laporte, 1840b: 12 type species *Cormyphora mandibularis* Laporte, 1840 (= *Laria dulcamarae* Scopoli, 1763)
Strychnobia Gistel, 1857: 573 type species *Nitidula dulcamarae* Illiger, 1798 (= *Laria dulcamarae* Scopoli, 1763)
compacta Cooper, 1982: 388 **A**: YE
dulcamarae Scopoli, 1763: 22 (*Laria*) **E**: AB AL AN AR AU BE BH BU BY CR CT CZ DE FR (Corse) GB GE GG GR HU IR IT(Sardegna, Sicilia) LA LS LT LU MC MR NL NR PL PT RO SK SL SP ST SV SZ TR UK YU **N**: AG CI EG MO **A**: CY ES FE IN IS JA JO SY TR WS YE **AFR**
breviuscula Kolenati, 1846b: 53 (*Nitidula*)
mandibularis Laporte, 1840b: 12 (*Cormyphora*)
olivacea Gyllenhal, 1813: 679 (*Nitidula*)
truncatella Marsham, 1802: 123 (*Silpha*)
gracilipes Cooper, 1982: 373 **A**: SA **AFR**
hirta Cooper, 1982: 385 **A**: YE **AFR**
indica Grouvelle, 1894b: 581 **A**: SD **ORR**
kolbei Grouvelle, 1909a: 103 **A**: SA **AFR**
raffrayi Grouvelle, 1909b: 135 **A**: SA **AFR**
tokarensis Nakane, 1959a: 55 **A**: JA (Tokara-Rettō)
transitoria Kirejtshuk, 1979c: 51 **E**: CT GG ST "Transcaucasia" **A**: ES WS
zenobia Jelínek, 1997: 126 **E**: GR **A**: IS TR

subfamily **Cybocephalinae** Jacquelin du Val, 1858

genus *Cybocephalus* Erichson, 1844: 441 type species *Anisotoma exigua* C. R. Sahlberg, 1834 (= *Nitidula polita* Gyllenhal, 1813)
Acribis C. O. Waterhouse, 1877a: 78 type species *Acribis serrativentris* C. O. Waterhouse, 1877
Dissia Chobaut, 1896a: 167 type species *Dissia albopilosa* Chobaut, 1896
Phantazomerus Jacquelin du Val, 1854: xxxvii type species *Phantazomerus aeneiceps* Jacquelin du Val, 1854
Stagonomorpha Wollaston, 1854: 482 type species *Stagonomorpha sphaerula* Wollaston, 1854
aegyptiacus Endrödy-Younga, 1968: 72 **N**: EG **A**: IS

afghanicus Endrödy-Younga, 1968: 58 **A**: AF
aksuensis Tian, 2000: 127 **A**: XIN
algiricus Endrödy-Younga, 1968: 61 **N**: AG
albopilosus Chobaut, 1896a: 167 (*Dissia*) **N**: AG EG
recticollis Abeille de Perrin, 1909a: 8 (*Dissia*)
amplus J. R. Sahlberg, 1908: 54 (*Dissia*) **N**: EG
atomus C. N. F. Brisout de Barneville, 1866: 369 **E**: SP
bernardi Peyerimhoff, 1948a: 22 **N**: LB
bicinctus Kirejtshuk, 1988: 92 **A**: FE
binotatus Grouvelle, 1908c: 394 **A**: HP JA UP
aeneonitens Champion, 1923c: 77
canariensis Endrödy-Younga, 1968: 87 **N**: CI (Tenerife, Gran Canaria)
chinensis G. Yu, 1995a: 31 **A**: GUA
chlorocephalus Erichson, 1844: 442 **A**: UP **ORR**
concavus G. Yu & Tian, 1995: 36 **A**: GUX
conduplicatus Tian, 2001: 43 **A**: GAN
diadematus Chevrolat, 1861: 265 **E**: FR IT (Sicilia) SP **N**: AG
smaragdifrons Reitter, 1885b: 121
dinghushanensis Tian, 1996: 92 **A**: GUA
dissectus G. Yu, 1994: 14 **A**: GUA GUX SCH
dunhuangensis Tian, 2001: 44 **A**: GAN
endroudii Tian, 1995: 42 **A**: HUB
explansus G. Yu, 1994: 15 **A**: GUA
festivus Erichson, 1845: 218 **E**: AU CR IT **N**: AG
clypeonitens Fiori, 1905b: 136
marginatus Fiori, 1905b: 134
flavicapitus Tian & G. Yu, 1994b: 209 **A**: GUA
fodori fodori Endrödy-Younga, 1965: 41 **E**: AL AU BH CR CZ FR GB GE GG GR HU IT MD PL RO SK TR YU "Caucasus" **A**: TM TR XIN
fodori minor Endrödy-Younga, 1968: 83 **N**: MO **A**: IS JO LE SA SY TR
freyi Endrödy-Younga, 1968: 70 **E**: IT (Sardegna) **N**: MO
gibbulus Erichson, 1844: 441 **A**: PA UP **ORR**
guilinensis G. Yu, 1995b: 278 **A**: GUX
heydeni Reitter, 1875f: 56 **E**: FR IT(Liguria) PT SP **N**: AG
hispanicus Reitter, 1885b: 121
intermedius G. Yu, 1997: 666 **A**: SCH
israelicus Endrödy-Younga, 1984: 1 **A**: IS
klapperichi Endrödy-Younga, 1968: 50 **A**: AF
krimicus Endrödy-Younga, 1968: 55 **E**: UK (Krym)
kuangi Tian, 2000: 128 **A**: GAN
kusnetzovi Kirejtshuk, 1988: 93 **A**: FE
malyi Obenberger, 1917a: 27 **E**: AB or AR "Araxestal" **A**: AF UZ
mashanus G. Yu, 1994: 13 **A**: GUA
mediterraneus Endrödy-Younga, 1968: 78 **E**: GR (Kríti) **A**: CY IS JO LE SA SY
membranaceus Reitter, 1874a: 9 **N**: EG **A**: CY LE SY
syriacus Reitter, 1878e: 91
metallicus Baudi di Selve, 1870: 51 **E**: GR (Kríti) **N**: AG EG LB MO **A**: CY IS
alutaceus Obenberger, 1917a: 27
ferrantei Reitter, 1908b: 41 (*Dissia*)
pectinatus Peyerimhoff, 1926a: 103
salome J. R. Sahlberg, 1908: 55 (*Dissia*)
micans Reitter, 1874a: 2 **E**: IT GR (Kríti) **N**: AG EG MO TU **A**: IS SY
aeneus Reiche, 1864: 237
chalcostoma Obenberger, 1917a: 27
dactylicus Peyerimhoff, 1949: 273
viridiaeneus Reitter, 1885b: 120

morio Endrödy-Younga, 1968: 93 **A**: SY
nanlingensis Tian & G. Yu, 1994a: 121, 122 **A**: HUN
nipponicus Endrödy-Younga, 1971: 244 **Ei**: IT **A**: BEI FUJ GAN GUA GUI GUX HAI HEB HEN HKG HUB HUN JA JIX PA SC SHA SCH UP YUN ZHE **AUR NAR**i
normandi Peyerimhoff, 1931a: 2 **N**: AG EG TU
pangi G. Yu & Tian, 1995: 37 **A**: GUI SCH ZHE
pharao Endrödy-Younga, 1968: 56 **N**: EG
pilosellus Endrödy-Younga, 1968: 64 **A**: TM UZ
planiceps planiceps Endrödy-Younga, 1968: 92 **E**: FR
planiceps platycephalus Endrödy-Younga, 1968: 93 **N**: AG
politissimus Reitter, 1898e: 117 **E**: FR GE IT ST **A**: ES FE WS
politus Gyllenhal, 1813: 680 (*Nitidula*) **E**: AU BE BH CR CT CZ DE EN FI FR GE GR HU IT LA NL NT PL RO SK SP SV SZ UK **A**: CY TR
exiguus C. R. Sahlberg, 1833: 473 (*Anisotoma*)
ruficeps C. R. Sahlberg, 1833: 472 (*Anisotoma*)
pulchellus Erichson, 1845: 218 **E**: AU CR CT CZ FR GE GR HU IT RO SK SP UK **N**: AG TU
aeneiceps Jacquelin du Val, 1854: xxxviii (*Phantazomerus*)
obliquestriatus Reitter, 1890a: 146
pullus Endrödy-Younga, 1968: 67 **N**: EG **A**: IS
reitterii Uhagón, 1879: 190 **E**: SP
rufifrons rufifrons Reitter, 1874a: 7 **E**: CR FR GE GR HU IT PT SP "Caucasus"
rufifrons flaviceps Reitter, 1874a: 6 **N**: AG EG MO TU **A**: IN IQ TR SA
seminulum aenescens J. R. Sahlberg, 1908: 52 (*Dissia*) **N**: EG **A**: AF
seminulum apterus Endrödy-Younga, 1968: 90 **A**: KZ TM
seminulum kandaharicus Endrödy-Younga, 1968: 90 **A**: AF
seminulum seminulum Baudi di Selve, 1870: 52 **A**: CY
similiceps Jacquelin du Val, 1858: 152 **E**: FR SP
sphaerula Wollaston, 1854: 484 (*Stagonomorpha*) **N**: CI (Tenerife) MR
rubriceps Har. Lindberg, 1951: 6
unicolor Wollaston, 1854: 485 (*Stagonomorpha*) **N**: CI MR
status Kirejtshuk, 1994a: 122 **A**: GUA
sublucidus Plavilshchikov, 1935: 298 (*Brachypterus*) **A**: XIN
taiwanensis Tian & Pang, 1994: 402 **A**: TAI
tetragonius G. Yu, 1995a: 32 **A**: GUA
tibialis Endrödy-Younga, 1968: 76 **A**: IS LE SY
torvus Endrödy-Younga, 1968: 53 **N**: AG MO
tricarinatus G. Yu, 1995b: 277 **A**: HEB
tryapitzini Kirejtshuk, 1984b: 530 **A**: TD
turanicus Endrödy-Younga, 1968: 58 **E**: AB or AR ("Araxestal") **A**: KZ MG TM
undatus Tian & Yu, 1994b: 211 **A**: SCH
wollastoni Har. Lindberg, 1951: 6 **E**: FR IT (Sardegna) SP **N**: AG CI MO TU
rabaticus Smirnoff, 1954: 100
subgenus *Theticephalus* Kirejtshuk, 1988: 95 type species *Cybocephalus aurocupreus* Reitter, 1900
aurocupreus Reitter, 1900f: 219 **E**: "Transcaucasia: Michalowo" **A**: TM MG
decamerus Endrödy-Younga, 1968: 100 **E**: "Caucasus" **A**: TD UZ
mesopotamicus Endrödy-Younga, 1968: 107 **A**: IQ JO
nigripes arab Endrödy-Younga, 1968: 104 **A**: YE
nigripes dexter Endrödy-Younga, 1968: 106 **N**: AG MO
nigripes heluanensis Endrödy-Younga, 1968: 106 **N**: EG
nigripes nigripes J. R. Sahlberg, 1908: 53 (*Dissia*) **N**: AG EG MO
nigriceps Grouvelle, 1913: 196 (*Dissia*)
nigripes sinister Endrödy-Younga, 1968: 105 **A**: AF SA TM
palmarum Peyerimhoff, 1931a: 2 **N**: AG EG **A**: SI SA
salsus Kirejtshuk, 1988: 95 **A**: UZ

genus *Hierronius* Endrödy-Younga, 1968: 108 type species *Cybocephalus laevis* Wollaston, 1864
angulosus Endrödy-Younga, 1968: 110 **N**: CI (Lanzarote)
laevis Wollaston, 1864: 117 (*Cybocephalus*) **N**: CI
nanus Endrödy-Younga, 1968: 111 **N**: CI

genus *Pastilodes* Endrödy-Younga, 1968: 112 type species *Cybocephalus agathidioides* Peyerimhoff, 1927
agathidioides Peyerimhoff, 1927: 238 (*Cybocephalus*) **N**: AG MO
areolatus Normand, 1940: 53 (*Cybocephalus*) **N**: MO

family Nitidulidae, nomina dubia
Cryptarcha sulcata Fischer von Waldheim, 1844: 41 **E**: ST
Cybocephalus politus Erichson, 1844: 441 [HN] **A**: IQ
Cychramus laticollis Heer, 1841: 407 **E**: SZ
Meligethes nigra Newman, 1834: 200 **E**: GB
Meligethes abdominalis Motschulsky, 1849: 49 **A**: ES
Ostoma ferruginea Laicharting, 1781: 104 **E**: AU

family MONOTOMIDAE Laporte, 1840
subfamily **Rhizophaginae** Redtenbacher, 1845

genus *Rhizophagus* Herbst, 1793: 18 type species *Lyctus bipustulatus* Fabricius, 1792
subgenus *Anomophagus* Reitter, 1907b: 28 type species *Rhizophagus cribratus* Gyllenhal, 1827
cribratus Gyllenhal, 1827: 637 (*Rhyzophagus*) **E**: AU BE BH BU BY CR CT CZ DE EN FI FR GE GB HU IR IT LT NL NR NT PL RO SK SL SP SV SZ UK **A**: FE
puncticollis C. R. Sahlberg, 1837: 179 (*Rhyzophagus*) **E**: BY CT FI NT PL SK UK **A**: FE JA
  wagae Wankowicz, 1865: 299
subgenus *Cyanostolus* Ganglbauer, 1899a: 558 type species *Rhizophagus aeneus* Richter, 1820
aeneus Richter, 1820: 9 **E**: AU BE BH BU BY CT CZ EN FI FR GG GE GB HU IT LA LT NL NR NT PL RO SK SP SZ UK YU
  coeruleipennis C. R. Sahlberg, 1837: 180 (*Rhyzophagus*)
  caeruleus Waltl, 1839: 225
  cyaneipennis Hardy, 1847: 379 (*Rhyzophagus*)
  nitidulus Duftschmid, 1825: 150 (*Synchita*)
  scoriaceus Gistel, 1857: 589
subgenus *Eurhizophagus* Méquignon, 1909: 105 type species *Rhyzophagus grandis Gyllenhal*, 1827
atticus Tozer, 1968: 57 **E**: GR
  atticus Nikolopoulos, 1969: 163 [HN]
depressus Fabricius, 1792b: 503 (*Lyctus*) **E**: AU BE BH BY CR CT CZ DE EN FI FR GG GE GB HU IR IT LA LT NL NR NT PL RO SK SL SP ST SV SZ UK **A**: WS **N**: AG CI
  subopacus Wollaston, 1864: 119
grandis Gyllenhal, 1827: 636 (*Rhyzophagus*) **E**: AU BE BY CT CZ DE EN FI FR GB GG GE IT NL NR NT PL SK SV SZ **A**: WS
subgenus *Rhizophagus* Herbst, 1793 type species *Lyctus bipustulatus* Fabricius, 1792
  Syringobidia Casey, 1916: 103 type species *Rhizophagus cylindricus* LeConte, 1866
beasoni Sengupta & De, 1988: 7 **A**: HP
bipustulatus Fabricius, 1792b: 503 (*Lyctus*) **E**: AU BE BH BU BY CR CT CZ DE EN FI FR GB GE GG GR HU IT LA LS LT LU MA NL NT NR PL RO SK SL SP ST SV SZ UK YU **A**: TR **N**: AG CI MO MR TU
  bipunctatus Herbst, 1793: 19 (*Ryzophagus*)
  gyllenhalii C. G. Thomson, 1885: 108
  longicollis Gyllenhal, 1827: 639 (*Rhyzophagus*)
  magniceps Reitter, 1897c: 123
  quercus Gistel, 1857: 534
  reductus Rey, 1889d: 28
brancsiki Reitter, 1905d: 312 **E**: AU BH BU CZ FR GG HU IT PL RO SK SL SV UK YU
championi Sengupta & De, 1988: 7 **A**: UP

dispar Paykull, 1800: 328 (*Lyctus*) **E**: AL AU BE BH BU BY CR CT CZ DE EN FI FRGB GE GG GR HU IR IT LA LS LT MC NL NR NT PL RO SK SL SP ST SV SZ UK YU **N**: AG
elongatus A. G. Olivier, 1790a: no 18: 8 (*Ips*)
punctulatus Guillebeau, 1897c: 226
fenestralis Linnaeus, 1758: 356 (*Dermestes*) **E**: AU BE BY CT CZ DE EN FI FR GB GE HU IT LT NL NR NT PL PT RO SK SV SZ UK **A**: ES FE WS
parvulus Paykull, 1800: 329 (*Lyctus*)
ferrugineus Paykull, 1800: 326 (*Lyctus*) **E**: AL AU BH BU BY CT CZ DE EN FI FR GB GE HU IR IT LT NR NT PL RO SK SL SP ST SV SZ UK YU **N**: AG CI **A**: MG
alpicola Baudi di Selve, 1889: 97
pinetorum Wollaston, 1864: 118
ghumus Sengupta & De, 1988: 12 **A**: SD
indicus Méquignon, 1913a: 90 **A**: UP
japonicus Reitter, 1884g: 271 [= 1885c: 101] **A**: FE JA
kali Sengupta & De, 1988: 12 **A**: UP
lineatus Sengupta & De, 1988: 12 **A**: NP
microps Jelínek, 1984a: 129 **A**: UZ
nitidulus Fabricius, 1798: 177 (*Lyctus*) **E**: AU BE BH BY CR CT CZ DE EN FI FR GB GE GR HU IT LA LT MC NL NR NT PL RO SK SL SP ST SV SZ UK YU
erythrocephalus Fabricius, 1801b: 557 (*Colydium*)
nobilis Lewis, 1893a: 83 (*Rhyzophagus*) **A**: FE JA
oblongicollis Blatch & Horner, 1892: 303 **E**: AB FR GB GE IT
lenkoranus Reitter, 1907c: 272
pahalgamus Sengupta & Biswas, 1977: 421 **A**: KA
parallelocollis Gyllenhal, 1827: 638 (*Rhyzophagus*) **E**: AU BE BH BY CT CZ DE EN FI FR GB GE HU IR IT LT MD NL NR NT PL RO SK SP ST SV SZ UK **NAR**
erichsonii C. G. Thomson, 1885: 108
hannenfeldii Tournier, 1872a: 251
robustus C. F. A. Schaeffer, 1911: 118
terebrans Stephens, 1839: 140
parviceps Reitter, 1884g: 271 [= 1885c: 101] **A**: JA
perforatus Erichson, 1845: 231 **E**: AB AU BE BH CR CT CZ DE FI FR GB GE HU IR IT NL NR PL RO SK SL SP ST SV SZ UK YU
doderoi Méquignon, 1925a: 13
fallax Rey, 1889d: 28
philippi Méquignon, 1925a: 14
picipes A. G. Olivier, 1790a: no 18: 7 (*Ips*) **E**: AU BE BH BY CR CT CZ DE EN FI FR GB GE HU IT NL NR NT PL RO SK SL SP SV SZ UK YU
bruckii Reitter, 1873b: 30
politus Hellwig, 1792: 404 (*Synchita*)
pratapi Sengupta & De, 1988: 8 **A**: KA
protensus Reitter, 1890c: 192 **E**: AR **A**: IN
striolatus Reitter, 1907c: 272 [HN]
similaris Reitter, 1876f: 289 **E**: AB GG ST
simplex Reitter, 1884g: 271 [= 1885c: 101] **A**: FE JA
singularis Sengupta & De, 1988: 10 **A**: KA
subvillosus Reitter, 1884g: 272 [= 1885c: 102] **A**: JA
suturalis Jelínek, 1965b: 385 **A**: SCH
unicolor P. H. Lucas, 1849: 475 **E**: FR GR IT MA PT SP **N**: AG MO TU **A**: CY TR
punctiventris Baudi di Selve, 1870: 53
serricollis Obenberger, 1917b: 76
ussuriensis Nikitsky, 1984: 65 **A**: FE

genus *Rhizophagus*, nomen dubium
intermedius Apfelbeck, 1917: 198 **E**: BH

subfamily **Monotominae** Laporte, 1840

genus *Europs* Wollaston, 1854: 149 type species *Europs impressicollis* Wollaston, 1854
subgenus *Europs* Wollaston, 1854: 149 type species *Europs impressicollis* Wollaston, 1854
Nomophloeus LeConte, 1861: 86 *Nomophloeus pallidipennis* LeConte
alutaceus Champion, 1924a: 168 **A**: UP
duplicatus Wollaston, 1862a: 146 **N**: CI (La Gomera, Gran Canaria, Tenerife)
impressicollis impressicollis Wollaston, 1854: 150 **N**: CI MR
impressicollis hierroensis Palm, 1975a: 49 **N**: CI (Hierro)
impressicollis palmensis Palm, 1975a: 49 **N**: CI (La Palma)
indicus Grouvelle, 1903a: 124 **A**: SD
temporis Reitter, 1884g: 272 [= 1885c: 103] **A**: FE JA
subgenus *Monotopion* Reitter, 1884g: 272 [= 1885a: 103] type species *Monotopion ferrugineum* Reitter, 1884
ferrugineum Reitter, 1884g: 272 [= 1885c: 103] (*Monotopion*) **A**: JA
todo Kôno, 1940: 61 (*Europs*)

genus *Indoleptipsius* Pal, 1996a: 2 type species *Indoleptipsius ushae* Pal, 1996
ushae Pal, 1996a: 4 **A**: AP

genus *Malinica* Sengupta, 1988: 24 type species *Malinica ranjana* Sengupta, 1988
ranjana Sengupta, 1988: 26 **A**: SD

genus *Mimemodes* Reitter, 1876g: 296 type species *Bactridium monstrosum* Reitter, 1874
carenifrons Grouvelle, 1914: 62 **A**: TAI
cribratus Reitter, 1874c: 515 (*Bactridium*) **A**: JA
emmerichi Mader, 1937: 179 **A**: JA
harmandi Grouvelle, 1903a: 123 (*Europs*) **A**: SD
japonus Reitter, 1874c: 515 (*Bactridium*) **A**: JA
megalocephalus Champion, 1924a: 169 **A**: UP
monstrosus Reitter, 1874c: 514 (*Bactridium*) **A**: GUX HUN JA SCH TAI
nigratus Sengupta, 1979: 69 **A**: SD
proximus Grouvelle, 1914: 63 **A**: TAI

genus *Monotoma* Herbst, 1793: 22 type species *Monotoma picipes* Herbst, 1793
Gyrocecis C. G. Thomson, 1859: 93 type species *Cerylon angusticolle* Gyllenhal, 1827
Monotomina Nikitsky, 1986: 1628 type species *Monotoma quadrifoveolata* Aubé, 1837
affinis Nikitsky, 1986: 1629 **A**: FE
aegyptiaca Motschulsky, 1868: 199 **N**: EG
americana Aubé, 1837: 461 **Ei**: CR IT **NAR**
corpulenta Motschulsky, 1868: 200
rhodeana Casey, 1916: 91
angusticeps Reitter, 1911b: 52 **A**: KI
angusticollis Gyllenhal, 1827: 634 (*Cerylon*) **E**: AU BE BY CT CZ DE FI FR GB GE IR IT LA LT NL NR NT PL SK SL ST SV SZ UK **A**: ES FE
formicetorum C. G. Thomson, 1863: 213 (*Gyrocecis*)
sulcicollis Trella, 1923: 112
bicolor A. Villa & G. B. Villa, 1835: 49 **E**: AB AU AZ BE BH BU BY CR CT CZ DE FI FR GB GE GG GR IR IT LA LS MA NL NR NT PT RO SK SP ST SV SZ UK YU **N**: AG CI EG MO MR TU **A**: AF CY ES FE GAN HEB IS JIA LIA NMO SHN SHX SY TD TR WS **N**: CI **AUS** **NAR**
elongata Rey, 1889g: 54
expansa Rey, 1889g: 54
obtusicollis Motschulsky, 1845a: 96
parallela LeConte, 1855: 305
parallela C. G. Thomson, 1870b: 334 [HN]
quisquiliarum L. Redtenbacher, 1849: 203
seriata Reitter, 1901i: 196
thomsoni Reitter, 1891a: 168 [RN]

brevicollis brevicollis Aubé, 1837: 460 **E**: AU BE BH BU BY CR CZ DE FI FR GB GE GG GR HU IT LS MA MC MD NL NR PL RO SK SP SV SZ UK YU **N**: "North Africa" **A**: KZ TR UZ **NAR**

blaivii Guérin-Méneville, 1839: 140
punctata Ragusa, 1892: 201
quadridentata C. G. Thomson, 1870b: 333
trapezicollis Chaudoir, 1845: 210

brevicollis lebedevi Roubal, 1929b: 47 **A**: TM

conicicollis Chevrolat, 1837: pl. 41 **E**: AU BE BY CR CT CZ DE FI FR GB GE IR IT LA LT NL NR NT PL RO SK SV SZ UK **A**: ES FE KZ

conicithorax Reitter, 1891b: 23 **E**: AB **A**: KZ TM

diecki Reitter, 1877b: 6 **E**: CR FR (Corse) IT

gotzi Holzschuh & Lohse, 1981: 175 **E**: AU DE LS

inseriata Reitter, 1901i: 197 **E**: AB

johnsoni Bousquet & Laplante, 1999: 78 **E**: PT SP **N**: CI MR **AFR AUR NAR NTR**

longicollis Gyllenhal, 1827: 635 (*Cerylon*) **E**: AU AZ BY CR CT CZ DE FI FR GB GE GG GR HU IR IT NR NT PL PT RO SK SP ST SV SZ UK **N**: CI (Tenerife, Grand Canaria) MR **A**: FE HEB HEI JA JIA TM **AFR AUS NAR**

angustata Stephens, 1830: 102
flavipes Kunze, 1839: 386
gracilis Curtis, 1840: 277

malyi Obenberger, 1914: 106 **A**: TD

perplexa Rey, 1889g: 54 **E**: FR

picipes Herbst, 1793: 24 **E**: AU AZ BY CR CT CZ DE EN FI FR GB GE GG GR IR IT LA LS LU MA SK SL SP ST SV SZ UK YU **N**: AG CI MO MR **A**: CY ES FE GAN HEN JA KZ NMO QIN TAI TD TM TR UZ WS "Korea" **COS**

brevipennis Kunze, 1839: 383
cavicula Reitter, 1901i: 194
congener Wollaston, 1857a: 68
contracta Marsham, 1802: 110
famelica Casey, 1916: 89
foveata LeConte, 1855: 305
fulvipennis Motschulsky, 1868: 199
fulvipes Melsheimer, 1844: 111
obsolescens Casey, 1916: 88
opaca Ziegler, 1845: 271
pallida Stephens, 1830: 103
robustula Motschulsky, 1868: 200
scabra Kunze, 1839: 384
subnigra Motschulsky, 1868: 199

punctaticollis Aubé, 1843a: 73 **E**: AB FR IT NL PL SP UK "Caucasus" **N**: AG MO TU

brevicornis Motschulsky, 1845a: 97

quadricollis Aubé, 1837: 465 **E**: AU BU CR CZ FI FR GB GE IT NL PT RO SL SZ YU **A**: KZ

quadrifoveolata Aubé, 1837: 468 **E**: AB AU AZ CR CZ DE FR GB GE GG GR HU IT NL PL PT RO SK SP ST SV SZ UK **N**: AG CI (Tenerife) MR **A**: FE HEB JA NMO SHN SHX TM TR UZ **COS**

quadrifoveolata Motschulsky, 1837a: 115 [HN]

specialis Nikitsky, 1985a: 47 **E**: ST

spinicollis Aubé, 1837: 463 **E**: AU AZ BE CR CT CZ DE EN FI FR GB GE GG GR IR IT HU MA NL NT PL PT RO SK SP SV SZ UK **N**: AG CI EG MO MR **A**: JA TD TR **COS**

spinifera Wollaston, 1857a: 67

testacea Motschulsky, 1845a: 97 **E**: AU CR CZ DE FI FR GB GE IT NL NR PL RO SK SP SV UK **N**: AG **A**: KZ TM WS "Caucasus" **NAR**

ferruginea C. N. F. Brisout de Barneville, 1863: 62
quadraria Casey, 1916: 90
quadriimpressa Motschulsky, 1845a: 94
rufa L. Redtenbacher, 1849: 203
subquadrifoveolata G. R.Waterhouse, 1858: 105 [RN]

genus *Monotomopsis* Grouvelle, 1896a: 38 type species *Monotomopsis monotomoides* Grouvelle, 1896
andrewesi Grouvelle, 1908c: 494 **A**: UP **ORR**

genus *Renuka* Sengupta, 1988: 30 ype species *Renuka rita* Sengupta, 1988
rita Sengupta, 1988: 31 **A**: SD

genus *Rhizophagoides* Nakane & Hisamatsu, 1963: 48 type species *Rhizophagoides kojimai* Nakane & Hisamatsu, 1963
kojimai Nakane & Hisamatsu, 1963: 48 **A**: FE JA

genus *Shoguna* Lewis, 1889b: 274 [NP] type species *Shoguna rufotestacea* Lewis, 1889
Holocephala Fairmaire, 1886b: 32 [RN, NO] type species *Pachycephala termitiformis* Fairmaire, 1883
Pachycephala Fairmaire, 1883a: 4 [HN] type species *Pachycephala termitiformis* Fairmaire, 1883
Shogunopsis Fauvel, 1903: 305 [RN] type species *Pachycephala termitiformis* Fairmaire, 1883
rufotestacea Lewis, 1889b: 274 (*Shoguna*) **A**: JA

genus *Tarunius* Sengupta, 1977: 532 type species *Tarunius punctatus* Sengupta, 1977
punctatus Sengupta, 1977b: 533 **A**: SD

family HELOTIDAE Reitter, 1876 / Chapuis, 1876

genus *Helota* W. S. MacLeay, 1825: 42 type species *Helota vigorsii* W. S. MacLeay, 1825
Neohelota Ohta, 1929c: 68 type species *Neohelota tumaaka* Ohta, 1929 (= *Helota helleri* Ritsema, 1911)
acutipennis Ritsema, 1914b: 166 **A**: YUN
attenuata Ritsema, 1894: 112 **A**: BT SD
bhutanensis M. Chûjô, 1975: 287 **A**: BT
boulei Ritsema, 1915: 247 **A**: BT
boysii Ritsema, 1889b: 189 **A**: BT SD
bretaudeaui Ritsema, 1894: 116 **A**: BT SD
caudata Ritsema, 1889a: 102 **A**: CH
cereopunctata Lewis, 1881b: 255 **A**: HUB JA JIL
chinensis Mader, 1955: 63 **A**: FUJ
consanguinea Ritsema, 1915: 248 **A**: BT SD
culta Olliff, 1883a: 55 **A**: SD **ORR**
curvipes Oberthür, 1883: 60 **A**: "Himalaya" **ORR**
desgodinsi Ritsema, 1893: 131 **A**: BT SD
distincta Ritsema, 1914a: 59 **A**: GUI SCH
donckieri Ritsema, 1907: 213 **A**: YUN
durelii Ritsema, 1905: 127 **A**: BT **ORR**
fairmairei Ritsema, 1889a: 101 **A**: BT SD "Himalaya" **ORR**
fulvitarsis Ritsema, 1889a: 107 **A**: BT SD **ORR**
fulviventris Kolbe, 1886: 182 **A**: FE HUB JA LIA SC
japonica Ohta, 1929d: 109
gemmata Gorham, 1874e: 448 **A**: HEI JA JIL SHG
gorhami Olliff, 1883a: 53 **A**: SC SHN **ORR**
guerinii Hope, 1840: 188 **A**: "India" **ORR**
guttata Ritsema, 1893: 134 **A**: BT SD
helleri Ritsema, 1911: 51 **A**: TAI
tumaaka Ohta, 1929c: 68 (*Neohelota*)
hopei Ritsema, 1915: 245 **A**: BT
immaculata Ritsema, 1891: 895 **A**: BT **ORR**
intermedia Ritsema, 1905: 125 **A**: BT
jordanii Ritsema, 1899: 252 **A**: NP **ORR**
klapperichi Mader, 1955a: 63 **A**: GUA
kolbei Ritsema, 1889a: 103 **A**: JIX SHG

krugeri Ritsema, 1900: 29 **A**: BT
laevigata Oberthür, 1883: 59 **A**: BT SD
lesnei Ritsema, 1906: 196 **A**: GUI
lewisi Ritsema, 1915: 244 **A**: YUN
longipes Ritsema, 1889a: 101 **A**: BT SD **ORR**
margarethae Ritsema, 1900: 27 **A**: BT
mariae Ritsema, 1899: 249 **A**: BT
montana Ohta, 1929c: 67 **A**: TAI
moutonii Ritsema, 1905: 129 **A**: FUJ
oberthueri Ritsema, 1889a: 100 **A**: BT FUJ SD TAI **ORR**
pusilla Oberthür, 1883: 60 **A**: SD
pustulata Ritsema, 1893: 133 **A**: BT **ORR**
renati Ritsema, 1905: 123 **A**: SHN
ritsemana Heller, 1923: 68 **A**: SCH
servillei Hope, 1840: 187 **A**: "Himalaya" **ORR**
severini Ritsema, 1893: 138 **A**: BT SD
sharpi Ritsema, 1915: 249 **A**: BT
sinensis Olliff, 1883a: 54 **A**: SHN TAI
sonani Ohta, 1929c: 67 **A**: TAI
miwai Ohta, 1931a: 136 (*Neohelota*)
taiwana Ohta, 1929c: 66 **A**: TAI
thibetana Westwood, 1842: 123 **A**: "Punjab" "Thibet"
mellii Westwood, 1848: 86
thoracica Ritsema, 1895: 49 **A**: SCH TAI XIZ **ORR**
tibialis Ritsema, 1893: 136 **A**: BT SD **ORR**
tonkinensis Ritsema, 1905: 119 **A**: SHN **ORR**
westwoodii Ritsema, 1907: 216 **A**: BT
yezoana Kôno, 1939: 158 **A**: JA

family PHLOEOSTICHIDAE Reitter, 1911

genus *Phloeostichus* W. Redtenbacher, 1842: 15 type species *Phloeostichus denticollis* W. Redtenbacher, 1842
denticollis W. Redtenbacher, 1842: 16 **E**: AU CZ DE FR GE IT PL SK "Carpathians" **A**: FE

family SILVANIDAE Kirby, 1837

subfamily **Brontinae** Erichson, 1845

tribe Brontini Erichson, 1845

genus *Dendrophagus* Schönherr, 1809: 50 type species *Cucujus crenatus* Paykull, 1799
crenatus Paykull, 1799: 168 (*Cucujus*) **E**: AU BY CR CT CZ EN FI FR GB GE IT LA LT NR NT PL RO SK SV SZ UK **A**: FE "Siberia"
longicornis Reitter, 1889r: 316 **A**: JA

genus *Macrohyliota* Thomas, 2004: 19 type species *Hyliota gracilicollis* Arrow, 1901
gracilicornis Arrow, 1901: 598 (*Hyliota*) **A**: JA TAI **ORR**
spinicollis Guérin-Méneville, 1831c: pl. 42 (*Brontes*) **A**: AP **ORR**
atratus Reitter, 1877e: 24 (*Brontes*)

genus *Parahyliota* Thomas, 2004: 22 type species *Uleiota costicollis* Reitter, 1876
alticola Pal, Sengupta & Crowson, 1985: 221 (*Uleiota*) **A**: AP SD **ORR**
costicollis Reitter, 1876c: 44 (*Uleiota*) **A**: TAI **ORR**
crenicollis Grouvelle, 1914: 57 (*Uleiota*)

genus *Uleiota* Latreille, 1797: 46 type species *Cucujus flavipes* Fabricius, 1792
Brontes Fabricius, 1801b: 97 type species *Cerambyx planatus* Linnaeus, 1761
Hyleota Seidlitz, 1888a: 55 [Gattungen] type species *Cerambyx planatus* Linnaeus, 1761
Hyliota Reitter, 1880d: 80 [unjustified emendation]
algericus Mařan, 1929: 35 **N**: AG MO
arboreus Reitter, 1889r: 317 (*Hyliota*) **A**: FE JA
planatus Linnaeus, 1761: 186 (*Cerambyx*) **E**: AB AR AU BE BH BU BY CR CT CZ DE EN FI FR GBi GE GG GR HU IT LA LS LT NL NR NT PL PT RO SK SL SP SV SZ UK **N**: AG TU **A**: FE JA IN TR
arabs Fabricius, 1792a: 97 (*Scarites*)
compressus Geoffroy, 1785: 76 (*Cerambyx*)
flavipes Fabricius, 1792b: 95 (*Cucujus*)
humeralis Faldermann, 1837: 259
pallens Fabricius, 1792b: 96 (*Cucujus*)

tribe Telephanini LeConte, 1861

genus *Cryptamorpha* Wollaston, 1854: 156 type species *Cryptamorpha musae* Wollaston, 1854
Parabrontes L. Redtenbacher, 1868: 40 type species *Parabrontes silvanoides* L. Redtenbacher, 1868
Pseudophanus LeConte, 1859b: 84 type species *Pseudophanus signatus* LeConte, 1859
bhutanensis Pal & Sengupta, 1979: 80 **A**: BT
desjardinsii Guérin-Méneville, 1844a: 196 (*Psammoecus*) **Ei**: AZ FI FR GB GE HU IT SV SZ **N**: CI MR **A**: JA NMO **COS**
fasciata L. Redtenbacher, 1868: 41 (*Telephanus*)
hubbardi Casey, 1885: 167
musae Wollaston, 1854: 157
signata LeConte, 1859b: 85 (*Pseudophanus*)
kaszabi Pal & Sengupta, 1979: 78 **A**: SD **ORR**
sculptifrons Reitter, 1889r: 320 **A**: BT JA SD YUN ZHE
opacifrons Grouvelle, 1908c: 474
punctifrons Grouvelle, 1908c: 474

genus *Megapsammoecus* Karner, 1995: 4 type species *Megapsammoecus christinae* Karner, 1995
christinae Karner, 1995: 5 **A**: YUN

genus *Psammoecus* Latreille, 1829b: 135 type species *Notoxus bipunctatus* Fabricius, 1792
Crypta Stephens, 1830: 103 type species *Notoxus bipunctatus* Fabricius, 1792
andrewesi Grouvelle, 1908c: 476 **A**: NP **ORR**
bambusae Pal, 1985a: 33 **A**: SD
bipunctatus Fabricius, 1792a: 212 (*Notoxus*) **E**: AB AU BE BH BY CR CT CZ DE FI FR GB GE HU IR IT LS NL PL RO SK SL SP ST SV SZ UK YU **A**: FE IN TM
algiricus Pic, 1920d: 9
boudieri P. H. Lucas, 1843a: 147
latior Pic, 1894c: 44
nigrinus Pic, 1894c: 44
pallidus Pic, 1894c: 44
bhutanicus Pal, 1985a: 27 **A**: BT
complexus Pal, Sengupta & Crowson, 1985: 23 **A**: SD
fasciatus Reitter, 1874c: 525 **A**: FE JA "Korea" **ORR**
harmandi Grouvelle, 1912g: 413 **A**: AP NP SD
lepidus Grouvelle, 1908c: 483 **A**: BH UP **ORR**
personatus Grouvelle, 1919a: 34 **E**: AZ **N**: CI MR
quadrimaculatus Reitter, 1874c: 525 **A**: JA
stultus Grouvelle, 1912b: 81 **A**: TAI
triguttatus Reitter, 1874c: 524 **A**: FE FUJ HEI HUN JA SCH "Korea"
trilochana Pal, 1985a: 29 **A**: HP UP NP
trimaculatus Motschulsky, 1858c: 45 (*Psammaechus*) **A**: AP BT JA KA NP UP **AUR ORR**
cephalotes Grouvelle, 1919a: 20
cruciger C. O. Waterhouse, 1876b: 125 (*Telephanus*)

incommodus Walker, 1859a: 53 (*Cucujus*)
upsilon Blackburn, 1903: 155
wittmeri Pal & Sengupta, 1979: 76 **A**: SD
x-notatus Grouvelle, 1912b: 84 **A**: TAI

subfamily **Silvaninae** Kirby, 1837

genus *Ahasverus* Gozis, 1881b: cxxvii type species *Cryptophagus advena* Waltl, 1834
advena Waltl, 1834: 169 (*Cryptophagus*) **E**: AL AU AZ BE BH BY CR CT CZ DE FI FR GB GE HU IT LS LT PL PT RO SK SL SP SVi SZ **N**: CI EG LB MO MR **A**: FE FUJ GAN GUA GUI GUX HEN HUB JA JIA JIL JIX NIN SCH TAI YUN ZHE **COS**
angustatus P. H. Lucas, 1846: 221 (*Cryptophagus*)
guerinii Allibert, 1847: 12 (*Cryptophagus*)
musaeorum Ziegler, 1844: 270 (*Lathridius*)
striatus Rouget, 1877: ccvii (*Cryptophagus*)
excisus Reitter, 1876e: 128 (*Cathartus*) **E**i: GE IT NL **NTR**

genus *Airaphilus* L. Redtenbacher, 1858: 999 type species *Cryptophagus elongatus* Gyllenhal, 1813
abeillei Grouvelle, 1889: 107 **A**: "Syria"
arcadius Reitter, 1884c: 67 **E**: GR IT (Sicilia)
calabricus Obenberger, 1914: 105 **E**: IT
carpetanus Heyden, 1870: 108 **E**: SP
chotanicus Semenov, 1893: 494 **A**: CH "Turkestan"
corsicus Grouvelle, 1874: xxix **E**: FR (Corse) IT (Elba, Sardegna)
depressus Reitter, 1889m: 278 **E**: AB **N**: EG
doramus Wurst & Lange, 1996: 59 **N**: CI (Grand Canaria)
elongatus Gyllenhal, 1813: 677 (*Cryptophagus*) **E**: AL AU BE BH BU CR CZ EN FI FR GE GR HU IT PL SK SL SV SZ **N**: EG **A**: MG NMO TM WS
frater Kraatz, 1862b: 127 (*Aeraphilus* [sic])
geminus Kraatz, 1858: 137 (*Silvanus*)
ruthenus Solsky, 1866b: 180 (*Aeraphilus*)
fallax Gouvelle, 1877: 296 **N**: MO **AFR**
ferrugineus Kraatz, 1862b: 129 (*Aeraphilus*) **E**: IT SP **N**: AG MO
filiformis Rosenhauer, 1856: 104 (*Silvanus*) **E**: SP SZ **N**: MO
grouvellei Reitter, 1880d: 83 **E**: FR SP
hellenicus Obenberger, 1922: 24 **E**: GR
hirtulus Reitter, 1884c: 66 **E**: GR
kaszabi Ratti, 1972a: 100 **A**: MG
montisatri Peyerimhoff, 1931b: 54 **N**: AG (Hoggar)
nasutus balearicus Español, 1943: 100 **E**: FR SP (Baleares)
nasutus nasutus Chevrolat, 1860: 211 **E**: GR IT MA (Gozo) SL **N**: AG MO
fibulatus Kraatz, 1862b: 128 (*Aeraphilus*)
natividadei Peyerimhoff, 1937: 89 **E**: PT
nubigena Wollaston, 1863: 217 (*Sylvanus*) **N**: CI
paganettii Obenberger, 1918: 60 **E**: GR (Kríti)
perangustus Lindberg, 1943: 54 **E**: FI PL SZ
peyerimhoffi Cobos, 1950a: 586 **E**: SP
seabrai Luna de Carvalho, 1951: 166 **E**: PT
semenowi Reitter, 1901d: 71 **A**: UZ
siculus Reitter, 1880d: 82 **E**: IT (Sicilia) SL
subferrugineus Reitter, 1884: 69 **E**: FR (Pyrénées)
syriacus Grouvelle, 1874: xxix **A**: "Syria"
talpa Kraatz, 1862b: 128 (*Aeraphilus*) **E**: IT
vaulogeri Grouvelle, 1913c: 327 **N**: TU

genus *Astilpnus* Perris, 1866: 184 type species *Astilpnus multistriolatus* Perris, 1866
Derotoma Reitter, 1872b: 170 type species *Derotoma lederi* Reitter, 1872 (= *Astilpnus multistriolatus* Perris, 1866)
Nilina Motschulsky, 1868: 201 type species *Nilina reflexicollis* Motschulsky, 1868
multistriolatus Perris, 1866: 185 **E**: IT (Sicilia) **N**: AG
lederi Reitter, 1872b: 171 (*Derotoma*)
sericellus Rottenberg, 1871: 237 (*Monotoma*)
reflexicollis Motschulsky, 1868: 201 (*Nilina*) **N**: EG

genus *Cathartus* Reiche, 1854b: 77 type species *Cathartus cassiae* Reiche, 1854 (= *Sylvanus quadricollis* Guérin-Méneville, 1844)
quadricollis Guérin-Méneville, 1844a: 198 (*Sylvanus*) **E**i: CR FR GB GE IT **A**: JIA SHG **COS**
annectens Sharp, 1899: 559
cassiae Reiche, 1854b: 78
gemellatus Jacquelin du Val, 1857: 250

genus *Monanus* Sharp, 1879: 85 type species *Monanus crenatus* Sharp, 1879
Emporius Ganglbauer, 1899b: 586 type species *Silvanus signatus* Frauenfeld, 1867
subgenus *Monanops* Grouvelle, 1913c: 334 type species *Monanus bouchardi* Grouvelle, 1913
rambicus Pal, 1982: 251 **A**: SD
subgenus *Monanus* Sharp, 1879: 85 type species *Monanus crenatus* Sharp, 1879
antennatus Grouvelle, 1913c: 368 **A**: TAI
concinnulus Walker, 1858: 207 (*Monotoma*) **E**: FR GE HU IT **A**: ANH FUJ GUA GUX HEB JA JIX SD SHG UP TAI YUN ZHE **COS**
fasciatus Wollaston, 1874: 169 (*Cryptomorpha*)
fascipennis Reitter, 1876e: 129 (*Cathartus*)
signatus Frauenfeld, 1867: 438 (*Silvanus*)
discoidalis Grouvelle, 1913c: 366 **A**: GUX TAI **ORR**

genus *Nausibius* Lentz, 1857: 75 type species *Corticaria dentata* Marsham, 1802 (= *Cucuius clavicornis* Kugelann, 1794)
Nausibius L. Redtenbacher, 1858: 998 [HN] type species *Corticaria dentata* Marsham, 1802 (= *Cucuius clavicornis* Kugelann, 1794)
clavicornis Kugelann, 1794: 571 (*Cucuius*) **E**i: AZ BE DE FR GB GE IR IT NR PL PT SP SV SZ **N**: CI MR **A**: CH JA **COS**
dentatus Marsham, 1802: 108 (*Corticaria*)
denticulatus W. S. MacLeay, 1825: 46 (*Sylvanus*)
intermedius Smith, 1851: 16 (*Silvanus*)
latus Fairmaire, 1850: 53 (*Silvanus*)
parallelus Walker, 1858: 206 (*Rhyzophagus*)

genus *Oryzaephilus* Ganglbauer, 1899b: 584 type species *Silvanus surinamensis* Linnaeus, 1858
abeillei Guillebeau, 1890: 222 (*Silvanus*) **E**: GR **A**: IS SY TR
fauveli Reitter, 1890g: 256 (*Silvanus*) **A**: IS TR
mercator Fauvel, 1889: 132 **E**: AU AZ BY CT CZ DE EN FI FR GB GE HU LA LT NR PL SK SL SV SZ **N**: CI EG MO MR **A**: ANH GAN HEN HUB FE FUJ GUA GUI GUX JA JIA SA SHG SHN YUN ZHE **COS**
gossypii Chittenden, 1897: 13 (*Silvanus*)
surinamensis Linnaeus, 1758: 357 (*Dermestes*) **E**: AU BH BY CR CT CZ EN FI FR DE GB GE GR HU IC IR IT LA LT MC NL NR PL PT RO SK SL SP SV SZ UK **N**: CI EG JO LB MO TU **A**: ANH FE FUJ GUA GUI HEN HKG HUB IS JA JIA JIX SA SCH SHA TAI TR WS ZHE "Korea" **COS**
bicornis Erichson, 1846: 337 (*Silvanus*)
cursor Linnaeus, 1758: 418 (*Tenebrio*)
frumentarius Fabricius, 1775: 62 (*Anobium*)
sexdentatus Herbst, 1783: 31 (*Ips*)
sexdentatus Fabricius, 1792a: 232 (*Dermestes*) [HN]

genus *Protosilvanus* Grouvelle, 1913c: 336 type species *Silvanus lateritius* Reitter, 1879
dehradunicus Pal & Sengupta, 1985: 240 **A**: UP

granosus Grouvelle, 1897: 396 (*Silvanus*) **A**: SD **AUR ORR**
lateritius Reitter, 1879b: 194 (*Silvanus*) **A**: AP BT CH HP JA NP SD TAI UP YUN "Korea" **ORR**
genus *Pseudonausibius* Halstead, 1980: 337 type species *Oryzaephilus maximus* Grouvelle, 1913
maximus Grouvelle, 1913c: 319 (*Oryzaephilus*) **A**: SHG HAI JA

genus *Silvanoides* Halstead, 1973: 87 type species *Silvanoides cheesmanae* Halstead, 1973
cribricollis Grouvelle, 1897: 397 (*Silvanus*) **A**: AP SD **ORR**
cheesmanae Halstead, 1973: 87

genus *Silvanolomus* Reitter, 1912b: 108 [RN] type species *Silvanus denticollis* Reitter, 1876
Silvanopsis Reitter, 1911a: 44 [HN] [type species not designated]
denticollis Reitter, 1876c: 56 (*Silvanus*) **A**: CH FE **ORR**
inermis Reitter, 1876c: 63 (*Silvanus*) **A**: JA
pullus Reitter, 1898d: 56 (*Silvanus*) **N**: MR MO **A**: YE **AFR**

genus *Silvanopsis* Grouvelle, 1892b: 286 type species *Silvanopsis simoni* Grouvelle, 1892
nepalensis Pal & Sengupta, 1979: 72 **A**: NP
simoni Grouvelle, 1892b: 286 **A**: CH JA TAI UP **ORR**
iyeri Imms & Chatterjee, 1915: 37
mysorensis Mhadihassan, 1955: 418 (*Silvanus*)

genus *Silvanoprus* Reitter, 1911a: 45 type species *Silvanus fagi* Guérin-Méneville, 1844
angusticollis Reitter, 1876c: 59 (*Silvanus*) **A**: BT CH JA KA NC NP SD TAI UP **NAR**i **ORR**
birmanicus Grouvelle, 1892c: 863 (*Silvanus*) **A**: TAI **ORR**
cephalotes Reitter, 1876c: 62 (*Silvanus*) **A**: BT HEB HKG JA MNO NC NP SCH SD TAI UP YUN **ORR**
fagi Guérin-Méneville, 1844a: 197 (*Sylvanus*) **E**: AU BE BY CR CT CZ DE EN FI FR GB GE HU IT LA LT NL NR NT PL RO SK SL ST SV SZ UK **A**: ES FE JA TAI WS
similis Erichson, 1846: 339 (*Silvanus*)
grouvellei Reitter, 1876c: 58 (*Silvanus*) **A**: JA UP **ORR**
prolixicornis Sengupta & Pal, 1996: 86
indicus Pal & Sengupta, 1979: 71 **A**: BT SD **ORR**
longicollis Reitter, 1876c: 60 (*Silvanus*) **A**: GUX JA SD SHG TAI UP **AFR ORR**
javanicus Grouvelle, 1885: 48 (*Silvanus*)
parallelocollis Reitter, 1876c: 59 (*Silvanus*) **A**: JA
scuticollis Walker, 1859a: 53 (*Silvanus*) **E**i: FR **N**: MR **A**: HUN JA SCH SD TAI UP YUN **AFR AUR NAR NTR ORR**
triangulus Reitter, 1876c: 60 (*Silvanus*)
sitkhoensis Krivolutskaya, 1992: 240 **A**: FE

genus *Silvanus* Latreille, 1804d: 158 [= 1804b: 242] type species *Ips unidentata* A. G. Olivier, 1790
Leptus Duftschmid, 1825: 156 type species *Ips unidentatus* A. G. Olivier, 1790
Microsilvanus Grouvelle, 1913c: 332 type species *Silvanus pumilus* Grouvelle, 1913
bhutanicus Sengupta & Pal, 1996: 67 **A**: BT
bidentatus Fabricius, 1792a: 233 (*Dermestes*) **E**: AU BY CR CT CZ DE EN FI FR GB GE HU IR IT LA LT NL NR NT PL RO SK ST SV SZ **A**: ES FE GUX HEI HEN HUB JA MNO TAI "Korea" **AFR AUR NAR ORR**
affinis Reitter, 1876c: 58
sulcatus Fabricius, 1792b: 495 (*Colydium*)
westerhauseri Gistel, 1857: 23 [= 1857: 535]
difficilis Halstead, 1973: 73 **A**: AP HP SD TAI UP **AFR AUR NAR ORR**
gibbus Pal & Sengupta, 1977: 275
ruficorpus Pal & Sengupta, 1977: 276
lateritius Broun, 1880: 222 (*Cryptamorpha*) **A**: AZ **N**: MRi **AUR AFR**i
lewisi Reitter, 1876c: 57 **E**i: DE **A**: GUA JA SD TAI UP "Korea" **AFR AUR ORR**
nigrans Pal & Sengupta, 1977: 277 **A**: SD
proximus Grouvelle, 1904: 183 **E**i: DE **AFR**
recticollis Reitter, 1876c: 61 **E**: CR IT **A**: CH JA TAI **AFR ORR**

minimus Grouvelle, 1913c: 334
pumilus Grouvelle, 1913c: 332
reflexus Reitter, 1880d: 85
vitulus Grouvelle, 1883: 294
robustus Halstead, 1973: 71 **A**: AP **ORR**
curvispinus Pal & Sengupta, 1977: 278
rossi Halstead, 1973: 54 **A**: HP UP **ORR**
unidentatus A. G. Olivier, 1790a: no 18: 9 (*Ips*) **E**: AB AU AZ BH BY CR CT CZ DE EN FI FR GB GE GR HU IT LA LS LT NL NR NT PL PT RO SK SL SP ST SV SZ **N**: CI MO **A**: CI ES FE MG NE TR WS "Korea" **COS**
gratiosus Motschulsky, 1863: 501
planus Herbst, 1797: 285 (*Colydium*)
siculus Stierlin, 1864: 147

family PASSANDRIDAE Blanchard, 1845 / Erichson, 1845

genus *Ancistria* Erichson, 1845: 305 type species *Colydium retusa* Fabricius, 1801
Anancistria Grouvelle, 1913e: 436 type species *Ancistria beccarii* Grouvelle, 1883
Bostrichoides Montrouzier, 1855: 17 type species *Bostrichoides angustatus* Montrouzier, 1855
Obancistria Grouvelle, 1913e: 436 type species *Ancistria longior* Grouvelle, 1912
Prionophora Westwood, 1848: 85 type species *Prionophora cylindrica* Westwood, 1848
apicalis Reitter, 1889r: 314 **A**: JA SCH **ORR**
fouqueti Grouvelle, 1913e: 481
bostrychoides Grouvelle, 1908c: 454 **A**: SD
brancuccii Burckhardt & Ślipiński, 1995a: 1014 **A**: NP
costata Burckhardt & Ślipiński, 1995a: 1018 **A**: SD
emarginata Grouvelle, 1914: 56 **A**: SCH TAI
kurosawai Sasaji, 1993: 22 **A**: JA
lewisi Reitter, 1889r: 315 **A**: JA
nepalensis Burckhardt & Ślipiński, 1995a: 1028 **A**: NP
pilosa Burckhardt & Ślipiński, 1995a: 1032 **A**: TAI
retusa Fabricius, 1801b: 555 (*Colydium*) **A**: HAI TAI **AFR AUR NTR ORR**
africana Grouvelle, 1919b: 56
albertisi Grouvelle, 1913e: 446
angustatus Montrouzier, 1855: 17 (*Bostrichoides*)
atra Grouvelle, 1913e: 445
brevis Grouvelle, 1913e: 469
cylindrica Westwood, 1848: 85 (*Prionophora*)
excavata Grouvelle, 1913e: 464
filiformis Grouvelle, 1913e: 458
gestroi Grouvelle, 1913e: 449
leai Grouvelle, 1913e: 462
magna Grouvelle, 1913e: 470
propingua Grouvelle, 1913e: 469
sauteri Grouvelle, 1914: 54

genus *Aulonosoma* Motschulsky, 1858c: 44 type species *Aulonosoma tenebrioides* Motschulsky, 1858
Asana Olliff, 1885: 71 type species *Trogosita rhyzophagoides* Walker, 1859
Laemotmetus Gerstaecker, 1871: 45 type species *Laemotmetus ferrugineus* Gerstaecker, 1871
Oryzoecus Reitter, 1876c: 37 type species *Oryzoecus carthartoides* Reitter, 1876
basalis Grouvelle, 1916a: 21 (*Laemotmetus*) **A**: SA **AFR**
insignis Grouvelle, 1891: 238 (*Laemotmetus*) **A**: BH NP UP **ORR**
tenebrioides Motschulsky, 1858c: 44 **A**: HUB TAI YUN **AFR AUR ORR**
carthartoides Reitter, 1876c: 38 (*Oryzoecus*)
ferrugineus Gerstaecker, 1871: 45 (*Laemotmetus*)
rhyzophagoides Walker, 1859a: 53 (*Trogosita*)

genus *Passandra* Dalman, 1817: 146 type species *Passandra sexstriata* Dalman, 1817
heros Fabricius, 1801b: 92 (*Cucujus*) **A**: CH TAI **AUR** **ORR**
atra Ślipiński, 1983: 83
autralicum C. O. Waterhouse, 1876b: 119 (*Hectarthrum*)
bistriatum Laporte, 1840b: 384 (*Hectarthrum*)
brevifossum Nerman, 1839: 392 (*Hectarthrum*)
cylindricum Smith, 1851: 12 (*Hectarthrum*)
depressum Smith, 1851: 20 (*Hectarthrum*)
orientale Schaufuss, 1887: 115 (*Hectarthrum*)
lineicollis Reitter, 1880a: 31 (*Hectarthrum*) **A**: YE **AFR**
modestum Fairmaire, 1884f: cxxi (*Hectarthrum*)
oblongicollis Fairmaire, 1886a: 441 (*Hectarthrum*) **A**: SA YE **AFR**
puncticollis Fairmaire, 1891g: cclxxxiii (*Hectarthrum*) **A**: SA **AFR**
tenuicornis Grouvelle, 1914: 53 (*Hectarthrum*) **A**: TAI **ORR**
trigemina Newman, 1839: 393 (*Hectarthrum*) **A**: "India" **ORR**
dejectum C. O. Waterhouse, 1876b: 120 (*Hectarthrum*)
latum Grouvelle, 1874: 27 (*Hectarthrum*)
minor C. O. Waterhouse, 1876b: 119 (*Hectarthrum*)
sociale C. O. Waterhouse, 1876b: 119 (*Hectarthrum*)

family CUCUJIDAE Latreille, 1802

genus *Cucujus* Fabricius, 1775: 204 type species *Cucujus depressus* Fabricius, 1775 (= *Meloe cinnaberina* Scopoli, 1763)
Earophilus Gistel, 1856: 375 type species *Cantharis sanguinolentus* Linnaeus, 1767 (= *Meloe cinnaberina* Scopoli, 1763)
Porphyrus Gozis, 1886: 20 type species *Cucujus depressus* Fabricius, 1775 (= *Meloe cinnaberina* Scopoli, 1763)
bicolor Smith, 1851: 2 [RN] **A**: NP
sanguinolentus Hope, 1831: 27 [HN]
cinnaberinus Scopoli, 1763: 60 (*Meloe*) **E**: AU BY CT CZ EN FI GE IT LT NR NT PL SK SP SV
depressus Fabricius, 1775: 204
geniculatus Reitter, 1893a: 110
sanguinolentus Linnaeus, 1767: 647 (*Cantharis*)
clavipes Fabricius, 1781: 257 (*Cucujus*) **Ei**: IT **NAR**
puniceus Mannerheim, 1843b: 303
siculus Tournier, 1894: 71
coccinatus Lewis, 1881a: 198 **A**: FE JA TAI
grouvellei Reitter, 1877b: 24 **A**: "Himalaya"
haematodes Erichson, 1845: 308 (*Cucuius* [sic!]) **E**: AR BY CT CZ EN FI IT LT PL SK "Caucasus" **A**: ES FE
caucasicus Motschulsky, 1845: 91
sibiricus Mannerheim, 1845: 91
imperialis Lewis, 1879c: 234 (*Cucujus*) **A**: SCH TAI YUN **ORR**
davidi Grouvelle, 1881: 89 (*Cucujus*)
kempi Grouvelle, 1913a: 107 **A**: UP
mniszechi Grouvelle, 1874: xxviii **A**: JA TAI
opacus Lewis, 1888: 84 **A**: JA TAI

genus *Pediacus* Shuckard, 1839: 150 type species *Cucujus dermestoides* Fabricius, 1792
bhutanicus Sengupta, 1978c: 221 **A**: AP BT
depressus Herbst, 1797: 286 (*Colydium*) **E**: BY CT CZ FI DE GE IT NR NT PL SK SV SZ "Caucasus" **NAR**
subglaber LeConte, 1854a: 73
dermestoides Fabricius, 1792b: 96 (*Cucujus*) **E**: AR CZ DE GE IT PL PT SK SZ "Caucasus" **A**: TR
elongatus Sengupta, 1978c: 219 **A**: BT
fuscus Erichson, 1845: 313 **E**: EN FI GE IT LA NR NT PL SK SV "European Russia" **A**: FE JA MG "Siberia" **NAR**
planus LeConte, 1850: 223 (*Silvanus*)
subcarinatus Mannerheim, 1852a: 363
japonicus Reitter, 1874c: 516 **A**: JA

kurosawai Sasaji, 1983: 18 **A**: JA
montivagus Champion, 1923c: 78 **A**: UP **ORR**
rufipes Grouvelle, 1908c: 461 **A**: AP **ORR**
smirnovi Nikitsky & Belov, 1979: 853 **E**: AB
tabellatus Wollaston, 1864: 131 **N**: CI (Tenerife)

family LAEMOPHLOEIDAE Ganglbauer, 1899

subfamily **Laemophloeinae** Ganglbauer, 1899

genus *Caulonomus* Wollaston, 1862a: 147 type species *Caulonomus rhizophagoides* Wollaston, 1862
rhizophagoides Wollaston, 1862a: 149 **N**: CI

genus *Cryptolestes* Ganglbauer, 1899b: 606 [RN] type species *Cucujus ferrugineus* Stephens, 1831
Fractophloeus Kessel, 1921: 28 type species *Laemophloeus fractipennis* Motschulsky, 1845
Leptus C. G. Thomson, 1863: 92 [HN] type species *Cucujus ferrugineus* Stephens, 1831
Tripsolestes Iablokoff-Khnzorian, 1978g: 341 type species *Cucujus pusillus* Schönherr, 1817
abietis Wankowicz, 1865: 298 (*Laemophloeus*) **E**: AU BY CT FI GE GR IT LA LT NR NT PL SK RU SV
puncticollis J. T. Fleischer, 1829: 72 (*Cucujus*)
candius Lefkovitch, 1965: 17 (*Cryptolestes*) **A**: YE
capensis Waltl, 1834: 169 (*Cucujus*) **Ei**: AU AZ CZ FR GB GE IT PL "Central Soviet Asia" **Ni**: AG MR **COS**
clavicollis Wollaston, 1854: 163 (*Laemophloeus*)
elongatulus P. H. Lucas, 1849: 479 (*Laemophloeus*)
obtusus Rey, 1889e: 35 (*Laemophloeus*)
rotundicollis Casey, 1884: 89 (*Laemophloeus*)
vermiculatus Wollaston, 1854: 161 (*Laemophloeus*)
corticinus Erichson, 1846: 327 (*Laemophloeus*) **E**: BY CZ DEi FI GE GR IT LT NR NT PL SK SP SV TR **A**: TR
krueperi Reitter, 1880d: 87 (*Laemophloeus*)
curus Lefkovitch, 1965: 18 **Ai**: YE **NAR**
dissimulatus Thomas, 1988: 47 (*Cryptolestes*)
divaricatus Grouvelle, 1893: 389 (*Laemophlaeus*) **A**: TIA **ORR**
duplicatus Waltl, 1839: 225 (*Cucujus*) **E**: BE BY CZ DE FIi GE IT KZ PL SK SP SV SZ UK **A**: FE
ferrugineus Stephens, 1831: 223 (*Cucujus*) **E**: AR AZ BEi CZ DE FI GB GE GR IT LT NL NRi PL SK SP SV SZ **N**: CI MR **A**: AE AF AHN FE FUJ GUA GUI GUX HAI HEB HEN HKG HUB HUN JIA JIX NMO SCH SHN SHX TAI YE YUN ZHE"India" **AFR** **NAR**i
alluaudi Grouvelle, 1906a: 120 (*Laemophloeus*)
carinulatus Wollaston, 1877: 44 (*Laemophloeus*)
concolor Smith, 1851: 6 (*Laemophloeus*)
emgei Reitter, 1887g: 286 (*Laemophloeus*)
exilis Rey, 1889e: 36 (*Laemophloeus*)
monilicornis Stephens, 1831: 223 (*Cucujus*)
obsoletus Smith, 1851: 7 (*Laemophloeus*)
perplexus Rey, 1889e: 35 (*Laemophloeus*)
testaceus Paykull, 1799: 168 (*Cucujus*) [HN]
fractipennis Motschulsky, 1845: 91 (*Loemophloeus* [sic!]) **E**: AB IT FR "Caucasus" **N**: "North Africa" **NAR**i
dufourii Laboulbène, 1848: 297 (*Laemophloeus*)
fursovi Iablokoff-Khnzorian, 1978g: 341 (*Cryptolestes*) **A**: UZ
halevyae Thomas, 1993: 113 (*Cryptolestes*) **A**: CY IS
klapperichi Lefkovitch, 1962b: 287 (*Cryptolestes*) **E**: GB **A**: AF HKG YE "China" **COS**
pusilloides Steel & Howe, 1952: 86 (*Laemophloeus*) **Ei**: AU GB GE IT PT SP **Ai**: HKG YUN **COS**
pusillus Schönherr, 1817: 55 (*Cucujus*) [RN] **E**: AZ BEi BY CZ DEi FIi GB GE IT NRi PL SK SP SVi **A**: AE AP BT FUJ HAI HUN NMO UP YE **COS**
brevis Fairmaire, 1850: 56 (*Loemophloes* [sic!])
crassicornis Waltl, 1839: 225 (*Cucujus*)
fuscus Lefkovitch, 1967: 312
longicornis Mannerheim, 1843b: 303 (*Laemophloeus*)
minutus A. G. Olivier, 1791: 243 (*Cucujus*) [HN]

parallelus Smith, 1851: 7 (*Laemophloeus*)
pauper Sharp, 1899: 530 (*Laemophloeus*)
testaceus Stephens, 1831: 224 (*Cucujus*)
spartii Curtis, 1834: pl. 50 (*Cucujus*) **E**: AB AZ BE CZ DEi FIi GB GE IT LAi SP SVi SZ UK "Caucasus" **N**: AG CI MR TU
ater A. G. Olivier, 1795b: no 74bis: 9 (*Cucujus*) [HN]
piceus Stephens, 1831: 223 (*Cucujus*) [HN]
rufipes P. H. Lucas, 1849: 478 (*Laemophloeus*)
suffuscus Wollaston, 1871: 244 (*Laemophloeus*)
turcicus Grouvelle, 1876a: xxxiii (*Laemophloeus*) **Ei**: AR AZ BE CZ DE FI FR GB GE GR IT NR PL SP SV SZ TR **Ni**: "North Africa" **Ai**: FUJ HAI HEN HKG HUB HUN IN JA JIX LIA NMO SC XIZ **COS**
immundus Reitter, 1874c: 519 (*Laemophloeus*)
truncatus Casey, 1884: 93 (*Laemophloeus*)
ugandae Steel & Howe, 1955: 107 (*Cryptolestes*) **Ei**: GB **AFR**
weisei Reitter, 1880d: 78 (*Laemophloeus*) **E**: GE IT PL RO

genus *Heterojinus* Sengupta & Mukhopadhyay, 1978: 142 type species *Laemophloeus semilaetaneus* Grouvelle, 1913
semilaetaneus Grouvelle, 1913a: 108 (*Laemophloeus*) **A**: AP

genus *Laemophloeus* Dejean, 1835: 340 type species *Cucujus muticus* Fabricius, 1781
biguttatus Say, 1827: 267 (*Cucujus*) **Ei**: IT **NAR**
bisignatus Guérin-Méneville, 1844a: 205
eremicus Peyerimhoff, 1948a: 23 **N**: LB
fasciatus Melsheimer, 1844: 113 (*Cucujus*) **Ei**: IT **NAR**
formosianus Grouvelle, 1914: 60 **A**: TAI
foveolatus Reitter, 1876c: 52 **A**: TAI **ORR**
harmandi Grouvelle, 1903a: 122 **A**: SD
humeralis Grouvelle, 1883: 283 **A**: TAI **ORR**
kraussi Ganglbauer, 1897b: 569 **E**: AU CT CZ FR GE IT PL SK SZ
monilis Fabricius, 1787a: 166 (*Cucujus*) **E**: AB AR BE CT CZ DEi FR GB GE GG IT NR PL SK SP SV SZ UK "Caucasus" **N**: AG TU **A**: UZ "South Siberia"
bimaculatus A. G. Olivier, 1795b: no 74bis: 6 (*Cucujus*)
bipustulatus Panzer, 1792c: 13 (*Cucujus*)
bucephalus Ganglbauer, 1899b: 610
denticulatus Preyssler, 1791: 117 (*Silpha*)
talyschensis Fursov, 1935a: 91
tunisiana Dajoz, 1981: 208
muticus Fabricius, 1781: 257 (*Cucujus*) **E**: CT DEi EN FR GE FI IT NR NT PL SK SV SZ WS **A**: ES FE WS
infuscatus Motschulsky, 1845a: 91 (*Loemophloeus* [sic!])
labiatus Kugellan, 1794: 572 (*Cucujus*)
piceus A. G. Olivier, 1795b: 74bis: 6 (*Cucujus*)
neglectus Grouvelle, 1883: 282 **A**: AP **ORR**
nigricollis P. H. Lucas, 1849: 478 **E**: BE FR GG GR IT SZ "Caucasus" "Balkan Peninsula" **N**: AG TU
mendax Sahlberg, 1913c: 61
ribbei Reitter, 1879e: 220 **A**: FE
singularius Hetschko, 1930: 41 [RN] **E**: ST **AUR**
insularis Kessel, 1921: 29 (*Laemophloeus*) [HN]
singularis Hetschko, 1928: 142 [RN] [HN]
submonilis Reitter, 1889r: 317 **A**: FE JA
subtilis Rey, 1889e: 35 **E**: GR

genus *Lathropus* Erichson, 1845: 307 type species *Trogosita sepicola* P. W. J. Müller, 1821
sepicola P. W. J. Müller, 1821: 208 (*Trogosita*) **E**: BY CT CZ FR GE IT NT PL SK SZ UK **A**: WS

genus *Leptophloeus* Casey, 1916: 119 type species *Lemophaleus angustulus* LeConte, 1866
Truncatophloeus Kessel, 1921: 28 type species *Laemophloeus mobilis* Grouvelle, 1908

abei Sasaji, 1986c: 684 **A**: JA
alternans Erichson, 1846: 325 (*Laemophloeus*) **E**: CT CZ FI FR GE GR HU IT LT NR NT PL SK SP SV SZ **A**: FE WS **NARi**
brevicornis C. G. Thomson, 1863: 97 (*Laemophloeus*)
delicatulus Obenberger, 1917: 26 (*Laemophloeus*)
axillaris Wollaston, 1854: 164 (*Laemophloeus*) **N**: MR
azoricus Ratti, 1972b: 283 (*Leptophloeus*) **E**: AZ
bupleuri Peyerimhoff, 1948c: 98 (*Laemophloeus*) **N**: AG
clematidis Erichson, 1846: 326 (*Laemophloeus*) **E**: AU BE CZ GB GE FR IT PL SK SP SZ UK
convexiusculus Grouvelle, 1877a: 213 (*Laemophloeus*) **Ei**: "Europe" **A**: JA
femoralis Sasaji, 1983: 25 **A**: JA
foveicollis Sasaji, 1986c: 681 **A**: JA
hypobori Perris, 1855: lxxvii (*Loemophloeus* [sic!]) **E**: AU FR IT **N**: AG
incertus Grouvelle, 1888: 627 (*Laemophlaeus* [sic!]) **A**: AP **ORR**
janeti Grouvelle, 1899b: 177 (*Laemophloeus*) **Ei**: BE GB GE **A**: AP **AFR ORR**
juniperi Grouvelle, 1874: xxviii (*Laemophloeus*) **E**: AU FR GE IT RO SK SP SZ UK **A**: SY
suspectus Rey, 1889e: 35 (*Laemophloeus*)
linearis Grouvelle, 1909b: 170 (*Laemophlaeus*) **Ai**: YE **AFR**
perrisi Grouvelle, 1876b: cxxii (*Laemophloeus*) **E**: FR IT **N**: AG
stenoides Wollaston, 1854: 165 (*Laemophloeus*) **N**: MR

genus *Microbrontes* Reitter, 1874c: 520 type species *Microbrontes laemophloeoides* Reitter, 1874
Tularthrum Broun, 1893: 1102 type species *Tularthrum lineatum* Broun, 1893
laemophloeoides Reitter, 1874c: 520 **A**: JA

genus *Microlaemus* Lefkovitch, 1962a: 196 type species *Microlaemus turneri* Lefkovitch, 1962
interceptus Grouvelle, 1903a: 122 (*Laemophloeus*) **A**: BT SD
picipennis Grouvelle, 1908c: 468 (*Laemophloeus*) **A**: BT **ORR**

genus *Nipponophloeus* Sasaji, 1983: 21 type species *Laemophloeus dorcoides* Reitter, 1874
boninensis Nakane, 1991a: 1 **A**: JA (Bonin Is.)
dorcoides Reitter, 1874c: 517 (*Laemophloeus*) **A**: FE JA **ORR**
prostomoides Reitter, 1874c: 516 (*Laemophloeus*)

genus *Notolaemus* Lefkovitch, 1959: 102 type species *Cucujus unifasciatus* Latreille, 1804
castaneus Erichson, 1845: 318 (*Laemophloeus*) **E**: AU CZ FR GE IT PL SK SV SZ
cribratus Reitter, 1889r: 317 (*Laemophloeus*) **A**: FE JA
lewisi Reitter, 1889r: 319 (*Laemophloeus*) **A**: JA TAI
nigroornatus Reitter, 1889r: 319 (*Laemophloeus*) **A**: JA
unifasciatus Latreille, 1804b: 256 (*Cucujus*) **E**: BE CZ DEi FR GB GE IT PL SK SV SZ UK **N**: TU
bimaculatus Paykull, 1801: 121 (*Cucujus*) [HN]
fasciatus Stephens, 1839: 142 (*Laemophloeus*)
ussuriensis Iablokoff-Khnzorian, 1977c: 619 (*Notolaemus*) **A**: FE

genus *Passandrophloeus* Kessel, 1921: 28 type species *Laemophloeus spinosus* Grouvelle, 1892
spinosus Grouvelle, 1892c: 861 (*Laemophloeus*) **Ei**: GB **AFR ORR**
divergens Grouvelle, 1908a: 17 (*Laemophloeus*)
glabriculus Grouvelle, 1897c: 395 (*Laemophloeus*)

genus *Placonotus* W. J. MacLeay, 1871: 168 type species *Placonotus longicornis* W. J. MacLeay, 1871
Silvanophloeus Sharp, 1899: 537 type species *Cucujus testaceus* Fabricius, 1787
admotus Grouvelle, 1914: 59 (*Laemophloeus*) **A**: TAI
africanus Lefkovitch, 1962a: 182 (*Placonotus*) **A**: AE YE **AFR**
debanus Mukhopadhyay, 1985: 186 **A**: AP
donacioides Wollaston, 1854: 159 (*Laemophloeus*) **N**: MR

fenestratus Reitter, 1889r: 318 (*Laemophloeus*) **A**: FE JA
granulatus Wollaston, 1854: 160 (*Laemophloeus*) **E**: AZ **N**: CI MR
hilleri Reitter, 1878j: 176 (*Laemophloeus*) **A**: FE JA
himalaicus Mukhopadhyay & Sengupta, 1977: 563 (*Placonotus*) **A**: SD
majus Lefkovitch, 1962a: 179 (*Placonotus*) **Ei**: IT **A**: AE **AFR NARi NTRi**
politissimus Wollaston, 1867: 67 (*Laemophloeus*) **Ei**: GB IT **Ai**: YE AE **COS**
commixtus Grouvelle, 1912f: 304 (*Laemophloeus*)
mirus Grouvelle, 1905b: 142 (*Laemophloeus*)
victus Kessel, 1926: 69 (*Laemophloeus*)
proximus Grouvelle, 1908c: 466 (*Laemophloeus*) **A**: AP **ORR**
saudicus Ślipiński, 1984: 257 **A**: AE
subtestaceus Grouvelle, 1883: 284 (*Laemophloeus*) **A**: TAI **ORR**
testaceus Fabricius, 1787a: 166 (*Cucujus*) **E**: AB AR AZ BE CZ DE IT FR GE GG KZ PL SK SZ SP "Caucasus" **N**: AG TU **A**: AP BT FE JA KI NE TAI UZ "NE China" **NAR ORR**
degener Rey, 1889e: 35 (*Laemophloeus*)
notatithorax Obenberger, 1917a: 25 (*Laemophloeus*)
stepensis Motschulsky, 1845a: 91 (*Loemophloeus* [sic!])
suberis P. H. Lucas, 1849: 479 (*Laemophloeus*)
zimmermanni LeConte, 1854a: 75 (*Laemophloeus*) **Ei**: IT **NAR**
apertus Casey, 1916: 128 (*Silvanophloeus*)
illustris Casey, 1916: 129 (*Silvanophloeus*)
sobrinus Casey, 1916: 129 (*Silvanophloeus*)

genus *Planolestes* Lefkovitch, 1958: 271 type species *Planolestes laevicornis* Lefkovitch, 1958
biskrensis Grouvelle, 1899b: 178 (*Laemophloeus*) **N**: AG **A**: AF **AFR**

genus *Pseudophloeus* Iablokoff-Khnzorian, 1977c: 620 type species *Laemophloeus fuscicornis* Reitter, 1874
fuscicornis Reitter, 1874c: 518 (*Laemophloeus*) **A**: FE JA

genus *Xylolestes* Lefkovitch, 1962a: 187 type species *Laemophloeus unicolor* Grouvelle, 1908
laevior Reitter, 1874c: 518 (*Laemophloeus*) **A**: JA TAI
calognathus Grouvelle, 1908c: 464 (*Laemophloeus*)
ovalis Grouvelle, 1883: 281 (*Laemophloeus*) **A**: AP BT **ORR**

genus *Xylophloeus* Lefkovitch, 1962a: 189 type species *Laemophloeus patens* Grouvelle, 1908
darjeelingensis Mukhopadhyay & Sengupta, 1978: 227 (*Xylophloeus*) **A**: SD
nainitalensis Mukhopadhyay & Sengupta, 1978: 226 (*Xylophloeus*) **A**: BT SD UP

subfamily **Propalticinae** Crowson, 1952

genus *Propalticus* Sharp, 1879: 88 type species *Propalticus oculatus* Sharp, 1879
Microloma John, 1939: 294 type species *Microloma acupinctum* John, 1939
indicus Sengupta, 1978a: 81 **A**: SD
japonicus Nakane, 1966b: 64 **A**: JA
kiuchii Sasaji, 1971b: 37 **A**: JA
morimotoi Kamiya, 1964: 281 **A**: JA (Irimote-jima)
ryukyuensis Kamiya, 1964: 283 **A**: JA (Irimote-jima, Ishigaki-jima)

family PHALACRIDAE Leach, 1815

subfamily **Phaenocephalinae** Matthews, 1899

genus *Phaenocephalus* Wollaston, 1873c: 167 type species *Phaenocephalus castaneus* Wollaston, 1873
castaneus Wollaston, 1873c: 168 **A**: FE JA **ORR**

genus *Sphaerostilbus* Champion, 1924a: 164 type species *Sphaerostilbus dilatatus* Champion, 1924
dilatatus Champion, 1924a: 164 **A**: UP **ORR**

subfamily **Phalacrinae** Leach, 1815

genus *Augasmus* Motschulsky, 1858c: 35 type species *Augasmus ligatus* Motschulsky, 1858
Heterolitus Guillebeau, 1893b: 375 type species *Heterolitus humilis* Guillebeau, 1893
Parischius Guillebeau, 1896c: 297 type species *Parischius auluaudi* Guillebeau, 1896
Liocrus Flach, 1889b: 271 type species *Litocrus coronatus* Flach, 1889
comptulus Lyubarsky, 2004: 62 **A**: NP
coronatus Flach, 1889b: 271 (*Litochrus*) **A**: JA TAI
gilbus Lyubarsky, 2004: 63 **A**: NP
luridus Lyubarsky, 2004: 62 **A**: NP
nigromaculatus Hisamatsu, 1985: 273 (*Heterolitus*) **A**: JA TAI
nipponicus Hisamatsu, 1985: 273 (*Heterolitus*) **A**: JA
perpolitus Lyubarsky, 2004: 64 **A**: NP
shirozui Hisamatsu, 1959d: 59 (*Heterolitus*) **A**: JA FE
sinuatus Motschulsky, 1858c: 36 **A**: JA TAI **ORR**
subflavus Lyubarsky, 2004: 64 **A**: NP

genus *Heterostilbus* Champion, 1924a: 165 type species *Heterostilbus marginatus* Champion, 1924
kobensis Champion, 1925e: 620 **A**: JA TAI
laevigatus Champion, 1924a: 166 **A**: UP
marginatus Champion, 1924a: 166 **A**: UP

genus *Litochrus* Erichson, 1845: 108 type species *Phalacrus brunneus* Erichson, 1842
subgenus *Lichrotus* Lyubarsky, 1993a: 17 type species *Litochrus strigosus* Reitter, 1899
strigosus Reitter, 1899b: 199 **A**: KZ
subgenus *Litochrus* Erichson, 1845: 108 type species *Phalacrus brunneus* Erichson, 1842
championi Hetschko, 1929: 156 [RN] **A**: CH JA UP **ORR**
brunneus Champion, 1924b: 240 [HN]
minutus Hisamatsu, 1985: 272 **A**: JA
rufoguttatus Champion, 1925e: 618 **A**: JA
ryukyuensis Hisamatsu, 1985: 272 **A**: JA

genus *Megistopalpus* Guillebeau, 1896a: 27 [RN] type species *Megapalpus simoni* Guillebeau, 1893
Megapalpus Guillebeau, 1893a: 297 [HN] type species *Megapalpus simoni* Guillebeau, 1893
simoni Guillebeau, 1893a: 297 (*Megapalpus*) **A**: YE

genus *Merobrachys* Guillebeau, 1896a: 26 [RN] type species *Stilbus koltzei* Reitter, 1887
Micromerus Guillebeau, 1892: 148 [HN] type species *Stilbus koltzei* Reitter, 1887
bimaculatus Matsumura, 1915a: 192 (*Phalacrus*) **A**: JA
rufosignatus Champion, 1925e: 620
fuscoguttatus Champion, 1924b: 241 **A**: UP
grouvellei Guillebeau, 1894a: 296 (*Micromerus*) **distr.**: "Sunésie"
koltzei Reitter, 1887h: 509 (*Stilbus*) **A**: ES FA
fenestratus Reitter, 1887h: 509 (*Stilbus*)
rufosignatus Champion, 1925e: 620 **A**: JA

genus *Nematolibrus* J. R. Sahlberg, 1913a: 21 type species *Nematolibrus filitarsis* J. R. Sahlberg, 1913
filitarsis J. R. Sahlberg, 1913a: 22 **A**: TR

genus *Olibrosoma* Tournier, 1889: 83 type species *Olibrosoma testacea* Tournier, 1889
Helectrus Guillebeau, 1892: 147 type species *Helectrus brisouti* Guillebeau, 1892
Litochroides Guillebeau, 1892: 148 type species *Litochroides sharpi* Guillebeau, 1892
Pyracoderus Guillebeau, 1892: 148 type species *Pyracoderus lemoroi* Guillebeau, 1892
sinuaticolle Guillebeau, 1892: 188 (*Litochroides*) **N**: EG MO
dorsale Guillebeau, 1892: 188 (*Litochroides*)
testacea Tournier, 1889: 84 **E**: SP **N**: AG EG MO TU **A**: JO SA

brisouti Guillebeau, 1892: 187 (*Helectrus*)
lemoroi Guillebeau, 1892: 187 (*Pyracoderus*)
sharpi Guillebeau, 1892: 187 (*Litochroides*)

genus *Olibrus* Erichson, 1845: 113 type species *Sphaeridium bicolor* Fabricius, 1792
Idiobius Gistel, 1856: 383 type species *Phalacrus flavicornis* Sturm, 1807
aenescens Küster, 1852: 60 **E**: FR IT PT SP **N**: AG MO
anthemidis Perris, 1869: 9
lucidus Guillebeau, 1892: 175
subniger Guillebeau, 1892: 175
aeneus Fabricius, 1792a: 83 (*Sphaeridium*) **E**: AU BH BU CZ DE EN FI FR GB GE HU IR IT LA LT NL NR NT PL SK SL SP SV SZ UK **N**: AG **A**: FE KZ TR
cognatus Stephens, 1829: 160 (*Phalacrus*)
multistriatus Zetterstedt, 1838: 233 (*Phalacrus*)
ovatus Marsham, 1802: 76 (*Dermestes*)
affinis Sturm, 1807: 76 (*Phalacrus*) **E**: AL AN AU AZ BH BU BY CR CZ DE EN FR GB GE GR HU IR IT LA LS LT MA MR NL PL PT RO RU SK SP SV SZ TR UK "Caucasus" **N**: AG EG LB MO MR TU **A**: ES FE IS JA MG SY TR
algericus Tournier, 1889: 92
aratus Gozis, 1889: 24
corcyrensis Flach, 1888a: 13
discoideus Küster, 1852: 61
hypocrites Tournier, 1889: 90
lederi Flach, 1888a: 13
rufus Guillebeau, 1892: 179
rufus Tournier, 1899: 87 [HN]
similaris Tournier, 1889: 91
stephensii Stephens, 1829: 164 (*Phalacrus*)
syriacus Tournier, 1889: 122
baudueri Flach, 1888a: 19 **E**: AU BE BY CZ DE FI FR GE GR IT NL PL NL NT SP SV **N**: EG **A**: SY
baudueri Tournier, 1889: 92 [HN]
perrisi Guillebeau, 1892: 170
siettii Guillebeau, 1897c: 224
bedeli Guillebeau, 1892: 180 **N**: AG MO TU
bicolor Fabricius, 1792a: 82 (*Sphaeridium*) **E**: AL AN AU BH BU BY CR CZ DE EN FI FR GB GE GR HU IT LA LT MA MR NL NR NT PL RO SK SP SV SZ YU UK **N**: MO TU **A**: ES FE KI KZ MG TM TR **NAR**
dohrni Flach, 1888a: 23
gentiles Guillebeau, 1892: 184
leachiellus Stephens, 1829: 163 (*Phalacrus*)
bimaculatus Küster, 1848a: 26 **E**: AU BH CR CZ EN FI FR GE GR HU IT MA NR NT PL PT RO SK SP SV SZ UK YU **N**: AG MO **A**: ES FE TR
abeillei Guillebeau, 1892: 185
biplagiatus Guillebeau, 1892: 184
maculifer Waltl, 1835: 84 (*Phalacrus*)
viennensis Guilebeau, 1892: 182
bisignatus Ménétriés, 1849: 271 (*Phalacrus*) **E**: AL AU BH CR CZ FR GE GR IT PL PT RO SK SP UK **N**: AG AR TU **A**: ES KZ TR
coccinella Flach, 1888a: 23
fauveli Guillebeau, 1892: 178
flavomaculatus Tournier, 1889: 89
calvosus Lyubarsky, 2004: 61 **A**: NP
camptoides Guillebeau, 1892: 184 **A**: "Turkestan"
castaneus Baudi di Selve, 1870: 49 **E**: BH CR FR GR (Corfu) IT MA SP **N**: AG MO **A**: CY LE SY TR UZ
caucasicus Tournier, 1889: 88 **E**: FR (Corse) IT (Sardegna) "Caucasus" **A**: TR
cinerariae Wollaston, 1854: 112 **N**: MR
congener Wollaston, 1864: 107 **N**: CI
consanguineus Flach, 1889b: 272 **A**: JA TAI

corticalis Panzer, 1797a: 11 (*Anisotoma*) **E**: AB AU BU BY CR CT CZ DE FR GB GE IR IT LA LT NL PT PL RO SK SP ST SV SZ UK YU (Crna Gora) **N**: AG CI EG TU **A**: SY TR
florum Wollaston, 1864: 106
stercoreum Fabricius, 1792a: 81 (*Sphaeridium*)
delicatulus Tournier, 1889: 121 **E**: ST
demarzoi Švec & Angelini, 1996: 207 **E**: IT **A**: TR
fallax Flach, 1888a: 20 **E**: AU IT
firmus Lyubarsky, 2004: 60 **A**: NP
flachi Reitter, 1891b: 22 **A**: KZ UZ
flavicornis Sturm, 1807: 78 (*Phalacrus*) **E**: AU CR CZ DE FI FR GB GE GR HU IT NL NR NT PL SK SL SP SV SZ UK **N**: AG MO **A**: ES IS TR
bonnarei Guillebeau, 1892: 181
convexus Rey, 1889a: 3
helveticus Tournier, 1876: Rye: 1876: 177
perfidus Flach, 1888a: 14
subsulcatus Rey, 1889a: 3
gerhardti Flach, 1888a: 22 **E**: AU BU CZ GE GR (Kerkyra) HU PL RO SK ST
bulgaricus Reitter, 1899a: 158
freyi Tournier, 1889: 97
heydeni Flach, 1888a: 21
globiformis Pic, 1894o: 111 **A**: TR
guttatus Tournier, 1889: 89 **E**: SZ **N**: MO **A**: TR
lepidus Tournier, 1889: 89
jelineki Švec & Ponel, 1999: 240 **A**: TR
judaicus J. R. Sahlberg, 1913a: 23 (*Olibus* [sic]) **A**: IS
kaszabi L. N. Medvedev, 1971: 222 [= 1972: 482] **A**: MG
koltzei Flach, 1888a: 19 **E**: AB BU GG SZ **A**: KZ
distinctus Roubal, 1946: 42
posticalis Reitter, 1913c: 125 [RN]
reitterianus Hetschko, 1928: 142 [RN]
laevisternus Guillebeau, 1897a: 22 **A**: SY
latisternus Guillebeau, 1894b: 459 **A**: UP **ORR**
liquidus Erichson, 1845: 117 **E**: AN AU AZ BE BH BU CR CZ FI FR GE GB GR HU IR IT LS MA MR NL NT PL PT RO SK SP SZ UK **N**: AG EG MO MR TU **A**: CY IQ UZ
sulcatus Tournier, 1889: 91
sulcipennis Guillebeau, 1892: 180
unicolor Tournier, 1889: 87
metallescens Flach, 1888a: 18 **A**: ES MG
millefolii Paykull, 1800: 439 (*Phalacrus*) **N**: AG **E**: AU AN BY CZ DE EN FI FR GB GE HU IT LA LT NL NR NT PL SK SL SP SV SZ UK **A**: ES KZ UZ
achillaeae Stephens, 1829: 162 (*Phalacrus*)
aeneopiceus Stephens, 1829: 163 (*Phalacrus*)
ulicis Gyllenhal, 1813: 430 (*Phalacrus*)
multesimus Luybarsky, 1993: 41 **A**: NP **ORR**
nainiensis Champion, 1924b: 238 **A**: UP **ORR**
norvegicus Munster, 1901: 34 **E**: BU DE EN FI GE GR LA LT NL NR NT PL SK SV SZ "Caucasus" **N**: AG **A**: FE IS TM TR
obscuricornis Guillebeau, 1894b: 459 **A**: HP **ORR**
obscurus Guillebeau, 1892: 184 **E**: IT SK
ovalis Iablokoff-Khnzorian, 1959b: 70 **E**: AR
particeps Mulsant & Rey, 1861d: 127 **E**: ES FR GR IT MC PT SP SZ **N**: AG EG MO TU **A**: CY ES IS JA LE
castanopterus Guillebeau, 1892: 174
demaisoni Flach, 1888a: 20
politus Tournier, 1889: 90
permicans Reitter, 1913c: 126 **A**: CH
pygmaeus Sturm, 1807: 84 (*Phalacrus*) **E**: AU BH CR CZ DE FI FR GB GE GR (Corfu) IT LA MA NL PL PT RU SK SP SV SZ UK **N**: AG EG MO TU **A**: TR SY

binderi Roubal, 1913c: 111
rasilis Lyubarsky, 2004: 60 **A**: NP
reitteri Flach, 1888a: 20 **E**: BU CR FR GR IT MA SP SZ **N**: AG LB EG MO TU **A**: TR
baudii Flach, 1888a: 21
bifenestratus Guillebeau, 1892: 176
desbrochersi Guillebeau, 1892: 171
ornatus Guillebeau, 1892: 176
raffrai Tournier, 1889: 91
similis Tournier, 1889: 98
reyi Guillebeau, 1892: 181 **E**: GR
seidlitzii Flach, 1888a: 21 **E**: ST **A**: MG
selvei Guillebeau, 1892: 177 **A**: CY
singularis Tournier, 1889: 100 **E**: SP **N**: MO
stictus Lyubarsky, 1993b: 43 **A**: CH **ORR**
stierlini Flach, 1888a: 22 **E**: BU CR FR GR IT PT RO SP SZ **N**: AG LB TU **A**: TR
apicatus Guillebeau, 1892: 184
atratus Guillebeau, 1892: 183
bilunulatus Guillebeau, 1892: 183
maximus Tournier, 1889: 89
striatissimus Reitter, 1899e: 282 **E**: AB "Caucasus" **A**: IN
subaereus Wollaston, 1864: 107 **N**: CI
tangerianus Tournier, 1889: 92 **N**: MO
turcicus Švec & Ponel, 1999: 243 **A**: TR

genus *Olibrus*, nomina dubia
aestuosus Tournier, 1889: 91 distr. not given
difficilis Tournier, 1889: 90 distr. not given
lederi Tournier, 1889: 91 distr. not given
minusculus Tournier, 1889: 91 distr. not given
minutus Tournier, 1889: 90 distr. not given
nigritus Tournier, 1889: 90 distr. not given
pyrenaeus Tournier, 1889: 90 distr. not given
subreptus Tournier, 1889: 92 distr. not given
subrugosus Tournier, 1889: 89 distr. not given

genus *Phalacrus* Paykull, 1800: 438 type species *Anisotoma corrusca* Panzer, 1797
atticus Guillebeau, 1894d: cxcii **E**: GR
borealis Lafer, 1992: 228 **E**: NT **A**: ES
brevidens Champion, 1925e: 606 **A**: JA
brunnipes C. N. F. Brisout de Barneville, 1863: 45 **E**: FR GB IT SP **N**: AG MO **A**: TR
mandibularis Tournier, 1889: 75
caricis Sturm, 1807: 80 **E**: AU BH BY CR CT CZ DE EN FI FR GB GE HU IR IT LT NL NT PL SK SV SZ UK **A**: TR MG FE "Siberia"
bonnairei Guillebeau, 1892: 157
delabyi Guillebeau, 1892: 157
flachi Schilski, 1891: 154
nigrinus Stephens, 1829: 146 (*Cercyon*)
nigrinus stephens, 1829: 162
championi Guillebeau, 1892: 158 **E**: AU BU CT GE HU NL PL SV SZ UK
suecicus Palm, 1947b: 182
corruscus Panzer, 1797a: 10 (*Anisotoma*) **E**: AL AN AR AU AZ BH BU CR CT CZ DE EN FI FR GB GE GR HU IR IT LT MA MD NL NR PL PT RO SK SL SP ST SV SZ UK YU **N**: AG CI EG MO **A**: CH CY FE IN KZ MG SY TD TM TR UZ "Tian Shan"
assimilis P. Rossi, 1794: 82 (*Sphaeridium*)
ater Herbst, 1792: 86 (*Tetratoma*)
humbertii E. C. Rye, 1872b: 37

intermedius Hochhut, 1872: 230
nigrinus Stephens, 1829: 162
picipes Stephens, 1829: 161
punctipennis J. R. Sahlberg, 1913d: 90
quercus Tournier, 1889: 32
rufipes Tournier, 1889: 33
corvinus Guillebeau, 1894b: 458 **A**: UP
exaluminatus Lyubarsky, 2004: 67 **A**: NP
fimetarius Fabricius, 1775: 68 (*Sphaeridium*) **E**: AU BE BU CZ DE FR GB GE GR HU IT LA MA NL PL RO SK SV SZ UK **N**: AG **A**: CY TR
brisouti E. C. Rye, 1872: 8
confusus Guillebeau, 1892: 153
genei Tournier, 1889: 27
hipponensis Guillebeau, 1892: 154
hybridus Flach, 1888a: 18
striatopunctatus Tournier, 1889: 53
subseriatus Rey, 1889a: 3
frater Flach, 1888: 27 **A**: TR
grossus Erichson, 1845: 111 **E**: AU BH CR CZ FR GE GR HU IT NL PL RO SK ST SV SZ **A**: ES KZ MG MO TR XIZ
dieckmanni Vogt, 1967: 161
grouvellei Guillebeau, 1892: 156 **A**: TU
immarginatus Champion, 1925e: 604 **A**: UP **ORR**
incommodus Flach, 1888a: 18 **E**: IT (Sicilia) SP UK **A**: AG TU
siculus Tournier, 1889: 52
indus Motschulsky, 1858c: 34 **A**: CH **ORR**
insularis Guillebeau, 1892: 156 **E**: GR (Kerkyra)
kuznetzovi Lafer, 1992: 228 **A**: FE JA
lateralis Guillebeau, 1893a: 298 **A**: YE
luteicornis Champion, 1924b: 238 **A**: CH JA UP
maspalomensis Palm, 1975a: 49 **N**: CI
maximus Fairmaire, 1852: 77 **E**: FR (Corse) IT SP **N**: AG LB MO TU
ambiguous Guillebeau, 1892: 151
reitteri Flach, 1888a: 8
mayeti Guillebeau, 1892: 158 **E**: SP **N**: AG MO
perfusorius Lyubarsky, 2004: 67 **A**: NP
punctatus Champion, 1925e: 605 **A**: HKG JA ZHE
seriepunctatus C. N. F. Brisout de Barneville, 1863: 44 **E**: FR IT SP **N**: AG MO **A**: CY SY
baudii Tournier, 1889: 76
striatulus Tournier, 1868: 143
substriatus Gyllenhal, 1813: 428 **E**: AU BY CR CZ DE EN FI FR GB GE HU IR IT LA LT MA NL NR NT PL SK SP SV SZ UK
fortestriatus J. R. Sahlberg, 1889: 82
minutus Tournier, 1868: 143
punctatostriatus Waltl, 1839: 226
trichopus Waltl, 1835: 84
tenuicornis Champion, 1925e: 607 **A**: JA ZHE **ORR**

genus *Podocesus* Guillebeau, 1894a: 281 type species *Podocesus semirufus* Guillebeau, 1894
Dolerus Guillebeau, 1894a: 282 [HN] type species *Dolerus limbatus* Guillebeau, 1894
Ledorus Guillebeau, 1896a: 27 [RN] type species *Dolerus limbatus* Guillebeau, 1894
interpositus Švec, 1992: 439 (*Stilbus*) **A**: JA

genus *Stilbomimus* Champion, 1924b: 242 type species *Stilbomimus polymorphus* Champion, 1924: 243
diluticollis Champion, 1924b: 244 **A**: UP
v-flavum Champion, 1924b: 244 **A**: UP

genus *Stilbus* Seidlitz, 1872: 35 type species *Anisotoma testacea* Panzer, 1797
subgenus *Stilbus* Seidlitz, 1872: 35 type species *Anisotoma testacea* Panzer, 1797
Eustilbus Sharp, 1888: 253 [HN] type species *Anisotoma testacea* Panzer, 1797
Microstilbus Guillebeau, 1894a: 283 type species *Phalacrus nitidus* Melsheimer, 1844
Olistherus Seidlitz, 1872: 157 [HN] type species *Silpha atomaria* Linnaeus, 1767
Stilboides Guillebeau, 1894a: 282 type species *Stilboides sublineatus* Guillebeau, 1894
ambagiosus Lyubarsky, 2004: 66 **A**: NP
atomarius Linnaeus, 1767: 574 (*Silpha*) **E**: AU BH CZ DE EN FI FR GE GB GR (Kríti) HU IT NL NT PL SK SP SV SZ UK **N**: AG MO **A**: FE IS JA TR SY **NAR**
picatus Flach, 1888a: 16
piceus Stephens, 1829: 165 (*Phalacrus*)
reitteri Flach, 1888a: 24
sulcatus Gerhardt, 1909: 418
avunculus Flach, 1889b: 272 **A**: JA
bipustulatus Champion, 1925e: 621 **A**: JA
coxalis Švec, 1992: 440 **A**: JA
ferrugienus Švec, 1992: 445 **E**: AB
japonicus Švec, 1992: 447 **A**: JA
merkli Švec, 1992: 445 **A**: FE
oblongus Erichson, 1845: 121 (*Olibrus*) **E**: AU CR CZ DE EN FI FR GB GE GR HU IT NL NT PL SK SP SV SZ UK **N**: EG **A**: AF KZ TR SY IS UZ
lebedevi Roubal, 1913a: 1
rauterbergi Reitter, 1911a: 78
uniformis Flach, 1888a: 17
olearis Lyubarsky, 2004: 66 **A**: NP
orbicularis Lyubarsky, 2004: 66 **A**: NP
pannonicus Franz, 1968: 169 **E**: AU BU CZ FR GE GR HU IT MC PL SK SP SZ UK **A**: AF KZ TR
sharpi Guillebeau, 1892: 191 (*Eustilbus*) **N**: EG **A**: SY **AFR**
obliquus Champion, 1925a: 52
testaceus Panzer, 1797a: 12 (*Anisotoma*) **E**: AB AL AU AZ BH BY CR CZ DE EN FI FR GE GB GR HU IT LA MA MR NL NR PL PT RO SK SP SV SZ UK **N**: AG AZ CI LB MO MR TU **A**: AF CY IN IS KI TM SY TR UZ **AFR**
consimilis Marsham, 1802: 75 (*Dermestes*)
rufipes Stephens, 1829: 164 (*Phalacrus*)
unicolor Flach, 1888a: 16
truncatus Švec, 1992: 448 **N**: MO
yezoensis Hisamatsu, 1985: 274 **A**: JA

genus *Tinodemus* Guillebeau, 1894a: 282 type species *Tinodemus grouvellei* Guilebeau, 1894
Astenulus Guillebeau, 1896c: 299 type species *Astenulus micropus* Guillebeau, 1896
bifurcus Švec, 1992: 434 (*Stilbus*) **A**: JA
chinensis Švec, 1992: 438 (*Stilbus*) **A**: FUJ
confusus Švec, 1992: 436 (*Stilbus*) **A**: JA
meridianus Švec, 1992: 432 (*Stilbus*) **A**: AF JA
mifsudi Švec, 2000: 61 **E**: MA
polygramma polygramma Flach, 1888a: 24 [= 1889a: 75] (*Stilbus*) **E**: FR (Corse) GR IT SP (Gibraltar) **N**: AG EG **A**: IS SY TR
polygramma rubidus Flach, 1889b: 272 (*Stilbus*) **A**: JA
similus Švec, 1992: 436 (*Stilbus*) **A**: JA SCH

genus *Tolyphus* Erichson, 1845: 108 type species *Phalacrus granulatus* Guérin-Méneville, 1834
subgenus *Pharcisinus* Guillebeau, 1894a: 278 type species *Tolyphus punctulatus* Rosenhauer, 1856
bimaculatus L. N. Medvedev, 1963: 147 **A**: KZ
jankovskii Skopin, 1951: 123 **A**: KZ
punctulatus Rosenhauer, 1856: 94 **E**: GR IT SP **N**: AG LB MO TU **A**: SA
flavipes A. Fleischer, 1928: 27

opacicollis Peyerimhoff, 1926c: 331
punctatostriatus Kraatz, 1858: 132
subsulcatus Fairmaire, 1860: 166
transcaspicus Reitter, 1913c: 125 **A**: TM
subgenus *Tolyphus* Erichson, 1845: 108 type species *Phalacrus granulatus* Guérin-Méneville, 1834
dubius Gridelli, 1930: 133 **N**: EG LB
granulatus Guérin-Méneville, 1834: pl. 50 (*Phalacrus*) **E**: FR GR IT PT SP **N**: AG EG MO LB TU **A**: IS LE SY TR
algiricus Guillebeau, 1892: 162
chalybaeus Guillebeau, 1892: 162
robustus Guillebeau, 1892: 162
simoni Flach, 1888a: 17 [= 1889a: 68]
striatipennis P. H. Lucas, 1849: 551 (*Phalacrus*)
subopacus J. R. Sahlberg, 1913d: 89
syriacus Reitter, 1884e: 251
rufescens Pic, 1914l: 51 **E**: IT **N**: EG
sedilloti Guillebeau, 1892: 163 **N**: LB TU

family Phalacridae, nomina dubia
Phalacrus nitens Marsham, 1802: 79 (*Dermestes*) **E**: GB
Phalacrus piceorhaeus Marsham, 1802: 78 (*Dermestes*) **E**: GB
Phalacrus striatus Stephens, 1835: 402 **E**: GB

family CRYPTOPHAGIDAE Kirby, 1837
subfamily **Cryptophaginae** Kirby, 1837
tribe Caenoscelini Casey, 1900

genus *Caenoscelis* C. G. Thomson, 1863: 266 type species *Cryptophagus ferrugineus* C. R. Sahlberg, 1820
Macrodea Casey, 1924: 194 type species *Macrodea antennalis* Casey, 1924
angelinii C. Johnson & Bowestead, 2003: 897 **E**: IT
bertiae C. Johnson & Bowstead, 2003: 894 **A**: PA
besucheti C. Johnson & Bowestead, 2003: 893 **A**: PA
ferruginea C. R. Sahlberg, 1820a: 58 (*Cryptophagus*) **E**: AU BE CT CZ DE EN FI FR GB GE GG HU IR IT LS LT NL NR NT PL RO SK SP SV SZ UK YU **A**: ES FE **NAR**
brisoutii Seidlitz, 1888a: 256
ochreosa Casey, 1900: 105
pallida Wollaston, 1846: 452 (*Atomaria*)
franzi Sengupta & Basak, 1985: 41 **A**: NP SD
jakesi C. Johnson, 1973: 262 **A**: AF
kurosai Sasaji, 1987: 26 **A**: JA
loebli C. Johnson & Bowestead, 2003: 892 **A**: PA
mussardi C. Johnson & Bowestead, 2003: 898 **N**: MO
sibirica Reitter, 1889o: 309 **E**: AU CT DE EN FI FR GB GE NR PL RO SK SV SZ UK **A**: FE JA MG NC
fleischeri Reitter, 1889o: 309
grandis C. G. Thomson, 1892: 1773
smetanai C. Johnson & Bowestead, 2003: 896 **A**: NP
subdeplanata C. N. F. Brisout de Barneville, 1882: xxix **E**: BE BH BY CT CZ DE EN FI FR GB GE HU LA NL NR NT PL RO SK ST SV SZ UK **A**: ES FE **NAR**
macilenta Casey, 1900: 106
tenerifensis C. Johnson & Bowestead, 2003: 890 **N**: CI
wangi C. Johnson & Bowestead, 2003: 894 **A**: TAI

genus *Dernostea* Sasaji, 1984: 27 type species *Dernostea tanakai* Sasaji, 1984
tanakai Sasaji, 1984: 28 **A**: JA

genus *Himascelis* Sengupta, 1978d: 275 type species *Himascelis brunneus* Sengupta, 1978

brunneus Sengupta, 1978d: 277 **A**: HP KA NP SD
gracilis Lyubarsky, 1999: 15
chinensis Nikitsky, 1996: 1368 (*Sternodea*) **A**: SHX
kashmirensis Sengupta & Pal, 1980b: 43 **A**: KA PA
nigratus Sengupta, 1978d: 279 **A**: NP SD
nepalensis Lyubarsky, 1999: 14
similis Nikitsky, 1996: 1369 (*Sternodea*) **A**: SCH
turgidus Lyubarsky, 1999: 15 **A**: NP

genus *Sternodea* Reitter, 1875g: 78 type species *Sternodea baudii* Reitter, 1875
baudii Reitter, 1875g: 79 **E**: AL BH CR GR IT UK YU
weisei Reitter, 1875l: 361
grilati Reitter, 1888a: 54 **N**: AG MO
japonica Sasaji, 1984: 25 **A**: JA
lederi Reitter, 1876f: 292 **E**: AB GG ST (Kavkaz) **A**: TR
dichroa Reitter, 1896f: 285
haroldi Reitter, 1876b: 292
miki Reitter, 1888d: 173 **E**: ST (Kavkaz)
raddei Reitter, 1876f: 292 **E**: AB GG ST (Kavkaz)

tribe Cryptophagini Kirby, 1837

genus *Antherophagus* Dejean, 1821: 45 type species *Tenebrio pallens* sensu Fabricius, 1781 (= *Antherophagus similis* Curtis, 1835)
caucasicus Reitter, 1878a: 169 **E**: AB AR GG ST (Kavkaz) UK
fursovi Lyubarsky, 1991b: 64 **E**: ST (Kavkaz) **A**: ES KI KZ MG
himalaicus Champion, 1922a: 68 **A**: HP KA NP UP
microphthalmus Grouvelle, 1919d: 182 **A**: SD
nigricollis Champion, 1925b: 108 **A**: UP
pallens Linné, 1758: 418 (*Tenebrio*) **E**: AU BE BY CT CZ DE EN FI FR GB GE HU IR IT LA LS LT NL NR NT PL RO SK SL SP SV SZ UK **A**: ES FE JA
fasciatus Roubal, 1935a: 54
fasciferus Roubal, 1935a: 54
maculipennis A. Villa & J. B. Villa, 1833: 33
nigricornis Fabricius, 1787a: 46 (*Ips*)
silaceus Herbst, 1792: 169 (*Ips*) **E**: AU BE BH BY CT CZ DE EN FR GB GE HU IR IT LT NL NT PL RO SK SL SP SV SZ UK YU **A**: KZ
canescens Grouvelle, 1916b: 28 [RN]
similis Curtis, 1835: pl. 546 **E**: AU BE BH BY CT CZ DE EN FI FR GB GE HU IR IT LA LT NK NL NR NT PL RO SK SL SP SV SZ UK YU **A**: KZ
[*pallens* sensu auctorum]
spaceki Roubal, 1935a: 53 **A**: FUJ SCH

genus *Asternodea* Leschen, 1996: 589 type species *Asternodea loebli* Leschen, 1996
loebli Leschen, 1996: 589 **A**: NP

genus *Catopochrotus* Reitter, 1889n: 290 type species *Catopochrotus crematogastri* Reitter, 1889
crematogastri Reitter, 1889n: 290 **E**: ST (Kavkaz)

genus *Cryptophagus* Herbst, 1792: 172 type species *Dermestes cellaris* Scopoli, 1763
Cryptophagistes Crotch, 1873a: 44 type species *Dermestes cellaris* Scopoli, 1763
Mnionomus Wollaston, 1864: 138 type species *Mnionomus ellipticus* Wollaston, 1864
acutangulus Gyllenhal, 1827: 285 **E**: AB AU BE BH BY CT CZ DE EN FI FR GB GE GG GR HU IC IR IT LA LT NL NR NT PL RO SK SL ST (Kavkaz) SV SZ UK YU **N**: EG MO **A**: AF BEI FE IN JA KI KZ NC NMO PA SA TD TM UZ **AFR NAR NTR**
heydeni Reitter, 1875b: 36
impressicollis Tournier, 1872a: 253

prominens Casey, 1924: 193
restrictus Casey, 1924: 193
subcylindricus Rey, 1889e: 36
waterhousei E. C. Rye, 1866: 101
anatolicus Bruce, 1963a: 30 **A**: TR
anitidus Lyubarsky, 2000: 324 **A**: NMO
anxiosus Grouvelle, 1916c: 67 **A**: NP SD
araxicola Reitter, 1889a: 21 **E**: AB AR
ariadne Otero & González, 1980: 177 **E**: SP
atratus Champion, 1922a: 69 **A**: HP NP UP
auropubens Grouvelle, 1916c: 75 **A**: HP NP SD
subnudus Grouvelle, 1916c: 76
aurovestitus Bruce, 1944: 3 **A**: NP UP **ORR**
axillaris Reitter, 1875b: 14 **E**: AL MD RO UK
jailensis Breit, 1913b: 294
posticus Reitter, 1888g: 425
bactrianus Solsky, 1876: 260 **A**: KZ TM
badius Sturm, 1845: 96 **E**: AL AR AU BE BH BU BY CR CT CZ DE EN FI FR GB GE GR IT LA LT NL NR NT PL RO SK SL SP ST (Kavkaz) SV SZ UK YU **N**: TU
demarzoi Otero & Angelini, 1981a: 139
irretusus Rey, 1889e: 36
lamellicornis C. N. F. Brisout de Barneville, 1882: xxviii
nescius Rey, 1889e: 36
rufus C. N. F. Brisout de Barneville, 1863: 64 [HN]
subarmatus Rey, 1889e: 36
subobtusus Rey, 1889e: 36
baldensis Erichson, 1846: 353 **E**: IT
besucheti Otero, 1997: 211 **A**: TR
bidentatus Mäklin, 1853: 205 **E**: NT **A**: ES FE WS **NAR**
kamtschaticus Lyubarsky, 1992b: 83
punctatissimus Mäklin, 1853: 209
blasi Otero & González, 1982: 161 **E**: SP
brisouti Reitter, 1875b: 27 **E**: AN FR SP
roubali Falcoz, 1929: 41
bruckii Reitter, 1875b: 26 **E**: AB AR BH GG SP ST (Kavkaz) **A**: AF FE IN IS JO PA SY TM TR
castanecens Grouvelle, 1916c: 65 **A**: AHN FUJ HP NP SD UP ZHE **ORR**
bimacularis Grouvelle, 1916c: 71
binotatus Grouvelle, 1916c: 72
heteroclitus Lyubarsky, 1997a: 52
vicinus Grouvelle, 1916c: 74 [DA]
caucasicus Motschulsky, 1845a: 111 (*Dapsa*) **E**: ST (Kavkaz)
osseticola Lyubarsky, 1992a: 68
cellaris Scopoli, 1763: 16 (*Dermestes*) **E**: AB AU BE BH BY CR CT CZ DE FI FR GB GE GG GR HU IR IT LA MA MD NL NR NT PL PT RO SL SP SK ST (Kavkaz) SV SZ UK **N**: AG AZ CI EG MO MR TU **A**: AF FE IN IQ IS JA JO KZ NC SA SD SY UP UZ **COS**
angustatus Casey, 1924: 192 [HN]
crenatus Herbst, 1792: 177
lomus Sengupta, 1980: 34
mascarensis Reitter, 1875b: 19
rupimontis Casey, 1924: 190
subfasciatus Grouvelle, 1896b: 90
subtilis C. G. Thomson, 1873: 530
cellaroides Reitter, 1921b: 90 **E**: AB ST (Kavkaz) **A**: IN
circassicus Reitter, 1888d: 171 **E**: GG ST (Kavkaz)
carinatus Bruce, 1936: 140
podolkai Roubal, 1916: 47
clavator Champion, 1924a: 166 **A**: HP UP

confusus Bruce, 1934b: 58 **E**: AR AU CT CZ EN FI GB GE GG HU NR NT PL SK SP SV SZ UK **A**: FE UZ
conicollis Reitter, 1896c: 67 **A**: ES WS
altaicus Bruce, 1938: 9
corticinus C. G. Thomson, 1863: 260 **E**: AU BE BH BY CT DE FI FR GB GE LA NL NR NT SV **A**: ES FE **NAR**
depressulus Casey, 1900: 98
cristatus Lyubarsky, 2000: 322 **A**: FE
croaticus Reitter, 1880b: 51 **E**: AU BH CR GE HU IT LS SL SZ
curtus Grouvelle, 1916c: 69 **A**: NP SD **ORR**
malaisei Bruce, 1944: 4
cylindrellus Johnson, **sp. nov**. [see New acts] **E**: AU BH CR CZ DE EN FR GE IT MD NL NT PL PT RO SK SL SP SV SZ UK **A**: TR
[*cylindrus* sensu auctorum]
daghestanicus Lyubarsky, 2000: 325 **E**: ST (Kavkaz)
decoratus Reitter, 1874b: 379 **A**: FE FUJ GUI HKG HUB JA JIA SCH SHX TAI YUN ZHE
callosipennis Grouvelle, 1919d: 191
elegans Grouvelle, 1919d: 188
longipennis Grouvelle, 1919d: 196
varians Grouvelle, 1919d: 190
dentatus Herbst, 1793: 15 (*Kateretes*) **E**: AB AN AR AU BE BH BU BY CR CT CZ DE EN FI FR GB GE GG GR HU IR IT LA LS LT MD NL NR NT PL RO SK SL SP ST SV SZ TR UK YU **N**: AG AZ CI EG LB MO MR TU **A**: FE IN TM TR UZ **NAR**
fortidens Rey, 1889f: 43
fumatus Marsham, 1802: 110 (*Corticaria*)
quadridentatus Mannerheim, 1843b: 256
desertus Lyubarsky, 1989: 61 **A**: KZ MG TM XIZ
dilatipennis Reitter, 1888a: 29 **E**: AR GG ST (Kavkaz) **A**: KZ TR
bipedes Bruce, 1963a: 31
delicatulus Roubal, 1913b: 47
schuberti Bruce, 1963a: 32
dilutus Reitter, 1874b: 380 **E**: AB AR AU BH BU CR CT GB GE GG GR HU IT MD PL PT SL SP ST (Kavkaz) TR UK YU **N**: AG MO TU **A**: AF FE IN IQ GUX HEI KI KZ JA JO MG NMO NC NP SA SHA SY TD TM TR UZ YUN **NAR**
deceptor Grouvelle, 1919d: 193
exilicornis Grouvelle, 1919d: 200
[*hexagonalis* sensu auctorum]
polonicus Roubal, 1927b: 33
subinermis Grouvelle, 1919d: 194 [HN]
distinguendus Sturm, 1845: 90 **E**: AB AR AU BE BH BY CR CT CZ DE EN FI FR GB GE GG GR HU IC IR IT LT MD NL NR NT PL RO SK SL ST (Kavkaz) SV SZ UK YU "Baltic" **N**: CI MO TU **A**: AF CY ES FE IN KI KZ **AFR NAR**
convexus Rey, 1889f: 43
keeni Casey, 1924: 191
lucifugus Falcoz, 1922: 189
rugicollis Rey, 1889f: 43
spelaeus Falcoz, 1922: 189
stramenti Donisthorpe, 1935: 110
subhirtus Rey, 1889b: 43
subinermis Rey, 1889f: 43
umbratus Erichson, 1846: 355
verrucifer Rey, 1889f: 43
vestitus Rey, 1889f: 43 [HN]
dorsaliformis Reitter, 1897e: 212 **E**: CT **A**: AF ES FE NMO WS
dorsalis C. R. Sahlberg, 1819: 56 **E**: AU BE CR CT CZ DE EN FI FR GE GG IT LT NL NR NT PL RO SK SP ST SV SZ **A**: UZ
angustus Ganglbauer, 1899b: 684 [RN]
parallelus C. G. Thomson, 1871: 387 [HN]
reichardti Bruce, 1936: 97

durus Reitter, 1878e: 93 **E**: IT SP
ludovici Breit, 1908: 62
dzungaricus Lyubarsky, 2000: 323 **A**: KZ
eliae Otero, 2002: 87 **A**: NP UP
ellipticus Wollaston, 1864: 138 (*Mnionomus*) **N**: CI
enormis Hisamatsu, 1962: 1 **A**: JA
falcozi Roubal, 1927b: 31 **E**: DE EN GB IT SK SL
westi Bruce, 1943c: 368
fallax Balfour-Browne, 1953: 258 **E**: AU BE BH BU CR CT CZ DE EN FI FR GB GE HU IT LT NL NR NT PL RO SK SL ST (Kavkaz) SV SZ UK YU **N**: EG **A**: KZ UZ **NAR**
fasciatus Kraatz, 1852: 229 **E**: AB BU CR FR GR IT MA MC PT SP ST (Kavkaz) UK **N**: MO TU **A**: KI KZ TM TR
gonzalezi Otero & Angelini, 1981b: 99
signatus H. Brisout de Barneville, 1862: 23
fusciclavis Bruce, 1943a: 159 **A**: FUJ SCH SHA ZHE
fuscicornis Sturm, 1845: 97 **E**: AU BE BH CT CZ FI FR GB GE HU IT LT NL NR PL RO SK SL SP SV SZ UK YU
fusiformis Wollaston, 1862a: 156 **N**: CI
gracilis Reitter, 1875b: 17 **E**: FR IT
bedeli Bruce, 1936: 139
grouvellei Bruce, 1938: 250 [RN] **A**: JA
convexus Grouvelle, 1919d: 187 [HN]
hauseri Reitter, 1890i: 358 **E**: AR CT ST (Kavkaz) UK **A**: AF IN KI KZ TD TM UZ XIN
amfidoxus Bruce, 1963b: 38
szekessyi Bruce, 1938: 5
hexagonalis Tournier, 1872b: 444 **E**: BY CT FI NT ST (Kavkaz) **A**: FE KZ NMO TM TR WS
subvittatus Reitter, 1888a: 29
himalaicus Bruce, 1952: 184 **A**: AF HP NP PA UP
robustus Bruce, 1959: 19 [HN]
honoratus Lyubarsky, 1999: 5 **A**: NP UP
humeridens Bruce, 1943a: 161 **A**: FUJ
hupalupae Israelson, 1985: 149 **N**: CI
impressus Wollaston, 1865: 22 **N**: CI
inaequalis Reitter, 1878b: 53 **E**: AU BH GR RO SL UK
infuscus Grouvelle, 1916c: 70 **A**: SD
insulicola Roubal, 1919: 36 **E**: GB GE GR SZ **A**: AF IN TD TM UZ
acuminatus Coombs & Woodroffe, 1955: 257
vagus Bruce, 1938: 6
intermedius Bruce, 1934b: 59 **E**: AB AR AU DE EN FR GB GE GG IT NR PL RO SK ST (Kavkaz) SV SZ
involatilis Lyubarsky, 1997b: 55 **A**: KI KZ UZ
jakowlewi Reitter, 1888g: 424 **E**: AU CT FI IT NR NT SV SZ **A**: ES FE KI KZ MG NMO TM UZ **NAR**
archangelicus J. R. Sahlberg, 1926: 118
confertus Casey, 1900: 94
uralensis Bruce, 1963b: 37
japonicus Reitter, 1874b: 380 **A**: ES JA
sculpturatus Bruce, 1941: 985
johnsoni Sengupta, 1980: 35 **A**: NP SD
bengalensis Sengupta, 1980: 36
martensi Sengupta, 1980: 37
kirgisicus Lyubarsky, 1989: 54 **A**: KI KZ UZ
klapperichi Bruce, 1943a: 160 **A**: FUJ SHA ZHE
labilis Erichson, 1846: 359 **E**: AU BE BH BY CT CZ DE EN FI FR GB GE GG GR HU IT LA LT NL NR NT PL RO SK SL SV SZ UK YU **A**: ES
lapidicola Reitter, 1880e: 471 **E**: AB AR GG ST (Kavkaz) **A**: TR
corpulentus Reitter, 1888d: 170

lapponicus Gyllenhal, 1827: 286 **E**: AU BH CT CZ EN FI FR GB GE GR IT NR NT PL SK SL SV SZ UK YU **A**: ES KZ UZ WS **NAR**
alpestris Rey, 1889f: 43
beringensis J. R. Sahlberg, 1885: 29
clavatus Zetterstedt, 1838: 96
rufulus J. R. Sahlberg, 1926: 123
suturalis Rey, 1889f: 43
validus Kraatz, 1856: 240
latangulus Reitter, 1889p: 303 **A**: ES FE JA NC TAI
laterangulus Reitter, 1891d: 197 **E**: ST (Kavkaz) **A**: IQ IN KZ TM
laticollis P. H. Lucas, 1846: 222 **E**: AB AR AU BE BH CR CT CZ DE FI FR GB GE GG GR HU IC IR IT LA LT MA NL NR NT PL PT RO SK SP SV SZ UK YU **N**: AG AZ CI EG LB MO MR TU **A**: AF FE HP IN IS IQ JO KA KI LE NP SY TM TR UP UZ XIN **AFR AUR NAR ORR**
affinis Sturm, 1845: 79 [HN]
cribricollis Casey, 1900: 97
galilei Otero, 1997: 208
inscitus Casey, 1900: 97
milleri Reitter, 1875b: 20
montemurroi Otero & Angelini, 1984: 83
obesulus Wollaston, 1864: 136
lemonchei Otero & González, 1985: 151 **E**: SP
lewisi Reitter, 1874b: 379 **A**: JA
lomus Sengupta, 1980: 34 **A**: UP
longinotus Lyubarsky, 1989: 53 **A**: KZ
longior Grouvelle, 1919d: 184 **A**: JA
longiventris Grouvelle, 1919d: 195 **A**: JA
lycoperdi Scopoli, 1763: 18 (*Dermestes*) **E**: AN AU BE BH BU BY CR CT CZ DE FI FR GB GE GR HU IR IT LA LT MD NL NR NT PL RO SK SL SP ST (Kavkaz) SV SZ UK YU **N**: TU **A**: TR **NAR**
bituberculatus Stephens, 1830: 74
fungorum Panzer, 1797c: 14 (*Dermestes*)
marshami Stephens, 1830: 75
rufus Marsham, 1802: 111 (*Corticaria*)
lysholmi Munster, 1932: 53 **E**: AU BH CZ FI NR SK SV
lyubarskyi Johnson, **sp. nov**. [see New acts] **A**: NP
robustus sensu Lyubarsky, 1999: 9 [HN]
micaceus Rey, 1889f: 44 **E**: AN BH BU CR CZ DE FR GB GE GR HU IT LA MD NR PL RO SK SL SP SV SZ TR UK **A**: TR
lovendali Ganglbauer, 1899b: 678
minor Lyubarsky, 1997b: 55 **A**: KI KZ
montanus C. N. F. Brisout de Barneville, 1863: 67 **E**: AL AU BH CR CZ FR GE HU IT MC PL RO SK SZ UK YU
arctomyos Sainte-Claire Deville, 1927: 44
bosniensis Obenberger, 1914: 107
deubeli Ganglbauer, 1897a: 562
madadorum Roubal, 1945: 120
okalii Reška, 1982: 460
silesiacus Ganglbauer, 1899b: 698
transsilvanicus Ganglbauer, 1899b: 698
tridentatus Roubal, 1936a: 170
monticola Grouvelle, 1916c: 73 **A**: SD
nitiduloides Wollaston, 1854: 618 **N**: MR
rutae Bruce, 1940: 4
nitidulus L. Miller, 1858: 93 **E**: AB AR AU BH CR CT CZ GE HU IT RO SK ST (Kavkaz) UK YU **A**: KZ
seriatus Reitter, 1888a: 20
obsoletus Reitter, 1879e: 221 **E**: GB GE LT SV **A**: ES FE JA MG NMO **NAR**
opacosericeus Reitter, 1896c: 67 **E**: AB ST (Kavkaz) **A**: KI KZ TD
opacus Reitter, 1898b: 19 **A**: KZ TM UZ

oromii Otero, 1989: 205 **N**: CI
pallidus Sturm, 1845: 69 **E**: AB AR AU BE BH BU CR CT CZ FR GE GG GR HU IT MD NL PL RO SK SL ST (Kavkaz) SZ TR UK **N**: AG LB TU **A**: IN IS LE TR
convexus Rey, 1889f: 43
rugicollis Rey, 1889f: 43
subhirtus Rey, 1889f: 43
subinermis Rey, 1889f: 43
thomsoni Reitter, 1875b: 32
verrucifer Rey, 1889f: 43
parallelus C. N. F. Brisout de Barneville, 1863: 65 **E**: AB BE CZ DE EN FI FR GB GE GR IT LA NR NT PL SK SV
trapezoidalis J. R. Sahlberg, 1912: 92
pedarius Lyubarsky, 1999: 8 **A**: NP
pilosus Gyllenhal, 1827: 287 **E**: AB AR AU BE BH BU BY CR CT CZ DE EN FI FR GB GE GG GR IR HU IT LS MC MD NL NR NT PL RO SK SL SP ST (Kavkaz) SV SZ TR UK **N**: MR **A**: ES FE IN IS TR UZ **AFR AUR NAR**
albipilus Rey, 1889f: 43
corticum Rey, 1889f: 43
denticulatus Heer, 1841: 426
obtusidens Rey, 1889f: 43
pseudodentatus Bruce, 1934f: 168
substriatus Rey, 1889f: 43
tamaricis Rey, 1889f: 43
populi Paykull, 1800: 355 **E**: AU BU CT CZ DE EN FI FR GB GE IT LA MD NL NR NT PL RO SK SL SV SZ UK **A**: UZ
grandis Kraatz, 1856: 240
pseudopacus Lyubarsky, 1991a: 145 **E**: KZ (Uralskaya Oblast')
pseudoschmidti Woodroffe, 1970: 9 **A**: CT ES MG NMO NT SHX WS
pubescens Sturm, 1845: 103 **E**: AU BE BH BU BY CR CT CZ DE FI FR GB GE GR HU IR IT LT MD NL NR PL PT RO SL SP SK ST (Kavkaz) SV SZ UK YU **N**: AG MO
domesticus Hubenthal, 1920: 241
pumilus Reitter, 1874b: 380 **A**: JA
puncticollis P. H. Lucas, 1846: 221 **E**: BH CR FR GB GE HU IT NL SP SZ YU **N**: AG
rotundatus Coombs & Woodroffe, 1955: 258
subelongatus Rey, 1889f: 43
punctipennis C. N. F. Brisout de Barneville, 1863: 63 **E**: AB AL AN AR AU BE BH BU BY CR CT CZ DE EN FI FR GB GE GG GR HU IC IR IT LA LS LT MD NL NR NT PL PT RO SK SL SP ST (Kavkaz) SV SZ UK **N**: AG AZ CI MO MR TU **A**: AF ES FE IN JA KI KZ MG NC TM TR **AUR NAR**
cicatricosus Casey, 1900: 95
infuscatus Casey, 1900: 95
parvinoda Casey, 1900: 24
peyerimhoffi Falcoz, 1921: 42
[*pilosus* sensu auctorum]
plectrum Casey, 1900: 95
praetermissus Hubenthal, 1920: 241
productus Rey, 1889e: 36
sparsus Rey, 1889e: 36
subconvexus Rey, 1889e: 36
thoracicus J. R. Sahlberg, 1903a: 16
vaulogeri Grouvelle, 1912a: 44
quadrihamatus Mäklin, 1853: 208 **E**: FI NR NT SV **A**: AF ES FE MG **NAR**
plagiatus Poppius, 1900: 189 [HN]
quadrimaculatus Reitter, 1877h: 293 **E**: AB AR CT FI GG ST (Kavkaz) UK **A**: ES FE IQ KI KZ MG NMO SY TD TM UZ WS **NAR**
kochi Bruce, 1943b: 295
ophthalmicus Bruce, 1938: 7
quercinus Kraatz, 1852: 228 **E**: AU BU BY CR CT CZ EN FI FR GE HU IT LA LT NR PL RO SK SV ST (Kavkaz) SZ UK **A**: FE KZ UZ

hamatus Ganglbauer, 1897a: 560
recticollis Solsky, 1876: 260 **A**: KZ UZ
reflexicollis Reitter, 1877d: 21 **E**: PL RO SK UK YU
reflexus Rey, 1889f: 43 **E**: AB AL AR AU BE BH BU BY CR CT CZ DE EN FI FR GB GE GG GR HU IR IT LT MC MD NL NR NT PL RO SK SL SP ST (Kavkaz) SV SZ TR UK YU **N**: LB MO TU **A**: FE IN IS KI LE MG NMO TD TR WS
fowleri Joy, 1910: 205
freyi Bruce, 1941: 986
lateralis Rey, 1889f: 43
[*pallidus* sensu auctorum]
robustus Grouvelle, 1919d: 186 **A**: JA
ruficornis Stephens, 1830: 78 **E**: AB BU FR GB GE IR IT RO SP YU **A**: SY
ater Baudi di Selve, 1870: 56 [DA]
lugens Petri, 1926: 184 [DA]
nigritulus Reitter, 1888a: 22
rufomaculus Lyubarsky, 2000: 325 **A**: FE
saginatus Sturm, 1845: 88 **E**: AB AU BE BH BU BY CR CT CZ DE EN FI FR GB GE GG GR HU IR IT LA LT MC MD NL NR NT PL RO SK SL SP ST SV SZ UK YU **N**: AG AZ CI MR TU **A**: ES FE TR **AUR NAR NTR**
agnatus Rey, 1889f: 43
amplus Grouvelle, 1912a: 46
bohemicus Roubal, 1924a: 24
escolai Otero & González, 1983a: 79
germaini Grouvelle, 1905a: 129
heyrovskyi Obenberger, 1917a: 24
solidus Rey, 1889f: 43
scanicus Linnaeus, 1758: 357 (*Dermestes*) **E**: AB AL AR AU BE BH BU BY CR CT CZ DE EN FI FR GB GE GG GR HU IC IR IT LA LT MD NL NR NT PL RO SK SL ST SV SZ UK YU **N**: MR TU **A**: FE IN TR WS **AFR NAR**
niger C. N. F. Brisout de Barneville, 1863: 66
nigrinus Rey, 1889f: 43
patruelis Sturm, 1845: 80
piceus J. R. Sahlberg, 1926: 124
subdentatus Rey, 1889f: 43
suturalis Rey, 1889f: 43
schaitan Lyubarsky, 2000: 321 **E**: ST (Kavkaz)
schmidtii Sturm, 1845: 93 **E**: AU BE CT CZ FR GB GE GG HU IR IT LA NL PL RO SK ST (Kavkaz) SZ UK **A**: FE KZ SY
schroetteri Reitter, 1912c: 252 **E**: BH CR IT SL YU
trebinjensis Roubal, 1919: 37
scutellatus Newman, 1834: 202 **E**: AR AU BE BH BY CR CT CZ DE EN FI FR GB GE GG HU IC IT IR LS LT MD NL NR NT PL PT RO SK SL SP ST (Kavkaz) SV SZ UK **N**: MO **A**: IN TR **NAR**
bicolor Sturm, 1845: 07
pumilio Ganglbauer, 1897a: 561
rufipennis Sturm, 1845: 107
setulosus Sturm, 1845: 92 **E**: AB AU BE BH BY CR CT CZ DE EN FI FR GB GE HU IR IT LT MD NL NR NT PL RO SK SL SP ST (Kavkaz) SZ UK **N**: AG MO **A**: AF ES KI KZ TR **NAR**
amplicollis J. R. Sahlberg, 1926: 116 [HN]
antennatus Casey, 1900: 98
crassicornis J. R. Sahlberg, 1900: 13
dilaticollis Tournier, 1872b: 443
grossulus Casey, 1924: 192
mazetieri Falcoz, 1929: 42
norvegicus J. R. Sahlberg, 1926: 119
sahlbergi Hellén, 1930: 78
vulpinus Reitter, 1888a: 30
simplex L. Miller, 1858: 94 **E**: AB AR AU CZ GB GE HU NL ST **A**: KI UZ
simulator Grouvelle, 1916c: 64 **A**: HP NP PA SD UP

skalitzkyi Reitter, 1875b: 12 **E**: AB AR CR CZ FR GE GG GR IT SP ST (Kavkaz) **A**: AF ES IN KA KI KZ PA TD TM TR UZ
anticus Rey, 1889e: 36
obtusus Rey, 1889e: 36
outereloi Otero & González, 1983b: 209
spadiceus Falcoz, 1925: 252 **E**: SP
jelineki Reška, 1982: 457
sporadum Bruce, 1934c: 160 **E**: AU BH BU CR CZ FR GE GR HU IT PL RO SK SP SZ **A**: IS TR
straussi Ganglbauer, 1897a: 564 **E**: AU FR
subdepressus Gyllenhal, 1827: 287 **E**: AU BE BH BU BY CR CT CZ DE EN FI FR GB GE GG HU IT LA MD NL NR NT PL RO SK SL ST (Kavkaz) SV SZ UK **N**: AG **A**: FE
depressus C. G. Thomson, 1863: 259
helveticus Tournier, 1872a: 254
subfumatus Kraatz, 1856: 241 **E**: AN AU BE BH CR CT CZ DE EN FI FR GB GE HU IC IR IT LA LT MD NL NR NT PL SK SL ST SV SZ UK **A**: AF ES FE KI TM UZ **NAR NTR**
kraatzii Reitter, 1875b: 34
tuberculosus Mäklin, 1852: 338 **E**: FI NR NT SV **A**: FE MG NMO **NAR**
boulderanus Casey, 1924: 190
brevipilis Casey, 1900: 97
debilis LeConte, 1858: 64
instabilis Bruce, 1936: 81
lepidus Casey, 1900: 97
turcicus Gillerfors & Otero, 1993: 229 **E**: ST (Kavkaz) **A**: KA KI TR
uncinatus Stephens, 1830: 75 **E**: AL AU BE BH BU CR DE FR GB GE GR HU IT LT MA MD NL PL SL SP SV SZ TR UK YU **N**: AG **A**: TM UZ
curtulus Rey, 1889f: 43
hirtulus Kraatz, 1858: 138
immixtus Rey, 1889f: 43
postpositus J. R. Sahlberg, 1903a: 15
retusus Rey, 1889e: 36
varus Woodroffe & Coombs, 1961: 190 **A**: FE **NAR**
versicolor Har. Lindberg, 1951: 6 **N**: CI
vseteckai Reška, 1977: 394 **Ei**: CZ
zonatus Lyubarsky, 1995: 46 **A**: ANH SCH

genus *Cryptophagus*, nomina dubia
cylindrus Kiesenwetter, 1858b: 138
minimus Falcoz, 1929: 42
flavipennis Faldermann, 1839: 6
erichsoni Reitter, 1888a: 22

genus *Henoticus* C. G. Thomson, 1868: 67 type species *Cryptophagus serratus* Gyllenhal, 1808
Coniophthalma Kangas, 1936: 70 type species *Henoticus germanicus* Reitter, 1906 (= *Cryptophagus californicus* Mannerheim, 1843)
Cryptophagops Grouvelle, 1919c: 67 type species *Cryptophilus alluaudi* Grouvelle, 1896
Glyptophorus Park, 1929: 430 type species *Glyptophorus mycetoecus* Park, 1929
bhutanicus Sengupta & Pal, 1980a: 39 **A**: BT
californicus Mannerheim, 1843b: 256 (*Cryptophagus*) **E**: BE DE FI FR GB GE IR NL SV **A**: NC **AFR NAR NTR**
germanicus Reitter, 1906d: 231
loberinus Sharp, 1900: 598
decoratus Bruce, 1943a: 162 **A**: FUJ
flavipennis C. Johnson, 1975b: 38 **A**: NP
indicus Grouvelle, 1916c: 78 **A**: HP NP SD UP **ORR**
pilifer Reitter, 1888a: 35 **A**: BEI ES FE HEI
regificus C. Johnson, 1975b: 38 **A**: NP
serratus Gyllenhal, 1808a: 171 (*Cryptophagus*) **E**: AU BE BH BY CR CT CZ DE EN FI FR GB GE GG IR IT

LA LT NL NR NT PL RO SK SL ST SV SZ UK YU **A**: ES FE
denticulatus LeConte, 1850: 223 (*Cryptophagus*) [HN]
reitteri Seidlitz, 1888a: 256
sinensis Bruce, 1943a: 163 **A**: FUJ JA NP SD TAI ZHE
japonicus Nakane & Hisamatsu, 1963: 50
nepalensis C. Johnson, 1975b: 38

genus *Henotiderus* Reitter, 1877e: 25 type species *Henotiderus centromaculatus* Reitter, 1877
Crosimus Casey, 1900: 89 type species *Crosimus obesulus* Casey, 1900 (= *Henotiderus centromaculatus* Reitter, 1877)
Henoticoides Hatch, 1962: 217 type species *Henoticoides lorna* Hatch, 1962
centromaculatus Reitter, 1877e: 25 **A**: FE JA JIA SHA **NAR**
hirtus Casey, 1900: 90 (*Crosimus*)
obesulus Casey, 1900: 90 (*Crosimus*)
parsonsi C. Johnson, 1972c: 274 (*Henoticoides*)

genus *Henotimorphus* Lyubarsky, 1987b: 951 type species *Henotimorphus belovi* Lyubarsky, 1987
belovi Lyubarsky, 1987: 952 **A**: FE

genus *Micrambe* C. G. Thomson, 1863: 263 type species *Dermestes abietis* Paykull, 1798
subgenus *Micrambe* C. G. Thomson, 1863: 263 type species *Dermestes abietis* Paykull, 1798
abietis Paykull, 1798: 291 (*Dermestes*) **E**: AU BE BH BY CT CZ DE EN FI FR GB GE GR HU IT LA LT NL NR NT PL PT RO SK SL SP SV SZ UK YU **N**: AG **A**: ES FE MG NC WS
pfefferi Roubal, 1944: 123 (*Cryptophagus*)
alatauensis Lyubarsky, 2000: 327 **A**: KZ
gonzaloi Otero, 2004: 247 **A**: NP
hesperia Wollaston, 1863: 217 (*Cryptophagus*) **N**: CI
wollastoni Bruce, 1941: 983 (*Cryptophagus*)
loebli Otero, 1997: 207 **A**: IS
longitarsis J. R. Sahlberg, 1900: 14 (*Cryptophagus*) **E**: AU FI GE NR NT PL SL SV
sericollis J. R. Sahlberg, 1887b: 33 (*Cryptophagus*) [HN]
mediterranica Otero & Johnson, 2001: 164 **E**: GR MA **A**: JO
micramboides Reitter, 1874b: 381 (*Cryptophagus*) **A**: JA
morula Bruce, 1943a: 158 (*Cryptophagus*) **A**: FE FUJ
nigricollis Reitter, 1876f: 290 (*Cryptophagus*) **E**: AB AR GG **A**: ES IN KI KZ MG NMO TD
occidentalis Wollaston, 1863: 217 (*Cryptophagus*) **N**: CI
perrisi C. N. F. Brisout de Barneville, 1882: xxviii (*Cryptophagus*) **E**: BU FR IT PT SL SP YU **N**: TU
pilosula Erichson, 1846: 373 (*Paramecosoma*) **E**: AU CZ EN IT FI FR GB GE HU LU NT PL SK SL SZ **A**: TR
lindbergorum Bruce, 1934a: 36 (*Cryptophagus*)
punctata Grouvelle, 1909b: 199 **Ei**: GB **AFR**
aubrooki Donisthorpe, 1939: 37
silvanoides Reitter, 1878e: 91 (*Cryptophagus*) **E**: ST
translata Grouvelle, 1916b: 27 (*Cryptophagus*) [RN] **E**: AB **A**: MG
substriata Reitter, 1898b: 19 (*Cryptophagus*) [HN]
ulicis Stephens, 1830: 76 (*Cryptophagus*) **E**: AB BE BH CR CT DE FR GB GE GG GR IR IT MC MD NL NT PL PT RO SL SP ST (Kavkaz) SZ UK YU **N**: AG MO MR TU **AFR**
vini Panzer, 1797d: 14 (*Dermestes*) [HN]
umbripennis Reitter, 1888a: 34 (*Cryptophagus*) **E**: FR IT SZ
woodroffei Johnson, **sp. nov**. [see New acts] **E**: AL AU BE BH CR CZ DE FR GB GE GR IR IT LA MC MD NL NR NT PL PT RO SK SL SP ST (Kavkaz) SV TR UK YU **N**: AG LB MO TU **A**: CY IN IS LE TR
[*villosus* sensu auctorum]
subgenus *Micrambinus* Reitter, 1906a: 332 type species *Dermestes bimaculatus* Panzer, 1798
baneti Otero, 2004: 246 **A**: NP SD
bimaculata Panzer, 1798a: 7 (*Dermestes*) **E**: AU BH CT CZ EN FI GB GE GR HU IR IT LA NR NT PL RO SK SL ST SV UK YU **A**: BEI FE GUI HEB HUB MG SCH SHA YUN ZHE
duclouxi Grouvelle, 1910b: 253 (*Micrambe*)
duclouxi Grouvelle, 1916c: 63 (*Pteryngium*) [HN]

sinensis Grouvelle, 1910b: 255 (*Micrambe*)
villosula J. R. Sahlberg, 1926: 113 (*Cryptophagus*)
reverenda Lyubarsky, 1995: 48 **A**: SCH

genus *Micrambe*, nomina dubia
obcordata Marsham, 1802: 112 (*Corticaria*)
ocularis Reitter, 1872b: 172 (*Paramecosoma*)
villosa Heer, 1841: 425 (*Cryptophagus*) [HN]

genus *Mnionomidius* Reitter, 1888a: 35 type species *Cryptophagus serricollis* Reitter, 1880
serricollis Reitter, 1880g: 515 (*Cryptophagus*) **E**: GG

genus *Paramecosoma* Curtis, 1833: 186 type species *Paramecosoma bicolor* Curtis, 1833 (= *Latridius melanocephalum* Herbst, 1793)
melanocephalum Herbst, 1793: 9 (*Latridius*) **E**: AU BE BH BY CT CZ DE EN FI FR GB GE GG HU IR IT LT NL NR NT PL RO SK SL SP ST SV SZ UK YU
bicolor Curtis, 1833: 186
fungorum Gyllenhal, 1808a: 172 (*Cryptophagus*) [HN]
infuscatum Halbert, 1910: 66
univestre Reitter, 1877h: 294

genus *Pteryngium* Reitter, 1888a: 34 type species *Dermestes crenatus* Fabricius, 1798
crenatum Fabricius, 1798: 72 (*Dermestes*) **E:** AU BH BU BY CR CT CZ DE DI EN FR GE IT LA LT NR NT PL RO SK SL ST SV SZ UK YU **NAR**
crenulatum Erichson, 1846: 369 (*Cryptophagus*)
malacum Casey, 1900: 102

genus *Salebius* Casey, 1900: 90 type species *Salebius sexdentatus* Casey, 1900 (= *Cryptophagus octodentatus* Mäklin, 1852)
octodentatus Mäklin, 1852: 327 (*Cryptophagus*) **A**: FE **NAR**
sexdentatus Casey, 1900: 90
tarsalis Casey, 1900: 91

genus *Serratomaria* Nakane & Hisamatsu, 1963: 49 type species *Serratomaria tarsalis* Nakane & Hisamatsu, 1963
tarsalis Nakane & Hisamatsu, 1963: 50 **A**: JA YUN
vulgaris Sasaji, 1984: 22 **A**: JA

genus *Spaniophaenus* Reitter, 1875b: 4 type species *Cryptophagus laticollis* L. Miller, 1858 (= *Catopochrotides termitophilus* Kieseritzky & Reichardt, 1936)
Catopochrotides Kieseritzky & Reichardt, 1936: 693 type species *Catopochrotides termitophilus* Kieseritzky & Reichardt, 1936
caucasicus Reitter, 1888b: 105 **E**: AB
fraudulentus Lyubarsky, 1997a: 54 (*Micrambe*) **A**: HP
lapidarius Fairmaire, 1862b: 648 (*Cryptophagus*) **E**: FR SP
amplicollis C. N. F. Brisout de Barneville, 1866: 369 (*Cryptophagus*)
termitophilus Kieseritzky & Reichardt, 1936: 693 (*Catopochrotides*) **E**: AU CT GG HU RO ST (Kavkaz) UK **A**: JO PA
laticollis L. Miller, 1858: 99 (*Cryptophagus*) [HN]

genus *Spavius* Motschulsky, 1844: 51 type species *Cryptophagus glaber* Gyllenhal, 1808
Emphylus Erichson, 1846: 346 type species *Cryptophagus glaber* Gyllenhal, 1808
glaber Gyllenhal, 1808a: 178 (*Cryptophagus*) **E**: AU BE BH BY CT CZ DE EN FI FR GE HU IT LA LT NL NR NT PL RO SK SL SV SZ UK YU **A**: ES

genus *Striatocryptus* Leschen, 1996: 602 type species *Striatocryptus wilkinsoni* Leschen, 1996
polyglandis Leschen, 1996: 603 **A**: UP
wilkinsoni Leschen, 1996: 603 **A**: NP

genus *Telmatophilus* Heer, 1841: 417 type species *Cryptophagus typhae* Fallén, 1802
Hydrophytophagus Shuckard, 1839: 173 type species *Cryptophagus typhae* Fallén, 1802
Limnesius Sturm, 1843: 93 type species *Cryptophagus sparganii* Ahrens, 1812
brevicollis Aubé, 1862: 72 **E**: AB AU BE BH BY CR CZ DE FR GB GE HU IT NL PL PT RO SL SP SK SV SZ UK YU **A**: KZ
balcanicus Karaman, 1960: 74 (*Thelmatophilus*)
parens Rey, 1889e: 36
rufus Reitter, 1875j: 226
caricis A. G. Olivier, 1790a: no 18: 15 (*Ips*) **E**: AB AU BE BY CT CZ DE EN FI FR GB GE GR HU IR IT LA LT NL NR NT PL RO SK SL SP ST SV SZ UK **N**: MO
longicollis Reitter, 1875j: 226
obscurus Fabricius, 1792a: 233 (*Dermestes*)
substriola Rey, 1889e: 36
orientalis Sasaji, 1987: 24 **A**: JA
schonherrii Gyllenhal, 1808a: 173 (*Cryptophagus*) **E**: AU BE CR CZ DE FI FR GB GE HU IR IT LA MC NL NR NT PL SK SP SV ST SZ **N**: MO TU **A**: KZ
bicolor Reitter, 1913b: 123
sparganii Ahrens, 1812: 21 (*Cryptophagus*) **E**: AB AU BE BH BU CR CZ FR GB GE GR HU IT NL PL PT RO SK SP SZ UK YU **N**: AG MO TU **A**: UZ
typhae Fallén, 1802: 16 (*Cryptophagus*) **E**: AU BE BY CT CZ DE EN FI FR GB GE GR HU IR IT LA LS LT MC NL NR NT PL SK SP ST SV SZ UK **A**: FE HEB JA NC **NAR**
prothetelie Kemner, 1914: 87
pumilus Reitter, 1875j: 226

subfamily **Atomariinae** LeConte, 1861

tribe Atomariini LeConte, 1861

genus *Atomaria* Stephens, 1829b: 83 type species *Atomaria linearis* Stephens, 1830
subgenus *Anchicera* C. G. Thomson, 1863: 274 type species *Atomaria munda* Erichson, 1846
Anathilopus Falcoz, 1921: 43 type species *Anathilopus theryi* Falcoz, 1921 (= *Dermestes pusillus* Paykull, 1798)
Microum Wollaston, 1854: 177 type species *Microum alternans* Wollaston, 1854
accola Lyubarsky, 1999: 21 **A**: NP YUN **ORR**
alternans Wollaston, 1854: 177 (*Microum*) **N**: MR
analis Erichson, 1846: 398 **E**: AB AU BE BH BY CR CT CZ DE EN FR GE HU IC IT LA LT PL RO SK SL ST (Kavkaz) SV SZ UK YU **A**: IN TR
cognata Erichson, 1846: 392
apicalis Erichson, 1846: 394 **E**: AB AR AU BE BH BU BY CR CT CZ DE EN FI FR GB GE GG GR HU IC IR IT LA LS LU MC NL NR NT PL RO SK SP ST SV SZ UK YU **N**: AG AZ CI MO MR TU **A**: ES FE FUJ IN IS KI KZ MG TR WS **NAR**
clavicornis Baudi di Selve, 1870: 57
myrmecophila Reitter, 1895f: 310
atra Herbst, 1793: 15 (*Kateretes*) **E**: AU BE BY CT CZ FI FR GB GE HU IT LA PL SK SZ UK **A**: FE WS
glabricula Rey, 1889g: 53
atricapilla Stephens, 1830: 66 **E**: AB AU BE BH BU BY CR CT CZ DE EN FI FR GB GE GG GR HU IR IT LA LT MA MC NL NR NT PL RO SK SP SR ST (Kavkaz) SV SZ TR UK YU **N**: MO TU **A**: CY IN IS JO SY TR
angulicollis Kangas, 1973: 69
nigriceps Erichson, 1846: 396
rufa Heer, 1841: 431 [HN]
attila Reitter, 1878b: 56 **E**: AB AU CT CZ DE GE GG HU PL RO SK ST SZ UK YU **A**: TR
banatica Reitter, 1888a: 48
flavicolor Roubal, 1931b: 184
sternodeoides Reitter, 1883c: 113

basalis Erichson, 1846: 391 **E**: AU BE BY CT CZ DE EN FI FR GB GE IR IT LA LT NL NT PL SK SV SZ UK **A**: ES FE WS
basicornis Reitter, 1888a: 51 **E**: AU BU FR GE GG HU RO SL ST (Kavkaz) **A**: TR
frondicola Reitter, 1889s: 372
singularis Reitter, 1888d: 172
bulbosa Wollaston, 1865: 24 **N**: CI
cephennoides Reitter, 1888a: 53 **E**: AR GG ST (Kavkaz) **A**: TR
circassica Reitter, 1888d: 173 **E**: ST (Kavkaz) **A**: TR
clavigera Ganglbauer, 1899a: 726 **E**: AU BH CT CZ DE FI GB GE HU IR NL NR NT PL SK SV SZ UK
convexiuscula Reitter, 1888a: 43 **A**: FE HEB HEI JIA NC SHA
deubeli Holdhaus, 1903: 369 **E**: BU RO
fasciata Kolenati, 1846b: 55 **E**: AB AG AR FR GG IT SP ST (Kavkaz) UK **N**: AG CI MO **A**: AF KA KI KZ NP TD XIN
pilosula Wollaston, 1864: 142
flava C. Johnson, 1968: 229 **A**: FE JA
fulvipennis Mannerheim, 1846: 515 **E**: FI NR NT SV **A**: ES FE MG **NAR**
semitestacea Reitter, 1888a: 50
fuscata Schönherr, 1808: 100 (*Cryptophagus*) **E**: AB AR AU BE BH BY CR CT CZ DE EN FI FR GB GE GG GR HU IR IT LA LS LT NL NR NT PL RO SK SL SP ST SV SZ UK **A**: BEI ES FE IS JA TU WS XIZ **NAR**
agnita Kangas, 1961: 112
castanea Stephens, 1830: 66
lacustris Casey, 1900: 121
lata Grimmer, 1841: 39 (*Cryptophagus*)
ochracea C. C. A. Zimmermann, 1869: 258
redtenbacheri Gistel, 1857: 522 (*Cryptophagus*)
rufipes Stephens, 1830: 68
xeniella Reitter, 1888a: 44
fuscipes Gyllenhal, 1808a: 182 (*Cryptophagus*) **E**: AR AU BE BH CT CZ DE EN FI FR GB GE GG HU IC IR IT LA LT NL NR NT PL RO SK SP ST (Kavkaz) SV SZ UK
carbonaria Stephens, 1830: 68
concolor Märkel, 1844: 244 (*Cryptophagus*)
dichroa Reitter, 1888a: 46
gibbula Erichson, 1846: 393 **E**: AU BE BH CR CT CZ FR GE GG HU IT NL PL SK SL SZ UK
hiemalis Baudi di Selve, 1870: 56
thorictoides Reitter, 1875b: 77
godarti Guillebeau, 1885a: 3 **E**: FR IT SP **N**: AG MO TU
graeseri Reitter, 1888a: 49 **A**: FE
grandicollis C. N. F. Brisout de Barneville, 1882: xxviii **E**: AU CR FR IT RO SZ
straussi Ganglbauer, 1899a: 721
gravidula Erichson, 1846: 395 **E**: AB AU BH BU CT CZ EN FR GE GG GR HU IT PL RO SK ST (Kavkaz) SZ UK **A**: IN TR
castanoptera Reitter, 1884a: 5
formosa Reitter, 1875b: 62
gutta Newman, 1834: 202 **E**: AU BE BH CR CT CZ DE EN FI FR GB GE GR HU IR IT LA MC MD NL NT PL PT RO SK SP ST SV SZ TR UK YU **A**: TR
csikii Kaszab & Székessy, 1953: 196
sellata Heer, 1841: 429
hislopi Wollaston, 1857b: 77 **E**: CT EN FI GB GE LA NR NT RO SV UK
horridula Reitter, 1877f: 112 **A**: BEI FE HEB HEI JA JIA NC SHG
incertula C. Johnson, 1971b: 228 **A**: GUX HP NP PA SD UP
insecta Wollaston, 1857a: 61 **N**: MR
maderensis Jansson, 1940: 31
jonica Reitter, 1888a: 50 **E**: AL AU BH BU FR GE GR HU IT MD RO SK SL YU
jureceki Obenberger, 1917a: 23
mehadiensis Ganglbauer, 1899a: 735
kamtschatica Motschulsky, 1845b: 362 **A**: ES FE MG WS **NAR**

globicollis Reitter, 1896c: 69
kaszabi C. Johnson, 1970c: 271 **A**: ES MG
khumbuensis C. Johnson, 1971d: 437 **A**: KA NP
mentita Lyubarsky, 1999: 22
klapperichi C. Johnson, 1971b: 229 **A**: AF NP
laevis Reitter, 1884c: 252 **A**: IS LE
laticollis Wollaston, 1865: 22 **N**: CI
brevis Har. Lindberg, 1951: 6
lederi C. Johnson, 1970a: 114 **E**: CT NT **A**: ES FE LIA MG NC WS **NAR**
lewisi Reitter, 1877f: 112 **E**: AB AU BE BU BY CT CZ DE EN FI GB GE HU IR IT LA LS LT MD NL NR PL SL SR ST (Kavkaz) SV SZ **A**: AF BEI BT FE FUJ GAN GUA GUI HEB HEI HUB IN JA KA KI KZ MG NC NMO NP SD SCH SHA SHG SHN TAI TD UP UZ YUN ZHE **COS**
herbigrada Reitter, 1896c: 69
psallioticola Hinton, 1941: 133
lucida Reitter, 1889p: 304 **A**: FE JA
laferi Lyubarsky, 1987: 953
lundbergi C. Johnson, 1978: 28 **E**: FI SV
marginicollis Reitter, 1888a: 47 **N**: CI
oromii Israelson, 1985: 113
rubricollis Wollaston, 1865: 144 [HN] [RN]
ruficollis Wollaston, 1864: 143 [HN]
mesomela Herbst, 1792: 143 (*Dermestes*) **E**: AU BE BH BY CR CT CZ DE EN FI FR GB GE HU IR IT LA LS LT MD NL NR NT PL RO SK SP ST (Kavkaz) SV SZ UK **N**: AG **A**: WS
dimidiata Marsham, 1802: 112 (*Corticaria*)
guttula Mannerheim, 1844b: 184
imitata Reitter, 1888a: 43
phaeorrhaea Marsham, 1802: 124 (*Silpha*)
mongolica C. Johnson, 1970c: 268 **E**: CT FI NT **A**: ES FE MG WS
montenegrina Reitter, 1888a: 50 **E**: BH CR YU
morio Kolenati, 1846b: 55 **E**: AB AU BY CT CZ DE FI GB GE HU LA LT NR NT PL SK SV SZ UK **A**: FE IS **NAR**
castanea C. G. Thomson, 1863: 276 (*Anchicera*)
munda Erichson, 1846: 388 **E**: AB AR AU AZ BE CR CT CZ DE EN FI FR GB GE GR HU IC IR IT LT MA NL NR NT PL SK SP SV SZ UK **N**: CI MO MR TU **A**: AF IS KZ TR
nigripennis Kugelann, 1794: 578 (*Latridius*) **E**: AU BE CT CZ DE EN FI FR GB GE GG GR HU IC IT LT NL NR NT PL RO SK SP ST (Kavkaz) SV SZ UK **N**: TU
pulchella Heer, 1841: 427
ruficollis Panzer, 1805: 13 (*Cryptophagus*)
nikitskyi Johnson, **nom. nov**. [see New acts] **A**: BEI FE MG SCH
lateralis Reitter, 1888a: 45 [HN]
nitidula Marsham, 1802: 123 (*Silpha*) **E**: AU CT DE EN FI GB GE HU IR LA LS LT NL NR NT RO SV SZ **A**: WS
borealis Sjöberg, 1947: 115
obliqua C. Johnson, 1971b: 230 **A**: AF FUJ JIA NP SCH SD TAI UP YUN ZHE **ORR**
angellata Lyubarsky, 1995: 50
obscura C. Johnson, 1970c: 266 **A**: HEI MG
ornata Heer, 1841: 431 **E**: AU BE CZ DE FI FR GB GE GR IT LS NL NR NT PL SK SV SZ **NAR**
contaminata Erichson, 1846: 385
pallidipennis Holdhaus, 1903: 370 **E**: FR IT SP **N**: AG MO TU
palmi C. Johnson, 1975a: 35 **N**: CI
parvula Reitter, 1875b: 77 **E**: IT
peltata Kraatz, 1853: 95 **E**: AB AR AU BE BH BY CT CZ DE EN FI FR GB GE GG HU IR IT NL NR NT PL RO SK ST SV SZ UK YU **A**: AF ES FE IN KZ MG UZ XIN WS
nitidula Heer, 1841: 429 [HN]
peltataeformis Sjöberg, 1947: 110 **E**: EN FI NR NT SV
peltatula Reitter, 1888a: 40 **A**: ES FE MG

plecta Lyubarsky, 1995: 50 **A**: HP KA NP SHA UP
frugi Lyubarsky, 1997a: 56
plicata Reitter, 1875b: 60 **E**: AU BU CZ FR GE IT PL RO SK SZ UK
amplipennis Reitter, 1875b: 63
plicatoides C. Johnson, 1970a: 115 **A**: ES FE MG WS
pseudatra Reitter, 1888a: 43 **E**: AU BY CT DE EN FI GB GE HU IT LA LT NT SV SZ
reitteri Lövendal, 1893: 252
pudica C. Johnson, 1971b: 229 **A**: FUJ GUA HKZ NP SD TAI UP
pusilla Paykull, 1798: 295 (*Dermestes*) **E**: AB AR AU BE BH BU BY CR CT CZ DE EN FI FR GB GE GG GR HU IC IR IT LA LS LT MC MD NL NR NT PL PT RO SK SL SP ST (Kavkaz) SV SZ TR UK YU **N**: AG CI MO MR TU **A**: AF FE IN KZ MG TR **NAR**
basella Stephens, 1830: 65
dimidiatipennis Mannerheim, 1844b: 185
fulvicollis Stephens, 1830: 64
minutissima Tournier, 1872b: 447
phaeogaster Marsham, 1802: 125 (*Silpha*)
theryi Falcoz, 1921: 43 (*Anathilopus*)
thoracica Stephens, 1830: 65
rhenonum Kraatz, 1853: 96 **E**: BE CT DE FI FR GB GE NL NR PT SP SV SZ **N**: AG
elevata Allen, 1938: 82
rhenana Kraatz, 1853: 96 [incorrect original spelling]
rubella Heer, 1841: 431 **E**: AU BE BH BU BY CT CZ DE FI FR GB GE HU IR IT LA LT NL NR NT PL PT SK ST (Kavkaz) SV SZ **A**: TR WS
berolinensis Kraatz, 1853: 94
carinula Rey, 1889g: 53
rubida Reitter, 1875b: 74 **E**: AU BH CR CT CZ FI FR GB GE GG HU IT PL SK SL SP ST (Kavkaz) SV SZ UK YU
viennensis Reitter, 1875b: 74
rubricollis C. N. F. Brisout de Barneville, 1863: 68 **E**: AB AU BH CR CT DE EN FI FR GB GE GG GR HU IR IT LA LS MD NL NT PL RO ST (Kavkaz) SV SZ UK **A**: CY IN IS JO KZ ST SY TM TR
atripennis Reitter, 1888a: 42
divisa E. C. Rye, 1876: 178
impubens Reitter, 1888a: 52
morula Reitter, 1875b: 75
scutellaris Motschulsky, 1849: 88 **E**: AU BH CR FR GB GR IT LA MA NL PT SP SK SV **N**: AG CI MO MR TU **A**: IS SY TR
canariensis Wollaston, 1864: 142
humeralis Kraatz, 1853: 98
semifuscata C. Johnson, 1970a: 112 **E**: CT **A**: ES
semusta C. Johnson, 1969: 78 **A**: CY IS JO
slavonica C. Johnson, 1971a: 3 **E**: AU BH BU CR CZ EN FR HU IT RO SK SZ YU
sodermani Sjöberg, 1947: 111 **E**: DE EN FI GE NR NT SV **A**: ES FE
sparsula Reitter, 1888a: 52 **E**: GG ST (Kavkaz) **A**: TR
subapicalis Reitter, 1888a: 46 **A**: ES FE
suffusa C. Johnson, 1971a: 1 **N**: MO
testacea Stephens, 1830: 66 **E**: AB AR AU BE BH BU BY CR CT CZ DE EN FI FR GB GE GG GR HU IR IT LA LS LT NR NT PL RO SK SP ST SV SZ TR UK YU **N**: AG MO TU **A**: KZ MG TR WS **NAR**
delicatula Tournier, 1872b: 447
dorsalis Stephens, 1830: 69
ruficornis Marsham, 1802: 125 (*Silpha*) [HN]
terminata Comolli, 1837: 20 (*Cryptophagus*)
torrida C. Johnson, 1971b: 231 **A**: JA NP SD TAI UP
tristis C. Johnson, 1971b: 226 **A**: HP NP SD UP YUN
turgida Erichson, 1846: 397 **E**: AU BE BH CR CT CZ DE EN FI FR GB GE GG HU IT LA LT NL NR NT PL SK SL ST (Kavkaz) SV SZ UK YU **A**: ES FE JA NC TR
dilatata Reitter, 1889p: 303
unifasciata Erichson, 1846: 385 **E**: AU BU CZ HU SK UK

venusta Wollaston, 1865: 23 **N**: CI
versicolor Erichson, 1846: 399 **E**: AU CZ FR GE GR IT NL PL SK
cretica Reitter, 1888a: 48
zetterstedti Zetterstedt, 1838: 97 (*Cryptophagus*) **E**: AU CR CT CZ DE EN FI FR GB GE IR IT LA NL NR NT PL SK ST (Kavkaz) SV SZ **A**: ES WS
rufa Waltl, 1838b: 271 (*Cryptophagus*)
salicicola Kraatz, 1853: 97
subgenus *Atomaria* Stephens, 1829b: 83 type species *Atomaria linearis* Stephens, 1830
Agathengis Gozis, 1886: 10 type species *Kateretes fimetarius* Herbst, 1793 (= *Dermestes fimetarius* Fabricius, 1792)
Grobbenia Holdhaus, 1903: 354 type species *Kateretes fimetarius* Herbst, 1793 (= *Dermestes fimetarius* Fabricius, 1792)
abietina J. R. Sahlberg, 1888: 38 **E**: AU FI IT PL SV
norica Ganglbauer, 1899b: 719
affinis R. F. Sahlberg, 1834b: 275 (*Cryptophagus*) **E**: BH BY CT CZ FI GE GG HU IT LA LT NR NT PL RO SK ST (Kavkaz) SV UK **A**: FE WS **NAR**
planiuscula Reitter, 1875b: 54
pseudaffinis C. Johnson & A. Strand, 1968: 94
alpina Heer, 1841: 430 **E**: AB AU BH BU CR CT CZ EN FI FR GE GG IT LA NR NT PL SK ST SV SZ UK **A**: FE HEI TR **NAR**
atrata Reitter, 1875b: 56 **E**: AU BE BH CT CZ EN FI GE HU IT LA LT NR NT PL RO SK SV SZ UK **A**: FE JA MG **NAR**
melanica Hatch, 1962: 220
badia Erichson, 1846: 381 **E**: AU BH BY CT CZ FI GB GE HU NR NT PL RO SK SV SZ UK
sahlbergi Sjöberg, 1947: 107
barbara Reitter, 1875b: 70 **E**: GR SP **N**: AG MO
sparsutula Reitter, 1888a: 37
bella Reitter, 1875b: 56 **E**: AU BE BH CT CZ DE EN FI GB GE HU IT NL NR NT PL SK ST SV SZ UK
bescidica Reitter, 1888a: 40 **E**: CZ FI NR NT PL RO SV UK **A**: TR
bicolor Erichson, 1846: 390 **E**: AU CR FR GE IT SK
acutifrons Ganglbauer, 1899b: 715
carpathica Reitter, 1875l: 360 **E**: UK
cribrella Reitter, 1888a: 39 **E**: GG
diluta Erichson, 1846: 380 **E**: AU BE BH CR CZ DE FI FR GB GE HU IT LT NR PL RO SK SL ST (Kavkaz) SV SZ UK YU
edithae Reitter, 1888a: 39 **A**: BEI ES FE HEI HUB JA NC
elongatula Erichson, 1846: 383 **E**: AU CT CZ FI GE HU LA PL RO SV SZ UK **A**: FE TR
herminea Reitter, 1875b: 55
fimetarius Fabricius, 1792a: 316 (*Dermestes*) **E**: AU BE BU CT CZ DE FI FR GB GE HU IR IT LT NL NR NT PL RO SK SP ST SV SZ UK **N**: AG **A**: FE **NAR**
abeillei Tournier, 1872b: 446
fimetarii Herbst, 1793: 14 (*Kateretes*) [HN]
parallelipipeda Waltl, 1838b: 271 (*Cryptophagus*)
statuaria Gistel, 1857: 523 (*Paramecosoma*) [DA]
gottwaldi C. Johnson, 1971c: 231 **E**: AU GE IT SK
gracilicornis Reitter, 1888a: 40 **E**: FI GG ST (Kavkaz) **A**: ES FE MG NP SD UP WS
grossepunctata Reitter, 1896c: 68 **E**: FI NT **A**: ES FE
ihsseni C. Johnson, 1978: 27 **E**: AU CZ FI GE
impressa Erichson, 1846: 389 **E**: AU BE BH CT CZ EN FI FR GB GE GG HU IT LA NL NT NR PL RO SK SL SV SZ UK **A**: TR **NAR**
lapponica C. Johnson, 1978: 26 **E**: FI NT SV
linearis Stephens, 1830: 71 **E**: AU BE BH BU BY CT CZ DE EN FR GB GE GG GR HU IR IT LA MD NL NR NT PL RO SK ST SV UK **N**: TU **A**: FE HEB HEI KZ TR **NAR**
bettae Macquart, 1831: 95 (*Cryptophagus*)
coriacea Petri, 1926: 184
distincticornis Roubal, 1913d: 122
pygmaea Heer, 1841: 430
lohsei C. Johnson & A. Strand, 1968: 94 **E**: CZ DE GB GE PL SP SZ

longicornis C. G. Thomson, 1863: 269 **E**: AU BE BY CT CZ DE EN FI GB GE IT LA LT NL NR NT PL RO SK SV SZ UK **A**: ES FE JA
[*procerula* sensu auctorum]
marginata A. Fleischer, 1909b: 271 **A**: KZ MG
nigrirostris Stephens, 1830: 69 **E**: AB AL AU BE BH BU BY CT CZ DE EN FI FR GB GE GG GR HU IR IT LA LS LT NL NR NT PL PT RO SK SL SP ST SV SZ UK YU **N**: AG **A**: ES FE IN MG TR UZ WS **NAR**
dilutella Solsky, 1876: 261
fumata Erichson, 1846: 378
fuscicollis Mannerheim, 1852a: 338
plicicollis Mäklin, 1863: 142 [RN]
talyschensis Reitter, 1888a: 41
nigriventris Stephens, 1830: 69 **E**: AU BE BU CR CT CZ DE EN FI FR GB GE GG GR HU IR IT LA MC NL NR NT PL RO SK SP ST (Kavkaz) SV SZ UK YU **N**: MO TU **A**: CY IS TR
inconstricta Rey, 1889g: 53
nana Erichson, 1846: 379
parvicollis Rey, 1889g: 53
pilifera Reitter, 1877f: 111 **A**: JA SHG
plicibasis Reitter, 1896c: 68 **A**: ES FE
pulchra Erichson, 1846: 383 **E**: AU BE BY CT CZ DE EN FI FR GB GE HU IR IT NL PL RO SK SV SZ UK YU **A**: AF
barani C. N. F. Brisout de Barneville, 1863: 69
pilosella Reitter, 1875b: 51
punctatissima Reitter, 1877f: 112 **A**: JA
puncticollis C. G. Thomson, 1868: 69 **E**: AU BE CT CZ DE FI GB GE IT NR NT PL SK SV UK
punctipennis Reitter, 1888a: 37 **A**: ES FE JIL WS
nigroscutellata Reitter, 1888a: 38
punctithorax Reitter, 1888a: 37 **E**: CR CZ DE FR GB GE HU PL SK SP SV
consanguinea C. Johnson, 1976b: 181
cribrata Rey, 1889c: 53
soror Ganglbauer, 1899b: 718 **E**: BU PL RO UK
sparreschneideri Munster, 1928: 294 **E**: FI NR NT SV
strandi C. Johnson, 1967: 45 **E**: AU DE FI GB IR NR RO SP SZ UK
subangulata J. R. Sahlberg, 1926: 132 **E**: CT DE FI GE LA NR NT SV UK **A**: ES **NAR**
trapeziformis C. Johnson, 1970c: 260 **A**: ES MG
umbrina Gyllenhal, 1827: 291 (*Cryptophagus*) **E**: AU BE BU BH BY CT CZ DE EN FI FR GB GE GG HU IR IT LA LS LT NL NR NT PL RO SK SV SZ UK **A**: FE HEI
nova Roelofs, 1939: 33
pumila Reitter, 1875b: 59
sjobergi Palm, 1949b: 241
suboblonga Rey, 1889g: 53
vespertina Mäklin, 1853: 164 **E**: AB AU BE BH BY CT CZ EN FI FR GB GE GG HU IT LA LT NL NR NT PL RO SK SP ST (Kavkaz) SV SZ UK **N**: CI **A**: ES FE IN JA MG NP TR UP WS **N**: CI **NAR**
[*prolixa* sensu auctorum]
sharpi Jakobson, 1915: 946 [RN]
wollastoni Sharp, 1867: 435 **E**: CT DE EN FI FR GB GE IR NL NR NT SV **NAR**
lineola Notman, 1920: 29 (*Agathengis*)
subfasciata Reitter, 1888a: 37

genus *Atomaria*, nomina dubia
dactylia Gistel, 1857: 523
lateralis Gistel, 1857: 523
parasitica Gistel, 1857: 598
procerula Erichson, 1846: 382
prolixa Erichson, 1846: 381
tumulorum A. Villa & G. B. Villa, 1835: 48 (*Cryptophagus*)
uhagoni Reitter, 1876b: 10

genus *Atomaroides* Lyubarsky, 1989: 51 type species *Atomaria ussurica* Lyubarsky, 1987
ussurica Lyubarsky, 1987: 953 (*Atomaria*) **A**: FE SCH

genus *Curelius* Casey, 1900: 127 type species *Ephistemus dilutus* Reitter, 1883
dilutus Reitter, 1883c: 114 (*Ephistemus*) **E**: AB GG **A**: AF KI KZ UZ
exiguus Erichson, 1846: 403 (*Epistemus*) **E**: AB AL AU BH BU BY CR CZ EN FR GE GG GR HU IT LA MA PL RO SK SP SZ TR UK **N**: EG MO TU **A**: CY IN IQ IS JO TD TR
lepidus Hochhuth, 1873: 305 (*Ephistemus*)
moczarskii Reitter, 1908b: 44 (*Ephistemus*)
ventrosus Baudi di Selve, 1870: 57 (*Ephistemus*)
japonicus Reitter, 1878j: 181 (*Ephistemus*) **E**: GE IT MA SP **N**: CI **A**: BEI BT HEI HKG HUB JA JIA NC NP SD SHN UP YE **AFR NAR NTR ORR**
punctatus Blatchley, 1925: 165 (*Ephistemus*)

genus *Ephistemus* Stephens, 1829a: 167 type species *Dermestes gyrinoides* Marsham, 1802 (= *Dermestes globulus* Paykull, 1798)
Elachistus A. Villa & G. B. Villa, 1833: 14 type species *Dermestes globulus* Paykull, 1798
Psychidium Heer, 1841: 432 type species *Dermestes globulus* Paykull, 1798
globulus Paykull, 1798: 295 (*Dermestes*) **E**: AB AL AU BE BH BU BY CR CT CZ DE EN FI FR GB GG GR HU IR IT LA LS LT MD NL NR NT PL PT RO SK SP ST (Kavkaz) SV SZ UK YU **N**: AG AZ CI MO TU **A**: FE HEI IN KZ MG **AUR NAR**
apicalis LeConte, 1863: 72
confinis Stephens, 1829a: 169
dimidiatus Sturm, 1807: 85 (*Phalacrus*)
dubius Fowler, 1889: 345
fungifex Gistel, 1857: 574
gyrinoides Marsham, 1802: 77 (*Dermestes*)
histeroides Stephens, 1835: 402
nitens Marsham, 1802: 79 (*Dermestes*)
nitidus Stephens, 1835: 402
ovulum Erichson, 1846: 402 (*Epistemus*)
piceorhaeus Marsham, 1802: 78 (*Dermestes*)
pulchellus Marsham, 1802: 72 (*Dermestes*)
reitteri Casey, 1900: 128 **E:** AB AL AU BH BU CZ GB GE GR HU IT MA PL PT SP ST (Kavkaz) SZ TR **A**: CY IN IS TR
splendens C. Johnson, 1971b: 232 **A**: GUA NP PA SD UP YUN **ORR**

genus *Ootypus* Ganglbauer, 1899b: 737 type species *Cryptophagus globosus* Waltl, 1838
globosus Waltl, 1838b: 271 (*Cryptophagus*) **E**: AU BE CT CZ DE EN FI FR GB GE HU IR IT LA LT NL NT NR PL RO SK SV UK
palustris Wollaston, 1846: 452 (*Ephistemus*)

tribe Cryptafricini Leschen, 1996

genus *Anitamaria* Leschen, 1996: 625 type species *Anitamaria thayerorum* Leschen, 1996
Scytomaria Lyubarsky, 1998: 2 type species *Scytomaria himalaica* Lyubarsky, 1998
himalaica Lyubarsky, 1998: 5 (*Scytomaria*) **A**: NP

tribe Hypocoprini Reitter, 1880

genus *Alfieriella* Wittmer, 1935b: 129 type species *Alfieriella rabinovitchi* Wittmer, 1935
Cyprogenia Baudi di Selve, 1870: 54 [HN] type species *Cyprogenia denticulata* Baudi di Selve, 1870
Obenbergerodes E. Strand, 1936: 168 [RN] type species *Cyprogenia denticulata* Baudi di Selve, 1870
denticulata Baudi di Selve, 1870: 54 (*Cyprogenia*) **A**: CY LE SY
laticollis Reitter, 1891d: 197 (*Cyprogenia*) **A**: UZ
naxiana Reitter, 1884c: 65 (*Cyprogenia*) **E**: GR IT
rabinovitchi Wittmer, 1935b: 131 **N**: EG

genus *Hypocoprus* Motschulsky, 1839: 72 type species *Hypocoprus latridioides* Motschulsky, 1839
Myrmecinomus Chaudoir, 1845: 206 type species *Myrmecinomus hochhuthi* Chaudoir, 1845
latridioides Motschulsky, 1839: 73 **E**: AU BU CR CT CZ DE EN FI FR GB GE GR HU IT LA LT LU MA MC NL NR NT PL PT SP ST SV UK YU **N**: CI EG MR TU **A**: ES KI **NAR**
caucasicus Kolenati, 1846a: 43 (*Monotoma*)
formicetorum Motschulsky, 1839: 72
hochhuthi Chaudoir, 1845: 207 (*Myrmecinomus*)
motschulskii Wollaston, 1854: 174
pici Lavagne, 1920: 111
quadricollis Reitter, 1878j: 180
tenuis Casey, 1916: 137

genus *Hypophagus* Lyubarsky, 1989: 49 type species *Hypophagus epipedus* Lyubarsky, 1989
epipedus Lyubarsky, 1989: 49 **A**: KZ

family EROTYLIDAE Latreille, 1802

subfamily **Xenoscelininae** Sengupta & Crowson, 1971

genus *Cathartocryptus* Sharp, 1886: 392 type species *Cathartocryptus obscurus* Sharp, 1886 (*Paramecosoma maculosa* Broun, 1881)
Xenoscelinus Grouvelle, 1910a: 143 type species *Xenoscelinus malaicus* Grouvelle, 1910
hiranoi Sasaji, 1989b: 226 (*Xenoscelinus*) **A**: JA

subfamily **Cryptophilinae** Casey, 1900

tribe Cryptophilini Casey, 1900

genus *Chinophagus* Lyubarsky, 1997c: 112 type species *Chinophagus mirabilis* Lyubarsky, 1997
mirabilis Lyubarsky, 1997c: 112 **A**: HUB

genus *Cryptophilus* Reitter, 1874b: 381 type species *Cryptophagus integer* Heer, 1841
Tomarops Grouvelle, 1903c: 343 type species *Tomarops punctatus* Grouvelle, 1903
cryptophagoides Grouvelle, 1916c: 55 **A**: JA TAI
hiranoi Sasaji, 1984: 30 **A**: JA
integer Heer, 1841: 426 (*Cryptophagus*) **E**: AB AU AZ CR CZ FR GB GE GG GR HU IT MA MD SK SP SZ "Benelux" "Caucasus" **N**: AG EG LB MR TU **A**: AF AP CY FE FUJ HEN HUB IN IS JA KI LE NC NMO SD SY TAI TD TM UP UZ **COS**
balearicus Schaufuss, 1869: 14 (*Paramecosoma*)
barnevillei Tournier, 1872b: 445 (*Cryptophagus*)
ceylonicus Motschulsky, 1866b: 396 (*Cryptophagus*)
debilis Sharp, 1885b: 145
frater Grouvelle, 1898: 43
muticus C. N. F. Brisout de Barneville, 1863: 67 (*Cryptophagus*)
simplex Wollaston, 1857a: 50 (*Paramecosoma*)
minimus Grouvelle, 1914: 72 **A**: TAI
mirus Grouvelle, 1914: 73 **A**: TAI
obliteratus Reitter, 1874b: 382 **Ei**: AU DE FR GE **A**: FE HEI JA NMO
propinquus Reitter, 1874b: 382 **A**: JA TAI
quadrisignatus Motschulsky, 1861a: 11 (*Cryptophagus*) **A**: JA FE SHA
reitteri Iablokoff-Khnzorian, 1975: 246 **A**: JA

tribe Toramini Sengupta, 1967

genus *Atomarops* Reitter, 1889p: 302 type species *Atomarops lewisi* Reitter, 1889
curvitibialis Sasaji, 1991b: 41 **A**: TAI
dharma Sasaji, 1989a: 19 **A**: JA
lewisi Reitter, 1889p: 302 **A**: JA

genus *Toramus* Grouvelle, 1916b: 26 [RN] type species *Tomarus pulchellus* LeConte, 1863
Itomarus Reitter, 1920: 220 [RN] type species *Tomarus pulchellus* LeConte, 1863
Tomarus LeConte, 1861: 99 [HN] type species *Tomarus pulchellus* LeConte, 1863
formosianus Grouvelle, 1914: 66 (*Tomarus*) **A**: JIX TAI
glisonothoides Reitter, 1874b: 382 (*Cryptophilus*) **A**: JA
pilifer Reitter, 1885h: 315 (*Tomarus*) **E**: AB "Caucasus" **A**: TD TR UZ
pulchellus LeConte, 1863: 71 (*Tomarus*) **Ei**: GE **NAR**
quadriguttatus Sasaji, 1989a: 11 **A**: JA
uenoi Sasaji, 1989a: 14 **A**: JA

subfamily **Languriinae** Crotch, 1873

tribe Hapalipini Leschen, 2003

genus *Bolerus* Grouvelle, 1919d: 92 type species *Croatchia* [sic!] *minuta* Fleutiaux, 1887
Thallisellodes Arrow, 1925: 256 type species *Thallis transversus* Gorham, 1895
flavipectus M. Chûjô, 1975: 286 (*Thallisellodes*) **A**: BT
lateralis Arrow, 1925: 257 (*Thallisellodes*) **A**: SD **ORR**

tribe Languriini Crotch, 1873

genus *Anadastus* Gorham, 1887a: 362 type species *Languria cambodiae* Crotch, 1876
Neolanguria Gorham, 1887a: 361 type species *Trogosita filiformis* Fabricius, 1801
Perilanguria Fowler, 1908: 19 type species *Languria monticola* Fowler, 1885
Stenodastus Gorham, 1887a: 362 type species *Languria melanosterna* Harold, 1879
amabilis Arrow, 1925: 216 **A**: GUI **ORR**
analis Fairmaire, 1889a: 376 (*Languria*) **A**: FUJ GUI HAI HKG YUN "India" **ORR**
angustior Fowler, 1913: 137 **A**: TAI
apicalis Zia, 1934a: 33 **A**: YUN **ORR**
asiaticus M. Chûjô, 1967a: 2 **A**: GUA **ORR**
atriceps Crotch, 1873c: 185 (*Languria*) **A**: JA
bhutanensis M. Chûjô, 1975: 283 **A**: BT
bifasciatus Motschulsky, 1860b: 241 (*Languria*) **A**: HKG SD UP **ORR**
jucundus Gorham, 1903: 343 (*Stenodastus*)
virgatus Fowler, 1893: 73 (*Languria*)
bocae Villiers, 1945: 107 **A**: FUJ HEN HUB YUN **ORR**
brevilinea Arrow, 1925: 219 **A**: UP
cambodiae Crotch, 1876: 388 (*Languria*) **A**: FUJ GUI GUA HAI JIX YUN ZHE **ORR**
chinensis M. Chûjô, 1967a: 4 **A**: GUA
convexus M. Chûjô, 1967a: 5 **A**: JA (Iriomote, Ishigaki)
densus Arrow, 1925: 223 **A**: AP SD **ORR**
distinctus Arrow, 1925: 222 [RN] **A**: HAI **ORR**
nigriventris Gorham, 1896: 275 [HN]
dohertyi Arrow, 1925: 216 **A**: SD **ORR**
filiformis Fabricius, 1801a: 152 (*Trogosita*) **A**: AP FUJ HAI HKG JA SCH SD TAI YUN **ORR**
nigripes Crotch, 1873c: 184 (*Languria*)
rufotestaceus Motschulsky, 1860b: 242 (*Languria*)
testaceus W. S. MacLeay, 1825: 45 (*Languria*)
flavimanus Arrow, 1925: 225 **A**: SD **ORR**
formosanus Fowler, 1913: 136 **A**: TAI YUN "South-West China"
fulvipes Motschulsky, 1860b: 241 (*Languria*) **A**: "Indies Orientales"
gratus Gorham, 1896: 274 **A**: SD **ORR**
hainanensis M. Chûjô, 1967a: 14 **A**: HAI
harmandi Villiers, 1945: 142 **A**: JA
kankaui Fowler, 1913: 138 **A**: TAI
laticornis Arrow, 1925: 232 **A**: SD **ORR**
latus Villiers, 1945: 98 **A**: YUN **ORR**

longior Arrow, 1925: 214 **A**: FUJ GUA GUI GUX HAI HUB JIA SCH SD YUN ZHE **ORR**
loricatus Arrow, 1925: 213 **A**: SD
matsuzawai M. Chûjô, 1955: 49 **A**: JA TAI
melanosternus Harold, 1879a: 82 (*Languria*) **N**: EGi **A**: GUA JA (Ryukyus) TAI (Lan Yu) **ORR**
amamiensis Miwa, 1936: 17
menetriesii Motschulsky, 1860b: 240 (*Languria*) **A**: FUJ FE JA JIL SC (Quelpart)
fucosa Lewis, 1884: 358 (*Languria*)
nigriventris Motschulsky, 1860b: 241 (*Languria*) **A**: SD **ORR**
ochreipennis Fowler, 1890: 107 (*Languria*) **A**: AP **ORR**
parallelus Arrow, 1925: 229 **A**: UP **ORR**
parvulus Wiedemann, 1823: 47 (*Languria*) **A**: UP **ORR**
popovi Zia, 1959a: 242 **A**: YUN
praetermissus Janson, 1873: 186 (*Languria*) **A**: JA
praeustus Crotch, 1873c: 185 (*Languria*) **A**: FUJ GUA GUI HAI HKG JA JIX YUN ZHE **ORR**
pulchelloides Nakane, 1982: 44 **A**: JA
pulchellus Arrow, 1925: 232 **A**: UP
punctatus M. Chûjô, 1967a: 18 **A**: HAI
robustior Fowler, 1913: 136 **A**: TAI
rubripes Zia, 1933: 33 **A**: FUJ GUI JIX
ruficeps Crotch, 1873c: 185 (*Languria*) **A**: FUJ GUA HAI JA JIX SCH ZHE
sakaii M. Chûjô, 1977: 78 **A**: TAI
sauteranus Fowler, 1913: 135 (*Perilanguria*) **A**: HAI TAI
scutellatus Crotch, 1876: 388 (*Languria*) **A**: AP BT FUJ GUI GUX HAI JIX YUN **ORR**
subtriangularis Zia, 1959a: 2242 **A**: SCH
tonkinensis Villiers, 1945: 142 **A**: YUN **ORR**
triangularis Villiers, 1945: 105 **A**: SCH YUN "South-West China" **ORR**
troncatus Zia, 1934a: 34 **A**: FUJ HAI YUN "South-West China" **ORR**
ventralis Gorham, 1896: 273 **A**: FUJ **ORR**
wiedemanni Gorham, 1896: 271 **A**: HAI **ORR**
vicinus Arrow, 1925: 226 **A**: YUN **ORR**
yunnanensis Zia, 1959a: 243 **A**: YUN

genus *Basulanguria* Sengupta & Mukherjee, 1977: 1 type species *Basulanguria lavanica* Sengupta & Mukherjee, 1977
lavanica Sengupta & Mukherjee, 1977: 4 **A**: SD

genus *Caenolanguria* Gorham, 1887a: 361 type species *Languria coarctata* Crotch, 1876
Acrolanguria Kolbe, 1897: 116 type species *Acrolanguria aeneonigra* Kolbe, 1897
Gurilana Heller, 1918: 31 type species *Gurilana ascedens* Heller, 1918
acutangula Zia, 1934b: 357 **A**: FUJ SCH ZHE
aeneipennis Fairmaire, 1889a: 377 (*Languria*) **A**: HAI YUN **ORR**
assamensis Fowler, 1886a: 24 (*Languria*) **A**: UP YUN **ORR**
birmanica Harold, 1879a: 74 (*Languria*) **A**: BT NP SD UP YUN "South-West China" **ORR**
insularis Miwa & M. Chûjô, 1937: 8 **A**: JA (Iriomote, Ishigaki, Okinawa, Amami-Ôshima) TAI
ruficornis Zia, 1934b: 356 **A**: FUJ HAI HUB JIA SCH ZHE
shibatai Maeda, 1972: 21 **A**: TAI (Lanyu: Botel-Tobago)
sinensis Zia, 1933: 37 **A**: GUA GUI

genus *Doubledaya* White, 1850a: 13 [= 1850b: 2834] type species *Doubledaya viator* White, 1850
Coptolanguria Gorham, 1896: 261 type species *Coptolanguria dilatipes* Gorham, 1896
Cosmolanguria Kraatz, 1900b: 352 type species *Cosmolanguria ruficollis* Kraatz, 1900
Glyphilanguria Fowler, 1908: 14 type species *Glyphilanguria andrewesi* Fowler, 1908
Languriosoma Crotch, 1876: 379 type species *Languriosoma muhoti* [sic!] Crotch, 1876 (= *Languriosoma mouhoti* Crotch, 1876)
atripennis Arrow, 1925: 189 **A**: AP SD **ORR**

bucculenta Lewis, 1884: 355 **A**: GUI JA
castanea Zia, 1933: 28 **A**: HAI YUN **ORR**
cribricollis Gorham, 1896: 260 (*Callilanguria*) **A**: YUN **ORR**
dilatipes Gorham, 1896: 262 (*Coptolanguria*) **A**: SD **ORR**
forcipata Arrow, 1925: 201 **A**: FUJ SD YUN "South-West China" **ORR**
monticola Villiers, 1945: 193 **A**: BT SD "Himalaya"
mouhoti Crotch, 1876: 379 (*Languriosoma*) **A**: SD YUN "South-West China" **ORR**
pieli Zia, 1935: 695 **A**: GUX
quadricollis Fowler, 1908: 14 (*Coptolanguria*) **A**: YUN "South-West China" **ORR**
ruficollis Kraatz, 1900b: 351 (*Cosmolanguria*) **A**: AP SD **ORR**
kraatzi Fowler, 1908: 13 (*Cosmolanguria*)
schenklingi Fowler, 1908: 21 (*Callilanguria*) **A**: SD **ORR**
severini Fowler, 1893: 74 **A**: BT FUJ SD
ustulata Arrow, 1925: 191 **A**: BT FUJ SCH SD XIZ YUN **ORR**
viator White, 1850a: 13 [= 1850b: 2834] **A**: SD TAI YUN **ORR**
lewisii Kraatz, 1895c: 208
varians Fowler, 1886a: 24
whitii Crotch, 1876: 382

genus *Epilanguria* Fowler, 1908: 18 type species *Epilanguria tenuicornis* Fowler, 1908
Leptolanguria Fowler, 1908: 35 type species *Languria longicollis* Fowler, 1887
gracilis M. Chûjô, 1940c: 218 **A**: TAI
tenuicornis Fowler, 1908: 18 **A**: "India"

genus *Languriomorpha* Gorham, 1887a: 361 type species *Languria lewisii* Crotch, 1873
kunashiriana Kôno, 1936: 150 **A**: FE (Kuril Islands)
lewisii Crotch, 1873c: 184 (*Languria*) **A**: FE JA
columella Lewis, 1884: 357 (*Languria*)
geniculata Harold, 1879a: 59 (*Languria*)
nara Lewis, 1884: 357 (*Languria*) **A**: JA
nigritarsis C. O. Waterhouse, 1873: 186 (*Languria*) **A**: JA
convexicollis Waterhouse, 1873: 187 (*Languria*) [HN]
ingens Lewis, 1884: 356 (*Languria*)
pectoralis C. O. Waterhouse, 1873: 187 (*Languria*)
sodalis C. O. Waterhouse, 1875: 213 (*Languria*) [RN]
unicolor Motschulsky, 1866a: 176 (*Languria*)
waterhousii Crotch, 1876: 390 (*Languria*) [RN]
yamamotoi Miwa & M. Chûjô, 1937: 7 (*Pachylanguria*) **A**: JA

genus *Languriophasma* Arrow, 1925: 203 type species *Languria cyanea* Hope, 1835
cyanea Hope, 1835: 94 (*Languria*) **A**: BT NP UP
cyanea Motschulsky, 1860b: 243 (*Languria*) [HN]
discoidalis Fowler, 1908: 17 (*Coenolanguria*) **A**: AP **ORR**
laeta Arrow, 1925: 204 **A**: AP "North-Western India"

genus *Megalanguria* Arrow, 1925: 168 type species *Pachylanguria metasternalis* Crotch, 1876
chinensis Mader, 1939a: 41 **A**: FUJ HEN
felix Arrow, 1925: 171 **A**: SD YUN **ORR**
gravis Arrow, 1929b: 319 **A**: FUJ GUI YUN **ORR**
metasternalis Crotch, 1876: 378 (*Pachylanguria*) **A**: SD **ORR**
producta Arrow, 1925: 170 **A**: YUN **ORR**

genus *Microlanguria* Lewis, 1884: 348 type species *Languria jansoni* Crotch, 1873
Microcladoxena Fowler, 1886c: 312 [RN] type species *Languria jansoni* Crotch, 1873
Platycladoxena Kraatz, 1900a: 312 type species *Platycladoxena castanea* Kraatz, 1900
jansoni jansoni Crotch, 1873c: 185 (*Languria*) **A**: HAI JA TAI "South China" **ORR**

borneensis Kraatz, 1900a: 313 (*Platycladoxena*)
jansoni takarana Nakane, 1959a: 59 **A**: JA

genus *Neanadastus* Zia, 1959b: 368 type species *Neanadastus gracilis* Zia, 1959
gracilis Zia, 1959b: 368 **A**: SCH

genus *Neocladoxena* Maeda, 1974: 6 type species *Neocladoxena hisamatsui* Maeda, 1974
hisamatsui Maeda, 1974: 7 **A**: TAI
chujoi Nakane, 1982: 43 (*Microlanguria*)

genus *Pachylanguria* Crotch, 1876: 377 type species *Languria paivae* (sic!) Wollaston, 1859 (= *Languria paivai* Wollaston, 1859)
aeneovirens Mader, 1937: 185 **A**: YUN
paivai Wollaston, 1859: 430 **A**: GUA GUI TAI YUN ZHE **ORR**

genus *Paederolanguria* Mader, 1939a: 44 type species *Paederolanguria holdhausi* Mader, 1939
Sinolanguria Zia, 1959b: 366 type species *Sinolanguria alternata* Zia, 1959
alternata Zia, 1959b: 366 (*Sinolanguria*) **A**: ZHE
bicoloripennis M. Chûjô, 1967a: 29 (*Sinolanguria*) **A**: GUA
cyanea M. Chûjô, 1967a: 30 (*Sinolanguria*) **A**: JA (Okinawa)
elegans M. Chûjô, 1967a: 31 (*Sinolanguria*) **A**: TAI
formosana M. Chûjô, 1967a: 32 (*Sinolanguria*) **A**: TAI
hisamatsui Maeda, 1972: 23 **A**: TAI
holdhausi Mader, 1939a: 44 **A**: HEN
klapperichi Mader, 1955a: 64 **A**: FUJ
montana Miwa & M. Chûjô, 1936b: 411 (*Labidolanguria*) **A**: TAI
okinawana M. Chûjô, 1970: 38 (*Tetralanguria*) **A**: JA (Okinawa)
oshimana Miwa, 1936: 17 (*Pachylanguria*) **A**: GUI JA (Amami-Ôshima)
tuberculata Zia, 1959b: 367 (*Sinolanguria*) **A**: FUJ

genus *Pentelanguria* Crotch, 1876: 380 *Pentelanguria elateroides* Crotch, 1876
elateroides Crotch, 1876: 380 **A**: GUI NP SCH SD XIZ YUN **ORR**
notopedalis Crotch, 1876: 380 **A**: SD

genus *Tetraphala* Sturm, 1843: 306 type species *Languria splendens* Wiedemann, 1823
Metabelus Gorham, 1887a: 361 type species *Pachylanguria borrei* Fowler, 1886
Tetralanguria Crotch, 1876: 378 type species *Languria splendens* Wiedemann, 1823
Tetralanguroides Fowler, 1886c: 318 type species *Tetralanguroides fryi* Fowler, 1886
apicata Zia, 1959a: 241 (*Tetralanguria*) **A**: SCH
collaris Crotch, 1876: 377 (*Pachylanguria*) **A**: FUJ GUI HAI HP JA JIX SC SCH SD TAI YUN XIZ ZHE "Mandjuria» **ORR**
punctata Harold, 1879a: 58 (*Languria*)
tripunctata Kraatz, 1900b: 347 (*Pachylanguria*)
yunnana Fairmaire, 1887a: 136 (*Languria*)
cuprea Arrow, 1925: 176 (*Pachylanguria*) **A**: BT SD UP **ORR**
elongata Fabricius, 1801a: 152 (*Trogosita*) **A**: AP BT FUJ GUA GUI HAI SCH SD TAI UP YUN ZHE "Himalaya" **ORR**
crucicollis Kraatz, 1900b: 348 (*Tetralanguria*)
cyanipennis Kraatz, 1899c: 218 (*Tetralanguria*)
micans Harold, 1875: 185 (*Languria*) [RN]
pyramidata W. S. MacLeay, 1825: 44 (*Languria*)
ruficollis Kraatz, 1900b: 348 (*Tetralanguria*)
splendens Motschulsky, 1860b: 242 (*Languria*) [HN]
splendida Crotch, 1876: 378 (*Tetralanguria*) [RN]
triplagiata Kraatz, 1900b: 348 (*Tetralanguria*)
tripunctata Wiedemann, 1823: 46 (*Languria*)

fraterna Zia, 1959a: 241 (*Tetralanguria*) **A**: FUJ SCH ZHE
fryi Fowler, 1886c: 319 (*Tetralanguroides*) **A**: FUJ JA (Tsushima) SC ZHE
humeralis Arrow, 1925: 177 (*Pachylanguria*) **A**: SW YUN **ORR**
miles Fowler, 1913: 134 (*Tetralanguria*) **A**: TAI
omeica Zia, 1959a: 242 (*Tetralanguria*) **A**: SCH
parallela Zia, 1933: 23 (*Pachylanguria*) **A**: SCH YUN **ORR**
sauteri Fowler, 1913: 133 (*Tetralanguroides*) **A**: TAI
simplex Fowler, 1913: 134 (*Tetralanguroides*) **A**: TAI
tienmuensis Zia, 1959b: 369 (*Tetralanguria*) **A**: FUJ HUB HUN SCH ZHE
tumidicollis Kraatz, 1900b: 349 (*Tetralanguria*) **A**: AP **ORR**
variventris Kraatz, 1900b: 348 (*Tetralanguria*) **A**: HP SD YUN **ORR**

subfamily **Xenoscelinae** Ganglbauer, 1899
tribe Xenoscelini Ganglbauer, 1899

genus *Henoticonus* Reitter, 1878i: 127 type species *Henoticus triphylloides* Reitter, 1878
triphylloides Reitter, 1878i: 127 (*Henoticus*) **A**: JA

genus *Leucohimatium* Rosenhauer, 1856: 179 type species *Leucohimatium angustum* Rosenhauer, 1856 (= *Tenebrio arundinaceus* Forskål, 1775)
alatum Reitter, 1884c: 62 (*Leucohivatium* [sic!]) **E**: GR "South Russia" **A**: KZ
arundinaceum Forskål, 1775: 78 (*Tenebrio*) **E**: AB AUi CR FR GEi GG GR HU IT MD PT SP "Caucasus" **N**: AG CI EG MR **A**: CY SY TR "Zakaspija" **AFR AUR**
  angustum Rosenhauer, 1856: 181
  elongatum Erichson, 1846: 371 (*Paramecosoma*)
breve Wollaston, 1874: 170 **A**: JA
jakowlewi Semenov, 1902: 255 **E**: HU UK **A**: KZ
  nigrosuturale Reitter, 1909c: 101
langii Solsky, 1866a: 90 (*Paramecosoma*) **E**: CT GG HU MD NT RO ST UK "South Russia" **A**: KZ MG NC TM
  brevicolle Reitter, 1878e: 93

genus *Loberogosmus* Reitter, 1876f: 291 type species *Engis fasciata* Kolenati, 1846
fasciatus Kolenati, 1846b: 54 (*Engis*) **E**: AB BH GG HU SL "Caucasus"

genus *Macrophagus* Motschulsky, 1845a: 51 type species *Macrophagus robustus* Motschulsky, 1845
  Haplolophus I. Frivaldszky, 1865: 190 type species *Haplolophus neglectus* I. Frivaldszky, 1865
albellus Reitter, 1889o: 313 (*Haplolophus*) **A**: UZ
robustus Motschulsky, 1845a: 51 **E**: AB AU AR AU GG HU SK ST UK "South Russia" **A**: KI KZ
  faustii Seidlitz, 1888a: 263
  frivaldszkyi I. Frivaldszky, 1865: 192 (*Cryptophagus*)
  neglectus I. Frivaldszky, 1865: 191 (*Haplolophus*)

genus *Pharaxonotha* Reitter, 1875b: 44 type species *Pharaxonotha kirschii* Reitter, 1875
  Planismus Casey, 1890: 500 type species *Planismus floridanus* Casey, 1890
discimaculata Mader, 1939a: 46 **A**: HEN
kirschii Reitter, 1875b: 44 **Ei**: CZ FR GE NL PT "Benelux" **Ai**: TAI **NAR**i **NTR** **ORR**i
  conradti Gorham, 1898: 249 (*Thallisella*)
yunnanensis Grouvelle, 1916c: 61 **A**: YUN

genus *Setariola* Jakobson, 1915: 941 [RN] type species *Setaria sericea* Mulsant & Rey, 1863
  Setaria Mulsant & Rey, 1863b: 1 [HN] type species *Setaria sericea* Mulsant & Rey, 1863
  Setarella Reitter, 1920: 220 [RN] type species *Staria sericea* Mulsant & Rey, 1863
sericea Mulsant & Rey, 1863b: 2 (*Setaria*) **E**: FR GR IT MA PT

genus *Xenoscelis* Wollaston, 1864: 132 [RN] type species *Pristoscelis deplanatus* Wollaston, 1862
Diochares Reitter, 1872b: 168 [HN] type species *Diochares depressus* Reitter, 1872 (= *Pediacus costipennis* Fairmaire, 1852)
Pediaphloeus Tournier, 1869: 11 type species *Pediacus costipennis* Fairmaire, 1852
Pristoscelis Wollaston, 1862a: 151 [HN] type species *Pristoscelis deplanatus* Wollaston, 1852
costipennis Fairmaire, 1852: 78 (*Pediacus*) **E**: GR FR IT SP (Mallorca) **N**: AG MO TU
depressus Reitter, 1872b: 169 (*Diochares*)
deplanatus Wollaston, 1862a: 152 (*Pristoscelis*) **N**: CI (Teneriffa, Palma, Hierro)
lauricola Israelson, 1980: 186 **N**: CI

genus *Zavaljus* Reitter, 1880g: 544 type species *Zavaljus fausti* Reitter, 1880 (= *Cryptophagus brunneus* Gyllenhal, 1808)
Eicolyctus J. R. Sahlberg, 1919: 4 type species *Cryptophagus brunneus* Gyllenhal, 1808
brunneus Gyllenhal, 1808a: 174 (*Cryptophagus*) **E**: CT FI LA NT SK SV
fausti Reitter, 1880g: 544

subfamily **Loberinae** Bruce, 1951

genus *Loberus* LeConte, 1861: 98 type species *Loberus impressus* LeConte, 1863
Glisonotha Motschulsky, 1863: 430 type species *Glysonotha* [sic!] *setosa* Motschulsky, 1863
excisus Grouvelle, 1919d: 126 **A**: JA
marginicollis Grouvelle, 1914: 64 (*Glisonotha*) **A**: TAI

subfamily **Erotylinae** Latreille, 1802

tribe Dacnini Crotch, 1876

genus *Combocerus* Bedel, 1868: 12 type species *Ips sanguinicollis* Fabricius, 1787 (= *Silpha glabra* Schaller, 1783)
glaber Schaller, 1783: 255 (*Silpha*) **E**: AB AU BE BH BU BY CR CT CZ DE EN FI FR GE GG HU IT LT NL NR NT PL RO RU SK SV SZ UK
quadrimaculatus Herbst, 1793: 148 (*Triplax*)
quadripustulatus Panzer, 1793a: 6 (*Dermestes*)
sanguinicollis Fabricius, 1787a: 46 (*Ips*)

genus *Cyrtengis* Reitter, 1900a: 51 type species *Cyrtengis testacea* Reitter, 1900
testacea Reitter, 1900a: 51 **A**: UZ

genus *Dacne* Latreille, 1797: 12 type species *Ips humeralis* Fabricius, 1787 (= *Dermestes bipustulatus* Thunberg, 1781)
subgenus *Dacne* Latreille, 1797: 12 type species *Ips humeralis* Fabricius, 1787 (= *Dermestes bipustulatus* Thunberg, 1781)
Engis Paykull, 1800: 349 type species *Ips humeralis* Fabricius, 1787 (= *Dermestes bipustulatus* Thunberg, 1781)
Cnecosophagus Reitter, 1875b: 42 type species *Cnecosophagus jekeli* Reitter, 1875 (= *Dermestes bipustulatus* Thunberg, 1781)
akitai Narukawa, 1992: 67 **A**: JA
bipustulata Thunberg, 1781: 6 (*Dermestes*) **E**: AL AR AU BE BH BY CR CT CZ DE EN FI FR GB GE GG GR HU IR IT LA LT NL NR NT PL RO SK SL SP SV SZ UK YU "Caucasus" **A**: FE "Siberia"
humeralis Fabricius, 1787a: 45 (*Ips*) [HN]
jekeli Reitter, 1875: 43 (*Cnecosophagus*)
scanica Fabricius, 1775: 58 (*Dermestes*) [HN]
fungorum fungorum Lewis, 1887a: 56 **A**: FE JA
binaeva Reitter, 1897c: 123 (*Engis*)
fungorum nigrocephala Mt. Chûjô, M. Chûjô & Lee, 1993b: 99 **A**: SC
indica Crotch, 1876: 401 (*Thallis*) **A**: HP UP
japonica Crotch, 1873c: 188 **A**: FE JA GUI GUX HUB SHG YUN **AUSi**
kidoi Nakane, 1982: 47 **A**: JA
minima Nikitsky & Kompantzev, 1995: 83 **A**: FE

notata Gmelin, 1790: 1617 (*Silpha*) **E**: AU BY CR CZ EN FR GE GG GR HU IT PL RO SK SL SV SZ UK "Caucasus" **A**: FE WS "Ural" "Siberia"

bipustulata Fabricius, 1787: 45 (*Ips*) [HN]

osawai Ashida & Kim, 1999: 381 **A**: SC (Chejudo)

picta Crotch, 1873c: 188 **Ei**: CZ GE **A**: FE GUA HUB JA SC ZHE

pontica Bedel, 1868: 9 (*Engis*) **E**: AB AU BH GG HU IT RO SK SL ST "Caucasus" **A**: IN TR

rufifrons Fabricius, 1775: 69 (*Tritoma*) **E**: AB AU BE BH CR CZ DE FR GB GE GG HU IT NL PL RO SK SP SV SZ UK YU "South and Middle Russia" "Caucasus" **A**: ES FE

angustata Kirby, 1830: 92 (*Engis*)

flava Marsham, 1802: 122 (*Silpha*)

fowleri Joy, 1905: 274

reitteri Schilsky, 1892: 195 (*Engis*)

rufa Marsham, 1802: 136 (*Nitidula*)

semirufula Reitter, 1897c: 123 (*Engis*) **E**: AB AR GG UK

subgenus *Xenodacne* Boyle, 1956: 84 type species *Dacne picea* LeConte, 1875

maculata M. Chûjô, 1940a: 78 **A**: FE JA

minor M. Chûjô, 1941b: 20

zonaria taiwana M. Chûjô, 1976: 1 **A**: TAI

zonaria zonaria Lewis, 1887a: 56 **A**: FE JA SC

jureceki Pic, 1921a: 2 (*Engis*)

genus *Dromodacne* Arrow, 1925: 35 type species *Dromodacne vivax* Arrow, 1925

vivax Arrow, 1925: 35 **A**: UP

genus *Microsternus* Lewis, 1887a: 3 type species *Megalodacne ulkei* Crotch, 1873

bhutanensis Chûjô, 1975: 280 **A**: BT

crotchi Lewis, 1887b: 57 **A**: FE JA

higonius Lewis, 1887b: 58 **A**: JA

hisamatsui Nakane, 1982: 46 **A**: JA (Iriomote)

nakanoi Narukawa, 2004: 277 **A**: JA (Tokunoshima) **ORR**

perforatus Lewis, 1883b: 140 (*Episcapha*) **A**: JA TAI

taiwanus M. Chûjô, 1976: 2 **A**: TAI

tokioensis Nakane, 1961a: 5 (*Microsternus*) **A**: JA NC SC

tricolor taiwanicus Nakane, 1966a: 59 **A**: TAI

tricolor tricolor Lewis, 1887b: 58 **A**: FE JA **ORR**

yamadai M. Chûjô & Shibata, 1963: 1 **A**: JA

genus *Xestus* Wollaston, 1864: 420 type species *Xestus throscoides* Wollaston, 1864

fungicola Wollaston, 1865: 57 (Appendix) **N**: CI (Gomera)

throscoides Wollaston, 1864: 421 **N**: CI (Tenerife)

tribe Encaustini Crotch, 1876

subtribe Encaustina Crotch, 1876

genus *Aulacochilus* Chevrolat, 1837b: 429 type species *Erotylus quadripustulatus* Fabricius, 1801

chevrolati P. H. Lucas, 1846: 220 **N**: AG

algerinus Reiche, 1868b: 47 (*Aulacochilus*)

episcaphoides Gorham, 1883b: 84 (*Aulacochilus*) **A**: SD **ORR**

grouvellei Achard, 1923: 48 (*Aulacochilus*) **A**: YUN

indicus Gorham, 1895a: 328 (*Aulacochilus*) **A**: UP **ORR**

issikii M. Chûjô, 1936a: 211 (*Aulacochilus*) **A**: TAI

janthinus Lacordaire, 1842: 250 **A**: AP HAI YUN **ORR**

japonicus Crotch, 1873c: 189 (*Aulacochilus*) **A**: JA "Korea"

javanus Guérin-Méneville, 1841: 155 (*Erotylus*) **A**: "Assam" **ORR**

klapperichi Mader, 1941c: 927 (*Aulacochilus*) **A**: FUJ

luniferus decoratus Reitter, 1879e: 223 (*Aulacochilus*) **A**: FE FUJ JA SC **ORR**

luniferus helleri Deelder, 1942: 78 **A**: GUI HAI TAI **ORR**
luniferus punctatellus Heller, 1920: 4 **A**: YUN **ORR**
nigrorufus Pal, 1993: 206 (*Aulacochilus*) **A**: AP SD
oblongus Arrow, 1925: 89 (*Aulacochilus*) **A**: SD **ORR**
quadripustulatus Fabricius, 1801b: 6 (*Erotylus*) **A**: HAI SD **ORR**
 celebensis Schaufuss, 1887: 141 (*Aulacohilus*)
 subrotundatus W. S. MacLeay, 1825: 42 (*Engis*)
reitteri Mader, 1939a: 41 (*Aulacochilus*) **A**: FUJ HEN
sericeus Bedel, 1871: 282 (*Aulacochilus*) **A**: CH **ORR**
shinoharai Miwa & M. Chûjô, 1939: 76 (*Aulacochilus*) **A**: TAI
sibiricus sibiricus Reitter, 1879e: 224 (*Aulacochilus*) **A**: FE
sibiricus bedeli Harold, 1880: 170 **A**: JA TAI
 hukuokensis Miyahara, 1942: 531
taliensis Achard, 1923: 48 (*Aulacochilus*) **A**: FUJ SCH YUN
violaceus Germar, 1824: 616 (*Triplax*) **E**: BH CR IT (Sicilia) RO

genus *Encaustes* Lacordaire, 1842: 33 type species *Engis verticalis* W. S. MacLeay, 1825
 Engis W. S. MacLeay, 1825: 41 [HN] type species *Engis vericalis* W. S. MacLeay, 1825
brevicollis Arrow, 1925: 76 **A**: "Assam"
cruenta formosana M. Chûjô, 1964b: 414 **A**: TAI
 formosensis Nakane, 1966a: 60
cruenta praenobilis Lewis, 1883b: 138 **A**: FE HEN JA "Korea"
marginalis birmanica Gorham, 1896: 276 **A**: SD **ORR**
nipponica Shinohara, 1941: 613 (*Eneaustes* [sic!]) **A**: TAI
verticalis gigantea Boheman, 1868: 211 **A**: SD **ORR**

genus *Hornerotylus* M. Chûjô, 1941b: 11 type species *Hornerotylus sauteri* M. Chûjô, 1941 (= *Aulacochilus abdominalis* Csiki, 1910)
abdominalis Csiki, 1910b: 444 (*Aulacochilus*) **A**: TAI **ORR**
 sauteri M. Chûjô, 1941b: 12

genus *Micrencaustes* Crotch, 1876: 498 type species *Engis lunulata* W. S. MacLeay, 1825
subgenus *Micrencaustes* Crotch, 1876: 498 type species *Engis lunulata* W. S. MacLeay, 1825
decipiens Arrow, 1925: 81 **A**: SD
liturata W. S. MacLeay, 1825: 42 (*Engis*) **A**: HAI **ORR**
maxima Bedel, 1871: 276 (*Aulacochilus*) **A**: "India" **ORR**
subgenus *Mimencaustes* Heller, 1920: 10 type species *Engis dehaanii* Laporte, 1840
dehaanii Laporte, 1840b: 15 (*Engis*) **A**: SD **ORR**
michioi Osawa & Mt. Chûjô, 1990: 69 **A**: TAI
taiwana Araki, 1941b: 31 **A**: TAI

genus *Satelia* Lewis, 1887b: 73 type species *Satelia scitula* Lewis, 1887
scitula Lewis, 1887b: 73 **A**: JA

subtribe Megalodacnina Sengupta, 1969

genus *Episcapha* Dejean, 1836: 137 type species *Engis quadrimacula* Wiedemann, 1823
subgenus *Ephicaspa* M. Chûjô, 1969: 97 type species *Megalodacne asahinai* M. Chûjô, 1936
asahinai asahinai M. Chûjô, 1936a: 212 (*Megalodacne*) **A**: TAI
asahinai amamiana M. Chûjô, 1969: 98 **A**: JA (Amami-Ôshima)
asahinai yakushimensis Araki, 1941a: 29 (*Megalodacne*) **A**: JA (Yakushima)
lewisi hayashii Nakane, 1955a: 35 (*Megalodacne*) **A**: JA (Hokkaido)
lewisi lewisi Nakane, 1950a: 12 (*Megalodacne*) **A**: JA (Tokara, Amami-Ôshima)
subgenus *Episcapha* Dejean, 1836: 137 type species *Engis quadrimacula* Wiedemann, 1823
flavofasciata cheni Nakane, 1995a: 26 **A**: TAI

flavofasciata flavofasciata Reitter, 1879e: 223 (*Megalodacne*) **A**: ES FE FUJ GUI HEN JA (Hokkaido) NO SC SCH
hamata Lewis, 1879d: 465
flavofasciata lucidofasciata M. Chûjô, 1969: 111 **A**: JA (Tsushima)
fortunii consanguinea Crotch, 1876: 408 **A**: HEI JA (Tokara, Amami-Ôshima) SHG SHN TAI "India" **ORR**
tapferi Mader, 1936d: 112
fortunii fortunii Crotch, 1873c: 188 **A**: CH JA "Himalaya" "Korea"
gorhami Lewis, 1879d: 465 **A**: GUI HUN JA
psiloides Bedel, 1918: 119 **A**: GUX **ORR**
quadrimacula Wiedemann, 1823: 132 (*Engis*) **A**: BT NP SD UP **ORR**
septentrionis Heller, 1920: 68 (*Episcapha*) **A**: HAI TAI SD **ORR**
hainanensis Araki, 1949: 5
subgenus *Psiloscapha* Heller, 1920: 55 type species *Engis glabra* Wiedemann, 1823
indica Crotch, 1876: 408 **A**: AP BT SD **ORR**
chapuisi Dohrn, 1878: 449
insignita Bedel, 1918: 120 **A**: YUN
lugubris Bedel, 1918: 119 **A**: YUN
mausonica Heller, 1920: 80 **A**: GUA **ORR**
morawitzi magna Araki, 1949: 4 **A**: TAI
morawitzi morawitzi Solsky, 1871: 266 (*Dacne*) **A**: ES FE HUB JA NC NE NO SC
taishoensis Lewis, 1874: 79
opaca Heller, 1920: 79 **A**: YUN
takasagona M. Chûjô, 1941d: 84 **A**: TAI

genus *Episcapha*, species incertae sedis
nocte Pal, 1993: 204 **A**: AP
xanthopustulata Gorham, 1890a: 47 **A**: AP **ORR**

genus *Episcaphula* Crotch, 1876: 409 type species *Episcaphula amboinensis* Crotch, 1876
subgenus *Episcaphula* Crotch, 1876: 409 type species *Episcaphula amboinensis* Crotch, 1876
andamanensis Gorham, 1888: 141 **A**: SD **ORR**
affinis Arrow, 1922: 289 (*Megalodacne*)
consimilis Arrow, 1925: 53 (*Megalodacne*) **A**: AP NP UP SD
hainanensis M. Chûjô, 1968b: 541 **A**: HAI
laevis Arrow, 1925: 54 (*Megalodacne*) **A**: NP SD
variolosicollis Mader, 1937: 184 **A**: FUJ SCH
subgenus *Isoscaphula* Heller, 1920: 57 type species *Episcapha elongata* Guérin-Méneville, 1841
elongata Guérin-Méneville, 1841: 160 (*Episcapha*) **A**: "India" **ORR**
kojimai Nakane, 1995a: 26 **A**: TAI
miwai M. Chûjô, 1941d: 87 (*Megalodacne*) **A**: TAI
singalensis Csiki, 1901a: 26 **A**: SD UP **ORR**
tonkinensis Heller, 1920: 83 **A**: BT HAI SD **ORR**
fratria Heller, 1920: 83
major Arrow, 1922: 288 (*Megalodacne*)
vitalisi Arrow, 1922: 287 (*Megalodacne*) **A**: AP BT FUJ SD **ORR**
subgenus *Lanugodacne* Heller, 1920: 57 type species *Episcaphula trifasciata* Heller, 1920
helleri Mader, 1934a: 40 [RN] **A**: NP SD **ORR**
pubescens Heller, 1920: 109 [HN]
matsumurai M. Chûjô, 1941d: 86 (*Megalodacne*) **A**: TAI

genus *Megalodacne* Crotch, 1873d: 352 type species *Ips fasciata* Fabricius, 1777
bellula Lewis, 1883b: 139 **A**: FE JA SC
chinensis Crotch, 1876: 417 **A**: SCH
immaculata M. Chûjô & Kiuchi, 1963: 31 **A**: JA
marginata Arrow, 1925: 47 **A**: PA

genus *Neodacne* M. Chûjô, 1976: 3 type species *Megalodacne okunii* M. Chûjô, 1936
okunii M. Chûjô, 1936a: 212 (*Megalodacne*) **A**: TAI

genus *Oretylus* Heller, 1920: 56 type species *Episcaphula hislopi* Crotch, 1876
carinicollis Heller, 1920: 107 (*Episcaphula*) **A**: TAI **ORR**
punctatissimus Heller, 1920: 106 (*Episcaphula*) **A**: TAI **ORR**

genus *Paracoptengis* Heller, 1920: 54 type species *Triplotoma lineata* Gorham, 1895 (= *Coptengis nigrolineatus* Allard, 1895)
brahminicus Gorham, 1883b: 80 (*Triplatoma*) **A**: AP **ORR**

genus *Triplatoma* Cuvier, 1832: 795 type species *Triplatoma variegata* Westwood, 1832
Pantheropterus J. Thomson, 1856: 323 type species *Pantheropterus pfeifferae* J. Thomson, 1856 (= *Triplatoma macleayi* Lacordaire, 1842)
Triplotoma Bedel, 1882: 439 [unjustified emendation]
davidi Fairmaire, 1889b: 12 (*Pantheropterus*) **A**: ES SCH

tribe Tritomini Curtis, 1834

genus *Amblyopus* Chevrolat, 1837b: 429 type species *Triplax vittata* A. G. Olivier, 1807
cinctipennis Lacordaire, 1842: 199 **A**: NP UP XIZ **ORR**
praepositus Walker, 1859a: 259 (*Tritoma*)
flaviventris Arrow, 1925: 99 **A**: CH **ORR**
interruptus Miwa, 1929a: 123 **A**: TAI
csikii M. Chûjô, 1964b: 414
pallidicornis Arrow, 1925: 98 **A**: NP UP
rusticoides Mader, 1937: 181 **A**: YUN
substriatus Gorham, 1896: 281 **A**: SD UP **ORR**
concolor Kuhnt, 1910: 232
triplacoides Gorham, 1895a: 326 **A**: UP **ORR**
vittatus A. G. Olivier, 1807: 490 (*Triplax*) **A**: FUJ SD **ORR**

genus *Camptotritoma* Heller, 1920: 28 type species *Camptotritoma fulva* Heller, 1920
amamiensis M. Chûjô, 1969: 169 (*Pseudotritoma*) **A**: JA (Amami-Ôshima)

genus *Cyrtomorphus* Lacordaire, 1842: 240 type species *Cyrtomorphus pantherinus* Lacordaire, 1842
chinensis Mader, 1955a: 66 **A**: FUJ
duodecimmaculatus Araki, 1941c: 55 **A**: TAI
dux Arrow, 1925: 113 **A**: SD
duxoides Mader, 1955a: 65 **A**: FUJ
liui M. Chûjô, 1967b: 69 **A**: TAI
yunnanus Mader, 1937: 182 **A**: YUN

genus *Dactylotritoma* Arrow, 1925: 105 type species *Triplax apicata* Crotch, 1876
apicata Crotch, 1876: 467 (*Triplax*) **A**: NP UP **ORR**
atricapilla Lewis, 1887b: 71 (*Triplax*) **A**: FE JA
chinensis Araki, 1942: 313 **A**: SHA
formosana Araki, 1942: 312 **A**: TAI

genus *Eutriplax* Lewis, 1887c: 342 [RN] type species *Eudaemonius tuberculifrons* Lewis, 1887
Eudaemonius Lewis, 1887b: 72 [HN] type species *Eudaemonius tuberculifrons* Lewis, 1887
tuberculifrons Lewis, 1887b: 72 (*Eudaemonius*) **A**: FE JA

genus *Neotriplax* Lewis, 1887b: 60 type species *Neotriplax atrata* Lewis, 1887
arisana Miwa, 1929a: 124 **A**: FUJ HEN TAI

reitteri Mader, 1939b: 71
atrata Lewis, 1887b: 61 **A**: FE JA
atratula Iablokoff-Khnzorian, 1975: 241 (*Tritoma*) [RN]
delkeskampi Nakane, 1961a: 4 **A**: JA
lewisii Crotch, 1873c: 189 (*Cyrtotriplax*) **A**: JA SC
miwai Nakane, 1966a: 63 **A**: TAI
rubens Hope, 1831: 22 (*Erotylus*) **A**: AP BT FUJ HP NP SCH SD UP **ORR**

genus *Ortitma* Heller, 1920: 27 type species *Cyrtotriplax cebana* Gorham, 1896 (= *Triplax nigripennis* Motschulsky, 1858)
cebana Gorham, 1896: 286 (*Cyrtotriplax*) **A**: UP **ORR**
nigripennis Motschulsky, 1858c: 114 (*Triplax*) [HN]
sumatrensis Gorham, 1901: 193 (*Triplacidea*)

genus *Pselaphandra* Jakobson, 1905: xxxv type species *Triplax cinnabarina* Reitter, 1879
Tetratriplax M. Chûjô, 1969: 210 type species *Dactylotritoma inornata* Chûjô, 1941
cinnabarina Reitter, 1879e: 222 (*Triplax*) **A**: FE JA NC
nigriceps Reitter, 1888a: 9 (*Triplax*)
inornata atrocephala Mt. Chûjô, M. Chûjô & Lee, 1993b: 102 **A**: SC
inornata inornata M. Chûjô, 1941b: 14 (*Dactylotritoma*) **A**: JA

genus *Pseudamblyopus* Araki, 1941d: 364 type species *Amblyopus palmipes* Lewis, 1889
palmipes Lewis, 1889c: 397 (*Amblyopus*) **A**: JA
similis Lewis, 1887b: 67 (*Amblyopus*) **A**: FE JA

genus *Pseudotritoma* Gorham, 1888: 147 type species *Tritomidea nigrocruciata* Crotch, 1876
Aporotritoma Arrow, 1925: 103 type species *Aporotritoma jucunda* Arrow, 1925
arakii arakii Nakane, 1954: 12 (*Tritoma*) **A**: JA FE
arakii fuscocephala Mt. Chûjô, M. Chûjô & Lee, 1993b: 101 (*Aporotritoma*) **A**: SC
consobrina consobrina Lewis, 1874: 78 (*Cyrtotriplax*) **A**: ES FE JA MG SC
chibana Nakane, 1991b: 115 (*Aporotritoma*) **A**: JA
solivaga Lewis, 1887b: 66 (*Cyrtotriplax*)
jucunda Arrow, 1925: 104 (*Aporotritoma*) **A**: HP UP
laetabilis Lewis, 1887b: 70 (*Triplax*) **A**: JA FE
atripes Araki, 1943: 561 (*Tritoma*)
kirishimensis Araki, 1943: 558 (*Tritoma*)
maedai Kiyoyama, 1979: 47 (*Aporotritoma*) **A**: TAI
nigrovariegata intersecta Mt. Chûjô, M. Chûjô & Lee, 1993b: 101 (*Aporotritoma*) **A**: SC
nigrovariegata nigrovariegata M. Chûjô, 1968a: 73 **A**: JA (Tsushima)
ruficornis Lewis, 1887b: 68 (*Cyrtotriplax*) **A**: JA
tsushimana M. Chûjô, 1969: 161 **A**: JA (Tsushima)
ussuriensis Krivolutskaya, 1992: 298 **A**: FE
valida Reitter, 1884a: 5 (*Triplax*) **E**: AB GG "Caucasus"
yasumatsui Nakane, 1948: 13 (*Tritoma*) **A**: JA

genus *Renania* Lewis, 1887b: 60 type species *Renania atrocyanea* Lewis, 1887
atrocyanea Lewis, 1887b: 60 **A**: JA

genus *Rhodotritoma* Arrow, 1925: 115 type species *Triplax coccinea* Crotch, 1876
albofasciata Nakane, 1982: 49 **A**: TAI
coccinea Crotch, 1876: 467 (*Triplax*) **A**: FUJ HP
fulva Reitter, 1879e: 223 (*Triplax*) **A**: FE JA
sufflava Lewis, 1887b: 70 (*Triplax*)
nigricornis Arrow, 1925: 117 **A**: SD **ORR**
rubicunda Araki, 1941e: 367 **A**: TAI

sanguinea Crotch, 1876: 468 (*Triplax*) **A**: HP UP

genus *Scelidopetalon* Delkeskamp, 1957: 82 [RN] type species *Petaloscelis instabilis* Gorham, 1896
Petaloscelis Gorham, 1896: 282 [HN] type species *Petaloscelis instabilis* Gorham, 1896
similis Arrow, 1925: 102 (*Petaloscelis*) **A**: "Assam Valley"

genus *Spondotriplax* Crotch, 1876: 469 type species *Spondotriplax endomychoides* Crotch, 1876 (= *Cyrtotriplax quadrimaculata* Kirsch, 1875)
bisbimaculata Mader, 1941c: 931 **A**: FUJ
flavofasciata M. Chûjô, 1941b: 12 **A**: JA TAI
flavomaculata M. Chûjô, 1941b: 13 **A**: TAI
fulviceps Arrow, 1925: 144 **A**: BT SD
horioi Nakane & Nobuchi, 1955: 15 **A**: JA
japonica Nakane & Nobuchi, 1955: 16
inornata Shibata, 1966: 38 **A**: JA (Ishigaki-jima)
obscura M. Chûjô, 1968b: 546
sawadai Narukawa, 2003: 31 **A**: JA (Iriomote-jima, Ishigaki-jima)

genus *Tetratritoma* Arrow, 1925: 145 type species *Tetratritoma miliaris* Arrow, 1925
angustata Arrow, 1925: 146 **A**: KA
chui Nakane, 1966a: 61 **A**: TAI
longicornis Arrow, 1925: 147 **A**: FUJ **ORR**
militaris Arrow, 1925: 146 **A**: HP NP SD

genus *Triplacidea* Gorham, 1901: 191 type species *Triplax motschulskyi* Bedel, 1873 (= *Triplax melanocephala* Motschulsky, 1860)
motschulskyi Bedel, 1873: 408 (*Triplax*) [RN] **A**: FUJ HP YUN **ORR**
melanocephala Motschulsky, 1860a: 107 (*Triplax*) [HN]

genus *Triplax* Herbst, 1793: 146 type species *Silpha russica* Linnaeus, 1758
Ogcotriplax Heller, 1920: 29 type species *Triplax pseuda* Heller, 1920
Peudotriplax Heller, 1920: 29 type species *Triplax tabayasi* Heller, 1920
Platichna C. G. Thomson, 1859: 96 type species *Erotylus rufipes* Fabricius, 1787
aenea Schaller, 1783: 254 (*Silpha*) **E**: AL AU BE BY CR CT CZ DE EN FI FR GB GE GG HU IT LA LT NR NT PL RO RU SK SL SP ST SV SZ UK YU "Balkan Peninsula" "Caucasus" **A**: FE RU
bicolor Marsham, 1802: 122 (*Silpha*) [HN]
ainonia Lewis, 1887b: 69 **A**: FE JA
akbesiana Escalera, 1925d: 457 [RN] **A**: SY
escalerai Pic, 1926a: 1 [RN]
syriaca Escalera, 1925a: 155 [HN]
amoena Solsky, 1871: 269 **A**: FE JA
analis Reitter, 1892a: 133 **E**: AB AR
andreinii Pic, 1930d: 5 **E**: IT
apicipennis Reitter, 1901e: 74 **E**: AR GG
atripennis Iablokoff-Khnzorian, 1975: 230 **A**: SY
atroguttata Araki, 1943: 556 (*Tritoma*) **A**: TAI
bedeli Reitter, 1888a: 6 **N**: MO
bicolorata M. Chûjô, 1941b: 14 (*Tritoma*) **A**: TAI
breviscutata Fairmaire, 1868: 502 **N**: MO
canalicollis Lewis, 1887b: 70 **A**: FE JA
carpathica Reitter, 1890j: 388 **E**: PL SK
castanea Peyerimhoff, 1925a: 8 **N**: AG MO
caucasica Reitter, 1892c: 257 **E**: "West Caucasus"
collaris Schaller, 1783: 256 (*Silpha*) **E**: AB AL AU BH BU CR CZ FR GE HU IT PL RO SK ST UK "Caucasus" **A**: IN

capistrata Lacordaire, 1842: 217
sulphuricollis Reitter, 1888a: 8
cyanescens Bedel, 1868: 26 [RN] **E**: FR PT SP **N**: AG
melanocephala Lacordaire, 1842: 211 [HN]
cyanipennis Motschulsky, 1858c: 114 **A**: "India"
devia Lewis, 1887b: 69 **A**: JA SC
discicollis Lewis, 1887b: 71 **A**: JA
elongata Lacordaire, 1842: 210 **E**: AU CZ GE HU IT PL RO SK **A**: ES FE WS
elongatoides Mader, 1941c: 930 **A**: FUJ
emgei Reitter, 1885i: 391 **E**: GR
flaviventris Tôyama & Kinugasa, 1986: 225 **A**: JA
fukudai M. Chûjô, 1941a: 1 (*Tritoma*) **A**: FE JA
gracilenta Solsky, 1871: 271 **A**: ES FE
bicoloripes Pic, 1926e: 5
horni M. Chûjô, 1941b: 16 (*Tritoma*) **A**: JA (Amami-Ôshima) TAI
hurusyoi M. Chûjô, 1941d: 90 (*Tritoma*) **A**: TAI
japonica Crotch, 1873c: 189 **A**: FE JA SC
lacordairii Crotch, 1870: 7 **E**: AU BH CR FR GB GE HU IT SK SL SP SZ **N**: AG MO
ragusae Reitter, 1892c: 257
lepida Faldermann, 1837: 395 (*Ischyrus*) **E**: AL AR AU BE BH BU BY CR CZ FR GE HU IT PL RO RU SK SP ST SV SZ UK YU "Caucasus" **A**: TR
marsueli Bedel, 1868: 24 **E**: AB AR BH CR FR GG IT PT SP YU "Caucasus" **N**: AG
discicollis Reitter, 1904e: 156 [HN]
reitteri Jakobson, 1915: 933 [RN]
meghriensis Iablokoff-Khnzorian, 1978a: 58 **E**: AR
melanocephala Latreille, 1804c: 39 (*Tritoma*) **E**: CR CT FR GB GE GG GR HU IT RO SK SP SZ "Caucasus" **N**: AG EG MO **A**: UZ "Asia Minor"
collaris Fabricius, 1801b: 572 (*Tritoma*) [HN]
nigriceps Lacordaire, 1842: 213
ruficollis Stephens, 1830: 90
nagaoi Nakane, 1977: 98 **A**: JA
nakanei M. Chûjô, 1941a: 3 (*Tritoma*) **A**: JA
hakonensis Nakane, 1980a: 99
yatsuensis Nakane, 1956: 34
nigrina Reitter, 1879e: 222 **A**: FE
nikritini Nikitsky & Kompanzev, 1997: 32 **A**: KZ
paganetti Obenberger, 1914: 108 **A**: TR (Taurus Mts.)
peyerimhoffi Escalera, 1925b: 240 **N**: MO
pseudambliopoides Nikitsky & Kompantzev, 1995: 88 **A**: FE
pygmaea Kraatz, 1872: 145 **E**: AU BH HU RO SK SL
quadrimaculata Motschulsky, 1858c: 115 **A**: "India"
quadrinotata Araki, 1943: 559 (*Tritoma*) **A**: TAI
rubrica Reitter, 1891b: 21 **A**: KZ TM UZ "Central Asia" "Altai Mts."
rudis Reitter, 1888a: 6 **E**: SP **N**: MO
rufipes Fabricius, 1787a: 92 (*Erotylus*) **E**: AU BE BH BY CR CT CZ DE FI FR GE GG HU IT LT LU NR NT PL RO RU SK SL SP SV SZ UK YU "Caucasus"
clavata Lacordaire, 1842: 217
swanetica Reitter, 1892a: 134
rufiventris Gebler, 1823: 125 **A**: ES (Altai) FE "Siberia"
amurensis Reitter, 1879e: 222
russica Linnaeus, 1758: 360 (*Silpha*) **E**: AL AU BE BH BY CR CT CZ DE EN FI FR GB GE HU IT LA LS LT NL NR NT PL RO RU SK SP SV SZ UK YU WS "Caucasus" **N**: AG **A**: IN **NAR**i
castanea Marsham, 1802: 122 (*Silpha*)
nigripennis Fabricius, 1792b: 513 (*Ips*)
nigripennis Latreille, 1804c: 39 (*Tritoma*) [HN]
palliata Schranck, 1798: 535 (*Chrysomela*)

rubra DeGeer, 1775: 283 (*Anthribus*)
scutellaris Charpentier, 1825: 244 **E**: AU BH BY CR CT CZ DE EN FI FR GB GE GG HU IT LA LT NR NT PL RO RU SK SP ST SV SZ UK YU "Caucasus" **A**: ES "Middle Asia" "West Siberia"
gyllenhalii Crotch, 1870: 7
seminigra Reitter, 1879e: 222 **A**: FE
sibirica sibirica Crotch, 1876: 466 **A**: ES FE HEI NE NO "Korea"
sibirica connectens Lewis, 1887b: 68 (*Cyrtotriplax*) **A**: JA SC
signaticollis Reitter, 1879e: 221 **A**: FE FUJ
longior Mader, 1941c: 930
sinica M. Chûjô, 1968b: 547 **A**: FUJ
subcylindrica Reitter, 1895a: 81 **A**: SY
syriaca Pic, 1908d: 43
subtilissima Reitter, 1901e: 75 **A**: KZ "Altai Mts"
tadzhica Nikitsky & Kompanzev, 1997: 31 **A**: TD
takashii Nakane, 1966a: 61 **A**: TAI
tanigutii M. Chûjô, 1940a: 80 (*Tritoma*) **A**: JA (Honshu)
tatjanae Nikitsky & Kompantzev, 1995: 86 **A**: FE
tergestana Reitter, 1881d: 229 **E**: CR IT
trimaculata M. Chûjô, 1941b: 19 (*Tritoma*) **A**: TAI
unifasciata Motschulsky, 1858c: 115 **A**: "India" **ORR**
yakushimana Nakane, 1959b: 71 **A**: JA (Yakushima)
yatoi Nakane, 1956: 33 (*Tritoma*) **A**: JA

genus *Tritoma* Fabricius, 1775: 68 type species *Tritoma bipustulata* Fabricius, 1775
subgenus *Rotitma* M. Chûjô & Mt. Chûjô, 1990: 46 type species *Cyrtotriplax pallidiventris* Lewis, 1887
circumcincta Lewis, 1887b: 66 (*Cyrtotriplax*) **A**: JA
discalis Lewis, 1887b: 65 (*Cyrtotriplax*) **A**: JA
pallidiventris Lewis, 1887b: 64 (*Cyrtotriplax*) **A**: JA
rufipennis Lewis, 1887b: 65 (*Cyrtotriplax*) **A**: FE JA
subgenus *Tritoma* Fabricius, 1775: 68 type species *Tritoma bipustulata* Fabricius, 1775
Cyrtotriplax Crotch, 1873c: 189 [RN] type species *Tritoma bipustulata* Fabricius, 1775
acuminata Mader, 1941c: 929 **A**: FUJ
andoi Narukawa, 2003: 33 **A**: JA (Ishigaki-jima)
asahinai Nakane, 1950a: 10 **A**: JA
biplagiata Lewis, 1887b: 61 (*Neotriplax*) **A**: JA
bipustulata Fabricius, 1775: 68 **E**: AU BE BH BY CR CZ DE EN FI FR GB GE GG GR HU IT LA LS NL NR NT PL RU SK SP SL SV SZ UK "Caucasus" **A**: FE "Siberia" "Ural"
bimaculata Herbst, 1783: 30 (*Sphaeridium*)
binotata Reitter, 1888a: 9 (*Cyrtotriplax*)
humeralis Marsham, 1802: 67 (*Dermestes*)
incerta P. Rossi, 1790: 48
cenchris Lewis, 1887b: 64 (*Cyrtotriplax*) **A**: JA SC
centralis Lewis, 1887b: 62 (*Cyrtotriplax*) **A**: JA (Honshu)
discaloides Nakane, 1986: 133 **A**: JA
excellens Arrow, 1925: 124 **A**: KA PA
fasciata M. Chûjô, 1941b: 15 **A**: TAI
gressitti M. Chûjô, 1968b: 548 (*Cyrtotriplax*) **A**: TAI
indica Crotch, 1876: 467 (*Triplax*) **A**: "India" **ORR**
infanta Arrow, 1925: 135 **A**: SD
kanekoi Araki, 1943: 557 **A**: TAI
karahutonis M. Chûjô, 1941b: 16 **A**: FE (Sakhalin)
kensakui M. Chûjô, 1955: 43 **A**: JA
kubotai Narukawa, 1994: 155 **A**: JA
kurinoensis Nakane, 1983b: 145 **A**: JA
laferi Krivolutskaya, 1992: 298 (*Pseudotritoma*) **A**: FE

lateripunctata Mader, 1936d: 113 **A**: GAN
latifasciata Lewis, 1887b: 63 (*Cyrtotriplax*) **A**: JA
basalis Lewis, 1887b: 67 (*Cyrtotriplax*) [HN]
lewisiana Nakane, 1983b: 144 [RN]
laxicornis Arrow, 1925: 123 **A**: SD
loochooana M. Chûjô, 1969: 247 (*Cyrtotriplax*) **A**: JA (Amami-Ôshima)
maackii Crotch, 1876: 460 (*Cyrtotriplax*) **A**: FE "Siberia"
maculifrons Lewis, 1887b: 65 (*Cyrtotriplax*) **A**: JA
pseudodiscalis Nakane, 1961a: 5 (*Tritoma*)
metasobrina M. Chûjô, 1941d: 91 **A**: TAI
michitakai M. Chûjô, 1960: 1 **A**: JA (Tsushima)
nigripes Heller, 1923: 69 **A**: SCH
nigropunctata Lewis, 1887b: 64 (*Cyrtotriplax*) **A**: JA
niponensis Lewis, 1874: 78 (*Cyrtotriplax*) **A**: FE JA
octonotata Bedel, 1874: cxcvi (*Cyrtotriplax*) **E**: GG ST "Transcaucasia"
osawai Nakane, 1982: 48 **A**: JA
otaitoensis Nakane, 1961a: 4 **A**: FE JA
pallidicincta Lewis, 1887b: 62 (*Neotriplax*) **A**: JA
pantherina Lewis, 1887b: 63 (*Cyrtotriplax*) **A**: JA
pulchella Fabricius, 1798: 177 **A**: "India"
quinquemaculata Mader, 1941c: 928 **A**: FUJ
recurrens Arrow, 1925: 127 **A**: UP
sasajii Narukawa, 2001: 275 **A**: JA (Okinawa)
scheerpeltzi Mader, 1955a: 67 **A**: FUJ
scutigera Arrow, 1925: 140 **A**: AP **ORR**
shibatai M. Chûjô, 1969: 249 (*Cyrtotriplax*) **A**: JA (Amami)
shimoyamai M. Chûjô & Nakane, 1950: 15 **A**: JA
shirakii M. Chûjô, 1936a: 213 **A**: TAI
sobrina Lewis, 1887b: 62 (*Cyrtotriplax*) **A**: JA
subbasalis Reitter, 1896f: 265 (*Cyrtotriplax*) **E**: AU BY CT HU IT LT NT PL SK ST SZ UK **A**: ES FE MG NC NMO WS "Siberia"
jakowlewi Semenov, 1898b: 550 (*Cyrtotriplax*)
sibirica Semenov, 1898b: 553 (*Cyrtotriplax*)
sungkangensis Nakane, 1966a: 62 **A**: TAI
taiwana M. Chûjô, 1936a: 214 **A**: JA (Ryukyus) TAI
takahashii Nakane, 1966b: 65 **A**: JA
takakurai Nakane, 1980b: 112 **A** JA
takasagona M. Chûjô, 1941b: 18 **A**: TAI
towadensis M. Chûjô, 1952: 17 **A**: FE JA
yachiensis Takeuchi, 1940: 761
tripartiaria Lewis, 1887b: 67 (*Cyrtotriplax*) **A**: JA
uniplagiata Nakane, 1983b: 144 **A**: JA
viridipennis Arrow, 1925: 131 **A**: SD
wittmeri M. Chûjô, 1975: 281 (*Cyrtotriplax*) **A**: BT
yamazii M. Chûjô, 1941a: 5 **A**: TAI
nanshanchica Nakane, 1966a: 63
yeiei Nakane, 1966a: 62 **A**: TAI

family BYTURIDAE Jacquelin du Val, 1858

genus *Haematoides* Fairmaire, 1878b: 116 type species *Haematoides davidis* Fairmaire, 1878
Byturodes Barber, 1942: 13 type species *Byturodes grahami* Barber, 1942
atriceps Pic, 1910l: 45 **A**: YUN
davidis Fairmaire, 1878b: 117 **A**: SCH YUN **ORR**
grahami Barber, 1942: 13 (*Byturodes*)

genus *Byturus* Latreille, 1797: 69 type speies *Dermestes tomentosus* DeGeer, 1774
Horticola Lindemann, 1865: 148 type species *Horticola urbana* Lindemann, 1865
Terobyturus Ohta, 1930a: 102 type species *Terobyturus shibakawai* Ohta, 1930
affinis Reitter, 1874c: 525 **A**: ES FE FUJ GUX HEI HUB JA JIA MG SC TAI ZHE
fallax Ohta, 1930a: 103
ferrugineus Reitter, 1874c: 526
japonicus Ohta, 1930a: 102
ohtai Barber, 1942: 11
olivaceus Fournel, 1840: 524
shibakawai Ohta, 1930a: 102
atricollis Reitter, 1874c: 526 **A**: FE JA
oblongulus Fairmaire, 1891f: cxcii **A**: HUB
ochraceus Scriba, 1790b: 153 (*Dermestes*) **E**: AU BE BH BU BY CR CT CZ DE EN FI FR GB GE HU IR IT LA LT NL NR NT PL SK SL ST SV SZ TR UK YU "Caucasus" **A**: AF CH (northeast) ES FE KZ MG WS TR "Korea"
bicolor Reitter, 1911a: 4
fleischeri Plavilshchikov, 1924: 232
tomentosus DeGeer, 1774: 199 (*Dermestes*) **E**: AU BE BH BU BY CR CT CZ DE EN FI FR GB GE GG HU IR IT LA LT LU NL NR NT PL SK SL ST SV SZ UK YU "Caucasus" **A**: AF CH (northeast) ES FE IN JA KZ MG NP TAI TR WS "Korea"
bhutanicus Sengupta, 1978b: 168
formosanus Hatta, 1970: 108
flavescens Marsham, 1802: 64 (*Dermestes*)
grisescens A. Fleischer, 1910a: 146
oakanus Ohta, 1930a: 101
obscurus Reitter, 1911a: 4
tomentosus Fabricius, 1775: 57 (*Dermestes*)
urbana Lindemann, 1865: 149 (*Horticola*)
wittmeri Sengupta, 1978b: 168 **A**: NP UP

genus *Xerasia* Lewis, 1895a: 120 type species *Xerasia variegata* Lewis, 1895
Byturellus Barber, 1942: 14 type species *Byturus grisescens* Jayne, 1882
Satorystia Reitter, 1905c: 243 type species *Satorystia meschniggi* Reitter, 1905
meschniggi Reitter, 1905c: 243 (*Satorystia*) **E**: HU SK
punica Goodrich & Springer, 1988: 346 **A**: IS LE SY TR
variegata Lewis, 1895a: 120 **A**: JA

family BIPHYLLIDAE LeConte, 1861

genus *Biphyllus* Dejean, 1821: 102 type species *Dermestes lunatus* Fabricius, 1787
Diphyllus Berthold, 1827: 393 type species *Dermestes lunatus* Fabricius, 1787
Thallestus Wollaston, 1862a: 153 type species *Thallestus semiellipticus* Wollaston, 1862
aequalis Reitter, 1889p: 300 (*Diphyllus*) **A**: JA (Honshu, Kyushu)
complexus Sasaji, 1983: 31 **A**: JA (Honshu, Kyushu, Shikoku)
euphorbiae Peyerimhoff, 1923: 51 (*Diphyllus*) **N**: MO
flexuosus Reitter, 1889p: 300 (*Diphyllus*) **A**: FE JA (Hokkaido , Honshu, Kyushu) TAI
formosianus Grouvelle, 1914: 67 (*Diphyllus*) **A**: TAI
frater Aubé, 1850: 330 (*Diphyllus*) **E**: AB AU BH CR CZ FR HU IT RO SK SL ST
humeralis Reitter, 1889p: 299 (*Diphyllu s*) **A**: JA (Honshu, Kyushu, Shikoku)
inaequalis Reitter, 1889p: 299 (*Diphyllus*) **A**: JA (Honshu, Shikoku)
japonicus Sasaji, 1983: 34 **A**: JA (Honshu)
kasuganus Nakane, 1988: 81 **A**: JA (Honshu)
kolosovi Nikitsky, 1983: 703 **A**: FE
kuzurius Sasaji, 1985: 11 **A**: JA (Honshu)
lewisi Reitter, 1889p: 299 (*Diphyllus*) **A**: JA (Honshu, Kyushu, Shikoku, Tsushima)

loochooanus Sasaji, 1991a: 12 **A**: JA (Ryukyus-Shotō)
lunatus Fabricius, 1787b: 378 (*Dermestes*) **E**: AU BE BH BY CR CT CZ DE FI FR GE GB GR HU IT NL NR PL PT RO SK SL SP ST SV SZ UK YU **N**: AG CI (La Palma, La Gomera, Gran Canaria) LB MR **A**: IN
nornatus Reitter, 1888d: 170 (*Diphyllus*)
marmoratus Reitter, 1889p: 300 (*Diphyllus*) **A**: FE JA (Honshu, Kyushu)
medius Grouvelle, 1914: 72 (*Diphyllus*) **A**: TAI
oshimanus Nakane, 1988: 80 **A**: JA (Kyushu, Ryukyus-Shotō)
rufopictus Wollaston, 1874: 172 (*Thallestus*) **A**: JA (Honshu, Kyushu, Shikoku)
satsumanus Nakane, 1988: 79 **A**: JA (Kyushu)
sauteri Grouvelle, 1914: 70 (*Diphyllus*) **A**: TAI
schenklingi Grouvelle, 1914: 68 (*Diphyllus*) **A**: TAI
subellipticus Wollaston, 1862a: 155 (*Thallestus*) **N**: CI (Tenerife)
suffusus Wollaston, 1874: 171 (*Thallestus*) **A**: JA (Honshu, Kyushu)
throscoides Wollaston, 1874: 171 (*Thallestus*) **A**: JA (Honshu, Kyushu, Shikoku, Tsushima)
typhaeoides Wollaston, 1862a: 155 (*Thallestus*) **N**: CI (La Gomera)
obscuricollis Wollaston, 1865: 158 (*Thallestus*)
ussuriensis Nikitsky, 1983: 703 **A**: FE

genus *Diplocoelus* Guérin-Méneville, 1844a: 196 type species *Triphyllus fagi* Chevrolat, 1837
Marginus LeConte, 1861: 105 type species *Marginus rudis* LeConte, 1863
humerosus Reitter, 1876f: 293 **E**: GG ST **A**: TR
fagi Chevrolat, 1837: pl. 41 (*Biphyllus*) **E**: AB AU BE BH BU BY CR CZ DE GG GE GB GR HU IT NR PL RO SK SL SV SZ UK YU **N**: AG **A**: IN

family BOTHRIDERIDAE Erichson, 1845

subfamily **Anommatinae** Ganglbauer, 1899

genus *Kocherius* Coiffait, 1984: 44 type species *Kocherius perpusillus* Coiffait, 1984
perpusillus Coiffait, 1984: 44 **N**: MO

genus *Abromus* Reitter, 1876d: 50 type species *Abromus brucki* Reitter, 1876
abeillei Belon, 1885a: 244 **E**: SP
algarvensis Dajoz, 1977: 256 **E**: PT
brucki Reitter, 1876d: 51 **E**: FR SP
besucheti Dajoz, 1968: 264 **N**: MO
coiffaiti Dajoz, 1968: 265 **E**: FR
convexus Dajoz, 1968: 265 **E**: SP
elongatus Dajoz, 1968: 263 **E**: SP
franzi Dajoz, 1969b: 377 **E**: SP
gigas Coiffait, 1984: 47 **N**: MO
comellinii Coiffait, 1984: 47 **E**: SP
elongatus Coiffait, 1984: 49 **N**: MO
hervei Dajoz, 1968: 264 **E**: FR
insularis Dajoz, 1968: 264 **E**: SP
lusitanicus Dajoz, 1969b: 378 **E**: SP
maroccanus Dajoz, 1969b: 380 **N**: MO
meridionalis Coiffait, 1984: 49 **N**: MO
minhoensis Coiffait, 1984: 47 **E**: PT
lusoensis Coiffait, 1984: 46 **E**: PT
palaui Español, 1945: 96 **E**: SP
rabatensis Coiffait, 1984: 48 **N**: MO
rifensis Coiffait, 1984: 48 **N**: MO
sintrensis Coiffait, 1984: 46 **E**: PT
vicinus Coiffait, 1984: 48 **N**: MO

zaerensis Coiffait, 1984: 48 **N**: MO
zariquieyi Dodero, 1916: 349 **E**: SP
    doderoi Español, 1945: 94

genus *Anommatus* Wesmael, 1835: 338 type species *Anommatus terricola* Wesmael, 1835
    Anommatoides Dudich, 1922: 124 type species *Anommatus reitteri* Ganglbauer, 1899
    Neanommatus Depoli, 1917: 11 type species *Anommatus thoracicus* Depoli, 1917
    Paranommatus J. Müller, 1913: 122 type species *Anommatus matzenaueri* J. Müller, 1913
anabelae Luna de Carvalho, 1985: 89 **E**: PT
balachowskyi Dajoz, 1969b: 384 **E**: SP
besucheti Dajoz, 1973b: 148 **E**: IT SZ
beszedesi Reitter, 1915a: 41 **E**: IT YU
biharicus Breit, 1911: 172 **E**: RO
boiteli Normand, 1950b: 106 **N**: AG
bosnicus Kaszab, 1947: 271 **E**: BH
cantabricus Dajoz, 1969b: 385 **E**: SP
catalunicus Dajoz, 1969b: 382 **E**: SP
coiffaiti Dajoz, 1968: 256 **E**: GR
confusus Dajoz, 1971: 89 **E**: CZ
convexus Breit, 1913a: 17 **E**: IT
curtii cansiglioensis Dajoz, 1973b: 153 **E**: IT
curtii curtii Reitter, 1910: 266 **E**: BH IT
curtii forojulensis Pace, 1979: 457 **E**: IT
curtii prealpinus Pace, 1979: 455 **E**: IT
curtii tarvisianus Pace, 1979: 456 **E**: IT
dalmatinus Reitter, 1922: 45 **E**: BH IT YU
davidi Luna de Carvalho, 1985: 88 **E**: PT
delamarei Dajoz, 1969b: 387 **E**: SP
dentatus Dajoz, 1973b: 151 **E**: IT SZ
dictyoderus Dodero, 1900a: 419 **E**: FR IT
diecki Reitter, 1875k: 310 **E**: AU DE FR GB IT SV SZ
    baudi Reitter, 1877e: 27
    sinuans Rey, 1889h: 55
distinctus Dodero, 1900a: 418 **E**: FR IT
    devillei Dodero, 1900a: 418
doderoi Binaghi, 1941b: 138 **E**: IT
duodecimstriatus P. W. J. Müller, 1821: 190 (*Lyctus*) **E**: AL AU AZ BE BH BU CR CT CZ DE FR GB GE GR HU IR IT NL NT PL RO RU SK SL SP ST SV SZ UK YU **N**: AG CI MR **A**: TR **AFR**i **NAR**i **NTR**i
    basalis Reitter, 1883e: 197
    corcyranus Dajoz, 1968: 249
    pusillus Schaufuss, 1862b: 49
    obsoletus Stephens, 1830: 98 (*Tribolium*)
    terricola Wesmael, 1835: 339
franzi Dajoz, 1969b: 383 **E**: SP
fodori Kaszab, 1947: 272 **E**: BH
foveolatus Všetečka, 1948: 42 **E**: IT
gallicus Reitter, 1922: 46 **E**: FR
grandis Dajoz, 1969b: 389 **E**: PT
herculis Všetečka, 1948 **E**: RO
    banaticus Dajoz, 1968: 252
herkulis Kaszab, 1947: 264 **E**: RO
hispanicus Dajoz, 1969b: 386 **E**: SP
hungaricus Dudich, 1922: 123 **E**: HU RO SK
    maramarosensis Kaszab, 1947: 263
    sorneri Roubal, 1925a: 57

trapezicollis Kaszab, 1947: 263
istrianus Reitter, 1910: 265 **E**: IT YU
jelinecki Dajoz, 1971: 90 **E**: HU
julianus Pace, 1979: 457 **E**: IT
kaszabi Dajoz, 1968: 258 **E**: MC
kiesenwetteri Reitter, 1881b: 47 **E**: FR SP
laevis Kaszab, 1947: 267 **E**: RO
libanicus Dajoz, 1968: 249 **A**: LE
littoralis Dajoz, 1969b: 388 **E**: SP
lombardus Pace, 1979: 442 **E**: IT
maderensis Coiffait, 1984: 45 **N**: MR
malisorum Všetečka, 1948: 43 **E**: AL
matzenaueri J. Müller, 1913: 122 **E**: BH CR YU
minutus Dajoz, 1969b: 386 **E**: SP
mixtus Dajoz, 1984a: 83 [= 1987: 187] **E**: YU
montandoni Dajoz, 1968: 252 **E**: RO
monsforensis Pace, 1979: 445 **E**: IT
mostarensis Reitter, 1915a: 40 **E**: BH
nanulus Obenberger, 1938: 92 **E**: CR YU
nonveilleri Dajoz, 1984b: 361 [= 1984a: 82] **E**: YU
novaki Dajoz, 1977: 238 **E**: CR YU
occidentalis Dajoz, 1968: 256 **E**: SP
octostriatus Dajoz, 1969b: 386 **E**: SP
osellai Pace, 1974: 113 **E**: IT
pannonicus Kaszab, 1947: 265 **E**: AU HU PL RO SK
paradoxus Breit, 1913a: 18 **E**: IT YU
pasubianus Pace, 1979: 451 **E**: IT
planicollis Fairmaire, 1869a: 232 **E**: FR IT
linderi Reitter, 1875k: 310
plicatus Reitter, 1894i: 299 **E**: IT YU
plicatoides Kaszab, 1947: 270 **E**: BH YU
politipennis Kaszab, 1947: 269 **E**: IT YU
quatordecemstriatus Coiffait, 1984: 45 **E**: CR
ramiroi Dajoz, 1969b: 389 **E**: PT
reitteri Ganglbauer, 1899b: 896 **E**: AL AU BH CR CZ HU IT RO SK SL SZ YU
apfelbecki Ganglbauer, 1899b: 896
roubali Všetečka, 1948: 39 **E**: CZ
rugifer Reitter, 1894i: 300 **E**: BH GR IT
pseudoparadoxus Obenberger, 1938: 91
scherleri Dajoz, 1973b: 150 **E**: IT SZ
schroetteri Reitter, 1910: 265 **E**: BH
semistriatus Coiffait, 1984: 46 **E**: FR
septemstriatus Reitter, 1922: 41 **E**: BH IT YU
serbicus Dajoz, 1984a: 84 [= 1987: 186] **E**: YU
stilleri Kaszab, 1947: 263 **E**: HU
sublaevis Kaszab, 1947: 268 **E**: RO
thoracicus Depoli, 1917: 11 **E**: IT YU
ticinensis Dajoz, 1973b: 152 **E**: IT SZ
titanus Reitter, 1901a: 23 **E**: CR
valombrosae Dieck, 1870: 353 **E**: IT **NARi**
venetus Pace, 1975: 14 **E**: IT
lessiniensis Pace, 1975: 17
veronensis Pace, 1975: 18 **E**: IT
vesulanus Binaghi, 1941b: 142 **E**: IT
vicentinus Pace, 1979: 448 **E**: IT

vienensis Coiffait, 1984: 45 **E**: AU
weiratheri Roubal, 1934b: 62 **E**: GR
zoufali Reitter, 1915a: 41 **E**: BH

subfamily **Xylariophilinae** Pal & Lawrence, 1986

genus *Xylariophilus* Pal & Lawrence, 1986: 185 type species *Xylariophilus comatus* Pal & Lawrence, 1986
comatus Pal & Lawrence, 1986: 198 **A**: AP

subfamily **Bothriderinae** Erichson, 1845

genus *Antibothrus* Sharp, 1885b: 124 type species *Antibothrus carinatus* Sharp, 1885
Antibothroides Nikitsky, 1985: 10 type species *Antibothrus fatalis* Nikitsky, 1985
fatalis Nikitsky, 1985b: 625 **A**: FE
ichihashii Narukawa, 2002: 123 **A**: JA
morimotoi Sasaji, 1997: 111 **A**: JA

genus *Ascetoderes* Pope, 1961: 104 type species *Bothrideres foveicollis* Reitter, 1877
Aeschyntelus C. O. Waterhouse, 1876a: 113 [HN] type species *Aeschyntelus ater* C. O. Waterhouse, 1876
bifossulatus Grouvelle, 1914: 51 **A**: TAI
popei Nakane, 1978: 157, 160 **A**: JA SW
sauteri Grouvelle, 1914: 51 (*Bothrideres*) **A**: TAI
takeii Nakane, 1968a: 75 (*Bothrideres*) **A**: JA

genus *Asosylus* Grouvelle, 1908c: 430 type species *Asosylus filiformis* Grouvelle, 1908
sauteri Grouvelle, 1914: 50 **A**: TAI
taiwanensis Heinze, 1943: 111 **A**: TAI

genus *Bothrideres* Dejean, 1835: 312 type species *Bothrideres contractus* Dejean, 1835
bipunctatus Gmelin, 1790: 1634 (*Opatrum*) **E**: AU BE BH BU BY CR CT CZ FI FR GB GE HU LA LT NL NR NT PT RO SK SV SZ UK "Caucasus" **A**: FE SY
contractus Dejean, 1835: 312
interstitialis Heyden, 1870: 107 **E**: FR SP **N**: AG MO
massanae Marquet, 1875: 511

genus *Cylindromicrus* Sharp, 1885a: 73 type species *Cylindromicrus gracilis* Sharp, 1885
gracilis Sharp, 1885a: 73 **A**: JA

genus *Dastarcus* Walker, 1858: 209 type species *Dastarcus porosus* Walker, 1858
Pathodermus Fairmaire, 1881: 79 type species *Pathodermus indicus* Fairmaire, 1881
cyprianus Dajoz, 1989: 35 **A**: CY
helophoroides Fairmaire, 1881: 81 (*Pathodermus*) **A**: SHG **ORR**
libanicus Fairmaire, 1881: 80 (*Pathodermus*) **A**: IS LE SA SY
kurosawai Sasaji, 1986b: 244 **A**: JA (Ryukyus)
longulus Sharp, 1885a: 76 **A**: FE JA NC TAI
turcicus Dajoz, 1973b: 146 **A**: TR

genus *Machlotes* Pascoe, 1863: 36 type species *Machlotes porcatus* Pascoe, 1863
costatus Sharp, 1885a: 75 (*Erotylathris*) **A**: JA TAI
incisus Pascoe, 1866b: 135 **A**: SD **ORR**

genus *Leptoglyphus* Sharp, 1885a: 75 type species *Leptoglyphus vittatus* Sharp, 1885
vittatus Sharp, 1885a: 75 **A**: JA

genus *Ogmoderes* Ganglbauer, 1899b: 899 type species *Bothrideres angusticollis* C. N. F. Brisout de Barneville, 1862 *angusticollis* C. N. F. Brisout de Barneville, 1862: 600 (*Bothrideres*) **E**: FR SP YU

genus *Pseudobothrideres* Grouvelle, 1908c: 438 type species *Pseudobothrideres neglectus* Grouvelle
velatus Grouvelle, 1908c: 441 **A**: SD **ORR**

genus *Sosylus* Erichson, 1845: 287 type species *Colydium rufipes* Fabricius, 1801
Metopiestes Pascoe, 1863: 38 type species *Metopiestes hirtifrons* Pascoe, 1863
striolatus Grouvelle, 1914: 49 (*Metopiestes*) **A**: TAI

genus *Triboderus* Grouvelle, 1894a: 463 type species *Triboderus andrewesi* Grouvelle, 1894
andrewesi Grouvelle, 1894a: 463 **A**: SD **ORR**

subfamily **Teredinae** Seidlitz, 1888

genus *Oxylaemus* Erichson, 1845: 275 type species *Lyctus cylindricus* Creutzer, 1796
cylindricus Creutzer, 1796a: 18 (*Lyctus*) **E**: AU BE BH BY CZ FR GR HU IT NL PL RO SK SP SZ UK YU
variolosus Dufour, 1843a: 94 (*Rhyzophagus*) **E**: FR GE GB IT PL SV SZ
caesus Erichson, 1845: 284

genus *Teredolaemus* Sharp, 1885a: 74 type species *Teredus politus* Lewis, 1879
guttatus Sharp, 1885a: 74 **A**: JA
politus Lewis, 1879d: 462 (*Teredus*) **A**: JA
major Champion, 1919c: 241 **A**: SD UP

genus *Teredus* Dejean, 1835: 313 type species *Lyctus nitidus* Fabricius, 1792
Teredosoma Curtis, 1840: 227 type species *Lyctus nitidus* Fabricius, 1792
cylindricus A. G. Olivier, 1790a: no 18: 9 (*Ips*) **E**: AU BE CR CZ GB GE FR IT PT SP SZ YU **N**: AG TU
nitidus Fabricius, 1792b: 505 (*Lyctus*)
opacus Habelmann, 1854: 29 **E**: AU CR CZ GE GR IT PT RO SK

family CERYLONIDAE Billberg, 1820

subfamily **Euxestinae** Grouvelle, 1908

genus *Euxestus* Wollaston, 1858c: 411 type species *Euxestus parkii* Wollaston, 1858
Neoplotera Belon, 1881: 29 type species *Neoplotera peregrina* Belon, 1881
Tritomidea Motschulsky, 1860a: 104 type species *Tritomidea translucida* Motschulsky, 1860 (= *Olibrus erithacus* Chevrolat, 1864)
erithacus Chevrolat, 1864: 599 (*Olibrus*) **Ei**: FR **Ni**: MR **A**: CH JA SD TAI
oblongus Motschulsky, 1860a: 107 (*Tritomoidea*)
peregrinus Belon, 1881: 30 (*Neoplotera*)
parkii Wollaston, 1858c: 413 **E**: AZ **N**: CI MR

genus *Euxestoxenus* Arrow, 1925: 155 type species *Euxestoxenus striatus* Arrow, 1925
striatus Arrow, 1925: 155 **A**: UP

genus *Cycloxenus* Arrow, 1925: 152 type species *Cycloxenus hispidus* Arrow, 1925
hispidus Arrow, 1925: 153 **A**: SD UP

genus *Hypodacne* J. L. LeConte, 1875: 170 type species *Hypodacne punctata* J. L. LeConte, 1875
Pachyochthes Reitter, 1897g: 248 type species *Pachyochthes edithae* Reitter, 1897
edithae Reitter, 1897g: 249 (*Pachyochthes*) **E**: AR

subfamily **Murmidiinae** Jacquelin du Val, 1857

genus *Murmidius* Leach, 1822: 41 type species *Mumidius ferrugineus* Leach, 1822
Ceutocerus Germar, 1824: 85 type species *Ceutocerus advena* Germar, 1824

ovalis Beck, 1817: 7 (*Hister*) **Ei**: AU FR DE GB GE NL PL SV SZ **Ai**: CH FE JA **COS**
ferrugineus Leach, 1822: 41
advena Germar, 1824: 85 (*Ceutocerus*)
segregatus C. O. Waterhouse, 1876a: 114 **Ei**: GB **AFR**

subfamily **Ceryloninae** Billberg, 1820

genus *Afrorylon* Ślipiński, 1980b: 469 type species *Cerylon substriatum* Grouvelle, 1919
shibatai Sasaji, 1983: 41 (*Philothermus*) **A**: JA (Tsushima)
takasago Sasaji, 1983: 38 (*Philothermus*) **A**: TAI

genus *Cautomus* Sharp, 1885a: 82 type species *Cautomus hystriculus* Sharp, 1885
Leptoxycheilus Besuchet, 1972: 127 type species *Cautomus sugerens* Besuchet, 1972
hystriculus Sharp, 1885a: 82 **A**: JA
myops Besuchet, 1972: 130 **A**: NE

genus *Cerylon* Latreille, 1802: 205 type species *Lyctus histeroides* Fabricius, 1792
Aphardion Gozis, 1886: 11 type species *Lyctus histeroides* Fabricius, 1792
bescidicum Reitter, 1911a: 119 **E**: CZ BU
caspicum Jelínek, 1981b: 124 **A**: IN
deplanatum Gyllenhal, 1827: 636 **E**: AU BU BY CT DE EN FI FR GB GE HU IT LA LT NL NR PL RO SK SL SV SZ UK **A**: FE
fagi C. N. F. Brisout de Barneville, 1867b: 176 **E**: AU BE BH BU BY CR CT CZ DE FI FR GB GE GR HU IR IT LA LT MC NR PL PT RO SK SL SP ST SV SZ TR UK YU
forticorne Mulsant & Rey, 1870: 119
laticolle Schweiger, 1949: 43
ferrugineum Stephens, 1830: 98 **E**: AU BE BH BU BY CR CT CZ DE EN FI FR GB GE GR HU IR IT LA LT MC NL NR PL RO SK SL SP SV SZ TR UK YU **A**: ES FE
arcuatum Roubal, 1925c: 76
angustatum Erichson, 1845: 295
atratulum Reitter, 1875m: 360
primroseae Donisthorpe, 1942: 2
grandicolle Reitter, 1888d: 176 **E**: AB ST (Kavkaz)
hazara Ślipiński, 1988b: 481 **A**: PA
histeroides Fabricius, 1792b: 295 (*Lyctus*) **E**: AB AL AR AU BE BH BU BY CR CT CZ DE EN FI FR GB GE GG GR HU IT KZ LA LT MC MD NL NR NT PL RO SK SL SP SV SZ UK YU **A**: ES FE ST
biimpressum Pic, 1894h: 71
caucasicum Reitter, 1876j: 389
conjunctum Rey, 1889e: 35
cordicolle Petri, 1926: 206
graecum Obenberger, 1917a: 26
nigripes Reitter, 1901j: 200
excavatum Fowler, 1886b: 74
reticulatum Roubal, 1925c: 76
ruficorne Rey, 1889e: 35
minimum Sharp, 1885a: 81 **A**: FE JA
impressum Erichson, 1845: 295 **E**: AU BY CT CZ EN FI FR IT LA LT NL NR PL SK SV SZ UK **A**: ES FE
sharpi Nakane, 1967d: 74 **A**: FE JA
takara Nakane, 1967d: 73 **A**: JA

genus *Ectomicrus* Sharp, 1885a: 78 type species *Ectomicrus rugicollis* Sharp, 1885
rugicollis Sharp, 1885a: 79 **A**: JA

genus *Gyrelon* Hinton, 1942b: 146 type species *Gyrelon mila* Hinton, 1942
rugosus Ślipiński, 1982: 53 (*Pseudocerylon*) **A**: SE

genus *Mychocerus* Erichson, 1845: 292 type species *Mychocerus ferrugineus* Erichson, 1845
Lapecautomus Sengupta & Crowson, 1973: 409 type species *Lapecautomus dybasi* Sengupta & Crowson, 1973
Lapethus Casey, 1890: 317 type species *Lapethus discretus* Casey, 1890
orientalis Sasaji, 1983: 42 (*Lapecautomus*) **A**: JA

genus *Paracerylon* Ślipiński, 1990: 116 type species *Cerylon biroi* Heinze, 1944
bhutanensis Dajoz, 1975: 309 (*Cerylon*) **A**: BT
curticolle Sharp, 1885a: 81 (*Cerylon*) **A**: JA
emarginatum Ślipiński, 1982: 32 (*Cerylon*) **A**: SW
globosum Ślipiński, 1981b: 430 (*Cerylon*) **A**: SD
humeridens Grouvelle, 1903a: 121 (*Cerylon*) **A**: BT SD
katmandu Dajoz, 1979a: 448 (*Cerylon*) **A**: NE
klapperichi Schweiger, 1952: 134 (*Cerylon*) **A**: SE
montanum Ślipiński, 1981b: 431 (*Cerylon*) **A**: SD
popei Ślipiński, 1980a: 399 (*Cerylon*) **A**: NE
yeti Ślipiński, 1982: 36 (*Cerylon*) **A**: KA

genus *Philothermopsis* Heinze, 1944: 30 type species *Cerylon alluaudi* Grouvelle, 1892
Cerylonopsis Heinze, 1944: 29 type species *Cerylon trifoveolatum* Heinze, 1944
abnormis Ślipiński, 1988b: 482 **A**: PA
crassipes Sharp, 1885a: 81 (*Cerylon*) **A**: JA

genus *Philothermus* Aubé, 1843b: 93 type species *Philothermus montandoni* Aubé, 1843
Pseudophilothermus Dajoz, 1973b: 154 type species *Cerylon semistriatum* Perris, 1865
brancucci Ślipiński, 1981b: 434 **A**: HP KA
clavipes Iablokoff-Khnzorian, 1956: 136 (*Cerylon*) **E**: AR
conicicollis Reitter, 1882b: 138 (*Cerylon*) **E**: AB ST (Kavkaz) **A**: IN
depressus Sharp, 1885a: 78 **A**: FE JA
elongatus Nakane, 1984: 590 (*Philothermes*) **A**: JA
evanescens Reitter, 1876j: 387 (*Cerylon*) **E**: AU BH CR CZ FR GE HU IT PL RO SK TR YU
japonicus Ślipiński, 1988a: 147 **A**: JA
magnicollis Reitter, 1882b: 137 (*Cerylon*) **E**: AB ST (Kavkaz)
micrus Ślipiński, 1988b: 480 **A**: PA
montandoni Aubé, 1843b: 94 **Ei**: FR IT RO
myops Ślipiński, 1988b: 479 **A**: PA
nepalenis Dajoz, 1979a: 447 **A**: NE
pakistanicus Ślipiński, 1988b: 481 **A**: PA
pubens Sharp, 1885a: 80 (*Ectomicrus*) **A**: JA
semistriatus Perris, 1865: 507 (*Cerylon*) **E**: AB FR GR HU IT SP TR YU **N**: AG MO
aetolicus Reitter, 1876i: 313 (*Cerylon*)
attenuatus Fairmaire, 1866a: 21 (*Cerylon*)
hellenicus Dajoz, 1973b: 154
spissicornis Fairmaire, 1877b: lxvi (*Cerylon*)

genus *Ploeosoma* Wollaston, 1854: 147 type species *Ploeosoma ellipticum* Wollaston, 1854
ellipticum Wollaston, 1854: 148 **N**: MR

genus *Spinocerylon* Ślipiński, 1988b: 483 type species *Spinocerylon apterum* Ślipiński, 1988
apterum Ślipiński, 1988b: 484 **A**: PA
mirabile Ślipiński, 1982: 34 (*Cerylon*) **A**: HP
wittmeri Ślipiński, 1981b: 428 (*Cerylon*) **A**: KA

genus *Thyroderus* Sharp, 1885a: 82 type species *Thyroderus porcatus* Sharp, 1885
porcatus Sharp, 1885a: 83 **A**: JA

family ALEXIIDAE Imhoff, 1856

genus *Sphaerosoma* Samouelle, 1819: 394 type species *Sphaerosoma quercus* Samouelle, 1819
Alexia Stephens, 1833: 23 type species *Tritoma piliferum* P. W. J. Müller, 1821 (= *Sphaerosoma quercus* Samouelle, 1819)
Arthrosphaerula Apfelbeck, 1915: 425 type species *Alexia bosnica* Reitter, 1885
Hygrotophila Champion, 1887: 227 [RN] [HN]
Lamprosphaerula Apfelbeck, 1910: 41 type species *Phalacrus globosus* Sturm, 1807
Neosphaerula Apfelbeck, 1910: 41 type species *Alexia laevicollis* Reitter, 1883
algiricum Reitter, 1889q: 304 (*Alexia*) **N**: AG
alutaceum Reitter, 1883f: 239 (*Alexia*) **E**: "Caucasus"
antennarium Apfelbeck, 1909: 500 **E**: BH HU
narentinum Apfelbeck, 1909: 501
apuanum Reitter, 1909d: 211 **E**: IT
bicome Peyerimhoff, 1917: 141 **N**: AG
bosnicum Reitter, 1885a: 35 (*Alexia*) **E**: BH
formaneki Reitter, 1902a: 5
carniolicum Apfelbeck, 1915: 427 **E**: SL
carpathicum Reitter, 1883f: 239 (*Alexia*) **E**: BU HU PL RO SK "Carpathians"
circassicum Reitter, 1888d: 169 (*Alexia*) **E**: ST (Kavkaz)
clamboides Reitter, 1888d: 170 (*Alexia*) **E**: ST (Kavkaz)
compressum Reitter, 1901f: 99 (*Alexia*) **E**: TR
corcyreum Reitter, 1883g: 394 (*Alexia*) **E**: AL GR
hellenicum Reitter, 1885a: 36 (*Alexia*)
valonense Apfelbeck, 1915: 423
csikii Apfelbeck, 1915: 421 **E**: BH BU
fiorii Ganglbauer, 1899b: 917 **E**: IT
globosum Sturm, 1807: 82 (*Phalacrus*) **E**: AU BE BH BU CR CZ FR GE HU IT LS LU NL PL RO SK SZ YU
glabrum Reitter, 1885d: 204
hemisphaericum Ganglbauer, 1899b: 916 **E**: BH
hemisphaericum Csiki, 1910a: 5 [HN]
laevicolle Reitter, 1883f: 240 (*Alexia*) **E**: AL AU BH BU CR IT YU
latitarse Apfelbeck, 1915: 427 **E**: IT
lederi Reitter, 1888d: 169 (*Alexia*) **E**: ST (Kavkaz)
leonhardi Apfelbeck, 1916: 494 **E**: GR
maritimum Reitter, 1904b: 45 (*Alexia*) **E**: IT
merditanum Apfelbeck, 1915: 418 **E**: AL MC
meridionale Reitter, 1883f: 239 (*Alexia*) **E**: SP **N**: AG
nevadense Reitter, 1883f: 241 (*Alexia*) **E**: SP
normandi Peyerimhoff, 1917: 141 **N**: AG
obsoletum Reitter, 1883f: 242 (*Alexia*) **E**: "Caucasus"
paganetti Obenberger, 1914: 110 **E**: IT
rotundatum Obenberger, 1914: 110
pilosellum Reitter, 1877h: 296 (*Alexia*) [RN] **E**: "Caucasus"
hirtulum Reitter, 1876f: 294 (*Alexia*) [HN]
pilosissimum Frivaldszky, 1881: 266 (*Alexia*) **E**: GR TR
pilosum Panzer, 1793b: 8 (*Tritoma*) **E**: AU BE BH BU CR CZ DE EN FI FR GB GE HU IT LS LU NL NR PL RO SK SP SV SZ YU
ignorans Reitter, 1883g: 393 (*Alexia*)
lunigerum Reitter, 1883f: 241 (*Alexia*)
piliferum Jacquelin du Val, 1858: pl. 55 (*Alexia*) [HN]
pubescens Frivaldszky, 1881: 265 (*Alexia*) **E**: TR
punctatum punctatum Reitter, 1878b: 63 (*Alexia*) **E**: AU HU IT PL RO SK YU
punctatum tatricum Reitter, 1897a: 47 (*Alexia*) **E**: "Tatra Mountains"
puncticolle Reitter, 1883f: 242 (*Alexia*) **E**: "Caucasus"
quercus Samouelle, 1819: 394 **E**: AU BE CZ FR GB GE IT LU NL SK SZ
infimum Baudi di Selve, 1889: 88 (*Alexia*)
piliferum P. W. J. Müller, 1821: 207 (*Tritoma*)

piligerum P. W. J. Müller, 1824: 617 (*Tritoma*)
rambouseki Apfelbeck, 1916: 490 **E**: YU
reitteri Ormay, 1888: 166 (*Alexia*) **E**: AL BH BU HU IT MC PL RO SK YU
albanicum Apfelbeck, 1906: 1670
aspromontanum Reitter, 1909a: 21
pilosissimum Reitter, 1883f: 241 (*Alexia*) [HN]
scymnoides Reitter, 1885a: 35 (*Alexia*) **E**: GR
seidlitzi Reitter, 1889o: 307 (*Alexia*) **E**: AU GE IT SZ
shardaghense Apfelbeck, 1915: 418 **E**: BU
solarii Reitter, 1904e: 155 **E**: IT
sparsum Reitter, 1909a: 22 **E**: IT
rotundum Obenberger, 1913: 110
sturanyi Apfelbeck, 1909: 500 **E**: BH BU
subglabrum Peyerimhoff, 1917: 140 **N**: AG TU
punctulatum Normand, 1936b: 94
sublaeve Reitter, 1883f: 239 (*Alexia*) **E**: GG "Caucasus"
tingitanum Peyerimhoff, 1917: 141 **N**: MO
vallombrosae Reitter, 1885a: 35 (*Alexia*) **E**: IT
winneguthi Apfelbeck, 1915: 425 **E**: AL

family DISCOLOMATIDAE Horn, 1878

genus *Aphanocephalus* Wollaston, 1873a: 278 type species *Aphanocephalus hemisphaericus* Wollaston, 1873
annamensis John, 1956: 30 **A**: SE **ORR**
asoanus John, 1965c: 53 **A**: JA
awashimanus John, 1968: 15 **A**: JA
besucheti Schawaller, 1989: 51 **A**: PA
birmanus Dodero, 1900b: 565 **A**: YUN **ORR**
chujoi John, 1955a: 3 **A**: JA
convexus Pal, 1996b: 12 **A**: SD
crinitus John, 1968: 17 **A**: JA (Ryukyus)
dahli John, 1971: 140 **A**: TAI
fallax John, 1955b: 17 **A**: TAI
formosanus John, 1955b: 17 **A**: TAI **ORR**
ganganicus Schawaller, 1989: 53 **A**: UP
gorkhus Schawaller, 1987: 4 **A**: NP
hemisphaericus Wollaston, 1873a: 278 **A**: CE JA **ORR**
hilleri John, 1971: 142 **A**: JA
himalayanus Schawaller, 1989: 53 **A**: NP SD
johni Pal, 1992: 48 **A**: AP
klapperichi John, 1956: 8 **A**: SE
nigriventris John, 1955a: 1 **A**: JA
pubescens Grouvelle, 1912e: 207 **A**: SE TAI **ORR**
pulinus John, 1967: 134 **A**: SE
quadrimaculatus Matthews, 1887: 114 **A**: CH **ORR**
satoi John, 1968: 14 **A**: JA
sauteri John, 1971: 139 **A**: TAI
secretus John, 1957: 252 **A**: SE
shibatai M. Chûjô & John, 1961: 1 **A**: JA
shirozui John, 1958: 61 **A**: JA
sikkima Pal, 1996b: 10 **A**: SD
superbus Pal, 1996b: 15 **A**: SD
taiwanus John, 1967: 134 **A**: TAI
takei John, 1963a: 8 **A**: JA
tamaii John, 1968: 18 **A**: JA (Ryukyus)

tonkinensis John, 1956: 13 **A**: YUN **ORR**
tsushimanus John, 1961: 1 **A**: JA
vitreus Matthews, 1887: 5 **A**: CH **ORR**
wollastoni E. C. Rye, 1873: 167 **A**: JA **ORR**

genus *Parafallia* Arrow, 1939: 7 type species *Parafallia castanea* Arrow, 1939
dissimilis Matthews, 1887: 115 (*Aphanocephalus*) **A**: CH
johni Geisthardt, 1975: 39 **A**: NP UP

genus *Parmaschema* Heller, 1912: 106 type species *Parmaschema nodimargo* Heller, 1912
indica Pal, 1992: 51 **A**: AP

family ENDOMYCHIDAE Leach, 1815
subfamily **Merophysiinae** Seidlitz, 1872
tribe Merophysiini Seidlitz, 1872

genus *Displotera* Reitter, 1887a: 10 type species *Displotera simoni* Reitter, 1887
maderae Wollaston, 1854: 180 (*Cholovocera*) **N**: MR
  oculata Belon, 1891: 878 (*Colovocera*)

genus *Cholovocera* Motschulsky, 1838: 177 type species *Cholovocera formicaria* Motschulsky, 1838
  Coluocera Gemminger & Harold, 1868: 905 [unjustified emendation]
afghana C. Johnson, 1977a: 123 **A**: AF
attae Kraatz, 1858: 140 (*Choluocera*) **E**: FR GR HU IT YU **N**: AG MO TU
brevicornis C. Johnson, 1977a: 124 **A**: AF
fleischeri Reitter, 1902a: 5 **E**: IT MA YU
formicaria Motschulsky, 1838: 179 **E**: AL FR GR IT MA SP SZ **N**: AG MO TU **A**: TR
  gallica Schaufuss, 1876: 398 (*Coluocera*)
  subterranea Motschulsky, 1845a: 111 (*Merophysia*)
major Reitter, 1887a: 10 **A**: TR
punctata Märkel, 1845: 255 **E**: FR IT SP SZ **N**: AG TU
  formiceticola Rosenhauer, 1856: 355
sardoa Reitter, 1911c: 70 **E**: IT (Sardegna)

genus *Reitteria* Leder, 1872: 137 type species *Reitteria lucifuga* Leder, 1872
balcanica Karaman, 1936: 131 **E**: MC
escherichi Wasmann, 1896: 238 **E**: IT **A**: CY TR
lucifuga Leder, 1872: 138 **N**: AG

genus *Merophysia* P. H. Lucas, 1852: 29 type species *Merophysia formicaria* P. H. Lucas, 1852
  Merophysiops Semenov, 1899: 15 type species *Merophysia bicarinata* Reitter, 1897
andreinii Gridelli, 1930: 142 **N**: LB
baudueri Reitter, 1877a: 6 **E**: SP **N**: AG MO TU
  acuminata Fairmaire, 1879a: 168
bicarinata Reitter, 1897e: 212 **A**: UZ
biplicata Reitter, 1884e: 253 **E**: GR **A**: IS SY
bisinuata Reitter, 1908f: 213 **A**: IS JO TR
bistriata Reitter, 1889j: 255 **E**: GR
carinulata Rosenhauer, 1856: 354 **E**: GB SP **N**: AG **A**: IS JO SY
carmelitana Saulcy, 1864: 423 **E**: GR **N**: AG MO TU **A**: CY IS LE SY
  minor Baudi di Selve, 1870: 59
cretica Kiesenwetter, 1872a: 163 **E**: GR **A**: CY PA TR
cyrenaica Gridelli, 1930: 139 **N**: LB
formicaria P. H. Lucas, 1852: xxix **E**: GR IT MA SZ **N**: AG MO TU **A**: CY TR
  ragusae Belon, 1895: 78

foveolata Baudi di Selve, 1870: 59 **A**: CY IS JO SY
lata Kiesenwetter, 1872a: 164 **E**: GR MA **A**: TR
letourneuxi Pic, 1914l: 49 **N**: EG **A**: IS LE SY **AFR**
longicornis Reitter, 1887a: 12 **N**: AG EG MO TU **A**: CY IS JO SY
lusitanica Reitter, 1913e: 654 **E**: PT
madoni Belon, 1885a: 241 **E**: IT (Sardegna) **A**: CY
nana Reitter, 1887a: 12 **A**: IS SY
oblonga Kiesenwetter, 1872a: 164 **E**: GR IT MA **A**: TR
orientalis Saulcy, 1864: 422 **E**: GR UK (Krym) **N**: AG **A**: IS JO LE SY TR
ovalipennis Coye, 1870: 376
procera Reitter, 1875k: 304 **A**: IS JO
sekerae Reitter, 1908g: 244 **E**: IT
sicula Kiesenwetter, 1872a: 166 **E**: FR GR IT **N**: TU
silvestrii Gridelli, 1927: 237 **N**: LB
striatella Reitter, 1890j: 388 **E**: GR IT UK (Krym)
uniplicata Reitter, 1884e: 253 **N**: MO

tribe Holoparamecini Seidlitz, 1888

genus *Holoparamecus* Curtis, 1833: 186 type species *Holoparamecus depressus* Curtis, 1833
subgenus *Calyptobium* Aubé, 1843c: 242 type species *Calyptobium caularum* Aubé, 1843
amabilis Sagaji, 1991: 22 **A**: JA
atomus Ragusa, 1888: 37 **E**: IT
caularum Aubé, 1843c: 244 (*Calyptobium*) **E**: AU AZ BH BE BU CT CZ GE DE FR GB GR HU IT MA NL PL SK SP SV SZ YU **N**: AG CI MO TU **A**: GUX HAI SY **COS**
beaucheni Croissandeau, 1891: 154
obtusicornis Motschulsky, 1867: 101 (*Calyptobium*)
tuberculum Motschulsky, 1867: 102 (*Calyptobium*)
contractus Wollaston, 1874: 203 **A**: JA
lyratus Reitter, 1891b: 22 **A**: UZ
microps Belon, 1895: 79 **N**: MO
nigriceps Belon, 1897: 122
niger Aubé, 1843c: 246 (*Calyptobium*) **E**: FR GR IT MA SP SZ **N**: AG CI EG LB MO MR TU **AFR**
argodi Croissandeau, 1891: 153
lowei Wollaston, 1879: 233
occultus Leder, 1872: 139
saulcyi Baudi di Selve, 1870: 60 **A**: CY
truquii Baudi di Selve, 1861: 182 **A**: CY LE
subgenus *Holoparamecus* Curtis, 1833: 186 type species *Holoparamecus depressus* Curtis, 1833
beloni Reitter, 1884c: 63 **E**: BH GR IT YU
depressus Curtis, 1833: 186 **E**: BE DE FR GB GR IT NL SZ YU **N**: MR **A**: HAI JA **COS**
integer Rey, 1889h: 54
kunzei Aubé, 1843c: 245 (*Calyptobium*)
ellipticus Wollaston, 1874: 201 **A**: ANH GAN FUJ GUA GUI GUX HEB HEN HUB JA JIL LIA MNO QIN SCH SHA SHN SHX YUN ZHE
punctulatus Reitter, 1908g: 244 **E**: IT
ragusae Reitter, 1875k: 309 **E**: BE CZ FR HU IT NL SZ **A**: JA **NAR NTR**
signatus Wollaston, 1874: 202 **A**: GUX HEN HUB JA
capitatus Wollaston, 1874: 202
singularis Beck, 1817: 14 (*Silvanus*) **E**: AZ BE FR GB GR IT NL SP SZ **N**: AG CI EG LB MO MR TU **A**: CY IS LE TR **AFR NAR ORR**
azoricus Méquignon, 1942b: 10
longipennis Motschulsky, 1867: 100
populi Motschulsky, 1867: 99
obtusus Reitter, 1884c: 63
villae Aubé, 1843c: 243 (*Calyptobium*)
subgenus *Tomyrium* Reitter, 1881b: 49 type species *Holoparamecus bertouti* Aubé, 1861

bertouti Aubé, 1861b: 198 **E**: FR IT (Sardegna, Sicilia) MA **N**: TU **A**: AE YE **AFR**
similis Belon, 1885b: xxxix

subfamily **Anamorphinae** Strohecker, 1953

genus *Aclemmysa* Reitter, 1904a: 41 type species *Aclemmysa solarii* Reitter, 1904
algirica Capra, 1924: 139 **N**: AG
solarii Reitter, 1904a: 42 **E**: IT

genus *Asymbius* Gorham, 1896: 299 type species *Asymbius crinipes* Gorham, 1896
formosanus Csiki, 1937: 173 **A**: TAI
foveicollis M. Chûjô & Kiuchi, 1974: 1 **A**: JA
hamulatus Arrow, 1926: 355 **A**: SD UP **ORR**
indicus Strohecker, 1982: 308 **A**: UP
rufus Arrow, 1925: 401 **A**: SD
sinensis Strohecker, 1982: 308 **A**: FUJ

genus *Bystodes* Strohecker, 1953: 19 type species *Bystodes paulus* Strohecker, 1953
flavoapicalis Sasaji, 1990: 70 **A**: JA
kidoi Sasaji, 1990: 71 **A**: JA
orbicularis Gorham, 1887b: 652 (*Symbiotes*) **A**: JA
taiwanensis Sasaji, 1970a: 1 **A**: TAI
yaeyamensis Sasaji, 1990: 72 **A**: JA

genus *Clemmus* Hampe, 1850: 353 type species *Clemmus troglodytes* Hampe, 1850
Clemmys Seidlitz, 1888a: 62 [RN]
Clemnus L. Redtenbacher, 1858: 975 [RN]
abbreviatus Reitter, 1889m: 277 (*Clemnus*) **A**: JA
troglodytes Hampe, 1850: 353 **E**: AU BU CR CZ HU IT RO SK SL

genus *Dexialia* Sasaji, 1970a: 4 type species *Dexialia ovalis* Sasaji, 1970
minor M. Chûjô, 1941c: 82 (*Symbiotes*) **A**: FE JA
mirabilis Sasaji, 1995: 22 **A**: JA
ovalis Sasaji, 1970a: 5 **A**: TAI
spectabilis Sasaji, 1984: 32 **A**: JA

genus *Dialexia* Gorham, 1891a: 146 type species *Dialexia setulosa* Gorham, 1891
hisanoi Sasaji, 1978: 12 **A**: JA

genus *Idiophyes* Blackburn, 1895: 234 type species *Idiophyes brevis* Blackburn, 1895
boninensis Sasaji, 1978: 19 **A**: JA
indicus Arrow, 1925: 397 (*Exysma*) **A**: UP
niponensis Gorham, 1874c: 225 (*Symbiotes*) **A**: JA
uenoi Sasaji, 1990: 73 **A**: JA

genus *Mychothenus* Strohecker, 1953: 24 [RN] type species *Mychophilus minutus* Frivaldszky, 1877
Mychophilus Frivaldszky, 1877: 19 [HN] type species *Mychophilus minutus* Frivaldszky, 1877
asiaticus Sasaji, 1978: 2 **A**: JA
caspicus Reitter, 1883c: 113 (*Mychophilus*) **E**: "Caucasus"
hirashimai Sasaji, 1990: 68 **A**: JA
minutus Frivaldszky, 1877: 20 (*Mychophilus*) **E**: AU CR HU IT SK SL YU

genus *Symbiotes* L. Redtenbacher, 1849: 198 type species *Symbiotes latus* L. Redtenbacher, 1849
Eponomastus Buysson, 1891b: xcv [RN]
Microchondrus Wollaston, 1854: 196 type species *Microchondrus domuum* Wollaston, 1854

armatus Reitter, 1881e: 227 **E**: AU CR GE IT SL SZ
gibberosus P. H. Lucas, 1846: 222 (*Cryptophagus*) **E**: AU BE CZ CR DE FR GE GR HU IT LU LS NL PL SK SL SP SZ **N**: AG CI MO MR **NAR**
domuum Wollaston, 1854: 197 (*Microchondrus*)
duryi Walton, 1912: 461 [HN]
minutus Schaufuss, 1882a: 544
montanus Casey, 1916: 140
pygmaeus Hampe, 1850: 353
latus L. Redtenbacher, 1849: 198 **E**: AU BE DE GB GE FR IT LU LS NL SZ
rubiginosus Reitter, 1880d: 95

subfamily **Danascelinae** Tomaszewska, 2000

genus *Danascelis* Tomaszewska, 1999: 279 type species *Danascelis elongata* Tomaszewska, 1999
elongata Tomaszewska, 1999: 279 **A**: PA

subfamily **Endomychinae** Leach, 1815

genus *Bolbomorphus* Gorham, 1887b: 647 type species *Bolbomorphus gibbosus* Gorham, 1887
chinensis Mader, 1955a: 69 [DA] **A**: FUJ
gibbosus Gorham, 1887b: 647 **A**: JA TAI
flavofasciatus M. Chûjô, 1940b: 216 **A**: TAI
mediojunctus Pic, 1921e: 1 (*Encymon*) **A**: YUN
disjunctus Mader, 1936b: 65 (*Eucteanus*)
mediojunctus Mader, 1936b: 64 (*Eucteanus*)
undulatus Pic, 1925e: 2
sexpunctatus Arrow, 1920a: 69 **A**: JIA JIX SHG
quadriguttatus Mader, 1938: 48
theryi Gorham, 1897: 461 **A**: CH

genus *Cyclotoma* Mulsant, 1851a: 71 type species *Cyclotoma testudinaria* Mulsant, 1851
Niteta J. Weise, 1890a: 21 type species *Niteta quatuordecimpunctata* J. Weise, 1890
Panomoea Gerstaecker, 1857: 241 type species *Panomoea coccinellina* Gerstaecker, 1857
conica Tomaszewska, 2000b: 354 **A**: TAI
formosana M. Chûjô, 1938: 403 [DA] **A**: TAI
indiana Gorham, 1897: 462 (*Panomoea*) **A**: SD TAI **ORR**

genus *Endomychus* Panzer, 1795c: 175 type species *Chrysomela coccinea* Linnaeus, 1758
Caenomychus Lewis, 1893b: 153 [RN] type species *Cyanauges gorhami* Lewis, 1874
Cyanauges Gorham, 1874a: 54 [HN] type species *Cyanauges gorhami* Lewis, 1874
armeniacus Motschulsky, 1835: 321 **E**: AR GG ST (Kavkaz) UK (Krym) **A**: IN
scovitzi Falderman, 1837: 411
atripes Pic, 1921e: 2 **A**: XIZ **ORR**
curtus Pic, 1927d: 10
bicolor Gorham, 1875: 22 **A**: NP SD **ORR**
chinensis Csiki, 1937: 174 **A**: FUJ
coccineus Linnaeus, 1758: 371 (*Chrysomela*) **E**: AL AU BE BH BU BY CR CZ DE EN FI FR GB GE GR HU IR IT LA LT LS LU MC MD NL NR NT PL RO SK SL SP SV SZ UK YU
biehli Reitter, 1888g: 424
krynickii Ganglbauer, 1899b: 940
quadrimaculatus de Geer, 1775: 301 (*Chrysomela*)
quadripunctatus Gorham, 1873a: 64
flavus Strohecker, 1943: 390 **A**: SCH
foveolatus Tomaszewska, 2002a: 151 **A**: YUN **ORR**
gorhami Lewis, 1874a: 55 (*Cyanauges*) **A**: FE JA
kyushuensis Sasaji, 1978: 25
violaceipennis Mader, 1941b: 170
humeralis Pic, 1922b: 10 (*Caenomychus*) **A**: KA PA

jureceki Mader, 1936c: 69 **A**: FE
micrus Tomaszewska, 1997a: 231 **A**: PA
mroczkowskii Tomaszewska, 1997a: 227 **A**: NP
nigricapitatus Tomaszewska, 2002a: 152 **A**: YUN
nigriceps M. Chûjô, 1938: 404 **A**: FUJ TAI
muelleri Mader, 1955a: 71 (*Caenomychus*)
nigricornis M. Chûjô, 1938: 405 **A**: TAI
nigripes Mader, 1955a: 70 **A**: FUJ
nigropiceus Gorham, 1887b: 651 (*Cyanauges*) **A**: JA SC
hiranoi Sasaji, 1978: 27
pakistanicus Tomaszewska, 1997a: 231 **A**: PA
plagiatus Gorham, 1887b: 650 (*Cyanauges*) **A**: JA
interruptus Nakane, 1994b: 82
quadra Gorham, 1887b: 651 (*Cyanauges*) **A**: JA
kojimai Nakane, 1994b: 83
ohbayashi Nakane, 1951: 116
quadratus Ohta, 1931c: 235 (*Caenomychus*)
shirahatai Nakane, 1951: 117
sasajii Tomaszewska, 1998: 301 **A**: TAI
sauteri M. Chûjô, 1938: 405 **A**: TAI
thoracicus Charpentier, 1825: 245 **E**: AU IT RO
tomishimai Nakane, 1994b: 82 **A**: JA
yunnani Tomaszewska, 1997a: 227 **A**: YUN

genus *Eucteanus* Gerstaecker, 1857: 240 type species *Eucteanus coelestinus* Gerstaecker, 1857
Homalosternus Guérin-Méneville, 1857: 581 type species *Eumorphus hardwickei* Hope, 1831
coelestinus Gerstaecker, 1857: 240 **A**: SD **ORR**
hardwickei Csiki, 1910a: 54 [HN]
cruciger Gorham, 1897: 460 **A**: AP **ORR**
dohertyi Gorham, 1897: 461 **A**: YUN **ORR**
elegans Arrow, 1937: 110 **A**: XIZ **ORR**
hardwickei Hope, 1831: 22 (*Eumorphus*) **A**: NP
marseuli Gorham, 1873a: 56 **A**: HP NP UP
mussuricus Arrow, 1925: 354 **A**: UP
subovatus Pic, 1925e: 2 (*Bolbomorphus*) **A**: YUN

genus *Meilichius* Gerstaecker, 1857: 240 type species *Meilichius nigricollis* Gerstaecker, 1857
Gibbiger Csiki, 1900c: 375 type species *Milichius fasciatus* Heller, 1898
Milichius Gemminger & Harold, 1874: 3737 [RN]
Thelgetrum Gorham, 1875: 314 type species *Thelgetrum ampliatum* Gorham, 1875
erotyloides Strohecker, 1951: 166 **A**: HAI
klapperichi Mader, 1941c: 933 **A**: FUJ TAI
multimaculatus Sasaji, 1970a: 15 **A**: TAI

subfamily **Eupsilobiinae** Casey, 1895

genus *Eidoreus* Sharp, 1885b: 146 type species *Eidoreus minutus* Sharp, 1885
Eupsilobius Casey, 1895: 454 type species *Eupsilobius politus* Casey, 1895
Pseudalexia Kolbe, 1910: 34 type species *Pseudalexia sechellarum*, Kolbe, 1910
japonicus Sasaji, 1991a: 18 **A**: JA

subfamily **Leiestinae** C. G. Thomson, 1863

genus *Leiestes* Chevrolat, 1836: 440 type species *Cryptophagus seminiger* Gyllenhal, 1808
Liesthes Ganglbauer, 1899b: 925 [RN]
fines Sasaji, 1995: 24 (*Liesthes*) **A**: JA
menetriesi Falderman, 1837: 396 [DA] (*Triplax*) **E**: "Caucasus"

seminiger Gyllenhal, 1808a: 178 (*Cryptophagus*) **E**: AU BE BH BY CR CZ EN GE HU FI FR IT LA LT LU NL NR NT PL RO SK SL SV SZ YU "Caucasus" **A**: IN JA

genus *Panaleies* Tomaszewska, 2000a: 80 type species *Panamomus decoratus* Gorham, 1887
decoratus Gorham, 1887b: 649 (*Panamomus*) **A**: JA

genus *Panamomus* Gorham, 1873b: 207 type species *Panamomus lewisi* Gorham, 1873
brevicornis Gorham, 1887b: 649 **A**: JA
lewisi Gorham, 1873b: 207 **A**: JA
sawadai Nakane, 1983b: 145 **A**: JA
yoshidai Kiuchi, 1972: 160 **A**: JA

subfamily **Lycoperdininae** L. Redtenbacher, 1844

genus *Amphisternus* Germar, 1843b: 85 type species *Amphisternus tuberculatus* Germar, 1843
coralifer Gerstaecker, 1857: 216 **A**: YUN **ORR**
laotianus Achard, 1926: 141

genus *Amphistethus* Strohecker, 1964: 337 type species *Amphistethus superbus* Strohecker, 1964
astrate Strohecker, 1951: 159 (*Amphisternus*) **A**: TAI

genus *Ancylopus* A. Costa, 1850: 13 type species *Endomychus melanocephalus* A. G. Olivier, 1808
melanocephalus A. G. Olivier, 1808: 1073 (*Endomychus*) **E**: BU HU IT SP
nigrothoracica Pal, 2003: 113 **A**: SD
phungi borealior Strohecker, 1972: 708 **A**: JIA
phungi phungi Pic, 1926c: 10 **A**: FUJ JA JIX XIZ ZHE
pictus asiaticus Strohecker, 1972: 706 **A**: CE JA SE SW TAI **ORR**
pictus indianus Strohecker, 1972: 706 **A**: AF NP SD UP **ORR**
villiersi Dajoz, 1972: 1046 **A**: JA **ORR**

genus *Avencymon* Strohecker, 1971b: 38 type species *Ancylopus concolor* Strohecker, 1951
ruficephalus Ohta, 1931c: 221 (*Encymon*) **A**: FUJ JIX TAI **ORR**
concolor Strohecker, 1951: 159 (*Ancylopus*)

genus *Beccariola* Arrow, 1943: 129 [RN] type species *Beccaria papuensis* Gorham, 1885
Beccaria Gorham, 1885b: 521 [HN] type species *Beccaria papuensis* Gorham, 1885
cardoni Gorham, 1897: 464 (*Beccaria*) **A**: NP **ORR**
fulgurata M. Chûjô, 1938: 399 (*Beccaria*) **A**: TAI
sikkimensis Tomaszewska, 2002b: 374 **A**: SD

genus *Cacodaemon* J. Thomson, 1857: 153 type species *Eumorphus satanas* J. Thomson, 1856
laotinus yunnanensis Kryzhanovskij, 1960: 880 (*Amphisternus*) **A**: YUN

genus *Cymbachus* Gerstaecker, 1857: 233 type species *Cymbachus pulchellus* Gerstaecker, 1857
koreanus Mt. Chûjô, M. Chûjô & Lee, 1993a: 95 **A**: SC
spilotus Arrow, 1925: 281 **A**: SD

genus *Dapsa* Latreille, 1829b: 159 type species *Endomychus denticollis* Germar& Kaulfuss, 1816
Phylira Mulsant, 1846: 23 type species *Dapsa trimaculata* Motschulsky, 1835
adami Tomaszewska, 1997b: 218 **A**: NP SD
barbara P. H. Lucas, 1849: 552 **N**: AG
subpunctata Marseul, 1868a: 107
bertiana Audisio & De Biase, 1996: 109 **N**: AG
celata Arrow, 1925: 337 **A**: NP SD
curta Franz, 1996: 131 **N**: CI

curvipes Audisio & De Biase, 1996: 104 **N**: AG
denticollis Germar & Kaulfuss, 1816: 8 (*Endomychus*) **E**: AU BU CR CZ GE HU IT PL RO SK SL UK YU
binotata Trella, 1923: 115
nigricollis Gerstaecker, 1858: 198
rufa Porta, 1929: 241
unimaculata Trella, 1930: 36
edentata Wollaston, 1864: 432 **N**: CI
fodori Csiki, 1907: 578 (*Phylira*) **E**: HU SK
gemina Audisio & De Biase, 1996: 108 **N**: AG
graeca Pic, 1902l: 71 **E**: GR
grancanariensis Palm, 1974: 32 **N**: CI
horvathi Csiki, 1901b: 106 (*Phylira*) **A**: ES FE KZ WS
rodiana Semenov, 1904: 289
indica Tomaszewska, 1997b: 218 **A**: SD
inornata Gorham, 1873a: 43 **A**: LE
syriaca Pic, 1902e: 33
kabylica Audisio & De Biase, 1996: 105 **N**: AG (Kabylia)
lederi Reitter, 1880d: 97 **E**: GG "Caucasus"
limbata Motschulsky, 1835: 323 **E**: AR GG TR "Caucasus"
nigripennis Reitter, 1879a: 89
obscurissima Pic, 1902a: 10 **E**: IT **N**: AG TU
tyrrhena Audisio, De Biase & Zampetti, 1983: 87
motschulskyi Nikitsky & Semenov, 2001: 61 **A**: MG FE
nikolajevi Nikitsky & Semenov, 2001: 65 **A**: UZ
opuntiae Reitter, 1884d: 117 **E**: AL CR GR IT YU
intermedia J. R. Sahlberg, 1913b: 13
sahlbergi Reitter, 1900f: 219
oxiana Iablokoff-Khnzorian, 1964: 167 **A**: TD
palaestinensis Audisio & De Biase, 1996: 82 **A**: IS "Palestina"
pallescens Marseul, 1868a: 109 **N**: AG MO
pardoi Peyerimhoff, 1949: 274 **N**: MO
palmaensis Franz, 1979: 91 **N**: CI
sellata Marseul, 1868a: 108 **N**: AG
spinicollis Fairmaire, 1868: 501 **E**: SP **N**: AG MO
acuticollis Reitter, 1882d: 231
trimaculata Motschulsky, 1835: 322 **E**: AB AL BH BU CR FR GG GR IT MC RO SL SP ST TU UK YU **A**: IN IS KZ LE SY TM UZ
testacea A. Costa, 1882: 36 (*Ancylopus*)
ussuriensis Nikitsky, Semenov & Audisio, 2001: 67 **A**: FE

genus *Encymon* Gerstaecker, 1857: 232 type species *Encymon violaceus* Gerstaecker, 1857
Cranterophorus Blackburn, 1895: 232 type species *Mycella clavicornis* Blackburn, 1890 (= *Eumorphus immaculatus* Montruzier, 1855)
cinctipes Gorham, 1897: 459 **A**: AP SD **ORR**
cyanipennis M. Chûjô, 1938: 396 **A**: TAI **ORR**
valgus Strohecker, 1951: 161
violaceus Gerstaecker, 1857: 233 **A**: YUN **ORR**
cupreatus Mader, 1936b: 63

genus *Eumorphus* Weber, 1801: 31 type species *Erotylus quadriguttatus* Illiger, 1800
Enaisimus Guérin-Méneville, 1858: 16 type species *Eumorphus quadrinotatus* Gerstaecker, 1857
Eumorphoides Guérin-Méneville, 1858: 12 type species *Eumorphus tetraspilotus* Hope, 1832
Haplomorphus Guérin-Méneville, 1858: 18 type species *Eumorphus bipunctatus* Perty, 1831
Heterandrus Guérin-Méneville, 1858: 26 type species *Eumorphus confusus* Guérin-Méneville, 1857
assamensis assamensis Gerstäker, 1857: 229 **A**: AP YUN **ORR**
austerus indianus Strohecker, 1968: 88 **A**: GUI YUN

bicoloripedoides Mader, 1955a: 68 (*Engonius*) **A**: FUJ SD YUN **ORR**
ocellatus Arrow, 1920b: 327 **A**: YUN **ORR**
panfilovi Kryzhanovskij, 1960: 877 **A**: YUN
quadriguttatus pulchripes Gerstaecker, 1857: 228 **A**: AP GUI HAI JA NP SD TAI YUN **ORR**
baibaranus Ohta, 1931c: 219 (*Engonius*)
formosanus Pic, 1930c: 8
taitoensis Ohta, 1931c: 219 (*Engonius*)
sanguinipes Guérin-Méneville, 1858: 20 (*Haplomorphus*) **A**: SD **ORR**
simplex erythromerus Kryzhanovskij, 1960: 880 **A**: YUN **ORR**
simplex simplex Arrow, 1920b: 324 **A**: YUN **ORR**
sybarita sybarita Gerstaecker, 1857: 229 **A**: YUN **ORR**
popovi Kryzhanovskij, 1960: 876

genus *Gerstaeckerus* Tomaszewska, 2005: 42 [RN] type species *Engonius sexguttatus* Gerstaecker, 1857
Engonius Gerstaecker, 1857: 220 [HN] type species *Engonius sexguttatus* Gerstaecker, 1857
chensicieni Kryzhanovskij, 1960: 874 (*Engonius*) **A**: YUN
gratus Gorham, 1891b: 399 (*Engonius*) **A**: YUN **ORR**
effractus Achard, 1922: 28
kamengensis Vazirani & Saha, 1972: 77 (*Engonius*) **A**: AP
sikkimensis Strohecker, 1971a: 18 (*Engonius*) **A**: SD

genus *Hylaia* Chevrolat, 1836: 440 type species *Lycoperdina rubricollis* Germar, 1844
Ceramis Gerstaecker, 1858: 220 [RN]
Hylaia Guérin-Méneville, 1857: 273 [HN] type species *Lycoperdina rubricollis* Germar, 1844
cardiophora Wurst, 1997b: 108 **E**: AB
dalmatina Kaufmann, 1883: 10 **E**: BH CR YU
dryadella Wurst, 1997b: 98 **E**: GR (Makedonia)
elongata Apfelbeck, 1912: 657 **E**: AL YU
podagrica Guérin-Méneville, 1857: 274 **E**: GG TR **A**: IN
reissi Csiki, 1911: 74 **E**: BU GR
rambouseki Obenberger, 1917a: 29
rubricollis Germar, 1844a: pl. 18 (*Lycoperdina*) **E**: CR MD RO

genus *Indalmus* Gerstaecker, 1858: 185 type species *Eumorphus kirbyanus* Latreille, 1807
Mycella Chapuis, 1876: 104 type species *Mycella lineella* Chapuis, 1876
coomani sinensis Strohecker, 1979: 289 **A**: GUA JIX
distinctus Arrow, 1923: 485 **A**: YUN **ORR**
formosanus Csiki, 1937: 7 (*Phaeomychus*) **A**: TAI
formosanus Strohecker, 1945: 1
hanchunglinensis Li, 1992: 134 **A**: JIL
indicus Gorham, 1875: 312 (*Ancylopus*) **A**: UP
liuchungloi Kryzhanovskij, 1960: 871 **A**: HAI YUN **ORR**
kirbyanus Latreille, 1807b: 72 (*Eumorphus*) **A**: HAI NP SD TAI **ORR**
atricornis Pic, 1921e: 2 (*Ancylopus*)
latefasciatus Pic, 1929c: 15
pubescens Arrow, 1925: 311 (*Engonius*) **A**: GUA YUN **ORR**
quadripunctatus Ohta, 1931c: 222 (*Encymon*) **A**: JA TAI

genus *Lycoperdina* Latreille, 1807b: 73 type species *Galleruca bovistae* Fabricius, 1792
Falsoylaia Pic, 1945: 11 type species *Falsoylaia obscuresuturalis* Pic, 1945
Golgia Mulsant, 1846: 17 type species *Silpha succincta* Linnaeus, 1767
Lycoperdinella Arrow, 1920a: 23 [HN] type species *Lycoperdinella morosa* Arrow, 1920
Lycoperdinodes Arrow, 1923: 485 [RN] type species *Lycoperdinella morosa* Arrow, 1920
banatica Ganglbauer, 1899b: 935 **E**: RO
bovistae Fabricius, 1792b: 20 (*Galleruca*) **E**: BE CZ DE FR GB GE HU IT NL PL SK SV SZ
immaculata Latreille, 1807b: 73

leprieuri Pic, 1943b: 9
rufescens Pic, 1943b: 9
canariensis Gillerfors, 1991: 24 **N**: CI
castaneipennis Gorham, 1874b: 151 **A**: JA
crassicornis Reitter, 1880d: 97 **E**: RO
dux Gorham, 1873b: 205 **A**: JA TAI
gomerae Franz, 1979: 87 **N**: CI
humeralis Wollaston, 1864: 432 **N**: CI (Tenerife)
koltzei Reitter, 1887h: 508 **A**: ES FE NC SC WS
mandarinea Gerstaecker, 1858: 212 **A**: CH FE JA MG NC SC TAI **ORR**
maritima Reitter, 1884c: 61 **E**: FR (Alpes-Maritimes) IT
pallida Gebler, 1841b: 623 **A**: ES WS
penicillata Marseul, 1868a: 96 **E**: SP **N**: AG MO
pulvinata Reitter, 1884c: 59 **E**: CR GR MC YU
obliqua Reitter, 1884c: 59
rifensis Franz, 1979: 88 **N**: MO
sanchezi Oromí & García, 1987: 87 **N**: CI (La Palma)
succincta Linnaeus, 1767: 573 (*Silpha*) **E**: AU CZ DE EN FI GB GE HU IT LA LT NL NR PL RO SK SP SV SZ **A**: ES WS
disca Gerstaecker, 1858: 213
fasciata Fabricius, 1801a: 505 (*Endomychus*)
humeralis Schilsky, 1888: 184
jakovlevi Semenov, 1904: 290
quadripustulata Fabricius, 1792b: 33 (*Galleruca*)
validicornis Gerstaecker, 1858: 217 **E**: FR (Corse) IT
subpubescens Reitter, 1884c: 60

genus *Mycetina* Mulsant, 1846: 15 type species *Chrysomela cruciata* Schaller, 1783
Mycetinina Pic, 1929h: 35 type species *Pseudindalmus testaceitarsis* Pic, 1926
Phaeomychus Gorham, 1887b: 649 type species *Endomychus rufipennis* Motschulsky, 1861
amabilis Gorham, 1873b: 206 **A**: JA
humeropicta Nakane, 1958a: 33
takahashii Nakane, 1958a: 33
ancoriger Gorham, 1873b: 206 **A**: JA
apicalis Motschulsky, 1835: 321 (*Endomychus*) **E**: AB AR GG "Caucasus" **A**: IN
bistripunctata Mader, 1938: 43 **A**: YUN
compacta Fairmaire, 1891f: ccxix **A**: CH
cruciata Schaller, 1783: 273 (*Chrysomela*) **E**: AU BE BH BU BY CR CZ DE EN FI FR GE HU IT LA LT LU LS MD NL NR PL RO SK SL SP SV SZ UK YU
balanica Csiki, 1900d: 401
binotata A. Costa, 1850: 10 (*Lycoperdina*)
calabra O. G. Costa, 1839a: 158 (*Eumorphus*)
fussi Csiki, 1900b: 150
interrupta Gredler, 1870: 17
lithophila Herbst, 1783: 37 (*Opatrum*)
cyanescens Strohecker, 1943: 382 **A**: SCH
emmerichi Mader, 1938: 43 **A**: SCH
fulva M. Chûjô, 1938: 394 **A**: TAI
humerosignata Nakane, 1968b: 135 **A**: TAI
inapicalis Pic, 1904m: 90 **A**: SY
kamikochiana Nakane, 1982: 50 **A**: JA
karin Sasaji, 1995: 26 **A**: JA
laticollis Gorham, 1887b: 644 **A**: JA
laticollis satsumana Nakane, 1984: 146 **A**: JA
maderi Strohecker, 1982: 30 **A**: FUJ
marginalis Gebler, 1830: 227 (*Lycoperdina*) **A**: ES FE WS

konumaensis Ohta, 1931c: 220 (*Engonius*)
obliquesignata Gorham, 1873a: 44
montivaga Csiki, 1900d: 401 **A**: HP SD UP **ORR**
rufipennis koyana Nakane, 1958a: 33 **A**: JA
rufipennis rufipennis Motschulsky, 1861a: 18 (*Endomychus*) **A**: FE JA TAI
sasajii Strohecker, 1982: 309 **A**: TAI
similis M. Chûjô, 1938: 395 (*Phaeomychus*) **A**: TAI
stackelbergi Kryzhanowskij, 1977: 39 **A**: FE
superba Mader, 1941c: 932 **A**: FUJ
variicornis Arrow, 1925: 317 (*Engonius*) **A**: AP SD **ORR**

genus *Ohtaius* M. Chûjô, 1938: 398 type species *Brachytrycherus mushanus* Ohta, 1931
laticollis Achard, 1922: 28 (*Engonius*) **A**: FUJ
mushanus Ohta, 1931c: 216 (*Brachytrycherus*) **A**: FUJ JIX TAI
signifer Gorham, 1875: 311 (*Engonius*) **A**: AP NP SD **ORR**

genus *Parindalmus* Achard, 1922: 29 type species *Parindalmus tonkineus* Achard, 1922
Pedanus Gerstaecker, 1857: 230 [HN] type species *Pedanus quadrilunatus* Gerstaecker, 1857
sinensis Strohecker, 1971a: 25 **A**: SCH
tonkineus Achard, 1922: 29 **A**: FUJ HAI YUN **ORR**
westermanni Gerstaecker, 1857: 231 (*Pedanus*) **A**: YUN **ORR**

genus *Pseudindalmus* Arrow, 1920b: 330 type species *Pseudindalmus tonkinensis* Arrow, 1920
longicornis Tomaszewska, 2001: 136 **A**: TAI

genus *Sinocymbachus* Strohecker & M. Chûjô, 1970: 511 type species *Engonius excisipes* Strohecker, 1943
angustefasciatus Pic, 1940d: 11 (*Engonius*) **A**: SCH
decorus Strohecker & M. Chûjô, 1970: 516 **A**: YUN
excisipes Strohecker, 1943: 383 (*Engonius*) **A**: SCH
humerosus Mader, 1938: 40 (*Cymbachus*) **A**: FUJ GUA GUX HAI TAI
quadrinotatus M. Chûjô, 1938: 396 (*Amphisternus*)
luteomaculatus Pic, 1921e: 1 (*Engonius*) **A**: YUN
parvimaculatus Mader, 1938: 40 (*Cymbachus*) **A**: YUN
politus Strohecker & M. Chûjô, 1970: 515 **A**: TAI
quadrimaculatus Pic, 1927d: 11 (*Amphisternus*) **A**: FUJ **ORR**
quadriundulatus M. Chûjô, 1938: 397 (*Amphisternus*) **A**: TAI

genus *Spathomeles* Gerstaecker, 1857: 218 type species *Spathomeles anaglyptus* Gerstaecker, 1857
Rhachidophorus Guérin-Méneville, 1857: 581 type species *Cacodaemon hopei* J. Thomson, 1857 (=*Eumorphus hopei* Guérin, 1857)
decoratus Gerstaecker, 1857: 219 **A**: AP SD **ORR**

subfamily Lycoperdininae, genus incertae sedis
genus *Polymus* Mulsant, 1846: 10 type species *Polymus nigricornis* Mulsant, 1846
nigricornis Mulsant, 1846: 10 **E**: FR

subfamily **Mycetaeinae** Jacquelin du Val, 1857

genus *Agaricophilus* Motschulsky, 1838: 175 type species *Agaricophilus reflexus* Motschulsky, 1838
dieneri Reitter, 1911e: 294 **E**: SK
reflexus Motschulsky, 1838: 177 **E**: ST (Kavkaz)
glabratus Kolenati, 1846b: 56 (*Hygrothophila*)
subaeneus Reitter, 1876f: 294

genus *Mycetaea* Stephens, 1829: 87 type species *Mycetaea fumata* Stephens, 1829 (= *Dermestes subterraneus* Fabricius, 1801)

Silpha Marsham, 1802: 124 [HN] type species *Silpha hirta* Marsham, 1802

subterranea Fabricius, 1801a: 318 (*Dermestes*) **E**: AU AZ BE BY CR CT CZ DE EN FI FR GB GE HU IC IR IT LA LT LU NL NR NT PL RO SK SL SV SP SZ "Caucasus" **N**: CI MR **NAR**

dubia Lea, 1921: 367 (*Idiophyes*)
fumata Stephens, 1829: 87 (*Mycetaea*)
hirta Marsham, 1802: 124 (*Silpha*)

subfamily **Pleganophorinae** Jacquelin du Val, 1858

genus *Pleganophorus* Hampe, 1855a: 97 type species *Pleganophorus bispinosus* Hampe, 1855
bispinosus bispinosus Hampe, 1855a: 98 **E**: BU CR CZ GR HU RO SK
bispinosus slovenicus Olexa, 1954a: 48 **E**: SK

genus *Trochoideus* Westwood, 1833: 673 type species *Paussus cruciatus* Dalman, 1825

Pseudopaussus Schultze, 1916 type species *Pseudopaussus monstrosus* Schultze, 1916

desjardinsi Guérin-Méneville, 1838c: 22 **A**: JA TAI **AFR AUR ORR**

amphora Cantor, 1846: 282
monstrosus Schultze, 1916: 292
particularis Pic, 1922a: 8
rouyeri Pic, 1922a: 8
termitophilus Roepke, 1919: 34

subfamily **Stenotarsinae** Chapuis, 1876

genus *Atrichonota* Arrow, 1925: 386 type species *Atrichonota flavipennis* Arrow, 1925
flavipennis Arrow, 1925: 386 **A**: TAI **ORR**

genus *Chondria* Gorham, 1887b: 651 type species *Chondria lutea* Gorham, 1887
agilis Arrow, 1943: 134 **A**: JA
araneola Arrow, 1925: 366 **A**: SD
formosana Sasaji, 1970a: 9 **A**: TAI
longipes Arrow, 1943: 134 **A**: JA
lutea Gorham, 1887b: 651 **A**: JA
nigropunctata Nakane, 1951: 113 **A**: TAI
tympanizans Arrow, 1943: 135 **A**: JA

genus *Danae* Reiche, 1847: 408 type species *Danae rufula* Reiche, 1847

Coniopoda Gorham, 1873b: 205 type species *Coniopoda orientalis* Gorham, 1873
Oediarthrus Gerstaecker, 1858: 344 type species *Oediarthrus natalensis* Gerstaecker, 1858
Rhabduchus Gorham, 1873c: 257 type species *Rhabduchus denticornis* Gorham, 1873

castanea Sasaji, 1978: 23 **A**: JA
chinensis Strohecker, 1951: 162 **A**: JIX
clauda Arrow, 1925: 390 **A**: SD UP
denticornis Gorham, 1873c: 257 (*Rhabduchus*) **A**: JA
orientalis Gorham, 1873b: 205 (*Coniopoda*) **A**: JA TAI
ornata Arrow, 1925: 392 **A**: UP
sericea Arrow, 1925: 391 **A**: SD **ORR**
shibatai Nakane, 1958a: 35 **A**: JA
taiwana M. Chûjô, 1938: 401 **A**: TAI

genus *Ectomychus* Gorham, 1887b: 646 type species *Ectomychus basalis*, Gorham, 1887
basalis Gorham, 1887b: 646 **A**: FE JA
musculus Gorham, 1887b: 645 (*Stenotarsus*) **A**: JA
nakanei M. Chûjô, 1941c: 81 **A**: JA
nepalensis Strohecker, 1982: 310 **A**: NP

nigriclavis Gorham, 1873b: 206 (*Stenotarsus*) **A**: JA
nigrofasciatus Sasaji, 1988a: 19 **A**: JA
ovatus Arrow, 1925: 369 **A**: SD **ORR**
sakaii M. Chûjô & Kiuchi, 1974: 4 **A**: JA
tappanus Ohta, 1931c: 230 **A**: TAI
yukihikoi Sasaji, 1995: 28 **A**: JA

genus *Perrisina* E. Strand, 1921: 106 [RN] type species *Lycoperdina brevis* Perris, 1864
Perrisia Marseul, 1868a: 113 [HN] type species *Lycoperdina brevis* Perris, 1864
brevis Perris, 1864: 303 (*Lycoperdina*) **N**: AG
theresae Pic, 1944a: 1 (*Perrisia*)

genus *Saula* Gerstaecker, 1858: 223 type species *Saula nigripes* Gerstaecker, 1858
chujoi Sasaji, 1970a: 7 **A**: TAI
japonica Gorham, 1874c: 224 **A**: JA TAI
longior Strohecker, 1951: 164 **A**: UP
taiwana M. Chûjô, 1938: 399 **A**: TAI

genus *Stenotarsus* Perty, 1832: 112 type species *Stenotarsus brevicollis* Perty, 1832
Quirinus J. Thomson, 1857: 157 type species *Quirinus sulcithorax* J. Thomson, 1857
Stenotarsoides Csiki, 1900d: 401 type species *Stenotarsoides quadrimaculatus* Csiki, 1900
Systaecha Gorham, 1890b: 132 type species *Systaecha cyanoptera* Gorham, 1890
albertisii Gorham, 1885b: 527 **A**: JA **ORR**
aokii Chûjô, 1942: 79 **A**: TAI
bicolor Gerstaecker, 1858: 343 **A**: UP **ORR**
alfierii Pic, 1922b: 9 (*Stenotarsoides*)
indianus Gorham, 1874d: 446
medianus Pic, 1922b: 9 (*Stenotarsoides*)
chrysomelinus Gorham, 1887b: 644 **A**: JA
chujoi Strohecker, 1957: 262 **A**: TAI
internexus Gorham, 1887b: 645 **A**: JA
kurosai Nakane, 1981a: 55 **A**: JA
maculosus Fairmaire, 1888b: 158 **A**: CH
nakanoshimensis Nakane, 1959a: 60 **A**: JA
oshimanus Nakane, 1981a: 55 **A**: JA
peguensis Gorham, 1896: 297 **A**: UP **ORR**
ryukyuensis M. Chûjô & Kiuchi, 1974: 5 **A**: JA
secticollis Strohecker, 1974: 136 **A**: UP
seminalis Arrow, 1925: 384 **A**: UP
yoshionis M. Chûjô, 1938: 402 **A**: TAI

family COCCINELLIDAE Latreille, 1807
subfamily **Sticholotidinae** J. Weise, 1901
tribe Limnichopharini Miyatake, 1994

genus *Limnichopharus* Miyatake, 1994: 269 type species *Limnichopharus hainanensis* Miyatake, 1994
hainanensis Miyatake, 1994: 271 **A**: HAI

tribe Microweiseini Leng, 1920

genus *Paracoelopterus* Normand, 1936c: 152 type species *Paracoelopterus rufinus* Normand, 1936
Diloponis Pope, 1962: 633 type species *Diloponis inconspicuus* Pope, 1962
berytensis J. Weise, 1884: 166 (*Pharus*) **E**: GR **N**: MO TU **A**: IN IS LE
fuerschi Yazdani & Ahmadi, 1992: 13 (*Diloponis*)
rufinus Normand, 1936c: 153

tribe Plotinini Miyatake, 1994

genus *Ballida* Mulsant, 1850b: 1042 type species *Ballida brahamae* Mulsant, 1850
Eoneda Iablokoff-Khnzorian, 1985: 142 type species *Eoneda sumatrensis* Iablokoff-Khnzorian, 1985
Palaeoeneis Crotch, 1874a: 205 type species *Palaeoeneis aurantiaca* Crotch, 1874
brahamae Mulsant, 1850b: 1042 **A**: GUA **ORR**
aurantiaca Crotch, 1874a: 205 (*Palaeoeneis*)
sumatrensis Iablokoff-Khnzorian, 1985: 142 (*Eoneda*)

genus *Plotina* Lewis, 1896a: 35 type species *Plotina versicolor* Lewis, 1896
muelleri Mader, 1955a: 73 **A**: FUJ GUI JIX **ORR**
versicolor Lewis, 1896a: 35 **A**: JA FUJ

genus *Sphaeroplotina* Miyatake, 1969: 202 type species *Sphaeroplotina hainanensis* Miyatake, 1969
hainanensis Miyatake, 1969: 204 **A**: HAI

tribe Serangiini Blackwelder, 1945

genus *Delphastus* Casey, 1899: 111 type species *Oeneis pussilus* LeConte, 1852
Lioscymnus Champion, 1913: 125 type species *Lioscymnus diversipes* Champion, 1913
catalinae Horn, 1895: 83 (*Cryptognatha*) **N**: CI **A**: FUJ IS YUN **AUR**i **NAR NTR**
occidentalis Juarez & Zaragoza, 1990: 298

genus *Serangiella* E. A. Chapin, 1940: 271 type species *Serangiella flavescens* E. A. Chapin, 1940
Microserangium Miyatake, 1961a: 127 type species *Microserangium shikokense* Miyatake, 1961
hainanense Miyatake, 1961b: 144 (*Microserangium*) **A**: HAI
okinawense Miyatake, 1961a: 131 (*Microserangium*) **A**: JA (Ryukyus) TAI
shikokense Miyatake, 1961a: 130 (*Microserangium*) **A**: JA (Shikoku)

genus *Serangium* Blackburn, 1889a: 187 type species *Serangium mysticum* Blackburn, 1889
Catana E. A. Chapin, 1940: 266 type species *Catana clauseni* E. A. Chapin, 1940
Catanella Miyatake, 1961b: 146 type species *Catanella formosana* Miyatake, 1961
Semichnoodes J. Weise, 1892b: 15 type species *Semichnoodes kunowi* J. Weise, 1892
buettikeri Fürsch, 2000: 233 **A**: SA
chapini Kapur, 1956a: 189 (*Catana*) **A**: UP
cincta J. Weise, 1898: 121 (*Semichnoodes*) **A**: YE
clauseni E. A. Chapin, 1940: 267 (*Catana*) **A**: HAI **NTR**i **ORR**
drepanicum Xiao, 1993: 368 **A**: HUB
formosanum Miyatake, 1961b: 138 (*Catanella*) **A**: TAI
japonicum E. A. Chapin, 1940: 269 **A**: FUJ GUA GUI GUX HEN HUB JA JIA SC SCH SHA SHG TAI YUN ZHE
suffusum Miyatake, 1961b: 143
tokaranum Nakane & M. Araki, 1959: 48
lygaeum Iablokoff-Khnzorian, 1972a: 118 **A**: FE
montazerii Fürsch, 1995: 20 **E**: FR GG **A**: IN IS PA SY UP
parcesetosum Sicard, 1929b: 184 **A**: UP **NAR ORR**
punctum Miyatake, 1963b: 13 **A**: JA
yezoense Miyatake, 1963b: 14
ruficolle Kamiya, 1965b: 114 **A**: JA (Ryukyus)
ryukyuense Kamiya, 1965b: 115 **A**: JA (Ryukyus)
yasumatsui Sasaji, 1967a: 9 (*Catana*) **A**: TAI

tribe Shirozuellini Sasaji, 1967

genus *Ghanius* Ahmad, 1973: 449 type species *Ghanius kiarachiensis* Ahmad, 1973
karachiensis Ahmad, 1973: 451 **A**: PA
schawalleri Canepari, 1997: 6 **A**: NP UP

genus *Medamatento* Sasaji, 1989a: 21 type species *Shirozuella ocularis* Sasaji, 1988
ocularis Sasaji, 1988a: 14 (*Shirozuella*) **A**: JA (Ryukyus)
secunda Sasaji, 1989a: 22 **A**: JA (Ryukyus)

genus *Sasajiella* Miyatake, 1994: 276 type species *Sasajiella amamiana* Miyatake, 1994
amamiana Miyatake, 1994: 277 **A**: JA (Ryukyus)

genus *Shirozuella* Sasaji, 1967a: 24 type species *Shirozuella mirabilis* Sasaji, 1867
alishanensis G.-Y. Yu & X.-F. Pang, 1992b: 39 **A**: TAI
appendiculata G.-Y. Yu & X.-F. Pang, 1992b: 38 **A**: TAI
bimaculata G.-Y. Yu, 1999b: 63 **A**: HEN
mirabilis Sasaji, 1967a: 25 **A**: TAI
nibagou G.-Y. Yu, 2000b: 188 **A**: SCH
parenthesis G.-Y. Yu, 1999b: 62 **A**: HEN
quadrimacularis G.-Y. Yu, 2000b: 187 **A**: YUN

tribe Sticholotidini J. Weise, 1901

genus *Coelopterus* Mulsant & Rey, 1852b: 224 [= 1853: 89] type species *Coelopterus salinus* Mulsant & Rey, 1852
armeniacus J. Weise, 1894: 144 **E**: AR **A**: IS
desertorum Dobrzhanskiy, 1928: 122 **A**: KZ
salinus Mulsant & Rey, 1852b: 224 [= 1853: 89] **E**: FR IT PT SP **N**: AG MO TU **A**: IN **AFR**

genus *Jauravia* Motschlusky, 1858c: 117 type species *Jauravia pallidula* Motschulsky, 1858
Clanis Mulsant, 1850b: 999 [HN] type species *Coccinella pubescens* Fabricius, 1798
Paraclitostethus Ohta, 1929a: 1 type species *Paraclitostethus ovatus* Ohta, 1929 (= *Jauravia limbata* Motschulsky, 1858)
hanifi Afroze, 1998: 131 **A**: UP
limbata Motschulsky, 1858c: 118 **A**: HAI NP TAI YUN **ORR**
ovata Ohta, 1929a: 2 (*Paraclitostethus*)
quadrinotata Kapur, 1946: 85 **A**: BT NP SD TAI YUN **ORR**

genus *Pharoscymnus* Bedel, 1906b: 93 [RN] species *Coccinella sexguttata* Gyllenhal, 1808
Gymnopharus Sicard, 1909: 157 type species *Pharoscymnus eburifer* Sicard, 1909
Pharus Mulsant, 1850b: 948 [HN] type species *Coccinella sexguttata* Gyllenhal, 1808
angohranensis Duverger, 1983: 75 **A**: IN
arabicus Fürsch, 1979: 238 **A**: AE IN SA
armenus Iablokoff-Khnzorian, 1970a: 68 **E**: AR
assamensis Kapur, 1963: 71 **A**: YUN **ORR**
auricomus Savoyskaya, 1971: 55 **A**: KI KZ
balkhashensis Savoyskaya, 1964: 360 **A**: KZ
brunneosignatus Mader, 1949: 223 **A**: MG XIN
c-luteum Sicard, 1907d: 417 (*Pharus*) **A**: SA YE
decemplagiatus Wollaston, 1857a: 137 (*Scymnus*) **N**: CI MR
grancanariensis Uyttenboogaart, 1934: 163
fleischeri J. Weise, 1883a: 67 (*Pharus*) **E**: GR **A**: IN TR
discopterus Smirnoff, 1956: 9
moestus J. Weise, 1885a: 64 (*Pharus*)
flexibilis Mulsant, 1853: 271 (*Scymnus*) **A**: AF IN KA PA SD UP **ORR**
kashmirensis Kapur, 1956b: 267
heptapotamicus Dobzhanskiy, 1927c: 241 **A**: KZ
horni J. Weise, 1901b: 434 (*Pharus*) **A**: UP **ORR**
koenigi Iablokoff-Khnzorian, 1970d: 254 **E**: AR **A**: TR
nubicus Mader, 1949: 19 **N**: EG
numidicus Pic, 1900k: 91 (*Pharus*) **N**: AG EG LB TU **A**: IS JO SA **AFR**
eichleri Mader, 1936a: 26

junctus Smirnoff, 1956: 9
unicolor Smirnoff, 1953: 148
ovoideus Sicard, 1929a: 61 **N**: AG MO TU **A**: IN JO SY **AFR**
deserti Smirnoff, 1956: 15
guttatus Smirnoff, 1956: 16
hamifer Smirnoff, 1953: 148
pharoides Marsuel, 1868: 215 (*Scymnus*) **N**: EG LB **A**: IN IS SA SY TR **AFR**
villosulus J. Weise, 1883a: 68 (*Pharus*)
pilosus Savoyskaya, 1971a: 61 **A**: KZ UZ
priesneri Mader, 1936a: 26 [DA]
setulosus Chevrolat, 1861: 269 (*Pharus*) **E**: SP **N**: AG EG LB MO TU **A**: AE IN IS JO SA
anchorago Fairmaire, 1884d: lxx (*Pharus*)
bardus Mulsant & Godart, 1872b: 222 (*Pharus*)
bifasciatus Pic, 1926a: 2 (*Pharus*)
brevemaculatus Pic, 1925f: 10 (*Pharus*)
brunneonotatus Pic, 1926a: 2 (*Pharus*)
isidis Kirsch, 1871: 395 (*Scymnus*)
letourneuxi Pic, 1926a: 2 (*Pharus*)
varius Kirch, 1871: 396 (*Scymnus*)
simmondsi Ahmad, 1970: 233 **A**: PA
smirnovi Dobzhanskiy, 1927c: 240 **E**: AB/GG (Eldar Steppe)
taoi Sasaji, 1967a: 21 **A**: FUJ GUA GUX TAI
tristiculus Sicard, 1907d: 418 (*Pharus*) **A**: YE **AFR**
tsharinensis Savoyskaya, 1965: 32 **A**: KZ

genus *Pharoscymnus*, incertae sedis
sexguttatus semijunctus Pic, 1926a: 2 (*Pharus*) **N**: AG

genus *Sticholotis* Crotch, 1874a: 200 type species *Sticholotis substriata* Crotch, 1874
Gymnoscymnus T. Blackburn, 1892: 241 type species *Gymnoscymnus quadrimaculatus* T. Blackburn, 1892
Nesolotis Miyatake, 1966a: 47 type species Nesolotis impunctata Miyatake, 1966
Paranesolotis Hoang, 1982a: 104 type species *Paranesolotis tamdaoensis* Hoang, 1982
amabilis Kamiya, 1965b: 111 (*Stictobura*) **A**: JA (Ryukyus)
amator Kapur, 1958: 334 **A**: NP
azumai Sasaji, 1967b: 177 (*Nesolotis*) **A**: JA (Ryukyus)
besucheti Canepari, 1986: 23 **A**: SD
bilineata J. Weise, 1910a: 50 **A**: FUJ SD **ORR**
decempunctata Sicard, 1910: 387 **A**: NP **ORR**
duodecimmaculata Canepari, 1997: 6 **A**: NP
duodecimpunctata J. Weise, 1910a: 50 **A**: NP SD
formosana J. Weise, 1923: 185 **A**: GUA HAI TAI
hilleri J. Weise, 1885b: 238 **A**: JA
hirashimai Sasaji, 1967a: 14 **A**: GUI TAI
impunctata Miyatake, 1966a: 49 (*Nesolotis*) **A**: JA (Ryukyus)
klapperichi Mader, 1955a: 74 **A**: FUJ
kumatai Miyatake, 1985a: 6 **A**: NP
loebli Canepari, 1986: 22
linguiformis G.-Y. Yu, 1995a: 124 **A**: GUA TAI
maculata Korschefsky, 1944: 51 **A**: TAI
madagassa J. Weise, 1909: 124 **A**: UP **AFR**
marginalis Kapur, 1956b: 270 **A**: KA **ORR**
morimotoi Kamiya, 1965b: 110 **A**: JA (Ryukyus) TAI
nepalensis Miyatake, 1985a: 4 **A**: NP
petila G. Yu, H. Pang & X.-F. Pang, 1993: 467 **A**: GUA
pictipennis Lewis, 1896a: 32 **A**: JA (Kyushu)
punctata Crotch, 1874a: 201 **A**: FUJ GUI HUB JA **ORR**

rufosignata J. Weise, 1885b: 239
punctifrons Miyatake, 1966a: 50 (*Nesolotis*) **A**: JA (Ryukyus)
ruficeps J. Weise, 1902b: 511 **A**: ANH GUA HEN **AUR ORR**
shirozui Sasaji, 1967a: 18 (*Nesolotis*) **A**: TAI
substriata Crotch, 1874a: 200 **A**: JA
taiwanensis Miyatake, 1982: 26 **A**: TAI
tsunekii Sasaji, 1967a: 17 (*Nesolotis*) **A**: TAI
undecimpunctata Miyatake, 1982: 23 **A**: TAI

genus *Synonychimorpha* Miyatake, l994: 246 type species *Synonychimorpha sexpunctata* Miyatake, 1994
chittagongi Vazirani, 1983: 301 (*Sticholotis*) **A**: NP **ORR**
rufolimbata Canepari, 1997: 6 (*Sticholotis*)
sexpunctata Miyatake, 1994: 247 **A**: HAI

tribe Sukunahikonini Kamiya, 1960

genus *Hikonasukuna* Sasaji, 1967a: 4 type species *Hikonasukuna monticola* Sasaji, 1967
monticola Sasaji, 1967a: 5 **A**: TAI

genus *Scymnomorphus* J. Weise, 1897: 303 type species *Scymnomorphus rotundatus* J. Weise, 1897
Scotoscymnus J. Weise, 1901c: 458 [RN] type species *Scymnomorphus rotundatus* J. Weise, 1897
Sukunahikona Kamiya, 1960: 22 type species *Sukunahikona japonica* Kamiya, 1960 (= *Alexia japonica* Reitter, 1889)
bicolor Kamiya, 1965b: 113 (*Sukunahikona*) **A**: JA (Ryukyus) TAI
japonicus Reitter, 1889m: 277 (*Alexia*) **A**: JA
japonicus Kamiya, 1960: 24 (*Sukunahikona*)

subfamily **Coccidulinae** Mulsant, 1846

tribe Coccidulini Mulsant, 1846

genus *Coccidula* Kugelann, 1798: 421 type species *Chrysomela scutellata* Herbst, 1783
Cacidula Dejean, 1821: 132 type species *Chrysomela pectoralis* Fabricius, 1792 (= *Dermestes rufus* Herbst, 1783)
Cacidula Curtis, 1826: 144 [HN] type species *Chrysomela scutellata* Herbst, 1783
lithophiloides Reitter, 1890b: 176 **E**: AR **A**: IN
reitteri Dodge, 1938: 127 [RN] **A**: MG ES
suturalis Reitter, 1897c: 127 [HN]
rufa Herbst, 1783: 22 (*Dermestes*) **E**: AL AN AU BH BU BY CR CZ DE EN FI FR GB GE GR HU IR ITLA LI LT LU NL NR NT PL RO SK SP SV SZ TR UK YU **N**: MO **A**: AF ES FE IN KZ MG TR UZ WS
conferta Reitter, 1890b: 176
nigropunctata Reitter, 1900f: 220
pectoralis Fabricius, 1792a: 328 (*Chrysomela*)
plagiata Gerhard, 1910: 556
rosea Marsham, 1802: 123 (*Silpha*)
unicolor Reitter, 1890b: 176
scutellata Herbst, 1783: 58 (*Chrysomela*) **E**: AN AU BH BU BY CR CT DE FI FR GB GE HU IT LI LU NL NR PL RO SK SL ST SV SZ UK YU **N**: MO **A**: KZ WS
aethiops Krauss, 1902b: 92
arquata J. Weise, 1879c: 131
bipunctata Gmelin, 1790: 1630 (*Nitidula*)
melanophthalma Gmelin, 1790: 1627 (*Silpha*)
quinquepunctata Fabricius, 1787a: 52 (*Nitidula*)
subrufa J. Weise, 1879c: 131

genus *Lindorus* Casey, 1899: 162 type species *Scymnus lophanthae* Blaisdell, 1892
Nothorhyzobius Brèthes, 1925a: 173 type species *Nothorhizobius ruficollis* Brèthes, 1925 (= *Scymnus lophantae* Blaisdell, 1892)
forestieri Mulsant, 1853: 286 (*Platyomus*) **E**: FR GR **AUR NAR**
circularis Sharp, 1889: 365 (*Scymnus*)

lophantae Blaisdell, 1892: 51 (*Scymnus*) **E**i: AL CR FR GR IT SP YU UK **N**i: AG CI MO MR TU **A**: IN TR UP **AFR**i **AUR NAR**i **NTR**i
coeruleipennis Sicard, 1909: 139 (*Pullus*)
ruficollis Brèthes, 1925a: 173 (*Nothorhyzobius*)
toowoombae T. Blackburn, 1892: 254 (*Rhizobius*)
unquicularis J. Weise, 1922: 39 (*Rhizobius*)

genus *Rhyzobius* Stephens, 1829: 239 type species *Nitidula litura* Fabricius, 1787
Axius Mulsant, 1850b: 1002 [HN] type species *Rhizobius burmeisteri* Mulsant, 1850
Nundina Dejean, 1837: 462 type species *Nitidula litura* Fabricius, 1887
Rhizobiellus Oke, 1951: 21 [RN]
Rhizobius Agassiz, 1846: 325 [RN]
bipartitus Fuente, 1918: 44 (*Rhizobius*) **E**: AN FR SP
bassus Normand, 1938a: 132 (*Rhizobius*)
chrysomeloides Herbst, 1792: 180 (*Strongylus*) **E**: AN AU CZ DE FR GE HU LU NL PL PT SK SP SZ UK **N**: AG EG LB MO **A**: TR
lineatellus Mulsant, 1846: 263 (*Rhizobius*)
subdepressus Seidlitz, 1872: 193 (*Rhizobius*)
litura Fabricius, 1787a: 52 (*Nitidula*) **E**: AL AU AZ BH BU CR CZ DE FR GE GR IR IT LU NL PL PT RO SK SP SV SZ TR YU **N**: AG CI EG LB MO MR TU **A**: TR
aurora Creutzer, 1796b: 5 (*Coccinella*)
coadunatus Marsham, 1802: 76 (*Dermestes*)
discimacula Mulsant, 1846: 266 (*Rhizobius*)
fasciatus Fabricius, 1798: 74 (*Nitidula*)
hypomelanus Marsham, 1802: 77 (*Dermestes*)
lividus A. G. Olivier, 1791: 161 (*Anthribus*)
maura O'Mahony, 1927: 208
marinus Heyden, 1882a: 44 (*Rhizobius*) [HN]
nigriventris C. G. Thomson, 1866: 331 (*Rhizobius*)
pallidulus Mulsant, 1850b: 1006 (*Rhizobius*)
pallidus Marsham, 1802: 79 (*Dermestes*)
testaceus Fabricius, 1794: 446 (*Nitidula*)

genus *Cryptolaemus* Mulsant, 1853: 268 type species *Cryptolaemus montrouzieri* Mulsant, 1853
montrouzieri montrouzieri Mulsant, 1853: 140 **E**: AE IT FR GR SP **N**: EG **A**: FUJ GUA IN HAI HKG JA SCH TAI **AFR AUR NAR NTR**

tribe Tetrabrachini Kapur, 1948

genus *Tetrabrachys* Kapur, 1948c: 320 [RN] type species *Tritoma connata* Creutzer, 1796 (= *Lithophilus ruficollis* J. A. Frölich, 1799)
Lithophilus J. A. Frölich, 1799: 11 [HN] type species *Tritoma connata* Creutzer, 1796 (= *Lithophilus ruficollis* J. A. Frölich, 1799)
arabicus Kapur, 1948c: 329 **A**: SA YE
araxis Reitter, 1897e: 226 (*Lithophilus*) **E**: AR **A**: TR
anatolicus Pic, 1901l: 81 (*Lithophilus*)
triplagiatus Reitter, 1897e: 227 (*Lithophilus*)
arnoldii Iablokoff-Khnzorian, 1969: 248 (*Lithophilus*) **E**: ST
berlandi Peyerimhoff, 1942: 216 **N**: MO **AFR**
bipustulatus armeniacus Iablokoff-Khnzorian, 1974b: 177 (*Lithophilus*) **E**: AR IN
bipustulatus bipustulatus Barovskij, 1909b: 258 **A**: AF IN TM
caucasicus J. Weise, 1878: 97 (*Lithophilus*) **E**: GG **A**: TR
coloratus Fürsch, 1960b: 64 **A**: IN
connatus Creutzer, 1796b: 18 (*Tritoma*) **E**: AU BH BU CR CZ HU RO SK UK YU
polkorabi Roubal, 1932c: 169 (*Lithophilus*)
ruficollis Frölich, 1799: 13 (*Lithophilus*)
rufolimbatus Roubal, 1932c: 169 (*Lithophilus*)
cordatus Rosenhauer, 1856: 213 (*Lithophilus*) **E**: PT SP **N**: AG MO

cordicollis Guérin-Méneville, 1844a: 318 (*Lithophilus*) **N**: AG EG LB MO TU
creticus Reitter, 1889j: 255 (*Lithophilus*) **E**: GR **A**: TR
cribratellus Fairmaire, 1876b: 50 (*Lithophilus*) **N**: AG EG SI SY TU
cruciatus J. R. Sahlberg, 1913a: 85 (*Lithophilus*)
henoni Pic, 1901l: 82 (*Lithophilus*)
latus Reitter, 1884e: 251 (*Lithophilus*)
limbatus Pic, 1903r: 170 (*Lithophilus*)
peyerimhoffi Pic, 1903c: 89 (*Litophilus* [sic])
subseriatus Reitter, 1883d: 63 (*Lithophilus*)
unicolor Reitter, 1883d: 64 (*Lithophilus*)
deserticola Wollaston, 1864: 431 (*Lithophilus*) **N**: CI MO
atlanticus Escalera, 1914: 128 (*Lithophilus*)
kochi Schatzmayr, 1939: 225 (*Lithophilus*)
festae Dodero, 1925: 9 (*Lithophilus*) **N**: LB
gandhara Kapur, 1948c: 325 **A**: PA (North-West Frontier Province)
gibbosus J. R. Sahlberg, 1913a: 87 (*Lithophilus*) **A**: TR
glazunovi Barovskij, 1909b: 259 **A**: UZ
graecus Reitter, 1880d: 94 (*Lithophilus*) **E**: GR TR **A**: IS TR
humeralis J. R. Sahlberg, 1913d: 267 (*Lithophilus*)
turcicus Roubal, 1932c: 169 (*Lithophilus*)
grandis Pic, 1901l: 82 (*Lithophilus*) **A**: IS SY
haemorrhous Solsky, 1882: 83 (*Lithophilus*) **A**: AF KZ TD UZ
cryptophagoides Kapur, 1948c: 326
hissariensis Barovskij, 1910: 40 (*Lithophilus*) **A**: TD
hoberlandti Günther, 1958: 22 **A**: TR
insculptus J. R. Sahlberg, 1913a: 86 (*Lithophilus*) **A**: TR
jacobsoni Barovskij, 1909a: 96 (*Lithophilus*) **A**: UZ
kalawrytus Reitter, 1883d: 63 (*Litophilus*) **E**: GR TR
kalavrytus Reitter, 1884c: 61 (*Lithophilus*)
kaszabi Iablokoff-Khnzorian, 1974b: 178 (*Lithophilus*) **A**: TR (Monan Göl)
khnzoriani Hernando & Ribes, 1990: 81 (*Lithophilus*) **E**: SP
klapperichi Iablokoff-Khnzorian, 1974b: 206 (*Lithophilus*) **A**: AF
kozlovi Barovskij, 1909b: 255 (*Lithophilus*) **A**: MG NMO
kiritshenkoi Bielawski, 1961: 275 (*Lithophilus*)
krasnowi Semenov, 1891: 381 (*Lithophilus*) **A**: KZ
latus Reitter, 1884e: 251 (*Lithophilus*) **A**: IS LE SY
lindemannae Fürsch, 1960b: 64 **A**: AF PA **ORR**
major Crotch, 1874a: 302 (*Lithophilus*) **A**: IN (Kurdistan) TR
marginatus Reitter, 1884e: 252 (*Lithophilus*) **E**: GR **A**: CY IS LE
silvestrii Schatzmayr, 1939: 224 (*Lithophilus*)
minimus Barovskij, 1910: 39 (*Lithophilus*) **A**: IN
minutus Pic, 1903r: 170 (*Lithophilus*) **N**: TU **A**: SA
naviauxi Duverger, 1983: 83 (*Lithophilus*) **A**: IN
nigripennis Reitter, 1891e: 222 (*Lithophilus*) **A**: UZ "Turkmenien"
opacus Korotyaev, 1989: 97 (*Lithophilus*) **A**: KZ
ovipennis Crotch, 1874a: 302 **N**: EG **A**: IS LE SI SY TR
pallidus Crotch, 1874a: 302 (*Lithophilus*) **A**: IN
pellucidus Barovskij, 1909b: 256 (*Lithophilus*) **A**: IN
rifensis Iablokoff-Khnzorian, 1974b: 210 (*Lithophilus*) **N**: MO
robusta Kapur, 1948c: 324 **A**: PA UP
roddi Barovskij, 1909a: 98 (*Lithophilus*) **A**: WS
semenovi Barovskij, 1909a: 97 (*Lithophilus*) **A**: TM
smarti Iablokoff-Khnzorian, 1974b: 208 (*Lithophilus*) **A**: IN
subobscurus Pic, 1901l: 82 (*Lithophilus*) **A**: IS LE
sumakovi Barovskij, 1909b: 260 (*Lithophilus*) **A**: KZ
tauricus Semenov, 1902: 254 (*Lithophilus*) **E**: UK

tenebrosus J. Weise, 1910a: 53 (*Lithophilus*) **A**: SA YE **AFR**
diversicollis Pic, 1922a: 9 (*Lithophilus*)
tenuepunctatus Iablokoff-Khnzorian, 1974b: 180 (*Lithophilus*) **A**: TR (Brussa)
tinerfensis Hodgson Torres, 1987: 117 **N**: CI (Tenerife)
trimaculatus Reitter, 1883d: 63 (*Lithophilus*) **A**: SY
subobliteratus J. R. Sahlberg, 1913d: 267 (*Lithophilus*)
villosus Faldermann, 1835b: 457 (*Cacidula*) **A**: BEI HEB SHX
volkonskyi Peyerimhoff, 1943: 10 (*Lithophilus*) **N**: AG **AFR**
weisei Reitter, 1880d: 94 (*Lithophilus*) **E**: GG "Caucasus"
yemenensis Kapur, 1948c: 328 **A**: YE
zarudnyi Barovskij, 1909b: 257 (*Lithophilus*) **A**: IN (Khorassan)

genus *Tetrabrachys*, nomen dubium
osculatii Marietti, 1844: 72 (*Lithophilus*) **E**: TR

subfamily **Exoplectrinae** Crotch, 1874

genus *Sumnius* J. Weise, 1892a: 29 type species *Sumnius cardoni* J. Weise, 1892
babai Sasaji, 1994: 236 **A**: TAI
brunneus Jing, 1983: 403 **A**: SCH YUN
cardoni J. Weise, 1892a: 30 **A**: NP **ORR**
hoangi Canepari, 1997: 47 **A**: NP
nigrofuscus Jing, 1983: 404 **A**: FE YUN
petiolimaculatus Jing, 1983: 405 **A**: YUN
thanhhoaensis Hoang, 1980: 8 **A**: NP **ORR**
vestitus Mulsant, 1850b: 934 (*Aulis*) **A**: AF FUJ NP PA SCH UP YUN **ORR**
hoematica Gorham, 1894b: 211 (*Aulis*)
renardi J. Weise, 1892a: 29
yunnanus Mader, 1955b: 978 **A**: SHA TAI YUN

subfamily **Scymninae** Mulsant, 1846

tribe Aspidimerini Mulsant, 1850

genus *Aspidimerus* Mulsant, 1850b: 944 type species *Aspidimerus spencii* Mulsant, 1850
blandus Mader, 1933c: 130 (*Cryptogonus*) **A**: YUN **ORR**
decemmaculatus X.-F. Pang & Mao, 1979a: 56 **A**: YUN
esakii Sasaji, 1968: 16 **A**: ANH GUX TAI
matsumurai Sasaji, 1968: 17 **A**: GUX HEN HUB SCH TAI YUN ZHE
mouhotii Crotch, 1874a: 202 **A**: YUN **ORR**
nigritus X.-F. Pang & Mao, 1979a: 61 (*Cryptogonus*) **A**: GUA YUN **ORR**
dongpaoensis Hoang, 1982a: 165
ruficrus Gorham, 1895b: 690 **A**: SCH YUN **ORR**
sexmaculatus X.-F. Pang & Mao, 1979a: 54

genus *Cryptogonus* Mulsant, 1850b: 945 type species *Coccinella orbicula* Gyllenhal, 1808
angusticarinatus Sasaji, 1968: 13 **A**: GUA TAI
ariasi Mulsant, 1853: 265 (*Aspidimerus*) **A**: SD UP **ORR**
bimaculatus Kapur, 1948a: 100 **A**: BT NP XIZ YUN **ORR**
brachylobius H. Pang, 1998: 188 **A**: GUA
complexus Kapur, 1948a: 110 **A**: BT GAN GUA GUX SCH SD TAI YUN **ORR**
forficulae C. Cao & Xiao, 1984: 120 **A**: YUN
fractemaculatus H. Pang, 1998: 189 **A**: GUA
guangdongiensis X.-F. Pang & Mao, 1979a: 67 **A**: GUA GUX HAI JIX
hainanensis X.-F. Pang & Mao, 1979a: 65 **A**: HAI
himalayensis Kapur, 1948a: 108 **A**: BT GUA SD YUN **ORR**
hingstoni Kapur, 1948a: 103 **A**: SD

horishanus Ohta, 1929a: 12 (*Scymnus*) **A**: FUJ GUA HUB JA JIX SCH TAI ZHE **ORR**
japonensis Nakane & M. Araki, 1958: 291 (*Pseudaspidimerus*)
kurosawai Sasaji, 1969: 11 **A**: TAI
lobulus Xiao, 1993: 374 **A**: SCH
loebli Canepari, 1986: 27 **A**: UP
lijiangiensis X.-F. Pang & Mao, 1979a: 66 **A**: GAN HUB SHA YUN
linguilatus H. Pang, 1998: 190 **A**: FUJ
nepalensis bhutanensis Bielawski, 1979: 109 **A**: BT
nepalensis nepalensis Bielawski, 1972: 294 **A**: NP XIZ YUN
octoguttatus Mader, 1933c: 129 **A**: SCH YUN
ohtai Sasaji, 1968: 8 **A**: TAI
orbiculus Gyllenhal, 1808b: 205 (*Coccinella*) **A**: ANH FUJ GUA GAN GUI GUX HAI HEB HKG HUB HUN JA JIX NC NP SC SCH SHA SHN TAI UP YUN ZHE **AUR ORR**
apicalis J. Weise, 1901b: 425
centroguttatus Boheman, 1859: 206
futahoshii Ohta, 1929a: 9 (*Diomus*)
japonicus J. Weise, 1901b: 425
klapperichi Mader, 1955a: 73 (*Scymnus*)
lunatus Kapur, 1948a: 93
lunigerus Thunberg, 1808: 207 (*Coccinella*)
malasiae Crotch, 1874a: 203
mitsuhashii Takizawa, 1917: 222 (*Scymnus*)
nigripennis J. Weise, 1895a: 137 (*Aspidimerus*)
occulatus [sic!] Motschulsky, 1858c: 118 (*Platynaspis*)
sellatus J. Weise, 1901b: 425
postmedialis Kapur, 1948a: 95 **A**: BT FUJ GUA HUB SCH SD TAI UP **ORR**
sushila Kapur, 1848a: 97
qianjiangensis Xiao, 1993: 374 **A**: SCH
quadriguttatus J. Weise, 1895f: 326 (*Aspidiphorus*) **A**: BT SD TAI UP YUN **ORR**
confluens Kapur, 1948a: 99
nigrescens Kapur, 1948a: 99
quinjiangensis Xiao, 1993: 374 **A**: SCH
robustus G.-Y. Yu, 1995a: 140 **A**: TAI
sagittiformis X.-F. Pang & Mao, 1979a: 63 **A**: YUN
schraiki Mader, 1933a: 80 **A**: HUB SCH YUN
urbani Mader, 1933a: 80
trifurcatus X.-F. Pang & Mao, 1979a: 62 **A**: YUN
trioblitus Gorham, 1895b: 690 **A**: NP SD UP YUN **ORR**
wuzhishanus X.-F. Pang & Mao, 1979a: 59 **A**: HAI HEN SCH SHA
xiushanensis Xiao, 1993: 373 **A**: SCH
yunnanensis C. Cao & Xiao, 1984: 121 **A**: YUN

genus *Pseudaspidimerus* Kapur, 1948a: 117 type species *Platynaspis circumflexa* Motschulsky, 1858 (= *Coccinella trinotata* Thunberg, 1781)
hoangi Kovář, **nom. nov**. [see New acts] **A**: HAI **ORR**
limbatus Hoang, 1982a: 178 [HN]

tribe Diomini R. D. Gordon, 2000

genus *Diomus* Mulsant, 1850b: 951 type species *Coccinella thoracica* Fabricius, 1801
anemicus Fürsch, 1960c: 300 **N**: EG
gillerforsi Fürsch, 1987a: 9 **N**: CI
brunsuturalis X.-F. Pang & R. D. Gordon, 1986: 192 **A**: BEI FUJ GUA HAI SCH SHA TAI ZHE **ORR**
brachysiphonicus X.-F. Pang & Huang, 1986: 59
lantauensis G.-Y. Yu, 2001b: 152 **A**: HKG
longilobus G.-Y. Yu, 1996b: 279 **A**: HEB

rubidus Motschulsky, 1837b: 418 (*Scymnus*) **E**: AR CR FR GR IT **N**: AG EG LB MO TU **A**: IN IQ IS LE SA SY YE
adenensis Kapur, 1959: 286 (*Nephus*)
cincticollis J. Weise, 1929: 33 (*Scymnus*)
fleischeri Duverger, 1998: 128 [RN]
inconspicuus Wollaston, 1867: 166 (*Scymnus*)
morio A. Fleischer, 1900a: 120
morio Fürsch, 1966b: 88 (*Scymnus*) [HN]
tumefactus G.-Y. Yu, 1996b: 280 **A**: GUA

tribe Hyperaspidini Mulsant, 1846

genus *Hyperaspis* Chevrolat, 1836: 435 type species *Coccinella reppensis* Herbst, 1783
subgenus *Hyperaspis* Chevrolat, 1836: 435 type species *Coccinella reppensis* Herbst, 1783
Cleothera Mulsant, 1850b: 541 type species *Cleothera buquetii* Mulsant, 1850
Oxynychus LeConte, 1850: 238 type species *Oxynychus moerens* LeConte, 1850
algirica Crotch, 1874a: 236 **E**: FR IT **N**: AG MO TU
amurensis amurensis J. Weise, 1887b: 187 **A**: FE HEN NC SC ZHE
amurensis babai Kamiya, 1963a: 82 **A**: JA
asiatica Lewis, 1896a: 33 **A**: ES FE HEB HEI JA JIA JIL LIA MG NC SC SCH SHA SHN ZHE
campestris Herbst, 1783: 48 (*Coccinella*) **E**: AL AU BE BH BU BY CT CZ FR GE GG GR HU IT LT LU NL NT PL RO SK ST SV SZ UK YU **N**: TU **A**: WS
lateralis Panzer, 1794b: 9 (*Coccinella*)
caucasica Crotch, 1874a: 37 [RN] **E**: AB GG **A**: TD
assimilis Zaslavskij, 1966: 60
sexpustulata Motschulsky, 1837b: 421 (*Coccinella*) [HN]
concolor Suffrian, 1843a: 93 (*Coccinella*) **E**: AU BE BU CZ FI FR GE HU IT PL RO SK SV SZ
inexpectata Günther, 1959: 257
dellabeffai Canepari, 1979b: 134 **E**: IT
desertorum J. Weise, 1885a: 58 **E**: ST **A**: KZ
duvergeri Fürsch, 1985: 233 **E**: CR FR HU IT PT SP **N**: AG MO
effusa J. Weise, 1885a: 58 **E**: KZ ST
erythrocephala Fabricius, 1787a: 61 (*Coccinella*) **E**: CT CZ DE GG HU PL RO SK ST UK **A**: ES HEI JIL KZ MG NC WS
bimacula Dobrzhanskiy, 1926a: 2
fabricii Mulsant, 1846: 182 [RN]
fausti J. Weise, 1885a: 60
guillardi Mulsant, 1853: 230
sexpustulata Mulsant, 1853: 229 (*Coccinella*)
femorata Motschulsky, 1837b: 421 (*Coccinella*) **E**: AB AR BU ST UK **A**: TM TR
collaris A. Fleischer, 1900a: 119
inaudax Mulsant, 1853: 231
guttulata Fairmaire, 1870: 404 **E**: SP **N**: AG MO TU **A**: MG TD
sexguttata C. N. F. Brisout de Barneville, 1866: 425
gyotokui Kamiya, 1963a: 83 **A**: FE HEB HEN JA NC SC SCH SHA
histeroides Faldermann, 1837: 408 (*Scymnus*) **E**: AB AR **A**: IN KZ
vicaria Zaslavskij, 1964: 153
hoffmannseggi Gravenhorst, 1809: 127 (*Coccinella*) **E**: FR IT SP
illecebrosa Mulsant, 1846: 181 **E**: PT SP
kamtchaticus Kuznecov & Ren, 1996: 426 **A**: FE
kansui Fürsch, 1981a: 15 **A**: TR
khnzoriani Fürsch, 1985: 249 **N**: EG
klapperichi Fürsch, 1985: 250 **A**: TR
leechi Miyatake, 1961c: 151 **A**: ANH ES FE FUJ HEB HEI HEN HUB JA JIA LIA MG NC SCH SHA SHG SC ZHE
marginaloides Canepari, 1997: 18 **A**: NP
marmottani Fairmaire, 1868: 501 (*Coccinella*) **N**: AG EG LB TU **A**: IN

albidiceps Walker, 1871: 19
camillae Bedel, 1885: 90
minois Fürsch, 1985: 241 **E**: BU GR
peezi Fürsch, 1976: 49 **E**: CR FR IT
persica Duverger, 1983: 79 **A**: IN
polita J. Weise, 1885a: 60 **N**: EG **A**: AF CY IS SY TD TM TR **AFR**
decemguttata A. Fleischer, 1900a: 119
pseudopustulata Mulsant, 1853: 232 **E**: AU BH CZ DE FR GB GE HU IT LU RO SK SL SP ST SV TR UK "Caucasus" **N**: AG **A**: TR WS KZ
pumila pumila Mulsant, 1850b: 655 **N**: LB **A**: SA YE **AFR**
quadrimaculata L. Redtenbacher, 1843a: 16 **E**: AL AU BU CZ HU GR RO SK **A**: LE TR SY
reppensis Herbst, 1783: 48 (*Coccinella*) **E**: AU BE BH BY CT CZ FI FR GB GE GR HU IR IT LT MD NT PL RO SK ST SV SZ UK **N**: TU **A**: WS KZ
bipustulata Thunberg, 1784a: 13 (*Chrysomela*)
galliae Duverger, 1989: 144 [RN]
marginella Quensel, 1790: 9 (*Dermestes*)
occidentalis Fürsch, 1967b: 261 [HN]
reppensis Thunberg, 1795: 111 (*Coccinella*) [HN]
subconcolor J. Weise, 1879c: 136
xanthocephala Quensel, 1790: 10 (*Dermestes*)
schatzmayri Canepari, 1985: 236 **E**: "Eastern Pyrenean"
sinensis Crotch, 1874a: 203 (*Cryptogonus*) **A**: ANH BEI ES FE FUJ GAN GUA GUI GUX HUB HEN JA JIA JIX NC SC SCH ZHE
japonica Crotch, 1874a: 203 (*Cryptogonus*)
kuwanai Ohta, 1921: 182 (*Scymnus*)
testaceicornis J. Weise, 1879a: 149
stigma A. G. Olivier, 1808: 1043 (*Coccinella*) [RN] **E**: FR GR IT PT SP **N**: MO
chevrolati Canepari, 1985: 235 [RN]
marginella Fabricius, 1801a: 378 (*Coccinella*) [HN]
syriaca J. Weise, 1885a: 57 **A**: IN IS LE SY TR
teinturieri Mulsant & Godart, 1869: 102 [= 1872a: 126] **N**: AG TU
terrea Zaslavskij, 1964: 152 **A**: KI
transversoguttata J. Weise, 1878: 94 **E**: AR **A**: IN KI KZ MG NMO TD TM UZ
alexandrae J. Weise, 1890c: 488 (*Oxynychus*)
turanica Iablokoff-Khnzorian, 1971: 181 **A**: UZ (Iskander-Darja)
uhligi Fürsch, 1985: 238 **E**: GR
vinciguerrae Capra, 1929: 241 **N**: LB **A**: SA YE **AFR**
vittifera Mulsant, 1850b: 680 **A**: KZ MG WS
vittata Gebler, 1844: 106 (*Coccinella*) [HN]
weisei Iablokoff-Khnzorian, 1971: 178 **A**: TR
wellschmiedi Fürsch, 1987c: 53 **A**: TR

genus *Hyperaspis*, nomina dubia
apicalis Dufour, 1843a: 116 (*Coccinella*) **E**: FR (Vallée de Ossau)
motschulskii Mulsant, 1850b: 689 [distribution unknown]
nigra Gmelin, 1790: 1662 (*Coccinella*) **E**: "Gallia"

tribe Scymnini Mulsant, 1846

genus *Acoccidula* Barovskij, 1931b: 130 type species *Acoccidula reichhardti* Barovskij, 1931
reichhardti Barovskij, 1931b: 130 **A**: TD

genus *Axinoscymnus* Kamiya, 1963b: 127 type species *Axinoscymnus beneficus* Kamiya, 1963
beneficus Kamiya, 1963b: 128 **A**: GUA JA (Ryukyus) HAI TAI
cardilobus Ren & X.-F. Pang, 1992: 20 **A**: GUA
glotticus Ren & X.-F. Pang, 1992: 22 **A**: GUI
navicularis Ren & X.-F. Pang, 1992: 21 **A**: GUI

nigripennis Kamiya, 1965b: 100 **A**: GUA GUX JA (Ryukyus) TAI
rai Kamiya, 1965b: 101 **A**: JA (Ryukyus)

genus *Clitostethus* J. Weise, 1885a: 65 type species *Coccinella arcuata* P. Rossi, 1794
Nephaspis Casey, 1899: 168 type species *Nephaspis gorhami* Casey, 1899
acutisiphonicus Peng, Ren & X.-F. Pang, 1998: 196 **A**: HAI
arcuatus P. Rossi, 1794: 88 (*Coccinella*) **E**: AL AR AU AZ BH CR CZ FR GB GE GG GR HU IT NL RO SK ST SZ UK YU **N**: CI MR **A**: IN IQ SY TR **AFR NAR**
abeillei J. Weise, 1884: 165 (*Scymnus*)
hausmanni J. Weise, 1885a: 74 (*Scymnus*) [HN]
heegeri Ganglbauer, 1899b: 972
nigrocinctus Ganglbauer, 1899b: 972
bawanglingensis Peng, Ren & X.-F. Pang, 1998: 197 **A**: HAI
brachylobus Peng, Ren & X.-F. Pang, 1998: 198 **A**: HAI
luteoniger Canepari, 1997: 8 (*Pseudoscymnus*) **A**: NP
nigrifrons G.-Y. Yu, 1997b: 715 **A**: SCH
sternalis X.-F. Pang & R. D. Gordon, 1986: 184 (*Pseudoscymnus*) **A**: GUA GUI HAI
wenbishanus G.-Y. Yu, 2000b: 158 **A**: YUN

genus *Horniolus* J. Weise, 1901b: 442 type species *Horniolus dispar* J. Weise, 1901
amamensis Miyatake, 1963a: 10 **A**: JA
fortunatus Lewis, 1896a: 38 (*Scymnus*) **A**: JA SHA
hisamatsui Miyatake, 1976: 34 **A**: HKG
kyushuensis Miyatake, 1963a: 9 **A**: JA
novempunctatus Miyatake, 1979: 105 **A**: TAI
okinawensis Chûjô & Miyatake, 1963: 11 **A**: JA

genus *Keiscymnus* Sasaji, 1971a: 95 type species *Keiscymnus tosaensis* Sasaji, 1971
securiformis G.-Y. Yu & X.-F. Pang, 1992a: 10 **A**: TAI
taiwanensis C.-T. Yang & R.-H. Wu, 1972: 118 **A**: FUJ GUI GUX HAI TAI
sternitus X.-F. Pang & R. D. Gordon, 1986: 170 (*Scymnus*)
tosaensis Sasaji, 1971a: 95 **A**: JA (Honshu, Shikoku)

genus *Leptoscymnus* Iablokoff-Khnzorian, 1978d: 115 type species *Leptoscymnus salsolae* Iablokoff-Khnzorian, 1978
salsolae Iablokoff-Khnzorian, 1978d: 116 **A**: KZ

genus *Nephus* Mulsant, 1846: 237 type species *Sphaeridium quadrimaculatum* Herbst, 1783
subgenus *Bipunctatus* Fürsch, 1987d: 66 type species *Scymnus bipunctatus* Kugelann, 1794
aguilari Gourreau, 1975: 21 **E**: FR
asiaticus Iablokoff-Khnzorian, 1970b: 120 **A**: TD
bicinctus Mulsant & Godart, 1872b: 223 (*Scymnus*) **E**: FR IT
bipunctatus Kugelann, 1794: 547 (*Scymnus*) **E**: BE BY CT CZ DE EN FI HU LA LT LU NT PL RO SK SV UK **A**: AF ES FE KZ WS
affinis Paykull, 1798: 153 (*Coccinella*) [HN]
bipustulatus Thunberg, 1795: 105 (*Scymnus*)
biverrucatus Panzer, 1795c: 149 (*Coccinella*)
bisignatus bisignatus Boheman, 1850: 76 (*Scymnus*) **E**: AZ CR CZ DE FI FR GB GR NR PT SK SV
ater Fürsch, 1958c: 101 (*Scymnus*) [HN]
etesiacus Fürsch, 1965: 183 (*Scymnus*)
bisignatus claudiae Fürsch, 1984a: 117 (*Scymnus*) **E**: AU GE HU SV
conjunctus Wollaston, 1870: 248 (*Scymnus*) **E**: IT PT SP **N**: AG CI EG MO TU **A**: IS SA TR **AFR**
confluens J. R. Sahlberg, 1913d: 266 (*Scymnus*) [HN]
includens Kirsch, 1871: 395 (*Scymnus*)
sannio J. Weise, 1883b: 137 (*Scymnus*)
deserticola Sicard, 1912a: 22 (*Diomus*) **N**: EG **A**: IS (Tiberias)

incinctus Mulsant, 1850b: 959 (*Scymnus*) **A**: BEI KZ MG NMO XIN
fedtshenkoi Savoyskaya, 1969c: 143 (*Scymnus*)
illiensis Savoyskaya, 1969c: 143 (*Scymnus*)
kaszabi Bielawski, 1975: 963 (*Scymnus*) **A**: MG
kiesenwetteri Mulsant, 1850b: 963 (*Scymnus*) **E**: FR IT MA **N**: AG EG LB MO TU **A**: IS UZ
corsicus J. Weise, 1879c: 153 (*Scymnus*)
siculus J. Weise, 1879c: 153 (*Scymnus*)
zigzag A. Costa, 1883: 87 [= 1883: 332] (*Scymnus*)
kompirasanus Kamiya, 1961a: 285 (*Scymnus*) **A**: JA (Honshu, Shikoku)
koreanus Fürsch, 1965: 189 (*Scymnus*) **A**: FE NC SC
kreticus Fürsch, 1965: 187 (*Scymnus*) **E**: GR (Kríti)
neixiangicus G.-Y. Yu, 1999b: 64 **A**: HEN
nigricans nigricans J. Weise, 1879c: 154 (*Scymnus*) **E**: AL CR FR GR IT TR **A**: TR
binaevulus Stenius, 1952: 154 (*Scymnus*)
pallidus Fürsch, 1980: 115 [HN]
nigricans niloticus Canepari, 1978: 168 **N**: EG
oshimaensis Sasaji, 1976: 53 **A**: JA (Honshu, Ryukyu)
peyerimhoffi Sicard, 1923: 79 (*Scymnus*) **E**: PT SP **N**: AG CI MO TU **A**: YE **AFR**
c-luteum Sicard, 1931: 231 (*Scymnus*)
incisus Har. Lindberg, 1951: 7 (*Scymnus*)
sinuatomaculatus J. R. Sahlberg, 1913d: 266 (*Scymnus*) **A**: AF TR
wittmeri Fürsch, 1979: 245 **A**: SA **AFR**
wrasei Pütz & Kusnetzov, 1995: 189 **A**: FE
subgenus *Geminosipho* Fürsch, 1987d: 66 type species *Scymnus bielawskii* Fürsch, 1965
arcuatus Kapur, 1959: 287 **A**: IN SA YE **AFR**
bielawskii Fürsch, 1965: 192 (*Scymnus*) **A**: AF
bilinearis G.-Y. Yu, 1997b: 718 **A**: GUA
dichosiphonicus Ren & X.-F. Pang, 1994b: 327 **A**: GUI SCH
dilepismoides X.-F. Pang & Pu, 1988: 240 **A**: FUJ GUA GUX HAI SCH TAI
ancyroides X.-F. Pang & Pu, 1988: 239
eurypodus G.-Y. Yu, 2001b: 155 **A**: HKG
fenestratus J. R. Sahlberg, 1913a: 83 (*Scymnus*) **N**: EG **A**: AE CY IN IS JO SA **AFR**
koltzei J. Weise, 1887b: 213 (*Scymnus*) **A**: BEI FE GAN GUI GUX HEB HEN LIA SCH
reunioni Fürsch, 1974: 275 (*Scymnus*) **E**: AB GG PT SP **A**: IS UZ **AFR**
sudanicus J. Weise, 1926: 228 **N**: MO **AFR**
triblulus Ren & X.-F. Pang, 1994b: 329 **A**: HUB HUN
wushanus G.-Y. Yu, 1997b: 717 **A**: SCH
ziguiensis G.-Y. Yu, 1997b: 716 **A**: HUB
subgenus *Nephus* Mulsant, 1846: 237 type species *Sphaeridium quadrimaculatum* Herbst, 1783
bicinctus Mulsant & Godart, 1870: 199 [= 1872b: 223] (*Scymnus*) **E**: IT **N**: AG TU
binotatus C. N. F. Brisout de Barneville, 1863: 122 (*Scymnus*) **E**: FR
boninensis Sasaji, 1982: 65 **A**: JA (Bonin) **ORR**
canepariii Fürsch & Uygun, 1980: 109 **A**: IN TR
crucifer A. Fleischer, 1900b: 235 **A**: IN SA TD UZ KZ SA **AFR**
intermedius Savoyskaya, 1968: 145 (*Scymnus*)
fijiensis Sicard, 1922: 353 (*Scymnus*) **A**: GUA JA **AUR ORR**
flavopictus Wollaston, 1854: 469 (*Scymnus*) **E**: AZ **N**: CI MR
fractus Wollaston, 1867: 163 (*Scymnus*)
maculosus Wollaston, 1864: 428 (*Scymnus*)
oblongior Wollaston, 1864: 427 (*Scymnus*)
ganglbaueri Kovář, **nom. nov.** [see New acts] **A**: IS JO LE SY
pictus Ganglbauer, 1899b: 970 [HN]
jacobsoni Barovskij, 1906: 8 (*Scymnus*) **E**: AU FI NT SK
deletomaculatus Fürsch, 1997: 11
rutaneni Fürsch, 1986: 107
kuznetsovi Fürsch, 2002: 125 **A**: FE

kaiensis Kamiya, 1961a: 286 (*Scymnus*) **A**: JA
limonii Donisthorpe, 1903: 287 (*Scymnus*) **E**: DE GB GE IC NL NR
ludyi J. Weise, 1879b: 153 (*Scymnus*) **E**: AR CR FR GR IT UK **N**: AG **A**: SY TR
italicus J. Weise, 1895d: 212 (*Scymnus*)
ponticus Iablokoff-Khnzorian, 1970b: 118
merkli Fürsch, 1994: 43 **A**: IS
parenthesis J. Weise, 1923: 187 (*Scymnus*) **A**: TAI
phosphorus Lewis, 1896a: 37 (*Scymnus*) **A**: JA SCH TAI
galloisi Sicard, 1907c: 211
quadrimaculatus Herbst, 1783: 30 (*Sphaeridium*) **E**: AL AU BE BY CR CZ DE FR GB GE GR HU IT MA NR PL SK SV YU **A**: KZ TAI TR
bisbipustulatus Panzer, 1793d: 5 (*Coccinella*) [HN]
colon Stephens, 1832a: 393 (*Scymnus*) [HN]
norvegicus Munster, 1928: 295 (*Scymnus*)
obliquus Comoli, 1837: 52 (*Scymnus*)
obliquus J. Weise, 1879c: 154 (*Scymnus*) [HN]
pulchellus Herbst, 1797: 345 (*Scymnus*)
quadrilunulatus Illiger, 1798: 416 (*Coccinella*)
redtenbacheri Mulsant, 1846: 240 (*Scymnus*) **E**: AU BE BH BU BY CR CT CZ DE EN FI FR GB GE GR HU IR IT LA LS LT NL NR NT PL SK ST SV UK YU **N**: AG **A**: ES FE IN TR WS
lividus Bold, 1872: 109 (*Scymnus*)
mulsanti G. R. Waterhouse, 1862: 133 (*Scymnus*)
nigrofemoratus Fürsch, 1958c: 101 (*Scymnus*)
ochraceipennis Fürsch, 1958c: 102 (*Scymnus*)
rufipennis Munster, 1930: 357 (*Scymnus*) [HN]
unicolor J. Weise, 1879c: 155 (*Scymnus*)
regularis Sicard, 1929b: 183 (*Scymnus*) **A**: PA UP **ORR**
roepkei Fluiter, 1939: 49 (*Scymnus*) **A**: GUA HAI JA (Bonin) **AUR ORR**
bipunctatus Fluiter, 1939: 51 (*Scymnus*) [HN]
niger Fluiter, 1939: 51 (*Scymnus*)
ryuguus Kamiya, 1961a: 289 (*Scymnus*) **A**: FUJ GUA GUI GUX HAI JA JIA SCH SHA UP
sauteri J. Weise, 1923: 187 **A**: TAI
schatzmayri Canepari & Tedeschi, 1977: 103 **E**: AN CR FR IT PT
tamaricis Capra, 1929: 242 **N**: EG LB MO **AFR**
stramineus Capra, 1929: 242
ulbrichi Fürsch, 1977: 652 **E**: GR IT PT SP
bucciarellii Canepari & Tedeschi, 1977: 101
yotsumon Kamiya, 1961a: 287 (*Scymnus*) **A**: JA NC SC
ziegleri Fürsch, 2001: 82 **E**: GR (Korfu)
subgenus *Sidis* Mulsant, 1850: 975 type species *Scymnus binaevatus* Mulsant, 1850
Pullosidis Fürsch, 1987d: 70 type species *Pullosidis ugandensis* Fürsch, 1987
besucheti Fürsch, 1977: 647 **A**: LE
caucasicus J. Weise, 1929: 33 (*Scymnus*) **E**: AR **A**: IN
plagiatus J. Weise, 1878: 95 (*Scymnus*)
depressiusculus Wollaston, 1867b: 167 (*Scymnus*) **N**: CI **AFR**
hiekei Fürsch, 1965: 201 (*Scymnus*) **E**: IT GR SP **N**: AG EG MO TU **A**: AE IN IQ IS LE SA TR
riyadhensis Fürsch, 1979: 247
jucundus Fürsch, 1977: 651 **A**: IN
kreissli Fürsch & Uygun, 1980: 115 **A**: TR
levaillanti Mulsant, 1850b: 964 (*Scymnus*) **E**: GR IT **N** EG **A**: AF FE IN IS JA JO LE PA SHA TAI **AFR ORR**
brevilineatus Fürsch, 1977: 654
montanus Hoang, 1982a: 133
patagiatus Lewis, 1896a: 39 (*Scymnus*)
seychellensis Sicard, 1912c: 363
voeltzkowi J. Weise, 1910e: 512
lindbergi Bielawski, 1959: 104 (*Scymnus*) **A**: AF PA

minqinensis Mao & Li, 1998: 169 **A**: GAN
pooti Fürsch, 1999: 75 **E**: PT
tagiapatus Kamiya, 1965b: 104 (*Scymnus*) **A**: GUA GUX JA (Ryukyus) PA TAI UP **ORR**
roonwali Kapur, 1967: 163

genus *Pseudoscymnus* E. A. Chapin, 1962: 50 type species *Scymnus hareja* J. Weise, 1879
amplus C.-T. Yang & R.-H. Wu, 1972: 122 **A**: GUX TAI
ancistroides X.-F. Pang & Huang, 1986: 65 **A**: FUJ
anmashanus C.-T. Yang, 1971: 88 **A**: TAI
bivalvis G.-Y. Yu, 1997b: 720 **A**: HUB
changi C.-T. Yang, 1971: 92 **A**: TAI
curvatus G.-Y. Yu, 1999a: 17 **A**: HEN
dapae Hoang, 1978a: 15 **A**: FUJ GUA HEN **ORR**
disselasmatus X.-F. Pang & Huang, 1986: 65 **A**: FUJ GUA HEN
fulvihumerus C.-T. Yang & R.-H. Wu, 1972: 123 **A**: TAI
funerarius Canepari, 1997: 7 **A**: NP
fuscus C.-T. Yang, 1971: 87 **A**: TAI
gibbosus G.-Y. Yu, 1997b: 720 **A**: SCH
hareja J. Weise, 1879a: 150 (*Scymnus*) **A**: FE GUI JA NC SC TAI
heijia G.-Y. Yu & Montgomery, 2000: 183 **A**: YUN
kurohime Miyatake, 1959: 136 (*Scymnus*) **A**: FUJ GUA GUI GUX HAI JA TAI YUN **ORR**
lamellatus C.-T. Yang & R.-H. Wu, 1972: 124 **A**: TAI
lancetapicalis X.-F. Pang & R. D. Gordon, 1986: 169 (*Nephus*) **A**: GUA
lewisi Kamiya, 1961a: 279 (*Clitostethus*) **A**: JA
nakanei M. Araki, 1964: 39 (*Clitostethus*) [HN]
luteoniger Canepari, 1997: 8 **A**: NP
montanus C.-T. Yang, 1971: 91 **A**: TAI
murreensis Ahmad, 1968: 377 **A**: PA **ORR**
nagasakiensis Kamiya, 1961a: 280 (*Clitostethus*) **A**: JA TAI
nakanei M. Araki, 1963: 253 (*Scymnus*) **A**: JA SC
nepalicus Miyatake, 1985a: 7 **A**: NP GUA
hamatus G. Yu, H. Pang & X.-F. Pang, 1993: 472
ocellatus G.-Y. Yu, 2000b: 184 **A**: SHA SCH
ocelliferus Canepari, 1997: 7 **A**: NP
ohtai C.-T. Yang & R.-H. Wu, 1972: 121 **A**: GUA HUB TAI
ohtsukai Sasaji, 1982: 63 **A**: JA (Kyushu)
orbiculatus C.-T. Yang, 1971: 90 **A**: TAI
paltatus X.-F. Pang & Huang, 1986: 64 **A**: FUJ
pilicrepus Lewis, 1896a: 36 (*Scymnus*) **A**: JA SC
pronotus X.-F. Pang & Huang, 1986: 64 **A**: FUJ GUA GUI
quinquepunctatus okinawanus Kamiya, 1965b: 103 **A**: JA (Ryukyus)
quinquepunctatus quinquepunctatus J. Weise, 1923: 188 (*Nephus*) **A**: GUI HUN JA (Ryukyus) TAI
seboshii Ohta, 1929a: 11 (*Scymnus*) **A**: FUJ GUI JA
seminigrinus G. Yu, H. Pang & X.-F. Pang, 1993: 471 **A**: GUA
shixingiensis H. Pang, 1993b: 493 **A**: GUA
simmondsi E. A. Chapin & Ahmad, 1966: 213 **A**: PA
sylvaticus Lewis, 1896a: 36 (*Scymnus*) **A**: HAI JA NC SC TAI
tainanensis Ohta, 1929a: 7 (*Pullus*) **A**: TAI
truncatulus G.-Y. Yu, 1997a: 435 **A**: YUN
tsugae Sasaji & McClure, 1997: 563 **A**: JA

genus *Scymniscus* Dorzhanskiy, 1928: 117 type species *Scymniscus zaisanensis* Dobzhanskiy, 1928
anomus Mulsant & Rey, 1852b: 222 (*Scymnus*) **E**: GR FR IT SP
infimior J. Weise, 1885a: 75 (*Scymnus*)

armeniacus Canepari, 1979: 1 (*Nephus*) **E**: AR
biflammulatus Motschulsky, 1837b: 418 (*Scymnus*) **E**: AR GG **A**: AF IN TM UZ "Kirgizen Steppe"
biguttatus Mulsant, 1850b: 991 (*Scymnus*) **A**: AF IN KZ
bipustulatus Motschulsky, 1837b: 417 (*Scymnus*)
quadriguttatus A. Fleischer, 1900a: 120 (*Scymnus*)
tetrastictus J. Weise, 1929: 33 (*Scymnus*)
castanopterus Mulsant, 1850b: 962 (*Scymnus*) **A**: KZ MG XIZ
depressus Savoyskaya, 1969c: 137 (*Scymnus*)
obsoletus J. Weise, 1890c: 491 (*Scymnus*)
elbursi Fürsch, 1981b: 137 (*Nephus*) **A**: IN
fuerschi Plaza, 1981: 239 (*Nephus*) **E**: PT SP
helgae Fürsch, 1965: 202 (*Scymnus*) **E**: AZ FR GR IT MA SP **N**: MO
horioni Fürsch, 1965: 199 (*Scymnus*) **E**: AL AU BH CR CZ GR HU IT SK **A**: TR
kahleni Fürsch, 1997: 15 (*Nephus*) **E**: AU
macilentus Stenius, 1952: 155 (*Nephus*) **A**: LE
mainanderi Fürsch, 1965: 208 (*Scymnus*) **E**: GR
marikovskii Savoyskaya, 1969c: 136 (*Scymnus*) **A**: KZ
semirufus J. Weise, 1885a: 75 (*Scymnus*) **E**: FR SP
biliottii Gourreau, 1974: 201 (*Scymnus*)
spilotus J. Weise, 1901b: 441 (*Scymnus*) **A**: ES MG
splendidulus Stenius, 1952: 155 (*Nephus*) **E**: IT **N**: AG MO TU
demarzoi Canepari, 2000: 60 (*Nephus*)
tristiculus J. Weise, 1929: 34 (*Sidis*) [RN] **E**: IT SP
tristis J. Weise, 1885a: 76 (*Scymnus*) [HN]
zaisanensis Dobrzhanskiy, 1928: 118 **A**: KZ MG
changajensis Bielawski, 1965b: 154 (*Scymnus*)
elongatus Savoyskaya, 1968: 139 [= 1971: 109] (*Scymnus*)

genus *Scymnus* Kugelann, 1794: 545 type species *Scymnus nigrinus* Kugelann, 1794
subgenus *Didion* Casey, 1899: 137 type species *Didion longulum* Casey, 1899
Nipponopullus Kamiya, 1961a: 284 type species *Scymnus pirikamenoko* Kamiya, 1961
pirikamenoko Kamiya, 1961a: 284 **A**: JA (Hokkaido)
subgenus *Mimopullus* Fürsch, 1987d: 72 type species *Scymnus mediterraneus* Iablokoff-Khnzorian, 1972 (= *Scymnus marinus* Mulsant, 1850)
epistemoides Wollaston, 1867: 276 **N**: MR
fennicus J. R. Sahlberg, 1886: 157 **E**: FI IT NR NT SV
flagellisiphonatus Fürsch, 1970: 112 (*Nephus*) **E**: AL AU BH BU CZ CR GR HU MC RO SK TR YU **A**: CY IS JO LE SY TR
fulvicollis Mulsant, 1846: 259 **E**: FR IT **N**: AG MO TU
limnichoides Wollaston, 1854: 470 **N**: MR
marinus Mulsant, 1850b: 1006 (*Rhizobius*) **E**: FR IT (Sicilia) PT SP **N**: AG CI EG MO MR TU
mediterraneus Iablokoff-Khnzorian, 1972b: 179 [RN]
mulsantianus Sicard, 1907b: 136 [RN]
nanus Mulsant, 1860: 169 [HN]
pharaonis Motschulsky, 1851: 657 **E**: AR FR GR **N**: EG **A**: IN IQ JO TR
araraticus Iablokoff-Khnzorian, 1969: 247
jordanensis Fürsch, 1970: 111 (*Nephus*)
sacium Roubal, 1927d: 137 (*Pullus*) **E**: BU HU IT MD RO UK
subgenus *Neopullus* Sasaji, 1971a: 177 type species *Scymnus hoffmanni* J. Weise, 1879
Caledonus Bielawski, 1973: 397 type species *Scymnus angusticollis* Fauvell, 1903
ater Kugelann, 1794: 548 **E**: BE BH BU BY CZ DE GE GR LA LT NL NR NT PL SK SV UK
celer J. Weise, 1891b: 391 [RN]
babai Sasaji, 1971a: 188 **A**: BEI FUJ HEB HEI HEN HUB HUN JA JIA JIL LIA NMO SC SCH SHA SHN ZHE
camptodromus G.-Y. Yu & Liu, 1997: 434 **A**: SCH YUN
fuscatus Boheman, 1859: 208 **A**: FE FUJ GUA JA NC NP SC SCH TAI YUN **ORR**

brunnescens Motschulsky, 1866b: 425
niponicus Lewis, 1896a: 37
picescens Gorham, 1894a: 207
vittipennis Boheman, 1859: 210
haemorrhoidalis Herbst, 1797: 342 **E**: AB AR AU BE BH BU BY CR CT CZ DE EN FI FR GB GE GG GR HU IT LA LS LT LU NL NR NT PL PT RO SK ST SV SZ UK YU **N**: MR **A**: ES FE IN KZ WS
analis P. Rossi, 1794: 88 (*Coccinella*) [HN]
bilituratus Marsham, 1802: 166 (*Coccinella*)
oculatissimus Wollaston, 1857a: 139 (*Rhizobius*)
hoffmanni J. Weise, 1879a: 152 (*Scymnus*) **A**: BEI FUJ GUX HEB HEI HEN HUN JA JIA JIL JIX NC SC SCH SHA SHG SHN TAI UP YUN ZHE
lijiangensis G.-Y. Yu, 2000b: 164 **A**: YUN
limbatus Stephens, 1832a: 395 **E**: AL AR AU BU BY CZ DE EN FI FR GB GE HU LA LT MC NR NT PL RO SK SV **N**: MR **A**: ES FE IN WS
atricapillus C. N. F. Brisout de Barneville, 1863: 124
cognatus J. Weise, 1887b: 213
pauper J. Weise, 1889a: 261
scutellaris Mulsant, 1850c: 220
lycotropus G.-Y. Yu, 2000b: 165 **A**: YUN
minisculus G.-Y. Yu, H. Pang & X.-F. Pang, 1993: 485 **A**: GUA
nakaikemensis Sasaji & Kishimoto, 1996: 15 **A**: JA (Honshu)
nigromarginalis G.-Y. Yu, 2000b: 166 **A**: YUN
ningshanensis G.-Y. Yu & Yao, 2000: 162 **A**: SHA
ohtai Sasaji, 1971a: 184 **A**: JA **ORR**
otohime Kamiya, 1961b: 314 **A**: JA
paralleus G.-Y. Yu & X.-F. Pang, 1994: 210 **A**: HEB
sinuanodulus G. Yu & Yao, 1997: 434 **A**: YUN **NAR**
tachengicus G.-Y. Yu, 1998a: 129 **A**: XIN
taishuensis Sasaji, 1971a: 186 **A**: JA (Tsushima) NC SC
testaceus Motschulsky, 1837b: 419 **E**: AB **A**: KI TD
thecacontus Ren & X.-F. Pang, 1993: 6 **A**: HUB YUN
yamato Kamiya, 1961b: 310 **A**: BEI FUJ GUA HEB HEN JA JIA SCH SHG **ORR**
subgenus *Orthoscymnus* Canepari, 1997: 16 type species *Scymnus smetanai* Canepari, 1997
rhododendri Canepari, 1997: 17 **A**: NP
smetanai Canepari, 1997: 17 **A**: NP
subgenus *Parapullus* C.-T. Yang, 1978a: 27 type species *Scymnus secula* C. T. Yang, 1978
Paucus Duverger, 1990: 67 type species *Coccinella abietis* Paykull, 1798
abietis Paykull, 1798: 154 (*Coccinella*) **E**: AU BE BU BY CT CZ DE EN FI FR GE HU IT LA LT NR NT PL RO SK SV SZ UK YU **A**: ES FE MG WS
alishanensis X.-F. Pang & G. Yu, 1993: 228 **A**: TAI
secula C. T. Yang, 1978a: 27 **A**: TAI
tsugae G.-Y. Yu & Yao, 2000: 186 **A**: YUN
subgenus *Pullus* Mulsant, 1846: 241 type species *Coccinella subvillosa* Goeze, 1777 (= *Coccinella fasciata* Geoffroy, 1785)
accampus X.-F. Pang & Pu, 1990: 336 **A**: GUX
agrumi Fürsch, l970: 109 **A**: SA
akonis Ohta, 1929a: 5 (*Pullus*) **A**: TAI
alocasia G.-Y. Yu, 2001b: 159 **A**: HKG
ambonoidea X.-F. Pang, 1988: 386 **A**: GUA GUX IS
ancontophyllus Ren & X.-F. Pang, 1993: 7 **A**: HUB SHA SCH YUN
arabicus Fürsch, 1989c: 116 **A**: SA
argutus Mulsant, 1850b: 980 **E**: AB AR GG **A**: AF IN IQ IS TM UZ
auritus Thunberg, 1795: 105 **E**: AL AU BH BU BY CR CT CZ DE EN FI FR GB GE GG GR HU LA LT LU MC MD NR NT PL PT RO SK SL SP ST SV SZ TR UK YU **A**: ANH CY ES FE FUJ GUA IN SA SCH SY TR WS

anatolicus J. Weise, 1905c: 220 (*Pullus*)
brisouti Crotch, 1874a: 248 [RN]
capitatus Fabricius, 1798: 79 (*Coccinella*)
fulvifrons Marsham, 1802: 168 (*Coccinella*)
lukesi Fürsch, 1958b: 87
moraviacus Fürsch, 1958b: 87
obenbergeri Fürsch, 1958b: 87
rufipes C. N. F. Brisout de Barneville, 1863: 123
vermionensis Fürsch, 1958b: 87
baoxingensis G.-Y. Yu, 2000b: 175 **A**: SCH
bengalicus Canepari, 1986: 25 **A**: SD
besucheti Canepari, 1986: 23 **A**: NP SD
bhaumiki Chakraborty & Biswas, 2003: 123 **A**: SD
bistortus G.-Y. Yu, 1995a: 137 **A**: TAI
bourdilloni Kapur, 1958: 335 (*Pullus*) **A**: NP SD
canariensis Wollaston, 1864: 426 **N**: CI
persimillis Har. Lindberg, 1951: 7
centralis Kamiya, 1965a: 81 **A**: FUJ GUA GUI HAI HEN TAI **ORR**
acidatus X.-F. Pang & Huang, 1985: 38
prosericatus X.-F. Pang, 1988: 385
cercyonides Wollaston, 1864: 428 **N**: CI
chujoi Sasaji, 1982: 69 **A**: JA
cladocerus Ren & X.-F. Pang, 1995b: 219 **A**: HUN
cnidatus X.-F. Pang & Pu, 1990: 336 **A**: GUX
coccivora Ramakrishna Ayyar, 1925: 491 **A**: PA UP **ORR**
comperei X.-F. Pang & R. D. Gordon, 1986: 171 **A**: GUA
compoceratus X.-F. Pang & Huang, 1986: 63 **A**: FUJ
contemtus J. Weise, 1923: 186 (*Pullus*) **A**: FUJ GUA JA TAI **ORR**
dorytatus X.-F. Pang, 1988: 387
corporosus Motschulsky, 1866b: 425 **A**: YUN **ORR**
cristiformis G. Yu, 1993: 293 **A**: YUN
cryphaconicus Ren & X.-F. Pang, 1994a: 257 **A**: GUI
curvus C.-T. Yang, 1978b: 111 **A**: TAI
dactylicus X.-F. Pang, 1988: 388 **A**: GUA
dichorionicus X.-F. Pang & Huang, 1986: 63 **A**: FUJ
dicorycus X.-F. Pang & Huang, 1986: 61 **A**: FUJ GUA HAI **ORR**
dorcatomoides J. Weise, 1879a: 151 **A**: FE FUJ HUB JA NC SC SCH TAI **ORR**
ebneri J. Weise, 1926: 228 **N**: EG **A**: AE IN **AFR**
exilis Canepari, 1997: 14 **A**: NP
exocorycus X.-F. Pang & Huang, 1986: 61 **A**: FUJ HAI
facetus Canepari, 1986: 24 **A**: UP
fanjingicus Ren & X.-F. Pang, 1995a: 151 **A**: GUI TAI
ferrugatus Moll, 1785: 183 (*Coccinella*) **E**: AU BE BH BU BY CT CZ GE FI FR HU IT LU NL NT PL SK SV SZ **A**: ES FE FUJ HUN KZ MG SHA WS
analis Fabricius, 1787a: 60 (*Coccinella*)
ruficollis A. G. Olivier, 1791: 81 (*Coccinella*)
formicarius Mulsant, 1850b: 996 **A**: ES FE
formosanus J. Weise, 1923: 188 (*Rhizobius*) **A**: GUA GUI GUX SCH TAI YUN ZHE **ORR**
habaciensis Hoang, 1982a: 140
fraxini Mulsant, 1850b: 991 **E**: AB AR BH BU CR FR GG GR HU IT SK TR UK YU **N**: EG **A**: AF IN SY TR
globosus J. Weise, 1879c: 145
piceus J. Weise, 1885a: 72
wichmanni Fürsch, 1960a: 55
fruticis Canepari, 1997: 14 **A**: NP
fujianensis X.-F. Pang & R. D. Gordon, 1986: 168 **A**: FUJ
geminus G.-Y. Yu & Montgomery, 2000: 177 **A**: YUN

giganteus Kamiya, 1961b: 321 **A**: FUJ GUA JA **ORR**
godavariensis Miyatake, 1985a: 10 **A**: NP
gucheng G.-Y. Yu, 2000b: 176 **A**: YUN
hatomensis Kamiya, 1965b: 108 **A**: JA (Ryukyus) TAI
hepaticus Canepari, 1997: 12 **A**: NP
heptaspilicus Ren & X.-F. Pang, 1995a: 152 **A**: GUI
heyuanus G.-Y. Yu, 2000b: 179 **A**: YUN
hingstoni Kapur, 1963a: 22 (*Pullus*) **A**: FUJ GUA NP SD UP **ORR**
hoabinhensis Hoang, 1982a: 150
hirsutus G.-Y. Yu, 1997b: 725 **A**: HUB
hoocalis X.-F. Pang & R. D. Gordon, 1986: 182 **A**: GUA GUX HAI TAI **ORR**
formosanus Fürsch, 1960d: 443
notus X.-F. Pang & Pu, 1990: 337
pangi Fürsch, 1989a: 34 [RN]
huashansong G.-Y. Yu, 1999c: 282 **A**: YUN
impexus Mulsant, 1850b: 979 **E**: AU CZ FR GE IT PL SK SP **A**: GUA **NAR**
jaculatorius G.-Y. Yu, 2000b: 181 **A**: YUN
janetscheki Bielawski, 1971: 5 **A**: NP
japonicus J. Weise, 1879a: 151 **A**: FUJ GUI HAI HEB HEN HUB HUN JA JIA JIL JIX LIA SC SCH SHA SHN XIZ YUN ZHE
jigongshan G.-Y. Yu, 1999a: 18 **A**: HEN
kabakovi Hoang, 1982a: 135 **A**: GUA YUN **ORR**
kaguyahime Kamiya, 1961b: 313 **A**: FUJ JA ZHE
kapuri Kovář, **nom. nov**. [see New acts] **A**: YE
rubicundus Kapur, 1951: 283 (*Pullus*) [HN]
kaszabianus Iablokoff-Khnzorian, 1976b: 307 **A**: ES
kawamurai Ohta, 1929a: 8 (*Pullus*) **A**: FUJ GUA JA SCH UP YUN ZHE
kamiyai M. Araki, 1964: 40
kimotoi Kamiya, 1965b: 106 **A**: JA (Ryukyus)
klapperichi X.-F. Pang & R. D. Gordon, 1986: 174 **A**: FUJ GUX TAI
klinosiphonicus Ren & X.-F. Pang, 1995a: 154 **A**: GUI
koebelei X.-F. Pang & R. D. Gordon, 1986: 188 (*Keiscymnus*) **A**: FUJ GUA GUX HAI
xanthostethus X.-F. Pang & Pu, 1990: 338
kosianus Canepari, 1997: 15 **A**: NP
latemaculatus Motschulsky, 1858c: 121 **N**: LB **A**: AF FUJ GUA GUX IN PA TAI ZHE **ORR**
belophallus Capra, 1925: 137 (*Pullus*)
quadrillum Motschulsky, 1858c: 120 [HN]
quadrinotatus Boheman, 1859: 210
taiwanus Ohta, 1929a: 6 (*Pullus*)
transversoplagiatus Motschulsky, 1858c: 120
leo C.-T. Yang, 1978: 113 **A**: TAI
linanicus G.-Y. Yu & X.-F. Pang, 1995: 475 **A**: ZHE
lonchiatus X.-F. Pang & Huang, 1985: 33 **A**: FUJ HUB SCH YUN
loxiphyllus Ren & X.-F. Pang, 1994a: 259 **A**: GUI
luciolus Canepari, 1997: 15 **A**: NP
luxorensis Fürsch, 1989c: 116 **N**: EG
martensi Canepari, 1997: 14 **A**: NP
mastigoides Ren & X.-F. Pang, 1995b: 219 **A**: HUN
mellinus Canepari, 1997: 11 **A**: NP
miyatakei Kamiya, 1961b: 320 **A**: JA (Ryukyus)
mongolicus J. Weise, 1890c: 490 **A**: GAN MG NMO
nankunicus X.-F. Pang, 1986: 308 **A**: FUJ GUA HAI
nepalensis Bielawski, 1971: 3 **A**: NP SD
nephrospilus Ren & Pang, 1995b: 221 **A**: HUN
nigrobasalis G.-Y. Yu, 2000b: 174 **A**: YUN

novenus G.-Y. Yu, 1995a: 134 **A**: TAI
oblongulus Normand, 1936c: 150 [DA] **N**: TU
oertzeni J. Weise, 1886: 25 **E**: CR FR GR
melanarius Gourreau, 1970: 153 (*Pullus*)
muelleri Penecke, 1907: 15
oestocraerus X.-F. Pang & Huang, 1985: 35 **A**: FUJ GUI GUX HAI TAI **ORR**
oncosiphonos C. Cao & Xiao, 1984: 122 **A**: YUN
orientalis Mader, 1955a: 73 **A**: YUN
osakaensis M. Araki, 1963: 251 **A**: JA
ovimaculatus Sasaji, 1968: 124 **A**: GUA **ORR**
paraperdere G. Yu, H. Pang & X.-F. Pang, 1993: 476 **A**: FUJ GUA
paratenuis Ren & X.-F. Pang, 1994a: 260 **A**: GUI
pelecoides X.-F. Pang & Huang, 1985: 35 **A**: FUJ ZHE
penicilliformis G.-Y. Yu, 1999b: 67 **A**: HEN
perdere C.-T. Yang, 1978b: 109 **A**: FUJ GUA TAI ZHE
nepenthus X.-F. Pang & Huang, 1985: 32
petalinus G.-Y. Yu, 1995a: 135 **A**: TAI
phylloides G.-Y. Yu, 1995a: 133 **A**: TAI
podoides G.-Y. Yu & X.-F. Pang, 1992c: 116 **A**: GUA GUI
posticalis Sicard, 1914a: 503 **A**: BT FUJ GUA GUI GUX HEN HUB JA NP SC SCH SD SHA TAI UP YUN **ORR**
ishidai M. Araki, 1963: 252
prostylotus X.-F. Pang & Huang, 1985: 31 **A**: FUJ
puellaris M. Araki, 1964: 42 **A**: JA
pyrocheilus Mulsant, 1853: 281 **A**: NP **ORR**
rectoides Sasaji, 1971a: 156 **A**: JA
rectus Ohta, 1929a: 4 (*Pullus*) **A**: JA
rhachiatus X.-F. Pang & Huang, 1985: 36 **A**: FUJ
rhamphiatus X.-F. Pang & Huang, 1985: 29 **A**: FUJ GUI HUB HUN SCH SHA ZHE
rhombicus G. Yu, H. Pang & X.-F. Pang, 1993: 482 **A**: GUA
robustibasalis G.-Y. Yu, 2000b: 180 **A**: YUN
ruficeps Ohta, 1929a: 5 (*Pullus*) **A**: JA SC TAI
kiotoensis M. Araki, 1963: 251
rufoniger Canepari, 1997: 13 **A**: NP
runcatus G.-Y. Yu & X.-F. Pang, 1995: 476 **A**: ZHE
sapporensis Ohta, 1929a: 7 (*Pullus*) **A**: JA
scalpratus G.-Y. Yu, 1993: 294 **A**: HAI
schawalleri Canepari, 1997: 13 **A**: NP
scrobiculatus G.-Y. Yu, 1993: 296 **A**: HAI
shirozui Kamiya, 1965a: 77 **A**: FUJ GUA HAI JIX TAI
shixingicus G.-Y. Yu & X.-F. Pang, 1992c: 118 **A**: GUA
silesiacus J. Weise, 1902a: 240 **E**: CR BU SK SL PL RO SV UK **A**: TR
lipicensis G. Müller, 1924: 75
stiglundbergi Fürsch, 1969: 55
singularis Mader, 1955a: 72 **A**: FUJ
sodalis J. Weise, 1923: 186 (*Pullus*) **A**: FUJ GUA GUX HEN HUB JA (Ryukyus) JIA NP SCH TAI UP XIZ ZHE **ORR**
spirosiphonicus Ren & X.-F. Pang, 1995a: 154 **A**: GUI
subvillosus Goeze, 1777: 247 (*Coccinella*) **E**: AB AL AN AR AU AZ BH BU CR CZ FR GE GG GR HU IT MC PT RO SK SL SP ST SZ TR UK YU **N**: AE AG CI EG LB MO MR TU **A**: AE AF CY IN IQ IS JO KI KU KZ LE PA QA SA SI SY TD TR UZ YE **AFR**
aurantiacus Panzer, 1813: 112 (*Coccinella*)
dorsalis Waltl, 1839: 226
durante Wollaston, 1854: 465
fasciatus Geoffroy, 1785: 149 (*Coccinella*) [HN]

fulvicollis J. R. Sahlberg, 1913d: 264 [HN]
juniperi Motschulsky, 1837b: 419
magnocapsularis Fürsch, 1966c: 161
meridionalis Fürsch, 1958b: 87
minutissimus Schrank, 1781a: 65 (*Coccinella*)
minutissimus P. Rossi, 1790: 71 (*Coccinella*) [HN]
nigropictus Wollaston, 1867b: 160
pubescens Panzer, 1794b: 13 (*Coccinella*) [HN]
quadrillum L. Redtenbacher, 1843a: 19
sahlbergianus Fürsch, 1958a: 87
transitus Fürsch, 1958b: 84
transversepustulatus Mulsant, 1850b: 976
villosus Contarini, 1847: 190 (*Coccinella*)

suturalis Thunberg, 1795: 106 [NP] **E**: AR AU AZ BE BH BU BY CR CZ DE EN FR GB GE GG GR HU IR IT LA LS LT LU MC MD NL NT PL PT RO SK SL SP SV SZ UK YU **N**: AG MR TU **A**: ES FE IS MG TR WS **NAR**
atriceps Stephens, 1832a: 396 (*Scymnus*)
gerhardti Fürsch, 1958b: 88
gratiosulus Fürsch, 1958b: 88
discoideus Illiger, 1798: 418 (*Coccinella*)
inlimbatus Fürsch, 1958b: 88
nigricans Gerhardt, 1898b: 336 [HN]
pilosus Herbst, 1797: 343
pini Marsham, 1802: 102 (*Byrrhus*)
plagiatus Beck, 1817: 16, pl. 4 (*Coccinella*) [HN]
pusillus O. F. Müller, 1776: 65 (*Coccinella*) [NO]
quercus Mulsant, 1850b: 982
testaceicolor Roubal, 1927d: 137 (*Pullus*)

suturaloides Canepari, 1997: 16 **A**: NP

syoitii Sasaji, 1971a: 146 **A**: FUJ JA **ORR**

syriacus Marsuel, 1868: 216 (*Nephus*) **N**: EG **A**: CY IN IQ IS JO LE SI SY
amoenus J. Weise, 1885a: 71

takabayashii Ohta, 1929a: 4 (*Pullus*) **A**: FUJ GUI HUN JA SCH

takaraensis Nakane & M. Araki, 1959: 46 **A**: JA (Ryukyus)

takasago Kamiya, 1965a: 80 **A**: TAI

tenuis C.-T. Yang, 1978b: 110 **A**: FUJ GUA HAI TAI YUN **ORR**
longisiphonulus C. Cao & Xiao, 1984: 123
longisiphonulus X.-F. Pang & R. D. Gordon, 1986: 169 [HN]

testacecollis Kapur, 1963a: 21 (*Pullus*) **A**: SD

toxosiphonius X.-F. Pang & Huang, 1986: 62 **A**: FUJ GUI YUN

trimaculatus G.-Y. Yu & X.-F. Pang, 1994: 209 **A**: YUN **ORR**

tschungi G.-Y. Yu, 1995b: 146 **A**: FUJ

tympanus G.-Y. Yu & X.-F. Pang, 1992c: 117 **A**: GUA

urgensis Jakobson, 1915: 974 [RN] **A**: KZ MG
arenarius J. Weise, 1929: 33
dorsalis A. Fleischer, 1900b: 235 [RN]
mongolicus A. Fleischer, 1900a: 119 [HN]

vencoxus Kamiya, 1965c: 316 [RN] **A**: JA
convexus Kamiya, 1961b: 315 [HN]

victoris Crotch, 1874a: 253 [RN] **A**: NP SD **ORR**
dorsualis J. Weise, 1892a: 28
ephippiatus Gorham, 1894a: 207
limbatus Motschulsky, 1858c: 118 [HN]
obsignatus J. Weise, 1901b: 435 (*Pullus*)
unimaculata Korschefsky, 1934: 5

villis J. Weise, 1923: 186 (*Pullus*) **A**: TAI **ORR**

vinhphuensis Hoang, 1982a: 152 **A**: FUJ GUA GUI HAI HUN TAI ZHE **ORR**
bicolor C.-T. Yang, 1978b: 114 [HN]

endocorycus X.-F. Pang & Huang, 1986: 62
yangi G.-Y. Yu & X.-F. Pang, 1993: 479 [RN]
xanthostethus X.-F. Pang & Pu, 1990: 338 **A**: GUX
yaling G.-Y. Yu, 1999b: 66 **A**: HEN
yemenensis Kapur, 1959: 281 (*Pullus*) **A**: OM SA YE
yungshanpingensis G.-Y. Yu, 1997a: 433 **A**: YUN
subgenus *Scymnus* Kugelann, 1794: 545 type species *Scymnus nigrinus* Kugelann, 1794
Anisoscymnus Crotch, 1874a: 273 type species *Coccinella rufipes* Fabricius, 1798
apetzi Mulsant, 1846: 228 **E**: AB AL AR AU BH BU CR CT CZ FR GE GG GR HU IT LI MA MC MD PL PT RO SK SP ST SZ TR UK YU **A**: IN IQ IS JO KI KZ LE SY TD TM TR WS
corpulentus Mulsant, 1850b: 967
megastigma Fürsch, 1958b: 84
apiciflavus Motschulsky, 1858c: 119 **A**: GUI GUX HUN SCH SHA **ORR**
andamanensis Kapur, 1967: 161
axinoides Ren & X.-F. Pang, 1995c: 469 **A**: GUI SCH
bifurcatus G.-Y. Yu, 1995a: 130 **A**: TAI
bivulnerus Baudi di Selve, 1894: 12 **E**: CR FR GR IT PT SP **N**: AG MO TU **A**: CY IS LE SY TR
bivulnerus Capra & Fürsch, 1967a: 212 [HN]
bogdoensis Bielawski, 1965b: 152 **A**: KZ MG
caprai Canepari, 1983: 189 **E**: IT
cedricolus Fürsch, 1977: 648 **A**: IN LE TR
crinitus Fürsch, 1966a: 40 **A**: FE GAN SCH
decemmaculatus G. Yu, H. Pang & X.-F. Pang, 1993: 474 **A**: GUA
damryi J. Weise, 1879c: 142 **E**: FR IT
dipterygicus Ren & X.-F. Pang, 1995c: 467 **A**: GUI
dissolobus X.-F. Pang & Huang, 1985: 41 **A**: FUJ GUA HUB
prosphykontus X.-F. Pang & Huang, 1985: 43
dolichonychus G.-Y. Yu & X.-F. Pang, 1995: 474 **A**: GUI HUN SCH ZHE
doriae Capra, 1924: 199 **E**: AU (Tyrol) BU IT YU **A**: KZ MG ES
dorotae Bielawski, 1982: 235 **A**: NC SC
femoralis Gyllenhal, 1827: 196 (*Coccinella*) **E**: AU CZ FI GE GB IT LS PL SK SV SZ
flavicollis L. Redtenbacher, 1843a: 13 **E**: AR **A**: CY IN IQ IS JO KV SY TD TR
araxicola A. Fleischer, 1900a: 120 (*Scymnus*)
rufithorax Baudi di Selve, 1894: 12
triangulifer A. Fleischer, 1900b: 235
folchinii Canepari, 1979a: 4 **A**: BEI HAI HEB ZHE
hainanensis X.-F. Pang & R. D. Gordon, 1986: 162
franzi Fürsch, 1967a: 214 **E**: SP (Mallorca)
frontalis Fabricius, 1787a: 60 (*Coccinella*) [NP] **E**: AB AL AR AU BE BH BU BY CR CT CZ DE EN FI FR GB GE GG GR HU IT LA LS LT LU NL NR NT PL PT RO SK SL SP ST SV SZ TR UK YU **A**: BEI ES FE FUJ IN HEB HEI HEN HUB JIX KI KZ SHA SHN TD TR UZ WS XIN
affinis L. Redtenbacher, 1843a: 19
alticus Schrank, 1781a: 84 (*Chrysomela*) [NO]
angulicinctus Fürsch, 1958b: 84
anticecinctus Fürsch, 1958b: 85
bimaculatus Marsham, 1802: 166 (*Coccinella*) [HN]
bimaculatus Thunberg, 1795: 105
bisbipustulatus Marsham, 1802: 167 (*Coccinella*) [HN]
bisbisignatus L. Redtenbacher, 1843a: 19
didymus Herbst, 1797: 341
flavellus Savoyskaya, 1969c: 127
flavilabris Paykull, 1798: 148 (*Coccinella*)
humeralis Panzer, 1795c: 149 (*Coccinella*) [HN]
immaculatus Suffrian, 1843b: 330
quadripustulatus Herbst, 1793: 344 (*Coccinella*)
quadrivulneratus Mulsant, 1850b: 970
suffriani J. Weise, 1879c: 150

gonatoides Kusnetzov & Ren, 1992: 228 **A**: YE
grammicus G.-Y. Yu, 1995a: 130 **A**: TAI
inderihensis Mulsant, 1850b: 971 **E**: IT UK **A**: GAN HEB HUB IN KI KZ MG SHA SHX TM TR UZ WS XIN
ornatus Savoyskaya, 1969c: 131
interruptus Goeze, 1777: 247 (*Coccinella*) **E**: AB AL AR AU BE BU BY CR CZ FR GE GG GR HU IT NL PL PT RO SK SP SZ TR UK **N**: AG EG LB MO MR TU **A**: IQ IS JO KZ MG TR SY YE
basalis L. Redtenbacher, 1843a: 20
basalis Kirsch, 1871: 394 (*Pharus*) [HN]
bimaculatus Herbst, 1792: 71 (*Hydrophilus*)
flexuosus J. Weise, 1879c: 152
inundatus J. Weise, 1879c: 152
pallidivestis Mulsant, 1853: 278
rufescens J. Weise, 1879c: 152
jakowlewi J. Weise, 1892d: 141 **E**: FI NR NT SV **A**: ES FE MG WS
sahlbergi Korschefsky, 1931: 136 [RN]
triangularis J. R. Sahlberg, 1914: 39 [HN]
laetificus J. Weise, 1879c: 152 **E**: IT PT **N**: AG MO TU
longmenicus X.-F. Pang, 1986: 308 **A**: GUA HAI
magnomaculatus Fürsch, 1958b: 84 **E**: AB AL AR AU BH BU CR FR GR IT MC RO SL TR **A**: IS LE SY TR
muelleri J. Weise, 1929: 33 [HN]
quadriguttatus Capra, 1924: 201 [HN]
quadriguttatus Fürsch, 1958b: 85 [HN]
quadriguttatus Fürsch & Kreissl, 1967: 215 [HN]
manipulus Fürsch & Kreissl, 1967: 219 **A**: GAN IN JIA MG TM
marginalis P. Rossi, 1794: 87 (*Coccinella*) **E**: AU (Tyrol) FR GR IT MA UK **N**: AG MO TU **A**: TR
incertus Mulsant, 1846: 228
morio Fabricius, 1801a: 380 (*Coccinella*) [HN]
marinellus Kovář, **nom. nov**. [see New acts] **A**: JA (Koshiki Okino-shima)
marinus Kamiya, 1961a: 299 [HN]
mesasiaticus Savoyskaya, 1971b: 109 [RN] **E**: BU CT GR **A**: KI KZ
gracilis Savoyskaya, 1969c: 129
miyamotoi Nakane & M. Araki, 1959: 46 **A**: JA
nigrinus Kugelann, 1794: 548 [NP] **E**: AU BE BU BY CZ DE EN FI FR GB GE GR HU IT LA LT NL NR NT PL RO SK SP ST SV SZ UK YU **A**: ES FE KZ MG WS
ater Thunberg, 1795: 105 [HN]
coelicolor Fürsch, 1958a: 84
decipiens J. Weise, 1885a: 76
minimus O. F. Müller, 1776: 65 (*Coccinella*) [NO]
morio Paykull, 1798: 151 (*Coccinella*)
simplocarioides Fürsch, 1958b: 84
najaformis G.-Y. Yu, 1997a: 432 **A**: YUN
notidanus X.-F. Pang & Huang, 1985: 37 **A**: FUJ
nubilus Mulsant, 1850b: 972 **E**: AZ GR IT PT SP **N**: CI EG MR **A**: AE AF IN IQ IS JA (Bonin, Ryukyus) JO KU LE OM NP PA SA SD SY TAI TR UP **AFR AUR ORR**
africanus J. Weise, 1895a: 141
amicus Mader, 1950: 20
curtisii Mulsant, 1850b: 973
floricolla Wollaston, 1867: 172
gracilis Motschulsky, 1866b: 426
lateralis Sicard, 1914a: 502
motschulskyi Crotch, 1874a: 253 [RN]
nigrosuturalis Kamiya, 1961a: 298
rudiventris Mader, 1950: 22
stabilis Motschulsky, 1966b: 425
suturalis Motschulsky, 1858c: 120 [HN]
togoensis Fürsch, 1960d: 444
paganus Lewis, 1896a: 38 (*Pullus*) **A**: FUJ HUB JA SCH TAI
pallipes Mulsant, 1850b: 966 **E**: ST **A**: KZ UZ TD TM XIN

paracrinitus G.-Y. Yu, 2000b: 172 **A**: YUN
pavesii Canepari, 1983: 189 **E**: IT **N**: AG
pinguis G.-Y. Yu, 1999b: 65 **A**: HEN
rubromaculatus Goeze, 1777: 247 (*Coccinella*) **E**: AL AN AU AZ BE BH BU BY CR CT CZ EN FR GE GR HU IT LU NL MC MD NT PL PT RO SK SL SP SV SZ TR UK YU **N**: MR **A**: ES FE KZ TR WS **AFR**
carbonarius Wollaston, 1867: 159
collaris Herbst, 1797: 345
flavilabris A. G. Olivier, 1808: 1041 (*Coccinella*)
flavipes Panzer, 1792a: 2 (*Tritoma*)
parvulus Fabricius, 1792a: 286 (*Coccinella*)
pubescens Herbst, 1797: 342 [HN]
pygmaeus Geoffroy, 1785: 150 (*Coccinella*)
sericeus Kugelann, 1794: 547
rufipennis Wollaston, 1864: 426 **N**: CI EG **AFR**
pallidulus Wollaston, 1867: 160
rubiginosus Mader, 1950: 35
rufipes Fabricius, 1798: 80 (*Coccinella*) **E**: FR IT PT SP **N**: AG TU MO
ahrensi Mulsant, 1846: 231
gallicus Capra & Fürsch, 1967a: 226
major A. Costa, 1882: 32
sangtanabiensis Bielawski, 1959: 103 **A**: AF
scapanulus X.-F. Pang & Huang, 1985: 39 **A**: FUJ ZHE
schmidti Fürsch, 1958a: 85 **E**: AU AZ BY CT CZ DE FI FR GB GE HU IT LU PL RO SK SV UK **A**: AF FUJ GAN GUI HEB HEI HUB JIX SHA SHN TR WS XIN ZHE
hispanicus Fürsch & Kreissl, 1967: 222
mimulus Fürsch, 1958b: 85
mimulus Capra & Fürsch, 1967a: 221 [HN]
schmidti Fürsch & Kreissl, 1967: 224 [HN]
sufflavus Canepari, 1997: 9 **A**: NP
suffrianioides apetzoides Capra & Fürsch, 1967a: 211 **E**: AU FR GE IT NT PL RO SK SL SV TR **A**: TR
suffrianioides suffrianioides J. R. Sahlberg, 1913d: 265 **A**: IQ IS JO LE SY TR
pallipediformis Günther, 1958: 25
tegminalis Hoang, 1985: 33 **A**: FUJ GUA **ORR**
tristigmaticus Gorham, 1895b: 692 **A**: NE **ORR**
trisulicus Canepari, 1997: 9 **A**: NP
tsushimaensis Sasaji, 1970b: 209 **A**: JA (Tsushima)
unciformis G.-Y. Yu, 2000b: 171 **A**: YUN

genus *Scydmus*, taxa incertae sedis
alpestris Mulsant & Rey, 1852b: 221 **E**: FR (Briançon)
arcuatus hausmanni Gredler, 1866: 457 **E**: IT (Bozen)
apetzi lindbergi Fürsch, 1959: 29 **E**: IT (Napoli)
stigmatopterus Faldermann, 1837: 409 **E**: AR
trojanus Mulsant & Godart, 1875: 16 **A**: TR

tribe Stethorini Dobrzhanskiy, 1924

genus *Stethorus* J. Weise, 1885a: 65 type species *Scymnus punctillum* J. Weise, 1891 (= *Scymnus pusillus* Herbst, 1797)
subgenus *Allostethorus* Iablokoff-Khnzorian, 1972a: 121 type species *Stethorus amurensis* Iablokoff-Khnzorian, 1972
amurensis Iablokoff-Khnzorian, 1972a: 121 **A**: BEI FE GAN HUB NC SHA
koreanus Bielawski, 1980: 232
shaanxiensis X.-F. Pang & Mao, 1975: 420
binchuangensis X.-F. Pang & Mao, 1975: 419 **A**: SCH YUN
chengi Sasaji, 1968: 4 **A**: GUI HEN HUB HUN JIA SHA SCH TAI
convexus G.-Y. Yu, 1996a: 33 **A**: BEI HEB HUB SCH

dongchuanensis C. Cao & Xiao, 1984: 124 **A**: YUN
emarginatus Miyatake, 1966b: 51 **A**: JA (Honshu, Shikoku)
hirashimai Sasaji, 1968: 5 **A**: TAI
klapperichi G.-Y. Yu, 1995a: 125 **A**: TAI
muriculatus G.-Y. Yu, 1995a: 126 **A**: TAI
parapauperculus X.-F. Pang, 1966: 79 **A**: FUJ GUA GUX HAI HEN YUN
pauperculus J. Weise, 1895c: 155 (*Scymnus*) **A**: SD YE **ORR**
vietnamicus Hoang, 1979: 11 **A**: GUX HAI **ORR**
yezoensis Miyatake, 1966b: 53 **A**: FE JA (Hokkaido)
subgenus *Parastethorus* X.-F. Pang & Mao, 1975: 421 type species *Stethorus yunnanensis* X.-F. Pang & Mao, 1975
baiyunshanensis Ren & X.-F. Pang, 1996: 326 **A**: GUA
dichiapiculus Xiao, 1993: 370 **A**: SCH
guangxiensis X.-F. Pang & Mao, 1975: 421 **A**: FUJ GUX HUB SCH SHA
indira Kapur, 1950: 148 **A**: GUX YUN **ORR**
malaicus Xiao, 1993: 371 **A**: SCH
truncatus Kapur, 1948b: 315 **A**: GUA **ORR**
yunnanensis X.-F. Pang & Mao, 1975: 421 **A**: YUN
subgenus *Stethorus* J. Weise, 1885a: 65 type species *Scymnus punctillum* J. Weise, 1891 (= *Scymnus pusillus* Herbst, 1797)
 Nephopullus Brèthes, 1925a: 167 type species *Nephopullus darwini* Brèthes, 1925
aptus aptus Kapur, 1948b: 314 **A**: FUJ GUA GUX HAI JA (Ryukyus) TAI ZHE **ORR**
aptus tsutsuii Nakane & M. Araki, 1959: 48 **A**: JA (Ryukyus)
balearicus Fürsch, 1987a: 2 **E**: SP (Mallorca) **N**: MO
cantonensis X.-F. Pang, 1966: 77 **A**: FUJ GUA GUX HAI HUB JIX SCH **ORR**
gilvifrons Mulsant, 1850b: 995 (*Scymnus*) **E**: AB AR CR BU CR FR GR IT **N**: EG **A**: AF CY IN IQ KA LE PA SA SY TR **ORR**
japonicus Kamiya, 1959b: 139 **A**: JA
loi Sasaji, 1968: 2 **A**: TAI
longisiphonulus X.-F. Pang, 1966: 78 **A**: FUJ GUA HAI **ORR**
pusillus Herbst, 1797: 346 (*Scymnus*) **E**: AB AL AN AR AU BE BH BU BY CR CT CZ DE EN FI FR GE GG GR HU LA LS LT LU MA MC MD NL NR NT PL RO SK SL ST SV TR UK YU **N**: EG MO **A**: BEI ES FE FUJ GAN GUI GUX HEB HEI HEN HUB IN IQ IS JIA JIX KI KZ LIA MG SCH SHA SHN TR UZ WS XIN ZHE **NAR**
 concolor J. Weise, 1887b: 213 (*Scymnus*)
 investitus Roubal, 1920: 79
 minimus P. Rossi, 1794: 89 (*Coccinella*) [HN]
 punctillum J. Weise, 1891b: 391 (*Scymnus*) [RN]
rani Kapur, 1948b: 313 **A**: FUJ GUA UP YUN **ORR**
sichuanensis Ren & X.-F. Pang, 1996: 320 **A**: SCH
siphonulus Kapur, 1948b: 314 **A**: FUJ GUA GUX HUB **ORR**
tenerifensis Fürsch, 1987a: 2 **N**: CI
wollastoni Kapur, 1948b: 305 **N**: CI MR
wulingicus Xiao, 1993: 370 **A**: SCH
yingjiangensis C. Cao & Xiao, 1984: 124 **A**: GUI YUN

subfamily **Chilocorinae** Mulsant, 1846

tribe Chilocorini Mulsant, 1846

genus *Brumoides* E. A. Chapin, 1965: 237 type species *Coccinella suturalis* Fabricius, 1798 (= *Brumus daldorfi* Crotch, 1874)
adenensis Fürsch, 1987b: 44 **A**: AE IN SA YE **AFR**
daldorfi Crotch, 1874a: 21 (*Brumus*) [RN] **A**: AF NP PA UP **ORR**
 suturalis Fabricius, 1798: 78 (*Coccinella*) [HN]
hainanensis Miyatake, 1970: 307 **A**: GUA GUX HAI YUN

lineatus J. Weise, 1885b: 229 (*Brumus*) **A**: GUA HAI JA (Ryukyus) NP TAI **ORR**
ruficollis J. Weise, 1885b: 230 (*Brumus*)
maai Miyatake, 1970: 307 **A**: FUJ GUX YUN
nigrosuturalis Kapur, 1951: 293 (*Brumus*) **A**: YE **AFR**
ohtai Miyatake, 1970: 309 **A**: JA TAI

genus *Chilocorus* Leach, 1815: 116 type species *Coccinella cacti* Linnaeus, 1758
Chilochorus Hope, 1840: 157 type species *Coccinella cacti* Linnaeus, 1758
alishanus Sasaji, 1968: 20 **A**: GUI TAI YUN
amamensis Kamiya, 1959a: 103 **A**: JA NC
bijugus Mulsant, 1853: 189 **A**: GUI HP HUB KA PA SD UP YUN **ORR**
infernalis Mulsant, 1853: 189
bipustulatus Linnaeus, 1758: 367 (*Coccinella*) **E**: AB AL AR AU AZ BE BH BY CZ CT DE EN FI FR GB GE GG GR HU IR IT LA LT LU MA MC MD NL NR NT PL PT RO SK SL SP ST SV SZ TR UK YU **N**: AG EG LB MO MR TU **A**: AF CY FE GAN IN IQ IS JO KI KZ LE MG SA SI SY TR UZ XIN **AFR NAR**
fasciatus O. F. Müller, 1776: 68 (*Coccinella*)
frontalis Thunberg, 1792: 105 (*Coccinella*) [HN]
minor J. R. Sahlberg, 1903b: 86
olivetorum O. G. A. Costa, 1839b: 104
strigatus Fabricius, 1798: 79 (*Coccinella*) [HN]
testudo Florencourt Chassot, 1796: 214 (*Cocccinella*)
transversoguttatus Boerner, 1776: 250 (*Coccinella*)
braeti J. Weise, 1895c: 154 **A**: SD
canariensis Crotch, 1874a: 185 **N**: CI
chalybeatus Gorham, 1892a: 84 **A**: FUJ GUA GUI GUX HUN JIX SCH SHA ZHE
chinensis Miyatake, 1970: 322 **A**: FUJ GUA GUX HAI YUN ZHE **ORR**
circumdatus Gyllenhal, 1808b: 152 (*Coccinella*) **A**: FUJ GUA GUX HAI HKG NP YUN ZHE **AUR ORR**
nigromarginatus Motschulsky, 1860c: 174
distigma Klug, 1835: 49 **A**: SA **AFR**
biplagiatus Walker, 1871: 19
ruficeps Mulsant, 1850b: 457
esakii Kamiya, 1959a: 102 **A**: GAN HEB JA (Kyushu) JIA LIA SCH SHN SHX
geminus Zaslavskij, 1962: 398 **A**: AF GAN KI MG NMO SCH TD TM XIN
hauseri J. Weise, 1895a: 135 **A**: FUJ GUA GIU HAI HUB SCH SD SHA YUN **ORR**
hupehanus Miyatake, 1970: 329 **A**: FUJ GAN GUI GUX HUB HUN SCH
infernalis Mulsant, 1853: 189 **A**: GAN GUI HP HUB JIA KA PA SCH SD UP XIZ YUN **ORR**
bijugus Mulsant, 1853: 189
inornatus J. Weise, 1887b: 211 **A**: FE NC SC
ishigakensis Kamiya, 1959a: 104 **A**: JA (Ryukyus)
kirgisicus Iablokoff-Khnzorian, 1976b: 305 **A**: KI
kuwanae Silvestri, 1909: 126 **Ei**: FR IT **A**: ANH BEI FE FUJ GUA GUI HEI HEN HUB HUN IS JA JIL JIX LIA NC SC SCH SHG SHN UP XIZ YUN ZHE **NAR ORR**
japonicus Sicard, 1907c: 211
matsumurai Miyatake, 1985a: 13 **A**: NP
melas J. Weise, 1898: 229 **A**: BT FUJ GUA GUX HAI HKG SCH SD YUN **ORR**
gressitti Miyatake, 1970: 330
nigritus Fabricius, 1798: 79 (*Coccinella*) **A**: FUJ GUA GUX HP NP PA SCH UP YUN **AFR ORR**
politus Mulsant, 1850b: 455 **A**: NP TAI XIZ YUN **AFR ORR**
renipustulatus L.G. Scriba, 1791: 105 (*Coccinella*) **E**: AU BE BH BY CR CT CZ DE EN FI FR GE GG HU IT LS LU NL NR NT PL RO SK SL ST SV SZ YU UK **A**: ES FE WS
abdominalis Thunberg, 1795: 111 (*Coccinella*)
mediopustulatus Schrank, 1798: 465 (*Coccinella*)
rubidus Hope, 1831: 31 **A**: ANH BEI ES FE FUJ GUI HAI HEB HEI HEN HUB HUN JA JIA JIL JIX KA LIA NIN NP MG NC NMO NP PA SC SCH SHN UP XIZ YUN ZHE **ORR**
fenestratus J. Weise, 1887b: 210
niger J. Weise, 1887b: 210

tristis Faldermann, 1835b: 452 (*Coccinella*)
rufitarsis Motschulsky, 1854c: 50 **A**: BEI FUJ GUA GUX HEN HKG HUN JIX SHG YUN ZHE
monachus Mulsant, 1856: 147
schirozui Sasaji, 1968: 22 **A**: TAI
similis P. Rossi, 1790: 68 (*Coccinella*) **E**: IT
takara Nakane & M. Araki, 1959: 52 **A**: JA (Ryukyus)
yunlongensis C. Cao & Xiao, 1984: 117 **A**: YUN

genus *Chujochilus* **Sasaji, 2005: 61** type species *Exochomus isenis* Kamiya, 1966
isensis Kamiya, 1966: 80 (*Exochomus*) **A**: JA SCH YUN

genus *Exochomus* L. Redtenbacher, 1843a: 11 type species *Coccinella quadripustulata* Linnaeus, 1758
Anexochomus Barovskij, 1922c: 292 type species *Exochomus undulatus* J. Weise, 1878
Brumus Mulsant, 1850b: 492 type species *Coccinella octosignata* Gebler, 1830
bifasciatus Barovskij, 1927b: 200 **A**: KZ IN XIN
cedri J. R. Sahlberg, 1913d: 262 **E**: AL BH BU CR CZ GR HU SK **A**: LE SY TR
muelleri Mader, 1930a: 134
discors Barovskij, 1927b: 200 (*Brumus*) **A**: KI TD
ericae Crotch, 1874a: 193 **N**: AG MO TU
anchorifer Bedel, 1885: 90
nigropictus Fairmaire, 1876c: 94 (*Chilocorus*)
picturatus Fairmaire, 1876c: 94 (*Chilocorus*)
gebleri J. Weise, 1885b: 55 **A**: AF IN TR
jacobsoni Barovskij, 1927b: 200 (*Brumus*) **A**: KZ MG NMO UZ
kiritshenkoi Barovskij, 1922c: 289 **A**: KI KZ MG TD
mongol Barovskij, 1922c: 291 **A**: ANH BEI ES FE HEB HEI JIA LIA MG NMO SC SHA SHN NC ZHE
freyi Fürsch, 1960c: 298 [HN]
georgi Fürsch, 1963: 52 [RN]
mongolicus A. Fleischer, 1900a: 118 (*Brumus*) **A**: ES KZ MG NMO
oblongus Weidenbach, 1859: 60 **E**: AU BH CZ GE IT
octosignatus Gebler, 1830: 225 (*Coccinella*) **E**: AB AR FR IT ST **A**: AF IN IQ KI KZ MG TD TM TR UZ
conjunctus A. Fleischer, 1900a: 118 (*Brumus*)
desertorum Gebler, 1841a: 376 (*Coccinella*)
desertus Motschulsky, 1840: 175 (*Coccinella*)
lasioides J. Weise, 1879c: 135 (*Brumus*)
quadriguttatus A. Fleischer, 1900a: 118 **E**: AR **A**: TR
cordiformis Roubal, 1926b: 245
illaesicollis Roubal, 1927c: 135
quadripustulatus Linnaeus, 1758: 367 (*Coccinella*) **E**: AB AL AR AU BE BH BU BY CR CT CZ DE EN FI FR GB GE GG GR HU IT LA LT LU MC MD NL NR NT PL PT RO SK SL SP ST SV SZ TR UK YU **N**: MO **A**: CY ES FE IN IQ IS KZ LE MG TR WS **NAR**
bilunulatus J. Weise, 1879c: 133
cassidoides Donovan, 1798: 74 (*Coccinella*)
distinctus Brullé, 1832: 273 (*Coccinella*) [HN]
floralis Motschulsky, 1837b: 423 (*Coccinella*) [HN]
haematideus A. Costa, 1849: 62
ibericus Motschulsky, 1837b: 422 (*Coccinella*)
koltzei J. Weise, 1879c: 134
lunulatus Gmelin, 1790: 1662 (*Coccinella*)
quadriverrucatus Fabricius, 1792a: 288 (*Coccinella*)
reitteri W. G. Schneider, 1881: 16
sexpustulatus Kraatz, 1873a: 192
unicolor Schaufuss, 1862b: 50
varius Schrank, 1798: 466 (*Coccinella*)
vittatus Fuente, 1910: 444
septemmaculatus J. Weise, 1885a: 53 **A**: KZ TJ UZ
undulatus J. Weise, 1878: 93 **E**: AB GG **N**: EG **A**: AF IN IQ IS LE SI SY TD

genus *Parexochomus* Barovskij, 1922c: 292 type species *Exochomus pubescens* Küster, 1848
bellus Wollaston, 1864: 425 (*Epilachna*) **N**: CI
melanocephalus Zubkov, 1833: 339 (*Coccinella*) **E**: AB AR BU GG TR **A**: IN KZ TD TM TR UZ
russicollis Mulsant, 1850b: 1033 (*Exochomus*)
nigripennis Erichson, 1843a: 267 (*Chilocorus*) **E**: IT PT SP **N**: AG CI EG LB MA MO TU **A**: AE AF IN IQ IS LE PA SA SI SY **AFR**
xanthoderus Fairmaire, 1864: 648 (*Exochomus*)
nigromaculatus Goeze, 1777: 248 (*Coccinella*) **E**: AB AL AR AU BE BH BU CR CZ DE EN FE FI FR GB GE GG GR HU IT KI KZ MC NL NT RO SK SL SP ST SV TR UK YU **A**: AF CY ES FE IN IQ IS KI KZ LE PA SY TD TR UZ WS XIN
auritus L. G. Scriba, 1791: 105 (*Coccinella*)
collaris Küster, 1849a: 100 (*Exochomus*)
humeralis Townson, 1800: 167 (*Coccinella*)
pyrenaeus Kraatz, 1873a: 194 (*Exochomus*)
rufipes Stephens, 1832a: 375 (*Chilocorus*)
testudinaris Geoffroy, 1785: 151 (*Coccinella*)
pubescens Küster, 1848a: 94 (*Exochomus*) **E**: FR GR IT SP **N**: AG EG LB MO TU **A**: AF IN IQ IS SA **AFR**
apicatus Fairmaire, 1884d: lxix (*Exochomus*)
circumcinctus J. R. Sahlberg, 1903b: 36 (*Exochomus*)
flavilabris Motschulsky, 1849: 155 (*Platynaspis*)
flavilabris Mulsant, 1850b: 947 (*Platynaspis*) [HN]
gestroi Fairmaire, 1875a: 540 (*Exochomus*)
lugubrivestis Mulsant, 1853: 194 (*Exochomus*)
saharae Sicard, 1929a: 60 (*Exochomus*)
quadriplagiatus Wollaston, 1864: 425 (*Epilachna*) **N**: CI MO
olcesi Crotch, 1874a: 195 (*Brumus*)
tetradymus Fairmaire, 1867: 416 (*Micraspis*)
semenowi J. Weise, 1887b: 211 (*Exochomus*) **A**: KI KZ MG XIN
kirgizorum Barovskij, 1922c: 290 (*Exochomus*)
sjoestedti J. Weise, 1910d: 260 (*Exochomus*) **A**: SA **AFR**
troberti troberti Mulsant, 1850b: 482 (*Exochomus*) **N**: CI **AFR**
wraniki Fürsch, 1989b: 37 (*Exochomus*) **A**: YE

genus *Phaenochilus* J. Weise, 1895a: 135 type species *Phaenochilus punctifrons* J. Weise, 1895
metasternalis Miyatake, 1970: 334 **A**: GUA GUX HAI HUN JIA JIX XIZ YUN **ORR**
mikado Lewis, 1896a: 32 (*Chilocorus*) **A**: JA (Kyushu)

genus *Priscibrumus* Kovář, 1997: 114 type species *Exochomus puniceipennis* Semenov, 1900
disjunctus Canepari, 1997: 45 **A**: NP
himalayaensis Kapur, 1958: 333 (*Exochomus*) **A**: NP
lituratus Gorham, 1894a: 203 (*Exochomus*) **A**: HP PA UP **ORR**
puniceipennis Semenov, 1900b: 684 (*Exochomus*) **A**: TD
trijunctus Kapur, 1973: 377 (*Exochomus*) **A**: KA
trubetzkoi Barovskij, 1927c: 251 (*Exochomus*) **A**: PA
uropygialis Mulsant, 1853: 196 (*Exochomus*) **A**: BT NP SD UP **ORR**

genus *Simmondsius* Ahmad & Ghani, 1966: 9 type species *Simmondsius pakistanensis* Ahmad & Ghani, 1966
pakistanensis Ahmad & Ghani, 1966: 9 **A**: PA

genus *Xanthocorus* Miyatake, 1970: 312 type species *Exochomus nigromarginatus* Miyatake, 1970
nigromarginatus Miyatake, 1970: 312 (*Exochomus*) **A**: FUJ GAN HUB HUN JIX SCH YUN ZHE

tribe Platynaspidini Mulsant, 1846

genus *Phymatosternus* Miyatake, 1961d: 167 type species *Platynaspis lewisii* Crotch, 1874
gressitti Miyatake, 1961d: 174 **A**: FUJ GUA HAI YUN
hainanensis Miyatake, 1961d: 172 **A**: HAI **ORR**

huangea C. Cao & Xiao, 1984: 119 (*Platynaspis*) **A**: YUN
khalaus Bielawski, 1979: 112 **A**: BT
lanyuanus Sasaji, 1991c: 49 **A**: TAI
lewisii Crotch, 1874a: 189 (*Platynaspis*) **A**: FUJ GUA GUI GUX HEN HUB JA JIA JIL JIX NC SC SHA SHG TAI YUN ZHE **ORR**
obscurus Sicard, 1907c: 211 (*Platynaspis*)
octoguttatus Miyatake, 1961d: 173 **A**: FUJ GUA
samchus Bielawski, 1979: 111 **A**: BT
saundersii Crotch, 1874a: 197 (*Platynaspis*) **A**: AF HP KA NP PA UP
tricolor Hoang, 1983a: 15 **A**: GUA **ORR**
wittmeri Bielawski, 1979: 110 **A**: BT

genus *Platynaspidius* Miyatake, 1961d: 161 type species *Platynaspis maculosa* J. Weise, 1910
Paraplatynaspis Hoang, 1983a: 8 type species *Paraplatynaspis bimaculatus* Hoang, 1983 (= *Platynaspis variegatus* Crotch, 1874)
angulimaculatus Mader, 1938: 51 (*Platynaspis*) **A**: GUA SCH XIZ YUN
babai Sasaji, 1988b: 41 **A**: TAI
maculosus J. Weise, 1910a: 48 (*Platynaspis*) **A**: FUJ GUA GUI GUX HAI HUB JIA JIX SCH SHA TAI ZHE **ORR**
quinquepunctatus Miyatake, 1961d: 165 **A**: TAI
sexmaculatus C. Cao & Xiao, 1984: 118 (*Platynspis*) **A**: YUN
variegatus Crotch, 1874a: 198 (*Platynaspis*) **A**: HAI **ORR**
bimaculatus X.-F. Pang & Mao, 1979: 94 (*Platynaspis*)
bimaculatus Hoang, 1983a: 9 (*Paraplatynaspis*) [HN]

genus *Platynaspis* L. Redtenbacher, 1843a: 11 type species *Coccinella bisbipustulata* Fabricius, 1792 (= *Coccinella luteorubra* Goeze, 1777)
Microrrhymbus Gerstaecker, 1871: 348 type species *Coccinella mesomela* Klug, 1833
luteorubra Goeze, 1777: 247 (*Coccinella*) **E**: AB AL AR AU BE BH BU BY CR CT CZ DE EN FI FR GE GG GR HU IT LA LT LU MC MD NL NR NT PL PT RO SK SL SP ST SV SY TR UK YU **N**: AG MO TU **A**: AF IN IQ IS JO KZ PA SY TM TR UZ WS
bisbipustulata Fabricius, 1792a: 288 (*Coccinella*)
bisbiverrucata Paykull, 1798: 147 (*Coccinella*)
confluens Mulsant, 1846: 217
karamani J. Weise, 1885a: 55
pubescens A. G. Olivier, 1791: 81 (*Coccinella*)
quadriguttata Brahm, 1790: 221 (*Coccinella*)
quadrimaculata P. Rossi, 1790: 71 (*Coccinella*) [HN]
quadrinotata Dufour, 1843a: 118 (*Coccinella*)
quadripustulata Kugelann, 1794: 547 (*Scymnus*) [HN]
rufimana Eschscholtz, 1818: 464 (*Coccinella*)
spectabilis Faldermann, 1837: 410 (*Scymnus*)
villosa Geoffroy, 1785: 149 (*Coccinella*)

tribe Platynaspidini, species incertae sedis
Platynaspis trimaculata J. Weise, 1910a: 49 **A**: SD

tribe Telsimiini Casey, 1899

genus *Telsimia* Casey, 1899: 165 type species *Telsimia tetrasticta* Casey, 1899
Lipernes Blackburn, 1899: 212 [HN] type species *Lipernes angulatus* Blackburn, 1889
Notolipernes Blackburn, 1900a: 68 [RN]
chujoi Miyatake, 1959: 140: **A**: JA (Ryukyus) TAI **AUR**
darjeelingensis Kapur, 1969: 50 **A**: SD
emarginata E. A. Chapin, 1926: 129 **A**: FUJ GUA GUX HEN HUB JIL SCH ZHE
huiliensis X.-F. Pang & Mao, 1979a: 102 **A**: SCH YUN
jinyangiensis X.-F. Pang & Mao, 1979a: 102 **A**: GUX SCH YUN

nagasakiensis Miyatake, 1978: 13 **A**: JA SC TAI
nigra J. Weise, 1879a: 149 (*Pentilia*) **A**: FUJ HUB JA SCH SHA TAI YUN **ORR**
 centralis X.-F. Pang & Mao, 1979a: 100
nitida E. A. Chapin, 1926: 131 **A**: JA (Bonin) **AUR**
scymnoides Miyatake, 1978: 15 **A**: TAI
shirozui Miyatake, 1965b: 55 **A**: TAI
sichuanensis X.-F. Pang & Mao, 1979a: 101 **A**: FUJ GUX SCH SHA

subfamily **Ortaliinae** Mulsant, 1850

tribe Noviini Mulsant, 1850

genus *Novius* Mulsant, 1846: [4, in Addenda et Errata] [RN] type species *Novius cruentatus* Mulsant, 1846
 Nomius Mulsant, 1846: 213 [HN] type species *Novius cruentatus* Mulsant, 1846
canariensis Korschefsky, 1935b: 2 **N**: CI
conicollis Korschefsky, 1935b: 1 **N**: CI
cruentatus Mulsant, 1846: 214 (*Nomius*) **E**: AU CR FR GE GR IT PL PT SP **N**: AG **A**: TR
 algiricus Crotch, 1874a: 283
 decempunctatus Kraatz, 1862c: 272

genus *Rodolia* Mulsant, 1850b: 902 type species *Rodolia ruficollis* Mulsant, 1850
 Macronovius J. Weise, 1885a: 63 type species *Novius limbatus* Motschulsky, 1866
breviuscula J. Weise, 1892a: 26 **A**: SD **ORR**
cardinalis Mulsant, 1850b: 906 (*Vedalia*) **E**: AL BU CR FR GR IT MA PT SP TR **N**: CI AG EG LB MO TU **A**: AE FUJ HAI GUA GUI HAI HEN HKG HUB IN IQ IS JA JIA JIX JO PA SC SCH SHA SHG TAI TR YUN ZHE **AFR AUR NAR NTR ORR**
 aegyptiaca Sicard, 1907a: 67
concolor Lewis, 1879d: 466 (*Novius*) **A**: FE GAN HEB HEN JA JIA NC SC
fausti J. Weise, 1885a: 63 (*Macronovius*) **A**: IN KI KZ UZ
formosana Korschefsky, 1935c: 255 **A**: TAI
fumida Mulsant, 1850b: 904 **A**: FUJ GUA PA SD UP YUN **AFR ORR**
 arethusa Mulsant, 1853: 254 (*Epilachna*)
 roseipennis Mulsant, 1850b: 904
 testicolor Mulsant, 1853: 255 (*Epilachna*)
hauseri Mader, 1930b: 181 (*Novius*) **A**: YUN SCH
limbata Motschulsky, 1866a: 178 (*Novius*) **A**: BEI ES FE HUB JA JIL NC SC "Mandzhuria"
 narae Lewis, 1896a: 40
octoguttata J. Weise, 1910a: 51 **A**: NP SCH SD UP **ORR**
pumila J. Weise, 1892a: 26 **A**: HKG JA (Bonin, Ryukyus) YUN TAI **AUR ORR**
 okinawensis Miyatake, 1959: 127
quadrimaculata Mader, 1939a: 48 **A**: GUA GUX HEI HEN HUB HUN JA LIA NMO SCH SHA TAI YUN
rubea Mulsant, 1850b: 903 **A**: FUJ GUA HAI HUB ZHE **ORR**
rufipennis Pic, 1926o: 230 (*Macronovius*) **N**: EG LB **A**: IN
rufocincta Lewis, 1896a: 40 **A**: BEI JA SC
rufopilosa Mulsant, 1850b: 903 **A**: ANH FUJ GAN GUA GUI GUX HAI HKG HUB HUN JIA JIX SCH SHA SHG XIN XIZ YUN ZHE **ORR**
 rowlandi Crotch, 1874a: 281
sexmaculata Korschefsky, 1940: 2 **A**: UP
sexnotata Mulsant, 1850b: 807 (*Epilachna*) **A**: BT FUJ GUA GUX NP SCH SD UP YUN ZHE **ORR**
 guerinii Crotch, 1874a: 282 (*Vedalia*)
 immsi J. Weise, 1912: 120
xianfengensis Xiao, 1993: 377 **A**: HUB

tribe Ortaliini Mulsant, 1850

genus *Amida* Lewis, 1896a: 34 type species *Scymnus tricolor* Harold, 1878
decemmaculata G.-Y. Yu, 2000a: 25 **A**: YUN
nigropectoralis X.-F. Pang & Mao, 1979a: 39 (*Ortalia*) **A**: YUN

quinquefasciata Hoang, 1982a: 185 **A**: HUN SCH **ORR**
tricolor formosana Kurisaki, 1920b: 407 **A**: TAI
formosana J. Weise, 1923: 185 [HN]
tricolor tricolor Harald, 1878: 87 (*Scymnus*) **A**: GUA JA **ORR**
vietnamica Hoang, 1982a: 189 **A**: GUX YUN **ORR**

genus *Anortalia* J. Weise, 1902b: 517 type species *Anortalia fleutiauxi* J. Weise, 1902
fleutiauxi J. Weise, 1902b: 518 **A**: YUN **ORR**
jinghongiensis X.-F. Pang & Mao, 1979a: 39 (*Ortalia*)
platyceps Hoang, 1982a: 187 (*Amida*)

genus *Ortalia* Mulsant, 1850b: 893 type species *Ortalia variata* Mulsant, 1850
bruneiana Bielawski, 1965a: 26 **A**: YUN **ORR**
menlungensis X.-F. Pang & Mao, 1979a: 40
chlorops Gorham, 1903: 347 **A**: UP **ORR**
horni J. Weise, 1901b: 433 **A**: GUA HAI SCH YUN **ORR**
yunnanensis X.-F. Pang & Mao, 1979a: 41

subfamily **Coccinellinae** Latreille, 1807
tribe Halyziini Mulsant, 1846

genus *Halyzia* Mulsant, 1846: 148 type species *Coccinella sedecimguttata* Linnaeus, 1758
maculata Jing, 1987a: 201 **A**: NP XIZ YUN
sanscrita Mulsant, 1853: 152 **A**: BT FUJ GAN GUI GUX HEB HEN HUN JIA JIX NP SD SCH SHA TAI XIZ YUN ZHE **ORR**
sedecimguttata Linnaeus, 1758: 367 (*Coccinella*) **E**: AB AL AR AU BE BH BU BY CR CT CZ DE EN FI FR GB GE GG GR HU IR IT LA LS LT LU MC NL NR NT PT RO SK SL SP UK YU **A**: HEB HEI JA JIL FE KZ MG TR SCH SHA WS XIN
pallasi Mulsant, 1850b: 1025
shirozui Sasaji, 1982: 36 **A**: TAI
straminea Hope, 1831: 31 (*Coccinella*) **A**: GAN NP SCH SD UP XIZ YUN **ORR**
tschitscherini Semenov, 1895: 142 **A**: AF KA KI KZ PA TD UZ
reitteri A. Fleischer, 1900a: 117

genus *Illeis* Mulsant, 1850b: 1026 [RN] type species *Coccinella cincta* Fabricius, 1798
subgenus *Illeis* Mulsant, 1850b: 1026 [RN] type species *Coccinella cincta* Fabricius, 1798
Egleis Mulsant, 1850a: 166 [HN] type species *Coccinella cincta* Fabricius, 1798
bistigmosa Mulsant, 1850a: 168 (*Egleis*) **A**: GUA GUX NP YUN **ORR**
simplex Mulsant, 1866: 128 (*Psyllobora*)
chinenis Iablokoff-Khnzorian, 1978e: 182 **A**: GUX HKG YUN
cincta Fabricius, 1798: 77 (*Coccinella*) **A**: UP **ORR**
confusa Timberlake, 1943: 44 **A**: FUJ HKG NP SHG
indica Timberlake, 1947: 61 **A**: HP PA SD UP **ORR**
kapuri Anand, Gupta & Ghani, 1990: 131 **A**: HP
koebelei amamiana Miyatake, 1959: 158 **A**: JA (Ryukyus)
koebelei koebelei Timberlake, 1943: 44 **A**: FUJ GUA GUI GUX HEB HEN HUB HUN JA JIA JIX SC SCH SHA SHX TAI YUN **AUR ORR**
shensiensis Timberlake, 1943: 61 **A**: FUJ GUX HEB SCH SHA TAI YUN

genus *Illeis*, species incertae sedis
nevili Dohrn, 1882: 374 (*Coccinella*) **A**: NP **ORR**

genus *Macroilleis* Miyatake, 1965b: 71 type species *Halyzia hauseri* Mader, 1930
Anchilleis Iablokoff-Khnzorian, 1979: 49 type species *Thea chapuisi* Crotch, 1874
hauseri Mader, 1930: 162 (*Halyzia*) **A**: BT HUB HUN SCH SHA TAI YUN **ORR**

genus *Psyllobora* Chevrolat, 1836: 434 type species *Coccinella lineola* Fabricius, 1792 (= *Psyllobora fabricii* Crotch, 1871
subgenus *Psyllobora Chevrolat*, 1836: 434 type species *Coccinella lineola* Fabricius, 1792
bisoctonotata Mulsant, 1850a: 204 (*Vibidia*) **N**: EG **A**: AE AF IN IQ IS JO KA PA SA SY UP YE **AFR ORR**
artemysiae Wollaston, 1867: 158 (*Coccinella*)
tricincta Sicard, 1931: 228 (*Vibidia*)
octodecimmaculata Thunberg, 1795: 109 (*Coccinella*) **A**: YE **AFR**
octodecimsignata Mulsant, 1866: 18
variegata Fabricius, 1781: 99 [HN]
vigintimaculata Say, 1824b: 96 (*Coccinella*) **A**: JAi **NAR**
obsoleta Casey, 1899: 101
separata Casey, 1899: 102
taedata LeConte, 1860: 70
subgenus *Thea* Mulsant, 1846: 159 type species *Coccinella vigintiduopunctata* Linnaeus, 1758
vigintiduopunctata Linnaeus, 1758: 366 (*Coccinella*) **E**: AB AL AR AU AZ BE BH BU BY CR CT CZ DE EN FI FR GB GE GG GR HU IR IT LA LS LT LU MA MC MD NL NR NT PL PT RO SK SL SP ST SV UK YU **N**: AG MO TU **A**: AF BEI ES FE HEN IN IQ IS KI KZ LE MG NC SC SCH SHG SY TD TR UZ WS XIN
bisdecempunctata Duméril, 1817: 495 (*Coccinella*)
circularis W. G. Schneider, 1881: 15 (*Halyzia*)
ellipsoidea Gradl, 1882: 328 (*Halyzia*)
flaviventris L.W. Schaufuss, 1862b: 50 (*Thea*)
lateripunctata J. Weise, 1879c: 125 (*Halyzia*)
signifera J. Weise, 1879c: 125 (*Halyzia*)
vigintipunctata Fabricius, 1775: 84 (*Coccinella*)
vigintiseptempunctata Motschulsky, 1849: 154 (*Coccinella*)
vulgaris Haworth, 1812: 282 (*Coccinella*) [HN]

genus *Vibidia* Mulsant, 1846: 150 type species *Coccinela duodecimguttata* Poda von Neuhaus, 1761
duodecimguttata Poda von Neuhaus, 1761: 25 (*Coccinella*) **E**: AR AU BE BH BU BY CR DE CZ FR GE GG GR HU IT LT LU NR NT PL PT RO SK SP SV SZ UK **A**: BEI ES FE FUJ GAN GUA GUI GUX HEB HEN HUB HUN IN JA JIL KZ MG NC QIN SC SCH TR WS XIZ YUN **ORR**
bissexguttata Fabricius, 1792a: 284 (*Coccinella*)
eburnea Bellier de la Chavignerie, 1870: xxviii (*Calvia*)
hispanica J. Weise, 1879c: 124 (*Halyzia*)
patricia J. Weise, 1879c: 124 (*Halyzia*)
tarnensis Sicard, 1891: 59
huiliensis X.-F. Pang & Mao, 1979b: 469 **A**: SCH
korschefskyi Mader, 1930: 161 (*Halyzia korschefskii*) **A**: FUJ SCH SHA YUN
luliangensis C. Cao & Xiao, 1984: 109 **A**: SCH YUN
nagayamai M. Araki, 1961: 153 **A**: FE JA YUN
xichangiensis X.-F. Pang & Mao, 1979b: 468 **A**: SCH
zhongdianensis Jing, 1992: 547 **A**: YUN

tribe Singhikaliini Miyatake, 1972

genus *Singhikalia* Kapur, 1963a: 16 type species *Singhikalia ornata* Kapur, 1963
Subepilachna Bielawski, 1963: 445 type species *Subepilachna latemarginata* Bielawski, 1963
duodecimguttata Xiao, 1993: 378 **A**: SCH
ornata Kapur, 1963a: 17 **A**: SD TAI **ORR**
subfasciata Miyatake, 1972: 92

tribe Tytthaspidini Crotch

genus *Anisosticta* Chevrolat, 1836: 432 type species *Coccinella novemdecimpunctata*: Fabricius, 1775 (=*Coccinella novemdecimpunctata* Linnaeus, 1758)
bitriangularis Say, 1824c: 296 **A**: FE
caucasica A. Fleischer, 1900a: 116 (*Adonia*) **E**: AR **A**: IN

kobensis Lewis, 1896a: 25 **A**: BEI FE HEN HUB HUN JA JIA JIX NC SC SHA SHN XIN
novemdecimpunctata Linnaeus, 1758: 366 (*Coccinella*) **E**: AL AR AU BE BH BU BY CR CT CZ DE EN FI FR GB GE GR HU IR IT LA LS LT LU NL NR NT PL PT RO SP SV UK YU **N**: EG TU **A**: ES FE IS WS
athesis J. Weise, 1881: 165
conjuncta J. Weise, 1879c: 94
egena J. Weise, 1887c: 335
flavicollis J. R. Sahlberg, 1903b: 85
gagliardii Della Beffa, 1913: 17
intermedia W. G. Schneider, 1881: 11
irregularis J. Weise, 1879c: 94
jakutensis Dobzhanskiy, 1926b: 3
maculata Cornelius, 1884: 55
pallida P. Rossi, 1790: 66 (*Coccinella*) [HN]
palustris Walter, 1882: 17
parvipunctata J. R. Sahlberg, 1913d: 260
pseudostrigata Palm, 1959: 107
thoracica J. Weise, 1879c: 94
tiesenhauseni J. Weise, 1881: 156
weisei G. Rossi, 1882: 214
westhoffi G. Rossi, 1882: 214
sibirica Bielawski, 1958: 92 **A**: ES FE MG WS
strigata Thunberg, 1795: 113 (*Coccinella*) **E**: FI NR NT SV **A**: ES FE WS
dohrniana Mulsant, 1856: 136
terminassianae Bielawski, 1959: 851 **A**: BEI ES FUJ GAN HEB HEN MG SHG

genus *Bulaea* Mulsant, 1850a: 69 type species *Coccinella lichatschovii* Hummel, 1827
lichatschovii Hummel, 1827b: 43 (*Coccinella*) **E**: AB AR BU CR CZi FR GG GR HU IT MD PLi RO ST TR UK **A**: AF IN KA KI KZ MG TD TM TR UZ XIN
coronata J. Weise, 1879c: 105
gebleri J. Weise, 1879c: 105
gilgitensis Fürsch, 1979: 242
novemdecimnotata Gebler, 1830: 225 (*Coccinella*)
salina J. Weise, 1879c: 105
lividula bocandei Mulsant, 1850a: 71 **N**: AG EG MO **A**: AE IN IQ IS JO PA SA SY YE **AFR**
albiventris Fürsch, 1971: 53 (*Thea*)
flavidula Mulsant, 1850a: 54
pallida Motschulsky, 1849: 154 (*Coccinella*) [HN]
pallida Mulsant, 1850a: 73 [HN]
quinquedecimpunctata Plaza, 1975: 104
suturella J. Weise, 1879: 105
lividula lividula Mulsant, 1850b: 1016 **E**: SP

genus *Coccinula* Dobrzhanskiy, 1925: 245 type species *Coccinella quatuordecimpustulata* Linnaeus, 1758
crotchi Lewis, 1897: 466 (*Coccinella*) **A**: JA
elegantula J. Weise, 1890c: 487 (*Coccinella*) **E**: ST **A**: IN KI KZ MG TM UZ
oresitropha Iablokoff-Khnzorian, 1972a: 116 **A**: KZ UZ
principalis J. Weise, 1887c: 335 (*Coccinella*) **A**: KZ
principalis Savoyskaya, 1969b: 104 [HN]
turkestanica A. Fleischer, 1900a: 116 (*Coccinella*)
quatuordecimpustulata Linnaeus, 1758: 368 (*Coccinella*) **E**: AB AL AR AU BE BH BU BY CR CT CZ DE EN FI FR GE GG GR HU IT LA LT LU MC MD NL NR NT PL RO SK SL ST SZ TR UK YU **A**: ES IN KI TR SY UZ WS XIN **AFR**
abundans J. Weise, 1879c: 116 (*Coccinella*)
bulaeoiedes Bedel, 1892: 36 (*Coccinella*)
cingulata J. Weise, 1879c: 115 (*Coccinella*)
colligata J. Weise, 1879c: 115 (*Coccinella*)
effusa J. Weise, 1879c: 116 (*Coccinella*)
leprieuri Pic, 1912r: 82 (*Coccinella*)

leucocephala Gmelin, 1790: 1662 (*Coccinella*)
mulsa J. Weise, 1903: 582 (*Coccinella*)
nigropicta J. Weise, 1879c: 116 (*Coccinella*)
quatuordecimguttata Donndorf, 1799: 34 (*Coccinella*) [HN]
quatuordecimmaculata Poda von Neuhaus, 1761: 26 (*Coccinella*)
taeniolata J. Weise, 1879c: 116 (*Coccinella*)
redimita J. Weise, 1885b: 234 (*Coccinella*) **A**: AF IN TM UZ
sinensis J. Weise, 1889c: 575 (*Coccinella*) **A**: BEI ES FE GAN HEB HEI JA JIL LIA MG NC NIN NMO SC SCH SHA SHN SHX
sinuatomarginata Faldermann, 1837: 404 (*Coccinella*) **E**: AB AL AR BU CZ FR HU GG IT MC MD PT RO SK SP ST UK **N**: MO **A**: AF ES (Tuva) IN IQ KI KZ LE SY TD TM TR UZ XIN
ambigua Gredler, 1857: 74 (*Coccinella*)
bellieri Chevrolat, 1866b: 324 (*Hyperaspis*)
ghilianii Bellier de la Chavignerie, 1870: xxvii (*Coccinella*)
trilobata A. Fleischer, 1900c: 245 (*Coccinella*)
vittula J. Weise, 1903: 582 (*Coccinella*)

genus *Isora* Mulsant, 1850a: 341 type species *Isora anceps* Mulsant, 1850
anceps Mulsant, 1850a: 341 **N**: EG **AFR**
circularis Mader, 1941a: 196

genus *Macronaemia* Casey, 1899: 76 type species *Coccinella episcopalis* Kirby, 1837
subgenus *Allonaemia* Iablokoff-Khnzorian, 1979: 44 type species *Coleomegilla paradoxa* Mader, 1947
paradoxa Mader, 1947: 1 (*Coleomegilla*) **A**: YUN
subgenus *Macronaemia* Casey, 1899: 76 type species *Coccinella episcopalis* Kirby, 1837
Micronaemia J. Weise, 1905c: 218 [RN]
hauseri J. Weise, 1905c: 217 (*Naeemia*) **A**: YUN

genus *Tytthaspis* Crotch, 1874a: 181 type species *Coccinella sedecimpunctata* Linnaeus, 1758
Barovskia Iablokoff-Khnzorian, 1979: 44 type species *Micraspis gebleri* Mulsant, 1850
gebleri Mulsant, 1850b: 1026 (*Micraspis*) [RN] **A**: KZ ST UZ WS
lineola Gebler, 1843: 39 (*Coccinella*) [HN]
lateralis A. Fleischer, 1900a: 118 **A**: MG NMO
phalerata A. Costa, 1849: 57 (*Micraspis*) **E**: IT **N**: AG TU
conjuncta Normand, 1936c: 152
sedecimpunctata Linnaeus, 1761: 156 (*Coccinella*) **E**: AL AU BE BH BU BY CR CT DE EN FI FR GB GR HU IR IT LA LT LU MC MD NL NR NT PL PT RO SK SP ST SZ UK YU **N**: MO **A**: SY TR XIN
communis J. Weise, 1879c: 127 (*Micraspis*)
duodecimpunctata Linnaeus, 1767: 581 (*Coccinella*)
flavida Bovie, 1897: 156 (*Micraspis*)
flavidula J. Weise, 1881: 166 (*Micraspis*)
haemorrhoidalis Della Beffa, 1913: 164 (*Micraspis*)
italica J. Weise, 1879c: 127 (*Micraspis*)
octodecimpunctata D. H. Schneider, 1787: 107 (*Coccinella*) [HN]
poweri J. Weise, 1879c: 127 (*Micraspis*)
primita Gabriel, 1905a: 142 (*Micraspis*)
suturata Goeze, 1777: 246 (*Coccinella*)
undecimpunctata Gmelin, 1790: 1652 (*Coccinella*) [HN]

tribe Coccinellini Latreille, 1807

genus *Aaages* Barovskij, 1926b: 69 type species *Aaages prior* Barovskij, 1926
prior Barovskij, 1926b: 69 **A**: GAN MG NMO QIN SCH YUN

genus *Adalia* Mulsant, 1846: [Addenda et errata, 2] type species *Coccinella bipunctata* Linnaeus, 1758
subgenus *Adalia* Mulsant, 1846 type species *Coccinella bipunctata* Linnaeus, 1758
Arrowella Brèthes, 1925a: 147 type species *Arrowella porteri* Brèthes, 1925 (= *Coccinella bipunctata* Linnaeus, 1758)
Idalia Mulsant, 1846: 44 [HN] type species *Coccinella bipunctata* Linnaeus, 1758

bipunctata Linnaeus, 1758: 364 (*Coccinella*) **E**: AB AL AN AR AU BE BH BU BY CR CT CZ DE EN FI FR GB GE GG GR HU IR IT LA LS LT LU NL NR NT PT RO SK SL SP ST SV SZ TR UK YU **N**: CI MR **A**: AF BEI FE FUJ GAN HEB HEI HEN HUB IN IQ IS JA JIA JIL JIX JO KI LE LIA MG NIN NMO SCH SHA SHN SHX SY TD TM TR UZ XIN XIZ YUN ZHE **AFR AUR NAR NTR**

anglicana J. Weise, 1879c: 99
annectens Crotch, 1873d: 371 (*Coccinella*)
annulata Voet, 1766: 2 pl. 45 (*Coccinella*)
annulata Linnaeus, 1767: 579 (*Coccinella*) [HN]
arctica Thunberg, 1795: 108 (*Coccinella*) [HN]
bioculata Say, 1824b: 94 (*Coccinella*)
bipunctata turanica Lusis, 1973: 105
bifasciata J. Weise, 1879c: 99
bisquadripustulata Haworth, 1812: 290 (*Coccinella*)
bistriverrucata Haworth, 1812: 287 (*Coccinella*)
borealis J. Weise, 1879c: 99
boreela J. Weise, 1879c: 99
bystricensis Roubal, 1929c: 56
cincta O. F. Müller, 1776: 68 (*Coccinella*)
coloradensis Casey, 1908: 401
concolor Wimmel, 1894: 11
conjuncta W. G. Schneider, 1881: 13
damryi J. Weise, 1879c: 102
daurica J. Weise, 1879c: 98
decipiens J. Weise, 1879c: 103
decempustulata Penecke, 1901: 21
disjuncta Randall, 1838a: 33 (*Coccinella*)
dispar D. H. Schneider, 1792: 173 (*Coccinella*)
distans J. Weise, 1879c: 98
doemmlingi Meier, 1899b: 101
duodecimpustulata Dobzhanskiy, 1927a: 47
faceta J. Weise, 1879c: 99
fasciata J. Weise, 1879c: 99
fasciatopunctata Faldermann, 1835b: 455 (*Coccinella*)
flava Haworth, 1812: 272 (*Coccinella*) [HN]
frigida D. H. Schneider, 1792: 172 (*Coccinella*)
glacialis J. Weise, 1879c: 99
hastata A. G. Olivier, 1791: 77 (*Coccinella*)
herbsti J. Weise, 1879c: 99
humeralis Say, 1824b: 95 (*Coccinella*) [HN]
humeralis R. H. Johnson, 1910: 71 [HN]
hyperborea Paykull, 1799: 38 (*Coccinella*)
illuminans Dobrzhanskiy, 1924: 204
inaequalis J. Weise, 1879c: 100
interpunctata Haworth, 1812: 272 (*Coccinella*)
interrupta Meier, 1899b: 101
jacobsoni Dobrzhansky, 1924b: 203
lineata Meier, 1899b: 100
ludovicae Mulsant, 1866: 36
lugubris J. Weise, 1879c: 101
lunigera J. Weise, 1879c: 101 (*Coccinella*)
melanoleura LeConte, 1859c: 286 (*Coccinella*)
munda J. Weise, 1879c: 99
obliquebinotata Roubal, 1936a: 222
ocellata R. H. Johnson, 1910: 71
octoguttata Sulzer, 1776: 32 (*Coccinella*)
octopustulata Penecke, 1901: 21
olivieri J. Weise, 1879c: 100
opthalmica Mulsant, 1850a: 56
ornatella Casey, 1899: 86
ottomana J. Weise, 1879c: 103

ovipennis Casey, 1899: 86
pantherina Linnaeus, 1758: 368 (*Coccinella*)
parvula J. Weise, 1879c: 98
perforata Marsham, 1802: 151 (*Coccinella*)
perplexa Meier, 1899b: 100
polyguttata Meier, 1899b: 101
postica Johnson, 1910: 68 (*Adalia annectens*)
primitiva Dobrzhanskiy, 1924: 204
pruni J. Weise, 1879c: 99
pulchella J. Weise, 1879c: 98
quadrimaculata Scopoli, 1763: 80 (*Coccinella*)
quadripunctata Donovan, 1813: 5 (*Coccinella*) [HN]
reiteri [sic!] Walter, 1882: 18
revelieri Mulsant, 1866: 40
rhombipunctata Haworth, 1912: 272 (*Coccinella*)
rubiginosa J. Weise, 1879c: 99
sardiniensis J. Weise, 1879c: 102
schoenherri J. Weise, 1879c: 99
schuetti Park, 1929: 429
semirubra J. Weise, 1879c: 101
septempustulata Marsham, 1802: 165 (*Coccinella*)
sesquipunctata Haworth, 1812: 272 (*Coccinella*)
sexpustulata Linnaeus, 1758: 367 (*Coccinella*)
sexpustulata R. H. Johnson, 1910: 71 [HN]
sibirica J. Weise, 1879: 98
similata Gabriel, 1905a: 141
simulatrix Dobrzhansky, 1927a: 47
simoni J. Weise, 1879c: 101
stephensi J. Weise, 1879c: 99
stictica Mulsant, 1850a: 50 (*Adonia*)
sublunata J. Weise, 1879c: 101
thunbergi J. Weise, 1879c: 99
transversalis Casey, 1899: 86
tripunctata J. J. Roemer, 1789: 5 (*Coccinella*) [HN]
tripustulata Gmelin, 1790: 1662 (*Coccinella*)
unifasciata Fabricius, 1777: 216 (*Coccinella*)
varia Schrank, 1798: 465 (*Coccinella*)
verticalis Fürsch, 1958a: 10
vulgaris Haworth, 1812: 272 (*Coccinella*) [HN]
westmani J. Weise, 1879c: 99

decempunctata Linnaeus, 1758: 366 (*Coccinella*) **E**: AB AL AR AU BE BH BU BY CR CT CZ DE EN FR GB GE GG GR HU IR IT LA LS LT LU NL MC MD NR NT PL PT RO SK SL SP SV SZ TR UK YU **N**: AG EG MO MR TU **A**: IN IQ SI SY TR WS

affinis Walter, 1882: 18 (*Coccinella*) [HN]
albipustulata Haworth, 1812: 292 (*Coccinella*)
arvensis J. Weise, 1879c: 114 (*Coccinella*)
austriaca Schrank, 1777: 62 (*Coccinella*)
autumnalis J. Weise, 1879c: 113 (*Coccinella*)
bella J. Weise, 1879c: 114 (*Coccinella*)
biguttata Fabricius, 1787a: 59 (*Coccinella*)
bimaculata Pontoppidan, 1763: 669 (*Coccinella*)
bimaculosa Herbst, 1786: 160 (*Coccinella*)
bina Haworth, 1912: 279 (*Coccinella*)
biscutellata Mulsant, 1866: 76 (*Coccinella*)
biocellata Gebler, 1830: 223 (*Coccinella*)
bipustulata Herbst, 1783: 48 (*Coccinella*)
buddebergi Heyden, 1904: 176 (*Coccinella*)
centromaculata J. Weise, 1879c: 113 (*Coccinella*)
clathrata Schaufuss, 1862c: 53 (*Coccinella*)
confluens Haworth, 1812: 278 (*Coccinella*)

conglobata Della Beffa, 1913: 68
conjuncta Gradl, 1880: 276 (*Coccinella*) [HN]
consita J. Weise, 1879c: 113 (*Coccinella*)
consolida J. Weise, 1879c: 113 (*Coccinella*)
curvipustulata Haworth, 1812: 287 (*Coccinella*)
decemguttata Haworth, 1812: 292 (*Coccinella*) [HN]
decempustulata Linnaeus, 1758: 368 (*Coccinella*)
didyma O. F. Müller, 1776: 68 (*Coccinella*)
dorsonotata J. Weise, 1879c: 112 (*Coccinella*)
ephippiata J. Weise, 1879c: 114 (*Coccinella*) [HN]
flava Goeze, 1777: 246 (*Coccinella*)
flavipustulata Haworth, 1812: 292 (*Coccinella*)
fulva Haworth, 1812: 289 (*Coccinella*)
fulvipustulata Haworth, 1812: 292 (*Coccinella*)
grossulariae Meier, 1899b: 101 (*Coccinella*)
guttatopunctata Linnaeus, 1758: 367 (*Coccinella*)
hildae Gradl, 1882: 328 (*Coccinella*)
humeralis Schaller, 1783: 266 (*Coccinella*)
humerata A. Villa & G. B. Villa, 1835: 50 (*Coccinella*)
inconstans Schaufuss, 1862c: 52 (*Coccinella*)
innotata Haworth, 1812: 279 (*Coccinella*)
larvata Roubal, 1936a: 221
lateralis J. Weise, 1879c: 112 (*Coccinella*) [HN]
lateripuncta Gradl, 1880: 276 (*Coccinella*)
limbata Gmelin, 1790: 1662 (*Coccinella*)
limbata Gradl, 1880: 275 (*Coccinella*) [HN]
limbella J. Weise, 1879c: 115 (*Coccinella*)
loricata J. Weise, 1879c: 112 (*Coccinella*)
lunaepustulata Haworth, 1812: 288 (*Coccinella*)
lunigera Brahm, 1790: 119 (*Coccinella*)
lunularis Marsham, 1802: 24 (*Coccinella*)
lutea P. Rossi, 1794: 89 (*Coccinella*)
lutea Schaufuss, 1862c: 53 (*Coccinella*) [HN]
maertensi Roubal, 1936a: 221
malleata Gradl, 1880: 276 (*Coccinella*)
marginata Thunberg, 1784a: 10 (*Coccinella*) [HN]
marginella Thunberg, 1795: 110 (*Coccinella*) [HN]
marginepunctata Marsham, 1802: 150 (*Coccinella*) [HN]
mediopunctata Penecke, 1901: 21 (*Coccinella*)
m-fuscum Mulsant, 1866: 35
mutabilis Gmelin, 1790: 1661 (*Coccinella*)
nassovica Heyden, 1904: 176 (*Coccinella*)
nigra Haworth, 1812: 288 (*Coccinella*)
nigrina J. Weise, 1879c: 115 (*Coccinella*)
obliquata Reiche, 1862: 300 (*Coccinella*)
obliterata Haworth, 1812: 293 (*Coccinella*) [HN]
obscura J. Weise, 1879c: 114 (*Coccinella*)
octoguttata Haworth, 1812: 293 (*Coccinella*)
octopunctata O. F. Müller, 1764: 6 (*Coccinella*)
octopunctata Fabricius, 1775: 81 (*Coccinella*) [HN]
octopustulata Della Beffa, 1913: 73
pantherina DeGeer, 1775: 382 (*Coccinella*)
parvipustulata Haworth, 1812: 292 (*Coccinella*)
pellucida J. Weise, 1879c: 112 (*Coccinella*)
quadripunctata Linnaeus, 1767: 580 (*Coccinella*) [HN]
quaterna Haworth, 1812: 279 (*Coccinella*)
quina Haworth, 1812: 280 (*Coccinella*)
recurva J. Weise, 1881: 165 (*Coccinella*)
relicta Heyden, 1883a: 53 (*Coccinella*)
rubellula J. Weise, 1879c: 112 (*Coccinella*)
rubripunctata Haworth, 1812: 293 (*Coccinella*)

rubripustulata Haworth, 1812: 292 (*Coccinella*)
saalmuelleri Heyden, 1883a: 53 (*Coccinella*)
salicis J. Weise, 1879c: 113 (*Coccinella*)
scribai J. Weise, 1879c: 115 (*Coccinella*) [RN]
sedula J. Weise, 1879c: 112 (*Coccinella*)
semicruciata Gradl, 1880: 276 (*Coccinella*)
semifasciata J. Weise, 1879c: 113 (*Coccinella*)
sena Haworth, 1812: 280 (*Coccinella*)
septena Haworth, 1812: 280 (*Coccinella*)
sexmaculata Thunberg, 1795: 106 (*Coccinella*) [HN]
sexpunctata Linnaeus, 1767: 580 (*Coccinella*)
silesiaca W. G. Schneider, 1881: 13 (*Coccinella*)
similata Thunberg, 1795: 112 (*Coccinella*)
subpunctata Schrank, 1781a: 51 (*Coccinella*)
superpunctata Gradl, 1880: 275 (*Coccinella*)
suturalis Haworth, 1812: 280 (*Coccinella*)
terna Haworth, 1912: 279 (*Coccinella*)
thoracica W. G. Schneider, 1881: 14 (*Coccinella*)
thunbergi Gmelin, 1790: 1666 (*Coccinella*) [RN]
transmutata Heyden, 1904: 176 (*Coccinella*)
tredecimmaculata Forster, 1771: 18 (*Coccinella*)
tredecimnotata Thunberg, 1784a: 10 (*Coccinella*)
triangularis J. Weise, 1879c: 114 (*Coccinella*)
tricuspis Thunberg, 1795: 107 (*Coccinella*)
trigemina J. Weise, 1879c: 113 (*Coccinella*)
trilunata Gradl, 1880: 277 (*Coccinella*)
troegneri Walter, 1882: 18 (*Coccinella*)
ulmi A. G. Olivier, 1791: 83 (*Coccinella*)
undecimmaculata Walter, 1882: 18 (*Coccinella*)
undecimnotata Marsham, 1802: 155 (*Coccinella*) [HN]
unifasciata L. G. Scriba, 1791: 107 (*Coccinella*) [HN]
variabilis Fabricius, 1777: 218 (*Coccinella*)
variabilis Herbst, 1783: 42 (*Coccinella*) [HN]
variabilis Illiger, 1798: 447 (*Coccinella*) [HN]
varians A. G. Olivier, 1791: 83 (*Coccinella*)
vulgaris Haworth, 1812: 277 (*Coccinella*) [HN]
weisei Sajó, 1880: 159

lenticula Gorham, 1892a: 82 **A**: JIX
puetzi Iablokoff-Khnzorian, 1986b: 80 **A**: PA
testudinea Wollaston, 1854: 463 (*Coccinella*) **N**: MR
tetraspilota Hope, 1831: 31 (*Coccinella*) **A**: AF HP IN NP SD TD TM UP UZ XIN **ORR**
dorsimacula Fabricius, 1798: 77 (*Coccinella*) [DA]
hopii Mulsant, 1850a: 57 [RN]
severini J. Weise, 1895c: 154
simmondsi Kapur & Sudha Rao, 1962: 47

subgenus *Adaliomorpha* Iablokoff-Khnzorian, 1979: 53 type species *Coccinella conglomerata* Linnaeus, 1758
conglomerata Linnaeus, 1758: 366 (*Coccinella*) **E**: AU BE BY CT CZ FI FR GE HU NR NT PL SK SV SZ UK **A**: ES FE JA MG SHA WS YUN
bicruciata W. G. Schneider, 1881: 12
botnica [sic!] Paykull, 1799: 34 (*Coccinella*)
bothnica Mulsant, 1846: 48
clausa Roubal, 1936a: 219
connexa W. G. Schneider, 1881: 11
crucifera J. Weise, 1879c: 97
decas Beck, 1817: 16 (*Coccinella*)
destituta J. Weise, 1879c: 97
encausta J. Weise, 1879c: 97
faldermanni Hummel, 1826: 28 (*Coccinella*)
immunda J. Weise, 1879c: 97
inconstans Roubal, 1936a: 219

intermedia W. G. Schneider, 1881: 12
reinecki Roubal, 1936a: 220
reticulum J. Weise, 1879c: 97
ronina Lewis, 1896a: 27 (*Coccinella*)
subcrucifera Roubal, 1936a: 219
tricruciata W. G. Schneider, 1881: 12
tricruciataeformis Roubal, 1936a: 219

genus *Aiolocaria* Crotch, 1871: 6 type species *Coccinella 6-spilota* Hope, 1931
Ithone Solsky, 1871: 275 [HN] type species *Leis mirabilis* Motschulsky, 1860
hexaspilota Hope, 1831: 31 (*Coccinella*) **A**: ES FE FUJ GAN GUI HEB HEI HP HUB HUN JA JIL KA LIA NMO NC NP PA QIN SC SCH SHA SHX SD TAI UP XIN XIZ YUN ZHE **ORR**
aethiops J. Weise, 1885a: 51 (*Ithone*)
degenerata Semenov, 1889: 226 (*Ithone*)
mirabilis Motschulsky, 1860b: 246 (*Leis*)
nigra Gadeau de Kerville, 1884: 70 (*Ithone*)
nigripennis Lewis, 1896a: 31 (*Ithone*) [RN]

genus *Alloneda* Iablokoff-Khnzorian, 1979: 49 type species *Coccinella dodecaspilota* Hope, 1831
callinotata Jing, 1988: 60 (*Lemnia*) **A**: FUJ GUA
quadricolor G. Yu, [H. Pang] & X.-F. Pang, 1993: 496
dentiformis G.-Y. Yu, 1998b: 284 **A**: HAI
dodecaspilota Hope, 1831: 31 (*Coccinella*) **A**: BT GUA HAI NP SD UP XIZ YUN **ORR**
duodecimspilota Mulsant, 1850a: 236 (*Caria*)
osawai Sasaji, 1986a: 4 **A**: HAI TAI

genus *Anatis* Mulsant, 1846: 133 type species *Coccinella ocellata* Linnaeus, 1758
halonis Lewis, 1896a: 28 **A**: JA FE NC SC
ocellata Linnaeus, 1758: 366 (*Coccinella*) **E**: AL AU BE BH CT CZ DE EN FI FR GB GE HU IT LA LS LT LU NL NR NT PL PT RO SK SL SP SV SZ UK YU **A**: ANH ES FE HEB HEI JIL KI KZ LIA MG NC NMO SC SHX UZ WS ZHE
badensis Heyden, 1883a: 53 (*Halyzia*)
bicolor J. Weise, 1879c: 121 (*Halyzia*)
biocellata J. Weise, 1879c: 121 (*Halyzia*)
bivittata J. Weise, 1879c: 121 (*Halyzia*)
boeberi Cederjhelm, 1798: 50 (*Coccinella*)
dominula Heyden, 1883a: 53 (*Halyzia*)
egrensis Gradl, 1880: 278 (*Halyzia*)
hebraea Linnaeus, 1758: 365 (*Coccinella*)
linnei Della Beffa, 1913: 171
mobilis Mulsant, 1850b: 1022
moscovica Gmelin, 1790: 1646 (*Coccinella*)
nana Gradl, 1880: 276 (*Halyzia*)
oblongopunctata Fabricius, 1787a: 54 (*Coccinella*)
octopunctata Walter, 1882: 19 (*Halyzia*)
pallida Marriner, 1929: 164
prava Heyden, 1892: 107
sexlineata Fabricius, 1781: 94 (*Coccinella*)
subfasciata J. Weise, 1879c: 121 (*Halyzia*)
tricolor J. Weise, 1879c: 121 (*Halyzia*)
vulgaris J. Weise, 1879c: 121 (*Halyzia*)

genus *Anegleis* Iablokoff-Khnzorian, 1982: 295 type species *Coelophora cardoni* J. Weise, 1892
cardoni J. Weise, 1892a: 19 (*Coelophora*) **A**: KA "Himalaya" **ORR**

genus *Aphidecta* J. Weise, 1893a: 106 type species *Coccinella obliterata* Linnaeus, 1758
Aphidecta J. Weise, 1900a: 375 [emendation]
Aphideita J. Weise, 1893a: 106 [unused orig. spelling] type species *Coccinella obliterata* Linnaeus, 1758

obliterata Linnaeus, 1758: 367 (*Coccinella*) **E**: AB AL AR AU BE BH BU BY CR CT CZ DE EN FI FR GB GE GG GR HU IR IT LA LS LT LU NL NR NT PL RO SK SL SP ST SV SZ UK YU **A**: TR
fenestrata J. Weise, 1879c: 96 (*Adalia*)
formosa Gravenhorst, 1807: 124 (*Coccinella*)
fumata J. Weise, 1879c: 96 (*Adalia*)
illigeri J. Weise, 1879c: 96 (*Adalia*)
livida DeGeer, 1775: 383 (*Coccinella*)
m-nigrum Fabricius, 1792a: 267 (*Coccinella*)
obsoleta D. H. Schneider, 1792: 147 (*Coccinella*)
pallida Thunberg, 1784a: 9 (*Coccinella*)
sexnotata Thunberg, 1784a: 9 (*Coccinella*)
sublineata J. Weise, 1879c: 96 (*Adalia*)

genus *Bothrocalvia* Crotch, 1874a: 143 type species *Coccinella albolineata* Gyllenhal, 1808
albolineata Gyllenhal, 1808b: 158 (*Coccinella*) **A**: FUJ JA GUA GUI GUX HEN HKG HUN JIX SCH TAI YUN ZHE **ORR**
circinatella Jing, 1992: 542 **A**: SCH
decemsignata Iablokoff-Khnzorian, 1982: 174 **A**: NP **ORR**
lewisii Crotch, 1874a: 143 (*Bothrocalvia*) **A**: BT FUJ GUA GUX **ORR**
pupillata Swartz, 1808: 184 (*Coccinella*) **A**: GUA GUX HAI HKG SCH YUN **ORR**
annectens Timberlake, 1943: 34

genus *Callicaria* Crotch, 1871: 6 type species *Caria superba* Mulsant, 1853
Pseudosynonycha Kurisaki, 1923: 101 type species *Synonycha japonica* Kurisaki, 1915
superba Mulsant, 1853: 159 (*Caria*) **A**: BT FUJ GAN GUI HP JA NP SCH SD SHA TAI UP XIZ YUN ZHE **ORR**
japonica J. Kurisaki, 1914: 443 (*Synonycha*) [HN]
japonica Sicard, 1907c: 211 (*Caria*)

genus *Calvia* Mulsant, 1846: 140 type species *Coccinella decemguttata* Linnaeus, 1767
Anisocalvia Crotch, 1873a: 4 type species *Coccinella quatuordecimguttata* Linnaeus, 1758
Eocaria Timberlake, 1943: 37 type species *Eocaria muiri* Timberlake, 1943
albida Bielawski, 1972: 308 **A**: NP SD UP
breiti Mader, 1932a: 6 **A**: BT HP KA NP PA SD UP
championorum Booth, 1997: 931 **A**: GAN SCH SHA TAI UP YUN **ORR**
chinensis Mulsant, 1850a: 142 (*Sospita*) **A**: FUJ GUA GUI GUX HAI HEB HEN HKG HUN JIA JIX SCH SHA YUN ZHE
decemguttata Linnaeus, 1767: 583 (*Coccinella*) **E**: AB AR AU BE BH BU BY CR CT CZ GB GE GG HU IT LT LU NL PL RO SK SL SP ST SV SZ TR UK YU **A**: ES FE HEI JA JIL MG NC SC SCH TR WS
deflorata Solsky, 1871: 272
fulva J. Weise, 1879c: 123 (*Halyzia*)
hololeuca Mulsant, 1850b: 1024
luigionii Della Beffa, 1913: 220
major Schaller, 1783: 264 (*Coccinella*)
flaveola Booth, 1997: 932 **A**: KA PA UP **ORR**
monosha Bielawski, 1979: 122 **A**: BT
muiri Timberlake, 1943: 38 (*Eocaria*) **A**: FUJ GUI GUX HEB HEN HUB JA JIX SC SCH SHA TAI YUN
parvinotata Miyatake, 1959: 149 (*Eocaria*) **A**: JA (Ryukyu)
punctata Mulsant, 1853: 143 (*Harmonia*) **A**: AF HP IN KA KI NP PA TD UP UZ
concolor Penecke, 1901: 21 (*Halyzia*)
duplicipuncta Semenov & Dobrzhanskiy, 1924: 101
nigripennis A. Fleischer, 1900a: 118 (*Halyzia*)
obversepunctata Mulsant, 1853: 156 (*Propylea*)
quadrivittata Miyatake, 1965b: 62 (*Sospita*) **A**: TAI
quatuordecimguttata Linnaeus, 1758: 367 (*Coccinella*) **E**: AB AL AN AR AU BE BH BU BY CR CT CZ DE EN FI FR GB GE GG GR HU IR IT LA LT LU NL NR NT MC MD NL NR NT PL PT RO SK SL SP ST SV

SZ UK YU **A**: BT ES FE HEB HP HUB JA JIL KZ MG NC NMO NP SC SCH SHN TAI TR UP WS XIZ YUN **NAR ORR**

bicordifera Casey, 1924: 159 (*Anisocalvia*)
cardiscae Randall, 1838a: 32 (*Coccinella*)
complexa Nunenmacher, 1912: 448 (*Agrabia*)
dorsonotata Mulsant, 1850a: 424 (*Oenopia*)
duodecimmaculata Gebler, 1832: 76 (*Coccinella*) [HN]
elliptica Casey, 1899: 97 (*Anisocalvia*)
exoleta J. Weise, 1897c: 122 (*Halyzia*)
fausti J. Weise, 1885a: 38 (*Coccinella*)
incarnata Kirby, 1837: 231 (*Coccinella*)
jakowlewi J. Weise, 1892d: 141 (*Coccinella*)
japonica Crotch, 1874a: 144 (*Anisocalvia*)
lacustris Casey, 1924: 158 (*Anisocalvia*)
obliqua Randall, 1838a: 33 (*Coccinella*)
ocelligera J. Weise, 1879c: 122 (*Halyzia*)
postplagiata Casey, 1924: 159 (*Anisocalvia*)
quadrisignata Casey, 1924: 159 (*Anisocalvia*)
rosti J. Weise, 1891a: 282 (*Propylaea*)
scutulata J. Weise, 1879c: 123 (*Halyzia*)
similis Randall, 1838b: 50 (*Coccinella*) [HN]
sicardi Nunenmacher, 1912: 448 (*Agrabia*)
uniformis Casey, 1924: 160 (*Anisocalvia*)
vancouveri Casey, 1924: 159 (*Anisocalvia*)
victoriana Casey, 1899: 96 (*Anisocalvia*)
yuasai Nakane, 1954: 13 (*Coccinella*)
zonata J. Weise, 1885a: 38 (*Coccinella*)

quindecimguttata Fabricius, 1777: 217 (*Coccinella*) **E**: AB BE BY CT CZ FI FR GE NT PL RO SK SZ UK **A**: ANH ES FE FUJ GAN GUA GUI GUX HEN HUN JA JIA JIX MG NC SC SCH SHA SHG SHN WS XIN YUN ZHE

albomarginata Goeze, 1777: 247 (*Coccinella*)
bisseptemguttata Schaller, 1783: 265 (*Coccinella*)
bisseptempunctata Sturm, 1826: 122 (*Coccinella*)
duodecimgemmata Herbst, 1793: 376 (*Coccinella*)
marginata Geoffroy, 1785: 147 (*Coccinella*) [HN]
murasei Ohta, 1929b: 59 (*Vibidia*)

septenaria Mulsant, 1866: 116 **A**: SCH

hauseri Mader, 1930: 163

schirozui Miyatake, 1965b: 68 (*Eocaria*) **A**: TAI

schiva Kapur, 1963a: 40 **A**: BT NP SD

durgae Kapur, 1963a: 42
connexa Miyatake, 1985a: 17
pasupati Kapur, 1963a: 45
pinaki Kapur, 1963a: 43
trilochana Kapur, 1963a: 42

sichuanica Kovář, **nom. nov**. [see New acts] **A**: GAN GUI HUB HUN SCH SHA YUN

sicardi Mader, 1930: 163 [HN]

sykesii Crotch, 1874a: 146 (*Anisocalvia*) **A**: NP SD UP **ORR**

vulnerata Hope, 1831: 31 (*Coccinella*) **A**: BT NP SD UP **ORR**

buddha Crotch, 1874a: 145 (*Anisocalvia*)
flaccida Mulsant, 1853: 151
krishna Crotch, 1874a: 145 (*Anisocalvia*)
uniramosa Hope, 1831: 31 (*Coccinella*)
vishnu Crotch, 1874a: 146 (*Anisocalvia*)

genus *Ceratomegilla* Crotch, 1873d: 365 type species *Ceratomegilla ulkei* Crotch, 1873

subgenus *Adaliopsis* Capra, 1926: 195 type species *Coccinella alpina* A. Villa & G. B. Villa, 1835

alpina alpina A. Villa & G. B. Villa, 1835: 50 (*Coccinella*) **E**: AU FR (Alpes) IT (north) LS SZ

primaria Della Beffa, 1913: 116 (*Adalia*)

pedemomtana J. Weise, 1879c: 98 (*Adalia*)
sabaudica J. Weise, 1897c: 98 (*Adalia*)
tyrolensis J. Weise, 1879c: 98 (*Adalia*)
alpina redtenbacheri Capra, 1928: 10 (*Adaliopsis*) [RN] **E**: AU (Alpen) BU GE (Beiern) PL (Tatra) RO SK (Tatra) UK (Karpaty)
alpina L. Redtenbacher, 1843a: 26 (*Coccinella*) [HN]
subgenus *Ceratomegilla* Crotch, 1873d: 365 type species *Ceratomegilla ulkei* Crotch, 1873
Asemiadalia Barovskij, 1931b: 128 type species *Semiadalia heydeni* J. Weise, 1892
Semiadalia Crotch, 1874a: 98 type species *Idalia inquinata* Mulsant, 1846 (= *Coccinella notata* Laicharting, 1781)
Spiladelpha Semenov & Dobrzhanskiy, 1923: 99 type species *Spiladelpha barovskii* Semenov & Dobrzhanskiy, 1923
apicalis J. Weise, 1879c: 104 (*Adalia*) **E**: AB AR GG **A**: ES FE TR
decimguttata Jing, 1986a: 73 **A**: XIZ
heydeni J. Weise, 1892c: 109 (*Semiadalia*) **A**: AF KA KI KZ TD XIZ
andrewesi Sicard, 1914a: 498 (*Semiadalia*)
przevalskii Savoyskaya, 1962: 316 (*Semiadalia*)
notata Laicharting, 1781: 119 (*Coccinella*) **E**: AR AU BH BU BY CR CT CZ EN FI FR GE GR HU IT LA LT NR NT PL PT RO SK SV SZ UK YU **A**: KI KZ MG TD TR
c-nigrum Della Beffa, 1913: 21 (*Semiadalia*)
elongata J. Weise, 1879c: 104 (*Adalia*)
inquinata Mulsant, 1846: 67 (*Idalia*)
quinquesignata Frivaldszky, 1884: 8 (*Adalia*)
obenbergeri Mader, 1933a: 82 (*Semiadalia*) **A**: SCH YUN
potanini J. Weise, 1889c: 650 (*Semiadalia*) **A**: GAN HUB SCH SHA
lixianensis Jing, 1986c: 206 (*Asemiadalia*)
spiculimaculata Jing, 1986c: 205 (*Asemiadalia*)
rickmersi J. Weise, 1901a: 281 (*Hippodamia*) **A**: KI KZ UZ TD XIZ
rufocincta doderoi Capra, 1944: 30 (*Semiadalia*) **E**: IT (Alpi Graies, Alpi Lepontine)
grundmanni Mandl, 1953: 129 (*Semiadalia*)
rufocincta rufocincta Mulsant, 1850a: 59 (*Adalia*) **E**: IT (Alpi) SZ
schelkovnikovi Dobzhansky, 1927b: 214 (*Semiadalia*) **E**: AR GG
schneideri J. Weise, 1878: 91 (*Coccinella*) **E**: AB AR GG ST
galuschkoii N. I. Nefedov, 1962: 93 (*Semiadalia*)
ulkei kiritschenkoi Barovskij, 1925: 206 (*Spiladelpha*) **A**: KI KZ TD (Hissar Pamir)
ulkei longula Barovskij, 1925: 207 (*Spiladelpha*) **A**: MG TD "Pamir"
ulkei ulkei Crotch, 1873d: 365 **A**: ES FE KZ WS **NAR**
barovskii Semenov & Dobrzhanskiy, 1923: 99 (*Spiladelpha*)
parva Watson, 1954: 45
undecimnotata D. H. Schneider, 1792: 379 (*Coccinella*) [NP] **E**: AB AL AR AU BE BH BU BY CR CT CZ FR GE GG GR HU IT MC MD NT PL PT RO SK SL ST SZ UK YU **A**: IN IQ KI KZ TD TM TR UZ WS
bargaglii Della Beffa, 1913: 33 (*Semiadalia*)
bipunctata Sajó, 1881: 273 (*Adalia*)
cardui Brahm, 1802: 142 (*Coccinella*)
distincta L. Redtenbcher, 1843: 25 (*Coccinella*) [HN]
etrusca Della Beffa, 1913: 34 (*Semiadalia*)
fulvimana Motschulsky, 1837b: 423 (*Coccinella*)
graminis J. Weise, 1879c: 104 (*Adalia*)
hungarica Sajó, 1881: 273 (*Adalia*)
saliana Faldermann, 1837: 399 (*Coccinella*)
maritima Ménétriés, 1832: 240 (*Coccinella*)
novempunctata Geoffroy, 1785: 144 (*Coccinella*)
parvula Walter, 1882: 18 (*Adalia*)
quinquenotata Walter, 1882: 18 (*Adalia*)
rugosa Fiori, 1913: 31 (*Semiadalia*)
schaefferi J. Weise, 1879c: 104 (*Adalia*)
undecimmaculata D. H. Schneider, 1792: 165 (*Coccinella*)
wingelmuelleri Ganglbauer, 1899b: 1017 (*Coccinella*)
weisei Frivaldszky, 1892: 120 (*Adonia*) **A**: QIN XIZ

genus *Cheilomenes* Chevrolat, 1836: 435 type species *Coccinella lunata* Fabricius, 1775
Chilomenes Crotch, 1874a: 179 [unjustified emendation]
Chilomenes J. Weise, 1879c: 135 [HN] type species *Chilomenes nilotica* Mulsant, 1850 (= *Cydonia vicina* Mulsant, 1850)
Cydonia Mulsant, 1850a: 430 type species *Cydonia quadrilineata* Mulsant, 1850 (= *Cydonia propinqua* Mulsant, 1850)
Cydoria J. Weise, 1926: 227 [RN]
Selenites Hope, 1840: 157 type species *Coccinella lunata* Fabricius, 1775
lunata lunata Fabricius, 1775: 86 (*Coccinella*) **N**: EG **AFR**
rivosa Thunberg, 1781: 22 (*Coccinella*)
lunata yemenensis Fürsch, 1989b: 38 **A**: SA YE
propinqua vicina Mulsant, 1850a: 441 (*Cydonia*) **N**: AG EG LB **A**: IQ IS LE SA YE **AFR**
isis Crotch, 1874a: 180 (*Chilomenes*)
cuppigera Mulsant, 1850a: 442 (*Cydonia*)
nilotica Mulsant, 1850a: 442 (*Cydonia*)
osiris Mulsant, 1858: 321 [= 1859: 63]
posticalis Fairmaire, 1893a: 49 (*Chilomenes*)
subsignata Pic, 1911a: 11 (*Cydonia*)
unicolor Crotch, 1874a: 180 (*Chilomenes*)

genus *Coccinella* Linnaeus, 1758: 364 type species *Coccinella septempunctata* Linnaeus, 1758
subgenus *Acoccinella* Iablokoff-Khnzorian, 1979: 53 type species *Coccinella barovskii* Iablokoff-Khnzorian, 1970
barovskii Iablokoff-Khnzorian, 1970d: 252 **A**: KZ UZ
subgenus *Chelonitis* J. Weise, 1879c: 126 type species *Chelonitis venusta* J. Weise, 1879
venusta adalioides Capra, 1944: 34 (*Chelonitis*) **E**: IT (Alpi)
venusta venusta J. Weise, 1879c: 127 (*Chelonitis*) **E**: FR (Alpes)
adalioides Porta, 1949: 225 [HN]
venustula Iablokoff-Khnzorian, 1982: 396 [RN]
subgenus *Coccinella* Linnaeus, 1758: 364 type species *Coccinella septempunctata* Linnaeus, 1758
Miliziella Fillipov, 1961: 381 type species *Semiadalia turkestanica* Semenov & Dobrzhanskiy, 1923
ainu Lewis, 1896a: 27 **A**: ES FE GAN HEI HUB JA LIA NC SC SHA XIN
algerica Kovář, 1977: 231 **E**: IT (Pantelleria) **N**: AG CI LB MO TU **AFR**
explanata Miyatake, 1963c: 49 **A**: JA SC
genistae Wollaston, 1854: 464 **N**: MR
hasegawai Miyatake, 1963c: 51 **A**: FE (Kuriles) JA CH
hieroglyphica hieroglyphica Linnaeus, 1758: 365 **E**: AU BE BY CT CZ DE EN FI FR GB GE HU IR IT LA LS LT LU NL NR NT PL RO SK SP SV SZ UK **A**: ES KZ WS
areata Panzer, 1794b: 7
bistriverrucata Haworth, 1912: 287
bracchiata Gradl, 1882: 328
brunnea J. Weise, 1879c: 111 [HN]
cervicornu Gradl, 1882: 327
colmarensis Pic, 1912r: 82
curva J. Weise, 1879c: 111
flammea Voet, 1806: 68
flexuosa Fabricius, 1777: 217
fuliginosa J. Weise, 1879c: 112
lineolata Marsham, 1802: 153
luctuosa J. Weise, 1879c: 112
marginemaculata Brahm, 1790: 30
octopustulata Thunberg, 1795: 112
quadrifasciata J. Weise, 1879c: 111
quinquepunctata J. Edwards, 1914: 140 [HN]
ribis L. G. Scriba, 1791: 104
robini Pic, 1912r: 82
rufocincta Haworth, 1912: 288
schneideri Gradl, 1882: 327 [HN]
sinuata Naezen, 1792: 171
sinuosa Marsham, 1802: 160

trilineata Herbst, 1783: 46 (*Coccinella*)
hieroglyphica mannerheimi Mulsant, 1850a: 106 **A**: ES FE HEI MG NC NMO SC **NAR**
hodeki Kovář, 2005: 138 **A**: TR
iranica Dobrzhanskiy, 1926b: 26 **A**: IN TR
lama Kapur, 1963a: 35 **A**: XIZ
leonina transversalis Fabricius, 1781: 97 **A**: BT FUJ GUA HAI NP SD TAI UP YUN **AUR ORR**
contempta Boisduval, 1835: 592
racemosa Germar, 1848: 245
repanda Thunberg, 1781: 18
signata Herbst, 1793: 304
tongataboae Boisduval, 1835: 595
tricincta Fabricius, 1787a: 55
varians Fabricius, 1798: 78 [HN]
longifasciata Liu, 1962: 265 **A**: ES (Tuva) FE GAN JIL NMO MG QIN SCH XIN XIZ
nigrovittata Kapur, 1963a: 37
trilineata J. Weise, 1889c: 652 (*Tytthaspis*) [HN]
luteopicta Mulsant, 1866: 45 (*Adalia*) **A**: BT NP SD UP XIZ YUN **ORR**
magnifica L. Redtenbacher, 1843a: 24 **E**: AB AR AU BE BU BY CT CZ DE EN FI GB GE GG HU LA LT LU NL NR NT PL RO SK SL SV ST SZ UK YU **A**: ES FE GAN HEB JIA KI KZ MG NMO SHA SHN WS XIN XIZ
distincta Faldermann, 1837: 404 [HN]
dobzhanskii C.-F. Wu, 1938: 581
domiduca J. Weise, 1879c: 108
intertexta J. Weise, 1879c: 109
labilis Mulsant, 1846: 84
magnopunctata Rybakow, 1889a: 289 **A**: GAN GUI KA MG QIN SCH XIN XIZ
semenowi J. Weise, 1889c: 651
marussii Kapur, 1973: 374 **A**: KA PA
nivicola Mulsant, 1850a: 109 **A**: ES FE MG NMO WS XIN
whitii Mulsant, 1850a: 108
quinquepunctata Linnaeus, 1758: 365 **E**: AB AL AR AU BE BH BU BY CT CZ DE EN FI FR GE GG HU IT LA LT LU NL NR NT PL RO SK SL SP ST SV SZ TR UK YU **N**: MO **A**: ES FE GAN KZ MG TR WS
arcuata J. Weise, 1879c: 110 [HN]
arthurica Jakobson, 1903: xii
cordia Jänichen, 1913: 552
elberti Flach, 1888b: 298
heraldica Schirmer, 1913: 317
jucunda J. Weise, 1879c: 110
kniephoffi Della Beffa, 1913: 128
minckwitzi Della Beffa, 1913: 126
multipunctata J. Weise, 1879c: 110
rossii J. Weise, 1879c: 109 [RN]
schreitteri A. Fleischer, 1886: 235
simplex J. Weise, 1879c: 109 [HN Thunberg, 1820: 363]
simulatrix J. Weise, 1879c: 110
subimpunctata Heyden, 1863: 50
tripunctata P. Rossi, 1790: 61 [HN]
undecimpunctatoides Munster, 1923: 241
reitteri J. Weise, 1891a: 282 **A**: AF KI KZ TD TM UZ XIN
sachalinensis Ohta, 1928: 112 **A**: FE HEI NC SC
saucerottii Mulsant, 1850a: 111 **E**: BY CZ CT PL NT RO SK ST UK **A**: ES KZ TD TR UZ WS XIN
bodemeyeri J. Weise, 1900d: 140 [= 1900e: 166]
klingenbergi Künnemann, 1917: 266
lutshniki Dobrzhanskiy, 1917: 46
sedakovii Mulsant, 1850b: 1020 **A**: AF ES KZ MG TD UZ
tianshanica Dobrzhanskiy, 1927a: 48

septempunctata Linnaeus, 1758: 365 **E**: AB AL AN AR AU BE BH BU BY CR CT CZ DE EN FI FR GB GE GG GR HU IR IT LA LS LT LU MC MD NL NR NT PL RO SK SL SP ST SV SZ TR UK YU **N**: MR **A**: AF ANH BEI BT CY FE FUJ GAN GUA GUI GUX HEB HEI HEN HUB HUN IN IQ IS JA JIA JIL JIX JO KA KI KU KZ LE LIA NIN NMO MG NC NP PA QIN SA SCH SC SD SHA SHG SHN SHX SI SY TAI TD TIA TM TR UP UZ XIN XIZ YUN ZHE **AFR NAR ORR**

anthrax J. Weise, 1879c: 108
atomaria J. Weise, 1879c: 107
brucki Mulsant, 1866: 90
confusa Wiedemann, 1823: 72
conspicua J. Weise, 1879c: 107
divaricata A. G. Olivier, 1808: 1001
externepunctata J. Weise, 1879c: 107
floricola J. Weise, 1879c: 107
graeca J. Weise, 1879c: 108
internepunctata J. Weise, 1879c: 107
lipsiensis J. Weise, 1879c: 108
lucida J. Weise, 1879c: 107
maculosa J. Weise, 1879c: 107
magnificoides Munster, 1930: 295
personata J. Weise, 1879c: 108
quinquenotata Haworth, 1912: 270
turcica J. Weise, 1879c: 107
vulgaris Haworth, 1812: 270
weisei Gradl, 1881: 303
zapluta J. Weise, 1879c: 107

tibetina Kapur, 1963a: 34 **A**: XIZ

transversoguttata transversoguttata Faldermann, 1835b: 454 **E**: **A**: ES FE GAN KA KI KZ MG NP SCH SHA WS YUN XIN XIZ

biinterrupta Mader, 1926: 151
geminopunctata Liu, 1962: 268

trifasciata trifasciata Linnaeus, 1758: 365 **E**: BY FI IT NR NT SV **A**: BEI ES FE GAN HEB HEI JIL LIA MG NIN NMO QIN SHA WS XIN XIZ

cimicifugae Pallas, 1773: 716
fennica J. Weise, 1879c: 109 [HN]

turkestanica Semenov & Dobrzhanskiy, 1923: 101 (*Semiadalia*) **A**: KI KZ UZ

weisei Rybakow, 1889a: 290 (*Micraspis*) **A**: QIN

subgenus *Spilota* Billberg, 1820: 61 type species *Coccinella undecimpunctata* Linnaeus, 1758

Dobzhanskia Iablokoff-Khnzorian, 1970a: 70 type species *Coccinella undecimpunctata* Linnaeus, 1758
Neococcinella Savoyskaya, 1969b: 104 type species *Coccinella undecimpunctata* Linnaeus, 1758

alpigrada Iablokoff-Khnzorian, 1957: 170 (*Adalia*) **E**: AR **A**: TR

miranda Wollaston, 1864: 422 **N**: CI

undecimpunctata boreolitoralis Donisthorpe, 1918: 123 **E**: IC

confluens Donisthorpe, 1902: 99 [HN]

undecimpunctata menetriesi Mulsant, 1850a: 104 **E**: GR IT ST **N**: AG AZ EG LB TU **A**: AE AF ES GAN GUI HEB IN IQ IS JO KI KU KZ MG NMO NIN PA SA SCH SHA SHX SI SY TR UP WS XIN

aegyptiaca Reiche, 1861c: 212 [RN]
arabica Mader, 1931a: 172
bagdadensis Roubal, 1932a: 62
undecimpunctata A. G. Olivier, 1808: 1017 [HN]

undecimpunctata tripunctata Linnaeus, 1758: 365 **E**: GR HU MD RO ST TR UK **A**: TR

pontica Dobrzhanskiy, 1927b: 212
tripunctata V. N. Nefedov & N. I. Nefedov, 1959: 162 [HN]

undecimpunctata undecimpunctata Linnaeus, 1758: 366 **E**: AZ BE CR CZ DE EN FI FR GB IR IT LA LU NL NR NT PL SK SP SV UK **A**: FE **AUR NAR**

brevifasciata J. Weise, 1879c: 111
collaris Paykull, 1799: 37
confluens Haworth, 1812: 274 [HN]
decempunctata Fabricius, 1775: 81 [HN]

impressa Verhoeff, 1891: 23
inscutellata Roubal, 1931a: 132
litoralis J. Weise, 1879c: 110
maculata Walter, 1882: 18
nigrofasciata P. Rossi, 1790: 62
novaezealandie Colenso, 1888: 40
novempunctata Linnaeus, 1758: 365
ocellata H. Piel de Churcheville & T. Piel de Churcheville, 1900: 26 [HN]
oculata Thunberg, 1795: 107 [HN]
oblonga Motschulsky, 1849: 153
peregrina J. Weise, 1879c: 110
pura J. Weise, 1885a: 29
salsolae J. Weise, 1879c: 110
tamaricis J. Weise, 1879c: 110
triangularis Thunberg, 1795: 108
variegata J. Weise, 1879c: 110 [HN]
vicina J. Weise, 1879c: 110

genus *Coccinella*, nomina dubia
biliturata Marsham, 1802: 166 **E**: GB
cervina Giorna, 1792: 109 **E**: IT
conspurcata Florencourt, 1796: 211 **E**: GE
exclamationis Eschscholtz, 1818: 464 "Livonia"
fuscomaculata Schrank, 1798: 455 **E**: GE (Bayern) [possibly syn. with *Coccidula scutellata*]
fuscopunctata Florencourt, 1796: 211 **E**: GE
humeralis Ponza, 1805: 49 [HN] **E**: IT (Nord Italien)
illigeri Gravenhorst, 1807: 127 distr. unknown
karpowkae Hummel, 1827a: 15 **E**: NT
obsoleta Ponza, 1805: 49 [HN] **E**: IT
ruficauda Gmelin, 1790: 1662 "Europa" [possible syn. with *Scymnus haemorrhoidalis*]
sedecimpustulata Linnaeus, 1758: 368 "Europa"
signata Gravenhorst, 1807: 124 [HN] "Europa"

genus *Coelophora* Mulsant, 1850a: 390 type species *Coccinella inaequalis* Fabricius, 1775
Artemis Mulsant, 1850a: 388 type species *Artemis circumusta* Mulsant, 1850
Autotela J. Weise, 1900c: 125 type species *Autotela picticollis* J. Weise, 1900
Cyphocaria Crotch, 1871: 6 type species *Caria duvaucelii* Mulsant, 1850
Lemnia Mulsant, 1850a: 376 type species *Lemnia fraudulenta* Mulsant, 1850 (= *Coccinella biplagiata* Swartz, 1808)
Microcaria Crotch, 1871: 7 type species *Coccinella mulsanti* Montrouzier, 1861
Osumia Kurisaki, 1923: 101 type species *Osumia bimaculata* Kurisaki, 1923 (= *Coccinella biplagiata* Swartz, 1808)
Spilocaria Timberlake, 1943: 58 type species *Coelophora bissellata* Mulsant, 1850
biplagiata Swartz, 1808: 196 **A**: FUJ GUX HAI HKG JA JIX NP SC TAI XIZ YUN ZHE **ORR**
bimaculata Kurisaki, 1923: 105 (*Osumia*)
fraudulenta Mulsant, 1850a: 379
minor Gorham, 1891b: 402
nepalensis Crotch, 1874a: 149
oblongata Thunberg, 1820: 368
personata J. Weise, 1910b: 147
subustulata Gorham, 1891b: 402
biquadripunctata Jing, 1988: 59 **A**: YUN
bissellata Mulsant, 1850a: 400 **A**: BT FUJ GUA GUI GUX HAI HUB JIX NP SC SCH SD TAI UP XIZ YUN **ORR**
gracilicornis J. Weise, 1902b: 505 (*Caria*)
nudipennis Sicard, 1910: 381
brunniplagiata Jing, 1988: 59 **A**: YUN
circumusta Mulsant, 1850a: 389 (*Artemis*) **A**: FUJ GUA GUX HAI HKG TAI YUN **ORR**
mandariana Mulsant, 1850a: 389 (*Artemis*)
rufula Mulsant, 1850a: 389 (*Artemis*)

westermanni Mulsant, 1850a: 391
circumvelata Mulsant, 1850a: 388 (*Lemnia*) [RN] **A**: ANH FUJ GUI GUX HAI HEN HKG HUB JIX NP SCH SHA TAI YUN ZHE
cincta Hope, 1831: 31 (*Coccinella*) [HN]
decimmaculata Jing, 1992: 541 **A**: YUN
duvaucelii Mulsant, 1850a: 233 (*Caria*) **A**: BT GUA GUX HKG SHG **ORR**
javana Mulsant, 1850a: 245 (*Leis*)
laichauana Hoang, 1983a: 95 (*Anisolemnia*)
novemmaculata C. Cao & Pu, 1985: 273 (*Alloneda*)
placens Mulsant, 1853: 182
ruficollis J. Weise, 1895a: 134 (*Cyphocaria*)
flavomarginata Sasaji, 1982a: 20 **A**: TAI
inaequalis Fabricius, 1775: 80 (*Coccinella*) **A**: JA (Bonin, Ryukyus) HKG TAI **AUR ORR**
artensis Montrouzier, 1861: 305 (Verania)
caledonica Crotch, 1874a: 153
circularis Thunberg, 1820: 365 (*Coccinella*) [HN]
comperei Timberlake, 1943: 32
consortula J. Weise, 1902b: 502
contortula J. Weise, 1903: 230 [RN]
desolata Mulsant, 1850a: 387 (*Lemnia*)
fasciata J. Weise, 1902b: 503
flachati Mulsant, 1866: 274
iridea Thunberg, 1781: 14 (*Coccinella*)
kochi J. Weise, 1913: 443
mendica Mulsant, 1850a: 407
mastersi T. Blackburn, 1892: 238
musae Montrouzier, 1855: 76 (*Coccinella*)
novemmaculata Fabricius, 1781: 97 (*Coccinella*) [RN]
novempunctata Fabricius, 1775: 81 (*Coccinella*) [HN]
novemsignata Thunberg, 1820: 365 (*Coccinella*) [HN]
octonotata Fabricius, 1801a: 366 (*Coccinella*)
patruelis Boisduval, 1835: 600 (*Coccinella*)
picta Voet, 1806: 61 (*Coccinella*)
psi Thunberg, 1781: 13 (*Coccinella*)
religiosa Lea, 1902: 488 (*Coccinella*)
reniplagiata Mulsant, 1850a: 392
ripponi Crotch, 1874a: 156
rouxi Heyden, 1911: 355
suturella J. Weise, 1913: 443
symbolica Mulsant, 1856: 146
veranioides T. Blackburn, 1894: 239
itoi Sasaji, 1982a: 23 **A**: TAI
jianfengensis Jing, 1988: 58 **A**: HAI
loi Sasaji, 1982a: 29 **A**: TAI
lushuiensis Jing, 1992: 564 **A**: YUN
saucia Mulsant, 1850a: 380 **A**: JA FUJ GAN GUA HEN HKG HUN JIA JIX NP SC SCH SHA SHN TAI YUN ZHE **ORR**
calypso Mulsant, 1856: 145 (*Leis*)
melanota Mulsant, 1850a: 381
mouhoti Crotch, 1874a: 149
ishidai Ohta, 1931a: 135 (*Chilomenes*)
swinhoii Crotch, 1874a: 149
takanonis Ohta, 1929c: 69 (*Chilomenes*)
sexguttata J. Weise, 1910c: 230 **A**: TAI **ORR**
tanoi Sasaji, 1982: 25 **A**: TAI

genus *Eoadalia* Iablokoff-Khnzorian, 1977b: 253 [= 1982: 450] type species *Adalia koltzei* J. Weise, 1887
koltzei J. Weise, 1887b: 209 (*Adalia*) **A**: FE
graeseri J. Weise, 1887b: 209 (*Adalia*)

genus *Harmonia* Mulsant, 1846: 108 type species *Coccinella marginepunctata* Schaller, 1783 (= *Coccinella quadripunctata* Pontoppidan, 1763)

Ballia Mulsant, 1853: 162 type species *Ballia christophori* Mulsant, 1853 (= *Ballia eucharis* Mulsant, 1853)
Callineda Crotch, 1871: 6 type species *Coccinella sedecimnotata* Fabricius, 1801
Callineda Crotch, 1874a: 161 type species *Callineda decussata* Crotch, 1874 (= *Harmonia manillana* Mulsant, 1866)
Leis Mulsant, 1850a: 241 type species *Coccinella dimidiata* Fabricius, 1781
Ptychanatis Crotch, 1874a: 122 type species *Coccinella axyridis* Pallas, 1773
Rhopaloneda Timberlake, 1943: 17 [RN]
Stictoleis Crotch, 1874a: 118 type species *Coccinella coryphaea* Guérin-Méneville, 1844

axyridis Pallas, 1773: 716 (*Coccinella*) **E**: CZiFI GBi GEi **A**: ES FE GAN GUA GUI GUX HKG HEB HEI HEN HKG HUB HUN JA JIA JIL JIX KZ LIA MG NC NIN NMO SC SCH SD SHA SHN SHX TAI WS XIN YUN ZHE **NAR ORR**

aulica Faldermann, 1835b: 451 (*Coccinella*)
besseri Faldermann, 1835b: 449 (*Coccinella*)
bissexpustulata Herbst, 1793: 303 (*Coccinella*)
circe Mulsant, 1856: 142 (*Anatis*)
conspicua Faldermann, 1835b: 448 (*Coccinella*)
duodecimsignata Frivaldszky, 1892: 121 (*Leis*)
frigida Mulsant, 1853: 161 (*Leis*)
henonii Mulsant, 1850a: 321 (*Daulis*)
henricae Mulsant, 1866: 253 (*Lemnia*)
insidiosa Mulsant, 1866: 254 (*Lemnia*)
multiplicata Mulsant, 1866: 78
novemdecimsignata Faldermann, 1835b: 456 (*Coccinella*)
octodecimspilota Hope, 1843a: 64 (*Coccinella*)
russica Herbst, 1783: 49 (*Coccinella*)
spectabilis Faldermann, 1835b: 450 (*Coccinella*)
succinea Hope, 1843a: 64 (*Coccinella*)

expalliata Sicard, 1914a: 499 **A**: HP KA UP **ORR**

breiti Mader, 1931b: 476
expallida Kapur, 1963c: 199 [HN]

dimidiata Fabricius, 1781: 94 (*Coccinella*) **A**: BT FUJ GUA GUI GUX HP HAI HEN HUN JA JIX KA NP PA SCH SD TAI UP XIZ YUN ZHE **ORR**

basalis L. Redtenbacher, 1844: 563 (*Coccinella*)
bicolor Hope, 1831: 31 (*Coccinella*) [HN]
dimidia Hope, 1831: 30 (*Coccinella*)
kikuchii Ohta, 1929: 70 (*Synonycha*)
quindecimmaculata Hope, 1831: 30 (*Coccinella*)
quindecimspilota Hope, 1931: 30 (*Coccinella*)

eucharis Mulsant, 1853: 167 (*Ballia*) **A**: KA NP PA SCH SD UP XIZ YUN **ORR**

bayadere Mulsant, 1866: 200 (*Neda*)
basalis L. Redtenbacher, 1844: 563 (*Coccinella*)
brahmae Mulsant, 1853: 164 (*Ballia*)
christophori Mulsant, 1853: 163 (*Ballia*)
dianae Mulsant, 1853: 295 (*Ballia*) [RN]
gustavii Mulsant, 1853: 165 (*Ballia*)
mayeti Mulsant, 1866: 189 (*Pelina*)
montivaga Mulsant, 1853: 167 (*Ballia*)
perplexa Crotch, 1874a: 128 (*Ballia*)
roubali Mader, 1929: 109
saundersi Crotch, 1874a: 128 (*Ballia*)
testacea Mulsant, 1853: 169 (*Ballia*)
zephirinae Mulsant, 1866: 190 (*Pelina*)

octomaculata Fabricius, 1781: 97 (*Coccinella*) **A**: JA FUJ GUA GUI GUX HAI HKG HUN SCH TAI YUN **AUR ORR**

arcuata Fabricius, 1787a: 55 (*Coccinella*)
cingulata Fabricius, 1787a: 55 (*Coccinella*)
duodecimnotata Fabricius, 1801a: 368 (*Coccinella*)
papuana J. Weise, 1902b: 496

postica Mulsant, 1850a: 79
rejiciens Walker, 1859b: 219 (*Coccinella*)
philippensis Mulsant, 1850a: 78
ungulata Herbst, 1793: 293 (*Coccinella*)
pardalina Gerstaecker, 1871: 346 (*Cheilomenes*) **A**: YE **AFR**
exiqueguttata Gorham, 1901: 407 (*Halyzia*)
pantherina Sicard, 1930: 174 (*Stictoleis*)
quadripunctata Pontoppidan, 1763: 669 (*Coccinella*) **E**: AL AU BU BY CR CT CZ DE FR GB GE GG GR HU IT LA LS LT LU NL NT PL RO SK SP ST SV UK **N**: AG **A**: ES FE NC TR SC SY WS YUN **NAR**
abieticola J. Weise, 1885a: 37 (*Coccinella*)
albida Gmelin, 1790: 1646 (*Coccinella*)
expallida J. Weise, 1907: 15 (*Coccinella*)
nebulosa J. Weise, 1879c: 119
marginella O. F. Müller, 1776: 65 (*Coccinella*)
marginepunctata Schaller, 1783: 260 (*Coccinella*)
multimacula Heyden, 1883a: 53 (*Coccinella*)
notata A. G. Olivier, 1791: 51 (*Coccinella*) [HN]
pinastri J. Weise, 1879c: 119
rustica J. Weise, 1879c: 119
sedecimpunctata Fabricius, 1781: 100 (*Coccinella*)
sordida J. Weise, 1879c: 119
sedecimnotata Fabricius, 1801a: 370 (*Coccinella*) **A**: FUJ GUA GUI GUX HAI HKG HUN JIX NP SCH YUN TAI XIZ ZHE **ORR**
shoichii Sasaji, 1988a: 17 **A**: TAI
yedoensis Takizawa, 1917: 220 (*Ptychanatis*) **A**: BEI FUJ GUA GUI GUX HEB HEN HUN JA NC SC SCH SHN TAI ZHE **ORR**
fallaciosa Mader, 1939a: 49 (*Leis*)
obscurosignata Liu, 1962: 263 (*Ballia*)

genus *Hippodamia* Chevrolat, 1836: 432 type species *Coccinella mutabilis* Scriba, 1791 (= *Coccinella variegata* Goeze, 1777)
subgenus *Hemisphaerica* Hope, 1840: 157 type species *Coccinella quinquesignata* Kirby, 1837
Parippodamia Iablokoff-Khnzorian, 1979: 51 type species *Coccinella arctica* D. H. Schneider, 1792
arctica D. H. Schneider, 1792: 148 (*Coccinella*) **E**: FE FI NR NT SV **A**: ES FE KA KI KZ MG NMO NP WS XIN **NAR**
amoena Faldermann, 1835b: 453 (*Coccinella*)
dubia F. Weber, 1801: 50 (*Coccinella*)
lapponica Thunberg, 1795: 113 (*Coccinella*)
scalaris Gebler, 1843: 33 (*Coccinella*)
septemmaculata DeGeer, 1775: 373 (*Coccinella*) **E**: AU BE BY CT CZ DE EN FI FR GE IT LA LT NC NR NT PL RO SK SV SZ UK **A**: ES FE KZ MG NC NMO SC WS
abbreviata Meier, 1897b: 365 [HN]
aestiva J. Weise, 1879c: 92
alligata Heyden, 1897: 97
axillaris J. Weise, 1879c: 91
baltica J. Weise, 1879c: 92
beccabungae Meier, 1897b: 365
bella Meier, 1897a: 317
berolinensis J. Weise, 1879c: 92
berulae J. Weise, 1879c: 92
calthae Meier, 1897b: 365
comari J. Weise, 1879c: 92
continua J. Weise, 1879c: 92
doemmlingi Meier, 1897a: 317
equiseti Meier, 1900: 78
ericae Schulz, 1899: 223
fontinalis J. Weise, 1879c: 92
frustrata Heyden, 1897: 97
germanica J. Weise, 1879c: 92

hamburgensis J. Weise, 1879c: 92
heydeni Meier, 1897b: 365 [RN]
heydeniana Meier, 1899b: 100 [RN]
holsatica J. Weise, 1879c: 92
hyperici Meier, 1897a: 317
imperfecta Heyden, 1897: 97
inchoata Heyden, 1897: 97
incompleta Schulz, 1899: 223
inornata J. Weise, 1879c: 91
lapponica J. Weise, 1879c: 92 [HN]
manca Heyden, 1897: 97
mendosa Heyden, 1897: 97
niemeyeri Schulz, 1899: 223
oblonga Herbst, 1783: 43 (*Coccinella*)
omostigma Heyden, 1897: 97
palustris J. Weise, 1879c: 92
paykulli J. Weise, 1879c: 92
pini Meier, 1897b: 365
prava Heyden, 1897: 97
quadrimaculata Schulz, 1899: 223
quinquemaculata Schulz, 1899: 223
republicana Heyden, 1897: 97
rubi Meier, 1897b: 365
rubra J. Weise, 1879c: 91
sanguisorbae Meier, 1897b: 365
sarothamni Meier, 1897a: 317
segetalis Naëzen, 1792: 170 (*Coccinella*)
sii J. Weise, 1879c: 92
submutabilis J. Weise, 1879c: 92
suecica J. Weise, 1879c: 92
supernumeraria Heyden, 1897: 98
tarda J. Weise, 1879c: 92
verna Meier, 1897b: 365
viadri J. Weise, 1879c: 92
vitiosa Heyden, 1897: 97
vorax J. Weise, 1879c: 92
wimmeli Meier, 1897b: 365
zetterstedti J. Weise, 1879c: 92

tredecimpunctata Linnaeus, 1758: 366 (*Coccinella*) **E**: AB AL AR AU BE BH BU BY CT CZ DE EN FI FR GB GE GR HU IR IT LA LS LT LU NL MC MD NR NT PL RO SK SL ST SV SZ TR UK YU **N**: AG MO **A**: AF ANH BEI JA ES FE GAN HEB HEI HEN HUB IN IQ JIA JIL JIX KI KZ LIA MG NC NIN QIN SC SHA SHN TD TM TR UZ XIN WS ZHE **NAR**

albingica Meier, 1897b: 366
borealis J. Weise, 1879c: 91
contorta J. Weise, 1879c: 91
c-nigrum J. Weise, 1879c: 91
fulvomarginata Wingelmüller, 1912: 183
gyllenhali J. Weise, 1879c: 91
impictipennis Fairmaire, 1876b: 50
incomta J. Weise, 1881: 165
laeta J. Weise, 1879c: 91
marchica J. Weise, 1879c: 91
octopunctata W. G. Schneider, 1881: 11
padana Comolli, 1837: 51 (*Coccinella*)
pallida Haworth, 1812: 276 (*Coccinella*) [HN]
quadripunctata G. Rossi, 1882: 214
scutellaris J. Weise, 1879c: 91
sellata J. Weise, 1879c: 91
signata Ménétriés, 1832: 240 (*Coccinella*) [HN]
signata Faldermann, 1837: 398 [HN]

spissa J. Weise, 1879c: 91
tibialis Say, 1824b: 94 (*Coccinella*)
tibialis timberlakei Capra, 1931: 17
triloba J. Weise, 1879c: 91
trinacris Geoffroy, 1785: 145 (*Coccinella*)
undecimmaculata G. A. Harrer, 1784: 98 (*Coccinella*)
vittata Goeze, 1777: 246 (*Coccinella*)
vulgaris Haworth, 1812: 276 (*Coccinella*) [HN]
xanthoptera Mulsant, 1850a: 10

subgenus *Hippodamia* Chevrolat, 1836: 432 type species *Coccinella mutabilis* L. G. Scriba, 1791 (= *Coccinella variegata* Goeze, 1777)
Adonia Mulsant, 1846: 39 type species *Coccinella mutabilis* L. G. Scriba, 1791 (= *Coccinella variegata* Goeze, 1777)

variegata Goeze, 1777: 247 (*Coccinella*) **E**: AB AE AL AN AR AU BE BH BU BY CR CT CZ DE EN FI FR GB GE GG GR HU IT LA LS LT LU MC MD NL NT PL PT RO SK SL SP ST SV SZ TR UK YU **N**: AG CI EG LB MO MR TU **A**: AE AF BEI BT FE FUJ GAN HEB HEN HP HUN IN IQ IS JIL JO KA KI KZ LE LIA MG NC NMO NIN NP PA SA SC SCH SHA SHX SD SI SY TD TM TR UP UZ WS YE YUN XIN XIZ **AFR NAR ORR**
abbreviata J. Weise, 1879c: 93 (*Adonia*)
affinis A. G. Olivier, 1791: 83 (*Coccinella*)
alacris J. Weise, 1879c: 93 (*Adonia*)
albicollis Chobaut, 1898: 87 (*Adonia*)
angulosa J. Weise, 1879c: 93 (*Adonia*)
arenaria Sajó, 1881: 273 (*Adonia*)
artemisiae J. Weise, 1879c: 93 (*Adonia*)
basilaris J. Weise, 1879c: 93 (*Adonia*)
bellieri Walter, 1882: 17 (*Adonia*)
betae Walter, 1882: 17 (*Adonia*)
biconstellata Sajó, 1881: 273 (*Adonia*)
bifurcata Mulsant, 1866: 28 (*Adonia*)
campestris J. Weise, 1879c: 93 (*Adonia*)
carpini Geoffroy, 1785: 114 (*Coccinella*)
conjuncta Walter, 1882: 17 (*Adonia*)
confluens Sajó, 1881: 273 (*Adonia*)
constellata Laicharting, 1781: 121 (*Coccinella*)
corsica Reiche, 1862: 299 (*Adonia*)
doubledayi Mulsant, 1850a: 38 (*Adonia*)
engelhardi B. G. Rye, 1893: 243
fennica Thunberg, 1795: 106 (*Coccinella*)
gradata Walter, 1882: 17 (*Adonia*)
hummeli J. Weise, 1879c: 92 (*Adonia*)
immaculata Gmelin, 1790: 1644 (*Coccinella*)
inhonesta J. Weise, 1879c: 92 (*Adonia*)
italica Walter, 1882: 17 (*Adonia*)
kriechbaumii Mulsant, 1866: 30 (*Adonia*)
laeta Fabricius, 1798: 78 (*Coccinella*)
limbata Fabricius, 1781: 497 (*Coccinella*)
litigiosa J. Weise, 1879c: 93 (*Adonia*)
lugubris Walter, 1882: 17 (*Adonia*)
lyrata J. Weise, 1900c: 123 (*Adonia*)
macrostigma Gabriel, 1905b: 142 (*Adonia*)
maculigera J. Weise, 1879c: 93 (*Adonia*)
mediopunctata Sajó, 1881: 273 (*Adonia*)
moraviaca Walter, 1882: 17 (*Adonia*)
mutabilis L. G. Scriba, 1791: 96 (*Coccinela*) [HN]
neglecta J. Weise, 1879c: 93 (*Adonia*)
novemnotata Haworth, 1912: 275 (*Coccinella*) [HN]
novempunctata Scopoli, 1763: 76 (*Coccinella*) [HN]
novempunctata Haworth, 1912: 275 (*Coccinella*) [HN]
obversepunctata Schrank, 1781a: 53 (*Coccinella*)
octopunctata Della Beffa, 1913: 9 (*Adonia*)

orientalis J. Weise, 1879c: 93 (*Adonia*)
portae Della Beffa, 1913: 13 (*Adonia*)
postica J. Weise, 1879c: 93 (*Adonia*)
quatuordecimnotata Donovan, 1804: 7 (*Coccinella*)
quinquemaculata Fabricius, 1787a: 56 (*Coccinella*)
quinquepunctata Walter, 1882: 17 (*Adonia*)
ripicola Mulsant, 1866: 13
scutellaris Della Beffa, 1913: 64 (*Adonia*) [HN]
septemnotata Fabricius, 1792a: 275 (*Coccinella*)
septempunctata Roubal, 1936a: 216 (*Adonia*)
sexpunctata Fabricius, 1781: 96
similis Schrank, 1781a: 56 (*Coccinella*) [HN]
simplex J. Weise, 1879c: 93 (*Adonia*)
thoracica J. Weise, 1889c: 572
transylvanica Penecke, 1901: 20 (*Adonia*)
tredecimpunctata Geoffroy, 1785: 144 (*Coccinella*) [HN]
tripunctata Haworth, 1812: 278 (*Coccinella*) [HN]
turkmenica Zubkov, 1833: 339 (*Coccinella*)
undecimpunctata Schrank, 1781a: 54 (*Coccinella*) [HN]
usambarica J. Weise, 1900c: 123 (*Adonia*)
ustulata J. Weise, 1879c: 93 (*Adonia*)
velox J. Weise, 1879c: 93 (*Adonia*)

genus *Hysia* Mulsant, 1850a: 47 type species *Coccinella endomycina* Mulsant, 1850
endomycina Boisduval, 1835: 603 (*Coccinella*) **A**: HP **AUR ORR**

genus *Megalocaria* Crotch, 1871: 6 type species *Neda reichii* Mulsant, 1850
Archaioneda Crotch, 1874a: 169 type species *Coccinella tricolor* Fabricius, 1787
dilatata Fabricius, 1775: 82 (*Coccinella*) **A**: FUJ GUA GUI GUX HKG SCH SHG TAI **ORR**
aucta Illiger, 1801c 179 (*Coccinella*)
suffusa Crotch, 1874a: 171 (*Caria*)
reichei pearsoni Crotch, 1874a: 170 **A**: HUN SD YUN **ORR**

genus *Menochilus* Timberlake, 1943: 40 [RN] type species *Coccinella sexmaculata* Fabricius, 1781
Cheilomenes Mulsant, 1850a: 443 [HN] type species *Coccinella sexmaculata* Fabricius, 1781
sexmaculata Fabricius, 1781: 96 (*Coccinella*) **A**: AE AF BT IN FUJ GAN GUA GUI GUX HAI HEN HKG HUB HUN JA JIA JIL JIX LIA NP OM PA SCH SD SHA SHG UP YUN ZHE **AFR AUR ORR**
aulata Heyden, 1911: 355 (*Chilomenes*)
australis J. Weise, 1908a: 12 (*Chilomenes*)
bifasciata Herbst, 1793: 274 (*Coccinella*)
flavofasciata Mulsant, 1850a: 446
flexuosa Thunberg, 1781: 17 (*Coccinella*) [HN]
hiugaensis Takizawa, 1917: 221 (*Chilomenes*)
inornata J. Weise, 1895c: 153 (*Chilomenes*)
insularis Nakane & M. Araki, 1959: 50 (*Menochilus*)
interrupta Fabricius, 1792a: 272 (*Coccinella*) [HN]
molipes Olliff, 1895: 30 (*Orcus*)
nigrofasciatus Miyatake, 1959: 146 (*Menochilus*)
quadriplagiata Swartz, 1808: 195 (*Coccinella*)
rufofasciata J. Weise, 1891a: 286 (*Chilomenes*)
saturata Gorham, 1894a: 202 (*Chilomenes*)
suffusa Gorham, 1894a: 202 (*Chilomenes*)
undulata Schaller, 1783: 262 (*Coccinella*) [HN]
unifasciata J. Weise, 1891a: 285 (*Chilomenes*)

genus *Micraspis* Chevrolat, 1836: 435 type species *Coccinella striata* Fabricius, 1792 (= *Coccinella lineata* Thunberg, 1781)
Alesia Mulsant, 1850a: 343 type species *Coccinella striata* Fabricius, 1792 (= *Coccinella lineata* Thunberg, 1781)
Cisseis Mulsant, 1850a: 129 [HN] type species *Coccinella furcifera* Guérin-Méneville, 1834

Cissela J. Weise, 1895c: 153 [RN]
Menevillidia Brèthes, 1923a: 228 [RN]
Pseudoverania Mader, 1941a: 193 type species *Pseudoverania nuda* Mader, 1941 (= *Micraspis sicardi* Mader, 1954)
Verania Mulsant, 1850a: 358 type species *Coccinella comma* Thunberg, 1781
allardi Mulsant, 1866: 249 (*Lemnia*) **A**: AF FUJ GUA HP KA PA UP YUN **ORR**
malaccensis Crotch, 1874a: 177 (*Verania*)
chinensis Mader, 1955a: 77 (*Verania*) **A**: FUJ
discolor Fabricius, 1798: 77 (*Coccinella*) **A**: FUJ GUA GUI GUX HAI HEN HKG HUB HUN JA JIA JIX SCH SHA SHG SHN TAI XIZ YUN ZHE **ORR**
bifurcata Gorham, 1895b: 687 (*Verania*)
bipartita Mulsant, 1850a: 371 (*Verania*)
crocea Mulsant, 1866: 58 (*Harmonia*)
licia Mulsant, 1856: 143 (*Cleis*)
nigrilabris Mulsant, 1866: 73 (*Harmonia*)
simplex Thunberg, 1820: 363 (*Coccinella*)
tabida Mulsant, 1866: 59 (*Harmonia*)
inops Mulsant, 1866: 286 (*Cydonia*) **A**: NP PA SD YUN **ORR**
vincta Gorham, 1895b: 686 (*Verania*)
kiotoensis Nakane & M. Araki, 1960: 117 (*Verania*) **A**: JA
kurosai Miyatake, 1977: 109 **A**: JA
satoi Miyatake, 1977: 111 **A**: JA
taiwanensis G.-Y. Yu, 2001a: 101 **A**: TAI
univittata Hope, 1831: 31 (*Coccinella*) **A**: NP XIZ **ORR**
inconsiderata Mulsant, 1866: 238 (*Alesia*)
yunnanensis Jing, 1985: 424 **A**: YUN

genus *Myrrha* Mulsant, 1846: 125 type species *Coccinella octodecimguttata* Linnaeus, 1758
subgenus *Metamhyrra* Capra, 1945: 91 type species *Thea thuriferae* Sicard, 1923
thuriferae Sicard, 1923: 78 (*Thea*) **E**: SP **N**: AG
subgenus *Myrrha* Mulsant, 1846: 125 type species *Coccinella octodecimguttata* Linnaeus, 1758
octodecimguttata Linnaeus, 1758: 367 (*Coccinella*) **E**: AL AR AU BH BU BY CT CZ DE EN FI FR GB GE GG GR HU IR IT LA LT LU NR NT PL RO SK SP ST SV UK YU **N**: AG CI MO MR TU **A**: ES FE KZ MG SY TR WS
andersoni Wollaston, 1862e: 337 (*Coccinella*)
discimacula Della Beffa, 1913: 208
formosa A. Costa, 1849: 38
inundata J. Weise, 1879c: 124 (*Halyzia*)
multiguttata Meier, 1899b: 102 (*Halyzia*)
ornata Herbst, 1783: 47 (*Coccinella*)
silvicola J. Weise, 1879c: 123 (*Halyzia*)

genus *Myzia* Mulsant, 1846: 277 [index] type species *Coccinella oblongoguttata* Linnaeus, 1758
Mysia Mulsant, 1846: 129 [HN] type species *Coccinella oblongoguttata* Linnaeus, 1758
Neomysia Casey, 1899: 98 type species *Coccinella pullata* Say, 1826
Paramysia Reitter, 1911a: 144 [RN] type species *Coccinella oblongoguttata* Linnaeus, 1758
gebleri Crotch, 1874a: 33 (*Mysia*) [RN] **A**: ES FE JA (Hokkaido) MG NMO WS
kasaii Kurisaki, 1920a: 231 (*Neomysia*)
ramosa Faldermann, 1833: 71 (*Coccinella*) [HN]
oblongoguttata nipponica Yuasa, 1963: 12 (*Neomysia*) **A**: JA NC SC
oblongoguttata oblongoguttata Linnaeus, 1758: 367 (*Coccinella*) **E**: AU BH BU BY CR CT CZ DE ES FI FR GB GE IR IT LA LT LU NR NT PL RO SK SP ST SV SZ TR UK YU **N**: MO **A**: ES FE KI KZ MG TR WS
abbreviata Walter, 1882: 19 (*Mysia*)
atrata Della Beffa, 1913: 168 (*Mysia*)
bisignata Fischer von Waldheim, 1842: 27 (*Coccinella*)
camschatsensis Gmelin, 1790: 1658 (*Coccinella*)
fleischeri Walter, 1882: 19 (*Mysia*)
leprieuri Pic, 1895h: 89 (*Mysia*)
maculata Walter, 1882: 19 (*Mysia*)

mixta Walter, 1882: 19 (*Mysia*)
mulsanti Schaufuss, 1862c: 52 (*Mysia*)
octoguttata Fabricius, 1787a: 59 (*Coccinella*) [HN]
vogeli Schaufuss, 1862b: 50 (*Mysia*)

genus *Oenopia* Mulsant, 1850a: 426 type species *Oenopia cinctella* Mulsant, 1850a (= *Coccinella cuneata* Thunberg, 1820)
Aza Mulsant, 1850a: 424 type species *Oenopia kirbyi* Mulsant, 1850
Gyrocaria Timberlake, 1943: 39 type species *Coelophora guttata* T. Blackburn, 1892 (= *Dysis excelens* Crotch, 1874)
Paramulsantina Hoang, 1982a: 9 type species *Paramulsantina gratiosa* Hoang, 1982
Protocaria Timberlake, 1943: 28 type species *Protocaria scalaris* Timberlake, 1943
Pseudoenopia Jablokoff-Khnzorian, 1986a: 59 type species *Pseudoenopia krausei* Iablokoff-Khnzorian, 1986
Pseudoharmonia Savoyskaya, 1969a: 37 type species *Pseudoharmonia montana* Savoyskaya, 1969
Synharmonia Ganglbauer, 1899b: 994 type species *Coccinella conglobata* Linnaeus, 1758
adelgivora Poorani, 2002: 107 **A**: SD
baoshanensis Jing, 1992: 559 **A**: YUN
billieti Mulsant, 1853: 144 (*Harmonia*) **A**: HP KA NP SHA SCH UP YUN SCH SD XIZ **ORR**
flava Fürsch, 1960c: 302 (*Synharmonia*)
flavidbrunna Jing, 1986b: 204
gonggarensis Jing, 1992: 560
indica Crotch, 1874a: 101 (*Adalia*)
indica J. Weise, 1903: 582 (*Coccinella*)
kehamae Crotch, 1874a: 158 (*Propylea*)
lanpingensis Jing, 1992: 561
picithoroxa Jing, 1992: 560
pomiensis Jing, 1987c: 413
pruthii Kapur, 1958: 327 (*Coccinella*)
testacea Kapur, 1958: 327 (*Coccinella*)
transgressa Mulsant, 1853: 145 (*Coccinella*)
bissexnotata Mulsant, 1850a: 269 (*Leis*) **A**: ES FE GAN GUI HEB HEI HUN JIL JIX LIA MG NC QIN SC SCH SHA SHN XIN YUN ZHE
cashmirense Korschefsky, 1935a: 299 (*Synharmonia*) **A**: KA
chinensis J. Weise, 1912: 113 (*Coelophora*) **A**: FUJ GAN GUA GUI GUX HUN JIA JIX SCH SHA SHG SHN YUN ZHE **ORR**
fallaciosa Mader, 1955a: 75 (*Coelophora*)
conglobata afghana Bielawski, 1968: 423 **A**: AF PA
conglobata conglobata Linnaeus, 1758: 366 (*Coccinella*) **E**: AB AL AR AU BE BH BU BY CR CT DE FR GE GG HU IT LA LS LT LU MD NL NT PL RO SK ST SV SZ TR UK YU **A**: ANH BEI ES FE GAN HEB HEI HEN IN JIA LE MG NC NIN NMO QIN SCH SHA SHN SHX SY TR WS ZHE
australis J. Weise, 1879c: 117 (*Coccinella*) [HN]
carmosina Voet, 1806: 67 (*Coccinella*)
conglobata major Dobrzhanskiy, 1927b: 216 (*Synharmonia*)
conjuncta Roubal, 1936a: 225 (*Synharmonia*)
dubia J. Weise, 1879c: 118 (*Coccinella*)
gemella Herbst, 1783: 44 (*Coccinella*)
houlberti Monnot, 1914: 69 (*Coccinella*)
juliana Gradl, 1882: 328 (*Coccinella*) [HN]
maculosa Gorham, 1891: 159 (*Coccinella*)
octodecimpunctata Scopoli, 1763: 77 (*Coccinella*)
pineti J. Weise, 1879c: 118 (*Coccinella*)
rosea DeGeer, 1775: 378 (*Coccinella*)
sedecimmaculata Fabricius, 1787a: 58 (*Coccinella*)
sedecimmaculata Thunberg, 1795: 109 (*Coccinella*) [HN]
specularis Bonelli, 1812: 182 (*Coccinella*)
vandalitiae J. Weise, 1879c: 117 (*Coccinella*)
conglobata contaminata Ménétriés, 1849: 272 (*Coccinella*) **A**: AF IN KZ MG NMO PA TD TM UZ XIN
buphthalmus Mulsant, 1850a: 90 (*Harmonia*)
desertorum Rybakow, 1889b: 574 (*Coccinella*)
quatuordecimplagiata Ballion, 1871: 353 (*Coccinella*)

dequenensis Jing, 1992: 558 **A**: SCH YUN
yunlongensis Jing, 1992: 561
diabolica Canepari, 1997: 54 **A**: NP
doublieri Mulsant, 1846: 118 (*Harmonia*) **E**: AL CR FR GR IT PT SP **N**: AG CI MO TU
gaditana J. Weise, 1879c: 119 (*Coccinella*)
scripta J. Weise, 1879c: 119 (*Coccinella*)
dracoguttata Jing, 1986b: 204 **A**: XIZ
emmerichi Mader, 1933b: 98 **A**: HUB JIX SCH YUN XIZ
formosana Miyatake, 1965b: 65 (*Gyrocaria*) **A**: TAI
hiekei Iablokoff-Khnzorian, 1977b: 250 (*Xanthadalia*) **A**: SCH YUN XIZ
hirayamai Yuasa, 1963: 12 (*Synharmonia*) **A**: JA
impustulata Linnaeus, 1767: 584 (*Coccinella*) **E**: AB AR AU BE BU CZ GE GG HU MD NL PL RO SK UK **A**: TR
caucasica Motschulsky, 1837b: 423 (*Coccinella*)
nigra Croissand, 1887: 279 (*Coccinella*)
vidua A. G. Olivier, 1808: 1040 (*Coccinella*)
viridula Hampe, 1880: 357 (*Coccinella*)
kirbyi Mulsant, 1850a: 425 (*Oenopia*) **A**: HUB SCH SD TAI XIZ YUN **ORR**
lyncea agnatha Rosenhauer, 1847: 64 (*Coccinella*) **E**: AR AU BH BU CZ FR GE GR IT HU PL RO SP (Salamanca) TR UK YU **A**: TR
pullata J. Weise, 1879c: 117 (*Coccinella*)
remota J. Weise, 1879c: 117 (*Coccinella*)
lyncea lyncea A. G. Olivier, 1808: 1056 (*Coccinella*) **E**: FR IT PT SP **N**: AG MO TU
castiliana J. Weise, 1903: 582 (*Coccinella*)
weisei Sicard, 1891: 59 (*Coccinella*) [HN]
mimica J. Weise, 1902b: 505 **A**: NP UP **ORR**
montana Savoyskaya, 1969a: 38 (*Pseudoharmonia*) **A**: IN KZ
oncina A. G. Olivier, 1808: 1048 (*Coccinella*) **E**: AB AR BU GR TR **N**: EG **A**: AE AF GAN IN IQ IS JO KA KI KU KZ LE MG PA SA SI SY TD TM TR UZ
addicta Mulsant, 1850a: 422
angelica J. Weise, 1903: 583 (*Coccinella*)
asiatica J. Weise, 1885a: 40 (*Coccinella*)
graeca J. Weise, 1903: 583 (*Coccinella*)
persica Faldermann, 1837: 406 (*Coccinella*)
sinaita J. Weise, 1903: 579 (*Coccinella*)
tenebrosa J. Weise, 1903: 583 (*Coccinella*)
quadripunctata Kapur, 1963a: 27 **A**: BT HUB JIA NP SD XIZ YUN **ORR**
sauzeti Mulsant, 1866: 281 **A**: BT GAN GUA GUI GUX HUN NP SCH SD SHA TAI XIZ YUN **ORR**
scalaris Timberlake, 1943: 29 (*Protocaria*) **A**: FUJ HEB GUA JA (Bonin, Honshu, Kyushu) NC SC TAI **ORR**
sexareata Mulsant, 1853: 181 (*Coelophora*) **A**: BT NP SD YUN **ORR**
sexmaculata Jing, 1986b: 206 **A**: SCH
smetanai Canepari, 1997: 53 **A**: NP UP
signatella Mulsant, 1866: 58 (*Harmonia*) **A**: BT GUX NP SD SHA UP XIZ YUN **ORR**
gratiosa Hoang, 1982b: 10 (*Paramulsantina*)
ornata Hoang, 1982b: 11 (*Paramulsantina*)
takasago Sasaji, 1982: 8 (*Chelonitis* [sic!]) **A**: TAI
yunlongensis Jing, 1992: 561 **A**: YUN
zonatus G.-Y. Yu, 2000b: 192 **A**: SCH

genus *Palaeoneda* Crotch, 1871: 5 type species *Neda auriculata* Mulsant, 1866
auriculata Mulsant, 1866: 193 (*Neda*) **A**: NP UP SD YUN **ORR**
miniata Hope, 1831: 30 (*Coccinella*) [HN]
genus *Phrynocaria* Timberlake, 1943: 34 type species *Coccinella congener* Billberg, 1808
Microlemnia Iablokoff-Khnzorian, 1982: 222 type species *Coelophora vidua* Mulsant, 1850
Pseudanegleis Iablokoff-Khnzorian, 1982: 296 type species *Coelophora perrotetti* Mulsant, 1850
congener Billberg, 1808: 197 (*Coccinella*) **A**: FUJ GUA HKG JA SCH TAI YUN **ORR**
approximans Crotch, 1874a: 153 (*Coelophora*)

consimilis M. Araki, 1962: 195 (*Lemnia*)
tereta Ohta, 1929c: 69 (*Chilomenes*)
decemguttata J. Weise, 1912: 114 (*Coelophora*) **A**: SD **ORR**
eberti Bielawski, 1972: 304 (*Coelophora*) **A**: NP
nigrilimbata Jing, 1986d: 307 **A**: YUN
piciella Jing, 1992: 556 **A**: YUN
shirozui Sasaji, 1982: 13 (*Propylea*) **A**: TAI
unicolor Fabricius, 1792a: 267 **A**: UP **ORR**
calliginosa Mulsant, 1850a: 414 (*Coelophora*)
cinctipennis J. Weise, 1892a: 20 (*Coelophora*)
dumortieri Mulsant, 1866: 261 (*Coelophora*)
*dupasquieri*Mulsant, 1866: 261 (*Coelophora*)
mitis J. Weise, 1912: 114 (*Coelophora*)
octosignata Mulsant, 1850a: 407 (*Coelophora*)
petrequini Mulsant, 1866: 276 (*Coelophora*)
quadripunctata J. Weise, 1892a: 21 (*Coelophora*)
romani Mulsant, 1866: 273 (*Coelophora*)
saundersii Crotch, 1874a: 160 (*Dysis*)
signata J. Weise, 1892a: 24 (*Coelophora*)
vidua Mulsant, 1850a 393 (*Coelophora*) **A**: TAI "Himalaya" **ORR**

genus *Propylea* Mulsant, 1846: 152 type species *Coccinella quatuordecimpunctata* Linnaeus, 1758
Pania Mulsant, 1850a: 420 type species *Oenopia luteopustulata* Mulsant, 1850
Vola Mulsant, 1850a: 377 type species *Lemnia dissecta* Mulsant, 1850
dissecta Mulsant, 1850a: 377 (*Lemnia*) **A**: NP SD SHG UP **ORR**
fallax Iablokoff-Khnzorian, 1977a: 61
feliciae Mulsant, 1866: 57 (*Harmonia*)
mystacea Mulsant, 1853: 178 (*Lemnia*)
japonica Thunberg, 1781: 12 (*Coccinella*) **A**: ANH BEI BT FE FUJ GAN GUA GUI GUX HAI HEB HEI HEN HUB HUN JA JIA JIL JIX LIA NC NIN NMO QIN SC SCH SHA SHN SHX TAI XIN YUN ZHE **ORR**
ambitiosa Mulsant, 1866: 72
ancora Wieise, 1880: 118
dionea Mulsant, 1856: 140
dentata Thunberg, 1781: 23 (*Coccinella*)
lewisi J. Weise, 1880: 119 (*Halyzia*)
tessellata J. Weise, 1892a: 18 (*Halyzia*)
tetraspilota Hope, 1843a: 64 (*Coccinella*) [HN]
tristis J. Weise, 1880: 119 (*Halyzia*)
virginalis J. Weise, 1892a: 17 (*Halyzia*)
luteopustulata Mulsant, 1850a: 421 (*Oenopia*) **A**: ANH BT FUJ GUA GUI GUX HEN HP HUB HUN JIX NP SHA SCH SD UP TAI XIZ YUN **ORR**
birmanica Gorham, 1895b: 685 (*Coelophora*)
bluethgeni Mader, 1933b: 97 (*Oenopia*)
hauseri Mader, 1930: 166 (*Oenopia*)
insularis Sicard, 1912b: 134 (*Coelophora*)
insularis J. Weise, 1923: 184 (*Coelophora*) [HN]
korschefskii Mader, 1930: 164
mariae Mulsant, 1853: 184 (*Coelophora*)
nigromaculata Mader, 1935: 340 (*Oenopia*)
pedicata Mulsant, 1853: 180 (*Coelophora*)
pracuae J. Weise, 1891a: 286 (*Oenopia*)
ramosa A. G. Olivier, 1808: 996 (*Coccinella*)
subpedicata Kapur, 1958: 331 (*Oenopia*)
thibetina Mulsant, 1853: 148 (*Anatis*)
thoracica J. Weise, 1891a: 283 (*Caria*)
thoracica J. Weise, 1895c: 152 (*Halyzia*)
victoriae Mulsant, 1866: 269 (*Coelophora*)
weisei Miwa, 1931a: 85 (*Coelophora*) [RN]

quatuordecimpunctata Linnaeus, 1758: 366 (*Coccinella*) **E**: AB AL AN AR AU BE BH BU BY CR CT CZ DE EN FI FR GB GE GG GR HU IT LA LS LT LU MC MD NL NR NT PL PT RO SK SL SP ST SV SZ TR UK YU **N**: MO **A**: CY ES FE FUJ GAN GUI HEB HEI HEN HKG HUB IN IQ IS JA JIA JIL JIX KZ LE LIA MG NC NMO PA SC SHA SY TR WS XIN **NAR**

angularis Sajó, 1881: 273 (*Halyzia*)
bifasciata Walter, 1882: 20 (*Halyzia*)
biflexuosa J. Weise, 1879c: 125 (*Halyzia*)
bilunata Walter, 1882: 20 (*Halyzia*)
bissexpustulata Fabricius, 1798: 80 (*Coccinella*)
brunnensis Walter, 1882: 20 (*Halyzia*)
decempustulata Fabricius, 1792a: 290 (*Coccinella*) [HN]
dentata Thunberg, 1795: 113 (*Coccinella*) [RN] [HN]
duodecimpunctata Walter, 1882: 19 (*Halyzia*)
duodecimpustulata Pontoppidan, 1763: 669 (*Coccinella*)
fasciata Walter, 1882: 20 (*Halyzia*)
figurata Walter, 1882: 20 (*Halyzia*)
fimbriata Sulzer, 1776: 32, pl. 3 (*Coccinella*)
frivaldszkyi Sajó, 1882: 158 (*Halyzia*)
herberi Heyden, 1904: 174 (*Coccinella*)
leopardina J. Weise, 1879c: 126 (*Halyzia*)
lunata Walter, 1882: 20 (*Halyzia*)
lyra Walter, 1882: 20 (*Halyzia*)
maeandra Walter, 1882: 20 (*Halyzia*)
moraviaca Walter, 1882: 20 (*Halyzia*)
ornata Voet, 1806: 67 (*Coccinella*) [HN]
palustris Sajó, 1881: 273 (*Halyzia*)
pannonica Sajó, 1882: 159 (*Halyzia*)
parumpunctata Sajó, 1882: 159 (*Halyzia*)
pedemontana Roubal, 1926b: 249
perlata J. Weise, 1879c: 126 (*Halyzia*)
picta Walter, 1882: 20 (*Halyzia*)
quatuordecimmaculata Fabricius, 1787a: 58 (*Coccinella*) [HN]
schaufussi Walter, 1882: 20 (*Halyzia*)
steineri Walter, 1882: 20 (*Halyzia*)
suturalis J. Weise, 1879c: 125 (*Halyzia*)
tessulata Scopoli, 1763: 79 (*Coccinella*)
tetragonata Laicharting, 1781: 125 (*Coccinella*)
trappi Walter, 1882: 20 (*Halyzia*)
walteri Walter, 1882: 20 (*Halyzia*)

genus *Protothea* J. Weise, 1898: 226 type species *Protothea firma* J. Weise, 1898
Nedina Hoang, 1983b: 150 type species *Nedina flavescens* Hoang, 1983
quadripunctata Mulsant, 1853: 155 (*Thea*) **A**: SD **ORR**
indica J. Weise, 1910a: 44

genus *Sospita* Mulsant, 1846: 137 type species *Coccinella tigrina* Linnaeus, 1758 (= *Coccinella vigintiguttata* Linnaeus, 1758)
vigintiguttata Linnaeus, 1758: 367 (*Coccinella*) **E**: AU BE BY CT CZ DE FR HU GE IT LT NL NT PL RO SK SL ST SV SZ TR UK **A**: TR
kochi Heyden, 1883a: 53 (*Halyzia*)
linnei J. Weise, 1879c: 122 (*Halyzia*)
tigrina Linnaeus, 1758: 368 (*Coccinella*)
trigaria J. Weise, 1879c: 122 (*Halyzia*)

genus *Synona* Pope, 1988: 660 [RN] type species *Synia melanaria* Mulsant, 1850
Synia Mulsant, 1850a: 375 [HN] type species *Synia melanaria* Mulsant, 1850
melanaria Mulsant, 1850a: 375 (*Synia*) **A**: FUJ GAN GUA GUI GUX HEN HUN JIX SCH SHA TAI XIZ YUN **ORR**
melanopepla Mulsant, 1850a: 376 (*Synia*)

rougeti Mulsant, 1866: 175 (*Leis*)
cassidoides Montrouzier, 1855: 77 (*Coccinella*) [HN]

genus *Synonycha* Chevrolat, 1836: 436 type species *Coccinella versicolor* Fabricius, 1792 (= *Coccinella grandis* Thunberg, 1781)
grandis Thunberg, 1781: 12 (*Coccinella*) **A**: FUJ GUA GUI GUX HEI HKG HUN JA JIL JIX LIA NP SD SHA TAI YUN **AUR ORR**
imperialis Herbst, 1793: 261 (*Coccinella*)
ursicolor Fabricius, 1787a: 58 (*Coccinella*)
versicolor Fabricius, 1792a: 279 (*Coccinella*)

genus *Xanthadalia* Crotch, 1874a: 99 type species *Harmonia rufescens* Mulsant, 1850
effusa rufescens Mulsant, 1850a: 76 (*Harmonia*) **N**: EG **A**: SA YE **AFR**
exsanquis Mader, 1941a: 195
miniata Mader, 1954: 152
medogensis Jing, 1987b: 301 [DA] **A**: XIZ

subfamily **Epilachninae** Mulsant, 1846

tribe Cynegetini C. G. Thomson, 1866

genus *Cynegetis* Chevrolat, 1836: 437 type species *Coccinella impunctata* Linnaeus, 1767
Cycnegetis Crotch, 1874a: [xi] type species *Coccinella impunctata* Linnaeus, 1767
impunctata Linnaeus, 1767: 579 (*Coccinella*) **E**: AU BE BH CT CZ DE FI FR GE HU IT NR PL RO SK SP ST SV SZ **A**: FE KZ NC
antalfyi K. Papp, 1943: 195
aptera Paykull, 1798: 154 (*Coccinella*)
atra Della Beffa, 1914: 19
elata Mulsant, 1846: 209
humeralis Della Beffa, 1914: 19
hungarica K. Papp, 1943: 195
maculigera K. Papp, 1943: 195
palustris Mulsant, 1846: 209
punctum Herbst, 1786: 161 (*Coccinella*)
striatopunctata K. Papp, 1943: 195
syriaca Mader, 1958b: 250 (*Subcoccinella*) **A**: IN SY TR
rapillyi Duverger, 1983: 90

tribe Epilachnini Mulsant, 1850

genus *Afidenta* Dieke, 1947: 109 type species *Afidenta mimetica* Dieke, 1947 (= *Epilachna misera* J. Weise, 1901)
misera J. Weise, 1901c: 420 (*Epilachna*) **A**: ANH FUJ GUA GUX HKG HUB HUN HP JIX NP SD SHN TAI UP XIZ **ORR**
mimetica Dieke, 1947: 110
simplex Dieke, 1947: 111
siamensis Dieke, 1947: 127 (*Afissa*) **A**: YUN **ORR**

genus *Afidentula* Kapur, 1958: 324 type species *Epilachna manderstjernae* Mulsant, 1853
bisquadripunctata Gyllenhal, 1808b: 186 (*Coccinella*) **A**: GUA GUX NP SCH YUN **ORR**
decimaculata C. Cao & H. Wang, 1992: 203 **A**: YUN
erberi Fürsch, 1984: 336 **A**: PA
himalayana Kapur, 1963a: 12 **A**: NP SD UP XIZ **ORR**
manderstjernae Mulsant, 1853: 256 (*Epilachna*) **A**: GUX NP SCH SD UP YUN **ORR**
minima Gorham, 1894a: 206 (*Epilachna*) **A**: UP **ORR**
quindecemguttata Dieke, 1947: 126 (*Afissa*) **A**: HUB SCH YUN
stephensi Mulsant, 1850b: 879 (*Epilachna*) **A**: NP **ORR**

genus *Afissula* Kapur, 1958: 319 type species *Afissula rana* Kapur, 1958
Afilachna Bielawski, 1967b: 793 type species *Afilachna kambaitana* Bielawski, 1967
expansa Dieke, 1947: 141 (*Afissa*) **A**: SCH YUN
henanica G.-Y. Yu, 1999b: 68 **A**: HEN
hydrangeae X.-F. Pang & Mao, 1979: 125 **A**: SCH YUN
kambaitana Bielawski, 1967: 793 (*Afilachna*) **A**: NP XIZ **ORR**
merkli Jadwiszczak, 1989: 741 **A**: SD
mysticoides Sicard, 1914b: 507 (*Solanophila*) **A**: NP SD XIZ **ORR**
nonggangensis T. Zeng & L.-F. Yang, 1996b: 271 **A**: GUX
ornithorrhyncha T. Zeng & L.-F. Yang, 1996b: 273 **A**: GUX
parvula Crotch, 1874a: 80 (*Epilachna*) **A**: NP SD
pronotonigeratus T. Singh & V. K. Singh, 1988: 217 **A**: UP
rana Kapur, 1958: 320 **A**: NP XIZ
sanscrita Crotch, 1874a: 82 (*Epilachna*) **A**: JIX NP SD SHN UP XIZ
spatulata C. Cao & Xiao, 1984: 110 **A**: YUN
uniformis X.-F. Pang & Mao, 1979a: 126 **A**: HUN SCH XIZ YUN
yunnanica C. Cao & H. Wang, 1992: 66 **A**: YUN

genus *Chnootriba* Chevrolat, 1836: 436 type species *Coccinella similis* Thunberg, 1781
similis Thunberg, 1781: 15 (*Coccinella*) **A**: YE **AFR**
assimilis Mulsant, 1850b: 699
atrocincta Sicard, 1930: 62
discordia Kapur, 1951: 278
erythromela Wiedemann, 1821: 182 (*Coccinella*)
primitiva Kapur, 1951: 278
quadrifasciata J. Weise, 1898b: 109 (*Epilachna*)
vitticollis J. Weise, 1898b: 108 (*Epilachna*)
tellinii J. Weise, 1905b: 139 (*Epilachna*)

genus *Epilachna* Chevrolat, 1836: 436 type species *Coccinella borealis* Fabricius, 1775
Afissa Dieke, 1947: 106 type species *Coccinella flavicollis* Thunberg, 1781
Aparodentata C. Cao & Wang, 1993: 126 type species *Epilachna yongshanensis* C. Cao & Xiao, 1993
Cleta Mulsant, 1850b: 866 type species *Epilachna eckloni* Mulsant, 1850
Hypsa Mulsant, 1850b: 860 type species *Epilachna nigrolimbata* C. G. Thomson, 1875
Solanophila J. Weise, 1898a: 99 type species *Epilachna gibbosa* Crotch, 1874
Uniparodentata C. Cao & Wang, 1993: 126 type species *Epilachna paramagna* X.-F. Pang & Mao, 1979
acuta J. Weise, 1900b: 384 (*Solanophila*) [RN] **A**: GAN JIA
acuminata J. Weise, 1889c: 648
acutula J. Weise, 1902b: 496 (*Solanophila*) [RN]
admirabilis Crotch, 1874a: 81 **A**: ANH FUJ GUA GUX HEB HEN HUB HUN JA JIA JIX SCH SHA TAI YUN ZHE **ORR**
continentalis Dieke, 1947: 118 (*Afissa*)
taiwanensis Miyatake, 1965b: 50
adscita Mader, 1930: 184 (*Solanophila*) **A**: SCH
ampliata X.-F. Pang & Mao, 1979a: 152 **A**: YUN
angusta Li, 1961: 79 **A**: TAI
anhweiana Dieke, 1947: 147 (*Afissa*) **A**: ANH
anthodea T. Zeng & L.-F. Yang, 1996a: 199 **A**: GUX SCH
apicilaris G.-Y. Yu, 1999b: 67 **A**: HEN
atypica Dieke, 1947: 133 (*Afissa*) **A**: BT SD **ORR**
bengalica Dieke, 1947: 130 (*Afissa*) **A**: BT NP SD XIZ **ORR**
besucheti Canepari, 1986: 29 **A**: SD
bhutanensis Bielawski, 1979: 91 **A**: BT NP
bicrescens Dieke, 1947: 142 (*Afissa*) **A**: SCH
bifibra Li, 1961: 82 **A**: TAI
brachyloba T. Zeng & L.-F. Yang, 1996a: 194 **A**: GUX

brivioi Bielawski & Fürsch, 1960: 70 (*Afissa*) **A**: YUN **ORR**
chayuensis X.-F. Pang & Mao, 1977: 324 **A**: XIZ
chinensis J. Weise, 1912: 112 (*Solanophila*) **A**: ANH FE FUJ GUA GUI HEN HUB HUN JA (Tsushima) JIX LIA NC SC SHN TAI ZHE **ORR**
separata Dieke, 1947: 150 (*Afissa*)
tsushimana Nakane & M Araki, 1960: 118 (*Afissa*)
chingjing G.-Y. Yu & H.-Y. Wang, 1999: 28 **A**: TAI
circummaculata X.-F. Pang & Mao, 1977: 325 **A**: XIZ
clematicola C. Cao & Xiao, 1984: 115 **A**: YUN
completa Dieke, 1947: 155 (*Afissa*) **A**: SCH
complicata Dieke, 1947: 159 (*Afissa*) SCH
sustenans Dieke, 1947: 161 (*Afissa*)
concuongensis Hoang, 1978b: 837 **A**: GUX **ORR**
confusa Li, 1961: 72 **A**: CH TAI
congener Gorham, 1895b: 693 **A**: SD **ORR**
convexa Dieke, 1947: 158 (*Afissa*) **A**: HUN SCH SHA
convextata T. Sing & V. K. Sing, 1990: 29 **A**: SD
crassimala Li, 1961: 68 **A**: TAI
crecentomaculata T. Sing & V.K. Sing, 1990: 33 **A**: HP
cuonaensis X.-F. Pang & Mao, 1977: 325 **A**: XIZ
decemguttata J. Weise, 1923: 183 (*Solanophila*) **A**: TAI
decemmaculata L. Redtenbacher, 1844: 564 **A**: BT HP KA NP UP XIZ **ORR**
deyrollii Crotch, 1874a: 78 **A**: SD **ORR**
dobzhanskyi Dieke, 1947: 163 (*Afissa*) **A**: HUB JIA SCH SHA
dorotae Bielawski, 1979: 86 **A**: BT
dumerili Mulsant, 1850b: 801 **A**: BT NP SD **ORR**
elvina Mulsant, 1853: 250 **A**: BT NP UP **ORR**
endomycina Gorham, 1903: 346 **A**: SD **ORR**
erythrotricha Hoang, 1978b: 833 **A**: GUX **ORR**
exornata Bielawski, 1965c: 223 **A**: YUN
fenchinica H. Pang, 1993a: 109 **A**: TAI
folifera X.-F. Pang & Mao, 1979a: 149 **A**: JIX YUN
formosana J. Weise, 1923: 183 (*Solanophila*) **A**: TAI
freyana Bielawski, 1965c: 219 **A**: HAI HUB HUN JIX SCH SHA YUN
fugongensis C. Cao & Xiao, 1984: 111 **A**: YUN
galerucinoides Korschefsky, 1934: 107 **A**: HAI TAI YUN **AUR ORR**
gedeensis Dieke, 1947: 129 (*Afissa*) **A**: YUN **ORR**
gibbera Crotch, 1874a: 80 **A**: NP SD **ORR**
glochinosa X.-F. Pang & Mao, 1979a: 138 **A**: GUX HUB HUN SCH YUN
glochisifoliata X.-F. Pang & Mao, 1979a: 139 **A**: SCH
gloiera Xiao, 1993: 386 **A**: SCH
globiera Xiao, 1993: 386
gorkhana Miyatake, 1985a: 26 **A**: NP
grayi Mulsant, 1850b: 774 **A**: NP SD TAI UP XIZ **ORR**
gressitti Li, 1961: 83 **A**: TAI
hauseri Mader, 1930: 182 (*Solanophila*) **A**: GUI SCH SHN YUN
boymi Jadwiszczak & Węgrzynowicz, 2003: 45 [RN]
hendecaspilota Mader, 1927a: 44 (*Solanophila*) **A**: NP SD TAI **ORR**
hingstoni Kapur, 1963a: 10 (*Afissa*) **A**: NP XIZ **ORR**
hopeiana Miyatake, 1985a: 28 **A**: GUX NP **ORR**
incauta Mulsant, 1850b: 803 **A**: TAI **ORR**
insignis Gorham, 1892: 84 **A**: ANH BEI FUJ GAN GUA GUI HEN HUB JIX SCH SHA YUN ZHE BEI FUJ GAN GUA GUI HEN HUB JIX SCH SHA YUN ZHE
fairmairei Frivaldszky, 1892: 121 [HN]
jianchuanensis C. Cao & Xiao, 1984: 113 **A**: YUN

laesicollis Mulsant, 1850b: 735 **A**: NP SD
langpingensis T. Zeng & L.-F. Yang, 1996a: 195 **A**: GUX
lata Li, 1961: 76 **A**: TAI
lichuaniensis Xiao, 1993: 387 **A**: HUB
longissima Dieke, 1947: 148 (*Afissa*) **A**: FUJ SCH TAI YUN
lugubris Dieke, 1947: 157 (*Afissa*) **A**: JIA SHN
macularis Mulsant, 1850b: 797 **A**: FUJ JIX NP SCH SD XIZ YUN **ORR**
maculicollis Sicard, 1922: 131 **A**: GUX TAI
maculivestis Mulsant, 1853: 251 **A**: BT NP XIZ **ORR**
magna Dieke, 1947: 167 (*Afissa*) **A**: FUJ SCH YUN **ORR**
malleforma Peng, H. Pang & Ren, 2002: 128 **A**: HUB
marginicollis Hope, 1831: 31 (*Coccinella*) **A**: BT NP SD UP XIZ **ORR**
maxima J. Weise, 1898: 236 (*Solanophila*) **A**: GUX SD TAI YUN **ORR**
media Li, 1961: 79 **A**: TAI
microgenitalia Li, 1961: 70 **A**: HUN TAI
militaris Dieke, 1947: 168 (*Afissa*) **A**: NP **ORR**
mirabiloides Dieke, 1947: 131 (*Afissa*) **A**: SCH **ORR**
mobilitertiae Li, 1961: 83 **A**: TAI
monsuna Bielawski, 1979: 100 **A**: BT NP
moorii Crotch, 1874a: 78 **A**: SD
mushana Li, 1961: 71 **A**: TAI
mystica Mulsant, 1850b: 841 **A**: BT NP SD UP XIZ **ORR**
nepalensis Kapur, 1958: 317 (*Afissa*) **A**: NP
nielamuensis X.-F. Pang & Mao, 1977: 323 **A**: NP XIZ
ocellataemaculata Mader, 1930: 183 (*Solanophila*) **A**: GUI HUB SCH TAI YUN **ORR**
ocreata T. Zeng & L.-F. Yang, 1996a: 196 **A**: GUX
paling G.-Y. Yu, 2001a: 99 **A**: TAI
parainsignis X.-F. Pang & Mao, 1979a: 133 **A**: GUX YUN
paramagna X.-F. Pang & Mao, 1979a: 135 **A**: HUN SCH YUN
pembertoni Crotch, 1874a: 80 **A**: BT **ORR**
pingbianensis X.-F. Pang & Mao, 1979a: 146 **A**: YUN
plicata J. Weise, 1889c: 649 **A**: GAN HUB SCH SHA YUN
provisoria Dieke, 1947: 151 (*Afissa*) **A**: SCH
quadricollis Dieke, 1947: 134 (*Afissa*) **A**: FUJ GUA GUX HEB JIA NC SHN TAI ZHE
rubiacis C. Cao & Xiao, 1984: 112 **A**: YUN
ryszardi Jadwiszczak & Węgrzynowicz, 2003: 112 [RN] **A**: BT
  sexsignata Bielawski, 1979: 99 [HN]
sauteri J. Weise, 1923: 182 (*Solanophila*) **A**: JA JIA TAI
schawalleri Canepari, 1997: 57 **A**: NP
septemnotata Bielawski, 1979: 104 **A**: BT
septemocellata T. Singh & V. K. Singh, 1990: 30 **A**: HP
sexpunctata Bielawski, 1979: 98 **A**: BT
sexpustulata Bielawski, 1979: 100 **A**: BT
shilliensis T. Singh & V. K. Singh, 1990: 27 **A**: HP
siphodenticulata Hoang, 1878: 27 **A**: GUX **ORR**
siphonechinulata Zeng & L.-F. Yang, 1996a: 197 **A**: GUX
sociolamina Li, 1961: 64 **A**: TAI
spiraloides C. Cao & Xiao, 1984: 114 **A**: YUN
subacuta Dieke, 1947: 165 (*Afissa*) **A**: FUJ GUX HUB JIA SCH ZHE
sureilica Kapur, 1963b: 136 (*Afissa*) **A**: SD **ORR**
suspiciosa J. Weise, 1901a: 281 **A**: SD
szechuana Dieke, 1947: 164 (*Afissa*) **A**: SCH
tianpingiensis X.-F. Pang & Mao, 1979a: 142 **A**: GUX HEN HUB HUN SCH
tridecimmaculosa X.-F. Pang & Mao, 1977: 326 **A**: XIZ

undecimspilota Hope, 1831: 31 (*Coccinella*) **A**: BT NP XIZ **ORR**
yongshanensis C. Cao & Xiao, 1984: 116 **A**: YUN

genus *Henosepilachna* Li, 1961: 35 type species *Coccinella sparsa* Herbst, 1786 (= *Coccinella vigintioctopunctata* Fabricius, 1775)
Elateria Fürsch, 1964: 182 type species *Coccinella elaterii* P. Rossi, 1794
aduncfolia Zeng, 2000: 136 **A**: ZHE
angusticollis Reiche, 1862: 545 (*Epilachna*) **E**: FR MA SP PT **N**: AG MO TU
enneastigma Fuente, 1907: 320 (*Epilachna*)
heptastigma Fuente, 1907: 320 (*Epilachna*)
undecimmaculata Fabricius, 1787a: 57 (*Coccinella*) [HN]
argus Geoffroy, 1785: 145 (*Coccinella*) **E**: AR AU BE BU CZ FR GB GE HU IT NL PL PT RO SK SP SZ YU **N**: AG MO **A**: LE TD TM TR UZ **AFR**
bedeli Sicard, 1892: 6 (*Epilachna*)
maroccana Kocher, 1953: 49 (*Epilachna*)
balajuensis Jadwiszczak & Pütz, 1994: 13 (*Epilachna*) **A**: NP
boisduvali Mulsant, 1850b: 765 (*Epilachna*) **A**: GUA JA (Ryukyus) TAI **AUR ORR**
chabuana Dieke, 1947: 80 (*Epilachna*)
fijiensis Crotch, 1874a: 89 (*Epilachna*)
montrouzieri Fauvel, 1862: 57 (*Epilachna*)
samoana Dieke, 1947: 81 (*Epilachna*)
tokarana Nakane & M. Araki, 1959: 45 (*Epilachna*)
dubiosa Dieke, 1947: 62 (*Epilachna*) **A**: SD **ORR**
assamensis Dieke, 1947: 63
elaterii elaterii P. Rossi, 1794: 85 (*Coccinella*) **E**: AL AU BH BU CR CZ FR GE GR HU IT LU PT RO SK SP SZ TR UK **N**: AG MO TU **A**: TR
costae J. Weise, 1879c: 128 (*Epilachna*)
furva J. Weise, 1879c: 128 (*Epilachna*)
hieroglyphica Sulzer, 1776: 31 (*Coccinella*) [HN]
nigrescens J. Weise, 1879c: 128 (*Epilachna*)
portugalensis Fürsch, 1964: 183
elaterii orientalis K. Zimmermann, 1939: 527 (*Epilachna*) **N**: EG LB **A**: AF CY IN IQ IS JO SI SA SY PA TR YE **AFR**
indica Mulsant, 1850b: 776 (*Epilachna*) **A**: BT NP SD TAI UP YUN **ORR**
ceylonica J. Weise, 1901b: 418 (*Epilachna*)
tertia Dieke, 1947: 66 (*Epilachna*)
introgibbera G. Yu, H. Pang & X.-F. Pang, 1993: 501 **A**: GUA
kaszabi Bielawski & Fürsch, 1960: 68 (*Epilachna*) **A**: YUN **ORR**
kathmanduensis Miyatake, 1985a: 23 **A**: NE
libera Dieke, 1947: 85 (*Epilachna*) **A**: SCH TAI YUN **ORR**
luteoguttata Mulsant, 1850b: 746 (*Epilachna*) **N**: EG **AFR**
maunsonica Jadwiszczak & Węgrzynowicz, 2003: 158 [RN] **A**: GUA **ORR**
freudei Fürsch, 1959: 7 (*Epilachna*) [HN]
niponica Lewis, 1896a: 23 (*Epilachna*) **A**: JA
ocellata L. Redtenbacher, 1844: 563 (*Epilachna*) **A**: BT HP KA NP PA SD UP XIZ **ORR**
oculea Mulsant, 1850b: 791 (*Epilachna*)
retexta Mulsant, 1850b: 793 (*Epilachna oculea*)
operculata Liu, 1950: 159 (*Epilachna*) **A**: BEI FUJ HUB JIX
processa Li, 1961: 45 **A**: TAI YUN **ORR**
pusillanima Mulsant, 1850b: 784 (*Epilachna*) **A**: FUJ NP SD TAI YUN **ORR**
dentulata Dieke, 1947: 46 (*Epilachna*)
languens Mulsant, 1850b: 786 (*Epilachna*)
parvinotata Dieke, 1947: 47 (*Epilachna*)
pustulosa Kôno, 1937: 99 (*Epilachna*) **A**: JA (Hokkaido)
pytharga Dieke, 1947: 74 (*Epilachna*) **A**: TAI **ORR**
quadriplagiata X.-F. Pang & Mao, 1977: 327 **A**: XIZ
septima Dieke, 1947: 58 (*Epilachna*) **A**: GUA GUX HAI **ORR**

kaiseri Bielawski, 1957: 73 (*Epilachna*)
sikkimica Kapur, 1963a: 7 (*Epilachna*) **A**: SD
subfasciata J. Weise, 1923: 182 (*Epilachna*) **A**: TAI
semifasciata Dieke, 1947: 103 (*Epilachna*)
tamdaoensis Hoang, 1977: 137 **A**: GUA **ORR**
umbonata X.-F. Pang & Mao, 1979a: 118 **A**: YUN
verriculata X.-F. Pang & Mao, 1979a: 117 **A**: YUN
vigintioctomaculata Motschulsky, 1858b: 40 (*Epilachna*) **A**: ANH BEI ES FE FUJ GAN GUI FUJ HEB HEI HUB JA JIA JIL LIA NC NP SC SCH SHA SHN SHX TAI XIZ YUN ZHE **ORR**
coalescens Dieke, 1947: 54 (*Epilachna*)
vigintioctopunctata Fabricius, 1775: 84 (*Coccinella*) **A**: AF ANH BT JA FE FUJ GUA GUI GUX HAI HEI HEB HKG HP HUB HUN JA JIA JIL JIX LIA MAC NC NP PA SCH SHA SHN SC SD TAI XIZ YUN ZHE **AUR ORR**
addita Mulsant, 1850b: 791 (*Epilachna*)
bengalensis Mulsant, 1850b: 789 (*Epilachna*)
chrysomelina Fabricius, 1775: 84 (*Coccinella*)
congressa Mulsant, 1850b: 791 (*Epilachna*)
dodecostigma Mulsant, 1850b: 789 (*Epilachna*)
egens Mulsant, 1850b: 836 (*Epilachna*)
fatalis Mulsant, 1850b: 788 (*Epilachna*)
gradaria Mulsant, 1850b: 789 (*Epilachna*)
indocilis Mulsant, 1850b: 788 (*Epilachna*)
infausta Mulsant, 1850b: 786 (*Epilachna*)
infuscata J. Weise, 1895c: 152 (*Epilachna*)
lacertosa Mulsant, 1850b: 838 (*Epilachna*)
lusoria Mulsant, 1850b: 789 (*Epilachna*)
multipunctata Mulsant, 1850b: 836 (*Epilachna*)
orientalis Dieke, 1947: 34 (*Epilachna*) [HN]
pardalis Boisduval, 1835: 596 (*Coccinella*)
pubescens Hope, 1831: 31 (*Coccinella*) [HN]
recta Mulsant, 1850b: 836 (*Epilachna*)
socors Mulsant, 1850b: 791 (*Epilachna*)
sparsa Herbst, 1786: 160 (*Coccinella*)
stolida Mulsant, 1850b: 791 (*Epilachna*)
territa Mulsant, 1850b: 787 (*Epilachna*)
vieta Mulsant, 1850b: 791 (*Epilachna*)
vigintioctopunctata Montrouzier, 1855: 75 (*Coccinella*) [HN]
vigintiquatuormaculata Fabricius, 1792a: 281 (*Coccinella*)
yasutomii Katakura, 1981: 361 **A**: JA (South Hokkaido, Honshu)

genus *Subcoccinella* Agassiz, 1845: 156 type species *Coccinella vigintiquatuorpunctata* Linnaeus, 1758
Lasia Hope, 1840: 157 [HN] type species *Coccinella globosa* D. H. Schneider, 1792 = *Coccinella vigintiquatuorpunctata* Linnaeus, 1758)
coreae Park & Yoon, 1991: 89 **A**: SC
vigintiquatuorpunctata Linnaeus, 1758: 366 (*Coccinella*) **E**: AB AL AR AU BH BU CR CT CZ DE EN FI GB GE GG GR IT IR HU LA LT LU MO MC NL NR NT PL PT RO SK SL SP ST SV SZ TR UK YU **N**: TU **A**: ES FE IN KI KZ LIA MG TR WS **NAR**
biundulata Pic, 1912r: 82
centrimaculata G. Rossi, 1882: 214
colchica Motschulsky, 1839: 51 (*Coccinella*)
colon Herbst, 1783: 42 (*Coccinella*)
gibbosa Duméril, 1817: 495 (*Coccinella*)
globosa D. H. Schneider, 1792: 149 (*Coccinella*)
haemisphaerica Schrank, 1798: 460 (*Coccinella*)
haemorrhoidalis Fabricius, 1777: 218 (*Coccinella*)
hungarica K. Papp, 1943: 195
immaculata P. Rossi, 1794: 86 (*Coccinella*) [HN]
inversa J. Weise, 1905a: 137
limbata Moll, 1784: 181 (*Coccinella*)

livida Herbst, 1783: 42 (*Coccinella*) [HN]
marmorata K. Papp, 1943: 195
meridionalis Motschulsky, 1837b: 420 (*Coccinella*)
micropunctata K. Papp, 1943: 195
nigra Della Beffa, 1912: 187
quadrinotata Fabricius, 1787a: 56 (*Coccinella*)
reticulata Della Beffa, 1912: 184
rufa Haworth, 1812: 283 (*Coccinella*)
saponariae J. Weise, 1879c: 130
vigintiduopunctata Haworth, 1812: 283 (*Coccinella*) [HN]
vigintiduopunctata Haworth, 1812: 284 (*Coccinella*) [HN]
vigintipunctata Haworth, 1812: 284 (*Coccinella*) [HN]
vigintiquinquepuncta Linnaeus, 1758: 366 (*Coccinella*)
vulgaris Haworth, 1812: 283 (*Coccinella*) [HN]
zonata Heyden, 1883a: 53

tribe Epivertini X.-F. Pang & Mao, 1979

genus *Epiverta* Dieke, 1947: 168 type species *Solanophila chelonia* Mader, 1933
chelonia Mader, 1933a: 79 (*Solanophila*) **A**: SCH XIZ YUN

family CORYLOPHIDAE LeConte, 1852

subfamily **Peltinodinae** Paulian, 1950

genus *Holopsis* Broun, 1883: 498 type species *Holopsis nigellus* Broun, 1883
Bathona Casey, 1900: 62 type species *Bathona carolinae* Casey, 1900
Corylophodes Matthews, 1885b: 160 type species *Corylophodes punctipennis* Matthews, 1899
Peltinodes Paulian, 1950: 19 type species *Peltinus gigas* Abeille de Perrin, 1894
gigas Abeille de Perrin, 1894b: 91 (*Peltinus*) **N**: AG TU
kirejtshuki Bowestead, 2003: 946 **A**: FE
kurilensis Bowestead, 2003: 947 **A**: FE
lewisi Bowestead, 2003: 945 **A**: FE JA
punctipennis Matthews, 1899: 150 (*Corylophodes*) **A**: JA

subfamily **Corylophinae** LeConte, 1852

tribe Corylophini LeConte, 1852

genus *Corylophus* Stephens, 1832b: 13 type species *Dermestes cassidoides* Marsham, 1802
Daubania Scott, 1917: 19 type species *Daubania seychellarum* Scott, 1917
Oligarthrum Matthews, 1887: 110 type species *Oligarthrum waterhousii* Matthews, 1887
cassidoides Marsham, 1802: 77 (*Dermestes*) **E**: AB AU BE BH BU CR CZ DE EN FI FR GB GE GG GR HU IR IT NL RO SV SZ UK **A**: FE IN JA KZ SY TM TR UZ
japonicus Matthews, 1899: 138
politus Reitter, 1898b: 18
sublaevipennis Jacquelin du Val, 1859a: 234 **E**: AZ GB GR FR IT SP SZ **N**: AG MO TU
nigripennis Rey, 1889a: 4
reitteri Matthews, 1890: 151
tectiformis Wollaston, 1854: 480 **N**: MR

genus *Microstagetus* Wollaston, 1861b: 103 type species *Microstagetus parvulus* Wollaston, 1861
parvulus Wollaston, 1861b: 106 **Ni**: AG CI EG MR **Ai**: IS
doderoi Peyerimhoff, 1917: 133

tribe Parmulini Poey, 1854

genus *Arthrolips* Wollaston, 1854: 475 type species *Clypeaster piceus* Comolli, 1837
Alloparmulus Paulian, 1950: 79 type species *Alloparmulus allaudi* Paulian, 1950
Lepadodes Matthews, 1872: 112 type species *Lepadodes chilensis* Matthews, 1887
Meioderus Matthews, 1899: 102 type species *Meioderus nitidus* Matthews, 1899: 10

Meizoderus Matthews, 1899: 105 type species *Meizoderus fuscus* Matthews, 1899
Pseudoparmulus Paulian, 1950: 100 type species *Sacium indescretum* Peyerimhoff, 1917
Sacina Broun, 1893: 1071 type species *Sacina oblonga* Broun, 1893
Sericoderistes Reitter, 1913e: 653 type species *Sericoderistes hetschkoi* Reitter, 1913

brittoni Paulian, 1948: 157 (*Sacium*) **A**: YE
convexiuscula Motschulsky, 1849: 90 (*Clypeaster*) **E**: AU AZ BU CR FR GE GR HU IT MA PT RO SP UK YU **N**: AG CI EG MO MR TU **A**: JO KZ TM TR UZ
aequale Wollaston, 1857a: 140
densatum Reitter, 1877c: 6 (*Sacium*)
doderoi Reitter, 1898b: 17
nitens Rey, 1889a: 4 (*Clypeaster*)
oblongum Matthews, 1899: 64 (*Sacium*)
curta Paulian, 1962: 91 **A**: JA
fasciata Erichson, 1842: 241 (*Corylophus*) **Ei**: FR IT **AUR**
pulchella Lea, 1891: 315 (*Clypeaster*)
fusca Matthews, 1899: 107 (*Meizoderus*) **A**: JA
galloisis Paulian, 1962: 82 (*Parmulus*) **A**: JA
hetschkoi Reitter, 1913e: 654 (*Sericoderistes*) **E**: FR (Corse) HU IT SP (Majorca, Menorca) RO UK **A**: IS KZ
humilis Rosenhauer, 1856: 347 (*Sericoderus*) **E**: FR IT SP **N**: AG EG MO MR TU **A**: AF IQ
marseulii Matthews, 1890: 151 (*Sacium*)
rufithorax Jacquelin du Val, 1859a: 232
vittatus Guillebeau, 1897a: 22
indescreta Peyerimhoff, 1917: 132 (*Sacium*) **E**: IT (Sardegna) **N**: AG
japonica Paulian, 1962: 102 (*Sacinops*) **A**: JA
klapperichi Bowestead, 1999: 54 **A**: AF
lewisii Matthews, 1899: 83 **A**: JA
longethorax Paulian, 1962: 93 **A**: JA
major Paulian, 1962: 103 (*Arthrolips*) **A**: JA
meionitida Bowestead, 1999: 42 (*Meioderus*) [RN] **A**: JA
nitidus Matthews, 1899: 105 [HN]
nana Mulsant & Rey, 1861c: 113 (*Clypeaster*) **E**: AU CR FR GE HU IT NL SZ
rhenanum Reitter, 1877c: 2 (*Sacium*)
oblonga Matthews, 1899: 93 **A**: JA
obscura C. R. Sahlberg, 1833: 474 (*Cossyphus*) **E**: AU BH CR FI FR GE GR HU IT LT NT RO ST SV SZ UK "Caucasus" **N**: AG **A**: CT IQ IS TR
bosnicus Obenberger, 1922: 22
corticinum C. G. Thomson, 1863: 118 (*Sacium*)
depressus Grimmer, 1841: 48 (*Clypeaster*)
discedens Jacquelin du Val, 1859a: 230 (*Sacium*)
polypori Fiori, 1915b: 73
sanctaebalmae Abeille de Perrin, 1895a: ccxlvi
picea Comolli, 1837: 51 (*Clypeaster*) **E**: AB AU BE BH BU CR CZ FR GG GR HU IT MA PL PT SL SP SZ UK **N**: AG CI MR MO TU **A**: AF CY IN IS ST TM TR **AFR**
ferrugatus Reitter, 1877c: 9
laevicollis Rey, 1889a: 4
longipes Matthews, 1899: 76 (*Sacium*)
mundum Matthews, 1899: 67 (*Sacium*)
parvum Matthews, 1886b: 432 (*Sacium*)
regularis Reitter, 1877c: 7
scaber Matthews, 1899: 101
sicardi Guillebeau, 1897b: 164
simoni Reitter, 1885f: 273
rufula Matthews, 1899: 72 (*Sacium*) **A**: JA
scotti Paulian, 1948: 158 **A**: YE
suffusus Matthews, 1887: 106 **A**: CH

genus *Clypastraea* Haldeman, 1842: 191 [RN] type species *Cossyphus pusillus* Gyllenhal, 1810
Aposacium Paulian, 1950: 103 type species *Aposacium major* Paulian, 1950
Aspidocha Gistel, 1848: 8 [RN]
Clypeastodes Casey, 1900: 70 type species *Sacium maderi* Kraatz, 1869
Clypeaster Dejean, 1821: 129 [HN]
Parmulus Gundlach, 1854: 323 [RN]
Sacinops Paulian, 1950: 101 type species *Sacinops macrothorax* Paulian, 1950
Sacium LeConte, 185: 144 [RN]
assimilis Matthews, 1899: 63 (*Sacium*) **A**: JA
brevicornis Matthews, 1899: 74 (*Sacium*) **A**: JA
brunnea C. N. F. Brisout de Barneville, 1863: 70 (*Sacium*) **E**: AU BH CR GR FR HU IT SP SK SZ **A**: TR
damryi Reitter, 1884c: 58
flava Matthews, 1899: 75 (*Sacium*) **A**: JA
flavicornis Matthews, 1899: 75 (*Sacium*) **A**: JA
lata Reitter, 1877c: 3 [= 1878: 149] (*Sacium*) **E**: AR GG ST (Kavkaz) **A**: IN TR
maderae Kraatz, 1869: 285 (*Sacium*) **N**: MR
major Paulian, 1950: 103 (*Aposacium*) **A**: JA
microthorax Paulian, 1962: 74 (*Parmulus*) **A**: JA
orientalis Reitter, 1877c: 4 (*Sacium*) **E**: CR HU GR IT **A**: CY IS TR
palmi Bowestead, 1999: 40 **N**: CI
picicornis Matthews, 1899: 73 (*Sacium*) **A**: JA
polita Reitter, 1878j: 192 (*Arthrolips*) **A**: JA
polita Matthews, 1899: 52 (*Sacium*) [HN]
pusilla Gyllenhal, 1810: 576 (*Cossyphus*) **E**: CT CZ FI FR GE HU LA LT NL NT PL SV SZ **A**: ES FE MG
reitteri Bowestead, 1999: 36 **E**: AU BH CR CZ HU IT RO SL YU **A**: TR
rufula Matthews, 1899: 72 (*Sacium*) **A**: JA
rugosa Matthews, 1899: 74 (*Sacium*) **A**: JA
yuasai Nakane, 1963a: 108 **A**: JA

tribe Sericoderini Matthews, 1888

genus *Aposericoderus* Paulian, 1950: 66 type species *Aposericoderus variegatus* Paulian, 1950
immigrans Israelson, 1987: 73 **Ei**: GR SP **Ni**: CI
revelierei Reitter, 1878h: 126 (*Sericoderus*) **E**: AB AL AR BH CR FR GG GR HU IT RO SP ST (Kavkaz) **N**: EG MO TU **A**: IS JO KZ TR
basalis Reitter, 1891b: 20 (*Sericoderus*) [HN]
chobauti Reitter, 1900b: 82 (*Sericoderus*)
flaviventris Reitter, 1891b: 21 (*Sericoderus*)
masoni Reitter, 1900d: 132 (*Sericoderus*) [RN]
variegatus Paulian, 1950: 67

genus *Sericoderus* Stephens, 1829: 187 type species *Sericoderus thoracicus* Stephens, 1829
Anisomeristes Matthews, 1886a: 225 type species *Anisomeristes sharpi* Matthews, 1886
Gryphinus L. Redtenbacher, 1849: 573 type species *Cossyphus lateralis* Gyllenhal, 1827
castaneus Reitter, 1877g: 126 **A**: JA
lateralis Gyllenhal, 1827: 516 (*Cossyphus*) **E**: AB AL AR AU AZ BE BH BU CR CT CZ DE FI FR GB GE GG GR IR IT HU LT MA MC NL NR PT RO SL SP SV SZ **N**: AG CI MO MR TU **A**: AF CY IN IS JA JO KI KZ LE SY TD TR UZ YE **COS**
blattina Gistel, 1857: 523 (*Paramecosoma*)
forticornis Matthews, 1899: 119
ptilioides Motschulsky, 1849: 91(*Clypeaster*) [DA]
thoracicus Stephens, 1829: 188
visnyae Csiki, 1940b: 918
pallidus Reitter, 1878j: 194 **A**: JA
pecirkanus Reitter, 1908b: 40 **E**: SP (Sicilia) **N**: EG **A**: AF ES IS JO YE

tribe Teplinini Pakaluk, Ślipiński & Lawrence, 1994

genus *Teplinus* Pakaluk, Ślipiński & Lawrence, 1994: 228 [RN] type species *Peltinus velatus* Mulsant & Rey, 1861

Peltinus Mulsant & Rey, 1861g: 121 [HN] type species *Peltinus velatus* Mulsant & Rey, 1861

matthewsi Reitter, 1885f: 273 (*Peltinus*) **E**: GR IT (Sardegna) SP **N**: AG EG CI JO MO **A**: CY IS

demaisoni Paulian, 1950: 28 (*Peltinus*)

intermedius Israelson, 1980: 184 (*Peltinus*)

peyerimhoffi Paulian, 1950: 28 (*Peltinus*)

walkeri Matthews, 1886a: 227 (*Peltinus*)

velatus Mulsant & Rey, 1861g: 121 (*Peltinus*) **E**: AL AU BH BU CR FR HU IT SP UK **A**: IS

alutaceus Reitter, 1885f: 273 (*Peltinus*)

subfamily **Orthoperinae** Jacquelin du Val, 1857

genus *Orthoperus* Stephens, 1829: 186 type species *Dermestes punctum* Marsham, 1802

Eutrilia Casey, 1900: 65 type species *Eutrilia brunnea* Casey, 1900: 66

Microsphaera L. Redtenbacher, 1845: 122 type species *Microsphaera corticalis* L. Redtenbacher, 1845

Pithophilus Heer, 1841: 433 type species *Pithophilus atomarius* Heer, 1841

acariformis Reitter, 1901d: 70 **A**: KZ

aequalis Sharp, 1885b: 128 **Ei**: CR FR GB IT SP SZ **Ni**: CI MR **ORR AUR**

nitidulus Allen, 1942: 89

anxius Mulsant & Rey, 1861c: 115 **E**: FR IT SP **N**: AG MO

perexiguus Rey, 1889a: 4

atomarius Heer, 1841: 433 (*Pithophilus*) **Ei**: AU BE FR GB GE HU IT NL SZ **Ni**: MR

atomus Gyllenhal, 1808a: 185 (*Cryptophagus*) **E**: AU BE BH CR CZ DE EN FI FR GB GE HU IR IT NL NR NT PL RO SV SZ **N**: MR **A**: AF TR

picea Stephens, 1829: 186

punctum Marsham, 1802: 80 (*Dermestes*) [DA]

brunnipes Gyllenhal, 1808a: 185 (*Cryptophagus*) **E**: AU BE CT DE EN FI GB GE LA NL NR NT SV UK **A**: ES WS **NAR**

kluki Wankowicz, 1865: 300

nigriclavis Stephens, 1829a: 168 (*Ephistemus*)

corticalis L. Redtenbacher, 1845: 155 (*Microsphera*) **E**: AU BE BH BU CR CT CZ DE FI FR GB GE GG HU IT NR NT PL RO SP ST (Kavkaz) SV SZ UK **A**: BY ES IN TM TR WS

improvisus Bruce, 1946: 106

punctatulus Matthews, 1885a: 108

intersitus Bruce, 1951b: 100 **E**: BH CR FR GE GR HU IT SP SZ YU

japonicus Matthews, 1899: 187 **A**: JA

loebli Bowestead, 1999: 109 **N**: AG EG MO **A**: IS

nigrescens Stephens, 1829: 187 **E**: BE CR CT CZ DE GB GE FI FR HU IT NL PL SV SZ UK **A**: TR

atomos Stephens, 1829: 187

coriaceus Mulsant & Rey, 1861d: 132

mundus Matthews, 1885a: 109

picatus Marsham, 1802: 80 (*Dermestes*) [DA]

truncatus Stephens, 1829: 187

nikitskyi Bowestead, 2000: 195 **E**: CT

pilosiusculus Jacquelin du Val, 1859a: 236 **E**: FR IT **N**: AG **A**: TR

borrei Matthews, 1885c: lxix

schneideri Reitter, 1902b: 137

punctatus Wankowicz, 1865: 299 **E**: AU CZ FI FR GE HU NR RO SK SL SV SZ **A**: ES FE

duplicatus Reitter, 1898b: 18

rogeri Kraatz, 1874: 122 **E**: AU CT FI FR HU NR LT NT PL RO SV SZ UK **A**: WS

punctulatus Reitter, 1876h: 312

sibiricus Bowestead, 2000: 195 **A**: FE

subfamily **Rypobiinae** Paulian, 1950

tribe Gloeosomatini Bowestead, 1999

genus *Catoptyx* Matthews, 1887: 111 type species *Catoptyx bowringii* Matthews, 1887
matthewsi Bowestead, 2003: 951 **A**: JA

genus *Gloeosoma* Wollaston, 1854: 480 type species *Gloeosoma velox* Wollaston, 1854
Gronevus Casey, 1900: 64 type species *Gronevus sticticus* Casey, 1900
Lewisium Matthews, 1899: 164 type species *Lewisium ceylonicum* Matthews, 1899
Moronillus Jacquelin du Val, 1854: xxxviii type species *Moronillus ruficollis* Jacquelin du Val, 1854
Sahlberginus Bruce, 1948 33 type species *Sahlberginus inexpectatus* Bruce, 1948
japonicum Matthews, 1899: 167 (*Lewisium*) **A**: JA
japonicus Matthews, 1899: 176 (*Rypobius*)
levantinum J. R. Sahlberg, 1913a: 12 (*Catoptyx*) **A**: IS LE TR
magnum Nakane, 1963c: 215 (*Lewisium*) **A**: JA
punctipenne Reitter, 1898b: 18 (*Corylophus*) **A**: UZ
sibericum Reitter, 1891b: 21 (*Moronillus*) **A**: ES FE JA NC
velox Wollaston, 1854: 482 **E**: FR GR IT SP "Caucasus" **N**: AG LB MO MR TU **A**: IS LE TR
discolor Rosenhauer, 1856: 348 (*Moronillus*)
inexpectatus Bruce, 1948: 34 (*Sahlberginus*)
obtusus Matthews, 1890: 151(*Rhypobius*)
pumilus Reitter, 1884d: 117 (*Moronillus*)
ruficollis Jacquelin du Val, 1854: xxxviii (*Moronillus*)

tribe Rypobiini Paulian, 1950

genus *Rypobius* LeConte, 1852c: 142 type species *Rypobius marinus* LeConte, 1852
dohrnii Matthews, 1899: 175 **A**: AF UZ
fiorianus Matthews, 1886b: 432 **E**: AL CR GR IT YU **A**: TR
praetermissus Bowestead, 1999: 123 **E**: AU BU FR GB GR HU IT SP SV YU **N**: AG

family LATRIDIIDAE Erichson, 1842

subfamily **Latridiinae** Erichson, 1842

tribe Latridiini Erichson, 1842

genus *Adistemia* Fall, 1899: 141 type species *Latridius watsoni* Wollaston, 1871
watsoni Wollaston, 1871: 253 (*Latridius*) **E**: BE CZ DE FI FR GB GE IR IT NL NR PT SP SV SZ **N**: AG EG MO MR **A**: IS JA **AFR NAR NTR**
bicostata Reitter, 1878j: 183 (*Cartodere*)
godarti Belon, 1881: 302 (*Cartodere*)

genus *Besuchetia* Dajoz, 1975: 202 type species *Besuchetia ceylanica* Dajoz, 1975
ceylanica Dajoz, 1975: 203 **A**: UP **ORR**

genus *Cartodere* C. G. Thomson, 1859: 93 type species *Latridius constrictus* Gyllenhal, 1827
subgenus *Aridius* Motschulsky, 1866c: 260 type species *Latridius nodifer* Westwood, 1839
Aridionomus Reitter, 1911a: 81 type species *Latridius nodifer* Westwood, 1839
Enicmoderes Reitter, 1911a: 81 type species *Lathridius apfelbecki* Reitter, 1901
apfelbecki Reitter, 1901g: 159 (*Lathridius*) **E**: TR **A**: TR
bifasciata Reitter, 1877g: 138 (*Coninomus*) **E**: BE DE FR GB GE IR NL PT SP SV SZ **N**: MR **A**: TR **AUR NAR**
dualis Broun, 1914: 181 (*Lathridius*)
nigromaculata Blackburn, 1888: 203 (*Lathridius*)
delamarei Dajoz, 1962: 287 (*Aridius*) **E**: FR **NTR**
lobli Dajoz, 1975: 205 (*Aridius*) **A**: NP SD UP YUN **ORR**
nodifer Westwood, 1839: 155 (*Latridius*) **E**: AU AZ BE BU BY CR CT CZ DE EN FI FR GB GE HU IR IT LA LS LT NL NR PL PT SP SK SL SV SZ **N**: CI MO MR **A**: JA **COS**
antipodum White, 1846: 18 (*Latridius*)

humile Rey, 1889h: 54 (*Lathridius*)
marginalis Broun, 1880: 233 (*Lathridius*)
nodulosa Motschulsky, 1866c: 261 (*Lathridius*)
rufescens Delahon, 1913: 530 (*Lathridius*)
sculpturata Broun, 1880: 233 (*Lathridius*)
satelles Blackburn, 1888: 202 (*Lathridius*) **E**: AZ FR GB GE NR SP **N**: EG MR **A**: JA **AUR**
norvegica A. Strand, 1940: 178 (*Lathridius*)
vietnamensis Rücker, 1979b: 161 (*Aridius*) **A**: YUN **ORR**
subgenus *Cartodere* C. G. Thomson, 1859: 93 type species *Latridius constrictus* Gyllenhal, 1827
Coninomus C. G. Thomson, 1863: 217 type species *Latridius constrictus* Gyllenhal, 1827
Latridulus Wollaston, 1877: 52 type species *Latridulus approximatus* Wollaston, 1877 (= *Latridius constrictus* Gyllenhal, 1827)
constricta Gyllenhal, 1827: 138 (*Latridius*) **E**: AU BE BY CT CZ DE EN FI FR GB GE GG HU IT LA LT NL NR NT PL RO SK SP ST (Kavkaz) SV SZ UK **N**: EG MO MR **A**: AF ES FE FUJ JA KA KZ MG NP PA SD UP WS YE **COS**
aegyptiaca Motschulsky, 1866c: 240 (*Lathridius*)
apicalis Blackburn, 1888: 204 (*Lathridius*)
approximata Wollaston, 1877: 52 (*Latridulus*)
carinata Gyllenhal, 1827: 173 (*Latridius*)
carinulata Mannerheim, 1844a: 81 (*Lathridius*)
castanea Broun, 1886: 834 (*Lathridius*)
incisa Mannerheim, 1844a: 80 (*Lathridius*)
limbata Förster, 1851: 38 (*Lathridius*)
monticola Mannerheim, 1844a: 82 (*Lathridius*)
nervosa Mannerheim, 1844a: 79 (*Lathridius*)
perpusilla Walker, 1858: 207 (*Lathridius*)
sculptile LeConte, 1855: 303 (*Lathridius*)
tenuis Belon, 1885a: 249 (*Lathridius*)
genus *Cartodere*, nomen dubium
ceylanicus Motschulsky, 1861c: 130 (*Lathridius*)

genus *Dienerella* Reitter, 1911a: 84 type species *Lathridius elegans* Aubé, 1850 (= *Dienerella vincenti* Johnson, 2007)
subgenus *Cartoderema* Reitter, 1911a: 84 type species *Latridius elongatus* Curtis, 1830
Cartoderopsis Dajoz, 1963: 65 type species *Cartoderopsis alinae* Dajoz, 1963
aequalis Reitter, 1877h: 295 (*Cartodere*) **E**: GG
anatolica Mannerheim, 1844a: 88 (*Lathridius*) **E**: AB BH CR GR IT YU **N**: EG **A**: IS TR
maritima Motschulsky, 1866c: 266 (*Aridius*)
angelinii Rücker, 1998: 1 **E**: IT
besucheti Vincent, 1994: 91 **N**: MO
clathrata Mannerheim, 1844a: 84 (*Lathridius*) **E**: AU BE BH BU CZ DE EN FI FR GB GE HU IT NL NR PL SK SL SV SZ
hexagonalis Motschulsky, 1866c: 254 (*Permidius*)
corsica Vincent, 1990a: 34 **E**: FR (Corse) IT
huguettae Vincent, 1991: 243 **E**: SP
kashmirensis Sengupta, 1976: 124 (*Microgamme*) **A**: KA NP
marginata Rücker, 1983b: 498 **A**: IS LE
ovata Sengupta, 1976: 125 (*Microgamme*) **A**: NP SD
parilis Rey, 1889h: 54 (*Cartodere*) **E**: AU FR IT
ruficollis Marsham, 1802: 111 (*Corticaria*) **E**: AU AZ BE BY CR CZ DE FI FR GB GE GG GR HU IC IR IT LS LT MA NL NR PL PT RO SK SP ST (Kavkaz) SV SZ UK **N**: CI EG MO MR **A**: CH IS JA LE JO TR **AFR NAR NTR**
atripennis Motschulsky, 1866c: 267 (*Aridius*)
collaris Mannerheim, 1844a: 86 (*Lathridius*)
concinna Mannerheim, 1844a: 88 (*Lathridius*)
exilis Mannerheim, 1844a: 86 (*Lathridius*)
liliputana A. Villa & V. G. Villa, 1833: 36 (*Latridius*)
nanula Mannerheim, 1844a: 87 (*Lathridius*)

pulicaria Melsheimer, 1844a: 115 (*Lathridius*)
separanda Reitter, 1887a: 25 (*Cartodere*) **E**: IT MA SP **N**: AG MO TU
siciliana Vincent, 1990b: 263 **E**: IT
vincenti Johnson, **sp. nov**. [see New acts] **E**: AU BE BH CR CT CZ DE EN FI FR GB GE HU IT LA LS LT NL NR PL PT RO SK SL SP SV SZ UK
elongata Curtis, 1830: 311 (*Latridius*) [HN]
subgenus *Dienerella* Reitter, 1911a: 84 type species *Lathridius elegans* Aubé, 1850
Microgramme Walkley, 1948: 150 type species *Latridius filiformis* Gyllenhal, 1827
argus Reitter, 1884b: 35 (*Cartodere*) **E**: AU BH CR CZ FI FR GB GE HU IT LS NL PL RO SK SL SP SZ UK **N**: AG EG MO MR TU **A**: AE IS JA TD **NAR**
beloni Reitter, 1882c: 164 (*Cartodere*) **E**: BE IT (Sardegna) NL **A**: HKG JA **AFR NTR**
costulata Reitter, 1877f: 114 (*Cartodere*) **E**: AU BE CR CZ DE FR GE GR HU IT LA NL PL SZ YU **A**: JA **NAR**
elegans Aubé, 1850: 334 (*Lathridius*) **E**: AR AU BY FR IT GB GR HU MA PL SP SV SZ UK **N**: AG MR **A**: AF FE IS JA TD **AFR NAR**
costipennis Reitter, 1877f: 114 (*Cartodere*)
delecta Wollaston, 1858c: 409 (*Lathridius*)
depilis Belon, 1895: 97 (*Cartodere*)
elegantula Reitter, 1890a: 147 (*Cartodere*)
forticostis Reitter, 1911a: 84 (*Cartodere*)
inflaticeps Motschulsky, 1866c: 255 (*Permidius*)
filiformis Gyllenhal, 1827: 143 (*Latridius*) **E**: AB AR AU BE BY CR CT CZ DE EN FI FR GB GE GR HU IR IT LA LS NL NT PL NR RO SK SP SV SZ TR UK **A**: AF JA UZ **NAR**
angustata Stephens, 1829: 94 (*Latridius*)
parallela Mannerheim, 1844a: 106 (*Lathridius*)
parallelipennis Solsky, 1876: 264 (*Lathridius*)
subcostella Reitter, 1915c: 66 (*Cartodere*)
tantilla Mannerheim, 1844a: 106 (*Lathridius*)
filum Aubé, 1850: 334 (*Lathridius*) **E**: AU BE BY CT CZ DE EN FI FR GB GE HU IC IR IT LA NL NR NT PL SK SP SV SZ UK YU **N**: AG **A**: FE FUJ JA **AUR NAR NTR**
laticeps Reitter, 1884e: 253 (*Cartodere*) **A**: IS
lurida Rücker, 1983b: 497 **Ei**: GB **A**: CH JA
tanakai Hisamatsu, 1985: 275
pilifera Reitter, 1875k: 334 (*Cartodere*) **E**: AU FR GR IT SP **N**: CI MO TU
schueppeli Reitter, 1881b: 57 (*Cartodere*) **E**: AB AU FR GB GE ST (Kavkaz) SZ UK **NAR**
strupii Hölzel, 1944: 81 (*Cartodere*)
genus *Dienerella*, nomen dubium
cordicollis Mannerheim, 1843b: 300 (*Latridius*)

genus *Enicmus* C. G. Thomson, 1859: 93 type species *Ips transversus* A. G. Olivier, 1790
Permidius Motschulsky, 1866c: 243 type species *Latridius rugosus* Herbst, 1793
alutaceus Reitter, 1885g: 305 **E**: RO ST (Kavkaz) **A**: ES FE IN MG TR WS
amici Lohse, 1981: 9 **E**: AU BU GE IT PL SK TR UK **A**: TR
geminatus Rücker, 1981: 166
apicalis J. R. Sahlberg, 1926: 92 **E**: FI NR NT SV **A**: FE
ussuricus Saluk, 1992: 380 [DA]
atriceps V. Hansen, 1962: 245 **E**: DE FR GE GR HU IT PL RO SV SZ
brevicornis Mannerheim, 1844a: 102 (*Lathridius*) **E**: AU BE BH CR CT CZ FR GB GE GG GR HU IT LA NL PL RO SK SP ST (Kavkaz) SV ST SZ UK YU **N**: AG MO TU **A**: IN TR
carbonarius Mannerheim, 1844a: 103 (*Lathridius*)
subterraneus Gistel, 1857: 527 (*Latridius*)
claviger C. Johnson, 1977a: 126 **A**: AF JO MG "Turkestan"
dubius Mannerheim, 1844a: 93 (*Lathridius*) **A**: ES KZ
fungicola C. G. Thomson, 1868: 336 **E**: AU BE BU BY CR CT CZ DE EN FI FR GB GE HU IT NL NT NR PL RO SK SL SV SZ UK
histrio Joy & Tomlin, 1910: 250 **E**: AB AU BE BH BU BY CR CT CZ DE EN FE FI FR GB GE HU IR IT LT

NL NT NR PL RO SK SP SV SZ UK **N**: AG MO MR **A**: AF BEI FE FUJ GUI JA KA KZ MG NC NP PA SD SHA SHN TR UP WS YE YUN **AUR ORR**

ventriosus Peyerimhoff, 1929d: 212

kaszabi Rücker, 1983a: 171 **A**: MG WS

lundbladi Palm, 1956: 24 **E**: FI SV

mannerheimii Kolenati, 1846a: 42 (*Lathridius*) **E**: GG ST (Kavkaz) UK

montanoasiaticus Saluk, 1995: 139 **A**: MG WS

planipennis A. Strand, 1940: 178 **E**: AU BY CT CZ FI GE NR NT SV SZ

rueckeri Johnson, **sp. nov**. [see New acts] **A**: ES FE MG

dubius sensu Rücker, 1992: 146

rugosus Herbst, 1793: 6 (*Latridius*) **E**: AU BE BY CR CT CZ DE EN FI FR GB GE GG GR HU IT LA LT NL NR NT PL PT RO SK SL SP SV SZ TR UK YU **N**: MO TU **A**: ES FE HEI IN MG WS

depressus Grimmer, 1841: 45 (*Latridius*)

ferrugineus Gerhardt, 1912: 538 [HN]

frater E. Weise, 1972: 161

planatus Mannerheim, 1844a: 93 (*Lathridius*)

ruficornis Kugelann, 1794: 574 (*Latridius*)

rugipennis Mannerheim, 1844a: 92 (*Lathridius*)

testaceus Stephens, 1830: 114 (*Latridius*) **E**: AU BE BY CZ DE EN FR GB GE GR HU IR IT LA NL NR NT PL RO SK SP SV SZ YU **N**: AG CI **A**: IN

cordaticollis Aubé, 1850: 332 (*Lathridius*)

crenicollis C. G. Thomson, 1868: 57

lederi Reitter, 1875k: 327 (*Enicmus*)

opacipennis Wollaston, 1864: 151 (*Latridius*)

pseudotestaceus Franz, 1974: 110

transversus A. G. Olivier, 1790a: no 18: 14 (*Ips*) **E**: AB AR AU BE BU BY CR CT CZ DE EN FI IR FR GB GE GG GR HU IT LA LS LT MA NL NR NT PL PT RO SK SP ST SV SZ UK **N**: AG EG MO MR TU **A**: AF IS JO NP SY WS

sculptilis Gyllenhal, 1827: 141 (*Latridius*)

ussuricus Saluk, 1992: 380 **A**: FE

varendorffi Reitter, 1903c: 234 **E**: GR **A**: TR

anatolicus Reitter, 1903c: 234

genus *Euchionellus* Reitter, 1908d: 133 type species *Metophthalmus albofasciatus* Reitter, 1891 (= *Metophthalmus zanzibaricus* Belon, 1887)

zanzibaricus Belon, 1887: 221 (*Metophthalmus*) **Ei**: GE PL **A**: JA **AFR AUR ORR**

albofasciatus Reitter, 1891b: 23 (*Metophthalmus*)

genus *Eufallia* Muttkowski, 1910: 161 [RN] type species *Cartodere unicostata* Belon, 1887 (= *Aridius seminiveus* Motschulsky, 1866)

Belonia Fall, 1899: 142 [HN] type species *Cartodere unicostata* Belon, 1887

seminivea Motschulsky, 1866c: 265 (*Aridius*) **E**: MA **A**: HKG JA **AUR NAR NTR**

unicostata Belon, 1887: 225 (*Cartodere*)

genus *Latridius* Herbst, 1793: 3 type species *Latridius porcatus* Herbst, 1793

Conithassa C. G. Thomson, 1859: 93 type species *Tenebrio minutus* Linnaeus, 1767

Isidius Motschulsky, 1866c: 257 type species *Latridius protensicollis* Mannerheim, 1843

amplus C. Johnson, 1977a: 124 **E**: AB FR GR SP **A**: AF CY IQ JO TR

creticus Rücker, 1986a: 129

latissimus Rücker, 1986b: 261

assimilis Mannerheim, 1844a: 98 (*Lathridius*) **E**: AU CR CZ DE FI FR GB GE IR IT LA NR NT PT RO SK SP SV SZ "Caucasus" **N**: TU **A**: LE **AFR AUR**

flavicornis Motschulsky, 1866c: 245 (*Permidius*)

herborisans Gistel, 1857: 527 (*Lathridius*) [DA]

pseudominutus A. Strand, 1958: 184 (*Enicmus*)

scitus Mannerheim, 1844a: 99 (*Lathridius*) [DA]

sulcifrons Broun, 1895: 196 (*Lathridius*) [DA]

brevicollis C. G. Thomson, 1868: 56 (*Conithassa*) **E**: AU BH BY CR CT CZ DE FI FR GE HU IT PL RO SK SV UK "Baltic" **A**: TR

carpathicus Reitter, 1875l: 361 (*Enicmus*)

canariensis Palm, 1972: 40 (*Enicmus*) **N**: CI

consimilis Mannerheim, 1844a: 99 (*Lathridius*) **E**: AU BE BH BU BY CT CZ DE FI FR GB GE GG HU IT LA LT NL NR NT PL RO SK SV SZ UK **A**: ES FE WS **AFR NAR**

basalis Motschulsky, 1866c: 246 (*Permidius*)

parallelocollis Mannerheim, 1844a: 101 (*Lathridius*)

subtestaceus Belon, 1881: 346 (*Enicmus*)

gemellatus Mannerheim, 1844a: 100 (*Lathridius*) **E**: AU BY CZ DE FI GE NR PL SP SV SZ

kolbei Wanka, 1929: 198 (*Enicmus*) [DA]

nidicola Palm, 1944b: 170 (*Enicmus*)

hirtus Gyllenhal, 1827: 139 **E**: AB AU BH BY CT CZ DE EN FI FR GE GR HU IT LA LT NL NR NT PL RO SK SL SV SZ UK **A**: FE WS **NAR**

distincticollis Roubal, 1933: 139 (*Enicmus*)

hirsutulus Stephens, 1829: 94

minutus Linnaeus, 1767: 675 (*Tenebrio*) **E**: AU BE BY CR CT CZ DE EN FI FR GB GE GG GR HU IC IR IT LA LS LT MA NL NR NT PL PT RO SK SP ST SV SZ UK **A**: ES FE MG TR WS

mongolicus Rücker, 1983a: 170 **A**: MG WS

peacockae Sengupta, 1976: 118 (*Lathridius*) **A**: AF KA NP PA UP

klapperichi C. Johnson, 1977a: 125

perminutus C. Johnson, 1977a: 125 **A**: AF

porcatus Herbst, 1793: 6 **E**: AR AU BE BY CT CZ DE FI FR GB GE HU IR IT LA LT NL NR PL PT RO SK SP ST (Kavkaz) SV SZ UK **N**: MR **A**: ES IS KZ **AUR NTR**

alius E. Weise, 1972: 159 (*Enicmus*)

anthracinus Mannerheim, 1844a: 97 (*Lathridius*)

marginatus Paykull, 1798: 300 (*Dermestes*) [DA]

minutissimus Motschulsky, 1858a: 24

pusillus Motschulsky, 1858b: plate

protensicollis Mannerheim, 1843b: 299 **A**: FE JA **NAR**

kamtschaticus Motschulsky, 1866c: 259 (*Isidius*)

quadricollis Mannerheim, 1843b: 299

recticollis Motschulsky, 1866c: 258 (*Isidius*)

sobrinus Mannerheim, 1852a: 362 (*Lathridius*)

suspectus Fall, 1899: 125 (*Enicmus*)

genus *Latridius*, nomina dubia

brunneus Grimmer, 1841: 45

exaratus Faldermann, 1837: 253

pulla Marsham, 1802: 111 (*Corticaria*)

genus *Lithostygnus* Broun, 1886: 950 type species *Lithostygnus costalus* Broun, 1886

serripennis Broun, 1914: 185 (*Lithostygnus*) **Ei**: GB GE **AUR**

genus *Metophthalmus* Motschulsky, 1850: 365 type species *Metophthalmus asperatus* Wollaston, 1854

subgenus *Bonvouloiria* Jacquelin du Val, 1859a: 245 type species *Bonvouloiria niveicollis* Jacquelin du Val, 1859

kabylianus Chobaut, 1906: 256 (*Bonvouloiria*) **N**: AG

niveicollis Jacquelin du Val, 1859a: 245 (*Bonvouloiria*) **E**: FR IT SP

intermedius Binaghi, 1946: 26

obesus Reitter, 1881b: 51

revelierei Belon, 1881: 264

subgenus *Metophthalmus* Motschulsky, 1850: 365 type species *Metophthalmus asperatus* Wollaston, 1854

Metatypus Belon, 1897: 124 type species *Metophthalmus bicolor* Belon, 1895

asperatus Wollaston, 1854: 193 **N**: MR

encaustus Wollaston, 1865: 154 **N**: CI

exiguus Wollaston, 1860a: 261 **N**: MR

ferrugineus Wollaston, 1865: 154 **N**: CI

hispanicus Reitter, 1908d: 134 **E**: SP **N**: MO
humeridens Reitter, 1884c: 65 **E**: AB ST (Kavkaz) **A**: TR
judaicus J. R. Sahlberg, 1913a: 33 **A**: IS JO
jordanensis J. R. Sahlberg, 1913a: 34
lacteolus Motschulsky, 1866c: 231 **E**: GR IT RO UK
brenskei Reitter, 1884c: 64
creticus Reitter, 1908d: 136
hungaricus Reitter, 1884c: 64
solarii Binaghi, 1946: 22
occidentalis Israelson, 1984b: 147 **E**: AZ
proximus Reitter, 1908d: 136 **E**: SP **N**: MO
raffrayi Belon, 1885b: xxxix **A**: YE
ragusae Reitter, 1875k: 315 **E**: GR IT (Sicilia) **N**: AG TU **A**: IS LE TR
convexiusculus J. R. Sahlberg, 1913a: 35
heiligi Reitter, 1908d: 135
siculus Binaghi, 1946: 25
syriacus Reitter, 1884c: 65
sculpturatus Wollaston, 1862e: 290 **N**: MR

genus *Revelieria* Perris, 1869: 12 type species *Revelieria spectabilis* Perris, 1869 (=*Lathridius genei* Aubé, 1850)
genei Aubé, 1850: 333 (*Lathridius*) **E**: FR GR IT SP **N**: AG MO TU **A**: IS TR
globosa J. R. Sahlberg, 1913a: 37
heydeni Reitter, 1875k: 339
spectabilis Perris, 1869: 12

genus *Stephostethus* LeConte, 1878b: 601 type species *Lathridius liratus* LeConte, 1863
alternans Mannerheim, 1844a: 76 (*Lathridius*) **E**: AU BE BH BY CZ DE EN FI FR GB GE HU IT LS LT NR RO SK SV SZ PL UK
altaicus Reitter, 1902c: 177 (*Lathridius*) **A**: WS
angusticollis Gyllenhal, 1827: 136 (*Latridius*) **E**: AB AU BE BU BY CR CT CZ DE EN FI FR GB GE GG GR HU IT LA LT NL NR NT PL RO SK SL SP ST SV SZ UK **A**: JA TR WS
angulatus Mannerheim, 1844a: 74 (*Lathridius*)
kokujevi Semenov, 1898c: 283 (*Lathridius*)
laticeps Belon, 1881: 271 (*Lathridius*)
obtusatus Roubal, 1951: 332 (*Lathridius*)
stricticollis Rey, 1889h: 54 (*Lathridius*)
undulatus Motschulsky, 1866c: 242 (*Lathridius*)
arunus Sengupta, 1983: 354 **A**: BT HP SD
attenuatus Mannerheim, 1844a: 69 (*Lathridius*) **E**: CT FI NT **A**: ES FE MG WS
jakovlevi Semenov, 1898c: 282 (*Lathridius*)
lenensis Poppius, 1903: 91 (*Lathridius*)
quadraticollis Semenov, 1898c: 282 (*Lathridius*)
semenowi Poppius, 1903: 87 (*Lathridius*)
barunus Sengupta, 1983: 356 **A**: HP
belonianus Reitter, 1889c: 65 (*Lathridius*) **E**: AR GG ST (Kavkaz)
carinatus Sengupta, 1976: 121 **A**: BT SD UP **ORR**
caucasicus Mannerheim, 1844a: 72 (*Lathridius*) **E**: AB AU CZ GB GE HU IT ST (Kavkaz) SZ UK **A**: TR
[*sinuatocollis* sensu auctorum]
chinensis Reitter, 1877f: 113 (*Lathridius*) **A**: CH FE JA
cinnamopterus Mannerheim, 1853: 213 (*Lathridius*) **E**: FI NR NT SV **A**: FE **NAR**
sahlbergi Reitter, 1894i: 300 (*Lathridius*)
indicus Motschulsky, 1866c: 241 (*Lathridius*) **A**: "East India"
kashmirensis Sengupta, 1983: 356 **A**: KA
lardarius DeGeer, 1775: 45 (*Tenebrio*) **E**: AU BE BY CT CZ DE EN FI FR GB GE HU IR IT LA LT NL NR NT PL RO SK SP SV SZ UK **A**: ES WS
acuminatus Paykull, 1798: 299 (*Dermestes*)

dilaticollis Motschulsky, 1866c: 236 (*Lathridius*)
pini Motschulsky, 1866c: 236 (*Lathridius*)
quadratus Herbst, 1793: 8 (*Dermestes*)
subbrevis Motschulsky, 1866c: 237 (*Lathridius*)
malabicus Sengupta, 1983: 351 **A**: KA
malinicus Sengupta, 1983: 359 **A**: SD
minaticus Sengupta, 1983: 350 **A**: HP
nepalensis Sengupta, 1983: 349 **A**: NP
nigratus Sengupta, 1976: 120 **A**: SD **ORR**
pandellei C. N. F. Brisout de Barneville, 1863: 71 (*Lathridius*) **E**: AU BE BH BY CT CZ EN FI FR GE GG HU IT LA LT NL NR NT PL RO SK SP ST (Kavkaz) SV SZ UK **A**: ES FE WS
poppiusi Semenov, 1904: 289 (*Lathridius*)
subnudus Reitter, 1889c: 66 (*Lathridius*)
tremulae C. G. Thomson, 1868: 335 (*Lathridius*)
paradoxus Sengupta, 1976: 122 **A**: HP SD UP **ORR**
productus Rosenhauer, 1856: 351 (*Lathridius*) **E**: FR IT PT SP **N**: AG CI MO TU
renukae Sengupta, 1983: 348 **A**: SD **ORR**
rugicollis A. G. Olivier, 1790a: no 18: 13 (*Ips*) **E**: AU BE BY CT CZ DE EN FI FR GE HU IT LA LT NL NR NT PL RO SK SL SV SZ UK YU **A**: ES HEI
campicola Gerhardt, 1911: 5 (*Lathridius*)
lignarius Gistel, 1857: 527 (*Latridius*)
ovatus Grimmer, 1841: 45 (*Latridius*)
rybinskii Reitter, 1894b: 14 (*Lathridius*) **E**: AR AU CZ GE HU PL SK ST UK
setosus Rücker, 2004: 151 **A**: BEI FUJ
tarunus Sengupta, 1983: 347 **A**: SD
variolosus Mannerheim, 1844a: 73 (*Lathridius*) **E**: FI NR NT SV **A**: ES FE MG WS
lapponum Mannerheim, 1844a: 75 (*Lathridius*)
genus *Stephostethus*, nomina dubia
sinuatocollis Faldermann, 1837: 254 (*Latridius*)
volgensis Mannerheim, 1844a: 77 (*Lathridius*)

genus *Thes* Semenov, 1910: 434 [RN] type species *Lathridius bergrothi* Reitter, 1881
Lar Semenov, 1904: 314 [HN] type species *Lathridius bergrothi* Reitter, 1881
bergrothi Reitter, 1881b: 53 (*Lathridius*) **E**: AU BE BY CT CZ DE EN FI FR GB GE HU IC IR IT LA LT NL NR NT PL SK SV SZ **AFR NAR**
africanus Dajoz, 1980: 286 (*Enicmus*)
microps Ericson, 1896: 266 (*Lathridius*)
tricarinatus Brown, 1934: 22 (*Enicmus*)

subfamily **Corticariinae** Curtis, 1829

genus *Corticaria* Marsham, 1802: 106 type species *Corticaria ferruginea* Marsham, 1802
Abothria Belon, 1897: 146 type species *Latridius elongatus* Gyllenhal, 1827
Adasia Belon, 1897: 147 type species *Dermestes serratus* Paykull, 1798
Brevina Belon, 1897: 140 type species *Corticaria sylvicola* C. N. F. Brisout de Barneville, 1863
Dapeda Belon, 1897: 197 [RN] type species *Corticaria cucujiformis* Reitter, 1881
Diarthrocera Broun, 1893: 1348 type species *Diarthrocera formicaephila* Broun, 1893
Epipeda Belon, 1897: 145 [HN] type species *Corticaria cucujiformis* Reitter, 1881
Epipedana E. Strand, 1942: 391 [RN] type species *Corticaria cucujiformis* Reitter, 1881
Parascheva Gozis, 1886: 10 type species *Latridius pubescens* Gyllenhal, 1827
abdominalis Dajoz, 1970c: 263 **E**: SP
aculeata C. Johnson, 1977c: 330 **A**: BT NP SD **ORR**
aequalis Reitter, 1898c: 29 **E**: AR **A**: IN TD TM
strandi Roubal, 1934d: 324
alleni C. Johnson, 1974: 101 **E**: AU BY CT CZ FI FR GB GE PL SK SV
alticola Lindberg, 1953: 8 **N**: CI
amurensis Reitter, 1879e: 224 **A**: FE
anatolica C. Johnson, 1989: 79 **A**: TR

aphictoides Reitter, 1898c: 31 **A**: MG
appenhageni Uyttenboogaart, 1930: 275 **N**: CI
armata Mannerheim, 1844a: 36 **A**: ES FE
bella L. Redtenbacher, 1849: 208 **E**: AU CZ FR GE GR HU IT PL SK SZ UK
beloni Reitter, 1889a: 21 **E**: AB AR GG IT **A**: IS TM TR
solarii Reitter, 1904e: 156
bengaliensis Rücker, 1978: 151 **A**: SD
brevilata Rücker, 1978: 152 **A**: NP SD
buddha C. Johnson, 1977b: 219 **A**: NP
canariensis C. Johnson, 1974: 106 **N**: CI
cognata Rücker, 1978: 152 **A**: SD
convexa Reitter, 1881b: 60 **E**: SP **N**: AG
corsica H. Brisout de Barneville, 1878: xcvi **E**: FR IT (Sardegna) **N**: AG TU
cotovillae Otero & Díaz Pazos, 1986: 167 **E**: SP
crenicollis Mannerheim, 1844a: 37 **E**: BH BY CT FI NR NT SV **A**: ES WS
lacerata Mannerheim, 1844a: 38
crenulata Gyllenhal, 1827: 125 (*Latridius*) **E**: AR AU BE CR CZ DE EN FI FR GB GE HU IR IT LA LT NL NR NT PT RO SK SP ST (Kavkaz) SV SZ UK **N**: AG TU **A**: ES IN TR WS
spectabilis Dajoz, 1970c: 266
cretica C. Johnson, 1989: 80 **E**: GR (Kriti)
cribricollis Fairmaire, 1863: 72 **E**: FR IT SP **N**: MO TU
angusta Aubé, 1866: 162
cucujiformis Reitter, 1881b: 66 **E**: FI FR GR IT SP SV **N**: AG MO TU **NAR**
planula Fall, 1899: 154
strandi Palm, 1949a: 109 [HN]
delgerchangaji Rücker, 1983a: 174 **A**: MG
dentiventris Poppius, 1903: 84 **E**: FI NR NT SV **A**: ES FE WS **NAR**
diecki Reitter, 1875o: 413 **E**: SP **N**: AG MO TU
kaufmanni Reitter, 1881b: 61
distincta Dajoz, 1970c: 271 **E**: SP
dubia Dajoz, 1970c: 269 **E**: FR GB GE HU SP YU
elongata Gyllenhal, 1827: 130 (*Latridius*) **E**: AB AR AU AZ BE BY CR CT CZ DE EN FR GB GE GG GR HU IC IR IT FI LA LT NL NT NR PL RO SK SP ST SV SZ UK **N**: MO TU **A**: AF CY FE JA KA NP PA SA TR WS **COS**
phoretica Dajoz, 1970a: 37
pseudoelongata Dajoz: 1970b: 239
villiersi Dajoz, 1970d: 433
fagi Wollaston, 1854: 188 **E**: AU AZ CZ DE FI FR GB GE HU NR PL RO SV UK **N**: MR **A**: FE **AFR AUR ORR**
aequidentata Allen, 1937: 52
pietschi Ganglbauer, 1899b: 806
fasciata Reitter, 1877f: 115 **A**: FE JA TAI
fastigata Rücker, 1978: 152 **A**: NP SD
fennica C. Johnson, 1974: 102 **E**: FI SV
ferruginea Marsham, 1802: 111 **E**: AB AU BE BH BY CT CZ DE EN FI FR GB GE GG GR HU IR IT LA LT NL NR NT PL RO SK SP SV SZ UK **A**: ES FE IN JIL MG WS **AUR NAR**
boreophila Motschulsky, 1867: 65
deleta Mannerheim, 1853: 212 (*Corticaria*)
denticulata Kirby, 1837: 110 [HN]
[*fenestralis* sensu auctorum]
kirbyi LeConte, 1855: 300
longula Broun, 1910: 26
nigriceps Waltl, 1839: 224 (*Latridius*)
nigricollis Zetterstedt, 1838: 199 (*Latridius*)
obtusa LeConte, 1855: 300
rufulus Zetterstedt, 1838: 199 (*Latridius*)
rugulosa LeConte, 1855: 300

subacuminata Mannerheim, 1844a: 46
foveola Beck, 1817: 14 (*Latridius*) **E**: AU BE CT CZ DE EN FI FR GE IT LT NL NR NT PL RO SV SZ UK **A**: FE TR
amplicollis Reitter, 1875o: 424
dilatipennis Reitter, 1878g: 96
foveolata Westerhauser, 1834: 154 (*Latridius*)
franzi Dajoz, 1969a: 96 **E**: SP
fulva Comolli, 1837: 39 (*Latridius*) **E**: AR AZ AU BE CR CT CZ DE EN FI FR GB GE GG GR HU IR IT LA LT NL NR NT PL SK SP SV SZ **N**: CI MO MR TU **A**: AF IQ KZ TD TM TR **COS**
attenuata Motschulsky, 1867: 67
cardiodera Fairmaire, 1875a: 505
ciliata Motschulsky, 1867: 55
concolor H. Brisout de Barneville, 1880: 236
flavescens C. G. Thomson, 1871: 383
hirtella C. G. Thomson, 1863: 232
metallica Reitter, 1874c: 526
normanna C. N. F. Brisout de Barneville, 1892: 68
unicarinulata Motschulsky, 1867: 76
fulvoides C. Johnson, 1989: 82 **N**: MO
geisha C. Johnson, 1989: 88 **A**: FUJ JA ZHE
gracilis Mannerheim, 1844a: 32 **A**: ES FE WS
hierroensis C. Johnson, 1985: 63 **N**: CI
illaesa Mannerheim, 1844a: 33 **E**: BU FR GR IT UK **N**: AG EG MO TU **A**: CY IN IQ IS JO SA TM TR YE
cypria Baudi di Selve, 1870: 61
monticola H. Brisout de Barneville, 1881: 388
pilosa Motschulsky, 1867: 48
punctatissima Motschulsky, 1867: 69
subparallela Fairmaire, 1875a: 505
villosa Motschulsky, 1867: 48
impressa A. G. Olivier, 1790a: no 18: 14 (*Ips*) **E**: AU BE BU BY CT CZ DE EN FI FR GB GE GG HU IR IT LA LT NL NR NT PL RO SK SP ST SV SZ UK YU **A**: BEI ES FE JP TR UZ WS **NAR**
atricollis Reitter, 1898c: 31
badia Mannerheim, 1844a: 25
campicola Mannerheim, 1844a: 26
denticulatus Gyllenhal, 1827: 126 (*Latridius*)
sculptipennis Faldermann, 1837: 252
validipes Motschulsky, 1867: 54
inconspicua Wollaston, 1860a: 260 **E**: BE CZ DE FR IR IT FI GB GE HU NL PL PT SK SP SV SZ **N**: CI MO MR TU **A**: CY
fagi H. Brisout de Barneville, 1881: 401 [HN]
siccana Normand, 1938b: 362
inflatipennis Champion, 1922c: 214 [RN] **A**: NP UP
inflata Champion, 1922a: 70 [HN]
interstitialis Mannerheim, 1844a: 21 **E**: CT CZ FI GE NR NT PL SV
japonica Reitter, 1877f: 116 **A**: CH JA **AFR AUR NTR**
adelaidae Blackburn, 1891a: 120
rudebecki Dajoz, 1970b: 236
kabakovi Saluk, 1992: 383 **A**: FE
khnzoriani C. Johnson, 1989: 85 **E**: AR **A**: TR
lapponica Zetterstedt, 1838: 199 (*Latridius*) **E**: AU BY CR CT CZ DE EN FI HU IT NR NT PL RO SK SV UK **A**: FE WS
robusta Ganglbauer, 1899b: 801
lateritia Mannerheim, 1844a: 44 **E**: AU BE CT CZ FI FR GE HU IT LA NR NT PL RO SV SZ **A**: FE
eppelsheimi Reitter, 1875o: 423
leileri C. Johnson, 1985: 64 **N**: CI
lineatoserrata C. Johnson, 1977b: 225 **A**: NP
longicollis Zetterstedt, 1838: 200 (*Latridius*) **E**: AU BE BH BY CR CT CZ DE EN FI FR GB GE GG HU IT LS LT NL NR NT PL PT RO SK SV SZ UK

formicetorum Mannerheim, 1843a: 85 (*Latridius*)
stigmosa Motschulsky, 1867: 79
longicornis Herbst, 1783: 23 (*Dermestes*) **E**: AU BE BY CT CZ DE EN FI FR GB GE IT NR NT PL RO SV SZ UK
abietorum Motschulsky, 1867: 71
depressa C. G. Thomson, 1871: 385
mannerheimi Reitter, 1875o: 427
maculosa lineata C. Johnson, 1974: 107 **N**: CI
maculosa maculosa Wollaston, 1858c: 408 **E**: AZ **N**: CI MR
magadanica Tsinkevich, 2001: 85 **A**: FE
martensi C. Johnson, 1977b: 222 **A**: NP SD
mongolica Rücker, 1983a: 173 **A**: MG WS
nebulosa Champion, 1922a: 71 **A**: BT HP NP PA UP
obscura C. N. F. Brisout de Barneville, 1863: 73 **E**: AB AU BE CZ FR GE GR HU IT NL PL RO SK SP SZ UK **N**: AG TU
subpicea Motschulsky, 1867: 70
obsoleta A. Strand, 1940: 179 **E**: FI NR NT SV **NAR**
olympiaca Reitter, 1875o: 417 **E**: BH BU CR GR UK
orbicollis Mannerheim, 1853: 212 **E**: BY FI NR NT PL SV **A**: MG WS **NAR**
munsteri A. Strand, 1937: 9
orientalis Champion, 1922a: 69 **A**: HKG HP KA NP SD UP
ornata Reitter, 1877f: 115 **A**: JA
ovicollis Reitter, 1887h: 509 **A**: MG WS
parvithorax Champion, 1922c: 214 [RN] **A**: HP NP PA SD UP **ORR**
parvicollis Champion, 1922a: 71 [HN]
peezi Johnson, 2007 **sp. nov**. [see New acts] **E**: AU BH BU GE GR IT RO ST
[*weisei* Reitter, 1875 sensu auctorum]
pharaonis Motschulsky, 1867: 74 **N**: EG **A**: AE CY
transversicollis Motschulsky, 1867: 76
pilosula Rosenhauer, 1856: 349 **E**: PT SP **N**: AG MO TU **A**: IS
pineti Lohse, 1960: 40 [RN] **E**: AL AR AU CZ DE EN FI GE GR IT NR SV SZ UK
pinicola Lohse, 1959: 33 [HN]
pinicola C. N. F. Brisout de Barneville, 1866: 370 **E**: SP
rufescens Reitter, 1875o: 420
polypori J. R. Sahlberg, 1900: 12 **E**: AU BY CZ EN FI GE HU LA NR NT PL RO SK SV SZ UK
porochini Johnson, 2007 **sp. nov**. [see New acts] **E**: AU CT CZ FI GE HU IT NL NT PL SP SV UK
[*longicornis* Herbst, sensu auctorum]
pubescens Gyllenhal, 1827: 123 (*Latridius*) **E**: AU BY CR CT CZ DE EN FI FR GE GG HU IT LA LT NR NT PL PT RO SK SP ST SV SZ UK **N**: CI MO MR TU **A**: ES FE TR WS **AFR AUR NAR**
brouni Hetschko, 1926: 48 [RN]
capensis Motschulsky, 1867: 51
concinnula Mannerheim, 1844a: 27
diluta Mannerheim, 1844a: 20
grossa LeConte, 1855: 299
intricata Mannerheim, 1844a: 20
robusta Broun, 1914: 183 [HN]
testacea Dalla Torre, 1879: 97
tincta Mannerheim, 1844a: 26
punctata Dajoz, 1970c: 261 **E**: SP
punctulata Marsham, 1802: 109 **E**: AL BE DE FR GB GE IR NL UK SP SZ **N**: MO
piligera Mannerheim, 1844a: 19
quadrimaculata Mannerheim, 1844a: 33 **E**: AB AR GG ST (Kavkaz) UK **A**: TM ZU
perpulchra Reitter, 1892a: 134
setosa Motschulsky, 1867: 49
relicta C. Johnson, 1977b: 223 **A**: NP
rhombifera Champion, 1922a: 70 **A**: KA NP SD UP **ORR**
rotundipennis C. Johnson, 1977b: 220 **A**: NP

rubripes Mannerheim, 1844a: 41 **E**: AU BE BY CT CZ DE EN FI FR GB GE HU IT LA NL NR NT PL RO SK SV SZ UK **A**: ES FE MG **NAR**

baicalica Mannerheim, 1844a: 41
limbicollis Motschulsky, 1867: 64
linearis Paykull, 1798: 302 (*Dermestes*) [HN]
obscuripes Reitter, 1898c: 32
spinulosa C. G. Thomson, 1871: 385 [HN]
thomsoni Reitter, 1881b: 66

rufa C. Johnson, 1977b: 223 **A**: BT NP
rugipennis Reitter, 1881b: 66 **N**: EG
saginata Mannerheim, 1844a: 24 **E**: AU BE BY CT CZ DE EN FI FR GE GR HU IT LA LT NR NT PL RO SK SV SZ UK **A**: ES JIL WS

curtipes Lövendal, 1893: 273

serrata Paykull, 1798: 300 (*Dermestes*) **E**: AB AU AZ BE BU BY CR CT CZ DE EN FI FR GB GE GG GR HU IR IT LA LS LT NL NR NT PL RO SK SP ST (Kavkaz) SV SZ UK **N**: AG CI MO MR TU **A**: AF CY IS JA KZ LE TR **COS**

adustipennis Motschulsky, 1867: 57
axillaris Mannerheim, 1844a: 30
chilensis Dajoz, 1967: 604
clairi H. Brisout de Barneville, 1881: 401 [DA]
flavifrons Motschulsky, 1867: 66
hispanica Dajoz, 1970c: 268
laticollis Mannerheim, 1844a: 29
melanophthalma Mannerheim, 184a4: 30
motschulskii Kolenati, 1846a: 41
octodentata Say, 1824a: 325
quadricollis Motschulsky, 1867: 59
rotulicollis Wollaston, 1854: 184
serrata Marsham, 1802: 109 [HN]
subcordata Rey, 1889h: 54
tenuicornis Motschulsky, 1867: 77
weisei Reitter, 1875o: 415

subamurensis Saluk, 1992: 386 **E**: FI **A**: FE
subpilosula Reitter, 1898c: 29 **E**: AR **A**: AE AF IN OM TM UZ

illaesiformis Reitter, 1898c: 30

suspecta C. Johnson, 1989: 84 **N**: MO
sylvicola C. N. F. Brisout de Barneville, 1863: 72 **E**: FR SP

pinguis Aubé, 1866: 162

tarragonensis Dajoz, 1970c: 260 **E**: SP

besucheti Dajoz, 1970c: 260

thea Reitter, 1894f: 101 **A**: TR
topali Rücker, 1978: 152 **A**: NP SD
tuberculata Dajoz, 1970c: 264 **E**: SP
tunisiensis H. Brisout de Barneville, 1884: lxxxi **N**: TU
umbilicata Beck, 1817: 13 (*Latridius*) **E**: AU BE BY CT CZ DE EN FI FR GB GE HU IR IT LA LS LT NL NR NT PL RO SK SP SV SZ UK **N**: CI MR **A**: ES WS

borealis Wollaston, 1855: ccvi
cylindrica Mannerheim, 1844a: 35
cylindripennis Motschulsky, 1867: 68
umbilicifera Mannerheim, 1844a: 37

genus *Corticarina* Reitter, 1881b: 70 type species *Corticaria truncatella* Mannerheim, 1844

Oropsime Gozis, 1881a: 161 type species *Melanophthalma carinulata* Motschulsky, 1867

acuta C. Johnson, 1975d: 102 **A**: YE **AFR**
alemannica Schiller, 1984: 103 **E**: GE SZ
bhutanensis C. Johnson, 1977c: 333 **A**: BT FUJ SD TAI YUN **ORR**
biharensis C. Johnson, 1979: 55 **A**: BT SD **ORR**

bruta C. Johnson, 1977c: 334 **A**: BT
carinifrons C. Johnson, 1990: 14 **A**: FUJ
cavicollis Mannerheim, 1844a: 56 (*Corticaria*) **Ei**: IT **NAR NTR**
angularis LeConte, 1855: 301 (*Corticaria*)
expansa LeConte, 1855: 301 (*Corticaria*)
laevis LeConte, 1855: 301 (*Corticaria*)
subfusca Sharp, 1902: 636 (*Corticaria*)
championi C. Johnson, 1972a: 103 **A**: NP SD UP
chinensis C. Johnson, 1972b: 189 **A**: FE HEI
clareae C. Johnson, 1972a: 102 **A**: BT HP NP SD TAI UP **ORR**
coei C. Johnson, 1972a: 105 **A**: NP SD UP **ORR**
cognata C. Johnson, 1972a: 104 **A**: NP SD UP **ORR**
curta Wollaston, 1854: 187 (*Corticaria*) **E**: AU AZ BE BU CR FR GB GG GR HU IT MA PL PT RO SP ST SZ UK YU **N**: AG CI EG MO MR TU **A**: CY TR UZ
carinulata Motschulsky, 1867: 84 (*Corticaria*)
[*fulvipes* sensu auctorum]
meridionalis Reitter, 1875o: 442 (*Melanophthalma*)
ooptera Fairmaire, 1875a: 506 (*Corticaria*)
pilitecta Motschulsky, 1867: 83 (*Corticaria*)
curta oblongipennis C. Johnson, 1981: 123 **A**: IN IS TR
delicatula Wollaston, 1871: 252 (*Corticaria*) [RN] **N**: CI
tenella Wollaston, 1864: 150 (*Corticaria*) [HN]
excavata C. Johnson, 1977a: 130 **A**: AF IN TD XIN
sororcula Rücker, 1980a: 93
fornicata Otto, 1978: 287 (*Cortarina*) **A**: IN
franzi C. Johnson, 1975d: 97 **A**: NP
fraudulenta C. Johnson, 1972a: 103 **A**: NP SD UP **ORR**
fukiensis C. Johnson, 1990: 13 **A**: FUJ YUN
gangolae C. Johnson, 1970b: 190 **A**: SD UP **ORR**
guptai C. Johnson, 1979: 55 **A**: NP SD
hammondi C. Johnson, 1972a: 102 **A**: SD UP **ORR**
irkutensis A. Strand, 1968: 40 **A**: ES FE
kabakovi Saluk, 1992: 391 **A**: FE
khnzoriani C. Johnson, 1976c: 197 **A**: FE
lambiana Sharp, 1910: 106 (*Corticaria*) **E**: AU BY DE FI GB GE HU LA LT NR SP SV SZ **A**: FE HEI JA MG WS
latipennis J. R. Sahlberg, 1871: 359 (*Corticaria*) **E**: CT CZ EN FI GB GE LA NR NT PL SV SZ **A**: FE **NAR**
fowleriana Sharp, 1910: 108 (*Corticaria*)
terricula Fall, 1899: 178 (*Melanophthalma*)
trinotata J. R. Sahlberg, 1926: 108 (*Melanophthalma*)
lutea Rücker, 1979a: 127 **A**: NC
minuta Fabricius, 1792a: 235 (*Dermestes*) **E**: AU BE BY CR CT CZ DE EN FI FR GB GE GG HU IR IT LA LT NL NR NT PL RO SK SP ST SV SZ UK **N**: CI **A**: AF ES FE HEB HEI KZ MG WS **NAR**
americana Mannerheim, 1844a: 50 (*Corticaria*)
brevicollis Mannerheim, 1844a: 52 (*Corticaria*)
compta LeConte, 1855: 301 (*Corticaria*)
fuscula Gyllenhal, 1827: 133 (*Latridius*)
melanocara Gistel, 1857: 585 (*Corticaria*)
ovalipennis Reitter, 1875o: 441 (*Melanophthalma*)
trifoveolata L. Redtenbacher, 1849: 211 (*Corticaria*)
nakanei C. Johnson, 1976c: 197 **A**: JA
nigerrima C. Johnson, 1975d: 97 **A**: NP
nigra C. Johnson, 1975d: 99 **A**: HP KA NP PA UP
paradoxa Saluk, 1992: 389 **A**: FE
parallela C. Johnson, 1972a: 104 **A**: UP

parvula Mannerheim, 1844a: 54 (*Corticaria*) **E**: AU BE BU BY CZ FI FR GE HU IT LA NR PL RO SK SL SV SZ
obfuscata A. Strand, 1937: 16
rotundipennis Wollaston, 1854: 186 (*Corticaria*) **N**: MR
saluki Johnson, **sp. nov**. [see New acts] **A**: FE
strandi sensu Saluk, 1992
seminigra C. Johnson, 1975d: 99 **A**: NP
similata Gyllenhal, 1827: 134 (*Latridius*) **E**: AB AU BE BU BY CR CT CZ DE EN FI FR GB GE GG GR HU IR IT LA LS LT NL NR NT PL RO SK SL SP ST SV SZ UK **N**: AG **A**: ES FE SHA TR
fulvipes Comolli, 1837: 39 (*Latridius*) [DA]
pusilla Rey, 1889h: 54 (*Melanophthalma*) [HN]
subtilis Mannerheim, 1844a: 57 (*Corticaria*)
strandi C. Johnson, 1972a: 105 **A**: UP
subcognata C. Johnson, 1979: 56 **A**: NP SD **ORR**
truncatella Mannerheim, 1844a: 59 (*Corticaria*) **E**: AU BE BU BY CR CT CZ DE EN FI FR GB GE GR HU IT LA LS LT NL NR PL RO SK SP SV SZ UK **N**: AG TU **A**: BEI ES FE HEI KZ MG NC SHA WS
fossulata Rey, 1889h: 54 (*Melanophthalma*)

genus *Cortinicara* C. Johnson, 1975e: 283 type species *Latridius gibbosus* Herbst, 1793
afghana C. Johnson, 1977a: 129 **A**: AF
bhutanica C. Johnson, 1977c: 333 **A**: BT SD UP **ORR**
gibbosa Herbst, 1793: 5 (*Latridius*) **E**: AB AU BE BU BY CR CT CZ DE EN FI FR GB GE GG GR HU IR IT LA LS LT NL NR NT PL PT RO SK SP ST SV SZ TR UK **N**: EG MR **A**: AF BEI BT ES FE FUJ HEB HEI HKG HP JA KA MG NC NP PA SD SHA TAI TR UP WS **AFR NAR ORR**
cylindricollis Motschulsky, 1866c: 289 (*Melanophthalma*)
fuscotestacea Motschulsky, 1861c: 128 (*Coricaria*)
gibbula Motschulsky, 1866c: 287 (*Melanophthalma*)
herbacea Gistel, 1857: 527 (*Latridius*)
intricata Rey, 1889h: 54 (*Melanophthalma*)
pallida Marsham, 1802: 109 (*Corticaria*)
resecta Walker, 1859a: 53 (*Corticaria*)
subgibbosa C. Johnson, 1972b: 101 (*Corticarina*) **A**: NP UP **ORR**

genus *Melanophthalma* Motschulsky, 1866c: 269 type species *Latridius transversalis* Gyllenhal, 1827
subgenus *Cortilena* Motschulsky, 1867: 95 type species *Corticaria pallens* Mannerheim, 1844
dewittei Dajoz, 1970a: 21 **A**: YE **AFR**
fuscipennis Mannerheim, 1844a: 62 (*Corticaria*) **E**: AU CR FR GG GR HU IT PT RO SL SP ST SZ TR **N**: EG MO MR TU **A**: CY TR
picipennis Mannerheim, 1844a: 63 (*Corticaria*)
pallens Mannerheim, 1844a: 58 (*Corticaria*) **E**: AR BU UK **A**: AF KZ
sakagutii C. Johnson, 1976c: 195 **A**: FE JA
subgenus *Melanophthalma* Motschulsky, 1866c: 269 type species *Latridius transversalis* Gyllenhal, 1827
Melanopsis Gozis, 1875: 18 [HN] type species *Corticaria hortensis* Mannerheim, 1844
aegyptiaca Otto, 1978: 289 **N**: EG
algirina Motschulsky, 1866c: 273 **E**: FR SP **N**: AG EG MO TU
angulata Wollaston, 1871: 251 (*Corticaria*) **N**: CI
angulicollis Motschulsky, 1866c: 278 **A**: NP UP **ORR**
ceylanica Dajoz, 1975: 206
hexagona Motschulsky, 1866c: 279
prominens C. Johnson, 1972a: 108
arabiensis Otto, 1979: 233 **A**: SA YE
argentea Rücker, 1984: 249 **A**: SA
cantabrica Otto, 1978: 288 **E**: FR MA PT SP **N**: MO
demoulini Dajoz, 1970a: 25 **A**: YE **AFR**
kaszabi Rücker, 1980b: 167
dentata Dajoz, 1970c: 240 **A**: NP SA YE **AFR**

distinguenda Comolli, 1837: 38 (*Latridius*) **E**: AU BE BY CR CZ DE FR GB GE GR HU IT LT MA NL PL PT RO SV SZ UK **N**: MR **A**: IQ TR
evansi C. Johnson, 1972a: 106 **A**: JA "North India"
extensa Rey, 1889h: 54 **E**: FR GR IT PT SP UK **A**: TR
franzi C. Johnson, 1972a: 109 **A**: BT HP NP PA SD UP **ORR**
japonica C. Johnson, 1976c: 195 **A**: JA
maura Motschulsky, 1866c: 271 **E**: AU CT CZ FR GE HU IT PL SP SV SZ UK **A**: AF ES FE TR WS
  austriaca Franz, 1967c: 107
ophthalmica Dajoz, 1970a: 29 **A**: SA **AFR**
parvicollis Mannerheim, 1844a: 62 (*Corticaria*) **E**: AB FR IT GR HU SP ST (Kavkaz) TR UK YU **N**: TU **A**: CY IN JO TR
phragmiteticola Franz, 1967c: 106 **E**: AU FR GE HU UK
sericea Mannerheim, 1844a: 61 (*Corticaria*) **E**: BU CR FR GR HU IT MA PT SP ST (Kavkaz) SZ TR **N**: EG MO TU **A**: AF CY IN LE TR
  albipilis Reitter, 1875o: 435
  parva C. Johnson, 1972b: 195
  rutila Motschulsky, 1866c: 288
similis Míka, 2000: 128 **A**: FUJ **ORR**
subvillosa C. Johnson, 1972a: 108 **A**: NP
suturalis Mannerheim, 1844a: 58 (*Corticaria*) **E**: AR BY CT CZ DE FI FR GB GE HU IT MC PL SP ST SV SZ UK **N**: MO **A**: AF KZ TR
  fallax Rey, 1889h: 54
taurica Mannerheim, 1844a: 51 (*Corticaria*) **E**: AB GE ST (Kavkaz) UK **A**: AF IN TD TR
  angulosa Motschulsky, 1849: 90 (*Corticaria*)
  asiatica Iablokoff-Khnzorian, 1970a: 67
transversalis Gyllenhal, 1827: 133 (*Latridius*) **E**: AR AU BE BU BY CR CT CZ DE EN FR GB GE HU IT FI LA LT NL NR NT PL RO SK ST (Kavkaz) SV SZ UK **A**: FE HEB HEI JA NC TR
  crocata Mannerheim, 1844a: 53 (*Corticaria*)
  curticollis Mannerheim, 1844a: 47 (*Corticaria*)
  hortensis Mannerheim, 1844a: 52 (*Corticaria*)
  moraviaca Reitter, 1875o: 435
  sinica C. Johnson, 1972b: 195
  wollastoni G. R. Waterhouse, 1859: 143 (*Corticaria*)
videns Otto, 1979: 232 **A**: SA **AFR**
wittmeri Otto, 1978: 289 **A**: JO TR

genus *Migneauxia* Jacquelin du Val, 1859a: 248 type species *Migneauxia serricollis* Jacquelin du Val, 1859 (= *Corticaria crassiuscula* Aubé, 1850)
crassiuscula Aubé, 1850: 331 (*Corticaria*) **E**: AB BU CR FR GG GR HU IT SP UK **A**: AF CY IQ IS JO TR UZ
  serricollis Jacquelin du Val, 1859a: 248
  villigera Motschulsky, 1867: 40 (*Mineauxia*)
lederi Reitter, 1875o: 444 **E**: AZ CR DE GB GE HU SP SV SZ **N**: AG CI EG MO TU **A**: AE AF BT IQ JA JIA JO SA SD UP YE **COS**
  orientalis Reitter, 1877g: 139
  parvicollis Peyerimhoff, 1929d: 212
ottoi Johnson, **sp. nov**. [see New acts] **E**: IT (Sardegna, Sicilia)
  [*lederi* sensu auctorum]
phili Johnson, **sp. nov**. [see New acts] **E**: SP **N**: AG MO TU
  [*inflata* sensu auctorum]
psammeticha Motschulsky, 1867: 80 (*Corticaria*) **N**: EG **AFR**
  subrugosa Motschulsky, 1867: 82 (*Corticaria*)
  sudanensis C. Johnson, 1976a: 31
subdola C. Johnson, 1977a: 128 **A**: AF
genus *Migneauxia*, nomen dubium
inflata Rosenhauer, 1856: 350 (*Corticaria*)

REFERENCES

Abeille de Perrin E. 1869a: Nouveaux coléoptères français. *Annales de la Société Entomologique de France* (4) **9**: 39-46.

Abeille de Perrin E. 1869b: Petites nouvelles. *Petites Nouvelles Entomologiques* **1** (1): 41-42.

Abeille de Perrin E. 1869c: Description de deux espèces nouvelles de coléoptères du bassin circum-méditerranéen et quelques observations sur divers Malachiaires, suivies de la description d'Anophthalmus nouveaux. *Annales de la Société Entomologique de France* (4) **9**: 401-409.

Abeille de Perrin E. 1872a: *Études sur les coléoptères cavernicoles suivies de la description de 27 coléoptères nouveaux français*. Marseille: M. Olive, 40 + [1] pp.

Abeille de Perrin E. 1872b: Études sur les coléoptères cavernicoles, suivies de la description de coléoptères nouveaux propres au midi de la France. *Bulletin de la Société Entomologique de France* **1872**: xli-xliv.

Abeille de Perrin E. 1875: Synopsis des Dryophilus du bassin de la Méditerranée. *Annales de la Société Entomologique de France* (5) **5**: 207-212.

Abeille de Perrin E. 1881: Contribution à la faune coléoptérologique d'Europe et des pays voisins. *Annales de la Société Entomologique de France* (6) **1**: 97-128.

Abeille de Perrin E. 1882a: Supplément à la monographie des Malachides d'Europe et des pays voisins. *Il Naturalista Siciliano* **1**: 110-115, 137-142, 145-149, 176-179.

Abeille de Perrin E. 1882b: Malachides nouveaux d'Algérie. *Revue d'Entomologie* **1**: 180-183.

Abeille de Perrin, E. 1882c: [Notes synonymiques]. *Bulletin de la Société Entomologique de France* **1882**: cxxiv-cxxv.

Abeille de Perrin E. 1883: Nouveau supplément à l'histoire des Malachides. *Revue d'Entomologie* **2**: 25-37, 49-57.

Abeille de Perrin E. 1885a: Nouveaux documents pour servir à l'histoire des Malachides. *Revue d'Entomologie* **4**: 4-28.

Abeille de Perrin E. 1885b: Malachides nouveaux. *Revue d'Entomologie* **4**: 139-153.

Abeille de Perrin E. 1885c: La description d'une nouvelle espèce de coléoptères. *Bulletin de la Société Entomologique de France* 1885: cxvi.

Abeille de Perrin E 1887a: [new taxa]. In: Heyden L. F. J. D. von.: XI. Beitrag zur Coleopteren-Fauna von Turkestan. *Deutsche Entomologische Zeitschrift* **31**: 305-336.

Abeille de Perrin E. 1887b: Anhang: Beschreibung der neuen Arten. Pp. 446-448. In: Heyden L. F. J. D. von.: Bericht über die von den Herrn Prof. Dr. Freiherrn von Fritsch und Prof. Dr. Rein in Morocco und dem Atlas-Gebirge gesammelten Käfer. *Deutsche Entomologische Zeitschrift* **31**: 432-448.

Abeille de Perrin E. 1888: [new taxa]. In: Heyden L. F. J. D. von. 1888. Käfer vom Cilicischen Taurus. *Deutsche Entomologische Zeitschrift* **32**: 68-71.

Abeille de Perrin E. 1890a: Études sur les Malachides. *Revue d'Entomologie* **9**: 35-55.

Abeille de Perrin E. 1890b: Descriptions de deux nouvelles espèces de Malachiides. *Annali del Museo Civico di Storia Naturale di Genova* **29**: 89-90.

Abeille de Perrin E. 1890c: Malachiidae. Malachides d'Europe et pays voisins. *Annales de la Société Entomologique de France* (6) **10**: 181-260.

Abeille de Perrin E. 1891a: Malachiidae. Malachides d'Europe et pays voisins. *Annales de la Société Entomologique de France* (6) **10** [1890]: 331-420, 567-680, pl. 4.

Abeille de Perrin E. 1891b: Malachiidae. Malachides d'Europe et pays voisins. *Annales de la Société Entomologique de France* **60**: 115-230, 405-446, pls. 6-7.

Abeille de Perrin E. 1892: Notices entomologiques. IV. Notices pour servir à l'histoire de Brachypterus (S.-G. Heterostomus Duv.). *Revue d'Entomologie* **11**: 66-67.

Abeille de Perrin E. 1893a: [Deux nouvelles espèces de Clérides méditerranées]. *Bulletin de la Société Entomologique de France* **1892**: ccxxxvi-ccxxxvii.

Abeille de Perrin E. 1893b: Les descriptions de quatre nouvelles espèces de Malachiidae. *Bulletin de la Société Entomologique de France* **1893**: clxii-clxv.

Abeille de Perrin E. 1893c: Description d'un nouveau Malachidae de Tunisie: Attalus cerastes. *Bulletin de la Société Entomologique de France* **1893**: clxxvi-clxxvii.

Abeille de Perrin E. 1893d: [rectification]. *Bulletin de la Société Entomologique de France* **1893**: clxxvii.

Abeille de Perrin E. 1894a: Notes pour servir à l'histoire des Malachides. *Anales de la Sociedad Española de Historia Natural* (2) **3** (23): 89-92.

Abeille de Perrin E. 1894b: Diagnoses de coléoptères réputés nouveaux. *L'Échange, Revue Linnéenne* **10**: 91-94.

Abeille de Perrin E. 1894c: [Description d'un genre nouveau et de quatre espèces nouvelles de coléoptères français: Ptinus, Coroebus et Troglophyes]. *Bulletin de la Société Entomologique de France* **63**: xxv-xxviii.

Abeille de Perrin E. 1894d: Sur trois Ptinus (Bruchus) du groupe de perplexus. *Bulletin de la Société Entomologique de France* **63**: cxxxi-cxxxii.

Abeille de Perrin E. 1895a: Descriptions de deux coléoptères du midi de la France et observation synonymique. *Bulletin de la Société Entomologique de France* **64**: ccxlvi-ccxlviii.

Abeille de Perrin E. 1895b: Deux Malachides paléarctiques nouveaux (Col.). *Bulletin de la Société Entomologique de France* **1895**: cccliii-cccliv.

Abeille de Perrin E. 1896a: Description de deux Malthiniens nouveaux (Col.): Podistrina Vaucheri et P. aptera. *Bulletin de la Société Entomologique de France* **1896**: 155-156.

Abeille de Perrin E. 1896b: Description de deux nouveaux Malachides d'Algérie et revision partielle des Axinotarsus (Col.). *Bulletin de la Société Entomologique de France* **1896**: 391-393.

Abeille de Perrin E. 1897: Diagnoses de deux Ptinus (Bruchus) français nouveaux (Col.). *Bulletin de la Société Entomologique de France* **1897**: 131-132.

Abeille de Perrin E. 1898a: Description d'un nouveau genre d'eucnémide de France. *Bulletin de la Société Entomologique de France* **1898**: 35-36.

Abeille de Perrin E. 1898b: Malthinus Devellei n. sp. *L'Abeille, Journal d'Entomologie* **29** [1896-1900]: 92.

Abeille de Perrin E. 1898c: Descriptions de coléoptères du Nord de l'Afrique. *Bulletin de la Société Entomologique de France* **1898**: 254-258.

Abeille de Perrin E. 1900a: Diagnoses de coléoptères présumés nouveaux. *Bulletin de l'Académie de Marseille* **27**: 1-23. [seen as separate].

Abeille de Perrin E. 1900b: Malachides recueillis par Eugène Simon au Cap de Bonne-Espérance. *Revue d'Entomologie* **19**: 163-177.

Abeille de Perrin E. 1903: Description de deux espèces de coléoptères d'Algérie. *Bulletin de la Société Entomologique de France* **1903**: 334-335.

Abeille de Perrin E. 1907a: Hypebaeus Peyerimhoffi n. sp. Pp. 18-19. In: Peyerimhoff P. de.: Liste des coléoptères du Sinai. *L'Abeille, Journal d'Entomologie* **31**: 1-47.

Abeille de Perrin E. 1907b: Sur les Dasytides nouveaux. *Annales de la Société des Sciences Naturelles de Provence, Comptes Rendus* **1**: xx-xxiii.

Abeille de Perrin E. 1909a: Description de deux coléoptères du Nord de l'Afrique. *Bulletin de la Société Linnéenne de Provence* **1** [1909-1913]: 8-9.

Abeille de Perrin E. 1909b: Coléoptères d'Algérie jugés nouveaux. *Bulletin de la Société Entomologique de France* **1909**: 179-181.

Achard J. 1922: Description de nouveaux endomychides (Col. Endomychidae). *Fragments Entomologiques* (Prague: Smichov) **1922**: 28-30.

Achard J. 1923: Aulacochilus nouveaux (Col. Erotylidae). *Fragments Entomologiques* (Prague: Smichov) **1923**: 48-51.

Achard J. 1926: Nouveaux Endomychidae. *Fragments Entomologiques* (Prague: Smichov) **1926** [1925]: 140-143.

Adams M. 1817: Descriptio insectorum novorum Imperii Rossici, imprimis Caucasi et Sibiriae. *Mémoires de la Société Impériale des Naturalistes de Moscou* **5**: 278-314.

Afroze S. 1998: [new taxon]. In: Afroze S. & Shujauddin 1998: A new species of Jauravia Motschulsky (Coleoptera: Coccinellidae) from India. *Shashpa* **5** (2): 131-132.

Agajev B. I. 1985a: Novyy rod zhukov-shchelkunov (Coleoptera, Elateridae) [A new click-beetle genus (Coleoptera, Elateridae)]. *Vestnik Zoologii* **1985** (4): 66-67.

Agajev B. I. 1985b: Novye vidy zhukov-shchelkunov (Coleoptera, Elateridae) iz vostochnogo Zakavkaz'ya [New species of click-beetles (Col. Elateridae) from Eastern Transcaucasia]. *Vestnik Zoologii* **1985** (6): 56-58.

Agajev B. I. 1985c: Novyy vid zhuka-shchelkuna (Coleoptera, Elateridae) iz Talysha [A new species of click-beetles (Coleoptera, Elateridae) from Talysch]. *Doklady Akademii Nauk Azerbaydzhanskoy SSR* **41** (7): 63-65.

Agajev B. I. 1988: *Zhestkokrylye shchelkuny v biocenozakh Azerbaydzhana* [*The click-beetles of the biocoenoses of Azerbayjan*]. Baku: Azerbaydzhanskoe gosudarstvennoe izdatel'stvo, 120 pp.

Agassis L. 1846a: *Nomina systematica generum Coleopterorum, tam viventium quam fossilium, secundum ordinem alphabeticum disposita, adjectis auctoribus, libris in quibus reperiuntur, anno editionis, etymologia et familiis ad quas pertinent*. Pp. xi +[i] + 170. In: Agassiz L.: *Nomenclatoris zoologici index universalis, continens nomina systematica classium, ordinum, familiarum et generum animalium omnium, tam viventium quam fossilium, secundum ordinem alphabeticum unicum disposita, adjectis homonymiis plantarum, nec non variis adnotationibus et emendationibus*. Soloduri: Jean et Gassmann.

Agassiz L. 1846b: *Nomenclatoris zoologici index universalis, continens nomina systematica classium, ordinum, familiarum et generum animalium omnium, tam viventium quam fossilium, secundum ordinem alphabeticum unicum disposita, adjectis homonymiis plantarum, nec non variis adnotationibus et emendationibus*. Soloduri: Jean et Gassmann, viii + 393 pp.

Ahmad R. 1968: A new species of Pseudoscymnus Chapin (Col., Coccinellidae) predaceous on scale insects in West Pakistan. *Entomophaga* **13**: 377-379.

Ahmad R. 1970: A new species of Pharoscymnus Bedel (Coleoptera: Coccinellidae) predaceous on scale insect in Pakistan. *Entomophaga* **15**: 233-235.

Ahmad R. 1973: A new tribe of the family Coccinellidae (Coleoptera). *Bulletin of Entomological Research* **62**: 449-452.

Ahmad R. & Ghani M. A. 1966: A new genus and species of Chilocorini (Coleoptera: Coccinellidae) from Pakistan. *Proceedings of the Royal Entomological Society of London* (B) **35** (1-2): 9-10.

Ahrens A. 1812: Beiträge zur Kenntnis deutscher Käfer. 8. (Mit Bemerkungen von Kunze). *Neue Schriften der Naturforschenden Gesellschaft zu Halle* **2** (2): 1-40, 2 pls.

Ahrens A. 1814: *Fauna Insectorum Europae. Fasciculus 2*. Halae: Kümmel, 25 pp., 25 pls.

Alfieri A. 1976: The Coleoptera of Egypt. *Mémoires de la Société Entomologique d'Egypte* **5**: xiv + 361 pp.

Allard E. 1869: Description de quelques coléoptères nouveaux et notes diverses. *L'Abeille Mémoires d'Entomologie* **5**: 465-478.

Allen A. A. 1937: Two species of Coleoptera new to science. *The Entomologist's Monthly Magazine* **73**: 51-54.

Allen A. A. 1938: Two British species of Atomaria (Col. Cryptophagidae) new to the European list. *The Entomologist's Monthly Magazine* **74**: 82-83.

Allen A. A. 1942: A new species of Orthoperus in Britain. *The Entomologist's Monthly Magazine* **78**: 89-90.

Allibert A. 1847: Notes sur divers coléoptères trouvés dans des graines de légumineuses rapportées de Canton par M. Yvan, médecin de l'ambassade française en Chine, et sur quelques autres espèces qui ont vécu dans les haricots venant du Brésil. *Revue Zoologique, par la Société Cuvierienne* **1847**: 11-19.

Amsel H. G. & Hering M. 1931: Beitrag zur Kenntnis der Minenfauna Palästinas. *Deutsche Entomologische Zeitschrift* **1931**: 113-152, 2 pls.

Anand R. K., Gupta A. K. & Ghani S. 1990: A new species of Illeis Muls. from North India. *Reichenbachia* **27**: 131-134.

Ancey C. M. F. 1870: Coléoptères nouveaux. *L'Abeille, Mémoires d'Entomologie* **7** [1869-1870]: 84-88.

Ancey C. M. F. 1879: Description d'une nouvelle espèce de coléoptère du genre Sinoxylon. *Le Naturaliste* **1**: 139-140.

Ancey C. M. F. 1881: Descriptions de coléoptères nouveaux. *Le Naturaliste* **3**: 509.

Apfelbeck V. 1894: Rezultati koleopterološkog ispitivanja iz alpinske regije južno-bosanskih planina. Enumeratio coleopterorum in alpium Bosniae meridionalis regione alpina collectorum. *Glasnik Zemaljskog Muzeja u Bosni i Hercegovini* (Sarajevo) **6**: 209-223.

Apfelbeck V. 1896: Fauna insectorum balcanica. II. Coleopterologische Forschungsergebnisse aus der alpinen Region der südbosnischen Hochgebirge. *Wissenschaftliche Mitteilungen aus Bosnien und Herzegowina* **4**: 539-559.

Apfelbeck V. 1906: Neue Koleopteren, gesammelt während einer im Jahre 1905 mit Subvention der Kaiserlichen Akademie der Wissenschaften in Wien durchgeführten Forschungsreise nach Albanien und Montenegro. *Sitzungsberichte der Matematisch-Naturwissenschaftlichen Classe der k. Akademie der Wissenschaften* **115** (Abt. 1): 1661-1674.

Apfelbeck V. 1909: Contributiones ad Coleopterorum faunam peninsulae balcanicae. *Glasnik Zemaljskog Muzeja u Bosni i Hercegovini* **21**: 495-504.

Apfelbeck V. 1910: Beiträge zur Kenntnis der Käferfauna der Balkanhalbinsel. Pp. 40-42. In: Bericht der Sektion für Koleopterologie. Versammlung am 3. February 1910. *Verhandlungen der Kaiserlich-Königlichen Zoologisch-Botanischen Gesellschaft in Wien* **60**: 40-49.

Apfelbeck V. 1911: Prinosi poznananju faune koleoptera balkanskog poluostrva. *Glasnik Zemaljskog Muzeja u Bosni i Hercegovini* **23**: 209-223.

Apfelbeck V. 1912: Fauna insectorum balcanica V. *Wissenschaftliche Mitteilungen aus Bosnien und der Herzegowina* **12**: 642-664.

Apfelbeck V. 1915: Revizija vrsta roda Sphaerosoma Leach balkanskog poluostrva (Col.). *Glasnik Zemaljskog Muzeja u Bosni i Hercegovini* **27**: 407-429.

Apfelbeck V. 1916: Zur Kenntnis der Gattung Sphaerosoma Leach (Col.). Revision der Arten von der Balkanhalbinseln. *Annales Historico-Naturales Musei Nationalis Hungarici* **14**: 471-500.

Apfelbeck V. 1917: K poznavanju balkanske faune koleoptera. Diversicornia Balkanskog Poluostrova s obzirom na faunu Male Azije. *Glasnik Zemaljskog Muzeja u Bosni i Hercegovini* **28** [1916]: 169-200.

Apfelbeck V. 1918: Diagnosen neuer Koleopteren von der Balkanhalbinsel. *Anzeiger der Akademie der Wissenschaften in Wien, Mathematisch-Naturwissenschaftlichen Klasse* **55**: 100-104.

Aragona L. A. 1830: *De quibusdam Coleopteris Italiae novis aut rarioribus*. Ticini Regii: Typographia Bizzomi, 31 pp.

Araki H. 1941a: On a new species of the genus Megalodacne Crotch from Yakushima Island, Japan. *Mushi* **14**: 29-30.

Araki H. 1941b: On a new species of the genus Micrencaustes Crotch from Formosa (Coleoptera, Erotylidae). *Mushi* **14**: 31-33.

Araki H. 1941c: Description of two species of Erotylidae from Formosa. *Transactions of the Kansai Entomological Society* **11**: 55-58.

Araki H. 1941d: Description of a new genus of Erotylidae (Col.) from Japan proper. Studies on the Erotylidae from the Japanese Empire (1). *Transactions of the Natural History Society of Formosa* **31**: 364-366.

Araki H. 1941e: Description of a new species of Erotylidae (Col.) from Formosa. Studies on the Erotylidae from the Japanese Empire (2). *Transactions of the Natural History Society of Formosa* **31**: 367-368.

Araki H. 1942: Descriptions of the two new species of the genus Dactylotritoma Arrow from Formosa and North China. *Transactions of the Natural History Society of Formosa* **32**: 312-314.

Araki H. 1943: Descriptions of five new species of Erotylidae from Japan proper and Formosa. Studies on the Erotylidae from the Japanese Empire (3). *Transactions of the Natural History Society of Taiwan* **33**: 556-562.

Araki H. 1949: A list of the genus Episcapha, with description of two new subspecies (Col. Erotylidae). *Transactions of the Kansai Entomological Society* **14**: 4-6.

Araki M. 1961: A new lady beetle from Japan, closely related to Vibidia duodecimguttata Poda (Coleoptera: Coccinellidae). *Scientific Reports of Kyoto Prefectural University* (*Natural Science and Living Science*) **3** (3) A: 153-154.

Araki M. 1962: On some lady-beetles from the Loochoo Islands. *Scientific Reports of Kyoto Prefectural University* (*Natural Science and Living Science*) **3** (4) A: 195-196 (in Japanese).

Araki M. 1963: Description of four new species of the genus Scymnus from Japan (Coleoptera: Coccinellidae). *Scientific Reports of Kyoto Prefectural University* (*Natural Science and Living Science*) **3** (5) A: 251-255.

Araki M. 1964: Three new species of the lady beetles of the tribe Scymnini from Japan (Insecta, Coleoptera) *Scientific Reports of Kyoto Prefectural University* (*Natural Science and Living Science*) **15** A: 39-42.

Arimoto H. 1987: A new species of the genus Yamatostrius Kishii from Amami-Ôshima Is. of the Ryukyus, Japan (Coleoptera, Elateridae, Negastriinae). *Entomological Review of Japan* **42**: 133-135.

Arimoto H. 1990: A new species of the genus Megapenthes (Coleoptera, Elateridae) from the Ryukyu Islands. *Elytra* **18**: 89-92.

Arimoto H. 1992: Two new elaterid beetles (Coleoptera, Elateridae) from Taiwan. *Elytra* **20**: 73-78.

Arimoto H. 2000: A new species of the genus Denticolloides (Coleoptera, Elateridae) from Hokkaido, Japan. *Elytra* **28**: 369-372.

Arimoto H. 2004: A new Ampedus-species (Coleoptera: Elateridae) from Ishigaki-jima Island of the Ryukyu Islands, Japan. *Entomological Review of Japan* **59**: 1-4.

Arimoto H. & Watanabe A. 1993: A new species of the genus Actenicerus (Coleoptera, Elateridae) from Mt. Ontake-san, Central Japan. *Elytra* **21**: 245-248.

Arnett R. H. Jr. 1952: A review of the Nearctic Adelocerina (Coleoptera: Elateridae, Pyrophorinae, Pyrophorini). *The Wasmann Journal of Biology* **10** (1): 103-126.

Arnett R. H. Jr. 1953: Notes on the genus Lacon (Elateridae). *The Coleopterists Bulletin* **7**: 5-7.

Arnett R. H. Jr. 1955: Supplement and corrections to J. A. Hyslop's genotypes of the elaterid beetles of the world. *Proceedings of the United States National Museum* **103** (3336): 599-619.

Arnol'di G. Ya. 1965: 40. Sem. Anobiidae – Tochil"shchiki. Pp. 244-257. In: Bey-Bienko G. Ya. (ed.): *Opredelitel" nasekomykh evropeyskoy chasti SSSR. Zhestkokrylye i Veerokrylye.* Moskva - Leningrad: Nauka, 668 pp.

Arrow G. J. 1901: The genus Hyliota of the coleopterous family Cucujidae, with descriptions of new forms and a list of the described species. *Transactions of the Entomological Society of London* **1901**: 593-601.

Arrow G. J. 1915a: A new species of Thaumaglossa bred from the egg-clusters of Mantidae. *Proceedings of the Entomological Society of London* **1915**: 112-113.

Arrow G. J. 1915b: Notes on the coleopterous family Dermestidae, and descriptions of some new forms in the British Museum. *The Annals and Magazine of Natural History* (8) **15**: 425-451.

Arrow G. J. 1917: The khapra beetle (Trogoderma khapra, sp. n.), an Indian grain-pest. *The Annals and Magazine of Natural History* (8) **19**: 481-482.

Arrow G. J. 1920a: A contributions to the classification of the coleopterous family Endomychidae. *Transactions of the Entomological Society of London* **1920**: 1-83, pl. 1.

Arrow G. J. 1920b: A list of the endomychid Coleoptera of Indo-China, with description of new species. *The Annals and Magazine of Natural History* (9) **5**: 321-336.

Arrow G. J. 1922: A list of the Erotylid Coleoptera of Indo-China, with descriptions of new species. *Transactions of the Entomological Society of London* **1921**: 285-306.

Arrow G. J. 1923: Notes on Endomychid Coleoptera and descriptions of new species in the British Museum. *Transactions of the Entomological Society of London* **1923**: 484-500.

Arrow G. J. 1925: *Coleoptera. Clavicornia. Erotylidae, Languriidae, and Endomychidae. The Fauna of British India, including Ceylon and Burma.* London: Taylor and Francis, xv + 416 pp., 1 pl., 1 map.

Arrow G. J. 1926: Notes on Oriental Endomychidae & Erotylidae (Col.). *Entomologische Mitteilungen* **15**: 354-358.

Arrow G. J. 1929a: A new genus of silphid Coleoptera from Persia. *Zoologischer Anzeiger* **82**: 96-99.

Arrow G. J. 1929b: On the families of Coleoptera related to the Erotylidae, with descriptions of a new family, two new genera, and a few new species. *The Annals and Magazine of Natural History* (10) **4**: 305-322.

Arrow G. J. 1937: Notes on some clavicorn Coleoptera and descriptions of a few new species and genera. *The Annals and Magazine of Natural History* (10) **20**: 101-113.

Arrow G. J. 1939: Entomological results from the Swedish expedition, 1934, to Burma and British India. Coleoptera: Clavicornia. Erotylidae, Languriidae, Endomychidae, Discolomidae and Georyssidae collected by René Malaise. *Arkiv för Zoologi* **31** (A) **17**: 1-9.

Arrow G. J. 1943: The endomychid Coleoptera of New Guinea and neighbouring Islands, with some new species. *The Annals and Magazine of Natural History* (11) **10**: 128-136.

Ashida H. & Kim Ch.-G. 1999: A new species of Dacne (Coleoptera, Erotylidae) from Chejudo Island of South Korea. *Elytra* **27**: 381-385.

Atamuradov Kh. I. 1990: Novyy vid zhuka-shchelkuna (Coleoptera, Elateridae) iz Kugitanga (Turkmen SSR, USSR) [A new click beetle species (Coleoptera, Elateridae) from Kugitang (Turkmen SSR, USSR)]. *Vestnik Zoologii* **1990** (3): 82-84.

Aubé C. 1837: Essai sur le genre Monotoma. *Annales de la Société Entomologique de France* **6**: 453-469.

Aubé C. 1843a: Description de deux coléoptères nouveaux appartenant à la faune parisienne. *Annales de la Société Entomologique de France* (2) **1**: 73-76.

Aubé C. 1843b: Note sur une nouvelle espèce de coléoptère tetramère qui devra servir de base à une coupe générique nouvelle. *Annales de la Société Entomologique de France* (2) **1**: 93-95, pl. 4.

Aubé C. 1843c: Note sur le genre Calyptobium. *Annales de la Société Entomologique de France* (2) **1**: 241-247, pl. 10.

Aubé C. 1850: Description de quelques insectes coléoptères appartenant à l'Europe et à l'Algérie. *Annales de la Société Entomologique de France* (2) **8**: 299-346.

Aubé C. 1861a: Description de quatre nouvelles espèces de coléoptères appartenant à un genre nouveau. *Annales de la Société Entomologique de France* (4) **1**: 93-96.

Aubé C. 1861b: Description de six espèces nouvelles de coléoptères d'Europe. *Annales de la Société Entomologique de France* (4) **1**: 195-199.

Aubé C. 1861c: [Note synonymique au sujet des Theca byrrhoides, elongata et raphaelensis]. *Bulletin de la Société Entomologique de France* **1861**: xlv.

Aubé C. 1862: Coléoptères nouveaux d'Europe et observations entomologiques. *Annales de la Société Entomologique de France* (4) **2**: 71-74.

Aubé C. 1866: Description de nouvelles espèces de coléoptères de France. *Annales de la Société Entomologique de France* (4) **6**: 161-164.

Audisio P. 1975: Note su alcune specie di Cleridae del Libano e della Turchia, comprehendenti la descrizione di Trichodes sbordonii n. sp. (Coleoptera). *Fragmenta Entomologica* **11**: 237-240.

Audisio P. 1976a: Note su alcune specie italiane del genere Meligethes Steph. (Coleoptera, Nitidulidae). *Bollettino dell'Associazione Romana di Entomologia* **30** [1975]: 2-16.

Audisio P. 1976b: Una nuova specie di Meligethes dell'Europa centro-meridionale (Coleoptera, Nitidulidae). *Fragmenta Entomologica* **12**: 299-307.

Audisio P. 1977a: Un nuovo Meligethes italiano del gruppo di specie di M. aeneus (F.) (Coleoptera, Nitidulidae). *Bollettino dell'Associazione Romana di Entomologia* **31** [1976]: 15-22.

Audisio P. 1977b: Una nuova specie di Meligethes dell'Anatolia (Coleoptera, Nitidulidae). *Bollettino dell'Associazione Romana di Entomologia* **31** [1976]: 23-28.

Audisio P. 1979: Note su alcuni Meligethes dell'Anatolia e del Caucaso, con descrizione di due specie nuove (Coleoptera, Nitidulidae). *Fragmenta Entomologica* **15**: 127-141.

Audisio P. 1980: Anamartus jelineki sp. n., a new species from Jordan (Coleoptera: Nitidulidae). *Folia Entomologica Hungarica* **41**: 21-23.

Audisio P. 1983: Tassonomia e distribuzione geografica di Epuraea latipes Grouvelle e specie correlate (Coleoptera, Nitidulidae). *Fragmenta Entomologica* **17**: 111-123.

Audisio P. 1984: [new taxon]. In: Audisio P., Angelici M. C. & Sbordoni V. 1984: Studio sistematico su Meligethes exilis Sturm, in base a dati elettroforetici, morfo-ecologici e biogeografici. *Fragmenta Entomologica* **17**: 359-372.

Audisio P. 1986a: Una nuova specie del genere Xenostrongylus Woll. delle Isole Canarie (Coleoptera, Nitidulidae). *Vieraea* **16**: 11-16.

Audisio P. 1986b: La posizione tassonomica di Meligethes fuscus (Olivier) e specie correlate (Coleoptera, Nitidulidae). *Bollettino del Museo di Scienze Naturali Torino* **4**: 543-556.

Audisio P. 1988: Tassonomia, ecologia e distribuzione geografica di alcuni Kateretidae e Nitidulidae ovest-paleartici (Coleoptera). *Fragmenta Entomologica* **20**: 189-231.

Audisio P. 1989a: Un nuovo Meligethes jugoslavo del gruppo di specie di M. difficilis (Heer) (Coleoptera, Nitidulidae). *Bollettino dell'Associazione Romana di Entomologia* **43** [1988]: 25-28.

Audisio P. 1989b: Un nuovo Meligethes anatolico del gruppo di M. erythropus (Marsham) (Coleoptera, Nitidulidae). *Fragmenta Entomologica* **21**: 179-182.

Audisio P. 1990: Nota sobre Nitidulidae ibero-marroquies (Col.). *Eos, Revista Española de Entomología* **66**: 25-27.

Audisio P. 1991: A new species of the genus Epuraea Erichson from Iran, and a replacement name for E. castanea (Duftschmid, 1825) (Coleoptera, Nitidulidae). *Revue Suisse de Zoologie* **98**: 517-520.

Audisio P. 1993a: Osservazioni su alcuni rappresentanti W-paleartici del genere Urophorus Murray, 1864, con descrizione di una nuova specie del Nord Africa (Coleoptera, Nitidulidae, Carpophilinae). *Fragmenta Entomologica* **24** [1992]: 173-179.

Audisio P. 1993b: Nuovi dati faunistici e bionomici sui Meligethes di Turchia orientale, con descrizione di una nuova specie del gruppo di M. difficilis (Heer) (Coleoptera, Nitidulidae, Meligethinae). *Fragmenta Entomologica* **24**: 181-193.

Audisio P. & De Biase A. 1996: Taxonomic revision, phylogeny and biogeography of the beetle genus Dapsa Latreille (Coleoptera: Endomychidae). *Memorie della Società Entomologica Italiana* **74**: 65-130.

Audisio P. & De Biase A. 1999: Bionomic and morphological evidence of a new southern European species of the Meligethes viridescens complex (Coleoptera: Nitidulidae). *Folia Heyrovskyana* **7**: 99-113.

Audisio P. & De Biase A. 2001: [new taxon]. In: Audisio P., De Biase A., Antonini G., Belfiore C. & Oliverio M.: Morphological, molecular and ecological evidence of a new Euro-Anatolian species of the Meligethes coracinus complex (Coleoptera: Nitidulidae). *Insect Systematics and Evolution* **31** [2000]: 361-385.

Audisio P., De Biase A. & Antonini G. 2003: A new exceptional Meligethes of the M. aeneus species-group from Western Alps and an updated key to identification of M. aeneus and allied species (Coleoptera: Nitidulidae: Meligethinae). *Insect Systematics and Evolution* **34**: 121-130.

Audisio P., De Biase A & Zampetti M. F. 1983: Una nuova specie di Dapsa (Coleoptera, Endomychidae) del litorale Tirrenico, con alcune considerazioni sistematiche sul genere. *Animalia* **8** [1981]: 87-93.

Audisio P. & Jelínek J. 1984: Nota tassonomica su Meligethes floribundus Reitter, 1877 e descrizione di M. dieckmanni sp. n. del Caucaso (Coleoptera, Nitidulidae). *Bollettino del Museo Regionale di Scienze Naturali Torino* **2**: 565-570.

Audisio P.& Jelínek J. 1990: Tassonomia e distribuzione geografica di Meligethes obscurus auct., con descrizione di una specie nuova (Coleoptera, Nitidulidae). *Fragmenta Entomologica* **22**: 75-85.

Audisio P. & Jelínek J. 2000 [new taxon]. In: Audisio P., Jelínek J., Mariotti A. & De Biase A.: The Coleoptera Nitidulidae and Kateretidae from Anatolian, Caucasian and Middle East regions. *Biogeographia* **21**: 241-354.

Audisio P.& Jelínek J. 2001: [new taxon]. In: Audisio P., Mariotti A., Jelínek J. & De Biase A.: Xenostrongylus levantinus, sp. n. from eastern Mediterranean, with notes on the generic classification of Xenostrongylus Wollaston, 1854 and Strongyllodes Kirejtshuk, 1993 (Coleoptera: Nitidulidae). *Annales de la Société Entomologique de France* (N. S.) **37**: 491-499.

Audisio P., Jelínek J. & Kirejtshuk A. G. 1990: [new taxon]. In: Kirejtshuk A. G.: Novye taksony zhukov-blestyanok vostochnogo polushariya. Chast′4 [New taxa of the Nitidulidae (Coleoptera) of the eastern Hemisphere, part 4]. *Trudy Zoologicheskogo Instituta Akademii Nauk SSSR* **211**: 84-103.

Audisio P., Jelínek J. & Stevanović M. 1999: Morphological and biogeographical reexamination of the Meligethes squamosus species complex (Coleoptera: Nitidulidae: Meligethinae). *Folia Heyrovskyana* **7**: 61-71.

Audisio P. & Kirejtshuk A. G. 1988: Novyy vid roda Meligethes Stephens gruppy difficilis s Kavkaza i obzor zapadnopalearkticheskikh vidov, blizkikh k M. viduatus (Heer) (Coleoptera, Nitidulidae). *Entomologicheskoe Obozrenie* **67**: 559-568.

Audisio P. & Kirejtshuk A. G. 1989: K sistematike i nomenklature nekotorykh palearkticheskikh zhukov-blestyanok (Coleoptera, Nitidulidae). *Entomologicheskoe Obozrenie* **67** [1988]: 790-797.

Audisio P. & Spornraft K. 1990: Taxonomie, Ökologie und Verbreitung von Meligethes coracinus auctt. mit Beschreibung einer neuen Art (Coleoptera: Nitidulidae). *Nachrichtenblatt der Bayerischen Entomologen* **39**: 70-75.

Audouin V. 1824: Recherches anatomiques sur la femelle du Drile jaunâtre, et sur le mâle de cette espèce. *Annales des Sciences Naturelles* **2**: 443-462.

Auroux L. 1967: Notas sobre coléopteros micetófilos. *Miscellànea Zoològica* **2**: 89-91.

Bach M. 1852: *Käferfauna für Nord- und Mitteldeutschland, mit besonderer Rücksicht auf die preussischen Rheinlande. Zweiter Band. Lieferung 3.* Coblenz: J. Hölscher, 6 + 148 pp.

Bach M. 1854: *Käferfauna fur Nord- und Mittel-Deutschland mit besonderer Rücksicht auf die preussischen Rheinlande. Zweiter Band. Lieferung 4.* Coblenz: J. Hoelscher, 149-392 pp.

Bach M. 1866: *Käferfauna für Nord- und Mitteldeutschland mit besonderer Rücksicht auf die preussischen Rheilande. Nachträge, Zusätze und Verbesserungen zum 1. Bande der Käferfauna.* Koblenz: J. Hölscher, pp. 415-523.

Bach M. 1872: Die Käfergattung Meligethes, mit besonderer Berücksichtigung von Meligethes aeneus und einer neuen Art, Meligethes Gresseri Bach. *Natur und Offenbarung* (Münster) **18**: 29-35.

Bach M. 1896: [new taxon]. In: Acloque A.: Faune de France, contenant la description des espèces indigènes disposées en tableaux analytiques et illustrées de figures représentant les types caractéristiques des genres et des sous-genres. Vol. 1 Coléoptères. Paris: J.-B. Baillière et Fils, [3] + 466 pp.

Bahillo de la Puebla P. & López-Colon J. I. 2004: Enicopus sarae, nueva especie de Enicopus Stephens, 1830 de España (Coleoptera, Dasytidae). *Boletín de la Sociedad Entomológica Aragonesa* **34**: 55-58.

Bahillo de la Puebla P., López-Colón J. I. & García-París M. 2003: Una especie nueva de Tillus Olivier, 1790 (Coleoptera, Cleridae) de la Península Ibérica. *Graellsia* **59**: 57-62.

Balfour-Browne J. 1953: Changes of nomenclature in the Cryptophagidae (Col.). *The Entomologist's Monthly Magazine* **89**: 258.

Ballion E. von. 1871: Eine Centurie neuer Käfer aus der Fauna des russischen Reiches. *Bulletin de la Société Impériale des Naturalistes de Moscou* **43** (2) [1870]: 320-353.

Ballion E. von. 1878: Verzeichniss der im Kreise von Kuldsha gesammelten Käfer. *Bulletin de la Société Impériale des Naturalistes de Moscou* **53** (1): 253-389.

Baranowski R. 1985: Central and northern European Dorcatoma (Coleoptera: Anobiidae), with a key and description of a new species. *Entomologica Scandinavica* **16**: 203-207.

Barber H. S. 1942: Rasberry fruitworms and related species. *Miscelaneous Publications of the U. S. Department of Agriculture* **468**, 32 pp.

Barovskij V. V. 1906: Novyy vid Scymnus iz S.-Peterburgskoy gubernii (Coleoptera, Coccinellidae). *Revue Russe d'Entomologie* **6**: 8-9.

Barovskij V. V. 1908a: Eine neue Aberration des Podabrus alpinus Payk. (1798), und Uebersicht der Aberrationen dieser Art (Coleoptera, Canthariididae). *Revue Russe d'Entomologie* **8**: 174-176.

Barovskij V. V. 1908b: Novyy vid Rhagonycha Esch. iz Vostochnoy Sibiri. (Coleoptera, Canthariididae) [Une espèce nouvelle de Rhagonycha Esch. de la Sibérie orientale (Coleoptera, Canthariididae)]. *Revue Russe d'Entomologie* **8**: 293-294.

Barovskij V. V. 1909a: Tri novykh vida Lithophilus iz aziatskoy Rossii (Coleoptera, Coccinellidae). *Revue Russe d'Entomologie* **9**: 96-99.

Barovskij V. V. 1909b: Novye aziatskye vidy Lithophilus Frhl. (Coleoptera, Coccinellidae). *Revue Russe d'Entomologie* **9**: 255-261.

Barovskij V. V. 1909c: Novye vidy Cantharis Linn. iz Aziatskoy Rossii (Coleoptera, Canthariididae) [New species of Cantharis Linn. from Asian Russia (Coleoptera, Canthariididae)]. *Revue Russe d'Entomologie* **9**: 325-327.

Barovskij V. V. 1910: Novye aziatskie vidy Lithophilus Frhl. (Coleoptera, Coccinellidae). *Revue Russe d'Entomologie* **10**: 39-41.

Barovskij V. V. 1922a: Species nova generis Drilus Oliv. e Persia septentrionali (Coleoptera: Cantharidae). *Russkoe Entomologicheskoe Obozrenie* **18**: 25-26.

Barovskij V. V. 1922b: Premier représentant du genre Ichthyurus Westw. (Coleoptera, Canthariididae) dans la faune de la Russie. *Annuaire du Musée Zoologique de l'Académie des Sciences de Russie* **23** [1918-1922]: 276-279.

Barovskij V. V. 1922c: Revisio specierum palaearticarum Coccinellidarum generis Exochomus Redtb. Obzor palearkticheskikh vidov roda Exochomus Redtb. (Coleoptera, Coccinellidae). *Annuaire du Musée Zoologique de l'Académie des Sciences de Russie* **23**: 289-303.

Barovskij V. V. 1925: De Coccinellidis centrasiatidis novis (Coleoptera) [Novye sredne-asiatskie korovki]. *Revue Russe d'Entomologie* **19**: 206-207.

Barovskij V. V. 1926a: Opisanie odnogo vida roda Silis Latr. (Coleoptera) [Description d'une nouvelle espèce du genre Silis Latr. (Coleoptera)]. *Materialy Komissii po Izucheniyu Jakutskoy A. S. S. Respubliki* **3**: 13-14.

Barovskij V. V. 1926b: De genere novo Coccinellidarum (Coleoptera). *Revue Russe d'Entomologie* **20**: 69-70.

Barovskij V. V. 1926c: Cantharidides asiatiques nouveaux (Coleoptera). *Revue Russe d'Entomologie* **20**: 235-244.

Barovskij V. V. 1927a: Une espèce nouvelle caucasique du genre Malchinus Kiesw. *Acta Societatis Entomologicae Stauropolitanae* **3** (1): 17.

Barovskij V. V. 1927b: Revisio specierum palaearticarum Coccinellidarum generis Brumus Muls. (Coleoptera). *Annuaire du Musée Zoologique de l'Académie de Sciences de l'URSS* **28**: 194-200.

Barovskij V. V. 1927c: Description d'une nouvelle espèce du genre Exochomus Redtb. *Revue Russe d'Entomologie* **21**: 251.

Barovskij V. V. 1928a: Cantharidides asiatiques nouveaux (Coleoptera). II. *Revue Russe d'Entomologie* **22**: 97-100.

Barovskij V. V. 1928b: Cantharidides asiatiques nouveaux (Coleoptera). III [Espèces nouvelles ou peu connues du genre Silis Latr.]. *Revue Russe d'Entomologie* **22**: 216-222.

Barovskij V. V. 1929a: Description d'une nouvelle espèce du genre Macrolycus Waterh. (Coleoptera, Lycidae). *Comptes Rendus de l'Académie des Sciences de l'URSS* **1929**: 231-232.

Barovskij V. V. 1929b: Cantharidides asiatiques nouveaux (Coleoptera). IV. *Revue Russe d'Entomologie* **23**: 266-269.

Barovskij V. V. 1929c: Une espèce nouvelle asiatique du genre Xeronthobius Mor. (Coleoptera). *Entomologicheskoe Obozrenie* **23**: 276.

Barovskij V. V. 1929d: Description d'une nouvelle espèce du genre Malthodes Kies. de l'Asie Centrale. *Comptes Rendus de l'Académie des Sciences de l'URSS* **1929**: 487-488.

Barovskij V. V. 1930a: Description d'un nouveau genre de la tribu Lycina, famille des Lycides (Coleoptera). *Comptes Rendus de l'Académie des Sciences de l'URSS* **1930**: 287-289.

Barovskij V. V. 1930b: Descriptio duarum novarum specierum generis Dictyoptera Latr. (Coleoptera, Lycidae) Asiae Orientalis. *Comptes Rendus de l'Académie des Sciences de l'URSS* **1930**: 359-362.

Barovskij V. V. 1930c: Description d'une nouvelle espèce du genre Macrolycus Waterh. (Coleoptera, Lycidae). *Comptes Rendus de l'Académie des Sciences de l'URSS* **1930**: 580-581.

Barovskij V. V. 1930d: Description d'une nouvelle espèce du genre Pyropterus Muls. provenant de la Transcaucasiae (Coleoptera). *Comptes Rendus de l'Académie des Sciences de l'URSS* **1930**: 615-616.

Barovskij V. V. 1931a: Description d'une nouvelle espèce du genre Malthodes Kies. de l'Asie orientale (Coleoptera, Canthariddidae). *Comptes Rendus de l'Académie des Sciences de l'URSS* **1931**: 101.

Barovskij V. V. 1931b: Coccinellidae (Coleoptera). In: Rat für Erforschung der produktiven Kräfte. *Abhandlungen über die Ergebnisse der Pamir-Expedition 1927* (Leningrad) **8**: 127-131.

Barovskij V. V. 1931c: Le premier représentant du genre Plateros Waterh. dans notre faune (Coleoptera, Lycidae). *Comptes Rendus de l'Académie des Sciences de l'URSS* **1931**: 160-161.

Barovskij V. V. 1932: Encore une nouvelle espèce des Lycides provenant de la Transcaucasiae (Coleoptera). *Comptes Rendus de l'Académie des Sciences de l'URSS* **1932**: 139-141.

Barovskij V. V. 1933: Un Trixagus Kug. (Throscus Latr.) nouveau de la région d'Ussuri (Coleoptera, Trixagidae). *Revue Russe d'Entomologie* **25**: 192.

Barr W. F. 1977: A new genus of European Cleridae. *The Pan-Pacific Entomologist* **53**: 32-33.

Barros J. M. C. de. 1929: Notas para o estudo das espécies da Fam. Cleridae existentes en Portugal. *Arquivos da Secçáo de Biología e Parasitologia do MuseoZoológico da Universidade de Coimbra* **1**: 1-25.

Bassi C. A. 1833: Description du genre Malacogaster. In: Guérin-Méneville F. E.: Insecta, Nr. 99, 1 pl. *Magasin de Zoologie, d'Anatomie Comparée et de Paléontologie.* Paris: Arthur Bertrand.

Bates H. W. 1866: On a collection of Coleoptera from Formosa, sent home by R. Swinhoe, Esq., H. B. M. Consul, Formosa. *Proceedings of the Scientific Meetings of the Zoological Society of London* **23**: 339-355.

Baudi di Selve F. 1859a: Malacodermatum quaedam novae species. *Berliner Entomologische Zeitschrift* **3**: 292-303.

Baudi di Selve F. 1859b: Kleine Mitteilungen. *Berliner Entomologische Zeitschrift* **3**: 321-342.

Baudi di Selve F. 1861: Coleoptera nonnulla nova-descripta. *Berliner Entomologische Zeitschrift* **5**: 182-183.

Baudi di Selve F. 1870: Coleopterorum messis in insula Cypro et Asia minore ab Eugenio Truqui congregatae recensitio: de Europaeis notis quibusdam additis. Pars tertia. *Berliner Entomologische Zeitschrift* **14**: 49-90.

Baudi di Selve F. 1871: Coleopterorum messis in insula Cypro et Asia minore ab Eugenio Truqui congregatae recensitio: de Europaeis notis quibusdam additis. Pars quarta. *Berliner Entomologische Zeitschrift* **15**: 49-71.

Baudi di Selve F. 1872: Europeae et circummediterraneae faunae Dascillidum et Malacodermatum specierum, quae Comes Dejean in suo Catalogo ed. 3 consignavit, ex ejusdem collectione in R. Taurinensi Musaeo asservata, cum auctorum hodierne recepta denominatione, collatio. *Berliner Entomologische Zeitschrift* **15** [1871]: 89-130.

Baudi di Selve F. 1873: Catalogo dei Dascilidi, Malacodermi e Teredili della fauna europea e circummediterranea appartenenti alla collezioni del Museo Civico di Genova per Flaminio Baudi. *Annali del Museo di Storia Naturale di Genova* **4**: 226-268.

Baudi di Selve F. 1874a: Europae et circummediterraneae faunae Dasytidum et Melyridum specierum, quae Comes Dejean in sou Catalogo ed. 3[a] consignavit, ex ejusdem collectione in R. Taurinensi Musaeo asservata, cum auctorum hodierne recepta denominatione, collatio. *Berliner Entomologische Zeitschrift* **17** [1873]: 293-316.

Baudi di Selve F. 1874b: Coleopterorum messis in insula Cypro et Asia minore ab Eugenio Truqui congregatae recensitio: de Europaeis notis quibusdam additis. Pars quinta. *Berliner Entomologische Zeitschrift* **17** [1873]: 317-338.

Baudi di Selve F. 1875: Lamprorhiza morio descrizione di una nuova specie italiana di Lampiridi. *Bullettino della Società Entomologica Italiana* **7**: 37.

Baudi di Selve F. 1889: Catalogo dei coleotteri del Piemonte. *Annali della Reale Accademia d'Agricoltura di Torino* **32**: 1-225.

Baudi di Selve F. 1893: [new taxa]. In: Ragusa E.: Catalogo regionale dei coleotteri di Sicilia (cont.). *Il Naturalista Siciliano* **13**: 27-46.

Baudi di Selve F. 1894: Viaggio del Dr. E. Festa in Palestina, nel Libano e regioni vicine. *Bollettino dei Musei di Zoologia e Anatomia Comparata della Reale Universitá di Torino* **9** (173): 1-13.

Bauduer P. 1874: Descriptions de quatre nouvelles espèces de coléoptères. *Bulletin de la Société Entomologique de France* **1874**: clxi-clxv.

Beal R. S. 1954: Biology and taxonomy of the Neartic species of Trogoderma (Coleoptera: Dermestidae). *University of California Publications in Entomology* (Berkeley and Los Angeles) **10**: 35-102.

Beal R. S. 1956: Synopsis of the economic species of Trogoderma occuring in the United States with description of a new species (Coleoptera: Dermestidae). *Annals of the Entomological Society of America* **49**: 559-566.

Beal R. S. 1961: Coleoptera: Dermestidae. *Insects of Micronesia* **16**: 109-131.

Beal R. S. 1967: A revisionary study of the North American dermestid beetles formerly included in the genus Perimegatoma (Coleoptera). *Miscellaneous Publications of Entomologica Americana* **5**: 281-312.

Beal R. S. 1992: A new genus and species of Dermestidae (Coleoptera) from Morocco. *The Coleopterist's Bulletin* **46**: 403-408.

Beck L. von. 1817: *Beiträge zur baierischen Insectenfaune, oder Beschreibung und Abbildung neuentdeckter Käfer, mit angehängtem Namensverzeichnisse der Eleuteraten des Landgerichtbezirks Zusmeshausen.* Augsburg: J. Wolff, 45 pp., 7 pls.

Becker A. 1864: Naturhistorische Mittheilungen. *Bulletin de la Société Impériale des Naturalistes de Moscou* **37** (2): 477-493.

Bedel L. 1868: Monographie des Erotyliens (Engides et Triplacides), d'Europe, du Nord de l'Afrique et de l'Asie Occidentale. *L'Abeille, Mémoires d'Entomologie* **5** [1868-1869]: 1-50.

Bedel L. 1871: Révision du genre Aulacochilus Lacordaire de la famille des Érotyliens. *Annales de la Société Entomologique de France* (5) **1**: 271-286.

Bedel L. 1873: Supplément à la révision du genre Aulacochilus Lacordaire et description de quatre espèces nouvelles. *Annales de la Société Entomologique de France* (5) **2** [1872]: 403-408.

Bedel L. 1874: [Description d'une espèce nouvelle du genre Cyrtotriplax Crotch (Tritoma Fabr. nec Geoffr.) de la famille des Érotylides]. *Bulletin de la Société Entomologique de France* **1874**: cxcvi-cxcvii.

Bedel L. 1882: Essai sur la Famille des Erotylidae. I. Révision des Triplotoma Westw. et genres voisins. *Annali del Museo Civico di Storia Naturale di Genova* **18** [1882-1883]: 435-444, pl. 10.

Bedel L. 1884: [Anthrenus miniopictus n. sp.]. *Annales de la Société Entomologique de France* (6) **4**: xxi.

Bedel L. 1885: Recherches sur les coléoptères du nord de l'Afrique. Recherche synonymiques (1). *Annales de la Société Entomologique de France* (6) **5**: 85-90.

Bedel L. 1899: Diagnoses et synonymie de plusieurs Clérides méditerranéens du groupe des Corynetini. *Bulletin de la Société Entomologique de France* **1899**: 115.

Bedel L. 1891: Observations synonymiques. *L'Abeille, Journal d'Entomologie* **27** [1890-1892]: 153-156.

Bedel L. 1892: [new taxon]. In: Weise J.: Coccinellidae d'Europe et du nord de l'Asie. *L'Abeille, Journal d'Entomologie* **28** [1892-1896]: 1-84. [Translated by A. Sicard].

Bedel L. 1900: Descriptions de deux coléoptères nouveaux, du Nord de l'Afrique. *Bulletin de la Société Entomologique de France* **1900**: 335-337.

Bedel L. 1906a: [new name]. In: Heyden L. F. J. D. von, Reitter E. & Weise J. (eds): *Catalogus Coleopterorum Europae, Caucasi et Armeniae rossicae. Editio secunda.* Berlin, Paskau, Caen: R. Friedländer & Sohn, E. Reitter, Revue d'Entomologie, [8] + 750 columns + pp. 750-775.

Bedel L. 1906b: Synonymies de coléoptères paléarctiques. *Bulletin de la Société Entomologique de France* **1906**: 91-93.

Bedel L. 1918: Cinq espèces nouvelles du genre Episcapha Lac. (Col. Erotylidae). *Bulletin de la Société Entomologique de France* **1918**: 118-120.

Bedel L. 1921: Coléoptères nouveaux du Maroc. Récoltes du Dr. M. Bedel dans le Moyen Atlas, 1916-1917. *Bulletin de la Société Entomologique de France* **1921**: 59-61.

Bedel L. & Bourgeois J. 1901: Telephorus geniculatus Luc. et Pygidia geniculata Mars. (Col.). *Bulletin de la Société Entomologique de France* **1901**: 176-178.

Bellés X. 1980: Une nouvelle espèce de Gibbinae (Coleoptera, Ptinidae): Gibbium einsteini n. sp. de Chine. *Bulletin de l'Académie Polonaise des Sciences, Série des Sciences Biologiques* **27**: 845-849.

Bellés X. 1981: Lapidoniptus nov. gen. pour Piarus hispanicus Pic, 1953 (Coleoptera, Ptinidae). *Spixiana* **4**: 69-72.

Bellés X. 1982: El primer representante de la familia Ptinidae (Col.) de la Islas Salvajes: Sphaericus bicolor n. sp. *Vieraea* **11**: 103-108.

Bellés X. 1991: Los géneros Kedirinus nov., Sundaptinus nov. y Hanumanus nov. en el Archipelago indo-australiano y sureste asiático, y nuevos datos sobre el género Maheoptinus Pic (Coleoptera, Ptinidae). *Graellsia* **47**: 71-96.

Bellés X. 1992: Cyphoniptus (Coleoptera: Ptinidae) a new genus from the Tibet upland and its southeastern counterparts. *Elytron* (*Supplementum*) **5**: 257-260.

Bellés X. 1994: El género Sphaericus Wollaston, 1854 (Coleoptera: Ptinidae). *Boletín de la Asociación Española de Entomología* **18**: 61-79.

Bellés X. 1995: Stereocaulophilus volcanius gen. n., sp. n. (Coleoptera: Ptinidae) from Lanzarote (Canary Islands). *Elytron* **8**: 43-47.

Bellés X. 1996a: Sinopsis del género Casapus Wollaston, 1862 (Coleoptera: Ptinidae). *Boletín de la Asociación Española de Entomología* **20**: 75-84.

Bellés X. 1996b: El género Dignomus Wollaston (Coleoptera, Ptinidae). *Bolletí de la Societat d'História Natural de les Balears* **39**: 209-228.

Bellés X. 2001: Description of Sphaericus selvagensis n. sp. from the Salvage Islands, and new data on Sphaericus bicolor Bellés (Coleoptera, Ptinidae). *Animal Biodiversity and Conservation* **24**: 9-13.

Bellés X. 2002: A synopsis of the subgenus Cyphoderes Mulsant & Rey, including the description of Ptinus (Cyphoderes) catalonicus sp. n. (Coleoptera: Ptinidae). *Elytron* **16**: 97-105.

Bellier de la Chavignerie J. B. E. 1870: Nouveaux coléoptères aux Basses Alpes. *Bulletin de la Société Entomologique de France* **1870**: xxvii-xxviii.

Belon M.-J. 1878: [Description d'une espèce nouvelle de Malachius]. *Bulletin de la Société Entomologique de France* **1878**: xxviii-xxix, cxv-cxvi.

Belon M.-J. 1881: Histoire Naturelle des coléoptères de France par M. Mulsant. Famille des Lathridiens 1re partie. Lyon: H. Georg & Paris: J.-B. Ballière & fils, [3] + 209 pp. [seen as reprint from the *Annales de la Société Linnéenne de Lyon* (N. S.) **26** : 1-209, 1879].

Belon M.-J. 1885a: Matériaux pour servir à l'étude des lathridiens. *Revue d'Entomologie* **4**: 239-254.

Belon M.-J. 1885b: [new taxa]. *Bulletin de la Société Entomologique de France* **1885**: xxxix.

Belon M.-J. 1891: Viaggio di Leonardo Fea in Birmania e regioni vicine XXXVIII. Lathridiidae. *Annali del Museo Civico di Storia Naturale di Genova* **30**: 877-880.

Belon M.-J. 1895: Nouvelle contribution à l'étude des lathridiens. *Annales de la Société Entomologiue de Belgique* **39**: 75-105.

Belon M.-J. 1897: Essai de classification générale des Lathridiidae. Avec le catalogue systématique et alphabétique de toute les espèces du globe. *Revue d'Entomologie* **16**: 105-221.

Bercedo-Páramo P. & López-Cólon J. I. 2003: Cebrio cantabricus sp. n. de la Península Ibérica (Coleoptera, Cebrionidae). *Boletín de la Sociedad Entomológica Aragonesa* **32**: 17-19.

Berg C. 1898: Substitución de nombres genéricos. *Comunicaciones del Museo Nacional de Buenos Aires* **1** [1898-1901]: 16-19.

Berthold A. A. 1827: *Latreille's natürliche Familien des Thierreichs. Aus dem französischen mit Anmerkungen und Zusätzen von Dr. Arnold Adolph Berthold.* Weimar: Industrie-Comptoirs, x + 602 pp.

Bessolitzina E. P. 1974: Fauna shchelkunov (Coleoptera, Elateridae) Irkutskoy oblasti [Fauna of the click beetles (Coleoptera, Elateridae) of the Irkutsk Region]. Pp. 77-104. In: Kulig S. A. (ed.): *Fauna nasekomykh Vostochnoy Sibiri i Dal"nego Vostoka.* Irkutsk: Izd-vo Irkutskogo gos. un-ta, 260 pp.

Bessolitzina E. P. & Dolin V. G. 1991: [new taxa]. In: Dolin V. G. & Bessolitzina E. P.: Novye vidy zhukov-shchelkunov roda Hypnoidus Steph. (Coleoptera, Elateridae) iz vostochnoy Palearktiki [New species of elaterid beetles of the genus Hypnoidus Steph. (Coleoptera, Elateridae) from East Palaearctic]. *Entomologicheskoe Obozrenie* **70**: 846-851.

Besuchet C. 1972: Les coléoptères aculagnathides. *Revue Suisse de Zoologie* **79**: 99-145.

Bethe 1868: Ein neuer europaïscher Throscus. *Entomologischen Zeitung* (Stettin) **29**: 36-37.

Bic V. 2002: Review of the genus Dilophotes Waterhouse (Coleoptera: Lycidae) of the Palaearctic Region and Indochina. *Zootaxa* **59**: 1-26.

Bickhardt H. 1906: Beitrag zur Coleopteren-Fauna von Korsika. *Wiener Entomologische Zeitung* **25**: 291-296.

Bickhardt H. 1908: Kleinere Beiträge zur Kenntnis der Coleopterenfauna von Korsika. *Entomologische Blätter* **4**: 201-202.

Bielawski R. 1958: A revision of the genus Anisosticta Duponch., with a description of a new species from Siberia (Coleoptera: Coccinellidae). *Annales Zoologici* **17** (7): 91-112.

Bielawski R. 1959: Beiträge zur Kenntnis der Coccinelliden von Afghanistan. II. (Coleoptera). *Entomologisk Tidskrift* **80**: 98-113.

Bielawski R. 1961: A new species of ladybeetle from Mongolia (Col., Coccinellidae). *Bulletin de l'Académie Polonaise des Sciences* Cl II *Série des Sciences Biologiques* **9**: 275-278.

Bielawski R. 1963: Monographie der Epilachninae (Coleoptera, Coccinellidae) der australischen Region. *Annales Zoologici* **21**: 295-461.

Bielawski R. 1965a: Über einige indomalaische Arten der Gattung Ortalia Muls. (Coleoptera, Coccinellidae). *Bulletin de l'Académie Polonaise de Sciences* Cl. II *Série des Sciences Biologiques* **13**: 21-28.

Bielawski R. 1965b: 50. Coccinellidae II. Ergebnisse der zoologischen Forschungen von Dr. Z. Kaszab in der Mongolei (Coleoptera). *Reichenbachia* **7**: 151-164.

Bielawski R. 1965c: Contribution to the knowledge of Coccinellidae (Coleoptera). IV. *Annales Zoologici* **23**: 211-236.

Bielawski R. 1966: Eine neue Gattung und Art der Epilachninae aus Burma (Coleoptera, Coccinellidae). *Bulletin de l'Académie Polonaise des Sciences* Cl. II *Série des Sciences Biologiques* **14**: 793-795.

Bielawski R. 1967: Drei neue Arten der Gattung Afissula Kapur aus Burma (Coleoptera: Coccinellidae). *Bulletin de l'Académie Polonaise des Sciences* Cl. II *Série des Sciences Biologiques* **15**: 151-157.

Bielawski R. 1968: Eine neue Unterart von Coccinelliden aus Afghanistan, Synharmonia conglobata afghana subsp. n. (Coleoptera, Coccinellidae). *Bulletin de l'Académie Polonaise des Sciences Cl. II. Série des Sciences Biologiques* **16**: 423-426.

Bielawski R. 1971: Über Coccinellidae (Coleoptera) aus Nepal. *Khumbu Himal: Ergebnisse des Forschungsunternehmens Nepal-Himalaya* **1**: 1-9.

Bielawski R. 1972: Die Marienkäfer (Coleoptera: Coccinellidae) aus Nepal. *Fragmenta Faunistica* **18**: 238-312.

Bielawski R. 1973: Rhyzobiini, Stethorini, Scymnini et Pharini (Coleoptera, Coccinellidae) de Nouvelle Calédonie. *Annales Zoologici* **30**: 387-409.

Bielawski R. 1979: Ergebnisse der Bhutan-Expedition 1972 des Naturhistorischen Museums in Basel Coleoptera: Fam. Coccinellidae. *Entomologica Basiliensia* **4**: 83-125.

Bielawski R. 1980: Die Marienkäfer (Coleoptera: Coccinellidae) der Koreanischen Volksdemokratischen Republik. *Fragmenta Faunistica* **25**: 221-254.

Bielawski R. 1984: Coccinellidae (Coleoptera) of Mongolia. *Annales Zoologici* **38**: 281-460.

Bielawski R. & Fürsch H. 1960: Zwei neue Coccinelliden aus Burma. *Mitteilungen der Münchener Entomologischen Gesellschaft* **50**: 68-71.

Bielz E. A. 1850: Entomologische Notizen. *Verhandlungen und Mitteilungen des Siebenbürgischin Vereins für Naturwissenschaften in Hermannstadt* **1**: 179-181.

Biggs H. E. J. 1947: A new species of Rhagonycha (Col., Cantharidae) from Marocco. *The Entomologist's Monthly Magazine* **83**: 9.

Bigliani L. 1907: Coleotteri nuovi o non segnati nel Catalogo del Bertolini. *Rivista Coleotterologica Italiana* **5**: 153-157.

Billberg G. J. 1808: [new names]. In: Schönherr C. J. *Synonyma Insectorum, oder: Versuch einer Synonymie aller bisher bekannten Insecten; nach Fabricii Systema Eleutheratorum &c. geordnet. Erster B. Eleutherata oder Käfer. Zweiter Theil. Sprechus---Cryptocephalus*. Stockholm: C. F. Marquard, ix + 424 pp., 1 pl.

Billberg G. J. 1820: *Enumeratio Insectorum in Museo Gust. Joh. Billberg*. Stockholm: Gadelianis, 138 + [2] pp.

Binaghi G. 1933: Un nuovo Hypnoidus italiano e nota sulle specie del gruppo del dermestoides Hbst. (Col. Elateridae). *Bollettino della Società Entomologica Italiana* **65**: 205-211.

Binaghi G. 1935: Appunti sugli elateridi italiani. II. *Bollettino della Società Entomologica Italiana* **67**: 22-26.

Binaghi G. 1936: Studio sul genere Adrastus Esch. (Col. Elateridae). *Bollettino di Zoologia Agraria e Bachicoltura* **7**: 217-227.

Binaghi G. 1940a: Idiotarmon, genere nuovo per Betarmon quadrivittatus Ragusa (Col. Elateridae). *Bollettino della Società Entomologica Italiana* **72**: 99-102.

Binaghi G. 1940b: Revisione delle Ctenicera s. str. europee con descrizione di due nuove specie e due varietá appartenenti alla fauna italiana (Col. Elateridae). *Memorie della Società Entomologica Italiana* **19**: 93-124.

Binaghi G. 1940c. Revisione degli Anostirus Thoms. europei (Col., Elateridae). *Memorie della Società Entomologica Italiana* **19**: 193-234.

Binaghi G. 1941a: Gli apparati genitali dell'Agrypnus notodonta Latr. e descrizione di una nuova specie di Agrypnus dell'Africa settentrionale (Col. Elateridae). *Bollettino della Società Entomologica Italiana* **73**: 68-74.

Binaghi G. 1941b: Due nuove specie di Anommatus della regione alpina (Col. Colydiidae). *Bollettino della Società Entomologica Italiana* **73**: 137-143.

Binaghi G. 1941c: Il Drasterius bimaculatus Rossi in Italia con note di sistematica sui Drasterius paleartici. Variazioni, geonimia, apparati genitali e loro pertinenza tra i Conoderini (Col. Elateridae). *Memorie della Società Entomologica Italiana* **20**: 162-183.

Binaghi G. 1946: Nuove forme di Metophthalmus Woll. con tabella di determinazione delle specie italiane. *Bollettino della Società Entomologica Italiana* **76**: 21-26.

Binaghi G. 1972: Notizie e discussioni su quattro specie di elateridi raccolte alle Isole Ponziane e descrizione di una nuova forma di Harminius castaneus (Coleoptera). *Fragmenta Entomologica* **8**: 143-148.

Bjerkander C. 1779: Beskrifning på et högst skadelig Rotmask. *Kungliga Svenska Vetenskapsakademiens Handlingar* **40**: 284-288, pl. X.

Biström O. 1977: Nomenclatoric notes on Coleoptera. *Notulae Entomologicae* **57**: 17-18.

Biström O. 1978: A replacement name for Ipidia quadrinotata (Coleoptera, Nitidulidae). *Notulae Entomologicae* **58**: 88.

Blackburn T. 1888: Further notes on Australian Coleoptera, with descriptions of new species. *Transaction and Proceedings and Report of the Royal Society of South Australia* **10**: 177-287.

Blackburn T. 1889a: Notes on Australian Coleoptera with descriptions of new species. *Transactions and Proceedings of the Royal Society of South Australia* **11**: 175-214.

Blackburn T. 1889b: Further notes on Australian Coleoptera with descriptions of new genera and species. *Proceedings of the Linnean Society of New South Wales* **3** (2): 1387-1506.

Blackburn T. 1890: Further notes on Australian Coleoptera, with descriptions of new genera and species. *Transactions and Proceedings of the Royal Society of South Australia* **13**: 82-93.

Blackburn T. 1891a: Further notes on Australian Coleoptera with descriptions of new genera and species. IX. *Transactions and Proceedings of the Royal Society of South Australia* **14**: 65-153.

Blackburn T. 1891b: Further notes on Australian Coleoptera, with descriptions of new genera and species. *Transactions and Proceedings of the Royal Society of South Australia* **14**: 292-345.

Blackburn T. 1892: Further notes on Australian Coleoptera, with descriptions of new genera and species, XII. *Transactions and Proceedings of the Royal Society of South Australia* **15**: 207-261.

Blackburn T. 1894: Further notes on Australian Coleoptera, with descriptions of new genera and species. *Transactions and Proceedings of the Royal Society of South Australia* **18**: 139-240.

Blackburn T. 1895: Further notes on Australian Coleoptera, with descriptions of new genera and species. *Transactions of the Royal Society of South Australia* **19**: 201-258.

Blackburn T. 1900a: Furthern notes on Australian Coleoptera with description of new genera and species. *Transactions and Proceedings of the Royal Society of South Australia* **24**: 35-68.

Blackburn T. 1900b: On some new genera and species of Australian Coleoptera. *Proceedings of the Royal Society of Victoria* **12**: 206-233.

Blackburn T. 1903: Further notes on Australian Coleoptera, with descriptions of new genera and species. XXXII. *Transactions of the Royal Society of South Australia* **27**: 91-182.

Blackburn T. & Sharp D. 1885: Memoirs on the Coleoptera of the Hawaiian Islands. *Scientific Transactions of the Royal Dublin Society* (2) **3** [1883-1887]: 119-290, pls. IV-V.

Blair K. G. 1937: Dasytidae. Pp. 181-182. In: Entomologische Sammelergebnisse der Deutschen Hindukusch-Expedition 1935 der Deutschen Forschungsgemeinschaft. II. Teil. *Arbeiten über Morphologische und Taxonomische Entomologie aus Berlin-Dahlem* **4**: 177-191.

Blair K. G. 1942: Trixagus seriatus sp. n. a new British beetle (Col. Trixagidae). *The Entomologist's Monthly Magazine* **78**: 287-288.

Blaisdell F. E. 1892: A new species of Coleoptera from California. *Entomological News* **3**: 51.

Blaisdell F. E. 1927: A blind beetle excavated from an Egyptian city's ruins dating between 117 and 235 A. D. *Proceedings of the Entomological Society of Washington* **29**: 121-125, pl. 6.

Blanchard E. 1853: Insectes. In: Hombrom J. B. & Jacquinot R. (eds): *Atlas d'histoire naturelle. Zoologie.* In: *Voyage au Pôle Sud et dans l'Océanie sur les corvettes l'Astrolabe et la Zélée, executé par l'ordre du Roi pendent les année 1837-1838-1839-1840 sous le commandement de M. J. Dumont-d'Urville, Capitaine de vaisseau. Tome Quatrième.* Paris: Gide et J. Baudry, [5] + 422 pp., 20 pls. [plates issued in 1847].

Blatch W. G. & Horner A. C. 1892: Description of a new species of Rhizophagus. *The Entomologist's Monthly Magazine* (2) **28**: 303-304.

Bleuse L. 1896: Nouvelle espèce de Danacaea. *Revue d'Entomologie* **15**: 250.

Bocák L. 1996: A new species of the genus Platycis from China (Coleoptera: Lycidae). *Folia Heyrovskyana* **4**: 1-2.

Bocák L. 1997: New and poorly known Ateliinae (Coleoptera: Lycidae). *Annales Zoologici* **47**: 175-178.

Bocák L. 1999: A revision and phylogenetic analysis of the genus Lyponia C.O.Watehouse, 1878 (Coleoptera, Lycidae). *Entomologica Basilensia* **21** [1998]: 59 -103.

Bocák L. & Bocáková M. 1987: A review of Lycidae (Coleoptera) from Taiwan with a description of a new species. *Acta Entomologica Bohemoslovaca* **84**: 273-286.

Bocák L. & Bocáková M. 1989: New tribe Lyropaeini, with a description of a new species of Lyropaeus (Coleoptera, Lycidae). *Polskie Pismo Entomologiczne* **58**: 717-723.

Bocák L. & Bocáková M. 1990: Revision of the supergeneric classification of the family Lycidae (Coleoptera). *Polskie Pismo Entomologiczne* **59**: 623-676.

Bocák L. & Bocáková M. 1991: Notes on some Palaearctic and Oriental representatives of the tribe Erotini (Coleoptera, Lycidae). *Acta Entomologica Bohemoslovaca* **88**: 313-326.

Bocák L. & Bocáková M. 1999: New taxa of Lycidae from China and Indonesia. *Acta Universitatis Palackianae Olomucensis, Facultas Rerum Naturalium, Biologica* **37**: 47-56.

Bocáková M. 1994: Notes on the genus Pyropterus with a description of a new species from Kashmir (Coleoptera, Lycidae). *Folia Heyrovskyana* **2**: 132-134.

Bocáková M. 1997a: New Melaneros-species from the Himalayas (Coleoptera, Lycidae). *Stuttgarter Beiträge zur Naturkunde* (A) **564**: 1-5.

Bocáková M. 1997b: Revision of the genus Melaneros from China with a note on Ditoneces (Coleoptera: Lycidae). *Acta Societatis Zoologicae Bohemicae* **61**: 175-190.

Bocáková M. 2000: Review of the genus Libnetis from China with descriptions of several species from Thailand (Coleoptera: Lycidae). *Acta Societatis Zoologicae Bohemicae.* **64**: 223-234.

Bocáková M. 2003: New Libnetini from China, Nepal, and Laos (Coleoptera, Lycidae). *Biologia* **58**: 175-179.

Bocáková M. 2004: Phylogenetic analysis of the tribe Libnetini and new genus Prolibnetis gen. n. (Coleoptera: Lycidae). *Deutsche Entomologische Zeitschrift* **51**: 53-64.

Bodemeyer B. von. 1929: *Ueber meine Entomologischen Reisen nach Kleinasien (1911), Ost-Sibirien, Schilka und Amur (1912), Tunis, Oasis Gafsa, Khroumerie (1913) und Iran, das Elbursgebirge (1914). Bd.III. Tunis, Oasis Gafsa und die Khroumerie. Band III. Tunis, Oasis Gafsa und die Khroumerie.* Stuttgart: Alfred Kernen Verlag, 79 pp., 1 pl.

Boheman C. H. 1850: Bildrag till Gottlands insekten-fauna. *Öfversigt af Kongliga Vetenskaps-Akademiens Förhandlingar* **7** (3): 70-76.

Boheman C. H. 1851: *Insecta Caffraria annis 1838 – 1845 a J. A. Wahlberg collecta, amici auxilio suffultus, descripsit Carolus H. Boheman. Pars I., Fasciculus 2. Coleoptera (Buprestides, Elaterides, Cebrionites, Rhipicerides, Cyphonides, Lycides, Lampyrides, Telephorides, Melyrides, Clerii, Terediles, Ptiniores, Palpatores, Silphales, Histeres, Scaphidilia, Nitidulariae, Cryptophagidae, Byrrhii, Dermestini, Parnidae, Hydrophilidae).* Holmiae: ex officina Norstedtiana, pp. 299-626, pl. 2.

Boheman C. H. 1854: Entomologiska anteckningar under en resa i Södra Sverige 1851. *Kongliga Vetenskaps-Akademiens Handlingar* **1852**: 55-210.

Boheman C. H. 1858-1859: Coleoptera. Species novas descripsit. Pp. 1-112, 113-217, pls I-II. In: Virgin C. (ed.): *Kongliga Svenska Fregatten Eugenies Resa omkring jorden under befäl af C. A. Virgin, ären 1851 - 1853. Vetenskapliga Iakttaagelser pa H. M. Konung Oscart den Försten befallning utgifna af K. Svenska Vetenskaps Akademien. II. Zoologi. 1. Insecta.* Uppsala et Stockholm: P. A. Norstedt & Söner, Almquist et Wiksells, 614 pp., 9 pls. [issued in parts, pp. 1-112 in 1858].

Boieldieu M. 1859: Descriptions d´espèces nouvelles de coléoptères. *Annales de la Société Entomologique de France* (3) **7**: 461-482.

Boisduval J. B. A. D. 1835: *Voyage de découvertes de l'Astrolabe exécuté par ordre du Roi, pendant les années 1826-1827-1828-1829, sous le commandement de M. J. Dumont D'Urville. Faune Entomologique de l'Océan Pacifique, avec l'illustration des insectes nouveaux recueillis pendant le voyage. Deuxième partie. Coléoptères et autres ordres.* Paris: Didot-Frères, J Tastu, éditeur-imprimeur, viii + 716 pp.

Boieldieu A. A. 1854: [Monographie des coléoptères de la tribu des Ptinoires]. *Bulletin de la Société Entomologique de France* (3) **2**: lxxvii-lxxxv.

Boieldieu M. 1856a: Monographie des Ptiniores. *Annales de la Société Entomologique de France* (3) **4**: 285-315.

Boieldieu M. 1856b: Monographie des Ptiniores. *Annales de la Société Entomologique de France* (3) **4**: 487-504.

Boieldieu M. 1856c: Monographie des Ptiniores. *Annales de la Société Entomologique de France* (3) **4**: 629-686, 5 pls.

Boieldieu M. 1859: Descriptions d'espèces nouvelles de coléoptères. *Annales de la Société Entomologique de France* (3) **7**: 461-482, pl. 8.

Bold T. J. 1872: A catalogue of the insects of Northumberland and Durham (Revision of Coleoptera). *Natural History Transactions of Northumberland and Durham and Newcastle-upon-Tyne.* London: Newcastle, **4** [1871-2]: 1-117.

Bolívar y Pieltain C. 1913: Especies nuevas del género "Cydistus" Bourg. (Drilidae). *Boletín de la Real Sociedad Española de Historia Natural* **13**: 316-318.

Bollow H. 1941: Zwei neue Amphicrossus-Arten und Bestimmungstabelle der pal. Arten (Col. Nitidulidae), sowie Berichtigung der Lage von Tienmuschan. *Mitteilungen der Münchener Entomologischen Gesellschaft* **31**: 234-238.

Bonneli F. A. 1812: Specimen Faunae subalpinae sistens Insecta Pedemontii hucusque inedita, aut rariora, aut ea quae commodi damnive gratia quod inferunt, prudentis agricolae magis interest cognoscere. Fasc. 1. Coleoptera plerumque iinedita comprehendens. *Memoire della Reale Sociedad Agraria di Torino* **9**: 149-183.

Bonvouloir H. A. de. 1859: *Essai monographique sur la famille des Throscides.* Paris: Deyrolle, 144 pp., 5 pls.

Bonvouloir H. A. de. 1862: Description de plusieurs espèces nouvelles de la famille des Throscides. *Annales de la Société Entomologique de France* (4) **1** [1861]: 349-360.

Bonvouloir H. A. de. 1871: Monographie de la Famille des Eucnémides, 1ère partie. *Annales de la Société Entomologique de France* (4) **10** (Suppl.): 5-288, pls. 1-21.

Booth R. G. 1997: A review of the species of Calvia (Coleoptera: Coccinellidae) from the Indian subcontinent, with descriptions of two new species. *Journal of Natural History* **31**: 917-934.

Börner I. C. H. 1776: Beschreibung eines neuen Insekts, des Dermestes 6-dentatus. *Der Ekonomische Nachricht der Gesellschaft in Schlesien* **4**: 78-80.

Börner I. C. H. 1776: Beschreibung eines neuen Insekts, des Coccinella transversepunctata. *Der Ekonomische Nachricht der Gesellschaft in Schlesien* **4**: 250.

Borowski J. 1995: On the genus Eurostoptinus Pic, 1895 (Coleoptera, Ptinidae). *Revue Suisse de Zoologie* **102**: 367-372.

Borowski J. 1999: Dignomus urbanus n. sp. from Saudi Arabia (Coleoptera, Ptinidae). *Doriana* **7** (231): 1-4.

Borowski J. 2000: Dignomus omanensis n. sp. from Oman (Coleoptera: Ptinidae). *Genus* **11**: 125-127.

Borowski J. 2002: Dignomus kroliki n. sp. from Turkey (Coleoptera, Ptinidae). *Baltic Journal of Colepterolgy* **2** (2): 155-158.

Boubers R. de & Leseigneur L. 1956: Un Athous nouveau appartenant à la faune des Alpes françaises (Col. Elateridae). *Bulletin de la Société Entomologique de France* **61**: 36-39.

Bourgeois J. 1879a: Lycides recueillis au Brésil par C. van Volxem. *Comptes Rendus de la Société Entomologique de Belgique* **1879**: xv-xix.

Bourgeois J. 1879b: Description d´une nouvelle espèce de coléoptère de la division des Lycides. *Bulletin de la Société Entomologique de France* 1879: xxxviii-xxxix.

Bourgeois J. 1882a: Monographie des Lycides de l'ancien-monde. *L'Abeille, Journal d'Entomologie* **20**: 1–117 + [3].

Bourgeois J. 1882b: Description d'une nouvelle espèce de Lycides. *Bulletin de la Société Entomologique de France* **1881**: cxliv-cxlv.

Bourgeois J. 1883a: Lycides nouveaux ou peu connus du Musée civique de Génes. *Annali del Musuo Civico di Storia Naturale di Genova* **18**: 621-648.

Bourgeois J. 1883b: Description d'une espèce nouvelle du genre *Dasytiscus* Kiesenw., trouvée en Palestine par M. Letourneux. *Bulletin de la Société Entomologique de France* **1883**: liii.

Bourgeois J. 1883c: [Communication]. *Bulletin de la Société Entomologique de France* **1883**: lix-lxii.

Bourgeois J. 1884a: Lycides nouveaux ou peu connus (4e partie, espèces des g. Calochromus, Calopteron, Lycus, Trichalus). *Annales de la Société Entomologique de France* (6) **4**: 63-68, pl. 4.

Bourgeois J. 1884b: Dascillides & Malacodermes de Nouvelle-Calédonie. *Revue d'Entomologie* **3**: 278-290.

Bourgeois J. 1884c: Description d'une espèce nouvelle du genre Henicopus provenant du Portugal. *Bulletin de la Société Entomologique de France* 1884: lxiv-lxv.

Bourgeois J. 1885a: Cébrionides, Malacodermes, 46 pp. In: Fauvel C. A. A.: *Faune Gallo-rhénane ou species des insectes qui habitent la France, la Belgique, la Hollande, le Luxembourg, la Prusse, le Nassau et le Valais – Coléoptères. Tome 4* [1884-88]. Caen.

Bourgeois J. 1885b: Diagnoses de Lycides nouveaux ou peu connus. 5e partie. *Annales de la Société Entomologique de France* (6) **5**: 71-84.

Bourgeois J. 1885c: Remarques sur le genre Dasytiscus et descriptions d'espèces nouvelles ou imparfaitement connues. *Annales de la Société Entomologique de France* (6) **5**: 253-271, 274, pl. 5.

Bourgeois J. 1885d: Description d'un genre nouveaux et d'une espèce nouvelle de Malacodermes de Syrie: Cydistus. *Bulletin de la Société Entomologique de France* **1885**: xxxvii-xxxviii.

Bourgeois J. 1885e: [new taxa]. *Bulletin de la Société Entomologique de France* **1885**: cxxvi-cxxvii.

Bourgeois J. 1885f: [new taxa]. *Bulletin de la Société Entomologique de France* **1885**: clxxvi-clxxvii.

Bourgeois J. 1886: Malacodermes (suite). In: Fauvel C. A. A.: Faune gallo-rhénane ou species des insectes qui habitent la France, la Belgique, la Hollande, le Luxembourg, la Prusse Rhénane, le Nassau et le Valais avec tableaux synoptiques et planches gravées. Coléoptères. 4. *Revue d'Entomologie, Supplement* **5**: 101-132.

Bourgeois J. 1887: Malacodermes (suite). In: Fauvel C. A. A.: Faune gallo-rhénane ou species des insectes qui habitent la France, la Belgique, la Hollande, le Luxembourg, la Prusse Rhénane, le Nassau et le Valais avec tableaux synoptiques et planches gravées. Coléoptères. 4. *Revue d'Entomologie, Supplement* **6**: 133-156.

Bourgeois J. 1888a: Synopsis du genre Henicopus Steph. *Annales de la Société Entomologique de France* (6) **8**: 5-34, pl. 2.

Bourgeois J. 1888b: Faune gallo-rhénane ou species des insectes qui habitent la France, la Belgique, la Hollande, le Luxembourg, la Prusse Rhénane, le Nassau et le Valais avec tableaux synoptiques et planches gravées. Coléoptères. 4. Malacodermes (suite). *Revue d'Entomologie, Supplement* **7**: 157-172.

Bourgeois J. 1889: [new taxa]. In: Heyden L. F. J. D. von: Insecta, a Cl. G. N. Potanin in China et in Mongolia novissime lecta. XII. Scarabaeidae. Cantharidae. Cleridae. Lagriidae. Meléndryidae. Pedilidae. Anthicidae. *Horae Societatis Entomologicae Rossicae* **23**: 654-677.

Bourgeois J. 1890a: Contributions à la faune Indo-Chinoise. 7e Mémoire. Rhipidoceridae, Dascillidae, Malacodermidae. *Annales de la Société Entomologique de France* (6) **10**: 169-176.

Bourgeois J. 1890b: Collection d'insects formée dans l'Indo-Chine par M. Pavie, consul de France au Cambodge. Coléoptères. Cebrionidae, Rhipidoceridae, Dascillidae, Malacodermidae. *Nouvelles Archives du Muséum* (3) **2**: 179-188.

Bourgeois J. 1891: Dascillides et Malacodermes du Bengale Occidentale. *Bulletin ou Comptes-Rendus des Séances de la Société Entomologique de Belgique* **35**: cxxxvii-cxli.

Bourgeois J. 1893: Faune gallo-rhénane. Coléoptères. Tom. 4. Supplément aux Malacodermes. *Revue d'Entomologie* **11** Supplement: 1-34.

Bourgeois J. 1894a: *Faune gallo-rhénane ou species des insectes qui habitent la France, la Belgique, la Hollande, le Luxembourg, la Prusse Rhénane, le Nassau et le Valais. Coléoptères. Tome 4*. Caen, 208 pp.

Bourgeois J. 1894b: Description d'un genre nouveau et d'une espèce nouvelle de Malacodermes d'Algérie. *Revue d'Entomologie* **13**: 121-123.

Bourgeois J. 1894c: Description d'une nouvelle espèce d'Amauronia, provenant d'Algérie. *Bulletin de la Société Entomologique de France* **1894**: cciv-ccv.

Bourgeois J. 1894d: [Description d'une nouvelle espèce de Malacoderme appartenant au genre Podistrina Fairm.]. *Bulletin de la Société Entomologique de France* **1893**: cclxxxix-ccxc.

Bourgeois J. 1895: [Description d'une nouvelle espèce de Malacoderme appartenant au genre Podistrina]. *Bulletin de la Société Entomologique de France* **1894**: cclxvi-cclxvii.

Bourgeois J. 1896: Note sur quelques espèces nouvelles ou peu connues de coléoptères de l'Inde et de la Birmanie. *Bulletin de la Société Entomologique de France* **1896**: 117-121.

Bourgeois J. 1897: Sur quelques Malthinides algériens. *Revue d'Entomologie* **16**: 39-44.

Bourgeois J. 1899a: Description de deux Podistrina nouvelles de la faune française (Col.). *Bulletin de la Société Entomologique de France* **1899**: 338-341.

Bourgeois J. 1899b: Notes sue quelques Malthinus paléarctiques et description d'une espèce nouvelle (Col.). *Bulletin de la Société Entomologique de France* **1899**: 368-371.

Bourgeois J. 1900a: [new taxa]. In: Bodemeyer B. von. 1900: *Quer durch Klein-Asien in den Bulgar-Dagh. Eine naturwissenschaftliche Studienreise*. Emmendingen: Verlangs-Aktiengesellschaft vormals Dölter, v + 169 pp.

Bourgeois J. 1900b: Note sur quelques Malthinus paléarctiques (suite) et description d'une espèce nouvelle (Col.). *Bulletin de la Société Entomologique de France* **1900**: 91-96.

Bourgeois J. 1900c: Description d'une variété nouvelle du Cantharis discoidalis et notes sur l'habitat de quelques autres Malacodermes (Col.). *Bulletin de la Société Entomologique de France* **1900**: 181-182.

Bourgeois J. 1900d: Notes sur quelques Podistrina et descriptions d'une espèce nouvelle d'Algérie (Col.). *Bulletin de la Société Entomologique de France* **1900**: 283-285.

Bourgeois J. 1900e: Description d'un Lampyride européen nouveau. *Bulletin de la Société Entomologique de France* **1900**: 337-338.

Bourgeois J. 1900f: Description d'une nouvelle espèce du genre Podistrina (Col.) et notes d'habitat. *Bulletin de la Société Entomologique de France* **1900**: 378-380.

Bourgeois J. 1901: Description d'une nouvelle espèce de Malthodes, d'Algérie (Col.). *Bulletin de la Société Entomologique de France* **1901**: 143-144.

Bourgeois J. 1902a: Notes sue quelques Podistrina du Nord de l'Afrique et description de deux espèces nouvelles (Col.). *Bulletin de la Société Entomologique de France* **1902**: 60-64.

Bourgeois J. 1902b: Malacodermes récoltés au Japon par M. J. Harmand (1900). *Bulletin du Muséum d'Histoire Naturele* (Paris) **8**: 89-92.

Bourgeois J. 1902c: Deux nouvelles espèces de Plateros de l'Himalaya. *Bulletin du Muséum d'Histoire Naturele* (Paris) **8**: 92-93.

Bourgeois J. 1902d: Description d'une nouvelle Podistrina (Col.). *Bulletin de la Société Entomologique de France* **1902**: 182-184.

Bourgeois J. 1902e: Trois nouvelles espèces de malacodermes de l'Himalaya. *Bulletin du Muséum d'Histoire Naturele* (Paris) **8**: 426-427.

Bourgeois J. 1903a: Notes sur quelques espèces de Malacodermes de la faune méditerranéenne (Col.). *Bulletin de la Société Entomologique de France* **1903**: 73-77.

Bourgeois J. 1903b: Diagnoses de trois Malthodes nouveaux de la faune méditerranéenne (Col.). *Bulletin de la Société Entomologique de France* **1903**: 152-154.

Bourgeois J. 1903c: Description d'une nouvelle espèce de Podistrina (Col.). *Bulletin de la Société Entomologique de France* **1903**: 210-211.

Bourgeois J. 1903d: Diagnoses de trois Malthodes nouveaux du Nord de l'Afrique (Col.). *Bulletin de la Société Entomologique de France* **1903**: 219-221.

Bourgeois J. 1903e: Description d'une nouvelle espèce de Malthodes de la faune française (Col.). *Bulletin de la Société Entomologique de France* **1903**: 247-248.

Bourgeois J. 1903f: Description d'une nouvelle espèce européenne du genre Malthodes (Col.). *Bulletin de la Société Entomologique de France* **1903**: 336-337.

Bourgeois J. 1905a: Description d'une nouvelle espèce de Malthodes de la faune française. *Bulletin de la Société Entomologique de France* **1905**: 21-23.

Bourgeois J. 1905b: Description de quatre espèces nouvelles de Lycides du Bengale. *Annales de la Société Entomologique de Belgique* **49**: 46-48.

Bourgeois J. 1905c: Description d'un nouveau Malacoderme d'Algerie appartenent au sous-genre Absidia (Col.). *Bulletin de la Société Entomologique de France* **1905**: 106-108.

Bourgeois J. 1905: Voyage du Dr. W. Horn à Ceylan: Malacodermes et Lymexylonides. *Annales de la Société Entomologique de France* **74**: 127-133.

Bourgeois J. 1905e: Description d une nouvelle espèce de Lycide (Col.) de la faune barbaresque. *Bulletin de la Société Entomologique de France* **1905**: 291-292.

Bourgeois J. 1906: Description d'une nouvelle espèce algérienne de sous-genre Absidia (Col.). *Bulletin de la Société Entomologique de France* **1906**: 97-98.

Bourgeois J. 1907: Sur quelques Malacodermes de l'Inde. *Annales de la Société Entomologique de Belgique* **51**: 291-293.

Bourgeois J. 1908a: Une espèce nouvelle du genre Cydistus (Col.). *Bulletin de la Société Entomologique de France* **1908**: 210-211.

Bourgeois J. 1908b: Description d une nouvelle espèce de Drilus (Col.). *Bulletin de la Société Entomologique de France* **1908**: 241-242.

Bourgeois J. 1909: Notes sur quelques espèces de coléoptères de la faune alpine. *Mitteilungen der Schweizerischen Entomologischen Gesellschaft* **11** [1903-1909]: 388-394.

Bourgeois J. 1911: Description d'un nouvel Haplocnemus marocain (Col. Melyridae). *Bulletin de la Société Entomologique de France* **1911**: 158-160.

Bousquet Y. 1989: A review of the North American genera of Cryptophaginae (Coleoptera: Cryptophagidae). *The Coleopterists Bulletin* **43**: 1-17.

Bousquet Y & Laplante S. 1999: Taxonomic review of the Canadian species of the genus Monotoma Herbst (Coleoptera: Monotomidae). *Proceedings of the Entomological Society of Ontario* **130**: 67-96.

Bouwer R. 1980: Revision der Ampedus ferrugatus-Gruppe nebst Beschreibung von vier neuen Arten (Insecta: Coleoptera: Elateridae). *Senckenbergiana Biologica* **60**: 321-336.

Bouwer R. 1984a: Ampedus carinthiacus n. sp., eine neue Art aus Österreich (Elateridae, Col.). *Koleopterologische Rundschau* **57**: 3-5.

Bouwer R. 1984b: Ampedus islamicus n. sp. aus Kirgizien (Elateridae, Col.). *Koleopterologische Rundschau* **57**: 7-8.

Bouwer R. 1984c: Zwei neue Ampedus-Arten fur die BRD (Coleoptera, Elateridae). *Deutsche Entomologische Zeitschrift* **31**: 83-87.

Bouwer R. 1990: Eine neue Ampedus-Art aus Frankreich: Ampedus gallicus n. sp. (Coleoptera: Elateridae). *Entomologische Zeitschrift* **100**: 59-61.

Bovie A. 1897: Les coccinelles de Belgique. *Annales de la Société Entomologique de Belgique* **41**: 133-162.

Bowestead S. 1999: A revision of the Corylophidae (Coleoptera) of the West Palaearctic Region. *Instrumenta Biodiversitatis* (Genève) **3**: 203 pp.

Bowestead S. 2000: Two new species of Orthoperus Stephens, 1829 (Coleoptera Corylophidae) from Russia. *Russian Entomological Journal* **9**: 195-197.

Bowestead S. 2003: A contribution to the knowledge of the Corylophidae of the Palaearctic Region, with 35 figs. Pp. 943-955. In: Cuccodoro G. & Leschen R. A. B. (eds): Systematics of Coleoptera: Papers Celebrating the Retirement of Ivan Löbl. *Memoirs on Entomology International* **17**: 1-955.

Boyle W. W. 1956: A revision of the Erotylidae of America north of Mexico (Coleoptera). *Bulletin of the American Museum of Natural History* **110**: 65-165.

Brahm N. J. 1790: *Insectenkalender für Sammler und Oekonomen. Handbuch der ökonomischen Insektengeschichte in Form eines Kalenders bearbeitet. Erster Theil.* Mainz: Kurfürstl. Privil. Universitätsbuchhandlung, lxvii + 248 pp.

Brahm N. J. 1797: Verzeichniss der von der Gattung Clerus Fabr. in den Gegenden von Mainz und Aschaffenburg einheinmischen Arten. Pp. 133-139. In: Hoppe D. H. (ed.): *Entomologisches Taschenbuch für die Anfänger und Liebhaber dieser Wissenschaft auf das Jahr 1797.* Regensburg: Montag und Weisse, 252 pp.

Brahm N. J. 1802: Bemerkungen über die von der Gattung Coccinella bei Mainz einheimischen Arten. *Naturforscher* **29**: 135-163.

Brancsik C. 1910: Species et varietates novae Coleopterorum exoticorum ac palaearcticorum. *Jahresheft des Naturwischenschaftlichen Vereins des Trencsiner Commitates* **31-33**: 176-190.

Brancsik K. 1914: Coleoptera nova. *Bericht des Museumvereines für das Comitat Trencsén* **1914**: 58-69.

Brancucci M. 1978: [new taxa]. In: Wittmer W. & Brancucci M.: Zur Kenntnis der Malthinini der indo-malaiischen Region (Col. Cantharidae). *Entomologica Basiliensia* **3**: 297-304.

Brancucci M. 1980: Morphologie comparée, évolution et systeématique des Cantharidae (Insecta: Coleoptera). *Entomologica Basiliensia* **5**: 215-388.

Brancucci M. 1981: Dritter Beitrag zur Kenntnis der Untergattung Malthinus (Indomalthinus) (Col. Cantharidae). *Mitteilungen der Entomologischen Gesellschaft Basel* (N. F.) **31**: 54-55.

Brancucci M. 1983: La famille des Cantharidae (Coleoptera) sur l'île de Taiwan. 4. La sous-famille des Chauliognathinae. *Entomologica Basiliensia* **8**: 256-309.

Brancucci M. 1985a: La famille des Cantharidae (Coleoptera) sur l'île de Taiwan. 4. La sous-famille des Chauliognathinae, supplément 1. *Entomologica Basiliensia* **10**: 239-244.

Brancucci M. 1985b: Trypheridium n. gen., un nouveau genre pour Trypherus nuristanicus Wittmer (Cantharidae, Coleoptera). *Entomologica Basiliensia* **10**: 245-249.

Brancucci M. 1985c: Revision du genre Trypherus Leconte (Coleoptera, Cantharidae). *Entomologica Basiliensia* **10**: 251-322.

Brancucci M. & Wittmer W. 1984: Zur Kenntnis der indo-malaiischen Malthininae (Coleoptera, Cantharidae). *Entomologica Basiliensia* **9**: 183-192.

Branham M. & Wenzel J. 2001: The evolution of bioluminescence in cantharoids (Coleoptera: Elateroidea). *Florida Entomologist* **84**: 565-586

Brébisson L. F. de. 1835: Catalogue des insectes de l'ordre de coléoptères qui se trouvent en Normandie, et notamment aux environs de Falaise, classés conformement aux derniers ouvrages de MM. Latreille et Dejean. *Mémoires de la Société Linnéenne de Normandie* **5**: 107-191.

Breit J. 1908: Eine koleopterologische Sammelreise auf Mallorca (Balearen). *Verhandlungen der Kaiserlich-Königlichen Zoologisch-Botanischen Gesellschaft in Wien* **58**: 52-67.

Breit J. 1908: [Beschreibung eines neuen Ptinus]. *Verhandlungen der Kaiserlich-Königlichen Zoologisch-Botanischen Gesellschaft in Wien* **58**: 160-162.

Breit J. 1911: Sechs neue mitteleuropäische Käferarten. *Wiener Entomologische Zeitung* **30**: 169-173.

Breit J. 1912: Beitrag zur Kenntnis der palaearktischen Käfer-Fauna. *Coleopterologische Rundschau* **1**: 5-7.

Breit J. 1913a: Beitrag zur Kenntnis der europäischen Blindkäferfauna. *Entomologische Mitteilungen* **2**: 12-19.

Breit J. 1913b: Beiträge zur Kenntnis der paläarktischen Coleopteren. *Entomologische Blätter* **9**: 292-299.

Brancsik C. 1899: Aliquot Coleoptera nova Russiae asiaticae. *Jahresheft des Naturwissenschaftlichen Vereins des Trencsiner Comitates* **21/22** [1898-1899]: 97-105.

Bremi-Wolf J. J. 1855a: Beschreibungen einiger muthmasslich neuer, oder doch noch nicht beschriebener schweizerischer Käfer. *Entomologische Zeitung* (Stettin) **16**: 196-200.

Bremi-Wolf J. J. 1855b: Ptinus Otti (Venetz.). *Entomologische Zeitung* (Stettin) **16**: 329-330.

Brèthes J. 1923a: Note sur un genre et une espèce de Coccinellides australiens passés inaperçus. *Bulletin de la Société Entomologique de France* **1923**: 227-229.

Brèthes J. 1923b: Catalogue synonymique des Coccinellides du Chili. *Rivista Chilena de Historia Natural Pura et Aplicata* **25** [1921]: 452-456.

Brèthes J. 1923c: Sur trois coléoptères chiliens. *Rivista Chilena de Historia Natural Pura et Aplicata* **25** [1921]: 457-461.

Brèthes J. 1925a: Sur une collection de coccinelides (et un Phalacridae) du British Museum. *Anales del Museo Argentine de Ciencia Natural* **33** [1923-1925]: 145-175.

Brèthes J. 1925b: Coccinellidae du British Muséum (avec une nouvelle famille de coléoptères). *Anales del Museo Nacional de Historia Natural "Bernardino Rivadavia"* **32**: 195-214.

Brisout de Barneville C. N. F. 1862: Espèces nouvelles de coléoptères français. *Annales de la Société Entomologique de France* (4) **1** [1861]: 597-606.

Brisout de Barneville C. N. F. 1863: [new taxa]. In: Grenier A. (ed.): *Catalogue des coléoptères de France et matériaux pour servir à la faune des coléoptères français*. Paris: L. Toinon, iv + 3-79 + 135 pp.

Brisout de Barneville C. N. F. 1866: Coléoptères nouveaux trouvés en Espagne pendant l'excursion de la Société en 1865. *Annales de la Société Entomologique de France* (4) **6**: 355-426.

Brisout de Barneville C. N. F. 1867a: [new names]. *Coleopterologische Hefte* **2**: 116-117.

Brisout de Barneville C. N. F. 1867b: [new taxa]. In: Grenier A. (ed.): *Matériaux pour servir à la faune des coléoptères de France, recueillis et publiés par le Dr. A. Grenier. 2e Cahier*. Paris: A. Grenier, [1] + iv + 131-194 pp.

Brisout de Barneville C. N. F. 1872: Synopse du genre Meligethes. *L'Abeille, Mémoires d'Entomologie* (2) **8** [1871]: 1-36.

Brisout de Barneville C. N. F. 1874: Descriptions suivantes de trois nouvelles espèces de coléoptères découvertes dans les environs de Paris. *Bulletin de la Société Entomologique de France* **1874**: lxx-lxxi.

Brisout de Barneville C. N. F. 1878: Descriptions de nouvelles espèces de coléoptères propres à la faune française. *Bulletin de la Société Entomologique de France* **1878**: lxii-lxiv.

Brisout de Barneville C. N. F. 1882: [Synonymies et descriptions relatives à quelques espèces de Cryptophagides et Nitidulides]. *Bulletin de la Société Entomologique de France* **1882**: xxvii-xxx.

Brisout de Barneville C. N. F. 1892: Description d'une Corticaria nouvelle de France. *Revue d'Entomologie* **11**: 68.

Brisout de Barneville H. 1862: Description de trois insectes coléoptères nouveaux qui se trouvent en France. *Revue et Magasin de Zoologie Pure et Appliquée* (2) **14**: 23-25.

Brisout de Barneville H. 1878: [new taxa]. *Bulletin de la Société Entomologique de France* **1878**: xcvi.

Brisout de Barneville H. 1880: Description d'une nouvelle espèce de Corticaria. *Annales de la Société Entomologique de France* (5) **10**: 236.

Brisout de Barneville H. 1881: Essai monographique des espèces d'Europe et des confins de la Mediterranée du genre Corticaria. *Annales de la Société Entomologique de France* (6) **1**: 375-422.

Brisout de Barneville H. 1884: [new taxa]. *Bulletin de la Société Entomologique de France* **1884**: lxxxi-lxxxii.

Brodský O. 1982: [new taxon]. In: Bílý S. & Brodský O.: Taxonomical, biological and faunistical notes on Buprestidae and Cleridae from east Mediterranean (Coleoptera). *Türkiye Bitki Koruma Dergisi* **6**: 185-194.

Brodský O. & Winkler J. R. 1988: Results of the Czechoslovak-Iranian Entomological Expeditions to Iran 1970, 1973 and 1977. Coleoptera, Cleridae. *Acta Entomologica Musei Nationalis Pragae* **42** [1987]: 293-305.

Broun T. 1880: *Manual of the New Zealand Coleoptera.* Wellington: James Hughes, xix + 651 pp.

Broun T. 1881: *Manual of the New Zealand Coleoptera. Part II.* Wellington: Georg Didsbury, viii + 653-744 + xxi-xxiii pp.

Broun T. 1882: Alternation of generic names. *The Annals and Magazin of Natural History* (5) **9**: 409.

Broun T. 1883: Revision of the New Zealand Cossonidae, with descriptions of new species. *The New Zealand Journal of Science* **1**: 498-499.

Broun T. 1886: *Manual of the New Zealand Coleoptera. Part IV.* Colonial Museum and Geological Survey Department. Wellington: G. Didsbury, pp. 817-973.

Broun T. 1893: *Manual of the New Zealand Coleoptera. Part V.* Wellington: Colonial Museum and Geological Survey Department, xvii + 975-1320 pp.

Broun T. 1895: Descriptions of new Coleoptera from New Zealand. (Continued). *The Annals and Magazine of Natural History* (6) **15**: 194-203.

Broun T. 1910: Descriptions of new genera and species of Coleoptera. *Bulletin of the New Zealand Institute* **1**: 3-78.

Broun T. 1914: Descriptions of new genera and species of Coleoptera. Part 3. *Bulletin of the New Zealand Institute* **1**: 143-266.

Brown W. J. 1929: Some new species of Coleoptera. *The Canadian Entomologist* **61**: 108-110.

Brown W. J. 1932: The North American species of Glischrochilus (Coleop.). *The Canadian Entomologist* **64**: 255-262.

Brown W. J. 1933: Studies in the Elateridae II. *The Canadian Entomologist* **65**: 173-182.

Brown W. J. 1934: New species of Coleoptera. 5. *Canadian Entomologist* **66**: 22-24.

Brown W. J. 1940: A key to the species of Ptinidae occurring in dwellings and warehouses in Canada (Coleoptera). *The Canadian Entomologist* **72**: 115-122.

Bruce N. 1934a: Eine neue Cryptophagus-Art aus Finnland. *Notulae Entomologicae* **14**: 35-36.

Bruce N. 1934b: Zwei neue Cryptophagen. *Entomologisk Tidskrift* **55**: 58-59.

Bruce N. 1934c: Eine neue Cryptophagus (Mnionomus) Art aus Südeuropa. *Entomologisk Tidskrift* **55**: 160-162.

Bruce N. 1934d: Studien über die Coleopterengattung Cryptophagus. *Entomologisk Tidskrift* **55**: 163-173.

Bruce N. 1936: Monographie der europäischen Arten der Gattung Cryptophagus Herbst mit besonderer Berücksichtigung der Morphologie des männlichen Kopulationsorgans. *Acta Zoologica Fennica* **20**: 1-167 pp., 17 pls., 8 maps.

Bruce N. 1938: Einige neue Cryptophagus-Arten aus dem Magyar Nemzeti Múzeum in Budapest. *Annales Historico-Naturales Musei Nationalis Hungarici. Pars Zoologica* **31**: 3-10.

Bruce N. 1940: Die Arthropodenfauna von Madeira nach den Ergebnissen der Reise von Prof. Dr. O. Lundblad Juli-August 1935. XXVIII. Coleoptera: Cryptophagidae: Gattung Cryptophagus. *Arkiv för Zoologi* **32** (23): 1-7.

Bruce N. 1941: Einige neue Cryptophagus-Arten aus der Sammlung Georg Frey-München. *Mitteilungen der Münchener Entomologischen Gesellschaft* **31**: 982-987.

Bruce N. 1943a: Cryptophagidae der J. Klapperichschen Fukien-Ausbeute. (Kol.). *Mitteilungen der Münchener Entomologischen Gesellschaft* **33**: 156-164, 2 pls.

Bruce N. 1943b: Ein neuer Cryptophagus aus der südlichen Ukraine (Col. Cryptoph.). *Mitteilungen der Münchener Entomologischen Gesellschaft* **33**: 295-297.

Bruce N. 1943c: Eine neue Cryptophagus-Art aus Dänemark. *Entomologiske Meddelelser* **22**: 367-368.

Bruce N. 1944: Entomological results from the Swedish Expedition 1934 to Burma and British India. *Arkiv för Zoo*logi **36** A (4): 1-5.

Bruce N. 1946: Orthoperus atomus Gyll. - En kollektivart. *Opuscula Entomologica* **11**: 106-107.

Bruce N. 1948: The Scandinavian species of the genus Orthoperus Steph. (Coleoptera). *Opuscula Entomologica Supplement* **9**: 1-34, 4 pls.

Bruce N. 1951a: *Cryptophagidae (Coleoptera Polyphaga). Exploration du Parc National Albert. Mission G. F. de Witte (1933-1935). Fascicule 75.* Bruxelles: Institut des Parcs Nationaux du Congo Belge, pp. 3-26.

Bruce N. 1951b: Eine neue Orthoperusart (Corylophidae). (Orthoperus intersitus n. sp.). *Entomologische Blätter* **45-46** [1949-1950]: 100-102.

Bruce N. 1952: XIX. Coleoptera Cryptophagidae in the British Museum -I. *The Annals and Magazine of Natural History* (12) **5**: 167-188.

Bruce N. 1959: Einige Cryptophagiden (Col.) aus Afghanistan nebst Beschreibung von Cryptophagus robustus n. sp. *Opuscula Entomologica* **24**: 18-20.

Bruce N. 1963a: Drei neue Cryptophagus-Arten aus Anatolien (Col. Cryptophagidae). *Opuscula Entomologica* **28**: 30-32.

Bruce N. 1963b: Beschreibung zweier neuer Cryptophagus-Arten uralensis n.sp. und amfidoxus n.sp. aus dem Ural. *Entomologische Blätter* **59**: 37-38.

Brullé G. A. 1832: IV.e Classe. Insectes. Pp. 1-240. In: Brullé G. A. & Guérin-Menéville F. M. (eds): *Expedition scientifique de Morée. Section des sciences physiques. Tome III. – 1.re partie. Zoologie. Deuxième Section.- Des animaux articulés.* Paris et Strasbourg: F. L. Levraut, 136 + 400 pp., 27-52 pls [note: pp. 1-240 issued in 1832, pp. 241-400 in 1833, plates issued in 1832-1836].

Brullé G. A. 1838: Entomologie. Pp. 53-95. In: Barker-Webb P. & Berthelot S.: *Histoire naturelle des Iles Canaries. (Animaux articulés recueillis aux Iles Canaries.). Vol. II. Pars 2.* Paris: Mellier, 119 pp., 8 pls. [issued in parts, from 1836 to 1840].

Büche B. 2001: A new fungivorous deathwatch-beetle from Europe (Coleoptera, Anobiidae). *Entomologische Blätter* **96**: 229-234.

Büche B. & Lundberg S. 2002: A new species of deathwatch beetle (Coleoptera: Anobiidae) discovered in Europe. *Entomologica Fennica* **13**: 79-84.

Buchholz L. & Dolin V. G. 1990: A new species of Melanotus Eschholz [sic!] (Coleoptera, Elateridae) from Tadjikistan (USSR). *Polskie Pismo Entomologiczne* **59**: 713-717.

Buquet J. B. L. 1840: Note sur quelques coléoptères nouveaux d'Algérie et particulièrement de Constantine. *Revue Zoologique* **3**: 240-243.

Burakowski B. & Buchholz L. 1991: Review of the European species of the genus Hylis Gozis (Coleoptera, Eucnemidae) with description of new species. *Annals of the Upper Silesian Museum, Entomology* **2**: 103-125.

Burchell W. J. 1822: *Travels in the interior of Southern Africa.* London: Longman, Hurst, Rees, Orme, and Brown, 7 pls., 1 map.

Burckhardt D. & Ślipiński S. A. 1995: A review of the Passandridae (Coleoptera, Cucujoidea) of the world. IV. Genus Ancistria. *Revue Suisse de Zoologie* **102**: 995-1044.

Burlini M. 1942: Forme nuove di coleotteri. *Bollettino della Società Entomologica Italiana* **74**: 13.

Buysson H. du. 1886a: [Description d'une nouvelle espèce d'élatérides]. *Bulletin de la Société Entomologique de France* **1885**: ccxvi-ccxvii.

Buysson H. du. 1886b: [Notes sur quelques élatérides d'Europe]. *Bulletin de la Société Entomologique de France* **1886**: xxvii-xxix.

Buysson H. du. 1887a: Révision des Silesis du bassin de la Méditerranée. *Revue d'Entomologie* **6**: 140-145.

Buysson H. du. 1887b: [Note sur quelques élatérides (1re partie)]. *Bulletin de la Société Entomologique de France* **1887**: lxxxix-xc.

Buysson H. du. 1887c: [Note sur quelques élatérides (2e partie)]. *Bulletin de la Société Entomologique de France* **1887**: xcv-xcvii.

Buysson H. du. 1887d: [Notes relatives à divers élatérides]. *Bulletin de la Société Entomologique de France* **1887**: cxxxii-cxxxiv.

Buysson H. du. 1888a: [Note au sujet de diverses élatérides]. *Bulletin de la Société Entomologique de France* **1887**: ccxi-ccxiii.

Buysson H. du. 1888b: [Note sur quelques élatérides (1re partie)]. *Bulletin de la Société Entomologique de France* **1888**: l-li.

Buysson H. du. 1889a: Rectification à la révision des Silesis du Bassin de la Méditerranée. *Revue d'Entomologie* **8**: 211-212.

Buysson H. du. 1889b: [Description d'une nouvelle espèce d'élatéride d'Europe]. *Bulletin de la Société Entomologique de France* **1888**: ccii.

Buysson H. du. 1890a: Aperçu dichotomique et observations sur quelques espèces d'élatérides. *Le Coléoptériste* **1890**: 4-5, 18, 34-36.

Buysson H. du. 1890b: [Descriptions de cinq espèces nouvelles d'élatérides (1re partie)]. *Bulletin de la Société Entomologique de France* **1889**: cciv-ccvii.

Buysson H. du. 1890c: [Espèces nouvelles d'élatérides (2e partie)]. *Bulletin de la Société Entomologique de France* **1889**: ccxiii-ccxv.

Buysson H. du. 1890d: [Description de deux espèces d'élatérides]. *Bulletin de la Société Entomologique de France* **1890**: clvii-clix.

Buysson H. du. 1891a: Aperçu dichotomique et observations sur quelques espèces d'élatérides. *Le Coléoptériste* **1890**: 241-245, 2 pls.

Buysson H. du. 1891b: [Communication: Le nom Symbiotes]. *Bulletin de la Société Entomologique de France* **1891**: xcv.

Buysson H. du. 1891c: [Description d'une nouvelle espèce d'élatérides: Cardiophorus Letourneuxi]. *Bulletin de la Société Entomologique de France* **1891**: lxviii-lxix.

Buysson H. du. 1891d: Descriptions de neuf espèces nouvelles d'élatérides, rapportée de la Haute Syrie et de l'Asie Mineure par M. C. Delagrange. *Bulletin de la Société Entomologique de France* **1891**: cxxxvii-cxliii.

Buysson H. du. 1893a: Coléoptères. Vol. 5: Elateridae. In: Fauvel C. A. A.: Faune gallo-rhénane ou description des insectes qui habitent la France, la Belgique, la Hollande, les Provinces Rhénanes et le Valais. *Revue d'Entomologie* **12** Supplement: 9-72.

Buysson H. du. 1893b: Description d'une espèce nouvelle d'élatéride. *Annali del Museo Civico di Storia Naturale di Genova* **33**: 417-418.

Buysson H. du. 1893c: [Description d'une nouvelle espèce d'élatérides d'Algérie]. *Bulletin de la Société Entomologique de France* **1893**: vii-viii.

Buysson H. du. 1894a: Coléoptères. Vol. 5: Elateridae. In: Fauvel C. A. A.: Faune gallo-rhénane ou description des insectes qui habitent la France, la Belgique, la Hollande, les Provinces Rhénanes et le Valais. *Revue d'Entomologie* **13** Supplement: 73-144.

Buysson H. du. 1894b: [Communications sur des élatérides]. *Bulletin de la Société Entomologique de France* 1893: cccxiv-cccxv.

Buysson H. du. 1895: [Sur des élatérides de Syrie, récoltés par M. C. Delagrange]. *Bulletin de la Société Entomologique de France* **1894**: cclxxxiii-cclxxxiv.

Buysson H. du. 1896a: Voyage de M. Charles Delagrange dans la Haute-Syrie en 1891. Élatérides. *Annales de la Société Entomologique de France* **65**: 38-42.

Buysson H. du. 1896b: Coléoptères. Vol. 5: Elateridae. In: Fauvel C. A. A.: Faune gallo-rhénane ou description des insectes qui habitent la France, la Belgique, la Hollande, les Provinces Rhénanes et le Valais. *Revue d'Entomologie* **15** Supplement: 145-224.

Buysson H. du. 1896c: Description d'un nouveau genre d'élatérides d'Espagne et rectification relative à l'Elater rubidus Cand. (Col.). *Bulletin de la Société Entomologique de France* **1895**: cdvi-cdviii.

Buysson H. du. 1897: Note sur le Cardiophorus discicollis Herbst var. Ganglbaueri var. nov. (Col.). *Bulletin de la Société Entomologique de France* **1897**: 259-260.

Buysson H. du. 1898a: Sur le Calostirus purpureus Poda var. atropilosus var. nov. (Col.). *Bulletin de la Société Entomologique de France* **1897**: 296-297.

Buysson H. du. 1898b: Description d'une nouvelle espèce d'élatérides (Col.). *Bulletin de la Société Entomologique de France* **1898**: 238-239.

Buysson H. du. 1898c: Description d'une espèce nouvelle d'élatérides (Col.). *Bulletin de la Société Entomologique de France* **1898**: 323.

Buysson H. du. 1899a: Note sur une variété de l'Agriotes lineatus L. et description d'une nouvelle espèce du genre Agriotes (Col.). *Bulletin de la Société Entomologique de France* **1899**: 20-22.

Buysson H. du. 1899b: Limonius turdus Cand. *Feuille des Jeunes Naturalistes* **29** [1898-1899]: 146.

Buysson H. du. 1899c: Description d'une espèce nouvelle d'élatérides (Col.). *Bulletin de la Société Entomologique de France* **1898**: 367-368.

Buysson H. du. 1899d: Remarques sur quelques élatérides (Col.). *Bulletin de la Société Entomologique de France* **1899**: 211-213.

Buysson H. du. 1899e: Description d'un élatéride nouveau du Nord de l'Afrique (Col.). *Bulletin de la Société Entomologique de France* **1899**: 213-214.

Buysson H. du. 1899f: Remarques sur quelques élatérides et description d'un genre nouveau du même groupe (Col.). *Bulletin de la Société Entomologique de France* **1899**: 279-282.

Buysson H. du. 1900a: Coléoptères. Vol. 5: Elateridae. In: Fauvel C. A. A.: Faune gallo-rhénane ou description des insectes qui habitent la France, la Belgique, la Hollande, les Provinces Rhénanes et le Valais. *Revue d'Entomologie* **19** Supplement: 225-264.

Buysson H. du. 1900b: Description d'une espèce et d'une variété nouvelles d'élatérides du Caucase (Col.). *Bulletin de la Société Entomologique de France* **1900**: 231-232.

Buysson H. du. 1901a: Description deux variétés nouvelles de coléoptères. *Revue d'Entomologie* **20**: 95.

Buysson H. du. 1901b: [new taxon, p. 101]. In: Warnier A.: *Catalogue des coléoptères de la faune gallo-rhénane*. Reims, 191 pp.

Buysson H. du. 1901c: Remarques sur quelques élatérides et description de deux espèces nouvelles (Col.). *Bulletin de la Société Entomologique de France* **1901**: 124-126.

Buysson H. du. 1902a: Notes sur le Cardiophorus argiolus Géné, var. (Col.). *Bulletin de la Société Entomologique de France* **1902**: 228.

Buysson H. du. 1902b: Coléoptères. Vol. 5: Elateridae. In: Fauvel C. A. A.: Faune gallo-rhénane ou description des insectes qui habitent la France, la Belgique, la Hollande, les Provinces Rhénanes et le Valais. *Revue d'Entomologie* **21** Supplement: 265-336.

Buysson H. du. 1902c: Remarques sur quelques élatérides du genre Cardiophorus et descriptions d'espèces nouvelles. *Annales de la Société Entomologique de France* **71** [1902-1903]: 422-439.

Buysson H. du. 1903: Description d'une nouvelle espèce d'élatéride du genre Cardiophorus (Col.). *Bulletin de la Société Entomologique de France* **1903**: 129-131.

Buysson H. du. 1904a: Notes sur quelques élatérides et description de deux espèces nouvelles. *Revue d'Entomologie* **23**: 5-8.

Buysson H. du. 1904b: Description d'un Agriotes nouveau. *Revue d'Entomologie* **23**: 42.

Buysson H. du. 1904c: Élatérides nouveaux et sous-genre nouveau (Col.). *Bulletin de la Société Entomologique de France* **1904**: 58-60.

Buysson H. du. 1905a: Sur quelques élatérides du Muséum. *Bulletin du Muséum National d'Histoire Naturelle* (Paris) **11**: 16-18.

Buysson H. du. 1905b: Description d'une espèce nouvelle d'élatérides (Col.). *Bulletin de la Société Entomologique de France* **1905**: 110.

Buysson H. du. 1905c: Descriptions d'espèces nouvelles d'élatérides (Col.). *Bulletin de la Société Entomologique de France* **1905**: 209-211.

Buysson H. du. 1905d: Coléoptères. Vol. 5: Elateridae. In: Fauvel C. A. A.: Faune gallo-rhénane ou description des insectes qui habitent la France, la Belgique, la Hollande, les Provinces Rhénanes et le Valais. *Revue d'Entomologie* **24** Supplement: 337-416.

Buysson H. du. 1906a: Note sur les élatérides récoltés en Perse par M. J. de Morgan en 1894. *Bulletin du Muséum National d'Histoire Naturelle* **12**: 17-21.

Buysson H. du. 1906b: Coléoptères. Vol. 5: Elateridae. In: Fauvel C. A. A.: Faune gallo-rhénane ou description des insectes qui habitent la France, la Belgique, la Hollande, les Provinces Rhénanes et le Valais. *Revue d'Entomologie* **25** Supplément: 417-494.

Buysson H. du. 1908: Vingt-deux ans après. Une promenade au Mont-Dore. *Revue Scientifique du Bourbonnais et du Centre de la France* **21**: 69-78.

Buysson H. du. 1911a: Tableaux analytiques des coléoptères de la faune franco-rhénane (France, Hollande, Belgique, Région rhénane, Valais). Familie LII, Elateridae. *Miscellanea Entomologica* **19**: 1-48.

Buysson H. du. 1911b: Documents sur quelques élatérides d'Égypte ou d'Afrique. *Bulletin de la Société Entomologique d'Égypte* **1910** (4): 134-145.

Buysson H. du. 1911c: Description d'une variété nouvelle de Cardiophorus. *Revue d'Entomologie* **28** [1909-1910]: 167-168.

Buysson H. du. 1912a: Matériaux pour servir à la faune des élatérides (Col.) de la Haute et Basse-Égypte. *Bulletin de la Société Entomologique d'Égypte* **1911**: 17-47.

Buysson H. du. 1912b: Les élatérides (recoltés en Perse par J. De Morgan). *Annales d'Histoire Naturelle* (Paris) **2** (1) *Entomologie*: 39-42.

Buysson H. du. 1912c: Note sur le Drasterius figuratus Germ. et un Cardiophorus nouveau. *Miscellanea Entomologica* **20**: 86-88.

Buysson H. du. 1912d: Cardiophorus bipunctatus Fabr. et ses variations. *Miscellanea Entomologica* **20**: 98-101.

Buysson H. du. 1912e: Élatérides nouveaux de l'Asie russe (Col.). *Bulletin de la Société Entomologique de France* **1912**: 128-131.

Buysson H. du. 1912f: Beschreibung neuer Elateriden-Arten und Varietäten. *Koleopterologische Rundschau* **1** [1911-1912]: 129-134.

Buysson H. du. 1912g: Contributions à la faune entomologique de Roumanie. Élatérides récoltés par M. A. L. Montandon dans les régions roumaines. *Buletinul Societății Romane de Ştiinţe* **21**: 133-143.

Buysson H. du. 1913a: Note sur le Cardiophorus ruficrus Brullé et description d'une espèce nouvelle. *Miscellanea Entomologica* **21**: 13-15.

Buysson H. du. 1913b: Tableaux analytiques des coléoptères de la faune franco-rhénane (France, Hollande, Belgique, Région rhénane, Valais). Familie LII, Elateridae. *Miscellanea Entomologica* **21**: 49-80.

Buysson H. du. 1914a: Drei neue Elateridenarten. *Koleopterologische Rundschau* **3**: 41-43.

Buysson H. du. 1914b: Tableaux analytiques des coléoptères de la faune franco-rhénane (France, Hollande, Belgique, Région rhénane, Valais). Familie LII, Elateridae. *Miscellanea Entomologica* **22**: 81-96.

Buysson H. du. 1915: Description d'un élatéride nouveau. *Miscellanea Entomologica* **23**: 9.

Buysson H. du. 1916: Tableaux analytiques des coléoptères de la faune franco-rhénane (France, Hollande, Belgique, Région rhénane, Valais). Familie LII, Elateridae. *Miscellanea Entomologica* **23**: 97-112.

Buysson H. du. 1917: Étude sur le Megathous algirinus Cand. (Col. Elateridae) et description d'une espèce nouvelle. *Bulletin de la Société d'Histoire Naturelle de l'Afrique du Nord* **8**: 11-15.

Buysson H. du. 1918a: Description d'un élatéride nouveau pour la faune française. *Miscellanea Entomologica* **24**: 27-28.

Buysson H. du. 1918b: Contributions à la faune du Maroc: Élatérides (Col.). *Bulletin de la Société d'Histoire Naturelle de l'Afrique du Nord* **9**: 109-112.

Buysson H. du. 1919: Contributions à la faune du Maroc et du Nord de l'Afrique Elatérides Col. (2e Note). *Bulletin de la Société d'Histoire Naturelle de l'Afrique du Nord* **10**: 114-116.

Buysson H. du. 1920a: Contribution à la faune du Maroc et du Nord de l'Afrique (3e Note). *Bulletin de la Société d'Histoire Naturelle de l'Afrique du Nord* **11**: 12-15.

Buysson H. du. 1920b: Melanotus castanipes var. nov. neopunctaticollis Buyss. (Col.). *Miscellanea Entomologica* **25**: 37-38.

Buysson H. du. 1924a: Tableaux analytiques des coléoptères de la faune franco-rhénane (France, Hollande, Belgique, Région rhénane, Valais). Familie LII, Elateridae. *Miscellanea Entomologica* **28**: 113-144.

Buysson H. du. 1924b: Indications relatives à la faune entomologique du Maroc (Élatérides). *Bulletin de la Société de Sciences Naturelles du Maroc* **4**: 206-210, 227.

Buysson H. du. 1925: Tableaux analytiques des coléoptères de la faune franco-rhénane (France, Hollande, Belgique, Région rhénane, Valais). Familie LII, Élatéridae. *Miscellanea Entomologica* **29**: 145-160.

Buysson H. du. 1926a: Note sur quelques élatérides de la collection Lubin. *Bulletin de la Société Linnéenne de Normandie* **2**: 47-48.

Buysson H. du. 1926b: Tableaux analytiques des coléoptères de la faune franco-rhénane (France, Hollande, Belgique, Région rhénane, Valais). Familie LII, Elateridae. *Miscellanea Entomologica* **29**: 161-208.

Buysson H. du. 1927: Tableaux analytiques des coléoptères de la faune franco-rhénane (France, Hollande, Belgique, Région rhénane, Valais). Familie LII, Elateridae. *Miscellanea Entomologica* **30**: 209-224.

Buysson H. du. 1928: Tableaux analytiques des coléoptères de la faune franco-rhénane (France, Hollande, Belgique, Région rhénane, Valais). Familie LII, Elateridae. *Miscellanea Entomologica* **31**: 225-240.

Buysson H. du. 1929: Tableaux analytiques des coléoptères de la faune franco-rhénane (France, Hollande, Belgique, Région rhénane, Valais). Familie LII, Elateridae. *Miscellanea Entomologica* **31**: 241-256.

Buysson H. du. 1930: Tableaux analytiques des coléoptères de la faune franco-rhénane (France, Hollande, Belgique, Région rhénane, Valais). Familie LII, Elateridae. *Miscellanea Entomologica* **32**: 257-272.

Cameron M. 1903: Description of a new European Soronia. *The Entomologist's Monthly Magazine* **39**: 97-98.

Caminero M. 1983a: Un nuevo Cebrionidae de la fauna española: Cebrio seguranus nov. sp. *Eos, Revista Española de Entomología* **59**: 27-30.

Caminero M. 1983b: Las especies de Cebrio de la región valenciana (Col.: Cebrionidae). Pp. 133-139. In: *Actas del I Congreso Ibérico de Entomología*, León.

Candèze E. 1857: Monographie des élatérides. I. *Mémoires de la Société Royale des Sciences de Liége* **12**: 8 + 400 pp., 7 pls.

Candèze E. 1859: Monographie des élatérides. II. *Mémoires de la Société Royale des Sciences de Liége* **14**: 543 pp., 7 pls.

Candèze E. 1860: Monographie des élatérides. III. *Mémoires de la Société Royale des Sciences de Liége* **15**: 512 pp., 5 pls.

Candèze E. 1861: Histoire des metamorphoses de quelques coléoptères exotiques. *Mémoires de la Société Royale des Science de Liége* **1861**: 325-410.

Candèze E. 1863: Monographie des Élatérides. IV. *Mémoires de la Société Royale des Sciences de Liége* **17**: 534 pp., 6 pls.

Candèze E. 1865: Élatérides nouveaux. (1). *Mémoires Couronnés et autres Mémoires publiés par l'Académie Royale des Sciences, des Lettres et des Beaux-Arts de Belgique* (*Classe Sciences*) **17** (1): 1-63.

Candèze E. 1869: Abänderungen vergebener Namen. *Coleopterologische Hefte* **5**: 122.

Candèze E. 1873: Insectes recueillis au Japon par M. G. Lewis pendant les années 1869-1871. Élatérides. *Mémoires de la Société Royale des Sciences de Liège* (2) **5**: [i]-ii + 32 pp.

Candèze E. 1874: Révision de la Monographie des élatérides. Premier fascicule. *Mémoires de la Société Royale des Sciences de Liège* (2) **4**, mémoire no. 1: viii + 218 pp.

Candèze E. 1878a: Relevé des élatérides recueillis dans les Iles Malaises, à la Nouvelle-Guinée et au Cap York, par MM. G. Doria, O. Beccari et L. M. D'Albertis. *Annali del Museo Civico di Storia Naturale di Genova* **12**: 99-143, 2 maps.

Candèze E. 1878b: Élatérides nouveaux. II. *Annales de la Société Entomologique de Belgique* **21**: 14, li-lxi, lxxv-lxxxv, cxxxv-cxliii, clxi-clxxii, clxxxix-cxcix.

Candèze E. 1879a: Élatérides de l'Amur. *Deutsche Entomologische Zeitschrift* **23**: 281-282, pl. II.

Candèze E. 1879b: [new taxa]. In: Candèze E. & Kraatz G.: Élatérides du Tarbagatai. *Deutsche Entomologische Zeitschrift* **23**: 283.

Candèze E. 1880: [new taxa]. In: Heyden L. F. J. D. von. 1880: Verzeichniss von Coleopteren aus Asturien, mit Beschreibungen neuer Arten von Candèze, v. Heyden, Kirsch, Kraatz , Stierlin. *Deutsche Entomologische Zeitschrift* **24**: 281-303.

Candèze E. 1882: Élatérides nouveaux. Troisième fascicule. *Mémoires de la Société Royale des Sciences de Liége* (2) **9**: ii + 117 pp.

Candèze E. 1885: Deux espèces nouvelles d'élatérides. *Notes from the Leyden Museum* **7**: 120-122, pl. 4.

Candèze E. 1886: [new taxa]. In: Heyden L. F. J. D. von & Kraatz G. 1886: Beiträge zur Coleopteren Fauna von Turkestan, namentlich des Alai-Gebirges. *Deutsche Entomologische Zeitschrift* **30** (1): 177-194.

Candèze E. 1888: Viaggio di Leonardo Fea in Birmania e regioni vicine XIV: Élatérides recueillis en Birmanie et au Ténasserim par M. L. Fea pendant les années 1885-1887. *Annali del Museo Civico di Storia Naturale di Genova* **26**: 667-689.

Candèze E. 1889a: Description de trois élaterides nouveaux du Musée de Leyde. *Notes from the Leyden Museum* **11**: 95-96.

Candèze E. 1889b: Élatérides nouveaux. Quatrième fascicule. *Annales de la Société Entomologique de Belgique* **33**: 67-123.

Candèze E. 1890: [Communication]. *Bulletin ou Comptes-Rendus des Séances de la Société Entomologique de Belgique* **1890**: cxx-cxxi.

Candèze E. 1891a: *Catalogue méthodique des élatérides connus en 1890*. Liége: H. Vaillant-Carmanne, [v]-xii + 246 pp.

Candèze E. 1891b: Viaggio di Leonardo Fea in Birmania e regioni vicine XXXVI. Élatérides recueillis en Birmanie en 1888 par M. L. Fea. 2e Article. *Annali del Museo Civico di Storia Naturale di Genova* **30**: 771-793.

Candèze E. 1892a: [Communication]. *Annales de la Société Entomologique de Belgique* **36**: 227-228.

Candèze E. 1892b: Insectes du Bengale. 24e Mémoire. Deuxième note sur les élatérides du Chota-Nagpore. *Annales de la Société Entomologique de Belgique* **36**: 480-495.

Candèze E. 1892c: Viaggio di Lamberto Loria nella Papuasia Orientale VI. Élatérides recueillis par M. Loria en 1889-1891, dans la Nouvelle-Guinée méridionale er régions voisines. *Annali del Museo Civico di Storia Naturale di Genova* **32**: 801-805.

Candèze E. 1893a: [new taxa]. In: Paulino d'Oliveira M.: *Catalogue des Insectes du Portugal. Coléoptères*. Coimbra: Imprensa da Universidade, [2] + 393 pp.

Candèze E. 1893b: Addition aux élatérides des Indes orientales. *Annales de la Société Entomologique de Belgique* **37**: 168-179.

Candèze E. 1893c: [Elater sanguinolentus var. paleatus]. *Annales de la Société Entomologique de Belgique* **37**: 286.

Candèze E. 1895: Élatérides nouveaux. Cinquième fascicule. *Mémoires de la Société Royale des Sciences de Liége* (2) **18** (4): 1-76.

Candèze E. 1897: Élatérides nouveaux. Sixième fascicule. *Mémoires de la Société Royale des Sciences de Liége* (2) **19** (2): 1-88.

Candèze E. 1900: Élatérides nouveaux. VII. *Annales de la Société Entomologique de Belgique* **44**: 77-101.

Canepari C. 1978: Alcuni Nephus della regione mediterranea e N. nigricans niloticus n. ssp. (Coleoptera Coccinellidae). *Bollettino della Società Entomologica Italiana* **110**: 166-170.

Canepari C. 1979a: Due nuove species di Scymnini paleartici: Nephus (Sidis) armeniacus e Scymnus (Pullus) folchinii (Coleoptera: Coccinellidae). *Doriana* **5** (232): 1-5.

Canepari C. 1979b: Hyperaspis dellabeffai n. sp., d'Italia (Coleoptera Coccinellidae). *Bollettino della Società Entomologica Italiana* **111**: 134-136.

Canepari C. 1985: [new taxa]. In: Canepari C., Fürsch H. & Kreissl E.: Die Hyperaspis-Arten Mittel-, West- und Südeuropa. Systematik und Verbreitung (Coleoptera Coccinellidae). *Giornale Italiano di Entomologia* **2**: 223-252.

Canepari C. 1986: Su alcuni Coccinellidi del'India e Nepal settentrionale del Museo di Storia Naturale di Ginevra (Coleoptera, Coccinellidae). *Revue Suisse de Zoologie* **93**: 21-36.

Canepari C. 1997: Coccinellidae (Coleoptera) from the Nepal Himalayas. *Stuttgarter Beiträge zur Naturkunde* (A) (Biologie), Stuttgart, Nr. 565, 65 pp.

Canepari C. 2000: The Nephus subgenus Sidis in Italy (Coleoptera Coccinellidae). *Bollettino della Società Entomologica Italiana* **132**: 49-62.

Canepari C. & Tedeschi M. 1977: Le sottospecie del Nephus quadrimaculatus (Herbst) (Coleoptera Coccinellidae). *Memoire della Società Entomologica Italiana* **55** [1976]: 101-105.

Cantor T. E. 1846: [new taxa]. In: Proceedings of Learned Societies. *The Annals and Magazine of Natural History* **17**: 270-293.

Cao C. & Pu T. 1985: A new species and a new record of Coccinellidae from Guangxi Zhuangzu Autonomous Region. *Entomotaxonimia* **7**: 273-275 (in Chinese).

Cao C. & Wang H. 1992: [new taxon]. In: Wang H. & Cao C.: A new species of the genus Afissula Kapur from Yunnan, China. *Entomotaxonomia* **14**: 66-68 (in Chinese).

Cao C. & Xiao N. 1984: New species of Coccinellidae from Yunnan, China. *Entomotaxonomia* **6**: 109-132 (in Chinese).

Cao Z. 1987: The species of Dermestes from Shaanxi. *Entomotaxonomia* **9**: 269-274.

Capra D. F. 1924: Nota sul genere Aclemmysa (Col. Endom.). *Bollettino della Società Entomologica Italiana* **56**: 138-141.

Capra F. 1924: Appunti sopra alcuni Coccinellidi Paleartici raccolti dal Marchese Giacomo Doria. *Annali del Museo Civico di Storia Naturale di Genova* **50**: 199-203.

Capra F. 1925: Appunti sistematici sui Coccinellidi. *Bollettino della Società Entomologica Italiana.* **57**: 136-139.

Capra F. 1926: Sulla posizione sistematica dell'Adalia alpina (Coleopt. Coccin.). *Annali del Museo Civico di Storia Naturale di Genova* **52**: 191-196.

Capra F. 1928: Le variazioni dell'Adaliopsis alpina (Villa) (Coleopt. Coccinellidae). *Bollettino della Società Entomologica Italiana* **60**: 6-10.

Capra F. 1929: Due nuovi Coccinellidi di Cirenaica. In: Risultati zoologici della Missione inviata della R. Società geografica italiana per l'esplorazione dell'oasi di Giarabub (1926-1927). *Annali del Museo Civico di Storia Naturale di Genova* **53** [1928-1930]: 241-242.

Capra F. 1931: Note sur alcuni Coccinellidi. *Bollettino della Società Entomologica Italiana* **63**: 15-20.

Capra F. 1945: Note sui Coccinellidi (Col.). II. Su Thea thurifera Sic. e sulla microfagia dei Coccinellidi. *Memoire della Società Entomologica Italiana* **24**: 91-96.

Capra F. & Fürsch H: 1967: [new taxa]. In: Fürsch H., Kreissl E. & F. Capra: Revision einiger europäischen Scymnus (s. str.)-Arten. *Mitteilungen der Abteilung für Zoologie und Botanik am Landesmuseum „Joanneum" in Graz* **28**: 207-259.

Carret A. 1904: Description d'un élatéride nouveau appartenant à la faune européenne (Col.). *Bulletin de la Société Entomologique de France* **1904**: 170-173.

Casey T. L. 1884: Revision of the Cucujidae of America North of Mexico. *Transactions of the American Entomological Society* **11**: 69-112.

Casey T. L. 1885: *Contribution to the descriptive and systematic coleopterology of North America. Part II.* Philadelphia 1-2: 61-198 [pp. 61-140 in 1884, pp. 141-198 in 1885].

Casey T. L. 1899: A revision of the American Coccinellidae. *Journal of the New York Entomological Society* **7**: 71-169.

Casey T. L. 1890: Coleopterological notices. II. *Annals of the New York Academy of Sciences* **5**: 307-504.

Casey T. L. 1891: Coleopterological notices. III. *Annals of the New York Academy of Science* **6**: 9-214.

Casey T. L. 1895: Coleopterological notices. VI. *Annals of the New York Academy of Sciences* **8**: 435-838.

Casey T. L. 1898: Studies in the Ptinidae, Ciodidae and Sphindidae of America. *Journal of the NewYork Entomological Society* **6**: 61-93.

Casey T. L. 1900: Review of the American Corylophidae, Cryptophagidae, Tritomidae and Dermestidae with other studies. *Journal of the New York Entomological Society* **8**: 51-172.

Casey T. L. 1908: Notes on the Coccinellidae. *The Canadian Entomologist* **40**: 393-421.

Casey T. L. 1916: Some random studies among the Clavicornia. Pp. 35-292. *Memoirs on the Coleoptera. VII.* Lancaster: The New Era Printing Company, 300 pp.

Casey T. L. 1924: Additions to the known Coleoptera of North America. *Memoirs on the Coleoptera. XI.* Lancaster: Lancaster Press, Inc., 347 pp.

Cate P. C. & Platia G. 1989: A new species of the genus Cidnopus Thomson from Greece: Cidnopus macedonicus n. sp. (Elateridae, Coleoptera). *Koleopterologische Rundschau* **59**: 185-188. [note: incomplete key corrected in *Koleopterologische Rundschau* **60**: 151].

Cate P. C. & Platia G. 1997: New species of Agriotes Eschscholtz (Coleoptera: Elateridae) from Greece, Turkey and Syria. *Zeitschrift der Arbeitsgemeinschaft Österreichischer Entomologen* **49**: 109-113.

Cate P. C., Platia G. & Schimmel R. 2002: New species and records of click beetles (Coleoptera: Elateridae) from Iran, with a checklist of known species. *Folia Heyrovskyana* **10**: 25-68.

Čechovský P. & Kubáň V. 1997: Hypoganus tibetis sp. nov. from China (Coleoptera: Elateridae: Ctenicerini). *Zeitschrift der Arbeitsgemeinschaft Österreichischer Entomologen* **49**: 115-117.

Čechovský P. & Platia G. 1991: A new species of the genus Anostirus Thomson, 1859, from Greece: Anostirus jarmilae n. sp. (Coleoptera, Elateridae, Ctenicerini). *Zeitschrift der Arbeitsgemeinschaft Österreichischer Entomologen* **43**: 13-16.

Cederhjelm N. J. 1798: *Faunae Ingricae Prodromus, exhibens methodicam descriptionem insectorum agri Petropolensis, &c.* Lipsiae: Hartknoch, 18 + 348 pp., 3 pls.

Chakraborty S. K. & Biswas S. 2003: Insecta: Coleoptera: Coccinellidae. Pp. 117-152. In: *State Fauna Series 9, Fauna of Sikkim (Part 3).* Kolkata: Zoological Survey of India, 411 pp.

Champenois A. 1898: Recherches sur la synonymie des Trichodes de l'ancien monde. *L'Abeille Journal d'Entomologie* **29**: 115-116.

Champenois A. 1900: Synopsis des espèces paléarctiques du genre Clerus Müller (Trichodes Herbst). *L'Abeille Journal d'Entomologie* **30**: 1-46.

Champion G. C. 1887: On the priority of various generic names in use in British Coleoptera. *The Entomologist's Monthly Magazine* **23**: 227-230.

Champion G. C. 1913: Notes on various Central American Coleoptera, with description of new genera and species. *Transactions of the Entomological Society of London* **1913** (1): 58-169.

Champion G. C. 1914: Revision of the Mexican and Central American Chauliognathinae (Fam. Telephoridae), based on the genital armature of the males. *Transactions of the Entomological Society of London* **1914**: 128-168, pls III-VIII.

Champion G. C. 1915: Revision of the Mexican and Central American Telephorinae (Fam. Telephoridae), with descrition of new species. *Transactions of the Entomological Society of London* **1915**: 16-146, pls III-IX.

Champion G. C. 1919a: The malacoderm genera Prionocerus and Idgia, and their sexual characters (Coleoptera). *The Annals and Magazine of Natural History* (9) **3**: 325-372, pls 11-12.

Champion G. C. 1919b: Notes on the African and Asiatic species of Melyris, Fab. (sensu lato), with an account of their sexual characters. *The Annals and Magazine of Natural History* (9) **4**: 157-219.

Champion G. C. 1919c: Some Indian Coleoptera 1. *The Entomologist's Monthly Magazine* **55**: 236-246.

Champion G. C. 1920a: Some Indian Coleoptera (2). *The Entomologist's Monthly Magazine* **56**: 68-77.

Champion G. C. 1920b: Some Indian Coleoptera (4). *The Entomologist's Monthly Magazine* **56**: 241-249.

Champion G. C. 1920c: Notes on various African and Asiatic species of Hapalochrus, Er., with an account of their accessory ♂-characters (Coleoptera). *The Annals and Magazine of Natural History* (9) **6**: 177-201, 249-266, 305-327, pl. 8.

Champion G. C. 1921a: Some Indian Coleoptera (5). *The Entomologist's Monthly Magazine* **57**: 70-78, 104-110, 145-146.

Champion G. C. 1921b: Notes on various African and Asiatic species of Laius, Guérin, with an account of their accessory ♂-characters (Coleoptera). *The Annals and Magazine of Natural History* (9) **7**: 322-343, pl. 8.

Champion G. C. 1921c: Revision of the African species of Hedybius, Er., and its allies, with an account of their accessory ♂-characters (Coleoptera). *The Annals and Magazine of Natural History* (9) **8**: 449-494, pls 13-14.

Champion G. C. 1922a: Some Indian Coleoptera (7). *The Entomologist's Monthly Magazine* **58**: 68-76.

Champion G. C. 1922b: Some Indian Coleoptera (8). *The Entomologist's Monthly Magazine* **58**: 126-131, 145-152.

Champion G. C. 1922c: Some Indian Coleoptera (9). *The Entomologist's Monthly Magazine* **58**: 214-220; 257-271.

Champion G. C. 1922d: Revision of the S. African species of Dinometopus, Troglops, Chalicorus, Colotes, Helcogaster, and the allied genera, with an account of their accessory ♂-characters (Coleoptera). *The Annals and Magazine of Natural History* (9) **10**: 309-358, pls 4-6.

Champion G. C. 1923a: A revision of the Malayan and Indian species of the melyrid subfamily Carphurinae represented in the Hope collection at Oxford and in the British Museum in London (Coleoptera). *The Annals and Magazine of Natural History* (9) **12**: 1-54, pl. 1.

Champion G. C. 1923b: Note on the Malacoderm- genus Amauronia Westwood. *The Entomologist's Monthly Magazine* **59**: 13.

Champion G. C. 1923c: Some Indian Coleoptera (10). *The Entomologist's Monthly Magazine* **59**: 77-80.

Champion G. C. 1923d: A revision of the eastern species of Nosodendridae (Coleoptera). *The Annals and Magazine of Natural History* (9) **12**: 578-591.

Champion G. C. 1924a: Some Indian Coleoptera (14). *The Entomologist's Monthly Magazine* **60**: 159-172.

Champion G. C. 1924b: Some Indian Coleoptera (15). *The Entomologist's Monthly Magazine* **60**: 234-247.

Champion G. C. 1924c: Some Indian Coleoptera. *The Annals and Magazine of Natural History* (9) **13**: 249-264.

Champion G. C. 1925a: Studies in Phalacridae. S. and E. African forms. *The Annals and Magazine of Natural History* (9) **15**: 35-53.

Champion G. C. 1925b: Some Indian (and Tibetan) Coleoptera (16). *The Entomologist's Monthly Magazine* **61**: 101-112.

Champion G. C. 1925c: Some Indian (and Tibetan) Coleoptera (17). *The Entomologist's Monthly Magazine* **61**: 169-181.

Champion G. C. 1925d: Some Indian (and Tibetan) Coleoptera (18). *The Entomologist's Monthly Magazine* **61**: 260-273.

Champion G. C. 1925e: Studies in Phalacridae (II.) – Asiatic and tropical American forms (Coleoptera). *The Annals and Magazine of Natural History* (9) **16**: 601-621.

Champion G. C. 1926a: Some Indian (and Tibetan) Coleoptera (19.). *The Entomologist's Monthly Magazine* **62**: 118-137.

Champion G. C. 1926b: Some Indian (and Tibetan) Coleoptera (20.). *The Entomologist's Monthly Magazine* **62**: 194-210.

Champion G. C. 1926c: Some Indian Coleoptera (21.). *The Entomologist's Monthly Magazine* **62**: 243-256.

Champion G. C. 1927: Some Indian Coleoptera (22). *The Entomologist's Monthly Magazine* **63**: 78-85.

Chao Y. & Lee H. 1966: A study of Chinese Trogoderma Berthold (Coleoptera, Dermestidae). *Acta Zootaxonomia Sinensia* **3**: 245-252, 2 pls.

Chapin E. A. 1919: New species of Coleoptera (Fam. Cleridae) from the Philippine and neighboring regions, collected by prof. Charles F. Baker. *Proceedings of the Biological Society of Washington* **32**: 225-234.

Chapin E. A. 1924: Classification of the Philippine components of the coleopterous family Cleridae. *The Philippine Journal of Science* **25**: 159-286, pls 1-5.

Chapin E. A. 1926: On some Coccinellidae of the tribe Telsimiini with descriptions of new species. *Proceedings of Biological Society of Washington* **39**: 129-134.

Chapin E. A. 1927: On some Asiatic Cleridae (Col.). *Proceedings of the Biological Society of Washington* **40**: 19-22.

Chapin E. A. 1940: New genera and species of lady-beetles related to Serangium Blackburn (Coleoptera: Coccinellidae). *Journal of the Washington Academy of Sciences* **30**: 263-272.

Chapin E. A. 1962: Pseudoscymnus, a new genus of Asiatic Scymnini (Coleoptera: Coccinellidae). *Psyche* **69**: 50-51.

Chapin E. A. 1965: The genera of the Chilocorini (Coleoptera: Coccinellidae). *Bulletin of the Museum of Comparative Zoology, Harvard University* **133** (4): 227-271.

Chapin E. A. & Ahmad R. 1966: A. new species of coccinellid feeding on scale insects in West Pakistan. *Entomophaga* **11**: 213-215.

Chapuis F. 1876: Famille des érotyliens, des endomychides et des coccinellides. In: Lacordaire T. & Chapuis F. (eds.): *Histoire naturelle des insectes. Genera des coléoptères ou exposé méthodique et critique de tous genres proposés jusqu'ici dans cet ordre d'insectes. Tome 12.* Paris: Roret, 424 pp.

Charpentier T. de. 1818: [new taxa]. In: Germar E. F.: Vermischte Bemerkungen über einige Käferarten. *Magazin der Entomologie* **3**: 228-260, pl. 3.

Charpentier T. de. 1825: *Horae entomologicae, adjectis tabulis novem coloratis*. Wratislaviae: A. Gosohorsky, xvi + 255 pp., 9 pls.

Chassain J. 1979a: Description de deux espèces nouvelles de Cardiophorus du Proche-Orient (Col. Elateridae). *L'Entomologiste* **35**: 70-77.

Chassain J. 1979b: Insects of Saudi Arabia. Coleoptera: Fam. Elateridae (1ère Partie). *Fauna of Saudi Arabia* **1**: 193-211.

Chassain J. 1983: Insects of Saudi Arabia. Coleoptera: Fam. Elateridae (2nd part). *Fauna of Saudi Arabia* **5**: 129-143.

Chassain J. 1984: La faune de l'île égéenne de Santorin. Part 4. Elateridae (Coleoptera). *Entomologische Blätter* **80**: 91-100.

Chassain J. 1985: Description d'une espèce nouvelle d'Athous du Sud-Ouest de la France (Col. Elateridae). *L'Entomologiste* **41**: 65-68.

Chassain J. 1985: Note homonymique. *Entomologische Blätter* **81**: 142.

Chassain J. 1991: Description de deux espèces paléarctiques nouvelles du genre Calais Castelnau (Coleoptera, Elateridae). *Nouvelle Revue d'Entomologie* (N. S.) **8**: 211-215.

Chassain J. 1992: Contribution à la connaissance du sous-genre Brachygonus Buysson, 1912 avec description d'une espèce nouvelle (Col. Elateridae). *L'Entomologiste* **48**: 323-335.

Chassain J. 1998: Note sur le genre Limoniscus avec désignation d'un lectotype et description d'une espèce nouvelle (Coleoptera, Elateridae). *Le Coléoptériste* **33**: 119-122.

Chaudoir M. 1845: Notices entomologiques sur le gouvernement et la ville de Kiew. *Bulletin de la Société Impériale des Naturalistes de Moscou* **18** (3): 158-213.

Chen Z.-L. & Yin H.-F. 2003: A new species of the genus Dysides from Indonesia (Coleoptera, Bostrichidae). *Acta Zootaxonomica Sinica* **28**: 113-115.

Cherepanov A. I. 1956: Nasekomye Tuvinskoy oblasti. Pp. 35-77. In: *Trudy Biologicheskogo Instituta Zap.-Sib. Filiala AN SSSR, Vol. 1, zoologicheskiy*. Novosibirsk. **Bitte Zitat überprüfen!! Bitte für mich kopieren!!**

Cherepanov A. I. 1957: *Zhuki-shchelkuny zapadnoy Sibiri* (*Coleoptera, Elateridae*). Novosibirsk: Knizhnoe Izdatel"stvo, 382 pp.

Cherepanov A. I. 1965: Novyy vid roda Agriotes Eschz. (Coleoptera, Elateridae) [A new species of the genus Agriotes Eschz. (Coleoptera, Elateridae)]. Pp. 35-38. In: *Novye i maloizvestnye vidy fauny Sibiri* [New and little known species of the Siberian fauna]. Novosibirsk: Nauka, 79 pp.

Cherepanov A. I. 1966a: Novyy vid shchelkuny roda Limonius (Coleoptera, elateridae) [A new species of the click-beetle genus Limonius (Coleoptera, Elateridae)]. Pp. 23-26. In: Cherepanov A. I. (ed.): *Novye vidy fauny Sibiri i prilegayushchikh regionov*. Novosibirsk: Nauka. 154 pp.

Cherepanov A. I. 1966b: Novyy vid shchelkuna roda Ischnodes (Coleoptera, Elateridae) [A new species of the click-beetle genus Ischnodes (Coleoptera, Elateridae)]. Pp. 27-28. In: Cherepanov A. I. (ed.): *Novye vidy fauny Sibiri I prilegayushchikh regionov* Novosibirsk: Nauka, 154 pp.

Chernyshev S. E. 1996: Novyy vid zhukov-krasnokrylov (Coleoptera, Lycidae) s Dalnego Vostoka Rossii [A new species of lycid beetles (Lycidae) from the Russian Far East. *Zoologicheskiy Zhurnal* **75**: 1092-1095.

Chernyshev S. E. 1998: Novyy vid myagkotelok roda Crudosilis (Coleoptera, Cantharidae) iz Tuvy [A new species of soldier-beetles of the genus Crudosilis (Coleoptera, Cantharodae) from Tuva]. *Zoologicheskiy Zhurnal* **77** (2): 240-242.

Chevrolat L. A. A. 1831: [new taxon, livr. 14: pl. 45]. In: Guérin-Méneville F. E.: [1829-1838]: *Iconographie du règne animal de G. Cuvier, ou représentation d'après la nature de l'une des espèces les plus remarquables et souvent non encore figurées, de chaque genre d'animaux. Avec un texte descriptif mis au courant de la science. Ouvrage pouvant servir d'atlas à tous les traités de zoologie. Iconographie du règne animal de G. Cuvier. II. Planches des animaux invertebrés. Insectes*. Paris: J. B. Bailière, 110 pls.

Chevrolat L. A. A. 1832: Description du genre Dryophilus. *Magasin de Zoologie* (F. E. Guérin-Méneville) (Cl. IX) pl. 3, [2] pp.

Chevrolat L. A. A. 1833: Description du genre Sphindus. *Revue Entomologique* (G. Silbermann) **1** (8): pl. 6. [pp. unnumb.].

Chevrolat L. A. A. 1834: *Coléoptères du Mexique*. 3e fascicule [species 49-70]. Strasbourg: G. Silbermann [unnumb.].

Chevrolat L. A. A. 1837: [new taxa, livr. 44, pl. 41]. In: Guérin-Méneville F. E.: [1829-1838]: *Iconographie du règne animal de G. Cuvier, ou représentation d'après la nature de l'une des espèces les plus remarquables et souvent non encore figurées, de chaque genre d'animaux. Avec un texte descriptif mis au courant de la science. Ouvrage pouvant servir d'atlas à tous les traités de zoologie. Iconographie du règne animal de G. Cuvier. II. Planches des animaux invertebrés. Insectes*. Paris: J. B. Bailière, 110 pls.

Chevrolat L. A. A. 1837b: [new names]. In: Dejean P. F. M. A.: *Catalogue des coléoptères de la Collection de M. le Comte Dejean. Troisième édition revue, corrigée et augmentée.* Livraison 5. Paris: Méquignon-Marvis Père et Fils, 385-503 pp.

Chevrolat L. A. A. 1840: Description de quelques coléoptères de la Galice et du Portugal provenant d'envois de M. Deyrolle fils. *Revue Zoologique par la Société Cuvierienne* **3**: 8-18.

Chevrolat L. A. A. 1842: Description de quelques térédiles de l'Afrique australe, du voyage de M. Drege. *Revue Zoologique par la Société Cuvierinne* **5**: 276-278.

Chevrolat L. A. A. 1843: Description de vingt-quatre nouvelles espèces de Terediles, pour faire suite à la monographie des Clairones, de M. le docteur Klug. *Annales de la Société Entomologique de France* (2) **1**: 31-42.

Chevrolat L. A. A. 1844: [new taxa]. In: Guérin-Méneville F. E.: [1829-1844]: *Iconographie du règne animal de G. Cuvier, ou représentation d'après la nature de l'une des espèces les plus remarquables et souvent non encore figurées, de chaque genre d'animaux. Avec un texte descriptif mis au courant de la science. Ouvrage pouvant servir d'atlas à tous les traités de zoologie. III. Texte explicatif. Insectes*. Paris: J. B. Baillière, 576 pp. [issued in parts].

Chevrolat L. A. A. 1845: [Hippodamia]. Pp. 622-623. In: D'Orbigny C.: *Dictionnaire universel d'Histoire Naturelle. Tome sixième.* Paris: Renard, Martinet et Cie, 792 pp.

Chevrolat L. A. A. 1852: Description de trois coléoptères. *Revue et Magasin de Zoologie Pure et Appliquée* (2) **4**: 578-579.

Chevrolat L. A. A. 1854: Coléoptères de Syrie. Premier Mémoire. – Suite. *Revue et Magasin de Zoologie Pure et Appliquée* (2) **8**: 432-437.

Chevrolat L. A. A. 1859: Description de dix coléoptères nouveaux d'Algérie. *Revue et Magasin de Zoologie Pure et Appliquée* (2) **11**: 380-390.

Chevrolat L. A. A. 1860: Description de coléoptères nouveaux d'Algérie. *Revue et Magasin de Zoologie Pure et Appliquée* (2) **12**: 75-82.

Chevrolat L. A. A. 1860: Description de coléoptères nouveaux d'Algérie. *Revue et Magasin de Zoologie Pure et appliquée* (2) **12**: 208-212.

Chevrolat L. A. A. 1861: Description de coléoptères nouveaux d'Algérie. *Revue et Magasin de Zoologie Pure et Appliquée* (2) **13**: 147-155, 264-270.

Chevrolat L. A. A. 1862: Observationes et notes synonymiques. *Annales de la Société Entomologique de France* (4) **1** [1861]: 389-392.

Chevrolat L. A. A. 1863: [new taxa]. In: Grenier A. J. F.: *Catalogue des coléoptères de France, et matériaux pour servir à la faune des coléoptères français.* Paris: L. Toinon, iv + 79 + 135 pp.

Chevrolat L. A. A. 1864: Coléoptères de l'île de Cuba. (Suite) (1). Notes, synonymies at descriptions d'espèces nouvelles. Quatriéme mémoire. Famille des Histériens, Phalacrides, Nitidulaires, Trogositaires, Colydiens, Rhyzodides, Cucujides, Mycétophagides, Dermestins, Byrrhiens et Chélonariides. *Annales de la Société Entomologiqu de France* (4) **3** [1863]: 589-620.

Chevrolat L. A. A. 1865: Descriptions de coléoptères d'Espagne, nouveaux ou peu connus. 1er Mémoire. *Revue et Magasin de Zoologie Pure et Appliquée* (2) **17**: 347-352.

Chevrolat L. A. A. 1866a: Descriptions de coléoptères nouveaux ou peu connus. *Revue et Magasin de Zoologie Pure et Appliquée* (2) **18**: 100-108.

Chevrolat L. A. A. 1866b: Description de coléoptères nouveaux ou peu connus. *Revue et Magasin de Zoologie Pure et Appliquée* (2) **18**: 321-326.

Chevrolat L. A. A. 1867: Coléoptères de l'île de Cuba. (Suite). Notes, synonymies et descriptions d'espèces nouvelles. Septième mémoire. Famille des Buprestides, Throscides, Eucnémides et Élatérides. *Annales de la Sociéte Entomologique de France* **4** (7): 571-616.

Chevrolat L. A. A. 1872: Description de plusieurs coléoptères d'Espagne et de deux Curculionites du Nord de l'Afrique. *Annales de la Société Entomologique de France* (5) **2**: 409-412.

Chevrolat L. A. A. 1873: Description de quelques espèces nouvelles de coléoptères de Syrie. *Annales de la Société Entomologique de France* (5) **3**: 203-206.

Chevrolat L. A. A. 1874a: Révision des Cébrionides. *Annales de la Société Entomologique de France* (5) **4**: 9-38.

Chevrolat L. A. A. 1874b: Catalogue des Clerides de la collection de M. A. Chevrolat. *Revue et Magasin de Zoologie Puré et Appliqué* **37**: 252-331.

Chevrolat L. A. A. 1874c: Révision des Cébrionides. 2. partie. *Annales de la Société Entomologique de France* (5) **4**: 363-426.

Chevrolat L. A. A. 1875: Révision des Cébrionides. 3e et dernière partie. *Annales de la Société Entomologique de France* (5) **4** [1874]: 507-540.

Chevrolat L. A. A. 1876a: *Mémoire sur la famille des Clerides.* Paris: L. Buquet, 51 pp.

Chevrolat L. A. A. 1876b: [Communication]. *Bulletin de la Société Entomologique de France*, **1876**: clxxiv-clxxxvii.

Chevrolat L. A. A. 1880: Taneroclerus [sic] girodi (sp. n.) Col. *Bulletin de la Société Entomologique de France* **43**: 31-32.

Chevrolat L. A. A. 1882: [Diagnoses de trois coléoptères nouveaux et des remarques sur une quatrième espèce]. *Bulletin de la Société Entomologique de France* **1882**: iv-v.

Chittenden F. H. 1897: Some little-known insects affecting stored vegetable products. A collection of articles detailing certain original observation made upon insects of this class. *Bulletin, Bureau of Entomology, United States Department of Agriculture* **8**: 45 pp.

Chitty A. J. 1893: A new variety of Telephorus figuratus. *The Entomologist's Monthly Magazine* **29**: 143.

Chobaut A. 1894: Description de Xyletinus Leprieuri, n. sp. *Annales de la Société Entomologique de France* **63**: 107-108.
Chobaut A. 1895: Notes sur quelques Danacaea et description d'une espèce nouvelle d'Anthicus (Col.). *Bulletin de la Société Entomologique de France* **1895**: clv-clvii.
Chobaut A. 1896a: Dissia albopilosa, genre nouveau et espèce nouvelle de la famille des Clambidae, tribu des Cybocephalini (Col.). *Bulletin de la Société Entomologique de France* **1896**: 167-168.
Chobaut A. 1896b: Descriptions de deux espèces nouvelles de Cebrio d'Algérie et de la femelle du C. gypsicola Graëlls (Col.). *Bulletin de la Société Entomologique de France* **1896**: 399-402.
Chobaut A. 1897a: Description de cinq variétés nouvelles de coléoptères de la faune circa-méditerranéenne. *Miscellanea Entomologica* **5**: 38-39.
Chobaut A. 1897b: Description de deux coléoptères nouveaux du Mzab (Sud-Algérien). *Bulletin de la Société Entomologique de France* **1897**: 170-171.
Chobaut A. 1897c: Description de deux espèces nouvelles de coléoptères du Mzab (Sud-algérien). *Bulletin de la Société Entomologique de France* **1897**: 204-206.
Chobaut A. 1897d: Descriptions de trois coléoptères nouveaux du Mzab (Sud-Algérien). *Bulletin de la Société Entomologique de France* **1897**: 239-242.
Chobaut A. 1897eb: Description de deux coléoptères nouveaux du sud de l'Algérie. *Bulletin de la Société Entomologique de France* **1897**: 277-279.
Chobaut A. 1898: Description de quelques espèces et variétés nouvelles de coléoptères algériens. *Revue d'Entomologie* **17**: 74-88.
Chobaut A. 1899a: Description d'un genre nouveau de cébrionides (Col.) du Sahara septentrional. *Bulletin de la Société Entomologique de France* **1899**: 22-23.
Chobaut A. 1899b: Description d'un Ernobius nouveaux (Col.) de la France méridionale. *Bulletin de la Société Entomologique de France* **1899**: 104-105.
Chobaut A. 1899c: Description d'un Ernobius nouveaux (Col.) de la France méridionale (2e note). *Bulletin de la Société Entomologique de France* **1899**: 117-119.
Chobaut A. 1901: Description d'un ptinide nouveau du midi de la France (Col.). *Bulletin de la Société Entomologique de France* **1901**: 298-299.
Chobaut A. 1904: Description de deux espèces nouvelles de coléoptères de l'Arabie. *Bulletin de la Société Entomologique de France* **1904**: 243-245.
Chobaut A. 1906: Description d'un lathridiide nouveau (Col.) du nord de l'Afrique. *Wiener Entomologische Zeitung* **25**: 256.
Chobaut A. 1914: [new taxa]. In: H. Caillol.: *Catalogue des coléoptères de Provence d'après des documents recueillis et groupés. Vol. 3.* Marseille: Société Linnéenne de Provence, 594 pp.
Chobaut A. 1924a: Une variété française de Lasioderma semirufulum Reitt. (Col., Anobiidae). *Bulletin de la Société Entomologique de France* **1924**: 103.
Chobaut A. 1924b: Quatre espèces nouvelles de Thorictus (Col. Thorictidae) du Maroc. *Bulletin de la Société des Sciences Naturelles du Maroc* **4**: 105-107.
Chobaut A. 1924c: Deux espèces nouvelles de Thorictus (Col. Thorictidae) de Tunisie. *Bulletin de la Société Entomologique de France* **1924**: 215-217.
Chûjô M. 1936a: Descriptions of five new species of Erotylidae from Formosa. *Transactions of the Natural History Society of Formosa* **26**: 211-214.
Chûjô M. 1936b: Description of a new species of Bostrychidae (Coleoptera) and notices of some others. *Transactions of the Natural History Society of Formosa* **26**: 407-410.
Chûjô M. 1938: Some additions and revisions to the Japanese Endomychidae (Coleoptera). *Transactions of the Natural History Society of Formosa* **28**: 394-406.
Chûjô M. 1940a: Descriptions of two new erotylid-beetles from Japan Proper. *Nippon no Kôchū* **3**: 78-81.
Chûjô M. 1940b: Description of a new endomychid-beetle from Formosa. *Transactions of the Natural History Society of Formosa* **30**: 216-217.
Chûjô M. 1940c: Description of a new languriid-beetle from Formosa. *Transactions of the Natural History Society of Formosa* **30**: 218-219.
Chûjô M. 1941a: Descriptions of three new erotylid-beetles from Japan proper and Formosa. *Transactions of the Kansai Entomological Society* **11**: 1-7.
Chûjô M. 1941b: Descriptions of some new Erotylidae (Coleoptera) the Japanese Empire. *Kontyû* **15**: 10-21.

Chûjô M. 1941c: Descriptions of two new species of the Japanese endomychid-beetles. *Mushi* **13**: 81-83.
Chûjô M. 1941d: Descriptions of six new erotylid-beetles from Formosa and the Marianna Islands. *Mushi* **13**: 84-92.
Chûjô M. 1942: Descriptions of a new endomychid-beetles from Formosa. *Mushi* **14**: 79-80.
Chûjô M. 1952: Erotylid-beetles of Aomori-Prefecture (the northern-end of Honshū, Japan), collected by Mr. K. Shimoyama (II). *The Entomological Review of Japan* **5**: 17-18.
Chûjô M. 1955: Erotylid-beetles of Aomori-Prefecture (the northern-end of Honshu, Japan), collected by Mr. K. Shimoyama (III). *The Entomological Review of Japan* **6**: 43-44.
Chûjô M. 1955: Studies on the languriid-beetles (VI). *Mushi* **28**: 49-51.
Chûjô M. 1959: Coleoptera of the Loo-choo Archipelago (I). *Memoirs of the Faculty of Liberal Arts and Education, Kagawa University* **2** (69): 1-15.
Chûjô M. 1960: Description of a new erotylid-beetle from the Island Tsushima, Japan. Studies on the erotylid-beetles (10). *Niponius* **1**: 1-4.
Chûjô M. 1964a: A systematic list of the erotylidae from Formosa (Coleoptera). *Folia Entomologica Hungarica (Series Nova)* **17**: 213-231.
Chûjô M. 1964b: On the erotylid beetles from Formosa preserved in the Zoological Department of the Hungarian Natural History Museum. *Annales Historico-Naturales Musei Nationalis Hungarici (Pars Zoologica)* **56**: 413-417.
Chûjô M. 1966: Some Bostrichidae, Languriidae and Ciidae (Col.) from the Philippine and Bismarck Islands collected by the Noona Dan Expedition. Noona Dan Papers No. 36. *Entomologiske Meddelelser* **34**: 521-532.
Chûjô M. 1967a: Languriid-beetles from Thailand, Laos, Viet-Nam, S. China, Is. Hainan, Formosa and Ryukyus. Studies on the languriid-beetles (12). *Memoirs of the Faculty of Education Kagawa University* **2** (149): 1-42.
Chûjô M. 1967b: Description of a new erotylid-beetle from Formosa. Studies on the erotylid-beetles (19). *Niponius* **2**: 69-72.
Chûjô M. 1968a: Description of a new erotylid-beetle from Japan. Studies on the erotylid-beetles (20). *Niponius* **2**: 73-75.
Chûjô M. 1968b: Erotylid beetles from South-China, Hainan, Taiwan and the Ryukyus. Studies on the erotylid beetles (20). *Pacific Insects* **10**: 539-550.
Chûjô M. 1969: *Erotylidae (Insecta: Coleoptera). Fauna Japonica.* Tokyo: Academic Press of Japan, xii + 316 pp.
Chûjô M. 1973: An unknown Bostrichidae collected from Tsushima, Japan. *Tôkaishibuhô, Japan Entomological Society, Nagoya* **26**: 10 (in Japanese).
Chûjô M. 1975: Ergebnisse der Bhutan-Expedition 1972 des Naturhistorischen Museums in Basel. Coleoptera: Fam. Erotylidae, Languriidae & Helotidae. *Entomologica Basiliensia* **1**: 279-292.
Chûjô M. 1976: Erotylid-beetles from Formosa (Taiwan) collected by Mr. M. Sakai, Ehime University, Japan, in 1972. Studies on the erotylid-beetles (24). *Transactions of the Shikoku Entomological Society* **13**: 1-8.
Chûjô M. 1977: Languriid-beetles from Formosa (Taiwan) collected by Mr. M. Sakai, Ehime University, Japan, in 1972. Studies on the languriid-beetles (15). *Transactions of the Shikoku Entomological Society* **13**: 77-82.
Chûjô M. & Chûjô Mt. 1990: A Catalogue of the Erotylidae (Coleoptera) from the Old World (excl. the Ethoipian Region) III. *Esakia* **29**: 1-67.
Chûjô Mt., Chûjô M. & Lee C. E. 1993a: Endomychidae from Korea (Insecta, Coleoptera). *Esakia* **33**: 95-98.
Chûjô Mt., Chûjô M. & Lee C. E. 1993b: Erotylidae from Korea (Insecta, Coleoptera). *Esakia* **33**: 99-108.
Chûjô M. & Hisamatsu S. 1964: 19. Family Nitidulidae. Pp. 217-219. In: Kira T. & Umesao T. (eds): *Nature and Life in Southeast Asia. Vol. 3.* Kyoto: Fauna and Flora Research Society.
Chûjô M. & John H. 1961: Eine neue Spezies von Aphanocephalus Woll. *Niponius* **1** (12): 1-2.
Chûjô M. & Kiuchi M. 1963: Description of a new erotylid-beetle from Japan. Studies on the erotylid-beetles (16) by M. Chûjô. *Niponius* **2**: 31-33.
Chûjô M. & Kiuchi M. 1974: Descriptions of three new species of endomychid-beetles from Loo-choo Archipelago. *Bulletin of the Japan Entomological Academy* **8**: 1-7.
Chûjô M. & Miyatake M. 1963: [new taxa]. Pp. 11-12. In: Miyatake M.: The genus Horniolus Weise of Japan and the Ryukyu Islands (Coleoptera: Coccinellidae). *Transactions of the Shikoku Entomological Society* **8**: 6-12.

Chûjô M. & Nakane T. 1950: [new taxon]. In: Chûjô M.: Erotylid-beetles of Aomori - Prefecture (the northern-end of Honshū, Japan), collected by Mr. K. Shimoyama. *The Entomological Review of Japan* **5**: 15-19.

Chûjô M. & Ôhira H. 1965: Elaterid- and dicronychid-beetles from Aomori Pref., Japan, collected by Mr. Kensaku Shimoyama. *Memoirs of the Faculty of Liberal Arts and Education, Kagawa University* **2** (132): 1-32 pp., 1 pl.

Chûjô M. & Satô M. 1970: Coleoptera of the Loo-Choo Archipelago (II). *Memoirs of Faculty of Liberal Arts & Education, Kagawa University, part II*, no. **192**: 27-32.

Chûjô M. & Satô M. 1972: Coleoptera of the Loo-Choo Archipelago (IV). 35. Family Lampyridae. *Memoirs of the Faculty of Liberal Arts & Education, Kagawa University, part II*, nr. **208**: 23-25.

Chûjô M. & Shibata T. 1963: Description of a new erotylid-beetle from the Island Amami-Oshima, Loochoos. Studies on the erotylid-beetles (14). *Niponius* **2**: 1-2.

Cobos A. 1950a: Datos para el catálogo de los coleópteros de España. Especies de los alrededores de Málaga. *Boletín de la Sociedad Española de Historia Natural* **47** [1949]: 563-609.

Cobos A. 1950b: Materiales para el catálogo de los Coleópteros de España: Famls. Buprestidae y Elateridae.- Especies de la provincia de Almeria, I. *Boletín de la Real Sociedad Española de Historia Natural* **48**: 151-157.

Cobos A. 1958: Primera contribución al estudio de los Elateridae (Coleoptera) del norte de África. Descripciones y rectificaciones diversas. *Bulletin de la Société des Sciences Naturelles et Physiques du Maroc* **38**: 51-57.

Cobos A. 1961: Sobre la posición sistemática del género Potergus Bonvouloir y revisión de las categorías supragénericas de la familia Throscidae (Coleoptera). *Bulletin de l'Institut Royal des Sciences Naturelles de Belgique* **37** (35): 1-6.

Cobos A. 1961: Exploración entomológica del cedral de Tizi-Ifri (macizo del Iguelmalet, Rif central, Marruecos). Coleópteros. *Archivos del Instituto de Aclimatación* **10**: 63-98, pl. 1.

Cobos A. 1963: Dos nuevas especies de "Throscus" paleárticos comunicados por el Naturhistorisches Museum (Viena), y nueva descripción de "T. orientalis" Bonvouloir, con comentarios acerca de su validez específica. *Miscelánea Zoológica* **1** (5): 49-54.

Cobos A. 1964: Materiales para el estudio de la familia Eucnemidae. Primera parte. *Eos, Revista Española de Entomología* **40**: 289-435.

Cobos A. 1966a: Estudios sobre Throscidae, II (Col. Sternoxia). *Eos, Revista Española de Entomología* **42**: 311-351.

Cobos A. 1966b: Nuevo género y especie de Agrypninae (Col., Elateridae). *Annales de la Société Entomologique de France* (N. S.) **2**: 651-653.

Cobos A. 1969: Tercer viaje entomológico a Marruecos. Coleópteros Sternoxia. *Archivos del Instituto de Aclimatación de Almería* **14**: 5-20.

Cobos A. 1969: Estudios sobre Throscidae, III (Col. Sternoxia). *Archivos del Instituto de Aclimatación* **14**: 39-63.

Cobos A. 1970: Ensayo monográfico sobre los Cardiophorus Esch., 1829 de Canarias (Col. Elateridae). *Eos, Revista Española de Entomología* **45** [1969]: 29-96.

Cobos A. 1978: Sobre Cardiophorus atlánticos extracanarios del subgénero Coptostethus (Coleoptera, Elateridae). Pp. 145-150. In: *Contribución al estudio de la Historia Natural de las Islas Salvajes. Resultados de la expedición científica 'Agamenon 76' (23 de febrero - 3 de marzo de 1976).* Tenerife: Museo de Ciencias Naturales del Cabildo Insular de Santa Cruz de Tenerife.

Cobos A. 1979: Una notable nueva especie de Cardiophorus (subgen. Coptostethus Wollaston) de Canarias (Col. Elateridae). *Boletín de la Asociación Española de Entomología* **3**: 199-202.

Cobos A. 1983: Tres nuevas especies de Cardiophorus Eschs. [sic] (Col. Elateridae) de Canarias. *Boletín de la Asociación Española de Entomología* **6** [1982]: 237-243.

Cobos A. 1985: Una nueva especie de Cebrio (Col., Cebrionidae) de Almería. *Boletín de la Asociación Española de Entomología* **9**: 331-333.

Coiffait H. 1984: Description d'un genre nouveau et de 11 espèces nouvelles d'Anommatidae. *Bulletin de la Société d'Histoire Naturalle de Toulouse* **120**: 43-45.

Cockerell T. D. A. 1900: Epismilia. *Science Gossip* **1900**: 606.

Cockerell T. D. A. 1903: The coccinellid genus Smilia, Weise. *The Canadian Entomologist* **35**: 38.

Colenso W. 1888: A description of a new species of Coccinella found in New Zealand. *Transactions and Proceedings of the New Zealand Institute* **2** [1887]: 40-42.

Comolli A. 1837: *De Coleopteris novis ac rarioribus minusve cognitis provinciae Novocomi.* Ticini: Fusi et Socii, vi + 7-54 pp.

Compte A. 1988: Una nueva especie de Cebrionidae: Cebrio elenacompteae nov. sp. (Insectos, Coleópteros). *Eos, Revista Española de Entomologia* **64**: 57-62.

Constantin R. 1971: Description de trois nouveaux Malthodes des Pyrenées-Occidentales et d'Espagne (Col. Cantharidae). *Annales de la Société Entomologique de France* (N. S.) **7**: 959-965.

Constantin R. 1975: Description de nouvelles espèces de Malthinus de France continentale, de Corse et d'Espagne (Col. Cantharidae). *L'Entomologiste* **31**: 80-88.

Constantin R. 1979: Révision des Malthinus d'Afrique du Nord (les espèces à élytres stries-ponctués) (Coleoptera - Cantharidae). *Annales de la Société Entomologique de France* (N. S.) **15**: 617-634.

Constantin R. 1983: Dasytides nouveaux de Crète et de Turquie (Coleoptera, Dasytidae). *Entomologica Basiliensia* **8**: 329-333.

Constantin R. 1988: Révision des Danacaea d'Afrique du Nord. I: Danacaea distincta (Lucas) et les espèces voisines. II. Étude anatomo-morphologique de l'appareil génital mâle (Coleoptera, Dasytidae). *Nouvelle Revue d'Entomologie* (N. S.) **4** [1987]: 341-358.

Constantin R. 1990: Danacaea coiffaiti, une nouvelle espèce de France, et note faunistique sur Danacaea longiceps Mulsant & Rey (Coleoptera, Melyridae). *Nouvelle Revue d'Entomologie* (N. S.) **7**: 63-68.

Constantin R. 1992: Description d'un Dasytes nouveau d'Espagne et notes faunistiques sur quelques Dasytinae aragonais (Coleoptera Melyridae). *Nouvelle Revue d'Entomologie* (N. S.) **8** [1991]: 399-406.

Constantin R. & Menier J. J. 1987: Étude d'un remarquable Melyridae aptère des Iles Canaries: Gietella fortunata, n. gen., n. sp., type d'une sous-famille nouvelle Gietellinae (Coleoptera, Cleroidea). *Revue Française d'Entomologie* (N. S.) **9**: 53-63.

Contarini N. B. 1847: Notizie sulla fauna terrestre e particolarmente sulla ornitologia del Veneto estuario con cenno sul passaggio degli uccelli e sulla caccia. Pp. 157-191. In: *Venezia e le sue lagune* 2 (1). Venezia: Antonelli.

Coombs C. W. & Woodroffe G. E. 1955: A revision of the British species of Cryptophagus (Herbst) (Coleoptera: Cryptophagidae). *Transactions of the Royal Entomological Society of London* **106**: 237-282.

Cooper M. C. 1982: The species of the genus Pria Stephens (Coleoptera: Nitidulidae). *Zoological Journal of the Linnean Society* **75**: 327-390.

Cooper W. 1865: Descriptions of new species of Canadian Coleoptera. *The Canadian Naturalist* (N. S.) **2**: 60-63.

Cornelius C. 1863: Entwicklungsgeschichten. *Entomologische Zeitung* (Stettin) **24**: 113-125.

Cornelius C. 1884: Verzeichnis der Käfer von Elberfeld und dessen Nachbarschaft, angeordnet in der Hauptgrundlage nach dem Catalogus Coleopterorum Europae et Caucasi, Auctoribus Dr. L. von Heyden, E. Reitter et J. Weise, Editio Tertia. *Jahresbericht des Naturwissenschaftlichen Vereins Elberfeld* **6**: 1-61.

Corporaal J. B. 1926: Some new species of Oriental Cleridae and remarks on known species. *Encyclopedie Entomologique Série B. Coleoptera* I [1925-1926]: 175-188, pl. II.

Corporaal J. B. 1926: On some Indian Cleridae (Coleoptera) (Part I). New species of Cleridae from British India and Burma. *Indian Forest Records, Entomology Series* **12** (8): 209-216.

Corporaal J. B. 1932: Notizen über palaearctische Cleriden (10te Mitteilung über Cleriden). *Tijdschrift voor Entomologie* **75** *Supplement*: 136-140.

Corporaal J. B. 1935: Two new Cleridae (Col.) from India. *Stylops* **4**: 34-35.

Corporaal J. B. 1939a: Some Cleridae from India, Burma, and Ceylon with descriptions of new species and notes on others. *Indian Forest Records* (*New Series*) *Entomology* **6** (2): 17-39.

Corporaal J. B. 1939b: Studies on Callimerus and allied genera (Col.). *Tijdschrift voor Entomologie* **82**: 182-195.

Corporaal J. B. 1939c: Revision of the Thaneroclerinae (Cleridae, Coll.). *Bijdragen tot de Dierkunde* **27**: 348-359.

Corporaal J. B. 1941: Notes on some Cleridae in the Hamburg Zoological Museum. *Tijdschrift voor Entomologie* **84**: 359-361.

Corporaal J. B. 1948: Nineteen notes on systematics and synonymy. *Entomologische Berichten* **12**: 242-246.

Corporaal J. B. 1949a: Fifth series of notes on systematics and synonymy. *Entomologische Berichten* **12**: 355-357.

Corporaal J. B. 1949b: Seventh series of notes on systematics and synonymy. *Entomologische Berichten* **12**: 420-422.

Corporaal J. B. 1950: Cleridae. In: *Coleopterorum Catalogus Supplementa. Pars 23*. Gravenhage: W. Junk, 373 pp.

Corporaal J. B. & Pic M. 1940: Les Callimerus du groupe de C. prasinatus Lewis (Col.). *Tijdschrift voor Entomologie* **83**: 189-192.

Corporaal J. B. & Vári L. 1946: Remarques sur quelques espèces paléarctiques du genre Trichodes Herbst (Col., Cleridae). *Miscellanea Entomologica* **43**: 77-86.

Corporaal J. B. & van der Wiel P. 1949: Studies in Cladiscus and allied genera (Col.). *Tijdschrift voor Entomologie* **91**: 180-198.

Costa A. 1847a: Specie nuove e rare d'insetti della montagne del Matese. *Annali del Accademia degli Aspiranti Naturalisti Napoli* (2) **1**: 89-131. [note: *Memorie Entomologiche*, pp. 9-51 is a reprinting. Citations of some authors list the pagination of the reprinting].

Costa A. 1847b: Descrizione di alcuni coleotteri del Regno di Napoli. *Annali del Accademia degli Aspiranti Naturalisti Napoli* (2) **1**: 134-162. [note: *Memorie Entomologiche*, pp. 52-80 is a reprinting. Citations of some authors list the pagination of the reprinting].

Costa A. 1849: Coleotteri Trimeri. Famiglia degli Coccinellidea. Pp. 1-112 pp., 7 pls. [foglio 33-46]. In: *Fauna dell regno di Napoli ossia enumerazione di tutti gli animali, che abitano le diverse regioni di questo regno e le acque che le bagnano contenente la descrizione de'nuovi o poco esattamente conosciuti noc figure ricavate da originali viventi e dipinte al nature di Oronzio-Gabrielle Costa. Coleotteri. Parte. I.a.* Napoli: Gaetano Sautto, xii + 364 pp. [note: families separately paginated].

Costa A. 1850: Coleotteri Trimeri. Famiglia degli Endomichidei – Endomichidea. Pp. 1-15 + [1]. [foglio 37-48]. In*: Fauna del Regno di Napoli ossia enumerazione di tutti gli animali che abitano le diverse regioni di questo regno e le acque che le bagnano contenente la descrizione de'nuovo o poco esattamente conosciuti con figure ricavate da originali viventi e dipinte al Naturale. Coleotteri. Parte. I.a.* Napoli: Gaetano Sautto, xii + 364 pp. [note: families separately paginated].

Costa A. 1857: De quibusdam novis Insectorum generibus iconibusque illustratis. *Memorie del Accademia di Scienze di Napoli* **2**: 219-233.

Costa A. 1858: *Ricerche entomologiche sopra i Monti Partenii nel principato ulteriore*. Napoli, 29 pp., 1 pl.

Costa A. 1875: Lampyroidea syriaca, sp. nov., Col. *Bulletin de la Société Entomologique de France* **1875**: clxix-clxx.

Costa A. 1882: Relazione di un viaggio nelle Calabrie per ricerche zoologiche fatto nella state del 1876. *Atti della Reale Accademia delle Scienze Fisiche e Matematiche Napoli* **9** (6) [1881]: 1-62 + [1] pp., 1 pl.

Costa A. 1883: Notizie ed osservazioni sulla Geo-fauna Sarda. Memoria Seconda. Risultamento di ricerche fatte in Sardegna nella primavera del 1882. *Atti della Reale Accademia Scienze Fisiche e Matematiche di Napoli* (2) **1** (2): 1-109 + [1].

Costa A. 1884: Diagnosi di nuovi Artropodi trovati in Sardegna. *Bullettino della Società Entomologica Italiano* **15** [1883]: 332-341.

Costa O. G. 1839a: Fauna di Aspromonte e sue adjacenze. *Atti della Reale Accademie delle Science Fisiche e Matematiche di Napoli* **4**: 61-173, 2 pls.

Costa O. G. 1839b: Monographia degli insecti dell'ulivo e delle olive. P. 104. In: *Corrispondenza Zoologica destinata a diffondere nel rogno delle due Sicilie tutto ciò che si va discuoprendo entro e fuori Europa risguardante la Zoologia in generale redatta. Vol. 1*. Napoli: Azzolino, 194 pp., 12 pls.

Coye C. H. 1870: Description de coléoptères nouveaux. *L'Abeille, Mémoires d'Entomologie* **6** [1869]: 368-389.

Creutzer C. 1796a: [new taxa]. In: Panzer G. W. F.: *Faunae Insectorum Germanicae initia oder Deutschlands Insecten. Heft 35*. Nürnberg: Felsecker, 24 pp. + 24 pls.

Creutzer C. 1796b: [new taxa]. In: Panzer G. W. F.: *Faunae Insectorum Germanicae initia oder Deutschlands Insecten. Heft 36*. Nürnberg: Felsecker, 24 pp. + 24 pls.

Creutzer C. 1797: [new taxa]. In: Panzer G. W. F.: *Faunae Insectorum Germanicae initia oder Deutschlands Insecten. Heft 39*. Nürnberg: Felsecker, 24 pp. + 24 pls.

Creutzer C. 1799: *Entomologische Versuche*. Wien: K. Schaumburg und Comp., 144 pp., 3 pls.

Croissandeau J. A. 1891: Holoparamecus argodi n. sp. *Le Coléoptériste* **1890-1891**: 153-154, pl. V.

Crotch G. R. 1867: On the Coleoptera of the Azores. *Proceedings of the Zoological Society of London* **1867**: 359-391, pl. XXIII.

Crotch G. R. 1870: Notes on British Coleoptera. *The Entomologist* **7**: 7-11.

Crotch G. R. 1871: *List of Coccinellidae.* Cambridge: Privately printed, 8 pp.

Crotch G. R. 1873a: *Check list of the Coleoptera of America, north of Mexico.* Salem: 136 pp.

Crotch G. R. 1873b: A list of Erotylidae collected by Edward M. Janson, in the vicinity of Santo Domingo, Chontales, Nicaragua, with descriptions of new genera and species. *Cistula Entomologica* **1** [1869-1876]: 141-150.

Crotch G. R. 1873c: A descriptive list of Erotylidae collected by Geo. Lewis, Esq., in Japan. *The Entomologist's Monthly Magazine* **9**: 184-189.

Crotch G. R. 1873d: Revision of the Coccinellidae of the United States. *Transactions of American Entomological Society* **4**: 363-382.

Crotch G. R. 1874a: *A revision of the coleopterous family Coccinellidae.* London: E. W. Janson, 311 pp.

Crotch G. R. 1874b: Descriptions of new species of Coleoptera from the Pacific Coast of the United States. *Transactions of the American Entomological Society* **5** [1874-1876]: 73-80.

Crotch G. R. 1876: A revision of the coleopterous family Erotylidae. *Cistula Entomologica* **1** [1869-1876]: 377-572.

Csiki E. 1899: Coleopterologische Notizen. *Természetrajzi Füzetek* **22**: 247-250.

Csiki E. 1900a: A magyarországi cleridák. *Rovartani Lapok* **7**: 117-124.

Csiki E. 1900b: Magyarország endomychidái. *Rovartani Lapok* **7**: 144-151.

Csiki E. 1900c: Die Endomychiden-Gattung Milichius Gerstaecker. *Természetrajzi Füzetek* **23**: 375-377.

Csiki E. 1900d: Coleoptera nova in collectione Musei Nationali Hungarici. *Természetrajzi Füzetek* **23**: 400-403.

Csiki E. 1901a: Endomychiden und Erotyliden aus Ceylon gesammelt von Dr. Walter Horn. *Természetrajzi Füzetek* **25** [1902]: 26-27.

Csiki E. 1901b: Bogarak. Coleopteren. Pp. 77-120. In: Horváth G. (ed.): *Zichy Jenö Gróf Harmadik Ázsiai Utazása. II. Kötet. Zichy Jenö Gróf Harmadik Ázsiai utazásának állattani Eredményei szerkesztette Dr. Horváth Géza. Dritte asiatische Forschungsreise des Grafen Eugen Zichy. Band II. Zoologische Ergebnisse der dritten asiatischen Forschungsreise des Grafen Eugen Zichy.* Budapest: Victor Hornyánszky, xli + 472 pp., 28 pls.

Csiki E. 1903: Magyarországi új bogarak. (Coleoptera nova ex Hungaria). *Annales Musei Nationalis Hungarici* **1**: 441-446.

Csiki E. 1906: Eucnemidae. Pp. 20-21. In: Heyden, L. von, Reitter E. & Weise J. (eds.): *Catalogus Coleopterorum Europae, Caucasi et Armeniae Rossicae. Editio secunda.* Berlin, Paskau, Caen: R. Friedländer & Sohn, [6] pp. + 750 columns + pp. 750-774.

Csiki E. 1907: Coleoptera nova in Museo Nationali Hungarico. *Annales Historico-Naturales Musei Nationalis Hungarici* **5**: 574-578.

Csiki E. 1910a: Endomychidae. Pars 12. In: Schenkling S. (ed.): *Coleopterorum Catalogus. Vol. 16.* Berlin: W. Junk, 68 pp.

Csiki E. 1910b: Coleoptera nova in Museo Nationali Hungarico. III. *Annales Histotico-Naturales Musei Nationalis Hungarici* **8**: 444.

Csiki E. 1911: Egy új Hylaia-faj a Balkánról. *Rovartani Lapok* **18**: 74-75.

Csiki E. 1937: Neue Endomychiden. *Entomologisches Nachrichtenblatt* (Troppau) **11**: 173-174.

Csiki E. 1940a: Csiki Ernö állattani kutatásal Albániában. - Explorationes zoologicae ab E. Csiki in Albania peractae. *Á Magyar Tudományos Akadémia Balkán-kutatásainak tudományos eredményei* **1**: 207-288.

Csiki E. 1940b: Coleoptera nova ex Hungaria. *Matematikai és Természettudományi Értesitö* **59**: 918-919.

Csiki E. 1951: Fauna zhukov gory Radna. *Acta Biologica Academiae Scientiarum Hungaricae* **2**: 119-168.

Curtis J. H. 1824: *British Entomology; being illustrations and descriptions of the Genera of Insects found in Great Britain and Ireland: containing Coloured Figures from Nature of the most rare and beautiful species, and in many instances of the plants upon which they are found. Vol. I.* London: J. Curtis, [8] pp. + 1-50 pls. and sheets.

Curtis J. H. 1826: *British Entomology; being illustrations and descriptions of the genera of Insects found in Great Britain and Ireland: containing coloured figures from naturee of the most rare and beautiful species, and in many instances of plants upon which they are found. Vol. III.* London: J. Curtis, [8] pp. + 99-146 pls. and sheets.

Curtis J. H. 1827: *British Entomology; being illustrations and descriptions of the genera of Insects found in Great Britain and Ireland: containing coloured figures from naturee of the most rare and beautiful species, and in many instances of plants upon which they are found. Vol. IV.* London: J. Curtis, [8] pp. + 147-194 pls. and sheets.

Curtis J. H. 1828: *British Entomology; being illustrations and descriptions of the genera of Insects found in Great Britain and Ireland: containing coloured figures from naturee of the most rare and beautiful species, and in many instances of plants upon which they are found. Vol. V.* London: J. Curtis, [8] pp. + 195-241 pls. and sheets.

Curtis J. H. 1829: *British entomology; being illustrations and descriptions of the genera of insects found in Great Britain and Ireland; containing coloured figures from nature of the most rare and beutiful species, and in many instances of the plants upon which they are found. Vol. VI.* London: J. Curtis, [8] pp. + 242-289 pls. and sheets.

Curtis J. H. 1830: *British Entomology; being illustrations and descriptions of the genera of insects found in Great Britain and Ireland: containing coloured figures from nature of the most rare and beautiful species, and in many instances of the plants upon which they are found. Vol. VII.* London: J. Curtis, [8] pp. + 290-337 pls. and sheets.

Curtis J. H. 1833: Characters of some undescribed genera and species, indicated in the "Guide to an Arrangement of British Insects". *The Entomological Magazine* **1**: 186-199.

Curtis J. H. 1835: *British Entomology; being illustrations and descriptions of the genera of insects found in Great Britain and Ireland; containing coloured figures from nature of the* most *rare and beautiful species, and in many instances of the plants upon which they are found. Vol XII.* London: J. Curtis, [8] pp. + 530-577 pls and sheets.

Curtis J. H. 1838: *British Entomology; being illustrations and descriptions of the genera of insects found in Great Britain and Ireland: containing coloured figures from nature of the most rare and beautiful species, and in many instances of the plants upon which they are found. Vol. XV.* London: J. Curtis, [4] pp. + 674-721 pls. and sheets + [2] pp.

Curtis J. H. 1839: Descriptions, &c. of the insects collected by Captain P. P. King, R.N.F.R.S. & L.S. in the Survey of the Straits of Magellan. *Transactions of the Linnaean Societ of London* **18**: 181-205, 1 pl.

Curtis J. H. 1840: Descriptions &c. of some rare or interesting indigenous insects. *The Annals of Natural History* **5**: 274-282.

Curtis J. H. 1854: Critical remarks upon the British Elateridae, with descriptions of some of the species. *Transactions of the Entomological Society of London* (N. S.) **3**: 10-17, pl. II.

Czempinski P. de 1778: *Dissertatio inauguralis zoologico-medica, sistens totius regni animalis genera, in classes et ordines Linnaeana methodo digesta, praefixa cuilibet classi terminorum explicatione, quam annuente inclyta facultate medica in antiquissima ac celeberrima Universitate Vindobonensi publicae disquisitioni submittit.* Viennae: Trautner, 16 + 122 pp.

Czwalina G. 1884: Malthodes lunifer n.sp. *Deutsche Entomologische Zeitschrift* **28**: 272.

Dahlgren G. 1968: Beiträge zur Kenntnis der Gattung Rhagonycha (Col. Cantharidae). *Entomologische Blätter* **64**: 93-124.

Dahlgren G. 1972: Beiträge zur Kenntnis der Gattung Rhagonycha (Col. Cantharidae) II. *Entomologische Blätter* **68**: 129-149.

Dahlgren G. 1974: Cantharis gemina n.sp. (Coleoptera: Cantharidae). *Mitteilungen der Entomologischen Gesellschaft Basel* (N. F.) **24**: 145-146.

Dahlgren G. 1975: Zur Taxonomie der Gattungen Rhagonycha, Pseudocratosilis und Cratosilis (Col. Cantharidae). *Entomologische Blätter* **71**: 100-112.

Dahlgren G. 1976: Zur Taxonomie der Gattungen Rhagonycha und Cantharis (Col. Cantharidae). *Mitteilungen der Entomologischen Gesellschaft Basel* (N. F.) **26**: 88-90.

Dahlgren G. 1978: Zwei neue Arten der Rhagonycha femoralis-Gruppe (Col. Cantharidae). *Mitteilungen der Entomologischen Gesellschaft Basel* (N. F.) **28**: 12-15.

Dahlgren G. 1979: Eine neue Art der Rhagonycha femoralis-Gruppe (Col. Cantharidae). *Mitteilungen der Entomologischen Gesellschaft Basel* (N. F.) **29**: 62-64.

Dahlgren G. 1985a: Beiträge zur Kenntnis der Gattung Rhagonycha (Col. Cantharidae) III. *Entomologische Blätter* **81**: 85-90.

Dahlgren G. 1985b: Zwei neue griechische Cantthariden. *Entomologische Blätter* **81**: 163-165.

Dajoz R. 1962: Notes sur les Lathridiidae (Coléoptères) 2. Description d'espèces nouvelles ou peu connues et localités nouvelles pour la faune d'Amérique du sud. *Biologie de l'Amérique Australe* **1**: 287-293.

Dajoz R. 1963: Description d'un nouveau coléoptère Lathridiidae des Indes: Cartoderopsis alinae n.g., n.sp. *Bulletin de la Société Entomologique de France* **68**: 65-67.

Dajoz R. 1967: Contribution à l'étude des coléoptères Lathridiidae du Chili. *Biologie de l'Amérique Australe* **3**: 587-609.

Dajoz R. 1968: Révision des Colydiidae anophtalmes de la faune paléarctique (Insectes, Coléoptères). Première partie: genres Anommatus et Abromus. *Revue d'Ecologie et Biologie du Sol* **5**: 239-269.

Dajoz R. 1969a: Étude des Corticaria du groupe sylvicola Ch. Brisout et description d'une espèce nouvelle (Coléoptères, Lathridiidae). *Revue d'Ecologie et Biologie du Sol* **6**: 93-97.

Dajoz R. 1969ba: Révision des Colydiidae anophthalmes de la faune Paléarctique II. - Notes complémentaires sur les genres Anommatus et Abromus. *Revue d'Ecologie et Biologie du Sol* **6**: 377-389.

Dajoz R. 1970a: Étude des coléoptères Lathridiidae de l'Afrique intertropicale. *Annales du Musée Royal de L'Afrique Centrale, Sciences Zoologiques* **184**: 1-49.

Dajoz R. 1970b: Coleoptera: Lathridiidae. *South African Animal Life* **14**: 230-244.

Dajoz R. 1970c: Contribution à l'étude des coléoptères Lathridiidae de la Peninsule Ibérique et du Maroc. *Revue d'Ecologie et Biologie du Sol* **7**: 255-275.

Dajoz R. 1970d: Contribution à la faune du Congo (Brazzaville) Mission A. Villiers et A. Descarpentries XCVII. Coléoptères Lathridiidae et Mycetophagidae. *Bulletin de l'I. F. A. N.* (A) **32**: 433-441.

Dajoz R. 1971: Coléoptères Colydiidae nouveaux ou peu connus de la région paléarctique. *L'Entomologiste* **27**: 83-101.

Dajoz R. 1972: Ancylopus melanocephalus (Oliv.) (Coléoptères, Endomychidae), nouveau pour la Bulgarie et description de deux espèces nouvelles. *Bulletin du Muséum National d'Histoire Naturelle* (Paris) (3) **64**: 1043-1048.

Dajoz R. 1973a: Les espèces balkaniques du genre Dima avec la description d'une espèce nouvelle (Coleoptera, Elateridae). *Cahiers des Naturalistes* **29**: 103-105.

Dajoz R. 1973b: Nouveaux coléoptères Colydiidae d'Europe et de Turquie. *L'Entomologiste* **29**: 146-156.

Dajoz R. 1973c: Coléoptères Lathridiides de Madagascar et Mascareignes. *Bulletin du Muséum National d'Histoire Naturelle* (Paris) (3) **64**: 1049-1056.

Dajoz R. 1975: Coleoptera: Lathridiidae de Ceylan. *Entomologica Scandinavica, Supplementa* **4**: 199-207.

Dajoz R. 1977: Coléoptères Colydiidae et Anommatidae Paléarctiques. *Faune de l'Europe et du Bassin Méditerranéen* 8. Paris: Masson, 275 pp.

Dajoz R. 1979: Coléoptères Cerylonidae nouveaux ou peu connus. *Bulletin Mensuel de la Société Linnéenne de Lyon* **48**: 441-452.

Dajoz R. 1980: Quelques coléoptères Lathridiidae du Cameroun. *Bulletin Mensuel de la Société Linnéenne de Lyon* **49**: 286-290.

Dajoz R. 1981: Note sur les coléoptères d'un champignon Ascomycète de Tunisie. *L'Entomologiste* **37**: 203-211.

Dajoz R. 1984a: Note sur les Anommatus de Yugolavie et description de trois especes nouvelles (coléoptères, Anommatidae). *Cahiers des Naturalistes* **40**: 81-84.

Dajoz R. 1984b: Description d'une nouvelle espèce d' Anommatus (Coleoptera, Anommatidae). Pp. 361-362. In: Nonveiller G. (ed.): *Fauna Durmitora.* Sveska 1. Prilozi endogejskoj fauni tvrdokrilaca Durmitora (Insecta, Coleoptera). Titograd: Crnogorska Akademija Nauka i Umjetnosti. Posebna Izdanja, Knjiga XVIII, Odjeljenje Prirodnih Nauka, Knjiga 11.

Dajoz R. 1987: La faune des coléoptères endogés et troglobionts de Serbie, 5. Note sur quelques espèces d'Anommatus de Serbie et descriptions de deux espèces nouvelles (Insecta, Coleoptera). *Zbornik Radova Fauni Srbije* **4**: 185-189.

Dajoz R. 1989: Le genre Dastarcus Walker (Col. Colydiidae) en Turquie et à Chypre. *L'Entomologiste* **45**: 35-38.

Dalla Torre K. W. von. 1879: Die Käferfauna von Oberösterreich. Systematisches Verzeichnis der in Oberösterreich bisher beobachteten Käfer. *Jahresberich des Vereins für Naturkunde in Oesterreich ob der Enns zu Linz* **10**: 1-125.

Dalla Torre K. W. von. 1880: Die Käferfauna von Oberösterreich. Systematisches Verzeichnis der in Oberösterreich bisher beobachteten Käfer. *Jahresbericht des Vereins für Naturkunde in Oesterreich ob der Enns zu Linz* **11**: 1-81.

Dalla Torre K. W. von. 1911a: Cebrionidae. Pars 25. In: Schenkling S. (ed.): *Coleopterorum Catalogus. Vol. 11*. Berlin: W. Junk, 18 pp.

Dalla Torre K. W. von. 1911b: Nosodendridae, Byrrhidae, Dermestidae. Pars 33. In: Schenkling S. (ed.): *Coleopterorum Catalogus. Vol. 14*. Berlin: W. Junk, 96 pp.

Dalman J. W. 1817: [new taxa]. In: Schönherr C. J. 1817: *Appendix ad Synonymiam Insectorum Bd I, Theil 3, sistens descriptiones novarum specierum*. Scaris: Lewerentz, 266 pp., 2 pls.

Dalman J. W. 1824: *Ephemerides Entomologicae*. Holmiae: Norstedt, 36 pp.

Dalman J. W. 1825: Om Insekter inneslutna i Kopal, jemte beskrifning på några deribland förekommande nya slägten och arter. *Kongelige Vetenskaps-Akademiens Handlingar* **46**: 375-410, 1 pl.

Daniel K. 1903: [new taxa]. In: Daniel K. & Daniel J.: Kleinere Abhandlungen, kritische und synonymische Bemerkungen. *Münchener Koleopterische Zeitschrift* **1** [1902-1903]: 249-255.

Daniel K. 1904: [new taxon]. In: Daniel K. & Daniel J.: Neue paläarktische Koleopteren. *Münchener Koleopterische Zeitschrift* **2**: 76-93.

Deelder C. L. 1942: Revison of the Erotylidae (Coleoptera) of the Leyden Museum. *Zoologische Mededeelingen* **24**: 49-115.

DeGeer C. de. 1774: *Mémoires pour servir à l'histoire des insectes. Tome quatrième*. Stockholm: P. Hesselberg, xii + 456 + [1] pp., 19 pls.

DeGeer C. de. 1775: *Mémoires pour servir à l'histoire des insectes. Tome cinquième*. Stockholm: Pierre Hesselberg, vii + [1] + 448 pp., 16 pls.

Dejean P. F. M. A. 1821: *Catalogue de la collection de coléoptères de M. le Baron Dejean*. Paris: Crevot, viii + 136 + [2] pp.

Dejean P. F. M. A. 1833: *Catalogue de la collection des coléoptères de M. le Comte Dejean. Deuxième édition*. [Livraison 1 (pp. 1-96), Livraison 2 (pp. 97-176)]. Paris: Méquignon-Marvis Père et Fils.

Dejean P. F. M. A. 1834: *Catalogue des coléoptères de la collection de M. le Comte Dejean. 3e Livraison*. Paris: Méquignon-Marvis Pères et Fils, pp. 177-256.

Dejean P. F. M. A. 1835: *Catalogue des coléoptères de la collection de M. le Comte Dejean. 4e Livraison*. Paris: Méquignon-Marvis Pères et Fils, pp. 257-360.

Dejean P. F. M. A. 1836: *Catalogue des coléoptères de la collection de M. le Comte Dejean. Troisième édition revue, corrigée et augmentée*. Livraisons 1-4. Paris: Méquignon-Marvis Père et Fils, 1-384 pp.

Dejean P. F. M. A. 1837: *Catalogue des coléoptères de la Collection de M. le Comte Dejean. Troisième édition revue, corrigée et augmentée*. Livraison 5. Paris: Méquignon-Marvis Père et Fils, 385-503 pp.

Delahon P. 1913: Nachträge zum Schilskyschen Verzeichniss der Käfer Deutschlands von 1909 und einige sonstige Bemerkungen über deutsche Käfer. *Deutsche Entomologische Zeitschrift* **1913**: 529-539.

Delkeskamp K. 1939: Cantharidae. Pars 165. In: Schenkling S. (ed.): *Coleopterorum Catalogus. Vol. 9*. Gravenhage: W. Junk, 357 pp.

Delkeskamp K. 1957: Neue Erotyliden aus dem Britischen Museum. 21. Beitrag zur Kenntnis der Erotyliden (Col.). *Bulletin of the British Museum (Natural History) (Entomology)* **6**: 79-90.

Della Beffa G. 1912: Revisione dei coccinellidi italiani. *Rivista Coleotterologica Italiana* **10**: 145-192, 117-132 [erroneously numeroted], 2 pls.

Della Beffa G. 1913: Revisione dei coccinellidi italiani. *Rivista Coleotterologica Italiana* **11**: 6-22, 29-44, 55-73, 83-98, 114-148, 149-180, 207-216, 218-233, 4 pls.

Della Beffa G. 1914: Revisione dei coccinellidi italiani. *Rivista Coleotterologica Italiana* **12**: 18-24, 88-101, 121-128, 1 pl.

Della Beffa G. 1931: Insetti del Kashmir raccolti dal Dott. Cesare Calciati nella spedizione Mario Piacenza (1913). Pp. 173-187. In: *Himálaia Cashmiriano; relazione del Dott. Cesare Calciat sulle note originali Borelli-Calciati-Piacenza*. Milano: Rizzoli & C.

Denisova L. A. 1948: Novye vidy Elateridae (triby Corymbitini i Elaterini). *Entomologicheskoe Obozrenie* **30**: 40-47.

Depoli G. 1912: Neue Käferformen aus dem Liburnischen Karst. *Wiener Entomologische Zeitung* **31**: 101-102.

Depoli G. 1913: Neue Käferformen aus dem Liburnischen Karst. *Wiener Entomologische Zeitung* **32**: 22.

Depoli G. 1917: Anommatus (subgen. nov. Neanommatus m.) thoracicus m. *Coleopterologische Rundschau* **6**: 11-12.

Depoli G. 1921: Die ostadriatischen Formen von Cantharis obscura L. *Wiener Entomologische Zeitung* **38**: 143-144.

Depoli G. 1928: I Coleotteri della Liburnia. III, Diversicornia (Malacodermata - Brachymera). *Fiumi, Rivista Semestrale della Società di Studi Fiumani* **6**: 208-243.

Desbrochers des Loges J. 1865: Remarques sur divers coléoptères. I. Description de deux nouvelles espèces de coléoptères propres au centre de la France. *Annales de la Société Entomologique de France* (4) **5**: 207-210.

Desbrochers des Loges J. 1869: [La description d'une assez grand nombre d'espèces nouvelles d'Europe et confins]. *Petites Nouvelles Entomologiques* **1** (8): 29-30, 33-34, 45-46.

Desbrochers des Loges J. 1870a: Description de coléoptères nouveaux algériens et remarques diverses. *Bulletin de l'Académie d'Hippone* **9**: 68-82.

Desbrochers des Loges J. 1870b: Descriptions de coléoptères nouveaux d'Europe et confins. *L'Abeille, Mémoires d'Entomologie* **7** [1869-1870]: 97-135.

Desbrochers des Loges J. 1871: Description de coléoptères nouveaux d'Europe et confins et rémarques diverses. *Mittheilungen der Schweizerischen Entomologischen Gesellschaft* **3** [1869-1872]: 337-376.

Desbrochers des Loges J. 1873a: [new taxa]. In: Marseul S. A. de: Répertoire des coléoptères d'Europe décrits isolément depuis 1863. (2ème partie). *L'Abeille, Mémoires d'Entomologie* **9** [1872-1873]: 161-448. [Note: pp. 1-160 issued in 1872].

Desbrochers des Loges J. 1873b: [Description de trois espèces nouvelles de coléoptères d'Europe]. *Bulletin de la Société Entomologique de France* **1873**: cxvii-cxx.

Desbrochers des Loges J. 1875a: Description d'un Throscus et de quelques élatérides nouveaux. *Opuscules Entomologiques* (*Coléoptères*) **1** [1874-1785]: 37-56.

Desbrochers des Loges J. 1875b: [Observations]. *Bulletin de la Société Entomologique de France* **1875**: cxxxvii-cxxxviii.

Desbrochers des Loges J. 1889: [Diagnoses de coléoptères nouveaux d'Algérie]. *Bulletin de la Société Entomologique de France* **1889**: clxxxv-clxxxvi.

Desbrochers des Loges J. 1896: Description d'un Corynetes nouveau d'Algérie. *Le Frelon* **5**: 167.

Desmarest E. 1860: Clériens. Pp. 226-279. In: Chenu J. C.: *Encyclopédie d'histoire naturelle ou traité de cette science d'après les travaux des naturalistes les plus éminents de tous les pays et de toutesles époques. Coléoptères. Deuxième partie.* Paris: Marescq et compagnie, 312 pp., 40 pls.

De Stefani-Perez T. 1911: Una specie inedita di Bostrychus (Coleottero) dell'Eritrea. *Giornale di Scienze Naturali ed Economiche* (Palermo) **28**: 61-63.

Desvignes T. 1842: New British Elater. *The Entomologist* **1**: 326.

Deyrolle H. 1878: [new taxon]. In: Deyrolle H. & Fairmaire L.: Descriptions de coléoptères recueillis par M. l'abbé David dans la Chine centrale. *Annales de la Société Entomologique de France* (5) **8**: 87-140.

Díaz de Castro J. & Sánchez-Ruiz A. 2002: Cardiophorus (Coptostethus) cobossanchezi, nom. nov. para Cardiophorus (Coptostethus) differens Cobos, 1983. *Boletín de la Sociedad Entomológica Aragonesa* **31**: 190.

Dieck G. 1869: Beiträge zur subterranen Käferfauna Südeuropas und Maroccos. (Erstes Stück). *Berliner Entomologische Zeitschrift* **13**: 337-360.

Dieck G. 1870: [new taxa]. In: Heyden L. F. J. D. von.: Entomologische Reise nach dem südlichen Spanien, der Sierra de Guadarrama und Sierra Morena, Portugal und den Cantabrischen Gebirgen, mit Beschreibungen der neuen Arten. *Berliner Entomologische Zeitschrift* **14** Beiheft: [1] + 218 pp., 2 pls.

Dieke G. H. 1947: Ladybeetles of the genus Epilachna (sens. lat.) in Asia, Europe and Australia. *Smithsonian Miscellaneous Collections* **106** (15): 1-183, pls. 1-25, iii.

Dietrich J. K. 1857: Einiges aus dem Gebiete der schweizerischen Käferfauna. *Entomologische Zeitung* (Stettin) **18**: 117-138.

Dillwyn L. W. 1829: *Memoranda relating to coleopterous insects, found in the neighbourhood of Swansea.* Swansea: W. C. Murray & D. Rees, 75 pp.

Dobrzhanskiy F. G. 1924: O geograficheskoy i individualnoy izmenchivosti Adalia bipunctata L. i A. decempunctata L. (Coleoptera, Coccinellidae) [Die geographische und individuelle Variabilität von Adalia bipunctata L. und A. decempunctata L. (Coleoptera, Coccinellidae)]. *Revue Russe d'Entomologie* **18**: 201-212.

Dobrzhanskiy F. G. 1925: Zur Kenntnis der Gattung Coccinella auct. *Zoologischer Anzeiger* **62**: 241-249.

Dobrzhanskiy F. G. 1926a: Nekotorye dannye po faune Coccinellidae (Coleoptera) Jakutii. *Materialy Komissii po izucheniyu Jakutskoy A. S. S. Respubliki* **3**: 1-10.

Dobrzhanskiy F. G. 1926b: Die paläearktischen Arten der Gattung Coccinella L. *Revue Russe d'Entomologie* **20**: 16-32.

Dobrzhanskiy F. G. 1927a: Materialy dlya fauny Coccinellidae Semirech'ya [Die Coccinelliden Heptapotamiens (Semiretchje)]. *Revue Russe d'Entomologie* **21**: 43-52.

Dobrzhanskiy F. G. 1927b: Neue und wenig bekannte Coccinelliden. *Revue Russe d'Entomologie* **21**: 212-217.

Dobrzhanskiy F. G. 1927c: Zwei neue Pharoscymnus-Arten nebst einem Beitrag zur Kenntniss der Morphologie der Coelopterina (Coleoptera, Coccinellidae). *Revue Russe d'Entomologie* **21**: 240-244.

Dobrzhanskiy F. G. 1928: Die Coccinelliden von Gouvernement Semipalatinsk. *Revue Russe d'Entomologie* **22**: 116-123.

Dobson R. M. 1954: A new species of Carpophilus Steph. (Col., Nitidulidae) found on stored produce. *The Entomologist's Monthly Magazine* **90**: 299-300.

Dobson R. M. 1956: A new species of Carpophilus Steph. (Col., Nitidulidae) associated with stored products. *The Entomologist's Monthly Magazine* **92**: 41-42.

Dobson R. M. 1969: A new species of Carpophilus Steph. (Col., Nitidulidae) from East Africa. *The Entomologist's Monthly Magazine* **105**: 99-100.

Dodero A. 1900a: Materiali per lo studio dei Coleotteri italiani con descrizioni di nuove specie. *Annali del Museo Civico di Storia Naturale di Genova* **40**: 400-419.

Dodero A. 1900b: Viaggio di Leonardo Fea in Birmania e regioni vicini LXXXII. *Annali del Museo Civico di Storia Naturale di Genova* **40**: 563-566.

Dodero A. 1916: Appunti Coleotterologici. II. *Annali del Museo Civico di Storia Naturale di Genova* **47**: 337-354.

Dodero A. 1922: Nuova specie e varietà inedite di Cyrtosus italiani (Col.). *Bollettino della Società Entomologica Italiana* **54**: 50-52.

Dodero A. 1925: Missione zoologica del Dr. E. Festa in Cirenaica. XIV. *Bollettino dei Musei di Zoologia ed Anatomia Comparata della R. Università di Torino* (N. S.) **39** (23) [1924]: 1-31.

Dodero A. 1926: Sulla esatta provenienza dell' Axinotarsus doderoi Esc. *Bollettino della Società Entomologica Italiana* **58**: 135-136. [Reference not seen].

Dodero A. 1929: Coleotteri di Valsavaranche (Valle d'Aosta). *Memoire della Società Entomologica Italiana* **6** [1927]: 223-236.

Dodge H. R. 1938: Coccidula suturalis synonymy (Coleop.: Coccinellidae). *Entomological News and Proceedings of the Entomological Section of the Academy of Natural Sciences of Philadelphia* **49**: 221-222.

Dohrn C. A. 1873: [new taxa]. In: Baudi di Selve F.: Catalogo dei Dascilidi, Malacodermi e teredili della Fauna europea e circummediterranea appartenenti alla collezioni del Museo Civico di Genova per Flaminio Baudi. *Annali del Museo di Storia Naturale di Genova* **4**: 226-268.

Dohrn C. A. 1878: Exotisches. *Entomologische Zeitung* (Stettin) **39**: 444-462.

Dohrn C. A. 1882: Exotisches. *Entomologische Zeitung* (Stettin) **43**: 363-374.

Doi H. 1931: A new species of Luciola from Korea. *Journal of Chosen Natural History Society* **12**: 54-55.

Doi H. 1932: A new species of Luciola from Korea. Luciola papariensis sp. nov. *Journal of Chosen Natural History Society* **14**: 63.

Dolin V. G. 1960: Neue und wenig bekannte Elateriden aus der Ukraine (Coleoptera: Elateridae). *Beiträge zur Entomologie* **10**: 189-201.

Dolin V. G. 1964: *Lichinki zhukov-shchelkunov (provolochniki) evropeyskoy chasti SSSR*. Kiev: Urozhay, 206 pp.

Dolin V. G. 1969: Einige neue Elateriden aus dem Krim-, Kaukasus- und Tjan-Schan-Gebirge. *Beiträge zur Entomologie* **19**: 123-135.

Dolin V. G. 1970: [new taxa]. In: Dolin V. G. & Gurjeva E. L.: Neue Ampedus-Arten (Coleoptera, Elateridae) aus der UdSSR. *Koleopterologische Rundschau* **48**: 13-25.

Dolin V. G. 1971: Novye vidy zhukov-shchelkunov (Coleoptera, elateridae) Sovetskogo Soyuza [New click-beetles (Coleoptera, Elateridae) from the Soviet Union]. *Entomologicheskoe Obozrenie* 50(3): 641-654.

Dolin V. G. 1973: [new taxon]. In Dolin V. G. & Ostafichuk V. G.: Novye lichinki zhukov-shchelkunov Moldavii [New larval forms of Elateridae from Moldavia]. Pp. 82-87. In: Yaroshenko M. F., Kiskin P. Kh.,

Plugaru I. G., Plugaru S. G. & Vereshchagin B. V. (eds): *Fauna i biologia zhukov Moldavii* [*Fauna and biology of the beetles of Moldavia*]. Kishinev: Shtiintsa, 186 pp.

Dolin V. G. 1975a: [new taxa]. In: Dolin V. G. & Kurcheva G. F.: Novye formy lichinok zhukov-shchelkunov (Coleoptera, Elateridae) s Dal'nego Vostoka [New larval forms of Elateridae (Coleoptera) from the Far East. *Entomologicheskie Issledovaniya na Dal'nem Vostoke, Trudy Biologo-Pochvennogo Instituta Dal'nevostochnyi Nauchnyi Tsentre Akademyi Nauk SSSR* Novaya seriya (Vladivostok) **28** (131): 113-118.

Dolin V. G. 1975b: [Wing venation of click-beetles (Coleptera, Elateridae) and its importance for taxonomy of the family]. *Zoologicheskiy Zhurnal* **54**: 1618-1633 (In Russian).

Dolin V. G. 1976: Novyy vid zhuka-shchelkuna (Coleoptera, Elateridae) s Yuzhnogo Sakhalina [A new species of click-beetle (Coleoptera, Elateridae) from southern Sakhalin]. *Zoologicheskiy Zhurnal* **55** (11): 1728-1729.

Dolin V. G. 1977a: Novyy vid zhukov-shchelkunov (Coleoptera, Elateridae) iz Kryma i Kavkaza [A new species of click-beetles (Coleoptera, Elateridae) from the Crimea and Caucasus]. Pp. 21-22. In: Skarlato O. A. (ed.): *Novye i maloizvestnye vidy nasekomykh evropeyskoy chasti SSSR*. Leningrad: Akademiya Nauk SSSR, 116 pp.

Dolin W. G. 1977b: Novi vidi zhukiv-kobalikiv (Coleoptera, Elateridae) z Turkmenij [New species of click beetles (Coleoptera, Elateridae) from Turkmenistan]. *Doklady Akademii Nauk Ukrainskoy SSR, Seriya B* **1977** (4): 354-357.

Dolin V. G. 1977c: Novye vidy zhukov-shchelkunov (Coleoptera, Elateridae) iz Turkmenii [New species of click beetles (Coleoptera, Elateridae) from Turkmenistan]. *Doklady Akademii Nauk Ukrainskoy SSR, Seriya B* **1977** (4): 357-361.

Dolin V. G. 1978: *Opredelitel' lichinok zhukov-shchelkunov fauny SSSR* [*Key to click-beetle larvae of the fauna of USSR*]. Kiev: Urozhay, 124 pp.

Dolin V. G. 1980: Novyy vid zhuka-shchelkuna roda Melanotus Esch. (Coleoptera, Elateridae) iz Kryma. *Vestnik Zoologii* **4**: 79-82.

Dolin V. G. 1982a: *Zhuki-kovaliki (Agrypnini, Negastriini, Dimiini, Athoini, Oestodini. Fauna Ukrayiny* **19** (3). Kiiv, 385 pp.

Dolin V. G. 1982b: Novye vidy roda Selatosomus (Coleoptera, Elateridae) iz gornykh raionov Kavkaza i Gissaro-Darvaza [New click-beetle species of the genus Selatosomus (Coleoptera, Elateridae) from the Caucasus and Ghissar-Darwaz Mountains]. *Vestnik Zoologii* **1982** (5): 20-25.

Dolin V. G. 1985a: [new taxa]. In: Dolin V. G. & Mardjanian M. A.: Novye vidy zhukov-shchelkunov roda Cardiophorus Eschz. fauny evropeyskoy chasti SSSR i Zakabkas'ya (Coleoptera, Elateridae) [New species of click-beetles of the genus Cardiophorus Eschz. of the fauna of the European part of the USSR and Transcaucasus]. *Doklady Akademii Nauk Armyanskoy SSR* **81** (3): 134-138.

Dolin V. G. 1985b: Novyy vid zhukov-shchelkunov (Coleoptera, Elateridae) iz Tundry Sibiri [A new species of click-beetles (Coleoptera, Elateridae) from the Sibirian tundra]. *Entomologicheskoe Obozrenie* **64**: 144-145.

Dolin V. G. 1986: Eine neue Athous (Orthathous) aus Ungarn (Coleoptera, Elateridae). *Annales Historico-Naturales Musei Nationalis Hungarici* **78**: 145-146.

Dolin V. G. 1987a: [new taxa]. In: Dolin V. G. & Suzuki W.: Vidy roda Penia (Coleoptera, Elateridae) gruppy P. opatroides gimalayskogo regiona [The species of the genus Penia (Coleoptera, Elateridae) of the group P. opatroides of the Himalayan region]. *Vestnik Zoologii* **1987** (1): 29-39.

Dolin W. G. 1987b: Novye vidy zhulov-shchelkunov (Coleoptera, Elateridae) s Dalnego Vostoka [New species of click-beetles (Coleoptera, Elateridae) from the Far East]. *Entomologicheskoe Obozrenie* **66**: 345-350.

Dolin V. G. 1988a: Zhuki-Shchelkuny. Cardiophoriny i Elateriny. *Fauna Ukrayiny* **19** (4). Kiiv, 204 pp.

Dolin V. G. 1988b: [new taxon]. In: Dolin W. G. & Latifi A. O. 1988a: Novye vidy zhukov-shchelkunov roda Melanotus Esch. (Coleoptera, Elateridae) iz Tadzhikistana [New species of click beetles of the genus Melanotus Esch. (Coleoptera, Elateridae) from Tadjikistan]. *Izvestiya Akademii Nauk Tadzhikskoy SSR, Otdelenie Biologicheskikh Nauk* **4** (113): 56-60.

Dolin V. G. 1988c: Dva novykh vida zhukov-shchelkunov roda Melanotus (Coleoptera, Elateridae) iz Tadzhikistana [Two new species of the elaterid genus Melanotus (Coleoptera, Elateridae) from Tadzhikistan]. *Vestnik Zoologii* **1988** (4): 78-81.

Dolin V. G. 1988d: [new taxon]. In: Dolin W. G. & Latifi A. O. 1988b: Dva novykh vida zhukov-shchelkunov roda Melanotus (Coleoptera, Elateridae) iz Tadzhikistana [Two new species of click beetles of the genus

Melanotus Esch. (Coleoptera, Elateridae) from Tadjikistan]. *Doklady Akademii Nauk Tadzhikskoy SSR* **31** (4): 825-827.

Dolin V. G. 1990: Neue Melanotus-Arten aus Zentralasien (Coleoptera: Elateridae). *Koleopterologische Rundschau* **60**: 117-124.

Dolin V. G. 1992a: Novye vidy zhukov – shchelkunov roda Oedostethus Lec. (Coleoptera, Elateridae, Negastriinae) iz Sibiri i s Dalnego Vostoka [New species of click-beetles of the genus Oedostethus Lec. (Coleoptera, Elateridae, Negastriinae) from Siberia and Far East]. Pp. 13-21. In: Amirkhanov A. M., Gabuzov O. S., Dezhkin V. V., Ilyinskiy V. O., Ilyshenko V. Yu., Nazarov A. A., Noskova N. K., Nukhimovskaya Yu. D. & Fertikov V. I.: Nasekomye Daurii I sopredelnykh territorij. Sbornik nauchnych trudov. *Vol. 1*. Moskva: Zentral'noy nauchno-issledovatel'skoy laboratoriy okhotnich'ego khozyaystva i zapovedinkov, 103 pp.

Dolin V. G. 1992b: Einige neue Cardiohypnus-Arten (Coleoptera, Elateridae) aus Himalaya und Zentral Asien und Stellung dieser Gattung im System der Elateridae. *Entomologica Basiliensia* **15**: 215-228.

Dolin V. G. 1993: Neue Quasimus-Arten (Col. Elateridae, Negastriinae) aus der Himalaya - Region. *Entomologica Basiliensia* **16**: 181-201.

Dolin V. G. 1994: [new taxa]. In: Dolin V. G. & Atamuradov Kh. I.: *Zhuky-shchelkuny* (*Elateridae*) *Turkmenistana. Click-beetles* (*Elateridae*) *of Turkmenistan*. Kiev: Institut zoologii NAN Ukrainy, 178 pp.

Dolin V. G. 1995: Neue Zorochrus-Arten (Coleoptera, Elateridae) aus dem Transkaukasus und Zentral-Asien. *Zeitschrift der Arbeitsgemeinschaft Österreichischer Entomologen* **47**: 19-24.

Dolin V. G. 1996: Neue Zorochrus-Arten aus der Sammlung G. Frey (Insecta, Coleoptera, Elateridae, Negastriinae). *Spixiana* **19**: 267-270.

Dolin V. G. 1997a: Neue und wenig bekannte Quasimus-Arten (Coleoptera: Elateridae, Negastriinae) aus Süd-Ost Asien. *Entomologica Basiliensia* **20**: 143-158.

Dolin V. G. 1997b: Novye vidy zhukov-shchelkunov roda *Melanotus* (Coleoptera, Elateridae) iz Sredney Azii [New species of the elaterid genus Melanotus Esch. (Coleoptera, Elateridae) from Middle Asia]. *Entomologicheskoe Obozrenie* **76**: 350-356.

Dolin V. G. 1998a: Neue Hypnoidus-Arten (Coleoptera, Elateridae, Hypnoidini) aus Mittelasien. *Journal of the Ukrainian Entomological Society* **4** (1-2): 3-9.

Dolin V. G. 1998b: Neue Schnellkäferarten (Coleoptera, Elateridae) aus Kyrgystan. *Vestnik Zoologii* **32** (5-6): 93-98.

Dolin V. G. 1999a: Neue Zorochrus-Arten mit ungekielten Halsschildhinterwinkeln aus Asien (Coleoptera: Elateridae: Negastriinae). *Zeitschrift der Arbeitsgemeinschaft Österreichischer Entomologen* **51**: 31-40.

Dolin V. G. 1999b: Drei neue Arten der Gattung Hypnoidus (Coleoptera, Elateridae) aus Tibet. *Vestnik Zoologii* **33** (4-5): 81-85.

Dolin V. G. 2000: Znachenie lichinochnykh priznakov i zhilkovaniya kryl'ev v sistematike Elateroidea (Coleoptera). The role of larval and wing venation characters in the systematics of Elateroidea (Coleoptera). In: Cheniya pamyati N. A. Kholodkovskogo. *Doklad na pyat'desyat vtorom ezhegodnom chtenin, 1 aprelya 1999 g. Meetings in memory of N. A. Cholodkovsky. Lecture at the 52nd annual meeting, 1 April, 1999*. St. Petersburg, 50 pp.

Dolin V. G. 2001: Zur Kenntnis der Quasimus-Arten Indiens (Coleoptera: Elateridae). *Koleopterologische Rundschau* **71**: 121-142.

Dolin V. G. 2002a: Zur Kenntnis der südostasiatischen Zorochrus Thomson, 1859 Arten nebst Beschreibung neuen Arten aus der Untergattung Thurana Stibick, 1971 (Coleoptera, Elateridae, Negastriinae). *Entomologica Basiliensia* **24**: 33-43.

Dolin V. G. 2002b: [new taxa]. In: Dolin V. G. & Mertlik J.: Drei neue klein-asiatischen und iranischen Zorochrus-Arten (Coleoptera, Elateridae, Negastriinae). *Vestnik Zoologii* **36** (6): 89-93.

Dolin V. G. 2003: Einige neue ostpalaearktischen Elateriden-Arten (Coleoptera, Elateridae). *Zeitschrift der Arbeitsgemeinschaft Österreichischer Entomologen* **55**: 29-38.

Dolin V. G. & Agajev B. I. 1974: Novyy vid zhuka-shchelkuna (Coleoptera, Elateridae) iz Talysha [A new species of click-beetle (Coleoptera, Elateridae) from Talysh]. *Doklady Akademii Nauk Azerbaydzhanskoy SSR* **30** (11): 54-56.

Dolin V. G. & Agajev B. I. 1983: Novye vidy zhukov-shchelkunov (Coleoptera, Elateridae) iz Azerbaydzhanskoy SSR [New click-beetle species (Coleoptera, Elateridae) from the Azerbaidjan SSR]. *Vestnik Zoologii* **1983** (6): 31-34.

Dolin V. G. & Agajev B. I. 1985: Novyy vid zhuka-shchelkuna roda Drapetes (Coleoptera, Elateridae) iz Talysha [A new species of the click beetle genus Drapetes (Coleoptera, Elateridae) from Talysh]. *Zoologicheskiy Zhurnal* **65**: 1746-1747.

Dolin V. G. & Agajev B. I. 1988: [new taxon]. P. 5. In: Agaev B. I. & Dolin V. G. 1988: Die Bedeutung der Larvenmerkmale in der Systematik des Tribus Athoini (Coleoptera, Elateridae). In: Dolin V. G., Zerova M. D., Nekrutenko Yu. P. & Filimonova N. S. (eds.): *XII. Mezhdunarodnyy Simpozium po Entomofaune Sredney Evropy. Kiev, 25 – 30 septembrya 1988 g. Tezisy Dokladov.* Kiev: Naukova Dumka, 601 pp.

Dolin V. G. & Agajev B. I. 1993: Novye vidy zhukov-shchelkunov (Coleoptera, Elateridae) iz Zakavkaz"ya [New click-beetle species (Coleoptera, Elateridae) from Transcaucasia]. *Vestnik Zoologii* **1993** (6): 65-68.

Dolin V. G. & Atamuradov Kh. I. 1980: Dva novikh vida zhukov-shchelkunov (Coleoptera, Elateridae) iz Badkhyza (Yuzhnaya Turkmeniya) [Two new species of click beetles (Coleoptera, Elateridae) from Badchyz (southern Turkmenistan)]. *Vestnik Zoologii* **1980** (2): 81-84.

Dolin V. G. & Atamuradov Kh. I. 1986: Novye vidy zhukov-shchelkunov (Coleoptera, Elateridae) iz yugo-vostochnogo Turkmenii [New species of click-beetles (Coleoptera, Elateridae) from Southeastern Turkmenistan]. *Izvestiya Akademii Nauk Turkmenskoy SSR, Seriya Biologicheskikh Nauk* **1986** (2): 69-71.

Dolin V. G. & Atamuradov Kh. I. 1987: Novye vidy zhukov-shchelkunov (Coleoptera, Elateridae) iz Kopetdaga [New species of click-beetles (Coleoptera, Elateridae) from Kopet-Dagh]. *Izvestiya Akademii Nauk Turkmenskoy SSR, Seriya Biologicheskikh Nauk* **1987** (4): 74-76.

Dolin V. G. & Atamuradov Kh. I. 1989: Novye vidy zhukov-shchelkunov (Coleoptera, Elateridae) iz Turkmenii [New species of click-beetles (Coleoptera, Elateridae) from Turkmenistan]. *Izvestiya Akademii Nauk Turkmenskoy SSR, Seriya Biologicheskikh Nauk* **1989** (4): 33-41.

Dolin V. G. & Atamuradov Kh. I. 1994: *Zhuki-shchelkuny* (*Elateridae*) *Turkmenistana* [*Click-beetles* (*Elateridae*) *of Turkmenistan*]. Kiev: Institut zoologii NAN Ukrainy, 178 pp.

Dolin V. G. & Bessolitzina E. P. 1982: Novye vidy zhukov-shchelkunov podsemeystva Negastriinae (Coleoptera, Elateridae) iz Charskoy kotloviny [New species of click-beetles of the subfamily Negastriinae (Coleoptera, Elateridae) from the Tshar Hollow]. *Zoologicheskiy Zhurnal* **62** (4): 631-634.

Dolin V. G. & Bessolitzina E. P. 1990: Novye vidy zhukv-shchelkunov roda Oedostethus (Coleoptera, Elateridae) s Dal"nego Vostoka [New click-beetle species of the genus Oedostethus (Coleoptera, Elateridae) from the Far East]. *Vestnik Zoologii* **1990** (6): 23-27.

Dolin V. G. & Bessolitzina E. P. 1992: Novye vidy zhukv-shchelkunov roda Hypnoidus Steph. (Coleoptera, Elateridae) iz vostochnoy Palearktiki [New species of elaterid beetles of the genus Hypnoidus Steph. (Coleoptera, Elateridae) from East Palaearctic]. *Entomologicheskoe Obozrenie* **70** [1991]: 846-851.

Dolin V. G. & Cate P. C. 1990: Eine neue Art der Gattung Melanotus Eschscholtz aus Tadzhikistan (Coleoptera: Elateridae). *Koleopterologische Rundschau* **60**: 113-115.

Dolin V. G. & Cate P. C. 1994: [new taxon]. In: Dolin V. G. & Atamuradov Kh. I.: *Zhuki-shchelkuny* (*Elateridae*) *Turkmenistana. Click-beetles* (*Elateridae*) *of Turkmenistan.* Kiev: Institut zoologii NAN Ukrainy, 178 pp.

Dolin V. G. & Cate P. C. 1995: Eine neue Art der Gattung Melanotus Eschscholtz aus dem westlichen Kopetdag Gebirge (Turkmenistan). *Koleopterologische Rundschau* **65**: 169-170.

Dolin V. G. & Cate P. C. 1998: Fünf neue Arten der Gattung Hypnoidus Dillwyn, 1829 (Coleoptera, Elateridae) aus Zentralasien. *Zeitschrift der Arbeitsgemeinschaft Österreichischer Entomologen* **50**: 45-51.

Dolin V. G. & Cate P. C. 1999: Revision der Gattung Denticolloides Gurjeva, 1963 (Coleoptera, Elateridae) sowie Beschreibungen zweier neuer Arten. *Zeitschrift der Arbeitsgemeinschaft Österreichischer Entomologen* **51**: 25-30.

Dolin V. G. & Cate P. C. 2001: Die Arten der Gattung Hypnoidus Dillwyn, 1829, aus dem Himalaya (Coleoptera: Elateridae). *Zeitschrift der Arbeitsgemeinschaft der Österreichischen Entomologen* **53**: 121-136.

Dolin V. G. & Cate P. C. 2002: Zur Kenntnis der Hypnoidus-Arten aus China (Elateridae: Athouinae: Hypnoidini). *Zeitschrift der Arbeitsgemeinschaft Österreichischer Entomologen* **54**: 61-76.

Dolin V. G. & Cate P. C. 2003: Nachtrag zur Kenntnis der Hypnoidus-Arten der Fauna Indiens und Chinas (Coleoptera: Elateridae). *Zeitschrift der Arbeitsgemeinschaft Österreichischer Entomologen* **55**: 39-43.

Dolin V. G. & Chantladze T. I. 1980: Novye vidy zhukov-shchelkunov (Coleoptera, Elateridae) iz vostochnoy Gryzii [New species of Elateridae beetles from eastern Georgia]. *Soobshcheniya Akademii Nauk Gruzinskoy* SSR **97** (1): 189-192.

Dolin V. G. & Chantladze T. I. 1982: Novyy vid zhuka-shchelkuna iz roda Athous Esch. (Coleoptera, Elateridae) s Malogo Kavkasa [A new species of the Elateridae beetle of the group Athous Esch. from the Lesser Caucasus]. *Soobshcheniya Akademii Nauk Gruzinskoy SSR* **105** (1): 161-164.

Dolin V. G. & Chantladze T. I. 1987: Dva novykh vida zhukov-shchelkunov (Coleoptera, Elateridae) iz vostochnoj Gruzii [Two new species of click-beetles (Col. Elateridae) from Eastern Georgia]. *Soobshcheniya Akademii Nauk Gruzinskoy SSR* **125** (3): 621-624.

Dolin V. G. & Chantladze T. I. 1989: Novyy vid zhuka-shchelkuna roda Ampedus (Coleoptera, Elateridae) iz vostochnoy Gruzii [A new species of click beetles of the genus Ampedus (Coleoptera, Elateridae) from Eastern Georgia]. *Soobshcheniya Akademii Nauk Gruzinskoy SSR* **133** (1): 169-172.

Dolin V. G. & Gurjeva E. L. 1976: Novyy vid zhukov-kovalikovikh z Talisha [A new species of click beetles from Talysh]. *Doklady Akademii Nauk Ukrainskoy SSR* (B) **1976** (7): 645-646.

Dolin V. G. & Gurjeva E. L. 1988: Novyy vid zhuka-shchelkuna roda Pseudanostirus Dolin (Coleoptera, Elateridae) iz Kazachstana [A new species of the click-beetle genus Pseudanostirus (Coleoptera, Elateridae) from Kazakhstan]. *Vestnik Zoologii* **1988** (2): 63-64.

Dolin V. G. & Gurjeva E. L. 1994: [new taxon]. In: Dolin V. G. & Atamuradov Kh. I.: *Zhuki-shchelkuny (Elateridae) Turkmenistana. Click-beetles (Elateridae) of Turkmenistan.* Kiev: Institut zoologii NAN Ukrainy, 178 pp.

Dolin V. G. & Katyukha S. A. 1977: Dva novykh vida zhukov-shchelkunov roda Negastrius Thoms. (Coleoptera, Elateridae) s yuzhnogo Sakhalina [Two new species of click-beetles from the genus Negastrius Thoms]. (Coleoptera, Elateridae) from South Sakhalin. *Doklady Akademii Nauk Ukrainskoy SSR* (B) **2**: 177-179 [in Ukrainian], 180-182.

Dolin V. G. & Katyukha S. A. 1987: Novye vidy zhukov-shchelkunov (Coleoptera, Elateridae) s yuzhnogo Sakhalina [New species of click-beetles (Coleoptera, Elateridae) from southern Sachalin]. Pp. 11-16. In: Ler P. A. & Storozheva M. A. (eds): *Novye dannye po sistematike nasekomykh Dal"nego Vostoka*. Vladivostok, DVO AN SSSR, 114 pp.

Dolin V. G. & Latifi A. O. 1988a: Novye vidy zhukov-shchelkunov roda Melanotus Esch. (Coleoptera, Elateridae) iz Tadzhikistana [New species of click beetles of the genus Melanotus Esch. (Coleoptera, Elateridae) from Tadjikistan]. *Izvestiya Akademii Nauk Tadzhikskoy SSR, Otdelenie Biologicheskikh Nauk* **4** (113): 56-60.

Dolin V. G. & Latifi A. O. 1988b: Dva novykh vida zhukov-shchelkunov roda Melanotus (Coleoptera, Elateridae) iz Tadzhikistana [Two new species of click beetles of the genus Melanotus Esch. (Coleoptera, Elateridae) from Tadjikistan]. *Doklady Akademii Nauk Tadzhikskoy SSR* **31** (4): 825-827.

Dolin V. G. & Latifi A. O. 1989: Novye vidy zhukov-shchelkunov (Coleoptera, Elateridae) iz Tadzhikskoy SSR [New species of click-beetles (Coleoptera, Elateridae) from the Tadjik SSR]. *Vestnik Zoologii* **1989** (6): 19-25.

Dolin V. G. & Latifi A. O. 1997: Neue Elateriden-Arten (Coleoptera) aus Tadzhikistan. *Vestnik Zoologii* **31** (3): 83-85.

Dolin V. G. & Medvedev A. A. 2002: Eine neue Oedostethus-Art (Coleoptera, Elateridae, Negastriinae) aus Nord des europäischen Teils Russlands mit der Übersicht der verwandten Arten. *Vestnik Zoologii* **36** (4): 95-97.

Dolin V. G. & Mertlik J. 2002: Drei neue klein-asiatischen und iranischen Zorochrus-Arten (Coleoptera, Elateridae, Negastriinae). *Vestnik Zoologii* **36** (6): 89-93.

Dolin V. G. & Ôhira H. 1976: Novye vidy zhukov-shchelkunov roda Ampedus Dej. (Coleoptera, Elateridae) c Dal"nego Vostoka [New species of click-beetles of the genus Ampedus Dej. (Coleoptera, Elateridae) from the Far East]. *Entomologicheskoe Obozrenie* **55**: 387-391.

Dolin V. G. & Penev L. D. 1988: Beitrag zur Taxonomie der mit Selatosomus latus (F.) verwandten Arten von der Krim und aus dem Kaukasus (Insecta, Coleoptera, Elateridae). *Reichenbachia* **26** (6): 25-30.

Dolin W. G. & Penev L. D. 2004: Revision der kaukasischen Athous-Arten aus der Untergattung Haplathous Reitter, 1905 (Coleoptera: Elateridae). *Linzer Biologische Beiträge* **36**: 89-104.

Dolin V. G. & Protsenko A. I. 1965: Dva novych vida zhulkov-shchelkunov (Coleoptera, Elateridae) iz Kirgizii [Two new species of click-beetles (Coleoptera, Elateridae) from Kirgizia]. *Sbornik Entomologicheskikh Rabot, Akademiya Nauk Kirgizskoy SSR* **4**: 21-25.

Dolin V. G. & Šauša O. 1997: Zwei neue Arten (Coleoptera: Elateridae) aus dem Fernen Osten Russlands. *Zeitschrift der Arbeitsgemeinschaft Österreichischer Entomologen* **49**: 41-44.

Dolin V. G. & Suzuki W. 1987: Vidy roda Penia (Coleoptera, Elateridae) gruppy P. opatroides gimalayskogo regiona [The species of the genus Penia (Coleoptera, Elateridae) of the group P. opatroides of the Himalayan region]. *Vestnik Zoologii* **1987** (1): 29-39.

Donisthorpe H. S. J. K. 1902: [new taxon]. *The Entomologist's Record and Journal of Variation* **14**: 99-100.

Donisthorpe H. S. J. K. 1903: Scymnus limonii, n. sp.- a species new to science. *The Entomologist's Record and Journal of Variation* **15**: 287.

Donisthorpe H. S. J. K. 1918: On the subspecies and aberrations of Coccinella 11-punctata L. *The Entomologist's Record and Journal of Variation* **30**: 121-128.

Donisthorpe H. S. J. K. 1935: Cryptophagus stramenti n. sp. (Cryptophagidae, Col.) a species of Coleoptera new to science, and a note on C. umbratus Er. and C. distinguendus Stm. *The Entomologist* **68**: 110-111.

Donisthorpe H. S. J. K. 1939: Micrambe aubrooki sp. n. (Cryptophagidae, Col.), a beetle new to science. *Entomologists Record & Journal of Variation* **60**: 37.

Donisthorpe H. S. J. K. 1942: Cerylon primroseae sp. n. (Col., Colydiidae) in Britain. *The Entomologist's Monthly Magazine* **78**: 2, pl. II.

Donndorff J. A. 1799: *Europäische Fauna oder Naturgeschichte der europäischen Thiere. Band VIII. Käfer.* Leipzig: Weidmann, 18 + 892 pp.

Donovan E. 1798: *The natural history of British Insects; explaining them in their several states, with the periods of their transformations, thei food, oeconomy &c. together wits the history of such minute insects as require investigations by the microscope. Vol. VII.* London: Rivington. 96 + 6 pp., pls 217-251.

Donovan E. 1804: *The natural history of British Insects; explaining them in their several states, with the periods of their transformations, thei food, oeconomy &c. together wits the history of such minute insects as require investigations by the microscope. Vol. XI.* London: Rivington. 100 + 7 pp., pls 361-396.

Donovan E. 1813: *The natural history of British Insects; explaining them in their several states, with the periods of their transformations, thei food, oeconomy &c. together wits the history of such minute insects as require investigations by the microscope. Vol. XVI.* London: Rivington. 91 + 10 pp., pls 541-576.

Dorn K. 1924: Elater Hjorti Rye und ferrugatus Lac. *Entomologisches Jahrbuch* **33/34** [1924/25]: 128-129.

Dowgailo K. 1997: Faunistische und systematische Übersicht der Ebaeus-Arten (Coleoptera, Malachiidae) aus Zentralasien. *Entomologica Basiliensia* **20**: 373-392.

Drapiez P. A. J. 1819: Description de huit espèces d'insectes nouveaux. *Annales Générales des Sciences Physiques* **1**: 130-138, pl. 8.

Drapiez A. 1820: Description de huit espèces d'insectes nouveaux. *Annales Générales des Sciences Physiques* (Bruxelles) **5**: 323-330, pl. LXXXIII [note: falsely numbered as pl. LXXVI].

Drury D. 1773: *Illustrations of Natural History. Wherein are exhibited upwards of two hundred and forty figures of exotic insects, according to their different genera; very few of which have hitherto been figured by any author, being engraved and coloured from nature,with the greatest accurancy, and under the author's own inspection, on fifty copper plates. With a particular description of each insect: interspersed with remarks and reflections on the nature and properties of many of them. Vol. II.* London: B. White, vii + 90 + [4] pp.

Dudich E. 1922: Eine neue Anommatus-Art aus Ungarn. *Annales Historico-Naturales Musei Nationalis Hungarici* **19**: 123-124.

Dufour L. J. M. 1843a: Excursion entomologique dans les montagnes de la vallée d'Ossau. *Bulletin des Sciences, Lettres et Arts de Pau* **8**: 1-118.

Dufour L. J. M. 1843b: Histoire des métamorphoses du Choragus Sheppardi et du Xyletinus hederae. *Annales de la Société Entomologique de France* (2) **1**: 313-328, 1 pl.

Dufour L. J. M. 1851 Des zones entomologiques dans nos Pyrénées occidentales et désignation des insectes qui les habitent. *Actes de la Société Linnéenne de Bordeaux* **17**: 303-364.

Dufour L. J. M. 1853: Excursion entomologique aux dunes de Biscarrosse et d'Arcachon, avec indication de quelques maneuvres insecticeptologiques, et réflexions. *Acta Societas Linneana Bordeaux* **19**: 283-317.

Duftschmid C. E. 1825: *Fauna Austriae. Oder Beschreibung der österreichischen Insecten, für angehende Freunde der Entomologie. Dritter Theil.* Linz: Priv. k. k. akademischen Kunst-, Muster- und Buchhandlung, 290 pp.

Duméril A. M. C. 1817: [Coccinelle (Entom.), Coccinella]. Pp. 490-496. In: *Dictionnaire des Sciences Naturelles, dans lequel on traite méthodiquement des différens êtres de la nature, considérés soit en eux-mèmes, d'après l'état actuel de nos connaissances, soit relativement à l'utilité qu'en peuvent retirer la médicine, l'agriculture, le commerce et les arts. Suivi d'une biographie des plus célèbres naturalistes. Ouvrage destiné aux médicins, aux agricultures, aux comerçans, aux artistes, aux manufacturiers, et à tous ceux qui ont intérêt à connaître les productions de la nature, leurs caractères génériques et spécifiques, leurs lieu naturel, leurs propriétés et leurs usages. Tome neuvième.* Strasbourg-Paris: F. G. Levraut, Le Normant, 560 pp.

Duverger C. 1983: Contribution à la connaissance des Coccinellidae d'Iran. *Nouvelle Revue d'Entomologie* **13**: 73-93.

Duverger C. 1989: Notes sur deux Coccinellidae peu connus de la faune de France. *Bulletin de la Sociètè Linnéenne de Bordeaux* **26**: 127-130.

Duverger C. 1990: Catalogue de coléoptères Coccinellidae de France continentale et de Corse. Essai de mise à jour critique. *Bulletin de la Société Linnéenne de Bordeaux* **28**: 61-87.

Duverger C. 2003: Phylogenie des Coccinellidae. *Bulletin de la Société Linnnéene de Bordeaux* **31**: 57-76.

Easton A. M. 1950: Some amendments of nomenclature in the genus Meligethes Stephens (Col., Nitidulidae). *The Entomologist's Monthly Magazine* **86**: 308-309.

Easton A. M. 1952: A review of the Meligethes (Col. Nitidulidae) associated with the labiate Marrubium vulgare L., including the description of a new species. *Bulletin de la Société des Sciences Naturelles du Maroc* **32**: 181-192.

Easton A. M. 1954a: 24. Coleoptera: (Nitidulidae) Meligethes Stephens. Pp. 337-350. In: *British Museum Expedition to South-West Arabia 1937-8. Vol. 1.* London: British Museum (Natural History).

Easton A. M. 1954b: A revision of the lugubris complex in the genus Meligethes Stephens (Col., Nitidulidae). *Transactions of the Royal Entomological Society of London* **105** (16): 373-392.

Easton A. M. 1955: The Palaearctic species of the subgenus Acanthogethes of the genus Meligethes Stephens (Coleoptera, Nitidulidae). *The Annals and Magazine of Natural History* (12) **8**: 225-240.

Easton A. M. 1956a: The Meligethes of North Africa (Coleoptera, Nitidulidae). *Mémoires de la Société des Sciences Naturelles et Physiques du Maroc* (N. S., Zoologie) **2** [1955]: 7-71.

Easton A. M. 1956b: The true identity of Meligethes flavicollis Reitter and of M. flavicollis Auctt. Japon. (Col., Nitidulidae). *The Annals and Magazine of Natural History* (12) **9**: 557-560.

Easton A. M. 1957a: A revision of the rotundicollis and lepidii species-groups in the genus Meligethes Stephens (Coleoptera: Nitidulidae). *The Annals and Magazine of Natural History* (12) **10**: 85-96.

Easton A. M. 1957b: The Meligethes (Col., Nitidulidae) of Afghanistan. *The Entomologist's Monthly Magazine* **92**: 385-401.

Easton A. M. 1957c: The Meligethes of Japan (Coleoptera: Nitidulidae). *Transactions of the Royal Entomological Society of London* **109**: 395-420.

Easton A. M. 1959a: A new Norvegian species of Meligethes Stephens (Col., Nitidulidae). *Norsk Entomologisk Tidsskrift* **11**: 50-53.

Easton A. M. 1959b: A new Nearctic species of Meligethes (Col., Nitidulidae). *The Entomologist's Monthly Magazine* **95**: 58-60.

Easton A. M. 1960: On the identity of Meligethes moestus Erichson and M. moestus Auctt. (Col., Nitidulidae). *Memorie del Museo Civico di Storia Naturale, Verona* **8**: 335-343.

Easton A. M. 1962: Meligethes stierlini Reitter, 1872 and M. stierlini Easton, 1955 (Col., Nitidulidae). *The Entomologist's Monthly Magazine* **98**: 74.

Easton A. M. 1964: Meligethes rosenhaueri Reitter, M. assimilis Sturm and related species (Coleoptera: Nitidulidae). *Proceedings of the Royal Entomological Society of London* (B) **33**: 1-6.

Easton A. M. 1968: The Meligethes of High Nepal (Coleoptera: Nitidulidae). *Khumbu Himal, Ergebnisse des Forschungsunternehmens Nepal Himalaya* **3**: 40-48.

Edwards J. 1914: On the variation in Britain of Coccinella hieroglyphica, L., with some collateral matter. *The Entomologist's Monthly Magazine* (2) **25**: 139-143.

Edwards J. 1921: On a second British species of Priobium: P. kiesenwetteri, nom. nov. = tricolor Kiesw. (nec Ol., nec Muls.). *The Entomologist's Monthly Magazine* **57**: 102-104.

Egorov A. B. 1992: 41. Sem. Throscidae (Trixagidae). In: Ler P. A. (ed.): *Opredelitel nasekomykh Dalnego Vostoka SSSR. Tom III. Zhestkokrylye, ili zhuki. Chast 2.* Sankt-Petersburg: Nauka, 704 pp.

Eichhoff W. 1863: Sammelberichte aus Hilchenbach bei Siegen. *Berliner Entomologische Zeitschrift* **7**: 436-437.

Emetz V. M. 1977: [new taxon]. In: Emetz V. M. & Logvinovskiy V. D.: Obzor zhukov-tochil"shchikov roda Xyletinus Latr. (Coleoptera, Anobiidae) fauny SSSR. *Entomologicheskoe Obozrenie* **56**: 409-419.

Endrödy-Younga S. 1965: Eine neue mitteleuropäische Art der Gattung Cybocephalus Er. (Coleoptera, Cybocephalidae). *Nachrichtenblatt der Bayerischen Entomologen* **14**: 41-43.

Endrödy-Younga S. 1968: Monographie der paläarktischen Arten der Familie Cybocephalidae (Coleoptera: Clavicornia). *Acta Zoologica Academiae Scientiarum Hungaricae* **14**: 27-115.

Endrödy-Younga S. 1971: Neue Ergebnisse bei der Bearbeitung der paläarktischen und orientalischen Cybocephaliden (Coleoptera: Clavicornia). *Acta Zoologica Academiae Scientiarum Hungaricae* **17**: 243-249.

Endrödy-Younga S. 1984: A new species of Cybocephalus (Coleoptera, Cucujoidea, Cybocephalidae) from Israel. *Israel Journal of Entomology* **18**: 1-2.

Erber D. & Hinterseher W. 2002: Revision der Gattung Sphaericus (Col. Ptinidae) der lauri-makaronesischen Region (Azoren, Madeira, Selvagens, Kanaren) einschliesslich Nordafrika und des europäischen Festlandes. *Coleoptera* **4** [2000]: 153-282.

Erichson W. F. 1839: Bemerkungen über einige Käferarten. *Germar's Zeitschrift für die Entomologie* **1** (2): 367-371.

Erichson W. F. 1840: *Entomographien, untersuchungen in dem Gebiete der Entomologie, mit besonderer Benutzung der Königl. Sammlung zu Berlin. Erstes Heft.* Berlin: F. H. Morin, 180 pp., 2 pls.

Erichson W. F. 1840: Ueber Elateriden ohne Bruststachel. (Cardiophorus). *Germar's Zeitschrift für die Entomologie* **2**: 279-341.

Erichson W. F. 1841a: Ueber die Elateriden mit kammförmig gezähnten Krallen. *Germar's Zeitschrift für die Entomologie* **3**: 88-129.

Erichson W. F. 1841b: Ueber die Insecten von Algier mit besonderer Berücksichtigung ihrer Verbreitung. Pp. 140-194. In: Wagner M. F.: *Reisen in der Regentschaft Algier in den Jahren 1836, 1837 und 1838 nebst einem naturhistorischen Anhang und einem Kupferatlas. Dritter Band.* Leipzig: L. Voss, xviii + 297 pp., 17 pls.

Erichson W. F. 1842: Beitrag zur Insecten-Fauna von Vandiemensland, mit besonderer Berücksichtigung der geographischen Verbreitung der Insecten. *Archiv für Naturgeschichte* **8**: 83-287, 2 pls.

Erichson W. F. 1843a: Beitrag zur Insecten-Fauna von Angola, in besonderer Beziehung zur geographischen Verbreitung der Insecten. *Archiv für Naturgeschichte* **9**: 199-267.

Erichson W. F. 1843b: Versuch einer systematischen Eintheilung der Nitidularien. *Zeitschrift für die Entomologie,* **4**: 225-361.

Erichson W. F. 1844: Einige Nachträge zu meinem Versuch einer systematischen Eintheilung der Nitidularien. *Germar's Zeitschrift für die Entomologie* **5**: 438-458.

Erichson W. F. 1845: [I., II. Lieferung, pp. 1-320]. In: *Naturgeschichte der Insecten Deutschlands. Erste Abtheilung. Coleoptera. Dritter Band.* Berlin: Nicolaischen Buchhandlung, vii + [2] + 968 pp., 1 pl. [issued in parts, pp. 321-480 in 1846, pp. 801-968 in 1848].

Erichson W. F. 1846: [III. Lieferung, pp. 321-480]. In: *Naturgeschichte der Insecten Deutschlands. Erste Abtheilung. Coleoptera. Dritter Band.* Berlin: Nicolaischen Buchhandlung, vii + [2] + 968 pp., 1 pl.

Erichson W. F. 1847: Conspectus Insectorum Coleopterorum, quae in Republica Peruana observata sunt. *Archiv für Naturgeschichte* **13**: 67-185.

Erichson W. F. 1871: [new taxon]. In: Baudi di Selve F.: Coleopterorum messis in insula Cypro et Asia minore ab Eugenio Truqui congregatae recensitio: de Europaeis notis quibusdam additis. Pars quarta. *Berliner Entomologische Zeitschrift* **15**: 49-71.

Ericson I. B. 1896: [Nya coleoptera: Acrotona curtipennis och Lathridius microps]. *Entomologisk Tidskrift* **17**: 265-266.

Escalera M. M. de la 1913a: Una campana entomologica en el Sus y description des los Coleópteros recogidas en elle. Coleópteros nuevos del Sus (Marruecos). *Trabajos del Museo Nacional de Ciencias Naturales Madrid, Serie Zoológica* **8**: 29-56.

Escalera M. M. de la 1913b: Un nuevo género de Malthínido (Cantharidae) de Marruecos. *Boletín de la Real Sociedad Española de Historia Natural* **13**: 322-323.

Escalera M. M. de la 1914: Los Coleópteros de Marruecos. *Trabajos del Museo Nacional de Ciencias Naturales. Serie Zoológica* **11**: 3-553.

Escalera M. M. de la 1917: Un nuevo Ptínido de España central (Col.). *Boletín de la Real Sociedad Española de Historia Natural* **17**: 234-235.

Escalera M. M. de la 1921: Especies nuevas de coleópteros de Tenerife. *Boletín de la Real Sociedad Española de Historia Natural* **21**: 296-308.

Escalera M. M. de la 1923: Especies nuevas de Thorictus (Col.) de Marruecos y riberas de Mediterráneo. *Boletín de la Real Sociedad Española de Historia Natural* **23**: 376-380.

Escalera M. M. de la 1925a: Una especie nueva oriental de Triplax (Col. Erotylidae). *Boletín de la Real Sociedad Española de Historia Natural* **25**: 155-156.

Escalera M. M. de la 1925b: Especies del género Triplax Payk. (Col. Erotyl.) de la Península Ibérica y Marruecos. *Eos, Revista Española de Entomología* **1**: 239-242.

Escalera M. M. de la 1925c: Especias nuevas de coleópteros del Mediano Atlas marroquí. *Boletín de la Real Sociedad Española de Historia Natural* **25**: 311-317.

Escalera M. M. de la 1925d: Nuevo nombre para un Triplax de Siria (Col. Erotyl.). *Boletín de la Real Sociedad Española de Historia Natural* **25**: 456-457.

Escalera M. M. de la 1926a: Avance para el concimiento de los Axinotarsus del Mediterráneo occidental (Col. Malachidae). *Eos, Revista Española de Entomología* **2**: 217-279.

Escalera M. M. de la 1926b: Coleópteros nuevos de las Sierras de Segura. *Boletín de la Real Sociedad Española de Historia Natural* **26**: 429-435.

Escalera M. M. de la 1927: Los Allotarsus Grlls. y géneros afines ibero-africanos (Col. Dasyt.). *Eos, Revista Española de Entomología* **3**: 5-28.

Escalera M. M. de la 1933a: Especies nuevas de Attalus Er. del Rif del grupo de A. convolvulus Ab. (Col. Malaq.). *Boletín de la Real Sociedad Española de Historia Natural* **33**: 219-220.

Escalera M. M. de la 1933b: Dos Sphinginus Rey del Norte de Marruecos (Col. Malach.). *Boletín de la Real Sociedad Española de Historia Natural* **33**: 253-255.

Escalera M. M. de la 1933c: Dos especies nuevas de Attalus Er. de Melilla del grupo de *A. paradoxus* Ab. (Col. Malach.). *Boletín de la Real Sociedad Española de Historia Natural* **33**: 359-360.

Escalera M. M. de la 1933d: Especies nuevas de Attalus Er. de Xauen y del Alto Atlas Marroquí del grupo de A. omophloides Ab. (Col. Malach.). *Boletín de la Real Sociedad Española de Historia Natural* **33**: 393-395.

Escalera M. M. de la 1936: Una especie nueva de Antholinus Rey (Attalus) del Rif central (Col. Malach.). *Boletín de la Real Sociedad Española de Historia Natural* **36**: 433-434.

Escalera M. M. de la 1937: Especies nuevas de Troglops de Marruecos. *Boletín de la Real Sociedad Española de Historia Natural* **37**: 47-49.

Escalera M. M. de la 1940: Especies nuevas de Maláchidos del Atlas y de Ifni (Col. Malach.). Pp. 255-262. *VI Congreso Internacional de Entomología. Madrid, 6-12, Septiembre 1935.*

Escalera M. M. de la 1942a: Especies nuevas de Attalus Er. concolores de España (Col. Malach.). *Boletín de la Real Sociedad Española de Historia Natural* **40**: 133-135.

Escalera M. M. de la 1942b: Especies nuevas de Thorictus Germ. (Col. Thorictidae) de Tripolitania y riberas occidentales del Mediterraneo de las colecciones A. Dodero y Museo Civico de Genova, revisadas por mi con enterioridad a 1934. *Eos, Revista Española de Entomología* **18**: 381-391, pls XI, XII.

Escherich K. L. 1892: Studien über die Cleriden-Gattung Trichodes Hbst. *Wiener Entomologische Zeitung* **11**: 226-230.

Escherich K. L. 1894: Zur Kenntnis der Coleopterengattung Trichodes Herbst. (Eine monographische Studie). *Verhandlunger der Kaiserlich-Königlichen Zoologisch-Botanischen Gesellschaft in Wien* **43** [1893]: 149-203, 2 pls.

Escherich K. L. 1897: Zoologische Ergebnisse einer von Dr. K. Escherich und Dr. L. Katharines nach Central-Kleinasien unternommenen Reise. IV. Theil. Coleopteren. Unter Mitwirkung von E. Brenske, L. Ganglbauer, Dr. L. von Heyden, E. Wasmann und J. Weise. *Entomologisches Zeitung* (Stettin) **58**: 3-63.

Eschscholtz J. F. von. 1818: Decades tres Eleutheratorum novorum. *Mémoires de l'Académie Impériale de Sciences St. Pétersburg* (5) **6**: 451-484.

Eschscholtz J. F. von. 1822: Description d'insectes nouveaux. *Entomographien: erste Lieferung.* Berlin: G. Reimer, iii + 128 pp., 2 pls.

Eschscholtz J. F. G. von. 1823: *Entomographien. Naturwissenschaftliche Abhandlungen* (Dorpat) **1**: 57-186 pp., 2. pls.

Eschscholtz J. F. G. von. 1829: Elaterites, Eintheilung derselben in Gattungen. *Thon's Entomologisches Archiv* **2** (1): 31-35.

Eschscholtz J. F. G. von. 1830a: Die Springkäfer Livlands, unter neuere Gattungen vertheilt. *Trautvetter, Quatember* (*Mitau*) **2** (3): 13-19.

Eschscholtz J. F. G. von. 1830b: Nova genera Coleopterorum faunae Europaeae. *Bulletin de la Société Impériale des Naturalistes de Moscou* **2**: 16-65.

Eschscholtz J. F. G. von. 1838: Classification des Elatérides. *Revue Entomologique* (G. Silbermann) **4**: [pl.].

Español F. 1943: Contribución al conocimiento de los Coleópteros de Ibiza y Formentera y un nuevo Heliophilus ibérico (Col. Tenebrionidae). *Anales de la Asociación Española para Progreso de las Ciencias, Madrid* **8**: 93-108.

Español F. 1945: Coleópteros nuevos o interesantes para la fauna Ibero-Balear. *Eos, Revista Española de Entomología* **21**: 83-105.

Español F. 1962: Notas sobre Anóbidos. 2-3. *Publicaciones del Instituto de Biología Aplicada* **33**: 57-69.

Español F. 1963a: Notas sobre Anóbidos. 4-6. *Publicaciones del Instituto de Biología Aplicada* **34**: 81-94.

Español F. 1963b: Notas sobre Anóbidos. VIII. Los Gastrallus Mediterráneos. *Publicaciones del Instituto de Biología Aplicada* **35**: 5-21.

Español F. 1964: Los Xestobium Motsch. paléarticos (Col. Anobiidae). *Eos, Revista Española de Entomología* **40**: 123-137.

Español F. 1965: Notas sobre Anóbidos (Coleoptera). XV. Precisiones sobre el género „Ptilinus" Geoffr. y otros elementos afines. *Eos, Revista Española de Entomología* **41**: 45-58.

Español F. 1967: Notas sobre Anóbidos (Coleoptera) XXVII.-XXIX. *Eos, Revista Española de Entomología* **43**: 27-56.

Español F. 1968: Notas sobre anóbidos (Coléopteros). XXXIII. – Nuevos datos sobre anóbidos de las Islas Canarias. *Miscelánea Zoológica* (Barcelona) **2** (3): 75-83.

Español F. 1969a: Notas sobre anóbidos (Col.). XXXVII.-XXXIX. *Publicaciones del Instituto de Biología Aplicada* **46**: 49-64.

Español F. 1969b: Notas sobre Anóbidos (Coleoptera). XXXV. – Los Stagetus Woll. del mediterráneo occidental. *Eos, Revista Española de Entomología* **44**: 103-119.

Español F. 1970a: Notas sobre Anóbidos (Coleoptera). LI. Complementos descriptivos. *Miscelánea Zoológica* (Barcelona) **2** (5): 97-98.

Español F. 1970b: Notas sobre Anóbidos. XXXVI. – Sobre la disparidad genérica del complejo Stagetus Woll. *Anales de la Escuela Nacional de Ciencias Biológicas* **17**: 157-168.

Español F. 1971: Ergebnisse der zoologischen Forschungen von Dr. Z. Kaszab in der Mongolei N.° 249 (Col. Anobiidae). *Eos, Revista Española de Entomología* **46**: 41-56.

Español F. 1972a: Notas sobre Anóbidos (Col.). LVI.-LVIII. *Publicaciones del Instituto de Biología Aplicada* **52**: 49-65.

Español F. 1972b: Notas sobre Anóbidos. LV. Contribución al conocimiento del género Metholcus Duv. *Miscelánea Zoológica* **3**: 61-74.

Español F. 1974: Sobre la presencia del género Deroptilinus Lea en el sudeste asiático (Col. Anobiidae, nota LXVI). *Miscellànea Alcobé* **1**: 57-60.

Español F. 1977a: Sobre algunos Anobiidae (Col.) de Turquía recogidos por el Dr. W. Wittmer (Nota 78). *Mediterránea* **1977** (No. 2): 5-11.

Español F. 1977b: Notas sobre anóbidos. LXXIV.-LXXVI. *Miscelànea Zoològica* **4**: 171-215.

Español F. 1977c: Ergebnisse der Bhutan-Expedition 1972 des Naturhistorischen Museum in Basel. *Entomologica Basiliensia* **2**: 305-317.

Español F. 1978a: A new species of Clada (Coleoptera: Anobiidae) from Israel. *Israel Journal of Entomology* **12**: 1-4.

Español F. 1978b: Sobre algunos Dorcatoma del mediterráneo occidental (Col. Anobiidae, Nota 86). *Miscelánea Zoológica* **4**: 77-85.

Español F. 1978c: Sobre la presencia del Género "Clada" Pascoe en las Islas Salvajes (Col. «Anobiidae», Nota 79). Pp. 151-154. In: Contribución al Estudio de la Historia Natural de las Islas Salvajes. Resultados de la Expedición Cientifica "Agamenón 76". Santa Cruz de Tenerife: Aula Cultura de tenerife, 209 pp.

Español F. 1979: Insects of Saudi Arabia. Coleoptera: Fam. Anobiidae. *Fauna of Saudi Arabia* **1**: 251-256.

Español F. 1981a: Un género y especie nuevos del norte de la India a colocar en la vecindad de Oligomerus Redt. (Col. Anobiidae, Nota 94). *Publicaciones del Departamento de Zoología de la Universidad de Barcelona* **6**: 45-55.

Español F. 1981b: Insects of Saudi Arabia. Coleoptera: Fam. Anobiidae (Part 2). *Fauna of Saudi Arabia* **3**: 266-272.

Español F. 1983: Nuevos Anobiidae (Col.) del Sudeste Asiatico. Nota 99. *Miscelánea Zoológica* **7**: 75-90.

Español F. 1990a: Un nuevo Anobiidae del Reino del Nepal (Coleoptera). *Sessió Conjunta de Entomologia. ICHN-SCL* No: **6**: 45-47.

Español F. 1990b: Nuevos datos sobre Anóbidos Ibéricos (Col., Anobiidae). *Miscelánea Zoológica* **12**: 171-175.

Español F. 1994: Anobiidae (Coleoptera) de la fauna Íbero-Marroquí, con una nueva especie de Stagetus Wollaston, 1861. *Miscelánea Zoológica* (Barcelona) **17**: 149-152.

Español F. & Bellés X. 1981: Los Dryophilinae Hispano-Marroquíes (Coleoptera, Anobiidae). *Fragmenta Entomologica* (Roma) **16**: 61-72.

Español F. & Bellés X. 1984: Dos nuevos Falsogastrallus Pic (Col. Anobiidae) asiáticos. *Entomologische Blätter* **80**: 26-28.

Español F. & Oromí P. 1984: Nuevos datos sobre los Anobiidae (Col.) de las Islas Canarias. *Vieraea* **13**: 83-92.

Evers A. M. J. 1944: Malachius viridis F. und bipustulatus L. 1. Beitrag zur Kenntnis der Malachiidae (Col.). *Entomologische Blätter* **40**: 138-139.

Evers A. M. J. 1948a: Neue Malachiidae (Col.). (5. Beitrag zur Kenntnis der Malachiidae). *Entomologische Blätter* **41-44** [1945-48]: 49-60.

Evers A. M. J. 1948b: Myrmecophasma fukienensis n. sp., a new malachiid-beetle from Fukien (South-East-China) (3rd contribution to the knowledge of Malachiidae). *Tijdschrift voor Entomologie* **89**: 155-157.

Evers A. M. J. 1948c: Synonymische Notizen. (8. Beitrag zur Kenntnis der Malachiidae). *Entomologische Blätter* **41-44** [1945-48]: 189-190.

Evers A. M. J. 1955: Einige von Herrn G. Fagel in Algerien gesammelte Malachiidae. Mit Beschreibung einer neuen Art. (10. Beitrag zur Kenntnis der Malachiidae). *Entomologische Blätter* **51**: 75-77.

Evers A. M. J. 1959: Die Malachiidae (Col.) der Kanarischen Inseln (12. Beitrag zur Kenntnis der Malachiidae). *Entomologische Blätter* **55**: 219-247.

Evers A. M. J. 1960b: Eine neu-entdeckte alte Ausbeute von den Kanarischen Inseln (14. Beitrag zur Kenntnis der Malachiidae). *Entomologische Blätter* **56**: 166-169.

Evers A. M. J. 1961a: Die Malachiidae (Col.) der Kanarischen Inseln. II. (16. Beitrag zur Kenntnis der Malachiidae). *Entomologische Blätter* **57**: 3-26.

Evers A. M. J. 1961b: Neue paläarktische Malachiidae (Col.) (17. Beitrag zur Kenntnis der Malachiidae). *Entomologische Blätter* **57**: 171-178.

Evers A. M. J. 1962: Die Malachiidae (Col.) von Marokko (18. Beitrag zur Kenntnis der Malachiidae). *Bulletin de la Société des Sciences Naturelles et Physiques du Maroc* **42**: 145-193.

Evers A. M. J. 1964: Neue palaearktische Malachiidae (Col.). (23. Beitrag zur Kenntnis der Malachiidae). *Entomologische Blätter* **60**: 15-17.

Evers A. M. J. 1965: 59. Malachiidae II. Ergebnisse der zoologischen Forschungen von Dr. Z. Kaszab in der Mongolei (Coleoptera) (29. Beitrag zur Kenntnis der Malachiidae). *Reichenbachia* **7**: 147-150.

Evers A. M. J. 1966a: Zwei neue Malachiidae (31. Beitrag zur Kenntnis der Malachiidae). *Entomologische Blätter* **62**: 44-46.

Evers A. M. J. 1966b: Die Malachiidae (Col.) von Marokko. II. (35. Beitrag zur Kenntnis der Malachiidae). *Bulletin de la Société des Sciences Naturelles et Physiques du Maroc* **46**: 87-120.

Evers A. M. J. 1966c: Eine neuer Troglocollops aus Turkestan (36. Beitrag zur Kenntnis der Malachiidae). *Notulae Entomologicae* **46**: 105-106.

Evers A. M. J. 1968a: 135. Malachiidae IV. Ergebnisse der zoologischen Forschungen von Dr. Z. Kaszab in der Mongolei (Coleoptera) (39. Beitrag zur Kenntnis der Malachiidae). *Reichenbachia* **11**: 25-34.

Evers A. M. J. 1968b: Malachiidae aus der Mongolei. Ergebnisse der Mongolisch-Deutschen Biologischen Expeditionen seit 1962, Nr. 37. (40. Beitrag zur Kenntnis der Malachiidae). *Mitteilungen aus dem Zoologischen Museum in Berlin* **44**: 123-124.

Evers A. M. J. 1969a: 182. Malachiidae V. Ergebnisse der zoologischen Forschungen von Dr. Z. Kaszab in der Mongolei (Coleoptera) (44. Beitrag zur Kenntnis der Malachiidae). *Reichenbachia* **12**: 187-198.

Evers A. M. J. 1969b: Neue Malachiidae aus dem Zoologischen Museum Berlin (41. Beitrag zur Kenntnis der Malachiidae). *Mitteilungen aus dem Zoologischen Museum in Berlin* **45**: 239-243.

Evers A. M. J. 1971a: Attalus oceanicus nov. spec. (Col. Malach.) (50. Beitrag zur Kenntnis der Malachiidae). *Entomologische Blätter* **66**: 175-177.

Evers A. M. J. 1971b: Über die paläarktischen Arten der Gattungen Sphinginus Rey und Fortunatius nov. gen. (49. Beitrag zur Kenntnis der Malachiidae). *Entomologische Blätter* **67**: 21-41.

Evers A. M. J. 1971c: Die Attalus pini-Gruppe der West-Kanaren. Eine neue Gruppe mit vier neuen Arten. (51. Beitrag zur Kenntnis der Malachiidae). *Entomologische Blätter* **67**: 65-73.

Evers A. M. J. 1971d: 206. Malachiidae VI. Ergebnisse der zoologischen Forschungen von Dr. Z. Kaszab in der Mongolei (Coleoptera). (45. Beitrag zur Kenntnis der Malachiidae). *Reichenbachia* **13**: 227-237.

Evers A. M. J. 1981: Ifnidius atlanticus n. sp. (Col., Malachiidae) von den Selvagens-Inseln. (53. Beitrag zur Kenntnis der Malachiidae). *Entomologische Blätter* **77**: 155-157.

Evers A. M. J. 1985: Aufteilung der paläarktischen Arten des Gattungkomplexes Malachius F. 54. Beitrag zur Kenntnis der Malachiidae. *Entomologische Blätter* **81**: 1-40.

Evers A. M. J. 1986a: Attalus remanei nov. spec. von Madeira. (60. Beitrag zur Kenntnis der Malachiidae). *Entomologische Blätter* **82**: 91-92.

Evers A. M. J. 1986b: Ein neuer Hypebaeus (Col. Malachiidae) aus Kashmir. 62. Beitrag zur Kenntnis der Malachiidae. *Entomologische Blätter* **82**: 101-102.

Evers A. M. J. 1986c: Die kanarischen Arten der Gattungen Cephaloncus Westw. und Cephalogonia Woll. mit Beschreibungen von vier neuen Arten. (Col. Malachiidae). *Entomologische Blätter* **82**: 129-144.

Evers A. M. J. 1987: Synopsis der Gattung Apalochrus Er. und der verwandten Gattungen der Welt (Col., Malachiidae). 63. Beitrag zur Kenntnis der Malachiidae. *Annalen Zoologische Wetenschappen* **253**: 1-73.

Evers A. M. J. 1988a: SW-paläarktische Malachiidae. (69. Beitrag zur Kenntnis der Malachiidae). *Entomologische Blätter* **84**: 72-78.

Evers A. M. J. 1988b: Die Gattungen Microlipus LeC. und Wittmerius gen. nov. (70. Beitrag zur Kenntnis der Malachiidae). *Entomologische Blätter* **84**: 165-170.

Evers A. M. J. 1988c: Synopsis des Monophylums um Attalus tristis Luc. mit Beschreibung einer neuen Art, Attalus santiagoi nov. spec. in der SW-Paläarktis. (71. Beitrag zur Kenntnis der Malachiidae). *Entomologische Blätter* **84**: 171-174.

Evers A. M. J. 1989b: Ein neuer schwarzer Troglops Er. aus Süd-Spanien. (72. Beitrag zur Kenntnis der Malachiidae). *Entomologische Blätter* **85**: 95-98.

Evers A. M. J. 1990: Fortunatius subextinctus nov. spec. von La Palma (Kanaren), eine neue, wahrscheinlich bereits ausgestorbene Art. (79. Beitrag zur Kenntnis der Malachiidae). *Entomologische Blätter* **86**: 161-165.

Evers A. M. J. 1991a: Synopsis der Kanaren-endemische Attalus tuberculatus-Gruppe. (80. Beitrag zur Kenntnis der Malachiidae). *Entomologische Blätter* **87**: 99-107.

Evers A. M. J. 1991b: Protocollops gen. nov. (Col. Malachiidae). (82. Beitrag zur Kenntnis der Malachiidae). *Entomologische Blätter* **87**: 108-109.

Evers A. M. J. 1993: Eine neue Ebaeus-Art aus Lappland. (86. Beitrag zur Kenntnis der Malachiidae). *Entomologische Blätter* **89**: 157-159.

Evers A. M. J. 1994: Anthocomus creticus nov. spec. von Kreta. (87. Beitrag zur Kenntnis der Malachiidae). *Entomologische Blätter* **90**: 77-79.

Evers A. M. J. 1995a: Juengerius nom. nov. für Wittmerius Evers (91. Beitrag zur Kenntnis der Malachiidae). *Entomologische Blätter* **91**: 2.

Evers A. M. J. 1995b: Neue SW-Palaarktische Malachiidae aus Marokko und von den Kanarischen Inseln. (90. Beitrag zur Kenntnis der Malachiidae). *Entomologische Blätter* **91**: 85-95.

Evers A. M. J. 1996: Eine neue Kuatunia-Art aus Nepal. (92. Beitrag zur Kenntnis der Malachiidae). *Entomologische Blätter* **92**: 90-91.

Evers A. M. J. & Plata P. 1990: Cephaloncus laureatus nov. spec. (Col. Malachiidae) von Gran Canaria. (78. Beitrag zur Kenntnis der Malachiidae). *Entomologische Blätter* **86**: 159-160.

Everts E. J. G. 1898: *Coleoptera Neerlandica. De Schildvleugelige insecten van Nederland en het aangrenzen gebied. Erste deel.* 'S Gravenhage: Martinus Nijhoff, [8] + 676 + [1] pp.

Everts E. J. G. 1920: Nieuwe vondsten voor de Niederlandische Coleopterenfauna, XXVII. *Entomologische Berichten* **5**: 226-231.

Everts E. J. G. 1921: Nieuwe vondsten voor de Niederlandische Coleopterenfauna, XXXII. *Entomologische Berichten* **5**: 302-309.

Fabricius J. C. 1775: *Systema entomologiae, sistens insectorum classes, ordines, genera, species, adiectis synonymis, locis, descriptionibus, observationibus*. Flensburgi et Lipsiae: Korte, xxxii + 832 pp.

Fabricius J. C. 1777: *Genera insectorum eorumoue characteres anturales secundum numerum, figuram, situm et proportionem omnium partium oris adiecta mantissa specierum nuper detectarum*. Chilonii: Mich. Friedr. Bartschii, xvi + 310 pp.

Fabricius J. C. 1781: *Species insectorum exhibentes eorum differentias specificas, synonyma auctorum, loca natalia, metamorphosin adiectis observationibus, descriptionibus. Tom 1*. Hamburgi et Kilonii: C. E. Bohn, viii + 552 pp.

Fabricius J. C. 1787a: *Mantissa insectorum sistens eorum species nuper detectas adiectis characteribus genericis, differentiis specificis, emendationibus, observationibus. Tom I*. Hafniae: Chist. Gottl. Proft, xx + 348 pp.

Fabricius J. C. 1787b: *Mantissa insectorum sistens eorum species nuper detectas adjectis characteribus genericis, differentiis specificis, emendationibus, observationibus. Tom II*. Haffniae: Christ. Gottl. Proft, 382 pp.

Fabricius J. C. 1792a: *Entomologia systematica emendata et aucta. Secundum classes, ordines, genera, species adjectis synonymis, locis, descriptionibus, observationibus. Tom I. Pars I*. Hafniae: Christ. Gottl. Proft, xx + 330 pp.

Fabricius J. C. 1792b: *Entomologia systematica emendata et aucta. Secundum classes, ordines, genera, species adjectis synonymis, locis, descriptionibus, observationibus. Tom I. Pars II*. Hafniae: Christ. Gottl. Proft, 538 pp.

Fabricius J. C. 1794: *Entomologia systematica emendata et aucta. Sécundum classes, ordines, genera, species adjectis synonimis, locis, observationibus, descriptionibus. Tom. IV*. [Appendix specierum nuper detectarum: pp. 435-462]. Hafniae: C. G. Proft, Fil. et Soc., [6] + 472 + [5] pp.

Fabricius J. C. 1798: *Supplementum entomologiae systematicae*. Hafniae: Proft et Storch, [2] + 572 pp.

Fabricius J. C. 1801a: *Systema Eleutheratorum secundum ordines, genera, species adiectis synonymis, locis, observationibus, descriptionibus.Tomus I*. Kiliae: Bibliopolii Academici Novi, xxiv + 506 pp.

Fabricius J. C. 1801b: *Systema Eleutheratorum secundum ordines, genera, species adiectis synonymis, locis, observationibus, descriptionibus. Tomus II*. Kiliae: Bibliopolii Academici Novi, 687 pp.

Fabricius J. C. 1803: *Index alphabeticus in J. C. Fabricii Systema Eleutheratorum genera et species continens*. Helmstadii: Fleckeisen, iv + 93 pp.

Fagniez C. 1946: Étude des Divales et Dasytes de France et de Corse (Col. Dasytidae). *Revue Française d'Entomologie* **13**: 19-27.

Fåhraeus O. L. 1871: Coleoptera Caffrariae, annis 1838-1854 a J.A. Wahlberg collecta. Heteromera. Fam.: Scolytidae, Paussidae, Bostrychidae et Cioidae. *Öfversigt af Finska Vetenskaps-Societetens Förhandlingar* **28** (6): 661-672.

Fairmaire L. 1849a: Insectes de Taïti des Marquises et des îles voisines. *Revue et Magasin de Zoologie Pure et Apliquée* (2) **1**: 34-36.

Fairmaire L. 1849b: Essai sur les coléoptères de la Polynésie. (Suite). *Revue et Magasin de Zoologie Pure et Appliquée* (2) **1**: 352-365.

Fairmaire L. 1849c: Description de quelques coléoptères nouveaux d'Europe et de France. *Annales de la Société Entomologique de France* (2) **7**: 419-427.

Fairmaire L. 1850: Essai sur les coléoptères de la Polynésie. *Revue et Magasin de Zoologie Pure et Appliquée* (2) **2**: 50-64.

Fairmaire L. 1851: Description de sept coléoptères nouveaux provenant du Midi de l'Europe et du Maroc. *Revue et Magasin de Zoologie Pure et Apliquée* (2) **3**: 527-528.

Fairmaire L. 1852: Quelques coléoptères nouveaux du Midi de l'Europe et du Nord de l'Afrique. *Annales de la Société Entomologique de France* (2) **10**: 69-93.

Fairmaire L. 1856a: Miscellanea entomologica. Première partie. *Annales de la Société Entomologique de France* (3) **4**: 517-548.

Fairmaire L. 1856b: Coleoptera maroccana. *Revue de Zoologie Pure et Appliquée* (2) **8**: 530-531.

Fairmaire L. 1857a: Rapport sur la session extraordinaire tenue à Montpellier. *Annales de la Société Entomologique de France* (3) **5**: 619-647.

Fairmaire L. 1857b: Miscellanea entomologica. *Annales de la Société Entomologique de France* (3) **5**: 725-745.

Fairmaire L. 1859a: Miscellanea entomologica. Troisième partie. *Annales de la Société Entomologique de France* (3) **7**: 21-64.

Fairmaire L. 1859b: Notice nécrologique sur Ph. Lareynie suivie d'observations sur les coléoptères de l'île de Corse. *Annales de la Société Entomologique de France* (3) **7**: 261-283.

Fairmaire L. 1859c: [Xyletinus sanguineo-cinctus]. *Bulletin de la Société Entomologique de France* **1859**: ciii-cv.

Fairmaire L.1859d: Descriptions de quelques espèces nouvelles de coléoptères récoltées par M. Gougelet pendant son voyage en Galice dans le courant de l'année 1859. *Bulletin de la Société Entomologique de France* **1859**: cl-clii.

Fairmaire L. 1860: [new taxa]. In: Fairmaire L. & Coquerel C.: Essai sur les coléoptères de Barbarie. Deuxième partie. *Annales de la Société Entomologique de France* (3) **8**: 145-176.

Fairmaire L. 1860a: Diagnose de nouvelles espèces de coléoptères. *Annales de la Société Entomologique de France* (3) **8**: 629-632.

Fairmaire L. 1862a: Miscellanea entomologica. Quartième partie. *Annales de la Société Entomologique de France* (4) **1** [1861]: 577-596.

Fairmaire L. 1862b: Description de coléoptères nouveaux recueillis pendant l'excursion en Savoie. *Annales de la Société Entomologique de France* (4) **1** [1861]: 647-650.

Fairmaire L. 1862c: Miscellanea entomologica. Cinquième partie. *Annales de la Société Entomologique de France* (4) **2**: 547-558.

Fairmaire L. 1863: [new taxon]. In: Grenier A.: *Catalogue des coléoptères de France, et materiaux pour servir à la faune des coléoptères Française*. Paris: L. Toinon, iv + 3-79 + 135 pp.

Fairmaire L. 1864: Coléoptères d'Algérie rapportés par M. Ogier de Baulny. *Annales de la Société Entomologique de France* (4) **3** [1863]: 637-648.

Fairmaire L. 1866a: [new taxa]. In: Fairmaire L. & Coquerel J. C.: Essai sur les coléoptères de Barbarie. Quatrième partie. *Annales de la Société Entomologique de France* (4) **6**: 17-74.

Fairmaire L. 1866b: Notice sur les coléoptères récoltés par M. J. Lédérer sur le Bosz-Dagh (Asie mineure). *Annales de la Société Entomologique de France* (4) **6**: 249-280.

Fairmaire L. 1867: Essai sur les coléoptères de Barbarie, Cinquième partie. *Annales de la Société Entomologiae de France* (4) **7**: 387-416.

Fairmaire L. 1868: Essai sur les coléoptères de Barbarie. Sixième partie. *Annales de la Société Entomologique de France* (4) **8**: 471-502.

Fairmaire L. 1869a: Coleoptera Europae nova. *Entomologische Zeitung* (Stettin) **1869**: 231-233.

Fairmaire L. 1869b: Notes sur les coléoptères recueillis par Charles Coquerel à Madagascar et sur les côtes d'Afrique. 1. Partie. *Annales de la Société Entomologique de France* (4) **8** [1868]: 753-820.

Fairmaire L. 1871: Essai sur les coléoptères de Barbarie. Septième Partie. *Annales de la Société Entomologique de France* (4) **10** [1870]: 369-404.

Fairmaire L. 1873: [new taxa]. In: Fairmaire L. & Raffray A.: Coléoptères du Nord de l'Afrique. Première partie. *Revue de Zoologie Pure et Appliquée* (3) **1**: 331-385.

Fairmaire L. 1874: Note sur quelques coléoptères des environs de Tuggurth. *Petites Nouvelles Entomologiques* **1** [1869-1875]: 407-408.

Fairmaire L. 1875a: Coléoptères de la Tunisie récoltés par M[r]. Abdul Kerim. *Annali del Museo di Storia Naturale in Genova* **7**: 475-540.

Fairmaire L. 1875b: Diagnose de quelques coléoptères nouveaux d'Algérie. *Petites Nouvelles Entomologiques* **1** [1869-1875]: 495-496.

Fairmaire L. 1876a: Diagnoses de coléoptères du nord de l'Afrique. *Petites Nouvelles Entomologiques* **2** [1876-1879]: 37-38.

Fairmaire L. 1876b: Diagnoses de coléoptères du nord de l'Afrique *Petites Nouvelles Entomologiques* **2** [1876-1879]: 49-50.

Fairmaire L. 1876c: Diagnoses de coléoptères du nord de l'Afrique. *Petites Nouvelles Entomologiques* **2** [1876-1879]: 93-94.

Fairmaire L. 1877a: Coléoptères nouveaux du nord de l'Afrique. *Petites Nouvelles Entomologiques* **2**: 141.

Fairmaire L. 1877b: [Descriptions de trois coléoptères nouveaux, propres à la faune européenne]. *Bulletin de la Société Entomologique de France* **1877**: lxv-lxvii.

Fairmaire L. 1878a: [new taxa]. In: Marseul S. A. de: Nouvelles et faits divers de l'Abeille. *L'Abeille Journal d'Entomologie* **21**: 81-84.

Fairmaire L. 1878b: [new taxa]. In: Deyrolle H. & Fairmaire L.: Descriptions de coléoptères recueillis par M. l'abbé David dans la Chine centrale. *Annales de la Société Entomologique de France* (5) **8**: 87-140, pls. 3,4.

Fairmaire L. 1878c: Coléoptères de Cochinchine recueillis par M. le docteur Morice. *Annales de la Société Entomolgique de France* (5) **8**: 269-274.

Fairmaire L. 1879a: Descriptions de coléoptères nouveaux du nord de l'Afrique. 1re partie. *Annales de la Société Entomologique de France* (5) **9**: 155-172.

Fairmaire L. 1879b: Descriptions de coléoptères nouveaux du nord de l'Afrique. 2 partie. *Annales de la Société Entomologique de France* (5) **9**: 245-258.

Fairmaire L. 1880a: Descriptions de coléoptères nouveaux du nord de l'Afrique. 3e Partie. *Annales de la Société Entomologique de France* (5) **10**: 5-32.

Fairmaire L. 1880b: Description d'un coléoptère nouveau et remarques synonymiques. *Le Naturaliste* **1**: 150.

Fairmaire L. 1880c: Diagnoses de Nitidulides d'Afrique. *Le Naturaliste* **1**: 181-182.

Fairmaire L. 1880d: Description de coléoptères d'Espagne et de Turquie. *Annales de la Société Entomologique de France* (5) **10**: 237-244.

Fairmaire L. 1880e: Diagnoses de coléoptères de Nossi-Bé. *Le Naturaliste* **2**: 246.

Fairmaire L. 1881: Descriptions de quelques coléoptères de Syrie. *Annales de la Société Entomologique de France* (6) **1**: 79-88.

Fairmaire L. 1882a: Description d'une espèce nouvelle du genre Atractocerus. *Notes from the Leyden Museum* **4**: 217-218.

Fairmaire L. 1882b: Trois nouvelles espèces de coléoptères appatenant au Musée Civique de Gênes. *Annali del Museo Civico di Genova* **18** [1882-83]: 445-447.

Fairmaire L. 1883a: Essai sur les coléoptères de l'archipel de la Nouvelle-Bretagne. *Annales de la Société Entomologique de Belgique* **27**: 1-58.

Fairmaire L. 1883b: Descriptions de coléoptères nouveaux ou peu connus récoltés par Mr. Raffray en Abyssinie. *Annales de la Société Entomologique de France* (6) **3**: 89-112.

Fairmaire L. 1883c: [Communication]. *Bulletin de la Société Entomologique de France* **1883**: xxxiv-xxxv.

Fairmaire L. 1883d: Descriptions de coléoptères recueillis par M. Bonnaire en Algérie. *Bulletin ou Comptes-Rendus des Séances de la Société Entomologique de Belgique* **1883**: cxi-cxiv, clvi-clix.

Fairmaire L. 1883e: Description d'un coléoptères nouveaux du nord de l'Afrique. *Bulletin de la Société Entomologique de France* **1882**: clxxxix-cxc.

Fairmaire L. 1884a: Liste des coléoptères recueillis par M. l'abbé David à Akbès (Asie-Mineure) et descriptions des espèces nouvelles. *Annales de la Société Entomologique de France* (6) **4**: 165-180.

Fairmaire L. 1884b: Description de coléoptères marocains. *Le Naturaliste* **6**: 446-447.

Fairmaire L. 1884c: Description d'une nouvelle espèce de coléoptère: Lampyris (Pelania) scutellata. *Bulletin de la Société Entomologique de France* **1884**: xxxv.

Fairmaire.L. 1884d: Descriptions de coléoptères recueillis par le Baron Bonnaire en Algérie. *Bulletin ou Comptes-Rendus des Séances de la Société Entomologique de Belgique* **1884**: lix-lxx.

Fairmaire L. 1884e: Descriptions de coléoptères nouveaux du Maroc. *Bulletin ou Comptes-Rendus des Séances de la Société Entomologique de Belgique* **1884**: cxvii-cxxi.

Fairmaire L. 1884f: Diagnoses de coléoptères de l'Afrique Orientale. *Bulletin ou Comptes-Rendus des Séances de la Société Entomologique de Belgique* **1884**: cxxi-cxxv.

Fairmaire L. 1885: Malacodermes, Lymexylonide et Hétéromères nouveaux receuillis par M. Deby à Sumatra et à Bornéo. *Bulletin ou Comptes-Rendus des Séances de la Société Entomologique de Belgique* **1885**: cv-cxii.

Fairmaire L. 1886a: Note sur les coléoptères recueillis par M. Laligant à Obock. *Annales de la Société Entomologique de France* (6) **5** [1885]: 435-462.

Fairmaire L. 1886b: Notes sur les coléoptères recueillis par M. Raffray, à Madagascar et descriptions des espèces nouvelles (2e partie). *Annales de la Société Entomologique de France* (6) **6**: 31-96, pl. 2.

Fairmaire L. 1886c: Descriptions de coléoptères de l'intérieur de la Chine (2. partie). *Annales de la Société Entomologique de France* (6) **6**: 303-356.

Fairmaire L. 1887a: Coléoptères de l'intérieur de la Chine (3e partie). *Annales de la Société Entomologique de Belgique* **31**: 87-136.

Fairmaire L. 1887b: Coléoptères nouveaux ou peu connus du Musée de Leyde. *Notes from the Leyden Museum* **9**: 145-162.

Fairmaire L. 1887c: [new taxa]. *Bulletin de la Société Entomologique de France* **1886**: clxii.

Fairmaire L. 1888a: Coléoptères de l'intérieur de la Chine. (Suite). *Annales de la Société Entomologique de Belgique* **32**: 7-46.

Fairmaire L. 1888b: Notes sur les coléoptères des environs de Pékin. *Revue d'Entomologie* **7**: 111-160.

Fairmaire L. 1889a: Descriptions de coléoptères de l'Indo-Chine. *Annales de la Société Entomologique de France* (6) **8** [1888]: 333-378.

Fairmaire L. 1889b: Coléptères de l'intérieur de la Chine. 5e partie. *Annales de la Société Entomologique de France* (6) **9**: 5-84.

Fairmaire L. 1889c: [Diagnoses de quelques coléoptères recueillis à Pékin par M. Provost.] *Bulletin de la Société Entomologique de France* **1888**: cc-cci.

Fairmaire L. 1891a: Notes sur quelques coléoptères de l'Afrique intertropicale et descriptions d'espèces nouvelles. *Annales de la Société Entomologique de France* **60**: 231-274, pl. 5.

Fairmaire L. 1891b: Coléoptères de l'intérieur de la Chine. (Suite, 6e partie). *Bulletin ou Comptes-Rendus des Séances de la Société Entomologique de Belgique* **1891**: vi-xxiv.

Fairmaire L.1891c: [Descriptions de deux coléoptéres des îles Seychelles]. *Bulletin de la Société Entomologique de France* **1891**: lxx-lxxi.

Fairmaire L. 1891d: Descriptions de coléoptères des montagnes de Kashmir. *Bulletin ou Comptes-Rendus des Séances de la Société Entomologique de Belgique* **1891**: lxxxviii-ciii.

Fairmaire L. 1891e: Descriptions de coléoptères des montagnes de Kashmir (Suite). *Bulletin ou Compte Rendus des Séances de la Société Entomologique de Belgique* **1891**: cxxi-cxxxiv.

Fairmaire L. 1891f: Coléoptères de l'intérieur de la Chine (Suite: 7e partie). *Bulletin ou Compte Rendus des Séances de la Société Entomologique de Belgique* **35**: clxxxvii-ccxxiii.

Fairmaire L. 1891g: Coléoptères de l'Afrique orientale. *Bulletin ou Compte Rendus des Séances de la Société Entomologique de Belgique* **35**: cclxxix-cccvii.

Fairmaire L. 1892a: Coléoptères d'Obock. Troisième partie. *Revue d'Entomologie* **11**: 77-127.

Fairmaire L. 1892b: Descriptions de coléoptères des environs d'Akbès (Syrie). *Annales de la Société Entomologique de Belgique* **36**: 144-159.

Fairmaire L. 1893a: Notes sur les coléoptères du Choa. *Annales de la Société Entomologique de Belgique* **37**: 9-50.

Fairmaire L.1893b: Un genre nouveau et espèces nouvelles de coléoptères. *Bulletin de la Société Entomologique de France* **1893**: lxviii-lxix.

Fairmaire L. 1894: Descriptions de coléoptères d'Algérie. *Annales de la Société Entomologique de Belgique* **38**: 310-313.

Fairmaire L. 1895a: Deuxième note sur quelques coléoptères des environs de Lang-Song. *Annales de la Société Entomologique de Belgique* **39**: 173-190.

Fairmaire L. 1895b: Coléoptères nouveaux d'Akbés, des dernières chasses de M. Charles Delagrange (Ipelates, Onthophagus, Capnodis, Malacogaster, Lampyris et Cteniopus). *Bulletin de la Société Entomologique de France* **1895**: cix-cxi.

Fairmaire L. 1896a: Coléoptères de l'Inde boréale, Chine et Malaisie. *Notes from the Leyden Museum* **18**: 81-129.

Fairmaire L. 1896b: Descriptions de coléoptères recueillis par le Dr Ch. Martin à Obock, Djibouti et Aden. *Bulletin de la Société Entomologique de France* **1896**: 223-225.

Fairmaire L. 1897a: Description de six coléoptères de l'île Ishigaki-Sima (Onthophagus, Eurylampyris, Amarygmus, Pachyrhynchus, Camtorhinus et Pachyosa). *Bulletin de la Société Entomologique de France* **1897**: 68-72.

Fairmaire L. 1897b: Coléoptères de l'Inde et de la Malaisie. *Notes from the Leyden Museum* **18** [1896/97]: 225-240.

Fairmaire L. 1897c: Description de coléoptères nouveaux de la Malaisie, de l'Inde et de la Chine. *Notes from the Leyden Museum* **19**: 209-233.

Fairmaire L. 1897d: Coléoptères du Szé-Tchouen et de Koui-Tchéou (Chine). *Notes from the Leyden Museum* **19**: 241-255.

Fairmaire L. 1898: Matériaux pour la faune coléoptérique de la région Malgache. *Annales de la Société Entomologique de Belgique* **42**: 222-260.

Fairmaire L. 1900: Description des coléoptères nouveaux recueillis en Chine par M. de Latouche. *Annales de la Société Entomologique de France* **68** [1899]: 616-649.

Falcoz L. 1921: Description d'un genre nouveau et de deux espèces nouvelles de Cryptophagidae (Col.) du nord de l'Afrique. *Bulletin de la Société Entomologique de France* **1921**: 42-44.

Falcoz L. 1922: Diagnoses de deux Cryptophagus nouveaux, de Sardaigne (Col., Cryptophaginae). *Bulletin de la Société Entomologique de France* **1922**: 189-190.

Falcoz L. 1925: Description d'un Cryptophagus nouveau d'Espagne (Col. Cryptophagidae). *Bulletin de la Société Entomologique de France* **1925**: 252-253.

Falcoz L. 1929: Description d'une espèce et de deux variétés nouvelles du genre Cryptophagus. (Col. Cryptophagidae). *Bulletin de la Société Entomologique de France* **1929**: 41-42.

Faldermann F. 1832: Lettre sur la faune d'insectes de Perse. *Bulletin de la Société Impériale des Naturalistes de Moscou* **4**: 201-207.

Faldermann F. 1835a: Addimenta entomologica ad faunam rossicam in itineribus Jussu Imperatoris Augustissimi annis 1827-1831 a Cl. Ménétriés et Szovitz susceptis collecta, in lucem edita. *Nouveaux Mémoires de la Société Impériale des Naturalistes de Moscou* (2) **4**: 1-310, 10 pls. [Separate 1836 edition has additional title: *Fauna entomologica Trans-caucasica. Coleoptera. Pars 1, Additamenta Entomologica ad Faunam Rossicam - Coleoptera Persico-Armeniaca*, 314 pp.].

Faldermann F. 1835b: Coleopterorum ab ill. Bungio in China boreale, Mongolia et montibus Altaicus collectorum nec non ab ill. Turczaninoffio et Stschukino e provincia Irkutzk missorum illustrationes. *Mémoires de l'Académie des Sciences de Saint Pétersbourg* **2**: 337-464, 5 pls.

Faldermann F. 1837: Fauna entomologica Trans-Caucasica. Coleoptera. Pars II. *Nouveaux Mémoires de la Société Impériale des Naturalistes de Moscou* **5**: 1-433 pp., 15 pls.

Faldermann F. 1839: Fauna Entomologica Trans-Caucasica. Pars III. Supplementum ad Faunam Trans-caucasicam. *Nouveaux Mémoires de la Société Impériale des Naturalistes de Moscou* **6**: 1-338 pp.

Fall H. C. 1899: Revision of the Lathridiidae of Boreal America. *Transactions of the American Entomological Society* **26**: 101-190, pls 3-5.

Fall H. C. 1905: Revision of the Ptinidae of Boreal America. *Transactions of the American Entomological Society* **31**: 97-296.

Fall H. C. 1910: Miscellaneous notes and descriptions of North American Coleoptera. *Transactions of the American Entomological Society* **36**: 89-197.

Fall H. C. 1932: New Coleoptera XV. *The Canadian Entomologist* **64**: 56-62.

Fallén C. F. 1802: *Observationes Entomologiciae. Part: 1.* Lundae: Litteris Belingianis, 20 pp.

Fallén C. F. 1807: *Monographia Cantharidum et Malachiorum Sueciae.* Lundae: Berlingianis, 26 pp.

Falzoni A. 1924: Sui Malthodes ganglbaueri Fiori e cognatus Costa. *Bollettino della Società Entomologica Italiana* **56**: 106-110.

Faust J. 1877a: Aeltere und einige neue Käfer der russischen Fauna. *Bulletin de la Société Impériale des Naturalistes de Moscou* **52**: 34-45.

Faust J. 1877b: Beiträge zur Kenntniss der Käfer des europäischen und asiatischen Russlands mit Einschluss der Küsten des Kaspischen Meeres. (Fortsetzung). *Horae Societatis Entomologica Rossicae* **12**: 300-332.

Fauvel A. 1885: Throscides et Eucnémides gallo-rhénans. Tableaux analytiques et catalogue. *Revue d'Entomologie* **4**: 330-351.

Fauvel A. 1889: Liste des coléoptères communs à l'Europe et à l'Amérique du Nord, d'après le Catalogue de M. J. Hamilton avec remarques et additions. *Revue d'Entomologie* **8**: 92-174.

Fauvel A. 1891: Essai sur l'entomologie de la Haute-Auvergne (Mont-Dore et Plomb du-Cantal). 2e Supplément. *Revue d'Entomologie* **10**: 50-60.

Fauvel A. 1903: Faune analytique des coléoptères de la Nouvelle Calédonie. *Revue d'Entomologie* **22**: 203-379.

Fender K. M. 1951: The Malthini of North America (Coleoptera - Cantharidae). *The American Midland Naturalist* **46**: 513-629.

Fender K. M. 1966: A new North American genus of soldier beetles. *The Pan-Pacific Entomologist* **42**: 321-323.

Ferrante G. 1908: Una nova specie di Malthinus d'Egitto. *Bulletin de la Société Entomologique d'Egypte* **1**: 26-28.

Ferrari J. A. G. 1869: Drei neue westasiatische Käfer. *Verhandlungen der Kaiserlich-Königlichen Zoologisch-Botanischen Gesellschaft in Wien* **19**: 193-200.

Fiori A. 1899a: Alcuni fatti di policromismo femminile nel genere Cantharis. *Rivista Italiana di Scienze Naturali* **19**: 138-143.

Fiori A. 1899b: Fauna entomologica italiana. Coleotteri nuovi o poco conosciuti. *Atti della Società dei Naturalisti di Modena* **31**: 153-164.

Fiori A. 1900: Nuove specie di Coleotteri. *Atti della Società dei Naturalisti e Matematici di Modena* (4) 1, **32** [1899]: 101-111, pls. V, VI.

Fiori A. 1904: Due nuove specie di Malthodes Kies. della Sicilia. *Il Naturalista Siciliano* **17**: 74-76.

Fiori A. 1905a: Descrizione di un Dasytidae nuovo e di un altro poco conosciuto in Italia. *Rivista Coleotterologica Italiana* **3**: 81-89.

Fiori A. 1905b: Descrizione di due specie credute nuove die dintorni di Bologna. *Rivista Coleotterologica Italiana* **3**: 131-139.

Fiori A. 1905c: Revisione delle specie italiane a me note del genere Malthodes Kiesw. *Rivista Coleotterologica Italiana* **3**: 221-252.

Fiori A. 1906a: Revisione delle specie italiane a me note del genere Malthodes Kiesw. (cont. e fine). *Rivista Coleotterologica Italiana* **4**: 1-26.

Fiori A. 1906b: Descrizione di quattro specie dell'Emilia credute nuove. *Rivista Coleotterologica Italiana* **4**: 257-268.

Fiori A. 1908: Ancora sui Malthodes italini. *Rivista Coleotterologica Italiana* **6**: 1-12.

Fiori A. 1909a: Aggiunte al Catalogo Heyden, Reitter e Weise, 1906. *Rivista Coleotterologica Italiana* **7**: 22-25.

Fiori A. 1909b: I Dasytes italiani del gruppo dell' alpigradus Kiesw. *Rivista Coleotterologica Italiana* **7**: 26-36.

Fiori A. 1912a: Dalle mie recenti caccie di Malthodes (con una figura). *Rivista Coleotterologica Italiana* **10**: 33-40.

Fiori A. 1912b: Indicazioni topografiche. *Rivista Coleotterologica Italiana* **10**: 127-139.

Fiori A. 1913: [new taxa]. In: Della Beffa G. Revisione dei Coccinellidi Italiani. *Rivista Coleopterologica Italiana* **11**: 31.

Fiori A. 1914a: Le Cantharis di Sicilia confrontate con quelle di altri paesi. *Rivista Coleotterologica Italiana* **12**: 1-18.

Fiori A. 1914b: Le Cantharis di Sicilia confrontate con quelle di altri paesi (contin. ved. N. 1). *Rivista Coleotterologica Italiana* **12**: 45-87.

Fiori A. 1915a: Nouve specie italiane dei generi Malthinus e Malthodes. *Rivista Coleotterologica Italiana* **13**: 45-56.

Fiori A. 1915b: Appunti sulla fauna coleotterologica dell'Italia meridionale e della Sicilia. *Rivista Coleotterologica Italiana* **13**: 57-84.

Fisher W. S. 1938: A new anobiid beetle (Coleoptera: Anobiidae) injurious to books. *Proceedings of the Entomological Society of Washington* **40**: 43-44.

Fisher W. S. 1950: A revision of the North American species of beetles belonging to the family Bostrychidae. *Miscellaneous Publications. United States Department of Agriculture, Washington,* No. **698**: 157 pp.

Fischer von Waldheim G. 1823: *Entomographie de la Russie, et genres des insectes. Entomographia Imperii Rossici, sue Caesareae Majestati Alexandro I dicata. Volume II.* Plates XVIII-L [Coleoptera] + VI-XI [Lepidoptera].

Fischer von Waldheim G. 1824: *Entomographie de la Russie, et genres des insectes. Entomographia Imperii Rossici, sue Caesareae Majestati Alexandro I dicata. Volume II.* Mosquae: Augusti Semen Typographi, xx + 264 pp.

Fischer von Waldheim G. 1829a: *Collectio Insectorum Steveniana / Museum histoire Universitatis caesareae mosquensis. Part 2. Insecta.* Pp. 3-147.

Fischer von Waldheim G. 1829b: Denops. Nouveau genre de coléoptères de la famille des clériens ou térédiles. *Bulletin de la Société Impériale des Naturalistes de Moscou* **1**: 65-68, pl. 2.

Fischer von Waldheim G. 1842: *Catalogus Coleopterorum in Sibiria orientali a Cel. Gregorio Silide Karelin collectorum.* Moscou, pp. 1-21 part 1, pp. 21-28 part 2.

Fischer von Waldheim G. 1844: Spicilegium Entomographiae Rossicae. *Bulletin de la Société Impériale des Naturalistes de Moscou* **17**: 3-144, pls 1-3.

Flach C. 1888a: *Bestimmungs Tabellen der europäischen Coleopteren. XVII. Heft. Enthaltend die Familie der Phalacridae.* Brünn: Ed. Reitter, 28 pp., 1 pl.

Flach K. 1888b: Eine neue Verietät von Coccinella 5-punctata L. *Wiener Entomologische Zeitung* **7**: 298.

Flach C. 1889a: Bestimmungs Tabellen der europäischen Coleopteren. Familie Phalacridae. *Verhandlungen des Naturforschenden Vereins in Brünn* **27** [1889]: 54-79.

Flach K. 1889b: Japanische Phalacriden. *Deutsche Entomologische Zeitschrift* **33**: 271-272.

Flach K. 1895: Neue Malachier. *Wiener Entomologische Zeitung* **14**: 45-46.

Flach K. 1907: Beitraege zur Käferfauna Calabriens. *Deutsche Entomologische Zeitschrift* **1907**: 15-17.

Fleischer A. 1886: Neue Käfer-Varietäten. *Wiener Entomologische Zeitung* **5**: 235-236.

Fleischer A. 1887: Bemerkenswerte neue Käfer-Varietäten. *Wiener Entomologische Zeitung* **6**: 237.

Fleischer A. 1889: Ein entomologischer Ausflug von Brünn an die Grenze von Siebenbürgen im Juni 1888. *Verhandlungen des Naturforschenden Vereins in Brünn* **27** [1888]: 81-94.

Fleischer A. 1900a: Neue Coccinelliden aus der Sammlung des Kais. Rathes Herrn Edmund Reitter. *Wiener Entomologische Zeitung* **19**: 116-120.

Fleischer A. 1900b: Über die Coleopt.-Gattung Scymnus Kugelann. *Wiener Entomologische Zeitung* **19**: 235.

Fleischer A. 1900c: Coleopterologische Notiz. *Wiener Entomologische Zeitung* **19**: 245.

Fleischer A. 1909a: Ein neuer Ebaeus. *Wiener Entomologische Zeitung* **28**: 244.

Fleischer A. 1909b: Eine neue Atomaria s.str. *Wiener Entomologische Zeitung* **28**: 271.

Fleischer A. 1910a: Neue Coleopteren. *Wiener Entomologische Zeitung* **29**: 146.

Fleischer A. 1910b: Zur Kenntnis der Coleopterenfauna von Mähren. *Wiener Entomologische Zeitung* **29**: 328-330.

Fleischer A. 1915: Ein neuer Tachys und ein neuer Ptinus vom Elbrus in Nordpersien (Col.). *Entomologische Mitteilungen* **4**: 130-131.

Fleischer A. 1919: Attalus chinensis n. sp. (Col. Canthar.). *Wiener Entomologische Zeitung* **37**: 211.

Fleischer A. 1928: Zajímavé nálezy brouků na Moravě a na Slovensku r. 1927. *Časopis Československé Společnosti Entomologické* **25**: 27-28.

Fleischer J. T. 1829: Trechus sericeus Coleopterorum species nova, descriptione illustrata. *Bulletin de la Société Impériale des Naturalistes de Moscou* **1**: 69-72.

Fleutiaux E. 1887: Descriptions de coléoptères nouveaux de l'Annam. Rapportés par M. le capitaine Delauney. *Annales de la Société Entomologique de France* (6) **7**: 59-68, pl. 4.

Fleutiaux E. 1889: Contributions à la faune indo-chinoise. Cicindelidae et Elateridae. 1e Mémoire. *Annales de la Société Entomologique de France* (6) **9**: 137-146.

Fleutiaux E. 1895: Contribution à la faune indo-chinoise. 15e Mémoire. Première addition aux Cicindelidae et Elateridae. *Annales de la Société Entomologique de France* **63** [1894]: 683-690.

Fleutiaux E. 1899: Eucnémides et Elatérides récoltés a l'île Nias par M. U. Raap en 1897 et 1898. *Annali del Museo Civico di Storia Naturale di Genova* **39**: 569-570.

Fleutiaux E. 1900: Liste des Cicindelidae, Elateridae et Eucnemidae recueillis dans le Japon Central par M. le Dr. J. Harmand de 1894 à 1897. *Bulletin du Muséum National d'Histoire Naturelle* (Paris) **6**: 356-361.

Fleutiaux E. 1902a: Deuxième liste des Cicindelidae, Elateridae et Melasidae (Eucnemidae) recueillis au Japon par M. J. Harmand. *Bulletin du Muséum National d'Histoire Naturelle* (Paris) **8**: 18-25.

Fleutiaux E. 1902b: Descriptions de quatre espèces nouvelles du genre Agrypnus (Col.). *Bulletin de la Société Entomologique de France* **1902**: 162-164.

Fleutiaux E. 1902c: Description de deux Adelocera nouvelles (Col. 9). *Bulletin de la Société Entomologique de France* **1902**: 213-214.

Fleutiaux E. 1902d: Essai d'une classification des Melasidae (Eucnemidae des auteurs). *Annales de la Société Entomologique de France* **70**: 636-664.

Fleutiaux E. 1903a: Contributions à la faune indo-chinoise (Indochine Française). 18e Mémoire. Deuxième addition aux Cicindelidae et Elateridae. *Annales de la Société Entomologique de France* **71** [1902]: 569-580.

Fleutiaux E. 1903b: Elateridae recueillis par le Dr. Achille Tellini pendant le voyage qu'il fit en Erythrée (octobre 1902 à fevrier 1903) (Col.). *Bulletin de la Société Entomologique de France* **1903**: 250-251.

Fleutiaux E. 1905: Voyage de M. Maurice Maindron dans l'Inde Méridionale (1901). 8e Mémoire. Elateridae. *Annales de la Société Entomologique de France* **74**: 319-330.

Fleutiaux E. 1908: Descriptions d'Élatérides nouveaux du Tonkin (Col.). *Bulletin de la Société Entomologique de France* **1908**: 164.

Fleutiaux E. 1911: Révision des Trixagidae, Melasidae et Elateridae (Col.) des Antilles Françaises. *Annales de la Société Entomologique de France* **80**: 235-264.

Fleutiaux E. 1914: Elateridae des Iles Philippines. *Philippine Journal of Science* (D) **9** (5): 441-449.

Fleutiaux E. 1916a: Descriptions de deux espèces nouvelles d'Elateridae appartenant au genre Dima (Col.). *Bulletin de la Société Entomologique de France* **1916**: 256-257.

Fleutiaux E. 1916b: Melasidae des îles Philippines récoltés par C. E. Baker. *Philippine Journal of Science* (D) **11**: 387-398.

Fleutiaux E. 1918a: Noms nouveaux pour quatre genres de Melasidae (Col.). *Bulletin de la Société Entomologique de France* **23**: 59.

Fleutiaux E. 1918b: Nouvelles contributions à la faune de l'Indo-Chine Française (Coleoptera: Serricornia). *Annales de la Société Entomologique de France* **87**: 175-278.

Fleutiaux E. 1918c: Coléopteres élatérides indochinois de la collection du Muséum d'Histoire Naturelle de Paris. Catalogue et description des espèces nouvelles. *Bulletin du Muséum National d'Histoire Naturelle* **24**: 205-236.

Fleutiaux E. 1919: Elateridae, Trixagidae et Melasidae. In: *Voyage de Ch. Alluaud et R. Jeannel en Afrique Orientale* (*1911-1912*). *Résultats Scientifiques. Insectes coléoptères XIII.* Paris: Léon Lhomme, 119 pp.

Fleutiaux E. 1921a: Études sur les Melasidae (Coleoptera Serricornia). Troisième partie. *Annales de la Société Entomologique de Belgique* **61**: 68-87.

Fleutiaux E. 1921b: Deux genres nouveaux de coléoptéres (Melasidae). *The Philippine Journal of Science* **18**: 71-72.

Fleutiaux E. 1921c: Deux espèces nouvelles de coléoptères (Elateridae). *Philippine Journal of Science* (D) **18** (1): 73-74.

Fleutiaux E. 1923a: Description de deux espèces nouvelles d'Elateridae du genre Diploconus Candèze. *Bulletin de la Société Entomologique de France* **1923**: 12-14.

Fleutiaux E. 1923b: Les Melasidae du Japon. *Annales de la Société Entomologique de France* **91**: 291-328.

Fleutiaux E. 1923c: Remarques sur les espèces du genre Pterotarsus Guérin (Galba aut.) (Coleoptera Melasidae). *Annales de la Société Entomologique de France* **91**: 301-309.

Fleutiaux E. 1925a: Descriptions du deux genres nouveaux d'Elateridae. *Bulletin de la Société Entomologique de France* **1925**: 101-103.

Fleutiaux E. 1925b: Description d'un Melasidae nouveau de la Chine orientale. *Bulletin de la Société Entomologique de France* **1925**: 177-178.

Fleutiaux E. 1925c: Elateridae nouveaux récoltés au Tonkin par le Père A. de Cooman (1re note). *Bulletin de la Société Entomologique de France* **1925**: 184-186.

Fleutiaux E. 1926a: Catalogue raissonné des Melasidae des Iles Philippines. *Annales de la Société Entomologique de France* **95**: 29-90.

Fleutiaux E. 1926b: Descriptions de deux espèces nouvelles appartenant au genre Campsosternus Latr. (Col. Elateridae). *Bulletin de la Société Entomologique de France* **1926**: 58-60.

Fleutiaux E. 1927a: Les élatérides de l'Indo-Chine Française (Catalogue raisonné). *Faune des Colonies Françaises* **1**: 53-122, 2 pls.

Fleutiaux E. 1927b: Remarques sur deux genres d'élatérides (Col.). *Bulletin de la Société Entomologique de* France **1927**: 90-92.

Fleutiaux E. 1927c: Description d'un genre nouveau d'élatéridae (Col.). *Bulletin de la Société Entomologique de France* **1927**: 104-105.

Fleutiaux E. 1927d: Addition aux Melasidae de l'Indochine française (Insectes Coleoptéres). *Bulletin de la Société Entomologique de France* **52**: 133-138.

Fleutiaux E. 1928a: Les élatérides de l'Indochine Française (Catalogue raisonné). Deuxième Partie. *Encyclopédie Entomologique, Série B. I., Coléoptères* **3** (3): 103-177.

Fleutiaux E. 1928b: Description d'une genre nouveau d'élatéridae de la sous-famille de Hypolithinae. *Bulletin de la Société Entomologique de France* **1928**: 252-254.

Fleutiaux E. 1928c: Note sur le genre Agrypnella Champion et description d'un genre nouveaux. *Bulletin de la Société Entomologique de France* **1928**: 282-287.

Fleutiaux E. 1929: Melasidae nouveaux (Col.). *Bulletin de la Société Entomologique de France* **1929**: 206-209.

Fleutiaux E. 1930a: A new genus and a new species of Melasidae and a new species of Elateridae. *The Indian Forest Records* **14**: 273-274.

Fleutiaux E. 1930b: Description d'un Campsosternus nouveau de la collection du Muséum National d'Histoire Naturelle de Paris. *Bulletin du Muséum National d'Histoire Naturelle* (2) **4**: 409.

Fleutiaux E. 1930c: Les élatérides de l'Indochine française. (Catalogue raisonné). Troisième partie. *Bulletin du Muséum National d'Histoire Naturelle* (2) **2**: 636-644.

Fleutiaux E. 1931: Les élatérides de l'Indochine Française (Catalogue raisonné). Quatrième partie. *Bulletin de la Société Zoologique de France* **56**: 306-334.

Fleutiaux E. 1932a: Sur quelques élatérides de Madagascar. Pp. 189-193. *Livre de Centenaire*. Paris: Société Entomologique de France, xii + 729 pp., 32 pls.

Fleutiaux E. 1932b: Descriptions de trois coléoptères nouveaux. *Bulletin de la Société Entomologique de France* **1932**: 228-230.

Fleutiaux E. 1933a: Entomological investigations on the spike disease of sandal (10). Melasidae and Elateridae (Col.). *The Indian Forest Records* (*Entomological Series*) **18** (7): 1-16.

Fleutiaux E. 1933b: Les élatérides de l'Indochine Française (Cinquième part ie). *Annales de la Société Entomologique de France* **102**: 205-235.

Fleutiaux E. 1934a: Descriptions d'élatérides nouveaux. *Bulletin de la Société Entomologique de France* **1934**: 178-185.

Fleutiaux E. 1934b: Quatre élaterides nouveaux. *Proceedings of the Hawaian Entomological* Society **8** (3): 475-476.

Fleutiaux E. 1934c: Eucnemididae et Elateridae nouveaux. *Bulletin de la Sociéte Entomologiques de Belgique* **74**: 363-370.

Fleutiaux E. 1935a: Essai d'un genera des Eucnemididae paléarctiques. *Revue Française d'Entomologie* **2**: 1-18.

Fleutiaux E. 1935b: Révision des genres Agraeus Candèze et Trachylacon Motschulsky (Coleop. Elater.). *Annales de la Société Entomologique de France* **104**: 11-29.

Fleutiaux E. 1935c: Elateridae. P. 296. In: Visser P. C. & Visser-Hooft J. (eds): *Wissenschaftliche Ergebnisse der niederländischen Expeditionen in den Karakorum und die angrenzenden Gebiete in den Jahren 1922, 1925 und 1929/30. Vol. 1.* Leipzig: F. A. Brockhaus, xviii + 499 pp., 1 map, 8 pls.

Fleutiaux E. 1936a: Schwedisch-chinesische wissenschaftliche Expedition nach den nordwestlichen Provinzen Chinas unter Leitung von Dr. Sven Hedin und Prof. Sü Ping-chang. Insekten gesammelt vom schwedischen Arzt der Expedition Dr. David Hummel 1927-1930. 39. Coleoptera, 7. Elateridae. *Arkiv för Zoologi* **27 A** (19): 15-21.

Fleutiaux E. 1936b: Les élateridae de l'Indochine Française (6e partie). *Annales de la Société Entomologique de France* **105**: 279-300.

Fleutiaux E. 1940a: Les élatérides de l'Indochine française. Septième partie. *Annales de la Société Entomologique de France* **108** [1939]: 121-148.

Fleutiaux E. 1940b: Les élatérides de l'Indochine française. Huitième et dernière partie. *Annales de la Société Entomologique de France* **109**: 19-40.

Fleutiaux E. 1940c: Élatérides nouveaux. *Bulletin et Annales de la Société Entomologique de Belgique* **80**: 89-104.

Fleutiaux E. 1940d: Élatérides nouveaux de l'Indochine française. *Bulletin de la Société Zoologique de France* **65**: 184-198.

Fleutiaux E. 1941: Sur la sous-famille des Tetralobitae et description d'une nouvelle espèce (Coleoptera, Elateridae). *Revue Française d'Entomologie* **7** [1940]: 105-107.

Fleutiaux E. 1942: Entomological Results from the Swedish Expedition to Burma and British India. Coleoptera: Elateridae, recueillis par Rene Malaise. *Arkiv för Zoologi* **33 A** (18): 1-24.

Fleutiaux E. 1944a: Affinité entre les genres Dima et Penia (Col. Elat.) et description d'une nouvelle espèce de Dima. *Revue Française d'Entomologie* **10**: 39-41.

Fleutiaux E. 1944b: Les Octocryptus Candèze (Col. Elateridae). *Revue Française d'Entomologie* **10**: 145-148.

Fleutiaux E. 1945: Révision des eucnémides Africains. *Abeille, Journal d'Entomologie* **34**: 149-274.

Fleutiaux E. 1947: Révision des élatérides (coléoptères) de l'Indo-Chine française. Première partie. *Notes d'Entomologie Chinoise* **11**: 233-420.

Fleutiaux E. & Sallé A. 1889: Liste des coléoptères de la Guadeloupe et descriptions d'espèces nouvelles. *Annales de la Société Entomologique de France* (6) **9**: 351-484.

Florencourt Chassot K. de. 1796: Verzeichniss der Insecten Göttinischer Gegend. Pp. 197-244. In: Meyer F. A. A.: *Zoologisches Archiv* (Leipzig) Band 1.

Fluiter H. J. de. 1939: Scymnus roepkei Nov. Spec., een roofvijand van de Witte luis van de Koffie (Pseudococcus citri). *Archief voor de Koffiecultuur in Nederlandisch-Indië* (Kolff-Malang) **12** [1938]: 49-56.

Ford E. J. 1958: Insects of Micronesia. Coleoptera: Anobiidae. *Insects of Micronesia* **16**: 75-83.

Forskål P. 1775: *Descriptiones Animalium. Avium, Amphibiorum, Piscium, Insectorum, Vermium; que in Itinere Orientali.* Hauniae: Mölleri, 20 + xxxiv + 2 + 164 pp.

Förster A. 1849: Uebersicht der Käfer-Fauna der Rheinprovinz. Erster Nachtrag. *Verhandlungen des Naturhistorischen Vereins der Preussischen Rheinlande und Westphalens* **6**: 1-45.

Förster A. 1851: Uebersicht der Käfer-Fauna der Rheinprovinz. *Verhandlungen des Naturhistorischen Vereins der Preussischen Rheinlande und Westphalens* **8**: 1-42, 1 pl.

Forster J. R. 1771: *Novae species Insectorum. Centuria I. Nam mihi contuenti se persuasit rerum naturalium, nihil incredible existimare de ea Plin. Lib. Xic. 2.* London: T. Davies & B White, viii + 100 pp.

Fournel D. 1840: *Faune de la Moselle, ou manuel de zoologie, contenant la description des animaux libres ou domestiques observés dans le département de la Moselle; ouvrage rédigé d'après la méthode de Cuvier. Vol. 1, Part 2.* Metz: Verronnais, 624 pp.

Fowler W. W. 1885: New species of Languriidae. *Transactions of the Entomological Society of London* **1885**: 381-388.

Fowler W. W. 1886a: On a small collection of Languriidae from Assam, with descriptions of two new species. *The Transactions of the Entomological Society of London* **1886**: 23-26.

Fowler W. W. 1886b: The European species of the genus Cerylon. *The Entomologists' Monthly Magazine* **23**: 71-76.

Fowler W. W. 1886c: New genera and species of Languriidae. *The Transactions of the Entomological Society of London* **1886**: 303-322.

Fowler W. W. 1886d: Descriptions of new species of Languriidae. *Bulletin ou Comptes-Rendus des Séances de la Société Entomologique de Belgique* **1886**: cvii-cxii.

Fowler W. W. 1887: A new genus and new species of Languriidae from the collection of the Leyden Museum. *Notes from the Leyden Museum* **9**: 121-125.

Fowler W. W. 1889: *The Coleoptera of the British Islands. A descriptive account of the families, genera, and species indigenousto Great Britain and Ireland, with notes as to localities, habitats, etc. Vol. III. Clavicornia* (*Leptinidae-Heteroceridae*). London: L.Reeve, 399 pp.

Fowler W. W. 1890: On a new species belonging to the genus Languria. *The Entomologist's Monthly Magazine* (2) **1**: 107.

Fowler W. W. 1893: Insectes du Bengale. 26e mémoire. Notes on a small collection of Languriidae from Bengal in the Royal Belgian Museum. *Annales de la Société Entomologique de Belgique* **37**: 73-75.

Fowler W. W. 1908: *Coleoptera. Fam. Erotylidae. Subfam. Languriinae.* In: P. Wytsman (ed.): *Genera Insectorum.* Fasc. 78. Bruxelles: P. Wytsman, **78**: 45 pp., 3pls.

Fowler W. W. 1913: H. Sauter's Formosa-Ausbeute: Languriidae. *Archiv für Naturgeschichte* (A) **79**: 132-138.

Franz H. 1967a: Zur Kenntnis der mitteleuropäischen Hypnoidus-Arten aus dem Subgenus Zorochrus Thoms. *Entomologische Blätter* **63**: 32-37.

Franz H. 1967b: Beiträge zur Systematik der europäischen und nordwestafrikanischen Agriotes s. str. (Elateridae). *Entomologische Blätter* **63**: 65-86.

Franz H. 1967c: Beitrag zur Kenntnis der mitteleuropaischen Arten der Gattung Melanophthalma Motsch. (Coleopt., Lathridiidae). *Nachrichtenblatt der Bayerischen Entomologen* **16**: 105-109.

Franz H. 1968: Eine neue mitteleuropäische Art der Gattung Stilbus Seidlitz. *Entomologische Blätter* **64**: 168-171.

Franz H. 1969a: Eine neue Art aus der Familie der Colydiidae von den Kanarischen Inseln (Coleoptera). *Eos, Revista Española de Entomologia* **44**: 135-139.

Franz H. 1969b: Eine neue Pocadius-Art aus Österreich (Col., Nitidulidae). *Entomologische Blätter* **65**: 146-148.

Franz H. 1974: Zur Kenntnis der Gattung Enicmus Thoms. Neue und ungenügend bekannte Arten von den Kanaren und der Iberischen Halbinsel (Col. Lathridiidae). *Eos, Revista Española de Entomología* **48**: 109-114.

Franz H. 1980: Weitere Beiträge zur Koleopterenfauna der Kanarischen Inseln. *Eos, Revista Española de Entomología* **54**: 111-115.

Franz H. 1982: Revision der Throscus-Arten (Coleoptera) der Makaroneischen Inseln. *Zeitschrift der Arbeitsgemeinschaft Österreichischer Entomologen* **34**: 49-58.

Franz H. 1996: Die Ergebnisse meiner langjahrigen Aufsammlungen der Coleopterenfauna auf der Insel Hierro (Kanarische Inseln). *Österreichische Academie der Wissenschaften Mathematisch-Naturwissenschaftliche Klasse Sitzungsberichte Abteilung I* **202** [1995]: 71-138.

Frauenfeld G. R. von. 1867: Zoologische Miscellen. XI. *Verhandlungen der Kaiserlich-Königlichen Zoologisch-Botanischen Gesellschaft in Wien* **17**: 425-502, pl. XII.

Friederichs K. 1901: Die Varietäten von Cryptohypnus pulchellus und sabulicola. *Allgemeine Zeitschrift für Entomologie* **6**: 81-83.

Frivaldszky J. 1865: Jellemzõ adatok Magyarország faunájához. *A Magyarország Tudományi Akademia Évkönyvei* (*Pest*) **11** (4), 274 pp., XIII pls.

Frivaldszky J. 1877: Coleoptera nova e Hungaria Meridionali. *Természetrajzi Füzetek* **1**: 17-22.

Frivaldszky J. 1881: Coleoptera nova in Europa orientali et Asia Minore. *Természetrajzi Füzetek* **4** [1880]: 260-266.

Frivaldszky J. 1882: Coleoptera Europaea nova. *Természetrajzi Füzetek* **5**: 26-29.

Frivaldszky J. 1884: Coleoptera nova ex Asia Minore. *Természetrajzi Füzetek* **8**: 1-8.

Frivaldszky J. 1892: Coleoptera in expeditione D. Comitis Belae Széchenyi in China, praecipue boreali, a dominis Gustavo Kreitner et Ludovico Lóczy anno 1879 collecta. (Pars secunda.). *Természetrajzi Füzetek* **15**: 114-125.

Frivaldszky von Frivald E. 1835: Közlések a Balkány vidékén tett természettudományi utazásrol. *A Magyar Tudós Társaság Évkönyvei* **2**: 235-276, 7 pls.

Frivaldszky von Frivald E. 1837: Balkány vidéki természettudományi utazás. Második közlés. *A Magyar Tudós Táraság Évkönyvei* (Buda) **3**: 156-184, 8 pls.

Frivaldszky von Frivald E. 1845: Rövid áttekintése egy természetrajzi utazásnak, az európai Törökbirodalomban, egyszersmind nehány a közben újdonnat fölfedezett állatnak leirása. *A' Király Magyar Természettudományi Társulat Évkönyvei* **1** [1841-1845]: 163-187, pls. i-iii.

Frölich J. A. 1792: Bemerkungen über einige seltene Käfer aus der Insectensammlung des Herrn Hofr. und Prof. Rudolph in Erlangen. *Der Naturforscher* **26**: 68-165.

Frölich J. A. 1799: Einige neue Gattungen und Arten von Käfern. *Der Naturforscher* **28**: 1-65, pl. 1.

Fuente J. M. de la. 1897: Datos para la fauna de la provincia de Ciudad Real. *Anales de la Sociedad Española de Historia Natural, Actas* (2) **6** (26): 129-132.

Fuente J. M. de la 1903: Datos para la fauna de la provincia de Ciudad Real. XVI. - Coleópteros. *Boletín de la Real Sociedad Española de Historia Natural* **3**: 342-348.

Fuente J. M. de la. 1907: Datos para la fauna de la provincia de Ciudad Real. XIX Coleópteros. *Boletín de la Real Sociedad Española de Historia Natural* **7**: 317-323.

Fuente J. M. de la 1909: Datos para la fauna de la provincia de Ciudad Real. XX. - Coleópteros. *Boletín de la Real Sociedad Española de Historia Natural* **9**: 306-310.

Fuente J. M. de la. 1910: Datos para la fauna de la provincia de Ciudad Real. XXI. *Boletín de la Real Sociedad Española de Historia Natural* **10**: 444-449.

Fuente J. M. de la. 1913: Datos para la fauna de la provincia de Ciudad Real. *Boletín de la Real Sociedad Española de Historia Natural* **13**: 473-475.

Fuente J. M. de la. 1918: Descripción de coléopteros nuevos españoles. *Boletín de la Sociedad Entomológica de España* **1**: 44-46.

Fuente J. M. de la 1920: Notas sobre coleópteros españoles. *Boletín de la Sociedad Ibérica de Ciencias Naturales* **19**: 149-151.

Fuessly J. C. 1775: *Verzeichnis der ihm bekannten Schweitzerischen Inseckten mit einer ausgemahlten Kupfertafel: nebst der Ankündigung eines Insecten Werks*. Zürich und Winterthur, xii + 62 pp., 1 pl.

Fürsch H. 1958a: Zwei für Deutschland neue Adalia-Arten? *Nachrichtenblatt der Bayerischen Entomologen* **7**: 9-11.

Fürsch H. 1958b: Die mitteleuropäischen Scymnini und deren Verbreitung mit besonderer Berücksichtigung Bayerns (Col. Cocc.). *Nachrichtenblatt der Bayerischen Entomologen* **7**: 75-91.

Fürsch H. 1958c: Die mitteleuropäischen Scymnini und deren Verbreitung mit besonderer Berücksichtigung Bayerns (Col. Cocc.) *Nachrichtenblatt der Bayerischen Entomologen* **7**: 100-102.

Fürsch H. 1960a: Eine neue Scymnus-Art aus Süditalien (Col. Cocc.). *Nachrichtenblatt der Bayerischen Entomologen* **9**: 55-56.

Fürsch H. 1960b: Zwei neue Tetrabrachys-Arten (Col. Cocc.). *Mitteilungen der Münchener Entomologischen Gesellschaft* **50**: 64-67.

Fürsch H. 1960c: Neue Coccinellidae aus dem Museum Frey. *Entomologische Arbeiten aus dem Museum G. Frey* **11**: 298-303.

Fürsch H. 1960d: Neue palaearktische und afrikanische Coccinellidae (Coleoptera). *Beiträge zur Entomologie* **10**: 442-451.

Fürsch H. 1962: Neues über die mittel-und südeuropäischen Arten der Scymnus-frontalis-Gruppe (Col. Cocc.). *Opuscula Zoologica* **65**: 1-9.

Fürsch H. 1964: Die Arten der Verwandtschaftsgruppe Henosepilachna elaterii (Rossi) (= E. chrysomelina auct.) (Coleoptera, Coccinellidae). *Reichenbachia* **3**: 181-208.

Fürsch H. 1965: Die palaearktischen Arten der Scymnus-bipunctatus-Gruppe und die europäischer Vertreter der Untergattung Sidis (Col. Cocc.). *Mitteilungen der Münchener Entomologischen Gesellschaft* **55**: 178-213.

Fürsch H. 1966a: Die Coccinelliden der Sven Hedin-Expedition nach Südkansu und Nordost Szechuan. *Entomologisk Tidskrift* **87**: 40-42.

Fürsch H. 1966b: Bemerkungen zur Systematik mitteleuropäischer Coccinelliden *Nachrichtenbatt der Bayerischen Entomologen* **15**: 85-90.

Fürsch H. 1966c: Die Scymnus-Arten Westafrikas (Col. Cocc.). *Entomologischen Arbeiten aus dem Museum G. Frey* **17**: 135-192.

Fürsch H. 1967a: Revision einiger europäischer Scymnus (s. str.)-Arten (Col., Coccinellidae). *Mitteilungen der Abteilung für Zoologie und Botanik am Landesmuseum „Joaneum" in Graz* **28**: 207-259.

Fürsch H. 1967b: Coccinellidae. Pp. 227-278. In: Freude H., Harde K. W., Lohse G. A. (eds): *Die Käfer Mitteleuropas. Band 7 Clavicornia*. Krefeld: Goecke & Evers, 310 pp.

Fürsch H. 1969: Eine neue Scymnus-Art aus Mittelschweden (Col. Cocc.). *Entomologisk Tidskrift* **90**: 55-56.

Fürsch H. 1970: Über Coccinellidenausbeuten aus dem Vorderen Orient. *Mitteilungen der Münchener Entomologischen Gesellschaft* **59** [1969]: 108-114.

Fürsch H. 1974: [new taxa]. Pp. 275-276. In: Chazeau J., Étienne J. & Fürsch H.: Les Coccinellidae de l'île de La Réunion (Insecta Coleoptera). *Bulletin du Muséum National d'Histoire Naturelle* (3) **210** (Zoologie 140): 265-297.

Fürsch H. 1976: Eine neue Hyperaspis-Art aus Südtirol. *Nachrichtenbatt der Bayerischen Entomologen* **25**: 49-51.

Fürsch H. 1977: Coccinellidenausbeuten aus Libanon und dem Iran im Museum Genf mit Beschreibung neuer Scymnini-Arten (Col. Cocc.). *Revue Suisse de Zoologie* **84**: 645-657.

Fürsch H. 1979: Insects of Saudi Arabia. Coleoptera: Fam. Coccinellidae. *Fauna of Saudi Arabia* **1**: 235-248.

Fürsch H. 1980: [new taxa]. In: Fürsch H. & Uygun N. 1980: Neue Scymnini aus der Türkei. *Nachrichtenblatt der Bayerischen Entomologen* **29**: 109-118.

Fürsch H. 1981a: [new taxon]. In: Uygun N. & Fürsch H.: Die Hyperaspis-Arten der Türkei (Coleoptera Coccinellidae). *Nachrichtenbatt der Bayerischen Entomologen* **30**: 12-15.

Fürsch H. 1981b: Eine neue Nephus-Art aus Iran (Coleoptera, Coccinellidae). *Revue Suisse de Zoologie* **88**: 137-139.

Fürsch H. 1984: Die Gattung Afidentula Kapur, 1958 (Coleoptera, Coccinellidae). *Entomofauna, Zeitschrift für Entomologie* **5**: 335-342.

Fürsch H. 1985: [new taxa]. In: Canepari C., Fürsch H. & Kreissl E.: Die Hyperaspis-Arten Mittel-, West- und Südeuropa. Systematik und Verbreitung (Coleoptera Coccinellidae). *Giornale Italiano di Entomologia* **2**: 223-252.

Fürsch H. 1986: Neue Coccinellidenart aus Finnland (Coleoptera). *Annales Entomologici Fennici* **52**: 107-108.

Fürsch H. 1987a: Die Scymninae der Kanaren, Azoren und Madeiras (Coleoptera Coccinellidae). *Acta Coleopterplogica* **3**: 1-14.

Fürsch H. 1987b: Neue Coccinelliden aus Ethiopien. *Folia Entomologica Hungarica* **48**: 39-44.

Fürsch H. 1987c: Eine neue Hyperaspis-Art aus Türkey (Coleoptera, Coccinellidae). *Nachrichtenblatt der Bayerischen Entomologen* **36**: 53-54.

Fürsch H. 1987d: Übersicht über die Genera und Subgenera der Scymnini mit besonderer Berücksichtigung der Westpalaearktis (Insecta, Coleoptera, Coccinellidae). *Entomologische Abhandlungen* (Dresden) **51**: 57-74.

Fürsch H. 1989a: Scymnus (Pullus) pangi nom. nov. *Coccinella* (Passau) **2**: 34.

Fürsch H. 1989b: New Coccinellidae from Yemen. *Coccinella* (Passau) **2**: 37-38.

Fürsch H. 1989c: The Arabian species of the Scymnus (Pullus) guimeti-group (Col. Cocc.). *Fauna of Saudi Arabia* **10**: 113-122.

Fürsch H. 1994: Neue Coccinelliden-Art aus Israel (Coleoptera). *Annales Historico-Naturales Musei Nationalis Hungarici* **86**: 43-44.

Fürsch H. 1995: A new Serangium-species from Iran. *Nachrichtenbatt der Bayerischen Entomologen* **44**: 20-23.

Fürsch H. 1997: Zwei Scymnini-Neufunde aus dem Burgenland. (Coleoptera, Coccinelldidae). *Veröffentlichungen des Tiroler Landesmuseums Ferdandeum* **75/76** [1995/96]: 15-22.

Fürsch H. 1999: Eine neue Nephus-Art aus Portugal (Coleoptera, Coccinellidae). *Mitteilungen Münchener Entomologischen Gesellschaft* **89**: 75-76.

Fürsch H. 2000: Annotated check-list of the Coccinellidae (Insecta: Coleoptera) of Yemen. *Fauna of Saudi Arabia* **18**: 211-243.

Fürsch H. 2002: Eine neue Nephus-Art aus dem russischen Ferner Osten (Coleoptera: Coccinellidae). *Entomologische Zeitschrift* (Stuttgart) **112**: 125-126.

Fürsch H. & Kreissl E. 1967: [new taxa]. In: Fürsch H., Kreissl E. & F. Capra: Revision einiger europäischen Scymnus (s.str.)-Arten. *Mitteilungen der Abteilung für Zoologie und Botanik am Landesmuseum „Joanneum" in Graz* **28**: 207-259.

Fursov N. I. 1935a: Tri novykh vida kavkazskikh zhukov iz semeystva Cucujidae, poleznikh v lesnom khozyaystve [Drei neue kaukasische Arten aus Familie Cucujidae nützlisch in der Forstwirschaft]. *Bulletin de la Société des Naturalistes de Moscou, Section Biologique* (N. S.) **44**: 90-91.

Fursov N. I. 1935b: Dva novykh vida vrediteley drevesiny sem. Lymexylonidae (Coleoptera) s Kavkaza i Dal'nevostochnogo Kraya [Zwei neue Arten aus Fam. Lymexylonidae (Coleoptera) von Kaukasus und Ost-Siberien schädlich in der Fortwirtschaft]. *Bulletin de la Société des Naturalistes de Moscou, Section Biologique* (N. S.) **44**: 92-93.

Fursov N. I. 1936a: Novye vidy zhukov, zhivushchikh v drevesnie (dva vida sem. Eucnemidae iz Dal'nebostochnogo kraya i oden bid sem. Bostrychidae iz Sredney Azii [Die neuen Arten der Holzkäfer (zwei aus Fam. Eucnemidae aus Ost-Sibirien und eine aus Fam. Bostrychidae aus Mittelasien]. *Bulletin de la Société des Naturalistes de Moscou, Section Biologique* (N. S.) **45**: 348-350.

Fursov N. I. 1936b: Novyy rod i 5 novykh vidov sem. Cleridae iz Vostochnoy Sibiri i Sredney Azii [Eine neue Gattung und 5 neue Arten auf Fam. Cleridae (Coleoptera) von Ost-Sibirien und Mittelasien]. *Bulletin de la Société des Naturalistes de Moscou, Section Biologique* (N. S.) **45**: 351-354.

Fursov N. I. 1939: Novyy rod i 6 novykh vidov zhukov iz Sredney Azii i Kavkaza. *Bulletin de la Société des Naturalistes de Moscou, Section Biologique* (N. S.) **48**: 88-90.

Fursov N. I. 1958: [new taxon]. In: Sinadskiy Yu. V.: Vrediteli lomonosa vostochnogo v tugayakh Amu-Dar"i. *Byulleten Moskovskogo Obshchestva Ispytateley Prirody, Otdel Biologicheskiy* **1958**: 143-144.

Fursov N. I. & Rikhter V. A. 1958: Novyy predstavitel roda Trichodes Herbst (Coleoptera, Cleridae) iz Sredney Azii. *Entomologicheskoe Obozrenie* **37**: 708-709.

Fuss G. 1850: Corticaria macularis, eine neue Art. *Verhandlungen und Mitteilungen des Siebenbürgischen Vereins für Naturwissenschaften zu Hermannstadt* **1**: 127-130.

Fuss C. 1863: Berichtigungen und Beiträge zur siebenbürgischen Käferfauna (Fortsetzung). *Verhandlungen und Mitteilungen des Siebenbürgischer Vereins für Naturwissenschaften zu Hermannstadt* **14** (4): 67-71.

Gabriel J. J. C. 1905a: Drei neue Varietäten. *Deutsche Entomologische Zeitschrift* **1905**: 141-142.

Gabriel J. J. C. 1905b: Adonia variegata var. macrostigma. *Deutsche Entomologische Zeitschrift* **1905**: 142.

Gadeau de Kerville H. 1884: Descriptions de quelques espèces nouvelles de la famille des Coccinellidae. *Annales de la Société Entomologique de France* (6) **4**: 67-72.

Gahan C. J. 1900a: The Expedition to Sokotra. XI. Descriptions of new genera and species of Coleoptera. *Bulletin of the Liverpool Museum* **3**: 8-13.

Gahan C. J. 1900b: [new taxon]. In: Waterhouse C. O., Gahan C. J. & Arrow G. J. 1900: Order 4. - Coleoptera. Pp. 89-127. In: Andrews C. W.: *A monograph of Christmas Island (Indian Ocean): Physical features and geology. With description of the fauna and flora.* London: printed by order of the Trustees, 13 + 338 + 22 pp.

Gahan C. J. 1903: Arthropoda. Insecta: Coleoptera. In: Natural History of Sokotra and Abd-el-Kuri. *Bulletin of the Liverpool Museum (Special Bulletin)* **1903**: 261-292 + 1 pl.

Gahan C. J. 1910: Notes of Cleridae and desriptions of some new genera and species of this family of Coleoptera. *The Annals and Magazine of Natural History* (8) **5**: 55-76.

Galeazzi G. 1854: *Coleoptera Europae dupleta, quae pro mutica commutatione offeri possunt.* Milano, 19 pp.

Gandolphe P.-L. 1865: Observations sur les mœurs des espèces algériennes du genre Cebrio et description des femelles de C. Barbarus (Lucas) et C. Melanocephalus (Leach). *Bulletin de l'Académie Hippone* **1**: 7-14.

Ganglbauer L. 1896: Sammelreisen nach Südungarn und Siebenbürgen. Coleopterologische Ergebnisse derselben. 1. Theil. *Annalen des K. K. Naturhistorischen Hoffmuseums* (Wien) **11**: 164-187.

Ganglbauer L. 1897a: Neue Cryptophagus aus dem mitteleuropäischen Faunengebiete. *Verhandlungen der Kaiserlich-Königlichen Zoologisch-Botanischen Gesellschaft in Wien* **47**: 560-564.

Ganglbauer L. 1897b: Einige neue Coleopteren des mitteleuropäischen Faunengebietes. *Verhandlungen der Kaiserlich-Königlichen Zoologisch-Botanischen Gesellschaft in Wien* **47**: 565-573.

Ganglbauer L. 1899a: *Die Käfer von Mitteleuropa. Die Käfer der österreichisch-ungarischen Monarchie, Deutschlands, der Schweiz, sowie des franzözischen und italienischen Alpengebietes. Familienreihe Clavicornia. Sphaeritidae, Ostomidae, Byturidae, Nitidulidae, Cucujidae, Erotylidae, Phalacridae, Thorictidae, Lathridiidae, Mycetophagidae, Colydiidae, Endomychidae, Coccinellidae. Volume III.* Wien: C. Gerold's Sohn, iii + 1046 pp.

Ganglbauer L. 1899b: Ueber einige, zum Theil neue mitteleuropäische Coleopteren. *Verhandlungen der Kaiserlich-Königlichen Zoologisch-Botanischen Gesellschaft in Wien* **49**: 526-535.

Ganglbauer L. 1900: [new taxa]. In: Bodemeyer E. von.: *Quer durch Klein-Asien in der Bulghar-Dagh. Eine naturwissenschaftliche Studien-Reise.* Emmendingen: Verlangs-Aktiengesellschaft vormals Dölter, v + 169 pp.

Ganglbauer L. 1904: *Die Käfer von Mitteleuropa. Die Käfer der österreichisch-ungarischen Monarchie, Deutschlands, der Schweiz, sowie des französischen und italienischen Alpengebietes. Vierter Band, erste Hälfte. Dermestidae, Byrrhidae, Nosodendridae, Georyssidae, Dryopidae, Heteroceridae, Hydrophilidae.* Wien: Karl Gerolds Sohn, 286 pp.

Ganglbauer L. 1905: Coleoptera. Ergebnisse einer naturwissenschaftlichen Reise zum Erdschias-Dagh (Kleinasien). Ausgeführt von Dr. Arnold Penther und Dr. Emerich Zederbauer auf Kosten der "Gesellschaft zur Förderung der naturhistorischen Erforschung des Orients in Wien im Jahre 1902. *Annalen des K. K. Naturhistorischen Hofmuseums* **20**: 246-290.

Ganglbauer L. 1906a: Malthinus marginicollis n.sp. *Il Naturalista Siciliano* **19** [1906-1907]: 49-50.

Ganglbauer L. 1906b: Die Malthodes aus der Gruppe des nigriceps Muls. *Il Naturalista Siciliano* **19** [1906-1907]: 50-55.

Ganglbauer L. 1911: [new taxa]. In: Reitter E.: *Fauna Germanica. Die Käfer des Deutsches Reiches. Nach der analytischen Methode bearbeitet. III. Band.* Stuttgart: K. G. Lutz Verlag, 436 pp., pls 81-128.

Ganglbauer L. 1912: Malthodes Moczarskii nov. spec. *Verhandlungen der Kaiserlich-Königlichen Zoologisch-Botanischen Gesellschaft in Wien* **62**: 181-182.

García R. de & Evers A. M. J. 1997: Descriptión de un nuevo Cephaloncus West., 1863 de la isla de La Palma (Islas Canarias). Cephaloncus palmensis n. sp. (Coleoptera, Malachiidae). *Entomologische Blätter* **93**: 87-89.

Garg P. & Saini M. S. 1996: Three new species of Ctenicera Latreille (Coleoptera: Elateridae: Ctenicerinae) from India. *Journal of Entomological Research* (New Dehli) **20**: 317-323.

Garg P., Saini M. S. & Vasu V. 1997a: New additions to genus Cardiophorus Eschscholtz (Coleoptera: Elateridae: Cardiophorinae) from India. *Journal of Entomological Research* (New Dehli) **21**: 65-70.

Garg P., Saini M. S. & Vasu V. 1997b: New additions to the Indian fauna of Agrypnus Eschscholtz (Coleoptera: Elateridae: Agrypninae). *Journal of Entomological Research* (New Dehli) **21**: 283-289.

Garg P., Saini M. S. & Vasu V. 1998: New species of Silesis Candeze (Coleoptera, Elateridae, Adrastinae) from India. *Journal of the Bombay Natural History Society* **95**: 460-464.

Garg P. & Vasu V. 1996: Two new species and a revised key to genus Neodiploconus Hyslop (Coleoptera, Elateridae: Melanotinae) from India. *Journal of the Bombay Natural History Society* **93**: 260-262.

Garg P. & Vasu V. 1997: New additions of genus Silesis Candèze (Insecta: Coleoptera: Elateridae: Adrastinae) to the indian Fauna. *The Rafflies Bulletin of Zoology* **45**: 53-59.

Garg P. & Vasu V. 1999: New species of Agonischius Candeze (Coleoptera, Elateridae: Ludiinae) from India. *Journal of the Bombay Natural History Society* **96**: 281-284.

Garg P., Vasu V. & Vats L. K. 1995: A new species of Monadicus Candeze (Coleoptera: Elateridae: Hypnoidinae) together with a key to Indian species. *Journal of Entomological Research* (New Dehli) **19**: 363-365.

Gardner J. C. M. 1935: A new Indian species of Atractocerus (Col. Lymexylonidae). *Stylops* **4**: 69-70.

Gardner A. E. & Classey E. W. 1962: Report on the insects collected by the E. W. Classey and A. E. Gardner expedition to Madeira in December 1957. Part II. Coleoptera. *Proceedings and Transactions of the South London Entomological & Natural History Society* **1961**: 149-159.

Gautier des Cottes C. de. 1872: [Coléoptères nouveaux]. *Petites Nouvelles Entomologiques* **1** [1869-1875]: 223-224.

Gebien H. 1899: Zwei neue deutsche Käfer. *Entomologische Nachrichten* **25**: 97.

Gebler F. A. von. 1817: Insecta Sibiriae rariora. *Mémoires de la Société Impériale des Naturalistes de Moscou* **5**: 315-333.

Gebler F. A. von. 1823: Chrysomelae Siberiae rariores. *Mémoires de la Société Impériale des Naturalistes de Moscou* **6**: 117-126.

Gebler F. A. von. 1824: [new taxon]. In: Fischer von Waldheim G.: *Entomographie de la Russie. Entomographia Imperii Rossici, sue Caesareae Majestati Alexandro I dicata. Volume II.* Mosquae: Augusti Semen Typographi, xx + 264 pp.

Gebler F. A. von. 1825: Coleopterorum Sibiriae species novae descriptae. *Essais Entomologique* (A. D. Hummel) **4**: 42-57.

Gebler F. A. von. 1830: Bemerkungen über die Insecten Sibiriens, vorzüglich des Altai. (Part 3). Pp. 1-228. In: C. F. Ledebour (ed.): *Reise durch das Altai-Gebirge und die songorische Kirgisen-Steppe. Auf Kosten der Kaiserlichen Universitä Dorpat im Jahre 1826 in Begleitung der Herren D. Carl Anton Miecherund D. Alexander von Bunge. Zweiter Theil.* Berlin: G. Reimer, 427 pp.

Gebler F. A. von. 1832: Notice sur les coléoptères qui se trouvent dans le district des mines de Nertschinsk, dans la Sibérie orientale, avec la description de quelques espèces nouvelles. *Nouveaux Mémoires de la Société Impériale des Naturalistes de Moscou* **2**: 23-78.

Gebler F. A. von. 1833: Notae et addidamenta ad catalogum coleopterorum Sibiriae occidentalis et confinis Tatariae operis, C. F. von Ledebours Reise in das Altaigebirge und die soongarische Kirgisensteppe (Zwyter Theil. Berlin 1830). *Bulletin de la Société Impériale des Naturalistes de Moscou* **6**: 262-309.

Gebler F. A. 1841a: Characteristik mehrerer neuen sibirischen Coleopteren. *Bulletin Scientifique de l'Académie Impériale des Sciences St. Péterbourg* **8**: 369-376.

Gebler F. A. von. 1841b: Notae et additamenta ad Catalogum Coleopterorum Sibiriae occidentalis et confinis Tatariae operis: C. F. Ledebour's Reise in das Altaigebirge und die soongarische Kirgisensteppe. Zweyter Theil. Berlin 1830. Fasciculus secundus. *Bulletin de la Société Impériale des Naturalistes de Moscou* **14** (4): 577-625.

Gebler F. A. von. 1843: Charakteristik der von Hn. Schrenk in den Jahren 1841 in den Steppen der Dsungarei gefundenen neuen Coleopterenarten. *Bulletin de la Classe Physico-Mathématique de l'Académie Impériale des Sciences de St.-Pétersbourg* **2** (1): 36-40 columns.

Gebler F. A. von. 1844: Charakteristik der von Hrn. Schrenk in den Jahren 1841-1843 in den Steppen der Dsungarei gefundenen neuen Coleopterenarten. *Bulletin de la Classe Physico-Mathématique de l'Académie Impériale des Sciences de St.-Pétersbourg* (2) **3** [1844-1845]: 97-106 columns.

Gebler F. A. von. 1847a: Verzeichnis der im Kolywano-Woskresenskischen Hüttenbezirke Süd-West-Sibiriens beobachteten Käfer mit Bemerkungen und Beschreibungen. *Bulletin de la Société Impériale des Naturalistes de Moscou* **20** (2): 263-361.

Gebler F. A. von. 1847b: Verzeichniss der im Kolywano-Woskresenskischem Hüttenbezirke Süd-West-Sibiriens beobachteten Käfer mit Bemerkungen und Beschreibungen. *Bulletin de la Société Impériale des Naturalistes de Moscou* **20** (4): 391-512.

Gebler F. A. von. 1848: Verzeihniss der in Kolywano-Woskresenskischen Huttenbezirke Süd-West-Sibiriens beobachteten Käfer mit Bemerkungen und Beschreibungen. *Bulletin de la Société Impériale des Naturalistes de Moscou* **21** (2): 317-423.

Geisthardt M. 1975: Parafallia johni n. sp. aus Nepal (Coleoptera: Discolomidae). *Senckenbergiana Biologica* **56**: 39-41.

Geisthardt M. 1982a: Zur Kenntnis von Nyctophila caucasica Motschulsky 1854) und Beschreibung neuer Subspecies und einer Species aus dem Kaukasus (Coleoptera: Lampyridae: Lampyrini). *Entomologische Blätter* **78**: 66-74.

Geisthardt M. 1982b: Beitrag zur Kenntnis der Gattung Nyctophila Olivier, 1884 (Coleoptera, Lampyridae). *Annales Historico-Naturales Musei Nationalis Hungarici* **74**: 115-128.

Geisthardt M. 1983a: Beitrag zur Revision der Gattung Lampyris Müller, 1764. I. Zur Kenntnis einiger nordafrikanischer und mediterraner Arten (Coleoptera, Lampyridae). *Mitteilungen der Münchener Entomologischen Gesellschaft* **73**: 23-44.

Geisthardt M. 1983b: Zur Kenntnis der Gattung Nyctophila Olivier 1884 und Aufstellung einer neuen Gattung Lychnobius gen. nov. (Coleoptera: Lampyridae: Lampyrinae). *Entomologische Blätter* **79**: 37-60.

Geisthardt M. 1985: Materialien zur Revision der Gattung Lampyris Müller 1764. II. Zur Kenntnis einiger europäischer und mediterraner Arten (Insecta: Coleoptera: Lampyridae). *Senckenbergiana Biologica* **65**: 279-294.

Geisthardt M. 1987: Materialien zur Revision der Gattung Lampyris Müller 1764. III. Bekannte und neue Arten der westmediterranen Inseln und Italiens. (Coleoptera: Lampyridae). *Mitteilungen des Internationalen Entomologischen Vereins* **11**: 89-110.

Geisthardt M. 1988: Zwei neue Lampyridae aus dem Iran, sowie Anmerkungen zu einigen bekannten Arten (Coleoptera). *Revue Suisse de Zoologie* **95**: 471-476.

Geisthardt M. 1990: Bestimmungsschlüssel der Arten von Nyctophila Olivier (Coleoptera: Lampyridae) mit Neubeschreibungen und Anmerkungen. *Stuttgarter Beiträge zur Naturkunde* A **447**: 1-15.

Geisthardt M. 1999: Neue paläarktische Lampyridae und Anmerkungen zu bekannten Arten (Coleoptera). *Mitteilungen des Internationalen Entomologischen Vereins* **24**: 95-109.

Geisthardt M. 2000: Lampyris monticola sp. nov. aus Griechenland (Coleoptera: Lampyridae, Lampyrinae). *Entomologische Zeitschrift* **110**: 175-176.

Geisthardt M. 2003: Lampyris satoi n. sp. aus Nepal (Coleoptera: Lampyridae: Lampyrinae). *Mitteilungen des Internationalen Entomologischen Vereins* **28**: 67-70.

Geisthardt M. 2003: Zwei neue Arten der Gattung Selasia Castelnau, 1836 aus Jemen. *Mitteilunden des Internationalen Entomologischen Vereins* **28**: 99-109.

Geisthardt M. 2004: New and known fireflies from Mount Emei (China) (Coleoptera: Lampyridae). *Mitteilungen des Internationalen Entomologischen Vereins* **29**: 1-10.

Geisthardt M. & Day J. 2004: Lampyroidea maculata (Coleoptera: Lampyridae): A new species of lampyrid from Iran. *Zootaxa* **427**: 1-6.

Gemminger M. 1869: [new name]. In: Gemminger M. & Harold E. von. (eds.): *Catalogus coleopterorum hucuscue descriptorum synonymicus et systematicus. Tomus VI.* Monachii: E. H. Gummi, pp. 1609-1800.

Gemminger M. 1870: [Abänderungen vergebener Namen]. *Coleopterologische Hefte* **6**: 119-124.

Gemminger M. 1872: [new names]. In: G. R. Crotch: Berichtungen und Zusätse zum Catalogus Coleopterorum synonymicus et systematicus. *Coleopterologische Hefte* **10**: 204-207.

Gemminger M. & Harold E. 1868: *Catalogus Coleopterorum hucusque descriptorum synonymicus et systematicus. Tom. III.* Monachii: E. H. Gummi, 753-978 + [6].

Gemminger M. & Harold E. 1869: *Catalogus Coleopterorum hucusque descriptorum synonymicus et systematicus. Tom. VI.* Monachii: E. H. Gummi, 1609-1800 + [6] pp.

Gemminger M. & Harold E. 1874: *Catalogus Coleopterorum hucusque descriptorum synonymicus et systematicusTom. XI. Chrysomelidae* (*Pars I.*). Monachii: G. Beck, 3233-3822 + [2] pp.

Gené J. [C. G.] 1836: De quisdam insectis Sardiniae novis aut minus cognitis. Fasciculus I. *Memoralia Regiae Scientiarum Accademiae Taurinensis* **39**: 161-199, 1 pl. [separate 39 pp., 1 pl.].

Gené J. [C. G.] 1839: De quisdam insectis Sardiniae novis aut minus cognitis. Fasciculus II. *Memoralia Regiae Scientiarum Accademiae Taurinensis* (2) **1**: 43-84, 2 pls. [separate, 44 pp., 2 pls].

Geoffroy E. L. 1762: *Histoire abrégée des insectes qui se trouvent aux environs de Paris; dans laquelle ces animaux sont rangés suivant un ordre méthodique. Tome premier.* Paris: Durand, [4] + xxviii + 523 + 1 pp., 11 pls.

Geoffroy E. L. 1785: [new taxa]. In: Fourcroy A. F. de.: *Entomologia parisiensis; sive Catalogus Insectorum quae in Agro Parisiensi reperiuntur; Secundum methodum Geoffraeanam in sectiones, genera et species distributus: cui addita sunt nomina trivialia et fere trecentae novae Species. Pars prima.* Parisiis: Privilegio Academiae, vii + [1] + 231 pp. [new names attributed to Geoffroy by Fourcroy].

Gerhardt J. 1898a: Neuheiten der schlesischen Käferfauna aus dem Jahre 1897. *Zeitschrift für Entomologie. Breslau* (N. S.) **23**: 12-16.

Gerhardt J. 1898b: Neuheiten der schlesischen Käferfauna von 1897. *Deutsche Entomologische Zeitschrift* **1898**: 334-336.

Gerhardt J. 1900: Neuheiten der schlesischen Käferfauna aus dem Jahre 1899. *Zeitschrift für Entomologie* (N. F.) **25**: 15-18.

Gerhardt J. 1905: Beitrag zur Kenntnis einiger Ernobius-Arten aus der Gruppe des nigrinus Er. *Deutsche Entomologische Zeitschrift* **1905**: 221-222.

Gerhardt J. 1908: Neuheiten der schlesischen Käferfauna aus dem Jahre 1907. *Deutsche Entomologische Zeitschrift* **1908**: 209-210.

Gerhardt J. 1909: Neuheiten der schlesischen Käferfauna aus dem Jahre 1908. *Deutsche Entomologische Zeitschrift* **1909**: 415-422.

Gerhardt J. 1910: Neuheiten der schlesischen Käferfauna aus dem Jahre 1909 (Col.). *Deutsche Entomologische Zeitschrift* **1910**: 554-557.

Gerhardt J. 1911: Eine neue schlesische Käferart. Lathridius campicola n. sp. *Jahresheft des Vereins für Schlesische Insektenkunde zu Breslau* **4**: 5-6.

Gerhardt J. 1912: Neuheiten der schlesischen Käferfauna aus dem Jahre 1911. *Deutsche Entomologische Zeitschrift* **1912**: 463.

Germain P. 1856: Descripcion de coléopteros de diversas especies que no se hallan en la obra del señor Gay. *Anales de la Universidad de Chile* **1855**: 386-407.

Germar E. F. 1813: Neue Insekten. *Magasin für Entomologie* (*Halle*) **1**: 114-133.

Germar E. F. 1817: *Reise nach Dalmatien und in das Gebiet von Ragusa.* Leipzig und Altenberg: F. A. Brockhaus, xii + 323 pp., 9 pls., 2 maps.

Germar E. F. 1818: VIII. Vermischte Bemerkungen über einige Käferarten. *Magazin der Entomologie* **3**: 228-260, pl. 3.

Germar E. F. 1823a: *Fauna Insectorum Europae. Fasciculus VI.* Halae: C. A. Kümmelii, 25 pls.

Germar E. F. 1823b: *Fauna Insectorum Europae. Fasciculus VII.* Halae: C. A. Kümmelii, 25 pls.

Germar E. F. 1824: *Insectorum species novae aut minus cognitae, descriptionibus illustratae. Volumen primum. Coleoptera.* Halae: J. C. Hendelii et Filii, xxiv + 624 pp., 2 pls.

Germar E. F. 1834: [Descriptions d'espèces nouvelles]. Description du Thorictus castaneus. *Revue Entomologique* (G. Silbermann) **2**: No. 15, pl. 17. [pp. unnumber.].

Germar E. F. 1836: *Fauna Insectorum Europae. Fasciculus XVIII.* Halae: C. A. Kümmelii, 25 pls.

Germar E. F. 1838: *Fauna Insectorum Europae. Fasciculus XX.* Halae: C. A. Kümmelii, 25 pls.

Germar E. F. 1839a: *Fauna Insectorum Europae. Fasciculus XXI.* Halae: C. A. Kümmelii, 25 pls.

Germar E. F. 1839b: Ueber die Elateriden mit häutigen Anhängen der Tarsenglieder. *Zeitschrift für die Entomologie* **1** (2): 193-236.

Germar E. F. 1840: Bemerkungen über Elateriden. *Zeitschrift für die Entomologie* **2** (2): 241-278, 1 pl.

Germar E. F. 1842: *Fauna insectorum Eropae. Fasciculus XXII.* Halae: C. A. Kümmelii, 25 pls. + 1 p.

Germar E. F. 1843a: Bemerkungen über Elateriden. *Zeitschrift für die Entomologie* **4**: 43-108.

Germar E. F. 1843b: Eumorphus. Pp. 84-86. In: Ersch J. S. & Gruber J. G. (eds): *Algemeine Encyclopaedie der Wissenschaften und Kuenste. Vol. 39.* Leipzig: Eugen - Ezzelino, 492 pp.

Germar E. F. 1844a: *Fauna Insectorum Europae. Fasciculus XXIII.* Halae: C. A. Kümmelii, 25 pls.

Germar E. F. 1844b: Bemerkungen über Elateriden. *Zeitschrift für die Entomologie* **5**: 133-192.

Germar E. F. 1846: Über die Elateridengattung Campylus. *Linnea Entomologica* **1**: 147-155.

Germar E. F. 1847: *Fauna Insectorum Europae. Fasciculus XXIV.* Halae: C. A. Kümmelii, 25 pls.

Germar E. F. 1848: Beiträge zur Insectenfauna von Adelaide. *Linnaea Entomologica* **3**: 153-247.

Germar E. F. & Kaulfuss F. 1816: *Fauna Insectorum Europae. Fasciculus III.* Halae: Car. Aug. Kümmelii, 25 pls.

Gerstaecker C. E. A. 1855: Diagnosen der von Peters in Mossambique gesammelten Käfer und Hymenoptera. Longicornia, Paussidae, Ptinoires. *Monatsbericht der Königlichen Akademie der Wissenschaften zu Berlin* **1855**: 265-268.

Gerstaecker C. E. A. 1857: Versuch einer systematischen Auseinandersetzung der Gattungen Eumorphus Web. und Endomychus Payk. *Archiv für Naturgeschichte* **23**: 211-243.

Gerstaecker C. E. A. 1858: Monographie der Endomychiden, einer Familie der Coleopteren. In: Gerstaecker C. E. A.: *Entomographien. Abhandlungen in Bereich der Gliederthiere, mit besonderer Benutzung der Koenigl. Entomologischen Sammlung zu Berlin. Erster Band.* Leipzig: W. Engelmann, xiv + 433 pp. + 3 pls.

Gerstaecker C. E. A. 1860: Die Arten der Gattung Lissomus Dalm., nach dem Material der Königl. Entomologischen Sammlung zu Berlin. *Linnaea Entomologica* **14**: 129-179.

Gerstaecker C. E. A. 1871: Beitrag zur Insektenfauna von Zanzibar. III. Coleoptera. *Archiv für Naturgeschichte* **37** (1): 42-86.

Gerstmeier R. 1985: Zwei neue Trichodes-Arten aus der ammios Artengruppe: Trichodes longicollis sp. n. aus dem Iran und Trichodes flavotarsis sp. n. aus dem Iraq (Coleoptera: Cleridae). *Mitteilungen der Münchener Entomologischen Gesellschaft* **75**: 127-136.

Gerstmeier R. 2002: Generic concept of clerid taxa related to Clerus (Coleoptera: Cleridae: Clerinae). *Entomological Problems* **32**: 99-111.

Gerstmeier R. & Kuff T. L. 1992: Revision der paläarktischen Arten der Gattungen Tillus Olivier, 1790, Tilloidea Castelnau, 1832, Falsotillus gen. n. und Flabelotilloidea gen. n. (Coleoptera, Cleridae, Tillinae). *Mitteilungen der Münchener Entomologischen Gesellschaft* **82**: 55-72.

Gestro F. 1874: Descrizione si tre specie nuove del genere Atractocerus appartenenti alle collezioni del Museo Civico di Genova. *Annali del Museo Civico di Storia Naturale di Genova* **6**: 544-547.

Gestro R. 1882: Sopra alcuni Coleotteri di Birmania raccolti dal Capitano G. B. Comotto. *Annali del Museo Civico di Storia Naturale di Genova* **18**: 297-317.

Gestro R. 1888: Viaggio di Leonardo Fea in Birmania e regioni vicine. IV. Nuove specie di Coleotteri. *Annali del Museo Civico di Storia Naturale di Genova* **26**: 87-132.

Gestro R. 1889: Viaggio ad Assab nel Mar Rosso dei signori G. Doria ed O. Beccari con il R. Avviso „Esploratore“ dal 16 novembre 1879 al 26 febbraio 1880. IV. Coleotteri. *Annali del Museo Civico di Storia Naturale Genova* **27** [1888-1889]: 5-72.

Gestro R. 1892: Appendice ai materiali per lo studio del genera Ichthyurus. *Annali del Museo Civico di Storia Naturale di Genova* **30**: 1022-1048.

Gestro R. 1906: Studii sugli Ichthyurus. *Annali del Museo Civico di Storia Naturale di Genova* **42** [1905-1906]: 266-307.

Gestro R. 1911: Alguni nuovi Ichthyurus. *Annali del Museo Civico di Storia Naturale di Genova* **45** [1911-1913]: 44-51.

Ghorpade K. D. 1976 An undescribed species of Illeis (Coleoptera: Coccinellidae) from South India. *Oriental Insects* **10**: 579-585.

Gillerfors G. 1991: Three new species of Coleoptera from the Canary Islands. *Vieraea* **20**: 21-26.

Gillerfors G. & Otero J. C. 1993: Contribución al conocimiento de los Cryptophagus Herbst de Turquia con descripción de una nueva especie (Coleoptera: Cryptophagidae). *Nouvelle Revue d'Entomologie* **10**: 229-232.

Gillogly L. R. 1962: Insects of Micronesia. Coleoptera: Nitidulidae. *Insects of Micronesia* **16** (4): 133-188.

Gillogly L. R. 1982: New species and a key to the genus Haptoncus (Coleoptera: Nitidulidae). *Pacific Insects* **24**: 281-291.

Giorna M. S. 1792: *Calendario entomologico ossia osservazioni sulla stagioni proprie agl' Insetti nell clima Piemontese, e particolarmente ne' contorni di Torino.* Torino: Nella Stamperia Reale, 146 pp.

Giraud J. E. 1851: Coleopterenfauna von Gastein. *Verhandlungen des Zoologisch-Botanischen Vereins in Wein* **1**: 132-140.

Gistel J. N. F. X. 1834: *Die Insecten-Doubletten aus der Sammlung des Herrn Grafen Rudolph von Jenison Walworth zu Regensburg, welche sowohl im Kauf als im Tausche abgegeben werden. Nro. I. Käfer.* München: G. Jaquet, 36 pp.

Gistel J. N. F. X. 1848: *Naturgeschichte des Thierreichs. Zür höhere Schulen.* Stuttgart: Scheitlin & Krais, xix + 216 + 4 pp., 32 pls. [seen 2nd ed. from 1851].

Gistel J. N. F. X. 1856: *Die Mysterien der europäischen Insectenwelt. Ein geheimer Schlüssel für Sammler aller Insecten-Ordnungen und Stände, behufs des Fangs, des Aufenthalts-Orts, der Wohnung, Tag- und Jahreszeit u. s. w., oder autoptische Darstellung des Insectenstaats in seinem Zusammenhange zum*

Bestehen des Naturhaushaltes überhaupt und insbesondere in seinem Einflusse auf die phanerogamische und cryptogamische Pflanzenbevölkerung Europa's. Zum ersten Male nach fünfundzwanzigjährigen eigenen Erfahrungen zusammengestellt und herausgegeben. Kempten: Tobias Dannheimer, xii + 530 + 2 pp.

Gistel J. N. F. X. 1857: Achthundert und zwanzig neue oder unbeschriebene wirbellose Thiere. Pp. 513-606. In: *Vacuna oder die Geheimnisse aus der organischen und leblosen Welt. Unterdruckte Originalien-Sammlung von grösstentheils noch lebenden und verstorbenen Gelehrten aus dem Gebiete sämmtlicher Naturwissenschaften, der Medizim, Litteraturgeschichte, des Forst- und Jagtwesens, der Oekonomie, Geschichte, Biographie, und der freien schönen Künste, herausgegeben von Professor Dr. Johannes Gistel. Zweiter Band.* Straubing: Schorner, 1031 pp. [Also issued as separate in 1857 by Schorner, 1-94 pp.]

Gmelin J. F. 1790: *Caroli a Linné Systema Naturae per regna tria naturae, secundum classes, ordines, genera, species, cum characteribus, differentiis, synonymis, locis. Editio decima tertia, aucta, reformata. Tomus I. Pars IV. Classis V. Insecta.* Lipsiae: Georg Enanuel Beer, 1517-2224 pp.

Goeze J. A. E. 1776: Geschichte der schädlichen Ptinen oder Bohrkäfer. *Naturforscher* **8**: 62-100, 1 pl.

Goeze J. A. E. 1777: *Entomologische Beytraege zu des Ritters Linné zwölften Ausgabe des Natursystems. Erster Theil.* Leipzig: Weidmanns Erben und Reich, [16] + 736 pp.

Golbach R. 1968: Pyrganus nov. gen. (Col. Elat.). *Acta Zoologica Lilloana* **22** [1967]: 197-199.

Golbach R. 1969: Anotaciones sobre el genero Lacon Cast. 1836 (Col. Elateridae). *Acta Zoologica Lilloana* **25**: 153-160.

Goodrich M. A. & Springer C. A. 1988: A new species of Xerasia (Coleoptera: Byturidae) from the Middle East, with key to the Xerasia of the world. *The Coleopterist's Bulletin* **42**: 345-351.

Gordon R. D. 1999: South American Coccinellidae (Coleoptera), Part VI: A systematic revision of the South American Diomini, new tribe (Scymninae). *Annales Zoologici* **49** (Supplement 1): 1-219 pp.

Gordon R. D. 2000: South American Coccinellidae (Coleoptera). Part VII: A systematic revision of South American Scymnus (Pullus) Mulsant (Scymninae: Scymnini). *Frustula Entomologica* (N. S.) **23**: 56-108.

Gorham H. S. 1873a: *A catalogue of the coleopterous group, Endomycici, with descriptions of new species, and notes. Endomycici Recitati.* London: Williams and Norgate, 64 pp., 1 pl.

Gorham H. S. 1873b: A list of Endomychidae collected in Japan by Geo. Lewis, Esq. with descriptions of new genera and species. *The Entomologist's Monthly Magazine* **9**: 205-207.

Gorham H. S. 1873c: Descriptions of a new genus and two new species of Coleoptera from Japan. *The Entomologist's Monthly Magazine* **9**: 257-258.

Gorham H. S. 1874a: [new taxa]. In: Gorham H. S. & Lewis G.: Description of a new genus and species of Coleoptera from Japan. *The Entomologist's Monthly Magazine* **11**: 54-55.

Gorham H. S. 1874b: Descriptions of new Japanese Lycoperdina. *The Entomologist's Monthly Magazine* **11**: 151.

Gorham H.S. 1874c: Descriptions of new species of Coleoptera from Japan. *The Entomologist's Monthly Magazine* **10**: 224-226.

Gorham H. S. 1874d: Descriptions of new species of Endomycici. *Transactions of the Entomological Society of London* **1874**: 437-446.

Gorham H. S. 1874e: Observations on the genus Helota, M'Leay, with description of a new species from Japan. *Transactions of the Entomological Society of London* **1874**: 447-449.

Gorham H. S. 1875: Descriptions of new species of Endomycici. *Transactions of the Entomological Society of London* **1875**: 311-314.

Gorham H. S. 1876: Notes on the coleopterous family Cleridae, with description of new genera and species. *Cistula Entomologica* **2** [1875-1882]: 57-106.

Gorham H. S. 1877: Descriptions of new species of Cleridae, with notes on the genera and corrections of synonymy. *Transactions of the Entomological Society of London* **1877**: 401-426.

Gorham H. S. 1878: Descriptions of new genera and species of Cleridae, with notes on the genera and corrections of synonymy. *Transactions of the Entomological Society of London* **1878**: 153-167.

Gorham H. S. 1880: Materials for a revision of the Lampyridae. *Transactions of the Entomological Society of London* **1880** (1): 1-36; 83-112.

Gorham H. S. 1881: Malacodermata. Fam. Lycidae. Pp. 1-29. In: Godman F. D. & Salvin O. (eds): *Biologia Centrali - Americana. Insecta. Coleoptera. Vol. III. Part 2.* [1880-1886]. London: Dulau & Co., B. Quaritsch, xii + 372 pp., 13 pls.

Gorham H. S. 1883a: Two new species of Lampyridae. *Notes from the Leyden Museum* **5**: 3-4.

Gorham H. S. 1883b: Descriptions of new species of beetles belonging to the family Erotylidae. *Proceedings of the Scientific Meetings of the Zoological Society of London* **1883**: 75-87.

Gorham H. S. 1883c: Malacodermata. Pp. 169-224. In: Godman F. D. & Salvin O. (eds): *Biologia Centrali – Americana. Insecta. Coleoptera. Vol. III. Part 2.* [1880-1886]. London: Dulau & Co., B. Quaritsch, xii + 372 pp., 13 pls.

Gorham H. S. 1883d: Revision of the genera and species of malacoderm Coleoptera of the Japanese fauna. Part.I. – Lycidae, Lampyridae. *Transactions of the Entomological Society of London* **1883**: 393-411, pl. xvii.

Gorham H. S. 1883e: Descriptions of Malacodermata in the Civic Museum of Natural History at Genua. *Annali del Museo Civico di Storia Naturale di Genova* **18**: 591-606.

Gorham H. S. 1885a: Description of a new genus of Bostrychidae. *Notes from the Leyden Museum* **7**: 51-53.

Gorham H. S. 1885b: Descriptions of some Endomychidae and Erotylidae in the Genoa Civic Museum. *Annali del Museo Civico di Storia Naturale di Genova* **22**: 517-530.

Gorham H. S. 1886: Insecta. Coleoptera. Malacodermata. Supplement. Pp. 313-372. In: Godman F. D. & Salvin O. (eds): *Biologia Centrali-America. Volume III. Part 2.* London: Dulau & Co., B. Quaritsch, xii + 372, 13 pls. [issued in parts, 1880-1886].

Gorham H. S. 1887a: On the classification of the Coleoptera of the subfamily Languriides. *Proceedings of the Scientific Meetings of the Zoological Society of London* **1887**: 358-362.

Gorham H. S. 1887b: Revision of the Japanese species of the coleopterous family Endomychidae. *Proceedings of the Scientific Meetings of the Zoological Society of London* **1887**: 642-653, pl. 53.

Gorham H. S. 1888: On Erotylidae of the Leyden Museum. *Notes from the Leyden Museum* **10**: 129-151.

Gorham H. S. 1889: Descriptions of new species and of a new genus of Coleoptera of the family Telephoridae. *Proceedings of the Zoological Society* **1889**: 96-111.

Gorham H. S. 1890a: A new species of the erotylid genus Episcapha. *Notes from the Leyden Museum* **12**: 47-48.

Gorham H. S. 1890b: Fam. Endomychidae [part]. Pp. 129-144. In: Godman F. & Salvin O. (eds): *Biologia Centrali-Americana. Insecta. Coleoptera. Vol. VII.* London: Taylor & Francis, xii + 276 pp., 13 pls. [issued in parts, 1887-1899].

Gorham H. S. 1890c: Notes on the species of the families Lycidae and Lampyridae, contained in the Imperial museum of Calcutta, with descriptions of new species, and list of the species at present described from India. *Transactions of the Entomological Society of London* **1890**: 541-550.

Gorham H. S. 1891a: Fam. Endomychidae [part]. Pp. 145-150. In: Godman F. & Salvin O. (eds): *Biologia Centrali-Americana. Insecta. Coleoptera. Vol. 7.* London: Taylor & Francis, xii + 276 pp., 13 pls. [issued in parts, 1887-1899].

Gorham H. S. 1891b: Contributions à la faune indo-chinoise. Malacodermata, Cleridae, Lyctidae, Erotylidae, Endomychidae, Cassididae, Coccinellidae. *Annales de la Société Entomologique de France* **60**: 397-404.

Gorham H. S. 1892a: Coleoptera of Central China and Korea. *The Entomologist* **25** (4) (Supplement): 81-85.

Gorham H. S. 1892b: Description of a new species of Elater. *The Entomologist* **25**: 304-305.

Gorham H. S. 1892c: Viaggio di Leonardo Fea in Birmania e regione vicine XLVIII. Cleridae. *Annali del Museo Civico di Storia Naturale di Genova* **32**: 718-746.

Gorham H. S. 1893: A list of the Coleoptera, of the family Cleridae collected by Mr. Doherty in Burmah and Nothern India, with descriptions of new species; and of some species from Borneo, Perak, &c., from the collection of Alexander Fry, Esq. *Proceedings of the Zoological Society of London* **1893**: 566-581.

Gorham H. S. 1894a: On the Coccinellidae from India in the collection of Mr. H. E. Andrewes of the Indian Forest Service. *Annales de la Société Entomologique de Belgique* **38**: 200-208.

Gorham H. S. 1894b: Insectes du Bengale (32e Mémoire). Observations on some species of the family Coccinellidae collected near Konbir and Mandar, India, Bengal, by P. Cardon. *Annales de la Société Entomologique de Belgique* **38**: 209-211.

Gorham H. S. 1895a: List of the Coleoptera in the collection of H. E. Andrews Esq. from India and Burma, with descriptions of new species and notes. Families: Malacodermata - Erotylidae - Endomychidae. *Annales de la Société Entomologique de Belgique* **39**: 293-330.

Gorham H. S. 1895b: Viaggio di Leonardo Fea in Birmania e regioni vicine. LXV. On the Coccinellidae collected by Mr. L. Fea in Birma. *Annali del Museo Civico di Storia Naturele di Genova* **34** [1894-1895]: 683-695.

Gorham H. S. 1896: Viaggio di Leonardo Fea in Birmania e regioni vicine. LXIX. Languriidae, Erotylidae and Endomychidae. *Annali del Museo Civico di Storia Naturale di Genova* **36**: 257-302.

Gorham H. S. 1897: Description of new species of Coleoptera of the family Endomychidae from the eastern hemisphere. *Proceedings of the Scientific Meetings of the Zoological Society of London* **1897**: 456-464, pl. 32.

Gorham H. S. 1898: Supplement. Pp. 247-252. In: Godman F. & Salvin O. (eds.): *Biologia Centrali-Americana. Insecta. Coleoptera. Vol. VII.* London: Taylor & Francis, xii + 276 pp., 13 pls. [issued in parts, 1887-1899].

Gorham H. S. 1901: Erotylidae, Endomychidae and Coccinellidae of Sumatra. *Entomologische Zeitung* (Stettin) **62**: 169-214.

Gorham H. S. 1903: On Coleoptera collected in India by MM. H.-E. and H.-L. Andrews. Families: Malacodermata, Erotylidae, Endomychidae and Coccinellidae. *Annales de la Société Entomologique de Belgique* **47**: 323-347.

Gory H. 1832: Taupin de Gory. E. Goryi. Duftschmid. *Magasin de Zoologie* (F. E. Guérin-Méneville) Classe IX. Insects. Pl. XI.

Gottwald J. 1971: Die mitteleuropäischen Arten aus der Verwandtschaft von Ernobius explanatus Mannerheim (Coleoptera, Anobiidae). *Ergebnisse der Wissenschaftlichen Untersuchungen im Sschweizerrischen Nationalpark* **11**: 168-173.

Gottwald J. 1977: Die paläarktischen Xyletinus-arten (Coleoptera, Anobiidae). *Acta Entomologica Bohemoslovaca* **74**: 158-177.

Gottwald J. 1983: Zur Taxonomie und Faunistik der paläarktischen Xyletinus Latreille, 1809-Arten (Coleoptera, Anobiidae). *Entomofauna, Zeitschrift für Entomologie* **4**: 133-137.

Gougelet J. S. 1859: [Descriptions de coléoptères nouveaux, de Galice et d'Algérie, faite conjointement avec M. Henry Brisout de Barbeville]. *Bulletin de la Société Entomologique de France* **1859**: ccxxxvii-ccxxxviii.

Gourreau J.-M. 1970: Pullus melanarius: une nouvelle espèce de Scymnini (Coléoptères, Coccinellidae) nommée mais non decrite par Rey. *Bulletin Mensuel de la Société Linnéenne de Lyon* **39**: 153-156.

Gourreau J.-M. 1974: Systématique de la tribu des Scymnini (Coccinellidae). *Annales de Zoologie, Ecologie Animale* Numéro Hors-Série **1974**: 221 + [2] pp.

Gozis M. des. 1875: *Catalogue des coléoptères de France et de la faune gallo-rhénane*. Monluçon: Crépin-Leblond, [4] + 108 pp.

Gozis M. des. 1881a: Révision des Latridiidae d'Europe par Edmond Reitter. Traduit de l'allemand par M. des Gozis accompagnée de généralitités sur l'histoire, les moeurs, la distribution géographique, la bibliographie de cette tribu avec addition des espèces extra européennes de l'ancien monde. *L'Abeille, Journal d'Entomologie* **18**: 1-178.

Gozis M. des. 1881b: Quelques rectifications synonymiques touchant différents genres et espèces de coléoptères français. *Bulletin des Séances de la Société Entomologique de France* **1881**: cxxvi-cxxvii, cxxxiv-cxxxv.

Gozis M. des. 1882: Notes et remarques pour le futur catalogue des coléoptères gallo-rhénans. *Revue d'Entomologie* **1**: 193-207.

Gozis M. des. 1886: *Recherche de l'espèce typique de quelques anciens genres. Rectification synonymiques et notes diverses*. Montluçon: Herbin, 36 pp.

Gozis M. des. 1889: Les Phalacrides d'Europe. *Revue d'Entomologie* **8**: 13-32.

Gradl H. 1880: Coccinelliden. *Entomologische Nachrichten* **6**: 253-256, 275-278.

Gradl H. 1881: Aus der Fauna des Egerlandes. Neue Beschreibungen von Insecten. *Entomologische Nachrichten* **7**: 294-309.

Gradl H. 1882: Aus der Fauna des Egerlandes. Zweite Folge. *Entomologische Nachrichten* **8**: 323-332.

Graells M. de la P. 1842: Description d'une nouvelle espèce du genre Dasytes. *Annales de la Société Entomologique de France* **11**: 221-223, pl. 10.

Graells M. de la P. 1846: Description de cinq espèces nouvelles de coléoptères d'Espagne. *Revue Zoologique par la Société Cuvierienne* **9**: 217-218.

Graells M. de la P. 1847: Description de cinq espèces nouvelles de coléoptères d'Espagne. *Annales de la Société Entomologique de France* (2) **5**: 305-308.

Graells M. de la P. 1851: Descripción de algunos insectos nuevos pertenecientes a la fauna central de España. *Memorias de la Academia de Ciencias de Madrid* **1** (2): 109-163.

Graells M. de la P. 1858: Insectos nuevos de España. Pp. 36-112, 7 pls. In: Luján F. De (ed.): *Memorias della Comisión del Mapa Geológico de España.* Año **1855**. Parte Zoológica. Madrid: Impr. Nacional, 1-111 + [1] pp., 7 pls.

Grandjean E. 1895: Description d'une nouvelle espèce de Ptinus. *Revue Scientifique du Bourbonnais et du Centre de la France* **8**: 70-71.

Gravenhorst J. L. C. 1807: *Vergleichende Uebersicht des Linnéischen und einiger neueren zoologischen Systeme, nebst dem eingeschalteten Verzeihniss der zoologischen Sammlung des Verfassers und den Beschreibungen neuer Thierarten, die in derselben vorhanden sind.* Goettingen: H. Dietrich, xx + 476 pp.

Gray G. R. 1832: Notices of new genera and species. In: Griffith E. (ed.): *The animal kingsdom arranged in conformity with its organization; by the Baron Cuvier, member of the Institute of France, &c. &c. &c. Volume the fourteenth. The class Insecta arranged by the Baron Cuvier. Volume the first.* London: Whittaker, Treacher, and Co., viii + 570 pp., 140 pls.

Gredler V. M. 1854: *Die Kaefer von Passeier 1. Heft.* Innsbruck: Wagner'sche Buchdruckerei, 20 pp.

Gredler V. M. 1857: *Die Käfer von Passeier. 2. Heft.* Innsbruck: Wagner'sche Buchdruckerei, 23-75 pp.

Gredler V. M. 1863: *Die Käfer von Tirol nach ihrer horizontalen und vertikalen Verbreitung. Mit mehren diagnostirten Novitäten.* Bozen: J. Eberle'sche Buchdruckerei, v + 235 pp.

Gredler V. M. 1866: *Die Käfer von Tirol nach ihrer horizontalen und vertikalen Verbreitung. 2. Teil. Dascillidae-Schluss.* Bozen: G. Ferrari (vorm. Eberle), [2] + 235-491 pp.

Gredler V. M. 1870: Zweite Nachlese zu den Käfern von Tirol. *Coleopterologische Hefte* **6**: 1-16.

Gredler V. M. 1878: Zur Käfer-Fauna Central Afrikas. *Verhandlungen der Kaiserlich-Königlichen Zoologisch-Botanischen Gesellschaft in Wien* **27** [1877]: 501-507.

Green J. W. 1951: The Lycidae of the United States and Canada. III. The tribe Platerodini (in part) (Coleoptera). *Transactions of the American Entomological Society* **78**: 1-22.

Gressitt J. L. 1939: Some Prionoceridae from China, Hainan, Formosa and Cochin-China (Coleoptera). *Lingnan Science Journal* (Canton) **18**: 187-196, 1 pl.

Gridelli E. 1927: Una nuova specie di Merophysia della Cirenaica. *Memoires de la Società Entomologica Italiana* **6**: 237-239.

Gridelli E. 1930: Rizultati zoologici della Missione invitata dalla R. Società Geografica Italiana per l'esplorazione dell' oasi di Giarabub (1926-1927). Coleotteri. *Annali del Museo Civico di Storia Naturale di Genova* **54**: 1-487.

Grimmer C. H. B. 1841: *Steiermarks Coleopteren mit Einhundertsechs neu beschriebenen Species.* Graz: Damian, iv +5-50 pp.

Grouvelle A. 1874: [Communications]. *Bulletin de la Société Entomologique de France* **1874**: xxvii-xxix.

Grouvelle A. 1876a: [Nouvelles espèces de cucujides.] *Bulletin de la Société Entomologique de France* **1876**: xxxii-xxxiii.

Grouvelle A. 1876b: [Diagnoses de deux nouvelles espèces de cucujides.] *Bulletin de la Société Entomologique de France* **1876**: cxxii-cxxiii.

Grouvelle A. 1877a: Cucujides nouveaux ou peu connus. 2e Mémoire. *Annales de la Société Entomologique de France* (5) **7**: 205-214.

Grouvelle A. 1877b: Descriptions d'espèces nouvelles de Cucujides. *Revue de Zoologie Pure et Appliquée* (3) **5**: 296-300.

Grouvelle A. 1881: Cucujides nouveaux ou peu connus. 6e Mémoire. *Annales de la Société Entomologique de France* (6) **1**: 89-96.

Grouvelle A. 1883: Cucujides nouveaux du Musée Civique de Génes. *Annali del Museo Civico di Storia Naturale di Genova* **18** [1882-1883]: 275-296.

Grouvelle A. 1885: Deux espèces nouvelles de Cucujides des îles de la Sonde. *Notes of the Leyden Museum* **7**: 47-48.

Grouvelle A. 1888: Viaggio de Leonardo Fea in Birmania e regioni vicine. XI. Cucujides. *Annali del Museo Civico di Storia Naturale di Genova* **26**: 624-629.

Grouvelle A. 1889: Cucujides nouveaux ou peu connus. 7e Mémoire. *Annales de la Société Entomologique de France* (6) **9**: 101-108, pl. 5.

Grouvelle A. 1890: Viaggio di Leonardo Fea in Birmania e regioni vicine. XXIII. Nitidulides. Premier mémoire. *Annali del Museo Civico di Storia Naturale di Genova* **29**: 120-126.

Grouvelle A. 1891: Clavicornes du Bengal occidentale. *Annales de la Société Entomologique Belgique* **35**: 237-238.

Grouvelle A. 1892a: Clavicornes du Bengale. 2me liste. *Annales de la Société Entomologique de Belgique* **36**: 60-61.

Grouvelle A. 1892b: Voyage de M. E. Simon a l'île de Luzon Philippines (Mars et Avril 1890). 5e Mémoire. Cucujides. *Annales de la Société Entomologique de France* **61**: 285-288.

Grouvelle A. 1892c: Viaggio de Leonardo Fea in Birmania e regioni vicine. L. Nitidulides, Cucujides et Parnides. 2.me Partie. *Annali del Museo Civico di Storia Naturale di Genova* **32**: 833-868.

Grouvelle A. 1893: Voyage de M. E. Simon à l'île de Ceylan (janvier-février 1892). 5e Mémoire. *Annales de la Société Entomologique de France* **62**: 383-390.

Grouvelle A. 1894a: [new taxa]. In: Grouvelle A. & Guillebeau F.: Clavicornes nouveaux récoltés dans l'Inde par Mr. H. E. Andrewes. *Annales de la Société Entomologique de Belgique* **38**: 458-465.

Grouvelle A. 1894b: Insectes de Bengale (35e mémoire). Clavicornes. 3me liste. *Annales de la Société Entomologique de Belgique* **38**: 578-587.

Grouvelle A. 1896a: Viaggio di Leonardo Fea in Birmania e regioni vicine. LXVIII: Colydiides et Monotomides. *Annali del Museo Civico di Storia Naturale di Genova* **36**: 29-42.

Grouvelle A. 1896b: Descriptions de clavicornes d'Afrique et de Madagascar. *Annales de la Société Entomologique de France* **65**: 71-94.

Grouvelle A. 1896c: Deux nouvelles espèces d'Epuraea (Col.) du bassin Méditerranéen. *Bulletin de la Société Entomologique de France* **1896**: 278-279.

Grouvelle A. 1897: Clavicornes nouveaux des Indes Orientales et pays voisins. *Annali del Museo Civico di Storia Naturale di Genova* **38**: 342-398.

Grouvelle A. 1898: Clavicornes de Grenada et de St. Vincent (Antilles) récoltés par M. H. H. Smith, et appartenant au Musée de Cambridge. *Notes from the Leyden Museum* **20**: 35-48.

Grouvelle A. 1899a: Nitidulides de l'Afrique occidentale (Cameroun) récoltés par M. L. Conradt. *Annales de la Société Entomologique de France* **68**: 125-135.

Grouvelle A. 1899b: Descriptions de clavicornes d'Afrique et de la Région Malgache. 2e Mémoire. *Annales de la Société Entomologique de France* **68**: 136-185.

Grouvelle A. 1901: Description d'un nouveau genre de Nitidulidae du nord et de l'est de l'Afrique. *Bulletin de la Société Entomologique de France* **1901**: 102-103.

Grouvelle A. 1902: Coléoptères (Silphidae, Nitidulidae, Rhysodidae, Cucujidae, Cryptophagidae), recueillis dans le Japon central par M. le Dr. Harmand. *Bulletin du Muséum d'Histoire Naturelle* (Paris) **8**: 15-17.

Grouvelle A. 1903a: Clavicornes de l'Inde septentrionale récoltés par M. Harmand. Nitidulidae.-Colydiidae.-Cucujidae.-Monotomidae.-Dryopidae. *Annales de la Société Entomologique de France* **72**: 108-124.

Grouvelle A. 1903b: Coléoptères de l'Inde récoltés par MM. H.-L. Andrewes et T. R. D. Bell. *Annales de la Société Entomologique de France* **72**: 125-127.

Grouvelle A. 1903c: Voyage de M. Maurice Maindron dans l'Inde méridionale (Mai-Novembre 1901). 2e Mémoire. Coléoptères clavicornes. *Annales de la Société Entomologique de France* **72** [1903-1904]: 340-347.

Grouvelle A. 1904: Descriptions de clavicornes nouveaux du Musée de Bruxelles. *Annales de la Société Entomologique de Belgique* **48**: 181-185.

Grouvelle A. 1905a: Quelques clavicornes nouveaux la République Argentine recueillis par M. Charles Bruch. *Revista Museum La Plata* **12**: 121-133.

Grouvelle A. 1905b: Description d'un Laemophloeus nouveau de Madagascar (Col.). *Bulletin de la Société Entomologique de France* **1905**: 142.

Grouvelle A. 1906a: Contribution à l'étude des coléoptères de Madagascar. Nitidulidae, Colydiidae, Cucujidae, Monotomidae, Cryptophagidae, Mycetophagidae, Dryopidae, Heteroceridae. *Annales de la Société Entomologique de France* **1906**: 67-168.

Grouvelle A. 1906b: Nitidulides nouveaux du British Museum. *Bulletin de la Société Entomologique de France* **1906**: 201-203.

Grouvelle A. 1906c: Clavicornes nouveaux du Musée Civique de Gènes. *Annali del Museo Civico di Storia Naturale di Genova* **42**: 308-333.

Grouvelle A. 1908a: Clavicornes récoltés dans l'Afrique Orientale Allemande par le Dr F. Eichelbaum (1903). (1re Liste). *Revue d'Entomologie* **27**: 5-22.

Grouvelle A. 1908b: Coléoptères récoltés dans l'Afrique Orientale Allemande par le Dr F. Eichelbaum (1903) (2-me liste). *Revue d'Entomologie* **27**: 97-206.

Grouvelle A. 1908c: Coléoptères de la région indienne. Rhysodidae, Trogositidae, Nitidulidae, Colydiidae, Cucujidae. (1er mémoire). *Annales de la Société Entomologique de France* **77** [1908-1909]: 315-495, pls 6-9.

Grouvelle A. 1909a: Coléoptères récoltés dans l'Afrique Orientale Allemande par le Dr. F. Eichelbaum (1903). (2e liste). *Revue d'Entomologie* **27** [1908]: 97-126.

Grouvelle A. 1909b: Coléoptères clavicornes de l'Afrique Australe et Orientale. *Revue d'Entomologie* **27** [1908]: 127-206.

Grouvelle A. 1910a: Description d'un nouveau genre et d'une nouvelle espèce d'Erotylidae (Col.). *Bulletin de la Société Entomologique de France* **1910**: 143-145.

Grouvelle A. 1910b: Nitidulides et Cryptophagides de l'Asie et des Indes orientales. *Notes from the Leyden Museum* **32**: 241-256.

Grouvelle A. 1912a: Description de deux Cryptophagus nouveaux d'Algérie (Col. Cryptophagidae). *Bulletin de la Société Entomologique de France* **1912**: 44-47.

Grouvelle A. 1912b: Psammoecus nouveaux du Musée de Leide. *Notes of the Leyden Museum* **34** [1911/1912]: 81-94.

Grouvelle A. 1912c: Description de deux Brachyleptus nouveaux d'Algérie et de Tunisie (Col. Nitidulidae). *Bulletin de la Société Entomologique de France* **1912**: 96-98.

Grouvelle A. 1912d: Description d'un Meligethes nouveau, de France (Col. Nitidulidae). *Bulletin de la Société Entomologique de France* **1912**: 168-169.

Grouvelle A. 1912e: Étude sur les Aphanocephalus et descriptions d'espèces nouvelles. *Notes from the Leyden Museum* **34**: 197-224.

Grouvelle A. 1912f: [new taxon]. In: Grouvelle A. & Raffray A.: Supplément à la liste des coléoptères de la Guadeloupe. 3e supplément. *Annales de la Société Entomologique de France* **81**: 289-312.

Grouvelle A. 1912g: Coléoptères cucujides nouveaux du genre Psammoecus faisant parti des collections du Muséum. *Bulletin du Muséum National d'Histoire Naturelle* (Paris) **18**: 409-414

Grouvelle A. 1912hg: Coléoptères du Tonkin récoltés par M. le colonel Bonifacy: Rhysodidae, Nitidulidae, Ostomidae. *Bulletin du Muséum National d'Histoire Naturelle* (Paris) **18**: 502-505.

Grouvelle A. 1913a: Results of the Abor Expedition, 1911-12. Part II. VIII. Coleoptera, I: Rhysodidae, Nitidulidae, Colydiidae, Cucujidae, Passandridae, Discolomidae, Cryptophagidae, Mycetophagidae, Dryopidae. Pp. 99-117. *Records of the Indian Museum* (Calcutta) **8** [1912-1922]: xxvii + 629 pp., 54 pls.

Grouvelle A. 1913b: Rectifications à la nomenclature des Nitidulidae (Col.). *Bulletin de la Société Entomologique de France* **1913**: 253.

Grouvelle A. 1913c: Notes sur les Silvanini. Synonymiques et descriptions de genres nouveaux et de nouvelles espèces. *Annales de la Société Entomologique de France* **81** [1912-1913]: 313-386.

Grouvelle A. 1913d: Famille des Nitidulidae. Notes synonymiques et rectifications à la nomenclature. *Annales de la Société Entomologique de France* **81** [1912-1913]: 387-400.

Grouvelle A. 1913e: Étude sur les Ancistria (Col. Passandridae). *Annales de la Société Entomologique de France* **81** [1912-1913]: 433-494.

Grouvelle A. 1914: H. Sauter's Formosa-Ausbeute. Rhysodidae, Nitidulidae, Ostomidae, Colydiidae, Passandridae, Cucujidae, Cryptophagidae, Diphyllidae, Lathridiidae, Mycetophagidae, Dermestidae. *Archiv für Naturgeschichte* (A) **79** (11) [1913]: 33-76.

Grouvelle A. 1915a: Clavicornes africains du Musée d'Histoire Naturelle de Luxembourg récoltés par M. Ed. Luja de Luxembourg. *Festschrift der Gesellschaft Luxemburger Naturfreunde* **25**: 103-123.

Grouvelle A. 1915b: Description d'un Colydiien nouveau. *Annales de la Société Entomologique de France* **84**: 521-522.

Grouvelle A. 1916a: Études sur les coléoptères. I. - Étude sur les Passandrella, Scalidia et Laemotmetus (Passandriidae). *Mémoires Entomologiques* **1**: 5-25.

Grouvelle A. 1916b: Études sur les coléoptères. II. - Famille des Cryptophagidae. Notes synonymiques et rectifications à la nomenclature. *Mémoires Entomologiques* **1**: 26-29.

Grouvelle A. 1916c: Études sur les coléoptères. III - Descriptions d'espèces nouvelles de Cryptophagidae. *Mémoires Entomologiques* **1**: 30-79.

Grouvelle A. 1916d: Description d´un Derodontus nouveau d'Italie (Coleoptera). *Bulletin de la Société Entomologique de France* **1916**: 296-297.

Grouvelle A. 1918: Coleoptera of the families Ostomidae, Monotomidae, Colydiidae and Notiophygidae from the Seychelles and Aldabra Islands. *Transactions of the Entomological Society of London* **1918**: 1-57.

Grouvelle A. 1919a: Descriptions d´espèces nouvelles du genre Psammoecus Latr. (Cucujidae). *Mémoires Entomologiques* **2** [1917]: 5-38.

Grouvele A. 1919b: Descriptions d'espèces nouvelles de clavicornes de l'Afrique australe. *Mémoires Entomologiques* **2** [1917]: 47-60.

Grouvelle A. 1919c: Études sur les coléoptères. V.– Famille des Cryptophagidae. Notes synonymiques et rectifications à la nomenclature. *Mémoires Entomologiques* **2** [1917]: 66-69.

Grouvelle A. 1919d: Études sur les coléoptères. VI. - Descriptions de genres nouveaux et d'espèces nouvelles de Cryptophagidae. *Mémoires Entomologiques* **2** [1917]: 70-203.

Grouvelle A. & Guillebeau F. 1894a: Clavicornes nouveaux récoltés dans l'Indie par Mr. H. E. Andrewes. *Annales Société Entomologique de Belgique* **38**: 458-465.

Grüss J. 1930: Ein altägyptischen blinder Biertrinker von 5000 Jahren. *Wochenschrift für Grauerei* **47**: 181-182.

Guérin E. 1892: Cardiophorus Reitteri, nova sp. (Coléoptères-Elatérides). *Feuille des Jeunes Naturalistes* **23** [1892-1893]: 31.

Guérin E. 1893: [Cardiophorus Reitteri (voir no. 266).] *Feuille des Jeunes Naturalistes* **23** [1892-1893]: 63.

Guérin-Menéville F. E. 1829: Livraison 4, pl. 12. In: *Iconographie du règne animal de G. Cuvier, ou représentation d'après la nature de l'une des espèces les plus remarquables et souvent non encore figurées, de chaque genre d'animaux. Avec un texte descriptif mis au courant de la science. Ouvrage pouvant servir d'atlas à tous les traités de zoologie. II. Planches des animaux invertébrés. Insectes*. [1829-1838]. Paris: J. B. Baillière, 110 pls.

Guérin-Menéville F. E. 1830: Livraison 6, pl. 13. In: *Iconographie du règne animal de G. Cuvier, ou représentation d'après la nature de l'une des espèces les plus remarquables et souvent non encore figurées, de chaque genre d'animaux. Avec un texte descriptif mis au courant de la science. Ouvrage pouvant servir d'atlas à tous les traités de zoologie. II. Planches des animaux invertébrés. Insectes*. [1829-1838]. Paris: J. B. Baillière, 110 pls.

Guérin-Méneville F. E. 1831a: Cladophorus. *Magasin de Zoologie* **1**: pl. II.

Guérin-Méneville F. E. 1831b: Crustacés, arachnides et insectes. Pls 3-12, 18-59. In: Duperrey L. I. (ed.): *Voyage autour du Monde. Exécuté par ordre du Roi, sur la Corvette de sa Majesté, La Coquille, pendant les années 1822, 1823, 1824 et 1825, sous le Ministère et conformément aux instructions de S. E. M. Le Marquis de Clermont-Tonnerre, Ministre de la Marine; et publié sous les auspices de son Excellence Mgr le Cre de Chabrol, Ministre de la Marine et des Colonies. Zoologie. Atlas*. Paris: Arthus Bertrand.

Guérin-Menéville F. E. 1831c: Livraison 13, pl. 42. In: *Iconographie du règne animal de G. Cuvier, ou représentation d'après la nature de l'une des espèces les plus remarquables et souvent non encore figurées, de chaque genre d'animaux. Avec un texte descriptif mis au courant de la science. Ouvrage pouvant servir d'atlas à tous les traités de zoologie. II. Planches des animaux invertébrés. Insectes*. [1829-1838]. Paris: J. B. Baillière, 110 pls.

Guérin-Méneville F. E. 1833: Mémoire sur deux nouveaux genres de l'ordre des coléoptères et description des espèces qui les composent. *Annales de la Société Entomologique de France* **2**: 155-160.

Guérin-Menéville F. E. 1834: Livraison 34, pl. 50. In: *Iconographie du règne animal de G. Cuvier, ou représentation d'après la nature de l'une des espèces les plus remarquables et souvent non encore figurées, de chaque genre d'animaux. Avec un texte descriptif mis au courant de la science. Ouvrage pouvant servir d'atlas à tous les traités de zoologie. II. Planches des animaux invertébrés. Insectes*. [1829-1838]. Paris: J. B. Baillière, 110 pls.

Guérin-Menéville F. E. 1835a: Livraison 37, pl. 15. In: *Iconographie du règne animal de G. Cuvier, ou représentation d'après la nature de l'une des espèces les plus remarquables et souvent non encore figurées, de chaque genre d'animaux. Avec un texte descriptif mis au courant de la science. Ouvrage pouvant servir d'atlas à tous les traités de zoologie. II. Planches des animaux invertébrés. Insectes*. [1829-1838]. Paris: J. B. Baillière, 110 pls.

Guérin-Menéville F. E. 1835b: Livraison 38, pl. 40. In: *Iconographie du règne animal de G. Cuvier, ou représentation d'après la nature de l'une des espèces les plus remarquables et souvent non encore figurées, de chaque genre d'animaux. Avec un texte descriptif mis au courant de la science. Ouvrage pouvant servir d'atlas à tous les traités de zoologie. II. Planches des animaux invertébrés. Insectes.* [1829-1838]. Paris: J. B. Baillière, 110 pls.

Guérin-Méneville F. E. 1838a: Crustacés, arachnides et insectes. Pp. i-xii, 1-319. In: Duperrey L. I. (ed.): *Voyage autour du Monde. Exécuté par ordre du Roi, sur la Corvette de sa Majesté, La Coquille, pendant les années 1822, 1823, 1824 et 1825, sous le Ministère et conformément aux instructions de S. E. M. Le Marquis de Clermont-Tonnerre, Ministre de la Marine; et publié sous les auspices de son Excellence Mgr le Cre de Chabrol, Ministre de la Marine et des Colonies. Zoologie. Tome Second. 2e Partie. 1re Div* (*1830*). Paris: Arthus Bertrand.

Guérin-Méneville F. E. 1838b: Genre phylocère, Phyllocerus, Latreille. *Revue Zoologique, par la Société Cuvierienne* **1**: 12-13.

Guérin-Méneville F. E. 1838c: Sur le genre trochoïde, Trochoideus, Westwood. *Revue Zoologique, par la Société Cuvierienne* **1**: 22-24.

Guérin-Méneville F. E. 1838d: Note sur le genre de coléoptères clavicornes, nommés par Latreille Globicornis, et description d'une espèce nouvelle de ce genre. *Revue Zoologique, par la Société Cuvierienne* **1**: 135-139.

Guérin-Méneville F. E. 1839: Description de deux coléoptères nouveaux découverts par M. le Vicomte de Lamote-Baracé, aux environs de Chinon (Indre-et-Loire). *Revue Zoologique, par la Société Cuvierienne* **2**: 139-141.

Guérin-Méneville F. E. 1840: Coléoptères nouveaux du plateau des Neelgheries, dans les Indes orientales, découverts par M. Adolphe Delessert. *Revue Zoologique, par la Société Cuvierienne* **3**: 37-42.

Guérin-Méneville F. E. 1841: Description de quelques nouvelles d'erotylides. Extraite du manuscript du texte explicatif de l'Iconographie du Règne Animal. *Revue Zoologique, par la Société Cuvierienne* **4**: 153-161.

Guérin-Méneville F. E. 1843: Revue critique de la tribu des eucnémides. *Annales de la Société Entomologique de France* **2** (1): 163-199, pls. 5-6.

Guérin-Méneville F. E. 1844a: *Iconographie du Règne Animal de G. Cuvier, ou représentation d'après nature de l'une des espèces les plus remarquables, et souvent non encore figurées, de chaque genre d'animaux. Avec un texte descriptif mis au courant de la science. Ouvrage pouvant servir d'atlas à tous les traités de zoologie. III. Texte explicatif. Insectes.* Paris: J. B. Baillière, 576 pp.

Guérin-Méneville M. F. E. 1844b: Note sur la découverte faite en Algérie, par M. le docteur Guyon, d'une nouvelle espèce du genre Cébrio. *Revue Zoologique, par la Société Cuvierienne* **7**: 403-404.

Guérin-Méneville F. E. 1845: [Sur quelques coléoptères trouvés dans la racine de squine (Smilax chine).] *Bulletin de la Société de Entomologique de France* **1845**: xvi-xvii.

Guérin-Méneville F. E. 1847: Décade entomologique. *Revue Zoologique, par la Société Cuvierienne* **10**: 2-11.

Guérin-Méneville F. E. 1850: Énumération des insectes qui consomment les tabacs. *Revue et Magasin de Zoologie Pure et Appliquée* (2) **2**: 426-442 pp., 1 pl.

Guérin-Méneville F. E. 1857: Matériaux pour une monographie des coléoptères du groupe des eumorphides, et plus spécialement du genre Eumorphus. *Revue et Magasin de Zoologie Pure et Appliquée* (2) **9**: 565-581.

Guérin-Méneville F. E. 1858: Matériaux pour une monographie des coléoptères du groupe des eumorphides, et plus spécialement du genre Eumorphus. *Revue et Magasin de Zoologie Pure et Appliquée* (2) **10**: 10-29.

Guérin-Méneville M. F. E. 1859: [new taxa]. *Bulletin de la Société Entomologique de France* **1859**: clxxxvi-cxci.

Guglielmi A. & Platia G. 1985: Contributo alla conoscenza degli elateridi di Grecia e Turchia (Coleoptera). *Fragmenta Entomologica* **18**: 169-224.

Guillebeau F. 1885a: Nouvelles espèces de coléoptères. *L'Échange, Revue Linnéenne* **1** (6): 2-3.

Guillebeau F. 1885b: Description de quelques nouvelles espèces de coléoptères (suite). *L'Échange, Revue Linnéenne* **1** (7): 2-3.

Guillebeau F. 1886: Description de trois espèces nouvelles de coléoptères. *L'Échange, Revue Linnéenne* **2** (19) [supplément]: 1-3.

Guillebeau F. 1890: Note sur les Silvanus du groupe du surinamensis. *Revue d'Entomologie* **9**: 220-224.

Guillebeau F. 1892: Révision des phalacrides de la faune Paléarctique. *Revue d'Entomologie* **11**: 141-197.

Guillebeau F. 1893a: Description de deux espèces de phalacrides recueillis à Aden par M. E. Simon. *Annales de la Société Entomologique de France* **62**: 297-298.

Guillebeau F. 1893b: Phalacridae. *Annales de la Société Entomologique de France* **62**: 374-379.

Guillebeau F. 1894a: Descriptions de quelques espèces de la famille de Phalacridae de la collection de M. Antoine Grouvelle. *Annales de la Société Entomologique de France* **63**: 275-310.

Guillebeau F. 1894b: Clavicornes nouveaux récoltés dans l'Indie. *Annales de la Société Entomologique de Belgique* **38**: 458-465.

Guillebeau F. 1894c: Sur Eledona agaricola Herbst et E. turcica Seidlitz et la description de quatre espèces nouvelles de coléoptères d'Algérie. *Bulletin de la Société Entomologique de France* **1893**: cccxxvi-cccxxviii.

Guillebeau F. 1894d: Description de deux espèces nouvelles de la coléoptères et une note sur Anthribus scapularis Gebler. *Bulletin de la Société Entomologique de France* **1894**: cxcii-cxciv.

Guillebeau F. 1896a: Note. *Bulletin de la Société Entomologique de France* **1895**: 25-27.

Guillebeau F. 1896b: Descriptions de quelques espèces nouvelles de coléoptères. *L'Échange, Revue Linnéenne* **12**: 48-50.

Guillebeau F. 1896c: Description de phalacridae recueillis par M. Ch. Alluaud dans le Nord de Madagascar, en 1893 (Col.). *Bulletin de la Société Entomologique de France* **1896**: 296-299.

Guilebeau F. 1897a: Description de quelques nouvelles espèces de coléoptères. *L'Échange, Revue Linnéenne* **13**: 22-25.

Guillebeau F. 1897b: Description de quelques espèces nouvelles de coléoptères. *Bulletin de la Société Entomologique de France* **1897**: 164.

Guilebeau F. 1897c: Descriptions de quelques espèces nouvelles de coléoptères. *Bulletin de la Société Entomologique de France* **1897**: 222-227.

Gundlach J. 1854: [new taxa]. P. 323. In: Poey F.: *Memorias sobre la historia natural de la Islas Cuba* (1) **5**.

Günther V. 1958: Ergebnisse der zoologischen Expedition des National-Museums in Prag nach der Turkei. 22. Coleoptera - Coccinellidae. *Acta Entomologica Musei Nationalis Pragae* **32**: 19-36.

Günther V. 1959: Vertreter des Tribus Hyperaspini (Col., Coccinellidae) aus der Tschechoslowakei. *Časopis Československé Společnosti Entomologické* **56**: 225-264.

Gurjeva E. L. 1954: Zhuki-shchelkuny (sem. Elateridae) rayonov srednego i nizhnego techeniya r. Urala i prilegayushchikh territoriy [Elaterid beetles (fam. Elateridae) from middle and lower flow of the r. Ural and neighbouring areas.] *Trudy Zoologicheskogo Instituta Akademii Nauk SSSR* **16**: 195-210.

Gurjeva E. L. 1957: Sistematicheskiy obzor vidov roda Elater L. (Coleoptera, Elateridae) fauny SSSR [A systematic review of the genus Elater L. (Coleoptera, Elateridae) in the fauna of the USSR.] *Entomologicheskoe Obozrenie* **36**: 451-475.

Gurjeva E. L. 1959: Sistematicheskiy obzor vidov roda Elater L. (Coleoptera, Elateridae) fauny SSSR. II [A systematic review of the genus Elater L. (Coleoptera, Elateridae) in the fauna of the USSR. II]. *Entomologicheskoe Obozrenie* **38**: 200-215.

Gurjeva E. L. 1961: Shzhelkuny roda Megapenthes Kiesenw. (Coleoptera, Elateridae) v faune SSSR [Click beetles of the genus Megapenthes Kiesenw. (Coleoptera, Elateridae) in the fauna of the USSR]. *Entomologicheskoe Obozrenie* **40**: 884-886.

Gurjeva E. L. 1962: Novyy vid shchelkuna (Coleoptera, Elateridae) iz Kirgizii [A new species of click-beetle (Coleoptera, Elateridae) from Kirghizia]. *Trudy Zoologicheskogo Instituta, Akademiya Nauk SSSR* **30**: 252-253.

Gurjeva E. L. 1963a: Hypnoidini (Coleoptera, Elateridae) Sredney Azii [Hypnoidini (Coleoptera, Elateridae) of Central Asia]. *Sbornik Entomologicheskikh Rabot Akademiya Nauk Kirgizskoy SSR* **2**: 20-32.

Gurjeva E. L. 1963b: Novyy rod shchelkuna (Coleoptera, Elateridae) s severnogo Tian-Shanya [A new Elaterid genus (Coleoptera, Elateridae) from the North Tien-Shan]. *Zoologicheskiy Zhurnal* **42**: 1410-1412.

Gurjeva E. L. 1964: Novye formy zhukov-shchelkunov (Coleoptera, Elateridae) iz Kazakhstana [New species of click beetles (Coleoptera, Elateridae) from Kazakhstan]. *Trudy Zoologicheskogo Instituta, Akademiya Nauk SSSR* **34**: 149-151.

Gurjeva E. L. 1966: Zhuki-shchelkuny podsemeystva Cardiophorinae (Coleoptera, Elateridae) Sredney Azii. *Trudy Zoologicheskogo Instituta* **37**: 62-96.

Gurjeva E. L. 1967a: Zhuki-shchelkuny (Coleoptera, Elateridae) Amuro-Zeiskogo plato [The click beetles (Coleoptera, Elateridae) of the Amur-Zeya plateau]. *Trudy Zoologicheskogo Instituta Akademiya Nauk SSSR* **41**: 134-138.

Gurjeva E. L. 1967b: Novyy vid shchelkuna roda Agriotes Esch. (Coleoptera, Elateridae), vredyashchiy bogarnym posevam v Sredney Azii [A new species of the click beetle genus Agriotes Esch. (Coleoptera, Elateridae) from Middle Asia]. *Entomologicheskoe Obozrenie* **46**: 201-204.

Gurjeva E. L. 1968: 129. Elateridae. Ergebnisse der zoologischen Forschungen von Dr. Z. Kaszab in der Mongolei (Coleoptera). *Reichenbachia* **9**: 291-297.
Gurjeva E. L. 1969: 181. Elateridae II. Ergebnisse der zoologischen Forschungen von Dr. Z. Kaszab in der Mongolei (Coleoptera). *Reichenbachia* **12**: 191-194.
Gurjeva E. L. 1970: [new taxon]. In: Dolin V. G. & Gurjeva E. L.: Neue Ampedus-Arten (Coleoptera, Elateridae) aus der UdSSR. *Koleopterologische Rundschau* **48**: 13-25.
Gurjeva E. L. 1971: Elateridae. Ergebnisse der zoologischen Forschungen von Z. Kaszab in der Mongolei (Coleoptera). *Faunistische Abhandlungen* (Dresden) **3**: 83-86.
Gurjeva E. L. 1972a: Novye vidy shchelkunov (Coleoptera, Elateridae) fauny SSSR i sopredel'nykh stran [New species of click beetles (Coleoptera, Elateridae) from the USSR and adjacent territories]. *Trudy Zoologicheskogo Instituta Akademiya Nauk SSSR* **52**: 299-308.
Gurjeva E. L. 1972b: Zhuki-shchelkuny (Coleoptera, Elateridae) Mongolskoy Narodnoy Respubliki [The click-beetles (Coleoptera, Elateridae) of the Mongolian People's Republic]. *Nasekomye Mongolii* **1**: 455-474.
Gurjeva E. L. 1972c: Obzor palearkticheskikh zhukv-shckelkunov roda Agriotes Esch. (Coleoptera, Elateridae) [A review of Palaearctic species click beetles of the genus Agriotes Esch. (Coleoptera, Elateridae)]. *Entomologicheskoye Obozrenie* **51**: 859-877.
Gurjeva E. L. 1973: Novaya triba zhukov-shchelkunov Megapentini tribus n. (Coleoptera, Elateridae) [A new tribe of click-beetles Megapenthini tribus n. (Coleoptera, Elateridae)]. *Zoologicheskiy Zhurnal* **52**: 448-451.
Gurjeva E. L. 1974a: Novye zhuki-shckelkuny roda Elater L. (Ludius Berth.) iz Talysha [New click beetles of the genus Elater L. (Ludius Berth.) (Coleoptera, Elateridae) from Talysh]. *Doklady Akademii Nauk Azerbaydzhanskoy SSR* **30**: 70-72.
Gurjeva E. L. 1974b: Zhuki-shzhelkuny (Coleoptera, Elateridae) vostochnoy Mongolii [The click beetles (Coleoptera, Elateridae) of eastern Mongolia]. *Nasekomye Mongolii* **2**: 144-147.
Gurjeva E. L. 1975a: Novye i maloizvestnye vidy shchelkunov roda Negastrius Thoms. (Coleoptera, Elateridae) iz Tuvy [New and little-known click-beetles of the genus Negastrius Thoms. (Coleoptera, Elateridae) from Tuva]. Pp. 47-49. In: Cherepanov A. I. (ed.): *Taksonomiya i ekologiya zhivotnykh Sibiri* (*Novye i maloizvestnye vidy fauny Sibiri,* **9**). Novosibirsk: Nauka, 122 pp.
Gurjeva E. L. 1975b: Zwei neue Elateriden-Arten (Coleoptera, Elateridae) aus Afghanistan. *Annales Historico-Naturales Musei Nationalis Hungarici* **66**: 169-170.
Gurjeva E. L. 1975c: Dva novykh vida zhukov-shchelkunov (Coleoptera, Elateridae) iz Kavkaza [Two new species of click beetles (Coleoptera, Elateridae) from the Caucasus]. *Doklady Akademii Nauk Armyanskoy SSR* **60**: 248-251.
Gurjeva E. L. 1975d: Novyy vid Anchastus (Coleoptera, Elateridae) s Dal'nego Vostoka [A new species of Anchastus (Coleoptera, Elateridae) from the Far East]. *Zoologicheskiy Zhurnal* **54**: 311-312.
Gurjeva E. L. 1976a: Novyy vid roda Silesis Cand. (Coleoptera, Elateridae) s Dalnego Vostoka [A new species of the genus Silesis Cand. (Coleoptera, Elateridae) from the Far East]. *Trudy Zoologicheskogo Instituta Akademiya Nauk SSSR* **62**: 36-37.
Gurjeva E. L. 1976b: Novyy rod i dva novykh vida shchelkunov (Coleoptera, Elateridae) iz triby Pleonomini [A new genus and two new species of click beetles (Coleoptera, Elateridae) in the Pleonomini tribe]. *Vestnik Zoologii* **1**: 77-80.
Gurjeva E. L. 1976c: Novye vidy zhukov-shchelkunov roda Adrastus Esch. (Coleoptera, Elateridae) s Kavkaza [New species of click beetles of the genus Adrastus Esch. (Coleoptera, Elateridae) from the Caucasus]. *Entomologicheskoe Obozrenie* **55**: 92-97.
Gurjeva E. L. 1976d: Novyy vid roda Pheletes Ksw. (Coleoptera, Elateridae) iz Mongolii [A new species of the genus Pheletes Ksw. (Coleoptera, Elateridae) from Mongolia]. *Nasekomye Mongolii* **4**: 195-197.
Gurjeva E. L. 1977a: Zhuki-shchelkuny roda Ampedus Dej. (Coleoptera, Elateridae) Primorskogo Kraya [The click beetles of the genus Ampedus Dej. (Coleoptera, Elateridae) from the Primorskij Area]. *Entomofauna Dal"nego Vostoka, Trudy Biologo-Pochvennogo Instituta, Novaya Seriya* **46** (149): 43-46.
Gurjeva E. L. 1977b: K sistematike zhukov-shchelkunov roda Ampedus Dej. (Coleoptera, Elateridae) [On the systematics of click beetles of the genus Ampedus (Coleoptera, Elateridae)]. *Entomologicheskoe Obozrenie* **56**: 795-804.
Gurjeva E. L. 1978a: Novye vidy shchelkunov (Coleoptera, Elateridea) iz Sredney Azii [New species of click beetles (Coleoptera, Elateridea) from Central Asia]. *Trudy Zoologicheskogo Instituta Akademiya Nauk SSSR* **61**: 138-146.

Gurjeva E. L. 1978b: Novyy vid roda Agriotes Esch. (Coleoptera, Elateridae) iz Pamira [A new species of click beetles of the genus Agriotes Esch. (Coleoptera, Elateridae) from Pamir]. *Trudy Zoologicheskogo Instituta Akademiya Nauk SSSR* **71**: 31-33.

Gurjeva E. L. 1979a: *Zhuki-shchelkuny (Elateridae) podsem. Elaterinae Triby Megapenthini, Physorhinini, Ampedini, Elaterini, Pomachiliini.* Fauna SSSR (N. S.) 118 Zhestkokrylye. Tom 12 (4). Leningrad: Nauka, 452 pp.

Gurjeva E. L. 1979b: Novyy vid zhukov-shzhelkunov roda Selatosomus Steph. Iz Sredney Azii [New species of click beetles of the genus Selatosomus Steph. (Coleoptera, Elateridae) from middle Asia]. *Trudy Zoologicheskogo Instituta Akademiya Nauk SSSR* **88**: 45-49.

Gurjeva E. L. 1979c: Novyy vid Elateridae roda Melanothus Esch. (Coleoptera) iz Sredney Azii [A new species of Elateridae of the genus Melanotus Esch. (Coleoptera) from central Asia]. *Trudy Vsesoyuznogo Entomologicheskogo Obshchestva* **61**: 70-71.

Gurjeva E. L. 1980: Dopolneniya k faune zhukov-shchelkunov (Coleoptera, Elateridae) Mongolskoy Narodnoy Respubliki [Supplementary notes on the fauna of click beetles (Coleoptera, Elateridae) of the Mongolian People's Republic]. *Nasekomye Mongolii* **7**: 92-96.

Gurjeva E. L. 1982a: Novyy vid zhukov roda Corymbitodes Buysson (Coleoptera, Elateridae) iz Tadzhikistana [A new species of beetles of the genus Corymbitodes Buysson (Coleoptera, Elateridae) from Tadzhikistan]. *Trudy Zoologicheskogo Instituta Akademiya Nauk SSSR* **110**: 20.

Gurjeva E. L. 1982b: Novyy vid zhukov roda Pseudanostirus Dol., gen. bonum (Coleoptera, Elateridae) iz Sredney Azii [A new species of beetles of the genus Pseudanostirus Dol., gen. bonum (Coleoptera, Elateridae) from middle Asia]. *Trudy Zoologicheskogo Instituta Akademiya Nauk SSSR* **110**: 21-28.

Gurjeva E. L. 1982c: K systematike i sinonimii zhukov-shchelkunov triby Ctenicerini (Coleoptera, Elateridae) [On the systematics and synonymy of the click-beetle tribe Ctenicerini (Coleoptera, Elateridae)]. *Entomologicheskoe Obozrenie* **61**: 542-547.

Gurjeva E. L. 1983: Novye rod i vid zhukov-shchelkunov (Coleoptera, Elateridae) iz Irana [A new genus and species of click-beetles (Coleoptera, Elateridae) from Iran]. *Vestnik Zoologii* **1983** (5): 81-82.

Gurjeva E. L. 1984a: Novyy vid zhuka-shchelkuna roda Pseudoanostirus Dol. (Coleoptera, Elateridae) s Altaya i iz severnoy Mongolii [A new species of click beetle of the genus Pseudanostirus Dol. (Coleoptera, Elateridae) from Altai and north Mongolia]. *Nasekomye Mongolii* **9**: 295-297.

Gurjeva E. L. 1984b: Palearkticheskie vidy shchelkunov roda Prosternon Latreille, 1834 (Coleoptera, Elateridae) [Palearctic species of the beetle genus Prosternon Latreille, 1834 (Coleoptera, Elateridae)]. *Entomologicheskoe Obozrenie* **63**: 552-555.

Gurjeva E. L. 1985a: Palearkticheskie vidy shchelkunov roda Selanostomus Steph. (Coleoptera, Elateridae) i blizkykh k nemu rodov [Palearctic species of the beetle genus Selatosomus (Coleoptera, Elateridae) and related genera]. *Entomologicheskoe Obozrenie* **64**: 563-579.

Gurjeva E. L. 1985b: Rod Corymbitodes (Coleoptera, Elateridae) vFaune SSSR [The genus Corymbitodes (Coleoptera, Elateridae) in the fauna of the USSR]. *Zoologicheskiy Zhurnal* **64**: 782-785.

Gurjeva E. L. 1986: Obzor zhukov-shchelkunov roda Liotrichus Ksw. (Coleoptera, Elateridae) mirovoy fauny [A review of the click beetle genus Liotrichus Ksw. (Coleoptera, Elateridae) of the world fauna]. *Entomologicheskoe Obozrenie* **65**: 551-558.

Gurjeva E. L. 1987a: Novye rod i vid zhukov-shchelkunov (Coleoptera, Elateridae) fauny SSSR [A new genus and species of click beetles (Coleoptera, Elateridae) of the fauna of the USSR]. *Trudy Zoologicheskogo Instituta Akademiya Nauk SSSR* **164**: 39-49.

Gurjeva E. L. 1987b: Novyy vid roda Tropihypnus Rtt. (Coleoptera, Elateridae) iz Tadzhikistana [A new species of the genus Tropihypnus Rtt. (Coleoptera, Elateridae) from Tadjikistan]. *Trudy Zoologicheskogo Instituta Akademiya Nauk SSSR* **170**: 50-51.

Gurjeva E. L. 1988a: Obzor palearkticheskikh vidov roda Anostirus C. Thoms. (Coleoptera, Elateridae) [Review of the palearctic species of the genus Anostirus C. Thoms. (Coleoptera, Elateridae)]. *Trudy Vsesoyuznogo Entomologicheskogo Obshchestva* **70**: 29-34.

Gurjeva E. L. 1988b: Novye vidyzhulov-shchelkunov roda Melanotus Esch. (Coleoptera, Elateridae) iz Sredney Azii [New species of click beetles of the genus Melanotus Esch. (Coleoptera, Elateridae) from Central Asia]. *Entomologicheskoe Obozrenie* **67**: 338-345.

Gurjeva E. L. 1989a: *Zhuki-shchelkuny (Elateridae). Podsemeystvo Athoinae. Triba Ctenicerini [Click beetles (Elateridae). Subfamily Athoinae. Tribe Ctenicerini]*. Fauna SSSR, Novaya Seriya, No. 136, Zhestkokrylye. Vol. 12 (3). Leningrad: Nauka, 295 pp.

Gurjeva E. L. 1989b: O faune zhukov-shchelkunov triby Melanotini (Coleoptera, Elateridae) iz Sredney Azii [On the click beetle fauna of the tribe Melanotini (Coleoptera, Elateridae) from Central Asia]. *Trudy Zoologicheskogo Instituta Akademiya Nauk SSSR* **208**: 23-42.

Gurjeva E. L. 1990: Novye vidy zhukov-shchelkunov roda *Melanotus* (Coleoptera, Elateridae) iz Afganistana [New species of click beetles of the genus Melanotus (Coleoptera, Elateridae) from Afghanistan]. *Trudy Zoologicheskogo Instituta Akademiya Nauk SSSR* **211**: 40-49.

Gurjeva E. L. & Dolin V. G. 1970: [new taxa]. In: Dolin V. G. & Gurjeva E. L.: Neue Ampedus-Arten (Coleoptera, Elateridae) aus der UdSSR. *Koleopterologische Rundschau* **48**: 13-25.

Gurjeva E. L. & Dolin W. G. 1972: Novyy vid zhuka-shchelkuna (Coleoptera, Elateridae) iz juzhnogo Tadzhikistana [A new genus of click beetles (Coleoptera, Elateridae) from southern Tadjikistan]. *Entomologicheskoe Obozrenie* **51**: 615-616.

Gurjeva E. L. & Penev L. D. 1983: Novyy vid zhukov-shchelkunov roda Penthelater Ôh. (Coleoptera, Elateridae) iz Bolgarii [A new species of the click-beetle genus Penthelater Oh. (Coleoptera, Elateridae) from Bulgaria]. *Entomologicheskoe Obozrenie* **62**: 736.

Gurjeva E. L. & Tugusheva R. S. 1967: Novyy vid zhukov-shchelkunov roda *Denticolloides* Gurjeva (Coleoptera, Elateridae) [A new click beetle species of the genus Denticolloides Gurjeva (Coleoptera, Elateridae)]. *Izvestiya Akademii Nauk Kazakhskoy SSR, Seriya Biologicheskaya* **1967**: 64-65.

Gurjeva E. L. & Valiakhmedow B. V. 1982: Description and biology of a new species of the beetle genus Phyllocerus Serv. (Coleoptera, Phylloceridae) from Tajikistan. Pp. 95-97. *Zoological Institute of the USSR Academy of Sciences, Leningrad and Institute of Soil Sciences, Dushanbe.*

Gyllenhal L. 1808a: *Insecta Suecica. Classis I. Coleoptera sive Eleuterata. Tomus 1.* Scaris: F.J. Leverentz, viii + [4] + 572 pp.

Gyllenhal L. 1808b: [new taxa]. In: Schönherr C. J.: *Synonyma Insectorum, oder: Versuch einer Synonymie aller bisher bekannten Insecten; nach Fabricii Systema Eleutheratorum &c. geordnet. Erster B. Eleutherata oder Käfer. Zweiter Theil. Sprechus---Cryptocephalus*. Stockholm: C. F. Marquard, ix + 424 pp., 1 pl.

Gyllenhal L. 1810: *Insecta Suecica, Classis 1. Coleoptera sive Eleuterata. Tom 1 Part 2.* Scaris: F. J. Leverentz, xix + [1] + 660 pp.

Gyllenhal L. 1813: *Insecta Suecica descripta. Classis I Coleoptera sive Eleuterata. Tom I Pars III.* Scaris: F. J. Leverenz, [4] + 730 + [2] pp.

Gyllenhal L. 1817: [new taxa]. In: Schönherr C. J.: *Appendix ad C. J. Schönherr Synonymiam Insectorum. Tom. 1. Pars 3. Sistens descriptiones novarum specierum*. Scaris: Officina Lewerentziana, 11 + 266 pp., pls. 5, 6.

Gyllenhal L. 1827: *Insecta Suecica. Classis I. Coleoptera sive Eleuterata. Tomi I. Pars IV. Cum appendice ad partes priores*. Lipsiae: F. Fleischer, viii + [2] + 761 + [1] pp.

Habelmann P. 1854: Eine neue Art der Käfergattung Teredus Dej. *Entomologische Zeitung (Stettin)* **15**: 27-29.

Haberfelner J. 1889: Ueber entomologische Zuchthäuser. *Deutsche Entomologische Zeitschrift* **1889**: 389-394.

Haenel K. 1940: [new taxon]. In: Ermisch K. (ed.): Kleine entomologische Mitteilungen. *Entomologische Blätter* **36**: 187-191.

Halbert J. N. 1910: Cryptophagus bimaculatus, Panz. and other Coleoptera at Lough Neagh. *The Entomologist's Monthly Magazine* **46**: 62-66.

Halbherr B. 1894: Elenco sistematico dei Coleotteri finora raccolti nella Valle Lagarina. Fascicolo VII. Cleridae inclusivo Pythidae. *Pubblicazione fatta per Cura del Museo Civico di Rovereto* **26**: 1-43.

Haldeman S. S. 1842: Remarks on Changes of Nomenclature in Natural History. *Proceedings of the Academy of Natural Sciences of Philadelphia* **1**: 190-191.

Halstead D. G. H. 1973: A revision of the genus Silvanus Latreille (s.l.) Coleoptera: Silvanidae). *Bulletin of the British Museum of Natural History, Entomology* **29**: 37-112.

Halstead D. G. H. 1980: A revision of the genus Oryzaephilus Ganglbauer, including description of related genera (Coleoptera: Silvanidae). *Zoological Journal of the Linnean Society* **69**: 271-374.

Halstead D. G. H. 1993: Keys for identification of beetles associated with stored products – II. Laemophloeidae, Passandridae and Silvanidae. *Journal of Stored Products Research* **29**: 99-197.

Halstead D. G. H. & Green M. 1979: Attagenus woodrofei sp. n., previously confused with Attagenus fasciatus (Thunberg) (Coleoptera: Dermestidae). *Notulae Entomologicae* **59**: 97-104.

Hampe C. 1850: Beschreibung einiger neuen Käfer-Arten. *Entomologische Zeitung* (Stettin) **11**: 346-358.
Hampe C. 1852: Beiträge zur Käfer-Fauna Österreichs. *Verhandlungen des Zoologisch-Botanischen Vereins in Wien* **1**: 158-160.
Hampe C. 1855a: Pleganophorus, nov. gen. *Verhandlungen und Mittheilungen des Siebenbürgischen Vereins für Naturwissenschaften zu Hermannstadt* **6**: 97-98.
Hampe C. 1855b: Neue Käfergattung. *Verhandlungen des Zoologisch-Botanischen Vereins in Wien* **5**: 255-256.
Hampe C. 1861: Einige neue Käfer aus Croatien und Siebenbürgen. *Wiener Entomologische Monatschrift* **5**: 65-69.
Hampe C. 1863: Ein kleiner Beitrag zur gross-österreicheischen Käferfauna. *Wiener Entomologische Monatschrift* **7**: 285-290.
Hampe C. 1864: Neue Käfer-Arten. *Wiener Entomologische Monatschrift* **8**: 190-193.
Hampe C. 1866: Beschreibung einiger neuen Käfer. *Berliner Entomologische Zeitschrift* **10**: 373-375.
Han M. T. 2000: *Taxonomy of the Negastriinae (Coleoptera: Elateridae) in Korea.* M. S. Thesis, Department of Biology, Graduate School, Sungshin Women's University, [5] + 42 + [10] pp.
Hansen V. 1962: Enicmus atriceps n. sp. (Coleoptera, Lathridiidae). *Entomologiske Meddelelser* **31**: 245-247.
Hardy J. 1847: Descriptions of new British Coleoptera, with additional notes. *The Annals and Magazine of Natural History* **19**: 379-383.
Harold E. von. 1868: Abänderungen vergebener Namen. *Coleopterologische Hefte* **4**: 104.
Harold E. von 1869: Abänderungen vergebener Namen. *Coleopterologische Hefte* **5**: 125.
Harold E. von. 1875: Geänderte Namen. *Coleopterologische Hefte* **13**: 185.
Harold E. von. 1877: Beiträge zur Käferfauna von Japan. (Zweites Stück). *Deutsche Entomologische Zeitschrift* **21**: 337-367.
Harold E. von. 1878: Beiträge zur Käferfauna von Japan. (Viertes Stück). Japanische Käfer des Berliner Königl. Museums. *Deutsche Entomologische Zeitschrift* **22**: 65-88.
Harold E. von. 1879a: Beiträge zur Kenntnis der Languria-Arten aus Asien und Neuholland. *Mittheilungen des Münchener Entomologischen Vereins* **3**: 46-94.
Harold E. von. 1879b: Neue Coleopteren aus dem Berliner Museum. *Entomologische Zeitung* (Stettin) **40**: 329-336.
Harold E. von. 1880: Einige neue Coleoptera. *Mittheilungen des Münchener Entomologischen Vereins* **3** [1879]: 148-171.
Harrer G. A. 1784: *Beschreibung derjenigen Insecten, welche D. Schaefer in CCLXXX ausgemalten Kupfertafeln unter dem Titel: Icones Insectorum circa Ratisbonam indigenorum in 3 Theilen herausgegeben hat. Theil I Hartschalige Insekten.* Regensburg: Kayser, 328 pp.
Hatch M. H. 1962: The beetles of the Pacific Northwest. Part III: Pselaphidae and Diversicornia I. *University of Washington Publications in Biology* **16** [1961]: ix + 503 pp.
Hatta K. 1970: Byturidae of Japan Archipelago. *Transactions of the Shikoku Entomological Society* **10**: 107-110.
Hauser F. 1894: Beitrag zur Coleopteren-Fauna von Transcaspien und Turkestan. Zusammengestellt unter Mitwirkung der Herrn Dr. Eppelsheim, Escherich, Faust, Dr. von Heyden, Kuvert, Reitter und Weise. *Deutsche Entomologische Zeitschrift* **1894**: 17-74, 1 pl.
Hausmann J. F. L. 1799: *Entomologische Bemerkungen.* Braunschweig u. Helmstedt, 4 + 64 pp.
Háva J. 1999a: Contribution to the knowledge of Old World Dermestidae (Coleoptera). Part 1: Genus Dermestes. *Folia Heyrovskyana* **7**: 133-140.
Háva J. 1999b: Contribution to the knowledge of Old World Dermestidae (Coleoptera). Part 2: Genus Dermestes. *Folia Heyrovskyana* **7**: 141-150.
Háva J. 2000a: Turcicornis kopeckyi gen. n., sp. n. (Coleoptera: Dermestidae: Megatominae) from Turkey. *Folia Heyrovskyana* **8**: 115-118.
Háva J. 2000b: New interesting Dermestidae (Coleoptera) from the world with descriptions of ten new species. *Veröffentlichungen Naturkundemuseum Erfurt* **19**: 161-171.
Háva J. 2000c: Globicornis (Dearthrus) kafkai sp. n. from Iran (Insecta: Coleoptera: Dermestidae). *Reichenbachia* **33**: 369-372.
Háva J. 2001a: Anthrenus (Nathrenus) jakli sp. n. (Coleoptera: Dermestidae) from the Sultanate of Oman, and distributional notes on some other Anthrenus species. *Acta Societatis Zoologicae Bohemiae* **65**: 1-3.
Háva J. 2001b: New and little known species of the genus Anthrenus O. F. Müller (Coleoptera: Dermestidae).

Entomologische Zeitschrift **111**: 58-61.

Háva J. 2001c: Anthrenus (Anthrenus) mesopotamicus sp. n. (Coleoptera: Dermestidae) from Iraq. *Folia Heyrovskyana* **9**: 65-66.

Háva J. 2001d: New and interesting Dermestidae from Nepal (Coleoptera) – I. *Veröffentlichungen Naturkundemuseum Erfurt* **20**: 209-213.

Háva J. 2002a: Five new species of Dermestidae (Coleoptera), with notes on synonymy and distribution of the family. *Acta Musei Moraviae, Scientiae Biologicae* **87**: 29-40.

Háva J. 2002b: A new species of the genus Anthrenus subgenus: Ranthenus (Coleoptera: Dermestidae) from Turkey. *Časopis Národního Muzea, Řada přírodovědná* **171**: 45-46.

Háva J. 2002c: New and interesting Dermestidae (Coleoptera) from Nepal – Part II. *Entomologische Zeitschrif* **112**: 157-158.

Háva J. 2002d: Contribution to the knowledge of Old World Dermestidae Part 3: Dermestes (Insecta: Coleoptera) *Reichenbachia* **34**: 303-307.

Háva J. 2002e: Taxonomical notes of Thaumaglossa Redtenbacher, 1867 from Oriental region (Insecta: Coleoptera: Dermestidae: Megatominae). *Reichenbachia* **34**: 309-315.

Háva J. 2002f: Descriptions of two new dermestid beetles (Coleoptera: Dermestidae) from Syria and Israel. *Revue Suisse de Zoologie* 109: 539-542.

Háva J. 2003a: Notes on Dermestidae (Coleoptera) with description of eight new species. *Annales Historico-Naturales Musei Nationalis Hungarici* **95**: 19-35.

Háva J. 2003b: Revision of the genus Hirtomegatoma (Coleoptera: Dermestidae: Megatominae). *Časopis Národního Muzea, Řada Přírodovědná* **172**: 173-175.

Háva J. 2003c: Descriptions of five new species and distributional notes on some other Thaumaglossa species (Coleoptera: Dermestidae). *Lambillionea* **103**: 205-213.

Háva J. 2003d: Anthrenus (Nanthrenus) zahradniki (Coleoptera: Dermestidae) from Turkey. *Klapalekiana* **39**: 225-227.

Háva J. 2003e: Distributional notes on some Oriental Nosodendridae (Insecta: Coleoptera) – III. Descriptions of a new subgenus and two new species from China and Nepal. Pp. 247-249. In: Hartmann M. & Baumach H. (eds.): *Biodiversität und Naturausstatung im Himalaya – Biodiversity and Natural Heritage of the Himalaya.* Erfurt: Verein der Freunde & Förder des Naturkundemuseums Erfurt e.V., 390 pp., 16 pls.

Háva J. 2003f: New and interesting Dermestidae (Coleoptera) from Nepal – III. Pp. 251-253. In: Hartmann M. & Baumach H. (eds.): *Biodiversität und Naturausstatung im Himalaya – Biodiversity and Natural Heritage of the Himalaya.* Erfurt: Verein der Freunde & Förder des Naturkundemuseums Erfurt e.V., 390 pp., XVI pls.

Háva J. 2004a: Anthrenus (Florilinus) sveci n. sp. from Greece. *Doriana* **7** (344): 1-5.

Háva J. 2004b: Nomenclatorial notes on Dermestidae and descriptions of new species (Coleoptera). *Anima.X* **7**: 1-11.

Háva J. 2004c: A new Phradonoma species and new distributional notes on some dermestid species from Yemen (Coleoptera: Dermestidae). *Deutsche Entomologische Zeitschrift* **51**: 77-79.

Háva J. 2004d: Notes on Dermestidae (Coleoptera) with new synonyms and description of three new species. *Annales Historico-Naturales Musei Nationalis Hungarici* **96**: 81-95.

Háva J. 2004e: World keys to the genera and subgenera of Dermestidae (Coleoptera), with descriptions, nomenclature and distributional records. *Acta Musei Nationalis Pragae* (*B, Natural History*) **60**: 149-164.

Háva J. 2004f: New and interesting Dermestidae (Coleoptera) from China. *Entomologische Zeitschrift* (Stuttgart) **114**: 225-232.

Háva J. & Farkač J. 2003: Distributional notes on some Oriental Nosodendridae (Insecta: Coleoptera) – II. Nosodendron nepalense sp. n. from Nepal. Pp. 245-246. In: Hartmann M. & & Baumach H. (eds): *Biodiversität und Naturausstatung im Himalaya – Biodiversity and Natural Heritage of the Himalaya.* Erfurt: Verein der Freunde & Förder des Naturkundemuseums Erfurt e.V., 390 pp., XVI pls.

Háva J. & Herrmann A. 2004: A new Orphinus species (Coleoptera: Dermestidae) from India: Sikkim. *Veröffentlichungen Naturkundemuseum Erfurt* **23**: 201-202.

Háva J. & Jelínek J. 1999: A new species of the genus Laricobius (Coleoptera, Derodontidae) from China. *Folia Heyrovskyana* **7**: 115-118.

Háva J. & Jelínek J. 2000: Laricobius schawalleri sp. nov. from Nepal, and the male of Laricobius mirabilis Háva & Jelínek, 1999, from China (Coleoptera: Derodontidae). *Entomologische Zeitschrift* **110**: 184-185.

Háva J. & Kalík V. 1999: The dermestid beetles (Coleoptera: Dermestidae) of Iran. Part 1: genus Dermestes. *Acta Societatis Zoologicae Bohemiae* **63**: 443-450.

Háva J. & Kovařík M. 2004: Attagenus lepidus sp. n. (Coleoptera: Dermestidae) from Morocco. *Acta Musei Reginaehradecensis* **30**: 83-84.

Háva J. & Tezcan S. 2004: A new species of Anthrenus O. F. Müller, 1764 (Coleoptera: Dermestidae) from Turkey. *Baltic Journal of Coleopterology* **4**: 53-54.

Havelka J. 1951: Dermestes maculatus pakistanicus ssp. n. introduced from Pakistan to Č.S.R. (3rd contribution to knowledge of Dermestes L.) (Coleoptera, Dermestidae). *Folia Entomologica* **10**: 143-144.

Havelka J. 1957: Contribution to the knowledge of Dermestes in China. Dermestes tesselatocollis Motsch. (Col. Dermestidae) (2nd communication on Dermestes). *Folia Entomologica Hungarica* **10**: 105-109.

Haworth A. H. 1812: An account of the genus Coccinella. *Transactions of the Entomological Society of London* **1**: 257-296.

Hayashi N. 1951: Studies of the Japanese species of the genus Dorcatoma II. The Classification of the larvae and pupae. (Studies of Dorcatominae, Anobiidae, I.). *Chûhô* **4**: 1-12, 3 pls.

Hayashi N. 1954: Taxonomic notes on Japanese species of the genus Dorcatoma Herbst, with descriptions of three new species (Studies on Dorcatominae, Anobiidae, I.). *Insecta Matsumurana* **18**: 87-93.

Hayashi N. 1955: Desriptions of a new genus and a new species of Dorcatominae from Japan. Studies on Dorcatominae, Anobiidae, II. *The Journal of New Entomologist* **4**: 21-23.

Hayek C. M. F. von 1973: A reclassification of the subfamily Agrypninae (Coleoptera: Elateridae). *Bulletin of the British Museum (Natural History) Entomology, Supplement* **20**: 1-309.

Heer O. 1841: *Fauna Coleopterorum Helvetica. Pars 1 (3)*. Turici: Orelii, Fuesslini et Sociorum, 361-652 pp. [Note: issued in parts, entire pagination is xii + 652 pp].

Heer O. 1856: Verzeichniss der Insecten. In: Hartung G.: Die geologischen Verhältnisse der Inseln Lanzarote und Fuerta Ventura. *Neue Denkschriften der Allgemeinen Schweizerischen Gesellschaft für die Gesamten Naturwissenschaften* **15** [1857]: 140-142.

Heinze E. 1943: Studien zur Kenntnis der Tribus Deretaphrini und deren Stellung im System. (Colydiidae.). *Entomologische Blätter* **39**: 85-93, 97-124.

Heinze E. 1944: Beiträge zur Kenntnis der Tribus Cerylini und Metacerylini (nov.) (Coleoptera: Colydiidae.). *Arbeiten über Morphologische und Taxonomische Entomologie aus Berlin-Dahlem* **11**: 19-32.

Hellén W. 1934: Koleopterologische Mitteilungen aus Finnland XII. *Notulae Entomologicae* **14**: 81-87.

Heller K. M. B. 1898: Neue Käfer von Celebes III. *Abhandlungen und Berichte des Königlichen Zoologischen, Anthropologischen und Ethnographischen Museums zu Dresden* **7**: 39-40.

Heller K. M. B. 1912: Eine neue Gattung der Discolomidae (Coleoptera) aus der orientalischen Region. *The Philippine Journal of Science* (D) **7**: 105-108.

Heller K. M. B. 1918: Philippinische Languriinae (Coleopt. Erotylidae). *Wiener Entomologische Zeitung* **37**: 25-33.

Heller K. M. B. 1920: Beitrag zur Kenntnis der Erotyliden der indo-australischen Region mit besonderer Berücksichtigung der philippinischen Arten. *Archiv für Naturgeschichte* (A) **84** [1918]: 1-121.

Heller K. M. B. 1922: New Philippine Coleoptera. *The Philippine Journal of Science* **19** [1921]: 523-625 + [1], 3 pls.

Heller K. M. B. 1923: Die Coleopterenausbeute der Stötznerschen Sze-Tschwan-Expedition (1913-1915). *Entomologische Blätter* **19**: 61-79.

Heller K. M. B. 1926: Fauna sumatrensis (Beitrag Nr. 29). Rhysodidae et familia nova Jacobsoniidae (propre Rhysodidae? Col.). *Supplementa Entomologica* **14**: 126-128.

Helliesen T. 1913: Nogle nye nordske Coleoptera. *Stavanger Museums Aarshefte* **1912** (3): 1-8.

Hellwig J. C. L. 1792: Dritte Nachricht von neuen Gattungen im entomologischen System. *Neuestes Magazin für die Liebhaber der Entomologie herausgegeben von D. H. Schneider* **1** (4): 385-408.

Henschel G. 1861: Zur Monographie der Psoa vinnensis; Tillus elongatus; Vorkommisse im unteren Mühlkreis. *Bericht über das Museum Francisco-Carolinum* (Linz) **21**: 41-46.

Herbst J. F. W. 1783: Kritisches Verzeichniss meiner Insektensammlung. *Archiv der Insectengeschichte* (Zürich: J. C. Fuessly) **4**: 1-72, pls. 19-23.

Herbst J. F. W. 1784a: Entomologische Bemerkungen aus verschiedenen akademischen Schriften. *Archiv der Insectengeschichte* (Zürich: J. C. Fuessly): **5**: 1-27.

Herbst J. F. W. 1784b: Kritisches Verzeichniss meiner Insektensammlung. *Archiv der Insectengeschichte* (Zürich: J. C. Fuessly): **5**: 73-151, pls 24-30.

Herbst J. F. W. 1786: Erste Mantisse zum Verzeichniss der ersten Klasse meiner Insektensammlung. *Archiv der Insectengeschichte* (Zürich & Winterthur: J. C. Fuessly) **7-8**: 153-182, pls 43-54.

Herbst J. F. W. 1790: *Natursystem aller bekannten in- und ausländischen Insekten, als eine Fortsetzung der von Büffonschen Naturgeschichte. Der Käfer dritter Theil.* Berlin, xiv + 325 pp.

Herbst J. F. W. 1792: *Natursystem aller bekannten in- und ausländischen Insecten, als eine Fortsetzung der von Büffonschen Naturgeschichte. Der Käfer vierter Theil.* Berlin: J. Pauli, viii + 197 pp., 12 pls.

Herbst J. F. W. 1793: *Natursystem aller bekannten in- und ausländischen Insecten, als eine Fortsetzung der von Büffonschen Naturgeschichte. Der Käfer fünfter Theil.* Berlin: Paulischen Buchhandlung, xvi + 392 pp., 16 pls.

Herbst J. F. W. 1797: *Natursystem aller bekannten in- und ausländischen Insekten, als eine Fortsetzung der von Büffonschen Naturgeschichte. Der Käfer siebenten Theil.* Berlin: Paulischen Buchhandlung, 346 + xi pp., 26 pls.

Herbst J. F. W. 1806: *Natursystem aller bekannten in- und ausländischen Insecten, als eine Fortsetung der von Büffonschen Naturgeschichte. Der Käfer zehnter Theil.* Berlin: Paulischen Buchhandlung, viii + 285 pp., 18 pls.

Hernando C. & Ribes E. 1990: Lithophilus khnzoriani sp. nov. (Coleoptera, Coccinellidae) de la fauna Ibèrica. *Miscellànea Zoològica* **14**: 81-83.

Hetschko A. 1913: Nomenclatorische Bemerkungen zu einigen Pselphiden- und Orthoperiden-Gattungen. *Wiener Entomologische Zeitung* **32**: 181-182.

Hetschko A. 1926: Lathridiidae. Pars 85. In : Schenkling S. (ed.): *Coleopterorum Catalogus. Vol. 15.* Berlin: W. Junk, 86 pp.

Hetschko A. 1928: Zur Nomenclatur einiger Colydiiden-, Cucujiden- und Phalacriden-Arten. *Wiener Entomologische Zeitung* **44**: 141-142.

Hetschko A. 1929: Nomenclatorisches über einige Lathridiiden- und Phalacriden-Arten. *Wiener Entomologische Zeitung* **46**: 156.

Hetschko A. 1930: Cucujidae, Thorictidae (Suppl.), Cossyphodidae (Suppl.). Pars 109. In: Schenkling S. (ed.): *Coleopterorum Catalogus. Vol. 15.* Berlin: W. Junk, 124 pp.

Heyden L. F. J. D. von. 1863: Beitrag zur Coleopterenfauna des Ober-Engadins, insbesondere der Umgebung von St. Moritz. *Jahresbericht der Naturforschenden Gesellschaft Graubündens* (N. F.) **8** [1861-1862]: 49-52.

Heyden L. von 1864: Nachtrag zum Beitrag zur Coleopterenfauna des Oberengadins, insbesondere der Umgebung von St. Moritz. *Jahresbericht der Naturforschenden Gesellschaft Graubündens* (N. F.) **9** [1862-1863]: 1-16.

Heyden L. F. J. D. von. 1867: Beiträge zur Kenntnis der deutschen Käferfauna. Erstes Stück. *Berliner Entomologische Zeitschrift* **11**: 376-383.

Heyden L. F. J. D. von. 1870: Entomologische Reise nach dem südlichen Spanien, der Sierra Guadarrama und Siera Morena, Portugal und Cantabrischen Gebirgen, mit Beschreibungen der neuen Arten. *Berliner Entomologische Zeitschrift* **14** Beiheft: [1] + 218 pp., 2 pls.

Heyden L. F. J. D. von. 1878: [new taxon]. In: Schneider O. & Leder H.: Beiträge zur Kenntniss der kaukasischen Käferfauna. *Verhandlungen des Naturforschenden Vereins in Brünn* **16** [1877]: 3-258, pls 1-4.

Heyden L. F. J. D. von. 1879: Die coleopterologische Ausbeute des Prof. Dr. Rein in Japan 1874-1875. *Deutsche Entomologische Zeitschrift* **23**: 321-365.

Heyden L. F. J. D. von. 1880a: *Catalog der Coleopteren von Sibirien mit Einschluss derjenigen der turanischen Länder, Turkestans und der chinensischen Grenzgebiete.* Berlin: A. W. Schade, xxiv + 224 pp.

Heyden L. F. J. D. von. 1880b: Verzeichniss von Coleopteren aus Asturien, mit Beschriebungen neuer Arten von Candèze, v. Heyden, Kirsch, Kraatz, Stierlin. *Deutsche Entomologische Zeitschrift* **24**: 281-303.

Heyden L. F. J. D. von. 1881: [new taxa]. In: Heyden L. F. J. D. von & Kraatz G. 1881: Beiträge zur Käferfauna von Turkestan. *Deutsche Entomologische Zeitschrift* **25**: 321-330.

Heyden L. F. J. D. von. 1882a: Beitrag zur Coleopteren-Fauna der Sierra de Cordoba. *Deutsche Entomologische Zeitschrift* **26**: 43-48.

Heyden L. F. J. D. von. 1882b: Neue und seltene Elateriden aus der Krimm. *Deutsche Entomologische Zeitschrift* **26**: 155.

Heyden L. F. J. D. von. 1883a: Phytodecta. Coccinella. Adoxus. *Entomologische Nachrichten* **9**: 53.

Heyden L. F. J. D. von. 1883b: [new taxa]. In: Heyden L. F. J. D. von & Weise J.: Neue südost-europäische und kleinasiatische Käfer. *Deutsche Entomologische Zeitschrift* **27**: 310-316.

Heyden L. F. J. D. von. 1883c: [new taxa]. In: Heyden L. F. J. D. von & Kraatz G. 1883: Käfer aus Tekke-Turcmenien. *Deutsche Entomologische Zeitschrift* **27**: 354-360.

Heyden L. F. J. D. von. 1884: Beitrag zur Coleopterenfauna der Insel Askold und anderer Theile des Amurgebietes. *Deutsche Entomologische Zeitschrift* **28**: 273-300.

Heyden L. F. J. D. von. 1885: [new taxa]. In: Heyden L. F. J. D. von & Kraatz G. 1885: Beiträge zur turkestanischen Coleopteren-Fauna. *Deutsche Entomologische Zeitschrift* **29**: 273-298.

Heyden L. F. J. D. von. 1886a: Zusammenstellung der von Herrn Dr. med. W. Kobelt von seiner Reise in den Provinzen Alger und Constantine, sowie von Tunis mitgebrachten Coleopteren. *Bericht über der Senckenbergischen Naturforschenden Gesellschaft in Frankfurt am Main* **1885-86**: 35-57.

Heyden L. F. J. D. von. 1886b: [new taxa]. In: Heyden L. F. J. D. von. & Kraatz G.: Beiträge zur Coleopteren-Fauna von Turkestan, namentlich des Alai-Gebirges. *Deutsche Entomologische Zeitschrift* **30**: 177-194.

Heyden L. F. J. D. von. 1886c: Die Coleopteren-Fauna des Suyfun-Flusses (Amur). *Deutsche Entomologische Zeitschrift* **30**: 269-277.

Heyden L. F. J. D. von. 1886d: Neue Käferarten aus Malatia (in südlichen Kleinasien). *Deutsche Entomologische Zeitschrift* **30**: 278-280.

Heyden L. F. J. D. von. 1887: Verzeichniss der von Herrn Otto Herz auf der chinesichen Halbinsel Korea gesammelten Coleopteren. *Horae Societatis Entomologica Rossicae* **21**: 243-273.

Heyden L. F. J. D. von. 1888a: Ueber die von L. Conradt aus dem Alai-Gebirge mitgebrachten Coleopteren. - Beitrag XII. *Deutsche Entomologische Zeitschrift* **32**: 39-44.

Heyden L. F. J. D. von. 1888b: Käfer vom Cilicischen Taurus. *Deutsche Entomologische Zeitschrift* **32**: 68-71.

Heyden L. F. J. D. von. 1889a: Descriptiones Coleopterorum novorum Regionis Palaearcticae. *Deutsche Entomologische Zeitschrift* **33**: 325-330.

Heyden L. F. J. D. von. 1889b: Insecta, a cl. G. N. Potanin in China et in Mongolia novissime lecta. *Horae Societatis Entomologicae Rossicae* **23**: 654-677.

Heyden L. F. J. D. von. 1890a: Aufzählung von Käfer-Arten aus Tunis und Tripolis aus Loosen von M. Quedenfeldt. *Deutsche Entomologische Zeitschrift* **1890**: 65-78.

Heyden L. F. J. D. von. 1890b: Beiträge zur Käferfauna von Turkestan.XIII. Beitrag zur Coleopteren-Fauna von Turkestan. (Unter Mitwirkung der Herrn Reitter und Weise). *Deutsche Entomologische Zeitschrif* **1890**: 353-367.

Heyden L. F. J. D. von. 1891: [new names]. In: Heyden L. F. J. D. von., Reitter E. & Weise J. *Catalogus Coleopterorum Europae, Caucasi et Armeniae Rossicae.* Berlin, Mödling, Caen: R. Friedländer & Sohn, E. Reitter, Revue d'Entomologie, viii + 420 pp.

Heyden L. F. J. D. von. 1892: Beitrag zur Coleopteren-Fauna von Turkestan, Turkmenien und Süd-West-Sibirien. Aus Sendungen des Herrn Dr. Staudinger. *Deutsche Entomogische Zeitschrift* **1892**: 105-110.

Heyden L. F. J. D. von. 1894: Ueber Meloë Olivieri Chevrolat und sechs neue Longicornen aus Klein-Asien. *Deutsche Entomologische Zeitschrift* **1894**: 85-89.

Heyden L. F. J. D. von. 1897: Zwölf neue Varietäten der Coccinellide: Hippodamia septemmaculata Degeer. *Entomologische Nachrichten* **23**: 97-98.

Heyden L. F. J. D. von. 1904: *Die Käfer von Nassau. II. Auflage*. Frankfurt am M., 425 pp.

Heyden L. F. J. D. von. 1906: [new names]. In: Heyden L. F. J. D. von, Reitter E. & Weise J. (eds): *Catalogus Coleopterorum Europae, Caucasi et Armeniae rossicae. Editio secunda.* Berlin, Paskau, Caen: R. Friedländer & Sohn, E. Reitter, Revue d'Entomologie, [8] + 750 columns + pp. 750-775.

Heyden L. F. J. D. von. 1911: Coleopteren der Aru- und Kei-Inseln. *Abhandlungen der Senckenbergischen Naturforschenden Gesellshaft* **33**: 349-372.

Hicker R. 1915a: In: Stolz H.: Über die Käferfauna des Monte Cavallo in den Venetianer Alpen. *Verhandlungen der Kaiserlich-Königlicher Zoologisch-Botanishen Gesellschaft in Wien* **65**: 238-254.

Hicker R. 1915b: Beschreibung einer neuen europaeischen Malthodes-Art. *Wiener Entomologische Zeitung* **34**: 380-381.

Hicker R. 1935: Malacodermata (Col.). Pp. 292-295. In: Visser C. & Visser-Hooft J. (eds): *Wissenschaftliche Ergebnissse der niederländischen Expeditionen in den Karakorum und die angrenzenden Gebiete 1922, 1925 und 1929/30.* Leipzig: F. A. Brockhaus, xviii + 499 pp., 1 map, 8 pls.

Hicker R. 1937: Entomologische Sammelergebnisse der Deutschen Hindukush-Expedition 1935 der Deutschen Forschungsgemeinschaft. Cantharidae. *Arbeiten über Morphologische und Taxonomische Entomologie aus Berlin-Dahlem* **4**: 180-181.

Hicker R. 1948: Zwei neue Laius-Arten aus China. *Entomologische Blätter* **41-44** [1945-48]: 60-62.

Hicker R. 1953: Ein neuer "Malthodes" aus dem Adamello-Gebiet. *Studi Trentini di Scienze Naturali* **30**: 130-132.

Hicker R. 1955: Neue palaearktische Arten aus der Familie Cantharidae. *Koleopterologische Rundschau* **33**: 55-60.

Hicker R. 1958: Fauna di Ronagna (collezione Zangheri). Eine neue Malthodes-Art (Coleopt. Cantharidae). *Bollettino della Società Entomologica Italiana* **87**: 26-27.

Hicker R. 1959: Neue Arten aus der Familie Malachiidae (Col.). *Entomologisches Nachrichtenblatt Österreichischer und Schweizer Entomologen* **11**: 69-73.

Hicker R. 1960: Neue Arten aus der Familie Cantharidae (Col.). *Zeitschrift der Arbeitsgemeinschaft Österreichischer Entomologen* **12**: 78-81.

Hinton H. E. 1941: A new Atomaria from mushroom-beds in South Africa (Col., Cryptophagidae). *Bulletin of Entomological Research* **32**: 133-134.

Hinton H. E. 1942a: A new leaf-mining nitidulid (Coleoptera). *The Entomologist* **75**: 126-129.

Hinton H. E. 1942b: A revision of the Cerylonini of Borneo (Coleoptera, Colydiidae). *The Annals and Magazine of Natural History* (11) **9**: 141-173.

Hinton H. E. 1943: Description and figure of a new Anthrenus (Col., Dermestidae). *Entomologist's Monthly Magazine* **79**: 14-16.

Hinton H. E. 1943: Stethomezium squamosum gen. et sp. n. infesting stored food in Britain, with notes on a south african ptinid not previously recorded in stored products (Coleoptera). *The Proceedings of the Royal Entomological Society of London (Series B Taxonomy)* **12**: 50-54.

Hinton H. E. 1944: A new nitidulid beetle from Burma, Carpophilus (Urophorus) prodicus sp. n. *The Entomologist* **77**: 172-173.

Hinton H. E. 1945: *A monograph of the beetles associated with stored products. Volume 1*. London: British Museum, viii + 443 pp.

Hintz E. 1902: Beitrag zur Kenntniss der paläarktischen Cleriden. *Deutsche Entomologische Zeitschrift* **1902**: 227-230.

Hintz E. 1904: Zur Kenntnis des Trichodes Kraatzi Reitter. *Deutsche Entomologische Zeitschrift* **1904**: 420-422.

Hintz E. 1905: Neue Cleriden aus Westafrika. *Deutsche Entomologische Zeitschrift* **1905**: 305-314.

Hisamatsu S. 1953: The insect fauna of Mt. Ishizuchi and Omogo valley, Iyo, Japan. The Nitidulidae. *Transactions of the Shikoku Entomological Society* **3**: 139-143.

Hisamatsu S. 1955: The Eucnemidae of Shikoku, Japan (Coleoptera). *Transactions of the Shikoku Entomological Society* **4**: 96-102.

Hisamatsu S. 1956: The Nitidulidae of the Amami Islands south of Kyushu, Japan (Coleoptera). *Memoirs of the Ehime University, Section VI. (Agriculture)* **1**: 51-57.

Hisamatsu S. 1956: The Eucnemidae of Shikoku, Japan (Coleoptera), II. *Memoirs of Ehime University, Sect. IV., Agriculture* **1**: 157-162.

Hisamatsu S. 1957: A new species of Eucnemidae from Japan (Coleoptera). *Akitu* **6**: 45-46.

Hisamatsu S. 1957: The Eucnemidae of Shikoku, Japan (Coleoptera). III. *Transactions of the Shikoku Entomological Society* **5**: 46-48.

Hisamatsu S. 1958: A revision of the genus Proxylobius Fleutiaux of Japan (Coleoptera: Eucnemidae). *Transactions of the Shikoku Entomological Society* **5**: 117-119.

Hisamatsu S. 1959a: The genus Prometopia in Japan (Coleoptera: Nitidulidae). *Memoirs of the Ehime University, Section VI. (Agriculture)* **4**: 21-23.

Hisamatsu S. 1959b: Notes on Japanese Bioxylus (Coleoptera: Eucnemidae). *Transactions of the Shikoku Entomological Society* **6**: 23-30.

Hisamatsu S. 1959c: A revision of the genus Pocadites of Japan and Ryukyu Islands (Coleoptera: Nitidulidae). *Transactions of the Shikoku Entomological Society* **6**: 30-32.

Hisamatsu S. 1959d: Notes on Japanese Clavicornia (Coleoptera). *Transactions of the Shikoku Entomological Society* **6**: 59-60.

Hisamatsu S. 1960: A revision of the genus Dirrhagus Latreille of Japan (Coleoptera: Eucnemidae). *Transactions of the Shikoku Entomological Society* **6**: 92-103.

Hisamatsu S. 1961: Four new species of Nitidulidae from Japan (Coleoptera). *Transactions of the Shikoku Entomological Society* **7**: 26-32.

Hisamatsu S. 1962: A new species of Cryptophagus (Col.: Cryptophagidae). *Niponius* **1** (20): 1-3.

Hisamatsu S. 1963: Six new species of Eucnemidae from Japan (Coleoptera). *Transactions of the Shikoku Entomological Society* **8**: 26-34.

Hisamatsu S. 1963: Carpophilus hemipterus and its Allied Species (Col., Nitidulidae). *Entomological Review of Japan* **15**: 59-62, pl. 8 (in Japanese, English summary).

Hisamatsu S. 1964: Four new species of Coleoptera from Japan. *Transactions of the Shikoku Entomological Society* **8**: 51-58.

Hisamatsu S. 1965: Some beetles from Formosa. *Special Bulletin of Lepidopterological Society of Japan* **1**: 130-140.

Hisamatsu S. 1968a: The genus Lasiodactylus Perty in Japan (Coleoptera: Nitidulidae). *Transactions of the Shikoku Entomological Society* **10**: 11-14.

Hisamatsu S. 1968b: A new species of Epuraea associated with eastern gall rust (Coleoptera: Nitidulidae). *Transactions of the Shikoku Entomological Society* **10**: 29-31.

Hisamatsu S. 1976: A new genus of Cateretinae from Taiwan (Coleoptera: Nitidulidae). *Transactions of the Shikoku Entomological Society* **13**: 19-23.

Hisamatsu S. 1978: Nitidulidae (Coleoptera) of Nepal collected by the Hokkaido University Expedition to the Himalaya. *Transactions of the Shikoku Entomological Society* **14**: 7-10.

Hisamatsu S. 1980: The genus Pterotarsus Guérin-Méneville in Japan (Coleoptera: Eucnemidae). *Transactions of the Shikoku Entomological Society* **15**: 27-30.

Hisamatsu S. 1982: Three new species of Ipidia Erichson (Coleoptera, Nitidulidae) from Southeast Asia. Pp. 55-61. *Special Issue to the Memory of Retirement of Emeritus Professor Michio Chûjô.* Nagoya: Association of the Memorial issue of Emeritus Professor M. Chûjô, 185 pp.

Hisamatsu S. 1985: [new taxa]. In: Kurosawa Y., Hisamatsu S. & Sasaji H. (eds.): *Colored illustrations of the Coleoptera of Japan. Vol. 3.* Osaka: Hoikusha Publishing Co., x + 500 pp., 72 pls (in Japanese).

Hisamatsu S. 1994: Record of Euryptychus ishiharai (Coleoptera, Eucnemidae) from Haha-jima Island of the Ogasawara Islands, Japan. *Transactions of the Shikoku Entomological Society* **20**: 151-153.

Hisamatsu S. & Sakai M. 1985: [new taxa]. In: Kurosawa Y., Hisamatsu S. & Sasaji H. (eds): *Colored illustrations of the Coleoptera of Japan. Vol. 3.* Osaka: Hoikusha Publishing Co., x + 500 pp., 72 pls (in Japanese).

Hisamatsu S. & Satô M. 1959: A new genus and new species of Eucnemidae from Japan. *Entomological Review of Japan* **10**: 15-17.

Hoang D. N. 1977: Kokcinellidy podsem. Epilachninae (Coleoptera, Coccionellidae) Severnogo Vietnama. I [Epilachninae (Coleoptera, Coccinellidae) of the northern Vietnam]. *Entomologicheskoe Obozrenie* **56**: 132-145.

Hoang D. N. 1978a: Genre Pseudoscymnus Chapin au Vietnam [Giong Pseudoscymnus Chapin (Coccinellidae, Coleoptera) o mien Bac Viet Nam]. *Tap San Sinh Vat Dia Hoc* **16** (4): 11-15.

Hoang D. N. 1978b: Kokcinellidy podsem. Epilachninae (Coleoptera, Coccionellidae) Severnogo Vietnama. II [Epilachninae (Coleoptera, Coccinellidae) of the northern Vietnam]. *Entomologicheskoe Obozrenie* **57**: 832-846.

Hoang D. N. 1979: Three new species of the subfamily Scymninae (Col. Cocc.) from the fauna of Vietnam. *Biological Review* **1**: 11-15.

Hoang D. N. 1980: Some new species of lady-beetles to the subfamily Coccidulinae (Coleoptera, Coccinellidae) in Viet Nam. *Tap Chi Sinh Vát Hoc* (Hanoi) **2** (1): 8-13 (in Vietnamese).

Hoang D. N. 1982a: *Coccinellidae of Vietnam* (*Insecta, Coleoptera*). I. Hanoi, 211 + [2] pp. (in Vietnamese).

Hoang D. N. 1982b: A new genus and some new species of subfamily Coccinellidae (Coleoptera, Coccinellidae) from Vietnam. *Tap Chi Sinh Vat Hoc* **4** (3): 10-15. (in Vietnamese).

Hoang D. N. 1983a: *Coccinellidae of Vietnam* (*Insecta: Coleoptera*). II. Hanoi, 159 + [1] pp. (in Vietnamese).

Hoang D. N. 1983b: Novyy rod Coccinellidae iz Vietnama [New genus of Coccinellidae from Vietnam]. Pp. 150-156. In: Medvedev L. N. (ed.): *Fauna i ekologiya zhivotnykh Vietnama* [*Fauna and ecology of the animals of Vietnam*]. Moskva: Nauka, 207 pp.

Hoang D. N. 1985: K izucheniyu Coccinellidae plato Teinguen v yuzhnom Vietname [On the study of Coccinellidae of Teinguen Plateau from South Vietnam]. Pp. 30-49. In: Medvedev L. N. (ed.): *Nasekomye Vietnama* [*Insects of Vietnam*]. Moskva, 182 pp.

Hochhut J. H. 1872: Enumeration der in den russischen Gouvernements Kiew und Volhynien bisher aufgefundenen Käfer. *Bulletin de la Société Impériale des Naturalistes de Moscou* **45** (2-3): 195-234.

Hochhuth J. H. 1873: Enumeration der in den russischen Gouvernements Kiew und Volhynien bisher aufgefundenen Käfer. *Bulletin de la Société Impériale des Naturalistes de Moscou* **45** (4): 283-322.

Hodgson Torres F. M. 1987: Lithophilus tinerfensis n. sp. de Coccinellide (Coleoptera) de Canarias. *Vieraea* **17**: 117-119.

Hodgson Torres F. M. & Plata Negrache P. 1987: Dos nuevas especies de la familia Malachiidae (Insecta, Coleoptera) de la Península Ibérica. *Entomologische Blätter* **83**: 173-180.

Hoffmann E. von. 1887: Eine neue Agonolia-Art. *Deutsche Entomologische Zeitschrift* **31**: 96.

Hoffmann J. J. 1803: Monografie der von den Verfassern dem Departemente vom Donnersberge, und den angrenzenden Gegenden der Departemente von der Saar, und von Rhein und Mosel einheimisch bemerkten Dorcatomen. *Entomologische Hefte enthaltend Beiträge zur weiteren Kenntniss und Aufklärung der Insektengeschichte* (Frankfurt am Main) **2**: 92-105.

Holdhaus K. 1903: Beiträge zur Kenntnis der Koleopterengattung Atomaria Steph. *Münchener Koleopterologische Zeitschrift* **1**: 350-382.

Holdhaus K. 1920: Koleopteren aus Mesopotamien. Ergebnisse der wissenschaftlichen Expedition nach Mesopotamien, 1910. *Annalen des Naturhistorischen Museum in Wien* **33** [1919]: 49-58.

Holdhaus K. 1923: Pp. 77-144. Elenco dei coleotteri dell'isola d'Elba, con studii sul problema della Tirennide. *Memorie della Società Entomologica Italiana* **2**: 77-175 [Pp. 145-175 were issued in 1824].

Hölzel E. 1944: Eine neue Art der Gattung Cartodere Thoms. aus Kärnten (Coleoptera, Lathridiidae), Mit 1 Abb. *Carinthia II* **134**: 81-83.

Hölzel E. 1960: Eine neue interessante Art der Gattung Epuraea Erichson (Col. Nitidulidae), Epuraea carinthiaca n. sp. aus den Kärntner Karawanken. *Nachrichtenblatt der Bayerischen Entomologen* **9**: 33-36.

Holzschuh C. & Lohse G. A. 1981: Eine neue Art der Gattung Monotoma Herbst aus Mitteleuropa: Monotoma gotzi n. sp. *Entomologische Blätter* **77**: 175-177.

Hope F. W. 1831: Synopsis of the new species of Nepaul insects in the collection of Major Gen. Hardwick. Pp. 21-32. In: Gray J. E. (ed.): *The Zoological Miscellany. Vol. 1.* London: Treuttel, Wurtz & Co., 40 pp., 4 pls.

Hope F. W. 1834: Dermestidae. In: Pettigrew T. J.: *History of Egyptian mummies, and an account of the worship and embalming of the sacred animals by the Egyptiams: with remarks on the funeral ceremonies of different nations, and observations on the mummies of the Canary islands, of ancient Peruvians, Burman priests.* London: Longman: xxi + 264 + [1] pp., 13 pls.

Hope F. W. 1835: Characters and descriptions of several new genera and species of coleopterous insects. *Transactions of the Zoological Society of London* **1**: 91-112.

Hope F. W. 1837: Description of various new species of insects found in Gum Animè. *Transactions of the Entomological Society of London* **2**: 52-57.

Hope F. W. 1840: *The Coleopterist's Manual, part the third, containing various families, genera, and species, of beetles, recorded by Linnaeus and Fabricius. Also, descriptions of newly discovered and unpublished Insects.* London: J. C. Bridgewater and Bowdery and Kerby, [5] + 191 pp., 3 pls.

Hope F. W. 1842a: A monograph on the coleopterous family Phyllophoridae. *Proceedings of the Zoological Society of London* **10**: 73-79.

Hope F. W. 1842b: Monograph of the Elaterideous genus Campsosternus Latr. *The Annals and Magazine of Natural History* **8**: 453-454.

Hope F. W. 1843a: Descriptions of the coleopterous insects sent to England by Dr. Cantor from Chusan and Canton, with observation on the entomology of China. *The Annals and Magazine of Natural History* **11**: 62-66.

Hope F. W. 1843b: Monograph on the genus Campsosternus of Latreille. *Transactions of the Entomological Society of London* **3**: 286-292.

Hope F. W. 1843c: Descriptions of new species of insects collected at Adelaide in South-Western Australia by Mr. Fortnum. *The Annals and Magazine of Natural History* **11**: 317-319.

Hope F. W. 1845a: On the entomology of China, with descriptions of the new species sent to England by Dr. Cantor from Chusan and Canton. *Transactions of the Entomological Society of London* **4** [1845-1847]: 4-17, 1 pl.

Hope F. W. 1845b: Descriptions of some new species of Coleoptera from Adelaide in New Holland. *Transactions of the Entomological Society of London* **4**: 100-113, pls 6-7.

Hormuzachi C. 1901: Cercetărĭ nouĭ asupra raporturilor faunistice din Bucovina, cu privire specială la clasa coleopterelor. Partea I. Caracterul fauneĭ şi vegetaţiuniĭ bucovinene. Impărţiuniĭ ţĕriĭ în regiunĭ naturale. Conclusiunĭ generale. *Buletinul Societăţii de Ştiinţe Bucaresti* **10**: 103.

Horn G. H. 1862: Monograph of the Trogosita inhabiting the United States. *Proceedings of the Academy of Philadelphia.* **1862**: 82-88.

Horn G. H. 1871: Descriptions of new species of Elateridae, of the United States. *Transactions of the American Entomological Society* **3**: 299-324, pl. IV.

Horn G. H. 1872: Synopsis of the Malachiidae of the United States. *Transactions of the American Entomological Society* **4**: 109-127.

Horn G. H. 1878: Revision of the species of the sub-family Bostrichidae of the United States. *Proceedings of the American Philosophical Society* **17**: 540-555.

Horn G. H. 1879: Revision of the Nitidulidae of the United States. *Transactions of the American Entomological Society and Proceedings of the Entomological Section of the Academy of Natural Sciences* **7** [1878-1879]: 267-336.

Horn G. H. 1890: Throscidae. Pp. 193-209. In: *Biologia Centrali-Americana. Insecta. Coleoptera. Vol. III. Part 1. Serricornia.* London: Taylor & Francis, xv + 690 pp., 27 pls. [issued in parts, 1882-1897].

Horn G. H. 1895: Studies in Coccinellidae. *Transactions of the American Entomological Society* **22**: 81-114.

Host N. T. 1789: Entomologica. Pp. 291-302, pl. 23. In: Jacquin N. J.: *Collectanea ad botanicam, chemiam, et historiam naturalem, spectantia, cum figuris coloatis. Vol. 3.* Vindobonae: Wappleriana, 306 pp., 23 pls.

Hubenthal W. 1920: Kleine coleopterologische Mitteilungen. *Entomologische Blätter* **16**: 241-247.

Hübner J. P. 1842: Mémoire pour servir à l'histoire de la Coccinella de la Saponaire. *Mémoires de la Société de Physique de Genève* **9**: 363-374.

Hummel A. D. 1826: Insectes de 1825. Pp. 29-51. *Essais Entomologiques* (St. Pétersbourg) **5**: 1-51 pp.

Hummel A. D. 1827a: Insectes de 1826. Pp. 1-20. *Essais Entomologiques* (St. Pétersbourg) **6**: 1-49 pp.

Hummel A. D. 1827b: Catalogus Insectorum quae itinere Petropoli in Chersonesum Tauricum et Iberiam anno 1825 collegit Benedictus Jaeger, membrum ordinarium Societatis Caesareae Natrae curiosorum Mosquensis. Pp. 27-43. *Essais Entomologiques* (St. Pétersbourg) **6**: 1-49.

Hummel A. D. 1829: Insectes de 1827 et 1828. Pp. 18-34. *Essais Entomologiques* (St. Pétersbourg) **7** (Tome second No. 1): 1-38.

Hyslop J. A. 1921: Genotypes of the elaterid beetles of the world. *Proceedings of the United States National Museum* **58**: 621-680.

Iablokoff A. K. 1937: Un nouvel Ampedus de France (Col. Elateridae). *Revue Française d'Entomologie* **4**: 64-67.

Iablokoff-Khnzorian S. M. 1956: Chetyre novykh vida zhestkokrylykh iz Armyanskoy SSR. *Doklady Akademii Nauk Armyanskoy SSR* **22**: 135-139.

Iablokoff-Khnzoryan S. M. 1959a: Novye vidy zhestkokrylyh iz Talysha. *Doklady Akademii Nauk Azerbayzhanskoy SSR* **3**: 58-59.

Iablokoff-Khnzorian S. M. 1959b: Novye vidy zhestkokrylych iz Armenskoy SSR i Nakhichevanskoy SSR II. *Zoologicheskiy Sbornik Akademii Nauk Armyanskoy SSR* **11**: 65-78.

Iablokoff-Khnzorian S. M. 1960: Coléoptères nouveaux de l'Armenie Sovietique. *Notulae Entomologicae* **40**: 140-153.

Iablokoff-Khnzorian S. M. 1962a: Novye vidy zhestkokrylykh iz Zakavkaz'ya. *Zoologicheskiy Sbornik Akademii Nauk Armyanskoy SSR* **12**: 99-124.

Iablokoff-Khnzorian S. M. 1962b: Die Gattungstypen von Latreille. *Folia Entomologica Hungarica* (Budapest) **15**: 419-426.

Iablokoff-Khnzorian S. M. 1964: Novye rody i vidy zhestkokrylych iz Zakavkaz'ya i Sredney Azii. *Zoologicheskiy Sbornik Akademii Nauk Armyanskoy SSR* **13**: 151-186.

Iablokoff-Khnzorian S. M. 1966a: Espèces nouvelles asiatiques du genre Chaetomalachius Er. (Coléoptères, Dasytides). *L'Entomologiste* **21** [1965]: 77-87.

Iablokoff-Khnzoryan S. M. 1966b: Dva novykh vida zhestkokrylykh iz Armyankoy SSR [Two new species of beetles from the Armenian SSR]. *Doklady Akademii Nauk Armyankoy SSR* **42**: 309-314.

Iablokoff-Khnzorian S. M. 1969: Dva novykh vida zhestokrylykh-kokcinellid s Kavkaza (Coleoptera, Coccinellidae). *Doklady Akademii Nauk Armyanskoy SSR* **48**: 247-250.

Iablokoff-Khnzorian S. M. 1970a: Novye vidy zhestkokrylych iz Armenii i drugich chastey SSSR (Insecta, Coleoptera). *Zoologicheskiy Sbornik Akademii Nauk Armyanskoy SSR* **15**: 50-80.

Iablokov-Khnzorian S. M. 1970b: Dva novykh vida iz roda Nephus Muls. iz SSSR (Coleoptera, Coccinellidae). *Doklady Akademii Nauk Armyanskoy SSR* **50**: 118-121.

Iablokoff-Khnzorian S. M. 1970c: Dva novykh bida shchelkunov iz SSSR. *Doklady Akademii Nauk Armyanskoj SSR* **50**: 177-181.

Iablokov-Khnzorian S. M. 1970d: Dva novykh vida kokcinellid iz Palearktiki (Coleoptera, Coccinellidae). *Doklady Akademii Nauk Armyansky SSR* **50**: 252-256.

Iablokoff-Khnzorian S. M. 1971: Synopsis des Hyperaspis paléartiques. *Annales de la Société Entomologique de France* (N. S.) **7**: 163-200.

Iablokoff-Khnzorian S. M. 1972a: Novye vidy zhestkokrylykh-kokcinellid iz SSSR (Coleoptera, Coccinellidae). *Doklady Akademii Nauk Armyanskoy SSR* **55**: 116-122.

Iablokoff-Khnzorian S. M. 1972b: Les types de Coccinellidae de la collection Motschulsky (Coléoptères: Coccinellidae). *Nouvelle Revue d'Entomologie* **2**: 163-184.

Iablokoff-Khnzorian S. M. 1974a: Novyy rod i vid zhestkokrylykh-shchelkunov iz Armenii (Coleoptera, Elateridae). *Doklady Akademii Nauk Armyanskoj SSR* **58**: 52-55.

Iablokoff-Khnzorian S. M. 1974b: Monographie der Gattung Lithophilus Froelich (Col. Coccinellidae) *Entomologische Arbeiten aus dem Museum G. Frey* **25**: 148-243.

Iablokoff-Khnzorian S. M. 1975: Étude sur les Erotylidae (Coleoptera) paléarctiques. *Acta Zoologica Cracoviensia* **20**: 201-265.

Iablokoff-Khnzorian S. M. 1976a: Zhestkokrylye-drevogryzy SSSR (Coleoptera, Lyctidae). *Zoologicheskiy Sbornik Akademii Nauk Armyanskoy SSR* **17**: 87-100.

Iablokoff-Khnzorian S. M. 1976b: Dva novykh vida zhestkokrylykh-kokcinellid iz aziatskoy chasti SSSR (Coleoptera, Coccinellidae). *Doklady Akademii Nauk Armyanskoy SSR* **62**: 305-309.

Iablokoff-Khnzorian S. M. 1976c: Novyy vid zhuka-tochil"shchika iz Darvaza. *Doklady Akademii Nauk Armyanskoy SSR* **63**: 124-127.

Iablokoff-Khnzorian S. M. 1977a: Novyy vid zhestkokrylykh kokcinellid iz Indii (Coleoptera, Coccinellidae). *Doklady Academii Nauk Armyanskoy SSR* **64**: 61-63.

Iablokoff-Khnzorian S. M. 1977b: Novyy vid zhestkokrylykh-kokcinellid iz Kitaya (Coleoptera, Coccinellidae). *Doklady Academii Nauk Armyanskoy SSR* **63**: 250-255.

Iablokoff-Khnzorian S. M. 1977c: Mukoyedy triby Laemophloeini (Coleoptera, Cucujidae) fauny SSSR. I. *Entomologicheskoe Obozrenie* **56**: 610-624.

Iablokoff-Khnzorian S. M. 1978a: Novyy vid zhestkokrylych-gribovikov iz Armenii (Coleoptera, Erotylidae). *Doklady Akademii Nauk Armianskoj SSR* **66**: 58-60.

Iablokoff-Khnzorian S. M. 1978b: Chetyre novykh vida zhestkokrylykh-malashek iz SSSR. (Coleoptera, Malachiidae). *Doklady Akademii Nauk Armyanskoy SSR* **66**: 172-176.

Iablokoff-Khnzorian S. M. 1978c: Novyy vid zhuka-tochil"shchika iz Tadzhikistana (Coleoptera, Anobiidae). *Doklady Akademii Nauk Armyanskoy SSR* **67**: 61-63.

Iablokoff-Khnzorian S. M. 1978d: Novyy rod i vid zhestkokrylykh-kokcinellid s Arala (Coleoptera, Coccinellidae). *Doklady Akademii Nauk Armyanskoy SSR* **67**: 115-117.

Iablokoff-Khnzorian S. M. 1978e: Dva novykh vida zhestkokrylykh-kokcinellid iz Vostochnoy Azii (Coleoptera, Coccinellidae). *Doklady Akademii Nauk Armyanskoy SSR* **67**: 180-183.

Iablokoff-Khnzorian S. M. 1978f: Novyy vid zhestkokrylykh-malashek iz Armyanskoy SSR (Coleoptera, Dasytidae). *Doklady Akademii Nauk Armyanskoy* SSR **67**: 243-245.

Iablokoff-Khnzorian S. M. 1978g: Mykoedy triby Laemophloeini (Coleoptera, Cucujidae) fauny SSSR. 2. *Entomologicheskoe Obozrenne* **57**: 337-352.

Iablokoff-Khnzorian S. M. 1979: Genera der paläarktischen Coccinellini (Coleoptera. Coccinellidae). *Entomologische Blätter* **75**: 37-75.

Iablokoff-Khnzorian S. M. 1982: *Les Coccinelles coléoptères-Coccinellidae, Tribu Coccinellini des régions Paléarctique et Orientale.* Paris: Boubée, 568 pp.

Iablokoff-Khnzorian S. M. 1985: Novyy rod i vid zhestkokrylykh-kokcinellid iz Sumatry (Coleoptera, Coccinellidae). *Doklady Akademii Nauk Armyanskoy SSR* **80**: 142-144.

Iablokoff-Khnzorian S. M. 1986a: Bemerkungen über einige Coccinellini der Alten Welt (Coleoptera, Coccinellidae). *Entomologische Abhandlungen* **50**: 59-80.

Iablokoff-Khnzorian S. M. 1986b: Adalia puetzi n. sp. nouvelle espèce du Pakistan (Col. Coccinellidae). *Nouvelle Revue d'Entomologie* **3**: 80.

Iablokoff-Khnzorian S. M. 1986c: Novyy vid zhestkokrylykh-malashek iz roda Malachius F. (Coleoptera, Malachiidae). *Doklady Akademii Nauk Armyanskoy SSR* **83**: 223-225.

Iablokoff-Khnzorian S. M. 1988: Novyy vid zhestkokrylykh-malashek s Altaya. *Doklady Akademii Nauk Armyanskoy SSR* **86**: 46-48.

Iablokoff-Khnzorian S. M. & Karapetyan A. P. 1986: Issledovaniya po rodu Ptinus L. (Coleoptera, Ptinidae). *Biologicheskiy Zhurnal Armenii* **39**: 512-513.

Iablokoff-Khnzorian S. M. & Karapetyan A. P. 1991: Ptinus-Studien (Coleoptera, Ptinidae). *Entomologische Blätter* **87**: 1-65.

Iga M. 1949: On the genus Stigmatium Gray (1832) of Japan, with description of a new species (Coleoptera: Cleridae). *The Entomological Review of Japan* **1**: 36-38.

Ihssen G. 1949: Beitrag zur Kenntnis der Gattung Anobium Fbr. Anobium hederae nov. spec. (Coleoptera, Anobiidae). *Koleopterologische Zeitschrift* **1**: 103-108.

Illiger J. K. W. 1794: Beschreibung einiger neuen Käferarten aus der Sammlung des Herrn Professor Hellwig in Braunschweig. *Neuestes Magazin für die Liebhaber der Entomologie herausgegeben von D. H. Schneider* **1** (5): 593-620.

Illiger J. K. W. 1798: *Verzeichniss der Käfer Preussens. Entworfen von Johann Gotlieb Kugelann, Apotheker in Osterode. Ausgearbeitet von Illiger, mit einer Vorrede von Hellwig und dem angehängten Versuch einer natürlichen Ordnung und Gattungsfolge der Insecten.* Halle: Johann Jacob Gebauer, xlii + 510 + [1] pp.

Illiger J. K. W. 1800: Vierzig neue Insekten aus der Helwigischen Sammlung in Braunschweig. Beschriben von Karl Illiger im Anfange des Februars 1800. Mit einigen Abbildungen. Pp. 103-150. In: Wiedemann C. P. W.: *Archiv für Zoologie und Zootomie.* 1 (2). Braunschweig: Reichard, iv + 318 pp., 2 pls.

Illiger J. K. W. 1801a: Nachtrag und Berichtigungen zum Verzeichnisse der Käfer Preussens. *Magazin für Insektenkunde* **1** (1-2): 1-94.

Illiger J. K. W. 1801b: Namen der Insekten-Gattungen, ihr Genitiv, ihr grammatisches Geschlecht, ihr Sittenmass, ihre Herleitung; zugleich mit den Deutschen Benennungen. *Magazin für Insektenkunde* **1** (1, 2): 125-155.

Illiger J. K. W. 1801c: Neue Insecten. *Magazin für Insektenkunde* **1** (1, 2): 163-208.

Illiger J. K. W. 1805: Zusätze, Berichtigungen und Bemerkungen zu Fabricii Systema Eleutheratorum, Tomus II. (Beschluss). *Magazin für Insektenkunde* **4**: 69-174.

Illiger J. K. W. 1806: Nachtrag zu den Zusätzen, Bemerkungen und berichtigungen zu Fabricii Systema Eleutheratorum. *Magazin für Insektenkunde* **5**: 221-246.

Illiger J. K. W. 1807a: Portugiesische Käfer. (Fortsetzung). *Magazin für Insektenkunde* **6**: 1-80.

Illiger J. K. W. 1807b: Zusätze, Berichtigungen und Bemerkungen zu Fabricii Systema Eleutheratorum. *Magazin für Insektenkunde* **6**: 296-317.

Illiger J. K. W. 1807c: Vorschlag zur Aufnahme im Fabricischen Systeme fehlende Käfergattungen. *Magazin für Insektenkunde* **6**: 318-348.

Imhoff L. 1843: Bericht über eine dem öffentlichen Museum vom Herrn Missionär Riis geschenkte Sammlung Guineensicher Käfer. *Bericht über die Verhandlungen der Naturforschenden Gesellschaft in Basel* **5**: 164-180.

Imms A. D. & Chatterjee N. C. 1915: On the structure and biology of Tachardia lacca Kerr, with observation on certain insects predaceous or parasitic upon it. *The Indian Forest Memoirs* **3** (1): 1-42.

Irish J. 1996: The southern African genera Stethomezium Hinton, Meziomorphum Pic and Lepimedozium Bellés (Coleoptera, Ptinidae, Gibbiinae). *Navorsinge van die Nasionale Museum Bloemfontein* **12**: 317-334.

Ishida K. 1986: A revision of the two genera, Athemus and Athemellus, of Japan (Coleoptera: Cantharidae). *Transactions of the Shikoku Entomological Society* **17**: 193-213.

Ishida K. 1989: Two new species of the cantharid beetles from Okinawa. *Entomological Review of Japan* **44**: 81-85.

Ishida K. & Satô M. 1993: A new species of Athemellus (Coleoptera, Cantharidae) from Japan. *Elytra* **21**: 47-50.

Israelson G. 1971a: Note on some North-European Coleoptera. *Entomologisk Tidskrift* **92**: 66-73.

Israelson G. 1971b: On the Stagetus species of the Canary Islands (Col., Anobiidae). *Nouvelles Revue d'Entomologie* **1**: 287-295.

Israelson G. 1974: New or poorly know Anobiidae from the Canarian Islands, with keys (Col.). *Miscellànea Zoològica* **3**: 71-89.

Israelson G. 1976: A key to the Palaearctic Mesocoelopus, with description of a new species (Col., Anobiidae). *Entomologisk Tidskrift* **97**: 9-14.

Israelson G. 1980: Taxonomical and nomenclatural notes on some Canarian Coleoptera. *Vieraea* **9** [1979]: 183-210.

Israelson G. 1984a: A new Sphaericus Wollaston from the Azores (Coleoptera, Ptinidae). *Bocagiana* **1984** (79): 1-3.

Israelson G. 1984b: Coleoptera from the Azores. *Bolletin Museu Municipal Funchal* **36**: 142-161.

Israelson G. 1985: A new Canarian Atomaria Stephens (Coleoptera, Cryptophagidae). *Vieraea* **15**: 113-115.

Israelson G. 1987: An Aposericoderus Paulian from the Canary Islands, with a note on the spermatheca of the Corylophidae (Coleoptera). *The Entomologist's Monthly Magazine* **123**: 73-76.

Israelson G. 1993: Taxonomical and ecological notes on the Canarian genera Casapus Wollaston and Piotes Wollaston (Coleoptera, Ptinidae), with description of a new species from La Gomera and observations on its pre-imaginal stages. *Vieraea* **22**: 113-118.

Istomina V. 1969: Novyy vid roda Malthodes (Coleoptera, Cantharidae) [A new species of the genus Malthodes (Coleoptera, Cantharidae)]. *Zoologicheskiy Zhurnal* **48**: 1104-1105.

Ivanov N. N. 1902: Elateridae (Coleoptera) click-beetles of the Moscow governatory. *Dnevnika Zoologischeskago Otdelanija Imperatorskago Obshchestva Ljubitelej Estestroznanija, Antropologii i Etnografii* **3**: 31-40.

Jacquelin du Val P. N. C. 1852: Description de deux genres nouveaux et de plusieurs espèces nouvelles de coléoptères propres à la faune française. *Annales de la Société Entomologique de France* (2) **10**: 695-718.

Jacquelin du Val P. N. C. 1854: [new taxa]. *Bulletin de la Société Entomologique de France* **1854**: xxxvi-xxxviii.

Jacquelin du Val P. N. C. 1857: Coleoptera. Pp. 1-136. In: Sagra R. de la. (ed.): *Historia física, política y natural de la isla de Cuba. Secunda parte: Historia natural. Vol. 7: Crustàceos, arácnidos e insectos*. Paris: Arthur Bertrand.

Jacquelin du Val P. N. C. 1858: Nitidulides. Pp. 134-160. In: *Manuel Entomologique. Genera des coléoptères d'Europe comprenant leur classification en familles naturelles, la description de tous les genres, des tableaux synoptiques destinés a faciliter l'étude, le catalogue de toutes les espèces, de nombreux dessins au trait de caractères. Tome deuxième.* Paris: A. Deyrolle, 285+ [3] + [53-122, Catalogue] + [2] pp., 67 pls. [issued in parts, Catalogue separately].

Jacquelin du Val P. N. C. 1859a: Corylophides, pp. 228-237; Lathridiides, pp.240-249; Dermestides, pp. 253-261. In: *Manuel Entomologique. Genera des coléoptères d'Europe comprenant leur classification en familles naturelles, la description de tous les genres, des tableaux synoptiques destinés a faciliter l'étude, le catalogue de toutes les espèces, de nombreux dessins au trait de caractères. Tome deuxième.* Paris: A. Deyrolle, 285+ (4) + [54-122, Catalogue] + [2] pp., 67 pls. [issued in parts, Catalogue separately].

Jacquelin du Val P. N. C. 1859b: Synopsis des espèces européennes du genre Lampyris. *Glanures Entomologiques* **1**: 5-22.

Jacquelin du Val P. N. C. 1859c: Description de quelques espèces nouvelles. *Glanures Entomologiques* **1**: 34-42.

Jacquelin du Val P. N. C. 1859d: Diagnoses de deux genres nouveaux du groupe des Dasytites. *Glanures Entomologiques* **1**: 42-43.

Jacquelin du Val P. N. C. 1859e: Note sur le genre Antidipnis Woll. et une coupe nouvelle voisine Homoeodipnis Jacq du Val. *Glanures Entomologiques* **1**: 43-48.

Jacquelin du Val C. 1860a: Eucnémides, pp. 112-121; Elatérides, pp. 122-144; Clérides, pp. 192-204; Lymexylonides, pp. 205-207. In: *Manuel Entomologique. Genera des coléoptères d'Europe comprenant leur Classification en Familles naturelles, la description de tous les genres, des tableaux synoptiques destinés à faciliter l'étude, le catalogue de toutes les espèces, de nombreux dessins au trait de caractères.*

Tome troisième. Paris: A. Deyrolle, 463 + [1] + [126-200, Catalogue] pp., 100 pls. [issued in parts, Catalogue appeared separately].

Jacquelin du Val P. N. C. 1860b: Essai monographique sur le genre Henicopus. *Glanures Entomologiques* **2**: 61-96.

Jacquelin du Val P. N. C. 1860c: Synopsis des espèces européennes du genre Lampyris. *Glanures Entomologiques* **2**: 97-103.

Jacquelin du Val P. N. C. 1860d: Synopsis des espèces européennes du genre Cebrio. *Glanures Entomologiques* **2**: 104-136.

Jacquelin du Val P. N. C. 1860e: Description de deux genres nouveaux et d'une espèce de la famille des Ptinides. *Glanures Entomologiques* **2**: 137-139.

Jacquelin du Val P. N. C. 1860f: Description de deux espèces nouvelles. *Glanures Entomologiques* **2**: 139-141.

Jacquelin du Val P. N. C. 1860g: Diagnoses provisoires de quelques genres nouveaux. *Glanures Entomologiques* **2**: 142-143.

Jacquet E. 1886b: Note sur le Montandonia catopoides. Nouveau genre appartenant à la famille Dermestidae. *Annales de la Société Linéenne de Lyon* **32**: 319-321.

Jadwiszczak A. S. 1989: Studies on the Oriental Epilachninae (Coleoptera, Coccinellidae). *Polskie Pismo Entomologiczne* **58**: 733-744.

Jadwiszczak A. S. & Pütz A. 1994: Descriptions of two new species of the genus Epilachna Dejean, 1837 from Nepal and India (Coleoptera, Coccinellidae). *Coccinella* (Dresden) **5** [1993/1994]: 13-17.

Jadwisyczak A. S. & Węgrzynowicz P. 2003: *World Catalogue of Coccinellidae. Part I- Epilachninae*. Olstyn: Mantis, 264 pp.

Jagemann E. 1934: Eine neue Silesis-Art aus dem palaearktischen Faunengebiete (Col. Elateridae). *Entomologisches Nachrichtenblatt* (Troppau) **8**: 124.

Jagemann E. 1939: Einige neue Elateriden-Formen aus dem palaearkt. Faunengebiet. *Entomologisches Nachrichtenblatt* (Troppau) **13**: 68-70.

Jagemann E. 1942a: De duobus novis elateridarum speciebus in China inventis. *Entomologické Listy* **5**: 29-30.

Jagemann E. 1942b: Ein neuer Elateridae aus dem Amurgebiet. (Col.). *Mitteilungen der Münchener Entomologischen Gesellschaft.* **32**: 596-598.

Jagemann E. 1943: Notulae elaterologicae (1. - 11.). *Entomologické Listy* **6**: 98-102.

Jagemann E. 1944: De elateridarum speciebus novis. *Sborník Entomologického Oddělení Zemského Musea v Praze* **21-22**: 332-335.

Jagemann E. 1945: Generis Heteroderes Latr. quattuor species novae. *Sborník Přírodovědeckého Klubu v Brně* **26**: 39-41.

Jakobson G. G. 1903: Spisok zhukov, naydennykh b 1899-1901 gg. Shtab-rotmistrom A. N. Gudzenko v yuzhnoy Manchzhurii i na kvantungskom poluostrove. *Annuaire du Musée Zoologique de l'Académie Impériale des Sciences de St.-Pétersbourg* **8**: xi-xvi.

Jakobson G. G. 1905: Localités de captures de quelques coléoptères présentants un certain intérêt. III. *Annuaire du Musée Zoologique de l'Académie Impériale des Sciences de St.-Pétersbourg* **9** [1904]: xxxiii-xxxiv.

Jakobson G. G. 1911a: *Zhuki Rossii i zapadnoy Evropy. Rukovodstvo k opredeleniyu zhukov. Vypusk 7*. St.-Pétersburg: A. F. Devrjen, 481-560 pp.

Jakobson G. G. 1911b: *Zhuki Rossii i zapadnoy Evropy. Rukovodstvo k opredeleniyu zhukov. Vypusk 9*. St.-Pétersburg: A. F. Devrjen, 641-720 pp.

Jakobson G. G. 1913: *Zhuki Rossii i zapadnoy Evropy. Rukovodstvo k opredeleniu zhukov. Vypusk 10*. St.-Péterbourg: A. F. Devrjen, 721-864 pp.

Jakobson G. G. 1915: *Zhuki Rossii i zapadnoy Evropy. Rukovodstvo k opredeleniu zhukov. Vypusk 11*. St.-Péterbourg: A. F. Devrjen, 865-1024 pp.

Jakovlev A. I. 1910: Enumeratio Coleopterorum a clarissimo L. Krulikovsky circa oppida Malmyzh (annis 1896-1899) et Urzhum (annis 1899-1908) provinciae Vjatkensis (Rossiae mediae orientalis) collectorum. *Horae Societatis Entomologicae Rossicae* **39**: 276-327.

Jänichen T. von. 1913: Eine neue Form von Coccinella quinquepunctata L. (Col.). *Deutsche Entomologische Zeitschrift* **1913**: 552.

Janson E. 1873: [new taxon]. In: Crotch G. R.: A descriptive list of Erotylidae collected by Geo. Lewis, Esq., in Japan. *The Entomologist's Monthly Magazine* **9**: 184-189.

Jansson A. 1936: Coleopterologiska bidrag 23-24. *Notulae Entomologicae* **16**: 39-43.

Jansson A. 1940: Die Arthropodenfauna von Madeira nach den Ergebnissen der Reise von Prof. Dr. O. Lundblad. Juli-August 1935. XXIX. Coleoptera: sämtliche Familien unter Ausschluss der Familien der Carabidae, Dytiscidae, Hydrophilidae und der Gattung Cryptophagus Herbst aus der Familie Cryptophagidae. *Arkiv för Zoologi* (A) **32** (24): 1-64, 2 pls.

Jansson A. 1942: De skandinaviska arterna inom Xyletinus ater-komplexet (Col., Anobiidae). Nebst Beschreibung des X. longitarsis nov. sp. *Opuscula Entomologica* (Lund) **7**: 22-27.

Jansson A. 1947: Zwei Namenänderungen (Coleoptera). *Opuscula Entomologica* (Lund) **12**: 118.

Jayne H. F. 1882: Revision of the Dermestidae of the United States. *Proceedings of the American Phil. Society, Philadelphia* **20**: 343-377, 1-4 pls.

Jelínek J. 1964: Nitidulidae of the Klapperich´s expedition in Afghanistan 1952-1953. *Annotationes Zoologicae et Botanicae* (Bratislava) **8**: 1-5.

Jelínek J. 1965a: 33. Nitidulidae. Ergebnisse der zoologischen Forschungen von Dr. Z. Kaszab in der Mongolei (Coleoptera). *Reichenbachia* **7**: 135-145.

Jelínek J. 1965b: A new Rhizophagus species from China (Coleoptera, Rhizophagidae). *Acta Entomologica Musei Nationalis Pragae* **36**: 385-389.

Jelínek J. 1967: 30th Result of the zoological Expedition of the National Museum in Prague to Turkey. Coleoptera - Nitidulidae. *Acta Entomologica Musei Nationalis Pragae* **37**: 23-30.

Jelínek J. 1968: New species of the genus Meligethes Steph. from Turkey (Coleoptera, Nitidulidae). *Annotationes Zoologicae et Botanicae* (Bratislava) **49**: 1-3.

Jelínek J. 1974: Generic classification of Oriental Cryptarchinae (Coleoptera, Nitidulidae). *Acta Entomologica Bohemoslovaca* **71**: 187-196.

Jelínek J. 1975a: New genus of Oriental Meligethinae with notes on supergeneric classification of Nitidulidae (Coleoptera, Nitidulidae). *Annotationes Zoologicae et Botanicae* (Bratislava) **102**: 1-9.

Jelínek J. 1975b: Revision of the genus Glischrochilus Reitter from the Oriental region and China (Coleoptera, Nitidulidae). *Acta Entomologica Bohemoslovaca* **72**: 127-144, 2 pls.

Jelínek J. 1976: Description and revision of the genus Anamartus gen. n. (Coleoptera, Nitidulidae). *Acta Entomologica Bohemoslovaca* **73**: 17-31.

Jelínek J. 1977: Revision of the genus Epuraea Er. from Africa with remarks to related genera (Col., Nitidulidae). *Acta Entomologica Musei Nationalis Pragae* **39**: 345-397.

Jelínek J. 1978a: Ergebnisse der Bhutan-Expedition 1972 des Naturhistorischen Museums in Basel. Coleoptera: Fam. Nitidulidae. *Entomologica Basiliensia* **3**: 171-218.

Jelínek J. 1978b: Two new species of the Meligethes elongatus species-group from the Middle East (Coleoptera, Nitidulidae). *Acta Entomologica Bohemoslovaca* **75**: 330-335.

Jelínek J. 1980: Zur Taxonomie der asiatischen Nitidulidae und Cerylonidae (Coleoptera). *Reichenbachia* **18**: 93-102.

Jelínek J. 1981a: Results of the Czechoslovak – Iranian entomological expeditions to Iran 1970 and 1973. Coleoptera: Nitidulidae. *Acta Entomologica Musei Nationalis Pragae* **40**: 105-119.

Jelínek J. 1981b: Results of Czechoslovak - Iranian Entomological Expedition to Iran 1970. Coleoptera: Erotylidae, Biphyllidae, Colydiidae, Cerylonidae, Endomychidae, Sphindidae. *Acta Entomologica Musei Nationalis Pragae* **40**: 121-126.

Jelínek J. 1982a: New and little known taxa of Nitidulidae (Coleoptera). *Acta Musei Nationalis Pragae* (B) **38**: 171-200.

Jelínek J. 1982b: New and little known species of the genus Meligethes from Turkey and neighbouring countries (Coleoptera, Nitidulidae). *Türkiye Bitki Koruma Dergesi* **5** [1981]: 201-214.

Jelínek J. 1982c: New palaearctic species of the genus Meligethes (Coleoptera, Nitidulidae). *Acta Entomologica Bohemoslovaca* **79**: 289-300.

Jelínek J. 1982d: Revision of the Meligethes obscurus species-group (Coleoptera, Nitidulidae). *Acta Entomologica Bohemoslovaca* **79**: 301-309.

Jelínek J. 1984a: New species of the genus Rhizophagus from Middle Asia (Coleoptera, Rhizophagidae). *Acta Faunistica Entomologica Musei Nationalis Pragae* **17**: 129-132.

Jelínek J. 1984b: Revision of the genus Stelidota from Asia, Australia and Pacific area (Coleoptera, Nitidulidae). *Acta Entomologica Bohemoslovaca* **81**: 132-156.

Jelínek J. 1986: A new species of Carpophilus from Asia related to C. delkeskampi (Coleoptera, Nitidulidae). *Acta Entomologica Bohemoslovaca* **83**: 455-464.

Jelínek J. 1988: Coleoptera: Nitidulidae of Saudi Arabia (Part 2). *Fauna of Saudi Arabia* **9**: 42-51.

Jelínek J. 1993: New Palaearctic species of the genus Amphicrossus Erichson (Coleoptera, Nitidulidae). *Folia Heyrovskyana* **1**: 36-40.

Jelínek J. 1995: New Asian species of the genus Aethina Erichson (Coleoptera, Nitidulidae). *Folia Heyrovskyana* **3**: 109-119.

Jelínek J. 1997: New descriptions and records of Brachypteridae and Nitidulidae from the Palaearctic region (Coleoptera). *Folia Heyrovskyana* **5**: 123-138.

Jelínek J. 1999a: New species of the genus Glischrochilus (Coleoptera: Nitidulidae. Cryptarchinae) from Asia. *Folia Heyrovskyana* **7**: 201-216.

Jelínek J. 1999b: Contribution to taxonomy of the beetle subfamily Nitidulinae (Coleoptera. Nitidulidae). *Folia Heyrovskyana* **7**: 251-281.

Jelínek J. & Audisio P. 1977: Tassonomia e distribuzione geografica di Meligethes coerulescens Kraatz e specie correlate (Coleoptera, Nitidulidae). *Bollettino del Museo di Zoologia dell'Università di Torino* **1**: 1-13.

Jelínek J. & Háva J. 2001: A new species of Laricobius (Coleoptera: Derodontidae) from Nepal. *Revue Suisse de Zoologie* **108**: 149-152.

Jelínek J. & Kirejtshuk A. G. 1986: [new taxa]. In: Kirejtshuk A. G.: Reviziya roda Aethina Er. (Coleoptera, Nitidulidae) fauny oriental'noy i palearkticheskoi oblastey [Revision of the genus Aethina Er. (Coleoptera, Nitidulidae) of the fauna of Palaearctic and Oriental regions]. *Trudy Zoologicheskogo Instituta AN SSSR* **140**: 44-82.

Jelínek J. & Marek O. 1966: Two new Palaearctic species of the genus Meligethes Steph. (Coleoptera, Nitidulidae). *Acta Entomologica Bohemoslovaca* **63**: 453-458.

Jelínek J. & Spornraft K. 1979: Die westpaläarktischen Arten der umbrosus-Gruppe der Gattung Meligethes Steph. (Coleoptera, Nitidulidae). *Mitteilungen der Münchener Entomologischen Geselschaft* **68**: 1-11.

Jeng M.-L. & Chang J.-C. 1998: [new taxon]. In: Jeng M.-L., Yang P.-S., Satô M., Lai J. & Chang J.-C.: The genus Curtos (Coleoptera, Lampyridae, Luciolinae) of Taiwan and Japan. *Japanese Journal of Systematik Entomology* **4**: 331-347.

Jeng M.-L. & Lai J. 1998: [new taxon]. In: Jeng M.-L., Yang P.-S., Satô M., Lai J. & Chang J.-C.: The genus Curtos (Coleoptera, Lampyridae, Luciolinae) of Taiwan and Japan. *Japanese Journal of Systematik Entomology* **4**: 331-347.

Jeng M.-L. & Lai J. 2003: [new taxon]. In: Jeng M.-L., Yang P.-S. & Lai J.: Notes on the genus Luciola Coleoptera, Lampyridae, Luciolinae) of Taiwan. *Special Bulletin of the Japanese Society of Coleopterology* **6**: 247-262.

Jeng M.-L., Lai J. & Yang P.-S. 2003: Lampyridae: A synopsis of aquatic fireflies with description of a new species (Coleoptera). Pp. 539-562. In: Jäch M. A. & Ji L. (eds.): *Water beetles of China. Volume III*. Wien: Zoologisch-botanische Gesellschaft und Wiener Coleopterologenverein, vi + 572 pp.

Jeng M.-L., Lai J., Yang P.-S. & Satô M. 1999: On the validity of the generic name Pyrocoelia Gorham (Coleoptera, Lampyridae, Lampyrinae) with a review of Taiwanese species. *Japanese Journal of Systematik Entomology* **5**: 347-362.

Jeng M.-L., Lai J., Yang P.-S. & Satô M. 2001: Revision of the genus Diaphanes Motschulsky (Coleoptera: Lampyridae: Lampyrinae) of Taiwan. *Japanese Journal of Systematik Entomology* **7**: 203-235.

Jeng M.-L. & Satô M. 1998: [new taxon]. In: Jeng M.-L., Yang P.-S., Satô M., Lai J. & Chang J.-C.: The genus Curtos (Coleoptera, Lampyridae, Luciolinae) of Taiwan and Japan. *Japanese Journal of Systematik Entomology* **4**: 331-347.

Jeng M.-L. & Yang P.-S. 1998: [new taxon]. In: Jeng M.-L., Yang P.-S., Satô M., Lai J. & Chang J.-C.: The genus Curtos (Coleoptera, Lampyridae, Luciolinae) of Taiwan and Japan. *Japanese Journal of Systematik Entomology* **4**: 331-347.

Jeng M.-L. & Yang P.-S. 2003: [new taxon]. In: Jeng M.-L., Yang P.-S. & Lai J.: Notes on the genus Luciola Coleoptera, Lampyridae, Luciolinae) of Taiwan. *Special Bulletin of the Japanese Society of Coleopterology* **6**: 247-262.

Jeng M.-L., Yang P.-S. & Satô M. 1998: The genus Cyphonocerus (Coleoptera, Lampyridae) from Taiwan and Japan, with notes on the subfamily Cyphonocerinae. *Elytra* **26**: 379-398.

Jeng M.-L., Yang P.-S. & Satô M. 1999: An additional new species of Cyphonocerus (Coleoptera, Lampyridae, Psilocladinae) from Taiwan. *Elytra* **27**: 405-408.

Jeng M.-L., Yang P.-S. & Satô M. 2002a: Notes on the morphology and systematics of the genus Pristolycus

Gorham (Coleoptera, Lampyridae). *Japanese Journal of Systematic Entomology* **8**: 87-108.
Jeng M.-L., Yang P.-S. & Satô M. 2002b: Checklist of Lycidae (Coleoptera: Elateriformia) of Taiwan. *Formosan Entomology* **22**: 171-187.
Jiang S.-H. 1990: New record of the genus Tetralobus Lepeletier and the species T. perroti Fleutiaux from China (Coleoptera: Elateridae). *Sichuan Journal of Zoology* **9** (1): 42.
Jiang S.-H. 1991: [Notes on the Chinese click beetles of the genus Campsosternus Latreille (Coleoptera: Elateridae), with descriptions of two new species]. *Entomotaxonomia* **13**: 275-280.
Jiang S.-H. 1993: *A Catalogue of the Chinese Elateridae*. Beijing: Beijing Agricultural University Publishing House.
Jiang S.-H. 1994: Notes on the genus Vuilletus from China (Coleoptera: Elateridae). *Sichuan Journal of Zoology* **13** (2): 84-85.
Jiang S.-H. 1995: Coleoptera: Elateridae. Pp. 220-222. In: Wu H. (ed.]. *Insects of Baishanzu Mountain, Eastern China*. China Beijing: Forestry Publishing House, i-xiii, 1-586 pp.
Jiang S.-H. 1999: [new taxa]. In: Jiang S.-H. & Wong S.-Y.: *Economic click beetle fauna of China (Coleoptera: Elateridae)*. Beijing: China Agricultural Press, i-x, 1-196.
Jiang S.-H., Wang S. & Li W. 1997: Coleoptera: Elateridae. Pp. 669-686. In: Yang X. (ed.): *Insects of the Three Gorge Reservoir area of Yangtze river. Part 1*. Chongqing: Chongqing Publishing House, i-xx, 1-974 pp.
Jing X.-L. 1983: Three new species of the genus Sumnius (Coleoptera: Coccinellidae). *Acta Zootaxonomica Sinica* **8**: 403-406 (in Chinese).
Jing X.-L. 1985: One new species and one new record of Micraspis from Chine (Coleoptera: Coccinellidae). *Acta Entomologica Sinica* **10**: 424-426 (in Chinese).
Jing X.-L. 1986a: A new species of the genus Semiadalia (Coleoptera: Coccinellidae). *Acta Entomologica Sinica* **29**: 73-74 (in Chinese).
Jing X.-L. 1986b: Three new species of Oenopia Mulsant (Coccinellidae: Coleoptera). *Acta Entomologica Sinica* **20**: 203-207 (in Chinese).
Jing X.-L. 1986c: Two new species of the genus Asemiadalia (Coleoptera: Coccinellidae). *Acta Zootaxonomica Sinica* **11**: 205–208 (in Chinese).
Jing X.-L. 1986d: A new species of Phrynocaria Timberlake (Coleoptera: Coccinellidae). *Acta Zootaxonomica Sinica* **11**: 307-308 (in Chinese).
Jing X.-L. 1987a: A new species of Halyzia Mulsant (Coleoptera: Coccinellidae). *Acta Entomologica Sinica* **30**: 201-202 (in Chinese).
Jing X.-L. 1987b: A new species of Chinese Xanthadalia (Coleoptera: Coccinellidae). *Acta Zootaxonomica Sinica* **12**: 301-302 (in Chinese).
Jing X.-L. 1987c: A new species and two new records of Oenopia Mulsant from China (Coleoptera: Coccinellidae). *Acta Zootaxonomica Sinica* **12**: 413-414 (in Chinese).
Jing X.-L. 1988: New species of Chinese Lemnia Mulsant (Coleoptera: Coccinellidae). *Acta Zootaxonomica Sinica* **13**: 57-62.
Jing X.-L. 1992: Coleoptera: Coccinellidae. Pp.541-574. In: Chen S. (ed.): *Insects of the Hengduan Mountains Region. Volume I*. Beijing: Science Press, 1992, xii + 865 pp. (in Chinese).
John H. 1955a: Zwei neue japanische Spezies von Aphanocephalus Wollaston (Notiophygidae, Col.). *Mushi* **28**: 1-3.
John H. 1955b: 5 neue Spezies von Aphanocephalus Woll. (Notiophygidae Col.) aus dem Staatl. Museum für Tierkunde Dresden. *Abhandlungen und Berichte aus dem Museum für Tierkunde Dresden* **22**: 13-21.
John H. 1956: Revision der Gattung Aphanocephalus Woll. (Notiophygidae Col.). *Entomologische Blätter* **52**: 1-35.
John H. 1957: Neue Discolomidae (= Notiophygidae, Col.). *Annalen des Naturhistorischen Museums in Wien* **61**: 249-253.
John H. 1958: Aphanocephalus shirozui spec. nov. (Notiophygidae, Col.). *Mushi* **31**: 61-62.
John H. 1961: Eine neue Spezies von Aphanocephalus Wollaston (Discolomidae = Notiophygidae, Col.). *Niponius* **1** (9): 1-2.
John H. 1962: Drei neue Spezies der Gattung Thorictus Germ. (Coleoptera, Thorictidae). *Annales Historico-Naturales Musei Nationalis Hungarici* **54**: 283-285.
John H. 1963a: Die japanischen Spezies im hemisphaericus-Kreis der Gattung Aphanocephalus Wollaston (Coleoptera: Discolomidae). *Niponius* **2** (2): 3-10.

John H. 1963b: Revision der Gattung Thorictus Germar (Thorictidae, Clavicornia Col.). Eine Bestandaufnahme, Kritik und Neubeschreibungen. *Entomologische Abhandlungen und Berichte aus dem Staatlichen Museum für Tierkunde in Dresden* **28**: 177-255.

John H. 1964a: Eine neue Spezies der Gattung Thorictus Germ. (Thorictidae Col.). *Notulae Entomologicae* **44**: 135-136.

John H. 1964b: Eine neue Spezies von Thorictus Germ. (Coleoptera, Thorictidae). *Atti della Società Italiana di Scienze Naturali e del Museo Civico di Storia Naturale in Milano* **103**: 163-164.

John H. 1964c: Zwei neue Spezies von Thorictes Germ. (Thorictidae Col.). *Entomologisk Tidskrift* **85**: 190-192.

John H. 1964d: Acht neue Spezies von Thorictus Germ. (Thorictidae, Coleoptera). *Annales Historico-Naturales Musei Nationalis Hungarici* **56**: 331-339.

John H. 1965a: Zwei neue Spezies von Thorictus Germ. (Thorictidae - Col.). *Bulletin de la Société des Sciences Naturelles et Physique du Maroc* **44** [1964]: 29-31.

John H. 1965b: Eine neue Spezies von Thorictus Germar (Coleoptera: Thorictidae). *Proceedings of the Royal Entomological Society of London* (B) **34**: 51-52.

John H. 1965c: Eine neue Spezies von Aphanocephalus Wollaston (Coleoptera - Discolomidae). *Niponius* **2** (10): 53-55.

John H. 1965d: Thorictus schatzmayri spec. n. (grazianii Schatzm. i. l.) (Coleoptera Thorictidae). *Atti della Società Italiana di Scienze Naturali e del Museo Civico di Storia Naturale in Milano* **104**: 67-68.

John H. 1965e: Revision der Gattung Thorictus Germar (Thorictidae, Clavicornia, Coleoptera). Eine Bestandaufnahme, Kritik und Neubeschreibungen. II. Teil. *Entomologische Abhandlungen und Berichte aus dem Staatlichen Museum für Tierkunde in Dresden* **31**: 319-395.

John H. 1965f: Neue Spezies der Gattung Thorictus Germ. (Col. Thorictidae). *Eos, Revista Española de Entomología* **40**: 507-513, pls VII, VIII.

John H. 1966: Neue Spezies von Thorictus Germ. (Col., Thorictidae). *The Entomologist's Monthly Magazine* **102**: 19-21.

John H. 1967: Pazifische Discolomidae (Coleoptera). *Pacific Insects* **9**: 111-137.

John H. 1968: 5. Familie Discolomidae. Neue Spezies von Aphanocephalus Wollaston. Pp. 13-20. In: Chûjô M. (ed.): Coleoptera from Southeast Asia (VIII). *Memoirs of the Faculty of Education Kagawa University, Part II,* No. 171.

John H. 1969: Drei neue Spezies von Thorictus Germ. (Thorictidae, Col.). *Archivos del Instituto de Aclimatación* (Almería) **14**: 125-128.

John H. 1971: Discolomidae und Thorictidae. *Entomologische Blätter* **66**: 137-152.

John A. & Andreae H. 1967: Revision der Gattung Thorictus Germar (Coleoptera, Thorictidae). III. Teil. *Entomologische Abhandlungen und Berichte aus dem Staatlichen Museum für Tierkunde in Dresden* **34**: 145-158.

Johnson C. 1967: Addditions and corrections to the British list of Atomaria s.str. (Col., Cryptophagidae), including a species new to science. *The Entomologist* **100**: 64-65.

Johnson C. 1968: Some interesting species of Atomaria Stephens (Col., Crytophagidae) in the British Museum. *The Entomologist* **101**: 228-231.

Johnson C. 1969: Atomariinae (Col., Cryptophagidae) from Jordan and Cyprus, with the description of a new species. *Norsk Entomologisk Tidsskrift* **16**: 77-80.

Johnson C. 1970a: Three new species of Atomaria Stephens (Coleoptera: Cryptophagidae) from the eastern Palaearctic. *Notulae Entomologicae* **50**: 112-116.

Johnson C. 1970b: Corticarina gangolae sp. nov. (Col., Lathridiidae) from Northern India. *The Entomologist* **103**: 190-191.

Johnson C. 1970c: Cryptophagidae: Atomariinae. Ergebnisse der Zoologischen Forschungen von Dr. Z. Kaszab in der Mongolei. *Reichenbachia* **12**: 259-276.

Johnson C. 1971a: Two new species of Atomaria Stephens from Europe and North Africa, and a re-description of A. versicolor Erichson (Col., Cryptophagidae). *Entomologische Blätter* **67**: 1-8.

Johnson C. 1971b: Atomariinae (Col. Cryptophagidae) from the northern parts of the Indian sub-continent, with descriptions of seven new species. *The Entomologist's Monthly Magazine* **107** [1970]: 224-232.

Johnson C. 1971c: Atomaria (s. str.) gottwaldi sp. n. (Coleoptera: Cryptophagidae) from Czechoslovakia and Italy. *Acta Entomologica Bohemoslovaca* **68**: 231-232.

Johnson C. 1971d: Cryptophagidae-Atomariinae from Nepal (Col.). *Senckenbergiana Biologica* **52**: 437-439.

Johnson C. 1972a: Some members of the genera Corticarina Reitter and Melanophthalma Motschulsky (Col., Lathridiidae) from India and Nepal. *The Entomologist* **105**: 97-110.

Johnson C. 1972b: Studies on the genera Corticarina Reitter and Melanophthalma Motschulsky (Col., Lathridiidae). *Nouvelle Revue d'Entomologie* **2**: 185-199.

Johnson C. 1972c: A new North American species of Henoticoides Hatch (Col., Cryptophagidae). *The Entomologist* **105**: 274-276.

Johnson C. 1973: Beiträge zur Kenntnis der Fauna Afghanistans, Atomariinae, Cryptophagidae, Coleoptera. *Časopis Moravského Musea* **56-57** [1971-72]: 261-266.

Johnson C. 1975a: The Atomaria species (Col., Cryptophagidae) of Madeira and the Canary Islands: a supplement. *Vieraea* **5**: 33-37.

Johnson C. 1975b: The Himalayan species of Henoticus Thomson (Coleoptera: Cryptophagidae). *Senckenbergiana Biologica* **56**: 37-38.

Johnson C. 1975c: A review of the Palaearctic species of the genus Ernobius Thomson. *Entomologische Blätter* **71**: 65-93.

Johnson C. 1975d: New species of Corticarina Reitter (Col., Lathridiidae) from Nepal and the Ethiopian Region. *The Entomologist's Monthly Magazine* **110** [1974]: 97-103.

Johnson C. 1975e: Cortinicara, a new genus of Corticariinae (Coleoptera: Lathridiidae). *Entomologica Scandinavica* **6**: 283-285.

Johnson C. 1976a: Coleoptera from North-East Africa, Lathridiidae (Zoological contribution from the Finnish expeditions to the Sudan No. 40). *Notulae Entomologicae* **56**: 31-32.

Johnson C. 1976b: Nine species of Coleoptera new to Britain. *The Entomologist's Monthly Magazine* **111** [1975]: 177-183.

Johnson C. 1976c: New Corticariinae (Coleoptera, Lathridiidae) from Japan and the Far East. *The Bulletin of the National Science Museum, Tokyo, Series A (Zoology)* **2**: 195-198.

Johnson C. 1977a: Beiträge zur Kenntnis der Fauna Afghanistans. Merophysiidae and Lathridiidae, Coleoptera. *Časopis Moravského Muzea. Vědy přírodní* **62**: 123-132.

Johnson C. 1977b: Studies on the genus Corticaria Marsham, Part 2 (Col., Lathridiidae). *The Entomologist's Monthly Magazine* **112** [1976]: 213-226.

Johnson C. 1977c: Ergebnisse der Bhutan-Expedition 1972 des Naturhistorischen Museums in Basel, Coleoptera: Fam. Lathridiidae Subfam. Corticariinae. *Entomologica Basiliensia* **2**: 329-335.

Johnson C. 1978: New or little-known species of Atomaria (Coleoptera: Cryptophagidae) from Northern and Central Europe. *Entomologica Scandinavica* **9**: 26-29.

Johnson C. 1979: New species of Corticarina Reitter (Col., Lathridiidae). *The Entomologist's Monthly Magazine* **114** [1978]: 55-62.

Johnson C. 1981: New species of Corticarina Reitter (Col., Lathridiidae), Part 2. *The Entomologist's Monthly Magazine* **117**: 117-124.

Johnson C. 1985: Two new species of Corticaria (Col.: Lathridiidae) from the Canary Islands. *Vieraea* **14**: 63-66.

Johnson C. 1989: Studies on the genus Corticaria Marsham (Coleoptera: Latridiidae), Part 3. *Entomologist's Gazette* **40**: 79-90.

Johnson C. 1990: New exotic species of Corticariinae (Col.: Latridiidae). *Entomologist's Record & Journal of Variation* **102**: 11-16.

Johnson C. & Bowestead S. 2003: A review of the apterous species of Caenoscelis Thomson, 1863 (Coleoptera, Cryptophagidae). Pp. 885-904. In: Cuccodoro G. & Leschen R. A. B. (eds): *Systematics of Coleoptera: Papers Celebrating the Retirement of Ivan Löbl*. Memoirs on Entomology, International, Volume **17**. Gainesville: Associated Publishers, 955 pp.

Johnson C. & Strand A. 1968: Notes on some species of Atomaria s.str. (Col., Cryptophagidae) with descriptions of two species new to science. *Norsk Entomologisk Tidsskrift* **15**: 93-96.

Johnson R. H. 1910: Determinate evolution in the color pattern of the lady-beetle. *Carnegie Institution* (Washington) No. **122**: 104 pp.

Jordan K. 1894: New species of Coleoptera from the Indo- and Austro-Malayan region, collected by Willliam Doherty. *Novitates Zoologicae* **1**: 104-138.

Joy N. H. 1905: Three species of Coleoptera new to Britain. *The Entomologist's Monthly Magazine* (2) **16**: 274-275.

Joy N. H. 1910: Cryptophagus fowleri sp. n. - a beetle new to Britain. *The Entomologist's Monthly Magazine* **46**: 205-206.

Joy N. H. & Tomlin J. R. le B. 1910: Enicmus histrio, sp. nov.: a beetle new to Britain. *The Entomologist's Monthly Magazine* **46**: 250-252.

Juarez Monroy. A. & Zaragoza Cabalero S. 1990: Una specie nueva de Delphastus Casey (Coleoptera: Coccinellidae). *Anales del Instituto de Biologia Universidad Nacional Autónoma de México, Serie Zoologia* **61**: 297-300.

Jurine L. 1814: [new taxon]. In: Champollion Figeac J. F.: Note sur une nouvelle espèce d'insecte du genre Corynetes Fabricius, trouvée dans une momie égyptienne, observée à Grenoble. *Philosophical Magazine and Journal* **44**: 131-133.

Kalík V. 1946: Noví Dermestidae palearktické fauny [Dermestidae nouveaux de la faune paléarctique]. *Folia Entomologica* **9**: 133-138.

Kalík V. 1950: Neue Dermestiden aus der Palaearktischen Fauna. (3. Beitrag). *Folia Entomologica Hungarica* **3**: 61-66.

Kalík V. 1951a: Noví Dermestidae palaearktické fauny [New Dermestidae of the Palaearctic fauna]. (2nd Contribution). *Sborník Entomologického Oddělení Národního Musea Praha* **26** (362) [1949]: 1-4.

Kalík V. 1951b: Poznámky o některých formách kožojedů rodu Dermestes L. [Notes sur quelques espèces et formes du genre Dermestes L.]. *Folia Entomologica* **14**: 39-46.

Kalík V. 1951c: Neue Dermestiden aus der Palaearktischen Fauna. (4. Beitrag). *Annales Historico-Naturales Musei Nationalis Hungarici* (S. N.) **1**: 152-156.

Kalík V. 1952a: New and interesting species of Dermestidae (Coleoptera). *Pan-Pacific Entomologist* **28**: 43-49.

Kalík V. 1952b: Neue Dermestiden aus der Paläarktischen Fauna. (5. Beitrag).). *Annales Historico-Naturales Musei Nationalis Hungarici* (S. N.) **2**: 75-77.

Kalík V. 1954: New and interesting Dermestidae (Coleoptera). *The Annals and Magazine of Natural History* (12) **7**: 367-370.

Kalík V. 1955: Dermestidae der aethiophischen Region aus der Sammlungen des Königl. Museum von Belgisch Kongo in Tervuren (Mit Bemerkungen über verwandte Arten). *Revue de Zoologie et Botanique Africaines* **52**: 294-322.

Kalík V. 1965: Coleoptera Dermestidae. Mission zoologique de l'I.R.A.C. en Afrique orientale (P. Basilewski et N. Leleup, 1957). XCII. *Annales du Musée Royale d'Afrique Centrale, Sciences Zoologiques* (Tervuren) **138**: 215-224.

Kalík V. & Ohbayashi N. 1982: A new dermestid beetle from Asia (Coleoptera). *Entomological Revue of Japan* **37**: 65-67.

Kalík V. & Ohbayashi N. 1985: Anthrenus nipponensis, a new dermestid beetle (Coleoptera, Dermestidae) from Japan, Korea and China. *Elytra* **13**: 75-79.

Kamiya H. 1959a: A revision of the tribe Chilocorini of Japan and the Loochoos (Coleoptera: Coccinellidae). *Kontyû* **27**: 99-104, pl. 9.

Kamiya H. 1959b: On the identity of Stethorus punctillum of Japanese authors (Coleoptera: Coccinellidae). *Kontyû* **27**: 139-143 (in Japanese).

Kamiya H. 1960: A new tribe of Coccinellidae (Coleoptera). *Kontyû* **28**: 22-26, pl. 3.

Kamiya H. 1961a: A revision of the tribe Scymnini from Japan and the Loochoos (Coleoptera: Coccinellidae). Part I. Genera Clitostethus, Stethorus and Scymnus (except subgenus Pullus). *Journal of the Faculty of Agriculture, Kyushu University* **11**: 275-301, pl. 38.

Kamiya H. 1961b: A revision of the tribe Scymnini from Japan and the Loochoos (Coleoptera: Coccinellidae). Part II. Genus Scymnus (subgenus Pullus). *Journal of the Faculty of Agriculture, Kyushu University*, **11**: 303-330, pl. 39.

Kamiya H. 1963a: A revision of the tribe Hyperaspini of Japan (Coleoptera: Coccinellidae). *The Memoirs of the Faculty of Liberal Arts, Fukui University Ser. II, Natural Science* **13**: 79-86.

Kamiya H. 1963b: On a new genus and species of Coccinellidae (Coleoptera) feeding on whiteflies. *Mushi* **37**: 127-130.

Kamiya H. 1964: Discovery of the family Propalticidae in the Ryukyus (Coleoptera, Cucujoidea-Clavicornia). *Kontyû* **32**: 281-285.

Kamiya H. 1965a: Tribe Scymnini (Coleoptera: Coccinellidae) from Formosa collected by Prof. T. Shirôzu. *Special Bulletin of the Lepidopterological Society of Japan* **1**: 75-82.

Kamiya H. 1965b: Coccinellid-Fauna of the Ryukyu Islands, south of the Amami-group (Coleoptera). *Kontyû* **33**: 97-122.

Kamiya H. 1965c: Scymnus (Pullus) vencoxus n. n. for Scymnus (Pullus) convexus H. Kamiya, 1961 nec Nunenmacher, 1937. *Kontyû* **33**: 316.

Kamiya H. 1966: On the Coccinellidae attacking the scale insects and mites in Japan and the Ryukyus. *Mushi* **39**: 65-93.

Kamiya H. 1979: A revision of the tribe Chilocorini of Japan and the Loochoos (Coleoptera: Coccinellidae). *Kontyû* **27**: 99-104.

Kang T.-H & Kim J.-I. 2000a: Taxonomic study of Korean Cantharidae (Coleoptera). IV. Subfamily Cantharinae: Genus Rhagonycha. *Korean Journal Entomology* **30**: 157-162.

Kang T.-H & Kim J.-I. 2000b: Taxonomic study of Korean Cantharidae (Coleoptera). II. Cantharinae: Genus Podabrus. *Insecta Koreana* **17** (3): 199-213.

Kang T.-H. & Okushima Y. 2003: Taxonomic study on Korean Cantharidae (Coleoptera). VI. Three new species from Is. Jejudo, Korea. *Elytra* **31**: 341-351.

Kangas E. 1936: Henoticus germanicus Reitt. als Vertreter einer neuen Cryptophagidengattung, Coniophthalma n. gen. (Col.). *Annales Entomologica Fennici* **2**: 67-73.

Kangas E. 1961: Über die Atomaria-Arten (Col., Cryptophagidae) Finnlands. *Annales Entomologica Fennici* **27**: 112-118.

Kangas E. 1973: Über die Coleopterenfauna in Spitzbergen. *Annales Entomologica Fennici* **39**: 68-70.

Kangas Y. 1955: Beschreibung einer finnischen Xyletinus-Art (Col., Anobiidae). *Annales Entomologici Fennici* **21**: 3-5.

Kangas Y. 1958: Weitere Beiträge zur Kenntnis der nordischen Xyletinus-Arten (Col., Anobiidae). *Annales Entomologici Fennici* **24**: 28-34.

Kapur A. P. 1946: A revision of the genus Jauravia Mots. (Coleoptera, Coccinellidae). *The Annals and Magazine of Natural History* (11) **13**: 73-93.

Kapur A. P. 1948a: A revision of the tribe Aspidimerini Weise (Coleoptera-Coccinellidae). *Transactions of the Royal Entomological Society of London* **99**: 77-128, pl. 1.

Kapur A. P. 1948b: On the Old World species of the genus Stethorus Weise (Coleoptera: Coccinellidae). *Bulletin of Entomological Research* **39**: 297-320.

Kapur A. P. 1948c: The genus Tetrabrachys (Lithophilus) with notes on its biology and a key to the species (Coleoptera, Coccinellidae). *Transactions of the Royal Entomological Society of London* **99**: 319-340.

Kapur A. P. 1950: A new species of Stethorus Weise from India (Coleoptera: Coccinellidae). *The Proceedings of the Royal Entomological Society of London* B **19**: 148-149.

Kapur A. P. 1951: 18. Coleoptera: Coccinellidae. Pp. 275-297, pl. xii-xiv. In: *British Museum (Natural History) Expedition to South-West Arabia 1937-8.* Norwich and London: Jarrold and sons, **1** (16-19): 179-305.

Kapur A. P. 1956a: A new species of Coccinellidae (Coleoptera) predaceous on the citrus white-fly in India. *Records of the Indian Museum* **52** [1954]: 189-193.

Kapur A. P. 1956b: Systematic and biological notes on the lady-beetles predaceous on the San Jose Scale in Kashmir with description of a new species (Coleoptera: Coccinellidae). *Records of the Indian Museum* **52** [1954]: 257-274.

Kapur A. P. 1958: Coccinellidae of Nepal. *Records of the Indian Museum* **53** [1955]: 309-338.

Kapur A. P. 1963a: The Coccinellidae of the third Mount Everest Expedition, 1924 (Coleoptera). *Bulletin of the British Museum of Natural History (Entomology)* **14**: 1-48.

Kapur A. P. 1963b: Some new or little known species of Coccinellidae (Insecta: Coleoptera). Part I - Four new species of Epilachninae from India and Burma. *Records of the Indian Museum* **59**: 131-141.

Kapur A. P. 1963c: The taxonomic status and further description of Harmonia expallida Weise (Col., Coccinellidae), feeding on Adelges species (Hem., Adelgidae) in north-western India. *Entomophaga* **8**: 199-203.

Kapur A. P. 1967: The Coccinellidae (Coleoptera) of the Andamans. *Proceedings of the National Institute of Sciences of India* **32** B [1966]: 148-189.

Kapur A. P. 1969: On some Coccinellidae of the tribe Telsimiini with description of new species from India. *Bulletin of Systematic Zoology* (Calcutta) **1**: 45-56.

Kapur A. P. 1973: The Coccinellidae (Coleoptera) of the Italian expeditions to the Karakoram and Hindu Kush. *Records of the Zoological Survey of India* **67** [1972]: 373-378.

Kapur A. P. & Sudha Rao V. 1962: A new species of lady-bird beetle (Col. Coccinellidae) feeding on Adelges spp. (Hem. Adelgidae) in North-West India. *Entomophaga* **7**: 47-51.

Karaman Z. 1936: Ein balkanischer Vertreter des Koleopteren-Genus Reitteria, Reit. balcanica n. sp. *Glasnik. Bulletin de la Société Scientifique de Skopje* **17**: 131-133.

Karaman Z. 1960: Kopulationsapparat des Genus Thelmatophilus (Cryptophagidae, Col.). Pp. 71-74. *XI. InternationalerKongress für Entomologie, Wien, 17. bis 25. August 1960. Verhandlungen Band I* (*Sektion I bis VI*).

Karapetyan A. P. 1980: Dva novykh vida zhukov-tochil"shchikov iz SSSR (Coleoptera, Anobiidae). *Doklady Akademii Nauk Armyanskoy SSR* **71**: 301-306.

Karner M. 1995: A new genus and two new species of Telephanini, with a redescription of Psammaechidius spinicollis Fairmaire and notes on the genus Psammoecus Latreille (Coleoptera, Silvanidae, Uleiotinae, Telephanini). *Coleoptera* **8**: 3-17.

Karsch F.A.F. 1881: Die Käfer der Rohlfs'schen Afrikanischen Expedition 1878-79. *Berliner Entomologische Zeitschrift* **25**: 41-50, 2 pls.

Kaszab Z. 1947: Beiträge zur Kenntnis der Gattung Anommatus Wesm., mit Beschreibung neuer Arten (Coleoptera, Colydiidae). *Annales Historico-Naturales Musei Nationalis Hungarici* **40**: 259-273.

Kawashima I. & Satou F. 2004: The lampyrid genus Stenocladius (Coleoptera, Lampyridae) from the Okinawa Islands, Middle Ryukyus, Southwest Japan, with description of two new local populations. *Elytra* **32**: 389-403.

Kazantsev S. V. 1988: Concerning red-wing beetles (Coleoptera, Lycidae) and the status of the species Dictyoptera erythroptera Baudi. Pp. 168-170. In: Medvedev L. N. & Striganova B. R. (eds): *Fauna i ekologiya nasekomykh Vietnama* [*The fauna and ecology of insects of Vietnam.*]. Moskva: Nauka, 198 pp.

Kazantsev S. V. 1989: To the knowledge of Palaearctic Cantharidae (Coleoptera). On the genera Bactrocantharis Barovsky, Ancistronycha Maerkel and Islamocantharis Wittmer & Magis of the USSR. *Entomologica Basiliensia* **13**: 239-245.

Kazantsev S. V. 1990: Two new subgenera and a new species of the genus Dictyoptera (Coleoptera, Lycidae) from East Asia. *Vestnik Zoologii* **1**: 10-15.

Kazantsev S. V. 1991: To the knowledge of Oriental Lycidae (Coleoptera). Remarks on Lycidae of Himalaya. *Entomologica Basiliensia* **14**: 153-182.

Kazantsev S. V. 1992a: Novye i maloizvestnye zhuki-myagkotelki (Coleoptera, Cantharidae) s Kavkaza i prilegayushchih territoriy [New and little known soldier-beetles (Coleoptera, Cantharidae) from the Caucasus and adjacent territories]. *Zoologicheskiy Zhurnal* **71** (1): 43-52.

Kazantsev S. V. 1992b: To the knowledge of Oriental Cantharoidea (Coleoptera). On the Lycidae of Himalaya. Part 2. *Entomologica Basiliensia* **15**: 261-266.

Kazantsev S. V. 1992c: Contribution to the knowledge of Palaearctic Cantharoidea (Coleoptera). Notes on Dichelotarsus Motschulsky. *Entomologica Basiliensia* **15**: 267-277.

Kazantsev S. V. 1993a: To the knowledge of Oriental Lycidae (Coleoptera). 3. Notes on Plateros Bourgeois and Ditoneces C. O. Waterhouse. *Russian Entomological Journal* **1** [1992]: 29-31.

Kazantsev S. V. 1993b: Lycides nouveaux ou peu connus de l'Indochine (Coleoptera). *Bulletin du Muséum National d'Histoire Naturelle* (Paris) A (4) **15**: 49-68.

Kazantsev S. V. 1993c: New species of the genus Libnetis (Coleoptera Lycidae) from Indochina and adjacent regions. *Russian Entomological Journal* **2** (3-4): 67-74.

Kazantsev S. V. 1993d: Novye i maloizvestnye vidy zhukov-krasnokrylov (Coleoptera, Lycidae) s Dal'nego Vostoka Rosii i Japonii [New and little known species of the lycid beetles (Coleoptera: Lycidae) from Far East of Russia and Japan]. *Entomologicheskoe Obozrenie* **72**: 100-105.

Kazantsev S. V. 1993e: Novyy rod zhukov-myagkotelok (Coleoptera, Cantharidae) iz Centralnoy Azii [A new genus of soldier-beetles (Coleoptera, Cantharidae) from Central Asia]. *Zoologicheskiy Zhurnal* **72** (7): 154-156.

Kazantsev S. V. 1994a: Dopolneniya k faune zhukov-myagkotelok (Coleoptera, Cantharidae) rossiyskogo Dalnego Vostoka [Supplementary remarks to the fauna of soldier-beetles (Coleoptera, Cantharidae) of the Russian Far East]. *Zoologicheskiy Zhurnal* **73** (3): 11-19.

Kazantsev S. V. 1994b: Obzor vidov Rhagonycha (Coleoptera, Cantharidae) aziatskoy chasti Rossii [A review of Rhagonycha species (Coleoptera, Cantharidae) of the Asian part of Russia]. *Zoologicheskiy Zhurnal* **73** (7-8): 71-100.

Kazantsev S. V. 1994c: The Palaearctic species of the genus Silis Charpentier with the description of Crudosilis n. gen. (Coleoptera, Cantharidae). *Elytron* **8**: 93-115.

Kazantsev S. V. 1995a: Novye dannye po rosiyskoy faune krasnokrylov (Coleoptera, Lycidae). New data on the Russian fauna of Lycidae (Coleoptera). *Zoologicheskiy Zhurnal* **74** (5): 35-44.

Kazantsev S. V. 1995b: A review of the genus Pristolycus Gorham with the description of a new species (Coleoptera, Lampyridae). *Russian Entomological Journal* **4**: 61-62.

Kazantsev S. V. 1995c: A key to Rhagonycha (Coleoptera, Cantharidae) east of the Ural Mountains with the description of a new subgenus. *Entomologica Basiliensia* **18**: 91-98.

Kazantsev S. V. 1995d: New taxons of Palaearctic Malthodes (Coleoptera, Cantharidae). *Entomologica Basiliensia* **18**: 99-108.

Kazantsev S. V. 1996a: A review of Pyropterus subgenus Helcophorus Fairmaire, 1891 with the description of two new Oriental species (Coleoptera, Lycidae). *Elytron* **10**: 81-88.

Kazantsev S. V. 1996b: Obzor myzgkotelok roda Podabrus (Coleoptera, Cantharidae) Rossii [Review of soldier-beetles of the genus Podabrus (Coleoptera, Cantharidae) of Russia]. *Zoologicheskiy Zhurnal* **75** (2): 200-211.

Kazantsev S. V. 1997a: A recent species of the fossil genus Kolibaceum Winkler (Coleoptera, Lycidae) from eastern China. *Entomologica Basiliensia* **20**: 159-164.

Kazantsev S. V. 1997b: Novye vidy Silis s zamechaniyami po zoogeografii roda (Coleoptera, Cantharidae) [New species of Silis with notes on the zoogeography of the genus (Coleoptera, Cantharidae)]. *Zoologicheskiy Zhurnal* **76**: 1214-1217.

Kazantsev S. V. 1998a: Novye palearkticheskye vidy Crudosilis (Coleoptera, Cantharidae) [New Palaearctic species of Crudosilis (Coleoptera, Cantharidae)]. *Zoologicheskiy Zhurnal* **77**: 285-294.

Kazantsev S. V. 1998b: Novye rossiyskie vidy Dichelotarsus Motschulsky (Coleoptera, Cantharidae) [New Russian species of Dichelotarsus Motschulsky (Coleoptera, Cantharidae)]. *Zoologicheskiy Zhurnal* **77**: 572-575.

Kazantsev S. V. 1998c: O komplekse Podistra Motschulsky (Coleoptera, Cantharidae) Rossii i prilegayushchihk territoriy [On Podistra Motschulsky complex (Coleoptera, Cantharidae) of Russia and adjacent territories]. *Zoologicheskiy Zhurnal* **77**: 689-695.

Kazantsev S. V. 1998d: Novyy vid zhukov-myagkotelok iz roda Bactrocantharis (Coleoptera, Cantharidae) [New species of soldier-beetles from the genus Bactrocantharis (Coleoptera, Cantharidae)]. *Zoologicheskiy Zhurnal* **77**: 865-866.

Kazantsev S. V. 1999: Revision of Lycocerus Gorham of Indochina and adjacent regions, with the description of Walteria, a new genus (Coleoptera, Cantharidae). *Entomologica Basiliensia* **21**: 115-134.

Kazantsev S. V. 2000a: New lycids from China (Coleoptera, Lycidae). *Miscellànea Zoològica* **23**: 79-92.

Kazantsev S. V. 2000b: On Dilophotes Waterhouse, 1879 and Mimodilophotes Pic, 1939 (Coleoptera, Lycidae). *Russian Entomological Journal* **9**: 329-332.

Kazantsev S. V. 2001a: New species of the genus Melaneros Fairmaire, 1879 from Nepal, China and the USA (Coleoptera, Lycidae). *Russian Entomolological Journal* **10**: 13-16.

Kazantsev S. V. 2001b: Macrolycus Waterhouse, 1878 (Coleoptera, Lycidae) of continental China. *Elytron* **14** [2000]: 99-109.

Kazantsev S. V. 2001c: New cantharid genus from North America and Far East Asia (Coleoptera: Cantharidae). *Elytron* **15**: 43-48.

Kazantsev S. V. 2001d: Novye taksony zhukov-myagkotelok (Coleoptera, Cantharidae) iz Srednej Azii i Kavkaza [New taxa of soldier-beetles (Coleoptera, Cantharidae) from Central Asia and the Caucasus]. *Zoologicheskiy Journal* **80** (9): 1076-1082.

Kazantsev S. V. 2002a: Supplementary notes to the revision of the genus Lyponia Waterhouse, 1878 (Coleoptera: Lycidae) with description of new taxa. *Russian Entomogical Journal* **11**: 197-206.

Kazantsev S. V. 2002b: New and little known species of Lycidae (Coleoptera) from China. *Russian Entomological Journal* **11**: 253-263.

Kazantsev S. V. 2003: A generic review of Duliticolinae, new subfamily (Coleoptera: Lycidae). *Elytron* **16** [2002]: 5-21.

Kazantsev S. V. 2004: Phylogeny of the tribe Erotini (Coleoptera: Lycidae), with descriptions of new taxa. *Zootaxa* **496**: 1-48.

Kazantsev S. V. &Takahashi K. 2001: Eight new species of the genus Rhagonycha (Coleoptera, Cantharidae) from Hokkaido, Japan. *Japanese Journal of Systematic Entomology* **7**: 269-277.

Kazantsev S. V. & Yang P. S. 1999: Taxonomic and synonymic notes on Lycidae (Coleoptera) with description of new species from Taiwan. *Chinese Journal of Entomology* **19**: 239-247.

Kaszab Z. 1955: Neue und wenig bekannte Malacodermata (Coleoptera) aus dem Karpatenbecken. *Acta Zoologica Academiae Scientiarum Hungaricae* **1**: 289-307.

Kaszab Z. & Székessy V. 1953: [new taxon]. Pp. 194-285. Bátorliget bogár-faunája Coleoptera. In: Székessy V. (ed.): *Bátorliget Élővilága*. Budapest: Akadémiai Kiando.

Katakura H. 1981: Classification and evolution of the phytophagous ladybirds belonging to Henosepilachna vigintioctomaculata complex (Coleoptera, Coccinellidae). *Journal of the Faculty of Science Hokkaido University, Series VI, Zoology* **22**: 301-378.

Kaufmann J. 1883: Eine neue Hylaia aus Dalmatien. *Wiener Entomologische Zeitung* **2**: 10.

Kawashima I. 1999: The lampyrid beetles of the genus Stenocladius (Coleoptera, Lampyridae) of the Ryukyu Islands, Southwest Japan, with description of two new species. *Elytra* **27**: 141-158.

Kawashima I. 2000: A redescription of Rhagophthalmus ingens (Coleoptera, Rhadophthalmidae) from northern Vietnam, with establishment of a new genus in the family. *Elytra* **28**: 131-140.

Kawashima I. & Satô M. 2001: Three new species of the genus Rhagophthalmus (Coleoptera, Rhagophthalmidae) from South East Asia. *Elytra* **29**: 423-434.

Kawashima I. & Satou F. 2004: The lampyrid genus Stenocladius (Coleoptera, Lampyridae) from the Okinawa Islands, Middle Ryukyus, Southwest Japan, with description of two new local populations. *Elytra* **32**: 389-403.

Kawashima I. & Sugaya H. 2003: An additional new species of the genus Rhagophthalmus (Coleoptera, Rhagophthalmidae) from Taiwan, with a key to the males of the Taiwanese and Japanese species. *Elytra* **31**: 353-359.

Kawashima I., Suzuki H. & Satô M. 2003: A check-list of Japanese fireflies (Coleoptera, Lampyridae and Rhagophthalmidae). *Japanese Journal of Systematic Entomology* **9**:241-261.

Kemner A. 1914: Ein Fall von Prothetelie nebst Bemerkungen über pränymphoide Studien in der Käfer Entwicklung. *Entomologisk Tidskrift* **35**: 87-95.

Kessel F. 1921: Neue Monotomiden, Cucujiden u. Passandriden aus der Sammlung des Deutschen Entomologischen Museums in Berlin. *Archiv für Naturgeschichte* (A) **87**: 25-33.

Kessel F. 1926: Synopse general do género Laemophloeus Cast. (Col.) com a descripção de algumas novas especies sul-americanas. *Archivos do Museu Nacional do Rio de Janeiro* **26**: 59-93.

Kiesenwetter E. A. H. von. 1850: Fünfzig Diagnosen unbeschriebener oder wenig bekannter europäischer Käfer. *Entomologische Zeitung* (Stettin) **11**: 217-225.

Kiesenwetter E. A. H. von. 1852: Énumération des coléoptères trouvés dans le midi de la France et en Catalogne. II. Partie. *Annales de la Société Entomologique de France* (2) **9** [1851]: 577-656, 1 pl.

Kiesenwetter E. A. H. von. 1858a: *Naturgeschichte der Insecten Deutschland. Erste Abtheilung. Coleoptera. Vierter Band.* Berlin: Nicolaische Verlagsbuchhandlung [1857-1863], vi + 746 + [1] pp. [issued in parts, pp. 1-178 in 1857; pp. 179-386 in 1858; pp. 387-570 in 1860; pp. 571-745 + [1] in 1863].

Kiesenwetter E. A. H. von. 1858b: [new taxa]. In: Kraatz G.: Beitrag zur Käferfauna Greichenlands. Drittes Stück: Staphylinidae (Schlufs), Trichopterygia, Histeridae, Phalacridae, Nitidulariae, Trogasitarii, Colydii, Cucujidae, Cryptophagidae, Thorictidae, Mycetophagidae, Dermestini, Byrrhii. *Berliner Entomologische Zeitschrift* **2**: 123-148.

Kiesenwetter E. A. H. von. 1859a: Beitrag zur Käferfauna Griechenlands. Fünftes Stück: Elateridae, Dascillidae, Malacodermata. *Berliner Entomologische Zeitschrift* **3**: 17-34.

Kiesenwetter E. A. H. von. 1859b: [new taxa]. In: Schaum H.: Beitrag zur europäischen Käferfauna. *Berliner Entomologische Zeitschrift* **3**: 42-59.

Kiesenwetter E. A. H. von. 1859c: Beitrag zur Käferfauna Griechenlands. Sechtes Stück: Malacodermata, Cleridae, Ptinidae, Anobiadae. *Berliner Entomologische Zeitschrift* **3**: 158-192, pls 2-3.

Kiesenwetter E. A. H. von. 1860: *Naturgeschichte der Insecten Deutschland. Erste Abtheilung. Coleoptera. Vierter Band.* Berlin: Nicolaische Verlagsbuchhandlung, vi + 746 + [1] pp. [issued in parts, pp. 1-178 in 1857; pp. 179-386 in 1858; pp. 387-570 in 1860; pp. 571-745 + [1] in 1863].

Kiesenwetter E. A. H. von. 1861: Eine entomologische Excursion in das Wallis und nach dem Monte Rosa in Sommer 1861. *Berliner Entomologische Zeitschrift* **5**: 360-395.

Kiesenwetter E. A. H. von. 1863a: *Naturgeschichte der Insecten Deutschland. Erste Abtheilung. Coleoptera. Vierter Band.* Berlin: Nicolaische Verlagsbuchhandlung, vi + 746 + [1] pp. [issued in parts, pp. 1-178 in 1857; pp. 179-386 in 1858; pp. 387-570 in 1860; pp. 571-745 + [1] in 1863].

Kiesenwetter E. A. H. von. 1863b: Weitere Beiträge zur Kenntnis der europäischen Malthiniden. *Berliner Entomologische Zeitschrift* **7**: 428-431.

Kiesenwetter E. A. H. von. 1864: Neue Arten der Gattungen Antidipnis und Dasytes. *Berliner Entomologische Zeitschrift* **8**: 387-389, pls 4-5.

Kiesenwetter E. A. H. von. 1866a: Eine entomologische Excursion nach Spanien in Sommer 1865. *Berliner Entomologische Zeitschrift* **9** [1865]: 359-396.

Kiesenwetter E. A. H. von. 1866b: Beiträge zur Käferfauna Spaniens. (Erstes Stück.) Malacodermata, Melyridae. *Berliner Entomoloische Zeitschrift* **10**: 241-274, pl. 1.

Kiesenwetter E. A. H. von. 1867a: [new taxa]. In: Seidlitz G.: Beitrag zur Käferfauna Siebenbürgens. *Verhandlungen und Mittheilungen des Siebenbürgischen Vereins für Naturwissenschaften zu Hermannstadt* **18**: 43-46.

Kiesenwetter E. A. H. von. 1867b: Beiträge zur Käferfauna Spaniens. (Zweiter Stück). Melyridae (Fortsetzung), Ptinidae. *Berliner Entomologische Zeitschrift* **11**: 109-134, 1 pl.

Kiesenwetter E. A. H von. 1867c: Revision der Dasytidengattung Dolichosoma. *Berliner Entomologische Zeitschrift* **11**: 136-140.

Kiesenwetter E. A. H von. 1870: [new taxa]. In: Heyden L. F. J. D. von: *Entomologische Reise nach dem Südlichen Spanien, der Sierra de Guadarrama und Sierra Morena, Portugal und den Cantabrischen Gebirgen, mit Beschreibungen der neuen Arten.* Berlin: Berliner entomologischer Verein (*Supplement, Berliner Entomologische Zeitschrift* **14**): [1] + 218 pp., 2 pls.

Kiesenwetter E. A. H. von. 1871: Beiträge zur Kenntniss der Malacodermen-Fauna von Corsica, Sardinien und Sicilien. *Beliner Entomologische Zeitschrift* **15**: 75-86.

Kiesenwetter H. A. H. von. 1872a: Übersicht der Arten der Gattung Merophysia. *Berliner Entomologische Zeitschrift* **16**: 163-166.

Kiesenwetter E. A. H. von. 1872b: Revision der Gattung Cerallus. *Berliner Entomologische Zeitschrift* **16**: 314-318.

Kiesenwetter E. A. H. von. 1872c: Revision der europäischen Arten der Gattung Malthodes (Erstes Stück). *Berliner Entomologische Zeitschrift* **16**: 369-392.

Kiesenwetter E. A. H. von. 1873: Zwei neue Aphyctus-Arten. *Berliner Entomologische Zeitschrift* **17**: 31-32.

Kiesenwetter H. von. 1874a: Revision der europäischen Arten der Gattung Malthodes (Zweites Stück). *Berliner Entomologische Zeitschrift* **18**: 45-70.

Kiesenwetter E. A. H. von. 1874b: Die Malacodermen Japans nach dem Ergebnisse der Sammlungen des Herrn G. Lewis während der Jahre 1869-1871. *Berliner Entomologische Zeitschrift* **18**: 241-288.

Kiesenwetter E. A. H. von. 1877: Anobiidae und Cioidae. Pp. 1-200. In: Kiesenwetter E. A. H. von & Seidlitz G. 1877-1898: *Naturgeschichte der Insecten Deutschlands. Erste Abtheilung. Coleoptera. Fünfter Band. Erste Hälfte.* Berlin: Nicolaische Verlags-Buchhandlung R. Stricker, 877 pp.

Kiesenwetter E. A. H. von. 1878: [new taxa]. In: Schneider O. & Leder H.: Beiträge zur Kenntniss der kaukasischen Käferfauna. *Verhandlungen des Naturforschenden Vereins in Brünn* **16** [1877]: 3-258, pls 1-4.

Kiesenwetter E. A. H. von. 1879a: Neue Amur-Käfer. *Deutsche Entomologische Zeitschrift* **23**: 145-146.

Kiesenwetter E. A. H. von. 1879b: Coleoptera japonica collecta a Domino Lewis et aliis. *Deutsche Entomologische Zeitschrift* **23**: 305-320.

Kieseritzky V. & Reichardt A. 1936: K voprosu o samostojatelnosti semeystva Catopochrotidae (Coleoptera) v sviazi s nakhozhdeniem novogo roda. *Trudy Zoologicheskogo Instituta Akademii Nauk SSSR* **3**: 693-697.

Kim J.-I., Han T. M. & Lee S. H. 2000: Three new species of the genus Oedostethus (Coleoptera, Elateridae, Negastriinae) in Korea. *The Korean Journal of Systematic Zoology* **16**: 69-77.

Kim J.-I. & Kang T.-H. 2000: Taxonomic study of Korean Cantharidae (Coleoptera). I. Silinae, Malthininae and Chauliognathinae. *Insecta Koreana* **17**: 111-120.

Kinoda T. & Nakane T. 1988: [new taxon]. In: Nakane T.: Notes on some little-known beetles (Coleoptera) in Japan. 3. *Kita-Kyūshū no Konchū* **35**: 77-79, pl. 9.

Kirby W. 1819: A century of insects, including several new genera described from his cabinet. *Transactions of the Linnean Society of London* **12** (2) [1818]: 375-453, pls.XXII-XXIII.

Kirby W. 1830: [new taxon]. In: Stephens J. F.: *Illustrations of british entomology or, a synopsis of indigenous insects: containing their generic and specific distinctions;with an account of their metamorphoses, times of appearance, localities, food, and economy, as far as practicable. Mandibulata. Volume III.* London: Baldwin and Cradock, 374 + 5 pp., pls XVI-XIX.

Kirby W. 1837: The Insects. Coleoptera. In: Richardson J. (ed.): *Fauna Boreali-Americana, or the Zoology Northern parts of British America: containing descriptions of the objects of Natural history collected on the late Northern Land Expeditions, under command of Captain Sir John Franklin, R. N. J. Fletcher, Part the fourth and last.* Norwich: Josian Fletcher, xxxix + 325 + [2] pp., 8 pls.

Kirby W. & Spence W. 1828: *An introduction to entomology or elements of the natural history of insects: comprising an account of noxious and useful insects, of their metamorphosis, food, stratagems, habitations, societies, motions, noises, hybernation, instinct, etc. etc. Edit. 5. Volume 2.* London: Longman, Orme, Brown and Green, 523 pp.

Kirejtshuk A. G. 1977a: Novye i maloizvestnye vidy blestyanok roda Meligethes Steph. (Coleoptera, Nitidulidae) iz Sredney Azii [New and little-known species of the genus Meligethes Stephens (Coleoptera, Nitidulidae) from Middle Asia]. *Trudy Zoologicheskogo Instituta Akademii Nauk SSSR* **1977**: 42-49.

Kirejtshuk A. G. 1977b: Novye i maloizvestnye blestyanki podsem. Meligethinae (Coleoptera, Nitidulidae) fauny palearktiki [New and little known species of subfamily Meligethinae (Coleoptera, Nitidulidae) in the Palaearctic fauna]. *Entomologicheskoe Obozrenie* **56**: 625-643.

Kirejtshuk A. G. 1978: Novye vidy zhukov-blestyanok roda Meligethes Stephens (Coleoptera, Nitidulidae) iz SSSR i obzor gruppy vidov, blizkikh k M. umbrosus Sturm [New species of the genus Meligethes Stephens (Coleoptera, Nitidulidae) from the USSR and review of the group of species related to Meligethes umbrosus Sturm]. *Entomologicheskoe Obozrenie* **57**: 578-595.

Kirejtshuk A. G. 1979a: Dva novykh vida zhukov-blestyanok roda Meligethes Stephens (Coleoptera, Nitidulidae) s Dal′nego Vostoka [Two new species of the genus Meligethes Stephens (Coleoptera, Nitidulidae) from the Far East]. *Trudy Zoologicheskogo Instituta Akademii Nauk SSSR* **81** [1978]: 66-75.

Kirejtshuk A. G. 1979b: Novyi vid zhukov-blestyanok roda Meligethes Stephens (Coleoptera, Nitidulidae) iz vostochnogo Kazakhstana [A new species of beetles of the genus Meligethes Stephens (Coleoptera, Nitidulidae) from eastern Kazakhstan]. *Trudy Zoologicheskogo Instituta Akademii Nauk SSSR* **83**: 39-42.

Kirejtshuk A. G. 1979c: Novye vidy zhukov-blestyanok podsem. Meligethinae (Coleoptera, Nitidulidae) iz Aziatskoi chasti SSSR i sopredel′nykh territorii [New species of subfamily Meligethinae (Coleoptera, Nitidulidae) from the Asian part of the USSR and adjacent territories]. *Trudy Zoologicheskogo Instituta Akademii Nauk SSSR* **88**: 50-68.

Kirejtshuk A. G. 1979d: Dva novykh roda i novye vidy zhukov-blestyanok podsem. Meligethinae (Coleoptera, Nitidulidae) iz V′etnama [Two new genera and new species of the subfamily Meligethinae (Coleoptera, Nitidulidae) from Vietnam]. *Entomologicheskoe Obozrenie* **58**: 355-368.

Kirejtshuk A. G. 1979e: Novyy vid roda Meligethes Stephens (Coleoptera; Nitidulidae) iz severo-vostochnoi Turcii [New species of the genus Meligethes Stephens (Coleoptera; Nitidulidae) from north-eastern Turkey]. *Doklady Akademii Nauk Ukrainskoi SSR, Seria B. Geologicheskie, khimicheskie i biologicheskie nauki* **1979** (8): 679-682.

Kirejtshuk A. G. 1980a: A new species of the erythropus group of the genus Meligethes Steph. from Algeria (Coleoptera, Nitidulidae, Meligethinae). *Fragmenta Entomologica* **15**: 353-355.

Kirejtshuk A. G. 1980b: Novye vidy zhukov-blestyanok podsem. Meligethinae (Coleoptera; Nitidulidae) iz Oriental′noi oblasti i sopredel′nykh territorii [New species of Meligethinae (Coleoptera, Nitidulidae) from the Oriental region and adjacent territories]. *Entomologicheskoe Obozrenie* **59**: 833-851.

Kirejtshuk A. G. 1982: Novyy vid zhukov-blestyanok roda Meligethes Stephens (Coleoptera; Nitidulidae) i reviziia gruppy planiusculus fauny Sredney Azii i Kazakhstana [New species of the genus Meligethes Stephens (Coleoptera; Nitidulidae) and revision of the planiusculus species group of the fauna of Middle Asia and Kazakhstan]. *Trudy Zoologicheskogo Instituta Akademii Nauk SSSR* **110**: 35-39.

Kirejtshuk A. G. 1984a: New taxa of Nitidulidae (Coleoptera) from the Indo-Malayan fauna. *Annales Historico-Naturales Musei Nationalis Hungarici* **76**: 169-195.

Kirejtshuk A. G. 1984b: Novye vidy zhukov semeistv Nitidulidae i Cybocephalidae (Coleoptera) fauny vostochnoy Palearktiki [New species of beetles of the families Nitidulidae and Cybocephalidae (Coleoptera) of the East Palaearctic fauna]. *Zoologicheskiy Zhurnal* **63**: 517-531.

Kirejtshuk A. G. 1985: Novye vidy Cyllodes Erichson i Viettherchnus gen. n. (Coleoptera, Nitidulidae) fauny V'etnama i sopredel'nykh territorii. Pp. 157-164. In : Medvedev L. N. (ed.): *Nasekomye V'etnama*. Moskva: Nauka, 182 pp.

Kirejtshuk A. G. 1986a: Analiz stroeniza genitalii dlya rekonstrukcii filogenii i obosnovaniya sistemy sem. Nitidulidae (Coleoptera) [Analysis of the morphology of genitalia for the reconstruction of phylogeny and the base of the system of family Nitidulidae (Coleoptera)]. *Trudy Vsesoyuznogo Entomologicheskogo Obshchestva* **68**: 22-28.

Kirejtshuk A. G. 1986b: Reviziya roda Aethina Er. (Coleoptera, Nitidulidae) fauny oriental'noi i palearkticheskoi oblastey [Revision of the genus Aethina Er. (Coleoptera, Nitidulidae) of the fauna of Oriental and Palaearctic regions]. *Trudy Zoologicheskogo Instituta Akademii Nauk SSSR* **140**: 44-82.

Kirejtshuk A. G. 1987a: Novye taksony zhukov-blestyanok (Coleoptera, Nitidulidae) vostochnogo polushariya (Chast'1) Omosita nearctica sp. n., vikariruyushchii s palearkticheskim O. colon (L.) [New taxa of nitidulid beetles (Coleoptera, Nitidulidae) of eastern hemisphere (Part 1). Omosita nearctica sp. n., vicariant of Palaearctic O. colon (L.)]. *Trudy Zoologicheskogo Instituta Akademii Nauk SSSR* **164**: 63-94.

Kirejtshuk A. G. 1987b: Obzor zhukov-blestyanok podsem. Cryptarchinae (Coleoptera, Nitidulidae) Indo-Malayskoy oblasti [Review of nitidulid beetles of subfamily Cryptarchinae (Coleoptera, Nitidulidae) of Indomalyan region]. *Trudy Zoologicheskogo Instituta Akademii Nauk SSSR* **170**: 62-95.

Kirejtshuk A. G. 1987c: Novye vidy kompleksa rodov, blizkikh k Cyllodes Erichson (Coleoptera, Nitidulidae) iz Indokitaya i sopredel'nykh stran [New species of the complex of genera related to Cyllodes Erichson (Coleoptera, Nitidulidae) from Indochina and adjacent countries]. Pp. 137-170. In: Medvedev L. N. (ed.): *Entomofauna V'etnama*. Moskva: Nauka, 192 pp.

Kirejtshuk A. G. 1988: Novye taksony zhukov-blestyanok (Coleoptera, Nitidulidae) vostochnogo polushariya. Chast'2 [New taxa of nitidulid beetles (Coleoptera, Nitidulidae) of Eastern Hemisphere. Part 2.]. *Trudy Zoologicheskogo Instituta Akademii Nauk SSSR* **178**: 62-97.

Kirejtshuk A. G. 1989a: Novye taksony zhukov-blestyanok (Coleoptera, Nitidulidae) vostochnogo polushariya (Chast' III) [New taxa of the Nitidulidae (Coleoptera) of the East Hemisphaera (Part III)]. *Trudy Zoologicheskogo Instituta Akademii Nauk SSSR* **208**: 64-89.

Kirejtshuk A. G. 1989b: Novye palearkticheskie rod i vidy semeistva Kateretidae (Coleoptera) i zamechaniya po sinonimii [New Palaearctic genus and species of the family Kateretidae (Coleoptera) and notes on the synonymy]. *Zoologicheskiy Zhurnal* **68**: 145-149.

Kirejtshuk A. G. 1990a: Novye vidy i zamechaniya po taxonomii zhukov-blestyanok (Coleoptera, Nitidulidae) Indokitaia i sopredel'nykh territorii. Chast'1 [New species and notes on the taxonomy of nitidulid beetles (Coleoptera, Nitidulidae) of Indochina and adjacent territories. Part 1.]. *Trudy Zoologicheskogo Instituta Akademii Nauk SSSR* **209**: 61-98.

Kirejtshuk A. G. 1990b: Novye taksony zhukov-blestyanok vostochnogo polushariya. Chast'4 [New taxa of the Nitidulidae (Coleoptera) of the eastern Hemisphere, part 4]. *Trudy Zoologicheskogo Instituta Akademii Nauk SSSR* **211**: 84-103.

Kirejtshuk A. G. 1992: 59., 61. Sem. Nitidulidae – Blestyanki. Pp. 114-209; 60. Sem. Kateretidae – Kateretidy. Pp. 210-216. In: Ler P. A. (ed.): *Opredelitel'nasekomykh Dal'nego Vostoka SSSR v shesti tomakh. Tom III. Zhestkokrylye, ili zhuki.* Sankt-Petersburg: Nauka, 704 pp.

Kirejtshuk A. G. 1994a: Novye vidy i zamechaniya po taxonomii zhukov-blestyanok (Coleoptera, Nitidulidae) Indokitaya i sopredel'nykh territorii. Chast'2 [New species and notes on taxonomy of nitidulid beetles (Coleoptera, Nitidulidae) of Indochina and adjacent territories. Part 2.]. *Trudy Zoologicheskogo Instituta Rossiyskoi Akademii Nauk* **257**: 92-127.

Kirejtshuk A. G. 1994b: Revision of the genus Neopallodes Reitter 1884 (Coleoptera, Nitidulidae) from the Palaearctic and Indo-Malayan regions. *Tropical Zoology* **7**: 225-253.

Kirejtshuk A. G. 1995a: Novye taksony zhukov-blestyanok (Coleoptera, Nitidulidae) vostochnogo polushariya. Chast'5 [New taxa of the Nitidulidae (Coleoptera) of the Eastern Hemisphere. Part 5]. *Trudy Zoologicheskogo Instituta Rossiyskoy Akademii Nauk* **258**: 3-50.

Kirejtshuk A. G. 1995b: Epuraea (Epuraea) dolosa sp. n. and notes on taxonomy of some Palaearctic species of the genus Epuraea (Coleoptera: Nitidulidae). *Zoosystematica Rossica* **3**: 279-282.

Kirejtshuk A. G. 1997: New Palaearctic nitidulid beetles, with notes on synonymy and systematic position of some species (Coleoptera: Nitidulidae). *Zoosystematica Rossica* **6**: 255-268.

Kirejtshuk A. G. 1998: *Nitidulidae (Coleoptera) of the Himalayas and northern Indochina*. Koenigstein: Koeltz Scientific Books, 489 pp.

Kirejtshuk A. G. & Gillerfors G. 1987: Brachypeplus pudicus sp. n., a new species of the genus Brachypeplus Erichson from the Azores (Coleoptera: Nitidulidae). *Bocagiana* **114**: 1-5.

Kirejtshuk A. G. & Lawrence J. F. 1999: Notes on the Aethina complex (Coleoptera: Nitidulidae: Nitidulinae), with a review of Aethina (Cleidorura) subgen. n. and Aethina (Idaethina Gemminger et Harold). *Annales Zoologici* **49**: 233-254.

Kirsch T. F. V. 1871: Neue Käfer-Arten aus Egypten, gesammelt von Dr. Schneider. *Berliner Entomologische Zeitschrift* **14** [1870-1871]: 389-396.

Kirsch T. F. V. 1874: Beiträge zur Kenntniss der Peruanischen Käferfauna auf Dr. Abendroth's Sammlungen basirt (II.). *Berliner Entomologische Zeitschrift* **17** [1873]: 339-418.

Kirsch T. F. V. 1875: Neue Käfer aus Malacca. *Mitteilungen aus dem Zoologischen Museum in Dresden* **1**: 25-58.

Kirsch T. F. W. 1889: Coleopteren gesammelt in den Jahren 1868-1877 auf einer Reise durch Süd Amerika von Alphons Stübel. *Abhandlungen und Berichte des Königlichen Zoologischen und Anthropologisch-Ethnographischen Museums zu Dresden* **4**: 1-58.

Kirchsberg O. V. von. 1898: Phyllocerus Ullmanni n. sp. *Verhandlungen der Kaiserlich-Königlichen Zoologisch-Botanischen Gesellschaft in Wien* **47** [1897]: 677-678.

Kishida K. 1934: Entomological World of Jehol, Manchoukus. Lampyridae of Jehol. *Kontyû* **8**: 116. (in Japanese).

Kishida K. 1936: Family Cantharidae. (5) **1**, Art. 46: 13-21. In: Insects of Jehol (VI) – Order Coleoptera (I). *Report of the 1st Scientific Expedition to Manchoukuo. Under the Leadership of Shigeyasu Tokunaga June October 1933*.

Kishii T. 1955a: Rare elaterid-beetles from Kibune and the adjacent area with descriptions of two new forms (I). *Akitu* **4**: 19-22.

Kishii T. 1955b: Some new forms of Elateridae in Japan (I). *Akitu* **4**: 77-82.

Kishii T. 1955c: Rare elaterid-beetles from Kibune and the adjacent Area (II). *Akitu* **4**: 109-110.

Kishii T. 1956: Some new forms of Elateridae in Japan (II). *Akitu* **5**: 17-20.

Kishii T. 1957a: Some new forms of Elateridae in Japan (III). *Entomological Revue of Japan* **8**: 10-12, pl. 3.

Kishii T. 1957b: [new taxa]. In: Baba K. & Kishii T.: Snappers of Niigata-Prefecture, on the collection in 1956, with he [sic] descriptions of some new forms. *Akitu* **6**: 67-75.

Kishii T. 1957c: Entomological results from the biological survey of the Nachi virgin forest. (II) Elateridae of Nachi, Japan, with description of a new species (Coleoptera). *Akitu* **6**: 84-89.

Kishii T. 1958: Snappers from Kyûshû-District, Japan, collected by Prof. Takashi Shirôzu. *Entomological Revue of Japan* **9**: 27-32.

Kishii T. 1959a: Elateridae of Is. Yakushima. The Snappers of Island (I). *Bulletin of the Heian High School, Kyôto* **3**: 1-24.

Kishii T. 1959b: Elateridae of Amami-Oshima collected by Prof. T. Shirôzu. *Akitu* **8**: 57-66.

Kishii T. 1961: Elateridae of Is. Tsushima. The Snappers of Island (II). *Bulletin of the Heian High School, Kyôto* **5**: 1-56.

Kishii T. 1962: Elateridae of Islands Rishiri-tô, Rebun-tô and Todo-jima. The Snappers of Islands (III). *Bulletin of the Heian High School, Kyôto* **7**: 1-31.

Kishii T. 1963: Snappers from Hyogo Prefecture, Japan, collected by Mr. Toshiro Takahashi. *Akitu* **11**: 9-13.

Kishii T. 1964: Elateridae of Islands Awa-shima, Hegura-jima and Nanatsu-jima. The Snappers of Islands (IV). *Bulletin of the Heian High School, Kyôto* **8**: 1-38.

Kishii T. 1966a: *Elateridae of Kyoto and its adjacent regions*. Kyôto: Biological Laboratory, Heian High School, 54 pp.

Kishii T. 1966b: Some new forms of Elateridae in Japan (IV). *Bulletin of the Heian High School, Kyôto* **10**: 1-11.

Kishii T. 1968: Some new forms of Elateridae in Japan (V). *Bulletin of the Heian High School, Kyôto* **13**: 1-15.

Kishii T. 1969: Some new forms of Elateridae in Japan (VI). *Bulletin of the Heian High School, Kyôto* **14**: 1-10.

Kishii T. 1970: On the genus Yukoana and Quasimus from Japan. Some new forms of Elateridae in Japan (VII). *Bulletin of the Heian High School, Kyôto* **15**: 1-28.

Kishii T. 1972: A list of elaterid-fauna from Ryûkyû Archipelago collected by the Society of Scientific Survey, Kinki University, with the descriptions of some new forms. The Snappers of Island (V). *Bulletin of the Heian High School, Kyôto* **16**: 1-12.

Kishii T. 1973: On the elaterid-beetles from Ryûkyû Archipelago collected by Mr. Isao Matoba. The Snappers of Island (VI). *Bulletin of the Heian High School, Kyôto* **17**: 1-24.

Kishii T. 1974: Elaterid-beetles from the Ryûkyû Archipelago, collected by Messrs. I. Matoba, O. Tamura and T. Takahashi in the spring of 1973, with some new forms and notes. The Snappers of Island (VII). *Bulletin of the Heian High School, Kyôto* **18**: 1-19.

Kishii T. 1975: Some new forms of Elateridae in Japan (VIII). *Bulletin of the Heian High School, Kyôto* **19**:1-7.

Kishii T. 1976a: Some new forms of Elateridae in Japan (IX). *The Scientific Reports of Kyôto Prefecture, Kyôto* **14**: 7-19.

Kishii T. 1976b: New Negastriinae with some notes. Some new forms of Elateridae in Japan (X). *Bulletin of the Heian High School, Kyôto* **20**: 17-46.

Kishii T. 1976c: Some new forms of Elateridae in Japan (XI). *Bulletin of the Heian High School, Kyôto* **20**: 47-56.

Kishii T. 1977a: Some new forms of Elateridae in Japan (XII). *The Scientific Reports of Kyôto Prefecture* **15**: 49-61.

Kishii T. 1977b: Elaterid beetles from Europe collected by Mr. A. Shinohara with descriptions of some new forms and notes. *Bulletin of the Heian High School* **21**: 19-34.

Kishii T. 1978: A study on the genus Actenicerus Kiesenwetter from Japan and its adjacent area (1). Some new forms of Elateridae in Japan (XIII). *Bulletin of the Heian High School* **22**: 17-29.

Kishii T. 1979a: Some new forms and records of Elateridae in Japan and its adjacent area. Some new forms of Elateridae in Japan (XIV). *Annual Report of Private Schools in Kyôto Prefecture* **17**: 1-21.

Kishii T. 1979b: Some new forms and notes of elateridae. Some new forms of Elateridae in Japan (XV). *Bulletin of the Heian High School* **23**: 31-43.

Kishii T. 1981: Notes on Elateridae from Japan and its adjacent area (1), with the description of a new species (Coleoptera). *Bulletin of the Heian High School* **25**: 17-25.

Kishii T. 1982: New species and new records of Elateridae from Japan (Coleoptera). Some new forms of Elateridae in Japan (XVI). *Bulletin of the Heian High School* **26**: 35-55.

Kishii T. 1983a: Some elaterid beetles from the Nansei Archipelago collected by Mr. T. Ogata in 1982 (Coleoptera, Elateridae). Notes on Elateridae of Japan and its adjacent area (2). *Entomological Review of Japan* **38**: 29-40.

Kishii T. 1983b: On some elaterid-species from Japan, with the descriptions of two new taxa (Coleoptera). Notes on Elateridae from Japan and its adjacent area (3). *Bulletin of the Heian High School* **27**: 47-64.

Kishii T. 1984: The elaterid-beetles of the tribe Agriotini from Japan (Coleoptera: Elateridae, Elaterinae). *Bulletin of the Heian High School* **28**: 11-77.

Kishii T. 1985: Some new forms of Elateridae in Japan (XVII). *Bulletin of the Heian High Schoo* **29**: 1-30.

Kishii T. 1986: Some new forms of Elateridae in Japan (XVIII). *Bulletin of the Heian High School* **30**: 37-56.

Kishii T. 1987: Some new forms of Elateridae in Japan (XIX). *Bulletin of the Heian High School* **31**: 1-20.

Kishii T. 1988a: A study of Ampedus (Ampedus) ivanovi (Jakobson, 1913) and its allied species from Japan, with descriptions of some new taxa (Some new forms of Elateridae in Japan, XXI). *Bulletin of the Heian High School* **32**: 11-20.

Kishii T. 1988b: A new Melanotus-species from Is. Mikura-jima in Izu Archipelago (Coleoptera: Elateridae). Some new forms of Elateridae in Japan (XX). *Entomological Review of Japan* **43**: 103-106.

Kishii T. 1988c: Some click-beetles from the Nansei Islands collected by Mr. T. Ogata (Coleoptera: Elateridae). Notes on Elateridae from Japan and its adjacent area (7). *Entomological Review of Japan* **43**: 123-134.

Kishii T. 1988d: On some elaterids collected by Mr. Shigeto Yamaya. *Entomological Review of Japan* **43**: 179-184.

Kishii T. 1989a: Elateridae from Taiwan, with descriptions of some new taxa (1) (Coleoptera). A study of the materials collected by Dr. Kintaro Baba in 1986. *Transactions of Essa Entomological Society Niigata* **67**: 19-34 (in Japanese with English descriptions).

Kishii T. 1989b: Elateridae from Taiwan, with descriptions of some new taxa (2) (Coleoptera). A study of the materials collected by Dr. Kintaro Baba in 1986. *Transactions of Essa Entomological Society Niigata* **67**: 35-63 (in Japanese with English descriptions).

Kishii T. 1989c: Elaterid-beetles from Mt. Houwau-zan and Gozaishi-kôsen Spa in Yamanashi Prefecture, collected by Mr. Kôichi Hosoda in 1987 and 1988 (Coleoptera: Elateridae). Notes on Elateridae from Japan and its adjacent area (8). *Bulletin of the Heian High School* **33**: 1-19, pls. 1-4.

Kishii T. 1989d: Elateridae from Taiwan, with Descriptions of some new Taxa (3) (Coleoptera). A study of the materials colected by Dr. Kintarô Baba in 1986 and 1987. *Transactions of Essa entomological Society Niigata* **68**: 27-58. (Japanese with English descriptions)

Kishii T. 1989e: New species of Kibunea Kishii and Gambrinus LeConte from Yamanashi Pref. collected by Mr. K. Hosoda (Coleoptera: Elateridae). Some new forms of Elateridae in Japan (XXII). *Entomological Review of Japan* **44**: 63-68, pls. 10-11.

Kishii T. 1989f: On a new subspecies of Hypolithus motschulskyi from Mt. Yamizo-san in Fukushima Prefecture (Coleoptera, Elateridae). *Entomological Review of Japan* **44**: 99-100.

Kishii T. 1990a: Taiwanese Elateridae collected by Mr. M. Yagi, with the descriptions of some new taxa (Coleoptera). *Entomological Review of Japan* **45**: 11-27.

Kishii T. 1990b: Notes on Elateridae from Japan and its adjacent area (9) (Coleoptera). *Bulletin of the Heian High School* **34**: 1-16.

Kishii T. 1990c: Elateridae from Taiwan, with descriptions of some new taxa (4) (Coleoptera). A study of the materials collected by Dr. Kintarô Baba from 1986 to 1989. *Transactions of Essa Entomological Society Niigata* **70**: 9-39 (in Japanese with English descriptions).

Kishii T. 1991a: Elateridae from Taiwan, with descriptions of some new taxa (5) (Coleoptera). A study of the materials collected by Dr. Kintarô Baba from 1986 to 1989. *Transactions of Essa Entomological Society* **71**: 3-32 (in Japanese with English descriptions).

Kishii T. 1991b: Elateridae from Taiwan, with Descriptions of some new Taxa (6) (Coleoptera). A study of the materials collected by Dr. Kintaro Baba from 1986 to 1989. *Transactions of Essa Entomological Society* **72**: 17-60 (in Japanese with English descriptions).

Kishii T. 1991c: Elateridae from Taiwan, with descriptions of some new taxa (7) (Coleoptera). A study of the materials collected by Dr. Kintaro Baba from 1986 to 1989. *Transactions of Essa Entomological Society* **73**: 5-27. (in Japanese with English descriptions).

Kishii T. 1991d: Notes on Elateridae from Japan and its adjacent area (10). *Bulletin of the Heian High School* **35**: 1-21, pls. 1-3.

Kishii T. 1992a: Two new elaterid-species from Taiwan collected by Mr. M. Yagi (Coleoptera). *Entomological Review of Japan* **47**: 5-10, pl. 1.

Kishii T. 1992b: Notes on Elateridae from Japan and its adjacent area (11). *Bulletin of the Heian High School* **36**: 1-30, pls. 1-5.

Kishii T. 1993a: Taiwanese Elateridae collected by Mr. M. Yagi in 1991, with the descriptions on some new taxa (Coleoptera). *Entomological Review of Japan* **48**: 15-34, pls. 1-5.

Kishii T. 1993b: Notes on Elateridae from Japan and its adjacent area (12). *Bulletin of the Heian High School* **37**: 1-19, pls. 1-5.

Kishii T. 1993c: A new Megapenthini in the generic and specific status (Coleoptera, Elateridae). Some new forms of Elateridae in Japan (XXIV). *Entomological Review of Japan* **48**: 87-91.

Kishii T. 1993d: A new species of Kibunea Kishii from Japan (Coleoptera, Elateridae). Some new forms of Elateridae in Japan (XXV). *Entomological Review of Japan* **48**: 151-153, pl. 11.

Kishii T. 1994a: Elateridae from Taiwan, with descriptions of some new taxa (8) (Coleoptera). A study of the materials collected by the late Dr. Kintaro Baba in 1986 to 1989. *Special Bulletin of Essa Entomological Society* **2**: 179-209. (in Japanese with English descriptions).

Kishii T. 1994b: Two new elaterid-species from Taiwan collected by the late Dr. Kintarô Baba, with a remainder of the deceased (Coleoptera, Elateridae). *Special Bulletin of Essa Entomological Society* **2**: 211-216.

Kishii T. 1994c: Notes on the Taiwanese Elateridae collected by Mr. M. Yagi in 1992, with the descriptions of three new species (Coleoptera). *Entomological Review of Japan* **49**: 5-15, pls. 1-2.

Kishii T. 1994d: Notes on Elateridae from Japan and its adjacent Area (13). *Bulletin of the Heian High School* **38**: 1-35.

Kishii T. 1995: A study on the Elaterid-beetles of Shibata Collection from Taiwan, I (Coleoptera, Elateridae). On the subfamilies Oxynopterinae and Agrypninae. *Entomological Review of Japan* **50**: 1-14, pls. 1-4.

Kishii T. 1996a: A study on the elaterid-beetles of Shibata collection from Taiwan, II (Coleoptera: Elateridae). On the subfamily Pyrophorinae. *Entomological Review of Japan* **50**: 95-108, pls. 7-9. [Corrigenda in *Entomological Review of Japan* **51**: 95-96].

Kishii T. 1996b: Notes on Elateridae from Japan and its adjacent area (14). *Bulletin of the Heian High School* **39**: 1-40.

Kishii T. 1996c: A new species of the genus Melanotus from Japan (Coleoptera: Elateridae, Melanotinae). Some new forms of Elateridae in Japan (XXVI). *Entomological Review of Japan* **51**: 71-73.

Kishii T. 1996d: A study on the elaterid-beetles of Shibata collection from Taiwan, III (Coleoptera: Elateridae). On the subfamily Conoderinae. *Entomological Review of Japan* **51**: 89-96.

Kishii T. 1996e: A study on the elaterid-beetles of Shibata collection from Taiwan, IV (Coleoptera, Elateridae). On the subfamily Pityobiinae. *Entomological Review of Japan* **51**: 103-106.

Kishii T. 1997a: Notes on Elateridae from Japan and its adjacent area (15). *Bulletin of the Heian High School* **40**: 1-11.

Kishii T. 1997b: A study on the elaterid-beetles of Shibata collection from Taiwan, V (Coleoptera: Elateridae). On the subfamily Denticollinae: Tribes Denticollini, Hemicrepidini and Dimini. *Entomological Review of Japan* **52**: 9-14.

Kishii T. 1998a: Notes on Elateridae from Japan and its adjacent area (16). *Bulletin of the Heian High School* **41**: 1-28, pls. I-V.

Kishii T. 1998b: Two new subspecies of *Homotechnus motschulskyi* (Fleutiaux, 1902) (Elateridae, Coleoptera) from Mie and Nara Prefectures. Some new forms of Elateridae in Japan (XXVII). *Entomological Review of Japan* **53**: 77-80.

Kishii T. 1998c: A new synaptine genus and species from Japan (Elateridae, Coleoptera). Some new forms of Elateridae in Japan (XXVIII). *Entomological Review of Japan* **53**: 81-85.

Kishii T. 1999a: A new species of the genus Ampedus Dejean (Elateridae Coleoptera) from Wakayama Pref. Some new forms of Elateridae, in Japan (XXIX). *Entomological Review of Japan* **54**: 1-4.

Kishii T. 1999b: A study on the elaterid beetles of Shibata collection from Taiwan. VII (Coleoptera, Elateridae). On the subfamily Elaterinae: Tribe Dicrepidiini. *Entomological Review of Japan* **54**: 5-9, pls. I-VIII.

Kishii T. 1999c: A new subspecies of Homotechnes Motschulskyi (Fleutiaux, 1902) (Elateridae, Coleoptera) from Tochigi Prefecture. Some new forms of Elateridae in Japan (30). *Entomological Review of Japan* **54**: 21-23.

Kishii T. 2001a: Revision of the genus Scutellathous from Japan, with descriptions of two new species and taxonomic notes on its relatives (Elateridae, Coleoptera). "Some new forms of Elateridae in Japan (XXIX)". *Special Publication of the Japan Coleopterological Society* **1**: 205-214.

Kishii T. 2001b: An addition to the genus Pectocera (Coleoptera: Elateridae) from Is. Kume-jima of the Nansei Archipelago. "Some new forms of Elateridae in Japan (XXXII)". *Entomological Revue of Japan* **56**: 41-44.

Kishii T. 2001c: Three new taxa of the genus Hemicrepidius (Coleoptera: Elateridae) from west Japan. "Some new forms of Elateridae in Japan (XXXIII)". *Entomological Revue of Japan* **56**: 45-50.

Kishii T. 2001d: Four new subspecies of Homotechnes motschulskyi (Fleutiaux, 1902) (Coleoptera: Elateridae) and a new locality of H. m. etchu Kishii, 1996. "Some new forms of Elateridae in Japan (XXXIV)". *Entomological Revue of Japan* **56**: 51-62.

Kishii T. 2002a: [new taxa]. In: Kishii T. & Paik J.-C. 2002: Elaterid-beetles of Is. Cheju (Quelpart Island) (Coleoptera: Elateridae). *Entomological Review of Japan* **57**: 1-27.

Kishii T. 2002b: Four new taxa of the genus Ampedus Dejean (Coleoptera: Elateridae) from Japan. Some new forms of Elateridae in Japan (XXXV). *Entomological Revue of Japan* **57**: 109-117.

Kishii T. 2002c: Two new species of the genus Limoniscus Reitter, 1905 (Coleoptera: Elateridae) from Hokkaido, Japan. Some new forms of Elateridae in Japan (XXXVI). *Entomological Revue of Japan* **57**: 129-138.

Kishii T. 2002d: A new subspecies of Homotechnes motschulskyi (Fleutiaux, 1902) (Coleoptera: Elateridae) from Shiga and Mie Prefectures. Some new forms of Elateridae in Japan (XXXVII). *Entomological Revue of Japan* **57**: 139-143.

Kishii T. 2004a: Two new species of the subfamily Dendrometrinae (Coleoptera: Elateridae) from Japan. *Entomological Review of Japan* **59**: 61-66.

Kishii T. 2004b: Revision of the species-group Ampedus kasugensis (Coleoptera: Elateridae) from Japan, with descriptions of three new species. *Entomological Review of Japan* **59**: 67-79.

Kishii T. 2004c: Three new subspecies of the subfamily Hypnoidinae (Coleoptera: Elateridae) from Japan. *Entomological Review of Japan* **59**: 161-165.

Kishii T. 2004d: A study on the click beetles collected by Mr. M. Yagi from Kume Island in the Okinawas (Coleoptera, Elateridae). *Entomological Review of Japan* **59**: 167-176.

Kishii T. & Jiang S.-H. 1994: Notes on the Chinese Elateridae, 1 (Coleoptera). *Entomological Review of Japan* **49**: 87-102, pls. 8-10.

Kishii T. & Jiang S.-H. 1996: Notes on the Chinese Elateridae (2) (Coleoptera). *Entomological Revue of Japan* **50**: 131-152, pls. 12-14.

Kishii T. & Jiang S.-H. 1999: Notes on the Chinese Elateridae (Coleoptera) (4). *Entomological Revue of Japan* **54**: 11-19.

Kishii T. & Ôhira H. 1956: Snappers of Niigata-Prefecture, especially on the collection by Dr. Kintaro Baba, with the descriptions of some new forms. *Akitu* **5**: 71-84 (in Japanese with English summary).

Kishii T. & Platia G. 1991: A new Melanotine-species from Taiwan (Coleoptera, Elateridae). *Entomological Review of Japan* **46**: 1-6.

Kishii T. & Platia G. 1993a: A new Melanotus from Is. Ishigaki-jima (Coleoptera, Elateridae). Some new forms of Elateridae in Japan (XXIII). *Entomological Review of Japan* **48**: 67-69.

Kishii T. & Platia G. 1993b: Two new species of the genus Melanotus from Taiwan (Coleoptera, Elateridae, Melanotinae). *Entomological Review of Japan* **48**: 117-122, pls. 9-10.

Kiuchi M. 1972: A new endomychid beetle from Japan. *Kontyû* **40**: 159-161.

Kiyoyama Y. 1979: A new species of Aporotritoma from Taiwan. (Col., Erotylidae). *The Entomological Review of Japan* **33**: 47-49.

Kleine R. 1924. Die Gattung Lyponia Waterhouse (4. Beitrag zur Kenntnis der Lycidae). *Entomologische Blätter* **20**: 170-175.

Kleine R. 1925a: Fauna Buruana. Coleoptera, Fam. Lycidae (2. Beitrag zur Kenntnis der Lycidae). *Treubia* **7**: 31-37.

Kleine R. 1925b: Drei neue Lycidengenera des orientalischen Gebietes. 8. Beitrag zur Kenntnis der Lycidae. *Stettiner Entomologische Zeitschrift* **87**: 97-110.

Kleine R. 1925c: Die Gattung Macrolycus C. O. Wtrh. (5. Beitrag zur Kenntnis der Lycidae). *Deutsche Entomologische Zeitschrift* **1925**: 325-331.

Kleine R. 1926a: Bemerkungen zu den Lyciden der Samlung des Forest Research Institutes zu Dehra Dun. *Societas Entomologica* **41**: 1-2.

Kleine R. 1926b: Bemerkungen zu den Lyciden der Samlung des Forest Research Institutes zu Dehra Dun. *Societas Entomologica* **41**: 6-8.

Kleine R. 1926c: Die Lyciden Formosas. *Stettiner Entomologische Zeitschrift* **87**: 97-110.

Kleine R. 1926d: Bestimmungstabelle der Gattung Cautires C. O. Waterhouse. *Archiv für Naturgeschichte* A **1926** (12): 118-145.

Kleine R. 1926e: Protaphes, a new Lycid-genus from Sarawak. *Sarawak Museum Journal* **10**: 363-365.

Kleine R. 1927a: Libnetomimus eine neue Lycidengattung von den Philippinen-Inseln. *Societas Entomologica* **42**: 1-2.

Kleine R. 1927b: Bestimmungstabelle der Gattung Xylobanus C. O. Waterhouse. *Tijdschrift voor Entomologie* **70**: 43-72.

Kleine R. 1928: Neue indische Lycidae nebst faunistischen Bemerkungen. *Indian Forest Records* **13** (6): 221-268.

Kleine R. 1930a: Neue Lycidae aus dem indischen Gebiet. *Stettiner Entomologische Zeitung* **91**: 94-103, 1 pl.

Kleine R. 1930b: Neue Lyciden der malaiischen Subregion. *Journal of the Federal Malay State Museum* **16** (1/2): 169-174.

Kleine R. 1930c: Three new species of Lycidae. *Indian Forest Records* **14**: 287-289.

Kleine R. 1933: Neue Lyciden und Bemerkungen zum Cat. Col. Junk-Schenkling Lycidae. *Stettiner Entomologische Zeitung* **94**: 1-20.

Kleine R. 1935: Neue Brenthiden und Lyciden aus Indien. *Indien Forest Records, Entomology* (N. S.) **1** : 73-78.

Kleine R. 1936a: Zwei neue Lycostomus aus China (Lycidae). *Lingnan Science Journal* (Canton) **15** (1): 13-15.

Kleine R. 1936b: Lycidae. P. 180. In: Entomologische Sammlergebnisse der Deutsch Hindukush-Expedition 1935 der Deutschen Forschungsgemeinschaft. *Arbeiten über Morphologische und Taxonomische Entomologie aus Berlin-Dahlem* **3**: 173-213.

Kleine R. 1936c: Zwei neue Plateros aus China nebst synonymischen Bemerkungen (Coleoptera: Lycidae). *Lingnan Science Journal* (Canton) **15** (2): 263-264.

Kleine R. 1939a: Entomological results from the Swedish expedition 1934 to Burma and British India. Coleoptera: Brenthidae und Lycidae, gesammelt von Herrn Rene Malaise. *Arkiv för Zoologi* A **31** (12): 1-23.

Kleine R. 1939b: Neue Lycini from der afrikanischen und orientalischen Region. *Sborník Entomologického oddělení Národního Musea v Praze* **17**: 19-22.

Kleine R. 1940: Klapperichs China-Ausbeute (Brenthidae und Lycidae). *Entomologische Blätter* **36**: 90-91.

Kleine R. 1942: Bestimmungstabelle der Lycidae. *Bestimmungstabellen der europäischen Coleopteren.* 123. Heft. Troppau: Edmund Reitter, 90 pp.

Kleine R. 1943: Neue Lyciden aus dem Ungarischen National-Museum (Coleopt.) *Annales Historico-Naturales Musei Nationalis Hungarici* **36**: 145-156.

Kleine R. 1950: Klapperichs China-Ausbeute. II. Teil (Lycidae und Brenthidae). *Entomologische Blätter* **45-46**: 22-24.

Klug J. C. F. 1829: *Preis-Verzeichniss vorräthiger Insectendoubleten des Königl. Zoologischen Museums der Universität.* Berlin, 18 pp.

Klug J. C. F. 1833: Bericht über eine auf Madagascar veranstaltete Sammlung von Insecten aus der Ordnung Coleoptera. *Abhandlungen der Königlichen Akademie der Wissenschaften in Berlin* **19** [1832-1833]: 91-223, 5 pls.

Klug J. C. F. 1842: *Versuch einer systematischen Bestimmung und Auseinandersetzung der Gattungen und Arten der Clerii, einer Insectenfamilie aus der Ordnung der Coleopteren.* Berlin: Königliche Akademie der Wissenschaften zu Berlin, 397 pp., 2 pls.

Klug J. C. F. 1855: Diagnosen neuer Coleoptera aus Mosambique. *Berichte und Verhandlungen der Akademie in Berlin* **1855**: 643-660.

Klug J. C. F. 1862: Coleoptera: Käfer. Pp. 145-348. In: Peters W. C. H. (ed.): *Naturwissenschaftliche Reise nach Mossambique auf Befehl Seiner Majestät des Königs Friedrich Wilhelm IV. in den Jahren 1842 bis 1848 ausgeführt. Zoologie V. Insecten und Myriopodes.* Berlin: G. Reimer, xxi + 566 pp., 34 pls.

Kniephof J. 1913: Neuheiten der pommerschen Käferfauna (Col.). *Deutsche Entomologische Zeitschrift* **1913**: 185-190.

Kocher L. 1952: Contribution à l'étude des Cebrio marocains (Col. Cebrionidae). *Bulletin de la Société des Sciences Naturelles et Physiques du Maroc* **32**: 301-323.

Kocher L. 1953: Localisations nouvelles ou intéressantes de coléoptères marocains. *Travaux de l'Institut Scientifique Chérifien, Série Zoologie* **7**: 7-142, 1 map.

Kocher L. 1954a: Note sur les Cantharis (téléphores) (Col. Cantharidae) de la région d'Ifrane. *Bulletin de la Société des Sciences Naturelles et Physiques du Maroc* **33** [1953]: 149-158.

Kocher L. 1954b: Observations sur quelques Malachius marocains (Col. Malachidae). *Comptes Rendus des Séances Mensuelles de la Société des Sciences Naturelles et Physiques du Maroc* **20**: 207-210.

Kocher L. 1954c: Prospection entomologique (coléoptères) de la moyenne Moulouya. *Bulletin de la Société des Sciences Naturelles et Physiques du Maroc* **33**: 263-286.

Kocher L. 1955a: Observations sur Clerus (Pseudoclerops) mutillarius F., avec description d'une sous-espèce nouvelle: africanus, s. sp. nov. (Col. Cleridae). *Comptes Rendu des Séances Mensuelles de la Société des Sciences Naturelles et Physiques du Maroc* **21**: 39.

Kocher L. 1955b: Description d'une nouvelle espèce d'anthrène (Col. Dermestidae). *Comptes Rendu des Séances Mensuelles de la Société des Sciences Naturelles et Physiques du Maroc* **21**: 40-41.

Kocher L. 1956a: Catalogue commenté des coléoptères du Maroc. Fasc. IV Clavicornes et groupes voisins. *Travaux de l'Institut Scientifique Chérifien* **11**: 1-136.

Kocher L. 1956b: Description d'un nouvel anobiide marocain. *Comtes Rendus de la Société des Sciences Naturelles et Physiques du Maroc* **22**: 43.

Kocher L. 1961a: Contribution à l'étude des Cebrio marocains (Col. Cebrionidae). 2e note. *Bulletin de la Société des Sciences Naturelles et Physiques du Maroc* **40** [1960]: 33-51.

Kocher L. 1961b: Une nouvelle espèce marocaine du genre Cantharis (Col. Malacod.) C. rifensis sp. nov. *Comptes Rendus des Séances Mensuelles de la Société des Sciences Naturelles et Physiques du Maroc* **27**: 147-149.

Kocher L. 1962: Coléoptères nouveaux ou intéressants du Maroc présaharien. *Bulletin de la Société des Sciences Naturelles du Maroc* **42**: 195-214.

Kocher L. 1963: Coléoptères nouveaux ou intéressants de la région rifaine. *Bulletin de la Société des Sciences Naturelles et Physiques du Maroc* **43**: 193-199.

Kocher L. 1964: *Catalogue commenté des coléoptères du Maroc. Fascicule 10: Addenda et corrigenda.* Rabat: Institut Scientifique Chérifien, Série Zoologie **30**, 200 pp.

Kocher L. 1967: Espèces ou formes nouvelles de Cebrio marocains (Col. Cebrionidae) (3e note). *Bulletin de la Société des Sciences Naturelles et Physiques du Maroc* **46**: 281-287.

Kocher L. 1969: Description de nouveaux coléoptères du Maroc. *Bulletin de la Société des Sciences Naturelles et Physiques du Maroc* **48**: 107-113.

Koenig E. 1887: Neue Elateriden und Bemerkungen über bekannte Arten. *Horae Societatis Entomologica Rossicae* **21**: 353-354.

Koenig E. 1889a: Neue Coleopteren aus Russland. *Horae Societatis Entomologicae Rossicae* **23**: 301-305.

Koenig E. 1889b: Insecta in itinere Cl. N. Przewalski in Asia centralis novissime lecta. XV. Elateridae. *Horae Societatis Entomologica Rossicae* **23**: 535-539.

Kofler A. 1970a: Zur Systematik und Verbreitung europäischer Xyletinus-Arten (Col., Anobiidae). *Mitteilungen der Abteilung für Zoologie und Botanik am Landesmuseum „Joanneum"* **35**: 61-74.

Kofler A. 1970b: Xyletinus cylindricus n. sp. aus der Türkey und Beschreibung der Allotype zu Xyletinus distinguendus Kofler. (Col., Anobiidae). *Zeitschrift der Arbeitsgemeinschaft Österreichischer Entomologen* **22**: 143-145.

Kofler A. 1970c: Die paläarktischen Xyletinus-Arten des Senckenberg-Museums (Col.: Anobiidae). *Senckenbergiana Biologica* **51**: 363-367.

Kolbe H. J. 1886: Beiträge zur Kenntniss der Coleopteren-Fauna Koreas, bearbeitet auf Grund der von Herrn Dr. C. Gottsche während der Jahre 1883 und 1884 in Korea veranstalteten Sammlung; nebst Bemerkungen über die zoogeographischen Verhälnisse dieses Faunengebietes und Untersuchungen über einen Sinnesapparat im Gaumen von Misolampidius morio. *Archiv für Naturgeschichte* **52** (1): 139-240, pls X, XI.

Kolbe H. J. 1897: *Coleopteren. Die Käfer Deutsch-Ost-Afrikas.* Berlin: Verlang von Dietrich Reimer, 368 pp.

Kolbe H. J. 1910: Die Coleopterenfauna der Seychellen. Nebst Betrachtungen über die Tiergeographie dieser Inselgruppe. *Mitteilungen aus dem Zoologischen Museum in Berlin* **5**: 1-49.

Kolbe H. J. 1924. Beiträge zur schlesischen Käferfauna. *Jahresheft des Vereins für Schlesische Insektenkunde* (Breslau) **14**: 40-55.

Kolenati F. 1846a: *Meletemata Entomologica. Fascicule III. Brachelytra caucasi cum distributione geographica. Adnexis Pselaphinis, Scydmaenis, Notoxidibus et Xylophagis.* Petropoli: Typis Imperialis Academiae Scientarum, [6] + 44 pp., pls 12-14.

Kolenati F. A. 1846b: *Meletemata Entomologica. Fascicule V. Insecta caucasi. Coleoptera, Dermaptera, Lepidoptera, Neuroptera, Mutillidae, Aphaniptera, Anoplura.* Petropoli: Typis Imperialis Academiae Scientarum, [8] + iii + 169 pp., pls. 17-19.

Kolibáč J. 1992: Revision of Thanerocleridae n. stat. (Coleoptera, Cleroidea). *Mitteilungen der Schweizerischen Entomologische Gesellschaft* **65**: 303-340.

Kolibáč J. 1998: Classification of the subfamily Hydnocerinae Spinola, 1944 (Coleoptera, Cleridae). *Acta Musei Moraviae, Science Biologica* **83**: 127-210.

Kollar V. 1844: [new taxa]. In: Kollar V. & Radtenbacher L.: Aufzählung und Beschreibung der von Freiher Carl von Huegel auf seiner Reise durch Kaschmir u. das Himalayagebirge gesammelten Insecten. Pp. 393-564. In: Hügel K. F. von. (ed.): *Kaschmir und das Reich der Siek. Vierter Band. Zweite Abtheilung.* Stuttgart: Hallbergerische Verlag, pp. 244-586 pp. [1844]; 587-865 + [6] pp., 31 pls, 1 map [1848].

Kôno H. 1930: Die Cleriden Japans (Col.). *Transactions of the Sapporo Natural History Society* **11**: 134-140.

Kôno H. 1932. Beitrag zur Lyciden-Fauna Japans. *Insecta Matsumurana* **7**: 54-64.

Kôno H. 1936: Die Clavicornien aus den Kurilen (Siebenter Beitrag zur Kenntnis der Käferfauna der Kurilen). *Insecta Matsumurana* **10**: 148-153.

Kôno H. 1937: Eine neue Epilachna-Art. *Insecta Matsumurana* **11**: 99.

Kôno H. 1938: Die Anobiidae und Ptinidae in Tannen- und Fichtenwaldung in Japan. *Insecta Matsumurana* **12**: 174-178.

Kôno H. 1938: Die Lymexyloniden, schädlich an Sachalintannen und Ezofichten in Sachalin und Hokkaido. *Transactions of the Sapporo Natural History Society* **15**: 199-201.

Kôno H. 1939: Helotidae of Japan, Korea, and Formosa (Coleoptera). *The Philippine Journal of Sciences* **69**: 157-160.

Kôno H. 1940: Die Nitiduliden und Cucujiden in Tannen- und Fichtenwald in Japan. *Insecta Matsumurana* **14** [1939-1940]: 56-62.

Kôno H. & Esaki T. 1937: Ein neuer, halophiler Käfer, Rotomalachius (n.g.) esakii (n.sp.) Kôno (Coleoptera, Malachiidae). *Annotationes Zoologicae Japonenses* **16**: 121-123.

Kôno H. & Kim H. K. 1937: Die Ptilininen Japans (Col.). *Insecta Matsumurana* **12**: 28-31.

Kôno H. & Tamanuki K. 1926: Die Käfer-Ausbeute aus Nordsachalin im Jahre 1922. *Dobutsugaku Zasshi* (Tokyo) **38**: 276-296.

Kopetz A. 2004: Zur Kenntnis der Gattung Themus Motschulsky, 1857 im Himalaya (Coleoptera, Cantharidae). *Entomologica Basiliensia* **26**: 113-153.

Korotyaev B. A. 1989: Materialy po faune i ecologii bozhiikh korovok roda Lithophilus Froel. (Coleoptera, Coccinellidae) v SSSR. *Trudy Zoologicheskogo Instituta* **170**: 96-98.

Korge H. 1960: Kleine Mitteilungen 1692. *Entomologische Blätter* **56**: 185-186.

Korschefsky R. 1931: Coccinellidae, I. Pars 118. In: Schenkling S. (ed.): *Coleopterorum Catalogus*. Berlin: W. Junk, 224 pp.

Korschefsky R. 1932: Coccinellidae, II. Pars 120. In: Schenkling S. (ed.): *Coleopterorum Catalogus*. Berlin: W. Junk, 225-659.

Korschefsky R. 1934: Entomological investigations on the spike disease of sandal, 16. Coccinellidae (Col.). *Indian Forestry Records* (Delhi) **19**(6): 1-9, 1 pl.

Korschefsky R. 1935a: Coccinelidae. P. 299 In: Visser P. C. & Visser-Hooft J. (eds): *Wissenschaftliche Ergebnisse der Niederlandishen Expedition in der Karakorum und die angrenzenden Gebiete 1922, 1925 und 1929-30. Vol. 1.* Leipzig: F. A. Brockhaus, xviii + 499 pp., 1 map, 8 pls.

Korschefsky R. 1935b: Zwei neue Novius-Arten von den Canaren. (9. Beitrag zur Kenntnis der Coccinellidae). *Commentationes Biologicae* **6** (3): 1-3.

Korschefsky R. 1935c: Neue Coccinelliden aus Africa, Brasilien und Formosa. *Arbeiten über Morphologische und Taxonomische Entomologie aus Berlin-Dahlem* **2**: 252-256.

Korschefsky R. 1940: Vier neue Coccinelliden der Alten Welt. *Entomologische Blätter* **36**: 1-3.

Korschefsky R. 1944: Neue altweltliche Coccinelliden. (Coleoptera: Coccinellidae). *Arbeiten über Morphologische und Taxonomische Entomologie aus Berlin-Dahlem* **11**: 47-56.

Kovář I. 1977: A new species of the genus Coccinella (Coleoptera) from North Africa. *Acta Entomologica Musei Nationalis Pragae* **39**: 231-235.

Kovář I. 1997: Revision of the genera Brumus Muls. and Exochomus Redtb. (Coleoptera, Coccinellidae) of the Palaearctic Region. *Acta Entomologica Musei Nationalis Pragae* **44** [1995]: 5-124.

Kozminykh V. O. 1997: Sostav fauny zhestkokrylych (Insecta: Coleoptera) semejstev Sphaeritidae, Histeridae i Dermestidae Permskoj oblasti. pp. 71-79. In: Esjunina S. L., Zinovjeva E. A., Litvinenko N. I. & Ovesnova S. A. (eds.): *Problemy regionalnoy Krasnoy knigi*: Mezhvedomstvennyy sbornik naucznych trudov, Permskij gosuniversitet, 152 pp.

Kraatz G. 1852: Bemerkungen über Cryptophagus. *Entomologische Zeitung* (Stettin) **13**: 227-230.

Kraatz G. 1853: Bemerkungen über Atomaria. *Entomologische Zeitung* (Stettin) **14**: 94-98.

Kraatz G. 1856: Bermerkungen über Cryptophagus. *Entomologische Zeitung* (Stettin) **17**: 239-241.

Kraatz G. 1858: Beitrag zur Käferfauna Griechenlands. Drittes Stück: Staphylinidae (Schluss), Trichopterygia, Histeridae, Phalacridae, Nitidulariae, Trogositarii, Colydii, Cucujidae, Cryptophagidae, Thorictidae, Mycetophagidae, Dermestini, Byrrhi. *Berliner Entomologische Zeitschrift* **2**: 123-148.

Kraatz G. 1862a: Von Dr. Krüper neue aufgefundene Arten. *Berliner Entomologische Zeitschrift* **6**: 121-126.

Kraatz G. 1862b: Ueber die Silvaniden-Gattungen Aeraphilus Redtb. und Cathartus Reiche und über Leucohimatium Rosenh. *Berliner Entomologische Zeitschrift* **6**: 127-134.

Kraatz G. 1862c: Beiträge zur europäischen Käfer-Fauna. *Berliner Entomologische Zeitschrift* **6**: 263-272.

Kraatz G. 1864: Zur kritischen Kenntniss europäeischer Käferarten. *Berliner Entomologische Zeitschrift* **8**: 131-142.

Kraatz G. 1869: Ueber die europäischen Arten der Gattung Sacium Le Conte und Arthrolips Woll. *Berliner Entomologische Zeitschrift* **12**: 283-287.

Kraatz G. 1872: Uebersicht der deutschen Triplax-Arten. *Berliner Entomologische Zeitschrift* **15** [1871]: 143-145.

Kraatz G. 1873a: Revision der europäischen Arten der Coccinelliden-Gattung Exochomus Redtb. *Berliner Entomologische Zeitschrift* **17**: 191-194.

Kraatz G. 1873b: Eine neue Art Trichodes Herbst, nebst Bemerkungen über einige früher Beschriebe. *Berliner Entomologische Zeitschrift* **17**: 239-242.

Kraatz G. 1874: Ueber ein interessantes Nomen in litteris. *Berliner Entomologische Zeitschrift* **18**: 138.

Kraatz G. 1875: Ueber Anthocomus fenestratus Linder und Verwandte. *Deutsche Entomologische Zeitschrift* **19**: 420.

Kraatz G. 1876: Ueber Ptinomorphus regalis Duftschmid. *Entomologische Monatsblätter* **1**: 149-151.

Kraatz G. 1879a: Neue Käfer vom Amur. *Deutsche Entomologische Zeitschrift* **23**: 121-144.

Kraatz G. 1879b: [new taxon]. In: Candèze E. & Kraatz G.: Elatérides du Tarbagatai. *Deutsche Entomologische Zeitschrift* **23**: 283.

Kraatz G. 1879c: Corymbites (Diacanthus) informis n. sp. *Deutsche Entomologische Zeitschrift* **23**: 287.

Kraatz G. 1881a: Einige für Deutschland neue Anobiidae. *Deutsche Entomologische Zeitschrift* **25**: 301-303.

Kraatz G. 1881b: Ueber die Gattung Pleonomus Ménétriés. In: Heyden L. F. J. D. von & Kraatz G.: Beiträge zur Käferfauna von Turkestan. *Deutsche Entomologische Zeitschrift* **25**: 324-235.

Kraatz G. 1882a: Zwei neue Melyrinen-Gattungen aus Margelan. *Deutsche Entomologische Zeitschrift* **26**: 96-98.

Kraatz G. 1882b: [new taxa]. In: Heyden L. F. J. D. von & Kraatz G.: Käfer um Margelan, gesammelt von Haberhauer. *Deutsche Entomologische Zeitschrift* **26**: 99-118.

Kraatz G. 1882c: [new taxa]. In: Heyden L. F. J. D. von & Kraatz G.: Käfer um Samarkand, gesammelt von Haberhauer. *Deutsche Entomologische Zeitschrift* **26**: 297-338.

Kraatz G. 1883: [new taxa]. In: Heyden L. F. J. D. von & Kraatz G.: Käfer aus Osch (in Turkestan). *Deutsche Entomologische Zeitschrift* **27**: 337-360.

Kraatz G. 1891: Ueber den griechischen Trichophorus Schaumi Candèze. *Deutsche Entomologische Zeitschrift* **35**: 125.

Kraatz G. 1893: Über die Amnios-Gruppe der Cleriden Gattung Trichodes. *Deutsche Entomologische Zeitschrift* **1893**: 78-80.

Kraatz G. 1894: Ergänzende Bemerkungen zu Escherich's monographisher Studie über Trichodes Herbst. *Deutsche Entomologische Zeitschrift* **1894**: 113-136.

Kraatz G. 1895a: Hadrocnemus nov. gen. Malachiidarum. *Deutsche Entomologische Zeitschrift* **39**: 59-62.

Kraatz G. 1895b: Nitidulidae von Togo. *Deutsche Entomologische Zeitschrift* **1895**: 145-153.

Kraatz G. 1895c: Die Languria-Arten von Togo. *Deutsche Entomologische Zeitschrift* **1895**: 201-208.

Kraatz G. 1899a: Verzeichniss der von Hrn. Conradt in Westafrica (Kamerun, Togo) gesammelten Cleriden. *Deutsche Entomologische Zeitschrift* **1899**: 81-107.

Kraatz G. 1899b: Verzeichniss der von Herrn Weyers in Südwest Sumatra gesammelten Cleridae. *Annales de la Société Entomologique de Belgique* **43**: 212-217.

Kraatz G. 1899c: Verzeichniss der von Herrn Weyers in Südwest Sumatra gesammelten Languridae [sic!]. *Annales de la Société Entomologique de Belgique* **43**: 218-219.

Kraatz G. 1900a: Ueber die Languriiden-Arten von Kamerun nebst einigen verwandten Formen. *Deutsche Entomologische Zeitschrift* **1899**: 307-315.

Kraatz G. 1900b: Einige Bemerkungen zu Gorham's Aufsatz von 1896: Languridae [sic!] in Birmania ex regione vicina a Leonardo Fea collecta. *Deutsche Entomologische Zeitschrift* **1899**: 345-352.

Kraatz G. & Kiesenwetter E. A. H. von. 1864: Ueber die zur Untergattungen Anthodytes Ksw. gehörenden Arten der Gattung Malachius. *Berliner Entomologische Zeitschrift* **8**: 305-312, pl. 4.

Kraus E. J. 1911: Revision of the powder-post beetles of the family Lyctidae of the United States and Europe. *United States Department of Agriculture Entomology. Technical Bulletin* **20**: 111-138.

Krauss H. 1894: Neue Absidien aus den Alpen. *Wiener Entomologische Zeitung* **13**: 180-185.

Krauss H. 1902a: *Bestimmungstabelle der europäischen Coleopteren. XLIX. Heft 49. Enthaltend: Cantharidae. III. Theil: Genus Malachius Fb. Uebersicht über die Arten der Käfer-Gattung Malachius Fabr., aus Europa und dem Caucasus*. Paskau: Edm. Reitter, 33 pp. + [Index].

Krauss H. 1902b: Coleopterologische Beiträge zur Fauna Austriaca. III. *Wiener Entomologische Zeitung* **21**: 89-92.

Krivolutskaya G. O. 1981: Novyy vid zhuka-pestryaka iz roda Thanasimus Latr. (Coleoptera, Cleridae) s Dalnego Vostoka SSSR. Pp. 34-35. In: Krivolutskaya G. O., Egorov A. B. & Fedikova V. S. (eds): *Novye svedeniya o nasekomykh Dalnego Vostoka.* Vladivostok: Akademiya Nauk, SSSR Dalnevostochnyy Nauchnyy Centr Biologo-Pochvennyy Institut, 125 pp.

Krivolutskaya G. O. 1992: 65. Sem. Cucujidae - Ploskotelki. Pp. 233-245. In: Ler P. A. (ed.): *Opredelitel nasekomykh Dalnego Vostoka SSSR v shestykh tomakh. Tom III Zhestkokrylye, ili zhuki. Chast 2.* Sankt-Petersburg: Nauka,704 pp.

Krivolutskaya G. O. 1992: 71. *Sem. Erotylidae - griboviki.* In: Ler P. A. (ed.): *Opredelitel" nasekomych Dal"nego Vostoka SSSR v shestykh tomah. Tom III. Zhestkokrylye, ili zhuki. Chast 2.* Sankt-Peterburg: Nauka, 704 pp.

Krynicki J. 1832: Enumeratio Coleopterorum Rossiae meridionalis et praecipue in Universitatis in Caesarea Charcoviensis circulo obvenientium, quae annorum 1827-1831 spatio observavit. *Bulletin de la Société Impériale des Naturalistes de Moscou* **5**: 65-179, pls. II-III.

Kryzhanovskij O. L. 1960: Nauchnye resul'taty Kitaysko-Sovetskikh zoologo-botanicheskikh ekspeditsiy 1955-1957 v yugo-zapadnyy Kitay. *Entomologicheskoe Obozrenie* **39**: 870-884.

Kryzhanovskij O. L. 1977: K faune i biologii Endomychidae (Coleoptera) Dal'nego Bostoka CCCP. *Trudy Zoologicheskogo Instituta* (Leningrad) **62**: 38-40.

Kugelann J. G. 1792: Verzeichniss der in einigen Gegenden Preussens bis jetzt entdeckten Käferarten, nebst kurzen Nachrichten von denselben. *Neuestes Magazin für die Liebhaber der Entomologie* **1** (4): 477-512.

Kugelann J. G. 1794: Verzeihniss der in einigen Gegenden Preussens bis jetzt entdeckten Käferarten, nebst kurzen Nachrichten von denselben. *Neuestes Magazin für die Liebhaber der Entomologie* **1** (5): 513-582.

Kuhnt P. 1908: Neue Erotyliden. (Col.). *Deutsche Entomologische Zeitschrift* **1908**: 626-633.

Kuhnt P. 1910: Neue Erotylidae. (Col.). *Deutsche Entomologische Zeitschrift* **1910**: 219-270.

Künnemann G. 1917: Eine neue Coccinella – Art (C. Klingebergi m.) aus den Karpaten. *Entomologische Blätter* **13**: 266-267.

Kunze G. 1839: Bemerkungen und Nachträge zu Aubé's Monotomen. Ann. de la Soc. ent. de France, 1837, 4. tri. p. 453 sqq. *Germar's Zeitschrift für die Entomologie* **1**: 383-387.

Kurisaki J. 1914: On two new species of Japanese Coccinellidae. [*Konchu Sekai*] *The Insect World* (Gifu) **18**: 441-443 (in Japanese).

Kurisaki M. 1920a: A new species of Coccinellidae in Japan. [*Konchu Sekai*] *The Insect World* (Gifu) **24**: 230-231 (in Japanese).

Kurisaki M. 1920b: On the genus Amida. [*Konchu Sekai*] *The Insect World* (Gifu) **24**: 405-407 (in Japanese).

Kurisaki M. 1923: Description of some interesting coccinellids. *Zoological Magazine* (*Dobutsugaku Zasshi*) **35** (413): 101-108 (in Japanese).

Kurosawa Y. 1949: Description of a new Lymexylonidae of the genus Lymexylon Fabricius from Japan. *Entomological Review of Japan* **1**: 26-28.

Kurosawa Y. 1956: A new genus and the species of the family Lymexylonidae with additional notes on some Japanese species (Coleoptera). *Bulletin of the National Science Museum* (Tokyo) **3**: 80-82.

Kurosawa Y. 1985: Revisional notes on the family Lymexylonidae (Coleoptera) in eastern and southeastern Asia. *Bulletin of the National Science Museum* (Tokyo) (A) (Zoology) **11**: 109-119.

Kuschakewitsch J. A. 1861: Eine neue Elateridae: Corymbites Baerii, J. Kusch. *Horae Societatis Entomologicae Rossicae* **1**: 55-56, pl. 1.

Küster H. C. 1844: *Die Käfer Europa's. Nach der Natur beschrieben. Mit Beiträgen mehrerer Entomologen. 1. Heft.* Nürnberg: Bauer und Raspe, [6] + 100 sheets, 2 pls.

Küster H. C. 1845: *Die Käfer Europa's. Nach der Natur beschrieben. Mit Beiträgen mehrerer Entomologen. 2. Heft.* Nürnberg: Bauer & Raspe, [6] + 100 sheets , 2 pls.

Küster H. C. 1846a: *Die Käfer Europa's. Nach der Natur beschrieben. Mit Beiträgen mehrerer Entomologen. 4. Heft.* Nürnberg: Bauer und Raspe, [6] + 100 sheets, 3 pls.

Küster H. C. 1846b: *Die Käfer Europa's. Nach der Natur beschrieben. 6. Heft.* Nürnberg: Bauer und Raspe, [4] + 100 sheets, 2 pls.

Küster H. C. 1847a: *Die Käfer Europa's. Nach der Natur beschrieben. Mit Beiträgen mehrerer Entomologen. 9. Heft.* Nürnberg: Bauer & Raspe, [4] + 100 sheets + 3 pl.

Küster H. C. 1847b: *Die Käfer Europa's. Nach der Natur beschrieben. Mit Beiträgen mehrerer Entomologen. 10. Heft.* Nürnberg: Bauer & Raspe, [4] + 100 sheets, 3 pls.

Küster H. C. 1848a: *Die Käfer Europa's. Nach der Natur beschrieben. Mit Beiträgen mehrerer Entomologen. 13. Heft*. Nürnberg: Bauer & Raspe, [4] + 100 sheets, 2 pls.

Küster H. C. 1848b: *Die Käfer Europa's. Nach der Natur beschrieben. Mit Beiträgen mehrerer Entomologen. 14. Heft*. Nürnberg: Bauer & Raspe, [4] + 100 sheets, 3 pls.

Küster H. C. 1848c: *Die Käfer Europa's. Nach der Natur beschrieben. Mit Beiträgen mehrerer Entomologen. 15. Heft* Nürnberg: Bauer & Raspe [4] + 100 sheets, 3 pls.

Küster H. C. 1849a: *Die Käfer Europa's. Nach der Natur beschrieben. Mit Beiträgen mehrerer Entomologen. 17. Heft*. Nürnberg: Bauer & Raspe, [4] + 100 sheets, 2 pls.

Küster H. C. 1849b: *Die Käfer Europa's. Nach der Natur beschrieben. Mit Beiträgen mehrerer Entomologen. 18. Heft*. Nürnberg: Bauer & Raspe, [4] + 100 sheets, 2 pls.

Küster H. C. 1849c: *Die Käfer Europa's. Nach der Natur beschrieben. Mit Beiträgen mehrerer Entomologen. 19. Heft*. Nürnberg: Bauer & Raspe, [4] + 100 sheets, 2 pls.

Küster H. C. 1850: *Die Käfer Europa's. Nach der Natur beschrieben. Mit Beiträgen mehrerer Entomologen. 21. Heft*. Nürnberg: Bauer & Raspe, [6] + 100 sheets, 2 pls.

Küster H. C. 1851b: *Die Käfer Europa's. Nach der Natur beschrieben. Mit Beiträgen mehrerer Entomologen. 22. Heft*. Nürnberg: Bauer & Raspe, [4] + 100 sheets, 2 pls.

Küster H. C. 1851c: *Die Käfer Europa's. Nach der Natur beschrieben. Mit Beiträgen mehrerer Entomologen. 23. Heft*. Nürnberg: Bauer & Raspe, [4] + 200 sheets, 2 pls.

Küster H. C. 1852: *Die Käfer Europa's. Nach der Natur beschrieben. Mit Beiträgen mehrerer Entomologen. 25. Heft*. Nürnberg: Bauer & Raspe, [4] + 100 sheets, 2 pls.

Küster H. C. 1853: *Die Käfer Europa's. Nach der Natur beschrieben. Mit Beiträgen mehrerer Entomologen. 27. Heft*. Nürnberg : Bauer & Raspe, [4] + 200 sheets, 2 pls.

Küster H. C. 1854: *Die Käfer Europa's. Nach der Natur beschrieben. Mit Beiträgen mehrerer Entomologen. 28. Heft*. Nürnberg: Bauer & Raspe, [4] + 100 sheets, 2 pls.

Kuwert A. F. 1893a: Neue und alte afrikanische Cleriden. *Annales de la Société Entomologique de Belgique* **37**: 467-479.

Kuwert A. F. 1893b: Einige neue indoautralische Cleriden. *Annales de la Société Entomologique de Belgique* **37**: 479-486.

Kuwert A. F. 1894a: Die Enopliinengattungen der Cleriden und einige amerikanische neue Arten derselben. *Annales de la Société Entomologique de Belgique* **38**: 6-13.

Kuwert A. F. 1894b: Revision der Cleridengattung Omadius Lap. *Annales de la Société Entomologique de Belgique* **38**: 63-97.

Kuwert A. F. 1894c: Revision des Genus Stigmatium und der diesem Genus verwandten Gattungen nebst Bestimmungstabelle der mir zur Kenntniss gekommenen und bisher beschriebenen Arten. *Annales de la Société Entomologique de Belgique* **38**: 398-457.

Kuznecov V. N. & Pang H. 1992: A new species of the genus Cryptogonus Mulsant (Coleoptera, Coccinellidae) from the Seychelle Islands. *Elytra* **20**: 231-233.

Kuznecov V. N. & Ren Shunxiang 1996: [new taxon]. In: Kuznecov V. N.: 77. Sem. Coccinellidae – bozh'i korovki. Dopolnenie 1. Rody Sumnius Ws. i Hyperaspis Chevr. Pp. 423-427. In: Ler P. A. (ed.): *Opredelitel' nasekomych Dal'nego Vostoka Rossii v shestykh tomach. Tom 3 - Zhestkokrylye, ili zhuki. Chast 3.* Vladivostok: Dal'nauka, 555 pp.

Kuznecova E. 1928: Ueber eine neue Dermestes-Art (Coleoptera, Dermestidae). *Entomologisches Nachrichtenblatt* (Troppau) **2**: 17-24.

Laboulbène A. 1848: Description d'une nouvelle espèce française de Laemophloeus. *Annales de la Société Entomologique de France* (2) **6**: 295-300.

Laco J. 1936: [new taxon]. In: Roubal J.: *Katalog coleopter (brouků) Slovenska a Podkarpatské Rusi na základě bionomickém a zoogeografickém a spolu systematický doplněk Ganglbauerových „Die Käfer von Mitteuleuropa" a Reitterovy „Fauna Germanica". Díl II. – Tome II.* Bratislava: Učená Společnost Šafaríka, [vii] +434 pp.

Lacordaire J. T. 1835: Coléoptères. In: Boisduval J. B. A. & Lacordaire J. T.: *Faune entomologique des environs de Paris; ou species général des insectes qui se trouvent dans un rayon de quinze à vingt lieues aux alentours de Paris. Tome premier.* Paris: Méquignon - Marvis, [6] + 1-17 + i-cii + 103-696 pp., 3 pls.

Lacordaire J. T. 1842: *Monographie des erotyliens, famille de l'ordre des coléoptères*. Paris: Roret, xi + 543 pp.

Lacordaire J. T. 1857: *Histoire naturelle des insectes. Genera des coléoptères ou exposé méthodique et critique de tous les genres proposés jusqu'ici dans cet ordre d'insectes. Tome quatrième contenant les familles des buprestides, troscides, eucnémides, élatérides, cébrionides, cérophytides, rhipicérides, dascyllides, malacodermes, clérides, lyméxylones, cupésides, ptiniores, bostrichides et cissides.* Paris: Libraire encyclopédique de Roret, 579 pp.

Lafer G. S. 1992: Phalacridae. Pp. 225-229. In: Ler P. A. (ed.): *Opredelitel' nasekomych Dal'nego Vostoka SSSR v shestykh tomach. Tom 3 - Zhestkokrylye, ili zhuki. Chast 2.* Sankt-Petersburg: Nauka, 704 pp.

Lai J., Satô M. & Yang P.-S.1998: Checklist of Lampyridae of Taiwan - Coleoptera: Polyphaga: Lampyridae. *Chinese Journal of Entomology* **18** (3): 207-215.

Laibner S. 1977: Limonius ruzenae sp. n. (Coleoptera, Elateridae) aus der Südslowakei. *Acta Entomologica Bohemoslovaca* **74**: 388-393.

Laibner S. 1985: Die bulgarischen Arten der Gattung Dicronychus (Coleoptera, Elateridae). *Acta Entomologica Bohemoslovaca* **82**: 129-137.

Laibner S. 1987: Adrastus montenegroensis sp. n. from Yugoslavia (Coleoptera, Elateridae). *Acta Entomologica Bohemoslovaca* **84**: 393-396.

Laibner S. 1991: Adrastus juditae sp. n. (Coleoptera, Elateridae) from Moravia. *Acta Entomologica Bohemoslovaca* **88**: 327-330.

Laicharting J. N. von. 1781: *Verzeichnis und Beschreibung der Tyroler – Insecten. I. Theil. Käferartige Insecten. I.* Zürich: Johann Gaspar Füessli, [4] + xii + [1]+ 248 pp.

Laporte [= de Castelnau] F. L. N. Caumont de. 1832: Mémoire sur cinquante espèces nouvelles ou peu connues d'insectes. *Annales de la Société Entomologique de France* **1**: 386-415.

Laporte [= de Castelnau] F. L. N. Caumont de. 1833: Essai d'une révision du genre Lampyre. *Annales de la Société Entomologique de France* **2**: 122-153.

Laporte [= de Castelnau] F. L. N. Caumont de. 1835a: *Études entomologiques ou descriptions des insectes nouveaux et observations sur leur synonymie.* Paris: Méquignon-Marvis, 159 pp., 4 pls.

Laporte [= de Castelnau] F. L. N. Caumont de. 1835b: Études entomologiques, ou descriptions d'insectes nouveaux et observations sur la synonymie. *Revue Entomologique* (G. Silberman) **3**: 157-181.

Laporte [= de Castelnau] F. L. N. Caumont de. 1838: Études entomologiques, ou descriptions d'insectes nouveaux et observations sur la synonymie. *Revue Entomologique* (G. Silbermann) **4** [1836]: 5-60, 1 pl.

Laporte [= de Castelnau] F. L. N. Caumont de. 1840a: *Histoire naturelle des insectes coléoptères. Avec une introduction renfermant l'anatomie et la physiologie des animaux articulés par M. Brullé. Tome Premier.* Paris: P. Duménil, i-cxxiv + 24 pls + 324 + [1] pp., 19 pls.

Laporte [= de Castelnau] F. L. N. Caumont de. 1840b: *Histoire naturelle des insectes coléoptères. Tome deuxième.* Paris: P. Duménil, 563 + [1] pp., 38 pls.

Lareynie P. 1852: [Description de trois nouvelles espèces de Ptiniores, Latr.]. *Bulletin de la Société Entomologique de France* **10**: xc-xci.

Latreille P. A. 1797: *Précis des caractères génériques des insectes, disposés dans ordre naturel.* Brive: F. Bourdeaux, xiv + 210 + 7 pp.

Latreille P. A. 1802: *Histoire naturelle, générale et particulière des crustacés et des insectes. Ouvrage faisant suite aux oeuvres de Leclerc de Buffon, et partie du Cours complet d'Histoire naturelle rédigé par C. S. Sonnini, membre de plusieurs Sociétés savantes. Tome troisième.* Paris: F. Dufart, x + 467 + [1] pp.

Latreille P. A. 1804a: *Histoire naturelle, générale et particulière des crustaces et des insectes. Ouvrage faisant suite aux oeuvres de Leclerc de Buffon, et partie du Cours complet d'histoire naturelle rédigé par C. S. Sonnini, membre de plusieurs Sociétés savantes. Tome neuvième.* Paris: F. Dufart, 400 + [16] pp., pls. 74-80.

Latreille P. A. 1804b: *Histoire naturelle, générale et particulière des crustacés et des insectes. Ouvrage faisant suite aux oeuvres de Leclerc de Buffon, et partie du Cours complet d'histoire naturelle rédigé par C. S. Sonnini, membre de plusieurs Sociétés savantes. Tome onzième.* Paris: F. Dufart, 422 pp., pls 91-93.

Latreille P. A. 1804c: *Histoire naturelle, générale et particulière des crustacés et des insectes. Ouvrage faisant suite aux oeuvres de Leclerc de Buffon, et partie du Cours complet d'histoire naturelle rédigé par C. S. Sonnini, membre de plusieurs Sociétés savantes. Tome douzième.* Paris: F. Dufart, 424 pp., pls 94-97.

Latreille P. A. 1804d: *Tableau méthodique des insectes. Classe huitième insectes, Insecta.* Pp. 129–200. In: Société des naturalistes et agriculteurs (ed.): *Nouveau Dictionnaire d'Histoire Naturelle, appliquée aux arts, principalement à l'agriculture et à l'économie rurale et domestique. Tome XXIV.* Paris: Deterville [each chapter with separate pagination].

Latreille P. A. 1806: *Genera crustaceorum et insectorum secundum ordinem naturalem in familias disposita, iconibus exemplisque plurimus explicata. Tomus primus.* Parisiis et Argentorati, Amand König, xvii + 302 + [1] pp., xvi pls.

Latreille P. A. 1807a: *Genera crustaceorum et insectorum secundum ordinem naturalem in familias disposita, iconibus exemplisque plurimis explicata. Tomus secundus.* Parisiis et Argentorati: Amand Koenig, Paris, 280 pp.

Latreille P. A. 1807b: *Genera crustaceorum et insectorum secundum ordinem naturalem in familia disposita, iconobis exemplisque plurimis explicta. Tomus tertius.* Parisii et Argentorati: Amand Koenig, 258 + [1] pp.

Latreille P. A. 1809: *Genera Crustaceorum et Insectorum secundum ordinem naturalem in familias disposita, iconibus exemplisque plurimis explicata. Tomus quartus et ultimus.* Parisiis et Argentorati: A. Koenig, 399 pp.

Latreille P. A. 1813: *Insectes de l'Amérique équinoxiale, recueillis pendand le voyage de MM. De Humboldt et Bonpland.* Pp. 65-96. In: Humboldt A. & Bonpland A.: *Voyage de Humbolt et Bonpland. Deuxième partie. Observations de zoologie et d'anatomie comparée, faite dans l'Océan atlantique, dans l'intérieur du Nouveau Continent et dans la Mer du Sud, pendant les années 1799, 1800, 1801, 1802 et 1803. Seconde volume.* Paris: F. Schoell et G. Dufour et comp., 138 pp., 13 pls.

Latreille P. A. 1817: *Les crustacés, les arachnides et les insectes.* In: Cuvier G. C. L. D.: *Le Règne Animal distribuè d'après son organisation, pour servir de base à l'histoire naturelle des animaux et d'introduction à l'anatomie comparée. Tom III.* Paris: Deterville, xxix + 653 pp., 2 pls.

Latreille P. A. 1827: Description des insectes. Pp. 271-292. In: Cailliaud F.: *Voyage à Méroé, au fleuve Blanc, au-delà de Fazoql dans le midi du royaume de Sennâr, à Syouah et dans cinq autres oasis; fait dans les années 1819, 1820, 1821 et 1822. Tome Quatrième.* Paris: Imprimerie Royale, 416 pp.

Latreille P. A. 1829a: Crustacés, arachnides et partie des insectes. In: Cuvier G. C. L. D.: *Le Règne animal distribué d'après son organisation pour servir de base à l'histoire naturelle des animaux et d'introduction à l'anatomie comparée. Avec figures dessinées d'après nature. Nouvelle édition, revue et augmentée. Tome IV.* Paris: Déterville, xxvii + 584 pp.

Latreille P. A. 1829b: Crustacés, arachnides et partie des insectes. In: Cuvier G. C. L. D.: *Le Règne Animal distribué d'après son organisation, pour servir de base à l'histoire naturelle des animaux et d'intoduction à l'anatomie comparée. Nouvelle édition, revue et augmentée. Tom V.* Paris: Déterville, xxiv + 556 pp.

Latreille P. A. 1834: Distribution méthodique et naturelle des genres de diverses tribus d'insectes coléoptères, de la famille des Serricornes. Ouvrage posthume. *Annales de la Société Entomologique de France* **3**: 113-170, pl. 2.

Lauffer G. 1905: Neue Arten und Varietäten von Coleopteren der pyrenäischen Halbinsel. *Boletín de la Real Sociedad Española de Historia Natural* **5**: 403-407.

Laurent L. 1967: La sous-famille Tetralobinae (Coleoptera, Elateridae). *Bulletin et Annales de la Société Royale Belge d'Entomologie* **103**: 83-109.

Lavagne H. 1914: Description d'un Ernobius nouveau (Col. Anobiidae). *Bulletin de la Société Entomologique de France* **1914**: 137-138.

Lavagne H. 1920: Description d'une variété nouvelle d'Hypocoprus quadricollis Reitter (Col. Cryptophagidae). *Bulletin de la Société Entomologique de France* **1920**: 111.

Laxmann E. 1770: Novae insectorum species. *Novi Commentarii Academiae Scientiarum Imperialis Petropolitanae* **14** (1): 593-604, pls XXIV-XXV.

Lea A. M. 1895: New species of Australian Coleoptera. *Proceedings of the Linnean Society of New South Wales* **10**: 308-319.

Lea A. M. 1898: Descriptions of new species of Australian Coleoptera. *Proceedings of the Linnean Society of New South Wales* **23**: 521-645.

Lea A. M. 1909: Revision of the Australian and Tasmanian Malacodermidae. *Transactions of the Entomological Society of London* **1909**: 45-111.

Lea A. M. 1919: Descriptions of new species of Australian Coleoptera. Part 16. *Proceedings of the Linnean Society of New South Wales* **43**: 715-746.

Lea A. M. 1921: Descriptions of new species of Australian Coleoptera. Part 17. *Proceedings of the Linnean Society of New South Wales* **46**: 351-369.

Lea A. M. 1924: On Australian Anobiides (Coleoptera). *Transaction and Proceeding of the Royal Society of South Australia* **48**: 15-64.
Leach W. E. 1815: Entomology. In: Brewster D. (ed.): *The Edinburgh Encyclopedia*. Volume 9. Edinburgh: Balfour, 384 pp.
Leach W. E. 1822: Characters of a new genus of coleopterous insects of the family Byrrhidae. *Transactions of the Linnaean Society of London* **13**: 41.
Leach W. E. 1824: Monograph on the Cebrionidae, a family of insects. *Zoological Journal* **1**: 33-46, 282-283.
Lechanteur F. 1959: Un genre nouveau de coléoptères Nitidulidae d'Afrique. *Bulletin et Annales de la Société Entomologique Belgique* **95**: 107-110.
LeConte J. L. 1850: *General remarks upon the Coleoptera of Lake Superior*. Pp. 201-242. In: Agassiz J. L. R. & Cabot J. E. (eds): *Lake Superior, its physical character, vegetation and animals, compared with those of other and similar regions. With a narrative of the tour, by J. Eliott Cabot. And contributions by other scientific gentlemen*. Boston: Gould, Kendall and Lincoln, x + 428 pp., 8 pls.
LeConte J. L. 1851: Synopsis of the Lampyridae of temperate North America. *Proceedings of the Academy of Natural Sciences of Philadelphia* **5**: 331-347.
LeConte J. L. 1852a: Synopsis of the eucnemides of temperate North America. *Proceedings of the Academy of Natural Sciences Philadelphia* **6**: 45-49.
LeConte J. L. 1852b: Descriptions of new species of Coleoptera from California. *Annals of the Lyceum of Natural History of New York* **5**: 125-216.
LeConte J. L. 1852c: Remarks upon the Coccinellidae of the United States. *Proceedings of the Academy of Natural Sciences of Philadelphia* **6**: 129-145.
LeConte J. L. 1852d: Catalogue of the Melyrides of the United States, with descriptions of new species. *Proceedings of the Academy of Natural Sciences of Philadelphia* **6**: 163-171.
LeConte J. L. 1853a: Descriptions of twenty new species of Coleoptera inhabiting the United States. *Proceedings of the Academy of Natural Sciences of Philadelphia* **6**: 226-235.
LeConte J. L. 1853b: Synopsis of the Silphales of America, North of Mexico. *Proceedings of the Academy of Natural Sciences of Philadelphia* **6**: 274-287.
LeConte J. L. 1853c: Revision of the Elateridae of the United States. *Transactions of the American Philosophical Society* (2) **10**: 405-508.
LeConte J. L. 1854a: Synopsis of the Cucuiides of the United States. *Proceedings of the Academy of Natural Sciences of Philadelphia* **7**: 73-79.
LeConte J. L. 1854b: Synopsis of the Dermestidae of the United States. *Proceedings of the Academy of the Natural Sciences of Philadelphia* **7**: 106-113.
LeConte J. L. 1855: Synopsis of the lathridiides of the United States, and northern contiguous territories. *Proceedings of the Academy of Natural Sciences of Philadelphia* **7** [1854-1855]: 299-305.
LeConte J. L. 1857: Reports upon insects collected on the Survey. *Reports of exploration and surveys for a railroad routr from the Mississippi river to the Pacific Ocean* **9** (1), 72 pp., 2 pls.
LeConte J. L. 1858: Description of new species of Coleoptera, chiefly collected by the United States and Mexican Boundary Commission, under Major W. H. Emory., U.S.A. *Proceedings of the Academy of Natural Sciences of Philadelphia* **10**: 59-89.
LeConte J. L. 1859a: The Coleoptera of Kansas and eastern New Mexico. *Smithsonian Contributions to Knowledge* **11**: 6 + 58 pp., 2 pls, 1 map.
LeConte J. L. 1859b: Catalogue of the Coleoptera of Fort Tejon, California. *Proceedings of the Academy of Natural Sciences of Philadelphia* **11**: 69-90.
LeConte J. L. 1859c: Additions to the coleopterous fauna of northern California and Oregon. *Proceedings of the Academy of Natural Sciences of Philadelphia* **11**: 281-292.
LeConte J. L. 1860: Reports of explorations and surveys for a railroad route from the Mississippi River to the Pacific Ocean. 9. 1. *Report upon Insects collected on the Survey* **4**: 1-72, 2 pls.
LeConte J. L. 1861: Classification of the Coleoptera of North America. Prepared for the Smithsonian Institution. Part I. *Smithsonian Miscellaneous Collections* **136**: xxv + 214 pp.
LeConte J. L. 1862: Classification of the Coleoptera of North America. Prepared for the Smithsonian Institution. Part I. *Smithsonian Miscellaneous Collections* **136**: 209-286.
LeConte J. L. 1863: New species of North American Coleoptera. Part I (1). *Smithsonian Miscellaneous Collections* **167** [1863-1866]: 1-92.

LeConte J. L. 1865: Prodromus of a monograph of the species of the tribe Anobiini, of the family Ptinidae, inhabiting North America. *Proceedings of the Academy of Natural Sciences of Philadelphia* **17**: 222-244.

LeConte J. L. 1866a: New species of North American Coleoptera. *Smithsonian Miscellaneous Collections* **6** (167): 87-168.

LeConte J. L. 1866b: Additions to the coleopterous fauna of the United States. No. 1. *Proceedings of the Academy of Natural Sciences of Philadelphia* **18**: 361-394.

LeConte J. L. 1869: List of Coleoptera collected in Vancouver's Island by Henry and Joseph Matthews, with descriptions of some new species. *The Annals and Magazine of Natural History* (4) **4**: 369-385.

LeConte J. L. 1874: Description of new Coleoptera chiefly from the Pacific Slope of North America. *Transactions of the American Entomological Society* **5**: 43-72.

LeConte J. L. 1875: Descriptions of new Coleoptera of the United States with notes on geographical distribution. *Transactions of the American Entomological Society* **5**: 169-176.

LeConte J. L. 1878a: Additional descriptions of new species. *Proceedings of the American Philosophical Society held at Philadelphia for Promoting useful Knowledge* **17**: 373-434.

LeConte J. L. 1878b: Descriptions of new species. Pp. 593-626. In: Hubbard H. G. & Schwarz E. A.: The Coleoptera of Michigan. *Proceedings of the American Philosophical Society* **17**: 593-699.

LeConte J. L. 1881: Synopsis of the Lampyridae of the United states. *Transactions of the American Entomological Society* **9** [1881-1882]: 15-72.

Leder H. 1872: Beschreibungen neuer Käfer aus Oran. *Berliner Entomologische Zeitschrift* **16**: 137-139.

Ledoux C. 1833: Nouvelle espèce du genre Enoplium. *Annales de la Société Entomologique de France* **2**: 474-475.

Lee S. H. 1998: Four new species of the genus Melanotus Eschscholtz (Coleoptera: Elateridae) from Korea I. *Korean Journal of Applied Entomology* **37**: 103-108.

Lee S. H. 1999: [new taxa]. In: Lee S. H., Woo K. S. & Kwo Y. J.: Systematic study of subfamily Melanotinae (Coleoptera, Melanotinae) from Korea I. The genus Melanotus Eschscholtz. *Insecta Koreana* **16**: 27-45.

Lee S. H. & Woo K. S. 1999: New species of the genus Melanotus Eschscholtz (Coleoptera: Elateridae) from Korea. *Journal of Asia-Pacific Entomology* **2**: 15-18.

Leesberg A. F. A. 1906: Insecten uit arachidennoten en Sesamzaad, in 1905 gevonden. *Entomologische Berichte* **2**: 78-81.

Lefebvre A. 1835: D'un coléoptère nouveau du genre Clerus sous ses divers états. *Annales de la Société Entomologique de France* **4**: 575-585.

Lefebvre A. 1838: Nouvelles diverses. *Bulletin de la Société Entomologique de France* **1838**: 13.

Lefkovitch L. P. 1958: A new genus of Laemophloeinae (Col., Cucujidae) from Africa. *The Entomologist's Monthly Magazine* **93**: 271-273.

Lefkovitch L. P. 1959: A revision of the European Laemophloeinae (Coleoptera: Cucujidae). *Transactions of the Royal Entomological Society of London* **111**: 95-118.

Lefkovitch L. P. 1962a: A revision of African Laemophloeinae (Coleoptera: Cucujidae). *Bulletin of the British Museum (Natural History), Entomology* **12**: 167-242.

Lefkovitch L. P. 1962b: First records and a new species of Cryptolestes Ganglbauer (Coleoptera: Cucujidae) from Afghanistan. *Annales Historico-Naturales Musei Nationalis Hungarici, Pars Zoologica* **54**: 287-288.

Lefkovitch L. P. 1965: Arabian Laemophloeinae (Coleoptera: Cucujidae). *The Proceedings of the Royal Entomological Society of London* **34**: 17-19.

Lefkovitch L. P. 1967: [new taxon]. In: Lefkovitch L. P. & Currie J. E.: Some morphological, biological and genetical differences between Cryptolestes pusillus fuscus ssp. n. and C. pusillus pusillus (Schönherr) (Coleoptera, Cucujidae). *Journal of Stored Products Research* **3**: 311-320.

Leiler T.-E. 1967: Studien über die Biologie und Beschreibung der Entwicklungsstadien einiger schwedischer Cardiophorus-Arten (Col., Elateridae). *Entomologisk Tidskrift* **88**: 89-109.

Leiler T.-E. 1969: Ptilinus banatensis Pic = Pseudoptilinus fissicollis Rtt. *Entomologisk Tidskrift* **90**: 57-58.

Leiler T.-E. 1979: Beschreibung einer neuen Mesocoelopus-Art (Col., Anobiidae) samt einer Bestimmungstabelle der paläarktischen Arten dieser Gattung. *Entomologische Blätter* **75**: 93-97.

Leiler T.-E. 1984: Zur Kenntnis alter und neuer Arten der Gattung Sphaericus Woll., Fam. Ptinidae, Diebskäfer. *Entomologische Blätter* **79**: 157-169.

Leinberg A. 1900: Über die Cateretes- (Cercus-) Varietäten. *Meddelanden af Societas pro Fauna et Flora Fennica* **26** [1899-1900]: 185-187.

Leinberg A. 1904: Über die finnischen Episernus-Arten. *Meddelanden af Societas pro Fauna et Flora Fennica* **30**: 16-22.

Leng C. W. 1908: Notes on Coccinellidae III. *Journal of the New York Entomological Society* **16**: 33-44.

Leng C. W. 1918: Notes on some changes in the list of Coleoptera. *Journal of the New York Entomological Society* **26**: 201-211.

Leng C. W. & Shoemaker E. 1915: A new genus and species of Lampyridae. *Journal of the New York Entomological Society* **23**: 55-56, pl. 5.

Lentz F. L. 1857: *Neues Verzeichniss der Preussischen Käfer.* Königsberg: Universitäts Buch- und Steindruckerei, 170 pp.

Leoni G. 1905: Due nuove varietà della Cantharis fusca L. *Rivista Coleotterologica Italiana* **3**: 129-131.

Leoni G. 1906a: Specie e varietà nuove o poco cognite di Coleotteri italiani. *Rivista Coleotterologia Italiana* **4**: 62-64.

Leoni C. 1906b: I Cebrio italiani. *Rivista Coleotterologica Italiana* **4**: 181-220, 1 pl.

Le Peletier de Saint-Fargeau A. L. M. & Audinet-Serville A. J. G. de. 1825: [new taxa]. In: Latreille P. A., Peletier de Saint-Fargeau A., Audinet-Serville A. J. G. de. & Guérin-Méneville F.: *Encyclopédie méthodique. Histoire naturelle. Entomologie, ou histoire naturelle des crustacées, des arachnides et des insectes. Tome Dixième. Par - Sca.* Paris: H. Agasse, [7] + 344 pp. [Pars 2, pp. 345-832, issued in 1828].

Lepesme P. 1939: Note synonymique sur les Dermestes (Coleoptera) et description d'une espèce et d'une variété nouvelles. *Bulletin de la Société Entomologique de France* **44**: 190-193.

Lepesme P. 1947: Bostrychoidea. Pp. 194-233. In: Fleutiaux E. J. B., Legros C., Lepesme P. & Paulian R.: *Coléoptères des Antilles. Vol. I., Faune de l'Empire Français* **7**: 1-239.

Lepesme P. 1950: Révision des Dermestes (Coleoptera, Dermestidae). *Annales de la Société Entomologique de France* **115** [1949]: 37-68.

Lepesme P. & Paulian R. 1944: Les Nemosoma et genres voisins (Col. Ostomatidae). *Revue Française d'Entomologie* **10**: 136-141.

Leschen R. A. B. 1996: Phylogeny and revision of the genera of Cryptophagidae (Coleoptera: Cucujoidea). *Kansas Science Bulletin* **55**: 549-634.

Leschen R. A. B. & Bowestead S. 2001: Synonymical notes for Corylophidae and Cryptophagidae. *The Coleopterists Bulletin* **55**: 312-316.

Leseigneur L. 1955: Coléoptères élatérides du Sahara nord-occidental. *Bulletin de la Société Entomologique de France* **60**: 86-92.

Leseigneur L. 1958: Description d'un Elateridae français nouveau (Col.). *Bulletin de la Société Entomologique de France* **63**: 88-91.

Leseigneur L. 1969: Contribution à l'étude du genre Adrastus Eschsch. (Coleoptera Elateridae). *Bollettino della Società Entomologica Italiana* **99-101**: 86-95.

Leseigneur L. 1970: Révision des Zorochrus européens (Col. Elateridae). *Bulletin Mensuel de la Société Linnéenne de Lyon* **39**: 19-44.

Leseigneur L. 1972: Coléoptères Elateridae de la faune de France continentale et de Corse. *Bulletin Mensuel de la Société Linnéenne de Lyon* (*Supplément au numéro de Février 1972*) **41**: 1-381.

Leseigneur L. 1974: Découverte en Corse d'un élatéridé nouveau, Adrastus temperi, nov. sp. *L'Entomologiste* **30**: 112-116.

Leseigneur L. 2004a: Cidnopus marginellus (Perris, 1864) (Coleoptera, Elateridae), espèce nouvelle pour la faune de France. Cidnopus marginellus subsp. brusteli nov. *Bulletin Mensuel de la Société Linnéenne* de Lyon **73**: 230-236.

Leseigneur L. 2004b: Description de Athous (Grypathous) lambeleti nov. sp. (Coleoptera: Elateridae) des Pyrenées-Atlantiques (France) et de Biscaye (Espagne). *Bulletin Mensuel de la Société Linnéenne* de Lyon **73**: 237-240.

Lesne P. 1895: Descriptions de genres nouveaux et d'espèces nouvelles de coléoptères de la famille des Bostrychides. *Annales de la Société Entomologique de France* **64**: 169-178.

Lesne P. 1896: Contributions à la faune indo-chinoise. 17e Mémoire. Bostrychidae. *Annales de la Société Entomologique de France* **65**: 511-512.

Lesne P. 1897a: Bostrychides indiens de la collection H.-E. Andrewes. *Annales de la Société Entomologique de Belgique* **41**: 18-22.

Lesne P. 1897b: Descriptions de deux espèces nouvelles de bostrychides algériens (Col.). *Bulletin de la Société Entomologique de France* **1897**: 235-237.

Lesne P. 1898a: Sur un coléoptère nouveau de la famille des lyctides. *Bulletin du Muséum National d'Histoire Naturelle* (Paris) **4**: 139-140.

Lesne P. 1898b: Révision des coléoptères de la famille des bostrychides. 2e Mémoire. Bostrychides hypocéphales. Dinoderinae. *Annales de la Société Entomologique de France* **66** [1897]: 319-350.

Lesne P. 1899a: Liste des bostrychides et lyctides recueillis sur le littoral de la baie de Tadjourah et description d'une espèce nouvelle. *Bulletin du Muséum National d'Histoire Naturelle* (Paris) **5**: 226-228.

Lesne P. 1899b: Révision des coléoptères de la famille des bostrychides. 3e Mémoire. Bostrychinae. *Annales de la Société Entomologique de France* **67**: 438-621.

Lesne P. 1901a: Diagnose d'un type générique nouveau de la tribu des Psoinae (Col.). *Bulletin de la Société Entomologique de France* **1901**: 348-350.

Lesne P. 1901b: Révision des coléoptères de la famille des bostrychides. 4e Mémoire. Bostrychinae sens. strict. – II. Les Xylopertha. *Annales de la Société Entomologique de France* **69**: 473-639.

Lesne P. 1902a: Synopsis des bostrychides paléarctiques. *L'Abeille, Journal d'Entomologie* **30**: 105-136.

Lesne P. 1902b: Voyage du Dr W. Horn à Ceylan. Bostrychidae et Anobiidae. *Annales de la Société Entomologique de France* **71**: 476-479.

Lesne P. 1904: Supplément au synopsis des bostrychides paléarctiques. *L'Abeille Journal d'Entomologie* **30**: 153-168 + 4 pl.

Lesne P. 1905: Un cléride saharien nouveau (Opilo desertorum nov. sp.). *Bulletin du Muséum d'Histoire Naturelle* (Paris) **11**: 388-390.

Lesne P. 1906a: Bostrychides nouveaux ou peu connus. *Annales de la Société Entomologique de France* **75**: 393-428.

Lesne P. 1906b: Révision des coléoptères de la famille des bostrychides. 5e Mémoire. Sinoxylinae. *Annales de la Société Entomologique de France* **75**: 445-561.

Lesne P. 1907: Un Lyctus Africain nouveau. (Col.). *Bulletin de la Société Entomologique de France* **1907**: 302-303.

Lesne P. 1908: Notes sur les coléoptères térédiles, un nouveau lyctide apparenté au Lyctus brunneus. *Bulletin du Muséum National d'Histoire Naturelle* (Paris) **15**: 356-358.

Lesne P. 1909: Révision des coléoptères de la famille des bostrychides. 6e Mémoire: Dinapatinae et Apatinae. *Annales de la Société Entomologique de France* **78**: 471-574 + 5 pl.

Lesne P. 1911a: Notes sur les coléoptères térediles. 6. Un Lyctidae paléarctique nouveau. *Bulletin du Muséum National d'Histoire Naturelle* (Paris) **17**: 48-50.

Lesne P. 1911b: Notes sur les coléoptères térediles. 8. Lyctides nouveaux du Mexique. *Bulletin du Muséum National d'Histoire Naturelle* (Paris) **17**: 534-538.

Lesne P. 1913a: Notes sur les coléoptères térediles. *Bulletin du Muséum National d'Histoire Naturelle* (Paris) **19**: 190-193.

Lesne P. 1913b: Un type nouveau d'Anobiidae appartenant à la faune Japonaise (Col.). *Bulletin de la Société Entomologique de France* **1913**: 292-296.

Lesne P. 1914: Un type nouveau de dinodériens. Variabilité du tarse chez les bostrychides (Col.). *Bulletin de la Société Entomologique de France* **1914**: 242-245.

Lesne P. 1916: Notes sur les coléoptères térediles. 15. Variabilité de certains lyctides de l'Amérique du Nord. Les formes typiques du genre Lyctus. *Bulletin du Muséum National d'Histoire Naturelle* (Paris) **22**: 92-100.

Lesne P. 1920: Notes sur les coléoptères térediles. 18. Un bostrychide nouveau de la faune yunnanaise. *Bulletin du Muséum National d'Histoire Naturelle* (Paris) **26**: 295-297.

Lesne P. 1921a: Un lycide nouveau de la faune française. *Bulletin de la Société Entomologique de France* **1921**: 180-182.

Lesne P. 1921b: Classification des coléoptères xylophages de la famille des bostrychides. *Association française pour l'avancement des sciences fusionnée avec l'Association scientifique de France, Strasbourg* **1920**: 285-289.

Lesne P. 1923: Notes sur les coléoptères térédiles. 19. Diagnose préliminaires de bostrychides nouveaux de l'Afrique tropicale. *Bulletin du Muséum National d'Histoire Naturelle* (Paris) **29**: 55-60.

Lesne P. 1924: Les coléoptères bostrychides de l'Afrique tropicale française. *Encyclopédie Entomologique* (Paris) **3**: 1-301, 1 pl.

Lesne P. 1925: Description d'un cléride nouveau du Maroc. *Bulletin de la Société des Sciences Naturelles du Maroc* **5**: 15-16.

Lesne P. 1925: Les coléoptères sphindides du bassin de l'Océan Indien. *Comptes Rendus du Congrès des Sociétés Savantes en 1925, Sciences,* pp. 510-513 [seen as reprint].

Lesne P. 1928: Deux clérides nouveaux de la région indo-chinoise. *Encyclopédie Entomologique, Série B Coleoptera* **3**: 35-37.

Lesne P. 1931: Notes sur les coléoptères térédiles. 20. Diagnoses de bostrychides nouveau faisant partie des collections du Muséum. *Bulletin du Muséum National d'Histoire Naturelle* (Paris) (2) **3**: 96-99.

Lesne P. 1932: Notes sur les coléoptères térédiles. 22. Diagnoses de bostrychides nouveaux de l'Asia orientale. *Bulletin du Muséum National d'Histoire Naturelle* (Paris) (2) **4**: 651-663.

Lesne P. 1933: Trois Dinoderus indo-malais nouveaux (Col. Bostrychidae). *Bulletin de la Société Entomologique de France* **38**: 257-260.

Lesne P. 1934a: Sur le genre Hendecatomus Mellié (Col. Bostrychidae). *Bulletin de la Société Entomologique de France* **39**: 174-175.

Lesne P. 1934b: Sur le Phonapate frontalis Fahr. (Col., Bostrychidae). *Bulletin de la Société Entomologique de France* **39**: 217-220.

Lesne P. 1935a: Les bostrychides de l'Arabie. *Revue Française d'Entomologie* **1** [1934]: 268-272.

Lesne P. 1935b: Lyctidae (Col.). Pp. 300-301. In: Visser P. C. & Visser-Hooft J. (eds): *Wissenschaftliche Ergebnisse der niederländischen Expeditionen in den Karakorum und die angrenzenden Gebiete in den Jahren 1922, 1925 und 1929/30. Vol. 1.* Leipzig: F. A. Brockhaus, xviii + 499 pp., 1 map, 8 pls.

Lesne P. 1936: Diagnoses préliminaires de lyctides et bostrychides nouveaux (Col.) de l'Indo et l'Austro-Malaisie. *Bulletin de la Société Entomologique de France* **41**: 131-137.

Lesne P. 1937: Xyloperthini paléarctiques peu connus ou nouveaux. *Bulletin de la Société Entomologique de France* **42**: 195-200.

Lesne P. 1941a: Sur les genres Xylogenes et Xylomedes. (Col. Bostrychidae). *Annales de la Société Entomologique de France* **109**: 131-136.

Lesne P. 1941b: Sur quelques bostrychides indiens. *Annales de la Société Entomologique de France* **109**: 137-152.

Letzner K. 1845: Bemerkungen zu Cantharis lemanoceros und denticollis Schummel. *Uebersicht der Arbeiten und Veränderungen der Schlesischen Gesellschaft für Vaterländische Kultur* (Breslau) **1844**: 71-72.

Letzner K. 1847: Zwei neue schlesische Arten der Gattung Cantharis. In: 2. Bericht über die Beschäftigung der Entomologischen Section im Jahre 1846. Erste Ordnung, Käfer. *Uebersicht der Arbeiten und Veränderungen der Schlesischen Gesellschaft für Vaterländische Kultur* (Breslau) **1846**: 73-85.

Léveillé A. 1889: Description de Temnochilides nouveaux. *Annales de la Société Entomologique de France* (6) **8** [1888]: 411-428.

Léveillé A. 1895: [Description de dix espèces nouvelles de coléoptères de la famille des temnochilides]. *Bulletin de la Société Entomologique de France* **1894**: cxliii-cxlviii.148.

Léveillé A. 1903: Diagnoses de deux espéces nouvelles de Temnochila et note sur une variété nouvelle de Tenebroides (Col.). *Bulletin de la Société Entomologique de France* **1903**: 190-191.

Léveillé A. 1905: Diagnoses de temnochilides nouveaux (Col.). *Bulletin de la Société Entomologique de France* **1905**: 92-93.

Léveillé A. 1907: Études sur la famille des temnochilides III. *Annales de la Société Entomologique de France* **76**: 309-411.

Léveillé A. 1908: Descriptions de deux formes nouvelles de temnochilides (Col.) et note rectificative. *Bulletin de la Société Entomologique de France* **1908**: 212-213.

Levrat J. N. B. G. 1857: Description de trois coléoptères nouveaux. *Annales de la Société Linnéenne de Lyon* (2) **4**: 417-419.

Levrat J. N. B. G. 1859: Description de quelques coléoptères nouveaux des environs de Tunis. *Études Entomologiques* (Lyon) **1**: 33-36.

Lewis G. 1874: Descriptions of three new species of Erotylidae. *The Entomologist's Monthly Magazine* **11**: 138-140.

Lewis G. 1879a: *A catalogue of Coleoptera from the Japanese Archipelago.* London: Tylor & Francis, 31 pp.

Lewis G. 1879b: Diagnoses of Elateridae from Japan. *The Entomologist's Monthly Magazine* **16**: 155-157.

Lewis G. 1879c: Description of new species of Cucujus from Assam, and of Ceratomyrrhina gemina from West Africa. *The Entomologist's Monthly Magazine* **15**: 234.

Lewis G. 1879d: On certain new species of Coleoptera from Japan. *The Annals and Magazine of Natural History* (5) **4**: 459-467.

Lewis G. 1881a: Notes on Cucujidae in Japan, with diagnosis of a new species. *The Entomologist's Monthly Magazine* **17**: 198-199.

Lewis G. 1881b: A new species of Helotidae from Japan. *The Entomologist's Monthly Magazine* **17**: 255-256.

Lewis G. 1883a: A new species of Peltastica Mannerh. (Trogositidae). *The Entomologist's Monthly Magazine* **20**: 79.

Lewis G. 1883b: On three new species of Japan Erotylidae, and notes of others. *The Entomologist's Monthly Magazine* **20**: 138-140.

Lewis G. 1884: Japanese Languriidae, with notes on their habits and external sexual structure. *Journal of the Linnaean Society, Zoology* **17**: 347-361.

Lewis G. 1887a: On a new genus of Erotylidae. *The Entomologist's Monthly Magazine* **24**: 3-4.

Lewis G. 1887b: A list of fifty Erotylidae from Japan, including thirty-five new species and four new genera. *The Annals and Magazine of Natural History* (5) **20**: 53-73.

Lewis G. 1887c: A list of the Japanese Silphidae. *The Annals and Magazine of Natural History* (5) **20**: 338-342.

Lewis G. 1888: Note on a species of Acrops from Japan. *The Entomologist* **21**: 220-221.

Lewis G. 1889a: On a species of Nosodendron from Japan. *The Entomologist's Monthly Magazine* **25**: 229.

Lewis G. 1889b: On a new genus of Coleoptera (Trogositidae). *The Annals and Magazine of Natural History* (6) **4**: 273-275.

Lewis G. 1889c: On a species of Amblyopus (Erotylidae) from Japan. *The Entomologist's Monthly Magazine* **25**: 397-398.

Lewis G. 1892: On the Japanese Cleridae. *The Annals and Magazine of Natural History* (6) **10**: 183-192.

Lewis G. 1893a: Japanese Coleoptera – two new species. *The Entomologist's Monthly Magazine* **29**: 83.

Lewis G. 1893b: A list of Coleoptera new to the fauna of Japan, with notices of unrecorded synonyms. *The Entomologist* **26**: 150-153.

Lewis G. 1894a: On the Elateridae of Japan. *The Annals and Magazine of Natural History* (6) **13**: 26-48, 182-201, 255-266, 311-320.

Lewis G. 1894b: On new species of Trogositidae from Japan. *The Entomologist's Monthly Magazine* **30**: 32-34.

Lewis G. 1895a: On the Dascillidae and malacoderm Coleoptera of Japan. *The Annals and Magazine of Natural History* (6) **16**: 98-122, pl. 6.

Lewis G. 1895b: On the Lamellicorn Coleoptera of Japan, and notices of some others. *The Annals and Magazine of Natural History* (6) **16**: 374-408.

Lewis G. 1896a: Notes on Japanese Coccinellidae. *The Annals and Magazine of Natural History* (6) **17**: 22-41.

Lewis G. 1896b: On a new species of Hylecoetus (Lymexylonidae) from Japan. *The Annals and Magazine of Natural History* (6) **17**: 65.

Lewis G. 1896c: On new species of Coleoptera from Japan, and notices of others. *The Annals and Magazine of Natural History* (6) **17**: 329-343.

Li J. 1992: *The Coleoptera fauna of northeast China*. Jilin: Jilin Education Publishing House, 205 pp. (in Chinese, English title).

Li J. 1993: [new taxa]. In: Li J. & Chen P.: *Studies on fauna and ecogeography of soil animal*. Chanchun: Northeast Normal University Press, [8] + 265 pp. (in Chinese, with English title).

Liberti G. 1963: Nota sul Dasytes plumbeus Muell. *Bollettino della Società Entomologica Italiana* **93**: 84-86.

Liberti G. 1979: Revisione delle specie italiane del genere Danacaea, primo gruppo (Coleoptera Dasytidae). *Memorie della Società Entomologica Italiana* **57** [1978]: 29-45.

Liberti G. 1981: Description of a new Danacaea from Calabria (Italy) (Col., Dasytidae). *Entomologica Basiliensia* **6**: 420-421.

Liberti G. 1982: Dasytes (Mesodasytes) of Taiwan (Col., Dasytidae). *Entomologica Basiliensia* **7**: 373-380.

Liberti G. 1984a: III Contributo alla conoscenza del genere Danacaea. Revisione della specie italiane del 4° gruppo. (Coleoptera Dasytidae). *Atti della Società Italiana di Scienze Naturali e del Museo Civico di Storia Naturale di Milano* **125**: 159-179.

Liberti G. 1984b: Dasytes (Mesodasytes) of Taiwan (Coleoptera, Dasytidae) (Part 2). *Entomologica Basiliensia* **9**: 324-327.

Liberti G. 1985: IV Contributo alla conoscenza del genere Danacaea Cast. (Col., Dasytidae). Descrizione del sottogenere nuovo Allodanacaea e revisione della specie italiane. *Annali del Museo Civico di Storia Naturale di Genova* **85**: 333-362.

Liberti G. 1986: Notes on some Dasytiscus Kiesw. (Col., Dasytidae) from Greece. *Giornale Italiano di Entomologia* **3**: 185-193.

Liberti G. 1989: V Contributo alla conoscenza del genere Danacaea Cast. (Coleoptera, Dasytidae). Revisione della specie italiane del II e del III gruppo. *Entomologica Basiliensia* **13**: 279-302.

Liberti G. 1995: Revisione della specie italiane del genere Aplocnemus Stephens (Coleoptera, Melyridae, Rhadalinae). *Memorie della Società Entomologica Italiana* **73** [1994]: 153-194.

Liberti G. 2004: Il genere Dasytes in Italia. Revisione e catalogo topografico, sinonimico e bibliografico delle specie italiane (Col. Dasytidae). *Annali del Museo Civico di Storia Naturale di Genova* **96**: 253-340.

Liberti G. & Baviera C. 2004: Le Danacea di Sicilia, con descrizione di D. zingara n. sp. (Coleoptera Dasytidae). *Bollettino della Società Entomologica Italiana* **136**: 145-156.

Liberti G. & Schembri S. 2002: Dasytidae of the Maltese archipelago (Coleoptera Cleroidea). *Bollettino della Società Entomologica Italiana* **134**: 175-186.

Liberto A. & Wurst C. 1999: A new species of Cardiophorus from La Palma (Canary Islands) (Coleoptera, Elateridae). *Fragmenta Entomologica* **31**: 333-338.

Lindberg Har. 1943: Nykomlingar till Finlands skalbaggsfauna III. *Notulae Entomologicae* **23**: 50-58.

Lindberg Har. 1951: Beitrag zur Kenntnis der Käferfauna der Kanarischen Inseln. *Commentationes Biologicae* **10** (18) [1950]: 1-20.

Lindberg Har. 1953: Zweiter Beitrag zur Kenntnis der Käferfauna der Kanarischen Inseln. *Commentationes Biologicae* **13** (12): 1-18.

Lindemann K. 1865: Zwei neue Käfer. *Bulletin de la Société Impériale des Naturalistes de Moscou* **38** (3): 147-149.

Linder J. 1863: [new taxa]. In: Grenier A. (ed.): *Catalogue des coléoptères de France et matériaux pour servir à la faune des coléoptères français*. Paris: L. Toinon, iv + 3-79 + 135 pp.

Linder J. 1864: Description de trois coléoptères européens nouveaux. *Annales de la Société Entomologique de France* (4) **4**: 250-252.

Linnaeus C. 1758: *Systema naturae per regna tria naturae, secundum classes, ordines, genera, species, cum characteribus, differentiis, synonymis, locis. Tomus I. Editio decima, reformata.* Holmiae: Laurentii Salvii, 823 pp.

Linnaeus C. 1761: *Fauna Suecica sistens Animalia Sueciae Regni: Mammalia, Aves, Amphibia, Pisces, Insecta, Vermes. Distributa per classes et ordines, genera et species, cum differentiis specierum, synonymis auctorum, nominibus incolarum, locis natalium, descriptionibus Insectorum. Editio altera, auctior.* Stockholmiae: Laurentii Salvii, 48 + 578 pp., 2 pls.

Linnaeus C. 1767: *Systema naturae, per regna tria naturae, secundum classes, ordines, genera. Species cum characteribus, differentiis, synonimis, locis. Tomus I. Editio duodecima. Tomus I., Pars II.* Holmiae: Laurentii Salvii, 2 + 533-1327 + [37] pp.

Liu C.-L. 1950: On the recognition of the so-called 28-spoted lady beetles of Nord China, with description of Epilachna operculata n. sp. *Scientia Sinica* **1**: 159-164 (in Chinese).

Liu C.-L. 1962: Description of new Chinese Coccinellids and observations on the external genitalia of ladybeetles. *Acta Entomologica Sinica* **11**: 261-268 (in Chinese).

Liu Y. & Zhang S. 1986: Chinese dermestid beetles associated with stored products. Beijing: Agriculture Publishing House, 170 pp. (in Chinese).

Logvinovskiy V. D. 1977a: K faune i sistematike zhukov-tochil"shchikov podsem. Ernobiinae (Coleoptera, Anobiidae) SSSR. *Entomologicheskoe Obozrenie* **56**: 121-131.

Logvinovskiy V. D. 1977b: Obzor zhukov-tochi"shchikov roda Lasioderma Stephens (Coleoptera, Anobiidae) fauny SSSR i Mongolii. *Nasekomyye Mongolii* **5**: 278-289.

Logvinovskiy V. D. 1977c: [new taxon]. In: Emetz V. M. & Logvinovskiy V. D.: Obzor zhukov-tochi"shchikov roda Xyletinus Latr. (Coleoptera, Anobiidae) fauny SSSR. *Entomologicheskoe Obozrenie* **56**: 409-419.

Logvinovskiy V. D. 1978: Novye rody i vidy zhukov-tochi"shchikov (Coleoptera, Anobiidae) iz Kazakhstana i Sredney Azii. *Trudy Zoologicheskogo Instituta AN SSSR* **71**: 26-30.

Logvinovskiy V. D. 1979a: Novyy rod i novye vidy zhukov-tochi"shchikov (Coleoptera, Anobiidae) iz Sredney Azii. *Trudy Zoologicheskogo Instituta AN SSSR* **88**: 42-44.

Logvinovskiy V. D. 1979b: Novyy vid zhuka-tochi"shchika (Coleoptera, Anobiidae) s Dal"nego Vostoka. *Trudy Vshesoyuznogo Entomologicheskogo Obshchestva AN SSSR* **11**: 69.

Logvinovskiy V. D. 1980: Obzor zhukov-tochi"shchikov roda Dorcatoma Herbst (Coleoptera, Anobiidae) fauny SSSR. *Entomologicheskoe Obozrenie* **59**: 148-153.

Logvinovskiy V. D. 1981: Novye vidy zhukov-tochi"shchikov (Coleoptera, Anobiidae) s Dal"nego Vostoka. Pp. 32-33. In: *Novye svedeniya o nasekomykh Dal"nego Vostoka.* Vladivostok: Dal'nevostochniy Nauchniy Tzentr. Biologo-Pochvenniy Institut, 125 pp.

Lohde R. 1899: Umtaufe einiger Cleriden-Arten. *Entomologische Nachrichten* **25**: 305.

Lohse G. A. 1954: Neuheiten der deutschen Käferfauna. *Deutscher Entomologentag in Hamburg* **1954**: 209-213.

Lohse G. A. 1957: Die mitteleuropaischen Arten der Gattung Xyletinus Latr. *Entomologische Blätter* **53**: 27-37.

Lohse G. A. 1959: Neuheiten der Deutschen Käferfauna 6. *Entomologische Blätter* **55**: 29-35.

Lohse G. A. 1960: Neuheiten der Deutschen Käferfauna 7. *Entomologische Blätter* **56**: 32-40.

Lohse G. A. 1970: Ernobius freudei n. sp., eine neue Anobiidae aus den Alpen (Col. Anob.). *Nachrichtenblatt der Bayerischen Entomologen* **18**: 99-100.

Lohse G. A. 1976: Elateriden-Studien. *Entomologische Blätter* **72**: 90-104.

Lohse G. A. 1977: Die mitteleuropaischen Arten der Gattung Haplocnemus Steph. sowie synonymische Bemerkungen zu anderen Melyriden. *Entomologische Blätter* **73**: 177-183.

Lohse G. A. 1978: Elateridenstudien II. *Entomologische Blätter* **74**: 21-28.

Lohse G. A. 1981: Enicmus amici n.sp., eine neue Lathridiide aus Mitteleuropa. *Entomologische Blätter* **77**: 9-10.

Lohse G. A. 1991: Ernobius angelinii n. sp., eine neue Anobiide aus Italien (Coleoptera, Anobiidae). *Acta Coleopterologica* **7**: 49-51.

Łomnicki J. 1923: Athous Mokrzecki nov. sp. Coleopt., Elateridae. *Polskie Pismo Entomologiczne* **2**: 53-54.

López-Colón J. I. 1997: Troglops alonsoi n. sp. de Maláquido de la Sierra de Gredos (Coleoptera, Malachiidae). *Lambillionea* **97**: 445-447.

Lövendal E. A. 1893: Fortegnelse over de i Danmark levende Cryptophagidae og Lathridiidae [contd.]. *Entomologisk Meddelelser* **3** [1892]: 241-276.

Lucas P. H. 1842: Sur une nouvelle espèce du genre Drilus qui habite le nord d'Afrique. *Comptes Rendus de l'Académie de Sciences Paris* **15**: 1187-1189.

Lucas P. H. 1843a: Note sur quelques nouvelles espèces d'insectes de la famille des Trachélides qui habitent les possessions françaises du nord de l'Afrique. *Revue Zoologique, par la Société Cuvierienne* **6**: 145-147.

Lucas P. H. 1843b: [Sur plusieurs coléoptères nouveaux, du nord de l'Afrique, nés à Paris]. *Revue Zoologique, par la Société Cuvieruenne* **6**: 158-159.

Lucas P. H. 1843c: [Note sur plusieurs coléoptères nouveaux du nord de l'Afrique, nés à Paris]. *Bulletin de la Société Entomologique de France* (2) **1**: xxiv-xxvi.

Lucas P. H. 1846: Pp. 1-360. In: *Exploration scientifique de l'Algérie pendant les années 1840, 1841, 1842 publiée par ordre du gouvernement et avec le concours d'une Commission Académique. Sciences Physiques Zoologie. Vol. II. Histoire naturelle des animaux articulés. Cinquième classe. Insectes. Premier Ordre. Les coléoptères.* Paris: Imprimerie Nationale [1849], 590 pp., 47 pls. (pp. 361-448 in 1847, the year 1849 is given for the rest).

Lucas P. H. 1847: Pp. 361-447. In: *Exploration scientifique de l'Algérie pendant les années 1840, 1841, 1842 publiée par ordre du gouvernement et avec le concours d'une Commission Académique. Sciences Physiques Zoologie. Vol. II. Histoire naturelle des animaux articulés. Cinquième classe. Insectes. Premier Ordre. Les coléoptères.* Paris: Imprimerie Nationale [1849], 590 pp., 47 pls. (pp. 1-360 issued in 1846, pp. 449-590 in 1849).

Lucas P. H. 1849: Pp. 449-590. In: *Exploration scientifique de l'Algérie pendant les années 1840, 1841, 1842 publiée par ordre du gouvernement et avec le concours d'une Commission Académique. Sciences Physiques Zoologie. Vol. II. Histoire naturelle des animaux articulés. Cinquième classe. Insectes. Premier Ordre. Les coléoptères.* Paris: Imprimerie Nationale [1849], 590 pp., 47 pls. (pp. 1-360 issued in 1846, pp. 361-448 in 1847).

Lucas P. H. 1850: Observations sur un nouveau genre de l'ordre des coléoptères (genus Stenomera) qui habite les possessions françaises du nord de l'Afrique. *Annales de la Société Entomologique de France* (2) **8**: 31-43.

Lucas P. H. 1852: [Communication]. *Bulletin de la Société Entomologique de France* **1852**: xxviii-xxix.

Lucas P. H. 1853: Essai sur les animaux articulés qui habitent l'île de Crête. *Revue et Magasin de Zoologie Pure et Appliquée* (2) **5**: 565-576.

Lucas P. H. 1870: Note sur le Malacogaster bassii (sp. nov.) vivant dans les Helix Dupoteti, Lucasi, etc. *Bulletin de la Société Entomologique de France* **1870**: lvii-lviii.

Lucas P. H. 1871: Description et figure de deux sexes d'une nouvelle espèce de Malacogaster, précédées de quelques remarques sur cette coupe générique de l'ordre des coléoptères et de la tribu des Malacodermes. *Annales de la Société Entomologique de France* (5) **1**: 19-28.

Lucas R. 1920: *Catalogus alphabeticus generum et subgenerum Coleopterorum orbis terrarum totius (famil., trib., subtrib., sect. incl.) I.* Berlin: Nicolaische Verlags-Buchhandlung R. Stricker, 696 pp.

Lucht W. 1982: Zwei neue Melasis-Arten aus Taiwan (Col. Eucnemidae). *Entomologische Blätter* **78**: 15-19.

Lucht W. 1984: Dirhagus klapperichi n. sp. eine neue Eucnemidenart aus China (Coleoptera, Eucnemidae). *Bonner Zoologische Beiträge* **35**: 289-294.

Lucht W. 1986: Calyptocerus favipunctatus n. sp. aus Taiwan (Col., Eucnemidae, Melasinae). *Entomologische Blätter* **82**: 103-106.

Lucht W. 1987: Die Gattungen Chapianus Fleutiaux und Pseudochapianus n. gen. (Col. Eucnemidae, Eucneminae) nebst Beschreibung je einer neuen Art. *Entomologische Blätter* **83**: 33-41.

Lucht W. 1989: Drei neue Arten der Gattung Otho Lacordaire, 1857 (Coleoptera, Eucnemidae, Melasinae). *Entomologica Basiliensia* **13**: 175-186.

Ludovici F. F. & Schaufuss L. W. 1869: *Beitrag zur Kenntniss der Coleopteren-Fauna der Balearen*. Prag: Heinr. Mercy, 31 pp.

Luigioni P. 1920: Contributo allo studio della fauna coleotterologica del Lazio. *Atti della Pontificia Accademia Romana delle Scienze - Nuovi Lincei* (Roma) **73**: 186-214.

Luigioni P. 1926: Una nuova specie italiana del genere „Enoplium" Latr. *Atti della Pontificia Accademia delle Scienze Nuovi Lincei* **79** [1925-1926]: 51-53.

Luna de Carvalho E. 1951: Notas coleopterológicas (IV nota). *Boletim da Sociedad Portuguesa de Ciências Naturais Lisboa* (2) **3**: 161-171.

Luna de Carvalho E. 1985: Anomatideos de Portugal (Coleoptera, Cucujoidea, Anommatidae). *Actas do Congresso Iberico de Entomologia* **2**: 85-91.

Lundberg S. 1969: Xyletinus vaederoeensis nov. sp. (Col. Anobiidae). *Entomologisk Tidskrift* **90**: 211-212.

Lundberg S. 1972: Xyletinus suecicus nov. sp. (Col. Anobiidae). *Entomologisk Tidskrift* **93**: 57-59.

Lundblad O. 1949: Die schwedischen Xyletinus-Arten. *Entomologisk Tidskrift* **70**: 1-5.

Lusis Ya. Ya. 1973: Formu taxonomiskas miattiecibas un geografiska izplate marisu ginti Adalia Mulsant [Taxonomical relationships and geographical distribution of forms in the ladybird genus Adalia Mulsant]. *Uchenye Zapiski, Problemy Genetiki i Evoljucii* (Riga) **1**: 101-123.

Lyubarsky [=Ljubarsky] G. Y. 1987: Novye taksony zhukov-skrytnoedov (Coleoptera, Cryptophagidae) fauny SSSR [New species of Cryptophagid beetles (Coleoptera: Cryptophagidae) in the fauna of the USSR]. *Zoologicheskiy Zhurnal* **66**: 951-954.

Lyubarsky [=Ljubarsky] G. Y. 1989: New taxa of Cryptophagidae beetles in fauna of the USSR. *Byulleten Moskovskogo Obshchestva Ispytateley Prirody, Otdel Biologicheskiy* (N. S.) **94**: 49-62.

Lyubarsky [=Ljubarsky] G. Y. 1991a: Novyy vid roda Cryptophagus (Coleoptera, Cryptophagidae) iz Prikaspiya [New species of the genus Cryptophagus (Coleoptera, Cryptophagidae) from the Caspian Sea Region]. *Zoologicheskiy Zhurnal* **70** (4): 145-146.

Lyubarsky [=Ljubarsky] G. Y. 1991b: Obzor palearkticheskikh vidov roda Antherophagus (Coleoptera, Cryptophagidae) [Review of palaearctic species of the genus Antherophagus (Coleoptera, Cryptophagidae)]. *Zoologicheskiy Zhurnal* **70** (7): 63-67.

Lyubarsky [=Ljubarsky] G. Y. 1992a: Kavkazkie zhuki-skrytnoedy roda Cryptophagus (Coleoptera, Cryptophagidae) [Beetles of the genus Cryptophagus (Coleoptera, Cryptophagidae) from the Caucasus]. *Zoologicheskiy Zhurnal* **71** (10): 68-82.

Lyubarsky [=Ljubarsky] G. Y. 1992b: Zhuki-skrytnoedy (Coleoptera, Cryptophagidae) severo-vostoka Rossii [Mould beetles (Coleoptera, Cryptophagidae) from the north-east of Russia]. *Zoologicheskiy Zhurnal* **71** (10): 83-89.

Lyubarsky [=Ljubarsky] G. Y. 1993a: Review of the genus Phalacrus Payk. (Coleoptera Phalacridae) of Oriental Region from V. Motschulsky's collection. *Russian Entomological Journal* **2** (1): 13-22.

Lyubarsky G. Y. 1993b: New species of the genus Olibrus Erichson, 1845 from the oriental region (Coleoptera, Phalacridae). *Russian Entomological Journal* **2** (5-6): 39-46.

Lyubarsky [=Ljubarsky] G. Y. 1995: Cryptophagidae and some Languriidae from palaearctic China. *Russian Entomological Journal* **4**: 45-53.

Lyubarsky [=Ljubarsky] G. Y. 1997a: Cryptophagidae and Languriidae from India (Coleoptera, Clavicornia). *Entomofauna Zeitschrift für Entomologie* **18**: 49-57.

Lyubarsky [=Ljubarsky] G. Y. 1997b: On new or poorly known Cryptophagidae and Languriidae from Middle Asia (Insecta: Coleoptera). *Reichenbachia* **32**: 53-57.

Lyubarsky [=Ljubarsky] G. Y. 1997c: Chinophagus mirabilis n. gen., n. sp. of Languriidae from China with new records of Cryptophagidae (Coleoptera, Clavicornia). *Miscellànea Zoològica* **20**: 111-117.

Lyubarsky [=Ljubarsky] G. Y. 1998: Scytomaria, a new genus of Cryptophagidae (Coleoptera) from the Nepal Himalayas. *Stuttgarter Beiträge zur Naturkunde* (A) (*Biologie*) **572**: 1-6.

Lyubarsky [=Ljubarsky] G. Y. 1999: Cryptophagidae (Coleoptera) from the Himalayas, with descriptions of new species, keys and remarks to some Languriidae. *Stuttgarter Beiträge zur Naturkunde* (A) (*Biologie*) **598**: 1-27.

Lyubarsky [=Ljubarsky] G. Y. 2000: New and rare species of the family Cryptophagidae from northern Palaearctic (former USSR) (Coleoptera: Clavicornia). *Russian Entomological Journal* **9**: 321-328.

Lyubarsky [=Ljubarsky] G. Y. 2003: Phalacridae from Nepal and other districts of Oriental Region. *Russian Entomological Journal* **12**: 59-69.

MacLeay W. J. 1871: Notes on a collection of insects from Gayndah. *Transactions of the Entomological Society of New South Wales* **2** [1869-1873]: 79-205.

MacLeay W. J. 1872: Notes on a collection of insects from Gaydah. *Transactions of the Entomological Society of New South Wales* **2** [1869-1873]: 239-318.

MacLeay W. J. 1886: The insects of the fly river, New Guinea, "Coleoptera". *Proceedings of the Linnaean Society of New South Wales* (2) **1**: 136-157.

MacLeay W. S. 1825: *Annulosa Javanica, or an attempt to illustrate the natural affinities and analogies of the insects collected in Java by Thomas Horsfield, M. D. F. L & G. S. and deposited by him in the Museum of the Honourable East-India Company. Number 1*. London: Kingsbury, Parbury, and Allen, xii + 50 pp., 1 pl.

MacLeay W. S. 1826: Annulosa. Catalogue of insects, collected by Captain King, R. N. Vol. 2. Appendix B. Pp. 438-473. In: King P. P.: *Narrative of a survey of the intertropical and western coasts of Australia*. London: John Murray.

Macquart J. 1831: Sur un insecte coléoptère qui dévore les betteraves. *Annales des Sciences Naturelles* **23**: 93-95.

Mader L. 1924: Coccinellidae, tribus Scymnini. In: Reitter E.: *Bestimmungs-Tabellen der europäischen Coleopteren. 94. Heft*. Troppau: Emmerich Reitter, 48 pp.

Mader L. 1926: Evidenz der paläarktischen Coccinelliden und ihrer Aberrationen in Wort und Bild. I. Teil: Epilachnini, Coccinellini, Halyziini, Synonychini. *Zeitschrift des Vereins der Naturbeobachter* (Wien) [1926–1934]: 1-4 (Mai); 9-12 (Juli); 13-16 (September); 17-20 (Dezember).

Mader L. 1927a: Evidenz der paläarktischen Coccinelliden und ihrer Aberrationen in Wort und Bild. I. Teil: Epilachnini, Coccinellini, Halyziini, Synonychini. *Zeitschrift des Vereins der Naturbeobachter* (Wien) [1926–1934]: 21-24 (Jänner); 25-28 (März); 29-32 (Mai); 33-36 (Juli); 37-40 (September); 41-44 (Oktober); 45-48 (Dezember).

Mader L. 1927b: Coccinellidae. Pp. 759-790. In: Winkler A. (ed.): *Catalogus Coleopterorum regionis palaearcticae*. Wien: A. Winkler [1924-1932], vi + [2] + 1698 pp.

Mader L. 1928: Evidenz der paläarktischen Coccinelliden und ihrer Aberrationen in Wort und Bild. I. Teil: Epilachnini, Coccinellini, Halyziini, Synonychini. *Zeitschrift des Vereins der Naturbeobachter* (Wien) [1926–1934]: 49-76 (note: issued in February).

Mader L. 1929: Evidenz der paläarktischen Coccinelliden und ihrer Aberrationen in Wort und Bild. I. Teil: Epilachnini, Coccinellini, Halyziini, Synonychini. *Zeitschrift des Vereins der Naturbeobachter* (Wien)

[1926–1934]: 77-84 (Jänner); 85-92 (April); 93-100 (Juni 1929); 101-108 (August); 109-116 (October); 117-124 (Dezember).

Mader L. 1930a: Evidenz der paläarktischen Coccinelliden und ihrer Aberrationen in Wort und Bild. I. Teil: Epilachnini, Coccinellini, Halyziini, Synonychini. *Zeitschrift des Vereins der Naturbeobachter* (Wien) [1926–1934]: 125-128 (Jänner); 129-136 (März); 137-144 (Mai); 145-152 (Juli); 153-160 (September); 161-168 (Dezember).

Mader L. 1930b: Neue Coccinelliden aus Yün-nan und Sze-tschwan (China). *Entomologischer Anzeiger* **10**: 161-166, 181-184.

Mader L. 1931a: Evidenz der paläarktischen Coccinelliden und ihrer Aberrationen in Wort und Bild. I. Teil: Epilachnini, Coccinellini, Halyziini, Synonychini. *Zeitschrift des Vereins der Naturbeobachter* (Wien) [1926–1934]: 169-176 (Jänner); 177-180 (Mai); 181-192 (Juli); 193-196 (August); 197-200 (November); 201-204 (Dezember).

Mader L. 1931b: Zur Kenntnis einiger Coccinelliden. *Entomologischer Anzeiger* **11**: 476-477.

Mader L. 1932a: Zur Kenntnis einiger Coccinelliden. *Entomologischer Anzeiger* **12**: 5-7.

Mader L. 1932b: Evidenz der paläarktischen Coccinelliden und ihrer Aberrationen in Wort und Bild. I. Teil: Epilachnini, Coccinellini, Halyziini, Synonychini. *Zeitschrift des Vereins der Naturbeobachter* (Wien) [1926–1934]: 205-212 (Jänner); 213-220 (April); 221-228 (Juni); 229-232 (August); 233-236 (Oktober); 237-244 (Dezember).

Mader L. 1933a: Über bekannte und neue Coccinelliden. *Entomologischer Anzeiger* **13**: 79-84.

Mader L. 1933b: Über neue und bekannte Coccinelliden. *Wiener Entomologische Zeitung* **50**: 93-99.

Mader L. 1933c: Weiteres über Coccinelliden an der Sammlung der National Museums. *Entomologischer Anzeiger* **13**: 129-131.

Mader L. 1933d: Evidenz der paläarktischen Coccinelliden und ihrer Aberrationen in Wort und Bild. I. Teil: Epilachnini, Coccinellini, Halyziini, Synonychini. *Zeitschrift des Vereins der Naturbeobachter* (Wien) [1926–1934]: pp. 245-252 (Jänner); 253-256 (Juni); 257-264 (Juli); 265-272 (August); 373-280 (Oktober); 281-288 (Dezember).

Mader L. 1934a: Coleopterologische Notizen II. *Entomologischer Anzeiger* **14**: 39-40.

Mader L. 1934b: Evidenz der paläarktischen Coccinelliden und ihrer Aberrationen in Wort und Bild. I. Teil: Epilachnini, Coccinellini, Halyziini, Synonychini. *Zeitschrift des Vereins der Naturbeobachter* (Wien) [1926–1934]: 289-296 (April); 297-316 (Mai); 317-324 (August); 325-336 (September).

Mader L. 1935: Evidenz der paläarktischen Coccinelliden und ihrer Aberrationen in Wort und Bild. I. Teil: Epilachnini, Coccinellini, Halyziini, Synonychini. *Entomologischer Anzeiger* **15**: 337-372.

Mader L. 1936a: Neue Pharoscymnus Bed. aus Aegypten (Coleoptera-Coccinellidae). *Bulletin de la Société Royale Entomologique d'Égypte* **1936**: 26-28.

Mader L. 1936b: Neue Coleopteren und Notizen. *Entomologische Rundschau* **54**: 61-66.

Mader L. 1936c: Neue Coleopteren und Notizen. *Entomologische Rundschau* **54**: 69-71.

Mader L. 1936d: Neue Coleopteren und Notizen. (Schluß.). *Koleopterologische Rundschau* **54**: 112-115.

Mader L. 1937: Neue Coleopteren aus China und Japan nebst Notizen. *Entomologisches Nachrichtenblatt* (Troppau) **11**: 179-186.

Mader L. 1938: Neue Coleopteren aus China und Japan nebst Notizen. *Entomologische Nachrichtenblatt* (Troppau) **12**: 40-61.

Mader L. 1939a: Neue Coleopteren aus China. *Entomologisches Nachrichtenblatt* (Troppau) **13**: 41-51.

Mader L. 1939b: Eine neue Erotylide (Coleopt.) aus China. *Entomologisches Nachrichtenblatt* (Troppau) **13**: 71-73.

Mader L. 1941a: Coccinellidae. In: *Exploration du Parc National Albert, mission G. F. de Witte (1933-1935).* Fascicule 34. Bruxelles, 208 pp.

Mader L. 1941b: Eine neue Endomychidae aus Japan. *Entomologische Blätter* **37**: 170.

Mader L. 1941c: Erotylidae und Endomychidae (Col.) von Fukien (China). *Mitteilungen der Münchener Entomologischen Gesellschaft* **31**: 927-933.

Mader L. 1947: Eine neue Coccinellide aus China, nebst einer Notiz. *Bulletin du Musée Royal d'Histoire Naturelle de Belgique* **23** (18): 1-4.

Mader L. 1949: Beitrag zur Kenntnis der Gattung Pharoscymnus Bedel. *Bulletin de la Société Fouad 1er d'Entomologie* **33**: 19-26.

Mader L. 1950: Coccinellidae. II. Teil. In: *Exploration du Parc National Albert. Mission de G.F. de Witte (1933-1935). Fascicule 34.* Bruxelles, 136 pp.

Mader L. 1954: Bemerkungen und Richtigstellungen zu einigen paläarktischen Coleopteren. *Entomologisches Nachrichtenblatt* (Wien) **1**: 48-53.

Mader L. 1955a: Neue Coleopteren aus Fukien (China). Helotidae, Languriidae, Erotylidae, Endomychidae, Coccinellidae. *Koleopterologische Rundschau* **33**: 62-78.

Mader L. 1955b: Evidenz der palaearktischen Coccinelliden und ihrer Aberrationen in Wort und Bild. II. Teil. *Entomologische Arbeiten aus dem Museum G. Frey* **6**: 764-1035, 1 pl.

Mader L. 1957: Weitere neue Coccinellidae aus Belgish Congo. *Revue de Zoologie et de Botanique Africaines* **55**: 101-124.

Mader L. 1958a: Neue Coccinelliden (Col.). *Entomologische Arbeiten aus dem Museum G. Frey* **9**: 178-183.

Mader L. 1958b: Subcoccinella syriaca nov. sp. (Col. Cocc.). *Annalen des Naturhistorischen Museums in Wien* **62**: 250.

Maeda Y. 1972: Two new species of the family Languriidae from Taiwan (Coleoptera). *The Entomological Review of Japan* **24**: 21-24.

Maeda Y. 1974: A new genus and a new species of tribe Cladoxenini from Taiwan (Col.). Reports on Languriidae from Taiwan, II. *The Entomological Review of Japan* **27**: 6-8.

Mahdissan S. 1935: Silvanus iyeri var. mysorensis new variety. *Current Science* (Bangalore) **24**: 418-419.

Majer K. 1982: Species of the genus Aplocnemus of Middle Europe. *Deutsche Entomologische Zeitschrift* (N. F.) **29**: 421-445.

Majer K. 1983: A new genus and species of Dasytinae from Central Asia (Coleoptera, Melyridae). *Acta Entomologica Bohemoslovaca* **80**: 385-389.

Majer K. 1984: A revision of the genus Divales Cast. (Coleoptera, Melyridae, Dasytinae). *Entomologica Basiliensia* **9**: 265-317.

Majer K. 1985: Supplementary notes to "Species of the genus Aplocnemus of Middle Europe". *Deutsche Entomologische Zeitschrift* (N. F.) **32**: 35-41.

Majer K. 1986: Dasytinae from Mongolia (Coleoptera: Melyridae). *Acta Zoologica Hungarica* **32**: 303-315.

Majer K. 1988a: A revision of the genus Dasytiscus Kiesenwetter, 1859 (Coleoptera, Melyridae, Dasytinae). *Mitteilungen der Münchener Entomologischen Gesellschaft* **78**: 127-154.

Majer K. 1988b: A rejection of the subgenus Trithrix Schilsky, 1896, and a further contribution to the systematics of the genus Chaetomalachius Kraatz, 1882 (Coleoptera, Melyridae, Dasytinae). *Entomologica Basiliensia* **12**: 419-430.

Majer K. 1988c: Kubanius wittmeri n. sp. from Pakistan (Coleoptera, Melyridae, Dasytinae). *Entomologica Basiliensia* **12**: 431-433.

Majer K. 1988d: A revision of the genus Bilyella n. gen. (Coleoptera, Melyridae, Dasytinae). *Entomologica Basiliensia* **12**: 435-467.

Majer K. 1989a: A revision of the tribe Danacaeomimini (Coleoptera, Melyridae, Dasytinae). *Deutsche Entomologische Zeitschrift* (N. F.) **36**: 1-31.

Majer K. 1989b: A revision of the genus Haplothrix Schilsky, 1896 (Coleoptera, Melyridae). *Mitteilungen der Münchener Entomologischen Gesellschaft* **79**: 31-46.

Majer K. 1989c: The genus Dasytidius Schilsky, 1896: Species from Turkey and the Balkans (Coleoptera, Melyridae). *Acta Entomologica Bohemoslovaca* **86**: 137-155.

Majer K. 1989d: A revision of the genus Chaetomalachius Kraatz (Coleoptera, Melyridae, Dasytinae). *Entomologica Basiliensia* **13**: 303-355.

Majer K. 1989e: Generic classification of the tribe Chaetomalachiini (Coleoptera, Melyridae). *Polskie Pismo Entomologiczne* **58**: 745-774.

Majer K. 1990a: A revision of the genus Parathrix Majer, 1989 (Col. Melyridae). *Deutsche Entomologische Zeitschrift* (N. F.) **37**: 15-21.

Majer K. 1990b: The genus Dasytidius Schilsky, 1896: Species from North Africa and Europe west of the Balkans. *Mitteilungen der Münchener Entomologische Gesellschaft* **80**: 39-63.

Majer K. 1990c: Results of the Czechoslovak-Iranian entomological expeditions to Iran, 1970, 1973, and 1977. Together with results of collections made in Anatolia. Coleoptera, Melyridae: Part 1 (Prionoceridae). *Acta Entomologica Musei Nationalis Pragae* **43**: 159-162.

Majer K. 1990d: Results of the Czechoslovak-Iranian entomological expeditions to Iran, 1970, 1973, and 1977. Together with results of collections made in Anatolia. Coleoptera, Melyridae: Part 2 (Melyridae, Cerallini). *Acta Entomologica Musei Nationalis Pragae* **43**: 163-181.

Majer K. 1990e: Results of the Czechoslovak-Iranian entomological expeditions to Iran, 1970, 1973, and 1977. Together with results of collections made in Anatolia. Coleoptera, Melyridae: Part III (Rhadalinae). *Acta Entomologica Musei Nationalis Pragae* **43**: 183-202.

Majer K. 1990f: A new tribe Listrini trib. n., including two new genera (Coleoptera, Melyridae). *Acta Entomologica Bohemoslovaca* **87**: 368-384.

Majer K. 1991a: The genus Dasytidius Schilsky, 1896 (Coleoptera, Melyridae): Species of Near and Middle East, chiefly west of Iranian and Turanian superprovince. Part 1. *Annotationes Zoologicae et Botanicae* **203**: 1-23.

Majer K. 1991b: The genus Dasytidius Schilsky, 1896 (Coleoptera, Melyridae): species of Near and Middle East, chiefly from Iranian and Turanian superprovince. Part 2. *Annotationes Zoologicae et Botanicae* **204**: 1-27.

Majer K. 1995: A revision of the genus Mimothrix Majer, 1989 (Insecta: Coleoptera: Dasytidae). *Entomologische Abhandlungen* (Dresden) **56** (10): 177-203.

Majer K. 1996a: A revision of the genus *Achaetomalachius* Majer, 1989 (Coleoptera, Dasytidae). *Annotationes Zoologicae et Botanicae* **219**: 3-26.

Majer K. 1996b: New species and synonymies of the genera Danacaeomimus, Dasytidius, Mimothrix and Haplothrix (Coleoptera: Dasytidae). *Entomological Problems* **27**: 85-91.

Majer K. 1996c: Internal classification of the genus Dasytidius Schilsky with descriptions of new species of the subfamily Chaetomalachiinae (Insecta, Coleoptera, Dasytidae). *Spixiana* **19**: 155-182.

Majer K. 1996d: Complementary generic revision of the subfamily Chaetomalachiinae (Insecta: Coleoptera: Dasytidae). *Annalen des Naturhistorischen Museums in Wien* **98B**: 435-500.

Majer K. 1997: A revision of the tribe Amauronioidini (Coleoptera, Dasytidae). *Acta Musei Moraviae Scientiae Naturales* **81**: 363-402.

Makino T. & Nakane T. 1981: A revision of the genus Podabrus Westwood in Japan (I) (Insecta, Coleoptera, Cantharidae). *Reports of the Faculty of Science, Kagoshima University,* (*Earth Sciences and Biology*) **14**: 55-63.

Mäklin F. W. 1845: Novae in fauna fennica Coleopterorum species. *Bulletin de la Société Impériale des Naturalistes de Moscou* **18** (2): 544-552.

Mäklin F. W. 1846: Coleoptera myrmecophila fennica. *Bulletin de la Société Impériale des Naturalistes de Moscou* **19** (1): 157-193.

Mäklin F. W. 1853: [new taxa]. In: Mannerheim C. G. G. von.: Dritter Nachtrag zur Käferfauna der Nord-Amerikanischen Länder des Russichen Reiches. *Bulletin de la Société Impériale des Naturalistes de Moscou* **26** (3): 95-273.

Mäklin F. W. 1863: Zur Synonymie einiger nordischer Käferarten. *Acta Societatis Scientiarum Fennicae* **7**: 133-144.

Mäklin F. W. 1878: Diagnosen öfver några nya siberiska insektarter. *Öfversigt af Finska Vetenskaps-Societetens Förhandlingar* **19** [1876-1877]: 15-32.

Mamaev B. M. 1976: Morphologicheskie typy lichinok zhukov-drevoedov (Coleoptera, Eucnemidae) i ikh evolutionnoe znachenie. Pp. 138-155. In: Mamaev B. M. (ed.): *Evolutionnaya morphologya lichinok nasekomykh.* Moskva, 202 pp.

Mandl K. 1953: Ergebnisse einer Koleopterologischen Sammelreise in das südwestliche Adamello-Gebiet. *Studi Trentini di Scienze Naturali* **30**: 128-135.

Mandl K. 1974: Eine neue Gattung, drei neue Arten und eine neue Unterart aus der Familie Karumiidae (Dascilloidea, Col.). *Verhandlungen der Naturforschenden Gesellschaft Basel* **84**: 635-645.

Mannerheim C. G. von. 1823: *Eucnemis, insectorum genus, Monographice tractatum iconibusque illustratum.* Petropoli ex officina directorii institutionis publicae, 36 pp., 2 pls.

Mannerheim C. G. von. 1825: Novae Coleopterorum species Imperii Rossici incolae. Pp. 19-41. In: *Essais Entomologiques* (A. D. Hummel), *Tome premier. No. IV.* St. Pétersbourg.

Mannerheim C. G. von. 1827: [new taxon]. In: Hummel A. D.: Insectes de 1826. Pp. 1-20. *Essais Entomologiques* (A. D. Hummel) No. VI, 49 pp.

Mannerheim C. G. von. 1843a: Mémoire sur la récolte d'insectes coléoptères, faite en 1842. *Bulletin de la Société Impériale des Naturalistes de Moscou* **16** (1): 70-87.
Mannerheim C. G. von. 1843b: Beitrag zur Kaefer-Fauna der Aleutischen Inseln, der Insel Sitkha und Neu-Californiens. *Bulletin de la Société Impériale des Naturalistes de Moscou* **16** (2): 175-314.
Mannerheim C. G. von. 1843c: Description de quelques autres nouvelles espèces de coléoptères de Finlande. *Bulletin de la Société Impériale des Naturalistes de Moscou* **16** (3): 88-99.
Mannerheim C. G. von. 1844a: Versuch einer monographischen Darstellung der Käfergattungen Corticaria und Lathridius. *Zeitschrift für die Entomologie* **5**: 1-112.
Mannerheim C. G. von. 1844b: Mémoire sur la récolté d'insectes coléoptères, faite en 1843. *Bulletin de la Société Impériale des Naturalistes de Moscou* **17** (1): 160-188.
Mannerheim C. G. von. 1845: [new taxon]. In: Motschulsky V. de.: Remarques sur la collection de coléoptères russes de Victor de Motschulsky. 1er Article. *Bulletin de la Société Impériale des Naturalistes de Moscou* **18** (1-2): 3-127, 3 pls.
Mannerheim C. G. von. 1846: Nachtrag zur Kaefer-fauna der aleutischen Inseln und der Insel Sitkha. *Bulletin de la Société Impériale des Naturalistes de Moscou* **19** (2): 501-516.
Mannerheim C. G. von. 1849: Insectes coléoptères de la Sibérie orientale nouveaux ou peu connus. *Bulletin de la Société Impériale des Naturalistes de Moscou* **22** (1): 220-249.
Mannerheim C. G. von. 1852a: Zweiter Nachtrag zur Kaefer-Fauna der nordamerikanischen Laender des russischen Reiches. *Bulletin de la Société Impériale des Naturalistes de Moscou* **25** (1) [1852-1853]: 283-387.
Mannerheim C. G. von. 1852b: Insectes coléoptères de la Sibérie orientale, nouveaux ou peu connus. Decades tertia, quarta et quinta. *Bulletin de la Société Impériale des Naturalistes de Moscou* **25** (2) [1852-1853]: 273-309.
Mannerheim C. G. von. 1853: Dritter Nachtrag zur Käferfauna der Nord-Amerikanischen Länder des Russischen Reiches. *Bulletin de la Société Impériale des Naturalistes de Moscou* **26** (3) [1853-1854]: 95-273.
Mao J.-L. & Li Y. 1998: A new species of Coccinellidae from Gansu Province (Coleoptera: Coccinellidae). *Acta Entomologica Sinica* **41**: 169-170.
Mařan J. 1929: Uleiota algerica m. nov. spec. *Časopis Československé Společnosti Entomologické* **26**: 33-36.
Mařan J. 1941: Nové druhy a formy palaearktických anobiidů. *Sborník Entomologického Oddělení Zemského Musea v Praze* **19**: 122-125.
Marcu O. 1936: Ein neuer Haplocnemus (H. Strandi nov.) aus der Bucovina. *Festschrift zum 60. Geburtstag von Professor Dr. Embrik Strand* **1**: 518-519.
Marcu O. 1939: Ein neuer Meligethes (Meligethes Peneckei n.) aus der Bucovina. *Folia Zoologica et Hydrobiologica* (Riga) **9** [1936-1939]: 374-375.
Marcu O. 1957: Contribuţii la cunoaşterea faunei coleopteleror Transilvaniei. *Bulletin University "Babeş Bolyai" Cluj, Ştiienţele Naturii* **2** (1-2): 527-544.
Mardjanian M. A. 1981: [new taxa]. In: Iablokoff-Khnzorian S. M. & Mardjanian M. A.: O processe sobokupleniya i stroenii edeagusa u zhukov-shchelkunov s opisaniem novogo v rode Cardiophorus Eschz. (Coleoptera, Elateridae) [On the process of copulation and structure of the aedeagus of the click beetles (Coleoptera, Elateridae) of the genus Cardiophorus Eschs.]. *Doklady Akademii Nauk Armyanskoy SSR* **73**: 244-249.
Mardjanian M. A. 1983: Dva novykh vida zhulkov-shchelkunov iz SSSR (Coleoptera, Elateridae) [Two new species of click-beetles from the U.S.S.R. (Coleoptera, Elateridae)]. *Doklady Akademii Nauk Armyanskoy SSR* **77**: 141-144.
Mardjanian M. A. 1985: [new taxon]. In: Dolin W. G. & Mardjanian M. A.: [New species of click-beetles of the genus Cardiophorus Eschz. of the fauna of the European part of the USSR and Transcaucasus]. *Doklady Akademii Nauk Armyanskoy SSR* **81** (3): 134-138.
Märkel J. C. F. 1845: Beiträge zur Kenntnis der unter Ameisen lebenden Insecten. *Zeitschrift für die Entomologie* **5** [1844]: 193-271.
Märkel J. C. F. 1847: Eine entomologische Excursion im Riesengebirge im Juli 1846. *Entomologische Zeitung* (Stettin) **8**: 73-91.
Märkel J. C. F. 1852: Rémarques. In: Kiesenwetter H.: Énumération des coléoptères trouvés dans le midi de la France et en Catalogne. *Annales de la Société Entomologique de France* (2) **9** [1851]: 577-656.

Märkel F. 1857: Zwei entomologische Oden. Hierzu ein Commentar. *Allgemeine Deutsche Naturhistorische Zeitung. Neue Folge* **3**: 161-180.

Marietti B. 1844: Coleopterorum enumertio, p. 72: Coleopterorum species novae. In: Osculati G.: *Note d'un viaggio nella Persia e nelle India orientali negli anni 1841, 1842.* Monza, 73 pp. 1 tab.

Marquet C. 1875: Description de deux coléoptères nouveaux de la France méridionale. *Petites Nouvelles Entomologiques* **1**: 511.

Marriner T. F. 1929: A pale Anatis ocellata. *The Entomologist's Record and Journal of Variation* (N. S.) **41**: 163-164.

Marseul S.-A. de. 1864: Telephorides. Tribu de la famille des malacodermes. *L'Abeille, Mémoires d'Entomologie* **1**: 1-108.

Marseul S. A. de. 1867: Descriptions d'espèces nouvelles. *L'Abeille, Mémoires d'Entomologie* **4**: xxxiii-xl.

Marseul S. A. de. 1868a: Monographie des Endomychides d'Europe et contrées limitrophes. *L'Abeille, Mémoires d'Entomologie* **5** [1868-1869]: 51-138.

Marseul S. A. de. 1868b: Descriptions des espèces nouvelles. *L'Abeille, Mémoires d'Entomologie* **5** [1868-1869]: 171-218.

Marsuel S. A. de. 1869: IV. Notes diverses. *L'Abeille, Journal d'Entomologie* **6**: 154-158.

Marseul S. A. de. 1870: Descriptions de coléoptères nouveaux. *L'Abeille, Mémoires d'Entomologie* **6** [1867-1869]: 368-389.

Marseul S. A. de. 1875: Mélanges. In: Nouvelles et faits diverses. *L'Abeille, Journal d'Entomologie* **14** [1874-1876]: xli-xliv.

Marseul S.-A. de. 1878a: Monographie des malthinides de l'Ancient-Monde. *L'Abeille, Journal d'Entomologie* **16**: 1-120.

Marseul S. A. de. 1878b: Mélanges coléoptérologiques. In: Nouveaux faits divers. *L'Abeille, Journal d'Entomologie* **16**: 54-60.

Marseul S. A. de. 1883: Mélanges (suite). *L'Abeille, Mémoires d'Entomologie* **21**: 182-184.

Marseul S. A. de. 1886: Catalogue synonymique et géographique des coléoptères de l'Ancien-Monde, Europe et contrées limitrophes en Afrique et en Asie. *L'Abeille, Journal d'Entomologie* **24**: 193-360.

Marseul S. A. de. 1890: [new taxon]. In: Abeille de Perrin E.: Malachiidae. Malachides d'Europe et pays voisins. *Annales de la Société Entomologique de France* (6) **10**: 181-260.

Marsham T. 1802: *Entomologia Britannica, sistens insecta britanniae indigena, secundum methodum linnaeanam disposita. Tomus I. Coleoptera.* Londini: Wilks et Taylor, J. White, xxxi + 548 pp.

Mateu J. 1954: Coleópteros de Sierra Nevada. Familias: Dytiscidae, Hydrophilidae, Dryopidae, Catopidae, Silphidae, Pselaphidae, Histeridae, Cleridae, Cantharidae, Dasytidae, Aderidae, Anthicidae. *Los Archivos del Instituto de Aclimatación* (Almería) **2**: 89-101, 2 pls.

Mathieu J. M. 1961: Revision of the genus Melanactes, with a proposed new genus (Coleoptera, Elateridae). *American Midland Naturalist* **65**: 459-480.

Matsuda K. 1992. Two new species of the genus Eropterus Green from Taiwan (Coleoptera, Lycidae). *Entomological Review of Japan* **47**: 91-98.

Matsumura S. 1910: *Die schädlichen und nützlichen Insecten vom Zuckerrohr Formosas.* Tokyo: The Keiseisha, 52 pp., 30 pls.

Matsumura S. 1911a: Erster Beitrag zur Insekten-Fauna von Sachalin. *Journal of the College of Agriculture, Tôhoku Imperial University (Sapporo)* **4**: 1-144, 2 pls.

Matsumura S. 1911b: Beschreibungen von an Zuckerrohr Formosas schädlichen oder nützlichen Insekten. *Mémoires de la Société Entomologique de Belgique* **18**: 129-150.

Matsumura S. 1915a: [Taxonomy of insects]. *Konchu Bunruigaku. Vol. II.* Tokyo: Keiseisha-Shoten, 316 pp., 5 pls (in Japanese).

Matsumura S. 1915b: *Dainippon Gaichu Zensho* **1915**: 183-184 (in Japanese).

Matsumura S. 1918: [The fireflies of Japan]. *Kyôiku-gahô [Education and Illustration Magazine]* **6** (3): 82-89 (in Japanese).

Matsumura S. 1928: [The fireflies]. Pp. 39-70, pls. 5-7. In: *Interesting insects for educational material* (in Japanese).

Matsumura S. 1935: *Dainippon Gaichu Zusetsu* **2**: 234 (in Japanese).

Matsumura S. & Yokoyama K. 1928a: An interesting dermested beetle from Japan. *Insecta Matsumurana* **2**: 130-132.

Matsumura S. & Yokoyama K. 1928b: New and hitherto unrecorded species of Dermestidae from Japan. *Insecta Matsumurana* **3**: 51-54.

Matthews A. 1885a: Synopsis of the British species of Orthoperus. *The Entomologist's Monthly Magazine* **22**: 107-110.

Matthews A. 1885b: On a new genus allied to Corylophus. *The Entomologist's Monthly Magazine* **22**: 160.

Matthews A. 1885c: [Description de deux Corylophides nouveaux]. *Bulletin ou Comptes-Rendus des Séances de la Société Entomologique de Belgique* **29**: lxviii-lxix.

Matthews A. 1886a: Description of a new genus, and some new species of Corylophidae. *The Entomologist's Monthly Magazine* **22**: 224-228.

Matthews A. 1886b: Corylophidarum species novae e musaeo fioriano. *Bullettino della Società Entomologica Italiano* **18**: 432.

Matthews A. 1887: New genera and species of Corylophidae in the collection of the British Museum. *The Annals and Magazine of Natural History* (5) **19**: 105-116.

Matthews A. 1890: Vier neue europäische Coleopteren-Arten aus der familie der Corylophidae. *Wiener Entomologische Zeitung* **9**: 151-152.

Matthews A. 1899: *A monograph of the coleopterous families Corylophidae and Sphaeriidae*. London: P. B. Mason, Janson & Son, 220 pp., 8 pls.

Mayer J. 1779: Bemerkungen über natürliche Gegenstände der Gegend um Schüttenhofen in Böhmen, und eines Theils der benachbarten Gebirge. *Abhandlungen einer Privatgesellschaft in Böhmen, zur Aufnahme der Mathematik, der vaterländischen Geschichte, und der Naturgeschichte. Band 4.* Prague.

Mayor A. J. 2004: Nomenclatorial corrections for Dasytidae and Malachiidae (Coleoptera). *Insecta Mundi* **17** [2003]: 85-96.

Mayor A. J. & Wittmer W. 1981: A new genus of Malachiidae from Asia (Coleoptera). *Entomologica Basiliensia* **6**: 416-419.

McDermott F. A. 1960: Lucidotopsis genus novum (Coleoptera; Lampyridae). *The Coleopterists Bulletin* **14**: 18.

McDermott F. A. 1964: The taxonomy of the Lampyridae (Coleoptera). *Transactions of the American Entomological Society* **90**: 1-72.

McDermott F. A. 1966: Lampyridae. In: Steel W. (ed.): *Coleopterorum Catalogus Supplementa. Pars 9* (*Editio Secunda*). Gravenhage: W. Junk, 149 pp.

Medvedev L. N. 1958: Eine neue Art der Gattung Collops aus Polarasien (Col. Malachiidae). *Deutsche Entomologische Zeitschrift* (N. F.) **5**: 469-470.

Medvedev L. N. 1963: Novyy vid roda Tolyphus (Coleoptera, Phalacridae) iz yugovostochnogo Kazachstana. *Akademia Nauk Kazachskoy SSR Trudy Instituta Zoologii* **21**: 147-148.

Medvedev L. N. 1964a: Novyy vid Ichthyurus (Coleoptera, Cantharidae) iz Tadzhikistana *Doklady Akademii Nauk Tadzhikskoy SSR* **7** (6): 44-46.

Medvedev L. N. 1964b: Novye vidy Malachiidae (Coleoptera) iz Centralnogo Kazakhstana. *Trudy Zoologicheskogo Instituta* (Leningrad) **34**: 155-158.

Medvedev L. N. 1966: Fauna Lycidae (Coleoptera) Kurilskikh ostrovov. Pp. 34-38. In: *Entomofauna lesov Kurilskikh ostrovov, poluostrova Kamchatki i Magadanskoy oblasti*. Moskva: L. Nauka.

Medvedev L. N. 1970: Novye i maloizvestnye vidy Lycidae i Melasidae (Coleoptera) Dalnego Vostoka. Pp. 163-168. In: Ivliev L. A. & Konovalova Z. A. (eds): *Entomologicheskiye Isledovania na Dalnem Vostoke*. Vladivostok: Akademia Nayk SSSR.

Medvedev L. N. 1971: Ergebnisse der Zoologischen Forschungen von Dr. Kaszab in der Mongolei. 254 Phalacridae (Coleoptera). *Annales Historico-Naturales Musei Nationalis Hungarici. Pars Zoologica* **63**: 217-224.

Medvedev L. N. 1972: Nitidulidae i Phalacridae (Coleoptera) Mongolskoy Narodnoy Respubliki [Nitidulidae and Phalacridae (Coleoptera) of the Mongolian People's Republic]. *Nasekomye Mongolii* **1**: 475-484.

Medvedev L. N. 1979: K poznaniyu Lycidae fauny SSSR. Pp. 239-243. In: Pravdin F. N. (ed.): *Nasekomye – razrushiteli drevesiny i ikh entomophagi*. Moskva: Nauka, 254 pp.

Medvedev L. N. & Kazantsev S. V. 1992: Novoe podsemejstvo i novyy rod zhukov-licid (Coleoptera, Lycidae) iz Jugo-Vostochnoy Asii. Pp. 55-60. In: Medvedev L. N. & Striganova B. R. (eds): *Fauna i ekologiya nasekomykh Vietnama*. Moskva: Nauka, 198 pp.

Medvedev L. N. & Ryvkin A. B.: 1989: Novye vidy Rhagonycha (Coleoptera, Cantharidae) iz Sibiri i Dalnego Vostoka [New species of Rhagonycha (Coleoptera, Cantharidae) from Siberia and Far East]. *Zoologicheskiy Zhurnal* **68** (10): 138-141.

Meier W. 1897a: Varietäten der Coccinellidae: Hippodamia septemmaculata Degeer. *Entomologische Nachrichten* **23**: 317.

Meier W. 1897b: Noch eine Bemerkungen über die Varietäten der Coccinellide Hippodamia septemmaculata Deg. *Entomologische Nachrichten* **23**: 365-366.

Meier W. 1899a: Xylotheca Meieri Reitter var. minor W. Mr. *Entomologische Nachrichten* **25**: 72.

Meier W. 1899b: Über Abänderungen einiger Coleopteren-Arten, welche bei Hamburg gefunden wurden. *Entomologische Nachrichten* **25**: 97-102.

Meier W. 1900: Timarcha v. Fracassii - Chrysomela sirentensis - Hippodamia v. equiseti. *Entomologische Nachrichten* **26**: 78.

Mellié J. 1847: Mélanges et nouvelles. *Revue Zoologique* **10**: 108-110.

Melnik I. V. 2004: A new species of the clerid genus Trichodes Herbst (Coleoptera, Cleridae) from Afghanistan. *Entomologicheskoe Obozrenie* **83**: 147-150.

Melsheimer F. E. 1844: Descriptions of new species of Coleoptera of the United States. *Proceedings of the Academy of Natural Sciences of Philadelphia* **2** [1844-1845]: 26-43, 98-118, 134-160.

Melsheimer F. E. 1845: Descriptions of new species of Coleoptera of the United States. *Proceedings of the Academy of Natural Sciences of Philadelphia* **2** [1844-1845]: 213-223.

Melsheimer F. E. 1846: Descriptions of new species of Coleoptera of United States. *Proceedings of the Academy of Natural Sciences of Philadelphia* **2** [1844-1845]: 302-318.

Ménétriés E. 1832: *Catalogue raisonné des objets de zoologie recueillis dans un voyage au Caucase et jusqu'aux frontières actuelles de la Perse.* St. Pétersbourgh: Académie Impériale des Sciences, 2 + xxxiii + 271 + iv + [1] pp.

Ménétriés E. 1836: Insectes nouveaux de la Turquie. *Bulletin Scientifique de l'Académie Impériale de St. Petersbourg* **1**: 149-151.

Ménétriés E. 1849: Catalogue des insectes recueillis par feu M. Lehmann avec les descriptions des nouvelles espèces. *Mémoires de l'Académie Impériale des Sciences de St.-Pétersbourg* (6) **8**: 217-328, pls 3-5.

Ménétriés E. 1851: Insecten. Pp. 43-76: In: *Reise in den äußersten Norden und Osten Sibiriens während der Jahre 1843 und 1844 mit allerhöchster Genehmigung auf Veranstaltung der Kaiserlichen Akademie der Wissenschaften zu St. Petersburg ausgeführt und in Verbindung mit vielen Gelehrten herausgegeben von Dr. A. Th. V. Middendorff. Zweiter Band. Theil 1: Tiere: Annulaten. Echinodermen. Insecten. Krebse. Mollusken. Parasiten.* St. Petersburg: Buchdruckerei der Kaiserlichen Akademie der Wissenschaften.

Menier J. J. 1981: Description de Sedlacekius poutianus, n. pp. Remarques sur la morphologie du genre (Coleoptera, Cleridae). *Revue Française d'Entomologie* (N. S.) **3**: 108-112.

Menier J. J. 1986: Coleoptera: Fam. Cleridae of Saudi Arabia. (Part 2). *Fauna of Saudi Arabia* **8**: 219-232.

Menier J. J. 1997: Description préliminaire de Bosquetoclerus arachnoides n. gen., n. sp., cleridae microptère de Chine (Coleoptera). *Revue Française d'Entomologie* (N. S.) [1996] **18**: 158.

Menier J. J. & Constantin R. 1988: Diagnose préliminaire de Gietella faialensis n. sp. (Coleoptera, Melyridae, Gietellinae). *Revue Française d'Entomologie* (N. S.) **10**: 30.

Menier J. J. & Villemant C. 1993: Description et biologie de Anthrenus (Peacockia n. sgen.) vladimiri n. sp. prédateur des pontes de Porthetria dispar (L.) au Maroc (Coleoptera, Dermestidae; Lepidoptera, Lymantriidae). *Revue Française d'Entomologique* (N. S.) **15**: 61-66.

Méquignon A. 1909: Révision des Rhizophagus paléarctiques. *L'Abeille. Journal d'Entomologie* **31** [1907-1923]: 103-119.

Méquignon A. 1913a: Description de trois Rhizophagus nouveaux (Col. Nitidulidae). *Bulletin de la Société Entomologique de France* **1913**: 90-92.

Méquignon A. 1913b: Synonymies et variétés nouvelles de Rhizophagus (Col. Nitidulidae). *Bulletin de la Société Entomologique de France* **1913**: 342-343.

Méquignon A. 1922: Étude synonymique sur les Melasidae (Coléoptères). *Annales de la Société Entomologique de France* **91**: 369-374.

Méquignon A. 1925a: Description de deux Rhizophagus nouveaux de l'Europe méridionale (Col. Nitidulidae). *Bulletin de la Société Entomologique de France* **1925**: 13-15.

Méquignon A. 1925b: Étude synonymique sur les Melasidae (Col.) (2e note). *Bulletin de la Société Entomologique de France* **1925**: 187-188.

Méquignon A. 1925c: Note rectificative. *Bulletin de la Société Entomologique de France* **1925**: 240.

Méquignon A. 1927: Variétés nouvelles ou peu connues de buprestides et d'élatérides gallo-rhénans. *Bulletin de la Société Entomologique de France* **1927**: 21-24.

Méquignon A. 1928: Sur quelques buprestides, mélasides et élatérides de France (Col.). *Bulletin de la Société Entomologique de France* **1928**: 130-134.

Méquignon A. 1929: Notes synonymiques sur quelques élatérides (Col.) (2e note). *Bulletin de la Société Entomologique de France* **1929**: 272-276.

Méquignon A. 1930a: Notes synonymiques sur quelques élatérides (Col.) (4e note). *Bulletin de la Société Entomologique de France* **1930**: 91-96.

Méquignon A. 1930b: Serricornia. In: Bedel L.: *Faune des coléoptères du bassin de la Seine*. Vol. 4 (3). Paris: Société Entomologique de France, 362 pp.

Méquignon A. 1932: Description d'un Harminius nouveaux du Portugal et remarques sur quelques Élatérides de même provenance (Col., Elateridae). *Bulletin de la Société Entomologique de France* **37**: 123-125.

Méquignon A. 1940: Coléoptères nouveaux ou rares pour la faune de France. *Bulletin de la Société Entomologique de France* **45**: 24-28.

Méquignon A. 1942a: Voyage de MM. L. Chopard et A. Méquignon aux Açores (Août-Septembre 1930). XIV. Catalogue des coléoptères açoréens. *Annales de la Société Entomologique de France* **111**: 1-66.

Méquignon A. 1942b: Voyage de MM. L. Chopard et A. Méquignon aux Açores (août-septembre 1930). XIII. Diagnoses de coléoptères nouveaux. *Bulletin de la Société Entomologique de France* **47**: 9-12.

Méquignon A. 1945: Sur quelques types d'Epuraea conservés au Muséum de Paris (Col. Nitidulidae). *Revue Française d'Entomologie* **12**: 14-21.

Méquignon A. 1948: Notes diverses sur des coléoptères de France. (Sixième série). Espèces et variétés rares, peu connues ou nouvelles pour la faune française. *Bulletin de la Société Entomologique de France* **53**: 75-80.

Méquignon A. 1949: Notes diverses sur les coléoptères de France. *Bulletin de la Société Entomologique de France* **54**: 114.

Mertlik J. 1996: Cidnopus platiai sp. n. (Coleoptera: Elateridae) from Slovakia. *Klapalekiana* **32**: 203-208.

Mertlik J. 1998: Zorochros merkli sp. n. (Coleoptera: Elateridae) from Greece. *Klapalekiana* **34**: 75-78.

Mertlik J. 2000: Contribution to the knowledge of Turkish species of the genus Selatosomus (Coleoptera: Elateridae). *Folia Heyrovskyana* **8**: 1-15.

Mertlik J. 2001: Ligmargus kalabi sp. n. (Coleoptera: Elateridae) from Pakistan. *Folia-Heyrovskyana* **9**: 157-160.

Meschnigg J. 1934: Beitrag zur Kenntnis der Coleopteren vom Balkan. *Entomologischer Anzeiger* **14**: 65-68.

Mielzinsky I. 1824: Mémoire sur une larve qui dévore les Helix nemoralis, et sur l'insecte auquel elle donne naisance. *Annales des Sciences Naturelles* **1**: 67-77.

Mignot E. C. 1969: [new taxon]. In: Arnett R. H. Jr., Mignot E. C. & Smith E. H. 1969: North American Coleoptera Fauna: Notes on Pyrophorinae, Elateridae. *The Coleopterists' Bulletin* **23**: 9-15.

Míka P. 2000: Corticariidae (Coleoptera) from southern Thailand; Part 1 – review of the genus Melanophthalma with description of some new species. Corticariidae (Coleoptera) jižního Thajska; 1.část – přehled druhů rodu Melanophthalma s popisy několika nových druhů. *Klapalekiana* **36**: 125-143.

Mikšić R. 1969: Contributo alla conoscenza delle specie italiane del genere Luciola. *Bollettino dell'Associazione Romana di Entomologia* **24**: 43-46.

Miller L. 1852 [new taxa]. *Verhandlungen des Zoologisch-Botanischen Vereins in Wien* **1**: 109-113.

Miller L. 1857: Malachius gracilis n. sp. *Wiener Entomoloigische Monatschrift* **1**: 138.

Miller L. 1858: Bemerkungen über die Gattung Cryptophagus. *Wiener Entomologische Monatschrift* **4**: 97-100.

Miller L. 1862: Ergebnisse einer entomologischen Reise nach Cephalonia. *Wiener Entomologische Monatschrift* **4**: 341-355.

Miller L. 1866: Neue Käfer-Arten. *Verhandlungen der Zoologisch-Botanischen Gesellschaft in Wien* **16**: 817-820.

Miller L. 1881: [new taxa]. In: Reitter E.: Neue und seltene Coleopteren, im Jahre 1880 in Süddalmatien und Montenegro gesammelt und beschrieben. *Deutsche Entomologische Zeitschrift* **25**: 177-230.

Milliron H. E. 1939: A parthenogenetic new species of the genus Perimegatoma Horn (Coleoptera:

Dermestidae). *Annals of the Entomological Society of America* **32**: 570-574.
Mink W. 1853: Beschreibung zweier neuer Käfer. *Entomologische Zeitung* (Stettin) **14**: 58-60.
Miwa Y. 1927a: Descriptions of new species of Japanese Elateridae. *Insecta Matsumurana* **2**: 12-22, pl. I.
Miwa Y. 1927b: New and some rare species of Elateridae from the Japanese Empire. *Insecta Matsumurana* **2**: 105-114.
Miwa Y. 1928a: New and some rare species of Elateridae from the Japanese Empire. *Insecta Matsumurana* **2**: 133-146, pl. V.
Miwa Y. 1928b: New and some rare species of Elateridae from the Japanese Empire. *Insecta Matsumurana* **3**: 36-51, pl. I.
Miwa Y. 1929a: On the Erotylidae of Japan, Formosa, Corea and Saghalien. *Transactions of the Natural History Society of Formosa* **19**: 120-128.
Miwa Y. 1929b: Elateridae of Formosa. (Contribution to the fauna of Formosan Coleoptera). *Transactions of the Natural History Society of Formosa* **19**: 225-246.
Miwa Y. 1929c: The elaterid-fauna of Loo-Choo. *Transactions of the Natural History Society of Formosa* **19**: 339-351.
Miwa Y. 1929d: On the elaterid-species from Saghalien and Kuriles.*Dobutsugaku Zasshi* (Tokyo) **41**: 441-454.
Miwa Y. 1929e: Elateridae of Formosa (II). *Transactions of the Natural History Society of Formosa* **19**: 485-495.
Miwa Y. 1930a: Elateridae of Formosa (III). *Transactions of the Natural History Society of Formosa* **20**: 1-12.
Miwa Y. 1930b: Elateridae of Formosa (IV). *Transactions of the Natural History Society of Formosa* **20**: 57-68.
Miwa Y. 1930c: H. Sauter's Formosa-Ausbeute (Elateridae I.). *Wiener Entomologische Zeitung* **47**: 91-97.
Miwa Y. 1931: A systematic catalogue of Formosan Coleoptera. *Report of the Department of Agriculture Government Research Institute* (Taihoku) **55**: 1-359.
Miwa Y. 1931b: Elateridae of Formosa (V). *Transactions of the Natural History Society of Formosa* **21**: 72-98.
Miwa Y. 1931c: H. Sauter's Formosa-Ausbeute (Elateridae II.) *Wiener Entomologische Zeitung* **47**: 205-208.
Miwa Y. 1931d: Supplementary notes on the elaterid-fauna of Loo-Choo (Coleoptera). *Transactions of the Natural History Society of Formosa* **22**: 259-261.
Miwa Y. 1933a: An enumeration of Coleoptera from the Island Iriomote in Loocho, with descriptions of new species. (The Coleopteran fauna of Loocho, I). *Transactions of the Natural History Society of Formosa* **23**: 4-15.
Miwa Y. 1933b: Elateridae in the collection of the Entomological Laboratory, Kyushu Imperial University (II). *Mushi (Fukuoka)* **6**: 66-73.
Miwa Y. 1934: The fauna of Elateridae in the Japanese Empire. *Department of Agriculture Government Research Institute Report* **65**: 1-289, 9 pls.
Miwa Y. 1935: Descriptions of two new species of Lymexylonidae from Formosa. *Transactions of the Natural History Society of Formosa* **25**: 456-458.
Miwa Y. 1936: Coleoptera from Amami Islands in Loo-Choo archipelago. *Transactions of the Kansai Entomological Society* **6** [1935]: 11-30, 2 pls. (in Japanese)
Miwa Y. 1939: Description of a new species belonging to the family Elateridae from Formosa. *Transactions of the Natural History Society of Formosa* **29**: 65-66.
Miwa Y. & Chûjô M. 1936: Description of a new species of Formosan Languriidae (Col.). *Transactions of the Natural History Society of Formosa* **26**: 411-412.
Miwa Y. & Chûjô M. 1937: Descriptions of two new species of Japanese Languriidae (Coleoktera) [sic!]. *Transactions of the Natural History Society of Formosa* **27**: 7-10.
Miwa Y & Chûjô M. 1939: Description of a new species of Erotylidae from Formosa (Coleoptera). *Mushi* **12**: 76-77.
Miyahara K. 1942: Studies on the Erotylidae of Fukuoka Prefecture, Kyushu, Japan. *Konchū-Kai* **10**: 529-534 (in Japanese).
Miyatake M. 1959: A contribution to the coccinellid-fauna of the Ryukyu Islands (Coleoptera). *Memoirs of the Ehime University, Sect. VI* (*Agriculture*) **4**: 125-161. 2 pl.
Miyatake M. 1961a: A new genus of the Coccinellidae from Japan and the Ryukyu Islands (Coleoptera). *Memoirs of the Ehime University, Sect. VI* (*Agriculture*) **6**: 127-133.
Miyatake M. 1961b: The East-Asian coccinellid-beetles preserved in the California Academy of Science, Tribe Serangiini. *Memoirs of the Ehime University, Sect. VI* (*Agriculture*) **6**: 135-146.

Miyatake M. 1961c: The East-Asian coccinellid beetles preserved in the California Academy of Science, tribe Hyperaspini. *Memoirs of the Ehime University, Sect. VI (Agriculture)* **6**: 147-155.

Miyatake M. 1961d: The East-Asian coccinellid-beetles preserved in the California Academy of Sciences, tribe Platynaspini. *Memoirs of the Ehime University, Sect. VI (Agriculture)* **6**: 157-176.

Miyatake M. 1963a: The genus Horniolus Weise of Japan and the Ryukyu Islands (Coleoptera: Coccinellidae). *Transactions of the Shikoku Entomological Society* **8**: 6-12.

Miyatake M. 1963b: Descriptions of two new species of the genus Serangium from Japan. *Transactions of the Shikoku Entomological Society* **8**: 13-14.

Miyatake M. 1963c: Description of two new Coccinella-species of Japan (Coleoptera: Coccinellidae). *Insecta Matsumurana* **26**: 49-53.

Miyatake M. 1965a: New species and records of the Cleridae from the Ryukyu Islands, I (Coleoptera) *The Entomological Review of Japan* **18**: 19-25, pl. 3.

Miyatake M. 1965b: Some Coccinellidae (excluding Scymnini) of Formosa (Coleoptera). *Special Bulletin of the Lepidopterological Society of Japan* **1**: 50-74.

Miyatake M. 1965c: Some species of Formosan Cleridae, collected by Prof. T. Shirôzu (Coleoptera). *Special Bulletin of the Lepidopterological Society of Japan* **1**: 157-166.

Miyatake M. 1966a: A new genus belonging to the tribe Pharini from the northern part of the Ryukyu Islands (Coleoptera: Coccinellidae). *Transactions of the Shikoku Entomological Society* **9**: 47-50.

Miyatake M. 1966b: Description of two new species of the genus Stethorus Weise from Japan (Coleoptera: Coccinellidae). *Transactions of the Shikoku Entomological Society* **9**: 51-54.

Miyatake M. 1969: The genus Plotina and related genera (Coleoptera: Coccinellidae). *Pacific Insects* **11**: 197-216.

Miyatake M. 1970: The East-Asian coccinellid-beetles preserved in the California Academy of Sciences, tribe Chilocorini. *Memoirs of the College of Agriculture, Ehime University* **14**: 303-340.

Miyatake M. 1972: A new Formosan species belonging to the genus Singhikalia Kapur, with proposal of a new tribe (Coleoptera: Coccinellidae). *Transactions of the Shikoku Entomological Society* **11**: 92-98.

Miyatake M. 1977: Two new species of the genus Micraspis Dejean from Japan (Coleoptera: Coccinellidae). *Tansactions of the Shikoku Entomological Society* **13**: 109-113.

Miyatake M. 1978: The genus Telsimia Casey of Japan and Taiwan (Coleoptera, Coccinellidae). *Transactions of the Shikoku Entomologcal Society* **14**: 13-19.

Miyatake M. 1979: A new species of the genus Horniolus Weise from Taiwan (Coleoptera, Coccinellidae). *Transactions of the Shikoku Entomological Society* **14**: 105-106.

Miyatake M. 1982: Two new species of the genus Sticholotis Crotch from Taiwan (Coleoptera: Coccinellidae). *Transactions of the Shikoku Entomological Society* **16**: 23-28.

Miyatake M. 1985a: Coccinellidae collected by the Hokkaido University expedition to Nepal Himalaya, 1968 (Coleoptera). *Insecta Matsumurana* (N. S.) **30**: 1-33.

Miyatake M. 1985b: Cleridae. Pp. 158-161. In: Kurosawa Y., Hisamatsu S. & Sasaji H. (eds): *Colored illustrations of the Coleoptera of Japan. Vol. 3.* Osaka: Hoikusha Publishing Co., x + 500 pp., 72 pls (in Japanese).

Miyatake M. 1994: Revisional studies on Asian genera of the subfamily Sticholotidinae (Coleoptera: Coccinellidae). *Memoirs of the College of Agriculture, Ehime University, Matsuyama, Entomological Series* No. 8, **38**: 223-293.

Montrouzier X. 1855: Essai sur la faune de l'île de Woodlark ou Moiou. *Annales de la Société d'Agriculture, Histoire Naturelle, et Arts utiles de Lyon* (2) **7** (1): 1-114.

Montrouzier X. 1860: Essai sur la faune entomologique de la Nouvelle-Calédonie (Balade) et des îles des Pins, Art, Lifu, etc. *Annales de la Société Entomologique de France* (3) **8**: 229-308, 1 pl.

Montrouzier X. 1861: Essai sur la faune entomologique de la Nouvelle-Calédonie (Balade) et des îles des Pins, Art, Lifu, etc. Coléoptères (Fin). *Annales de la Société Entomologique de France* (4) **1**: 265-306.

Moore B. P. 1957: The identity of Ptinus latro auct. (Coleoptera: Ptinidae). *Proceedings of the Royal Entomological Society of London* (*Serie B Taxonomy*) **26**: 199-202.

Morawitz F. F. 1861: Einige für die russisch-europäische Fauna neue Käfer beschrieben. *Bulletin de la Société Impériale des Naturalistes de Moscou* **34** (1) [1861-1862]: 284-294.

Morawitz F. F. 1862: Einige neue Melyridae. *Bulletin de la Société Impériale des Naturalistes de Moscou* **34** [1861-1862] (3): 314-320.

Morawitz F. F. 1863: Notiz über die russischen Xyletininae. *Trudy Russkago Entomologicheskago Obshchestva* **2**: 161-37.

Moscardini C. 1962: Tre nuove specie appenniniche del genere Cantharis del gruppo tristis Fabr. *Memorie del Museo Civico di Storia Naturale di Verona* **10**: 105-119.

Moscardini C. 1963: Osservazioni sulle Cantharis obscura L., C. obscura liburnica Dep. et C. csikii Still. et descrizione di una nuova specie (Coleoptera, Cantharidae). *Bollettino della Società Entomologica Italiana* **93**: 107-113.

Moscardini C. 1967: Cantharis puncticollis Levrat, C. erichsoni Bach et descrizione di una nuova specie (Coleoptera, Cantharidae). *Bollettino della Società Entomologica Italiana* **97**: 28-33.

Moscardini C. & Sassi F. 1970: Nuovo genere di Cantharidae (Coleoptera Malacodermata). *Bollettino della Società Entomologica Italiana* **102**: 192-196.

Motschulsky V. de. 1835: Description de quelques coléoptères, recueillis dans un voyage au Caucase et dans les provinces transcaucasiennes russes en 1834 et 1835. Fungicoles. *Nouveaux Mémoires de la Société Impériale des Naturalistes de Moscou* **4**: 321-323, pl. 11.

Motschulsky V. de. 1837a: Extrait d'une lettre adressée par M. V. Motschoulsky à M. B. Zoubkoff. *Bulletin de la Société Impériale des Naturalistes de Moscou* **10** (5): 97-124.

Motschulsky V. de. 1837b: Description de quelques coléoptères recueillis dans un voyage au Caucase et dans les provinces transcaucasiennes russes en 1834 et 1835. *Nouveaux Mémoires de la Société Impériale des Naturalistes de Moscou* **5**: 413-425, pl. 16.

Motschulsky V. de. 1838: Coléoptères du Caucase et des provinces transcaucasiennes, recueillis et décrit. Fungicoles. *Bulletin de la Société Impériale des Naturalistes de Moscou* **11**: 175-188, pl. 3.

Motschulsky V. de. 1839: Coléoptères du Caucase et des provinces transcaucasiennes. *Bulletin de la Société Impériale des Naturalistes* **12**: 68-93, pls. 5, 6.

Motschulsky V. de. 1840: Insectes du Caucase et des provinces transcaucasiennes recueillis et decrits (Continuation). *Bulletin de la Société Impériale des Naturalistes de Moscou* **13**: 169-180, pl. IV.

Motschulsky V. de. 1844: Bemerkungen zu dem im Vten Bande der Zeitschrift für die Entomologie p. 192 von H.rm. Mærkel gegebenen "Beitræge zur Kenntniss der unter Ameisen lebenden Insekten". *Bulletin de la Société Impériale des Naturalistes de Moscou* **17**: 812-823.

Motschulsky V. de. 1845a: Remarques sur la collection de coléoptères Russes de Victor de Motschulsky. 1er Article. *Bulletin de la Société Impériale des Naturalistes de Moscou* **18** (1-2): 3-127, 3 pls.

Motschulsky V. 1845b: Observations sur le Musée entomologique de l'Université impériale de Moscou. *Bulletin de la Société Impériale des Naturalistes de Moscou* **18** (4): 332-388.

Motschulsky V. de. 1849: Coléoptères reçus d'un voyage de M. Handschuh dans le Midi de l'Espagne, énumérés et suivis de notes. *Bulletin de la Société Impériale des Naturalistes de Moscou* **22** (3): 52-163.

Motschulsky V. de. 1850: Voyage de Motschulsky I. Lettres à Mr. Renard. *Bulletin de la Société Impériale des Naturalistes de Moscou* **23** (3): 351-368.

Motschulsky V. de. 1851: Correspondance. Lettre à Monsieur le Dr. Renard, secrétaire de la Société Impériale des naturalistes de Moscou. *Bulletin de la Société Impériale des Naturalistes de Moscou* **24** (2): 648-657.

Motschulsky V. de. 1853a: Malthinides. *Études Entomologiques* **1**: 1-15, 25-58.

Motschulsky V. de. 1853b: Notices. *Études Entomologiques* **1**: 21-24.

Motschulsky V. de. 1853c: Nouveautés. *Études Entomologiques* **1**: 77-79.

Motschulsky V. de. 1854a: Lampyrides. *Études Entomologiques* **2** [1853]: 1-14, 33-43.

Motschulsky V. de. 1854b: Nouveautés. *Études Entomologiques* **2** [1853]: 28-32.

Motschulsky V. de. 1854c: Diagnoses de coléoptères nouveaux, trouvés par M. M. Tatarinoff et Gasckéwitsch aux environ de Pékin. *Études Entomologiqes* **2** [1853]: 44-51.

Motschulsky V. de. 1854d: Nouveautés. *Études Entomologiques* **2** [1853]: 55-56.

Motschulsky V. de. 1854e: Lampyrides. *Études Entomologiques* **3**: 15-26, 47-62.

Motschulsky V. de. 1854f: Coléoptères du nord de la Chine (Shingai). *Études Entomologiques* **3**: 63-65.

Motschulsky V. de. 1858a: Voyages et excursions entomologiques. *Études Entomologiques* **6** [1857]: 4-24.

Motschulsky V. de. 1858b: Entomologie spéciale. Insectes du Japon. *Études Entomologiques* **6** [1857]: 25-41, 1 pl.

Motschulsky V. de. 1858c: Insectes des Indes orientales. 1:ière Série. *Études Entomologiques* **7**: 20-122, 2 pls.

Motschulsky V. de. 1859a: Coléoptères nouveaux de la Californie. *Bulletin de la Société Impériale des Naturaliste de Moscou* **32** (3): 122-185.

Motschulsky V. de. 1859b: Coléoptères du Gouvernement de Iakoutsk, recueillis par M. Pavlofski. *Mélanges Biologiques tirés du Bulletin Physiko-Mathématique de l'Académie Impérial des Sciences de St.-Pétersbourg* **3** (2): 221-238.

Motschulsky V. de. 1859d: Coléoptères du Gouvernement de Iakoutsk, recueillis par M. Pavlofski. *Bulletin de la Classe Physico-Mathématique de l'Académie Impériale des Sciences de St.-Pétersbourg* **17**: 567-574 (columns).

Motschulsky V. de. 1860a: Insectes des Indes orientales, et de contrées analogues. 2:me série. Coléoptères. *Études Entomologiques* **8** [1859]: 25-118, 2 pls.

Motschulsky V. de. 1860b: Coléoptères rapportés de la Sibérie orientale et notamment des pays situés sur les bords du fleuve Amour par MM. Schrenck, Maack, Ditmar, Voznessenski etc. Pp. 77-257 + [1] pp., pls 6-11, 1 map. In: Schrenck P. L.: *Reisen und Forschungen im Amur-Lande in den Jahren 1854-1856 im Auftrage der Kaiserl. Akademie der Wissenschaften zu St. Peterburg ausgeführt und in Verbindung mit mehreren Gelehrten herausgegeben von Dr.Leopold Schrenck. Band II. Zweite Lieferung. Coleopteren.* St. Peterburg: Kaiserliche Akademie der Wissenschaften, 976 pp.

Motschulsky V. de. 1860c: IV. Entomologie appliquée. Insectes utiles et nuisibles. *Études Entomologiques* **8** [1859]: 169-174.

Motschulsky V. de. 1860d: Coléoptères nouveaux de la Californie. *Bulletin de la Société Impériale des Naturalistes de Moscou* **32** (4) [1859]: 357-410.

Motschulsky V. de. 1860e: Catalogue des insectes rapportés des environs du fl. Amour, depuis le Schilka jusqu'à Nikolaevsk. *Bulletin de la Société Impériale des Naturalistes de Moscou* **32** (4) [1859]: 487-507.

Motschulsky V. de. 1860f: Coléoptères rapportés en 1859 par M. Sévertsef des Steppes méridionales des Kirghises. *Bulletin de l'Académie Impériale des Sciences de St.-Pétersbourg* **2**: 513-546.

Motschulsky V. de. 1861a: Insectes du Japon. *Études Entomologiques* **9** [1860]: 4-39.

Motschulsky V. de. 1861b: Diognoses [sic!] d'insectes nouveaux des rives du fl. Amour et de la Daourie méridionale. *Études Entomologiques* **9** [1860]: 39-41.

Motschulsky V. de. 1861c: Essai d'un catalogue des insectes de l'île Ceylan. *Bulletin de la Société Impériale des Naturalistes de Moscou* **34** (1): 95-155.

Motschulsky V. de. 1863: Essai d'un catalogue des insectes de l'île Ceylan. *Bulletin de la Société Impériale des Naturalistes de Moscou* **36** (2): 421-532, 1 pl.

Motschulsky V. de. 1866a: Catalogue des insectes reçus du Japon. *Bulletin de la Société Impériale des Naturalistes de Moscou* **39** (1): 163-200.

Motschulsky V. de. 1866b: Essai d'un catalogue des insectes de l'île de Ceylan. Supplément. *Bulletin de la Société Impériale des Naturalistes de* **39** (2): 393-446.

Motschulsky V. de. 1866c: Enumération des espèces de coléoptères rapportés de ses voyages. *Bulletin de la Société Impériale des Naturalistes de Moscou* **39** (3): 225-290.

Motschulsky V. de. 1867: Enumération des espèces de coléoptères rapportées de ses voyages. 5-ème article. *Bulletin de la Société Impériale des Naturalistes de Moscou* **40** (1): 39-103.

Motschulsky V. 1868: Enumération des nouvelles espèces de coléoptères rapportés de ses voyages. 6-ième Article. *Bulletin de la Société des Naturalistes de Moscou* **41** (2) [1868-1869]: 170-201, pl. VIII.

Motschulsky V. de. 1870: Genres et espèces d'insectes, publiés dans différentes ouvrages par Victor Motschulsky. *Horae Societatis Entomologicae Rossicae* **6** [1869] Supplément: [1] + 118 pp.

Mroczkowski M. 1950: A new species of Anthrenus Geoffr. from Poland (Coleoptera, Dermestidae). *Annales Musei Zoologici Polonici* **14**: 187-192., pl. 17.

Mroczkowski M. 1951: Anthrenus (Helocerus) polonicus sp. n. from Central and Eastern Europe with a description of a new subgenus of Anthrenus Geoffr. *Annales Musei Zoologici Polonici* **14**: 253-260, pl. 35.

Mroczkowski M. 1952: Contribution to the knowledge of the Dermestidae with description of a new species and a new subspecies (Coleptera). *Annales Musei Zoologici Polonici* **15**: 25-32.

Mroczkowski M. 1959: Zweiter Beitrag zur Kenntnis der Dermestiden von Afghanistan, nebst Beschreibung einer neuen Art (Coleoptera). *Kungl. Fysiografiska Sällskapets i Lund Förhandlingae* **29**: 99-101.

Mroczkowski M. 1960: Kozheedy (Coleoptera, Dermestidae) Turkmenii. *Trudy Zoologičeskogo Instituta Akademii Nauk SSSR* **27**: 208-219.

Mroczkowski M. 1961a: Vierter Beitrag zur Kenntnis der Dermestiden von Afghanistan nebst Beschreibung von zwei neuen Arten der Gattung Ctesias Steph. (Coleoptera). *Entomologisk Tidskrift* **82**: 191-196.

Mroczkowski M. 1961b: Two new species of the genus Anthrenus Geoffr. from the Tadzhik S.S.R. (Coleoptera,

Dermestidae). *Bulletin de l'Académie Polonaise des Sciences, Cl. II, Warszawa* **8**: 591-593.

Mroczkowski M. 1962: Dermestidae (Coleoptera) from the Kazakh SSR. *Annales Zoologici* **20**: 229-259.

Mroczkowski M. 1963a: Some remarks on the Dermestidae (Coleoptera) from the Zoological Museum in Hamburg. *Entomologische Mitteilungen* (Hamburg) **2**: 377-380.

Mroczkowski M. 1963b: Anthrenus sarnicus sp. n. (Coleoptera, Dermestidae) from the Island of Guernsey. *The Annals and Magazine of Natural History* (13) **5**: 697-700.

Mroczkowski M. 1964: Systematic and synonymic notes upon certain species of Dermestidae (Coleoptera). *Annales Zoologici* **22**: 179-187.

Mroczkowski M. 1966: Contribution to the knowledge of Silphidae and Dermestidae of Korea (Coleoptera). *Annales Zoologici* **23**: 433-443.

Mroczkowski M. 1967: The Palaearctic species of Megatoma Herbst (Coleoptera, Dermestidae). *Polskie Pismo Entomologiczne* **37**: 3-24.

Mroczkowski M. 1968: Distribution of the Dermestidae (Coleoptera) of the world with a catalogue of all known species. *Annales Zoologici* **26**: 15-191.

Mroczkowski M. 1980: Insects of Saudi Arabia. Coleoptera: Fam. Dermestidae. Part 2, Description of three new species. *Fauna of Saudi Arabia* **2**: 124-126.

Mukhopadhyay P. & Chakraborty P. 2003: Insecta: Coleoptera: Elateroidea: Elateridae. *State Fauna Series* **9**, *Fauna of Sikkim* **3**: 217-236.

Mukhopadhyay P. & Sengupta [= Sen Gupta] T. 1977: Studies on Placonotus (Coleoptera: Cucujidae: Laemophloeinae) from India. *Oriental Insects* **11**: 555-556.

Mukhopadhyay P. & Sengupta [= Sen Gupta] T. 1978: Ergebnisse der Bhutan-Expedition 1972 des Naturhistorischen Museums in Basel. Coleoptera: Fam. Cucujidae Subfam. Laemophloeinae. *Entomologica Basiliensia* **3**: 223-230.

Müller J. 1908: Beschreibungen neuer dalmatinischer Koleopteren. *Münchener Koleopterologische Zeitschrift* **3** [1906-1908]: 314-316.

Müller J. 1913: Ein neuer Annomatus aus der Herzegowina. *Wiener Entomologische Zeitung* 32: 122.

Müller J. 1916: Coleopterologische Beiträge zur Fauna der österreichischen Karstprovinzen und ihrer Grenzgebiete. *Entomologische Blätter* **12**: 73-109.

Müller J. [G.] 1924: Nuovi coleotteri Paleartici. *Bollettino della Società Entomologica Italiana* **56**: 68-80.

Müller J. [G.] 1925: Nuovi coleotteri della regione adriatica. *Studi Entomologici* **1**: 25-31.

Müller J. [G.] 1935: Risultati scientifici della spedizione Ravasini-Lona in Albania. IX. Carabidae, Cantharidae. *Bollettino della Società Entomologica Italiana* **67**: 152-163.

Müller J. [G.] 1946: Nuovi coleotteri della regione balcanica occidentale (Dalmazia, Montenegro, Albania e Grecie). *Redia* **31**: 107-122.

Müller J. [G.] 1960: Eine neue Käfergattung aus der Herzegowina. *Bulletin Scientifique* (Zagreb) **5**: 71-72.

Müller O. F. 1764: *Fauna insectorum Fridrichsdalina, sive methodica descriptio insectorum agri Fridrichsdalensis cum characteribus genericis et specificis, nominibus trivialibus, locis natulibus, iconibus allegatis, novisque pluribus speciebus additis.* Hafniae et Lipsiae: F. Gleditschi, xxiv + 96 pp.

Müller O. F. 1776: *Zoologiae Danicae prodromus, seu animalium Daniae et Norvegiae indigenarum characteres, nomina et synonyma imprimis popularium.* Hafniae: Hallageriis, xxxii + 282 pp.

Müller P. W. J. 1821: Neue Insecten. *Magazin der Entomologie* (Halle) **4**: 184-230.

Müller P. W. J. 1824: [new taxa]. In: Germar E. F.: *Insectorum species novae aut minus cognitae, descriptionibus illustratae, Vol. 1.* Halae: J. C. Hendelii et filii, xxiv + 624 pp., 2 pls.

Mulsant E. 1838: Tableau synoptique des Lycides, ou des espèces du genre Lycus, qui se rencontrent dans les environs de Lyon. *Annales des Sciences Physiques et Naturelles, d'Agriculture et d'Industrie* (Lyon) **1**: 77-81.

Mulsant E. 1846: *Histoire Naturelle des coléoptères de France. Sulcicolles - Sécuripalpes.* Paris: Maison, xxiv + 280 pp., 1 pl.

Mulsant E. 1850a: Species des coléoptères trimères sécuripalpes. *Annales des Sciences Physiques et Naturelles, d'Agriculture et d'Industrie de Lyon* (2) **2** (1): xv + 450 pp.

Mulsant E. 1850b: Species des coléoptères trimères sécuripalpes. *Annales des Sciences Physiques et Naturelles, d'Agriculture et d'Industrie de Lyon* (2) **2** (2): 451-1104.

Mulsant E. 1850c: Description d'une nouvelle espèce de coléoptère sécuripalpes. *Annales de la Société Linnéenne de Lyon* **2** [1847-1849]: 220-222.

Mulsant E. 1851a: Description de quelques espèces d'insectes devant servir à constituer une tribu nouvelle parmi les coléoptères trimères. *Mémoires de l'Académie Nationale des Sciences, Belles-Lettres et Arts de Lyon. Classe de Sciences Naturelle* **1**: 69-74.

Mulsant E. 1851b: Description d'une espèce nouvelle du genre Zygia. *Mémoires de l'Académie Impériale des Sciences, Belles-Lettres et Arts de Lyon. Classe de Sciences Naturelle* **1**: 190-191.

Mulsant E. 1851c: Note sur le Bostrichus trispinosus d'Olivier. *Mémoires de l'Académie Impériale des Sciences, Belles-Lettres et Arts de Lyon, Classe des Sciences Naturelle* **1** (2): 205-208.

Mulsant E. 1851d: Description d'un insecte inédit consistuant un genre nouveau parmi les coléoptères. *Mémoires de l'Académie Impériale des Sciences, Belles-Lettres et Arts de Lyon, Classe des Sciences Naturelle* **1** (2): 209-211.

Mulsant E. 1852: Description d'un coléoptère nouveau de genre Homalisus. *Opuscules Entomologiques* **1**: 79-81.

Mulsant E. 1853: Supplément à la monographie des coléoptères trimères sécuripapes [sic!]. *Annales de la Société Linnéenne de Lyon* (N. S.) (2) **1** [1852-1853]: 129-333.

Mulsant E. 1856: Additions et rectifications au Catalogue des coccinellides, publié en 1853. *Annales de la Société Linéenne de Lyon* (N. S.) **3**: 135-156.

Mulsant E. 1858: Description d'une espèce nouvelle de Coccinellidae. *Annales des Sciences Physiques et Naturelles, d'Agriculture et d'Industrie de Lyon* (3) **2**: 321-322.

Mulsant E. 1859: Description d'une espèce nouvelle de Coccinellidae. *Opuscules Entomologiques* **9**: 63-64.

Mulsant E. 1861: Descriptions d'une espèce nouvelle de coléoptères de la tribu des mollipennes. *Annales de la Société Linnéenne de Lyon* (N. S.) **7** [1860]: 188-192.

Mulsant E. 1862a: *Histoire Naturelle des coléoptères de France. Mollipennes*. Paris: Magnin, Blanchard et Co., [6] + 440 pp., 3 pls.

Mulsant E. 1862b: *Histoire Naturelle des coléoptères de France. Mollipennes. Annales de la Société Linnéenne de Lyon* (N. S.) **9**: 57-494.

Mulsant E. 1862c: [new name]. In: Reiche L. J.: Notes synonymiques sur divers coléoptères. *Annales de la Société Entomologique de France* (4) **2**: 79-80.

Mulsant E. 1866: *Monographie des Coccinellides*. Paris: Savy, [ii] + 292 pp.

Mulsant E. & Godart A. 1855a: Description de quelques espèces de coléoptères nouveaux ou peu connus. *Opuscules Entomologiques* **6**: 161-183. [pp. 168-183 are misnumbered as 268-283].

Mulsant E. & Godart A. 1855b: Description de quelques espèces de coléoptères nouveaux ou peu connus. *Annales de la Société Linnéenne de Lyon* **2**: 261-283.

Mulsant E. & Godart A. 1859: Description de trois espèces nouvelles de coléoptères. *Opuscules Entomologiques* **9**: 177-183.

Mulsant E. & Godart A. 1866: Description de quelques coléoptères nouveaux ou peu connus. *Annales de la Société Linnéenne de Lyon* (N. S.) **12** [1865]: 447-456.

Mulsant E. & Godart A. 1868: Description d'une espèce nouvelle de coléoptère du genre Athous. *Annales de la Société Linnéenne de Lyon* (N. S.) **15** [1867]: 409-410.

Mulsant E. & Godart A. 1869: Description d'une espèce nouvelle de Coccinellidae. *Annales de la Société Linéenne de Lyon* (N. S.) **16**: 102-103.

Mulsant E. & Godart A. 1870: Description de deux nouveaux scymniens (tribus des coccinellides). *Annales de la Société Linéenne de Lyon* (N. S.) **16**: 198-200.

Mulsant E. & Godart A. 1872a: Description d'une espèce nouvelle de Coccinellidae. *Opuscules Entomologiques* **14**: 126-127.

Mulsant E & Godart A. 1872b: Description de deux nouveaux scymniens (tribu du Coccinellides). *Opuscules Entomologiques* **14**: 222-224 + [errata].

Mulsant E. & Godart A. 1872c: Description d´une espèce nouvelle de coléoptères du genre Anthrenus. *Opuscules Entomologiques* **14**: 240-241.

Mulsant E. & Godart A 1875: Description d'un espèce nouvelle de Scymnus. *Opuscules Entomologiques* **16**: 167-168.

Mulsant E. & Guillebeau F. 1853: Description d'un coléoptère inédit constituant un genre nouveau parmi les élatérides. *Opuscules Entomologiques* **2**: 189-192.

Mulsant E. & Guillebeau F. 1855a: Description de quelques espèces d'élatérides. *Annales de la Société Linnéenne de Lyon* (2) **2**: 17-32. [reprinted in same year, same pp. in *Opuscules Entomologiques* 6].

Mulsant E. & Guillebeau F. 1855b: Description de deux nouvelles espèces d'élatérides. *Annales de la Société Linnéenne de Lyon* (2) **2**: 317-320.

Mulsant E. & Guillebeau F. 1856: Description de quelques élatérides nouveaux ou peu connus. *Annales de la Société Linnéenne de Lyon* (2) **3**: 60-99. [reprinted in same year, same pp. in *Opuscules Entomologiques* 7].

Mulsant E. & Revelière E. 1860: Description d'une espèce nouvelle du genre Lampyris. *Opuscules Entomologiques* **11**: 130-133.

Mulsant E. & Revelière E. 1861: Description d'une espèce nouvelle de coléoptère du genre Dasytes. *Opuscules Entomologiques* **12**: 10-12.

Mulsant E. & Rey C. 1852a: Description d'une espèce nouvelle de malachie. *Mémoires de l'Académie Impériale des Sciences, Belles-Lettres et Arts de Lyon* (2) **2**: 219-220. [reprinted in 1853 *Opuscules Entomologiques* **2**: 93-94].

Mulsant E. & Rey C. 1852b: Description de trois coléoptères nouveaux de la famille des scymniens. *Mémoires de Académie Impériale des Sciences, Belles-Lettres et Arts de Lyon* (2) **2**: 221-225. [reprinted in 1853 *Opuscules Entomologiqes* **2**: 86-90].

Mulsant E. & Rey C. 1853a: Description de quelques coléoptères nouveaux ou peu connus. *Annales de la Société Linnéenne de Lyon* (2) **1**: 2-9.

Mulsant E. & Rey C. 1853b: Description de quelques coléoptères nouveaux ou peu connus. *Opuscules Entomologiques* **2**: 4-11.

Mulsant E. & Rey C. 1853c: Description de trois espèces nouvelles du genre Anobium (Dryophilus), Chevrolat. *Opuscules Entomologiques* **2**: 14-20.

Mulsant E. & Rey C. 1859a: Description de quelques coléoptères nouveaux. *Opuscules Entomologiques* **9**: 158-176.

Mulsant E. & Rey C. 1859b: Description d'une espèce nouvelle du genre Scymnus (coléoptères sécuripalpes). Scymnus nanus. *Opuscules Entomologiqes* **11** [1859-1860]: 169-172.

Mulsant E. & Rey C. 1861a: Description d'un genre nouveau de la famille des anobides. *Opuscules Entomologiques* **12**: 40-46.

Mulsant E. & Rey C. 1861b: Description de quelques coléoptères nouveaux ou peu connus. *Entomologiques* **12**: 47-92.

Mulsant E. & Rey C. 1861cc: Description de quelques coléoptères nouveaux ou peu connus. *Annales de la Société Linnéenne de Lyon* (N. S.) **8**: 113-118, 121-122.

Mulsant E. & Rey C. 1861dc: Description de quelques coléoptères nouveaux ou peu connus et de deux genres nouveaux. *Opuscules Entomologiques* **12**: 120-134.

Mulsant E. & Rey C. 1861e: Etablissement d'un nouveau genre de téléphorides (coléoptères de la tribu des mollipennes). *Mémoires de la Société Impériale des Sciences Naturelle de Cherbourg* **8**: 190-192, pl. 3.

Mulsant E. & Rey C. 1861f: Description d'un genre nouveau de la famille des anobides. *Annales de la Société Linnéenne de Lyon* (N. S.) **7**: 293-299.

Mulsant E. & Rey C. 1861g: Description de quelques coléoptères nouveaux ou peu connus. *Annales de la Société Linnéenne de Lyon* (N. S.) **7**: 300-345.

Mulsant E. & Rey E. 1863a: *Histoire naturelle des coléoptères de France. Angusticolles*. Paris: Magnin, Blanchard et Cie, 134 pp., 2 pls.

Mulsant E. & Rey C. 1863b: Description d'un genre nouveau de la famille des cryptophagides. *Annales de la Société Linnéenne de Lyon* (N. S.) **10**: 1-3.

Mulsant E. & Rey C. 1863c: Description de quelques coléoptères nouveaux et peu connus. *Annales de la Société Linnéenne de Lyon* (N. S.) **10**: 4-29.

Mulsant E. & Rey C. 1863d: Essai sur la famille des anobides proprement dits. *Opuscules Entomologiques* **13**: 30-143.

Mulsant E. & Rey C. 1864: *Histoire naturelle des coléoptères de France. Térédiles*. Paris: F. Savy, [8] + 392 + 13 pp., 10 pls.

Mulsant E. & Rey C. 1867: Histoire naturelle des coléoptères de France. Vésiculifères. *Annales de la Société Agricoles de Lyon* (3) **11**: 625-943, pls 1-7. [Reprinted by F. Savy, Libraire-Éditeur, Paris. 308 pp., pls 1-7, plate explanations unnumbered].

Mulsant E. & Rey C. 1868a: *Histoire naturelle des coléoptères de France. Scuticolles*. Paris: F. Savy, 186 pp., 3 pls. [reprint from *Annales de la Société Linnéenne de Lyon* (N. S.) **15**: 1-188, 3 pls].

Mulsant E. & Rey C. 1868b: *Histoire naturelle des coléoptères de France. Gibbicolles.* Paris: Deyrolle, Naturaliste, 224 + 16 pp., 14 pls. [reprint from *Annales de la Société Agricoles de Lyon* (4) **1**: 179-421, 14 pls].

Mulsant E. & Rey C. 1868c: Histoire naturelle des coléoptères de France. Floricoles. *Annales de la Société Linnéenne de Lyon* (N. S.) **15**: 237-402; **16**: 83-231, pls 1-19.

Mulsant E. & Rey C. 1868c: Histoire naturelle des coléoptères de France. Floricoles. Paris: Deyrolle, 315 pp., pls 1-19 + [5] pp. [reprint from *Annales de la Société Linnéenne de Lyon* (N. S.) **15**: 237-402; **16**: 83-231, pls 1-19].

Mulsant E. & Rey C. 1870: Description de diverses espèces nouvelles de coléoptères. *Opuscules Entomologiques* **14**: 105-122.

Mulsant E. & Rey C. 1875a: Description d'une nouvelle espèce d'élatéride. *Opuscules Entomologiques* **16**: 81-83.

Mulsant E. & Rey C. 1875b: Description d'un genre nouveau de la tribu des Élatérides. *Annales de la Société Linnéenne de Lyon* (2) **21**: 405-408.

Mulsant E. & Rey C. 1875c: Description d'une nouvelle espèce d'élatéride. *Annales de la Société Linnéenne de Lyon* (N. S.) **21** [1874]: 416-418.

Mulsant E. & Wachanru A. 1852a: Première série de coléoptères nouveaux ou peu connus. *Mémoires de l'Académie Impériale des Sciences, Belles-Lettres et Arts de Lyon* (2) **2**: 1-17.

Mulsant E. & Wachanru A. 1852b: Première série de coléoptères nouveaux ou peu connus. *Opuscules Entomologiques* **1**: 161-177.

Munster T. G. 1901: Index Coleopterorum Norvegiae I. Phalacridae. *Christiania Videnskabs-Selskabs Forhandlinger* **1**: 32-35.

Munster T. G. 1923: Insecta, ex Sibiria meridionali et Mongolia in itinere Ørjan Olsen 1914 collecta. A. Coleoptera, a Fritz Jensen lecta. VI. Coccinellidae. *Norsk Entomologisk Tidsskrift* **1** [1920-1924]: 241-243.

Munster T. G. 1928: Tillaeg og bemaerkninger til Norges koleopterfauna. *Norsk Entomologisk Tidsskrift* **2** [1924-1930]: 262-298.

Munster T. G. 1930: Tillaeg og bemaerkninger til Norges koleopterfauna II. *Norsk Entomologisk Tidsskrift* **2** [1924-1930]: 353-357.

Munster T. G. 1932: Cryptophagus Lysholmi sp. n. fra Målselven. *Norsk Entomologisk Tidsskrift* **3**: 53.

Munster T. G. 1933: Tillaeg og bemaerkninger til Norges Koleopterfauna III. *Norsk Entomologisk Tidsskrift* **3** [1932-1935]: 267-278.

Muona J. 1991: A revision of the Indomalesian tribe Galbitini new tribe (Coleoptera, Eucnemidae). *Entomologica Scandinavica* **39** (suppl.): 67 pp.

Muona J. 2002: Trixagus leseigneuri n. sp. (Coleoptera, Throscidae). *Bulletin de la Société Entomologique de France* **107**: 187-190.

Murray A. 1864: Monograph of the family of Nitidulariae. Part I. *Transactions of the Linnean Society of London* **24**: 211-414, pls. 32-35.

Murray A. 1867a: List of Coleoptera received from Old Calabar, on the West Coast of Africa. *The Annals and Magazine of Natural History* (3) **19**: 167-180.

Murray A. 1867b: List of Coleoptera received from Old Calabar, on the West Coast of Africa. *The Annals and Magazine of Natural History* (3) **20**: 20-23, 83-95, 314-323.

Mutchler J. & Weiss H. B. 1927: The dermestid beetles of New Jersey including the carpet beetle and those which feed on animal products. *Circular New Jork Agriculture* **108**: 31 pp., 4 pls.

Muttkowski R. A. 1910: Eufallia, a new name for Belonia Fall (Coleoptera). *Bulletin of the Wisconsin Natural History Society* **8**: 161-162.

Naezén D. E. 1792: Beskrifning på några, vid Umeå; fundne, okände arter iblad Sklalbaggarne. *Kongelige Svenska Vetenskaps-Academiens nya Handlingar* (Stockholm) **13**: 167-174, pl. 7.

Nakane T. 1948: Description of a new species of the genus Tritoma from Japan (Coleoptera, Erotylidae). *Mushi* **19**: 13-14.

Nakane T. 1950a: New or little known Coleoptera from Japan and adjacent regions II. Erotylidae. *The Entomological Review of Japan* **5**: 6-13.

Nakane T. 1950b: New or little known Coleoptera from Japan and its adjacent regions. IV - Drilidae. *Mushi* **21**: 29-31.

Nakane T. 1951: New or little known Coleoptera from Japan and its adjacent regions. III. Endomychidae. *Insecta Matsumurana* **17**: 113-118.

Nakane T. 1954: New or little known Coleoptera from Japan and its adjacent regions. X. Descriptions of some new species from Shikoku, Japan. *Transactions of the Shikoku Entomological Society* **4**: 7-15.

Nakane T. 1955a: New or little-known Coleoptera from Japan and its adjacent regions XII. *The Scientific Reports of the Saikyo University* (*Natural Science and Living Science*) (A) **2**(1): 24-42, 3 pls.

Nakane T. 1955b: Marine insects of the Tokara Islands V. Three new species of the genus Laius Guerin from Kyushu and the Tokara Islands, with notes on a species from Mariana. *Publication of the Seto Mariana Biological Laboratory* **4** (2-3): 373-378, pl. 25.

Nakane T. 1956: New or little-known Coleoptera from Japan and its adjacent Regions, XIII. *The Scientific Reports of the Saikyo University* (*Natural Science and Living Science*) (A) **2**: 29-44, 159-174.

Nakane T. 1957: New or little-known Coleoptera from Japan and its adjacent regions, XIV. *The Scientific Reports of the Saikyo University* (*Natural Science and Living Science*) (A) **2**: 235-240.

Nakane T. 1958a: Notes on the Endomychidae of Japan (I). *Akitu* **7**: 33-36.

Nakane T. 1958b: On the Coleoptera of Shimokita Peninsula, northern end of Honshu Japan (Insecta), I. *Miscellaneous Reports of Research Institute for Natural Resources, Tokyo* **46/47**: 83-92.

Nakane T. 1959a: Entomological results from the scientific survey of the Tokara Islands. VII. Coleoptera: Clavicornia – Nitidulidae, Rhizophagidae, Languriidae, Erotylidae & Endomychidae. *The Scientific Reports of Kyoto Prefectural University* (*Natural Science and Living Science*) **3** (1): 53-61.

Nakane T. 1959b: The Coleoptera of Yakushima Island, Clavicornia (Nitidulidae, Cryptophagidae, Biphyllidae, Languriidae, Erotylidae, Endomychidae, Cisidae). *The Scientific Reports of Kyoto Prefectural University* (*Natural Science and Living Science*) **3**: 69-72.

Nakane T. 1959c: Notes on the Elaterid-beetles of Japan (I). *Akitu* **8**: 91-94.

Nakane T. 1961a: New or little-known Coleoptera from Japan and its adjacent regions, XV. *Fragmenta Coleopterologica* **1**: 1-5.

Nakane T. 1961b: The lycid-beetles from the Loochoo Islands, with descriptions of a few new forms (Coleoptera). *Entomological Review of Japan* **13**: 11-15.

Nakane T. 1963a: [new taxa]. In: Nakane T., Obhayashi K., Nomura S. & Kurosawa Y.: *Iconographia Insectorum Japonicum Vol. 2* (*Coleoptera*). Tokyo: Hokuryukan, 443 pp., 192 pls. (in Japanese).

Nakane T. 1963b: New or little-know Coleoptera from Japan and its adjacent Regions XVI.-XXII. *Fragmenta Coleopterologica* **10-12**: 18-48.

Nakane T. 1963c: New or little-known Coleoptera from Japan and its adjacent regions. XXV. *Scientific Reports of the Kyoto Prefecture University* (*Natural Sciences and Living Sciences*) (A) **3**: 221-226.

Nakane T. 1966a: Notes on the Erotylidae of Formosa (Taiwan), with descriptions of few new forms (Coleoptera). *Fragmenta Coleopterologica* **16**: 59-64.

Nakane T. 1966b: New or little-known Coleoptera from Japan and its adjacent regions, XXIV. *Fragmenta Coleopterologica* **16**: 64-66.

Nakane T. 1966c: Coleoptera of Afghanistan. *Results of the Kyoto University Scientific Expedition to the Karakoram and Hindukush, 1955* **8**: 231-248.

Nakane T. 1967a: On the genus Cyphonocerus Kiesenwetter in Japan and Formosa (Insecta, Coleoptera, Lampyridae). *Bulletin of the National Science Museum of Tokyo* **10**: 7-9.

Nakane T. 1967b: New and little known Coleoptera from Japan and its adjacent regions, XXIV. *Fragmenta Coleopterologica* **17**: 67-69.

Nakane T. 1967c: Notes on the genus Macrolycus Waterhouse in Japan and Formosa (Coleoptera: Lycidae). *Fragmenta Coleopterologica* **17-18**: 69-73, 1 pl.

Nakane T. 1967d: New or little-known Coleoptera from Japan and its adjacent regions. XXVI. *Fragmenta Coleopterologica* **18**: 73-74.

Nakane T. 1967e: Descriptions of three new species of Lycid beetles from Formosa, with notes on some others (Insecta, Coleoptera). *Bulletin of the National Science Museum* (Tokyo) **10**: 285-292.

Nakane T. 1968a: New or little-known Coleoptera from Japan and its adjacent regions. XXVI. *Fragmenta Coleopterologica* **19**: 75-76.

Nakane T. 1968b: On some species of Endomychidae from Formosa with description of a new species (Insecta, Coleoptera). *Bulletin of the National Science Museum* (Tokyo) **11**: 135-139.

Nakane T. 1968c: On some new species of Lycidae from Formosa (Insecta: Coleoptera). *Bulletin of the National Science Museum* (Tokyo) **11**: 217-224.

Nakane T. 1969a: *Lycidae (Insecta: Coleoptera). Fauna Japonica*. Tokyo: Academic Press of Japan, 224 pp., 8 pls.

Nakane T. 1969b: New species of Lycidae from Formosa, with notes on some known species (Insecta: Coleoptera). *Bulletin of the National Science Museum* (Tokyo) **12**: 9-16.

Nakane T. 1969c: Zwei neue Arten der Lyciden aus Formosa (Insecta: Coleoptera). *Bulletin of the National Science Museum* (Tokyo) **12**: 471-474.

Nakane T. 1970a: On the cantharoid Coleoptera found in the Islands of Tsushima (Insecta). *Memoirs of the National Science Museum of Tokyo* **3**: 285-288 (in Japanese).

Nakane T. 1970b: Japanese Lycidae in the collection of the California Academy of Sciences with notes on a few species from Manchuria (Insecta: Coleoptera). *Bulletin of the National Science Museum* (Tokyo) **13**: 357-361.

Nakane T. 1971: Formosan Lycidae in the collection of the California Academy of Sciences (Insecta: Coleoptera). *Bulletin of the National Science Museum* (Tokyo) **14**: 137-155.

Nakane T. 1975: On the Lycid-beetles of Kyushu (Coleoptera). *Kita-Kyūshū no Konchū* **21**: 29–32.

Nakane T. 1977a: New or little-known Coleoptera from Japan and its adjacent regions. XXVIII. *Fragmenta Coleopterologica* **22-24**: 87-98.

Nakane T. 1977b: Notes on some species of Coleoptera (Insecta) occuring in the Ogasaware and Volcano Islands, with desciptions of new species. *Memoirs of the National Science Museum* (Tokyo) **10**: 147-162.

Nakane T. 1978: On some interesting species of Coleoptera from Izu and Ogasawara Islands, with descriptions of new species. *Memoirs of the National Science Museum* (Tokyo) **11**: 155-161 (in Japanese with English descriptions).

Nakane T. 1979: New or little-known Coleoptera from Japan and its adjacent regions. XXIX. *Reports of the Faculty of Science, Kagoshima University (Earth Sciences and Biology)* **11** [1978]: 129-134.

Nakane T. 1980a: New or little-known Coleoptera from Japan and its adjacent regions. XXVIII. *Fragmenta Coleopterologica* **25-28** [1979]: 99.

Nakane T. 1980b: New or little-known Coleoptera from Japan and its adjacent regions. XXXI. *Fragmenta Coleopterologica* **25-28** [1979]: 112-114.

Nakane T. 1981a: Notes on the Endomychidae of Japan (II). *Kita-Kyūshū no Konchū* **28**: 55-58.

Nakane T. 1981b: New or litle-known Coleoptera from Japan and its adjacent regions. XXXIII. *Fragmenta Coleopterologica* **29-32**: 125-130.

Nakane T. 1981c: New or little-known Coleoptera from Japan and its adjacent regions. XXXII. *Reports of the Faculty of Science, Kagoshima University (Earth Sciences and Biology)* **13** [1980]: 127-130.

Nakane T. 1982: New or little-known Coleoptera from Japan and its adjacent regions. XXXIV. *Reports of the Faculty of Science, Kagoshima University (Earth Sciences and Biology)* **14** [1981]: 43-53.

Nakane T. 1983a: Coleoptera of Nepal collected by the members of the Hokkaido University scientific expedition to Nepal Himalaya: Lycidae. *Reports of the Faculty of Science, Kagoshima University (Earth Sciences and Biology)* **16**: 115-123.

Nakane T. 1983b: New or little known Coleoptera from Japan and its adjacent regions. XXXVI. *Fragmenta Coleopterologica* **35-37**: 139-150.

Nakane T. 1984: On the Coleoptera occurring in Yaku-shima Island. Pp. 587-631. In: *Conservation Reports of the Yaku-shima Wildness Area, Kyushu, Japan*, 59. Nature Conservation Bureau, Environment Agency of Japan, Tokyo.

Nakane T. 1985a: Notes on Japanese Coleoptera (1) Lymexylonidae. *Chô Chô* **8**: 27-30.

Nakane T. 1985b: Notes on some Japanese Coleoptera (3) Genus Drapetes Redtenbacher (Throscidae). *Chô Chô* **8**: 30-32 (in Japanese with English abstract).

Nakane T. 1985c: New or little known Coleoptera from Japan and its adjacent regions, XXXVIII. *Fragmenta Coleopterologica* **38/40**: 153-164.

Nakane T. 1986: Notes on some species of Erotylidae ocuring [sic!] in Japan, with description of a new species. *Kita-Kyūshū no Konchū* **33**: 131-134.

Nakane T. 1987a: Notes on Japanese Melyridae (Coleoptera) II. *Kita-Kyūshū no Konchū* **34**: 1-5, pl. 1 (in Japanese, new taxa in Japanese and English).

Nakane T. 1987b: Notes on the Eucnemidae of Japan (Coleoptera). *Gekkan-Mushi* **198**: 7-11 (in Japanese, with English description).

Nakane T. 1987c: New or little-known Coleoptera from Japan and its adjacent regions. XXXIX. *Review of the Social Science, Miyazaki Sangyo Keiei University* **1** (1): 171-177.

Nakane T. 1987d: Notes on some little-known beetles (Coleoptera) in Japan. *Kita-Kyūshū no Konchū* **34**: 171-177 (in Japanese).

Nakane T. 1988: Notes on some little-known beetles (Coleoptera) in Japan. 3. *Kita-Kyūshū no Konchū* **35**: 77-79, pl. 9.

Nakane T. 1989a: Notes on some little-know beetles (Coleoptera) in Japan. 4. *Kita-Kyūshū no Konchū* **36**: 1-10.

Nakane T. 1989b: [new taxa]. In: Nakane T. & Makino T.: A revision of the genus Podabrus Westwood in Japan (II) (Insecta, Coleoptera, Cantharidae). *Review of Miyazaki Sangyo-Keiei University* **1** (2): 1-18.

Nakane T. 1989c: Notes on some litle-known beetles (Coleoptera) in Japan. 5. *Kita-Kyūshū no Konchū* **36**: 143-152, pl. 1.

Nakane T. 1990: [new taxa]. In: Nakane T. & Makino T.: A revision of the genus Podabrus Westwood in Japan (III) (Insecta, Coleoptera, Cantharidae). *Fragmenta Coleopterologica* **45/48**: 183-198.

Nakane T. 1991a: Notes on some little-known beetles (Coleoptera) in Japan. 7. *Kita-Kyūshū no Konchū* **38**: 1-9.

Nakane T. 1991b: Notes on some little-known beetles Coleoptera) in Japan. 8. *Kita-Kyūshū no Konchū* **38**: 111-115, pl. 1.

Nakane T. 1992: Notes on some little-known beetles (Coleoptera) in Japan. 9. *Kita-Kyūshū no Konchū* **39**: 73-79.

Nakane T. 1993: Notes on some little-known beetles (Coleoptera) in Japan. 11. *Kita-Kyūshu no Konchū* **40**: 155-162.

Nakane T. 1994a: Revisional notes on the species of the genus Macrolycus Waterhouse in Japan (Insecta, Coleoptera, Lycidae). *Transactions of the Nagasaki Biological Society* No. 44, **1994**: 3-9.

Nakane T. 1994b: Notes on some little-known beetles (Coleoptera) in Japan. 12. *Kita-Kyūshū no Konchū* **41**: 81-86.

Nakane T. 1995a: The beetles collected by Drs. Keizo Kojima and Shingo Nakamura in Taiwn [sic!] (Insecta, Coleoptera) I. *Miscellaneous Reports of the Hiwa Museum for Natural History* **33**: 23-28.

Nakane T. 1995b: Notes on some little-known beetles (Coleoptera) in Japan. 13. *Kita-Kyūshu no Konchū* **42**: 41-46.

Nakane T. 1996: The beetles collected by Drs. Keizo Kojima and Shingo Nakamura in Taiwan (Insecta, Coleoptera) 2. *Miscellaneous Reports of the Hiwa Museum for Natural History* **34**: 129-140.

Nakane T. & Araki M. 1958: The Coleoptera of Yakushima Island. Coccinellidae. *Scientific Reports of the Saikyo University, Natural Science and Living Science* **2**: 291-193 (in Japanese).

Nakane T. & Araki M. 1959: Entomological results from the scientific survey of the Tokara Islands. VI. Coleoptera: Coccinellidae. *The Scientific Reports of the Kyoto Prefectural University* (*Natural Science and Living Science*) **3** A: 45-52.

Nakane T. & Araki M. 1960: Descriptions of two new forms of Coccinellidae from Japan (Insecta, Coleoptera). *Scientific Reports of the Kyoto Prefectural University* (*Natural Science and Living Science*) **3** A: 117-119.

Nakane T. & Baba K. 1961: In Nakane T. 1961: New or little-known Coleoptera from Japan and its adjacent regions, XV. *Fragmenta Coleopterologica* **1**: 1-5.

Nakane T. & Hisamatsu S. 1955: A new species of the genus Cyllodes Erichson from Japan (Coleoptera: Nitidulidae). *Akitu. Transactions of the Kyoto Entomological Society* **4** (3): 55-56.

Nakane T. & Hisamatsu S. 1963: Two new genera and three new species of clavicorn Coleoptera from Japan. *Fragmenta Coleopterologica* **12**: 48-50.

Nakane T. & Hisamatsu S. 1991: [new taxon]. In: Nakane T.: Notes on little-known bettles (Coleoptera) in Japan. 8. *Kita-Kyūshū no Konchū* **38**: 111-115, 1 pl.

Nakane T. & Kishii T. 1954a: [new taxon]. In: Nakane T.: New or little known Coleptera from Japan and its adjacent regions. X. Descriptions of some new species from Shikoku, Japan. *Transactions of the Shikoku Entomological Society* **4**: 7-15.

Nakane T. & Kishii T. 1954b: Elateridae. In: Nakane T.: A list of Coleoptera (Polyphaga) from Oze with descriptions of some new species. *Scientific Researches of the Ozegahara Moor* (Tokyo) pp. 727-740.

Nakane T. & Kishii T. 1955a: Descriptions of a few new forms of the elaterid-beetles from Japan, with notes on some others. *The Scientific Reports of the Saikyo University* (*Natural Science and Living Science*) (A) **2**: 43-46.

Nakane T. & Kishii T. 1955b: Entomological Results from the Scientific Survey of the Tokara Islands. I. Coleoptera: Elateridae. *Bulletin of the Osaka Municipal Museum of Natural History* **2**: 1-8. 2 pls.

Nakane T. & Kishii T. 1956: On the elaterid-beetles from Loochoo, collected by Mr. Kaku Sato (Coleoptera). *The Scientific Reports of the Saikyo University* (*Natural Science and Living Science*) (A) **2**: 157-158.

Nakane T. & Kishii T. 1958: The Coleoptera of Yakushima Island. Elateridae. *The Scientific Reports of the Saikyo University* (*Natural Science and Living Science*) (A) **2**: 294-302.

Nakane T. & Nobuchi A. 1955: Two new species of the genus Spondotriplax Crotch from Japan (Coleoptera: Erotylidae). *Akitu* **4**: 15-16.

Nakane T. & Ohbayashi K. 1949: Description of a new species of the genus Lucidina from Japan (Coleoptera: Lampyridae). *Entomological Review of Japan* **1**: 34-35.

Nakane T. & Ohbayashi K. 1955: Description of a new species of the genus Calochromus from Japan (Coleoptera: Lycidae). *Akitu* **4**: 29-30.

Nakane T. & Ohbayashi K. 1958: Two new species of the genus Benibotarus Kôno from Japan (Coleoptera: Lycidae). *Akitu* **7**: 79-80.

Nakane T. & Winkler J. R. 1952: Dictyoptera elegans n. sp. *Časopis Československé Společnosti Entomologické* **49**: 133-135.

Narukawa N. 1992: A new species of the genus Dacne from the Kii Peninsula, Central Japan (Coleoptera, Erotylidae). *The Entomological Review of Japan* **47**: 67-69.

Narukawa N. 1994: A new species of the genus Tritoma from Japan (Coleoptera, Erotylidae). *The Entomological Review of Japan* **49**: 155-158.

Narukawa N. 2000: A new species of the genus Tritoma (Coleoptera, Erotylidae) from Northeast Japan. *The Entomological Review of Japan* **55** (2): 1-4.

Narukawa N. 2001: Description of a new species of the genus Tritoma (Coleoptera, Erotylidae) from the Ryukyu Islands, Japan. *Sukunahikona. Special Publication of the Japan Coleopterological Society* **1**: 275-278.

Narukawa N. 2002: A new species of the genus Antibothrus (Coleoptera: Bothrideridae) from Japan. *Entomological Review of Japan* **57**: 123-126.

Narukawa N. 2003: Descriptions of two new species of Erotylidae (Coleoptera) from the Ryukyus, Japan. *Japanese Journal of Systematic Entomology* **9**: 31-35.

Narukawa N. 2004: Description of a new species of the genus Microsternus (Coleoptera, Erotylidae) from Tokunoshima Is., Japan. *Japanese Journal of Systematic Entomology* **10**: 277-279.

Nawa U. 1921: Research Outline of Luciola ibukiyamana. *Insect World, Gifu* (*Konchu Sekai*) **25**: 256-260 (in Japanese).

Nawa U. 1939: The History of the Studies on Luciola parvula Kiess. *Kontyû* **13**: 262-263 (in Japanese).

Nefedov N. I. 1962: Novyy vid korovki iz Kabardino-Balkarii. *Uchenye Zapisky Kabardino-Balkarskogo Gosudarstvennogo Universiteta. Seria sel'Skokhozyaystvenaya i Biologicheskay* **14**: 93-95.

Nefedov V. N. & Nefedov N. I. 1959: Novoe v faune kokcinellid. *Uchenye Zapisky Kabardino-Balkarskogo Gosudarstvennogo Universiteta. Seria Sel'skokhozyaystvenaya i Biologicheskaya* **11**: 161-164.

Newman E. 1833: Entomological notes (continuation). *Entomological Magazine* **1**: 505-514.

Newman E. 1834: Entomological notes. *The Entomological Magazine* **2**: 200-205.

Newman E. 1838: Entomological notes. *The Entomological Magazine* **5**: 372-402.

Newman E. 1839: On the synonymy of Passandra, with descriptions of all the old and of a new species. *Annals of Natural History* **2**: 388-398.

Newman E. 1840: Entomological notes. *The Entomologist* **1**: 1-16.

Nietner J. 1856: Entomological Papers - being descriptions of new Ceylon Coleoptera, with such observations on their habits as appear any way interesting. *Journal of the Asiatic Society of Bengal* **25**: 523-554.

Nigrin Z. 1984: Danacaea pickai n. sp., ein neuer Dasytinae aus Tunesien (Coleoptera, Melyridae). *Entomologica Basiliensia* **9**: 318-323.

Nigrin Z. 1986: Danacaea zahradniki sp. n. aus dem Nahen Osten (Coleoptera, Melyridae, Dasytinae). *Acta Entomologica Bohemoslovaca* **83**: 359-364.

Nikitsky N. B. 1983: Vidy roda Biphyllus (Coleoptera, Biphyllidae) vostochnoy palearktiki [Species of the genus Biphyllus (Coleoptera, Biphyllidae) in the East Palaearctic]. *Zoologicheskiy Zhurnal* **62**: 695-706.

Nikitsky N. B. 1984: Species of the genus Rhizophagus Hbst. (Coleoptera, Rhizophagidae) from the eastern Palaearctic region. *Byulleten Moskovskogo Obshchestva Ispytatelei Prirody. Otdel Biologicheskii* **89** (3): 63-69.

Nikitsky N. B. 1985a: Monotoma specialis n. sp. from the NW-Caucasus area (Insecta: Coleoptera: Rhizophagidae). *Senckenbergiana Biologica* **66**: 47-49.

Nikitsky N. B. 1985b: Novyy vid roda Antibothrus (Coleoptera, Colydiidae) iz yuzhnogo Primorya [A new species of the genus Antibothrus (Coleoptera, Colydiidae) from the south of the Soviet Far East]. *Zoologicheskiy Zhurnal* **64**: 625-628.

Nikitsky N. B. 1986: Zhestkokrylye podsemeistv Monotominae i Thioninae (Coleoptera, Rhizophagidae) Dal'nego Vostoka SSSR. *Zoologicheskiy Zhurnal* **65**: 1622-1630.

Nikitsky N. B. 1987: Coleopterans of the family Derodontidae (Coleoptera, Polyphaga) in eastern Palaearctic. *Byulleten Moskovskogo Obshchestva Ispytaleley Prirody. Otdel Biologicheskiy* **92**: 31-36.

Nikitsky N. B. 1992: [new taxon]. In: Nikitsky N. B. & Lafer G. S.: 47. Sem. Derodontidae. Pp. 40-44. In: Ler P. A. (ed.): *Opredelitel' nasekomykh Dal'nego Vostoka SSSR v shestykh tomakh. Tom III. Zhestkokrylye, ili zhuki. Chast' 2.* St. Petersburg: Nauka, 704 pp.

Nikitsky N. B. 1993: New species of Derodontidae and Biphyllidae (Coleoptera) from the Caucasus. *Zoologicheskiy Zhurnal* **72**: 152-156.

Nikitsky N. B. 1996: Novye zhuki iz Kitaya [New Coleoptera from China]. *Zoologicheskiy Zhurnal* **75**: 1366-1373.

Nikitsky N. B. & Belov V. V. 1979: Novyy i maloizvestnyy vidy Clavicornia (Coleoptera) iz Talysha [New and poorly known species of Clavicornia (Coleoptera) from Talysh]. *Zoologicheskiy Zhurnal* **58**: 849-854.

Nikitsky N. B. & Kompantzev A. V. 1995: Novye vidy zhukov-gribovikov (Coleoptera, Erotylidae) rossiyskogo Dalnego Vostoka s kommentariyami po rasprostraneniyu i biologii nekotorykh drugikh vidov [The new species of pleasing fungus beetles (Coleoptera, Erotylidae) from the Russian Far East with the comments on the distribution and biology of some other species]. *Zoologicheskiy Zhurnal* **74**: 83-92.

Nikitsky N. B. & Kompantzev A. V. 1997: Dva novykh vida roda Triplax Herbst (Coleoptera, Erotylidae) iz Sredney Azii i Kazakhstana [Two new species of the genus Triplax Herbst (Coleoptera, Erotylidae) from Middle Asia and Kazakhstan]. *Byulleten Moskovskogo Obshchestva Ispytaleley Prirody. Otdel Biologicheskiy* **102**: 31-33.

Nikitsky N. B. & Semenov W. B. 2001: O vidakh roda Dapsa Latreille, 1829 (Coleoptera, Endomychidae) byvshego USSR. *Byulleten Moskovskogo Obshchestva Ispytaleley Prirody. Otdel Biologicheskiy* **106**(6): 57-70.

Nikolopoulos C. N. 1969: Peri enos neou eidous koleopterou aneirothentos eis to elatodasos tís parnithos Rhizophagus atticus n. sp. *Nea Agrotiki Epitheorísis* **1969**: 163-166.

Nördlinger H. 1855: *Die kleinen Feinde der Landwirthschaft oder Abhalun der in Feld, Garten und Haus schädlichen oder lästigen Kerfe, sonstigen Gliederthierchen, Würmer, und Schnecken, mit besonderer Berücksichtigung ihrer natürlichen Feinde und der gegen sie anwendbaren Schutzmittel.* Stuttgart und Augsburg: J. G. Cotta, 636 pp.

Nördlinger H. 1880: *Lebensweise von Forstkerfen oder Nachträge zu Ratzeburg's Forstinsekten, zweite vermehrte Auflage.* Stuttgart: J. G. Cotta, v + 73 pp.

Normand H. 1935: Contribution au catalogue des coléoptères de Tunisie (6me Fascicule). *Bulletin de la Société d'Histoire Naturelle de l'Afrique du Nord* **26** [1933-1934]: 235-251.

Normand H. 1936a: Contribution au catalogue des coléoptères de la Tunisie (7e fascicule). *Bulletin de la Société d'Histoire Naturelle de l'Afrique du Nord* **26** [1935]: 286-304.

Normand H. 1936b: Contribution au catalogue des coléoptères de la Tunisie (8e fascicule). *Bulletin de la Société d'Histoire Naturelle Afrique du Nord* **27**: 81-100.

Normand H. 1936c: Contribution au catalogue des coléoptères de la Tunisie (8e fascicule) Clavicornia. *Bulletin de la Société d'Histoire Naturelle de l'Afrique du Nord* **27**: 136-155.

Normand H. 1938a: Coléoptères nouveaux de la faune française et remarques diverses. *Revue Française d'Entomologique* **5**: 131-135.

Normand H. 1938b: Contribution au catalogue des coléoptères de la Tunisie. Premier Supplément. *Bulletin de la Société d'Histoire Naturelle de l'Afrique du Nord* **29**: 337-371.

Normand H. 1940: Nouveaux coléoptères du Nord d'Afrique (24e note). *Bulletin de la Société Entomologique de France* **45**: 51-54, pl. 1.
Normand H. 1941: Remarques sur les espèces Nord-Africaines du groupe du Cardiophorus rufipes Goeze. *Bulletin de la Société d'Histoire Naturelle de l'Afrique du Nord* **32**: 331-334.
Normand H. 1948: Nouveaux Dasytes de la Tunisie. *Revue Française d'Entomologie* **15**: 88-93.
Normand H. 1950a: Contribution au catalogue des coléoptères de la Tunisie (Troisième supplément-fascicule 3). *Bulletin de la Société des Sciences Naturelles de Tunisie* **2** [1949]: 65-104, pls. VI-VII.
Normand H. 1950b: Nouveaux coléoptères du Nord de l'Afrique. *Revue Française d'Entomologie* **17**: 105-108.
Normand H. 1951: Nouveaux coléoptères de l'Afrique du nord. *Revue Française d'Entomologie* **18**: 24-30.
Notman H. 1920: Coleoptera collected at Schoharie, N.Y., June 9-14, 1918, with descriptions of new species. *Journal of the New York Entomological Society* **28**: 14-31.
Novoa F. & Baselga A. 2000: A new species of Ernobius Thomson (Coleoptera: Anobiidae: Ernobiinae) from the Cíes Islands, Spain. *The Coleopterists Bulletin* **54**: 403-407.
Nunenmacher F. W. 1912: Studies amongst the Coccinellidae, No. 4. (Coleoptera). *Entomological News* **23**: 448-451.
Obenberger J. 1914: Beitrag zur Kenntnis der palaearktischen Käferfauna. *Coleopterologische Rundschau* **3**: 97-115, 129-142.
Obenberger J. 1916: Ueber die europäischen Corynetesarten (Coleoptera, Cleridae). *Neue Beiträge zur Systematischer Insektenkunde* **1**: 22-23.
Obenberger J. 1917a: Beitrag zur Kenntnis der palaearktischen Käferfauna. *Archiv für Naturgeschichte* A **82** (4) [1916]: 9-45, pls 1, 2.
Obenberger J. 1917b: Einige neue palaearktische Käferarten. *Entomologische Blätter* **13**: 75-77.
Obenberger J. 1918: Vier neue palaearktische Coleopteren. *Entomologische Blätter* **14**: 58-61.
Obenberger J. 1922: II. Beitrag zur Kenntnis der palaearktischen Käferfauna. *Archiv für Naturgeschichte* A **82** (4) [1917]: 9-44, 2 pls.
Obenberger J. 1938: Dva nové druhy rodu Anommatus Wesm. (Col. Colydiidae). *Sborník Entomologického Oddělení Národního Musea v Praze* **16**: 91-92.
Oberthür R. 1883: Trois espèces nouvelles du genre Helota. *Coleopterorum Novitates* **1**: 59-61.
Ohbayashi K. 1949: A new species of the genus Malthinus from Japan (Coleoptera: Cantharidae). *Entomological Review of Japan* **1**: 21-22.
Ohbayashi K. 1954: Five new species of Lycidae from Japan (Coleoptera). *Mushi* **26** (6): 19-22.
Ohbayashi K. 1955: A new species of the genus Maltypus from Japan (Col. Cantharidae). *Entomological Review of Japan* **6**: 51-52.
Ohbayashi K. 1956: New Lycidae from Japan (Coleoptera). *Akitu* **5** (3): 57-59.
Ohbayashi N. 1977: Revisional study of Dermestidae (Coleoptera) from Japan I. *Kontyû* **45**: 349-359.
Ohbayashi N. 1982: A new dermestid beetle from Japan (Coleoptera, Dermestidae). In: Satô M., Hori Y., Arita Y. & Okadome T. (eds.): *Special Issue to the Memory of Retirement of Emeritus Prof. M. Chûjô*. Nagoya, 185 pp.
Ohbayashi N. 1985: Three new species of the genus Anthrenus (Coleoptera, Dermestidae) from Japan. *Elytra* **13**: 59-66.
Ohbayashi K. & Satô M. 1963: Description of a new species of the genus Pristolycus Gorham from Shikoku, Japan (Coleoptera: Lycidae). *Transactions of the Shikoku Entomological Society* **7**: 127-128.
Ohbayashi N. & Satô M. 1968: Description of a new species of cantharid-beetle from Amami-Ôshima Island. *Entomological Review of Japan* **20**: 33-34.
Ôhira H. 1954: New or little-known Elateridae from Japan. I. (Coleoptera). *Mushi* (Fukouka) **27**: 41-47, pl. 4.
Ôhira H. 1956. [new taxa]. In: Baba K. & Ôhira H.: Elateridae from some islands (Sado, Tobi and Awa Island) in the Japan Sea off the coast of Niigata, Honshu, Japan. *Kontyû* **24**: 8-18.
Ôhira H. 1960: New or little-known Elateridae from Japan II. *Akitu* **9**: 23-24.
Ôhira H. 1962a: New or little-known Elateridae from Japan. III. (Coleoptera). *Kontyû* **30**: 198-201.
Ôhira H. 1962b: New or little-known Elateridae from Japan. IV (Coleoptera). *Kontyû* **30**: 262-264, 1 pl.
Ôhira H. 1963a: New or little-known Elateridae from Japan, VI (Coleoptera). *Transactions of the Shikoku Entomological Society* **8**: 15-18.
Ôhira H. 1963b: New or little-known Elateridae from Japan. V. (Coleoptera). *Kontyû* **31**: 176-179.
Ôhira H. 1963c: Notes on some elaterid-beetles from Japan (Coleoptera) (I). *Kontyû* **31**: 270-275.

Ôhira H. 1964a: New or little-known Elateridae from Japan. IX. (Coleoptera). *Bulletin of the Japan Entomological Academy* **1**: 13-14.
Ôhira H. 1964b: New or little-known Elateridae from Japan VIII (Coleoptera). *Akitu* **11**: 35-36.
Ôhira H. 1964c: New or little-known Elateridae from Japan. VII (Coleoptera). *Kontyû* **32**: 233-235.
Ôhira H. 1966a: Notes on some elaterid-beetles from the Ryukyu Islands, II. *Bulletin of the Japan Entomological Academy* **2**: 7-12.
Ôhira H. 1966b: New or little-known Elateridae from Japan. X (Coleoptera). *Transactions of the Shikoku Entomological Society* **9**: 41-43.
Ôhira H. 1966c: Notes on some Elateridae-beetles from Formosa I. *Kontyû* **34**: 215-222.
Ôhira H. 1966d: Notes on some Elateridae-beetles from Formosa II. *Kontyû* **34**: 266-274.
Ôhira H. 1966e: Notes on some Elateridae-beetles from Formosa III. *Kontyû* **34**: 331-338.
Ôhira H. 1967a: Notes on some Elateridae-beetles from Formosa IV. *Kontyû* **35**: 55-59.
Ôhira H. 1967b: The Elateridae of the Ryukyu Archipelago, I (Coleoptera). *Transactions of the Shikoku Entomological Society* **9**: 95-106.
Ôhira H. 1967c: The Elateridae of the Ryukyu Archipelago, II. (Coleoptera). *The Entomological Review of Japan* **19**: 41-48.
Ôhira H. 1967d: The Elateridae of the Ryukyu Archipelago, III. *Bulletin of the Japan Entomological Academy* **3**: 27-38.
Ôhira H. 1968a: The Elateridae of the Ryukyu Archipelago, IV (Coleoptera). *Bulletin of Aichi University of Education* **17**: 129-134.
Ôhira H. 1968b: Some Elateridae-beetles collected by the late Mr. K. Ohbayashi in Japan. *The Entomological Review of Japan* **20**: 73-77.
Ôhira H. 1968c: The Elateridae of the Ryukyu Archipelago, V (Coleoptera). *Kontyû* **36**: 134-143.
Ôhira H. 1968d: New or little-known Elateridae from Japan. XI. (Coleoptera). *Bulletin of the Japan Entomological Academy* **4**: 9-12.
Ôhira H. 1968e: Notes on Hypolithus (Hypnoidus) motschulskyi (Fleutiaux, 1902) (I) (Coleoptera: Elateridae). *The Entomological Review of Japan* **21**: 25-32.
Ôhira H. 1968f: Notes on some Elateridae-beetles from Formosa V. *Kontyû* **36**: 363-368.
Ôhira H. 1969a: The Elateridae of the Ryukyu Archipelago, VI (Coleoptera). *Bulletin of Aichi University of Education* **18**: 89-103.
Ôhira H. 1969b: Notes on some species of Elateridae of the Izu Islands (Coleoptera). *Transactions of the Shikoku Entomological Society* **10**: 41-48.
Ôhira H. 1969c: The Elateridae of the Ryukyu Archipelago, VII (Coleoptera). *Bulletin of the Japan Entomological Academy* **4**: 27-32.
Ôhira H. 1970a: New or little-known Elateridae from Japan, XII (Coleoptera). *Bulletin of the Japan Entomological Academy* **5**: 9-13.
Ôhira H. 1970b: The Elateridae of the Ryukyu Archipelago, VIII (Coleoptera). *Bulletin of Aichi University of Education* **19**: 103-111.
Ôhira H. 1970c: Notes on some elaterid-beetles from Japan (Coleoptera) (V). *The Entomological Review of Japan* **22**: 75-88, pls. 8-11.
Ôhira H. 1970d: Elaterid-beetles from Lan-hsu I. of Formosa in 1968. *Bulletin of the Japan Entomological Academy* **6**: 17-21.
Ôhira H. 1970e: A new elaterid-beetle from Bonins (Coleoptera). *Gensei* **21**: 9-12.
Ôhira H. 1970f: A list of the elaterid-beetles from South Asia Preserved in the Hungarian Natural History Museum (Coleoptera) Parts I.-V. *Annales Historico-Naturales Musei Nationalis Hungarici, Pars Zoologica* **62**: 207-243.
Ôhira H. 1971a: New or little-known Elateridae from Japan. XIII. (Coleoptera). *Kontyû* **39**: 39-40.
Ôhira H. 1971b: On some Elaterid-beetles from the Ryukyu Archipelago (Coleoptera). *Pacific Insects* **13**: 531-543.
Ôhira H. 1971c: A List of the elaterid-beetles from South Asia preserved in the Hungarian Natural History Museum, Part VI. (Coleoptera). *Annales Historico-Naturales Musei Nationalis Hungarici, Pars Zoologica* **63**: 205-216.
Ôhira H. 1972a: New or little-known Elateridae from Japan, XIV (Coleoptera). *Bulletin of the Japan Entomological Academy* **7**: 19-21.

Ôhira H. 1972b: Elaterid-beetles from Taiwan in Bishop Museum (Coleoptera). *Pacific Insects* **14**: 1-14.
Ôhira H. 1972c: Notes on some elaterid-beetles from Japan (Coleoptera) (XIII). *New Entomologist* **21**: 45-48.
Ôhira H. 1973a: New or little-known Elateridae from Japan, XV (Coleoptera). *Kontyû* **41**: 97-98.
Ôhira H. 1973b: New or little-known Elateridae from Japan, XVI (Coleoptera). *Transactions of the Shikoku Entomological Society* **11**: 117-120.
Ôhira H. 1973c: Elaterid-beetles from the Ryukyu Archipelago collected by Mr. Makihara in 1968-1970 (Coleoptera). *Bulletin of the Japan Entomological Academy* **7**: 27-33.
Ôhira H. 1973d: Notes on some elaterid-beetles from Japan (Coleoptera) (XVII). *New Entomologist* **22**: 21-23.
Ôhira H. 1973e: Notes on some elaterid-beetles from Japan (Coleoptera) (XX). *New Entomologist* **22**: 35-40.
Ôhira H. 1973f: A list of the elaterid-beetles from South Asia preserved in the Hungarian Natural History Museum (Coleoptera). Part VII. *Folia Entomologica Rovartani Közlemények* (S. N.) **26** Suppl.: 317-334.
Ôhira H. 1974a: New or little-known Elateridae from Japan, XIX (Coleoptera). *Transactions of the Shikoku Entomological Society* **12**: 56-59.
Ôhira H. 1974b: New or little-known Elateridae (Coleoptera) from Japan, XVIII. *Kontyû* **42**: 24-28.
Ôhira H. 1976: New or little-known Elateridae (Coleoptera) from Japan, XX. *Transactions of the Shikoku Entomological Society* **13**: 47-50.
Ôhira H. 1977a: New or little-known Elateridae (Coleoptera) from Japan, XXII. *The Entomological Review of Japan* **30**: 11-12.
Ôhira H. 1977b: Notes on some elaterid-beetles from Formosa, VI. *Elytra* **5**: 33-35.
Ôhira H. 1978a: New or little-known Elateridae (Coleoptera) from Japan, XXI. *Transactions of the Shikoku Entomological Society* **14**: 31-34.
Ôhira H. 1978b: New or little-known Elateridae (Coleoptera) from Japan, XXIII. *Kontyû* **46**: 552-556.
Ôhira H. 1982: New or little-known Elateridae (Coleoptera) from Japan, XXIII [sic!]. *Special Issue to the Memory of Retirement of Emeritus Professor Michio Chûjô* **1982**: 21-24.
Ôhira H. 1984: [new taxon]. In: Ôhira H. & Shimoyama K.: Some elaterid-beetles from Aomori Pref., Honshu, Japan. *Annual Report of the Aomori Prefectural Museum* **9**: 53-61.
Ôhira H. 1986a: [new taxon]. In: Ôhira H. & Kusui Y.: Notes on some elaterid-beetles from Teuri Is. and Yagishiri Is., Hokkaido, Japan. *Jezoensis* **13**: 87-90.
Ôhira H. 1986b: Two new species and a new subspecies of Elateridae from Japan (Coleoptera). Papers on Entomology presented to Prof. Takehiko Nakane in commemoration of his retirement. *Special Bulletin of the Japanese Society of Colepterology* **2**: 213-218.
Ôhira H. 1986c: Notes on the Colaulon-species from Japan (Coleoptera: Elateridae). *Transactions of the Essa Entomological Society* **63**: 3-13.
Ôhira H. 1987a: Notes on some rare Elateridae from Kyushu (24). *Kita-Kyūshū no Konchū* **34**: 91-92, pl. 7.
Ôhira H. 1987b: New or little-known Elateridae (Coleoptera) from Japan, XXV. *Transactions of the Shikoku Entomological Society* **18**: 285-288.
Ôhira H. 1988: Notes on some rare Elateridae from Kyushu (28). *Kita-Kyūshū no Konchū* **35**: 155-157.
Ôhira H. 1989a: Notes on Ampedus (Ampedus) optabilis (Lewis, 1894) and its allied species (Coleoptera: Elateridae). *Transactions of the Essa Entomological Society* **67**: 3-17.
Ôhira H. 1989b: A new genus of the subfamily Cardiophorinae (Coleoptera: Elateridae) from Japan. *The Entomological Review of Japan* **44**: 79-80.
Ôhira H. 1989c: Notes on Dalopius-species from Japan (Coleoptera: Elateridae). *Transactions of the Essa Entomological Society* **68**: 59-79.
Ôhira H. 1990: Notes on the genus Silesis (Elateridae) from Japan. *Entomological Revue of Japan* **45**: 73-75.
Ôhira H. 1991: Notes on Oedostethus telluris and its allied species (Coleoptera: Elateridae). *Transactions of the Essa Entomological Society* **72**: 2-16 (Japanese with English summary).
Ôhira H. 1992a: New or little-known Elateridae (Coleoptera) from Japan, XXVI. *Elytra* **20**: 223-226.
Ôhira H. 1992b: New or little-known Elateridae (Coleoptera) from Japan, XXVII. *The Entomological Revue of Japan* **47**: 123-124.
Ôhira H. 1992c: Notes on Melanotus fortnumi and its allied species (Coleoptera, Elateridae). *Transactions of the Essa Entomological Society* **74**: 23-35.
Ôhira H. 1992d: Notes on some elaterid-beetles from Hokkaido and its adjacent regions (VII). *Jezoensis* **19**: 79-89.

Ôhira H. 1992e: Notes on some elaterid-beetles from Hokkaido and its adjacent regions (VIII). *Jezoensis* **19**: 91-94.

Ôhira H. 1993: New or little-known Elateridae (Coleoptera) from Japan, XXVIII. *Elytra* **21**: 249-254.

Ôhira H. 1994a: New or little-known Elateridae (Coleoptera) from Japan, XXIX. *Transactions of the Shikoku Entomological Society* **20**: 285-287.

Ôhira H. 1994b: Notes on the genera and species of Conoderinae (Coleoptera: Elateridae) from Japan. *Special Bulletin of Essa Entomological Society* **2**: 217-234 (in Japanese with English summary).

Ôhira H. 1994c: Notes on the morphological structure of Fleutiauxellus yezoensis (Coleoptera, Elateridae) from Japan. *Elytra* **22**: 153-157.

Ôhira H. 1994d: New or little-known Elateridae (Coleoptera) from Japan, XXX. *Elytra* **22**: 315-320.

Ôhira H. 1995a: New or little known Elateridae (Coleoptera) from Japan, XXXI. *Japanese Journal of Systematic Entomology* **1**: 95-97.

Ôhira, 1995b: [new taxa]. In: Ôhira H. & Yasuda N.: Notes on Oedostethus-species from Hokkaido (Coleoptera, Elateridae). *Bulletin of the Sounkyo Museum of Natural History* **15**: 1-12.

Ôhira H. 1995c: New or little-known Elateridae (Coleoptera) from Japan, XXXII. *Elytra* **23**: 25-30.

Ôhira H. 1995d: New or little-known Elateridae (Coleoptera) from Japan, XXXIII. *Elytra* **23**: 187-193.

Ôhira H. 1996a: New or little-known Elateridae (Coleoptera) from Japan, XXXIV. *Japanese Journal of Systematic Entomology* **2**: 25-27.

Ôhira H. 1996b: Notes on the Anchastus-species from Japan (Elateridae, Elaterinae, Physorhinini). *Transactions of the Essa Entomological Society* **76**: 1-14.

Ôhira H. 1996c: New or little-known Elateridae (Coleoptera) from Japan, XXXV. *Elytra* **24**: 311-316.

Ôhira H. 1997a: Notes on Paracardiophorus pullatus and its allied species from Japan (Coleoptera: Elateridae). *Miscellaneous Reports of the Hiwa Museum for Natural History* **35**: 1-16, pls. 1-14 (in Japanese with English summary).

Ôhira H. 1997b: New or little-known Elateridae (Coleoptera) from Japan, XXXVI. *Japanese Journal of Systematic Entomology* **3**: 25-27.

Ôhira H. 1997c: Notes on Orthostethus sieboldi and its allied species from Japan (Coleoptera, Elateridae). *Hibakagaku - Journal of the Hiba Society of Natural History* **182**: 37-44 (in Japanese with English summary).

Ôhira H. 1997d: New or little-known Elateridae (Coleoptera) from Japan, XXXVII. *Elytra* **25**: 343-348.

Ôhira H. 1998a: New or little-known Elateridae (Coleoptera) from Japan, XXXVIII. *Elytra* **26**: 161-166.

Ôhira H. 1998b: New or little-known Elateridae (Coleoptera) from Japan, XXXIX. *Elytra* **26**: 373-377.

Ôhira H. 1999: New or little-known Elateridae (Coleoptera) from Japan, XL. *Elytra* **27**: 409-416.

Ôhira H. 2000: New or little-known Elateridae (Coleoptera) from Japan, XLI. *Elytra* **28**: 365-368.

Ôhira H. 2001: Notes on the morphological structure of the female internal reproductive organs of the Melanotus species in Japan (Coleoptera: Elateridae). *Miscellaneous Reports of the Hiwa Museum for Natural History* **40**: 17-27. pls. I-VIII.

Ôhira H. 2002a: New or little-known Elateridae (Coleoptera) from Japan, XLII. *Special Bulletin of the Japanese Society of Coleopterology* **5**: 337-339.

Ôhira H. 2002b: New or little-known Elateridae (Coleoptera) from Japan, XLIII. *Elytra* **30**: 483-485.

Ôhira H. 2003a: New or little-known Elateridae (Coleoptera) from Japan, XLIV. *Special Bulletin of the Japanese Society of Coleopterology* **6**: 243-245.

Ôhira H. 2003b: New or little-known Elateridae (Coleoptera) from Japan, XLV. *Elytra* **31**: 361-365.

Ôhira H. 2004: Notes on the Morphological Structure of Agrypnus species from Japan (III) (Coleoptera: Elateridae, Agrypninae, Agrypnus, Colaulon group). *Miscellaneous Reports of the Hiwa Museum for Natural History* **43**: 67-89, pls. I-XVII.

Ôhira H. & Becker E. C. 1971: Elateridae (Coleoptera) from the Canadian Nepal Expedition (1967). 1. Descriptions of three new species of Silesis. *Oriental Insects* **5**: 577-581.

Ôhira H. & Becker E. C. 1972a: Elateridae (Coleoptera) from the Canadian Nepal Expedition (1967). 2. Descriptions of four new species of Glyphonyx and Parhemiops with a new record for Hemiops flava (Lap. de Cast.). *Oriental Insects* **6**: 135-140.

Ôhira H. & Becker E. C. 1972b: Elateridae (Coleoptera) from the Canadian Nepal Expedition (1967). 3. Descriptions of new species and records of Dima, Penia, and Neocsikia new genus. *Oriental Insects* **6**: 531-537.

Ôhira H. & Becker E. C. 1973a: Elateridae (Coleoptera) from the Canadian Nepal Expedition (1967). 4. Descriptions of new species and records of the subfamilies Negastriinae and Hypnoidinae. *Oriental Insects* **7**: 69-77.

Ôhira H. & Becker E. C. 1973b: Elateridae (Coleoptera) from the Canadian Nepal Expedition (1967). 5. Records and descriptions of Paracalais, Compsolacon, Meristhus, Pectocera and Octocryptus species. *Oriental Insects* **7**: 461-469.

Ôhira H. & Becker E. C. 1973c: Elateridae (Coleoptera) from the Canadian Nepal Expedition (1967). 6. Description of two new species of Acteniceromorphus of the subfamily Ctenicerinae. *Oriental Insects* **7**: 563-566.

Ôhira H. & Becker E. C. 1974a: Elateridae (Coleoptera) from the Canadian Nepal Expedition (1967). 7. Descriptions of three new species of Anchastus, Procraerus, and Megapenthes. *Oriental Insects* **8**: 385-390.

Ôhira H. & Becker E. C. 1974b: Elateridae (Coleoptera) from the Canadian Nepal Expedition (1967). Descriptions of new species and records of Zorochrus, Melanotus and Silesis. *Oriental Insects* **8**: 557-562.

Ôhira H. & Satô M. 1964: Notes on some elaterid-beetles from the Ryukyu Islands, I (Coleoptera). *Aomori Kenritsu Kyodokan Chosa Kenkyu Nenpo* **9**: 53-61.

Ôhira H. & Yasuda N. 1994: Notes on Hypnoidus rivalis (Lewis, 1894) (Col. Elateridae) from Hokkaido. *Bulletin of the Sounkyo Museum of Natural History* **14**: 1-12.

Ohta Y. 1928: Ueber einen neuen Coccinelliden und zwei Abberationen aus Sakhalin. *Insect World* (Gifu) **32**: 112-114 (in Japanese).

Ohta Y. 1929a: Scymninen Japans. *Insecta Matsumurana* **4**: 1-16.

Ohta Y. 1929b: Beitrag zur Coccinelliden-Fauna aus Sachalin und Kurilen. *Insecta Matsumurana* **3**: 57-62.

Ohta Y. 1929c: Einige neue Helotiden- und Coccinelliden-Arten aus Formosa. *Insecta Matsumurana* **4**: 66-70.

Ohta Y. 1929d: Beitrag zur Kenntnis der Helotiden-Fauna von Japan, Korea und Formosa. *Insecta Matsumurana* **3**: 108-110.

Ohta Y. 1930a: Byturiden Japans. *Insecta Matsumurana* **4**: 100-103.

Ohta Y. 1930b: Beitrag zur Kenntnis der Ptiniden Japans. *Insecta Matsumurana* **4**: 105-109.

Ohta Y. 1931a: Ueber einige Coelophoren- und Helotiden-Arten aus Formosa, mit Beschreibung von 2 neuen Arten. *Insecta Matsumurana* **5**: 134-136.

Ohta Y. 1931b: On a new species and new recorded species of the tribe Scymnini from Japan. *Insect World* (Gifu) **35**: 182-184 (in Japanese).

Ohta Y. 1931c: Beitrag zur Kenntnis der Endomychiden Japans. *Journal of the Faculty of Agriculture, Hokkaido Imperial University* **30**: 205-242, pl. 3.

Okushima Y. 1991a: A new species of the cantharid genus Athemus (Coleoptera, Cantharidae) from Okinawa Island, Southwest Japan. *Elytra* **19**: 35-38.

Okushima Y. 1991b: A new cantharid beetles of the genus Themus (Coleoptera, Cantharidae) from Ishigaki Island, Southwest Japan. *Elytra* **19**: 195-199.

Okushima Y. 1996: The genus Athemellus (Coleoptera, Cantharidae) of the Ryukyu Islands, Southwest Japan. *Elytra* **24**: 113-124.

Okushima Y. & Imasaka S. 2000: Three new species of the genus Athemus (Coleoptera, Cantharidae) from Japan. *Elytra* **28**: 373-385.

Okushima Y. & Satô M. 1997a: Two new species of the genus Stenothemus (Coleoptera, Cantharidae) from Taiwan. *Elytra* **25**: 85-91.

Okushima Y. & Satô M. 1997b: Cantharid beetles of the genus Prothemus (Coleoptera, Cantharidae) of the Ryukyu Islands, Southwest Japan. *Elytra* **25**: 333-342.

Okushima Y. & Satô M. 1999a: Cantharid beetles of the genus Stenothemus (Coleoptera, Cantharidae) from Taiwan. *Elytra* **27**: 131-140.

Okushima Y. & Satô M. 1999b: Cantharid beetles of the genus Habronychus (Coleoptera, Cantharidae) from Taiwan, with description of a new subgenus. *Elytra* **27**: 387-403.

Okushima Y. & Satô M. 2000a: A new species of the genus *Themus* (Coleoptera, Cantharidae) from Taiwan. *Japanese Journal of Systematic Entomology* **6**: 147-149.

Okushima Y. & Satô M. 2000: Note on a new subspecies of Athemus suturellus (Coleoptera, Cantharidae) from Central Honshu. *Japanese Journal of Systematic Entomology* **6**: 321-324.

Okushima Y. & Satô M. 2002: A new small species of the genus Prothemus (Coleoptera, Cantharidae) from Taiwan. *Special Bulletin of the Japanese Society of Coleopterology* (Tokyo) **5**: 347-350.

Olexa A. 1954a: Pleganophorus bispinosus Hampe var. slovenicus var. nov. (Col. Endomychidae). *Bulletin de la Société Entomologique de Mulhouse* **1954**: 48.

Olexa A. 1954b: Hypocoelus fleischeri sp. nov. (Col. Eucnemidae). *Bulletin de la Société Entomologique de Mulhouse* **1954**: 66-67.

Olexa A. 1960: Xylophilus humeralis Duf. et X. leseigneuri n. sp. (Col. Eucnemidae). *Bulletin de la Société Entomologique de Mulhouse* **1960**: 55-56.

Olexa A. 1962: Dirrhagus ferrugineus Reitt. et D. bodemeyeri n. sp. (Col. Eucnemidae). *Bulletin de la Société Entomologique de Mulhouse* **1962**: 43-44.

Olexa A. 1963: Dirhagus palmi n.sp., eine neue Art aus dem Gebiete der Ost-Karpathen. *Entomologische Blätter* **59**: 88-91.

Olexa A. 1969: Xylophilus corticalis (Payk. 1800) et X. talyshensis n. sp. Une nouvelle espèce des Monts Talysh (Coleoptera Eucnemidae). *Bulletin de la Société Entomologique de Mulhouse* **1969**: 61-63.

Olexa A. 1970: Hypocoelus simonae n. sp. Une nouvelle espèce de Bulgarie (Coleoptera Eucnemidae). *Bulletin de la Société Entomologique de Mulhouse* **1969**: 91-94.

Olexa A. 1975: Reklassifikation der Gattung Dirhagus und verwandter Gattungen (Coleoptera - Eucnemidae). *Entomologische Blätter* **71**: 155-164.

Oliveira M. Paulino de. 1893: *Catalogue des insectes du Portugal. Coléoptères.* Coimbra: Universidade, [2] + 393 pp.

Olivier A. G. 1790a: *Entomologie, ou histoire naturelle des insectes, aves leurs caractères génériques et spécifiques, leur description, leur synonymie, et leur figure enluminée. Coléoptères. Tome second.* Paris: de Baudouin, 485 pp., 63 pls. [Note: each genus with separat pagination].

Olivier A. G. 1790b: *Encyclopédie méthodique ou par ordre de matières; par une société de gens de lettres, de savans et d'artistes; précédée d'un vocabulaire universel, servant de table pour tout l'ouvrage, ornée des portraits de Mm. Diderot et d'Alembert, premiers éditeurs de l'Encyclopédie. Histoire Naturelle. Insectes. Tome quatrième.* Paris: C.-J. Panckoucke et Liége: Plomteux pp. 45-331. [pp. i-ccclxxiii issued in 1792, pp. 1-44 in 1789, following pp. in 1790].

Olivier A. G. 1790c: *Encyclopédie méthodique ou par ordre de matières; par une société de gens de lettres, de savans et d'artistes; précédée d'un vocabulaire universel, servant de table pour tout l'ouvrage, ornée des portraits de Mm. Diderot et d'Alembert, premiers éditeurs de l'Encyclopédie. Histoire Naturelle. Insectes. Tome cinquième.* Paris: C.-J. Panckoucke, 793 pp.

Olivier A. G. 1791: *Encyclopédie métodique, ou par ordre de matières; par une société de gens de lettres, de savans et d'artistes; précédée d'un vocabulaire universel, servant de table pour tout l'ouvrage, ornée des portraits de Mm. Diderot et d'Alembert, premiers éditeurs de l'Encyclopédie. Histoire Naturelle. Insectes. Tome sixième. Pars I.* Paris: Panckoucke 704 pp.

Olivier A. G. 1792: Sur quelques nouvelles espèces de coléoptères. *Journal d'Histoire Naturelle* (Paris) **1**: 262-268, pl. 14.

Olivier A. G. 1795a: *Entomologie, ou histoire naturelle des Insectes, avec leurs caractères génériques et spécifiques, leur description, leur synonymic et leur figure enluminée. Coléoptères. Tome troisième.* Paris: de Lanneau, 557 pp., 65 pls. [Note: each genus separatly pagination].

Olivier A. G. 1795b: *Entomologie, ou histoire naturelle des Insectes, aves leurs caractères génériques et spécifiques, leur description, leur synonymie, et leur figure enluminée. Coléoptères. Tome quatrième.* Paris: de Lanneau, 519 pp., 72 pls. [Note: each genus with separat pagination].

Olivier A. G. 1807: *Entomologie, ou histoire naturelle des insectes, avec leurs caractères génériques, leur description, leur synonymie, et leur figure enluminée. Coléoptères. Tome cinquième.* Paris: Desray, 2 + 612 pp. [Note: each genus with separat pagination].

Olivier A. G. 1808: *Entomologie, ou histoire naturelle des insectes, avec leur caractères génériques et spécifiques, leur description, leur synonymie, et leur figure eluminée. Coléoptères. Tome sixième.* Paris: Desray, 613-1104 pp., 41 pls

Olivier A. G. 1811: [new taxa]. In: Manuel B. E., Olivier A. G. & Latreille P. A.: *Encyclopédie méthodique ou par ordre de matières; par une société de gens de lettres, de savans et d'artistes; précédée d'un vocabulaire universel, servant de table pour tout l'ouvrage, ornée des portraits de Mm. Diderot et d'Alembert, premiers éditeurs de l'Encyclopédie. Histoire Naturelle. Insectes. Tome huitième.* Paris: H. Agasse, 722 pp.

Olivier E. 1883a: Lampyrides nouveaux ou peu connus II. *Revue d'Entomologie* **2**: 326-333.

Olivier E. 1883b: Deux nouvelles espèces de Lampyridae: Pelania et Lampyris. *Bulletin de la Société*

Entomologique de France **1883**: lxix-lxx.

Olivier E. 1884: Essai d'une révision des espèces européennes & circaméditerranéennes de la famille des lampyrides. *Abeille, Journal d'Entomologie* **22**: 1-54 + [2], 2 pls + Notes complémentaire, 1-4.

Olivier E. 1885: Catalogue des lampyrides faisant partie des collections du Musée civiques de Gênes. *Annali del Museo Civico di Storia Naturale di Genova* **22**: 333-374.

Olivier E. 1886: Révision du genre Pyrocoelia, Gorh. (Ordre des coléoptères famille des lampyrides). *Notes from the Leyden Museum* **8**: 195-208.

Olivier E. 1891: Lampyrides rapportés de Birmanie par M. L. Fea, avec descriptions des espèces nouvelles. *Annali del Museo Civico di Storia Naturale di Genova* **30**: 595-604.

Olivier E. 1894: Description d'un Lampyris nouveau d'Algérie. *L'Échange, Revue Linnéenne* **10**: 135-136.

Olivier E. 1895a: Une variété nouvelle de Lampyris. *Revue Scientifique du Bourbonnais et du Centre de la France* **8**: 212.

Olivier E. 1895b: Deux espèces nouvelles de lucioles. *Bulletin de la Société Entomologique de France* **1895**: cxlviii-cxlix.

Olivier E. 1899: Contribution à l'étude des lampyrides (Col.) descriptions et observations. *Bulletin de la Société Entomologique de France* **1899**: 86-93.

Olivier E. 1900: Contribution à l'étude de la faune entomologique de Sumatra /Côte Ouest: Indrapoera-Vice-Résidence Painan (Voyage de M. J.-L. Weyers). Lampyrides déterminés et décrits. *Annales de la Société Entomologique de Belgique* **44**: 234-238.

Olivier E. 1901: Variétés nouvelles de Lampyris noctiluca. *Miscellanea Entomologica* **9**: 129.

Olivier E. 1902a: Catalogue des espèces de «Luciola» et genre voisins décrits jusqu'à ce jour. *Revue Scientifique du Bourbonnais et du Centre de la France* **15**: 69-88.

Olivier E. 1902b: Coléoptères lampyrides recueillis aux environs de Tokio (Japon). *Bulletin du Muséum d'Histoire Naturelle* (Paris) **8**: 188-190.

Olivier E. 1903: Coléoptères lampyrides capturés à Dardjilling par M. le Dr Harmand. *Bulletin du Muséum d'Histoire Naturelle* (Paris) **9**: 19-20.

Olivier E. 1905: Description de lampyrides nouveaux. *Annales de la Société Entomologique de Belgique* **49**: 206-209.

Olivier E. 1907a: Collections recueillies en Perse par M. de Morgan. Coléoptères: lampyrides. *Bulletin du Muséum d'Histoire Naturelle* (Paris) **13**: 28-29.

Olivier E. 1907b: Coleoptera Fam. Lampyridae. In: Wytsman P. (ed.): *Genera Insectorum*. Fasc. 53, Brüssel, 74 pp., 3 pls.

Olivier E. 1907c: Descriptions de lampyrides nouveaux. *Revue Scientifique du Bourbonnais et du Centre de la France* **20**: 175-181.

Olivier E. 1908a: Le genre Ototreta E. Oliv. et descriptions d'espèces nouvelles (Col.). *Bulletin de la Société Entomologique de France* **1908**: 113-116.

Olivier E. 1908b: Lampyrides nouveaux du Musée de Genève. *Annales de la Société Entomologique de Belgique* **52**: 262-263.

Olivier E. 1909a: Coléoptères lampyrides rapportés de Kouy-Tcheou, Région de Pin Fa (Chine), par les P. Cavalerie et Fortunat. *Bulletin du Muséum d'Histoire Naturelle* (Paris) **15**: 247-250.

Olivier E. 1909b: Descriptions de lampyrides nouveaux. *Annales de la Société Entomologique de Belgique* **53**: 394-398.

Olivier E. 1910a: Lampyridae. Pars 9. In: Schenkling S. (ed.): *Coleopterorum Catalogus. Vol. 9*. Berlin: W. Junk, 68 pp.

Olivier E. 1910b: Descriptions de lampyrides nouveaux (Col.). *Bulletin de la Société Entomologique de France* **1910**: 282-287.

Olivier E. 1911a: Révison des lampyrides. *Revue Scientifique du Bourbonnais et du Centre de la France* **24**: 24-27, 39-58, 63-85.

Olivier E. 1911b: Lampyrides rapprotés de Formosa par M. Hans Sauter determinés et décrits par Ernest Olivier. *Annali del Museo Civico di Storia Naturale di Genova* **45**: 145-148.

Olivier E. 1911c: Révision du genre Rhagophthalmus et description d'espèces nouvelles. *Annales de la Société Entomologique de France* **82**: 467-472.

Olivier E. 1912: Descriptions de quelques lampyrides (Col.) nouveaux de ma collection. *Bulletin de la Société Entomologique de France* **1912**: 170-172.

Olivier E. 1913a: Description d'une nouvelle espèce de Drilidae. *Bulletin de la Société Entomologique de France* **1913**: 92-93.

Olivier E. 1913b: H. Sauter's Formosa-Ausbeute: Lampyridae. *Entomologische Mitteilungen* **2**: 269-272.

Olliff A. S. 1883a: Descriptions of two larvae and new genera and species of clavicorn Coleoptera, and a synopsis of the genus Helota, MacLeay. *Cistula Entomologica* **3**: 49-61.

Olliff A. S. 1883b: Remarks on a small collection of clavicorn Coleoptera from Borneo, with descriptions of new species. *Transactions of the Entomological Society of London* **1883**: 173-186.

Olliff A. S. 1884: Descriptions of two new species of Nitidulidae from Sumatra. *Notes from the Leyden Museum* **6**: 245-247.

Olliff A. S. 1885: Notes on certain Ceylonese Coleoptera (Clavicornia) described by the late Mr. Francis Walker. *Proceedings of the Linnean Society of New South Wales* **10**: 69-72.

Olliff A. S. 1890: New species of Lampyridae, including a notice of the Mt. Wilson fire-fly. *Proceedings of the Linnaean Society of New South Wales* (2) **4** [1889]: 643-653.

Olliff A. S. 1895: Entomological notes - A new friendly ladybird. *Agricultural Gazette of New South Wales* **6**: 30-31.

O'Mahony E. 1927: A new variety of Rhizobius litura F. from Ireland. *The Entomologist's Monthly Magazine* **63**: 208-209.

Orlov V. N. 1993: Novye vidy shchelkunov roda Cardiophorus Esch. (Coleoptera, Elateridae) s Kavkaza [New species of elaterid beetles of the genus Cardiophorus Esch. (Coleoptera, Elateridae) from the Caucasus]. *Entomologicheskoe Obozrenie* **72**: 96-99, 254.

Orlov V. N. 1994: Novye vidy zhukov-shchelkunov roda Athous Esch. (Coleoptera, Elateridae) c Kavkaza [New species of elaterid beetles of the genus Athous Esch. (Coleoptera, Elateridae) from the Caucasus]. *Entomologicheskoe Obozrenie* **73**: 682-689.

Orlov V. N. 1995: Novyy vid shchelkunov roda Odontoderus Schw. (Coleoptera, Elateridae) s Kavkaza [A new species of elaterids of the genus Odontoderus Schw. (Coleoptera, Elateridae) from the Caucasus]. *Entomologicheskoe Obozrenie* **74**: 619-621.

Orlov V. N. 1997a: Novye vidy zhukov-shchelkunov roda Cardiophorus Esch. (Coleoptera, Elateridae) s Kavkaza [New species of the elaterid genus Cardiophorus Esch. (Coleoptera, Elateridae) from the Caucasus]. *Entomologicheskoe Obozrenie* **76**: 172-175.

Orlov V. N. 1997b: A new species of Agriotes from the Caucasus (Coleoptera, Elateridae). *Zoosystematica Rossica* **5**: 275-276.

Ormay S. 1888: Coleoptera nova e Transsilvania. *Wiener Entomologische Zeitung* **7**: 165-168.

Oromí P. & García R. 1987: Sobre una nueva de Lycoperdina de las Islas Canarias (Coleoptera, Endomychidae). *Vieraea* **17**: 87-91.

Osawa S. & Chûjô Mt. 1990: A new species of Micrencaustes (Mimencaustes) from Taiwan, Republic of China (Coleoptera, Erotylidae). *Esakia* **29**: 69-71.

Osculati G. 1844: Coleopterorum species novae. In: *Note d'un viaggio nella Persia e nelle Indie orientali negli anni 1841-1842*. Monza, 73 pp., 1 pl.

Otero J. C. 1989: Cryptophagus oromii n. sp., de las Islas Canarias (Coleoptera, Cryptophagidae). *Nouvelle Revue d'Entomologie* (N. S.) **6**: 205-207.

Otero J. C. 1997: Three new species and distributional records of Micrambe C.G. Thomson, 1863 and Cryptophagus Herbst, 1792 (Coleoptera: Cryptophagidae) from Israel and Turkey. *Revue Suisse de Zoologie* **104**: 207-216.

Otero J. C. 2001: Records of Cryptophagus Herbst, 1792 from the Himalaya, with a new species from Thailand (Coleoptera: Cryptophagidae). *Revue Suisse de Zoologie* **108**: 987-992.

Otero J. C. 2002: A new species of Cryptophagus Herbst (Coleoptera: Cryptophagidae) from India and Nepal. *Oriental Insects* **36**: 87-91.

Otero J. C. 2004: Two new species of Micrambe Thomson (Coleoptera: Cryptophagidae) from India and Nepal. *Oriental Insects* **38**: 245-249.

Otero J. C. & Angelini F. 1981b: Su alcuni Cryptophagus italiani nuovi poco conosciuti (Coleoptera Cryptophagidae). *Bollettino della Società Entomologica Italiana* **113**: 96-100.

Otero J. C. & Angelini F. 1984: Contributo alla conoscenza dei Cryptophagidae italiana (Coleoptera). *Entomologica* **18**: 81-96.

Otero J. C. & Díaz Pazos J. 1986: Lathridiidae de la Península Ibérica. IV: Descripción de Corticaria cotovillae n.sp. (Coleoptera). *Nouvelle Revue d'Entomologie* (N. S.) **3**: 167-169.

Otero J. C. & González M. A. 1980: Cryptophagus ariadne n. sp. (Col. Cryptophagidae) de la Península Ibérica. *Nouvelle Revue d'Entomologie* **10**: 177-179.

Otero J. C. & González M. A. 1982: Sobre la presencia de una nueva especie del gen. Cryptophagus (Col. Cryptophagidae) de la Península Ibérica. *Nouvelle Revue d'Entomologie* **12**: 161-164.

Otero J. C. & González M. A. 1983a: Nueva especie de Cryptophagus (Col. Cryptophagidae) de la Región Catalana. *Speleon* **26-27**: 79-81.

Otero J. C. & González M. A. 1983b: Una nueva especie de Cryptophagus Herbst (Col. Cryptophagidae) de la Península Ibérica. *Nouvelle Revue d'Entomologie* **13**: 209-211.

Otero J. C. & González M. A. 1985: Criptofagidos de Leon (España). Descripción de Cryptophagus lemonchei n. sp. (Col. Cryptophagidae). *Boletim da Sociedade Portuguesa de Entomologia* (*Actas do II Congresso Ibérico de Entomologia*) Suplemento n.o **1**: 149-154.

Otto J. 1978: Neue Lathridiidae aus dem Naturhistorischen Museum Basel (Col.). *Entomologica Basiliensia* **3**: 287-290.

Otto J. 1979: Insects of Saudi Arabia. Coleoptera: Fam. Lathridiidae: Subfam. Corticariinae. *Fauna of Saudi Arabia* **1**: 232-234.

Pace R. 1974: Tre nuove specie di coleotteri della fauna Appeninica. *Bollettino del Museo Civico di Storia Naturale di Verona* **1**: 105-119.

Pace R. 1975: Due nuove species e una nuova sottospecie di Anommatus Wesmael del Veneto (Coleoptera, Colydiidae) (XI Contributo alla conoscenza della dauna endogea). *Atti e Memorie dell'Accademia di Agricoltura Scienze e Lettere di Verona* **26** [1974-1975]: 13-22.

Pace R. 1979: Gli Anommatus del Veneto e delle regioni vicine (Coleoptera Anommatidae). *Bollettino del Museo Civico di Storia Naturale di Verona* **5**: 439-463.

Pacher D. D. 1853: Über die Käfer in der Umgebung von Sagritz und Heiligenblut. *Jahrbuch des Naturhistorischen Landesmuseums von Kärnthen* **2**: 30-52.

Pakaluk J., Ślipiński S. A. & Lawrence J. F. 1994: Current classification and family-group names in Cucujoidea. *Genus* **5** (4): 223-268.

Pal T. K. 1982: On Monanus Sharp (Coleoptera: Silvanidae) from India. *Oriental Insects* **15** [1981]: 241-255.

Pal T. K. 1985a: A revision of Indian Psammoecus Latreille (Coleoptera: Silvanidae). *Records of the Zoological Survey of India, Miscellaneous Publications, Occasional Papers* **71**: 1-54.

Pal T. K. 1985b: Fauna of Namdapha: Arunachal Pradesh. Insecta: Coleoptera: Silvanidae. *Records of the Zoological Survey of India* [**1982**]: 195-200.

Pal T. K. 1992: Discolomidae, a beetle family new from India. *Hexapoda* (*Insecta Indica*) **4** (1): 45-56.

Pal T. K. 1993: On a collection of Clavicorn beetles (families Erotylidae, Endomychidae and Languriidae) from Arunachal Pradesh, India. *Records of the Zoological Survey of India* **90** [1992]: 203-210.

Pal T. K. 1996a: Indoleptipsius ushae, a new genus and species of Rhizophagidae (Coleoptera: Cucujoidea) from Arunachal Pradesh, India. *Hexapoda* (*Insecta Indica*) **8** (1): 1-7.

Pal T. K. 1996b: Three new species of Discolomidae (Coleoptera Clavicornia) from Sikkim, India. *Hexapoda* (*Insecta Indica*) **8** (1): 10-17.

Pal T. K. 1998: A new species of Sarothrias Grouvelle from Northeast India (Coleoptera; Jacobsoniidae). *Doriana* **7** (307): 1-7.

Pal T. K. 2003: Insecta: Coleoptera: Endomychidae. Pp. 107-115. In: *State Fauna Series* 9, *Fauna of Sikkim* (*Part 3*). Kolkata: Zoological Survey of India, 411 pp.

Pal T. K. & Lawrence J. F. 1986: A new genus and subfamily of mycophagous Bothrideridae (Coleoptera: Cucujoidea) from the Indo-Australian Region, with notes on related families. *Journal of the Australian Entomological Society* **25**: 185-210.

Pal T. K. & Sengupta T. 1977: A revision of Silvanus (Coleoptera: Silvanidae) from India. *Oriental Insects* **11**: 269-284.

Pal T. K. & Sengupta T. 1979: Ergebnisse der Bhutan-Expedition 1972 des Naturhistorischen Museums in Basel. Coleoptera: Fam. Silvanidae. *Entomologia Basiliensia* **4**: 69-82.

Pal T. K. & Sengupta T. 1985: On Protosilvanus (Coleoptera: Silvanidae) from India. *Oriental Insects* **18** [1984]: 235-249.

Pal T. K. & Sengupta T. 2003: On a collection of Sphindidae (Coleoptera: Clavicornia) from Western Himalaya, India. *Entomon* **28**: 97-104.

Pal T. K., Sengupta T. & Crowson R. A. 1985: Revision of Uleiota (Coleoptera: Silvanidae) from India and Sri Lanka and its systematic position. *Oriental Insects* **18** [1984]: 213-233.

Palisot de Beauvois A. M. F. J. 1801: Mémoire sur un nouveau genre d'insecte trouvé en Afrique. *La Décade Philosophique, Littéraire et Politique* **36**: 513-518, 1 pl.

Palisot de Beauvois A. M. F. J. 1805: Livraison 1, pp. i-xvi + 1-24. In: *Insectes recueillis en Afrique et en Amérique, dans les Royaumes d'Oware et de Bénin, à Saint-Dominque et dans les États Unis pendant les années 1786-1797.* Paris: Levrault, Schoedl et Cie, xvi + 276 pp., 90 pls. [entire work published 1805-1821].

Pallas P. S. 1772: *Spicilegia zoologica qiubus novae imprimis et obscurae animalium species iconibus, decriptionibus atque commentariis illustrantur. Fasciculus Nonus.* Berolini: G. A. Lange, 86 pp.

Pallas P. S. 1773: *Reise durch verschiedene Provinzen des russischen Reichs. Zweiter Theil. Zweytes Buch vom Jahr 1771.* St.-Peterburg: Kayserliche Akademie der Wissenschaften, pp. 371-744.

Palm T. 1940: Systematische Bemerkungen zu den schwedischen Anthrenus-Arten (Col. Dermestidae). *Entomologisk Tidschrift* **61**: 1-3.

Palm T. 1944a: De svenska arterna av Scymnus frontalis F.-gruppen (Col. Coccinellidae). *Entomologisk Tidskrift* **64**: 1-8.

Palm T. 1944b: For Sverige nya Coleoptera VIII. *Entomologisk Tidskrift* **65**: 166-172.

Palm T. 1947a: Systematiska studier över svenska Elater-arter (Col. Elateridae). *Entomologisk Tidskrift* **68**: 155-170.

Palm T. 1947b: Våra Phalacrus-Arter (Col. Phalacridae). *Entomologisk Tidskrift* **68**: 179-183.

Palm T. 1949a: Eine neue Corticaria aus Lappland (Col. Lathridiidae). *Entomologisk Tidskrift* **70**: 109-110.

Palm T. 1949b: Atomaria (s. str.) Sjöbergi n. sp. (Col. Cryptophagidae). *Entomologisk Tidskrift* **70**: 241-242.

Palm T. 1955: Weitere Beiträge zur Kenntnis der schwedischen Hypocoelus-Arten (Col. Eucnemidae). *Opuscula Entomologica* **20**: 211-217.

Palm T. 1956: Enicmus lundbladi n.sp. und verwandte Arten (Col., Lathridiidae). *Entomologisk Tidskrift* **77**: 25-28.

Palm T. 1958: Dirrhagus (Microrrhagus) Lindbergi n. sp., eine neue Käferart aus Finnland (Eucnemidae). *Entomologisk Tidskrift* **79**: 34-37.

Palm T. 1959: Bidrag till kännedomen om svenska skalbaggars biologi och systematik. 24-27. *Entomologisk Tidskrift* **79** [1958]: 104-114.

Palm T. 1972: Eine neue Art der Enicmus minutus-Gruppe (Col. Lathridiidae). *Entomologisk Tidskrift* **93**: 39-41.

Palm T. 1974: Zur Kenntnis der Käferfauna der Kanarischen Inseln. 1-3. *Entomologisk Tidskrift* **95**: 31-45.

Palm T. 1975a: Zur Kenntnis de Käferfauna der Kanarischen Inseln. 4-7. *Entomologisk Tidskrift* **96**: 37-50.

Palm T. 1975b: Zur Kenntnis der Käferfauna der Kanarischen Inseln. 8. Revision der Gattung Malthinus Latr. (Coleoptera: Cantharidae). *Entomologica Scandinavica* **6**: 119-130.

Palm T. 1976a: Zur Kenntnis der Käferfauna der Kanarischen Inseln 11-14. *Entomologisk Tidskrift* **97**: 23-38.

Palm T. 1976b: Zur Kenntnis der Käferfauna der Kanarischen Inseln, 15-16. *Entomologica Scandinavica* **7**: 71-74.

Palm T. 1976c: Ampedus suecicus nom. nov. for Elater borealis Palm, 1947 nec Paykull, 1800. *Notulae Entomologicae* **56**: 140.

Palm T. 1978: Zur Kenntnis der Käferfauna der Kanarischen Inseln 23-24. 23. Die Gattung Denops Fisch. (Coleoptera: Cleridae). 24. Die Gattung Criocephalus Muls. (Coleoptera: Cerambycidae). *Entomologica Scandinavica* **9**: 35-42.

Pandelle L. 1867: [new taxa]. In: Grenier A. (ed.): *Catalogue des coléoptères de France et matériaux pour servir à la faune des coléoptères français.* Paris: A. Grenier, [1] + iv + 3-79 + 135 pp.

Pang H. 1993a: The Epilachninae (Coleoptera: Coocinellidae [sic!]) from Taiwan collected by Klapperich in 1977 with description of a new species. *Journal of South China Agricultural University* (Guangzhou) **14**: 105-110.

Pang H. 1993b: A new species of Pseudoscymnus. *Acta Entomologica Sinica* **36**: 493-494.

Pang H. 1998: The new record and new species of Aspidimerinae (Coleoptera: Coccinellidae) from China. *Entomotaxonomia* **20**: 185-193 (in Chinese).

Pang X.-F. 1966: New species of Stethorus (Coleoptera, Coccinellidae) from Canton, China. *Acta Zootaxonomica Sinica* **3**: 76-81 (in Chinese).

Pang X.-F. 1986: New species of Scymnus from Guangdong Province (Coleoptera: Coccinellidae). *Acta Entomologica Sinica* **29**: 308-311 (in Chinese).

Pang X.-F. 1988: Descriptions of four new species of Scymnus from Guangdong Province (Coleoptera, Coccinellidae: Scymnini). *Acta Zootaxonomica Sinica* **13**: 385-391 (in Chinese).

Pang X.-F. & Gordon R. D. 1986: The Scymnini (Coleoptera: Coccinellida) of China. *The Coleopterists Bulletin* **40**: 157-199.

Pang X.-F. & Huang B. 1985: Description of twelve new species of ladybeetles from Fujian Province (Col. Cocc. Scymnini). *Wuyi Scientific Journal* **5**: 29-46 (in Chinese).

Pang X.-F. & Huang B. 1986a: A new species of Diomus from Fujian (Coleoptera: Coccinellidae). *Wuyi Science Journal* **6**: 59-60 (in Chinese).

Pang X.-F. & Huang B. 1986b: Six new species of Scymnus and four new species of Pseudoscymnus from Fujian (Col. Cocc. Scymnini). *Wuyi Science Journal* **6**: 61-73 (in Chinese).

Pang X.-F. & Mao J.-L. 1975: The natural enemies of the phytophagous mites – Stethorus Weise. *Acta Entomologica Sinica* **18**: 418-424 (in Chinese).

Pang X.-F. & Mao J.-L. 1977: Notes on the Epilachninae of Tibet Autonomous region (Coleoptera: Coccinellidae). *Acta Entomologica Sinica* **20**: 323-328 (in Chinese).

Pang X.-F & Mao J.-L. 1979a: *Economic insect fauna of China* (*Coleoptera: Coccinellidae-II*). Beijing: Science Press, 170 pp. (in Chinese).

Pang X.-F. & Mao J.-L. 1979b: Two new species of Vibidia Mulsant (Coleoptera: Coccinellidae). *Acta Entomologica Sinica* **22**: 468-470 (in Chinese).

Pang X.-F. & Pu T. 1988: Description of two new species of Nephus (Scymnini) from Guangxi (Coleoptera: Coccinellidae). *Entomotaxonomia* **10**: 239-242.

Pang X.-F. & Pu T. 1990: Description of four new species of Scymnus from Guangxi. *Acta Entomologica Sinica* **33**: 336-341.

Pang X.-F. & Yu G.-Y. 1993: Validity of Scymnus (Parapullus) Yang with description of a new species (Coleoptera, Coccinellidae) from Taiwan. *The Coleopterists Bulletin* **47**: 228-231.

Panzer W. G. F. 1789: Einige seltene Insecten beschrieben. *Naturforscher* **24**: 1-35, 1 pl.

Panzer W. G. F. 1792a: *Faunae insectorum Germanicae initia oder Deutschlands Insecten. Heft 2.* Nürnberg: Felsecker, 24 pp. + 24 pls.

Panzer G. W. F. 1792b: *Faunae insectorum Germanicae initia oder Deutschlands Insecten. Heft 3.* Nürnberg: Felsecker, 24 pp. + 24 pls.

Panzer W. G. F. 1792c: *Faunae insectorum Germanicae initia oder Deutschlands Insecten. Heft 4.* Nürnberg: Felsecker, 24 pp. + 24 pls.

Panzer W. G. F. 1793a: *Faunae insectorum Germanicae initia oder Deutschlands Insecten. Heft 6.* Nürnberg: Felsecker, 24 pp. + 24 pls.

Panzer G. W. F. 1793b: *Faunae insectorum Germanicae initia oder Deutschlands Insecten. Heft 7.* Nürnberg: Felsecker, 24 pp. + 24 pls.

Panzer G. W. F. 1793c: *Faunae insectorum Germanicae initia oder Deutschlands Insecten. Heft 8.* Nürnberg: Felsecker, 24 pp. + 24 pls.

Panzer G. W. F. 1793d: *Faunae insectorum Germanicae initia oder Deutschlands Insecten. Heft 13.* Nürnberg: Felsecker, 24 pp. + 24 pls.

Panzer G. W. F. 1793e: *Faunae insectorum Germanicae initia oder Deutschlands Insecten. Heft 15.* Nürnberg: Felsecker, 24 pp. + 24 pls.

Panzer G. W. F. 1794a: *Faunae insectorum germanicae initia oder Deutschlands Insecten. Heft 21.* Nürnberg: Felsecker, 24 pp. + 24 pls.

Panzer G. W. F. 1794b: *Faunae insectorum Germanicae initia oder Deutschlands Insecten. Heft 24.* Nürnberg: Felsecker, 24 pp. + 24 pls.

Panzer G. W. F. 1795a: *Faunae insectorum Germanicae initia oder Deutschlands Insecten. Heft 25.* Nürnberg: Felsecker, 24 pp. + 24 pls.

Panzer G. W. F. 1795b: *Faunae Insectorum Germanicae initia oder Deutschlands Insecten. Heft 26.* Nürnberg: Felsecker, 24 pp. + 24 pls.

Panzer G. W. F. 1795c: *Entomologia germanica exhibens insecta per Germaniam indigena, secundum classes, ordines, genera, species, adiectis, synonymis, locis, observationibus. I. Eleuterata. Cum tabulis aeneis. Deutschlands Insectenfaune oder entomologisches Taschenbuch für das Jahr 1795*. Norimbergae: Felsecker, [8] + 368 + [2] pp., 12 pls.

Panzer G. W. F. 1796a: *Faunae insectorum Germanicae initia oder Deutschlands Insecten. Heft 31*. Nürnberg: Felsecker, 24 pp. + 24 pls.

Panzer G. W. F. 1796b: *Faunae insectorum germanicae initia oder Deutschlands Insecten. Heft 34*. Nürnberg: Felsecker, 24 pp. + 24 pls.

Panzer G. W. F. 1796c: *Faunae insectorum Germanicae initia oder Deutschlands Insecten. Heft 35*. Nürnberg: Felsecker, 24 pp. + 24 pls.

Panzer G. W. F. 1797a: *Faunae insectorum Germanicae initia oder Deutschlands Insecten. Heft 37*. Nürnberg: Felsecker, 24 pp. + 24 pls.

Panzer G. W. F. 1797b: *Faunae insectorum Germanicae initia oder Deutschlands Insecten. Heft 38*. Nurnberg: Felsecker. 24 pp. +24 pls.

Panzer G. W. F. 1797c: *Faunae insectorum Germanicae initia oder Deutschlands Insecten. Heft 39*. Nürnberg: Felsecker, 24 pp. + 24 pls.

Panzer G. W. F. 1797d: 98: *Fauna insectorum Germaniae initia oder Deutschlands Insecten. Heft 40*. Nürnberg: Felsecker, 24 pp. + 24 pls.

Panzer G. W. F. 1798a: *Fauna insectorum Germaniae initia oder Deutschlands Insecten. Ed. 2. Heft 57*. Nürnberg: Flessecker, 24 pp. + 24 pls.

Panzer G. W. F. 1798b: *Voets Beschreibungen und Abbildungen hartschaaligter Insecten. Theil 4*. Nürnberg: Val. Bishoff und Erlangen: J. J. Palm, 16 + 112 pp., pls 23-48.

Panzer G. W. F. 1799a: *Faunae insectorum germanicae initia oder Deutschlands Insecten. Heft 61*. Nürnberg: Felsecker, 24 pp. + 24 pls.

Panzer G. W. F. 1799b: *Fauna insectorum Germaniae initia oder Deutschlands Insecten. Heft 67*. Nürnberg: Felsecker, 24 pp. + 24 pls.

Panzer G. W. F. 1800a: *Faunae insectorum germanicae initia oder Deutschlands Insecten. Heft 76*. Nürnberg: Flessecker, 24 pp. + 24 pls.

Panzer G. W. F. 1800b: *Faunae insectorum germanicae initia oder Deutschlands Insecten. Heft 77*. Nürnberg: Flessecker, 24 pp. + 24 pls.

Panzer G. W. F. 1803: *Faunae insectorum Germanicae initia oder Deutschland Insecten. Heft 90*. Nürnberg: Felsecker, 24 pp. + 24 pls.

Panzer G. W. F. 1805: *Faunae insectorum Germanicae initia oder Deutschland Insecten. Heft 99*. Nürnberg: Felsecker, 24 pp. + 24 pls.

Panzer G. W. F. 1806: *Faunae insectorum Germanicae initia oder Deutschland Insecten. Heft 101*. Nürnberg: Felsecker, 24 pp. + 24 pls.

Panzer G. W. F. 1813: *Index entomologicus, sistens omnes insectorum species in G. W. F. Panzeri Fauna Insectorum Germanica descriptas atque delineatus secundum methodum Fabricianam: adjectis emendationibus, observationibus. Pars I. Eleutherata*. Norimbergae: Felsecker, viii + 216 pp.

Papp K. 1943: *A Magyar bogárfauna határozója*. Budapest: Ruszkabányai Könyvkiadó Vállalat, 477 pp.

Pardo Alcaide A. 1945: Contribución al conocimiento de la fauna entomológica marroquí II. Malacodermos nuevos o interesantes del Marruecos español. *Boletín de la Real Sociedad Española de Historia Natural* **43**: 457-465.

Pardo Alcaide A. P. 1946: Contributión al conocimiento de la fauna entomológica marroquí. II. Malacodermos nuevos o interessantes del Marruecos español. *Boletín de la Sociedad Española de Historia Natural* **43**: 457-465.

Pardo Alcaide A. 1948: Notas de entomológica ibérica. Una nueva y singular especie de Attalus Er. de Cataluña (Col. Malach.). *Eos, Revista Española de Entomología* **24**: 25-28.

Pardo Alcaide A. 1950: *Contribución al conocimiento de la fauna entomológica marroquí III*. Tetuán: Publicaciones Instituto General Franco, 74 pp.

Pardo Alcaide A. 1952: Contribución al conocimiento de la fauna entomológica marroquí IV. Los Ebaeus (Col. Malach.) del grupo del thoracicus (Fourc.), con descripción de una nueva especie de Marruecos. *Eos, Revista Española de Entomología* **28**: 77-82.

Pardo Alcaide A. 1954a: Nota sobre algunos coleópteros euroafricanos. *Eos, Revista Española de Entomología* **30**: 81-87.

Pardo Alcaide A. 1954b: [new taxa]. In: Mateu J.: Coleópteros de Sierra Nevada. Familias: Dytiscidae, Hydrophilidae, Dryopidae, Catopidae, Silphidae, Pselaphidae, Histeridae, Cleridae, Cantharidae, Dasytidae, Aderidae, Anthicidae. *Archivos del Instituto de Aclimatación* (Almería) **2**: 89-101, 2 pls.

Pardo Alcaide A. 1954c: Coleópteros de Sierra Nevada "Malachiidae". *Archivos del Instituto de Aclimatación* (Almería) **2**: 105-109.

Pardo Alcaide A. 1955: Contribución al conocimiento de la fauna entomológica marroquí VI. Coleópteros del valle inferior del Uad Muluya. *Tamuda* **3**: 39-74, 1 map.

Pardo Alcaide A. 1956: Un género y tres especies nuevas de Cleroidea (Col.) de Andalucía Oriental. *Archivos del Instituto de Aclimatación* (Almería) **5**: 17-24.

Pardo Alcaide A. 1958: Diagnosis preliminares de maláquidos ibéricos nuevos (Col. Malach.). *Eos, Revista Española de Entomología* **34**: 169-170.

Pardo Alcaide A. 1960: Observaciones sobre Malachiidae ibéricos (Coleoptera). *Eos, Revista Española de Entomología* **36**: 119-136.

Pardo Alcaide A. 1961: [new taxon]. In: Cobos A.: Exploración entomológica del cedral de Tizi-Ifri (macizo del Iguelmatel, Rif central, Marruecos). Coleópteros. *Archivos del Instituto de Aclimatación* **10**: 63-98, 1 pl.

Pardo Alcaide A. 1962: Malachiidae del Meditarráneo occidental (Coleoptera). *Eos, Revista Española de Entomología* **38**: 263-298.

Pardo Alcaide A. 1966a: Notas sobre Malachiidae del Mediterráneo occidental (Coleoptera). *Graellsia* **22**: 33-45.

Pardo Alcaide A. 1966b: Notas sobre Dasytidae ibero-mauritanos (Coleoptera). *Graellsia* **22**: 177-189.

Pardo Alcaide A. 1966c: Notas sobre Malachiidae del Mediterráneo occidental (Coleoptera) IV. *Eos, Revista Española de Entomología* **42**: 527-534.

Pardo Alcaide A. 1966d: Nueva especie de Henicopus Stephens de Portugal (Col. Dasytid.). *Arquivos do Museu Bocage* (2) **1**(6): xxiii-xxvi.

Pardo Alcaide A. 1967: Notas sobre Malachiidae del Mediterráneo occidental (Col.) V. El grupo de Malachius parilis Erichson con descripcion de una nueva especie de Italia peninsular. *Memorie del Museo Civico di Storia Naturale Verona* **15**: 97-104.

Pardo Alcaide A. 1968: Malachiidae de la Isla de Sicilia (Col.). *Memorie del Museo Civico di Storia Naturale di Verona* **15** [1967]: 269-276.

Pardo Alcaide A. 1970: Notes sur Malachiidae (Col.) VI. Sur le Malachius falcifer Abeille et espèces voisines. *Memorie del Museo Civico di Storia Naturale di Verona* **17**: 349-358.

Pardo Alcaide A. 1971: Notas sobre Malachiidae (Coleoptera). VII. Dos nuevas especies de *Axinotarsus* Motschulsky de Marruecos y revisión de las especies ibero-marroquíes. *Eos, Revista Española de Entomología* **46**: 229-262.

Pardo Alcaide A. 1975a: Notas sobre Malachiidae (Col.) XI. Un género y dos especies nuevas para la ciencia y comentarios sobre otras especies de la fauna hispano-marroquí. *Archivos del Instituto de Aclimatación* (Almería) **20**: 89-104.

Pardo Alcaide A. 1975b: Notas sobre Malachiidae (Coleoptera). VIII-X. VIII. El verdadero status de los Malachius graecus Kraatz y elaphus Abeille con descripción de una especie ignorada de la fauna europea: Malachius agenjoi nov. sp. IX. Sobre la subordinación real del Malachius fallaciosus Baudi. X. Acerca de la existencia de un "excitador" tergo-abdominal en Attalus rufomaculatus Wittmer. *Eos, Revista Española de Entomología* **49**: 233-246.

Park H. C. & Yoon I. B. 1991: A taxonomic revision of subfamily Epilachninae in Korea (Coleoptera: Coccinellidae). *Entomological Research Buletin* **17**: 81-92.

Parsons C. T. 1943: A revision of Nearctic Nitidulidae (Coleoptera). *Bulletin of the Museum of Comparative Zoology at Harvard College* **92**: 121-278, pls. 1-13.

Parsons C. T. 1967: North American Nitidulidae (Coleoptera). IV. Epuraea associated with fungi of pine and oak. *The Canadian Entomologist* **99**: 734-737.

Pascoe F. P. 1860: Notices of new or little-know genera and species of Coleoptera. *Journal of Entomology* **1**: 36-64, pls. II, III.

Pascoe F. P. 1862: Notices of new or little known genera and species of Col. *Journal of Entomology* **1**: 319-370, pls XVI, XVII.

Pascoe F. P. 1863: Notices of new or little-known genera and species of Coleoptera. Part IV. *The Journal of Entomology* **2**: 26-56, pls. II, III.

Pascoe F. P. 1866a: List of the Colydiidae collected in the Amazons Valley by H.W. Bates, Esq., and descriptions of a new species. *The Journal of Entomology* **2**: 79-99.

Pascoe F. P. 1866b: List of the Colydiidae collected in the Indian Islands by Alfred R. Wallace, Esq., and descriptions of new species. *The Journal of Entomology* **2**: 121-143.

Pascoe F. P. 1866c: Notices of new or little-known genera and species of Coleoptera. *Journal of Entomology* **2**: 443-493, pls. XVIII-XIX.

Pascoe F. P. 1887: Notes on Coleoptera, with description of new genera and species. – Part VI. *The Annals and Magazine of Natural History* (5) **20**: 8-20 pp., 1 pl.

Paulian R. 1948: Coleoptera: Corylophidae. Pp. 157-159. In: *The British Museum Expedition to South-West Arabia* (1) part 13. London: H. Scott & E. B. Britton, 504 pp.

Paulian R. 1950: Les Corylophidae d'Afrique. *Memoires de l'Institut Français d'Afrique Noire* **12**: 1-126.

Paulian R. 1962: Observation sur les Saciinae d'Asie tropicale (Col. Corylophidae). *Reichenbachia* **1**: 73-105.

Paulus H. F. 1972: Die systematische und phylogenetische Stellung der Karumiidae, mit einer Beschreibung von Escalerina serraticornis n. sp. aus S-Persien (Ins.: Coleoptera: Cantharoidea). *Senckenbergiana Biologia* **53**: 37-54.

Paulus H. F. 1981: Insects of Saudi Arabia. Coleoptera fam. Cebrionidae. *Fauna of Saudi Arabia* **3**: 257-267.

Paulus H. F. 2004: Urtea graeca nov. gen. et nov. spec., der erster Vertreter der tropischen Atractocerinae in Europa sowie eine Beschreibung von Hymaloxylon aspoeki nov. spec. aus Yunnan (China) (Coleoptera, Cucujiformia, Lymexylidae, Atractocerinae nov. status). *Denisia* **13**: 277-290.

Paykull G. von. 1798: *Fauna suecica. Insecta. Tomus 1.* Upsaliae: J. F. Edman, [8] + 358 + [2] pp.

Paykull G. von. 1798: Anmårkningar vid genus Coccinella, och beskisning öfver de Svånska arter deraf som åro med fina hår beströdde. *Kongliga Vetenskaps Akademiens nya Handlingar* **19**: 144-156.

Paykull G. von. 1799: *Fauna suecica. Insecta. Tomus II.* Upsaliae: J. F. Edman, 234 pp.

Paykull G. von. 1800: *Fauna suecica. Insecta. Tomus III.* Upsaliae: J. F. Edman, 459 pp.

Paykull G. von. 1801: Beskrifning å nya och rättelser vid beskrifningarna af några förr kände Svänska insecter. Tredje Stycket. *Kongliga Svenska Vetenskaps-Akademiens Nya Handlingar* **22**: 116-122.

Peacock E. R. 1987: A review of the Rhadalinae (= Aplocneminae) (Coleoptera: Melyridae). *Bulletin of the British Museum of Natural History (Entomology)* **56** (3): 129-170.

Pečírka J. 1926: Ein neuer Cardiophorus (Elat.) aus Central-Spanien. *Coleopterologisches Centralblatt* **1**: 67-68.

Peez A. von. 1967: Lathridiidae. Pp. 168-190. In: Freude H., Harde K. W. & Lohse G. A. (eds): *Die Käfer Mitteleuropas Band 7 Clavicornia.* Krefeld: Goecke & Evers, 310 pp.

Pelerin J. 1829: Notice of a new species of Elater. *The Zoological Journal* **4**: 524.

Penecke K. A. 1901: Coleopterologische Miscellen. II. *Wiener Entomologische Zeitung* **20**: 11-21.

Penecke K. A. 1907: [new taxon]. In: Penecke K. A. & Müller J.: Koleopterologische Ergebnisse einer Sammelreise nach Dalmatien im Sommer 1905. *Verhandlungen der Zoologisch-Botanischen Gesellschaft in Wien* **57**: 1-19.

Penecke K. A. 1912: Zwei neue Käferarten aus Bulgarien. *Wiener Entomologische Zeitung* **31**: 240-244.

Penev L. D. 1983: Eine neue Art der Gattung Adrastus Esch. aus Bulgarien (Coleoptera, Elateridae). *Reichenbachia* **21**: 181-186.

Penev L. D. 1987: Novyy vid roda Haterumelater (Coleoptera, Elateridae) iz Bolgarii [A new species of the genus Haterumelater Oh. (Coleoptera, Elateridae) [A new species of the genus *Haterumelater* Oh. (Coleoptera, Elateridae) from Bulgaria]. *Zoologicheskiy Zhurnal* **66**: 1586-1589.

Peng Z.-Q., Pang H. & Ren S.-X. 2002: On new species of the genus Epilachna Chevrolat (Coleoptera: Coccinellidae from Hubei Province, China. *Entomotaxonomia* **24**: 128-130 (in Chinese).

Peragallo A. 1863: Seconde note pour servir à l'histoire des lucioles. *Annales de la Société Entomologique de France* (4) **3**: 661-665.

Perez Arcas L. 1865: *Insectos nuevos ó poco conocidos de la fauna española. Primera parte.* Madrid: D. Eusebio Aguado, 56 pp. [issued also in *Revista de los Progresos de las Ciencias Exactas, Físicas y Naturales* **15**].

Perez Arcas L. 1868: *Insectos nuevos ó poco conocidos de la fauna española. Tercera parte.* Madrid: D. Eusebio Aguado, pp. 57-86 [issued also in *Revista de los Progresos de las Ciencias Exactas, Físicas y Naturales* **15**: 413-444].

Perez Arcas L. 1872: Especies nuevas ó críticas de la fauna española. *Anales de la Sociedad Española de Historia Natural* **1**: 89-137, pls I-III.

Perez Arcas L. 1874: Especies nuevas ó críticas de la fauna española. *Anales de la Sociedad Española de Historia Natural* **1**: 111-155, pls I-III.

Péringuey L. 1888: Second contribution to the South African Coleopterous fauna. *Transactions of the South African Philosophy Society* **4**: 67-194.

Péringuey L. 1899: Fifth contribution to the South African Coleopterous fauna. *Annals of the South Africa Museum* **1** (2): 240-330.

Perris E. 1847: Lettre de M. Édouard Perris A M. M sur une excursion dans les Grandes Landes. *Mémoires de l'Académie Royale des Sciences, Belles-Lettres et Arts de Lyon. Section des Sciences* **2**: 433-506.

Perris E. 1853: Histoire des insectes du Pin Maritime. Suite. *Annales de la Société Entomologique de France* (3) **1**: 555-644, pls. 17-19.

Perris E. 1855: [new taxa]. *Bulletin de Société Entomologique de France* **1855**: lxxvii.

Perris E. 1857: Nouvelles excursions dans les Grandes Landes. *Annales de la Société Linnéenne de Lyon* (2) **4**: 83-180.

Perris E. 1862: Histoire des insectes du pin maritime. Supplément aux coléoptères et rectifications. *Annales de la Société Entomologique de France* (4) **2**: 173-243, pls. 5-6.

Perris E. 1864: Description de quelques espèces nouvelles de coléoptères et notes diverses. *Annales de la Société Entomologique de France* (4) **4**: 275-310.

Perris E. 1865: Descriptions de quelques nouvelles espèces de coléoptères, rectifications et notes. *Annales de la Société Entomologique de France* (4) **5**: 505-512.

Perris E. 1866: Descriptions de quelques insectes nouveaux. *Annales de la Société Entomologique France* (4) **6**: 181-196.

Perris E. 1869: Descriptions de quelques coléoptères nouveaux. Rectifications et notes. *L'Abeille, Mémoires d'Entomologie* **7** [1869-70]: 3-37.

Perris E. 1873: Nouvelles et faits diverses. Synonymies et rectifications. *L'Abeille, Journal d'Entomologie* **15** [1873-1877]: 5-7.

Perris É. 1875: Descriptions de quelques insectes jugés nouveaux. *L'Abeille, Journal d'Entomologie* **13**: 1-14.

Perroud B.-P. & Montrousier R. P. 1864: Essai sur la faune entomologique de Kanala (Nouvelle-Calédonie) et description de quelques espèces nouvelles ou peu connues. *Annales de la Société Linnéenne de Lyon* (N. S.) **11**: 46-257, 1 pl.

Persov M. P. 1964: Novyy vid zhuka-tochil"shchika roda Ptilinus (Coleoptera, Anobiidae) iz Primorskogo kraya. *Entomologicheskoe Obozrenie* **48**: 169.

Perty J. A. M. 1831: *Observationes nonnullae in Coleoptera Indiae orientalis. Dissertatio philosophico-entomologica, quam unacum praemissis thesibus auctoritate et consensu illustris philosophorum ordinis in Academia Ludovico-Maximilianea facultatem legendi rite adepturus.* Monachii: M. Lindauer, xxxxiv pp., 1 pl.

Perty J. A. M. 1832: Fasc. 2, pp. 61-124, pls 13-24. In: *Delectus animalium articulatorum, quae in itinere per Brasiliam annis MDCCCVII-MDCCCXX jussu et auspiciis Maximiliani Josephi I. Bavariae regis augustissimi peracto collegerunt Dr J. B. de Spix ... et Dr C. F. Ph. Martius. Accedit dissertatio de insectorum in America meridionali habitantium vitae, genere, moribus et distributione geographica.* Monachii: J. A. M. Perty, [6] + 44 + 244 pp., 40 pls.

Petagna V. 1787: *Specimen insectorum ulterioris Calabriae.* Francofurti and Moguntiae: Varrentrapp & Wenner, 46 pp., 1 pl.

Petri K. 1906: Malthodes serbotae n.sp., ein neuer Malthodes des Subgenus Podistrella Seidl. aus den transsilvanischen Alpen. *Wiener Entomologische Zeitung* **25**: 224-225.

Petri K. 1912: *Siebenbürgens Käferfauna auf Grund ihrer Erforschung bis zum Jahre1911.* Hermannstadt: Jos. Drotleff, xi + [1] pp., 376 columns.

Petri K. 1926: Ergänzungen und Berichtigungen zur Käferfauna Siebenbürgens, 1912. *Verhandlungen und Mitteilungen des Siebenbürgischen Vereins für Naturwissenschaft zu Hermannstadt* **75-76** [1925/1926]: 165-206.

Peyerimhoff P. M. de. 1907: Liste des coléoptères du Sinai. *L'Abeille, Journal d'Entomologie* **31** [1907-1923]: 1-47.

Peyerimhoff P. M. de. 1913: Nouveaux coléoptères du Nord-Africain. *Annales de la Société Entomologique de France* **81**: 515-521.

Peyerimhoff P. M. de. 1917: Nouveaux coléoptères du Nord-Africain. Vingt-cinquième note avec des renseignements synonymiques sur les Silphidae et les Histeridae. *Annales de la Société Entomologique de France* **86**: 117-148.

Peyerimhoff P. M. de. 1920: Nouveaux coléoptères du Nord-Africain. Trente-cinquième note: Fauna de l'Aurès. *Bulletin de la Société Entomologique de France* **1920**: 243-246.

Peyerimhoff P. M. de. 1920: Nouveaux coléoptères du Nord-Africain: Trente-sixième note. Anobiidae parasites du Cèdre et du Pin d'Alep. *Bulletin de la Société Entomologique de France* **1920**: 266-268.

Peyerimhoff P. M. de 1923: Les coléoptères des euphorbes dans le Maroc méridional. *Bulletin de la Société de Sciences Naturelles du Maroc* **3**: 43-51.

Peyerimhoff P. M. de. 1924: Nouveaux coléoptères du Nord-Africain. Quarante sixième note: Récoltes de M. C. Dumant dans la région d´El-Goléa. *Bulletin de la Société Entomologique de France* **1924**: 62-67.

Peyerimhoff P. M. de. 1925a: Nouveaux coléoptères du Nord-Africain. Cinquantième note. *Annales de la Société Entomologique de France* **94**: 1-29.

Peyerimhoff P. M. de. 1925b: Nouveaux coléoptères du Nord-Africain. Cinquante-et-unième note: Faune du Grand Atlas marocain. *Bulletin de la Société Entomologique de France* **1925**: 55-59.

Peyerimhoff P. M. de. 1925c: Nouveaux coléoptères du Nord-Africain. Cinquante-troisième note: Faune du Grand-Atlas marocain. *Bulletin de la Société Entomologique de France* **1925**: 267-269.

Peyerimhoff P. M. de. 1926a: Nouveaux coléoptères du Nord-Africain. Cinquante-sixième note: Récoltes de M. C. Dumont à El-Goléa (suite). *Bulletin de la Société Entomologique de France* **1926**: 103-108.

Peyerimhoff P. M. de. 1926b: Nouveaux coléoptères du Nord-Africain. Cinquante-septième note. Faune de la Zousfana et du Guir. *Bulletin de la Société Entomologique de France* **1926**: 193-196.

Peyerimhoff P. M. de. 1926c: Notes sur la biologie de quelques coléoptères phytophages du Nord Africain (quatrième série) aves les descriptions de quatre espèces nouvelles et de quatre sous-espèces. *Annales de la Société Entomologique de France* **95**: 319-390.

Peyerimhoff P. M. de. 1927: Nouveaux coléoptères du Nord-Africain. Cinquante-neuvième note. Matériaux récolté ou réunis par L. Bedel. *Bulletin de la Société Entomologique de France* **1927**: 237-240.

Peyerimhoff P. M. de. 1928: Nouveaux coléoptères du Nord-Africain. Soixantième note. Récoltes de M. le lieutenant Callies dans la région du Mzab. *Bulletin de la Société Entomologique de France* **1927**: 304-306.

Peyerimhoff P. M. de. 1929a: Nouveaux coléoptères du Nord-Africian. Soixante-troisième note. Récoltes de M. C. Dumont en Tunisie méridionale. *Bulletin de la Société Entomologique de France* **1929**: 37-41.

Peyerimhoff P. M. de. 1929b: Nouveaux coléoptères du Nord-Africian. Soixante-sixième note. Fauna du Hoggar et des massifs voisins. *Bulletin de la Société Entomologique de France* **1929**: 107-111.

Peyerimhoff P. M. de. 1929c: Nouveaux coléoptères du Nord-Africain. Soixante et onzième note: Faune du Hoggar et des massifs voisins (suite). *Bulletin de la Société Entomologique de France* **1929**: 191-194.

Peyerimhoff P. M. de. 1929d: Nouveaux coléoptères du nord-Africain. Soixante-douzième note. Faune du Hoggar et des massifs voisins (suite). *Bulletin de la Société Entomologique de France* **1929**: 209-213.

Peyerimhoff P. M. de. 1931a: Nouveaux coléoptères du Nord-African. Soixante-quinzième note (avec table et résumé). *Annales de la Société Entomologique de France* **100** [1930]: 1-20.

Peyerimhoff P. M. de. 1931b: Mission scientifique du Hoggar envoyée de Février à Mai 1928 par M. Pierre Bordes Gouverneur Général de l'Algérie. Coléoptères. *Mémoires de la Société d'Histoire Naturelle de l'Afrique du Nord* **2**: 1-172, 3 pls, 2 maps.

Peyerimhoff P. M. de. 1935: Description de quatre coléoptères malacodermes récemment découverts au Maroc. *Bulletin de la Société des Sciences Naturelles du Maroc* **14** [1934]: 191-196.

Peyerimhoff P. M. de. 1937: Un Airaphilus nouveau du Portugal (Col. Cucujidae). *Bulletin de la Société Entomologique de France* **42**: 89-92.

Peyerimhoff P. M. de 1942: Récoltes entomologiques faites par L. Berland à Villa-Cisneros (Rio de Oro) Coléoptères. *Bulletin du Muséum National d'Histoire Naturelle* (Paris) (2) **14**: 215-222.

Peyerimhoff P. M. de. 1943: Matériaux pour un Catalogue des coléoptères sahariens. II. - Descriptions d´espèces nouvelles. *Bulletin de la Société d'Histoire Naturelle de l'Afrique du Nord* **34**: 7-35.

Peyerimhoff P. M. de. 1948a: Insectes coléoptères. Pp. 7-84. In: Bernard F. & Peyerimhoff P. M. de. (eds.): *Mission scientifique du Fezzân (1944-1945). Tome V. Zoologie.* Alger: Institut de recherches sahariennes de l'Université d'Alger, 248 pp., 2 pls.

Peyerimhoff P. M. de. 1948b: Coléoptères nouveaux ou mal connus de Berbérie XIX. Le genre Calotroglops Ab. (Malachiidae). *Bulletin de la Société Entomologique de France* **53**: 72-75.

Peyerimhoff P. M. de. 1948c: Description d'un Laemophloeus (Col. Cucujidae) algérien commensal de Triotemnus Gangeri Peyerh. (Scolytidae). *Bulletin de la Société Entomologique de France* **53**: 97-98.

Peyerimhoff P. M. de. 1948d: Communications écrites. Malachiidae. *Bulletin de la Société des Sciences Naturelles du Maroc* **25-27**: 267-288.

Peyerimhoff P. M. de. 1949: Études et descriptions de coléoptères marocain. II. *Bulletin de la Société des Sciences Naturelles de Maroc* **25-27** [1945-1947]: 248-308.

Peyron E. 1857a: Note monographique sur le genre Thorictus de Germar. *Annales de la Société Entomologique de France* (3) **5**: 697-714.

Peyron E. 1857b: Description de quelques coléoptères nouveaux et observations diverses. *Annales de la Société Entomologique de France* (3) **5**: 715-723.

Peyron E. 1877: Étude sur les Malachiides d'Europe et du bassin de la Méditerranée *L'Abeille, Journal d'Entomologie* **15**: 1-312.

Philippi R. A. & Philippi F. 1864: Beschreibung einiger neuen chilenischen Käfer. *Entomologische Zeitung* (Stettin) **25**: 266-284.

Philippi R. A. 1866: Einige Insekten von Chile. *Entomologische Zeitung* (Stettin) **27**: 109-117.

Pic M. 1893a: [new taxa]. *Revue Scientifique du Bourbonnais et du Centre de la France* **6**: 33.

Pic M. 1893b: Sur des variétés de coléoptères. *Feuille des Jeunes Naturalistes* **23**: 46.

Pic M. 1893b: Description de coléoptères de la faune circa-européenne. *La Feuilles des Jeunes Naturalistes* **24**: 60-61.

Pic M. 1893c: Coléoptères nouveaux de la faune circa-européenne. *Feuille des Jeunes Naturalistes* **23**: 111.

Pic M. 1893d: Descriptions d'espèces nouvelles de coléoptères et notes synonymiques. *L'Échange, Revue Linnéenne* **9**: 111-112.

Pic M. 1893e: Notes sur quelques coléoptères, avec descriptions. *L'Échange, Revue Linnéenne* **9**: 122.

Pic M. 1893f: Descriptions d'espèces nouvelles de coléoptères et notes synonymiques. *Bulletin de la Société Entomologique de France* **62**: lxxxv-lxxxviii.

Pic M. 1893g: Deux coléoptères nouveaux. *Bulletin de la Société Entomologique de France* **1893**: clxi.

Pic M. 1893h: [new taxa]. *Bulletin de la Société Entomologique de France* **1893**: cclxxvi-cclxxviii.

Pic M. 1894a: Étude sur les Melyris proprement dits. *Feuille des Jeunes Naturalistes* **25**: 13-14.

Pic M. 1894b: Descriptions de coléoptères nouveaux. *Le Naturaliste* **16**: 27.

Pic M. 1894c: Descriptions de coléoptères de la faune circa-européenne. *Feuille des Jeunes Naturalistes* **24**: 44.

Pic M. 1894d: Description de deux coléoptères nouveaux. *L'Échange, Revue Linnéenne* **10**: 49-50.

Pic M. 1894e: Descriptions de coléoptères de la faune circa-européenne. *Feuille des Jeunes Naturalistes* **24**: 60-61.

Pic M. 1894f: Descriptions et notes entomologiques (coléoptères). *L'Échange, Revue Linnéenne* **10**: 66.

Pic M. 1894g: Descriptions de coléoptères du nord de l'Afrique. *Le Naturaliste* **16**: 71.

Pic M. 1894h: Quelques notes en passant sur plusieurs coléoptères de la collection Henri Tournier. *L'Échange, Revue Linnéenne* **10**: 71-72.

Pic M. 1894i: Trois coléoptères syriens. *L'Échange, Revue Linnéenne* **10**: 75-76.

Pic M. 1894j: Description de quatre nouvelles espèces de coléoptéres d'Algérie. *L'Échange, Revue Linnéenne* **10**: 77-78.

Pic M. 1894l: Sur le genre Ceralliscus Bourg. *L'Échange, Revue Linnéenne* **10**: 95-96.

Pic M. 1894m: Essai d'une étude sur les Danacaea Cast. de la faune d'Europe et circa. *L'Échange, Revue Linnéenne* **10**: 99-106.

Pic M. 1894n: Habitats de quelques coléoptères algériens et descriptions d'espèces nouvelles. *Annales de la Société Entomologique de France* **63**: 101-106.

Pic M. 1894o: Descriptions de deux coléoptères de la Turquie d'Asie. *L'Échange, Revue Linnéenne* **10**: 110-111.

Pic M. 1894p: Notes sur quelques Dasytides d'Algérie. *L'Échange, Revue Linnéenne* **10**: 111-112.

Pic M. 1894q: Contribution à l'étude des Amauronia West. *L'Échange, Revue Linnéenne* **10**: 113-116.

Pic M. 1894r: Notes entomologiques. *L'Échange, Revue Linnéenne* **10**: 116.
Pic M. 1894s: Description de trois coléoptères nouveaux du groupe des Dermestides. *L'Échange, Revue Linnéenne* **10**: 119.
Pic M. 1894t: Contribution à l'étude du genre Cychramus Kugl. *L'Échange, Revue Linnéenne* **10**: 132-134.
Pic M. 1894u: Addenda à l'essai d'une étude sur les Danacaea. *L'Échange, Revue Linnéenne* **10**: 134-135.
Pic M. 1894v: Contribution à la faune coléoptérologique d'Algérie. *Le Naturaliste* **16**: 180.
Pic M. 1894w: Descriptions de coléoptères. *Mémoires de la Société Zoologique de France* **7**: 203-207.
Pic M. 1894x: Contribution à la faune coléoptérologique d'Algérie. Description d'espèces nouvelles. *Le Naturaliste* **16**: 210-211.
Pic M. 1894y: [Ptinus (Bruchus) Theryi, n. sp. et Dorcadion griseolineatum, n. sp.]. *Bulletin de la Société Entomologique de France* **62**: cccxlviii-cccxlix.
Pic M. 1895a: Excursion entomologique dans la province d'Oran (Algérie). *Revue Scientifique du Bourbonnais et du Centre de la France* **8**: 10-13, 131-137, 173-180, 218-223.
Pic M. 1895b: Préliminaires d'une étude synoptique sur le genere Ptinus L. *Feuille des Jeunes Naturalistes* **25**: 26-29.
Pic M. 1895c: Ptinides d'Algérie. *L'Échange, Revue Linnéenne* **11**: 50-51.
Pic M. 1895d: Sur les Danacaea Lap. (Coléoptères malacodermes). *Miscellanea Entomologica* **3**: 66-69.
Pic M. 1895e: Descriptions de coléoptères d'Algérie. *L'Échange, Revue Linnéenne* **11**: 78-82.
Pic M. 1895f: A propos de variétés. *L'Échange, Revue Linnéenne* **11**: 87-89.
Pic M. 1895g: Exochomu nigromaculatus (flavipes Thunb.) Goeze var. hipponensis n. var. *L'Échange, Revue Linnéenne* **11**: 89.
Pic M. 1895h: Mysia oblongoguttata var. Leprieuri n. var. *L'Échange, Revue Linéenne* **11**: 89.
Pic M. 1895i: Notes diverses sur les Ptinides et descriptions d'espèces nouvelles. *L'Échange, Revue Linnéenne* **11**: 99-103.
Pic M. 1895j: A propos de variétés. 2me article. *L'Échange, Revue Linnéenne* **11**: 106-108.
Pic M. 1895l: Notes complementaires ou observations diverses à propos d'espèces et variétés omises ou cataloguées. *L'Échange, Revue Linnéenne* **11**: 110-120.
Pic M. 1895m: Diagnoses de Danacaea. *Miscellanea Entomologica* **3**: 121-122.
Pic M. 1895n: Deuxième étude sur les Melyrides. *L'Échange, Revue Linnéenne* **11**: 123-126.
Pic M. 1895o: Descriptions d'espèces et variétés de Colèoptéres asiatiques. *L'Échange, Revue Linnéenne* **11**: 142-144.
Pic M. 1895p: Notes sur des coléoptères rares ou nouveaux d'Algérie. *Bulletin de la Société Entomologique de France* **1895**: cxxvi-cxxx.
Pic M. 1895q: [Description de deux coléoptères nouveaux et une note sur Anthicus dromioides Pic]. *Bulletin de la Société Entomologique de France* **1895**: cxci-cxcii.
Pic M. 1895r: Notes entomologiques diverses (Col.). *Bulletin de la Société Entomologique de France* **1895**: cccl-cccliii.
Pic M. 1896a: Descriptions de quatre ptinides exotiques. *Miscellanea Entomologica* **4**: 6-7.
Pic M. 1896b: Remarques synonimiques et descriptives sur le genere Cercomorphus, sur divers phytophages, sur des Anthicides, etc. (Col.). *Bulletin de la Sociétét Entomologique de France* **1896**: 29-31.
Pic M. 1896c: Descriptions et notes sur divers coléoptères. *Miscellanea Entomologica* **4**: 41-43.
Pic M. 1896d: Préliminaires d'une étude synoptique sur le genre Ptinus L. (Fin). *Feuille des Jeunes Naturalistes* **26**: 42-44.
Pic M. 1896e: Diagnoses de dasytides divers (in collection Pic). *Miscellanea Entomologica* **4**: 47-48.
Pic M. 1896f: Sur le groupe des Hedobia Sturm (Col.). *Bulletin de la Société Entomologique de France* **1896**: 69-72.
Pic M. 1896g: Notes et diagnoses. *L'Échange, Revue Linnéenne* **12**: 87-88.
Pic M. 1896h: Coléoptères du Nord de l'Afrique. *Miscellanea Entomologica* **4**: 93-94.
Pic M. 1896i: Habitats de ptinides du nord de l'Afrique. *L'Échange, Revue Linnéenne* **12**: 107-110.
Pic M. 1896j: Description d'une Danacaea asiatique (Col.). *Bulletin de la Société Entomologique de France* **1896**: 148.
Pic M. 1896k: Notes et decriptions (Coléoptères). *Feuille des Jeunes Naturalistes* **26**: 201-202.
Pic M. 1897a: Diagnoses. *Miscellanea Entomologica* **5**: 2-3.
Pic M. 1897b: Descriptions de coléoptères. *L'Échange, Revue Linnéenne* **13**: 5-6.

Pic M. 1897c: Descriptions de coléoptères. *Miscellanea Entomologica* **5**: 41-43.

Pic M. 1897d: Descriptions de coléoptères. *Miscellanea Entomologica* **5**: 61-63.

Pic M. 1897e: Remarques et observations diverses sur le genre Danacaea Laporte et sur Käfer Europa's XXXIII. *Miscellanea Entomologica* **5**: 94-97.

Pic M. 1897f: Description d'une coupe nouvelle et de trois especès de ptinides. *Feuille des Jeunes Naturalistes* **27**: 102-103.

Pic M. 1897g: Descriptions de coléoptères. *Feuille des Jeunes Naturalistes* **27**: 119-120.

Pic M. 1897h: Descriptions de coléoptères. *Bulletin de la Société d'Histoire Naturelle d'Autun* **10**: 194-198.

Pic M. 1897i: Notes et descriptions diverses (Coléoptères). *Feuille des Jeunes Naturalistes* **27**: 202-204.

Pic M. 1897j: Descriptions de coléoptères nouveaux d'Algérie et d'Asie Mineure. *Bulletin de la Société Entomologique de France* **1897**: 219-221.

Pic M. 1897k: Descriptions de coléoptères. *Bulletin de la Société Histoire Naturelle de Autun* **10**: 295-302.

Pic M. 1898a: Variétés et nouvelles espèces de coléoptères. *Miscellanea Entomologica* **6**: 2-5.

Pic M. 1898b: Diagnoses de coléoptères malacodermes. *Feuille des Jeunes Naturalistes* **29**: 26-28.

Pic M. 1898c: Description de coléoptères nouveaux. *Le Naturaliste* **20**: 46, 63.

Pic M. 1898d: Notes descriptives sur plusieurs coléoptères et sur un Ichneumon (hyménoptère). *Miscellanea Entomologica* **6**: 73-75.

Pic M. 1898e: Descriptions d'espèces ou variétés de coléoptères de la faune d'Europe et circa. *Miscellanea Entomologica* **6**: 97-99.

Pic M. 1898f: Description de coléoptères. *Bulletin de la Société d'Histoire Naturelle d'Autun* **11**: 116-125.

Pic M. 1898g: Rectifications et renseignements entomologiques. 2 pp. In: Chobaut A.: Voyage chez les Beni-Mzab (Sud de la province d'Alger). Contribution à l'étude de la faune entomologique du Sahara algérien. *Mémoires de l'Académie de Vaucluse* (Avignon) **17**: 131-235, 1 pl., 1 map.

Pic M. 1898h: Diagnoses de coléoptères malacodermes et phytophages. *Miscellanea Entomologica* **6**: 137-140.

Pic M. 1898i: Description de coléoptères nouveaux. *Le Naturaliste* **20**: 141.

Pic M. 1898j: Diagnoses de coléoptères malacodermes et phytophages (2me article). *Miscellanea Entomologica* **6**: 153-154.

Pic M. 1898k: Coléoptères rares ou nouveaux récoltés par M. Maurice Pic cette année dans les Alpes. *Bulletin de la Société d'Histoire Naturelle d'Autun* **11**: 154-156.

Pic M. 1898l: Description de coléoptères nouveaux. *Le Naturaliste* **20**: 273.

Pic M. 1899a: St-Martin Vesubie. Addenda. *L'Échange, Revue Linnéenne* **15**: 18-20.

Pic M. 1899b: Coléoptères européens et exotiques nouveaux. *Bulletin de la Société Zoologique de France* **24**: 24-28.

Pic M. 1899c: Description d´un genre nouveau de coléoptères d´Algérie. *Bulletin de la Société Entomologique de France* **1899**: 28-29.

Pic M. 1899d: Diagnoses de ptinides et anthicides (Col.) des collections du Muséum de Paris. *Bulletin du National Muséum d'Histoire Naturelle* **5**: 28-31.

Pic M. 1899e: Notes sur les Ptinides d'Orient avec diagnoses. *Miscellanea Entomologica* **7**: 81-83.

Pic M. 1899f: Diagnoses de coléoptères communiqués par M. L. Bleuse. *Bulletin de la Société Entomologique de France* **1899**: 85-86.

Pic M. 1899g: Sur divers coléoptères de la faune paléarctique. *Miscellanea Entomologica* **7**: 113-116.

Pic M. 1899h: Descriptions d'Élatérides et Curculionides d'Europe et circa. *Miscellanea Entomologica* **7**: 139-142.

Pic M. 1899i: Nouvelle étude synoptique sur le genre Microjulistus Reitt. (Ceralliscus Bourg.). *Feuille des Jeunes Naturalistes* **29** [1898-99]: 169-171.

Pic M. 1899j: Diagnoses de coléoptères d'Orient. *Bulletin de la Société d'Histoire Naturelle d'Autun* **12**: 204-210.

Pic M. 1899k: Notes sur quelques coléoptères d'Orient. *Bulletin de la Société Entomologique de France* **1899**: 230-232.

Pic M. 1899l: Diagnoses de coléoptères d´Orient. Deuxième article. *Bulletin de la Société d´Histoire Naturelle d'Autun* **12**: 252-261.

Pic M. 1899m: Description d'un coléoptère malacoderme, d'Asie Mineure. *Bulletin de la Société Entomologique de France* **1899**: 357-358.

Pic M. 1899n: Essai d'une étude synoptique sur les Malthinus Latr. (Coléoptères malacodermes) d'Orient. *Annales de la Société Entomologique de Belgique* **43**: 370-377.
Pic M. 1900a: Notes diverses sur les coléoptères. *Le Frelon* **8**: 1-8.
Pic M. 1900b: Contribution à l'étude des coléoptères d'Europe et des regions voisines. *Bulletin de la Société Zoologique de France* **25**: 14-15.
Pic M. 1900c: Diagnoses de coléoptères d'Orient recoltés en 1899. *Bulletin de la Société d'Histoire Naturelle d'Autun* **13**: 25-32.
Pic M. 1900d: Notes sur divers coléoptères. *L'Échange, Revue Linnéenne* **16**: 37-38.
Pic M. 1900e: Description de trois Ptinus d'Orient (Col.). *Bulletin de la Société Entomologique de France* **1900**: 48-49.
Pic M. 1900f: Descriptions de coléoptères algériens et tunisiens. *L'Échange, Revue Linnéenne* **16**: 53-56.
Pic M. 1900g: Diagnoses de coléoptères américains et asiatiques. *Le Naturaliste* (2) **14**: 57.
Pic M. 1900h: Notes et diagnoses. *L'Échange, Revue Linnéenne* **16**: 61-63.
Pic M. 1900i: Notes et diagnoses 2e article. *L'Échange, Revue Linnéenne* **16**: 77-80.
Pic M. 1900j: Diagnoses de malacodermes et d'un Cryptocephalus. *L'Échange, Revue Linnéenne* **16**: 85-88.
Pic M. 1900k: Diagnoses de coléoptères du globe. *L'Échange, Revue Linnéenne* **16**: 89-91.
Pic M. 1900l: Description et habitats nouveaux de divers coléoptères d'Algérie et d'Orient. *Bulletin de la Société Entomologique de France* **1900**: 123-125.
Pic M. 1900m: Contribution à l'étude des coléoptères de la Tripolitaine et la Tunisie. *Bulletin de la Société Entomologique de France* **1900**: 164-166.
Pic M. 1900n: Coléoptères nouveaux de la faune paléarctique. *Bulletin de la Société Zoologique de France* **25**: 182-185.
Pic M. 1901a: Notes diverses et diagnoses. *L'Échange, Revue Linnéenne* **17**: 9-12.
Pic M. 1901b: Notes diverses et diagnoses (3e article). *L'Échange, Revue Linnéenne* **17**: 17-20.
Pic M. 1901c: Notes diverses et diagnoses (4e article). *L'Échange, Revue Linnéenne* **17**: 25-27.
Pic M. 1901d: Notes diverses et diagnoses (5e article). *L'Échange, Revue Linnéenne* **17**: 33-35.
Pic M. 1901e: Coléoptères d'Orient. *Bulletin de la Société d'Histoire Naturelle d'Autun* **14**: 33-39.
Pic M. 1901f: Description abrégées de coléoptères Provenant de Grèce. *L'Échange, Revue Linnéenne* **17**: 49-52.
Pic M. 1901g: Notes diverses et diagnoses (6e article). *L'Échange, Revue Linnéenne* **17**: 57-59.
Pic M. 1901h: Notes diverses et diagnoses (7e article). *L'Échange, Revue Linnéenne* **17**: 65-68.
Pic M. 1901i: Synopsis des Euanoma Reitt. et Pseudeuanoma Pic. *L'Échange, Revue Linnéenne* **17**: 74-75.
Pic M. 1901j: Espèces et variétés nouvelles de coléoptères Paléarctiques. *L'Échange, Revue Linnéenne* **17**: 79-80.
Pic M. 1901k: Coléoptères et ichneumoniens recueillis dans la Haute-Savoie. *L'Échange, Revue Linnéenne* **17**: 80-81.
Pic M. 1901l: Notes diverses et diagnoses *L'Échange, Revue Linnéenne* **17**: 81-83.
Pic M. 1901m: Coléoptères nouveaux d'Orient, Arménie et Tunisie. *L'Échange, Revue Linnéenne* **17**: 89-91.
Pic M. 1901n: Diagnoses ou descriptions abrégées de divers coléoptères exotiques. *L'Échange, Revue Linnéenne* **17**: 93-96.
Pic M. 1901o: Sur le group Podistrina. Col. Malacodermes. *La Feuille des Jeunes Naturalistes* **31**: 104-106.
Pic M. 1901p: Diagnoses preliminaires de deux Malthinus, du Nord de l'Afrique. *Bulletin de la Société Entomologique de France* **1901**: 111-112.
Pic M. 1901q: Coléoptères nouveaux de Amérique Méridionale. *Le Naturaliste* **23**: 278.
Pic M. 1901r: Liste des Ptinidae, Hedobiini et Anthicidae recueillis au Japon central par M. le docteur Harmand, ministre plénipotentiaire de France à Tokio. *Bulletin de Muséum National d'Histoire Naturelle* (Paris) **7**: 343.
Pic M. 1902a: Descriptions et notes diverses. *L'Échange, Revue Linnéenne* **18**: 9-10.
Pic M. 1902b: Descriptions et notes diverses (2e article). *L'Échange, Revue Linnéenne* **18**: 17-19.
Pic M. 1902c: Diagnoses de coléoptères de l'Ancien et du Nouveau Monde. *L'Échange, Revue Linnéenne* **18**: 23-24.
Pic M. 1902d: Coléoptères paléarctiques présumés nouveaux. *L'Échange, Revue Linnéenne* **18**: 25-27.
Pic M. 1902e: Nouvelles espèces et variétés de coléoptères paléarctiques. *L'Échange, Revue Linnéenne* **18**: 31-33.
Pic M. 1902f: Coléoptères intéressants, rares ou nouveaux pour la faune de France. *L'Échange, Revue*

Linnéenne **18**: 46.

Pic M. 1902g: Descriptions et notes diverses (3e article). *L'Échange, Revue Linnéenne* **18**: 47-49.

Pic M. 1902h: Sur quelques "Rhagonycha" de France. *L'Échange, Revue Linnéenne* **18**: 53-54.

Pic M. 1902i: Descriptions et notes diverses (4e article). *L'Échange, Revue Linnéenne* **18**: 55-57.

Pic M. 1902j: "Haplocnemus subinteger n. sp. *Journ. Ins. Nat.* **2**: 60" [reference unknown].

Pic M. 1902k: Coléoptères presumés nouveaux de la faune circamediterranéenne. *L'Échange, Revue Linnéenne* **18**: 63-65.

Pic M. 1902l: Descriptions et notes diverses (5e article). *L'Échange, Revue Linnéenne* **18**: 71-72.

Pic M. 1902m: Nouvelles espèces et variétés de coléoptères paléarctiques (2e article). *L'Échange, Revue Linnéenne* **18**: 79-80.

Pic M. 1902n: Note sur le genre Cardiophorus Esch. (Col.). *Bulletin de la Société Entomologique de France* **1902**: 305-307.

Pic M. 1903a: Notes entomologiques II. - Notes sur plusieurs Anobiidae d'Afrique. *Bulletin de la Société Entomologique de France* **1903**: 81.

Pic M. 1903b: Coléoptères exotiques nouveaux. *Le Naturaliste* **25**: 81.

Pic M. 1903c: Notes entomologiques diverses et descriptions. *L'Échange, Revue Linnéenne* **19**: 89-91.

Pic M. 1903d: Descriptions et notes complémentaires sur divers coléoptères paléarctiques. *L'Échange, Revue Linnéenne* **19**: 97-98.

Pic M. 1903e: Coléoptères exotiques nouveaux. *L'Échange, Revue Linnéenne* **19**: 98-100.

Pic M. 1903f: Coléoptères exotiques nouveaux. *L'Échange, Revue Linnéenne* **19**: 105-107.

Pic M. 1903g: Descriptions et habitats de divers coléoptères du Nord de l'Afrique. *L'Échange, Revue Linnéenne* **19**: 113-115.

Pic M. 1903h: Coléoptères nouveaux d'Asie et d'Afrique. *L'Échange, Revue Linnéenne* **19**: 119-120.

Pic M. 1903i: Notes entomologiques et descriptions. *L'Échange, Revue Linnéenne* **19**: 121-125.

Pic M. 1903j: Sur les Dasytes du s.-g. Metadasytes M. R. *L'Échange, Revue Linnéenne* **19**: 127-128.

Pic M. 1903k: Captures, notes diverses et diagnoses. *L'Échange, Revue Linnéenne* **19**: 129-131.

Pic M. 1903l: Coléoptères européens et "Clytus" syrien nouveaux. *L'Échange, Revue Linnéenne* **19**: 138-139.

Pic M. 1903m: Deux captures intéressantes. Diagnoses de divers coléoptères. *L'Échange, Revue Linnéenne* **19**: 145-147.

Pic M. 1903n: Quelques coléoptères rares ou nouveaux pour la Faune gallo-rhénane. *L'Échange, Revue Linnéenne* **19**: 153-155.

Pic M. 1903o: Note sur divers malacodermes du Nord de l'Afrique. Corrigenda (Col.). *Bulletin de la Société Entomologique de France* **1903**: 155-157.

Pic M. 1903p: Malthinides nouveaux. *L'Échange, Revue Linnéenne* **19**: 155-158.

Pic M. 1903q: Espèces et variétés nouvelles de coléoptères. *L'Échange, Revue Linnéenne* **19**: 161-163.

Pic M. 1903r: Espèces et variétés nouvelles de coléoptères (2me article). *L'Échange, Revue Linnéenne* **19**: 169-171.

Pic M. 1903s: Sur les genres "Micranobium" Gorham et "Rhadine" Baudi. *L'Échange, Revue Linnéenne* **19**: 171-172.

Pic M. 1903t: Espèces et variétés nouvelles de coléoptères. *L'Échange, Revue Linnéenne* **19**: 177-178.

Pic M. 1903u: Diagnoses génériques et spécifiques de divers coléoptères exotiques. *L'Échange, Revue Linnéenne* **19**: 182-183.

Pic M. 1904a: Coléoptères français nouveaux. *L'Échange, Revue Linnéenne* **20**: 2.

Pic M. 1904b: Notes et descriptions. *L'Échange, Revue Linnéenne* **20**: 2-5.

Pic M. 1904c: Notes et descriptions. *L'Échange, Revue Linnéenne* **20**: 9-11.

Pic M. 1904d: Diagnoses de coléoptères asiatiques provenant surtout de Sibérie. *L'Échange, Revue Linnéenne* **20**: 25-27.

Pic M. 1904e: Diagnoses de malacodermes africains et américains. *L'Échange, Revue Linnéenne* **20**: 27-29.

Pic M. 1904f: Diagnoses de coléoptères paléarctiques et exotiques. *L'Échange, Revue Linnéenne* **20**: 33-36.

Pic M. 1904g: Descriptions d'un Bryaxis et de plusieurs malacodermes ou rhyncophores. *L'Échange, Revue Linnéenne* **20**: 49-51.

Pic M. 1904h: Diagnoses de divers coléoptères d'Europe et Turquie d'Asie. *L'Échange, Revue Linnéenne* **20**: 57-58.

Pic M. 1904i: Quelques chasses faites avant et après le coucher du soleil. *L'Échange, Revue Linnéenne* **20**: 60-61.
Pic M. 1904j: A propos de quelques femelles brachyptères du genre Cantharis L. (Col.). *Bulletin de la Société Entomologique de France* **1904**: 71.
Pic M. 1904k: Coléoptères nouveaux de la Turquie d'Asie. *L'Échange, Revue Linnéenne* **20**: 73-74.
Pic M. 1904l: Nouvelles espèces et variétés de coléoptères paléarctiques. *L'Échange, Revue Linnéenne* **20**: 81-82.
Pic M. 1904m: Diagnoses ou descriptions abrégées de coléoptères paléarctiques. *L'Échange, Revue Linnéenne* **20**: 89-94.
Pic M. 1904n: Sur un certain nombre de variétés de coléoptères omises dans le "Catalogue" de Reitter. *Bulletin de la Société Entomologique de France* **1904**: 143-144.
Pic M. 1904o: Coléoptères asiatiques nouveaux. *Bulletin de la Société Entomologique de France* **1904**: 287-289.
Pic M. 1904p: Description d'un Cardiophorus Eschsch. de Syrie (Col.). *Bulletin de la Société Entomologique de France* **1904**: 298-299.
Pic M. 1905a: Descriptions de coléoptères exotiques nouveaux. *Le Naturaliste* **27**: 92-93.
Pic M. 1905b: Espèces et variétés nouvelles de coléoptères paléarctiques. *L'Échange, Revue Linnéenne* **21**: 97-100.
Pic M. 1905c: Descriptions abrégées et notes diverses. *L'Échange, Revue Linnéenne* **21**: 105-107.
Pic M. 1905d: Descriptions abrégées et notes diverses (2e article). *L'Échange, Revue Linnéenne* **21**: 113-115.
Pic M. 1905e: Descriptions abrégées et notes diverses (3e article). *L'Échange, Revue Linnéenne* **21**: 121-122.
Pic M. 1905f: Descriptions abrégées et notes diverses (4e article). *L'Échange, Revue Linnéenne* **21**: 129-131.
Pic M. 1905g: Diagnoses de coléoptères algériens. *L'Échange, Revue Linnéenne* **21**: 145-147.
Pic M. 1905h: Nouveaux coléoptères d'Algérie et d'Asie occidentale. *L'Échange, Revue Linnéenne* **21**: 153-156.
Pic M. 1905i: Coléoptères nouveaux. Provenant de France, Grèce, Algérie et Turquie d'Asie. *L'Échange, Revue Linnéenne* **21**: 161-163.
Pic M. 1905j: Capture diverses, noms nouveaux et diagnoses (Coléoptères). *L'Échange, Revue Linnéenne* **21**: 169-171.
Pic M. 1905k: Espèces et variétés nouvelles de coléoptères de France et Turquie d'Asie. *L'Échange, Revue Linnéenne* **21**: 177-179.
Pic M. 1905l: Sur le genre 'Phyllocerus' Lep. *L'Échange, Revue Linnéenne* **21**: 180-182.
Pic M. 1905m: Notes sur divers coléoptères. *Bulletin de la Société Entomologique de France* **1905**: 181-182.
Pic M. 1905n: Notes entomologiques et descriptions. *L'Échange, Revue Linnéenne* **21**: 185-187.
Pic M. 1906a: Nouveaux coléoptères d'Europe, Asie, Afrique et Amérique (1). *L'Échange, Revue Linnéenne* **22**: 1-5.
Pic M. 1906b: Nouveaux coléoptères d'Europe, Asie, Afrique et Amérique (Suite). *L'Échange, Revue Linnéenne* **22**: 9-12.
Pic M. 1906c: Coléoptères européens, algériens et asiatiques nouveaux. *L'Échange, Revue Linnéenne* **22**: 17-20.
Pic M. 1906d: Étude synoptique sur les "Malthinus" de France (Suite). *L'Échange, Revue Linnéenne* **22**: 22-24.
Pic M. 1906e: Habitats et descriptions de divers coléoptères paléarctiques. *L'Échange, Revue Linnéenne* **22**: 25-26.
Pic M. 1906f: Étude synoptique sur les "Malthinus" de France (Suite). *L'Échange, Revue Linnéenne* **22**: 30-32.
Pic M. 1906g: Habitats et descriptions de divers coléoptères paléarctiques (Suite). *L'Échange, Revue Linnéenne* **22**: 33-35.
Pic M. 1906h: Habitats et descriptions de divers coléoptères paléarctiques (Suite). *L'Échange, Revue Linnéenne* **22**: 41-42.
Pic M. 1906i: Coléoptères en partie nouveaux provenant de la Perse et des régions avoisinantes. *L'Échange, Revue Linnéenne* **22**: 49-51.
Pic M. 1906j: Sur divers coléoptères paléarctiques nouveaux, rares ou peu connus. *L'Échange, Revue Linnéenne* **22**: 65.
Pic M. 1906k: Noms nouveaux et diagnoses de "Cantharini" (Telephorides) européens et exotiques. *L'Échange, Revue Linnéenne* **22**: 81-85.

Pic M. 1906l: Noms nouveaux et diagnoses de "Cantharini" (Telephorides) européens et exotiques (Suite). *L'Échange, Revue Linnéenne* **22**: 89-93.
Pic M. 1906m: Notes entomologiques diverses (Suite). *L'Échange, Revue Linnéenne* **22**: 96.
Pic M. 1906n: Un nouveau malacoderme de Sicile. *Il Naturalista Siciliano* **18**: 197-198.
Pic M. 1906o: Deux nouveaux malacodermes de Chine (Col.). *Bulletin de la Société Entomologique de France* **1906**: 217-218.
Pic M. 1906p: Sur divers ptinides et anobiides (Col.). *Bulletin de la Société Entomologique de France* **1906**: 227-228.
Pic M. 1907a: Coléoptères paléarctiques nouveaux. *L'Échange, Revue Linnéenne* **23**: 97-100.
Pic M. 1907b: Notes entomologiques diverses (Suite). *L'Échange, Revue Linnéenne* **23**: 104.
Pic M. 1907c: Coléoptères paléarctiques nouveaux (Suite). *L'Échange, Revue Linnéenne* **23**: 105-106.
Pic M. 1907d: Notes entomologiques diverses (Suite). *L'Échange, Revue Linnéenne* **23**: 113-115.
Pic M. 1907e: Notes entomologiques diverses (Suite). *L'Échange, Revue Linnéenne* **23**: 121-123.
Pic M. 1907f: Coléoptères exotiques nouveaux ou peu connus (Suite). *L'Échange, Revue Linnéenne* **23**: 125-128.
Pic M. 1907g: Notes entomologiques diverses (Suite). *L'Échange, Revue Linnéenne* **23**: 129-130.
Pic M. 1907h: Coléoptères provenant de l'Afrique australe. *L'Échange, Revue Linnéenne* **23**: 130-134.
Pic M. 1907i: Coléoptères exotiques nouveaux ou peu connus (Suite). *L'Échange, Revue Linnéenne* **23**: 134-135.
Pic M. 1907j: Notes entomologiques diverses (Suite). *L'Échange, Revue Linnéenne* **23**: 137-139.
Pic M. 1907k: Notes entomologiques diverses (Suite). *L'Échange, Revue Linnéenne* **23**: 153-154.
Pic M. 1907l: Coléoptères exotiques nouveaux ou peu connus (Suite). *L'Échange, Revue Linnéenne* **23**: 158.
Pic M. 1907m: Notes entomologiques diverses (Suite). *L'Échange, Revue Linnéenne* **23**: 161-162.
Pic M. 1907n: Coléoptères exotiques nouveaux ou peu connus (Suite). *L'Échange, Revue Linnéenne* **23**: 165-166.
Pic M. 1907o: Notes entomologiques diverses (Suite). *L'Échange, Revue Linnéenne* **23**: 169-170.
Pic M. 1907p: Diagnose de coléoptères asiatiques nouveaux. *L'Échange, Revue Linnéenne* **23**: 171-174.
Pic M. 1907q: Coléoptères exotiques nouveaux ou peu connus (Suite). *L'Échange, Revue Linnéenne* **23**: 174-175.
Pic M. 1907r: Notes entomologiques diverses (Suite). *L'Échange, Revue Linnéenne* **23**: 177-179.
Pic M. 1907s: Notes entomologiques diverses. *L'Échange, Revue Linnéenne* **23**: 185-187.
Pic M. 1907t: Coléoptères exotiques nouveaux ou peu connus (Suite). *L'Échange, Revue Linnéenne* **23**: 190-191.
Pic M. 1907u: Un Silis et deux Anthicus nouveaux provenant des Indes (Col.). *Bulletin de la Société Entomologique de France* **1907**: 195-196.
Pic M. 1907v: Nouveaux coléoptères recueillis en Algérie par M. P. de Peyerimhoff. *Bulletin de la Société Entomologique de France* **1907**: 274-275.
Pic M. 1908a: Descriptions ou diagnoses et notes diverses (Suite). *L'Échange, Revue Linnéenne* **24**: 2.
Pic M. 1908b: Descriptions ou diagnoses et notes diverses (Suite). *L'Échange, Revue Linnéenne* **24**: 9.
Pic M. 1908c: Coléoptères nouveaux de l'Afrique orientale allemande et anglaise. *L'Échange, Revue Linnéenne* **24**: 25-28.
Pic M. 1908d: Descriptions ou diagnoses et notes diverses (Suite). *L'Échange, Revue Linnéenne* **24**: 41-46.
Pic M. 1908e: Sur divers coléoptères français rares ou nouveaux. *L'Échange, Revue Linnéenne* **24**: 46-47.
Pic M. 1908f: Descriptions ou diagnoses et notes diverses (Suite). *L'Échange, Revue Linnéenne* **24**: 49-51.
Pic M. 1908g: Descriptions ou diagnoses et notes diverses (Suite). *L'Échange, Revue Linnéenne* **24**: 57-58.
Pic M. 1908h: Coléoptères exotiques nouveaux ou peu connus. *L'Échange, Revue Linnéenne* **24**: 59-61.
Pic M. 1908i: Descriptions ou diagnoses et notes diverses (Suite). *L'Échange, Revue Linnéenne* **24**: 65-66.
Pic M. 1908j: Descriptions ou diagnoses et notes diverses (Suite). *L'Échange, Revue Linnéenne* **24**: 73-75.
Pic M. 1908k: Descriptions ou diagnoses et notes diverses (Suite). *L'Échange, Revue Linnéenne* **24**: 81-83.
Pic M. 1908l: Descriptions ou diagnoses et notes diverses (Suite). *L'Échange, Revue Linnéenne* **24**: 89-91.
Pic M. 1908m: Diagnoses ou descriptions abrégées de divers Malacodermes exotiques. *L'Échange, Revue Linnéenne* **24**: 94-96.
Pic M. 1908n: Coléoptères recueillis dans le département, en 1907-1908. *Bulletin de la Société d'Histoire Naturelle d'Autun* **21**: 116-120.

Pic M. 1908o: Description d'une nouvelle espèce d'Idgia Cast. de Chine (Col.). *Bulletin de la Société Entomologique de France* **1908**: 128.
Pic M. 1908p: Deux Haplocnemus Steph. algériens nouveaux (Col.). *Bulletin de la Société Entomologique de France* **1908**: 200.
Pic M. 1908q: Malacodermes et hétéromères nouveaux de l'Inde (Col.). *Bulletin de la Société Entomologique de France* **1908**: 228-230.
Pic M. 1909a: Sur plusieurs malacodermes paléarctiques et exotiques (Col.). *Bulletin de la Société Entomologique de France* **1908**: 328-329.
Pic M. 1909b: Nouveaux Coléoptères (malacodermes et hétéromères) des Indes. *Bulletin de la Société Entomologique de France* **1909**: 37-39.
Pic M. 1909c: Descriptions ou diagnoses et notes diverses. *L'Échange, Revue Linnéenne* **25**: 97-100.
Pic M. 1909d: Descriptions ou diagnoses et notes diverses. *L'Échange, Revue Linnéenne* **25**: 105-106.
Pic M. 1909e: Coléoptères exotiques nouveaux ou peu connus (Suite). *L'Échange, Revue Linnéenne* **25**: 108-110.
Pic M. 1909f: Descriptions ou diagnoses et notes diverses (Suite). *L'Échange, Revue Linnéenne* **25**: 113-115.
Pic M. 1909g: Coléoptères exotiques nouveaux ou peu connus. *L'Échange, Revue Linnéenne* **25**: 117-118.
Pic M. 1909h: Descriptions ou diagnoses et notes diverses (Suite). *L'Échange, Revue Linnéenne* **25**: 129-131.
Pic M. 1909i: Coléoptères exotiques nouveaux ou peu connus. *L'Échange, Revue Linnéenne* **25**: 133-134.
Pic M. 1909j: Descriptions ou diagnoses et notes diverses (Suite). *L'Échange, Revue Linnéenne* **25**: 137-139.
Pic M. 1909k: Coléoptères exotiques nouveaux ou peu connus (Suite). *L'Échange, Revue Linnéenne* **25**: 142-143.
Pic M. 1909l: Coléoptères exotiques nouveaux ou peu connus (Suite). *L'Échange, Revue Linnéenne* **25**: 146-148.
Pic M. 1909m: Descriptions ou diagnoses et notes diverses (Suite). *L'Échange, Revue Linnéenne* **25**: 153-156.
Pic M. 1909n: Coléoptères exotiques nouveaux ou peu connus (Suite). *L'Échange, Revue Linnéenne* **25**: 160.
Pic M. 1909o: Sur divers Cantharis (Telephorus). *L'Échange, Revue Linnéenne* **25**: 162-164.
Pic M. 1909p: Coléoptères exotiques nouveaux ou peu connus (Suite). *L'Échange, Revue Linnéenne* **25**: 166-167.
Pic M. 1909q: Descriptions ou diagnoses et notes diverses (Suite). *L'Échange, Revue Linnéenne* **25**: 169-172.
Pic M. 1909r: Coléoptères exotiques nouveaux ou peu connus (Suite). *L'Échange, Revue Linnéenne* **25**: 173-175.
Pic M. 1909s: Descriptions ou diagnoses et notes diverses (Suite). *L'Échange, Revue Linnéenne* **25**: 177-179.
Pic M. 1909t: Coléoptères exotiques nouveaux ou peu connus (Suite). *L'Échange, Revue Linnéenne* **25**: 179-181.
Pic M. 1909u: Descriptions ou diagnoses et notes diverses (Suite). *L'Échange, Revue Linnéenne* **25**: 185-186.
Pic M. 1909v: Coléoptères exotiques nouveaux ou peu connus (Suite). *L'Échange, Revue Linnéenne* **25**: 189.
Pic M. 1910a: Descriptions ou diagnoses et notes diverses (Suite). *L'Échange, Revue Linnéenne* **26**: 1-2.
Pic M. 1910b: Coléoptères exotiques nouveaux ou peu connus. *L'Échange, Revue Linnéenne* **26**: 5-7.
Pic M. 1910c: Descriptions ou diagnoses et notes diverses. *L'Échange, Revue Linnéenne* **26**: 9-11.
Pic M. 1910d: Coléoptères exotiques nouveaux ou peu connus. *L'Échange, Revue Linnéenne* **26**: 12-14.
Pic M. 1910e: Descriptions ou diagnoses et notes diverses (Suite). *L'Échange, Revue Linnéenne* **26**: 17-19.
Pic M. 1910f: Coléoptères exotiques nouveaux ou peu connus. *L'Échange, Revue Linnéenne* **26**: 20-22.
Pic M. 1910g: Descriptions ou diagnoses et notes diverses. *L'Échange, Revue Linnéenne* **26**: 25-27.
Pic M. 1910h: Coléoptères exotiques nouveaux ou peu connus. *L'Échange, Revue Linnéenne* **26**: 28-30.
Pic M. 1910i: Descriptions ou diagnoses et notes diverses. *L'Échange, Revue Linnéenne* **26**: 33-34.
Pic M. 1910j: Coléoptères malacodermes et hétéromères en partie nouveaux. *Bulletin de la Société d'Histoire Naturelle d'Autun* **14**: 37-39.
Pic M. 1910k: Descriptions ou diagnoses et notes diverses. *L'Échange, Revue Linnéenne* **26**: 41-42.
Pic M. 1910l: Coléoptères exotiques nouveaux ou peu connus. *L'Échange, Revue Linnéenne* **26**: 45-47.
Pic M. 1910m: Descriptions ou diagnoses et notes diverses (Suite). *L'Échange, Revue Linnéenne* **26**: 49-51.
Pic M. 1910n: Coléoptères exotiques nouveaux ou peu connus. *L'Échange, Revue Linnéenne* **26**: 53-54.
Pic M. 1910o: Descriptions ou diagnoses et notes diverses (Suite). *L'Échange, Revue Linnéenne* **26**: 57-58.
Pic M. 1910p: Coléoptères exotiques nouveaux ou peu connus. *L'Échange, Revue Linnéenne* **26**: 60-63.
Pic M. 1910q: Descriptions ou diagnoses et notes diverses (Suite). *L'Échange, Revue Linnéenne* **26**: 65-66.

Pic M. 1910r: Sur divers coléoptères intéressants de France. *L'Échange, Revue Linnéenne* **26**: 66-67.

Pic M. 1910s: Coléoptères exotiques nouveaux ou peu connus (Suite). *L'Échange, Revue Linnéenne* **26**: 69-71.

Pic M. 1910t: Coléoptères exotiques nouveaux ou peu connus. *L'Échange, Revue Linnéenne* **26**: 74-78, 86-87.

Pic M. 1910u: Descriptions ou diagnoses et notes diverses. (Suite). *L'Échange, Revue Linnéenne* **26**: 89-91.

Pic M. 1910v: [Elateridae]. In: Viturat C. & Fauconnet L.: Catalogue analytique et raisonné des coléoptères de Saône-et-Loire et des régions limitrophes. *Bulletin de la Société d'Histoire Naturelle d'Autun* **23**: 121-232.

Pic M. 1910w: Malacodermes et hétéromères nouveaux d'Afrique et d'Asie. *Le Naturaliste* **32**: 259.

Pic M. 1910x: Malacodermes et hétéromères nouveaux d'Afrique et d'Asie. *Le Naturaliste* **32**: 271-272.

Pic M. 1911a: Sur divers coléoptères d'Egypte plus ou moins intéressants et en partie nouveaux. *Bulletin de la Société Entomologique d'Egypte* **4**: 11-13.

Pic M. 1911b: Sur quelques Ototreta E. Oliv. *Revue Scientifique du Bourbonnais et du Centre de la France* **24**: 35-37.

Pic M. 1911c: Coléoptères égyptiens nouveaux ou plus ou moins rares. *Bulletin de la Société Entomologique d'Égypte* **4**: 71-76.

Pic M. 1911d: Notes sur divers coléoptères paléarctiques. *Bulletin de la Société Entomologique de France* **1911**: 91-92.

Pic M. 1911e: Descriptions ou diagnoses et notes diverses (Suite). *L'Échange, Revue Linnéenne* **27**: 97-98.

Pic M. 1911f: Coléoptères exotiques nouveaux ou peu connus. *L'Échange, Revue Linnéenne* **27**: 99-101.

Pic M. 1911g: Descriptions ou diagnoses et notes diverses (Suite). *L'Échange, Revue Linnéenne* **27**: 105-107.

Pic M. 1911h: Descriptions de plusieurs coléoptères malacodermes et hétéromères. *Bulletin de la Société d'Histoire Naturelle d'Autun* **15**: 107-109.

Pic M. 1911i: Quelques notes et renseignements généraux sur les Malachides de France. *L'Échange, Revue Linnéenne* **27**: 111-112.

Pic M. 1911j: Descriptions ou diagnoses et notes diverses (Suite). *L'Échange, Revue Linnéenne* **27**: 113-114.

Pic M. 1911k: Quelques notes et renseignements généraux sur les Malachides de France. *L'Échange, Revue Linnéenne* **27**: 118-119, 127.

Pic M. 1911l: Descriptions ou diagnoses et notes diverses (Suite). *L'Échange, Revue Linnéenne* **27**: 121-122.

Pic M. 1911m: Diagnoses préliminaires de 30 coléoptères exotiques. *L'Échange, Revue Linnéenne* **27**: 122-124.

Pic M. 1911n: Coléoptères exotiques nouveaux ou peu connus. *L'Échange, Revue Linnéenne* **27**: 127-128, 129-130, 132-134.

Pic M. 1911o: Descriptions ou diagnoses et notes diverses (Suite). *L'Échange, Revue Linnéenne* **27**: 136-139.

Pic M. 1911p: Coléoptères exotiques nouveaux ou peu connus. *L'Échange, Revue Linnéenne* **27**: 142-144.

Pic M. 1911q: Descriptions ou diagnoses et notes diverses (Suite). *L'Échange, Revue Linnéenne* **27**: 145-146.

Pic M. 1911r: Coléoptères exotiques nouveaux ou peu connus. *L'Échange, Revue Linnéenne* **27**: 149-152.

Pic M. 1911s: Descriptions ou diagnoses et notes diverses (Suite). *L'Échange, Revue Linnéenne* **27**: 153-154.

Pic M. 1911t: Coléoptères exotiques nouveaux ou peu connus. *L'Échange, Revue Linnéenne* **27**: 155-159, 164-167.

Pic M. 1911u: Descriptions ou diagnoses et notes diverses (Suite). *L'Échange, Revue Linnéenne* **27**: 169-170.

Pic M. 1911v: Coléoptères exotiques nouveaux ou peu connus (Suite). *L'Échange, Revue Linnéenne* **27**: 172-174.

Pic M. 1911w: Descriptions ou diagnoses et notes diverses (Suite). *L'Échange, Revue Linnéenne* **27**: 177-178.

Pic M. 1911x: H. Sauter's Formosa-Ausbeute. Cantharidae, Lampyridae, Mordellidae (Col.). *Deutsche Entomologische National-Bibliothek* **2**: 188-189.

Pic M. 1911y: Malacodermes et hétéromères nouveaux d'Afrique et d'Asie. *Le Naturaliste* **32** [1910]: 271-172.

Pic M. 1912a: Notes sur certains genres litigieux ou peu connus. *Mélanges Exotico-Entomologiques* **3**: 1-3. Lycidae!

Pic M. 1912b: Descriptions ou diagnoses et notes diverses. (Suite). *L'Échange, Revue Linnéenne* **28**: 1-3, 9-10.

Pic M. 1912c: Étude dichotomique du genre Drilosilis Pic. *Mélanges Exotico-Entomologiques* **2**: 12-14.

Pic M. 1912d: Notes sur certains genres litigieux. *Mélanges Exotico-Entomologiques* **3**: 1-5.

Pic M. 1912e: Descriptions d'espéces et variétés diverses. *Mélanges Exotico-Entomologiques* **3**: 6-20.

Pic M. 1912f: Descriptions ou diagnoses et notes diverses. (Suite). *L'Échange, Revue Linnéenne* **28**: 17.

Pic M. 1912g: Descriptions ou diagnoses et notes diverses (Suite). *L'Échange, Revue Linnéenne* **28**: 25-26.

Pic M. 1912h: Coléoptères exotiques nouveaux ou peu connus (Suite). *L'Échange, Revue Linnéenne* **28**: 29-32.

Pic M. 1912i: Descriptions ou diagnoses et notes diverses. (Suite). *L'Échange, Revue Linnéenne* **28**: 33-35, 41-43.

Pic M. 1912j: Coléoptères exotiques nouveaux ou peu connus. *L'Échange, Revue Linnéenne* **28**: 46-47.

Pic M. 1912k: Descriptions ou diagnoses et notes diverses. (Suite). *L'Échange, Revue Linnéenne* **28**: 49-51.

Pic M. 1912l: Descriptions ou diagnoses et notes diverses. (Suite). *L'Échange, Revue Linnéenne* **28**: 57-58.

Pic M. 1912m: Nouveaux « Malthodes » Ksw. paléarctiques. *L'Échange, Revue Linnéenne* **28**: 58-61.

Pic M. 1912nm: Descriptions ou diagnoses et notes diverses. (Suite). *L'Échange, Revue Linnéenne* **28**: 65-66.

Pic M. 1912on: Coléoptères égyptiens nouveaux ou plus ou moins rares. *Bulletin de la Société Entomologique d'Égypte* **1912**: 71-76.

Pic M. 1912po: Descriptions ou diagnoses et notes diverses. (Suite). *L'Échange, Revue Linnéenne* **28**: 73-74.

Pic M. 1912q: Captures de quelques coléoptères dans les Alpes françaises. *L'Échange, Revue Linnéenne* **28**: 74-75.

Pic M. 1912r: Anomalies, captures et nouveautés de coléoptères. *L'Échange, Revue Linnéenne* **28**: 81-83.

Pic M. 1912s: Synonymies de divers genres, espèces et variétés de coléoptères. *Bulletin de la Société Entomologique de France* **1912**: 83-85.

Pic M. 1912t: Nouveaux coléoptères paléarctiques. *L'Échange, Revue Linnéenne* **28**: 89-90.

Pic M. 1912u: Coléoptères du Maroc. *Bulletin du Muséum National d'Histoire Naturelle* (Paris) **18**: 219-221.

Pic M. 1912v: Renseignements bibliographiques et synonymiques sur divers anobiides. (Col.). *Bulletin de la Société Entomologique de France* **1912**: 265.

Pic M. 1913a: Un genre et divers malacodermes nouveaux. *Mélanges Exotico-Entomologiques* **5**: 3-6.

Pic M. 1913b: Espèces et variétés nouvelles, ou peu connues, rentrant dans diverses familles. *Mélanges Exotico-Entomologiques* **7**: 10-20.

Pic M. 1913c: Notes diverses, descriptions et diagnoses. (Suite). *L'Échange, Revue Linnéenne* **29**: 97-98.

Pic M. 1913d: Coléoptères exotiques en partie nouveaux. *L'Échange, Revue Linnéenne* **29**: 98-100.

Pic M. 1913e: Notes entomologiques. *Bulletin de la Société d'Histoire Naturelle d'Autun* **25**: 100-103.

Pic M. 1913f: Notes diverses, descriptions et diagnoses. *L'Échange, Revue Linnéenne* **29**: 105-106.

Pic M. 1913g: Notes diverses, descriptions et diagnoses. (Suite). *L'Échange, Revue Linnéenne* **29**: 113-114.

Pic M. 1913h: Notes diverses, descriptions et diagnoses (Suite). *L'Échange, Revue Linnéenne* **29**: 121-122.

Pic M. 1913i: Notes diverses, descriptions et diagnoses (Suite). *L'Échange, Revue Linnéenne* **29**: 137-139.

Pic M. 1913j: Notes diverses, descriptions et diagnoses (Suite). *L'Échange, Revue Linnéenne* **29**: 145-146.

Pic M. 1913k: Notes diverses, descriptions et diagnoses (Suite). *L'Échange, Revue Linnéenne* **29**: 153-154.

Pic M. 1913l: Notes diverses, descriptions et diagnoses (Suite). *L'Échange, Revue Linnéenne* **29**: 161-162.

Pic M. 1913m: Notes diverses, descriptions et diagnoses (Suite). *L'Échange, Revue Linnéenne* **29**: 169-171.

Pic M. 1913n: Notes diverses, descriptions et diagnoses (Suite). *L'Échange, Revue Linnéenne* **29**: 181-183.

Pic M. 1913o: Notes diverses, descriptions et diagnoses (Suite). *L'Échange, Revue Linnéenne* **29**: 185-187.

Pic M. 1913p: Une nouvelle espèce de Malacodermes des Indes. P. 199. In: Zoological results of Abor Expedition, 1911-12. XII. Coleoptera. VI. *Records of the Indian Museum* (Calcutta) **8** [1912-1922]: xxvii + 629 pp., 54 pls.

Pic M. 1914a: Contribution à l'étude des "Cantharis" L. du Turkestan. *L'Échange, Revue Linnéenne* **30**: 2-6.

Pic M. 1914b: Coléoptères divers du Tonkin et de l'Indochine. Malacodermes (Telephorides=Cantharides). *Mélanges Exotico-Entomologiques* **9**: 5-12.

Pic M. 1914c: Nouveaux coléoptères de diverses familles. *Mélanges Exotico-Entomologiques* **10**: 7-20.

Pic M. 1914d: Ptinidae et Anobiidae (Col.). *Supplementa Entomologica* **3**: 8-11.

Pic M. 1914e: Notes diverses, descriptions et diagnoses (Suite). *L'Échange, Revue Linnéenne* **30**: 10-13.

Pic M. 1914f: Coléoptères exotiques en partie nouveaux. *L'Échange, Revue Linnéenne* **30**: 15-16.

Pic M. 1914g: Étude dichotomique et biologique des malachides de France. *L'Échange, Revue Linnéenne* **30** [hors-texte]: 21-40.

Pic M. 1914h: Notes diverses, descriptions et diagnoses (Suite). *L'Échange, Revue Linnéenne* **30**: 25-27.

Pic M. 1914i: Notes diverses, descriptions et diagnoses (Suite). *L'Échange, Revue Linnéenne* **30**: 33-34.

Pic M. 1914j: Coléoptères exotiques en partie nouveaux (Suite). *L'Échange, Revue Linnéenne* **30**: 38-40.

Pic M. 1914k: Notes diverses, descriptions et diagnoses (Suite). *L'Échange, Revue Linnéenne* **30**: 41-42.

Pic M. 1914l: Notes diverses, descriptions et diagnoses (Suite). *L'Échange, Revue Linnéenne* **30**: 49-51.

Pic M. 1914m: Notes sur les Cantharidae paléarctiques et diagnoses de formes nouvelles. *L'Échange, Revue Linnéenne* **30**: 51-53.

Pic M. 1914n: Coléoptères exotiques en partie nouveaux (Suite). *L'Échange, Revue Linnéenne* **30**: 54-56.
Pic M. 1914o: Notes diverses, descriptions et diagnoses (Suite.). *L'Échange, Revue Linnéenne* **30**: 57-59.
Pic M. 1914p: Notes diverses, descriptions et diagnoses (Suite.). *L'Échange, Revue Linnéenne* **30**: 73-75.
Pic M. 1914q: Notes diverses, descriptions et diagnoses (Suite.). *L'Échange, Revue Linnéenne* **30**: 77-79.
Pic M. 1914r: Catalogue analytique et raisonné des coléoptères de Saône-et-Loire et des départements limitrophes. *Bulletin de la Société d'Histoire Naturelle d'Autun* **27**: 153-231.
Pic M. 1915a: Notes diverses, descriptions et diagnoses (Suite). *L'Échange, Revue Linnéenne* **31**: 1-3.
Pic M. 1915b: Nouveautés de diverses familles. *Mélanges Exotico-Entomologiques* **13**: 2-13.
Pic M. 1915c: Nouveautés rentrant dans diverses familles. *Mélanges Exotico-Entomologiques* **14**: 2-20.
Pic M. 1915d: Nouvelles espèces de diverses familles. *Mélanges Exotico-Entomologiques* **15**: 2-24.
Pic M. 1915e: Coléoptères exotiques en partie nouveaux (Suite). *L'Échange, Revue Linnéenne* **31**: 14-15.
Pic M. 1915f: Genres nouveaux, espèces et variétés nouvelles. *Mélanges Exotico-Entomologiques* **16**: 2-13.
Pic M. 1915g: Notes diverses, descriptions et diagnoses. (Suite.). *L'Échange, Revue Linnéenne* **31**: 17-18.
Pic M. 1915h: Notes diverses, descriptions et diagnoses (Suite). *L'Échange, Revue Linnéenne* **31**: 21-22.
Pic M. 1915i: Notes diverses, descriptions et diagnoses. (Suite.). *L'Échange, Revue Linnéenne* **31**: 25-27.
Pic M. 1915j: Notes diverses, descriptions et diagnoses (Suite). *L'Échange, Revue Linnéenne* **31**: 29-30.
Pic M. 1915k: Notes diverses, descriptions et diagnoses. (Suite.). *L'Échange, Revue Linnéenne* **31**: 45-46.
Pic M. 1915l: Deux nouveaux Lycocerus Gorh. du Thibet (Col. Malacodermata). *Bulletin de la Société Entomologique de France* **1915**: 78-79.
Pic M. 1915m: Nouveaux Malacodermes exotiques. *Bulletin de la Société Zoologique de France* **40**: 95-97.
Pic M. 1916a: Notes diverses, descriptions et diagnoses (Suite). *L'Échange, Revue Linnéenne* **32**: 1-2.
Pic M. 1916b: Contribution abrégée pour l'étude du genre "Ernobius" Thoms. *L'Échange, Revue Linnéenne* **32** [horse texte]: 1-8.
Pic M. 1916c: Notes et descriptions abrégées diverses (Suite). *Mélanges Exotico-Entomologiques* **17**: 2-8.
Pic M. 1916d: Diagnoses spécifiques. *Mélanges Exotico-Entomologiques* **17**: 8-20.
Pic M. 1916e: Diagnoses génériques et spécifiques. *Mélanges Exotico-Entomologiques* **18**: 2-20.
Pic M. 1916f: Diagnoses abrégées diverses. *Mélanges Exotico-Entomologiques* **21**: 2-20.
Pic M. 1916g: Notes diverses, descriptions et diagnoses (Suite). *L'Échange, Revue Linnéenne* **32**: 5-7.
Pic M. 1916h: Coléoptères exotiques en partie nouveaux (Suite). *L'Échange, Revue Linnéenne* **32**: 8.
Pic M. 1916i: Notes diverses, descriptions et diagnoses (Suite). *L'Échange, Revue Linnéenne* **32**: 9-11.
Pic M. 1916j: Notes diverses, descriptions et diagnoses (Suite). *L'Échange, Revue Linnéenne* **32**: 13-14.
Pic M. 1916k: Coléoptères exotiques en partie nouveaux (Suite). *L'Échange, Revue Linnéenne* **32**: 15-16.
Pic M. 1916l: Notes diverses, descriptions et diagnoses (Suite). *L'Échange, Revue Linnéenne* **32**: 21-23.
Pic M. 1917a: Notes diverses, descriptions et diagnoses (Suite). *L'Échange, Revue Linnéenne* **33**: 1-3.
Pic M. 1917b: Descriptions abrégées diverses. *Mélanges Exotico-Entomologiques* **22**: 2-20.
Pic M. 1917c: Descriptions abrégées diverses. *Mélanges Exotico-Entomologiques* **24**: 2-24.
Pic M. 1917d: Descriptions abrégées diverses. *Mélanges Exotico-Entomologiques* **25**: 2-24.
Pic M. 1917e: Descriptions abrégées diverses. *Mélanges Exotico-Entomologiques* **26**: 2-24.
Pic M. 1917f: Notes diverses, descriptions et diagnoses (Suite). *L'Échange, Revue Linnéenne* **33**: 5-6.
Pic M. 1917g: Notes diverses, descriptions et diagnoses (Suite). *L'Échange, Revue Linnéenne* **33**: 13-15.
Pic M. 1917h: Notes diverses, descriptions et diagnoses (Suite). *L'Échange, Revue Linnéenne* **33**: 17.
Pic M. 1917i: Notes diverses, descriptions et diagnoses (Suite). *L'Échange, Revue Linnéenne* **33**: 21-23.
Pic M. 1917j: Deux nouveaux Cantharis de Chine. *Bulletin de la Société Entomologique de France* **1917**: 160-161.
Pic M. 1917l: Nouveaux coléoptères exotiques. *Bulletin du Muséum National d'Histoire Naturelle* (Paris) **23**: 258-263.
Pic M. 1918a: Notes diverses, descriptions et diagnoses. *L'Échange, Revue Linnéenne* **34** (385): 1-3.
Pic M. 1918b: Courtes descriptions diverses. *Mélanges Exotico-Entomologiques* **27**: 1-24.
Pic M. 1918c: Courtes descriptions diverses. *Mélanges Exotico-Entomologiques* **28**: 1-24.
Pic M. 1918d: Contribution à l'étude des Dasytides. *L'Échange, Revue Linnéenne* **34** [hors-texte] (385-386): 5-8.
Pic M. 1918e: Notes diverses, descriptions et diagnoses (Suite). *L'Échange, Revue Linnéenne* **34**: 9-11.
Pic M. 1918f: Contribution à l'étude des Dasytides. *L'Échange, Revue Linnéenne* **34** [hors-texte] (385-386): 9-12.

Pic M. 1918g: Notes diverses, descriptions et diagnoses (Suite). *L'Échange, Revue Linnéenne* **34**: 13-15, 17-19.
Pic M. 1918h: Notes diverses, descriptions et diagnoses (Suite). *L'Échange, Revue Linnéenne* **34**: 21-24.
Pic M. 1919a: Nouveautés diverses. *Mélanges Exotico-Entomologiques* **30**: 1-21.
Pic M. 1919b: Notes diverses, descriptions et diagnoses (Suite). *L'Échange, Revue Linnéenne* **35**: 5-6.
Pic M. 1919c: Étude détaillée de divers insectes. *L'Échange, Revue Linnéenne* **34** [hors-texte] (392): 5-8.
Pic M. 1919d: Notes diverses, descriptions et diagnoses (Suite). *L'Échange, Revue Linnéenne* **35**: 9-11.
Pic M. 1919e: Nouveautés diverses. *Mélanges Exotico-Entomologiques* **31**: 11-24.
Pic M. 1919f: Notes diverses, descriptions et diagnoses (Suite). *L'Échange, Revue Linnéenne* **35**: 13-15.
Pic M. 1919g: Coléoptères exotiques en partie nouveaux (Suite). *L'Échange, Revue Linnéenne* **35**: 15.
Pic M. 1919h: Notes diverses, descriptions et diagnoses (Suite). *L'Échange, Revue Linnéenne* **35**: 17-20.
Pic M. 1919i: Coléoptères exotiques en partie nouveaux (Suite). *L'Échange, Revue Linnéenne* **35**: 24.
Pic M. 1920a: Nouveautés diverses. *Mélanges Exotico-Entomologiques* **32**: 1-28.
Pic M. 1920b: Notes diverses, descriptions et diagnoses (Suite). *L'Échange, Revue Linnéenne* **36**: 5-8.
Pic M. 1920c: Diagnoses et notes sur le genre Idgia Cast. *L'Échange, Revue Linnéenne* **36** [hors-texte]: 7-8, 12.
Pic M. 1920d: Notes diverses, descriptions et diagnoses. *L'Échange, Revue Linnéenne* **36**: 9-12.
Pic M. 1920e: Notes diverses, descriptions et diagnoses (Suite). *L'Échange, Revue Linnéenne* **36**: 13-15.
Pic M. 1920f: Notes diverses, descriptions et diagnoses (Suite). *L'Échange, Revue Linnéenne* **36**: 17-18.
Pic M. 1920g: Diagnoses de coléoptères exotiques (Suite). *L'Échange, Revue Linnéenne* **36**: 18-20.
Pic M. 1920h: Notes diverses, descriptions et diagnoses. *L'Échange, Revue Linnéenne* **36**: 21-22.
Pic M. 1920i: Sur divers coléoptères d'Egypte et Sinai en partie nouveaux. *Bulletin de la Société Entomologique d'Égypte* **6** [1919]: 44-56.
Pic M. 1921a: Notes diverses, descriptions et diagnoses. *L'Échange, Revue Linnéenne* **37**: 1-4.
Pic M. 1921b: Contribution à l'étude des Lycides. *L'Échange, Revue Linnéenne* **37** [hors texte] (404-406): 1-12.
Pic M. 1921d: Nouveautés diverses. *Mélanges Exotico-Entomologiques* **33**: 1-32.
Pic M. 1921e: Nouveautés diverses. *Mélanges Exotico-Entomologiques* **34**: 1-32.
Pic M. 1921f: Notes diverses, descriptions et diagnoses (Suite). *L'Échange, Revue Linnéenne* **37**: 5-6.
Pic M. 1921g: Diagnoses de coléoptères exotiques. *L'Échange, Revue Linnéenne* **37**: 6-8.
Pic M. 1921h: Notes diverses, descriptions et diagnoses. *L'Échange, Revue Linnéenne* **37**: 9-10.
Pic M. 1921i: Diagnoses de coléoptères exotiques. *L'Échange, Revue Linnéenne* **37**: 10-12.
Pic M. 1921j: Notes diverses, descriptions et diagnoses. *L'Échange, Revue Linnéenne* **37**: 13-15.
Pic M. 1921k: [new taxa]. In: Alfieri A.: Contribution à la faune entomologique du nord-est du Sinaï. *Bulletin de la Société Entomologique d'Egypte* **6** [1920]: 40-53.
Pic M. 1921l: Nouveaux coléoptères. *Bulletin de la Société Entomologique d'Égypte* **6** [1920]: 54-55.
Pic M. 1921m: Nouveaux malacodermes du Maroc (Insectes Coléoptères). *Bulletin de la Société des Sciences Naturelles du Maroc* **1**: 146-147.
Pic M. 1922a: Nouveautés diverses. *Mélanges Exotico-Entomologiques* **35**: 1-32.
Pic M. 1922b: Nouveautés diverses. *Mélanges Exotico-Entomologiques* **36**: 1-32.
Pic M. 1922c: Nouveautes diverses. *Mélanges Exotico-Entomologiques* **37**: 1-32.
Pic M. 1922d: Contribution à l'étude des Lycides. *L'Échange, Revue Linnéenne* **38**: [hors texte] (407-410): 13-28.
Pic M. 1922e: Notes diverses, descriptions et diagnoses (Suite). *L'Échange, Revue Linnéenne* **38**: 17-19.
Pic M. 1922f: Notes diverses, descriptions et diagnoses (Suite). *L'Échange, Revue Linnéenne* **38**: 21-23.
Pic M. 1922g: Notes diverses, descriptions et diagnoses (Suite). *L'Échange, Revue Linnéenne* **38**: 25-28.
Pic M. 1922h: Notes diverses, descriptions et diagnoses (Suite). *L'Échange, Revue Linnéenne* **38**: 29-30.
Pic M. 1922i: Coléoptères exotiques en partie nouveaux (Suite). *L'Échange, Revue Linnéenne* **38**: 31-32.
Pic M. 1922j: Coléoptères. Malacodermes nouveaux des collections du Muséum. *Bulletin du Muséum d'Histoire Naturelle* (Paris) **28**: 49-53.
Pic M. 1922k: Nouveaux coléoptères exotiques. *Annales de la Société Linnéenne de Lyon* **69**: 73-76.
Pic M. 1922l: Diagnoses de Dasytides asiatiques (Col. Cantharidae). *Bulletin de la Société Entomologique de France* **1922**: 155-156.
Pic M. 1922m: Coléoptères. Malacodermes nouveaux des collections du Muséum. *Bulletin du Muséum d'Histoire Naturelle* (Paris) **28**: 157-161.
Pic M. 1923a: Notes diverses, descriptions et diagnoses (Suite). *L'Échange, Revue Linnéenne* **39**: 1-3.
Pic M. 1923b: Nouveautés diverses. *Mélanges Exotico-Entomologiques* **38**: 1-32.

Pic M. 1923c: Nouveautés diverses. *Mélanges Exotico-Entomologiques* **39**: 3-32.
Pic M. 1923d: Nouveautés diverses. *Mélanges Exotico-Entomologiques* **40**: 3-32.
Pic M. 1923e: Coléoptères exotiques en partie nouveaux (Suite). *L'Échange, Revue Linnéenne* **39**: 4.
Pic M. 1923f: Coléoptères exotiques en partie nouveaux (Suite). *L'Échange, Revue Linnéenne* **39**: 7-8.
Pic M. 1923g: Étude des malacodermes de l'Indochine recueillis par M. R. Vitalis de Salvaza. *Faune Entomologique de l'Indochine Française* (Saigon) **6**: 7-63.
Pic M. 1923h: Notes diverses, descriptions et diagnoses (Suite). *L'Échange, Revue Linnéenne* **39** [hors-texte] (413): 9-10.
Pic M. 1923i: Notes diverses, descriptions et diagnoses (Suite). *L'Échange, Revue Linnéenne* **39**: 9-11.
Pic M. 1923j: Descriptions de deux malacodermes marocains (Insectes Coléoptères). *Bulletin de la Société des Sciences Naturelles du Maroc* **3**: 19.
Pic M. 1923k: Contribution à l'étude des Lycides. *L'Échange, Revue Linnéenne* **39** [hors-texte] (411): 29-40.
Pic M. 1923l: Prionocerus Perty et Idgia Cast. du Muséum National de Paris (Coléoptères, Malacodermes). *Bulletin du Muséum d'Histoire Naturelle* (Paris) **1923**: 72-73.
Pic M. 1923m: Nouveaux coléoptères exotiques. *Annales de la Société Linnéennel de Lyon* (N. S.) **69**: 73-74.
Pic M. 1923n: Notes diverses et corrections sur les malachides d'Afrique (Col.). *Bulletin de la Société Zoologique de France* **48**: 79-83.
Pic M. 1923o: Nouveaux coléoptères malachides, II. *Bulletin du Muséum d'Histoire Naturelle* (Paris) **29**: 372-379.
Pic M. 1924a: Malacodermes exotiques. *L'Échange, Revue Linnéenne*, hors-texte **39** (418): 1-4.
Pic M. 1924b: Notes diverses, descriptions et diagnoses (Suite). *L'Échange, Revue Linnéenne* **39**: 17-18.
Pic M. 1924c: Notes diverses, descriptions et diagnoses (Suite). *L'Échange, Revue Linnéenne* **39**: 21-23.
Pic M. 1924d: Notes diverses, descriptions et diagnoses (Suite). *L'Échange, Revue Linnéenne* **39**: 25-27.
Pic M. 1924e: Coléoptères exotiques en partie nouveaux (Suite). *L'Échange, Revue Linnéenne* **39**: 27-28.
Pic M. 1924f: Notes diverses, descriptions et diagnoses (Suite). *L'Échange, Revue Linnéenne* **39**: 29-30.
Pic M. 1924g: Catalogue analytique et raisonné des coléoptères de Saône-&-Loire et des départements limitrophes. *Bulletin de la Société d'Histoire Naturelle d'Autun* **28**: 51-114.
Pic M. 1924h: Un nouveau Dasytes F. de Sardaigne. *Bollettino della Società Entomologica Italiana* **56**: 80.
Pic M. 1924i: Nouveaux coléoptères asiatiques. *Bulletin de la Société Zoologique de France* **49**: 89-91.
Pic M. 1924j: Nouveaux Malacodermes asiatiques. *Bulletin du Muséum d'Histoire Naturelle* (Paris) **30**: 475-482.
Pic M. 1925a: Notes diverses, descriptions et diagnoses (Suite). *L'Échange, Revue Linnéenne* **41**: 1-3.
Pic M. 1925b Missione zoologica del Dr. E. Festa in Cirenaica. XII. Coléoptères nouveaux de Cyrénaïque. *Bollettino dei Musei di Zoologica et Anatomia Comparata di Torino* (N. S.) **39** (20) [1924]: 1-3.
Pic M. 1925c: Malacodermes exotiques. *L'Échange, Revue Linnéenne* **41** [hors texte] (418-422): 1-20.
Pic M. 1925d: Nouveautés diverses. *Mélanges Exotico-Entomologiques* **43**: 1-32.
Pic M. 1925e: Nouveautés diverses. *Mélanges Exotico-Entomologiques* **44**: 1-32.
Pic M. 1925f: Notes diverses, descriptions et diagnoses (Suite). *L'Échange, Revue Linnéenne* **41**: 5-6, 9-11.
Pic M. 1925g: Notes diverses, descriptions et diagnoses (Suite). *L'Échange, Revue Linnéenne* **41**: 13-14.
Pic M. 1925h: Malacodermes exotiques. *L'Échange, Revue Linnéenne* **41** [hors-texte] (421-422): 13-16, 17-20.
Pic M. 1925i: Nouveaux Malacodermes asiatiques. *Bulletin du Muséum National d'Histoire Naturelle* (Paris) **31**: 72-73.
Pic M. 1926a: Notes diverses, descriptions et diagnoses (Suite). *L'Échange, Revue Linnéenne* **42**: 1-2.
Pic M. 1926b: Entomologische Ergebnisse der Schwedischen Kamtchatka-Expedition 1920-1922. 7. Coleoptera (ex p.). *Arkiv för Zoologi* (B) **18** (3): 1-5.
Pic M. 1926c: Nouveautés diverses. *Mélanges Exotico-Entomologiques* **45**: 1-32.
Pic M. 1926d: Coléoptères exotiques en partie nouveaux. *L'Échange, Revue Linnéenne* **42**: 4.
Pic M. 1926e: Notes diverses, descriptions et diagnoses (Suite.). *L'Échange, Revue Linnéenne* **42**: 5-6.
Pic M. 1926f: Notes divers, descriptions et diagnoses (Suite). *L'Échange, Revue Linnéenne* **42**: 9-10.
Pic M. 1926g: Coléoptères exotiques en partie nouveaux (Suite). *L'Échange, Revue Linnéenne* **42**: 15-16.
Pic M. 1926h: Malacodermes exotiques. *L'Échange, Revue Linnéenne* **42** [hors-texte] (424-426): 21-36.
Pic M. 1926i: H. Sauter's Formosa Ausbeute. Mordellidae, Lycidae, Cantharidae. *Entomologische Mitteilungen* **15**: 67-69.

Pic M. 1926j: Sur le genre Lyponia Wat. (Col. Lycidae). *Bulletin Bi-Mensuel de la Société Linnéenne de Lyon* **5**: 68-70.

Pic M. 1926k: Mutations et descriptions de nouveaux coléoptères asiatiques. *Bulletin du Muséum National d'Histoire Naturelle* (Paris) **32**: 76-80.

Pic M. 1926l: Nouveaux coléoptères exotiques. *Bulletin de la Société Entomologique de France* **1926**: 143.

Pic M. 1926m: Sept coléoptères exotiques nouveaux. *Bulletin de la Société Entomologique de France* **1926**: 153-155.

Pic M. 1926n: Hetéromères et Malacodermes nouveaux. *Bulletin du Muséum d'Histoire Naturelle* (Paris) **32**: 211-212.

Pic M. 1926o: Coléoptères nouveaux. *Bulletin de la Société Royale Entomologique d'Egypte* **18** [1925]: 227-231.

Pic M. 1926p: Nouveaux coléoptères du globe. *Bulletin de la Société Zoologique de France* **51**: 235-238.

Pic M. 1926: Coléoptères asiatiques nouveaux. *Bulletin de la Société Entomologique de France* **1925**: 301-303.

Pic M. 1926r: Nouveaux coléoptères exotiques. *Bulletin du Muséum d'Histoire Naturelle* (Paris) **32**: 354-359.

Pic M. 1926s: Remarques critiques, synonymies et diagnoses (Col.). *Bulletin de la Société Zoologique de France* **51**: 451-455.

Pic M. 1927a: Notes diverses, descriptions et diagnoses (Suite). *L'Échange, Revue Linnéenne* **43**: 1-2.

Pic M. 1927b: Descriptions de deux coléoptères d'Egypte. *Bulletin de la Société Royale Entomologique d'Égypte* **1927**: 1-2.

Pic M. 1927c: Nouveautés diverses. *Mélanges Exotico-Entomologiques* **48**: 1-32.

Pic M. 1927d: Coléoptères de l'Indochine. *Mélanges Exotico-Entomologiques* **49**: 1-36.

Pic M. 1927e: Coléoptères du globe. *Mélanges Exotico-Entomologiques* **50**: 1-36.

Pic M. 1927f: Notes diverses, descriptions et diagnoses (Suite). *L'Échange, Revue Linnéenne* **43**: 5-7.

Pic M. 1927g: Notes diverses, descriptions et diagnoses (Suite). *L'Échange, Revue Linnéenne* **43**: 9-11.

Pic M. 1927h: Notes diverses, descriptions et diagnoses (Suite). *L'Échange, Revue Linnéenne* **43**: 13-14.

Pic M. 1927i: Malacodermes exotiques. *L'Échange, Revue Linnéenne* **43** [hors texte] (427-430): 37-52.

Pic M. 1927j: Fauna sumatrensis. (Beitrag Nr. 34). Byrrhidae, Dermestidae, Anobiidae (Coleoptera). *Supplementa Entomologica* **15**: 65-66.

Pic M. 1927k: Nouveaux coléoptères du globe II. *Bulletin de la Société Zoologique de France* **52**: 106-110.

Pic M. 1927l: Coléoptères asiatiques nouveaux. *Bulletin Bi-Mensuel de la Société Linnéenne de Lyon* **6**: 132-133.

Pic M. 1927m: Nouveaux coléoptères exotiques. *Bulletin de la Société Entomologique de France* **1927**: 139-140.

Pic M. 1927n: Coléoptères nouveaux de Chine et du Japon. *Bulletin de la Société Entomologique de France* **1927**: 152-154.

Pic M. 1927o: Sur quelques coléoptères nouveaux ou récemment décrits. *Bulletin de la Société Entomologique de France* **1927**: 167-168.

Pic M. 1927p: Nouveaux coléoptères exotiques. *Entomologische Mitteilungen* **16**: 246-255.

Pic M. 1927q: Travaux scientifiques de l'Armée d'Orient (1916-1918). Coléoptères malacodermata. *Bulletin du Muséum d'Histoire Naturelle* (Paris) **33**: 371-375.

Pic M. 1928a: Notes diverses, descriptions et diagnoses (Suite). *L'Échange, Revue Linnéenne* **44**: 1-2.

Pic M. 1928b: Notes et descriptions. *Mélanges Exotico-Entomologiques* **51**: 1-36.

Pic M. 1928c: Nouveautés diverses. *Mélanges Exotico-Entomologiques* **52**: 1-36.

Pic M. 1928d: Étude synoptique du genre Pristolycus Gorh. (Col.). *Entomologisches Nachrichtenblatt* (Troppau) **2**: 3-4.

Pic M. 1928e: Notes diverses, descriptions et diagnoses (Suite). *L'Échange, Revue Linnéenne* **44**: 5-6, 9-10.

Pic M. 1928f: Deuxième liste de coléoptères de la Somalie italienne avec description de nouvelles espèces. Missione Entomologica Paoli in Somalia - 1926. *Memoire della Società Entomologica Italiana* **6** [1927]: 37-43.

Pic M. 1928g: Malacodermes exotiques. *L'Échange, Revue Linnéenne* **44** [hors texte] (431-436): 53-68.

Pic M. 1928h: Nové druhy Koleopter z Brasilie. *Sborník Entomologického Oddělení Národního Musea v Praze* **6**: 74-76.

Pic M. 1928i: Coléoptères nouveaux de la Cyrénaique. *Bollettino della Società Entomologica Italiana* **60**: 103-105.

Pic M. 1928j: Communications entomologiques. *Bulletin de la Société Zoologique de France* **53**: 265-268.
Pic M. 1928k: Nouveaux coléoptères surtout tonkinois. *Bulletin de la Société Zoologique de France* **1928**: 377-379.
Pic M. 1929a: Notes diverses, nouveautés. *L'Échange, Revue Linnéenne* **45**: 1-2.
Pic M. 1929b: Nouveautés diverses. *Mélanges Exotico-Entomologiques* **53**: 1-36.
Pic M. 1929c: Nouveautés diverses. *Mélanges Exotico-Entomologiques* **54**: 1-36.
Pic M. 1929d: Notes diverses, nouveautés. *L'Échange, Revue Linnéenne* **45**: 5-7.
Pic M. 1929e: Coléoptères exotiques en partie nouveaux (Suite). *L'Échange, Revue Linnéenne* **45**: 7-8.
Pic M. 1929f: Notes diverses, nouveautés. *L'Échange, Revue Linnéenne* **45**: 9-10.
Pic M. 1929g: Notes diverses, nouveautés. *L'Échange, Revue Linnéenne* **45**: 13-14.
Pic M. 1929h: Notes sur divers Endomychides. *Revue Scientifique du Bourbonnais et du Centre de la France* **1929**: 32-36.
Pic M. 1929i: Malacodermes exotiques (Suite). *L'Échange* **65** [hors-texte] (437-438): 69-76.
Pic M. 1929j: Coléoptères nouveaux de la Cyrénaique. *Bollettino della Società Entomologica Italiana* **61**: 94-96.
Pic M. 1929k: Coléoptères d'Égypte et Sinaï. *Bulletin de la Société Royale Entomologique d'Égypte* **1929**: 108-110.
Pic M. 1929l: Sur le genre Triblius Fairm. (Col. Malacodermata). *Bulletin de la Société Zoologique de France* **1929**: 195-196.
Pic M. 1930a: Deux nouveaux coléoptères d'Egypte et Sinaï. *Bulletin de la Société Royale Entomologique d'Égypte*, **1930**: 1-2.
Pic M. 1930b: Notes diverses, nouveautés (Suite). *L'Échange, Revue Linnéenne* **46**: 1-3.
Pic M. 1930c: Nouveautés diverses. *Mélanges Exotico-Entomologiques* **55**: 1-36.
Pic M. 1930d: Notes diverses, nouveautés (Suite). *L'Échange, Revue Linnéenne* **46**: 5-7.
Pic M. 1930e: Notes diverses, nouveautés (Suite). *L'Échange, Revue Linnéenne* **46**: 9-11.
Pic M. 1930f: Notes diverses, nouveautés. *L'Échange, Revue Linnéenne* **46**: 13-14.
Pic M. 1930g: Nouveautés diverses. *Mélanges Exotique-Entomologique* **56**: 1-36.
Pic M. 1930h: Sur quelques coléoptères recueillis en Egypte. *Bulletin de la Société Royale Entomologique d'Egypte* **14**: 61-64.
Pic M. 1930i: Malacodermes exotiques (Suite). *L'Échange* **66** [hors-texte] (439): 77-80.
Pic M. 1931a: Nouveautés diverses. *Mélanges Exotico-Entomologiques* **57**: 1-36.
Pic M. 1931b: Nouveautés diverses. *Mélanges Exotico-Entomologiques* **58**: 1-36.
Pic M. 1931c: Notes diverses, nouveautés (Suite). *L'Échange, Revue Linnéenne* **47**: 5-6.
Pic M. 1931d: Notes diverses, nouveautés (Suite). *L'Échange, Revue Linnéenne* **47**: 9-10.
Pic M. 1931e: Notes diverses, nouveautés (Suite). *L'Échange, Revue Linnéenne* **47**: 13-14.
Pic M. 1931f: Coléoptères nouveaux du Maroc. *Bulletin de la Société Entomologique de France* **1931**: 60-61.
Pic M. 1931g: Neue exotische Coleopteren (Malacodermata). *Entomologischer Anzeiger* **11**: 77-78.
Pic M. 1931h: Malacodermes exotiques. *L'Échange, Revue Linnéenne* **47** [hors-texte] (443-445): 93-96, 97-100, 101-104.
Pic M. 1931i: Coléoptères intéressants ou nouveaux d'Egypte et du Sinaï. *Bulletin de la Société Royale Entomologique d'Egypte* **15**: 134-136.
Pic M. 1931j: Nouveaux coléoptères d'Italie. *Bollettino della Società Entomologica Italiana* **63**: 159.
Pic M. 1932a: Diagnoses préliminaires. *Mélanges Exotico-Entomologiques* **59**: 1-9.
Pic M. 1932b: Nouveautes diverses. *Mélanges Exotico-Entomologiques* **60**: 1-36.
Pic M. 1932c: [new taxa]. In: Pic M. & Lindberg Håk. 1932: Inventa entomologica, itineris Hispanici et Maroccani, quod a. 1926 fecerunt Harald et Håkan Lindberg. XII. Anobiidae, Cleridae, Malacodermata, Heteromera (ex parte). *Commentationes Biologicae* (Helsingfors) **3** (18): 1-37.
Pic M. 1932d: Notes diverses, nouveautés (Suite). *L'Échange, Revue Linnéenne* **48**: 17-18.
Pic M. 1932e: Notes diverses, nouveautés (Suite). *L'Échange, Revue Linnéenne* **48**: 21-23.
Pic M. 1932f: Notes diverses, nouveautés (Suite). *L'Échange, Revue Linnéenne* **48**(449): 25-26.
Pic M. 1932g: Diagnoses préliminaires. *L'Échange, Revue Linnéenne* **48**: 26-28.
Pic M. 1932h: Notes diverses, nouveautés (Suite). *L'Échange, Revue Linnéenne* **48**: 29-31.
Pic M. 1932i: Quatre nouveaux coléoptères de Chine et du Tonkin. *Bulletin de la Société Entomologique de France* **37**: 206-208.

Pic M. 1932j: Neue exotische Coleopteren (Malacodermata) II. *Entomologischer Anzeiger* **12**: 250-255.
Pic M. 1933a: Notes diverses, nouveautés (Suite). *L'Échange, Revue Linnéenne* **49**: 1-2.
Pic M. 1933b: Schwedisch-chinesische wissenschaftliche Expedition nach den nordwestlichen Provinzien Chinas unter Leitung von Dr. Sven Hedin und Prof. Sü Ping-chang. Insekten gesammelt vom schwedischen Arzt der Expedition Dr. David Hummel 1927-1930. 16. Coleoptera. 2. Helmidae, Dermestidae, Anobiidae, Cleridae, Malacodermata, Dascillidae, Heteromera (ex p.), Bruchidae, Cerambycidae, Phytophaga (ex p.). *Arkiv för Zoologi* **27 A** (2): 1-14.
Pic M. 1933c: Nouveautés diverses. *Mélanges Exotico-Entomologiques* **62**: 1-36.
Pic M. 1933d: Notes diverses, nouveautés (Suite) *L'Échange, Revue Linnéenne* **49**: 5-6.
Pic M. 1933e: Notes diverses, nouveautés (Suite). *L'Échange, Revue Linnéenne* **49**: 9-11.
Pic M. 1933f: Notes diverses, nouveautés (Suite). *L'Échange, Revue Linnéenne* **49**: 13-15.
Pic M. 1933g: Neue Coleopteren-Clavicornia. *Entomologisches Nachrichtenblatt* (Troppau) **7**: 71-72.
Pic M. 1933h: Malachides nouveaux du Maroc. *Bulletin de la Société des Sciences Naturelles du Maroc* **13**: 102-103.
Pic M. 1933i: Malacodermes exotiques. *L'Échange, Revue Linnéenne* **49** [hors-texte] (451-454): 109-112, 113-116, 117-120.
Pic M. 1934a: Notes diverses, nouveautés (Suite). *L'Échange, Revue Linnéenne* **50**: 21-22.
Pic M. 1934b: Notes diverses, nouveautés (Suite). *L'Échange, Revue Linnéenne* **50**: 25-27.
Pic M. 1934c: Notes diverses, nouveautés (Suite). *L'Échange, Revue Linnéenne* **50**: 29-31.
Pic M. 1934d: Neue exotische Coleopteren (Malacodermata). III. *Entomologischer Anzeiger* **14**: 45-47, 54-55.
Pic M. 1934e: Notes sur divers coléoptères. *Revue Française d'Entomologie* **1**: 132-135.
Pic M. 1934f: [new taxa]. *L'Échange, Revue Linnéenne* **50** [hors text] (456-458): 129, 133.
Pic M. 1934g: Nouveaux coléoptères de la Cyrénaïque. *Atti del Museo Civico di Storia Naturale Trieste* **12**: 147-150.
Pic M. 1934h: [new taxa]. In: Wittmer W.: Resultats scientifiques des expéditions entomologiques de S. A. le Prince Alexandre C. Della Torre e Tasso en Egypte et au Sinaï. VI. Malacodermata (Coleoptera). *Bulletin de la Société Royal Entomologique d'Egypte* **18**: 449-455.
Pic M. 1935a: Nouveautés diverses. *Mélanges Exotico-Entomologique* **66**: 1-36.
Pic M. 1935b: Notes diverses, nouveautés. (Suite.). *L'Échange, Revue Linnéenne* **51**: 3-4.
Pic M. 1935c: Notes diverses, nouveautés. (Suite.). *L'Échange, Revue Linnéenne* **51**: 5-8.
Pic M. 1935d: Notes diverses, nouveautés (Suite.). *L'Échange, Revue Linnéenne* **51**: 11-12.
Pic M. 1935e: Nouveaux coléoptères de l'île Maurice. *Bulletin de la Société Entomologique de France* **40**: 12-14.
Pic M. 1935f: Notes diverses, nouveautés (Suite.). *L'Échange, Revue Linnéenne* **51**: 13-15.
Pic M. 1935g: Contribution à l'étude des coléoptères malacodermes. *Annales de la Société Entomologique de France* **104**: 109-115.
Pic M. 1935h: Nouveaux coléoptères de la Cyrénaique. *Atti del Museo Civico di Storia Naturale di Trieste* **12**: 147-150.
Pic M. 1935i: Description de coléoptères paléarctiques. *Bulletin de la Société Entomologique de France* **40**: 158-160.
Pic M. 1935j: [new taxa]. In: Normand H.: Contribution au Catalogue des coléoptères de Tunisie. *Bulletin de la Société d'Histoire Naturelle de l'Afrique du Nord* **26**: 235-251.
Pic M. 1935k: Neue Malacodermata (Coleopteren). *Entomologischer Anzeiger* **15**: 253-256.
Pic M. 1935l: Mission scientifique de l'Omo. Coleoptera. VII. Lampyridae, Cantharidae, Melyridae, Geneceridae et Helodidae. *Mémoires du Muséum National d'Hisoire Naturelle* (Paris) **2**: 291-302.
Pic M. 1936a: Neue Coleoptera (Malacodermata). *Entomologische Anzeiger* **16**: 1-3.
Pic M. 1936b: Nouveaux coléoptères paléarctiques. *L'Échange, Revue Linnéenne* **51** [hors-texte] (463): 1-4.
Pic M. 1936c: Nouveaux coléoptères de Chine. *Notes d'Entomologie Chinoises, Musée Heude* **3**: 15-17.
Pic M. 1936d: Notes diverses, nouveautés (Suite). *L'Échange, Revue Linnéenne* **52**: 25-27.
Pic M. 1936e: Quelques coléoptères nouveaux de la presqu'île malaise (III.). *Journal of the Federated Malay States Museum* **18**: 54-59.
Pic M. 1936f: Deux nouveaux coléoptères néotropiques. (Coleoptera: Cantharidae et Ptinidae). *Arbeiten über Morphologische und Taxonomische Entomologie aus Berlin-Dahlen* **3**: 131.
Pic M. 1936g: Nouveaux coléoptères du Congo. *Revue de Zoologie et de Botanique Africaines* **28**: 419-420.

Pic M. 1937a: Dasytidae II. Dasytinae (Danacaeini, Henicopini, Dasytini). Pars 155. In: Schenkling S. (ed.): *Coleopterorum Catalogus. Vol. 10.* Berlin: W. Junk, 130 pp.
Pic M. 1937b: Notes diverses, nouveautés. *L'Échange, Revue Linnéenne* **53**: 1.
Pic M. 1937c: Nouveautés diverses. *Mélanges Exotico-Entomologiques* **67**: 1-36.
Pic M. 1937d: Nouveautés diverses. *Mélanges Exotico-Entomologiques* **69**: 1-36.
Pic M. 1937e: Coléoptères exotiques en partie nouveaux (Suite). *L'Échange, Revue Linnéenne* **53**: 4.
Pic M. 1937f: Notes diverses, nouveautés (Suite). *L'Échange, Revue Linnéenne* **53**: 5-6.
Pic M. 1937g: Notes diverses, nouveautés (Suite). *L'Échange, Revue Linnéenne* **53**: 9-11.
Pic M. 1937h: Nova Coleoptera Malacodermata. *Entomologische Blätter* **33**: 58-65.
Pic M. 1937i: Cis Latr. et Anobiides nouveaux des Indes (Coléoptères). *Indian Forest Records* **3**: 123-126.
Pic M. 1937j: Malacodermes exotiques. *L'Échange, Revue Linnéenne* **53** [hors texte] (466-470): 137-148.
Pic M. 1937k: Coléoptères nouveaux de Chine. *Notes d'Entomologie Chinoise* **4**: 169-176.
Pic M. 1937l: Contribution à l'étude du genre Laemoglyptus Fairm. (Coléoptères Malacodermes). *Bulletin de la Société Zoologique de France* **62**: 187-190.
Pic M. 1938a: Notes diverses, nouveautés (Suite). *L'Échange, Revue Linnéenne* **54**: 1-3.
Pic M. 1938b: [new taxa]. In: Sjöstedt Y.: Insekten aus China im Naturhistorischen Reichsmuseum zu Stockholm. Heimgebracht von Direktor Kjell Kolthoff und anderen Schwedischen Forschern und Reisenden. *Arkiv för Zoologi* **30 A** (13): 1-19.
Pic M. 1938c: Nouveautés diverses, mutations. *Mélanges Exotico-Entomologiques* **70**: 1-36.
Pic M. 1938d: Notes diverses, nouveautés (Suite). *L'Échange, Revue Linnéenne* **54**: 5-6.
Pic M. 1938e: Notes diverses, nouveautés (Suite). *L'Échange, Revue Linnéenne* **54**: 9-11.
Pic M. 1938f: Coléoptères exotiques en partie nouveaux (Suite). *L'Échange, Revue Linnéenne* **54**: 12.
Pic M. 1938g: Notes diverses, nouveautés (Suite). *L'Échange, Revue Linnéenne* **54**: 13-14.
Pic M. 1938h: Mutations et coléoptères nouveaux. *L'Échange, Revue Linnéenne* **54**: 14-16.
Pic M. 1938i: Nouveaux coléoptères d'Afrique et mutations. *Bulletin de la Société Zoologique de France* **63**: 82-85.
Pic M. 1938j: Descriptions des coléoptères et notes synonymiques. *Bulletin de la Société Entomologique de France* **43**: 121-124.
Pic M. 1938k: Malacodermes exotiques. *L'Échange, Revue Linnéenne* **54** [hors-texte] (472-474): 149-156, 157-160, 161-164.
Pic M. 1938l: Mission Robert Ph. Dollfus en Égypte. XIV Coleoptera II. Description d'un Colotes. *Mémoires de l'Institut d'Égypte* **37**: 251-252.
Pic M. 1938m: Divers coléoptères nouveaux de la Presqu'île Malaise (IV). *Journal of the Federal Malaysian State Museum* **18**: 279-286.
Pic M. 1939a: Mutations et nouveautes diverses. *Mélanges Exotico-Entomologiques* **71**: 1-36.
Pic M. 1939b: Notes diverses, nouveautés. *L'Échange, Revue Linnéenne* **55**: 17-19.
Pic M. 1939c: Nouveaux coléoptères de l'Indochine. *Bulletin de la Société Entomologique de France* **43**: 17-19.
Pic M. 1939d: Notes diverses, nouveautés. *L'Échange, Revue Linnéenne* **55**: 21-22.
Pic M. 1939e: Notes diverses, nouveautés. *L'Échange, Revue Linnéenne* **55**: 25-26.
Pic M. 1939f: Deux nouveaux coléoptères de Chine. *Notes d'Entomologie Chinoise* **6**: 135-136.
Pic M. 1939g: Coléoptères nouveaux d'Egypte et du Sinaï. *Bulletin de la Société Fouad Ier d'Entomologie* **23**: 143-149.
Pic M. 1939h: Malacodermes exotiques. *L'Échange, Revue Linnéenne* **55** [hors texte] (475-477): 165-172.
Pic M. 1939i: Coléoptères asiatiques nouveaux. *Bulletin de la Société Entomologique de France* **43** [1938]: 220-221.
Pic M. 1940a: Tenthredides [Hymén.] et coléoptères nouveaux. *L'Échange, Revue Linnéenne* **56**: 1-2.
Pic M. 1940b: Opuscula martialia. *L'Échange, Revue Linnéenne, Numéro spécial* **1**: 1-16.
Pic M. 1940c: Diagnoses de coléoptères exotiques. *L'Échange, Revue Linnéenne* **56**: 6-8.
Pic M. 1940d: Diagnoses de coléoptères exotiques. *L'Échange, Revue Linnéenne* **56**: 10-12.
Pic M. 1941a: Opuscula martialia. *L'Échange, Revue Linnéenne, Numéro spécial* **2**: 1-16.
Pic M. 1941b: Opuscula martialia. *L'Échange, Revue Linnéenne, Numéro spécial* **4**: 1-16.
Pic M. 1941c: Opuscula martialia. *L'Échange, Revue Linnéenne, Numéro spécial* **5**: 1-16.
Pic M. 1942a: Opuscula martialia. *L'Échange, Revue Linnéenne, Numéro spécial* **6**: 1-16.
Pic M. 1942b: Opuscula martialia. *L'Échange, Revue Linnéenne, Numéro spécial* **7**: 1-16.

Pic M. 1942c: Opuscula martialia. *L'Échange, Revue Linnéenne, Numéro Spécial* **8**: 1-16.
Pic M. 1942d: Coléoptères nouveaux de l'Adrar des Iforas. *Bulletin de la Société Entomologique de France* **47**: 77-79.
Pic M. 1943a: Opuscula martialia. *L'Échange, Revue Linnéenne, Numéro Spécial* **9**: 1-16.
Pic. M. 1943b: Coléoptères du globe. *L'Échange, Revue Linnéenne* **59**: 9-12.
Pic M. 1944a: Opuscula martialia. *L'Échange, Revue Linnéenne, Numéro Spécial* **12**: 1-16.
Pic M. 1944b: Opuscula martialia. *L'Échange, Revue Linnéenne, Numéro Spécial* **13**: 1-16.
Pic M. 1944c: Coléoptères du globe (suite). *L'Échange, Revue Linnéenne* **60**: 2-4, 5-8, 10-12.
Pic M. 1945: Coléoptères du globe (suite) *L'Échange, Revue Linnéenne* **61**: 10-12, 13-16.
Pic M. 1946: Coléoptères du globe (suite). *L'Échange, Revue Linnéenne* **62**: 5-8, 9-12.
Pic M. 1947: Coléoptères du globe (suite). *L'Échange, Revue Linnéenne* **63**: 5-8.
Pic M. 1948a: Coléoptères du globe (suite). *L'Échange, Revue Linnéenne* **64**: 1-4, 5-8, 13-16.
Pic M. 1948b: Captures et coléoptères nouveaux. *Diversités Entomologiques* **3**: 12-16.
Pic M. 1948c: Notes et nouveautés diverses. *Diversités Entomologiques* **4**: 12-16.
Pic M. 1949a: Coléoptères du globe. *L'Échange, Revue Linnéenne* **65**: 5-8, 13-16.
Pic M. 1949b: Nouveaux coléoptères du Maroc. *Bulletin de la Société des Sciences Naturelles du Maroc* **29**: 73-75.
Pic M. 1949c: Anobiidae de la mission de l'Omo. *Revue Française d'Entomologie* **16**: 87-89.
Pic M. 1949d: Nouveaux coléoptères de Madagascar. *Mémoires de l'Institut Scientifique de Madagascar* (A) **3**: 341-345.
Pic M. 1950ac: Nouveautés diverses, mutation. *Diversités Entomologiques* **6**: 15-16.
Pic M. 1950b: Description et notes variées. *Diversités Entomologiques* **7**: 1-16.
Pic M. 1950c: Coléoptères du globe. (suite). *L'Échange, Revue Linnéenne* **66**: 9-12.
Pic M. 1950d: Deux nouveaux coléoptères d'Egypte et Nubie. *Bulletin de la Société Fouad 1er d'Entomologie* **34**: 23-24.
Pic M. 1950e: Cinq nouveaux coléoptères du Maroc. *Comptes Rendus de la Société des Sciences Naturelles du Maroc* **30**: 92-94.
Pic M. 1950f: Coléoptères Cleridae, Anobiidae et Dermestidae. *Mémoires de l'Institut Français d'Afrique Noire* **10**: 180-182.
Pic M. 1951a: Descriptions diverses et notes. *Diversités Entomologiques* **8**: 7-16.
Pic M. 1951b: Notes et descriptions. *Diversites Entomologiques* **9**: 4-16.
Pic M. 1951c: Coléoptères du globe (suite). *L'Échange, Revue Linnéenne* **67**: 5-8, 13-16.
Pic M. 1952: Coléoptères du globe (suite). *L'Échange, Revue Linnéenne* **68**: 1-4.
Pic M. 1952b: Descriptions diverses, notes. *Diversités Entomologiques* **10**: 9-16.
Pic M. 1952c: Missione biologica Sagon-Omo diretta dal prof. Edoardo Zavattarii. Coléoptères nouveaux. *Rivista di Biologia Coloniale* **11**: 25-36.
Pic M. 1952d: Nouveaux coléoptères du Maroc. *Comptes Rendus de la Société des Sciences Naturalles du Maroc* **18**: 65-66.
Pic M. 1953a: Coléoptères du globe. *L'Échange, Revue Linnéenne* **69**: 2-4, 5-8, 9-12, 13-16.
Pic M. 1953b: Notes et descriptions. *Diversités Entomologiques* **12**: 5-16.
Pic M. 1953c: Quatre nouveaux coléoptères du Maroc. *Comtes Rendus de la Société des Sciences Naturalles du Maroc* **19**: 18-20.
Pic M. 1953d: Coléoptères nouveaux de la région marocaine. *Comptes Rendus de la Société des Sciences Naturelles du Maroc* **19**: 61-63.
Pic M. 1953e: Nouveaux Malacodermes du Maroc (Coléoptères). *Comptes Rendus de la Société des Sciences Naturelles du Maroc* **19**: 117.
Pic M. 1953f: Trois coléoptères nouveaux d'Espagne. *Archiv del Instituto Aclimatación* (Almería) **1**: 143-144.
Pic M. 1953g: Neue Cleriden aus der Sammlung des Deutschen Entomologischen Instituts (Coleoptera). *Beiträge zur Entomologie* **3**: 320-322.
Pic M. 1953h: Contribution à l'étude de la zone d'inondation du Niger (mission G. Remaudière, 1950). II. Coléoptères malacodermoidea. *Bulletin de l'Institut Français d'Afrique Noire* **15**: 1529-1534.
Pic M. 1954a: Coléoptères du globe (Suite). *L'Échange, Revue Linnéenne* **70**: 1-3.
Pic M. 1954b: Descriptions et notes. *Diversités Entomologiques* **13**: 4-12.
Pic M. 1954c: Coléoptères du globe (Suite). *L'Échange, Revue Linnéenne* **70**: 9-12, 13-16.

Pic M. 1954d: Coléoptères nouveaux de Chine. *Bulletin de la Société Entomologique Mulhouse* **1954**: 53-59.

Pic M. 1954e: Coléoptères nouveaux de Chine (Suite). *Bulletin de la Société Entomologique Mulhouse* **1954**: 61-64.

Pic M. 1954f: [new taxa]. In: Mateu J.: Coleópteros de Sierra Nevada. Familias: Dytiscidae, Hydrophilidae, Dryopidae, Catopidae, Silphidae, Pselaphidae, Histeridae, Cleridae, Cantharidae, Dasytidae, Aderidae, Anthicidae. *Archivos del Instituto de Aclimatación* (Almería) **2**: 89-101, 2 pls.

Pic M. 1955a: Coléoptères du globe (Suite). *L'Échange, Revue Linnéenne* **71**: 7-11.

Pic M. 1955b: Descriptions diverses. *Diversites Entomologiques* **14**: 7-20.

Pic M. 1955c: Coléoptères nouveaux de Chine. *Bulletin de la Société Entomologique de Mulhouse* **1955**: 21-23, 25-26, 29-32.

Pic M. 1955d: Coléoptère lampyride nouveau du Karakorum. *Atti del Museo Civico di Storia Naturale di Trieste* **20**: 47-48.

Pic M. 1955e: Coléoptères malacodermes nouveaux de la Cyrénaïque. *Atti del Museo Civico di Storia Naturale di Trieste* **20**: 131-132.

Pic M. 1956: Anobiidae, Ptinidae et Aderidae (Coleoptera) de l'expédition J. Klapperich en Afghanistan (1952-1953). *Bulletin de la Société Entomologique Mulhouse* **1956**: 5-7.

Pic M. & Pic T. 1901: Uebersicht der Arten der Coleopteren-Gattung Hedobia Latr. aus der palaearktischen Fauna. *Wiener Entomologische Zeitung* **20**: 169-174.

Piel de Churcheville H. & Piel de Churcheville T. 1900: Description d'une nouvelle variété de Coccinelle. *Miscellanea Entomologica* **8**: 26.

Piller M. & Mitterpacher L. 1783: *Iter per Posegnam Sclavoniae provinciam mensibus Junio, et Julio Anno MDCCXXXII suspectum*. Budae: Typis regiae universitatis, 147 pp., 16 pls.

Plata Negrache P. 1987a: Descripción de un nuevo Attalus del grupo tuberculatus Woll., de las Islas Canarias Attalus eversi n. sp. *Entomologische Blätter* **83**: 157-159.

Plata Negrache P. 1987b: [new taxon]. In: Plata Negrache P. & Evers A. M. J.: Revision del género Ifnidius, Escalera con descriptión de una especie nueva Ifnidius petricola n. sp. de las Islas Canarias. *Entomologische Blätter* **83**: 160-165.

Plata Negrache P. 1990a: Notas sobre Maláquidos canarios. I. (Coleoptera: Malachiidae). *Vieraea* **18**: 49-55.

Plata Negrache P. 1990b: Notas sobre Maláquidos canarios. II. (Coleoptera: Malachiidae). *Vieraea* **18**: 57-64.

Plata Negrache P. & Prendes-Ayala C. 1980: Amphotis tenorioi nov. sp. de Nitidulidae (Coleoptera) de las Islas Canarias. *Eos, Revista Española de Entomología* **55-56** [1979-1980]: 191-193.

Plata Negrache P. & Prendes-Ayala C. 1981: Los Ostomidae (Col.) del Archipiélago Canario. *Boletín de la Asociación Española de Entomología* **4** [1980]: 225-234.

Plata Negrache P. & Santiago Hernández C. T. 1987: *Revision de la familia Malachiidae Erichson* (*Insecta: Coleoptera*) *en Marruecos*. Krefeld: Goecke & Evers, 842 pp.

Platia G. 1983: Descrizione di un nuovo Anostirus Thomson dei Monti della Laga (Appennino Centrale) (Coleoptera, Elateridae). (I Contributo alla conoscenza degli elateridi). *Bollettino del Museo Civico di Storia Naturale di Verona* **9** [1982]: 143-150.

Platia G. 1984ab: Descrizione di una nuova specie di Athous dell'Isola d'Elba (IV contributo alla conoscenza dei Coleoptera Elateridae). *Doriana* **6** (252): 1-4.

Platia G. 1984ba: Descrizione di una nuova specie di Athous della Sardegna e considerazioni su Athous dorgaliensis Buysson (III contributo alla conoscenza dei Coleoptera Elateridae). *Giornale Italiano di Entomologia* **2**: 119-122.

Platia G. 1985a: Descrizione di una nuova specie di Athous Eschz. (Col. Elateridae) dell'Albania. *Doriana* **6** (255): 1-4.

Platia G. 1985b: Descrizione di due nuovi specie di elateridi paleartici (Coleoptera) (V contributo alla conoscenza dei Coleoptera Elateridae). *Bollettino della Società Entomologica Italiano* **117**: 93-96.

Platia G. 1986: Due nuovi generi e cinque nuove specie della sottofamiglia Melanotinae Candeze della fauna indomalese (Col. Elateridae). *Acta Coleopterologica* **2**: 1-8.

Platia G. 1988a: Reperti di elateridi (Coleoptera) dal Nord Pakistan con descrizione di due nuove specie. I contributo. *Giornale Italiano di Entomologia* **4**: 5-10.

Platia G. 1988b: Descrizione di due nuove specie di elateridi dall'Italia (Coleoptera). *Bollettino della Società Entomologica Italiano* **119**: 159-162.

Platia G. 1988c: Tre nuovi specie di Athous Eschscholtz di Italia e Jugoslavia (Coleoptera, Elateridae). *Fragmenta Entomologica* **20**: 303-307.

Platia G. 1989: Descrizione di nuove specie di elateridi della Grecia e della Turchia (Coleoptera Elateridae). *Lavori - Società Veneziana di Scienze Naturali* **14**: 79-86.

Platia G. 2001a: Description de trois nouvelles espèces du genre Ampedus Dejean de la région paléarctique, avec une synonymie nouvelle. (Coleoptera: Elateridae). *Biocosme Mésogéen* **18**: 65-73.

Platia G. 2001b: Descrizione di nuove specie di Athous Eschscholtz dell'Europa sud orientale con note sistematiche e geonemiche (Coleoptera, Elateridae). *Lambillionea* **101**: 522-527.

Platia G. 2003a: Descrizione di otto specie nuove di Agriotes del gruppo nuceus Fairmaire della Turchia con chiave di determinatione alle specie nota (Insecta, Coleoptera, Elateridae). *Il Naturalista Valtellinese - Atti del Museo Civico di Storia Naturale di Morbegno* **14**: 13-28.

Platia G. 2003b: Contribotu alla conscenza delle specie di Athous (Neonomopleus) Schenkling, 1927 della Penisola Iberica (Coleoptera: Elateridae: Dendrometrinae: Dendrometrini). *Boletín de la Sociedad Entomologica Aragonesa* **33**: 33-39.

Platia G. 2003c: Descrizione di nuove species di Athous Eschsholtz [sic] del sottogenere Orthathous Reitter della Turchia (Coleoptera, Elateridae). *Lambillionea* **103**: 263-268.

Platia G. 2004a: Nuovi elateridi d'Europa, Turchia e Qatar con note tassonomiche e geonomiche su alcune specie (Coleoptera: Elateridae). *Boletín de la Sociedad Entomológica Aragonesa* **35**: 39-45.

Platia G. 2004b: Description de deux nouvelles espèces du genre Peripontius Gurjeva de l'Asie Mineure avec un tableau de determination des especes bicolores de la region. (Coleoptera: Elateridae: Synaptini). *Biocosme Mésogéen* **20** [2003]: 85-92.

Platia G. 2004c: Descrizione di nuove elateridi della regione palearctica con note geonemiche e sinonimiche. *Lambillionea* **104**: 98-108.

Platia G. 2004d: Descrizione di un nuovo genere di elateridi della fauna indomalese con chiave alle specie note (Coleoptera, Elateridae, Elaterini). *Lambillionea* **104**: 701-708.

Platia G. & Bartolozzi L. 1988: Contributo alla conoscenza degli elateridi della Calabria (Coleoptera, Elateridae). *Redia* **71**: 61-97.

Platia G. & Borges P. A. V. 2002: Description of a new species of Athous and record of the female of A. azoricus Platia & Gudenzi from the Azores (Coleoptera: Elateridae). *Elytron* **16**: 91-95.

Platia G. & Cate P. C. 1990: Note sistematiche e sinonimiche su elateridi Paleartici (Coleoptera Elateridae). *Bollettino della Società Entomologica Italiana* **122**: 111-114.

Platia G. & Cate P. C. 2001: Description of a new species of the C. syriacus group of Cardiophorus Eschscholtz, 1829 (Coleoptera: Elateridae: Cardiophorinae). *Zeitschrift der Arbeitsgemeinschaft der Österreichischen Entomologen* **53**: 117-120.

Platia G., Furlan L. & Gudenzi I. 2002: Descrizione di sei nuove specie di elateridi dell'Iran (Insecta, Coleoptera, Elateridae). *Il Naturalista Valtellinese - Atti del Museo Civico di Storia Naturale di Morbegno* **13**: 65-77.

Platia G. & Gudenzi I. 1985: Descrizione di un nuovo Cidnopus Thomson (Col., Elateridae) della fauna europea. *Bollettino del Museo Civico di Storia Naturale di Verona* **10** [1983]: 21-29.

Platia G. & Gudenzi I. 1996a: Descrizione di tre nuove specie di elateridi di Grecia, Turchia ed Isola di Cipro con note geonemiche e sinonimiche (Coleoptera, Elateridae). *Bollettino del Museo Regionale di Scienze Naturali - Torino* **14**: 139-150.

Platia G. & Gudenzi I. 1996b: Descrizione di tre nuove specie di Athous Eschscholtz del sottogenere Orthathous della Turchia con chiave di determinazione dei maschie delle specie note. (Coleoptera, Elateridae). *Lambillionea* **96**: 235-241.

Platia G. & Gudenzi I. 1997a: Contributo alla conoscenza degli Agriotes del gruppo nuceus di Siria e Iraq (Coleoptera, Elateridae). *Bollettino della Società Entomologica Italiana* **128**: 201-211.

Platia G. & Gudenzi I. 1997b: Revisione delle specie del genere Drasterius Eschscholtz della regione orientale. (Coleoptera Elateridae, Conoderini). *Lambillionea* **97**: 402-416.

Platia G. & Gudenzi I. 1998a: Note tassonomiche e faunistiche su elateridi del vicino oriente (Coleoptera, Elateridae). *Bollettino dell'Associazione Romana di Entomologia* **53**: 49-62.

Platia G. & Gudenzi I. 1998b: Descrizione di una nuova specie di Athous Eschscholtz, 1829 e della femmina di Athous pedemontanus Platia, 1988, del Piemonte, con nuovi dati geonemici su alcuni elateridi della fauna italiana (Coleoptera, Elateridae). *Rivista Piemontese di Storia Naturale* **19**: 267-273.

Platia G. & Gudenzi I. 1998c: Descrizione di nuovi elateridi di Grecia e Turchia con note e chiave di determinazione per gli Agriotes del gruppo gurgistanus Faldermann (Coleoptera, Elateridae). *Lambillionea* **98**: 629-640.

Platia G. & Gudenzi I. 1999a: Descrizione di nuove specie di elateridi della regione palearctica con note geonemiche e sinonimiche (Insecta Coleoptera Elateridae). *Quaderno di Studi e Notizie di Storia Naturale della Romagna* **11**, suppl.: 17-31.

Platia G. & Gudenzi I. 1999b: New species and new or interesting records of click-beetles from Iran and Iraq (Coleoptera, Elateridae). *Entomologica Fennica* **10**: 117-125.

Platia G. & Gudenzi I. 2000a: Description de deux nouvelles espèces d'élatérides de France et d'Espagne. *Biocosme Mésogéen* **16** [1999]: 147-152.

Platia G. & Gudenzi I. 2000b: Descrizione di nuovi generi e specie di elateridi dell'Asia orientale. *Annali del Museo Civico di Storia Naturale "G. Doria"* **93**: 253-269.

Platia G. & Gudenzi I. 2000c: Descrizione di due nuove specie di elateridi della Romagna con note geonemiche su alcune specie della fauna italiana (Insecta Coleoptera Elateridae). *Quaderno di Studi e Notizie di Storia Naturale della Romagna* **13**: 11-16.

Platia G. & Gudenzi I. 2000d: Contributo alla conoscenza dei Cardiophorus della Grecia (Coleoptera Elateridae). *Bolletino della Società Entomologica Italiano* **132**: 135-155.

Platia G. & Gudenzi I. 2000e: Revisione dei Cardiophorus Eschscholtz della Turchia. I: Le specie bicolori. (Coleoptera, Elateridae, Cardiophorinae). *Lambillionea* **100**: 601-612.

Platia G. & Gudenzi I. 2000f: Descrizione di un nuovo genere e tredici nuove specie di elateridi della regione paleartica con note geonemiche. *Quaderno di Studi e Notizie di Storia Naturale della Romagna* **13**, suppl.: 65-83.

Platia G. & Gudenzi I. 2001: Descrizione di una nuova specie di Agriotes Eschscholtz dell'isola di Cefalonia (Grecia) (Coleoptera, Elateridae). *Atti della Società Italiana di Scienze Naturali e del Museo Civico di Storia Naturale di Milano* **141** [2000]: 309-312.

Platia G. & Gudenzi I. 2002a: Revisione dei Cardiophorus della Turchia. II. Le specie unicolori (Coleoptera Elateridae). *Bollettino della Società Entomologica Italiana* **134**: 27-55.

Platia G. & Gudenzi I. 2002b: Descrizione di venti nuove specie di elateridi della regione paleartica (Insecta Coleoptera Elateridae). *Quaderno di Studi e Notizie di Storio Naturale della Romagna* **16**, suppl.: 73-98.

Platia G. & Gudenzi I. 2003: Contributo alla conoscenza dei Dicronychus Brullé della Grecia con descrizione di nuove specie e chiave di determinazione per quelle note (Coleoptera: Elateridae: Cardiophorini). *Annali del Museo Civico di Storia Naturale di Ferrara* **5** [2002]: 27-39.

Platia G. & Gudenzi I. 2004: Contributo alla conoscenza dei Dicronychus Brullé, 1832 della Turchia con descrizione di nuove specie e chiave di determinazione per la specie note (Coleoptera: Elateridae: Cardiophorinae). *Boletín de la Sociedad Entomologica Aragonesa* **34**: 9-22.

Platia G. & Marini M. 1990: Gli elateridi (Coleoptera) della collezione Andrea Fiori, conservati nel Museo di Zoologia dell'Università di Bologna. *Giornale Italiano di Entomologia* **5** [1989]: 31-47.

Platia G. & Mertlik J. 1996: Anostirus turcicus sp. n. (Coleoptera: Elateridae) from Turkey. *Klapalekiana* **32**: 225-228.

Platia G. & Schimmel R. 1988: Descrizione di due nuove specie di elateridi della fauna italiana (Coleoptera Elateridae). *Bollettino del Museo Civico di Storia Naturale di Verona* **14** [1987]: 453-457.

Platia G. & Schimmel R. 1991a: Il genere Silesis Candèze nella regione indiana e cinese (Coleoptera, Elateridae, Adrastini). *Fragmenta Entomologica* **23**: 101-177.

Platia G. & Schimmel R. 1991b: Descrizione di nuove specie di elateridi di Grecia e Turchia con chiave per due gruppi di specie (Coleoptera Elateridae). *Lavori - Società Veneziana di Scienze Naturali* **16**: 131-142.

Platia G. & Schimmel R. 1992: Descrizione di nuove specie di elateridi di Grecia e Turchia (Coleoptera, Elateridae). *Lavori - Societa Veneziana di Scienze Naturali* **17**: 87-95.

Platia G. & Schimmel R. 1993a: Descrizione di nuove specie di elateridi di Grecia e Turchia con note su alcuni Agriotes Eschscholtz (Coleoptera, Elateridae). *Lavori - Societa Veneziana di Scienze Naturali* **18**: 51-57.

Platia G. & Schimmel R. 1993b: Nuove specie e nouvi reperti di Silesis Candèze della regione indiana e cinese (Coleoptera, Elateridae, Adrastini). *Fragmenta Entomologica* **25**: 117-127.

Platia G. & Schimmel R. 1994: Descrizione di tre nuove specie di elateridi di Grecia, Turchia e Siria, con note su Agriotes cribrithorax Pic, 1910 (Coleoptera, Elateridae). *Lavori - Società Veneziana di Scienze Naturali* **19**: 77-82.

Platia G. & Schimmel R. 1995: I Priopus della regione indiana (Coleoptera Elateridae, Melanotinae). *Lavori - Società Veneziana di Scienze Naturali* **20**: 53-65.

Platia G. & Schimmel R. 1996a: I Priopus Castelnau di Birmania, Thailandia, Laos, Cambodia, Vietnam, China (Coleoptera, Elateridae, Melanotinae). *Annali del Museo Civico di Storia Naturale "G. Doria"* **91**: 179-225.

Platia G. & Schimmel R. 1996b: Nuove specie e nuovi dati di Silesis Candèze della regione indiana e cinese (Coleoptera, Elateridae). *Fragmenta Entomologica* **28**: 245-255.

Platia G. & Schimmel R. 1997: Click-beetles (Coleoptera: Elateridae) from the Sultanate of Oman with additional new species and records from neighbouring regions. *Fauna of Saudi Arabia* **16**: 291-318.

Platia G. & Schimmel R. 2001: Revisione delle specie orientale (Giappone e Taiwan esclusi) del genere Melanotus Eschscholtz, 1829 (Coleoptera, Elateridae, Melanotinae). *Museo Regionale di Scienze Naturali, Torino, Monografie XXVII*:1-638.

Platia G. & Schimmel R. 2002: Revisione delle species orientali (Giappone e Taiwan esclusi) del genere Melanotus Eschscholtz, 1829 (Coleoptera, Elateridae, Melanotinae) I. Supplemento. *Bollettino del Museo Regionale di Scienze Naturali Torino* **19**: 305-388.

Platia G. & Schimmel R. 2004: Revisione delle specie della Regione Orientale (Giappone e Taiwan esclusi) del genere Melanotus Eschascholtz, 1829 (Coleoptera, Elateridae, Melanotinae). II. Supplemento. *Bollettino del Museo Regionale di Scienze Naturali Torino* **21**: 99-122.

Platia G. & Serrano, A. R. M. 2002: Contribution to the knowledge of the click beetles of Portugal (Coleoptera, Elateridae). *Boletim da Sociedade Portuguesa de Entomologia* **VII-26**: 299-326.

Platia G. & Tarnawski D.1998: Description of two new species of click-beetles from Turkey (Coleoptera: Elateridae). *Genus* **9**: 269-274.

Plavilshchikov N. N. 1924: Analecta Coleopterologica. *The Annals and Magazine of Natural History* (9) **13**: 230-233.

Plavilshchikov N. N. 1935: Nitidulidae (Col.). P. 298. In: Visser P. C. & Visser-Hooft J. (eds): *Wissenschaftliche Ergebnisse der niederländischen Expeditionen in den Karakorum und die angrenzenden Gebiete in den Jahren 1922, 1925 und 1929/30. Vol. 1.* Leipzig: F. A. Brockhaus, xviii + 499 pp., 1 map, 8 pls.

Plaza E. 1975: Acerca de la especie Bulaea lichatschovi (Humm., 1872) (Col., Cocc.). *Graellsia* **29**: 99-110.

Plaza E. 1981: Los representantes españoles de Nephus del subgénero Sidis Muls. 1851 (Col., Coccinellidae). *Boletín de la Asociación Española de Entomología* **4** [1980]: 235-242.

Pliginsky V. 1923: Novye palearkticheskye vidy zhestkokrylykh [Espèces nouvelles des coléoptères paléarctiques]. *Revue Russe d'Entomologie* **18**: 141-143.

Poda von Neuhaus N. 1761: *Insecta Musei Graecensis, que in ordines, genera et species juxta Systema Naturae Linnaei digessit.* Widmanstad: Graecii, 127 + 18 pp. 2 pls.

Pontoppidan E. L. 1763: *Den Danske atlas, eller Konge-Riget Dannemark, med dets naturlige egenskaber, elementer, inbyggere, vaexter, dyr og andre affødninger, dets gamle tildragelser og naervaerende omstaendigheder i alle provintzer, staeder, kirkzer, slotte og herre-gaarde. Forestillet ved en udførlig Lands-Beskrivelse, saa og oplyst med dertil forfaerdigede landkort over enhver provintz, samt ziret med staedernes prospecter, grund-ridser, og andre merkvaerdige kaaber-stykker. Efter Høy-Kongelig allernaadigst Befalning. Tomus I.* Kiøbenhavn: A.H. Godieche, xl + [iv] + 723 + [1] pp., 30 pls.

Ponza L. 1805: Coleoptera Salutiensia sive enumeratio methodica Coleopterorum, quae in agro Salutiensi reperiuntur, locorum natalium indicatione, observationibus, novisque aliquot speciebus aucta. *Mémoires de l'Académie Impériale des Sciences, Littérature et Beaux-Arts de Turin* **14**: 29-94, pls 1-2.

Poorani J. 2002: A review of the genus Oenopia Mulsant (Coleoptera: Coccinellidae) in the Indian subcontinent, with description of a new species. *Oriental Insects* **36**: 97-116.

Pope R. D. 1961: Colydiidae (Coleoptera Clavicornia). Pp. 115 pp. *Exploration du Parc National de la Garamba. Mission H. de Saeger.* Bruxelles: Institut des Parcs Nationaux du Congo Belge.

Pope R. D. 1962: A review of the Pharini (Coleoptera: Coccinellidae). *The Annals and Magazine of Natural History* (13) **4** [1961]: 627-640.

Pope R. D. 1988: A revision of the Australian Coccinellidae (Coleoptera). Part 1. Subfamily Coccinellinae. *Invertebrate Taxonomy* **2**: 633-735.

Poppius B. 1900: Eine neue Art der Gattung Cryptophagus Herbst. *Meddelanden af Societas pro Fauna et Flora Fennica* **26**: 189-192.

Poppius B. 1903: Über einige Lathridiiden. *Meddelanden af Societas pro Fauna et Flora Fennica* **29**: 84-91.

Poppius B. 1904: Neue paläarctische Coleopteren. *Öfversigt af Finska Vetenskaps-Societens Förhandlingar* **46** (16) [1903-1904]: 1-14.

Porta A. 1929: *Fauna Coleopterorum italica. Vol. III. – Diversicornia.* Piacenza: Tipografico Piacentino, 2 + 466 pp.

Portevin G. 1931: *Histoire naturelle des coléoptères de France. Tome II. Polyphaga: Lamellicornia, Palpicornia, Diversicornia.* In: *Encyclopédie Entomologique.* Paris: Paul Lechevalier et Fils, vii + 542 pp., pls. 6-10.

Preyssler J. D. E. 1790: *Verzeichniss böhmischer Insecten. Erstes Hundert.* Prague: Schoenfeld-Meissner, 108 pp., 2 pls.

Preiss R. 2003: A new Elathous Reitter, 1890, from Syria (Coleoptera: Elateridae). *Zeitschrift der Arbeitsgemeinschaft Österreichischer Entomologen* **55**: 55-57.

Preiss R. & Platia G. 2003: The click beetles of Cyprus with descriptions of two new species and notes on species of the genus *Haterumelater* Ôhira, 1968 (Coleoptera: Elateridae). *Zeitschrift der Arbeitsgemeinschaft Österreichischer Entomologen* **55**: 97-123.

Preyssler J. D. E. 1791: *Beschreibungen und Abbildungen derjenigen Insecten, welche in Saimmlungen nicht aufzubewahren sind, dann aller, die noch ganz neu, und solcher, von denen wir noch keine oder doch sehr schlechte Abbildung besitzen. Erste-Dritte Sammlung.* In: Mayer J. (ed.): *Sammlung physikalischer Aufsätze, besonders die Böhmische Naturgeschichte betreffetid. Band 1.* Dresden: J. Mayer, pp. 55-152.

Preyssler J. D. E. 1792: *Beschreibungen und Abbildungen derjenigen Insecten, welche in Saimmlungen nicht aufzubewahren sind, dann aller, die noch ganz neu, und solcher, von denen wir noch keine oder doch sehr schlechte Abbildung besitzen. Vierte Sammlung.* In: Mayer J. (ed.): *Sammlung physikalischer Aufsätze, besonders die Böhmische Naturgeschichte betreffetid. Band 2.* Dresden: J. Mayer, pp. 1-46.

Priesner H. 1951: Description of the common Egyptian Trogoderma (Coleoptera: Dermestidae). *Bulletin de la Société Fouad I Entomologie* **35**: 133-135.

Procházka J. 1894: *Bestimmungs-Tabelle der europäischen Coleopteren: Cantharidae, II. Theil: Genus Danacaea. XXX. Heft.* Brünn: Ed. Reitter, 35 pp., 1 pl.

Procházka J. 1895: Dasytiscus Ragusae Proch. nov. sp. *Il Naturalista Siciliano* **14**: 139.

Puppin O. 1974: Dictyoptera aurora caprai n. subsp. ed illustrazioni di altri Dictyopterini. *Bollettino della Società Entomologica Italiana* **106**: 40-46.

Puton A. 1865: Description d'une nouvelle espèce de coléoptères de la division des Malacodermes. *Annales de la Société Entomologique de France* (4) **5**: 131-132.

Pütz A. & Kusnetzov V. 1995: Nephus (Bipunctatus) wrasei sp. n. - eine neue Art von der Insel Sachalin (Coleoptera: Coccinellidae). *Koleopterologische Rundschau* **65**: 189-191.

Quedenfeldt F. O. G. 1886: Verzeichniss der von Herrn Major a. D. von Mechow in Angola und am Quango-Strom 1878-1881 gesammelten Anthribiden und Bostrychiden. *Berliner Entomologische Zeitschrift* **30**: 303-328.

Quelle F. 1932: Elateriden-Studien II. *Coleopterologisches Centralblatt* **5**: 202-216.

Quensel C. 1790: *Dissertatio historico-naturalis ignotas insectorum species continens, cujus partem primam, consent. ampliss. facult. philosoph. publico examini modeste subjiciunt praeses Conrad. Quensel, philosophiae magister, et. respondens C. P. Lundgård.* Lundae: Typis Berlingianis, 20 pp.

Raffray A. 1873: [new taxa]. In: Fairmaire L. & Raffray A.: Coléoptères du Nord de l'Afrique. *Revue et Magasin de Zoologie Pure et Appliquée* (3) **1**: 331-385.

Ragusa E. 1870: Descrizione di due nuove specie di coleotteri trovate in Sicilia. *Bullettino della Società Entomologica Italliana* **2**: 315-316.

Ragusa E. 1871: Descrizione di una nuova specie d'Attalus della Sicilia. *Bullettino della Società Entomologica Italiana* **3**: 282-283.

Ragusa E. 1872: Haplocnemus trinacriensis. *Bullettino della Società Entomologica Italiana* **4**: 83, 96, pl. 1.

Ragusa E. 1881: Coleotteri nuovi o poco conosciuto della Sicilia. *Il Naturalista Siciliano* **1**: 5-9, pl. 1.

Ragusa E. 1883: Lygistopterus anorachilus nov. sp. *Il Naturalista Siciliano* **2** [1882-83]: 251.

Ragusa E. 1886: Coleotteri nuovi o poco conosciuti della Sicilia. *Il Naturalista Siciliana* **5**: 138-142.

Ragusa E. 1888: Coleotteri nuovi o poco conosciuti della Sicilia. *Il Naturalista Siciliano* **8** [1888-1889]: 36-37.

Ragusa E. 1889: Coleotteri nuovi o poco conosciuti della Sicilia. *Il Naturalista Siciliano* **8** [1888-1889]: 234-236.

Ragusa E. 1892: Catologo ragionato dei coleotteri di Sicilia. *Il Naturalista Siciliano* **11** [1891-1892]: 185-209.

Ragusa E. 1893: Catalogo ragionato dei coleotteri di Sicilia. (Cont.). *Il Naturalista Siciliano* **12** [1892-1893]: 289-307.

Ragusa E. 1894: Catalogo ragionato dei coleotteri di Sicilia (Cont.). *Il Naturalista Siciliano* **13** [1893-1894]: 2-14, 37-47.

Ragusa E. 1896a: Coleotteri nuovi o poco conosciuti della Sicilia. *Il Naturalista Siciliano* **15**: 138-142.

Ragusa E. 1896b: Catalogo ragionato dei coleotteri di Sicilia (Cont.). *Il Naturalista Siciliano* (N. S.) **1**: 69-106.

Ragusa E. 1923: Coleotteri nuovi o poco conosciuti della Sicilia. *Atti Accademia di Scienze, Lettere e Arti di Palermo* (3) **12**: 1-18.

Rai K. 1966: Studies on the morphology and taxonomy of Indian Bostrychidae. VI. A new species of the genus Bostrychopsis Lesne from India. *Journal of the Bombay Natural History Society* **62** [1965]: 576-578.

Rai K. 1967: Studies on the morphology and taxonomy of Indian Bostrychidae. VIII. A new species of the genus Xylopsocus Lesne from India. *Science & Culture* **33** 140-141.

Rai K. & Chatterjee P. N. 1964: Studies on the morphology and taxonomy of Indian Bostrychidae. IV. A new species of Xylopertha Guerin-Meneville (Coleoptera: Bostrychidae) from North India. *Indian Forest* **90** (2): 122-124.

Ramakrishna Ayyar T. V. 1925: Notes on the status of some parasitic Hymenoptera in South India. *Journal of the Bombay Natural History Society* **30**: 487-491.

Randall J. W. 1838a: Descriptions of new species of coleopterous insects, inhabiting the State of Maine. *Boston Journal of Natural History* **2**: 1-33.

Randall J. W. 1838b: Description of new species of coleopterous insects inhabiting the State of Massachusetts. *Boston Journal of Natural History* **2**: 34-52.

Ratti E. 1972a: I Cucujidae (senzo lato) della Mongolia. *Bollettino dela Società Entomologica Italiano* **103** [1971]: 98-103.

Ratti E. 1972b: Considerazioni sui Laemophloeinae delle Azzorre, con descrizione di Leptophloeus azoricus n. sp. (Coleoptera Cucujidae). *Atti della Società Italiana di Scienze Naturali e del Museo Civico di Storia Naturale di Milano* **113**: 281-287.

Ratzeburg J. T. C. 1837: *Die Forst-Insecten oder Abbildungen und Beschreibungen der in den Wäldern Preussens und der Nachbarstaaten als schädlich oder nützlich bekannt gewordenen Insecten. Erster Theil. Die Käfer.* Berlin: Nicolai, x + 4 + 202 pp., 21 pls.

Razoumowsky G. de. 1789: *Histoire naturelle du Jorat et de ses environs; et celle des trois Lacs de Neuschatel, Morat et Bienne; précédées d'un essai sur le Climat, les Productions, le Commerce, les Animaux de la partie du Pays de Vaud ou de la Suisse Romande, qui entre dans le plan de cet Ouvrage.* Lausanne: Jean Mourer, 322 pp.

Rebmann O. 1940a: Zwei neue nordafrikanische Meligethes-Arten aus der Verwandtschaft des M. elongatus (Col. Nitid.). Mit Bestimmungstabelle der M. elongatus-Gruppe und einer Kritik der Untergattung Clypeogethes Scholz. (3. Beitrag zur Kenntnis der Nitiduliden). *Mitteilungen der Münchener Entomologischen Gesellschaft* **30**: 547-555.

Rebmann O. 1940b: Drei neue Meligethes-Arten aus dem Mittelmeergebiet, nebst Bemerkungen zu einigen anderen Arten der lugubris-Verwandtschaft (Col. Nitid.). (4. Beitrag zur Kenntnis der Nitiduliden). *Mitteilungen der Münchener Entomologischen Gesellschaft* **30**: 576-583.

Rebmann O. 1944: Zwei neue Nitiduliden-Gattungen aus China: Osotima nov. gen. und Meligethopsis nov. gen. (8. Beitrag zur Kenntnis der Nitiduliden). *Entomologische Blätter* **40**: 22-26.

Rebmann O. 1956a: Revision der Gattung Meligethes Subgenus Odontogethes (Col. Nitid.). (6. Beitrag zur Kenntnis der Nitiduliden). *Entomologische Blätter* **52**: 42-48.

Rebmann O. 1956b: Revision der Gattung Meligethes (Col. Nitid.). Die paläarktischen Arten der Gruppe des M. atratus Ol. (7. Beitrag zur Kenntnis der Nitiduliden). *Entomologische Blätter* **52**: 124-135.

Reclaire A. & van der Wiel P. 1932: 2e Korte mededeeling omtrent eenige op Vlieland waargenomen insecten. *Entomologische Berichten* (Amsterdam) **8**: 421-431.

Redtenbacher L. 1843a: *Tentamen dispositionis generum et specierum Coleopterorum Pseudotrimerorum Archiductus Austriae.* Vindobone: Disert. Inaug, 32 pp.

Redtenbacher L. 1843b: Coleopterorum Syriae genera et species novae. Pp. 971-990. In: Russegger J.: *Reisen in Europa, Asien und Afrika, mit besonderer Rücksicht auf die naturwissenschaftlichen Verhältnisse der betreffenden Länder, unternommen in den Jahren 1835 bis 1841. Erster Band. Reise in Griechenland, Unteregypten, im nördlichen Syrien und Kleinasien, mit besonderer Rücksicht auf die naturwissenschaftlichen Verhältnisse der betreffenden Länder, unternommen in dem 1836. Zweiter Theil.* Stuttgart: E. Schweitzerbart'sche Verlangshandlung.

Redtenbacher L. 1844: [new taxa]. In: Kollar V. & Radtenbacher L.: Aufzählung und Beschreibung der von Freiher Carl von Huegel auf seiner Reise durch Kaschmir u. das Himalayagebirge gesammelten Insecten. Pp. 393-564. In: Hügel K. F. von. (ed.): *Kaschmir und das Reich der Siek. Vierter Band. Zweite Abtheilung.* Stuttgart: Hallbergerische Verlag, pp. 244-586 pp. [1844]; 587-865 + [6] pp., 31 pls, 1 map [1848].

Redtenbacher L. 1845: *Die Gattungen der deutschen Käfer-Fauna nach der analytischen Method bearbeitet, nebst einem kurtz geffassten Leitfaden, zum Studium dieses Zweiges der Entomologie.* Wien: C. Ueberreuter, [13] + 177 + [1] pp., 2 pls.

Redtenbacher L. 1849: *Fauna Austriaca. Die Käfer. Nach der analytischen Methode bearbeitet.* Wien: Carl Gerold, xxvii + 883 pp., 2 pls.

Redtenbacher L. 1850: [new taxa]. In: Kollar V. & Redtenbacher L.: Ueber den Charakter der Insecten-Fauna von Südpersien. *Denkschriften der Kaiserlichen Akademie der Wissenschaften, Wien* **1**: 42-53.

Redtenbacher L. 1858: *Fauna Austriaca. Die Käfer. Nach der analytischen Methode bearbeitet. Zweite, gänzlich umgearbeitete, mit mehreren hunderten von Arten und mit der Charakteristik sämmlicher europäischen Käfergattungen vermehrte Auflage.* Wien: Carl Gerold's Sohn, cxxxvi + 1017 pp., 2 pls.

Redtenbacher L. 1868: Zoologischer Theil. Zweiter Band. I. Abtheilung A. 2. Coleopteren. In: *Reise der Öesterreichischen Fregatte Novara um die Erde in den Jahren 1857, 1858, 1859 unter den Befehlen des Commodore B. von Wüllerstorf-Urbair. Zoologischer Theil. Zweiter Band: Coleopteren.* 1867. Wien: Karl Gerold's Sohn, iv + 249 pp., 5 pls.

Redtenbacher L. 1874: *Fauna Austriaca. Die Käfer, nach der analytischen Methode bearbeitet. Dritte, gänzlich umgearbeitete und bedeutend vermehrte Auflage.* Wien: C. Gerold's Sohn, cliii + 564 + 725 + viii pp., 2 pls.

Redtenbacher W. 1842: *Quaedam genera et species Coleopterorum Archiducatus Austriae nondum descriptorum. Dissertatio inauguralis.* Vindobonae: Carolus Ueberreuter, 31 pp.

Reichardt H. 1966: Bostrychidae (Coleoptera) 7: A new Xylothrips from China. *The Coleopterists' Bulletin* **20**: 81-83.

Reichardt H. & Toskina I. N. 1973: [new taxon]. In: Toskina I. N.: O sistematicheskom polozhenii Coelostethus (Coleoptera, Anobiidae). *Zoologicheskiy Zhurnal* **52**: 1335-1340.

Reiche L. J. 1847: [new taxa]. In: Ferret P. V. A. & Galinier J. G. [1847-1848]: *Voyage en Abyssinie, dans les provinces du Tigre, du Samen et de l'Ahmara. Entomologie. Vol. 3.* Paris: Paulin, 536 pp.

Reiche L. J. 1854a: *Catalogue des espèces d'insectes coléoptères recueillies par M. F. de Saulcy pendant son voyage en Orient.* Paris: Gide, 19 pp.

Reiche L. J. 1854b: Observations entomologiques. *Annales de la Société Entomologique de France* (3) **2**: 77-81.

Reiche L. J. 1858: Description d'une nouvelle espèce du genre Lampyris, Lin. *Annales de la Société Entomologique de France* (3) **6**: 155-156.

Reiche L. J. 1860: [Note relative à la synonymie de deux espèces du genre Corymbites]. *Bulletin de la Société Entomologique de France* **1860**: cviii-cix.

Reiche L. J. 1861a: Species novae Coleopterorum descripta, quae in Syria invenit Dom. Kindermann. *Wiener Entomologische Monatschrift* **5**: 1-8.

Reiche L. J. 1861b: Coléoptères nouveaux recueillis en Corse par M. E. Bellier de la Chavignerie. *Annales de la Société Entomologique de France* (4) **1**: 201-210.

Reiche L. J. 1861c: Notes synonymiques. *Annales de la Société Entomologique de France* (4) **1**: 211-212.

Reiche L. J. 1862: Notes synonymiques sur divers coléoptères. *Annales de la Société Entomologique de France* (4) **2**: 79-80.

Reiche L. J. 1862: Espèces nouvelles de coléoptères appartenant à la faune Circa-Méditerranéenne (Suite). *Annales de la Société Entomologique de France* (4) **2**: 539-546.

Reiche L. J. 1863: Examen rapide de quelques pages du Cat. Col. Eur. de Schaum. *Annales de la Société Entomologique de France* (4) **3**: 121-132.

Reiche L. J. 1864: Espèces nouvelles de coléoptères d'Algérie. *Annales de la Société Entomologique de France* (4) **4**: 233-246.

Reiche L. J. 1868a: [new taxa]. In: Mulsant E. & Rey C.: *Histoire Naturelle des coléoptères de France. Tribu des Scuticolles.* Paris: F. Savy, 186 pp., 3 pls.

Reiche L. J. 1868b: [new taxon]. In: Bedel L. 1868: Monographie des Erotyliens (Engides et Triplacides), d'Europe, du Nord de l'Afrique et de l'Asie occidentale. *L'Abeille, Mémoires d'Entomologie* **5** [1868-1869]: 1-50.

Reiche L. J. 1869: Notes sur quelques espèces du genre Athous des coléoptères Élatérides. *Annales de la Société Entomologique de France* (4) **9**: 379-387.

Reiche L. J. 1878a: Espèces nouvelles de Téléphorides. *Annales de la Société Entomologique de France* (5) **8**: 383-384.

Reiche L. J. 1878b: Description de deux nouvelles espéces de coléoptères de la famille des Lycides. *Bulletin de la Société Entomolomologique de France* **1878**: xxvii-xxviii.

Reiche L. J. & Saulcy F. de. 1857a: Espèces nouvelles ou peu connues de coléoptères, recueillies par M. F. de Saulcy, membre de l'Institut, dans son voyage en Orient. *Annales de la Société Entomologique de France* (3) **4**: 385-422.

Reiche L. J. & Saulcy F. de. 1857b: Espèces nouvelles ou peu connues de coléoptères, recueillies par M. F. de Saulcy membre de l'Institut, dans son Voyage en Orient. *Annales de la Société Entomologique de France* (3) **5**: 169-276, pl. 5.

Reichenbach H. G. L. 1826: [new taxon]. In: Sturm J.: *Catalog meiner Insecten-Sammlung. Erster Theil. Käfer*. Nürnberg: Verfasser, viii + 207 + 16 pp., 4 pls.

Reichensperger A. 1913: Zur Kenntnis von Myrmekophilen aus Abessinien. *Zoologische Jahrbücher. Abteilung für Systematik, Geographie und Biologie der Tiere* **35**: 185-218, pls. 5-6.

Reichensperger A. 1926: Beobachtungen und Versuche mit Cataglyphis und Thorictus nebst dessen Metamorphose. Beschreibungen zweier neuer Myrmekophilen. *Verhandlungen des Naturhistorischen Vereins der Preussischen Rheinlande, Westfalen und des Regierungsbezirks Osnabrück* **82**: 73-110, 1 pl.

Reitter E. 1871: Revision der europäischen Meligethes-Arten. *Verhandlungen des Naturforschenden Vereins in Brünn* **9** [1870]: 39-169, pls. 1-6.

Reitter E. 1872a: Erster Nachtrag zur Revision der europäischen Meligethes-Arten. *Berliner Entomologische Zeitschrift* **16**: 125-134, pl. 7.

Reitter E. 1872b: Neue Käferarten von Oran, gesammelt von Hans Leder. (Erstes Stück). *Berliner Entomologische Zeitschrift* **16**: 167-186.

Reitter E. 1872c: Zweiter Nachtrag zur Revision der europäischen Meligethes-Arten. *Berliner Entomologische Zeitschrift* **16**: 265-269.

Reitter E. 1873a: Revision der europäischen Epuraea-Arten. *Verhandlungen des Naturforschenden Vereins in Brünn* **11** [1872]: 2-25, pl. 1.

Reitter E. 1873b: Die Rhizophaginen, monographisch bearbeitet. *Verhandlungen des Naturforschenden Vereins in Brünn* **11** [1872]: 27-48, pl. 1.

Reitter E. 1873c: Neue Meligethesarten. *Verhandlungen des Naturforschenden Vereins in Brünn* **11** [1872]: 49-52.

Reitter E. 1873d: Systematische Eintheilung der Nitidularien. *Verhandlungen des Naturforschenden Vereins in Brünn* **12** (1): 3-194.

Reitter E. 1874a: Diagnosen der bekannten Cybocephalus-Arten. *Verhandlungen des Naturforschenden Vereins in Brünn* **12** (2) [1873]: 1-10.

Reitter E. 1874b: Beitrag zur Kenntniss der japanesischen Cryptophagiden. *Verhandlungen der Kaiserlich-Königlichen Zoologisch-Botanischen Gesellschaft in Wien* **24**: 379-382.

Reitter E. 1874c: Beschreibungen neuer Käfer-Arten nebst synonymische Notizen. *Verhandlungen der Kaiserlich-Königlichen Zoologisch-Botanischen Gesellschaft in Wien* **24**: 509-529.

Reitter E. 1875a: Die europäischen Nitidularien mit kurzer Charakteristik der Gattungen und Bemerkungen über schwierige Arten verzeichnet. *Deutsche Entomologische Zeitschrift* **19** (3): 1-30.

Reitter E. 1875b: Revision der europäischen Cryptophagiden. *Deutsche Entomologische Zeitschrift* **19** (3): 1-86.

Reitter E. 1875c: Revision der Gattung Trogosita Oliv. (Temnochila Westw.). *Verhandlungen des Naturforschenden Vereins in Brünn* **13** [1874]: 3-44.

Reitter E. 1875d: Ueber eine neue Gattung der Thorictidae. *Coleopterologische Hefte* **14**: 45-46.

Reitter E. 1875e: Darstellung der mit Epuraea verwandten Gattungen. *Verhandlungen des Naturforschenden Vereins in Brünn* **13** [1874]: 53-64, 1 pl.

Reitter E. 1875f: Diagnosen neuer Cybocephalus-Arten als Nachtrag zu der gleichnamigen Abhandlung in den Verhandlungen des Naturhistorischen Vereins in Brünn 1873. *Coleopterologische Hefte* **13**: 55-56.

Reitter E. 1875g: Beitrag zur Kenntniss der aussereuropäischen Cryptophagiden. *Coleopterologische Hefte* **13**: 73-87.

Reitter E. 1875h: Nachtrag zu den europäischen Nitidularien. *Deutsche Entomologische Zeitschrift* **19** (3): 87-88.

Reitter E. 1875i: Beschreibungen neuer Nitidulidae aus der Sammlung der Herrn Deyrolle in Paris. *Verhandlungen des Naturforschenden Vereins in Brünn* **13** [1874]: 99-122.

Reitter E. 1875j: Die bekannten *Telmatophilus*-Arten, übersichtlich dargestellt. *Deutsche Entomologische Zeitschrift* **19**: 225-228.

Reitter E. 1875k: Revision der europäischen Lathridiidae. *Entomologische Zeitung* (Stettin) **36**: 297-340.

Reitter E. 1875l: [new taxa]. In: Putzeys J., Reitter E., Saulcy F. de & Weise J.: Neue Käferarten aus Ungarn. *Deutsche Entomologische Zeitschrift* **19**: 355-364.

Reitter E. 1875m: Cerylon atratulum Reitter n. sp. *Deutsche Entomologische Zeitschrift* **19**: 360.

Reitter E. 1875n: Meligethes prioides Reitter nov. spec. *Deutsche Entomologische Zeitschrift* **19**: 393.

Reitter E. 1875o: Revision der europäischen Lathridiidae. *Entomologische Zeitung* (Stettin) **36**: 410-445.

Reitter E. 1876a: Systematische Eintheilung der Trogositidae. (Familia Coleopterorum). *Verhandlungen des Naturforschenden Vereins in Brünn* **14** [1875]: 3-66.

Reitter E. 1876b: Atomaria uhagoni. *Entomologische Montsblätter* **1**: 10.

Reitter E. 1876c: Neue Gattungen und Arten aus der Familie der Cucujidae. *Coleopterologische Hefte* **15**: 37-64.

Reitter E. 1876d: Nachtrag zur Revision der europäischen Lathridiidae. *Entomologische Zeitung* (Stettin) **37**: 50-52.

Reitter E. 1876e: Die Arten der Gattung Cathartus Reiche. *Coleopterologische Hefte* **15**: 125-130.

Reitter E. 1876f: Neue transcaucasische Coleopteren, gesammelt von Hans Leder (Erstes Stück). *Deutsche Entomologische Zeitschrift* **20**: 289-294.

Reitter E. 1876g: Revision der Monotomidae (sensu Le Conte). *Deutsche Entomologische Zeitschrift* **20**: 295-301.

Reitter E. 1876h: Über Orthoperus punctatus. *Deutsche Entomologische Zeitschrift* **20**: 312.

Reitter E. 1876i: Uebersicht der europäischen Cerylon-Arten. *Deutsche Entomologische Zeitschrift* **20**: 313-314.

Reitter E. 1876j: Revision der Cerylon-Arten aus Europa und den angrezenden Ländern. *Deutsche Entomologische Zeitschrift* **20**: 385-394, pl. II.

Reitter E. 1877a: Ueber Merophysia Luc., Coluocera Motsch. und Reitteria Leder. *Mittheilungen des Münchener Entomologischen Vereins* **1**: 1-6.

Reitter E. 1877b: Bemerkungen über die Arten der Gattung Monotoma. *Zeitschrift für Entomologie* (Breslau) (N. F.) **6**: 1-7.

Reitter E. 1877c: Description des espèces d'Europe des genres Sacium Lec. et Arthrolips Woll. *L'Abeille, Journal d'Entomologie* **16** [1878]: 1-12.

Reitter E. 1877d: [new taxa]. In: Putzeys J. A. A. H., Reitter E., Saulcy F. de & Weise J.: Coleopterologische Ergebnisse einer Reise nach Südungarn und in die Transsylvanischen Alpen. *Verhandlungen des Naturforschenden Vereins in Brünn* **15** [1876]: 3-30.

Reitter E. 1877e: Neue Arten aus den Familien der Cucujidae, Nitidulidae, Colydiidae und Cryptophagidae. *Mittheilungen des Münchener Entomologischen Vereins* **1**: 22-28.

Reitter E. 1877f: [new taxa]. In: Putzeys J. A. A. H., Weise J., Kraatz G., Reitter E. & Eichhoff W.: Beiträge zur Käferfauna von Japan, meist auf R. Hiller's Sammlungen basirt (Erstes Stück). *Deutsche Entomologische Zeitschrift* **21**: 81-128.

Reitter E. 1877g: Beiträge zur Kenntnis aussereuropäischer Coleopteren. *Mittheilungen des Münchener Entomologischen Vereins* **1**: 126-140.

Reitter E. 1877h: Neue caucasische Coleopteren, gesammelt von Hans Leder (Zweites Stück). *Deutsche Entomologische Zeitschrift* **21**: 289-296.

Reitter E. 1877i: Beiträge zur Käferfauna von Japan. (Drittes Stück.). *Deutsche Entomologische Zeitschrift* **21**: 369-384.

Reitter E. 1878a: [new taxa]. In: Schneider O & Leder H.: Beiträge zur Kenntnis der kaukasischen Käferfauna. *Verhandlungen des Naturforschenden Vereins in Brünn* **16** [1877]: 1-258, pls 1-4.

Reitter E. 1878b: Beitrag zur Coleopteren-Fauna der Carpathen. *Deutsche Entomologische Zeitschrift* **22**: 33-64.

Reitter E. 1878c: Strongylus literatus n. sp. *Deutsche Entomologische Zeitschrift* **22**: 89.
Reitter E. 1878d: Caenocara rufitarsis n. sp. *Deutsche Entomologische Zeitschrift* **22**: 90.
Reitter E. 1878e: Neue Käfer-Arten aus Algier und dem Orient. *Deutsche Entomologische Zeitschrift* **22**: 91-95.
Reitter E. 1878f: Hadrotoma quadriguttata n. sp. *Deutsche Entomologische Zeitschrift* **22**: 95.
Reitter E. 1878g: Synonymische Bemerkungen etc. *Deutsche Entomologische Zeitschrift* **22**: 96.
Reitter E. 1878h: Sericoderus Revelierei n. sp. europea. *Deutsche Entomologische Zeitschrift* **22**: 126.
Reitter E. 1878i: Henoticonus nov. gen. Cryptophagidarum. *Deutsche Entomologische Zeitschrift* **22**: 127-128.
Reitter E. 1878j: Coleopterorum species novae. *Verhandlungen der Kaiserlich-Königlichen Zoologisch-Botanischen Gesellschaft in Wien* **27** [1877]: 165-194.
Reitter E. 1878k: Über die europäischen Orthoperus-Arten. *Deutsche Entomologische Zeitschrift* **22**: 199-201.
Reitter E. 1879a: Dapsa nigripennis n. sp. *Verhandlungen des Naturforschenden Vereins in Brünn* **17**: 89.
Reitter E. 1879b: Neue Cucujidae des königl. Museums in Berlin. *Verhandlungen der Kaiserlich-Königlichen Zoologisch-Botanischen Gesellschaft in Wien* **28** [1878]: 185-195.
Reitter E. 1879c: Beitrag zur Kenntniss der Lyctidae. *Verhandlungen der Kaiserlich-Königlichen Zoologisch-Botanischen Gesellschaft in Wien* **28** [1878]: 195-199.
Reitter E. 1879d: Die Arten der Gattungen Sphindus und Aspidophorus analytisch dargestellt. *Verhandlungen der Kaiserlich-Königlichen Zoologisch-Botanischen Gesellschaft in Wien* **28** [1878]: 200-202.
Reitter E. 1879e: Verzeichniss der von H. Christoph in Ost-Sibirien gesammelten Clavicornier etc. *Deutsche Entomologische Zeitschrift* **23**: 209-226.
Reitter E. 1880a: Einige neue Coleopteren. *Verhandlungen des Naturforschenden Vereins in Brünn* **18** [1879]: 29-33.
Reitter E. 1880b: Coleopterologische Ergebnisse einer Reise nach Croatien und Slavonien. Unter Mitwirkung der Herren Dr. Eppelsheim und Dr. von Heyden. *Verhandlungen der Kaiserlich-Königlichen Zoologisch-Botanischen Gesellschaft in Wien* **29** [1879]: 35-56.
Reitter E. 1880c: Descriptions of three new species of Coleoptera collected during the recent scientific Sumatra-expedition. *Notes from the Leyden Museum* **2**: 41-44.
Reitter E. 1880d: Bestimmungs-Tabellen der Europäischen Coleopteren. I. Enthaltend die Familien: Cucujidae, Telmatophilidae, Tritomidae, Mycetaeidae, Endomychidae, Lyctidae und Sphindidae. *Verhandlungen der Kaiserlich-Königlichen Zoologisch-Botanischen Gesellschaft in Wien* **29** [1879]: 71-100.
Reitter E. 1880e: [new taxa]. In: Leder H.: Beitrag zur kaukasischen Käfer-Fauna. Unter Mitwirkung hervorragender Fachgenossen. *Verhandlungen der Kaiserlich-Königlichen Zoologisch-Botanischen Gesellschaft in Wien* **29** [1879]: 451-488.
Reitter E. 1880f: Neue Nitidulidae des Museo Civico di Storia Naturale, beschrieben von Edmund Reitter, in Wien. II Serie. *Annali del Museo Civico di Storia Naturale di Genova* **15** [1879-1880]: 454-460.
Reitter E. 1880g: Neue Coleopteren aus dem südöstlichen Russland, aufgefunden von Herrn J. Faust aus Helsingfors. *Verhandlungen der Kaiserlich-Königlichen Zoologisch-Botanischen Gesellschaft in Wien* **29** [1879]: 543-546.
Reitter E. 1881a: Die aussereuropäischen Dermestiden meiner Sammlung. Mit 70 Diagnosen neuer Arten. *Verhandlungen des Naturforschenden Vereins in Brünn* **19** [1880]: 27-60.
Reitter E. 1881b: Bestimmungs-Tabellen der europäischen Coleopteren. III Heft. I. Auflage. Enthaltend die Familien: Scaphidiidae, Lathridiidae und Dermestidae. *Verhandlungen der Königlich-Kaiserlichen Zoologisch-Botanischen Gesellschaft in Wien* **30** [1880]: 41-94.
Reitter E. 1881c:Two new species of Coleoptera collected during the recent scientific Sumatra-expedition. *Notes from the Leyden Museum* **3**: 75-76.
Reitter E. 1881d: Neue und seltene Coleopteren im Jahre 1880 in Süddalmatien und Montenegro gesammelt und beschrieben. Unter Mitwirkung der Herren Dr. Eppelsheim in Gründstadt, Dr. Kraatz in Berlin, L. Miller und Fr. Wachtl in Wien. *Deutsche Entomologische Zeitschrift* **25**: 177-230, 2 pls.
Reitter E. 1881e: Coleopterologische Ergebnisse einer Reise nach Croatien, Dalmatien und der Herzegowina im Jahre 1879. *Verhandlungen der Kaiserlich-Königlichen Zoologisch-Botanischen Gesellschaft* **30** [1880]: 201-228.
Reitter E. 1881f: Bestimmungs-Tabellen der europäischen Coleopteren. IV. Cistelidae, Georyssidae und Thorictidae. *Verhandlungen der Königlich-Kaiserlichen Zoologisch-Botanischen Gesellschaft in Wien* **31**: 67-96, pl. II.
Reitter E. 1882a: Über Malthodes brachypterus Kiesw. *Wiener Entomologische Zeitung* **1**: 28-31.

Reitter E. 1882b: Bestimmungs-Tabellen der europäischen Coleopteren. VI. Enthaltend die Familien Colydiidae, Rhysodidae, Trogositidae. *Verhandlungen des Naturforschenden Vereins in Brünn* **20** [1881]: 113-149.

Reitter E. 1882c: Histoire Naturelle des coléoptères de France. (Par E. Mulsant). Famille des Lathridiens, par le R.P. Fr. Marie-Joseph Belon. Lyon 1881. *Deutsches Entomologisches Zeitschrift* **26**: 161-165.

Reitter E. 1882d: Dapsa acuticolle nov. sp. *Il Naturalista Siciliano* **1**: 231.

Reitter E. 1883a: [new names]. In: Heyden L. F. J. D. von, Reitter E. & Weise J.: *Catalogus Coleopterorum Europae et Caucasi. Editio tertia.* Londini: E. Janson, Berolini: Libraria Nicolai, Parisiis: L. Buquet, 2 + 228 pp.

Reitter E. 1883b: Zwei neue ost-indische Coleopteren. *Notes from the Leyden Museum* **5**: 9-10.

Reitter E. 1883c: Neue Coleopteren aus Russland und Bemerkungen über bekannte Arten. *Revue Mensuelle d'Entomologie* **1**: 40-44, 111-117.

Reitter E. 1883d: Uebersicht der bekannten Litophilus-Arten. *Deutsche Entomologische Zeitschrift* **27**: 61-64.

Reitter E. 1883e: Ueber die deutschen Anommatus-Arten. *Wiener Entomologische Zeitung* **2**: 195-197.

Reitter E. 1883f: Revision der Alexia-Arten. *Deutsche Entomologische Zeitschrift* **27**: 236-242.

Reitter E. 1883g: Weitere Mittheilungen über Alexia. *Deutsche Entomologische Zeitschrift* **27**: 393-394.

Reitter E. 1884a: Diagnosen neuer Coleopteren aus Lenkoran. *Verhandlungen des Naturforschenden Vereins in Brünn* **22** [1883]: 3-10.

Reitter E. 1884b: Coleopterologische Notizen. *Wiener Entomologische Zeitung* **3**: 33-37.

Reitter E. 1884c: [new taxa]. In: Brenske E. & Reitter E.: Neuer Beitrag zur Käferfauna Griechenlands. Unter Mitwirkung der Herren Dr. Eppelsheim und L. Ganglbauer. *Deutsche Entomologische Zeitschrift* **28**: 17-100, 2 pls.

Reitter E. 1884d: Resultate einer coleopterologischen Sammel-Campagne auf den jonischen Inseln. *Deutsche Entomologische Zeitschrift* **28**: 101-122.

Reitter E. 1884e: Neue Coleopteren aus Syrien und Marocco. *Deutsche Entomologische Zeitschrift* **28**: 251-254.

Reitter E. 1884f: Einige neue Coleopteren aus Süd-Europa. *Deutsche Entomologische Zeitschrift* **28**: 255-259.

Reitter E. 1884g: Die Nitiduliden Japans. *Wiener Entomologische Zeitung* **3**: 257-272, 299-302.

Reitter E. 1884h: Bestimmungs-Tabellen der europäischen Coleopteren. XI. Bruchidae (Ptinidae). *Verhandlungen des Naturforschenden Vereins in Brünn* **22** [1883]: 295-323.

Reitter E. 1885a: *Bestimmungs-Tabellen der europäischen Coleopteren. I. Enthaltend die Familien: Cucujidae, Telmatophilidae, Tritomidae, Mycetaeidae, Endomychidae, Lyctidae und Sphindidae. II. Auflage.* Mödling: E. Reitter, 45 pp.

Reitter E. 1885b: Bestimmungs-Tabellen der europäischen Coleopteren. XII. Necrophaga (Platypsyllidae, Leptinidae, Silphidae, Anisotomidae und Clambidae). *Verhandlungen des Naturforschenden Vereins in Brünn* **23** [1884]: 3-122.

Reitter E. 1885c: Die Nitiduliden Japans. *Wiener Entomologische Zeitung* **4**: 15-18, 39-44, 75-80, 101-104, 141-142, 173-176.

Reitter E. 1885d: Coleopterologische Ergebnisse einer Excursion nach Bosnien im Mai 1884. *Deutsche Entomologische Zeitschrift* **29**: 193-216.

Reitter E. 1885e: Uebersicht der bekannten Dasytiscus-Arten. *Entomologische Nachrichten* **11**: 241-247.

Reitter E. 1885f: Coleopterologische Notizen, XIII. *Wiener Entomologische Zeitung* **4**: 273.

Reitter E. 1885g: [new taxa]. In: Heyden L.von, Eppelsheim E. & Reitter E.: Weitere Beiträge zur Coleopteren-Fauna des Amur-Gebietes. *Deutsches Entomologisches Zeitschrift* **29**: 299-307.

Reitter E. 1885h: Coleopterologische Notizen, XIV. *Wiener Entomologische Zeitung* **4**: 313-317.

Reitter E. 1885i: Neue Coleopteren aus Europa und den angrenzenden Ländern, mit Bemerkungen über bekannte Arten. *Deutsche Entomologische Zeitschrift* **29**: 353-397.

Reitter E. 1886: Ueber die Coleopteren-Gattung Dendrodipnis, Woll. aus Sumatra. *Notes from the Leyden Museum* **8**: 215-218.

Reitter E. 1887a: *Bestimmungs-Tabellen der europäischen Coleopteren. III. Heft. II. Auflage. Enthaltend die Familien: Scaphidiidae, Lathridiidae und Dermestidae*. Mödling: Edmund Reitter [1886], 75 pp.

Reitter E. 1887b: Coleopterologische Notizen. XX. *Wiener Entomologische Zeitung* **6**: 28-29.

Reitter E. 1887c: Una nuova Epuraea di Sicilia. *Il Naturalista Siciliano* **6**: 83-84.

Reitter E. 1887d: Coleopterologische Notizen. XXIII. *Wiener Entomologische Zeitung* **6**: 171-173.

Reitter E. 1887e: Insecta in itinere cl. N. Przewalskii in Asia centrali novissime lecta. VI. Clavicornia,

Lamellicornia et Serricornia. *Horae Societatis Entomologicae Rossicae* **21**: 201-234.

Reitter E. 1887f: Ueber Elater-Arten aus der Verwandtschaft des E. ochtropterus Eschsch. *Wiener Entomologische Zeitung* **6**: 211-213.

Reitter E. 1887g: Neue Coleopteren aus Europa, den angrenzenden Ländern und Sibirien, mit Bemerkungen über bekannte Arten. Dritter Theil. *Deutsche Entomologische Zeitschrift* **31**: 241-288.

Reitter E. 1887h: Neue Coleopteren aus Europa, den angrenzenden Ländern und Sibirien, mit Bemerkungen über bekannte Arten. Vierter Theil. *Deutsche Entomologische Zeitschrift* **31**: 497-528.

Reitter E. 1888a: Bestimmungs-Tabellen der europäischen Coleopteren. XVI. Heft. Enthaltend die Familien: Erotylidae und Cryptophagidae. *Verhandlungen des Naturforschenden Vereins in Brünn* **26** [1887]: 1-56.

Reitter E. 1888b: Coleopterologische Notizen. XXIX. *Wiener Entomologische Zeitung* **7**: 104-105.

Reitter E. 1888c: Coleopteren aus Circasien, gesammelt von Hans Leder im Jahre 1887. III. Theil. *Wiener Entomologische Zeitung* **7**: 143-156.

Reitter E. 1888d: Coleopteren aus Circassien, gesammelt von Hans Leder im Jahr 1887. IV.Theil. *Wiener Entomologische Zeitung* **7**: 169-180.

Reitter E. 1888e: Coleopteren aus Circassien, gesammelt von Hans Leder im Jahre 1887. V. Theil. *Wiener Entomologische Zeitung* **7**: 207-214.

Reitter E. 1888f: Uebersicht der Arten der Coleopterengattung Alexia Steph. aus Europa und den angrenzenden Ländern. *Wiener Entomologische Zeitung* **7**: 322-327.

Reitter E. 1888g: Neue Coleopteren aus Europa, den angrenzenden Ländern und Sibirien, mit Bemerkungen über bekannte Arten. Fünfter Theil. *Deutsche Entomologische Zeitschrift* **1888**: 417-432.

Reitter E. 1889a: Neue Coleopteren aus Europa, den angrenzenden Ländern und Sibirien, mit Bemerkungen über bekannte Arten. Sechster Theil. *Deutsche Entomologische Zeitschrift* **1889**: 17-44.

Reitter E. 1889b: Analytische Tabelle zur Bestimmung der europäischen Throscus-Arten. *Wiener Entomologische Zeitung* **8**: 35-37.

Reitter E. 1889c: Coleoptera aus Circassien gesammelt vons Hans Leder in Jahre 1887. X. Theil. *Wiener Entomologische Zeitung* **8**: 63-70.

Reitter E. 1889d: Coleopterologische Ergebnisse der im Jahre 1886 und 1887 in Transcaspien von Dr. G. Radde, Dr. A. Walter und A. Konchin ausgeführten Expedition. *Verhandlungen des Naturforschenden Vereins in Brünn* **27** [1888]: 95-133.

Reitter E. 1889e: Neue Coleopteren aus Circassien, gesammelt von A. Starck. *Wiener Entomologische Zeitung* **8**: 97-104.

Reitter E. 1889f: Uebersicht der mir bekannten Elater-Arten der palaearktischen Fauna. *Entomologische Nachrichten* **15**: 110-116.

Reitter E. 1889g: Coleopterologische Notizen. XXXI. *Wiener Entomologische Zeitung* **8**: 125-128.

Reitter E. 1899h: Ueber einige Coleopteren aus der palaearctischen Fauna und aus Japan. *Entomologische Nachrichten* **14**: 216-221.

Reitter E. 1889i: Zwei neue Trogositiden aus Japan. *Wiener Entomologische Zeitung* **8**: 217.

Reitter E. 1889j: Berichte über die von E. v. Oetzen im Jahre 1887 in Griechenland u. Klein-Asien gesammelten Coleopteren. IX. Neue Arten aus verschiedenen Familien. *Deutsche Entomologische Zeitschrift* **1889**: 251-259.

Reitter E. 1889l: Bemerkungen zu europäischen Elater-Arten. *Deutsche Entomologische Zeitschrift* **1889**: 260.

Reitter E. 1889m: Neue Coleopteren aus Europa und den angrenzenden Ländern und Sibirien, mit Bemerkungen über bekannte Arten. Siebenter Theil. *Deutsche Entomologische Zeitschrift* **1889**: 273-288.

Reitter E. 1889n: Zwei neue Coleopteren-Gattungen aus Transkaukasien. *Wiener EntomologischeZeitung* **8**: 289-292.

Reitter E. 1889o: Bemerkungen und Berichtigungen zu den Clavicornen in der Fauna Baltica 2. Aufl. und Fauna Transsylvanica von Dr. G. Seidlitz. Ein Beitrag zur Lösung strittiger, coleopterologischer Fragen. *Deutsche Entomologische Zeitschrift* **1889**: 289-318.

Reitter E. 1889p: Die bekannten Cryptophagiden Japans, mit Beschreibungen neuer, in den Jahren 1880 und 1881 von George Lewis gesammelten Arten. *Wiener Entomologische Zeitung* **8**: 299-304.

Reitter E. 1889q: Eine neue Alexia aus Algier. *Wiener Entomologische Zeitung* **8**: 304.

Reitter E. 1889r: Verzeichniss der Cucujiden Japans mit Beschreibungen neuer Arten. *Wiener Entomologische Zeitung* **8**: 313-320.

Reitter E. 1889s: Neue Coleopteren aus Europa, den angrenzenden Ländern und Sibirien, mit Bemerkungen über bekannte Arten. Theil VIII. *Deutsche Entomologische Zeitschrift* **1889**: 369-376.

Reitter E. 1889t: Insecta, a Cl. G. N. Potanin in China et in Mongolia novissime lecta. VIII. Clavicornia. Hydrophilidae. Bruchidae. *Horae Societatis Entomologicae Rossicae* **23**: 555-559.

Reitter E. 1890a: Neue Coleopteren aus Europa, den angrenzenden Ländern und Sibirien, mit Bemerkungen über bekannte Arten. Neunter Theil. *Deutsche Entomologische Zeitschrift* **1890**: 145-164.

Reitter E. 1890b: Neue Coleopteren aus Europa, den angrenzenden Ländern und Sibirien, mit Bemerkungen über bekannte Arten. Zehnter Theil. *Deutsche Entomologische Zeitschrift* **34**: 165-176.

Reitter E. 1890c: Beschreibungen neuer Coleopteren aus Europa, dem Kaukasus, Russisch-Armenien und Turkestan. *Wiener Entomologische Zeitung* **9**: 189-198.

Reitter E. 1890d: Uebersicht der bekannten Arten der Coleopteren-Gattung Athous aus dem Kaukasus. *Entomologische Nachrichten* **16**: 241-247.

Reitter E. 1890e: Eine neue mit Athous verwandte Elateriden-Gattung aus russisch Armenien. *Entomologische Nachrichten* **16**: 247-248.

Reitter E. 1890f: Ein neuer Athous aus Montenegro. *Entomologische Nachrichten* **16**: 249.

Reitter E. 1890g: Ueber Silvanus surinamensis L. und Verwandte. *Wiener Entomologische Zeitung* **9**: 255-256.

Reitter E. 1890h: Coleopterologische Notizen. XXXIX. *Wiener Entomologische Zeitung* **9**: 264-267.

Reitter E. 1890i: [new taxa]. In: Heyden L. F. J. D. von: XIII. Beitrag zur Coleopteren-Fauna von Turkestan. Unter Mitwirkung der Herren Reitter und Weise. *Deutsche Entomologische Zeitschrift* **1890**: 353-367.

Reitter E. 1890j: Neue Coleopteren aus Europa und den angrenzenden Ländern und Sibirien, mit Bemerkungen über bekannte Arten. Elfter Theil. *Deutsche Entomologische Zeitschrift* **1890**: 385-396.

Reitter E. 1891a: [new names]. In: Heyden L. F. J. D. von, Reitter E. & Weise J.: *Catalogus Coleopterorum Europae, Caucasi et Armeniae Rossicae*. Mödling: Edmund Reitter, viii + 420 pp.

Reitter E. 1891b: Neue Coleopteren aus Europa und den angrenzenden Ländern und Sibirien, mit Bemerkungen über bekannte Arten. Zwölfter Theil. *Deutsche Entomologische Zeitschrift* **1891**: 17-36.

Reitter E. 1891c: Übersicht der Arten der Coleopteren-Gattung Aeolus Esch. aus Europa und Russisch-Asien. *Wiener Entomologische Zeitung* **10**: 145-148.

Reitter E. 1891d: Zweiter Beitrag zur Coleopteren-Fauna des russischen Reiches. *Wiener Entomologische Zeitung* **10**: 195-199.

Reitter E. 1891e: Dritter Beitrag zur Coleopterenfauna des russischen Reiches. *Wiener Entomologische Zeitung* **10**: 221-224.

Reitter E. 1891f: Coleopterologische Notizen. XLI. *Wiener Entomologische Zeitung* **10**: 226-228.

Reitter E. 1891g: Vierter Beitrag zur Coleopteren-Fauna des russischen Reiches. *Wiener Entomologische Zeitung* **10**: 233-240.

Reitter E. 1891h: Erster Beitrag zur Coleopteren-Fauna von Europa und den angrenzenden Ländern. *Wiener Entomologische Zeitung* **10**: 246-249.

Reitter E. 1891i: Zweiter Beitrag zur Coleopteren-Fauna von Europa und den angrenzenden Ländern. *Wiener Entomologische Zeitung* **10**: 259-262.

Reitter E. 1892a: Sechster Beitrag zur Coleopteren-Fauna des russischen Reiches. *Wiener Entomologische Zeitung* **11**: 133-136.

Reitter E. 1892b: Siebenter Beitrag zur Coleopterenfauna des russischen Reiches. *Wiener Entomologische Zeitung* **11**: 151-154.

Reitter E. 1892c: Zwei neue Triplax-Arten. *Il Naturalista Siciliano* **11**: 257.

Reitter E. 1893a: Achter Beitrag zur Coleopteren-Fauna des russischen Reiches. *Wiener Entomologische Zeitung* **12**: 109-114.

Reitter E. 1893b: Dritter Beitrag zur Coleopteren-Fauna von Europa und den angrenzenden Ländern. *Wiener Entomologische Zeitung* **12**: 172-176.

Reitter E. 1893c: Zwei neue Trichodes-Arten. *Wiener Entomologische Zeitung* **12**: 258.

Reitter E. 1893d: Ein neuer Trichodes aus der Verwandtschaft des Trichodes spectabilis. *Wiener Entomologische Zeitung* **12**: 303-304.

Reitter E. 1894a: *Bestimmungs-Tabellen der europäischen Coleopteren*: *Cantharidae. 1 Theil. Drilini. Heft XXIX.* Paskau: Edmund Reitter, 8 pp.

Reitter E. 1894b: Ein neuer Lathridius aus Ostgalizien. *Wiener Entomologische Zeitung* **13**: 14.

Reitter E. 1894c: Analytische Uebersicht der europäischen Arten der Coleopteren-Gattung Epuraea Er. *Verhandlungen des Naturforschenden Vereins in Brünn* **32** [1893]: 18-36.

Reitter E. 1894d: Bestimmungs-Tabelle der Coleopteren-Familie der Cleriden des palaearctischen Faunengebietes. *Verhandlungen des Naturforschenden Vereins in Brünn* **32** [1893]: 37-89.

Reitter E. 1894e: Die Coleopteren-Gattung Podistrina Fairm. und deren Arten. *Wiener Entomologische Zeitung* **13**: 46-48.

Reitter E. 1894f: Vierter Beitrag zur Coleopteren-Fauna von Europa und den angrenzenden Ländern. *Wiener Entomologische Zeitung* **13**: 101-107.

Reitter E. 1894g: Elfter Beitrag zur Coleopteren-Fauna des russischen Reiches. *Wiener Entomologische Zeitung* **13**: 238-248.

Reitter E. 1894h: Coleopterologische Notizen. LIII: *Wiener Entomologische Zeitung* **13**: 251-254.

Reitter E. 1894i: Fünfter Beitrag zur Coleopteren-Fauna von Europa und den angrenzenden Ländern. *Wiener Entomologische Zeitung* **13**: 299-306.

Reitter E. 1895a: Beschreibung neuer oder wenig gekannter Coleopteren aus der Umgebung von Akbes in Syrien. *Wiener Entomologische Zeitung* **14**: 79-88.

Reitter E. 1895b: Übersicht der Hypnoidus-Arten aus der Verwandtschaft des quadriguttatus Cast. der palaearctischen Fauna (Subgen. Zorochus Thoms.). *Entomologische Nachrichten* **21**: 87-91.

Reitter E. 1895c: Sechster Beitrag zur Coleopteren-Fauna von Europa und den angrenzenden Ländern. *Wiener Entomologische Zeitung* **14**: 199-204.

Reitter E. 1895d: Uebersicht der Elateriden- (Coleopteren-) Arten aus der Verwandtschaft des Limonius (Pheletes) aeneo-niger Deg. aus der palaearctischen Fauna. *Wiener Entomologische Zeitung* **14**: 256.

Reitter E. 1895e: Einiges über bekannte und neue Thorictus. *Deutsche Entomologische Zeitschrift* **1895**: 295-296.

Reitter E. 1895f: Atomaria myrmecophila n. sp. *Deutsche Entomologische Zeitschrift* **1895**: 310.

Reitter E. 1896a: Coleopterologische Notizen. LVI. *Wiener Entomologische Zeitung* **15**: 17-18.

Reitter E. 1896b: Beschreibungen neuer Coleopteren aus dem russischen Reiche. *Deutsche Entomologische Zeitschrift* **1896**: 33-48.

Reitter E. 1896c: Dreizehnter Beitrag zur Coleopteren-Fauna von Europa und den angrenzenden Ländern. *Wiener Entomologische Zeitung* **15**: 64-77.

Reitter E. 1896d: Uebersicht der bekannten Arten der Coleopteren-Gattung: Adrastus Esch. aus Europa und den angrenzenden Ländern. *Deutsche Entomologische Zeitschrift* **1896**: 157-159.

Reitter E. 1896e: Abbildungen mit Beschreibungen zehn neuer Coleopteren aus der palearctischen Fauna. *Wiener Entomologische Zeitung* **15**: 233-240, pl. III.

Reitter E. 1896f: Achter Beitrag zur Coleopteren-Fauna von Europa und den angrenzenden Ländern. *Wiener Entomologische Zeitung* **15**: 265-272.

Reitter E. 1896g: Vierzehnter Beitrag zur Coleopteren-Fauna des russischen Reiches. *Wiener Entomologische Zeitung* **15**: 285-291.

Reitter E. 1896h: Uebersicht der bekannten palaearctischen Arten der Coleopteren Gattung Brachyleptus Motsch. *Entomologische Nachrichten* **22**: 293-296.

Reitter E. 1897a: Neunter Beitrag zur Coleopteren-Fauna von Europa und den angrenzenden Ländern. *Wiener Entomologische Zeitung* **16**: 45-47.

Reitter E. 1897b: Coleopterologische Notizen LX. *Wiener Entomologische Zeitung* **16**: 77-80.

Reitter E. 1897c: Fünfzehnter Beitrag zur Coleopteren-Fauna des russischen Reiches. *Wiener Entomologische Zeitung* **16**: 121-127.

Reitter E. 1897d: Zehnter Beitrag zur Coleopteren-Fauna von Europa und den angrenzenden Ländern. *Wiener Entomologische Zeitung* **16**: 203-206.

Reitter E. 1897e: Dreissig neue Coleopteren aus russisch Asien und der Mongolei. *Deutsche Entomologische Zeitschrift* **1897**: 209-228.

Reitter E. 1897f: Coleopterologische Notizen LXI. *Wiener Entomologische Zeitung* **16**: 217-220.

Reitter E. 1897g: Abbildungen mit Beschreibungen acht neuer Coleopteren aus der palearctischen Fauna. *Wiener Entomologische Zeitung* **16**: 246-252, pl. 3.

Reitter E. 1897h: 2 neue russische Meligethes. *Deutsche Entomologische Zeitschrift* **1896**: 312.

Reitter E. 1898a: Eine Decade neuer Coleopteren aus der Buchara. *Wiener Entomologische Zeitung* **17**: 10-16.

Reitter E. 1898b: Sechzehnter Beitrag zur Coleopteren-Fauna des russischen Reiches. *Weiner Entomologische Zeitung* **17**: 17-22.

Reitter E. 1898c: Ueber die nächsten Verwandtenten der Corticaria fulva Com. und einige Arten aus anderen Gruppen. *Wiener Entomologische Zeitung* **17**: 28-32.

Reitter E. 1898d: Coleopterologische Notizen. LXIII. *Wiener Entomologische Zeitung* **17**: 54-56.

Reitter E. 1898e: Siebzehnter Beitrag zur Coleopteren-Fauna des russischen Reiches. *Wiener Entomologische Zeitung* **17**: 109-127.

Reitter E. 1898f: Die europäisch-kaukasischen Arten der Coleopteren-Gattung: Hedobia Strm. *Wiener Entomologische Zeitung* **17**: 137-139.

Reitter E. 1898g: Uebersicht der Coleopteren-Gattung Megapenthes Kiesw. aus Europa und dem Kaukasus. *Entomologische Nachrichten* **24**: 180-181.

Reitter E. 1898h: Neue Coleopteren aus Europa und den angrenzenden Ländern. *Deutsche Entomologische Zeitschrift* **1898**: 337-360.

Reitter E. 1899a: Elfter Beitrag zur Coleopteren-Fauna von Europa und der angrenzenden Ländern. *Wiener Entomologische Zeitung* **17**: 155-161.

Reitter E. 1899b: Beitrag zur Coleopteren-Fauna des russischen Reiches und den angrenzenden Ländern. *Deutsche Entomologische Zeitschrift* **1899**: 193-209.

Reitter E. 1899c: Ueber einige Coleopteren aus der palaearctischen Fauna und aus Japan. *Entomologische Nachrichten* **25**: 216-220.

Reitter E. 1899d: Neue Coleopteren aus Europa, den angrenzenden Ländern, russisch Asien und der Mongolei. *Deutsche Entomologische Zeitschrift* **1899**: 273-281.

Reitter E. 1899e: Abbildungen und Beschreibungen neuer oder wenig bekannter Coleopteren aus der palaearctichen Fauna. *Wiener Entomologische Zeitung* **18**: 282-287.

Reitter E. 1900a: Beitrag zur Coleopteren-Fauna des russischen Reiches. *Deutsche Entomologische Zeitschrift* **1900**: 49-59.

Reitter E. 1900b: Beitrag zur Coleopteren-Fauna von Europa und den angrenzenden Ländern. *Deutsche Entomologische Zeitschrift* **1900**: 81-88.

Reitter E. 1900c: Uebersicht der mir bekannten Arten der Coleopteren-(Elateriden-)Gattung Pleonomus Mén. aus Central-Asien. *Entomologische Nachrichten* **26**: 87-89.

Reitter E. 1900d: Coleopterologische Notizen, LXIX. *Wiener Entomologische Zeitung* **19**: 132.

Reitter E. 1900e: Coleoptera, gesammelt in Jahre 1898 in Chin. Central-Asien von Dr. Holderer in Lahr. *Wiener Entomologische Zeitung* **19**: 153-166, pl. 1.

Reitter E. 1900f: Neue, von Herrn Dr. John Sahlberg auf seinem Reisen in Corfu, Palästina und Centralasien gesammelte Coleopteren. *Wiener Entomologische Zeitung* **19**: 217-220.

Reitter E. 1900g: Beschreibung und Abbildung von neun Coleopteren der palaearctischen Fauna. *Wiener Entomologische Zeitung* **19**: 225-232, pl. 2.

Reitter E. 1900h: Neue Coleopteren aus Europa, der angrenzenden Ländern, russisch Asien und der Mongolei. *Deutsche Entomologische Zeitung* **1899**: 273-281.

Reitter E. 1901a: Analytische Uebersicht der palaearctischen Gattungen und Arten der Coleopteren-Familien: Byrrhidae (Anobiidae) und Cioidae. *Verhandlungen des Naturforschenden Vereins in Brünn* **40**: 3-64.

Reitter E. 1901b: Zwölfter Beitrag zur Coleopteren-Fauna von Europa und den angrenzenden Ländern. *Wiener Entomologische Zeitung* **20**: 22-23.

Reitter E. 1901c: Coleopterologische Notizen. *Wiener Entomologische Zeitung* **20**: 57-59.

Reitter E. 1901d: Ueber Throscus-Arten mit ganz ungetheilten Augen, aus der Gruppe des brevicollis Bonv. *Wiener Entomologische Zeitung* **20**: 60.

Reitter E. 1901e: Weitere Beiträge zur Coleopteren-Fauna des russischen Reiches. *Deutsche Entomologische Zeitschrift* **1901**: 65-84.

Reitter E. 1901f: Uebersicht der Arten der Coleopteren-Gattungen Triplax Payk. aus Europa und den angrenzenden Ländern. *Wiener Entomologische Zeitung* **20**: 73-76.

Reitter E. 1901g: Dreizehnter Beitrag zur Coleopteren-Fauna von Europa und den angrenzenden Ländern. *Wiener Entomologische Zeitung* **20**: 99-101.

Reitter E. 1901h: Abbildungen und Beschreibungen neuer Coleopteren aus den paläarctischen Fauna. *Wiener Entomologische Zeitung* **20**: 157-164, pl. 3.

Reitter E. 1901i: Eine Serie neuer Coleopteren aus dem russischen Reiche. *Deutsche Entomologische Zeitschrift* **1901**: 177-186.
Reitter E. 1901j: Die europäischen Arten der Coleopteren-Gattung Monotoma Hrbst. *Wiener Entomologische Zeitung* **20**: 193-197.
Reitter E. 1901l: Vierzehnter Beitrag zur Coleopteren-Fauna von Europa und den angrenzenden Ländern. *Wiener Entomologische Zeitung* **20**: 200-202.
Reitter E. 1902a: Neue und seltene Coleopteren, gesammelt im Jahre 1901, in der Herzegowina, in Dalmatien und Bosnien. *Wiener Entomologische Zeitung* **21**: 1-9.
Reitter E. 1902b: Neue Coleopteren der palaearktischen Fauna. *Wiener Entomologische Zeitung* **21**: 137-141.
Reitter E. 1902c: Eine Serie neuer Coleopteren aus dem russischen Reiche. *Deutsche Entomologische Zeitschrift* **1901**: 177-186.
Reitter E. 1902d: Beschreibungen neuer Coleopteren aus dem russischen Reiche. *Wiener Entomologische Zeitung* **21**: 185-190.
Reitter E. 1902e: Neue Coleopteren aus Europa und den angrenzenden Ländern. *Deutsche Entomologische Zeitschrift* **1901**: 187-188.
Reitter E. 1902f: Fünfzehnter Beitrag zur Coleopteren-Fauna von Europa und den angrenzenden Ländern. *Wiener Entomologische Zeitung* **21**: 191-196.
Reitter E. 1902g: Coleopterologische Studien. I. *Wiener Entomologische Zeitung* **21**: 203-217.
Reitter E. 1902h: Uebersicht der Arten der Gattung: Necrobinus. *Wiener Entomologische Zeitung* **21**: 212-213.
Reitter E. 1902i: Coleopterologische Notizen. *Wiener Entomologische Zeitung* **21**: 221-222.
Reitter E. 1902j: Uebersicht der mir bekannten Arten der Coleopteren-Gattung Cerallus Duv. *Wiener Entomologische Zeitung* **21**: 226-229.
Reitter E. 1902k: Uebersicht der Arten der Coleopteren-Gattung Trichoceble Thoms. (Iulistus Kiesw.). *Wiener Entomologische Zeitung* **21**: 257-260.
Reitter E. 1903a: Sechzehnter Beitrag zur Coleopteren-Fauna von Europa und den angrenzenden Ländern. *Wiener Entomologische Zeitung* **22**: 43-46.
Reitter E. 1903b: Vier neue Coleopteren der palaearktischen Fauna. *Wiener Entomologische Zeitung* **22**: 231-232.
Reitter E. 1903c: Uebersicht der Coleopteren-Arten aus der nächsten Verwandtschaft des Enicmus mannerheimi Kolen. *Wiener Entomologische Zeitung* **22**: 234.
Reitter E. 1903d: Dorcatoma Lomnickii n. sp. *Wiener Entomologische Zeitung* **22**: 239.
Reitter E. 1903e: Denticollis (Campylus) Jakobsoni n. sp. *Wiener Entomologische Zeitung* **22**: 280-281.
Reitter E. 1904a: Eine neue Mycetaeiden-Gattung aus Italien. *Wiener Entomologische Zeitung* **23**: 41-42.
Reitter E. 1904b: Alexia maritima n. sp. *Wiener Entomologische Zeitung* **23**: 45.
Reitter E. 1904c: Über vier Coleopteren aus der palaearktischen Fauna. *Wiener Entomologische Zeitung* **23**: 81-82.
Reitter E. 1904d: Drei neue Coleopteren aus Europa. *Wiener Entomologische Zeitung* **23**: 147-148.
Reitter E. 1904e: Sechzehn neue Coleopteren aus Europa und den angrenzenden Ländern. *Wiener Entomologische Zeitung* **23**: 151-160.
Reitter E. 1904f: Neue Arten des Coleopteren-Genus Athous aus Spanien. *Boletín de la Real Sociedad Española de Historia Natural* **4**: 236-239.
Reitter E. 1905a: Bestimmungs-Tabelle der palaearctischen, mit Athous verwandten Elateriden (Subtribus Athouina), mit einer Übersicht der verwandten Coleopteren-Familien: Sternoxia und mit einem Bestimmungsschlüssel der Gattungen der Elateriden. *Verhandlungen des Naturforschenden Vereins in Brünn* **43**: 1-122.
Reitter E. 1905b: Acht neue Coleopteren aus der palaearktischen Fauna. *Wiener Entomologische Zeitung* **24**: 90-94.
Reitter E. 1905c: Sechzehn neue Coleopteren aus palaearktischen Fauna. *Wiener Entomologische Zeitung* **24**: 241-251.
Reitter E. 1905d: Drei neue Coleopteren aus der palaearktischen Fauna. *Wiener Entomologische Zeitung* **24**: 311-312.
Reitter E. 1906ab: [new names]. In: Heyden L. F. J. D. von, Reitter E. & Weise J. (eds): *Catalogus Coleopterorum Europae, Caucasi et Armeniae Rossicae.* Berlin: R. Friedländer & Sohn; Paskau: Edmund Reitter; Caen: Revue d'Entomologie, [8] pp. +750 columns + pp. 750-774.

Reitter E. 1906b: Neue Coleopteren aus der palaearktischen Fauna. *Wiener Entomologische Zeitung* **25**: 31-37.
Reitter E. 1906c: Epuraea Ragusae nov. sp. *Il Naturalista Siciliano* **18** [1905-1906]: 219.
Reitter E. 1906d: Henoticus germanicus n. sp. (Col.). *Wiener Entomologische Zeitung* **25**: 231.
Reitter E. 1906e: Dreizehn neue Coleopteren aus der palaearktischen Fauna. *Wiener Entomologische Zeitung* **25**: 237-244.
Reitter E. 1906f: Übersicht der mir bekannten Campylus-Arten (Coleopteren), aus der Verwandtschaft des C. rubens Pill., aus Europa und den angrenzenden Ländern. *Wiener Entomologische Zeitung* **25**: 273-274.
Reitter E. 1906g: Ubersicht der Coleopteren-Arten aus der nächsten Verwandtschaft des Ptinus fur L., mit weissen Schuppenflecken auf den Flügeldecken. *Wiener Entomologische Zeitung* **25**: 281-284.
Reitter E. 1906h: Einige neue Coleopteren. *Deutsche Entomologische Zeitschrift* **50**: 449-451.
Reitter E. 1907a: Dictyoptera fiedleri sp. n. *Societas Entomologica* **22**: 25.
Reitter E. 1907b: Coleopterologische Notizen. *Wiener Entomologische Zeitung* **26**: 27-30.
Reitter E. 1907c: Rhizophagus striolatus n. sp. (Col.). *Deutsche Entomologische Zeitschrift* **1907**: 272.
Reitter E. 1908a: Siebzehn Coleopteren-Neuheiten aus der palaearktischen Fauna. *Wiener Entomologische Zeitung* **27**: 17-24.
Reitter E. 1908b: Beschreibung einiger neuen Käferarten von Egypten. *Bulletin de la Société Entomologique d'Egypte* **1**: 39-40, 41-56.
Reitter E. 1908c: Neue paläarktische Käfer. *Entomologische Blätter* **4**: 129.
Reitter E. 1908d: Analytische Übersicht der Arten der Coleopterengattung Metophthalmus Woll. aus der paläarktischen Fauna. (Col.). *Deutsche Entomologische Zeitschrift* **1908**: 133-136.
Reitter E. 1908e: Zwei neue Coleopteren aus Adana in Kleinasien. *Wiener Entomologische Zeitung* **27**: 133-136.
Reitter E. 1908f: Sieben neue Coleopteren aus Europa und den angrenzenden Ländern. *Wiener Entomologische Zeitung* **27**: 213-216.
Reitter E. 1908g: Acht neue Coleopteren aus Europa, Kleinasien und Turkestan. *Wiener Entomologische Zeitung* **27**: 243-247.
Reitter E. 1908h: Coleopterologische Notizen. *Wiener Entomologische Zeitung* **27**: 296.
Reitter E. 1909a: Due nuove specie di Sphaerosoma della Calabria (Sph. aspromontanum e Sph. sparsum). *Rivista Coleotterologica Italiana* **7**: 21-22.
Reitter E. 1909b: Einige neue Coleopteren aus der paläarktischen Fauna. *Wiener Entomologische Zeitung* **28**: 53-58.
Reitter E. 1909c: Neun neue Coleopterenarten und -Varietäten aus der paläarktischen Fauna. *Wiener Entomologische Zeitung* **28**: 99-103.
Reitter E. 1909d: Tre nuovi coleotteri italiani (Liodes calcaratoides, Agathidium labratum, Sphaerosoma apuanum). *Rivista Coleotterologica Italiana* **7**: 209-211.
Reitter E. 1909e: Neue Coleopteren aus der paläarktischen Fauna. *Wiener Entomologische Zeitung* **28**: 303-312.
Reitter E. 1909f: Übersicht der Arten der Coleopterengattung Thorictus Germ. aus den Kaukasusländern. *Wiener Entomologische Zeitung* **28**: 315-317.
Reitter E. 1910a: Zwei neue Thorictus aus Transkaspien. *Wiener Entomologische Zeitung* **29**: 128.
Reitter E. 1910b: Beschreibungen neuer Coleopteren aus der Familie der Elateriden und Bemerkungen über bekannte Arten. *Wiener Entomologische Zeitung* **29**: 165-178.
Reitter E. 1910c: Über die Arten der Lyciden-Gattung Lycostomus Motsch. (Col.). *Wiener Entomologische Zeitung* **29**: 204-205.
Reitter E. 1910d: Ueber eine wenig bekannte Gattung der Dermestiden. *Entomologische Zeitschrift* **23**: 239.
Reitter E. 1910e: Übersicht der europäischen Anommatus-Arten (Coleopt.). *Wiener Entomologische Zeitung* **29**: 264-266.
Reitter E. 1911a: *Fauna Germanica. Die Käfer des Deutschen Reiches. Nach der analytischen Methode bearbeitet. III. Band*. Stuttgart: K. G. Lutz, 436 pp., pls 81-128.
Reitter E. 1911b: Paläarktische Coleopterennovitäten. *Wiener Entomologische Zeitung* **30**: 47-55.
Reitter E. 1911c: Coluocera punctata Märk. subsp. sardoa nov. *Wiener Entomologische Zeitung* **30**: 70.
Reitter E. 1911d: Zwei neue Coleopteren-Varietäten. *Wiener Entomologische Zeitung* **30**: 291.
Reitter E. 1911e: Coleopterologische Notizen. *Wiener Entomologische Zeitung* **30**: 294-295.
Reitter E. 1912a: Zwei neue Coleopteren. *Wiener Entomologische Zeitung* **31**: 104.
Reitter E. 1912b: Namensänderung. *Wiener Entomologische Zeitung* **31**: 108.

Reitter E. 1912c: Cryptophagus schroetteri n. sp. (Col.). *Wiener Entomologische Zeitung* **31**: 252.

Reitter E. 1912d: Athous hilfi n. sp. (Col. Elateridae). *Wiener Entomologische Zeitung* **31**: 317.

Reitter E. 1913a: Gastrallus indicus n. sp. (Col. Anob.). *Wiener Entomologische Zeitung* **32**: 16.

Reitter E. 1913b: Coleopterologische Novitäten der palaearktischen Fauna. *Coleopterologische Rundschau* **2**: 121-125.

Reitter E. 1913c: Drei neue Phalacriden. *Coleopterologische Rundschau* **2**: 125-126.

Reitter E. 1913d: Neue paläarktische Coleopteren. *Entomologische Blätter* **9**: 289-292.

Reitter E. 1913e: Eine Serie neuer Coleopteren aus der paläarktischen Fauna. *Deutsche Entomologische Zeitschrift* **1913**: 649-668.

Reitter E. 1914a: Schematische Übersicht der Anostirus Thoms. (Calosirus Thoms.) der Elateriden-Gattung Corymbites Latr. (Col.). *Berliner Entomologische Zeitschrift* **58** [1913]: 194-196.

Reitter E. 1914b: Athous (Subg. Grypocarus Thoms.) sardiniensis n. sp. *Wiener Entomologische Zeitung* **33**: 266-267.

Reitter E. 1915a: Drei neue Coleopteren-Arten der Gattung Anommatus Wesmael. *Entomologische Blätter* **11**: 40-42.

Reitter E. 1915b: Zwei neue paläarktische Käferarten. *Entomologische Blätter* **11**: 42-43.

Reitter E. 1915c: Cartodere (Cart. s.str.) subcostella n.sp. *Wiener Entomologische Zeitung* **34**: 66.

Reitter E. 1915d: Theca sardoa n. sp. (Col. Anobiidae). *Wiener Entomologische Zeitung* **34**: 116.

Reitter E. 1915e: Coleopterologishe Notizen. *Wiener Entomologische Zeitung* **34**: 124-126.

Reitter E. 1915f: Neue Coleopteren aus Aegypten. *Bulletin de la Société Entomologique d'Égypte* **6** [1913]: 135-137.

Reitter E. 1916: Ein neuer Ptilinus vom Ural. *Coleopterologische Rundschau* **5**: 85-86.

Reitter E. 1918a: Einige neue Coleopteren aus Albanien. *Entomologische Blätter* **14**: 42-45.

Reitter E. 1918b: Bestimmungs-Tabelle der paläarktischen Elater-Arten. *Wiener Entomologische Zeitung* **37**: 81-105.

Reitter E. 1919: Bestimmungs-Tabelle der Coleopterenfamilien: Nitidulidae und Byturidae aus Europa und den angrenzenden Ländern. *Verhandlungen des Naturforschenden Vereins in Brünn* **56** [1918-1919]: 1-104.

Reitter E. 1920: Coleopterologische Notizen. *Entomologische Blätter* **15** [1919]: 220-221.

Reitter E. 1921a: Bestimmungs-Tabellen des europaeischen Coleopteren 90. Trixagidae, Eucnemidae, Cerophytidae und Phylloceridae. *Wiener Entomologische Zeitung* **38**: 65-90.

Reitter E. 1921b: Cryptophagus cellaroides n. sp. (Col.). *Wiener Entomologische Zeitung* **38**: 90.

Reitter E. 1922: *Bestimmungs-Tabellen der europaeischen Coleopteren. VI. Heft. Enhaltend die Familien: Colydiidae, Rhysodidae, Ostomidae. Zweite, ganzlich umgearbeitete und auf die palaearktische Fauna ausgedehnte Auflage*. Troppau: E. Reitter, 73 pp.

Ren S.-X. & Pang X.-F. 1992: The Axinoscymnus Kamiya (Col.: Coccinellidae) from China. *Coccinella* (Passau) **4**: 20-23.

Ren S.-X. & Pang X.-F. 1993: Two new species of Scymnus Kugelann from Hubei (Coleoptera: Cocinellidae). *Journal of South China Agricultural University* (Guangzhou) **14** (3): 6-9.

Ren S.-X. & Pang X.-F. 1994a: Three new species of Scymnus (Coleoptera: Coccinellidae) from Guizhou, China. *Entomotaxonomia* **16**: 257-262.

Ren S.-X. & Pang X.-F. 1994b: The genus Nephus Mulsant (Coleoptera, Coccinellidae) of China. *Elytra* **22**: 325-333.

Ren S.-X. & Pang X.-F. 1995a: Four new species of Scymnus from China (Insecta, Coleoptera, Coccinellidae). *Spixiana* **18**: 151-155.

Ren S.-X. & Pang X.-F. 1995b: Three new species of the genus Scymnus Kugelann from China (Coleoptera: Coccinellidae). *Acta Zootaxonomia Sinica* **20**: 219-224.

Ren S.-X. & Pang X.-F. 1995c: Two new species of Scymnus Kugelann from China (Coleoptera: Coccinellidae). *Acta Entomologica Sinica* **38**: 467-470.

Ren S.-X. & Pang X.-F. 1996: The genus Stethorus Weise (Coleoptera, Coccinellidae) of China. *Elytra* **24**: 317-329.

Reška M. 1977: Cryptophagus vseteckai sp. n. aus der Tschechoslowakei (Coleoptera, Cryptophagidae). *Acta Entomologica Bohemoslovaca* **74**: 394-397.

Reška M. 1982: Zwei neue europäische Arten der Gattung Cryptophagus (Coleoptera, Cryptophagidae). *Acta Entomologica Bohemoslovaca* **79**: 457-464.

Reška M. 1994: Bestimmungstabellen der mitteleuropäischen Arten der Gattungen Micrambe Thomson und Cryptophagus Herbst (Insecta: Coleoptera: Cryptophagidae). *Annalen des Naturhistorischen Museums in Wien* (B) **96**: 247-342.

Rey C. 1887: *Essai d'études sur certaines larves de coléoptères et descriptions de quelques espèces inédites ou peu connues*. Beaune (Côte-d'Or): Bibliothèque entomologique de Ed. André, 126 pp.

Rey C. 1889a: Remarques en passant (Suite). *L'Échange, Revue Linnéenne* **5** (50): 3-4.

Rey C. 1889b: Remarques en passant. Coléoptères (Suite) Familles des Phalacrides. Famille des Corylophides. *L'Échange, Revue Linnéenne* **5**: 23.

Rey C. 1889c: Remarques en passant. Famille des Nitidulides. *L'Échange. Revue Linnéenne* **5** (51): 3-4.

Rey C. 1889d: Remarques en passant. Famille des Nitidulides. *L'Échange. Revue Linnéenne* **5** (52): 27-29.

Rey C. 1889e: Remarques en passant. Famille des Colydides, Famille des Cucujides, Famille des Cryptophagides. *L'Échange, Revue Linnéenne* **5** (53): 35-36.

Rey C. 1889f: Remarques en passant. Cryptophagides (suite). *L'Échange, Revue Linnéenne* **5** (54): 43-44.

Rey C. 1889g: Remarques en passant. Cryptophagides (suite) Famille des Monotomides. *L'Échange. Revue Linnéenne* **5** (55): 53-54.

Rey C. 1889h: Remarques en passant. Famille des Lathridides. *L'Échange. Revue Linnéenne* **5** (55): 54-55.

Rey C. 1889i: Remarques en passant. Familles des Thorictides Famille des Dermestides. *L'Échange, Revue Linnéenne* **5** (56): 58-59.

Rey C. 1891a: Remarques en passant (suite). *L'Échange, Revue linnéenne* **7**: 26-27.

Rey C. 1891b: Remarques en passant. Elatérides (Suite). *L'Échange* **7**: 68.

Rey C. 1891c: Remarques en passant. Elatérides (Suite). *L'Échange* **7**: 85.

Rey C. 1891d: Remarques en passant. Familles des Lycides et Lampyrides. *L'Échange, Revue Linnéenne* **7**: 114.

Rey C. 1891e: Remarques en passant. Famille des Téléphorides. *L'Échange, Revue Linnéenne* **7**: 114-115.

Rey C. 1891f: Remarques en passant (Suite). *L'Échange, Revue Linnéenne* **7**: 130-131.

Rey C. 1892: Remarques en passant. Famille des Térédiles ou Anobides. *L'Échange, Revue Linnéenne* **8**: 2-3.

Reymond A. 1956: Presentation d'insectes (Coléoptères) nouveaux de la faune marocaine. *Comptes Rendus Mensuelles des Séances de la Société des Sciences Naturelles et Physique du Maroc* **22**: 40-42.

Richardson J. 1837: *Fauna Boreali-Americana; or the Zoology of the Northern parts of British America: containing of the objects of natural history collected on the late northern land expeditions, under command of Captain Sir John Franklin, R.N.* Norwich: J. Fletcher, 325 pp.

Richter C. F. W. 1820: *Supplementa faunae sectorum Europae. Teil 1.* Vratislaviae: Richter, 12 pls.

Ritsema C. 1881: Two new species of the coleopterous genus Helota, Mac Leay. *Notes from the Leyden Museum* **3**: 79-81.

Ritsema C. 1885: Description of a new species of the Nitidulid genus Platynema, Rits. (Orthogramma, Murray) nec Guenés). *Notes from the Leyden Museum* **7**: 29-30.

Ritsema C. 1889a: Preliminary descriptions of new species of the coleopterous genus Helota, MacLeay. *Notes from the Leyden Museum* **11**: 99-111.

Ritsema C. 1889b: On an undescribed species of the coleopterous genus Helota, MacLeay. *Notes from the Leyden Museum* **11**: 189-190.

Ritsema C. 1891: Viaggio di Leonardo Fea in Birmania e regioni vicine. The genus Helota, as represented in the Civic Museum of Natural History at Genoa, with descriptions of the new species collected in Burma by Mr. L. Fea. *Annali del Museo Civico di Storia Naturale di Genova* **30** [1890-1892]: 885-898.

Ritsema C. 1893: Five new species of the genus Helota from Sikkim and Darjeeling. *Notes from the Leyden Museum* **15**: 131-140.

Ritsema C. 1894: On a collection of Helotidae from Kurseong. *Notes from the Leyden Museum* **14**: 111-118.

Ritsema C. 1895: A new species of the genus Helota from Thibet. *Notes from the Leyden Museum* **17**: 49-50.

Ritsema C. 1899: Three new species of the genus Helota. *Notes from the Leyden Museum* **20**: 249-254.

Ritsema C. 1900: Two new species of the genus Helota from British Bhotan. *Notes from the Leyden Museum* **22**: 27-32.

Ritsema C. 1905: Eight new Asiatic species of the coleopterous genus Helota. *Notes from the Leyden Museum* **25**: 117-132.

Ritsema C. 1906: Une nouvelle espèce chinoise du genre Helota. *Bulletin de la Museum d'Histoire Naturelle du Paris* **12**: 196-198.

Ritsema C. 1907: Two new Asiatic species of the coleopterous genus Helota. *Notes from the Leyden Museum* **28**: 213-217.

Ritsema C. 1911: A species of the coleopterous genus Helota from Formosa. *Notes from the Leyden Museum* **34**: 51-53.

Ritsema C. 1914a: Two new Asiatic species of the coleopterous genus Helota. *Notes from the Leyden Museum* **36**: 57-59.

Ritsema C. 1914b: On three species of the coleopterous genus Helota. *Notes from the Leyden Museum* **36**: 165-169.

Ritsema C. 1915: Description of five new Asiatic species of the coleopterous genus Helota. *Tijdschrift voor Entomologie* **58**: 244-250.

Robert J. 1909: Meligethes capucinus, sp. nov. *Bulletin de la Société Linnéenne de Provence* **1** [1909-1912]: 42-43.

Roelofs P. J. 1939: Un Cryptophagidae nouveau de Belgique. *Bulletin et Annales de la Société Royale d'Entomologie de Belgique* **79**: 33.

Roelofs P. J. 1946: Coléoptères capturés en Belgique. *Bulletin et Annales de la Société Royale d'Entomologie de Belgique* **82**: 239-242.

Roemer J. J. 1789: *Genera Insectorum Linnei et Fabricii iconibus illustrata.* Vitoduri: Steiner, 86 pp., 37 pls.

Roepke W. 1919: Ein termitophile Trochoideine von Java. Trochoideus termitophilus n.sp. (Col. Endomychidae). *Treubia* **1**: 34-45.

Ronchetti V. 1910: Trichodes favarius var. nov. interruptefasciatus mihi. *Il Naturaliste Siciliano* **21**: 201.

Rosenhauer W. G. 1846: *Broscosoma und Laricobius, zwei neue Käfer-Gattungen, entdeckt, beschrieben und in Stahl abgebildet.* Erlangen: Blaesing, 8 pp., 1 pl.

Rosenhauer W. G. 1847: *Beiträge zur Insekten-Fauna Europas. Erstes Bändchen; enthält die Beschreibung von sechzig neuen Käfern aus Bayern, Tyrol, Ungarn etc., so wie die Käfer Tyrols nach dem Ergebnisse von vier Reisen.* Erlangen: Theodor Blaesing, x + 159 pp., 1 pl.

Rosenhauer W. G. 1856: *Die Thiere Andalusiens nach dem Resultate einer Reise zusammengestellt, nebst den Beschreibungen von 249 neuen oder bis jetzt noch unbeschrieben Gattungen und Arten.* Erlangen: Theodor Blaesing, viii + 429 pp., 3 pls.

Rossi G. de. 1882: Die Käfer der Umgegend von Neviges. *Verhandlungen des Naturhistorischen Vereins der Preussischen Rheinlande, Westfalens und des Regierungbezirk Osnabrück* **39**: 196-215.

Rossi P. 1790: *Fauna Etrusca, sistens Insecta, quae in provinciis Florentina et Pisana praesertim collegit. Tomus primus.* Liburni: Thomae Masi & Sociorum, xxiii + 272 pp.

Rossi P. 1792: *Mantisa insectorum, exhibens species nuper in Etruria collectas, adiectis faunae Etruscae illustrationibus, ac emendationibus. Tomus Primus.* Pisis: Typographia Polloni, 148 pp.

Rossi P. 1794: *Mantissa insectorum exhibens species nuper in Etruria collectas, adiectis faunae Etruscae illustrationibus, ac emendationibus. Tomus II.* Pisis: Typographia Prosperi, 154 pp., 8 pls.

Rossi P. 1795: *Fauna Etrusca sistens Insecta quae in provinciis Florentina et Pisana praesertim collegit. Tomus primus.* [*Ed. Sec.*]. Helmstadii: C. G. Fleckeisen, xxviii + 457 pp.

Rost C. 1892: Drapetes sulcatus Rost n. sp. *Deutsche Entomologische Zeitschrift* **1892**: 402.

Rost C. 1893: Neue oder wenig bekannte caucasische Coleopteren. *Entomologische Nachrichten* **19**: 338-344.

Roth J. R. 1851: Diagnosen neuer Coleoptera aus Abyssinien. *Archives für Naturgesschichte* **17**: 115-133.

Rottenberg A. L. A. M. F. von. 1871: Beiträge zur Coleopteren-Fauna von Sicilien (zweites Stück.). *Berliner Entomologische Zeitschrift* **14** [1870-1871]: 235-260, pl. 2.

Roubal J. 1909: Danacaea Rambouseki sp. nova (Col.). *Wiener Entomologische Zeitung* **28**: 341.

Roubal J. 1911: Koleopterologické výsledky mé cesty na Kavkaz, v červenci r. 1910 [Quid novi de Coleopterorum Caucasi ad orientem vergentis fauna in mio itinere Julio mense anni 1910 suscepto cognoverium]. *Časopis České Společnosti Entomologické* **8**: 101-104.

Roubal J. 1912: Koleopterologické výsledky mé cesty na Kavkaz v červenci r. 1910 [Quid novi de Coleopterorum Caucasi ad occidentem vergentis fauna in meo itinere Julio mense anni 1910 suscepto cogno-verim]. *Časopis České Společnosti Entomologické* **9**: 124-126.

Roubal J. 1913a: Phalacrus Lebedevi novyy vid iz okrestnostey Kieva [Phalacrus lebedevi, species nova de regione Kijevensi, Rossia]. *Entomologicheskiy Vestnik* **2**: 1.

Roubal J. 1913b: Cryptophagus (Mnionomus) delicatulus sp. n. m. (Col.). *Deutsche Entomologische Zeitschrift* **1913**: 47-48.

Roubal J. 1913c: Quinque Coleoptera nova de Bohemia. *Časopis České Společnosti Entomologické* **10**: 106-112. [issued with erroneous vol. nr. 9]

Roubal J. 1913d: Zwei neue Coleopteren aus dem Kaukasus. *Entomologische Blätter* **9**: 121-122.

Roubal J. 1914: Contribution à la faune des coléoptères de la Slovaquie occidentale avec la description de deux formes nouvelles. *Bulletin de la Société Entomologique de France* **1914**: 153-154.

Roubal J. 1915: Vier Coleopteren-Neuheiten. *Entomologische Blätter* **11**: 169-170.

Roubal J. 1916: Neue palaearktische Coleopteren. *Coleopterologische Rundschau* **5**: 45-48.

Roubal J. 1917a: Drei neue Xyletinus-Formen. *Societas Entomologica* **32** (3): 11.

Roubal J. 1917b: Wissenschaftliche Ergebnisse der Bearbeitung von O. Leonhards Sammlungen. 8. Eine neue Gattung, zwei neue Arten und eine neue Subspecies aus der Familie Anobiidae der europäischen Fauna. *Archiv für Naturgeschichte* A **82** (3): 50-53.

Roubal J. 1919: Sechs neue palaearktische Coleopteren. *Archiv für Naturgeschichte* A **83** (7) [1917]: 36-39.

Roubal J. 1920: Beschreibung von vier neuen paläarktischen Coleopteren. *Entomologische Mitteilungen* **9**: 78-79.

Roubal J. 1923: Description d'un Soronia nouveau d'Europe (Col. Nitidulidae). *Bulletin de la Société Entomologique de France* **1923**: 181.

Roubal J. 1924a: Cryptophagus (subg. Mnionomus) bohemicus sp. n. *Časopis Československé Společnosti Entomologické* **21**: 23-24.

Roubal J. 1924b: Einige Käferneuheiten. *Entomologischer Anzeiger* **4**: 80-81.

Roubal J. 1925a: Anommatus Šorneri sp. n. (Col.). *Entomologische Mitteilungen* **14**: 57-58.

Roubal J. 1925b: Contribution à la faune des coléoptères de la Slovaquie occidentale avec la description de deux formes nouvelles. *Bulletin de la Société Entomologique de France* **1925**: 61-63.

Roubal J. 1925c: Zwei neue palaearktische Cerylon. *Entomologische Blätter* **21**: 76-77.

Roubal J. 1926a: Zwei neue palaearktische Coleopteren. *Entomologischer Anzeiger* **6**: 171.

Roubal J. 1926b: Seieben neue Coleopteren aus der paläarktischen Region. *Coleopterologisches Centralblatt* **1**: 244-249.

Roubal J. 1927a: Sex nova Coleoptera palaearctica. *Boletín de la Real Sociedad Española de Historia Natural* **27**: 14-16.

Roubal J. 1927b: Zwei neue Cryptophagus-Arten aus den Carpathen. *Koleopterologische Rundschau* **13**: 31-33.

Roubal J. 1927c: Sex nova Coleoptera palaearctica. *Boletín de la Real Sociedad Española de Historia Natural* **27**: 134-136.

Roubal J. 1927d: Zwei neue gelbe europäische Pullus (Col. Coccinell.). *Entomologische Mitteilungen* **16**: 137-138.

Roubal J. 1929a: Beschreibung von einigen neuen palaearktischen Coleopteren. *Entomologische Nachrichtsblatt* **3**: 31-33.

Roubal J. 1929b: Vier neue Coleopteren aus SSSR. *Entomologische Blätter* **25**: 46-48.

Roubal J. 1929c: Beschreibung von sechs neuen Coleopteren. *Časopis Československé Společnosti Entomologické* **26**: 56-57.

Roubal J. 1931a: Beschreibungen von vier neuen Coleopteren. *Časopis Československé Společnosti Entomologické* **28**: 131-132.

Roubal J. 1931b: Einige neue palaearktische Koleopteren. *Entomologische Blätter* **27**: 183-185.

Roubal J. 1932a: Sur quelques coléoptères des environs de Bagdad. *Bulletin de la Société Entomologique de France* **1932**: 59-64.

Roubal J. 1932b: Addition à la liste des coléoptères de Monte Gargano. *Bollettino della Società Entomologica Italiana* **64**: 66.

Roubal J. 1932c: Beschreibungen drei neuer Lithophilus Fröll. (Col. Cocc.). *Časopis Československé Společnosti Entomologické* **29**: 169-170.

Roubal J. 1932d: Zwei neue Elateriden aus Bulgarien. *Wiener Entomologische Zeitung* **49**: 282-284.

Roubal J. 1933: Popisy několika nových Koleopter ze Slovenska. Beschreibung einig neuen Coleopterenformen aus der Slovakei. *Časopis Československé Společnosti Entomologické* **30**: 139.

Roubal J. 1934a: Popisy čtyř nových Coleopter z Československé republiky. *Časopis Československé Společnosti Entomologické* **31**: 36-38.

Roubal J. 1934b: Beschreibung einer neuen Art und drei neuer Formen paläarktischer Coleopteren. *Mitteilungen der Deutschen Entomologischen Gesellschaft, E.V.* **5**: 62-63.

Roubal J. 1934c: Descriptiones quattuor Coleopterorum regionis palaearcticae. *Sbornik Entomologického Oddělení Národního Musea v Praze* **12**: 206.

Roubal J. 1934d: Beschreibung zweien neuen Coleopteren nebst Bemerkungen. *Folia Zoologica et Hydrobiologica* **5**: 323-325.

Roubal J. 1935a: Popis nového Antherophaga: A. špačeki n. (Col., Cryptophagidae) a přehled palaearktických druhů tohoto rodu. *Sborník Entomologického Oddělení Národního Musea v Praze* **13**: 53-54.

Roubal J. 1935b: Athous hirtus Herbst v. Balthasari n. *Entomologische Blätter* **31**: 61.

Roubal J. 1936a: *Katalog coleopter (brouků) Slovenska a Podkarpatské Rusi na základě bionomickém a zoogeografickém a spolu systematický doplněk Ganglbauerových „Die Käfer von Mitteuleuropa" a Reitterovy „Fauna Germanica". Díl II. – Tome II.* Bratislava: Učená Společnost Šafaríka, vii +434 pp.

Roubal J. 1936b: Contribution à la connaissance des élatérides (Coleopt.) de la Bulgarie. *Bulletin des Institutions Royales d'Histoire Naturelle* (Sofia) **9**: 57-67.

Roubal J. 1936c: L'exploration zoologique dans le parc national limitrophe "Pieniny" sur la rive du Dunajec entre la Pologne at la Tchécoslovaquie. *Sborník Entomologického Oddělení Národního Musea v Praze* **14**: 180-199.

Roubal J. 1939: Popis nové Epuraey – Epuraea grandiclava n. z Vysokých Tater. Description d'une Epuraea nouvelle de V. Tatry: Epuraea grandiclava n. *Entomologické Listy* **2**: 5-7.

Roubal J. 1940: Elateridae z entomologické cesty Heyrovského do Syrie a Palestiny. *Časopis Československé Společnosti Entomologické* **37**: 51-52.

Roubal J. 1941a: Popisy několika nových Palaearktických coleopter. Faunae palaearcticae coleopterorum novorum descriptiones. *Časopis České Společnosti Entomologické* **38**: 10-14.

Roubal J. 1941b: Monografie evropských a kavkaských druhů rodu Denticollis Pill. et Mitterp. (Col., Elateridae). - Monographia Denticollis Pil. Et Mitterp. specierum ex Europa et Caucaso. (Col., Elateridae). *Sborník Entomologického Oddělení Zemského Musea v Praze* **19**: 26-59.

Roubal J. 1942: Xyletinus Latr. (Col., Anob.). Klíč palearktických druhů a nižších systematických jednotek a popisy dvou nových. *Folia Entomologica* **5**: 19-25.

Roubal J. 1943: Popis čtyř nových forem Meligethes Steph. palaearktické zvířeny. Quattuor novi Meligethes palaearctici. *Entomologické Listy* **6**: 65-67.

Roubal J. 1944: Seznam druhů rodu Cryptophagus Herbst (Col.) Čech, Moravy a přilehlého území s výčtem známých a biotopickým zhodnocením. Připojen popis nového druhu. Catalogus speciorum generis Coleopterum Cryptophagus Herbst Bohemiae, Moravieae confiniumque. Reigonum descriptione speciei novae adiuncta. *Sborník Entomologického Oddělení Národního Musea v Praze* **21-22**: 95-126.

Roubal J. 1945: Nový balkánský Cryptophagus Hrbst. z podrodu Mnionomus Woll.: C. madororum m. Cryptophagus (Mnionomus Woll.) madarorum sp. n. de Peninsula Balcanica. *Časopis Československé Společnosti Entomologické* **62**: 120-121.

Roubal J. 1946: Několik nových Coleopter z mé sbírky (Quelques coléoptères nouveaux de ma collection). *Časopis Československé Společnosti Entomologické* **43**: 41-44.

Roubal J. 1951: Miscellanea Coleopterologica Bohemica. *Časopis Československé Společnosti Entomologické* **48**: 32-33.

Rouget A. 1856: Catalogue des insectes coléoptères du Départment de la Côte-d'Or. *Mémoires de l'Académie Impériale des Sciences, Arts et Belles-lettres de Dijon* **2** (4): 113-195.

Rouget A. 1877: [new taxa]. *Bulletin de la Société Entomologique de France* **1876**: ccvii.

Rücker H. W. 1978: Lathridiiden aus India (Coleoptera, Lathridiidae). *Annales Historico-Naturales Musei Nationalis Hungarici* **70**: 151-154.

Rücker H. W. 1979: Lathridiiden aus Korea (Coleoptera: Lathridiidae). *Folia Entomologica Hungarica. Rovartani Kozlemenyek* **32**: 127-128.

Rücker H. W. 1980a: Eine neue Art der Gattung Corticarina aus Turkestan (Col. Lathridiidae). *Deutsche Entomologische Zeitschrift* **27**: 93-94.

Rücker H. W. 1980b: Merophysiidae und Lathridiidae (Coleoptera) aus der äthiopischen Region. *Annales Historico-Naturales Musei Nationalis Hungarici* **72**: 165-168.

Rücker H. W. 1981: Ein neuer Enicmus (s. str.) aus Mitteleuropa (Coleoptera, Lathridiidae). *Entomologische Blätter* **77**: 165-168.

Rücker H. W. 1983a: Lathridiidae (Coleoptera) aus der Mongolei. *Annales Historico-Naturales Musei Nationalis Hungarici* **75**: 169-176.

Rücker H. W. 1983b: Neue Lathridiiden der Gattung Dienerella Reitter (Coleoptera: Lathridiidae). *Revue Suisse de Zoologie* **90**: 497-499.

Rücker H. W. 1983c: Hexasternum und Lathrapion, zwei neue Gattungen und weitere neue Arten der Familie Merophysiidae (Coleoptera). *Revue Suisse de Zoologie* **90**: 679-687.

Rücker H. W. 1984: Insects of Saudi Arabia. A new species of Melanophthalma from Saudi Arabia (Coleoptera: Lathridiidae). *Fauna of Saudi Arabia* **6**: 249-250.

Rücker H. W. 1986a: Ein neuer Vertreter der Latridius minutus-Gruppe von Creta (Coleoptera: Lathridiidae). *Revue Suisse de Zoologie* **93**: 129-131.

Rücker H. W. 1986b: Ein neuer Vertretet der Latridius minutus-Gruppe aus Frankreich (Coleoptera, Lathridiidae). *Entomofauna, Zeitschrift für Entomologie* **7**: 261-268.

Rücker H. W. 1998: Una nuova specie Italiana di Latridiidae: Dienerella angelinii nov.sp. (Coleoptera). *Doriana* **7** (306): 1-7.

Rüschkamp F. 1928: Zur rheinischen Käferfauna III. *Entomologische Blätter* **24**: 145-154.

Rybakow G. 1889a: Insecta in itinere Cl. N. Przewalski in Asia Centrali novissime lecta. XIV. Chrysomelidae et Coccinellidae. *Horae Societatis Entomologicae Rossicae* **23**: 286-290.

Rybakow G. 1889b: [new taxon]. In: Weise J.: Insecta, a CL. G. N. Potanin in China et Mongolia novissime lecta. IX. Chrysomelidae et Coccinellidae. *Horae Societatis Entomologicae Rossicae* **23**: 560-653.

Rybiński M. 1897: Wykaz chrząszczów nowych dla fauny galicyjskiej. *Sprawozdanie Komisyi Fizyograficznej* (Krakow) **32**: 46-62.

Rybiński M. M. 1902: Coleopterorum species novae, minusve cognitae, in Galicia inventae. *Bulletin de l'Académie des Sciences de Cracovie* **42**: 10-13, pls II-III.

Rye B. G. 1893: [new variety]. *Entomological Record* **4**: 243-244.

Rye B. G. 1903: Notes on the genus Elater (Ampedus). *Entomologiske Meddelelser* **1**: 78-80.

Rye B. G. 1905: To nye arter af slaegten Elater (Ampedus). *Entomologiske Meddelelser* **2**: 355-358.

Rye E. C. 1866: Description of a new species of Cryptophagus; and note on the occurrence of another species of that genus new to Britain. *The Entomologist's Monthly Magazine* **3**: 101-102.

Rye E. C. 1867: Description of a new species of Elater. *The Entomologist's Monthly Magazine* **3**: 249-250.

Rye E. C. 1871: Description of a new species of Meligethes from Britain. *The Entomologist's Monthly Magazine* **8**: 74-76.

Rye E. C. 1872a: Additions &c., to the list of British Coleoptera, with descriptions of three new species. *The Entomologist's Monthly Magazine* **9** [1872-1873]: 5-11.

Rye E. C. 1872b: Notes on British Coleoptera, including three species new to our list. *The Entomologist's Monthly Mag*azine **9** [1872-1873]: 36-39.

Rye E. C. 1873: Description of a second species of Aphanocephalus, Wollaston, from Japan. *The Entomologist's Monthly Magazine* **10** [1873-1874]: 167.

Rye E. C. 1876: Notes on British Coleoptera with descriptions of three new species. *The Entomologist's Monthly Magazine* **12** [1875-1876]: 175-180.

Sáez-Bolaño J. A. 1989: Descripción de una nueva especie de Harminius de España (Coleoptera, Elateridae). *Boletín del Grupo Entomológico de Madrid* **4**: 23-27.

Sahlberg C. R. 1819: Pars 2-4: pp. 9-56. In: *Insecta Fennica enumerans, dissertatio entomologica, A. 1817-1834 editis. Tomus I.* Helsingfors: J. C. Frenckel, viii + 519 pp.

Sahlberg C. R. 1820a: Pars 5: pp. 57-72. In: *Insecta Fennica enumerans, dissertatio entomologica, A. 1817-1834 editis. Tomus I, pars V.* Helsingfors: J. C. Frenckel, viii + 519 pp.

Sahlberg C. R. 1820b: Pars 6: pp. 73-88. In: *Insecta fennica, dissertationibus academicis, A. 1817-1834 editis. Tomus I.* Helsingfors: J. C. Frenckel, viii + 519 pp.

Sahlberg C. R. 1822: Pars 8: pp. 105-120. In: *Insecta Fennica enumerans, dissertatio entomologica, A. 1817-1834 editis. Tomus 1.* Helsingfors: J. C. Frenckel, viii + 519 pp.

Sahlberg C. R. 1833: Pars 30-31: pp. 457-488. In: *Insecta fennica, dissertationibus academicis, A. 1817-1834 editis. Tomus I.* Helsingfors: J. C. Frenckel, viii + 519 pp.

Sahlberg C. R. 1837: *Dissertatio entomologica Insecta Fennica enumerans. Tomus II, pars XII.* Aboae: J. C. Frenckel, pp. 177-192.

Sahlberg J. R. 1871a: Entomologiska anteckningar från en resa i sydöstra Karelen sommaren 1866. *Notiser ur Sällskapets pro Fauna et Flora Fennica* **11** [1870]: 327-384.

Sahlberg J. R. 1871: Anteckningar till Lapplands Coleopter-fauna. *Notiser ur Sällskapets pro Fauna et Flora Fennica Förhandligar* **11** [1870]: 385-440.

Sahlberg J. R. 1883: Negastrius algidus, en ny högnordisk Elaterid. *Meddelanden af Societas pro Fauna et Flora Fennica* **9**: 98-99.

Sahlberg J. R. 1887a: Bidrag till Tschuktsch-Halföns insectfauna. Coleoptera och Hemiptera, insamlade under Vega-expeditionens vid halföns norra och östra kust 1878-1879. Pp. 1-42. In: Nordeskiöld N. A. E. von.: *Vega-Expeditionen Vetenskapliga Jakttagelser bearbetade af deltagare i resan och andra forskare. Fjerte Bandet.* Stockholm: Beijers, 582 pp.

Sahlberg J. R. 1887b: [new taxa]. In: Reitter E.: Bestimmungs-Tabellen der europäischen Coleopteren. XVI Heft. Enthaltend die Familien: Erotylidae und Cryptophagidae. *Verhandlungen des Naturforschenden Vereins in Brünn* **26** [1886]: 1-56.

Sahlberg J. R. 1889: Enumeratio Coleopterorum Clavicornium Fenniae. Systematisk förteckning öfver de inom Finlands naturaalhistoriska omrade hittills funnas Coleoptera Clavicornia jernte uppgift om arternas utbredning och beskrifningar af nya och mindre kända species. *Acta Societatis pro Fauna et Flora Fennica* **6** [1889-1890]: 1-152.

Sahlberg J. R. 1900: Coleoptera nova vel minus cognita faunae fennicae. *Acta Societatis pro Fauna et Flora Fennica* **19** (3): 1-23.

Sahlberg J. R. 1902: Coleoptera Levantina mensibus Februario et Martio 1896 in Palaestina et Aegypto inferiore collecta. *Oefversigt af Finska Vetenskaps-Societetens Förhandlingar* **45 A** (18) [1902-1903]: 1-36.

Sahlberg J. R. 1903a: Coleoptera mediterranea et rosso-asiatica nova et minus cognita, maxima ex parte itineribus annis 1895-1896 et 1898-1899 collecta. *Öfversigt Finska Vetenskaps-Societetens Förhandlingar* **45** (10) [1902-1903]: 1-40.

Sahlberg J. R. 1903b: Messis hiemalis Coleopterorum Corcyraeorum. Enumeratio Coleopterorum mensibus Novembri-Februario 1895-1896 et 1898-1899 nec non primo vere 1896 in insula Corcyra collectorum *Öfversigt af Finska Vetenskaps-Societetens Förhandlingar* **45** (11) [1902-1903]: 1-87.

Sahlberg J. R. 1903c: Coleoptera Levantina mensibus Februario et Martio 1896 in Palaestina et Egypto inferiora collecta. *Öfvesigt af Finska Vetenskaps-Societetens Förhandlingar* **45** (18) [1902-1903]: 1-36.

Sahlberg J. R. 1903d: Coleoptera Numido-Punica mensibus Martio, Aprili et Majo 1899 in Tunisia et Algeria orientali collecta. *Öfversigt af Finska Vetenskaps-Societetens Förhandlingar* **45** (19) [1902-1903]: 1-70.

Sahlberg J. R. 1908: Coleoptera mediterranea et rosso-asiatica nova et minus cognita, maxima et parte itineribus annis 1895-1896, 1898-1899 et 1903-1904 collecta. *Öfversigt af Finska Vetenskaps-Societetens Förhandlingar* **50** (7) [1907-1908]: 1-94.

Sahlberg J. R. 1911: Fyra för fauna nya Coleoptera, funna i trakten af Helsingfors i oktober 1910. *Meddelanden af Societas pro Fauna et Flora Fennica* **37** [1910-1911]: 42-52.

Sahlberg J. R. 1912: En ny finsk art av släktet Cryptophagus. *Meddelanden af Societatis pro Fauna etFlora Fennica* **38**: 92.

Sahlberg J. R. 1913a: Coleoptera mediterranea et rosso-asiatica nova et minus cognita maxima ex parte itineribus annis 1895-1896, 1898-1899 et 1903-1904 collecta. *Öfversigt af Finska Vetenskaps-Societetens Förhandlingar* (A) **55** (8) [1912-1913]: 1-88.

Sahlberg J. R. 1913b: Messis nova hiemalis Coleopterorum Corcyreorum. *Öfversigt af Finska Vetenskaps-Societetens Förhandlingar* (A) **55** (12): 1-28.

Sahlberg J. R. 1913c: Coleoptera balcanica quae mansibus Octobri et Decembri 1903 atque Martis et Aprili 1906 in peninsula balcanica collegerunt John Sahlberg et Unio Saalas, enumeravit John Sahlberg. *Öfversigt af Finska Vetenskaps-Societetens Förhandlingar* (A) **55** (15) [1912-1913]: 1-108.

Sahlberg J. R. 1913d: Coleoptera mediterranea orientalia quae in Aegypto, Palaestina, Syria, Caramania atque in Anatolia occidentali anno 1904 collegerunt John Sahlberg et Unio Saalas. *Öfversigt af Finska Vetenskaps-Societetens Förhandlingar* (A) **55** (19) [1912-1913]: 1-282.

Sahlberg J. R. 1914: Scymnus triangularis, en ny finsk Coleopter-art. *Acta Societatis pro Fauna et Flora Fennica* **40**: 39-41.

Sahlberg J. R. 1919: Vad är Cryptophagus brunneus Gyll? *Entomologisk Tidskrift* **40**: 1-8.

Sahlberg J. R. 1926: Enumeratio Coleopterorum Fenniae. Lamellicornes, Platysoma, Xylophagi, Fungicola. *Annales Societatis Zoologicae-Botanicae Fennicae Vanamo* **4**: 1-169.

Sahlberg R. F. 1834a: *Dissertatio Academica novas Coleopterorum Fennicorum species sistens.* Helsingforsiae: Typis Frenckelliorum, 12 pp.

Sahlberg R. F. 1834b: Novae Coleopterorum Fennicorum species. *Bulletin de la Société Impériale des Naturalistes de Moscou* **7**: 267-280.

Sainte-Claire Deville J. 1906: [new taxon]. In: Bickhardt H.: Beitrag zur Coleopteren-Fauna von Korsika. *Wiener Entomologische Zeitung* **25**: 291-296.

Sainte-Claire Deville J. 1908: [Pp. 137-248]. In: *Catalogue critique des coléoptères de la Corse* [1906-1914]. Caen: G. Poisson et Cie, 537 pp., 1 map. [published separately as supplement to *Revue d'Entomologie* **27**].

Sainte-Claire Deville J. 1910: [Pp. 315]. *Catalogue critique des coléoptères de la Corse*. Caen: G. Poisson et Cie, 537 pp., 1 map. [published separately as supplement to *Revue d'Entomologie* **27**].

Sainte-Claire Deville J. 1921: Catalogue critique des coléoptères de la Corse. 2e supplément. *Annales de la Société Entomologique de France* **89**: 377-404.

Sainte-Claire Deville J. 1927: Descriptions de quatre espèces de coléoptères découvertes dans le terriers de Marmottes (Col. Staphylinidae, Silphidae, Cryptophagidae). *Bulletin de la Société Entomologique de France* **1927**: 41-45.

Sajó K. 1881: Acht neue Coccinelliden-Varietäten aus Central-Ungarn. *Entomologische Nachrichten* (Stettin) **7**: 273-274.

Sajó K. 1882: Beiträge zu Weise's Bestimmungs-Tabellen der europäischen Coccinelliden. *Entomologische Monatsblattätter* **2**: 159-160.

Sakai M. 1975: The Xyletininae of Japan (Coleoptera: Anobiidae). *Transactions of the Shikoku Entomological Society* **12**: 95-110.

Sakai M. 1977: A new genus and new species of Anobiidae from Japan (Coleoptera). *Transactions of the Shikoku Entomological Society* **13**: 119-122.

Sakai M. 1978: Studies on the Anobiidae (Coleoptera) from Japan and neighboring countries. I. On the genus Hadrobregmus Thomson from Japan. *Transactions of the Shikoku Entomological Society* **14**: 35-38.

Sakai M. 1979: Studies on the Anobiidae (Coleoptera) from Japan and neighboring countries. II. Four new Japanese species. *Transactions of the Shikoku Entomological Society* **14**: 131-139.

Sakai M. 1982a: Studies on the Anobiidae (Coleoptera) from Japan and neighboring countries. III. A review of the genus Oligomerus Redtenbacher of Japan. *Special Issue to the Memory of Retirement of Emeritus Professor Michio Chûjô* **1982**: 43-49.

Sakai M. 1982b: Occurrence of the genus Eurypogon in Japan and Taiwan (Coleoptera: Artematopidae). *Transactions of the Shikoku Entomological Society* **16**: 51-61.

Sakai M. 1983: Notes on Nepalese Anobiidae (Coleoptera). *Bulletin of the National Science Museum Series A* (Zoology) **9**: 19-28.

Sakai M. 1984a: Studies on the Anobiidae (Coleoptera) from Japan and neighboring countries. IV. Three new species of Dorcatominae from Japan. *Transactions of the Shikoku Entomological Society* **16**: 11-17.

Sakai M. 1984b: Studies on the Anobiidae (Coleoptera) from Japan and neighboring countries. V. Two new Japanese species of the genus Gastrallus. *Transactions of the Shikoku Entomological Society* **16**: 19-24.

Sakai M. 1985: Studies on the Anobiidae (Coleoptera) from Japan and neighboring countries. VI. Description of a new species of the genus Trichodesma from the Ryukyus. *Transactions of the Shikoku Entomological Society* **17**: 27-29.

Sakai M. 1986: Studies on the Anobiidae (Coleoptera) from Japan and neighboring countries. VII. Pp. 211-214. In: Uéno S.-I. (ed.): *Entomological Papers Presented to Yoshihiko Kurosawa on the Occasion of his Retirement*. Tokyo: Coleopterists Association of Japan, 342 pp.

Sakai M. 1987: Studies on the Anobiidae (Coleoptera) from Japan and neighboring countries. VIII. The genus Clada Pascoe of Taiwan (Coleoptera: Anobiidae). *Entomological Review of Japan, Supplementum* **42**: 65-74.

Sakai M. 1991: Studies on the Anobiidae (Coleoptera) from Japan and neighboring countries. IX. Occurence of the genus Clada Pascoe in the Ryukyus. *Transactions of the Shikoku Entomological Society* **19**: 167-170.

Sakai M. 1993: Studies on the Anobiidae (Coleoptera) from Japan and neighboring countries. X. Discovery of the genus Rhamna Peyerimhoff from Japan with description of two new species. *Japanese Journal of Entomology* **61**: 569-575.

Sakai M. 1994: Studies on the Anobiidae (Coleoptera) from Japan and neighboring countries. XI. A new species of the genus Hadrobregmus Thomson, with a key to the Japanese species. *Transactions of the Shikoku Entomological Society* **20**: 289-292.

Sakai M. 1995a: Discovery of the dorcatomine genus Stagetodes Español (Coleoptera, Anobiidae) from Japan with descriptions of two new species. *Japanese Journal of Systematic Entomology* **1**: 243-248.
Sakai M. 1995b: Stagetus uenoi (Coleoptera, Anobiidae), a new dorcatomine species from Japan. *Special Bulletin of the Japanese Society of Coleopterology* **4**: 419-423.
Sakai M. 1996: Three new dorcatomine species (Coleoptera, Anobiidae) associated with Ganoderma fungi in Japan and Taiwan. *Japanese Journal of Systematic Entomology* **2**: 275-281.
Sakai M. 2001: Hyperisus as a distinct genus, with description of a new species from Japan (Coleoptera, Anobiidae). *Japanese Journal of Systematic Entomology* **7**: 327-332.
Sakai M. 2002: The genus Ernobius Thomson of Japan (Coleoptera, Anobiidae, Ernobiinae). *Japanese Journal of Systematic Entomology* **8**: 119-130.
Sakai M. 2003: Mimogastrallus hideoi, a new genus and a new species of the subfamily Anobiinae, with notes on the related genus Falsogastrallus (Coleoptera, Anobiidae). *Japanese Journal of Systematic Entomology* **9**: 333-338.
Saluk S. V. 1992: 79. Sem. Lathridiidae. Pp. 378-392. In: Ler P. A. (ed.): *Opredelitel nasekomykh Dalnego Vostoka SSSR v shestykh tomakh. Tom III. Zhestkokrylye, ili zhuki. Chast 2*. Sankt-Petersburg: Nauka, 704 pp.
Saluk S. V. 1995: K poznaniyu vidov roda Enicmus blizkikh k E. transversus (Olivier) (Coleoptera, Latridiidae) [Contribution to the knowledge of Enicmus species related to E. transversus (Olivier) (Coleoptera, Latridiidae)]. *Proceedings of the Zoological Museum of Byelorus University* (Minsk) **1**: 139-145.
Samouelle G. 1819: *The Entomologist's useful compendium; or an introduction to the knowledge of British insects, comprising the best means of obtaining and preserving them, and a description of the apparatus generally used; together with the genera of Linné, and the modern method of arranging the classes Crustacea, Myriapoda, spiders, mites, and insects from their affinities and structure, according to the views of Dr. Leach. Also an explanation of the term used in entomology; a calendar of the times of appearance, and usual situation of near 3000 species of British insects; with instructions for collecting and fitting up objects for the microscope.* London: Thomas Boys, 496 pp., 12 pls.
Sánchez-Ruiz A. 1996: Catálogo bibliográfico de las especies de la familia Elateridae (Coleoptera) de la Península Ibérica e Islas Baleares. *Documentos Fauna Ibérica* **2**: 1-265.
Sánchez-Ruiz Y. & Rosa J. J. de la. 2003: Melasis fermini sp. nov., un nuevo Eucnemidae (Coleoptera) para la fauna Europea. *Boletín de la Sociedad Entomológica Aragonesa* **32**: 1-4.
Santiago C. T. 1987: [new taxa]. In: Plata Negrache P. & Santiago Hernández C. T.: *Revision de la familia Malachiidae Erichson (Insecta: Coleoptera) en Marruecos*. Krefeld: Goecke & Evers Verlag, 842 pp.
Saraiva A. C. 1957: Insectos de madeira-em-ombra e seu combate. *Memorias Laboratorio Nacional de Engenharia Civil* **107**: 1-79.
Sasaji H. 1967a: A revision of the Formosan Coccinellidae (I), the subfamily Sticholotinae, with an establishment of a new tribe (Coleoptera). *Etizenia* **25**: 1-28.
Sasaji H. 1967b: Description of a new coccinellid species attacking the scale insects, with a taxonomic note on a known species (Coleoptera). *Mushi* **41**: 177-180.
Sasaji H. 1968: A revision of the Formosan Coccinellidae (II), tribes Stethorini, Aspidimerini and Chilocorini (Coleoptera). *Etizenia* **32**: 1-24.
Sasaji H. 1970a: Notes on the Formosan Endomychidae, with description of a new genus and several new species (Coleoptera). *Etizenia* **43**: 1-18, pls. 4.
Sasaji H. 1970b: Biogeographical consideration of the Coccinellidae (Coleoptera) of the Tsuchimas, Japan. *Bulletin of the National Science Museum* **13**: 207-217.
Sasaji H. 1971a: *Fauna Japonica: Coccinellidae (Insecta: Coleoptera)*. Tokyo: Academic Press of Japan, ix + 340 pp., 16 pls.
Sasaji H. 1971b: A new Propalticus-species of Japan (Coleoptera: Propalticidae). *Kontyû* **39**: 37-38.
Sasaji H. 1976: Insect fauna of Oshima Island, Fukui Pref., Central Japan, with description of a new coccinellid species. *The Memoirs of the Faculty of Education, Fukui University Ser. II (Natural Science)* **26** (2): 27-57 (in Japanese with English description).
Sasaji H. 1978: Notes on the Japanese Endomychidae, with an establishment of a new subfamily (Coleoptera). *The Memoirs of the Faculty of Education, Fukui University Series II (Natural Science)* **28** (1): 1-31.
Sasaji H. 1982: Additions to the Japanese fauna of the coccinellid tribe Scymnini (Coleoptera). Pp. 63-72. *Special Issue to the Memory of Retirement of Emeritus Professor Michio Chûjô*. Nagoya, 185 pp.

Sasaji H. 1983: Contribution to the taxonomy of the superfamily Cucujoidea (Coleoptera) of Japan and her adjacent districts I. *The Memoirs of the Faculty of Education, Fukui University, Series II.* (*Natural Science*) **33** (2): 17-52.

Sasaji H. 1984: Contribution to the taxonomy of the superfamily Cucujoidea (Coleoptera) of Japan and her adjacent districts, II. *The Memoirs of the Faculty of Education, Fukui University Series II* (*Natural Science*) **34**: 21-63.

Sasaji H. 1985: Descriptions of a new species of the genus Biphyllus from Japan (Coleoptera: Biphyllidae). *The Memoirs of the Faculty of Education, Fukui University, Series II.* (*Natural Science*) **35**: 11-14.

Sasaji H. 1986a: The Cucujoidea (Insecta: Coleoptera) collected by the Nagoya University Scientific Expedition to Formosa in 1984. *The Memoirs of the Faculty of Education, Fukui University Series II* (*Natural Science*) **36**(2): 1-14.

Sasaji H. 1986b: Notes on the Colydiidae (Coleoptera) of Japan and formosa. Pp. 243-249. In: Uéno S.-I. (ed.): *Entomological papers presented toYoshihiko Kurosawa on the occasion of his retirement.* Tokyo: The Coleopterists' Association of Japan, v + 342 pp.

Sasaji H. 1986c: A revision of the genus Leptophloeus (Coleoptera, Cucujidae) of Japan. *Kontyû* **54**: 681-687.

Sasaji H. 1987: Contribution to the taxonomy of the superfamily Cucujoidea (Coleoptera) of Japan and her adjacent districts, III. *The Memoirs of the Faculty of Education, Fukui University. Series II* (*Natural Science*) **37**: 23-55.

Sasaji H. 1988a: Contribution to the taxonomy of the superfamily Cucujoidea (Coleoptera) of Japan and her adjacent districts, IV. *The Memoirs of the Faculty of Education, Fukui University Series II* (*Natural Science*) **38**: 13-48.

Sasaji H. 1988b: The Formosan Coccinellidae collected by Dr. K. Baba in 1986. *Transactions of Essa Entomological Society* (Nigata) **65**: 37-52.

Sasaji H. 1989a: Contribution to the taxonomy of the superfamily Cucujoidea (Coleoptera) of Japan and her adjacent districts, V. *The Memoirs of the Faculty of Education, Fukui University, Series II* (*Natural Science*) **39** (2): 5-26.

Sasaji H. 1989b: Discovery of the curious beetle genus Xenoscelinus Grouvelle (Languriidae-Cryptophilinae) from Japan. *Kanagawa-Chûhô* **90**: 223-228.

Sasaji H. 1990: The family Mychothenidae of Japan (Coleoptera). *Esakia* (Special Issue) **1**: 65-75.

Sasaji H. 1991a: Contribution to the taxonomy of the superfamily Cucujoidea (Coleoptera) of Japan and her adjacent districts, VI. *The Memoirs of the Faculty of Education, Fukui University Series II* (*Natural Science*) **41**: 11-25.

Sasaji H. 1991b: Formosan Languriidae (Coleoptera: Cucujoidea) collected by Dr. K. Baba in 1987-1989. *Transactions of Essa Entomological Society* **71**: 37-47.

Sasaji H. 1991c: The Coccinellidae (Coleoptera) collected from the Island of Lan Yu, Formosa by Dr. K. Baba in 1987, with the description of a new species. *Transactions of Essa Entomological Society Niigata* **71**: 48-52.

Sasaji H. 1993: Contributions to the taxonomy of the superfamily Cucujoidea (Coleoptera) of Japan and her adjacent districs, VII. *The Memoirs of the Faculty of Education, Fukui University, Series II* (*Natural Science*) **44** (2): 17-25.

Sasaji H. 1994: The Formosan Coccinellidae (Coleoptera) collected by the Late Dr. K. Baba (Third Report). *Special Bulletin of the Essa Entomological Society* **2**: 235-240.

Sasaji H. 1995: Contribution to the taxonomy of the superfamily Cucujoidea (Coleoptera) of Japan and her adjacent districts, VIII. *The Memoirs of the Faculty of Education, Fukui University Series II* (*Natural Science*) **47**: 21-30.

Sasaji H. 1997: A new species of the genus Antibothrus (Boleoptera, Bothrideridae) from Japan, with notes on the Japanese Bothrideridae. *Esakia* **37**: 111-116.

Sasaji H. 1999: Discovery of the family Cerophytidae from Japan (Coleoptera: Elateroidea), with the description of a new species. *Entomological Review of Japan* **54**: 97-102.

Sasaji H. & Kishimoto O. 1996: Insect fauna of Nakaikemi Marsh, Fukui Prefecture, Central Japan and its natural environmental conservation - with description of a new species and its life history - (Insecta: Coleoptera: Coccinellidae). *The Memoirs of the Snowdraft Laboratory* (Fukui University) **3**: 15-36.

Sasaji H. & McClure M. S. 1997: Description and distribution of Pseudoscymnus tsugae sp. nov. (Coleoptera: Coccinellidae), an important predator of hemlock woolly adelgid in Japan. *Annals of the Entomological Society of America* **90**: 563-568.

Satô M. 1968: The lampyrid-beetles of Japan, I. *Bulletin of the Japan Entomological Academy* **4** (5): 19-26.

Satô M. 1970: Notes on the genus Cyphonocerus Kiesenwetter from Formosa (Col.: Lampyridae). *Bulletin of the Japan Entomological Academy* **5** (1): 22-24.

Satô M. 1976: New Cantharoidea from Japan (Coleoptera). *Transactions of the Shikoku Entomological Society* **13**: 51-60.

Satô M. 1982: The Coleoptera of the Ozegahara Moor. Pp. 379-408. In: Hara H. (ed.): *Ozegahara. Scientific researches of the high-moor in central Japan.* Tokyo: Japanese Society for the Promotion of Science, ix + 456 pp.

Satô M. 1985a: Family Lampyridae. Pp. 120-124. In: Kurosawa Y., Hisamatsu S. & Sasaji H. (eds): *Colored illustrations of the Coleoptera of Japan. Volume 3.* Osaka: Hoikusha Publishing Co., x + 500 pp., 72 pls (in Japanese).

Satô M. 1985b: Family Melyridae. Pp. 163-167, pl. 26. In: Kurosawa Y., Hisamatsu S. & Sasaji H. (eds): *Colored illustrations of the Coleoptera of Japan. Volume 3.* Osaka: Hoikusha Publishing Co., x + 500 pp., 72 pls (in Japanese).

Satô M. 1986a: The genus Attalus (Coleoptera, Melyridae) of the Ryukyu Archipelago. Pp. 233-242. In: Uéno S.-I. (ed.): *Entomological Papers Presented to Yoshihiko Kurosawa on the Occasion of his Retirement.* Tokyo: Coleopterists Association of Japan, 342 pp.

Satô M. 1986b: A new Dasytes from Hokkaido, Japan (Coleoptera: Melyridae). *Transactions of the Shikoku Entomological Society* **17**: 253-254.

Satô M. 1986c: New Cantharoidea from Japan, II (Coleoptera). *Transactions of the Shikoku Entomological Society* **17**: 255-261.

Satô M. 1989: Notes on Podabrus (Coleoptera, Cantharidae) from the Zaou Mountains, Miyagi Prefecture, Japan. *Elytra* **17**: 181-186.

Satô M. 1991: A new Cyphonocerus (Coleoptera, Lampyridae) from Japan. *Elytra* **19**: 191-193.

Satô S. 2002: Notes on the genus Iwawakia, with description of a new species from Central Japan. *Japanese Journal of Systematic Entomology* **8**: 109-114.

Satô M. & Ishida K. 1982: The genus Themus of Japan (Coleoptera, Cantharidae). Pp. 31-39. *Special Issue to the Memory of Retirement of Emeritus Professor Michio Chûjô.*

Satô M. & Kimura M. 1994: Discovery of a new firefly of the genus Luciola (Coleoptera, Lampyridae) from Kume-jima of the Ryukyu Islands. *Elytra* **22**: 159-162.

Satô M. & Ohbayashi N. 1968: Notes on the lycid-beetles of the Ryukyu Archipelago. *Entomological Review of Japan* **20**: 65-72.

Satô M. & Okushima Y. 1992: The genus Athemus (Coleoptera, Cantharidae) of the Ryukyu Islands, Southwest Japan. *Japanese Journal of Entomology* **60**: 857-865.

Satô M. & Okushima Y. 2001: New elegant species of the genus Lycocerus (Coleoptera, Cantharidae) from Iriomote-jima Island, Southwest Japan. *Japanese Journal of Systematic Entomology* **7**: 117-119.

Satô M., Okushima Y. & Ishida K. 2002: Japanese species of the genus Cantharis Linnaeus (Coleoptera: Cantharidae). *Entomological Revue of Japan* **57**: 205-217.

Satô M. & Takahashi K. 1989: A new species of the cantharid genus Themus (Coleoptera) from Japan. *Kanagawa-Chûhô* **90**: 215-217.

Satô M. & Wittmer W. 1989: Some notes on Japanese Melyridae (Coleoptera). *Elytra* 17: 57-60.

Saulcy F. H. C. de. 1864: Descriptions des espèces nouvelles de coléoptères recueillis en Syrie, en Égypte et en Palestine pendant les mois d'octobre 1863 à Janvier 1864. *Annales de la Société Entomologique de France* (4) **4**: 421-440.

Saulcy F. H. C. de. 1868: [new taxa]. In: Mulsant E. & Rey C.: *Histoire Naturelle des coléoptères de France. Tribu des Scuticolles*. Paris: F. Savy, 186 pp., 3 pls.

Savoyskaya G. I. 1962: Morfologia i sistematika lichinok roda Coccinella (Col., Coccinellidae). *Trudy Kazachskogo Instituta Zashchity Rasteniy* **7**: 316-322.

Savoyskaya G. I. 1963: Lichinki kokcinellid rodov Semiadalia Crotch, Hippodamia Muls., Adonia Muls., i Anisosticta Dup. (Col., Coccinellidae). In: Materialy po izucheniyu nasekomykh Kazakhstana: Alma Ata. *Trudy Instituta Zoologii* **21**: 74-95.

Savoyskaya G. I. 1964: O nekotorykh lichinkakh trib Coleopterini i Hyperaspini s opysaniem novogo vida, Pharoscymnus balkhashensis Savoyskaya sp. n. (Coleoptera, Coccinellidae). *Trudy Instituta Zachshchity Rosteniy* **8**: 358-370.

Savoyskaya G. I. 1965: O novom vide kokcinellid. *Vestnik Sel'skokhozyaystvenoy Nauki* (Alma-Ata) **1**: 31-37.

Savoyskaya G. I. 1969a: Novyy rod i vid kokcinellid iz Kazakhstana. *Vestnik Sel'skokhazyaystvenoy Nauki* (Alma-Ata) **6**: 37-43.

Savoyskaya G. I. 1969b: O vydelenii novikh taxonomicheskich kategoriy kokcinellid (Col., Coccinellidae). *Vestnik Sel'skokhozyaystvenoy Nauki* (Alma-Ata) **9**: 101-105.

Savoyskaya G. I. 1969c: K izucheniju Scymnin (Coleoptera, Coccinellidae) juzhnogo i vostochnogo Kazakhstana. *Trudy Kazakhskogo Nauchno-Issedovatel'skogo Instituta Zashchity Rosteniy* **10** [1968]: 118-157.

Savoyskaya G. I. 1971a: K izucheniju roda Pharoscymnus Bed. (Coleoptera, Coccinellidae). *Materialy po Chlenistonogim Entomofagam Kirgizii* (Frunze) **7**: 54-68.

Savoyskaya G. I. 1971b: Kokcinellidy triby Hyperaspini (Col., Coccinellidae) iz yuzhnogo i yugo-vostochnogo Kazakhstana. In: Fauna i Biologiya nasekomykh Kazakhstana. *Trudy Instituta Zoologii* (Alma-Ata) **32**: 98-110.

Saxesen W. 1838: Elater heyeri n. sp. *Isis* (Leipzig) **1838**: 805.

Say T. 1824a: Descriptions of coleopterous insects collected in the late expedition to the Rocky Mountains, preformed by order of Mr. Calhoun, Secretary of War, under command of Major Long. Continuation. *Journal of the Academy of Natural Sciences of Philadelphia* **3** [1823]: 298-331.

Say T. 1824b: Descriptions of coleopterous insects collected in the late expedition to the Rocky Mountains, performed by order of Mr. Calhoun, Secretary of War, under the command of Major Long. *Journal of the Academy of Natural Sciences of Philadelphia* **4**: 83-99.

Say T. 1824c: Narrative of an expedition to the source of the St Peter's River, etc. under the command of Stephan H. Long, Major, U.S.T.E. *Journal of the Academy of Natural Sciences of Philadelphia* [Appendix] **4**: 268-378.

Say T. 1825a: Descriptions of new species of coleopterous insects inhabiting the United States. *Journal of the Academy of Natural Sciences of Philadelphia* **5**: 160-204.

Say T. 1825b: Descriptions of new American species of the genera Buprestis, Trachys and Elater. *Annals of the Lyceum of Natural History of New York* **1**: 249-268.

Say T. 1826: Descriptions of new species of coleopterou insects inhabiting the United States. *Journal of the Academy of Natural Sciences of Philadelphia* **5**: 293-304.

Say T. 1827: Descriptions of new species of coleopterous insects inhabiting United States. *Journal of the Academy of Natural Sciences of Philadelphia* **5** (2): 237-284, 293-304.

Say T. 1835: Descriptions of new North American coleopterous insects, and observations on some already described. *Boston Journal of Natural History* **1**: 151-203.

Schaeffer C. F. A. 1911: New Coleoptera and miscellaneous notes. *Journal of the New York Entomological Society* **19**: 113-126.

Schaeffer C. F. A. 1931: On a few new and known Coleoptera. *Bulletin of the Brooklyn Entomological Society* (Lancaster) **26**: 174-176.

Schaeffer J. C. 1766a: *Icones Insectorum circa Ratisbonam indigenorum coloribus naturam referentibus expressae. Natürlich ausgemalte Abbildungen Regensburgischen Insecten. Vol. 1, pars 1.* Ratisbone: Heinrich Gottfried Zungel, [60] pp., pls. I-L.

Schaeffer J. C. 1766b: *Elementa Entomologica. CXXXV tabulae aere excusae floridisque coloribus distinctae.* Ratisbonae: Weiss, [9] + 135 pls, 135 pp.

Schaeffer J. C. 1771: *Icones insectorum circa Ratisbonam indigenorum coloribus naturam referentibus expressae. Volume III.* Regenburg: Heinrich Gottfried Zunkel, pls cci-cclxxx, + [80] pp. [seen as Edition nova, 1804].

Schaeffer J. C. 1777: *Elementorum entomologicorum appendix quinque insectorum nova genera exhibens. Fünf neue Insectengeschlechter zur Einleitung in die Insectenkenntnis. Tabulae V. vivis coloribus expressae.* Ratisbonae, [6] pp. + pls 136-140.

Schaller J. G. 1783: Neue Insecten. *Abbhandlungen der Hallischen Naturforschenden Gesellschaft* **1**: 217-328.

Schatzmayr A. 1939: Nuovi contributi alla conoscenza della fauna delle isole Italiane dell'Egeo X. Coccinellidae et Oedemeridae (Col.). *Bollettino del Laboratorio di Zoologia Generale e Agraria della Facolta Agraria di Portici* **31**: 222-229.

Schaufuss L. W. 1862a: Ueber drei neue oesterreichische Käfer. *Verhandlungen der Kaiserlich-Königlichen Zoologisch-Botanischen Gesellschaft in Wien* **12**: 18-20.

Schaufuss L. W. 1862b: Beschreibungen von neuen Käferarten. *Sitzungsberichte und Abhandlungen der Naturwissenschaftlichen Gesselschaft Isis zu Dresden* **1861**: 47-51.

Schaufuss L. W. 1862c: Über Monotropus Normanni und über Coccinellen mit vier neuen Arten. *Sitzungsberichte und Abhandlungen der Naturwissenschaftlichen Gesselschaft Isis zu Dresden* **1861**: 52-53.

Schaufuss L. W. 1862d: Diagnoses de coléoptères nouveaux. *Annales de la Société Entomologique de France* (4) **2**: 309-312.

Schaufuss L. W. 1863: [new taxon]. *Sitzungsberichte und Abhandlungen der Naturwissenschaftlichen Gesselschaft Isis zu Dresden* **1862**: 185-186.

Schaufuss L. W. 1863: Beitrag zur Käfer-Fauna Spaniens. *Sitzungs-Berichte der Naturwissenschaftlichen Gesellschaft Isis zu Dresden* **1862**: 189-204.

Schaufuss L. W. 1866: Coleopterorum europaeorum species novae. *Revue et Magasin de Zoologie Pure et Appliquée* (2) **18**: 180-181.

Schaufuss L. W. 1867: Beitrag zur Gruppe der Malacodermata. *Entomologische Zeitung* (Stettin) **28**: 81-86.

Schaufuss L. W. 1869: [new taxa]. In: Ludovici F. F. & Schaufuss L. W.: *Beitrag zur Kenntniss der Coleopteren-Fauna der Balearen*. Prag: Selbstverlag, 31pp.

Schaufuss L. W. 1870: Phosphaenopterus. Nov. gen. Malacodermata. *Nunquam Otiosus* **1** [1870-1871]: 60-61.

Schaufuss L. W. 1871: Vermischtes vom Herausgaber. *Nunquam Otiosus* **1** [1870-1871]: 156-160.

Schaufuss L. W. 1874: Die bisherigen Arbeiten des Herausgebers. Fortsetzung. *Nunquam Otiosus* **2** [1872]: 361-384.

Schaufuss L. W. 1876: Notizen zu Gem. & Harolds Catalog Coleopt. *Nunquam Otiosus* **2** [1872]: 384.

Schaufuss L. W. 1876: Ueber Merophysia Luc., Coluocera Mtsch. und Reitteria Led. *Nunquam Otiosus* **2** [1872]: 393-400.

Schaufuss L. W. 1882a: Zoologische Ergebnisse von Excursionen auf den Balearen. III. (Addenda und Fortsetzung). *Nunquam Otiosus* **3**: 527-552.

Schaufuss L. W. 1882b: Zoologische Ergebnisse von Excursionen auf den Balearen. I. *Verhandlungen der Kaiserlich-Königlichen Zoologisch-Botanischen Gesellschaft in Wien* **31** [1881]: 619-624, pl. XXI.

Schaufuss L. W. 1887: Beitrag zur Fauna der Niederlandischen Besitzungen auf den Sunda-Inseln. *Horae Societatis Entomologicae Rossicae* **21**: 102-147.

Schaum H. R. 1852: *Catalogus Coleopterorum Europae. Vierte Auflage.* Berlin: Entomologischer Verein in Stettin, v + 98 + 12 pp.

Schaum H. R. 1858: [new taxa]. In: Kraatz G.: Beitrag zur Käferfauna Griechenlands. Drittes Stück: Staphylinidae (Schluss), Trichopterygia, Histeridae, Phalacridae, Nitidulariae, Trogositarii, Colydii, Cucujidae, Cryptophagidae, Thorictidae, Mycetophagidae, Dermestini, Byrrhi. *Berliner Entomologische Zeitschrift* **2**: 123-148.

Schaum H. R. 1859a: Beitrag zur europäischen Käferfauna. *Berliner Entomologische Zeitschrift* **3**: 42-59.

Schaum H. R. 1859b: [new taxa]. In: Kraatz G.: Ueber die Gattungen Micropeplus, Thorictus, Antidipnis, Cnemeplatia und Foucartia. *Berliner Entomologische Zeitschrift* **3**: 65-78.

Schawaller W. 1987: Discolomidae aus dem Nepal-Himalaya (Coleoptera). *Stuttgarter Beiträge zur Naturkunde* (A) **410**: 1-9.

Schawaller W. 1989: New Discolomidae (Coleoptera) from the Himalayas. *Tropical Zoology* **2**: 45-58.

Schenkling S. 1898: Zehn neue Cleriden nebst Bemerkungen über schon beschriebene Arten. *Deutsche Entomologische Zeitschrift* **1898**: 361-368.

Schenkling S. 1899ab: Neue Cleriden des Museums zu Genua nebst Bemerkungen über bereits beschriebene Arten. *Annali del Museo Civico di Storia Naturale di Genova* **40** [1899-1901]: 331-346.

Schenkling S. 1899ba: Cleriden aus Somali-Land. *Annali del Museo Civico di Storia Naturale di Genova* **39** [1898-1899]: 545-550.

Schenkling S. 1900: Neue Cleriden des Hamburger Museums. *Mitteilungen aus dem Naturhistorischen Museum in Hamburg* **17**: 11-20.

Schenkling S. 1901: Neue Cleriden des Königl. Museums zu Brüssel. *Annales de la Société Entomologique de Belgique* **45**: 104-108.

Schenkling S. 1902a: Neue Cleriden des Museum zu Leyden. *Notes from the Leyden Museum* **23**: 123-130.

Schenkling S. 1902b: Cleridés nouveax du Muséum d'Histoire naturelle de Paris. *Bulletin du Muséum d'Histoire Naturelle* **8**: 317-333.

Schenkling S. 1903a: Coleoptera Malacodermata Fam. Cleridae. In: Wytsman P. (ed.): *Genera Insectorum*. Fasc. 13. Bruxelles: P. Wytsman, 124 pp., 5 pls.

Schenkling S. 1903b: Zur Systematik der Cleriden. *Deutsche Entomologische Zeitschrift* **1903**: 9-16.

Schenkling S. 1903c: Einige neue Cleriden des Stettiner Museums. *Entomologische Zeitung* (Stettin) **64**: 16-21.

Schenkling S. 1906: Die Cleriden des Deutschen Entomologischen National-Museums, nebst Beschreibungen neuer Arten. *Deutsche Entomologische Zeitschrift* **1906**: 241-320, pl. II.

Schenkling S. 1908a: Die Cleriden des Deutschen Entomolog. National-Museums (Col.). Nachtrag I. *Deutsche Entomologische Zeitschrift* **1908**: 361-367.

Schenkling S. 1908b: Die Cleriden des Deutschen Entomolog. National-Museums (Col.). Nachtrag II. *Deutsche Entomologische Zeitschrift* **1908**: 477-487.

Schenkling S. 1908c: Die Cleriden des Deutschen Entomolog. National-Museums (Col.). Nachtrag III. *Deutsche Entomologische Zeitschrift* **1908**: 701-707.

Schenkling S. 1912a: Ein neuer Phloeocopus aus Unteraegypten (Col.). *Bulletin de la Société Entomologique de Egypte* **5**: 38-39.

Schenkling S. 1912b: H. Sauter's Formosa-Ausbeute. Cleridae (Col.). *Entomologische Mitteilungen* **1**: 321-330.

Schenkling S. 1915: Neue Beiträge zur Kenntnis der Cleriden (Col.) I. *Entomologische Mitteilungen* **4**: 107-114.

Schenkling S. 1916: H. Sauter's Formosa-Ausbeute: Cleridae II (Col.). *Archiv für Naturgeschichte* A **82** [1917]: 117-118.

Schenkling S. 1926: Ein neuer Opetiopalpus von Spanien (Col., Cler.). *Coleopterologisches Centralblatt* **1**: 102.

Schenkling S. 1927: Elateridae II. Pars 88. In: Schenkling S. (ed.): *Coleopterorum Catalogus*. Berlin: W. Junk, 265-636 pp.

Schenkling S. 1928: Throscidae, Cerophytidae, Perothopidae. Pars 101. In: Schenkling S. (ed.): *Coleopterorum Catalogus*. Berlin: W. Junk, 26 pp.

Schenkling S. 1932: Cleriden von Yunnan (Col.). *Miteillungen der Deutschen Entomologische Gesellschaft* **3**: 24-25.

Schiller W. 1984: Corticarina alemannica n. sp., ein neuer Schimmelkäfer aus Sudbaden (Col. Lathridiidae). *Entomologische Blätter* **80**: 103-106.

Schilling P. S. 1843: Salz-Bohrkäfer (Ptinus salinus mihi). *Arbeiten und Veränderungen der Schlesischen Gesellschaft für Vaterländische Kultur im Jahre* **1843**: 175.

Schilsky J. 1888: Beitrag zur Kenntnis der deutschen Käferfauna. *Deutsche Entomologische Zeitschrift* **32**: 177-190.

Schilsky J. 1889: Berichtungen und Ergänzungen zum Verzeichniss der deutschen Käfer. *Deutsche Entomologische Zeitschrift* **33**: 337-344.

Schilsky J. 1890: Beiträge zur deutschen Fauna. V. *Deutsche Entomologische Zeitschrift* **34**: 177-199.

Schilski J. 1891: VI. Beitrag zur Kenntnis der deutschen Käferfauna. *Deutsche Entomologische Zeitschrift* **35**: 153-157.

Schilsky J. 1892: VII. Beitrag zur Kenntniss der deutschen Käferfauna. *Deutsche Entomologische Zeitschrift* **1892**: 193-208.

Schilsky J. 1893: VIII. Beitrag zur Kenntniss der deutschen Käferfauna. *Deutsche Entomologische Zeitschrift* **1893**: 353-357.

Schilsky J. 1894a: *Die Käfer Europa's. Nach der Natur beschrieben. Heft 30*. Nürnberg: Bauer & Raspe, viii pp. + 100 sheets.

Schilsky J. 1894b: Un nuovo Dasytes di Sicilia. *Il Naturalista Siciliano* **13**: 15-16.

Schilsky J. 1894c. Beitrag zur Kenntniss der Dasytinen. *Deutsche Entomologische Zeitschrift* **1894**: 225-236.

Schilsky J. 1894d: IX. Beitrag zur deutschen Käferfauna. *Deutsche Entomologische Zeitschrift* **1894**: 329-332.

Schilsky J. 1895: *Die Käfer Europa's. Nach der Natur Beschrieben. Heft 31*. Nürnberg: von Bauer und Raspe (E. Küster), v pp. + 100 sheets, 1 pl.

Schilsky J. 1896a: *Die Käfer Europa's. Nach der Natur Beschrieben. Heft 32*. Nürnberg: von Bauer und Raspe (E. Küster), viii pp. + 100 sheets.
Schilsky J. 1896b: Ein neuer Acanthocnemus aus Guinea (West-Afrika). *Deutsche Entomologische Zeitschrift* **40**: 361-363.
Schilsky J. 1897a: *Die Käfer Europa's. Nach der Natur Beschrieben. Heft 33*. Nürnberg: von Bauer und Raspe (E. Küster), nr 33A – 33R [= 18 pp.] + iv + pp. + 100 nr. [= 1 or 2 sheets] + [3] pp.
Schilsky J. 1897b: *Die Käfer Europa's. Nach der Natur Beschrieben. Heft 34*. Nürnberg: von Bauer und Raspe (E. Küster), viii pp. + 100 nr. [each on 1 or 2 sheets] + nr. 34A – 34G [= 54 pp.].
Schilsky J. 1897c: Trichoceble heydeni n. sp. aus China. *Deutsche Entomologische Zeitschrift* **41**: 155.
Schilsky J. 1898: *Die Käfer Europa's. Nach der Natur Beschrieben. Heft 35*. Nürnberg: von Bauer und Raspe (E. Küster), viii pp. + 100 nr. [each on 1 to 3 sheets] + [43] pp.
Schilsky J. 1899: *Die Käfer Europa's. Nach der Natur Beschrieben. Heft 36*. Nürnberg: von Bauer und Raspe (E. Küster), iv + 100 nr. [each on 1 to 3 sheets] + [54] pp.
Schilsky J. 1900: *Die Käfer Europa's. Nach der Natur Beschrieben. Heft 37*. Nürnberg: von Bauer und Raspe (E. Küster), iv pp. + 100 nr. [each on 1 to 3 sheets] + [22] pp.
Schilsky J. 1903: *Die Käfer Europa's. Nach der Natur Beschrieben. Heft 40*. Nürnberg: von Bauer und Raspe (E. Küster), viii pp. + 100 nr. [each on1 or 2 sheets] + [22] pp.
Schilsky J. 1906a: *Die Käfer Europa's. Nach der Natur Beschrieben. Heft 42*. Nürnberg: von Bauer und Raspe (E. Küster), vi pp. + 119 nr. [each on 1 or 2 sheets].
Schilsky J. 1906b: *Die Käfer Europa's. Nach der Natur Beschrieben. Heft 43*. Nürnberg: von Bauer und Raspe (E. Küster), viii + cxix pp. 30 nr. [each on 1 or 2 sheets] + [1] p.
Schilsky J. 1907: *Die Käfer Europa's. Nach der Natur Beschrieben. Heft 44*. Nürnberg: von Bauer und Raspe (E. Küster), iv pp. + 100 + 2 sheets, 30 pls. [106 pp.].
Schilsky J. 1908: Neue märkische Käfer und Varietäten aus der Gegend von Luckenwalde. *Deutsche Entomologische Zeitschrift* **1908**: 599-604.
Schilsky J. 1910: *Die Käfer Europa's. Nach der Natur beschrieben. Heft 46*. Nürnberg: Bauer und Raspe (E. Küster), 100 sheets.
Schimmel R. 1982: Ein neuer Ampedus (Elateridae) aus Niedersachsen. *Entomologische Blätter* **78**: 183-186.
Schimmel R. 1984: Ein neuer Ampedus aus der Verwandtschaft des Ampedus balteatus (L.) aus Deutschland (Elateridae, Col.). *Koleopterologische Rundschau* **57**: 9-12.
Schimmel R. 1987: Zwei neue Elateridae aus Griechenland und aus Jugoslawien. *Entomologische Blätter* **83**: 42-46.
Schimmel R. 1990: Neue Schnellkäfer, sowie Bemerkungen über bekannte Arten dieser Familie aus der Türkei, aus Spanien und aus Deutschland (Coleoptera, Elateridae). *Entomologische Blätter* **86**: 71-84.
Schimmel R. 1993a: Revision, phylogenetische und zoogeographische Analyse des supraspezifischen Taxons Ampedus Dejean, 1833, aus dem Himalaya und angrenzenden Regionen (Coleoptera: Elateridae). *Entomologica Basiliensia* **16**: 79-180.
Schimmel R. 1993b: Neue Arten sowie eine neue Gattung der Unterfamilie Diminae Candeze, 1863 aus Ostasien und dem Balkan (Coleoptera: Elateridae). *Koleopterologische Rundschau* **63**: 245-259.
Schimmel R. 1994: Die Arten der supraspezifischen Taxa Campsosternus Latreille, 1834 und Pectocera Hope, 1872, aus dem Himalaya (Coleoptera: Elateridae). *Entomologische Blätter* **90**: 45-68.
Schimmel R. 1995: Neue Elateriden aus Zypern und aus Sudostasien (Insecta: Coleoptera: Elateridae). *Reichenbachia* **31**: 49-56.
Schimmel R.1996a: Neue und wenig bekannte Elateriden aus Griechenland und Ostasien (Coleoptera: Elateridae). *Koleopterologische Rundschau* **66**: 161-177.
Schimmel R. 1996b: Neue Arten der Tribus Ampedini aus Griechenland und Zypern (Coleoptera: Elateridae). *Mitteilungen des Internationalen Entomologischen Vereins e. V.* **21**: 63-67.
Schimmel R. 1996c: Neue Arten aus den- Elater- und Megapenthes-Gruppen aus Indien und Neuseeland (Coleoptera, Elateridae). *Entomologische Blätter* **92**: 121-128.
Schimmel R. 1996d: Eine neue Peripontius-Art aus dem Troodos-Gebirge in Zypern (Coleoptera: Elateridae). *Mitteilungen des Internationalen EntomologischenVereins e. V.* **21**: 121-124.
Schimmel R. 1996e: Das Monophylum Diminae Candèze, 1863 (Insecta, Coleoptera, Elateridae). *Pollichia-Buch* **33**: 1-227.

Schimmel R. 1997: Neue Pectocera-Arten aus Südostasien und systematische Bemerkungen zu Ceropectus Fleutiaux, 1927 (Coleoptera: Elateridae, Oxynopterini). *Entomologische Blätter* **93**: 43-49.

Schimmel R. 1998a: Neue Elateriden aus Ostasien und dem Iran (Coleoptera: Elateridae). *Koleopterologische Rundschau* **68**: 217-226.

Schimmel R. 1998b: Neue Schnellkäfer-Arten aus Südostasien (Coleoptera, Elateridae). *Entomologische Blätter* **94**: 101-109.

Schimmel R. 1998c: Neue und wenig bekannte Schnellkäfer aus Südostasien (Coleoptera: Elateridae). *Entomological Problems* **29**: 149-155.

Schimmel R. 1998d: Neue und wenig bekannte Elateriden sowie eine neue Gattung, Acumenator n. gen., aus Südostasien (Insecta: Coleoptera, Elateridae). *Mitteilungen der Pollichia* **85**: 235-259.

Schimmel R. 1999: Die Megapenthini-Arten Süd- und Südostasiens. Erster Teil: Procraerus, Ectamenogonus, Xanthopenthes, Dolinolus n. gen., Girardelater n. gen. und Preusselater n. gen. *Pollichia-Buch* **38**: 1-299.

Schimmel R. 2002: Neue und wenig bekannte Diminae-Arten aus China, Indien, Laos, Myanmar, Nepal und Vietnam (Insecta, Coleoptera: Elateridae). *Mitteilungen der Pollichia* **88**: 205-231.

Schimmel R. 2003a: Die Megapenthini-Arten Süd- und Südostasiens. Zweiter Teil: Acumenator, Anchastelater, Dicarinator, Gamepenthes, Ganoxanthus, Martiniana und Simodactylus. *Pollichia-Buch* **42**: 1-261.

Schimmel R. 2003b: Neue Ampedini-, Physorhinini-, Pectocerini-, Elaterini- und Diminae-Arten aus Südostasien (Coleoptera: Elateridae). *Mitteilungen der Pollichia* **90**: 256-292.

Schimmel R. 2004a: Zwei neue Gattungen, Platianulus n. gen. und Rieseulus n. gen. sowie weitere neue Arten aus der Tribus Physorhinini aus Suedostasien (Insecta: Coleoptera, Elateridae). *Veröffentlichungen des Naturkundemuseums Erfurt* **23**: 207-215.

Schimmel R. 2004b: Die Megapenthini-Arten Süd- und Südostasiens. Dritter Teil: Abelater, Amamipenthes, Cateanus n. gen., Friedrichiellus n. gen., Hayekpenthes, Megapenthes, Melanoxanthus, Pengamenthes, Platianeus n. gen., Sawadapenthes und Wallaceus n. gen. (Insecta: Coleoptera, Elateridae). *Pollichia-Buch* **45**: 1-504.

Schimmel R. & Cate P. 1991: Eine neue Art der Gattung Dima Charpentier 1825, sowie ein Wiederfund von Dima yunnana Fleutiaux aus China (Coleoptera: Elateridae). *Koleopterologische Rundschau* **61**: 137-138.

Schimmel R. & Platia G. 1988: Eine neue Art der Gattung Ampedus Germar, 1844 (Coleoptera: Elateridae) aus Nord-Afrika. *Entomologische Blätter* **84**: 38-40.

Schimmel R. & Platia G. 1991a: Revision der Subtribus Dimina Candèze, 1863 aus dem Himalaya, mit Bestimmungstabellen der Gattungen und Arten (Coleoptera: Elateridae). *Entomologica Basiliensia* **14**: 261-382.

Schimmel R. & Platia G. 1991b: Die Arten der Gattung Neoathousius n. gen. aus Nordindien, Pakistan und Afghanistan (Coleoptera: Elateridae). *Revue Suisse de Zoologie* **98**: 879-895.

Schimmel R. & Platia G. 1992a: Neue Arten des supraspezifischen Taxons Dimina Candeze, 1863 aus Sudostasien (Coleoptera: Elateridae). *Beiträge zur Entomologie* **42**: 293-315.

Schimmel R. & Platia G. 1992b: Die Arten des supraspezifischen Taxons Senodoniinae Schenkling, 1927 (Coleoptera: Elateridae). *Entomologica Basiliensia* **15**: 229-254.

Schimmel R. & Platia G. 1994: Eine neue Art der Gattung Neoathousius Schimmel & Platia, 1991, aus China. *Koleopterologische Rundschau* **64**: 233-234.

Schimmel R., Platia G. & Martin O. 1992: Eine neue Ampedus-Art aus dem Atlas-Gebirge (Coleoptera: Elateridae). *Koleopterologische Rundschau* **62**: 173-176.

Schiødte J. M. C. 1870: De metamorphosi Eleutheratorum observationes: Bidrag til insekternes udviklingshistorie. *Naturhistorisk Tidsskrift* (3) **6** [1869-1870]: 467-536, pls. III-X.

Schiødte J. M. C. 1872: Fortegnelse over de i Danmark levende Malacordermi, Cleri, Ptini og Anobia. *Naturhisorisk Tidsskrift* (3) **8** [1872-1873]: 1-22.

Schirmer C. von. 1913: Coccinella quinquepunctata L. var. heraldica Schirmer (Col.). *Deutsche Entomologische Zeitschrift* **1913**: 317-318.

Schmidt F. J. 1832: Graf's Schnellkäfer (Elater Grafii). *Faunus* **1**: 85-86.

Schmidt R. 1904: Ueber eine neue Thorictus-Art aus Ost-Buchara (Th. kaznakovi sp.n.; Coleptera, Thorictidae). *Annuaire du Musée Zoologique de l'Académie Impériale des Sciences de St.-Petersbourg* **8** [1903]: 389-391.

Schneider D. H. 1787: Einige Berichtigungen und Ergänzungen der aus Schäffers Icones Insectorum Ratisbonsium in Fabricii Species Insectorum angeführten Allegaten verglichen mit Harres Beschreibung der von Schäffer abgebildeten Insecten. *Neues Magazin für die Liebhaber der Entomologie* **3** (2): 97-140.

Schneider D. H. 1791: Nachrichten von neu angenommenen Gattungen (Generibus) im entomologischen System. *Neuestes Magazin für die Liebhaber der Entomologie* **1**: 14-31.

Schneider D. H. 1791: Rezensionen. *Neuestes Magazin für die Liebhaber der Entomologie* **1**: 90-109.

Schneider D. H. 1792: Verzeichniss und Beschreibung der in der Sammlung des Herausgebers befindlichen zur Gattung Coccinella gehörigen europäischen Käfer (Sonnen-Käfer oder Blattlaus-Käfer). *Neuestes Magazin für die Liebhaber der Entomologie* **2**: 128-256.

Schneider W. G. 1881: Einige neue Varietäten von Coccinellidaen. *Zeitschrift für Entomologie* (N. F.) (Breslau) **8**: 10-16.

Scholz M. F. R. 1932: Ein neuer Meligethes aus Südeuropa und Bemerkungen zu einigen Arten. *Entomologische Blätter* **28**: 97-100.

Scholz M. F. R. 1933: Drei neue Käferarten aus Europa. *Entomologische Blätter* **29**: 74-78.

Schönherr C. J. 1806: *Synonymia Insectorum, oder: Versuch einer Synonymie aller bisher bekannten Insecten: nach Fabricii Systema Eleutheratorum geordnet, mit Berichtigungen und Anmerkungen, wie auch Beschreibungen neuer Arten und illuminirten Kupfern. Erster Band. Eleutherata oder Käfer. Erster Theil.* Stockholm: Heinr. A. Nordström, xxii + 289 + [2] pp., 3 pls.

Schönherr C. J. 1808: *Synonymia Insectorum, oder: Versuch einer Synonymie aller bisher bekannten Insecten; nach Fabricii Systema Eleutheratorum geordnet. Mit Berichtigungen und Anmerkungen, wie auch Beschreibungen neuer Arten und illuminirten Kupfern. Erster Band. Eleutherata oder Käfer. Zweiter Theil.* Stockholm: Carl Friedr. Marquard, ix + [1] + 424 pp., 4 pls.

Schönherr C. J. 1809: Entomologiska Anmärkningar och beskrifningar på några för Svenska Fauna nya Insekter. *Kongelige Svenska Vetenskaps-Akademiens nya Handlingar* **30**: 48-58.

Schönherr C. J. 1817: *Synonymia Insectorum, oder: Versuch einer Synonymie aller bisher bekannten Insecten; nach Fabricii Systema Eleutheratorum etc. geordnet. Mit Berichtigungen und Anmerkungen, wie auch Beschreibungen neuer Arten und illuminirten Kupfern. Erster Band. Eleutherata oder Käfer. Dritter Theil.* Upsala: Em. Bruzelius, xi + [1] + 506 pp.

Schrank F. von Paula. 1776: *Beyträge zur Naturgeschichte.* Augsburg: Gebrüder Beith, [7] + 130 pp., 7 pls.

Schrank F. von Paula. 1781a: *Enumeratio insectorum Austriae indigenorum.* Augustae Vindelicorum: V. E. Klett et Franck, [24] + 548 + [4] pp., 4 pls.

Schrank F. von Paula. 1781b: Kleine entomologische Anmerkungen. *Schriften der Berlinischen Gesellschaft Naturforschender Freunde* **2**: 307-318.

Schrank F. von Paula. 1785: Verzeichniss beobachteter Insecten im Fürstenthume Berchtesgaden. *Neues Magazine für die Liebhaber der Entomologie* **2**: 313-345.

Schrank F. von Paula. 1786: *Baierische Reise.* München: Johann Baptist Strobl, 276 pp., 2 pls.

Schrank F. von Paula. 1789: Entomologische Beobachtungen. (Coleoptera). *Naturforscher* (Halle) **24**: 60-90.

Schrank F. von Paula. 1796: *Sammlung naturhistorischer und physikalischer Aufsätze.* Nürnberg: Raspe, 456 pp., 7 pls.

Schrank F. von Paula. 1798: *Fauna Boica. Durchgedachte Geschichte der in Baiern einheimischen und zahmen Thiere. Erster Band, zweyte Abtheilung.* Nürnberg: Stein'sche Buchhgandlung, 293-720 pp.

Schulz H. 1899: Neue Varietäten von Hippodamia septemmaculata. *Entomologische Nachrichten* **25**: 223.

Schulze W. 1916: Beitrag zur Coleopteren Fauna der Philippinen. *The Philippine Journal of Science* **11**, D (4): 291-299, 2 pls.

Schummel T. E. 1844: [In den ersten Tagen des August 1843 auf dem Gesenke gefundenen Insekten einiger Ordnungen]. *Uebersicht der Arbeiten und Veränderungen der Schlesischen Gesellschaft für Vaterländische Kultur* (Breslau) [1843]: 184-199.

Schwarz O. 1891a: Revision der paläarktischen Arten der Elateriden-Gattung Agriotes Eschsch. *Deutsche Entomologische Zeitschrift* **1891**: 81-114, pls. I-II.

Schwarz O. 1891b: Zwei neue Elateriden von Syrien und Madagascar. *Deutsche Entomologische Zeitschrift* **1891**: 151-152.

Schwarz O. 1892a: Spheniscosomus, eine neue, bisher verkannte Elateriden-Gattung. *Wiener Entomologische Zeitung* **11**: 132.

Schwarz O. 1892b: Neue Melanotus aus der paläarktischen Zone. *Deutsche Entomologische Zeitschrift* **1891**: 365-366.

Schwarz O. 1892c: Revision der paläarktischen Arten der Elateriden-Gattung Melanotus Eschsch. *Deutsche Entomologische Zeitschrift* **1892**: 145-164.

Schwarz O. 1892d: Neue Cardiophorus-Arten (Elateriden). *Deutsche Entomologische Zeitschrift* **1892**: 383-384.

Schwarz O. 1893a: Cardiophorus albofasciatus nov. sp. *Il Naturalista Siciliano* 13: 1.

Schwarz O. 1893b: [new taxon]. In: Ragusa E.: Catalogo ragionato die coleotteri di Sicilia. *Il Naturalista Siciliano* **13**: 2-14.

Schwarz O. 1893c: Eine neue europäische Art der Elateriden-Gattung Elathous. *Deutsche Entomologische Zeitschrift* **1893**: 191.

Schwarz O. 1893d: Adelocera Kraatzi Schwarz nov. spec. *Deutsche Entomologische Zeitschrift* **1893**: 192.

Schwarz O. 1894: Neue Elateriden. *Deutsche Entomologische Zeitschrift* **1894**: 145-150.

Schwarz O. 1895a: Bemerkungen zu Cardiophorus und Beschreibung einer neuen Art. *Deutsche Entomologische Zeitschrift* **1895**: 37-38.

Schwarz O. 1895b: Ueber Cardiophorus musculus Er. als Vertreter einer besonderen Gattung. *Deutsche Entomologische Zeitschrift* **1895**: 39-40.

Schwarz O. 1896a: Elateriden aus Ost-Afrika, gesammelt von R. v. Bennigsen. *Deutsche Entomologische Zeitschrift* **1896**: 89-93.

Schwarz O. 1896b: Neue Elateriden-Arten. *Deutsche Entomologische Zeitschrift* **1896**: 97-104.

Schwarz O. 1896c: Heterostethus novum genus Elateridarum. *Deutsche Entomologische Zeitschrift* **1896**: 147-148.

Schwarz O. 1897a: Neue palaearctischen Elateriden-Arten. *Wiener Entomologische Zeitung* **16**: 128-136.

Schwarz O. 1897b: Ueber Athus [sic] Henoni Ab. *Deutsche Entomologische Zeitschrift* **1897**: 63-64.

Schwarz O. 1897c: Eine neue Ludius-Art (Steatoderus Eschs.) aus dem cilicischen Taurus. *Deutsche Entomologische Zeitschrift* **1897**: 400.

Schwarz O. 1898a: Beschreibung neuer Elateriden. *Deutsche Entomologische Zeitschrift* **1898**: 129-156.

Schwarz O. 1898b: Elateriden aus Usambara (Ost-Afrika). *Deutsche Entomologische Zeitschrift* **1898**: 183-192.

Schwarz O. 1898c: Aenderungen der Gattungsnamen Enoploderes und Craspedonotus Schwarz. *Deutsche Entomologische Zeitschrift* **1898**: 414.

Schwarz O. 1900: Neue paläarktische Elateriden. *Deutsche Entomologische Zeitschrift* **44**: 98-111.

Schwarz O.1902a: Neue Elateriden. *Stettiner Entomologische Zeitung* **63**: 194-316.

Schwarz O. 1902b: Neue Elateriden aus dem tropischen Asien, den malayischen Inseln und den Inseln der Südsee. *Deutsche Entomologische Zeitschrift* **1902**: 305-350.

Schwarz O. 1903: Drei neue Elathous-Arten (Coleoptera, Elateridae). *Wiener Entomologische Zeitung* **22**: 207-208.

Schwarz O. 1906: Fam. Elateridae. Fascicles 46 A, B. In: Wytsman P. (ed.): *Genera Insectorum, Coleoptera*. Bruxelles: P. Wytsman, 1-224 pp.

Schwarz O. 1907: Fam. Elateridae. Fascicle 46 C. In: Wytsman P. (ed.): *Genera Insectorum, Coleoptera*. Bruxelles: P. Wytsman, 225-370 pp., 6 pls.

Schweiger H. 1949: Eine neue Art der Gattung Cerylon aus Griechenland (Col. Colydiidae). *Entomon* (München) **1**: 43-45.

Schweiger H. 1952: Eine neue Cerylon-Art aus Fukien (Col. Colydiidae). *Entomologische Blätter* **17/18**: 134-136.

Scopoli J. A. 1763: *Entomologia Carnioloca exhibens Insecta Carnioloae indigena et distributa in ordines, genera, species, varietates. Methodo Linnaeana*. Vindobonae: Joannis Thomae Trattner, xxxii + 420 + [4] pp.

Scopoli J. A. 1772: *Annus V. Historico-Naturalis*. Lipsiae: Christ. Gottlob Hilscher, 128 pp.

Scopoli J. A. 1777: *Introductio ad historiam naturalem, sistens genera lapidum, plantarum et animalium hactenus detecta, characteribus essentialibus donata, in tribus divisa, subinde ad leges naturale.* Pragae: W. Gerle, 506 pp. + [34].

Scott H. 1922: The Percy Sladen Trust expedition to the Indian Ocean in 1905, under the leadership of Mr. J. Stanley Gardiner, M. A. Vol. vii., No. iv – Coleoptera: Scydmaenidae, Scaphidiidae, Phalacridae, Cucujidae (supplement), Lathridiidae, Mycetophagidae (including Propalticus), Bostrychidae, Lyctidae. *Transactions of the Linnean Society of London* (2) *Zoology* **18**: 195-260, 4 pls.

Scott H. 1924: Coleoptera: Ptinidae and Anobiidae of the Seychelles. *The Annals and Magazine of Natural History* (9) **14**: 345-426.

Scriba L. G. 1790a: Verzeichniss der Insekten der Darmstädter Gegend. *Journal für die Liebhaber der Entomologie* **1** (1): 41-73.

Scriba L. G. 1790b: Erste Fortzetzung des Verzeichnisses der Insekten der Darmstädter Gegend. *Journal für die Liebhaber der Entomologie* **1** (2): 151-192.

Scriba L. G. 1791: *Beiträge zu der Insecten-Geschichte. Zweites Heft.* Frankfurt: Varrentrapp & Wenner, pp. 69-194, pls 7-12.

Scriba L. G. 1793: Fortsetzung der Beschreibung verschiedener Käfer. *Beiträge zur Insekten-Geschichte herausgegeben von Ludwig Gottlieb Scriba* **3**: 219-230, pl. xiv.

Seidlitz G. K. M. von. 1867: Einige entomologische Excursionen in den Castilischen Gebirgen im Sommer 1865. *Berliner Entomologische Zeitschrift* **11**: 167-193.

Seidlitz G. K. M. von. 1872-1875: *Fauna Baltica. Die Käfer (Coleoptera) der Ostseeprovinzen Russlands. Lieferung 1, 2.* Dorpat: Dorpader Naturforschende-Gesellschaft, xlii + 142 + 560 pp. [issued in parts, xx + 48 + 208 pp., Fam. 1-35 in 1872; Fam. 36-42 in 1874; following in 1875].

Seidlitz G. K. M. von. 1887a-1891a: *Fauna Baltica. Die Kaefer (Coleoptera) der deutschen Ostseeprovinzen Russlands. Zweite Auflage.* Königsberg: Hartugsche Verlagsdruckerei, 10+lvi+192+818 pp., 1 pl. [issued in parts: pp. i-xl, Gattungen 1-16, Arten 1-96: 1887; pp. xli-xlviii, Gattungen 17-80, Arten 97-336: 1888; pp. Gattungen 81-128, Arten 337-512: 1889; pp. Gattungen 129-160, Arten 513-608: 1890; pp. xlix-lvi, Gattungen 161-192, Arten 609-818: 1891].

Seidlitz G. K. M. von. 1888b-1891b: *Fauna Transsylvanica. Die Kaefer (Coleoptera) Siebenbürgens.* Königsberg: Hartungsche Verlagsdruckerei, lvi + 192 + 914 pp., 1 pl. [issued in parts: pp. i-xl, Gattungen 1-48, Arten 1-240: 1888; pp. xli-lvi, Gattungen 49-128, Arten 241-544: 1889; pp. xlix-lvi, Gattungen 129-192, Arten 545-914: 1891].

Semenov A. P. 1890: Diagnoses coleopterorum novorum ex Asia centrali et orientali. II. *Horae Societatis Entomologicae Rossicae* **24** [1890-1891]: 193-226.

Semenov A. P. 1891: Diagnoses coleopterorum novorum ex Asia centrali et orientali. III. *Horae Societatis Entomologicae Rossicae* **25** [1890-1891]: 262-382.

Semenov A. P. 1893: Coleoptera asiatica nova. *Horae Societatis Entomologicae Rossicae* **27**: 494-507.

Semenov A. P. 1895: Coleoptera asiatica nova. III. *Horae Societatis Entomologicae Rossicae* **29** [1894-1895]: 125-144.

Semenov A. 1898a: Zamyatki o zhestkokrylykh (Coleoptera) evropeyskoy Rossii i Kavkaza. *Bulletin de la Société Impériale des Naturalistes de Moscou* (N. S.) **12**: 68-115.

Semenov A. P. 1898b: Coleoptera nova Rossiae europaeae caucasique. *Horae Societatis Entomologicae Rossicae* **31** [1897-1898]: 542-554.

Semenov A. P. 1898c: Coleoptera nova Rossiae europaeae caucasique. *Horae Societatis Entomologicae Rossicae* **32**: 280-290.

Semenov A. 1899: Neskolko soobrazheniy o proshlom fauny i flory Kryma po novodu nakhozhdenia tam gornoy kuropatky (Caccabis Chukar G. R. Gray). Quelque considérations sur le passé de la faune et flore de la Crimée en rapport avec le habitat de la perdrix de montagne (Caccabis Chukar G. R. Gray). *Mémoires de l'Académie Impériale des Sciences de St.-Pétersbourg* (8) **8** (6): 1-19.

Semenov A. P. 1900a: Coleoptera asiatica nova. X. *Horae Societatis Entomologica Rossicae* **34** [1899-1900]: 577-605.

Semenov A. P. 1900b: Coleoptera asiatica nova. XI. *Horae Societatis Entomologicae Rossicae* **34** [1899-1900]: 676-688.

Semenov A. P. 1902: Coleoptera nova Rossiae europaeae caucasique. *Horae Societatis Entomologicae Rossicae* **35**: 253-265.

Semenov A. P. 1903: Coleoptera nova faunae turanicae. *Revue Russe d'Entomologie* **3**: 353-357.

Semenov A. P. 1904: Novae coleopterorum formae e Sibiria. *Revue Russe d'Entomologie* **4**: 289-291.

Semenov A. P. 1910: Analecta coleopterologica. XVI. *Revue Russe d'Entomologie* **9** [1909]: 433-495.

Semenov A. P. & Dobrzhanskiy F. G. 1924: Tres novae coccinellidarum species e fauna Rossiae asiaticae (Coleoptera). *Revue Russe d'Entomologie* **18**: 99-102.

Semenov A. P. & Pjatakova V. 1936: Sur un genre inédit, représentant une nouvelle tribu dans la famille Elateridae (Coleoptera). - Nový neznámý rod, zástupce nové skupiny v čeledi Elateridae. *Časopis Československé Společnosti Entomologické* **33**: 101-103.

Sengupta [= Sen Gupta] T. 1967: A new subfamily of Languriidae (Coleoptera) based on four genera, with a key to the species of Toramus. *Proceedings of the Royal Entomological Society of London* (B) **36**: 167-176.

Sengupta [= Sen Gupta] T. 1976: Lathridiinae (Coleoptera: Lathridiidae) from India. *Oriental Insects* **10**: 113-135.

Sengupta [= Sen Gupta] T. 1977: A new genus and species of Rhizophagidae (Clavicornia: Coleoptera) from Sikkim. *Oriental Insects* **11**: 531-535.

Sengupta [= Sen Gupta] T. 1978a: Discovery of the family Propalticidae (Coleoptera) from India, with description of a new species of Propalticus. *Oriental Insects* **12**: 81-83.

Sengupta [= Sen Gupta] T. 1978b: Ergebnisse der Bhutan-Expedition 1972 des Naturhistorischen Museums in Basel. Coleoptera: Fam. Byturidae. *Entomologia Basilensia* **3**: 163-170.

Sengupta [= Sen Gupta] T. 1978c: Ergebnisse der Bhutan-Expedition 1972 des Naturhistorischen Museums in Basel. Coleoptera: Fam. Cucujidae. *Entomologica Basiliensia* **3**: 219-222.

Sengupta [= Sen Gupta] T. 1978d: A new genus of Atomariinae (Coleoptera: Cryptophagidae) with descriptions of two new species from India. *Oriental Insects* **12**: 275-279.

Sengupta [= Sen Gupta] T. 1979: On the genus Mimemodes (Rhizophagidae: Coleoptera), with descriptions of new species from India. *Journal of the Zoological Society of India* **28** [1976]: 65-73.

Sengupta [= Sen Gupta] T. 1980: On the genus Cryptophagus Herbst (Cryptophagidae: Coleoptera) with descriptions of four new species from India and Nepal. *Proceedings of the Zoological Society of Calcutta* **31**: 33-38.

Sengupta [= Sen Gupta] T. 1983: Revision of the Indian Stephostethus LeConte (Coleoptera, Lathridiidae). *Entomologica Basiliensia* **8**: 342-368.

Sengupta [= Sen Gupta] T. 1988: Review of the genera of the family Rhizophagidae (Clavicornia: Coleoptera) of the world. *Memoirs of the Zoological Survey of India* **17** (1): 1-58, pls I-XXIII.

Sengupta [= Sen Gupta] T. & Basak P. K. 1985: Discovery of the genus Caenoscelis Thomson (Coleoptera: Cryptophagidae) from Indian region and description of a new species from Nepal. *Bulletin of the Zoological Survey of India* **7**: 41-43.

Sengupta [= Sen Gupta] T. & Biswas D. N. 1977: On the genus Rhizophagus Herbst (Coleoptera: Rhizophagidae) and descrition of a new species from India. *Records of the Zoological Survey of India* **72**: 419-423.

Sengupta [= Sen Gupta] T. & Crowson R. A. 1973: A review of the classification of Cerylonidae (Coleoptera, Clavicornia). *Transactions of the Royal Entomological Society of London* **124**: 365-446.

Sengupta [= Sen Gupta] T. & De M. 1988: Rhizophagus Herbst from the Indian region with description of seven new species (Coleoptera, Rhizophagidae). *Elytron* **2**: 3-14.

Sengupta [= Sen Gupta] T. & Mukherjee A. K. 1977: A new genus of Languriidae (Coleoptera) from India. *Oriental Insects* **11**: 1-4.

Sengupta [= Sen Gupta] T. & Mukhopadhyay P. 1978: The systematic position of Laemophloeus semilaetaneus Grouvelle (Coleoptera: Cucujidae: Laemophloeinae). *Bulletin of the Zoological Survey of India* **1**: 141-145.

Sengupta [= Sen Gupta] T. & Pal T. K. 1980a: Ergebnisse de Bhutan Expedition 1972 des Naturhistorischen Museum in Basel. Coleoptera: Fam. Cryptophagidae, genus Henoticus. *Entomologica Basiliensia* **5**: 31-41.

Sengupta [= Sen Gupta] T. & Pal T. K. 1980b: Ergebnisse de Bhutan Expedition 1972 des Naturhistorischen Museum in Basel. Coleoptera: Fam. Cryptophagidae, genus Himascelis. *Entomologica Basiliensia* **5**: 42-44.

Sengupta [= Sen Gupta] T. & Pal T. K. 1982: Three new species of Sphindidae (Coleoptera: Clavicornia) from India and Sri Lanka. *Entomologica Basiliensia* **7**: 387-393.

Sengupta [= Sen Gupta] T. & Pal T. K. 1996: *Fauna of India and the adjacent countries. Calvicornia* [sic]: *Coleoptera Family Silvanidae*. Calcutta: Zoological Survey of India, [10] + 262 pp.

Sharp D. 1867: [new taxa]. In: Crotch G. R. & Sharp D.: Additions to the catalogue of British Coleoptera, with descriptions of new species. *Transactions of the Entomological Society of London* **1867**: 435-451.

Sharp D. 1877: On the Elateridae of New Zealand. *The Annals and Magazine of Natural History* (4) **19**: 396-413, 469-487.

Sharp D. 1878: On some Nitidulidae from the Hawaiian Islands. *Transactions of the Entomological Society of London* **1878**: 127-140.

Sharp D. 1879: On some Coleoptera from the Hawaiian Islands. *Transactions of the Entomological Society of London* **1879**: 77-105.

Sharp D. 1881: On some new Coleoptera from the Hawaiian Islands. *Transactions of the Entomological Society of London* **1881**: 507-534.

Sharp D. 1885a: On the Colydiidae collected by Mr. G. Lewis in Japan. *The Journal of the Linnaean Society. Zoology* **19**: 58-84, pl. 3.

Sharp D. 1885b: [new taxa]. In: Blackburn T. & Sharp D.: Memoirs on the Coleoptera of the Hawaiian Islands. *The Scientific Transactions of the Royal Dublin Society* (2) **3**: 119-289, 300 [=290]. pls. 4, 5.

Sharp D. 1886: On New Zealand Coleoptera with descriptions of new genera and species. *Transactions of the Royal Dublin Society* (2) **3**: 351-454, pls. 12, 13.

Sharp D. 1888: Fam. Phalacridae. Pp. 244-264. In: In: Godman F. D. & Salvin O. (eds): *Biologia Centrali Americana. Insecta. Coleoptera. Vol. II. Part 1.* London: Taylor and Francis, xii + 717 pp., 19 pls. [issued in parts, 1887-1905].

Sharp D. 1889: Two new species of Scymnus. *Insect Life* **1**: 364-365.

Sharp D. 1890: Fam. Nitidulidae. Pp. 305-336, pls 8-10. In: Godman F. D. & Salvin O. (eds): *Biologia Centrali-Americana. Insecta. Coleoptera. Vol. II. Part 1.* London: Dulau and Co., xii + 717 + 1 pp., 19 pls. [issued in parts 1887-1905].

Sharp D. 1891: Fam. Nitidulidae. Pp. 337-440, pls 10-12. In: Godman F. D. & Salvin O. (eds): *Biologia Centrali Americana. Zoologia. Insecta. Coleoptera. Vol. II. Part 1.* London: Dulau & Co., xii + 717 + 1 pp., 19 pls. [issued in parts, 1887-1905].

Sharp D. 1899: Fam. Cucujidae. Pp: 499-563, pls 16, 17. In: Godman F. D. & Salvin O. (eds): *Biologia Centrali-Americana. Insecta. Coleoptera. Vol. II. Part 1.* London: Dulau and Co., xii + 717 + [1] pp., 19 pls. [issued in parts, 1887-1905].

Sharp D. 1900: Fam. Cryptophagidae. Pp. 579-624, pl. 18. In: Godman F. & Salvin O. (eds): *Biologia Centrali-Americana. Insecta. Coleoptera. Vol. II. Part 1.* London: Dulau and Co., xii + 717 + [1] pp., 19 pls. [issued in parts, 1887-1905].

Sharp D. 1902: Lathridiidae, Mycetophagidae, Dermestidae. Pp. 625-688. In: Godman F. & Salvin O. (eds): *Biologia Centrali-Americana, Insecta, Coleoptera. Vol II. Part 1* London: Dulau and Co., xii + 717 + [1] pp., 19 pls. [issued in parts, 1887-1905].

Sharp D. 1908: [new taxa]. In: Sharp D. et Scott H.: *Fauna Hawaiiensis or the zoology of the Sandwich (Hawaiian) Isles: being result of the exploration by the Joint Committee appointed by the Royal society of London for promoting natural knowledge and the British Association for the advancement of science and carried on with the assistance of those bodies and of the trustees of the Bernice Papuahi Bishop Museum at Honolulu. Volume III. Part V. Coleoptera. III.* Cambridge: University Press, 367-579 pp., 13-26 pls.

Sharp D. 1910: Notes on Corticaria, with descriptions of two new species. *The Entomologist's Monthly Magazine* **46**: 105-108.

Sharp D. 1916a: Additional British species of Ernobius. *The Entomologist's Monthly Magazine* **52**: 178-180.

Sharp D. 1916b: Additions and corrections in the genus Ernobius, with notes on the copula. *The Entomologist's Monthly Magazine* **52**: 219-224.

Shaw G. 1792: *Vivarium naturae, or the Naturalist's Miscellany.* London, pls. 88-125.

Shibata T. 1966: Some reports on the pleasing fungus-beetles from Japan, I. (Col. Erotylidae). *The Entomological Review of Japan* **18**: 37-40.

Shinohara T. 1941: Description of a new species of Eneaustes (sic!) from Formosa. *Konchū-Kai* **9**: 613-614 (in Japanese, with English title).

Shuckard W. E. 1839: *Elements of British Entomology. Containing a general introduction to the science, a systematic description of all the genera, and a list of all the species of British Insect, with a history of their transformation, habits, economy, and distribution, with outline figures of the families and their larvae and pupae, an explanation of the technical terms, and full directions. Vol. 1.* London: Hippolyte Baillière, 240 pp.

Shuckard W. E. 1840. *The British Coleoptera delineated, consisting of figures of all the genera of British Beetles, drawn in outline by W. Spry, M.E.S.* London, 83 pp., 94 pls.

Sicard A. 1891: Habitat de Coccinellides et variétés nouvelles. *Le Coléoptériste* (Paris) **1890**: 58-60.

Sicard A. 1892: Scymnus Kiesenwetteri et ses variétés. *Miscelanea Entomologica* **1**: 151-152, 156-157.

Sicard A. 1907a: Description d'une nouvelle espèce de coccinellide paléartique (Col.). *Bulletin de la Société Entomologique de France* **1907**: 67-68.

Sicard A. 1907b: Notes sur quelques coccinellides paléartiques (Col.). *Bulletin de la Société Entomologique de France* **1907**: 132-136.

Sicard A. 1907c: Coléoptères coccinellides du Japon, recueillis par MM. Harmand et Gallois. Liste et description d'espèces nouvelles. *Bulletin du Muséum National d'Histoire Naturelle* (Paris) **13**: 210-212.

Sicard A. 1907d: Espèces nouvelles de coccinellidae d'Afrique (Col.). *Annales de la Société Entomologique de France* **76**: 412-418.

Sicard A. 1909: Revision de coccinellides de la Faune Malgache. II. Coccinellidae Aphidiphagae. *Annales de la Société Entomologique de France* **78** [1909-1910]: 63-165.

Sicard A. 1910: Espèces et variétés nouvelles de Coccinellidae de l'Inde (Coleopt.) appartenant à la collection de M. Andrewes (de Londres). *Annales de la Société Entomologique de France* **79**: 377-389.

Sicard A. 1912a: Description d'un Diomus (Coccinellidae) nouveau. *L'Échange, Revue Linnéenne* **28**: 22.

Sicard A. 1912b: Description d'espèces et variétés nouvelles de coccinellides de la collection de Deutsches Entomologisches Museum de Berlin-Dahlem. *Archiv für Naturgeschichte* A **28** (6): 129-138.

Sicard A. 1912c: Coleoptera, Coccinellidae [of the Seychelles]. *Transactions of the Linnean Society of London* **15**: 361-366.

Sicard A. 1914a: Notes sur quelques coccinellides de l'Inde et de Birmanie appartenant à la collection de M. Andrewes, de Londres et description d'espèces et de variétés nouvelles. *Annales de la Société Entomologique de France* **81** [1912-1913]: 495-506.

Sicard A. 1914b: Coccinellides nouveaux de M. Walter, de Rägern (Moravie). *Annales de la Société Entomologique de France* **81** [1912-1913]: 507-513.

Sicard A. 1922: Descriptions de variétés, espèces et genres nouveaux appartenant à la famille des coccinelides. *The Annals and Magazine of Natural History* (9) **9**: 349-360.

Sicard A. 1923:Coccinellides nouveaux de l'Afrique du Nord. *Bulletin de la Société d'Histoire Naturelle de l'Afrique du Nord* **14**: 78-80.

Sicard A. 1929a: Description de deux espèces nouvelles de coccinellides paléarctiques. *Bulletin de la Société d'Histoire Naturelle de l'Afrique du Nord* **20**: 60-62.

Sicard A. 1929b: Descriptions de quelques espèces nouvelles de coccinellides de la faune Indo-malaise. *The Annals and Magazine of Natural History* (10) **3**: 179-184.

Sicard A. 1929c: Description d'espèces et variétés nouvelles de Coccinellidae appartenant au genre Caria Muls. *Annales de la Société Entomologique de France* **98**: 188-192.

Sicard A. 1930: Étude sur les coccinellides du Congo Belge. *Revue de Zoologie et Botanique Africaines* **19**: 56-78.

Sicard A. 1931: Descriptions d'espèces nouvelles appartenant à la famille des coccinellides (Col.). *The Annals and Magazine of Natural History* (10) **8**: 228-234.

Silfverberg H. 1977: Nomenclatoric notes on Coleoptera Polyphaga. *Notulae Entomologicae* **57**: 91-94.

Silvestri F. 1909: Nuovo Coccinellide introdotto in Italia. *Rivista Coleotterologica Italiana* **7**: 126-129.

Singh T. & Singh V. K. 1988: Afissula pronotonigeratus (Coleoptera: Coccinellidae) a new species from India. *Uttar-Pradesh Journal of Zoology* **8**: 217-219.

Sjöberg O. 1939a: Insecta, ex Sibiria meridionali et Mongolia, in itinere Orjan Olsen 1914 collecta. A. Coleoptera, a Fritz Jensen lecta. VII. Nitidulidae. *Norsk Entomologisk Tidsskrift* **5**: 97-99.

Sjöberg O. 1939b: Beitrag zur Kenntnis der Gattung Epuraea Er. (Col. Nitidulidae). Bestimmungstabelle der paläarktischen Arten. *Entomologisk Tidskrift* **60**: 108-126.

Sjöberg O. 1947: Die Ergebnisse einer Genitaluntersuchung der nordischen Arten der Gattung Atomaria Steph. (Col., Cryptophagidae). *Entomologisk Tidskrift* **68**: 91-119.

Sjöberg O. 1950: Drei neue Kaefer aus der borealen Region Schwedens. *Entomologisk Tidskrift* **71**: 159-164.

Sjöstedt Y. 1938: Insecten aus China im Naturhistorischen Reichsmuseum zu Stockholm. *Arkiv för Zoologi* **30** A (13): 1-19.

Skopin N. G. 1951: Novye i maloizvestnye vidy saranchovykh i zhestkokrylykh iz pustyni Betpak-Dala. *Uchenye Zapiski Kazakhovo Gosudarstvenogo Instituta* **13**: 121-126.

Ślipiński S. A. 1980a: New species of Cerylonidae (Coleoptera). *Polskie Pismo Entomologiczne* **50**: 397-411.

Ślipiński S. A. 1980b: Two new species of Philothermopsis Heinze (Coleoptera, Cerylonidae) from Ivory Coast. *Revue Suisse de Zoologie* **87**: 467-471.

Ślipiński S. A. 1981b: Studies on the Cerylonidae (Coleoptera, Clavicornia). Part III. On the Cerylonini from the Himalaya. *Entomologica Basiliensia* **6**: 427-436.

Ślipiński S. A. 1982: New and little known species of Cerylonidae (Coleoptera). *Polskie Pismo Entomologiczne* **52**: 31-58.

Ślipiński S. A. 1983: A review of the Ethiopian species of Passandra Dalman, with notes on the constitution of the Cucujidae (Coleoptera). *Polskie Pismo Entomologiczne* **59**: 85-129.

Ślipiński S. A. 1984: Insects of Saudi Arabia. Coleoptera: Fam. Silvanidae and Cucujidae. *Fauna of Saudi Arabia* **6**: 255-259.

Ślipiński S. A. 1988a: On the Cerylonidae (Coleoptera; Cucujoidea) of Japan. *Revue Suisse de Zoologie* **95**: 145-149.

Ślipiński S. A. 1988b: Cerylonidae (Coleoptera, Cucujoidea) of Pakistan. *Revue Suisse de Zoologie* **95**: 477-485.

Ślipiński S. A. 1990: A monograph of the world Cerylonidae (Coleoptera; Cucujoidea). Part I - Introduction and higher classification. *Annali del Museo Civico di Storia Naturale di Genova* **88**: 1-273.

Ślipiński S. A. 2004: Revision of the Australian Coccinellidae (Coleoptera). Part 2. Tribe Sticholotidini. *Annales Zoologici* **54**: 389-402.

Slosson A. T. 1903: A coleopterous conundrum. *Canadian Entomologist* **35**: 183-187.

Smirnoff W. A. 1953: Les Pharoscymnus (Col. Coccinellidae) d'Afrique du Nord prédateurs de Parlatoria blanchardi Targ. *Revue de Pathologie Végétale et d'Entomologie Agricole de France* **32**: 143-160.

Smirnoff W. A. 1954: Les Cybocephalus (Col. Cybocephalidae) d'Afrique du Nord, prédateurs de Parlatoria blanchardi Targ. (Homoptera, Coccoidea) parasite du palmier-dattier. *Revue de Pathologie Végétale et d'Entomologie Agricole de France* **33**: 84-101, pls I-II.

Smirnoff W. A. 1956: Les Pharoscymnus (Coléoptères Coccinellidae). *Travaux originaux. Service de la Défense des Végétaux, Rabat Maroc* (No 9): 1-38, pl. A, B.

Smith E. H. 1969: [new taxon]. In: Arnett R. H. Jr., Mignot E. C. & Smith E. H.: North American Coleoptera Fauna: Notes on Pyrophorinae, Elateriadae [sic]. *The Coleopterist's Bulletin* **23**: 9-15.

Smith F. 1851: *List of the coleopterous insects in the collection of the British Museum. Part I. - Cucujidae, &c.* London: Trustees, 4 + 25 pp.

Snellen van Vollenhoven S. C. & Sélys Longchamps M. E. de. 1869: *Recherches sur la faune de Madagascar et de ses dépendances, d'après les découvertes de François P. L. Pollen et D. C. van Dam. 5. Insectes.* Leyden: J. K. Steenhoff, 25 pp., 2 pls.

Sokolov E. A. 1972: Materialy k faune i ekologii kozheedov (Coleoptera, Dermestidae) Kazakhstana. *Trudy Kazakhskogo Nauchno-Issledovatelskogo Instituta* **11**: 115-137.

Sokolov E. A. 1974: Novye vidy kozheedov (Coleoptera, Dermestidae) iz Kazakhstana i Sredney Azii. *Trudy Instituta Zoologii Akademii Nauk Kazakhskoy SSR* **35**: 91-96.

Sokolov E. A. 1979: Dva novykh vida kozheedov (Coleoptera, Dermestidae) iz Sredney Azii. *Zoologicheskiy Zhurnal* **58**: 923-925.

Sokolov E. A. 1980: Kozheedy (Coleoptera, Dermestidae) Aksu-dzhabaglinskovo zapovednika. *Trudy Instituta Zoologii Akademii Nauk Kazakhskoy SSR* **39**: 61-63.

Sokolov E. A. 1983: Noviy vid kozheeda (Coleoptera, Dermestidae) iz Nuratau. *Zoologicheskiy Zhurnal* **62**: 803-804.

Solervicens J. & Estrada P. 1996: Coleópteros de Follaje de la Reserva Nacional Río Clarillo (Chile Central). *Acta Entomologica Chilena* **20**: 29-44.

Solier A. J. J. 1849: Orden III. Coleoptera. In: Gay C.: *Historia fisica y politica de Chile segun documentos adquiridos en esta republica durante doce años de residencia en ella y publicada bajo los auspicios del supremo gobierno. Tomo quarto. Zoologia (Coleoptera: Pentamera – Heteromera I., exl. Cleroideos).* Paris-Santiago: Museo de Historia Natural de Santiago, pp. 105-511.

Solsky [= Solskij] S. M. 1866a: Matériaux pour servir à l'étude des insectes de la Russie. I. Notes sur quelques coléoptères nouveaux ou peu connus. *Horae Societatis Entomologicae Rossicae* **4** [1866-1867]: 79-96.

Solsky [= Solskij] S. M. 1866b: Matériaux pour servir à l'étude des insectes de la Russie. II. Insectes nouveaux et remarques sur des espèces connues. *Horae Societatis Entomologicae Rossicae* **4** [1866-1867]: 179-185.

Solsky [= Solskij] S. M. 1868: Coléoptères nouveaux. *Horae Societatis Entomologicae Rossicae* **5**: 29-37.

Solsky [= Solskij] S. M. 1871: Coléoptères de la Sibérie orientale. *Horae Societatis Entomologicae Rossicae* **7** [1870]: 334-406.

Solsky [= Solskij] S. M. 1872: Coléoptères de la Sibérie orientale. *Horae Societatis Entomologicae Rossicae* **8** [1870-1872]: 232-277.

Solsky [= Solskij] S. M. 1876: Zhestkokrylya (Coleoptera). Tetrad' vtoraya. In: Puteshestvie v Turkestan chlena-osnovatelya obshchestva A. P. Fedchenko, sovershennoe ot imperatorskago obshchestva lyubiteley estestvoznaniya po porucheniyu Turkestanskago General-Gubernatora K. P. von-Kaufmana. *Izvestiya Imperatorskago Obshetshvo Lyubiteley Yestestvoznaniya, Antropologii i Entografii* **21** (1): 223-398 + xxiv pp., 2 pls.

Solsky [= Solskij] S. M. 1881: Novyya i maloizve'stnyya zhestkokrylyya okrain Rossiyckoy Imperii i prilegayushchikh k ney stran. *Trudy Russkago Entomologicheskabo Obshchestva* **12** [1880-81]: 231-265.

Solsky [= Solskij] S. M. 1882: Novyya ili maloizve'stnyya zhestkokrylyya okrain Rossiyckoy Imperii i prilegayushchikh k ney stran. *Horae Societatis Entomologicae Rossicae* **13** [1881-82]: 31-84.

Sowerby J. 1806: The British Miscellany: or coloured figures of new, rare or little known animal subjects; Many not before ascertained to be inhabitants of the British Isles; and chiefly in the possession of the author. London, vi + 131 + 31 + [133]-137 pp., 76 pls.

Spaeth F. 1898: Homalisus (Phaeopterus) flavangulus nov. spec. *Verhandlungen der Kaiserlich-Königlichen Zoologisch-Botanischen Gesellschaft in Wien* **48**: 657-658.

Spinola M. 1838: Note sur le Phyllocerus de Sicile. *Annales de la Société Entomologique de France* **7**: 41-42.

Spinola M. 1841: Monographie des térédiles. *Revue Zoologique, par la Société Cuvierienne* **4**: 70-76.

Spinola M. 1844: *Essai monographique sur les clérites insectes coléoptères. Tome premier.* Gênes: Frères Ponthenier, ix + 386 pp.

Spinola M. 1844: *Essai monographique sur les clérites insectes coléoptères. Tome second.* Gênes: Frères Ponthenier, 216 +[4] pp., xlvii pls.

Spornraft K. 1966: Kurze Beiträge zur Kenntnis der mitteleuropäischen Nitiduliden (Coleoptera). *Nachrichtenblatt der Bayerischen Entomologen* **15**: 15-16.

Spornraft K. 1975: Eine neue Meligethes-Art der difficilis-Gruppe aus Italien. *Nachrichtenblatt der Bayerischen Entomologen* **24**: 13-15.

Stebbing E. P. 1914: *Indian Forest Insects of Economic importance. Coleoptera.* Dehra Dun: S. Bishen Singh Mahendra Pal Singh, xvi + 648 pp.

Steel W. O. & Howe R. W. 1952: A new species of Laemophloeus (Col.: Cucujidae) associated with stored products. *Proceedings of the Royal Entomological Society of London, Series B. Taxonomy* **21**: 86-88.

Steel W. O. & Howe R. W. 1955: A new species of Cryptolestes (Coleoptera: Cucujidae) associated with stored products in Africa. *Proceedings of the Royal Entomological Society of London, Series B. Taxonomy* **24**: 107-109.

Steinheil E. W. 1873: Symbolae ad historiam Coleopterorum Argentinae meridionalis, ossia enumerazione dei coleotteri dal prof. P. Strobel nell' Argentinia meridionale, e descrizione delle specie nuove. *Atti della Società Italiana di Scienze Naturalle* **15**: 554-578.

Stenius G. 1952: Paläarktische Arten aus der Gruppe Scymnus (Nephus) bipunctatus Kugel. (Col. Cocc.). *Notulae Entomologicae* **32**: 147-155.

Stepanov E. M. 1930a: Neue und seltene Elateriden aus SSSR. *Entomologisches Nachrichtenblatt* (Troppau) **4**: 86-93.

Stepanov E. M. 1930b: Přehled druhů podrodu Calambus Motsch. - Übersicht der Arten der Selatosomus-Untergattung Calambus Motsch. (Col., Elateridae). *Časopis Československé Společnosti Entomologické* **27**: 134-136.

Stepanov E. M. 1935: Neue und wenig bekannte Elateridae aus S.S.S.R. *Entomologisches Nachrichtenblatt* (Troppau) **9**: 187-200.

Stephens J. F. 1829: *The nomenclature of British Insects; being a compendious list of such species as are contained in the systematics catalogue of British Insects, and forming a guide to their classification, &c &c.* London: Baldwin & Cradock, 68 pp.

Stephens J. F. 1829: *A systematic catalogue of British Insects: Being an attempt to arrange all the hitherto discovered indigenous insects in accoradance with their natural affinities. Containing also the references to every English writer on entomology, and to the principal foreign authors. With all the published British genera to the present time. Part 1.* London: Baldwin and Cradock, xxxiv + 416 + 388 pp.

Stephens J. F. 1829: *Illustrations of British entomology or, a synopsis of indigenous insects: containing their generic and specific distinctions;with an account of their metamorphoses, times of appearance, localities, food, and economy, as far as practicable. Mandibulata. Volume II.* London: Baldwin and Cradock, 200 pp., pls. X-XV.

Stephens J. F. 1830: *Illustrations of British entomology or, a synopsis of indigenous insects: containing their generic and specific distinctions;with an account of their metamorphoses, times of appearance, localities, food, and economy, as far as practicable. Mandibulata. Volume III.* London: Baldwin and Cradock, 447 + [1] pp., pls. XVI-XIX.

Stephens J. F. 1831: *Illustrations of British entomology or, a synopsis of indigenous insects: containing their generic and specific distinctions;with an account of their metamorphoses, times of appearance, localities, food, and economy, as far as practicable. Mandibulata. Volume IV.* London: Baldwin and Cradock, 413 + [1] pp., pls. XX-XXIII. [published in parts, pp. 1-366 in 1831; pp. 367-413 in 1832].

Stephens J. F. 1832a: [Pp. 367-413]. In: *Illustrations of British entomology or, a synopsis of indigenous insects: containing their generic and specific distinctions;with an account of their metamorphoses, times of appearance, localities, food, and economy, as far as practicable. Mandibulata. Volume IV.* London: Baldwin et Cradock, 413 + [1] pp., pls. XX-XXIII. [published in parts, pp. 1-366 in 1831].

Stephens J. F. 1832b: [Pp. 1-240]. In: *Illustrations of British entomology or, a synopsis of indigenous insects: containing their generic and specific distinctions;with an account of their metamorphoses, times of appearance, localities, food, and economy, as far as practicable. Mandibulata. Volume V.* London: Baldwin and Cradock, 448 pp., pls. XXIV-XXVII.. [published in parts, pp. 241-304 in 1833; pp. 305-368 in 1834; pp. 369-448 (Appendix) in 1835].

Stephens J. F. 1833: *The nomenclature of British Insects; together with their synonymes: being a compendious list of such species as are contained in the systematic catalogue of British insects, and of those discovered subsequently to its publication; forming a guide to their classification, &c. &c.* London: Baldwin and Cradock, [2 pp] + 135 columns.

Stephens J. F. 1835: [Appendix, pp. 369-448]. In: *Illustrations of British entomology or, a synopsis of indigenous insects: containing their generic and specific distinctions;with an account of their metamorphoses, times of appearance, localities, food, and economy, as far as practicable. Mandibulata. Volume V.* London: Baldwin and Cradock, 448 pp., pls. XXIV-XXVII.. [published in parts, pp. 1-240 in 1832; pp. 241-304 in 1833; pp. 305-368 in 1834].

Stephens J. F. 1839: *A manual of British Coleoptera or beetles; containing a brief description of all species of beetles hitherto ascertained to inhabit Great Britain and Ireland; together with a notice of their chief localities, times and places of appearences, etc.* London: Longman, Orme, Brown, Green, and Longmans, xii + 443 pp.

Steven C. von. 1808: [new taxa]. In: Schönherr C. J.: *Synonymia Insectorum, oder: Versuch einer Synonymie aller bisher bekannten Insecten; nach Fabricii Systema Eleutheratorum geordnet. Mit Berichtigungen und Anmerkungen, wie auch Beschreibungen neuer Arten und illuminirten Kupfern. Erster Band. Eleutherata oder Käfer. Zweiter Theil.* Stockholm: Carl Friedr. Marquard, ix + [1] + 424 pp., 4 pls.

Steven C. 1829: Notice sur quelques insectes de la collection de C. Steven. *Bulletin de la Société Impériale des Naturalistes de Moscou* **1**: 284-288.

Steven C. 1830: Description de l'elater Parreyssi et de quelques nouveaux Buprestes. *Bulletin de la Société Impériale des Naturalistes de Moscou* **2**: 153-172.

Steven C. 1832: Description de l'Elater parreysii et de quelques nouveaux Buprestes. *Nouveaux Mémoires de la Société Impériale des Naturalistes de Moscou* **2**: 81-94, pls. 2-3.

Stibick J. N. L. 1968: A preliminary revision of the genus Tropihypnus (Coleoptera, Elateridae). *Entomological News* **79**: 169-187.

Stibick J. N. L. 1971: The generic classification of the Negastriinae (Coleoptera: Elateridae). *Pacific Insects* **13**: 371-390.

Stibick J. N. L. 1976: A revision of the Hypnoidinae of the World (Col., Elateridae). Part I. Introduction, phylogeny, biogeography. The Hypnoidinae of North and South America. The genera Berninelsonius and Ligmargus. *Eos, Revista Española de Entomología* **51** [1975]: 143-223.

Stibick J. N. L. 1979: A revision of the Hypnoidinae of the world (Col. Elateridae). Part 3. The Hypnoidinae of Eurasia. *Eos, Revista Española de Entomología* **53** [1977]: 223-307.

Stibick J. N. L. 1980: A revision of the Hypnoidinae of the world (Col. Elateridae). Part 4. The Hypnoidinae of India. *Eos, Revista Española de Entomología* **54** [1979]: 247-273.

Stierlin W. G. 1861: Beitrag zur Insekten-Fauna von Epirus. *Wiener Entomologische Monatschrift* **5**: 216-226.

Stierlin W. G. 1862: Über einiger neue oder wenig bekannte Arten der schweizerischen Käfer-Fauna. *Mitteilungen der Schweizerischen Entomologischen Gesellschaft* **1**: 35-38.

Stierlin W. G. 1863a: Verzeichniss der während einer entomologischen Exkursion nach dem Engadin im Juni 1862 gesammelten Käfer. *Mitteilungen der Schweizerischen Entomologischen Gesellschaft* **1**: 57-66.

Stierlin W. G. 1863b: Ein neuer europäischer Athous. *Mittheilungen der Schweizerischen Entomologischen Gesellschaft* **1**: 91-92.

Stierlin W. G. 1864: Ueber einige neue und wenig bekannte sicilianische Käferarten. *Berliner Entomologische Zeitschrift* **8**: 144-153.

Stierlin W. G. 1868: Beschreibung zweier neuer Käferarten. *Mitteilungen der Schweizerischen Entomologischen Gesellschaft* **2**: 345-347.

Stierlin W. G. 1875: Beschreibung einiger neuer Käferarten. *Mittheilungen der Schweizerischen Entomologischen Gesellschaft* **4**: 473-493.

Stierlin W. G. 1879a: Einige neue schweizerische Käferarten. *Mittheilungen der Schweizerischen Entomologischen Gesellschaft* **5**: 438-441.

Stierlin W. G. 1879b: Beschreibung einiger neuer Käferarten. *Mittheilungen der Schweizerischen Entomologischen Gesellschaft* **5**: 510-515.

Stierlin W. G. 1880: Ein neuer europäischer Athous. *Mittheilungen der Schweizerischen Entomologischen Gesellschaft* **5**: 590-591.

Stierlin W. G. 1884: Über eine seltene Varietät des Corymbites melancholicus F. *Mittheilungen der Schweizerischen Entomologischen Gesellschaft* **7** [1884-1887]: 43-44.

Stierlin W. G. 1896: Beschreibung einiger neuen Rüsselkäfer. *Mittheilungen der Schweizerischen Entomologischen Gesellschaft* **9** [1893-1897]: 326-329.

Stierlin W. G. 1898: *Fauna coleopterorum helvetica. Die Käfer-Fauna der Schweiz nach analytischen Methode bearbeitet. II. Theil.* Schaffhausen: Balli & Böcherer, 662 pp.

Stierlin W. G. 1902: Beschreibung einiger neuen Species von Coleoptera. *Mittheilungen der Schweizerischen Entomologischen Gesellschaft* **10**: 425-427.

Stiller V. 1926: A. fiumei tengermellék és a mögöttes Karszt-hegységet összekötötáj bogárfaunája [Die Käferfauna der Kontanktzone zwischen dem Litorale von Fiume und dem dahinterliegenden Karstgebiete]. *Rovartani Lapok* **26**: 131-140.

Stiller V. 1942: Cantharis obscura L. var. szegedensis nov. *Mitteilungen der Münchener Entomologischen Gesellschaft* **32**: 223-225.

Strand A. 1937: Bemerkungen über nordische Corticaria- und Corticarina-Arten, nebst Beschreibung von zwei neuen Arten. *Norsk Entomologisk Tidsskrift* **5**: 7-23.

Strand A. 1938: Eine neue Dorcatoma-Art (Col. Anobiidae) aus Norwegen. *Norsk Entomologisk Tidsskrift* **5**: 80-82.

Strand A. 1940: Drei neue Lathridien (Col.) aus Fennoscandien. *Norsk Entomologisk Tidsskrift* **5**: 178-180.

Strand A. 1958: Enicmus pseudominutus n.sp. (Col., Lathrid.). *Norsk Entomologisk Tidsskrift* **10**: 184-186.

Strand A. 1968: Corticarina irkutensis n.sp. (Col., Lathridiidae). *Norsk Entomologisk Tidsskrift* **15**: 40.

Strand E. 1921: Neue Namen dreier Insektengattungen. *Entomologische Zeitschrift* **34**: 106.

Strand E. 1928: Nomenclatorische Bemerkungen über einige Coleopteren-Gattungen. *Entomologische Nachrichtsblatt* **2**: 2-3.

Strand E. 1936: Miscellanea nomenclatorica zoologica et paleontologica. *Folia Zoologica et Hydrobiologica* **9**: 167-170.

Strand E. 1942: Miscellanea nomenclatorica zoologica et palaeontoligica. XI. *Folia Zoologica et Hydrobiologica* **11**: 386-402.

Strassen R. zur. 1954: Eine Käfer-Ausbeute aus Sardinien. Mit zwei Neubeschreibungen (Malthodes sassariensis n. sp., Amphimallon montanum n. sp.) und vielen Neunachweisen. *Senckenbergiana* **34**: 259-289, 1 map.

Strohecker H. F. 1943: Some fungus beetles of the family Endomychidae in the United States National Museum, mostly from Latin America and the Philippine Islands. *Proceedings of the United States National Museum* **93**: 381-392.

Strohecker H. F. 1945: Three new species of Endomychidae (Coleoptera). *American Museum Novitates* **1275**: 1-3.

Strohecker H. F. 1951: New species of oriental Endomychidae, with remarks on the previously known species. *The Pan-Pacific Entomologist* **27**: 157-167.

Strohecker H. F. 1953: Coleoptera Fam. Endomychidae. In: Wytsman P. (ed.): *Genera Insectorum*. Bruxelles: Desmet-Verneuil, 140 pp., 5 pls.

Strohecker H. F. 1957: New species of Endomychidae (Coleoptera) in the collections of the Hungarian National Museum and other institutions. *Annales Historico-Naturales Musei Nationalis Hungarici* **8**: 259-283.

Strohecker H. F. 1964: A synopsis of the Amphisternini (Coleoptera: Endomychidae). *Pacific Insects* **6**: 319-357.

Strohecker H. F. 1968: A synopsis of the genus Eumorphus (Coleoptera: Endomychidae). *Pacific Insects* **10**: 79-112.

Strohecker H. F. 1971a: The genera Engonius and Parindalmus (Coleoptera: Endomychidae). *Pacific Insects* **13**: 13-25.

Strohecker H. F. 1971b: Synopsis of genus Encymon and Avencymon, new genus (Coleoptera: Endomychidae). *Pacific Insects* **13**: 27-39.

Strohecker H. F. 1972: The genus Ancylopus in Asia and Europe (Coleoptera: Endomychidae). *Pacific Insects* **14**: 703-708.

Strohecker H. F. 1974: Two new Stenotarsus from India in Museum G. Frey (Col., Endomychidae). *Entomologische Arbeiten aus dem Museum G. Frey* **25**: 136-138.

Strohecker H. F. 1979: The genus Indalmus in Asia, New Guinea and Australia, with description of a new genus Platindalmus (Coleoptera: Endomychidae). *Pacific Insects* **20**: 279-292.

Strohecker H. F. 1982: New Endomychidae (Coleoptera) from Asia. *Revue Suisse de Zoologie* **89**: 307-311.

Strohecker H. F. & Chûjô M. 1970: Sinocymbachus, n. gen. from the Orient (Coleoptera: Endomychidae). *Pacific Insects* **12**: 511-518.

Sturm J. 1807: *Deutschlands Fauna in Abbildungen nach der Natur mit Beschreibungen. V. Abteilung. Die Insecten. Zweites Bändchen. Käfer*. Nürnberg: Verfasser, iv + 279 pp., pls XXI-LII.

Sturm J. 1826: *Catalog meiner Insecten-Sammlung. Erster Theil. Käfer*. Nürnberg: Verfasser, viii + 207 + 16 pp., 4 pls.

Sturm J. 1837a: *Deutschlands Fauna in Abbildungen nach der Natur mit Beschreibungen. V. Abtheilung, Die Insecten. Neuntes Bändchen. Käfer*. Nürnberg: Verfasser, 148 pp., 15 pls.

Sturm J. 1837b: *Deutschland Fauna in Abbildungen nach der Natur mit Beschreibungen. V. Abtheilung. Die Insekten. Deutschland Insekten, Käfer. Elftes Bändchen*. Nürnberg: Bersasserg, 148 pp., 16 pl.

Sturm J. 1837c: *Deutschland Fauna in Abbildungen nach der Natur mit Beschreibungen. V. Abtheilung. Die Insekten. Zwölftes Bändchen. Käfer*. Nürnberg: Bersasserg, 88 pp., 15 pl.

Sturm J. 1839: *Deutschlands Fauna in Abbildungen nach der Natur mit Beschreibungen. V. Abtheilung. Die Insecten. Vierzehntes Bändchen. Käfer*. Nürnberg: Verfasser, 123 pp., pls. cclxxii-cclxxxvii.

Sturm J. 1843: *Catalog der Kaefer-Sammlung*. Nürnberg: J. Sturm, xii + 386 pp., 6 pls.

Sturm J. 1844: *Deutschlands Fauna in Abbildungen nach der Natur mit Beschreibungen. V. Abtheilung. Die Insecten. Fünfzehntes Bändchen. Käfer*. Nürnberg: Verfasser, xii + 140 pp., pls. cclxxxviii-ccciii.

Sturm J. 1845: *Deutchslands Fauna in Abbildungen nach der Natur mit Beschreibungen. V. Abtheilung Die Insecten. Zweites Bändchen. Käfer*. Nürnberg: Verfasser, 114 pp., pls ccciv-cccix.

Suffrian C. W. L. E. 1843a: Entomologische Bemerkungen. *Entomologische Zeitung* (Stettin) **4**: 91-95.

Suffrian C. W. L. E. 1843b: Entomologische Bemerkungen. *Entomologische Zeitung* (Stettin) **4**: 330-337.

Suffrian C. W. L. E. 1851: [new taxa]. In: Küster H. C.: *Die Käfer Europa's. Nach der Natur beschrieben. 22. Heft*. Nürnberg: Bauer & Raspe, 2 pp. + 100 nrs, 2 pls.

Sulzer J. H. 1776: *Abgekürzte Geschichte der Insecten. Nach dem Linnaeischen System. Erster Theil*. Winterthur: H. Steiner und Comp., xxviii + 274 + 72 pp., 32 pls.

Sumakow G. 1903: Diagnosen neuer Coleopteren aus dem Kaukasus. *Wiener Entomologische Zeitung* **22**: 47-48.

Sumakow G. 1907: Novyy vid iz roda Trogoderma Latr. (Coleoptera, Dermestidae) i obzor vidov etogo roda. *Russkoe Entomologicheskoe Obozrenie* **7**: 15-19.

Suzuki H. 1997: Molecular phylogenetic studies of Japanese fireflies and their mating systems (Coleoptera: Cantharoidea). *Tokyo Metropolitan University, Bulletin of Natural History* (**3**): 1-53.

Suzuki W. 1976: A new elaterid beetle of the genus Pectocera from the Ryukyu Islands. *Kontyû* **44**: 263-266.

Suzuki W. 1978: Notes on elaterid beetles from Taiwan I. The genus Thaumastiellus. *Kontyû* **46**: 439-444.

Suzuki W. 1979: A New elaterid beetle of the Genus Dima from Central Taiwan. *Bulletin of the National Science Museum Series A (Zoology)* **5**: 207-210.

Suzuki W. 1982a: Notes on the genus Neocsikia Ôhira and Becker (Coleoptera: Elateridae). *Transactions of the Shikoku Entomological Society* **16**: 77-82.

Suzuki W. 1982b: A new denticolline genus, Parapenia, from the Indo-Chinese subregion (Coleoptera: Elateridae). *Transactions of the Shikoku Entomological Society* **16**: 83-94.

Suzuki W. 1984: Elaterid beetles from the Island of Yonaguni, collected by Mr. and Mrs. Shusei Saito. *Transactions of the Shikoku Entomological Society* **16**: 37-41.

Suzuki W. 1985a: On some Elateridae of the Far East (Coleoptera). *Transactions of the Shikoku Entomological Society* **17**: 79-89.

Suzuki W. 1985b: A new species of the genus Dima (Coleoptera, Elateridae) from Northeastern Pakistan. *Entomologische Blätter* **81**: 183-186.

Suzuki W. 1986: Two elaterid beetles of the genus Magapenthes (Coleoptera, Elateridae) from Japan and Taiwan. Pp. 205-210. In: Uéno S.-I. (ed.): *Entomological Papers presented to Yoshihiko Kurosawa on the Occasion of his Retirement*. Tokyo: Coleopterists Association of Japan, 342 pp.

Suzuki W. 1999: Catalogue of the family Elateridae (Coleoptera) of Taiwan. *Miscellaneous Reports of the Hiwa Museum for Natural History* **38**: 1-348.

Suzuki W. 2002: A revisional study of the genus Campsosternus (Coleoptera, Elateridae) from Taiwan, with description of a new species. *Special Bulletin of the Japanese Society of Coleopterology* **5**: 323-335.

Suzuki W. & Dolin W. G. 1984: Eine neue Art der Gattung Penia Cast. (Coleoptera, Elateridae) aus Darjeerling, Indien. *Entomologica Basiliensia* **9**: 168-172.

Suzuki W. & Dolin W. G. 1987: [new taxon]. In: [The species of the genus Penia (Coleoptera, Elateridae) of the group P. opatroides of the Himalayan region]. *Vestnik Zoologii* **1987** (1): 29-39 (in Russian).

Suzuki W. & Lucht W. 1983: A new species of the genus Melasis Olivier from Sachalin (Coleoptera, Eucnemidae). *Entomological Review of Japan* **38**: 41-44.

Švec Z. 1992: A review of the Palaearctic species of the genus Stilbus (Coleoptera, Phalacridae). *Acta Entomologica Bohemoslovaca* **89**: 429-450.

Švec Z. 2000: Tinodemus mifsudi sp. n. (Coleoptera, Phalacridae) from Malta. *The Central Mediterranean Naturalist* **3**: 61-64.

Švec Z. & Angelini F. 1996: The Italian species of Phalacridae with the description of Olibrus demarzoi sp. n. *Bollettino della Società Entomologica Italiana* **127**: 199-212.

Švec Z. & Ponel P. 1999: Taxonomical and biogeographical notes on the Phalacridae of Turkey (Coleoptera, Phalacridae). *Nouvelle Revue d'Entomologie* (N. S.) **16**: 231-247.

Švihla V. 1977: New Rhagonycha species from the Caucasus and Lebanon (Coleoptera, Cantharidae). *Acta Entomologica Bohemoslovaca* **74**: 178-183.

Švihla V. 1980a: New species of Malithinini (Coleoptera, Cantharidae) from the Palaearctic region. *Acta Entomologica Bohemoslovaca* **77**: 242-251.

Švihla V. 1980b: New species of Malachiinae (Coleoptera, Melyridae) from Soviet Central Asia. *Acta Entomologica Bohemoslovaca* **77**: 394-399.

Švihla V. 1983: New species of the family Cantharidae (Coleoptera) from the west Palaearct. *Annotationes Zoologicae et Botanicae* **156**: 1-10.

Švihla V. 1984: Cephaloncus kubani sp. n. from Soviet Central Asia (Coleoptera, Melyridae). *Acta Faunistica Entomologica Musei Nationalis Pragae* **17**: 251-252.

Švihla V. 1987: Results of the Czechoslovak-Iranian entomological expeditions to Iran 1973 and 1977. Coleoptera, Melyridae, Malachiinae. *Acta Entomologica Musei Nationalis Pragae* **42**: 277-286.

Švihla V. 1990a: Notes on Cantharidae (Coleoptera) from the western Palaearctic region, with description of new species. *Acta Entomologica Bohemoslovaca* **87**: 195-203.

Švihla V. 1990b: Revision of Ebaeus flavicornis species-group in Central Europe (Coleoptera, Malachiidae). *Acta Entomologica Bohemoslovaca* **87**: 355-359.

Švihla V. 1992: Revision of the subfam. Cantharinae without Podabrus (Coleoptera: Cantharidae) from Soviet Central Asia, Afghanistan and Chinese Turkestan. *Entomologica Basiliensia* **15**: 279-332.

Švihla V. 1993: Contribution to the knowledge of the genus Rhagonycha Eschscholtz (Coleoptera: Cantharidae) from Eastern Mediterranean. *Entomologica Basiliensia* **16**: 255-277.

Švihla V. 1994: New taxa of Palaearctic Cantharidae (Coleoptera). *Folia Heyrovskyana* **2**: 116-123.

Švihla V. 1995: Contribution to the knowledge of the genus Rhagonycha Eschscholtz (Coleoptera: Cantharidae) II. *Entomologica Basiliensia* **18**: 71-90.

Švihla V. 1997a: Malthinus moravicus sp. n. (Coleoptera: Cantharidae) surprising discovery of new species in Central Europe. *Klapalekiana* **33**: 101-102.

Švihla V. 1997b: Contribution to the knowledge of the subfamily Malthininae (Coleoptera: Cantharidae) from the Western Palaearctic. *Časopis Národního Muzea, Řada Přírodovědná* **166**: 123-126.

Švihla V. 1998: Contribution to the knowledge of the subfamily Malthininae (Coleoptera: Cantharidae) of Turkey. *Klapalekiana* **34**: 227-232.

Švihla V. 1999: Contribution to the knowledge of the genus Cantharis L. and related genera from Turkey and adjacent regions (Coleoptera: Cantharidae). *Entomologica Basiliensia* **21**: 135-170.

Švihla V. 2002a: A contribution to the knowledge of the subfamily Malthininae (Coleoptera: Cantharidae) from the western Palaearctic. *Folia Heyrovskyana* **10**: 119-154.

Švihla V. 2002b: A contribution to the knowledge of the genus Rhagonycha Eschscholtz, 1830 (Coleoptera, Cantharidae). III. *Entomologica Basiliensia* **24**: 305-319.

Švihla V. 2004a: Contribution to the knowledge of the family Cantharidae (Coleoptera) from the western Palaearctic. *Journal of National Museum, Natural History Series* **173** (1-4): 77-88.

Švihla V. 2004b: New taxa of the subfamily Cantharinae (Coleoptera, Cantharidae) from southeastern Asia with notes on other species. *Entomologica Basiliensia* **26**: 155-238.

Swartz O. 1808: [new taxa]. In: Schönherr C. J.: *Synonyma Insectorum, oder: Versuch einer Synonymie aller bisher bekannten Insecten; nach Fabricii Systema Eleutheratorum &c. geordnet. Band I. Eleuterata oder Käfer. Theil 2.* Stockholm: C. F Marquard, x + 424 pp., 1 pl.

Szombathy K. 1909: Notes sur quelques élaterides exotiques. *Annales Musei Nationalis Hungarici* **7**: 118-124.

Szombathy K. 1910a: Egy magyarországi új elateridáról. *Rovartani Lapok* **17**: 190-191.

Szombathy K. 1910b: Élatérides nouveaux ou peu connus appartenent au Musèe National Hongrois. I. *Annales Historico-Naturales Musei Nationalis Hungarici* **8**: 353-360.

Szombathy K. 1910c: Élatérides nouveaux ou peu connus appartenant au Musée National Hongrois. II. *Annales Historico-Naturales Musei Nationalis Hungarici* **8**: 445-448.

Szombathy K. 1910d: Beiträge zur Kenntnis der ungarischen Elateriden. *Annales historico-naturales Musei Nationalis Hungarici* **8**: 575-589.

Szombathy K. 1911: Egy új Corymbites Boszniából. *Rovartani Lapok* **18**: 182-183.

Takahashi K. 1992: The fauna of family Cantharidae (Coleoptera) in Kanagawa Prefecture, Japan. *Kanagawa-Chûhô* **100**: 71-124.

Takahashi K. 1996: Additional notes on the fauna of Cantharidae (Coleoptera) in Kanagawa Prefecture, Japan. *Kanagawa-Chûhô* **113**: 19-37.

Takahashi K. 1999a: Three new blackish species of the genus Podabrus (Coleoptera, Cantharidae) from central and northeastern Honshu, Japan. *Elytra* **27**: 70-75.

Takahashi K. 1999b: A new Caccodes (Coleoptera, Cantharidae) from the Ogasawara Islands, Japan. *Japanese Journal of Systematic Entomology* **5**: 203-205.

Takahashi K. 2002a: Eight new species of the genus Podabrus (Coleoptera, Cantharidae) mainly from Eastern Honshu, Japan. *Japanese Journal of Systematic Entomology* **8**: 49-58.

Takahashi K. 2002b: A new species of the genus Asiopodabrus (Coleoptera, Cantharidae) from Eastern Honshu, Japan. *Elytra* **30**: 195-201.

Takahashi K. 2002c: Three new species and a new subspecies of the genus Malthodes (Coleoptera, Cantharidae) from Eastern Honshu, Japan. *Japanese Journal of Systematic Entomology* **8**: 273-280.

Takahashi K. 2003: A revision of the Japanese genus Yukikoa (Coleoptera, Cantharidae) with reference to the systematic position and speciation. *Japanese Journal of Systematic Entomology* **9** (1): 81-100.

Takahashi N. & Imasaka S. 1997: A new species of the genus Rhagonycha (Coleoptera, Cantharidae) from the Koshiki-jima Islands, of southern Kyushu, Japan. *Elytra* **25**: 79-83.

Takahashi K. & Kiriyama I. 2000: Eighteen new species and two new subspecies of the genus Podabrus (Coleoptera, Cantharidae) mainly from Gifu Prefecture, Central Honshu, Japan. *Japanese Journal of Systematic Entomology* **6**: 121-146.

Takahashi K. & Kiriyama I. 2003: Two remarkable new species of the genus Asiopodabrus (Coleoptera, Cantharidae) from Central Honshu, Japan. *Special Bulletin of the Japanese Society of Coleopterology* **6**: 289-297.

Takahashi K., Okushima Y. & Ichita T. 1999: Cantharid beetles (Coleoptera, Cantharidae) of Aomori Prefecture, northern Honshu, Japan, III. *Bulletin of the Kurashiki Museum of Natural History* **14**: 17-48.

Takakura Y. 1987: Description of a new species of the genus Athemellus from Kyushu, Japan (Coleoptera, Cantharidae). *Kita-Kyūshū no Konchū* **34**: 179-180.

Takakura Y. 1988a: Description of a new species of the genus Hypebaeus from Kyushu, Japan (Coleoptera, Melyridae). *Kita-Kyūshū no Konchū* **35**: 151.

Takakura Y. 1988b: Description of a new species of the genus Podabrus from Kyushu, Japan (Coleoptera, Cantharidae). *Kita-Kyūshū no Konchū* **35**: 152.

Takakura Y. 1988c: Description of two new species of the genus Malthodes from Kyushu, Japan (Coleoptera, Cantharidae). *Kita-Kyūshū no Konchū* **35**: 153-154, pl. 16.

Takakura Y. 1989: On a new subspecies and a new species of Cantharidae from Kyushu, Japan (Coleoptera). *Kita-Kyūshū no Konchū* **36**: 157-158.

Takizawa M. 1917: Some new species of Coccinellidae in Japan. *Transactions of the Sapporo Natural History Society* **6**: 220-224.

Tanzer P. 1929: Eine neue Varietät des Elater aethiops Lacordaire. *Entomologisches Nachrichtenblatt* (Troppau) **3**: 76-77.

Tanzer P. 1934: Harminius undulatus Deg. und v. bifasciatus Gyll. *Entomologisches Nachrichtenblatt* (Troppau) **8**: 125-127.

Tarnawski D. 1995: A revision of the genus Selatosomus Stephens, 1830 (Coleoptera: Elateridae: Athoinae: Ctenicerini). *Genus, Supplement*, 183 pp.

Tarnier F. 1860: Coléoptères des îles Açores. Pp. 87-96. In: Morelet A.: *Notice sur l'histoire naturelle des Açores, suivie d'une description des molusques terrestres de cet Archipel*. Paris: J.-B. Baillière et Fils.

Tenenbaum S. 1915: Un nouveau coléoptère de Baléares. *Pracownia Biologiczna* **1915**: 1-3.

Thomas M. C. 1988: A revision of the New World species of Cryptolestes Ganglbauer (Coleoptera: Cucujidae: Laemophloeinae). *Insecta Mundi* **2**: 43-65.

Thomas M. C. 1993: A new species of Cryptolestes Ganglbauer from the Middle East and a new synonym (Coleoptera: Laemophloeidae [Cucujidae sensu lato]). *Israel Journal of Entomology* **27**: 113-118.

Thomas M. C. 2004: The Brontini of the world: A generic review of the tribe (Coleoptera: Silvanidae: Brontinae). *Insecta Mundi* **17** [2003]: 1-31.

Thomson C. G. 1859: *Skandinaviens Coleoptera, synoptiskt bearbetade. Tom I.* Lund: Berlingska, [6] + 290 pp.

Thomson C. G. 1862: *Skandinaviens Coleoptera, synoptiskt Bearbetade. Tom IV.* Lund: Lundbergska Boktryckeriet, 269 pp.

Thomson C. G. 1863: *Skandinaviens Coleoptera, synoptisk bearbetade. Tom V.* Lund: Lundbergska Boktryckeriet, 340 pp.

Thomson C. G. 1864: *Skandinaviens Coleoptera, synoptiskt bearbetade. Tom VI.* Lund: Lundbergska Boktryckeriet, 386 pp.

Thomson C. G. 1866: *Scandinaviens Coleoptera, Synoptiskt bearbetade. Tom. VIII.* Lund: Lundbergska Boktryckeriet, 409 + lxxv pp.

Thomson C. G. 1868: *Skandinaviens Coleoptera, Synoptiskt Bearbetade. Tom. X.* Lund: Lundbergska Boktryckeriet, [4] + 420 pp.

Thomson C. G. 1870a: *Opuscula Entomologica* **2**: [1] + 83-304.

Thomson C. G. 1870b: XVIII. Bidrag till Sveriges Insect-Fauna. (1) (a) Coleoptera. *Opuscula Entomologica* **3**: 305-358, 2 pls.

Thomson C. G. 1871: XX. Bidrag till Sveriges Insect-faunae (2) (a) Coleoptera. *Opuscula Entomologica* **4**: 359-452.

Thomson C. G. 1873: XXII: Några för Sveriges fauna nya Coleoptera. *Opuscula Entomologica* **5**: 453-532, 1 pl.

Thomson C. G. 1874: XXIII. Bidrag till Skandinaviens Coleoptera. *Opuscula Entomologica* **6**: 533-612.

Thomson C. G. 1885: Communications. *Bulletin de la Société Entomologique de France* **1885**: cviii-cix.

Thomson C. G. 1892: XLVI. Bidrag till Sveriges Insectfauna. *Opuscula Entomologica* **16**: 1657-1774.

Thomson J. 1856: Description de dix-sept coléoptères. *Revue et Magasin de Zoologie* (2) **8**: 472-483.

Thomson J. 1857: Description d'un genre nouveau de la familie des Eumorphides, et de plusieurs espèces qui rentrent dans cette division. *Archives Entomologique* **1**:153-157.

Thomson J. 1858: Deuxième Partie. Insects. I. Ordre Coléoptères. Pp. 30-343. In: Voyage au Gabon. Histoire naturelle des insectes et des arachnides recueillis pendant un voyage fait au Gabon en 1856 et en 1857 par M. Henry C. Deyrolle sous les auspices de MM. Le Comte de Mniszech et James Thomson, précédée de l'histoire du voyage par J. Thomson; Arachnide par H. Lucas. *Archives Entomologiques* **2**: 1-469 + 14 pp., XIV pls.

Thomson J. 1860 [1860-1868]: *Musée scientifique ou recueil d'histoire naturelle*. Paris: J. Thomson, 96 pp., 9 pls.

Thunberg C. P. 1781: *Dissertatio entomologica novas Insectorum species, sistens cujus partem primam*. Upsaliae: J. Edman, 28 pp., 1 pl.

Thunberg C. P. 1784a: Novae insectorum species. *Nova Acta Regiae Societatis Scientarum Upsaliensis* **4**: 1-28, 1 pl.

Thunberg C. P. 1784b: Dissertatio entomologica novas Insectorum species, sistens, cujus partem quartam, cons. exper. Upsaliae: Facult. Med. Publice Ventilandam Exhibent, pp. 69-85.

Thunberg C. P. 1787: *Donation. Thunbergianae 1785. Continuat. II.* Museum naturalium Academiae Upsaliensis. Pars. 4. Upsaliae: Joh. Edman, pp. 43-58, 1 pl.

Thunberg C. P. 1791: *Characteres generum Insectorum, variis cum adnotationibus, denuo edidit Frid. Alb. Ant. Meyer*. Gottingae: Dieterich, 48 pp.

Thunberg C. P. 1792: Descriptiones insectorum svecicorum. *Nova Acta Regiae Societatis Scientiarum Upsaliensis* **5**: 85-119.

Thunberg C. P. 1794: *Dissertatio entomologica sistens Insecta Svecica. Quorum partem Quintam. Cons. Exp. facult. Med. Ups. praeside Carol. Pet. Thunberg publico examini subjicit Isaacus Haij, stip. Kahreanus Westrogoth*. Upsaliae: litteris viduae direct. Johann. Edman, pp. 63-72.

Thunberg C. P. 1795: *Dissertatio entomologica, sistens Insecta Suecica. IX.* Exam. Steno Edvardus Westman. Upsaliae, pp. 105-113.

Thunberg C. P. 1798: *Donationes Thunbergiana - Museum Naturalium Academiae Upsaliensis, Appendix part 6*. Upsaliae, pp. 111-117.

Thunberg C. P. 1805: Göttingische gelehrte Anzeigen **29**: 281-282.

Thunberg C. P. 1808: [new taxa]. In: Schönherr C. J.: *Synonyma Insectorum, oder: Versuch einer Synonymie aller bisher bekannten Insecten; nach Fabricii Systema Eleutheratorum &c. geordnet.Erster Band. Eleutherata oder Käfer. Zweiter Theil*. Stockholm: C. F. Marquard, x + 424 pp., 1 pl.

Thunberg C. P. 1815: Anthreni monographia. *Nova Acta Regiae Societatis Scientiarum Upsaliensis* **7**: 150-156.

Thunberg C. P. 1820: Coleoptera Capensis Antennarum clava solida et perfoliata. *Memoires de l'Académie des Sciences de St Pétersburg* **7**: 362-372.

Tian M.-Y. 1995: Description of one new species of genus Cybocephalus from China (Coleoptera: Cybocephalidae). *Journal of South China Agricultural University* **16**: 42-43.

Tian M.-Y. 1996: A new species of the genus Cybocephalus Erichson from Dinghushan Nature Reserve, Guangdong, China (Coleoptera: Cybocephalidae). *Acta Zootaxonomica Sinica* **21**: 92-94 (in Chinese, with English title and summary).

Tian M.-Y. 2000: Two new species of Cybocephalidae (Coleoptera) from Northwest China. *Entomologica Sinica* **7**: 127-131.

Tian M.-Y. 2001: Two new species of Cybocephalidae fram Gansu province of China (Coleoptera: Cybocephalidae). *Acta Zootaxonomica Sinica* **26**: 43-45 (in Chinese, with English summary).

Tian M.-Y. & Pang H. 1994: A new species of Cybocephalus (Coleoptera: Cybocephalidae) from Taiwan. *Chinese Journal of Entomology* **14**: 401-403.

Tian M.-Y. & Yu G. 1994a: Descriptions of two known and one new species of the genus Cybocephalus Erichson of China. *Natural Enemies of Insects* **16** (3): 119-122 (in Chinese, with English summary).

Tian M.-Y. & Yu G. 1994b: Two new species of the genus Cybocephalus from China (Coleoptera: Cybocephalidae). *Entomologica Sinica* **1**: 209-213.

Timberlake P. H. 1919: Notes on the American species of Hippodamia (Coleoptera). *Journal of the New York Entomological Society* **27**: 162-174.

Timberlake P. H. 1943: The Coccinellidae or ladybeetles of the Koebele-collection. Part 1 *Bulletin of the Experiment Station of the Hawaiian Sugar Planters' Association. Entomological Series, Bulletin no: 22. The Hawaiian Planters Record* **47**: 1-67. 2 pls.

Tomaszewska K. W. 1997a: A review of the genus Endomychus Panzer (Coleoptera: Endomychidae), with descriptions of seven new species. *Annales Zoologici* **47**: 215-241.

Tomaszewska K. W. 1997b: Two new species of Dapsa Latreille from Nepal and India (Coleoptera: Endomychidae). *Polskie Pismo Entomologiczne* **66**: 217-221.

Tomaszewska K.W. 1998: A supplement to the recent review of the genus Endomychus Panzer (Coleoptera: Endomychidae), with descriptions of two new species. *Annales Zoologici* **48**: 299-304.

Tomaszewska K. W. 1999: A new genus and species of Epipocinae (Coleoptera: Endomychidae) from Pakistan. *Revue Suisse de Zoologie* **106**: 277-284.

Tomaszewska K. W. 2000a: A review and a phylogenetic analysis of the genera of Leiestinae (Coleoptera, Endomychidae). *Deutsche Entomologische Zeitschrift* **47**: 65-86.

Tomaszewska K. W. 2000b: A review of the genus Cyclotoma Mulsant (Coleoptera: Endomychidae) with descriptions of three new species. *Annales de la Société Entomologique de France* **36**: 345-369.

Tomaszewska K. W. 2000c: Morphology, phylogeny and classification of adult Endomychidae (Coleoptera: Cucujoidea). *Annales Zoologici* **50**: 449-558.

Tomaszewska K. W. 2001: On the Oriental genus Pseudindalmus Arrow, 1920 with descriptions of four new species (Coleoptera: Endomychidae). *Annales Zoologici* **51**: 133-141.

Tomaszewska K. W. 2002a: Two new species of Endomychus Panzer (Coleoptera: Endomychidae), from China and Burma. *Annales Zoologici* **52**: 151-153.

Tomaszewska K. W. 2002b: A contribution to the Oriental genus Beccariola Arrow, with description of a new species from India (Coleoptera: Endomychidae). *Annales Zoologici* **52**: 373-378.

Tomaszewska K. W. 2003: A review of the genus Avencymon Strohecker, 1971 (Coleoptera: Endomychidae). *Annales Zoologici* **53**: 705-710.

Tomaszewska K.W. 2005: Phylogeny and generic classification of the subfamily Lycoperdininae with a re-analysis of the family Endomychidae (Coleoptera: Cucujoidea). *Annales Zoologici* **55** (supplement 1): 1-172.

Toskina I. N. 1973: O sistematicheskom polozhenii Coelostethus (Coleoptera, Anobiidae). *Zoologicheskiy Zhurnal* **52**: 1335-1340.

Toskina I. N. 1976: Dva novych vida tochil"shchikov (Coleoptera, Anobiidae). *Zoologicheskiy Zhurnal* **55**: 1730-1732.

Toskina I. N. 1995: Wood-borers of the genus Ptilinus (Coleoptera Anobiidae) of Russia and adjacent countries. *Russian Entomological Journal* **4**: 15-34.

Toskina I. N. 1998: Rare and new Anobiidae (Coleoptera) from the Arabian Peninsula. *Fauna of Saudi Arabia* **17**: 51-91.

Toskina I. N. 1999: New and rare species of the genus Lasioderma Stephens, 1860 (Coleoptera: Anobiidae) of Russia and adjacent countries. *Russian Entomological Journal* **8**: 23-46.

Toskina I. N. 2001: Three new Palaearctic species of the genus Ptinomorphus Mulsant & Rey, 1868 (Coleoptera: Anobiidae). *Russian Entomological Journal* **10**: 143-151.

Toskina I. N. 2002a: A new species of the genus Xyletinus (Coleoptera: Anobiidae) from Arabia and a note about Mesocoelopus. *Russian Entomological Journal* **11**: 207-208.

Toskina I. N. 2002b: Some new palaearctic species of wood-borers from subfamily Ernobiinae (Coleoptera: Anobiidae). *Russian Entomological Journal* **11**: 387-400.

Toskina I. N. 2003: The genera Gastrallus Jacquelin du Val, 1860, and Falsogastrallus Pic, 1914 (Coleoptera: Anobiidae) of Palaearctic Eurasia. *Russian Entomological Journal* **12**: 187-197.

Toskina I. N. 2004a: About genera Anobium Fabricius, 1775, and Cacotemnus LeConte, 1861 (Coleoptera: Anobiidae). *Russian Entomological Journal* **13**: 53-68.

Toskina I. N. 2004b: A new species of Clada Pascoe, 1887 (Coleoptera: Anobiidae) from the Socotra Archipelago. *Fauna of Saudi Arabia* **20**: 459-462.

Toskina I. N. & Nikitsky N. B. 2003: Novyy vid zhukov tochil"shchikov (Coleoptera, Anobiidae) roda Episernus s severo-vostoka evropeyskoy chasti Rossii. *Zoologicheskiy Zhurnal* **82**: 1126-1128.

Tournier H. 1860: Drei neue europäische Käferarten. *Entomologische Zeitschrift* **4**: 317-319.

Tournier H. 1868: Coléoptères nouveaux d'Europe, d'Algérie et d'Egypte. *L'Abeille, Mémoires d'Entomologie* **5**: 141-148.

Tournier H. 1869: [new taxa]. *Petites Nouvelles Entomologiques* **1**: 11.

Tournier H. 1872a: Nouveaux coléoptères d'Europe. *Revue et Magasin de Zoologie Pure et Appliquée* (2) **23** [1871-1872]: 250-256.

Tournier H. 1872b: Coléoptères européens et circumeuropéens. Descriptions d'espèces nouvelles. *Mitteilungen der Schweizerischen Entomologische Gesellschaft* **3**: 436-448.

Tournier H. 1874: Tableau synoptique des espèces européennes du genre Lyctus Fabr. *Petites Nouvelles Entomologiques* **6**: 411-412.

Tournier H. 1889: Coléoptères. Famille des phalacrides. Essai monographique des espèces européennes et des contrées linitrophes. *L'Entomologiste Genevois* **1889**: 2-10; 25-34; 49-55; 73-92, 97-101, 121-124.

Tournier H. 1890: Coléoptères. Famille des phalacrides. Essai monographique des espèces européennes et des contrées linitrophes. *L'Entomologiste Genevois* **1890**: 185-188.

Tournier H. 1894: [new taxon]. In: Pic M.: Quelques notes en passant sur plusieurs coléoptères de la collection Henri Tournier. *L'Échange, Revue Linnéenne* **10**: 71-72.

Tozer E. R. 1968: A new species of Rhizophagus Herbst (Coleoptera: Rhizophagidae) from Greece. *Proceedings of the Royal Entomological Society of London* (B) **37**: 57-61.

Trella T. 1923: Wykaz chrząszczów okolic Przemyśla. *Polskie Pismo Entomologiczne* **2**: 110-123.

Trella T. 1925: Wykaz chrząszczów okolic Przemyśla. Wzerki – Teredilia, Miekopokrowe – Malacodermata. *Polskie Pismo Entomologiczne* **3**: 122-127.

Trella T. 1930: Ergänzungen su den bisher erschienen Verzeichnissen und die Familien: Lathridiidae, Anthribidae, Nemonychidae u. Ipidae. *Polskie Pismo Entomologiczne* **9**: 33-39.

Trella T. 1932: Wykaz chrząszczów okolic Przemyśla. Uzupelnienie dotychczasowych wykazóv. *Polskie Pismo Entomologiczne* **10**: 223-229.

Trost P. 1801: *Kleiner Beytrag zur Entomologie in einem Verzeichnisse der Eichstettischen bekannten und neuentdekten Insekten mit Anmerkungen für Kenner und Liebhaber. Erster Heft.* Erlangen: Johann Jakob Palm, viii + 71 pp.

Tshernyshev S. E. 1994: A revision of the genus Troglocollops Wittmer, 1965 (Coleoptera, Melyridae, Malachiinae). *Russian Entomological Journal* **3**: 51-80.

Tshernyshev S. E. 1997: New data on the genus Cephaloncus (Coleoptera, Malachiidae) from central Asia. *Zoologicheskiy Zhurnal* **76**(6): 676-683.

Tshernyshev S. E. 1999a: Towards the knowledge of the Malachius group soft-winged flower beetles (Coleoptera: Malachiidae) of the fauna of Russia and adjacent countries. Part 1. *Russian Entomological Journal* **7** [1998]: 129-146.

Tshernyshev S. E. 1999b: A new species of the genus Hypebaeina Wittm. (Coleoptera, Malachiidae) from Kirghizia. *Species Diversity* **4**: 367-370.

Tshernyshev S. E. 2000: Contributions to the knowledge of the Malachius group of soft-winged flower beetles (Coleoptera, Malachiidae) of the fauna of Russia and adjacent countries. Part II. *Bulletin de l'Institut Royal des Sciences Naturelles de Belgique, Entomologie* **70**: 199-218.

Tshernyshev S. E. 2001: A new species of the genus Cephaloncus Westwood from Turkmenistan (Insecta: Coleoptera: Malachiidae). *Reichenbachia* **34**: 155-160.

Tshernyshev S. E. 2002: Novyy vid zhukov dazitid poda Chaetomalachius Er. gruppy C. acutipennis Majer, 1989 (Coleoptera, Dasytidae) iz Kazakhstana [New species of the genus Chaetomalachius Er. in C. acutipennis Majer, 1989 species group (Coleoptera, Dasytidae) from Kazakhstan]. *Euroasian Entomological Journal* **1**: 211-214.

Tshernyshev S. E. 2003: A review of soft-winged flower beetles (Coleoptera, Malachiidae) of Russia and the adjacent countries: genus Ebaeus Erichson, 1840. Part I. *Euroasian Entomological Journal* **2**: 281-299.

Tsinkevich V. A. 2001: A new species and some new records of Palaearctic Lathridiidae (Coleoptera). *Zoosystematica Rossica* **10**: 85-86.

Türkay M. 1974: Zur Kenntnis der Gattung Pelania (Coleoptera, Lampyridae). *Entomologische Zeitschrift* **84**: 21-28.

Uhagón S. de. 1879: Coleópteros de Badajoz. Segunda parte. *Anales de la Sociedad Española de Historia Natural* **8**: 187-216, pl. 4.

Uhagón S. de. 1898: Diagnosis de especies nuevas de maláquidos de España. *Anales de la Sociedad Española de Historia Natural* **27**: 41-42.

Uhagón S. de. 1900: Ensayo sobre los maláquidos de España. *Anales de la Sociedad Española de Historia Natural* **29**: 5-63, 291-361.

Uhagón S. de. 1901a: Ensayo sobre los maláquidos de España. *Anales de la Sociedad Española de Historia Natural* **30**: 5-102.

Uhagón S. de. 1901b: Notas sobre malaquidos de España. *Boletín de la Real Sociedad Española de Historia Natural* **1901**: 359-360.

Uyttenboogaart D. L. 1929: VII. Description of a new species of Troglops Er. (*Cephalogonia* Woll.) (Col. Cantharidae). *Tijdschrift voor Entomologie* **72**: 158-161. [In: Contributions to the knowledge of the fauna of the Canary-Islands, edited by Dr. D. L. Uuyttenboogaart (Renkum), pp. 154-168, 341-355].

Uyttenboogaart D. L. 1930: Contributions to the knowledge of the fauna of the Canary Islands. *Tijdschrift voor Entomologie* **73**: 275-278.

Uyttenboogaart D. L. 1934: Contributions to the knowledge of the fauna of the Canary Islands. XVII. Remarks concerning collections of canarian coleoptera in the zoological museum at Hamburg and in the museo Pietro Rossi at Duino. *Tijdschrift voor Entomologie* **77**: 162-166.

Uyttenboogaart D. L. 1935: Report on Canarian Coleoptera collected by R. Frey and R. Storå in 1931 for the Museum Zoologicum Universitatis Helsingfors. *Commentationes Biologicae, Societas Scientiarum Fennica* **6** (2): 1-17.

Van Dyke E. C. 1923: New species of Coleoptera from California. *Bulletin of the Brooklyn Entomological Society* **18**: 37-53.

Van Dyke E. C. 1932: Miscellaneous studies in the Elateridae and related families of Coleoptera. *Proceedings of the California Academy of Sciences* (4) **20**: 291-465.

Van Zwaluwenburg R. H. 1932: Elateridae of the Marquesas, with a new species from Fiji. *Bulletin of the Bernice P. Bishop Museum* **98**: 129-143, pl. 4.

Van Zwaluwenburg R. H. 1957: Insects of Micronesia. Coleoptera: Elateridae. *Insects of Micronesia* **16**: 1-66.

Vats L. K. & Chauhan R. L. 1991a: Species of Melanotus Eschscholtz (Elateridae: Coleoptera). *Journal of Entomological Research* (New Dehli) **15**: 187-196.

Vats L. K. & Chauhan R. L. 1991b: Species of Hypnoidinae from North India (Coleoptera: Elateridae). *Journal of Entomological Research* (New Dehli) **15**: 294-300.

Vats L. K. & Chauhan R. L. 1991c: The species of Cardiophorus from North India (Elateridae: Coleoptera). *Research Bulletin (Science) of the Panjab University Science* **42**: 11-29.

Vats L. K. & Chauhan R. L. 1992a: Two new species of Ampedinae (Coleoptera: Elateridae). *Journal of Entomological Research* (New Dehli) **16**: 20-23.

Vats L. K. & Chauhan R. L. 1992b: New species of Ctenicera Latreille from northern India (Coleoptera: Elateridae). *Journal of Entomological Research* (New Dehli) **16**: 24-32.

Vats L. K. & Chauhan R. L. 1992c: Species of Conoderus Esch. with notes on genus (Conoderinae, Elateridae: Coleoptera). *Uttar Pradesh Journal of Zoology* **12**: 28-36.

Vats L. K. & Chauhan R. L. 1992d: Two new species of Adiaphorus Candeze from North India (Coleoptera: Elateridae). *Journal of Entomological Research* (New Dehli) **16**: 177-180.

Vats L. K. & Chauhan R. L. 1992e: Two new species of Adrastus Eschscholtz (Coleoptera: Elateridae). *Journal of Entomological Research* (New Dehli) **16**: 181-184.

Vats L. K. & Chauhan R. L. 1992f: New species of Ludiinae (Coleoptera: Elateridae). *Journal of Entomological Research* (New Dehli) **16**: 189-192.

Vats L. K. & Chauhan R. L. 1992g: New species of Glyphonyx Candeze (Elateridae: Coleoptera) from North India. *Indian Journal of Forestry* **15**: 203-206.

Vats L. K. & Chauhan R. L. 1993a: Four new species of Elateridae (Coleoptera) from north India. *Journal of Entomological Research* (New Dehli) **17**: 35-41.

Vats L. K. & Chauhan R. L. 1993b: Species of Silesis Candeze from North India (Coleoptera: Elateridae). *Uttar Pradesh Journal of Zoology* **13**: 37-43.

Vats L. K. & Chauhan R. L. 1993c: Species of Heteroderes Latreille from North-West India (Conoderinae, Elateridae: Coleoptera). *Entomon* **18**: 169-173.

Vats L. K. & Chauhan R. L. 1994: Species of Agonischius Candeze from north India (Elateridae: Coleoptera). *Research Bulletin (Science) of the Panjab University Science* **43** [1993]: 143-150.

Vats L. K. & Kashyap S. L. 1992a: Seven new species of Adelocera Latreille from North-West India (Coleoptera: Elateridae). *Journal of Entomological Research* (New Dehli) **16**: 1-12.

Vats L. K. & Kashyap S. L. 1992b: Four new species of Lacon Castelnau from North India (Coleoptera: Elateridae). *Journal of Entomological Research* (New Delhi) **16**: 13-19.

Vats L. K. & Kashyap S. L. 1992c: Forty-eight species of Agrypnus Eschscholtz including forty new additions from North-West India (Coleoptera: Elateridae). *Journal of Entomological Research* (New Delhi) **16**: 87-154.

Vats L. K. & Kashyap S. L. 1992d: Four new species of subfamily Hemirrhipinae (Coleoptera: Elateridae). *Journal of Entomological Research* (New Delhi) **16**: 193-200.

Vats L. K. & Kashyap S. L. 1992e: Description of three species including two new species of subfamily Denticollinae (Coleoptera: Elateridae) from north India. *Journal of Entomological Research* (New Delhi) **16**: 247-251.

Vats L. K. & Kashyap S. L. 1992f: A new genus with a description of a new species of Crepidomeninae (Coleoptera: Elateridae) from north India. *Journal of Entomological Research* (New Delhi) **16**: 252-254.

Vats L. K. & Kashyap S. L. 1993: Species of Lanelater Arnett from North India (Elateridae: Coleoptera). *Indian Journal of Forestry* **16**: 214-222.

Vats L. K. & Kashyap S. L. 1995a: A new species of the genus Tetralobus Lepel. and Serv. (Coleoptera : Elateridae). *Journal of Entomological Research* (New Delhi) **19**: 171-173.

Vats L. K. & Kashyap S. L. 1995b: A new species of Octocryptus Candeze (Coleoptera: Elateridae: Octocryptinae) with notes on the genus. *Journal of Entomological Research* (New Delhi) **19**: 215-217.

Vats L. K. & Kashyap S. L. 1996: Species of Meristhus Candeze from North India (Coleoptera: Elateridae: Agrypninae). *Journal of Entomological Research* (New Delhi) **20**: 67-71.

Vazirani T. G. 1983: Sticholotis chittagongi sp.n. (Coleoptera: Coccinellidae) a predator on aphids on Ziziphus in Bangladesh. *Bulletin of Entomological Research* **73**: 301-303.

Vazirani T. G. & Saha G. N. 1972: Notes on a collection of Endomychidae (Insecta: Coleoptera) from NEFA with description of a new species. *Records of the Zoological Survey of India* **66**: 75-79.

Verhoeff C. 1891: Ein Beitrag zur Coleopteren-Fauna der Inseln Norderney. *Entomologische Nachrichten* **17**: 17-26.

Vijay Veer, Chauhan N. & Singh S. 2004: A new species of Thaumaglossa (Coleoptera: Dermestidae) from India. *Oriental Insects* **38**: 389-394.

Villa A: 1837: [new taxon]. In: Comolli A.: *De Coleopteris novis ac rarioribus minusve cognitis provinciae Novocomi.* Ticini: Fusi et Socii, vi + 7-54 pp.

Villa A. & Villa G. B. 1833: Coleopterorum species novae in catalogo dupletorum extantes. Diagnosibus, adumbrationibus atque observationibus illustratae. Pp. 32-36. In: *Coleoptera europae dupleta in collectione Villa quae pro muttua commutatione offerri possunt.* Mediolani: Villa, 36 pp.

Villa A. & Villa G. B. 1835: Coleopterorum species novae in supplemento salutatae. Diagnosibus, adumbrationibus atque observationibus illustratae. Pp. 47-50. In: *Supplementum Coleopterorum europae dupletorum catalogo collectionis Villa id est species aliae, quae nunc pro mutua commutatione itidem offerri possunt; nec non emendationes aliquarum specierum in catalogo anni 1833 extantium.* Mediolani: Villa, 37-50 pp.

Villa A. & Villa G. B. 1837: [new taxon]. In: Comolli A. 1837: *De Coleopteris novis ac rarioribus minusve cognitis provinciae Novocomi.* Ticini: Fusi et Socii, vi + 7-54 pp.

Villa A. & Villa G. B. 1838: Coleopterorum species novae in hoc altero supplemento commemoratae. Diagnosibus atque observationibus illustratae. Pp. 61-63. In: *Alterum supplementum Coleopterorum Europae sive additio ad Catalogum et Supplementum I. Dupletorum Collectionis Villa continens species alias, nunc pro mutua commutatione adhuc offerrendas; nec non aliquarum specierum emendatioones et synonima quae in catalogo anni 1833 et supplemento anni 1835 oblatae fuerunt.* Mediolani: Villa, 51-66 pp.

Villaret F. de. 1833: Description d'une nouvelle espèce du genre Lampyris. *Annales de la Société Entomologique de France* **2**: 352-356.

Villers C. J. de. 1789: *Caroli Linnaei entomologia faunae suecicae descriptionibus aucta; DD. Scopoli, Geoffroy, de Geer, Fabricii, Schrank, &c. speciebus vel in systemate non enumeratis, vel nuperrime detectis, vel speciebus Galliae australis locupletata, generum specierumque rariorum iconibus ornata. Tomus primus.* Lugduni: Piestre et Delamolliere, xxiv + 765 pp., 3 pls.

Villiers A. 1945: Révision des languriides de l'Ancien Monde. *L'Abeille* **37**: 1-317.

Vincent R. 1990a: Contribution à l'étude de quelques espèces françaises du genre Dienerella Reitter, 1911 (Coleoptera, Latridiidae). *Revue Française d'Entomologie* (N. S.) **12**: 29-36.

Vincent R. 1990b: Nouvelle contribution à l'étude du genre Dienerella Reitter 1911 (Coleoptera, Latridiidae). *Nouvelle Revue d'Entomologie* (N. S.) **7**: 263-266.

Vincent R. 1991: Nouveau statut de Dienerella separanda (Reitter) 1887 et description d'une espèce nouvelle d'Espagne D. huguettae n. sp. (Coleoptera, Lathridiidae). *L'Entomologiste* **47**: 241-248.

Vincent R. 1994: Nouvelle contribution à l'étude du genre Dienerella Reitter 1911, D. besucheti espèce nouvelle du Maroc (Coleoptera, Latridiidae). *Nouvelle Revue d'Entomologie* (N. S.) **11**: 91-93.

Tomaszewska K. W. 2000c: A review of the genus Cyclotoma Mulsant (Coleoptera: Endomychidae) with descriptions of three new species. *Annales de la Société Entomologique de France* **36**: 345-369.

Voet J. E. 1791: *Beschreibungen und Abbildungen hartschaaligter Insecten Coleoptera Linn. Aus dem Original getreu übersetzt mit der in selbigem fehlenden Synonymie und beständigen Kommentar versehen von Dr. Georg Wolfgang Franz Panzer. Zweiter Theil.* Nürnberg, [viii]+134 pp., 48 pls.

Voet J. E. 1806: *Catalogus systematicus Coleopterorum. Tomus secundus.* La Haye, 82 pp., 50 pls.

Vogt H. 1967: Phalacridae. Pp. 158-166. In: Freude H., Harde K. W. & Lohse G. A. (eds.): *Die Käfer Mitteleuropas. Band 7. Clavicornia.* Krefeld: Goecke und Evers, 310 pp.

Vosyka V. 1957: Příspěvek k poznání československých druhů páteřníkových (Col. Cantharidae) [Beitrag zur Kenntnis der tschechoslovakischen Arten der Familie Cantharidae]. *Časopis Československé Společnosti Entomologické* **54**: 123-125.

Vrydagh J.-M. 1961: Contribution à l'étude des Bostrychidae. N° 28. – Étude des types de Fåhraeus, désignation de lectotypes. *Bulletin de l'Institut Royal des Sciences Naturelles de Belgique* **37** (7): 1-10.

Všetečka K. 1948: O nových i známých druzích rodu Anommatus Wesm. *Časopis Československé Společnosti Entomologické* **45**: 38-44.

Walckenaer C. A. 1802: *Faune parisienne, insectes. Ou histoire abrégée des insectes des environs de Paris, classés d'après le système de Fabricius. Tome premier.* Paris: Dentu, Imprimeur-Libraire, xxx + 303 pp.

Walker F. 1858: Characters of some apparently undescribed Ceylon insects. *The Annals and Magazine of Natural History* (3) **2**: 202-209, 280-286.

Walker F. 1859a: Characters of some apparently undescribed Ceylon Isects. *The Annals and Magazine of Natural History* (3) **3**: 50-56, 258-265.

Walker F. 1859b: Characters of some apparently undescibed Ceylon Insects. *The Annals and Magazine of Natutal History* (3) **4**: 217-224.

Walker F. 1871: *List of Coleoptera collected by J. K. Lord, Esq. in Egypt, Arabia and near the African shore of the Red Sea. With characters of the undescribed species.* London: E. W. Janson, 19 pp.

Walkley L. M. 1948: Notes on nomenclature in the Lathridiini (Coleoptera, Lathridiidae). *Proceedings of the Entomological Society of Washington* **50**: 149-150.

Walter A. 1882: Coccinellen-Varietäten. *Entomologische Nachrichten* **8**: 17-20.

Waltl J. 1834: Ueber das Sammeln exotischer Insecten. *Faunus. Zeitschrift für Zoologie und Vergleichende Anatomie* **1** (3): 166-170.

Waltl J. 1835: *Reise durch Tyrol, Oberitalien und Piemont nach dem südlichen Spanien. Zweiter Theil.* [*Ueber die Tiere Andalusiens*]. Passau: Pustetschen Buchhandlung, 247 pp. [+ 120 pp., zweiter Theil].

Waltl J. 1838a: Beiträge zur nähern naturhistorischen Kenntniss des Unterdonaukreises in Bayern. *Isis von Oken* **4**: 250-263.

Waltl J. 1838b: Verzeichniss der um Passau vorkommenden seltenen Käfer nebst Beschreibung der neuen Arten. *Isis von Oken* **4**: 263-273.

Waltl J. 1838c: Malachius coccineus n. sp. *Isis von Oken* **4**: 455-457.

Waltl J. 1839: Verzeichniss der um Passau vorkommenden seltener Käfer nebst Beschreibung der neuen Arten. *Isis von Oken* **5**: 221-227.

Walton L. 1912: Symbiotes duryi, a new species of Endomychidae. *Ohio Naturalist* **12**: 461-464.

Wang H. & Cao C. 1993: Studies on the systematic position and genetic relationships of the tribe Epilachnini (Coleoptera: Coccinellidae) from Yunnan. *Zoological Research* **14**: 118-127 (in Chinese).

Wang S. & Yang C. 1990: Two new Cantharidae species (Coleoptera: Cantharidae) from Nei Mongol autonomous region. *Entomotaxonomia* **12**: 240-242.

Wang S. & Yang J. 1993: Coleoptera: Cantharidae. Pp. 264-267. In: Huang F. (ed.): *Insects of Wuling mountains area, southwestern China.* Beijing:Science Press, x + 777 pp.

Wanka T. 1929: Zwei neue Käferarten aus Schlesien. *Coleopterologisches Centralblatt* **3** [1928-1929]: 197-198.

Wankowicz J. 1865: Description de quelques coléoptères nouveaux de Lithuanie. *Annales de la Société Entomologique de France* (4) **5**: 297-300.
Wankowicz J. 1867: Descriptions de trois coléoptères nouveaux trouvés en Lithuanie. *Annales de la Société Entomologique de France* (4) **7**: 249-254.
Wankowicz J. 1869: Notices entomologiques. *Annales de la Société Entomologique de France* (4) **9**: 411-422.
Wasmann E. 1894: *Kritisches Verzeichniss der myrmekophilen und termitophilen Arthropoden. Mit Angabe der Lebensweise und mit Beschreibung neuer Arten.* Berlin: Felix L. Dames, xiii + 231 pp.
Wasmann E. 1895a: Zur Kenntniss einiger schwieriger Thorictus-Arten. *Deutsche Entomologische Zeitschrif* **1895**: 41-44.
Wasmann E. 1895b: Zur Kenntniss einer Thorictus-Arten. Zweites Stück. *Deutsche Entomologische Zeitschrift* **1895**: 291-294.
Wasmann E. 1896: Zoologische Ergebnisse einer von Dr. K. Escherich und Dr. L. Kathariner nach Central-Kleinasien unternommenen Reise. – Myrmekophilen. *Deutsche Entomologische Zeitschrift* **1896**: 237-241.
Wasmann E. 1897: Zur Kenntniss einiger Thorictus-Arten. Drittes Stück. *Deutsche Entomologische Zeitschrift* **1896**: 242-243.
Wasmann E. 1902: Coléoptères myrmecophiles recueillis par M. le prof. A. Lameere en Algérie. *Annales de la Société Entomologique de Belgique* **46**: 159.
Wasmann E. 1916: Wissenschaftliche Ergebnisse einer Forschungsreise nach Ostindien, ausgeführt im Auftrage der Kgl. Preuß. Akademie der Wissenschaften zu Berlin von H. v. Buttel-Reepen. V. Termitophile und myrmecophile Coleopteren. *Zoologische Jahrbücher* **39**: 169-210, pls. 4-5.
Wasmann E. 1925: Die Ameisenmimikry. Ein Exakter Beitrag zum Mimikryproblem und zur Theorie der Myrmekophilen. *Abhandlungen zur Theoretischen Biologie* (Berlin) **19**: xii + 164 pp., 3 pls.
Waterhouse C. O. 1873: [new taxa]. In: Crotch G. R.: A descriptive list of Erotylidae collected by Geo. Lewis, Esq., in Japan. *The Entomologist's Monthly Magazine* **9**: 184-189.
Waterhouse C. O. 1875: [Geänderte Namen]. *Coleopterologische Hefte* **14**: 213.
Waterhouse C. O. 1876a: New species of Coleoptera from the Island of Rodriguez, collected by the naturalists accompanying the Transit of Venus Expedition. *The Annals and Magazine of Natural History* (4) **18**: 105-121.
Waterhouse C. O. 1876b: Descriptions of new Cucujidae and Cleridae. *The Entomologist's Monthly Magazine* **13** [1876-1877]: 118-126.
Waterhouse C. O. 1877a: VII. Coleoptera. Pp. 77-82. In: Günther A.: Account of the zoological collection made during the visit of H. M. S. 'Peterel'to the Galapagos Islands. *Proceedings of the Scientific Meetings of the Zoological Society of London* **1877**: 64-93.
Waterhouse C. O. 1877b: Monograph of the coleopterous genus Calochromus of the family Lycidae. *Cistula Entomologica* **2**: 195-202.
Waterhouse C. O. 1878: On the different forms occurring in the Coleopterous family Lycidae, with description of new genera and species. *Transactions of the Entomological Society of London* **1878**: 95-118.
Waterhouse C. O. 1879: *Illustrations of typical specimens in the collection of the British Museum part I - Lycidae.* British Museum of Natural History, London. pp. 1-98.
Waterhouse C. O. 1881: On the coleopterous insects collected by Prof. I. Bailey Balfour in the island of Socotra. *Proceedings of the Zoological Society of London* **1881**: 469-478.
Waterhouse C. O. 1883: Description of a new species of Anthrenus from India (Coleoptera, Dermestidae). *The Annals and Magazine of Natural History* (5) **11**: 61.
Waterhouse C. O. 1884a: On the coleopterous insects collected by H. O. Forbes in the Timor-Laut Islands. *Proceedings of the Zoological Society of London* **16**: 213-219.
Waterhouse C. O. 1884b: Coleoptera collected during the expedition of HMS "Challenger". *The Annals and Magazine of Natural History* (5) **13**: 276-283.
Waterhouse C. O. 1888: Some observations on the coleopterous family Bostrichidae. *The Annals and Magazine of Natural History* (6) **1**: 348-350.
Waterhouse G. R. 1849: [Descriptions of Tricorynus zeae and Cryptorhynchus batatae]. *Proceedings of the Entomological Society of London* **1849**: lxvii-lxx.
Waterhouse G. R. 1858: *Catalogue of British Coleoptera.* London: Taylor and Francis, 117 pp.
Waterhouse G. R. 1859: A revision of the British species of Corticaria. *Transactions of the Entomological Society of London* (N. S.) **5**: 133-145.

Waterhouse G. R. 1862: Description of Scymnus discoideus (family Coccinellidae) and two allied species, and description of a new species of Bryaxis (fam. Pselaphidae). *Transactions of the Entomological Society of London* (3) **1** [1862-1864]: 132-136.

Watson W. Y. 1954: Two new species of Coccinellidae (Coleoptera). *The Canadian Entomologist* **86**: 45-47.

Weber F. 1801: *Observationes entomologicae, continentes novorum quae condidit generum characteres, et nuper detectarum specierum descriptiones.* Kiliae: Impensis Bibliopolii Academici Novi, xii + 116 + [1] pp.

Weidenbach C. von & Petry A. 1859: Systematische Ubersicht der Käfer um Augsburg. *Jahresbericht der Naturhistorischen Vereins in Augsburg* **12**: 33-82.

Weise E. 1972: Zwei neue Arten der Gattung Enicmus Thoms. (Col., Lathridiidae). *Entomologische Blätter* **68**: 159-163.

Weise J. 1876: Coleopterologische Ergebnisse einer Bereisung der Czernahora. *Verhandlungen des Naturforschenden Vereins in Brünn* **14** [1875]: 85-114.

Weise J. 1877: [new taxon]. In: Putzeys J. A. A. H., Reitter E., Saulcy F. de & Weise J.: Coleopterologische Ergebnisse einer Reise nach Südungarn und in die Transsylvanischen Alpen. *Verhandlungen des Naturforschenden Vereins in Brünn* **15** [1876]: 3-30.

Weise J. 1878: [new taxa]. In: Schneider O. & Leder H.: Beiträge zur Kenntniss der kaukasischen Käferfauna. *Verhandlungen des Naturforschenden Vereins in Brünn* **16** [1877]: 1-359, 3 pls.

Weise J. 1879a: Beitäge zur Käferfauna von Japan. (Fünftes Stück). *Deutsche Entomologische Zeitschrift* **23**: 147-152.

Weise J. 1879b: Ceutorrhynchus Venedicus n. sp. und Scymnus Ludyi n. sp. *Deutsche Entomologische Zeitschrift* **23**: 153-154.

Weise J. 1879c: Bestimmungs-Tabellen der europaischen Coleopteren II. Coccinellidae. *Zeitschrift für Entomologie* (N. F.) (Breslau) **7**: 88-156.

Weise J. 1880: Ueber Halyzia japonica. *Entomologische Monatsblätter* **2**: 117-119.

Weise J. 1881: Nachträge zu den Bestimmungstabelen II der Coccinellidae. *Deutsche Entomologische Zeitschrift* **25**: 165-166.

Weise J. 1883a: Zwei neue Pharus-Arten. *Wiener Entomologische Zeitung* **2**: 66-68.

Weise J. 1883b: Scymnus (Nephus) sannio n. sp. *Il Naturalista Siciliano* **2**: [1882-1883]: 137-138.

Weise J. 1884: Einige neue Chrysomeliden und Coccinelliden. *Deutsche Entomologische Zeitschrift* **28**: 161-166.

Weise J. 1885a: Coccinellidae. II. Auflage. Mit Berücksichtigung der Arten aus dem nördlichen Asien. In: Reitter E. *Bestimmungs-Tabellen der europäischen Coleopteren. II. Heft.* Mödling: H. Büsing, 83 pp.

Weise J. 1885b: Beschreibung einiger Coccinelliden. *Entomologische Zeitung* (Stettin) **46**: 227-241.

Weise J. 1886: Scymnus Oertzeni. *Deutsche Entomologische Zeitschrift* **30**: 25-26.

Weise J. 1887a: Mitteilungen über das Sammeln von Käfern und über die Fangstellen im Glatzer Gebirge. *Zeitschrift für Entomologie* (N. F.) (Breslau) **12**: 47-60.

Weise J. 1887b: Neue sibirische Chrysomeliden und Coccinelliden nebst Bemerkungen über früher beschriebene Arten. *Archiv für Naturgeschichte* **53** (1): 164-214.

Weise J. 1887c: [new taxa]. In: Heyden L. F. J. D. von.: Vierter Beitrag zur Kenntniss der Coleopteren-Fauna der Amurländer. *Deutsche Entomologische Zeitschrift* **31**: 299-336.

Weise J. 1889a Neue Chrysomeliden und Coccinelliden aus dem Kaukasus. *Wiener Entomologische Zeitung* **8**: 259-262.

Weise J. 1889b: Synonymische Bemerkungen. *Deutsche Entomologische Zeitschrift* **1889**: 384.

Weise J. 1889c: Insecta, a CL. G. N. Potanin in China et Mongolia Novissime Lecta. IX. Chrysomelidae et Coccinellidae. *Horae Societatis Entomologicae Rossicae* **23**: 560-653.

Weise J. 1890a: Einige neue Coccinelliden. *Deutsche Entomologische Zeitschrift* **1890**: 21-22.

Weise J. 1890b: Malthodes liegeli nov. spec. *Deutsche Entomologische Zeitschrift* **1890**: 348.

Weise J. 1890c: Insecta, a CL. G. N. Potanin in China et in Mongolia novissime lecta. XVI. Chrysomelidae et Coccinelidae. *Horae Societatis Entomologicae Rossicae* **24** [1889-1890]: 477-492.

Weise J. 1891a: Neue Coccinelliden. *Deutsche Entomologische Zeitschrift* **1891**: 282-288.

Weise J. 1891b: [new names]. In: Reitter E. (ed.): *Catalogus Coleopterorum Europae, Caucasi et Armeniae rossicae.* Berlin: R. Friedländer & Sohn, Mödlling: Edmund Reitter, Caen: Revue entomologique: viii + 420 pp.

Weise J. 1892a: Les:Coccinellides du Chota-Nagpore. *Annales de la Société Entomologique de Belgique* **36**: 16-30.

Weise J. 1892b: Kleine Beiträge zur Coccinelliden-Fauna Ost-Afrika s. *Deutsche Entomologische Zeitschrift* **1892**: 15-16.

Weise J. 1892c: [new species]. In: Heyden L. F. J. D. von.: XV. Beitrag zur Coleopteren-Fauna von Turkestan, Turkmenien und Süd-West-Sibirien. Aus Sendungen des Herrn Dr. Staudinger. *Deutsche Entomologische Zeitschrift* **1892**: 105-110.

Weise J. 1892d: Coccinelliden aus Ostsibirien. *Deutsche Entomologische Zeitschrift* **1892**: 140-141.

Weise J. 1892e: Malthodes brevicollis Payk., Typ einer Gattung. *Deutsche Entomologische Zeitschrift* **1892**: 407-410.

Weise J. 1893a: Nouvelle répartition des tribus et des genres de Coccinellides paléartiques. *L'Abeille, Journal d'Entomologie* **28** [1892-1896]: 105-108.

Weise J. 1893b: Beitrag zur weiteren Zerlegung der Malthodes und Beschreibung einer neuen Art. *Deutsche Entomologische Zeitschrift* **1893**: 364-365.

Weise J. 1894: Coelopterus armeniacus. *Deutsche Entomologische Zeitschrift* **1894**: 144.

Weise J. 1895a: Neue Coccinelliden, sowie Bemerkungen zu bekannten Arten. *Annales de la Société Entomologique de Belgique* **39**: 120-146.

Weise J. 1895b: Über die mit Novius Muls. verwandten Gattungen. *Annales de la Société Entomologique de Belgique* **39**: 147-150.

Weise J. 1895c: Insectes du Bengale 36 mémoire, Coccinellidae. *Annales de la Société Entomologique de Belgique* **39**: 151-157.

Weise J. 1895d: Luperus Fiorii n. spec. *Deutsche Entomologische Zeitschrift* **1895**: 212.

Weise J. 1895e: Bekannte und neue Malthiniden. *Deutsche Entomologische Zeitschrift* **1895**: 213-216.

Weise J. 1895f: Aspidophorus 4-guttatus n. sp. *Deutsche Entomologische Zeitschrift* **1895**: 326.

Weise J. 1897: Coccinellen aus Ostafrika (Usambara). *Deutsche Entomologische Zeitschrift* **1897**: 289-304.

Weise J. 1898: Über bekannte und neue Coccinelliden. *Archiv für Naturgeschichte* **64** (1): 225-238.

Weise J. 1900a: Bemerkungen zu den neuesten Bearbeitungen der Coccinelliden. *Deutsche Entomologische Zeitschrift* **1899**: 369-378.

Weise J. 1900b: Synonymische Bemerkungen. *Deutsche Entomologische Zeitschrift* **1899**: 384.

Weise J. 1900c: Kurze Mittheilungen über ostafrikanische Coccinelliden und Beschreibungen neuer Arten. *Deutsche Entomologische Zeitschrift* **1900**: 113-131.

Weise J. 1900d: Neue Coleopteren aus Kleinasien. *Deutsche Entomologische Zeitschrift* **1900**: 132-140.

Weise J. 1900e: [new taxon]. P. 166. In: Bodemeyer E. von.: *Quer durch Klein-Asien in den Bulghar-Dagh. Eine naturwissenschaftliche Studien-Reise*. Emmendingen: Verlangs-Aktiengesellschaft vormals Dölter, v + 169 pp.

Weise J. 1901a: Neue Coccinelliden. *Annales de la Société Entomologique de Belgique* **45**: 273-286.

Weise J. 1901b: Coccinelliden aus Ceylon gesammelt von Dr. Horn. *Deutsche Entomologische Zeitschrift* **1900**: 417-445.

Weise J. 1901c: Beschreibung africanischer [sic!] Chrysomeliden nebst synonymischen Bemerkungen. *Deutsche Entomologische Zeitschrift* **1900**: 446-459.

Weise J. 1902a: Ein neuer deutscher Scymnus. *Deutsche Entomologische Zeitschrift* **1902**: 240.

Weise J. 1902b: Coccinelliden aus der Sammlung des ungarischen National-Museums. *Termmészetrajzi Füzetek* **25**: 489-520.

Weise J. 1903: [Coccinellidae]. Pp. 579-583 In: Kneucker A.: Zoologische Aubeute einer botanischen Studienreise durch die Sinai-Halbinseln im März und April 1902. *Verhandlungen der Kaiserlich-Königlichen Zoologisch-Botanischen Gesellschaft in Wien* **53**: 575-587.

Weise J. 1905a: Ueber Coccinellen. *Deutsche Entomologische Zeitschrift* **1905**: 137-138.

Weise J. 1905b: Coccinelliden aus Erythrea. *Deutsche Entomologische Zeitschrift* **1905**: 139-140.

Weise J. 1905c: Ueber Coccinelliden. *Deutsche Entomologische Zeitschrift* **1905**: 217-220.

Weise J. 1907: Kleine Mitteilungen und synonymische Bemerkungen. *Wiener Entomologische Zeitung* **26**: 13-15.

Weise J. 1808a: Chrysomelidae und Coccinellidae. Pp. 1-13. In: Michaelson W. & Hartmeyer R. (eds): *Die Fauna Südwest-Australiens. Ergebnisse der Hamburger südwest-australischen Forschungsreise 1905. Band II, Lieferung 1*. Jena: G.Fischer.

Weise J. 1908b: Coleopteren aus Ostindien. *Entomologische Zeitung* (Stettin) **69**: 214-230.

Weise J. 1909: Hispinen und Coccinelliden aus Madagaskar. *Archiv für Naturgeschichte* **75** (1): 111-125.

Weise J. 1910a: Chrysomeliden und Coccinelliden. *Verhandlungen des Naturforschenden Vereins in Brünn* **48** [1909]: 25-53.

Weise J. 1910b: Verzeichnis von Coleopteren aus den Philippinen, nebst zwei neuen Arten aus Niederlandisch Ostindien. *The Philippine Journal of Science* **5**: 139-148.

Weise J. 1910c: Über Chrysomeliden und Coccinelliden der Philippine: I. Theil. *The Philippine Journal of Science* **5**: 223-232.

Weise J. 1910d: Coleoptera. 12. Coccinellidae. Pp. 248-265. In: Sjöstedt Y. (ed.): *Wissenschaftliche Ergebnisse der Schwedischen Zoologischen Expedition nach dem Kilimandjaro, dem Meru und den umgebenden Massaisteppen Deutsch-Ostafrikas 1905-1906 unter der Leitung von Prof. Dr. Yngve Sjörstedt. 1. Band.* Stockholm: Königl. Schwedisches Akademie der Wissenschaften, 442 pp.

Weise J. 1910e: Coccinellidae von Madagascar, den Comoren und den Inseln Ostafrikas. Pp. 507-520. In: Voeltzkow A.: *Reise in Ostafrika*. Stuttgart: E. Schweizerbart.

Weise J. 1912: Über Hispinen und Coccinelliden. *Archiv für Naturgeschichte* A **78** (2): 100-120.

Weise J. 1913: Chrysomeliden und Coccinelliden. *Nova Guinea* **9**: 423-446.

Weise J. 1922: Coleoptera e Collectione Bruchiana. *Annales de la Sociedad Cientifica Argentina* **94**: 30-40.

Weise J. 1923: H. Sauter's Formosan-Ausbeute: Coccinellide. *Archiv für Naturgeschichte* A **89** (2): 182-188.

Weise J. 1926: XXIII Coleoptera B. Chrysomelidae et Coccinellidae. In: Wissenschaftliche Ergebnisse der von Werner unternommenen zoologischen Expedition nach dem Anglo-Ägyptischen Sudan (Kordofan) 1914. *Denkschrift der Kaiserlichen Akademie der Wissenschaften. Mathematisch-Naturwissenschaftliche Classe* **100**: 223-228.

Weise J. 1929: Westindische Chrysomeliden und Coccinelliden. *Zoologische Jahrbücher, Abteilung Systematik, Ökologie und Geographie der Tiere, Supplement* (Jena) **16**: 11-34.

Wellschmied K. 1978: [new taxon]. In: Dolin W. G. 1978: *Opredelitel' lichinok zhukov-shchelkunov fauny SSSR* [*Key to click-beetle larvae of the fauna of USSR*]. Kiev: Urozhay, 124 pp.

Wesmael C. 1835: Description d'un nouveau genre d'insectes coléoptères, de la famille des Xylophages, tribu des Bostrichins. *Bulletins de l'Académie Royal de Bruxelles* **2**: 338-340, pl. 4 [part].

Westhoff F. 1881: *Die Käfer Westfalens zusammengestellt. I. Abtheilung. Supplement zu den Verhandlungen des naturhistorischen Vereins der preussischen Rheinlande und Westfalens. Achtunddreissigster Jahrgang. Vierte Folge: 8. Jahrgang.* Bonn: Max Cohen & Sohn (Fr. Cohen), xxvii + 328 pp.

Westwood J. O. 1832: [new taxon]. In: Cuvier G. L. C. F. D.: *The animal kingdom arranged in conformity with its organization, with supplementary additions to each order, by Edward Griffith and others. Volume the Fifteenth.* London: Whittaker, Treacher, and Co., 796 pp.

Westwood J. O. 1833: On the Paussidae, a family of coleopterous insects. *Transactions of the Linnean Society of London* **16**: 607-684, pl. 33.

Westwood J. O. 1838: *Synopsis of the genera of British insects.* Pp. 1-48. In: *An introduction to the modern classification of insects; founded on the natural habits and corresponding organization of the different families. Vol. II.* London: Orne, Brown, Green and Longmans, 158 pp. [issued in parts, 1838-1840].

Westwood J. O. 1839: Description of a new genus of coleopterous insects from Corfu. *Transactions of the Entomological Society of London* **2**: 174-175, pl. 14.

Westwood J. O. 1842: Insectorum novorum Centuria. *The Annals and Magazine of Natural History* **8**: 120-130.

Westwood J. O. 1843: Dodecatoma bicolor. In: Guérin-Méneville F. C.: *Spécies et iconographie générique des animaux articulés ou représentation des genres avec leur descriptions et celle de toutes les espèces de cette grande division du Règne animal. I. Coléoptères.* Paris: Auteur.

Westwood J. O. 1848: *The cabinet of Oriental entomology; being a selection of some of the rarer and more beautiful species of insects, natives of India and the adjacent islands, the greater portion of which are now for the first time described and figured.* London: William Smith, 2 + 88 pp., 42 pls. [issued in parts, in 1847 and 1848].

Westwood J. O. 1852: Descriptions of new species of Cleridae, from Asia, Africa, and Australia. *The Proceedings of the Zoological Society of London* **20**: 34-55, pls XXXIV-XXVII.

Westwood J. O. 1855: Descriptions of some new species of Cleridae, collected at Singapore by Mr. Wallace. *The Proceedings of the Zoological Society of London* **23**: 19-26, pl. XXXVIII.

Westwood J. O. 1863: Two new Col. from the Canaries. *The Proceedings of the Entomological Society of London* **1863**: 178-179.

Westwood J. O. 1875: Descriptions of a new genus of Cleridous Coleoptera, from the Malayan Archipelago. *Transactions of the Entomological Society of London* **1875**: 241-242.

White A. 1849: *Nomenclature of Coleopterous Insects in the British Museum. Part IV. Cleridae*. London: Trustees, [4] + 68 pp.

White A. 1850a: [no title]. *Transactions of the Entomological Society of London* (*New Series*) **1** [1850-1851]: 13.

White A. 1850b: [no title]. *The Zoologist* **8**: 2834.

White R. E. 1966: Cryptoramorphus floridanus, New Genus, New Species (Coleoptera: Anobiidae), with Generic Reassignments. *Annals of the Entomological Societa of America* **59**: 959-961.

White R. E. 1973: A new genus, two new species, and a species key for Byrrhoides (Coleoptera: Anobiidae). *The Proceeding of the Entomological Society of Washington* **75**: 48-54.

White R. E. 1974: Type-species for world genera of Anobiidae (Coleoptera). *Transactions of the American Entomological Society* **99**: 415-475.

White A. & Butler A. G. 1846: Insects. Pp. 1-24. In: Richardson J. & Gray J. E.: *The Zoology of the Voyage of H.M.S. Erebus et Terror under the command of Captain Sir James Clark Ross, R.N., F.R.S. during the years 1839 to 1843*. London: E. W. Janson, 51 pp.

Wiedemann C. R. W. 1821: [new taxa]. In: Wiedemann C. R. W. & Germar E. F.: Neue exotische Käfer. *Magazin der Entomologie* **4**: 107-183.

Wiedemann C. R. W. 1823: Zweihundert neue Käfer von Java, Bengalen und dem Vorgebirge der guten Hoffnung. *Zoologisches Magazin* (Ultona) **2** (1): 3-133.

Wimmel T. 1894: [new taxa]. In: Timm R. & Wimmel T.: Neue und seltener Käferder Hamburger Gegend. In: Beiträge zur Fauna und Flora der Niederelb-Gegend. *Verhandlungen des Vereins für Naturwissenschaftlichen Unterhaltung zu Hamburg* **8** [1891-1893]: 1-11.

Wingelmüller A. 1912: Diagnosen neuer Coccinelliden-Aberrationen. *Verhandlungen der Kaiserlich-Königlichen Zoologisch-Botanisch Geselscheft in Wien* **62**: 182-185.

Winkler J. R. 1953: Doplňkové poznámky ke R. Kleineovým "Bestimmungs-Tabellen" s popisy nových druhů východoasijských lycidů. Supplementary remarks to R. Kleine's „Bestimmungstabellen" with the descriptions of new species of east Asiatic Lycidae. *Acta Entomologica Musei Nationalis Pragae* **28** [1952]: 401-410.

Winkler J. R. 1958: Tillopilo corporaali n. gen. n. sp., new genus and species of the checkered beetles from China (Col., Cleridae). *Časopis Československé Společnosti Entomologické* **55**: 244-249.

Winkler J. R. 1959: Trichodes crabroniformis hidalgo nueva subespecie del Clerido de España. *Graellsia* **17**: 69-75.

Winkler J. R. 1963: Trichodes israelicus, eine neue Buntkäferart aus Palestina (Coleoptera: Cleridae). *Beiträge zur Entomologie* **13**: 890-893.

Winkler J. R. 1972: Sedlacekius rex n. gen. n. sp. (Col. Cleridae). Un nouveau Cléride aptère de Somalie. *Bulletin de la Société Entomologique de Mulhouse* **1972**: 38-41.

Winkler J. R. 1974: Cleridae (Coleoptera) of Afghanistan, with description of new genera and species. *Acta Universitatis Carolinae – Biologica* **1972**: 375-408, 4 pls.

Winkler J. R. 1978: Sinobaneus sedlaceki gen. n., sp. n. (Coleoptera: Cleridae) from Tienmuschan Mountains, Oriental, not Palaearctic area of China. *Acta Universitatis Carolinae – Biologica* **1975-1976** (5-6): 247-255.

Winkler J. R. 1981: Insects of Saudi Arabia Coleoptera: Fam. Cleridae. *Fauna of Saudi Arabia* **3**: 251-256.

Winkler J. R. 1982a: New subspecies of Tilloidea unifasciata (F., 1787) from Iran (Col. Cleridae). *Deutsche Entomologische Zeitschrift* (N. F.) **29**: 7-9.

Winkler J. R. 1982b: A new genus established for reception of Clerus paivae Wollaston, 1862 (Col., Cleridae). *Deutsche Entomologische Zeitschrift* (N. F.) **29**: 475-478.

Winkler J. R. 1984a: Enopliinae of Palaearctis, Part 2. Syriopelta gen.n., new genus established for Syrian Orthopleura funebris Fairmaire, 1892 (Col., Cleridae). *Deutsche Entomologische Zeitschrift* (N. F.) **31**: 177-185.

Winkler J. R. 1984b: Opilo hladirorum n. sp., new species from Greece (Coleoptera, Cleridae). *Entomologica Basiliensia* **9**: 193-203.

Winkler J. R. 1985: Über einige sizilianische Cleriden, mit Beschreibung der neuen Unterart Tillus pallidipennis espinosai subsp. n. (Coleoptera, Cleridae). *Mitteilungen den Münchener Entomologischen Gesellschaft* **75**: 117-126.

Winkler J. R. 1987: Ekisius vitreus gen. n., sp. n., the first representative of apterous Cleridae in Oriental region (Coleoptera). *Deutsche Entomologische Zeitschrift* (N. F.) **34**: 169-177.

Winkler J. R. & Žirovnický J. 1980: Trichodes Herbst, 1792, series „leucopsideus" - a revision of Pontic-Mediterranean group of species (Coleoptera: Cleridae). *Acta Universitatis Carolinae – Biologica* **1978**: 457-484.

Wittmer W. 1934: Résultats scientifiques des expéditions entomologiques de S. A. le Prince Alexandre C. della Torre e Tasso en Égypte et au Sinai. VI. Malacodermata (Coleoptera). *Bulletin de la Société Royale Entomologique d'Égypte* **18**: 449-455.

Wittmer W. 1935a: Risultati Scientifici delle caccie entomologiche di S.A.S. il Principe Alessandro della Torre e Tasso in Italia IV. Malacodermata von M.te Pollino (Lucania). (2. Beitrag zur Kenntnis der palaearkt. Malac.). *Bollettino de la Società Entomologica Italiana* **67**: 72-76.

Wittmer W. 1935b: Ein neues Genus aus der Familie der Cucujidae. *Bulletin de la SociétéEntomologique d'Egypte* **19**: 129-131.

Wittmer W. 1935c: Wissenschaftliche Ergebnisse der entomologischen Expeditionen seiner Durchlaucht des Fürsten Alessander C. della Torre e Tasso nach Aegypten und der Sinaihalbinsel. VII. Malacodermata (Coleoptera). (5. Beitrag zur Kenntnis der palaearktischen Malacodermata). *Eos, Revista Española de Entomología* **11**: 181-189.

Wittmer W. 1935d: Risultati Scientifici delle cacce entomologiche di S. A. S. il Principe Alessandro della Torre e Tasso Nelle Isole Dell'egeo. II. Malacodermata. (3. Beitrag zur Kenntnis der palaearktischen Malacodermata). *Bollettino del Laboratorio di Zoologia Generale ed Agraria* **28**: 247-256.

Wittmer W. 1935e: Zwei neue palaearktishe Malthodes Kies. (6. Beitrag zur Kenntnis der palaearkt. Malacodermata). *Boletín de la Sociedad Española de Historia Natural* **35**: 459-462.

Wittmer W. 1936a: Zur Kenntnis der Malacodermata Aegyptens (8. Beitrag zur Kenntnis der palaearktischen Malacodermata) (Coleoptera - Malachiidae). *Bulletin de la Société Royale Entomologique d'Égypte* **29**: 152-154.

Wittmer W. 1936b: Un nouveau Cephaloncus tripolitain (Coleopt. Malachiidae). (4e contribution à la connaissance des Malacodermes paléarctiques). Pp. 375-376. In: *Livre Jubilaire de M. E. L. Bouvier*, Paris.

Wittmer W. 1936c: Malthinus axillaris Kies. und verwandte Formen. (7. Beitrag zur Kenntnis der palaearkt. Malacodermata). *Mitteilungen der Schweizerischen Entomologischen Gesellschaft* **16**: 676-679.

Wittmer W. 1937: Zur Kenntnis der Malacodermata Aegyptens (9. Beitrag zur Kenntnis der palaearktischen Malacodermata). *Bulletin de la Société Royale Entomologique d'Égypte* **30**: 167-170.

Wittmer W. 1938a: 10. Beitrag zur Kenntnis der palaearktischen Malacodermata. *Bollettino della Società Entomologica Italiana* **70**: 86-87.

Wittmer W. 1938b: 3. Beitrag zur Kenntnis der indo-malayischen Malacodermata(Col.). *Treubia* **16**: 301-306.

Wittmer W. 1939a: Neues aus der Gattung Malthodes Kies. (Canth.) (11. Beitrag zur Kenntnis der palaearktischen Malacodermata). *Mitteilungen der Münchener Entomologischen Gesellschaft* **29**: 1-4.

Wittmer W. 1939b: 5. Beitrag zur Kenntnis der indo-malayischen Malacodermata (Col.). *Treubia* **17**: 21-32.

Wittmer W. 1940a: Neue Malacodermata aus der Sammlung des Museum Heude, Shanghai. 13. Beitrag zur Kenntnis der palaearktischen Malacodermata. *Musée Heude, Notes d'Entomologie Chinoise* **7**: 99-104.

Wittmer W. 1940b: Neue Malacodermata aus der Sammlung des Herrn Georg Frey, Müchen (Col. Malacod.). 12. Beitrag zur Kenntnis der palaearktischen Malacodermata. *Mitteilungen der Münchener Entomologischen Gesellschaft* **30**: 544-547.

Wittmer W. 1941: Ergänzungen zum Katalog der Malachiidae von J. Greiner (Pars 159 des Coleopterorum Catalogus). *Mitteilungen der Münchener Entomologischen Gesellschaft* **31**: 1122-1127.

Wittmer W. 1944: Catalogue des Drilidae E. Oliv. (Coleoptera-Malacodermata). *Revista de la Sociedad Entomológica Argentina* **12**: 203-221.

Wittmer W. 1948: Notas sinonimicas y sistematicas sobre Malacodermata (1ª nota). *Anales de la Sociedad Científica Argentina* **145**: 167-173.

Wittmer W. 1951: Neue Cantharidae aus Herrn Joh. Klapperichs' Südchina Ausbeute. *Entomologische Blätter* **47/48** [1951/1952]: 96-103.

Wittmer W. 1953a: 15. Beitrag zur Kenntnis der palaearktischen Malacodermata, Col. *Entomological Review of Japan* **6**: 35-37.

Wittmer W. 1953b: Malachiidae (Col.) aus Shansi (17. Beitrag zur Kenntnis der palaearktischen Malacodermata). *Mushi* **25**: 71-73.

Wittmer W. 1954a: Cantharidae, Malachiidae und Prionoceridae (Coleopt.) von Sumbaund Flores. Wissenschaftliche Ergebnisse der Sumba-Expedition des Museums für Völkerkunde und des Naturhistorischen Museums in Basel, 1949. *Verhandlungen der Naturforschenden Gesellschaft in Basel* **65**: 24-28.

Wittmer W. 1954b: Zur Kenntnis der Malthodes Japans (Col. Cantharidae). 18. Beitrag zur Kenntnis der palaearktischen Malacodermata. *Mushi* **26** (9): 47-52.

Wittmer W. 1954c: 20. Beitrag zur Kenntnis der palaearktischen Malacodermata (Col.). *Mitteilungen der Schweizerischen Entomologischen Gesellschaft* **27**: 109-114.

Wittmer W. 1954d: Zur Kenntnis der Cantharidae und Malachidae der Insel Formosa. *Revue Suisse de Zoologie* **61**: 271-282.

Wittmer W. 1954e: 21. Coleoptera: Malachiidae und Dasytidae. Pp. 315-323, pl. 15. *British Museum Expedition to South-West Arabia, 1937-8. Part 1*. London: British Museum (Natural History).

Wittmer W. 1955: 21. Beitrag zur Kenntnis der palaearktischen Malacodermata (Coleoptera). *Mushi* **29**: 39-45, pls 4, 5.

Wittmer W. 1956a: Afghanistan Ausbeute von Herrn J. Klapperich, 1952/53 (22. Beitrag zur Kenntnis der palaearktischen Malacodermata, Col.). *Revue Suisse de Zoologie* **63**: 115-161.

Wittmer W. 1956b: Neue Malacodermata aus der Sammlung der California Academy of Sciences (16. Beitrag zur Kenntniss der palaearktischen Malacodermata, Col.). *Mitteilungen der Schweizerischen Entomologischen Gesellschaft* **29**: 303-313.

Wittmer W. 1956c: 11. Beitrag zur Kenntnis der Malacodermata Afrikas (Col.). *Entomologischen Arbeiten aus dem Museum G. Frey* **7**: 1072-1088.

Wittmer W. 1957a: 20. Beitrag zur Kenntnis der indo-malayischen Malacodermata (Col.). *Mitteilungen der Schweizerischen Entomologischen Gesellschaft* **30**: 25-47.

Wittmer W. 1957b: 18. Beitrag zur Kenntnis der indo-malayischen Malacodermata (Col.) (2. Teil). *The Annals and Magazine of Natural History* (12) **10**: 567-578.

Wittmer W. 1958: 23. Beitrag zur Kenntnis der palaearktischen Malacodermata (Col.). *Entomologischen Arbeiten aus dem Museum G. Frey* **9**: 114-130.

Wittmer W. 1960a: 22. Beitrag zur Kenntnis der indo-malayischen Malacodermata (Col.). *Mitteilungen der Schweizerischen Entomologischen Gesellschaft* **33**: 101-106.

Wittmer W. 1960b: 24. Beitrag zur Kenntnis der palaearktischen Malacodermata (Col.). *Mitteilungen der Schweizerischen Entomologischen Gesellschaft* **33**: 107-110.

Wittmer W. 1961: 25. Beitrag zur Kenntnis der palaearktischen Malacodermata (Col.) *Niponius* **1**: 1-2.

Wittmer W. 1963: 26. Beitrag zur Kenntnis der palaearktischen Malacodermata (Col.) *Entomologischen Arbeiten aus dem Museum G. Frey* **14**: 114-117.

Wittmer W. 1964: 27. Beitrag zur Kenntnis der palaearktischen Malacodermata (Col.). *Entomologischen Arbeiten aus dem Museum Gg. Frey* **15**: 100-105.

Wittmer W. 1965a: 29. Beitrag zur Kenntnis der palaearktischen Malacodermata (Col.). *Mitteilungen der Schweizerischen Entomologischen Gesellschaft* **38**: 82-87.

Wittmer W. 1965b: Ergebnisse der zoologischen Forschungen von Dr. Z. Kaszab in der Mongolei. 43. Cantharidae (Col.). *Reichenbachia* **7**: 83-85.

Wittmer W. 1965c: Mitteilungen über Cantthariden un Malachiiden aus Nepal (Coleoptera) 28. Beitrag zur Kenntnis der palaearktischen Malacodermata. *Khumbu Himal - Ergebnisse des Forschungsunternehmen Nepal Himalaya* **1**: 85-87.

Wittmer W. 1965d: Kommt die Gattung Collops Er. (Col. Malachiidae) in der Palaearktis vor? *Mitteilungen der Schweizerischen Entomologischen Gesellschaft* **38**: 142-144.

Wittmer W. 1966a: 32. Beitrag zur Kenntnis der palaearktischen Malacodermata (Col.). *Israel Journal of Zoolology* **15**: 1-8.

Wittmer W. 1966b: 26. Beitrag zur Kenntnis der indo-malayischen Malacodermata. *Mitteilungen der Schweizerischen Entomologischen Gesellschaft* **38**: 229-237.

Wittmer W. 1966c: 33. Beitrag zur Kenntnis der palaearktischen Malacodermata (Col.). *The Annals and Magazine of Natural History* (13) **9**: 491-503.

Wittmer W. 1967: Contribution à la faune de l'Iran 3. Coléoptères Drilidae, Cantharidae et Malachiidae. *Annales de la Société Entomologique de France* (N. S.) **3**: 1087-1091.

Wittmer W. 1968a: Ergebnisse der mongolisch-tschechoslowakischen entom.-bot. Expeditionen (1965, 1966) in der Mongolei Nr. 15: Coleoptera - Cantharidae und Malachiidae. *Acta Faunistica Entomologica Musei Nationales Pragae* **13**: 21-22.

Wittmer W. 1968b: 37. Beitrag zur Kenntnis der palaearktischen Cantharidae und Malachiidae (Col.). *Mitteilungen der Schweizerischen Entomologischen Gesellschaft* **40**: 276-280.

Wittmer W. 1968c: Die Ebaeus-Arten aus der Gruppe turkestanicus-caspius (38. Beitrag zur Kenntnis der paläarktischen Malachiidae, Col.). *Mitteilungen der Schweizerischen Entomologischen Gesellschaft* **41**: 205-210.

Wittmer W. 1969a: Cantharidae und Malachiidae (Coleoptera) aus der Mongolei. *Annales Zoologici* (Warszawa) **27**: 54-58.

Wittmer W. 1969b: Zur Kenntnis der Gattung Metacantharis Bourg. (Col. Cantharidae). *Verhandlungen der Naturforschenden Gesellschaft in Basel* **80**: 70-93.

Wittmer W. 1969c: Synonymische und systematische Notizen über Coleopteren. *Mitteilungen der Schweizerischen Entomologischen Gesellschaft* **42**: 126-134.

Wittmer W. 1969d: Über die Gattung Malachidius Motsch. (Progeutes Ab.). *Mitteilungen der Schweizerischen Entomologischen Gesellschaft* **42**: 169-180.

Wittmer W. 1969e: Resultate der Expedition des Zoologischen Institutes in Leningrad nach d. Mongolei 1967. Coleoptera: Cantharidae und Malachiidae (42. Beitrag zur Kenntnis der Paläarktischen Cantharidae und Malachiidae). *Entomologischen Arbeiten aus dem Museum G. Frey* **20**: 510-517.

Wittmer W. 1970a: Zur Kenntnis der Gattung Malthodes (Coleoptera, Cantharidae) (49. Beitrag zur Kenntnis der palaearktischen Cantharidae). *Entomologischen Arbeiten aus dem Museum G. Frey* **21**: 13-107.

Wittmer W. 1970b: Coleoptera of the Loo-Choo Archipelago (II). Cantharidae und Malachiidae von den Inseln Okinawa und Ryukyu, Gesammelt durch Herrn Prof. Dr. Michio Chûjô (46. Beitrag zur Kenntnis der palaearktischen Cantharidae und Malachiidae.). *Memoirs of the Faculty of Education Kagawa University* **2** (192): 22-27.

Wittmer W. 1970c: 47. Beitrag zur Kenntnis der palaearktischen Cantharidae und Malachiidae. *Mitteilungen der Schweizerischen Entomologischen Gesellschaft* **43**: 73-98.

Wittmer W. 1970d: 19. Beitrag zur Kenntnis der Cantharidae und Malachiidae Afrikas. *Mitteilungen der Schweizerischen Entomologischen Gesellschaft* **43**: 149-155.

Wittmer W. 1971a: Coleoptera of the Loo-Choo Archipelago (III). Cantharidae und Malachiidae von den Inseln Okinawa und Ryukyu, Gesammelt durch Herrn Prof. Dr. M. Chûjô (46. Beitrag zur Kenntnis der palaearktischen Cantharidae und Malachiidae.). *Memoirs of the Faculty of Education Kagawa University* **2** (202): 16-17.

Wittmer W. 1971b: Ein weiterer Beitrag zur Kenntnis der Gattung Malthinus (50. Beitrag zur Kenntis der paläarktischen Cantharidae/Coleoptera). *Mitteilungen der Schweizerischen Entomologischen Gesellschaft* **43**: 169-188.

Wittmer W. 1971c: Ergebnisse der zoologischen Forschungen von Dr. Z. Kaszab in der Mongolei. Cantharidae der V. & VI. Expeditionen. (49. Beitrag zur Kenntnis der palaearktischen Cantharidae). *Annales Historico-Naturales Musei Nationalis Hungarici* **63**: 189-203.

Wittmer W. 1971d: Zur Kenntnis der Gattung Cantharis (51. Beitrag zur Kenntnis der palaearktischen Cantharidae/Col.). *Entomologischen Arbeiten aus dem Museum G. Frey* **22**: 226-239.

Wittmer W. 1971e: 53. Beitrag zur Kenntnis der palaearktischen Cantharidae und Malachiidae (Coleoptera). *Mitteilungen der Schweizerischen Entomologischen Gesellschaft* **44**: 363-373.

Wittmer W. 1971f: Col. Cantharidae. Genus Malthodes. Ergebnisse der Albanien-Expedition 1961 des Deutsches Entom. Instituts. *Beiträge zur Entomologie* **21**: 517-519.

Wittmer W. 1972a: 56. Beitrag zur Kenntnis der palaearktischen Cantharidae (Col.). *Mitteilungen der Schweizerischen Entomologischen Gesellschaft* **45**: 61-77.

Wittmer W. 1972b: 55. Beitrag zur Kenntnis der palaearktischen Cantharidae und Malachiidae (Col.). *Entomologischen Arbeiten aus dem Museum G. Frey* **23**: 122-141.

Wittmer W. 1972c: Zur Kenntnis der Cantharidae Irans (54. Beitrag zur Kenntnis der palaearktischen Cantharidae Col.). *Verhandlungen der Naturforschenden Gesellschaft in Basel* **82**: 193-204.

Wittmer W. 1973a: Zur Kenntnis der Gattung Pseudoabsidia. *Mitteilungen der Entomologischen Gesellschaft Basel* (N. F.) **23**: 117-120.

Wittmer W. 1973b: Zur Kenntnis der Gattung Themus Motsch. (Col. Cantharidae). *Entomologischen Arbeiten aus dem Museum G. Frey* **24**: 186-228.

Wittmer W. 1974a: 57. Beitrag zur Kenntnis der palaearktischen Cantharidae und Malachiidae (Coleoptera). *Fragmenta Entomologica* **10**: 1-20.

Wittmer W. 1974b: Zwei neue Malthodes aus dem Libanon. *Fragmenta Entomologica* **10**: 21-24.

Wittmer W. 1974c: Zur Kenntnis der Gattung Stenothemus Bourg. (Col. Cantharidae). *Mitteilungen der Schweizerischen Entomologischen Gesellschaft* **47**: 49-62.

Wittmer W. 1974d: Zur Kenntnis der Gattung Malthinus Latr. II. (Col. Cantharidae). *Entomologischen Arbeiten aus dem Museum G. Frey* **25**: 358-427.

Wittmer W. 1974e: Ergebnisse der Bhutan-Expedition 1972 des Naturhistorisches Museums in Basel. Col. Fam. Cantharidae (1. Teil). *Verhandlungen der Naturforschenden Gesellschaft in Basel* **84**: 627-632.

Wittmer W. 1975a: Ergebnisse der Bhutan-Expedition 1972 des Naturhistorisches Museums in Basel. Col.: Fam. Cantharidae (2. Teil). *Entomologica Basiliensia* **1**: 249-278.

Wittmer W. 1975b: 61. Beitrag zur Kenntnis der palaearktischen Fauna Cantharidae und Malachiidae. *Mitteilungen der Schweizerischen Entomologischen Gesellschaft* **48**: 259-267.

Wittmer W. 1976a: Coleoptera: Fam. Cantharidae und Malachiidae auf den Sovietisch-Mongolischen biologischen Expeditionen aufgesammelt (62. Beitrag zur Kenntnis der palaearktischen Cantharidae und Malachiidae). *Insects of Mongolia* **4**: 170-184.

Wittmer W. 1977a: Ergebnisse der Bhutan-Expedition 1972 des Naturhistorisches Museums in Basel. Coleoptera: Fam. Cantharidae (3. Teil). Gleichzeitig ein Beitrag zur Kenntnis der indo-malaiischen Silini. *Entomologica Basiliensia* **2**: 273-303.

Wittmer W. 1977b: Canthariden und Malachiiden (Coleoptera, Insecta) aus der Mongolei. *Mitteilungen aus dem Zoologischen Museum in Berlin* **53**: 281-283.

Wittmer W. 1977c: Ergebnisse der Bhutan-Expedition 1972 des Naturhistorischen Museums in Basel. Coleoptera: Fam. Malachiidae (1. Teil), Colotes gorhami Champ. und verwandte Arten. *Entomologica Basiliensia* **2**: 319-324.

Wittmer W. 1978a: Neue Coleoptera aus dem Orient, Neuguinea und Australien. *Pacific Insects* **18**: 61-66.

Wittmer W. 1978b: Neue Ichthyurini (Coleoptera: Cantharidae-Chauliognathinae). *Transactions of the Shikoku Entomological Society* **14**: 85-87.

Wittmer W. 1978c: Ergebnisse der Bhutan-Expedition 1972 des Naturhistorisches Museums in Basel. Coleoptera: Fam. Cantharidae (4. Teil) und Bemerkungen zu einigen Arten aus angrenzenden Gebieten. *Entomologica Basiliensia* **3**: 151-161.

Wittmer W. 1978d: [new taxa]. In: Wittmer W. & Brancucci M.: Zur Kenntnis der Malthinini der indo-malaiischen Region (Col. Cantharidae). *Entomologica Basiliensia* **3**: 297-304.

Wittmer W. 1978e: Beitrag zur Kenntnis der palaearktischen und indo-malaiischen Cantharidae und Malachiidae (Col.). *Entomologica Basiliensia* **3**: 347-376.

Wittmer W. 1979a: Zur Kenntnis der Gattung Malachius F. sensu lato (Col. Malachiidae) (66. Beitrag zur Kenntnis der palaearktischen Cantharidae & Malachiidae). *Entomologischen Arbeiten aus dem Museum G. Frey* **28**: 1-8.

Wittmer W. 1979b: Zur Kenntnis der Lampyridae der orientalischen Region (Coleoptera). *Entomologische Arbeiten aus dem Museum G. Frey* **28**: 83-92.

Wittmer W. 1979c: 65. Beitrag zur Kenntnis der palaearktischen Cantharidae und Malachiidae (Col.). *Entomologischen Arbeiten aus dem Museum G. Frey* **28**: 133-156.

Wittmer W. 1979d: Insects of Saudi Arabia Coleoptera: Fam. Malachiidae, Dasytidae, Karumiidae. *Fauna of Saudi Arabia* **1**: 186-192.

Wittmer W. 1979e: 64. Beitrag zur Kenntnis der palaearktischen Cantharidae, Phengodidae und Malachiidae (Col.). *Entomologica Basiliensia* **4**: 327-346.

Wittmer W. 1980a: Fauna of Saudi Arabia. Coleoptera: Fam. Drilidae, Malachiidae, Prionoceridae (2. Beitrag). *Fauna of Saudi Arabia* **2**: 114-118.

Wittmer W. 1980b: 67. Beitrag zur Kenntnis der palaearktischen Cantharidae (Col.). *Entomologica Basiliensia* **5**: 389-414.

Wittmer W. 1980c: Zur Kenntnis der Gattungen Indiebaeus und Tropiebaeus (Col. Malachiidae). *Entomologica Basiliensia* **5**: 467-474.

Wittmer W. 1981a: Eine neue Malthodes-Art aus der Schweiz (Coleoptera, Cantharidae). *Entomologische Berichte Luzerne* **6**: 95-97.

Wittmer W. 1981b: Zur Kenntnis der Malachiidae (Col.) des Himalaja und angrenzender Gebiete. *Entomologischen Arbeiten aus dem Museum G. Frey* **29**: 181-233.

Wittmer W. 1981c: 33. Beitrag zur Kenntnis der indo-malaiischen Cantharidae (Col.). *Entomologica Basiliensia* **6**: 387-398.

Wittmer W. 1981d: Die Gattung Habronychus (Col. Cantharidae) auf dem indischen Subcontinent (34. Beitrag zur Kenntnis der indo-malaiischen Cantharidae). *Entomologica Basiliensia* **6**: 399-405.

Wittmer W. 1981e: 68. Beitrag zur Kenntnis der palaearktischen Cantharidae und Malachiidae (Coleoptera). *Entomologica Basiliensia* **6**: 406-415.

Wittmer W. 1982ab: Die Arten der Gattung Themus Motschulsky aus der Verwandschaft von versicolor (Gorham) und davidis (Fairmaire) (Coleoptera, Cantharidae). Pp. 25-30. *Special Issue to the Memory of Retirement of Emeritus Professor Michio Chûjô*. Nagoya: Association of the Memorial Issue of Emeritus Professor M. Chûjô, 185 pp.

Wittmer W. 1982b: Zur Kenntnis der Malachiidae (Coleoptera) des Ryukyu Archipels (70. Beitrag zur Kenntnis der palaearktischen Malachiidae). Pp. 51-54. *Special Issue to the Memory of Retirement of Emeritus Professor Michio Chûjô*. Nagoya: Association of the Memorial Issue of Emeritus Professor M. Chûjô, 185 pp.

Wittmer W. 1982c: Insects of Saudi Arabia Coleoptera: Fam. Malachiidae (Part 3). *Fauna of Saudi Arabia* **4**: 108-110.

Wittmer W. 1982d: Die Familie Cantharidae (Col.) auf Taiwan (1. Teil). *Entomological Review of Japan* **37**: 119-140.

Wittmer W. 1982e: 71. Beitrag zur Kenntnis der palaearktischen Cantharidae. *Entomologica Basiliensia* **7**: 340-347.

Wittmer W. 1982f: Zur Kenntnis der Malachiidae von Taiwan (Coleoptera). *Entomologica Basiliensia* **7**: 348-372.

Wittmer W. 1983a: Die Familie Cantharidae (Col.) auf Taiwan (2. Teil). *Entomological Review of Japan* **38**: 147-172.

Wittmer W. 1983b: Beitrag zur einer Revision der Gattung Themus Motsch. Coleoptera: Cantharidae. *Entomologischen Arbeiten aus dem Museum G. Frey* **31/32**: 189-239.

Wittmer W. 1983c: Die Gattung Micropodabrus Pic im Himalaja (Coleoptera, Cantharidae). (35. Beitrag zur Kenntnis der indo-malaiischen Fauna). *Entomologica Basiliensia* **8**: 233-255.

Wittmer W. 1983d: 36. Beitrag zur Kenntnis der Malachiidae der indomalaiischen Region. *Entomologica Basiliensia* **8**: 310-328.

Wittmer W. 1984a: 72. Beitrag zur Kenntnis der palaearktischen Fauna (Coleoptera: Fam. Cantharidae und Malachiidae). *Mitteilungen der Entomologischen Gesellschaft Basel* **34**: 1-5.

Wittmer W. 1984b: 36. Beitrag zur Kenntnis der indo-malaiischen Cantharidae (Coleoptera). *Mitteilungen der Schweizerischen Entomologischen Gesellschaft* **57**: 95-109.

Wittmer W. 1984c: Die Familie Cantharidae (Col.) auf Taiwan (3. Teil). *Entomological Review of Japan* **39**: 141-166.

Wittmer W. 1984d: Insects of Saudi Arabia Coleoptera: Fam. Malachidae (Part 4). *Fauna of Saudi Arabia* **6**: 243-248.

Wittmer W. 1984e: Die schwarzen Troglops Er. (Coleoptera, Malachiidae) des west-palaearktischen Faunengebietes (73. Beitrag zur Kenntnis der palaearktischen Fauna). *Entomologica Basiliensia* **9**: 244-253.

Wittmer W. 1984f: Die Gattung Nepachys Thoms. (Coleoptera, Malachiidae) in Kaschmir und dem nördlichen Pakistan. (74. Beitrag zur Kenntnis der palaearktischen Fauna). *Entomologica Basiliensia* **9**: 254-264.

Wittmer W. 1985a: Zur Kenntnis der mit Hedybius Er. verwandten Gattungen unter besonderer Berücksichtigung der Fauna des südlichen Afrikas. *Entomologischen Arbeiten aus dem Museum G. Frey* **33/34**: 291-347.

Wittmer W. 1985b: 38. Beitrag zur Kenntnis der indo-malaiischen Fauna. *Entomologischen Arbeiten aus dem Museum G. Frey* **33/34**: 381-390.
Wittmer W. 1985c: Malachiidae (Coleoptera) aus dem südlichen Mittelmeergebiet. *Entomologica Basiliensia* **10**: 397-419.
Wittmer W. 1986a: 76. Beitrag zur Kenntnis der palaearktischen Fauna. *Mitteilungen der Entomologischen Gesellschaft Basel* **36**: 100-122.
Wittmer W. 1986b: Nachtrag zu den Fam. Cantharidae und Malachiidae (Col.) auf Taiwan. *Entomological Review of Japan* **41**: 131-135.
Wittmer W. 1986c: Laius plagiatus (Walker) und verwandte Arten (Coleoptera, Malachiidae) (39. Beitrag zur Kenntnis der indo-malaiischen Fauna). Pp. 215-232. In: S.-I. Uéno (ed.): *Entomological Papers Presented to Yoshihiko Kurosawa on the Occasion of his Retirement*. Tokyo: Coleopterists Association of Japan, 342 pp.
Wittmer W. 1987a: Zur Kenntnis der Gattung Prothemus Champion (Coleoptera: Cantharidae). *Mitteilungen der Entomologischen Gesellschaft Basel* **37**: 69-88.
Wittmer W. 1987b: 40. Beitrag zur Kenntnis der indo-malaiischen Fauna (Coleoptera: Cantharidae, Malachiidae). *Mitteilungen der Entomologischen Gesellschaft Basel* **37**: 101-115.
Wittmer W. 1987c: Die Gattung Colotes Erichson in Aethiopien, Somalia und Eritrea und beschreibung einer neuen Gattung (Coleoptera - Malachiidae) (26. Beitrag zur Kenntnis der Fauna Afrikas). *Annali del Museo Civico di Storia Naturale di Genova* **86**: 627-655.
Wittmer W. 1988a: 78. Beitrag zur Kenntnis der palaearktischen Fauna (Coleoptera: Malachiidae). *Mitteilungen der Entomologischen Gesellschaft Basel* **38**: 21-30.
Wittmer W. 1988b: Zur Kenntnis der Malachiidae (Coleoptera) von Jemen. *Mitteilungen der Entomologischen Gesellschaft Basel* **38**: 47-52.
Wittmer W. 1988c: Zur Kenntnis der Cantharidae (Coleoptera) Chinas und angrenzenden Länder. *Entomologica Basiliensia* **12**: 343-372.
Wittmer W. 1988d: Clanoptilus kruperi (Pic) und verwandte Arten (Coleoptera, Malachiidae). (77. Beitrag zur Kenntnis der palaearktischen Fauna). *Entomologica Basiliensia* **12**: 397-406.
Wittmer W. 1988e: Neue Gattungen und neue Arten der Familie Malachiidae (Coleoptera) häuptsachlich aus dem südlichen Afrika. *Annals of the Transvaal Museum* **34**: 557-619.
Wittmer W. 1989: 42. Beitrag zur Kenntnis der indo-malaiischen Cantharidae und Malachiidae (Coleoptera). *Entomologica Basiliensia* **13**: 207-237.
Wittmer W. 1992a: Neue Arten aus der Gattung Podosilis Wittmer (Coleoptera: Cantharidae: Silinae) aus der Ausbeute von L. Fea (44. Beitrag zur Kenntnis der indo-malaiischen Fauna (Coleoptera: Cantharidae, Malachidae). *Doriana* **6** (276): 1-9.
Wittmer W. 1992b: 78. Beitrag zur Kenntnis der palaearktischen Malachiidae (Coleoptera). *Mitteilungen der Entomologischen Gesellschaft Basel* **42**: 11-34.
Wittmer W. 1992c: Die Malthininae (Coleoptera Cantharidae) des Caucasus. *Russian Entomological Journal* **1**: 17-35.
Wittmer W. 1992d: 45. Beitrag zur Kenntnis der indo-malaiischen Cantharidae (Col.). *Mitteilungen der Entomologischen Gesellschaft Basel* **42**: 98-111.
Wittmer W. 1993a: Zur Kenntnis der palaearktischen und indo-malaiischen Cantharidae (Coleoptera). *Entomologica Basiliensia* **16**: 203-253.
Wittmer W. 1993b: 79. Beitrag zur Kenntnis der palaearktischen Fauna (Coleoptera - Cantharidae). *Entomologica Basiliensia* **16**: 279-305.
Wittmer W. 1994a: Neue Malachiidae (Coleoptera) aus dem palaearktischen und indomalayischen Faunengebiet. *Mitteilungen der Entomologische Gesellschaft Basel* **44**: 2-26.
Wittmer W. 1994b: [new taxa]. In: Wittmer W. & Ohba N.: Neue Rhagophthalmidae (Coleoptera) aus China und benachbarten Ländern. *Japanese Journal of Entomology* **62**: 341-355.
Wittmer W. 1995a: Lampyridae und Omethidae aus dem indo-malaiischen Faunengebieten (Coleoptera). *Mitteilungen der Entomologischen Gesellschaft Basel* **45**: 106-117.
Wittmer W. 1995b: Neue Cantharidae (Col.) aus dem indo-malaiischen und palaearktischen Faunengebiet mit Mutationen. *Entomologica Basiliensia* **18**: 109-169.
Wittmer W. 1995c: Zur Kenntnis der Gattung Athemus Lewis (Col. Cantharidae). *Entomologica Basiliensia* **18**: 171-286.

Wittmer W. 1995d: Zur Kenntnis der Familie Malachiidae (Coleoptera). Teil I. *Mitteilungen der Schweizerischen Entomologischen Gesellschaft* **68**: 263-295.

Wittmer W. 1995e: Zur Kenntnis der Familie Malachiidae (Coleoptera) 2. Beitrag. *Entomologica Basiliensia* **18**: 287-391.

Wittmer W. 1995f: 46. Beitrag zur Kenntnis der indo-malaiischen Cantharidae (Coleoptera). *Special Bulletin of the Japanese Society of Coleopterology* (Tokyo) **4**: 407-418.

Wittmer W. 1996: Zur Kenntnis der Familie Malachiidae (Coleoptera). Teil II. *Mitteilungen der Schweizerischen Entomologischen Gesellschaft* **69**: 297-327.

Wittmer W. 1997a: Neue Cantharidae (Col.) aus China und Vietnam der Ausbeuten von Prof. Dr. Masataka Satô in den Jahren 1995 und 1996. *Japanese Journal of Systematic Entomology* **3**: 33-42.

Wittmer W. 1997b: Zur Kenntnis einiger fernöstlicher Vertreter der Gattungen Nepachys Thomson und Attalus Erichson (Coleoptera, Malachiidae). *Japanese Journal of Systematic Entomology* **3**: 43-48.

Wittmer W. 1997c: Neue Cantharidae und Malachiidae (Coleoptera) hauptsächlich aus der Sammlung des Staatlichen Museums für Naturkunde Stuttgart. *Entomologische Blätter* **93**: 152-162.

Wittmer W. 1997d: Zur Kenntnis der Gattungen Intybia Pascoe und Stenolaius Wittmer (Coleoptera, Malachiidae). *Japanese Journal of Systematic Entomology* **3**: 181-211.

Wittmer W. 1997e: Neue Cantharidae (Col.) aus dem indo-malaiischen und palaearktischen Faunengebiet mit Mutationen. 2. Beitrag. *Entomologica Basiliensia* **20**: 223-365.

Wittmer W. 1997f: Neue Rhagophthalmidae und Phengodidae (Coleoptera). *Mitteilungen der Schweizerischen Entomologischen Gesellschaft* **70**: 257-263.

Wittmer W. 1999: Zur Kenntnis der Familie Malachiidae (Coleoptera). 3. Beitrag. *Entomologica Basiliensia* **21**: 171-252.

Wittmer W. & Magis N. 1978: Zur Kenntnis einiger mit Cantharis L. verwandter Gattungen (Coleoptera, Cantharidae). *Bulletin et Annales de la Société d'Entomologie de Belgique* **114**: 133-139.

Wollaston T. V. 1854: *Insecta Maderensia; being an account of the insects of the Islands of the Madeiran Group.* London: J. van Voorst, xliv + 634 pp., 13 pls.

Wollaston T. V. 1857a: *Catalogue of the coleopterous insects of Madeira in the collection of the British Museum.* London: printed by order of the trustees, xvi + 234 pp., 1 pl.

Wollaston T. V. 1857b: A revision of the British Atomariae; with observations on the genus. *Transactions of the Entomological Society of London* (N. S.) **4**: 64-82.

Wollaston T. V. 1857c: On certain coleopterous insects from the Cape Verde Islands. *The Annals and Magazine of Natural History* (2) **20**: 503-506.

Wollaston T. V. 1858a: Description of a coleopterous insect from the Canary Islands. *The Annals and Magazine of Natural History* (3) **2**: 196-197.

Wollaston T. V. 1858b: On a new genus of European Coleoptera. *The Annals and Magazine of Natural History* (3) **2**: 337-339.

Wollaston T. V. 1858c: On additions to the Madeiran Coleoptera. *The Annals and Magazine of Natural History* (3) **2**: 407-415.

Wollaston T. V. 1859: Descriptions of two coleopterous insects from the North of China. *The Annals and Magazine of Natural History* (3) **4**: 430-431.

Wollaston T. V. 1860a: On addition to the Madeiran Coleoptera. *The Annals and Magazine of Natural History* (3) **5**: 252-267.

Wollaston T. V. 1860b: On addition to the Madeiran Coleoptera. *The Annals and Magazine of Natural History* (3) **5**: 358-366.

Wollaston T. V. 1861a: On the Anobiidae of the Canary Islands. *The Annals and Magazine of Natural History* (3) **7**: 11-19.

Wollaston T. V. 1861b: Additions to the Madeiran Coleoptera. *The Annals and Magazine of Natural History* (3) **8**: 103-106.

Wollaston T. V. 1862a: On the Euphorbia-infesting Coleoptera of the Canary Islands. *Transactions of the Entomological Society of London* (3) **1** [1862-1864]: 136-189.

Wollaston T. V. 1862b: On the Ptinidae of the Canary Islands. *Transactions of the Entomological Society of London* (3) **1**: 190-214, 1 pl.

Wollaston T. V. 1862c: On the Canarian malacoderms. *Journal of Entomology* **1**: 421-450, pl. 20.

Wollaston T. V. 1862d: Brief diagnostic characters of new Canarian Coleoptera. *The Annals and Magazine of Natural History* (3) **9**: 437-442.

Wollaston T. V. 1862e: On additions to the Madeiran Coleoptera. *The Annals and Magazine of Natural History* (3) **10**: 287-293, 331-342.

Wollaston T. V. 1863: Diagnostic notices of new Canarian Coleoptera. *The Annals and Magazine of Natural History London* (3) **11**: 214-221.

Wollaston T. V. 1864: *Catalogue of the coleopterous insects of the Canaries in the collection of the British Museum.* London: Trustees, xiv + 648 pp.

Wollaston T. V. 1865: *Coleoptera Atlantidum, being an enumeration of the coleopterous insects of the Madeiras, Salvages, and Canaries.* London: J. van Voorst, xlvii + 526 pp. + Appendix and Index 140 pp.

Wollaston T. V. 1867: *Coleoptera Hesperidum, being an enumeration of the coleopterous insects of the Cape Verde Archipelago.* London: J. van Voorst, xxxix + 285 pp., 1 map.

Wollaston T. V. 1870: On addition to the coleopterous fauna of the Cape-Verde Islands. *The Annals and Magazine of Natural History* (4) **5**: 245-251.

Wollaston T. V. 1871: On additions to the Atlantic Coleoptera. *Transactions of the Entomological Society of London* **1871**: 203-314.

Wollaston T. V. 1873a: On a new coleopterous genus from Japan. *The Entomologist's Monthly Magazine* **9**: 278-281.

Wollaston T. V. 1873b: On a genus of the Byrrhidae from Japan. *The Entomologist's Monthly Magazine* **10**: 33-34.

Wollaston T. V. 1873c: On new Coleoptera from Japan. *The Entomologist's Monthly Magazine* **10** [1873-1874]: 167-168.

Wollaston T. V. 1874: On new Coleoptera from Japan. *The Entomologist's Monthly Magazine* **10** [1873-1874]: 169-172, 200-203.

Wollaston T. V. 1875: [new taxa]. In: Bates H. W. & Wollaston T. V.: Description of new species referred to in Mr. Blackmore's notes on the Coleoptera from south Morocco. *The Entomologist's Monthly Magazine* **11**: 217-220.

Wollaston T. V. 1877: *Coleoptera Sanctae-Helenae.* London: Van Voorst, xxv + 256 pp.

Wollaston T. V. 1879: [new taxon]. In: Belon M.-J.: Histoire Naturelle des coléoptères de France par M. Mulsant. Famille des Lathridiens 1re partie. *Annales de la Société Linnéenne de Lyon* (N. S.) **26**: 157-365.

Woodroffe G. E. 1970: A new species of Cryptophagus, and the male of Micrambe translatus Grouvelle from Mongolia. Ergebnisse der zoologischen Forschungen von Dr. Z. Kaszab in der Mongolei (Coleoptera, Cryptophagidae). *Reichenbachia* **13**: 9-13.

Wu C.-F. [= Hu J.] 1938: [Coleoptera-Coccinellidae]. Pp. 553-587. In: *Catalogus Insectorum Sinensium (Catalogue of Chinese Insects). Volume III.* Peiping: Fan Memorial Institute of Biology [1937], 1312 pp.

Wurst C. 1994a: Neue Schnellkäfer der Unterfamilie Athoinae aus dem Taurusgebirge, Südtürkei (Coleoptera: Elateridae). *Mitteilungen Entomologischer Verein Stuttgart 1869 e. V.* **29**: 59-63.

Wurst C. 1994b: Elater asmodaius n. sp., ein neuer Schnellkäfer aus Griechenland. (Coleoptera: Elateridae). *Entomologische Zeitschrift* **104**: 118-121.

Wurst C. 1995a: Ampedus tingitanus n. sp. und Brachygonus hadullanus n. sp., zwei neue Schnellkäfer aus Nordafrika und Kleinasien (Coleoptera, Elateridae). *Mitteilungen Entomologischer Verein Stuttgart 1869 e. V.* **30**: 30-34.

Wurst C. 1995b: Anostirus zenii binaghianus subsp. nov. und Athous silicensis Laibner, zwei bemerkenswerte Neufunde für die Schnellkäferfauna Sloweniens (Coleoptera: Elateridae). *Acta Entomologica Slovenica* **3**: 87-90.

Wurst C. 1995c: Vier neue Schnellkäfer aus Europa und Kleinasien (Coleoptera: Elateridae). *Entomologische Zeitschrift* **105**: 147-156.

Wurst C. 1995d: Neues von Anostirus Thomson (Pseudostirus Binaghi) reissi (Reitter) mit Beschreibung einer neuen Art aus Kärnten: Anostirus (Pseudostirus) lauianus n. sp. (Col., Elateridae). *Entomologische Nachrichten und Berichte* **38** [1994]: 261-266.

Wurst C. 1996: Zur Kenntnis der Phylloceriden Kleinasiens mit Beschreibung einer neuen Art aus der Sudtürkei: Phyllocerus oreadis n. sp. (Coleoptera: Elateroidea: Phylloceridae). *Mitteilungen des Internationalen Entomologischen Vereins* **21**: 51-61.

Wurst C. 1997a: Ampedus minos sp. n. und Dima raineri sp. n., zwei neue Schnellkäfer aus Griechenland, mit Bemerkungen zur Synonymie von Ampedus ocellatus (Buysson, 1891) (Coleoptera, Elateridae). *Entomologische Blätter* **93**: 1-7.

Wurst C. 1997b: Zur Taxonomie, Biogeographie und Bionomie der Gattung Hylaia mit Beschreibung zweier neuer Arten aus Griechenland und Vorderasien (Coleoptera: Endomychidae). *Folia Heyrovskyana* **5**: 97-114.

Wurst C. & Cate P. 1994: Neue Schnellkäfer der Gattung Cardiophorus Eschscholtz (Untergattung Coptostethus Wollaston) von den Kanarischen Inseln (Coleoptera: Elateridae). 1. Teil: Gran Canaria. *Entomologische Zeitschrift* **104**: 419-424.

Wurst C. & Lange F. 1996: Airaphilus doramas n. sp., ein neuer Plattkäfer von den Kanarischen Inseln (Coleoptera, Silvanidae). *Mitteilungen des Entomologischen Vereins in Stuttgart* **31**: 59-64.

Wurst C. & Schimmel R. 1995: Anostirus (Parastirus) nubilosus n. sp., ein neuer Schnellkäfer aus dem türkischen Kaukasus (Coleoptera, Elateridae). *Mitteilungen Entomologischer Verein Stuttgart 1869 e. V.* **30**: 25-29.

Wurst C., Schimmel R. & Platia G. 2001: Contribution to the fauna of click-beetles from Yemen (Coleoptera: Elateridae). Part 1: Description of species new to science with remarks on genera Monadicus Stibick, Buffeventris Fleutiaux, and Optitarynus Stibick (plate 27). *Esperiana, Buchreihe zur Entomologie* **8**: 541-558.

Xambeu P. V. 1909: Anomalies, variétes, aberrations, cas particuliers. *L'Échange, Revue Linnéenne* **25**: 1-6.

Xiao N. 1993: [new taxa]. In: Xiao N. & Li H.: [Coleoptera: Coccinellidae]. Pp. 368-390. In: Huang F. et al. (eds.): *Insects of Wuling mountains area, southwestern China.* Beijing: Science press, x + 777 pp.

Yajima T. & Nakane T. 1969a: Two new species of Cleridae from the Loochoos (Insecta, Coleoptera). *Bulletin of the National Science Museum* **12**: 5-7.

Yajima T. & Nakane T. 1969b: On the species of the genus Athemus Lewis in Japan, with descriptions of a few new forms (Insecta, Coleoptera, Cantharidae). *Bulletin of the National Science Museum* **12**: 183-189.

Yajima T. & Nakane T. 1988: [new taxon]. In: Nakane T.: Notes on some little-known beetles (Coleoptera) in Japan. 3. *Kita-Kyūshū no Konchū* **35**: 77-79, pl. 9.

Yang C. 1980: Niptinus nom. nov. for Eurostus Brown 1940 (Coleoptera: Ptinidae) nec Dallas 1851 (Hemiptera: Pentatomidae). *Entomotaxonomia* **2** (1): 26.

Yang C.-T. 1971: Notes on the species of genus Pseudoscymnus from Taiwan (Coleoptera: Coccinellidae). *Journal of Agriculture and Forestry* **20**: 85-96.

Yang C.-T. 1978a: A new subgenus and species of Coccinellidae. *Bulletin of the Society of Entomology* (Taichung) **13**: 27-28.

Yang C.-T. 1978b: Scymnus (Subgenus Pullus) (Coleoptera, Coccinellidae) of Taiwan. *Plant Protection Bulletin* (Taiwan) **20**: 106-116.

Yang C.-T. & Wu R.-H. 1972: Notes on some Coccinellidae of Taiwan. *Journal of Agriculture and Forestry* **21**: 115-128.

Yazdani A. & Ahmadi A. A. 1992: Four Coccinellid species of subfamily Sticholotidinae in Fars (Coleoptera: Coccinellidae). *Journal of Entomological Society of Iran* (Tehran) **11**: 13-18 (51-60) (in English and in Farsi).

Yu G. 1994: Three new species of Cybocephalidae (Coleoptera) from China. *Zoological Research* **15** (3): 13-17 (in Chinese, with English abstract and title).

Yu G. 1995a: Two new species of Cybocephalidae (Coleoptera) from Guangdong, China. *Entomotaxonomia* **17**: 31-34.

Yu G. 1995b: Two new species of Cybocephalidae (Coleoptera) from China. *Entomotaxonomia* **17**: 277-280.

Yu G. 1997: Coleoptera: Cybocephalidae. Pp. 665-668. In Yang X. (ed.): *Insects of the Three Gorge Reservoir of Yangtze river. Part I.* Chongqing: Chongqing Publishing House, xx + 974 pp. (in Chinese, English summary).

Yu G. & Tian M. 1995: Notes on the genus Cybocephalus Erichson from China (Coleoptera: Cybocephalidae). *Entomologica Sinica* **2**: 35-38.

Yu G.-Y. 1993: Three new species of Scymnus Kugelann from China (Coleoptera: Coccinellidae). *Zoological Research* **14**: 293-298.

Yu G.-Y. 1995a: The Coccinellidae (excluding Epilachninae) collected by J. Klapperich in 1977 on Taiwan (Insecta, Coleoptera). *Spixiana* **18**: 123-144.

Yu G.-Y. 1995b: Coccinellid beetles from Fujian, China, preserved in the Zoologische Staatsammlung Muenchen, Germany (Insecta, Coleoptera, Coccinellidae). *Spixiana* **18**: 145-150.

Yu G.-Y. 1996a: A list of Stethorus Weise (Coleoptera: Coccinellidae) from China with description of a new species. *Entomotaxonomia* **18**: 32-35.

Yu G.-Y. 1996b: The Chinese species of Diomus Mulsant (Coleoptera: Coccinellidae). *Entomotaxonomia* **18**: 276-282.

Yu G.-Y. 1997a: [new taxa]. In: Yu G.-Y., Yao D. & H. Liu: The Coccinellidae collected from Tsuga with Adelges tsugae Annand (Homoptera: Adelgidae). *Scientia Silvae Sinicae* **33**: 432-440 (in Chinese).

Yu G.-Y. 1997b: Coleoptera: Coccinellidae: Scymninae. Pp. 714-730. In: Yang X. (ed.): *Insects of the Three Gorge Reservoir area of Yangtze river. Part 1.* Chongqing: Chongqing Publishing House, i-xx, 1-974 pp.

Yu G-Y. 1998a: [new taxa]. In: Yu G.-Y. & Liang H.-B.: Coccinellidae collected in the wheat fields of Xinjiang, China with description of a new species (Coleoptera). *Entomotaxonomia* **20**: 127-132 (in Chinese).

Yu G.-Y. 1998b: Notes on the genus Alloneda Iablokoff-Khnzorian with description of a new species (Coleoptera, Coccinellidae). *Zoological Studies* **37**: 282-285.

Yu G.-Y. 1999a: New species of family Coccinellidae from Mt. Jigong in Henan (Coleoptera: Coccinellidae). *Fauna and Taxonomy of Insects in Henan* 1999 **3**: 17-20.

Yu G.-Y. 1999b: Eigth new species of lady beetles from Henan, China (Coleoptera, Coccinellidae). *Fauna and Taxonomy of Insects in Henan* **4**: 62-73 (in Chinese).

Yu G.-Y. 1999c: Lady beetles (Coleoptera: Coccinellidae) from Pinus armandii Franchet infested with Pineus sp. *Entomotaxonomia* **21**: 281-287.

Yu G.-Y. 2000a: Amida Lewis, with description of a new species (Coleoptera: Coccinellidae) from China. *Zoological Studies* **39**: 23-27.

Yu G.-Y. 2000b: [new taxa]. In: Yu G.-Y., Montgomery M. E. & Yao D.: Lady beetles (Coleoptera: Coccinellidae) from Chinese hemlocks infested with the hemlock wooly adelgid, Adelges tsugae Anand (Homoptera: Adelgidae). *The Coleopterists Bulletin* **54**: 154-199.

Yu G.-Y. 2001a: Two new species of ladybeetles (Coleoptera: Coccinellidae) from Taiwan. *Bulletin of the National Science Museum* (Taichung) **14**: 99-103.

Yu G.-Y. 2001b: [new taxa]. In: Yu G.-Y. & Lau C.-S.: A contribution to knowledge of the ladybirds of Hong Kong with description of three new species. *Memoirs of the Hong-Kong Natural History Society* **24**: 147-179.

Yu G.-Y. & Liu H. 1997: [new taxa]. In: Yu G.-Y., Yao D. & Liu H.: The Coccinellidae collected from Tsuga with Adelges tsugae Annand (Homoptera: Adelgidae). *Scientia Silvae Sinicae* **33**: 432-440 (in Chinese).

Yu G.-Y. & Montgomery M. E. 2000: [new taxa]. In: Yu G.-Y., Montgomery M. E. & Yao D.: Lady Beetles (Coleoptera: Coccinellidae) from Chinese hemlocks infested with the hemlock wooly adelgid, Adelges tsugae Anand (Homoptera: Adelgidae). *The Coleopterists Bulletin* **54**: 154-199.

Yu G.-Y. & Pang X.-F. 1992a: A new species of Keiscymnus from Taiwan. *Coccinella* (Passau) **4**: 10-11.

Yu G.-Y. & Pang X.-F. 1992b: Description of a male of Shirozuella mirabilis Sasaji with two additional new species from Taiwan. *Journal of South China Agricultural University* **13**: 39-47.

Yu G.-Y. & Pang X.-F. 1992c: Descriptions of three new species of Scymnus (Pullus) Mulsant (Coleoptera: Coccinellidae) from Guangdong province. *Entomotaxonomia* **14**: 116-122 (in Chinese)

Yu G.-Y., Pang H. & Pang X.-F. 1993: Coccinellidae collected from Chebaling National nature reserve. Pp. 467-511. In: Xu Y. (ed.): *Collected papers for investigation in National Chebaling Nature Reserve.* Guangzhou: Guangdong Scient Technol Publ House.

Yu G.-Y. & Pang X.-F. 1994: Description of two new species of Scymnus Kugelann from China (Coleoptera: Coccinellidae). *Acta Entomologica Sinica* **37**: 209-212 (in Chinese).

Yu G.-Y. & Pang X.-F. 1995: Description of three new species of Scymnus Kugelann from Zheijiang Province, China (Coleoptera: Coccinellidae). *Acta Zootaxonomica Sinica* **19**: 474-479.

Yu G.-Y. & Wang H.-Y. 1999: A new species and a new record of lady beetles (Coleoptera: Coccinellidae) from Taiwan. *Journal of Taiwan Museum* **52**: 27-31.

Yu G.-Y. & Yao D. 1997: [new taxa]. In: Yu G.-Y., Yao D. & Liu H.: The Coccinellidae collected from Tsuga with Adelges tsugae Annand (Homoptera: Adelgidae). *Scientia Silvae Sinicae* **33**: 432-440 (in Chinese).

Yu G.-Y. & Yao D. 2000: [new taxa]. In: Yu G.-Y., Montgomery M. E. & Yao D.: Lady beetles (Coleoptera: Coccinellidae) from Chinese hemlocks infested with the hemlock wooly adelgid, Adelges tsugae Anand (Homoptera: Adelgidae). *The Coleopterists Bulletin* **54**: 154-199.

Yuasa H. 1937: Description of a new genus of the family Lampyridae (Coleoptera). *Kontyû* **11**: 107-109.

Yuasa H. 1963: Two new species of Coccinellidae from Japan. *Fragmenta Coleopterologica* (Tokyo) **3**: 12.

Zahradník P. 1993: New species of the genus Dorcatoma from Central Europe (Coleoptera, Anobiidae). *Folia Heyrovskyana* **1**: 80-83.

Zahradník P. 1996a: Four new species of Anobiidae (Coleoptera) from Turkey and Rhodes Island. *Folia Heyrovskyana* **4**: 93-101.

Zahradník P. 1996b: New species of Gastrallus from Europe (Coleoptera: Anobiidae). *Klapalekiana* **32**: 267-269.

Zahradník P. 1997: Six new Species of Anobiidae (Coleoptera) from the East Palaearctic Region. *Entomologica Basiliensia* **20**: 101-114.

Zahradník P. 1998: Anobiidae of Turkey (Coleoptera). *Klapalekiana* **34**: 263-286.

Zahradník P. 2000: Ernobius besucheti sp.n. (Coleoptera: Anobiidae) from Switzerland. *Folia Heyrovskyana* **8**: 133-136.

Zaitzev P. A. 1915: Materialia ad cognitionem faunae coleoptorum Caucasi 1. Fam. Cleridae et Dermestoidae (seu Corynetidae). *Otchet po Kavkazkomu Muzeyu i Riflisskoy Publichnoy Bibliotek. Bulletin du Musée du Caucase* (Tiffis) **9** [1913]: 36-52.

Zaitzev P. A. 1919: Materialy k faune zhestkokrylych Kavkazskovo kraya. VIII. Dermestidae et Bostrychidae. *Izvestija Kavkazskovo Muzea* (Tiflis) **12**: 161-186.

Zaslavskij V. A. 1962: Novyy palearkticheskiy vid Chilocorus (Coleoptera, Coccinellidae). *Entomologicheskoe Obozrenie* **41**: 398-401.

Zaslavskij V. A. 1964: Novye i maloizvestnye Hyperaspis (Coleoptera, Coccinellidae) iz Kazakhstana i Srednej Azii. *Trudy Zoologichekogo Instituta* (Leningrad) **34**: 152-154.

Zaslavskij V. A. 1966: K poznaniyu palearkticheskikh Hyperaspis (Coleoptera Coccinellidae). *Trudy Zoologichekogo Instituta* (Leningrad) **37**: 60-61.

Zeising M. 1981: Ein neuer Ampedus (Elateridae) aus den Spanischen Pyrenäen. *Entomologische Blätter* **77**: 17-18.

Zeising M. & Sieg J. 1983: Anmerkungen zur Ampedus pomorum-Gruppe (Coleoptera: Elateridae) nebst Beschreibung einer neuen Art (Ampedus ziegleri sp. n.) aus der sanguinolentus-Gruppe. *Entomologische Blätter* **79**: 3-29.

Zeising M. & Brunne G. 2003: Neue Elateriden aus Portugal, Marokko und der Türkei (Coleoptera, Elateridae). *Entomologische Blätter* **98**: 189-198.

Zeising M. & Sieg J. 1992: Beschreibung von neuen Arten aus der Verwandschaft des Ampedus preaustus Fabricius, 1792 und des Ampedus rufipennis Stephens, 1830 nebst Anmerkungen zur Taxonomie. *Entomologische Blätter* **88**: 103-141.

Zeng T. 2000: A new species and two new records of Henosepilachna (Coleoptera: Coccinellidae). Pp. 136-139. In: Zhang Y. (ed.): *Systematic and faunistic research of Chinese insects*. Proceedings of the 5th National Congress of Insect Taxonomy. China Agriculture Press, 2000: xvi + 331 pp. (in Chinese).

Zeng T. & Yang L.-F. 1996a: Five new species of Epilachna (Coleoptera: Coccinellidae) from Guanxi, China. *Entomotaxonomia* **18**: 193-200 (in Chinese).

Zeng T. & Yang L.-F. 1996b: Two new species of Afissula from Guanxi, China. *Entomotaxonomia* **18**: 271-275 (in Chinese).

Zetterstedt J. W. 1824: Nya Svenska Insect-arter fundne och beskrifne af Joh. Wilh. Zetterstedt. *Kongliga Vetenskaps-Akademiens Handlingar* **1824**: 149-159.

Zetterstedt J. W. 1828: *Fauna Insectorum Lapponica. Pars I.* Hammone: Libraria Schulziana, xx + 563 pp.

Zetterstedt J. W. 1838: Coleoptera. Columns 7-868. In: *Insecta Lapponica. Sectio prima. Coleoptera, Orthoptera et Hemiptera.* Lipsiae: Leopoldi Voss, vi + 1140 columns [issued in parts: 1838-1840].

Zhang S. & Liu Y. 1988: A new species of Thorictodes from Yunnan China (Coleoptera, Dermestidae). *Entomotaxonomia* **10**: 27-29 (in Chinese).

Zhang S., Liu Y. & Jing X. 1999: A new species of Trogoderma (Coleoptera: Dermestidae) found damaging stored grain in China. *Proceedings 7th International Working Conference on Stored product Protection* **2**: 1765-1767.

Zhantiev R. D. 1963a: Kozheedy roda Attagenus Latr. (Coleoptera, Dermestidae) fauny sovetskogo Soyuza. [Dermestid beetles of the genus Attagenus Latr. (Coleoptera, Dermestidae) from the USSR]. *Entomologicheskoe Obozrenie* **42**: 415-424.

Zhantiev R. D. 1963b: Novye vidy kozheedov roda Attagenus Latr. (Coleoptera, Dermestidae) iz Sredney Azii. [New species of dermestids belonging to the genus Attagenus Latr. (Coleoptera, Dermestidae) from Central Asia]. *Zoologicheskiy Zhurnal* **42**: 454-457.

Zhantiev R. D. 1967: Opyt taksonomycheskovo analyza roda Dermestes L. (Coleoptera, Dermestidae). *Zoologicheskiy Zhurnal* **46**: 1350-1356.

Zhantiev R. D. 1973a: Ergebnisse der zoologischen Forschungen von Dr. Z. Kaszab in der Mongolei. Nr. 321. Dermestidae (Coleoptera). *Annales Historico-Naturales Musei Nationalis Hungarici* **65**: 187-194.

Zhantiev R. D. 1973b: Novye i maloizvestnye kozheedy (Coleoptera, Dermestida) fauny SSSR. [New and little known dermestids (Coleoptera) in the fauna of the USSR]. *Zoologicheskiy Zhurnal* **52**: 282-284.

Zhantiev R. D. 1975: Novye vidy kozheedov (Coleoptera, Dermestidae) iz Kazakhstana i Sredney Azii [New species of Dermestidae (Coleoptera) from Kazakhstan and Central Asia]. *Zoologicheskiy Zhurnal* **54**: 308-311.

Zhantiev R. D. 1976: Zhuki-kozheedy Fauny SSSR. Moskva: Izdatelstvo Moskovskogo Universiteta, 180 pp.

Zhantiev R. D. 1979: Novyy vid roda Anthrenus Schaeffer (Coleoptera: Dermestidae) iz Uzbekistana. *Trudy Vshesoyuznovo Entomologicheskogo Obshchestva* **61**: 68-69.

Zhantiev R. D. 1988: Novye vidy zhukov-kozheedov (Coleoptera, Dermestidae) s Dal'nego Vostoka SSSR. [New species of Dermestid beetles (Coleoptera) from the Far East of the USSR]. *Zoologicheskiy Zhurnal* **67**: 1420-1423.

Zhantiev R. D. 2001: Zhuki-kozheedy roda Orphilus Er. (Coleoptera, Dermestidae) fauny Palearktiki [Dermestids of the genus Orphilus Er.]. (Coleoptera, Dermestidae) of the Palaearctic Fauna]. *Entomologicheskoe Obozrenie* **80**: 611-620.

Zhantiev R. D. 2002: New and little-known dermestid-beetles (Coleoptera, Dermestidae) from Transcaucasia. *Zoologicheskiy Zhurnal* **81**: 1-4.

Zhantiev R. D. 2004a: Novye vidy zhukov-kozheedov roda Anthrenus (Coleoptera, Dermestidae) iz Sredney Azii. [New species of dermestid beetles of the genus Anthrenus (Coleoptera, Dermestidae) from Central Asia]. *Zoologicheskiy Zhurnal* **83** (5): 1-5.

Zhantiev R. D. 2004b: Novye vidy zhukov-kozheedov roda Anthrenus (Coleoptera, Dermestidae) iz Sredney Azii i Mongolii. [New species of dermestid beetles of the genus Anthrenus (Coleoptera, Dermestidae) from Central Asia and Mongolia]. *Zoologicheskiy Zhurnal* **83** (6): 1-4.

Zia Y. 1933: On the Languridae (sic!) of the Provinces Kweichow and Yunnan. *Sinesia* **4**: 15-37.

Zia Y. 1934a: Étude morphologique et systématique des Languriidae du Tonkin. *Mémoire présenté à la Faculté des Sciences de l'Université de Paris pour obtenir le Diplôme d'Études Supérieures* **554**: 1-41.

Zia Y. 1934b: On some species of Languriidae of Hangchow. *Sinesia* **4**: 353-358.

Zia Y. 1935: Catalogue of Chinese Languriidae with description of a new species. *Sinesia* **6**: 689-697.

Zia Y. 1959a: Results of the Zoologico-Botanical expedition of Academia Sinica and Academy of Sciences USSR to Southwest China 1955-1957 (Coleoptera, Languriidae). *Acta Entomologica Sinica* **9**: 234-243.

Zia Y. 1959b: New genera and species of Chinese Languriidae. *Acta Entomologica Sinica* **9**: 366-372.

Ziegler D. R. 1844: Descriptions of new North American Coleoptera. *Proceedings of the Academy of Natural Sciences of Philadelphia* **2**: 266-272.

Ziegler D. R. 1845: Descriptions of new North American Coleoptera. *Proceedings of the Academy of Natural Sciences of Philadelphia* **2**: 226-272.

Zimmermann C. C. A. 1869: Synonymical notes on Coleoptera of the United States, with descriptions of new species from the mss. of the late Dr. C. Zimmermann (ed. by J. L. LeConte). *Transactions of the American Entomological Society* **2**: 243-259.

Zimmermann K. 1939: Die geographischen Rassen von Epilachna chrysomelina F., und ihre Beziehungen zu Epilachna capensis Thunbg. *Zeitschrift für Induktive Abstammungs- und Vererbungslehre* (Leipzig) **71**: 527-537.

Zimmermann S. 1971: Ergebnisse zoologischer Sammelreisen in die Turkei. Gattung Trichodes Herbst (Cleridae, Coleoptera). *Annalen des Naturhistorisches Museum Wien* **75**: 591-625.

Zimmermann S. 1973: Ergebnisse zoologischer Sammelreisen in der Türkei. Gattung Trichodes Herbst (Cleridae, Coleoptera). *Annalen des Naturhistorisches Museum Wien* **77**: 409-419.

Žirovnický J. 1974: Records of Cleridae from Algeria, including the description of Trichodes hofferi sp. n. (Coleoptera). *Acta Entomologica Bohemoslovaca* **71**: 315-317.
Žirovnický J. 1976: Trichodes winkleri n. sp. de Grèce (Col., Cleridae). *Bulletin de la Société Entomologique de Mulhouse* **1976**: 15-16.
Žirovnický J. 1977: Eurenoplium genre nouveau pour Orthopleura sanguinicollis (F.) (Col., Cleridae). *Bulletin de la Société Entomologique de Mulhouse* **1977**: 45-46.
Zoufal V. 1894: Bestimmungs-Tabelle der Bostrychidae aus Europa und angrenzenden Ländern. *Wiener Entomologische Zeitung* **13**: 33-42.
Zoufal V. 1897: Zwei neue Gastrallus-Arten. *Wiener Entomologische Zeitung* **16**: 206.
Zubkov [= Zoubkoff] B. 1833: Nouveaux coléoptères recueillis en Turcménie. *Bulletin de la Société Impériale des Naturalistes de Moscou* **6**: 310-340.
Zurcher C. 1911a: Description d'une espèce nouvelle de Danacaea de l'île de Chypre. *Bulletin de la Société Entomologique de France* **1911**: 242-243.
Zurcher C. 1911b: Nouvelles espèces de coléoptères de Chypre et d'Asie Mineure. *Bulletin de la Société Entomologique de France* **1911**: 253-256.

INDEX

Family-group names

INDEX

Genus-group names
(synonyms are in *italics*)

www.ingramcontent.com/pod-product-compliance
Lightning Source LLC
La Vergne TN
LVHW020955300926
849160LV00007B/155